COMPREHENSIVE BIOTECHNOLOGY

*The Principles, Applications and Regulations
of Biotechnology in Industry,
Agriculture and Medicine*

EDITOR-IN-CHIEF
MURRAY MOO-YOUNG
University of Waterloo, Ontario, Canada

Volume 4

The Practice of Biotechnology: Speciality Products and Service Activities

VOLUME EDITORS
CAMPBELL W. ROBINSON
University of Waterloo, Ontario, Canada

and
JOHN A. HOWELL
University College of Swansea, UK

PERGAMON PRESS
OXFORD · NEW YORK · TORONTO · SYDNEY · FRANKFURT

U.K.	Pergamon Press Ltd., Headington Hill Hall, Oxford OX3 0BW, England
U.S.A.	Pergamon Press Inc., Maxwell House, Fairview Park, Elmsford, New York 10523, U.S.A.
CANADA	Pergamon Press Canada Ltd., Suite 104, 150 Consumers Road, Willowdale, Ontario M2J 1P9, Canada
AUSTRALIA	Pergamon Press (Aust.) Pty. Ltd., P.O. Box 544, Potts Point, N.S.W. 2011, Australia
FEDERAL REPUBLIC OF GERMANY	Pergamon Press GmbH, Hammerweg 6, D-6242 Kronberg-Taunus, Federal Republic of Germany

First edition 1985

Library of Congress Cataloging in Publication Data

Main entry under title:
Comprehensive biotechnology.
Includes bibliographies and index.
Contents: v. 1. The principles of biotechnology – scientific fundamentals / volume editors, Alan T. Bull, Howard Dalton –
v. 2. The principles of biotechnology – engineering considerations / volume editors, A.E. Humphrey, Charles L. Cooney – [etc.] –
v. 4. The practice of biotechnology – speciality products and service activities / volume editors, C.W. Robinson, John A. Howell.
1. Biotechnology.   I. Moo-Young, Murray.
TP248.2.C66   1985   660'.6   85–6509

British Library Cataloguing in Publication Data

Comprehensive biotechnology: the principles, applications and regulations of biotechnology in industry, agriculture and medicine.
1. Biotechnology
I. Title   II. Moo-Young, Murray
660'.6   TP248.3

ISBN 0–08–032512–2 (vol. 4)
ISBN 0–08–026204–X (4–vol. set)

Printed in Great Britain by A. Wheaton & Co. Ltd., Exeter

Contents

Preface

In his recent book, entitled 'Megatrends', internationally-celebrated futurist John Naisbitt observed that recent history has taken industrialized civilizations through a series of technology-based eras: from the chemical age (plastics) to an atomic age (nuclear energy) and a microelectronics age (computers) and now we are at the beginning of an age based on biotechnology. Biotechnology deals with the use of microbial, plant or animal cells to produce a wide variety of goods and services. As such, it has ancient roots in the agricultural and brewing arts. However, recent developments in genetic manipulative techniques and remarkable advances in bioreactor design and computer-aided process control have founded a 'new biotechnology' which considerably extends the present range of technical possibilities and is expected to revolutionize many facets of industrial, agricultural and medical practices.

Biotechnology has evolved as an ill-defined field from inter-related activities in the biological, chemical and engineering sciences. Inevitably, its literature is widely scattered among many specialist publications. There is an obvious need for a comprehensive treatment of the basic principles, methods and applications of biotechnology as an integrated multidisciplinary subject. *Comprehensive Biotechnology* fulfils this need. It delineates and collates all aspects of the subject and is intended to be the standard reference work in the field.

In the preparation of this work, the following conditions were imposed. (1) Because of the rapid advances in the field, it was decided that the work would be comprehensive but concise enough to enable completion within a set of four volumes published simultaneously rather than a more encyclopedic series covering a period of years to complete. In addition, supplementary volumes will be published as appropriate and the work will be updated regularly via *Biotechnology Advances*, a review journal, also published by Pergamon Press with the same executive editor. (2) Because of the multidisciplinary nature of biotechnology, a multi-authored work having an international team of experts was required. In addition, a distinguished group of editors was established to handle specific sections of the four volumes. As a result, this work has 10 editors and over 250 authors representing 15 countries. (3) Again, because of the multidisciplinary nature of the work, it was virtually impossible to use a completely uniform system of nomenclature for symbols. However, provisional guidelines on a more unified nomenclature of certain key variables, as provided by IUPAC, was recommended. (4) According to our definition, aspects of biomedical engineering (such as biomechanics in the development of prosthetic devices) and food engineering (such as product formulations) are not included in this work. (5) Since the work is intended to be useful to both beginners as well as veterans in the field, basic elementary material as well as advanced specialist aspects are covered. For convenience, a glossary of terms is supplied. (6) Since each of the four volumes is expected to be fairly self-contained, a certain degree of duplication of material, especially of basic principles, is inevitable. (7) Because of space constraints, a value judgement was made on the relative importance of topics in terms of their actual rather than potential commercial significance. For example, 'agricultural biotechnology' is given relatively less space compared to 'industrial biotechnology', the current raison d'être of biotechnology as a major force in the manufacture of goods and services. (8) Finally, a delicate balance of material was required in order to meet the objective of providing a comprehensive and stimulating coverage of important practical aspects as well as the intellectual appeal of the field. Readers may wish to use this work for initial information before possibly delving deeper into the literature as a result of the critical discussions and wide range of references provided in it.

Comprehensive Biotechnology is aimed at a wide range of user needs. Students, teachers, researchers, administrators and others in academia, industry and government are addressed. The requirements of the following groups have been given particular consideration: (1) chemists, especially biochemists, who require information on the chemical characteristics of enzymes, metabolic processes, products and raw materials, and on the basic mechanisms and analytical techniques involved in biotechnological transformations; (2) biologists, especially microbiologists and molecular biologists, who require information on the biological characteristics of living organisms involved in biotechnology and the development of new life forms by genetic engineering techniques; (3) health scientists, especially nutritionists and toxicologists, who require information on biohazards and containment techniques, and on the quality of products and by-products of biotechnological processes, including the pharmaceutical, food and beverage industries; (4) chemical engineers, especially biochemical engineers, who require information on mass and energy balances and rates of processes, including fermentations, product recovery and feedstock pretreatment, and the equipment for carrying out these processes; (5) civil engineers, especially environmental engineers, who require information on biological waste treatment methods and equipment, and on contamination potentials of the air, water and land within the ecosystem, by industrial and domestic effluents; (6) other engineers, especially agricultural and biomedical engineers, who require information on advances in the relevant sciences that could significantly affect the future practice of their professions; (7) administrators, particularly executives and legal advisors, who require information on national and international governmental regulations and guidelines on patents, environmental pollution, external aid programs and the control of raw materials and the marketing of products.

No work of this magnitude could have been accomplished without suitable assistance. For guidance on the master plan, I am indebted to the International Advisory Board (J. D. Bu'Lock, T. K. Ghose, G. Hamer, J. M. Lebault, P. Linko, C. Rolz, H. Sahm, B. Sikyta and H. Taguchi). For structuring details of the various sections, the invaluable assistance of the section editors is gratefully acknowledged, especially Alan Bull, Charles Cooney, Harvey Blanch and Campbell Robinson, who also acted as coordinators for each of the four volumes. For the individual chapters, the 250 authors are to be commended for their hard work and patience during the two years of preparation of the work. For checking the hundreds of literature references cited in the various chapters, the many graduate students are thanked for a tedious but important task well done. A special note of thanks is due to Jonathan and Arlene Lamptey, who acted as editorial assistants in many diverse ways. At Pergamon Press, I wish to thank Don Crawley for originally suggesting this project and Colin Drayton for managing it. Finally, I am pleased to note the favourable evaluations of the work by two distinguished authorities, Sir William Henderson and Nobel Laureate Donald Glaser, who provided a foreword and a guest editorial, respectively, to the treatise.

MURRAY MOO-YOUNG
Waterloo, Canada
December 1984

Foreword

This very comprehensive reference work on biotechnology is published ten years after the call by the National Academy of Sciences of the United States of America for a voluntary worldwide moratorium to be placed on certain areas of genetic engineering research thought to be of potential hazard. The first priority then became the evaluation of the conjectural risks and the development of guidelines for the continuation of the research within a degree of containment. There had hardly been a more rapid response to this type of situation than that of the British Advisory Board for the Research Councils. The expression of concern by Professor Paul Berg and the committee under his chairmanship, and the call for the moratorium, was published in *Nature* on 19 July 1974. The Advisory Board agreed at their meeting on the 26 July to establish a Working Party with the following terms of reference:

'To assess the potential benefits and potential hazards of the techniques which allow the experimental manipulation of the genetic composition of micro-organisms, and to report to the Advisory Board for the Research Councils.'

Because of the conviction of those concerned that recombinant DNA techniques could lead to great benefits, the word order used throughout the report of the Working Party (Chairman, Lord Ashby) always put 'benefits' before 'hazards'. The implementation of the recommendations led to the development of codes of practice. This was followed by the establishment of the Genetic Manipulation Advisory Group as a standing central advisory authority operating within the framework of the Health and Safety at Work, *etc.* Act 1974 and, later, more specifically within the framework of the Health and Safety (Genetic Manipulation) Regulations 1978. Similar moves took place in many other countries but the other most prominent and important activity was that of the US National Institutes of Health. This resulted in the adoption by most countries of the NIH or the UK guidelines, or the use of practices based on both.

The significant consequence of the debates, the discussions and of the recommendations that emerged during these early years of this decade (1974–1984) was that research continued, expanded and progressed under increasingly less restriction at such a pace that now makes it possible and necessary to devote the first Section of Volume 1 of this work to genetic engineering. Many chapters of the subsequent Sections and Volumes are of direct relevance to the application of genetic engineering.

The reason for identifying today's genetic engineering for first mention in this foreword is its novelty. It was being conceived barely more than ten years ago. Ten years by most standards is a short time. Although in the biological context it represents at least 10^4 generation times of the most vigorous viruses, it is less than one of man even for the most precocious. The current developments in biotechnology, whether they be in recombinant DNA, monoclonal antibodies, immobilized enzymes, *etc.* are mostly directed towards producing a better product, or a better process. This is commendable and is supportable by the ensuing potential commercial benefits. The newer challenge is the application of the new biotechnology to achieve what previously could scarcely have been contemplated. Limited biological sources of hormones, growth regulators, *etc.* are being, and will be increasingly, replaced by the use of transformed microorganisms, providing a vastly increased scale of production. Complete safety of vaccines by the absence of ineffectively inactivated virus is one of the great advantages of the genetically engineered antigen. This is quite apart from the ability to prepare products for which, at present, there is a technical difficulty or which is economically not feasible by standard methods.

A combination of advances in recombinant DNA research, molecular biology and in blastomere manipulation has provided the technology to insert genetic material into the totipotent animal cell. The restriction on the application of this technology for improved animal production is the lack of knowledge on the genetic control of desirable biological characteristics for transfer from one breed line to another.

There are probably greater potential benefits to be won in the cultivation of the domesticated plants than in the production of the domesticated animals. In both cases, the objectives are to increase the plant's or the animal's resistance to the prejudicial components of its environment and to increase the yield, quality and desired composition of the marketable commodities. These include the leaf, the tuber, the grain, the berry, the fruit or the milk, meat and other products of animal origin. This is not taking into account the other valuable products of horticulture, of oil or wax palms, rubber trees and forestry in general. Genetic engineering should be able to provide short-cuts to reach objectives attainable by traditional procedures, for example by by-passing the sequential stages of a traditional plant breeding programme by the transfer of the genetic material in one step. Examples of desirable objectives are better to meet user specifications with regard to yield, quality, biochemical composition, disease and pest resistance, cold tolerance, drought resistance, nitrogen fixation, *etc*. One of the constraints in this work in plants is the scarcity of vectors compared with the many available for the transformation of microorganisms. The highest research priority on the plant side is to determine by one means or another how to increase the efficiency of photosynthesis. The photosynthetic efficiency of temperate crop plants is no more than 2–2.5% in terms of conversion of intercepted solar energy. These plants possess the C_3 metabolic pathway with the energy loss of photorespiration. Tropical species of plants with the C_4 metabolic pathway have a higher efficiency of photosynthesis in that they do not photorespire. One approach for the breeder of C_3 plants is to endow them with a C_4 metabolism. If this transformation is ever to be achieved, it is most likely to be by genetic engineering. Such an advance has obvious advantages with regard, say, to increased wheat production for the ever-increasing human population. Nitrogen fixation as an agricultural application of biotechnology is given prominence in Section 1 of Volume 4. Much knowledge has been acquired about the chemistry and the biology of the fixation of atmospheric nitrogen. This provides a solid foundation from which to attempt to exploit the potential for transfer of nitrogen-fixing genes to crop plants or to the symbiotic organisms in their root systems. If plants could be provided with their own capability for nitrogen fixation, the energy equation might not be too favourable in the case of high yielding varieties. Without an increase in the efficiency of photosynthesis, any new property so harnessed would have to be at the expense of the energy requirements of existing characteristics such as yield.

Enzymes have been used for centuries in the processing of food and in the making of beverages. The increasing availability of enzymes for research, development and industrial use combined with systems for their immobilization, or for the immobilization of cells for the utilization of their enzymes, is greatly expanding the possibilities for their exploitation. Such is the power of the new biotechnology that it will be possible to produce the most suitable enzymes for the required reaction with the specific substrate. An increasing understanding at the molecular level of enzyme degradation will make it possible for custom-built enzymes to have greater stability than those isolated from natural sources.

The final section of Volume 4 deals with waste and its management. This increasingly voluminous by-product of our society can no longer be effectively dealt with by the largely empirical means that continue to be practised. Biological processes are indispensable components in the treatment of many wastes. The new biotechnology provides the opportunity for moving from empiricism to processes dependent upon the use of complex biological reactions based on the selection or the construction of the most appropriate cells or their enzymes.

The very comprehensive coverage of biotechnology provided by this four-volume work of reference reflects that biotechnology is the integration of molecular biology, microbiology, biochemistry, cell biology, chemical engineering and environmental engineering for application to manufacturing and servicing industries. Viruses, bacteria, yeasts, fungi, algae, the cells and tissues of higher plants and animals, or their enzymes, can provide the means for the improvement of existing industrial processes and can provide the starting points for new industries, for the manufacture of novel products and for improved processes for management of the environment.

SIR WILLIAM HENDERSON, FRS
Formerly of the *Agricultural Research Council*
and *Celltech Ltd., London, UK*

Guest Editorial

Since 1950, the new science of molecular biology has produced a remarkable outpouring of new ideas and powerful techniques. From this revolution has sprung a new discipline called genetic engineering, which gives us the power to alter living organisms for important purposes in medicine, agriculture and industry. The resulting biotechnologies span the range from the ancient arts of fermentation to the most esoteric use of gene splicing and monoclonal antibodies. With unprecedented speed, new scientific findings are translated into industrial processes, sometimes even before the scientific findings have been published. In earlier times there was a more or less one-way flow of new discoveries and techniques from scientific institutions to industrial organizations where they were exploited to make useful products. In the burgeoning biotechnology industry, however, developments are so rapid that there is a close intimacy between science and technology which blurs the boundaries between them. Modern industrial laboratories are staffed with sophisticated scientists and equipped with modern facilities so that they frequently produce new scientific discoveries in areas that were previously the exclusive province of universities and research institutes, and universities not infrequently develop inventions and processes of industrial value in biotechnology and other fields as well.

Even the traditional flow of new ideas from science to application is no longer so clear. In many applications, process engineers may find that the most economical and efficient process design requires an organism with new properties or an enzyme of previously unknown stability. These requirements often motivate scientists to try to find in nature, or to produce through genetic engineering or other techniques of molecular biology, novel organisms or molecules particularly suited for the requirements of production. A recent study done for the United States Congress* concluded that "in the next decade, competitive advantage in areas related to biotechnology may depend as much on developments in bioprocess engineering as on innovations in genetics, immunology, and other areas of basic science."

These volumes bring together for the first time in one unified publication the scientific and engineering principles on which the multidisciplinary field of biotechnology is based. Following accounts of the scientific principles is a large set of illustrations of the diverse applications of these principles in the practice of biotechnology. Finally, there are sections dealing with important regulatory aspects of the potential hazards of the growing field and of the need for promoting biotechnology in developing countries.

Comprehensive Biotechnology has been produced by a team of some of the world's foremost experts in various aspects of biotechnology and will be an invaluable resource for those wishing to build bridges between 'academic' and 'commercial' biotechnology, the ultimate form of any technology.

DONALD A. GLASER
University of California, Berkeley
and *Cetus Corp., Palo Alto, CA, USA*

*"Commercial Biotechnology: An International Analysis," Office of Technology Assessment Report, U.S. Congress, Pergamon Press, Oxford, 1984.

Executive Summary

In this work, biotechnology is interpreted in a fairly broad context: the evaluation and use of biological agents and materials in the production of goods and services for industry, trade and commerce. The underlying scientific fundamentals, engineering considerations and governmental regulations dealing with the development and applications of biotechnological processes and products for industrial, agricultural and medical uses are addressed. In short, a comprehensive but concise treatment of the principles and practice of biotechnology as it is currently viewed is presented. An outline of the main topics in the four volumes is given in Figure 1.

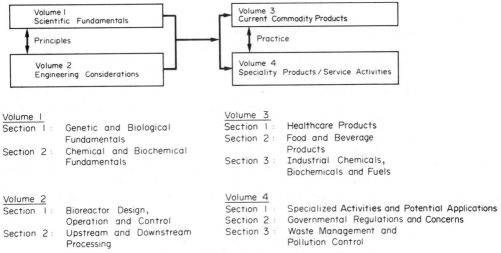

Volume I		Volume 3	
Section 1 :	Genetic and Biological Fundamentals	Section 1 :	Healthcare Products
Section 2 :	Chemical and Biochemical Fundamentals	Section 2 :	Food and Beverage Products
		Section 3 :	Industrial Chemicals, Biochemicals and Fuels

Volume 2		Volume 4	
Section 1 :	Bioreactor Design, Operation and Control	Section 1 :	Specialized Activities and Potential Applications
Section 2 :	Upstream and Downstream Processing	Section 2 :	Governmental Regulations and Concerns
		Section 3 :	Waste Management and Pollution Control

Figure 1 Outline of main topics covered

As depicted in Figure 2, it is first recognized that biotechnology is a multidisciplinary field having its roots in the biological, chemical and engineering sciences leading to a host of specialities, *e.g.* molecular genetics, microbial physiology, biochemical engineering. As shown in Figure 3, this is followed by a description of technical developments and commercial implementation,

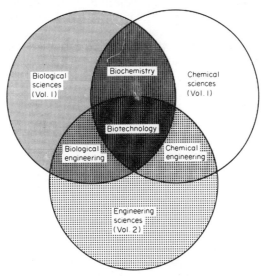

Figure 2 Multidisciplinary nature of biotechnology

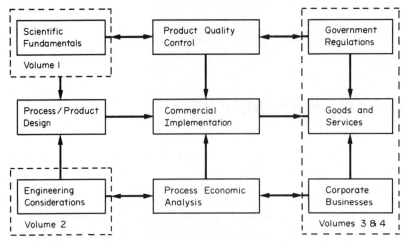

Figure 3   Interrelationships between biotechnology principles and applications

the ultimate form of any technology, which takes into account other important factors such as socio-economic and geopolitical constraints in the marketplace.

There are two main divisions of the subject matter: a pedagogical academic coverage of the disciplinary underpinnings of the field (Volumes 1 and 2) followed by a utilitarian practical view of the various commercial processes and products (Volumes 3 and 4). In the integration of these two areas, other common factors dealing with product quality, process economics and government policies are introduced at appropriate points throughout all four volumes. Since biotechnological advances are often ahead of theoretical understanding, some process descriptions are primarily based on empirical knowledge.

The four volumes are relatively self-contained according to the following criteria. Volume 1 delineates and integrates the unifying multidisciplinary principles in terms of relevant scientific fundamentals. Volume 2 delineates and integrates the unifying multidisciplinary principles of biotechnology in terms of relevant engineering fundamentals. Volume 3 describes the various biotechnological processes which are involved in the manufacture of bulk commodity products. Volume 4 describes various specialized services, potential applications of biotechnology and related government concerns. In each volume, a glossary of terms and nomenclature guideline are included to assist the beginner and the non-specialist.

This work takes into account the relative importance of the various topics, primarily in terms of current practice. Thus, bulk commodity products of the manufacturing industries (Volume 3) are accorded more space compared to less major ones and for potential applications (part of Volume 4). This proportional space distribution may be contrasted with the expectations generated by the recent news media 'biohype'. For example, virtually no treatment of 'biochips' is presented. In addition, since the vast majority of commercial ventures involve microbial cells and cell-derived enzymes, relatively little coverage is given to the possible use of whole plant or animal cells in the manufacturing industries. As future significant areas of biotechnology develop, supplementary volumes of this work are planned to cover them. In the meantime, on-going progress and trends will be covered in Pergamon's complementary review journal, *Biotechnology Advances*.

M. Moo-Young
University of Waterloo, Canada

Contributors to Volume 4

Professor D. R. Absolom
Department of Mechanical Engineering, University o Toronto, Toronto, Ontoario M5S 1A4, Canada

Dr D. W. Agar
ZAV/T-M 300, BASF, D-6700 Ludwigshafen, Federal Republic of Germany

Mr K. Allner
Environmental Microbiology Safety Reference Laboratory, Centre for Applied Microbiology and Research, Public Health Laboratory Service, Porton Down, Salisbury, Wiltshire SP4 0JG, UK

Professor R. Bakke
Civil Engineering Department, College of Engineering, Montana State University, Bozeman, MT 59717, USA

Dr K. O. Bayer
Law Offices of Pluymen & Bayer, 8140 Mopac Expressway, Westpark 2, Suite 150, Austin, TX 78759, USA

Ms S. Berch
Département d'Ecologie et Pedologie, Faculté de Foresterie et Géodesie, Université Laval, Quebec, Quebec G1K 7P4, Canada

Dr A. J. Biddlestone
Department of Chemical Engineering, University of Birmingham, PO Box 363, Edgbaston, Birmingham B15 2TT, UK

Dr A. G. Boon
Water Research Centre, Elder Way, Stevenage, Herts SG1 1TH, UK

Dr R. Briggs
23 Deards Wood, Knebworth, Herts SG3 6PG, UK

Dr J. D. Bryers
Department of Biochemical Engineering, Duke University, Durham, NC 27706, USA

Dr J. O. Carroll
NOVO Laboratories Inc., 59 Danbury Road, Wilton, CT 06897, USA

Professor T. M. S. Chang
Artificial Cells and Organs Research Centre, McGill University, 3655 Drummond Street, Montreal, Quebec H3G 1Y6, Canada

Professor W. G. Characklis
Civil Engineering Department, College of Engineering, Montana State University, Bozeman, MT 59717, USA

Mr P. F. Cooper
Water Research Centre, Elder Way, Stevenage, Herts SG1 1TH, UK

Dr B. Danielsson
Department of Pure and Applied Biochemistry, Lund Institute of Technology, Chemical Centre, PO Box 740, S-220 07 Lund, Sweden

Dr E. J. DaSilva
Division of Scientific Research and Higher Education, UNESCO, 7 place de Fontenoy, 75700 Paris, France

Dr A. L. Downing
Binnie and Partners, Artillery House, Artillery Row, London SW1P 1RX, UK

Dr S. J. Duff
Rideau Falls Laboratory, Division of Biological Sciences, National Research Council of Canada, Ottawa, Ontario K1A 0R6, Canada

Dr G. Eisenbarth
Joslin Diabetes Center, Harvard Medical School, 1 Joslin Place, Boston, MA 02215, USA

Professor D. W. Emerich
Department of Biochemistry, University of Missouri-Columbia, Columbia, MO 65211, USA

Professor K.-E. Eriksson
Swedish Forest Products Research Laboratory, Box 5604, S-114 86 Stockholm, Sweden

Dr S. Esumi
Research and Development, Kaken Pharmaceutical Co Ltd, 2-28-8 Honkomazome, Bunkyo-ku, Tokyo 113, Japan

Professor A. H. Fraenkel
Department of Preventive Medicine, Ohio State University, Columbus, OH 43210, USA

Dr K. R. Gray
Department of Chemical Engineering, University of Birmingham, PO Box 363, Edgbaston, Birmingham B15 2TT, UK

Dr G. Gregoriadis
MRC Group, Academic Department of Medicine, Royal Free Hospital School of Medicine, Pond Street, London NW3 2QG, UK

Dr J. Gregory
Civil Engineering Department, University College London, Gower Street, London WC1E 6BT, UK

Professor G. Hamer
ETH/EAWAG, CH-8600 Dübendorf-Zürich, Switzerland

Mr H. A. Hawkes
Department of Civil Engineering, University of Birmingham, PO Box 363, Edgbaston, Birmingham B15 2TT, UK

Dr C. G. Heden
Microbiological Resources Centre (MIRCEN), Karolinska Institute, Fack, S-10401 Solnavagen, Stockholm, Sweden

Dr E. C. Hill
E. C. Hill and Associates, Unit M22, Cardiff Workshops, Lewis Road, East Moors, Cardiff CF1 5EG, UK

Professor J. A. Howell
Department of Biochemical Engineering, University of Bath, Claverton Down, Bath BA2 7AY, UK

Professor R. L. Irvine
Department of Civil Engineering, University of Notre Dame, Notre Dame, IN 46556, USA

Dr T. R. Jack
NOVA Husky Research Corporation, 1411 25th Avenue NE, Calgary, Alberta T2E 7L6, Canada

Dr B. E. Jank
Environment Canada, Wastewater Technology Centre, PO Box 5050, 867 Lakeshore Drive, Burlington, Ontario L7R 4A6, Canada

Dr D. W. M. Johnstone
Sir William Halcrow and Partners, Burderop Park, Swindon, Wilts, SN4 0QD, UK

Dr G. L. Jones
Water Research Centre, Elder Way, Stevenage, Herts SG1 1TH, UK

Dr K. Kano
Department of Immunology, Institute of Medical Science, University of Tokyo, 4-6-1 Shiro-ganedai, Minatoku, Tokyo 108, Japan

Professor B. Kendrick
Department of Biology, University of Waterloo, Waterloo, Ontario N2L 3G1, Canada

Dr K. J. Kennedy
Rideau Falls Laboratory, Division of Biological Sciences, National Research Council of Canada, Ottawa, Ontario K1A 0R6, Canada

Dr T. K. Kirk
Forest Products Laboratory, PO Box 5130, Madison, WI 53705, USA

Dr J. S. Knapp
Department of Microbiology, University of Leeds, Leeds LS2 9JT, UK

Professor S. Krimsky
Department of Urban & Environmental Policy, Tufts University, Medford, MA 02155, USA

Dr R. P. Lanzilotta
Codon, 430 Valley Drive, Brisbane, CA 94005, USA

Dr P. J. Larkin
Division of Plant Industry, CSIRO, PO Box 1600, Canberra City, ACT 2601, Australia

Professor V. Linek
Department of Chemical Engineering, Prague Institute of Chemical Technology, Suchbatarova 5, 166 28 Praha 6, Czechoslovakia

Professor R. G. McDaniel
Department of Plant Sciences, University of Arizona, Tucson, AZ 85721, USA

Professor T. O. McGarity
School of Law, University of Texas, Austin, TX 78701, USA

Dr H. Melcer
Biological Processes Section, Environment Canada, Wastewater Technology Centre, Canada Centre for Inland Waters, PO Box 5050, 867 Lakeshore Drive, Burlington, Ontario L7R 4A6, Canada

Professor B. S. Montenecourt
Department of Biology, Williams Hall No 31, Lehigh University, Bethlehem, PA 18015, USA

Dr E. R. Nestmann
Mutagenesis Section, Environmental and Occupational Toxicology Division, Health Protection Branch, Environmental Health Centre 308, Tunney's Pasture, Ottawa, Ontario K1A 0L2, Canada

Professor A. W. Neumann
Department of Mechanical Engineering, University of Toronto, Toronto, Ontario M5S 1A4, Canada

Mr N. E. Nituch
Ayerst Laboratories, 1025 Laurentian Blvd, Saint-Laurent, Quebec H4R 1J6, Canada

Dr M. Okada
Laboratory of Freshwater Environment, National Institute for Environmental Studies, PO Box Tsukuba-Gakuen, Tsukuba, Ibaraki 305, Japan

Dr G. Olsson
Department of Automatic Control, Lund Institute of Technology, S-22007 Lund, Sweden

Dr H. A. Painter
Water Research Centre, Elder Way, Stevenage, Herts SG1 1TH, UK

Dr J. M. Parry
Department of Genetics, University College of Swansea, Singleton Park, Swansea SA2 8PP, UK

Dr E. B. Pike
Water Research Centre, Henley Road, Medmenham, PO Box 16, Marlow, Bucks SL7 2HD, UK

Professor K. E. Porter
Department of Chemical Engineering, University of Aston in Birmingham, Gosta Green, Birmingham B4 7ET, UK

Professor B. J. Ralph
School of Metallurgy, University of New South Wales, PO Box 1, Kensington, NSW 2033, Australia

Dr J. F. Rees
Biotechnica Ltd, 5 Chiltern Close, Cardiff Industrial Park, Llanishen, Cardiff CF4 5DL, UK

Professor C. W. Robinson
Department of Chemical Engineering, University of Waterloo, Waterloo, Ontario N2L 3G1, Canada

Professor H. W. Rossmoore
Department of Biological Sciences, Wayne State University, Detroit, MI 48202, USA

Mr R. Saliwanchik
8753 Merrimac, Richland, MI 49083, USA

Professor E. D. Schroeder
Civil Engineering Department, University of California, Davis, CA 95616, USA

Dr W. R. Scowcroft
Division of Plant Industry, CSIRO, PO Box 1600, Canberra City, ACT 2601, Australia

Professor N. S. Scrimshaw
The United Nations University, Hunger, Health & Society Sub-programme, MIT 20-A 201, Cambridge, MA 02139, USA

Dr S. E. Shumate II
Engenics Inc, 3760 Haven Avenue, Menlo Park, CA 94025, USA

Dr J. Sinkule
Department of Chemical Engineering, Prague Institute of Chemical Technology, Suchbatarova 5, 166 28 Praha 6, Czechoslovakia

Dr L. A. Slotin
Policy, Planning and Program Development, Medical Research Council of Canada, Jeanne Mance Building, Tunney's Pasture, Ottawa, Ontario K1A 0W9, Canada

Professor J. E. Smith
Department of Bioscience and Biotechnology, University of Strathclyde, 204 George Street, Glasgow G1 1XW, UK

Dr R. Stace-Smith
Plant Pathology Section, Agriculture Canada, Research Station, 6660 NW Marine Drive, Vancouver, British Columbia V6T 1X2, Canada

Dr D. A. Stafford
CLEAR Ltd, Unit M40, Lewis Road, East Moors, Cardiff CF1 5EG, UK

Dr E. I. Stentiford
Department of Civil Engineering, University of Leeds, Leeds LS2 9JT, UK

Dr J. P. Stephenson
Canviro Consultants Ltd, 178 Louisa Street, Kitchener, Ontario N2H 5M5, Canada

Dr G. W. Strandberg
Oak Ridge National Laboratory, PO Box X, Oak Ridge, TN 37830, USA

Dr D. A. Stringer
UNICELPE, Rue du Collège St.-Michel 10, 1150 Brussels, Belgium

Dr R. Sudo
Laboratory of Freshwater Environment, National Institute for Environmental Studies, PO Box Tsukuba-Gakuen, Tsukuba, Ibaraki 305, Japan

Professor M. S. Switzenbaum
Department of Civil Engineering, University of Massachusetts, Amherst, MA 01003, USA

Professor M. G. Trulear
Civil Engineering Department, College of Engineering, Montana State University, Bozeman, MT 59717, USA

Dr V. Vacek
Department of Chemical Engineering, Prague Institute of Chemical Technology, Suchbatarova 5, 166 28 Praha 6, Czechoslovakia

Dr L. van den Berg
Rideau Falls Laboratory, Division of Biological Sciences, National Research Council of Canada, Ottawa, Ontario K1A 0R6, Canada

Professor J. D. Wall
Department of Biochemistry, University of Missouri-Columbia, Columbia, MO 65211, USA

Professor H. Y. Wang
Department of Chemical Engineering, University of Michigan, Ann Arbor, MI 48109, USA

Dr E. M. Waters
Department of Genetics, University College of Swansea, Singleton Park, Swansea SA2 8PP, UK

Dr M. J. D. White
Process Engineering Research and Technology Fasson (Nederland) bV, Lammenschansweg 140, Leiden, Netherlands

Professor L. B. Wingard, Jr.
Department of Pharmacology, School of Medicine, University of Pittsburgh, Pittsburgh, PA 15261, USA

Professor W. Zingg
Institute of Biomedical Engineering, University of Toronto, Toronto, Ontario M5S 1A4, Canada

Contents of All Volumes

Section 2 Upstream and Downstream Processing

Volume 3 The Practice of Biotechnology: Current Commodity Products
Section 1 Healthcare Products

SPECIALIZED ACTIVITIES AND POTENTIAL APPLICATIONS

1

Introduction

C. W. ROBINSON
University of Waterloo, Ontario, Canada

The first three volumes of this treatise deal, for the most part, with fundamental biotechnologies and their applications to well-established or recently-implemented commercial processes. **Volume 4, Section 1** deals with a number of biotechnologies and bioprocesses which, while neither yet having reached the commercial production stage nor yet having found widespread industrial or medical application, nonetheless show considerable promise of doing so within the next few years.

Our choice of topics for inclusion in the sub-sections on *Biomedical and Chemotherapeutic Applications, Agricultural and Forestry Applications* and *Process Applications* in each case should not be considered as necessarily being a complete, definitive list of all biotechnologies currently under development which have a high probability of eventual commercial success or more extensive application. Given the rapid pace of new developments in biotechnology today, undoubtedly other useful processes, products and applications are on the horizon, or may even have become public knowledge while this treatise was going to press. On the other hand, due to the economic uncertainties existent worldwide today, some of the processes, products or applications now perceived as being promising may turn out to be economically impractical under future conditions. However, we do believe that our choices of the specific biotechnology-based processes, products and applications for inclusion in **Section 1** are sound in view of present-day knowledge, societal needs and technico-economic trends.

Section 1 also provides the reader with fundamental knowledge of many application techniques for those *Analytical Methods and Instruments* which find common use throughout the biotechnologies, both on the experimental bench or pilot-plant scale (biological kinetics and transport processes evaluation and modelling) and on the production scale (process monitoring and control).

At both the developmental (bench or pilot-plant scale research) and commercial implementation stages of a new biotechnological process, due care and attention must be given to the identification and containment of known or potential biohazards in order to protect the health and safety of both employees and the general public. Background information on these important areas of biotechnology and recommended plant or laboratory design and operating procedures are given in the *Detection and Containment of Biohazards* sub-section. Readers should consult **Section 2**, Chapters 27–30, for details on some specific biohazards of particular concern in bioproducts and bioprocesses; Chapters 31–33 of that section cover various governmental regulations pertaining to the manufacturing use, commercial applications and disposal of potentially hazardous or actually hazardous biological substances or organisms. Chapters 23 and 24 in this section describe and discuss the present technological bases for hazard assessment and, where necessary, hazard containment in order to accommodate the concerns and regulations described in **Section 2**.

The reader's attention is also drawn to Chapter 26 on *Patenting* in **Section 2**. The procedures associated with the filing of a patent application are established as governmental 'regulations'. However, successful attainment (and, if necessary, defence) of a patent is highly dependent upon the careful implementation of several in-house actions right from the start of the project, which actions can be viewed as a type of specialized activity.

2

Pharmacokinetics

L. B. WINGARD, Jr.
University of Pittsburgh, PA, USA

2.1 INTRODUCTION

Pharmacokinetics is the quantitative description of how the concentration of a compound of exogenous origin in specific body fluids or tissues varies with time. At any instant the concentration of the compound in a particular body fluid depends on several factors; these include competing rate processes for input and disappearance of a drug, relative volumes of different body fluids and tissues in which the compound distributes, and relative solubility of the compound in each of these fluids and tissues. The quantitative relationships between these factors are discussed in detail in Sections 2.3.2.1 and 2.3.3.1. The application of these quantitative relationships makes use of several pharmacokinetic parameters; the main parameters are described in Sections 2.2 and 2.3.2.2.

The concepts of pharmacokinetics apply to any exogenous material, whether it be a therapeutically beneficial drug or a toxicologically harmful poison. Pharmacokinetic modelling normally is applied to mammals, although many of the concepts could be applied also to fish, reptiles and even plants. In the therapeutic use of drugs, the objective usually is to maintain the plasma drug concentration roughly between a so-called 'minimum effective' level and an upper 'toxic' level. A knowledge of the pharmacokinetic parameters for a given drug aids markedly in selecting the rate of administration for individual patients so as to achieve the desired concentration. Reduced input rates are especially important with patients having impaired capabilities for eliminating the drug from the body. With toxic agents the objective is reversed, with the goal being to reduce the plasma concentration of the toxic material to less than the minimum effective level as quickly as possible. A knowledge of the pharmacokinetic parameters for the toxic agent provides a firm basis for predicting how long the detoxification will take for alternative modes of treatment.

Drugs as well as toxic materials produce pharmacologic or toxic effects, with the intensity of the effect often related to the concentration of the drug. The variation in the intensity of effect with time is known as pharmacodynamics. Some people use the term pharmacokinetics to mean the time variation of both concentration and pharmacologic effect. Other people follow the more traditional nomenclature, wherein pharmacokinetics refers to concentrations and pharmacodynamics to effects. This chapter is concerned only with concentrations and thus uses the traditional definitions.

The overall sequence of processes that drug molecules may undergo from administration to eventual elimination is shown schematically in Figure 1. In plasma many drugs bind reversibly to albumin and globulins, thus markedly reducing the plasma concentration of unbound ('free') drug. This is an important concept, since only the free drug can cross capillary walls and reach cellular tissues or undergo metabolism and renal elimination. The term metabolism here means the chemical conversion of the drug to another compound known as a metabolite. Some metabolites have high pharmacologic or toxic activity, while others are essentially inert. Unless otherwise stated, experimentally measured concentrations of drug mean the free unbound plus the bound drug. The experimental determination of the concentration of unbound drug is very difficult to do and for some highly bound drugs requires sensitivities beyond the capabilities of present day methodology.

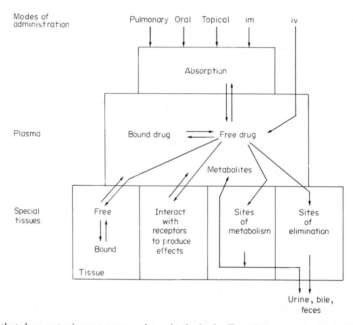

Figure 1 Processes that drug or toxic agent may undergo in the body. Free drug means not bound; im and iv are intramuscular and intravenous injection modes of administration. Gaseous anesthetic agents and air pollutants enter through the pulmonary system

Several reviews are available that cover pharmacokinetics from the standpoints of clinical use (Rowland and Tozer, 1980), mathematical development (Gibaldi and Perrier, 1975), applications in the design of sustained release preparations (Kwan, 1978), and applications in industrial and environmental toxicology (Gehring *et al.*, 1976).

2.2 MAIN PARAMETERS AND AREAS OF APPLICATION

In applying pharmacokinetic concepts to the therapeutic use of drugs or elimination of toxicologic agents, there are three parameters that are particularly useful. These are the elimination half-life, the apparent volume of distribution, and the clearance. These parameters are described here in physical terms and used later in Section 2.3.2.2.

The elimination half-life is the length of time it takes for the plasma concentration to be cut in half, following a single dose or after stopping multiple dosing of a drug. For many drugs the half-life is independent of the concentration, while for others the half-life may increase as the concentration of the drug decreases. These points are described further in Section 2.3.2.2. Actual half-lives range from a few minutes to a couple of days; a knowledge of the half-life is very useful in deciding on a practical dosing schedule for a specific drug.

The apparent volume of distribution is the second parameter of interest. There are several techniques for mathematical analysis of experimental pharmacokinetic data to give values which have the units of volume. These values are called the apparent volumes in which the drug appears to be distributed. Actual apparent volumes of distribution range from 5–40 000 l. In a standard size (70 kg) adult human, the plasma occupies about 3 l and the total body water about 45 l. The explanation for how the apparent volume of distribution can greatly exceed the total body volume is discussed in Section 2.3.2.2.

The third parameter, and possibly the most useful, is called the total body clearance. Clearance is defined as the rate of drug elimination (mg min^{-1}) divided by the plasma concentration of drug (mg ml^{-1}). Thus, clearance has the units of ml min^{-1}. Conceptually it represents the volume of plasma that would need to be totally voided of drug each minute in order to attain the actual number of milligrams of drug removed per minute. This is a term borrowed from physiology, where it is used to define the ability of the kidney to remove materials from the blood. For example, a drug eliminated primarily by passage through the renal glomeruli (filters) and not undergoing subsequent reabsorption has a clearance of about 120 ml min^{-1} for a normal functioning pair of human kidneys.

For many drugs pharmacokinetic considerations are used only during the initial introduction of the drug into clinical use, and then mainly to determine a suitable dosing schedule (how much and how often) for the average sized patient. This schedule then can be adjusted, based on body weight or body surface area, for larger or smaller patients. These drugs are relatively easy to use because there is considerable latitude between the dose or concentration needed to produce the therapeutic effect and the dose or concentration that causes toxicity. However, there are several dozen drugs where the ratio of the toxic to the therapeutic dose or concentration is only 1.2–2.0. With so narrow a range of usable concentrations, pharmacokinetic considerations need to be applied more rigorously in developing safe and effective dosing schedules for individual patients. Examples of drugs with a narrow range of useful concentrations are the cardiac agents digoxin, lidocaine and propranolol, and the antiasthmatic bronchiodilator theophylline. Another important area of application for pharmacokinetics is in cases of impaired renal function, where the glomerular filtration clearance has dropped from the normal 120 ml min^{-1} to perhaps only 20 ml min^{-1}. Here the normal dose of 50 mg every 6 hours of a renal eliminated drug having a normal half-life of 3 hours must be reduced to 50 mg every 24 hours (Schönebeck *et al.*, 1973). When two or more drugs are given concurrently, there is a significant possibility for an interaction to produce a magnified effect that can be lethal if adjustments are not made in the dosing schedule. Pharmacokinetic considerations often are a major factor in deciding what dosing adjustments to make (Rowland and Tozer, 1980). Several informative examples of how pharmacokinetics can aid in evaluation of the toxicology and hazardous nature of industrial chemicals, *e.g.* 2,4-dinitrophenol, vinyl chloride and dioxane, are described by Gehring *et al.* (1976). Pharmacokinetics are also important in assessing the toxicologic and teratogenic hazards in administering drugs during certain periods of pregnancy, since many drugs are transferred to the fetus. This transfer is exemplified by the muscle relaxant pancuronium bromide that is often used during caesarean section delivery in humans (Abouleish *et al.*, 1980).

2.3 METHODS IN PHARMACOKINETICS

2.3.1 Available Experimental Samples

The key elements that govern the successes as well as the limitations of pharmacokinetics are the type, accuracy and quantity of experimental data that can be obtained. Values for the parameters mentioned in the previous section cannot be obtained without experimental data. The suggestion that a particular mathematical expression should be especially appropriate for describing the plasma concentration X of drug *versus* time means little unless the expression can be tested against experimental data. Herein lies the most difficult aspect of pharmacokinetics, namely devising methodology for obtaining the needed experimental measurements. This is a major limitation in human pharmacokinetic applications since sampling sites are limited to urine and plasma, with very infrequent justification for cerebrospinal fluid taps, bone marrow smears and liver or other tissue samples. Even the taking of plasma samples needs to be considered from the patient's standpoint and cannot be done simply to get a few more data points.

Pharmacokinetic parameters can vary 10- to 20-fold among different individuals for a given drug. Therefore, in the experimental determination of half-life and other pharmacokinetic parameters, it is highly preferable to get a complete set of concentration *versus* time samples from each subject. The alternative method of using one group of subjects for the concentrations at the first time point and other groups of subjects for concentrations at additional time points is much less desirable, but often must be used because of limitations in methodology.

2.3.2 Compartment Models Without Anatomical Significance to Compartments

2.3.2.1 *Mathematical formulation of problems and solutions*

The major approaches that are currently used in pharmacokinetic modelling are deterministic rather than stochastic. The approach described in this section can be applied to mammals, reptiles, fish, plants and essentially any type of organism where samples can be obtained to show experimentally how the concentration varies with time. Another approach, described in Section 2.3.3, can be applied only to organisms that have a circulatory system.

In the approach outlined in this section, the organism is described by one, two, or perhaps three well-mixed open compartments. In engineering terminology such compartments are known as well-mixed stirred tank reactors. The compartments are fitted with inlet (drug administration) and outlet (drug elimination and metabolism) streams and with reversible interconnections, as shown in Figure 2. First order processes are assumed to describe elimination, metabolism and transfer of the drug between compartments, although the case with mixed order non-linear Michaelis–Menton metabolism also will be described. A material balance equation is written for the drug for each compartment for a finite element of time. In the following examples, the input is assumed to be a single dose added instantaneously at time = zero to compartment number 1.

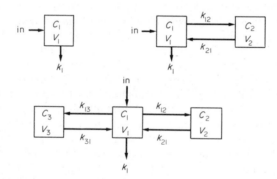

Figure 2 One-, two- or three-compartment models. C is concentration, V is volume of the compartment and the k symbols represent first order rate constants for transfer out of or between compartments

1 compartment:

$$V_1(\mathrm{d}C_1/\mathrm{d}t) = -k_1C_1V_1 \tag{1}$$

2 compartments:

$$V_1(\mathrm{d}C_1/\mathrm{d}t) = -k_1C_1V_1 - k_{12}C_1V_1 + k_{21}C_2V_2 \tag{2}$$
$$V_2(\mathrm{d}C_2/\mathrm{d}t) = k_{12}C_1V_1 - k_{21}C_2V_2 \tag{3}$$

3 compartments:

$$V_1(\mathrm{d}C_1/\mathrm{d}t) = -k_1C_1V_1 - k_{12}C_1V_1 + k_{21}C_2V_2 - k_{13}C_1V_1 + k_{31}C_3V_3 \tag{4}$$
$$V_2(\mathrm{d}C_2/\mathrm{d}t) = k_{12}C_1V_1 - k_{21}C_2V_2 \tag{5}$$
$$V_3(\mathrm{d}C_3/\mathrm{d}t) = k_{13}C_1V_1 - k_{31}C_3V_3 \tag{6}$$

where C = concentration in compartment 1, 2 or 3 (mg ml^{-1}); V = volume of compartment 1, 2 or 3 (ml); k = first order rate constant for elimination-metabolism (k_1) or for transfer between compartments (example k_{12}; min^{-1}); D = dose (mg); C_1^0 = concentration in compartment 1 at time = zero.

For each case the equations are solved simultaneously, for example by Laplace transforms, to give the concentrations as functions of time. For humans and other mammals, the pathways and organs for drug elimination and metabolism as well as the circulatory system are assumed to be in compartment number 1. Thus, $C_1(t)$ is really the only concentration of interest since the anatomical meanings of C_2 or C_3 are not known.

The analytical solutions for $C_1(t)$ in the first two cases are as follows:

1 compartment:

$$C_1(t) = C_1^0 e^{-k_1 t} \tag{7}$$

$$\text{or } \log[C_1(t)] = \log C_1^0 - (k_1 t/2.303) \tag{8}$$

2 compartments:

$$C_1(t) = Ae^{-\alpha t} + Be^{-\beta t} \tag{9}$$

$$A = C_1^0(\alpha - k_{21})/(\alpha - \beta) \tag{10}$$

$$B = C_1^0(k_{21} - \beta)/(\alpha - \beta) \tag{11}$$

$$\alpha + \beta = k_1 + k_{12} + k_{21} \tag{12}$$

$$\alpha\beta = k_1 k_{21} \tag{13}$$

The solution to the 3-compartment case is more complex. The general form of the solution is shown in equation (14); the reader is referred to Gibaldi and Perrier (1975) for detailed definitions of A, B, G, α, β and γ.

$$C_1(t) = Ae^{-\alpha t} + Be^{-\beta t} + Ge^{-\gamma t} \tag{14}$$

2.3.2.2 *Parameter evaluation and comparison between simulated and experimental results*

The general method for obtaining parameter values is by curve-fitting the equations against experimental data for plasma concentration *versus* time. In curve-fitting, the parameter values are adjusted to give the best fit between the experimental data and the curves from the equations.

An example is shown in Figure 3 to demonstrate how the fit can be improved by adding more compartments (Wingard and Cook, 1976). Three compartments are usually the maximum number employed with pharmacokinetic studies in humans because the limited accuracy of the experimental data cannot justify additional compartments. In this example, the neuromuscular blocking agent (+)-tubocurarine was administered in a single intravenous dose of 0.30 mg kg^{-1} of body weight to five human subjects. Blood samples were taken at the times indicated and assayed for (+)-tubocurarine. The experimental data were then fit, using the PROPHET computer system and the MLAB modelling routine, against one-, two- and three-compartment models, *i.e.* against equations (7), (9) and (14). In the fitting the data points were weighted as the inverse of their variance. The fitted equations were as follows for the 1-, 2- and 3-compartment models:

$$C_1(t) = 0.821e^{-0.00244t} \tag{15}$$

$$C_1(t) = 5.24e^{-0.334t} + 1.11e^{-0.00701t} \tag{16}$$

$$C_1(t) = 7.17e^{-0.741t} + 1.66e^{-0.0704t} + 0.808e^{-0.00456t} \tag{17}$$

2.3.2.3 *Additional aspects of one-compartment models*

The concepts covered in this section apply equally well to multi- and single-compartment models. However, the mathematics are much more complex with multicompartment models, so the simpler case is used so as not to lose sight of the ideas due to the more complex mathematics.

If the data fit the one-compartment model (equation 7) then there is no point in using the more complex models. From equation (7), values for the elimination-metabolism half-life, an apparent volume of distribution and the total body clearance can be calculated. The anticoagulant dicumarol in rats follows single-compartment pharmacokinetics at low doses, as shown in Figure 4. The half-life can be read directly from a plot of log C_1 *versus* t (equation 8; about 7 hours from Figure 4), or it can be calculated from the slope of the same plot by the expression

$$t_{1/2} = 0.693/k_1 \tag{18}$$

Note that in Figures 3 and 4 the plot gives a straight line for the one-compartment model. With such a linear relationship, the half-life becomes independent of the plasma concentration of drug. For the two- and three-compartment models the plots of log C_1 *versus* time are curved; in these cases the half-life increases as the concentration of drug diminishes. This is attributed to the drug

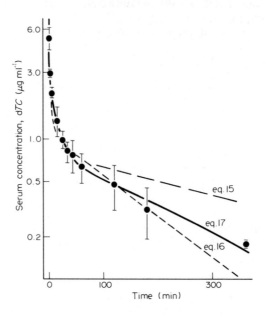

Figure 3 Serum concentration of (+)-tubocurarine at different times following intravenous injection of 0.30 mg kg^{-1} in five human subjects. Datum point of 0.030 μg ml^{-1} at 1440 min not shown but was used in the fitting. The data are shown as means ± SD. The lines represent the best fits: 1-compartment, equation (15); 2-compartment, equation (16); 3-compartment, equation (17) (Wingard and Cook, 1976)

returning, from poorly perfused tissue beds or from tissues to which the drug is strongly bound, to be metabolized and eliminated in the hepatic–renal system.

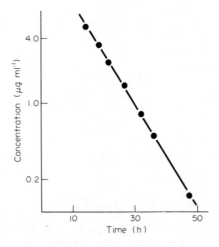

Figure 4 Disappearance of ^{14}C-dicumarol after intraperitoneal injection of 20 mg kg^{-1} in rat No. 6-3 from the study by Wingard and Levy (1973)

The calculation of the apparent volume of distribution is most easily described for the one-compartment model. The apparent volume of distribution V_1 can be defined as the amount of drug in the body divided by the plasma concentration of drug. By extrapolating the log C_1 *versus t* plot to time zero, the one-compartment model intercept gives a value of 26.0 μg ml^{-1}. At zero time all of the dose is in the plasma, so that $V_1 = D/C_1^0$. From Figure 4, 20 mg kg^{-1} divided by 0.026 mg ml^{-1} gives $V_1 = 0.77$ ml kg^{-1}. It was mentioned earlier that some drugs have apparent volumes of distribution greatly in excess of the actual total body volume. This arises when a drug is highly bound to tissue sites. On extrapolation of the log C_1 *versus* time plot (equation 8) back to time zero, a much smaller C_1^0 value is obtained than when the drug is not bound to tissue. When the dose is divided by C_1^0, the result is an unrealistically large apparent volume of distribution for highly tissue-bound drugs.

Total body clearance was defined earlier as the rate of removal of drug from the plasma divided by the plasma concentration of drug. From equation (1) the term $k_1V_1C_1$ also stands for the rate of disappearance of drug from the plasma. Thus, the clearance must be the same as k_1V_1 and can be calculated if values are available for those two parameters. The clearance can also be calculated directly by rearranging the above definition to give:

$$dX/dt = (\text{clearance})_p C_1(t) \tag{19}$$

where X is the amount of drug and the subscript p refers to total clearance from the plasma. Rearrangement of equation (19) and integration gives:

$$\int_0^D dX = D = (\text{clearance})_p \int_0^\infty C_1(t)dt \tag{20}$$

The integral in equation (20) is evaluated by taking the area under the curve of C_1 *versus* time.

The elimination of drug by renal mechanisms and the metabolism of drug by enzymatic processes generally follow first order kinetics. However, the special case of drug metabolism at very high drug concentrations, such that the drug metabolizing enzymes are saturated, deserves special mention. With Michaelis–Menton drug metabolism kinetics, and neglecting any elimination by first order kinetics, equation (1) becomes:

$$V_1(dC_1/dt) = -v_m C_1/(K_m + C_1) = V_1 v \tag{21}$$

where v_m is the maximum enzyme-catalyzed reaction velocity (mg min^{-1}), K_m is the Michaelis constant (mg ml^{-1}) and v denotes the rate expression dC_1/dt. Rearrangement of equation (21) results in a linear relationship, when $C_1 v^{-1}$ is plotted against C_1, from equation (22).

$$K_m + C_1 = v_m C_1/(V_1 v) \tag{22}$$

An example of equation (22) is shown by the saturation of salicylate metabolism in humans (Levy *et al.*, 1972). When the enzymes are saturated, the disappearance of drug occurs according to zero order kinetics.

The above pharmacokinetic modelling concepts are all presented for intravenous administration of a single dose of drug. Oral administration and continuous or multiple dosing are also quite important, although space does not allow for development of the equations in each case.

For oral administration with single-compartment kinetics, the material balance equation becomes:

$$V_1(dC_1/dt) = -k_1 V_1 C_1 + k_a C_a \tag{23}$$

where k_a is the first order rate constant for absorption and C_a the concentration of drug at the site of absorption. Actual absorption kinetics often approximate to first order kinetics, so the use of that assumption for equation (23) is realistic. Equation (23) can be solved, with the initial condition that C_1 is zero at time zero, to give equation (24). The fraction absorbed, F, is termed the bioavailability. A typical graph of equation (24) is shown in Figure 5.

$$C_1(t) = k_a FD/[V_1(k_a - k_1)][e^{-k_1 t} - e^{-k_a t}] \tag{24}$$

Multiple dosing is the more usual therapeutic situation as compared to single acute doses. However, the mathematical complexity is much greater for multiple than single dosing, therefore the principles described earlier for acute dosing will now be summarized for multiple dosing. If the interval between doses is long compared to the elimination-metabolism half-life, then each dose behaves pharmacokinetically like an isolated single dose. This is because there is no drug remaining from the previous dose when the next one is administered. At the other extreme is the situation where the time interval between doses is negligible, as with a drug solution being infused, *e.g.* pumped or dripped through an indwelling needle, continuously into the patient. The material balance equation and its solution for a single-compartment void of drug at time zero and with a constant rate of infusion, k_0, are, respectively:

$$V_1(dC_1/dt) = k_0 - k_1 V_1 C_1 \tag{25}$$
$$C_1(t) = (k_0/V_1 k_1)(1 - e^{-k_1 t}) \tag{26}$$

Equation (26) is shown in Figure 6a. If the dosing interval T is less than the disappearance half-life, then the curve of C_1 *versus* time looks like that in Figure 6b. In both cases depicted in Figure 6 the concentration C_1 builds up until the average rate of drug input equals the average rate of elimination-metabolism. This occurs because the rate of elimination-metabolism increases with concentration. The detailed pharmacokinetic calculations for the Figure 6b case are given else-

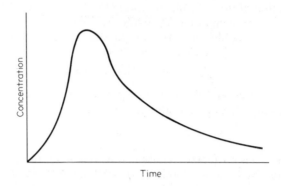

Figure 5 Representative time course of plasma drug concentration following oral administration of drug, as per equation (24)

where (Gibaldi and Perrier, 1975). The magnitude of the steady-state or plateau level is determined by the ratio of the rate of input divided by the clearance, as indicated in Figure 6. Thus, any change in the dose, the dosing interval, or the clearance will cause eventual establishment of a new plateau level, unless a compensating change is made in one of the other variables. Another important consideration is the length of time needed to reach the plateau level. It turns out that the time to reach steady-state depends only on the half-life for the drug. The fractional attainment of the steady-state concentration, C_{ss}, is presented in terms of the number of half-lives, n, as follows:

n	C_1/C_{ss}
1	0.50
2	0.75
3	0.875
4	0.938
5	0.969

Therefore, a drug with an elimination-metabolism half-life of 14 hours will take 42 hours to reach 88% of the plateau level if no loading dose is given. In clinical situations where such a long wait would be dangerous for the patient, then a loading dose is administered to get the plasma concentration hopefully to 50–75% of the plateau level in only a few minutes.

2.3.3 Compartment Models With Anatomical Significance to Compartments

2.3.3.1 *Mathematical formulation of problems and solutions*

This approach to compartmental pharmacokinetic modelling is based on giving each compartment specific anatomical significance and then connecting the compartments by the circulatory system. In this modelling system drug molecules move from one compartment to another as a result of convective blood flow. The actual blood or tissue volumes of each compartment can be determined by independent measurements. One of the earliest attempts to apply this 'flow' or 'physiological' modelling concept to humans was done for the fast acting thiobarbiturate, thiopental (Bischoff and Dedrick, 1968). A recent review specifically of flow-type pharmacokinetic modelling is by Himmelstein and Lutz (1979).

A simplified flow model is shown in Figure 7, with blood, visceral tissue, lean tissue and adipose tissue as the four compartments. The material balance equations for the Figure 7 model are:

blood: $\quad V_B(dC_B/dt) = Q_V C_V + Q_L C_L + Q_A C_A - (Q_V + Q_L + Q_A)C_B + DI(t)$ $\quad\quad$ (27)

viscera: $\quad V_V(dC_V/dt) = Q_V C_B - Q_V C_V - k_1 V_V C_V$ $\quad\quad\quad\quad\quad\quad\quad\quad\quad\quad\quad\quad$ (28)

lean: $\quad V_L(dC_L/dt) = Q_L C_B - Q_L C_L$ $\quad\quad\quad\quad\quad\quad\quad\quad\quad\quad\quad\quad\quad\quad\quad\quad$ (29)

adipose: $\quad V_A(dC_A/dt) = Q_A C_B - Q_A C_A$ $\quad\quad\quad\quad\quad\quad\quad\quad\quad\quad\quad\quad\quad\quad\quad$ (30)

where Q represents blood flow rate in ml min^{-1} and $I(t)$ the input function in min^{-1}. The above equations (27–30) can be solved using iterative computer techniques to give each of the concentrations as a function of time.

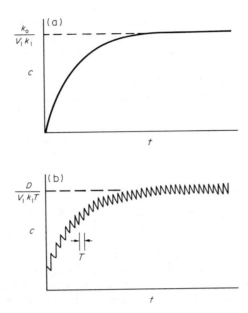

Figure 6 Change in plasma concentration of drug with time following intravenous administration: (a) continuous infusion at a rate of k_0 mg min^{-1}; (b) multiple dosing with dose D mg given every T minutes (dosing interval). In both (a) and (b) the mean plateau-level concentration depends on the ratio of the average input rate divided by the average clearance of drug

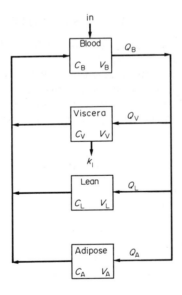

Figure 7 Flow type of pharmacokinetic model; Q represents blood flow rate

A variety of complexities can be added, for example (1) differentiation between plasma-protein bound and unbound drug; (2) addition of tissue beds to the compartments with equilibrium partitioning of drug between plasma and tissue; and (3) small time-delay volumes between major compartments to account for the transit times between major organs.

2.3.3.2 Parameter evaluation and comparison between simulated and experimental results

The flow model approach has the advantage that the parameters can be evaluated by separate experiments, so that none of the parameter values are adjusted in order to have the simulated results agree with the experimental data. Although this major advantage of flow models holds true in principle, it is normally a much more difficult task in practice to develop valid flow models. Average blood and tissue weights (volumes) for specific organs or tissue types and average blood flow rates to different organs or body regions can be obtained for animals and with lower accuracy for humans. But the question of what volumes and flows to use for an individual human subject has no ready answer, since the measurement of these parameter values in individual patients cannot be justified. As the flow model gets more features added, the ability to determine meaningful values for the added parameters becomes exceedingly difficult if not impossible. This difficulty is especially noticeable when blood:tissue partitioning of the drug or even plasma–protein binding relationships, many of which are nonlinear (concentration dependent), are required.

2.3.4   Comparison of the Two Modelling Approaches

On the one hand we have the more anatomical realistic flow models, where in simple cases the parameter values can be determined by separate experiments. However, parameter values are not normally available for individual subjects and binding or partitioning parameter values are extremely difficult to estimate. On the other hand, there are the one-, two-, or three-compartment models wherein the parameter values are adjusted to give the best agreement between the concentrations predicted from the model and those measured experimentally. There is no magic guideline to say when to use one approach and when to use the other, except to suggest that flow models are closer to reality. In most cases, the justification for pharmacokinetic modelling studies is to learn something that can be applied eventually to a better understanding of therapeutics or toxicology in humans. Pharmacokinetic parameter values obtained in animal species in general cannot be applied to calculations for humans. Thus, although the flow model gives a more realistic situation, its more difficult application to humans often is a major limitation.

2.4   FUTURE ROLE OF PHARMACOKINETICS IN BIOTECHNOLOGY

Pharmacokinetics is a very useful concept in quantifying the relative internal exposure of organisms to drugs, toxins, industrial chemicals and environmental additives, and in so doing to help assess the benefits or hazards of such exposure. Since many of the expected products or by-products of biotechnology are expected to possess biological activity either of a direct or indirect nature, it seems prudent to expect that many of these materials will need to undergo pharmacokinetic testing in animals and in some cases also in humans before these materials can be released for widespread use.

2.5   LIST OF SYMBOLS

A	constant in equation (9) (mg ml^{-1})
B	constant in equation (9) (mg ml^{-1})
C	concentration (mg ml^{-1})
C_a	concentration at adsorption site (mg ml^{-1})
C_{ss}	steady-state concentration (mg ml^{-1})
D	drug dose (mg)
F	fraction of drug adsorbed
G	constant in equation (14) (mg ml^{-1})
$I(t)$	input function (min^{-1})
k_0	rate of drug infusion (mg min^{-1})
k_1	rate constant for elimination-metabolism (min^{-1})
k_a	first order rate constant for adsorption (min^{-1})
k_{ij}	rate constant for species transfer between compartments i and j (min^{-1})
K_m	Michaelis constant (mg ml^{-1})
Q	blood flow rate (ml min^{-1})
t	time (min)
$t_{1/2}$	half-life (min)
v	reaction rate (mg ml^{-1} min^{-1})

V_m maximum enzyme-catalyzed reaction velocity (mg min^{-1})
V compartment volume (ml)
X amount of drug (mg)
α constant in equation (9)
β constant in equation (9)
γ constant in equation (14)
T dosing interval (min)
Subscripts
1, 2, 3 compartment number
A adipose tissue
B blood
L lean tissue
V viscera
Superscript
0 initial value (at $t = 0$)

2.6 REFERENCES

Abouleish, E., L. B. Wingard, Jr., S. D. L. Vega and N. Uy (1980). Pancuronium in caesarean section and its placental transfer. *Br. J. Anaesth.*, **52**, 531–536.
Bischoff, K. B. and R. L. Dedrick (1968). Thiopental pharmacokinetics. *J. Pharm. Sci.*, **57**, 1346–1351.
Gehring, P. J., G. E. Blau and P. G. Watanabe (1976). Pharmacokinetic studies in evaluation of the toxicological and environmental hazard of chemicals. In *Advances in Modern Toxicology*, ed. M. Mehlman, vol. 1, pp. 195–270. Hemisphere, Washington, DC.
Gibaldi, M. and D. Perrier (1975). *Pharmacokinetics*. Dekker, New York.
Himmelstein, K. J. and R. J. Lutz (1979). A review of the applications of physiologically based pharmacokinetic modeling. *J. Pharmacokinet. Biopharm.*, **7**, 127–145.
Kwan, K. C. (1978). Pharmacokinetic considerations in the design of controlled and sustained release drug delivery systems. In *Sustained and Controlled Release Drug Delivery Systems*, ed. J. R. Robinson, pp. 595–629. Dekker, New York.
Levy, G., T. Tsuchiya and L. P. Amsel (1972). Limited capacity for salicyl phenolic glucuronide formation and its effect on the kinetics of salicylate elimination in man. *Clin. Pharmacol. Ther.*, **13**, 258–273.
Rowland, M. and T. N. Tozer (1980). *Clinical Pharmacokinetics. Concepts and Applications*. Lea and Febiger, Philadelphia, PA.
Schönebeck, J., A. Polak, M. Fernex and H. J. Scholer (1973). Pharmacokinetic studies on the oral antimycotic agent 5-fluorocytosine in individuals with normal and impaired kidney function. *Chemotherapy*, **18**, 321–336.
Wingard, L. B., Jr. and D. R. Cook (1976). Pharmacodynamics of tubocurarine in humans. *Br. J. Anaesth.*, **48**, 839–845.
Wingard, L. B., Jr. and G. Levy (1973). Kinetics of anticoagulant effect of dicumarol in rats. *J. Pharmacol. Exp. Ther.*, **184**, 253–260.

3

Use of Liposomes as a Drug Delivery System

G. GREGORIADIS*
Clinical Research Centre, Harrow, Middlesex, UK

3.1 INTRODUCTION

Targeting of drugs to tissues, cells or subcellular organelles by carrier systems is now widely accepted as a means to improve drug selectivity (Gregoriadis, 1981a). Systems proposed to date include certain macromolecules, cells, viruses and a variety of synthetic polymers and particles (Gregoriadis *et al.*, 1982). Most of these, however, are limited in the quantity and range of drugs they can incorporate or are often unable to prevent contact of the drug they carry with the normal biological milieu or to promote the drug's entry into areas in need of pharmacological intervention. Additional limitations include toxicity of the carrier's components, unavailability of the latter, and technical problems such as the preparation of the drug–carrier complex. Efforts in many laboratories therefore have been focused on the development of the ideal drug carrier. This should be able to direct a wide variety of agents selectively to the area of action and provoke no ill effects in the process.

A carrier system that has recently attracted much attention is liposomes (Gregoriadis and Allison, 1980). These are microscopic vesicles formed from natural constituents such as phospholipids and were originally used as a model for cellular membranes with which they share the bimolecular leaflet structure unit (Bangham *et al.*, 1974). It was soon appreciated, however, that some of the properties of liposomes, notably entrapment and retention of virtually any pharmacologically active agent, structural versatility and an apparently innocuous nature, were attributes of a potentially powerful tool for the control of drug action. In the decade that has elapsed since the liposome drug carrier concert was put forward (Gregoriadis and Ryman, 1972a, 1972b), progress in the technology of the liposomes (Gregoriadis, 1983), in understanding their behaviour

*Now at Royal Free Hospital School of Medicine, London, UK.

within the biological milieu and the multitude of uses that such behaviour would entail, has been remarkable (Gregoriadis and Allison, 1980; Tom and Six, 1980).

3.2 TECHNICAL ASPECTS

Most of the success of liposomes in experimental drug delivery can probably be accounted for by the great number of forms they can assume. Made by an ever increasing variety of techniques (Gregoriadis, 1984), liposomes are composed of one, a few, or many concentric bilayer membranes which alternate with aqueous spaces, and have a size that can be as small as 25 nm in diameter or as large as several microns (Figure 1). Depending on their solubility and the method adopted, antimicrobial, cytotoxic and other conventional drugs, enzymes, antigens, hormones, genetic material, viruses and bacteria can be incorporated, often quantitatively, in the aqueous or the lipid phase of liposomes. Furthermore, the wide array of natural and synthetic phospholipids available can, in conjunction with cholesterol and other lipids, contribute to the design of vesicles endowed with the required structural and biological properties. Thus, vesicle membrane fluidity and permeability, surface charge, stability in biological fluids and tissues, and clearance from the circulation or injected tissue can all be arranged to comply with particular needs. Prepared under defined conditions, liposomes will retain both structural integrity and drug contents after long-term storage, even at temperatures below zero or can be lyophilized and then reconstituted before use. The preparation of sterile and pyrogen-free liposomes is also possible under certain conditions (Gregoriadis, 1984; Leserman and Barbet, 1982).

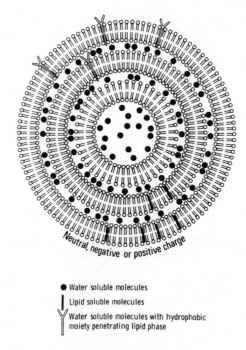

● Water soluble molecules

| Lipid soluble molecules

Y Water soluble molecules with hydrophobic moiety penetrating lipid phase

Figure 1 Diagrammatic representation of a liposome in which three bilayer phospholipid membranes alternate with aqueous spaces

3.3 INTERACTION OF LIPOSOMES WITH THE BIOLOGICAL MILIEU

Initial work (Gregoriadis and Ryman, 1972a, 1972b; Gregoriadis, 1973; Rahman, 1980) on the fate of liposomes administered intravenously into animals established some of the principles governing liposomal behaviour *in vivo*. For instance, latency of liposomal enzymes was found to be largely retained in the circulating blood, suggesting that the structural integrity of the carrier in the intravascular space could be maintained, at least to some extent. Examination of a number of tissues also showed that liposomes and their contents are taken up mostly by the liver- and spleen-fixed macrophages through endocytosis to end up in the lysosomes. Such findings supported the notion (Gregoriadis, 1976; Poste, 1980) that the system was suitable as a means of drug delivery into the intracellular environment, particularly the lysosomes. Drugs could, in addition, diffuse through the lysosomal membrane to enter other cell compartments. During the last decade, this laboratory and many others (Gregoriadis and Allison, 1980; Ryman and Tyrrell, 1980; Papahadjopoulos, 1978) have revealed an intriguing variety of aspects relating to the drug transport potential of liposomes. Some of the basic facets of the behaviour of liposomes within living systems, possible ways of manipulating such behaviour to our advantage and potential applications in the treatment or prevention of disease will be briefly discussed here.

3.3.1 Delivery of Liposomal Drugs into Cells *in vitro*

The realization that enzymes and other agents could be delivered by liposomes into the lysosomes of cells raised the question as to whether agents transported into these organelles could liberate themselves from the carrier and be able to survive in an active form in the intralysosomal milieu. It was established that cultured cells (*e.g.* macrophages or fibroblasts) deficient in a specific enzyme were capable of taking up liposomes, the contents of which (in this case the missing enzyme) were delivered into the lysosomes. After the disintegration of the carrier's membranes by lysosomal hydrolases, the enzyme was set free to perform its task (Gregoriadis and Buckland, 1973). The fate and effect of liposomal agents in the lysosomes were found to depend on the physical characteristics of such agents. For instance, hydrolytic enzymes, metal chelating agents and certain antimicrobial drugs that are relatively stable in the lysosomal milieu could act within it or cross lysosomal membranes to reach, and act in, other cellular regions (Poste, 1980).

Fusion of cellular membranes with those of liposomes has been suggested by a number of workers (Poste, 1980) as an alternative mechanism of cell–liposome interaction. This has led to the possibility of introducing agents entrapped in the aqueous space of liposomes directly into the cytoplasm. Alternatively, agents incorporated into the lipid phase of liposomes could, on the basis of this mechanism, incorporate themselves into the cellular membrane. Although there has been a considerable amount of indirect experimental evidence to support cell–liposome fusion, there are some doubts as to its actual occurrence, or at least its extent (Poste, 1980). Another mechanism by which agents may be introduced into cells is based on the destabilization of both cellular and liposomal membranes when in contact and subsequent diffusion of liposomal contents into the cell. A scheme of the variety of ways by which liposomes are thought to interact with cells is given in Figure 2.

The ability of liposomes to introduce drugs into distinct intracellular sites *in vitro* has led to the development of techniques for the study of cell behaviour in cell biology, pharmacology, immunology and genetics (Gregoriadis *et al.*, 1982).

3.3.2 Fate of Liposomes *in vivo*

The behaviour of liposomes after *in vivo* administration varies according to the particular environment with which they come into contact. Most of our knowledge with regard to the influence that liposomes and biological milieu exert on each other and to ways that such influence can be modified have come from studies by the intravenous route (Gregoriadis, 1981a; Ryman and Tyrrell, 1980). The events that occur between initial contact of liposomes with blood and eventual drug liberation from its carrier and action are complex and depend both on the type of preparation used (including its drug contents) and the physiological state of the animal. To begin with, immediately after their injection liposomes are attacked by plasma high density lipoproteins (HDL) (Krupp *et al.*, 1976; Scherphof *et al.*, 1978; Kirby *et al.*, 1980a). This leads to the transfer of liposomal phospholipid molecules to HDL followed by disruption of liposomal membrane continuity and the leakage of entrapped drugs into the circulation (Kirby and Gregoriadis, 1980). Liposomes, at various stages of damage, are then removed by the fixed macrophages of the liver and spleen. A portion of liposomes will also interact with phagocytes of other tissues and, to a small extent, with circulating monocytes. Small liposomes (less than 100 nm in diameter) will almost certainly reach the hepatic parenchymal cells. However, liposome passage, even for the smallest vesicles, through normal capillaries to extravascular areas is in doubt, although some passage may occur in the case of inflamed or damaged tissues. Liposomes interacting with phago-

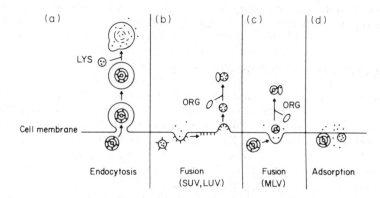

Figure 2   Possible mechanisms for cell–liposome interaction. (a) Endocytosis of the multilamellar (or monolamellar) liposome is followed by fusion of the endocytic vacuole containing the liposome with a lysosome (LYS). Lysosomal phospholipases or other factors (×) disrupt the lipid bilayers of liposomes and free the entrapped agent (·) which can then act either within the lysosome or, after its diffusion, in other cell compartments. (b) Fusion of a monolamellar liposome with the cellular membrane is followed by the entrance of water-soluble agents (·) into the cell's cytoplasm, from which they can reach other organelles. Agents incorporated in the lipid bilayers of the liposome (−) are transferred onto the membrane of the cell. The latter agents can be internalized by the process of endocytosis. Endocytic vacuoles thus formed can interact with other organelles (ORG). (c) Fusion of a multilamellar liposome with the cellular membrane is followed by the entrance of agents (·) in the outer aqueous space of the liposome and of the inner core into the cell's cytoplasm. Free agents and the core can then interact with other organelles (ORG). (d) Adsorption of a multilamellar or monolamellar liposome onto the cellular membrane can be of no consequence or can induce changes in the permeability of both liposomal and cellular membranes. Agents (·) diffusing from liposomes can thus penetrate the cell's interior (Gregoriadis, 1979)

cytic cells are assimilated (together with any drug contents that are still retained by the carrier) by endocytosis to end up in the lysosomes. As already mentioned, transported drugs that are stable in the lysosomal milieu will either act within it or, if capable of diffusing through the lysosomal membrane, escape to other cellular regions. Such lysosomotropic action of liposomes is instrumental in facilitating the entry of drugs to areas in the cell that are often inaccessible to them and thus forms the basis for several important applications in medicine (see later).

Using alternative routes of administration, liposomes can reach a variety of tissues. After intraperitoneal, intramuscular or subcutaneous injection, liposomes and their drug contents enter the lymphatics to localize in the lymph nodes draining the injected sites (Segal *et al.*, 1975; Osborne *et al.*, 1979; Jackson, 1981; Parker *et al.*, 1981). A portion of the administered liposomes flows further *via* the lymph ducts to join the blood stream and eventually reaches the liver and spleen. On the other hand, liposomes of large size are retained at the site of injection and release their drug content locally at rates which are related to the carrier's lipid composition and its vulnerability to the liposome-disrupting efficiency of the site (Segal *et al.*, 1975). Several days' delay in the clearance of liposomes from the site of injection is observed (Mauk *et al.*, 1980) when the vesicle surface is coated with certain amino sugar derivatives of cholesterol which adhere onto nearby cells. However, intraperitoneally given liposomes will be, regardless of size, largely removed into the lymphatics (Dapergolas *et al.*, 1976; Ellens *et al.*, 1981; Parker *et al.*, 1981).

The intratracheal, intra-articular and topical (*e.g.* skin and eyes) routes have also been used to show that liposomes and entrapped antitumour agents, corticosteroids, antibiotics, *etc.* reach and act in relevant tissues and cells more efficiently than the same drugs given by conventional means (Gregoriadis, 1981a). Further, attempts have been made (Dapergolas and Gregoriadis, 1976; Patel and Ryman, 1976) to use liposomes by the oral route but with little success so far. Because of their lipid nature, liposomes are rapidly disrupted in the gut by phospholipases and by the detergent effect of bile salts. There is evidence, however, that when insulin-containing liposomes are administered to diabetic rats or dogs intragastrically or intraduodenally, a small proportion of the hormone reaches the periphery to induce a significant fall in blood glucose (Dapergolas and Gregoriadis, 1976; Patel *et al.*, 1982). Others have claimed similar findings of absorption of liposomal Factor VIII in haemophiliacs (Hemker *et al.*, 1980). Facilitation of protein entry into the periphery by liposomes may be related to the temporary protection they offer against protease attack until the protein reaches less hostile areas (Patel *et al.*, 1982). It is also possible that at the same time a small portion of surviving liposomes are absorbed as such (Dapergolas and Gregoriadis, 1976). Indeed, it has been proposed (Dapergolas and Gregoriadis, 1977) that liposomes prepared from phospholipids (such as distearoylphosphatidylcholine) which are likely to protect them from gross disruption in the gut may contribute to a more efficient absorption of proteins and other drugs regardless of the underlying mechanism.

3.4 OPTIMIZATION OF LIPOSOME BEHAVIOUR

Among the advantages offered by liposomes with regard to drug delivery, versatility in structural characteristics is most prominent. For instance, appropriate choice of lipid composition, size and surface charge of liposomes will all influence the carrier's fate and, in many instances, contribute to the optimization of drug action. Because a major objective in the *in vivo* use of liposomes is drug delivery to cells accessible to the carrier (*i.e.* cells in the blood, vascular endothelium and also in extravascular areas separated from the circulating blood by membranes permeable to liposomes), drugs must be retained by liposomes for periods of time sufficient to ensure effective access to, and association with the target. Some of the factors influencing retention of drugs by liposomes *in vitro* and in the circulating blood and rates of liposome clearance after intravenous injection are discussed here. All liposomes used in the studies to be described were neutral, small, unilamellar and prepared by sonication followed by molecular sieve chromatography (Senior and Gregoriadis, 1982a). The choice of size was based (Gregoriadis, 1981a) on both the well-known property of small liposomes to assume half-lives which are longer than those of large vesicle versions and their ability to pass through membranes with pores of a diameter greater than about 25 nm, which is the size of the smallest liposome. As a model solute in bilayer permeability studies we chose carboxyfluorescein (CF) because, when entrapped at high concentrations (*e.g.* 0.1 M), the dye self-quenches. Upon leakage and dilution in the large volume of the surrounding medium, CF fluoresces and thus provides a simple and direct method for the measurement of its leakage. Latent (non-leaked) CF is estimated from $100 (Dye_t - Dye_f)/Dye_t$ where t and f denote total (measured in the presence of Triton X-100) and free dye respectively (Kirby *et al.*, 1980b).

3.4.1 Retention of Solutes by Liposomes in the Presence of Blood Plasma

It has been shown that drugs which are quantitatively retained by liposomes following their entrapment, leak out in the presence of whole blood or plasma *in vitro* and in the circulation of injected animals (Gregoriadis, 1973). As already mentioned, such leakage results from the loss of liposomal phospholipid to plasma HDL (Krupp *et al.*, 1976; Scherphof *et al.*, 1978; Kirby *et al.*, 1980a). Using liposomes as models of cell membranes, other workers had already found that the presence of cholesterol in liposomes above the liquid–crystalline phase transition temperature (T_c) of their phospholipid component condenses phospholipid molecules, thus reducing bilayer permeability to solutes (Papahadjopoulos, 1978). On this basis it was reasoned (Gregoriadis and Davis, 1979) that bilayer packing could also diminish or prevent altogether removal of phospholipid molecules by HDL. As discussed below, this was proved of importance in controlling liposomal stability in the presence of blood and vesicle clearance from the circulation.

Work from this laboratory (Kirby *et al.*, 1980b; Senior and Gregoriadis, 1982a) has shown that, irrespective of the liposomal content of cholesterol, CF entrapped in small unilamellar liposomes made of egg phosphatidylcholine (PC) retains most of its latency in the presence of buffer at 37 °C. In the presence of serum, however, and when liposomes are devoid of cholesterol, CF latency rapidly decreases to low values, presumably because of phospholipid loss to HDL and subsequent release of the dye. Incorporation of some cholesterol (PC:cholesterol molar ratio 1:0.28) results, in the presence of serum, in the retention of much of the CF latency and, with liposomes composed of equimolar phospholipid and cholesterol, CF latency is retained fully. A similar action of cholesterol is also observed (Senior and Gregoriadis, 1982a) with dioleoylphosphatidylcholine (DOPC) and sphingomyelin (SM), although there are quantitative differences: in the presence of serum, for instance, leakage of CF from SM and DOPC cholesterol-rich liposomes is less drastic compared with that seen in cholesterol-rich PC liposomes, probably because removal of liposomal SM and DOPC by serum HDL is less efficient (Allen, 1981). Cholesterol also reduces the permeability of liposomes to CF when these are composed of a variety of other phospholipids (Senior and Gregoriadis, 1982a; Allen, 1981). However, for phospholipids with saturated fatty acid esters of more than 16 carbon atoms and a T_c above or at 37 °C, *e.g.* distearoylphosphatidylcholine, DSPC; dipalmitoylphosphatidylcholine, DPPC, incorporation of some cholesterol (1:0.28 phospholipid : cholesterol molar ratio) enhances the loss of CF latency at 37 °C, presumably because bilayers become fluid and are thus vulnerable to HDL attack (Senior and

Gregoriadis, 1982a). On the other hand, incorporation of equimolar cholesterol in liposomes composed of such phospholipids (*e.g.* DSPC) brings about the opposite effect with the bilayers becoming virtually impermeable to CF (Senior and Gregoriadis, 1982b; see later).

The proposition (Kirby *et al.*, 1980b) that cholesterol reduces leakage of solutes by preventing liposomal phospholipid loss to HDL, was investigated (Kirby *et al.*, 1980a) with CF-containing ^{3}H-PC labelled liposomes. On incubation in serum, and chromatography, it was found that with cholesterol-free liposomes only a fraction of the ^{3}H-PC and CF markers eluted from the column together as intact vesicles. The remainder was recovered as ^{3}H-PC in fractions containing HDL and as free CF in subsequent fractions. Increasing amounts of cholesterol in the liposomal membrane progressively reduced the proportion of ^{3}H-PC appearing with HDL and of CF eluted as free (Kirby *et al.*, 1980a). Since the mode by which HDL removes phospholipids from liposomes is believed to involve insertion of HDL hydrophobic regions into the bilayer prior to associating with the phospholipid (Chobanian *et al.*, 1979; Allen, 1981), packing of bilayers by the sterol is likely to obstruct such insertion.

It has been suggested (Scherphof *et al.*, 1978) that HDL-induced destabilization of liposomes leads to their disintegration and total release of solutes. On the other hand, quantitative solute release in the presence of serum could also occur through formed pores in otherwise intact vesicles (Gregoriadis and Davis, 1979). These pores are expected to vary in size in direct proportion to the amount of phospholipid removed, thus allowing discrimination of solute release according to molar mass. Our attempts to establish which of the two events occurs (*i.e.* vesicle destruction or leaky vesicles) were two-fold (Kirby and Gregoriadis, 1981). (a) Radiolabelled solutes of decreasing molar mass were entrapped in PC liposomes and their release in the presence of plasma was then measured. In the event of complete disintegration of membranes, release of contents into the medium should be independent of solute size. If, however, release of solutes preferentially favours those of smaller size, this can only be accounted for by pore formation. (b) 'Empty' (buffer-loaded) PC liposomes were mixed with plasma containing CF at a concentration of 0.2 M. As already mentioned, at this concentration CF self-quenches. Assuming that there is pore formation in the vesicles, the dye would diffuse into their interior and, upon re-isolation, liposomes would be expected to contain self-quenched CF.

As shown in Table 1, release of sucrose, inulin and poly(vinylpyrrolidone) (PVP) after incubation of the liposome-entrapped markers in the presence of plasma is non-uniform and confirms that solute release is dependent on size. For instance, with cholesterol-free liposomes 80.6, 68.9 and 26.9% of sucrose, inulin and PVP, respectively, are released (Table 1). Such preferential loss of the smaller sucrose and inulin molecules seems incompatible with vesicle disruption and consistent with the maintenance of a closed but leaky structure. Because of the similarity in the loss of the two smaller solutes (Table 1), formed pores are likely to be large enough to allow relatively free diffusion of sucrose and inulin but not of PVP. It could be argued that such pores, with hydrophobic fatty acid chains of phospholipids exposed on the pore's surface, cannot exist in a stable form. Further, since the small size of the vesicles would not allow further condensation of the bilayer to close the pores, the vesicles should disintegrate to smaller fragments. On the other hand, it is possible that pore stability is maintained by the interaction of the hydrophobic regions of the phospholipids on the surface of the pore channels with the hydrophobic regions of components present in the plasma.

Table 1   Release of Solutes from Liposomes in the Presence of Plasma[a]

Liposomes	Sucrose	Inulin	Poly (vinyl-pyrrolidone)
Cholesterol-free	80.6 ± 10.4 (3)	68.9 ± 6.9 (3)	26.9 ± 3.7 (3)
Cholesterol-poor	42.2 ± 2.8 (3)	31.1 ± 7.2 (3)	26.1 ± 3.0 (3)
Cholesterol-rich	4.1 ± 2.1 (3)	7.7 ± 0.9 (3)	6.6, 7.7

[a] Small unilamellar liposomes containing radiolabelled sucrose, inulin and poly(vinylpyrrolidone) were incubated in the presence of mouse plasma at 37 °C for 30 min. Samples were than chromatographed on Ultrogel AcA 34 columns. Radioactivity released from liposomes and eluted in corresponding fractions was pooled and expressed as a percentage of the total applied. Results are means ± SD for the numbers of experiments indicated in parentheses (Kirby and Gregoriadis, 1981).

Further examination of Table 1 shows that enrichment of liposomes with some cholesterol (PC:cholesterol molar ratio 1:0.28) reduces pore size so that loss of inulin and PVP is now similar, whereas that of sucrose still remains relatively high (Table 1). The size of pores in such cholesterol-poor liposomes is, apparently, small enough to restrict the passage of inulin and PVP more than of sucrose. With cholesterol-rich liposomes, release of all three solutes is low and pores present must be too small even for the release of sucrose (Table 1). In the reverse experiment, in which buffer-loaded liposomes were incubated with plasma containing quenched CF, it was shown (Kirby and Gregoriadis, 1981) that after chromatography of the mixture, fractions expected to contain liposomes also contained latent (quenched) CF. This suggests entry of dye into vesicles, presumably through pores formed by the action of plasma.

3.4.2   Retention of Solutes by Liposomes in the Circulating Blood

Cholesterol-induced reduction of solute leakage in the presence of plasma is also observed in liposomes circulating in the blood (Kirby and Gregoriadis, 1980). With PC liposomes containing equimolar cholesterol, for instance, clearance of entrapped CF is linear reflecting the clearance of its carrier (Kirby and Gregoriadis, 1980; Kirby et al., 1980b). This was confirmed using CF-containing liposomes labelled with radioactive lipids: ratios of CF: lipid radioactivity in the blood at time intervals after injection were very similar to the ratio in the injected preparation and, as with the *in vitro* experiments, cholesterol was found to diminish phospholipid loss to HDL (Kirby and Gregoriadis, 1980).

In recent work (Kirby and Gregoriadis, 1983), we have studied the effect of cholesterol presence in liposomes containing melphalan and vincristine on the release of drugs from vesicles circulating in the blood of injected animals. As expected, the rate of clearance of drugs given as free was reduced considerably when these were given with (cholesterol-free) liposomes and a further reduction was achieved with cholesterol-rich liposomes (phospholipid : cholesterol molar ratio 1:1). However, in contrast to findings with CF, clearance of the drugs entrapped in cholesterol-rich liposomes was more rapid than that of the carrier. It is possible that vincristine and melphalan, being lipophilic, leak through the circulating liposomes to a considerable extent and are then removed rapidly into the extravascular space.

3.4.3   Clearance of Liposomes and Entrapped Solutes from the Circulation

Recent experiments (Senior and Gregoriadis, 1982b; Gregoriadis and Senior, 1980) have established a relationship between leakiness of cholesterol-rich liposomes and rates of their clearance from the circulation after intravenous injection. For instance, quenched CF clearance patterns from the blood of injected mice correspond to half-lives ranging from 0.1 [cholesterol-rich dilauroylphosphatidylcholine (DLPC) liposomes] to 16 h (cholesterol-rich SM liposomes; Figure 3). A nearly identical half-life for small, unilamellar, cholesterol-rich SM liposomes has also been reported by others (Hwang et al., 1980).

Such phospholipid-dependent differences in liposome clearance cannot be explained on the basis of vesicle surface charge or size variations, because all preparations used were neutral and of similar average vesicle size (about 30–60 nm in diameter). We have, therefore, looked into the possibility of a relationship between the varying resistance of phospholipids to being removed from bilayers by plasma HDL (Allen, 1981) and vesicle clearance rates. The quenched CF-containing liposomes used in these studies (Senior and Gregoriadis, 1982b) were composed of varying proportions of PC and SM (hybrid liposomes) and of cholesterol equimolar to total phospholipid. Upon incubation of the preparations with mouse plasma, CF loss was only slight for liposomes composed of SM only, but was augmented considerably in the hybrids and in proportion to PC present (Figure 4). Interestingly, after injection of mice with the SM liposomes (Figure 5), the rate of clearance of quenched CF was, as expected (Gregoriadis and Senior, 1980; Hwang et al., 1980), slow and remained so when some (23%) of the liposomal SM was replaced by PC (Figure 5). However, clearance rates were increased progressively with 53, 77 and 100% substitution with PC. Thus, half-lives as estimated from the clearance patterns of quenched CF were reduced from 11 (SM and 77% SM:23% PC liposomes) to 2.1 h (PC liposomes), with intermediate values (7 and 4.5 h) for the hybrid preparations (47% SM:53% PC and 23% SM:77% PC) reflecting the effect of PC concentration. The validity of these half-lives was confirmed by the use of radiolabelled phospholipid markers of the liposomal lipid phase and has been discussed extensively elsewhere (Senior and Gregoriadis, 1982b).

The possibility that the relationship between clearance of liposomes and the nature of their phospholipid component could also derive from the structural characteristics of the latter

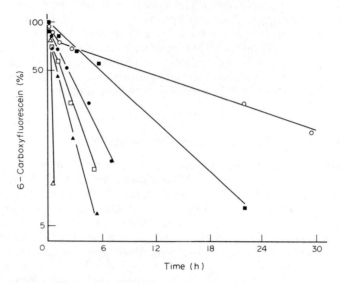

Figure 3 The effect of the phospholipid component of liposomes on the clearance of entrapped CF from the blood of injected mice. Mice were bled at time intervals after intravenous injection with CF entrapped in small unilamellar liposomes (3 mg phospholipid) composed of DLPC ($\triangle$), DOPC ($\blacktriangle$), DSPC ($\square$), PC ($\bullet$), DMPC ($\blacksquare$) or SM ($\bigcirc$). With the exception of liposomes made of DSPC, all preparations contained cholesterol (1:1 molar ratio). Latent CF values (means from three animals) in total mouse plasma are expressed as percentages of latent CF injected. Half-lives estimated from slopes were: DLPC, 0.1; DOPC, 1; DSPC, 1.5; egg PC, 2; DMPC, 6; SM, 16 h (Gregoriadis and Senior, 1980)

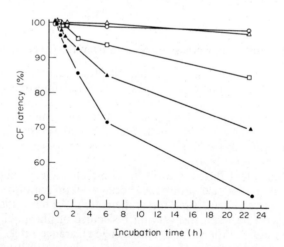

Figure 4 The effect of phospholipid composition of liposomes on their permeability in plasma. In a typical experiment shown here, small unilamellar liposomes containing quenched CF and composed of SM ($\triangle$), 77% SM:23% PC ($\bigcirc$), 47% SM:53% PC ($\square$), 23% SM:77% PC ($\blacktriangle$) and PC ($\bullet$) were incubated in the presence of mouse plasma at 37 °C. All liposomal preparations contained cholesterol, equimolar to the total phospholipid. CF latency values at time intervals are percentages of total CF present (Senior and Gregoriadis, 1982b)

expressed on the liposomal surface (for instance through specific association of phospholipids with certain plasma proteins and/or recognition by tissues) was also investigated (Senior and Gregoriadis, 1982b). If this were so, liposomal preparations (*e.g.* cholesterol-rich and cholesterol-free DSPC liposomes) sharing the same phospholipid would be expected to show similar clearance kinetics. However, observed half-life values (20 and 1.3 h, respectively (Figure 6b) could not support the above proposition and were too diverse to be attributed to different phospholipid densities on the bilayer (Senior and Gregoriadis, 1982b). On the other hand it is of interest to note that half-lives in Figure 6b correspond to complete impermeability and partial permeability of the cholesterol-rich and cholesterol-free liposomes in the presence of plasma (Figure 6a). A

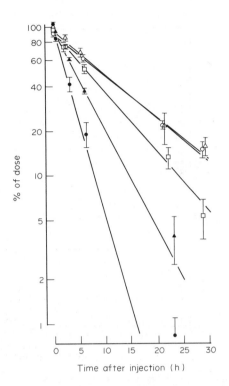

Figure 5 The effect of phospholipid composition of liposomes on their clearance from the circulation. Liposomes as described in Figure 4 were injected intravenously into mice. Latent CF values (mean ± SD; 3–6 animals) at time intervals are percentages of injected latent CF per total mouse plasma. For other details see legend to Figure 4 (Senior and Gregoriadis, 1982b)

scheme correlating bilayer leakiness of cholesterol-rich small unilamellar liposomes and their rate of removal from the circulation appears in Figure 7.

The effect of the phospholipid component of liposomes on entrapped CF clearance from the circulation (Figures 3 and 5) may not be as straightforward with lipophilic drugs. For instance, whereas CF clearance is linear throughout for most types of cholesterol-rich small unilamellar liposomes studied (Figures 3 and 5), clearance of melphalan and vincristine was found to be biphasic showing a rapid initial phase followed by a slower linear one (Kirby and Gregoriadis, 1983). The rapid loss of drugs during the initial phase may, as discussed elsewhere (Kirby and Gregoriadis, 1983), reflect dissociation of drugs absorbed onto the liposomal membrane and their subsequent rapid redistribution in the extravascular space. This and an increased leakage (and ensuing clearance) of the lipophilic drugs from liposomes during the second slower phase may have contributed to overall drug clearance rates that were substantially greater than those of CF in liposomes of corresponding phospholipid composition (Kirby and Gregoriadis, 1983).

Phospholipid-induced variations in the half-lives of liposomes are reflected in the hepatic and splenic uptake of entrapped solutes. Thus, administration of liposomes of increasing half-lives into mice results in the gradual reduction of the uptake of the liposomal marker (^{111}In) by the liver and spleen (Gregoriadis and Senior, 1980). For instance, hepatic uptake of ^{111}In was 32% for PC, 23% for dimyristoylphosphatidylcholine (DMPC), 16% for SM and 10% (unpublished data) for DSPC (cholesterol-rich) liposomes.

3.5 POTENTIAL APPLICATIONS IN MEDICINE

The transport of drugs by liposomes to phagocytic cells has made apparent a number of possible applications in therapeutic and preventive medicine (Gregoriadis, 1981a, 1982; Ryman and Tyrrell, 1980). Before the liposome-drug carrier concept was formed and even during its development, other carrier systems that are 'foreign' and therefore easy prey for phagocytes were or have been made available for a similar role (Breimer and Speiser, 1981; Gregoriadis *et al.*, 1982; Gregoriadis, 1979). However, the limited range of drugs that such carriers can accommodate, difficulties in the preparations of carrier–drug conjugates or complexes that could retain drugs

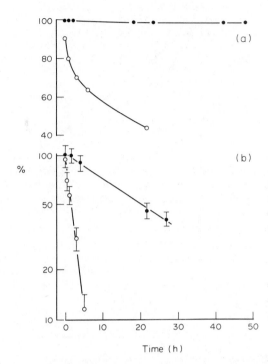

Figure 6 Permeability of DSPC liposomes in mouse plasma and clearance from the circulation. Small unilamellar liposomes containing quenched CF and composed of DSPC (○) or equimolar DSPC and cholesterol (●) were incubated in the presence of mouse plasma at 37 °C (a) or injected intravenously into four mice (b). Values at time intervals are percentages of latent CF of total CF present or percentages (mean ± SD) of latent CF of injected latent CF per total mouse plasma (Senior and Gregoriadis, 1982b)

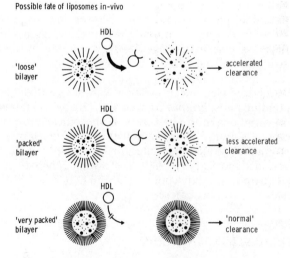

Figure 7 Correlation between porosity of cholesterol-rich small unilamellar liposomes and clearance from the circulation. In the proposed scheme, the extent of bilayer porosity (and diffusion of solutes) attained in blood is dependent on the facility with which high density lipoproteins (HDL) remove phospholipid molecules. Vesicles that remain virtually intact (full solute retention) will exhibit a 'normal' rate of clearance which, for a given dose of liposomal lipid, should be similar for all stable vesicles of the same size and charge, regardless of their composition. With vesicles that become leaky (because of HDL action for instance), clearance rates will increase according to the extent of porosity. No mechanism is provided here for the accelerated clearance of the porous cholesterol-rich liposomes. For discussion see Senior and Gregoriadis, 1982b

effectively and preserve their activity and resistance to biodegradation have prevented their widespread adoption. As it happens, liposomes have none of these impediments and, furthermore, their structural versatility allows for modifications that are not attainable with other systems.

Among diseases affecting the reticuloendothelial system in which microorganisms invade and thrive, phagocytes are prominent. Although there are a large number of antimicrobial drugs available, many of these drugs are either unable to penetrate the cell's interior or are toxic. The extent to which liposomes can optimize antimicrobial drug action is illustrated by the successful treatment of experimental visceral and cutaneous leishmaniasis with relevant drugs entrapped in liposomes at doses which are much smaller than those for free drugs used as such (Alving, 1982). Further, several antibiotics (*e.g.* dihydrostreptomycin, gentamycin, amicacin and kanamycin) presented in liposomes of appropriate designs can gain access and kill intracellular organisms such as *Staphylococcus aureus*, *Brucella abortus* and *Listeria monocytogenes* much more effectively than the free drugs do (Gregoriadis, 1982). Intracellular transport of drugs for the removal of unwanted materials stored in the lysosomes is also facilitated by liposomes. Such materials, varying from metabolic products,the breakdown of which to smaller units is blocked by the absence of a relevant enzyme (Belchetz *et al.*, 1977), to metals accumulating in the tissues as a result of disease (*e.g.* iron loading in haemochromatosis) or poisoning (*e.g.* plutonium, gold, mercury, *etc.*) (Rahman, 1980), are removed with varying degrees of efficiency by the transported drugs.

The phagocyte–liposome interaction has recently been put to use in circumventing the resistance of tumours to the action of macrophages and other killer cells. Encouraging evidence (Poste *et al.*, 1982) has shown that systemically administered liposome-entrapped lymphokines or muramyl dipeptide activate the tumouricidal properties of macrophages in animal models to the extent that established tumour metastases are destroyed by an immunologically non-specific mechanism which is independent of tumour cell heterogeneity or cycle. It would, therefore, appear that macrophage activation with liposomal immunomodulators may prove superior as a modality for cancer therapy (Poste *et al.*, 1982).

The use of liposomes as immunological adjuvants in vaccines is another promising development with important implications in medicine (Gregoriadis, 1981b). It is well recognized that human and animal immunization programmes would benefit greatly from a safe and efficient adjuvant which reduces the amounts of antigens required for vaccination or enables relatively weak antigens to induce stronger immune responses. Unfortunately, presently available adjuvants (*e.g.* Freund's adjuvant, bacterial endotoxins, mineral absorbents, polyelectrolytes, *etc.*) are either toxic, lack efficiency or have short-lived effects. Several laboratories (Gregoriadis, 1981b) have now shown that liposomes can (without the presence of side effects seen with other adjuvants) act as powerful immunological adjuvants to induce both humoural and cellular immunity for a variety of bacterial and viral antigens relevant to disease. There include diphtheria toxoid, *Plasmodium falsiparum* antigens, hepatitis B surface antigens, rubella haemagglutinin, cholera toxin and influenza virus subunits (Gregoriadis, 1981b). The mechanism by which liposomes induce antibody response, although still unknown, is certain to involve the macrophages.

Recent studies in this laboratory indicate that liposomes hold promise as an adjuvant for a hepatitis B vaccine (Gregoriadis and Manesis, 1980). HB_sAg can be readily incorporated into multilamellar liposomes and immunization of guinea pigs with liposomal HB_sAg as such or in conjunction with killed *Bordetella pertussis* or saponin produces earlier conversion rates and much higher antibody responses than with the free antigen, alone or in association with the two other adjuvants (Manesis *et al.*, 1979). Delayed hypersensitivity tests in these animals show that the immune response is cell-mediated as well. This should be an advantage since there is strong evidence that cell-mediated immunity plays a major role in the protection against most viral infections. Encouraging results (Gregoriadis and Manesis, 1980) have also been obtained in inbred mice where a single injection of the liposomal HB_sAg produced plateau antibody values lasting for at least four months. From comparative studies (Gregoriadis and Manesis, 1980), it appears that liposomal HB_sAg can be as or more effective than HB_sAg administered with a variety of other adjuvants, including alum oxide, muramyl dipeptide and complete Freund's adjuvant. Although in recent clinical trials (cited in Gregoriadis, 1981b) immunization with HB_sAg using alum as an adjuvant has proven successful, the tendency of antibody titres to decline several months after the first injection is likely to affect long-term protection from the disease. In addition, individuals in these trials were injected three times, a practice with obvious socioeconomic disadvantages. On the other hand, liposomes could prolong further HB_sAb antibody titres in immunized subjects with smaller amounts of antigen and fewer injections.

Among a variety of synthetic adjuvants, liposomes are unique in their flexibility in composition and structure. This should, therefore, enable us to control their behaviour *in vivo* so as to fulfil requirements compatible with an effective immunological adjuvant. As already discussed (Gregoriadis, 1982), manipulation of size, surface charge and phospholipid composition can all influence the rate of clearance of liposomes from the injected site or the blood and their uptake by tissues. Such modifications, in conjunction with control of antigen diffusion from liposomes *in situ* may lead to patterns of antigen distribution and fate that would lead to augmented immunogenicity and thus help in producing 'single shot' vaccines.

Although there is little doubt that liposomes will play an important role in interfering with macrophage function in the treatment or prevention of disease, there are indications that other domains in medicine may also benefit. For instance, several reports suggest that liposomes can serve as diagnostic agents when carrying radioactive tracers or radiodense molecules. In particular, liposomes made of radio-opaque iodinated or brominated phospholipids have been used for contrast image enhancement of the liver and spleen so as to improve the diagnostic accuracy of computed tomography (Gregoriadis, 1983). Similar success has been obtained in lymph node imaging or treatment, after subcutaneous or intramuscular injection of liposomes containing appropriate agents (Segal *et al.*, 1975; Osborne *et al.*, 1979).

3.6 TARGETING OF LIPOSOMES

Effective use of liposomes as drug carriers *in vivo* will obviously require access to, and interaction with, cell targets. As already mentioned, with the exception of cells associated with the reticuloendothelial system, which have an avidity for injected liposomes, other cells interact with liposomes only poorly and non-selectively. In our early attempts (Gregoriadis and Neerunjun, 1975) to target liposomes to specific cells we were able to show that IgG raised against tumour cells and subsequently incorporated on the surface of drug-containing liposomes mediates uptake of the liposomal carrier and its contents by the respective cells. Using liposomes bearing desialylated fetuin (Gregoriadis and Neerunjun, 1975) or certain glycosides (Ghosh *et al.*, 1981) which exhibit specific affinity for the liver parenchymal and Kupffer cells, respectively, it was possible to target liposomes *in vivo* after intravenous injection. Results from these experiments supported the view that targeting of antibody-bearing liposomes *in vivo* could be realistically achieved for cells accessible to the carrier. However, in spite of convincing evidence (Leserman *et al.*, 1980; Huang *et al.*, 1980; Martin *et al.*, 1981) that liposomes coated with antibodies by a variety of methods interact *in vitro* with target cells selectively, understanding of factors that influence such selective interaction is poor. It would be of interest, for instance, to know the relative importance in targeting of antibody purity, stability of antibody-coated liposomes, availability of the antigen-recognizing regions of IgG on the liposomal surface and ability of these regions to recognize and interact with the respective antigens, particularly within the biological milieu (Gregoriadis and Meehan, 1981). It is widely appreciated that optimal targeting *in vivo* can only be achieved by liposomes that are stable in the blood, circulate for extended periods of time and can, if needed, cross membranes. Evidence discussed here indicates that proper adjustments of the lipid composition, size and surface charge can produce liposomes with defined characteristics in terms of solute retention and vesicle survival in the circulation. This and acquired target-oriented properties could influence liposomal drug fate and, hopefully, efficacy.

3.7 ABBREVIATIONS

CF	carboxyfluorscein
DLPC	dilaurylphosphatidylcholine
DMPC	dimyristoylphosphatidylcholine
DOPC	dioleoylphosphatidylcholine
DPPC	dipalmitoylphosphatidylcholine
DSPC	distearoylphosphatidylcholine
HDL	high-density lipoproteins
PC	egg phosphatidylcholine
PVP	poly(vinylpyrrolidone)
SM	sphingomyelin
T_c	liquid-crystalline phase transition temperature

ACKNOWLEDGEMENTS

This work was supported in part by an NIH National Cancer Institute contract (NO I-CM-87171). I thank Mrs. M. Moriarty for unrivalled secretarial work.

3.8 REFERENCES

Allen, T. M. (1981). A study of phospholipid interactions between high-density lipoprotein and small unilamellar vesicles. *Biochim. Biophys. Acta*, **640**, 385–397.

Alving, C. R. (1982). Therapeutic potential of liposomes as carriers in leishmaniasis, malaria and vaccines. In *Targeting of Drugs*, ed. G. Gregoriadis, J. Senior and A. Trouet, pp. 337–353. Plenum, New York.

Bangham, A. D., M. W. Hills and N. G. A. Miller (1974). Preparation and use of liposomes as models of biological membranes. In *Methods in Membrane Biology*, ed. E. D. Korn, vol. 1, pp. 1–68. Plenum, New York.

Belchetz, P. E., I. P. Braidman, J. C. W. Crawley and G. Gregoriadis (1977). Treatment of Gaucher's disease with liposome-entrapped glucocerebroside β-glucosidase. *Lancet*, **2**, 116–117.

Breimer, D. D. and P. Speiser (eds.) (1981). *Topics in Pharmaceutical Sciences*. Elsevier/North Holland, Amsterdam.

Chobanian, J. V., A. R. Tall and P. I. Brecher (1979). Interaction between unilamellar egg yolk lecithin vesicles and human high density lipoprotein. *Biochemistry*, **18**, 180–187.

Dapergolas, G. and G. Gregoriadis (1976). Hypoglycaemic effect of liposome-entrapped insulin administered intragastrically into rats. *Lancet*, **2**, 824–827.

Dapergolas, G. and G. Gregoriadis (1977). The effect of liposomal lipid composition on the fate and effect of liposome-entrapped insulin and tubocurarine. *Biochem. Soc. Trans.*, **5**, 1383–1386.

Dapergolas, G., E. D. Neerunjun and G. Gregoriadis (1976). Penetration of target areas in the rat by liposome-associated bleomycin, glucose oxidase and insulin. *FEBS Lett.*, **63**, 235–239.

Ellens, H., H. Morselt and G. Scherphof (1981). *In-vivo* fate of large unilamellar sphingomyelin–cholesterol liposomes after intraperitoneal and intravenous injection into rats. *Biochim. Biophys. Acta*, **674**, 10–18.

Ghosh, P., P. K. Das and B. K. Bachhawat (1981). Selective uptake of liposomes by different cell types of liver through the involvement of liposomal surface glycosides. *Biochem. Soc. Trans.*, **9**, 512–514.

Gregoriadis, G. (1973). Drug entrapment in liposomes. *FEBS Lett.*, **36**, 292–296.

Gregoriadis, G. (1976). The carrier potential of liposomes in biology and medicine. *New Engl. J. Med.*, **295**, 704–710 and 765–770.

Gregoriadis, G. (ed.) (1979). *Drug Carriers in Biology and Medicine*. Academic, New York.

Gregoriadis, G. (1981a). Targeting of drugs : Implications in medicine. *Lancet*, **2**, 241–247.

Gregoriadis, G. (1981b). Liposomes: a role in vaccines? *Clin. Immunol. Newslett.*, **2**, 33–36.

Gregoriadis, G. (1982). Use of monoclonal antibodies and liposomes to improve drug delivery. *Drugs*, **24**, 261–266.

Gregoriadis, G. (ed.) (1984). *Liposome Technology*. CRC Press, Roca Baton, FL.

Gregoriadis, G. and A. C. Allison (eds.) (1980). *Liposomes in Biological Systems*. Wiley, New York.

Gregoriadis, G. and R. A. Buckland (1973). Enzyme-containing liposomes alleviate a model for storage disease. *Nature (London)*, **244**, 170–172.

Gregoriadis, G. and C. Davis (1979). Stability of liposomes *in vivo* and *in vitro* is promoted by their cholesterol content and the presence of blood cells. *Biochem. Biophys. Res. Commun.*, **89**, 1287–1293.

Gregoriadis, G. and E. K. Manesis (1980). Liposomes as immunological adjuvants for hepatitis B surface antigens. In *Liposomes and Immunobiology*, ed. B. H. Tom and H. R. Six, pp. 271–283. Elsevier/North Holland, Amsterdam.

Gregoriadis, G. and A. Meehan (1981). Interaction of antibody-bearing small unilamellar liposomes with antigen-coated cells: The effect of antibody and antigen concentration on the liposomal and cell surface respectively. *Biochem. J.*, **200**, 211–216.

Gregoriadis, G. and D. E. Neerunjun (1975). Homing of liposomes to target cells. *Biochem. Biophys. Res. Commun.*, **65**, 537–544.

Gregoriadis, G. and B. E. Ryman (1972a). Lysosomal localization of β-fructofuranosidase-containing liposomes injected into rats. *Biochem. J.*, **129**, 123–133.

Gregoriadis, G. and B. E. Ryman (1972b). Fate of protein-containing liposomes injected into rats. An approach to the treatment of storage diseases. *Eur. J. Biochem.*, **24**, 485–491.

Gregoriadis, G. and J. Senior (1980). The phospholipid component of small unilamellar liposomes controls the rate of clearance of entrapped solutes from the circulation. *FEBS Lett.*, **119**, 43–46.

Gregoriadis, G., J. Senior and A. Trouet (eds.) (1982). *Targeting of Drugs*. Plenum, New York.

Hemker, H. C., W. Th. Hermens, A. D. Muller and R. F. A. Zwaal (1980). Oral treatment of haemophilia A by gastrointestinal absorption of Factor VIII entrapped in liposomes. *Lancet*, **1**, 70–71.

Huang, A., L. Huang and S. J. Kennel (1980). Monoclonal antibody covalently coupled with fatty acid. *J. Biol. Chem.*, **255**, 8015–8018.

Hwang, K. J., K.-F. S. Luk and P. L. Baumier (1980). Hepatic uptake and degradation of unilamellar sphingomyelin/cholesterol liposomes: a kinetic study. *Proc. Natl. Acad. Sci. USA*, **77**, 4030–4034.

Jackson, A. J. (1981). Intramuscular absorption and regional lymphatic uptake of liposome-entrapped insulin. *Drug Metab. Dispos.*, **9**, 535–540.

Kirby, C. and G. Gregoriadis (1980). The effect of the cholesterol content of small unilamellar liposomes on the fate of their lipid components *in vivo*. *Life Sci.*, **27**, 2223–2230.

Kirby, C. and G. Gregoriadis (1981). Plasma-induced release of solutes from small unilamellar liposomes is associated with pore formation in the bilayers. *Biochem. J.*, **199**, 251–254.

Kirby, C. and G. Gregoriadis (1983). The effect of lipid composition of small unilamellar liposomes containing melphalan and vincristine on drug clearance after injection into mice. *Biochem. Pharmacol.*, **32**, 609–615.

Kirby, C., J. Clarke and G. Gregoriadis (1980a). The cholesterol content of small unilamellar liposomes controls phospholipid loss to high density lipoproteins in the presence of serum. *FEBS Lett.*, **111**, 324–328.

Kirby, C., J. Clarke and G. Gregoriadis (1980b). Effect of the cholesterol content of small unilamellar liposomes on their stability *in vivo* and *in vitro*. *Biochem. J.*, **186**, 591–598.

Krupp, L., A. V. Chobanian and I. P. Brecher (1976). The *in-vivo* transformation of phospholipid vesicles to a particle resembling HDL in the rat. *Biochem. Biophys. Res. Commun.*, **72**, 1251–1258.

Leserman, L. and J. Barbet (eds.) (1982). *Liposome Methodology*. Inserm, Paris.

Leserman, L. D., J. Barbet, F. Kourilsky and J. N. Weinstein (1980). Targeting to cells of fluorescent liposomes covalently coupled with monoclonal antibody or protein A. *Nature (London)*, **288**, 602–604.

Manesis, E. K., C. H. Cameron and G. Gregoriadis (1979). Hepatitis B surface antigen-containing liposomes enhance humoral and cell-mediated immunity to the antigen. *FEBS Lett.*, **102**, 107–111.

Martin, F. J., W. L. Hubbell and D. Papahadjopoulos (1981). Immunospecific targeting of liposomes to cells: a novel and efficient method for covalent attachment of Fab′ fragments *via* disulfide bonds. *Biochemistry*, **20**, 4229–4238.

Mauk, M. R., R. C. Gamble and T. D. Baldeschwieler (1980). Targeting of lipid vesicles: Specificity of carbohydrate receptor analogues for leucocytes in mice. *Proc. Natl. Acad. Sci. USA*, **77**, 4430–4434.

Osborne, M. P., V. J. Richardson, K. Jeyasingh and B. E. Ryman (1979). Radionuclide-labelled liposomes. A new lymph node imaging agent. *Int. J. Nucl. Med. Biol.*, **6**, 75–83.

Papahadjopoulos, D. (ed.) (1978). Liposomes and their uses in biology and medicine. *Ann. N.Y. Acad Sci.*, **308**.

Parker, R. J., K. D. Hartman and S. M. Sieber (1981). Lymphatic absorption and tissue disposition of liposome-entrapped [^{14}C]adriamycin following intraperitoneal administration of rats. *Cancer Res.*, **41**, 1311–1317.

Patel, H. M. and B. E. Ryman (1976). Oral administration of insulin encapsulated within liposomes. *FEBS Lett.*, **62**, 60–63.

Patel, H. M., R. W. Stevenson, J. A. Parsons and B. E. Ryman (1982). Use of liposomes to aid intestinal absorption of entrapped insulin in normal and diabetic dogs. *Biochim. Biophys. Acta*, **716**, 188–193.

Poste, G. (1980). Interaction of liposomes with cultured cells and their use as carriers for drugs and macromolecules. In *Liposomes in Biological Systems*, ed. G. Gregoriadis and A. C. Allison, pp. 101–151. Wiley, New York.

Poste, G., C. Bucana and I. J. Fidler (1982). Stimulation of host response against metastatic tumours by liposome-encapsulated immunomodulators. In *Targeting of Drugs*, ed. G. Gregoriadis, J. Senior and A. Trouet, pp. 261–284. Plenum, New York.

Rahman, Y.-E. (1980). Liposomes and chelating agents. In *Liposomes in Biological Systems*, ed. G. Gregoriadis and A. C. Allison, pp. 265–298. Wiley, New York.

Ryman, B. E. and D. A. Tyrrell (1980). Liposomes — bags of potential. *Essays Biochem.*, **16**, 49–98.

Scherphof, G., F. Roerdink, M. Waite and J. Parks (1978). Disintegration of phosphatidylcholine liposomes in plasma as a result of interaction with high density lipoproteins. *Biochim. Biophys. Acta*, **542**, 296–307.

Segal, A. W., G. Gregoriadis and C. D. V. Black (1975). Liposomes as vehicles for the local release of drugs. *Clin. Sci. Mol. Med.*, **49**, 99–106.

Senior, J. and G. Gregoriadis (1982a). Stability of small unilamellar liposomes in serum and clearance from the circulation: The effect of the phospholipid and cholesterol components. *Life Sci.*, **30**, 2123–2136.

Senior, J. and G. Gregoriadis (1982b). Is half-life of circulating liposomes determined by changes in their permeability? *FEBS Lett.*, **145**, 109–114.

Tom, B. H. and H. R. Six (eds.) (1980). *Liposomes and Immunobiology*. Elsevier/North Holland, Amsterdam.

4
Monoclonal Antibodies

G. S. EISENBARTH
Harvard Medical School, Boston, MA, USA

4.1 INTRODUCTION

'Biotechnology' utilizing antibodies for therapy of human disease has a long history and antiserum therapy for selected human illnesses continues today, including administration of antivenoms, antitoxins, globulin replacement therapy in immunodeficient individuals, antiserum to Rh antigens given at the time of childbirth and immunosuppression with antithymocyte globulin. In addition, antibody reagents (*e.g.* radioimmunoassays, immunocytophathology) are essential in the diagnosis and monitoring of plant, animal and human diseases. The importance of currently available antiserums highlights areas where monoclonal antibody technology has obvious potential. Furthermore, the unique properties of monoclonal antibodies and the essentially unlimited quantities of these antibodies should allow monoclonal antibodies to be used in novel ways. In particular, monoclonal antibodies are proving to be extremely important research reagents because they usually react with a single molecular species despite immunization with impure antigens. The production of specific monoclonal antibodies to cell surface differentiation antigens following immunization with whole cells is an extreme example of the use of 'impure antigens'. This review will discuss the general methodology for the production of monoclonal antibodies, recent developments in this technology, specific applications of monoclonal antibody techniques, and, finally, potential applications of hybridoma antibodies.

4.2 ANTIBODY PRODUCTION

The basic technology for the production of monoclonal antibodies derives from the studies of Köhler and Milstein (1975). A series of reviews and books discussing monoclonal antibody techniques are available (Milstein, 1980; Melchers *et al.*, 1978; Kennett *et al.*, 1980; Fellows and Eisenbarth, 1981; Oi and Hertzenberg, 1980; Köhler, 1980; Eisenbarth, 1981; Eisenbarth and Jackson, 1982; McKay *et al.*, 1981). Since 1975, the number of investigators using this technology has increased exponentially and many monoclonal antibodies can be obtained either commercially, through cell distribution centers such as the American Type Tissue Collection, or from individual investigators. With such widespread use of monoclonal antibody techniques, application of this technology to new areas has become easier. It takes relatively little time to fuse cells to produce hybridoma cell lines. Major time investments in the production of monoclonal antibodies usually occur only after potentially important antibodies are derived. At this time, cell lines are cloned to ensure their permanency and the antibody is studied in detail.

Figure 1 illustrates the general methodology for the production of monoclonal antibodies. A myeloma parental cell line (usually murine in origin) which lacks the enzyme hypoxanthine phosphoribosyltransferase is incubated with spleen cells from a mouse (the same strain as the parental myeloma) immunized with the desired antigen. To produce hybrid cells, the incubation occurs in the presence of poly(ethylene glycol). After several minutes of incubation, the poly(ethylene glycol) is diluted with medium and the cells are pelleted and the supernatant poured off. The cell mixture is then suspended in a culture medium containing fetal calf serum and aliquoted into microculture wells. Following removal of the poly(ethylene glycol), the cell mixture is usually distributed into ten 96-well microtiter plates (Köhler, 1980). Approximately one week after the fusion, the cultures are fed by adding one drop of additional medium. The culture medium contains aminopterin and hypoxanthine. Because the parental myeloma is deficient in the enzyme hypoxanthine phosphoribosyltransferase (HPRT), in the presence of aminopterin parental myeloma cells die. The normal spleen cells cannot proliferate in culture and thus only colonies of used cells develop. By approximately two weeks, colonies are readily visible. Culture supernatants are then assayed for the presence of antibody reacting with the immunizing antigen. Colonies producing the desired antibody are transferred to larger culture wells (0.5 to 1 ml of medium). When these cultures produce enough cells, their supernatants are assayed. At this stage, the supernatant is assayed for antibody specificity (Eisenbarth *et al.*, 1980, 1981a). For example, if antibodies reacting with islet cells are sought, the supernatant is tested for reaction with islet cells and with fibroblasts of the same species. If the culture supernatant contains antibodies which react only with islets, the culture is transferred to a flask (5 ml medium) and cloned by limiting dilution using a spleen cell feeder layer and 96-well plates. It is very important in the production of stable hybridomas to clone early. In addition, when sufficient cells are available, they are frozen viably in dimethyl sulfoxide-containing medium and ascites tumors are induced in mice. Culture supernatants produce approximately 10 to 20 μg ml^{-1} of antibody, while ascites fluid often contains more than 10 mg ml^{-1} of antibody.

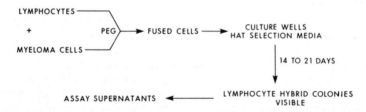

Figure 1 General scheme for producing monoclonal antibodies

There are several critical stages where cultures can be lost. In particular, it is useful to grow interesting hybrid colonies in different batches of medium for fear of contamination and when vials of frozen cells are obtained, these vials should be stored in more than one liquid nitrogen freezer. Finally, the cells from ascites tumors should be viably frozen, as they can be used to propagate ascites tumors and the culture.

There have been several recent improvements in the methodology for the production and utilization of monoclonal antibodies. These improvements include: (1) the development of new parental myeloma cell lines, such as a rat myeloma line (Galfrè *et al.*, 1979) and 'non-secretor' mouse lines (Schulman *et al.*, 1978): the initial cell line, P3X63 used by Köhler and Milstein (1975), secretes its own antibody, and when hybrid cells are produced with normal spleen cells, each hybrid cell has the potential to produce a series of 'hybrid' antibody molecules, containing antibody chains from both parents; (2) the development and utilization of human parental myeloma cell lines (Olsson and Kaplan, 1980; Croce *et al.*, 1980; Eisenbarth *et al.*, 1982b): by fusing human lymphocytes with human myeloma cells it is possible to produce human antibodies of defined specificity and investigators have also succeeded in fusing human lymphocytes with mouse myeloma cell lines to produce human immunoglobulins; (3) development of a serum-free medium to propagate murine hybridoma cell lines (Murakami *et al.*, 1982); (4) commercial availability of a series of immunologic reagents for the assaying of monoclonal antibodies, such as radioactivity labelled protein A, antimouse antibody, enzyme linked antiantibodies for use in ELISA assays and monoclonal antibodies; and (5) establishment of a cell bank for hybridomas and myeloma parents at the American Type Tissue Collection in Rockville, MD.

Will monoclonal antibodies replace current antiserums which are raised in a variety of animals from guinea pigs to horses? Monoclonal antibodies will then find use as antitoxins, antivenoms, 'antidotes', antibiotics, antilymphocyte reagents in the therapy of autoimmune disease, antitumor reagents, and as diagnostic reagents in 'radioimmunoassays' and in immunopathology. The remainder of this review will discuss specific applications of monoclonal antibody biotechnology.

4.3 SPECIFIC APPLICATIONS

4.3.1 Antidotes

Murine monoclonal antibodies have been produced which react with digoxin and tetanus toxin. Prior to the development of monoclonal antidigoxin antibodies, rabbit antibodies reacting with digoxin were produced (Smith *et al.*, 1976) and animal and clinical trials were conducted to test the ability of such antibodies to reverse potentially fatal cardiac arrhythmias induced by overdosage of this drug. Following antibody injection, the amount of circulating digoxin increases, apparently by removing digoxin bound to its receptor. Almost immediately the cardiac toxicity is reversed. The digoxin in circulation is inactive when coupled to the antibody and is cleared in the urine. A major drawback of digoxin antibodies produced in rabbits is the limited supply of such antibodies and the variability of antiserums produced in different rabbits. Monoclonal antibodies are homogeneous and can potentially be produced in unlimited quantity. Thus, monoclonal antidigoxin antibodies have now been produced (Haber *et al.*, 1981). A potential difficulty in the therapeutic use of these mouse antibodies is that repeated use of such antibodies may be limited by allergic phenomena or blocking by patient anti-antibodies. Future clinical trials will answer the question as to whether this will be a significant problem.

Several monoclonal antibodies to tetanus toxin have been generated (Haber *et al.*, 1981). In addition to potential clinical utility, antibodies to tetanus toxin have been used *in vitro* to aid in the localization of cells and tissues reacting with tetanus toxin (Figure 2). Binding of tetanus toxin to a cell surface can be detected by incubation with the monoclonal antitetanus toxin antibody followed by fluorescein coupled antimouse antibody. In addition to murine monoclonal antibodies, Butler and coworkers have recently reported the production of human monoclonal antibodies reacting with tetanus toxoid (Butler *et al.*, 1982).

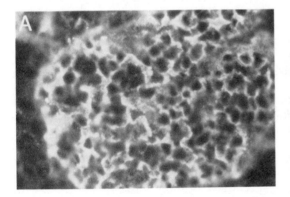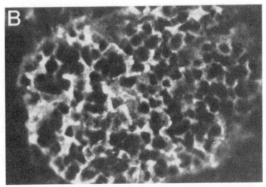

Figure 2 Binding of tetanus toxin (A) and monoclonal antibody A2B5 (B) specifically to islet cells of sectioned pancrease (detected with indirect immunofluorescence)

4.3.2 Antibiotics

A large number of murine monoclonal antibodies reacting with viruses, bacteria, fungi and other parasites are now available. One reason for this plethora of antibodies is that fusions following immunization with infectious agents result in a high percentage of antibodies reacting with the infecting organism. These antibodies are being used to identify and subtype viruses and bacteria. Probably of more import, a number of studies suggest that specific antibodies may block infection. A striking example of this antibiotic effect is the ability of two monoclonal antibodies reacting with herpes simplex virus glycoproteins to block fatal viral induced neurological disease (Dix *et al.*, 1981). There are other reports of the protective effect of monoclonal antibodies including activity against rabies virus (Wiktor and Kaprowski, 1978) and Group B streptococcal infection (Shigeoka *et al.*, 1982).

4.3.3   Antitumor Applications

Many investigators have begun to explore the use of monoclonal antibodies for cancer chemotherapy and tumor localization. The most impressive case report describing this form of therapy is by Miller and coworkers (1982). They describe the treatment of a B-cell lymphoma with monoclonal anti-idiotype antibody. The patient, after intravenous injection of murine monoclonal antibody directed at his malignant lymphoma, went into remission which lasted for more than six months (duration of follow-up) without further therapy of any kind. The form of tumor expressed by this patient, a B-cell lymphoma, allowed Miller and colleagues to circumvent a major question concerning immunotherapy of tumors. Namely, do tumor 'specific' antigens exist? Since the surface immunoglobulin of the B-cell lymphoma is monoclonal with a unique antibody variable region, the production of antibodies specific for this variable region results by definition in a tumor specific antibody. The effect of the injection of the anti-idiotype antibody was long lasting, and raises the possibility that the antibody perhaps stimulated the body's own immune mechanisms (anti-idiotype) to continually suppress the tumor. Animal studies suggest that anti-idiotype antibodies may indeed activate a suppressive 'immunoregulatory circuit' (Lynch, 1982).

The antitumor effect of other monoclonal antibodies administered to cancer patients has been limited (Ritz *et al.*, 1981; Ritz and Schlossman, 1982; Miller *et al.*, 1981; Sears *et al.*, 1982; Vogel and Müller-Eberhard, 1981). A number of factors (reviewed by Ritz and Schlossman, 1982) including the presence of circulating antigen, loss of specific antigen by tumor cells (antigenic modulation), the development of 'antimouse' anti-antibodies, and limited destruction of cells binding antibodies appear to hinder successful therapy. A number of investigators have coupled monoclonal antibodies to toxins (Youle and Neville, 1982; Neville and Youle, 1982). In particular, the A chain of ricin has been coupled to monoclonal antibodies. The A chain of ricin in the absence of the B chain, which binds to cell surfaces, is much less toxic than the intact ricin molecule. Monoclonal–ricin A chain conjugates are selectively cytotoxic to cells reacting with the monoclonal antibody. In addition to proposals for use of such conjugates in cancer therapy, such conjugates have been used to deplete bone marrow of mature T cells (Vallera *et al.*, 1982a, 1982b) prior to bone marrow transplantation.

In addition to use as chemotherapeutic agents, the potential use of monoclonal antibodies to image tumors has been studied in animals (Ballou *et al.*, 1979; Shimizu *et al.*, 1982) and in humans (Stuhmiller *et al.*, 1982). In typical studies, monoclonal antibodies are labelled with ^{131}I or radioactive metal chelates (Scheinberg *et al.*, 1982) and administered intravenously. Localization of the labelled monoclonal antibody is detected with a gamma camera. Most of the monoclonal antibodies studied to date have been selected because the antibody is 'tumor specific'. We have recently discovered that antibody A2B5, which reacts with a complex (GQ) cell surface ganglioside expressed on a series of normal cells (islet, neurons, other neuroendocrine) and 'neuroendocrine' tumors, when administered intravenously specifically localizes in tumor cells (Figure 3). An advantage of studying imaging techniques with this particular antibody is that it reacts with cells from many vertebrate species (rat, mouse, dog and human) and thus scanning techniques can be developed for studying a rat tumor in a rat with an antigen also expressed by human cells. The majority of monoclonal antibodies (usually reacting with proteins) react only with cells from the species used for immunization. The specificity of binding of monoclonal antibody A2B5 to the rat insulinoma and absence of binding to brain and islet cells (Eisenbarth *et al.*, 1981b; Shimizu *et al.*, 1982) suggests that tumor localization may depend on other factors in addition to antibody affinity, perhaps altered circulation within the tumor facilitating access of this IgM antibody to cells. Differential access *in vivo* to tumor cells may facilitate development of useful monoclonal antibodies for tumor imaging, since tumors express a series of 'activation' antigens [*e.g.* 4F2, 5E9 (transferrin receptor)] which are also expressed on selective normal cells (Eisenbarth *et al.*, 1980; Haynes *et al.*, 1981c).

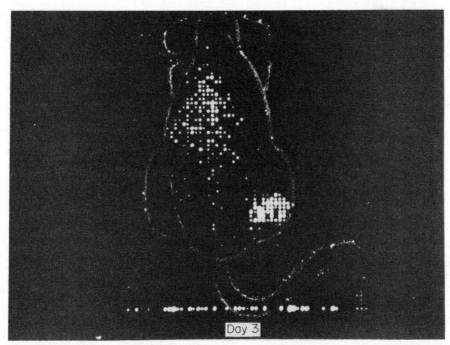

Figure 3 External scan of a rat bearing a syngenic insulinoma in its right hindquarters following injection of radioactive monoclonal antibody A2B5. The concentration of the antibody in the tumor is apparent. The central radioactive area corresponds to heart, liver and lungs

4.3.4 Lymphocyte Phenotyping and Monoclonal Antibody Therapy of Autoimmune Diseases

A large number of monoclonal antibodies reacting with leukocyte subsets, and in particular, T-lymphocyte subsets of man, mouse and rat, are now available (Haynes, 1981). These antibodies, particularly when coupled with fluorescent cell sorter analysis, allow investigators to rapidly quantitate circulating subsets of cells which could not be detected seven years ago. It is likely that there will not be a simple correlation between former broad definitions of 'helper' and 'suppressor' T lymphocytes and monoclonal antibody defined subsets, such as 3A1, T4, T8, *etc.* (Haynes, 1981). The complexity of the immune system and the complexity of specific *in vitro* functional assays make precise correlation unlikely. In addition, a series of monoclonal antibodies has been discovered whose antigens are absent on resting T cells but appear when T cells are triggered to divide, including the transferrin receptor (antibody 5E9; Haynes *et al.*, 1981a), antigen 4F2, a 120 000 dalton glycoprotein (Haynes *et al.*, 1981b), and the Ia antigen (Winchester and Kunkel, 1979). These reagents now permit a much more precise study of T cell subsets and T cell activation and their correlation with disease. Even in illnesses not classically considered 'autoimmune' in etiology, such as diabetes mellitus, these reagents are proving useful. For example, an increasing body of data indicate that one form of diabetes mellitus, Type I diabetes, is autoimmune in origin (Cahill and McDevitt, 1981). Important supporting evidence of this autoimmune theory is the ability of immunotherapy to prevent or 'cure' the 'genetic' diabetes of the BB rat (Like *et al.*, 1979). Approximately 50% of BB rats develop an acute form of diabetes characterized by β-cell destruction with lymphocytic infiltration of the islets. We and other investigators (Jackson *et al.*, 1981b; Poussier *et al.*, 1982; Kadison *et al.*, 1982) have discovered a severe T cell lymphopenia of the BB rats which is characterized by a selective deficiency of circulating monoclonal antibody defined W3/25 cells. These W3/25 cells in normal rats comprise approximately 80% of circulating T cells, whereas in the BB rat there are less than 10% W3/25 positive cells (Table 1). The relationship of this severe immunodeficiency to the development of diabetes and other autoimmune diseases of these animals is unknown, but it is likely that this severe disorder of immunoregulation predisposes animals to the development of diabetes. The abnormalities in the BB rat are not directly analogous to the T cell abnormalities we have found in Type I diabetes mellitus (Jackson *et al.*, 1982).

Table 1 Comparison of Circulating White Blood Cells

	Control strain	*BB Wistar*	*Difference*
Experiment 1			
White blood count	12 612[a] ± 1 677	7 422 ± 805	5 190[b]
Polymorphonuclear leucocytes	1 982 ± 419	2 600 ± 560	−618
Monocytes	418 ± 64	283 ± 68	135
Lymphocytes	10 290 ± 1 220	4 549 ± 409	5 741[b]
B lymphocytes	1 338 ± 171	1 581 ± 224	−243
Non-B lymphocytes	8 953 ± 1 058	2 968 ± 246	5 985[b]
Experiment 2			
B lymphocytes	1 569 ± 242	1 495 ± 595	74
Non-B lymphocytes	6 254 ± 765	2 412 ± 451	3 833[b]
W3/13 + non-B	5 929 ± 727	2 111 ± 396	3 818[b]
W3/25 + non-B	4 996 ± 612	169 ± 32	4 827[b]
W3/25 − non-B	1 249 ± 153	2 243 ± 419	−994

[a] In experiment 1, mean ± SEM of four control rats and five BB Wistar rats; experiment 2, mean ± SEM of five control rats and three BB Wistar rats.   [b] $p < 0.01$ by Student's t test. From Jackson *et al.* (1981b).

In human Type I diabetes there is no lymphopenia (Table 2). There is, however, a marked elevation of T cells bearing the Ia antigen (Figure 4) recognized with monoclonal antibody L243. The Ia antigen of humans is coded for by the HLA-D region of chromosome 6 and is similar to the Ia antigens of mice and other species. This antigen is present on B lymphocytes and some monocytes; small quantities may be expressed on normal circulating T lymphocytes, but activated T cells express large amounts of the Ia antigen.

Using indirect immunofluorescence and monoclonal antibody reagents (anti-Ia, monoclonals

Table 2 T-Lymphocyte Subsets in Patients with Type I Diabetes Mellitus and in Controls, as Defined by Monoclonal Antibodies[a]

Group *According to no. of cells/μl*	White cells	Lymphocytes	3A1	T4	T8	4F2	5E9	Ia[b]
Patients (n=11)	6366±500	2756±380	1781±216	1143±216	588±138	32±11	18±6.6	235±48[c]
Controls (n=8)	6215±620	2406±445	1822±360	834±67	524±67	23±9	13±6	24±12

According to % of T cells	T Cells	B Cells	3A1	T4	T8	4F2	5E9	Ia[b]
Patients (n=11)	2299±294	320±74	84±2	53±3.5	28±1.8	1.8±0.5	1.2±0.5	10.1±1.5[c]
Controls (n=8)	2142±425	241±35	85±4.3	54±4.3	35±13	1.4±0.5	0.5±0.2	0.9±0.3

[a] Data are expressed as mean ± SEM. [b] Determined by anti-Ia antibody L243. [c] $p < 0.01$ by Student's t test. [d] Reproduced with permission from Jackson *et al.* (1982).

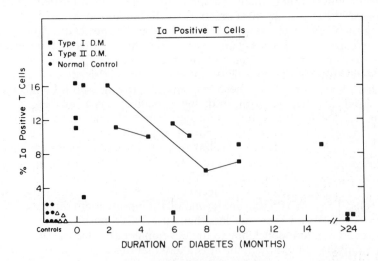

Figure 4 Activated T cells in patients with Type I diabetes mellitus, an autoimmune endocrine disease. These T cells are quantitated by the binding of a monoclonal antibody to the Ia antigen. The percentage of positive cells is shown on the x-axis (Reproduced from Jackson *et al.*, 1982, with permission of the New England Journal of Medicine)

4F2 or 5E9), T cell activation is now being studied in a series of autoimmune diseases. In Type I diabetes only the Ia antigen is detected, whereas in other illnesses such as lupus erythematosus T cells bearing the Ia antigen and transferrin receptor (5E9) are found (Cohen *et al.*, 1982). Monoclonal antibody defined T cell abnormalities in a number of human diseases have been defined (Veys *et al.*, 1981; Reinherz *et al.*, 1981a; Sridama *et al.*, 1982). In addition, the T cells directly infiltrating tissue are being studied using monoclonal antibodies (Jackson *et al.*, 1981a; Rowe *et al.*, 1981).

A somewhat natural development following the identification and characterization of T cell subsets with monoclonal antibodies is the use of these same antibodies to treat autoimmune disease. In particular, a number of investigators have now used monoclonal antibodies in animals and man to block 'graft-*versus*-host' disease. When mature bone marrow cells are given to an immunodeficient animal, these lymphocytes grow in their new host and can react with the host tissue and cause a fatal disease, graft-*versus*-host disease. Elimination *in vitro* of mature bone marrow cells with anti-T cell monoclonal antibodies [anti-Lyt 1.2 in the mouse (Vallera *et al.*, 1982a); OKT3 (Prentice *et al.*, 1982); and T16 (Reinherz *et al.*, 1982); monoclonal antibodies in man] has been used to block the development of, or to treat graft-*versus*-host disease. The monoclonal antibody is incubated *in vitro* in the presence of complement with donor marrow cells, prior to marrow tranplantation, thereby killing mature T cells. Initial studies suggest that elimination of these mature T cells can reduce (Prentice *et al.*, 1982) or prevent (Reinherz *et al.*, 1982) graft-*versus*-host disease. In the study of a single patient reported by Reinherz and co-workers (1982) in addition to *in vitro* marrow treatment, the monoclonal antibody was injected *in vivo*. *In vivo* administration of monoclonal antibody OKT3 has also been used to treat acute rejection of renal transplants (Cosimi *et al.*, 1981).

Another area of human transplantation in which monoclonal antibodies are beginning to be used is for histocompatibility typing. Because most xenogenic monoclonal antibodies react with non-polymorphic regions of histocompatibility antigens, relatively few tissue typing monoclonal antibodies are available (Parham and Bodmer, 1978; Lampson *et al.*, 1978; Herrmann and Mescher, 1979; Haynes *et al.*, 1982). Nevertheless, with time, more of these stable and well characterized reagents will become available and they will hopefully replace current serums from multiparous women for routine tissue typing.

4.3.5 Radioimmunoassays

Monoclonal antibodies have the potential to replace or supplement conventional antibodies in a series of assays (Staehelin *et al.*, 1981; Deverill *et al.*, 1981; Stenman *et al.*, 1981b; Greene *et al.*, 1980; Wards *et al.*, 1982; Slovin *et al.*, 1982; Gheuens and McFarlin, 1981; Schneider and Eisenbarth, 1979; Bundesen *et al.*, 1980; Green and Jensen, 1981). In particular, availability of more than one monoclonal antibody reacting with different sites of single molecules allows investigators to produce highly specific radioassays in which one of the monoclonal antibodies is labelled, rather than the antigen. A number of the antibodies generated have the requisite specificity and sensitivity for use in radioimmunoassays.

4.3.6 Novel Applications of Monoclonal Antibodies

In addition to replacing conventional antisera, the availability of large amounts of monoclonal antibodies has led to a number of novel uses for these antibodies. For example, they have been used to purify molecules such as Mullerian inhibiting substance (Mudgett-Hunter *et al.*, 1982) and α-fetoprotein (Stenman *et al.*, 1981a); to prepare Factor IX deficient human plasma by immunoaffinity chromatography (Goodall *et al.*, 1982); to stimulate growth (Schreiber *et al.*, 1981) or inhibit the growth (Trowbridge and Lopez, 1982) of cells in culture; to isolate subsets of lymphocytes using bulk techniques (Reinherz *et al.*, 1981b); to isolate islet cells from pancreatic digests with a fluorescent cell sorter (Eisenbarth *et al.*, 1981b; Eisenbarth *et al.*, 1982b; Eisenbarth *et al.*, 1979) or plating techniques (Kortz *et al.*, 1982).

4.4 CONCLUSIONS

This review has dealt in large part with medical aspects of the use of monoclonal antibodies. By analogy to the areas discussed, these techniques will find application in many fields. The most interesting uses for monoclonal antibodies are probably yet to come as investigators from all disciplines exploit this technology. The ability to patent hybrid cell lines (Raub, 1981) and the novel processes developed with such antibodies (Hoscheit, 1981) will undoubtedly contribute to the development of this biotechnology.

ACKNOWLEDGEMENTS

Research supported by grants from the Kroc Foundation, National Foundation, National Institutes of Health. G. Eisenbarth is the recipient of a Career Development Award from the Juvenile Diabetes Foundation.

4.5 REFERENCES

Becton-Dickenson, Monoclonal Antibody Source Book. Becton-Dickenson, Sunnyvale, CA.
Ballou, B., G. Levine, T. R. Hakala and D. Solter (1979). Tumor location detected with radioactively labeled monoclonal antibody and external scintigraphy. *Science*, **206**, 844–847.
Bundesen, P. G., R. G. Drake, K. Kelly, I. G. Worsley, H. G. Friesen and A. H. Sehon (1980). Radioimmunoassay for human growth hormone using monoclonal antibodies. *J. Clin. Endocrinol. Metab.*, **51**, 1472–1474.
Butler, J. L., H. C. Lane and A. S. Fauci (1982). Delineation of optimal conditions for producing hybridomas from human peripheral blood B lymphocytes of immunized subjects. *Clin. Res.*, **30**, 344A (abstract).
Cahill, G. F. and H. O. McDevitt (1981). Insulin-dependent diabetes mellitus: the initial lesion. *New Engl. J. Med.*, **304**, 1454–1456.
Cohen, P. L., D. A. Litvin, G. S. Eisenbarth and J. B. Winfield (1982). Transferrin receptors are released by stimulated human T cells. *Clin. Res.*, **30**, 345A (abstract).
Cosimi, A. B., R. B. Colvin, R. C. Burton, R. H. Rubin, G. Goldstein, P. C. Kung, W. P. Hansen, F. L. Delmonico and P. S. Russell (1981). Use of monoclonal antibodies to T-cell subsets of immunologic monitoring and treatment in recipients of renal allografts. *New Engl. J. Med.*, **305**, 308–314.
Croce, C. M., A. Linnenbach, W. Hall, Z. Steplewski and H. Koprowski (1980). Production of human hybridomas secreting antibodies to measles virus. *Nature (London)*, **288**, 488–489.
Deverill, I., R. Jeffries, N. R. Ling and W. G. Reeves (1981). Monoclonal antibodies to human IgG: reaction characteristics in the centrifugal analyzer. *Clin. Chem.*, **27**, 2044–2047.
Dix, R. D., L. Pereira and J. R. Baringer (1981). Use of monoclonal antibody directed against herpes simplex virus glycoproteins to protect mice against acute virus-induced neurologic disease. *Infect. Immun.*, **34**, 192–199.
Eisenbarth, G. S. (1981). Application of monoclonal antibody techniques to biochemical research. *Anal. Biochem.*, **111**, 1–16.
Eisenbarth, G. S. and R. A. Jackson (1982). Application of monoclonal antibody techniques to endocrinology. *Endocr. Rev.*, **3**, 26–39.
Eisenbarth, G. S., B. Haynes, J. Schroer and A. Fauci (1980). Production of monoclonal antibodies reacting with peripheral blood mononuclear cell surface differentiation antigens. *J. Immunol.*, **124**, 1237–1244.
Eisenbarth, G. S., F. S. Walsh and M. Nirenberg (1979). Monoclonal antibody to a plasma membrane antigen of neurons. *Proc. Natl. Acad. Sci. USA*, **76**, 4913–4917.
Eisenbarth, G. S., H. Oie, A. Gazdar, W. Chick, J. A. Schultz and R. M. Scearce (1981a). Production of monoclonal antibodies reacting with rat islet cell membrane antigens. *Diabetes*, **30**, 226–230.
Eisenbarth, G. S., K. Shimizu, B. Mittler, M. Conn and S. Wells (1981b). Monoclonal antibody F12A2135: expression on neuronal and endocrine cells. In *Monoclonal Antibodies to Neural Antigens, Cold Spring Harbor Symposium*, ed. R. McKay *et al.*, pp. 209–218. Cold Spring Harbor Laboratory.
Eisenbarth, G. S., A. Linnenbach, H. Chopra, R. A. Jackson, R. Scearce and C. Croce (1982a). Human hybridomas secreting anti-islet autoantibodies. *Nature (London)*, **300**, 264–267.
Eisenbarth, G. S., K. Shimizu, M A. Bowring and S. Wells (1982b). Expression of receptors for tetanus toxin and monoclonal antibody A2B5 by pancreatic islet cells. *Proc. Natl. Acad. Sci. USA*, **79**, 5066–5070.
Fellows, R. E. and G. S. Eisenbarth (eds.) (1981). *Monoclonal Antibodies in Endocrine Research*. Raven Press, New York.
Galfrè, G., C. Milstein and B. Wright (1979). Rat × rat hybrid myelomas and a monoclonal anti-Fd portion of mouse IgG. *Nature (London)*, **277**, 131–133.
Gheuens, J. and D. E. McFarlin (1981). A multivalent antibody radioimmunoassay (MARIA) for screening specific antibody secretion by lymphocyte hybridoma cultures. *J. Immunol. Methods*, **47**, 183–189.
Goodall, A. H., G. Kemble, D. P. O'Brien, E. Rawlings, F. Rotblat, G. C. Russell, G. Janossy and E. G. D. Tuddenham (1982). Preparation of factor IX deficient human plasma by immunoaffinity chromatography using a monoclonal antibody. *Blood*, **59**, 664–670.
Greene, G. L., F. W. Fitch and E. V. Jensen (1980). Monoclonal antibodies to estrophilin: probes for the study of estrogen receptors. *Proc. Natl. Acad. Sci. USA*, **77**, 157–161.
Greene, G. S. and E. V. Jensen (1981). Monoclonal antibodies to estrophilin: probes for the study of steroid hormone action. In *Monoclonal Antibodies in Endocrine Research*, ed. R. Fellows and G. S. Eisenbarth, pp. 143–155. Raven Press, New York.
Haber, E., P. Donahoe, P. Ehrlich, J. Hurrell, H. Katus, B. A. Khaw, M. N. Margolies, M. Mudgett-Hunter and V. R. Zurawski (1981). Resolving antigenic sites and purifying protein with monoclonal antibodies. In *Monoclonal Antibodies in Endocrine Research*, ed. R. Fellows and G. S. Eisenbarth, pp. 1–11. Raven Press, New York.
Haynes, B. F. (1981). Human T lymphocyte antigens as defined by monoclonal antibodies. *Immunol. Rev.*, **57**, 127–161.

Haynes, B. F., M. Hemler, T. Cotner, D. L. Mann, G. S. Eisenbarth, J. L. Strominger and A. S. Fauci (1981a). Characterization of a monoclonal antibody (5E9) that defines a human cell surface antigen of cell activation. *J. Immunol.*, **127**, 347–351.

Haynes, B. F., M. E. Hemler, D. L. Mann, G. S. Eisenbarth, J. Shelhamer, H. S. Mostowski, C. A. Thomas, J. L. Strominger and A. S. Fauci (1981b). Characterization of a monoclonal antibody (4F2) that binds to human monocytes and to a subset of activated lymphocytes. *J. Immunol.*, **126**, 1409–1414.

Haynes, B. F., M. E. Hemler, D. L. Mann, G. S. Eisenbarth, J. L. Strominger and A. S. Fauci (1981c). Characterization of a monoclonal antibody (4F2) which binds to human PB monocytes and activated lymphocytes. In *Heterogeneity of Mononuclear Phagocytes*, ed. O. Foster, pp. 53–59. Academic, London.

Haynes, B. F., E. G. Reisner, M. E. Hemler, J. L. Strominger and G. S. Eisenbarth (1982). Description of a monoclonal-antibody defining an HLA allotypic determinant that include specificities within the B5 cross-reacting group. *Human Immunol.*, **4**, 273–285.

Herrmann, S. H. and M. F. Mescher (1979). Purification of the H-2K^k molecule of the murine major histocompatibility complex. *J. Biol. Chem.*, **254**, 8713–8716.

Hoscheit, D. H. (1981). United States patent requirements. *In Vitro*, **17**, 1084–1085.

Jackson, R., M. Bowring, M. Morris, B. Haynes and G. S. Eisenbarth (1981a). Increased circulating Ia positive T cells in recent onset Graves' disease and insulin-dependent diabetes. *63rd Annual Meeting of the Endocrine Society*, p. 195.

Jackson, R., N. Rassi, T. Crump, B. Haynes and G. S. Eisenbarth (1981b). The BB diabetic rat: profound T cell lymphocytopenia. *Diabetes*, **30**, 887–889.

Jackson, R. A., M. A. Morris, B. F. Haynes and G. S. Eisenbarth (1982). Increased circulating Ia-antigen-bearing T cells in Type I diabetes mellitus. *New Engl. J. Med.*, **306**, 785–788.

Kadison, P., R. Jackson, N. Rassi, B. Haynes, B. Jegasothy and G. S. Eisenbarth (1982). The BB rat: selective W3/25 T cell lymphocytopenia and profound depression of mitogen responsiveness. *Clin. Res.*, **30**, 396A (abstract).

Kennett, R. H., T. J. McKearn and K. B. Bechtol (eds.) (1980). *Monoclonal Antibodies*. Plenum, New York.

Köhler, G. (1980). *Hybridoma Techniques*. EMBO, SKMB Course, Basel.

Köhler, G. and C. Milstein (1975). Continuous cultures of fused cells secreting antibody of predefined specificity. *Nature (London)*, **256**, 495–497.

Kortz, W. J., T. H. Reiman, R. R. Bollinger and G. S. Eisenbarth (1982). Identification and isolation of rat and human islet cells using monoclonal antibodies. *Surg. Forum*, **33**, 354–356.

Lampson, L. A., R. Levy, F. C. Grumet, D. Ness and D. Pious (1978). Production *in vitro* of murine antibody to a human histocompatibility alloantigen. *Nature (London)*, **271**, 461–462.

Like, A. A., A. A. Rossini, D. L. Guberski, M. C. Appel and R. M. Williams (1979). Spontaneous diabetes mellitus: reversal and prevention in the BB/W rat with antiserum to rat lymphocytes. *Science*, **206**, 1421–1423.

Lynch, R. (1982). Immunoregulation of malignant lymphoma. *New Engl. J. Med.*, **306**, 543–544.

McKay, R., M. C. Raff and L. F. Reichardt (1981). *Monoclonal Antibodies to Neural Antigens*. Cold Spring Harbor Laboratory, Cold Spring Harbor, NY.

Melchers, F., M. Potter and N. L. Warner (eds.) (1978). Lymphocyte hyridomas. Second workshop on functional properties of tumors of T and B lymphocytes. Preface. In *Current Topics in Microbiology and Immunology*, vol. 81, ix–xxii. Springer-Verlag, Berlin.

Miller, R. A., D. G. Maloney, J. McKillop and R. Levy (1981). *In vivo* effects of murine hybridoma monoclonal antibody in a patient with T-cell leukemia. *Blood*, **58**, 78–86.

Miller, R. A., D. G. Maloney, R. Warnhe and R. Levy (1982). Treatment of B cell lymphoma with monoclonal anti-idiotype antibody. *N. Engl. J. Med.*, **306**, 517–522.

Milstein, C. (1980). Monoclonal antibodies. *Sci. Am.*, **243**, 66–74.

Mudgett-Hunter, M., G. P. Budzik, M. Sullivan and P. K. Donahoe (1982). Monoclonal antibody to Mullerian inhibiting substance. *J. Immunol.*, **128**, 1327–1333.

Murakami, H., H. Masui, G. H. Sato, N.Sueoko, T. P. Chow and T. Kano-Sueoka (1982). Growth of hybridoma cells in serum-free medium: ethanolamine is an essential component. *Proc. Natl. Acad. Sci. USA*, **79**, 1158–1162.

Neville, D. M., Jr. and R. J. Youle (1982). Monoclonal antibody–ricin or ricin A chain hybrids: kinetic analysis of cell killing for tumor therapy. *Immunol. Rev.*, **62**, 75–91.

Oi, V. T., and L. A. Herzenberg (1980). Immunoglobulin-producing hybrid cell lines. In *Selected Methods in Cellular Immunology*, ed. B. B. Mishell and S. M. Shiligi. Freeman Press, San Francisco.

Olsson, L. and H. S. Kaplan (1980). Human–human hybridomas producing monoclonal antibodies of predefined antigenic specificity. *Proc. Natl. Acad. Sci. USA*, **77**, 5429–5431.

Parham, P. and W. F. Bodmer (1978). Monoclonal antibody to a human histocompatibility alloantigen, HLA-A2. *Nature (London)*, **276**, 397–399.

Poussier, P., A. F. Nakhooda, J. A. Falk, C. Lee and E. B. Marliss (1982). Lymphopenia and abnormal lymphocyte subsets in the 'BB' rat: relationship to the diabetic syndrome. *Endocrinology*, **110**, 1825–1827.

Prentice, H. G., G. Janossy, D. Skeggs, H. A. Blacklock, H. F. Bradstock, G. Goldstein and A. V. Hoffbrand (1982). Use of anti-T-cell monoclonal antibody OKT3 to prevent acute graft-*versus*-host disease in allogenic bone-marrow transplantation for acute leukemia. *Lancet*, March, 700–703.

Raub, W. F. (1981). NIH policies on hybridomas. *In Vitro*, **17**, 1089–1090.

Reinherz, E. L., M. D. Cooper and S. F. Schlossman (1981a). Abnormalities of T cell maturation and regulation in human beings with immunodeficiency disorders. *J. Clin. Invest.*, **68**, 699–705.

Reinherz, E. L., A. C. Penta, R. E. Hussey and R. F. Schlossman (1981b). A rapid method for separating functionally intact human T lymphocytes with monoclonal antibodies. *Clin. Immunol. Immunopathol.*, **21**, 257–266.

Reinherz, E. L., R. Geha, J. M. Rappeport, M. Wilson, S. F. Schlossman and F. S. Rosen (1982). Immune reconstitution in severe combined immunodeficiency following transplantation with T lymphocyte depleted HLA mismatched bone marrow. *Clin. Res.*, **30**, 515A (abstract).

Ritz, J. and S. F. Schlossman (1982). Utilization of monoclonal antibodies in the treatment of leukemia and lymphoma. *Blood*, **59**, 1–11.

Ritz, J., J. M. Pesando, S. E. Sallan, L. A. Clavell, J. Notis-McConarty, P. Rosenthal and S. F. Schlossman (1981). Serotherapy of acute lymphoblastic leukemia with monoclonal antibody. *Blood*, **58**, 141–152.

Rowe, D. J., D. A. Isenberg, J. McDougall and P. C. L. Beverly (1981). Characterization of polymyositis infiltrates using monoclonal antibodies to human leukocyte antigens. *Clin. Exp. Immunol.*, **45**, 290–298.

Scheinberg, D. A., M. Strand and O. A. Gansow (1982). Tumor imaging with radioactive metal chelates conjugated to monoclonal antibodies. *Science*, **215**, 1511–1513.

Schneider, M. D. and G. S. Eisenbarth (1979). Transfer plate radioassay using cell monolayers to detect anti-cell surface antibodies synthesized by lymphocyte hybridomas. *J. Immunol. Methods*, **29**, 331–342.

Schreiber, A. B., I. Lax, Y. Yarden, Z. Eshhar and J. Schlessinger (1981). Monoclonal antibodies against receptor for epidermal growth factor induce early and delayed effects on epidermal growth factor. *Proc. Natl. Acad. Sci. USA*, **78**, 7535–7539.

Schulman, M., C. D. Wilde and G. Köhler (1978). A better cell line for making hybridomas secreting specific antibodies. *Nature (London)*, **276**, 268–269.

Sears, H. F., J. Mattis, D. Herlyn, P. Häyry, B. Atkinson, C. Ernst, Z. Steplewski and H. Koprowski (1982). Phase-I clinical trial of monoclonal antibody in treatment of gastrointestinal tumors. *Lancet*, ii, 762–765.

Shigeoka, A. O., S. H. Pincus and H. R. Hill (1982). Protective effect of hybridoma type specific antibody for experimental group B streptococcal infection. *Clin. Res.*, **30**, 127A (abstract).

Shimizu, K., D. Reingten, E. Colman, W. Briner, H. Seigler, R. Rowley and G. S. Eisenbarth (1982). *In vivo* and *in vitro* binding of iodinated-monoclonal antibody A2B5 to RIN insulinoma cells. *Hybridoma*, **2**, 69–77.

Slovin, S. F., D. M. Frisman, C. D. Tsoukas, I. Royston, S. M. Baird, S. B. Wormsley, D. A. Carson and J. H. Vaughn (1982). Membrane antigen on Epstein–Barr virus infected human B cells recognized by a monoclonal antibody. *Proc. Natl. Acad. Sci. USA*, **79**, 2649–2653.

Smith, T. W., E. Haber, L. Yeatman and V. P. Butler (1976). Reversal of advanced digoxin intoxication with Fab fragments of digoxin-specific antibodies. *New Engl. J. Med.*, **294**, 797–800.

Sridama, V., F. Pacini and L. Degroot (1982). Decreased suppressor T-lymphocytes in autoimmune thyroid diseases detected by monoclonal antibodies. *J. Clin. Endocr. Metab.*, **54**, 316–319.

Staehelin, T., C. Stähli, D. S. Hobbs and S. Pestka (1981). A rapid quantitative assay of high sensitivity for human leukocyte interferon with monoclonal antibodies. *Methods Enzymol.*, **79**, 589–595.

Stenman, U.-H., M.-L. Sutinen, R.-K. Selander, K. Tontti and J. Schröder (1981a). Characterization of a monoclonal antibody to human alpha-fetoprotein and its use in affinity chromatography. *J. Immunol. Methods*, **46**, 337–345.

Stenman, U.-H., P. Tanner, T. Ranta, J. Schroder and M. Seppala (1981b). Monoclonal antibodies to charionic gonadotropin: Use in a rapid radioimmunoassay for gynecological emergencies. *Obst. Gynecol.*, **59**, 375–377.

Stuhlmiller, G. M., M. J. Borowitz, B. P. Croker and H. F. Seigler (1982). Multiple assay characterization of murine monoclonal antimelanoma antibodies. *Hybridoma*, **1**, 447–460.

Trowbridge, I. S. and F. Lopez (1982). Monoclonal antibody to transferrin receptor blocks transferrin binding and inhibits human tumor cell growth *in vitro*. *Proc. Natl. Acad. Sci. USA*, **79**, 1175–1179.

Vallera, D. A., C. C. B. Soderling and J. H. Kersey (1982a). Bone marrow transplantation across major histocompatibility barriers in mice. III. Treatment of donor grafts with monoclonal antibodies directed against Lyt determinants. *J. Immunol.*, **128**, 871–875.

Vallera, D. A., R. J. Youle, D. M. Neville and J. H. Kersey (1982b). Bone marrow transplantation across major histocompatability barriers. Protection of mice from lethal graft-*vs.*-host disease by pretreatment of donor cells with monoclonal Anti-Thy-1.2 coupled to the toxin ricin. *J. Exp. Med.*, **155**, 949–954.

Veys, E. M., P. Hermanns, G. Goldstein, P. Kung, J. Schindler and J. VanWauwe (1981). Determination of T lymphocyte subpopulations by monoclonal antibodies in rheumatoid arthritis. Influence of immunomodulating agents. *Int. J. Immunopharmacol.*, **3**, 313–319.

Vogel, C.-W. and H. J. Müller-Eberhard (1981). Induction of immune cytolysis: tumor cell killing by complement is initiated by covalent complex of monoclonal antibody and stable C3/C5 convertase. *Proc. Natl. Acad. Sci. USA*, **78**, 7707–7711.

Wards, J. R., R. R. Bruns, R. I. Carlson, A. Ware, J. E. Menitove and K. J. Isselbacher (1982). Monoclonal IgM radioimmunossay for hepatitis B surface antigen: High binding activity in serum that is unreactive with conventional antibodies. *Proc. Natl. Acad. Sci. USA*, **79**, 1277–1281.

Wiktor, T. J. and H. Koprowski (1978). Monoclonal antibodies against rabies virus produced by somatic cell hybridization: Detection of antigenic variants. *Proc. Natl. Acad. Sci. USA*, **75**, 3938–3942.

Winchester, R. J. and H. G. Kunkel (1979). The human Ia system. *Adv. Immunol.*, **28**, 221–292.

Youle, R. J. and D. M. Neville, Jr. (1982). Kinetics of protein synthesis inactivation by Ricin-anti-thy 1.1 monoclonal antibody hybrids. *J. Biol. Chem.*, **257**, 1598–1601.

5
Transplantation Immunology

K. KANO
University of Tokyo, Japan

5.1 INTRODUCTION

The objective of organ and tissue transplantation in clinical medicine is rather straightforward, namely the replacement of a diseased, injured or lost organ by a graft from another individual. Such a graft would replace the vital function of the organ that cannot adequately be restored by other means. This situation is quite analogous to that of an old car requiring the replacement of various parts by new ones to maintain its normal function. The ancient dream involving the creation of an individual carrying a part of the body of another or even another species, was expressed in the form of mythical beasts such as mermaids or chimeras with supernatural forces. The latter, according to Greek mythology, consisted of a lion's head, goat's body and dragon's tail. In modern terminology, a chimera denotes a mixture of tissues or cells originating from genetically different individuals in the same organism.

Although sporadic attempts to rectify physiologic deficits by transplantation of appropriate grafts had been made many centuries ago, in modern biology transplantation methods have served as one of the key procedures of research in the fields of embryology and tumor biology. Notably, as early as in 1903, it was observed that tumor cells grafted from a mouse to another mouse usually were destroyed by the hosts (see Snell *et al.*, 1976a; rejection of the graft). Grafts from a member of the same species are called homografts, whereas grafts from different species are called heterografts or xenografts. However, even the former types of tumor grafts were accepted in exceptional cases. Little and Tyzzer (1916) demonstrated that acceptance of a tumor was dependent on the sharing of dominant genes by the donor and the host. This rule of transplantation, described above, that homografts are usually rejected (with rare exceptions) holds true not only for transplantation of malignant cells, but also for transplantation of normal tissues, such as skin grafts.

Our understanding of the underlying mechanisms leading to homograft rejection as well as its genetic basis had to wait for the development of genetically uniform experimental animals. In the 1930s, groups of mice fulfilling the above-mentioned criteria became available: each group of mice is called an inbred strain resulting from repeated sister–brother matings (inbreeding) for many generations. Members of an inbred strain are identical for almost all genetic traits, similar to identical twins, and they are homozygous at almost all genetic loci. By using these inbred strains of mice, Gorer (1937) was the first to discover the immunological basis of homograft rejection and the presence of strong transplantation antigens, the H-2 system in mice. Upon transplantation of a homograft, such as skin from a member of an inbred strain of mice to another strain of mice, antigens in the graft are recognized as foreign substances by the host's immune system. The immune response of the host then results in the generation of effector cells and effector molecules which react *in vivo* with the antigens of the graft. The *in vivo* reactions of the host's immune effectors with the antigens of the graft trigger pathologic processes leading to graft rejection.

The antigens of the graft foreign to the host which induce immune response are collectively called histocompatibility (transplantation) antigens. All of them are under genetic control and the fate of a homograft, whether it is accepted or rejected, is dependent on the genetic relationship between the donor and the host. Therefore, grafts can be classified on the basis of the genetic relationship: (i) hetero (xeno) graft, grafts from different species which are very quickly rejected within minutes due to 'natural' immunity against antigens of foreign species; (ii) autograft, grafts from the host himself; since there is no genetic disparity involved, such a graft survives permanently; (iii) homografts, grafts from different individuals in the same species. Homografts can be further divided into two groups, syngeneic (iso) and allogeneic grafts. Syngeneic grafts are those from genetically-identical individuals, such as grafts between identical twins, or between members of the same inbred strain of mice. Allogeneic homografts (allografts) are those from genetically non-identical individuals.

Autografts are frequently utilized in plastic surgery. For example, burned skin tissues of a patient can be rectified by transplantation of a piece of healthy skin taken from the patient himself. Allogeneic homografts, however, are the most important grafts in clinical transplantation, since only very few patients who require organ transplantation are fortunate enough to have an identical twin as the donor. For this reason, the following sections are mostly devoted to the discussion of allogeneic homografts.

5.2   HOMOGRAFT REJECTION

As mentioned in the preceding section, all histocompatibility antigens are under genetic control and, therefore, rejection of a homograft depends on genetic disparity and the host's immune response to the incompatible antigens of the donor (Milgrom *et al.*, 1979). In this respect, it is essential for the readers to familiarize themselves with genetic principles involved in homograft rejection in order to understand the immunological basis of graft rejection.

5.2.1   Genetic Laws of Transplantation

These laws were established in the 1940s on the basis of experiments performed on inbred strains of mice (Snell *et al.*, 1976b). (i) Syngeneic homografts are always accepted. Grafts between members within an inbred strain are accepted because there are no genetic disparities between the donor and the host, analogous to grafts between identical twins. (ii) Allogeneic homografts are usually rejected. Members of two different inbred strains AA and BB differ at many genetic loci, including those of histocompatibility antigen systems. Therefore, immune response of the host against histocompatibility antigens of the donor leads to rejection of the homograft. (iii) Grafts from either inbred parent to the F_1 hybrid are always accepted, but grafts in the reverse direction, *i.e.* from the F_1 hybrid to either parent, are always rejected. As shown below, the genetic makeup of the F_1 hybrid (AB) consists of exactly one half of genes from AA and BB parents. No histocompatibility antigens of either parent are foreign to the immune system of the F_1 hybrid and, therefore, no immune response occurs after transplantation of grafts from the parents and they are accepted permanently. On the other hand, for the immune system of the parents AA or BB, histocompatibility antigens of grafts from the F_1 hybrid (AB) coded by the genes of the opposite parent are foreign, *i.e.*, for the parent AA, the B part of the F_1 hybrid (AB) is foreign, and, therefore, they are always rejected (Figure 1). (iv) Since all genes of members of the F_2 generation resulting from mating of F_1 hybrids are present in the F_1 hybrid, grafts from F_2 and subsequent generations to the F_1 hybrid are always accepted. (v) Grafts from an inbred parent are accepted by some but not all members of the F_2 generation (resulting from $F_1 \times F_1$ mating) or by the backcross generation. Backcross denotes mating of a F_1 hybrid with one of the parents. This fifth law is the key to understanding the genetic rules of transplantation, since genes of parents segregate into members of these generations. As shown below, 3/4 of the members of an F_2 generation, *i.e.* AA and 2 AB accept grafts from the parents AA, but the remaining 1/4 (BB) reject the graft. Among members of the backcross (AB × BB) generation, 2 AB accept grafts from AA, but 2 BB reject the grafts (Figure 2). This, however, is based on the assumption that histocompatibility antigens are controlled by genes A and B at a single locus. If n number of independent loci are present, $(3/4)^n$ members of the F_2 generation and $(1/2)^n$ of the BC generation accept grafts from the parent AA. By using this simple formula, it was estimated that 4–13 independent histocompatibility loci, depending on the kinds of grafts, are most probably involved in acceptance or rejection of the grafts by inbred strains of mice. In other words, an allogeneic homograft originating from a donor sharing the same genes at the histocompatibility loci with the host will be accepted, even though the donor and the host possess different genes at other loci.

Figure 1 Rejection of grafts between parents and F_1 hybrids

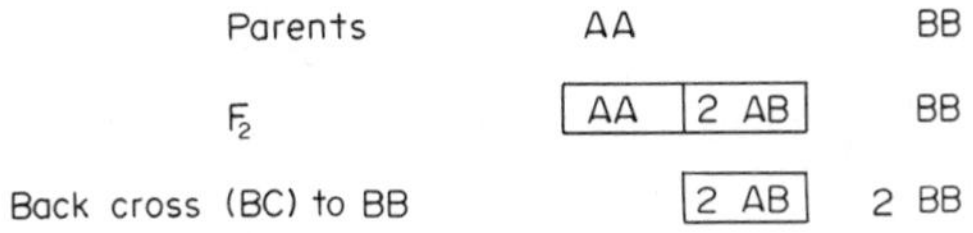

	Parents	AA		BB
	F_2	AA	2 AB	BB
Back cross (BC) to BB			2 AB	2 BB

Figure 2 Rejection of grafts between parents and F_2 and F_1 backcross to parent (BB)

5.2.2 Host *versus* Graft Rejection (HvGR)

The primary function of the immune system is to recognize foreign substances (antigens), such as bacteria and viruses, invading the body and to generate immune effectors, effector T cells and antibodies which are specific for the antigens. These effectors combine with the antigens *in vivo* and initiate non-specific pathologic processes which often result in destruction of antigen-carrying invaders or the elimination of them. Therefore, if a graft has histocompatibility antigens foreign to the host, the fate of such a graft is not an exception to this rule. The histocompatibility antigens are recognized by the immune system of the host and immune effectors react with the antigens of the graft resulting in destruction (rejection) of the graft (Snell *et al.*, 1976a; Milgrom *et al.*, 1979).

A skin allograft transplanted to a non-immune host is accepted for a few days as if it were an autograft. Mononuclear leukocytes, including effector T cells, start to invade the graft 5 to 6 days after transplantation; infiltration of mononuclear cells increases progressively and by the 10–12th day the graft will be destroyed completely by the necrotizing processes (Snell *et al.*, 1976b).

Once the host rejects an allograft, then the host becomes immune to the histocompatibility antigens of the initial donor. When the second graft from the same donor is grafted to the immune host, the graft will be rejected much more quickly and violently by the host. This type of rapid rejection of the second graft by the immune host is called second set rejection, whereas rejection of a graft by a non-immune host is called first set rejection.

The homograft immunity has the following characteristics: (i) it is a systemic immunity specific for histocompatibility antigens to which the host was exposed; (ii) it is a long-lasting immunity for months and years; (iii) it is a donor-specific immunity, but not organ- or tissue-specific immunity; (iv) it can be induced not only by transplantation, but also by a proper immunization with the donor's cells and tissues.

Transfer experiments have demonstrated the crucial role played by lymphocytes of the immune host for homograft immunity. Figure 3 illustrates such an experiment: mouse *b* received a skin allograft from mouse *a* of a different strain and rejected the graft 10 days later. Spleen cells, mostly lymphocytes and serum, were obtained from the immune mouse *b* and transferred separately to two syngeneic mice which were then challenged by skin grafts from the same donor *a*. Six days later, the mouse which received immune lymphocytes rejected the graft in a second set fashion, whereas another mouse which received immune serum (antibodies) could not reject the graft at this time. Obviously, the latter mouse eventually will reject the skin allograft in a first set fashion. This transfer experiment clearly indicates that homograft immunity is a cell-mediated rather than an antibody-mediated immunity.

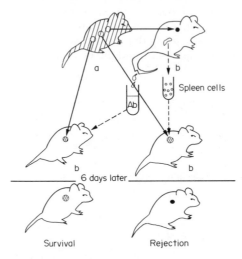

Figure 3 Adoptive transfer of homograft immunity

Recently, cellular events taking place in the host following allografting have been elucidated. Figure 4 illustrates such events. Histocompatibility antigens are carried to the host immune centers, lymph nodes and spleen, where macrophages take up the antigens and present them in an immunogenic form to T cells and B cells with specific receptors for the antigens. At least two kinds of T cells recognize different aspects of the antigens; one rather quickly goes through blast transformation, proliferates and becomes helper T cells (Th). Another kind of T cell, with the help of Th, proliferates and becomes effector T cells and memory T cells. Meanwhile, B cells with receptors specific for the antigenic determinants are stimulated and, under the help of Th, differentiate to mature plasma cells which secrete antibodies.

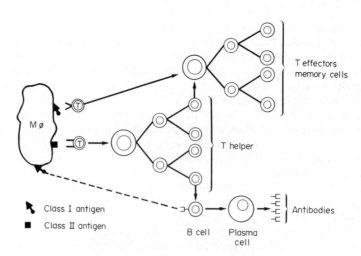

Figure 4 Immune response in HvGR

Figure 5 illustrates the kinetics of generation of cytotoxic T cells (Tc), which are one of the effector T cells, and formation of antibodies after transplantation of a skin allograft. As seen in Figure 5, Tc cells start to appear in the host spleen 4 days after transplantation, increase progressively in number and reach maximal on day 8 just before rejection. On the other hand, antibody formation starts 8 days after transplantation and reaches its peak a few weeks later. Therefore, in the first set rejection, the major effector is the T cells and, most probably, antibodies are not involved. In the second set rejection, however, both T cells and antibodies most probably are involved.

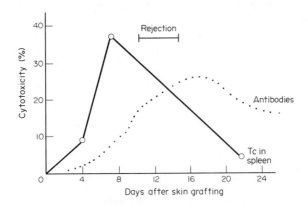

Figure 5   Kinetics of generation of cytotoxic lymphocytes and antibodies (Modified from Canty and Wunderlich, Jr., 1971)

5.2.3   Graft *versus* Host Rejection (GvHR)

If an allograft consisted of immunologically competent cells, such as bone marrow graft, and the host is unable to reject the graft, then immune response in a reversed direction, *i.e.* graft *versus* host rejection (GvHR), would take place. GvHR is essentially the same immunological phenomenon as HvGR. In GvHR, grafted lymphocytes recognize histocompatibility antigens of the host and become immune effector cells which attack the host and cause various clinical symptoms, such as retarded growth (runting syndrome), hemolytic anemia, dermatitis, enteritis and immunosuppression.

5.2.4   General Characteristics of Histocompatibility Antigens

As mentioned in Section 5.1, histocompatibility antigens are defined as antigens present in the graft and missing from tissues of the host which are capable of inducing homograft immunity. They share the following general characteristics (Lafferty, 1980; Milgrom *et al.*, 1979; Morris, 1980). (i) They are responsible for induction of homograft immunity. (ii) They are under genetic control. Almost all of them, except sex-linked antigens, are inherited as simple Mendelian condominant traits. In other words, there are no amorph or recessive genes coding for histocompatibility antigens. (iii) They are T-dependent antigens. As mentioned for HvGR (Section 5.2.2) immune response to histocompatibility antigens requires generation of Th. There are no 'natural' antibodies directed against histocompatibility antigens. (iv) They are present on the cell membrane (cell surface antigens) and distributed in practically all tissues. Their concentrations, however, vary from one tissue to another. The antigenic determinants reside in protein or glycoprotein molecules of the cell membrane. (v) Within a species, there are many genetically independent histocompatibility antigen systems. However, the immunogenic strength of each system varies. The strongest one is called 'major' and the remaining 'minor' histocompatibility antigen systems. The complex of genes coding for major histocompatibility antigens are called major histocompatibility complexes (MHC). Every mammalian species studied thus far possesses one MHC. (vi) They are usually detected by a variety of serologic tests such as agglutination (Dausset, 1958; van Rood and Van Leeuwen, 1963), cytotoxicity (Terasaki and McClelland, 1964), complement fixation (Ceppellini and van Rood, 1974) and other more complicated procedures (Milgrom *et al.*, 1979). Some antigens, however, are detectable at the present time only by mixed lymphocyte reaction (MLR; Ceppellini and van Rood, 1974).

There are significant differences between major and minor histocompatibility antigens besides the difference in their immunogenic potency in transplantation. Table 1 lists important differences between them.

Table 1   Comparison of Major and Minor Histocompatibility Antigens

Characteristics	Major	Minor
Genetic determination	Complex (MHC)	Single locus
MST[a]	<2–3 weeks	>3 weeks
Effect of immunosuppression	Weak	Strong
Effect of pre-immunization	Not impressive	Significant
Antibody formation	Common	Irregular
Tolerance induction	Difficult	Easy
MLR	Always	Variable
Association with Ir[b]	Yes	No

[a] Mean survival time of allografts.   [b] Antibody responses to certain simple antigens.

5.3   MAJOR HISTOCOMPATIBILITY COMPLEX (MHC)

Gorer (1937) was the first to discover one of gene products of the murine MHC (H-2 system). He demonstrated H-2 antigens by means of the hemagglutination test using sera of an inbred strain of mice which rejected skin allografts from another strain. Since the discovery of the H-2 system, the presence of MHC has been confirmed in all mammalian species studied thus far. The names of MHC of representative species are: mouse, H-2; rat, Rt-1 (previously called AgB); guinea pig, GPLA; dog, DLA; pig, SLA; and rheusus monkey, RhLA.

Unlike the situation in the discovery of MHC in experimental animals, the human MHC (the HLA system) was discovered in the 1950s during studies on leukocyte antibodies in patients with various blood dyscrasia. In 1952, Dausset and Nenna (cited in Ceppellini and van Rood, 1974) found leukocyte alloantibodies in sera of leukopenic patients. Subsequently, Dausset (1958) using sera of patients with multiple transfusions, discovered the first HLA antigen, Mac (A2).

The history of the HLA system is quite unique in that international collaborative efforts, in the form of periodic workshops starting with the first one in 1964, have been devoted to the establishment of this highly polymorphic system (Ceppellini and van Rood, 1974).

In this section, the H-2 system will be described first, followed by a description of the HLA system since the murine MHC has been studied more thoroughly than the MHC of any other species and represents the prototype of the MHC in other species.

5.3.1   The H-2 System

Figure 6 illustrates the H-2 complex. The H-2 complex resides on the 17th chromosome and consists of several separate but closely linked regions: K, I, S and D. The I region is subdivided into 5 subregions: A, B, J, E and C. On each of these regions or subregions of the complex, there exists an allele (an alternative form of a gene) that codes for a specific molecule of a given strain. Gene products of the complex can be divided into three different classes: I, II and III. As shown in Table 2, class I molecules (K and D) carry histocompatibility antigens of a given strain and act as strong immunogens in transplantation. Class II molecules are collectively called *Ia* (*I* region associated) antigens, and are also strong histocompatibility antigens. *Ia* antigens are known to induce *in vitro* proliferation of T cells of another strain when they are cultured together (MLR). Class III molecules represent polymorphism of serum proteins, such as C2 and C4 of complement components.

Physicochemical structures of class I and class II molecules of the complex have been elucidated. Figure 7 illustrates their structures. The class I molecule consists of a glycoprotein of 45 000 daltons and a smaller polypeptide β_2 microglobulin (β_2M) of 11 000 daltons. The antigenic specificity of the class I molecule depends on the primary structure of a variable portion close to the N terminal of the glycoprotein molecule. There are no alloantigenic specificities on the β_2M molecule.

Class II molecules consist of two glycoprotein molecules of 35 000 daltons and 28 000 daltons. The primary structure of the variable portion of the larger molecule is responsible for the specificity of the *Ia* antigen of a particular strain.

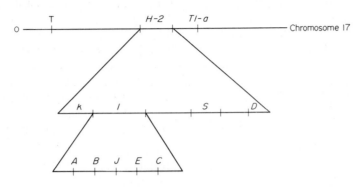

Figure 6 H-2 complex

Table 2 Different Classes of H-2 Molecules

	I	II	III
Specific molecules	H-2*K* H-2*D*	*Ia*1 *Ia*5/*Ia*3 *Ia*4	Ss S1p
Biochemistry	Glycoprotein (45 000 daltons), associated with β_2M	Glycoprotein α (35 000 daltons) β (28 000 daltons) non-convalently bound	Protein α (95 000 daltons) β (70 000 daltons) γ (30 000 daltons)

Antigenicity in allogenic
system
	Multiple specificities
	1. Private
	  (specific for an allele)
	2. Public
	  (shared by different alleles)
			Two specificities S1p and
			H-2.7 (on chain)

Cellular distribution			
T cells	All	Some	None
B cells	All	All	None
Macrophages	All	All	Some (in cytoplasm)
Epithelial cells	All	Some	None
Others	All but at different concentrations	Sperm	Hepatocytes (in cytoplasm) Erythrocytes (passive adsorption)
Causing allograft rejection	+	+	−
Major biological functions	Restricting specificity of immune responses (cellular)	Restricting cell collaboration, restricting specificity of immune responses (humoral)	C4 of complement (Ss)

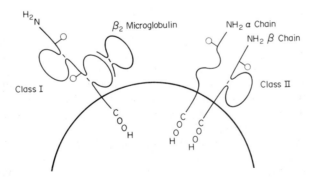

Figure 7 Physiocochemical structure of H-2 antigens (Courtesy of Dr. M. Zaleski, State University of New York, Buffalo)

The class I molecules are found in all cells and tissues, but their concentration varies from one cell type to another. Lymphoid cells are the richest source of the molecules and brain tissues contain the least amount. Distribution of class II molecules is more restricted than that of class I molecules. They are predominantly expressed on B cells and macrophages, but only very small amounts or even none on normal T cells and erythrocytes (Table 2).

When a graft is transplanted across the MHC barrier, the donor and the host differ at the MHC. The HvGR is a complex phenomenon, as shown in Figure 4, since the immune response is directed against multiple antigens of different classes. This may be one of the reasons why the immunogenic strength of major histocompatibility antigens is far greater than that of minor histocompatibility antigens.

It should also be stressed that primary response to MHC antigens seen in the first set rejection is quite uniform and unusually strong, regardless of the direction of a graft, from A strain to B strain or from B to A. This would indicate that clones directed against MHC antigens are developed surprisingly more uniformly and solidly than the clones against minor histocompatibility antigens, suggesting a fundamental biological importance of the MHC and corresponding clones beyond their roles in homograft rejection, which is undoubtedly a very artificial event.

Both class I and II molecules not only play an important role in induction of homograft immunity, but also play fundamental roles in cellular cooperation in antibody responses, genetic controls of antibody responses to certain antigens and cell-mediated immune reactions (see Table 2).

5.3.2   The HLA System

The HLA resides on the short arm of the sixth chromosome and consists of four separate but closely linked loci: A, B, C and D (DR) (Figure 8). As seen in Table 3, at each locus of the complex there exist multiple alleles, 20 for A, 40 for B, 8 for C, 12 for D, and 10 for DR. Each of these alleles is designated by its locus and number, such as A1; w stands for workshop nomenclature. The alleles without w are approved by the World Health Organization (WHO) committee, whereas those with w are not yet approved by the committee. The multiple loci with multiple alleles make the HLA system the most polymorphic system known in man. The question whether D and DR are the same or separate loci cannot be answered at the present time, since no recombinants between them have been found.

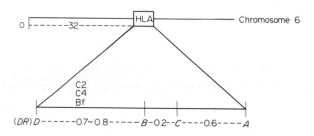

Figure 8   HLA complex

Gene products of the HLA can also be divided into three classes: I, II and III. Gene products of A, B and C belong to class I, analogous to murine K and D products, whereas D (DR) gene products belong to class II, analogous to murine Ia molecules. Distribution of class I and II molecules in different cells and tissues is quite similar to that of the H-2 molecules.

Class I HLA antigens are routinely demonstrated by means of the micro-lymphocytotoxicity test (Terasaki and McClelland, 1964). In this test, peripheral blood lymphocytes are separated and incubated with individual alloimmune sera followed by incubation with rabbit complement. When antibodies combine with HLA antigens on lymphocytes and fix complement, membrane damage occurs. This is demonstrated by dye exclusion, such as eosin staining. Antisera to class I and II HLA antigens are selected from sera of multiparous women, or sera of patients with transfusion, or sera of recipients of renal grafts. They are standardized and distributed to various HLA tissue-typing laboratories for their clinical use.

Genes at the HLA complex are inherited together as a set, except in the case of recombinants due to crossover between the loci. As seen in Figure 8, the distance between A and B is 0.8 centi-

Table 3 World Health Organization Recognized HLA Specificities (1980)

HLA-A locus	HLA-B locus		HLA-C locus	HLA-D locus	HLA-DR locus
A1	B5	Bw46	Cw1	Dw1	DR1
A2	B7	Bw47	Cw2	Dw2	DR2
A3	B8	Bw48	Cw3	Dw3	DR3
A9	B12	Bw49(Bw21)	Cw4	Dw4	DR4
A10	B13	Bw50(Bw21)	Cw5	Dw5	DR5
A11	B14	Bw51	Cw6	Dw6	DRw6
Aw19	B15	Bw52	Cw7	Dw7	DR7
Aw23(A9)	Bw16	Bw53	Cw8	Dw8	DRw8
Aw24(A9)	B17	Bw54		Dw9	DRw9
A25(A10)	B18	Bw55(Bw22)		Dw10	DRw10
A26(A10)	Bw21	Bw56(Bw22)		Dw11	
A28	Bw22	Bw57(B17)		Dw12	
A29(Aw19)	B27	Bw58(B17)			
Aw30(Aw19)	Bw35	Bw59			
Aw31(Aw19)	B37	Bw60(B40)			
Aw32(Aw19)	Bw38(Bw16)	Bw61(B40)			
Aw33(Aw19)	Bw39(Bw16)	Bw62(B15)			
Aw34	B40	Bw63(B15)			
Aw36	Bw41				
Aw43	Bw42	Bw4			
	Bw44(B12)	Bw6			
	Bw45(B12)				

Morgans (cM), *i.e.* the crossover rate between these two loci is 0.8%. The distance between *B* and *D* is 0.7–0.8 cM. The *C* locus is present between *A* and *B*, but is closer to *B*.

A set of genes at the HLA on a single sixth chromosome is called HLA haplotype. If one assigns a,b for paternal and c,d for maternal HLA haplotypes, then children's haplotypes in this family would be ac, ad, bc, bd as shown below:

Parents   a,b   x   c,d

Children   ac, ad, bc, bd

Figure 9 Heredity of HLA haplotypes

In other words, the likelihood for any two siblings being HLA identical is 25%.

Some haplotypes appear quite frequently in a given population due to linkage disequilibrium. For example, the frequency of HLA *A1-B8* haplotype in the white population is far greater than the value that is calculated from gene frequencies of *A1* gene and *B8* gene in the entire population. The reason for the existence of linkage disequilibrium in the HLA system is not known and is a matter of speculation. An attractive hypothesis would be that individuals with certain HLA haplotypes might have survival advantages during the evolution of the human race over those individuals who do not have the haplotypes. Perhaps hypothetical resistance genes or Ir genes for certain infectious agents were linked closely to the HLA complex and they were in linkage disequilibria with HLA genes of a particular haplotype.

Frequencies of some HLA genes are quite different from one race to another, as shown in Table 4. For example, the HLA *A1* gene is frequent among white populations, whereas it is extremely rare among Orientals. Furthermore, both *Aw23* and *DRw3* genes are frequently found in white as well as negroid populations but absent from Orientals. *Aw43* gene is only found in the African negroid population.

Physicochemical structures of class I and II HLA molecules are quite similar to those of H-2 molecules. The class I HLA molecule consists of a glycoprotein carrying a given antigenic determinant and a β_2M molecule, whereas class II HLA molecules consist of two glycoprotein molecules and the larger one possesses the antigenic determinant. Class II (*D*) antigens are shown to induce strong *in vitro* MLR when the lymphocytes are cocultured with allogeneic lymphocytes.

There have been several reports which indicate that class I and II HLA molecules play important roles in cellular cooperation in antibody responses as well as in *in vitro* cell-mediated immune reactions (Thorsby, 1979; Lafferty, 1980; Morris, 1980).

Table 4   Distribution of HLA Genes[a]

Genes		Caucasian	Negroid	Oriental
A1		15.8	3.9	1.2
A2		27.0	9.4	25.3
A3		12.6	6.4	0.7
A9	Aw23	2.4	10.8	0.0
	Aw24	8.8	2.4	36.7
Aw43		0.0	4.0	0.0
B5		5.9	3.0	20.9
B8		9.2	7.1	0.2
B18		6.2	2.0	0.0
DRw1		6.2	0.0	4.5
DRw2		11.2	8.7	16.5
DRw3		8.9	11.7	0.0

[a] % of population.

5.4   CLINICAL SIGNIFICANCE OF MHC

The formal evidence that HLA antigens are indeed major histocompatibility antigens has been obtained from the results of experimental skin grafting on volunteers. The mean survival time of skin grafts between ABO blood group matched siblings showed a clearcut bimodal distribution; 1/4 of such grafts showed a mean survival time of 22 days, whereas that of the remaining grafts was 10 days (Ceppellini and van Rood, 1974; Milgrom *et al.*, 1979). Those sibling combinations with the longer survival time were found to be HLA identical, and the remainder with the shorter survival time HLA nonidentical combinations.

It should be stressed that besides the HLA system, A and B antigens of the major blood group system are also very strong histocompatibility antigens. This was shown in experimental skin grafting as well as experiments in which group O individuals were immunized with purified blood group substances who rejected skin grafts from the donors with corresponding blood groups in the second set fashion.

Renal transplantation now became one of the potential therapeutic means to save patients at the end stage of renal failure. Every year transplantation of several thousands of renal grafts is now being performed all over the world and about half of them are carried out in the USA. This achievement in the clinical application of transplantation has been based on the matching of donor and recipient for HLA antigens and the development of immunosuppressive agents.

Transplantation of other organs such as heart, lung, pancreas and liver has been attempted. However, the success rate of these organ grafts is still far below that of renal grafts. Bone marrow transplantation has been applied for treatment of immunological and hematological disorders. Because of the problems of GvHR, clinical bone marrow transplantation has thus far been limited to HLA identical siblings.

5.4.1   Renal Transplantation

As pointed out in the preceding section, blood group A and B antigens were shown to be potent histocompatibility antigens and, therefore, the donor and the recipient have to be matched at first for the ABO blood group. Effects of HLA matching on the clinical outcome of renal grafts have been well appreciated in related donor–recipient combinations. Grafts from ABO compatible, HLA identical siblings, have a 3 year survival rate of over 90% under immuno-suppressive therapy. Many of the recipients were able to return to their normal lives. Grafts from HLA haploidentical siblings or from parents enjoy the second best clinical outcome. Because of obvious reasons, transplantation of grafts from ABO incompatible or HLA nonidentical siblings should not be considered.

The situation of HLA matching on renal grafts from unrelated (cadaveric) donors is still not quite clear. At the present time, it can be stated that the survival rate of grafts from unrelated donors perfectly matched for HLA *A* and *B* antigens is significantly better than that of very poorly matched grafts. In this respect, it should be mentioned that the number of pretransplant blood transfusions influences significantly the survival of cadaveric renal grafts (Morris, 1980).

The exact mechanism(s) of the beneficial effect of blood transfusions on the clinical outcome of the renal grafts is still unknown.

Prior to transplantation, crossmatch tests just like the major crossmatch test for blood transfusion have to be performed to ascertain that the recipient's serum does not contain antibodies directed against any of the donor's antigens. If the crossmatch tests were positive, the recipient's antibodies would combine *in vivo* with the antigens of the graft and may cause 'hyperacute rejection' which occurs within a matter of minutes after transplantation.

Finally, the recipient having successfully received a well-matched graft has to be treated with immunosuppressive agents. This is simply because of our ignorance of minor histocompatibility antigen systems in man. As discussed in the preceding section, minor histocompatibility antigens are also able to induce HvGR, which eventually leads to graft rejection. Currently, a combination of synthetic corticosteroid such as prednin and immuran, a derivative of 6-mercaptopurine, is routinely used.

When a cytotoxic drug such as immuran is given immediately after transplantation, it blocks DNA synthesis of rapidly proliferating, antigen-stimulated lymphocytes of the recipient, resulting in suppression of the triggered immune responses. On the other hand, synthetic corticosteroids act mainly as an anti-inflamatory agent by suppressing the activities of inflamatory cells, including immune effector cells generated by the recipient.

5.4.2 HLA and Disease Association

Recently much interesting data have been accumulated on the association of various diseases with certain HLA antigens. General features of the associations are the following: (i) there are no HLA and disease associations which confer 100% association; (ii) almost all HLA and disease associations are for susceptibility rather than resistance to the disease and the susceptibility effect appears to be dominant; (iii) many HLA and disease associations appear to be due to genes at the *D* locus (Figure 8) or genes showing strong linkage disequilibrium with genes at the *D* locus; (iv) most of the HLA-associated diseases do not affect the individuals until after the peak reproductive period; (v) only a few HLA-associated diseases were studied for the linkage with the HLA.

Table 5 lists some examples of the HLA and disease associations. The most striking association thus far demonstrated is that between ankylosing spondilitis and HLA-B27. The frequency of B27 in patients with this disease is 90%, whereas frequency of this antigen in a general population is less than 10% (Ceppellini and van Rood, 1974; Bodmer, 1978). This, however, does not imply that all those individuals with B27 develop the disease and, therefore, such information would only serve as an adjunct to the clinical diagnosis.

Table 5   HLA and Disease Association[a]

Disease	Antigen	Frequency (%)		Relative risk
		Patients	*Controls*	
Coeliac disease	Dw3	96	27	64.5
Chronic active hepatitis	DRw3	41	17	3.4
Myasthenia gravis: caucasian	DRw3	32	17	2.3
Graves' disease	Dw3	53	18	5.1
Juvenile-onset diabetes	DRw3	27	17	1.8
	B8	32	16	2.5
	DRw4	39	15	3.6
Rheumatoid arthritis	DRw4	56	15	7.2
Myasthenia gravis: Japanese	DRw4	59	35	2.7
Ankylosing spondylitis	B27	90	8	103.5
Reiter's disease	B27	80	9	40.4
Haemochromatosis	A3	72	21	9.7
Psoriasis: Japanese	B13	18	1	22
	B37	35	2	26
	A1	30	2	21

[a] Modified from Bodmer (1978).

Nevertheless, studies on HLA and disease association have contributed to mapping of some disease-related genes within or close to the HLA and also to our better understanding of the heterogeneity of some diseases which otherwise have been believed to be a single disease entity.

For example: the gene presumably responsible for 21-hydroxylase deficiency which causes congenital adrenal hyperplasia is mapped within the HLA. The first example of the separation of a disease into two subgroups based on its HLA association would be psoriasis; psoriasis vulgaris, but not pustular psoriasis, is associated with the HLA antigen which, in turn, confirmed the clinical distinction between these two types of psoriasis.

Finally, it should be mentioned that HLA typing provides a powerful tool for paternity testing which, in the past, has been based on only the blood group determination. Because of the extreme polymorphism of the HLA, the exclusion rate of paternity by HLA typing went up from 15% of the blood group determination to over 95%. Accordingly, many states in the USA made HLA typing mandatory for paternity testing.

5.5 REFERENCES

Bodmer, W. F. (ed.) (1978). The HLA system. *Br. Med. Bull.*, **34** (3).
Canty, T. G. and R. Wunderlich, Jr. (1971). Quantitative assessment of cellular and humoral responses to skin and tumor allografts. *Transplantation*, **11**, 111.
Cepellini, R. and J. J. van Rood (1974). The HLA system I and II. *Semin. Hematol.*, **11**, 233–252.
Dausset, J. (1958). Iso-leuco-anticorps. *Acta Hematol.*, **20**, 156–166.
Gorer, P. A. (1937). The genetic and antigenetic basis of tumor transplantation. *J. Pathol. Bacteriol.*, **44**, 691–697.
Lafferty, K. J. (1980). Immunogenicity of foreign tissues. *Transplantation*, **29**, 179–181.
Little, C. C. and E. E. Tyzzer (1916). Further studies on inheritance of susceptibility to transplantable tumor of Japanese waltzing mice. *J. Med. Res.*, **33**, 393–425.
Milgrom, F., K. Kano and M. B. Zaleski (1979). Transplantation immunology. In *Principles of Immunology*, 2nd edn., pp. 376–401. Macmillan, New York.
Morris, P. J. (1980). Suppression of rejection of organ allografts by alloantibody. *Immunol. Rev.*, **49**, 93–125.
Snell, G. P., J. Dausset and S. Nathenson (1976a). *Histocompatibility*, pp. 1–9. Academic, New York.
Snell, G. P., J. Dausset and S. Nathenson (1976b). *Histocompatibility*, pp. 357–385. Academic, New York.
Terasaki, P. I. and J. D. McClelland (1964). Microdroplet assay of human serum cytotoxin. *Nature (London)*, **204**, 998–1000.
Terasaki, P. I., D. Gjertson, D. Bernoco, M. R. Mickey, J. Bond and G. Cornacchione (1980). HLA in paternity testing. In *Immunobiology of Major Histocompatibility Complex*, pp. 270–276. Karger, Basel.
Thorsby, E. (1979). The human major histocompatibility complex HLA. *Transpl. Proc.*, **11**, 616–623.
van Rood, J. J. and A. Van Leeuwen (1963). Leucocyte grouping. A method and its applications. *J. Clin. Invest.*, **42**, 1382–1390.

6

Biotechnology of Artificial Cells Including Application to Artificial Organs

T. M. S. CHANG
McGill University, Montreal, Quebec, Canada

6.1 PRINCIPLE OF ARTIFICIAL CELLS

Artificial cells are prepared making use of some of the simpler properties of biological cells (Chang, 1957, 1964, 1965, 1977b). Each artificial cell consists of a spherical ultrathin semipermeable membrane of cellular dimensions, enveloping biologically active material. The semipermeable membrane of each artificial cell separates the contents from the external environment. Each standard artificial cell has an ultrathin membrane of 0.02 μm thickness, an equivalent pore radius of about 1.8 nm, and a large surface/volume relationship (2.5 cm^2 in 10 ml of 20 μm diameter microcapsules). This allows for an extremely rapid equilibration of external permeant molecules at a rate which is 400 times faster than that of a standard hemodialysis machine. In this artificial cell system (Figure 1), the semipermeable membrane prevents external proteins, antibodies or cells from entering, but external permeant molecules can equilibrate rapidly to come in contact with the enclosed materials (Chang, 1964, 1965, 1972a, 1972b; Chang and Poznansky, 1968b). The membrane composition, permeability and characteristics can be varied over a wide range. Furthermore, it is possible to enclose almost any combination of enzymes, multienzyme systems, cofactors regenerating enzyme systems, cell extracts, whole cells, proteins, adsorbents, magnetic materials, multicompartmental systems and others (Figure 2). Artificial cells are being used in detoxifiers, artificial kidney, artificial liver, immunosorbents, blood substitutes and other areas. The other possible applications of artificial cells, including biotechnology, have just begun to be seriously investigated. Detailed reviews in the areas of artificial cells are available (Chang, 1972b, 1975a, 1976c, 1976e, 1977d, 1978, 1979a, 1980c, 1981a, 1984a, 1984b).

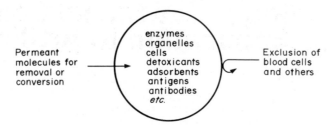

Figure 1   Artificial cell membranes: examples of possible variations in membrane composition, permeability and surface properties (Chang, 1978). Reproduced with permission from Plenum Press, New York

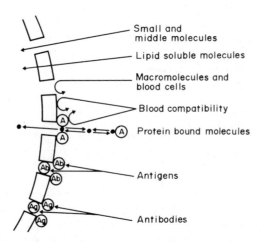

Figure 2   Contents of artificial cells: examples of materials which have been enclosed within artificial cells (Chang, 1978). Reproduced with permission from Plenum Press, New York

6.2   TECHNICAL ASPECTS OF ARTIFICIAL CELLS

6.2.1   Principle of Methods of Preparation

As emphasized earlier (Chang, 1972b), the basic methods used for the preparation of artificial cells are, in fact, physical examples for demonstrating the principle of artificial cells and encapsulation of enzymes. It has been suggested that many new physical systems could be developed to demonstrate the same principle (Chang, 1972b). An increasing number of new model systems have been developed since then (Figure 3).

The following synthetic polymer membrane systems have been used. Spherical ultrathin polymer membranes can be formed using emulsification followed by interfacial polymerization (Chang, 1957, 1964, 1965, 1972a, 1976c, 1977a). Numerous chemical reactions available for interfacial polymerization have been adapted for the same end (Mori *et al.*, 1973; Shiba *et al.*, 1970). Multiple-compartment membrane systems consisting of smaller artificial cells enveloped within larger artificial cells have also been formed (Chang, 1965, 1972b). Another way is to use silastics, cellulose acetate and other polymers for the formation of artificial cell membranes by secondary emulsion (Chang, 1965, 1966, 1972b, 1976c, 1977a). Liquid membranes can also be used in the

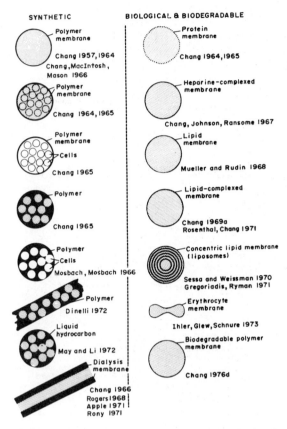

Figure 3 Artificial cells: example of possible variations in membrane compositions and configurations (Chang, 1977a). Reproduced with permission from Plenum Press, New York

secondary emulsion approach (May and Li, 1972). Polyethyleneimine nylon microcapsules have been used for proteolytic enzymes (Aisina *et al.*, 1976).

Biological and biodegradable membranes can also be used to microencapsulate enzymes and other biologically active materials. The following are examples: spherical ultrathin cross-linked protein membranes (Chang, 1964, 1965, 1972b, 1976c, 1977a; Chang *et al.*, 1966), heparin-complexed polymer membranes (Chang *et al.*, 1967), spherical ultrathin lipid membranes (Mueller and Rudin, 1968), lipid-complexed membranes (Chang, 1969a, 1972b; Rosenthal and Chang, 1980), liposomes (Gregoriadis and Ryman, 1971; Sessa and Weissman, 1970), erythrocyte-encapsulated enzymes (Ihler *et al.*, 1973) and biodegradable polymer membranes (Chang, 1976d). A cross-linked protein system can also be formed (Chang, 1964, 1965, 1971a, 1972b).

6.2.2 Contents of Artificial Cells

Almost any biologically active material can be enclosed within artificial cells (Figure 4). The following have been enclosed: enzymes, multienzyme systems, cell extracts and other proteins (Chang, 1957, 1964, 1972b, 1976c, 1977a; Kitajima and Kondo, 1971; Østergaard and Martiny, 1973), granules of enzyme systems and proteins (Chang, 1957, 1972b), combined enzyme and adsorbent systems (Chang, 1966, 1972b; Gardner *et al.*, 1971; Sparks *et al.*, 1969) and biological cells (Chang, 1965, 1972b, 1977a; Chang *et al.*, 1966; Mosbach and Mosbach, 1966). Magnetic materials have also been included within artificial cells to allow external magnetic fields to direct the movements of the artificial cells (Chang, 1966, 1977a). Other materials include radioisotope-labelled enzymes and proteins (Chang, 1965, 1972b), insolubilized enzymes (Chang, 1969b, 1972b), cofactor recycling systems (Campbell and Chang, 1975, 1976; Chang, 1977b; Chang and Malouf, 1979; Chang *et al.*, 1979b; Cousineau and Chang, 1977; Grunwald and Chang, 1978; Yu and Chang, 1981a, 1981b, 1982) and antigens, antibodies, vaccines and hormones (Chang, 1976c, 1976d, 1977a).

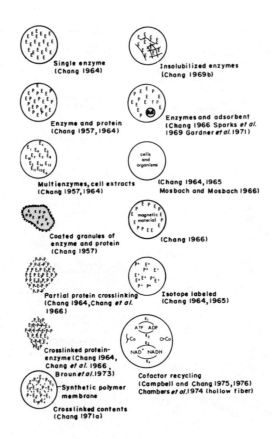

Figure 4   Artificial cells: examples of variations in compositions and contents of artificial cells (Chang, 1977a). Reproduced with permission from Plenum Press, New York

6.3   ARTIFICIAL CELLS CONTAINING ENZYME AND MULTIENZYME SYSTEMS

Enzymes and proteins in the body are mostly located in the intracellular environment. Enzyme systems immobilized within the intracellular environment carry out their function by acting sequentially on substrates, including those which cross the cell membranes by passive movement or by special transport mechanisms. If one were to ignore this natural scheme of things and inject enzymes in solution into the body, the foreign proteins in the free form may result in hypersensitivity reactions, production of antibodies and rapid removal and inactivation. Free enzymes in solution cannot be kept at the sites where the action is desired. Enzymes in free solution are not stable, especially at a body temperature of 37 °C. Furthermore, multienzyme systems require the enzymes and substrates to be in closer proximity and also for cofactor recycling preferably in an intracellular environment. These problems also apply to other areas of application of enzyme systems, though to a lesser extent.

The above problems have resulted in extensive research into the possible uses of artificial cells to immobilize enzymes and proteins (Chang, 1972b, 1976c, 1977a). Unlike enzymes immobilized to a solid support, enzymes immobilized in artificial cells remain in free solution. This way they function more like enzyme systems in red blood cells, especially in complex multienzyme systems. There is no limit to the number of different enzyme systems that can be enclosed together in one artificial cell.

6.3.1   Artificial Cells Containing Simple Enzyme Systems

By selecting one of the many methods available (Chang, 1972b, 1976c, 1977a), all single enzymes tested so far can be successfully enclosed within artificial cells. A 10 g dl^{-1} hemoglobin solution is usually present in the standard artificial cells. This gives an intracellular environment somewhat comparable to red blood cells. Thus, the enzymes enclosed in the artificial cells are stabilized by the high concentration of protein. Further stabilization can be obtained by cross-linking with glutaraldehyde (Chang, 1971a). In most cases the enzyme retained at least 90% of its original activity following disruption of the artificial cells to release the enzymes for analysis. However, because of the high concentrations of enzymes in the artificial cells and the membrane restriction to free diffusion of substrates, the assayed activity of the artificial cells containing the enzymes is usually about 30% of that of the enzymes in free solution.

6.3.2 Artificial Cells Containing Multienzyme Systems

Most metabolic functions are carried out in cells by complex multienzyme systems with cofactor requirements. As a result, research has been carried out for the microencapsulation of multienzyme systems with cofactor regeneration incorporated. Artificial cells containing hexokinase and pyruvate kinase could recycle ATP for the continuous conversion of glucose into G-6-P and phosphoenol pyruvate into pyruvate; similarly, artificial cells containing alcohol dehydrogenase and malic dehydrogenase can recycle NADH making use of NAD^+ (Campbell and Chang, 1976, 1977). A multienzyme system consisting of urease, glutamate dehydrogenase and glucose-6-phosphate dehydrogenase all within each artificial cell can convert urea into ammonia which is then incorporated to α-ketoglutarate in the presence of NADPH to form an amino acid, glutamate, with glucose-6-phosphate dehydrogenase to recycle the cofactor (Cousineau and Chang, 1977). In order to allow for the use of blood glucose instead of glucose 6-phosphate, a new system of artificial cells containing urease, glutamine dehydrogenase and glucose dehydrogenase has been developed using glucose as an energy source to recycle the cofactor (Chang and Malouf, 1978). Optimization has resulted in an artificial cell system which can function using glucose at concentrations normally present in the blood (Chang and Malouf, 1979). Further development of this approach has been carried out to convert glutamate formed from urea or ammonia into other amino acids. The first system tested involved the conversion of urea into ammonia then to glutamic acid and then to alanine (Chang *et al.*, 1979b). This was done using artificial cells each containing a multienzyme system of urease, glutamate dehydrogenase, alcohol dehydrogenase and transaminase. Artificial cells containing multienzyme systems have also been studied for galactose removal (Chang and Kuntarian, 1978).

These studies show that it is feasible to prepare artificial cells containing multienzyme systems for the sequential conversion of substrates into products and at the same time to recycle the required cofactors. This principle is applicable to the conversion of waste metabolites in liver failure, renal failure or other metabolic disorders into useful or removable products.

The above studies demonstrate the feasibility of using multienzyme systems in artificial cells for the conversion of substrate which could not be done previously with single enzyme systems. This has been facilitated by the ability of multienzyme systems in artificial cells to recycle the required cofactors. This way, a very low external concentration of cofactor is required to carry out these types of reactions. However, for *in vivo* applications, although it is possible to introduce external cofactors for recycling, it would be much more desirable to retain the cofactors within the artificial cells so that it is more analogous to biological artificial cells where cofactors can be retained inside the artificial cells for continuous recycling. One approach, carried out in our laboratory, is to link cofactors to dextran to form soluble macromolecules. This way the cofactors linked to the macromolecules can be retained within the artificial cells and do not cross the semipermeable membrane (Grunwald and Chang, 1978, 1979). Semipermeable nylon polyethyleneimine was prepared to contain ethanol dehydrogenase and malic dehydrogenase together with a soluble dextran–NAD^+ (Grunwald and Chang, 1981). In the presence of the substrate ethanol and oxaloacetic acid, dextran–NAD^+ was successfully recycled within the artificial cells by the sequential reactions of the included enzymes. NAD^+ deactivation in the presence of crude hemoglobin is caused by enzymes such as NAD^+-glycohydrolases and poly(ADP–ribose) polymerase which are known to be present in hemolysate. The deactivation of NAD^+ is eliminated after purification of hemoglobin by affinity chromatography on NAD^+–Sepharose. After this the stability of NAD^+ and dextran–NAD^+ in the presence of purified hemoglobin is greatly improved (Grunwald and Chang, 1979, 1981). The stability of the microcapsules can be further improved by cross-linking the microcapsules with glutaraldehyde, but this treatment causes a considerable decrease in the recycling activity. These artificial cells have been retained in a column, to allow substrate solution to circulate through the column. In this self-sufficient continuous flow shunt system, 65% of the oxaloacetic acid is converted into malic acid in one passage through the column. The dextran–NAD^+ present within the microcapsules is regenerated twice each minute during the reaction. The shunt showed good stability, the reaction rate remaining constant for the first hour and 83% of the original activity being retained even after 3 h of continuous reaction. The cross-linked microcapsules showed complete stability during 3 h of continuous reaction, although their activity was only about 10% of that of the untreated microcapsules. The same shunt could be stored and reused several times. The microcapsules are suitable for use in continuous flow reactors without the need to supply external cofactors (Grunwald and Chang, 1979). In those studies where one wants to avoid the use of hemoglobin in artificial cells, polyethyleneimine, a macromolecule, can replace hemoglobin. The recycling of dextran–NAD^+ within the artificial cells in this form was measured in a continuous flow column. The recycling activity remained stable with no significant decrease in activity during the first 3 h of reaction. After this reaction, each column was stored at 4 °C. In reusing this after 7 and 12 days of storage, 63% of the original activity was retained after 7 days and 41% was still retained after 12 days (Grunwald and Chang, 1981). Artificial cells could be further stabilized by cross-linking with glutaraldehyde, but this cross-linking caused a sharp decrease in the initial recycling activity within the microcapsules.

Another approach for retaining cofactors inside artificial cells for continuous recycling is being studied for a system which acts on lipid-soluble substrates. Previously, it had not been possible to retain cofactors inside artificial cells because the artificial cell membrane, being made of polymeric material, is highly permeable. A lipid–polymer complex membrane artificial cell system (Chang, 1972b) has been analyzed in detail (Rosenthal and Chang, 1980). In this system, the artificial cells with their polymer membranes are complexed with a lipid material similar to lipid in biological cells. This way the membrane has very low permeability to nonlipid-soluble materials, but lipid-soluble substrates can cross the membrane very freely. The permeability coefficients for the unmodified semipermeable microcapsules to Na^+ and Rb^+ were $(13.0 \pm 0.6) \times 10^{-6}$ cm s^{-1} and $(14.3 \pm 0.6) \times 10^{-6}$ cm s^{-1}, respectively. For lipid-incorporated membranes the permeability coefficients for sodium and rubidium were $(2.6 \pm 0.2) \times 10^{-6}$ cm s^{-1} and $(1.3 \pm 0.2) \times 10^{-6}$ cm s^{-1}, respectively. The influx of sodium across lipid-incorporated microcapsule membranes was $(3.8 \pm 0.3) \times 10^{-10}$ mol cm^{-2} s^{-1}, which is comparable to fluxes across biological cell membranes. The addition of valinomycin did not increase the sodium permeability of lipid-incorporated membranes, but selectively increased the rubidium permeability coefficient to a value of $(4.1 \pm 0.5) \times 10^{-6}$ cm s^{-1}. The membrane of the artificial cells could also be modified by the incorporation of Na^+–K^+-ATPase. After being incorporated into the polymer membrane, the Na^+–K^+-ATPase retained 50–100% of its original ouabain-sensitive activity.

Studies were carried out using ultrathin lipid–polymer membrane artificial cells containing multienzymes, cofactors and substrates for multistep enzyme reactions (Yu and Chang, 1981a, 1981b, 1982). The ultrathin lipid–nylon membrane microcapsules can retain enzymes, $NADP^+$, NADPH and α-ketoglutarate. External ammonia and alcohol can cross the lipid membrane to take part in the multistep reactions as shown in Figure 5. Two grams of artificial cells contained 0.25 mmol of NAD^+, alcohol dehydrogenase, glutamic dehydrogenase, α-ketoglutarate, $MgCl_2$, KCl and ADP. When these artificial cells were added to the substrate solution, the change in ammonia levels when compared to the control indicated that 10 mmol ammonia was converted into glutamate. The 0.25 mmol of NAD^+ retained in 2 g of artificial cells with cofactor recycling multienzyme system could be recycled 40 times to convert 10 mmol of ammonia at a rate comparable to 2 g of artificial cells containing 10 mmol of NADH with no multienzyme systems for cofactor recycling. The supernatant when analyzed showed no leakage of enzymes, $NADP^+$ or α-ketoglutarate. The cofactor $NADP^+$ and α-ketoglutarate did not leak out from the microcapsules despite extensive washings. These results show that cofactors and substrates required for multienzyme reactions can be retained within ultrathin lipid–nylon microcapsules to act on permeant external substrates.

6.3.3 Role in Immobilized Enzyme and Enzyme Engineering—Applications in Medicine, Industry and Biology

Artificial cells containing enzymes and proteins have been used in a number of experimental and therapeutic conditions (Chang, 1972b, 1976e, 1977b, 1977c, 1977d, 1979a; Chang *et al.*, 1982c). At the beginning, the simpler single enzyme systems had been first tested. Some of these are briefly summarized below.

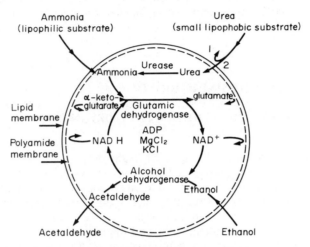

Figure 5   Schematic representation of artificial cells containing multienzyme system with cofactor recycling (Yu and Chang, 1982). Reproduced with permission from *J. Microbial Enzyme Technol.*

Artificial cells with red blood cell hemolysate have been tested for use as red blood cell substitutes (Chang, 1957, 1964, 1972a, 1980c; Sekiguchi and Kondo, 1966). We have also experimented on artificial blood cells in the form of polyhemoglobin (Keipert *et al.*, 1982). These will be described later.

Microencapsulated urease has been used as a model immobilized enzyme system for experimental therapy (Chang, 1964, 1965, 1966, 1972a; Chang *et al.*, 1966). The basic results obtained paved the way for other types of enzyme replacement therapy. The first demonstration of using immobilized enzymes for replacement in hereditary enzyme-deficiency conditions was the use of microencapsulated catalase to effectively replace a hereditary catalase deficiency in acatalasemic mice (Chang, 1972b; Chang and Poznansky, 1968a; Poznansky and Chang, 1974). Repeated injections did not result in the production of immunological reactions to the heterogeneous enzyme in the artificial cells (Poznansky and Chang, 1974). Subsequently, liposome-microencapsulated enzymes have been used for replacement in hereditary enzyme-deficiency conditions related to storage diseases (Gregoriadis, 1979). More recently, red blood cell entrapped enzymes (Ihler *et al.*, 1973) and entrapped enzymes have also been tested for possible use in storage diseases. Artificial cells containing tyrosinase have been used in extracorporeal hemoperfusion to lower tyrosine levels in rats (Shu and Chang, 1980, 1981). Research into the therapeutic applications of artificial cells containing enzymes has also included the use of asparaginase for tumor suppression. Having demonstrated the effectiveness of artificial cells containing asparaginase for experimental tumor suppression (Chang, 1971b, 1972b), more detailed studies were carried out on the various aspects of artificial cells containing asparaginase (Chang, 1973; Mori *et al.*, 1973; Siu Chong and Chang, 1974).

Artificial cells have been used for the construction of artificial kidneys, artificial livers and detoxifiers (Chang, 1966, 1969b, 1972a; Chang and Malave, 1970; Chang *et al.*, 1971). Ten milliliters of microencapsulated urease in an extracorporeal hemoperfusion chamber lowered the systemic blood urea of dogs by 50% within 45 min (Chang, 1966). The ammonium formed was removed by microencapsulated ammonium adsorbent (Chang, 1966) in the hemoperfusion chamber. This principle of urea removal using urease and ammonium demonstrated by us (Chang, 1966) was later successfully adapted into the Redy system for urea removal (Gordon *et al.*, 1969). Artificial cells containing a combined urease and ammonia adsorbent system have also been investigated for the possible removal of urea by oral administration (Chang, 1972b, 1976b; Chang and Loa, 1970; Gardner *et al.*, 1971). In initial studies, artificial cells containing tyrosinase were prepared and tested *in vitro* and found to act effectively on tyrosine (Shu and Chang, 1980). Further studies were then carried out to use artificial cells containing tyrosinase for hemoperfusion in galactosamine-induced fulminant hepatic failure (FHF) rats. In these studies it was found that hemoperfusion through tyrosinase artificial cells resulted in a significant lowering of tyrosine in the systemic circulation (Shu and Chang, 1981). Thus, 60 min after hemoperfusion, the plasma tyrosine concentration was 6.84 ± 2.03 mg dl^{-1}. The tyrosine level decreased to 45.7% of the original level of 14.94 ± 3.72 mg dl^{-1} for hepatic coma rats. However, during this 1 h of hemoperfusion, the tyrosine concentration in the brain remained unchanged (Shi and Chang, 1982). These hemoperfusion studies show that immobilized tyrosinase artificial cells can effectively remove tyrosine.

6.4 ARTIFICIAL CELLS CONTAINING BIOLOGICAL CELLS AND OTHER BIOLOGICAL MATERIALS—APPLICATIONS IN MONOCLONAL ANTIBODY PRODUCTION, INTERFERON PRODUCTION, DIAGNOSTICS AND ARTIFICIAL PANCREAS

Artificial cells containing biological cells have been prepared (Chang, 1965, 1972b, 1977a; Chang *et al.*, 1966). The reason for doing this is to place the cells in an intracellular environment and to separate them from the external environment. It was proposed that in this form cells can be implanted and still retain function, since they will be prevented from rejection by the immunological system of the body (Chang, 1965, 1972b). Thus, endocrine cells from heterogeneous sources could be enclosed within artificial cells and implanted. In this form the cells inside the artificial cells can respond to external substrate concentrations (*e.g.* blood glucose), and the required hormone (*e.g.* insulin) can be secreted into the systemic circulation (Chang, 1965, 1972b). This potential has already been demonstrated in the implantation of artificial cells containing enzymes (*e.g.* catalase and asparaginase) for successful *in vivo* action without causing immunological reactions (Poznansky and Chang, 1974; Siu Chong and Chang, 1974). Further recent development elsewhere has resulted in actual *in vivo* experiments in animals demonstrating that this proposal is possible. Thus rat-islet cells have been microencapsulated and then implanted intraperitoneally into diabetic rats (Lim and Sun, 1980). This way the microencapsulated islet cells can avoid rejection and can function to maintain normal glucose levels in the diabetic animals. However, the biocompatibility of the membrane of the artificial cells requires further improvement to avoid deposition of fibrin and cells, resulting in a decrease in permeability a few weeks after implantation. Artificial cells containing fibroblasts or plasma cells have also been used in *in vitro* cultures for the production of interferon and monoclonal antibodies, respectively (Damon Corporation, 1981). Other biologically active materials such as hormones, antigens, antibodies and vaccines can be enclosed within artificial cells (Chang, 1972b, 1975c, 1975d, 1976d, 1977a). Artificial cells containing antibodies to hormones have also been studied in clinical laboratories to measure plasma hormone levels (Ashkar *et al.*, 1980).

6.5 THE USE OF ARTIFICIAL CELLS IN ARTIFICIAL ORGANS

Artificial cells have a very large surface to volume relationship. Thus 33 ml of 0.1 mm diameter artificial cells have a total surface area of 2 m^2, double the total surface area of a standard hemodialysis machine. Furthermore, the membrane thickness of artificial cells is only 0.02 μm, which is 1000 times thinner than the hemodialysis membrane (20 μm) being used in standard dialysis machines. If we look at this simplistically from a theoretical point of view, the transport rate across the membranes contained in 33 ml of 0.1 mm diameter artificial cells could be up to 2000 times higher than that across the standard hemodialyzer (Chang, 1964, 1966, 1972a, 1974a). The very small volume of artificial cells required would result in a miniaturized artificial organ for the removal or conversion of substances in the body. Enzymes, adsorbents or other materials can be placed inside these artificial cells to remove or to convert solutes diffusing into the artificial cells. Thus in one experiment urease was placed in 10 ml of artificial cells, which were retained in a small container through which blood from dogs was recirculated continuously from the femoral arteries and returned to the femoral veins. This way it was found that within 45 min the systemic blood urea level in the dog was decreased to 50% of its original level by enzymatic conversion (Chang, 1966). Tyrosinase artificial cells used this way could also lower systemic tyrosine to less than 50% of its original level (Shu and Chang, 1981). Similar studies were carried out using artificial cells containing activated charcoal or ion exchange resin to remove other toxic materials (Chang, 1966). Thus, both theoretical analysis and animal experimental studies demonstrated the feasibility of using the principle of artificial cells to form extremely compact, efficient and simple artificial organs. Four examples of this approach already being used in patients will be briefly described: blood detoxifier, artificial liver, artificial kidney and immunosorbents.

6.5.1 Blood Detoxifier

Artificial cells containing activated charcoal can be prepared in a number of ways for the construction of a novel detoxifier. A simple approach is to coat an ultrathin membrane directly on charcoal granules. The microencapsulating membrane prevents harmful charcoal powder from being released into the patients in the form of emboli. If the permeability and the membrane thickness are adjusted properly, toxins, drugs and waste metabolites can enter the artificial cells rapidly and are removed by the enclosed activated charcoal. By using blood-compatible polymers which do not adversely affect blood cells, artificial cells containing activated charcoal would not adversely affect platelets or other formed elements of blood (Chang, 1966, 1969c, 1972b, 1974b, 1976b) (Figure 6). A simple and inexpensive system is the use of 0.05 μm thick collodion membrane coating on activated charcoal granules. We found that to increase the blood compatibility further, especially in certain conditions like liver failure where the platelets are especially sensitive, an albumin coating could be applied to the collodion membrane to make the surface even more blood compatible (Figure 6) (Chang, 1969c, 1972b, 1974b, 1976b). The albumin coating, in addition to making the surface more blood compatible, also acts as a carrier mechanism to facilitate transport of loosely protein-bound substances in the bloodstream (Chang, 1978). The artificial cells containing activated charcoal are placed in a container with a screen at both ends of the container. The screen (0.1 mm openings) retained the artificial cells (1 mm diameter in these cases) inside the container, but allowed blood to circulate freely through the container and to come in direct contact with the artificial cells. This way, blood from patients can be recirculated through the detoxifier to be purified. Dogs which had received lethal doses of barbiturate, salysalate and Doridan recovered rapidly when treated with this artificial cell detoxifier device, as the drugs were rapidly removed from the circulation (Chang, 1969c, 1972b). This led us to apply this in patients for the treatment of acute drug intoxication (Chang *et al.*, 1973a, 1973b).

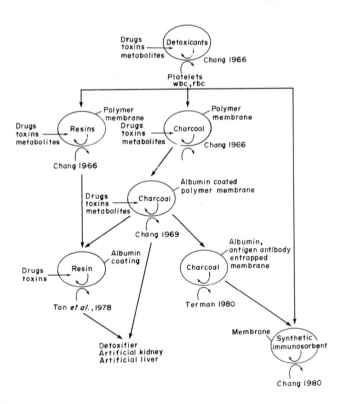

Figure 6 Development of artificial cells with adsorbents and immunosorbents (Chang, 1981b). Reproduced with permission from Springer-Verlag, Berlin

The initial clinical trial carried out in patients with suicidal doses of sleeping pills involved three drugs that are normally not removed well by the standard hemodialysis machine. These are Doridan, methylprylone and methylqualone. The patients recovered rapidly as the drugs are rapidly removed (Chang *et al.*, 1973a, 1973b). The artificial cells are much more effective than the large and bulky standard hemodialyzer in removing these medications. This led us and others to extend and treat more patients (Agishi *et al.*, 1980; Bonomini and Chang, 1982; Chang, 1975b, 1976a, 1978, 1980a; Gelfand *et al.*, 1977; Sideman and Chang, 1980; Vale *et al.*, 1975), including infants (Chang *et al.*, 1980; Chavers *et al.*, 1980; Papadopoulou and Novello, 1982). The use of artificial cells containing activated charcoal initiated in our laboratories is now a routine procedure in many large medical centers around the world for the treatment of patients with acute drug poisoning involving drugs with a small volume distribution (Chang, 1978, 1980a, 1981a; Sideman and Chang, 1980). Further investigations are required to decide its effectiveness for drugs with large volume distribution located mostly intracellularly in the body, for instance digoxin (Prichard *et al.*, 1977; Sideman and Chang, 1980).

The results described stimulated a number of other centers and industries to develop blood detoxifiers based on artificial cells (Andrade *et al.*, 1971; Klinkmann *et al.*, 1979; Kolff, 1970; Leber *et al.*, 1976; Siemsen *et al.*, 1978). In coating activated charcoal with polymer membranes, the thickness, permeability and blood compatibility of the membranes are extremely important factors. Those artificial cells which use thicker membranes or blood-incompatible membranes are not as effective. One of the most effective blood detoxifiers now consists of 80 g of spherical petroleum-based charcoal (1 mm diameter) coated by an ultrathin collodion membrane (Figure 7) (Chang *et al.*, 1982b). This is only a fraction of the size of the washing machine size standard hemodialyzer.

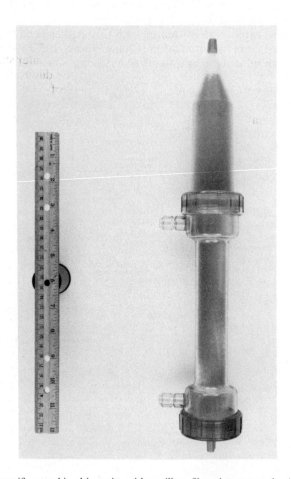

Figure 7 Artificial cell detoxifier combined in series with capillary fibers into one unit. Artificial cells containing activated charcoal (80 g) are in the conical container. Blood passes through the capillary fibers and then enters directly into the artificial cells. If no dialysis or ultrafiltration is required, the artificial cell detoxifier is used by itself without the capillary fibers section; this is the case in the treatment of acute intoxication

We have also used the principle of artificial cells to microencapsulate ion exchange resin inside artificial cells (Chang, 1966, 1972b). This prevents the ion exchange resin from adversely affecting blood cells. Another approach is to apply the albumin coating technique developed by us for improving blood compatibility (Chang, 1969c, 1972b) to resins like amberlites (Sideman *et al.*, 1981; Ton *et al.*, 1979). Thus amberlites (Rosenbaum, 1978) can now be used for blood detoxification (Figure 6).

6.5.2 Artificial Liver

Unlike kidney failure, in liver failure there is at present no artificial liver support system available. The first partial success with an artificial liver support system was carried out in 1972 when a patient with grade IV hepatic coma was treated by hemoperfusion with the artificial cell detoxification device (Chang, 1972c; Chang and Migchelson, 1973). With each treatment of 2 h the unconscious and unresponsive grade IV hepatic coma patient regained full consciousness lasting a few hours. Repeated treatment returned the patient's consciousness each time. These results stimulated a number of centers to use this approach to treat patients with hepatic coma. A total of about 500 cases have been treated around the world and in 60–80% of the patients treated there was temporary recovery of consciousness (Agishi *et al.*, 1980; Bonomini and Chang, 1982; Chang, 1975b, 1976a, 1978; Gelfand *et al.*, 1978; Kikolaef and Strelko, 1979; Odaka *et al.*, 1978; Sideman and Chang, 1980; Williams and Murray-Lyon, 1975). However, the effect on significantly improving the survival rate in grade IV coma patients has not been conclusive.

A detailed study carried out by us using the galactosamine-induced FHF rat model has demonstrated the following findings. The artificial cell detoxifier for artificial liver support significantly increased the long-term survival of rats in the earlier stages of hepatic coma (from 30% in control to 70% in treated) (Chirito *et al.*, 1977). However, this increase in survival was possible when the rats were treated in the earlier stages of hepatic coma, but not in the later stage of coma (Chang *et al.*, 1978; Niu *et al.*, 1980; Tabata and Chang, 1980). This may be due to the fact that in the later stage of hepatic coma irreversible damage may have occurred. Perfusion of donor livers also gives the same results, *i.e.* an increase in survival rate happened only if the rats were treated in the earlier stages of hepatic coma, but not in the later stages of coma (Mohsini *et al.*, 1980). These results from our animal studies have prompted another center to start treating patients in the earlier stages of coma using the detoxification device. Their results showed that this treatment improved the survival rates of 23 paracetomal-induced grade III FHF coma patients from 30% to 70%. Treatment of 23 patients in grade IV FHF coma did not improve the survival rate (Gimson *et al.*, 1982).

These results are very promising. This is the first successful artificial liver support system available. However, further studies will be required to gain further knowledge and to devise a more complete artificial liver support system for the later grades of coma and also for chronic terminal liver failure patients (Chang, 1981b, 1982b). Along this line, further studies are being carried out as follows. Artificial cells are being used to carry out some metabolic functions of the liver in experimental studies. Ammonia, tyrosine and some other amino acids are elevated in liver failure. There have been some proposals that tyrosine may contribute as precursor for false neurotransmitters in the brain and contribute to hepatic coma. Ammonia could also contribute to hepatic coma. Studies are being carried out here to prepare artificial cells containing tyrosinase (Shi and Chang, 1982; Shu and Chang, 1980, 1981). Tyrosinase artificial cells retained in extracorporeal shunts perfused by blood can effectively lower the systemic tyrosine levels in the blood of liver failure rats. Artificial cells to carry out other metabolic functions of the liver are also being studied. For instance, artificial cells containing multienzyme systems with cofactor recycling are being studied *in vivo* for the sequential conversion of ammonia into different amino acids (Chang *et al.*, 1979b). When further basic knowledge of the toxins responsible for hepatic coma (Chang and Lister, 1980, 1981) is available, artificial cells with more specific adsorbent or enzymes can be constructed.

6.5.3 Artificial Kidney

Artificial cells are being developed as the basis of an artificial kidney to replace the bulky and expensive hemodialysis machine (Chang, 1966, 1972b, 1974a, 1975b, 1976b, 1978, 1979b, 1981a; Chang and Malave, 1970; Pişkin and Chang, 1982, 1983). The blood detoxifier, consisting of artificial cells containing activated charcoal, can maintain terminal renal failure patients alive and eliminate their uremic symptoms of nausea, vomiting, fatigue, bleeding and other problems

(Chang, 1972b, 1976b, 1979b, 1981a; Chang *et al.*, 1971, 1972). This can be accomplished by using the blood detoxifier for two hours, twice a week, instead of six hours, three times a week, using standard hemodialysis. This is due to the much more efficient removal of uremic toxins by artificial cells (Chang and Migchelsen, 1973; Chang *et al.*, 1974). However this blood detoxifier based on artificial cells containing activated charcoal does not remove water, electrolytes or urea.

Since 1975 the detoxifier was used in our laboratories in series with the hemodialysis machine (Chang, 1976a, 1978, 1979b; Chang *et al.*, 1975, 1981, 1982a, 1982b). Hemodialysis is used by us only to remove water, electrolytes and urea. It was found that with this combination the time of treatment could be cut down from the standard six hours of hemodialysis to two hours. This decrease of treatment time means that the patients would have more time for their work and other responsibilities. It also means that hospitals could reduce the cost of space and staff requirements. These results are supported by other centers (Bonomini and Chang, 1982; Bonomini *et al.*, 1982; Odaka *et al.*, 1978, 1980; Stefoni *et al.*, 1980; Winchester *et al.*, 1976). In patients who develop uremic symptoms despite standard hemodialysis treatment, the artificial cell approach can eliminate the complications of pericarditis, nausea, vomiting, peripheral neuropathy or others (Bonomini and Chang, 1982; Chang *et al.*, 1972, 1974; Martin *et al.*, 1979; Odaka *et al.*, 1978, 1980; Stefoni *et al.*, 1980). The effectiveness of combining the blood detoxifier based on artificial cells and hemodialysis is thus demonstrated. At present a new development involves the combination of this artificial cell detoxifier in series with a capillary membrane system into one single unit (Figure 7) (Chang *et al.*, 1981, 1982a, 1982b).

Thus, the artificial cells blood detoxifier is much more effective in removing uremic waste metabolites and toxins. This has shortened the treatment time and has been effective in treating patients who have complications while on standard hemodialysis. However, in this approach one still needs the hemodialysis machine which is used in combination with artificial cells only to remove water, salt and urea. Thus, the following studies were carried out by us to eliminate the need of the expensive and bulky hemodialysis machine. The artificial cell blood detoxifier was combined in series with a small ultrafiltrator (Chang, 1976a, 1978; Chang *et al.*, 1975, 1977, 1979a). In this way the artificial cells can remove the uremic waste metabolites and toxins while the ultrafiltrator would remove sodium chloride and water extremely effectively (Chang *et al.*, 1975, 1977, 1979a). By using this approach the hemodialysis machine is no longer required and the smallest artificial kidney based on artificial cells is now available. This has been tested successfully on a long-term basis (Chang *et al.*, 1979a). However, there is still the problem of removal of potassium, phosphate and urea. Potassium could be removed by the oral administration of potassium adsorbent. Means for phosphate removal and urea removal are being studied to complete this system (Chang, 1980b).

The original approach of urea removal was the use of artificial cells containing urease for hemoperfusion in dogs (Chang, 1966). This lowered the systemic urea level of the dogs to 50% within 45 min. Urea is converted into ammonia which is toxic to the body. Therefore, microencapsulated ammonia adsorbent was used for the removal of ammonia (Chang, 1966). Another approach we tested was to administer urease artificial cells and ammonia adsorbent into the intestinal tract (Chang, 1972b, 1976a; Chang and Loa, 1970; Chang and Poznansky, 1968a). This way, urea from blood diffusing into the intestine was converted into ammonia which is removed by the ammonia adsorbent. This study was extended further by other laboratories (Gardner *et al.*, 1971), and is now being tested clinically (Kjellstrand *et al.*, 1981). Unfortunately, the ammonia adsorbent does not have sufficient capacity and a large volume is required. An ammonia adsorbent with a much higher adsorbing capacity has to be developed. Another attempt which is still not yet conclusive is to prepare urease artificial cells with a liquid membrane which is permeable to urea but impermeable to ammonium ions (Asher *et al.*, 1975; May and Li, 1972). We are studying another approach using artificial cells containing urease and a complex enzyme system for cofactor recycling and for the conversion of urea into ammonia, which is then sequentially converted into different types of amino acids (Chang and Malouf, 1978; Chang *et al.*, 1979b; Cousineau and Chang, 1977; Yu and Chang, 1981b, 1982). In the meantime, oxystarch (Giordano, 1979; Giordano *et al.*, 1968) has been modified into a more acceptable form for oral administration for the removal of urea (Espinosa-Melendez *et al.*, 1982). However, the problem here is the large amount required.

While waiting for the development of an optimal urea removal system, the very compact artificial kidney system consisting of the artificial cells blood detoxifier device with a small ultrafiltrator has already been demonstrated in preliminary studies to be able to substitute one of the three weekly hemodialysis treatments on a long-term basis (Chang *et al.*, 1979a). The artificial cell blood detoxifier has been made into one unit in series with an ultrafiltrator (Chang *et al.*, 1981,

1982a, 1982b). With this approach, no hemodialysis machine nor dialysate fluid is required. This is the simplest and most compact artificial kidney available (Figure 7).

6.5.4 Immunosorbent

The artificial cells containing activated charcoal being used in the detoxification device were prepared in some cases with albumin adsorbed on the surface of collodion to make the surface blood compatible (Chang, 1969c, 1974b, 1976b, 1978). This albumin coating also serves to interact with different substances in the blood to facilitate the transport of loosely protein-bound molecules in the blood into the artificial cells (Chang *et al.*, 1973a, 1973b). It was found later by Terman that albumin on the surface of the ACAC could also be used to remove antibodies to albumin in the circulating blood of animals (Terman, 1980; Terman *et al.*, 1977). This has led to a new line of research whereby different antibodies or antigens could be incorporated into the collodion-coated charcoal and used for the removal of specific antibody or antigen from the circulating blood (Terman, 1980). In more recent studies, albumin was substituted by protein A on the collodion-activated charcoal (Terman *et al.*, 1981). Plasma which has been hemoperfused over this column, when given to patients, was found to initiate an immunological reaction to stimulate the rejection of breast cancer in patients (Terman *et al.*, 1981). These results are preliminary (Chang, 1982a).

Synthetic immunosorbent is available for the removal of antibodies to blood groups. We find that by applying the principle of artificial cells with an albumin–collodion coating on the surface, the adverse effects on platelet removal and particulate embolism are eliminated (Chang, 1980d). Albumin coating on this immunosorbent has now been applied in an initial clinical trial to remove blood group antibodies before bone marrow transplantation (Bensinger *et al.*, 1981a, 1981b).

6.6 ARTIFICIAL CELLS USED FOR BLOOD SUBSTITUTES

The use of donor blood for transfusion is associated with a number of possible problems. These include the availability of blood donors, transmission of hepatitis, requirements for cross-matching and short duration of storage. It is, therefore, not surprising that there has been a continuing search for red blood cell substitutes. The two systems being investigated as possible red blood cell substitutes at present are artificial cells with hemoglobin and artificial cells with organic material (Chang, 1980c; Jamieson and Greenwalt, 1978; Mitsuno and Naito, 1979).

Research was started in our laboratories (Chang, 1957, 1964, 1965, 1972b) to investigate the feasibility of hemoglobin artificial cells. Since then, extensive studies have been carried out in a number of centers to study the preparation of different types of hemoglobin artificial cells (Chang, 1980c). These studies might be divided into two major approaches.

(a) One approach is to envelop microdroplets of hemolysate solution with spherical ultrathin membranes of cellular dimensions (Chang, 1957, 1964, 1965, 1972b, 1980c) (Figure 8). Collodion, nylon polystyrene, heparin-complexed membranes, cellulose acetate and other synthetic polymers have been studied. Lipids have also been used to form spherical membranes, although their stability and strength are insufficient (Mueller and Rudin, 1968). Lipids have been complexed to polymer or protein to form the membranes of hemolysate artificial cells (Chang, 1969a, 1972b; Rosenthal and Chang, 1980).

(b) A second major approach is the use of cross-linking (Chang, 1964, 1965, 1971a, 1972b, 1980c) (Figure 8). The hemoglobin of the hemolysate microdroplets can be cross-linked at the surface to form an ultrathin cross-linked hemoglobin membrane. Intermolecular cross-linking has also been used to cross-link part or all of the hemoglobin in the hemolysate microdroplet. The size of each cross-linked hemoglobin complex depends on the total number of hemoglobin molecules. The smaller the total number of hemoglobin molecules, the smaller would be the cross-linked complex. Further extension of this principle involves the cross-linking of hemoglobin with dextran to result in a larger molecule which is not as rapidly excreted by the kidney. Intramolecular cross-linking of the hemoglobin molecule itself has also been carried out to prevent the conversion of tetramers into dimers. In the intermolecular cross-linkage of hemoglobin described earlier there is also intramolecular cross-linkage.

There was no agglutination when the above two types of hemoglobin artificial cells were placed in contact with plasma (Chang, 1972b), even when the artificial cells were prepared using hemoglobin obtained from blood of a heterogeneous blood group. Depending on the type of procedure

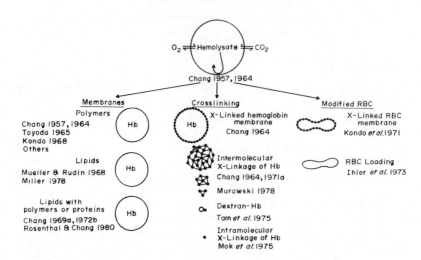

Figure 8 Development of artificial red blood cells based on hemoglobin. At present the most promising approach is the cross-linking of hemoglobin into polyhemoglobin as shown in the middle section (Chang, 1980c). Reproduced with permission from the American Society for Artificial Internal Organs

used, hemoglobin, carbonic anhydrase and catalase in these artificial cells retained different degrees of activities. Normal red blood cells contain complicated multienzyme systems for the Embden–Meyerhof pathway and the pentose phosphate pathway for the reduction of methemoglobin by the methemoglobin reductase system, for the recycling of oxidized glutethione to the reduced form, for decreasing the affinity of oxyhemoglobin for oxygen, and for other functions. Artificial red blood cells have been prepared by us containing multienzyme systems required for the recycling of these cofactors (Campbell and Chang, 1975, 1976, 1977; Chang, 1980c; Grunwald and Chang, 1978, 1979; Yu and Chang, 1981a). Studies demonstrated that artificial cells containing multienzyme systems can effectively regenerate cofactors such as ATP, NADH and NADPH.

The major problem still to be solved is that the larger hemoglobin artificial cells when injected intravenously are removed rapidly by the reticuloendothelial system for the circulation. Some progress is being made in this area using small size cross-linked hemoglobin complexes. Our studies show that these cross-linked hemoglobin artificial cells, each consisting of soluble polyhemoglobin, survive significantly longer in the circulation when compared to free hemoglobin (Keipert *et al.*, 1982; Keipert and Chang, 1983, 1984, 1985).

The development of artificial cells based on organic materials went through a number of stages (Chang, 1980c) (Figure 9). Since silicone rubber is excellent for oxygen transport, 5 μm diameter silicone rubber microspheres consisting of two parts of silicone rubber and one part of hemolysate were first prepared (Chang, 1966). This type of silicone rubber microsphere was found to carry oxygen well. However, being solid silicone rubber microspheres of 5 μm diameter, they do not survive well in the circulation. At about the same time, another group was using first silicone oil and then fluorocarbon oil for organ perfusions (Jamieson and Greenwalt, 1978). While silicone oil and fluorocarbon oil work well for O_2 carriage, the organic liquids do not contain water-soluble nutrients required by the perfused organs. The problems of solid silicone rubber microspheres and the problems of silicone oil and fluorocarbon oils were solved when a fine emulsion of fluorocarbon was used. These fine fluorocarbon emulsions were effective in O_2 carriage when tested in rats. However, problems related to blood compatibility and also emulsion stability resulted in the need to use albumin coating or lipid coating on these microdroplets. These coatings form a membrane over each fluorocarbon emulsion. One of the systems being tested clinically at present involves a lipid-coated fluorocarbon system (Mitsuno and Naito, 1979).

The major problems related to fluorocarbon artificial cells are: (1) short survival in the circulation and (2) accumulation of fluorocarbon in the body. Nevertheless, the fluorocarbon artificial cells are the first artificial blood cells ready for clinical trial as a red blood cell substitute. These major problems will prevent its routine clinical application. One of the advantages of hemoglobin artificial cells is the biodegradability of the hemoglobin component. The polyhemoglobin system being studied by us may prove to be a much better system than fluorocarbon (Chang, 1980c, 1985a; Keipert *et al.*, 1982; Keipert and Chang, 1983, 1984, 1985).

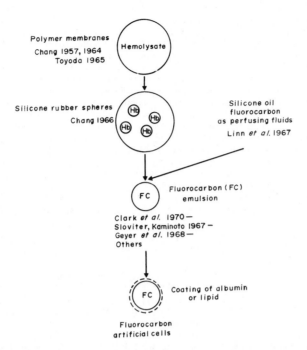

Figure 9 Development of artificial red blood cells from microencapsulated hemolysate, to silicone rubber microspheres, to fluorocarbon emulsion and finally to phospholipid coated fluorocarbon emulsion (Chang, 1980c). Reproduced with permission from the American Society for Artificial Internal Organs.

6.7 CONCLUDING REMARKS

Cells are the basic and fundamental units of all organisms. Therefore, artificial cells can be prepared to fulfill a large number of potential functions. Up to now, only comparatively simple artificial cells have been developed to a stage for actual application. The present extensive research into biotechnology will no doubt contribute directly to the further development of artificial cells.

6.8 ADDENDUM

This is a rapidly progressing field. Since the submission of the present manuscript, much new progress has been made. This has been updated in more recent reviews (Chang, 1984a, 1984b). A number of recent research reports from this laboratory include results on polyhemoglobin (Chang, 1985a; Keipert and Chang, 1983, 1984, 1985); removal of trace metals (Chang and Barre, 1983); and artificial cells immobilized enzyme systems (Bourget and Chang, 1984, 1985; Chang, 1985b, 1985c; Ergan *et al.*, 1984; Ortmanis *et al.*, 1984; Neufeld *et al.*, 1984).

6.9 REFERENCES

Agishi, T., N. Yamashita and K. Ota (1980). Clinical results of direct charcoal hemoperfusion for endogenous and exogenous intoxication. In *Hemoperfusion: Part 1—Kidney and Liver Support and Detoxification*, ed. S. Sideman and T. M. S. Chang, pp. 255–263. Hemisphere, Washington, DC.

Aisina, R. B., N. F. Kazanskata, E. V. Lukasheva and V. Berezin (1976). Microcapsules by interfacial polymerization. *Biokhimiya*, **41**, 1656–1661.

Andrade, J. D., K. Kunitomo, R. van Wagenon, B. Kastigir, D. Gough and W. J. Kolff (1971). Coated adsorbents for direct blood perfusion: HEMA activated carbon. *Trans. Am. Soc. Artif. Intern. Organs*, **17**, 222–228.

Apple, M. (1971). Hemodialysis against enzymes as a method of 'gene replacement' in cases of inherited metabolic diseases. *Proc. West. Pharmacol. Soc.*, **14**, 125.

Asher, W. J., K. C. Bovée, J. W. Frankenfeld, R. W. Hamilton, L. W. Henderson, P. G. Holtzapple and N. N. Li (1975). Liquid membrane system directed toward chronic uremia. *Kidney Int.*, **7**, S409–S412.

Ashkar, F. S., R. J. Buehler, T. Chan and M. Hourani (1980). Radioimmunoassay of free thyroxine with prebound anti-T₄ microcapsules. *J. Nucl. Med.*, **20**, 956–960.

Bensinger, W. I., D. A. Baker, C. D. Buckner, R. A. Clift and E. D. Thomas (1981a). Immunoadsorption for removal of A and B blood-group antibodies. *New Engl. J. Med.*, **304**, 160–162.

Bensinger, W. I., D. A. Baker, C. D. Buckner, R. A. Clift and E. D. Thomas (1981b). *In vitro* and *in vivo* removal of anti-A erythrocyte antibody by adsorption to a synthetic immunoadsorbent. *Transfusion*, **21**, 335–342.

Bonomini, V. and T. M. S. Chang (eds.) (1982). *Hemoperfusion*. Karger, Basel.

Bonomini, V., S. Stefoni, C. U. Casciani, M. Taccone Gallucci, A. Albertazzi, P. Cappelli, V. Mioli, R. Boggi, F. Mastrangelo and S. Rizzelli (1982). Multicentric experience with combined hemodialysis/hemoperfusion in chronic uremia. In *Hemoperfusion*, ed. V. Bonomini and T. M. S. Chang, pp. 133–142. Karger, Basel.

Bourget, L. and T. M. S. Chang (1984). Artificial cells immobilized phenylalanine ammonia lysase. *Appl. Biochem. Biotechnol.*, **10**, 57–59.

Bourget, L. and T. M. S. Chang (1985). Experimental enzyme replacement for phenylketouria using phenylalanine ammonia-lysase immobilized by microencapsulation with artificial cells. *FEBS Lett.*, in press.

Broun, G., D. Thomas, G. Gellf, D. Domurado, A. M. Berjonneau and T. Guillon (1973). New methods for binding enzyme molecules into a water-insoluble matrix: properties after insolubilization. *Biotechnol. Bioeng.*, **15**, 359–375.

Brunner, G. and F. W. Schmidt (eds.) (1981). *Artificial Liver Support*. Springer-Verlag, Berlin.

Campbell, J. and T. M. S. Chang (1975). Enzymatic recycling of coenzymes by a multi-enzyme system immobilized within semipermeable collodion microcapsules. *Biochim. Biophys. Acta*, **397**, 101–109.

Campbell, J. and T. M. S. Chang (1976). The recycling of NAD$^+$ (free and immobilized) within semipermeable aqueous microcapsules containing a multienzyme system. *Biochem. Biophys. Res. Commun.*, **69**, 562–569.

Campbell, J. and T. M. S. Chang (1977). Immobilized multienzyme systems and coenzyme requirements: Perspectives in biomedical applications. In *Biomedical Applications of Immobilized Enzymes and Proteins*, ed. T. M. S. Chang, vol. 2, pp. 281–302. Plenum, New York.

Chambers, R. P., J. R. Ford, J. H. Allender, W. H. Baricos and W. Cohen (1974). Continuous processing with cofactors requiring enzymes, coenzyme retention, and regeneration. In *Enzyme Engineering*, ed. E. K. Pye and J. B. Wingard, Jr., p. 195. Plenum, New York.

Chang, T. M. S. (1957). Hemoglobin corpuscles. Report of research project for B.Sc. Honours, McGill University, Montreal, Canada.

Chang, T. M. S. (1964). Semipermeable microcapsules. *Science*, **146**, 524–525.

Chang, T. M. S. (1965). Semipermeable aqueous microcapsules. Ph.D. Thesis, McGill University, Montreal, Canada.

Chang, T. M. S. (1966). Semipermeable aqueous microcapsules ('artificial cells'): with emphasis on experiments in an extracorporeal shunt system. *Trans. Am. Soc. Artif. Intern. Organs*, **12**, 13–19.

Chang, T. M. S. (1969a). Lipid-coated spherical ultrathin membranes of polymer or cross-linked protein as possible cell membrane models. *Fed. Proc., Fed. Am. Soc. Exp. Biol.*, **28**, 461.

Chang, T. M. S. (1969b). Clinical potential of enzyme technology. *Sci. Tools*, **16**, 33–39.

Chang, T. M. S. (1969c). Removal of endogenous and exogenous toxins by a microencapsulated absorbent. *Can. J. Physiol. Pharmacol.*, **47**, 1043–1045.

Chang, T. M. S. (1971a). Stabilisation of enzymes by microencapsulation with a concentrated protein solution or by microencapsulation followed by cross-linking with glutaraldehyde. *Biochem. Biophys. Res. Commun.*, **44**, 1531–1536.

Chang, T. M. S. (1971b). The *in vivo* effects of semipermeable microcapsules containing L-asparaginase on 6C3HED lymphosarcoma. *Nature (London)*, **229** (528), 117–118.

Chang, T. M. S. (1972a). A new approach to separation using semipermeable microcapsules (artificial cells): Combined dialysis, catalysis and absorption. In *Recent Development in Separation Science*, ed. N.N.Li, vol. 1, pp. 203–216. CRC Press, Cleveland, OH.

Chang, T. M. S. (1972b). *Artificial Cells*. Thomas, Springfield.

Chang, T. M. S. (1972c). Haemoperfusions over microencapsulated adsorbent in a patient with hepatic coma. *Lancet, ii*, 1371–1372.

Chang, T. M. S. (1973). L-Asparaginase immobilized within semipermeable microcapsules: *in vitro* and *in vivo* stability. *Enzyme*, **14** (2), 95–104.

Chang, T. M. S. (1974a). A comparison of semipermeable microcapsules and standard dialysers for use in separation. *Sep. Purif. Methods*, **3**, 245–262.

Chang, T. M. S. (1974b). Platelet–surface interaction: Effect of albumin coating or heparin complexing on thrombogenic surfaces. *Can. J. Physiol. Pharmacol.*, **52** (2), 275–285.

Chang, T. M. S. (1975a). Artificial cells. *Chem. Tech.*, **5**, 80–85.

Chang, T. M. S. (1975b). Microencapsulated adsorbent hemoperfusion for uremia, intoxication and hepatic failure. *Kidney Int.*, **7**, S387–S392.

Chang, T. M. S. (1975c). The one shot vaccine. In *Socio-Economic and Ethical Implications of Enzyme Engineering*, ed. C.-G. Hedén, pp. 17–18. International Federation of Institutes for Advanced Studies, Stockholm.

Chang, T. M. S. (1975d). Artificial cells as carriers for biologically active materials in therapy. *Clin. Pharmacol.*, **5**, 81–90.

Chang, T. M. S. (1976a). Hemoperfusion alone and in series with ultrafiltration or dialysis for uremia, poisoning and liver failure. *Kidney Int.*, **10**, S305–S311.

Chang, T. M. S. (1976b). Microcapsule artificial kidney: including updated preparative procedures and properties. *Kidney Int.*, **10**, S218–S224.

Chang, T. M. S. (1976c). Microencapsulation of enzymes and biologicals. In *Methods in Enzymology: Immobilized Enzymes*, ed. K. Mosbach, vol. XLIV, pp. 201–218. Academic, New York.

Chang, T. M. S. (1976d). Biodegradable semipermeable microcapsules containing enzymes, hormones, vaccines and biologicals. *J. Bioeng.*, **1**, 25–32.

Chang, T. M. S. (1976e). Methods for the therapeutic applications of immobilized enzymes. In *Methods in Enzymology: Immobilized Enzymes*, ed. K. Mosbach, vol. XLIV, pp. 676–698. Academic, New York.

Chang, T. M. S. (1977a). Encapsulation of enzymes, cell contents, cells, vaccines, antigens, antiserum, cofactors, hormones and proteins. In *Biomedical Applications of Immobilized Enzymes and Proteins*, ed. T. M. S. Chang, vol. 1, pp. 69–90. Plenum, New York.

Chang, T. M. S. (1977b). Rationale and strategies for the therapeutic applications of immobilized enzymes. In *Biomedical Applications of Immobilized Enzymes and Proteins*, ed. T. M. S. Chang, vol. 1, pp. 93–104. Plenum, New York.

Chang, T. M. S. (1977c). Experimental therapy using semipermeable microcapsules containing enzymes and other biologically active material. In *Biomedical Applications of Immobilized Enzymes and Proteins*, ed. T. M. S. Chang, vol. 1, pp. 147–162. Plenum, New York.

Chang, T. M. S. (ed.) (1977d). *Biomedical Applications of Immobilized Enzymes and Proteins*, vols. 1 and 2. Plenum, New York.

Chang, T. M. S. (1978). *Artificial Kidney, Artificial Liver, and Artificial Cells*. Plenum, New York.

Chang, T. M. S. (1979a). Artificial cells as drug carriers in biology and medicine. In *Drug Carriers in Biology and Medicine*, ed. G. Gregoriadis, pp. 271–285. Academic, New York.

Chang, T. M. S. (1979b). Assessments of clinical trials of charcoal hemoperfusion in uremic patients. *Clin. Nephrol.*, **11**, 111–119.

Chang, T. M. S. (1980a). Clinical experience with ACAC coated charcoal hemoperfusion in acute intoxication. *Clin. Toxicol.*, **17**, 529–542.

Chang, T. M. S. (1980b). New approaches using immobilized enzymes for the removal of urea and ammonia. In *Enzyme Engineering*, ed. H. H. Weetall and G. P. Royer, vol. 5, pp. 225–229. Plenum, New York.

Chang, T. M. S. (1980c). Artificial red blood cells. *Trans. Am. Soc. Artif. Intern. Organs*, **26**, 354–357.

Chang, T. M. S. (1980d). Blood compatible coating of synthetic immunoadsorbents. *Trans. Am. Soc. Artif. Intern. Organs*, **26**, 546–549.

Chang, T. M. S. (1981a). Current status of sorbent microencapsulation. In *Advances in Basic and Clinical Nephrology*, ed. W. Zurukzoglu, M. Papadimitriou, M. Pyrpasopoulos, M. Sion and C. Zamboulis, pp. 400–406. Karger, Basel.

Chang, T. M. S. (1981b). Hemoperfusion, exchange transfusion, cross circulation, liver perfusion, hormones and immobilized enzymes. In *Artificial Liver Support*, ed. G. Brunner and F. W. Schmidt, pp. 126–133. Springer-Verlag, Berlin.

Chang, T. M. S. (1982a). Plasma perfused over immobilized protein A for breast cancer. *New Engl. J. Med.*, **306**, 936.

Chang, T. M. S. (1982b). Earlier haemoperfusion in fulminant hepatic failure. *Lancet*, **ii**, 1039.

Chang, T. M. S. (1984). Artificial cells: biotechnology and medical applications. *Appl. Biochem. Biotechnol.*, **10**, 5–24.

Chang, T. M. S. (1985a). Blood substitute. *Trans. Am. Soc. Artif. Intern. Organs*, in press.

Chang, T. M. S. (1985b). Artificial cells containing multienzyme systems. In *Methods in Enzymology Series: Immobilized Enzymes and Cells*, ed. K. Mosbach. Academic, New York, in press.

Chang, T. M. S. (1985c). Methods in the medical applications of immobilized proteins, enzymes and cells. In *Methods in Enzymology Series: Immobilized Enzymes and Cells*, ed. K. Mosbach. Academic, New York, in press.

Chang, T. M. S. and P. Barre (1983). Effect of desferrioxamine on removal of aluminium and iron by coated charcoal haemoperfusion and haemodialysis. *Lancet*, Nov, 1051–1053.

Chang, T. M. S. and N. Kuntarian (1978). Galactose conversion using a microcapsule immobilized multienzyme cofactor recycling system. In *Enzyme Engineering*, ed. G. B. Broun, G. Manecke and L. B. Wingard, Jr., vol. 4, pp. 193–197. Plenum, New York.

Chang, T. M. S. and C. Lister (1980). Analysis of possible toxins in hepatic coma including the removal of mercaptan by albumin–collodion charcoal. *Int. J. Artif. Organs*, **3** (2), 108–112.

Chang, T. M. S. and C. Lister (1981). Middle molecules in hepatic coma and uremia. *Artif. Organs*, **4**, S169–S172.

Chang, T. M. S. and S. K. Loa (1970). Urea removal by urease and ammonia adsorbent in the intestine. *Physiologist*, **13**, 70.

Chang, T. M. S. and N. Malave (1970). The development and first clinical use of semipermeable microcapsules (artificial cells) as a compact artificial kidney. *Trans. Am. Soc. Artif. Intern. Organs*, **16**, 141–148.

Chang, T. M. S. and C. Malouf (1978). Artificial cells microencapsulated multienzyme system for converting urea and ammonia to amino acid using α-ketoglutarate and glucose as substrate. *Trans. Am. Soc. Artif. Intern. Organs*, **24**, 18–20.

Chang, T. M. S. and C. Malouf (1979). Effects of glucose dehydrogenase in converting urea and ammonia into amino acid using artificial cells. *Artif. Organs*, **3** (1), 38–41.

Chang, T. M. S. and M. Migchelsen (1973). Characterization of possible 'toxic' metabolites in uremia and hepatic coma based on the clearance spectrum for larger molecules by the ACAC microcapsule artificial kidney. *Trans. Am. Soc. Artif. Intern. Organs*, **19**, 314–319.

Chang, T. M. S. and M. J. Poznansky (1968a). Semipermeable microcapsules containing catalase for enzyme replacement in acatalasaemic mice. *Nature (London)*, **218**, 243–245.

Chang, T. M. S. and M. J. Poznansky (1968b). Semipermeable aqueous microcapsules (artificial cells): V. Permeability characteristics. *J. Biomed. Mater. Res.*, **2**, 187–199.

Chang, T. M. S., F. C. MacIntosh and S. G. Mason (1966). Semipermeable aqueous microcapsules: I. Preparation and properties. *Can. J. Physiol. Pharmacol.*, **44**, 115–128.

Chang, T. M. S., L. J. Johnson and O. Ransome (1967). Semipermeable aqueous microcapsules: IV. Nonthrombogenic microcapsules with heparin-complexed membranes. *Can. J. Physiol. Pharmacol.*, **45**, 705–715.

Chang, T. M. S., A. Gonda, J. H. Dirks and N. Malave (1971). Clinical evaluation of chronic, intermittent, and short term hemoperfusions in patients with chronic renal failure using semipermeable microcapsules (artificial cells) formed from membrane-coated activated charcoal. *Trans. Am. Soc. Artif. Intern. Organs*, **17**, 246–252.

Chang, T. M. S., A. Gonda, J. H. Dirks, J. F. Coffey and T. Lee-Burns (1972). ACAC microcapsule artificial kidney for the long term and short term management of eleven patients with chronic renal failure. *Trans. Am. Soc. Artif. Intern. Organs*, **18**, 465–472.

Chang, T. M. S., J. F. Coffey, P. Barré, A. Gonda, J. H. Dirks, M. Levy and C. Lister (1973a). Microcapsule artificial kidney: treatment of patients with acute drug intoxication. *Can. Med. Assoc. J.*, **108**, 429–433.

Chang, T. M. S., J. F. Coffey, C. Lister, E. Taroy and A. Stark (1973b). Methaqualone, methyprylon, and glutethimide clearance by the ACAC microcapsules artificial kidney: *in vitro* and in patients with acute intoxication. *Trans. Am. Soc. Artif. Intern. Organs*, **19**, 87–91.

Chang, T. M. S., M. Migchelsen, J. F. Coffey and A. Stark (1974). Serum middle molecule levels in uremia during long term intermittent hemoperfusions with the ACAC (coated charcoal) microcapsule artificial kidney. *Trans. Am. Soc. Artif. Intern. Organs*, **20**, 364–371.

Chang, T. M. S., E. Chirito, P. Barré, C. Cole and M. Hewish (1975). Clinical performance characteristics of a new combined system for simultaneous hemoperfusion–hemodialysis–ultrafiltration in series. *Trans. Am. Soc. Artif. Intern. Organs*, **21**, 502–508.

Chang, T. M. S., E. Chirito, P. Barré, C. Cole, C. Lister and E. Resurreccion (1977). Clinical evaluation of the clearance profiles of a portable, compact, dialysate-free system incorporating microencapsulated charcoal hemoperfusion for blood purification with ultrafiltration for fluid removal. *J. Dialysis*, **1** (3), 239–259.

Chang, T. M. S., C. Lister, E. Chirito, P. O'Keefe and E. Resurreccion (1978). Effects of hemoperfusion rate and time of initiation of ACAC charcoal hemoperfusion on the survival of fulminant hepatic failure rats. *Trans. Am. Soc. Artif. Intern. Organs*, **24**, 243–245.

Chang, T. M. S., E. Chirito, P. Barré, C. Cole, C. Lister and E. Resurreccion (1979a). Long-term clinical assessment of combined ACAC hemoperfusion–ultrafiltration in uremia. *Artif. Organs*, **3**, 127–131.

Chang, T. M. S., C. Malouf and E. Resurreccion (1979b). Artificial cells containing multienzyme systems for the sequential conversion of urea into ammonia, glutamate, then alanine. *Artif. Organs*, **3**, S284–S287.

Chang, T. M. S., E. Espinosa-Melendez, T. E. Francoeur and N. R. Eade (1980). Albumin–collodion activated charcoal hemoperfusion in the treatment of severe theophylline intoxication in a 3-year-old patient. *Pediatrics*, **65**, 811–814.

Chang, T. M. S., Y. Lacaille, X. Picart, E. Resurreccion, A. Loebel, D. Messier and N. K. Man (1981). Composite artificial kidney CAK: A single unit combining hemodialysis and hemoperfusion. *Artif. Organs*, **5**, S200–S203.

Chang, T. M. S., P. Barré, S. Kuruvilla, N. K. Man, Y. Lacaille, D. Messier, M. Messier and E. Resurreccion (1982a). Hemoperfusion–hemodialysis in a single unit: Composite artificial kidney. In *Artificial Support Systems*, ed. J. Belinger, pp. 63–67. Saunders, London.

Chang, T. M. S., P. Barré, S. Kuruvilla, D. Messier, N. Man and E. Resurreccion (1982b). Phase one clinical trial of a new composite artificial kidney: A single unit combining dialysis with hemoperfusion. *Trans. Am. Soc. Artif. Intern. Organs*, **28**, 43–48.

Chang, T. M. S., C. D. Shu, Y. T. Yu and J. Grunwald (1982c). Artificial cells immobilized enzymes for metabolic disorders. In *Advances in the Treatment of Inborn Errors of Metabolism*, ed. M. Crawford, D. Gibbs and R. W. E. Watts, pp. 175–184. Wiley, London.

Chavers, B. M., C. M. Kjellstrand, C. Wiegand, J. Ebben and S. M. Mauer (1980). Techniques for use of charcoal hemoperfusion in infants: experience in two patients. *Kidney Int.*, **18**, 386.

Chirito, E., B. Reiter, C. Lister and T. M. S. Chang (1977). Artificial liver: the effect of ACAC microencapsulated charcoal hemoperfusion on fulminant hepatic failure. *Artif. Organs*, **1** (1), 76–83.

Clark, L. C., Jr., S. Kaplan, F. Becattini and G. Benzing, III (1970). Perfusion of whole animals with perfluorinated liquid emulsions using the Clark bubble–defoam heart–lung machine. *Fed. Proc.*, **29**, 1764–1770.

Cousineau, J. and T. M. S. Chang (1977). Formation of amino acid from urea and ammonia by sequential enzyme reaction using a microencapsulated multienzyme system. *Biochem. Biophys. Res. Commun.*, **79** (1), 24–31.

Damon Corporation (1981). *Bulletin on Tissue Microencapsulation*, Needham Heights, MA.

Dinelli, D. (1972). Fibre-entrapped enzymes. *Process Biochem.*, **7**, 9.

Ergan, F., D. Thomas and T. M. S. Chang (1984). Microencapsulation of micro-organisms and 3α-hydroxysteroid dehydrogenase for NAD recycling and stereospecific steroid oxidation. *Appl. Biochem. Biotechnol.*, **10**, 61–71.

Espinosa-Melendez, E., M. Zelman, P. Barré, C. Lister and T. M. S. Chang (1982). Oxystarch modified by boiling: effects of oral administration in stable hemodialysis patients. In *Hemoperfusion and Artificial Organs*, ed. E. Pişkin and T. M. S. Chang, pp. 83–85. Artificial Organs Society, Ankara.

Gardner, D. L., R. D. Falb, B. C. Kim and D. C. Emmerling (1971). Possible uremic detoxification via oral-ingested microcapsules. *Trans. Am. Soc. Artif. Intern. Organs*, **17**, 239.

Gelfand, M. C., J. F. Winchester, J. H. Knepshield, K. M. Hansen, S. L. Cohan, B. S. Strauch, K. L. Geoly, A. C. Kennedy and G. E. Schreiner (1977). Treatment of severe drug overdose with charcoal hemoperfusion. *Trans. Am. Soc. Artif. Intern. Organs*, **23**, 599–603.

Gelfand, M. C., J. F. Winchester, J. H. Knepshield, S. L. Cohan and G. E. Schreiner (1978). Biochemical correlates of reversal of hepatic coma coated with charcoal hemoperfusion. *Trans. Am. Soc. Artif. Intern. Organs*, **24**, 239–242.

Geyer, R. P., R. G. Monroe and K. Taylor (1968). Survival of rats totally perfused with a fluorocarbon–detergent preparation. In *Organ Perfusion and Preservation*, ed. J. C. Norman, J. Folkman, W. G. Hardison, L. E. Rudolf and F. J. Veith, pp. 85–96. Appleton-Century-Crofts, New York.

Gimson, A. E. S., S. Brande, P. J. Mellon, J. Canalese and R. Williams (1982). Earlier charcoal hemoperfusion in fulminant hepatic failure. *Lancet*, September, 681–683.

Giordano, C. (ed.) (1979). *Sorbents and Their Clinical Applications*. Academic, New York.

Giordano, C., R. Esposito and G. Demma (1968). Possibilita' di ridurve l'azotemia nell'uomo mediante somministrazione di una polialdeide. *Boll. Soc. Ital. Biol. Sper.*, **44**, 2232–2234.

Gordon, A., M. A. Greenbaum, L. B. Marantz, M. S. McArthur and M. D. Maxwell (1969). A sorbent based low volume recirculating dialysate system. *Trans. Am. Soc. Artif. Intern. Organs*, **15**, 347–349.

Gregoriadis, G. (ed.) (1979). *Drug Carriers in Biology and Medicine*. Academic, New York.

Gregoriadis, G. and B. E. Ryman (1971). Enzyme-entrapment in liposomes. *FEBS Lett.*, **14**, 95.

Grunwald, J. and T. M. S. Chang (1978). Nylon polyethyleneimine microcapsules for immobilizing multienzymes with soluble dextran-NAD$^+$ for the continuous recycling of the microencapsulated dextran-NAD$^+$. *Biochem. Biophys. Res. Commun.*, **81** (2), 565–570.

Grunwald, J. and T. M. S. Chang (1979). Continuous recycling of NAD$^+$ using an immobilized system of collodion microcapsules containing dextran-NAD$^+$, alcohol dehydrogenase, and malic dehydrogenase. *J. Appl. Biochem.*, **1**, 104–114.

Grunwald, J. and T. M. S. Chang (1981). Immobilization of alcohol dehydrogenase, malic dehydrogenase and dextran-NAD$^+$ within nylon-polyethyleneimine microcapsules: Preparation and cofactor recycling. *J. Mol. Catal.*, **11**, 83–90.

Ihler, G. M., R. H. Glew and F. W. Schnure (1973). Enzyme loading of erythrocytes. *Proc. Natl. Acad. Sci. USA*, **70**, 2663.

Jamieson, G. A. and T. J. Greenwalt (1978). *Blood Substitutes and Plasma Expanders*. Liss, New York.

Keipert, P. E. and T. M. S. Chang (1983). In vivo assessment of pyridoxylated cross-linked polyhemoglobin as an artificial red cell substitute in rats. *Trans. Am. Soc. Artif. Intern. Organs*, **29**, 329–333.

Keipert, P. E. and T. M. S. Chang (1984). Preparation and in-vitro characteristics of pyridoxylated polyhemoglobin as blood substitute. *Appl. Biochem. Biotechnol.*, **10**, 133–141.

Keipert, P. E. and T. M. S. Chang (1985). Pyridoxylated polyhemoglobin as a blood substitute for resuscitation of lethal hemorrhagic shock in conscious rats. *Biomat. Med. Dev., Artif. Organs*, in press.

Keipert, P. E., J. Minkowitz and T. M. S. Chang (1982). Cross-linked stroma-free polyhemoglobin as a potential blood substitute. *Int. J. Artif. Organs*, **5**, 383–385.

Kikolaef, V. G. and V. V. Strelko (1979). *Hemoperfusion*. Naukova Dumka, Kiev.

Kitajima, M. and A. Kondo (1971). Fermentation with multiplication of cells using microcapsules that contain zymase complex and muscle enzyme extract. *Bull. Chem. Soc. Jpn.*, **44**, 3201.

Kjellstrand, C., H. Borges, C. Pru, D. Gardner and D. Fink (1981). On the clinical use of microencapsulated zirconium phosphatase-urease for the treatment of chronic anemia. *Trans. Am. Soc. Artif. Intern. Organs*, **27**, 24–30.

Klinkmann, H., D. Falkenhagen and J. M. Courtney (1979). Blood purification by hemoperfusion. *Int. J. Artif. Organs*, **2**, 296.

Kolff, W. J. (1970). Artificial organs in the seventies. *Trans. Am. Soc. Artif. Intern. Organs*, **16**, 534–540.

Kondo, A. (1968). Personal communication.

Kondo, A., M. Kitajima and W. Sekiguchi (1971). A modification of red blood cells by isocyanates. *Bull. Chem. Soc. Jpn.*, **44**, 139.

Leber, H. W., M. Neuhauser and G. Goubeaud (1976). Hemoperfusion *versus* hemodialysis in uremia: biochemical analysis. *Eur. Soc. Artif. Organs*, **3**, 202.

Lim, F. and A. M. Sun (1980). Microencapsulated islets as bioartificial endocrine pancreas. *Science*, **210**, 908.

Linn, B. S., W. Canaday, Jr., M. Berton and F. Gollan (1967). Kidney preservation by perfusion with organic liquids. *Surg. Forum*, **18**, 278–279.

Martin, A. M., T. K. Gibbins, T. Kimmit and F. Rennie (1979). Hemodialysis and hemoperfusion in the treatment of uremic pericarditis. A study of 13 cases. *Dial. Transplant.*, **8**, 135.

May, S. W. and N. N. Li (1972). The immobilization of urease using liquid-surfactant membranes. *Biochem. Biophys. Res. Commun.*, **47**, 1179–1185.

Miller, I. F. (1978). Discussion 6 in *Blood Substitutes and Plasma Expanders*, ed. G. A. Jamieson and T. J. Greenwalt, p. 225. Liss, New York.

Mitsuno, T. and R. Naito (1979). *Perfluorochemical Blood Substitutes*, p. 469. Excerpta Medica, Amsterdam.

Mohsini, K., C. Lister and T. M. S. Chang (1980). The effects of homologous cross-circulation and *in situ* liver perfusion on fulminant hepatic failure rats. *Artif. Organs*, **4**, 171–175.

Mok, W., D. E. Chen and A. Mazur (1975). Covalent linkage of subunits of hemoglobin. *Fed. Proc.*, **34**, 1458.

Mori, T., T. Tosa and I. Chibata (1973). Enzymatic properties of microcapsules containing asparaginase. *Biochem. Biophys. Acta*, **321**, 653–661.

Mosbach, K. and R. Mosbach (1966). Entrapment of enzymes and microorganisms in synthetic cross-linked polymers and their applications in column techniques. *Acta Chem. Scand.*, **20**, 2807–2810.

Mueller, P. and D. O. Rudin (1968). Resting and action potentials in experimental bimolecular lipid membranes. *J. Theor. Biol.*, **18**, 222–258.

Murawski, K. (1978). Discussion 6 in *Blood Substitutes and Plasma Expanders*, ed. G. A. Jamieson and T. J. Greenwalt, p. 224. Liss, New York.

Neufeld, R. J., M. Arbeloa and T. M. S. Chang (1984). Design of a fluidized bed reactor for microencapsulated urease. *Appl. Biochem. Biotechnol.*, **10**, 109–119.

Niu, Z., S. R. Jia, D. Y. Zhang, C. X. Xu, X. J. Tang, W. K. Fan, Y. P. Luo and Z. M. Li (1980). The effects of hemoperfusion in galactosamine induced FHF rats. *Chungking Medical College Bull.*, November, 1–6.

Odaka, M., Y. Tabata, H. Kobayashi, Y. Nomura, H. Soma, H. Hirasawa and H. Sato (1978). Clinical experience of bead-shaped charcoal haemoperfusion in chronic renal failure and fulminant hepatic failure. In *Artificial Kidney, Artificial Liver, and Artificial Cells*, ed. T. M. S. Chang, pp. 79–88. Plenum, New York.

Odaka, M., H. Hirasawa, H. Kobayashi, M. Ohkawa, K. Soeda, Y. Tabata, M. Soma and H. Sato (1980). Clinical and fundamental studies of cellulose coated bead-shaped charcoal haemoperfusion in chronic renal failure. In *Hemoperfusion: Part 1—Kidney and Liver Support and Detoxification*, ed. S. Sideman and T. M. S. Chang, pp. 45–55. Hemisphere, Washington, DC.

Ortmanis, A., R. J. Neufeld and T. M. S. Chang (1984). Study of microencapsulated urease in a continuous feed, stirred tank reactor. *Enzyme Microb. Technol.*, **6**, 135–139.

Østergaard, J. C. W. and S. C. Martiny (1973). The immobilization of β-galactosidase through encapsulation in water-insoluble microcapsules. *Biotechnol. Bioeng.*, **15**, 561–563.

Papadopoulou, Z. L. and A. C. Novello (1982). The use of hemoperfusion in children. *Pediatric Clinics of North America*, **26**, 1039–1052.

Pişkin, E. and T. M. S. Chang (1982). *Hemoperfusion and Artificial Organs*. Turkish Artificial Organs Society, Ankara, Turkey.

Pişkin, E. and T. M. S. Chang (1983). *Past, Present and Future of Artificial Organs*. Turkish Artificial Organs Society, Ankara, Turkey.

Poznansky, M. J. and T. M. S. Chang (1974). Comparison of the enzyme kinetics and immunological properties of catalase immobilized by microencapsulation and catalase in free solution for enzyme replacement. *Biochim. Biophys. Acta*, **334**, 103–115.

Prichard, S., E. Chirito, T. M. S. Chang and A. D. Sniderman (1977). Microencapsulated charcoal hemoperfusion: A possible therapeutic adjunct in digoxin toxicity. *J. Dialysis*, **1** (4), 367–377.

Rogers, S. (1968). Dialysis against enzymes. *Nature (London)*, **220**, 1321.

Rony, P. R. (1971). Multiphase catalysis: II. Hollow fibre catalysts. *Biotechnol. Bioeng.*, **13**, 431–447.

Rosenbaum, J. L. (1978). Experience with resin hemoperfusion. In *Artificial Kidney, Artificial Liver, and Artificial Cells*, ed. T. M. S. Chang, pp. 217–224. Plenum, New York.

Rosenthal, A. M. and T. M. S. Chang (1971). The effect of ialinomycin on the movement of rubidium across lipid coated semipermeable microcapsules. *Proc. Can. Fed. Biol. Soc.*, **14**, 44.

Rosenthal, A. M. and T. M. S. Chang (1980). The incorporation of lipid and Na^+-K^+-ATPase into the membranes of semipermeable microcapsules. *J. Membrane Sci.*, **6** (3), 329–338.

Sekiguchi, W. and A. Kondo (1966). Studies of microencapsulated hemoglobin. *J. Jpn. Soc. Blood Transfusion*, **13**, 153–154.

Sessa, G. and G. Weissman (1970). Incorporation of lysozyme into liposomes. *J. Biol. Chem.*, **245**, 3295–3301.

Shi, Z. Q. and T. M. S. Chang (1982). The effects of hemoperfusion using coated charcoal on tyrosinase artificial cells on middle molecules and tyrosine in brain and serum of hepatic coma rats. *Trans. Am. Soc. Artif. Intern. Organs*, **28**, 205–209.

Shiba, M., S. Tomioka, M. Koishi and T. Kondo (1970). Studies on microcapsules: V. Preparation of polyamide microcapsules containing aqueous protein solution. *Chem. Pharm. Bull.*, **18**, 803.

Shu, C. D. and T. M. S. Chang (1980). Tyrosinase immobilized with artificial cells for detoxification in liver failure. I. Preparation and *in vitro* studies. *Int. J. Artif. Organs*, **3** (5), 287–291.

Shu, C. D. and T. M. S. Chang (1981). Tyrosinase immobilized within artificial cells for detoxification in liver failure: II. *In vivo* studies in fulminant hepatic failure rats. *Int. J. Artif. Organs*, **4**, 82–84.

Sideman, S. and T. M. S. Chang (1980). *Hemoperfusion: Kidney and Liver Support and Detoxification*. Hemisphere, Washington, DC.

Sideman, S., L. Mor, L. S. Fishler, I. Thaler and J. M. Brandes (1981). Bilirubin removal by sorbent hemoperfusion from jaundice blood. In *Artificial Liver Support*, ed. G. Brunner and F. W. Schmidt, pp. 103–109. Springer-Verlag, Berlin.

Siemsen, A. W., G. Dunea, B. H. Mamdani and G. Guruprakash (1978). Charcoal hemoperfusion for chronic renal failure. *Nephron*, **22**, 386–390.

Siu Chong, E. D. and T. M. S. Chang (1974). *In vivo* effects of intraperitoneally injected L-asparaginase solution and L-asparaginase immobilized within semipermeable nylon microcapsules with emphasis on blood L-asparaginase, 'body' L-asparaginase, and plasma L-asparagine levels. *Enzyme*, **18**, 218–239.

Sloviter, H. A. and T. Kaminoto (1967). Erythrocyte substitute for perfusion of brain. *Nature (London)*, **216**, 458.

Sparks, R. E., R. M. Salemme, P. M. Meier, M. H. Litt and O. Lindan (1972). Removal of waste metabolites in uremia by microencapsulated reactants. *Trans. Am. Soc. Artif. Intern. Organs*, **15**, 353–358.

Stefoni, S., L. Coli, G. Feliciangeli, L. Baldrati and V. Bonomini (1980). Regular hemoperfusion in regular dialysis treatment. A long-term study. *Int. J. Artif. Organs*, **3**, 348.

Tabata, Y. and T. M. S. Chang (1980). Comparisons of six artificial liver support regimes in fulminant hepatic coma rats. *Trans. Am. Soc. Artif. Intern. Organs*, **26**, 394–399.

Tam, S. C., J. Blumenstein and J. T. F. Wong (1975). Dextran–hemoglobin complex as a potential blood substitute. *Proc. Natl. Acad. Sci. USA*, **73**, 2128–2131.

Terman, D. S. (1980). Extracorporeal immunoadsorbents for extraction of circulating immune reactants. In *Sorbents and Their Clinical Applications*, ed. C. Giordano, pp. 469–490. Academic, New York.

Terman, D. S., T. Tavel, D. Petty, M. R. Racic and G. Buffaloe (1977). Specific removal of antibody by extracorporeal circulation over antigen immobilized in collodion–charcoal. *Clin. Exp. Immunol.*, **28**, 180–188.

Terman, D. S., J. B. Young, W. T. Shearer, C. Ayus, D. Lehane, C. Mattioli, R. Espada, J. F. Howell, T. Yamamoto, H. I. Zaleski, L. Miller, P. Frommer, L. Feldman, J. F. Henry, R. Tillquist, G. Cook and Y. Daskal (1981). Preliminary observations of the effect on breast adenocarcinoma of plasma perfused over protein A collodion charcoal. *New Engl. J. Med.*, **305**, 1195–1200.

Ton, H.-Y., R. D. Hughes, D. B. A. Silk and R. Williams (1979). Albumin-coated amberlite XAD-7 resin for hemoperfusion in acute liver failure. *Artif. Organs*, **3**, 20.

Toyoda, T. (1965). Blood replacement. *Kagaku*, **36**, 7.

Vale, J. A., A. J. Rees, B. Widdop and R. Goulding (1975). Use of charcoal hemoperfusion in the management of severely poisoned patients. *Br. Med. J.*, **1**, 5.

Williams, R. and I. M. Murray-Lyon (1975). *Artificial Liver Support*. Pitman, London.

Winchester, J. F., M. T. Apiliga and A. C. Kennedy (1976). Short term evaluation of charcoal hemoperfusion combined with dialysis in uremic patients. *Kidney Int.*, **10**, S315–S319.

Yu, Y. T. and T. M. S. Chang (1981a). Ultrathin lipid–polymer membrane microcapsules containing multienzymes, cofactors and substrates for multistep enzyme reactions. *FEBS Lett.*, **125** (1), 94–96.

Yu, Y. T. and T. M. S. Chang (1981b). Lipid–nylon membrane artificial cells containing multienzyme systems, cofactors and substrates for the sequential conversion of ammonia and urea into glutamate. *Trans. Am. Soc. Artif. Intern. Organs*, **27**, 535–538.

Yu, Y. T. and T. M. S. Chang (1982). Lipid–polyamide membrane microcapsules immobilized multienzymes and cofactors for sequential conversion of lipophilic and lipophobic substrates. *J. Microbial Enzyme Technol.*, **4**, 327–331.

7

Nitrogen Fixation

D. W. EMERICH and J. D. WALL
University of Missouri, Columbia, MO, USA

7.1 INTRODUCTION

A logarithmic expansion in our understanding of the biochemistry, physiology and genetics of nitrogen fixation has occurred in the last 20 years. Although few agricultural or industrial uses have resulted to date, considerable effort is being expended on model systems as a prerequisite to commercial application. This article will first review nitrogen fixation and then explore how some of this recently acquired knowledge may be applied. The rapid advances that have been witnessed in the field of nitrogen fixation lead us to believe that direct application of this knowledge will be forthcoming.

7.2 BIOCHEMISTRY

7.2.1 Historical Review

Ancient peoples acknowledged the benefits of annually rotating leguminous and non-leguminous crop plants (Burris, 1974). The first explanation for this effect was made in the 1830s by the French scientist Boussingault, who suggested that leguminous plants could utilize atmospheric nitrogen for growth. Boussingault's work met the harsh criticism of the distinguished German organic chemist, Liebig, and consequently Boussingault's claim fell into disregard with the scientific elite of the period. However, during the next 50 years others sought to verify Boussingault's results (Lawes *et al.*, 1861; Atwater, 1885). The debate ended when Hellriegel and Wilfarth (1888) published a comprehensive paper firmly establishing the biological fixation of atmospheric nitrogen.

The three decades following the experiments of Hellriegel and Wilfarth (1888) found researchers seeking to identify not only the organisms responsible for symbiotic nitrogen fixation but also free-living N_2-fixing microbes. Winogradsky (1893) demonstrated fixation by the strictly anaerobic *Clostridium* spp., Beijerinck (1901) showed the obligately aerobic *Azotobacter* spp. were capable of nitrogen fixation and Drewes (1928) reported fixation by the algae *Nostoc* and *Anabaena*. Fred, Baldwin and McCoy (1932) thoroughly reviewed the work on the leguminous symbioses in their treatise. That monograph describes the morphological, cultural and physiological characteristics of the rhizobia, cross-inoculation groups, the nodule formation process, inoculation methods, the quantification of fixation rates under various conditions, and so on.

In the late 1920s, investigations on the biochemistry of the nitrogen fixation process began in earnest (Burris, 1974). Progress was made in several major areas: (a) the effects of different partial pressures of N_2 on the fixation process were investigated; (b) H_2 was reported to be a specific competitive inhibitor of nitrogen fixation; (c) ureides were found to be the major nitrogen assimilate in certain leguminous plants; and (d) the inter-relationships of carbon and nitrogen metabolism during the N_2 fixation process were actively investigated.

It was during this period, before active, cell-free extracts or pure enzyme became available, that the first stable product of nitrogen fixation was identified. Virtanen (1938) supported hydroxylamine, whereas Burris and Wilson (1946) considered ammonia to be the first stable product. In the early 1940s, a new technique was introduced which aided in establishing ammonia as the first stable product of nitrogen fixation. Burris (1974) developed the use of the stable isotope, $^{15}N_2$, for nitrogen fixation research. Not only did this technique resolve the hydroxylamine *versus* ammonia controversy, but it continues to provide the most definitive and accurate method of quantifying nitrogen fixation. By the early 1950s, it was firmly established from many additional lines of evidence that ammonia was the first stable product of nitrogen fixation.

In 1960, both the group at Wisconsin (Schneider *et al.*, 1960) and at Dupont (Carnahan *et al.*, 1960) reported reproducible, active, cell-free extracts from blue-green algae and *Clostridium pasteurianum*, respectively. These reports were quickly followed by others describing active extracts from *Azotobacter vinelandii*, *Klebsiella pneumoniae*, *Bacillus polymyxa*, *Azotobacter chroococcum*, *Rhodospirillum rubrum*, *Chromatium* spp. and *Rhizobium* bacteroids (Burris, 1969). These extracts were used to define the requirements for the nitrogen-fixing reaction and also pointed out why most previous attempts had failed. Nitrogenase (EC1.18.2.1), the enzyme catalyzing the reduction of N_2 to NH_4^+, was found to require large amounts of ATP, to be strongly inhibited by ADP, to require a constant supply of low potential electrons and to be irreversibly inactivated by O_2.

In these early studies on the cell-free systems, the ATP was supplied *via* the phosphoroclastic metabolism of pyruvate and was later replaced by an exogenous ATP-regenerating system. The utilization of $Na_2S_2O_4$ as an electron donor, as suggested by Bulen *et al.* (1965), coupled with the

ATP-regenerating system, provided a readily available and reliable *in vitro* method for measuring nitrogenase activity. These surrogate systems permitted investigation of the ATP and electron requirements, the H_2 evolution activity of nitrogenase, the discovery of alternative substrates and the purification of nitrogenase (Burris, 1969, 1974, 1975).

In the 1970s, highly purified nitrogenase component proteins suitable for protein chemistry and enzymological characterization were obtained from *C. pasteurianum, K. pneumoniae, A. vinelandii, A. chroococcum, B. polymyxa* and *R. rubrum* (Emerich and Burris, 1978; Eady, 1980). The investigations on the nature of nitrogenase structure and function advanced rapidly, utilizing such classical tools as enzyme kinetics and protein chemistry, and such state-of-the art physical methodologies as electron paramagnetic resonance (EPR), Mössbauer spectroscopy and extended X-ray absorption fine structure.

Nitrogenase requires the functioning of two distinct proteins for the reduction of atmospheric nitrogen. Neither of these component proteins separately displays any activity characteristic of nitrogenase itself. The smaller of the two proteins, dinitrogenase reductase (also called Fe protein, component 2), is reduced by an appropriate reductant and binds two molecules of ATP. Dinitrogenase reductase then transfers one electron to dinitrogenase (also called MoFe protein, component 1) with the concommitant hydrolysis of both ATP molecules. The electrons are transferred within dinitrogenase to the various iron–sulfur centers and the iron–molybdenum cofactor (FeMoCo). The latter is believed to be the substrate-reduction site.

Hageman and Burris (1978a, 1978b, 1980) recently proposed a nomenclature for nitrogenase based on a functional role of the component proteins (dinitrogenase and dinitrogenase reductase), rather than on a physical property or an arbitrary designation. This nomenclature, though not accepted by all, will be utilized here since it does describe the major functional role for these proteins (as far as our present understanding permits) and it provides the reader unfamiliar with the field with a way of more easily comprehending the biochemistry of the process.

7.2.2 Molecular Properties of Nitrogenase

7.2.2.1 *Dinitrogenase*

Dinitrogenase is the larger and more complex component of nitrogenase. Dinitrogenase is an $\alpha_2\beta_2$ tetramer ranging between 210 000 daltons for *C. pasteurianum* and 245 000 daltons for *A. vinelandii* (Mortenson and Thorneley, 1979; Harker and Wullstein, 1981). The molar masses for the two subunits are approximately 50 000 and 60 000 daltons. Metal analyses indicate two molybdenum atoms and 24 to 32 iron atoms per molecule (Orme-Johnson and Davis, 1977). The acid-labile sulfur content is approximately equivalent to the iron content. The metals appear to be arranged as four Fe_4S_4 clusters, two iron–molybdenum (FeMo) cofactors and possibly an Fe_2S_2 center (Huynh *et al.*, 1979, 1980; Zimmermann *et al.*, 1978). Dinitrogenase as normally isolated yields a characteristic electron paramagnetic resonance (EPR) spectrum with g values of approximately 4.3, 3.7 and 2.01 (Mortenson and Thorneley, 1979) which originate from the FeMo cofactor. Reduction of the paramagnetic form of dinitrogenase by dinitrogenase reductase results in the loss of the EPR signal. The EPR signal returns upon catalytic reoxidation (substrate reduction). Orme-Johnson *et al.* (1977, 1981) monitored the change in the EPR signal during oxidative titration of reduced *A. vinelandii* dinitrogenase. Removal of four electrons, one from each of the Fe_4S_4 centers as monitored by Mössbauer spectroscopy, did not alter the EPR signal. Removal of the fifth and sixth electrons (from the FeMo cofactor) resulted in the disappearance of the signal. These two distinct phases are not observed with dinitrogenase from either *C. pasteurianum* or *K. pneumoniae*, implying subtle differences among these nitrogenases in the spatial arrangements of these metal centers (Orme-Johnson *et al.*, 1981).

7.2.2.2 *FeMo cofactor*

Shah and Brill (1977, 1981) were able to extract the labile FeMo cofactor of dinitrogenase utilizing *N*-methylformamide and have subsequently purified the molecule. The isolated FeMo cofactor (i) contains 7–8 iron atoms per molybdenum atom, (ii) contains no amino acids, (iii) possesses an EPR signal with g values of 4.6, 3.4 and 2.0, (iv) is able to reconstitute FeMo cofactor-less mutants of *A. vinelandii* and *K. pneumoniae*, and (v) is capable of binding and reducing acetylene in the presence of an appropriate reductant (Burgess *et al.*, 1981b; Shah *et al.*, 1978). X-ray absorption spectroscopy (XAS) and extended X-ray absorption fine structure (EXAFS) have been used to probe the environment around the molybdenum atoms in the FeMo cofactor and in dinitrogenase. Data from both materials indicate that three to four sulfur atoms and two to three iron atoms are at an average distance of 0.235 and 0.266 nm, respectively, from the molybdenum atoms. The only discrepancy is the appearance of oxygen or perhaps nitrogen in the FeMo cofactor, but not in the dinitrogenase spectra. This difference may be an artifact of the isolation procedure (Burgess *et al.*, 1981b). The isolation and analysis of the FeMo cofactor has initiated considerable activity in synthesizing inorganic iron–molybdenum model compounds.

7.2.2.3   *Dinitrogenase reductase*

Dinitrogenase reductase is composed of two identical subunits, containing a single Fe_4S_4 cluster, with a molar mass between 55 000 and 67 000 daltons depending upon the source of the protein (Emerich and Evans, 1980). Reported molar masses of the subunits range between 27 500 and 34 600 daltons. Dinitrogenase reductase from *C. pasteurianum*, which has been completely sequenced, has 273 amino acid residues per subunit, yielding a molar mass of 57 674 daltons (Mortenson and Thorneley, 1979). The amino acid sequence had no sequence homology with other known iron–sulfur proteins. *C. pasteurianum* dinitrogenase reductase has a higher content of glycine–glycine sequences than any other protein previously examined. The twelve cysteine residues were randomly distributed, unlike the clustered sequences of ferredoxins. Thus, it was not possible to deduce the liganding structure of the Fe_4S_4 cluster between the two subunits. The Fe_4S_4 cluster can be extruded readily with the aid of thiol ligands and characterized (Mortenson and Thorneley, 1979).

Dinitrogenase reductase possesses two catalytically active binding sites for MgATP (Mortenson and Thorneley, 1979). MgADP, which strongly inhibits catalysis, binds strongly to only one of the MgATP sites (Ljones and Burris, 1978). The binding of MgATP causes a number of changes in dinitrogenase reductase indicative of a conformational change of the protein. Mg ATP binding (i) decreases the midpoint potential from about -250 mV to -400 mV, (ii) increases the accessibility of the Fe_4S_4 center to iron chelators, (iii) increases the sensitivity to O_2 and (iv) alters the —SH titer of the protein in the presence of dithionitrobenzoate or iodoacetamide (Mortenson and Thorneley, 1979). Dinitrogenase reductase has an EPR signal with g values of approximately 2.04, 1.94 and 1.88. The binding of MgATP to dinitrogenase reductase induces a transition of the EPR spectrum from a rhombic to an axial-type signal (Orme-Johnson *et al.*, 1977). Orme-Johnson and Davis (1977) have questioned the significance of the MgATP-induced EPR spectral change for *C. pasteurianum* dinitrogenase reductase because the spectral change can be observed only at pH values above 7.5 and not at values close to the pH optimum for substrate reduction (pH 6.5).

7.2.3   Substrates

Nitrogenase is capable of reducing a large number of doubly- and triply-bonded molecules (Burris and Orme-Johnson, 1976; Burns and Hardy, 1975; McKenna *et al.*, 1980). The substrates are reduced in two-electron steps or multiples thereof:

Nitrogen: $N_2 + 6 H^+ + 6 e^- \rightarrow 2 NH_3$

Acetylene: $C_2H_2 + 2 H^+ + 2 e^- \rightarrow C_2H_4$

Protons: $2 H^+ + 2 e^- \rightarrow H_2$

Cyanide: $HCN + 6 H^+ + 6 e^- \rightarrow CH_4 + NH_3$
and $HCN + 4 H^+ + 4 e^- \rightarrow CH_3NH_2$

Nitrous oxide: $N_2O + 2 H^+ + 2 e^- \rightarrow N_2 + H_2O$

Azide: $N_3^- + 3 H^+ + 2 e^- \rightarrow N_2 + NH_3$
and $N_3^- + 7 H^+ + 6 e^- \rightarrow N_2H_4 + NH_3$
and $N_3^- + 9 H^+ + 8 e^- \rightarrow 3 NH_3$

Hydrazine: $N_2H_4 + 2 H^+ + 2 e^- \rightarrow 2 NH_3$

Cyclopropene: $C_3H_4 + 2 H^+ + 2 e^- \rightarrow C_3H_6$ (cyclopropane)
and $C_3H_4 + 2 H^+ + 2 e^- \rightarrow C_3H_6$ (propane)

N_2 is the natural substrate for the enzyme, but the acetylene reduction technique is commonly employed as an index of nitrogenase activity. Although rapid and convenient, and acceptably adequate for routine applications, the acetylene reduction method may yield misleading results if not used judiciously (Thorneley and Eady, 1977). Michaelis and inhibitor constants for substrates and inhibitors (competition for electrons in the presence of two or more substrates) have been calculated from steady-state kinetics with nitrogenases from *C. pasteurianum, A. vinelandii, A. chroococcum* and *K. pneumoniae* (Winter and Burris, 1976; Zumft and Mortenson, 1975). These studies are the basis of a hypothesis predicting five distinct substrate-reducing sites. The physical relationship between these kinetically determined sites and FeMo cofactor or the other metal sites is not known.

Hydrazine, long suspected as a bound intermediate of N_2 reduction, was shown by Bulen (1976) to be a substrate for nitrogenase. The reaction was dependent upon pH, indicating that the protonated form of hydrazine ($N_2H_5^+$) was the preferred substrate. The reactivity of hydrazine provided the first evidence for its existence as a bound intermediate. Further evidence was reported by the ARC unit at Sussex, UK (Thorneley *et al.*, 1978). By quenching an actively N_2-reducing preparation of nitrogenase with ethanolic HCl, Thorneley *et al.* (1978) captured a hydrazine-like intermediate with *p*-dimethylaminobenzaldehyde. The time course for the production of this compound is consistent with its existence as an intermediate of N_2 reduction. Recently, Dilworth and Thorneley (1981) while investigating the reduction of azide to ammonia discovered that hydrazine was a product of azide reduction. Addition of $^{15}N_2H_4$ during azide reduction did not result in $^{15}NH_3$, indicating lack of equilibration between enzyme-bound N_2H_4-like intermediates and added N_2H_4.

Recently, two new substrates have been reported that show promise as active site probes. McKenna *et al.* (1979, 1980) have demonstrated that cyclopropene is reduced by nitrogenase to a mixture of propene (1/3 of total) and cyclopropane (2/3 of total). Orme-Johnson *et al.* (1981) reported that diazirine is a substrate for nitrogenase.

During the reduction of N_2, HD is produced from reaction mixtures containing either H_2 + D_2O or D_2 + H_2O (Burgess *et al.*, 1981a). Burgess *et al.* (1981a) have proposed a mechanism whereby HD formed by the N_2-dependent process originates from a bound, reduced N_2 intermediate. Thus, H_2 inhibition of N_2 reduction and N_2-dependent HD formation are believed to arise from the same molecular process.

Steric effects are a major factor determining which unsaturated molecules will serve as substrates and also their effectiveness in competing for available electrons (Burns and Hardy, 1975; McKenna *et al.*, 1979, 1980). For example, the Michaelis constants for $CH_2CHCHCN$, CH_2CN and CH_3CHCCH_3 are three orders of magnitude higher than N_2, C_2H_2 and HCN. Rates of substrate reduction have been reported to be equivalent for all substrates, although these have primarily been comparisons between N_2 and C_2H_2. Recently, Hageman and Burris (1980) have demonstrated that the electron flux through dinitrogenase controls the relative effectiveness and thus the rates at which various substrates are reduced by nitrogenase. Furthermore, the relative electron allocation is temperature dependent (Thorneley and Eady, 1977).

7.2.4 Energy Requirements

The nitrogen–nitrogen triple bond is exceedingly strong requiring 941 kJ mol^{-1} to completely dissociate the molecule. The large energy requirement for the reduction to the double-bonded intermediate (523 kJ mol^{-1}) may explain the difficulty in reducing dinitrogen even though the free energy for the overall reaction is negative (Rossini *et al.*, 1952):

$$N_2(gas) + 3\,H_2(gas) \rightleftharpoons 2\,NH_3(aq), \quad 25\,°C \quad \Delta G = -53.34\,kJ\,mol^{-1}$$

The energy to drive biological nitrogen fixation is supplied by ATP. Since two ATPs are hydrolyzed for each electron transferred from dinitrogenase reductase to dinitrogenase, a minimum of 12 ATPs are needed to reduce N_2. However, for each mole of N_2 reduced, a minimum of one mole of H_2 is evolved from nitrogenase (Emerich and Evans, 1980). The evolution of H_2 requires the same energy input (two ATPs per electron transferred) as other substrates, thus raising the minimum number of ATPs to 16. Also, the eight low potential electrons (six for N_2; two for H^+) can be assumed to possess energy as they could alternatively be used for oxidative phosphorylation or some other energy yielding reactions. Evans *et al.* (1980) assumed each pair of electrons is equivalent to three ATPs and, therefore, the apparent minimum requirement for N_2 reduction is approximately 28 ATPs.

The ATP for biological nitrogen fixation must originate from the metabolism of carbon compounds (although photosynthetic microorganisms may derive energy directly from light). Theoretically, 0.11 mol of glucose is needed to produce 1 mol of ammonia (Gutschick, 1980).

Commonly, in the leguminous symbioses, 12 g of glucose must be metabolized to produce the energy needed to reduce 1 g of N_2. This results in an overall efficiency of 12%. The efficiencies of the free-living heterotrophs are considerably lower.

7.2.5 Electron Donors

The efficacy of $Na_2S_2O_4$ as an *in vitro* reductant has deferred the need to identify the endogenous electron carriers to nitrogenase. Thus, many donors have not been identified nor have the metabolic pathways from which the electrons originate.

The *in vivo* reductants for dinitrogenase reductase that have been identified are ferredoxins or flavodoxins (Burns and Hardy, 1975; Emerich and Evans, 1980). Ferredoxins reduced by the phosphoroclastic pathway have been shown to be the source of electrons for dinitrogenase reductase from *C. pasteurianum* and *B. polymyxa*. Carter *et al.* (1980) reported that a ferredoxin purified from *R. japonicum* bacteroids could supply electrons to nitrogenase utilizing an assay system containing 5-deazariboflavin or heat-treated chloroplasts. Hageman and Burris (1978a, 1978b) have demonstrated that *Azotobacter vinelandii* flavodoxin is the likely *in vivo* reductant because of the significant stimulation of activity compared to the *in vitro* reductant, $Na_2S_2O_4$. A flavodoxin reduced *via* pyruvate metabolism has been shown to be the electron donor in *K. pneumoniae via* biochemical and genetic analysis of the *nifJ* and *nifF* gene products (Hill and Kavanagh, 1980).

7.2.6 Catalytic Mechanism

Currently the most detailed mechanisms of nitrogenase catalysis are those offered by Thorneley and Lowe (1981) and Burris *et al.* (1981). The flow of electrons through nitrogenase is well established; reduced flavodoxin/ferredoxin transfers an electron to dinitrogenase reductase, then the single electron is transferred to dinitrogenase with the hydrolysis of two molecules of MgATP to MgADP. The mechanism by which the electrons are transferred or stored within dinitrogenase and the manner by which the electrons are allocated to substrates is poorly understood.

The oxidation/reduction of dinitrogenase reductase can be monitored by EPR, Mössbauer, potentiometry and visible absorption spectroscopy (Mortenson and Thorneley, 1979). The Fe_4S_4 center operates between the $[Fe_4S_4(Cys)_4]^{2-}$ state and the $[Fe_4S_4(Cys)_4]^{3-}$ state. The binding of MgATP can be followed by EPR, thiol group reactivity, O_2 sensitivity, chelation susceptibility and, of course, ligand binding methods.

It is not known if there is an obligatory *in vivo* order in which dinitrogenase reductase is charged with MgATP and then reduced or if these processes occur independently. The rate of reduction by $Na_2S_2O_4$ of oxidized dinitrogenase reductase in the absence of MgATP is approximately 1000 times faster than the rate observed during catalytic turnover. Thus, dissociation of the MgADP-oxidized dinitrogenase reductase complex may greatly affect the rate of re-reduction (Mortenson and Thorneley, 1979). However, $Na_2S_2O_4$, which is normally utilized in these investigations, is not a very effective reductant for dinitrogenase reductase and the use of the natural electron donors may yield different results (Hageman and Burris, 1978a, 1978b). Theoretically, reduced dinitrogenase reductase, free in solution, may transfer electrons to a molecule of oxidized dinitrogenase reductase that is complexed to dinitrogenase.

The reduced, MgATP-complexed dinitrogenase reductase rapidly associates with dinitrogenase. The rate constant for this association has been estimated at greater than 10^7 m^3 mol^{-1}s^{-1} with a dissociation constant of the order of 0.5 μmol m^{-3} (Mortenson and Thorneley, 1979). Ratios of dinitrogenase reductase to dinitrogenase of both 2 to 1 and 1 to 1 have been reported as optimal for nitrogenase activity. Hageman and Burris (1978a) reported that a 1:1 ratio of dinitrogenase reductase:dinitrogenase produced full activity of the ATP hydrolysis and electron transfer reactions for *A. vinelandii* nitrogenase. A report on an heterologous nitrogenase complex (*A. vinelandii* dinitrogenase plus *C. pasteurianum* dinitrogenase reductase) indicates that *A. vinelandii* dinitrogenase does have two binding sites for dinitrogenase reductase (Emerich and Burris, 1976). *C. pasteurianum* nitrogenase requires a 2 to 1 ratio of dinitrogenase reductase to dinitrogenase for full catalytic activity (Emerich *et al.*, 1981). These differences in binding ratios may reflect subtle distinctions in the electron transfer reactions between various nitrogenases.

Although nitrogenase is frequently referred to as a complex, there is little evidence to suggest a given dinitrogenase reductase molecule remains bound to a particular dinitrogenase molecule for a complete catalytic cycle (in terms of substrate reduction). Recently, Hageman and Burris (1978b) have presented persuasive evidence that during proton reduction dinitrogenase reductase and dinitrogenase associate and dissociate with each electron transfer event. Mortenson and Thorneley (1979) suggest that perhaps for acetylene or N_2 reduction a longer-lived complex may be necessary.

7.2.7 Inhibitors

7.2.7.1 *Classical inhibitors*

The so-called 'classical' inhibitors are those compounds that also serve as substrates for nitrogenase. These inhibitors can be classified into five different groups depending upon their inhibition patterns *versus* the other substrates (Burris and Orme-Johnson, 1976). (a) H_2, N_2O and cyclopropene: these compounds are competitive *versus* N_2. (b) CN^-, N_3 and CH_3NC: these three are mutually competitive but non-competitive *versus* N_2. (c) C_2H_2: acetylene is non-competitive *versus* N_2 and N_3^-. N_2 is competitive *versus* acetylene. H_2 evolution is completely suppressed by the presence of substrate levels of acetylene. (d) CO: carbon monoxide is competitive with all substrates but is unable to block H^+ reduction. (e) H_2 evolution: H^+ reduction is not inhibited by the presence of H_2 nor blocked by CO.

Originally these results were interpreted as five different sites or perhaps modification of a single site (Burris and Orme-Johnson, 1976). There is no active site data to support this interpretation and these results may simply reflect differences in the allocation of electrons to substrates (Hageman and Burris, 1980).

Recently, several new substrates have been investigated to pursue the conformation of the substrate-reducing active site(s) of dinitrogenase. McKenna *et al.* (1980, 1981) have synthesized and utilized cyclopropene and 3,3-difluorocyclopropene as substrates and active site probes. 3,3-Difluorocyclopropene inhibits H_2 evolution, acetylene reduction and N_2 reduction. Diazurine, a strained-ring diazene analog, is a substrate and also an inhibitor of acetylene reduction (McKenna *et al.*, 1981; Orme-Johnson *et al.*, 1981). Dilworth and Thorneley (1981) reported that N_2, CO and N_2O inhibited the reduction of azide to hydrazine. H_2 was not an inhibitor of hydrazine formation from azide.

7.2.7.2 *Regulatory inhibitors*

The regulation of nitrogenase is governed by energy, nitrogenous compounds and oxygen. Normally, nitrogen fixation activity is expressed when it is required for the growth of organisms. Once the nitrogen fixing system has been genetically turned on, it can either be genetically or biochemically turned off. In this section we will deal only with the biochemical considerations (see Section 7.3).

(i) Energy

Nitrogen fixation imposes a considerable energy drain on the energy metabolism of an organism. Nitrogenase requires ATP for catalysis and is inhibited by ADP. The ratio of ADP/ATP affects electron transfer and thus allocation of electrons to substrates as well as total activity. *In vivo*, N_2-fixing cells have ADP/ATP ratios of 0.3–0.5, whereas organisms under non-N_2-fixing conditions have ratios of 0.8–0.9 (Mortenson and Upchurch, 1981). The *in vivo* data, if considered exclusively, would imply nitrogenase is 80% inhibited under normal conditions. Mortenson and Upchurch (1981) utilized $p/2e^-$ ratios as a measure of the efficiency of nitrogenase *in vitro* and found that the best apparent efficiency is at an ADP/ATP ratio of 0.3–0.5. Thus many other factors in addition to ADP/ATP ratios contribute to optimize rates of N_2 fixation *in vivo*.

(ii) Ammonia

The effect of added nitrogenous compounds, particularly ammonia, to cultures of actively fixing cultures is perhaps the most often studied aspect of nitrogenase regulation. Addition of ammonia to the culture media causes a rapid 'switch-off' of nitrogenase, a simple repression of nitrogenase synthesis or a combination of these effects (Eady, 1981). The observed effects depend upon the organism, the carbon and fixed nitrogen sources and the degree of membrane energization. The mechanism by which ammonia represses nitrogenase synthesis is not known. However, during derepression of *K. pneumoniae*, the *nif* mRNA possess a half-life of about 20 min. After the addition of ammonia to a derepressed culture, the half-life of these mRNAs was reduced to approximately 9 min (Eady, 1981). Apparently, there is no need for post-transcriptional control of nitrogenase by ammonia in *K. pneumoniae* (see Section 7.3).

(iii)   Covalent modification

The nitrogenases of the photosynthetic bacteria *Rhodospirillum rubrum*, *Rhodopseudomonas capsulata* and *Rhodopseudomonas palustris* exist in one of two different forms depending upon the nitrogen source (Ludden, 1980; Zumft *et al.*, 1981). The two different forms are actually interchangeable by the addition or removal of a group covalently bound to dinitrogenase reductase containing phosphate, a sugar (presumably ribose) and an adenine-like molecule. Cultures grown on glutamate or N_2 are poised so that modification occurs during harvesting and express little or no nitrogenase activity upon isolation. Cultures grown under NH_4^+-limited conditions lack this potential and possess active cell-free N_2-fixing preparations. The inactive dinitrogenase reductase can be converted into the active form by removal of the covalently linked adenine-like molecule. The deactivation is not well understood, but during this process ribose, phosphate and the adenine-like molecule are attached to dinitrogenase reductase. The inactive form of dinitrogenase reductase is incapable of transferring electrons to dinitrogenase (Ludden *et al.*, 1982). Glutamine synthetase may play a role in the interconversions of dinitrogenase reductase (Eady, 1981; Alef and Zumft, 1981).

(iv)   Oxygen

In vitro, oxygen irreversibly inactivates both nitrogenase components. The half-lives of dinitrogenase and dinitrogenase reductase in air are approximately 10 min and less than one min, respectively. *In vivo*, it is not known if O_2-denatured nitrogenase proteins can be reactivated or must be completely resynthesized. Based on investigations of putaredoxin and other iron–sulfur proteins, O_2 denaturation of the nitrogenase proteins is most probably due to the oxidation of the labile sulfide of the Fe_4S_4 clusters and/or FeMo cofactor to the zero oxidation state (Petering *et al.*, 1971).

The physiology of each nitrogen-fixing organism must have the capacity to maintain the activity of the O_2-sensitive nitrogenase proteins as well as provide adequate sources for energy and reducing equivalents. The physiology of each N_2-fixing organism must be compatible with its respective ecological niches (see Section 7.4.1). For example, the *Azotobacter* possess a unique mechanism to protect nitrogenase components during O_2 stress. A complex is formed between an Fe_2S_2 protein and dinitrogenase reductase (Veeger *et al.*, 1980). This Fe_2S_2 protein has a molar mass of 14 000 daltons in *A. chroococcum* and 23 000 in *A. vinelandii*. O_2 stress may initially oxidize the Fe_2S_2 protein, thereby increasing its affinity for dinitrogenase reductase. Reduction of the Fe_2S_2 protein (removal of O_2 stress) causes dissociation of the complex and restores nitrogenase activity (Eady, 1981).

Conversely, sub-optimal partial pressures of O_2 can severely affect nitrogenase activity in those organisms which generate ATP *via* oxidative phosphorylation. Low O_2 tensions reduce ATP levels and thus lower the nitrogenase activity.

7.2.8   Ammonia Assimilation

Ammonia, the first stable product of nitrogen fixation, must be assimilated for transport to symbionts and/or for conversion into proteins. Classical enzyme studies, ^{15}N and more recently ^{13}N (Wolk, 1980) have been employed to trace the assimilatory pathways of N_2-fixing organisms. Several reviews appeared recently describing the ammonia assimilation process in prokaryotes (Tyler, 1978; Magasanik, 1982), plants (Miflin and Lea, 1976) and legume nodules (Boland *et al.*, 1980).

There are two possible primary routes for the assimilation of ammonia into amino acids: (1) glutamate dehydrogenase and (2) glutamine synthetase–glutamate synthase. Glutamate dehydrogenase provides the major assimilatory pathway when ammonia is abundant. However, during nitrogen fixation when nitrogen-limiting conditions prevail, it operates predominately in a catabolic mode rather than a synthetic one. Glutamate dehydrogenase has a high Michaelis constant for ammonia and furthermore the specific activities in crude extracts of nitrogen limited cells are usually quite low. Conversely, Lees *et al.* (1981) and Kennedy *et al.* (1981c) have reported the kinetic parameters for soybean and lupine glutamate dehydrogenase, respectively, and suggest that this enzyme may have a more important role in ammonia assimilation than implied by the Michaelis constant for ammonia. Other amino acid dehydrogenase activities, such as alanine or aspartate, are too low to be of physiological importance (Folkes and Sims, 1974; Brown *et al.*, 1974).

The glutamine synthetase–glutamate synthase pathway is the major assimilatory pathway under nitrogen-fixing conditions in prokaryotes. The Michaelis constants for all the substrates are at physiological levels and the specific activities can account for the observed rates of ammonia assimilation. This pathway requires ATP in addition to reductant and thus may represent a mechanism for regulatory control in free-living organisms (Tyler, 1978).

Once glutamine and glutamate are formed, all the other amino acids and other nitrogenous compounds can be formed by transamination or similar processes. However, in some symbiotic associations more complex pathways are required. In nodulated leguminous plants, the ammonia produced *via* nitrogen fixation within the *Rhizobium* bacteroids is excreted into the plant cytosol. Although two glutamine synthetases have been reported in free-living cultures of rhizobia (Darrow and Knotts, 1977; Ludwig, 1980b) only one has been reported within bacteroids (Bishop *et al.*, 1976). Present evidence suggests glutamine synthetase of *Rhizobium* bacteroids performs an insignificant role in symbiotic ammonia assimilation.

Glutamine synthetase of the plant cytosol is the major pathway of ammonia assimilation in nodulated leguminous plants (Boland *et al.*, 1980). In certain leguminous plant species such as lupine, the principal form of nitrogen assimilates that are translocated to other plant tissues are amides. In other leguminous species, for example soybeans, allantoin and allantoic acid, commonly referred to as the ureides, are the major transport forms of nitrogen (Matsumoto *et al.*, 1977a, 1977b, 1977c). The ureides are synthesized entirely in the plant host cell. Schubert (1981) has shown that the synthesis of the purines takes place entirely within the plant cytosol, whereas the catabolism of the purines involves sequential steps within the cytosol, peroxisome and endoplasmic reticulum.

7.3 GENETICS

7.3.1 Introduction

Until the 1970s, the molecular genetics of N_2 fixation was essentially unexplored. Earlier progress was confined to the isolation and biochemical characterization of mutants unable to use N_2 as a nitrogen source (for reviews see Burns and Hardy, 1975; Schwinghamer, 1977). The primary impediment to further analysis was the absence of genetic exchange systems for those diazotrophs which had received the most biochemical attention, *i.e. Clostridium* and *Azotobacter*. Because many of the genetic tools available for *Escherichia coli* could be applied to *K. pneumoniae*, it became clear that this nitrogen-fixing cousin of *E. coli* offered the greatest advantages for genetic analysis. As a result of this recognition, two major advances were made. First, Streicher *et al.* (1971) reported on the intrastrain transfer of *K. pneumoniae* genes for nitrogen fixation (*nif*) by bacteriophage Pl and showed their linkage to genes for histidine biosynthesis. Almost simultaneously, Dixon and Postgate (1972) demonstrated that genes for nitrogen fixation could be transferred from *K. pneumoniae* to *E. coli* by conjugation and, amazingly, expressed there to form a functional nitrogenase system. These advances established the approximate chromosomal location of *nif*, identified an easily-selected, linked marker and demonstrated that some *nif* genes were clustered on the chromosome. These successes, a political climate favorable to the support of this research and major advances in molecular technology, provided the basis for the exponential acquisition of genetic and molecular information which has taken place in the ensuing decade.

The genetic tools and current results of analyses will be outlined for several types of diazotrophs after a general discussion of the available approaches and methods.

7.3.2 Approaches and Techniques Available

An excellent review of the techniques used to explore the *nif* genes of *K. pneumoniae* has recently been published (Cannon, 1980). Because it is written for use in the laboratory and is part of a more comprehensive methods volume, it is certainly a critical reference for those working in the area of nitrogen fixation. Here we will attempt to point out approaches which can be used with diazatrophs not as genetically accessible as *K. pneumoniae* and initially confine our remarks to free-living organisms.

First, a bank of independent mutations blocking the ability to fix dinitrogen can be generated. Although NTG (*N*-methyl-*N*-nitro-*N*-nitrosoguanidine), a powerful mutagen, is often used, it readily results in multiple lesions which may complicate the interpretation of pleiotropic phenotypes. Other mutagens such as alkylating agents, hydroxylamine, nitrous acid or UV (Miller, 1972; Meynell and Meynell, 1965; Carlton and Brown, 1981) may offer fewer problems. Because all strains are not equally sensitive to these agents, a dose–response curve should be established for each before use (Cannon, 1980).

After mutagenesis of an appropriately genetically marked strain, Nif⁻ mutants are isolated as those unable to grow with N_2 but able still to grow well with ammonium salts. Next some phenotypic characterization of the *nif* mutant strains can be made. If the wild-type phenotype(s) is restored spontaneously at a frequency of 10^{-9} or higher, the original lesion is likely to be a point mutation, although the possibility of suppression should be considered. Generally loss of a plasmid, deletions or multiple mutations do not revert at these frequencies. In addition, any pleiotropic effects of the mutations should be examined, *e.g.* altered growth rates on nitrogen sources other than N_2.

A biochemical and physical analysis of the *nif* components of the mutants can proceed without regard to the genetic capabilities of the organism. However, the establishment of the number of genes represented among the mutations greatly decreases the amount of biochemistry which must be done. Residual *in vivo* and *in vitro* nitrogenase activity (Cannon, 1980) should be measured to determine suitability for further biochemical or genetic studies. Furthermore, if activity is present *in vitro* but not *in vivo*, a block in endogenous electron flow to the nitrogenase complex is likely. Additional functional characterization can be made if antibodies to purified protein components of the nitrogenase complex are available. Purified components can sometimes be used to restore activity to inactive extracts of mutants, thus identifying the altered protein.

Because the three polypeptides of the dinitrogenase and the dinitrogenase reductase can be easily visualized by a comparison of the protein patterns of repressed and derepressed cells on denaturing polyacrylamide gels (SDS-PAGE), a rapid screen of mutants representative of each linkage group or gene will assess the gross integrity of the nitrogenase complex. Additional *nif*-specific polypeptides can be identified by pulse-labelling proteins during derepression of nitrogenase, separation by SDS-PAGE, followed by autoradiography (Cannon, 1980). Finally, as a further refinement, pulse-labelled extracts can be subjected to two-dimensional gel electrophoresis and subsequent autoradiography (Roberts and Brill, 1980). These experiments can give gene–protein relationships as well as operon structure when used to analyze the appropriate mutations.

If a genetic exchange mechanism such as transduction, transformation or conjugation is available, mapping can proceed. For Gram-negative organisms lacking indigenous transferrable plasmids, conjugation has often been obtained with the promiscuous drug resistance plasmids of the Pl incompatibility group. Two factor crosses of *nif-1* × *nif-2* can be made to determine the relative linkage of the *nif* lesions. This procedure, called the ratio test (Hayes, 1968), can be performed using an unlinked marker to standardize the results of each cross (Streicher *et al.*, 1971; Bishop *et al.*, 1977). Thus the closer two mutations are in the chromosome, the fewer prototrophic recombinants will be obtained relative to the number of recombinants for the unlinked marker. If a linked marker is available, such as *hisD* in *K. pneumoniae*, three point crosses can be carried out in a similar manner (*e.g.* Kennedy, 1977). It should be remembered that the size of the DNA transferred, *e.g.* the capacity of the phage head in transduction, determines the physical limits for establishing linkage between markers. In addition, fine structure mapping requires a relatively small genetic vector.

If a linked marker is not available for mapping purposes, it is possible to generate such a mutation by the use of NTG (Dixon *et al.*, 1977). A Nif⁻ strain is mutagenized with NTG and plated for Nif⁺ revertants. Most of these revertants will have been the result of the action of NTG and, therefore, will have a high probability of having an additional mutation(s) nearby. Temperature sensitive mutations can be isolated and used, even without a knowledge of the genes altered, or specific phenotypic alterations can be sought.

To investigate the number of genes represented by the mutational clusters, complementation analysis is required. The complementation test simply determines whether two mutations affect the same gene. For this analysis it must be possible to establish a relatively stable merodiploid, *i.e.* a strain with two copies of the chromosomal segment covering the mutations. This is generally brought about by the introduction (conjugation or transformation) of a plasmid containing the appropriate region of DNA into a recombination-deficient recipient strain (Rec$^-$). Although complementation has been successfully performed in a Rec$^+$ background (Dixon *et al.*, 1977), an absolute interpretation of results is questionable (Roberts and Brill, 1981).

The construction of a plasmid containing *nif* may be accomplished either *in vivo* or *in vitro*. To form an F-prime or an R-prime *in vivo* carrying *nif*, conjugation into a Nif$^-$ Rec$^-$ recipient can be carried out and a Nif$^+$ recipient selected. Since recombination is prevented, the incoming chromosomal DNA can be stably maintained only if it is a part of the plasmid replicon. Other ways of restricting recombination are to use a recipient with a large deletion of the area coding for *nif* proteins, or to use another, closely related species as recipient so that DNA homology is low but one in which *nif* expression can be monitored.

In vitro construction of plasmids carrying *nif* genes can be carried out using recombinant DNA techniques (*e.g.* Pühler and Klipp, 1981), which have been adequately reviewed elsewhere (Davis *et al.*, 1980; Schleif and Wensink, 1981; Maniatis *et al.*, 1982). The highly conserved sequence of *nifH* and *nifD* genes among diazotrophs allows the identification of nitrogenase structural genes from almost any species with an appropriately labelled probe, *e.g.* pSA30 from *K. pneumoniae* (Ruvkun and Ausubel, 1980).

Once a plasmid containing the *nif* genes has been obtained, mutations can be introduced either by mutagenesis of the plasmid, by homogenotization with chromosomal mutations or by cotransduction of *nif* lesions with a selectable marker into the plasmid. Following the derivation of a bank of plasmid and chromosomal mutations, all pair-wise crosses can be constructed and the Nif phenotypes observed (see Roberts and Brill, 1981, for references).

Information concerning operon structure can be obtained from insertion mutations (Cannon, 1980; Roberts and Brill, 1981) induced by transposable genetic elements such as drug resistance transposons (Kleckner, 1981; Davis *et al.*, 1980) or bacteriophage Mu (Miller, 1972). These insertions have been shown to produce strongly polar effects on genes operator-distal to the point of the insertion. Therefore, noncomplementarity between an insertion mutation and point mutations in several clustered genes suggests an operon or transcriptional unit.

Another advantage of insertion mutations is their usefulness in the construction of deletions. Since these elements excise imprecisely, deletions are often observed among strains selected for loss of the insert (Cannon, 1980).

A third type of genetic insertion resulting in the fusion of the gene for β-galactosidase to the operator-promoter region of the gene of interest has proved to be especially valuable in studying *nif* regulation (Dixon *et al.*, 1980; MacNeil *et al.*, 1981). Because of the extreme O$_2$ sensitivity of some *nif* products and the lack of discriminatory assay systems for most of these products, observations of the control of expression are quite difficult. By fusing the gene for β-galactosidase (which is easily assayed) to the control region of the various *nif* operons, fluctuations in expression are readily followed. The limitation to this procedure, as for the other transposable elements, is to obtain a suitable vector for introducing the element into the organism of choice. Unless the bacterium of interest is sensitive to λ or μ bacteriophages (Casadaban and Cohen, 1979), a conjugable plasmid or transformation must be used which may have limited efficiency.

7.3.3 *nif* Genes in *Klebsiella pneumoniae*

The results of numerous investigations aimed at the elucidation of the number, arrangement and control of the genes essential for nitrogen fixation by *K. pneumoniae* have been reviewed well and frequently in the recent literature (Ausubel and Cannon, 1981; Eady, 1981; Kennedy *et al.*, 1981a; Roberts and Brill, 1981). Here we will attempt to summarize briefly the information with updates from the primary journals.

Figure 1 shows the order, operon organization, direction of transcription and approximate molar mass of the products of the genes presently identified in the *nif* regulon of *K. pneumoniae*.

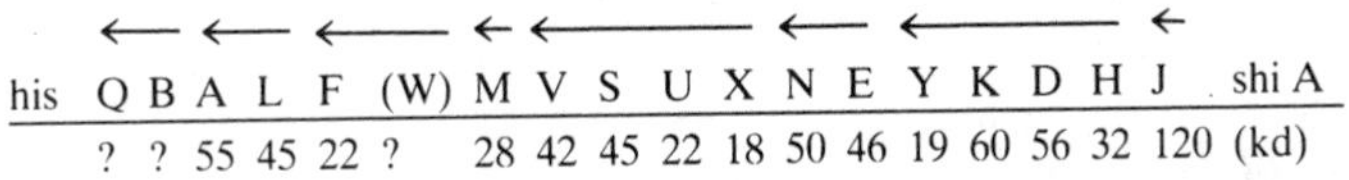

$$\text{← ← ←——— ← ← ——— ← ←——— ←}$$

his	Q	B	A	L	F	(W)	M	V	S	U	X	N	E	Y	K	D	H	J	shi A
	?	?	55	45	22	?	28	42	45	22	18	50	46	19	60	56	32	120	(kd)

Figure 1 Order, operon organization, direction of transcription and approximate molar mass of the products of the *nif* regulon genes of *Klebsiella pneumoniae*

The order was established relative to *hisD* by three-point crosses, deletion mapping and physical analysis with cloned fragments. The additional use of transposon mutagenesis of the cloned fragments has allowed a more detailed physical map to be established. It is now apparent that the *nif* regulon is contained in a 23 kb segment and that all the essential genes in this segment have been identified (Roberts and Brill, 1981).

The number of operons and the direction of transcription have been identified by complementation between insertion mutations and point mutations. Present data suggest seven or eight transcriptional units read in the leftward direction toward the genes for histidine biosynthesis (Figure 1).

Although the protein products of only five of the *nif* genes have been purified to homogeneity, the functions of most are known but not the actual enzymatic reactions catalyzed. The dinitrogenase reductase and the dinitrogenase, which have been purified, are coded by *nifH* and *nifKD*, respectively.

The protein products of both genes involved in electron transport to the nitrogenase complex, *nifF* and *nifJ*, have been purified. *nifF* codes for a flavoprotein which is essential for physiological electron transport (Nieva-Gómez *et al.*, 1980). In contrast, the J protein is a dimer of *ca.* 245 000 daltons containing 30 mol iron and 24 mol labile sulfur/mol protein and is probably an oxido-reductase (Bogusz *et al.*, 1981).

Three of the genes whose products have not been purified are involved in the FeMo cofactor synthesis, *nifB*, *nifN* and *nifE*.

Because extracts of strains with mutations in *nifM* which have inactive nitrogenase can be restored to activity by the addition of dinitrogenase reductase, it appears that M protein is involved in a post-transcriptional modification of the product of *nifH*. A recent analysis of mutations in *nifV* (McLean and Dixon, 1981) indicated that the V product may also be involved in post-transcriptional modification but of the dinitrogenase protein. Although dinitrogenase in a Nif V⁻ background reduces some substrates, it is not capable of reducing N_2 with physiological electron fluxes (McLean and Dixon, 1981). Two other gene products, those of *nifS* and *nifU*, also appear to be involved in the maturation of dinitrogenase since strains containing Mu insertions in *S* or *U* were shown to lack 'normal' levels of this protein.

The last two genes for which functions are known are those comprising the *nifLA* operon. The *nifA* product is required for the expression of all *nif* genes except for its own operon (Buchanan-Wollaston *et al.*, 1981a). In contrast, the *nifL* product is not essential for *nif* expression and acts as a repressor for all transcripts (except its own operon) in response to O_2 (Hill *et al.*, 1981), NH_4^+ and temperature (Merrick *et al.*, 1982).

The remaining genes, *nifQ*, *nifX*, *nifY* and *nifW* have not been assigned a specific function. The gene designated *nifQ* was inferred from the observation that a deletion from *hisD* which recombined with all known *nif* mutations resulted in a leaky Nif⁻ phenotype. In addition, the strain containing this deletion was not dramatically altered in acetylene reduction, suggesting that protein may influence substrate selection. By analysis of the protein products of cloned fragments of the *nif* regulon, genes *X* (Pühler and Klipp, 1981) and *Y* (Cannon *et al.*, 1979; Pühler and Klipp, 1981) were identified. These two genes have not been shown to be necessary for nitrogen fixation.

Only limited analysis has been carried out on *nifW* (Roberts and Brill, 1980). It was assigned on the basis of reversion of a Mu induced mutation which was polar onto *nifF* and may represent an operator proximal region of *nifF* that is non-essential.

Finally a mutation between *nifJ* and *nifH* which appeared to complement a *nifJ* mutation was assigned the designation *nifC* (Roberts and Brill, 1981). Recently additional mapping studies have demonstrated that this mutation lies between well characterized *nifJ* lesions and that intracistronic complementation occurs which confused the interpretation of earlier results (Stacey *et al.*, 1982).

In summary, 18 to 19 gene assignments have been made in the *nif* regulon. All except *nifW* have been shown to produce a protein product, 15 by genetic experiments and two to three by examination of cloned fragments. It is clear that the commitment to nitrogen fixation is a major investment for a bacterium.

7.3.4 Regulation of *nif*

A large number of environmental factors are involved in the regulation of expression of nitrogen fixing activity, among which are ammonia concentration, dissolved oxygen tension, presence of amino acids, availability of molybdenum and temperature. It appears that control may operate at transcription, the stability of transcripts and the stability of the enzyme complex. In most cases, evidence for regulatory effectors still rests with the physiological descriptions rather than detailed molecular mechanisms.

Ammonia, the end product of nitrogen fixation, is assimilated by the successive action of glutamine synthetase (GS) and glutamate synthase. As a result, the investigations of the mechanism of ammonia repression of nitrogenase have focussed on these enzymes. Only recently has it been directly proven that this control occurs at the level of transcription (Roberts and Brill, 1981; Buchanan-Wollaston *et al.*, 1981b; Drummond *et al.*, 1983).

Because the levels of GS and the state of its covalent modification were seen to fluctuate in response to the supply of fixed nitrogen available to the cell (see reviews by Magasanik, 1977; Tyler, 1978), attention began to focus on the importance of this enzyme in a generalized nitrogen control system. Concomitant with the changes in GS activity and content, changes in the levels of other enzymes capable of supplying NH_4^+ or glutamate to the cell were also observed and a cause and effect relationship was proposed. The description of two types of mutants mapping in the region of the structural gene for GS, (a) those leading to glutamine auxotrophy, Gln$^-$, which also resulted in a Nif$^-$ phenotype and (b) those designated Gln^c which were constitutive for *nif* expression in the presence of NH_4^+, confirmed the involvement of the assimilatory system (see also Shanmugam *et al.*, 1978). The model which grew out of these studies suggested that the activation of the synthesis of the enzymes of nitrogen fixation and nitrogen metabolism required an increase in the cellular level of GS and its conversion to the unadenylylated form, which could then function as a positive regulatory element for transcription (Magasanik, 1977; Tyler, 1978).

Since this model was proposed, there has been an accumulation of evidence showing that there is no apparent correlation between the adenylylation state of GS and the level of other enzymes of nitrogen metabolism (Leonardo and Goldberg, 1980). The discovery of additional genes involved in nitrogen regulation tightly linked to *glnA* has made a re-evaluation of nitrogen regulation necessary. Genetic and physical data presently support the model proposed by McFarland and coworkers (1981) for *Salmonella* and elaborated on by Merrick (1982) and Ow and Ausubel (1983), illustrated in Figure 2 in a modified form showing an interaction with *nif* genes.

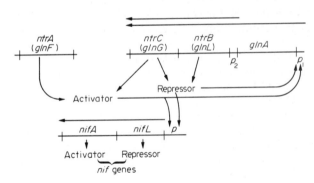

Figure 2 Model for regulation of nitrogen metabolism. Gene designations in parentheses are those used in *Escherichia coli*. The *ntr* designation derives from nitrogen regulation ◄— indicate transcriptional units and directionality ← indicate regulatory functions

Loss of *ntrA* or *ntrC* results in loss of ability to express nitrogen controlled genes at high levels. Thus, when nitrogen is limiting, *ntrA* protein or an enzymatic product of that protein interacts with *ntrC* protein to form an activator. The *ntrC* product also can interact with the *ntrB* protein to form a repressor when ammonia levels are high. As a consequence of the loss of *ntrC*, both repression and activation of *glnA* are eliminated resulting in a low *glnA* expression insensitive to nitrogen availability. In contrast, mutations in *ntrB* allow a high constitutive level of expression of *glnA*. This model now appears to hold for several enteric bacteria (Pahel *et al.*, 1982; Rothman *et al.*, 1982).

In *E. coli* (Magasanik, 1982) and *K. pneumoniae* (Ow and Ausubel, 1983) the P_{II} protein, a component of the adenylylation system for glutamine synthetase, has been reported to be a corepressor capable of interacting with the *ntrBC* complex. Mutants incapable of converting P_{II} to its uridylylated form in response to low levels of ammonia were unable to activate nitrogen assimilation genes including nitrogenase.

Mutations occurring within the *nif* regulon which render the expression of that regulon insensitive to fixed nitrogen are at the operator–promoter site of *nifLA* (Roberts and Brill, 1981) or within the *nifLA* operon (Hill *et al.*, 1981). These results suggest that the transcription of *nifLA* is controlled by the general nitrogen-regulatory system and the products of that operon in turn regulate the remaining *nif* operons.

Recent experiments (Drummond *et al.*, 1983; Ow and Ausubel, 1983) have pointed out the remarkable similarities between the *nifA* and *ntrC* protein products. Both are approximately 55 000 daltons, have similar isoelectric points, and require a functional *ntrA* gene for regulatory activity. Both have been shown to be capable of activity with the *nifLA* operon, *glnAntrBntrC* operon, and the *Rhizobium meliloti nifH* promoter in addition to those genes governing the catabolism of certain amino acids. These similarities have led to the suggestion that the two genes are evolutionarily related (Ow and Ausubel, 1983). Significantly the *K. pneumoniae nifH* promoter is responsive only to *nifA* protein.

As yet the physiological meaning of the ability of *nifA* protein to substitute for *ntrC* protein is unclear. Perhaps it is simply a sparing effect for *ntrC* or the vestiges of the evolutionary antecedent.

The *nifL* product which is the Nif regulon repressor affords a more immediate response to fixed nitrogen than is possible through nitrogen regulation. Since *nifL* and *nifA* coding for the activator protein are transcribed as a single operon, the *nifL* protein must be maintained in an inactive state during active nitrogen fixation and be converted into a repressor state by molecular signals sensing fixed nitrogen or O_2 (Kennedy *et al.*, 1981a; Merrick *et al.*, 1982).

Synthesis of most *nif* polypeptides in *K. pneumoniae* has been shown to be temperature sensitive. It is as yet unresolved whether A protein (Buchanan-Wollaston *et al.*, 1981a) and/or L protein (Merrick *et al.*, 1982) are responsible for thermosensitivity.

The mechanisms involved in the control mediated by amino acids or by nitrate remain to be elucidated. However, both repressive effectors are still functional in mutants of *K. pneumoniae* derepressed for nitrogenase biosynthesis in the presence of ammonium (see discussion in Eady, 1981).

The involvement of molybdenum (Mo) in the regulation of biosynthesis of nitrogenase in *K. pneumoniae* has been brought into question with results from more recent studies (Kahn *et al.*, 1982). Under Mo deprivation some dinitrogenase apoprotein is made and can be activated by the addition of molybdate to the cells (Kahn *et al.*, 1982). Thus Mo must not be essential for *nif* polypeptide synthesis in this bacterium. Only in *A. vinelandii* does Mo appear to be required for conventional dinitrogenase synthesis (Eady, 1981).

A form of control at the level of mRNA destabilization operates in *Klebsiella* for *nif* expression. Under nitrogen fixing conditions, *nif* mRNA appears to be remarkably stable with a half-life of 18 min (Kaluza and Hennecke, 1981) or longer (Roberts and Brill, 1981). When these diazotrophs were shifted to repressive conditions, the half-life became significantly shorter (Kaluza and Hennecke, 1981). However after addition of rifampicin plus NH_4^+, mRNA hybridization experiments suggested that *nif* mRNA was detectable well after *nif* protein synthesis was terminated (J. Collins cited in Roberts and Brill, 1981). This result would suggest that a translational control was superimposed on the transcriptional regulation and mRNA stability in *Klebsiella* or that a specific nuclease was required to shorten the half-life of *nif* mRNA upon repression.

7.3.5 *Azotobacter* Species

The members of the Gram-negative *Azotobacter* genus are nitrogen-fixing obligate aerobes. They are widely distributed in soil and some species have been found associated with the rhizosphere of tropical grasses (van Berkum and Bohlool, 1980). There are reports of increased plant growth following inoculation with *Azotobacter*, possibly due to the production of plant growth hormones (van Berkum and Bohlool, 1980). Because of the potential for coupling ammonia excretion from N_2 reduction with plant-hormone production in an organism that has an associative growth mode, this genus deserves considerable attention.

Although *A. vinelandii* has been reported to have 10 times the amount of DNA that *E. coli* contains (Sadoff *et al.*, 1979), mutants lacking the ability to fix nitrogen were obtained rather readily (see discussion in Roberts and Brill, 1981). While purine and pyrimidine auxotrophs of *A. vinelandii* have been isolated, amino acid auxotrophs have not been reported even though considerable effort has been made to obtain them (Sadoff *et al.*, 1979). Therefore, a very limited array of genetic markers are presently available for extensive mapping efforts in this genus.

Several *nif* mutations have been characterized with respect to activities for dinitrogenase and dinitrogenase reductase, antigenic cross-reacting material (CRM) and electron paramagnetic resonance signals (refs. in Roberts and Brill, 1981). Among the mutants described were all the predictable classes for a system of two proteins which could be assayed separately. Two additional classes of special interest were found: one which was Nif⁻ but hyperproduced the reductase and one which produced nitrogenase in the presence of ammonia.

Although a transformation system was described 10 years earlier, a reliable procedure was not established until 1976 (Page and Sadoff, 1976). Refinements of this procedure have been published (Page and von Tigerstrom, 1979). The plate-transformation system (Page and Sadoff, 1976) was used for the determination of a rough linkage map for several Nif⁻ mutant strains (Bishop *et al.*, 1977; Bishop and Brill, 1977). Because the size of DNA transferred was not determined and congression (simultaneous transfer of markers on separate fragments of DNA) was relatively high (Bishop *et al.*, 1977), no physical interpretation of distance could be made from the ratio test crosses (Bishop and Brill, 1977). However, the results did demonstrate that the *nif* genes in *A. vinelandii* do not fall into one cluster (Bishop and Brill, 1977). Subsequent DNA–DNA hybridization studies with cloned *K. pneumoniae nif* genes have shown that the highly-conserved structural genes for nitrogenase *nif H* and *D* do occur on a single, small fragment (Ruvkun and Ausubel, 1980).

Regulation of the nitrogen-fixing complex of *Azotobacter* is similar to other free-living diazotrophs (see Eady, 1981; Roberts and Brill, 1981) in being ammonia repressible with the involvement of the ammonia assimilatory system. Two classes of regulatory lesions have been described: (1) presumed point mutations resulting in the loss of activity and CRM for both nitrogenase proteins, as well as (2) a mutation overproducing the dinitrogenase reductase and simultaneously not producing dinitrogenase. The existence of these phenotypes has been interpreted to mean that a common regulatory gene is required for expression of the structural genes for the nitrogenase complex (Roberts and Brill, 1981).

Surprisingly, an extended analysis of *A. vinelandii* Nif⁻ mutants which included two-dimensional PAGE of the proteins of several Nif⁺ revertants revealed that the 'conventional' dinitrogenase polypeptides were missing and four additional, ammonia-repressible proteins were present. Thus the Nif⁺ revertants were pseudorevertants (Bishop *et al.*, 1980). Further investigation has led to the conclusion that this diazotroph may possess an alternative nitrogen fixing system which is Mo repressed (Bishop *et al.*, 1980; Bishop *et al.*, 1982; Riddle *et al.*, 1982). While the description of this alternative system awaits further confirmation, the survival potential for a bacterium forced to fix nitrogen in an environment depleted of Mo is clear. The biochemistry of N_2 reduction without the involvement of Mo should prove extremely interesting.

7.3.6 Cyanobacteria

The cyanobacteria can be roughly divided into three main groups, the unicellular, filamentous nonheterocystous and filamentous heterocystous forms. Although nitrogen fixing species occur in all classes, all the filamentous heterocystous species have been shown to possess this capacity (Stewart, 1977, 1980). The heterocyst is the differentiated cell that functions as an anaerobic site specifically designed for N_2 reduction and release of fixed nitrogen to vegetative cells (Haselkorn, 1978; Hazelkorn *et al.*, 1980).

The only reliable and efficient genetic exchange system for the cyanobacteria to date has been developed for *Anacystis nidulans*, a non-N_2 fixing species (see Sherman and van de Putte, 1982, and refs. therein). The consequences of the absence of a genetic exchange system in the diazotrophic cyanobacteria was brought into clear focus by their omission from a recent review of the genetics of nitrogen fixation by Roberts and Brill (1981). However, progress is being made towards genetic analysis through the generation of defined mutations and cloning technology.

Although a few mutant strains had been isolated in several species of filamentous cyanobacteria, until the work of Currier and coworkers (1977) there had not been a concerted effort to obtain well characterized mutants in one species. Now a bank of mutant strains has been isolated in *A. variabilis* (Currier *et al.*, 1977), including auxotrophs and those altered in nitrogen fixation and heterocyst development. To accomplish mutant isolation from a filamentous organism, it was necessary to separate the rare mutant cell from the filament (Wolk, 1980). To rupture cells randomly and fragment the filaments, cavitation was used and the short filaments or single cells were then subjected to penicillin enrichment (Wolk and Wojciuch, 1971), and selection for auxotrophs (Currier *et al.*, 1977).

The strongly conserved nature of the structural genes for nitrogenase, *nif H* and *D*, was demonstrated by heterologous hybridization of *K. pneumoniae nif* genes carried on the plasmid pSA30 with fragments of *Anabaena* DNA (Mazur *et al.*, 1980). In addition, homology between *nif* genes other than the structural genes for the nitrogenase was reported for DNA from *Anabaena* and *K. pneumoniae* (Mazur *et al.*, 1980) which may be a unique feature among the diazotrophs (Ruvkun and Ausubel, 1980). These studies have indicated that the *nif* structural genes in *Anabaena* are rearranged from the order found in *K. pneumoniae* and are probably separated on the chromosome (Mazur *et al.*, 1980; Rice *et al.*, 1982).

During the course of the hybridization experiments, what appears to be a second copy of the *nifH* gene was found (Rice *et al.*, 1982). Because the restriction pattern was different from that of the *nifH* gene located near *nifD*, it is not known whether this is a second copy of *nifH*, a nonfunctional pseudogene, or another gene which by chance has homology. A role for such a gene in a hypothetical alternative nitrogen-fixing system such as that found in *A. vinelandii* was also suggested (Rice *et al.*, 1982).

The regulation of the capacity for nitrogen fixation by cyanobacteria differs from other free-living diazotrophs in a few areas and is discussed in several reviews (Stewart, 1977; Haselkorn *et al.*, 1980; Wolk, 1980). Ammonia repression (and probably O_2 repression) operates in the filamentous cyanobacteria (Stewart, 1977). Supporting the involvement of a generalized nitrogen regulatory system in nitrogenase control were results which showed that the GS inhibitor MSX (methionine-*SR*-sulfoximine) caused a relief of inhibition of heterocyst formation by exogenous ammonia and the excretion of newly fixed ammonia from cells of N_2-fixing *Anabaena cylindrica* (Stewart, 1977). The majority of the nitrogenase in filamentous cyanobacteria occurs within the heterocysts of a photosynthetically active culture (Wolk and Wojciuch, 1971). The elegant experiments of Wolk and coworkers have demonstrated that ammonia generated by nitrogenase in *A. cylindrica* is assimilated *via* the glutamine synthetase–glutamate synthase enzyme couple (reviewed in Wolk, 1980). Therefore, as might have been expected, the GS activity, which is the result of a single enzyme in *Anabaena* (Stacey *et al.*, 1979), is slightly higher in heterocysts, although the glutamate synthase activity is lower as compared to vegetative cells (Haselkorn *et al.*, 1980). These findings support the contention that glutamine is produced from newly fixed N_2 within the heterocyst and exported to the vegetative cell where glutamate is made (Wolk, 1980).

The GS in the cyanobacteria does not enjoy large fluctuations in activity in response to nitrogen source nor is it covalently modified (Stacey *et al.*, 1979; Orr *et al.*, 1981). In addition, the excretion of ammonia from *Anabaena azollae* when in association with the water fern *Azolla caroliniana* has been shown to be the result of the absence of GS activity and antigen from the cyanobacterium (Haselkorn *et al.*, 1980). Thus some question arises concerning the role of GS or *gln* associated genes in the regulation of nitrogenase. As a result, molecular biological techniques have been applied to the problem and the *glnA* gene from *Anabaena* 7120 has been cloned into *E. coli* where it complements a *glnA* deletion (Fisher *et al.*, 1981). Interestingly the *Anabaena* gene is not repressible in *E. coli* (Fisher *et al.*, 1981). Additional studies are under way to elucidate the extent of the *glnA* region cloned and the regulatory properties encoded.

7.3.7 Photosynthetic Bacteria

The purple, non-sulfur phototrophs have been most intensively studied among the free-living photosynthetic bacteria, excluding the cyanobacteria. Within the Rhodospirillaceae, the nitrogenase system of *R. rubrum* has received most biochemical attention, while genetic transfer systems have been established in *R. capsulata* and *Rhodopseudomonas sphaeroides*. Although some metabolic and morphological differences exist, it is assumed here that the nitrogen fixing function is conserved in these organisms, an assumption which has received support from a number of physiological studies (see discussion by Yoch, 1978).

Interest in the nitrogen fixing capacity of these bacteria has been rekindled by the demonstration of regulation of nitrogenase activity by covalent modification of the dinitrogenase reductase. This mechanism of inactivation and subsequent activation (discussed here earlier, see Ludden, 1980; Eady, 1981; Zumft *et al.*, 1981) suggest that the number of genes essential for nitrogen fixation may be increased in these organisms.

Several genetic tools are available in the Rhodospirillaceae and a recent review (Marrs, 1982) has an excellent description of each. Here we will briefly summarize the transfer mechanisms and the strains in which they have been successfully applied.

Conjugation mediated by Pl incompatibility group plasmids appears to be ubiquitous among these Gram-negative phototrophs. Plasmid transfer was first used to promote chromosomal transfer and demonstrate linkage in *R. sphaeroides*. Subsequently, transfer of several P and W group plasmids into this species has been shown as well as low chromosomal mobilization. Similar experiments of conjugational transfer have been reported for *R. capsulata*. The introduction of RP4::Mu *cts* has been accomplished in both species. Although Mu was not thermoinducible in *R. sphaeroides*, phages were produced. In contrast, *R. capsulata* strains differ in Mu expression; 37b4 containing RP4::Mu *cts* was reported to be thermoinducible (Yu *et al.*, 1981) while no Mu expression was observable in B100 (Wall and Lee, unpublished). These experiments suggest that *in vivo* engineering with Mu may be practical in *R. sphaeroides* and *R. capsulata* 37b4. In addition, the introduction of R751::Mudlac (Ap^R, *lac*) into *R. sphaeroides* has been accomplished and should allow operon fusions in this organism (Nano and Kaplan, 1982).

By the introduction of the mercury resistance transposon Tn501 into RP1, Pemberton and Bowen (1981) have been able to demonstrate that this plasmid promoted a high frequency of chromosomal transfer in *R. sphaeroides*. As a result, they have published the first map of auxotrophic markers in the photosynthetic bacteria, which opens the way for additional mapping and strain construction. A single *nif* mutation has subsequently been mapped to a position very near the genes for the photosynthetic apparatus (Pemberton, personal communication).

A derivative of RP1, pBLM2, which has enhanced chromosome mobilization ability in *R. capsulata* was isolated by Marrs (1981) by screening rare exconjugant clones for donor ability. A frequency of chromosome transfer of 6×10^{-4}/donor was obtained for some markers and R-primes bearing the genes for photosynthesis were found. Similar constructs should be possible for the *nif* genes of the organism.

These preceding genetic tools have been adaptations of systems first described in the enteric bacteria. *R. capsulata* enjoys, in addition, an elegant endogenous system of generalized gene transfer discovered by Marrs in 1974. The agent (GTA) that serves as the vector for gene transfer appears to be a small phage-like particle. DNA extracted from the GTA particles is linear, double-stranded of *ca.* 3×10^6 daltons and is randomly packaged from the chromosome (Marrs, 1982). Mapping results with the GTA system have yielded map distances that are remarkably close to the physical distances obtained by restriction endonuclease mapping (Marrs, 1982).

This transfer agent has been used to show *nif* transfer between *R. capsulata* mutant strains (Wall *et al.*, 1975) and to begin to construct a linkage map of *nif* genes. By applying the ratio test (Hayes, 1968), markers of the same phenotype less than 2500 base pairs (bp) apart can be reliably demonstrated to be linked with this agent (Wall, unpublished). Although 13 Nif$^-$ mutations have been shown to fall into five linkage groups (Wall and Braddock, unpublished), the question of overall clustering must be settled by different exchange techniques employing larger pieces of the chromosome. Preliminary indications from cloning studies using the heterologous pSA30 plasmid to identify *R. capsulata nif* structural genes indicated that these genes are clustered (P. Scolnik, personal communication).

Regulation of the repression of synthesis of nitrogenase in the Rhodospirillaceae is assumed to be similar to that in other diazotrophs. No activity is measurable in cultures grown in ammonia, complex medium or air. In addition, the polypeptides corresponding to the nitrogenase proteins are absent under these conditions (Wall and Braddock, unpublished observations).

Enzymatic studies with *R. capsulata* and *Rhodopseudomonas palustris* demonstrated that GS and glutamate synthase are the key enzymes of ammonia assimilation (Johansson and Gest, 1976; Alef and Zumft, 1981). The involvement of a generalized nitrogen regulatory system in the control of nitrogenase was indicated by the derepression of the enzyme complex in the presence of the GS inhibitor, MSX (Alef and Zumft, 1981, and refs. therein). Support for the involvement of common regulatory elements for *gln* and *nif* has been derived from the isolation of glutamine auxotrophs lacking GS activity which are derepressed for nitrogenase in the presence of ammonia (Wall and Gest, 1979).

Recently, an interesting class of spontaneous Nif$^-$ mutants has been isolated after serial subculture of *R. capsulata* cells in derepressing medium (Wall and Love, unpublished). Because the only substrate for nitrogenase was protons, these cells produced large quantities of H$_2$. After 10 to 12 subcultures, a substantial portion of the population was Nif$^-$. This result points out one of the difficulties which may be encountered with applications of diazotrophs for commerical H$_2$ or fertilizer production.

7.3.8   *Rhizobium* Species

The agricultural economic importance of the symbiotic associations involving the rhizobial species has focussed attention on these organisms for many years. The mechanisms of recognition, infection and nodulation have been investigated extensively; however, the innate difficulties of a developmental symbiosis has slowed the acquisition of knowledge which might be used to improve the process. Renewed interest has arisen because of the recent advances in understanding nitrogen fixation in the free-living bacterial systems. Much of the genetic research with *Rhizobium* has been thoroughly reviewed in the recent literature (see Beringer, 1980, 1981; Heumann, 1981; Dénarié *et al.*, 1981; Kondorosi and Johnston, 1981). Earlier genetic studies were critically summarized by Schwinghamer (1977), Kuykendall (1981) and Roberts and Brill (1981). Here we will summarize the mechanisms available for analysis and some of the recent findings.

7.3.8.1   *Fast growing species*

Because of the faster doubling times, *R. meliloti*, *R. leguminosarum*, *R. trifolii* and *R. phaseoli* have been easier to manipulate by standard genetic techniques than the slower growing species. Mutants have been isolated after chemical mutagenesis and penicillin enrichment techniques *via* classical procedures (Kondorosi and Johnston, 1981). Although transformation procedures have been available for most species for many years, little linkage data has resulted from its use (see summary in Schwinghamer, 1977). In addition, generalized transduction in *R. meliloti* was reported as early as 1967; however, only one instance of cotransduction of markers has been demonstrated with this system (see Schwinghamer, 1977). It is expected that these genetic exchange processes will begin to be re-investigated since the establishment of circular linkage maps of the chromosomes has been accomplished *via* conjugation. Transductional analysis will be essential for fine structure mapping and the ability to transform opens the way for additional cloning manipulations. Indeed, the recent literature reflects this interest since transduction has now been reported for *R. leguminosarum* and *R. trifolii* (Buchanan-Wollaston, 1979) and improved methods for plasmid transformation of *R. meliloti* have been developed (Selvaraj and Iyer, 1981).

Systems of conjugation have played the most prominent role in the genetic investigations of the rhizobial species. Early endogenous conjugational systems have either not been rigorously pursued or the *Rhizobium* species used has been questioned (see discussion in Kondorosi and Johnston, 1981). More recently, the existence of conjugative plasmids in three strains of *R. leguminosarum* which code for bacteriocin production and have chromosome mobilization ability (Cma) has been demonstrated.

In contrast with the endogenous plasmids which are interesting because of their own genetic content, the plasmids of the Pl incompatibility group originally from *Pseudomonas aeruginosa* have been the most productive for genetic analysis. As a result of their use, chromosome linkage maps now exist for *R. meliloti* 2011, for *R. meliloti* 41 and for *R. leguminosarum*. Recombination and linkage between markers in crosses of *R. leguminosarum* with *R. trifolii* or *R. phaseoli* were essentially the same as results obtained from crosses within *R. leguminosarum*. Therefore, the map derived for *R. leguminosarum* is believed to represent all three species. A recent comparison of these maps shows the similarities and complementation obtained among the *R. meliloti* strains and *R. leguminosarum* (Kondorosi *et al.*, 1980).

Use of the promiscuous plasmids has also made the introduction of transposons into *Rhizobium* a fairly straightforward procedure (reviewed in Kondorosi and Johnston, 1981). Although Pl plasmids are generally stable in *Rhizobium* when bacteriophage Mu is present in the plasmid, the plasmid is no longer stably maintained after its introduction by conjugation. When a transposon conferring drug resistance is also included on such a plasmid, selection for that drug resistance after transfer of the plasmid to *Rhizobium* selects for those cells in which transposition has occurred. Two such 'suicide' plasmids, one containing Tn7 and one containing Tn5, have been constructed and used.

An additional procedure for transposon mutagenesis has been described by Ruvkun and Ausubel (1981). Any cloned gene can be introduced into *E. coli* and there mutagenized with transposons. If it is not in a transmissable vector, it may be inserted into a broadhost-range cloning vehicle (Ditta *et al.*, 1980) and conjugated back into the original host. The cloning vehicle can then be 'chased' from the bacterial cytoplasm by the introduction of an incompatible plasmid while continuous drug resistance selection is made. The majority of drug resistant exconjugants will have undergone a homologous recombinational event, or homogenotization, such that the transposon now resides in the chromosomal gene (Ruvkun and Ausubel, 1981).

In *R. trifolii* (Prakash *et al.*, 1981), *R. phaseoli* and *R. leguminosarum* (Hombrecker *et al.*, 1981; Prakash *et al.*, 1982), the genes for nodulation lost specificity and nitrogen fixation have been reported to be present on large plasmids. In contrast, experiments with *R. meliloti* 102F34 designed to examine the location of *nif* genes, suggested both a chromosomal position and that at least three discrete units necessary for nitrogen fixation were contained in an 11.2 kb fragment (Corbin *et al.*, 1982). Most intriguing was the recent observation of the occurrence of reiterated *nifD* and *nifH* genes of *R. phaseoli* (Quinto *et al.*, 1982) with at least one copy on a larger plasmid.

7.3.8.2 *Slow growing species*

Included in the slower growing group of rhizobial strains are *R. lupini*, *R. japonicum* and *Rhizobium* sp. or cowpea rhizobia. The genetic analysis of these species lags behind that of the faster growers primarily because of the greater difficulty encountered in the microbiological manipulations. In contrast to the fast growers, these strains express high nitrogenase activities in defined culture; therefore, these bacteria may ultimately be more amenable to the analysis of *nif* functions. Although most of the mutants isolated early were drug resistant, more recently Nif⁻ mutants of *R. japonicum* have been obtained after mutagenesis and screening *in planta* or *ex planta* (Noel *et al.*, 1982). Glutamine auxotrophs of cowpea strain 32H1 showing impaired nitrogenase activity have also been isolated. Few other auxotrophs have been reported in these strains (see Kondorosi and Johnston, 1981; Kuykendall, 1981).

Although an endogenous conjugation system was reported for a non-nodulating *R. lupini* strain as early as 1968, methods for gene transfer in other strains of slow growers have only now begun to be developed (see Kondorosi and Johnston, 1981). Using an uncharacterized bacteriophage, Shah *et al.* (1981) have now generated a linkage map of *R. japonicum* by transduction. This map can now be used to locate and manipulate the genes essential for nodulation and nitrogen fixation.

In preparation for a more extensive genetic analysis, Kennedy and coworkers (1981b) have reassessed the ability of several strains of cowpea rhizobia to transfer, maintain and express P group plasmids. Strains which have a reasonable frequency of transfer and good plasmid stability during nodulation were identified but no Cma was reported. Obviously Cma is the next step in the analysis of the slow growers.

7.3.8.3 *Regulation*

From studies with cloned DNA fragments containing *nif* genes, it has been shown that the control for expression of these genes in the fast-growing species operates at the level of transcription during nodulation (Corbin *et al.*, 1982). Similar experiments with slow growers have not yet been done.

The involvement of the ammonia assimilatory system in the regulation of nitrogenase of *Rhizobium* remains an open question. In contrast to *K. pneumoniae*, results with bacteroids have suggested that the assimilation system of the bacteria is essentially non-functional during greatest N_2-fixing activity (Planqué *et al.*, 1978). Studies with *R. japonicum* bacteroids (Bergerson and Turner, 1967) and free-living bacteria (O'Gara and Shanmugam, 1976) supported this finding in that the ammonia produced from N_2 was found in the medium. In addition, mutants lacking glutamate synthase (Ali *et al.*, 1981) and strains naturally lacking glutamate dehydrogenase were normal in their abilities to nodulate and fix N_2.

On the other hand, studies with glutamine auxotrophs showed that nitrogenase activity was lacking in the mutant bacteroids (see Roberts and Brill, 1981). These results must be interpreted with care, since it is now known that all *Rhizobium* species so far examined have two distinct glutamine synthetase enzymes (Fuchs and Keister, 1980). The two enzymes have different physical properties; GSI undergoes adenylylation in response to a nitrogen signal while GSII does not (Darrow and Knotts, 1977). The description of an auxotrophic revertant of *Rhizobium* sp. 32H1 in which (1) GSI was constitutively adenylylated, (2) GSII was still missing and (3) nitrogenase was constitutively synthesized implies common regulatory elements for GSI and nitrogenase (Ludwig, 1980a). Because the auxotrophy of GSII mutants could be satisfied by either glutamine or purines, a role for this enzyme in assimilation has been questioned (Ludwig, 1980b).

7.4 APPLICATIONS

7.4.1 Physiology of Organisms

Application of the biological nitrogen fixation process requires a knowledge of the physiology of these organisms. The broad range of the ecological adaptations of nitrogen-fixing organisms permits an even broader range of applications. The capacity to fix atmospheric nitrogen occurs in a larger number of diverse bacteria (Wilson, 1958; Rubenchik, 1963; Mishustin and Shil'nikova, 1966; Jensen, 1965a; Postgate, 1971) but does not naturally occur in eukaryotes except within symbiotic associations.

7.4.1.1 Aerobes

Aerobic nitrogen-fixing bacteria are found mainly in the family Azotobacteriaceae, which includes the taxonomically similar genera *Azotobacter*, *Azomonas*, *Azotomonas*, *Azotococcus*, *Beijerinckia* and *Derxia*. Members of the genera *Azotobacter*, *Beijerinckia* and *Derxia* have been demonstrated to fix N_2, but Parejko and Wilson (1968) could find no evidence for fixation in strains of *Azotomonas* that were examined. Most of these aerobic bacteria actually fix nitrogen only under microaerophilic conditions. It has been proposed that these organisms adapt physiologically by producing copious amounts of polysaccharide which hinders diffusion of oxygen thereby permitting the functioning of nitrogenase (Postgate, 1971). Of these organisms, the *Azotobacter* is best adapted to aerobic nitrogen fixation. *Azotobacter* possess very high rates of respiration; indeed, the highest respiratory rates ever measured (Q_{0_2}) are those of *Azotobacter*. These high respiratory rates have been called 'respiratory protection' since they maintain intracellular partial pressures of oxygen that are compatible with the functioning of nitrogenase. However, even the *Azotobacter* strains are susceptible to oxygen at partial pressures beyond 20% O_2, or when forced into phosphate limitation (Postgate, 1971). Under these conditions, the nitrogenase is 'switched off' or protected by an iron–sulfur protein that forms a complex with dinitrogenase reductase. Nitrogenase is switched on when favorable conditions return.

Although *Azotobacter* is generally thought of as a free-living nitrogen fixing organism, several species have been reported to form associations with the roots of plants. For example, *Azotobacter paspali* forms a loose association with the roots of *Paspalum notatum* and other grasses (Döbereiner and Day, 1976).

Azospirillum (Tarand *et al.*, 1978) fixes atmospheric nitrogen microaerophilically forming characteristic pellicles in culture. The discovery of associations between *Azospirillum* and many plant roots, including some agriculturally important plants, has spawned much promise and controversy.

Aerobic nitrogen-fixing organisms also include iron- and methane-oxidizing bacteria. Three strains of *Thiobacillus ferrooxidans* have been shown to incorporate ^{15}N when growing in culture media at pH values of 2 (MacKintosh, 1978). Also nitrogen fixation among the methane-oxidizing bacteria is common (Postgate, 1981). Dalton (1981) has shown that the methane mono-oxygenase enzyme of *Methylococcus capsulata* provides respiratory protection for the organism such that cultures can be conditioned to withstand elevated partial pressures of oxygen without increased respiration.

7.4.1.2 Facultative anaerobes

The majority of the physiological and biochemical investigations of nitrogen fixation have been performed with facultatively anaerobic bacteria. This class includes species of *Bacillus*, *Klebsiella*, *Rhodopseudomonas*, *Rhodomicrobium* and *Rhodospirillum*. These bacteria, although able to grow aerobically on fixed nitrogen, cannot fix nitrogen in air. Only under strict anaerobic, or extremely low oxygen tensions are these organisms capable of nitrogen fixation. The non-sulfur phototrophic bacteria are able to utilize light energy directly for the fixation of atmospheric nitrogen anaerobically and recently Madigan *et al.* (1979) have demonstrated that *Rhodopseudomonas capsulata* fixes nitrogen microaerophilically in the dark.

Certain cyanobacteria, most notably the filamentous *Anabaena* species fix atmospheric nitrogen aerobically. This would appear to be a dilemma since the cyanobacteria conduct plant-type photosynthesis and thereby evolve O_2. *Anabaena* has overcome this incongruity by morphologically altering approximately every tenth cell along its length into a thick walled 'heterocyst' devoid of photosystem II but possessing the nitrogen-fixing apparatus (Haselkorn, 1978; Haselkorn *et al.*, 1980). The heterocyst supplies fixed nitrogen to the vegetative cells, which in turn provide reduced carbon compounds for the fixation process.

The coccoid cyanobacteria *Gleocapsa* also fixes nitrogen aerobically. *Gleocapsa* forms large dense colonies in which the outermost cells restrict O_2 from diffusing into cells at the center of the colony mass. Thus, a diffusion gradient is formed and cells at the center of the colony are able to fix N_2 under microaerophilic conditions.

7.4.1.3   Anaerobes

The most notable anaerobic diazotrophs are in the genus *Clostridium*. *Clostridium* is one of the most primitive of organisms and thus its physiology has attracted considerable attention (Mishustin and Shil'nikova, 1966). These organisms are strict anaerobes, but will tolerate oxygen to varying degrees in culture without irreversible denaturation of their nitrogenase. By comparison the nitrogenase of the sulfate-reducing bacteria is impaired by any level of oxygen present in the culture.

7.4.1.4   Symbionts

The complexity of many of the symbioses has hindered progress toward elucidating the physiology and biochemistry of the prokaryotic symbiont. Although the free-living state of these organisms is well defined, in many cases the symbiotic state is poorly understood.

The *Rhizobium* species are defined as heterotrophs able to grow both aerobically and anaerobically in culture. Recently, Hanus *et al.* (1979) have demonstrated certain *R. japonicum* and cowpea rhizobia are capable of chemolithotrophic growth if provided with a fixed nitrogen source. *Rhizobium*, the endophytes of leguminous plants, have been demonstrated to fix measurable levels of N_2 when grown in culture on well-controlled, low partial pressures of oxygen. Whereas nitrogen fixation can be obtained on defined media in the slow-growing rhizobia (*R. japonicum*, *R. lupin*; cowpea miscellany), the fast growers (*R. melitoti*, *R. leguminosarum*, *R. trifolii*, *R. phaseoli*) can fix nitrogen *ex planta* only in media containing liquid extract from plant callus growth media (Postgate, 1981). The rates of fixation are extremely low compared to those obtained when these organisms are participating in their respective symbioses.

The genus *Frankia* has been designated to encompass those endophytes of symbioses between non-leguminous plants and the actinorhizii. Callaham *et al.* (1978) were the first to successfully isolate the endophyte of these associations, obtaining the organism from *Comptonia peregrina* nodules in pure culture. Subsequently, the endophytes from *Alnus* and *Eleagnus* have also been isolated (Quispel and Burggraaf, 1980). These organisms are actinomycetes with a characteristic sporangia. The doubling time of these strains on defined media is of the order of several days.

The cyanobacteria participate as the nitrogen-fixing endophyte in lichens and with the water fern *Azolla*. The lichens are associations between cyanobacteria and a fungus. In the case of *Peltigera aphthosa*, a complex symbiosis occurs consisting of an ascomycetous fungus, a species of the cyanobacteria *Nostoc* and a green alga *Coccomyxa* sp. The *Nostoc* in the symbiosis contains a much higher number of heterocysts than is normally found in the free-living culture (20% *versus* 5–10%; Stewart *et al.*, 1981). *Anabaena azollae*, the endophyte found in the leaf cavities of the water fern *Azolla*, also has an increased number of heterocysts (Peters *et al.*, 1980). The factors controlling the increased heterocyst frequency in these cyanobacterial symbionts have not been determined. The *Anabaena–Azolla* symbiosis is a major contributor to rice production in China and Southeast Asia. Talley *et al.* (1981) report nitrogen accretion rates between 30 and 100 kg N ha^{-1} year^{-1} in replicated experiments conducted in California.

7.4.2   Agronomic Applications

The most grandiose application will be the integration of the Nif genes directly into plant cells such that the transformed eukaryotes now are capable of providing their own fixed nitrogen directly from the atmosphere. Whereas the transfer of the genes may certainly be possible, the expression of the genes at a significant level, provision of sufficient energy for nitrogen fixation without severely affecting plant growth or development, protection of the O_2 labile nitrogenase proteins and so on, represent intractable barriers that will require extraordinary effort and time to surmount. The most logical intermediate step to achieve nitrogen fixing eukaryotes is *via Agrobacterium. Agrobacterium*, taxonomically related to the *Rhizobium*, is able to infect a wide variety of plants, rather than being restricted to the legumes as are the rhizobia, and form crown galls (Giles and Altherly, 1981). This infection and host–bacteria relationship require the transfer of a plasmid from the bacteria into the plant genome. Thus, *Agrobacterium* may be a very useful model system to solve the problems that the nitrogen fixation process may impart upon the host plant. This section will address some of the more practical, short term applications.

7.4.2.1 Rhizobium

The *Rhizobium* spp. have attracted considerable attention as the primary approach to increasing nitrogen fixation potential in crop legumes. The features that make *Rhizobium* symbioses attractive are (1) their symbiotic relationship with major economic/agricultural crops and (2) that the *Rhizobium* and many of the important legumes can be manipulated genetically. Compared to the associative systems, the intimate symbiosis between *Rhizobium* and leguminous plants afford a more definitive approach toward improvement of its efficiency and effectiveness.

During the early stages of plant development, a period of nitrogen limitation is endured before the required nodule mass has been synthesized (Fred *et al.*, 1932; Phillips and coworkers, 1981). The high energy demand of nitrogen fixation then imposes a carbon stress upon the metabolic processes of the plant. This and a multitude of other inter-relationships govern the yield potential of the leguminous crops (Vincent, 1980). To increase yields in the short term, the proper *Rhizobium* strain–plant cultivar combination (Burton, 1976; Seetin and Barnes, 1977) as well as the necessary disease resistance traits should be utilized. The use of the proper *Rhizobium* strain–plant cultivar can markedly increase yields (Burton, 1976), but the farmer often does not have access to the desired information. Moreover, new cultivars are introduced more rapidly than *Rhizobia* strain–plant profiles can be completed.

Typically, fields in which leguminous crops are planted contain large indigenous populations of rhizobia. For example, the majority of these indigenous *R. japonicum* in midwestern soils consist of the 123 serotype, a group which is not conducive to optimal yields with most soybean cultivars (Ham, 1980). This serogroup is quite competitive *versus* applied strains and thus forms a high percentage of the bacteroids found within the nodules of plants grown in these soils. Applying strains of rhizobia possessing superior biochemical and genetic traits is futile if they eventually form only a minority of the nodule bacteroid population. Thus, improved strains of rhizobia need to contain the necessary characteristics for competitiveness as well as any beneficial biochemical or genetic characteristics. Thus, strains of rhizobia should be selected for competitiveness before genetic/biochemical alterations are made or, conversely, the genes for competitiveness should be transferred to improved strains of rhizobia. However, the parameters controlling competitiveness are not understood at this time, although lectins are believed to be a major factor (Dazzo, 1980; Napoli *et al.*, 1980; Bauer, 1980).

The problem of competitiveness may be circumvented by sterilization of the soil or by producing rhizobia strains resistant to those agrichemicals deleterious to rhizobia. At present, soil sterilization methods are impractical on a large scale. A short term solution may be 'reagent-selective' inoculants. Legume seeds can be coated with various agents deleterious to the indigenous rhizobia. A desired *Rhizobium* strain which can be made resistant to this agent also can be incorporated into the seed coating. The nodules resulting from this treatment contain a high proportion of the desired *Rhizobium*. Thus, full expression of the beneficial traits are obtained.

The list of desired traits is endless and with the development of *Rhizobium* genetics the near future promises an abundance of improved strains. Among these traits is the hydrogen recycling mechanism as exemplified by certain strains of *R. japonicum*. During the nitrogenase reaction, a minimum of 1 mol of hydrogen is evolved for each mole of atmospheric nitrogen reduced. Under less favorable conditions, more hydrogen is evolved than nitrogen reduced. In the extreme case, that is in the absence of nitrogen, only hydrogen is evolved. Evans and associates (Schubert *et al.*, 1977; Carter *et al.*, 1978) have identified strains that do not evolve hydrogen during the fixation of atmospheric nitrogen. These strains possess an uptake hydrogenase which consumes all of the hydrogen evolved, *via* nitrogenase, in an oxygen-dependent reaction and concomitantly produces

ATP. This ATP can then be recycled for the fixation of nitrogen. Use of these strains in field tests has increased the nitrogen content of soybeans by 10–13% (Evans *et al.*, 1981). The transfer of this trait to other rhizobia possessing additional beneficial parameters will be possible in the very near future.

At the present time, identification of the biochemical processes that limit symbiotic nitrogen fixation is more difficult than the genetic transfers. The biochemistry and physiology of the symbioses have not been adequately characterized, so that the limitations of the symbiotic fixation process can be defined. During the peak of nitrogen fixation activity, the symbiotic process is thought to be carbon limited. There is considerable physiological information supporting this concept and it has been calculated that if the peak of nitrogen fixation activity could be lengthened by several days the nitrogen content of soybean seed theoretically could be doubled. The interdependence of carbon metabolism and nitrogen metabolism in *Rhizobium*–leguminous plant symbioses suggest that any increase in the availability of carbon compounds to the root nodules, or alternatively more efficient utilization of available carbon compounds within root nodules, will increase the nitrogen fixation potential. As an index of photosynthate availability many workers have reported differences between cultivars in the rate of CO_2 uptake in soybeans (Moss, 1976). However, the rate of CO_2 uptake is not a reliable indicator of net photosynthate production nor photosynthate transport to root nodules. More systematic methodologies for screening cultivars and cultivar–*Rhizobium* strain combinations are required.

Mutations affecting the metabolism of the citric acid cycle or the uptake of these metabolites significantly alters the nitrogen fixation capacity of the symbiosis (Ronson *et al.*, 1981; Glenn and Brewin, 1981). Compared to other carbon compounds, the citric acid cycle intermediates support greater rates of nitrogen fixation in suspensions of anaerobically isolated bacteroids. However, it is not known how the photosynthetic compounds supplied to the root nodule, primarily glucose or sucrose, are metabolized (Embden–Meyerhof *vs.* Entner–Doudoroff *vs.* pentose shunt) or in which compartment of the symbiotic tissue (infected plant cell *vs.* non-infected plant cell *vs.* bacteroid) the metabolism occurs.

The genetics and biochemistry of the host plant must also be considered and a number of investigators are focusing on this area. For example, in *Pisum sativum*, where two genes have been found that control nodule number, a correlation exists between nodule number and seed yield in peas (Gelin and Blixt, 1964). Abu-Shakra *et al.* (1978) have described a soybean plant line that does not undergo senescence like normal plants. Hopefully these plants will continue to fix N_2 for prolonged periods and thus lead to increased yields.

Perhaps the greatest application of *Rhizobium* research will be in the tropics and subtropics where the majority of the 20 000 species of leguminous plants thrive (Allen and Allen, 1981; Anon, 1981). These regions coincide with the majority of the world's underdeveloped population. Because of the poor soils in many tropical regions, applications of nitrogen fixation technology have the greatest potential for helping mankind. However, agricultural research on legumes has been conducted primarily in the temperate climates of the more developed countries.

7.4.2.2 Azospirillum

A large number of diazotrophs have been reported in association with plants, usually found near, on or within the roots. These associations are not nearly as specialized in their symbiotic morphology as the *Rhizobium*, but their biochemistry may be equally complex. These associations have generated considerable excitement, since their wider range of association indicates they may be adaptable to many non-leguminous, agriculturally important plants. In the forseeable future, the expected benefits in terms of plant productivity are considerably less than the *Rhizobium*–leguminous plant symbioses, but they may provide significant crop improvement in soil with poor or moderate nitrogen fertility.

Azospirillum has been the most widely studied associative organism because (1) it can form associations on a large variety of plants often in high numbers (10^7 per g of root) and (2) some reported rates of fixation can significantly affect crop yields. *Azospirillum* has been reported associated with maize, sorghum, sugar cane, rice, millet, oats, rye, barley, forage grasses and the water plant *Spartina alternifora* (Döbereiner, 1978; Döbereiner and De-Polli, 1980). Döbereiner and Day (1976) have reported rates of fixation of *Azospirillum* on *Digitaria decumbens* as high as 1 kg N ha^{-1} d^{-1}. Although some of the early nitrogen fixation measurements were overestimates due to experimental conditions, more careful work by Okon and others has shown that in Israel,

Azospirillum association can contribute the majority of the nitrogen required by the plant (Kapulnik *et al.*, 1981a, 1981b). In the more temperate regions, particularly the midwestern and northern United States, it appears that *Azospirillum* may contribute little to the nitrogen economy of typical crop plants (Klucas and Pedersen, 1980). However, it may be practical to select cell lines (mutants or genetically improved strains) of *Azospirillum* as inoculum for specific cultivars to form productive associations under a particular set of environmental or ecological conditions (Ela *et al.*, 1982).

7.4.2.3 Cyanobacteria

In the United States, the cyanobacteria are usually associated with the eutrophication of lakes, rivers and streams and thus are treated as a scourge. However, they are treated as a boon in the rice growing regions of the world since the cyanobacteria are beneficial to the crop (Watanabe, 1981). The cyanobacteria are nurtured and subsequently used as a green manure to fertilize the rice field after it is drained. The extent of the contribution of cyanobacteria to rice is not yet known (Watanabe, 1981), but a typical rice crop will remove about 50 kg N ha^{-1}. Wetland rice can be grown continuously with reasonable yield levels without N-fertilizer additions if adequate cyanobacteria populations are maintained.

Cyanobacteria frequently dominate the phytoplankton population of eutrophic lakes (Peterson *et al.*, 1977) and account for the major source of fixed nitrogen. In non-eutrophic lakes and rivers and in the open ocean (Burris, 1976), N$_2$-fixation rates may be imperceptible. Burris (1976) has reported that cyanobacteria attached to rocks in the intertidal zone of the Great Barrier Reef could fix 6.8 to 30.6 kg N ha^{-1} of rock surface year^{-1}. Stewart (1977) reported rates of fixation of 25 kg N ha^{-1} year^{-1} on the supralittoral fringe of temperate shores. Much higher rates of fixation have been reported in association with coral reefs (Burris, 1976). Stacey *et al.* (1977) have isolated a rapid growing blue-green alga from the warm shallow coastal area of Port Aransas, Texas capable of fixing 40 kg N ha^{-1} year^{-1}.

The contribution of cyanobacteria to soils has not been estimated. Although contributions to the nitrogen economy may be quite substantial in certain environments, the contribution on a global scale is rather small.

7.4.2.4 Cyanobacterial associations

Associations of cyanobacteria with eukaryotic cells are rare (Stewart *et al.*, 1980). The two primarily studied cyanobacterial associations are *Azolla* and the lichens. *Azolla* spp. are some of the few vascular plants capable of forming an intimate association with a blue-green alga. *Azolla* spp. are fast-growing water ferns that contain the cyanobacteria, *Anabaena azollae*, as an endophyte within their leaf cavities. In the symbiotic state, the *Anabaena* express an elevated number of nitrogen-fixing heterocysts, as high as 20% of the total cell. In southeast Asia, *Azolla* are nurtured simultaneously with rice or as a green manure crop in fallow paddies (Talley and Rains, 1980). When grown simultaneously with rice, some species of *Azolla* serve only as a future green source of nitrogen, but other *Azolla* species may leak or excrete ammonia into the aquatic environment which may then be available to the rice crop immediately. Thus, the nitrogen input of *Azolla* blooms toward a particular rice crop is difficult to determine; when *Azolla* are considered solely as a green manure crop, values as high as 250 kg N ha^{-1} have been estimated, but values between 30 to 100 may be more common (Talley *et al.*, 1981; Watanabe *et al.*, 1977). *Azolla* have provided a readily available inexpensive source of fertilizer nitrogen to underdeveloped regions of the world, where the cost and technology of applying commercial fertilizer would be prohibitive.

Lichens have little direct agronomic importance; however, their ability to fix atmospheric nitrogen and occupy habitats not suitable for other plants make them important long term investments of fixed nitrogen (Stewart *et al.*, 1980; Jordan, 1981). Lichens can colonize poor and/or acidic soils, survive prolonged periods of desiccation and fix N$_2$ under severe temperature extremes. Lichens are composed of diverse classes of organisms. The fungal partner is usually an ascomycete but may also be a basidiomycete or in the Fungi Imperfecti group. The cyanobacterial partner may have to provide the photosynthetically fixed carbon in addition to the fixed nitrogen although this depends upon the composition of the lichen. The reported ranges for fixation by lichens is 0.2 to 12.0 kg N ha^{-1} year^{-1} (Jordan, 1981).

7.4.2.5 Photosynthetic bacteria

The photosynthetic bacteria, although ubiquitous, probably do not provide meaningful agronomic levels of fixed nitrogen. The photosynthetic bacteria contribute in anaerobic and microaerophillic environments, especially those rich in H_2S but nutrient-poor and illuminated. They have been reported to contribute fixed nitrogen in rice paddies, salt marshes, estuarine muds and sulfur-spring ditches (Jordan, 1981; Habte and Alexander, 1980).

7.4.2.6 New associations

The discovery of *Azospirillum* in association with the roots of grasses (Döbereiner and Day, 1976) initiated a renewed interest in associative systems. Although most of this interest has centered around *Azospirillum*, a number of other associative systems have been reported. The best defined new associations are (1) sugar cane–*beijerinckia*; (2) wheat–*Bacillus* spp. and/or *Erwinia herbicola*; (3) rice–*Achromobacter*; (4) *Paspalum notatum–Azotobacter paspali*; and (5) *Azospirillum*. Because of the importance of wheat as a major agronomic crop, the reports of associations with *Bacillus* spp. and *E. herbicola* have received particular interest and scepticism. Klucas and Pedersen (1980) have used *K. pneumoniae* or *E. herbicola* as inoculum on winter wheat and grain sorghum. They reported significant differences between plant cultivars and inoculants, pointing out that plant genotype, species of microorganism and environmental conditions all play a major role in determining yield. Depending upon the various treatments, inoculation produced increases of up to 44.4 g dry wt. or decreases of as much as 24.7 g dry wt. Rennie and Larson (1979) have shown that inoculation of wheat with *Bacillus* in sterile leonard jars resulted in increased plant nitrogen. As discussed by Jensen (1965a) and Burris (1977) a great number of difficulties remain before the associative systems become agronomically feasible. Although the yields of most of these associative systems appear small when judged against yields in typical agricultural soils, their benefits are significant when compared to the poor soils (and crop yields) in many underdeveloped countries.

7.4.3 Industrial Applications

7.4.3.1 Chemical catalysts

The Haber–Bosch process, the principal industrial process for ammonia production since 1913, requires pressures of 350–1000 atm, temperatures of around 350 °C plus elementary hydrogen for the reduction of N_2. The usual catalyst is composed of iron. Ruthenium and osmium catalysts are available that perform at lower temperature and which push the equilibrium further toward ammonia, but their greater costs discourage their use. The high energy costs of the Haber process means an entirely different procedure must be developed to provide cheaper ammonia. Even the biological process requires large inputs of energy, but unlike the industrial process it occurs at room temperature and atmospheric pressure. Hopefully, new catalysts and procedures will not only provide cheaper sources of ammonia but will require a minimum amount of equipment and capital investment so that ammonia can be produced where it is needed and the costs of transportation can be reduced.

The three types of chemical reactions being considered as possible alternative procedures are (1) nitriding reactions, (2) reactions with transition metals in aqueous or alcoholic solutions, and (3) formation of coordination compounds with concomitant reduction to hydrazine or ammonia (Chatt, 1981). None of these reactions has as yet yielded a suitable new method. Some of the metal–nitrogen complexes have yielded nearly quantitative yields of hydrazine or ammonia, but have not been catalytic. Those methods that have been capable of turnover or catalysis exhibit rates that are extremely low and not suitably reproducible. A large majority of the research on nitrogen complexes has been done in non-aqueous solutions due to the hydrophobic nature of the metal ligands. In these solutions, production of ammonia has been favored when tungsten or vanadium are used as the coordinating metal rather than molybdenum. The most promising systems utilizing aqueous or alcoholic media have been those based on vanadium. The most effective metal in the nitriding systems has been titanium.

Recently, the emphasis has switched to molybdenum-containing compounds modelled after the FeMo cofactor of dinitrogenase. Shah *et al.* (1978) have demonstrated that isolated FeMoCo is capable of catalytically reducing acetylene to ethylene in the presence of sodium borohydride.

Also, Thorneley *et al.* (1978) have demonstrated a hydrazine-like intermediate which they believe is bound end-on to the molybdenum of dinitrogenase. A number of cubane iron–sulfur clusters of the sum formula $[Mo_2Fe_6S_9(SC_2H_5)_8]^{3-}$ (Wolff *et al.*, 1978, 1979) or $[Mo_2Fe_6S_8(SC_2H_5)_9]^{3-}$ (Christou *et al.* 1978a, 1978b; Cramer *et al.*, 1978) have been synthesized in which one Mo atom replaces one Fe atom in a corner of a typical Fe_4S_4 cluster. Analysis of these clusters by EXAFS show that they closely resemble native dinitrogenase. A large number of other structures which bear resemblance to the FeMo cofactor of dinitrogenase have also been reported (Newton *et al.*, 1981). At present there have been no reports that these compounds function catalytically.

Nitrogenase, in addition to reducing nitrogen to ammonia, can reduce acetylene to ethylene, evolve H_2 from protons and reduce a number of other triple-bonded compounds. Developments toward creating better procedures for producing ammonia could also be applied for other oxidation–reduction reactions. Although the same catalyst may not be appropriate for all the various reactions needed, the variety of new Mo–Fe compounds synthesized indicates there should be no shortage of specialized catalysts. Possible applications include producing methylamine from cyanide, removing nitrogen oxides from emission gases, forming doubly- or triply-bonded carbon and nitrogen reagents, preparing benzene or unsaturated heterocyclic ring systems and catalyzing specific substitutions in unsaturated ring systems. A large number of chemical feedstocks will be prepared with these newly developed catalysts.

7.4.3.2 *Ammonia production*

Ammonia production from continuously cultured organisms or organisms bound to solid supports has been widely discussed (Shanmugam and Valentine, 1975; Shanmugam *et al.*, 1977; Karube *et al.*, 1981), but application has been difficult. The requirements for microbial production of ammonia are (i) derepression of the genes for nitrogen fixation, (ii) repression of ammonia assimilation enzymes, (iii) availability of large quantities of energy for nitrogen fixation and cell viability, (iv) export of the ammonia produced by nitrogenase into the surrounding media, and (v) extraction of ammonia from undesirable components in the effluent. There has been considerable variability and instability of genetic alterations described in points (i) and (ii) above during continuous culture experiments. The best strains in terms of longevity of ammonia export are those capable of maintaining low rates of protein synthesis (Shanmugam *et al.*, 1977).

The energy sources utilized most frequently for culturing nitrogen fixing organisms are expensive reduced carbon sources such as sucrose, glucose, mannitol and/or organic acids. Thus, provision of an inexpensive energy source has been a major limitation. The transfer of cellulase genes into a nitrogen-fixing organism has not yet been reported, but utilization of cellulose or any other cheap energy source (industrial or commercial wastes) is a necessary prerequisite for microbial production of ammonia to be cost-effective.

If anhydrous ammonia is needed, extraction of ammonia from the media would require available technologies to achieve the desired product purity. This step may require considerable expense. Should the desired, stable genetic alteration of a nitrogen-fixing organism be achieved, the extraction procedure may become the major cost limitation. If an effective extraction procedure can be devised with minimal capital investment and furthermore can be automated along with the culture production, then on-site ammonia generation may be available for the average farmer or small industrial consumer.

The overall cost-effectiveness depends upon the rate of ammonia production per culture, the density at which the organisms can be cultured or attached to a matrix and the extent or type of extraction procedure needed. The actual ammonia production per culture or per microbe depends upon the genetic alteration achieved, assuming the rate of ammonia production from genetically altered cells can be equated with the doubling time of the faster growing nitrogen-fixing organisms. The rapid advances of gene cloning and gene regulation coupled with the rising costs and dwindling supplies of fossil fuels indicate biological ammonia production may be cost-effective within the next few decades.

7.4.3.3 *Hydrogen production*

Like biological ammonia production, hydrogen production *via* nitrogen-fixing organisms requires derepressed strains capable of utilizing cheap, available energy sources. Unlike ammonia production, cellular export and extraction of H_2 from the media is not a problem. Considerable volumes of H_2 are generated under certain culture conditions. For example, the build-up of hydrogen from *Rhodopseudomonas capsulata* can rupture sealed gas culture tubes. H_2 evolution from *Rhodospirillum rubrum* has been reported at rates of 20 ml $h^{-1}g^{-1}$ dry wt. of culture (Zürrer and Bachofen, 1979).

The culturing conditions required for H_2 production (continuous or matrix) and enzyme source are similar to those described above for ammonia production. The same genetically altered diazotroph could probably be utilized for both ammonia production and hydrogen production, since in both cases nitrogenase regulation and cloning are required plus utilization of a cheap energy source. Culturing problems for both products would be very similar. The major technological difference would be product extraction. Since the hydrogen gas will be mixed with whatever gas (or gas mixture) has been used to support cellular metabolism, the extraneous gases will have to be removed. However, the methodologies for purification by liquefaction are available. The extraction requirement would limit small site applications and add the cost of transportation for the consumer. Again the cost-effectiveness depends upon the genetically altered rate of production that can be achieved as well as capital purchases, maintenance and other production and transportation costs.

7.4.3.4 Biomass conversion

The nitrogen fixation process has often been linked with biomass conversion (Sasaki *et al.*, 1981; Benemann, 1979). Since nitrogen is quite often limiting in diets of people in the less developed countries, biomass conversion with nitrogen supplementation (*via* nitrogen fixation) has been a popular theme. Utilization of neutralized pulp mill wastes, spoiled grains and wastes from food processing could serve as energy and carbon sources for nitrogen-fixing organisms. Coculturing of diazotrophs and other microbes has been suggested as a means of providing additional vitamins or minerals not necessarily found in diazotrophs. The resulting biomass could then be processed as food for humans and/or animals. Biomass production rates of blue-green algae in waste water treatment ponds have been reported as high as 100 tons ha^{-1} $year^{-1}$ with the average centering around 50 tons ha^{-1} $year^{-1}$ (Benemann, 1979). Major drawbacks have been production costs, nutritional quality, palatability and, when dealing with less developed countries, transportation. The technology required would limit its availability in those areas where these products would be most needed. Alternatively, the biomass produced could be converted into nitrogen fertilizers, either by utilizing the biomass directly or by extraction of the nitrogen-containing components (Benemann, 1979).

7.4.3.5 Timber production

Nitrogen is the most common limiting nutrient for timber production as well as for agronomic crops (Gordon *et al.*, 1979). The contribution of nitrogen fixation to forests can occur *via* lichens–cyanobacteria, leguminous plants–*Rhizobium* symbioses and woody plant–actinomycete symbioses. The most often used biological method of forest fertilization has been the use of actinomycete-nodulated plants, such as *Myrica*, *Alnus*, *Purshia* and others. Actinomycetous root nodules have been reported in more than 140 species of plants, most of which are woody shrubs or trees. Many of these trees provide excellent lumber for building materials, furniture, pulp wood, or fuel.

In the Pacific Northwest, red alder has the highest rate of nitrogen fixation (up to 480 kg ha^{-1}) followed by snowbush (up to 100 kg ha^{-1}), Scotch broom (160 kg ha^{-1}) and lupines (100 kg ha^{-1}) (Cromack *et al.*, 1979). Most reports show mixed stands of alder and Douglas fir increase total timber production, in some cases up to 100%, compared to pure stands of Douglas fir (DeBell, 1979). However, management of these mixed forests can require additional labor and capital. Alder and other native woody nitrogen-fixing plants are considered nuisances since they will outgrow the Douglas fir during the first several years after planting and consequently shade and reduce the growth of the fir. Prohibiting the growth of these nitrogen-fixing plants is difficult, since many are pioneer species and rapidly colonize selective or clean-cut areas. Proper management requires intercropping of alder several years after the fir have been planted to avoid shading. This demands cultivation of the Douglas fir for the first few years, perhaps additions of chemical nitrogen fertilizers and then planting of the alder. The uses of alder as a second lumber product from this intercropping system are pulp, furniture and fuel. With regard to the application of biological nitrogen fixation to forests, DeBell (1979) has stated 'In many instances the most limiting factors are *not* the need for more scientific research or breakthroughs regarding N_2-fixation *per se*, but rather some fundamental biological information as well as appropriate demonstration of benefits.'

7.4.3.6 Phytochemical production

Many of the legumes and non-leguminous plants produce valuable biochemicals such as dyes, fibers, flavorings, odors, substitutes for chocolate, citrus, coffee, garlic, licorice, tea, tobacco and vanilla, medicines, pharmaceuticals, oils for fragrances, cooking and machinery and a host of other chemicals (Allen and Allen, 1981; May, 1981). Cultivation of these plants on a commercial scale would require large initial expenditures. However, since the symbionts for many of these plants are known or are obtainable, biological nitrogen fixation is the logical alternative to chemical fertilizer to reduce the cost of applying nitrogen. Bio-Alionetics has recently announced that it will use an algal strain to produce commercially 3.8 to 7.6 million l of ethanol per year. The algae residue after processing will be sold as fertilizer (Anon., 1983).

ACKNOWLEDGEMENTS

We gratefully appreciate the typing skills of Linda Winfrey and Sandra Graziano.

7.5 REFERENCES

Abu-Shakra, S. S., D. A. Phillips and R. C. Huffaker (1978). Nitrogen fixation and delayed leaf senescence in soybeans. *Science*, **199**, 973–975.
Alef, K. and W. G. Zumft (1981). Regulatory properties of glutamine synthetase from the nitrogen-fixing phototrophic bacterium. *Rhodopseudomonas palustris. Z. Naturforsch., Teil C*, **36**, 784–789.
Ali, H., C. Niel and J. B. Guillaume (1981). The pathways of ammonium assimilation in *Rhizobium meliloti. Arch. Microbiol.*, **129**, 391–394.
Allen, O. N. and E. K. Allen (1981). *The Leguminosae. A Source Book of Characteristics, Uses and Nodulation.* University of Wisconsin Press, Madison.
Anon. (1981). Tropical legumes: Resource for the future. National Academy of Sciences Report. Washington, DC.
Anon. (1983). Ethanol produced from algae. *Chem. Eng. News*, **61**, 23.
Atwater, W. O. (1885). On the acquisition of atmospheric nitrogen by plants. *Am. Chem. J.*, **6**, 365–388.
Ausubel, F. M. and F. C. Cannon (1981). Molecular genetic analysis of *Klebsiella pneumoniae* nitrogen fixation (*nif*) genes. *Cold Spring Harbor Symp. Quant. Biol.*, **45**, 487–499.
Bauer, W. D. (1980). Role of soybean lectin in the soybean–*Rhizobium japonicum* symbiosis. In *Nitrogen Fixation*, ed. W. E. Newton and W. H. Orme-Johnson, vol. II, pp. 205–214. University Park Press, Baltimore, MD.
Beijerinck, M. W. (1901). Über oligonitrophile Mikrobew. *Zentralbl. Bakteriol. Parasitenk. Abstr. 2*, **7**, 561–582.
Benemann, J. R. (1979). Production of nitrogen fertilizer with nitrogen-fixing blue-green algae. *Enzyme Microb. Technol.*, **1**, 83–90.
Bergerson, F. J. and G. L. Turner (1967). Nitrogen fixation by the bacteroid fraction of breis of soybean root nodules. *Biochim. Biophys. Acta*, **141**, 507–515.
Beringer, J. E. (1980). The development of *Rhizobium* genetics. The fourth Fleming lecture. *J. Gen. Microbiol.*, **116**, 1–7.
Beringer, J. E. (1981). The identification, location and manipulation of genes in *Rhizobium*. In *Genetic Engineering of Symbiotic Nitrogen Fixation and Conservation of Fixed Nitrogen*, ed. J. M. Lyons, R. C. Valentine, D. A. Phillips, D. W. Rains and R. C. Huffaker, pp. 55–63. Plenum, New York.
Bishop, P. E. and W. J. Brill (1977). Genetic analysis of *Azotobacter vinelandii* mutant strains unable to fix nitrogen. *J. Bacteriol.*, **130**, 954–956.
Bishop, P. E., J. G. Guevara, J. A. Engelke and H. J. Evans (1976). Relation between glutamine synthetase and nitrogenase activities in the symbiotic association between *Rhizobium japonicum* and *Glycine max. Plant Physiol.*, **57**, 542–546.
Bishop, P. E., J. K. Gordon, V. K. Shah and W. J. Brill (1977). Transformation of nitrogen fixation genes in *Azotobacter*. In *Genetic Engineering for Nitrogen Fixation*, ed. A. Hollaender, pp. 67–80. Plenum, New York.
Bishop, P. E., D. M. L. Jarlenski and D. R. Hetherington (1980). Evidence for an alternative nitrogen fixation system in *Azotobacter vinelandii. Proc. Natl. Acad. Sci. USA*, **77**, 7342–7346.
Bishop, P. E., D. M. L. Jarlenski and D. R. Hetherington (1982). Expression of an alternative nitrogen fixation system in *Azotobacter vinelandii. J. Bacteriol.*, **150**, 1244–1251.
Bogusz, D., J. Houmard and J. P. Aubert (1981). Electron transport to nitrogenase in *Klebsiella pneumoniae*: purification and properties of the *nifJ* protein. *Eur. J. Biochem.*, **120**, 421–426.
Boland, M. J., K. J. F. Farnden and J. G. Robertson (1980). Ammonia assimilation in nitrogen-fixing legume nodules. In *Nitrogen Fixation*, ed. W. E. Newton and W. H. Orme-Johnson, vol. II, pp. 33–52. University Park Press, Baltimore, MD.

Brown, C. M., D. S. MacDonald-Brown and J. L. Meers (1974). Physiological aspects of microbial inorganic nitrogen metabolism. *Adv. Microb. Physiol.*, **11**, 1–52.

Buchanan-Wollaston, V. (1979). Generalized transduction in *Rhizobium leguminosarum*. *J. Gen. Microbiol.*, **112**, 135–142.

Buchanan-Wollaston, V., M. C. Cannon, J. L. Beynon and F. C. Cannon (1981a). Role of the *nifA* gene product in the regulation of *nif* expression in *Klebsiella pneumoniae*. *Nature (London)*, **294**, 776–778.

Buchanan-Wollaston, V., M. C. Cannon and F. C. Cannon (1981b). The use of cloned *nif* (nitrogen fixation) DNA to investigate trancriptional regulation of *nif* expression in *Klebsiella pneumoniae*. *Mol. Gen. Genet.*, **184**, 102–106.

Bulen, W. A. (1976). Nitrogenase from *Azotobacter vinelandii* and reactions affecting mechanistic interpretations. In *Proceedings of the First International Symposium on Nitrogen Fixation*, ed. W. E. Newton and C. J. Nyman, vol. 1, pp. 177–186. Washington State University Press, Pullman, WA.

Bulen, W. A., R. C. Burns and J. R. LeComte (1965). Nitrogen fixation: hydrosulfite as electron donor with cell-free preparations of *Azotobacter vinelandii* and *Rhodospirillum rubrum*. *Proc. Natl. Acad. Sci. USA*, **53**, 532–539.

Burgess, B. K., S. Wherland, W. E. Newton and E. I. Stiefel (1981a). Nitrogenase reactivity: insight into the nitrogen-fixing process through hydrogen inhibition and HD-forming reactions. *Biochemistry*, **20**, 5140–5146.

Burgess, B. K., S.-S. Yang, C.-B. You, J.-G. Li, G. D. Friesen, W.-H. Pan, E. I. Stiefel and W. E. Newton (1981b). Iron–molybdenum cofactor and its complementary protein from *Azotobacter vinelandii* UW 45. In *Current Perspectives in Nitrogen Fixation*, ed. A. H. Gibson and W. E. Newton, pp. 71–74. Australian Academy of Science, Canberra.

Burns, R. C. and R. W. F. Hardy (1975). *Nitrogen Fixation in Bacteria and Higher Plants*. Springer-Verlag, New York.

Burris, R. H. (1969). Progress in the biochemistry of nitrogen fixation. *Proc. R. Soc. London, Ser. B*, **172**, 339–354.

Burris, R. H. (1974). Biological nitrogen fixation, 1924–1974. *Plant Physiol.*, **54**, 443–449.

Burris, R. H. (1975). Preparation and properties of nitrogenase proteins. In *Nitrogen Fixation by Free-living Micro-organisms*, ed. W. D. P. Stewart, International Biological Programme, vol. 6, pp. 333–349. Cambridge University Press, London.

Burris, R. H. (1976). Nitrogen fixation by blue-green algae of the Lizard Island area of the Great Barrier Reef. *Aust. J. Plant Physiol.*, **3**, 41–51.

Burris, R. H. (1977). A synthesis paper on non-leguminous N_2-fixing systems. In *Recent Developments in Nitrogen Fixation*, ed. W. E. Newton, J. R. Postgate and C. Rodriguez-Barrueco, pp. 487–511. Academic, New York.

Burris, R. H. and W. H. Orme-Johnson (1976). Mechanism of biological dinitrogen fixation. In *Proceedings of the First International Symposium on Nitrogen Fixation*, ed. W. E. Newton and C. J. Nyman, vol. 1, pp. 208–233. Washington State University Press, Pullman, WA.

Burris, R. H. and P. W. Wilson (1946). Ammonia as an intermediate in nitrogen fixation by *Azotobacter*. *J. Bacteriol.*, **52**, 505–512.

Burris, R. H., D. J. Arp, R. V. Hageman, J. P. Houchins, W. J. Sweet and M.-Y Tso (1981). Mechanism of nitrogenase action. In *Current Perspectives in Nitrogen Fixation*, ed. A. H. Gibson and W. E. Newton, pp. 56–66. Australian Academy of Science, Canberra.

Burton, J. C. (1976). Pragmatic aspects of the *Rhizobium*: leguminous plant association. In *Proceedings of the First International Symposium on Nitrogen Fixation*, ed. W. E. Newton and C. J. Nyman, vol. 1, pp. 429–446. Washington State University Press, Pullman, WA.

Callaham, D., P. Del Tredici and J. G. Torrey (1978). Isolation and cultivation *in vitro* of the actinomycete causing root nodulation in *Comptonia*. *Science*, **199**, 899–902.

Cannon, F. C. (1980). Genetic studies with diazotrophs. In *Methods for Evaluating Biological Nitrogen Fixation*, ed. F. J. Bergersen, pp. 367–413. Wiley, New York.

Carlton, B. C. and B. J. Brown (1981). Gene mutation. In *Methodology for General Bacteriology*, ed. P. Gerhardt, chap. III, Genetics, ed. E. W. Nester, pp. 222–242. American Society for Microbiology, Washington, DC.

Carnahan, J. E., L. E. Mortenson, H. F. Mower and J. E. Castle (1960). Nitrogen fixation in cell-free extracts of *Clostridium pasteurianum*. *Biochim. Biophys. Acta*, **38**, 188–189.

Carter, K. R., N. T. Jennings, J. Hanus and H. J. Evans (1978). Hydrogen evolution and uptake by nodules of soybeans inoculated with different strains of *Rhizobium japonicum*. *Can. J. Microbiol.*, **24**, 307–311.

Carter, K. R., J. Rawlings, W. H. Orme-Johnson, R. R. Becker and H. J. Evans (1980). Purification and characterization of a ferredoxin from *Rhizobium japonicum* bacteroids. *J. Biol. Chem.*, **255**, 4213–4223.

Casadaban, M. J. and S. N. Cohen (1979). Lactose genes fused to exogenous promoters in one step using a Mu-*lac* bacteriophage; *in vivo* probe for transcriptional control sequences. *Proc. Natl. Acad. Sci. USA*, **79**, 4530–4533.

Chatt, J. (1981). Towards new catalysts for nitrogen fixation. In *Current Perspectives in Nitrogen Fixation*, ed. A. H. Gibson and W. E. Newton, pp. 15–21. Australian Academy of Science, Canberra.

Christou, G., C. D. Garner, F. E. Mabbs and T. J. King (1978a). Crystal structure of tris(tetra-*n*-butylammonium)tri-μ-benzenethiolato-bis{tris-μ-sulphido-[μ_3-sulphido-tris(benzenethiolatoiron)] molybdenum } [Bu$_4^n$N]$_3$[{(PhSF$_3$)$_3$MoS$_4$}$_2$(SPh)$_3$]; an Fe$_3$MoS$_4$ cubic cluster dimer. *J. Chem. Soc., Chem. Commun.*, **17**, 740–741.

Christou, G., C. D. Garner and F. E. Mabbs (1978b). A molybdenum derivative of a four-iron ferredoxin type centre. *Inorg. Chim. Acta*, **28**, L189–L190.

Corbin, D., G. Ditta and D. R. Helinski (1982). Clustering of nitrogen fixation (*nif*) genes in *Rhizobium meliloti*. *J. Bacteriol.*, **149**, 221–228.

Cramer, S. P., K. O. Hodgson, E. I. Stiefel and W. E. Newton (1978). A systematic X-ray absorption study of molybdenum complexes. The accuracy of structural information from extended X-ray absorption fine structure. *J. Am. Chem. Soc.*, **100**, 2748–2761.

Cromack, Jr., K., C. C. Delwiche and D. H. McNabb (1979). Prospects and problems of nitrogen management using symbiotic nitrogen fixers. In *Symbiotic Nitrogen Fixation in the Management of Temperature Forests*, ed. J. C. Gordon, C. T. Wheeler and D. A. Perry, pp. 210–223. Forest Research Laboratory, Oregon State University, Corvallis, OR.

Currier, T. C., J. F. Haury and C. P. Wolk (1977). Isolation and preliminary characterization of auxotrophs of a filamentous cyanobacterium. *J. Bacteriol.*, **129**, 1556–1562.

Dalton, H. (1981). Nitrogen fixation. *Annu. Proc. Phytochem. Soc. Eur.*, **18**, 177–195.

Darrow, R. A. and R. R. Knotts (1977). Two forms of glutamine synthetase in free-living root-nodule bacteria. *Biochem. Biophys. Res. Commun.*, **78**, 554–559.

Davis, R. W., D. Botstein and J. R. Roth (1980). *Advanced Bacterial Genetics*. Cold Spring Harbor Laboratory, Cold Spring Harbor, NY.

Dazzo, F. B. (1980). Determinants of host specificity in the *Rhizobium*-clover symbiosis. In *Nitrogen Fixation*, ed. W. E. Newton and W. H. Orme-Johnson, vol. II, pp. 165–187. University Park Press, Baltimore, MD.

DeBell, D. S. (1979). Future potential for use of symbiotic nitrogen fixation in forest management. In *Symbiotic Nitrogen Fixation in the Management of Temperate Forests*, ed. J. C. Gordon, C. T. Wheeler and D. A. Perry, pp. 451–466. Forest Research Laboratory, Oregon State University, Corvallis, OR.

Dénarié, J., P. Boistard, F. Casse-Delbart, A. G. Atherly, J. O. Berry and P. Russell (1981). Indigenous plasmids of *Rhizobium*. In *Biology of the Rhizobiaceae*, ed. Y. L. Giles and A. G. Atherly, pp. 225–246. Academic, New York.

Dilworth, M. J. and R. N. F. Thorneley (1981). Nitrogenase of *Klebsiella pneumoniae*. Hydrazine is a product of azide reduction. *Biochem. J.*, **193**, 971–983.

Ditta, G., S. Stanfield, D. Corbin and D. R. Helinski (1980). Broad host range DNA cloning system for Gram-negative bacteria: construction of a gene bank of *Rhizobium meliloti*. *Proc. Natl. Acad. Sci. USA*, **77**, 7347–7351.

Dixon, R. A. and J. R. Postgate (1972). Genetic transfer of nitrogen fixation from *Klebsiella pneumoniae* to *Escherichia coli*. *Nature (London)*, **237**, 102–103.

Dixon, R., C. Kennedy, A. Kondorosi, V. Krishnapillai and M. Merrick (1977). Complementation analysis of *Klebsiella pneumoniae* mutants defective in nitrogen fixation. *Mol. Gen. Genet.*, **157**, 189–198.

Dixon, R., R. R. Eady, G. Espin, S. Hill, M. Iaccarino, D. Kahn and M. Merrick (1980). Analysis of regulation of *Klebsiella pneumoniae* nitrogen fixation (*nif*) gene cluster with gene fusions. *Nature (London)*, **286**, 128–132.

Döbereiner, J. (1978). Potential for nitrogen fixation in tropical legumes and grasses. In *Limitations and Potentials for Biological Nitrogen Fixation in the Tropics*, ed. J. Döbereiner *et al.*, pp. 13–24. Plenum, New York.

Döbereiner, J. and J. M. Day (1976). Associative symbioses in tropical grasses: characterization of microorganisms and dinitrogen-fixing sites. In *Proceedings of the First International Symposium on Nitrogen Fixation*, ed. W. E. Newton and C. J. Nyman, vol. 2, pp. 518–538. Washington State University Press, Pullman,WA.

Döbereiner, J. and H. De-Polli (1980). Diazotrophic Rhizococenoses. Nitrogen fixation. *Annu. Proc. Phytochem. Soc. Eur.*, **18**, 301–333.

Drewes, K. (1928). Über die assimilation des Luftstickstoffs durch Blaualgen. *Zentralbl. Bakteriol. Parasitenk. Abstr. 2*, **76**, 88–101.

Drummond, M., J. Clements, M. Merrick and R. Dixon (1983). Positive control and autogenous regulation of the *nifA* promotor in *Klebsiella pneumoniae*. *Nature (London)*, **301**, 302–307.

Eady, R. R. (1980). Isolation and characterization of various nitrogenases. *Methods Enzymol.*, **69**, 753–778.

Eady, R. R. (1981). Regulation of nitrogenase activity. In *Current Perspective in Nitrogen Fixation*, ed. A. H. Gibson and W. E. Newton, pp. 172–181. Australian Academy of Science, Canberra.

Ela, S. W., M. A. Anderson and W. J. Brill (1982). Screening and selection of maize to enhance associative bacterial nitrogen fixation. *Plant Physiol.*, **70**, 1564–1567.

Emerich, D. W. and R. H. Burris (1976). Interactions of heterologous nitrogenase components that generate catalytically inactive complexes. *Proc. Natl. Acad. Sci. USA*, **73**, 4369–4373.

Emerich, D. W. and R. H. Burris (1978). Purification of nitrogenase. *Methods Enzymol.*, **53**, 314–329.

Emerich, D. W. and H. J. Evans (1980). Biological nitrogen fixation with an emphasis on the legumes. In *Biochemical and Photosynthetic Aspects of Energy Production*, ed. A. San Pietro, pp. 117–145. Academic, New York.

Emerich, D. W., R. V. Hageman and R. H. Burris (1981). Interactions of dinitrogenase and dinitrogenase reductase. *Adv. Enzymol.*, **52**, 1–22.

Evans, H. J., D. W. Emerich, T. Ruiz-Argüeso, R. J. Maier and S. L. Albrecht (1980). Hydrogen metabolism in the legume-*Rhizobium* symbiosis. In *Nitrogen Fixation*, ed. W. E. Newton and W. H. Orme-Johnson, vol. II, pp. 69–86. University Park Press, Baltimore, MD.

Evans, H. J., K. Purohit, M. A. Cantrell, G. Eisbrenner, S. A. Russell, F. J. Hanus and J. E. Lepo (1981). Hydrogen losses and hydrogenases in nitrogen-fixing organisms. In *Current Perspectives in Nitrogen Fixation*, ed. A. H. Gibson and W. E. Newton, pp. 84–96. Australian Academy of Science, Canberra.

Fisher, R., R. Tuli and R. Haselkorn (1981). A cloned cyanobacterial gene for glutamine synthetase functions in *Escherichia coli*, but the enzyme is not adenylylated. *Proc. Natl. Acad. Sci. USA*, **78**, 3393–3397.

Folkes, B. F. and A. P. Sims (1974). The significance of amino acid inhibition of NADP-linked glutamate dehydrogenase in the physiological control of glutamate synthesis in *Candida utilis*. *J. Gen. Microbiol.*, **82**, 77–95.

Fred, E. B., I. L. Baldwin and E. McCoy (1932). *Root Nodule Bacteria and Leguminous Plants*. University of Wisconsin Press, Madison, WI.

Fuchs, R. L. and D. L. Keister (1980). Comparative properties of glutamine synthetases I and II in *Rhizobium* and *Agrobacterium* spp. *J. Bacteriol.*, **144**, 641–648.

Gelin, O. and S. Blixt (1964). Root nodulation in peas. *Agric. Hort. Genet.*, **22**, 149–159.

Giles, K. L. and A. G. Atherly (1981). *Biology of the Rhizobiaceae*. *International Review of Cytology*. Academic, New York.

Glenn, A. R. and N. J. Brewin (1981). Succinate-resistant mutants of *Rhizobium leguminosarum*. *J. Gen. Microbiol.*, **126**, 237–242.

Gordon, J. C., C. T. Wheeler and D. A. Perry (eds.) (1979). *Symbiotic Nitrogen Fixation in the Management of Temperate Forests*. Forest Research Laboratory, Oregon State University, Corvallis, OR.

Gutschick, V. P. (1980). Energy flows in the nitrogen cycle, especially in fixation. In *Nitrogen Fixation*, ed. W. E. Newton and W. H. Orme-Johnson, vol. I, pp. 17–28. University Park Press, Baltimore, MD.

Habte, M. and M. Alexander (1980). Nitrogen fixation by photosynthetic bacteria in lowland rice cultures. *Appl. Environ. Microbiol.*, **39**, 342–347.

Hageman, R. V. and R. H. Burris (1978a). Kinetic studies on electron transfer and interactions between nitrogenase components from *Azotobacter vinelandii*. *Biochemistry*, **17**, 4117–4124.

Hageman, R. V. and R. H. Burris (1978b). Nitrogenase and nitrogenase reductase associate and dissociate with each catalytic cycle. *Proc. Natl. Acad. Sci. USA*, **75**, 2699–2702.

Hageman, R. V. and R. H. Burris (1980). Electron allocation to alternative substrates of *Azotobacter* nitrogenase is controlled by the electron flux through dinitrogenase. *Biochim. Biophys. Acta*, **591**, 63–75.

Ham, G. E. (1980). Inoculation of legumes with *Rhizobium* in competition with naturalized strains. In *Nitrogen Fixation*, ed. W. E. Newton and W. H. Orme-Johnson, vol. II, pp. 131–138. University Park Press, Baltimore, MD.

Hanus, F. J., R. J. Maier and H. J. Evans (1979). Autotrophic growth of H_2-uptake-positive strains of *Rhizobium japonicum* in an atmosphere supplied with hydrogen gas. *Proc. Natl. Acad. Sci. USA*, **76**, 1788–1792.

Harker, A. R. and L. H. Wullstein (1981). Resolution of two subunits from the molybdenum–iron protein of *Azotobacter vinelandii* nitrogenase. *J. Biol. Chem.*, **256**, 11 981–11 983.

Haselkorn, R. (1978). Heterocysts. *Annu. Rev. Plant Physiol.*, **29**, 319–344.

Haselkorn, R., B. Mazur, J. Orr, D. Rice, N. Wood and R. Rippka (1980). Heterocyst differentiation and nitrogen fixation in cyanobacteria (blue-green algae). In *Nitrogen Fixation*, ed. W. E. Newton and W. H. Orme-Johnson, vol. II, pp. 259–278. University Park Press, Baltimore, MD.

Hayes, W. (1968). *The Genetics of Bacteria and Their Viruses*, 2nd edn., pp. 138–139. Wiley, New York.

Hellriegel, H. and H. Wilfarth (1888). Untersuchungen über die Stickstoff-nahrung der Gramineen und Leguminosen. *Beilageheft zu der Ztscher. Ver. Rübenzucker-Industrie Deutschen Reichs.*

Heumann, W. (1981). *Rhizobium* genetics. In *Biology of Inorganic Nitrogen and Sulfur*, ed. H. Bothe and A. Trebst, pp. 87–102. Springer-Verlag, Berlin.

Hill, S. and E. Kavanagh (1980). Roles of *nifF* and *nifJ* gene products in electron transport to nitrogenase in *Klebsiella pneumoniae*. *J. Bacteriol.*, **141**, 470–475.

Hill, S., C. Kennedy, E. Kavanagh, R. B. Goldberg and R. Hanau (1981). Nitrogen fixation gene (*nifL*) involved in oxygen regulation of nitrogenase synthesis in *K. pneumoniae*. *Nature (London)*, **290**, 424–426.

Hombrecker, G., N. J. Brewin and A. W. B. Johnston (1981). Linkage of genes for nitrogenase and nodulation ability on plasmids in *Rhizobium leguminosarum* and *R. phaseoli*. *Mol. Gen. Genet.*, **182**, 133–136.

Huynh, B. H., M. T. Henzl, J. A. Christner, R. Zimmerman, W. H. Orme-Johnson and E. Munck (1980). Nitrogenase XII: Mossbauer studies of the MoFe protein from *Clostridium pasteurianum* W-5. *Biochim. Biophys. Acta*, **623**, 124–138.

Huynh, B. H., E. Münck and W. H. Orme-Johnson (1979). Nitrogenase XI: Mössbauer studies on the cofactor centers of the MoFe protein from *Azotobacter vinelandii* OP. *Biochim. Biophys. Acta*, **576**, 192–203.

Jensen, H. L. (1965a). Nonsymbiotic nitrogen fixation. In *Soil Nitrogen Agronomy Monograph 10*, ed. W. V. Bartholomew and F. E. Clark, pp. 436–480. American Society of Agronomy Inc., Madison, WI.

Jensen, H. L. (1965b). Soil nitrogen. In *Soil Nitrogen Agronomy Monograph 10*, ed. W. V. Bartholomew and F. E. Clark, p. 440. American Society of Agronomy, Inc., Madison, WI.

Johansson, B. C. and H. Gest (1976). Inorganic nitrogen assimilation by the photosynthetic bacterium *Rhodopseudomonas capsulata*. *J. Bacteriol.*, **128**, 683–688.

Jordan, D. C. (1981). Nitrogen fixation by selected free-living and associative micro-organisms. In *Current Perspectives in Nitrogen Fixation*, ed. A. H. Gibson and W. E. Newton, pp. 317–320. Australian Academy of Science, Canberra.

Kahn, D., M. Hawkins and R. R. Eady (1982). Nitrogen fixation in *Klebsiella pneumoniae*: nitrogenase levels and the effect of added molybdate on nitrogenase derepressed under molybdenum deprivation. *J. Gen. Microbiol.*, **128**, 779–787.

Kaluza, K. and H. Hennecke (1981). Regulation of nitrogenase messenger RNA synthesis and stability in *Klebsiella pneumoniae*. *Arch. Microbiol.*, **130**, 38–43.

Kapulnik, Y., J. Kigel, Y. Okon, I. Nur and Y. Henis (1981a). Effect of *Azospirillum* inoculation on some growth parameters and N-content of wheat, sorghum and panicum. *Plant and Soil*, **61**, 65–70.

Kapulnik, Y., Y. Okon, J. Kigel, J. Nur and Y. Henis (1981b). Effects of temperature, nitrogen fertilization and plant age on nitrogen fixation by *Setaria italica* inoculated with *Azospirillum brasilense* (strain cd). *Plant Physiol.*, **68**, 340–343.

Karube, I., T. Matsunaga, Y. Otomine and S. Suzuki (1981). Nitrogen fixation by immobilized *Azotobacter chroococcum*. *Enzyme Microbiol. Technol.*, **3**, 309–313.

Kennedy, C. (1977). Linkage map of the nitrogen fixation (*nif*) genes in *Klebsiella pneumoniae*. *Mol. Gen. Genet.*, **157**, 199–204.

Kennedy, C., F. Cannon, M. Cannon, R. Dixon, S. Hill, J. Jensen, S. Kumar, P. McLean, M. Merrick, R. Robson and J. Postgate (1981a). Recent advances in the genetics and regulation of nitrogen fixation. In *Current Perspectives in Nitrogen Fixation*, ed. A. H. Gibson and W. E. Newton, pp. 146–156. Australian Academy of Science, Canberra.

Kennedy, C., B. Dreyfus and J. Brockwell (1981b). Transfer, maintenance and expression of P plasmids in strains of cowpea rhizobia. *J. Gen. Microbiol.*, **125**, 233–240.

Kennedy, I. R., L. Copeland and S. R. Stone (1981c). Involvement of glutamate dehydrogenase in assimilation of fixed nitrogen in legume nodules: complementary roles for GDH and glutamine synthetase. In *Current Perspectives in Nitrogen Fixation*, ed. A. H. Gibson and W. E. Newton, p. 386. Australian Academy of Science, Canberra.

Kleckner, N. (1981). Transposable elements in prokaryotes. *Annu. Rev. Genet.*, **15**, 341–404.

Klucas, R. V. and W. Pedersen (1980). Nitrogen fixation associated with roots of sorghum and wheat. In *Nitrogen Fixation*, ed. W. E. Newton and W. H. Orme-Johnson, vol. II, pp. 243–255. University Park Press, Baltimore, MD.

Kondorosi, A. and A. W. B. Johnston (1981). The genetics of *Rhizobium*. In *Biology of the Rhizobiaceae*, ed. K. L. Giles and A. G. Atherly, pp. 191–224. Academic, New York.

Kondorosi, A., E. Vincze, A. W. B. Johnston and J. E. Beringer (1980). A comparison of three *Rhizobium* linkage maps. *Mol. Gen. Genet.*, **178**, 403–408.

Kuykendall, L. D. (1981). Mutants of *Rhizobium* that are altered in legume interaction and nitrogen fixation. In *Biology of the Rhizobiaceae*, ed. K. L. Giles and A. G. Atherly, pp. 299–309. Academic, New York.

Lawes, J. B., J. H. Gilbert and E. Pugh (1861). On the sources of nitrogen in vegetation; with special reference to the question whether plants assimilate free or uncombined nitrogen. *Philos. Trans. R. Soc. London, Ser. B*, **151**, 431–577.

Lees, E. M., L. Copeland and E. A. McKenzie (1981). Soybean glutamate dehydrogenases: their role in ammonia assimilation. In *Current Perspectives in Nitrogen Fixation*, ed. A. H. Gibson and W. E. Newton, p. 387. Australian Academy of Science, Canberra.

Leonardo, J. M. and R. B. Goldberg (1980). Regulation of nitrogen metabolism in glutamine auxotrophs of *Klebsiella pneumoniae*. *J. Bacteriol.*, **142**, 99–110.

Ljones, T. and R. H. Burris (1978). Nitrogenase: the reaction between the Fe protein and bathophenanthrolinedisulfonate as a probe for interactions with MgATP. *Biochemistry*, **17**, 1866–1871.

Ludden, P. W. (1980). Nitrogen fixation by photosynthetic bacteria: properties and regulation of the enzyme system from *Rhodospirillum rubrum*. In *Nitrogen Fixation*, ed. W. E. Newton and W. H. Orme-Johnson, vol. I, pp. 139–156. University Park Press, Baltimore, MD.

Ludden, P. W. and R. H. Burris (1976). Activating factor for the iron protein of nitrogenase from *Rhodospirillum rubrum*. *Science*, **194**, 424–426.

Ludden, P. W., R. V. Hageman, W. H. Orme-Johnson and R. H. Burris (1982). Properties and activities of 'inactive' Fe protein from *Rhodospirillum rubrum*. *Biochim. Biophys. Acta*, **700**, 213–216.

Ludwig, R. A. (1980a). Regulation of *Rhizobium* nitrogen fixation by the unadenylylated glutamine synthetase I system. *Proc. Natl. Acad. Sci. USA*, **77**, 5817–5821.

Ludwig, R. A. (1980b). Physiological roles of glutamine synthetases I and II in ammonium assimilation in *Rhizobium* sp. 32H1. *J. Bacteriol.*, **141**, 1209–1216.

MacKintosh, M. E. (1978). Nitrogen fixation by *Thiobacillus ferroxidans*. *J. Gen. Microbiol.*, **105**, 215–218.

MacNeil, D., J. Zhu and W. J. Brill (1981). Regulation of nitrogen fixation in *Klebsiella pneumoniae*: isolation and characterization of strains with *nif–lac* fusions. *J. Bacteriol.*, **145**, 348–357.

Madigan, M. T., J. D. Wall and H. Gest (1979). Dark anaerobic dinitrogen fixation by a photosynthetic microorganism. *Science*, **204**, 1429–1430.

Magasanik, B. (1977). Regulation of bacterial nitrogen assimilation by glutamine synthetase. *Trends Biochem. Sci.*, **2**, 9–12.

Magasanik, B. (1982). Genetic control of nitrogen assimilation in bacteria. *Annu. Rev. Genet.*, **16**, 135–168.

Maniatis, T., E. F. Smith and J. Sambrook (1982). *Molecular Cloning. A Laboratory Manual*. Cold Spring Harbor Laboratory, Cold Spring Harbor, NY.

Marrs, B. (1981). Mobilization of the genes for photosynthesis from *Rhodopseudomonas capsulata* by a promiscuous plasmid. *J. Bacteriol.*, **146**, 1003–1012.

Marrs, B. L. (1982). Genetics and molecular biology. In *The Anoxygenic Phototrophic Bacteria*, ed. J. Ormerod, chap. 10. Blackwell Scientific Publications, Oxford.

Mazur, B. J., D. Rice and R. Haselkorn (1980). Identification of blue-green algal nitrogen fixation genes by using heterologous DNA hybridization probes. *Proc. Natl. Acad. Sci. USA*, **77**, 186–190.

Matsumoto, T., M. Yatazawa and Y. Yamamoto (1977a). Distribution and change in the contents of allantoin and allantoic acid in developing nodulating and nonnodulating soybean plants. *Plant Cell Physiol.*, **18**, 353–359.

Matsumoto, T., M. Yatazawa and Y. Yamamoto (1977b). Incorporation of ^{15}N into allantoin in nodulated soybean plants supplied with ^{15}N$_2$. *Plant Cell Physiol.*, **18**, 459–462.

Matsumoto, T., M. Yatazawa and Y. Yamamoto (1977c). Effects of exogenous nitrogen-compounds on the concentration of allantoin and various constituents in several organs of soybean plants. *Plant Cell Physiol.*, **18**, 613–624.

May, R. M. (1981). Useful tropical legumes. *Nature (London)*, **294**, 516–517.

McFarland, N., L. McCarter, S. Artz and S. Kustu (1981). Nitrogen regulatory locus '*glnR*' of enteric bacteria is composed of cistrons *ntrB* and *ntrC*: identification of their protein products. *Proc. Natl. Acad. Sci. USA*, **78**, 2135–2139.

McKenna, C. E., M.-C. McKenna and C. W. Huang (1979). Low stereoselectivity in methylacetylene and cyclopropene reductions by nitrogenase. *Proc. Natl. Acad. Sci. USA*, **76**, 4773–4777.

McKenna, C. E., C. W. Huang, J. B. Jones, M.-C. McKenna, T. Nakajima and H. T. Nguyen (1980). Cyclopropenes: new chemical probes of nitrogenase active site interactions. In *Nitrogen Fixation*, ed. W. E. Newton and W. H. Orme-Johnson, vol. I, pp. 223–236. University Park Press, Baltimore, MD.

McKenna, C. E., H. Eran, T. Nakajima and A. Osumi (1981). Active site probes for nitrogenase. In *Current Perspectives in Nitrogen Fixation*, ed. A. H. Gibson and W. E. Newton, p. 358. Australian Academy of Science, Canberra.

McLean, P. A. and R. A. Dixon (1981). Requirement of *nifV* gene for production of wild-type nitrogenase enzyme in *Klebsiella pneumoniae*. *Nature (London)*, **292**, 655–656.

Merrick, M. J. (1982). A new model for nitrogen control. *Nature (London)*, **297**, 362–363.

Merrick, M., S. Hill, H. Hennecke, M. Hahn, R. Dixon and C. Kennedy (1982). Repressor properties of the *nifL* gene product in *Klebsiella pneumoniae*. *Mol. Gen. Genet.*, **185**, 75–81.

Meynell, G. G. and E. Meynell (1965). *Theory and Practice in Experimental Bacteriology*, 2nd edn. Cambridge University Press, Cambridge.

Miflin, B. J. and P. J. Lea (1976). The pathway of nitrogen assimilation in plants. *Phytochemistry*, **15**, 873–885.

Miller, J. H. (1972). *Experiments in Molecular Genetics*. Cold Spring Harbor Laboratory, Cold Spring Harbor, NY.

Mishustin, E. N. and N. B. Shil'nikova (1966). *Biological fixation of atmospheric nitrogen*. USSR Academy of Sciences, Moscow.

Mortenson, L. E. and R. N. F. Thorneley (1979). Structure and function of nitrogenase. *Annu. Rev. Biochem.*, **48**, 387–418.

Mortenson, L. E. and R. G. Upchurch (1981). Effect of adenylates on electron flow and efficiency of nitrogenase. In *Current Perspectives in Nitrogen Fixation*, ed. A. H. Gibson and W. E. Newton, pp. 75–78. Australian Academy of Science, Canberra.

Moss, D. N. (1976). Studies on increasing photosynthesis in crop plants. In *CO$_2$ Metabolism and Plant Productivity*, ed. R. H. Burris and C. C. Black, pp. 31–41. University Park Press, Baltimore, MD.

Nano, F. and S. Kaplan (1982). Introduction of R751::Mud*l* (Ap^r, *lac*) into *Rhodopseudomonas sphaeroides*. *Abstracts of the Annual Meeting of the American Society for Microbiology, 1982*, p. 127.

Napoli, C., R. Sanders, R. Carlson and P. Albersheim (1980). Host–symbiont interactions: recognizing *Rhizobium*. In *Nitrogen Fixation*, ed. W. E. Newton and W. H. Orme-Johnson, vol. II, pp. 189–203. University Park Press, Baltimore, MD.

Newton, W. E., J. W. McDonald, G. D. Friesen and B. K. Burgess (1981). Molybdenum–iron–sulfur complexes and their relevance to the molybdenum site of nitrogenase. In *Current Perspectives in Nitrogen Fixation*, ed. A. H. Gibson and W. E. Newton, pp. 30–39. Australian Academy of Science, Canberra.

Nieva-Gómez, D., G. P. Roberts, S. Klevickis and W. J. Brill (1980). Electron transport to nitrogenase in *Klebsiella pneumoniae*. *Proc. Natl. Acad. Sci. USA*, **77**, 2555–2558.

Noel, K. D., G. Stacey, S. R. Tandon, L. E. Silver and W. J. Brill (1982). *Rhizobium japonicum* mutants defective in symbiotic nitrogen fixation. *J. Bacteriol.*, **152**, 485–494.

O'Gara, F. and K. T. Shanmugam (1976). Regulation of nitrogen by fixation rhizobia. Export of fixed N_2 and NH_4^+. *Biochim. Biophys. Acta*, **437**, 313–321.

Orme-Johnson, W. H. and L. C. Davis (1977). Current topics and problems in the enzymology of nitrogenase. In *Iron–Sulfur Proteins*, ed. W. Lovenberg, vol. 4, pp. 15–60. Academic, New York.

Orme-Johnson, W. H., L. C. Davis, M. T. Henzl, B. A. Averill, N. R. Orme-Johnson, E. Münck and R. Zimmermann (1977). Components and pathways in biological nitrogen fixation. In *Recent Developments in Nitrogen Fixation*, ed. W. E. Newton, J. R. Postgate and C. Rodriguez-Barrueco, pp. 131–178. Academic, New York.

Orme-Johnson, W. H., P. Lindahl, J. Meade, W. Warren, M. Nelson, S. Groh, N. R. Orme-Johnson, E. Münck, B. H. Huynh, M. Emptage, J. Rawlings, J. Smith, J. Roberts, B. Hoffman and W. B. Mims (1981). Nitrogenase: prosthetic groups and their reactivities. In *Current Perspectives in Nitrogen Fixation*, ed. A. H. Gibson and W. E. Newton, pp. 79–83. Australian Academy of Science, Canberra.

Orr, J., L. M. Keefer, P. Keim, T. D. Nguyen, T. Wellems, R. L. Heinrikson and R. Haselkorn (1981). Purification, physical characterization, and NH$_2$-terminal sequence of glutamine synthetase from the cyanobacterium *Anabaena* 7120. *J. Biol. Chem.*, **256**, 13 091–13 098.

Ow, D. W. and F. Ausubel (1983). Regulation of nitrogen metabolism genes by *nifA* gene product in *Klebsiella pneumoniae*. *Nature (London)*, **301**, 307–313.

Page, W. J. and H. L. Sadoff (1976). Physiological factors affecting transformation of *Azotobacter vinelandii*. *J. Bacteriol.*, **125**, 1080–1087.

Page, W. J. and M. von Tigerstrom (1979). Optimal conditions for transformation of *Azotobacter vinelandii*. *J. Bacteriol.*, **139**, 1058–1061.

Pahel, G., D. M. Rothstein and B. Magasanik (1982). Complex *glnA–glnL–glnG* operon of *Escherichia coli*. *J. Bacteriol.*, **150**, 202–213.

Parejko, R. A. and P. W. Wilson (1968). Taxonomy of *Azotomonas* species. *J. Bacteriol.*, **95**, 143–146.

Pemberton, J. M. and A. R. St. G. Bowen (1981). High-frequency chromosome transfer in *Rhodopseudomonas sphaeroides* promoted by broad-host-range plasmid RP1 carrying mercury transposon *Tn501*. *J. Bacteriol.*, **147**, 110–117.

Petering, D., J. A. Fee and G. Palmer (1971). The oxygen-sensitivity of spinach ferredoxin and other iron–sulfur proteins. *J. Biol. Chem.*, **246**, 643–653.

Peters, G. A., T. B. Ray, B. C. Mayne and R. E. Toia, Jr. (1980). *Azolla–Anabaena* association: morphological and physiological studies. In *Nitrogen Fixation*, ed. W. E. Newton and W. H. Orme-Johnson, vol. II, pp. 293–309. University Park Press, Baltimore, MD.

Peterson, R. B., E. F. Friberg and R. H. Burris (1977). Diurnal variation in N_2 fixation and photosynthesis by aquatic blue-green algae. *Plant Physiol.*, **59**, 74–80.

Phillips, D. A., T. M. DeJong and L. E. Williams (1981). Carbon and nitrogen limitations on symbiotically-grown soybean seedlings. In *Current Perspectives in Nitrogen Fixation*, ed. A. H. Gibson and W. E. Newton, pp. 117–120. Australian Academy of Science, Canberra.

Planqué, K., G. E. de Vries and J. W. Kijne (1978). The relationship between nitrogenase and glutamine synthetase in bacteroids of *Rhizobium leguminosarum* of various ages. *J. Gen. Microbiol.*, **106**, 173–178.

Postgate, J. R. (1971). Fixation by free-living microbes: physiology. In *The Chemistry and Biochemistry of Nitrogen Fixation*, ed. J. R. Postgate, chap. 5, pp. 161–190. Plenum, London.

Postgate, J. R. (1981). Microbiology of the free-living nitrogen-fixing bacteria, excluding cyanobacteria. In *Current Perspectives on Nitrogen Fixation*, ed. A. H. Gibson and W. E. Newton, pp. 217–228. Australian Academy of Science, Canberra.

Prakash, R. K., R. A. Schilperoort and M. P. Nuti (1981). Large plasmids of fast-growing rhizobia: homology studies and location of structural nitrogen fixation (*nif*) genes. *J. Bacteriol.*, **145**, 1129–1136.

Prakash, R. K., R. J. M. van Veen and R. A. Schilperoort (1982). Restriction endonuclease mapping of a *Rhizobium leguminosarum* sym plasmid. *Plasmid*, **7**, 271–280.

Pühler, A. and W. Klipp (1981). Fine-structure analysis of the gene region for N_2-fixation (*nif*) of *Klebsiella pneumoniae*. In *Biology of Inorganic Nitrogen and Sulphur*, ed. H. Boethe and A. Trebst, pp. 276–286. Springer-Verlag, Berlin.

Quinto, C., H. de la Vega, M. Flores, L. Fernández, T. Ballado, G. Soberón and R. Palacios (1982). Reiteration of nitrogen fixation gene sequences in *Rhizobium phaseoli*. *Nature (London)*, **299**, 724–726.

Quispel, A. and A. J. P. Burggraaf (1980). Frankia, the diazotrophic endophyte from Actinorhiza's. In *Current Perspectives in Nitrogen Fixation*, ed. A. H. Gibson and W. E. Newton, pp. 229–236. Australian Academy of Science, Canberra.

Rennie, R. J. and R. I. Larson (1979). International Workshop on Associative N_2 Fixation. Piracicaba, Brazil.

Rice, D., B. J. Mazur and R. Haselkorn (1982). Isolation and physical mapping of nitrogen fixation genes from the cyanobacterium *Anabaena* 7120. *J. Biol. Chem.*, **257**, 13 157–13 163.

Riddle, G. D., J. G. Simonson, B. J. Hales, and H. D. Braymer (1982). Nitrogen fixation system of tungsten—resistant mutants of *Azotobacter vinelandii*. *J. Bacteriol.*, **152**, 72–80.

Roberts, G. P. and W. J. Brill (1980). Gene-product relationships of the *nif* regulon of *Klebsiella pneumoniae*. *J. Bacteriol.*, **144**, 210–216.

Roberts, G. P. and W. J. Brill (1981). Genetics and regulation of nitrogen fixation. *Annu. Rev. Microbiol.*, **35**, 207–235.

Ronson, C. W., P. Lyttleton and J. G. Robertson (1981). C_4-dicarboxylate transport mutants of *Rhizobium trifolii* form ineffective nodules on *Trifolium repens*. *Proc. Natl. Acad. Sci. USA*, **78**, 4284–4288.

Rossini, F. D., D. D. Wagman, W. H. Evans, S. Levine and I. Jaffe (1952). Selected values of chemical thermodynamic properties. Circular of the National Bureau of Standards, 500, U.S. Department of Commerce, pp. 52 and 55. United States Government Printing Office, Washington, DC.

Rothman, N., D. Rothstein, F. Foor and B. Magasanik (1982). Role of *glnA*-linked genes in regulation of glutamine synthetase and histidase formation in *Klebsiella aerogenes*. *J. Bacteriol.*, **150**, 221–230.

Rubenchik, L. I. (1963). *Azotobacter* and its use in agriculture. Office of Technical Services, U.S. Department of Commerce, Washington, DC.

Ruvkun, G. B. and F. M. Ausubel (1980). Interspecies homology of nitrogenase genes. *Proc. Natl. Acad. Sci. USA*, **77**, 191–195.

Ruvkun, G. B. and F. M. Ausubel (1981). A general method for site-directed mutagenesis in prokaryotes. *Nature (London)*, **289**, 85–88.

Sadoff, H. L., B. Shimei and S. Ellis (1979). Characterization of *Azotobacter vinelandii* deoxyribonucleic acid and folded chromosomes. *J. Bacteriol.*, **138**, 871–877.

Sasaki, K., N. Noparatnaraporn, M. Hayashi, Y. Nishizawa and S. Nagai (1981). Single-cell protein production by treatment of soybean wastes with *Rhodopseudomonas gelatinosa*. *J. Ferment. Technol.*, **59**, 471–477.

Schleif, R. F. and P. C. Wensink (1981). *Practical Methods in Molecular Biology*. Springer-Verlag, New York.

Schneider, K. C., C. Bradbeer, R. N. Singh, L. C. Wang, P. W. Wilson and R. H. Burris (1960). Nitrogen fixation by cell-free preparations from microorganisms. *Proc. Natl. Acad. Sci. USA*, **46**, 726–733.

Schwinghamer, E. A. (1977). Genetic aspects of nodulation and dinitrogen fixation by legumes: the microsymbiont. In *A Treatise on Dinitrogen Fixation, Section III Biology*, ed. R. W. F. Hardy and W. S. Silver, pp. 577–622. Wiley, New York.

Schubert, K. R. (1981). Enzymes of purine biosynthesis and catabolism in *Glycine max*. I. Comparison of activities with N_2 fixation and composition of xylem exudate during nodule development. *Plant Physiol.*, **68**, 1115–1122.

Schubert, K. R., J. A. Engelke, S. A. Russell and H. J. Evans (1977). Hydrogen reactions of nodulated leguminous plants. I. Effect of rhizobial strains and plant age. *Plant Physiol.*, **60**, 651–654.

Seetin, M. W. and D. K. Barnes (1977). Variation among alfalfa genotypes for rate of acetylene reduction. *Crop Sci.*, **17**, 783–787.

Selvaraj, G. and V. N. Iyer (1981). Genetic transformation of *Rhizobium meliloti* by plasmid DNA. *Gene*, **15**, 279–283.

Shah, K., C. Patel and V. V. Modi (1983). Linkage mapping of *Rhizobium japonicum* D211 by phage M-1 mediated transduction. *Can. J. Microbiol.*, **29**, 33–38.

Shah, V. K. and W. J. Brill (1977). Isolation of an iron–molybdenum cofactor from nitrogenase. *Proc. Natl. Acad. Sci. USA*, **74**, 3249–3253.

Shah, V. K. and W. J. Brill (1981). Isolation of a molybdenum–iron cluster from nitrogenase. *Proc. Natl. Acad. Sci. USA*, **78**, 3438–3440.

Shah, V. K., J. R. Chisnell and W. J. Brill (1978). Acetylene reduction by the iron–molybdenum cofactor from nitrogenase. *Biochem. Biophys. Res. Commun.* **81**, 232–235.

Shah, K., S. de Sousa and V. V. Modi (1981). Studies on transducing phage M-1 for *Rhizobium japonicum* D211. *Arch. Microbiol.*, **130**, 262–266.

Shanmugam, K. T. and R. C. Valentine (1975). Microbial production of ammonium ion from nitrogen. *Proc. Natl. Acad. Sci. USA*. **72**, 136–139.

Shanmugam, K. T., F. O'Gara, K. Anderson, C. Morandi and R. C. Valentine (1977). Genetic control of nitrogen fixation (*nif*). In *Recent Developments in Nitrogen Fixation*, ed. W. E. Newton, J. R. Postgate and C. Rodriquez-Barrueco, pp. 321–330. Academic, London.

Shanmugam, K. T., F. O'Gara, K. Anderson and R. C. Valentine (1978). Biological nitrogen fixation. *Annu. Rev. Plant Physiol.*, **29**, 263–276.

Sherman, L. A. and P. van de Putte (1982). Construction of a hybrid plasmid capable of replication in the bacterium *Escherichia coli* and the cyanobacterium *Anacystis nidulans*. *J. Bacteriol.*, **150**, 410–413.

Stacey, G., C. van Baalen and F. R. Tabita (1977). Isolation and characterization of a marine *Anabaena* sp. capable of rapid growth on molecular nitrogen. *Arch. Microbiol.*, **114**, 197–201.

Stacey, G. A., G. van Baalen and F. R. Tabita (1979). Nitrogen and ammonia assimilation in the cyanobacteria: regulaton of glutamine synthetase. *Arch. Biochem. Biophys.*, **194**, 457–467.

Stacey, G., J. Zhu, V. K. Shah, S.-C. Shen and W. J. Brill (1982). Intragenic complementation by the *nifJ*-coded protein of *Klebsiella pneumoniae*. *J. Bacteriol.*, **150**, 293–297.

Stewart, W. D. P. (1977). Blue-green algae. In *A Treatise on Dinitrogen Fixation, Section III, Biology*, ed. R. W. F. Hardy and W. S. Silver, pp. 63–123. Wiley, New York.

Stewart, W. D. P. (1980). Systems involving blue-green algae (cyanobacteria). In *Methods for Evaluating Biological Nitrogen Fixation*, ed. F. J. Bergersen, pp. 583–635. Wiley, New York.

Stewart, W. D. P., P. Rowell and A. N. Rai (1980). Symbiotic nitrogen-fixing cyanobacteria. Nitrogen fixation. *Annu. Proc. Phytochem. Soc. Eur.*, 239–277.

Stewart, W. D. P., A. N. Rai, R. H. Reed, E. Creach, G. A. Codd and P. Rowell (1981). Studies on the N_2-fixing lichen *Peltigera aphthosa*. In *Current Perspectives in Nitrogen Fixation*, ed. A. H. Gibson and W. E. Newton, pp. 237–243. Australian Academy of Science, Canberra.

Streicher, S., E. Gurney and R. C. Valentine (1971). Transduction of the nitrogen-fixation genes in *Klebsiella pneumoniae*. *Proc. Natl. Acad. Sci. USA*, **68**, 1174–1177.

Talley, S. N. and D. W. Rains (1980). *Azolla* as a nitrogen source for temperate rice. In *Nitrogen Fixation*, ed. W. E. Newton and W. H. Orme-Johnson, vol. II, pp. 311–320. University Park Press, Baltimore, MD.

Talley, S. N., E. Lim and D. W. Rains (1981). Application of *Azolla* in crop production. In *Genetic Engineering of Symbiotic Nitrogen Fixation and Conservation of Fixed Nitrogen*, ed. J. M. Lyons, R. C. Valentine, D. A. Phillips, D. W. Rains and R. C. Huffaker, pp. 363–384. Plenum, New York.

Tarand, J. J., N. R. Kreig and J. Döbereiner (1978). A taxonomic study of the *Spirillum lipoferum* group, with descriptions of a new genus, *Azospirillum* gen. nov., and two species, *Azospirillum lipoferum* (Beijerinck) comb. nov. and *Azospirillum brasilense* sp. nov. *Can. J. Microbiol.*, **24**, 967–980.

Thorneley, R. N. F. and D. J. Lowe (1981). Pre-steady state kinetic studies with nitrogenase from *Klebsiella pneumoniae* using quenched-flow to monitor NH_3 and H_2 formation. In *Current Perspectives in Nitrogen Fixation*, ed. A. H. Gibson and W. E. Newton, p. 360. Australian Academy of Science, Canberra.

Thorneley, R. N. F. and R. R. Eady (1977). Nitrogenase of *Klebsiella pneumoniae*. Distinction between proton-reducing and acetylene-reducing forms of the enzyme: effect of temperature and component protein ratio on substrate reducing kinetics. *Biochem. J.*, **167**, 457–461.

Thorneley, R. N. F., R. R. Eady and D. J. Lowe (1978). Biological nitrogen fixation by way of an enzyme-bound dinitrogen-hydride intermediate. *Nature (London)*, **272**, 557–558.

Tyler, B. (1978). Regulation of the assimilation of the nitrogen compounds. *Annu. Rev. Biochem.*, **47**, 1127–1162.

van Berkum, P. and B. B. Bohlool (1980). Evaluation of nitrogen fixation by bacteria in association with roots of tropical grasses. *Microbiol. Rev.*, **44**, 491–517.

Veeger, C., C. Laane, G. Scherings, L. Matz, H. Haaker and L. van Zeeland-Wolbers (1980). Membrane energization and nitrogen fixation in *Azotobacter vinelandii* and *Rhizobium leguminosarum*. In *Nitrogen Fixation*, ed. W. E. Newton and W. H. Orme-Johnson, vol. I, pp. 111–137. University Park Press, Baltimore, MD.

Vincent, J. M. (1980). Factors controlling the legume–*Rhizobium* symbiosis. In *Nitrogen Fixation*, ed. W. E. Newton and W. H. Orme-Johnson, vol. II, pp. 103–129. University Park Press, Baltimore, MD.

Virtanen, A. I. (1938). *Cattle Fodder and Human Nutrition*. Cambridge University Press, London.

Wall, J. D. and H. Gest (1979). Derepression of nitrogenase activity in glutamine auxotrophs of *Rhodopseudomonas capsulata*. *J. Bacteriol.*, **137**, 1459–1463.

Wall, J. D., P. F. Weaver and H. Gest (1975). Genetic transfer of nitrogenase–hydrogenase activity in *Rhodopseudomonas capsulata*. *Nature (London)*, **258**, 630–631.

Watanabe, I. (1981). Biological nitrogen fixation associated with wetland rice. In *Current Perspectives in Nitrogen Fixation*, ed. A. H. Gibson and W. E. Newton, pp. 313–316. Australian Academy of Science, Canberra.

Watanabe, I., C. R. Espinas, N. S. Berja and C. B. Almiagno (1977). Utilization of the *Azolla–Anabaena* complex as a nitrogen fertilizer for rice. *Int. Rice Res. Inst. Res. Paper Ser.*, **11**, 1.

Wilson, P. W. (1958). Asymbiotic Nitrogen Fixation. In *Encyclopedia of Plants*, ed. W. Ruhland, pp. 9–47. Springer-Verlag, Berlin.

Winogradsky, S. (1893). Sur l'assimilation de l'azole gazeux de l'atmosphere par les microbes. *C. R. Hebd. Seances Acad. Sci.*, **116**, 1385–1388.

Winter, H. C. and R. H. Burris (1976). Nitrogenase. *Annu. Rev. Biochem.*, **45**, 409–426.

Wolff, T. E., J. M. Berg, C. Warrick, K. O. Hodgson, R. H. Holm and R. B. Frankel (1978). The molybdenum–iron–sulphur cluster complex $[Mo_2Fe_6S_9(SC_2H_5)_8]^{3-}$. A synthetic approach to the molybdenum site in nitrogenase. *J. Am. Chem. Soc.*, **100**, 4630–4632.

Wolff, T. E., J. M. Berg, K. O. Hodgson, R. B. Frankel and R. H. Holm (1979). Synthetic approaches to the molybdenum site in nitrogenase. Preparation and structural properties of the molybdenum–iron–sulfur-'double-cubane' cluster complexes $[Mo_2Fe_6S_8(SC_2H_5)_9]^{3-}$ and $[Mo_2Fe_6S_9(SC_2H_5)_8]^{3-}$. *J. Am. Chem. Soc.*, **101**, 4140.

Wolk, C. P. (1980). Heterocysts, ^{13}N, and N_2-fixing plants. In *Nitrogen Fixation*, ed. W. E. Newton and W. H. Orme-Johnson, vol. II, pp. 279–292. University Park Press, Baltimore, MD.

Wolk, C. P. and E. Wojciuch (1971). Photoreduction of acetylene by heterocysts. *Planta*, **97**, 126–134.

Yoch, D. C. (1978). Nitrogen fixation and hydrogen metabolism by photosynthetic bacteria. In *The Photosynthetic Bacteria*, ed. R. K. Clayton and W. R. Sistrom, pp. 657–676. Plenum, New York.

Yu, P.-L., J. Cullum and G. Drews (1981). Conjugational transfer systems of *Rhodopseudomonas capsulata* mediated by R plasmids. *Arch. Microbiol.*, **128**, 390–393.

Zimmerman, R., E. Munck, W. J. Brill, V. K. Shah, M. T. Henzl, J. Rawlings and W. H. Orme-Johnson (1978). Nitrogenase X. Mössbauer and EPR studies on reversibly oxidized MoFe protein from *Azotobacter vinelandii* OP. Nature of the iron centers. *Biochim. Biophys. Acta*, **537**, 185–207.

Zumft, W. G. and L. E. Mortenson (1975). The nitrogen-fixing complex of bacteria. *Biochim. Biophys. Acta*, **416**, 1–52.

Zumft, W. G., K. Alef and S. Mummier (1981). Regulation of nitrogenase activity in Rhodospirillaceae. In *Current Perspectives in Nitrogen Fixation*, ed. A. H. Gibson and W. E. Newton, pp. 190–193. Australian Academy of Science, Canberra.

Zürrer, H. and R. Bachofen (1979). Hydrogen production by the photosynthetic bacterium *Rhodospirillum rubrum*. *Appl. Environ. Microbiol.*, **37**, 789–793.

8

Mycorrhizae: Applications in Agriculture and Forestry

B. KENDRICK
University of Waterloo, Ontario, Canada
and
S. BERCH
Université Laval, Québec, Canada

8.1 INTRODUCTION

The evolution of the eukaryotic cell, which made possible the great plethora of living things that now inhabit the earth, is thought to have resulted from the repeated development of mutualistic symbioses between prokaryotes (Margulis, 1981). This ancient tradition of mutualism has been maintained among the eukaryotes, and when green plants first invaded the land more than four hundred million years ago, their ultimate success in taking this giant step may well have been due to their establishment of an intimate and mutually beneficial relationship with fungi (Pirozynski and Malloch, 1975).

While at first sight this may seem improbable, on reflection it is not such an unlikely partnership: the early land plants could photosynthesize effectively even before the evolution of leaves, but they undoubtedly had difficulty in accumulating the water and mineral nutrients they needed to sustain this activity. The filamentous fungi, which had themselves only recently emerged from the water, were perfectly adapted for ramifying through the soil and scavenging for those very things, but had an absolute requirement for energy-rich carbon compounds of the kind produced by the plants. We now have some proof of the antiquity of this relationship. Fossils of Devonian plants have been found to contain well-preserved fungal structures virtually indistinguishable from those that can be seen today in the roots of many healthy modern plants (Kidston and Lang, 1921). Similar structures occur in the rhizomes of a Carboniferous fern (Andrews and Lenz, 1943), and have in fact been recorded sporadically in the underground parts of Palaeozoic, Mesozoic and Cenozoic plants.

Man first became aware of this long tradition of plant–fungus interaction about a century ago, when several biologists noticed that some plant roots, while extensively invaded by fungi, did not become diseased. The name 'mycorrhiza', literally translated 'fungus root', was coined by Frank in 1885. Although the beneficial nature of the relationship was not established until much later, there is now an overwhelming body of evidence showing that in many situations (particularly in infertile soils) mycorrhizal plants grow better than non-mycorrhizal plants (Gerdemann, 1968; Kleinschmidt and Gerdemann, 1972; Mosse, 1973; Sanders *et al.*, 1975; Hayman, 1978; Mikola, 1980; Schoenbeck, 1980; Schoenbeck and Dehne, 1981). It has been demonstrated that the hyphae of the fungal symbionts permeate the soil and obtain scarce and relatively immobile elements, especially phosphorus, but also nitrogen, potassium, copper and zinc, more effectively than the root hairs on a non-mycorrhizal plant (Gerdemann, 1968; Kleinschmidt and Gerdemann, 1972; Mosse, 1973).

But long before the true nature of the mycorrhizal association was suspected, the constant association of particular fungi with certain trees had been noted. In the third century BC, Theophrastus commented on fungi which grew from the roots of oaks and other trees (Kelley, 1950), and more recently, unsung naturalists coined such suggestive names as *dubovik* or 'oak mushroom' for *Boletus luridus* in Russia, and *Larchenmilchling*, or 'larch milky cap' for *Lactarius porinsis* in Germany. From the early 1800s the French had been encouraging the growth of truffles, which were (and still are) in great demand for *haute cuisine*, by planting oak trees in particular kinds of soil.

Interest in this symbiosis has escalated dramatically in recent years, partly because of what we have learned about the benefits of mycorrhizae, and partly because of economic and geopolitical events.

On the geopolitical and economic fronts, several factors emerged which stimulated interest in mycorrhizae. (1) The world-wide energy crisis: the large-scale manufacture, transportation and application of fertilizers are energy-intensive processes, so drastic increases in the price of oil have made fertilizers more expensive. (2) The population explosion: this has necessitated increased agricultural production, which can be achieved either through the 'green revolution', using new races of crop plants that require heavy and repeated applications of fertilizer, or by bringing marginal land into cultivation, which is again often assumed, according to conventional wisdom, to call for extensive fertilization. (3) The enormous requirements of industrial societies for raw materials: these have resulted in widespread destruction of natural soils and vegetation. (4) The vast quantities of waste products generated by industrial and domestic consumption: these have led to massive pollution of various kinds. Among other things, these undesirable aspects of unregulated private (and also, sometimes, of government) enterprise eventually engendered the environmental movement, which has successfully lobbied for efforts at restoration of many damaged ecosystems. Each of these phenomena has helped to increase interest in mycorrhizae.

On the academic front, we learned that some plants cannot become established or grow normally without an appropriate mycobiont (Rhodes, 1980), and that as compared to non-mycorrhizal plants, those of the same species but with a suitable fungal partner need less fertilizer (Powell, 1977); withstand heavy metal and acid rain pollution better; grow better on the infertile soils of marginal lands (Nicolson, 1960), or on mine spoils and other areas in need of revegetation (Schramm, 1966; Daft and Nicolson, 1974; Daft *et al.*, 1975), and at high elevations (Crush, 1973; Powell, 1975); are more resistant to soil-borne diseases; withstand high soil temperatures better; grow better in soils of high salinity; can tolerate greater extremes of soil pH; and survive transplant shock better. We will explore each of these areas in some detail in the appropriate section of the chapter. As these positive features emerged, the rapidly accumulating data permitted other generalizations. It became apparent that mycorrhizal fungi are ubiquitous, and it is now estimated that over 90% of all higher plants are normally mycorrhizal.

The dramatic increase in research effort being devoted to mycorrhizae can be demonstrated by two simple statistics. At the first North American Conference on Mycorrhizae in 1969, 26 communications were presented. At the fifth NACOM in 1981, the number had risen to 263. In 1970, the abstracting service of the Commonwealth Agricultural Bureaux listed six papers on mycorrhizae. By 1979, the CAB listed 314 mycorrhizal references (Nemec, 1982).

There are several different kinds of mycorrhiza, but this chapter will consider only the two most important and widespread forms, which are those constantly found in association with our agricultural and forest crops. These mycorrhizae differ widely both in their structure and in the systematic position of the fungi involved. By far the commoner of the two is the vesicular–arbuscular endotrophic mycorrhiza (VAM). This kind of symbiosis involves what appears to be a relatively small number of fungi which will grow only in association with plant roots (*i.e.* they are obligate biotrophs) and seem not to reproduce sexually. These attributes made experimental work with VAM difficult, and left the fungi in something of a taxonomic limbo, although they are grouped in a family called the Endogonaceae, which is tentatively placed in the phylum Zygomycota.

Nevertheless, they do form mycorrhizae with hundreds of thousands of plant species, including almost all our field crops, with such exceptions as sugar beet and the brassicas, so investigators persisted. At first spores for experimental work had to be painstakingly sieved and picked out by hand from naturally infected soil. Later a variety of flotation and density-gradient techniques were devised in order to speed up the collection process. Now even more efficient ways of producing inoculum have been found, and are described in the appropriate section of this chapter. We anticipate that these will make the application of VA mycorrhizal fungi in agriculture and revegetation a practical process.

The second kind of mycorrhiza is the ectotrophic mycorrhiza. This kind of symbiosis involves a fairly large number of fungi which, although they are usually found associated with tree roots, can in most cases be grown in pure culture, and are almost all known to produce sexual fructifications *in vivo*, if not universally *in vitro*. These fungi are mainly Basidiomycetes, though a few are Ascomycetes: both belong to the phylum Dikaryomycota. Since most of them can be grown in axenic culture, they are much easier to work with, and it is therefore not too surprising that their application in forestry is further advanced than that of the VA mycorrhizae in agriculture.

The mycorrhizal symbiosis, whether ectotrophic or endotrophic, must have three basic functioning components: (1) fungal mycelium exploring large volumes of soil and retrieving mineral nutrients; (2) a fungus–plant interface where the exchange of chemicals occurs; and (3) plant tissues which produce and store carbohydrates. The success of mycorrhizae is probably due to three factors: (1) their long evolutionary history: they predate the root hair and have had a long time to beome finely tuned to their environment; (2) their economy: a delicate but extensive fungal mycelium requires less investment than a macroscopic system of fine roots and root hairs (Sanders and Tinker, 1973); (3) their efficiency: the fungi may have surface phosphatases that enable them to obtain soil phosphates more rapidly than non-mycorrhizal roots can (Bartlett and Lewis, 1973). What is abundantly clear is that we are just now on the threshold of important developments in the exploitation of both VA mycorrhizae and ectomycorrhizae. We hope this chapter will make more scientists aware of that potential.

We note that the synergistic attributes of mycorrhizal fungi and their hosts cannot be pooled by gene cloning and transformation. The properties of mycorrhizal fungi that make them valuable to their hosts are inherent in the morphology of the fungi, especially in their diffuse, tubular, nutrient-absorbing thalli.

8.2 COMPARISON OF ECTOTROPHIC AND VESICULAR–ARBUSCULAR MYCORRHIZAE

The two kinds of mycorrhizae have essentially the same functions. Acting as an interface between plant and fungus, they are the site of an ongoing exchange. Plant photosynthates are translocated to the fungus in return for phosphorus and other inorganic nutrients obtained by the fungal hyphae from a larger volume of soil than that to which the roots of non-mycorrhizal plants have access. But once we begin to consider other aspects of these phenomena, their numbers, their relationships, their morphology and development, their host ranges, we find little congruence between them.

The physical contrast is exemplified in Figure 1, which shows a vertically sectioned root with diagrammatic representations of characteristic endo- and ecto-mycorrhizal structures. Vesicular–arbuscular mycorrhizal (VAM) fungi cause no macroscopic changes in the roots they inhabit, and obviously represent a relatively minor investment on the part of their host plants. Even their fructifications are on a microscopic scale, with the exception of those species that form sporocarps. The ectomycorrhizal (EM) fungi, on the other hand, produce macroscopic changes in the morphology of their host roots, and their visible mantles and their numerous, large and often colourful fruit bodies just as obviously represent a considerable investment on the part of their host plants. This is partly explained by the fact that the roots which become ectomycorrhizal are usually perennial. Since they are expected to be a long-term, re-usable investment they can involve much more biomass than a short-lived phenomenon like the VAM arbuscule, which is often associated with the roots of annual herbaceous plants. We have evidence that ectotrophic mantles act as nutrient sinks for both fungus and host, but no such function is ascribed to the VA mycorrhiza.

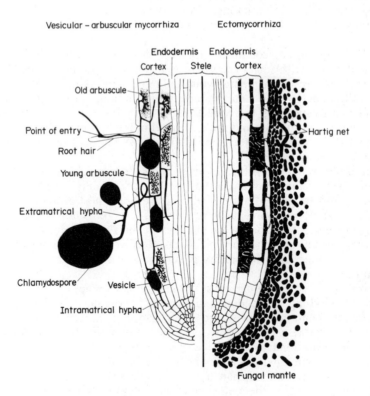

Figure 1 Vertical section of a root with a schematic representation of vesicular–arbuscular endotrophic myccorrhiza (left) and ectomycorrhiza (right)

Most EM fungi produce sexual fructifications only during a short season, which is determined by the climate rather than the time of infection, though asexual sclerotia are produced by some species throughout the growing season. VAM fungi may produce spores or sporocarps whenever their host plants are growing, but there tends to be a population peak toward the end of the season. VAM spores may germinate at any time if ambient conditions are suitable and, in some species, if dormancy has been broken, so we must assume that they are also being continuously replaced.

While both kinds of mycorrhiza involve filamentous fungi, the two groups of fungi are taxonomically unrelated. All VAM fungi are still treated as members of a single family having affinity with the phylum Zygomycota, though this is by no means fully established. Most EM fungi are members of a completely different evolutionary line, the phylum Dikaryomycota, mostly hymenomycetous or gasteroid holobasidiomycetes with a few operculate, unitunicate ascomycetes, though a few Zygomycetes form ectomycorrhizae with conifers and eucalypts. Individual hyphae of the VAM fungi penetrate the root cells of their host plant, and the interfaces at which exchange of materials takes place, the finely branched arbuscules, develop within individual cells of the root cortex. This is why VA mycorrhizae are often called endomycorrhizae. By definition, hyphae of EM fungi mass around the feeder roots of their host to form a macroscopic sheath, and grow between and around, but never within, the cells of the root cortex to form the Hartig net, which is the functional interface between host and fungus. The extracellular nature of this interface gives the mycorrhiza the name ectomycorrhizae (though under some conditions, limited intracellular penetration may occur).

The different degrees of intimacy inherent in the two types of mycorrhiza may be reflected in the growth requirements of the two kinds of fungi involved. VAM fungi are obligately biotrophic, and can be grown only in dual culture, with a suitable host plant. Many, probably most, EM fungi can be grown in axenic culture, though a variety of vitamins and other growth substances must often be added to the culture medium.

Trees in boreal forests are usually ectomycorrhizal, while those in tropical rainforests usually have vesicular–arbuscular mycorrhizae. The reasons for this contrast are worth considering. The tree flora in tropical rainforests is extremely diversified. It is common to find 100 different species of trees per hectare (Wyatt-Smith, 1953), and the total number of woody species may approach 400 ha^{-1} (Longman and Jenik, 1974). As many as 208 species of trees have been recorded in 1000 m^2 (White, 1983), and individual specimens of the same tree species may be 800 m apart. This diversity may not be reflected in the VA mycorrhizal fungi associated with the roots of the trees. However, the VAM fungi of tropical forests are largely unknown, and recent observations suggest that many taxa await discovery and description. New techniques may be necessary to study these, since turnover is high and production may be seasonal. The described world mycota comprises about 100 species at present, and even if, as Trappe (1982) predicts, this number doubles by the year 1990 (largely as a result of the renascent interest in these fungi), they will still not even remotely approach the diversity of fungi involved in ectotrophic mycorrhizae.

The tree flora of the boreal forest is extremely restricted. This great band of forest that stretches completely across North America, Europe and Asia is essentially made up of representatives of only six genera; spruce, hemlock, fir, pine, willow and birch. Sometimes, pure stands of a single coniferous species extend for many kilometres. Within this extremely uniform plant community, there are hundreds of different species of ectomycorrhizal fungi. We can emphasize this comparison by giving approximate world totals for the number of species involved on each side of the mycorrhizal equations: VA mycorrhizae, 200 fungi and 300 000 plants; ectomycorrhizae, 5000 fungi and 2000 plants. This seems to show which partner has undergone adaptive radiation to cope with habitat variability. It has been suggested that a tropical rainforest is like a boreal forest standing on its head. In the tropical forest the diversity is above ground, in the boreal forest it is below ground. If this is true, it represents an enormous reversal of roles, and is worthy of further study for what it will tell us about the evolution of mycorrhizae and of trees.

Vesicular–arbuscular mycorrhizae are, as we mentioned in the introduction, by far the oldest-established and commonest form of mycorrhiza. They have been detected in some of the oldest land plants, *Rhynia* and *Asteroxylon*, they occur scattered throughout the subsequent fossil record, and are present in the roots of the great majority of living plants. From the sample of plants so far examined, it is estimated that they will be found to occur in about 90% of all extant vascular plants. In fact, the most logical way of discussing their host range is to list the groups of plants that do not normally have VA mycorrhizae (see Section 8.4.1).

The only woody family among the entirely non-mycorrhizal plants is the Proteaceae, which is characterized by finely branched roots and extensive roothairs. Many of the rest are weeds, vigorous pioneer herbaceous annuals with highly opportunistic lifestyles. They germinate quickly in disturbed or depauperate soils, grow rapidly, and in some cases flower and set seed in a phenomenally short period. This means that they have no time to sit around and wait for the local VAM fungi to find and colonize their roots — indeed in some of the habitats occupied by weeds there may be little or no VAM inoculum — so the plants have evolved systems of roothairs that enable them to get along without mycobionts. The endomycorrhizal relationship is extremely old, and it is hardly surprising that some plants may now, after hundreds of millions of years, be evolving alternate strategies. That this should happen mainly among the herbaceous annuals is even more understandable when we consider that they are the most recently evolved group of plants. In addition to their weedy habit, members of the Brassicaceae and related families may have evolved chemical defences to repel herbivorous animals, and thereby discouraged their now inessential mycorrhizal fungi. Glenn (1982) suggested that these families lack functional mycorrhizae because of the presence of glucosinolates and their hydrolysis products, isothiocyanates, in and around their roots.

This leads to some speculations on the reasons for the emergence of the ectotrophic mycorrhiza. Pirozynski (personal communication) suggests that, as land plants and insects evolved, the effects of leaf-eating insects became more and more severe. This selection pressure favoured those plants which contained unpalatable or toxic substances such as tannins, phenols and resins. These substances, however, if they became disseminated throughout the plant, as they did in the Pinaceae, Myrtaceae, *etc.*, could be inimical to the continued presence of endomycorrhizal fungi. The less intimate association involved in ectomycorrhizae may well have evolved to fill the void left when the VAM fungi were effectively expelled from some of their hosts.

It is well known that EM fungi produce plant growth hormones, notably auxins, in axenic culture. In addition, they produce phenols that have been demonstrated to have a regulatory action on auxins. This suggests that the EM fungus may have a vital role in controlling root development in its host, a function presumably absent from the VAM symbiosis. This may correlate with the obviously modified structure of ectomycorrhizal roots, and the lack of such modification in VA mycorrhizal roots.

Another instructive comparison can be drawn between the distributions of the two kinds of mycorrhiza. Although our knowledge of the agarics is far from complete, we have considerable biogeographical data on them, information that has been supplemented by our attempts to plant conifers in parts of the world to which they were not indigenous. In many cases, these attempts failed initially, simply, we now know, because no appropriate ectomycorrhizal fungi were present. The dependence of the fungi on their tree hosts is so complete that despite their long-range airborne propagules, they were obviously unable to establish themselves on many islands, or in many areas of the southern hemisphere, in advance of their hosts. Thus the only way to establish either partner is to introduce both simultaneously. The VAM fungi suffer under no such disability. Although they are obligately biotrophic, their host range, at least *in vitro*, is so enormous, and their history so long (predating, for example, the breakup of Pangaea) that wherever there is soil capable of supporting plant growth, there are likely to be VAM fungi.

Ectomycorrhizal hosts tend to produce small, dry, wind-dispersed fruits, which are most suited to short-range dispersal. This is in marked contrast to the large, fleshy and appetizing fruits of many tropical endomycorrhizal trees, which stand a good chance of becoming established even if transported over fairly long distances by animal or bird vectors.

Ectomycorrhizal tree species tend to occur at timberline in both Northern and Southern Hemispheres (*viz.* the *Nothofagus* forests in the Southern Alps of New Zealand) and to grow better than endomycorrhizal plants on soil that is poor or disturbed. The disturbance may be anthropogenic, due to mining, or natural, due to landslides or even glaciation. The implication is that ectomycorrhizal trees have a competitive edge in marginal conditions, and that major climatic shifts in the recent geologic past (especially the ice ages) have been instrumental in the spread of ectomycorrhizal trees, while the endomycorrhizal trees retreated to areas of warmer climate. However, this generalization cannot be applied to the herbaceous plants. In the recolonization of areas devastated by the 1980 eruption of Mount St. Helens, the first mycorrhizal plants were endomycorrhizal species.

8.3 ECTOMYCORRHIZAE

8.3.1 Systematics of Ectomycorrhizal Fungi and their Hosts

In the preceding section, we pointed out that the species of ectomycorrhizal fungi outnumber those of their hosts. An estimated 5000 fungi can establish ectotrophic mycorrhizae with about 2000 woody hosts. We obviously cannot give any detailed coverage of the very large number of fungi so far known to be ectomycorrhizal. We think it is appropriate, however, to sketch in the broad taxonomic outlines. All ectomycorrhizal fungi, with only one or two exceptions, belong to the mainstream fungal phylum, Dikaryomycota. The great majority are Basidiomycetes. Miller

(1982) records mycorrhiza-forming activity in representatives of 73 basidiomycete genera distributed among 27 families in nine orders. These orders, families, and the number of mycorrhizal genera known for each family are recorded in Table 1. These fungi are holobasidiomycetes: agarics (mushrooms, toadstools); agarics which have evolved, or are evolving, into a gasteroid habit and have in many cases become hypogeous (puffballs, false truffles); and non-gilled hymenomycetes from among the club fungi, chanterelles, tooth fungi and what are called the resupinate hymenomycetes (paint fungi). There are also ectomycorrhizal representatives of 16 unitunicate ascomycete genera from eight families and two orders. All but one of the ascomycete genera involved are hypogeous, and all but one are related to the operculate Discomycetes (cup fungi).

Table 1 Taxonomic Distribution of Ectomycorrhizal Fungi[a]

Phylum: Dikaryomycota			Melanogastrales	
Subphylum: Basidiomycotina			Melanogastraceae	(2)
Order: Agaricales			Leucogastraceae	(2)
Family: Amanitaceae	(2)[b]		Sclerodermatales	
Hygrophoraceae	(1)		Sclerodermataceae	(2)
Tricholomataceae	(6)		Astraceae	(1)
Entolomataceae	(1)		Aphyllophorales	
Cortinariaceae	(5)		Cantharellaceae	(3)
Paxillaceae	(2)		Clavariaceae	(?)
Gomphidiaceae	(5)		Corticiaceae	(3)
Boletaceae	(13)		Hydnaceae	(?)
Strobilomycetaceae	(3)		Thelephoraceae	(2)
Russulales			Subphylum: Ascomycotina	
Russulaceae	(5)		Order: Pezizales	
Elasmomycetaceae	(4)		Family: Pezizaceae	(1)
Gautieriales			Balsamiaceae	(3)
Gautieriaceae	(1)		Geneaceae	(1)
Hymenogastrales			Helvellaceae	(1)
Octavianinaceae	(4)		Pyronemataceae	(3)
Hymenogastraceae	(1)		Terfeziaceae	(4)
Rhizopogonaceae	(2)		Tuberaceae	(2)
Hydnangiaceae	(1)		Elaphomycetales	
Phallales			Elaphomycetaceae	(1)
Hysterangiaceae	(1)			
Lycoperdales				
Mesophelliaceae	(1)			

[a] Data from Miller (1982).   [b] Number of confirmed ectomycorrhizal genera.

Some of the evidence that particular fungi are mycorrhizal is rather circumstantial, based on field observations, though enough is now known about the patterns of growth and fruiting of ectomycorrhizal fungi to make these putative connections fairly reliable. For example, although no representatives of the Clavariaceae or Hydnaceae have yet been confirmed as EM fungi by synthesizing mycorrhizae in culture, field observations suggest that this confirmation may not be long delayed.

Most ectomycorrhizal fungi can be accurately identified only from their macroscopic fruit bodies (basidiomata or ascomata), which are produced during a relatively short season each year. General technical works for the identification of ectomycorrhizal fungi are those of Singer (1975) (the most comprehensive work on the genera of Agaricales yet published), Moser (1978; excellent, but in German), Ainsworth *et al.* (1973a, 1973b), Smith *et al.* (1979), Smith and Smith (1973), Miller and Farr (1975) and Watling and Watling (1980). Good illustrations can be found in Pomerleau (1980), Smith (1975), Miller (1980), Miller and Miller (1980), Dahncke and Dahncke (1979) and Cetto (1978, 1979a, 1979b).

Some of the groups have been covered in greater detail. Schaeffer (1933) and Romagnesi (1967) gave extensive treatments of *Russula*, Hesler and Smith (1979) monographed *Lactarius* in North America and Pegler and Young (1979) dealt with the gasteroid Russulales. Miller (1973) revised the Gomphidiaceae and their gasteroid relatives. North American boletes have been extensively treated by Smith and Thiers (1971), Snell and Dick (1971) and Thiers (1975). Thiers and Trappe (1969) dealt with *Gastroboletus*, and Watling and Watling (1980) list other bolete literature. Singer and Smith (1960) covered the gasteroid agarics. Jülich (1972) and Eriksson and Ryvarden (1973) compiled many of the Corticiaceae. The hypogeous Ascomycetes have been revised by Gilkey (1939), Hawker (1954) and Trappe (1979).

Many EM fungi belong to cosmopolitan, highly speciated agaric genera such as *Russula*, *Lactarius*, *Cortinarius*, *Amanita*, *Tricholoma*, *Inocybe* and *Laccaria*, which, despite much attention from agaricologists, are still incompletely known. Moser (1978) estimates that *Cortinarius* has 2000 species. One of the features of the great expansion in our knowledge of EM fungi that has taken place in recent years is the increasing number of ectomycorrhizal hypogeous Basidiomycetes, which are frequently, and perhaps logically, gasteroid relatives of the conspicuous epigeous agarics. Some families are almost entirely ectomycorrhizal — Boletaceae, Gomphidiaceae, Russulaceae, Strobilomycetaceae, Cantharellaceae — as are all or most species of the genera *Amanita*, *Armillaria*, *Astraeus*, *Cortinarius*, *Hebeloma*, *Inocybe*, *Laccaria*, *Pisolithus*, *Ramaria*, *Rozites*, *Scleroderma*, *Thelephora* and *Tricholoma*, as well as almost all hypogeous Basidiomycetes.

At the other end of the scale, although a few scattered members of the Aphyllophorales are ectomycorrhizal, the vast majority of the members of this diverse order are vigorous saprobes, with the enzymatic capability to degrade lignin and/or cellulose. In the Ascomycetes, members of the hypogeous Tuberaceae are probably all mycorrhizal. One phylum-wide generalization may be in order: if a fungus produces macroscopic, hypogeous fruit bodies, then it is very likely to be mycorrhizal, whether it is a basidiomycete or an ascomycete.

So many different fungi are involved in ectomycorrhizal partnerships, and ectomycorrhizal hosts are so relatively few, that distributional patterns of these interdependent organisms are bound to be rich in biogeographic significance, and important to would-be mycorrhiza synthesizers. It is unlikely that the conifers, for example, could spontaneously make the leap across wide water barriers to oceanic islands, since they would have to be accompanied by their mycobiont(s), and the evidence is that they have not been able to achieve this (Pirozynski, personal communication). Since many of the EM fungi coevolved with their hosts, they will be found only within the distribution range of the hosts. They may not occur throughout that range, however, because they may have been replaced in some areas by other EM fungi.

The Gomphidiaceae and Rhizopogonaceae were originally restricted to the Northern Hemisphere, with their conifer hosts. Species of the bolete genera, *Suillus*, *Fuscoboletinus* and *Boletopsis* are associated only with Northern Hemisphere Pinaceae, but other bolete genera are more cosmopolitan. *Mesophellia*, the sole genus of the Mesophelliaceae, and the only known mycorrhizal member of the Lycoperdales, is restricted to Australia, where its phytosymbiont is *Eucalyptus*. The group of Agaricales which evolved mycorrhizae with the Pinaceae, Betulaceae, Fagaceae and Salicaceae in the Northern Hemisphere seem also to have done so with the ancestors of *Nothofagus* before these reached the Southern Hemisphere. Conceivably, the mycobionts associated with *Eucalyptus* and other Myrtaceae in the tropics and the Southern Hemisphere are derived from those which arrived there with *Nothofagus* (Malloch *et al.*, 1980).

The uniformity of phytobionts in the boreal forest is reflected also in forests of Caesalpinioid legumes in the tropics, and of *Eucalyptus* in Australia (Malloch *et al.*, 1980). In each case, the diversity lies in the mycobionts. The dominance of ectomycorrhizal trees over vast areas of the Northern Hemisphere was promoted by various episodes of cold or dry climate during the Tertiary or the Cenozoic, since the ectomycorrhizal symbionts are often well adapted to climatic stress and to poor soils, as explained in Section 8.2.

Some EM fungi have a wide host range. Typical examples of this group are *Amanita muscaria*, *Boletus edulis*, *Cantharellus cibarius*, *Cenococcum geophilum*, *Laccaria laccata*, *Pisolithus tinctorius* and *Thelephora terrestris*. These would appear to be likely candidates for exploitation. Others appear to be more selective, and some are virtually host-specific, *e.g.* *Suillus grevillei* fruits only under *Larix*, *Suillus lakei* only with *Pseudotsuga menziesii*, *Lactarius obscuratus* with *Alnus*, *Gomphidius vinicolor* with *Pinus* and *Cortinarius hemitrichus* with *Betula*.

A single tree may have several or many different ectomycorrhizal partners on its roots at the same time; Trappe (1977) gives a range of from five to dozens of different fungi, and these mycobionts may give place to others as the tree ages. Thus, a single tree species may have a very large number of potential ectomycorrhizal partners. Trappe (1977) estimated that Douglas fir (*Pseudotsuga menziesii*) might be able to form ectomycorrhizae with as many as 2000 different species of fungi. The potential diversity in these partnerships is even greater than that number would indicate. Different isolates of the same fungus may also behave very differently as mycobionts with the same tree species. Marx (in Trappe, 1977) reported that of 20 isolates of *Pisolithus tinctorius*, one was much better than all the rest as a mycorrhizal partner of southern pines, while some isolates did not form mycorrhizae at all with these hosts.

This leads us to a consideration of the full host spectrum of the ectomycorrhizal fungi. Briefly, this can be described as follows: Gymnosperms—Pinaceae and some Cupressaceae; Angiosperms—a few monocots; *Kobresia bellardi* (Fontana, 1963), *Kobresia myosuroides*

(Haselwandter and Read, 1980), *Euterpe globosa* (Edmisten, 1970) and *Festuca rubra* (Read and Haselwandter, 1981); all or some members of 21 dicot families; all members of the Fagaceae, Betulaceae, Salicaceae, and Dipterocarpaceae subfamily Dipterocarpoideae, and most Myrtaceae; also the tribes Amherstieae and Detarieae of the family Caesalpinioideae, some Mimosoideae and Papilionoideae, *Coccoloba* of the Polygonaceae, *Neea* and *Pisonia* of the Nyctaginaceae, scattered representatives of the Aceraceae, Bignoniaceae, Combretaceae, Euphorbiaceae, Juglandaceae, Rhamnaceae, Rosaceae, Sapindaceae, Sapotaceae, Tiliaceae, Ulmaceae, and even some ferns.

The 2000 plants comprising the above list have several features in common. They are almost all woody and perennial. Many of them grow in extensive pure stands. Many are indigenous to the Northern Hemisphere, and some are the main constituents of the boreal forest. The Pinaceae are the single most important ectomycorrhizal family, since they cover vast areas of the globe, and are harvested and replanted in astronomical numbers each year. It is the hope of enriching that harvest, and improving the success of the replanting, that has kindled our interest in the biotechnological exploitation of ectomycorrhizal fungi.

8.3.2 Morphology and Development of Ectomycorrhizae

In describing the development of vesicular–arbuscular mycorrhizae (Section 8.4.2) we are basically concerned only with the activities of the fungus, since there is little visible change in the roots. But the formation of ectomycorrhizae involves morphological responses by both host and fungus, so this process must be described in terms of both partners. The various morphological types of ectomycorrhizae have been described in detail by Chilvers (1968) and Zak (1969; 1971a,b,c; 1973). Zak (1973) gave a number of excellent photographs which should make it easy for anyone to recognize an ectomycorrhiza on sight.

Ectomycorrhizae normally begin to develop 1 to 3 months after the tree seed germinates. Although EM fungi may colonize the long roots, they occur most commonly on the 'short' or 'feeder' roots that develop from the sides of those long roots in the upper, humic layers of the soil. In many cases the fungus is spreading through the soil from a nearby mycorrhizal root and can call upon ample reserves of energy derived from its host. If, however, isolated inoculum of spores or mycelium is involved, it can apparently subsist in the rhizosphere (the area around the roots) on root exudates before actual infection occurs (Bowen and Theodorou, 1973). Mycorrhization occurs when an EM fungus encounters a feeder root of a hospitable tree species. Colonization will occur only in a specific unsuberized zone of the root, behind the root tip, and before the region in which the primary cortex has begun to deteriorate. Although the root tip is not invaded, it eventually becomes completely covered by the fungal mantle.

Fortin *et al.* (1980), using the root pouch technique, observed that mycelial inoculum of *Pisolithus tinctorius* initiated the mycorrhization of short root primordia of *Pinus strobus* within 5 days. This involved: (1) the penetration of hyphae between the cells of the root cortex and the formation of the characteristic Hartig net, in which hyphae completely surround cortical cells (Figure 1); and (2) the establishment of a mantle of hyphae around the outside of the root. A *Cenococcum geophilum* isolate took 2 weeks to reach a comparable stage. The growth rate of the short roots was increased, and they eventually branched dichotomously. We should point out that changes in growth rate and pattern are typical of ectomycorrhizae, and are caused by the liberation of growth hormones from the fungus (Slankis, 1973). These changes will occur if auxins are applied in the absence of an EM fungus.

Marks and Foster (1973) described two patterns of mycorrhization. In primary colonization, typical of an uncolonized root in colonized soil, the mantle (a mass of fungal hyphae surrounding the root) begins to form at about the time the first leaves appear. When the xylem vessels begin to differentiate in the vascular cylinder at the centre of the root, hyphae penetrate between the cells of the cortex and form a Hartig net (a single layer of closely packed fungal hyphae). In most cases, this net spreads slowly inward until it reaches the endodermis, which effectively bars any further penetration, though in some angiosperms the Hartig net may not penetrate beyond the first layer of cortical cells. In secondary colonization, typically derived from mycelia already established on other parts of the same root system, the mantle often simply spreads along a long root and rapidly envelops any emerging short roots, sending hyphae into their cortex to set up the vital Hartig net interface.

The penetration of hyphae between the cells of the root cortex does not lead to plasmolysis or other deleterious cytological alterations in the host cells. As the hyphae insinuate themselves between the cortical cells, the latter simply separate at the middle lamella, and an almost complete single layer of fungal hyphae eventually separates and virtually encapsulates each cell (Figure 1), though plasmodesmatal connections may remain between cortical cells. Far from being deleterious, the presence of this fungal net actually prolongs the life of the cortical cells and of the root as a whole. The fungal mantle, which develops concomitantly around the outside of the root, varies from a relatively loose weft of hyphae to a thick, dense, pseudoparenchymatous layer which may account for nearly half the biomass of the mycorrhiza. Mycelial strands or hyphae often extend from the mantle into the surrounding soil, while the formation of roothairs by the plant is suppressed.

It has been shown that sugars are translocated from the root *via* the Hartig net to the fungal mantle, where they tend to accumulate. As sugars pass from plant to fungus, they are converted into trehalose (a disaccharide), mannitol (a polyhydric alcohol) and glycogen, all three being typical fungal carbohydrates. The glycogen is insoluble, and therefore unavailable for possible reabsorption by the plant. More surprisingly, although the mannitol and trehalose remain in solution in the fungus, the plant is incapable of reabsorbing them. Thus, the fungal mantle acts as a sink where reserves of carbohydrates derived from the phytobiont are stored. This fact is emphasized when, as autumn approaches, many of the fungi mobilize the stored carbohydrates and produce flushes of their large, fleshy basidiomata near the tree. If we add up the various parts of the fungus, the conspicuous fruit bodies, the extensive but usually inconspicuous mycelium ramifying through the soil, and the rootlet mantles, we can calculate that the tree often invests at least 10% of its total production of photosynthates in its mycorrhizal fungus (Harley, 1975). This investment is clearly more than compensated for by the increased efficiency of mineral absorption provided by the EM fungus.

Visual inspection will show that, compared to non-mycorrhizal roots, mycorrhizal roots are: (1) a different colour, because of the enveloping fungal mantle; (2) thicker, because of the presence of the mantle and because their cells are larger and less liable to collapse; and (3) branched much more often, pinnately and racemosely in *Abies*, *Fagus* and *Eucalyptus*, and dichotomously in *Pinus*. Thus, the ectomycorrhiza is characterized by its form, colour and texture; by the presence of a sheathing fungal mantle (sometimes not very well developed); and by the presence in the outer layer(s) of the root cortex of a Hartig net, however restricted in development, which is the truly diagnostic structure. Although several different morphologies have been described among ectomycorrhizae, and the black monopodial mycorrhizae of *Cenococcum geophilum* are unique, it is widely thought impossible to determine the identity of the mycobiont unless this could be persuaded to produce fruit bodies (basidiomata or ascomata). Zak (1973) gives a photograph of a basidioma of *Hysterangium separabile* attached by a mycelial strand to a mycorrhiza of *Pseudotsuga menziesii*. This kind of evidence is unusual, but has been used to establish the mycorrhizal nature of a number of fungi (Woodroof, 1933; Zak and Marx, 1964; Schramm, 1966; Chilvers, 1968; Laiho, 1970; Zak, 1969, 1973). Today, there is renewed interest in the taxonomic value of mycorrhizal morphology.

Ectomycorrhizae remain active for periods ranging from several months to 3 years (Harley, 1969). Roots and mantles often extend at the same rate, but root extension sometimes outpaces mantle extension, and the root breaks through and extends beyond the mantle. The root may then be colonized by other EM fungi.

8.3.3 Sources of Ectomycorrhizal Inoculum

Ectomycorrhizae may be initiated by several different kinds of inoculum, which can be categorized as: (1) natural inoculum in the form of airborne spores (usually basidiospores, but in some cases also conidia); (2) soil already colonized by an EM fungus or fungi; (3) seedlings already colonized by an EM fungus or fungi, that is, bearing mycorrhizal roots; (4) fungal sporomata, spores or sclerotia specifically collected for the purpose; and (5) fungal mycelium produced in axenic culture. It is worth comparing the merits of these different kinds of inoculum.

8.3.3.1 Natural airborne spore inoculum

This is, of course, one of the prime dispersal mechanisms for these fungi in nature, but there are a variety of reasons why it is basically unsuited to forestry applications. (a) It is available only during a relatively short period of the year, since most agarics fruit in late summer or early fall.

(b) Many ectomycorrhizal fungi are hypogeous, and their spores may not be aerially transmitted in significant numbers. (c) Even when spore inoculum is being produced, its availability in a specific area where tree seedlings are being produced may be sporadic or quantitatively inadequate, especially if the nursery is a long way from the nearest stand of ectomycorrhizal trees. (d) We have no effective control over the nature of the fungal partners being introduced. (e) If the seedlings are being started at a low elevation for high elevation outplanting, they may acquire local mycobionts unsuited to conditions at the ultimate growth site. Such erratic, patchy and possibly unsuitable inoculation is basically unacceptable in modern forestry practice. Since spontaneous inoculation, even where it is a possibility, seems to be so unreliable, we will consider all available alternatives, dealing with them in increasing order of complexity and cost.

8.3.3.2 Soil already colonized by an EM fungus or fungi

This has been fairly widely employed, especially when attempts were being made to establish conifers in new areas. But it has a number of drawbacks. (a) Soil inoculum is very bulky and heavy, and although it can be used in small-scale operations, it is unsuited for any large-scale afforestation projects. (b) Soil is a complex system, usually containing a large number of living organisms, some of which may be pests or pathogens. It would be most unfortunate to introduce any of these to a new area. (c) As in the case of spontaneous inoculation, the nature of the mycobiont(s) in the soil will often be uncertain or unknown. The desirability of using precisely identified fungal partners is clearly established in Section 8.3.4.

8.3.3.3 The introduction of seedlings with established mycorrhizae

Although this implies that the mycobiont(s) being introduced are compatible with the desired tree species, this method shares several disadvantages with the previous one. (a) Pests or pathogens may be introduced with the mycorrhizal seedlings. (b) The nature of the fungal symbiont(s) will often be uncertain. (c) The spread of infection within a nursery or plantation may be slow and possibly uneven, unless inordinately large numbers of infected seedlings can be used.

8.3.3.4 The deliberate introduction of spores, sporocarps or sclerotia

This would seem to be an obvious way of improving upon nature, and there is no doubt that it has potential, especially since the identity of the desired mycobiont could be established at the outset. But several difficulties stand in the way of its widespread adoption. (a) Since the only major source of this kind of inoculum is currently the naturally occurring fruit bodies of the fungi concerned, it must be apparent that the availability of most of these structures is seasonal. (b) This also means that in most cases the quantities of inoculum available will be limited, and that they will inevitably fluctuate from year to year as the fruiting of the fungi is affected by the climate. (c) Since the basidiomata of most agarics, although they may be locally abundant, are usually sporadic in occurrence and scattered over large areas, the collection of the quantities that would be required for large-scale forestry applications would be extremely labour-intensive (but see the case of *Pisolithus tinctorius*, below).

Also, in most instances, a certain level of taxonomic expertise would be required of the collectors, since they would have to be able to discriminate between desirable and undesirable species. (d) Even if an appropriate quantity of basidiomata of a suitable fungus could be collected, the question of how these could be stored would still remain. Although it is conceivable that in some cases the inoculum would be applied immediately after collection, this is likely to be the exception rather than the rule, since most fumigation and seeding are carried out in spring. The basidiomata of agarics are a notoriously perishable commodity, and if they are not immediately lyophilized, dried, frozen, or treated in some other way to preserve the viability of the inoculum, the natural processes of decay will frustrate the entire operation. At best, the long-term storage of such inoculum is problematical. (e) Initiation of mycorrhizae by basidiospore inoculum takes 3 to 4 weeks longer than when mycelial inoculum is used (see below). This delay gives pathogens a greater chance to attack the roots, and the later-developing mycorrhizae also provide less growth stimulation during the crucial early stages of growth.

There is another fungal reproductive structure that is much less perishable than either spores or basidiomata, since it has obviously evolved as a long-term survival mechanism. We refer, of course, to the sclerotium. Unfortunately, most agarics do not produce these structures, but one widespread mycobiont is characterized by them. The sclerotia of *Cenococcum geophilum* occur naturally in the soil in huge numbers, making up a not inconsiderable biomass that might be harvested and used as inoculum. Fogel and Hunt (1979) estimated that the A_0 and A_1 horizons of soil, *i.e.* the top 5 cm, in a 35 to 50 year old Douglas fir stand (*Pseudotsuga menziesii*) contained 2785 kg ha^{-1} dry wt. of *Cenococcum geophilum* sclerotia.

8.3.3.5 *Mycelial inoculum derived from pure cultures of known mycobionts*

Here the identity of the chosen fungus will be known, pests and pathogens will be absent, inoculum will be relatively compact, and it should be available on a year-round basis. However, it too has some inherent problems, since it must be easily the most expensive of the alternatives. (a) Some ectomycorrhizal fungi are difficult to isolate in pure culture, the process also calling for highly trained personnel. (b) Cultures are expensive to maintain, and they tend to grow relatively slowly, taking a long time to produce the quantity of biomass required for large-scale applications. (c) We still do not know how well such inoculum survives in the soil in face of predation and competition from indigenous organisms. (d) We have not yet defined the best possible fungus–host combinations for many soil–climate combinations. It is hardly worth going to the expense of mass-producing mycelium of single species until we are sure that the results will be economically worthwhile.

Pure cultures of ectomycorrhizal fungi can be derived from fruit body tissue, surface-sterilized mycorrhizal roots and sclerotia. Isolation from rhizomorphs or mycelial strands is also possible. Isolation of EM fungi from basidiospores is difficult and is rarely attempted.

Isolation from fruit bodies is usual, because this allows precise identification of the fungus at the outset. Members of some genera are often fairly easy to isolate. Among these are *Amanita*, *Astraeus*, *Boletus*, *Cortinarius*, *Fuscoboletinus*, *Hebeloma*, *Hymenogaster*, *Hysterangium*, *Laccaria*, *Lactarius*, *Leccinum*, *Melanogaster*, *Paxillus*, *Pisolithus*, *Rhizopogon*, *Scleroderma*, *Suillus* and *Tricholoma* (Molina and Palmer, 1982). Fortunately, these include some of the better mycorrhizal partners with the broadest host ranges, *e.g. Pisolithus*. Other genera are more recalcitrant; only a few species of *Russula*, and none of *Gomphidius*, have been cultured. We believe that most EM fungi can ultimately be grown in axenic culture when the rather stringent nutritional requirements they have developed as a result of their more or less obligately biotrophic lifestyle have been worked out.

Cultures should be made from young basidiomata or ascomata, though mature specimens should also be collected in order to facilitate identification and the preparation of voucher specimens. When collected, fruit bodies should be kept in waxed or brown paper bags until isolation can be attempted, since storage in plastic bags often causes rapid loss of viability. Extensive field notes, photographs and spore prints should be taken and the advice or assistance of an experienced fungal taxonomist may well be invaluable, since the accurate identification called for by mycorrhizal research is no easy matter for the uninitiated. If there is any uncertainty about the name to be applied to the organism, voucher specimens should be sent to a competent taxonomist, preferably one who has published recently on the group in question. If the fungus is to be used in research, voucher specimens should be deposited in an established, internationally recognized herbarium where they will be available for study by other researchers. Such herbaria are listed in the Index Herbariorum (Holmgren *et al.*, 1981), and the accession number and the code letters of the Herbarium (*e.g.* DAOM 129643) should be cited in all publications concerning the fungus. If this is not done, the culture is in many ways an orphan, since essential information concerning its identity, and some of the value of subsequent research, will have been permanently lost.

Isolations should be carried out in still, and preferably sterile, air. A laminar-flow bench is ideal, but any draught-free place may be adequate, and isolations can even be made in the field, if necessary, in a portable isolation chamber. The work area should be sterilized with chlorine bleach; we have found that this is much more effective than alcohol. The isolation is made with flame-sterilized scalpels. Shallow slits are made in the surface tissues of the fruit body, then these are opened up to expose the inner tissues by gently bending the tissue on each side away from the slit. Now, small cubes of the freshly exposed tissue can be cut out with another flamed scalpel and transferred to modified Melin–Norkrans nutrient agar (Molina and Palmer, 1982). Tubes containing agar ('slants') are usually much less susceptible to random atmospheric contamination than

Petri plates, especially if the neck is flamed before and after the tissue is introduced. Ten to twenty isolations should be attempted from several different locations in the fruit body. Molina and Palmer (1982) give many useful hints for dealing wth particular genera, but experience is the best teacher.

The isolates can be incubated at room temperature for 3 to 4 days, then examined under a dissecting microscope for signs of mycelial growth. Note that some fungi are slow starters and may require up to 6 weeks to establish themselves. Fast-growing fungi should be transferred after 2 to 4 weeks; slow-growing species not for up to 4 months. Really slow-growing isolates will probably not be of much use, and may well fail to survive subculturing. Contaminating bacteria and molds (usually hyphomycetes) will be encountered, but with a little experience these will be easily recognized and rejected. The ultimate test of any cultures obtained is their ability to establish ectomycorrhizae with aseptically grown tree seedlings.

Cultures may also be derived from freshly collected ectomycorrhizae, if basidiomata or ascomata are not available. The mycorrhizal roots are washed in tapwater, shaken vigorously in a dilute detergent solution (*e.g.* Tween 20), washed again, soaked in a surface-sterilizing agent (*e.g.* 100 mg l^{-1} mercuric chloride for 4 min, or 30% hydrogen peroxide for 5–20 s), rinsed with sterile water, and finally transferred aseptically to nutrient agar (Zak, 1971a, 1973).

The sclerotia of *Cenococcum geophilum*, a possible anamorph of *Elaphomyces* (Miller, 1982), are present in very large numbers in the soil under many conifers. They will be readily extracted from the soil (Trappe, 1969) and used to initiate cultures. Large, clean sclerotia are washed, surface sterilized with 30% hydrogen peroxide, rinsed in sterile water, then aseptically transferred to nutrient agar. The wide, black hyphae of this fungus are unmistakeable.

Stock cultures of EM fungi should be maintained at 3 to 4 °C and transferred three or four times a year. Marx and Daniel (1976) found that many cultures of EM fungi lost their ability to induce mycorrhizae after prolonged culturing, though some could be successfully stored as plugs cut from growing cultures and kept in cold sterile water in darkness. We must emphasize that not all culturable EM fungi will behave in the same way, and that some experimentation may be necessary to obtain the best response from many isolates. Carefully kept records will prevent loss of hard-won information.

Melin (1921, 1922, 1923, 1936) pioneered the techniques which permitted the synthesis of ectomycorrhizae *in vitro*, and thus established the true identities of many EM fungi. As a result of his work, we can now use field observations to predict many functional host–fungus combinations. Mycorrhizae synthesized in this way have also permitted investigations of fungal and host physiology and interactions, particularly as these concern the uptake and translocation of water and nutrients by the fungus, the movement of photosynthate from plant to fungus, the protection from pests and pathogens afforded the plant by the fungus, temperature effects, and the various degrees of host–fungus compatibility. Molina and Palmer (1982) give details of the various steps necessary to establish aseptic tree seedlings and to inoculate them with EM fungi.

We must emphasize that before the considerable expense of producing large quantities of mycelial inoculum *in vitro* is undertaken, as many as possible of the factors just listed should be investigated, so that the final choice from among so many potential mycobionts may be made intelligently. In fact, since we know that the average ectomycorrhizal tree has several to many different symbiotic fungi on its roots at any one time, and may also change partners over the years, there is almost certainly no single ideal mycobiont.

Nevertheless, it is also certain that inoculation of young conifer seedlings with an appropriate fungal partner will greatly increase their chances of survival during the first year or so of life. The subsequent replacement of that partner by others may be regarded as part of the normal course of events.

Therefore, attempts are being made to mass-produce and market mycelial inoculum of several broad-spectrum EM fungi, most notably *Pisolithus tinctorius*, but also several others, including *Thelephora terrestris* and *Cenococcum geophilum*. These fungi not only meet most mycorrhiza-forming criteria, but are also easy to isolate, in most cases grow relatively quickly, and can withstand the various manipulations involved in preparation, storage, transportation and application of inoculum. *Pisolithus tinctorius* can grow at 42 °C, and also at 7 °C, can tolerate a pH range of 2.6–8.4, and can survive prolonged freezing. It also forms abundant mycelial strands. *Cenococcum geophilum* is extremely drought tolerant and will form ectomycorrhizae from pH 3.4 to 7.5.

It is easy to produce enough mycelial inoculum for small-scale research projects, but much more difficult to generate enough to inoculate the many millions of seedlings routinely produced each year, while preventing losses due to microbial contamination.

Moser (1958b) grew *Suillus plorans* in liquid medium in small flasks, then added this starter inoculum to 10 l tanks of the same medium. These tanks were aerated for 2–3 h daily over 3–4 months. The mycelium thus generated was used to inoculate 5 l flasks containing sterile peatmoss soaked with liquid medium. Over the next few months the mycelium ramified throughout this substrate. This inoculum was then packaged in sterile plastic bags, taken to the nursery, and used within 3 days. As a result of the various manipulations involved, inoculum often became contaminated by common molds and bacteria. Moser (1958a,b,c) went to all this trouble because he found that neither mycelium growing on solid agar media, nor a mycelial suspension in the liquid medium, made effective inocula. Moser used the same technique to produce inoculum of seven other EM fungi from the genera *Suillus* (3), *Paxillus* (1), *Phlegmacium* (1), *Amanita* (1) and *Lactarius* (1).

Göbl (1975) grew mycelia in sterilized cereal grains (wheat, millet), often with an addition of calcium sulfate, in 1 l bottles. Shaken weekly, these substrates became thoroughly permeated by mycelium after about 4 weeks, and could be stored at 4 °C for up to 9 months. The actual inoculum was produced by adding the colonized grain to peatmoss variously amended with inorganic nutrients plus ammonium tartrate, asparagine, soybean meal, blood meal, malt extract and glucose, depending on the requirements of the species at hand. One or more bottles of the grain were added to plastic bags containing 10–15 l of peatmoss, the bags were plugged with cotton and periodically shaken over 3–6 weeks.

Takacs (1961, 1964, 1967) used a similar approach to produce inoculum of seven EM fungi for the establishment of conifers on formerly treeless land in Argentina, which had no indigenous EM fungi. Inocula of four different fungi were mixed with soil or litter at the nursery site, and incubated for a few weeks in small heaps. In this way, twenty 200 ml starter cultures were used to generate 100–200 kg of 'inoculum', though in the absence of ectomycorrhizal host plants and special nutrients, it is unlikely that the starter cultures can have done any more than survive, in a considerably diluted form.

Similar work has been done by Hatch (1936) and Park (1971) in North America, and by Theodorou and Bowen (1970) in Australia. Hacskaylo and Vozzo (1967) used a modification of Moser's (1958a) technique to grow *Cenococcum geophilum*, *Corticium bicolor*, *Rhizopogon roseolus* and *Suillus cothurnatus* for introduction into areas in Puerto Rico which lacked EM fungi, and which were to be afforested with *Pinus caribaea*. Inoculum grown in Maryland was successfully transported to Puerto Rico, and all species but one were eventually demonstrated to form ectomycorrhizae.

Marx (1980) tried various methods of producing mycelial inoculum of *Pisolithus tinctorius*, *Thelephora terrestris* and *Cenococcum geophilum*. Functional inoculum could not be produced on sterilized grains, but vermiculite plus peatmoss moistened with modified Melin–Norkrans (MMN) nutrient solution gave good results. The MMN solution contained: 0.05 g $CaCl_2$, 0.025 g NaCl, 0.5 g KH_2PO_4, 0.25 g $(NH_4)_2PO_4$, 0.15 g $MgSO_4\cdot7H_2O$, 1.2 ml 1% $FeCl_3$, 100 µg thiamine HCl, 3 g malt extract, 10 g glucose and distilled water to make 1 l.

Peatmoss–vermiculite in a ratio of 28:1 is moistened with half its volume of MMN solution (*e.g.* 1400 ml vermiculite, 50 ml peatmoss, 750 ml MMN solution). The initial inoculum can be in the form of plugs cut from an agar culture, or blended mycelium grown in liquid culture. Using blended mycelium mixed throughout the substrate, *Thelephora terrestris* and *Pisolithus tinctorius* will thoroughly colonize the substrate in 1–2 months at room temperature. *Cenococcum geophilum* may take 4–5 months to achieve the same result. If this inoculum is mixed directly with fumigated nursery soil, it will soon be overgrown by saprophytic microorganisms which exploit the unused nutrients. This problem can be reduced or eliminated by leaching the inoculum in tapwater for 2–3 min to remove those nutrients. This leaves a bulky, sticky paste with 90% water. Much of the water is normally removed by drying the inoculum on wooden frames at 20–26 °C and 35–45% relative humidity for about 4 days, mixing every few hours to reduce excessive surface drying. The final inoculum weighs about 400 g and its water content is 20–65% (Marx and Kenney, 1982). Inoculum of *Pisolithus tinctorius* can be stored at 5 °C for 9 weeks without much loss of activity, but the sooner it is used, the better.

The USDA and Abbott Laboratories have developed a commercial formulation of *Pisolithus tinctorius* mycelial inoculum, called 'MycoRhiz'. This is grown in the vermiculite–peatmoss–MMN medium. The starter mycelium is grown in MMN solution in large fermenters, with continuous agitation and aeration, at 28–32 °C for 7–14 days. The vermiculite–peatmoss–MMN substrate is steam-sterilized in deep-tank or drum fermenters, then inoculated with the starter culture at a rate of 5–20% of fermenter volume. The medium is mixed, and then incubated with or without agitation for 1–4 weeks. The final inoculum is harvested by flotation on water, then dried in an Aeromatic fluid bed dryer (Aeromatic Ltd., Switzerland) to a moisture content of

20–25%. 'MycoRhiz' weighs about 250–300 g l^{-1}, is packaged in 50 l units, and is stored at 5 °C, when it has a shelf life of 5–6 weeks. A tractor-drawn combined seeder-inoculater has recently been developed (Cordell *et al.*, 1981).

However 'MycoRhiz' was withdrawn from the market in September 1983, due to quality control problems. Abbott Laboratories anticipated that it would return in 1984, but at the time of publication this had not occurred. The Butler Mushroom Farm in Pennsylvania is at present willing to produce inoculum for any viable EM under contract. Mycelial inoculum of *Suillus*, *Thelephora* and *Laccaria* is being field tested in Oregon. Sylvan Spawn Laboratories, PA are now producing mycelial inoculum of *Pisolithus tinctorius* in breathable plastic bags. One factor in the acceptance of inoculum such as 'MycoRhiz' is the high total cost of application at around $15 m^{-2}.

8.3.4 Evaluation and Selection of Ectomycorrhizal Fungi

The thousands of different ectomycorrhizal fungi are probably necessary evolutionary responses to the diverse needs of many hosts in a multiplicity of habitats. Trappe (1977) addressed the question of how many host–fungus–soil–climate combinations we may expect to find. He reported that a 250 km transect running east from the coast of Oregon or Washington, USA, can pass through 17 major forest zones, hundreds of habitat types and at least 10 genera of ectomycorrhizal hosts that are grown in nurseries for forestation, erosion control or as ornamentals.

How, then, are we to select the best possible mycobionts for our trees? The first step is to define the range of fungal symbionts available for the chosen tree species. Even this may be difficult, since some trees are compatible with a wide spectrum of mycobionts, and the success of the symbiosis often varies with the provenance or strain of either partner. It will often be found, however, that one or two of the characteristics discussed below will override the rest in importance. For example, if a fungus cannot be grown in pure culture for the large-scale production of inoculum, it will be effectively excluded from consideration for biotechnological applications.

All potential host–fungus pairs should ideally be tested for all of the following characteristics: (1) rapidity and extent of mycorrhization; (2) host response; (3) inorganic nutrient uptake; (4) water relations (keeping in mind the conditions under which the pair must operate after outplanting); (5) tolerance of extremes of temperature (*cf.* field conditions); (6) tolerance of extremes of pH (*cf.* field conditions); (7) tolerance of natural or anthropogenic soil toxicity; (8) stability of the partnership (one expression of the competitive ability of the fungus); (9) disease resistance (this might be tested only for diseases known to exist in the outplanting site); (10) mycelial strand formation by the fungus; (11) ease of isolating the fungus in pure culture; (12) ease and rapidity with which large quantities of inoculum can be produced; and (13) edibility of the fruit bodies of the fungus.

The potential range of mycobionts for a given tree in a given area may initially be checked by field observations of basidiomata associated with that tree, combined with estimations of the degree of mycorrhizal infection on the roots, though this should not rule out the possibility of introducing new and efficient EM partners to the area. Field observations showed Moser (1958b) that *Suillus plorans* was the predominant EM fungus associated with *Pinus cembra* at treeline in the Alps. Other obvious pairs are listed in Section 8.3.1, but the choice is not usually so simple. The process of testing host–fungus pairs involves the isolation of the fungi in pure culture, and their inoculation onto seedlings grown individually and aseptically in tubes (Hacskaylo, 1953) or in root pouches (Fortin *et al.*, 1980), where the development of mycorrhizae can be visually checked. Molina (1980) inoculated containerized conifer seedlings with 15 potential mycobionts. Only two of the fifteen fungi formed abundant mycorrhizae with all the conifers tested.

It is important to design experiments with the inherent variability of the material in mind. In the host plant, much less variability may be expected in material derived from cuttings than from seed (though conifers are not usually propagated by cuttings), while the geographic origin of the host can be important (Lundeberg, 1968). In the mycobiont, geographic origin, original host and age of the culture may all introduce variation (Molina, 1980; Marx, 1981). Marx (1981) suggested that before the mycorrhizal potential of any fungal species can be properly assessed, several isolates representing the range of variation expected should be tested. Appropriate conditions for experiments with EM fungi and their hosts are discussed in detail by Reid and Hacskaylo (1982). We will discuss our 13 criteria in sequence.

8.3.4.1   Rapidity and extent of mycorrhization

Knowledge of the rate at which, and the degree to which, a tree's root system is colonized by EM fungi is an essential part of any comprehensive understanding of ectomycorrhizae. Ectomycorrhizae are distinguishable by the naked eye, and therefore require no special treatment before they can be quantified in various ways. Entire root systems of seedlings can be examined, but in older plants only a representative sample obtained by soil coring or local excavation can be studied. Soil cores reported in the literature vary in diameter from 1.2–10 cm, and in depth from 7.5 to 90 cm, with 30 cm as the norm. The number of samples required must be determined statistically for each study. Soil cores are soaked for up to 36 h before being agitated in water to clean the roots of soil particles. The percentage of mycorrhizal short roots can be determined visually (Richards and Wilson, 1963). Grand and Harvey (1982) discuss other assessment techniques, such as direct counts of entire root systems, of selected roots and of ectomycorrhizal tips, in tandem with determinations of the weights of these structures. Results may be expressed as number and weight of ectomycorrhizal structures per unit area, or per unit volume of soil.

If the amount of mycorrhizal material in core samples is very large, subsampling may be necessary. Marks *et al.* (1968) used a 10 inch-square plexiglass tray divided into 1 in squares and 0.1 in subdivisions. Sieved material from cores was spread out evenly over the tray in 200 ml water, and the ectomycorrhizae and tips counted in seven randomly selected squares under a dissecting microscope.

8.3.4.2   Host response

The response of the host seedling or tree to mycorrhizal colonization can be measured in various ways. The non-destructive, and therefore easiest, method is to determine seedling survival, expressed as percentages of the initial uninoculated and inoculated populations. Such data can be gathered at various ages, before and after outplanting. Other non-destructive measures are height of plant, thickness of stalk at ground level, number of leaves, leaf length and leaf area. More definitive measurements involve determining the dry weight of the whole plant, or of separate root and shoot systems. Roots, of course, will often have a significant fungal component, but this is usually acknowledged or assessed without measurement, though it can be determined by a glucosamine assay (Wu and Stahmann, 1975). Details of these parameters are given by Sinclair and Marx (1982). Measurements of stem height and stem diameter at soil line are eventually replaced by diameter at breast height (1.4 m) in older trees.

Sinclair and Marx (1982) propose a 'mycorrhizal influence value' (MIV), which can be determined for any of the parameters mentioned above by expressing the mean value for non-inoculated plants as 100, and calculating the value for mycorrhizal plants as an integer relative to that 100. Thus the MIV would be a percentage of the control value in each case.

8.3.4.3   Inorganic nutrient uptake

Phosphorus uptake, and levels achieved in the mantle and in the plant (determined using radiotracer techniques by Mejstřik, 1970, and Mejstřik and Krause, 1973; for methodology see Rhodes and Hirrel, 1982) are among the most important reflections of the effects of EM fungi on their hosts. Ectomycorrhizal plants also absorb many other minerals, *e.g.* calcium, potassium, copper, molybdenum, magnesium and zinc, from the soil much more efficiently than non-mycorrhizal plants (Hatch, 1937; Mejstřik, 1970; Bowen *et al.*, 1974). The fungal mantle can store inorganic nutrients, *e.g.* chloride (Smith, 1972), ammonium (Carrodus, 1967) and especially phosphate (Harley and McCready, 1952) and release them to the plant during periods of deficiency or active growth. *Pisolithus tinctorius* thrives in soils of extremely low fertility, such as mine spoils (Marx, 1977), while *Paxillus involutus* does well only on sites with relatively abundant available nitrogen (Laiho, 1970). But since it is in the uptake of phosphorus, often a limiting nutrient in poor soils, that EM fungi make their greatest contribution to the symbiosis, evaluation of rate and amount of phosphorus accumulation must be one of the most important criteria in selection.

8.3.4.4   Water relations

Theodorou (1978) found that isolates of some mycorrhizal fungi were better able than others to survive drought while still contributing directly to the welfare of the plant partner. To assess the response of isolates of *Cenococcum geophilum*, *Suillus luteus* and *Thelephora terrestris* to induced water stress, Mexal and Reid (1973) grew them in artificial media with the water potential controlled at various levels by the osmoticum, poly(ethylene glycol) 4000. Of the three fungi, *Cenococcum geophilum* was found to be especially tolerant of low water potential, which correlates well with its propensity for forming ectomycorrhizae in dry areas. In fact, because *Cenococcum* grows best at a water potential of − 1.5 MPa, it can be difficult to establish this fungus in irrigated nurseries, where it may be replaced by *Thelephora terrestris* (Marx *et al.*, 1978). Other work demonstrating the superior performance of ectomycorrhizal plants under conditions of water stress has been carried out by Dixon *et al.* (1980), and Sands and Theodorou (1978).

8.3.4.5 Temperature tolerance

Moser (1958c) was careful to inoculate *Pinus cembra* destined for high-altitude outplanting with a cold-adapted strain of *Suillus plorans*. Other fungi, especially *Pisolithus tinctorius*, have been found to be adapted to high temperatures (Marx *et al.*, 1970; Marx and Bryan, 1971; Marx, 1977). *Cenococcum geophilum* appears to tolerate both extremes relatively well (Theodorou, 1978). Hacskaylo *et al.* (1965) examined the temperature responses of six EM fungi commonly associated with *Pinus virginiana*.

8.3.4.6 pH tolerance

Marx and Artman (1979) reported that pine seedlings with *Pisolithus tinctorius* ectomycorrhizae survived and grew better on acid coal spoils than did non-mycorrhizal seedlings. Dale and McComb (1955) found that ectomycorrhizae improved the growth of pines in an alkaline soil. Marx and Zak (1965) investigated the effects of pH on formation of ectomycorrhizae on pine in aseptic culture.

8.3.4.7 Tolerance to soil toxicity

Zak (1971c) demonstrated the destruction of heat-formed phytotoxins in the soil by EM fungi. Bowen (1973) pointed out that in view of the selective absorption of various ions by mycorrhizal fungi, and their capacity for storing ions in the mantle, they may be active in ameliorating marginal soil toxicities. There is still little published work in this area, but we are aware of studies in progress on the spoils derived from nickel mining at Sudbury, Ontario, which indicate the ability of some EM fungi to tolerate fairly high levels of some heavy metals in the substrate. Powell *et al.* (1968) treated soil around pecan trees with a variety of nematocides and fungicides, and observed a large increase in mycorrhiza formation by *Scleroderma bovista*.

8.3.4.8 Stability of the partnership

The stability of the partnership need only be established in the short term, since the choice of a mycorrhizal partner for a tree should probably be based on the immediate benefits it bestows. Although the initial mycobiont has in many cases been shown to be supplanted or supplemented by other EM fungi after the seedling has been outplanted, its presence in the early days may well make the difference between death and survival of very young seedlings. Once again, selection of a mycobiont adapted to the conditions of the outplanting site, and preferably already established there as determined by the occurrence of its basidiomata or ascomata, may produce the best results.

8.3.4.9 Disease resistance

The presence of EM fungi on the roots of trees has repeatedly been shown to confer some protection against the effects of several important root-pathogenic fungi. Hyppel (1968) found that *Boletus bovinus* helped to protect *Picea abies* from *Fomes annosus*. Wingfield (1968) observed that *Pisolithus tinctorius* increased the survival rate of *Pinus taeda* seedlings exposed to *Rhizocto-*

nia solani. Corte (1969) reported that mycorrhizae formed by *Suillus granulatus* seemed to protect seedlings of *Pinus excelsa* from a root-rotting *Rhizoctonia* sp. Marx (1969) reviewed the antagonism of mycorrhizal fungi to root-pathogenic fungi. Ross and Marx (1972) found that seedlings of *Pinus clausa* were protected against *Phytophthora cinnamomi* by mycorrhizae of *Pisolithus tinctorius*, while Marx (1973) found that *Cenococcum geophilum* or *Pisolithus tinctorius* reduced the impact of the same pathogen on *Pinus echinata*. The effects of the pathogen *Mycelium radicis atrovirens* on *Picea mariana* and *Pinus resinosa* were markedly reduced by the presence of *Suillus granulatus* (Richard *et al.*, 1971; Richard and Fortin, 1975).

8.3.4.10　Strand formation

Mycelial strands, associations of parallel hyphae, serve both as effective agents for the spread of the fungi through the soil and in the long-distance translocation of phosphate and other nutrients to the mycorrhizae (Skinner and Bowen, 1974). Different species, and different isolates of the same species of EM fungus, may have different capacities for mycelial strand formation. Other things being equal, it would seem reasonable to choose a strand-forming fungus, such as *Pisolithus tinctorius*, over one that did not produce these structures.

8.3.4.11　Ease of pure culture formation

Many of the fungi responsible for ectotrophic mycorrhizae can be isolated in axenic culture without much difficulty, but most will not fruit in culture, grow slowly, and are heterotrophic for vitamins like thiamine, for some amino acids, and for other normally root-derived substances, as well as for simple carbohydrates. Most of them are incapable of degrading cellulose or lignin, though these substances are the principal diet of many other basidiomycetes. Techniques and media for isolating EM fungi, and the amenability of many genera to culturing, are dealt with in Section 8.3.3.

8.3.4.12　Ease and rapidity of production

The various attempts to produce inoculum of ectomycorrhizal fungi on a large scale are fully documented in Section 8.3.3. Since *Pisolithus tinctorius* has been shown to establish mycorrhizae with almost 50 different tree species, thrives over a wide range of soil pH, tolerates high temperatures well, and can establish mycorrhizae in the poorest soils, it has been touted as a panacea for all our ectomycorrhizal problems. It is also the first EM fungus to be made available in the form of commercially produced mycelial inoculum. We applaud this initiative, but we would encourage researchers to continue working on other fungi, because it seems unlikely to us that *Pisolithus* can be all things to all EM trees. It is probably at its best coping with heat and drought stress. It has, to the best of our knowledge, been collected only once or twice in Canada, and we suspect that it will not turn out to be the perfect partner for boreal conifers.

8.3.4.13　Edibility of the fruit bodies

If it should transpire that several potential partners are more or less equivalent, then the ultimate choice may be dictated by secondary, though not negligible, factors such as the edibility or otherwise of the sporomata of the fungi being considered. For example, if a hypothetical choice lay between a species of *Amanita* known to have highly toxic basidiomata, and another agaric whose basidiomata are edible and choice, the decision would be obvious. Less obvious, but also important, is the caution that species known to have toxic fruit bodies should not be introduced to new areas as mycorrhizal partners, even if they might seem otherwise desirable. One of the most toxic of all agarics, *Amanita phalloides*, was inadvertently introduced into South America as a mycobiont of oak seedlings imported from Europe during the early part of this century (Herter, 1934; Martinez, 1945). The cyclopeptide toxins (amanitins) in this fungus have caused many fatalities. One example in which the conscious dissemination of poisonous agarics has been avoided is provided by the Australian government, which refused to allow the importation of cultures of *Amanita pantherina*, a good mycorrhizal partner, but a species producing basidiomata containing dangerous levels of ibotenic acid (Trappe, 1977).

At the other end of the scale are the French experiments with 'trufficulture', the deliberate use of *Tuber melanosporum* as a mycorrhizal partner with an eye to the production of truffles, an extremely valuable crop. The first steps toward the culture of other choice edible fungi have been made, again by the French (Poitou *et al.*, 1981). Using pure cultures of the famous 'cepe' (*Boletus edulis*) and three other boletes, as well as *Lactarius deliciosus* and *Tricholoma flavovirens*, ecto-mycorrhizae have been established on *Pinus pinaster* and *Pinus radiata* in tubes and in green-house pots. It remains to be seen whether outplanted seedlings bearing mycorrhizae of these species will ultimately produce basidiomata, thereby providing an interesting and perhaps valu-able byproduct of afforestation.

8.3.5 Application of Ectomycorrhizal Inoculum

We will now consider actual forestry research and practice, using the headings for different kinds of inoculum which we introduced in Section 8.3.3.

8.3.5.1 *Natural inoculum: airborne spores*

Although spontaneous inoculation by airborne spores is still operative in some cases, this is largely by default, in the absence of any active human intervention. It will usually be regarded only as a supplementary source when mycorrhizal fungi are already known or presumed to be present in the soil, and will play a role chiefly in conditions where natural regeneration is being allowed or encouraged, and significant investment of time or money is unavailable or is con-sidered to be unnecessary.

8.3.5.2 *Soil colonized by EM fungi*

In Western Australia, pine seeds planted in 14 new nurseries germinated and produced seed-lings that grew relatively normally for a few months, then began to decline and die. A few seed-lings remained healthy and when these were examined it was found that they had developed ectomycorrhizae, structures conspicuous by their absence from the sickly seedlings. When soil from around the healthy seedlings was used to inoculate other seedbeds, the seedlings in those beds began to recover, and ultimately thrived (Mikola, 1973). Soil from beneath established ecto-mycorrhizal trees is a fairly reliable source of inoculum. About 10% by volume of infected soil is often added to a new nursery bed. Although the mycorrhiza-forming fungi are often unknown, and there is some risk of introducing pests and pathogens, this has rarely caused difficulty, because the location from which the soil is derived can be carefully scrutinized for such problems. Infected soil has been used to establish exotic pines in various parts of the Southern Hemisphere, and soil transfer is still a regular procedure in many third-world countries (Mikola, 1970).

8.3.5.3 *Seedlings colonized by EM fungi*

The planting of 'nurse' seedlings that carry established mycorrhizae has also worked relatively well (Mikola, 1970, 1973) despite the drawbacks we mentioned earlier. This technique was first applied on a large scale in Indonesia, and is apparently still in use there. Mycorrhizal seedlings are planted in seedbeds at 1–2 m intervals. When the plants are lifted, some are left behind to infect the next crop. Attempts to extend the range of the indigenous *Pinus insularis* from the Philippine highlands to lowland sites in the Philippines, Hawaii and South Africa failed comple-tely until seedlings potted in highlands soil were introduced to the lowland sites. This was a very clear-cut example, because there were no indigenous potential ectomycorrhizal partners in the lowland soil to confuse the issue.

8.3.5.4 *Fungal sporomata or sclerotia*

Spores or chopped basidiomata or ascomata have also been used with some success, though only experimentally (Donald, 1975; Marx, 1976; Theodorou, 1971; Theodorou and Bowen,

1973). On occasion it can be feasible to collect sufficient quantities of basidiomata of particular species for large scale inoculation. Over 450 kg of basidiomata of *Pisolithus tinctorius* were collected on mine spoils in about 75 man-days. If it is assumed that 1 mg of spores is required for successful inoculation of a single plant, this collection represented enough inoculum for 225 million pine seedlings (Marx and Kenney, 1982). Trappe (1984) suggests that in most years he could collect enough basidiomata of *Rhizopogon* spp. to inoculate many millions of conifer seedlings. The fruit bodies of gasteromycetous fungi are a particularly concentrated form of spore inoculum, and it would clearly be impossible to collect spores of any Agaricales or Aphyllophorales on this scale.

8.3.5.5 *Mycelial inoculum*

Axenically grown mycelial inoculum of EM fungi. This is now considered to have the greatest potential for general application. Nevertheless, this kind of inoculum presents some problems. Since EM fungi are, to all intents and purposes, obligately biotrophic in the natural habitat, mycelial inoculum will not be able to grow far through the soil to find and colonize a hospitable root. Therefore, an initial inoculum must be delivered in a very precise manner—it must be placed in contact with, or in the path of, the young roots. Several methods of application have been tried.

(i) *Broadcast inoculation*

A known quantity of inoculum is spread out over a given area of seedbed and mixed into the top 10–20 cm of soil before the bed is seeded (Marx *et al.*, 1976, 1978). Inoculum of *Pisolithus tinctorius* broadcast at a rate of $1 \ l \ m^{-2}$ gave results equivalent to those obtained with higher levels of inoculum. Marx *et al.* (1982) found that inoculum incorporated in container growth media at a rate of 6% by volume produced effective mycorrhization in many tree species. With containerized seedlings, inoculation and container filling processes can be combined (Molina, 1977). Marx and Bryan (1975) obtained good results by inoculating *Pinus taeda* nursery beds with cultures of *Pisolithus tinctorius*. Laboratory grown inoculum was leached under running tapwater, cool-dried to about 20% moisture and kept cold, but not frozen, until used. Nursery beds previously fumigated with methyl bromide–chloropicrin were inoculated with about 100 cm³ of the dried preparation, dug into the top 7–10 cm of soil. The bed was then machine sown.

(ii) *Banding of inoculum below seeds*

This concentrates inoculum in a zone that will be penetrated by the growing roots. Seeds and inoculum can now be simultaneously dispensed by a modified nursery seed planter. One important advantage of this method over the broadcast technique is that it needs only about a third as much inoculum. Riffle and Maronek (1982) reviewed recent developments in this field.

(iii) *Slurry inoculum*

With slurry inoculum, bare-root or container-grown seedlings can be inoculated by dipping before transplanting (Marx, 1976; Marx *et al.*, 1982).

Shaw and Molina (1980) showed that mycelial inoculum of EM fungi can be safely transported over long distances without loss of viability. Mycelial inoculum was grown in Oregon, leached, placed in plastic bags and cooled to 5 °C, air freighted in an ice chest to Juneau, Alaska, and stored at 5 °C for up to 3 weeks. Container-grown Sitka spruce seedlings were successfully inoculated with this material when it was mixed with the growth medium.

8.4 ENDOMYCORRHIZAE (VESICULAR–ARBUSCULAR MYCORRHIZAE)

8.4.1 *Systematics of Vesicular–Arbuscular Mycorrhizal Fungi and their Hosts*

In the light of what we say elsewhere in this chapter about the enormous differences in efficiency among the various fungus–host–soil–climate combinations, it should be apparent that the identity of the mycobionts to be used in agricultural systems must be established with the greatest precision possible.

The taxonomy of the VA mycorrhizal fungi is still in a state of active ferment. Where thirty species of VAM fungi were recognized in 1974, seventy-eight species had been described by 1982 (Trappe, 1982), and this number is expected to more than double in the next decade. None of these apparently obligately biotrophic fungi have ever been seen to undergo sexual reproduction, and they are sufficiently different from all other known fungi to constitute something of a taxonomic enigma. They are all grouped in a single family, the Endogonaceae, which has usually been tentatively placed in the phylum Zygomycota of Kingdom Fungi. Of this family, only the non-endomycorrhizal genus *Endogone* has been seen to form zygosporangia (the kind of meiosporangia peculiar to the Zygomycota).

The disposition of the VA mycorrhizal fungi in the Zygomycota was challenged by Pirozynski and Malloch (1975), who suggested that these fungi might well be members of the phylum Oomycota in the Kingdom Protoctista. That challenge led to some investigations of wall chemistry by Weijman and Meuzelaar (1979), who found that the non-chitinous glucans which characterize the walls of oomycotan fungi are absent from the walls of the Endogonaceae. This effectively ruled out the Oomycota as a home for the VA mycorrhizal fungi, and strengthened the likelihood of a connection with the Zygomycota. Recent observations of *Endogone pisiformis*, the type species of *Endogone*, type genus of the Endogonaceae, by Berch and Fortin (1983a) have revealed arbuscule-like and vesicle-like structures in this non-VAM, saprophytic fungus, and have reinforced the cohesion of the Endogonaceae, making it even more likely that at least some VAM fungi are zygomycetous.

The Endogonaceae is composed of six genera. Four of these are unequivocally mycorrhizal—*Acaulospora*, *Gigaspora*, *Glomus* and *Sclerocystis*. Another little-known genus, *Entrophospora*, may also be endomycorrhizal, but is too infrequently collected to concern us here. *Glaziella*, formerly placed in the Endogonaceae, now has been demonstrated to be an ascomycete with uni-spore asci (Gibson *et al.*, 1984). *Acaulospora* and *Gigaspora* form what have been called 'azygospores', which are supposed to be parthenogenetic zygospores, though there is little or no karyological evidence to support this hypothesis. *Glomus* and *Sclerocystis* produce what are called chlamydospores: these are supposed to be asexual resting spores, a perfectly reasonable assumption in our opinion. For the sake of simplicity, we usually refer to all these structures as spores.

Mosse and Bowen (1968) produced the first modern key to nine spore types associated with VA mycorrhizae. Although their coverage was limited, they attempted to focus on diagnostic features of size, shape and wall structure of spores, and on the mycelium giving rise to the spores. They gave helpful line drawings and photomicrographs of these features.

The first modern monograph of the Endogonaceae was produced by Gerdemann and Trappe (1974), who gave detailed descriptions of 30 taxa in 4 mycorrhizal genera. Hall and Fish (1979) produced a dichotomous key to many named, and some unnamed, species. Trappe (1982) gave synoptic keys to a total of 78 species and varieties in the four proven VA mycorrhizal genera. *Acaulospora* had 9 distinct taxonomic entities; *Gigaspora*, 18; *Glomus*, 43; and *Sclerocystis*, 8. By 1984, over 100 taxa had been described.

The literature is, in general, very inadequately illustrated: for example, the Gerdemann and Trappe (1974) monograph contains 28 illustrations, but while 30 VA mycorrhizal taxa were dealt with in the monograph, reasonably diagnostic illustrations were provided for only two, and one new genus remained unillustrated. This kind of substitution of words for pictures is fairly typical of the literature on VA mycorrhizal fungi, and has made it very difficult for the uninitiated to identify VA endophytes. Hall and Abbott (1979) tried to remedy the situation by compiling a set of nearly 250 photographic transparencies illustrating diagnostic features of the Endogonaceae. Complete sets can be found in the Farlow Herbarium of Harvard University; the Department of Botany, Oregon State University, Corvallis; the Department of Biology, University of Waterloo, Waterloo, Ontario, Canada; the Plant Diseases Division, DSIR, Auckland, New Zealand; the Herbarium, Royal Botanic Gardens, Kew; and the Invermay Agricultural Research Centre, Mosgiel, New Zealand. Unfortunately, not all of the species in this collection are illustrated from type material. As our understanding of the taxonomic concepts of these species matures, the names of many of the fungi illustrated will have to be changed. In the meantime, this collection should not be used as an authoritative guide to identification.

The main features of VA mycorrhizal fungi are neatly summarized by Trappe and Schenck (1982), with black-and-white and coloured figures of spores and sporocarps. We are optimistic enough to discern a trend toward better illustration of VA fungi, in terms of both scope and quality, in some recent papers. Although we have neither space, nor a mandate, to make our own contribution toward improving the situation, good illustrations should be a priority for taxonomists working in this group.

Relative to the number of genera of ectomycorrhizal fungi, there are very few genera of VAM fungi, which allows us to characterize them here.

Glomus, the most common genus of VAM fungi, has over 50 species which form globose, ellipsoid or rather irregularly shaped spores that range from 20–400 μm in their largest dimension. These spores are thick-walled (up to 30 μm), and hyaline, yellow, red-brown, brown or black. They are attached to a single subtending hypha (or to two, three or more in exceptional species). They are produced in the soil of the rhizosphere (near plant roots), at the soil surface, or occasionally in roots, and they may form singly, in groups of a few to many, or in large aggregates called sporocarps, which may range from 1 mm to more than 20 mm in their largest dimension. The sporocarps of certain species, such as *Glomus versiforme* (= *Glomus epigaeum*), are formed at the surface of the soil (epigeous), while those of other species, such as *Glomus macrocarpum* are formed in the soil or in leaf litter (hypogeous). Sporocarps are most commonly found in undisturbed forest communities with perennial plants and accumulated organic material. Sporocarpic species, when grown on plants in pot culture, usually produce spores singly or in small aggregates. Some *Glomus* species (*e.g. Glomus lacteum*) are not known to form spores in sporocarps, while others form spores outside the plant root only infrequently (*e.g. Glomus intraradices*).

Sclerocystis species produce chlamydospores that are very similar to those of *Glomus*, except that they tend to be clavate rather than globose. All *Sclerocystis* spp. form spores only in sporocarps that are up to 700 μm in diameter. The spores, which may be up to 125 μm long, develop in a single layer, radially arranged around a central plexus of sterile and sporogenous hyphae. The sporocarps may group together in masses up to several cm in diameter on the soil surface, or on leaves, twigs and mosses.

The so-called 'azygospores' of *Acaulospora* spp. form laterally on thin-walled terminal vesicles. When mature, the spore does not display a subtending hypha because the vesicular structure that gives rise to it loses its cytoplasm and collapses during development of the spore. The spores are globose or ellipsoid, from under 100 to over 400 μm in diameter, and hyaline, yellow or reddish-brown. The surface of the spore wall may be ornamented with pits, projections of various shapes, folds, spines or reticulations, and the wall is up to 12 μm thick. One species of *Acaulospora* is now known to produce sporocarps, but in the other species, spores are normally found in the soil of the rhizosphere.

Individual spores of *Gigaspora* spp. can be over 600 μm in diameter, and the smallest ones are just under 200 μm. Spores develop terminally on a bulbous hyphal suspensor, which remains attached at maturity and may bear short lateral projections that are all that remain of collapsed fine hyphal branches. Spores are hyaline, white, yellow, pinkish, grey-green, brown or black, and the wall may be ornamented with minute pits, warts, spines or reticulations and is up to 20 μm thick. These spores form singly in rhizosphere soil. Additional structures that have some importance in the identification of species of *Gigaspora*, and are not reported to be formed by any other fungi, are called 'ornamented vesicles' or 'accessory cells'. These are usually 20–50 μm in diameter, borne singly or in clusters of 12 or more, and are typically formed on spiralling hyphae. Vesicle walls may be smooth, or have rounded lobes or knobs, digitate or coralloid projections, or spines. These structures have not been seen to germinate, so their function is obscure.

Another fungus, or group of fungi, that forms endomycorrhizae has been referred to in the literature as 'the fine endophyte', since its hyphae are less than 4 μm wide, compared to 10–20 μm for typical VAM fungi. This anomalous endophyte has been described as *Glomus tenue*, though it does not resemble the other species of *Glomus*. Its spores are 10–15 μm in diameter, and when young are hyaline and indistinguishable from the vesicles that form in the roots: when mature, however, they are pigmented. Because of their small size when compared to those of other VAM fungi, the spores of '*Glomus tenue*' are not often recorded in nature, though mycorrhizae of this type are common, particularly in dry regions.

Trappe and Fogel (1977) pointed out that VA mycorrhizae have actually been observed in about 1000 genera of plants representing some 200 families. This is a relatively small sample of the 350 000 extant species of higher plants, but, as in opinion polls, a small sample is often perfectly adequate to establish general trends, sometimes with surprising accuracy. In this case, we now confidently forecast that about 90% of vascular plants normally establish symbiotic relationships with VAM fungi.

The taxonomy of the plants with which the VAM fungi form symbioses is a vast subject which lies far beyond the scope of this chapter. In fact, relatively few families are known to be largely non-VA mycorrhizal; the Brassicaceae, Commelinaceae, Cyperaceae, Juncaceae and Proteaceae, as well as some members of the Capparaceae, Polygonaceae, Resedaceae, Urticaceae and herbaceous members of the Caryophyllales (Amaranthaceae, Caryophyllaceae, Chenopodiaceae, Portulacaceae), plus, of course, most of the 2000 woody plant species that are ectomycorrhizal.

Since Cronquist (1981) lists just under 400 families of higher plants, we estimate that all or most members of over 380 of these families form VA mycorrhizae. In addition, there are about 50 families of non-flowering vascular plants (Gymnosperms, Pteridophytes, *etc.*), many of which are also VA mycorrhizal. The total number of plant species involved is astronomical. It is estimated that there are about 350 000 species of higher plants, so we suggest that over 300 000 of these will be receptive hosts for the incomparably smaller number of VA mycorrhizal fungi (78 taxa recognized as of February 1983). The implications of these figures for the host range of the VAM fungi are profound. If the hosts were divided up evenly among the fungi, with no overlap in host range, each fungus would have over 3 000 potential partners. In fact, we know that host ranges overlap very extensively, and we can suggest that some individual VAM fungi may well have access to tens of thousands of hosts.

Present information suggests that the distribution of VAM fungi is essentially world-wide, since they could occur wherever more than 300 000 host plants grow. This forecast may well change, as host ranges and ecological specificities of individual fungal species are investigated in the field. A few distributional patterns are already emerging. A recent taxonomic redefinition of *Glomus macrocarpum* has reduced the apparent distribution of this species (Berch and Fortin, 1983b). The same study showed that *Glomus australe*, a segregate from *Glomus macrocarpum* sensu lato, may be restricted to Tasmania and mainland Australia.

8.4.2 Morphology and Development of Vesicular–Arbuscular Mycorrhizae

The vesicular–arbuscular mycorrhiza is a much more subtle and less obvious phenomenon than the ectotrophic mycorrhiza. The presence of a VA mycobiont in a root is usually undetectable by the naked eye. There is no overt morphological change in the roots, no mycelial sheath, no exuberant eruption of macroscopic fungal fructifications. Yet, as appropriate clearing and staining will demonstrate, the absorbing roots may be extensively colonized. The technique most frequently used is that described by Phillips and Hayman (1970) or some modification thereof (for example, Berch and Kendrick, 1982). Since anyone who wishes to work with VA mycorrhizae will have to examine colonized roots for various purposes, we will give a complete staining technique.

Roots, fresh or fixed in FAA, are heated in 10% (w/v) KOH at 90 °C for 1–2 h to clear the cytoplasm from the cortical cells. If roots are darkly pigmented, it may be necessary to immerse them in an alkaline solution of 10 vol. H_2O_2 at 20 °C for 10–60 min until bleached. The roots are then washed in several changes of water, the last change acidified by addition of a few drops of lactic acid or HCl. Roots are then stained in 0.5% acid fuchsin or 0.05% trypan blue in lactoglycerol heated to fuming. Excess stain is subsequently removed by heating the material in lactoglycerol. Segments (0.5–1.0 cm) of the stained roots are mounted in lactoglycerol or poly(vinyl alcohol)–lactophenol. Kormanik *et al.* (1980) described a technique for handling large numbers of samples simultaneously.

We may outline the life cycle of the vesicular–arbuscular mycorrhizal fungi as follows. Spores in the soil germinate even in the absence of a suitable partner, and often in the absence of any apparent stimulus, but usually in conditions that are also appropriate for plant seed germination and root growth. If a receptive root is not encountered, the cytoplasm may be retracted from the germ tube and the spore may retain the capacity to germinate on several subsequent occasions. This multiple germination may reduce the vulnerability of the fungus to grazing by mycophagous soil microfauna (Koske, 1981), as well as increasing the likelihood that the fungus will successfully colonize a plant.

The interaction of fungus and plant may begin well before the hypha and the root make physical contact. Germ tubes of *Gigaspora gigantea* were attracted through the air toward roots of bean or corn, presumably by volatile substances that were active over at least 10 mm (Koske, 1982). Response of the VAM fungus to volatile attractants emitted by roots would increase the probability of eventual mycorrhiza formation. In addition, germinating spores of *Glomus mosseae* liberated substances having the same kind of biological activity as gibberellins and cytokinins (Barea and Azcón-Aguilar, 1982). The role of these substances in recognition and colonization under natural conditions deserves investigation.

If the germ tube, or any part of the network of hyphae ramifying through the soil, encounters a receptive root or root hair, an appressorium forms at the point of contact, and penetration of the root cortex follows. Penetration occurs into or between epidermal cells in the zone of differentiation and elongation of the fine feeder roots. As is the case in other root symbioses, such as ectomycorrhizae and actinorrhizae, the symbiosis is initiated in juvenile tissues or cells. Hussey and Peterson (1982) observed that the preferred site of penetration of *Asparagus vulgaris* roots by *Glomus* sp. was the 'short cells', cells which had not yet developed the secondary wall thickening characteristic of mature epidermal cells.

Within the epidermal and outermost cortical cells, the VAM fungus may form loops of hyphae, or it may traverse these cells to penetrate the next deeper layer of cells. Subsequently, the hyphae grow between or within the cortical cells, although the meristematic cells and the endodermal cells are never entered. Specialized branches enter individual cortical cells and form finely arborescent structures called arbuscules, which remain enclosed by the host plasmalemma, and are apparently the sites of exchange between the fungus and the plant. After some time, the arbuscules gradually break down and their fine branches disappear. It was formerly assumed that the plant was digesting the fungal structure, and that phosphorus passed to the plant as a result of this digestion. However, Cox and Tinker (1976) demonstrated that digestion of the arbuscule could not provide the plant with more than 0.065% of the phosphorus that enters the mycorrhiza (Sanders and Tinker, 1973). This seems to indicate that there must be an active mechanism transferring phosphorus to the plant throughout the duration of the mycorrhizal relationship. Polyphosphate granules, involved in P transport within the fungal cytoplasm, have been seen in vacuoles in the inter- and intra-cellular hyphae (Cox *et al.*, 1975), but are no longer observed in the finest branches of the arbuscules (Callow *et al.*, 1978). Those branches have been found to contain acid and alkaline phosphatases (Gianinazzi *et al.*, 1979).

Individual arbuscules seem to function for a period of 4–15 days (Carling and Brown, 1982), and the cytoplasm of the colonized host cell remains alive, and seems in fact to be greatly stimulated, throughout this period. The nucleus, in particular, of the arbuscule-containing cells is much larger than that of non-colonized cells, and the volume of the cytoplasm is also greatly increased. Ultimately, retraction septa form in the arbuscule as the cytoplasm is drawn back, beginning with the fine branches and progressing toward the main trunk (Kaspari, 1975). The fine branches then collapse and the cytoplasm of the root cell eventually returns to normal, though traces of the fungal wall may persist long after the active fungus has decamped.

Many, though not all, endomycorrhizal fungi that are characterized by the formation of arbuscules also form terminal and/or intercalary vesicles within the mycorrhiza. These are thin-walled, expanded structures which are not delimited by a septum, and often contain large quantities of lipids. We have recorded up to 500 vesicles cm^{-1} of root in leek, the root cortex taking on the appearance of an almost solid mass of vesicles. This occurs when the root zone concerned is no longer as actively involved in absorption as in the earlier stages of infection, and it seems probable that the cortex is no longer functional as a pathway for nutrient uptake by the plant. Since the stele is not infected, it can of course still translocate substances to and from the active feeder roots.

Vesicles, which may be involved in temporary storage of modified photosynthates received from the plant, are not formed by the species of *Gigaspora*. Daft and Nicolson (1974) proposed that the term 'arbuscular mycorrhiza' or 'AM' be adopted to describe the kind of endomycorrhiza formed by members of the Endogonaceae, since the arbuscule is the only structure common to them all. Although we agree that 'arbuscular mycorrhiza' is indeed more accurate and therefore more appropriate than 'vesicular–arbuscular mycorrhiza' as a unifying descriptive term for this phenomenon, we have generally followed the established convention in this chapter, to permit easier access to the literature, which is usually keyed to the traditional term, vesicular–arbuscular mycorrhiza or 'VAM'.

In conjunction with the development of the fungus inside the root, which is generally described as the intramatrical phase, the fungus proliferates in the soil outside the root, this being described as the extramatrical phase. A network of extramatrical hyphae ramifies through the soil, extending as far as 8 cm from the root (Rhodes and Gerdemann, 1975). This means that a mycorrhizal plant can exploit several times the volume of soil available to a non-mycorrhizal plant, since active translocation of minerals along the extramatrical hyphae to the host plant has been convincingly demonstrated (Cooper and Tinker, 1981; Hattingh *et al.*, 1973; Pearson and Tinker, 1975; Rhodes and Gerdemann, 1975).

In exchange, the fungus receives photosynthates, some of which are transported to the extramatrical hyphae, and to the spores that develop on those hyphae, as has been established through ^{14}C tracer studies (Bevege *et al.*, 1975; Cox *et al.*, 1975; Ho and Trappe, 1973).

The large, rounded, presumably asexual spores of the fungus are formed in the soil singly, or in aggregations up to 1 cm in diameter called sporocarps. They are also occasionally found in roots, empty seed coats (Taber, 1982), insect carapaces (Rabatin and Rhodes, 1982), rhizomes (Taber and Trappe, 1982) and other available, relatively protected spaces. These spores range in size from 50 to 600 μm, and may be very thick-walled (up to 30 μm), darkly pigmented, and filled with storage lipids, all features that emphasize their role in long-term survival when host plants are absent or dormant. These spores may be dispersed in flood waters or in wind-driven soil, and are well equipped to survive desiccation and UV radiation (Tommerup, 1981). Soil fauna, including earthworms and small rodents, may contribute to the dispersal of the spores, which are known to remain viable after passage through the gut of such animals (Diehl, 1939; Maser *et al.*, 1978).

The spores will eventually germinate, producing hyphae which will once more grow through the soil and perhaps encounter another plant. The identity of the plant may not matter much, since VA mycorrhizal fungi can usually relate successfully to a very large number of host species, though there is no doubt that some pairings are more efficient than others, as has been demonstrated by Fairweather and Parbery (1982). This aspect of the relationship is discussed elsewhere in the chapter.

8.4.3 Sources of VAM Inoculum

The inoculation of experimental plants can be carried out with colonized soil, or colonized roots, or spores, or a mixture of all three. The first two kinds of inocula have some profound disadvantages. Soil may contain more than one mycorrhizal fungus, and it may also contain pathogenic organisms. Root inoculum can be used if it has been grown under more or less aseptic conditions, and if the original inoculum used to infect the roots was of a single, named species. The structures developing in roots, although diagnostic of VAM fungi as a group, are not usually diagnostic of a particular species. The features diagnostic of individual species are present only in the spores, which develop primarily on extramatrical hyphae, so these are perhaps the best inoculum for laboratory experiments.

Since the fungi cannot be grown in axenic culture, the only source of spores is the soil near the roots of infected plants. Agricultural soil has been found to contain varying numbers of VAM fungal spores. Sutton and Barron (1972) recorded up to 70 ± 17 spores g^{-1} dry soil under maize, and up to 86 ± 11 under strawberry, all in the uppermost 24 cm of soil. Hayman (1970) and Kruckelmann (1975) recorded much lower numbers, possibly because their extraction techniques were less effective. Kessler and Blank (1972) estimated that the upper 10 cm of soil in an undisturbed *Acer*-dominated hardwood forest in Michigan contained nearly 7 000 000 sporocarps ha^{-1} of VAM species.

Although inoculum for experiments could be derived from naturally colonized soil, this kind of inoculum will usually contain several different VAM fungi, and is also present at a fairly low level. It was discovered that individual VAM fungi could be propagated in a dual culture system, fungus plus host plant, in what is called pot culture. Furlan and Fortin (1975) noted that soil in which onions inoculated with *Gigaspora calospora* had been grown for 19 months contained 325 spores g^{-1} dry soil.

Gerdemann (1955) devised the first useful technique for extracting spores from soil. A soil sample was suspended in four times its volume of water, heavier particles were allowed to settle for a few seconds, then the liquid was decanted through a sieve with 1 mm mesh. Whatever passed through this sieve was then poured through another sieve with 0.25 mm (250 μm) mesh. Material retained by this sieve was washed and transferred to a Petri dish, and the spores picked out by hand under a dissecting microscope. This technique was slightly refined by Gerdemann and Nicolson (1963) who used the following series of sieves: 1 mm, 710 μm, 420 μm, 250 μm, 149 μm, 105 μm, 74 μm and 44 μm. They found that most of the desired spores fell in the 420–149 μm range, and they used this fraction for their study. Since it is recommended that from 50 to 500 spores are required for inoculation of a single pot, and a well-designed experiment might need scores of pots, an investigator was condemned to spend an excessive amount of time simply to collect enough spores for each experiment. The initial sieving could, however, be speeded up by nesting the set of sieves in a vertical series.

Ohms (1957) reported that Gerdemann's technique yielded only about 100 spores per operator per day, and tried to make spore recovery less laborious by centrifuging the organic fraction of soil on a sucrose gradient. He sieved the soil initially to obtain the fraction retained by a 177 μm mesh, then suspended this material in water. Lighter material was decanted after the spores and heavier soil particles had settled, or had been centrifuged for 3 min at 3100 r.p.m. (presumably in a clinical centrifuge). A suspension of the spores and the remaining debris was added to a 50 ml centrifuge tube containing a sucrose gradient established by placing 10 ml of 50% sucrose at the bottom, 15 ml of 25% sucrose above it, and 10 ml of water at the top. The upper layers were added very carefully with a hypodermic syringe equipped with a curved extension that allowed the liquid to be placed directly on the side wall of the vertically held tube. After centrifuging for 5 min at 3100 r.p.m., the spores were largely to be found in the middle layer, while the soil particles had settled to the bottom layer. The spores in the middle layer were pipetted into a beaker, and the sucrose solution diluted with water to decrease the specific gravity of the solution and allow the spores to sink to the bottom. The solution was then decanted and the spores were washed several times with water. They could then be readily hand-sorted. This method yielded over 1500 spores per man-day. One disadvantage of this technique was the osmotic pressure exerted by the sugar solution, which could apparently damage young spores.

To avoid this problem, Mosse and Jones (1968) separated the spores by differential sedimentation on gelatin columns. Gelatin was dissolved in distilled water to make solutions of 5%, 15% and 20% (w/v), which had specific gravities of 1.02, 1.04 and 1.05, respectively, at 32 °C. Liquid 20% gelatin at 30 °C was poured into glass tubes 30 cm long and with an internal diameter of 37 mm to a depth of about 5 cm and allowed to solidify at 4 °C. A 5 cm layer of 15% gelatin was then added and allowed to solidify, then a similar layer of the 5% solution, and finally a 10 cm layer of water. Columns were stored at 4 °C until used. For spore separation, the columns were placed upright in a water bath at 32 °C and allowed to liquefy. Then the fraction of a 50 g soil sample that passed through a 250 μm sieve but was retained on a 100 μm sieve was washed into the water layer at the top of a column. After 30 min of sedimentation, the columns were cooled to 4 °C again. Spores were extracted from the columns as follows. The water layer was discarded. The column was then held under a hot water tap until the gelatin became free from the wall of the glass tube, and could be slid out of the tube. The core could now be cut into segments. It was found that VAM spores tended to sediment near the 5–15% interface. A 1 cm segment of the column (0.5 cm on either side of the interface) was cut out, gently melted, then strained through a piece of finely woven material supported on a 100 μm sieve. This gave good recovery of four different kinds of VAM spores. Mosse and Jones (1968) claimed that their gelatin columns had the following advantages over the sucrose solutions used by Ohms (1957): (1) high viscosity and relatively high specific gravity; (2) low osmotic pressure; (3) solidification below 30 °C, which makes it easy to set up columns, to maintain sharp boundaries between layers, even in very wide columns, and to segregate portions of the resolidified columns for recovery of the material they contain.

Sutton and Barron (1972) extracted spores from soil by a flotation–adhesion technique that depended on the rapid flotation of spores on water, the hydrophobic nature of their surface layer, and their tendency to adhere to glass surfaces. A 10 g subsample of soil was placed in a 50 ml cylinder which was then filled with water to within about 2 cm of the brim, and shaken for 10–15 s. The suspension was allowed to settle for about 2 min, while the VAM fungal spores accumulated in the scum floating at the meniscus. Most of the aqueous layer was then decanted into a separatory funnel, carrying most of the spores with it. Further water was added to the cylinder, and the shaking, sedimentation and decanting processes repeated twice more. The accumulated liquid in the separatory funnel was then allowed to settle for 2–3 min. This allowed the spores to collect around the margin of the meniscus in the funnel. The water was then drained at 75–100 ml min^{-1} from the funnel into a second 250 ml separatory funnel, where it was allowed to settle once more before being drained at a similar rate. The scum and spores adhering to the insides of the two funnels were then washed onto a filter paper. Preliminary experimentation showed that this technique recovered 94–98% of the spores in any given soil sample, unfortunately including spores that are dead.

In an effort to simplify and speed up the process of collecting the thousands of spores needed for experimental purposes, Furlan and Fortin (1975) introduced a flotation–bubbling collection technique. The soil was first sieved to concentrate the spores, then the material was added to a column containing a 50% aqueous solution of glycerol, which was found to have a higher specific gravity (1.13) than the spores without exerting any deleterious osmotic effects on them. Compressed air was injected at the bottom of the column through a fritted disc, so that the small bubbles would both agitate the suspension, helping to separate the spores from soil particles, and propel the spores toward the surface, where they could be recovered. Furlan and Fortin (1975) found that the 74–500 μm soil fraction was most appropriate for this technique. The apparatus consisted of a glass tube of 6 cm internal diameter and 55 cm long, held firmly in the vertical position, and with a fritted disc with pore size 4–5.5 μm fused in position 5 cm above the bottom. Compressed air (monitored by a pressure gauge) was supplied to the bottom of the column at approximately 83 kPa. A litre of 50% glycerol solution was added to the tube, the air pressure raised to 103.4 kPa, and 10 g of soil fraction added. After the bubbling had continued for 1–2 min the air was turned off, the foam allowed to subside, soil particles adhering to the wall of the tube above the liquid were carefully washed down, and the suspension was allowed to settle. After 30 min the supernatant was drawn off, the tip of the aspiration tube being held just below the surface of the liquid to pick up all the spores. All supernatant was aspirated to within 1 cm of the deposit at the bottom. The supernatant was now sieved, and the spores were washed thoroughly in tap water and stored in Ringer's solution (adjusted to a pH of 7.4 with NaOH, filtered and autoclaved). This technique was found to give 94.5% recovery of spores. If virtually total recovery was required, a second treatment would increase the yield to 99.5%.

Mertz *et al.* (1979) used discontinuous sucrose gradients to recover large numbers of spores from massive soil samples. They decanted and wet-sieved 18 kg of soil with cold water, and found that most spores were present in the 425–250 μm fraction. Spores were separated from most of the remaining debris using discontinuous 30% (w/v) aqueous sucrose gradients. The sieved material was layered on 600 ml water over 200 ml sucrose in a 1 l beaker. After settling, the spores and debris that collected at the interface were removed by vacuum aspiration, rinsed in cold water, and centrifuged for 1–5 min at 1600 $\times$ g in a clinical centrifuge on a second gradient (15 ml water over 20 ml sucrose in a 50 ml tube), the duration of centrifugation being determined by the kind and amount of debris present. Fresh pot culture material needed 5 min of centrifugation to ensure complete separation of the finest root fragments. The spores were then aspirated from the gradient interface, rinsed, and stored in cold water. In this manner, up to 10 000 spores d^{-1} could be recovered by a single operator. The authors claimed that their method was faster than the bubbling–glycerol technique of Furlan and Fortin (1975).

Furlan *et al.* (1980) used several radiopaque media (substances designed to be injected intravenously into humans for a variety of radiographic diagnostic procedures), instead of sucrose, to establish density gradients. These media are free from the deleterious osmotic effects associated with prolonged immersion in strong sucrose solutions. Also, being innocuous in human tissues, they should have no harmful effects on the fungal spores. This assumption turned out to be correct. The radiopaque media used were Renografin-60 (E. R. Squibb & Sons Ltd., Montreal, Canada), Telebrix-30 and Telebrix-38 (Laboratoires Guerbet, France), and Radioselectan (Laboratoires Schering, France). These substances are of generally similar composition: Renografin-60 contains 52.6% of the meglumine (*N*-methylglucamine) salt and 7.5% of the sodium salt of diatrizoic (3,5-diacetamido-2,4,6-triiodobenzoic) acid. It also contains 0.32% sodium citrate as a buffer, and 0.04% disodium edetate as a sequestering agent. It has a pH of 7.0–7.6.

The 53–250 μm or 53–500 μm soil fraction was initially subjected to the flotation–bubbling technique of Furlan and Fortin (1975), then the spores and soil particles derived from the supernatant were centrifuged in concentration gradients of the radiopaque media. Each of the gradients used contained a series of four dilutions of the medium, 60%, 40%, 20% and 10% (v/v). These gradients were assembled in 50 ml centrifuge tubes by placing 10 ml of 60% solution at the bottom of the tube, followed by 5 ml of each of the 40%, 20% and 10% solutions, and lastly 5 ml of water or Ringer's solution. Two ml of the spore concentrate was added to the top of the gradient and the tube centrifuged at 500 $\times$ g for 10 min. The spores usually concentrated just above the various gradient interfaces, with most above the 10% and 20% step gradients. After extraction, spores were rinsed in tap water and stored in Ringer's solution at 4 °C. This technique is very effective for cleaning up the spore concentrate derived from the flotation–bubbling process. Furlan *et al.* (1980) also found that another medium, Percoll (Pharmacia), could be used, and that a continuous gradient could be formed by centrifugation at high speed, 40 000$\times$ g for 20 min.

An alternative approach to the separation of VAM spores from soil has been developed by Tommerup and Carter (1982) from a technique designed originally to measure the particle size of powders. Soils, collected dry, were fractionated in a dry particle size analysis elutriator. The basic principles are: (1) mobilization of particles into an airstream; (2) regulation of the airstream velocity to suspend particles of equivalent diameters; (3) collection of the fractions on filter papers; and (4) separation of spores in a sucrose gradient. Propagules of fungi remained viable after undergoing this treatment, as evidenced by their ready germination. This technique greatly reduces the exposure of the spores to wetting, and the sucrose gradient could probably be eliminated for rough bulk separation of spores. Since wetting may induce germination of many VAM spores, the older wet-sieving and decanting technique could result in some loss of inoculum potential. The dry particle size analysis elutriator is a simple and inexpensive piece of equipment that can be adapted to either small- or large-scale spore separation.

It must be noted that spores collected by the methods described above are not sterile. Mertz *et al.* (1979) devised a gentle but effective surface sterilization process as a logical corollary of their spore-collecting technique. Within 24 h of being collected, spores were washed twice for 1 min by shaking in an 0.05% (w/v) aqueous solution of Tween 20, a surfactant, and rinsed in sterile water after each washing. Then the spores were immersed twice in 2% (w/v) Chloramine T for 10 min, 5 min of each treatment being under a vacuum of up to 80 kPa. Spores were rinsed in sterile water after each treatment. More rigorous sterilization was achieved by storing the spores for at least a week in the dark at 4 °C in a filter-sterilized solution of 200 mg l^{-1} streptomycin and 100 mg l^{-1} gentamicin. Just before use as inoculum, spores were treated for 20 min in Chloramine T, rinsed with sterile water, then rinsed in sterile streptomycin–gentamicin solution. This procedure yielded spores that were 100% surface sterile, yet still 90% germinable. Spores could be stored for up to 3 months (and possibly longer) at 4 °C without loss of viability.

All the techniques discussed above recover spores from naturally or artificially inoculated soil in which suitable host plants have been grown. Some effort is now being made to maximize the yield of spores, not simply by efficient methods of recovery, but by discovering what factors, if any, can stimulate the fungus to produce larger numbers of spores per unit volume of soil. Sylvia and Schenck (1982) tried six approaches: (1) pruning the host shoot; (2) withholding water; (3) subjecting the symbionts to cold and darkness; (4) spraying with the herbicide, paraquat; (5) drenching the soil with P_2O_5; and (6) drenching the soil with ethazole. Unfortunately, some of these treatments sharply reduced the number of spores produced, and none increased the yield. But Ferguson (1981) found that plants grown in 15 000 cm^3 pots produced 90 times as many spores as those grown in 750 cm^3 pots, although the soil volume in the larger pots was only 20 times greater. Ferguson (1981) also noted that limited daily watering, and a temperature regime most suited to the host plant, were stimulatory to the VAM fungi. Clearly, these avenues deserve further exploration.

Starter cultures of at least 15 different named VA mycorrhizal fungi were until recently commercially available from Abbott Research Center, Long Grove, Illinois, USA. At $25 per culture, this was probably the cheapest and easiest way of initiating a research program on these fungi. Each starter culture consisted of living fungal spores, root tissue of the previous host and soil. According to the vendor, each culture weighed about 200 g, and contained enough spores to inoculate at least five pot cultures. Unfortunately, apparently because of quality control problems, these cultures are no longer available. Mixed inoculum of several VAM fungi was recently available under the trade name 'Biofert' from Biofertec Inc., Ancienne Lorette, Quebec, Canada, but quality control problems have led to its withdrawal from the market, at least for the present.

We will now discuss the two phenomena among the VAM fungi which seem to hold out the greatest promise for production of inoculum on a large scale. The first is exhibited by a recently rediscovered species, *Glomus epigaeum* (Daniels and Trappe, 1979) (= *G. versiforme* fide; Berch and Fortin, 1983b) which is most unusual among VAM fungi in producing macroscopic sporocarps on the surface of the soil in pot cultures. Each of these sporocarps may contain millions of spores. They develop without the usual admixture of soil found in the hypogeous sporocarps of other species, and they can be collected without disturbing the pot culture, which can thus go on producing further sporocarps for years.

The commercial potential of *Glomus versiforme* has recently been examined by Daniels and Menge (1981). These authors investigated several characteristics of the fungus. (1) Sporocarp production in pot culture with several hosts under different water regimes. The greatest number of sporocarps, 4.5 pot^{-1} $month^{-1}$, was produced on Sudan grass (*Sorghum vulgare*), watered manually. Sporocarps ranged from 3 to 12 mm diam, and individual sporocarps contained from 1 × 10^5 to 7.5 × 10^6 spores. (2) Germinability of spores following 3 month storage under 13 different conditions, and with or without surface sterilization spores failed to germinate, with the exception of those stored in wet loamy sand. Non-surface sterilized spores stored under 6 of the 13 conditions germinated. Germination rates were as follows. Stored in wet loamy sand, 1%; in streptomycin–gentamycin solution, 13%; in dry loamy sand, 33%; in sterile water, 37%; in sterile bentonite, 80%; and as entire sporocarps, 83%. Daniels and Menge (1981) also found that *Glomus versiforme* infected six of seven potential hosts tested, and produced significant increases in growth ($P = 0.05$) of four of five hosts tested.

In summary, *Glomus versiforme* can produce up to 1.8 × 10^7 spores $month^{-1}$ 10 cm pot^{-1} culture over an extended period. These spores can be easily harvested, stored, and used as inoculum when required. The fungus probably has a wide host range, and usually produces marked beneficial effects in its hosts under phosphorus-deficient conditions. Paradoxically, however, Daniels *et al.* (1982) reported difficulty in even detecting infection caused by an isolate of *Glomus versiforme* in Sudan grass. Although the isolate number was, unfortunately, not mentioned by Daniels and Menge (1981), it seems possible that the radically different performances reported in the two studies were due to isolate differences. We cannot stress too strongly the need for fully documenting all isolates used in mycorrhizal research and for depositing voucher specimens in a recognized herbarium.

The second method of producing VAM inoculum we shall mention may be the one with the most potential for generating the amounts of material required for large-scale agricultural applications. We refer to infected root inoculum. Jalraji-Hare is at present working (personal communication) with what is believed to be a species of *Glomus* that rarely produces extramatrical spores when grown in pot cultures. It does, however, produce very large numbers of vesicles in infected roots, and in the absence of spores she was forced to use root inoculum in her experiments. Since she obtains consistently high levels of infection with this material, it occurred to us (and to Fortin and his coworkers, from whom we had obtained our original isolate) that root inoculum was, in fact, far easier to obtain in large quantities than spores. Assuming that the dual cultures have been established using a sterile soil or soil-substitute, and that reasonably clean conditions have been maintained in the growth chambers to reduce the possibility of contamination by other unwanted VAM or non-mycorrhizal fungi, it is possible to harvest a very large amount of inoculum in a short time. All that is necessary is to pull up the infected plants, wash the root system, then cut off and use or store the infected roots, presumably refrigerated in Ringer's solution or an appropriate aqueous solution of glycerol. Of course, some kind of quality control in the form of regular staining and microscopic observations of root samples would be necessary to ensure that roots were not colonized by pathogens. But we see little to prevent the commercial development of root inoculum in the near future.

Mosse and coworkers at Rothamsted have been studying the potential for producing commercial quantities of infected root inoculum in nutrient film culture (NFT). The infected host roots are bathed in a shallow film of recirculating nutrient solution in which low levels of phosphorous are maintained by passing the solution over relatively unavailable sources of phosphorus. Although no special precautions were taken to exclude airborne spores, contamination appears to be minimal. Since tomatoes are already produced commercially by NFT, the potential exists for a crop of infected plants to be raised and the roots to be used to infect the next crop. Also Mugnier (personal communication) has developed a system for producing mycorrhizae in genetically transformed root organ tissue culture. This holds promise for large scale inoculum production and for detailed studies of spore development.

8.4.4 Evaluation and Selection of VAM Fungi

VA mycorrhizal fungi may be detectable as spores or sporocarps in the soil, or as hyphae, arbuscules, and in three of the four genera, vesicles, within the roots. We discussed the recovery of spores from soil in Section 8.4.3. Spores are certainly one expression of the success of the symbiosis, but although some of the methods elaborated for their recovery are capable of giving almost 100% retrieval, spore production is not usually employed in quantitative assessments of VA mycorrhizae. There are a number of reasons for this. (1) Spores may not be produced until the mycorrhizal relationship has been established for several months. (2) Spores may be eaten by arthropods, they may be parasitized by fungi, and they may disappear or lose their viability for a variety of reasons. (3) Spores may be of very different sizes, both among species and within them. They are sometimes too small to be readily separated from soil, as in *Glomus tenue*, or are found concentrated in occasional sporocarps, as in *Glomus fasciculatum* (Abbott and Robson, 1982). (4) Some VAM fungi rarely if ever produce spores.

If it is desired to follow the development of the colonization during the course of an experiment, or to assess the level of colonization present in indigenous plants collected in the field, it is logical to study stained roots. Giovannetti and Mosse (1980) compared several different methods which had been developed by various authors to estimate the amount of colonization in roots. They found that the gridline intersect method was the most accurate of those tested, so we will describe this method here.

Cleared and stained roots are spread out evenly over the bottom of a 10 cm Petri dish which is marked with a grid of lines delimiting 0.5 inch (13 mm) squares. The gridlines are systematically scanned under a dissecting microscope, and the presence or absence of colonization is recorded at each point at which a root intersects a line. If three sets of 100 intersects are recorded, and the mean value determined, the percentage of the root length colonized will be estimated with a fair degree of accuracy. This technique also allows the total length of roots in the dish to be determined. All grid-intersects are counted, and the final total equals the total length of the roots in cm. Newman (1966) related the length of roots in a given area to the number of times they intersect a number of straight lines placed randomly in that area. Marsh (1971) refined the method by arranging the lines in a regular grid, and found that if the distance between lines was 14/11 of the chosen measuring unit, then the number of intersects equalled the length of the roots, expressed in that unit. Conveniently, 14/11 cm = 0.5 in.

The responses of host plants to VAM can be measured in several ways. One of the most universally accepted parameters is dry weight production (recorded after tissues have been dried at 70 °C for 72 h). It is often instructive to record dry weight of the root and shoot systems separately, since the response of the two systems to the presence of the fungus may differ. Other possible measures of response are differences in plant height, stem diameter, shoot volume and leaf number and area, all of which, since they are non-destructive, can be measured repeatedly on the same plants at intervals during their growth (Linderman and Hendrix, 1982). Crop yield is also a valid measure of mycorrhizal success. Transplant survival and disease reaction are others that may be important in particular situations. Mineral analyses or tracer studies may reflect physiological changes resulting from VAM infection. VAM inoculation of *in vitro*-produced raspberry plants (*Rubus occidentalis*) and apple trees (*Malus pumila*) facilitated their recovery from transplant shock and also accelerated their subsequent growth (Granger *et al.*, 1983).

Chilvers and Daft (1982) demonstrated that onion plants (*Allium cepa*) inoculated with *Glomus caledonicum* were better able than non-mycorrhizal plants to withstand and recover from drops in temperature (in this instance to 5 °C). This indicates that when host response to mycorrhization by particular VAM fungi is being evaluated, normal climatic conditions can be allowed to play their usual role in the selection of both plant and fungal survivors. Since the VAM fungus is often essentially parasitic on its host during the early phase of colonization, it is essential to allow a growth period long enough to reflect the positive effects that accrue after the fungus has developed an extensive extramatrical mycelium and is contributing to the mineral nutrition of the plant. Alternatively, multiple sequential harvests can be employed. Selection of an appropriately compatible host is also important. In Section 8.4.1 we listed the families of plants that are not hospitable to VAM fungi, but every host–fungus combination must be tested individually.

The extent of colonization of the root and the development of the extramatrical mycelium can normally be directly related to the enhancement of plant growth in response to mycorrhization. However, Graham *et al.* (1982) found that sometimes growth is not improved despite extensive root colonization. This can occur because the extramatrical mycelium fails to develop properly, perhaps because of the presence of some inhibitory influence(s) in the soil. Among these may be certain collembola, mites and perhaps also nematodes which can feed on the extramatrical hyphae (Hayman, 1982). Particularly in the greenhouse, arthropods can be persistent pests, and their impact on the outcome of experiments cannot be discounted.

Safir and Duniway (1982) give a discussion of the interplay between a number of environmental factors (phosphorus, pH, nitrogen, water and temperature) and the host–fungus relationship.

Significant differences exist among the relative efficiencies of VAM fungi as mycorrhizal partners, both among different species (Mosse, 1972, 1973, 1977; Powell, 1975), and among different strains of what appears to be single morphologically defined species (Abbott and Robson, 1978). Although the mechanisms underlying these differentials are not well understood, there is some suggestion that much of the effect may well be due to differences in ability to infect host roots quickly and extensively (Abbott and Robson, 1982). Other factors, such as the extent of the extramatrical mycelium, and the efficiency of nutrient absorption and translocation, may also be important. Extramatrical hyphae have been shown to retrieve phosphorus 27 mm from a root, phosphorus that was completely unavailable to non-mycorrhizal roots at similar distances (Hattingh *et al.*, 1973). Owusu-Bennoah and Wild (1979) showed that the zone of depletion around mycorrhizal roots was twice that around non-mycorrhizal roots. Barrow *et al.* (1977) showed that VAM fungi were equally adept at accumulating soluble and insoluble phosphorus.

Abbott and Robson (1982) have stressed the importance of knowing the response curves of mycorrhizal and non-mycorrhizal plants to P application when evaluations of the impact of VAM fungi on the growth, P uptake and mycorrhiza dependency of different species of plants are being made. No stimulation normally results from mycorrhization by even the most efficient VAM fungi if available phosphorus is present at luxury levels; in fact, depression of growth can occur under these conditions. To standardize the expression of 'relative mycorrhiza dependency' for a given plant at a known level of available P, Menge *et al.* (1978) used the ratio of the dry weight of the mycorrhizal plant to that of the non-mycorrhizal plant, expressing it as a percentage. This method of calculation sometimes gives a rather inappropriate 'mycorrhiza dependency' of well over 100%. Plenchette *et al.* (1983) proposed that the expression of mycorrhiza dependency be shifted into the normal 0–100% range by application of the formula:

$$100 \times \frac{\text{dry wt of mycorrhizal plant } - \text{ dry wt of non-mycorrhizal plant}}{\text{dry wt of mycorrhizal plant}}$$

to be calculated for any given level of P availability. Thus, Plenchette (1982) reports that, at 100 μg g^{-1} available P, the relative mycorrhiza dependency of carrot (*Daucus carota* var. Nantaise) is 99.2%, and that of wheat (*Triticum aestivum* var. Glenlea) is 0%. No value was calculated for garden beets (*Beta vulgaris* var. Dark Red), since VAM were not formed in this species. While satisfying Abbott and Robson's (1982) requirement that mycorrhiza dependency be expressed in relation to the level of available P, this calculation also permits comparisons of the responses of any species of plant to different sources of P, and to different mycorrhizal fungi.

As yet we know little about the contributions to plant growth made by VAM fungi occurring naturally in field soils (although we extrapolate a good deal from the results of our laboratory experiments). This is partly because any treatment designed to kill all spores (and thus allow comparisons to be made) is sufficiently drastic that it will also change many other factors in the chemical makeup of the soil, and probably alter its nutrient status (Bowen and Rovira, 1969). However, Plenchette *et al.* (1981) found that young apple trees preinoculated with VAM fungi grew better than uninoculated trees, even when planted out in unsterilized field soil which contained an indigenous VAM flora.

It is now well established that the presence of VA mycorrhizal fungi in the roots of plants tends to reduce the incidence and the severity of soil-borne diseases in those plants. Dehne (1982) lists 56 reports of interactions between VAM and 18 soil-borne plant pathogenic fungi, 8 plant parasitic nematodes, and 3 viruses, involving a total of 21 different crop plants. The majority of these investigations showed that mycorrhizal plants had less disease, except in the case of the viruses, when symptoms were always more severe. Dehne advanced several reasons for the reduction in damage caused by fungi. (1) The mycorrhizal plants were healthier and more able to repel or resist the attacks of the pathogenic fungi. (2) As an apparently normal part of the nutrient exchange process, the cells of the mycorrhizal plants may partly digest the arbuscules of the fungus. This implies chitinolytic activity on the part of the plant, a talent that may well be turned against other invading fungi. (3) Possible infection sites on the surface of the plant roots may be preferentially occupied by the VA fungus, to which the plant is probably more susceptible anyway. We refer our readers to other accounts of VAM–pathogen interactions by Schenck and Kellam (1978), Schoenbeck (1980), Schenck (1981), and Schoenbeck and Dehne (1981).

A number of factors must be taken into account when the suitability of specific VAM fungi as mycorrhizal partners for specific plants in specific areas is being considered. These include geographic distribution, frequency of occurrence, host range, soil type, pH, fertility, moisture and organic matter, persistence of inoculum in the soil, rapidity of germination and infection, and phosphorus efficiency, which is a function of the soil volume a fungus explores, the rate at which it takes up phosphorus, and the rate at which that phosphorus is translocated to the plant. Equally important are considerations of inoculum production: how easy it is to persuade the fungus to produce propagules on a large scale under controlled conditions and to store, transport and deliver that inoculum without loss of viability.

The global distribution of most VAM fungi is incompletely known. Although herbaria contain increasing numbers of collections from many parts of the world, their number is still inadequate to permit proper distributions to be extracted from them, and even that distributional data inherent in existing collections has not yet been collated. This is largely because the taxonomy of these fungi is still in a state of flux, new taxa being described every year and old ones being reassessed (see Section 8.4.1). Whatever the reason for the dearth of knowledge, it is most unfortunate that we have so little, because it would be very valuable in the selection of appropriate VAM partners for specific areas. In the absence of such background information we must approach each new problem by conducting extensive sampling in the field, to determine the nature of the indigenous flora, and perhaps to accumulate sufficient inoculum both to carry out an initial experimental program, and to establish pot cultures which will multiply the inoculum for future use.

Despite the apparent lack of host specificity under the artificial conditions of pot culturing in a greenhouse or growth chamber, there is some evidence that, in the field, plants select certain indigenous VAM fungi in preference to others. Schenck and Kinloch (1980) assessed the numbers of spores associated with monocultures of annual crop plants grown for seven years on a newly cleared woodland site. They concluded that the major influence on spore production was host plant specificity for particular VAM fungi. Nemec *et al.* (1981) found that in Florida and California citrus orchards, production of spores by certain VAM fungi was favoured by high soil P, B, Ca and Mg, or salinity, while that of other species was not. As they point out, much of the apparent disagreement over the effects of local conditions on VAM fungi could probably be eliminated if species of VAM fungi were considered individually, instead of as a group. In fact, certain VAM fungi have been linked to particular kinds of soil: *Glomus mosseae* with fine-textured, fertile, high pH soils; *Acaulospora laevis* with coarse-textured, acid soils; *Gigaspora* species with sand dune soil.

For it to be worthwhile to produce and deliver VAM inoculum on a large scale, the organism involved must have been shown to be a more efficient scavenger of phosphorus than other VAM fungi present in the local soil, or to boost the soil population to a level at which it can bring about improvements in the nutrition of the desired crop. Unfortunately, phosphorus efficiency by itself is no guarantee of success for a VAM fungus. In a non-sterilized field of birdsfoot trefoil (*Lotus corniculatus*), introduced strains of VAM fungi improved yield, but only at the first harvest (Lambert *et al.*, 1980). In subsequent years the indigenous strains, presumably better adapted to local conditions, apparently replaced the introduced, more P-efficient strain, and yields returned to normal. It seems that the improvement in plant nutrition produced by the presence of VAM results mainly from the effective increase in the volume of soil that is being exploited by the plant. Therefore, a measure of the amount and extent of the extramatrical mycelium would be of considerable value. Unfortunately, such a measure is difficult to obtain with any degree of accuracy, since when the root–fungus–soil system is disturbed for sampling purposes, the fine network of extramatrical hyphae is usually extensively disrupted. Removal of a root system from soil normally results in almost complete loss of the extramatrical mycelium. A few estimates have been made, using such indicators as the degree to which the extramatrical mycelium binds together the soil around the roots (Sutton and Sheppard, 1976). Fortunately, when Sanders *et al.* (1977) carefully compared the level of root infection with the amount of extramatrical mycelium in four VAM fungi infecting onions, they found a strong positive correlation between these two parameters. Thus the actual density of colonization of plant roots by a VAM fungus appears to be an appropriate measure of its efficiency in exploring the soil.

Nevertheless, we have little or no comparative data on the respective abilities of various VAM fungi to absorb phosphorus from the soil, or to transport it to the host plant. These are two areas in which differentials may well exist. The relationship between what we may call phosphorus efficiency and speed of infection is also important. The effectiveness of a VAM fungus depends on both factors. Since plants normally have higher phosphorus requirements when young, the faster an infection can be established, the better. Thus a fungus which is phosphorus-efficient may still be an unsuitable mycorrhizal partner because it establishes infection too slowly. Once again, comparative studies are lacking.

As Abbott and Robson (1982) have pointed out, it may be possible to simplify the screening process to some extent. For example, it is obviously much easier to test fungi for the rapidity with which they can cause infection than for their ability to increase phosphorus uptake. Such simplification could be justified only if it was determined that early infection was an appropriate index of mycorrhizal suitability. Soil pH is sometimes a limiting factor in the germination of certain VAM fungal spores (Green *et al.*, 1976; Daniels and Trappe, 1980), so it must be noted that the prevailing pH may render some VAM fungi unsuitable for inoculation into certain soils.

Gildon and Tinker (1981) isolated a strain of *Glomus mosseae* from mycorrhizae of clover growing naturally on sites heavily contaminated with Zn and Cd. This VAM strain tolerated high concentrations of these two metals. The existence of such strains suggests that when metal-tolerant plants capable of revegetating such sites are being sought, the tolerance of their endophytes should also be considered.

Fairweather and Parbery (1982) recently discovered a very efficient VA mycorrhiza on tomatoes growing in an extremely phosphate-deficient soil in south-east Australia. They attempted to reconstitute the partnership experimentally, using four kinds of VAM spores recovered from the soil concerned. In view of their striking results, it is worth detailing the experimental procedures they used.

Sand with no Olsen-extractable phosphorus was sterilized by γ-irradiation. Tomato seeds surface-sterilized in 1% sodium hypochlorite for 1–2 min were sown in the sterile sand. When the first leaf emerged, the roots were washed with distilled water and 30 spores of one of the VAM fungi were placed on the roots of each seedling. Control plants were not inoculated, and remained free from infection throughout. The plants were watered as required and 50 ml of Hoagland's solution 2, containing only 25% of the normal level of phosphate, and 50% of the normal levels of the other nutrients, was added weekly to each 6 in pot. Plants were harvested after 49 days.

The control plants had a mean dry weight of 6.9 mg. Those infected with *Acaulospora levis*, *Gigaspora gigantea*, an unnamed VAM fungus and *Gigaspora margarita* had mean dry weights of 13.3 mg, 18.2 mg, 24.0 mg and 82.3 mg, respectively. *Gigaspora margarita* is obviously by far the most suitable fungal symbiont among those tested for tomatoes grown under the chosen conditions. We believe that much more of this type of research will be needed to ensure that the host–fungus combinations used in agricultural applications are optimized, as far as is possible.

8.4.5 Applications of VAM Inoculum

This section examines the problems related to the actual delivery or application of the inoculum, first in the laboratory, then in the greenhouse and finally in the field.

8.4.5.1 Laboratory experiments

For complete precision in the establishment of aseptic dual cultures (host + fungus) the methods described by Hepper (1981) are appropriate, and the inoculum usually takes the form of spores. (1) The agar slant method. A surface-sterilized seed is germinated at the top of the slant. Pregerminated spores are then put on the roots, and a little free water added to the bottom of the slant. (2) The paper substrate method. A seed is germinated between a paper strip and a slide standing in a tube with its lower end immersed in a nutrient solution. Spores are then placed on the young roots. (3) The Fahraeus slide method. A seed is germinated between a slide and a coverslip held parallel to, but at a distance of 800 μm from, the slide. The lower part of the slide is again immersed in nutrient solution in a tube, and spores are added to the young roots.

It is possible to infect a root with a single spore, and Daft and Nicolson (1969) showed that inocula of 3 spores seed^{-1} and 250 spores seed^{-1} produced similar levels of infection in mature tomato plants. However, we would not suggest using very low spore numbers, since other workers have found that the level of inoculum used can affect the rate of infection (Carling *et al.*, 1979). Higher levels of inoculum seem to shorten the lag phase in mycorrhiza development (Sutton, 1973) or hasten the achievement of the extensive infection desired, the plateau phase (Tinker, 1975). When spores are used as inoculum, the number applied usually ranges from 50 to 500 plant^{-1}.

8.4.5.2 Greenhouse crops

Crops that are grown in the greenhouse, or that are usually transplanted to field situations, can be infected before planting out without involving significant extra investment of time in an already labour-intensive process. This is appropriate for all container-grown seedlings, and has particular relevance to many horticultural crops. Since relatively small volumes of soil are involved, it should not be difficult to incorporate a reasonable concentration of inoculum in this soil during its preparation. The chances for early and widespread infection are presumably increased as the number of spores or other infective propagules increases, up to a point. The speed of infection can presumably be increased by ensuring that the propagules are in close juxtaposition with the roots. Inoculum can be mixed with the soil, or placed below or to one side of the seed or the roots of the seedling, or even pipetted onto the roots. But at least one worker has suspended the spores in a viscous, non-toxic medium containing carboxymethylcellulose, then dipped the roots of seedlings in this medium before planting.

8.4.5.3 Field-sown crops

Little experimental work has yet reached the field trial stage, but adapting to this challenge will inevitably be the next major phase of endomycorrhizal research. Inoculum for field use may be granulated, pelletized or in slurry form. (1) Granular inoculum. Inoculum derived from pot cultures is roughly milled or chopped up, then dried to 5–10% moisture content. It consists of a peat, perlite, vermiculite or surface substrate with an admixture of spores and infected root fragments.

(2) Pelletized inoculum. This has been prepared (Hall, 1979, 1980) by mixing 20 parts pot culture inoculum with 1 part clay (mean particle size 16 μm) and 1 part clay (mean particle size 2–3 μm). Water was mixed with these constituents until the medium could be rolled into pellets weighing 1–2 g. Jung *et al.* (1981) made pellets with a rather complex recipe involving mycorrhizal inoculum, 150 ml phosphate buffer, 50 ml buffered acrylamide solution, 50 ml buffered N,N^1-methylenebisacrylamide solution, 1 ml freshly prepared ammonium persulfate solution, and 76 μl N,N,N,N^1-tetramethylethylenediamine, all of which form a gel within 10 min. Pellets are prepared by cutting up the gel into small pieces. The inoculum is 'polymer-entrapped', and apparently does not lose viability or infectivity after several months at room temperature.

For annual, field-sown crops, it is apparent that the inoculum must be introduced with, or at about the same time as, the seeds. The technique cited by Hayman *et al.* (1981) as being most widely used is the addition of an infective soil inoculum to the furrow, below the seed, at planting time. This soil comes from around the roots of plants deliberately infected with the VA endophyte some months previously. Owusu-Bennoah and Mosse (1979) used this method of inoculation on an experimental scale for onions, lucerne and barley, but concluded that for field application, about 2.5 tons infected soil ha^{-1} would be needed. Clearly, this would be an impracticably large amount.

Ross and Harper (1970) successfully inoculated soybean with *Glomus macrocarpum*, using about 500 g inoculum m^{-2}, with an estimated spore inoculum of about 30 000 m^{-2}. Black and Tinker (1977) used a similar approach, as did Timmer and Leyden (1978) and Ferguson (1981), to inoculate citrus seedlings with *Glomus fasciculatum*. Five studies placed layers, pads or pellets of inoculum beneath the seeds. Kleinschmidt and Gerdemann (1972) inoculated citrus in this way, using 300 g inoculum m^{-2}. Black and Tinker (1977) did the same with potatoes, applying 3000 g inoculum m^{-2}, with over 30 000 spores m^{-2}. LaRue *et al.* (1975) placed pads of *Glomus fasciculatum* inoculum under peaches, at a rate of 5 g plant^{-1}. Owusu-Bennoah and Mosse (1979) inoculated alfalfa, barley and onions in this way, using larger pads (10–20 g plant^{-1}). Clarke and Mosse (1981) used 20 g pads of *Glomus mosseae*, *Glomus caledonicum* and *Glomus fasciculatum* inoculum under barley seeds. Hall (1980) placed soil pellets containing *Glomus fasciculatum* inoculum under seeds of *Lotus pedunculatus*. His pellets had to be placed with great accuracy, because each weighed less than 2 g and contained fewer than 30 spores. Obviously, this technique is not suited to large scale, mechanized applications.

Seeds pelleted with *Glomus fasciculatum* inoculum were used by Crush and Pattison (1975), Hattingh and Gerdemann (1975) and Gaunt (1978). Powell (1977) used soil inoculum pellets containing many seeds, but on a field scale this still translated into about 2 t of soil inoculum ha^{-1}. A more concentrated and less bulky inoculum composed of spores, mycelium and infected root fragments concentrated from soil by wet-sieving (or any of the other techniques described in the section on recovery of VAM inoculum from soil) has been used as a seed-coating, when formulated with a suitable binding agent (Hattingh and Gerdemann, 1975; Mosse, 1977; Gaunt, 1978). This method is applicable to crops with large seeds, but is unsuited to small-seeded crops, because the inoculum is still bulky relative to the seeds; a single spore of a VAM fungus is commonly 200–500 μm in diameter.

Witty and Hayman (1978) adapted the fluid-drilling technique described by Salter (1978), and 'injected' pots with a slurry containing pre-germinated seeds of white clover, *Rhizobium trifolii*, and wet-sievings of the VAM fungus, *Glomus fasciculatum*, all suspended in 4% methyl cellulose. This method gave 20% infection after 6 weeks, and eliminated the extra manipulations required by several other procedures.

Hayman *et al.* (1981) compared five treatments. (1) VAM inoculum and seeds broadcast over plot and raked in. (2) VAM inoculum applied in 10 g portions immediately below seeds sown in furrows. (3) VAM inoculum and pre-germinated seeds suspended in 4% methyl cellulose and applied to the furrows as a slurry at the rate of 250 ml m^{-2}. (4) VAM inoculum made into 1 cm diam pellets which were rolled in seeds until 3–4 seeds adhered to each pellet. Pellets were broadcast and then thinly covered with soil. (5) Non-inoculated controls. They found that treatments (2) and (3) above were very efficient: over 60% of the plant roots were colonized after 9 weeks. Multi-seeded pellets (treatment 4) were less effective: only a third of the seedlings became extensively colonized (more than 40% of root length infected). Treatment (1), the broadcasting of inoculum, was essentially ineffective, since plants in plots given this treatment had no more infection than those in control plots.

Of the treatments compared by Hayman *et al.* (1981), fluid drilling appeared to be the method of choice, especially for legumes, since *Rhizobium* as well as the VA endophyte can be incorporated in the slurry. Hayman *et al.* (1981) did not use chopped-up infected roots as inoculum, but mentioned that this type of inoculum, which loses much of its viability during air drying and subsequent storage (Elmes and Mosse, 1980), might survive better in slurry form.

Menge *et al.* (1977) compared several methods of inoculation, and found that placing a layer of inoculum below citrus seed, or as a band alongside the seed, was better than pelleting seed with inoculum. It was also more efficient to have inoculum 2.5 cm below the seed than 5 cm below it.

Most of these experiments, although carried out in the field, involved hand sowing or the use of manual planters. The development work necessary to mechanize these techniques for large-scale application still lies before us, though the requisite technical expertise undoubtedly exists. Ferguson (1981) was exceptional in his use of a modified tractor-drawn fertilizer bander to apply a band of *Glomus fasciculatum* inoculum to one side of citrus seedlings, subsequently recording 48% infection. Band placement of inoculum did not appear to be effective in corn and soybean (Jackson *et al.*, 1972). It is clear that the inoculum must be placed in the zone of root growth if it is to bring about the requisite degree of infection.

8.5 PROSPECTS

Since, as Trappe and Fogel (1977) so felicitously put it, 'Most woody plants require mycorrhizae to survive, and most herbaceous plants need them to thrive', we envisage increasing exploitation of both VAM and EM fungi. Man, after all, requires food to survive, and lumber to thrive.

The world's endomycorrhizal rainforests are rapidly being destroyed by extraction of timber, slash and burn agriculture, and other processes, all, unhappily, anthropogenic. Since these forests do not regenerate easily or quickly, it is very likely that in many areas they will be replaced by plantations of exotic conifers which often grow much faster than indigenous trees. Such plantations may also be initiated on many other currently treeless areas. And existing forests need to be re-established after harvesting. In every case, our biotechnological investments in ectomycorrhizal fungi will be well repaid by the ease and speed with which the trees become established, and by the gains in growth of the maturing forests.

Tree nurseries and field crops are often heavily fertilized. This effectively discourages many mycorrhizal fungi, which are highly adapted to infertile soils. It will be some time before we successfully wean forest nurserymen and farmers away from the chemicals that have served them so well. In the long run, increasing energy costs may provide the required impetus.

Horticulturists commonly use their own mixture of sand, peat and wood shavings as a propagating medium for cuttings, often adding lime and fertilizer. Needless to say, the mix does not contain VAM inoculum. The addition of VAM fungi would confer all the usual benefits upon these plants when they were outplanted, and would have the added advantage of reducing the amount of fertilizer that must be added to the mixture. Plenchette (1982) suggested that VAM fungi could be added to any substrate used in the greenhouse to produce stock for transplanting to the field. At present, tomatoes raised in greenhouses have their root systems plunged in fertilizer before outplanting. This step may eventually be replaced by plunging in VAM inoculum.

Starter cultures of VAM fungi have been available in the past from universities and from industry, but the difficulties associated with maintaining the purity of the inoculum have been major obstacles to their commercialization. Although pure inoculum of a single species is often used in laboratory experiments, this might not be necessary, and might not indeed be desirable, for field applications. As we mentioned in Section 8.4.3, a special mixture of selected VAM fungi was recently placed on the market by Biofertec Inc. in Canada. The producers claimed that this mixed inoculum can make most species of garden plants, fruit trees and bushes grow better with less phosphorus fertilizer. We have no experience of this product, and we can neither confirm nor refute the claims made for it. It is certainly possible that a mixture of several fungi might constitute a more generally applicable inoculum by increasing the chance that at least one of the potential mycobionts would be well suited to the particular host–soil–climate combination. We have learned that Biofert has been temporarily taken off the market because of quality control problems.

What we know about the much greater natural diversity of EM fungi suggests that a similar approach could be fruitful there, too, at least in the many cases where the ideal mycobiont has not been pinpointed. So far, mycelial inoculum of only one EM fungus, *Pisolithus tinctorius*, is being produced on a commercial scale. We expect that others, including but not restricted to, *Suillus granulatus*, *Rhizopogon luteolus*, *Thelephora terrestris* and *Cenococcum geophilum*, will prove worthy of similar biotechnological development. We also anticipate breakthroughs in the axenic culturing of many hitherto recalcitrant EM fungi, such as members of the genera *Russula* and *Gomphidius*, which appear to be excellent mycorrhizal partners. Even more important and exciting would be the successful axenic culturing of the VAM fungi, which would remove one of the major barriers to research and development of these enigmatic fungi.

Certain agricultural, forestry and management practices might be modified to encourage, rather than disturb, existing populations of mycorrhizal fungi. Since soil factors such as pH, addition of manure, chemical fertilizers and fungicides can favour particular mycorrhizal fungi, Hayman (1982) has suggested that it may be possible to manipulate soil populations of these fungi both qualitatively and quantitatively by applying or withholding certain treatments.

Read *et al.* (1976) suggested that the root-based hyphal network in soil, rather than the resting spores, is the primary inoculum for seedlings that become established on natural grasslands. This hypothesis is supported by the observation that the infectivity of a forest soil was greatly reduced by passing it through a 9.5 mm sieve (Clark, 1964). This treatment would have damaged the hyphae, but not the spores or infected roots. Hayman (1982) suggested that direct drilling and minimum tillage might reduce disruption of the mycelial network and favour established populations of VAM fungi. It seems to us that this knowledge can be turned to good account in another way. When we need to introduce highly efficient strains, thorough cultivation should actually help them to become established by cutting down competition from indigenous organisms.

Another way of doing this, or of eliminating known pathogens, is to fumigate the soil with methyl bromide. Indeed, in California the law requires that this be done when plants destined for sale are to be produced. This treatment is extremely effective in eradicating soil microorganisms, and once the soil has been detoxified for the recommended time, introduced mycorrhizal fungi can proliferate without competition. Soil fumigation does, however, require equipment, time and money, and may not always be necessary.

In other situations, the soil may contain persistent toxic substances: mine tailings are a good example. Many of the plants that colonize this habitat are non-mycorrhizal. Yet the re-establishment of a more complex and more normal vegetational pattern depends on the advent of mycorrhizal species. This cannot happen until appropriate inoculum has been introduced (Reeves *et al.*, 1979). Accordingly, one of the prime strategies for reclamation of mine tailings should be the selection, introduction and maintenance of suitable mycorrhizal fungi.

For many years, substances with specific toxicity toward fungi have been deliberately introduced to the plant or to the soil in order to control pathogens. Most early fungicides were toxic to a wide range of fungi, and also to many non-target organisms, not excluding the very plants they were designed to protect. Members of the second generation of fungicides are far less phytotoxic, and are also systemic. They enter the plant but remain apoplastic, moving only upward in the transpiration stream. Third generation fungicides are not only systemic, but symplastic, moving upward in the transpiration stream and downward in the phloem. The early fungicides posed a threat to mycorrhizal fungi only when deliberately applied to the seed or the soil, or when washed down into the soil. The most recent fungicides will reach the interior of the roots no matter where the application is made. Obviously, there is concern that these newer fungicides, in addition to controlling the target pathogen, may deter, damage or displace the desirable mycorrhizal fungi present in or near the roots. Fortunately, some of the most recently introduced fungicides are highly selective in their action, and have negligible toxicity to many non-target organisms. In fact, it has recently been demonstrated that foliar applications of a selectively anti-oomycete fungicide, Aliette, actually stimulate the colonization of the roots of onion by a species of *Glomus* (Jabaji-Hare and Kendrick, unpublished). The possible impact on indigenous or introduced mycorrhizal fungi of the wide range of biocides currently used in agriculture and forestry has not usually been considered in the past, but we believe it should be a subject for ongoing research, and a concern in the choice of strategies for the control of pests and pathogens.

The possible role of mycorrhizae in the management of marginally productive lands, such as arid and semi-arid rangelands where herd animals graze, has been considered by Trappe (1981). Overgrazing is deleterious to the mycorrhizal system, perhaps because the animals selectively eat forage plants that are mycorrhizal annuals, leaving behind unpalatable, non-mycorrhizal perennials. Non-disruptive management practices, such as the introduction of efficient mycorrhizal fungi, removal of non-mycorrhizal perennials and planting of mycorrhizal perennials, might result in higher productivity from such marginal lands.

Surveying the VAM fungi of citrus orchards in California and Florida, Nemec *et al.* (1981) found that plants of different ages tended to have different VAM symbionts. They concluded that the age of the plant plays a role in determining which VAM fungus will be the dominant mycorrhizal partner. It seems quite likely that under natural conditions, over the long term, a succession of VAM fungi will associate with a single perennial plant. Since the source of inoculum for the later stages of this succession will in all probability be the community of VAM fungi in the surrounding soil, programs aiming at introducing efficient VAM fungi to perennials should concentrate on those best adapted to the earliest stages of growth: those which will bring about quick and extensive infection, and increase the probability of the young plant's survival. A similar sequence of events occurs with ectomycorrhizal plants, and our conclusions are essentially the same.

The apparent succession of EM fungi could be investigated by outplanting, into sterilized soil, trees bearing specific EM partners, and following the succession of fungi that would presumably result from natural airborne spore inoculation. A more controlled experimental program could be undertaken in the greenhouse, using seedlings with various initial mycobionts which could later be challenged by a range of potential secondary and tertiary EM fungi. This kind of research, which is now being done by Fortin and his coworkers in Quebec, may give valuable insights into the process of ectomycorrhizal succession as it occurs in nature. This succession, and the factors that determine it, may play an important role in the success or failure of reforestation programs.

Successions, however, are not always inevitable, and in some cases they would be counterproductive, at least from man's point of view. Since 1973, more than 300 000 oak and hazelnut seedlings bearing ectomycorrhizae of *Tuber melanosporum* (the black truffle) have been produced and marketed by a private company in France (Chevalier and Frochot, 1981). These seedlings are used to establish 'truffle orchards' which come into production within 3–5 yr of planting. From 71 hazelnut trees in one orchard, 2.3 kg of truffles were harvested in 1981. Considering that fresh truffles now cost about $1100 per kilogram in Toronto, Canada, this yield represents a good return, and the replacement of the *Tuber* by a local EM fungus is the last thing the trufficulturist wants.

8.6 REFERENCES

Abbott, L. K. and A. D. Robson (1978). Growth of subterranean clover in relation to the formation of endomycorrhizas by introduced and indigenous fungi in a field soil. *New Phytol.*, **81**, 575–585.

Abbott, L. K. and A. D. Robson (1982). The role of vesicular–arbuscular mycorrhizal fungi in agriculture and the selection of fungi for inoculation. *Aust. J. Agric. Res.*, **33**, 389–408.

Ainsworth, G. C., F. K. Sparrow and A. S. Sussman (eds.) (1973a). *The Fungi*, vol. 4A. Academic, New York.

Ainsworth, G. C., F. K. Sparrow and A. S. Sussman (eds.) (1973b). *The Fungi*, vol. 4B. Academic, New York.

Andrews, H. N. and L. W. Lenz (1943). A mycorrhizome from the carboniferous of Illinois. *Bull. Torrey Bot. Club*, **70**, 120–125.

Barea, J. M. and C. Azcón-Aguilar (1982). Production of plant growth-regulating substances by the vesicular–arbuscular mycorrhizal fungus *Glomus mosseae*. *Appl. Environ. Microbiol.*, **43**, 810–813.

Barrow, N. J., N. Malajczuk and T. C. Shaw (1977). A direct test of the ability of vesicular–arbuscular mycorrhizas to help plants take up fixed soil phosphate. *New Phytol.*, **78**, 269–276.

Bartlett, E. M. and D. H. Lewis (1973). Surface phosphatase activity of mycorrhizal roots of beech. *Soil Biol. Biochem.*, **5**, 249–257.

Berch, S. M. and J.-A. Fortin (1983a). *Endogone pisiformis*: axenic culture and associations with *Sphagnum*, *Pinus sylvestris*, *Allium porrum* and *Allium cepa*. *Can. J. Bot.*, **61**, 899–905.

Berch, S. M. and J.-A. Fortin (1983b). Lectotypification of *Glomus macrocarpum* and proposal of *Glomus australe*, *Glomus versiforme* and *Glomus tenebrosum* (Endogonaceae). *Can. J. Bot.*, **61**, 2608–2617.

Berch, S. M. and B. Kendrick (1982). Vesicular–arbuscular mycorrhizae of Southern Ontario ferns and fern-allies. *Mycologia*, **74**, 769–776.

Black, R. L. B. and P. B. Tinker (1977). Interaction between effects of vesicular–arbuscular mycorrhiza and fertiliser phosphorus on yields of potatoes in the field. *Nature (London)*, **267**, 510–511.

Bowen, G. D. (1973). Mineral nutrition of ectomycorrhizae. In *Ectomycorrhizae: their Ecology and Physiology*, ed. G. C. Marks and T. T. Kozlowski, pp. 151–205. Academic, New York.

Bowen, G. D. and A. D. Rovira (1969). The influence of micro-organisms on growth and metabolism of plant roots. In *Root Growth*, ed. W. J. Whittington, pp. 170–201. Butterworths, London.

Bowen, G. D., M. F. Skinner and D. I. Bevege (1974). Zinc uptake by mycorrhizal and uninfected roots of *Pinus radiata* and *Araucaria cunninghamii*. *Soil Biol. Biochem.*, **6**, 141–144.

Bowen, G. D. and C. Theodorou (1973). Growth of ectomycorrhizal fungi around seeds and roots. In *Ectomycorrhizae: their Ecology and Physiology*, ed. G. C. Marks and T. T. Kozlowski, pp. 107–150. Academic, New York.

Callow, J. A., L. C. M. Capaccio, G. Parish and P. B. Tinker (1978). Detection and estimation of polyphosphate in vesicular–arbuscular mycorrhizas. *New Phytol.*, **80**, 125–134.

Carling, D. E. and M. F. Brown (1982). Anatomy and physiology of vesicular–arbuscular and non-mycorrhizal roots. *Phytopathology*, **72**, 1108–1114.

Carling, D. E., M. F. Brown and R. A. Brown (1979). Colonization rates and growth responses of soybean plants infected by vesicular–arbuscular mycorrhizal fungi. *Can. J. Bot.*, **57**, 1769–1772.

Carrodus, B. B. (1967). Absorption of nitrogen by mycorrhizal roots of beech. II. Ammonium and nitrate as sources of nitrogen. *New Phytol.*, **66**, 1–4.

Cetto, B. (1978). *Der Grosse Pilzefuhrer*, vol. 1. Arti Grafiche Saturnia, Trento.

Cetto, B. (1979a). *Der Grosse Pilzefuhrer*, vol. 2. Arti Grafiche Saturnia, Trento.

Cetto, B. (1979b). *Der Grosse Pilzefuhrer*, vol. 3. Arti Grafiche Saturnia, Trento.

Chevalier, G. and H. Frochot (1981). Truffle production from artificially mycorrhizal plants: first results. *Abstracts of the 5th North American Conference on Mycorrhizae, Quebec*.

Chilvers, G. A. (1968). Some distinctive types of eucalypt mycorrhiza. *Aust. J. Bot.*, **16**, 49–70.

Chilvers, M. T. and M. J. F. Daft (1982). Effects of low temperature on development of the vesicular–arbuscular mycorrhizal association between *Glomus caledonicum* and *Allium cepa*. *Trans. Br. Mycol. Soc.*, **79**, 153–157.

Clark, F. B. (1964). Microorganisms and soil structure affect yellow poplar growth. *USDA For. Serv. Res. Pap. CS-9.*

Clarke, C. and B. Mosse (1981). Plant growth responses to vesicular–arbuscular mycorrhiza. XII. Field inoculation responses of barley at two soil P levels. *New Phytol.*, **87**, 695–703.

Cooper, K. M. and P. B. Tinker (1981). Translocation and transfer of nutrients in vesicular–arbuscular mycorrhizas. IV. Effect of environmental variables on movement of phosphorus. *New Phytol.*, **88**, 327–339.

Cordell, C. E., J. P. Conn, D. H. Marx and D. S. Kenney (1981). Practical machine application of ectomycorrhizal fungus inoculum in forest tree nurseries. *Abstracts of the 5th North American Conference on Mycorrhizae, Quebec*, p. 74. Université Laval, Québec.

Corte, A. (1969). Research on the influence of the mycorrhizal infection on the growth, vigor, and state of health of three *Pinus* species. *Arch. Bot. Biogeogr. Ital.*, **45**, 1.

Cox, G., F. E. Sanders, P. B. Tinker and J. A. Wild (1975). Ultrastructural evidence relating to host-endophyte transfer in a vesicular–arbuscular mycorrhiza. In *Endomycorrhizas*, ed. F. E. Sanders, B. Mosse and P. B. Tinker, pp. 297–312. Academic, London.

Cox, G. and P. B. Tinker (1976). Translocation and transfer of nutrients in vesicular–arbuscular mycorrhizas. I. The arbuscule and phosphorus transfer: a quantitative ultrastructural study. *New Phytol.*, **77**, 371–378.

Cronquist, A. (1981). *An Integrated System of Classification of Flowering Plants*. Columbia University Press, New York.

Crush, J. R. (1973). Significance of endomycorrhizas in tussock grassland in Otago, New Zealand. *N.Z. J. Bot.*, **11**, 645–660.

Crush, J. R. and A. C. Pattison (1975). Preliminary results on the production of vesicular–arbuscular mycorrhizal inoculum by freeze-drying. In *Endomycorrhizas*, ed. F. E. Sanders, B. Mosse and P. B. Tinker, pp. 485–493. Academic, London.

Daft, M. J., E. Hacskaylo and T. H. Nicolson (1975). Arbuscular mycorrhizas in plants colonising coal spoils in Scotland and Pennsylvania. In *Endomycorrhizas*, ed. F. E. Sanders, B. Mosse and P. B. Tinker, pp. 561–580. Academic, London.

Daft, M. J. and T. H. Nicolson (1969). Effect of *Endogone* mycorrhiza on plant growth. II. Influence of soluble phosphate on endophyte and host in maize. *New Phytol.*, **68**, 945–952.

Daft, M. J. and T. H. Nicolson (1974). Arbuscular mycorrhizas in plants colonizing coal wastes in Scotland. *New Phytol.*, **73**, 1129–1138.

Dahncke, R. M. and S. M. Dahncke (1979). *700 Pilze in Farbfotos*. AT Verlag, Aarau.

Dale, J. and A. L. McComb (1955). Chlorosis, mycorrhizae, and the growth of pines on a high lime soil. *For. Sci.*, **1**, 148–157.

Daniels, B. A., P. M. McCool and J. A. Menge (1982). Comparative inoculum potential of spores of six vesicular–arbuscular mycorrhizal fungi. *New Phytol.*, **89**, 385–391.

Daniels, B. A. and J. A. Menge (1981). Evaluation of the commercial potential of the vesicular–arbuscular mycorrhizal fungus, *Glomus epigaeus*. *New Phytol.*, **87**, 345–354.

Daniels, B. A. and J. M. Trappe (1979). *Glomus epigaeus* sp. nov., a useful fungus for vesicular–arbuscular mycorrhizal research. *Can. J. Bot.*, **57**, 539–542.

Daniels, B. A. and J. M. Trappe (1980). Factors affecting spore germination of the vesicular–arbuscular mycorrhizal fungus *Glomus epigaeus*. *Mycologia*, **72**, 457–471.

Dehne, H. W. (1982). Interaction between vesicular–arbuscular mycorrhizal fungi and plant pathogens. *Phytopathology*, **72**, 1115–1119.

Diehl, W. W. (1939). *Endogone* as animal food. *Science*, **90**, 442.

Dixon, R. K., G. M. Wright, G. T. Behrns, R. O. Teskey and T. M. Hinckley (1980). Water deficits and root growth of ectomycorrhizal white oak seedlings. *Can. J. For. Res.*, **10**, 545–548.

Donald, D. G. M. (1975). Mycorrhizal inoculation for pines. *S. Afr. For. J.*, **92**, 27–29.

Edmister, J. (1970). Survey of mycorrhiza and nodules in the El Verde forest. In *Tropical Rain Forest — A Study of Irradiation and Ecology at El Verde, Puerto Rico*, ed. H. T. Odum, pp. F15–20. UCAEC, Oak Ridge, TN.

Elmes, R. and B. Mosse (1980). Vesicular–arbuscular mycorrhizae: nutrient film technique. *Rothamsted Exp. Station Rep. 1979.*, Part 1, 188.

Eriksson, J. and L. Ryvarden (1973). *The Corticiaceae of North Europe*, **2**, 60–286.

Fairweather, J. V. and D. G. Parbery (1982). Effects of four vesicular–arbuscular mycorrhizal fungi on growth of tomato. *Trans. Br. Mycol. Soc.*, **79**, 151–153.

Ferguson, J. J. (1981). Inoculum production and field application of vesicular–arbuscular mycorrhizal fungi. Ph.D. Thesis, University of California, Riverside.

Fogel, E. and G. Hunt (1979). Fungal and arboreal biomass in a western Oregon Douglas fir ecosystem: distribution patterns and turnover. *Can. J. For. Res.*, **9**, 245–256.

Fontana, A. (1963). Micorrize ectotrifiche in una ciperacea: *Kobresia bellardi* Degl. *Giorn. Bot. Ital.*, **70**, 639–641.

Fortin, J.-A., Y. Piché and M. Lalonde (1980). Technique for the observation of early morphological changes during ectomycorrhiza formation. *Can. J. Bot.*, **58**, 361–365.

Frank, A. B. (1885). Über die auf Wurzelsymbiose beruhende Ernahrung gewisser Baume durch unterirdische Pilze. *Ber. Dtsch. Bot. Ges.*, **3**, 128–145.

Furlan, V., H. Bartschi and J.-A. Fortin (1980). Media for density gradient extraction of endomycorrhizal spores. *Trans. Br. Mycol. Soc.*, **75**, 336–338.

Furlan, V. and J.-A. Fortin (1975). A flotation–bubbling system for collecting Endogonaceae spores from sieved soil. *Naturaliste Canadien*, **102**, 663–667.

Gaunt, R. E. (1978). Inoculation of vesicular–arbuscular mycorrhizal fungi on onion and tomato seeds. *N.Z. J. Bot.*, **16**, 69–71.

Gerdemann, J. W. (1955). Relation of a large soil-borne spore to phycomycetous mycorrhizal infection. *Mycologia*, **47**, 619–632.

Gerdemann, J. W. (1968). Vesicular–arbuscular mycorrhiza and plant growth. *Annu. Rev. Phytopathol.*, **6**, 397–418.

Gerdemann, J. W. and T. H. Nicolson (1963). Spores of mycorrhizal *Endogone* species extracted from soil by wet sieving and decanting. *Trans. Br. Mycol. Soc.*, **46**, 235–244.

Gerdemann, J. W. and J. M. Trappe (1974). The Endogonaceae in the Pacific Northwest. *Mycologia Mem.*, **5**, 1–76.

Gianinazzi, S., V. Gianinazzi-Pearson and J. Dexheimer (1979). Enzymatic studies on the metabolism of vesicular–arbuscular mycorrhiza. III. Ultrastructural localization of acid and alkaline phosphatase in onion roots infected by *Glomus mosseae* (Nicol. & Gerd.). *New Phytol.*, **82**, 127–132.

Gibson, J. L., J. W. Kimbrough and G. L. Benny (1984). Ultrastructural evidence for the Ascomycete-like nature of *Glaziella aurantiäca. Proc. 6th N. Am. Conf. Mycorrhizae, Bend, Oregon.*

Gildon, A. and P. B. Tinker (1981). A heavy metal-tolerant strain of a mycorrhizal fungus. *Trans. Br. Mycol. Soc.*, **77**, 648–649.

Gilkey, H. M. (1939). *Tuberales of North America.* Oregon State University, Corvallis.

Giovannetti, M. and B. Mosse (1980). An evaluation of techniques for measuring vesicular–arbuscular mycorrhizal infection in roots. *New Phytol.*, **84**, 489–500.

Glenn, M. G. (1982). VA mycorrhizal infection in *Brassica*, a non-host. *Phytopathology*, **72**, 951.

Göbl, F. (1975). Erfahrungen bei der Anzucht von Mykorrhiza–Impfmaterial. *Centralbl. Ges. Forstw.*, **92**, 227–237.

Graham, J. H., R. G. Linderman and J. A. Menge (1982). Development of external hyphae by different isolates of mycorrhizal *Glomus* spp. in relation to root colonization and growth of Troyer citrange. *New Phytol.*, **91**, 183–189.

Grand, L. F. and A. E. Harvey (1982). Quantitative measurement of ectomycorrhizae on plant roots. In *Methods and Principles of Mycorrhizal Research*, ed. N. C. Schenck, pp. 157–164. Am. Phytopath. Soc., St. Paul.

Granger, R., C. Plenchette and J.-A. Fortin (1983). Effect of a vesicular–arbuscular (VA) endomycorrhizal fungus (*Glomus epigaeus*) on the growth and leaf mineral content of two apple clones propagated *in vitro. Can. J. Plant Sci.*, **63**, 551–555.

Green, N. E., S. O. Graham and N. C. Schenck (1976). The influence of pH on the germination of vesicular–arbuscular mycorrhizal spores. *Mycologia*, **68**, 929–934.

Hacskaylo, E. (1953). Pure culture synthesis of pine mycorrhizae in terra-lite. *Mycologia*, **45**, 971–975.

Hacskaylo, E., J. G. Palmer and J. A. Vozzo (1965). Effect of temperature on growth and respiration of ectotrophic mycorrhizal fungi. *Mycologia*, **57**, 748–756.

Hacskaylo, E. and J. A. Vozzo (1967). Inoculation of *Pinus caribaea* with pure cultures of mycorrhizal fungi in Puerto Rico. *Proc. 14th Int. Union For. Res. Org. Munich.*, **5**, 139–148.

Hall, I. R. (1979). Soil pellets to introduce vesicular–arbuscular mycorrhizal fungi into soil. *Soil Biol. Biochem.*, **11**, 85–86.

Hall, I. R. (1980). Growth of *Lotus pedunculatus* Cav. in an eroded soil containing soil pellets infested with endomycorrhizal fungi. *N.Z. J. Agric. Res.*, **23**, 103–105.

Hall, I. R. and L. Abbott (1979). Photographic slide collection illustrating features of the Endogonaceae. See p. 129 for locations.

Hall, I. R. and B. J. Fish (1979). A key to the Endogonaceae. *Trans. Br. Mycol. Soc.*, **73**, 261–270.

Harley, J. L. (1969). *The Biology of Mycorrhiza*, 2nd edn. Leonard Hill, London.

Harley, J. L. (1975). Problems of mycotrophy. In *Endomycorrhizas*, ed. F. E. Sanders, B. Mosse and P. B. Tinker, pp. 1–24. Academic, London.

Harley, J. L. and C. C. McCready (1952). The uptake of phosphate by excised mycorrhizal roots of the beech. II. Distribution of phosphorus between host and fungus. *New Phytol.*, **51**, 56–64.

Haselwandter, K. and D. J. Reed (1980). Fungal associations of roots of dominant and sub-dominant plants in high-alpine vegetation systems with special reference to mycorrhiza. *Oecologia (Berlin)*, **45**, 57–62.

Hatch, A. B. (1936). The role of mycorrhizae in afforestation. *J. For.*, **34**, 22–29.

Hatch, A. B. (1937). The physical basis of mycotrophy in the genus *Pinus. Black Rock Forest Bull.*, **6**, 1–168.

Hattingh, M. J. and J. W. Gerdemann (1975). Inoculation of Brazilian sour orange seed with an endomycorrhizal fungus. *Phytopathology*, **65**, 1013–1016.

Hattingh, M. J., L. E. Gray and J. W. Gerdemann (1973). Uptake and translocation of ^{32}P-labelled phosphate to onion roots by endomycorrhizal fungi. *Soil Sci.*, **116**, 385–387.

Hawker, L. E. (1954). British hypogeous fungi. *Philos. Trans. R. Soc. London, Ser. B*, **237**, 429–546.

Hayman, D. S. (1970). *Endogone* spore number in soil and vesicular–arbuscular mycorrhiza in wheat as influenced by season and soil treatment. *Trans. Br. Mycol. Soc.*, **54**, 53–63.

Hayman, D. S. (1978). Endomycorrhizae. In *Interactions Between Non-pathogenic Soil Microorganisms and Plants*, ed. Y. R. Dommergues and S. V. Krupa, pp. 401–442. Elsevier, Amsterdam.

Hayman, D. S. (1982). Influence of soils and fertility on activity and survival of vesicular–arbuscular mycorrhizal fungi. *Phytopathology*, **72**, 1119–1125.

Hayman, D. S., E. J. Morris and R. J. Page (1981). Methods for inoculating field crops with mycorrhizal fungi. *Ann. Appl. Biol.*, **99**, 247–253.

Hepper, C. M. (1981). Techniques for studying the infection of plants by vesicular–arbuscular mycorrhizal fungi under axenic conditions. *New Phytol.*, **88**, 641–647.

Herter, W. G. (1934). La aparicion del hongo venenoso *Amanita phalloides* en Sudamerica. *Rev. Sud-Am. Bot.*, **1**, 111–119.

Hesler, L. R. and A. H. Smith (1979). *North American Species of Lactarius.* University of Michigan Press, Ann Arbor.

Ho, I. and J. M. Trappe (1973). Translocation of ^{14}C from *Festuca* plants to their endomycorrhizal fungi. *Nature New Biol.*, **244**, 30–31.

Holmgren, P. K., W. Keuken and E. K. Schofield (1981). *Index Herbariorum, Part 1. The Herbaria of the World*, 7th edn., Regnum Vegetabile, vol. 106. W. Junk, The Hague.

Hyppel, A. (1968). Effect of *Fomes annosus* on seedlings of *Picea abies* in the presence of *Boletus bovinus*. *Stud. For. Suec.*, **66**, 1–16.

Jackson, N. E., R. E. Franklin and R. H. Miller (1972). Effects of VA mycorrhizae on growth and phosphorus content of three agronomic crops. *Proc. Soil. Sci. Soc. Am.*, **36**, 64–67.

Jülich, W. (1972). Monographie der Athelieae (Corticiaceae, Basidiomycetes). *Willdenowia*, **7**, 1–283.

Jung, G., G. Mungier, D. Diem Hoang and Y. R. Dommergues (1981). Polymer-entrapped symbiotic microorganisms as inoculants for legumes and non-legumes. *Abstracts of the 5th North American Conference on Mycorrhizae, Quebec*, p. 54. Université Laval, Québec.

Kaspari, H. (1975). Fine structure of the host–parasite interface in endotrophic mycorrhiza of tobacco. In *Endomycorrhizas*, ed. F. E. Sanders, B. Mosse and P. B. Tinker, pp. 325–334. Academic, London.

Kelley, A. P. (1950). *Mycotrophy in Plants*. Chronica Botanica Co., Waltham.

Kessler, K. J., Jr. and R. W. Blank (1972). *Endogone* sporocarps associated with sugar maple. *Mycologia*, **64**, 634–638.

Kidston, R. and W. H. Lang (1921). On Old Red Sandstone plants showing structure from the Rhynie chert bed, Aberdeenshire. Part V. The Thallophyta occurring in the peat bed; the succession of the plants throughout a vertical section of the bed, and the conditions of accumulation and preservation of the deposit. *Trans. R. Soc. Edinburgh*, **52**, 855–902.

Kleinschmidt, G. D. and J. W. Gerdemann (1972). Stunting of citrus seedlings in fumigated nursery soils related to the absence of endomycorrhizae. *Phytopathology*, **62**, 1447–1453.

Kormanik, P. P., W. C. Bryan and R. C. Schultz (1980). Procedures and equipment for staining large numbers of plant roots for endomycorrhizal assay. *Can. J. Microbiol.*, **26**, 536–538.

Koske, R. E. (1981). Multiple germination by spores of *Gigaspora gigantea*. *Trans. Br. Mycol. Soc.*, **76**, 328–330.

Koske, R. E. (1982). Evidence for a volatile attractant from plant roots affecting germ tubes of a VA mycorrhizal fungus. *Trans. Br. Mycol. Soc.*, **79**, 305–310.

Kruckelmann, H. W. (1975). Effects of fertilizers, soil tillage, and plant species on the frequency of *Endogone* chlamydospores and mycorrhizal infection in arable soils. In *Endomycorrhizas*, ed. F. E. Sanders, B. Mosse and P. B. Tinker, pp. 511–525. Academic, London.

Laiho, O. (1970). *Paxillus involutus* as a mycorrhizal symbiont of forest trees. *Acta For. Fenn.*, **106**, 1–73.

Lambert, D. H., H. Cole, Jr. and D. E. Baker (1980). Variation in the response of alfalfa clones and cultivars to mycorrhizae and phosphorus. *Crop Sci.*, **20**, 615–618.

LaRue, J. H., W. D. McClellan and W. L. Peacock (1975). Mycorrhizal fungi and peach nursery nutrition. *Calif. Agric.*, **29**, 7–9.

Linderman, R. G. and J. W. Hendrix (1982). Evaluation of plant response to colonization by vesicular–arbuscular mycorrhizal fungi. A. Host variables. In *Methods and Principles of Mycorrhizal Research*, ed. N. C. Schenck, pp. 69–76. Am. Phytopath. Soc., St. Paul.

Longman, K. A. and J. Jenik (1974). *Tropical Forest and its Environment*. Longman, London.

Lundeberg, G. (1968). The formation of mycorrhizae in different provenances of pine (*Pinus sylvestris* L.). *Sv. Bot. Tidskr.*, **62**, 269.

Malloch, D. W., K. A. Pirozynski and P. H. Raven (1980). Ecological and evolutionary significance of mycorrhizal symbioses in vascular plants (a review). *Proc. Natl. Acad. Sci. USA*, **77**, 2113–2118.

Margulis, L. (1981). *Symbiosis in Cell Evolution*. W. H. Freeman, San Francisco.

Marks, G. C., N. Ditchburne and R. C. Foster (1968). Quantitative estimates of mycorrhiza populations in radiata pine forests. *Aust. For.*, **32**, 26–38.

Marks, G. C. and R. C. Foster (1973). Structure, morphogenesis, and ultrastructure of ectomycorrhizae. In *Ectomycorrhizae: their Ecology and Physiology*, ed. G. C. Marks and T. T. Koslowski, pp. 2–41. Academic, New York.

Marsh, B. B. (1971). Measurement of length in random arrangements of lines. *J. Appl. Ecol.*, **8**, 256.

Martinez, A. (1945). La presencia en la Argentina del hongo venenoso *Amanita phalloides*. *Notas Mus. La Plata*, **10**, 93–98.

Marx, D. H. (1969). The influence of ectotrophic mycorrhizal fungi on the resistance of pine roots to pathogenic infections. I. Antagonism of mycorrhizal fungi to root pathogenic fungi and soil bacteria. *Phytopathology*, **59**, 153–163.

Marx, D. H. (1973). Mycorrhizae and feeder root diseases. In *Ectomycorrhizae, their Ecology and Physiology*, ed. G. C. Marks and T. T. Kozlowski, pp. 351–382. Academic, New York.

Marx, D. H. (1976). Synthesis of ectomycorrhizae on loblolly pine seedlings with basidiospores of *Pisolithus tinctorius*. *For. Sci.*, **22**, 13–20.

Marx, D. H. (1977). Tree host range and world distribution of the ectomycorrhizal fungus *Pisolithus tinctorius*. *Can. J. Microbiol.*, **23**, 217–223.

Marx, D. H. (1980). Ectomycorrhizal fungus inoculations: a tool for improving forestation practices. In *Tropical Mycorrhiza Research*, ed. P. Mikola, pp. 13–71. Oxford University Press, London.

Marx, D. H. (1981). Variability in ectomycorrhizal development and growth among isolates of *Pisolithus tinctorius* as affected by source, age, and reisolation. *Can. J. For. Res.*, **11**, 168–174.

Marx, D. H. and J. D. Artman (1979). The significance of *Pisolithus tinctorius* ectomycorrhizae to survival and growth of pine seedlings on coal spoils in Kentucky and Virginia. *Reclam. Rev.*, **2**, 23–31.

Marx, D. H. and W. C. Bryan (1971). Influence of ectomycorrhizae on survival and growth of aseptic seedlings of loblolly pine at high temperatures. *For. Sci.*, **17**, 37–41.

Marx, D. H. and W. C. Bryan (1975). Growth and ectomycorrhizal development of pine seedlings in nursery soils infested with the fungal symbiont *Pisolithus tinctorius*. *For. Sci.*, **22**, 91–100.

Marx, D. H. and W. J. Daniel (1976). Maintaining cultures of ectomycorrhizal and plant pathogenic fungi in sterile water cold storage. *Can. J. Microbiol.*, **22**, 338–341.

Marx, D. H. and D. S. Kenney (1982). Production of ectomycorrhizal fungus inoculum. In *Methods and Principles of Mycorrhizal Research*, ed. N. C. Schenck, pp. 131–146. Am. Phytopath. Soc., St. Paul.

Marx, D. H. and B. Zak (1965). Effect of pH on mycorrhizal formation of slash pine in aseptic culture. *For. Sci.*, **11**, 66–75.

Marx, D. H., W. C. Bryan and C. B. Davey (1970). Influence of temperature on aseptic synthesis of ectomycorrhizae of *Thelephora terrestris* and *Pisolithus tinctorius* on loblolly pine. *For. Sci.*, **16**, 424–431.

Marx, D. H., W. C. Bryan and C. E. Cordell (1976). Growth and ectomycorrhizal development of pine seedlings in nursery soils infested with the fungal symbiont *Pisolithus tinctorius*. *For. Sci.*, **22**, 91–100.

Marx, D. H., W. G. Morris and J. G. Mexal (1978). Growth and ectomycorrhizal development of loblolly pine seedlings in fumigated and non-fumigated soil infested with different fungal symbionts. *For. Sci.*, **24**, 193–203.

Marx, D. H., C. E. Cordell, D. S. Kenney, J. G. Mexal, J. D. Artman, J. W. Riffle and J. R. Molina (1982). Commercial vegetative inoculum of *Pisolithus tinctorius* and inoculation techniques for development of ectomycorrhizae on bare-root seedlings. *For. Sci.*, **28**, 1–101.

Maser, C., J. M. Trappe and R. A. Nussbaum (1978). Fungal–small mammal interrelationships with emphasis on Oregon coniferous forests. *Ecology*, **59**, 799–809.

Mejstřik, V. K. (1970). The uptake of ^{32}P by different kinds of ectotrophic mycorrhiza of *Pinus*. *New Phytol.*, **69**, 295–298.

Mejstřik, V. K. and H. H. Krause (1973). Uptake of ^{32}P by *Pinus radiata* roots inoculated with *Suillus luteus* and *Cenococcum graniforme* from different sources of available phosphate. *New Phytol.*, **72**, 137–140.

Melin, E. (1921). Über die Mykorrhizenpilze von *Pinus sylvestris* L., und *Picea abies* (L.) Karst. *Sv. Bot. Tidskr.*, **15**, 192–203.

Melin, E. (1922). Untersuchungen über die *Larix* Mykorrhiza. I. Synthese der Mykorrhiza in Reinkultur. *Sv. Bot. Tidskr.*, **16**, 161–196.

Melin, E. (1923). Experimentelle Untersuchungen über die Birken und Espenmykorrhizen und ihre Pilzsymbionten. *Sv. Bot. Tidskr.*, **17**, 479–520.

Melin, E. (1936). Methoden der experimentellen Untersuchung Mykotropher Pflanzen. *Handb. Biol. Arbeitsmeth.*, **11**, 1015–1108.

Menge, J. A., H. Lembright and E. L. V. Johnson (1977). Utilization of mycorrhizal fungi in citrus nurseries. *Proc. Int. Soc. Citricult.*, **1**, 129–132.

Menge, J. A., E. L. V. Johnson and R. G. Platt (1978). Mycorrhizal dependency of several citrus cultivars under three nutrient regimes. *New Phytol.*, **81**, 553–559.

Mertz, S. M., J. J. Heithaus and R. L. Bush (1979). Mass production of axenic spores of the endomycorrhizal fungus *Gigaspora margarita*. *Trans. Br. Mycol. Soc.*, **72**, 167–169.

Mexal, J. and C. P. P. Reid (1973). The growth of selected mycorrhizal fungi in response to induced water stress. *Can. J. Bot.*, **51**, 1579–1588.

Mikola, P. (1970). Mycorrhizal inoculation in afforestation. *Int. Rev. For. Res.*, **3**, 123–196.

Mikola, P. (1973). Application of mycorrhizal symbioses in forestry practice. In *Ectomycorrhizae: Their Ecology and Physiology*, ed. C. G. Marks and T. T. Kozlowski, pp. 383–411. Academic, New York.

Mikola, P. (ed.) (1980). *Tropical Mycorrhiza Research*. Oxford University Press, Oxford.

Miller, O. K., Jr. (1973). A new gastroid genus related to *Gomphidius*. *Mycologia*, **65**, 226–229.

Miller, O. K., Jr. (1980). *Mushrooms of North America*. E. P. Dutton, New York.

Miller, O. K., Jr. (1982). Taxonomy of ecto- and ectendomycorrhizal fungi. In *Methods and Principles of Mycorrhizal Research*, ed. N. C. Schenck, pp. 91–101. Am. Phytopath. Soc., St. Paul.

Miller, O. K., Jr. and D. F. Farr (1975). *An Index of the Common Fungi of North America*. Cramer, Vaduz.

Miller, O. K., Jr. and H. H. Miller (1980). *Mushrooms in Color*. E. P. Dutton, New York.

Molina, R. (1977). Ectomycorrhizal fungi and forestry practice. In *Mushrooms and Man, an Interdisciplinary Approach to Mycology*, ed. T. Walters. USDA. Forest Service.

Molina, R. (1980). Ectomycorrhizal inoculation of containerized western conifer seedlings. *Pac. N.W. For. Range Exp. Station Res.*, Note 357, 1–10.

Molina, R. and J. G. Palmer (1982). Isolation, maintenance and pure culture manipulation of ectomycorrhizal fungi. In *Methods and Principles of Mycorrhizal Research*, ed. N. C. Schenck, pp. 115–129. Am. Phytopath. Soc., St. Paul.

Moser, M. (1958a). Die kunstliche Mykorrhizaimpfung an Forstpflanzen. I. Erfahrungen bei der Reinkultur von Mykorrhizapilzen. *Forstwiss. Centralbl.*, **77**, 32–40.

Moser, M. (1958b). Die kunstliche Mykorrhizaimpfung an Forstpflanzen. II. Die Torfstreukultur von Mykorrhizapilzen. *Forstwiss. Centralbl.*, **77**, 257–320.

Moser, M. (1958c). Die Einfluss tiefer Temperaturen auf das Wachstum und die Lebenstätigkeit höherer Pilze mit spezieller Berücksichtigung von Mykorrhizapilzen. *Sydowia*, **12**, 386–399.

Moser, M. (1978). *Die Rohrlinge und Blätterpilze (Polyporales, Boletales, Agaricales, Russulales)*. Gustav Fischer Verlag, Stuttgart.

Mosse, B. (1972). The influence of soil type and *Endogone* strains on the growth of mycorrhizal plants in phosphate-deficient soils. *Rev. Ecol. Biol. Sol.*, **9**, 529–537.

Mosse, B. (1973). Advances in the study of vesicular–arbuscular mycorrhiza. *Annu. Rev. Phytopathol.*, **11**, 171–196.

Mosse, B. (1977). Plant growth responses to vesicular–arbuscular mycorrhiza. X. Responses of *Stylosanthes* and maize to inoculation in unsterile soils. *New Phytol.*, **78**, 277–288.

Mosse, B. and G. D. Bowen (1968). A key to the recognition of some *Endogone* spore types. *Trans. Br. Mycol. Soc.*, **51**, 469–483.

Mosse, B. and G. W. Jones (1968). Separation of *Endogone* spores from organic soil debris by differential sedimentation on gelatin columns. *Trans. Br. Mycol. Soc.*, **51**, 604–608.

Nemec, S. (1982). Aspects of vesicular–arbuscular mycorrhizae in plant disease research. *Phytopathology*, **72**, 1102.

Nemec, S., J. A. Menge, R. G. Platt and E. L. V. Johnson (1981). Vesicular–arbuscular mycorrhizal fungi associated with citrus in Florida and California and notes on their distribution and ecology. *Mycologia*, **73**, 112–127.

Newman, E. I. (1966). A method of estimating the total length of root in a sample. *J. Appl. Ecol.*, **3**, 139.

Nicolson, T. H. (1960). Mycorrhiza in the Gramineae. II. Development in different habitats. particularly sand dunes. *Trans. Br. Mycol. Soc.*, **43**, 132–145.

Ohms, R. E. (1957). A flotation method for collecting spores of a phycomycetous mycorrhizal parasite from soil. *Phytopathology*, **47**, 751–752.

Owusu-Bennoah, E. and B. Mosse (1979). Plant growth responses to vesicular–arbuscular mycorrhiza. XI. Field inoculation responses in barley, lucerne and onion. *New Phytol.*, **83**, 671–679.

Owusu-Bennoah, E. and A. Wild (1979). Autoradiography of the depletion zone of phosphate around onion roots in the presence of vesicular–arbuscular mycorrhiza. *New Phytol.*, **82**, 133–140.

Park, J. Y. (1971). Preparation of mycorrhizal grain spawn and its practical feasibility in artificial inoculation. *USDA For. Serv. Misc. Publ.*, No. 1189, 239–240.

Pearson, V. and P. B. Tinker (1975). Measurement of phosphorus fluxes in the external hyphae of endomycorrhizas. In *Endomycorrhizas*, ed. F. E. Sanders, B. Mosse and P. B. Tinker, pp. 277–287. Academic, London.

Pegler, D. N. and T. W. K. Young (1979). The gasteroid Russulales. *Trans. Br. Mycol. Soc.*, **72**, 353–388.

Phillips, J. M. and D. S. Hayman (1970). Improved procedures for clearing roots and staining parasitic and vesicular–arbuscular mycorrhizal fungi for rapid assessment of infection. *Trans. Br. Mycol. Soc.*, **55**, 158–160.

Pirozynski, K. A. and D. W. Malloch (1975). The origin of land plants: a matter of mycotrophism. *BioSystems*, **6**, 153–164.

Plenchette, C. (1982). Les endomycorrhizes à vesicules et arbuscules (VA): un potentiel à exploiter en agriculture. *Phytoprotection*, **63**, 86–104.

Plenchette, C., V. Furlan and J.-A. Fortin (1981). Growth stimulation of apple trees in unsterilized soil under field conditions with VA mycorrhiza inoculation. *Can. J. Bot.*, **59**, 2003–2008.

Plenchette, C., J.-A. Fortin and V. Furlan (1983). Growth responses of several plant species to mycorrhizae in a soil of moderate P-fertility. I. Mycorrhizal dependency under field conditions. *Plant and Soil*, **70**, 199–209.

Poitou, N., M. Mamoun and J. Delmas (1981). Quelques resultats obtenus concernant la mycorhization de plantes-hôtes par des champignons mycorhiziens comestibles. *Abstracts of the 5th North American Conference on Mycorrhizae*, Quebec, p. 61. Université Laval, Québec.

Pomerleau, R. (1980). *Flore des Champignons au Quebec*. Les Editions La Presse, Montreal.

Powell, C. L. (1975). Plant growth responses to vesicular–arbuscular mycorrhiza. VIII. Uptake of P by onion and clover infected with different *Endogone* spore types in ^{32}P labelled soil. *New Phytol.*, **75**, 563–566.

Powell, C. L. (1977). Efficient mycorrhizal fungi reduce fertilizer requirements. *Proceedings of the Ruakura Farmers' 29th Conference, Hamilton, N.Z.*, pp. 22–47. Ministry of Agric. and Fish., Wellington.

Powell, W. M., F. F. Hendrix and D. H. Marx (1968). Chemical control of feeder root necrosis of pecans caused by *Pythium* species and nematodes. *Plant Dis. Rep.*, **52**, 577.

Rabatin, S. C. and L. H. Rhodes (1982). *Acaulospora bireticulata* inside oribatid mites. *Mycologia*, **74**, 859–861.

Read, D. J. and K. Haselwandter (1981). Observations on the mycorrhizal status of some alpine plant communities. *New Phytol.*, **88**, 341–352.

Read, D. J., H. K. Koucheki and J. Hodgson (1976). Vesicular–arbuscular mycorrhiza in natural vegetation systems. I. The occurrence of infection. *New Phytol.*, **77**, 641–653.

Reeves, F. B., D. Wagner, T. Moorman and J. Kiel (1979). The role of endomycorrhizae in revegetation practices in the semi-arid west. I. A comparison of incidence of mycorrhizae in severely disturbed vs. natural environments. *Am. J. Bot.*, **66**, 6–13.

Reid, C. P. P., and E. Hacskaylo (1982). Evaluation of plant response to inoculation. B. Environmental variables. In *Methods and Principles of Mycorrhizal Research*, ed. N. C. Schenck, pp. 175–187. Am. Phytopath. Soc., St. Paul.

Rhodes, L. H. (1980). The use of mycorrhizae in crop production systems. *Outlook Agric.*, **10**, 275–281.

Rhodes, L. H. and J. W. Gerdemann (1975). Phosphate uptake zones of mycorrhizal and non-mycorrhizal onions. *New Phytol.*, **75**, 555–561.

Rhodes, L. H. and M. C. Hirrel (1982). Radiotracer methods for mycorrhizal research. In *Methods and Principles of Mycorrhizal Research*, ed. N. C. Schenck, pp. 189–200. Am. Phytopath. Soc., St. Paul.

Richard, C. and J.-A. Fortin (1975). Role protecteur de *Suillus granulatus* contre le *Mycelium radicis atrovirens* sur les semis de *Pinus resinosa*. *Can. J. For. Res.*, **5**, 452–456.

Richard, C., Fortin, J.-A. and Fortin, A. (1971). Protective effect of an ectomycorrhizal fungus against the root pathogen *Mycelium radicis atrovirens*. *Can. J. For. Res.*, **1**, 246–251.

Richards, B. N. and G. L. Wilson (1963). Nutrient supply and mycorrhiza development in Caribbean pine. *For. Sci.*, **9**, 405.

Riffle, J. W. and D. M. Maronek (1982). Ectomycorrhizal inoculation procedures for greenhouse and nursery studies. In *Methods and Principles of Mycorrhizal Research*, ed. N. C. Schenck, pp. 147–155. Am. Phytopath. Soc., St. Paul.

Romagnesi, H. (1967). *Les Russules d'Europe et d'Afrique du Nord*. Bordas, Paris.

Ross, E. W. and Marx, D. H. (1972). Susceptibility of sand pine to *Phytophthora cinnamomi*. *Phytopathology*, **62**, 1197.

Ross, J. P. and J. A. Harper (1970). Effect of *Endogone* mycorrhiza on soybean yields. *Phytopathology*, **60**, 1552–1556.

Safir, G. R. and J. M. Duniway (1982). Evaluation of plant response to colonization by vesicular–arbuscular mycorrhizal fungi. B. Environmental variables. In *Methods and Principles of Mycorrhizal Research*, ed. N. C. Schenck, pp. 77–80, Am. Phytopath. Soc., St. Paul.

Salter, P. J. (1978). Fluid drilling — a new approach to crop establishment. In *Advances in Agriculture*, pp. 16–23. University of Aston, Birmingham.

Sanders, F. E., B. Mosse and P. B. Tinker (eds.) (1975). *Endomycorrhizas*. Academic, London.

Sanders, F. E. and P. B. Tinker (1973). Phosphate flow into mycorrhizal roots. *Pestic. Sci.*, **4**, 383–395.

Sanders, F. E., P. B. Tinker, R. L. B. Black and S. M. Palmerley (1977). The development of endomycorrhizal root systems: I. Spread of infection and growth-promoting effects with four species of vesicular–arbuscular endophyte. *New Phytol.*, **78**, 257–268.

Sands, R. and C. Theodorou (1978). Water uptake by mycorrhizal roots of radiata pine seedlings. *Aust. J. Plant Physiol.*, **5**, 301–309.

Schaeffer, J. (1933). *Russula* — Monographie. *Ann. Mycol.*, **31**, 305–516.

Schenck, N. C. (1981). Can mycorrhiza control root disease? *Plant Dis.*, **65**, 230–234.

Schenck, N. C. and M. K. Kella (1978). The influence of vesicular–arbuscular mycorrhizae on disease development. *Florida Agric. Exp. Station Bull.*, **799**.

Schenck, N. C. and R. A. Kinloch (1980). Incidence of mycorrhizal fungi on six field crops in monoculture on a newly cleared woodland site. *Mycologia*, **72**, 445–456.

Schoenbeck, F. (1980). Endomycorrhiza: ecology, function and phytopathological aspects. *Forum Microbiol.*, **3**, 90–96.

Schoenbeck, F. and H.-W. Dehne (1981). Mycorrhiza and plant health. *Gesunde Pflanzen*, **33**, 186–190.

Schramm, J. R. (1966). Plant colonization studies on black wastes from anthracite mining in Pennsylvania. *Trans. Am. Phil. Soc.*, **47**, 331.

Shaw, C. G. and R. Molina (1980). Formation of ectomycorrhizae following inoculation of containerized Sitka spruce seedlings. *Pacific N. W. For. Range Exp. Station Res.*, Note 351, 1–8.

Sinclair, W. A. and D. H. Marx (1982). Evaluation of plant reponse to inoculation. A. Host variables. In *Methods and Principles of Mycorrhizal Research*, ed. N. C. Schenck, pp. 165–174. Am. Phytopath. Soc., St. Paul.

Singer, R. (1975). *The Agaricales in Modern Taxonomy*. Cramer, Vaduz.

Singer, R. and A. H. Smith (1960). Studies on Secotiaceous fungi. IX. The Astrogastraceous series. *Bull. Torr. Bot. Club*, **21**, 1–112.

Skinner, M. F. and G. D. Bowen (1974). The uptake and translocation of phosphate by mycelial strands of pine mycorrhizas. *Soil Biol. Biochem.*, **6**, 53–56.

Slankis, V. (1973). Hormonal relationships in mycorrhizal development. In *Ectomycorrhizae: their Ecology and Physiology*, ed. G. C. Marks and T. T. Kozlowski, pp. 231–298. Academic, New York.

Smith, A. H. (1975). *A Field Guide to Western Mushrooms*. University of Michigan Press, Ann Arbor.

Smith, A. H. and H. Thiers (1971). *The Boletes of Michigan*. University of Michigan Press, Ann Arbor.

Smith, A. H., H. V. Smith and N. S. Weber (1979). *How to Know the Gilled Mushrooms*. W. C. Brown, Dubuque.

Smith, F. A. (1972). A comparison of the uptake of nitrate, chloride and phosphate by excised beech mycorrhizas. *New Phytol.*, **71**, 875–882.

Smith, H. V. and A. H. Smith (1973). *How to Know the Non-gilled Fleshy Fungi*. W. C. Brown, Dubuque.

Snell, W. and E. Dick (1971). *The Boleti of Northeastern North America*. Cramer, Weinheim.

Sutton, J. C. (1973). Development of vesicular–arbuscular mycorrhizae in crop plants. *Can. J. Bot.*, **51**, 2487–2493.

Sutton, J. C. and G. L. Barron (1972). Population dynamics of *Endogone* spores in soil. *Can. J. Bot.*, **50**, 1909–1914.

Sutton, J. C. and B. R. Sheppard (1976). Aggregation of sand-dune soil by endomycorrhizal fungi. *Can. J. Bot.*, **54**, 326–333.

Sylvia, D. M. and N. C. Schenck (1982). Effect of post-colonization treatments on sporulation of vesicular–arbuscular mycorrhizal fungi. *Phytopathology*, **72**, 950.

Taber, R. A. (1982). Occurrence of *Glomus* spores in weed seeds in soil. *Mycologia*, **74**, 515–520.

Taber, R. A. and J. M. Trappe (1982). Vesicular–arbuscular mycorrhiza in rhizomes, scale-like leaves, roots, and xylem of ginger. *Mycologia*, **74**, 156–161.

Takacs, E. A. (1961). Inoculacion de especies de pinos con hongos formadores de micorrizas. *Silvicultura*, **15**, 5–17.

Takacs, E. A. (1964). Inoculacion artificial de pinos de regiones subtropicales con hongos formadores de micorrizas. *Idia. Supl. For.*, **2**, 41–44.

Takacs, E. A. (1967). Produccion de cultivos puros de hongos micorrizogenos en el Centro Nacional de Investigaciones Agropecuarias, Castelar. *Idia. Supl. For.*, **4**, 83–87.

Theodorou, C. (1971). Introduction of mycorrhizal fungi into soil by spore inoculation of seed. *Aust. For.*, **35**, 23–26.

Theodorou, C. (1978). Soil moisture and the mycorrhizal association of *Pinus radiata* D. Don. *Soil Biol. Biochem.*, **10**, 33–37.

Theodorou, C. and G. D. Bowen (1970). Mycorrhizal responses of radiata pine in experiments with different fungi. *Aust. For.*, **34**, 183–191.

Theodorou, C. and G. D. Bowen (1973). Inoculation of seeds and soil with basidiospores of mycorrhizal fungi. *Soil Biol. Biochem.*, **5**, 765–771.

Thiers, H. (1975). *California Mushrooms. A Field Guide to the Boletes*. Macmillan, New York.

Thiers, H. and J. M. Trappe (1969). Studies in the genus *Gastroboletus*. *Brittonia*, **21**, 244–254.

Timmer, L. W. and R. F. Leyden (1978). Stunting of citrus seedlings in fumigated soil in Texas and its correction by phosphorus fertilization and inoculation with mycorrhizal fungi. *J. Am. Soc. Hort. Sci.*, **103**, 533–537.

Tinker, P. B. (1975). Soil chemistry of phosphorus and mycorrhizal effects on plant growth. In *Endomycorrhizas*, ed. F. E. Sanders, B. Mosse and P. B. Tinker, pp. 353–371. Academic, London.

Tommerup, I. (1981). Survival mechanisms of VA mycorrhizal fungi. *Abstracts of the 5th North American Conference on Mycorrhizae*, p. 16. Université Laval, Québec.

Tommerup, I. and D. J. Carter (1982). Dry separation of microorganisms from soil. *Soil Biol. Biochem.*, **14**, 69–71.

Trappe, J. M. (1969). Studies on *Cenococcum graniforme*. I. An efficient method for isolation from sclerotia. *Can. J. Bot.*, **47**, 1389–1390.

Trappe, J. M. (1977). Selection of fungi for ectomycorrhizal inoculation in nurseries. *Annu. Rev. Phytopathol.*, **15**, 203–222.

Trappe, J. M. (1979). The orders, families and genera of hypogeous Ascomycotina (truffles and their relatives). *Mycotaxon.*, **9**, 297–340.

Trappe, J. M. (1981). Mycorrhizae and productivity of arid and semiarid rangelands. In *Advances in Food Producing Systems for Arid and Semiarid Lands*, pp. 581–599. Academic, New York.

Trappe, J. M. (1982). Synoptic keys to the genera and species of zygomycetous mycorrhizal fungi. *Phytopathology*, **72**, 1102–1108.

Trappe, J. M. and R. D. Fogel (1977). Ecosystematic functions of mycorrhizae. In *The Belowground Ecosystem: A Synthesis of Plant-Associated Processes*. Colorado State Univ. Range Sci. Dept. Sci. Ser., **26**, 205–214.

Trappe, J. M. and N. C. Schenck (1982). Taxonomy of the fungi forming endomycorrhizae. A. Vesicular–arbuscular mycorrhizal fungi (Endogonales). In *Methods and Principles of Mycorrhizal Research*, ed. N. C. Schenck, pp. 1–9. Am. Phytopath. Soc., St. Paul.

Watling, R. and E. Watling (1980). *A Literature Guide for Identifying Mushrooms*. Mad River Press, Eureka.

Weijman, A. C. M. and H. L. C. Meuzelaar (1979). Biochemical contributions to the taxonomic status of the Endogonaceae. *Can. J. Bot.*, **57**, 284–291.

White, P. T. (1983). Nature's dwindling treasures — rainforests. *Natl. Geogr.*, **163**, 2–46.

Wingfield, E. B. (1968). Mycotrophy in loblolly pine. Ph.D. Thesis, Virginia Polytechnic Institute, Blacksburg.

Witty, J. F. and D. S. Hayman (1978). Slurry inoculation of VA mycorrhiza. *Rothamsted Exp. Station Rept. 1977.*, Part 1, 239–240.

Woodroof, N. (1933). Pecan mycorrhizas. *Georgia. Agric. Exp. Station Bull.*, **178**.

Wu, L. and M. A. Stahmann (1975). Chromatographic estimation of fungal mass in plant materials. *Phytopathology*, **65**, 1032–1034.

Wyatt-Smith, J. (1953). A note on the vegetation of some islands in the Malacca Straits. *Malay. For.*, **16**, 191–205.

Zak, B. (1969). Characterization and classification of mycorrhizae of Douglas fir. I. *Pseudotsuga menziesii + Poria terrestris* (blue- and orange-staining strains). *Can. J. Bot.*, **47**, 1833–1840.

Zak, B. (1971a). Characterization and classification of Douglas-fir mycorrhizae. In *Mycorrhiza*, ed. E. Hacskaylo, USDA Misc. Publ., No. 1189, 38–53.

Zak, B. (1971b). Characterization and classification of mycorrhizae of Douglas fir. II. *Pseudotsuga menziesii + Rhizopogon vinicolor. Can. J. Bot.*, **49**, 1079–1084.

Zak, B. (1971c). Detoxification of autoclaved soils by a mycorrhizal fungus. *Pacific N. W. For. Range Exp. Station Res.*, Notes PNW-159, 1–4.

Zak, B. (1973). Classification of ectomycorrhizae. In *Ectomycorrhizae: their Ecology and Physiology*, ed. G. C. Marks and T. T. Kozlowski, pp. 43–78. Academic, New York.

Zak, B. and D. H. Marx (1964). Isolation of mycorrhizal fungi from roots of individual slash pines. *For. Sci.*, **10**, 214.

9

Somaclonal Variation, Cell Selection and Genotype Improvement

W. R. SCOWCROFT and P. J. LARKIN
CSIRO, Canberra, ACT, Australia

9.1 INTRODUCTION

Plant improvement employs genetic solutions to maximize crop productivity. Whether the motives be humanitarian or directed by the market place, plant improvement continues to be an essential investment for both public and private sectors. The testimony of the recent past, where varietal improvement has directly accounted for up to 50% of the yield increase of crops such as wheat, maize and barley, is witness to this compulsion (Bingham, 1981). The prognosis that future crop productivity requires a sustained average annual rate of increase of more than 2% (Anon., 1980) justifies continued and even enhanced investment.

The challenge of mitigating the effects of factors which perturb crop productivity, such as diseases and pests, climatic and edaphic influences, changing consumer demands and economic and political constraints, must involve the plant breeder. To be increasingly effective, the conventional armoury of plant improvement will need to be augmented by more efficient technology.

The genetic enhancement of crop productivity depends on two vital components. The extent of genetic variability available for the traits that need to be improved is of prime importance. While the natural germ-plasm of the crop species in question is often quite vast, much of it might not be 'available'. For example, modern cultivars of wheat and maize are highly sophisticated energy converters. By introducing genetic variability from wild near relatives to upgrade a particular

153

trait, much of the preceding plant breeding effort may be very difficult to recover from such a hybrid gene pool. Sequential improvement of an already crop-production-efficient gene pool is arguably more desirable.

The second component is the ability to select efficiently genotypes which are specific improvements over pre-existing cultivars. To this end, gene effect specificity and numbers are the operative parameters. The more accurately the phenotypic selection scheme reflects heritable gene effects, the more rapid the progress towards defined objectives. Genetic changes in specific traits, particularly those which lead to an improvement, will be rare. Therefore, the larger the population of progeny that can be screened, the greater the probability of recovering desirable variants.

For these two crucial elements of plant improvement, the *in vitro* genetic manipulation of plants is already proving very useful. Rapid developments have been made in the application of cellular selection to recover useful genetic variants: haploidy, both to hasten the attainment of homozygosity and to provide more efficient screening among recombinants; embryo culture for recovering inviable hybrids; somatic hybridization for recombining genomes of asexually incompatible species and the use of cell culture to generate genetic variability (Scowcroft, 1977; Thomas *et al.*, 1979; Chaleff, 1981; Larkin and Scowcroft, 1981). Further downstream is the possibility of specific gene addition or modification by recombinant DNA technology.

Here we wish to develop the thesis that tissue culture generated variability (somaclonal variation) and cell selection offer immediate prospects for use in varietal improvement. The focus of this article will be on the accessibility and application of cell culture in an agricultural context.

9.2 SOMACLONAL VARIATION

9.2.1 Historical Perspective

Efficient cell culture is relatively recent among agricultural species. We define this as the relative ease whereby cell lines can be established, proliferated under *in vitro* conditions with reasonable speed and, most importantly, plants regenerated in reasonably large numbers. Many non-agriculturally important plants meet this criterion (Tisserat *et al.*, 1980; Evans *et al.*, 1981), but for many years legume and cereal species remained recalcitrant. Particular efforts by the Hawaiian Sugar Planters Association Experiment Station (Heinz, 1973) with sugar-cane, by Green and Phillips (1975) with maize, and Bingham *et al.* (1975) with lucerne (alfalfa) gave reality to the vision of a Rockefeller working party convened in 1969 to contemplate the application of plant cell culture in plant improvement. In the few years since the mid-1970s, efficient cell cycles have been established with several important legumes (though soybean is still proving difficult) and important grain crops such as maize, oats, rice, barley, sorghum and, most recently, wheat.

The ingress of geneticists and plant breeders into this research area did not at first dispel the historically-based assumption that cell culture was simply a rapid method of cloning a particular genotype. Potential use of cell culture was seen in the context of providing a more efficient selection scheme. Genetic changes, either spontaneous or induced, were thought to occur at rates comparable to those found in conventional sexual cycles. An advantage of cell culture was seen as providing a more efficient means of selection to enhance the rate of recovery of desirable variants (Scowcroft, 1977).

Phenotypic variants among plants regenerated from cell cultures, if anything, were considered annoying 'artefacts of tissue culture' arising from epigenetic events. They were relegated to the dust bin as being unworthy of any use or of genetic interest. Arguably there were a few exceptions to this. Researchers in Hawaii had observed variation among plants regenerated from cell cultures of sugar-cane as early as 1969 and saw this as a potential tool (Heinz, 1973) in the improvement of this species. Skirvin and Janick (1976) observed considerable variation among plants (calliclones) regenerated from cell cultures of scented geranium (*Pelargonium* spp.) and concluded 'that tissue culture techiques may have special uses for the plant breeder, and should provide an additional method to increase intraclonal variation'.

In both sugar-cane and *Pelargonium*, tissue culture generated variation was seen as circumventing the considerable constraints imposed on the use of conventional breeding practice for improving asexually propagated species. Greater credence to this approach was subsequently provided by Shepard *et al.* (1980), who observed extensive agronomically useful variation among protoplast derived plants (protoclones) of potato. They chose to work with an old, but widely grown cultivar, Russett Burbank, which was excluded from conventional improvement because of its sterility.

Thus, tissue culture as a source of useful genetic variability—somaclonal variation—gained currency as a potential adjunct to plant improvement (Larkin and Scowcroft, 1981).

9.2.2 The Manifold Incidence of Somaclonal Variation

Table 1 lists the species in which somaclonal variation has been reported to date. Qualitative details of many of these reports were assembled by Larkin and Scowcroft (1981) and Scowcroft and Larkin (1982) and will not be repeated here except where relevant to more recent findings.

The explant refers to the plant organ(s) from which the cell line or *in vitro* culture was initiated. The R mode column describes the culture conditions from which plants were regenerated. A single or two word description is admittedly inadequate. Callus, for example, can be disorganized as in the case of tobacco, or have organized meristems as is found with most of the monocots. The adventitious shoot mode may involve a more or less discernible callus phase. Androgenesis implies that the somaclones evaluated were either haploids or dihaploids, but diploid plants can also be derived from diploid tissue of the anther pedicel, *i.e.* the cell wall. The column on transmission is provided only to give a general notion that at least some of the variant traits observed in R_0 plants are transmitted through an asexual or sexual cycle. From this, some appreciation of the genetic credibility of somaclonal variation can be deduced. In many cases, the transmission or otherwise of variant characters either through sexual or asexual cycles has not been reported (nr). Citations usually refer to the original reports, but in some cases, *e.g.* Heinz *et al.* (1977), a review is more appropriate. For species of particular interest, readers should consult the orginal references(s).

In referring to plants derived from *in vitro* culture and their progeny, we will use the notation of Chaleff (1981), namely, R (or R_0) is the primary regenerated plant and R_1, R_2, *etc.* the seed derived progeny of R and R_1, respectively, and so on.

We will not discuss details of somaclonal variation in each species (see Larkin and Scowcroft, 1981), but rather look at the general features and interesting facets which emerge from the reports in Table 1 about the phenomenon as a whole.

9.2.2.1 *Range of species*

Where a number of plants regenerated from cell culture have been observed, somaclonal variation appears to be the general rule. There is a range of diligence with which character variation has been observed, from great detail, as in potatoes (Secor and Shepard, 1981), to little more than cursory observations, as in peanuts (Bajaj *et al.*, 1981). Seed propagated species are no less prone to somaclonal variation than asexually propagated ones. This is particularly relevant because it not only eases the genetic understanding of the phenomenon, but also adds a new dimension to the creation of variability in conventional plant breeding programmes.

Many earlier reports describing plants regenerated from cell culture failed to mention phenotypic variation with any conviction. This was probably due to the entrenched assumption that *in vitro* propagation of plants was truly clonal propagation. Several authors have made passing reference to the apparent homogeneity among regenerated plants of species such as celery (Williams and Collins, 1976), asparagus and iris (Reuther, 1977), potato (Wenzel *et al.*, 1979) and *Pennisetum purpureum* (Haydu and Vasil, 1981). In contrast to variation displayed among plants regenerated from root adventitious meristems in cauliflower (Grout and Crisp, 1980), in earlier studies of over 4000 plants derived from culture of the cauliflower curd no variants were observed (Crisp and Walkey, 1974).

9.2.2.2 *Characters displaying variation*

Many of the cell culture studies where variants have been observed have been more or less related to plant improvement. In consequence, variations in agronomically important characters comprise a large proportion of cases listed in Table 1. In tobacco, in particular, emphasis has been on characters which relate to the marketable quality of the leaf. The results of Burk and Matzinger (1976), where they found as much variation for yield, grade index, flowering time, *etc.* among somaclonally derived double haploids as among a segregating F_2 population between two cultivars, are most telling. The 46 dihaploids which were analyzed in a replicated block exper-

Table 1　Species Displaying Somaclonal Variation[a]

Species	Explant	R mode	Variant character in R or R_n plants	Transmission[b] Asexual	Sexual	Ref.
MONOCOTS						
Avena sativa	(a) Immature embryo, apical meristem	Callus	Plant ht, heading date, leaf striping, twin culms, awns		+	Cummings et al. (1976)
	(b) Immature embryo	Callus	Cytogenetic abnormalities, e.g. chromosome loss, interchanges, trisomy, monosomy		nr	McCoy et al. (1982)
Hordum spp.	(a) Immature embryo	Callus	Plant ht, tillering, albinism		+	Deambrogio and Dale (1980)
	(b) Anthers	Callus	Cytological abnormalities		nr	Mix et al. (1978)
	(c) Immature ovaries	Callus	Growth habit and rate, size, head morphology, auricles		nr	Orton (1980)
Triticum aestivum	(a) Immature embryo	Callus	Plant ht, stem thickness, leaf size, spike shape, pollen fertility		nr	Ahloowahlia (1982)
	(b) Immature embryo	Callus	Plant and head morphology, awns, chlorophyll deficiency, plant ht		+	Scowcroft and Larkin (unpublished)
Oryza sativa	(a) Seedling	Callus	Tiller no., fertile tiller no., panicle length, seed fertility		nr	Henke et al. (1978)
	(b) Dihaploid seed	Callus	Chlorophyll content, seed fertility, plant ht, flowering date, grain no., kernel wt.		+	Oono (1981)
Saccharum officinarum	(a) Parenchyma inflorescence, young leaf	Callus	Eyespot, Fiji virus, downy mildew diseases	+		Heinz et al. (1977)
	(b) Meristems inflorescence, young leaf	Callus	Auricle length, dewlap shape, top leaf attitude, chr no., esterase isozymes, cane yield, sugar yield, stalk no., smut resistance	+		Liu (1981)
	(c) Leaf base	Callus	Eyespot disease, leaf scald	+		Larkin and Scowcroft (1982)
Zea mays	(a) Immature embryo	Callus	Abphyll syndrome, pollen fertility		nr	Green (1977)
	(b) Immature embryo	Callus	Drechslera maydis race T toxin resistance, male fertility, mtDNA sequence rearrangement		+	Brettell et al. (1980); Kemble et al. (1982); Gengenbach and Umbeck (1982)
	(c) Immature embryo	Callus	Endosperm and seedling mutants	+		Edallo et al. (1981)
Sorghum bicolor	Immature embryos, seedlings	Callus	Fertility, leaf morphology growth habit		nr	Gamborg et al. (1977)
Allium sativa	Shoot tips, leaf base	Callus	Bulb size and shape, clove no., aerial bulbil germination	+		Novák (1980)
Haworthia	Flower bud	Callus	Chr. no. and meiotic behaviour, vigour, leaf shape, leaf colour, stomata no.,		nr	Ogihara (1981)

Species	Explant	Culture	Characters			Reference
DICOTS						
Solanum tuberosum	(a) Leaf	Protoplast	Tuber shape, yield, maturity date, photoperiod, plant morphology, early and late blight resistance, numerous field traits	+		Shepard et al. (1980); Secor and Shepard (1981)
	(b) Leaf	Protoplast	Leaf colour and morphology vigour, ht, anthocyanin pigment	+		Thomas et al. (1982)
	(c) Leaf	Adv. shoots	Stem, leaf, flower morphology, skin colour	+		van Harten et al. (1981)
Nicotiana tabacum	(a) Anther	Androgenesis	Plant ht, stem diameter, leaf size, yield		+	Oinuma and Yoshida (1974)
	(b) Anther	Androgenesis	Yield, grade index, flowering time, plant ht, leaf number, length and width, alkaloids, reducing sugars		+	Burk and Matzinger (1976)
	(c) Leaf	Protoplast, callus, adv. shoots	2 specific leaf colour loci, leaf shape and size		+	Barbier and Dulieu (1980)
N. sylvestris	Anther	Androgenesis	Crumpled leaf		+	De Paepe et al. (1981)
Medicago sativa	Immature ovaries	Suspension/callus	Multifoliate leaves, elongated petioles, growth habit, primary branch no., plant ht, dry matter yield	+		Reisch and Bingham (1981)
Pelargonium	Stem	Callus	Plant, leaf and flower, morphology, essential oils, fasciation, anthocyanin pigmentation, pubescence	+	+	Skirvin and Janick (1976)
Begonia × *hiemalis*	Leaf	Adv. shoot	Colour, size, form of flowers and leaves		nr	Roest et al. (1981)
Chrysanthemum	(a) Nodes, shoot tip	Adv. shoot callus	Flower colour, flower induction temperature	+		Jung-Heiliger and Horn (1980)
	(b) Shoot tip	Callus	Shoot and leaf morphology	+	nr	Sutter and Langhans (1981); Schwartz et al. (1981)
Fragaria ananassa	Stolon tips	Adv. shoots	Vigour, runnering, chlorosis, compact habit, yield			
Anasas comosus	Syncarp, slip, crown, axillary bud	Callus	Spine and leaf colour, waxiness, foliage density, leaf width and spines		nr	Wakasa (1979)
Brassica napus	(a) Anther	Androgenesis	Flowering time, glucosinolates, growth habit		nr	Hoffmann et al. (1982)
	(b) Androgenic embryo	Callus	Resistance/tolerance to *Phoma lingam*		+	Sacristán (1982)
Brassica oleracea	Root crown meristem	Adv. shoots	Waxiness, stem branching, precocious flower formation	+		Grout and Crisp (1980)
Lactuca sativa	Cotyledons	Callus	Leaf wt, length, width, flatness & colour, bud no.		+	Sibi (1976)
Arachis hypogaea	Anther	Callus	Chr no.		nr	Bajaj et al. (1981)

[a] See text for heading descriptions. [b] nr = not reported.

iment were derived by anther culture from a 16-generation inbred line. The expectation of phenotypic identity between the dihaploids and between the dihaploid somaclones and the parent was, surprisingly, not fulfilled.

In sugar-cane, variants have been observed for disease resistance (eyespot disease, Fiji virus disease, downy mildew, smut resistance), stalk number, sugar content, sugar yield and leaf traits, characters which are of concern in plant improvement. In Taiwan, Liu and Chen (1976) found that one particular somaclonal variant had a substantial improvement over its donor parent for both cane and sugar yield and, in addition, out-yielded Taiwan's best cultivar by 16%.

The most extensive evaluation of somaclonal variation was conducted on 65 somaclones derived from potato by protoplast culture (Secor and Shepard, 1981). These 65 remained from an original population of 1700 which had been so screened in two seasons to eliminate grossly aberrant types. They possessed acceptable vigour, vine and tuber characteristics. The 65 somaclones were evaluated relative to the parent, Russet Burbank, in a replicated field trial for 35 characters, ranging from pre-plant sprouting through 19 separately measured vine traits, maturity and morphology and yield of tubers after harvest. Statistically significant variation was found for 22 characters. Every somaclone varied for at least one character and one clone differed from the parent, Russet Burbank, in 17 characters. The modal class of 15 somaclones differed from the parent in four characters. Shepard (1981) points out, however, that although these somaclones did vary both with respect to each other and to the parent, the plants 'closely resembled Russet Burbank in morphology and general tuber characteristics and therefore had retained most traits of the (parental) cultivar'.

It is particularly significant that somaclonal variants can be recovered which have enhanced tolerance to pathogenic diseases. In addition to the four sugar-cane diseases mentioned before, this also is true for late blight (*Phytophthora infestans*) and early blight (*Alternaria solani*) in potatoes (Shepard *et al.*, 1980), southern corn leaf blight (*Drechslera maydis*) in maize (Brettell *et al.*, 1980), and blackleg (*Phoma lignam*) in oilseed rape (Sacristán, 1982). The frequency of recovering such disease tolerant variants is relatively high. In our own research on eyespot disease in sugar-cane, based on sensitivity of somaclones to *H. sacchari* toxin, 15–20% of plants were judged resistant. In potato, 5/500 somaclones were more resistant to *A. solani* toxin than the parent and four of these displayed field resistance. Among 800 somaclones screened for reaction to late blight 20 (2.5%) were resistant, some of which were resistant to multiple races of the pathogen.

In a study to confirm earlier work (Gengenbach *et al.*, 1977) on the use of *D. maydis* race-T pathotoxin to select southern corn leaf blight resistant genotypes of maize, Brettell *et al.* (1980) also screened plants regenerated from cell lines (derived from a toxin sensitive, male sterile T-cytoplasm line) which had not been exposed to toxin selection. Surprisingly, more than half (31/50) of the plants reverted to toxin resistant, male-fertile types. In contrast, the T-cytoplasm characteristic is extremely stable among seed grown plants.

While we are enthusiatic for the idea that somaclonal variation may be an important source of variability for disease resistance, sufficient data has not yet been accumulated to generalize the notion. In breeding for disease resistance, cell culture and somaclonal variation is only one of several equally worthy and efficient strategies.

9.2.2.3 *Genetic nature of somaclonal variants*

The question is often put, does somaclonal variation lead to completely novel genotypes? For many species and for even more characters it is still too early to claim that variants are even genetically determined. The general appreciation, however, is that the variants observed among regenerated plants and their progeny are phenotypically similar to those mutant (or possibly recombinant) phenotypes which have been previously described for that species. A tentative proposition can be advanced, namely that somaclonal variation generates mutant genotypes of a class similar to that found as a consequence of spontaneous or induced mutation, but at a much higher frequency.

Edallo *et al.* (1981) adopted the classical description of maize mutants in screening for somaclonal variants among the progeny of 77 fertile maize regenerants. They observed 12 classes of defective endosperm and six mutant seedling types, respectively. To avoid the confounding effect of cell culture induced phenocopies, only R_0 regenerants which gave R_2 progeny that segregated in Mendelian proportions were classed as true genetic variants. On average, each regenerated plant carried at least one recognizable Mendelian mutation. No mutations were observed among plants grown from seed of the parent.

Among 1666 tobacco somaclones screened for mutations at two specific loci (a_1, yg) which affect chlorophyll synthesis, Barbier and Dulieu (1980) found that the frequency of mutations was approximately 3.5% at each locus. Their genetic testing of a sample of these altered genotypes (reversions to a wild-type allele or deletions) indicated that such changes were similar in kind to those which occur spontaneously.

In maize, the difference between T-cytoplasm (male-sterile, *D. maydis* race T toxin susceptible) and N-cytoplasm (male-fertile, toxin resistant) is encoded in the mitochondrial genome. The reversion of T-cytoplasm from the toxin susceptible male-sterile form to a toxin resistant male-fertile type among maize somaclones is also associated with genetic changes in the mitochondrial genome (Brettell *et al.*, 1982; Kemble *et al.*, 1982). The mtDNA from such reverted somaclones has been subjected to restriction endonuclease analysis to determine whether the change to toxin resistance and male-fertility is a true reversion to N-cytoplasm type mitochondria. None of the reverted somaclones gave a fragmentation pattern identical to that of the N-cytoplasm mitochondria. However, 14/15 of the Gengenbach and Umbeck (1982) lines and all four of the Kemble *et al.* (1982) lines, many of which had been regenerated from cell cultures subjected to toxin selection, had changes associated with a 6.6 kb XhoI fragment. The major point is that regardless of toxin selection during cell culture at least one common change appears to characterize mtDNA among reverted lines.

Recently, Dixon *et al.* (1982) have screened the *in vitro* translation products of mitochondria isolated from the progeny of male fertile, T-toxin resistant plants regenerated from cell lines with and without selective treatment with T-toxin. There appeared to be a close correlation between the revertant genotypes and qualitative and/or quantitative changes in a 13 000 M_r mitochondrial polypeptide. In male-fertile, toxin-resistant somaclone progeny this polypeptide is either not expressed or maximally expressed at only 33% of that of the male sterile, toxin susceptible types.

In this maize example the molecular analysis is approaching a resolution of the causal relationship between mtDNA changes and the male-sterile, toxin-susceptible reversion to fertility and resistance. The somaclonal variants and their progeny which have been analysed appear to be not unusual with respect to the expected types of changes which have occurred.

While there seems to be a concurrence, at least phenotypically, between variants found in cell culture regenerated plants and those found in nature or as a consequence of induced mutation, cytogenetic evidence indicates that there might be subtle differences. In *Haworthia*, as well as morphological and esterase isozyme differences among regenerated plants, Ogihara (1981) found a number of somaclones with chromosomal interchanges. None of the break points of these interchanges occurred in the centromere region. Similarly, non-centric chromosomal breaks predominate in plants regenerated from *Crepis* (Sacristan and Wendt-Gallitelli, 1971) and *N. sylvestris* (Maliga *et al.*, 1979) cell cultures. In contrast, in naturally occurring populations of *Haworthia*, *Haplopappus* and *Gibasis* (see Ogihara, 1981) centric interchanges are the rule. This contrast could mean that cytological variants in cell culture may be different to those which occur in nature. Alternatively, non-centric interchanges are at a selective disadvantage relative to centric interchanges under natural conditions.

9.2.2.4 *Pre-existing or culture induced variation*

Somaclonal variation may simply reflect genetic changes that have occurred in the somatic cells of the donor explant prior to culture. This view was put forward by Wenzel *et al.* (1979) and Brettell and Ingram (1979). While it is difficult, and indeed unnecessary, to discount the contribution of pre-existing changes to somaclonal variation, the evidence suggests that the majority of genetic changes leading to variation occurred during the culture phase.

Shepard (1981) found that under the influence of different growth regulators (2,4-D *vs.* NAA) during protoplast culture not only were protoplast plating efficiencies affected, but the frequency of variant regenerants was greater with NAA as against 2,4-D. Such a result is difficult to relate to pre-existing variation. Thomas *et al.* (1982) in confirming somaclonal variation in a different cultivar of potato, Maris Bard, provide more compelling evidence that variation is largely culture induced. Variation existed between plants regenerated from a single callus, which in turn derived from a single protoplast. Hence, variation arose during callus growth.

The relatively detailed analysis of somaclonal variants in maize (Edallo *et al.*, 1981) can be interpreted as supporting the notion that variation arose during the culture phase. Many of the regenerated maize plants which carried different mutants were pedigree related in that they derived from the same cultured embryo. In one instance, the progeny of 11 plants regenerated from subcultures each derived from the same embryo contained 12 mutants, of which nine were different.

The extensive data of Barbier and Dulieu (1980) on somaclonal variation in tobacco tends to be somewhat equivocal, however. They observed the frequency of variants among plants regenerated directly from cotyledon explants (30 days in culture), from cotyledon derived cells cultured for either 135, 180 or 255 days prior to plantlet initiation, and among plants derived from cotyledon protoplasts (80 days from protoplast culture to plants). The frequency of mutations at the specific a_1 and yg loci was at a maximum in plants derived from protoplasts and early callus culture. This was interpreted as pre-existing mutations in the cotyledon cells either as a functional mutation or as a lesion in DNA of the latent cotyledon cell, which on dedifferentiation during culture was 'repaired' and expressed following mitotic segregation. Conversely, unstable chlorophyll variants and leaf shape and size morphological variants did increase in frequency with prolonged culture.

In our own research utilizing somaclonal variation to recover eyespot disease (*H. sacchari*) resistant sugar-canes, we have found that the incidence of toxin resistant somaclones is much greater among plants regenerated from old (30 months) cultures than from younger cultures. Similarly, McCoy *et al.* (1982) in screening oat somaclones for meiotic abnormalities (heteromorphic bivalents, monosomics, trisomics, translocations) found a dramatic increase in frequency with duration in culture.

In several species, such as tobacco (see Table 1 and Arcia *et al.*, 1978; Collins *et al.*, 1973; Deaton *et al.*, 1982), studies have observed an unexpectedly high level of variation among dihaploids derived from inbred lines. The analysis of this variation relates to the issue of pre-existing or culture induced variation. It was found that some dihaploids in tobacco (De Paepe *et al.*, 1981), rape (Hoffmann *et al.*, 1982) and rice (Oono, 1981) derived by androgenesis from inbred lines proved to be heterozygous when progeny tested. The mutations could only have arisen subsequently to microspore culture initiation and spontaneous chromosome doubling, *i.e.* during the culture phase.

Thus, in broad terms, the evidence is overwhelmingly in favour of the concept that somaclonal variation is generated during the culture phase *per se*. Pre-existing somatic variation in the explant material will of course amplify the level of variation observed.

9.2.2.5 *Genetic and explant source effects*

Any procedure which examines the occurrence of genetic variation is also likely to uncover a genetic predisposition to elaborate more or less genetic variation. Such is the origin of transposable genetic elements in prokaryotes, which have parallels in the mutator genes and transposing DNA sequences described for higher organisms such as yeast, *Drosophila* and several plant species (Finnegan *et al.*, 1982). Though only limited conclusions can be drawn at this stage, several studies have shown that the level of somaclonal variation can be cultivar, *i.e.* genotype, dependent. McCoy *et al.* (1982) in a detailed examination of chromosomal abnormalities among plants regenerated from cell cultures of two oat cultivars (Tippecanoe and Lodi), found that the latter cultivar consistently gave a higher frequency of chromosomal variants than Tippecanoe. An evaluation of variants among meristem tip derived plants of strawberry showed that one cultivar 'Earligrow' produced more than twice the number of crown, leaf and fruit variants than either 'Redchief' or 'Guardian' subclones (Schwartz *et al.*, 1981). In *Pelargonium*, somaclones (calliclones) derived from callus were more variable from some cultivars, *e.g.* 'Attar of Rose', than others, *e.g.* 'Old Fashioned Rose' (Skirvin and Janick, 1976). One cultivar, 'Snowflake', which is inherently unstable through conventional stem cutting propagation, is also highly unstable through calliclonal propagation. In *Begonia* × *hiemalis*, as part of a study to evaluate mutation breeding, Roest *et al.* (1981) found that the frequency of variants (colour, size and form of leaves and flowers) in non-irradiated controls was much higher in one cultivar (43%) than in the other (7%).

It can be seen from Table 1 that quite diverse explants can give rise to somaclonal variants. Some explant sources and modes of producing *in vitro* plants, as for example adventitious shoots from petiole or leaf explants, minimize the duration of the callus phase. While such a procedure does reduce the extent of somaclonal variation, it does not obviate it. Barbier and Dulieu (1980) found somaclonal variants not only among plants regenerated from callus and cultured proto-plasts, but also among plants rapidly derived from adventitious shoots on leaf explants.

By inference it was considered that the extensive variation identified in potatoes by Shepard (1981) required the use of a protoplast culture system. While this is desirable from the standpoint that individual somaclones have a unique single cell origin, it is not essential for the elaboration of somaclonal variation. Thomas *et al.* (1982) found variation between different plants regenerated from a single callus which itself had been derived from cultured protoplasts of the cultivar 'Maris Bard'. They also produced plants by stem embryogenesis from a different cultivar 'Majestic' and found relatively low levels of variability which could reflect varietal differences as discussed earlier. In contrast, from a different cultivar of potato, 'Desiree', van Harten *et al.* (1981) found substantial variation for stem, leaf, flower and skin colour variants among plants regenerated by the adventitious shoot technique from rachis, petiole and leaf-disc explants. In the case of plants produced from rachis and petiole explants, not previously exposed to X-rays (as were other explants), 50% of the plants were variant relative to the donor parent.

The knowledge that there is a cultivar (genotypic) and explant influence on the level of somaclonal variation observed is significant. Where clonal identity is desirable, as in the use of *in vitro* techniques for the rapid propagation of horticultural varieties, a judicious choice of both cultivar and explant source may obviate some of the expense in rogueing the resultant plants. Conversely, where additional genetic variability is required, as in a plant breeding program, optimizing the level of variability will depend not only on cultivar and explant source but also on the period of time the cell line is held in culture prior to regenerating somaclones.

9.3 THE ORIGIN OF SOMACLONAL VARIATION

An understanding of the genetic and/or molecular events which underlie somaclonal variation not only has intrinsic value, but also a pragmatic consequence. The plant propagator in seeking clonal uniformity would wish to mitigate against such variation, while the plant breeder may wish to enhance or, if it were possible, produce specific types of genetic variation.

We have examined some of the possible mechanisms which might give rise to somaclonal variation (Larkin and Scowcroft, 1981) as have Skirvin (1978), Chaleff (1981) and Shepard (1981). These include ploidy changes, nuclear fragmentation, inter- and intra-chromosomal interchanges, somatic crossing over and sister chromatid exchanges, gene amplification and diminution and transposable genetic elements. One of the impacts of the molecular genetic analysis of eukaryotes is the realization that the eukaryotic genome exists in a much greater state of genetic flux than classical genetics might lead us to believe. It is thus useful to offer some additional thoughts on the events which might give rise to somaclonal variation.

9.3.1 Chromosomal Abnormalities

There is an extensive literature documenting a wide range of chromosomal abnormalities which occur in cultured plant cells (see Baylis, 1980; Larkin and Scowcroft, 1981). Notwithstanding the fact that chromosomal abnormalities do occur in plants regenerated from cell culture, the frequency of gross karyotypic changes is much reduced relative to that observed in cell cultures *per se* (Ogihara, 1981). It is highly likely that a reasonably intact and balanced genome is necessary for plant regeneration, so that many of the gross karyotypic changes that occur during cell culture will be selected against during the regeneration process. Karyotype studies in potato, *Pelargonium*, lucerne and maize indicate that chromosome number changes seem not to be very prevalent among regenerated plants which display somaclonal variation. Further, there is only a very limited causal relationship between phenotypic variation among somaclones and ploidy changes. Ogihara (1981) also made the interesting observation that chromosomal variability in cell cultures of *Haworthia* can be carried over to the somatic cells (root tips) of the regenerates, but appear to be eliminated from the meiotic cycle.

Mitotic analysis of regenerated plants gives only limited information about cytological abnormalities. Chromosomal changes such as deletions, duplications, translocations and inversions can usually be detected only by meiotic analysis. Detailed meiotic analysis of regenerated plants in barley (Orton, 1980), *Haworthia* (Ogihara, 1981) and oats (McCoy *et al.*, 1982) have revealed extensive chromosomal rearrangements such as reciprocal and non-reciprocal translocations, inversions, sub-chromatid aberrations and partial chromosome loss. It seems, therefore, that chromosome breakage and interchanges during cell culture is a relatively frequent event.

As pointed out by Larkin and Scowcroft (1981) and McCoy *et al.* (1982), somatic chromosome interchanges may favour transgression of desirable genes from alien species into commercial cultivars. Transgression often fails to occur in conventional interspecific hybridization because of the lack of recombination between disparate genomes. Several reports which describe the phenotype of plants regenerated from cell cultures of intespecific hybrids of *Lolium* species (Ahloowahlia, 1976) between *Lolium multiflorum* and *Festuca arundinaceae* (Kasperbauer *et al.*, 1979) and between *Hordeum* species (Orton, 1980) indicate that transgression does occur at an appreciable frequency. Increasingly wide hybridization either by conventional means or by somatic hybridization through protoplast fusion is being viewed as a plant breeding option. The potential of cell culture as a means of enhancing gene exchange could play an important role in this strategy.

9.3.2 Molecular Possibilities

The molecular analysis of eukaryotic genomes is revealing a luxuriant, almost prodigal, variety of ways in which gene expression can be regulated and modified (see Brown, 1981). Several mechanisms can cause either gene amplification or diminution, gene rearrangements and the alteration of specific methylation patterns leading to altered gene activity can be inherited. Transposable DNA sequences appear to abound in eukaryotes and an increasing number of sequences coding for a functional gene product are being recognized as multigene families. We would like to consider just two of these possibilities in the control of somaclonal variation.

9.3.2.1　*Gene amplification and diminution*

The quantity of a gene product can be increased or decreased simply by differential gene amplification or diminution. The dramatic amplification of specific genes during development, *e.g. Drosophila* chorion genes, is documented by Chisholm (1982). With appropriate selection, stable amplification can result. For example, selection for resistance to the drug methotrexate, which inhibits dihydrofolate reductase activity, can lead to a 200-fold amplification of the gene coding for this enzyme (Schimke *et al.*, 1977). Similarly, the gene complex which includes the coding sequence for aspartate transcarbamylase is amplified by up to 190-fold when hamster cell lines are selected for resistance to a transition state analogue (Wahl *et al.*, 1979).

In plants, ribosomal RNA gene amplification and depletion has been documented in wheat, rye, hyacinth, maize, *Vicia*, melon and tobacco. The DNA of flax is known to change in response to environmental pressure (Cullis and Goldsborough, 1980). Reassociation kinetics showed that one plant form possessed a class of moderately repeated DNA not present in the other form.

If such amplification (or depletion) of DNA sequences can occur in plant cell cultures it may account at least in part for somaclonal variation. This possibility requires a mechanism to amplify the specific genes. Somatic crossing over and sister chromatid exchange can cause duplications and deletions given asymmetry between homologous sequences. Molecular analysis has shown that eukaryotic genomes contain a number of different repeated DNA sequences distributed either as tandem arrays or dispersed throughout the genome (Jeffreys and Harris, 1982). Often such families are flanked by short tandem or inverted repeats. Thus a ready means is provided for asymmetric association of homologous sequences (chromosomes or chromatids) and when followed by an exchange, duplication results. Given one duplication, further amplification can readily occur.

9.3.2.2　*Transposable elements*

Transposable genetic elements are DNA sequences which can move from one position in the genome to another independently of extensive sequence homology (Calos and Miller, 1980). A transposition may or may not have a detectable effect, but if inserted into a coding sequence, that gene is likely to be inactivated. Imprecise excision of the transposable elements may generate rearrangements (deletions, inversions) of adjacent chromosomal sequences.

In eukaryotes, genetic evidence suggests that certain unstable mutants may be explained by transposable elements (McClintock, 1980). Eukaryotes, such as *Drosophila*, contain a large proportion of moderately repetitive DNA organized as a number of families of transposable elements located at 20 to 40 sites in the genome. Typically these transposable sequences are flanked by long direct repeats, each of which contains short flanking inverted repeats (Finnegan *et al.*, 1982). In yeast, the transposition of a particular element has been directly associated with a mutational event at the his-4 locus and in *Drosophila* mutations at the white and bithorax loci are known to contain insertions. It is yet to be proven, however, that in higher eukaryotes mutation and transpositional events are causally related.

9.4 CELL SELECTION

Where there is a correlation between cellular and whole plant response, then it becomes possible to view cell culture as a means of enriching for specific variants. Progress in the development, success and understanding of the limitations of cell culture selection systems have been well documented (Thomas *et al.*, 1979; Chaleff, 1981). Most success has been achieved in selecting for resistance/tolerance to antimetabolites, such as amino acid analogues, base analogues and antibiotic drugs. Biochemical auxotrophs can also be recovered, though this involves the relatively tedious total isolation and individual screening approach (Gebhardt *et al.*, 1982).

We will be concerned here with the use of cell culture selection to isolate mutants which are more or less directly relevant to an agricultural end use. Several of the foregoing aspects of somaclonal variation are also relevant and complementary to mutant selection.

Table 2 contains some details of several instances where deliberate selection was imposed at the cell proliferation phase to recover potentially useful agricultural variants. In general, the plants regenerated from selected cell lines showed enhanced tolerance to the particular antimetabolite or stress imposed, but this did not always result in genotypes with a greater level of tolerance/resistance than can be achieved by conventional means.

9.4.1 Disease Resistance

Details of the *H. maydis* toxin resistance maize regenerants have already been covered (Sections 9.2.2, 9.2.3). As in the case of *H. sacchari* toxin selection in sugar-cane, it appears that the only advantage of toxin selection over simply screening somaclones for tolerance is to enhance the frequency of tolerant types recovered. The evidence from the mtDNA analysis of toxin resistant maize plants indicates that those regenerated from toxin selected cell lines do not have any distinguishing features relative to those recovered without cell selection.

The success of using cell selection systems to recover late blight (*P. infestans*) tolerant potato and blackleg (*P. lingam*) tolerant rape plants adds a new and potentially valuable dimension. In both these cases, the respective causal pathogens do not produce a host-specific toxin as does *H. maydis* and *H. sacchari*. Behnke (1979) and Sacristán (1982) used a culture filtrate of the respective fungal pathogens as the selective agent; callus growth was inhibited in both cases. Plants regenerated from selected cell lines when challenged with the pathogen had an enhanced level of tolerance. Sacristán (1982) argued that such a non-selective toxin was a codeterminant of pathogenicity. The *P. lingam* toxin necrotized leaf tissue and arguably conditioned it for invasion by the parasite.

As pointed out by Brettel and Ingram (1979), there is a limited but significant number of plant diseases which produce host-specific toxins and these would seem amenable to cell culture selection strategies. In view of the results of Behnke (1979) and Sacristán (1982), there is a case for an integrated effort by plant pathologists and somatic cell geneticists to evaluate the use of cell culture systems to select for resistance to pathogens which do not necessarily elaborate a host-specific toxin.

9.4.2 Herbicide Tolerance

Herbicide tolerant crop plants are desirable for several reasons. For example, the available cost-effective herbicides are not always sufficiently selective and the increasing emphasis on conservation tillage options to minimize input costs and to conserve top soil demands effective weed control. Cell culture could provide a means of selecting for herbicide tolerance. Though several cases exist where selection has produced cell lines tolerant to herbicides such as 2,4-D, 2,4,5-T,

Table 2 Cell Culture Selection for Agronomically Useful Genotypes

Selected trait	Species	Plant regeneration and expression	Inheritance	Ref.
DISEASE (TOXIN) TOLERANCE				
Pseudomonas tabaci[a]	Nicotiana tabacum	+	Semi-dominant	Carlson (1973)
Helminthosporium maydis	Zea mays	+	Cytoplasmic	Gengenbach et al. (1977)
				Brettell et al. (1980)
H. sacchari	Saccharum officinarum	+	?[b]	Larkin and Scowcroft (in prep.)
Phytophthora infestans	Solanum tuberosum	+	?	Behnke (1979)
Phoma lingam	Brassica napus	+	Nuclear?[c]	Sacristán (1982)
HERBICIDE TOLERANCE				
Picloram	N. tabacum	+	Dominant or semi-dominant	Chaleff and Parsons (1978)
PHYSIOLOGICAL TOLERANCE				
NaCl	N. tabacum	+	Nuclear?	Nabors et al. (1980)
NUTRITIONAL QUALITY				
Enhanced threonine	Zea mays	+	Dominant	Hibberd and Green (1982)

[a] A pseudo-analogue of wildlife toxin. methionine sulfoximine, was used as the selective agent. [b] Stable for at least 5 normal sett generations. [c] Nuclear transmission but inheritance pattern not determined.

aminotriazole, anulam, and propham, there is only one case where herbicide tolerant plants have been produced following cell selection.

Chaleff and Parsons (1978) selected several tobacco cell lines for resistance to lethal concentrations of picloram (4-amino-3,5,6-trichloropicolinic acid). Plants regenerated from some of these lines were resistant and transmitted the resistance to progeny. Resistance was shown to be controlled by a dominant allele in three cases and a semi-dominant allele in a fourth case. In a subsequent study, Chaleff and Keil (1981) found that three of five picloram resistance lines were also resistant to hydroxyurea, which normally inhibits cell culture growth. Hydroxyurea resistance was controlled by a single dominant gene and in two of the three cases assorted independently of picloram resistance. This surprising result, which begged explanation, may be a case of extreme somaclonal variation.

9.4.3 Nutritional Quality

In cereals in particular it is desirable to upgrade the levels of certain amino acids such as lysine, threonine and tryptophan. Several selection schemes are based on the principle that overproduction of specific amino acids might occur in cell lines, and plants could be regenerated therefrom which have been selected for resistance to growth inhibition by false feedback inhibitors (Thomas *et al.*, 1979; Chaleff, 1981). Hibberd and Green (1982) have significantly upgraded the threonine content of maize kernels by selecting maize cell lines for resistance to growth inhibition occasioned by equimolar concentrations of threonine and lysine. The selection procedure was based on the commonality of the aspartate pathway for lysine, threonine, isoleucine and methionine synthesis. The lysine/threonine resistance was inherited as a single dominant gene and overproduction of threonine in the kernels of the resistant plants increased threonine content by 33–59%. Levels of free methionine, serine and proline were also marginally increased.

9.4.4 Other Cell Selection Systems

Cell selection to recover salt tolerant genotypes, though often advocated, has only been successfully achieved once (Nabors *et al.*, 1980). This was in tobacco, where a stepwise selection regime with increasing salt concentrations was applied to cell cultures. Plants regenerated from selected cell lines survived watering with 0.57 kmol m^{-3} NaCl, whereas only a few plants from non-selected cultures survived 0.26 kmol m^{-3} NaCl. The tolerance was seed transmitted, but the inheritance pattern was unclear. Some physiological adaptation was apparent because F_2 progeny are more salt tolerant if the respective F_1 plants were also exposed to salt.

There are several reports where cell selection has resulted in cell lines with enhanced tolerance to imposed stress or antimetabolites such as aluminum, salt and chilling stress (see Chaleff, 1981). With chilling stress selection, regenerated plants had no greater tolerance than unselected plants and the response of plants from cells selected for metal tolerance has not been reported.

An interesting consequence of selection, *vis-à-vis* somaclonal variation, involved the use of ethionine selection in cell culture in an attempt to recover lucerne which overproduced methionine (Reisch *et al.*, 1981; Reisch and Bingham, 1981). Ethionine resistant cell lines could be recovered with higher levels of methionine, but there was a concomitant increase in total amino acids. Among plants regenerated from ethionine resistant cell lines, there was a staggering amount of variability with respect to leaf and growth habit, plant height, shoot length, number of primary branches and field plot yield. Plants were also regenerated from non-ethionine selected cell lines. Though the variation among these was much reduced relative to those from selected cell lines, the highest yielding variant was derived from an unselected cell line. The inheritance of these changes is yet to be determined. The morphological abnormalities were attributed largely (but not exclusively) to the ethionine treatment, possibly as a consequence of transethylation of nucleic acids which generate mutations because of impaired DNA replication and/or repair.

9.5 CONCLUSIONS

There is heightened, albeit tempered, enthusiasm for the use of cell culture technology in plant improvement. The enthusiasm derives from the continual quest for technology which will speed the development of improved cultivars. The use of somaclonal variation in cultivar improvement has already advanced to the stage of field testing of promising variants in sugar-cane, potatoes and pasture species. Some horticultural cultivars of *Pelargonium* and *Begonia* have been derived from cell culture variants.

Temperance must apply lest an unwarranted panacea be promulgated. The conventional wisdom and practice of plant improvement can never be obviated. It can however be amplified, extended and made more efficient. Plant biotechnology will become an integral part of plant improvement within the immediate future.

9.6 ADDENDUM

This review was completed in August 1982. A few additional references (p. 168) should assist the reader to gain a better grasp of the current state of the art. In particular, somaclonal variation has now been established as a significant way to generate useful variation in a number of additional crop species (Orton, 1984; Scowcroft and Ryan, 1985) including tomato (Evans and Sharp, 1983), lettuce (Engler and Grogan, 1984), alfalfa (Groose and Bingham, 1984) and wheat (Larkin *et al.*, 1984). In addition, we now have a better understanding of the genetic events which give rise to somaclonal variation (Scowcroft and Ryan, 1985). *In vitro* selection has also become a very powerful method to recover specific useful genetic variants (Chaleff, 1983; Maliga, 1984), some of which are now known to result from gene amplification (Donn *et al.*, 1984).

9.7 REFERENCES

Ahloowahlia, B. S. (1976). Chromosomal changes in parasexually produced ryegrass. In *Current Chromosome Research*, ed. K. Jones and P. E. Brandham, pp. 115–122. Elsevier/North Holland Biomedical Press, Amsterdam.

Ahloowahlia, B. S. (1982). Plant regeneration from callus culture in wheat. *Crop Sci.*, **22**, 405–410.

Anon. (1980). The Global 2000 Report to the President. *Entering the Twenty First Century*, vol. 2. Council on Environmental Quality and the Dept. of State, Government Printing Office, Washington, DC.

Arcia, M. A., E. A. Wernsman and L. G. Burk (1978). Performance of anther-derived dihaploids and their conventionally inbred parents as lines, in F_1 hybrids, and in F_2 generations. *Crop Sci.*, **18**, 413–418.

Bajaj, Y. P. S., A. K. Ram, K. S. Labana and H. Singh (1981). Regeneration of genetically variable plants from the anther-derived callus of *Arachis hypogaea* and *Arachis villosa*. *Plant Sci. Lett.*, **23**, 35–39.

Barbier, M. and H. L. Dulieu (1980). Effets génétiques observés sur des plantes de Tabac régénérées à partir de cotyledons par culture *in vitro*. *Ann. Amélior. Plant*, **30**, 321–344.

Bayliss, M. W. (1980). Chromosomal variation in plant tissues in culture. *Int. Rev. Cytol. Suppl.*, **11A**, 113–144.

Behnke, M. (1979). Selection of potato callus for resistance to culture filtrates of *Phytophthora infestans* and regeneration of resistant plants. *Theor. Appl. Genet.*, **55**, 69–71.

Bingham. J. (1981). The achievements of conventional plant breeding. *Philos. Trans. R. Soc. London, Ser. B*, **292**, 441–455.

Bingham, E. T., L. V. Hurley, D. M. Kaatz and J. W. Saunders (1975). Breeding alfalfa which regenerates from callus tissue in culture. *Crop Sci.*, **15**, 719–721.

Brettell, R. I. S. and D. S. Ingram (1979). Tissue culture in the production of novel disease-resistant crop plants. *Biol. Rev.*, **54**, 329–345.

Brettell, R. I. S., E. Thomas and D. S. Ingram (1980). Reversion of Texas male-sterile cytoplasm maize in culture to give fertile, T-toxin resistant plants. *Theor. Appl. Genet.*, **58**, 55–58.

Brettell, R. I. S., M. F. Conde and D. R. Pring (1982). Analysis of mitochondrial DNA from four fertile maize lines obtained from a tissue culture carrying Texas cytoplasm. *Maize Genet. Coop. Newslett.*, **56**, 13–14.

Brown, D. D. (1981). Gene expression in eukaryotes. *Science*, **211**, 667–674.

Burk, L. G. and D. F. Matzinger (1976). Variation among anther-derived doubled haploids from an inbred line of tobacco. *J. Hered.*, **67**, 381–384.

Calos, M. P. and J. H. Miller (1980). Transposable elements. *Cell*, **20**, 579–595.

Carlson, P. (1973). Methionine sulfoximine-resistant mutants of tobacco. *Science*, **180**, 1366–1368.

Chaleff, R. S. (1981). *Genetics of Higher Plants*. Cambridge University Press, Cambridge.

Chaleff, R. S. and R. L. Keil (1981). Genetic and physiological variability among cultured cells and regenerated plants of *Nicotiana tabacum*. *Mol. Gen. Genet.*, **181**, 254–258.

Chaleff, R. S. and M. F. Parsons (1978). Direct selection *in vitro* for herbicide-resistant mutants of *Nicotiana tabacum*. *Proc. Natl. Acad. Sci. USA*, **75**, 5104–5107.

Chisholm, R. (1982). Gene amplification during development. *TIBS*, **7**, 161–162.

Collins, G. B., P. D. Legg and C. C. Litton (1973). The use of anther-derived haploids in *Nicotiana*. 2. Comparisons of doubled-haploid lines with lines obtained by conventional breeding methods. *Tob. Sci.*, **18**, 40–42.

Crisp, P. and D. G. A. Walkey (1974). The use of aseptic meristem culture in cauliflower breeding. *Euphytica*, **23**, 305–313.

Cullis, C. A. and P. B. Goldsborough (1980). Variation in the ribosomal RNA gene number — a model system for DNA changes in flax genotrophs. In *The Plant Genome*, ed. D. R. Davies and D. A. Hopwood, pp. 91–97. The John Innes Charity, Norwich.

Cummings, D. P., C. E. Green and D. D. Stuthmam (1976). Callus induction and plant regeneration in oats. *Crop Sci.*, **16**, 465–470.

Deambrogio, E. and P. J. Dale (1980). Effect of 2,4-D on the frequency of regenerated plants in barley and on genetic variability between them. *Cereal Res. Commun.*, **8**, 417–423.

Deaton, W. R., P. D. Legg and G. B. Collins (1982). A comparison of barley tobacco double-haploid lines with their source inbred cultivars. *Theor. Appl. Genet.*, **62**, 69–74.

De Paepe, R., D. Bleton and F. Gnangbe (1981). Basis and extent of genetic variability among doubled haploid plants obtained by pollen culture in *Nicotiana sylvestris*. *Theor. Appl. Genet.*, **59**, 177–184.

Dixon, L. K., C. J. Weaver, R. I. S. Brettell and B. G. Gengenbach (1982). Mitochondrial sensitivity to *Drechslera maydis* T-toxin and the synthesis of a variant mitochondrial polypeptide in plants derived from maize tissue cultures in Texas male-sterile cytoplasm. *Theor. Appl. Genet*, **63**, 75–80.

Edallo, S., C. Zuccinali, M. Perenzin and F. Salamini (1981). Chromosomal variation and frequency of spontaneous mutation associated with *in vitro* culture and plant regeneration in maize. *Maydica*, **26**, 39–56.

Evans, D. A., W. R. Sharp and C. E. Flick (1981). Plant regeneration from cell cultures. *Hortic. Rev.*, **3**, 214–314.

Finnegan, D. J., B. H. Will, A. A. Bayev, A. M. Bowcock and L. Brown (1982). Transposable DNA sequences in eukaryotes. In *Genome Evolution*, ed. G. A. Dover and R. B. Flavell, pp. 29–40. Academic, London.

Gamborg, O. L., J. P. Shyluck, D. S. Brar and F. Constab (1977). Morphogenesis and plant regeneration from callus of immature embryos of sorghum. *Plant Sci. Lett.*, **10**, 67–74.

Gebhardt, C., V. Schnebli and P. J. King (1982). Isolation of biochemical mutants using haploid mesophyll protoplasts of *Hyoscyamus muticus*. II. Auxotrophic and temperature-sensitive clones. *Planta*, **153**, 81–89.

Gengenbach, B. G. and P. Umbeck (1982). Characteristics of T-cytoplasm revertants from tissue culture. *Maize Genet. Coop. Newslett.*, **56**, 140–142.

Gengenbach, B. G., C. E. Green and C. M. Donovan (1977). Inheritance of selected pathotoxin resistance in maize plants regenerated from cell cultures. *Proc. Natl. Acad. Sci. USA*, **74**, 5113–5117.

Green, C. E. (1977). Prospects of crop improvement in the field of cell culture. *Hortic. Sci.*, **12**, 7–10.

Green, C. E. and R. L. Phillips (1975). Plant regeneration from tissue cultures of maize. *Crop Sci.*, **15**, 417–421.

Grout, B. W. W. and P. Crisp (1980). The origin and nature of shoots propagated from cauliflower roots. *J. Hortic. Sci.*, **55**, 65–70.

Haydu, Z. and I. K. Vasil (1981). Somatic embryogenesis and plant regeneration from leaf tissues and anthers of *Pennisetum purpureum* Schum. *Theor. Appl. Genet.*, **59**, 269–273.

Heinz, D. J. (1973). Sugar-cane improvement through induced mutations using vegetative propagules and cell culture techniques. In *Induced Mutations in Vegetatively Propagated Plants*, pp. 53–59. International Atomic Energy Agency, Vienna.

Heinz, D. J., M. Krishnamurthi, L. G. Nickell and A. Maretzki (1977). Cell, tissue and organ culture in sugarcane improvement. In *Applied and Fundamental Aspects of Plant Cell, Tissue and Organ Culture*, ed. J. Reinert and Y. P. S. Bajaj, pp. 3–17. Springer-Verlag, Berlin.

Henke, R. R., M. A. Mansur and M. J. Constantin (1978). Organogenesis and plantlet formation from organ- and seedling-derived calli of rice (*Oryza sativa*). *Physiol. Plant*, **44**, 11–14.

Hibberd, K. A. and C. E. Green (1982). Inheritance and expression of lysine plus threonine resistance selected in maize tissue culture. *Proc. Natl. Acad. Sci. USA*, **79**, 559–563.

Hoffmann, F., E. Thomas and G. Wenzel (1982). Anther culture as a breeding tool in rape. II. Progeny analyses of androgenetic lines and induced mutants from haploid cultures. *Theor. Appl. Genet.*, **61**, 225–232.

Jeffreys, A. J. and S. Harris (1982). Processes of gene duplication. *Nature (London)*, **296**, 9–10.

Jung-Heiliger, H. and W. Horn (1980). Variation nach mutagener Behandlung von Stecklingen und in vitro-Kulturen bei *Chrysanthemum*. *Z. Pflanzenzücht.*, **85**, 185–199.

Kasperbauer, M. J., R. C. Buckner and L. P. Bush (1979). Tissue culture of annual ryegrass × tall fescue F_1 hybrids: callus establishment and plant regeneration. *Crop Sci.*, **19**, 457–460.

Kemble, R. J., R. B. Flavell and R. I. S. Brettell (1982). Mitochondrial DNA analyses of fertile and sterile maize plants derived from tissue culture with the Texas male sterile cytoplasm. *Theor. Appl. Genet.*, **62**, 213–217.

Larkin, P. J. and W. R. Scowcroft (1981). Somaclonal variation — a novel source of variability from cell culture for plant improvement. *Theor. Appl. Genet.*, **60**, 197–214.

Larkin, P. J. and W. R. Scowcroft (1982). Somaclonal variation and crop improvement. In *Genetic Engineering of Plants*, ed. A. Hollaender, C. P. Meredith and T. Kosuge. Plenum, New York.

Liu, M.-C. (1981). *In vitro* methods applied to sugar cane improvement. In *Plant Tissue Culture*, ed. T. A. Thorpe, pp. 299–323. Academic, New York.

Liu, M.-C. and W.-H. Chen (1976). Tissue and cell culture as aids to sugarcane breeding. I. Creation of genetic variation through callus culture. *Euphytica*, **25**, 393–403.

Maliga, P., Z. R. Kiss, P. J. Dix and G. Lázár (1979). A streptomycin-resistant line of *Nicotiana sylvestris* unable to flower. *Mol. Gen. Genet.*, **172**, 13–15.

McCoy, T. J., R. L. Phillips and H. W. Rines (1982). Cytogenetic analysis of plants regenerated from oat (*Avena sativa*) tissue cultures; High frequency of partial chromosome loss. *Can. J. Genet. Cytol.*, **24**, 37–50.

McClintock, B. (1980). Modified gene expressions induced by transposable elements. In *Mobilization and Reassembly of Genetic Information*, ed. W. A. Scott, R. Werner, D. R. Joseph and J. Schutz, pp. 11–19. Academic, New York.

Mix, G., H. M. Wilson and B. Foroughi-Wehr (1978). The cytological status of plants of *Hordeum vulgare* L. regenerated from microspore callus. *Z. Pflanzenzücht.*, **80**, 89–99.

Nabors, M. W., S. E. Gibbs, C. S. Bernstein and M. E. Meis (1980). NaCl-tolerant tobacco plants from cultured cells. *Z. Pflanzenphysiol.*, **97**, 13–17.

Novák, F. J. (1980). Phenotype and cytological status of plants regenerated from callus cultures of *Allium sativum* L. *Z. Pflanzenzuecht.*, **84**, 250–260.

Ogihara, Y. (1981). Tissue culture in *Haworthia*. Part 4: Genetic characterization of plants regenerated from callus. *Theor. Appl. Genet.*, **60**, 353–363.

Oinuma, T. and T. Yoshida (1974). Genetic variation among doubled haploid lines of barley tobacco varieties. *Jpn. J. Breeding*, **24**, 211–216.

Oono, K. (1981). *In vitro* methods applied to rice. In *Plant Tissue Culture*, ed. T. A. Thorpe, pp. 273–298. Academic, New York.

Orton, T. J. (1980). Chromosomal variability in tissue cultures and regenerated plants of *Hordeum*. *Theor. Appl. Genet.*, **56**, 101–112.

Reisch, B. and E. T. Bingham (1981). Plants from ethionine-resistant alfalfa tissue cultures: variation in growth and morphological characteristics. *Crop Sci.*, **21**, 783–788.

Reisch, B., S. H. Duke and E. T. Bingham (1981). Selection and characterization of ethionine-resistant alfalfa (*Medicago sativa* L.) cell lines. *Theor. Appl. Genet.*, **59**, 89–94.

Reuther, G. (1977). Adventitious organ formation and somatic embryogenesis in callus of *Asparagus* and *Iris* and its possible application. *Acta Hortic.*, **78**, 217–224.

Roest, S., M. A. E. Van Berkel, G. S. Bokelmann and C. Broertjes (1981). The use of an *in vitro* adventitious bud technique for mutation breeding of *Begonia* × *hiëmalis*. *Euphytica*, **30**, 381–388.

Sacristán, M. D. (1982). Resistance responses to *Phoma lingam* of plants regenerated from selected cell and embryogenic cultures of haploid *Brassica napus*. *Theor. Appl. Genet.*, **61**, 193–200.

Sacristán, M. D. and M. F. Wendt-Gallitelli (1971). Transformation to auxin-autotrophy and its reversibility in a mutant line of *Crepis capillaris* callus culture. *Mol. Gen. Genet.*, **110**, 355–360.

Schimke, R. T., F. W. Alt, R. E. Kellems and J. R. Bertino (1977). Amplification of dihydrofolate reductase gene in methotrexate-resistant variants of cultured murine cells. *Cold Spring Habor Symp. Quant. Biol.*, **42**, 649–657.

Schwartz, H. J., G. J. Galletta and R. H. Zimmerman (1981). Field performance and phenotypic stability of tissue culture-propagated strawberries. *J. Am. Soc. Hortic. Sci.*, **106**, 667–673.

Scowcroft, W. R. (1977). Somatic cell genetics and plant improvement. *Adv. Agron.*, **29**, 39–81.

Scowcroft, W. R. and P. J. Larkin (1982). Somaclonal variation: a new option for plant improvement. In *Plant Improvement and Somatic Cell Genetics*, ed. I. K. Vasil, W. R. Scowcroft and K. J. Frey, pp. 159–178. Academic, New York.

Secor, G. R. and J. F. Shepard (1981). Variability of protoplast-derived potato clones. *Crop Sci.*, **21**, 102–105.

Shepard, J. F. (1981). Protoplasts as sources of disease resistance in plants. *Annu. Rev. Phytopathol.*, **19**, 145–166.

Shepard, J. F., D. Bidney and E. Shahin (1980). Potato protoplasts in crop improvement. *Science*, **208**, 17–24.

Sibi, M. (1976). La notion de programme génétique chez les végétaux supérieurs. II. Aspect expérimental. Obtention de variants par culture de tissus *in vitro* sur *Lactuca sativa* L. apparition de vigueur chez les croisements. *Ann. Amélior. Plant.*, **26**, 523–547.

Skirvin, R. M. (1978). Natural and induced variation in tissue culture. *Euphytica*, **27**, 241–266.

Skirvin, R. M. and J. Janick (1976). Tissue culture-induced variation in scented *Pelargonium* spp. *J. Am. Soc. Hortic. Sci.*, **101**, 281–290.

Sutter, E. and R. W. Langhans (1981). Abnormalities in *Chrysanthemum* regenerated from long term cultures. *Ann. Bot.*, **48**, 559–568.

Thomas, E., P. J. King and I. Potrykus (1979). Improvement of crop plants *via* single cells *in vitro* — an assessment. *Z. Pflanzenzücht.*, **82**, 1–30.

Thomas, E., S. W. J. Bright, J. Franklin, V. A. Lancaster, B. J. Miflin and R. Gibson (1982). Variation amongst protoplast-derived potato plants (*Solanum tuberosum* cv. 'Maris Bard'). *Theor. Appl. Genet.*, **62**, 65–68.

Tisserat, B., E. B. Esan and T. Murashige (1980). Somatic embryogenesis in angiosperms. *Hortic. Rev.*, **1**, 1–78.

van Harten, A. M., H. Bouter and C. Broertjes (1981). *In vitro* adventitious bud techniques for vegetative propagation and mutation breeding of potato (*Solanum tuberosum* L.). II. Significance for mutation breeding. *Euphytica*, **30**, 1–8.

Wahl, G. M., R. A. Padgett and G. R. Stark (1979). Gene amplification causes overproduction of the first three enzymes of UMP synthesis in *N*-(phosphonacetyl)-L-aspartate-resistant hamster cells. *J. Biol. Chem.*, **254**, 8679–8689.

Wakasa, K. (1979). Variation in the plants differentiated from the tissue culture of pineapple. *Jpn. J. Breeding*, **29**, 13–22.

Wenzel, G., O. Schieder, T. Przewozny, S. K. Sopory and G. Melchers (1979). Comparison of single cell culture derived *Solanum tuberosum* L. plants and a model for their application in breeding programs. *Theor. Appl. Genet.*, **55**, 49–55.

Williams, L. and H. A. Collins (1976). Growth and cytology of celery plants derived from tissue cultures. *Ann. Bot.*, **40**, 333–338.

ADDITIONAL REFERENCES

Chaleff, R. S. (1983). Isolation of agronomically useful mutants from plant cell cultures. *Science*, **219**, 676–682.

Donn, G., E. Tischer, J. A. Smith and H. M. Goodman (1984). Herbicide-resistant alfalfa cells: an example of gene amplification in plants. *J. Mol. Appl. Genet.*, **2**, 621–635.

Engler, D. E. and R. G. Grogan (1984). Variation in lettuce plants regenerated from protoplasts. *J. Hered.*, **75**, 426–430.

Evans, D. A. and W. R. Sharp (1983). Single gene mutations in tomato plants regenerated from tissue culture. *Science*, **221**, 949–951.

Groose, R. W. and E. T. Bingham (1984). Variation in plants regenerated from tissue culture of tetraploid alfalfa heterozygous for several traits. *Crop Sci.*, **24**, 655–658.

Larkin, P. J., S. A. Ryan, R. I. S. Brettell and W. R. Scowcroft (1984). Heritable somaclonal variation in wheat. *Theor. Appl. Genet.*, **67**, 443–455.

Maliga, P. (1984). Isolation and characterization of mutants in plant cell culture. *Annu. Rev. Plant Physiol.*, **35**, 519–542.

Oono, K. (1985). Putative homozygous mutations in regenerated plants of rice. *Mol. Gen. Genet.*, **198**, 377–384.

Orton, T. J. (1984). Somaclonal variation: theoretical and practical considerations. In *Gene Manipulation in Plant Improvement*, ed. J. P. Gustafson, p. 427. Plenum, New York.

Scowcroft, W. R. (1986). Somaclonal variation: the myth of clonal uniformity. In *Plant Gene Research*, vol. 2, *Genetic Flux in Plants*, ed. B. Hohn and E. S. Dennis. Springer Verlag, Berlin (in press).

Scowcroft, W. R. and S. A. Ryan (1985). Tissue culture and plant breeding. In *Plant Cell Culture Technology*, ed. M. Yeoman. Blackwell Scientific, Oxford (in press).

10

Virus-free Clones Through Plant Tissue Culture

R. STACE-SMITH

Agriculture Canada, Vancouver, British Columbia, Canada

10.1 INTRODUCTION

Virus diseases differ from other infectious diseases in that they have no independent metabolism but depend on their host both for sustenance and for the mechanism whereby they are synthesized. They enter into such a close association with their hosts that they essentially become part of them, with the virus nucleic acid adding a new pattern of synthesis that leads to the cell producing virus nucleic acid and protein. This intimate association makes the cure of virus diseases by chemicals much more difficult than the cure of bacterial or fungal diseases. Antibiotics or other pesticides owe their success to their ability to stop specific and essential processes of the pathogen but, since viruses appear to use the same mechanisms and processes that make nucleic acids and proteins of the host cells, complete eradication of viruses from an infected plant by chemical means is virtually impossible. The only chemicals that could be expected to interfere with virus multiplication are those that affect the synthesis of nucleic acids and these must also be expected to interfere with the synthesis of normal nucleic acids. These expectations have not been realized in practice: although there are examples of some analogues of purines and pyrimidines reducing the virus content of infected plants, there are none yet of a plant being made virus-free.

Viruses are economically important only in plants in which they become systemic, *i.e.* plants in which the virus moves from a single point of infection to invade all or most of the vegetative parts. Once systemically infected, plants remain so for as long as they live; consequently, virus diseases are much more important in perennial plants, or those that are propagated vegetatively, than in crops raised annually from true seed. Although some viruses do enter the seed set by infected plants, most do not, and seedlings usually start life virus-free. By contrast, viruses are consistently transmitted to offspring whenever plants are propagated vegetatively, whether by bulbs, tubers, cuttings, or by grafting or budding, and the new plants raised by these methods will almost all be infected.

The symptoms of virus diseases are characteristically more severe in the initial phases of infection than later on, but full recovery from a virus infection is rare under conditions in which plants normally grow. Most trees and other perennial plants that become infected thrive indefinitely, yielding less each year than they would if they were not infected. Another problem with perennial plants is that they do not remain infected with a single virus, but with time and field exposure they tend to acquire a complex of two or more viruses. Such accumulation of viruses is common in many of the popular plant cultivars, and useful clonal material has often become so crippled by viruses that their cultivation has had to be abandoned.

To supply healthy planting material, certification schemes are now operating for several kinds of plants, and the raising of stocks for propagation is often separated from the growing of a crop for its main purpose. Special stocks that form the basis of certification schemes are built up by propagating from single virus-free plants. Until recently, the production of virus-free stocks of plants that are grown as clonal varieties depended upon finding an uninfected plant of the variety to start the stock. Fortunately, this is no longer so for methods have been found whereby many plants can be freed from some or all of the viruses that commonly infect them. The state has now been reached where it can be said that no useful clone need be abandoned because the whole clone is virus-infected. The method that has found the widest application is meristem tip culture, usually taken from clonal material that has been subjected to a prolonged period of growth at a temperature that is considerably higher than would normally be used for plant production.

10.2 DISTRIBUTION OF VIRUSES IN PLANTS

It has already been noted that most viruses do not enter the seed of infected plants. Likewise, many fail to enter the meristematic tissue at the growing points of stems. Different viruses seem to differ in their ability to approach this region of actively dividing cells. Some are remote enough from it so that tip cuttings several millimeters long may be virus-free; these are pieces long enough to be rooted directly or to be grafted on to healthy plants and later rooted as cuttings. Other viruses approach the growing points so closely that little other than the apical meristem itself is uninfected. The technique of developing virus-free plants by propagating meristems excised from infected plants was based originally on the theory that viruses are unevenly distributed in their host, so that a sufficiently small piece of tissue might be virus-free. While this is indeed true for some viruses, the majority still appear to be able to penetrate into the meristematic cells. Infectivity tests to determine the extent of viral penetration into the apical region were inconclusive until recently. Kassanis (1967) first suggested that electron microscopy of apical meristems might detect the presence of virus particles and indeed this proved to be the case. Appiano and Pennazio (1972), using thin section electron microscopy, detected potato virus X (PVX) particles in the cytoplasm of apical dome cells of potato. Further, Pennazio and Redolfi (1974) were able to detect infective particles of PVX in potato meristems by direct inoculation using meristematic tips as a source of inoculum.

The maximum heat that the host can tolerate will result in a drastic reduction of virus concentration. Indeed, for a few viruses that are particularly labile, keeping the plant at about 37 °C for a few weeks has been sufficient to eradicate the virus from the whole plant. More commonly, however, plant viruses are not completely eradicated and once a treated plant is returned to normal growing conditions the virus replication process resumes and no permanent cure has been achieved. Thus, to be effective, heat treatment must be coupled with some process whereby the treated plant is subdivided into explants which are capable of being established on their own. This may be done by rooting cuttings, culturing meristematic tips, or grafting excised buds on to healthy plants.

Heat therapy is the most successful and widely used therapeutic method. The number of viruses that have been successfully inactivated has steadily increased; 15 were reported by 1950, 75 by 1960, 100 by 1966 and 120 by 1969 (Nyland and Goheen, 1969). To put these figures in perspective, it must be noted that successful eradication did not depend on heat therapy alone, but usually on heat therapy combined with meristem-tip culture.

There is little agreement in the literature as to how high a temperature the various plant species can tolerate on an indefinite basis. The reason for disparity in reported results is that a number of factors contribute to plant survival and the equipment and techniques vary. Important considerations would include the physiological state of the plant at the beginning of treatment, the watering regime, the fluctuations in temperature, and the differences between root temperature and top temperature.

10.3 TECHNIQUES FOR ERADICATION

Three methods have been used in attempts to isolate virus-free tissue that will regenerate into a rooted plant. The first is to subject the infected plant to temperatures near 37 °C, which retards or inhibits the multiplication of some viruses. Meristematic tips excised from heat-treated plants are often free from viruses that are not easily eliminated by meristem culture alone. The second is the use of metabolites, either applied to the infected plant before bud excision, or incorporated into the nutrient medium. As first noted by Norris (1954), some metabolites decrease the concentration of virus without irreparable harm to the plant tissue. The third is based on the observation that virus concentration diminished from older to younger vegetative parts of an infected plant. By isolating the apical meristems (Morel and Martin, 1952) from infected plants and growing them in a nutrient medium, virus-free tissue cultures have been established, some of which may develop into rooted plants.

For the purpose of discussion, these three techniques will be considered separately, although it must be pointed out that in practice the techniques are mutually compatible and the three may be employed simultaneously to eradicate those viruses that do no readily respond to any single eradication process.

10.3.1 Heat Treatment

Most viruses multiply less well in plants held above 30 °C than below and keeping virus-infected plants continuously near the maximum temperature they can tolerate will often free them from infection. A few weeks at about 36 °C has often been enough to cure plants of some virus diseases. More frequently, however, the virus is not completely eradicated and, although the virus concentration is drastically reduced, virus replication resumes when the plant is returned to normal growing conditions. Nevertheless, heat treatment is still a valuable technique in establishing a virus-free clone. After long periods at high temperature, some parts of the plant may be virus-free, even though others are still infected. New shoots produced during treatment are especially likely to be uninfected, so when cuttings are taken from plants and rooted, one or more may produce a virus-free plant.

Although applicable to some viruses, heat therapy followed by excision and rooting of shoots does not work with most viruses, hence alternative techniques must be used. The smaller the tissue unit used to establish an explant, the less likelihood of it being infected. Carried to the extreme, the minute shoot tip constitutes the ideal tissue unit. The small size required to be virus-free is incapable of surviving in an exposed environment, but it is able to survive in the protected environment of a culture tube. For this reason, the most common technique for obtaining virus-free clones is to subject the mother plant to a prolonged period of heat therapy and, immediately after treatment, excise and culture meristematic tips from the plant. Thus heat treatment is frequently combined with meristem culture to eradicate those viruses that are not readily eliminated by either treatment alone.

Heat treatment is perhaps the most useful method of freeing plants from viruses. Hot water treatment was used a few decades ago, but hot air is now universally accepted as the preferred method. The standard method involves growing the infected plant at about 37 °C for several weeks. A useful innovation is to establish first an *in vitro* culture of the infected clone and then subject the culture to heat treatment, excising meristematic tips at intervals during treatment. This technique has proved particularly effective in obtaining virus-free grapevine clones (Harris and Stevenson, 1982).

For a more complete discussion on heat therapy, the reader is referred to review articles by Kassanis (1965), Hollings (1965) and Nyland and Goheen (1969).

10.3.2 Chemotherapy

There are many reports in the literature of substances that suppress symptoms or reduce the virus concentration in plants and it is logical to assume that some of these substances may be usefully applied to a virus-infected plant to obtain virus-free explants. However, the results to date are disappointing. Some of the analogues of purines and pyrimidines, used at concentrations that do not destroy plant tissue, inhibit the multiplication of virus. Some authors have actually claimed virus eradication, but the reports have been based on small numbers of plants or the results have been open to other interpretations (Hollings, 1965; Mellor and Stace-Smith, 1977). Despite the apparent lack of success in virus eradication, the antiviral compound could augment the effectiveness of virus eradication by meristem-tip culture, particularly with those viruses that are difficult to eradicate by the standard heat treatment–meristem culture combination.

A recent report (Cassels and Long, 1980) suggests that one of the antiviral agents used by animal virologists may be more effective than the antimetabolites that have been used previously. Virazole, when incorporated into a differentiation medium, resulted in the eradication of potato virus Y and cucumber mosaic virus from a proportion of the explants. This is the most encouraging report to date on the possible beneficial effects of including an antiviral agent in culture media.

10.3.3   Meristem Culture

Meristematic buds from apices of stems or from leaf axils are usually selected for tissue culture. Their advantage is that the incipient shoot has already differentiated; to establish an independent plant, only elongation and root differentiation are required. It is of little practical value to culture other tissue, but it should be recognized that almost any part of a plant has the potential of producing propagules indistinguishable from the mother plant. Non-meristematic tissue must undergo a development process involving the formation of callus tissue and subsequent differentiation into embryos and plantlets. For this reason, meristematic buds have become the tissue of choice used by virtually all workers as the starting material in the production of virus-free clones.

Meristem culture involves the dissection of a portion of the meristematic region and culturing it on a nutrient medium for plantlet regeneration. The explant sometimes used is the apical dome, although in most cases one to several leaf primordia of the subapical region are also included (Figure 1). The number of leaf primordia included depends upon the plant species, the stability of the virus (or viruses) being eradicated, and the experience of the operator with respect to rooting success. Some workers refer to meristem culture as the culturing of the apical dome only, while others include a portion of the stem beyond the meristematic subapical region of their explant. The latter procedure is usually referred to as 'shoot-tip culture'. Other terms, such as 'meristem-tip culture', 'tip culture' and 'axillary bud culture' frequently appear in the literature. In this article I shall refer to meristem culture as a unit including the meristematic dome plus one or two pairs of leaf primordia, the unit being from 0.2 to 0.5 mm in length. (Figure 2).

10.3.3.1   Culture media

No aspect of meristem culture is as contentious as the composition of the culture medium that should be used to culture a particular species. The literature is rife with contradictions. This is understandable considering the fact that most research on the subject has been done by plant physiologists studying growth responses. Their findings have been adapted by applied workers wanting tissue differentiation and root development. As a result, there is little agreement as to what constitutes the ideal meristem culture medium for a particular plant species. White's medium (White, 1943) was the most widely used during the early days of meristem culture, but since then many improvements in media have been made. A useful basic medium is the one devised by Murashige and Skoog (1962) and most workers today favour this medium or variations of this medium. It contains major and minor elements, sucrose as a carbon source, and a variety of vitamins and growth substances.

As a generalization, meristem tips excised from herbaceous plants develop and root more readily than comparable tissue from woody plants. However, this may reflect the state of the art with respect to tissue culturing and, as techniques improve, there may be little or no intrinsic difference between the two groups. With recent reports of success in culturing species that were once considered difficult or impossible to culture, it appears that virtually any plant species of economic significance can be cultured, provided expertise and effort are applied. An indication of how the field is advancing can be gained by noting the rate of expansion of lists of species for which virus-free plants have been obtained by meristem culture. The first report of the use of

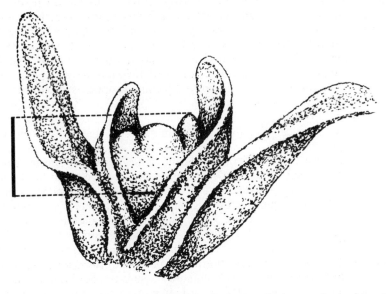

Figure 1 Drawing of axillary bud of potato, with rudimentary leaves spread to expose the meristematic dome. Dotted line indicates portion to be excised for culture. Bar represents 0.5 mm

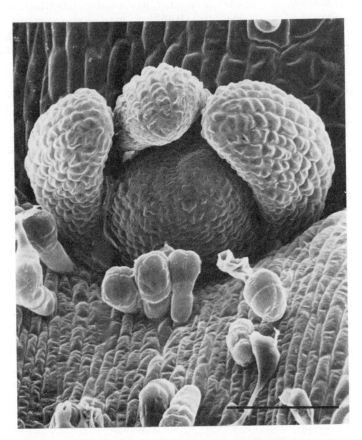

Figure 2 Scanning electron micrograph of the meristematic tissue in a potato leaf axil. Photograph shows the meristematic dome with the initials of a leaf primordium plus three leaf primordia. Bud was exposed by bending back the leaf petiole on a slender shoot raised *in vitro*. Bar represents 0.1 mm

meristem culture for virus eradication was with dahlia (Morel and Martin, 1952). By 1965, the technique had been used to eradicate one or more viruses from carnation, chrysanthemum, cymbidium, narcissus, potato, strawberry and sweet potato (Hollings, 1965). The application steadily increased and by 1977 the list had reached 38 species (Quak, 1977) and, in the most recent compilation (Wang and Hu, 1980), it had reached 48 species. It is evident from this that a suitable culture medium can be devised for virtually any commercial clone for which virus-free material is unavailable and for which no alternative eradication technique appears applicable.

10.3.3.2 *Factors affecting development and rooting*

The choice of the culture medium for use in culturing a particular species is of the utmost importance. Fortunately, along with successes, many workers have published their failures, so that for many species information is available as to concentration of components that may inhibit development and rooting. However, even though limited success has been reported with a species having fastidious requirements, there is no guarantee that another worker in a different laboratory will achieve the same degree of success. If work is required on a new clone, about all one can do in designing preliminary experiments aimed at devising a suitable culture medium is be guided by previous experience. It does not have to be the optimal medium, since this would require considerable time, but it must be adequate to assure the development and rooting of a proportion of the explants to yield at least one virus-free plant. Some species, like potato, offer a broad range of media possibilities (Mellor and Stace-Smith, 1977); others, such as raspberry, are more exacting in their requirements (Donnelly *et al.*, 1980).

While recognizing that the composition of the culture medium is the single most important factor affecting development and rooting of excised buds, it is by no means the only factor. The size of the excised meristem influences rooting, but the presence of leaf primordia, rather than the volume of tissue, appears to determine development. Stace-Smith and Mellor (1968) reported that potato buds 1 mm or longer rooted more readily than smaller buds. However, bud size had a marked influence on the elimination of viruses. Buds less than 0.3 mm long were usually rejected because they were unlikely to root, and buds more than 0.7 mm long were seldom used because they were liable to be infected. Within this narrow size range, there was little difference in rooting.

The environmental conditions under which cultures are held also has an influence. Most workers hold their cultures at room temperature (20–28 °C) illuminated by fluorescent lights of 2000–4000 lux for a 16 hour day. This is not necessarily optimum, but specific requirements for individual species have generally not been determined. Pennazio and Redolfi (1973) studied the effect of three light regimes on potato cultures: lamps which were rich in red light, lamps which approximated daylight and a 2:1 mixture of both. Rooting under each regime was 18, 34 and 67%, respectively. Moreover, the plantlets under mixed light were the most vigorous. They found no difference in percentage rooting between light intensity of 2000 and 4000 lux, but leaf and root development were better at 4000 lux. The effect of temperature on the growth of meristems of several potato cultivars was investigated by Mellor and Stace-Smith (1977). At 29 °C some cultivars grew quickly and rooted plantlets developed within 5 weeks. At 26 °C growth was slower, but shoot growth usually preceded root growth, whereas at 23 °C growth was even slower but shoots and roots usually developed simultaneously.

Mellor and Stace-Smith (1977) noted other factors affecting the development of potato meristems, factors which may be important with other plant species. Cultivar differences may determine the rate of development. With many cultivars, a few buds develop a shoot and root within 2 months of excision while others, under similar cultural conditions, require 4–6 months, and one or more transfers to fresh media, to reach the same degree of development. Buds from a single shoot also vary and there appears to be no predictable pattern for rate of development with a single group of buds. Position on shoot and bud size, within the range 0.3–0.7 mm, appears to have little influence on early rooting.

10.4 VIRUS ERADICATION

A frustrating aspect of the literature on virus eradication is that the virus or viruses present in a treated plant or the cultured explants may not be adequately identified. Such reports are of dubious value. In a compilation prepared by Murashige (1978), plants claimed to have been freed of viruses, but where the tests and the pathogens eliminated remained unidentified, were deliberately excluded from a list of specific pathogen-free plants attained *in vitro* from infected plants. He further notes that a serious error among plant tissue culturists has been the claim of disease-free or pathogen-free plants without having appropriately tested to confirm and identify the excluded pathogens.

Since plants that are vegetatively propagated are subject to infection by many unrelated viruses, detection and identification may be difficult. Sap inoculations, aphid inoculations, graft inoculation, serology, electron microscopy, or electrophoresis may be required. If the only objective is to produce a virus-free clone, it might seem unnecessary to know which viruses were present before treatment, but indexing the source plant will reduce the extent of indexing required to determine the virus status of the regenerants. Use of a single source plant, and knowledge of its virus content, indicates which tests must be used to detect any surviving viruses in possible numerous plantlets. The single source plant should be subdivided prior to treatment so that an untreated control can serve as a check when regenerants are indexed. Adequate controls and rigorous indexing are a necessary prerequisite to justify claims of virus freedom in the end product.

A simplified version of strawberry meristem culture is given in Figure 3. What is not included is such steps as prior heat treatment and the extensive indexing required to determine the virus status of the source plant and progeny.

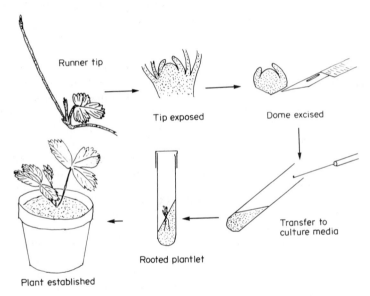

Figure 3 Schematic drawing of the procedure used to establish a strawberry plant *via* meristem culture. Steps involving heat treatment to assist virus eradication and indexing for virus detection are omitted from the sketch

With regard to the mechanism of eradication, there is conclusive evidence with a number of viruses, as noted earlier, that infectious particles are present in the meristematic tissue and that excised tips contain virus at the time they are placed on the culture medium. Electron microscopic evidence is limited to relatively few viruses, but the fact that particles have been detected whenever a determined search has been made suggests that the phenomenon is the rule rather than the exception. Further, a portion of these tips develop into plantlets that are free from virus, providing conclusive evidence that eradication takes place during culture.

The mechanism of virus eradication during culture is not known but a number of explanations have been proposed. Hollings and Stone (1964) suggested that an inactivation system exists within the apex, whose action is helped by removing the mature portion of the plant. Ingram (1973) suggested that it may be due either to some inactivating factor produced by the explant, or to the effect of some constituent in the culture medium. Quak (1977) speculated that the disappearance of virus particles may be attributed to contact of the meristem with the culture medium. Mellor and Stace-Smith (1977) believe that viral replication requires enzymes that are normally available to the cells near the meristematic dome and that excision of the tip temporarily disorganizes the growth processes, with the result that enzymes required for one or more steps of viral replication are unavailable. Without viral replication, the normal process of viral degradation would continue but no new particles would be formed.

At present no general explanation is available for the fact that some viruses are readily inactivated during culture and others are not. Possibly the various viruses differ in the rate at which they are naturally degraded in excised tips. Once could also speculate that those viruses that are not inactivated during culture were initially present in the meristematic cells at a much higher concentration than those viruses that are more readily eliminated. This may explain why meristem culture alone is unsuccessful whereas, if the plant is subjected to prior heat treatment, the virus concentration in the meristematic tissue is reduced to the point where there are relatively few particles remaining at the time of excision and these are unable to survive in culture.

10.5   MAJOR USE OF VIRUS-FREE CLONES

Considerable effort has been directed toward acquiring virus-free clones of many vegetatively propagated plants. These efforts have succeeded to the point where all major cultivars are either available in a virus-free state or, if the demand arose, could be rendered virus-free. It is pertinent to ask the question, is the effort worthwhile and should it be continued? There is little doubt that it has been worthwhile in specific instances, but of dubious value in other instances. At least it has provided researchers with material for three basic uses, namely (1) to undertake controlled experiments on the effects of a specific virus or virus combination on a cultivar, (2) to provide the basic clonal stock for replacement of infected material and (3) to provide material for mass *in vitro* propagation.

10.5.1   Study Effect of Virus Infection

The question as to the effect of a specific virus on the yield of a particular clone is in theory easy to answer. In practice, many reports of increased yield are unable to withstand close scrutiny. First, consideration must be confined to the so-called 'latent' viruses, viruses which exhibit no obvious symptoms of infection. The reason for this is that, even in the absence of virus-free material, propagation schemes based upon the use of best available stock have usually resulted in the selection of clones that appear healthy. The introduction of virus-free clones is therefore an added refinement of schemes that are already in operation. Second, the fact that those viruses being eradicated are latent is an indication that they are responsible for minimal damage, which means that it is more difficult to design experiments that will reliably detect minor differences in yield. Third, since virus-free clones can readily be reinfected, it would be essential to ascertain the virus status of both the infected plants and the virus-free plants at the conclusion of the experiment. Fourth, since the virus-free clone may not be genetically identical to the original untreated clone, it is essential that a portion of the virus-free clone be used for control plots and that another portion of the virus-free clone be inoculated with a typical field isolate to provide planting material for the infected plot. Technically, all of these criteria should be met in order to obtain statistically sound data. In actuality, it is the rare report on the effect of virus on yield that meets all or most of the criteria. For this reason, the persistent question that is raised as to whether increased yields are realized when clones are freed from latent virus infections will continue to be debated.

When reliable comparisons have been made, be it on carnation, narcissus, geranium, rhubarb or potato, the virus-free clones outyielded infected clones by a significant margin. Results vary with the crop and the virus involved but yield increases generally range from 10–20% (Quak, 1977; Mellor and Stace-Smith, 1977). These results are to be expected since previous yield comparisons using naturally available virus-free material consistently show yield reductions associated with virus infection. To my knowledge, no worker has ever demonstrated that a plant virus can be benign or beneficial. It is possible that, under certain cultural conditions, virus-infected plots may outyield virus-free plots. Also, there is suggested evidence that latent viruses may influence either susceptibility or resistance to other types of disease. This emphasizes the requirement that cultural conditions may have to be altered to achieve the full potential of the virus-free clones.

10.5.2   Source for Clonal Propagation

The major use of virus-free clones is to provide a source for propagation. This is based upon three assumptions, namely (1) that the virus-free clone is genetically identical to the infected clone, (2) that the virus-free clone can be propagated to the point of commercial production in an essentially virus-free state and (3) that the increased costs associated with production of virus-free planting-stock will be justified in terms of increased yields. These three assumptions, although partially accepted, have not been universally accepted. Ineptitude, or poor planning on the part of those involved with clonal propagation, has contributed to the general reluctance to accept the virus-free clones.

One can not assume that the progeny from a plant selected for virus eradication will be genetically identical to the mother plant. There is little evidence to support the frequently expressed view that heat treatment and/or meristem culture increases the mutation rate of a particular species, but it must be recognized that, since mutations are sometimes detected in cuttings, they should be expected to occur in plantlets derived from meristem-tips. As noted by Quak (1977), it is generally believed that meri-clones (*i.e.* clones derived by meristem-tip culture), with an increased mutation rate, lead to genetic aberrations although no statistical data on mutation frequency are available. I prefer to accept the view expressed by Walkey (1978) that, in general, plants regenerated by meristem-tips show no more genetic variation than those produced by conventional macro-methods of vegetative propagation.

The need for attention to clonal selection can not be overemphasized. There is absolutely no point in releasing a virus-free clone without the assurance that it is genetically equivalent to the best available clone of a particular cultivar. It is essential that the regenerated virus-free clones be monitored for changes in their horticultural characteristics before being used for large scale propagation.

The assumption that virus-free clones can be propagated to the point of commercial production in an essentially virus-free state is sometimes difficult to fulfill. With apple clones it appears to be easy since apple viruses rarely, if ever, spread naturally. Field infections can always be traced to the planting of infected trees and not to spread from tree to tree. Most of the increase in virus occurrence comes about by the use of clonal understock that are thoroughly infected (Hansen and Denby, 1978). In contrast, potato clones freed from the latent viruses (potato virus X and potato virus S) are exceedingly difficult to multiply on a commercial scale without becoming recontaminated (Shepard and Claflin, 1975). Between these two extremes are such crops as raspberry, strawberry, chrysanthemum and carnation, where re-infection is a constant threat but where satisfactory programs have evolved.

Will the increased costs associated with the production of virus-free planting stock be justified in terms of increased yields? The evidence suggests that it is definitely justified in those crops that are not readily reinfected, but there is still some doubt in those crops where recontamination is almost inevitable. It is evident that attempts to incorporate certification requirements for latent viruses on the potato certification schemes that are in place in the US have not worked. Technically, there were no latent virus-free potato certification programs in the US in 1975 and only three states were engaged in large scale certification for potato virus X (Shepard and Claflin, 1975). The fact that a virus-free potato program in British Columbia has been effective (Wright, 1981) suggests that similar programs may eventually be cost-effective in the US.

10.5.3 Source for *In Vitro* Mass Propagation

As noted above, there are increased costs associated with production of virus-free planting stock and, with some crops, recontamination with virus is almost inevitable. It is obvious that reinfection is fully excluded as long as the plantlets remain under aseptic conditions. Carried one step further, it is advantageous as well to perform micropropagation *in vitro*, provided of course that the process is economically viable. Tissue culture is just another method of cloning, but it differs from traditional methods by enabling faster rates of plant multiplication and protection against contamination. Rapid multiplication is particularly advantageous when a newly released cultivar is in high demand. It has other advantages: pathogens are often unknowingly excluded, and the tissue-culture derived plant can often lead to a commodity with superior qualities. The most extensive application of plant tissue culture has been in rapid clonal propagation, especially with ornamental crops (Murashige, 1978).

While virus-free clones are routinely used for tissue cultured clones, there are relatively few instances where *in vitro* cloning has been used specifically to avoid the virus problems that plague traditional multiplication methods. One example of *in vitro* cloning being used primarily to ensure the health of the plant is strawberry. Boxus *et al.* (1977) pioneered the industrial multiplication of strawberry plants by *in vitro* methods and reported that it is possible to obtain several million plants from a single mother plant within one year. They note that the main drawbacks of micropropagation are economic and psychological in nature—the cost of producing *in vitro* strawberry plants is at present rather high, and strawberry growers will have to accept a new type of plant and adapt themselves to a few new requirements.

10.6 CONCLUDING REMARKS

Many proponents of virus eradication through heat treatment and meristem culture are enthusiastic about the improvement in plant quality and yield. They project increased popularity of plant tissue cultures, with more and more scientists all over the world devoting their efforts to the *in vitro* regeneration of pathogen-free crop plants. They further predict that the number of pathogen-free clonal plants is expected to increase at an accelerated rate (Wang and Hu, 1980). In my reading of trends, I have arrived at a different conclusion. My conclusion is that adequate procedures of virus eradication have in fact already been developed and there is little need for scientists all over the world to devote their efforts to regeneration of pathogen-free plants. Taking the potato as an example, virtually every cultivar of economic significance is already available as a virus-free clone. All that remains to be done now is to index any new cultivars of potential value and, for those determined to be infected, undertake eradication procedures. This does not require the effort of a scientist; a competent technician can do the job. About the only circumstance where a scientific input may be required is in devising a suitable culture medium to meet a requirement in those crops which to date have defied attempts to culture.

Technology that is already available should be exploited in devising schemes to mass produce virus-free clones without extensive reinfection. Again, little new research is necessary, only ingenuity and cooperation between researchers, certification agencies, nurserymen and growers. In circumstances where virus spread can not be kept within reasonable bounds, it may well be decided that the extra effort required is not commensurate with possible increased yields. Alternative control measures such as cross protection may be more satisfactory. An exciting example of the application of cross protection to control citrus tristeza has recently been reported (Costa and Müller, 1980). In the future, this technique may be utilized to a greater extent to complement virus-free schemes.

10.7 REFERENCES

Appiano, A. and S. Pennazio (1972). Electron microscopy of potato meristem tips infected with potato virus X. *J. Gen. Virol.*, **14**, 273–276.
Boxus, P., M. Quoirin and J. M. Laine (1977). Large scale propagation of strawberry plants from tissue culture. In *Plant Cell, Tissue, and Organ Culture*, ed. J. Reinert and Y. P. S. Bajaj, pp. 130–143. Springer-Verlag, Berlin.
Cassels, A. C. and R. D. Long (1980). The regeneration of virus-free plants from cucumber mosaic virus- and potato virus Y-infected tobacco explants cultured in the presence of virazole. *Z. Naturforsch., Teil B*, **35**, 350–351.
Costa, A. S. and G. W. Müller (1980). Tristeza control by cross protection: a US–Brazil cooperative success. *Plant Dis.*, **64**, 538–541.
Donnelly, D. J., R. Stace-Smith and F. C. Mellor (1980). *In vitro* culture of three *Rubus* species. *Acta Hortic.*, **112**, 69–75.
Hansen, A. J. and L. Denby (1978). Apple viruses—an endangered species? *Canada Agric.*, **23** (4), 8–11.
Harris, R. E. and J. H. Stevenson (1982). Tissue culture of grapes. *Canada Agric.*, **27** (1), 29–33.
Hollings, M. (1965). Disease control through virus-free stock. *Annu. Rev. Phytopathol.*, **3**, 367–396.
Hollings, M. and O. M. Stone (1964). Investigation of carnation viruses. 1. Carnation mottle *Ann. Appl. Biol.*, **53**, 103–118.
Ingram, D. S. (1973). Growth of plant parasites in tissue culture. In *Plant Tissue and Cell Culture*, ed. H. E. Street, pp. 392–421. Blackwell, Oxford.
Kassanis, B. (1965). Therapy of virus-infected plants. *J. R. Agric. Soc. Engl.*, **126**, 105–114.
Kassanis, B. (1967). Plant tissue culture. In *Methods in Virology*, ed. K. Maramorosch and H. Koproski, vol. 1, pp. 537–566. Academic, New York.
Mellor, F. C. and R. Stace-Smith (1977). Virus-free potatoes by tissue culture. In *Plant Cell, Tissue and Organ Culture*, ed. J. Reinert and Y. P. S. Bajaj, pp. 616–635. Springer-Verlag, Berlin.
Morel, G. and C. Martin (1952). Guérison de dahlias atteints d'une maladie à virus. *C. R. Hebd. Seances Acad. Sci.*, **235**, 1324–1325.
Murashige, T. (1978). The impact of plant tissue culture on agriculture. In *Frontiers of Plant Tissue Culture*, ed. T. A. Thorpe, pp. 15–26. University of Calgary, Calgary, Alberta.
Murashige, T. and F. Skoog (1962). A revised medium for rapid growth and bio assays with tobacco tissue cultures. *Physiol. Plant.*, **15**, 473–497.
Norris, D. O. (1954). Development of virus-free stock of Green Mountain by treatment with malachite green. *Aust. J. Agric. Res.*, **5**, 658–663.

Nyland, G. and A. C. Goheen (1969). Heat therapy of virus diseases of perennial plants. *Annu. Rev. Phytopathol.*, **7**, 331–354.

Pennazio, S. and P. Redolfi (1973). Factors affecting the culture *in vitro* of potato meristem tips. *Potato Res.*, **16**, 20–29.

Pennazio, S. and P. Redolfi (1974). Potato virus X eradication in cultured potato meristem tips. *Potato Res.*, **17**, 333–335.

Quak, F. (1977). Meristem culture and virus-free plants. In *Plant Cell, Tissue, and Organ Culture*, ed. J. Reinert and Y. P. S. Bajaj, pp. 598–615. Springer-Verlag, Berlin.

Shepard, J. F. and L. Claflin (1975). Critical analyses of the principles of seed potato certification. *Annu. Rev. Phytopathol.*, **13**, 271–293.

Stace-Smith, R. and F. C. Mellor (1968). Eradication of potato viruses X and S by thermotherapy and axillary bud culture. *Phytopathology*, **58**, 199–203.

Walkey, D. G. A. (1978). *In vitro* methods for virus elimination. In *Frontiers of Plant Tissue Culture*, ed. T. A. Thorpe, pp. 245–254. University of Calgary, Calgary, Alberta.

Wang, P. J. and C. Y. Hu (1980) Regeneration of virus-free plants through *in vitro* culture. In *Advances in Biochemical Engineering*, ed. A. Fiechter, vol. 18, pp. 61–99. Springer-Verlag, Berlin.

White, P. R. (1943). *A Handbook of Plant Tissue Culture*. Jaques Cattell Press, Lancaster, PA.

Wright, N. S. (1981). A strategy for PVX and PVS control on seed farms. *Am. Potato J.*, **58**, 525.

11

Metabolites from Recombinant DNA Modified Plants

R. G. McDANIEL
University of Arizona, Tucson, AZ, USA

11.1 INTRODUCTION

Higher plant species exhibit myriad metabolic pathways which have evolved for the production of special chemical products, many of which are both characteristic of, and unique to, these species. The diversity of such plant-produced metabolites rivals and probably exceeds the number of metabolic products found in the cells of multicellular animals. The alternation of generations of higher plants has no doubt compelled the synthesis of numerous metabolites. Differentiation and development of specialized plant cells and organs within the framework of structures of sexual reproduction have resulted in an array of different organizational approaches by plants to metabolite production during evolution of gymnosperms and angiosperms.

This chapter considers the factors which have contributed to the production of diverse metabolites by higher plants, and some examples of metabolic pathways involved. Also discussed is the manipulation of the genes controlling these pathways using the tools of recombinant DNA biotechnology. The most valuable, useful, as well as the most feasible, metabolites which may be produced *via* recombinant DNA plant modification will be reviewed.

11.2 EUKARYOTIC BIOCHEMICAL GENETICS

11.2.1 Plant Biochemistry in Relation to Cell Differentiation and Tissue Development

Eukaryotic cells operate under the constraints and controls imposed upon their enzymatic machinery by the regulatory properties of the nuclear genome. The present conceptualization of the eukaryotic genome considers gene transcription to be closely controlled both by the higher order structure of the nucleosome and by gene products themselves (Nagl and Fusenig, 1979). Nucleosome structure can be simplistically described as the ordered winding of the DNA polymer around successive spaced beads of nucleoproteins. These beads are octamers of histones — basic proteins of the nucleus, one of whose functions it is to assist chromatin 'packing' in the nucleus. Associated with the DNA in the spacer regions between the 'core' histone octamers is another histone, the highly reactive H1. This H1 histone has the capability to severely repress gene transcription and additionally is readily phosphorylated, acetylated or methylated. These chemical reactions modify the activity of histone H1 in processes which are closely linked with phases of the cell cycle.

Implicit in the structure and function of eukaryotic chromosome structure is the orderly, developmental activation and deactivation of genes, eliciting an ebb and flow of messenger RNAs which, in ways too subtle for present understanding, mold the ontogeny of the organism. The net result of such temporal control of genetic message production is the construction of the inherent physicochemical properties of the organism, which serve to distinguish it, as a species, from all other organisms.

As a cell divides and becomes surrounded by its fellows, chemical metabolites produced by the surrounding cells elicit the feedback of information which activates the appropriate genes and enables the cells to form tissues, organs and, ultimately, the entire organism. This occurs simply through the appropriate sequential activation of the proper genetic templates, in essence, the message of the species. Prime among the metabolites directing the differentiation and development of higher plants are undoubtedly plant hormones. Manipulation and alteration of the course of development, and indeed the choice of developmental pathways, followed by higher plants can be shown experimentally to be readily achieved by adding exogenous plant growth hormones.

The synthesis of plant growth hormones is closely associated with the meristematic activity of the growing apices of the shoots and roots during plant development. Such hormones are divided into several classes, each of which may be represented by one or more structural analogues which function essentially like the parent compound. Plant hormones include auxins, characterized by an indole ring; gibberellins, first isolated from fungi; cytokinins, closely related to kinetin, a nucleic acid degradation product first described in fish sperm; abscisic acid; brassins; several classes of short and long chain alcohols; and ethylene, unique among plant growth hormones because it exists in biological systems as a gas.

Common to these hormones are three attributes closely linked with the developmental biochemistry of plants. First, they are ubiquitous in their presence in nearly all plant tissues. Secondly, the most prominent effect attributed to these hormones normally occurs at a site in the plant remote to their site of synthesis. Thirdly, each of the plant hormones is characterized by a multiplicity of observed effects on plant growth, cell division and differentiation. Effects can be measured on intermediary metabolic pathways, organelle functions and nucleic acid metabolism. Evidence has indicated that plant hormones often may be found in 'bound' form and as the 'free' hormone in the same tissues. The bound forms exist as a pool of hormone which could serve to supplement depleted free hormone, without the necessity to await the transport of a fresh supply of hormone through the vascular system from a remote organ. Most importantly, numerous researchers utilizing radiotracer techniques or sensitive bioassays have described gradients of hormone concentrations throughout the plant. For example, a micromolar concentration of indoleacetic acid may be detected in root tissues, while a thousand-fold higher concentration might exist in the meristematic tissues at the shoot apex.

This point is emphasized here, not only because hormones are of paramount importance in the *in vitro* culture and growth of plant cells, but also because their concentrations may change as a result of any alteration of the genetic structure of plants. One must consider that an extremely slight change in the hormone balance of these living systems is likely to have a profound influence on the resultant product, be it a whole plant or a single secondary metabolite of interest.

In addition to the concentration gradient which is expected from one plant tissue to the next, numerous studies have revealed an interaction between plant hormones with regard to the effects of a given concentration of one hormone upon the equilibrium concentrations of other hormones in the same tissues. In essence, this means that the addition of a given concentration of exogenous hormone to one tissue may elicit a completely different response than the equivalent amount of the same hormone added to a different tissue which possesses a distinctly different balance of other plant hormones. Some of the problems inherent to growth hormones will be discussed later with regard to the regeneration of genetically-engineered plants from tissue culture.

Germane to any discussion of plant hormones and development is the evidence coming from scattered experiments in several laboratories, in which plant hormones are being found to associate closely with DNA or RNA. Low molar mass plant hormones have the ability to intercalate with nucleic acids, and thus conceivably may also be capable of acting as modulators of gene transcription or translation processes (Kline *et al.*, 1982). The extremely rapid responses of tissues to exogenous growth hormones may point to the actual activation of genes within cells of treated tissues (Higgins *et al.*, 1976; Melanson and Trewavas, 1982). The structure of the genes which code for the biosynthetic steps in the synthesis of growth hormones has been little studied. It would be of considerable importance if it could be proven that the gene products themselves are capable of regulation by acting directly upon the genetic material itself.

Sequential activation and inactivation of genes within the cells of developing plant tissues and organs is universally considered to be the catalytic factor in the function of the specialized metabolic pathways which are responsible for the synthesis of secondary plant products. Plant taxonomists have long relied on the presence of secondary metabolites to confirm or disconfirm the degree of relatedness of plant taxa (Bell, 1980). Indeed, an entire branch of plant biology, chemosystematics, has arisen from the use of plant chemicals to establish such relationships (Crawford and Giannasi, 1982). Secondary metabolites, especially such molecules as flavonoids and terpenoids, have been shown to exhibit slight structural differences depending on which species, genus or broader taxonomic group is compared. Thus, at the same time there are both genetic components which irrevocably set the pattern of the secondary metabolites produced, and temporal and spatial ones which provide the cue to activate the genes in question in a specific tissue at a specific time (Luckner, 1980).

Table 1 illustrates some of the better known secondary metabolites, the plant tissues or organs in which they are found in the highest concentrations and the plant sources recognized as producers of these metabolites. It is noteworthy that numerous studies of metabolic pathways have revealed that the basic enzymatic machinery necessary for the synthesis of the majority of these secondary products is present across an array of plant genera (Harborne, 1980; Stephen, 1980). Additionally, very low concentrations of a number of secondary metabolites occasionally may be found in tissues of plants which had not been considered to produce such secondary products at all. It would appear in many instances in higher plants that the basic message necessary for the synthesis of secondary products does exist, albeit in an inactive form within the genome of these plant genera.

One of the first perceived goals of plant genetic engineers then should be to learn more about factors such as hormones which have profound influences on the differentiation and development of plant tissues. At the same time, geneticists should seek a more complete understanding of the mechanistic properties of the plant genes which control the synthesis of these metabolic pathways.

11.2.2 Plant Genome Structure and Function

11.2.2.1 Organization of the genome at the chromosomal level

As multicellular eukaryotes, higher plants are in no way unique among the assemblage of living organisms. Although plant genetic systems have given researchers a few surprises, chief among

Table 1 A List of Higher Plant Anatomical Features and Selected Examples of Metabolites Obtained from Each[a]

Plant structure	Metabolite produced	Representative plant sources
Reproductive structures		
Seeds	Oils	Soybean, sunflower
Berries	Storage proteins	Corn, cotton
	Fatty acids	Castorbean
	Plastics	Corn, peanuts
	Steroids	Fenugreek
	Coffee	*Coffea*
Flowers, floral organs	Saffron	Crocus
	Carotene	Marigold
	Pyrethrins	Chrysanthemum
	Opium	Poppy
	Vanillin	*Vanilla*
Above-ground structures		
Leaves	Tea	*Camellia*
Leaflets	Chlorophyll	Various plants
	Digoxin	Foxglove
	Vincristine	Periwinkle
Stems	Latex, chicle	*Achras*
	Rubber	Guayule
	Nicotine	Tobacco
	Essential oils	Citrus
	Sterols	Potato
	Alcohols	Corn
	Phenolics	Cotton
	Cellulose	Pines, hardwoods
	Resins	Hardwoods, pines
	Gums	*Acacia*
	Terpenes	Pines
	Glucomannans	Pines, *Larix*
	Lignins	Various plants
Bark	Tannins	Oak, walnut
	Cellulose	Hardwoods, pines
	Rubber	*Hevea*
	Quinine	*Cinchona*
Subterranean structures		
Roots	Phytoalexins	Various plants
	Psoralens	Parsnip
	Rotenone	*Amorpha*
	Ginseng	Ginseng
Tubers	Starch	Cassava, potato
	Sugar	Sugar beet, sugar cane
	Steroidal sapogenins	*Dioscorea*, agave
Rhizomes	Alkaloids	Water lily
Bulbs	Colchicine	Crocus

[a] Blunden *et al.* (1975), Bell (1980), Ivie *et al.* (1981), Princen (1977), Watowich and Schultz (1982), McDaniel (1981b).

these being transposable elements or 'jumping genes', for the most part plant chromosome structure and gene function have been most amenable to analysis by conventional means.

Plant genomes consist of multiples of distinct types of paired chromosomes, at least some of which are thought to have arisen from ancestral chromosomes or pieces of chromosomes by polyploidization or gene duplication mechanisms. Classical genetic analyses often have been able to show that genes with the same function may reside on two or more nonhomologous chromosomes within the same organism. For example, genes eliciting the expression of male sterility, that is failure of the plant to form functional male gametes, are located at various positions on each of the seven chromosomes of barley. Although not all of these genes have the same mode of action, that is they do not block formation of viable pollen at the same metabolic step, state of the art analyses do indicate that the actions of several of these male sterility genes are identical. It is commonly speculated that one or two of the seven pairs of chromosomes of modern barley arose from one or more of about five 'original' ancestral chromosomes. McDaniel and Ramage (1970) have presented biochemical evidence for such chromosome organizational changes during the evolution of barley.

Plant chromosomes can be stained with the characteristic tools of the cytogeneticist. Giemsa-stained plant chromosomes show much the same generality of banding patterns as do mammalian chromosomes. The chromosomes of corn, barley, wheat and other crop plants have been well mapped and the relative positions of hundreds of genes of potential economic importance have been ascertained. Flow cytospectrofluorometry has been utilized in a number of instances to separate and purify individual plant chromosomes which may be useful in chromosome transfer from one plant cell to another. Unlike the normal situation in mammals and some lower animals, for the most part plants are quite tolerant of polyploidy. Many crop plants are tetraploid, or exhibit even higher ploidy levels. The combination of genomes or portions of genomes contributed from wild progenitor species has aided these plants in gaining the characteristics which have potentiated their domestication. Some tree species may have multiples of chromosome numbers greater than 10 or 20 times the diploid or 2C chromosome number. Alfalfa, for example, has diploid through octopoloid representatives which resemble each other quite closely. Walbot and Goldberg (1978) have attempted to tie together the historical cytogenetic concepts of the plant genome with the present concepts of plant genes based upon findings of molecular biologists.

11.2.2.2 *Macromolecular gene structure*

Until quite recently, the analysis of the fine structure of plant genes has been nearly impossible. It has simply not been worthwhile to go to the enormous trouble necessary to obtain the large quantities of plant DNA, greatly enriched for specific genes, which are needed to carry out the studies of the base sequences of these genes (Bendich and Ward, 1980; Griesbach *et al.*, 1981). The advent of restriction endonucleases as powerful tools in the analysis of base sequences of DNA in prokaryotes found rapid utility in the analysis of mammalian gene structure (Roberts, 1977).

Simplistically, restriction endonucleases are enzymes found in the cytoplasm of some types of bacteria. These enzymes serve as watchdogs to protect the bacteria from the egress of alien nucleic acids. These enzymes recognize specific base sequences in the DNA of invading viruses, and cleave, that is make a double-stranded cut in, the DNA of the invader, thus preventing infection and subsequent replication of the foreign DNA within the host bacterium. The DNA of the bacterium itself is protected from degradation by an extensive modification/methylation system. Hundreds of restriction enzymes have now been purified; most recognize uniquely different sets of several base pair sequences of DNA. Thus, each enzyme will make a unique series of cuts at different points along the length of a DNA molecule and thereby produce a set of distinct molar mass fragments of DNA. Use of restriction enzymes thereby enables researchers to break the DNA into populations of identically sized fragments. Each population of fragments contains a great many copies of identical genes. Smith and Birnstiel (1976) have discussed these applications of restriction enzymes in detail.

The cloning of such DNA fragments in bacteria has been accomplished *via* the use of plasmid vectors to insert these fragments so that the genetic machinery of the bacterium recognizes them and replicates them continuously. Unfortunately, it has been difficult to obtain reasonable yields of plant genes by the use of the cloning methods normally utilized in the analysis of animal genes. Ironically, some of the difficulties in the isolation and cloning of plant genomic DNA have been considered to result from high concentrations of secondary metabolites in the plant tissues which interfere with purification of the DNA by conventional methods (Sung and Slightom, 1981).

Use of tissue culture cells as a source of plant DNA appears to obviate some of these difficulties (Yadav *et al.*, 1980). Additionally, new, more sophisticated methodology for plant DNA isolation should serve to increase the cloning efficiency of plant genes. Recent papers have detailed such developments in methodology which have enabled the isolation of genes coding for zein seed protein in maize (Lewis *et al.*, 1981; Pedersen *et al.*, 1982), phaseolin genes from French beans (Chee *et al.*, 1982), soybean seed protein genes (Goldberg *et al.*, 1981) and tobacco 'structural' genes (Kamalay and Goldberg, 1980). Some structural aspects of these genes are presented in Table 2. In addition to these and other efforts to characterize and clone specific genes within the nucleus of plant cells, substantial efforts are in progress to delineate the structure and function of organelle genes which are to be found in the DNA of cytoplasmic organelles, *i.e.* mitochondria and chloroplasts, of plants (Leaver and Forde, 1980). Fox and Leaver (1981), for example, have reported the characterization (by base sequence analysis) of a plant mitochondrial gene which codes for subunit II of cytochrome oxidase. Bogorad (1979) has discussed the isolation and cloning of chloroplast genes. Genetic mapping and cloning of chloroplast genes is underway in a number of laboratories, with emphasis on ribulosebisphosphate carboxylase genes (Palmer, 1982).

Table 2   Structural Aspects of Some Plant Genes Judged by Base Sequence Analysis

Gene	Structural distinctions	References
Leghemoglobin	Contained in 7.5 kb soybean nuclear DNA fragment Three intervening sequences Consensus sequences present One of a family of several genes Gene base sequence known	Marcker *et al.* (1981) Brisson and Verma (1982)
Cytochrome oxidase	Subunit II mitochondrial gene Contains a central intervening sequence Gene base sequence known Contained in a 1.1 kb DNA segment	Fox and Leaver (1981)
Ribulose-1,5- bisphosphate carboxylase	The large subunit of fraction I protein Contained in a 20.5 kb chloroplast DNA fragment Lies within the single copy region of the chloroplast genome	Palmer (1982)
Zein protein	Seed storage protein gene Contained in a 7.7 kb nuclear DNA fragment in corn Contains no intervening sequences One of a family of several genes Gene base sequence known	Pedersen *et al.* (1982) Lewis *et al.* (1981)

11.2.2.3   *Mechanisms of gene transcription, processing and translation*

Perhaps the most striking feature of eukaryotic gene structure, in comparison with prokaryotic genes, is the inclusion of intervening sequences, 'introns', in some genes of plants and animals. These introns are surrounded by 'exons', the regions of the gene which encode a functional messenger RNA molecule. Evidence gathered from studies of the amino acid sequence similarities of related proteins suggests that these introns may have served as archetypal 'building blocks' for what have now evolved, for the most part, into non-functional 'relict' genes, through a process of gene duplication *via* tandem repeats.

These relict genes must be 'edited out' of the RNA message in order for a fully functional protein to be synthesized. As only a single functional gene may have been necessary at any one time during the evolution of a given plant, additional copies may have gone through an error accumulation process, with the relict sequences gradually becoming surrounded by functional genes or portions of read-through sequences. Alternatively, introns could conceivably have been inserted into eukaryotic genes during evolution. The discovery of introns in mitochondrial genes might lend support to this contention (Fox and Leaver, 1981). In general, however, the proportion of total base sequences within a gene which are introns is a function of the total size, or number of bases, of the exons comprising the gene (Naora and Deacon, 1982). As prokaryote gene transcription mechanisms do not recognize and edit out introns, as do those of eukaryotes, introns must be cut out before plant genes may be expressed in bacteria. Alternatively, a mature plant mRNA which has been processed in order to remove introns may be used to construct a cDNA (complementary DNA) clone as has been done with soybean mRNA (Goldberg *et al.*, 1981) and bean seed mRNA (Hall *et al.*, 1980, Sun *et al.*, 1980). Brisson and Verma (1982) have detailed some of the ramifications of gene evolution with regard to pseudo and truncated genes in the soybean leghemoglobin gene family.

If one wishes to insert such a gene sequence into either a plant or a bacterial recipient in order to potentiate increased secondary metabolite synthesis, another aspect of genetic fine structure must be considered. In addition to editing out any introns which may be present in the cloned gene, one must also include a promoter sequence or insert the gene into the host DNA at a site adjacent to a strong appropriate promoter sequence. The promoter DNA sequence serves the function of binding the RNA polymerase, thereby initiating transcription, and thus insuring that the inserted gene is read. If a strong promoter sequence is not present, the inserted gene may not be transcribed frequently, and little if any gene product may be recovered.

11.3   RECOMBINANT DNA TECHNIQUES APPLICABLE TO PLANTS

11.3.1   Expression of Plant Genes in Yeasts and Bacteria

Initial efforts to insert plant genes into bacteria failed for a number of reasons. First, many plant genes contain introns, and the bacteria, as prokaryotes, were unable to process out the introns prior to transcription. Secondly, the presence of a strong bacterial promoter sequence at the proper site adjacent to the inserted gene was lacking, and the polymerase may not have been able to read the inserted gene often enough to make a detectable amount of gene product. Genetic engineers have been able to insert numerous examples of mammalian genes which also included introns into the genomes of bacteria and achieve production of insulin, human growth hormone and other peptide hormones. This was accomplished by *in vitro* synthesis of complementary DNA from mRNA transcripts of the desired gene in which the introns were absent. The cDNA could then be linked to an appropriate bacterial promoter sequence and vectored into the bacteria *via* a recombinant DNA plasmid.

Although word of similar successes with plant genes, as judged by actual product formation as a result of insertion of the plant gene, has been little heard, there is no reason to expect that bacteria would not transcribe plant genes in an identical manner to mammalian cDNA copies.

In fact, plant genes are being routinely cloned as cDNA copies by workers in a number of laboratories at the present time. Numerous fragments of chloroplast and mitochondrial DNA have been cloned into *E. coli* and a number of genomic libraries of cloned organelle genes are now available.

11.3.2 Plasmid Vectors of Utility

Plasmids, the small circular DNAs which are found to coexist with the genomic DNA in many types of bacteria, have proven to be nearly ideal vectors to transfer genes from one species or type of bacterium to another. In essence, the plasmid DNA is linearized by cleavage with an appropriate restriction endonuclease and a gene fragment which is desired to be transferred is annealed to one end of the plasmid DNA. The two ends of the linear DNA 'hybrid' then are ligated, or rejoined together enzymatically to produce a circular recombinant DNA plasmid. This plasmid is capable of 'infecting' and potentially transforming a bacterium with which it is incubated. The bacterium then synthesizes the product for which the new gene codes, using the bacterial protein synthesis machinery. Additionally, the plasmid DNA is normally self-replicating in the bacterium, thereby resulting in a permanent man-made genetic alteration of a living organism, a genetically-engineered bacterium. Such genetically-engineered plasmids are being utilized to insert new genes into animal cells.

Plant molecular biologists have reported that plasmid vectors appear to be useful tools to vector genes into plant cells. Plasmid pBR313, for example, has been reported to be taken up by plant protoplasts, to associate to some extent with nuclear DNA (Hughes *et al.*, 1977), and to be stable for several days (Fernandez *et al.*, 1978). There is a question, however, of whether such bacterial plasmids are truly self-replicating in plant cells (Lurquin and Kado, 1977).

The vector of proven utility in plant cell culture, as well as in intact plants themselves, is the Ti plasmid. This is a DNA plasmid found in strains of *Agrobacterium tumefaciens*, which is considered to be the causal agent of crown gall disease. Since the crown gall organism has a very wide host range, showing pathogenesis across many hundreds of dicotyledenous species, if genes could be engineered into the Ti plasmid found within the *A. tumefaciens* bacterium, then there would be a strong possibility of success in transferring such genes into a wide array of host plants (Ream and Gordon, 1982). Importantly, many of the plants subject to crown gall disease are crop plants of considerable significance in world agriculture. Kemp and Hall have cloned a gene for a seed storage protein from the French bean, placed this gene into the Ti plasmid from *A. tumefaciens*, and infected sunflower tissue with bacteria containing this genetically engineered plasmid (Check, 1982). The plan of this study was to see if the French bean gene, vectored into a dissimilar dicot (sunflower) *via* the Ti plasmid would be self-replicating in the cells of the host plant, and whether the protein product of the transferred gene would, in fact, be expressed. Available evidence points to the successful recognition of the French bean gene in the cells of the sunflower recipient; however, presence of the seed protein coded for by the bean gene has not been reported at the time of writing. Chilton *et al.* (1977) and Schell and collaborators (Zambryski *et al.*, 1980; Otten *et al.*, 1981) have reported similar successes in engineering different genes by use of the Ti plasmid using several test plant species. *Agrobacterium rhizogenes* likewise appears to be most promising as a plant gene transfer vector. Recent work by Chilton *et al.* (1982) has clearly demonstrated that this plant pathogen, a close relative of *Agrobacterium tumefaciens*, contains virulence plasmids. Infection of plant roots (within dicots) elicits the over-production of roots by the host plant. These plasmids have been termed Ri or root inducing pathogenic plasmids, and like the plasmids associated with *A. tumefaciens*, the crown gall organism, a portion of the T-

DNA is integrated into the plant genome. A comparison of possibly useful plant vectors is presented in Table 3.

Table 3	Vectors of Possible Utility for Genetic Engineering of Higher Plants *via* Recombinant DNA

Vector nomenclature	*Type and source*	*Status*	*References*
Ti plasmid	Circular DNA plasmid from *Agrobacterium tumefaciens*	Reports of successful transfer of plant genes into host cells in culture	Zambryski *et al.* (1980), Otten *et al.* (1981), Check (1982), Ream and Gordon (1982)
Ri plasmid	Circular DNA plasmid from *Agrobacterium*	Report of successful incorporation of T-DNA into host plant roots	Chilton *et al.* (1982)
Cauliflower mosaic virus	Plant DNA virus, cloned in *E. coli*, infects crucifers	Yield of virus has been increased *via* cloning and infectivity retained	Howell *et al.* (1980), Shepherd *et al.* (1981)
Dahlia mosaic virus, potato leafroll virus, gemini virus and others	Plant DNA viruses	Little relevant work reported; low yield of virus	Kado (1979), Kado and Kleinhofs (1980)
E. coli plasmids: pBR313, pBR322 and others	Circular DNA plasmids of bacterium; restriction mapped	Proven vectors for recombinant prokaryote and animal genes; promising vectors for plant genes	Hughes *et al.* (1977), Lurquin and Kado (1977), Fernandez *et al.* (1978)
Phage mutants: Lambda, M 13	Single-stranded, small DNA molecules; restriction mapped	Considered potentially useful vectors for small DNA segments	Kado (1979), Kado and Kleinhofs (1980)
S–Plasmids	Small DNA molecules found associated with mitochondrial DNA of some monocots	Status as a vector uncertain, as plasmid may be unstable in some cytoplasms	Pring *et al.* (1977)

Another vector of potential utility in plant genetic engineering is the cauliflower mosaic virus (CMV). This is a double-stranded DNA plant virus, infecting members of the crucifer family, which is of low complexity and small molar mass. Recent evidence indicates that cauliflower mosiac virus is capable of vectoring in a certain amount of DNA in recombinant form (Howell *et al.*, 1980; Shepherd *et al.*, 1981). The limitations to use of this virus as a plant vector are that it has a somewhat limited host range as it exists in nature, and that experimental data indicate that it has a very limited 'carrying capacity' with regard to the length of DNA fragment which can be annealed to the virus without loss of infectivity or of functionality (Hull, 1981). Kado (1979) has discussed attributes of CMV and similar viruses in detail.

Yet another potential vector which is being closely scrutinized is the S-plasmid which has been reported to occur associated with the mitochondria in a few lines of maize and possibly sorghum. These are low molar mass, apparently linear DNA molecules which copurify with the mitochondrial DNA of the S-type cytoplasmic sources of maize (Pring *et al.*, 1977). These lines of maize are characterized by an unstable cytoplasmic male sterility response. With time, S-type plants which appear sterile may revert to fully fertile forms. The reversion to fertility has been associated with the loss of, or perhaps the reintegration of, the S-type plasmids into the mitochondrial genome in a manner analogous to transposable elements.

### 11.3.3	Microinjection and Encapsulation

Micromanipulation of isolated protoplasts for the purpose of microinjection of recombinant DNA fragments appears to be gaining acceptance as a potentially usable tool in genetic engineering of plant cells. Cell microinjection apparatus is available at a large number of universities, and several laboratories are presently evaluating the effectiveness with which naked liposome encapsulated DNA may be injected into protoplasts. In some systems, it appears to be possible to 'deliver' the DNA with protoplasts (Matthews and Cress, 1981; Fraley *et al.*, 1982; Lurquin and Sheehy, 1981). These approaches would avoid the complexities of constructing a recombinant DNA vector system altogether. Schaefer-Ridder *et al.* (1981) have reported transformation efficiencies as high as 10% using liposomes to transfer recombinant plasmids into mouse L cells.

11.3.4 Protoplast and Organelle Fusion

In the same vein, numerous successes have been reported in isolating protoplasts (Vasil and Vasil, 1980), fusing them together, and regenerating the resultant hybrid plant. Advantages to such fusion systems include the ability to recognize and micromanipulate the hybrid cells, use of selective markers to distinguish true hybrids, and a relatively high efficiency of hybridization (Schieder and Vasil, 1980). One major disadvantage is the selective loss of chromosomes which may take place when cells from genetically diverse plants are fused (Kao, 1977). Mitochondria and chloroplasts when incubated with protoplasts under appropriate conditions can be taken up by the protoplasts and maintained in a fairly stable manner for several cell generations at least. As a considerable number of secondary metabolites are formed in these organelles (Bogorad, 1979) the potential for organelle transplantation appears most promising. McDaniel (1981a) recently has reviewed the status of organelle transfer in relation to plant genetic engineering.

11.4 ACTION OF INTRODUCED GENES

11.4.1 Gene Integration and Expression

Probably the most disturbing difficulty for plant genetic engineers has been the problem encountered in insuring the integration, maintenance and expression of a transferred gene. Although added DNA does not have to be integrated into the host genome, it stands to reason that this would be the most positive assurance of expression of the gene through future generations. Our present concepts of gene integration suggest that the higher the degree of sequence homology of some part of the added DNA with a portion of the host DNA, the better the likelihood of integration of the added segment. At the same time it is considered advantageous to be able to insert multiple copies of a desired gene into a single host genome. This in effect increases the copy number of the inserted gene, and helps to insure that the polymerase will read multiple copies of this gene, thus potentially amplifying the genetic message to cause the increased synthesis of the gene product.

11.4.2 Directed Synthesis of Gene Products

Integration of genes into the genome of higher plants may still be inadequate to insure gene transcription. The locus at which the added gene is placed may be of paramount importance. The added gene must be sited with a strong promoter region the proper distance away for polymerase directional reading. It also appears that amplification sequences are necessary 'upstream' from the added gene. If one wants to control the expression of the gene in a temporal manner, then it may be necessary to insert the gene into a developmentally regulated gene 'cluster' in order to activate or inactivate the gene at the proper time.

11.4.3 Examples of Plasmid-induced Metabolite Synthesis

Very few examples of plasmid-induced metabolite synthesis in plant tissues can be found in the literature at this juncture. Those that have been reported are germane, as they serve as models for the expected results of a recombinant DNA-induced metabolite synthesis when work in progress in many laboratories is reported. A list of the best studied samples of metabolites from plasmid infection of plant nuclei is presented in Table 4. Probably the prime examples are the synthesis of novel amino acid derivatives, opines, nopalines and octopines (Zambryski *et al.*, 1980). Such distinctive biosynthetic products should be useful as markers to show transformation has in fact taken place.

11.5 PLANT METABOLITES OF IMPORTANCE

11.5.1 Amino Acids, Peptides and Proteins

The most obvious candidate metabolites which can be manipulated *via* recombinant DNA technology are the primary products of gene action, *i.e.* proteins. Just as human insulin and

Table 4 Examples of Plasmid-induced Metabolite Synthesis in Higher Plants

Plasmid type	Host plant	Metabolite synthesized	References
Ti plasmid (of *A. tumefaciens*)	Dicots: tobacco, sunflower, others	Amino acid derivatives: octopine, nopalines, agrocinopine, lysopine dehydrogenase, auxins, cytokinins	Otten *et al.* (1981), Ream and Gordon (1982)
Ri plasmid (of *A. rhizogenes*)	Dicots, carrots	Mannopine	Chilton *et al.* (1982)

bovine rennin were manufactured in *E. coli* using recombinant DNA vectors, so too should plant proteins or useful amino acids be among the first results of recombinant DNA technology applied to plants. One of the first goals in a number of laboratories is to increase the amount of limiting amino acids, such as lysine, methionine and tryptophan, for human nutrition in some of our food crops.

One approach to improve protein quality is to increase the number of genes whch act to regulate the production of amino acids or of specific proteins in that portion of the plant used as food. The hope is to overproduce a given amino acid by having genes which effect the synthesis of that amino acid transcribed with greater frequency. Recent investigations concerning the fine structure of genes which code for seed storage proteins are elucidating not only the structure and temporal function of these genes, but also are suggesting ways in which these genes can be engineered to produce higher protein quality in legumes (Goldberg *et al.*, 1981).

11.5.2 Drugs, Toxins and Antimetabolites

A substantial proportion of the drugs listed in the US Pharmacopoeia is derived from plants. Some of the pharmaceutical products from plants are among the most valuable substances in the world, costing over a million dollars a kilogram. Many of these drugs are secondary metabolites (Tabata, 1977). Some toxic products and antimetabolites are among the most effective treatments for human diseases, and are especially active against several forms of cancer. The narcotic opiates, digoxins and reserpines for cardiovascular disease, colchicine for gout, diosgenin for steroid biosynthesis, and marijuana as a pain deadener, to name but a few, are all derived from plant sources. Farnsworth (1977) has reviewed the role of plants as sources of drugs. Farnsworth estimated that over 25% of the prescriptions in the United States contained one or more of a total of 76 active compounds derived from higher plants. This would not include many over-the-counter formulations containing caffeine, citrus biflavonoids, or camphor. A number of such examples, their plant source, and references for further reading have been listed in Table 5.

Applezweig (1977) has traced the history of diosgenin, a botanical drug source of enormous present and future importance. Diosgenin is the steroidal sapogenin obtained from the tubers of *Dioscorea composita*, the Mexican yam. Diosgenin was found to serve as an ideal steroid raw material for the synthesis or conversion of testosterone, progesterone and estrogens. With the development of a fermentative oxidation of progesterone, diosgenin could also be used for the synthesis of corticosteroids. The ever-growing demand for steroids and oral contraceptives has fired the development of competitive sources and chemosynthetic processes for synthesis of these corticosteroids from steroidal sapogenins, phenomena which may dilute the market share of *Dioscorea*-derived chemicals. The possible application of recombinant DNA technology to improve the fermentative processes, or to increase the yield of steroidal sapogenins from plant sources, could greatly enhance the efficiency of future steroid derivative production.

11.5.3 Isoprenoids and Other Polymers

There are a number of potentially valuable, high molar mass plant metabolites whose synthesis should prove amenable to improvement *via* recombinant DNA technology. Princen (1977) has enumerated many of these compounds, and has discussed their higher plant sources and economic utility. Such polymers include starch, waxes, gums, cellulose, lignins, tannins and polyisoprene rubber. There has been considerable renewed interest in the native American shrub guayule. Guayule produces in its stem and root tissues a latex which is chemically identical to the

Table 5	A List of Selected Higher Plant Metabolites of Economic Importance

Class of metabolite	Specific examples	References
Alkaloids	Atropine	Constabel *et al.* (1981)
	Cocaine	Fodor (1980)
	Colchicine	Gröger (1980)
	Dioscorine	Kutney *et al.* (1981)
	Ephedrine	Leete (1980)
	Indole alkaloids	Shiio and Ohta (1973)
	Morphine	
	Nicotine	
Amines	Cadaverine	Smith (1980)
	Dopamine	
	Putrescine	
Carbohydrates	Cellulose	Luckner (1980)
	Gums	Princen (1977)
	Pectins	Stephen (1980)
	Polysaccharides	
Cinnamic acid	Various lignins and	Böhm (1980)
	flavonoids	Luckner (1980)
Cyanogenic glycosides	Antimetabolites	Conn (1980)
Glucosinolates	Essential oil producing	Underhill (1980)
	compounds	
Lipids	Alcohols and ketones	Princen (1977)
	Cutins	Thompson (1980)
	Fatty acids	Watowich and Shultz (1984)
	Hydrocarbons	Yamaoto *et al.* (1981)
	Waxes	
Pigments	Chlorophyll	Luckner (1980)
	Ubiquinone	Threlfall (1980)
Phenolics	Anthocyanins	Böhm (1980)
	Bioflavonoids	Conn (1980)
	Coumarins	Harborne (1980)
	Flavonoids	Ivie *et al.* (1981)
	Gossypol	
	Lignins	
	Psoralens	
	Tannins	
Sterols	Cholesterol	Applezweig (1977)
	Diosgenin	Biesboer and Mahlberg (1980)
		Blunden *et al.* (1975)
		Grunwald (1980)
Terpenoids	Carotenoids	Archer (1980)
	Isoprenoids	Goodwin (1980, 1981)
	Polyisoprene rubber	Radin *et al.* (1982)
	Xanthophylls	Spurgeon and Porter (1981)

Hevea rubber from Asian and South American rubber trees. If the 'rubber' genes could be found and amplified, then the production of rubber within the borders of the United States would be commercially feasible, and could help lift the dependence on petroleum feedstocks for this strategic resource.

Genetic manipulation of starch crops to increase the quantity and quality of starch, as well as reducing losses during starch purification would seem to be goals which may be achieved nearly as quickly as the improvements in amino acid and protein quality are realized. Most importantly, many of these polymers result as the end products of metabolic pathways which have common intermediates. These include mevalonic acid, a common intermediate in the synthesis of isoprenoids, steroids and terpenes; shikimic acid, a precursor in phenolic, aromatic amino acid, and coumarin synthesis; and cinnamic acid, a key intermediate in lignin, flavonoid and tannin synthesis. Figure 1 illustrates some of the metabolic pathways important in secondary metabolite synthesis, and shows the close relationship and possible coordination of metabolite production through the specific activation of 'finishing enzymes' in individual plant species for the production of each species-specific final metabolite.

The point here is to suggest that the genetic engineering of secondary metabolite production may not be as difficult as has been envisioned. Instead of the necessity to activate or to isolate and transfer a coordinated array of plant genes specifying a biochemical pathway, it is entirely possible that all or nearly all of the necessary enzymatic steps are in place within the genetic code of

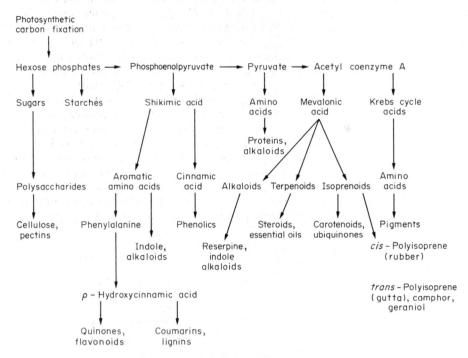

Figure 1 Examples of key metabolic pathways which elicit the production of secondary plant products (Mooney, 1972; Bush and Grunwald, 1973; Downing and Mitchell, 1975; Walsh, 1979; Goodwin, 1980, 1981; Archer, 1980; Harborne, 1980; Spurgeon and Porter, 1981)

the cell. What remains is a means of activating them, or perhaps adding *via* recombinant DNA technology the last enzyme or two in the biosynthetic pathway to coordinate with the already present major metabolic sequences in the cell. Although conjectural, this hypothesis is eminently testable, and if proven would set the stage for a much more rapid commercial bioproduction of plant metabolites than can now be envisioned. Paramount to the effective production of such metabolites is the development of a means to insure the transcription of the added genes coding for the final steps in the metabolic sequences, or an effective chemical or genetic means to activate these genes if they are present in inactive form.

11.5.4　Alcohols and Other Industrial Chemicals

There are rather serious doubts as to whether metabolite production from transformed higher plant cells in a chemostat culture can really compete economically with the production of alcohols, fatty acids and other industrial biochemicals *via* fermentation. Plant fatty acids are widely used as feedstuffs for the industrial production of plastics, nylons, epoxides and lubricants, however, the present sources of most of these fatty acids are the seed oils of crop plants (Watowich and Schultz, 1982; Princen, 1977). A number of pesticides are also derived from plant sources. Pyrethrins are effective insecticides derived from chrysanthemums. Antijuvenile insect hormone-like chemicals are also obtained from plants (Brady, 1982), as is the insect poison rotenone. It is difficult to envision a competitive *in vitro* production of these chemicals in a fermentation-type plant cell culture system, however, addition of genes *via* recombinant DNA methodology to enhance the production of such chemicals in seeds would appear to be of considerable economic potential.

11.6　COORDINATION OF METABOLITE PRODUCTION

11.6.1　Selection of Adapted Cell Variants

It is not enough to perfect the ways and means to insert new and useful genes into plant cells. One must at the same time devise techniques to select, propagate and regenerate these transformed plant cells into their final useful form, namely a recombinant DNA engineered plant capable of high efficiency production of a pure desirable metabolite. Some plant cells are more amenable to tissue culture than others. With the present state of the art in plant cell biology, it is necessary to enzymatically dissolve the pectin (the intracellular cement) and to strip away the plant cell wall *via* hydrolytic cellulose-degrading enzymes. The resultant cell is a protoplast devoid of a cell wall, but still membrane-bound and fully viable. The protoplast is the recipient cell in nearly all of the genetic engineering experiments underway with plants.

After transformation by vectoring in a recombinant DNA, the foreign DNA must be integrated with the plant genome or established in an independently self-replicating plasmid in the daughter cells of the protoplast. Cell walls must be reformed, cell division must have taken place, and an *in vitro* clone of the transformed cell formed. This group of cells, probably in the form of callus, must be induced to redifferentiate the organs of the whole plant in order to be taken from culture and once again established as an agronomically viable competitive strain. Vigorous selection programs are underway in a number of laboratories to derive lines of plant cell cultures which are capable of high incidence of plant regeneration. The expected steps in the transformation of plant cells and subsequent metabolite production are diagrammed in Figure 2.

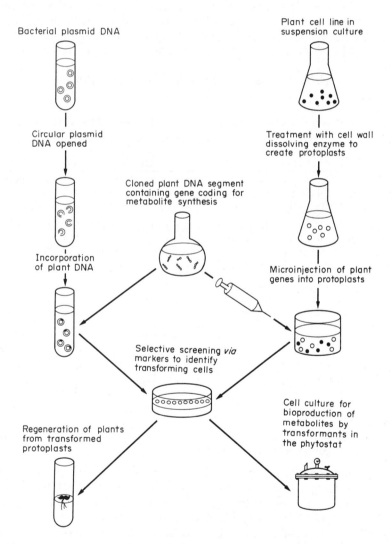

Figure 2 Steps in the transformation of plant cells *via* recombinant DNA technology

One intriguing aspect of protoplast regeneration is the phenomenon of 'somaclonal variation', where different regenerated cell lines from a single plant exhibit genetic variation. Shepard and colleagues (Secor and Shepard, 1981) have pioneered studies of this unexpected type of variation. Although the genetic basis for such variation within cells of a single organism has not been fully explained, several examples of mitochondrial variation in maize tissue culture and regenerated plants show evidence for possible mitotic segregation of organelle DNA which could help explain some instances of somaclonal variation (Brettell *et al.*, 1979).

11.6.2 Screening of Mutants

Many of the plant tissue culture strains currently studied have been subjected to mutagenic chemicals, ionizing radiation or other means to elicit potentially useful mutant strains which are capable of being utilized in positive selection schemes. These experiments are somewhat analogous to the temperature-sensitive thimadine kinase mutant screening system developed for mammalian tissue culture. In essence, all the mutant cells will survive when challenged with a toxic intermediate, or when some normally essential component of the culture medium is omitted. If the mutant gene which confers this ability is isolated and is ligated with the gene for metabolite production which one wishes to transfer, then any plant cell which is transformed by a recombinant vector which contains these two genes will be capable of surviving when challenged, whereas untransformed cells would not. Plant tissue culture researchers have encountered considerable difficulties in selecting usable mutants. Because of the multiplicity of metabolic pathways in plant tissues, almost every mutant found is 'leaky', and is capable of making the 'blocked' product in at least small quantities, thus circumventing any metabolic block imposed by mutation. Widholm (1977) has discussed these problems of plant mutant isolation, and has cited some successful examples.

A list of mutant plant cell lines in culture is presented in Table 6, along with their distinguishing chemosynthetic properties.

Table 6 Metabolic Characteristics of Some Mutant Plant Cell Lines in Culture

Cell line	Mutagenic agent or selection method	Metabolic characteristic	References
Tobacco callus	Repression by plant hormones	Nicotine production	Shiio and Ohta (1973)
Periwinkle suspension culture	Spontaneous/anther explants	Alkaloid production	Kutney *et al.* (1981)
Periwinkle suspension culture	Somaclonal variation	Alkaloid production	Constabel *et al.* (1981)
Maize callus	*Drechslera maydis* T-toxin	Resistance to T-toxin	Brettell *et al.* (1979)
Tobacco callus	5-Bromodeoxyuridine	Resistance to the chemical	P.S. Carlson reviewed by Widholm (1977)
Carrot suspension culture	Ethylmethane sulfonate or UV radiation	Resistance to 5-methyltryptophan	Widholm (1977)
Tobacco suspension culture	Methionine sulfoximide	Resistance to the chemical	P.S. Carlson reviewed by Widholm (1977)
Corn callus	*Helminthosporium maydis* T-Toxin	Resistance to T-toxin	Gengenbach reviewed by Maliga (1980)
Tobacco callus	Streptomycin	Resistance to streptomycin	Maliga (1980)
Carrot suspension culture	Ethylmethane sulfonate or nitrosoguanidine	Resistance to cyclo-heximide	Sung reviewed by Maliga (1980)

11.6.3 Release of Metabolites from Cultured Cells

As recently as the mid-1970s, the release of metabolites into the medium by cultured plant cells was considered to be a difficult phenomenon to elicit. More recent work has indicated that this should not be a problem with most of the recombinant DNA engineered plant cell cultures presently envisioned. Dougall (1977) has presented detailed data concerning the relative yields of numerous secondary plant metabolites in tissue cultured cells, and has reported that in a large number of instances the production of secondary products in cell cultures equalled or exceeded the concentration of such compounds in the tissues of the intact plant sources. Alfermann and Reinhard (1980) have discussed the secretion of secondary metabolites into the culture medium, a necessary attribute if the cultures are to be maintained continuously in a non-destructive fashion. They reported that over 90% of β-methyldigoxin, a steroid precursor, produced was excreted into the media by cell cultures of *Digitalis purpurea*. Although one must anticipate unique difficulties with every proposed plant cell biotransformation reaction, there is an obvious enthusiasm for the scale-up and production of a number of plant metabolites.

11.7 LIMITATIONS ON METABOLITE PRODUCTION BY CULTURED PLANT CELLS

11.7.1 Cell Growth Constraints

One of the greatest hurdles to metabolite production in recombinant DNA modified plant cell cultures is the slow growth of the cells. One must consider that plant cells grow and divide slowly relative to bacteria and that the aseptic procedures and expense of media required will be economic impediments to plant cell bioproduction on a commercial scale. Obviously, the most valuable metabolites are most likely to be synthesized in this manner. There are reports of more rapidly growing cell cultures, in which the cell doubling time has been dramatically increased in comparison to wild-type cells (F. H. K. Katterman, personal communication, 1982). Another potential cell growth constraint is the inability to regulate the addition of a precise optimum blend of plant growth hormones, one which exactly complements the bound and free hormone flux of the tissues in culture. The possibility of continuously monitoring cell hormone status in tissue culture is becoming stronger as more and more sensitive instrumental methods for analysis of minute hormone quantities are being perfected.

A third major hurdle in plant metabolite production is that in the process of culturing and maintaining plant cells, one must guard against spontaneous genetic variation in the genomes of the plant cells. It is well known that length of time in culture, exposure to some phytohormones and the presence of high plant hormone concentrations can elicit changes in the chromosomal constitution of plant cells (Bayliss, 1980). Most damaging is the appearance of aneuploid or polyploid cells in the cultures. Such chromosomal variants may differ substantially from the wild-type cells with regard to levels of metabolite production, or, indeed, may produce no metabolite at all. It is apparent that the chromosome status of plant cells being cultured for metabolite production will need to be monitored.

11.7.2 Equipment and Technical Constraints

It has become clear that much of the equipment presently developed for the culture of microorganisms will need to be re-engineered before being eminently suitable for plant cell culture. At the same time there may have to be some 'redesigning' of plant cells *via* recombinant DNA technology to render them more suitable for the modified chemostat or phytostat operations.

Plant cells can be more fragile than their bacterial and yeast counterparts, which will necessitate some redesign of the methods for suspending, stirring or otherwise agitating the cells. Osmolality of culture media will have to be optimized, perhaps with a different osmolality for each phase of the culture process, in addition to variations needed to accommodate cells from different species of plants.

Effective means to separate out dead or unproductive aging cells will have to be designed, for plant cells cannot always be removed with the methodology utilized in conventional fermentations. Some plant cells in culture show a tendency to begin to redifferentiate if conditions are not exactly maintained, however, it should be possible to select out cell lines which do not show this behavior.

11.8 POTENTIAL AND ECONOMICS OF FUTURE COMMERCIALIZATION

Probably the greatest impediment to the successful restructuring of plant genes and the utilization of such restructured plants for metabolite production, or for any other purpose, has been the historic lack of funding substantial enough to impel sufficient numbers of well trained plant molecular biologists to attack these problems with vigor. It will take some time before the application of recombinant DNA technology to metabolite production in plants reaches fruition.

One can envision that the present era of biotechnological advancements will enable the commercial exploitation of plant cell cultures on a very large scale if our breakthroughs in the plant sciences keep pace with those in the biomedical area (Geissbühler *et al.*, 1982). A number of experts envision commercial successes in plant metabolite production *via* recombinant DNA in the near future. In this present decade, at least, the economics of metabolite production will probably continue to be closely linked to the cost of petroleum supplies.

11.9 SUMMARY

All aspects of genomes of higher plants which are relevant to the manipulation of plants *via* recombinant DNA technology have been explored. The alternative methods by which DNA could be introduced into the genome of higher plants have been considered, as well as the means by which the expression of such alien genes could be insured and optimized. The array of plant secondary metabolites which should be amenable to manipulation and enhanced synthesis *via* recombinant DNA technology have been discussed, and some strategies for insuring the production of such metabolites investigated. Parameters of plant cell culture necessary to accommodate the manipulation of such cells in order to add genes for metabolite synthesis were delineated, and overall limiting factors were detailed. Conclusions in regard to the future potential of metabolite production by genetically altered plants were positive; future commercialization of processes for plant metabolite production are fully expected.

ACKNOWLEDGEMENTS

The author wishes to express his sincere appreciation to Dr. James Bonner for his thoughtful critique of the overall chapter outline, and to Sedley Josserand for her contribution to the graphics and editing of the manuscript.

11.10 REFERENCES

Alfermann, A. W. and E. Reinhard (1980). Biotransformation by plant tissue cultures. In *Plant Cell Cultures: Results and Perspectives*, ed. F. Sala, B. Parisi, R. Cella and O. Cifferri, pp. 399–404. Elsevier/North Holland Biomedical Press, Amsterdam.

Applezweig, N. (1977). *Dioscorea*—the pill crop. In *Crop Resources*, ed. D. S. Seigler, pp. 149–163. Academic, New York.

Archer, B. L. (1980). Polyisoprene. In *Secondary Plant Products*, ed. E. A. Bell and B. V. Charlwood, pp. 309–327. Springer-Verlag, Berlin.

Bayliss, M. W. (1980). Chromosomal variation in plant tissues in culture. In *International Review of Cytology*, ed. I. K. Vasil, Suppl. 11A, pp. 113–144. Academic, New York.

Bell, E. A. (1980). The possible significance of secondary compounds in plants. In *Secondary Plant Products*, ed. E. A. Bell and B. V. Charlwood, pp. 11–21. Springer-Verlag, Berlin.

Bendich, A. D. and B. L. Ward (1980). On the evolution and functional significance of DNA sequence organisation in vascular plants. In *Genome Organization and Expression in Plants*, ed. C. J. Leaver, pp. 17–30. Plenum, New York.

Biesboer, D. D. and P. G. Mahlberg (1980). Sterol synthesis and identification in cultures of *Euphorbia tirucalli* L. In *Plant Cell Cultures: Results and Perspectives*, ed. F. Sala, B. Parisi, R. Cella and O. Ciferri, pp. 351–357. Elsevier/North-Holland Biomedical Press, Amsterdam.

Blunden, G., M. C. Culling and K. Jewers (1975). Steroidal sapogenins: a review of actual and potential plant sources. *Trop. Sci.*, **17**, 139–154.

Bogorad, L. (1979). The chloroplast, its genome and possibilities for genetically manipulating plants. In *Genetic Engineering: Principles and Methods*, ed. J. K. Setlow and A. Hollaender, vol. I, pp. 181–203. Plenum, New York.

Böhm, H. (1980). The formation of secondary metabolites in plant tissue and cell cultures. In *Perspectives in Plant Cell and Tissue Culture*, ed. K. Vasil, Suppl. 11B, pp. 183–208. Academic, New York.

Brady, N. C. (1982). Chemistry and world food supplies. *Science*, **218**, 847–853.

Brettell, R. I. S., B. V. D. Goddard and D. S. Ingram (1979). Selection of Tms-cytoplasm maize tissue cultures resistant to *Drechslera maydis* T-toxin. *Maydica*, **24**, 203–213.

Brisson, N. and D. P. S. Verma (1982). Soybean leghemoglobin gene family: normal, pseudo, and truncated genes. *Proc. Natl. Acad. Sci. USA*, **79**, 4055–4059.

Bush, P. B. and C. Grunwald (1973). Effect of light on mevalonic acid incorporation into the phytolsterols of *Nicotiana tabaccum* L. seedling. *Plant Physiol.*, **51**, 110–114.

Check, W. A. (1982). Engineering the botanical gene. *Mosaic*, **13**, 19–25.

Chee, P. O., D. R. Ersland, L. M. Hoffman and J. Slightom (1982). Analysis of French bean storage protein 'phaseolin' gene family. *DNA*, **1**, 197.

Chilton, M.-D., M. H. Drummond, D. J. Merlo, D. Sciaky, A. L. Montoya, M. P. Gordon and E. W. Nester (1977). Stable incorporation of plasmid DNA into higher plant cells: the molecular basis of crown gall tumorigenesis. *Cell*, **11**, 263–271.

Chilton, M.-D., D. A. Tepfer, A. Petit, C. David, F. Casse-Delbart and J. Tempé (1982). *Agrobacterium rhizogenes* inserts T-DNA into the genomes of host plant root cells. *Nature (London)*, **295**, 432–434.

Conn, E. E. (1980). Cyanogenic glycosides. In *Secondary Plant Products*, ed. E. A. Bell and B. V. Charlwood, pp. 461–492. Springer-Verlag, Berlin.

Constabel, F., S. Rambold, K. B. Chatson, W. G. M. Kurz and J. P. Kutney (1981). Alkaloid production in *Catharanthus roseus* (L.) G. Don. VI. Variation in alkaloid spectra of cell lines derived from one single leaf. *Plant Cell Rep.*, **1**, 3–5.

Crawford, D. J. and D. E. Giannasi (1982). Plant chemosystematics. *BioScience*, **32**, 114–124.

Dougall, D. K. (1977). Factors affecting the yields of secondary products in plant tissue cultures. In *Plant Cell and Tissue Culture Principles and Applications*, ed. W. R. Sharp, P. O. Larsen, E. F. Paddock and V. Raghavan, pp. 728–743. Ohio State University Press, Columbus.

Downing, M. R. and E. D. Mitchell (1975). Mevalonate-activating enzymes in callus culture cells from *Nepeta cataria*. *Phytochemistry*, **14**, 369–371.

Farnsworth, N. R. (1977). The current importance of plants as a source of drugs. In *Crop Resources*, ed. D. S. Seigler, pp. 61–73. Academic, New York.

Fernandez, S. M., P. F. Lurquin and C. I. Kado (1978). Incorporation and maintenance of recombinant-DNA plasmid vehicles pBR313 and pCRI in plant protoplasts. *FEBS Lett.*, **87**, 277–282.

Fodor, G. B. (1980). Alkaloids derived from phenylalanine and tyrosine. In *Secondary Plant Products*, ed. E. A. Bell and B. V. Charlwood, pp. 92–127. Springer-Verlag, Berlin.

Fox, T. D. and C. J. Leaver (1981). The *Zea mays* mitochondrial gene coding cytochrome oxidase subunit II has an intervening sequence and does not contain TGA codons. *Cell*, **26**, 315–323.

Fraley, R. T., S. L. Dellaporta and D. Papahadjopoulos (1982). Liposome-mediated delivery of tobacco mosaic virus RNA into tobacco protoplasts: a sensitive assay for monitoring liposome–protoplast interactions. *Proc. Natl. Acad. Sci. USA*, **79**, 1859–1863.

Geissbühler, H., P. Brenneisen and H.-P. Fischer (1982). Frontiers in crop production: chemical research objectives. *Science*, **217**, 505–510.

Goldberg, R. B., G. Hoschek, G. S. Ditta and R. W. Breidenbach (1981). Developmental regulation of cloned superabundant embryo mRNAs in soybean. *Dev. Biol.*, **83**, 218–231.

Goodwin, T. W. (1980). Carotenoids. In *Secondary Plant Products*, ed. E. A. Bell and B. V. Charlwood, pp. 257–287. Spring-Verlag, Berlin.

Goodwin, T. W. (1981). Biosynthesis of plant sterols and other triterpenoids. In *Biosynthesis of Isoprenoid Compounds*, ed. J. W. Porter and S. L. Spurgeon, pp. 443–480. Wiley, New York.

Griesbach, R. J., P. J. Koivuniemi and P. S. Carlson (1981). Extending the range of plant genetic manipulation. *BioScience*, **31**, 754–756.

Gröger, D. (1980). Alkaloids derived from trypotophan and anthranilic acid. In *Secondary Plant Products*, ed. E. A. Bell and B. V. Charlwood, pp. 128–159. Springer-Verlag, Berlin.

Grunwald, C. (1980). Steroids. In *Secondary Plant Products*, ed. E. A. Bell and B. V. Charlwood, pp. 221–256. Springer-Verlag, Berlin.

Hall, T. C., S. M. Sun, B. U. Buchbinder, J. W. Pyne, F. A. Bliss and J. D. Kemp (1980). Bean seed globulin mRNA: translation, characterization, and its use as a probe towards genetic engineering of crop plants. In *Genome Organization and Expression in Plants*, ed. C. J. Leaver, pp. 259–272. Plenum, New York.

Harborne, J. B. (1980). Plant phenolics. In *Secondary Plant Products*, ed. E. A. Bell and B. V. Charlwood, pp. 329–402. Springer-Verlag, Berlin.

Higgins, T. J. V., J. A. Zwar and J. V. Jacobsen (1976). Gibberellic acid enhances the level of translatable mRNA for alpha-amylase in barley aleurone layers. *Nature (London)*, **260**, 166–169.

Howell, S. H., L. L. Walker and R. K. Dudley (1980). Cloned cauliflower mosaic virus DNA infects turnips (*Brassica rapa*). *Science*, **208**, 1265–1267.

Hughes, B. G., F. G. White and M. A. Smith (1977). Fate of bacterial plasmid DNA during uptake by barley protoplasts. *FEBS Lett.*, **79**, 80–84.

Hull, R. (1981). Cauliflower mosaic virus DNA as a possible gene vector for higher plants. In *Genetic Engineering in the Plant Sciences*, ed. N. J. Panopoulos, pp. 99–109, Praeger, New York.

Ivie, G. W., D. L. Holt and M. C. Ivey (1981). Natural toxicants in human foods: psoralens in raw and cooked parsnip-root. *Science*, **213**, 909–910.

Kado, C. I. (1979). Host–vector systems for genetic engineering of higher plant cells. In *Genetic Engineering: Principles and Methods*, ed. J. K. Setlow and A. Hollaender, vol. 1, pp. 223–239. Plenum, New York.

Kado, C. I. and A. Kleinhofs (1980). Genetic modification of plant cells through uptake of foreign DNA. In *Perspectives in Plant Cell and Tissue Culture*, ed. I. K. Vasil, Suppl. 11B, pp. 47–80. Academic, New York.

Kamalay, J. C. and R. B. Goldberg (1980). Regulation of structural gene expression in tobacco. *Cell*, **19**, 935–946.

Kao, K. N. (1977). Chromosomal behavior in somatic hybrids of soybean–*Nicotiana glauca*. *Mol. Gen. Genet.*, **150**, 225–230.

Kline, E. L., L. B. Hendry, E. D. Bransome and M. Chernin (1982). Transcription initiation at pregene sequences in *E. coli* with responses to small molecules (camp and indoles). In *Promoters: Structure and Function*, ed. R. L. Rodriguez and M. S. Chamberlin. Praeger, New York.

Kutney, J. P., L. S. Choi, P. Kolodziejczyk, S. K. Sleigh, K. L. Stuart, B. Worth, W. G. W. Kurz, K. B. Chatson and F. Constabel (1981). Alkaloid production in *Cathanthus roseus* cell cultures. V. Alkaloids from the 176G, 299Y, 340Y and 951G cell lines. *J. Nat. Prod.*, **44**, 536–540.

Leaver, C. J. and B. G. Forde (1980). Mitochondrial genome expression in higher plants. In *Genome Organization and Expression in Plants*, ed. C. J. Leaver, pp. 407–425. Plenum, New York.

Leete, E. (1980). Alkaloids derived from ornithine, lysine and nicotinic acid. In *Secondary Plant Products*, ed. E. A. Bell and B. V. Charlwood, pp. 65–91. Springer-Verlag, Berlin.

Lewis, E. D., G. Hagen, J. I. Mullins, P. N. Mascia, W. D. Park, W. D. Benton and I. Rubenstein (1981). Cloned genomic segments of *Zea mays* homologous to zein mRNAs. *Gene*, **14**, 205–215.

Luckner, M. (1980). Expression and control of secondary metabolism. In *Secondary Plant Products*, ed. E. A. Bell and B. V. Charlwood, pp. 23–63. Springer-Verlag, Berlin.

Lurquin, P. F. and C. I. Kado (1977). *Escherichia coli* plasmid pBR313 insertion into plant protoplasts and into their nuclei. *Mol. Gen. Genet.*, **154**, 113–121.

Lurquin, P. F. and R. E. Sheehy (1981). Binding of large liposomes to plant protoplasts and delivery of encapsulated DNA. *Plant Sci. Lett.*, **25**, 133–146.

Maliga, P. (1980). Isolation, characterization, and utilization of mutant cell lines in higher plants. In *International Review of Cytology*, ed. I. K. Vasil, Suppl. 11A, pp. 225–248. Academic, New York.

Marcker, K. A., K. Gausing, B. Jochimsen, P. Jorgensen, K. Paludan and E. Truelsen (1981). The cloning and organization of soybean leghemoglobin genes. In *Genetic Engineering in the Plant Sciences*, ed. N. J. Panopoulos, pp. 63–71. Praeger, New York.

Matthews, B. F. and D. E. Cress (1981). Liposome-mediated delivery of DNA to carrot protoplasts. *Planta*, **153**, 90–94.

McDaniel, R. G. and R. T. Ramage (1970). Genetics of a primary trisomic series in barley: identification by protein electrophoresis. *Can. J. Genet. Cytol.*, **12**, 490–495.

McDaniel, R. G. (1981a). Plant genetic engineering: possibilities for organelle transfer. In *Genetic Engineering for Crop Improvement*, pp. 185–205. The Rockefeller Foundation.

McDaniel, R. G. (1981b). Recombinant DNA for plant genetic improvement. In *Recombinant DNA, Proceedings of the Third Cleveland Symposium on Macromolecules, Cleveland, Ohio*, ed. G. A. Walton, pp. 245–259. Elsevier, Amsterdam.

Melanson, D. and A. J. Trewavas (1982). Changes in tissue protein pattern in relation to auxin induction of DNA synthesis. *Plant, Cell Environ.*, **5**, 53–64.

Mooney, H. A. (1972). The carbon balance of plants. In *Annual Review of Ecology and Systematics*, vol. 3, pp. 315–346. Annual Reviews Inc., Palo Alto, CA.

Nagl, W. and H.-P. Fusenig (1979). Types of chromatin organization in plant nuclei. In *Genome and Chromatin: Organization, Evolution, Function*, ed. W. Nagl, V. Hemleben and F. Ehrendorfer, pp. 221–233. Springer-Verlag, New York.

Naora, H. and N. J. Deacon (1982). Relationship between the total size of exons and introns in protein-coding genes of higher eukaryotes. *Proc. Natl. Acad. Sci. USA*, **79**, 6196–6200.

Otten, L., H. De Greve, J. P. Hernalsteens, M. Van Montagu, O. Schieder, J. Straub and J. Schell (1981). Mendelian transmission of genes introduced into plants by the Ti plasmids of *Agrobacterium tumefaciens*. *Mol. Gen. Genet.*, **183**, 209–213.

Palmer, J. D. (1982). Physical and gene mapping of chloroplast DNA from *Atriplex triangularis* and *Cucumis sativa*. *Nucleic Acids Res.*, **10**, 1593–1605.

Pedersen, K., J. Devereu, D. R. Wilson, E. Sheldon and B. A. Larkins (1982). Cloning and sequence analysis reveal structural variation among related zein genes in maize. *Cell*, **29**, 1015–1026.

Princen, L. H. (1977). Potential wealth in new crops: research and development. In *Crop Resources*, ed. D. S. Seigler, pp. 1–15. Academic, New York.

Pring, D. R., C. S. Levings, W. W. L. Hu and D. H. Timothy (1977). Unique DNA associated with mitochondria in the 'S'-type cytoplasm of male-sterile maize. *Proc. Natl. Acad. Sci. USA*, **74**, 2904–2908.

Radin, D. N., H. M. Behl, P. Proksch and E. Rodriquez (1982). Rubber and other hydrocarbons produced in tissue cultures of guayule (*Parthenium argentatum*). *Plant Sci. Lett.*, **26**, 301–310.

Ream, L. W. and M. P. Gordon (1982). Crown gall disease and prospects for genetic manipulation of plants. *Science*, **218**, 854–858.

Roberts, R. J. (1977). The role of restriction endonucleases in genetic engineering. In *Recombinant Molecules: Impact on Science and Society*, ed. R. F. Beers, Jr. and E. G. Basset, pp. 21–32. Raven, New York.

Secor, G. A. and J. F. Shepard (1981). Variability of protoplast-derived potato clones. *Crop Sci.*, **21**, 102–105.

Schaefer-Ridder, M., Y. Wang and P. H. Hofschneider (1981). Liposomes as gene carriers: efficient transformation of mouse L cells by thymidine kinase gene. *Science*, **215**, 166–168.

Schieder, O. and I. K. Vasil (1980). Protoplast fusion and somatic hybridization. In *Perspectives in Plant Cell and Tissue Culture*, ed. I. K. Vasil, Suppl. 11B, pp. 21–46. Academic, New York.

Shepherd, R. J., B. Gronenborn, R. Gardner and S. D. Daubert (1981). Recent developments: molecular cloning of foreign DNA in plants using cauliflower mosaic virus as recombinant vector. In *Genetic Engineering in the Plant Sciences*, ed. N. J. Panopoulos, pp. 255–257. Praeger, New York.

Shiio, I. and S. Ohta (1973). Nicotine production by tobacco callus tissues and effect of plant growth regulators. *Agric. Biol. Chem.*, **37**, 1857–1864.

Smith, H. O. and M. L. Birnstiel (1976). A simple method for DNA restriction site mapping. *Nucleic Acids Res.*, **3**, 2387–2398.

Smith, T. A. (1980). Plant amines. In *Secondary Plant Products*, ed. E. A. Bell and B. V. Charlwood, pp. 433–460. Springer-Verlag, Berlin.

Spurgeon, S. L. and J. W. Porter (1981). Introduction. In *Biosynthesis of Isoprenoid Compounds*, ed. J. W. Porter and S. L. Spurgeon, pp. 2–46. Wiley, New York.

Stephen, A. M. (1980). Plant carbohydrates. In *Secondary Plant Products*, ed. E. A. Bell and B. V. Charlwood, pp. 555–584. Springer-Verlag, Berlin.

Sun, S. M., J. L. Slightom and T. C. Hall (1980). Intervening sequences in a plant gene — comparison of the partial sequence of cDNA and genomic DNA of French bean phaseolin. *Nature (London)*, **289**, 37–41.

Sung, M. T. and J. L. Slightom (1981). Methods for preparation of plant nucleic acids optimally suited for restriction endonuclease digesting and cloning: the construction of jack bean and soybean phage libraries in charon 4A. In *Genetic Engineering in the Plant Sciences*, ed. N. J. Panopoulos, pp. 39–61. Praeger, New York.

Tabata, M. (1977). Recent advances in the production of medicinal substances by plant cell cultures. In *Plant Tissue Culture and its Bio-Technical Application*, ed. W. Barz, E. Reinhard and M. H. Zenk, pp. 3–16. Springer-Verlag, Berlin.

Thompson, G. A., Jr. (1980). Plant lipids of taxonomic significance. In *Secondary Plant Products*, ed. E. A. Bell and B. V. Charlwood, pp. 535–553. Springer-Verlag, Berlin.

Threlfall, D. R. (1980). Polyprenols and terpenoid quinones and chromanols. In *Secondary Plant Products*, ed. E. A. Bell and B. V. Charlwood, pp. 288–308. Springer-Verlag, Berlin.

Underhill, E. W. (1980). Glucosinolates. In *Secondary Plant Products*, ed. E. A. Bell and B. V. Charlwood, pp. 492–511. Springer-Verlag, Berlin.

Vasil, I. K. and V. Vasil (1980). Isolation and culture of protoplasts. In *Perspectives in Plant Cell and Tissue Culture*, ed. I. K. Vasil, Suppl. 11B, pp. 1–19. Academic, New York.

Walbot, V. and R. Goldberg (1978). Plant genome organization and its relationship to classical plant genetics. In *Nucleic Acids in Plants*, ed. J. Davis and T. Hall, pp. 3–40. CRC Press, Bocas Raton, FL.

Walsh, C. (1979). Enzyme-catalyzed alkylations involving prenyl-group transfer. In *Enzymatic Reaction Mechanisms*, pp. 867–879. W. H. Freeman, San Francisco, CA.

Watowich, A.-M. V. and E. B. Shultz, Jr. (1984). Potential for polymers from novel oilseeds in the Southeastern United States. In *Fuels and Chemicals from Oilseeds: Technology and Policy Options*, ed. E. B. Shultz, pp. 103–123. AAS/ Westview Press, Boulder, CO.

Widholm, J. M. (1977). Isolation of biochemical mutants of cultured plant cells. In *Molecular Genetic Modification of Eucaryotes*, ed. I. Rubenstein, R. L. Phillips, C. E. Green and R. Desnick, pp. 57–65. Academic, New York.

Yadav, N. S., K. Postle, R. K. Saiki, M. F. Thomashow and M.-D. Chilton (1980). T-DNA of a crown gall teratoma is covalently joined to host plant DNA. *Nature (London)*, **287**, 458–461.

Yamaoto, Y., R. Mizuguchi and Y. Yamada (1981). Chemical constituents of cultured cells of *Euphorbia tirucalli* and *E. millii*. *Plant Cell Rep.*, **1**, 29–30.

Zambryski, P., M. Holsters, K. Kruger, A. Depicker, J. Schell, M. Van Montagu and H. M. Goodman (1980). Tumor DNA structure in plant cells transformed by *A. tumefaciens*. *Science*, **209**, 1385–1391.

12

Biotechnology Applied to Raw Minerals Processing

B. J. RALPH
University of New South Wales, Kensington, NSW, Australia

12.1 INTRODUCTION

Early evidence for the ability of some microbial species to influence the course of inorganic chemical reactions was provided by the work of Winogradsky (1888), which indicated that the growth of certain filamentous bacteria in ferruginous water was accompanied by the accumulation of oxidized and insoluble forms of iron. Rudolfs and Helbronner (1922) later reported the oxidation of zinc sulfides by soil microorganisms but no real understanding of the microbial role in geochemical phenomena appeared until Colmer and Hinkle (1947) isolated and described bacteria with the capacity for enormously accelerating the oxidation of ferrous iron. The past three decades have seen a blossoming of interest in fundamental biogeochemical phenomena and a slow but steady increase of sophistication in the derived technologies. The earliest comprehensive account of these areas was published by Silverman and Ehrlich (1964); the state of the art with respect to the basic science and the applications in bacterial leaching and related areas has been progressively reviewed in recent years (Tuovinen and Kelly, 1974a, 1974b, 1974c; Torma, 1977; C. L. Brierley, 1978; Kelly *et al.*, 1979; Torma and Bosecker, 1982). The proceedings of international conferences in recent years provide evidence of the lively and increasing interest in both fundamental and applied aspects of mineral biodegradation phenomena (Schwartz, 1977; Murr *et al.*, 1978; Trudinger *et al.*, 1980; Czegledi, 1980; Rossi and Torma, 1983). The biogeochemical phenomena underlying both the genesis and breakdown of minerals have also been comprehensively described (Trudinger and Swaine, 1979).

The most extensively practised applications of phenomena of these kinds have been the recovery of copper by treatment of the drainage waters from abandoned mines and from dumps of overburden material containing low percentages of copper sulfide minerals. Such methods are said to have been practised as far back as the Soong Dynasty (960–1279 AD) in China (Yao Dun Pu, 1982) and were certainly well-developed at the Rio Tinto mines in Spain by the 17th century (Taylor and Whelan, 1942; Avery, 1974). The empirical technology evolved during these early developments has been progressively rationalized during the past 30 years, with greater understanding of the basic phenomena involved, in particular the essential role of the associated microbial populations. The recovery of copper by dump and heap leaching and by biologically-assisted solution mining is currently widely practised in a number of countries on a substantial scale, for example in the USA these methods are said to account for more than 10% of total copper production (Brierley, 1982). Similar bacterial leaching techniques have been applied to the commercial recovery of uranium.

In spite of the relative slowness of recovery of metals by biologically-assisted leaching methods as compared to chemical hydrometallurgical procedures, there is likely to be a continuing and increasing usage of dump, heap and solution mining techniques in the future. Solution mining, in particular, might provide access to deep ore bodies (Bhappu, 1982) and current research suggests that microorganisms with the capabilities for the facilitation of mineral biodegradation at elevated temperatures and under substantial hydrostatic pressures could be available. The inevitable exhaustion of high-grade mineral deposits, the sheer bulk of dilute metal sources, the lower energy consumptions and more complete recoveries possible with biodegradative processes, and the steadily increasing understanding of the basic phenomena involved, suggest a continuing role for these modes of recovery of copper, uranium and some other metals. The greater usage of biohydrometallurgical processes is also likely to be stimulated by the unamenability of some complex and fine-grained sulfidic ores to orthodox metallurgical treatment and perhaps by the inherent inability of some common procedures to meet increasingly strict standards for control of environmental pollution.

The well-established bacterial leaching techniques have been exclusively applied to sulfidic ores, or to ores containing the ubiquitous iron sulfide minerals. Such ores include their own energy sources, in the form of oxidizable sulfur and iron, for the driving of the degradation reactions or the generation of essential chemical reagents. In contrast, the microbial degradation of silicate minerals requires the availability of an external energy source. In spite of considerable investigation and encouraging results at laboratory and pilot scales, economically viable biodegradative procedures for silicate minerals have not yet emerged; such methods for the recovery of nickel from lateritic ores and for the beneficiation of low-grade bauxite would greatly extend the availability of sources for nickel and aluminum.

New directions in biohydrometallurgical research and development are becoming evident from the literature and the proceedings of recent conferences (Ralph, 1982a, 1982b; Rossi and Torma, 1983). Reactor leaching is beginning to emerge as a viable alternative to some orthodox metallurgical procedures, not only for the direct solubilization of metals such as copper (chalcopyrite concentrate leaching) and iron (desulfurization of coal, beneficiation of glass-making sand), but also for the release of high-value components (gold, silver, tin minerals) by dissolution or modification of associated low-value ore components. Such procedures lend themselves to closer control of the biological processes and to the use of superior biological agents derived from genetic engineering techniques. Considerable attention is being paid to the basic genetics of microbial species known to be involved in leaching processes, and to their amenability to genetic manipulation. Also at the fundamental level, substantial progress continues to be made with the assessment of the roles in mineral biodegradation of the lesser known components of the complex microbial associations which develop in leaching dumps and other field situations, and some understanding has begun to emerge on the interactions of mixed microbial populations during the leaching processes. The recent findings on the role of anaerobic and micro-aerophilic processes in leaching situations have implications for both novel processes and for the better control of environmental pollution from mining and metallurgical wastes.

12.2 BIOGEOCHEMICAL REACTIONS

The flux of elements and compounds in the earth's crust, the water bodies and the atmosphere occurs *via* a series of physical and chemical changes, the rates of which are affected by biological agencies. As far as the economically important minerals are concerned, biologically-mediated reactions of both synthetic and degradative kinds have been identified over the past 100 years and some of these are of central importance in biohydrometallurgical processes. The most significant reactions are degradative, that is they facilitate the destruction of mineral structures and the solubilization or freeing of valuable components. The biogenesis of minerals, insofar as it has been delineated (Trudinger and Mendelsohn, 1976) involves processes of accumulation of specific metallic ions, separation and reprecipitation. Some of these processes involve biological agencies and some have considerable relevance to the ancillary processes of biohydrometallurgy such as the recovery of metals from admixture in dilute leaching solutions. Major attention has, however, been focussed on biogeochemical reactions of a biodegradative type.

12.2.1 Chemical Mechanisms and Controlling Factors

The metallic sulfides are the most studied minerals in respect of mineral biodegradation and of these the most common, iron pyrites (FeS_2), and the economically-important copper sulfides, chalcopyrite ($CuFeS_2$) and chalcocite (Cu_2S), have attracted most attention. About 250 metallic sulfides are known to occur as stable, crystalline phases in the geological environment, but less than 50 of these have been examined with respect to their amenability to biodegradation and limited information is available on all but an economically-significant handful (principally, iron pyrites, arsenopyrites, marcasite, pyrrhotite, chalcopyrite, bornite, chalcocite, covellite, enargite, pentlandite, galena, sphalerite, marmatite, stibnite and molybdenite). The gaps in basic physicochemical information about mechanisms of chemical breakdown of even some common metallic sulfides are surprising. Useful compilations of the general characteristics of metallic sulfides (*e.g.* Jellinek, 1968) and of semiconductor properties (Shuey, 1975) are available. The recent comprehensive reviews on physicochemical properties and the mechanisms of oxidation of pyrite in aqueous media (Hiskey and Schlitt, 1982; Lowson, 1982) are model compilations and evaluations of relevant data.

The metallic sulfide minerals which have been most intensively studied have generally been shown to be degraded *via* multiple pathways and, in some cases, individual steps in pathways have been identified. Some breakdown routes are chemical; others are electrochemical, since the metallic sulfides have semiconductor properties. Some steps in pathways are amenable to biological intervention and when this occurs overall rates of breakdown, and indeed the nature of the final products, can be affected. An understanding of chemical and electrochemical breakdown reactions greatly facilitates the generation of experimentally-examinable hypotheses for the elucidation of the intimate details of biodegradation mechanisms.

A number of factors influence the rates of the degradation reactions of sulfide minerals; these include the morphology and electronic structure of the mineral surface (Tributsch and Bennett, 1981a, 1981b), the surface area, the temperature, the pH and redox potential of the milieu (Garrels and Christ, 1965; Pourbaix, 1966; Majima and Peters, 1968), the presence or absence of catalysts, the partial pressure of oxygen, and, in undersaturated systems, the relative humidity (Howie, 1977a, 1977b; Ralph, 1980). These factors have been discussed in detail by Lowson (1982). Some degradation reactions of sulfide minerals have been studied only under hydrometallurgical conditions, that is at temperatures above 100 °C and at above atmospheric pressures, and it has been assumed that biological interventions could not occur under such circumstances. This assumption may need re-examination in the light of the recent discovery of biogenic reactions at temperatures in excess of 250 °C (Baross and Deming, 1983). A number of the factors mentioned above are also influential in the determination of the activity of microbial and biochemical systems and may influence biodegradation in a less direct manner.

A number of the chemical reactions which have been identified as occurring in the chemical oxidation of pyrite are also significant in the biodegradation of sulfide and other minerals, as are some in which sulfur or sulfur species participate. Some of these reactions are affected by biological catalysis (Kelly *et al.*, 1979).

The basic chemistry of the degradation of silicates and aluminosilicates has been less studied than that of the metallic sulfides, insofar as a relevance to mineral biodegradation is concerned. Loughnan (1969) has summarized concepts of the chemical breakdown of silicates in the generalized equation:

$$M^+ (Mineral)^- + H^+OH^- \rightleftharpoons H^+ (Mineral)^- + M^+OH^- \tag{1}$$

The equilibrium reaction represented by the equation indicates that any process that changes the concentrations of any of the reactants or products will affect the extent of the reaction. The chemical studies of Huang and Keller (1972) on the leaching activities of different organic acids on various silicate minerals suggests that while hydrogen ions are effective agents in the depolymerization of the silicate lattice, the metal chelation characteristics of the acid anion are influential in determining the course of the overall degradation reaction.

12.2.2　Microbial Interventions

By relatively simple experimental procedures, it has been possible over the past 40 years to demonstrate unequivocally that the rate of dissolution in aqueous media of a number of sulfide and silicate minerals is significantly enhanced by the presence of various microbial populations, as compared with the breakdown rate in sterile controls. Chemical analysis of dissolved metallic ions and of the leached residues clearly establishes the overall stoichiometry of such biodegradation processes. The reaction patterns for the sulfides of antimony, arsenic, bismuth, cadmium, cobalt, copper, gallium, germanium, iron, lead, mercury, molybdenum, nickel and zinc, and the oxides of manganese and uranium have been reviewed by Torma and Bosecker (1982). General aspects of the biodegradation of sulfide minerals have been described by Ralph (1979).

The numerous studies on the mechanisms of biodegradation of iron pyrites, which were initiated by the pioneering work of Colmer and Hinkle (1947), and the intensive investigations of the bacterial leaching of the economically-important copper iron sulfide mineral, chalcopyrite, have paved the way for a better understanding of the various modes by which the biological intervention in geochemical breakdown processes can occur. In the case of iron pyrites, the overall oxidation is described by equation (2), and two electron acceptors can function in the process, molecular oxygen and ferric ions.

$$4\,FeS_2 + 15\,O_2 + 2\,H_2O \rightarrow 2\,Fe_2(SO_4)_3 + 2\,H_2SO_4 \qquad (2)$$

At pH values above 2.5, ferric ions increasingly react with water to form insoluble ferric oxides (equation 3).

$$Fe^{3+} + 3\,H_2O \rightarrow Fe(OH)_3 + 3\,H^+ \qquad (3)$$

The initiating reaction, when pyrite is exposed to oxygen and water, is a slow spontaneous reaction (equation 4) which can be catalysed by some bacteria.

$$FeS_2\ (solid) + 3.5\,O_2 + H_2O \xrightarrow{\text{bacteria}} Fe^{2+} + 2\,SO_4^{2-} + 2\,H^+ \qquad (4)$$

The generation of hydrogen ions in this reaction leads to the development of acidic conditions in the immediate milieu, under which conditions the auto-oxidation of ferrous ions is very slow (Singer and Stumm, 1970). The rate of this reaction (equation 4) is only significant when the bacteria are in close proximity to the mineral surface, and is a function of their numbers. The ferrous ions generated in this reaction (equation 4), in spite of their stability at low pH in the absence of catalysts, can be rapidly oxidized by oxygen in the presence of various bacteria, the best known of which is *Thiobacillus ferro-oxidans*. Accelerations of the reaction under these conditions can be of the order of 10^6 times (Lacey and Lawson, 1970).

$$Fe^{2+} + 0.25\,O_2 + H^+ \xrightarrow{\text{bacteria}} Fe^{3+} + 0.5\,H_2O \qquad (5)$$

The ferric ions resulting from this reaction (equation 5) can chemically degrade sulfides, including pyrite, by the following reaction:

$$FeS_2 + 2\,Fe^{3+} + 3\,O_2 + 2\,H_2O \rightarrow 3\,Fe^{2+} + 2\,SO_4^{2-} + 4\,H^+ \qquad (6)$$

There is some evidence that the reaction described by equation (6) proceeds by two steps, one of which is bacterially catalysed:

$$FeS_2 + 2\,Fe^{3+} \rightarrow 3\,Fe^{2+} + 2\,S^0 \qquad (7)$$

$$2\,S^0 + 3\,O_2 + 2\,H_2O \xrightarrow{\text{bacteria}} 4\,H^+ + 2\,SO_4^{2-} \qquad (8)$$

It was originally suggested by Silverman (1967) and further expanded by Smith and Shumate (1970) that two mechanisms operate concurrently for the bacterial oxidation of pyrite, one requiring close physical contact between the bacterium and the mineral particles (direct mechanism) and the other, in which ferric ions are regenerated by bacterial catalysis, not requiring such close proximity (indirect mechanism). The numerous studies carried out on other sulfide minerals such as chalcopyrite ($CuFeS_2$), chalcocite (Cu_2S), covellite (CuS) and sphalerite (ZnS) suggest that similar direct and indirect mechanisms operate concurrently in the biodegradation of other sulfide minerals.

The evidence for a direct mechanism of bacterial oxidation of sulfide minerals has been closely examined by a number of authors (Gaidarjiev *et al.*, 1975; Groudev and Genchev, 1978a; C. L. Brierley, 1978; Ralph, 1979; Kelly *et al.*, 1979). Direct observational studies by electron microscopy (Corrans, 1970; Brierley *et al.*, 1973; Karavaiko and Pivovarova, 1973; Berry and Murr, 1978) clearly demonstrate that physical attachment occurs selectively to the sulfide phases in prepared mineral specimens. Further observations by Bennett and Tributsch (1978) have indicated characteristic patterns of bacterially etched pits on the surface of pyrite crystals exposed to *T. ferro-oxidans*; these authors suggest that the bacterial distribution on the surface (and hence leaching sites) is critically dependent on deviations in the crystal structure such as fracture lines and dislocations.

A considerable amount of attention has been paid to ascertaining whether sulfide mineral biodegradation can occur in the absence of iron, which is an ingredient of the indirect mechanism. The leaching of the copper minerals chalcocite and covellite, in the presence of *T. ferro-oxidans* under iron-free conditions, has been demonstrated by several workers (Bryner *et al.*, 1954; Corrans *et al.*, 1972; Beck, 1977; Imai, 1978). In other studies, the nature of the enzymic profiles of various strains of *T. ferro-oxidans* has been examined (Tuovinen *et al.*, 1976) and throws light on the specific enzymic catalysts which are likely to be involved in the biochemical mechanisms for the oxidation of sulfur species, present or arising from prior chemical processes, on the sulfide mineral surface. Gaidarjiev *et al.* (1975), and Groudev and Genchev (1978a) have studied the leaching activity of some 120 strains of *T. ferro-oxidans* towards chalcopyrite, iron and elemental sulfur and have noted that the most active strains are characterized by strong sulfur-oxidizing activity and high activity levels of those enzymes which participate in the transformations of inorganic sulfur compounds. A high level of rhodanese activity appeared particularly significant. Further evidence on the occurrence of direct mechanisms for the biodegradation of covellite has been reported by Groudev (1980a) and for sphalerite by Khalid and Ralph (1977) and Groudev (1980b).

Other investigations on the extent of direct oxidation mechanisms have employed selective inhibition of iron- and sulfur-oxidizing enzymatic activities. The early investigations of Landesman *et al.* (1966a, 1966b) clarified the effects of the various conditions controlling the optimum oxidation rates of ferrous iron, sulfur and reduced sulfur compounds by *T. ferro-oxidans* and experiments on soluble iron, sulfur and iron-containing sulfide minerals (chalcopyrite, bornite and iron pyrite) established that iron and sulfur can be oxidized simultaneously. On the basis of these experiments, Duncan *et al.* (1967) established the differential susceptibility of the bacterial oxidation of ferrous iron and sulfur to *N*-ethylmaleimide and sodium azide and determined the effect of these inhibitors on pyrite and chalcopyrite oxidation. The evidence obtained by these workers for bacterial oxidation (independent of ferric iron) of the sulfur moiety in mineral sulfides which contain iron, together with that on the oxidation of iron-free synthetic sulfides (Torma, 1971; Silver and Torma, 1974; Sakaguchi *et al.*, 1976), provides strong support for direct attack mechanisms on mineral sulfides by bacteria.

The models devised to accommodate the observed leaching and bacterial growth rates during the continuous microbiological leaching of zinc sulfide concentrates by *T. ferro-oxidans* (Gormely *et al.*, 1975; Sanmugasunderam, 1981) assume the operation of a direct mechanism and the attachment of the bacteria to the mineral surfaces. A recent study on the leaching of iron pyrite by *T. ferro-oxidans* in a continuous system (Chang and Myerson, 1982) has yielded results which are consistent with the concept of preferential bacterial attachment to certain sites on the mineral surface. The concentration of bacteria suspended in solution was related to the concentration of bacteria attached to the pyrite surface by a Langmuir-type adsorption–desorption relation and the fitting of experimental data to this relation yielded a value for the area occupied on the pyrite surface per bacterium of 86 μm^2.

From early in the study of sulfide mineral biodegradation there has been considerable evidence supporting an indirect attack mechanism involving microbially-regenerated ferric ions as the leaching agent (Kuznetsov *et al.*, 1963; Silverman and Ehrlich, 1964; Zajic, 1969). As pointed out by Ralph (1979), 'the ubiquitous occurrence of iron in natural situations and the wide range of reactions with sulfide minerals that can occur, combine to make the ferric ion, either as such or in complex form, the most important chemical species involved in indirect attack mechanisms.' The general form (equation 9) of the ferric attack mechanism on iron pyrite (equation 6) is known to be applicable to a number of sulfide minerals under aerobic conditions (Bryner *et al.*, 1954; Ivanov *et al.*, 1961) and may be expressed as:

$$MeS + 2 Fe^{3+} + H_2O + 1.5 O_2 \rightarrow Me^{2+} + 2 Fe^{2+} + SO_4^{2-} + 2 H^+ \qquad (9)$$

Silverman (1967) has shown that in the case of iron pyrites, under aerobic conditions and in the absence of microorganisms, elemental sulfur may be produced (equation 7) during the ferric leaching reaction, and Berry and Murr (1977) have detected, by electron microscopical techniques, sulfur crystals on freshly leached pyrite surfaces. These authors have speculated that such accumulation of elemental sulfur occurs during the very early stages of the leaching of pyrite at a point where bacterial activity is not sufficient to completely convert the sulfur to sulfuric acid (equation 8). In spite of the formal similarities between abiotic acid ferric sulfate leaching of pyrite and other sulfides and the postulated biologically-assisted indirect ferric attack mechanism, there is evidence to suggest that the two processes are fundamentally different (Keller and Murr, 1982).

There is the further possibility that other indirect attack mechanisms, analogous to the microbially-regenerated ferrous/ferric system, may be operative in the biodegradation of some mineral sulfides in which the metallic component is oxidizable. There is evidence that the Cu(I)/Cu(II) oxidation can contribute to the energy-yielding metabolism of *T. ferro-oxidans* (Nielsen and Beck, 1972; Golding *et al.*, 1974). It is relevant to note that cupric ions, stabilized in acidified aqueous acetonitrile solution, are more potent oxidants than ferric ions in acid solution. Muir *et al.* (1976) and Kelly *et al.* (1979) have drawn attention to the similar reaction systems which occur in the *T. ferro-oxidans*-mediated oxidation of copper selenide (Torma and Habashi, 1972) and the antimony sulfide, stibnite (Lyalikova, 1972; Torma and Gabra, 1977).

It is clear that both direct and indirect attack mechanisms contribute to the biodegradation of sulfide minerals and that the relative contributions of each degradative mode will be determined by a number of factors which will include the physical and chemical characteristics of individual minerals, the particular microbial agent(s) and the nature of the physicochemical environment. Murr (1980) has extensively reviewed the leaching of copper ores and has concluded that the pyrite which is commonly associated with the copper sulfide minerals is degraded predominantly by the indirect mechanism. The studies by Smith and Shumate (1970) similarly suggest that the pyrite oxidation responsible for acid coal mine drainage is generally indirect. Hiskey and Schlitt (1982) have usefully discussed other aspects of the biodegradation of pyrite. In the case of other sulfides, however, such as chalcopyrite, the evidence is strong that the direct mechanism predominates (Landesman *et al.*, 1966b; Duncan *et al.*, 1967).

The classification of microbial interventions in mineral biodegradation processes into 'direct' and 'indirect' types, in spite of its widespread usage in relation to the leaching of sulfide minerals, is over-simplistic and not particularly illuminating; it does not accommodate the unfolding complexity and diversity of the degradation modes than can occur. The term 'direct', for example, implies only that the microbial cells are in close proximity to the mineral surface. The elegant experiments and deductions of Vanselow (1976; summarized by Ralph, 1979) lead to the general conclusion that the behaviours of microbial populations do not reflect a mechanistic necessity for actual physical contact between cells and mineral surface, but that evolutionary adaptations which favoured the adsorption of cells to surfaces would be advantageous when the concentration of intermediates, reactants or biocatalysts was so low that their diffusion through a liquid film might be rate-limiting. Further, the terms 'direct' and 'indirect' throw no light on the intimate mechanisms involved, except to imply that the first type must occur in close proximity to the mineral surface, whereas the second type may operate elsewhere.

A number of modes of biological intervention have been demonstrated by various workers. They include the following categories.

(i) The facilitation of the mass transfer of reactants by more effective wetting of the mineral surface by surface active agents excreted by the microorganisms. A number of such agents have been identified in the case of *Thiobacillus* species (Starkey *et al.*, 1956; Schaeffer and Umbreit, 1963; Shively and Benson, 1967; Barridge and Shively, 1968; Beebe and Umbreit, 1971; Dees and Shively, 1982; Gupta and Mishra, 1983). Similar effects have been noted with synthetic detergents (Duncan *et al.*, 1964; Trussell *et al.*, 1964; Martinez Garcia *et al.*, 1981).

(ii) The removal of passivation layers on the mineral surface arising from chemical or electrochemical reactions (Corrans *et al.*, 1972; Golding *et al.*, 1977; Rickard and Vanselow, 1978; Tomizuka and Yagisawa, 1978). In particular, sulfur layers can arise on the surface of sulfide minerals by chemical or electrochemical processes and limit further degradation; data on this occurrence in the case of the copper sulfide mineral covellite have been reported by Corrans *et al.*

(1972), Golding *et al.* (1977), and Rickard and Vanselow (1978), and in the case of the lead sulfide mineral galena by Tomizuka and Yagisawa (1978).

(iii) The excretion of enzymes and cofactors and their adsorption to the mineral surface in a manner analogous to that suggested by Nickerson (1969) for the biodegradation of solid, organic polymeric substrates. There is little evidence to date for this mode of biological intervention in the case of solid sulfide mineral substrates, even though a great deal is known of the biochemistry of inorganic sulfur oxidation by *Thiobacilli*, the principal organisms involved (Suzuki, 1974; Aleem, 1975). Most of the reduced sulfur compounds studied have been soluble compounds $(S^{2-}, H_2S, S_2O_3^{2-}, SO_3^{2-}, S_4O_6^{2-}, n(S-S)^{2-})$; most of the known reactions can be accommodated in the scheme shown in Figure 1.

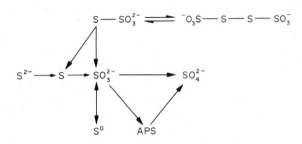

Figure 1 Reactions of soluble reduced-sulfur compounds

It is probable that sulfide oxidation occurs in two stages; sulfide first loses two electrons mediated by a sulfide oxidase and subsequently polymerization of the resulting sulfur atoms occurs (Silver, 1978). Moriarty and Nicholas (1969) have shown that the oxidation of short-chain polysulfides, thought to be associated with membranes, to polymeric sulfur compounds can then occur. Membrane fractions prepared from acidophilic sulfur-grown *T. ferro-oxidans* oxidize soluble sulfide (Tano and Lundgren, 1978). Further, polythionates can oxidize sulfide ion (Szekeres, 1974; equation 10).

$$S^{2-} + S_nO_6^{2-} \rightarrow 2S_2O_3^{2-} + (n-3)S \qquad (10)$$

It has recently been suggested that the biodegradation of covellite by some strains of *T. ferro-oxidans* may involve trithionate and thiosulfate as soluble intermediates (Golding *et al.*, 1983). The localization of intermediates and associated catalysts in such cases is not yet clear.

(iv) The shifting of dissociation equilibria on sulfide mineral surfaces. It was suggested by Torma and Sagaguchi (1978), on the basis of the direct correlation between the rates of bacterially-mediated oxidation of a number of synthetic metal sulfides and their solubility products, that the first step in metal sulfide oxidation is the dissociation of the substrate (equation 11).

$$MS \rightarrow M^{2+} + S^{2-} \qquad (11)$$

The released sulfide moiety is envisaged as being immediately captured by the enzymatic system of the bacteria, its removal shifting the dissociation reaction towards completion. The concept has been further examined by Tributsch and Bennett (1981a, 1981b) in the context of exploring the influence of the semiconductor–electrochemical characteristics of metal sulfides on their bacterial leaching behaviour. These authors conclude that the most critical rate-determining parameter (in the absence of any electron acceptors, *e.g.* Fe^{3+}) was the solubility product of the sulfide. They present experimental evidence consistent with a mechanism in which *T. ferro-oxidans* utilizes H^+ and Fe^{3+} as catalytic agents which break surface bonds by chemical and electrochemical mechanisms, respectively. They are subsequently recycled while the bacterium oxidizes the generated surface products, $HS^{\ominus}$, S^0, *etc.*, to sulfate. It is suggested on the basis of the derived mechanism that it should be possible to estimate and predict the suitability of sulfides for bacterial oxidation and to analyse the influence of crystalline quality and impurities on the rate of oxidation. It has been noted earlier that the biodegradation of silicate minerals may be based on biological intervention in an equilibrium reaction (equation 1).

(v) Production of chelating agents by the microbial agent. There is sufficient published evidence on the kinetics of the bacterial oxidation of ferrous ions to suggest that the complexing of iron with components of the leaching solutions (*e.g.* Na^+, PO_4^{3-}, Cl^-) or products of the oxidation reactions and of the organism's metabolism (*e.g.* Fe^{3+}, SO_4^{2-}) may have a significant effect on the rate of reaction (Dugan and Lundgren, 1965; Smith and Shumate, 1970; Tuovinen and Kelly, 1972; Steiner and Lazaroff, 1974; Tuovinen *et al.*, 1978; Golding *et al.*, 1983). Possible metal-chelating capabilities and roles for the extracellular materials excreted by *T. ferro-oxidans* (Agate *et al.*, 1969; Goodman *et al.*, 1980) do not appear to have been further investigated.

The production of organic compounds with strong chelating or binding capacity for metals and other cations is probably of central significance in the biodegradation of silicate minerals. Biogenic acids such as citric, gluconic and 2-ketogluconic have been identified as playing a role in silicate mineral breakdown. The topic has been extensively reviewed by Silverman (1979). It should be noted that metallo-organic complexes can themselves be microbial substrates (Silverman and Ehrlich, 1964; Brantner, 1970; Berthelin *et al.*, 1977; Marshall, 1979) and as such participate in processes by which the metallic component is solubilized or deposited.

(vi) *Modification of physicochemical parameters*. The principal physicochemical factors which influence the rate of chemical degradation reactions and the level of microbial activities are temperature, pH and oxidation–reduction potential (*Eh*), together with other factors such as particle size and surface area of the mineral substrate, and the concentrations of nutrients and reactants. The effects of these parameters on the rate of mineral biodegradation have been exhaustively discussed by Torma (1977) and Lundgren and Silver (1980).

The metabolic activities of microbial populations can profoundly affect the magnitudes of these factors. For example, microbial processes are frequently accompanied by heat generation (Goodman and Ralph, 1980; Soljanto and Tuovinen, 1980) and if the milieu does not permit ready heat dissipation, the temperature of the microenvironment will rise. Similarly, acidity may increase due to biogenic production of hydrogen ions and the oxidation–reduction potential may be varied by the changing of ferric/ferrous ion ratios or the exhaustion of oxygen by aerobic processes. Further, essential nutrients may by immobilized by, for example, jarosite formation as a result of iron oxidation (Duncan and Walden, 1972; Sulligoi, 1972). The inhibitory effects of particulate materials (DiSpirito *et al.*, 1981) may be enhanced by the secondary effects of low pH and mineral dissolution on the physical breakdown of gangue materials. That microbial populations influence these parameters is well demonstrated in field situations by the succession of microorganism types which arises (Lundgren and Silver, 1980), each favoured by the particular combinations of conditions which arise in a temporal continuum. Similar situations have been very frequently observed in laboratory leaching column experiments where the natural populations are permitted to develop, the changes in dominant species reflecting the change in levels of physicochemical factors (Madgwick *et al.*, 1981). In laboratory experiments under monoculture conditions, the changes in factors such as pH or level of essential nutrients can be such as to virtually eliminate the operation of the microbial agent.

The biokinetic ranges and the optimum values for the various activities relevant to mineral biodegradation have been determined for the principal organisms currently known to be involved (Torma, 1977; Lundgren and Silver, 1980). It should be noted, however, that most of the published data have been secured by simple factorial design experimentation; more sophisticated techniques which accommodate the interdependence of parameters can lead to substantially different values in some cases (L.A.V. Sulligoi, personal communication).

12.2.3 Nature and Characteristics of Biogeochemically-important Microorganisms

A diversity of microbial types have been identified as playing primary or secondary roles in the processes of mineral biodegradation and some of these have been studied in considerable detail. The most common source of such microorganisms has been field situations in which mineral breakdown is occurring; such milieu commonly arise as the result of the disturbance of ore bodies during mining operations, in the waste dumps or tailings dams for the disposal of the low-value by-products of metallurgical processing, or in the drainage waters from mined ore bodies or residual waste materials. The methodologies for the isolation, purification, identification and characterization of such microorganisms have been described by a number of authors (*e.g.* Silverman and Lundgren, 1959; Collins, 1969; Tuovinen and Kelly, 1973; Manning, 1975; Mishra and Roy, 1979; C. L. Brierley *et al.*, 1980; Goodman *et al.*, 1981a; Marsh and Norris, 1983).

It is characteristic of both fractured ore bodies and the wastes from their processing that the degree of compositional heterogeneity is likely to be very high and that a wide range of physicochemical conditions will occur in microenvironments. It is not surprising that ecological studies of the microbial populations occurring in such situations have revealed most complex associations of microorganisms which show considerable variation in kind and in relative magnitude according to location and time (Goodman *et al.*, 1981a, 1981b). However, circumstances can occur which favour the predominance of a very limited number of species. Ecological successions commonly develop over time in locations such as sulfidic waste dumps, the earlier-developing populations modifying the microenvironments in respect of factors such as acidity, temperature and availability of biomatter, and paving the way for their own decline and the ascendancy of species better adapted to the new conditions. In the example quoted, the colonizing microbial populations are frequently species which can utilize the sulfur moiety of sulfide minerals as an energy source (*e.g. Thiobacillus delicatus* and *T. rubellus*) (Mizoguchi *et al.*, 1976; Mizoguchi and Okabe, 1980), with concomitant generation of hydrogen ions, but which themselves have low tolerance to hydrogen ions or metallic ions in solution. Such species may substantially modify the composition of sulfide minerals without release of soluble metallic ions, but increase the acidity of the microenvironments and generate biomass. At these higher pH levels, conditions are not favourable for the solution of some metallic ions and there is little information available on the details of the mechanisms by which mineral structure is modified without metal release (Silver and Torma, 1974). The continuation of processes of this kind leads to a succession of increasingly acidophilic, sulfur-oxidizing species and eventually to conditions under which the stability of ferrous ions is sufficiently high for them to be available as energy sources for iron-oxidizing species such as *Metallogenium* and *T. ferro-oxidans*. The accumulation of dead biomass from autotrophic species earlier in the successions may stimulate the development of heterotrophic species (*e.g.* some nitrogen-fixing types) and encourage mixotrophic modes of nutrition.

The complex microbial communities which are associated with sulfide mineral degradation in field situations are dominated by organisms with sulfur- and iron-oxidizing abilities, and with tolerance to high levels of acidity and metal ion concentrations (Tuovinen *et al.*, 1971a, 1971b; Tuovinen and Kelly, 1974a, 1974b, 1974c; Imai *et al.*, 1973). The taxonomic characteristics of the *Thiobacilli* have been tabulated by Ralph (1979) and the growth characteristics of the principal sulfur-oxidizing and iron-oxidizing bacteria involved in sulfide mineral degradation surveyed by C. L. Brierley (1978), Ralph (1979), Lundgren and Silver (1980), Goodman *et al.* (1981a) and Torma and Bosecker (1982). The sulfur-oxidizing and iron-oxidizing microflora are associated with a diverse array of metal- and acid-tolerant heterotrophic bacteria, fungi, yeasts, algae and protozoa (Joseph, 1953; Marchlewitz and Schwartz, 1961; Kuznetsov *et al.*, 1963; Ehrlich, 1963; Moss and Andersen, 1968; Tuttle *et al.*, 1968; Arrietta and Grez, 1971; Wenberg *et al.*, 1971; Lundgren *et al.*, 1972; Updegraf and Duoros, 1972; Bhurat *et al.*, 1973; Belly and Brock, 1974; Dugan, 1975; Madgwick and Ralph, 1977; Groudev *et al.*, 1978). The role of such mixed microbial communities in metal recovery has been recently reviewed (Norris and Kelly, 1982).

The principal acid-generating microorganism associated with the degradation of sulfide ores is *Thiobacillus ferro-oxidans*; its physiology and biochemistry have been intensively studied and extensively reviewed (Tuovinen and Kelly, 1972; Torma, 1977; Lundgren and Silver, 1980). A number of authors have noted the population heterogeneity of the organism as it occurs in leaching dumps and other mineral degradation situations (Groudev *et al.*, 1978) and as manifested by isolates in colony morphology on solid media (Manning, 1975; Goodman *et al.*, 1980), by the different capacities of adapted strains to oxidize particular substrates (Silver and Torma, 1974), by their different DNA base compositions (Guay *et al.*, 1976) and by the varying enzymic profiles of isolated and mutated strains (Groudeva and Groudev, 1980; Groudeva *et al.*, 1981). It is unlikely that the full range of sulfide mineral degrading capabilities of the numerous biochemical and physiological variants of the species *T. ferro-oxidans* have yet been fully explored. A steady stream of new information continues to emerge on such aspects as the lower limits of pH tolerance in respect of iron-oxidizing ability (Atkins, 1978; Bruyensteyn *et al.*, 1980b), tolerance to elevated hydrostatic pressures (Bosecker *et al.*, 1979; Davidson *et al.*, 1981) and the ability of the organism to fractionate sulfur isotopes during leaching processes (Karavaiko *et al.*, 1980). Other recent studies have been concerned with the tolerance of the organism to metallic ions (Kovalenko and Karavaiko, 1981; Sugio *et al.*, 1981; Wong *et al.*, 1982), with its fine structure (Pereversev *et al.*, 1981; DiSpirito *et al.*, 1982) and with novel chemical activities such as the volatilization of mercury compounds (Olsen *et al.*, 1982) and the direct oxidation of uranyl ions (DiSpirito and Tuovinen, 1982a, 1982b). The range of useful abilities of this organism is likely to be further extended when a more detailed understanding of its genetic patterns emerges from the considerable volume of recent and current studies (Chakrabarty, 1978; Martin *et al.*, 1981; Harrison, 1982; Holmes *et al.*, 1983; Rawlings *et al.*, 1983).

The potential for mineral degradation processes of organisms other than the mesophilic *Thiobacilli* has received considerable attention during the past decade. A number of thermophilic, thiobacillus-like bacteria have been isolated and their applicability to the bacterial leaching of sulfide minerals demonstrated (Le Roux *et al.*, 1977; Brierley and Lockwood, 1977; J. A. Brierley, 1978; Brierley and Brierley, 1978; Brierley *et al.*, 1978; Le Roux and Wakerley, 1980; Brierley, 1980). Of great interest are the thermophilic organisms isolated from hot springs and identified as *Sulfolobus* species (*S. acidocaldarius* and related types) which are capable of growth at temperatures of 45 to 80 °C and at pH values ranging from 1 to 6 (Brock, 1969; Darland *et al.*, 1970; Brock *et al.*, 1972; Brierley and Brierley, 1973; De Rosa *et al.*, 1975; Millonig *et al.*, 1975; Mosser *et al.*, 1974; Bohlool, 1975; Marsh *et al.*,1983). *Sulfolobus*-type microorganisms have been shown to be capable of oxidizing ferrous iron and metal sulfides (Brierley and Murr, 1973; Berry and Murr, 1978; Brierley and Brierley, 1978). The abilities of these higher temperature organisms have provided a stimulus for the development of high-temperature bacterial leaching processes and for attempts to achieve greater leaching rates by inoculation with thermophilic organisms (Murr and Brierley, 1978). These developments are likely to be pursued with even greater vigour with the isolation of microorganisms from 350 °C waters emanating from marine sulfide chimneys ('black smokers') and their cultivation at temperatures of at least 250 °C and at 265 atm pressure (Baross and Deming, 1983).

There are circumstances in which non-sulfur-oxidizing, iron-oxidizing microorganisms could have some utility in biohydrometallurgical processes. Some studies have established a limited ferrous iron-oxidizing role for the filamentous bacterium *Metallogenium* at pH values near 4.5 (Walsh and Mitchell, 1972a, 1972b). Above pH 5, where the autooxidation of ferrous iron proceeds at a significant rate and the possibility of energy harvesting from the ferrous–ferric reaction is limited, heterotrophic or mixotrophic organisms such as *Gallionella* spp. may be extremely useful (Hanert, 1973, 1974) for facilitating the oxidation of iron sulfides such as pyrrhotites. In very low pH situations (*ca.* pH 1), the vibrioid, iron-oxidizing bacterium *Leptospirillum ferro-oxidans* (Balashova *et al.*, 1974; Norris and Kelly, 1982; Norris, 1983) is of considerable interest.

While it has been noted by a number of workers during the assessment of the biodegradability of sulfide ore materials that higher rates are frequently achieved with mixed cultures from natural sources than with monocultures of the individual components from such mixtures, the systematic study of mixed culture phenomena in mineral biodegradation situations is of comparatively recent advent. Norris and Kelly (1978) have discussed the role of different *Thiobacilli* in pyrite oxidation when cultured alone and in mixed culture; the general question of the use of mixed microbial cultures in metal recovery has been examined in more detail by the same authors (Norris and Kelly, 1982) and a number of examples of different types of interaction reviewed. Of interest are potentially mutualistic associations, involving the exchange of organic material and fixed nitrogen, between organisms such as *T. ferro-oxidans* and *Beijerinckia lacticogenes* (Tsuchiya *et al.*, 1974; Tsuchiya, 1977), between nitrogen-fixing and non-nitrogen-fixing strains of *T. ferro-oxidans* (Mackintosh, 1978), and nitrogen-fixing strains of *T. rubellus* and other *Thiobacilli* (Goodman and Ralph, 1978).

Norris and Kelly (1978, 1982) have also pointed out that interactions can be envisaged between different bacterial types during sulfide mineral breakdown which result in inorganic material recycling between the bacteria catalysing the reactions. The oxidations of sulfur (with concomitant reduction of ferric ions) by *T. ferro-oxidans* and *T. thio-oxidans* (Brock and Gustafson, 1976; Brock, 1977), and of chalcopyrite (Babij *et al.*, 1981) and marmatite (Goodman *et al.*, 1983) under anaerobic conditions, provide mechanisms by which metal sulfide degradation can proceed even when oxygen is limiting. Cyclic leaching mechanisms of this kind probably occur in flooded opencuts with residual sulfide ore exposures or in which sulfidic tailings have been dumped (Goodman *et al.*, 1981b). A number of possible mixed culture reaction cycles, involving chemical attack on sulfide minerals by ferric ions, mineral dissolution by iron-oxidizing organisms such as *Leptospirillum*, and sulfur removal by sulfur-oxidizing organisms such as *T. thio-oxidans*, *T. acidophilus* (Arkesteyn, 1980) and *T. organoparus* (Markosyan, 1973), under both aerobic and anaerobic conditions, can be envisaged.

The leaching of sulfides and various metal-containing wastes by *T. ferro-oxidans* and *T. thio-oxidans* has been reported; for example, zinc sulfide minerals (Khalid and Ralph, 1977); nickel sulfide ores (Bosecker, 1977); fly-ash, lead slag, zinc-rich jarosite (Ebner, 1977, 1978); oil sands cokes (Zajic *et al.*, 1977); carbonate-rich copper shale (Bosecker *et al.*, 1978); carbonate-bearing uranium ores (Bosecker and Wirth, 1980); petroleum cokes (Sullivan *et al.*, 1980); tin-rich pyrrhotites (Sawe, 1980); metalliferous industrial wastes (Bosecker, 1983; Schäfer, 1983; pyritic cinders (Dogan *et al.*, 1983). However, systematic studies on the effect of mixed cultures on commercially-important minerals such as chalcopyrite do not appear to have been attempted. The mixed culture study of Norris and Kelly (1978) on pyrite degradation would serve as an excellent model for such experimentation.

There is growing interest in microorganisms other than the *Thiobacilli*, with mineral-degrading capabilities. Lyalikova (1972, 1974) has reported the isolation and characterization of the autotrophic organism *Stibiobacter senarmontii* which can utilize the Sb(III)/Sb(V) oxidation as an energy source. Understanding of the role of heterotrophic organisms in mineral degradation continues to expand. Torma and Bosecker (1982) have drawn attention to the reports on the solubilization of gold by *Bacillus* spp., *Pseudomonas fluorescens*, *Serratia marcescens* and *Agrobacterium tumefaciens*. The involvement of other heterotrophic species in the degradation of carbonate and silicate ores of copper (Kiel, 1977) and copper–nickel sulfide concentrates (Le Roux *et al.*, 1978) has been described. Heterotrophic microorganisms capable of leaching manganese ores have been reviewed by Marshall (1979) and Ehrlich (1980) and the possibility of microbially-mediated processes for the refining of manganese dioxide slimes continues to attract attention (Mercz and Madgwick, 1982; Holden and Madgwick, 1983). The role of heterotrophic organisms in the biodegradation of silicate minerals has been reviewed by Silverman (1979) and referred to earlier.

12.3 ENGINEERING CONSIDERATIONS

The translation of information on the mechanisms of mineral biodegradation or the possibility of novel methods for accomplishing such breakdowns into viable large-scale processes presents more difficulties than those commonly encountered in the scale-up of biotechnological processes based on organic raw materials. Firstly, the scale of material handling in mining and metallurgical operations is orders of magnitude greater than that in the fermentation industries and the valuable product content of the primary raw materials may be only a fraction of one percent. Secondly, mineral ores, as extracted by mining procedures, have particular characteristics as substrates for biological modification. They are commonly extremely heterogeneous in terms of chemical composition and physical form, and may vary in particle size from boulders, metres in diameter, to particles of submicron dimensions. They are usually highly insoluble in water, very dense and often highly abrasive.

These considerations apply also to orthodox metallurgical processes, the economic success of which depends upon the balancing of the value of extracted metal or of a more or less uniform mineral concentrate against the costs of primary extraction, reduction in particle size, physical separation of components by a range of techniques including gravity separation or flotation, and treatment by pyrometallurgical or hydrometallurgical procedures. Amongst a number of other factors, the profitability of the overall process will be determined by the grade of the primary raw material and the efficiency of the recovery processes. There will normally be a cut-off point below which the concentration of the valuable component of the primary raw material is too low to sustain the costs of its extraction; the actual level of the cut-off grade will vary with the market price of the valuable component and the overall efficiency of the technology by which it is recovered.

In the majority of mining and metallurgical operations, two wastes of large dimensions occur; firstly, low-grade ore whose valuable mineral content lies below the cut-off grade and which is uneconomic to process by orthodox technology, and secondly, the residues from metallurgical processes (process tailings) which frequently contain significant quantities of valuable components unamenable to recovery by usual procedures. The first waste, low-grade ore material, separated as far as possible from waste rock and higher grade ore material at the time of mining extraction, may be separately stockpiled against the possibility of future processing, but is commonly dumped together with valueless overburden material. The process tailings which are normally finely divided are pumped as slurries to tailings dams for dewatering and stabilization.

Low-grade waste ore has been and continues to be the principal raw material for biohydrometallurgical processing in the case of copper and uranium ores. Process tailings represent a raw material resource of impressive magnitude for the treatment of which biologically-mediated hydrometallurgical procedures probably have an important role to play. A further step in the development of biohydrometallurgical processes would involve them as viable interlocking processes with established pyrometallurgical and hydrometallurgical procedures for the treatment of mineral concentrates secured by physical separation methods.

The successful operation or otherwise of bacterial leaching processes for the treatment of low-grade ores is heavily conditioned by the engineering considerations which govern their disposal. The absolute magnitude of this material together with the overburden with which it is commonly mixed can be of impressive dimension; for example, it is estimated that at the end of mine life of the Bougainville Copper Ltd. operation at Panguna in Papua New Guinea, the dumps will contain 400 to 500 million tonnes of waste rock (Piercy, 1982). The magnitude of the task of removal demands the limitation of transport distance and the most expeditious handling procedures. The nature of the surrounding terrain frequently limits the choice of sites for dumping and it may be difficult to ensure that drainage from subsequent irrigation can be collected without substantial loss into permeable underlying strata. Of prime importance to the mining engineer are modes of construction and the geometry of the dump which give assurance of its long-term physical stability in face of the ravages of climatic conditions and seismic activity. These legitimate engineering imperatives are frequently in conflict with the requirements for subsequent leaching operations and lead to configurations in dumps which impose severe limitations on the efficiency of biologically-mediated processes.

The common forms of waste rock dumps make use of natural features and involve the filling of valleys, hillside tipping or the construction of free-standing, truncated pyramidal heaps. These geometries have the advantage of compactness and minimum usage of available ground area, but their lower surface area/volume ratios are an impediment to the free access of the air necessary for aerobic breakdown processes. Further, the practice of over-end tipping from heavy wheeled transport vehicles, often at successive levels, may lead to highly compacted and impermeable strata within the heap. Such construction practices can lead to irregular flow patterns and severe permeation problems when the dump is subsequently irrigated with leaching solutions. Many of these problems arise because the possibility of metal recovery from low-grade material by leaching is not seriously considered at an early enough stage in overall planning and development. If subsequent leaching is seen not only as a mode for more complete metal recovery but as some degree of insurance against the high cost of control of environmental pollution arising from autogenous leaching, the necessary compromises between minimum cost dump construction practices and those yielding a dump geometry more suited to subsequent leaching might be achievable.

The foregoing points have been laboured since it is still true that many dump leaching operations appear to have been initiated as after-thoughts in overall planning, and their implementation to have taken little cognisance of the progress made in design and management of leaching heaps since the pioneering studies of Malouf and Prater (1962) over 20 years ago. As a generalization, the rate and extent of metal recovery in a dump leaching operation are determined less by the inherent biochemical abilities of the complex microbial populations present than by the physicochemical constraints which arise from the nature of the raw material and the manner in which its assemblage has been engineered. Recent developments will be considered in a later section.

The large-scale handling of process tailings for the recovery of valuable metal content by biohydrometallurgical processes has not yet been developed to any significant extent, but has considerable potential as more readily treated sources of metals become depleted. The development of suitable procedures presents different problems to those associated with the dump leaching of low-grade ores. In general, tailings are finely divided as a result of prior processing and have a particle size distribution extending from 300 to 400 microns to submicron dimensions. The 'slime' fraction (less than about 5 micron particle size) may contain substantial metal values which have eluded recovery by gravity separation or flotation procedures. The larger particle size fractions frequently consist of complex and intimate mineral associations whose separation is not practicable by physical methods. It should be noted that this latter fraction is similar to the material obtainable by bulk flotation from high-grade, complex ores composed of fine-grained mineral components unamenable to separation by selective flotation procedures. Biohydrometallurgical procedures for the treatment of process tailings could have a direct application to the primary processing of such high-grade material.

The leaching of tailings by percolation of static masses is not practicable by virtue of the impermeability of the consolidated fine-particle material to water and to gases, and agitated reactor systems appear to be the appropriate alternative. Considerable attention has been paid to the design of mineral bioreactor systems at laboratory and pilot-scale levels and the various investigators have drawn largely on the sophisticated technology evolved for the design of reactors for the fermentation industries. The design of reactor systems for mineral materials has much in common with that of fermenters for organic biotransformations, in the requirements for adequate agitation, efficient transfer of gases, and the monitoring and control of process variables such as temperature and pH. Special problems arise from the dense, insoluble and abrasive character of the mineral substrate and the low pH at which some leaching reactions must be carried out.

The main preference to date has been for stirred tank systems; a flexible batch reactor system for mineral leaching investigations has been described by Babij *et al.* (1980a). The development of continuous flow reactor systems, some with recycle, has received some attention (Moss and

Andersen, 1968), and equipment for the continuous reactor leaching of zinc and copper sulfide concentrates (Gormely *et al.*, 1975; McElroy and Bruynesteyn, 1978; Sanmugasunderam, 1981), and of gold-bearing arsenopyrite/pyrite ore (Livesey-Goldblatt *et al.*, 1983) has been described. Other types of reactor systems for the treatment of finely-divided mineral materials have been investigated; the mass transfer characteristics of an airlift fermenter with internal loop for the leaching of chalcopyrite concentrate have been studied by Yukawa (1975) and similar reactors have been designed and evaluated by Ebner (1980) and Kiese *et al.* (1980). Comparison of the efficacy of different mineral bioreactors has been made by Atkins and Pooley (1983). A continuous flow loop reactor for use with thermophilic and bathyphilic organisms at elevated temperatures and pressures is extant (B. Bubela, Baas Becking Geobiological Laboratory, Canberra; personal communication). Procedures for the design and scale-up of reactors for the microbial desulfurization of coal have been developed (Huber *et al.*, 1983) and should have wider application.

The likelihood that the capacity of bioreactors for use in mineral treatment would need to be considerably larger than that of those commonly used in fermentation or chemical industry has stimulated some workers to examine the possibilities of modification of established metallurgical processing equipment. Andersen (1971) has suggested modifications to thickener systems and to pachucas, and has speculated on the use of standard flotation equipment and the Davcra flotation system (Gauci and Cusack, 1968). The use of modified pachucas as air lift reactors has also been advocated by Torma and Bosecker (1982). There would appear to be scope for the adaptation of other reactor systems such as the loop fermenter developed by ICI Ltd., UK for the production of biomass from methanol and the deep shaft aeration systems proposed for organic waste treatment.

12.4 APPLICATIONS OF BIOGEOCHEMICAL PROCESSES IN MINING AND METALLURGY

Over the past 50 years, opencut mining techniques have enabled economies of scale to be achieved in the extraction of metalliferous and other ores and the refinement of flotation and other physical separation technologies has opened the way for the economic processing of lower grade ore resources. In spite of these successes, however, a number of circumstances suggest that other extractive methods and novel mineral processing procedures may be necessary to meet the growing needs for metals. The depletion of high-grade ore deposits in near-surface locations, together with the improvement of exploration methods for the identification of ore bodies at depth, suggest that greater attention will be focused on solution mining methods. Alternative methods are needed for the processing of the very-low-grade ore deposits which are still relatively abundant in a number of countries, but which are uneconomic to treat with existing technology. In spite of the general efficiency of metallurgical processes, physical methods for mineral concentration and separation may well have reached a plateau, and be inherently incapable of achieving higher metal recovery rates or of dealing with the more complex mineral assemblages. Hydrometallurgical methods may have the potential to overcome some of these problems, whether in the solution mining context or in the extension of mineral processing procedures, and seem likely to be more extensively used in the future.

Biohydrometallurgical methods, based on an understanding of biogeochemical processes in the natural environment and extended by precise information on the capabilities of the biological executive agents, the possibilities for their genetic improvement and their responses to optimized physicochemical factors, may well provide one of the new tools necessary to greatly extend the exploitation of metalliferous ore resources and to improve the efficiency of extractive recovery procedures. A further circumstance which may favour the wider usage of biological mining and biometallurgical processes could be the potential of these procedures for the reduction of environmental damage and pollution.

As has been pointed out by Bhappu (1982), it is not inconceivable that eventually ore reserves will consist largely of higher grade deposits at depth, low-grade zones near previously worked deposits, caved ore and stopes filled with submarginal ore, waste dumps, tailings ponds and slag heaps. Solution mining involving both chemical and biologically-mediated processes may well prove to be the most economic method of recovery of metals from these sources (Ralph, 1982b). Information on modern developments in a number of areas of solution mining technology has been published recently (Schlitt, 1982).

12.4.1 Dump, Heap and *In Situ* Leaching

Dump, heap and *in situ* leaching are the most common modes of solution mining and have been applied to the recovery of copper and uranium from low-grade ores on a very extensive scale. It is estimated (Solozhenkin, 1980) that these methods are practised in about 20 countries and account for approximately 20% of recovered copper and a substantial part of uranium production in the USA, the USSR, Canada, Mexico, Spain, Yugoslavia, Bulgaria and other countries. In general terms, the three modes are distinguished by the following features. As mentioned earlier (Section 12.3), dump leaching is practised on submarginal ore material, assembled into large dumps (up to 4×10^9 tonnes of material; C. L. Brierley, 1978) which are characteristically highly hetero-geneous in composition. Dump leaching is relatively slow and inefficient; nevertheless production by this means yields about 200 000 tonnes of copper per year in the USA at the present time, or 15% of total production (Bhappu, 1982). The term heap leaching is frequently used synony-mously with dump leaching, but the description is preferably restricted to systems in which low-grade or mixed oxide–sulfide ores are leached under controlled conditions in relatively small heaps ($<10^5$ tonnes) on impermeable pads or prepared surfaces. Some operators leach to com-pletion, remove the exhausted ore, reheap the pad and repeat the leaching operation on success-ive batches of ore. Similar procedures are used for the chemical leaching of gold ores (Hickson, 1982). In heap leaching, the ore may be crushed to less than 10 cm particle size and is usually more homogeneous in composition than dump material. The surface/volume ratio may be greatly increased for better aeration and distribution of leaching solutions by arranging the ore in rela-tively thin layers or in 'finger' heaps (Robinson, 1972). Leaching in heaps is usually more rapid and more complete than in dumps. In *in situ* leaching, the ore material, either naturally perme-able or suitably fractured to ensure permeability, is irrigated in place with appropriate leaching solutions. Wadsworth (1975, 1977) has described three types of *in situ* leaching situations: near surface deposits above the water table, near surface deposits below the water table and deep deposits below the water table.

Descriptions of dump and heap leaching operations are plentiful in the literature (Woodcock, 1976; Malouf and Prater, 1962; Anderson *et al.*, 1966; Andersen and Allman, 1968; Sheffer and Evans, 1968; Stirkov *et al.*, 1975; Piercy, 1982). Most of the problems associated with dump leaching arise from poor construction and inadequate knowlege of the internal conditions and reactions occurring in the dump. Considerable progress has been made in the understanding of the mechanisms of breakdown of individual sulfide mineral species and of the roles played by the various organisms in the facilitation of such reactions. Secondary reactions can also be significant (Stanczyk and Rampacek, 1963; Wadsworth, 1972; Equations 12 and 13).

$$\begin{array}{cccc} Cu_5FeS_4 \rightarrow & CuFeS_2 & + & Cu_2S \\ \text{(bornite)} & \text{(chalcopyrite)} & & \text{(chalcocite)} \end{array} \tag{12}$$

$$\begin{array}{cccc} CuFeS_2 \rightarrow & CuS & + & FeS \\ & \text{(covellite)} & & \text{(troilite)} \end{array} \tag{13}$$

Galvanic interactions between pyrite and chalcopyrite have been demonstrated (Berry *et al.*, 1978). Other secondary reactions of importance result in the formation of various iron oxide species (Bhappu *et al.*, 1969) and of jarosites (Lazaroff *et al.*, 1982).

$$3\,Fe^{3+} + A^+ + 2\,SO_4^{2-} + 6\,H_2O \rightarrow A{\cdot}Fe_3(SO_4)_2{\cdot}(OH)_6 + 6\,H^+ \tag{14}$$
$$\text{(jarosite)}$$

where A^+ is a hydrogen or other monovalent ion. Duncan and Walden (1972) have noted the buffering effect in the range pH 2.0 to 2.2 of the precipitation and re-solution of ferric iron; buf-fering may also occur with the formation of jarosite–$Fe(OH)_3$ couples (Miller, 1980). In the dump situation, these reactions can maintain a degree of acidity favourable to the proliferation and activity of *Thiobacillus ferro-oxidans* and other acidophilic species. An unfavourable conse-quence of the formation of insoluble iron complexes may be the impeding of the flow of leaching solutions, the development of anaerobic zones and the 'blinding' of reactive mineral surfaces (Madgwick and Ralph, 1981).

When one considers the percentage of the gangue minerals in the composition of dumps, it is surprising that more attention has not been paid to their influence on the reactions involved in the degradation of the ore minerals and subsequently upon the leachates. The gangue minerals play a major role in the determination of the microenvironment of the sulfide mineral surfaces. Physical screening effects may limit access of water, dissolved nutrients, gases and the microbial populations; the rate-limiting factors in oxidative degradations in dumps may in fact lie in the diffusion characteristics of the gangue material (Auck and Wadsworth, 1973) rather than in the chemical characteristics of the ore minerals or the biochemical abilities of the microbial populations. Further limitations may be imposed by the chemical nature of the gangue in respect of its influence on pH and redox potential and the adsorption of essential reactants. Some silicates have marked ion exchange capacities and their lability under acid conditions may lead to saturation concentrations of silicic acid in the leachates (Madgwick and Ralph, 1980) with the possibility of reprecipitation of solubilized metallic ions such as copper as complex silicates. The gangue minerals also play a major role in their contribution to the matrix component of the water potential of the microenvironments, a parameter which is known to affect the rate of microbial leaching (Brock, 1975). The temperature profiles of the dump mass are likely to affect the rates of both chemical and microbial reactions and to have considerable influence upon the nature of the microbial populations (Brierley and Lockwood, 1977; C. L. Brierley, 1978; Brierley *et al.*, 1978; Marsh and Norris, 1983). The thermal properties of the gangue material may be such as to facilitate rises in temperature to a point at which chemical reactions are initiated (Bryner *et al.*, 1967). There have been few systematic studies of the profiles of the physicochemical parameters in the interior of waste and leaching dumps. The long-term studies on the microenvironments within waste rock dumps in which autogenous leaching of pyrite and other sulfides is occurring, have yielded a great deal of information on water movements, temperature profiles and the composition of the gases within the pore spaces (Daniel *et al.*, 1980a, 1980b, 1981, 1982; Harries and Ritchie, 1981, 1982, 1983a, 1983b). Similar information has been obtained from the use of very large-scale, column test facilities which simulate sections through dumps and which also permit correlation of physicochemical parameters with microbial activity (Murr and Brierley, 1978; Murr, 1980; Murr *et al.*, 1982).

The modelling of the dump leaching process has been attempted by a number of workers (Tayler and Whelan, 1942; Harris, 1969; Wadsworth, 1975; Madsen and Wadsworth, 1977). The majority of these models are shrinking core-diffusional models in which the reacting particles are assumed to be spheres enclosing a shrinking core of unreacted material surrounded by the reaction product. The topic has been comprehensively reviewed by Murr (1980), who has noted that 'while the shrinking core-diffusional models .... have demonstrated considerable success on a laboratory scale or in small pilot-scale leaching situations, they have not been tested directly in an actual (industrial scale) leach dump .... Few dumps have been devised on the basis of such models because experimental data on such a large scale has been lacking'. Modelling attempts continue (Liddell and Bautista, 1981), but the overall complexity of the dump leaching process is such that it has so far defied efforts to construct models which accommodate all the observed phenomena. However, the commonly used empirical laboratory test procedures, based principally on the use of percolation columns (2 to 6 m in length), can give extremely useful evaluations of the likely leaching behaviour of ores (Bruynesteyn *et al.*, 1976).

The possibility of improving dump leaching processes by the introduction and establishment of superior strains of leaching organisms, obtained by genetic manipulation techniques, has led to cautious speculation (C. L. Brierley, 1978; Ralph, 1982a) and some experimentation. As indicated earlier, some progress has been made in identifying strains of *Thiobacillus ferro-oxidans* with superior performance in the breakdown of sulfide minerals. In the mixed culture situations which occur in leaching dumps, such improved strains appear to have great difficulty in establishing themselves and are rapidly displaced by wild strains (Groudev *et al.*, 1978; Groudev, 1980a, 1980b, 1980c). It is clear that if genetically-modified leaching organisms are to play a useful role in dump leaching processes, a range of characteristics other than those concerned with the leaching mechanisms *per se* will need to be modified if stable populations are to be established in the complex environment of a leaching dump.

During the past decade a very considerable body of information on the mechanisms of dump leaching processes and upon the factors governing optimum construction and operation has accumulated. The most important aspects, together with some continuing areas of obscurity requiring further investigation, have been succinctly summarized by Murr (1980).

The three principal types of *in situ* leaching situations have been mentioned earlier. The first type, in which naturally permeable or suitable fractured ores above the water table are irrigated with leaching solutions, have close similarities to dump and heap leaching operations and a number of successful ventures of this kind have been described (Imai, 1971; Fletcher, 1971; Ruderhausen, 1974; Ward, 1974; Catanach, 1976; Butler *et al.*, 1982). Such operations greatly extend the ore resources available by enabling recovery of metal from submarginal ore zones and areas inaccessible to conventional mining practices. The second type of situation envisages the leaching of shallow deposits (<300 m from the surface) and which are under the water table. Such deposits may need to be fractured in place and drained before leaching cycles are initiated. Leaching and recovery may be effected by bore hole mining technology and this has been extensively used in the USA and elsewhere for the *in situ* leaching of uranium deposits. The bore hole mining technique involves the drilling of wells for the injection of leaching solution and the recovery of metal-bearing liquors on a designed pattern such as the common five-spot programme (four holes at corner points for injection and one central hole for recovery).

The leaching of deep-seated deposits (>300 m depth) and under the water table (third situation type) is still a developing technology. The leaching of copper sulfide deposits at such depths after fracturing of the ore body with nuclear devices has been proposed (Lewis, 1970). Alternate methods of rubblization by modified block caving techniques have also been proposed, and involve the use of conventional explosives to shatter the ore into a pattern of free space generated by mining appropriate lamellae every 30–40 m of depth (Wadsworth and Pitt, 1980).

Bhappu (1982) has outlined the advantages and disadvantages of *in situ* leaching and has discussed the controlling parameters which apply. The advantages of *in situ* mining include less environmental disturbance, lower capital and operating costs, lower total energy expenditure, extension of ore resources, recovery of metal in forms that may be more attractive than the usual sulfide and oxide forms, and flexibility of interlocking with conventional mining and milling processes. Disadvantages are seen in physical and chemical restrictions which may limit the possible recovery of some metals, the possibility of ground water contamination, the difficulty of prior testing of feasibility and the current lack of information on the effects of the various physical, chemical and microbiological factors in actual large-scale operation.

Considerable current attention is being paid to the fragmentation of ore bodies at depth (Porter, 1973; McKee *et al.*, 1982); hydrofracing is said to show considerable promise (Bhappu, 1982). Environmental monitoring techniques have been described by Canterford *et al.* (1982), and Vandell (1982) has considered the ground water problems that can arise. The modelling of the *in situ* leaching of primary copper ores has been attempted (Gao *et al.*, 1982) as has that of a complex zinc–copper sulfide ore (Derry and Whittemore, 1983). A most interesting study on the feasibility of the *in situ* leaching of metallic ores other than copper and uranium has been reported (Potter *et al.*, 1982).

While the physicochemical and engineering aspects of *in situ* leaching are under rapid development, biological aspects have not escaped attention. The effect of pH and *Eh* on the chemical and biological leaching of uranium ores has been studied and the results demonstrate the superior performance of biological leaching when these factors are appropriately controlled (Bruynesteyn *et al.*, 1980a, 1980b, 1980c); Bruyensteyn and Viszoli, 1982). Of related interest is the demonstration that uranium(IV) may be directly oxidizable by *T. ferro-oxidans* (Soljanto and Tuovinen, 1980; DiSpirito and Tuovinen, 1981, 1982a, 1982b). Little information exists on the ability of leaching bacteria to withstand the elevated hydrostatic pressures and temperatures that would prevail during the solution mining of a deeply-situated ore body. The results of recent investigations (Bosecker *et al.*, 1979; Davidson *et al.*, 1981) suggest a high degree of resistance by iron-oxidizing bacteria to hydrostatic pressure alone and a capability for activity at depths as great as 3000 m. The data suggest that hyperbaric oxygen might prove to be a more important limitation to a bacterially-assisted solution mining operation than elevated hydrostatic pressure. It has been pointed out that temperatures generally increase with depth below the earth's surface and this, coupled with the heat generated by exothermic leaching reactions, might also restrict bacterial activity. Extreme thermophiles such as *Sulfolobus* species can function at temperatures as high as 85 °C, but their resistance to elevated pressure and oxygen tensions is at present unknown. The high temperature/high pressure organisms recently reported (Baross and Deming, 1983) might have a role in this connection. It might also be noted that the bacterial oxidation of sulfides can occur under anaerobic conditions (Goodman *et al.*, 1983); the effects of high temperatures and pressures on these reactions are not known. Current studies on microbiological processes involved in enhanced oil recoveries are likely to be highly relevant to the biological aspects of deep solution mining (Bubela, 1983).

Acidic ferric sulfate solution is a potent oxidizing agent which has found wide application in hydrometallurgy and is of considerable importance in the solution mining of uranium and copper ores. Its capacity for the oxidative degradation of metallic sulfides and other minerals is well known. During the dissolution reactions, it is reduced to ferrous sulfate and the regeneration of the spent reagent is necessary if it is to be recycled. The spent reagent can be reoxidized chemically (*e.g.* by hydrogen peroxide and various peracids). The autooxidation of ferrous ions at the low pH of the reagent is extremely slow in the absence of catalysts (Singer and Stumm, 1970), but rapid in the presence of the bacterium *T. ferro-oxidans*. The kinetics of the reaction with particular strains of the organism have been evaluated by a number of workers (*e.g.* Lacey and Lawson, 1970). During leaching processes, the reoxidation of the reagent can occur at the site of attack on the mineral substrate or at some other point in the leaching circuit. In the latter case, special equipment is generally employed such as stirred reactor systems (Görög *et al.*, 1980), or a fixed film reactor such as the Bacfox system (Livesey-Goldblatt *et al.*, 1977). The kinetics of ferrous iron oxidation using fixed films of iron-oxidizing bacteria on a rotating contactor have been evaluated (Wichlacz and Unz, 1981). Whichever mode is employed, it is important that the bacterial strain has biochemical and physiological characteristics which are attuned to the particular leaching operation. In the case of the use of *T. ferro-oxidans* as the catalyst for reagent regeneration in the biologically-assisted leaching of uranium ores, it is clear that improvements are needed in the organism's capabilities in respect of: rate of ferrous iron oxidation; greater tolerance to low pH (*ca.* pH 1 to 1.6) and to sudden change in pH round the leaching circuit; better tolerance to high concentrations of uranium, thorium and various common heavy metals; ability to function under saline conditions; and thermophilic and barophilic tolerances for deep mine operation. Gene technology is probably well enough advanced for these objectives to be achieved in the near future.

12.4.2 Reactor Leaching

The suitability of reactor leaching for the further recovery of metals from process tailings has been pointed out earlier (Section 12.3). Although such processes have not yet been developed on a commercial scale to any significant extent, the pressures arising from environmental control legislation and the need to improve overall recovery rates from complex ores may well provide the necessary stimulus in the near future. There is considerable current interest in another area to which reactor leaching is applicable, namely, the possibility of the development of viable hydrometallurgical processes based on the biologically-assisted reactor leaching of mineral concentrates. Biological leaching in the form of solution mining has been applied principally to low-grade primary ores and mining wastes which cannot be economically treated by conventional technologies. The raw material in these cases is predominantly of large particle size and the rate of leaching is severely limited by a number of factors, including gaseous and aqueous diffusion rates and the diluting effects of the large preponderance of gangue minerals. The large increases in leaching rates which can be achieved by diminution of the mineral particle size are well established; the available data have been reviewed by Torma (1977). The prior concentration of the valuable mineral components of an ore, by physical separation techniques such as flotation, of necessity involves reductions in particle size and a concomitant reduction in bulk compatible with the handling capacity of feasible reactor systems. The concentration of the valuable mineral components also facilitates the possibility of securing high metal tenors in the leachate with the likely easing of the problems of metal recovery from solution.

The principal factors which have stimulated inquiry into the feasibility of reactor leaching of concentrates have been: increases in smelting, refining and freight costs; the pressures of environmental pollution control legislation; the development of attrition grinding techniques (Gerlach *et al.*, 1973; Beckstead *et al.*, 1976); the technical improvements in continuous-flow, multistage, recycling reactor systems with precise monitoring and control of process variables (see Section 12.3); the possibility of usage as process catalysts, monocultures or simple mixed cultures of genetically-improved microorganisms and organisms with the thermophilic and barophilic tolerances; the likelihood of achieving rates of mineral dissolution approaching those of conventional hydrometallurgical processes at lower levels of energy consumption overall; and improvements in the technologies for the recovery of metals from solution.

The possibilities of reactor techniques for the microbial leaching of sulfide concentrates of a number of common metals have been examined. Considerable attention has been devoted to zinc sulfide concentrates (Torma *et al.*, 1970, 1972; Gormely *et al.*, 1975; Torma and Guay, 1976; Sanmugasunderam, 1981) and to chalcopyrite concentrates (Bruynesteyn and Duncan, 1971; Torma *et al.*, 1979; McElroy and Bruynesteyn, 1978; Torma and Rozgoni, 1980; Bruynesteyn *et al.*, 1983). Pyrite/arsenopyrite concentrates have received some attention (Pinches, 1975; Babij and Ralph, 1979; Livesey-Goldblatt *et al.*, 1983) as have nickel iron sulfides (Corrans, 1974) and a lead sulfide concentrate (Torma and Subramanian, 1974). The reduction of particle size to submicron dimensions by attrition grinding adds a further dimension to reactor leaching; a general scheme for the leaching of chalcopyrite concentrate using attrition-ground material, reactor leaching with an unimproved strain of *T. ferro-oxidans* as the biological catalyst, and recovery of copper by solvent extraction/electrowinning has been proposed by Torma and Rozgonyi (1980). The rate of extraction achieved by pilot runs was $1.65 \text{ g Cu } l^{-1} h^{-1}$.

Economic feasibility studies of the reactor leaching of chalcopyrite concentrates have been made (B. C. Research, 1978; McElroy and Bruynesteyn, 1978). The following advantages for the process have been claimed: negligible atmospheric emissions; production of refined copper (*i.e.* potentially lower transport cost and increased market flexibility); feasibility for small-scale mine-site operations; reduced costs for concentrates containing appreciable bornite, chalcocite, *etc.*, regardless of grade or pyrite content; potentially increased returns for silver, gold and (possibly) molybdenite in concentrates; and production of dilute sulfuric acid suitable for leaching of oxide ore and/or mill tailing and/or acid-consuming waste dumps. The authors suggest that where any of these factors are significant, biological reactor leaching should be considered as an option.

Reactor leaching is likely to find major application in the development of integrated flotation/ biological leaching procedures for the treatment of complex sulfide ores. Conventional treatment of mixed sulfide ores employs selective flotation processes which can yield clean, single metal concentrates with recoveries in excess of 90%. In some ores, however, generally composed of copper, lead and zinc sulfides, together with iron pyrites and minor sulfides such as those of arsenic and cadmium, intergrowth of the sulfides occurs and the quantitative recovery of the individual minerals becomes extremely difficult by physical separation processes. The fine mineralization requires reduction to small particle size if physical separation processes are to be successful and differences in grindability of individual sulfides may add to the technical difficulties, as may the association of some gangue minerals such as talcs. The conventional treatment of such complex sulfide ores generally yields 'dirty' concentrates (*i.e.* containing more than one metallic sulfide mineral) and recoveries may be as low as 70%. Alternate treatment approaches have advocated the biological leaching of composite concentrates obtained by bulk flotation after relatively coarse grinding, or of 'dirty' concentrates obtained by conventional selective flotation (Torma and Subramanian, 1974). Torma (1978) has reviewed the earlier attemps to remove copper selectively from a lead blast furnace matte and arsenic and copper from oxidized tin concentrates by microbiological means, and has described and assessed a bacterial leaching process for the removal of zinc, copper and cadmium from an off-grade lead sulfide concentrate. The leach residue from the process is an upgraded lead sulfide concentrate which is partially oxidized to lead sulfate and which can be used directly in the classical smelting process for the recovery of metallic lead (equation 15).

$$PbS + PbSO_4 \rightarrow 2 Pb^0 + 2 SO_2 \qquad (15)$$

A general process for the beneficiation of complex ores of lead, zinc and copper has been described by Carta *et al.* (1980a). The process consists of a bulk flotation step, preceded by a relatively coarse grind, with the production of a mixed sulfide concentrate at a recovery of more than 80%, followed by a regrinding step. The reground product is microbially leached in a reactor system with adapted strains of *T. ferro-oxidans*, and yields a virtually zinc-free lead concentrate in which the lead content is three-fold upgraded. Only minor removal of copper occurs in the example cited. While this sort of process has the potential for recovery of all metals present in the bulk concentrate, the results indicate that the relative abundances of different sulfides have an influence on the rates of leaching and on the metal content of the leach liquor. Further investigations of interactions of this kind between sphalerite and chalcopyrite (Carta *et al.*, 1980b) and chalcopyrite and pyrrhotite (Rossi *et al.*, 1983) have been reported.

Another area in which reactor leaching is likely to be significant is in the microbial desulfurization of coal. The topic has been reviewed by Dugan and Apel (1978) and Kargi (1982a), who have described a process for the removal of pyritic sulfur from pulverized coal slurries by mixed enrichment cultures of acidophilic microorganisms. These were more effective than mixtures of pure cultures of *T. ferro-oxidans* and *T. thio-oxidans*. Approximately 97% of the pyritic sulfur is removed within five days from coal which had an initial pyritic sulfur content of 3.1%. Kinetic data on the desulfurization of various coals have been determined (Olsen *et al.*, 1980). The bacterial inhibition observed during the desulfurization of some coals (Rossi and Salis, 1977) and the role of *T. thio-oxidans* in coal pyrite oxidation have been further examined by Groudev and Genchev (1979). The superior performance of thermophilic organisms for coal desulfurization has been reported (Murr and Mehta, 1982) and enhancement of pyritic sulfur removal has been achieved with concentrated cell suspensions of *T. ferro-oxidans* and an external carbon dioxide supply (Kargi, 1982b). The composition of the mineral salts medium (Kos *et al.*, 1983) and the removal of organic sulfur by mixed cultures (Mishra *et al.*, 1983) have been investigated. Kempton *et al.* (1980) have proposed a novel process for the removal of pyrite from finely-divided coal (up to 90% removal) in which conditioning with cells of *T. ferro-oxidans* is followed by oil agglomeration. This process has recently been evaluated (Doddema, 1983) and its applicability to different types of coal and different coal particle sizes investigated. The modelling, scale-up and design of a bioreactor has been reported (Huber *et al.*, 1983) and a comparison made of dump leaching and process techniques for coal desulfurization (Pooley and Atkins, 1983).

12.4.3 Beneficiation of Ores and Process Residues

Most of the processes which have been proposed or developed for the beneficiation of ores and process residues, including some of those referred to in the previous section, involve what has been loosely termed 'reverse' leaching, that is the leaching process is directed to the solubilization or modification of the components of the ore, concentrate or process residue, which are of low value and whose removal upgrades the value of the residue. The removal may be immediate, by solubilization, or the leach treatment may so modify the low value component that its separation by conventional techniques is facilitated. This concept has considerable utility for the disruption of intimate mineral associations recalcitrant to physical separation methods and commonly requiring the use of pyrometallurgical or drastic chemical procedures for their treatment. Such applications normally involve the microbial catalysis of oxidative degradation processes, but some promote the dissolution of silicates, and others are concerned with the microbial sulfidization of oxide minerals for improved flotation separation of their components. Very few of the processes of this type which have been proposed appear to have been implemented on a commercial scale, but the potential of the approach would seem to be considerable. The following examples will illustrate these types of biohydrometallurgical process.

12.4.3.1 Recovery of gold and silver

Very intimate associations of gold and silver with iron sulfide minerals are common as are solid solutions of these metals in other sulfides. Such complexes resist the complete removal of the precious metal by cyanidation or treatment with other solvents. The association can be broken by roasting but this procedure is sometimes technically difficult and increasingly the pyrometallurgical procedure generates intractable environmental problems. Some mineral assemblages of this kind can be disrupted by microbial oxidation of sulfides with acclimatized strains of *T. ferro-oxidans* (B. C. Research, 1969; Pinches, 1975; Babij and Ralph, 1979; Solozhenkin and Lyubavina, 1980; Groudev, 1981). The bacterial pretreatment of an auriferous pyrites ore is claimed to increase the recovery of gold from approximately 40% to 85% (Bruynesteyn, 1983). A full process for gold recovery from arsenopyrite/pyrite ore by bacterial leaching and cyanidation has been described and its economics assessed (Livesey-Goldblatt *et al.*, 1983).

12.4.3.2 Beneficiation of sulfidic tailings from tin processing

Some processes for the recovery of tin from massive ores yield process tailings which are predominantly iron sulfide concentrates (pyrite or pyrrhotite) containing a small but significant and valuable percentage of cassiterite. The upgrading of such material to a tin grade of 8–10% at an acceptable recovery level would provide a feedstock for fuming procedures and allow the devising of a more fully integrated recovery/smelting process. At laboratory level, it has been demonstrated that bacterial leaching at low pH (2.5) with iron-oxidizing bacteria will yield a residue with a tin grade in excess of 12% at an overall recovery level of 46% (Sawe, 1980). Subsequent experimentation at higher pH levels with a mixed enrichment culture has yielded a treated residue with a tin grade of 10% at a recovery of 65% for the +6 micron fraction of the waste sulfide concentrate (Harris *et al.*, 1983).

12.4.3.3 Purification of ferruginous sands

Sand for glass-making or foundry use is commonly beneficiated by removal of iron minerals by strong acid leaching. If, as is frequently the case, the iron mineral is pyrite or marcasite, substantially complete removal of iron can be effected by leaching at room temperature at pH 2.5 with *T. ferro-oxidans* as the catalyst (Babij and Ralph, 1976). Oxidized iron minerals may require leaching with heterotrophic acid-producing bacteria and fungi; a two-stage process of this kind by which the iron content of sand can be lowered to below 0.01% has been described (Groudev *et al.*, 1983).

12.4.3.4 Beneficiation of bauxite

The removal of both iron and silica from bauxite has been attempted in the beneficiation of low-grade bauxite. The former component can be removed by processes similar to that described above for the purification of sand (Groudev *et al.*, 1983; J. C. Nixon, Comalco Ltd., personal communication). A considerable number of microorganisms are known which can solubilize silicates and silica; some of these have been applied to the leaching of lateritic ores and to the beneficiation of low-grade bauxite (Groudev and Genchev, 1978b; Silverman, 1979; Torma and Bosecker, 1982). A good deal of attention has been devoted to the leaching of aluminosilicate minerals (Rossi, 1978; Groudev and Genchev, 1979; Heydeman *et al.*, 1981), but the mechanisms of aluminum concentration are still not clear. It has been suggested that aluminum concentration can be achieved by one or more of three routes: the selective extraction of silica which results in solid residues enriched in aluminum oxide, the selective 'solubilization' of aluminum itself, or the selective removal of iron from aluminosilicates. It is not clear whether aluminum is actually solubilized or exists as an oxide during the leaching processes. It should be noted that all the organisms implicated in these processes are heterotrophic bacteria or fungi and as such require reduced carbon compounds as an energy source. The need for energy sources of this kind has been a major impediment to the development of such leaching processes but palaeontological evidence related to the genesis of bauxite deposits suggests that, by the use of mixed culture systems, they could be driven by solar energy (Ralph, 1977). In some locations, agricultural and industrial wastes have been used as energy sources for such leaching processes (Kiel, 1977).

12.4.3.5 Applications of sulfate-reducing bacteria

It has been demonstrated by Lyalikova *et al.* (1977), Solozhenkin *et al.* (1979) and Solozhenkin and Lyubavina (1980) that improved extraction of oxide ores of antimony and bismuth can be achieved by prior treatment with sulfate-reducing bacteria. Sulfidization occurs and significant improvement in the flotation of mixed concentrates is claimed. There appear to be some effects on the flotation process *per se* in addition to sulfidization (Kupeyeva *et al.*, 1977). An effect on the flotation properties of sulfide minerals has been reported by Kim *et al.* (1981). The possibility of recovery of metals from leaching solutions and acid mine drainage by the use of sulfate-reducing bacteria has been demonstrated (Tomizuka and Yagisawa, 1978).

12.4.4 Environmental Pollution Control

The sheer scale of mining operations and metallurgical processing leads to considerable disruption of natural landforms and the visual impact of opencuts, spoil heaps, waste dumps and tailings ponds is aesthetically disturbing. Further, the exposure of sulfidic and other mineral material to air and water usually leads to some form of autogenous leaching and the concomitant contamination of groundwater systems with acid and heavy metal ions at concentration levels damaging or toxic to the indigenous flora and fauna. Until comparatively recently, the decommissioning of minesites has not been regarded as more than an incidental part of the total operation and such measures as have been undertaken have been in the main empirically based. However, in most developed countries at the present time, increasingly strict legislation governs the termination of mining operations and indeed their ongoing practices. Such legislation imposes close control on measures to ensure the ultimate restoration of the landscape to an acceptable standard, the establishment of vegetative cover and the minimization of environmental pollution.

The increasing understanding of the basic mechanisms involved in the breakdown of minerals and the roles played by microbial populations in facilitating such degradations has provided a more rational geochemical basis for management strategies aimed at minimizing the long-term impact of environmental contaminants (Pidgeon, 1985). A problem of long standing and one to which complete solutions are not yet available is that of acid mine drainage (Lundgren *et al.*, 1972; Lundgren, 1975; Dugan, 1975), arising principally from the microbially-catalysed, oxidative degradation of pyrites and other sulfide minerals. The primary degradation reactions can be inhibited by a number of factors, the manipulation of some of which are employed as control measures. Limitation of oxygen access to mineral sites can slow down the degradation reactions but, as noted earlier, some oxidation of sulfides can occur in the absence of oxygen if alternate electron acceptors such as ferric ions are available. The oxidation of pyrite is also inhibited by ions such as chromate and slow release systems such as calcium chromate in poly(vinyl acetate) have been suggested as control agents (Hartford, 1970). Biocidal or bacteriostatic agents have been frequently suggested for the elimination of the catalytic role of the microbial populations but have not found acceptance due in the main to their cost, their likely rapid biodegradation in field situations, and the certainty that microbial tolerances would be rapidly established. The breaking of ecological successions by ferrous iron inhibition of iron-oxidizing bacteria has also been proposed as a control measure (Walsh and Mitchell, 1975).

Of paramount importance in mineral degradation and the development of catalysing microbial populations is the availability of water; even when liquid water is not present, the oxidation of most sulfide minerals will occur if the relative humidity of the microenvironment exceeds about 50% (Ralph, 1980). The overwhelming importance of oxygen and water as the basic determinants of sulfide oxidation in geologically disturbed situations has led to the considerable emphasis in control strategies on the management of water movement and where practicable, on sealing techniques to limit ingress of these two reagents. Successful employment of these two approaches in a particular situation has been described by Craze (1980) and aspects of the revegetation of mine dumps and other parts of minesites have been discussed by Andersen (1980). The total control of ingress of oxygen and water is in some situations almost impossible to achieve, and in such circumstances there is no alternative to the treatment of drainage waters prior to discharge into natural waterways. In the case of the drainage waters from high sulfur waste dumps and abandoned underground workings, the principal pollutional constituents are ferrous ions and sulfuric acid and if this effluent is untreated, the subsequent oxidation of the iron leads to depletion of oxygen in receiving streams and the precipitation of the iron as gelatinous, hydrated ferric oxides. Common modes of treatment involve chemical neutralization of the acid, followed by oxidation and precipitation of the iron (Lovell, 1973). More recent procedures have favoured oxidation of the iron prior to neutralization and precipitation (Ishikawa *et al.*, 1983) and good results have been obtained with rotating biological contactors (Olem and Unz, 1980; Wichlacz and Unz, 1981).

In operating mines, waste process waters frequently contain undesirable levels of metallic contaminants. Brierley and Brierley (1980) have described biological ponding methods for the removal of inorganic pollutants from uranium mine waste water. The use of live or dead biomass as adsorbents for metals appears to have considerable promise for the treatment of metal-containing waste waters (Shumate, II *et al.*, 1980; Strandberg *et al.*, 1981; Khummongkol *et al.*, 1981; McEntee *et al.*, 1983). In some circumstances, the legislating authorities may require a mine operator to totally contain all process water; in the recycle situation which results, water treatment processes are required which can cope with the biodegradation of flotation reagents in the presence of substantial levels of heavy metal ions. Such processes have been described (Carta *et al.*, 1980c).

Sulfide tailings dumps and ponds often present special problems in respect of environmental pollution. On occasion, such material can be returned as back fill to exhausted underground workings and hopefully can be isolated from air and water access. Another common practice has been their return to abandoned opencuts where 2 m of water cover has been believed to be sufficient to ensure anaerobic conditions and a minimization of subsequent sulfide mineral degradation. Recent investigations have shown this practice to be of dubious utility (Babij *et al.*, 1980b; Goodman *et al.*, 1981b). The chemical processes in abandoned tailings dumps have been described in some detail (Boorman and Watson, 1976) and the use of physical properties such as grain size stratification and the shaping of tailings piles has been advocated by Pidgeon (1982) as a means of achieving long-term chemical stability. The thickened discharge method for tailings disposal (Robinsky, 1979) illustrates some of these points. The particular problems of the mobility of residual radium in uranium ore processing tailings have been examined by Silver and Andersen (1979) and its removal by salt washing described.

In spite of considerable progress in techniques for the rehabilitation of mined areas, the control of pollution-generating mechanisms, and the treatment of unavoidable, polluted drainage outflows, the need remains for continuing comprehensive studies of selected situations. In many ways the environmental problems arising from the mining and processing of sulfide-associated uranium ores are typical of the general problems found in other areas of the mining industry, and the continuing studies in a number of countries (Davy, 1975) have yielded extremely significant information not only on the specific objectives of pollution control but on matters relevant to the further development of biohydrometallurgical processes.

12.5 CONCLUDING REMARKS

It can be deduced from this brief survey of the current level of understanding of biogeochemically-significant processes affecting minerals of industrial importance, and the technologies which stem from them, that further developments of considerable significance are imminent. The new directions in basic research and in application are becoming clearer and are summarized in the following comments.

Information on the characteristics and roles of individual components of the microbial associations which develop in biogeochemical situations is still comparatively rudimentary. Considerable progress has been made with the identification and characterization of lesser-known species such as *Leptospirillum*, *Metallogenium*, the less acid-tolerant *Thiobacilli*, the acid- and metal-tolerant heterotrophs, and those heterotrophs effective in the breakdown of silicates. It is evident, however, that even isolation and identification of the full range of species involved are incomplete let alone the elucidation of particular roles and the interactions in mixed populations. With those species which have been characterized, a good deal more information is needed on the limits of their abilities and the extent of their capabilities under optimized physicochemical conditions.

Further impetus is needed in the study of thermophilic and barophilic organisms and on the behaviour of mesophilic organisms under conditions of elevated temperature and pressure. It is evident that this area of research is currently inhibited by the slow evolution of techniques for the study of organisms under conditions of elevated temperature and pressure. Fundamental studies on the maintenance of metabolic function under such conditions have scarcely begun.

A good deal of filling out remains to be done on the fine details of mineral degradation mechanisms and in particular the relationships between the fine structure of mineral surfaces and their amenability or otherwise to biodegradation. A formidable arsenal of physical, chemical, electrochemical and mineralogical techniques is currently available for such investigations.

The mapping of the genomes of organisms important in biohydrometallurgical processes has begun and accelerating progress in the application of genetic manipulation technology to such organisms can be anticipated, with extremely significant implications for the improvement of existing processes and the design of novel procedures.

The on-going studies on water and gas movements and the physicochemical profiles in rock masses, and the correlation of this information with the activities of the resident microbial communities are likely to affect profoundly the engineering design of waste rock dumps and leaching heaps. Similar information will be essential for the proper design and management of *in situ* leaching operations.

The full exploitation of solution mining possibilities will lean heavily upon the availability of more extensive microbiological information, the details of mineral–microorganism interactions, the devising of effective and economical methods for the securing of permeability, and the understanding of water and gas flows.

The modelling of processes in dumps, heaps and in solution mining situations will undoubtedly continue to attract considerable attention and will facilitate the design, operation and control of more highly optimized configurations.

A bottle-neck in the development of novel biohydrometallurgical processes could develop unless substantial progress is made in the design of large-scale reactor systems, suitable for the handling of mineral materials and permitting the optimization of the rates of biologically-catalysed processes. Designs will need to accommodate the possibility of usage of novel organism systems with genetically-enhanced capabilities and tolerances for operation at elevated temperatures and pressures.

The selective bioleaching of complex mineral ores and concentrates demands more extensive investigation, particularly in respect of the influence of variations in the relative abundances of the constituent minerals. Similar attention is required to elucidate the intricacies of the mechanisms involved in the biodegradation of aluminosilicates and other metallic silicates.

The attractive possibilities of the complementary interlocking of biologically-assisted procedures and conventional hydrometallurgical and pyrometallurgical processes will need the most serious attention, and the development of integrated processes may provide significant relief to existing problems of low recoveries and subgrade products in the treatment of complex and finely-mineralized ores. The more extensive usage of solution mining procedures and of biohydrometallurgical processes could make a large contribution to the reduction of environmental damage from mining and metallurgical operations.

Finally, the need is urgent for the continuation and extension of comprehensive investigations of all aspects of the environmmental damage caused by past and existing mining and metallurgical operations, and continued monitoring of the various control measures which have been implemented.

12.6 ADDENDUM

The foregoing topics have been further discussed by Ralph (1985), with particular reference to recent progress in microbially-enhanced oil recovery and the unravelling of the genetic mechanisms of the biogeochemically-important microorganisms.

12.7 REFERENCES

Agate, A. D., M. S. Korczynski and D. G. Lundgren (1969). Extracellular complex from the culture filtrate of *Ferrobacillus ferro-oxidans*. *Can. J. Microbiol.*, **15**, 259–264.
Aleem, M. I. H. (1975). Biochemical reaction mechanisms in sulfur oxidation by chemosynthetic bacteria. *Plant Soil*, **43**, 587–607.
Andersen, J. E. (1971). Studies on bacterial mineral leaching. Ph.D. Thesis. University of New South Wales.
Andersen, J. E. (1980). Mine dump vegetation: practice and problems. In *Biogeochemistry of Ancient and Modern Environments*, ed. P. A. Trudinger, M. R. Walter and B. J. Ralph, pp. 713–718. Australian Academy of Science, Canberra.
Andersen, J. E. and M. B. Allman (1968). Some operational aspects of heap leaching at Rum Jungle. *Proc. Australas. Inst. Min. Metall.*, **225**, 27–31.
Andersen, J. E., G. L. Herwig and R. B. Moffitt (1966). Heap leaching at Rum Jungle. *Aust. Min.*, **58**, 35–40.
Arkesteyn, G. J. M. W. (1980). Contribution of microorganisms to the oxidation of pyrite. Doctoral thesis, University of Wageningen.
Arrietta, L. and R. Grez (1971). Solubilisation of iron-containing minerals by soil micro-organisms. *Appl. Microbiol.*, **22**, 487–490.
Atkins, A. S. (1978). Studies on the oxidation of sulphide minerals (pyrite) in the presence of bacteria. In *Metallurgical Applications of Bacterial Leaching and Related Microbiological Phenomena*, ed. L. E. Murr, A. E. Torma and J. A. Brierley, pp. 403–426. Academic, New York.
Atkins, A. S. and F. D. Pooley (1983). Comparison of bacterial reactors employed in the oxidation of sulphide concentrates. In *Recent Progress in Biohydrometallurgy*, ed. G. Rossi and A. E. Torma, pp. 111–125. Associazione Mineraria Sarda-09016 Iglesias-Italy.
Auck, Y. T. and M. E. Wadsworth (1973). Physical and chemical factors in copper dump leaching. In *International Symposium on Hydrometallurgy, Chicago*, ed. D. J. Evans and R. S. Shoemaker, pp. 645–700. American Institute of Mining Engineering, New York.
Avery, D. (1974). *Not on Queen Victoria's Birthday — the Story of the Rio Tinto Mines*. Collins, London.
Babij, T. and B. J. Ralph (1976). Unpublished information.
Babij, T. and B. J. Ralph (1979). Assessment of metal recovery and metal pollution potentiality of sulphidic mine wastes. *Proc. GIAM V 1979*, 476–487.
Babij, T., R. B. Doble and B. J. Ralph (1980a). A reactor system for mineral leaching investigations. In *Biogeochemistry of Ancient and Modern Environments*, ed. P. A. Trudinger, M. R. Walter and B. J. Ralph, pp. 563–572. Australian Academy of Science, Canberra.
Babij, T., A. Goodman, A. M. Khalid and B. J. Ralph (1980b). Environmental studies of flooded opencuts. In *Biogeochemistry of Ancient and Modern Environments*, ed. P. A. Trudinger, M. R. Walter and B. J. Ralph, pp. 637–649. Australian Academy of Science, Canberra.
Babij, T., A. Goodman, A. M. Khalid, and B. J. Ralph (1981). Microbial ecology of Rum Jungle III. Leaching behaviour of sulphidic waste material under controlled conditions. Australian Atomic Energy Commission, AAEC/E520.
Balashova, V. V., I. Ya. Vedenina, G. E. Markosyan and G. A. Zavarzin (1974). The autotrophic growth of *Leptospirillum ferro-oxidans*. *Mikrobiologiya*, **43**, 581–585 (Engl. transl.) pp. 491–494.
Baross, J. A. and J. W. Deming (1983). Growth of 'black smoker' bacteria at temperatures of at least 250 °C. *Nature (London)*, **303**, 423–426.
Barridge, J. K. and J. M. Shively (1968). Phospholipids of the Thiobacilli. *J. Bacteriol.*, **95**, 2182–2185.
B. C. Research, Vancouver, B. C., Canada (1969). Leaching of an arseno-pyrite-gold concentrate. In *Microbiological Leaching of Sulfide Ores*, Progress Report No. 28.
B. C. Research, Vancouver, B. C., Canada (1978). Concentrate leaching: A new look at a pollution-free method to treat sulfide concentrate. April, 1978.
Beck, J. V. (1977). Chalcocite oxidation by concentrated cell suspensions of *Thiobacillus ferro-oxidans*. In *GBF Monograph Series*, No. 4 (August, 1977), *Conference Bacterial Leaching 1977*, ed. W. Schwartz, pp. 119–128. Verlag Chemie, Weinheim.

Beckstead, L. W., P. B. Munoz, J. L. Sepulvada, J. A. Herbst, J. D. Miller, F. A. Olsen and M. E. Wadsworth (1976). Acid ferric sulfate leaching of attrition-ground chalcopyrite concentrates. In *Extractive Metallurgy and Electrowinning*, ed. J. C. Yannapoulos and J. C. Agarval, pp. 611–632. Port City Press, Baltimore, MD.

Beebe, J. L. and W. Umbreit (1971). Extracellular lipid of *Thiobacillus thiooxidans. J. Bacteriol.*, **108**, 612–614.

Belly, R. T. and T. D. Brock (1974). Ecology of iron-oxidizing bacteria in pyritic materials associated with coal. *J. Bacteriol.*, **117**, 726–732.

Bennett, J. C. and H. Tributsch (1978). Bacterial leaching patterns on pyrite crystal surfaces. *J. Bacteriol.*, **134**, 310–317.

Berry, V. K. and L. E. Murr (1977). An SEM study of bacterial catalysis and its effect on surface reactions at sulfide phases in the leaching of low-grade ore. In *Scanning Electron Microscopy/1977*, ed. O. Johari, vol. 1, pp. 137–146. IIT Research Institute, Chicago.

Berry, V. K. and L. E. Murr (1978). Direct observations of bacteria and quantitative studies of their catalytic role in the leaching of low-grade, copper-bearing waste. In *Metallurgical Applications of Bacterial Leaching and Related Microbiological Phenomena*, ed. L. E. Murr, A. E. Torma and J. A. Brierley, pp. 103–136. Academic, New York.

Berry, V. K., L. E. Murr and J. B. Hiskey (1978). Galvanic interaction between chalcopyrite and pyrites during bacterial leaching of low-grade waste. *Hydrometallurgy*, **3**, 309–326.

Berthelin, J., G. Belgy and R. Magne (1977). Some aspects of the mechanisms of solubilization and insolubilization of uranium from granites by heterotrophic microorganisms. In *GBF Monograph Series*, No. 4 (August, 1977), *Conference Bacterial Leaching 1977*, ed. W. Schwartz, pp. 251–260. Verlag Chemie, Weinheim.

Bhappu, R. B. (1982). Past, present and future of solution mining. In *Interfacing Technologies in Solution Mining, Proceedings of the 2nd SME–SPE International Solution Mining Symposium, Denver, Colorado*, November 18–20, 1981, ed. W. J. Schlitt, pp. 1–9. Lucas Guinn Co., Hoboken, N.J.

Bhappu, R. B., P. H. Johnson, J. A. Brierley and D. H. Reynolds (1969). Theoretical and practical studies on dump leaching. *Trans. Soc. Min. Eng. AIME*, **244**, 307–320.

Bhurat, M. C., K. K. Dwivedy, K. M. V. Jayaram and K. K. Dar (1973). Some results of microbial leaching of uranium ore samples from Narwapahar, Bhatin and Keruadungri, Singhbhum District, Bihar. *Natl. Metall. Lab. Tech. J.*, **15**, 47–51.

Bohlool, B. B. (1975). Occurrence of *Sulfolobus acidocaldarius*, an extremely thermophilic bacterium in New Zealand hot springs. *Arch. Mikrobiol.*, **106**, 171–174.

Boorman, R. S. and D. M. Watson (1976). Chemical processes in abandoned sulphide tailings dumps and environmental implication for northeastern New Brunswick. *CIM Bull.*, August, 86–96.

Bosecker, K. (1977). Studies on the bacterial leaching of nickel ores. In *GBF Monograph Series*, No. 4 (August, 1977), *Conference Bacterial Leaching 1977*, ed. W. Schwartz, pp. 139–144. Verlag Chemie, Weinheim.

Bosecker, K. (1983). Microbial recycling of industrial waste products. In *Recent Progress in Biohydrometallurgy*, ed. G. Rossi and A. E. Torma, pp. 331–345. Associazione Mineraria Sarda-09016 Iglesias-Italy.

Bosecker, K. and G. Wirth (1980). Bacterial leaching of a carbonate-bearing uranium ore. In *Biogeochemistry of Ancient and Modern Environments*, ed. P. A. Trudinger, M. R. Walter and B. J. Ralph, pp. 577–582. Australian Academy of Science, Canberra.

Bosecker, K., D. Neuschütz and U. Scheffler (1978). Microbiological leaching of carbonate-rich German copper shale. In *Metallurgical Applications of Bacterial Leaching and Related Microbiological Phenomena*, ed. L. E. Murr, A. E. Torma and J. A. Brierley, pp. 389–401. Academic, New York.

Bosecker, K., A. E. Torma and J. A. Brierley (1979). Microbiological leaching of a chalcopyrite concentrate and the influence of hydrostatic pressure on the activity of *Thiobacillus ferro-oxidans. Eur. J. Appl. Microbiol. Biotechnol.*, **7**, 85–90.

Brantner, H. (1970). Investigations on biological oxidation of iron and manganese. *Zentralbl. Bakteriol. Parasitenkd., Infektionskr. Hyg., Abt. II*, **124**, 412–426 (from *Microbiol. Abstr.*, 1970, No. B7755).

Brierley, C. L. (1978). Bacterial leaching. *CRC Crit. Rev. Microbiol.*, 207–262.

Brierley, C. L. (1982). Microbiological mining. *Sci. Am.*, **247**, 42–49.

Brierley, C. L. and J. A. Brierley (1973). A chemoautotrophic and thermophilic microorganism isolated from an acid hot spring. *Can. J. Microbiol.*, **19**, 183–188.

Brierley, C. L. and J. A. Brierley (1980). Biological methods to remove selected inorganic pollutants from uranium mine waste water. In *Biogeochemistry of Ancient and Modern Environments*, ed. P. A. Trudinger, M. R. Walter and B. J. Ralph, pp. 661–667. Australian Academy of Science, Canberra.

Brierley, C. L. and L. E. Murr (1973). Leaching: Use of a thermophilic and chemoautotrophic microbe. *Science*, **179**, 488–490.

Brierley, C. L., J. A. Brierley and L. E. Murr (1973). Using the SEM in mining research. *Res. Dev.*, **24**, 24–28.

Brierley, C. L., J. A. Brierley, P. R. Norris and D. P. Kelly (1980). Metal-tolerant micro-organisms of hot, acid environments. In *Microbial Growth and Survival in Extremes of Environment*, Society of Applied Bacteriology Technical Series No. 15, ed. G. W. Gould and J. E. L. Corry, pp. 39–51. Academic, London.

Brierley, J. A. (1978). Thermophilic iron-oxidizing bacteria found in copper leaching dumps. *Appl. Environ. Microbiol.*, **36**, 523–525.

Brierley, J. A. (1980). Facultative thermophilic *Thiobacillus*-like bacteria in metal leaching. In *Biogeochemistry of Ancient and Modern Environments*, ed. P. A. Trudinger, M. R. Walter and B. J. Ralph, pp. 445–450. Australian Academy of Science, Canberra.

Brierley, J. A. and C. L. Brierley (1978). Microbial leaching of copper at ambient and elevated temperatures. In *Metallurgical Applications of Bacterial Leaching and Related Microbiological Phenomena*, ed. L. E. Murr, A. E. Torma and J. A. Brierley, pp. 477–490. Academic, New York.

Brierley, J. A. and S. J. Lockwood (1977). The occurrence of thermophilic iron-oxidizing bacteria in a copper leaching system. *FEMS Microbiol. Lett.*, **2**, 163–165.

Brierley, J. A., P. R. Norris, D. P. Kelly and N W. Le Roux (1978). Characteristics of a moderately thermophilic and acidophilic iron-oxidising *Thiobacillus. Eur. J. Appl. Microbiol. Biotechnol.*, **5**, 291–299.

Brock, T. D. (1969). Microbial growth under extreme conditions. *Symposia of the Society for General Microbiology, No. 19. Microbial Growth*, pp. 15–41.

Brock, T. D. (1975). Effect of water potential on growth and iron oxidation by *Thiobacillus ferro-oxidans*. *Appl. Microbiol.*, **29**, 495–501.

Brock, T. D. (1977). Ferric iron reduction by sulphur- and iron-oxidizing bacteria. In *GBF Monograph Series*, No. 4 (August, 1977), *Conference Bacterial Leaching 1977*, ed. W. Schwartz, p. 47. Verlag Chemie, Weinheim.

Brock, T. D. and J. Gustafson (1976). Ferric iron reduction by sulphur- and iron-oxidizing bacteria. *Appl. Environ. Microbiol.*, **32**, 567–571.

Brock, T. D., K. M. Brock, R. T. Belly and L. R. Weiss (1972). *Sulfolobus* : A new genus of sulphur-oxidising living at low pH and high temperature. *Arch. Mikrobiol.*, **84**, 54–68.

Bruynesteyn, A. (1983). Effect of a bacterial pre-treatment on the recovery of gold from a pyritic ore. B. C. Research Council, Vancouver, B. C., Canada. Unpublished information.

Bruynesteyn, A. and D. W. Duncan (1971). Microbiological leaching of sulphide concentrates. *Can. Metall. Quart.*, **10**, 57–63.

Bruynesteyn, A. and A. Viszolyi (1982). The effect of pH and Eh on the chemical and biological leaching of a pyritic uranium ore. In *Interfacing Technologies in Solution Mining, Proceedings of the 2nd SME–SPE International Solution Mining Symposium, Denver, Colorado*, November 18–20, 1981, ed. W. J. Schlitt, pp. 159–174. Lucas Guinn Co., Hoboken, N.J.

Bruynesteyn, A., D. W. Duncan and J. K. Ballard (1976). An evaluation of the leaching characteristics of Butte leaching ore. *Hydrometallurgy*, **2** (3) (1976/1977), 235–248.

Bruynesteyn, A., A. Vizsolyi and R. Vos (1980a). Effect of low pH on the rate of ferrous iron oxidation by *Thiobacillus ferro-oxidans*. In *Proceedings of the International Conference on the Use of Micro-organisms in Hydrometallurgy, Pecs*, 4–6 December, 1980, ed. B. Czegledi, pp. 155–160. Hungarian Academy of Sciences.

Bruynesteyn, A., A. Vizsolyi and R. Vos (1980b). The ability of *Thiobacillus ferro-oxidans* to withstand changes in pH. In *Proceedings of the International Conference on the Use of Micro-organisms in Hydrometallurgy, Pecs*, 4–6 December, 1980, ed. B. Czegledi, pp. 151–154. Hungarian Academy of Sciences.

Bruynesteyn, A., A. Vizsolyi and C. C. Walden (1980c). Effect of pH and Eh on the chemical and biological leaching of a low grade uranium ore. In *Proceedings of the International Conference on the Use of Micro-organisms in Hydrometallurgy, Pecs*, 4–6 December, 1980, ed. B. Czegledi, pp. 85–93. Hungarian Academy of Sciences.

Bruynesteyn, A., R. W. Lawrence, A. Vizsolyi and R. Hackl (1983). An elemental sulphur producing biohydrometallurgical process for treating sulphide concentrates. In *Recent Progress in Biohydrometallurgy*, ed. G. Rossi and A. E. Torma, pp. 151–168. Associazione Mineraria Sarda-09016 Iglesias-Italy.

Bryner, L. C., R. B. Walker and R. Palmer (1967). Some factors influencing the biological and non-biological oxidation of sulphide minerals. *Trans. Soc. Min. Eng.*, **238**, 56–61.

Bryner, L. C., J. F. Beck, D. B. Davis and D. G. Wilson (1954). Micro-organisms in leaching sulphide minerals. *Ind. Eng. Chem.*, **46**, 2578–2592.

Bubela, B. (1983). Microbiological processes involved in enhanced oil recoveries. In *Microbiologically Enhanced Oil Recovery*, eds. J. E. Zajic and T. R. Jack, pp. 1–8. Penn Well, Tulsa.

Butler, J. E., M. C. Ackland and P. C. Robinson (1982). Development of *in situ* leaching by Gunpowder Copper Limited, Queensland, Australia. In *Interfacing Technologies in Solution Mining, Proceedings of the 2nd SME–SPE International Solution Mining Symposium, Denver, Colorado*, November 18–20, 1981, ed. W. J. Schlitt, pp. 251–260. Lucas Guinn Co., Hoboken, N.J.

Canterford, J. H., A. N. Smith, K. F. Bampton and F. Bologiannis (1982). Pilot-plant solution mining and environmental monitoring at the Mutooroo copper mine, South Australia. In *Interfacing Technologies in Solution Mining, Proceedings of the 2nd SME–SPE International Solution Mining Symposium, Denver, Colorado*, November 18–20, 1981, ed. W. J. Schlitt, pp. 261–270. Lucas Guinn Co., Hoboken, N.J.

Carta, M., M. Ghiani and G. Rossi (1980a). Beneficiation of a complex sulphide ore by an integrated process of flotation and bioleaching. *Proceedings of the Complex Sulphide Ores Conference, Rome*, 5–8 October, 1980. pp. 178–185. The Institution of Mining and Metallurgy, London.

Carta, M., M. Ghiani and G. Rossi (1980b). Complex sulphides concentrates bioleaching performance as related to feed composition. In *Proceedings of the International Conference on the Use of Micro-organisms in Hydrometallurgy, Pecs*, 4–6 December, 1980, ed. B. Czegledi, pp. 203–209. Hungarian Academy of Sciences.

Carta, M., M. Ghiani and G. Rossi (1980c). Biochemical beneficiation of mining industry effluents. In *Biogeochemistry of Ancient and Modern Environments*, ed. P. A. Trudinger, M. R. Walter and B. J. Ralph, pp. 669–688. Australian Academy of Science, Canberra.

Catanach, C. B. (1976). Development and in place leaching of Mountain City chalcocite ore body. In *Extractive Metallurgy of Copper*, ed. J. C. Yannapoulos and J. C. Agarwal, vol. 2, p. 849. The American Institute of Mining, Metallurgical and Petroleum Engineers, New York.

Chakrabarty, A. M. (1978). Genetic mechanism in metal-microbe interactions. In *Metallurgical Applications of Bacterial Leaching and Related Microbiological Phenomena*, ed. L. E. Murr, A. E. Torma and J. A. Brierley, pp. 137–149. Academic, New York.

Chang, Y. C. and A. S. Myerson (1982). Growth models of the continuous bacterial leaching of iron pyrite by *Thiobacillus ferro-oxidans*. *Biotechnol. Bioeng.*, **24**, 889–902.

Collins, V. G. (1969). Isolation, cultivation and maintenance of autotrophs. *Methods Microbiol.*, **38**, 1–52.

Colmer, A. R. and M. E. Hinkle (1947). The role of microorganisms in acid mine drainage. *Science*, **106**, 253–256.

Corrans, I. J. (1970). Studies on the bacterial leaching of natural and synthetic iron copper sulphide minerals. M.Sc. Thesis, University of New South Wales.

Corrans, I. J. (1974). Kinetic and mechanistic studies of the biological and chemical leaching of nickel from sulphide ores. Ph.D. Thesis, University of Natal.

Corrans, I. J., B. Harris and B. J. Ralph (1972). Bacterial leaching: An introduction to its application and theory and a study of mechanism of operation. *J. South Afr. Inst. Min. Metall.*, **72**, 221–230.

Craze, B. (1980). Mine waste pollution control at Captains Flat, New South Wales. In *Biogeochemistry of Ancient and Modern Environments*, ed. P. A. Trudinger, M. R. Walter and B. J. Ralph, pp. 705–712. Australian Academy of Science, Canberra.

Czegledi, B. (ed.) (1980). *Proceedings of the International Conference on the Use of Micro-organisms in Hydrometallurgy, Pecs*, 4–6 December, 1980. Hungarian Academy of Sciences.

Daniel, J. A., J. R. Harries and A. I. M. Ritchie (1980a). Temperature distributions in an overburden dump undergoing pyritic oxidation. In *Biogeochemistry of Ancient and Modern Environments*, ed. P. A. Trudinger, M. R. Walter and B. J. Ralph, pp. 630–636. Australian Academy of Science, Canberra.

Daniel, J. A., J. R. Harries and A. I. M. Ritchie (1980b). Water movement caused by monsoonal rainfall in an overburden dump undergoing pyritic oxidation. In *Biogeochemistry of Ancient and Modern Environments*, ed. P. A. Trudinger, M. R. Walter and B. J. Ralph, pp. 623–629. Australian Academy of Science, Canberra.

Daniel, J. A., J. R. Harries and A. I. M. Ritchie (1981). The transport of oxygen into a waste rock dump undergoing pyritic oxidation. In *Physics for Australia's Development*, 2nd. *Applied Physics Conference of the Australian Institute of oPhysics*, vol. 1, pp. 185–188. Royal Melbourne Institute of Technology.

Daniel, J. A., J. R. Harries and A. I. M. Ritchie (1982). Runoff and seepage from waste rock dumps containing pyritic material. In *Proceedings of Hydrology and Water Resources Symposium, Melbourne*, 11–13 May, 1982, pp. 28–32. The Institution of Engineers, Australia.

Darland, G., T. D. Brock, W. Samsonoff and S. F. Conti (1970). A thermophilic, acidophilic mycoplasma isolated from a coal refuse pile. *Science*, **170**, 1416–1418.

Davidson, M. S., A. E. Torma, J. A. Brierley and C. L. Brierley (1981). Effects of elevated pressures on iron- and sulphur-oxidising bacteria. *Biotechnol Bioeng. Symp.*, **11**, 603–618.

Davy, D. R. (ed.) (1975). *Rum Jungle Environmental Studies*, Australian Atomic Energy Commission Report AAEC/E365.

Dees, C. and J. M. Shively (1982). Localization and quantitation of the ornithine lipid of *Thiobacillus thio-oxidans*. *J. Bacteriol.*, **149**, 798–799.

De Rosa, M., A. Gambacorta and J. D. Bu'lock (1975). Extremely thermophilic acidophilic bacteria convergent with *Sulfolobus acidocaldarius*. *J. Gen. Microbiol.*, **86**, 156–164.

Derry, R. and R. G. Whittemore (1983). Modelling the *in-situ* bioleaching of zinc copper sulphide ore from Avoca, Ireland. In *Recent Progress in Biohydrometallurgy*, ed. G. Rossi and A. E. Torma, pp. 243–263. Associazione Mineraria Sarda-09016 Iglesias-Italy.

DiSpirito, A. A. and O. H. Tuovinen (1981). Oxygen uptake coupled with uranous sulphate oxidation by *Thiobacillus ferro-oxidans* and *T. acidophilus*. *Geomicrobiol. J.*, **2**, 275–291.

DiSpirito, A. A. and O. H. Tuovinen (1982a). Uranous ion oxidation and carbon dioxide fixation by *Thiobacillus ferro-oxidans*. *Arch. Microbiol.*, **133**, 28–32.

DiSpirito, A. A. and O. H. Tuovinen (1982b). Kinetics of uranous ion and ferrous iron oxidation by *Thiobacillus ferro-oxidans*. *Arch. Microbiol.*, **133**, 33–37.

DiSpirito, A. A., P. R. Dugan and O. H. Tuovinen (1981). Inhibitory effects of particulate materials in growing cultures of *Thiobacillus ferro-oxidans*. *Biotechnol. Bioeng.*, **23**, 2761–2769.

DiSpirito, A. A., M. Silver, L. Voss and O. H. Tuovinen (1982). Flagella and pili of iron-oxidizing Thiobacilli isolated from a uranium mine in Northern Ontario, Canada. *Appl. Environ. Microbiol.*, **43**, 1196–1200.

Doddema, H. J. (1983). Partial microbial oxidation of pyrite in coal followed by oil-agglomeration. In *Recent Progress in Biohydrometallurgy*, ed. G. Rossi and A. E.Torma, pp. 467–478. Associazione Mineraria Sarda-09016 Iglesias-Italy.

Doğan, Z. M., C. F. Gökçay and E. Atabey (1983). Bacterial leaching of Turkish pyrite cinders. In *Recent Progress in Biohydrometallurgy*, ed. G. Rossi and A. E. Torma, pp. 693–704. Associazione Mineraria Sarda-09016 Iglesias-Italy.

Dugan, P. R. (1975). Bacterial ecology of strip mine areas and its relationship to the production of acidic mine drainage. *Ohio J. Sci.*, **75**, 266–279.

Dugan, P. R. and W. A. Apel (1978). Microbiological desulphurization of coal. In *Metallurgical Applications of Bacterial Leaching and Related Microbiological Phenomena*, ed. L. E. Murr, A. E. Torma and J. A. Brierley, pp. 223–250. Academic, New York.

Dugan, P. R. and D. G. Lundgren (1965). Energy supply for the chemoautotroph *Ferrobacillus ferro-oxidans*. *J. Bacteriol.*, **89**, 825–834.

Duncan, D. W. and C. C. Walden (1972). Microbiological leaching in the presence of ferric iron. *Dev. Ind. Microbiol.*, **13**, 66–75.

Duncan, D. W., P. C. Trussell and C. C. Walden (1964). Leaching of chalcopyrite with *Thiobacillus ferro-oxidans*: Effect of surfactants and shaking. *Appl. Microbiol.*, **12**, 122–126.

Duncan, D. W., J. Landesman and C. C. Walden (1967). Role of *Thiobacillus ferro-oxidans* in the oxidation of sulphide minerals. *Can. J. Microbiol.*, **13**, 397–403.

Ebner, H. G. (1977). Metal extraction from industrial waste with *Thiobacilli*. In *GBF Monograph Series*, No. 4 (August, 1977), *Conference Bacterial Leaching 1977*, ed. W. Schwartz, pp. 217–222. Verlag Chemie, Weinheim.

Ebner, H. G. (1978). Metal recovery and environmental protection by bacterial leaching of inorganic waste materials. In *Metallurgical Applications of Bacterial Leaching and Related Microbiological Phenomena*, ed. L. E. Murr, A. E. Torma and J. A. Brierley, pp. 195–206. Academic, New York.

Ebner, H. G. (1980). Bacterial leaching in an airlift reactor. In *Proceedings of the International Conference on the Use of Micro-organisms in Hydrometallurgy, Pecs*, 4–6 December 1980, ed. B. Czegledi, pp. 211–217. Hungarian Academy of Sciences.

Ehrlich, H. L. (1963). Micro-organisms in acid drainage from a copper mine. *J. Bacteriol.*, **86**, 350–352.

Ehrlich, H. L. (1980). Bacterial leaching of manganese ores. In *Biogeochemistry of Ancient and Modern Environments*, ed. P. A. Trudinger, M. R. Walter and B. J. Ralph, pp. 609–614. Australian Academy of Science, Canberra.

Fletcher, J. B. (1971). In-place leaching at Miami Mine, Miami, Arizona. *Trans. Soc. Min. Eng. AIME*, **250**, 310–314.

Gaidarjiev, S., S. Groudev and F. Genchev (1975). Direct mechanism of bacterial oxidation of sulphide minerals. Paper 35 presented at *11th International Mineral Processing Congress, Cagliari*, April, 1975.

Gao, H. W., H. Y. Sohn and M. E. Wadsworth (1982). A mathematical model for the *in situ* leaching of primary copper ore. In *Interfacing Technologies in Solution Mining, Proceedings of the 2nd SME–SPE International Solution Mining Symposium, Denver, Colorado*, November 18–20, 1981, ed. W. J. Schlitt, pp. 197–208. Lucas Guinn Co., Hoboken, N.J.

Garrels, R. S. M. and C. L. Christ (1965). *Solutions, Minerals and Equilibria*. Harper and Row, New York.

Gauci, G. and B. L. Cusack (1968). DAVCRA flotation cell. Preliminary report on the operation of a 60 tph. circuit at the Zinc Corporation mill. *Aust. Inst. Min. Metall. Convention*, Broken Hill, 1968.

Gerlach, J. K., E. D. Gock and S. K. Ghosh (1973). Activation and leaching of chalcopyrite concentrates with dilute sulphuric acid. In *Extractive Metallurgy of Copper. Hydrometallurgy and Electrowinning*, ed. D. J. I. Evans and R. S. Shoemaker, pp. 403–416. American Institute of Mining, Metallurgical and Petroleum Engineers, Inc., New York.

Golding, R. M., B. Harris, B. J. Ralph, P. A. D. Rickard and D. G. Vanselow (1977). The nature of the passivation film on covellite exposed to oxygen. In *GBF Monograph Series*, No. 4 (August, 1977), *Conference Bacterial Leaching 1977*, ed. W. Schwartz, pp. 191–200. Verlag Chemie, Weinheim.

Golding, R. M., B. Harris, B. J. Ralph, P. A. D. Rickard and D. G. Vanselow (1984). Mechanisms of oxidation of copper sulphide (CuS) by *Thiobacillus ferro-oxidans*. Unpublished information.

Golding, R. M., A. D. Rae, B. J. Ralph and L. Sulligoi (1974). A new series of polynuclear copper dithiocarbamate–copper halide polymers. *Inorg. Chem.*, **13**, 2499–2504.

Goodman, A. and B. J. Ralph (1978). Suppression and control of microbial leaching, Progress Report No. 6, The isolation and characterisation of organisms representative of some ecologically important bacterial species in the Rum Jungle area. *Australian Atomic Energy Commission Research Contract 74/F/40*.

Goodman, A. and B. J. Ralph (1980). A microcalorimetric study of the metabolic activity of two *Thiobacillus* species. In *Biogeochemistry of Ancient and Modern Environments*, ed. P. A. Trudinger, M. R. Walter and B. J. Ralph, pp. 477–483. Australian Academy of Science, Canberra.

Goodman, A. E., T. Babij, A. I. M. Ritchie (1983). Leaching of a sulphide ore by *Thiobacillus ferro-oxidans* under anaerobic conditions. In *Recent Progress in Biohydrometallurgy*, ed. G. Rossi and A. E. Torma, pp. 361–376. Associazione Mineraria Sarda-09016 Iglesias-Italy.

Goodman, A., A. M. Khalid and B. J. Ralph (1980). A scanning electron microscopy study of the colony morphology of *Thiobacillus ferro-oxidans*. In *Biogeochemistry of Ancient and Modern Environments*, ed. P. A. Trudinger, M. R. Walter and B. J. Ralph, pp. 459–467. Australian Academy of Science, Canberra.

Goodman, A. E., A. M. Khalid and B. J. Ralph (1981a). Microbial ecology of Rum Jungle. Part I. Environmental study of sulphidic overburden dumps, experimental heap-leach piles and tailings dam area Australian Atomic Energy Commission. AAEC/E531.

Goodman, A. E., A. M. Khalid and B. J. Ralph (1981b). Microbial ecology of Rum Jungle. II. Environmental study of two flooded opencuts and smaller, associated water bodies. Australian Atomic Energy Commission. AAEC/E527.

Gormely, L. S., D. W. Duncan, R. M. R. Branion and K. L. Pinder (1975). Continous culture of *Thiobacillus ferro-oxidans* on a zinc sulphide concentrate. *Biotechnol. Bioeng.*, **17**, 31–49.

Görög, J., G. Pap, J. Hollo and T. Lakatos (1980). An efficient method for the production of leaching medium. In *Biogeochemistry of Ancient and Modern Environments*, ed. P. A. Trudinger, M. R. Walter and B. J. Ralph, pp. 601–607. Australian Academy of Science, Canberra.

Groudev, S. N. (1980a). Differences between *Thiobacillus ferro-oxidans* strains with respect to their ability to oxidise covellite. *C. R. Acad. Bulg. Sci.*, **33**, 659–662.

Groudev, S. N. (1980b). Leaching of sphalerite with different strains of *Thiobacillus ferro-oxidans*. *C. R. Acad. Bulg. Sci.*, **33**, 1119–1120.

Groudev, S. N. (1980c). Leaching of copper-bearing mineral substrates with wild microflora and with laboratory-bred strains of *Thiobacillus ferro-oxidans*. In *Biogeochemistry of Ancient and Modern Environments*, ed. P. A. Trudinger, M. R. Walter and B. J. Ralph, pp. 485–503. Australian Academy of Science, Canberra.

Groudev, S. N. (1981). Oxidation of arseno-pyrite by pure and mixed cultures of *Thiobacillus ferro-oxidans* and *T. thiooxidans*. *C. R. Acad. Bulg. Sci.*, **34**, 1139–1142.

Groudev, S. N. and F. Genchev (1978a). Mechanisms of bacterial oxidation of chalcopyrite. *Mikrobiologiya*, **15**, 139–152.

Groudev, S. N. and F. Genchev (1978b). Bioleaching of bauxites by wild and laboratory-bred microbial strains. *Proceedings of the 4th International Congress for the Study of Bauxites, Alumina and Aluminium*, Athens, October, 1978, vol. 1, pp. 271–278.

Groudev, F. N. and F. Genchev (1979). Microbial coal desulphurization: Effect of the cell adaptation and mixed cultures. *C. R. Acad. Bulg. Sci.*, **32**, 353–355.

Groudev, S. N., F. N. Genchev and S. S. Gaidarjiev (1978). Observations on the microflora in an industrial copper dump leaching operation. In *Metallurgical Applications of Bacterial Leaching and Related Microbiological Phenomena*, ed. L. E. Murr, A. E. Torma and J. A. Brierley, pp. 253–274. Academic, New York.

Groudev, S. N., F. N. Genchev, V. I. Groudeva, E. C. Petrov and D. J. Mochey (1983). Removal of iron from sands by means of micro-organisms. In *Recent Progress in Biohydrometallurgy*, ed. G. Rossi and A. E. Torma, pp. 441–450. Associazione Mineraria Sarda-09016 Iglesias-Italy.

Groudeva, V. and S. Groudev (1980). Strain improvement of *Thio-bacillus ferro-oxidans* for the purpose of leaching of copper sulphide minerals. *Proceedings of the 12th Meeting of Miners and Metallurgists*, Technical Faculty — Bor (Yugoslavia) and Institute for Copper—Bor; Bor, October, 1980, vol. II, pp. 354–364.

Groudeva, V. I., S. N. Groudev and K. I. Markov (1981). Selection of *Thiobacillus ferro-oxidans* mutants tolerant to high concentrations of copper ions. *C. R. Acad. Bulg. Sci.*, **34**, 375–378.

Guay, R., M. Silver and A. E. Torma (1976). Base composition of DNA isolated from *Thiobacillus ferro-oxidans* grown on different substrates. *Rev. Can. Biol.*, **35**, 61–67.

Gupta, M. D. and A. K. Mishra (1983). Profile of phospholipids during growth of *Thiobacillus ferro-oxidans*. In *Recent Progress in Biohydrometallurgy*, ed. G. Rossi and A. E. Torma, pp. 1–10. Associazione Mineraria Sarda-09016 Iglesias-Italy.

Hanert, H. (1973). Quantifizierung der Massenentwicklung des Eisenbakteriums *Gallionella ferruginea* unter natürlichen Bedingungen. *Arch. Microbiol.*, **88**, 225–243.

Hanert, H. (1974). Untersuchungen zur individuellen Entwicklungskinetik von *Gallionella ferruginea* in statischer Mikrokultur. *Arch. Microbiol.*, **96**, 59–74.

Harries, J. R. and A. I. M. Ritchie (1981). The use of temperature profiles to estimate the pyritic oxidation rate in a waste rock dump from an opencut mine. *Water Air Soil Pollut.*, **15**, 405–423.

Harries, J. R. and A. I. M. Ritchie (1982). Measurement of oxygen concentration and temperature in overburden dumps at a mine site. *Proceedings of Soil Science Conference*, Canberra, May, 1982. Australian Society for Soil Science (A.C.T. Branch), Canberra.

Harries, J. R. and A. I. M. Ritchie (1983a). Runoff fraction and pollution levels in runoff from a waste rock dump undergoing pyritic oxidation. *Water Air Soil Pollut.*, **19**, 155–170.

Harries, J. R. and A. I. M. Ritchie (1983b). The microenvironment within waste rock dumps undergoing pyritic oxidation. In *Recent Progress in Biohydrometallurgy*, ed. G. Rossi and A. E. Torma, pp. 377–392. Associazione Mineraria Sarda-09016 Iglesias-Italy.

Harris, J. A. (1969). Development of a theoretical approach to the heap leaching of copper sulphide ores. *Proc. Australas. Inst. Min. Metall.*, **230**, 81–92.

Harris, B., A. M. Khalid, B. J. Ralph and R. Winby (1983). Biohydrometallurgical beneficiation of tin process tailings. In *Recent Progress in Biohydrometallurgy*, ed. G. Rossi and A. E. Torma, pp. 595–616. Associazione Mineraria Sarda-09016 Iglesias-Italy.

Harrison, A. P., Jr. (1982). Genomic and physiological diversity amongst strains of *Thiobacillus ferro-oxidans* and genomic comparison with *Thiobacillus thio-oxidans*. *Arch. Microbiol.*, **131**, 68–76.

Hartford, W. H. (1970). Control of acidic mine drainage. *Science*, **169**, 504.

Heydeman, M. T., A. M. Button and H. D. Williams (1981). Preliminary investigation of micro-organisms occurring in some open blanket lateritic silicate bauxites. Personal communication.

Hicksan, R. J. (1982). Heap leaching practices at Ortiz gold mine, Santa Fe County, New Mexico. In *Interfacing Technologies in Solution Mining, Proceedings of the SME–SPE International Solution Mining Symposium, Denver, Colorado*, November 18–20, 1981, ed. W. J. Schlitt, pp. 209–222. Lucas Guinn Co., Hoboken, N.J.

Hiskey, J. B. and W. J. Schlitt (1982). Aqueous oxidation of pyrite. In *Interfacing Technologies in Solution Mining, Proceedings of the SME–SPE International Solution Mining Symposium, Denver, Colorado*, November 18–20, 1981, ed. W. J. Schlitt, pp. 55–74. Lucas Guinn Co., Hoboken, N.J.

Holden, P. J. and J. C. Madgwick (1983). Mixed culture bacterial leaching of manganese dioxide. *Proc. Australas. Inst. Min. Metall.*, **286**, 61–63.

Holmes, D. S., J. H. Lobos, L. H. Bopp and G. C. Welch (1983). Setting up a genetic system *de novo* for studying the acidophilic thiobacillus *T. ferro-oxidans*. In *Recent Progress in Biohydrometallurgy*, ed. G. Rossi and A. E. Torma, pp. 541–554. Associazione Mineraria Sarda-09016 Iglesias-Italy.

Howie, F. M. P. (1977a). Pyrite and conservation, Part 1 : Historical aspects. *Technical Report, Palaeontology Dept., British Museum (Natural History)*, pp. 457–463.

Howie, F. M. P. (1977b). Pyrite and conservation, Part 2. *News letter of the Geological Curators' Group*, No. 10 (September 1977), pp. 497–512. British Museum (Natural History).

Huang, W. H. and W. D. Keller (1972). Organic acids as agents of chemical weathering of silicate minerals. *Nature (London), Phys. Sci.*, **239**, 149–151.

Huber, T. F., N. W. F. Kossen, P. Bos and J. G. Kuenen (1983). Modelling design and scale-up of a reactor for microbial desulphurization of coal. In *Recent Progress in Biohydrometallurgy*, ed. G. Rossi and A. E. Torma, pp. 279–289. Associazione Mineraria Sarda-09016 Iglesias-Italy.

Imai, K. (1971). Bacterial leaching in the mining industry. In *Biochemical and Industrial Aspects of Fermentation*, ed. K. Sakaguchi, T. Uemura and S. Kinoshita, pp. 329–336. Kodansha Ltd., Tokyo.

Imai, K. (1978). On the mechanism of bacterial leaching. In *Metallurgical Applications of Bacterial Leaching and Related Microbiological Phenomena*, ed. L. E. Murr, A. E. Torma and J. A. Brierley, pp. 275–295. Academic, New York.

Imai, K., H. Sakaguchi, T. Sugio and T. Tano (1973). On the mechanism of chalcocite oxidation by *Thiobacillus ferro-oxidans. J. Ferment. Technol.*, **51**, 865–870.

Ishikawa, T., T. Murayama, I. Kawahara and T. Imaizumi (1983). A treatment of acid mine drainage utilizing bacterial oxidation. In *Recent Progress in Biohydrometallurgy*, ed. G. Rossi and A. E. Torma, pp. 393–407. Associazione Mineraria Sarda, 09016 Inglesias, Italy.

Ivanov, V. I., F. I. Nagirnyak and B. A. Stepanov (1961). Bacterial oxidation of sulphide ores. I. Role of *Thiobacillus ferro-oxidans* in the oxidation of chalcopyrite and sphalerite. *Mikrobiologiya*, **30**, 688–692. (English translation, 575–578.)

Jellinek, F. (1968). Sulphides. In *Inorganic Sulphur Chemistry*, ed. G. Nickess, pp. 669–747. Elsevier, Amsterdam.

Joseph, J. N. (1953). Microbiological study of acid mine waters: Preliminary report. *Ohio J. Sci.*, **53**, 125–127.

Karavaiko, G. I. and T. A. Pivovarova (1973). Oxidation of elementary sulphur by *Thiobacillus thio-oxidans. Mikrobiologiya*, **42**, 389–395. (English translation, 345–350.)

Karavaiko, G. I., Y. U. M. Miller, O. A. Kapustin and T. A. Pivovarova (1980). Fractionation of stable isotopes of sulphur during its oxidation by *Thiobacillus ferro-oxidans. Microbiology*, **46**, 667–671.

Kargi, F. (1982a). Microbiological coal desulphurization. *Enzyme Microb. Technol.*, **4**, 13–19.

Kargi, F. (1982b). Enhancement of microbial removal of pyritic sulphur from coal using concentrated cell suspension of *Thiobacillus ferro-oxidans* and an external carbon dioxide supply. *Biotechnol. Bioeng.*, **24**, 749–752.

Keller, L. and L. E. Murr (1982). Acid-bacterial and ferric sulphate leaching of pyrite single crystals. *Biotechnol. Bioeng.*, **24**, 83–96.

Kelly, D. P., P. R. Norris and C. L. Brierley (1979). Microbiological methods for the extraction and recovery of metals. In *Microbial Technology, Society for General Microbiology Symposium 29*, ed. A. T. Bull, D. C. Ellwood and C. Ratledge, pp. 263–308. Society for General Microbiology Ltd.

Kempton, A. G., N. Moneib, R. G. L. McCready and C. E. Capes (1980). Removal of pyrite from coal by conditioning with *Thiobacillus ferro-oxidans* followed by oil agglomeration. *Hydrometallurgy*, **5**, 117–125.

Khalid, A. M. and B. J. Ralph (1977). The leaching behaviour of various zinc sulphide minerals with three *Thiobacillus* species. In *GBF Monograph Series*, No. 4 (August, 1977), *Conference Bacterial Leaching 1977*, ed. W. Schwartz, pp. 165–173. Verlag Chemie, Weinheim.

Khummongkol, D., G. S. Canterford and C. Fryer (1981). Accumulation of heavy metals in unicellular algae. Paper presented at *Engineering Foundation* (New York) *Conference on Advances in Fermentation Recovery Process Technology*, Banff, 7–12 June, 1981.

Kiel, H. (1977). Laugung von Kupferkarbonat und Kupfersilikatersen mit heterotrophen Mikroorganismen. In *GBF Monograph Series*, No. 4 (August, 1977), *Conference Bacterial Leaching 1977*, ed. W. Schwartz, pp. 261–270. Verlag Chemie, Weinheim.

Kiese, S., H. G. Ebner and U. Onken (1980. A simple laboratory airlift fermentor. *Biotechnol. Lett.*, **2**, 345–350.

Kim, D. K. H., N. G. Klimenko, G. I. Karavaiko and N. D. Klyueva (1981). Effect of *Thiobacillus ferro-oxidans* on flotation properties of sulphide minerals. *Appl. Biochem. Microbiol.*, **17**, 210–213.

Kos, C. H., W. Bijleveld, T. Grotenhuis, P. Bos, J. G. Kuenen and R. P. E. Poorter (1983). Composition of mineral salts medium for microbial desulphurization of coal. In *Recent Progress in Biohydrometallurgy*, ed. G. Rossi and A. E. Torma, pp. 479–490. Associazione Mineraria Sarda-09016 Iglesias-Italy.

Kovalenko, T. V. and G. I. Karavaiko (1961). Effect of temperature on the resistance of *Thiobacillus ferro-oxidans* to bivalent copper ions. *Mikrobiologiya*, **50**, 913–918. (English translation, 690–695.)

Kupeyeva, R. D., P. M. Solozhenkin and G. A. Khan (1977). Use of bacteria for xanthogenate desorption of collective concentrate from the mineral surface. *Ore Flotation*, Irkutsk, N5, 158–164.

Kuznetsov, S. I., M. V. Ivanov and N. N. Lyalikova (1963). *Introduction to Geological Microbiology*. McGraw-Hill, New York.

Lacey, D. T. and F. Lawson (1970). Kinetics of the liquid-phase oxidation of acid ferrous sulphate by the bacterium *Thiobacillus ferro-oxidans*. *Biotechnol. Bioeng.*, **12**, 29–50.

Landesman, J., D. W. Duncan and C. C. Walden (1966a). Iron oxidation by washed cell suspensions of the chemoautotroph, *Thiobacillus ferro-oxidans*. *Can. J. Microbiol.*, **12**, 25–33.

Landesman, J., D. W. Duncan and C. C. Walden (1966b). Oxidation of inorganic sulphur compounds by washed cell suspensions of *Thiobacillus ferro-oxidans*. *Can. J. Microbiol.*, **12**, 957–964.

Lazaroff, N., W. Sigal and A. Wasserman (1982). Iron oxidation and precipitation of ferric hydroxysulphates by resting *Thiobacillus ferro-oxidans* cells. *Appl. Environ. Microbiol.*, **43**, 924–938.

Le Roux, N. W. and D. S. Wakerley (1980). The leaching of sulphide ores by a thermophilic bacterium. In *Biogeochemistry of Ancient and Modern Environments*, ed. P. A. Trudinger, M. R. Walter and B. J. Ralph, pp. 451–457. Australian Academy of Science, Canberra.

Le Roux, N. W., D. S. Wakerley and S. D. Hunt (1977). Thermophilic Thiobacillus-type bacteria from Icelandic thermal areas. *J. Gen. Microbiol.*, **100**, 197–201.

Le Roux, N. W., D. S. Wakerley and V. S. Perry (1978). Leaching of minerals using bacteria other than *Thiobacilli*. In *Metallurgical Applications of Bacterial Leaching and Related Microbiological Phenomena*, ed. L. E. Murr, A. E. Torma and J. S. Brierley, pp. 167–191. Academic, New York.

Lewis, A. E. (1970). Chemical mining of primary copper ores by use of nuclear technology. *Proceedings of the Symposium of Engineering with Nuclear Explosives*, *Las Vegas, Nevada*, January 14–16, 1970. Conf. 700101, **21**, 907–917.

Liddell, K. C. and R. G. Bautista (1981). A partial equilibrium chemical model for the dump leaching of chalcopyrite. *Metall. Trans. B*, **128**, 627–637.

Livesey-Goldblatt, E., T. H. Tunley and I. F. Nagy (1977). Pilot-plant bacterial film oxidation (Bacfox Process) of recycled acidified uranium plant ferrous sulphate leach solution. In *GBF Monograph Series*, No. 4 (August, 1977), *Conference Bacterial Leaching 1977*, ed. W. Schwartz, pp. 175–190. Verlag Chemie, Weinheim.

Livesey-Goldblatt, E., P. Norman and D. R. Livesey-Goldblatt (1983). Gold recovery from arsenopyrite/pyrite ore by bacterial leaching and cyanidation. In *Recent Progress in Biohydrometallurgy*, ed. G. Rossi and A. E. Torma, pp. 627–641. Associazione Mineraria Sarda-09016 Iglesias-Italy.

Loughnan, F. C. (1969). *Chemical Weathering of the Silicate Minerals*. Elsevier, New York.

Lovell, H. L. (1973). EPA-670/2-73-093, US Environmental Protection Agency, Washington, DC.

Lowson, R. T. (1982). Aqueous oxidation of pyrite by molecular oxygen. *Chem. Rev.*, **82**, 461–497.

Lundgren, D. G. (1975). Microbiological problems in strip mine areas: Relationship to the metabolism of *Thiobacillus ferro-oxidans*. *Ohio J. Sci.*, **75**, 280–287.

Lundgren, D. G. and M. Silver (1980). Ore leaching by bacteria. *Annu. Rev. Microbiol.*, **34**, 263–283.

Lundgren, D. G., J. R. Vestall and F. R. Tabita (1972). The microbiology of mine drainage. In *Water Pollution Microbiology*, ed. R. Mitchell, pp. 69–88. Wiley-Interscience, New York.

Lyalikova, N. N. (1972). Oxidation of trivalent antimony to higher oxides as an energy source for the development of a new autotrophic organism *Stibiobacter* gen. nov. *Dokl. Akad. Nauk SSSR*, **205**, 1228–1229.

Lyalikova, N. N. (1974). *Stibiobacter senarmontii* — A new antimony-oxidizing microorganism. *Mikrobiologiya*, **43**, 941–943. (English translation, 799–805.)

Lyalikova, N. N., L. L. Lyubavina and P. M. Solozhenkin (1977). Application of sulphate-reducing bacteria for the enrichment of ores. In *GBF Monograph Series*, No. 4 (August, 1977), *Conference Bacterial Leaching 1977*, ed. W. Schwartz, pp. 93–100. Verlag Chemie, Weinheim.

McEntee, J. D., S. F. Minney and A. V. Quirk (1983). Adsorption of cadmium by a cadmium-resistant organism isolated from sewage sludge. In *Recent Progress in Biohydrometallurgy*, ed. G. Rossi and A. E. Torma, pp. 617–626. Associazione Mineraria Sarda-09016 Iglesias-Italy.

McElroy, R. O. and A. Bruynesteyn (1978). Continuous biological leaching of chalcopyrite concentrates: demonstration and economic analysis. In *Metallurgical Applications of Bacterial Leaching and Related Microbiological Phenomena*, ed. L. E. Murr, A. E. Torma and J. A. Brierley, pp. 441–462. Academic, New York.

McKee, C. R., R. H. Jacobson, S. C. Way, M. E. Hanson and K. Chong (1982). Design criteria for *in situ* mining of hard rock ore deposits. In *Interfacing Technologies in Solution Mining, Proceedings of the 2nd SME–SPE International Solution Mining Symposium*, *Denver, Colorado*, November 18–20, 1981, ed. W. J. Schlitt, pp. 103–121.

Mackintosh, M. E. (1978). Nitrogen fixation by *Thiobacillus ferro-oxidans*. *J. Gen. Microbiol.*, **105**, 215–218.

Madgwick, J. C. and B. J. Ralph (1977). The metal-tolerant alga *Hormidium fluitans* (Gay) Heering from acid mine drainage waters in Northern Australia and Papua-New Guinea. In *GBF Monograph* Series, No. 4 (August, 1977). *Conference Bacterial Leaching 1977*, ed. W. Schwartz, pp. 85–91. Verlag Chemie, Weinheim.

Madgwick, J. C. and B. J. Ralph (1980). Laboratory studies on mineral leaching. In *Biogeochemistry of Ancient and Modern Environments*, ed. P. A. Trudinger, M. R. Walter and B. J. Ralph, pp. 589–600. Australian Academy of Science, Canberra.

Madgwick, J. C. and B. J. Ralph (1981). Removal of rate-limiting iron precipitates from low-grade copper ore during bacterial leaching. *Proc. Australas. Inst. Min. Metall.*, **279**, 33–35.

Madgwick, J. C., B. J. Ralph and W. R. Sadler (1981). Sulphide oxidising bacteria of copper porphyry ore. *Proc. Australas. Inst. Min. Metall.*, **279**, 37–40.

Madsen, B. W. and M. E. Wadsworth (1977). The applicatin of a mixed kinetics leaching model to an ore containing an assemblage of different copper minerals. In *Energy and Mineral Resource Recovery*, American Nuclear Society Topical meeting, 1977. Washington, D. C. Technology Information Center Publication CONF-770440, 1977, pp. 619–630.

Majima, H. and E. Peters (1968). Electrochemistry of sulphide dissolution in hydrometallurgical systems. *Proceedings of the VIII International Mineral Processing Congress, Leningrad.*

Malouf, E. E. and J. D. Prater. (1962). New technology of leaching waste dumps. *Min. Congr. J.*, **48**, 82–107.

Marchlewitz, B. and W. Schwartz (1961). Investigation of microbial associations in acid mine waters. *Z. Allg. Mikrobiol.*, **1**, 100–114.

Markosyan, G. E. (1973). A new mixotrophic sulphur bacterium developing in acid media, *Thiobacillus organoparus*, sp. n. *Dokl. Akad. Nauk SSSR*, **211**, 1205–1208. (Engl. transl., pp. 318–320.)

Manning, H. L. (1975). New medium for isolating iron-oxidising and heterotrophic acidophilic bacteria from acid mine drainage. *Appl. Microbiol.*, **30**, 1010–1016.

Marsh, R. M. and P. R. Norris (1983). The isolation of some thermophilic, autotrophic, iron- and sulphur-oxidizing bacteria. *FEMS Microbiol. Lett.*, **17**, 311–315.

Marsh, R. M., P. R. Norris and N. W. Le Roux (1983). Growth and mineral oxidation studies with *Sulfolobus*. In *Recent Progress in Biohydrometallurgy*, ed. G. Rossi and A. E. Torma, pp. 71–81. Associazione Mineraria Sarda-09016 Iglesias-Italy.

Marshall, K. C. (1979). Biogeochemistry of manganese minerals. In *Biogeochemical Cycling of Mineral-forming Elements*, ed. P. A. Trudinger and D. J. Swaine, pp. 253–292. Elsevier, Amsterdam.

Martin, P. A. W., P. R. Dugan and O. H. Tuovinen (1981). Plasmid DNA in acidophilic, chemolithotrophic thiobacilli. *Can. J. Microbiol.*, **27**, 850–853.

Martinez Garcia, J. P., G. Gimenez Gimenez, E. Garay Auban, P. Matinez Germes and E. Hernandez Gimenez (1981). Growth of aciduric and non-aciduric species of *Thiobacillus* in the presence of two anionic detergents. *Zentralbl. Bakteriol. Parasitenkd. Infektionskr. Hyg. Zweite Naturwiss. abt. Mikrobiol. Landwirtsch.Technol. Umweltschutzes*, **136**, 682–691.

Mercz, T. I. and J. C. Madgwick (1982). Enhancement of bacterial manganese leaching by microalgal growth products. *Proc. Australas. Inst. Min. Metall.*, **283**, 43–46.

Miller, S. D. (1980). Sulphur and hydrogen ion buffering in pyritic strip-mine spoil. In *Biogeochemistry of Ancient and Modern Environments*, ed. P. A. Trudinger, M. R. Walter and B. J. Ralph, pp. 537–543. Australian Academy of Science, Canberra.

Millonig, G., M. de Rosa, A. Gambacorta and J. D. Bu'lock (1975). Ultrastructure of an extremely thermophilic acidophilic micro-organism. *J. Gen. Microbiol.*, **86**, 165–173.

Mishra, A. K. and P. Roy (1979). A note on the growth of *Thiobacillus ferro-oxidans* on solid medium. *J. Appl. Bacteriol.*, **47**, 289–292.

Mishra, A. K., P. Roy, S. S. Roy Mahapatra and D. Chandra (1983). The role of *Thiobacillus ferro-oxidans* and other microbes in biohydrometallurgy with special reference to desulphurization of coal. In *Recent Progress in Biohydrometallurgy*, ed. G. Rossi and A. E. Torma, pp. 491–510. Associazione Mineraria Sarda-09016 Iglesias-Italy.

Mizoguchi, T. and T. Okabe (1980). Isolation of iron- and sulphur-oxidizing bacteria from mine water in Japan and some investigations on the leaching of sulphide ores. In *Biogeochemistry of Ancient and Modern Environments*, ed. P. A. Trudinger, M. R. Walter and B. J. Ralph, pp. 505–513. Australian Academy of Science, Canberra.

Mizoguchi, T., T. Sato and T. Okabe (1976). New sulphur-oxidizing bacteria capable of growing heterotrophically, *Thiobacillus rubellus* nov. sp. and *Thiobacillus delicatus* nov. sp. *J. Ferment. Technol.*, **54**, 181–191.

Moriarty, D. J. W. and D. J. D. Nicholas (1969). Enzymic sulphide oxidation by *Thiobacillus concretivorus*. *Biochim. Biophys. Acta*, **184**, 114–123.

Moss, F. J. and J. E. Andersen (1968). The effects of environment on bacterial leaching rates. *Proc. Australas. Inst. Min. Metall.*, **225**, 15–25.

Mosser, J. L., B. B. Bohlool and T. D. Brock (1974). Growth rates of *Sulfolobus acidocaldarius* in nature. *J. Bacteriol.*, **118**, 1075–1082.

Muir, D. M., A. J. Parker and D. W. Giles (1976). Cuprous hydrometallurgy, Part IV. Rates and equilibria in the reaction of copper sulpides with Cu(II) sulfate in aqueous acetonitrile. *Hydrometall.*, **2**, 127–140.

Murr, L. E. (1980). Theory and practice of copper sulphide leaching in dumps and *in situ*. *Miner. Sci. Eng.*, **12**, 121–189.

Murr, L. E. and J. A. Brierley (1978). The use of large-scale test facilities in studies of the role of microorganisms in commercial leaching operations. In *Metallurgical Applications of Bacterial Leaching and Related Microbiological Phenomena*, ed. L. E. Murr, A. E. Torma and J. A. Brierley, pp. 491–520. Academic, New York.

Murr, L. E., A. E. Torma and J. A. Brierley (eds.) (1978). *Metallurgical Applications of Bacterial Leaching and Related Microbiological Phenomena*. Academic, New York.

Murr, L. E., W. J. Schlitt and L. M. Cathles (1982). Experimental observations of solution flow in the leaching of copper-bearing waste. In *Interfacing Technologies in Solution Mining, Proceedings of the 2nd SME–SPE International Mining Symposium, Denver, Colorado*, November 18–20, 1981, ed. W. J. Schlitt, pp. 271–290. Lucas Guinn Co., Hoboken, N.J.

Murr, L. E. and A. P. Mehta (1982). Coal desulfurization by leaching involving acidophilic and thermophilic microorganisms. *Biotechnol. Bioeng.*, **24**, 743–748.

Nickerson, W. J. (1969). Microbial transformations of naturally-occurring polymers. In *Fermentation Advances*, ed. D. Perlman, pp. 631–634. Academic, New York.

Nielsen, A. M. and J. V. Beck (1972). Chalcocite oxidation and coupled carbon dioxide fixation by *Thiobacillus ferro-oxidans*. *Science.*, **175**, 1124–1126.

Norris, P. R. (1983). Iron and mineral oxidation with *Leptospirillum*-like bacteria. In *Recent Progress in Biohydrometallurgy*, ed. G. Rossi and A. E. Torma, pp. 83–96. Associazione Mineraria Sarda-099016 Iglesias-Italy.

Norris, P. R. and D. P. Kelly (1978). Dissolution of pyrite (FeS$_2$) by pure and mixed cultures of some acidophilic bacteria. *FEMS Microbiol. Lett.*, **4**, 143–146.

Norris, P. R. and D. P. Kelly (1982). The use of mixed microbial cultures in metal recovery. In *Microbial Interactions and Communities*, ed. A. T. Bull and J. H. Stater, pp. 443–474. Academic, London.

Olem, H. and R. F. Unz (1980). Rotating disc biological treatment of acid mine drainage. *J. Water Pollut. Control Fed.*, **52**, 257–269.

Olsen, T. M., P. R. Ashman, A. E. Torma and L. E. Murr (1980). Desulphurization of coal by *Thiobacillus ferro-oxidans*. In *Biogeochemistry of Ancient and Modern Environments*, ed. P. A. Trudinger, M. R. Walter and B. J. Ralph, pp. 693–703. Australian Academy of Science, Canberra.

Olsen, G. J., F. D. Porter, J. Rubinstein and S. Silver (1982). Mercuric reductase enzyme from a mercury-volatilizing strain of *Thiobacillus ferro-oxidans*. *J. Bacteriol.*, **151**, 1230–1236.

Pereverzev, N. A., L. A. Gromova, G. I. Karavaiko and A. A. Manykin (1981). Preparation of *Thiobacillus ferro-oxidans* cell wall fragments and their ultrastructural organization. *Microbiology*, **50**, 509–513.

Pidgeon, R. T. (1985). Non-radiological contaminants in the long-term management of uranium mining and milling. Environment Paper of Department of Arts, Heritage and Environment. Australian Government Publishing Service, Canberra.

Piercy, P. (1982). Waste dump leaching in a tropical environment. In *Interfacing Technologies in Solution Mining, Proceedings of the 2nd SME–SPE International Solution Mining Symposium, Denver, Colorado*, November 18–20, 1981, ed. W. J. Schlitt, pp. 241–250. Lucas Guinn Co., Hoboken, N.J.

Pinches, A. (1975). Bacterial leaching of an arsenic-bearing sulfide concentrate. In *Leaching and Reduction in Hydrometallurgy*, ed. A. R. Burkin, pp. 28–35. The Institute of Mining and Metallurgy, London.

Pooley, F. D. and A. S. Atkins (1983). Desulphurization of coal using bacteria by both dump and process plant techniques. In *Recent Progress in Biohydrometallurgy*, ed. G. Rossi and A. E. Torma, pp. 511–526. Associazione Mineraria Sarda-09016 Iglesias-Italy.

Porter, D. D. (1973). Blast design for *in situ* leaching. Presented at Southwest Mineral Industry Conference Phoenix, AZ, April 27, 1973, 7 pp. Available from E. I. duPont de Nemours and Co. Inc., Wilmington, DE 19898.

Potter, G. M., C. K. Chase and P. G. Chamberlain (1982). Feasibility of *in situ* leaching of metallic ores other than copper and uranium. In *Interfacing Technologies in Solution Mining, Proceedings of the 2nd SME–SPE International Solution Mining Symposium, Denver, Colorado*, November 18–20, 1891, ed. W. J. Schlitt, pp. 123–130. Lucas Guinn Co., Hoboken, N.J.

Pourbaix, M. J. N. (ed.) (1966). *Atlas of Electrochemical Equilibria in Aqueous Solutions*. Pergamon, Oxford.

Ralph, B. J. (1977). Photosynthetic approaches to ore leaching. *Proceedings of the Symposium on Practical Applications of Photosynthesis*, 26 August, 1977, University of New South Wales and International Solar Energy Society (ANZ Section).

Ralph, B. J. (1979). Oxidative reactions in the sulfur cycle. In *Biogeochemical Cycling of Mineral-forming Elements*, ed. P. A. Trudinger and D. J. Swaine, pp. 369–400. Elsevier, Amsterdam.

Ralph, B. J. (1980). Weathering of sulfur in rocks. In *Sulfur in Australia*, ed. J. R. Freney and A. J. Nicholson, pp. 146–157. Australian Academy of Science, Canberra.

Ralph, B. J. (1982a). The implications of genetic engineering for biologically-assisted hydrometallurgical processes. In *Proceedings of the Symposium on Genetic Engineering — Commercial Opportunities in Australia*, 18–20 November 1981, Sydney. pp. 82–102. Australian Government Publishing Service, Canberra.

Ralph, B. J. (1982b). Biological mining and biohydrometallurgy — prospects and opportunities. *The Australian I.M.M. Conference*, August, Melbourne, pp. 403–409. Aust. Inst. Min. Metall., Melbourne.

Ralph, B. J. (1985). Geomicrobiology and the new biotechnology. ONR Lecture, Society for Industrial Microbiology, Fort Collins, CO, August 1984. *Dev. Ind. Microbiol.*, **26**, 411–447.

Rawlings, D. D., C. Gawith, A. Petersen and D. R. Woods (1983). Characterisation of plasmids and potential genetic markers in *Thiobacillus ferro-oxidans*. In *Recent Progress in Biohydrometallurgy*, ed. G. Rossi and A. E. Torma, pp. 555–570. Associazione Mineraria Sarda-09016 Iglesias-Italy.

Rickard, P. A. D. and D. G. Vanselow (1978). Investigations into the kinetics and stoichiometry of bacterial oxidation of covellite (CuS) using a polarographic oxygen probe. *Can. J Microbiol.*, **24**, 998–1003.

Robinsky, E. I. (1979). Tailing disposal by the thickened discharge method for improved economy and environmental control. In *Tailing Disposal Today*, vol. 2, *Proceedings of the 2nd International Tailing Symposium, Denver, Colorado*, May, 1978. Miller Freeman, San Francisco.

Robinson, W. I. (1972). Finger dump preliminaries promise improved copper leaching at Butte. *Min. Eng.*, **N9**, 24.

Rossi, G. (1978). Potassium recovery through leucite bioleaching: possibilities and limitations. In *Metallurgical Applications of Bacterial Leaching and Related Microbiological Phenomena*, ed. L. E. Murr, A. E. Torma and J. A. Brierley, pp. 297–319. Academic, New York.

Rossi, G. and E. Salis (1977). Possibilities for large-scale microbial leaching processes. Joint Meeting, University of Manchester Institute of Science and Technology, December 1977.

Rossi, G. and A. E. Torma (eds.) (1983). *Recent Progress in Biohydrometallurgy*. Associazione Mineraria Sarda-09016 Iglesias-Italy. (*Proceedings of the International Symposium in Biohydrometallurgy, Cagliari, Italy*, May, 1983).

Rossi, G., A. E. Torma and P. Trois (1983). Bacteria-mediated copper recovery from a cupriferous pyrrhotite ore: chalcopyrite/pyrrhotite interactions. In *Recent Progress in Biohydrometallurgy*, ed. G. Rossi and A. E. Torma, pp. 185–200. Associazione Mineraria Sarda-09016 Iglesias-Italy.

Ruderhausen, C. G. (1974). Copper solution mining at Old Reliable. Presented at the National Meeting of the American Institute of Chemical Engineers, Salt Lake City, 18–21 August, 1974.

Rudolfs, W. and A. Helbronner (1922). Oxidation of zinc sulphide by microorganisms. *Soil Sci.*, **14**, 459–464.

Sakaguchi, H., A. E. Torma and M. Silver (1976). Microbiological oxidation of synthetic chalcocite and covellite by *Thiobacillus ferro-oxidans*. *Appl. Environ. Microbiol.*, **31**, 7–10.

Sanmugasunderam, V. (1981). The continuous microbiological leaching of zinc sulphide concentrate with recycle. Ph.D. Thesis, University of British Columbia.

Sawe, A. V. (1980). The beneficiation of Renison sulphide concentrate by bacterial leaching. M.Sc. (Biotech.) Thesis, 1980, University of New South Wales.

Schaeffer, W. I. and W. W. Umbreit (1963). Phosphatidyl inositol as a wetting agent in sulphur oxidation by *Thiobacillus thio-oxidans*. *J. Bacteriol.*, **85**, 492–493.

Schäfer, W. (1983). Leaching of industrial waste products with thiobacilli and heterotrophic micro-organisms. In *Recent Progress in Biohydrometallurgy*, ed. G. Rossi and A. E. Torma, pp. 427–440. Associazione Mineraria Sarda-09016 Iglesias-Italy.

Schlitt, W. J. (ed.) (1982). *Interfacing Technologies in Solution Mining. Proceedings of the 2nd SME–SPE International Solution Mining Symposium, Denver, Colorado*, November 18–20, 1981. Lucas Guinn Co., Hoboken, N.J.

Schwartz, W. (ed.) (1977). *Bacterial Leaching Conference 1977. GBF Monograph Series*, No. 4 (August, 1977). Verlag Chemie, Weinheim.

Sheffer, H. W. and L. G. Evans (1968). Copper leaching practices in the Western United States. Washington DC US Bureau of Mines, I. C. '341.

Shively, J. M. and A. A. Benson (1967). Phospholipids of *Thiobacillus thio-oxidans*. *J. Bacteriol.*, **94**, 1679–1683.

Shuey, R. T. (1975). *Semiconducting Ore Minerals. Developments in Economic Geology*, vol. 4. Elsevier, Amsterdam.

Shumate II, S. E., G. W. Strandberg, D. A. McWhirter, J. R. Parrott, Jr., G. M. Bogaski and B. R. Locke (1980). Separation of heavy metals from aqueous solutions using biosorbents — development of contacting devices for uranium removal. *Biotechnol. Bioeng. Symp.*, **10**, 27–34.

Silver, M. (1978). Metabolic mechanisms of iron-oxidizing *Thiobacilli*. In *Metallurgical Applications of Bacterial Leaching and Related Microbiological Phenomena*, ed. L. E. Murr, A. E. Torma and J. A. Brierley, pp. 3–17. Academic, New York.

Silver, M. and J. E. Andersen (1979). Removal of radium from Elliott Lake uranium tailings by salt washing. Presented at the 15th Canadian Symposium on Water Pollution Research in Canada, University of Sherbrooke, December 7, 1979.

Silver, M. and A. E. Torma (1974). Oxidation of metal sulphides by *Thiobacillus ferro-oxidans* grown on different substrates. *Can. J. Microbiol.*, **20**, 141–147.

Silverman, M. P. (1967). Mechanism of bacterial pyrite oxidation. *J. Bacteriol.*, **94**, 1046–1051.

Silverman, M. P. (1979). Biological and organic chemical decomposition of silicates. In *Biogeochemical Cycling of Mineral-forming Elements*, ed. P. A. Trudinger and D. J. Swaine, pp. 445–465. Elsevier, Amsterdam.

Silverman, M. P. and H. L. Ehrlich (1964). Microbial formation and degradation of minerals. *Adv. Appl. Microbiol.*, **6**, 153–206.

Silverman, M. P. and D. G. Lundgren (1959). Studies on the chemoautotrophic bacterium *Ferrobacillus ferro-oxidans*. I. An improved medium and a harvesting procedure for securing high cell yields. *J. Bacteriol.*, **77**, 642–647.

Singer, P. C. and W. Stumm (1970). Acidic mine drainage: the rate-determining step. *Science*, **167**, 1121–1123.

Smith, E. E. and K. S. Shumate (1970). The sulphide to sulphate reaction mechanism; a study of the sulphide to sulphate reaction mechanism as it relates to the formation of acid mine waters. *Water Pollution Control Research Series*. Ohio State University Research Foundation.

Soljanto, P. and O. H. Tuovinen (1980). A microcalorimetric study of U(IV)-oxidation by *Thiobacillus ferro-oxidans* and ferric iron. In *Biogeochemistry of Ancient and Modern Environments*, ed. P. A. Trudinger, M. R. Walter and B. J. Ralph, pp. 469–475. Australian Academy of Science, Canberra.

Solozhenkin, P. M. (1980). Brief report on bacterial leaching of non-ferrous metals. In *Proceedings of the International Conference on the Use of Micro-organisms in Hydrometallurgy, Pecs*, 4–6 December 1980, ed. B. Czegledi, pp. 11–25. Hungarian Academy of Sciences.

Solozhenkin, P. M. and L. L. Lyubavina (1980). The bacterial leaching of antimony- and bismuth-bearing ores and the utilization of sewage waters. In *Biogeochemistry of Ancient and Modern Environments*, ed. P. A. Trudinger, M. R. Walter and B. J. Ralph, pp. 615–621. Australian Academy of Science, Canberra.

Solozhenkin, P. M., L. L. Lyubavina, L. F. Samokhvalova and V. S. Pupkov (1979). Studies of flotation properties of sulphate-reducing bacteria at ore treatment. *Izv. Vuzov, Tsvetn. Metall.*, **3**, 13–21.

Starkey, R. L., G. E. Jones and L. R. Frederick (1956). Effects of medium agitation and wetting agents on oxidation of sulphur by *Thiobacillus thio-oxidans*. *J. Gen. Microbiol.*, **15**, 329–334.

Steiner, M. and N. Lazaroff (1974). Direct method for continuous determination of iron oxidation by autotrophic bacteria. *Appl. Microbiol.*, **28**, 872–880.

Stirkov, P., M. Semkov, P. Nestorova, G. Merazchiev, M. Michaelov and P. Popov (1975). Technological and economical aspects of the bacterial and chemical leaching of waste ores and refractory-to-dress copper ores in the People's Republic of Bulgaria. *Proceedings of the XI International Mineral Processing Congress, Cagliari*, April, 1975. Universita Instituto di Arte Mineraria e Preparazione dei Minerali.

Stanczyk, M. H. and C. Rampacek (1963). Oxidation leaching of copper sulphides in acidic pulps at elevated temperatures and pressures. Washington DC, *US Bureau of Mines, RI 6193*, 1963, pp. 1–15.

Strandberg, G. W., S. E. Shumate II, and J. R. Parrott, Jr. (1981). Microbial cells as biosorbents for heavy metals: accumulation of uranium by *Saccharomyces cerevisiae* and *Pseudomonas aeruginosa*. *Appl. Environ. Microbiol.*, **41**, 237–245.

Sugio, T., T. Tano and K. Imai (1981). Isolation and some properties of silver ion resistant iron oxidising bacterium, *Thiobacillus ferro-oxidans*. *Agric. Biol. Chem.*, **45**, 2037–2052.

Sulligoi, L. A. V. (1972). Unpublished information on the reactor leaching of a nickeliferous pyrrhotite. School of Biotechnology, University of New South Wales, Kensington, NSW, Australia.

Sullivan, E. A., J. E. Zajic and T. R. Jack (1980). The effect of chemical and biological redox reactions on the growth of *Thiobacillus thio-oxidans*. In *Biogeochemistry of Ancient and Modern Environments*, ed. P. A. Trudinger, M. R. Walter and B. J. Ralph, pp. 521–528. Australian Academy of Science, Canberra.

Suzuki, I. (1974). Mechanisms of inorganic oxidation and energy coupling. *Annu. Rev. Microbiol.*, **28**, 85–101.

Szekerces, L. (1974). Analytical chemistry of the sulphur acids. *Talanta*, **21**, 1–44.

Tano, T. and D. G. Lundgren (1978). Sulphide oxidation by spheroplasts of *Thiobacillus ferro-oxidans*. *Appl. Environ. Microbiol.*, **35**, 1198–1205.

Taylor, J. H. and P. F. Whelan (1942). The leaching of cupreous pyrites and the precipitation of copper at Rio Tinto, Spain. *Trans. Inst. Min. Metall.*, **52**, 35–71.

Tomizuka, N. and M. Yagisawa (1978). Optimum conditions for leaching of uranium and oxidation of lead sulphide with *Thiobacillus ferro-oxidans* and recovery of metals from bacterial leaching solution with sulphate-reducing bacteria. In *Metallurgical Applications of Bacterial Leaching and Related Microbiological Phenomena*, ed. L. E. Murr, A. E. Torma and J. A. Brierley, pp. 321–344. Academic, New York.

Torma, A. E. (1971). Microbiological oxidation of synthetic cobalt, nickel and zinc sulphides by *Thiobacillus ferro-oxidans*. *Rev. Can. Biol.*, **30**, 209–216.

Torma, A. E. (1977). The role of *Thiobacillus ferro-oxidans* in hydrometallurgical processes. In *Advances in Biochemical Engineering*, ed. T. K. Ghose, A. Fiechter and N. Blakebrough, pp. 1–37. Springer Verlag, Berlin.

Torma, A. E. (1978). Complex lead sulphide concentrate leaching by micro-organisms. In *Metallurgical Applications of Bacterial Leaching and Related Microbiological Phenomena*, ed. L. E. Murr, A. E. Torma and J. A. Brierley, pp. 375–387. Academic, New York.

Torma, A. E. and K. Bosecker (1982). Bacterial leaching. *Prog. Ind. Microbiol.*, **16**, 77–118.

Torma, A. E. and G. G. Gabra (1977). Oxidation of stibnite by *Thiobacillus ferro-oxidans*. *Antonie van Leeuwenhoek J. Microbiol. Serol.*, **43**, 1–6.

Torma, A. E. and R. Guay (1976). Effect of particle size on the biodegradation of a sphalerite concentrate. *Nat. Can. (Que.)*, **103**, 133–138.

Torma, A. E. and F. Habashi (1972). Oxidation of copper (II) selenide by *Thiobacillus ferro-oxidans*. *Can. J. Microbiol.*, **18**, 1780–1781.

Torma, A. E. and T. Rozgonyi (1980). Influence of attrition grinding on the recovery of copper from a high-grade chalcopyrite concentrate by the BSE-process. In *Biogeochemistry of Ancient and Modern Environments*, ed. P. A. Trudinger, M. R. Walter and B. J. Ralph, pp. 583–588. Australian Academy of Science, Canberra.

Torma, A. E. and H. Sakaguchi (1978). Relation between the solubility product and the rate of metal sulphide oxidation by *Thiobacillus ferro-oxidans*. *J. Ferment. Technol.*, **56**, 173–178.

Torma, A. E. and K. N. Subramanian (1974). Selective bacterial leaching of a lead sulphide concentrate. *Int. J. Miner. Process.*, **1**, 125–134.

Torma, A. E., P. R. Ashman, T. M. Olsen and K. Bosecker (1979). Microbiological leaching of chalcopyrite concentrate and recovery of copper by solvent extraction and electrowinning. *Metallurgy*, **33**, 479–484.

Torma, A. E., C. C. Walden and R. M. R. Branion (1970). Microbiological leaching of a zinc sulfide concentrate. *Biotechnol. Bioeng.*, **12**, 501–517.

Torma, A. E., C. C. Walden, D. W. Duncan and R. M. R. Branion (1972). The effect of carbon dioxide and particle surface area on the microbiological leaching of a zinc sulphide concentrate. *Biotechnol. Bioeng.*, **14**, 777–786.

Tributsch, H. and J. C. Bennett (1981a). Semiconductor–electrochemical aspects of bacterial leaching. I. Oxidation of metal sulphides with large energy gaps. *J. Chem. Technol. Biotechnol.*, **31**, 565–577.

Tributsch, H. and J. C. Bennett (1981b). Semiconductor–electrochemical aspects of bacterial leaching. II. Survey of rate-controlling sulphide properties. *J. Chem. Technol. Biotechnol.*, **31**, 627–635.

Trudinger, P. A. and F. Mendelsohn (1976). Biological processes and mineral deposition. In *Stromatolites, Developments in Sedimentology 20*, ed. M. R. Walter, pp. 663–672. Elsevier, Amsterdam.

Trudinger, P. A. and D. J. Swaine (eds.) (1979). *Biogeochemical Cycling of the Mineral-forming Elements*. Elsevier, Amsterdam.

Trudinger, P. A., M. R. Walter and B. J. Ralph (eds.) (1980). *Biogeochemistry of Ancient and Modern Environments*. Australian Academy of Science, Canberra.

Trussell, P. C., D. W. Duncan and C. C. Walden (1964). Biological mining. *Can. Min. J.*, **83**, 46–49.

Tsuchiya, H. M. (1977). Leaching of Cu-Ni sulphide concentrate from the Duluth Gabbro. In *GBF Monograph Series*, No. 4 (August, 1977), *Conference Bacterial Leaching 1977*, ed. W. Schwartz, pp. 101–106. Verlag Chemie, Weinheim.

Tsuchiya, H. M., N. C. Trivedi and M. L. Schuler (1974). Microbial mutualism in ore leaching. *Biotechnol. Bioeng.*, **16**, 991–995.

Tuovinen, O. H. and D. P. Kelly (1972). Biology of *Thiobacillus ferro-oxidans* in relation to the microbiological leaching of sulphide ores. *Z. Allg. Mikrobiol.*, **12**, 311–346.

Tuovinen, O. H. and D. P. Kelly (1973). Studies on the growth of *Thiobacillus ferro-oxidans*. I. Use of membrane filters and ferrous iron agar to determine viable numbers, and comparison with $^{14}CO_2$-fixation and iron oxidation as measures of growth. *Arch. Mikrobiol.*, **88**, 285–298.

Tuovinen, O. H. and D. P. Kelly (1974a). Studies on the growth of *Thiobacillus ferro-oxidans*. II. Toxicity of uranium to growing cultures and tolerance conferred by mutation, other metal cations and EDTA. *Arch. Microbiol.*, **95**, 153–164.

Tuovinen, O. H. and D. P. Kelly (1974b). Studies on the growth of *Thiobacillus ferro-oxidans*. III. Influence of uranium, other metal ions and 2,4-dinitrophenol on ferrous iron oxidation and carbon dioxide fixation by cell suspensions. *Arch. Microbiol.*, **95**, 165–180.

Tuovinen, O. H. and D. P. Kelly (1974c). Studies on the growth of *Thiobacillus ferro-oxidans*. IV. Influence of monovalent metal cations on ferrous iron oxidation and uranium toxicity in growing cultures. *Arch. Microbiol.*, **98**, 167–174.

Tuovinen, O. H., B. C. Kelley and D. J. D. Nicholas (1976). Enzymic comparisons of the inorganic sulphur metabolism in autotrophic and heterotrophic *Thiobacillus ferro-oxidans*. *Can. J. Microbiol.*, **22**, 109–113.

Tuovinen, O. H., S. I. Niemelä and H. G. Gyllenberg (1971a). Tolerance of *Thiobacillus ferro-oxidans* to some metals. *Antonie van Leeuwenhoek J. Microbiol. Serol.*, **37**, 489–496.

Tuovinen, O. H., S. I. Niemelä and H. G. Gyllenberg (1971b). Effect of mineral nutrients and organic substances on the development of *Thiobacillus ferro-oxidans*. *Biotechnol. Bioeng.*, **13**, 517–527.

Tuovinen, O. H., D. P. Kelly, C. S. Dow and M. Eccleston (1978). Metabolic transitions in cultures of acidophilic *Thiobacilli*. In *Metallurgical Applications of Bacterial Leaching and Related Microbiological Phenomena*, ed. L. E. Murr, A. E. Torma and J. A. Brierley, pp. 61–81. Academic, New York.

Tuttle, J. H., C. I. Randles and P. R. Dugan (1968). Activity of micro-organisms in acid mine water. I. Influence of acid water on aerobic heterotrophs of a normal stream. *J. Bacteriol.*, **95**, 1495–1503.

Updegraff, D. M. and D. J. Duoros (1972). The relationship of micro-organisms to uranium deposits. *Dev. Ind. Microbiol.*, **13**, 76–90.

Vandell, T. D. (1982). Groundwater problems at uranium *in situ* leach mining operations in the Powder River Basin, Wyoming. In *Interfacing Technologies in Solution Mining, Proceedings of the 2nd SME–SPE International Solution*

Mining Symposium, Denver, Colorado, November 18–20, 1981. ed. W. J. Schlitt, pp. 299–309. Lucas Guinn Co., Hoboken, N.J.

Vanselow, D. G. (1976). Mechanisms of bacterial oxiation of the copper sulphide mineral covellite. Ph.D. Thesis, University of New South Wales.

Wadsworth, M. E. (1972). Advances in the leaching of sulphide minerals. *Miner. Sci. Eng.*, **4**, 36–47.

Wadsworth, M. E. (1975). Physico-chemical aspects of solution mining. In *Symposium on In Place Leaching and Solution Mining, Reno, Nevada*, ed. Y. S. Kim, Mackay School of Mines, 1975.

Wadsworth, M. E. (1977). Interfacing technologies in solution mining. *Min. Eng.*, **29**, 30–33.

Wadsworth, M. E. and C. H. Pitt (1980). An assessment of energy requirements in proven and new copper processes. USDOE Contract No. EM-78-S-07-1743, December, 1980.

Walsh, F. and R. Mitchell (1972a). A pH-dependent succession of iron bacteria. *Environ. Sci. Technol.*, **6**, 809–812.

Walsh, F. and R. Mitchell (1972b). An acid-tolerant iron-oxidising *Metallogenium*. *J. Gen. Microbiol.*, **72**, 369–376.

Walsh, F. and R. Mitchell (1975). Mine drainage pollution reduction by inhibition of iron bacteria. *Water Res.*, **9**, 525–528.

Ward, M. H. (1974). Surface blasting followed by *in situ* leaching at the Big Mike mine. *Proceedings of the Solution Mining Symposium, AIME Annual Meeting, Dallas, Texas*, February 25–27, 1974, pp. 243–251. A.I.M.E., New York.

Wenberg, G. M., F. H. Erbisch and M. E. Violin (1971). Leaching of copper by fungi. *Society of Mining Engineers, AIME, Trans.*, **250**, 207–212.

Wichlacz, P. and R. F. Unz (1981). Fixed film biokinetics of ferrous iron oxidation. *Biotechnol. Bioeng. Symp.*, **11**, 493–504.

Winogradsky, S. N. (1888). Über Eisenbacterin. *Bot. Z.*, **46**, 262–270.

Wong, C., M. Silver and D. J. Kushner (1982). Effects of chromium and manganese on *Thiobacillus ferro-oxidans*. *Can. J. Microbiol.*, **28**, 536–544.

Woodcock, J. T. (1967). Copper waste dump leaching. *Proc. Australas. Inst. Min. Metall.*, **58**, 47–66.

Yao Dun Pu (1982). The history and present status of practice and research work on solution mining in China. In *Interfacing Technologies in Solution Mining, Proceedings of the 2nd SME–SPE International Solution Mining Symposium, Denver, Colorado*, November 18–20, 1981, ed. W. J. Schlitt, pp. 13–20. Lucas Guinn Co., Hoboken, N.J.

Yukawa, T. (1975). Mass transfer studies in microbial systems. Ph.D. Thesis, University of New South Wales.

Zajic, J. E. (1969). *Microbial Biogeochemistry*. Academic, New York.

Zajic, J. E., T. R. Jack and E. A. Sullivan (1977). Chemical and microbially-assisted leaching of Athabasca oil sands coke. In *GBF Monograph Series*, No. 4 (August, 1977), *Conference Bacterial Leaching 1977*, ed. W. Schwartz, pp. 233–242. Verlag Chemie, Weinheim.

13

Accumulation of Metals by Microbial Cells

S. E. SHUMATE II
Engenics Inc., Menlo Park, CA, USA

and

G. W. STRANDBERG
Oak Ridge National Laboratory, Oak Ridge, TN, USA

13.1 INTRODUCTION

Certain species of microorganisms have been found to accumulate surprisingly large quantities of important metals. Important metals in this case include metals involved in toxicity to humans (*e.g.* the transport of cadmium through food chains) and metals of commercial economic value (*e.g.* the recovery of silver from industrial waste solutions).

Through determination of the physical and chemical mechanisms which regulate microbial metal accumulation phenomena, additional value could be realized. Such information would provide the basis for effective design and control of metal recovery process systems. Design information currently available is both meager and empirical. Ultimately, one might capture the biomimetic value of mechanistic knowledge and create more effective synthetic sorbents or metal complexing agents.

The use of microorganisms to treat aqueous streams for the removal, concentration and recovery of toxic and valuable heavy metals, although receiving increased attention in the last 5 to 10 years, was considered years earlier. Ruchloft (1949) observed that activated sludge efficiently removed plutonium-239 from contaminated water. He found that 96% removal could be accomplished in a single-stage treatment and pointed out that a two-stage process could be used if a greater degree of decontamination was required. Also, he described the decontamination process as the propagation of a microbial population 'having gelatinous matrices with tremendous surface areas that are capable of adsorbing radioactive materials'. The removal of heavy metals from municipal and industrial wastes by biological treatment systems has continued to be of interest (Cheng *et al.*, 1975; Eden, 1960; Neufeld and Hermann, 1975).

Although the microbial populations in waste treatment systems can effect heavy metal removal, there is always a real danger that these metals may poison the system, stopping biological activity and microbial growth. More recently, greater attention has been given to decoupling the propagation of the microbial sorbent from contact with the metal-contaminated stream to circumvent the toxicity problem. Jilek *et al.* (1975) described the preparation and pilot-scale use of a mycelial preparation from *Penicillium chrysogenum*, polymerized by urea–formaldehyde treatment, to remove uranium and radium from mine water. A patent was issued to Heide *et al.* (1975) for the use of a matrix of green algae to extract uranium from seawater. As will be discussed in detail later, interest in the use of organisms specifically propagated for their metal accumulating properties has continued. This includes not only the use of microorganisms as sorbents, but also the development of microbial populations that are able to grow in the presence of, and at the same time accumulate, heavy metals.

An important consideration for the practical utilization of microorganisms for the accumulation/separation/recovery of metals is the amount of metal that can be accumulated by the cell mass. The amounts of metals which microbial cells can accumulate vary from a few micrograms per gram of cells to several percent of the dry cell weight. Table 1 presents several examples where substantial amounts of certain metals have been observed to be accumulated.

Table 1   Levels of Metal Accumulation by Microorganisms[a]

Metal	Organism	Metal uptake (g metal g^{-1} cells dry wt)	Ref.
Ag	Mixed culture	0.32	Charley and Bull (1979)
	Pseudomonas maltophilia	0.18	Charley and Bull (1979)
	Thiobacillus ferrooxidans and *T. thiooxidans*	0.25	Pooley (1982)
Co	*Proteus vulgaris*	0.08	Neyland *et al.* (1952)
	Zoogloea sp.	0.25	Friedman and Dugan (1968)
Cu	*Zoogloea* sp.	0.34	Friedman and Dugan (1968)
Ni	*Zoogloea* sp.	0.13	Friedman and Dugan (1968)
Pb	*Citrobacter* sp.	0.35	Aickin *et al.* (1979)
U	*Rhizopus arrhizus*	0.18	Tsezos and Volesky (1982a)
	Saccharomyces cerevisiae	0.15	Strandberg *et al.* (1981)
	Pseudomonas aeruginosa	0.15	Strandberg *et al.* (1981)
	Penicillium chrysogenum	0.08	Jilek *et al.* (1975)
Th	*Rhizopus arrhizus*	0.17	Tsezos and Volesky (1982b)

[a] These examples were extracted from the references indicated and are presented solely to indicate the magnitudes of metal accumulation observed.

13.2   PHENOMENA/MECHANISTIC CONSIDERATIONS

An understanding of the mechanisms by which microorganisms accumulate metals is important to the development of microbial processes for the concentration, removal and recovery of metals from aqueous solutions. For example, a knowledge of the chemical or physiological reactions that occur during metal uptake might enable specification and control of process parameters to increase the rate, quantity and specificity of metal accumulation. Also, by knowing those inherent properties or activities of an organism that are responsible for metal uptake, it is possible to enhance the microorganism's ability to accumulate metals through environmental (*i.e.* growth conditions) or genetic manipulation.

The ability to accumulate a particular metal varies among microbial species; certain metals are also accumulated to a greater extent than others. Thus the mechanistic bases for observed metal accumulation phenomena must be considered with respect to the properties of both the microorganisms and the metals.

We have arranged this discussion first on the basis of the physical site of metal accumulation and then, where evidence exists, the explanations that have been given as to how and why metals accumulate at these sites. Metal accumulation has been observed to occur in association with extracellular products (*e.g.* polymers), at or within the cell membrane, and intracellularly. There are several processes which can lead to metal accumulation at these sites. Biosorption is the nondirected, physical–chemical complexation reaction between dissolved metal species and charged cellular components, akin in many respects to ion exchange. The precipitation or crystallization of metals can take place at or near the cell and possibly subsequent to initial biosorptive complexation. Also, insoluble metal species can be physically entrapped in microbially produced extracellular polymers. Metabolically mediated metal uptake involves specialized cell functions such as ion transport systems, enzymes and energy expenditure. Additionally, metals may accumulate within microbial cells by undefined nonmetabolically mediated processes.

Although we do not consider the detailed chemical or physiological reactions that actually occur, it is important to realize that the physiological condition of the cells, the chemical state of the reactive sites on the cells or cell products and the chemical state of the metals are strongly influenced by the environment. Consequently, any interaction that takes place between microbial cells and metal ions may be influenced accordingly.

Finally, we have restricted our coverage to examples of the accumulation of metals having environmental and industrial significance.

13.2.1 Extracellular Accumulation

The removal of metals from aqueous streams in biological waste treatment systems has been the subject of numerous reports. Both metabolically mediated and biosorptive phenomena can and do occur in such systems. However, most studies have concentrated on the role of extracellular polymers in effecting metal removal (Brown and Lester, 1979; Dugan and Pickrum, 1972; Friedman and Dugan, 1968). Brown and Lester (1979) reviewed the mechanisms of metal removal by microbial polymers such as that produced by *Zoogloea*. Included were the physical entrapment of precipitated metals by the polymer matrix (possibly mediated by microbially induced changes in the environment, pH for example) and the complexation of soluble metal species by charged constituents of the polymers. Although many extracellular microbial polymers consist of neutral polysaccharides, others frequently contain compounds such as uronic acids, hexosamines and organically-bound phosphates that are capable of complexing metal ions. The complexation of metal ions by charged constituents may be likened to an ion-exchange type reaction and thus is affected by the chemical environment and the presence of other metals or ligands. These types of complexation reactions are considered in greater detail in the following section.

The composition and, hence, the metal binding properties of extracellular polymers differ among microbial species and can depend on the conditions in which the organisms are grown. For example, Strandberg *et al.* (1981) demonstrated that the uranium-complexing capacities of extracellular phosphomannans isolated from two different species of yeast were directly related to the phosphate content of the individual polymers. Although metal accumulation by extracellular polymers is generally considered a passive phenomenon requiring no directed microbial activity, it has been suggested that bacteria produce large amounts of extracellular organic material in the presence of toxic metal ions (Jones, 1967). However, during a study of the metal-binding properties of extracellular, surface-active polymers of marine bacteria, Corpe (1975) found no evidence for this.

13.2.2 Cell Surface Accumulation

There have been numerous studies in which metal accumulation was either demonstrated or implied to occur at the cell surface or within the cell-wall matrix. It has been generally assumed that surface accumulation is the result of complexation reactions between metal ions and the charged, receptive constituents of the cell walls. However, in only a few instances has there been a detailed investigation of the phenomena involved. The composition of cell walls of microorganisms is highly species-dependent and differs considerably among Gram-negative and Gram-positive bacteria, yeasts, filamentous fungi and algae. Without elaborating on the structural details, it appears that there are constituents in these cell walls which have the potential to complex metal ions. As in the formation of extracellular polymers, the wall composition is not only species-dependent but, to some extent, is subject to the conditions under which the organisms are grown. For example, the protein and phosphate contents of yeast cell walls are significantly greater for organisms cultured at a slow growth rate (McMurrough and Rose, 1967).

Studies by Matthews *et al.* (1979), Beveridge and Murray (1976, 1980), and Doyle *et al.* (1980) have provided strong evidence that the carboxyl groups (of glutamic acid) in the peptidoglycans present in cell walls of the Gram-positive *Bacillus subtilis* are the primary sites of divalent metal complexation. This is in contrast to earlier studies (Heptinstall *et al.*, 1970; Huges *et al.*, 1973), which suggested that the phosphate-containing teichoic acid in the cell wall was responsible for metal binding. More recently, Beveridge *et al.* (1982) demonstrated that teichoic and teichuronic acids are the prime sites for metal deposition in cell walls of the Gram-positive *B. licheniformis*. Thus, the metal binding sites seem to differ even among related species of bacteria. From studies with the structurally and chemically more complex cell envelope of the Gram-negative bacterium *Escherichia coli*, Beveridge and Koval (1981) concluded that the primary sites of metal binding are the polar head groups of the phospholipids, the available anionic sites of the lipopolysaccharides and the acidic groups of exposed polypeptides.

The biosorption of uranium by yeasts has been examined to some extent. Rothstein and Meier (1951) suggested that the surfaces of yeast cells contain reactive groups that are chemically similar to the high molecular weight polyphosphates which are responsible for uranium accumulation. The uranium-complexing ability of phosphomannans which can be present in yeast cell walls was noted in Section 13.2.1 (Strandberg *et al.*, 1981). However, it has also been established that the carboxyl groups of proteins can effectively complex uranium (Dounce and Flagg, 1949), and proteins are present in yeast cell walls (Farkas, 1979). Strandberg *et al.* (1981) found that the treatment of *Saccharomyces cerevisiae* cells with formaldehyde increased the rate of uranium uptake, presumably by decreasing the repulsive force of the positively charged amino groups of surface proteins. These same authors found evidence that the dicarboxylic amino acids will more effectively complex uranium than will the monocarboxylic amino acids. They suggested that the positive charge on the amino group may prevent uranium from complexing with the proximal carboxyl group, while the distant carboxyl group of a dicarboxylic amino acid is free to complex the metal.

Several filamentous fungi have been shown effective as biosorbents. As with bacteria and yeasts, it has been generally assumed that reactive groups on the mycelial surface are responsible for metal binding. However, Tsezos and Volesky (1982a) recently determined, by studies using electron microscopy, energy dispersive X-ray analysis and IR spectroscopy, that uranium was complexed in the stratified layers of chitin in cell walls of *Rhizopus arrhizus* (Figure 1). They suggested that uranium biosorption includes at least three processes, including uranium coordination with the amine nitrogen of chitin, the adsorption of additional uranium in the cell-wall chitin structure, and the precipitation of uranyl hydroxide within the cell-wall matrix. In contrast, thorium (Figure 2) accumulated primarily at the outer cell surface (Tsezos and Volesky, 1982b). The proposed mechanism included the coordination of thorium with the nitrogen of the cell-wall chitin and the adsorption of hydrolyzed thorium ions at the cell surface, apparently by other cell-wall constituents.

As defined earlier, biosorption is the physical/chemical reaction between positively charged, dissolved metal species and charged, reactive cellular components. This process has been compared to an ion-exchange phenomenon. Therefore, it is not unexpected that biosorptive metal uptake is subject to environmental conditions that affect the reaction chemistry of both the receptive sites and the metals. For example, pH and temperature can significantly affect biosorption (Neyland *et al.*, 1952; Strandberg *et al.*, 1981; Tsezos and Volesky, 1981, 1982a,b). Also as expected, certain metals are preferentially complexed. Marquis *et al.* (1976), studying three Gram-positive strains of bacteria, reported selective binding of metals in the following order: $La^{3+} \gg Cd^{2+} > Sr^{2+} > Ca^{2+} > Mg^{2+} > K^+ > Na^+$. Nakajima *et al.* (1981a) found slightly differing affinity series for the heat-killed and living cells of *Chlorella regularis*. Metal uptake decreased in the order: $UO_2^{2+} \gg Cu^{2+} \gg Zn^{2+} \gtrsim Ba^{2+} \simeq Mn^{2+} \gtrsim Co^{2+} \simeq Cd^{2+} \gtrsim Ni^{2+} \simeq Sr^{2+}$ for living cells and in the order: $UO_2^{2+} \gg Cu^{2+} \gg Mn^{2+} \gtrsim Ba^{2+} > Zn^{2+} \gtrsim Co^{2+} \gtrsim Cd^{2+} \gtrsim Ni^{2+} > Sr^{2+}$ for heat-killed cells. Strandberg *et al.* (1981) found that Ca^{2+} could partially displace uranium during uptake by *S. cerevisiae*. Tsezos and Volesky (1982a,b) observed that Fe^{2+} and Zn^{2+} interfered with uranium biosorption by *R. arrhizus*, but did not affect thorium binding.

Generally, metal uptake experiments are carried out by contacting the microorganism with an aqueous solution containing a soluble salt of the metal. However, depending upon the solution environment, the chemistry of the metals may be quite complex. For example, in addition to free metal ions, both hydroxide and carbonate complexes may exist. A quantitative knowledge of the metal species present in solution and of those associated with the cells is important for proper interpretation of the behavior being observed. Unfortunately, these data are difficult to obtain under actual conditions of the experiment and are seldom reported.

In our experiments on uranium biosorption (Shumate *et al.*, 1978, 1980; Strandberg *et al.*, 1981) the optimal pH range for uranium uptake by *S. cerevisiae* was pH 3.0 to 4.5. In this pH range, soluble uranium exists as UO_2^{2+} (Baes and Mesmer, 1976) and as other hydrolysis products $[(UO_2)_2(OH)_2^{2+}, UO_2(OH)^+, (UO_2)_3(OH)_5^+]$; carbonate complexes are also possible. In our experiments, as soluble uranium disappeared from solution and was bound to the cell, the solution pH increased from ~4.0 to 5.5–6.0, indicating a release of free hydroxyl ions. This suggests that UO_2^{2+} may have been the form of the bound metal. A similar conclusion was reached by Tsezos and Volesky (1981) for uranium deposition in cell walls of *R. arrhizus*. They also sug-

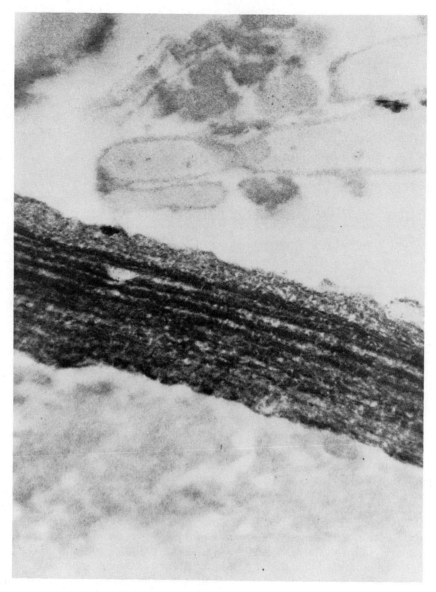

Figure 1 *Rhizopus arrhizus* cell wall following uranium biosorption. Electron micrograph (80 000 ×). From Tsezos and Volesky, 1982a. Reproduced by permission of Wiley, New York

gested that hydrolysis products precipitated within the chitin cell-wall matrix (Tsezos and Volesky, 1982a).

In many instances, the amount of metal accumulated by an apparent biosorptive complexation reaction reaches several percent of the dry weight of the biomass. If biosorptive complexation were strictly an ion-exchange phenomenon, it would be difficult to conceive of sufficient reactive sites existing on the cell to account for the quantity of metal actually accumulated. There have thus been suggestions that associated phenomena must occur. Beveridge (1978) observed that greater-than-stoichiometric amounts of metals such as gold (as metallic gold) accumulated on cell walls of *B. subtilis*. He proposed that the metal ions react initially with the carboxyl groups of the peptidoglycans, occupying the reactive sites. These bound molecules then act as nucleation sites for the crystallization of additional molecules. Strandberg *et al.* (1981) observed what appeared to be crystalline deposits of uranium on the surface of *S. cerevisiae* cells (Figure 3). These yeast cells accumulated 10–15% of their dry cell weight as uranium, on a total population basis. However, electron micrographs revealed that only 32% of the cells in a population had visible uranium deposits. Thus, for those cells accumulating uranium, the concentration approached 50% of their dry weight.

Figure 2 *Rhizopus arrhizus* cell wall following thorium biosorption at pH 4. Electron micrograph (80 000 ×). From Tsezos and Volesky (1982b). Reproduced by permission of Wiley, New York

The precipitation of metals may also explain the greater-than-stoichiometric amounts of metals accumulated. Tsezos and Volesky (1982a) proposed that a significant fraction of uranium accumulated within the wall structure of *R. arrhizus* by the precipitation of uranyl hydroxide. The surface accumulation of silver by *Thiobacillus ferrooxidans* and *T. thiooxidans* has been reported by Pooley (1982). Electron microscopic examination of these organisms recovered from experimental silver-leaching systems showed crystalline silver sulfide granules adhering to the cell surface. Aickin *et al.* (1979) found evidence that lead accumulation by *Citrobacter* involved the formation of firmly bound $PbHPO_4$ on the cell surface, with the phosphate moiety being contributed by the cells. Although biosorption, by definition, does not require cell metabolism, it is possible that a cell-mediated microenvironment may enhance the deposition of metals (see review by Kelly *et al.*, 1979).

13.2.3 Intracellular Metal Uptake

Microorganisms encounter both essential and toxic metals in the environment and, therefore, require mechanisms for controlling the intracellular concentrations and the chemical states of metals. For example, there are highly specific, active transport systems which are utilized for the uptake and removal of essential metal ions. Although the transport mechanisms for essential elements such as sodium, potassium and calcium have been studied extensively, much less is known about the systems for regulating the intracellular concentrations of other metals. Because of the ecological and toxicological significance of metal ions in the environment, the mechanisms by which microorganisms (and other biological systems) respond to metals are receiving increased attention.

The intracellular accumulation of many metals (including Cd, Ag, Zn, Cu, Cr, Ni, U, Pb, Hg, As, Tl, Pt, Pd and Au) has been found to occur in bacteria, fungi and algae. It has been inferred, in several instances, that the accumulation of a metal results from the lack of specificity in a nor-

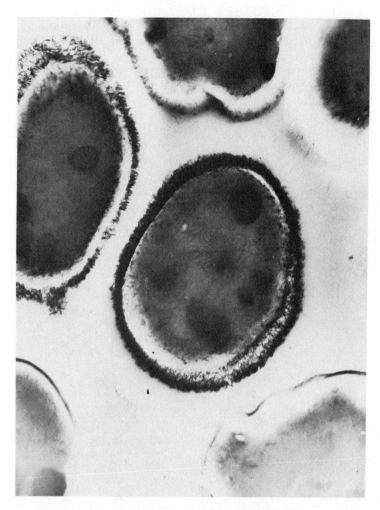

Figure 3 Electron micrograph of *Saccharomyces cerevisiae* showing surface accumulation of uranium (35 000 ×). From Strandberg *et al* (1981)

mal metal-transport system and that, at high concentrations, metals may act as competitive substrates in a transport system. These inferences are based on experiments measuring the competitive effects of specific metal-ion uptake (see review by Kelly *et al.*, 1979).

Although metals such as Ag, As, Hg, Zn, Pb and Cd are generally toxic, certain microorganisms show resistance to them. One mechanism of resistance has been shown to be a rapid, energy-dependent efflux system, such as exists for arsenate (Silver, 1981). However, toxic metals may also be rendered innocuous by systems that lead to their intracellular deposition and accumulation. The specific mechanisms by which these metals are taken up by the cells are essentially unknown, but in many cases their uptake has been shown to be under genetic control and, more specifically, to be controlled by plasmid-linked genes (Chakrabarty, 1976, 1978; Silver, 1981). Chakrabarty (1976) described microbial strains capable of accumulating Hg, Au, Pt and Pd. The capability for mercury accumulation was found to be transferable *via* a plasmid from a resistant strain to a normally sensitive strain of *Pseudomonas*. Charley and Bull (1979) established a community of three distinct species in chemostat culture that could grow in the presence of up to 0.1 mol l^{-1} Ag$^+$. Silver resistance was shown to reside in the ability of one member, *Pseudomonas maltophilia*, to accumulate the metal. Circumstantial evidence was found for the existence of a plasmid specifying silver resistance in this organism.

Intracellular metal deposition occurs by nonmetabolically mediated processes as well. Heldwein *et al.* (1977) concluded that lead accumulated intracellularly in *S. cerevisiae* by diffusion, while in this same organism Cd and Co uptake was energy dependent. Nakajima *et al.* (1981b) observed that U, Cu and Cd accumulated intracellularly in *Chlorella regularis*. Strandberg *et al.* (1981) demonstrated, by electron micrography, the intracellular deposition of uranium by

Pseudomonas aeruginosa (Figure 4) and by a nondefined, mixed culture of denitrifying pseudomonads. Uranium uptake by washed, resting cells of *P. aeruginosa* was very rapid (<10 s) and was unaffected by metabolic inhibitors, interfering metal ions, or environmental conditions (*e.g.* temperature, pH). Estimates of the mass-transfer coefficient for uranium uptake indicated that the process occurred at a slower rate than would be expected based on diffusion through the bulk solution (Strandberg and Shumate, 1982), and that the transfer of the metal through the cell envelope was the rate-limiting step.

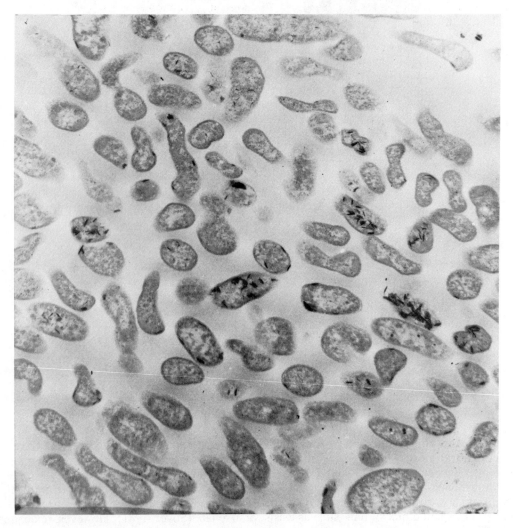

Figure 4   Electron micrograph of *Pseudomonas aeruginosa* showing intracellular accumulation of uranium (27 000 ×). From Strandberg *et al.* (1981)

Although there are many cellular constituents capable of forming complexes with intracellularly accumulated metals, the nature of the complexes is largely unknown. The metals may even be deposited in a metallic form. However, it is not unreasonable to expect that inducible, specific, metal-binding proteins are produced by those organisms that purposefully accumulate metals. These types of proteins are well known in higher organisms (see review by Simkiss, 1979). Recently, the metal-binding protein metallothionein, frequently found in higher organisms, was identified in a blue-green bacterium (Olafson *et al.*, 1979).

13.3   APPLICATIONS

There appear to be no instances of technically sophisticated processes based on microbiological metal separation phenomena in use on a large scale. One can, however, obtain an appreciation for the range of possibilities by reviewing several cases.

Gale and Wixson (1979) described the use of tailings ponds and artificial stream-meander systems using photosynthetic organisms to treat lead mining and milling effluents in Missouri's New Lead Belt. Lead mine-mill operators encountered problems with escaping Pb, Zn, Cu and Mn, and one company constructed an artificial stream-meander system to encourage biological removal of metals prior to discharge of the stream to the receiving body. During its operation from 1972 to 1979, the system reduced the number and severity of heavy metal release episodes. Filamentous algae (*e.g. Cladophora*) were shown to have contributed significantly to heavy metal binding. Field specimens of algae collected from the system did not release the metals when washed extensively with stream or tap water. Since washing with 0.01–0.1 kmol m^{-3} ethylenediaminetetracetate solution did rapidly remove the Pb, Zn, Cu and Mn without extensive cell lysis, it is likely that the metals were complexed with ionic groups on the outer cell surface. Figure 5 illustrates the concentration of Pb, Mn and Zn associated with biomass (*Cladophora*) at points along the water course downstream from the tailings dam.

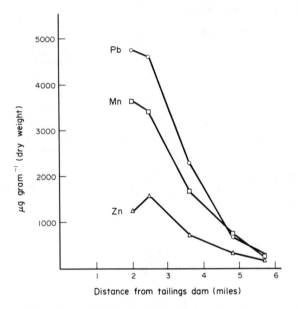

Figure 5 Heavy metals accumulation in algae (*Cladophora*) downstream from lead mine-mill tailings pond. From Gale and Wixson (1979). Reproduced by permission of the Society for Industrial Microbiology

Spisak (1979) proposed a more sophisticated process for biological removal of arsenic from an aqueous effluent produced in the refining of copper. Pyrometallurgical smelter flue dusts are collected and treated in a hydrometallurgical process to recover valuable metals. This Kosaka-type process typically discharges 300 l min^{-1} of aqueous solution containing 1 mg l^{-1} of arsenic. This arsenic concentration must be reduced to 50 μg l^{-1} or less before being released to the environment.

Spisak (1979) recommended that an algae-based arsenic removal process be employed, as depicted in Figure 6. Arsenic assimilation would be accomplished in two photosynthetic reactor vessels containing 1300 kg of algae in 24 700 kg of aqueous solution. Arsenic-laden algae would be harvested from the reactors at a rate of 430 kg d^{-1}, filtered, dried and incinerated at a temperature >300 °C to vaporize the As$_2$O$_3$. The discharge liquor would contain 40 μg l^{-1} of arsenic.

For a copper production operation that produces 60 tonnes of copper per day, Spisak (1979) estimated that the capital investment for the algae-based arsenic removal plant would be $834 240 US (1978 dollars). Based on 1978 economics, this would have cost 0.20 cents per kilogram of copper produced. No estimate of the operating cost was offered.

A third example illustrates an interesting approach to the simultaneous removal of metallic and non-metallic pollutants from industrial wastewater streams. Certain processing operations for nuclear materials generate effluents which must be treated to remove both nitrate and uranium. A mixed culture of bacteria, predominantly pseudomonads, was developed which was very effective in converting nitrate to gaseous molecular nitrogen (Hancher *et al.*, 1979). Cells formed as a

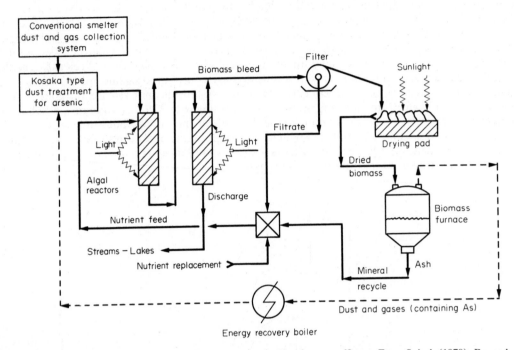

Figure 6 Supplemental arsenic scrub circuit for treatment of a Kosaka-type effluent. From Spisak (1979). Reproduced by permission of the Society for Industrial Microbiology

byproduct of the denitrification reaction were used to remove uranium from the stream in a unit operation separate from the denitrification bioreactor (Shumate *et al.*, 1980). A simplified flow-sheet is shown in Figure 7.

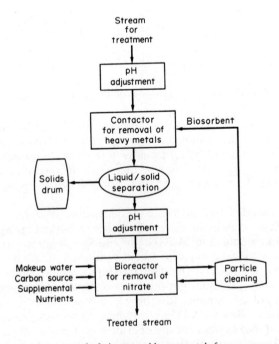

Figure 7 Flowsheet for the removal of nitrate and heavy metals from an aqueous waste stream

The denitrification bioreactor was operated in a continuous mode with the microbial cells as a film growing on and adhering to the surface of anthracite coal particles, which were fluidized by passage of the liquid stream upward through the bioreactor. An optimum microbial film thickness was maintained by continuously removing particles with excess cell mass (which accumulated at the upper boundary of the fluidized bed), passing the particles through a Sweco vibrating screen, and recycling the particles to the lower region of the fluidized bed. Excess cells, stripped from the particles by passage through the vibrating screen, were used as a biosorbent in a stirred-tank reactor for removal of uranium.

An alternate mode of biosorbent–liquid contacting was developed. The unit operation is depicted schematically in Figure 8 (Shumate *et al.*, 1980). Microbial cells were grown on, and left attached to, the surface of anthracite coal particles as described previously. The cell-laden particles were harvested from the denitrification bioreactor and placed in a vibrating particle feeder. The biosorbent particles were continuously metered into a column and settled counter-current to the flow of the uranium solution being treated. As indicated in Figure 8, the soluble uranium concentration could be reduced from 25.0 to 0.5 g m^{-3} in a mean liquid residence time of only 8 min.

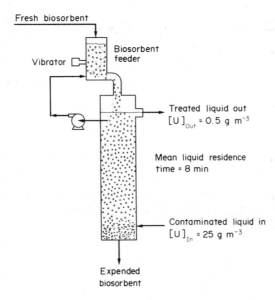

Figure 8 Schematic diagram of countercurrent contactor used for continuous separation of uranium by a film of denitrifying microorganisms attached to coal particles. From Shumate *et al.* (1980)

This mode of biosorbent–liquid contacting should be attractive in cases where cells reversibly adsorb or complex a metal on the outer cell surface. In this instance, the metal could be removed from the biosorbent in a concentrated form and the biosorbent recycled for removal of additional metal.

13.4 SUMMARY

Accumulation of metals by microbial cells has been observed to occur by metabolic and nonmetabolic processes. Nonmetabolically mediated accumulation may result from biosorption and precipitation of metals extracellularly, at or within the cell-wall matrix, and intracellularly. Metabolically mediated accumulation is usually intracellular and, in some cases, is under the control of plasmid-linked genes.

Accumulation at the cell surface is strongly influenced by the local physical and chemical environment. Important parameters include temperature, pH and concentrations of other metal ions. Several investigations have discovered specificity (preferred complexation) in metal-ion accumulation. Cell-associated metal concentrations of up to 50% of dry cell weight have been reported.

Commercial utilization of these phenomena has been rare. In a few instances, photosynthetic microorganisms have been utilized in stream-meander systems to prevent the discharge of environmentally objectionable metals. The development of more technically sophisticated process strategies and unit operations using microbial accumulation of metals has been reported.

13.5 REFERENCES

Aickin, R. M., A. C. R. Dean, A. K. Cheetham and A. J. Skarnulis (1979). Electron microscopic studies on the uptake of lead by a *Citrobacter* species. *Microbios Lett.*, **9**, 7–15.

Baes, C. F., Jr. and R. E. Mesmer (1976). *The Hydrolysis of Cations*. Wiley, New York.

Beveridge, T. J. (1978). The response of cell walls of *Bacillus subtilis* to metals and to electron-microscopic stains. *Can. J. Microbiol.*, **24**, 89–104.

Beveridge, T. J. and S. F. Koval (1981). Binding of metals to cell envelopes of *Escherichia coli* K-12. *Appl. Environ. Microbiol.*, **42**, 325–335.

Beveridge, T. J. and R. G. E. Murray (1976). Uptake and retention of metals by cell walls of *Bacillus subtilis*. *J. Bacteriol.*, **127**, 1502–1518.

Beveridge, T. J. and R. G. E. Murray (1980). Sites of metal deposition in the cell wall of *Bacillus subtilis*. *J. Bacteriol.*, **141**, 876–887.

Beveridge, T. J., C. W. Forsberg, and R. J. Doyle (1982). Major sites of metal binding in *Bacillus licheniformis* walls. *J. Bacteriol.*, **150**, 1438–1448.

Brown, M. J. and J. N. Lester (1979). Metal removal in activated sludge: The role of bacterial extracellular polymers. *Water Res.*, **13**, 817–837.

Chakrabarty, A. M. (1976). Which way genetic engineering? Recent trends in genetic engineering and its potential applications. *Ind. Res.*, January, 45–50.

Chakrabarty, A. M. (1978). Genetic mechanisms in metal–microbe interaction. In *Metallurgical Applications of Bacterial Leaching and Related Microbiological Phenomena*, ed. L. E. Murr, A. E. Torma, and J. A. Brierley, pp. 137–149. Academic, New York.

Charley, R. C. and A. T. Bull (1979). Bioaccumulation of silver by a multispecies community of bacteria. *Arch. Microbiol.*, **123**, 239–244.

Cheng, M. H., J. W. Patterson and R. A. Minear (1975). Heavy metals uptake by activated sludge. *J. Water Pollut. Control Fed.*, **47**, 362–376.

Corpe, W. A. (1975). Metal-binding properties of surface materials from marine bacteria. *Dev. Ind. Microbiol.*, **16**, 249–255.

Dounce, A. J. and J. F. Flagg (1949). The chemistry of uranium compounds. In *Pharmacology and Toxicology of Uranium Compounds*, ed. C. Voegtlin and H. C. Hodge, part I, pp. 55–145. McGraw-Hill, New York.

Doyle, R. J., T. A. Matthews and U. N. Streips (1980). Chemical basis for selectivity of metal ions by the *Bacillus subtilis* cell wall. *J. Bacteriol.*, **143**, 471–480.

Dugan, P. R. and H. M. Pickrum (1972). Removal of mineral ions from water by microbially produced polymers. *Purdue Univ. Eng. Ext. Series Eng. Bull.*, **141**, 1019–1038.

Eden, G, E. (1960). Biological concentration of radioactivity and its application to the treatment of liquid effluents. In *Radioactive Wastes, Their Treatment and Disposal*, ed. J. C. Collins, pp. 141–165. E. and F. N. Spon, London.

Farkaš, V. (1979). Biosynthesis of cell walls of fungi. *Microbiol. Rev.*, **43**, 117–144.

Friedman, B. A. and P. R. Dugan (1968). Concentration and accumulation of metallic ions by the bacterium *Zoogloea*. *Dev. Ind. Microbiol.*, **9**, 381–388.

Gale, N. L. and B. G. Wixson (1979). Removal of heavy metals from industrial effluents by algae. *Dev. Ind. Microbiol.*, **20**, 259–273.

Hancher, C. W., P. A. Taylor and J. M. Napier (1979). Operation of a fluidized-bed bioreactor for denitrification. *Biotechnol. Bioeng. Symp.*, **8**, 361–378.

Heide, E.-A., M. Paschke, K. Wagener and M. Wald (1975). Aus kultivierbaren Mutanten von einzelligen Grünalgen bestehende Matrix sowie Verfahren zur Urangewinnung mittels dieser Matrix. *Ger. Pat.* 2 345 430.

Heldwein, R., H. W. Tromballa and E. Broda (1977). Aufnahme von Cobalt, Blei un Cadmium durch Bäckerhefe. *Z. Allg. Mikrobiol.*, **17**, 299–308.

Heptinstall, S., A. R. Archibald and J. Baddiley (1970). Teichoic acids and membrane function in bacteria. *Nature (London)*, **225**, 519–521.

Hughes, A. H., I. C. Hancock and J. Baddiley (1973). The function of teichoic acids in cation control in bacterial membranes. *Biochem. J.*, **132**, 83–93.

Jilek, R., H. Prochazka, K. Stamberg, J. Katzer and P. Nemec (1975). Some properties and development of cultivated biosorbent (Nekteré vlastnosti a výroj nakultivovaneho biologickeho sorbentu). *Rudy*, **23**, 282–286.

Jones, G. E. (1967). Growth of *Escherichia coli* in heat- and copper-treated synthetic seawater. *Limnol. Oceanogr.*, **12**, 167–172.

Kelly, D. P., P. R. Norris and C. L. Brierley (1979). Microbiological methods for the extraction and recovery of metals. In *Microbial Technology: Current State, Future Prospects*, ed. A. T. Bull, C., Ellwood and C. Ratleds, 29th Symp. Soc. Gen. Microbiol., pp. 263–308. Cambridge University Press, Cambridge.

Marquis, R. E., K. Mayzel and E. L. Carstensen (1976). Cation exchange in cell walls of gram-positive bacteria. *Can. J. Microbiol.*, **22**, 975–982.

Matthews, T. H., R. J. Doyle and U. N. Streips (1979). Contribution of peptidoglycan to the binding of metal ions by cell wall of *Bacillus subtilis*. *Current Microbiol.*, **3**, 51–53.

McMurrough, I. and A. H. Rose (1967). Effect of growth rate and substrate limitation on the composition and structure of the cell wall of *Saccharomyces cerevisiae*. *Biochem. J.*, **105**, 189–203.

Nakajima, A., T. Horikoshi and T. Sakaguchi (1981a). Studies on the accumulation of heavy metal elements in biological systems. XVII. Selective accumulation of heavy metal ions by *Chlorella regularis*. *Eur. J. Appl. Microbiol. Biotechnol.*, **12**, 76–83.

Nakajima, A., T. Horikoshi and T. Sakaguchi (1981b). Distribution and chemical state of heavy metal ions absorbed by *Chlorella* cells. *Agric. Biol. Chem.*, **45**, 903–908.

Neufeld, R. D. and E. R. Hermann (1975). Heavy metal removal by acclimated activated sludge. *J. Water Pollut. Control Fed.*, **47**, 310–329.

Neyland, M., P. Dunkel and A. L. Schade (1952). The uptake of cobalt by *Proteus vulgaris*. *J. Gen. Microbiol.*, **7**, 409–416.

Olafson, R. W., K. Abel and R. G. Sim (1979). Prokaryotic metallothionein: Preliminary characterization of a blue-green alga heavy-metal-binding protein. *Biochem. Biophys. Res. Commun.*, **89**, 36–43.

Pooley, F. D. (1982). Bacteria accumulate silver during leaching of sulphide ore minerals. *Nature (London)*, **296**, 642–643.

Rothstein, A. and R. Meier (1951). The relationship of the cell surface to metabolism. VI. The chemical nature of uranium-complexing groups of the cell surface. *J. Cell. Comp. Physiol.*, **38**, 245–270.

Ruchloft, C. C. (1949). The possibilities of disposal of radioactive wastes by biological treatment methods. *Sewage Works J.*, **21**, 877–883.

Shumate, S. E., II, G. W. Strandberg, D. A. McWhirter, J. R. Parrott, Jr., G. M. Bogacki and B. R. Locke (1980). Separation of heavy metals from aqueous solutions using 'biosorbents'—Development of contacting devices for uranium removal. *Biotechnol. Bioeng. Symp.*, **10**, 27–34.

Shumate, S. E., II, G. W. Strandberg and J. R. Parrott, Jr. (1978). Biological removal of metal ions from aqueous process streams. *Biotechnol. Bioeng. Symp.*, **8**, 13–20.

Silver, S. (1981). Mechanisms of bacterial resistances to toxic heavy metals: Arsenic, antimony, silver, cadmium, and mercury. In *Environmental Speciation and Monitoring Needs for Trace-Metal-Containing Substances from Energy-Related Processes*, ed. F. E. Brinckman and R. H. Fish, pp. 301–311. NBS Publication 618.

Simkiss, K. (1979). Metal ions in cells. *Endeavour*, **3**, 2–6.

Spisak, J. F. (1979). Metallurgical effluents — growing challenges for second generation treatment. *Dev. Ind. Microbiol.*, **20**, 249–257.

Strandberg, G. W., S. E. Shumate, II and J. R. Parrott, Jr. (1981). Microbial cells as biosorbents for heavy metals: Accumulation of uranium by *Saccharomyces cerevisiae* and *Pseudomonas aeruginosa*. *Appl. Environ. Microbiol.*, **41**, 237–245.

Strandberg, G. W. and S. E. Shumate, II (1982). *Accumulation of Uranium, Cesium, and Radium by Microbial Cells Bench-Scale Studies*. ORNL/TM-7599, Oak Ridge National Laboratory, Oak Ridge. TN.

Tsezos, M. and B. Volesky (1981). Biosorption of uranium and thorium. *Biotechnol. Bioeng.*, **23**, 583–604.

Tsezos, M. and B. Volesky (1982a). The mechanism of uranium biosorption by *Rhizopus arrhizus*. *Biotechnol. Bioeng.*, **24**, 385–401.

Tsezos, M. and B. Volesky (1982b). The mechanism of thorium biosorption by *Rhizopus arrhizus*. *Biotechnol. Bioeng.*, **24**, 955–969.

14

Microbial Degradation of Water-based Metalworking Fluids

H. W. ROSSMOORE
Wayne State University, Detroit, MI, USA

14.1 INTRODUCTION

Metalworking fluids (MWFs) have three distinct functions: (1) cooling tools, dies, forms and workpieces; (2) lubricating tools, dies, forms and workpieces; and (3) carrying off chips, fines and swarf. The processes all involve a change in or a reduction of net shape from some metal removal or forming operation.

If an application primarily involves lubrication, the base of the metalworking fluid (MWF) is essentially non-aqueous with a viscosity greater than water (*e.g.* may contain anticorrosion and extrapressure addition). If an application primarily involves cooling, the MWF is essentially water with similar additives for heavy duty operations and corrosion protection and, in addition, antimicrobial agents.

The purposeful use or addition of water to metalworking operations has created a more than favorable environment for a variety of microorganisms. They can directly or indirectly cause occupational hazard, economic loss and environmental insult. This is the *raison d'être* of this chapter.

14.2   NATURE OF METALWORKING FLUIDS

14.2.1   Classification by Chemistry

Until recently, the dominant scheme for classifying MWFs has been based on fluid chemistry (American Society for Testing and Materials, 1978), rather than fluid function. The following is simplified from the American Society for Testing and Materials (ASTM) Designation D 2881-73 for water-containing fluids:

   I   Oil Emulsions
     A.   Water in oil — Primarily deformation
     B.   Oil in water
   II   Chemical Solution
     A.   True solution
     B.   Colloidal solution
   III   Pre-Formed Emulsion, Oil in Water Emulsions in Concentrated Form Containing Additives

The general chemistry of all of the above fluid types has been well documented in a number of patents (Andlid and Linden, 1980; Compton and Sucher, 1965; Koh, 1980; Piotrowski and Davis, 1981). However, the vast majority of formulations are hidden in trade secrets. The emulsifiable oils are either based on paraffin or naphthenic (70–85%) structures with sulfonated derivatives (petroleum sulfonate) as emulsifiers, with or without extrapressure (EP) additives. These include fats, fatty acids, other esters, sulfurized fats and mineral oils, and chlorinated fats and mineral oils. The so-called 'heavy duty products' have increasing amounts of EP additives at the expense of the mineral oil content. The oil film of a stable emulsion usually is sufficient anticorrosion protection. Thus corrosion of tools and workpieces is one result of emulsion separation, whether it be from chemical or biological causes.

Chemical solutions are just that. The concentrated product contains about 75% water (none in the emulsion) by weight. The advantage to these products is their greater receptivity to a variety of water-soluble additives, including anticorrosion and antimicrobial agents. The number of justifiable permutations in formulations varies with the process (*ca.* 15) and the formulation (*ca.* 300+) (Table 1). Primarily, these include amine-based wetting agents, fatty acids and alcohols, waxes, borates, phosphates and some corrosion protection additives. Until the mid 1970s, sodium nitrite was the most widely used corrosion inhibitor. However the discovery of nitrosamines in MWF containing amines and nitrite (Centers for Disease Control, 1976; Fan *et al.*, 1977), has resulted in the gradual elimination of nitrite from formulations.

The preformed emulsions are dilute oil-in-water emulsions containing 10% or more mineral oil in the concentrate. They have the advantage of chemical solutions and possibly the extra lubrication offered by the emulsified oil.

14.2.2   Classification by Use

Currently, a new approach to MWF classification is under consideration by the International Standards Organization (ISO). This is a functional scheme based on cooling and lubricant properties with the accessory attributes such as EP and corrosion inhibition extending the classification.

This approach seems more reasonable. Among the use modes treated in this new outline are operations involving deformation of metals (*e.g.* stamping, rolling, and drawing) and net shape reduction of metals (*e.g.* grinding, machining, drilling, gear cutting).

14.3   DELIVERY SYSTEMS

The size and type of delivery system play a major role in the prevention, detection and treatment of microbial problems.

Sizes of unit fluid systems vary from under 40 l to a behemoth, in some cases exceeding 10^6 l.

Table 1 Relationship of Metalworking Function to Fluid Selection

Severity and type of operation	Metal machineability	Performance level	Fluids recommended	
			Fluid type	Ingredients
Most severe Examples: Broaching Tapping Threading Hobbing, gear cutting, shaving	Low	High (heavy duty)	Soluble oils	20–50% EP additives, 10–30% mineral oil, rest emulsifiers
			Semi-synthetics (possible)	10–40% EP additives, 20–50% water, rest emulsifiers
Medium severity Examples: Deep hole drilling, reaming, gun drilling Boring Multiple-spindle, automatic screw machines and turret lathes High-speed, light feed screw machining	Medium	Medium	Soluble oils	15–25% fat, 50–70% mineral oil, rest emulsifiers
			Semi-synthetics (possible)	10–40% EP additives, 20–50% water, rest emulsifiers
			Synthetics	50–70% water, rest synthetic EP additives, etc.
Relatively easy Examples: Drilling Milling Sawing Planing Turning	High	Low	Soluble oils	70–85% mineral oil, rest emulsifiers
			Semi-synthetics	10–40% mineral oil, 20–50% water, 20–30% emulsifiers
			Synthetics	50–70% water, rest rust inhibitors, biocides, etc.

The former are linked to individual machines, while the latter are reservoirs for a recirculating central delivery system. These can service a large number of diverse metalworking operations and especially those involved with transfer lines. The central systems contain as much as 25% of the fluid not actually involved in the engineering operation. This is the in-transit and reservoir fluid.

There are many trade-offs in the maintenance of a central system when compared to individual machine sumps. Intrinsically, by design, the central system imposes more impressive microbiological problems. Recirculation of large volumes of fluid (*ca.* 10 min system^{-1}) is ideally accomplished with the concomitant removal of metal chips and fine, so-called 'tramp oil', and other miscellaneous particulate contamination (Brandt, 1974). Except for the most recently built facilities, the capability of preventing sludge accumulation is limited, and frequently there are dead spots (*i.e.* regions of minimal flow) all resulting in the potential for maximizing microbial growth and, indeed, minimizing its treatment. In these large systems (10^4–10^6 l), the physical half-life of the fluid tends to be longer than for individual sumps. That is, the turnover time due to the manufacturing operation can vary from one to three months. Fluid is lost on the workpiece by splashing and by misting. All types of operations lose an additional volume from water evaporation. The establishment of the appropriate fluid concentration is thus a resultant of these two kinds of losses.

Many operations have calculated from experience the average concentration change from fluid loss and water evaporation and add make-up at a lower concentration than the original. With some fluids, analytical chemical procedures permit a more exact determination of fluid concentration. For example, soluble oil emulsions can be monitored by refractivity while some synthetic fluids can be evaluated by infra red (IR), high-performance liquid chromatography (HPLC) analyses, or by titration of alkaline reserve. Obviously these procedures depend upon the sophistication and mission of the metallurgical or chemical laboratory.

Make-up rate and concentration maintenance have a dual impact on microbiological activity. The longer the physical half-life of the fluid in the system, the greater the potential for chemical breakdown and subsequent microbial growth (Rossmoore and Holtzman, 1974). Any innate bioresistance exhibited by some fluids would be decreased with dilution below optimal engineering levels (Bennett, 1974b). Frequently, antimicrobial agents are included in the concentrated MWF to yield the optimal level when diluted for use. Thus, any reduction in the working concentration of the fluid would likewise result in the reduction of biocide levels. In addition, even with maintenance of the desired fluid concentration, both the physical and chemical half-life of the fluid may exceed the chemical and biological half-life of the biocide (Rossmoore and Holtzman, 1974).

This latter effect is a separate issue and will be dealt with in a later section. However, monitoring biocide and microbial levels are both accomplished more or less routinely. The cost effectiveness of this surveillance is directly proportional to the size of the system. A large central system gone bad not only poses horrendous logistical problems with respect to disposal (the potential interruption of a very large-scale production), but also the cost of fluid replacement. With care, large central systems can last for years without dumping (McCoy, 1978; Quanstrom, 1977).

Individual machine sumps are frequently at the whim of the operator since they can more readily manipulate fluid concentration. This can be microbiologically disastrous. Monitoring and treating microbial problems in small systems is economically prohibitive as a standard operating procedure. The time and materials for testing a 25 l sump is the same as for 1000 l sumps. In some instances, the cost of surveillance exceeds the value of biocide and the fluid. In these systems, there is a greater reliance on a premixed biocide in the fluid which is frequently diluted in a central location for a one-way trip to machine sumps.

14.4 FACTORS CONTRIBUTING TO BIODEGRADATION

14.4.1 Type of Operation

The operation is inextricably tied to the selection of the fluid and in this sense there can be a two-fold impact from fluids which, based on engineering demands, have their own biosusceptibility profile and from the mechanics of the process, including the metal being worked.

The higher the speed of the operation, the greater the heat generated and usually the greater the water content (Table 1). This latter factor is a major contributor to biodeterioration.

During deformation and production of net shape, waste metal and grinding swarf accumulate.

The smaller the particles, the greater their reactivity with fluid commponents, biocides and pioneering microbial species. The former fact, increasing growth rate by increasing surface area for attachment of sessile organisms, is an established theorem of microbial ecology.

In the drawing and ironing (D and I) formation of two-piece aluminum cans, particulate aluminum is partially solubilized; the aluminate reacts with the fatty acid of the fluid resulting in aluminum soaps which are an important contributor to lubricity. Filtration of can-forming fluids is obligatory and both aluminum fine and the soaps contribute to cake formation and filter registering. Growth on the filter cake with soap hydrolysis and free fatty acid precipitation results in earlier filter plugging (King, 1982). Intrinsic to all operations are the inevitable, inescapable hydraulic leaks from presses, tools and transfer lines and contamination from way lubricant. These intrusive fluids can adversely affect microbiological resistance (Bennett, 1976), primarily by sequestering biocide from the MWF.

The metal worked can play a significant role in fluid bioresistance. Some metals, especially the amphoteric species, dissolve significantly at fluid pH. These include aluminum and tin (also bronze). These can intrinsically inhibit some organisms (Rossmoore, 1979; Table 2). In addition, there is the significant danger of metal ion reducing biocide activity (Bennett *et al.*, 1982; Rossmoore, 1979).

Table 2 Effect of Metal Machined on Activity of Hexahydro-1,3,5-tris(2-hydroxyethyl)-*s*-triazine in 5% Oil-in-Water Cutting Fluid[a,b]

	Cast iron	Steel	Bronze	Brass	Aluminum	Copper
Microbiocide (0.1%)	B−/F+	B−/F+	B−/F+	B−/F+	B+/F−	B−/F+
No microbiocide	B+/F−	B+/F−	B+/F−	B+/F−	B+/F−	B+/F−

[a] Modified from Rossmoore, 1979. [b] B+ = bacterial growth (10^7 ml^{-1}) after 6 weeks; B− = no bacterial growth; F+ = countable fungal colonies and/or visible fungal growth in system; F− = no fungal growth; inoculum: spoiled cutting fluid 10% with 10^8 bacteria ml^{-1} and 10^3 fungi ml^{-1}.

All operations are dependent upon water for dilution of fluid. There is no doubt that the quality of the water affects the functional stability of the fluid. Hard water (so classified on the basis of water containing more than 200 mg l^{-1} Ca^{2+}/Mg^{2+}) may in fact contribute to emulsion separation and fatty acid precipitation.

The effect of hardness on microbial growth in fluids is equivocal. There is evidence for indirect but permissive stimulation by biocide neutralization (Bennett, 1973a; Rossmoore, 1979). However, in continued experiments in the laboratory, both stimulation (Feisal and Bennett, 1961) and inhibition (Byrom and Hill, 1971) of *Pseudomonas aeruginosa* were reported.

14.4.2 Effect of Fluid Formulation

Although formulation of a bioresistant fluid remains an idealistic goal, primary concerns should involve engineering requirements, occupational safety and disposability. Fluid formulations may show an increase in biosusceptibility after dilution beyond the most popular use concentration. (For example, at 5% a fluid may exhibit a practical level of resistance, but at 2.5% this resistance is lost.) Fluids formulated for aluminum need not be at the relatively high pH values (*ca.* 9–9.5) used in ferrous metalworking. Although high pH is not sufficiently inhibitory to prevent growth, a lower beginning pH permits a more rapid increase in bacterial populations (Rossmoore, unpublished data and 1962).

There is also a difference among fluids with regard to their relative resistance to bacteria and fungi. Bacteria do not apparently grow as well in certain synthetic fluids containing amine borate esters, but the trade-off is usually luxuriant fungal growth (Bennett, 1972; Rossmoore and Holtzman, 1974; Rossmoore *et al.*, 1976). Despite the preferential appetite of fungi for synthetic fluids and, indeed, what appears to be the converse with bacteria and the soluble emulsions, these are not absolutes. Bacteria do indeed grow in these synthetic fluids, and fungi do grow in the emulsions. However, it would be safe to say that in each fluid type, one microbial group is the primary and initial cause of biodeterioration.

An ancillary role of the fluid can be related to lack of compatibility with a selected biocide, a fact that is frequently not learned, sadly, until it is too late. Incompatibilities between fluids can also lead to biodeterioration. In a particular machining operation involving nodular iron with a 4% soluble oil, a decision was made to switch to a synthetic fluid. The central system capacity was *ca.* 200 000 l. The new fluid passed all engineering and disposability criteria. Disposal of existing fluid was not considered an economical option. Thus, the loss of 'old' fluid during operation was made up daily with 'new' fluid. When approximately 90% of the old had been replaced (about six weeks), residual oil coalescence had become foci for fungal growth (Rossmoore and Treusch, 1975). The black masses on the weirs contained 4×10^4 fungal colony-forming units (c.f.u.) per gram (Figure 1).

Figure 1 Central system settling tank (57 000 US gal) heavily contaminated with fungi (4×10^4 c.f.u. g^{-1})

14.4.3 Housekeeping

Although most fluids can and do support the growth of large numbers of organisms (as many as 10^9 aerobic bacteria ml^{-1}) and there is evidence for the utilization independently of many of the components of the fluids (hydrocarbons, emulsifiers, fatty acids, waxes, amines *etc.*) (Bennett, 1974b; Hill, 1968; Sabina and Pivnick, 1956), frequently there is no evidence of overt degradation of the fluid. A major contributor to the onset of degradation is growth on substrates extrinisic to the fluid's components. In many instances, central systems become recipients of floor sweepings, food, cigarettes, and human and animal waste. Certainly these increase the organic burden, add needed nitrogen for the microbial diet and stimulate growth (Bennett, 1972; Rossmoore, 1974).

No fluid can maintain resistance for long periods of time unless by design and intent extraneous additions are kept from the fluid. This means good housekeeping.

14.4.4 Evaluation of Bioresistance

At best, this measurement is relative to the needs of the system. If the fluid residence time is long, then bioresistance should be concomitantly long; however, if turnover is rapid, then bioresistance becomes a two-edged sword. The waste treatment facility relies on biodegradability to dispose of used fluids. In practical terms, this is becoming a problem even with fluids no longer functional. This will be discussed later under the heading of disposability. Currently, bioresistance is only measured by the ability of the fluid to inhibit or control the growth of a selected microbial inoculum. For specific sites, the inoculum should be selected from that site, while for more general evaluation, a repository battery of organisms will suffice (ASTM, 1980; Hill, 1978; Rossmoore and Rossmoore, 1980; Rossmoore *et al.*, 1977).

14.5 MICROBIAL SUCCESSION

14.5.1 Aerobic Bacteria

With the purposeful addition of water to MWFs, the microbial problems begin. From the first definitive study, members of the genus *Pseudomonas* were recognized as the dominant group (Duffett *et al.*, 1943) and subsequently a new type species, *Pseudomonas oleovorans*, was described (Lee and Chandler, 1941). Although numerous species have since been isolated (Bennett and Wheeler, 1954; Pivnick *et al.*, 1954), the *Pseudomonas* cluster remains dominant (Bennett, 1974a; Rossmoore and Rossmoore, 1980; Table 3).

Table 3 Frequency of Isolation of Bacterial Metalworking Fluid Species[a]

	Fluid type	
Species[b]	*Soluble oil*	*Synthetic and semi-synthetic*
Pseudomonas aeruginosa	16	19
Pseudomonas fluorescens	2	1
Pseudomonas cepatia	0	2
Pseudomonas stutzeri	1	0
Pseudomonas alcaligenes	0	1
Pseudomonas pseudomallei	0	1
Pseudomonas putida	1	0
Aeromonas hydrophila	1	1
Proteus mirabilis	2	3
Proteus vulgaris	1	1
Proteus rettgeri	1	1
Enterobacter cloacae	3	7
Enterobacter agglomerans	2	1
Enterobacter gergoviae	1	3
Citrobacter freundii	2	12
Escherichia coli	1	4
Klebsiella pneumoniae (oxytoca)	6	8
Klebsiella ozaenae	3	0
Serratia liquefaciens	1	0

[a] Modified from Rossmoore and Rossmoore, 1980. [b] Total number of isolates: 45 in soluble oil; 65 in synthetic and semi-synthetic fluids.

Water-miscible metalworkings are decidedly a Gram-negative milieu with only occasional isolation of Gram-positive organisms. In laboratory studies (Tant and Bennett, 1958), they survive poorly when added to fluids. The pseudomonads have a generic penchant for nutritional adaptation and are structurally suited to be the pioneers in the metalworking ecosystem. External polysaccharide fibrils (McCoy *et al.*, 1981) allow them to initiate attachment to tool surfaces, metal fines and begin biofilm formation; they become dispersed throughout the system benefitting from the extensive aeration produced by fluid flow. During growth they can utilize all the organic components of the fluid, oxidizing alkanes to fatty acids, β-oxidizing fatty acids, deaminating amines, hydrolyzing petroleum sulfonates and degrading corrosion inhibitors. By their activities they discolor the fluids, cause separation of emulsions, and drop the pH as well as the eH. They prepare the way for the anaerobic sulfate reducers (Isenberg and Bennett, 1959) by either metabolizing an inhibitor or reducing the redox potential.

14.5.2 Sulfate Reducers

It has never been reported that sulfate reducers are capable of initiating growth in new fluids before prior aerobic growth in the system by aerobic species. Once contaminated with sulfate reducers, they are difficult to eliminate. The reasons have more to do with engineering than microbiology. The appearance of these organisms suggests that the fluid has become anaerobic, perhaps during periods of shutdown (third shift) or on weekends when central systems stop circulating. This lack of aeration allows the subsequent drop in dissolved oxygen (DO) to encourage resident sulfate reducers (see aerobic activity above). The reservoirs of sulfate reducers are the critical factor, since these may be dead ends either at the bottom of the sump or at some points in the delivery system (Figure 2).

Figure 2   Progressive blackening with increasing levels of sulfate-reducing bacteria. Notice that the sulfate reduction, as indicated by iron(II) sulfide formation, begins at the bottom where the dissolved oxygen levels are the lowest

Without a doubt, these organisms were the first related to microbial degradation of MWFs. They are easily recognized organoleptically. White emulsions turn grey-black due to the formation of FeS, and free H_2S is readily detectable by smell. Often, large amounts of H_2S are released suddenly as fluid aeration begins after a weekend shut-down. The ensuing olfactory insult is euphemistically referred to as 'Monday morning stink'.

These organisms are representative of larger groups comprising two genera, all of which are dissimilatory sulfate reducers; sulfate is used as a terminal electron acceptor. The taxonomy and nomenclature currently accepted (Postgate, 1967; Postgate and Campbell, 1966) have basically separated the sulfate reducers into two groups, spore formers and non-spore formers. It is to the latter descriptive, *Desulfovibrio*, that MWFs accommodate. They are unique anaerobes; they can use sulfate or pyruvate as an electron acceptor, oxidize hydrogen, lactate (Postgate, 1979) or paraffins (Senez, 1962), and survive in aerated and non-aerated MWFs (Rossmoore, 1962). They have relatively long generation times in MWFs (Rossmoore and Brazin, 1968; Rossmoore *et al.*, 1964). They apparently survive desiccation since they have been recovered by air sampling near a working system with known sulfate reducer contamination (Vedder, 1985). Perhaps this is partially explained by the recent findings of their oxygen tolerance (Hardy and Hamilton, 1981) and especially catalase production, a metabolic rarity for obligative anaerobes (Hardy, 1981).

14.5.3   Fungi

This ubiquitous group has achieved more recent notoriety as spoilers, primarily due to their differential resistance to both the newer generation of fluids and biocides. Their growth produces its own set of sights and smells. In the extreme, fungal (mycelial) growth blocks flumes, clogs sprays, and inhibits settling (weirs). It has been aptly described as 'oily Spanish moss' and more crudely as 'black snot', and it it this sobriquet that causes a significant level of revulsion among the workforce. This can be coupled to the strong musty, locker-room odor which accompanies growth.

It would appear that in the absence of evidence for the significant breakdown of fluid components, the physical effects of growth can be considered the primary degradative effects. However, fungi have never been reported from fluids where neither the fluid nor the system had previous aerobic bacterial growth. The corollary appears to be true; you can expect fungal growth in some systems following a permanent or temporary die-off of pioneering bacterial species (Bennett, 1974a; Rossmoore and Holtzman, 1974; Rossmoore *et al.*, 1972, 1976).

In a contrived MWF ecology, fungal inocula survived in the presence of a large population of *Pseudomonas*, grew in sterile fluid in which *Pseudomonas* previously grew, but died off in fresh MWF. Because of the sessile nature of the growth, the total fungal biomass is difficult to assess from c.f.u. data generated from sampling the fluid. A case in point is seen in Figure 1. The fungal count g^{-1} was 4×10^4, while the count in the fluid was only *ca.* 10 ml^{-1}. Thus, fluid plate counts are only the tip of a very widely distributed 'iceberg'. Either smelling or seeing fungal growth is certainly more diagnostic than a negative or low plate count.

Despite the accessibility of MWF systems to a large variety of airborne and waterborne fungal species, only a very few have consistently made this environment their home. These are all (so-called) Fungi Imperfecti, including *Cephalosporium* sp. and *Fusarium* sp. and, to a lesser extent, *Aspergillus*, *Penicillium* and *Cladosporium*. Additionally, in some synthetic fluids a number of yeast species are also found, primarily *Candida* and *Trichosporon* sp.

14.6   DETECTION, DETERIORATION AND DISEASE

14.6.1   Selection and Detection

With incipient biodeterioration, the questions asked about the microbial population are: (1) How many are there?, (2) What are they?, and (3) How soon can you tell (1) and (2)?

Any method used must give meaningful answers to the above. Detection methods currently available for aquatic systems include a variety of microscopic and analytic techniques that do not distinguish living from dead microorganisms (Buck, 1979). Others rely on some metabolic activity which can be detected more rapidly than visible growth (Tilton, 1982).

Three metabolic detection systems have been evaluated in MWF. In soluble oil emulsions, methylene blue reduction was proportional to *Pseudomonas* and/or aerobic populations, and breakpoint results were obtained within five hours of test initiation. In those systems with detectable sulfate reducer populations, the correlation of methylene blue reduction and operator assessment of spoilage was highly positive. No fungi were included or evaluated in the study (Rossmoore, 1972).

The detection of ATP levels in MWF can only reflect total viable biomass since no distinction can be made between bacteria and fungi (Jankowskus, personal communication) and, furthermore, can lead to faulty conclusions since a yeast cell may have 10^3 times more ATP than a bacterial cell. Similarly, equating O_2 release from H_2O_2 by cell catalase (present in members of the dominant population) with viable cell levels does not separate bacteria from fungi. In addition, the type and condition of each fluid demands the preparation of separate nomographs for each (Gannon and Bennett, 1981). This latter method has not been explored in depth in MWFs and may still have other shortcomings preventing its universal acceptance (Supavej *et al.*, 1972, 1973).

Other methods, including radiorespirometric, conductometric and calorimetric, have not been tried in MWFs. Moreover, the current available instrumentation is not sufficiently sensitive to give early warnings during the working day except with larger levels of contamination. Furthermore, they do not differentiate between bacteria and fungi.

There is some real doubt about the practicality of routine chemical tests that do not separate bacteria from fungi, especially since choice of type and time of biocide addition should be based on the total microbial picture. Indeed, even if rapid methods were microbially selective, the operator response would have to be equally rapid in making biocide addition decisions. Certainly systems that are checked weekly will not suffer if 24 hours intervene between sampling and dosing. An exception would be a completely automated system in which monitoring of microbial populations is coupled to biocide injection. This may be the system of the future, but for now, reliance on more traditional culturing methods based on the plate count and simplified on-site derivatives thereof is the case.

After the first awareness that microbial growth in MWFs contributed to deterioration, many manufacturing plants either developed some minimal microbiological competence or sent samples to an outside laboratory (within the corporate structure or independent thereof). Often the fluid vendor serves as the extramural laboratory, offering not only microbial results, but the biocide as well. These resources are not always timely. More often than not, results that should have been available (based on the microbiology) in 48 hours, sometimes took one week. This is frequently too late, the system in question being so far gone that it is already dumped.

During the past 10 years, a number of convenience methods have been introduced based on either agar (Genner, 1976; Hill, 1975) or paper (ASTM, 1980; Rossmoore *et al.*, 1977) and compare favorably with standard methodology (Bailey and May, 1979). These devices do not require microbiological facilities or specialized training, can be used on-site, and are relatively inexpensive. Without dilutions, the level of sensitivity of these devices called dip-slides or dip-strips covers four orders of magnitude (*e.g.* 10^2–10^6 c.f.u. ml^{-1}). Currently, kits are available for detecting total aerobic bacteria, fungi and sulfate reducing bacteria. The media in these systems are both selective and differential and were developed by microbiologists knowledgeable about MWF microflora.

There is the danger of misinterpretation arising from the use of improper media. For example, the production of extracellular sulfide by microorganisms can proceed by three pathways (reactions 1–3). Almost all bacteria can carry out the assimilatory reduction of sulfate (reaction 4).

$$S_2O_3^{2-} \xrightarrow{\text{[H]}} SO_3^{2-} + S^{2-} \text{(extracellular)} \tag{1}$$

$$\downarrow \text{[H]}$$

$$S^{2-} \text{(assimilated)}$$

$$H-S-\underset{\underset{H}{|}}{\overset{\overset{H}{|}}{C}}-\underset{\underset{NH_2}{|}}{\overset{\overset{H}{|}}{C}}-COOH \longrightarrow H-\underset{\underset{H}{|}}{\overset{\overset{H}{|}}{C}}-\underset{\underset{NH_2}{|}}{\overset{\overset{H}{|}}{C}}-COOH + S^{2-} \text{(extracellular)} \tag{2}$$

$$\downarrow$$

$$\text{(assimilated)}$$

$$SO_4^{2-} \xrightarrow{\text{[H]}} S^{2-} \text{(extracellular)} \tag{3}$$

$$SO_4^{2-} \xrightarrow{\text{[H]}} SO_3^{2-} \xrightarrow{\text{[H]}} S^{2-} \text{(assimilated)} \tag{4}$$

Reaction (1) above likewise is a specialized case of assimilatory reduction with thiosulfate as the source of sulfur. This ability is found in a number of enteric organisms, including *Salmonella* sp., *Proteus* sp. and *Citrobacter* sp. Only the anaerobic sulfate reducers are capable of using SO_4^{2-} as an electron acceptor with S^{2-} as the end product. The use of organic-based media containing thiosulfate will detect reactions (1), (2) and (3); however, only a selective medium for sulfate reducers will detect reaction (3). Thus, organisms capable of reactions (1) and (2) are not sulfate reducers *sensu strictu*. These media should not be used for detecting sulfate reducers (Russ, 1980). Pseudomonads, especially *Pseudomonas putrefaciens*, can split H_2S from appropriate amino acids.

Inadvertently, a recovery medium can give an incorrect assessment of the microbial population. Most media routinely used for fungal isolation are selective solely based on low pH. With very large bacterial populations, sufficient numbers are acidophilic and overgrow and completely mask fungal growth (Rossmoore and Treusch, 1975).

14.6.2 Biodeterioration and Function

There is general agreement that continuous and extensive growth of any of the three microbial groups will result in biodeterioration of the fluid. However, there appears to be no real consensus concerning how many of which groups constitute either a biological early warning or spoilage.

Aerobic bacteria c.f.u. from 10^5 to 10^7 ml^{-1} have been used as warning levels by different groups. Ostensibly, some operations (*e.g.* two-piece aluminum can formation and aluminum rolling) are affected by low levels of organisms which, when ashed, could cause surface pitting. In one instance, 10^4 ml^{-1} coliforms was established as the deterioration breakpoint level (Brandeberry and Myers, 1960). This was in no way intended for sanitary significance, but to the sponsoring group was a better predictor of aerobic degradation than the total count.

However, many systems have been found with up to 10^9 c.f.u. ml^{-1} without apparent functional failure. It is possible that the organisms have utilized nutrients extrinsic to the fluids. Several studies have been reported attempting to relate growth to chemical changes and functional failure of the fluid. Byrom and Hill (1971) examined the effect on an aluminum rolling emulsion; growth increased emulsion instability, as reflected by capacitance increase in the contaminated fluid when compared to sterile controls. However, a performance test, in which torque was applied until system seizure, showed no differences between contaminated and sterile emulsions.

Another report (Mattison *et al.*, 1975) also failed to relate biodegradation to performance using the rolling four ball test. More recently (Almen *et al.*, 1982; Holtzman *et al.*, 1982), IR and gas chromatography measurements of heavily contaminated emulsions showed decided reduction in hydrocarbon levels. Nonetheless, there was no significant difference in tapping torque values except in one emulsion sample that was visibly split. In this instance, the tap seized.

Certainly, sophisticated analytical and engineering devices are not needed to predict failures in such cases where the appearance of the fluid precludes the need for further testing. In all fairness, it should be stated that no methods have been designed for water-based fluids and, perhaps with some refinements, that the tapping torque procedure could give better correlative data with microbial growth.

14.6.3 Disease

There are numerous reports about frank pathogens in MWFs. They have been isolated from (Tant and Bennett, 1956), survive in (Bennett and Wheeler, 1954), grow in (Pivnick *et al.*, 1954) and coexist with non-pathogens (Maharajah *et al.*, 1956) in MWFs. The worker is exposed to fluids through inhalation, ingestion and contact. Considering the numbers of exposures and very high levels of contamination, one could anticipate occasional reports of illness attributable to the fluid microflora. Up to 1981, there had been no reports of any infectious disease attributed to MWFs.

The primary source of contact, and the most obvious, is the skin. Staphylococcal folliculitis has been reported among tool workers exposed to non-aqueous MWF, but even in these cases the source of organisms was not the fluid but the skin itself. No coagulase positive staphylococci could be found after continuous centrifugation of 5 l of soluble oil. A hospital isolate of *Staphylococcus aureus* died off rapidly when added to sterile used emulsion (Rossmoore and Williams, 1971). Several reports had disavowed the probability of infection of the skin from MWF microflora (Key *et al.*, 1966; Menter *et al.*, 1975; Rycroft, 1980). The dominant population, the pseudomonads, are frequently referred to as opportunistic pathogens and are certainly capable of causing infections under special conditions (Glick *et al.*, 1978), but not one case has either appeared in print or been verified verbally in occupational health circles.

During metalworking operations, there is extensive aerosolization of the fluid. Air samples were found to contain organisms from the fluid (Rossmoore *et al.*, 1976) and, subsequently, a significant number of droplets were found to be of respirable size (Hill, 1979). It was presumed from animal studies (Hill, 1979; Hill and Al-Zubaidy, 1979) that the inhalation of large numbers of *Pseudomonas* sp. poses a potential threat to the immunological responsiveness of the exposed individual.

Does the exception make the rule? On August 28, 1978 the following headline appeared in *The Detroit Free Press* on page 3: 'Ford Workers Stricken With Legionnaire's Bacillus.' More than 250 workers in the Windsor, Ontario engine plant of Ford Motor Company were clinically and epidemiologically diagnosed as having had Pontiac Fever. This is a high morbidity, non-fatal, short-term febrile myalgic condition first attributed to *Legionella pneumophila* (Glick *et al.*, 1978). Diagnostic conclusions were aided by an initial finding of suspect organisms by direct fluorescent antibody (DFA) in two systems distal to the outbreak site. The agglutination titers from recovery sera neither reacted with the types detected by DFA nor with any of the known *Legionella* serotypes. Subsequently, an isolate was found (Herwaldt, personal communication) in a sample of soluble oil from a piston machine operating proximal to the outbreak that possessed *Legionella*-like characteristics. This organism was agglutinated by affected workers' serum to a statistically greater level than non-affected control serum.

Although a great deal of information concerning this case is in written reports directed to all interested parties (Canadian Ministry of Health, Ford Motor Company and United Auto Workers), it has not been reported in the *Morbidity and Mortality Weekly Report* of the Centers for Disease Control or in the open refereed literature. The documented account, including the description of the new isolate, is anticipated in an appropriate peer review journal (Herwaldt, *et al.*, 1984). Under most conditions, the above report would not be included in a technical publication like this chapter without a verifiable citation. However, this author was professionally involved in the Windsor site during the course of the investigation evaluating the non-*Legionella* microbiology and so is privy to the sequence stated.

Many of the publications dealing with MWFs and health imply that the lack of positive reports of infectious disease is the result of poor reporting. Let not this be the case with the Windsor Ford–Pontiac Fever, despite the fact that prudence would dictate waiting for a formal publication before accepting a presumed cause and effect relationship between a newly-described organism and an infectious disease.

Studies (Hill, 1979; Rossmoore *et al.*, unpublished data) on the exposure of workers to organisms in fluids and mists revealed only transient colonization of hands and throat and practically none in nasal passages. However, these results should not imply a lack of infectivity, but rather a high innate host resistance. In a study on survival of fluid isolates, the graduate student involved contracted typhoid fever from the laboratory growth system (Bennett and Wheeler, 1954).

Water-based fluids may pose health concerns not directly related to the infectious process. These have included reports of generalized dermatitis from excessive use of biocides (Rossmoore, unpublished observations) as well as more specific contact dermatitis among hypersensitive members of the workforce to selected biocides (Keczkes and Brown, 1976). The finding of nitrosamines in fluids (Fan *et al.*, 1977) from the nitrosation of amines by nitrite has resulted in the replacement of nitrite by other corrosion inhibitors. However, the continuing surveillance for nitrosamines in MWF suggests that the presence of very low levels (p.p.m.) may be difficult to control. They may be from chemical contaminants or from microbial metabolism or from a breakdown product of a fluid component (Keefer and Hansen, 1982).

It is because of concerns of potential acute and chronic toxicity that two groups have established task forces to outline problem areas and establish standards. These are, respectively, the MWF Problem Committee of the International Biodeterioration Research Group (IBRG) (Minutes of the Metalworking Fluid Task Group, Biodeterioration Research Group, Aberdeen, March 1981) and the Health and Safety Committee of the Society of Manufacturing Engineers (SME) (SME, 1977). Because of legal constraints, this committee and its activities have been discontinued. The work is continuing in the American Society for Testing and Materials (ASTM) under Committee E34 (Occupational Health and Safety) as Subcommittee E34.50 (Health and Safety of Metalworking Fluids).

14.7 PREVENTION OF DETERIORATION

14.7.1 Biocides

Chemical antimicrobial agents have become part of MWF formulations since the first notice of microbial activity.

The use and selection of biocides for MWF is to control biodeterioration and not potential infectious disease. However, it has become universal practice to legally sanction biocides for specific applications and to demand proof of toxicological and environmental safety as well as some indication of effectiveness (Environmental Protection Agency, 1975, 1978).

It is important that compounds be evaluated in the fluids at their use concentrations (Carlson and Bennett, 1960) and challenged by representative organisms from a contaminated source (Rossmoore *et al.*, 1977). Many methods have been prepared for laboratory testing (ASTM, 1980; Bennett, 1974a; Hill, 1969; Himmelfarb and Scott, 1968; Pivnick and Fabian, 1953; Rogers *et al.*, 1975; Rossmoore and Williams, 1971; Yanis and Wolfe, 1960). All of the currently used methods involve some form of aeration, reinoculation, and the addition of appropriate metal fines or chips. There is disagreement on the severity of the challenge (10^6 to 10^8 bacterial c.f.u. ml^{-1}) and on the level consistently indicating biocide failure (10^5 c.f.u. ml absolute to 3 logs difference between control and treated sample). There is no consensus on challenge or survival levels for sulfate reducer bacteria or fungi. To many, the presence of any of these organisms after treatment poses a biodeterioration threat. Since both of these groups tend to be more greatly localized in MWF systems than aerobic bacteria, they are inherently more difficult to treat (Ruseska *et al.*, 1982). Thus, planktonic survivors in a simulated laboratory test would be an indication of resistance of sessile organisms in the MWF system.

Biocides are added directly into the diluted MWF at the recommended dose or are incorporated into the MWF concentrate. Some biocides cannot be compounded at the levels needed due to insolubility or chemical reactivity at those concentrations. However, even when it is physically and chemically possible to include the biocide in the MWF concentrates, there can be problems since the fluids are diluted to different unpredictable levels due to the varying demands of the metalworking operations. This makes biocide levels impossible to predict. This fact, coupled with operational problems such as fluid turnover rates, metals involved and accumulated organic load, makes tankside addition of biocides a logical direction for microbial control, and should supplement the biocide already in the fluid with prior testing to ensure compatibility if a new compound is involved.

Antagonism (Bachenheimer and Bennett, 1961; Rossmoore, 1974) as well as synergism (Rossmoore *et al.*, 1979) has been reported between two fluids and between two biocides in the same fluid. No one biocide currently available is most effective in all fluids, with all metals and against all types of organisms. Rossmoore (1979) found that highly electropositive metals are antagonistic to the activities of 1,3,5-tris(β-hydroxyethyl) hexahydrotriazine (HTH), while Bennett and his coworkers (1982) reported on a large array of antagonisms and synergisms using 15 metals and 13 biocides. In addition, Bennett and his coworkers examined a large number of putative corrosion inhibitors for biocidal activity (Bennett, 1978, 1979; Bennett *et al.*, 1979) with mixed results. More promising is the use of EDTA as a synergist especially with HTH where as much as a 10-fold increase in activity has been reported (Izzat and Bennett, 1979; Rossmoore, 1979; Table 2).

Lists of biocides have appeared continually during the last 25 years as well as conditions governing use and effectiveness (Bennett, 1956; Hill, 1972; Pivnick and Fotopoulos, 1957; Rossmoore, 1981; Smith, 1969). Toxicological and environmental imperatives and the introduction of new MWF formulations have continued to change the members of the list or markedly restrict their use. The high cost of development, including hazard evaluation, has severely limited the number of new compounds introduced to the metalworking industry. Also, legal approval for use (*i.e.* registration) is not always worldwide. In Table 4 are some of the more popular biocides approved for use in the US. It is not unusual to find compounds or mixtures included in formulation concentrates as stabilizers or corrosion inhibitors which may have antimicrobial activity that need not be claimed.

One interesting compound based on copper monooxycitrate was patented (Maurer and Shringarpurey, 1977) for its reported antimicrobial activity and subsequently complemented by a claim for soluble oil stabilization (Maurer and Shringarpurey, 1978). It is also reported to control odors from microbial activity.

14.7.2 Physical Methods

There are several reports on the effectiveness of some form of thermal pasteurization (Heinrichs and Rossmoore, 1971; Hill, 1982; Porter and Rowe, 1982; Rossmoore and Heinrichs, 1970). Temperatures from 63 to 70 °C for relatively short periods (20–60 s) produced significant reduction in bacterial populations. In one report (Rossmoore and Heinrichs, 1970), organisms from a stagnant soluble oil system were more resistant to heat than those taken from an actively aerated

Table 4 Biocides EPA Registered for Use in Metalworking Fluids[a]

Manufacturer	Trade name and active ingredients	Dose range (% of formulated biocide)	EPA Reg. No.	Comments
Buckman Laboratories, Inc.	*Busan 85* potassium dimethyldithiocarbamate 50%	0.03–0.2	1448–70	Compatible with most concentrates (exceptions usually oil/water emulsion)
The Dow Chemical Co.	*Dowicil 75* 1-(3-chloroallyl)-3,4,7-triaza-1-azoniaadamantane chloride (67.5% sodium bicarbonate inert stabilizer)	0.015–0.2	464–403	Not recommended for fluid concentrates, usually added on site; hygroscopic powder
	Dowicide I O-phenylphenol 98%	0.1–0.15	464–70	Compatible with most concentrates; restrictions on discharge levels of phenols —inhibits use for many
	Dowicide A sodium O-phenylphenate 97%	0.1–0.15	464–70	Compatible with most concentrates; restrictions on discharge levels of phenols —inhibits use for many
	XD-8254 DBNPA 2,2-dibromo-3-nitrilopropionamide 10%	0.02–0.04 daily	464–500AA	Cannot be used in concentrates; unstable in alkaline pH; usually added continually — degrades easily
Givaudan Corp.	*Givgard DXN* 6-acetoxy-2,4-dimethyl-*m*-dioxane 100%	0.05–0.02	824–7	Incompatible with most concentrates
ICI United States, Inc.	*Proxel CRL* 1,2-benzisothiazolin-3-one 30–35%	0.001–0.06	10182–2	Stable in many concentrates; not being aggressively marketed by manufacturer for metalworking fluids at this time
IMC Chemical Group, Inc.	*Bioban P-1487* 4-(2-nitrobutyl)morpholine 70%; 4,4'-(2-ethyl-2-nitrotrimethylene)-dimorpholine 20%	0.01–0.3	271–30	Good oil solubility; stable in oil/water emulsion concentrates
	Tris Nitro tris(hydroxymethyl)nitromethane; aqueous 50%, powder 100%	0.2 aqueous; 0.1 powder	271–261 271–18	Poorly compatible with most concentrates; recommended for tankside use
Lehn & Fink Industrial Products, Div. of Sterling Drug, Inc.	*Grotan HD2* 2-chloro-N-(hydroxymethyl)acetamide 39%; sodium tetraborate 41%; potassium iodide 0.39%	0.1–0.15	10000–3	Incompatible with concentrates; recommended tankside: soluble oil needs higher dose

Manufacturer	Product and composition	Concentration (%)	Code no.	Properties
	Grotan BK[b] hexahydro-1,3,5-tris(2-hydroxyethyl)-s-triazine 78%	0.15	10000-1	Stable in many concentrates; less effective *vs.* fungi at low-dose levels
Merck Chemical Div. of Merck & Co., Inc.	*Tektomer 38* 1,2-dibromo-2,4-dicyanobutane; liquid 25%, powder 98%	1000–4000 p.p.m. liquid; 250–1000 p.p.m. powder	10445-22 10445-56; 10445-33	Suitable for use in some concentrates as well as for tankside addition
Olin Corp.	*Zinc Omadine* zinc 2-pyridinethiol 1-oxide powder 95%; aqueous dispersion 48%	0.0075	1258-840 powder; 1258-841 aqueous	Not soluble in concentrates
	Sodium Omadine sodium 2-pyridinethiol 1-oxide powder 90%; aqueous solution 40%	0.0025–0.015	1258-842 power; 1258-843 aqueous	Soluble in most concentrates except with difficulty in oil/water emulsion concentrates; more effective *vs.* fungi and yeasts than bacteria
	Triadine-10 hexahydro-3,5-tris(2-hydroxyethyl)-s-triazine 63.6%; sodium 2-pyridinethiol 1-oxide 6.4%	0.07 in synthetic fluids; 0.1 in oil-containing fluids	1258-990	Used in oil/water emulsion concentrates with care; compatible with most other concentrates
Onyx-Millmaster	*Onyxide 200* hexahydro-1,3,5-tris(2-hydroxyethyl)-s-triazine 78%	0.15	1839-59	Stable in many concentrates; less effective *vs.* fungi at low-dose levels
Rohm and Haas Co.	*Kathon 886 MW* 5-chloro-2-methyl-4-isothiazolin-3-one 8.6%; 2-methyl-4-isothiazolin-3-one 2.6%	0.0025–0.0125	707-129	Stable in some concentrates; not particularly in oil/water emulsion concentrates — possibly inactivated by some amines and sulfides
R. T. Vanderbilt Co., Inc.	*Vancide TH* hexahydro-1,3,5-triethyl-s-triazine 95%	0.05–0.1	1965-55	Good oil and water solubility; stable in most concentrates; strong amine odor
	Vancide 51 sodium dimethyldithiocarbamate 27.6%; sodium 2-mercaptobenzothiazole 2.4%	4% in water as bactericide; 2% in water as fungicide	1965-8	Stable in some concentrates but recommended at use-dilution by user

[a] Modified from Rossmoore, 1981.   [b] Product sold as Grotan in the United States

system, while the necessity of reheating every five hours to maintain control is emphasized in another report (Hill, 1982).

Most assuredly, there are many advantages to thermal microbial control, such as no fluid incompatibility, toxicity, or environmental problems. However, there is no residual control, and, short of sterilizing the fluid, there is a recurring need to reheat. Thus far, studies have been done only on small systems. The energy input needed for large control systems even on a bypass could be costly. Eventually, with intermittent pasteurization the operating temperature of the fluid is sure to be raised unless there is intervening cooling. This too will add to the cost. The continual heating would increase evaporation rate, ambient temperature, and humidity of the air, and would reduce the cooling capacity of a fluid whose function is cooling.

Other more exotic methods of physical control have been evaluated, including gamma and ultraviolet (UV) radiation, ultrasonication, and microwave radiation. Rossmoore and Brazin (1968) found LD_{50}s of aerobic bacteria and sulfate reducers in deteriorated soluble oil were 2 and 10 krad, respectively. Intermittent irradiation did not produce increases in radioresistance in survivors. Cost extrapolations to large systems (e.g. 10^5 l) suggest that flow rates of 10% of total volume minute^{-1} past an appropriately shielded ^{137}Cs source could control the fluid microbial population. Further, it was found (Heinrichs and Rossmoore, 1971) that a demonstrated synergism between gamma radiation and certain chemical inhibitors might make it possible to reduce the source size, or at least account for loss of radioactivity through decay. Observations made on fresh and deteriorated fluid after 100 krad radiation showed apparent increases in stability in a soluble oil emulsion.

On a small scale (2 l), ultrasound (25 kHz) effectively reduced population levels (*e.g.* 10^9–10^4 ml^{-1}) with a concomitant rise in fluid temperature from 20 to 45 °C (Rossmoore, 1974), but the practicality of in-plant use remains questionable.

All of the physical methods suffer the single drawback of having no residual protection. In addition, gamma radiation, sonication and microwave radiation have not been sufficiently tested to remove doubts of occupational hazard in the metalworking environment.

14.7.3 Mechanical Methods

Filtration and centrifugation are both used routinely for separating particulate contaminants and tramp oils from a variety of MWF operations. The filtration media and porosities vary (Brandt, 1973; Nehls, 1977) but none are constructed for microbial removal; downstream fluids were almost as contaminated as before filtration (Rossmoore, unpublished data). Only one filter unit has been specifically designed for microbial removal (Symes and Cowap, 1976) and although apparently successful for small systems (200 l, 8 l min^{-1} flow) has never been developed for larger systems.

Bowl centrifuges similar to cream separators of varying capacities certainly remove some microbial contamination, especially organisms adsorbed on larger particulates. This may account for as much as 50% of the attached population and could ostensibly remove fungal slime as well. Some centrifuges have in-line heating units to raise the fluid temperature to 65–70 °C during centrifugation. This facilitates separation of tramp oil. Manufacturers of these units claim the additional benefits of pasteurization. Obviously, there is some reduction in viable counts. However, there are no data available relating frequency of use to microbial regrowth.

14.7.4 Significance of Recycling

The cost of replacement and disposal suggests that efforts to maintain and recover fluids should be a universal goal. Some newer plants have designed their central systems with additional holding reservoirs for cleaning and decontaminating fluids. In addition, several companies have unit packages that are designed for recycling fluid from smaller systems (*e.g.* up to 2500 l; Figures 3 and 4). These successively provide appropriately diluted fluid to machines; dirty fluid then is pumped out into a holding reservoir and then centrifuged. Solids are disposed of, waste oil can be re-refined or reblended, and the cleaned fluid is returned.

14.8 DISPOSAL AND DISPOSABILITY

There are severe legal restrictions (Dzindzeleta, 1981) on dumping used MWF either into natural waterways or indirectly into municipal waste treatment facilities. MWFs contain excessive

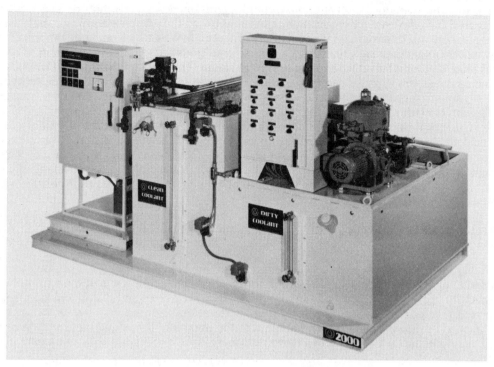

Figure 3 Photograph of XYBEX™ 2000, a unitized recycling system, courtesy of Master Chemical Corporation (see Figure 4 for description)

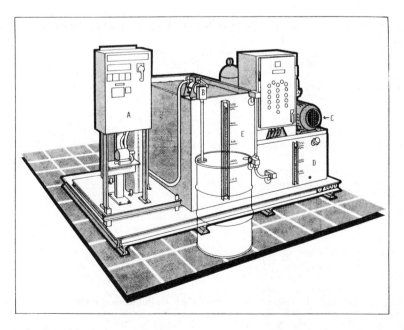

Figure 4 An example of a unitized recycling system, XYBEX™ 2000, courtesy of Master Chemical Corporation. Parts labeled A–E are: A, automatic deionizer; B, automatic coolant proportioner; C, disk bowl centrifuge, high speed for tramp oil removal; D, 350 US gal dirty fluid tank linked with automatic controls to centrifuge; E, 500 US gal clean fluid tank linked with automatic controls to centrifuge and to coolant proportioner

levels of oil, suspended and dissolved metals, biocides, corrosion inhibitors and products of microbial metabolism. Hauling costs frequently exceed the initial cost of some fluid concentrates.

Soluble oils and synthetic fluids offer different problems of disposal (Bennett, 1973b; Coursey, 1968). Before current restrictive legislation, emulsions posed the greater effort since they constituted a much greater environmental impact than synthetic MWFs which could be dumped without any treatment under some situations. Currently, emulsions are split into oil and water phases, which is readily accomplished with mineral acid and alum (Harlow and Doran, 1981). After pH adjustment, the aqueous phase can be discharged into the sanitary sewer. The oil is recovered and usually sold to reclaimers who bring it back to original specifications for MWFs.

A more direct method for dewatering soluble oil emulsions is by ultrafiltration (Dick, 1982). It is more economical than conventional chemical methods for flow rates below $400\ 000\ \mathrm{l\ day}^{-1}$. Synthetic fluids by design have all their substituents in solution. Spent fluids may have biochemical oxygen demands (BODs) in excess of $10\ 000\ \mathrm{mg\ l}^{-1}$. This level should be reduced before entering a sanitary system. Many users have installed biological treatment systems to handle this fluid. An alternative is reverse osmosis which can effectively reduce the level of dissolved constituents.

14.9   SUMMARY

There has been a trend recently to increasing use of synthetic oil-free MWF, although not to the exclusion of soluble oil emulsions. Shifts in formulations have resulted in new microbiological problems with greater yeast and mold involvement. Increased costs and restrictions on disposal have placed greater emphasis on control of biodeterioration resulting in a greater demand for more effective biocides. The current awareness of the microbial levels and types in MWFs has focused on potential health hazards. However, there are few reports to substantiate these concerns.

14.10   REFERENCES

Almen, R., G. Mantelli, P. McTeer and S. Nakayama (1982). Application of high-performance liquid chromatography to the study of the effect of microorganisms in emulsifiable oils. *Lubr. Eng.*, **38**, 99–103.
American Society for Testing and Materials (1978). Standard classification of metalworking fluids and related materials. ANSI/ASTM Designation D 2881-73 (Reapproved 1978). In *1978 Annual Book of ASTM Standards, Part 24*, pp. 750–751. ASTM, Philadelphia, PA.
American Society for Testing and Materials (1980). Standard method for evaluating the bioresistance of water-soluble metal-working fluids. ASTM Designation D 3946-80. In *1982 Annual Book of ASTM Standards, Part 25*, pp. 544–547. ASTM, Philadelphia, PA.
Andlid, S. and L. Linden (1980). Metal working emulsion. *US Pat.* 4 237 021.
Bachenheimer, A. G. and E. O. Bennett (1961). The sensitivity of mixed populations of bacteria to inhibitors. I. The mechanism by which *Desulfovibrio desulfuricans* protects *Pseudomonas aeruginosa* from the toxicity of mercurials. *Antonie Van Leeuwenhoek (J. Microbiol. Serol.)*, **27**, 180–188.
Bailey, C. A. and M. E. May (1979). Evaluation of microbiological test kits for hydrocarbon fuel systems. *Appl. Environ. Microbiol.*, **37**, 871–877.
Bennett, E. O. (1956). Control of bacterial spoilage of emulsion oils. *Soap Chem. Spec.*, Oct.–Nov., 1–7.
Bennett, E. O. (1972). The biology of metalworking fluids. *Lubr. Eng.*, **28**, 237–247.
Bennett, E. O. (1973a). Microbiological aspects of metalworking fluids. *SME Technical Paper* MR73-826. Society of Manufacturing Engineers.
Bennett, E. O. (1973b). The disposal of metal cutting fluids. *Lubr. Eng.*, **29**, 300–307.
Bennett, E. O. (1974a). The biological testing of cutting fluids. *Lubr. Eng.*, **30**, 128–135.
Bennett, E. O. (1974b). The deterioration of metal cutting fluids. *Prog. Ind. Microbiol.*, **13**, 121–149.
Bennett, E. O. (1976). Effect of dragout and hydraulic fluid contamination on rancidity control in cutting fluids. *American Society of Lubrication Engineers* (ASLE) *Preprint* No. 76-AM-1B-1.
Bennett, E. O. (1978). The antimicrobial properties of diethylene triamines in metal working fluids. *Int. Biodeterior. Bull.*, **14**, 21–29.
Bennett, E. O. (1979). Corrosion inhibitors as preservatives for metal-working fluids — ethanolamines. *Lubr. Eng.*, **35**, 137–144.
Bennett, E. O., M. C. Adams and G. Tavana (1979). Antimicrobial properties of butanolamines and propanolamines in metal working fluids. *J. Gen. Appl. Microbiol.*, **25**, 63–69.
Bennett, E. O., J. E. Gannon and I. U. Onyekwelu (1982). The effects of metals upon the inhibitory activities of cutting fluid preservatives. *Int. Biodeterior. Bull.*, **18**, 7–12.
Bennett, E. O. and H. O. Wheeler (1954). Survival of bacteria in cutting oil. *Appl. Microbiol.*, **2**, 368–371.
Brandeberry, L. J. and H. V. Myers (1960). Test procedures for compounds used as preservatives in industrial coolants. *Lubr. Eng.*, **16**, 161–164.
Brandt, R. (1973). Bacteria–tramp oil problems in central coolant system — a unit solution. *SME Technical Paper* MR73-119. Society of Manufacturing Engineers.

Brandt, R. H. (1974). Fluid longevity and central clarification systems. *SME Technical Paper* MR74-171. Society of Manufacturing Engineers.

Buck, J. D. (1979). The plate count in aquatic microbiology. In *Native Aquatic Bacteria: Enumeration, Activity, and Ecology — ASTM STP 695*, ed. J. W. Costerton and R. R. Colwell, pp. 19–28. ASTM, Philadelphia, PA.

Byrom, D. and E. C. Hill (1971). The microbiology of rolling mills. *Microbiol. 1971, Proc. Conf., London, Jan. 17–28, 1971*, 42–59.

Carlson, V. and E. O. Bennett (1960). The relationship between the oil–water ratio and the effectiveness of inhibitors in oil soluble emulsions. *Lubr. Eng.*, **16**, 572–574.

Centers for Disease Control (1976). *Current Intelligence Bulletin: Nitrosamines in Cutting Fluids, October 6*. National Institute for Occupational Safety and Health, Rockville, MD.

Compton, M. and E. Sucher (1965). Cutting composition. *US Pat.* 3 215 630.

Coursey, W. M. (1968). The application, control, and disposal of cutting fluids. *Lubr. Eng.*, **25**, 200–204.

Dick, R. M. (1982). Ultrafiltration for oily wastewater treatment. *Lubr. Eng.*, **38**, 219–222.

Duffett, N. D., S. H. Gold and C. L. Weirich (1943). Normal bacterial flora of cutting oil emulsions. *J. Bacteriol.*, **45**, 37–38.

Dzindzeleta, M. (1981). An overview of EPA regulations regarding petroleum products. Paper presented at SME Symposium on Alternatives to the Disposal of Industrial Fluids, November 17–19, 1981.

Environmental Protection Agency (1975). Guidelines for registering pesticides in United States. *Fed. Regist.*, **40** (II), 26 810, 26 832–26 833.

Environmental Protection Agency (1978). Proposed guidelines for registering pesticides. *Fed Regist.*, **43**, 29 696, 37 336.

Fan, T. Y., J. Morrison, D. P. Rounbehler, D. H. Fine, R. Ross, W. Miles and N. P. Sen (1977). *N*-Nitrosodiethanolamine in synthetic cutting fluids: a part-per-hundred impurity. *Science*, **196**, 70–71.

Feisal, E. V. and E. O. Bennett (1961). The effect of water hardness on the growth of *Pseudomonas aeruginosa* in metal cutting fluids. *J. Appl. Bacteriol.*, **24**, 125–130.

Gannon, J. E. and E. O. Bennett (1981). A rapid technique for determining microbial loads in metalworking fluids. *Trib. Int.*, **14**, 3–6.

Genner, C. (1976). Evaluation of the 'dipslide' technique for microbiological testing of industrial fluids. *Process Biochem.*, **11** (6), 39–40, 48.

Glick, T. H., M. B. Gregg, B. Berman, G. Mallison, W. W. Rhodes Jr. and I. Kassanoff (1978). Pontiac fever, an epidemic of unknown etiology in a health department: I. Clinical and epidemiologic aspects. *Am. J. Epidemiol.*, **107**, 149–160.

Hardy, J. A. (1981). The enumeration, isolation and characterization of sulfate-reducing bacteria from North Sea waters. *J. Appl. Bacteriol.*, **51**, 505–516.

Hardy, J. A. and W. A. Hamilton (1981). The oxygen tolerance of sulfate-reducing bacteria isolated from North Sea waters. *Curr. Microbiol.*, **6**, 259–262.

Harlow, B. D. and T. M. Doran (1981). Chrysler upgrades oily wastes treatment in Kokomo. *Ind. Wastes*, **27** (6), 24–27.

Heinrichs, T. F. and H. W. Rossmoore (1971). Effects of heat, chemicals, and radiation on cutting fluid flora. *Dev. Ind. Microbiol.*, **12**, 341–345.

Herwaldt, L. A., G. W. Gorman, T. McGrath, S. Toma, B. Brake, A. W. Hightower, J. Jones, A. L. Reingold, P. A. Boxer, P. W. Tang, C. W. Moss, H. Wilkinson, D. J. Brenner, A. G. Steigerwalt and C. V. Broome (1984). A new *Legionella* species, *Legionella feeleii* species *nova*, causes Pontiac Fever in automobile plant. *Ann. Intern. Med.*, **100**, 333–338.

Hill, E. C. (1968). Microbial degradation of lubricant oils and emulsions and its engineering significance. In *Biodeterioration of Materials, Microbiological and Allied Aspects*, ed. H. A. Walters and J. J. Elphick, pp. 381–385. Elsevier, New York.

Hill, E. C. (1969). Microbiological examination of petroleum products. *Tribology*, Feb., 5–10.

Hill, E. C. (1972). Biocides for petroleum products. *J. Inst. Pet.*, **58**, 248–253.

Hill, E. C. (1975). Biodeterioration of petroleum products. In *Microbial Aspects of the Deterioration of Materials*, ed. R. J. Gilbert and D. W. Lovelock, pp. 127–136. Academic, New York.

Hill, E. C. (1978). Biodegradation of hyrocarbon-based products in industrial use. In *Developments in Biodegradation of Hydrocarbons-1*, ed. R. J. Watson, chap. 7, pp. 201–225. Applied Science Publishers, London.

Hill, E. C. (1979). Microbial infection of lubricating and other oils and its significance. *SME Technical Paper* MF79-391. Society of Manufacturing Engineers.

Hill, E. C. (1982). The current state of the art on chemical and physical anti-microbial measures. In *Lubrication in Metal Working*, ed. W. J. Bartz and J. Wolff, Proceedings of the Third International Colloquium of the Esslingen Technical Academy, Jan. 12–14, 1982, vol. 1, pp. 83.1–83.6. Esslingen Technical Academy, West Germany.

Hill, E. C. and T. Al-Zubaidy (1979). Some health aspects of infections in oil and emulsions. *Trib. Int.*, 161–164.

Himmelfarb, P. and A. Scott (1968). Simple circulating tank test for evaluation of germicides in cutting fluid emulsion. *Appl. Microbiol.*, **16**, 1437–1438

Holtzman, G. H. M., H. W. Rossmoore, E. Holodnik and M. Weintraub (1982). Interrelationships between biodeterioration, chemical breakdown, and function in soluble oil emulsions. *Dev. Ind. Microbiol.*, **23**, 207–216.

Isenberg, D. L. and E. O. Bennett (1959). Bacterial deterioration of emulsion oils. II. Nature of the relationship between aerobes and sulfate-reducing bacteria. *Appl. Microbiol.*, **7**, 121–125.

Izzat, I. N. and E. O. Bennett (1979). Potentiation of the antimicrobial activities of cutting fluid preservatives by EDTA. *Lubr. Eng.*, **35**, 153–159.

Keczkes, K. and P. M. Brown (1976). Hexahydro-1,3,5-tris(2-hydroxyethyl)triazine, a new bacteriocidal agent as a cause of allergic contact dermatitis. *Contact Dermatitis*, **2**, 92–98.

Keefer, L. K. and I. J. Hansen (1982). Primary amine use and other strategies for preventing human exposure to *N*-nitroso compounds: application to cutting fluids. In *The Proceedings of the Conference on N-Nitroso Compounds: Occurrence and Biological Effects*, ed. H. Bartsch, I. K. O'Neill, and M. Castegnaro-Mokaida. International Agency for Research on Cancer Publication #41, Geneva, Switzerland.

Key, M. M., E. J. Ritter and K. A. Arndt (1966). Cutting and grinding fluids and their effects on the skin. *Am. Ind. Hyg. Ass. J.*, **27**, 423–427.

King, R. D. (1982). Harmful effects and control of microbial populations encountered in metalworking fluids. In *Lubrication in Metal Working*, ed. W. J. Bartz and J. Wolff, Proceedings of the Third International Colloquium of the Esslingen Technical Academy, Jan. 12–14, 1982, vol. 1. pp. 83.1–83.6. Esslingen Technical Academy, West Germany.

Koh, K.-W. (1980). Cooling and lubricating fluid for metal working. *US Pat.* 4 218 329.

Lee, M. and A. C. Chandler (1941). A study of the nature, growth and control of bacteria in cutting compounds. *J. Bacteriol.*, **41**, 373–386.

Maharajah, R. S., H. Pivnick, W. E. Engelhard and S. Templeton (1956). The coexistence of pathogens and pseudomonads in soluble oil emulsions. *Appl. Microbiol.*, **4**, 293–299.

Mattison, R., G. I. Lloyd and J. Schofield (1975). The fatigue effect on En 31 steel balls of microbially degraded soluble oil emulsion. *Trib. Int.*, Dec., 253–255.

Maurer, G. L. and S. K. Shringarpurey (1977). Complexes of heavy metal ions and polyfunctional organic ligands as antimicrobial agents. *US Pat.* 4 055 655.

Maurer, G. L. and S. K. Shringarpurey (1978). Metalworking fluid compositions and methods of stabilizing same. *US Pat.* 4 129 509.

McCoy, J. S. (1978). A practical approach to central system control. *Lubr. Eng.*, **34**, 180–186.

McCoy, W. F., J. D. Bryers, J. Robbins and J. W. Costerton (1981). Observations of fouling biofilm formation. *Can. J. Microbiol.*, **27**, 910–917.

Menter, P., W. Harrison and W. G. Woodin (1975). Patch testing of coolant fractions. *J. Occup. Med.*, **17**, 565–568.

Nehls, B. L. (1977). Particulate contamination in metalworking fluids. *Lubr. Eng.*, **33**, 179–183.

Piotrowski, A. B. and R. H. Davis (1981). Metal working lubricant containing an alkanolamine and a cycloaliphatic acid. *US Pat.* 4 259 206.

Pivnick, H., W. E. Engelhard and T. L. Thompson (1954). The growth of pathogenic bacteria in soluble oil emulsions. *Appl. Microbiol.*, **2** 140–142.

Pivnick, H. and F. W. Fabian (1953). Methods for testing the germicidal value of chemical compounds for disinfecting soluble oil emulsions. *Appl. Microbiol.*, **1**, 204–207.

Pivnick, H. and C. K. Fotopoulos (1957). Disinfection of soluble oil emulsions. *Lubr. Eng.*, **13**, 151–156.

Porter, T. M. and G. W. Rowe (1982). Maintenance of cutting and grinding fluid quality. In *Lubrication in Metal Working*, Proceedings of the Third International Colloquium of the Esslingen Technical Academy, Jan. 12–14, 1982, ed. W. J. Bartz and J. Wolff, vol. 1, pp. 91.1–91.8. Esslingen Technical Academy, W. Germany.

Postgate, J. R. (1967). Report of the subcommittee on sulfate-reducing bacteria (1962–1966) to the international committee on nomenclature of bacteria. *Int. J. Syst. Bacteriol.*, **17**, 111–112.

Postgate, J. R. (1979). *The Sulphate-Reducing Bacteria*. Cambridge University Press, Cambridge.

Postgate, J. R. and L. L. Campbell (1966). Classification of *Desulfovibrio* species, the non-sporulating sulfate-reducing bacteria. *Bacteriol. Rev.*, **30**, 732–738.

Quanstrom, L. (1977). Central coolant systems — closing the loop on metalworking fluids. *Lubr. Eng.*, **33**, 14–19.

Rogers, M. R., A. M. Kaplan, and E. Beaumont (1975). A laboratory in-plant analysis of a test procedure for biocides in metalworking fluids. *Lubr. Eng.*, **31**, 301–310.

Rossmoore, H. W. (1962). Correlation of coliform activity with anaerobic sulfate reduction with deterioration of cutting fluids. *Lubr. Eng.*, **18**, 226–229.

Rossmoore, H. W. (1972). Methylene blue reduction for rapid inplant detection of coolant breakdown. *Int. Biodeterior. Bull.*, **7**, 147–154.

Rossmoore, H. W. (1974). Microbiological causes of cutting fluid deterioration. *SME Technical Paper* MR74-169. Society of Manufacturing Engineers.

Rossmoore, H. W. (1979). Heterocyclic compounds as industrial biocides. *Dev. Ind. Microbiol.*, **20**, 41–71.

Rossmoore, H. W. (1981). Antimicrobial agents for water-based metalworking fluids. *J. Occup. Med.*, **23**, 247–254.

Rossmoore, H. W. and J. G. Brazin (1968). Control of cutting oil deterioration with gamma radiation. In *Biodeterioration of Materials, Microbiological and Allied Aspects*, ed. H. A. Walters and J. J. Elphick, pp. 386–402. Elsevier, New York.

Rossmoore, H. W. and T. F. Heinrichs (1970). Physical and chemical combinations for the control of microbial breakdown of cutting fluids. *SME Technical Paper* MR70-267. Society of Manufacturing Engineers.

Rossmoore, H. W. and G. H. Holtzman (1974). Growth of fungi in cutting fluids. *Dev. Ind. Microbiol.*, **15**, 273–280.

Rossmoore, H. W., G. H. Holtzman and L. Kondek (1976). Microbial ecology with a cutting edge. In *Proceedings of the Third International Biodegradation Symposium*, ed. J. M. Sharpley and A. M. Kaplan, pp. 221–232. Applied Science Publishers, London.

Rossmoore, H. W., G. H. Holtzman and L. A. Rossmoore (1972). Microbial succession in cutting fluid milieu. *Abstracts of the Annual Meeting — 1972*, p. 3. American Society for Microbiology, Washington, DC.

Rossmoore, H. W. and L. A. Rossmoore (1980). The identification of a defined microbial inoculum for the evaluation of biocides in water-based metalworking fluids. *Lubr. Eng.*, **36**, 16–20.

Rossmoore, H. W., P. Sceszny and L. A. Rossmoore (1977). Evaluation of source of bacterial inoculum in development of a cutting fluid test procedure. *Lubr. Eng.*, **33**, 372–377.

Rossmoore, H. W., M. E. Shearer and C. Shearer (1964). Growth studies on *Desulfovibrio desulfuricans*. *Dev. Ind. Microbiol.*, **5**, 334–342.

Rossmoore, H. W., J. F. Sieckhaus, L. A. Rossmoore and D. DeFonzo (1979). The utility of biocide combinations in controlling mixed microbial populations in metalworking fluids. *Lubr. Eng.*, **35**, 559–563.

Rossmoore, H. W. and P. J. Treusch (1975). Comparison of several media for enumeration of cutting fluid fungi. *Dev. Ind. Microbiol.*, **16**, 475–482.

Rossmoore, H. W. and B. W. Williams (1971). An evaluation of a laboratory and plant procedure for preservation of cutting fluids. *Int. Biodeterior. Bull.*, **7**, 55–60.

Ruseska, I., J. Robbins, J. W. Costerton and E. S. Lashen (1982). Biocide testing against corrosion-causing oil field bacteria. *Oil Gas J.*, Mar. 8, 253–264.

Russ, G. A. (1980). Coolant control of large central systems. *Lubr. Eng.*, **36**, 21–24.

Rycroft, R. J. G. (1980). Bacteria and soluble oil dermatitis. *Contact Dermatitis*, **6**, 7–9.

Sabina, L. R. and H. Pivnick (1956). Oxidation of soluble oil emulsions and emulsifiers by *Pseudomonas oleovorans* and *Pseudomonas formicans*. *Appl. Microbiol.*, **4**, 171–175.

Senez, J. C. (1962). Some considerations on the energetics of bacterial growth. *Bacteriol. Rev.*, **26**, 95–107.

Smith, T. H. F. (1969). Toxicological and microbiological aspects of cutting fluid preservatives. *Lubr. Eng.*, **25**, 313–320.

Society of Manufacturing Engineers (1977). Developing a workable standard for cutting fluids. *Manuf. Eng.*, Nov., 48.

Supavej, S., E. Meyer and B. Wurtz (1972). Recherches sur les variations de l'activité catalasique dans les suspensions non-proliférantes de *Pseudomonas fluorescens* (type R). I. Relation avec la lyse. Société de Biologie de Strasbourg Séance du 22 Avril 1972. *Biologie. Comptes Rendus*, **166**, 703–709.

Supavej, S., E. Meyer and B. Wurtz (1973). Recherches sur les variations de l'activité catalasique dans les suspensions non-proliférantes de *Pseudomonas fluorescens* (type R). III. Influence des groupes SH libres. Société de Biologie de Strasbourg Séance du 16 Juin 1973. *Biologie. Comptes Rendus*, **167**, 1043–1047.

Symes, W. R. and D. Cowap (1976). A new approach to the control of microbial degradation of metal-working coolants and lubricants. In *Proceedings of the Third International Biodegradation Symposium*, ed. J. M. Sharpley and M. Kaplan, pp. 233–242. Applied Science Publishers, London.

Tant, C. O. and E. O. Bennett (1956). The isolation of pathogenic bacteria from used emulsion oils. *Appl. Microbiol.*, **4**, 332–338.

Tant, C. O. and E. O. Bennett (1958). The growth of aerobic bacteria in metal-cutting fluids. *Appl. Microbiol.*, **6**, 388–392.

Tilton, R. C. (ed.) (1982). *Rapid Methods and Automation in Microbiology*. American Society for Microbiology (ASM), Washington, DC.

Vedder, K. (1985). Recovery of airborne sulfate-reducing bacteria from metalworking fluid systems. Masters Thesis, Wayne State Univ., Dept. of Biological Sciences, Detroit, MI.

Yanis, R. J. and G. F. Wolfe (1960). Test procedures for the evaluation of cutting fluids. *Lubr. Eng.*, **16**, 164–170.

15

Biopulping, Biobleaching and Treatment of Kraft Bleaching Effluents with White-rot Fungi

K.-E. ERIKSSON
Swedish Forest Products Research Laboratory, Stockholm, Sweden
and
T. K. KIRK
Forest Products Laboratory, Madison, WI, USA

15.1 BIOPULPING

15.1.1 Properties of Wood-rotting Fungi and Lignin Degradation by these Fungi

In nature there is a continuous degradation of dead plant materials by saprophytic microorganisms. Degradation of woody materials is mainly accomplished by higher fungi; these organisms are the most damaging to wood and are present in most of the environments where wood has found use. Fungi infect wood by the aid of spores or hyphae which grow into sound wood from infected material. The strong wood-degrading effect that fungi have depends in part upon the organization of their hyphae, which gives the organisms a penetrating capacity. The wood is then discolored and/or decayed. Decay involves specific enzymes; these are excreted from the fungal hyphae and attack the cell walls of the wood fibers. Different types of fungi cause different types of decay; one normally distinguishes between soft-rot, brown-rot and white-rot fungi. The 'blue-staining' fungi are also associated with wood damage, but do not cause wood decay.

The term soft-rot emanates from the fact that there is a softening of the surface layer when wood is attacked by this group of fungi. Although the soft-rot fungi slowly degrade lignin, it can be concluded that they are more successful in degrading wood polysaccharides than lignin. Soft-rot fungi are *Ascomycetes* or *Fungi Imperfecti*, whereas brown-rot and white-rot fungi are *Basidiomycetes*.

The hyphae of the brown-rot and white-rot fungi are normally localized in the wood cell lumens and penetrate from one cell to another through openings or by producing bore holes through the cell walls. Brown-rot fungi mainly degrade cellulose and hemicelluloses in wood. In an early stage of degradation, they depolymerize cellulose faster than the degradation products are utilized.

The white-rot group of fungi is a rather heterogeneous group of organisms. They have in common a capacity to degrade lignin as well as the other wood components. They also have in common the ability to produce extracellular enzymes which oxidize phenolic compounds including many related to lignin. The relative amounts of lignin and polysaccharides degraded and utilized by these fungi vary and so does the order of preferential attack. The normal method of degradation by white-rot fungi is for the cellulose and the lignin to be attacked simultaneously. A totally specific attack on the lignin by white-rot fungi probably does not occur. Evidence indicates that energy to degrade lignin must be derived from more easily accessible energy sources, including polysaccharides and low molecular weight sugars (Ander and Eriksson, 1975; Hiroi and Eriksson, 1976; Kirk *et al.*, 1976; Jeffries *et al.*, 1981).

Physiological requirements for lignin metabolism have been studied in some detail in the white-rot fungus *Phanerochaete chrysosporium* Burds. In addition to a requirement for a cosubstrate, the ligninolytic system: (a) is produced even in the absence of lignin; (b) is expressed only during secondary ('maintenance') metabolism; (c) is triggered by carbon-, sulfur- or nitrogen-limitation; (d) is strongly repressed by glutamate and certain other amino acids; (e) is sensitive to the balance of trace metals supplied; (f) has a relatively narrow pH optimum (pH 4–5); and (g) is markedly affected by O_2 concentration (Keyser *et al.*, 1978; Fenn *et al.*, 1981; Jeffries *et al.*, 1981; Kirk, 1984; Kirk and Shimada, 1985).

With *P. chrysosporium* (*Sporotrichum pulverulentum* is the imperfect stage of *P. chrysosporium*), the highest rates of lignin degradation occur in nitrogen-limited culture. Nutrient nitrogen depletion triggers secondary metabolism, including the ligninolytic system (Keyser *et al.*, 1978). Nitrogen concentration does not affect all white-rot fungi in this way; lignin degradation by some is quite insensitive to nitrogen (Janshekar *et al.*, 1982; Leatham and Kirk, 1982). Ander *et al.* (1983) showed recently in *Sporotrichum pulverulentum* that whereas lignin degradation is repressed by high nitrogen, degradation of vanillic and syringic acids is stimulated under these conditions.

The biochemistry of lignin degradation by white-rot fungi has been reviewed recently (Kirk, 1984; Kirk and Shimada, 1985).

15.1.2 Production of Cellulase-less Mutants

The normal way for white-rot fungi to degrade wood is to degrade the cellulose and lignin simultaneously. As pointed out above, it seems as if the need for energy in lignin degradation demands that a further, easily metabolized energy source be available. The polysaccharides of the wood function, therefore, as cosubstrates.

Even if an absolutely specific attack on the lignin cannot be achieved, it has been shown that enough lignin can be degraded to give rise to a decreased need for energy in mechanical pulping if the wood chips are pretreated with cellulase-less mutants of white-rot fungi or if glucose is added to wild types to repress cellulolysis (Eriksson and Vallander, 1982; Bar-Lev *et al.*, 1982).

In the first genetic experiments to produce cellulase-less mutants, a spore suspension of the fungus *Polyporus adustus* was irradiated with UV light until approximately 97% of the spores were killed (Eriksson and Goodell, 1974). The spores were then spread over a cellulose agar plate to which small amounts of glucose had been added to allow growth of cellulase-less mutants (Figure 1). In these first experiments to produce cellulase-less mutants, colonial growth was obtained simply by decreasing the pH value to 3.2. In experiments with *S. pulverulentum* it has been necessary to add chemicals, such as saponin (Ander and Eriksson, 1975) or desoxycholate (Gold and Cheng, 1978), in order to obtain such growth. After one to two weeks, clear zones were formed around most of the colonies, *i.e.* where the opaque cellulose had been degraded. Colonies without clear zones were further investigated for their eventual cellulase production by transferring these organisms to test tubes (Figure 1). In this way, several cellulase-less mutants have been isolated with different white-rot fungi. Particularly, the mutants from the white-rot fungus *Sporotrichum pulverulentum* have been of interest. Most of our work so far has been done with Cel 44, a *Cel⁻* mutant from *S. pulverulentum*. However, Cel 26, a cellulase-less mutant of *Phlebia radiata*, has also been used in wood-chip delignification experiments (Samuelson *et al.*, 1980).

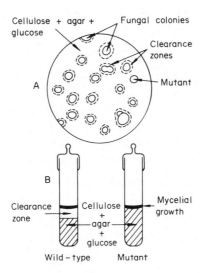

Figure 1 Selection of cellulase-less mutants of white-rot fungi: A, cellulose agar plate with normal fungal colonies and a mutant without clearance zone; B, cellulose agar tube with normal (wild-type) fungus and a mutant (Reproduced with the permission of Elsevier Scientific Publ. Co. from Ander and Eriksson, 1978)

A project, 'biological pulping', at the Swedish Forest Products Research Laboratory has been directed at eventual energy savings in mechanical pulping after treatment of wood chips with cellulase-less mutants of white-rot fungi. The project has been divided into three parts as follows. (a) Studies of growth conditions in wood have been conducted with fungi and their cellulase-less mutants (Eriksson *et al.*, 1980a). (b) Scanning and transmission electron microscopy studies have been undertaken to determine how wild-type fungi and cellulase-less mutants grow in wood and attack the wood cell walls (Eriksson *et al.*, 1980b; Ruel *et al.*, 1981). (c) Mechanical pulp has been made on a laboratory scale from fungus-pretreated and reference wood chips. Energy input has been measured and the properties of the produced pulp have been studied (Eriksson and Vallander, 1982).

15.1.3 Growth Conditions in Wood for Wild-type Fungi and Cellulase-less Mutants

To study growth conditions in wood, three white-rot fungi and their cellulase-less mutants were used (Eriksson *et al.*, 1980a); *Sporotrichum pulverulentum* and its cellulase-less mutant Cel 44; *Phlebia radiata* and Cel 26; and *Phlebia gigantea* and Cel 50. The pH and temperature optima as well as the growth rates in wood were studied. For the mutants the optimum nitrogen source and C/N ratios for treatment of birch, spruce and pine woods were determined. Wood blocks 20 and 60 mm long were inoculated on one of the cross-sections and stored at the respective optimum temperatures for growth of the different strains. After different times the opposite cross-section was pressed against the surface of a malt agar plate. These plates were incubated at 25 °C and by observing when fungal growth first appeared on the plates the time required for growth through the wood was estimated for the different strains.

By cultivating the fungi on malt agar plates the growth rates on this medium were estimated. In one day, *S. pulverulentum* and Cel 44 grew 20 mm, whereas the other strains needed 3.5–4 days to grow as much. The lag time is included in these results. The maximal growth rate was obtained after 2–3 days. All fungal strains grew slower in wood than on a malt agar surface. In wood, the growth rate of *S. pulverulentum* and Cel 44 decreased to approximately half of their growth rates on malt agar, to 0.5 mm h⁻¹. The decrease in growth rate of the other strains was less pronounced. The values for growth in wood indicate that in the *Cel⁻* mutants this will not be a rate-limiting factor in biopulping.

The optimum source of nitrogen as well as the optimum C/N conditions for growth on wood flour were evaluated for these cellulase-less mutants (Figure 2). A proportion of 10% urea and 90% $NH_4H_2PO_4$ was found to be optimum for Cel 44 on both pine and spruce wood. The optimal C/N ratio was found to be 160/1 for pine and 200/1 for spruce. Cel 26 and Cel 50 were very similar in their demand for nitrogen, but they differed from Cel 44. Notably, the optimum C/N ratios were higher for these mutants than for Cel 44.

urea / $NH_4H_2PO_4$

C:N	0/100	10/90	20/80
160:1		x	
200:1	x	x	x
250:1	x	x	
400:1		x	

	Pine		Spruce	
	urea		urea	
Mutant	$NH_4H_2PO_4$	C:N	$NH_4H_2PO_4$	C:N
Cel 44 (*S.p.*)	10/90	160:1	10/90	200:1

C:N	10/90	20/80	25/75	30/70
160:1		x		
250:1		x		
400:1	x	x	x	x
650:1		x		

Mutant	urea $NH_4H_2PO_4$ (Pine)	C:N	urea $NH_4H_2PO_4$ (Spruce)	C:N
Cel 26 (*P. rad*)	20/80	250:1	20/80	400:1
Cel 50 (*P.gig*)	20/80	250:1	25/75	400:1

Figure 2 Optimal nitrogen sources for the degradation of wood meal by three cellulase-less mutants of white-rot fungi. These are determined from weight loss after three weeks (Reproduced with permission of Wiley, Inc. from Eriksson *et al.*, 1980a)

It was also demonstrated that both water- and acetone-extractable substances in wood support the growth of the cellulase-less mutants. When some glucose was added to the wood, the weight-loss caused by Cel 44 increased. All of these observations support earlier findings that lignin in wood cannot be degraded by white-rot fungi unless a more easily metabolizable carbon source is used simultaneously.

The pH optima for all of the studied strains were found to be very similar. *Sporotrichum pulverulentum* and Cel 44 had optimum growth rates at pH 4.7, while *Phlebia radiata* and Cel 26 grew maximally at pH 4.9–5.0 and *P. gigantea* and Cel 50 at pH 4.8. All the investigated strains grew at pH values up to 6.5 and on the acidic side of the optimum. *S. pulverulentum* and Cel 44 grew at pH values as low as 3.4.

The studied strains differed considerably in optimum growth temperature. *S. pulverulentum* and Cel 44 were shown to have optimum growth rates at 38–39 °C and *P. radiata* and Cel 26 at 26 °C and *P. gigantea* and Cel 50 at 28 °C.

15.1.4   Growth in Wood of Wild-type Fungi and Cellulase-less Mutants

15.1.4.1   *Studies by scanning electron microscopy*

To understand better how the attack on the wood cell wall is carried out by wild types and *Cel⁻* mutants of *S. pulverulentum*, studies have been undertaken both on the micromorphological level and on the ultrastructural level using scanning electron microscopy (SEM) and transmission electron microscopy (TEM). Two important questions to answer in these studies have been: what is the growth pattern in wood of wild-type fungi and their cellulase-less mutants and how can a cellulase-less mutant which cannot degrade the cellulose in the wood fiber wall grow from one wood cell to another?

The SEM technique has been used to study fungal growth in wood (Blanchette, 1980; Eriksson *et al.*, 1980b). In Figures 3 and 4 it can be seen that Cel 44 grows from one wood cell to another by utilizing existing holes. Branching of the fungal hypha within the cells is common for both wild-type and mutants. In birch (Figure 3) the mutant grows from one cell to another by penetration of the perforated plates between the vessels. In pine (Figure 4) the mutant spreads through the pits between ray cells and between the ray cells and adjacent tracheids. Penetration by the wild-type hypha directly through the cell wall has also been observed (Figure 5). In contrast, the *Cel⁻*

mutants grow through pores in the wood cell (Figure 6). That the fungus does not need immediate support of the fiber cell wall but can also grow at some distance from it is illustrated in Figure 7. The hypha seems to be stiff enough to keep itself up at least for short distances.

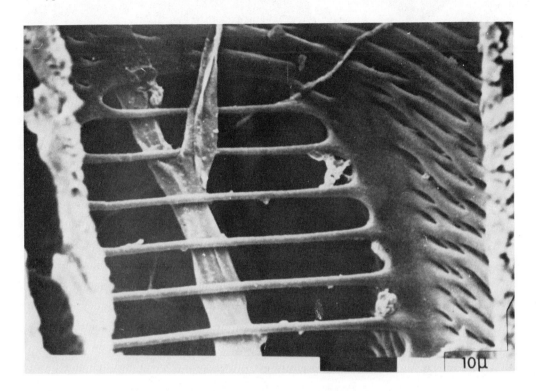

Figure 3 Growth of Cel 44 (a *Cel⁻* mutant of *Sporotrichum pulverulentum*) through the perforated plate between vessels in birch (Reproduced with permission from Eriksson, 1981)

Another characteristic which is very obvious for the wild type of the white-rot fungi is that they can cause thinning of the wood cell walls (Figure 8). However, such thinning has not been observed with the cellulase-less mutants. Instead, these mutants produce cavities and fissures in the fiber walls, particularly in connection with the middle lamella. Apparently, a rather specific attack by certain fungi on the middle lamella can take place (Figure 9); in this case the fungus is a wild-type strain of *Phellinus isabellinus*.

In Figure 10, an interesting observation can be made, namely that the surface layer itself seems to have been removed by the fungus. It seems obvious from this picture that enzymes attacking the wood polymers diffuse from the fungal cell wall. From this photograph, the diffusion distance appears to be approximately 3 to 6 μm.

15.1.4.2 Studies by transmission electron microscopy

While the use of the SEM technique has given valuable information on the mode of growth of white-rot fungi in the wood, TEM offers further possibilites. Ruel *et al.* (1981) have used this technique to study changes in spruce wood on attack by either the wild-type *S. pulverulentum* or its cellulase-less mutant Cel 44. These studies have given information both at the micromorphological level and at the ultrastructural level, where changes in the macromolecular organization of the three main fiber wall components, cellulose, hemicelluloses and lignin, have been investigated. It has been confirmed in these studies that the wild-type *S. pulverulentum* can grow through the fiber cell walls, mainly the S_2 layer, while Cel 44 cannot. The capacity to bore holes through the fiber cell wall must therefore be related to the power of the wild-type fungus to attack the cellulose. In Figure 11, it can be seen that the mutant mainly attacked the space between the S_1 and S_2 layers. In the transition layers between S_1 and S_2 regions, the cellulose microfibrills change direction, which seems to make these layers more susceptible to attack by the mutant. The transition layers are easier to degrade for both wild-type fungi and cellulase-less mutants.

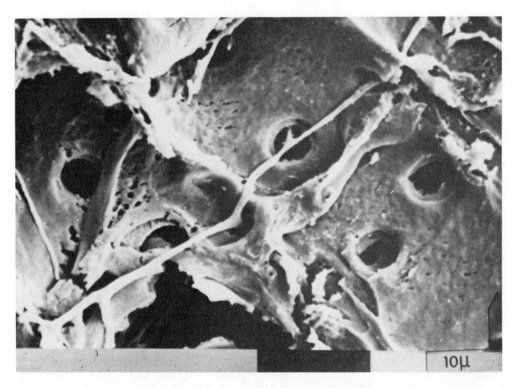

Figure 4 Growth of Cel 44 (a *Cel⁻* mutant of *Sporotrichum pulverulentum*) through the pits between ray cells and between a ray cell and a tracheid cell in pine (Reproduced with permission from Eriksson, 1981)

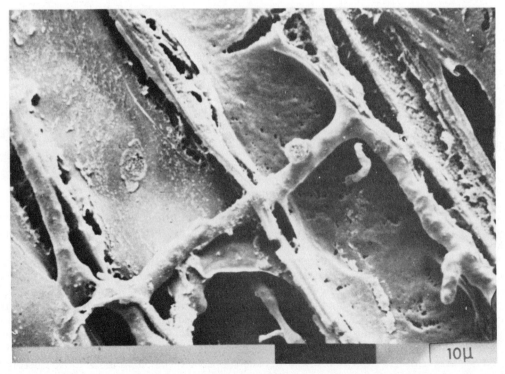

Figure 5 The wall of a ray cell penetrated by *Sporotrichum pulverulentum* (Reproduced with permission from Eriksson, 1981)

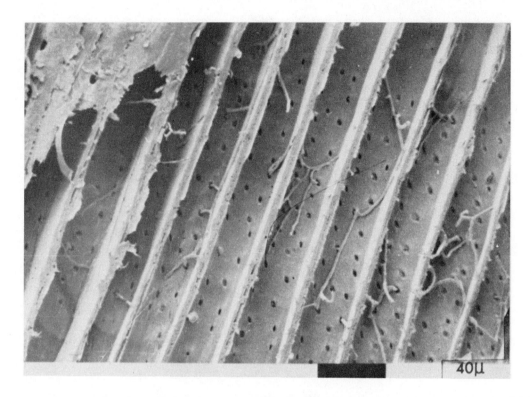

Figure 6   Growth of Cel 50 (a *Cel⁻* mutant of *Phlebia gigantea*) in spruce. A concentration of fungal hyphae can be found
in pit regions (Reproduced with permission of Walter de Gruyter and Co. from Eriksson *et al.*, 1980b)

The transition layer between S_2 and S_3 is also easily degraded, whereas the S_3 layer is surprisingly resistant to microbial attack (Figure 12). The studies by Ruel *et al.* (1981) have clearly demonstrated that the most resistant parts to microbial degradation are the S_3 layer and the middle lamella.

The average distance from the fungal cell wall where lignin was visibly modified was of the order of 2–3 μm for wild-type decay, but only around 0.2–0.3 μm for the mutant. In the areas around the fungal hypha, it seems as if the hemicelluloses were the first wood components to be degraded. This has been shown using cytochemical tests and is also in agreement with the findings illustrated in Figure 10. The substance removed from the tracheid cell wall in this figure was most likely hemicellulose. However, it is also clear that the lignin can be changed at least morphologically at a distance from the fungal cell wall. Diffusion of enzymes from the fungal cell wall apparently caused the lignin to aggregate in the form of granules instead of being layered in the wood in the form of long barriers (Figures 13 and 14). To judge from these pictures, it seems plausible that lignin modification is not to be considered as purely a contact phenomenon, *i.e.* modification can take place at some distance from the fungal wall. This requires further study.

15.1.5   Energy Savings and Properties of Pulp from Fungus-treated Wood Chips

The fungi which have been used in the wood pretreatment experiments are Cel 44 (a *Cel⁻* mutant of *Sporotrichum pulverulentum*) and Cel 26 (a *Cel⁻* mutant of *Phlebia gigantea*). The Cel 44 mutant has been used primarily.

Treatment of wood chips has been carried out in a bench-scale apparatus containing four 20 l steel cylinders (Eriksson and Vallander, 1982). Optimum quantities of urea and $NH_4H_2PO_4$ (Eriksson *et al.*, 1980a) were added to the wood chips before inoculation with mycelium. In some experiments, the wood chips were vacuum-pressure impregnated with a glucose solution. Sterile humid air was continuously passed through the cylinder and the concentration of carbon dioxide in the outgoing air analyzed and recorded. The temperature was kept at optimum growth temperature for the mutant used (for Cel 44 at 39 °C); treatment has normally been for 14 days.

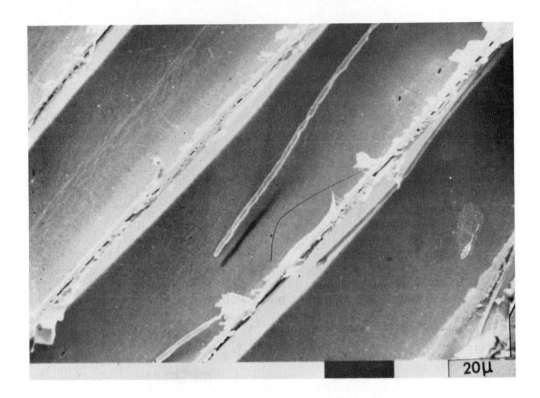

Figure 7   Growth of Cel 26 (a *Cel*⁻ mutant of *Phlebia radiata*) in spruce. The fungus hypha does not need to be mechanically supported against the cell wall (tracheid) along its whole length (Reproduced with permission of Walter de Gruyter
and Co. from Eriksson *et al.*, 1980b)

The defibration, sheet formation, paper properties and bleaching evaluations were carried out
as described by Eriksson and Vallander (1982). In the first experiments, ordinary wood chips
were used. In later experiments the chips were impregnated with glucose before the fungal treatment. The purpose of impregnating wood chips with glucose was two-fold. Firstly, addition of
glucose promoted the growth and rapid establishment of the fungus within the wood. Secondly,
as discussed above, white-rot fungi need an additional, easily metabolizable carbon source to
degrade lignin. Thus, if the wood chips were impregnated with glucose, the cellulase-less mutant
would preferentially metabolize the added glucose and naturally-occurring low molecular mass
sugars instead of degrading the wood cell wall hemicelluloses during delignification, thereby protecting the fibers. The protective effects of glucose had earlier been demonstrated by Yang *et al.*
(1980). Wood weight losses, which were small in all the experiments and never exceeded 3%,
decreased in the presence of glucose. It was shown by Eriksson and Hamp (1978) that cellulose-
degrading enzymes, *endo*-1,4-β-glucanases, are repressed at glucose concentrations lower than 50
mg l⁻¹ in liquid cultures of *S. pulverulentum*. One can anticipate that mannan- and xylan-degrad-
ing enzymes are also repressed at the same glucose level.

In the absence of glucose, there was a more intense fungal attack on the wood cell-wall components, rendering the fibers more susceptible to mechanical action. Thus it can be seen in Figure 15
that the longer the fungal treatment (increased weight loss), the smaller the amount of energy
needed, at freeness levels from 200 ml and upwards, to refine the pulp to a certain freeness level.
(The greater the beating (refining), the slower the water drains from a pulp sheet. A rapidly
draining pulp has a high freeness, a more slowly draining pulp has a low freeness.) At a freeness
of around 100 ml no difference was distinguished.

Where glucose-impregnated wood chips were refined, the scatter of values obtained with fungus-treated samples did not allow them to be distinguished from the values obtained with the
reference sample if the energy–freeness relationship is plotted as in Figure 15.

In Figure 16, the tensile index *versus* energy is plotted for pulp made from Cel 44-treated, glucose-impregnated (1.7% glucose calculated on wood weight) spruce wood. The fungus-treated
sample used 1250 kW h tonne⁻¹ and the reference 1600 kW h tonne⁻¹ to obtain a tensile index
(tensile strength) of 33 N m g⁻¹, corresponding to a 23% energy reduction in the refining stage.

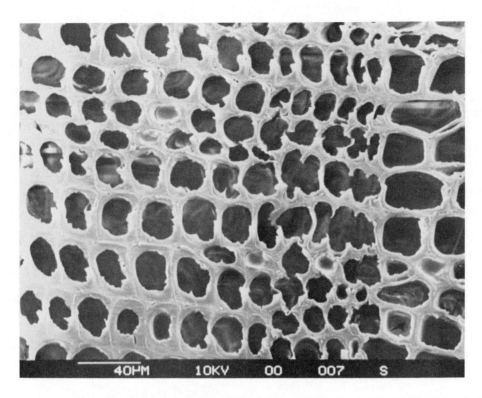

Figure 8 Growth of *Phlebia radiata* in pine. The fungus has caused a thinning of the cell wall (Reproduced with permission of Walter de Gruyter and Co. from Eriksson *et al.*, 1980b)

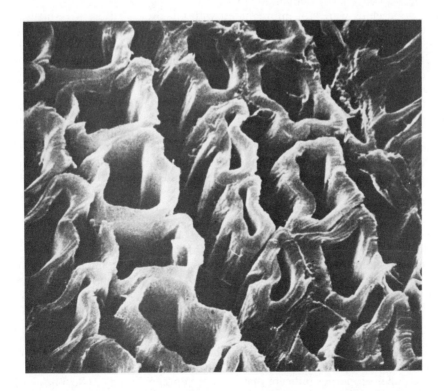

Figure 9 Growth of *Phellinus isabellinus* in birch wood. It can be seen that the middle lamella is heavily attacked (Reproduced with permission of Academic Press from Käärik, 1974; courtesy T. Nilsson)

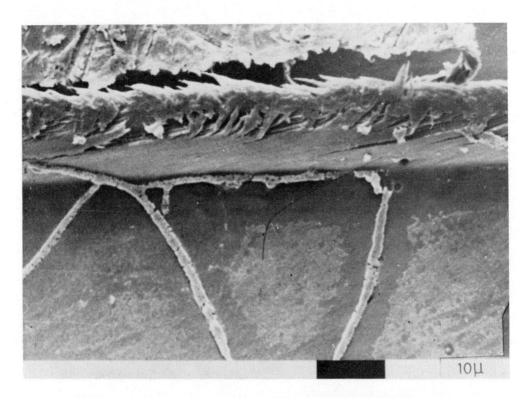

Figure 10 Growth of *Phlebia radiata* in spruce. Diffusion of enzymes from the fungus cell wall degrades a substance covering the tracheid cell wall (Reproduced with permission of Walter de Gruyter and Co. from Eriksson *et al.*, 1980b)

Defibration energy was 745 and 715 kW h tonne^{-1}, respectively. (Defibration of wood into mechanical pulp is a very energy-demanding process.)

When pine wood chips were not impregnated with glucose, a different result was obtained (Figure 17). Compared with the reference pulp more energy was needed at high tensile index, less energy at low values of the tensile index. For the cellulase-less mutant used in this experiment it can thus be concluded that pretreatment of wood not containing added glucose is of little interest. Bar-Lev *et al.* (1982) reached a similar conclusion with TMP (thermomechanical pulp) and a wild-type strain of *Phanerochaete chrysosporium*.

It has been observed, when the tensile index–density relationship is considered, that test samples from glucose-impregnated wood tend to have greater strength at density values above 400 kg m^{-3} than their reference samples (Figure 16). In contrast, with wood which was not glucose-impregnated, the reference pulps gave higher tensile index values than the test pulps.

How can the observed difference between pulps made from impregnated and non-impregnated chips be explained? The tensile index–density relationship is influenced by pulp yield, degree of fiber collapse and fiber length distribution.

The differences between weight loss values for glucose-impregnated and non-impregnated samples were small. Consequently, differences in tensile index due to glucose impregnation can hardly be explained by variations in pulp yield.

It is possible that the degree of fiber collapse is affected by glucose impregnation and fungal action. The glucose might, as pointed out above, influence the extent and chemical nature of the fungal attack on the wood and hence the collapsibility of the fibers. This has not been studied.

Andersson and Mohlin (1980) found tensile index to be approximately proportional to the long fiber content and to the z-strength (internal bond strength). A decrease in long fiber content was observed for fungus-treated samples without glucose compared to the reference, whereas in samples with glucose no such difference between reference and test samples was observed. This may partly explain the higher tensile index obtained with glucose-impregnated samples. The influence of z-strength has not been determined.

The pulp and paper properties indicate that glucose impregnation has a beneficial effect on the ability of the fibers to bond to each other in the paper sheet, which could be due to a repression of hemicellulose degradation.

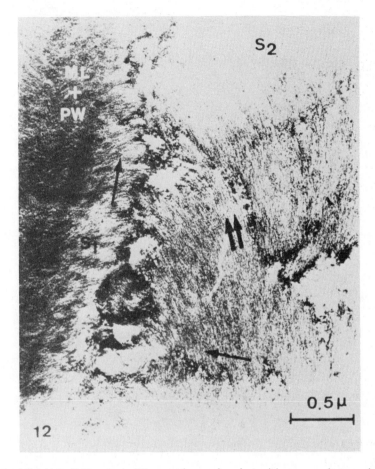

Figure 11 Growth of Cel 44 (a *Cel⁻* mutant of *Sporotrichum pulverulentum*) in spruce; photograph shows vertical progression of decay through the S₁/S₂-layer (Reproduced with permission of Walter de Gruyter and Co. from Ruel *et al.*, 1981)

Attack on wood by white-rot fungi leads to the formation of colored substances, and a darker pulp is obtained. This effect is more pronounced with cellulase-less mutants than with wild types of the white-rot fungi. The explanation is most likely to be found in the enzyme systems. The wild type, as opposed to the cellulase-less mutant, produces an enzyme, cellobiose:quinone oxidoreductase which, in the presence of cellobiose, eliminates quinones formed during lignin degradation (Westermark and Eriksson, 1974).

To characterize the optical properties of the pulps, brightness, light scattering and light absorption were determined. Initially, great losses in brightness were observed in the experiments. Pulps from non-impregnated wood had a reduction in the order of 10 units of ISO brightness compared to the reference pulps. (For an explanation of ISO brightness see International Standards Organization, 1977.) In subsequent experiments, the difference was 4 units. The fungal treatment also leads to lower light-scattering coefficients than for the reference pulps at constant density.

In order to study how pulp from treated wood responded to bleaching, pulps from glucose-impregnated fungus-treated chips and the reference pulp were treated with 4% hydrogen peroxide. Table 1 shows a change in brightness of the reference pulp from 55 to 73% (% ISO) and for the fungus-treated pulp from 43 to 66%. Although the gap in brightness between the pulps was reduced from 12 to 7% due to the bleaching, the figures clearly demonstrate the poor optical properties of pulps from fungus-treated wood. It is therefore necessary to study how this undesirable discoloration is caused and how it can be avoided.

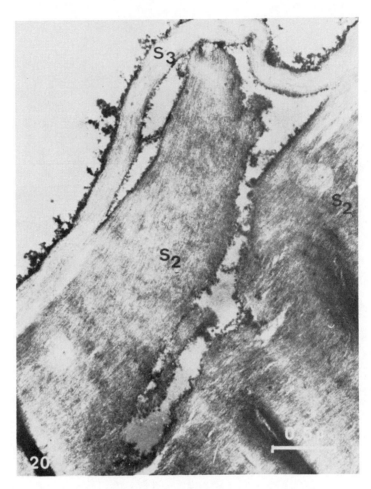

Figure 12 Growth of *Sporotrichum pulverulentum* in spruce, demonstrating the relative resistance of the S_3-layer (Reproduced with permission of Walter de Gruyter and Co. from Ruel *et al.*, 1981)

15.1.6 Possibilities for Improvements in Biopulping and Microbial Delignification in General

To be able to successfully run a large scale biotechnical process, such as fungal pretreatment of wood chips for energy reduction in mechanical pulping, it is necessary to have a thorough knowledge of the physiological requirements of the fungus for lignin degradation, to know at least in some detail the enzyme mechanisms for lignin degradation, and particularly to know more about the primary attack on the lignin polymer.

The mutants so far used in the wood chip treatment experiments have had only one virtue, namely to be cellulase-less. This is obviously not enough, particularly since the *Cel*⁻ property often is combined with a low phenol oxidase production and undoubtedly there are other differences from the wild type. It is not known in any detail which enzymes are necessary and important for lignin degradation; evidence indicates that phenol oxidases are necessary (Ander and Eriksson, 1976). For this reason, the Swedish Forest Products Research Laboratory has recently begun to screen for *Cel*⁻, *phenol oxidase*⁺ (*Po*⁺) mutants. These mutants have proved to be more efficient degraders of ¹⁴C-labelled dimers and synthetic lignin than the previously used *Cel*⁻ mutants (Eriksson *et al.*, 1983). Treatment of wood chips and pulp production with these new mutants is now under way.

The project 'biopulping' in Stockholm is thus in many ways back on the basic research level where we particularly aim to identify which additional important qualifications a *Cel*⁻ mutant should have.

Since the work was begun, good progress has been made toward understanding the physiology,

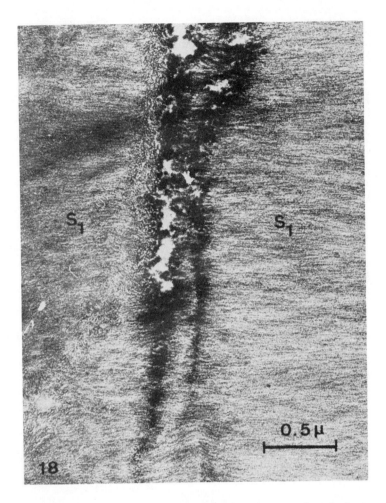

Figure 13 Growth of *Sporotrichum pulverulentum* in spruce. Diffusion of enzymes from the fungus cell wall tears down 'the lignin lamella' in the wood (Reproduced with permission of Walter de Gruyter and Co. from Ruel *et al.*, 1981)

chemistry and biochemistry of the degradation of lignin by white-rot fungi (Ander and Eriksson, 1978; Kirk 1984; Kirk and Shimada, 1985; Higuchi, 1982; Crawford, 1981). Application of this new knowledge will help transform the idea of biomechanical pulping into a practical process.

15.2 BIOBLEACHING OF KRAFT PULP

15.2.1 Nature and Importance of Kraft Pulping

The kraft process is the world's major pulping method and is likely to remain so into the foreseeable future. It has evolved over a period of 100 years, and has become highly refined. Currently about 70% of the world's annual pulp output of approximately 100 million tonnes is produced by the kraft process. Despite some shortcomings, it is the most cost effective, versatile and efficient wood delignification method available. Because of this fact and the large amount of capital already invested in kraft pulping, it is unlikely that the process will be replaced in the near future.

The kraft process results in the degradation and solubilization of lignin. Wood chips are cooked in a solution of $Na_2S/NaOH$ at about 170 °C for about 2 h to degrade and solubilize the lignin. (Details of the process are given by Rydholm, 1965; Sanyer and Chidester, 1963; Sjöström, 1981.) The lignin undergoes a variety of reactions, but the most important of these result in partial depolymerization and formation of ionizable (mainly phenolic hydroxyl) groups; these changes lead to the dissolution of the lignin in the alkaline pulping liquor. The lignin reactions involved in kraft pulping have been studied extensively (Gierer, 1970, 1981; Ljunggren, 1980; Marton, 1971).

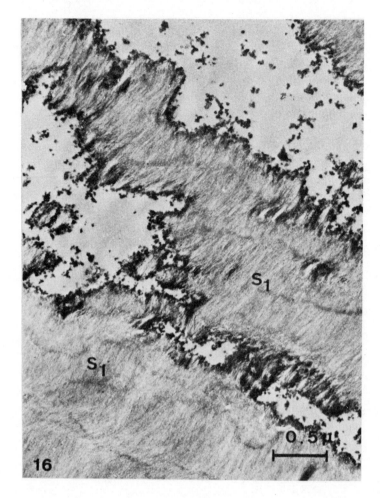

Figure 14 Growth of *Sporotrichum pulverulentum* in spruce. After prolonged rotting, the lignin aggregates and forms aggregates 40–80 nm in size (Reproduced with permission of Walter de Gruyter and Co. from Ruel *et al.*, 1981)

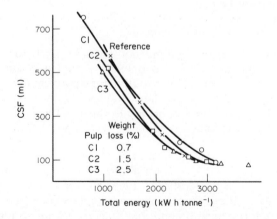

Figure 15 Total energy needed for defibration and refining to different freeness levels. Pine chips, without glucose impregnation, pretreated with Cel 44 (a *Cel⁻* mutant of *Sporotrichum pulverulentum*) (Reproduced with permission from Eriksson and Vallander, 1982)

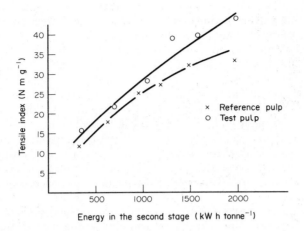

Figure 16 Energy demand in refining of glucose-impregnated spruce wood chips, pretreated with Cel 44 (a *Cel⁻* mutant of *Sporotrichum pulverulentum*) to different tensile indices (Reproduced with permission from Eriksson and Vallander, 1982)

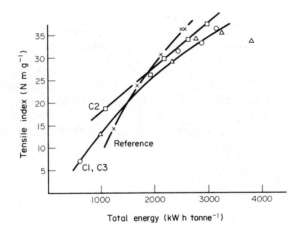

Figure 17 Energy demand in defibration and refining of wood chips to different tensile indices. The wood chips were pretreated with Cel 44 (a *Cel⁻* mutant of *Sporotrichum pulverulentum*) (Reproduced with permission from Eriksson and Vallander, 1982)

Table 1 Peroxide Bleaching[a] of Spruce Pulp
Pretreated With Cell 44

Sample	Brightness[b] Before	After
Test sample (Cel 44)	43	66
Reference sample	55	73

[a] 4% H_2O_2. [b] Brightness as % ISO (International Standards Organization, 1977).

15.2.2 Chemical Bleaching of Kraft Pulp

In kraft pulping about 90% of the lignin is removed. The 10% or so remaining in the pulp is primarily responsible for the brown color characteristic of kraft pulp and papers. The lignin remaining in the pulp has been heavily modified, almost as much as the 'kraft lignin' that is removed. Its persistence in the pulp probably reflects covalent binding to the hemicelluloses

(Yamasaki *et al.*, 1981). The brown color is due to various conjugated structures, including quinones, complexed catechols, chalcones and stilbenes, which absorb visible light (Falkehag *et al.*, 1966; Gierer, 1970). Natural lignin is nearly colorless.

Almost half of the kraft pulp produced annually is bleached before use. Because the bleaching process is costly and results in some weight loss, bleached kraft pulp sells for about 10–20% more than unbleached pulp. Bleaching is done by processes employing mainly chlorine and its oxides. The residual lignin is degraded and dissolved in various sequences of bleaching and extraction steps in which chlorine, hypochlorite, chlorine dioxide, oxygen and hydrogen peroxide are used. Perhaps the most common sequence is 'CEDED' (the initials stand for *c*hlorination, alkali *e*xtraction, chlorine *d*ioxide treatment, alkali *e*xtraction, and chlorine *d*ioxide treatment), which is followed by washing to remove residual chemicals. The effluents produced during bleaching of kraft pulp are the major contributors to waste water pollution from the pulp industry and have to be treated before discharge; they are of increasing environmental concern (Trubacek and Wiley, 1979; see Section 15.3 below).

Newer bleaching procedures based on either oxygen or ozone have been developed; the former are already being used commercially. These newer methods, however, cannot yet replace chlorine-based procedures because they are not fully effective. They are being developed to reduce the use of chlorine and the associated environmental concerns and also to alleviate corrosion problems. The oxygen-based bleaching steps produce effluents that are not corrosive, permitting them to be recycled within the mill (Singh and Dillner, 1979). The chlorine-based effluents must be treated, and they put a heavy burden on the waste water treatment plant due both to the organic load and to the toxic chlorinated organics.

15.2.3 Preliminary Experiments on Bleaching with White-rot Fungi

Biological bleaching processes have been explored only in a very preliminary way, but with results that suggest the desirability of additional study. The potential for biobleaching was recognized when it was discovered that kraft lignin is degraded by white-rot fungi (Hiroi and Eriksson, 1976; Lundquist *et al.*, 1977). These fungi have been shown to reduce quinones to their corresponding phenols and thereby probably cause bleaching in part by the action of the enzymes cellobiose:quinone oxidoreductase (Westermark and Eriksson, 1974), and NAD(P)H:quinone oxidoreductase (Buswell *et al.*, 1979).

The biochemical mechanism by which the white-rot fungi degrade lignin is rapidly being elucidated (Kirk and Shimada, 1985). After many years of underpinning research, the first lignin-degrading enzyme was discovered in late 1982 (Tien and Kirk, 1983). The enzyme, ligninase, is now known to be a type of peroxidase, requiring H_2O_2 for its reactions (Tien and Kirk, 1984); its basic catalytic mechanism is oxidation of the aromatic nuclei of lignin to unstable cation radicals which decompose *via* several different reactions (Kersten *et al.*, 1985). The enzyme is nonspecific in the substrates it attacks, and its action leads to formation of a variety of products, in accord with the known chemistry of lignin degradation by the white-rot fungi (Chen and Chang, 1985).

The nonspecificity of the ligninolytic system of the white-rot fungi is clearly evidenced in the fact that kraft lignin is readily metabolized (Hiroi and Eriksson, 1976; Lundquist *et al.*, 1977); in fact, kraft pulping appears to increase the degradability of lignin (Lundquist *et al.*, 1977; Figure 18).

The finding that kraft lignin is metabolized by white-rot fungi prompted a preliminary investigation into biobleaching (Kirk and Yang, 1979). Pulp was amended with nutrients required for fungal growth, and inoculated with *Phanerochaete chrysosporium*. Incubation was in shallow stationary cultures at 39 °C. Various experiments showed that the kappa number (a measure of lignin content) could be reduced 50 to 75% in 6 to 8 days. Up to 50% of the pulp cellulose was depleted in 7 days, but this could be retarded by addition of glucose (or malt extract, cane molasses, corn syrup, starch, *etc.*), to repress the cellulolytic system. Delignification was enhanced two-to-three-fold by maintaining the cultures under >50% O_2.

Exploratory experiments demonstrated that other white-rot fungi gave similar results, although *P. chrysosporium* was the most rapid.

Figure 19 gives the time course of decrease in kappa number of kraft pulp in *P. chrysosporium* cultures containing 0.2% N (as NH_4NO_3 + L-asparagine) and glucose and incubated under ~100% O_2. No decrease in kappa number was observed prior to day 2. The subsequent change did not follow first order kinetics, but was biphasic; an initial rapid decrease was followed after 4 days by a much slower decrease.

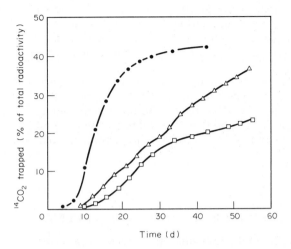

Figure 18 $^{14}CO_2$ evolution from cultures of *Phanerochaete chrysosporium* containing synthetic [ring-^{14}C]lignins: □, untreated control lignin; △, kraft-pulped; ●, kraft-pulped and 'CE' treated. Cellulose served as growth substrate in the nitrogen-limited, shallow stationary cultures which were incubated in air (Lundquist *et al.*, 1977)

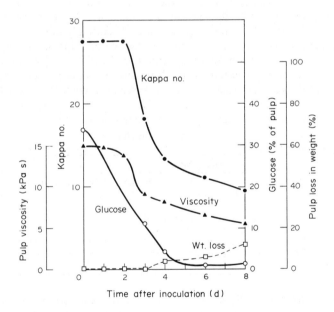

Figure 19 Changes in kappa number, viscosity and weight loss of kraft pulp, and in concentration of added glucose, in cultures of *Phanerochaete chrysosporium*. Shallow, stationary cultures were grown under O_2 (Reproduced with permission from Kirk and Yang, 1979)

The 2-day lag seen before the start of the decrease in kappa number reflects depletion of nutrient nitrogen. This depletion triggers the development of ligninolytic activity in *P. chrysosporium* (Keyser *et al.*, 1978), and is therefore necessary for bleaching. Mutants not repressed by nutrient nitrogen might not exhibit such a lag period; in fact, certain white-rot fungi such as *Lentinus edodes* apparently do not exhibit nitrogen repression, and do not exhibit the lag period (Leatham and Kirk, 1982).

Evaluation of chlorine consumption in a standard 'CEDED' bleaching sequence to a pulp brightness of 85 showed a decrease of 27% for a fungus-treated pulp sample in which the kappa number had been reduced by 29% or from kappa 24 to 17.

These results demonstrated that ligninolytic fungi can partially bleach kraft pulp. The observed concomitant reduction in the amount of chlorine required for subsequent chemical bleaching was expected, because chlorine requirements generally parallel kappa number.

In this experiment pulp viscosity loss, an indicator of pulp strength loss, paralleled extent of bleaching, despite the presence of glucose. Use of cellulase-less mutants, discussed in Section 15.1, would be expected to preclude serious loss in pulp strength.

The presence of fungal mycelium in pulp can have various influences on pulp properties, depending on the fungal species. By proper choice of fungus, the mycelium might be beneficial (Johnson and Carlson, 1978).

As is clear from the above, research to date on biobleaching has only been cursory. The primary problems would appear to be control of cellulose degradation, the technical aspects of scale-up, and avoidance of the initial lag phase. Better understanding of the basic mechanisms of bleaching (*i.e.* of lignin degradation) by the white-rot fungi should help solve these problems. The recent discovery of ligninase opens the door to possible use of isolated enzymes for bleaching (Farrell, 1985).

15.3   TREATMENT OF KRAFT BLEACH PLANT EFFLUENTS

15.3.1   Origin and Nature of the Effluents and Current Treatment

As mentioned above, kraft pulp is commonly bleached with chlorine and its oxides, and, to an increasing extent, with a prior oxygen bleaching step. A variety of effluents from the several stages of chlorine bleaching are produced which must be treated before release into receiving waters.

Table 2 summarizes the major components and properties of the two principal types of chlorine bleach plant wastewater: the first chlorination and the first extraction-stage effluents. Effluents from subsequent stages in the bleaching sequence are of relatively minor concern because of their much lower contents of dissolved organics and the absence of chlorinated organics. Chlorination-stage effluents, although toxic and mutagenic (Ander *et al.*, 1977; Eriksson *et al.*, 1979; Nazar and Rapson, 1980), are successfully treated by alkalinization (Eriksson *et al.*, 1982), and indeed are largely removed by well-operated waste treatment systems (Leach *et al.*, 1978).

Table 2   Some Components and Properties of Chlorination- and Extraction-stage Effluents
from Kraft Pulp Bleach Plants

Effluent	Major components	Properties
Chlorination-stage	Chlorinated phenols and resin and fatty acids, chlorinated hydrocarbons, hypochlorite, chloride, carbohydrates and their degradation products	Low molecular weight, low color, high BOD, toxic, bacteriostatic, mutagenic
Extraction-stage	Chlorinated and oxidized kraft lignin, hemicelluloses and their degradation products	Mostly polymeric, high color, high BOD and COD.

Bleaching effluents are generally sewered with pulping and other mill wastes and the combined effluent stream is first treated in a primary clarifier to remove lost pulp fibers and other insolubles. The resulting effluent is then taken through some type of biological treatment such as the activated sludge process or an aerated lagoon. Finally, this effluent is put through a secondary clarifier to remove residual biological solids. The final effluent is discharged to receiving waters. The primary clarifier solids (primary sludge) are dewatered, and, together with secondary clarifier solids, are either landfilled or burned to recover energy. These procedures remove BOD, toxicity, mutagenicity, and some of the COD and TOC (total organic carbon). Color with its associated COD, however, is only partially reduced. In Japan, color is removed from the colored E_1-stage effluent at several mills by alum coagulation followed by air (foam) flotation or by ultrafiltration, and burned in the chemical recovery furnace (H.-m. Chang, personal communication, 1982). Color is removed in only a handful of European mills, and not in US or Canadian mills.

Effluent color, therefore, remains an incompletely solved problem in the treatment of kraft bleach plant effluents. Over 90% of the color from a chlorine bleach plant is in the first extraction-stage ('E_1') effluent. Over half of the color from an entire 'bleached kraft' mill is in this relatively small-volume effluent, and the contribution of E_1 effluent is increasing as mills increasingly recycle process water and recover wastes. Thus it is mainly the E_1 effluent that must be decolorized if mill color is to be controlled. Although several physical/chemical processes for color removal or reduction have been developed (coagulation/flotation, reverse osmosis, ultrafiltration, massive lime precipitation, ozonolysis, alum precipitation, ion exchange, *etc.*), there is a need for less expensive methods.

15.3.2 The Chromophoric Material in Extraction-stage Effluent

Investigations have shown that the chromophoric materials in the E_1 effluent from kraft chlorine bleach plants are water-soluble, polymeric, chlorine-containing oxidized lignin fragments with a low aromatic content (Bennett *et al.*, 1971; Hardell and deSousa, 1977; Pfister and Sjöström, 1979; Sundman *et al.*, 1981). The chromophoric component can be isolated readily by membrane filtration, and further purified by physical methods. In comparison to kraft lignin, the isolated material is found to have a low methoxyl content and a high oxygen content. It also contains a substantial amount (6 to 10%) of organically bound chlorine (Hardell and deSousa, 1977; Pfister and Sjöström, 1979; Sundman *et al.*, 1981; Collins *et al.*, 1969). The chromophores exhibit an absorption of ~ 21 g^{-1} cm^{-1} at 465 nm (pH 7.6; Sundman *et al.*, 1981).

15.3.3 Decolorization with Ligninolytic Fungi

In 1977 two reports described the degradation of the extraction-stage effluent from the chlorine bleaching of kraft lignin by ligninolytic fungi. Fukuzumi and coworkers (1977) showed that several white-rot fungi can decolorize the effluent, and that some species are superior. Lundquist *et al.* (1977) demonstrated that ^{14}C-labelled 'effluent' (prepared by kraft-cooking, chlorine bleaching and alkali — 'CE' — treatment of synthetic ^{14}C-lignins) is decomposed to $^{14}CO_2$ by the white-rot fungus *P. chrysosporium* (Figure 18). Subsequent investigations have optimized culture conditions (Eaton *et al.*, 1980) and demonstrated that the chromophores are both bleached and decomposed (Fukuzumi, 1980; Sundman *et al.*, 1981).

Based on laboratory and bench-scale experiments, a process for decolorization by *P. chrysosporium* (or other suitable fungus) has been outlined (Eaton *et al.*, 1982). Several factors are important. (1) The fungi require a growth substrate for decolorization (Eaton *et al.*, 1980) just as they do for lignin degradation (Ander and Eriksson, 1975; Hiroi and Eriksson, 1976; Kirk *et al.*, 1976). Investigations have demonstrated that the cellulose-rich primary sludge serves this purpose well. This waste is probably present in quantities sufficient to support decolorization, and in fact disposal of the sludge itself is currently a waste management problem. (2) The fungus requires a fixed surface, or pregrowth without agitation in shallow medium, for efficient color reduction. A fixed-film reactor of the rotating biological contactor (RBC) design (Figure 20) has proved to be effective (Campbell *et al.*, 1982). Such reactors are commercially available and are already used in waste treatment (Antonie, 1976). (3) Color removal is greatly stimulated by oxygen-enriched atmospheres. The RBC provides good aeration and can be enclosed for addition of supplemental oxygen. (4) A growth stage is necessary before decolorization begins. During this stage nutrient nitrogen is depleted and the fungus becomes ligninolytic (Keyser *et al.*, 1978), and able to decolorize.

Based on these considerations, the procedure depicted in Figure 21 can be outlined; this method has been termed the FPL/NCSU MyCoR process (Eaton *et al.*, 1982; Campbell *et al.*, 1982). The *My*celial *Co*lor *R*emoval process resulted from cooperative research between the US Forest Products Laboratory (FPL), and North Carolina State University (NCSU). A fixed film MyCoR reactor is charged with growth nutrients, which can include primary sludge as the carbon source, and inoculated with a suitable fungus. Depending on the mill, the sludge will provide certain of the required mineral nutrients and trace elements as well as carbon. Nitrogen-rich secondary sludge might even be used to supply the nitrogen required for growth. After the mycelium has grown over the reactor surfaces, depleted the available nitrogen, and become ligninolytic ('pregrowth' stage; 2 to 4 days), the reactor is ready for use. The active life of the mycelium and the effect of periodic additions of nutrient to give continuous operation must still be examined in detail. Operation for over 60 days has been achieved in bench reactors in a batch mode.

Although it might be necessary or desirable to steam-treat the waste sludge to reduce the number of contaminating microorganisms, further aseptic procedures will probably not be required with fungi such as *P. chrysosporium*, which grows vigorously and is aggressive. Further-

Figure 20 Bench-scale rotating biological contactor used for studying fungal decolorization of first extraction-stage effluent from the chlorine-based bleaching of kraft pulp: the 'FPL/NCSU MyCoR' method. The mycelium of *Phanerochaete chrysosporium* covers the disk surfaces in a thin film. Slow rotation of the disks provides aeration and mixing (Eaton *et al.*, 1982)

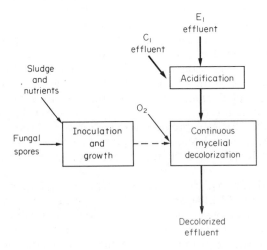

Figure 21 Flow diagram for a fungal decolorization operation. Inoculation and growth are carried out in the same vessel as decolorization in the rotating biological contactor (Figure 20)

more, the conditions of temperature, pH, and nutritional limitation in the MyCoR reactor will restrict or totally inhibit most other fungi and bacteria. Results with the bench scale apparatus described above, which has not been used aseptically, support this contention. E_1 effluent, at 28 to 40 °C and adjusted to approximately pH 4.5 with C_1 effluent or other acid source, passes into the reactor, where color, BOD and COD are reduced by fungal action. Oxygenation is accomplished as the disk surfaces contact an O_2-enriched atmosphere.

The laboratory investigations with the bench-scale (2.5 l) RBC reactors have given promising results, indicating that larger scale testing is needed. The bench-scale studies have shown that decolorization is first order below about 10 000 NCASI color units per liter and zero order above that concentration, and that decolorization rates exceed 65 000 color units per day per m^2 of mycelium surface. Because E$_1$ effluents usually contain fewer than 10 000 color units per liter, concentrating the effluents prior to decolorization can be considered (T. W. Joyce, H.-m. Chang and T. K. Kirk, unpublished). A color unit is defined by the US National Council for Air and Stream Improvement (NCASI) in terms of the weight in mg/liter of a standard chromophore in aqueous solution (NCASI, 1971).

Figure 22 shows the reduction of color from E$_1$ effluent in a MyCoR bench-scale reactor. The reactor was operated in a 'fill-and-draw' mode (A. Campbell *et al.*, unpublished). Continuous operation has also given promising results.

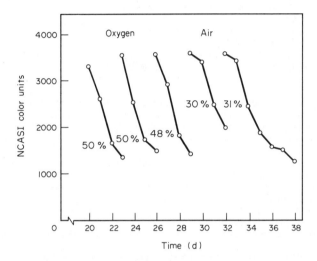

Figure 22   Color reduction of an extraction-stage effluent in a rotating biological contactor of the type pictured in Figure 20, operated with *Phanerochaete chrysosporium* (Campbell *et al.*, 1982). The unit was repeatedly drained and refilled with colored effluent. Percentages of color reduction are given for each cycle. Color reduction is substantially faster in an oxygen-enriched atmosphere

Livernoche *et al.* (1981) have reported that white-rot fungal mycelium (*Coriolus versicolor*) immobilized in calcium alginate gel can be used to decolorize E$_1$ effluent. Such immobilization can permit an entirely different approach to reactor design. Undoubtedly other types of immobilization methods can also be used.

15.3.4   Treatment of C-stage Effluents: Removal of Toxic Chlorinated Phenols and Mutagenic Substances

Bleaching effluents from pulp mills are potentially hazardous to the environment due to their high contents of toxic chlorinated organic compounds (Leach *et al.*, 1976; Holmbom, 1980; Lindström *et al.*, 1981). The C-stage effluents in particular exhibit mutagenic effects (Ander *et al.*, 1977; Eriksson *et al.*, 1979; Nazar and Rapson, 1980).

Aerated ponds have been used to remove chlorophenolic compounds but such treatments have not been very successful (Holmbom, 1980). However, other techniques such as the Billerud–Uddeholm ion exchange method and the Enso–Fenox biooxidation method have been developed on a full scale (Hakulinen and Salkinoja-Salonen, 1982).

White-rot fungi have so far found no technical use for degradation of chlorinated aromatic compounds in bleaching effluents. However, it has been demonstrated in laboratory studies that the white-rot fungus *Phanerochaete chrysosporium* can decompose the low molecular weight chlorinated aromatics found in kraft bleach plant effluents; compounds decomposed include 2,4,6-trichlorophenol, 6-chlorovanillin, and tri- and tetra-chloroguaiacols (Chang *et al.*, 1983; Huynh *et al.*, 1985). Some of the chlorinated guaiacol compounds are transformed by the fungus to veratryl compounds (*i.e.* they are *O*-methylated), which might be a disadvantage because of the greater lipophilicity of such compounds. Further study is required. Mutagenic substances can be eliminated from C-stage effluents by both chemical and microbial treatments of the spent liquor (Eriksson *et al.*, 1979, 1982); the microbial treatments have not been conducted with white-rot fungi, but with the mixed bacterial floras from a community waste purification plant and from a biological purification plant at a papermill. Interestingly, *P. chrysosporium* grown under conditions optimized for E_1 color reduction also decomposes chlorinated compounds considered to be resistant to bacterial treatments; such compounds include polychlorinated biphenyls and DDT (Eaton, 1985; Bumpus *et al.*, 1985). Obviously, therefore, further research with the white-rot fungi is warranted.

15.4 REFERENCES

Ander, P. and K.-E. Eriksson (1975). Influence of carbohydrates on lignin degradation by the white-rot fungus *Sporotrichum pulverulentum*. *Sven. Papperstidn.*, **78**, 643–652.

Ander, P. and K.-E. Eriksson (1976). The importance of phenol oxidase activity in lignin degradation by the white-rot fungus *Sporotrichum pulverulentum*. *Arch. Microbiol.*, **109**, 1–8.

Ander, P., K.-E Eriksson, M.-C. Kolar, K. Kringstad, U. Rannug and C. Ramel (1977). Studies on the mutagenic properties of bleaching effluents. Part 1. *Sven. Papperstidn.*, **80**, 454–459.

Ander, P. and K.-E. Eriksson (1978). Lignin degradation and utilization by microorganisms. *Prog. Ind. Microbiol.*, **14**, 1–58.

Ander, P., K.-E. Eriksson and H.-s. Yu (1983). Physiological requirements for degradation of lignin and lignin-related substances by *Sporotrichum pulverulentum*. *Eur. J. Appl. Microbiol. Biotechnol.*, **18**, 347–380.

Andersson, M. and U.-B. Mohlin (1980). Z-strength of mechanical pulp. *Pap. Puu*, **62**, 583–586.

Antonie, R. L. (1976). *Fixed Biological Surface Wastewater Treatment. The Rotating Biological Contactor*. CRC Press, Boca Raton, FL.

Bar-Lev, S. S. and T. K. Kirk (1981). Effects of molecular oxygen on lignin degradation by *Phanerochaete chrysosporium*. *Biochem. Biophys. Res. Commun.*, **99**, 373–378.

Bar-Lev, S. S., H.-m. Chang and T. K. Kirk (1982). Evidence that fungal treatment can reduce the energy requirement for secondary refining of thermomechanical pulp. *Tappi*, **65**, 111–113.

Bennett, D. J., C. W. Dence, F.-L. Jung, P. Luner and M. Ota (1971). The mechanism of color removal in the treatment of spent bleaching liquors with lime. *Tappi*, **54**, 2019–2026.

Blanchette, R. A. (1980). Wood decomposition by *Phellinus* (*Fomes*) *pini*: a scanning electron microscopy study. *Can. J. Bot.*, **58**, 1496–1503.

Bumpus, J. A., M. Tien, D. Wright and S. D. Anst (1985). Oxidation of persistent environmental pollutants by a white rot fungus. *Science*, **228**, 1434–1436.

Buswell, J. A., S. Hamp and K.-E. Eriksson (1979). Intracellular quinone reduction in *Sporotrichum pulverulentum* by a NAD(P)H: quinone oxidoreductase: Possible role in vanillic acid catabolism. *FEBS Lett.*, **108**, 229–232.

Campbell, A. G., E. D. Gerrard, T. W. Joyce, H.-m. Chang and T. K. Kirk (1982). The MyCoR process for color removal from bleach plant effluent: bench scale studies. *Proceedings of the 1982 Tappi Research and Development Conference, Asheville, North Carolina*, pp. 209–214. Tappi Press, Atlanta.

Chang, H.-m., T. W. Joyce, A. G. Campbell, E. D. Gerrard, V. B. Huynh and T. K. Kirk (1983). Fungal decolorization of bleach plant effluents. In *Recent Advances in Lignin Biodegradation Research*, ed. T. Higuchi, H.-m. Chang and T. K. Kirk, pp. 257–268. Uni Publishers, Tokyo.

Chen, C.-L. and H.-m. Chang (1985). Chemistry of lignin biodegradation. In *Biosynthesis and Biodegradation of Wood Components*, ed. T. Higuchi, pp. 535–556. Academic, San Diego, CA.

Collins, J. W., A. A. Webb, H. P. Didwania and B. F. Lueck (1969). Components of wood pulp bleach effluents. *Environ. Sci. Technol.*, **3**, 371–377.

Crawford, R. L. (1981). *Lignin Biodegradation and Transformation*. Wiley, New York.

Eaton, D. C. (1985). Mineralization of polychlorinated biphenyls by *Phanerochaete chrysosporium*: a ligninolytic fungus. *Enzyme Microb. Technol.*, **7**, 194–196.

Eaton, D., H.-m. Chang and T. K. Kirk (1980). Fungal decolorization of kraft bleach plant effluents. *Tappi*, **63** (10), 103–106.

Eaton, D. C., H.-m. Chang, T. W. Joyce, T. W. Jeffries and T. K. Kirk (1982). Reducing the color of extraction-stage kraft bleach effluents: the FPL/NCSU MyCoR method. *Tappi*, **65**, 89–92.

Eriksson, K.-E. (1981). Fungal degradation of wood components. *Pure Appl. Chem.*, **53**, 33–43.

Eriksson, K.-E. and E. W. Goodell (1974). Pleiotropic mutants of the wood-rotting fungus *Polyporus adustus* lacking, cellulase, mannanase and xylanase. *Can. J. Microbiol.*, **20**, 371–378.

Eriksson, K.-E. and S. G. Hamp (1978). Regulation of *endo*-1,4-β-glucanase production in *Sporotrichum pulverulentum*. *Eur. J. Biochem.*, **90**, 183–190.

Eriksson, K.-E. and L. Vallander (1982). Properties of pulps from thermomechanical pulping of chips pretreated with fungi. *Sven. Papperstidn.*, **85**, R33–R38.

Eriksson, K.-E., M.-C. Kolar and K. Kringstad (1979). Studies on the mutagenic properties of bleaching effluents. Part 2. *Sven. Papperstidn.*, **82**, 95–104.

Eriksson, K.-E., A. Grünewald and L. Vallander (1980a). Studies on growth conditions in wood for three white-rot fungi and their cellulase-less mutants. *Biotechnol. Bioeng.*, **22**, 363–376.

Eriksson, K.-E., A. Grünewald, T. Nilsson and L. Vallander (1980b). A scanning electron microscopy study of the growth and attack on wood by three white-rot fungi and their cellulase-less mutants. *Holzforschung*, **34**, 207–213.

Eriksson, K.-E., M.-C. Kolar, K. Kringstad, F. de Sousa and L. Strömberg (1982). Studies on the mutagenic properties of spent bleaching liquors. Elimination of mutagenicity by treatment with alkali or sodium bisulfite. *Sven. Papperstidn.*, **85**, R73–R76.

Eriksson, K.-E., S. C. Johnsrud and L. Vallander (1983). Degradation of lignin and lignin model compounds by various mutants of the white-rot fungi *Sporotrichum pulverulentum*. *Arch. Microbiol.*, **135**, 161–168.

Falkehag, I., J. Marton and E. Adler (1966). Chromophores in kraft lignin. *Adv. Chem. Ser.*, **59**, 75–89.

Farrell, R. (1984). Biocatalysts hold promise of better pulp quality. *Tappi J.*, **67**, 31–33.

Fenn, P., S. Choi and T. K. Kirk (1981). Ligninolytic activity of *Phanerochaete chrysosporium*: physiology of suppression by NH_4^+ and L-glutamate. *Arch. Microbiol.*, **130**, 66–71.

Fukuzumi, T. (1980). Microbial decolorization and defoaming of pulping waste liquors. In *Lignin Biodegradation: Microbiology, Chemistry and Potential Application*, ed. T. K. Kirk, H.-m. Chang and T. Higuchi, vol. 2., pp. 161–177. CRC Press, Boca Raton, FL.

Fukuzumi, T., A. Nishida, K. Aoshima and K. Minami (1977). Decolourization of kraft waste liquor with white-rot fungi. I. Screening of fungi and culturing condition for decolourization of kraft waste liquor. *Mokuzai Gakkaishi*, **23**, 290–298.

Gierer, J. (1970). The reactions of lignin during pulping: A description and comparison of conventional pulping processes. *Sven. Papperstidn.*, **73**, 571–596.

Gierer, J. (1981). Chemical aspects of delignification. In *The Ekman-Days 1981, International Symposium on Wood and Pulping Chemistry*, vol. 2, pp. 12–17. Swedish Forest Products Laboratory, Stockholm, Sweden.

Gold, M. H. and T. M. Cheng (1978). Induction of colonial growth and replica plating of the white rot basidiomycete *Phanerochaete chrysosporium*. *Appl. Environ. Microbiol.*, **35**, 1223–1225.

Hakulinen, R. and M. S. Salkinoja-Salonen (1982). Treatment of pulp and paper industry waste water in an anaerobic fluidised bed reactor. *Process Biochem.*, **17** (2), 18–22.

Hardell, H.-L. and F. de Sousa (1977). Characterization of spent bleaching liquors. I. Spent liquors from the chlorine and alkali extraction stages in the prebleaching of kraft pulp. *Sven. Papperstidn.*, **80**, 110–120.

Higuchi, T. (1982). Biodegradation of lignin: biochemistry and potential application. *Experientia*, **38**, 159–166.

Hiroi, T. and K.-E. Eriksson (1976). Microbiological degradation of lignin. Part 1. Influence of cellulose on the degradation of lignins by the white-rot fungus *Pleurotus ostreatus*. *Sven. Papperstidn.*, **79**, 157–161.

Holmbom, B. R. (1980). Studies on resin acids and chlorinated phenolics in Finnish pulp mill effluents and on their bioaccumulation in fish. 89th Natl. AIChE Meeting, Portland, Oregon, August 17–20.

Huynh, V.-B., H.-m. Chang, T. W. Joyce and T. K. Kirk (1985). Dechlorination of chloro-organics by a white-rot fungus. *Tappi J.*, **68**, 98–102.

International Standards Organization (1977). Paper and board measurement of diffuse blue reflectance factor (ISO Brightness). ISO Standard 2470.

Janshekar, H., T. Haltmeier and C. Brown (1982). Fungal degradation of pine and straw alkali lignins *Eur. J. Appl. Microbiol. Biotechnol.*, **14**, 174–181.

Jeffries, T. W., S. Choi and T. K. Kirk (1981). Nutritional regulation of lignin degradation by *Phanerochaete chrysosporium*. *Appl. Environ. Microbiol.*, **42**, 290–296.

Johnson, M. A. and J. A. Carlson (1978). Mycelial paper: a potential resource recovery process? *Biotechnol. Bioeng.*, **20**, 1063–1084.

Käärik, A. (1974). Decomposition of wood. In *Biology of Plant Litter Decomposition 1*, ed. C. H. Dickinson and G. J. F. Pugli, p. 129. Academic, New York.

Kersten, P. J., M. Tien, B. Kalyanaraman and T. K. Kirk (1985). The ligninase of *Phanerochaete chrysosporium* generates cation radicals from methoxybenzenes. *J. Biol. Chem.*, **260**, 2609–2612.

Keyser, P., T. K. Kirk and J. G. Zeikus (1978). Ligninolytic enzyme system of *Phanerochaete chrysosporium*: synthesized in the absence of lignin in response to nitrogen starvation. *J. Bacteriol.*, **135**, 790–797.

Kirk, T. K. (1983). Degradation of lignin. In *Biochemistry of Microbial Degradation*, ed. D. T. Gibson. Dekker, New York.

Kirk, T. K. and M. Shimada (1985). Lignin biodegradation: the microorganisms involved and the physiology and biochemistry of degradation by white-rot fungi. In *Biosynthesis and Biodegradation of Wood Components*, ed. T. Higuchi, pp. 579–605. Academic, San Diego, CA.

Kirk, T. K. and H.-H. Yang (1979). Partial delignification of unbleached kraft pulp with ligninolytic fungi. *Biotechnol. Lett.*, **1**, 347–352.

Kirk, T. K., H.-m. Chang and T. Higuchi (eds.) (1980). *Lignin Biodegradation: Microbiology, Chemistry and Potential Application*. CRC Press, Boca Raton, FL.

Kirk, T. K., W. J. Connors and J. G. Zeikus (1976). Requirement for a growth substrate during lignin decomposition by two wood-rotting fungi. *Appl. Environ. Microbiol.*, **32**, 192–194.

Leach, J. M., J. C. Mueller and C. C. Walden (1976). Identification and removal of toxic materials from kraft and groundwood pulp mill effluent. *Process Biochem.*, **11** (1), 7–10.

Leach, J. M., J. C. Mueller and C. C. Walden (1978). Biological detoxification of pulp mill effluents. *Process Biochem.*, **13** (1), 18–26.

Leatham, G. F. and T. K. Kirk (1983). Regulation of lignolytic activity by nutrient nitrogen in white-rot basidiomycetes. *FEMS Lett.*, **16**, 65–67.

Lindström, K., J. Nordin and F. Österberg (1981). In *Advances in the Identification and Analysis of Organic Pollutants in Water*, ed. L. H. Keith, pp. 1039–1058. Ann Arbor Science, Ann Arbor, MI.

Livernoche, D., L. Jurasek, M. Desrochers and I. A. Veliky (1981). Decolorization of a kraft mill effluent with fungal mycelium immobilized in calcium alginate gel. *Biotechnol. Lett.*, **3**, 701–706.

Ljunggren, S. (1980). The significance of aryl ether cleavage in kraft delignification of softwood. *Sven. Papperstidn.*, **83**, 363–369.

Lundquist, K., T. K. Kirk and W. J. Connors (1977). Fungal degradation of kraft lignin and lignin sulfonates prepared from synthetic ^{14}C-lignins. *Arch. Microbiol.*, **112**, 291–296.

Marton, J. (1971). Reactions of lignin in alkaline pulping. In *Lignins: Occurrence, Formation, Structure and Reactions*, ed. K. V. Sarkanen and C. H. Ludwig, pp. 639–694. Wiley-Interscience, New York.

Nakatsubo, F., I. D. Reid and T. K. Kirk (1982). Fungal metabolism of a lignin substrate model compound of the β-1 type: incorporation of $^{18}O_2$ and absence of stereospecificity in product formation. *Biochim. Biophys. Acta*, **719**, 284.

Nazar, M. A. and W. H. Rapson (1980). Elimination of the mutagenicity of bleach plant effluents, *Pulp Pap. Mag. Can.*, **T 191**, 75.

NCASI (1971). An investigation of improved procedures for measurements of mill effluent and receiving water color. *Technical Bulletin 253*, National Council for Air and Stream Improvement, Inc., New York.

Pfister, K. and E. Sjöström (1979). Characterization of spent bleaching liquors. 3. Composition of material dissolved during alkali extraction (CEH sequence). *Pap. Puu*, **61**, 367–370.

Rydholm, A. A. (1965). *Pulping Processes*. Interscience, New York.

Ruel, K., F. Barnoud and K.-E. Eriksson (1981). Micromorphological and ultrastructural aspects of spruce wood degradation by wild-type *Sporotrichum pulverulentum* and its cellulase-less mutant Cel 44. *Holzforschung*, **35**, 157–171.

Samuelsson, L., P. J. Mjöberg, N. Hartler, L. Vallander and K.-E. Eriksson (1980). Influence of fungal treatment on the strength versus energy relationship in mechanical pulping. *Sven. Papperstidn.*, **83**, 221–225.

Sanyer, N. and C. H. Chidester (1963). Manufacture of wood pulp. In *The Chemistry of Wood*, ed. B. L. Browing, pp. 441–534. Interscience, New York.

Singh, R. P. and B. C. Dillner (1979). Oxygen bleaching. In *The Bleaching of Pulp*, ed. R. P. Singh, pp. 159–209. Tappi Press, Atlanta, GA.

Sjöström, E. (1981). *Wood Chemistry: Fundamentals and Applications*. Academic, New York.

Sundman, G., T. K. Kirk and H.-m. Chang (1981). Fungal decolorization of kraft bleach plant effluent: fate of the chromophoric material. *Tappi*, **64**, 145–148.

Tien, M. and T. K. Kirk (1983). Lignin-degrading enzyme from the hymenomycete *Phanerochaete chrysosporium* Burds. *Science*, **221**, 661–663.

Tien, M. and T. K. Kirk (1984). Lignin-degrading enzyme from *Phanerochaete chrysosporium*: Purification, characterization and catalytic properties of a unique H_2O_2-requiring oxygenase. *Proc. Natl. Acad. Sci. USA*, **81**, 2280–2284.

Trubacek, I. and A. Wiley (1979). Bleaching and pollution. I *The Bleaching of Pulp*, ed. R. P. Singh, pp. 423–461. Tappi Press, Atlanta, GA.

Westermark, U. and K.-E. Eriksson (1974). Carbohydrate-dependent enzymic quinone reduction during lignin degradation. *Acta Chem. Scand., Ser. B*, **28**, 204–208.

Yamasaki, T., S. Hosoya, C.-L. Chen, J. S. Gratzl and H.-m. Chang (1981). Characterization of residual lignin in kraft pulp. In *The Ekman-Days 1981, International Symposium on Wood and Pulping Chemistry*, vol. 2, pp. 34–42. Swedish Forest Products Laboratory, Stockholm, Sweden.

Yang, H. H., M. J. Effland and T. K. Kirk (1980). Factors influencing fungal degradation of lignin in a representative lignocellulosic, thermomechanical pulp. *Biotechnol. Bioeng.*, **22**, 65–77.

16

Microbially Enhanced Oil Recovery

T. R. JACK

NOVA Husky Research Corporation, Calgary, Alberta, Canada

16.1 INTRODUCTION

Microbially enhanced oil recovery (MEOR) is not a single new technology. Rather it is the application of microorganisms in the unique environment of the underground oil reservoir that defines MEOR as a new and challenging area of endeavour.

The organisms are actually used as underground factories to generate various agents or conditions *in situ* which can stimulate, extend or increase the recovery of oil. Since quite different strategies may be required depending on the nature of the oil and reservoir, there is no universal approach to increase oil recovery. In effect the actual biotechnology required is largely site specific. Consequently, MEOR research may be best viewed as the adaptation of a chosen biotechnological process such as polymer production to a given reservoir problem rather than as a new and separate discipline.

16.2 OIL RESERVOIRS

The oil reservoir is a geological formation which has trapped petroleum over geological time. Such formations hold the oil in the pore space of a permeable host rock, commonly sandstone or limestone.

In the petroleum industry, the permeability of the porous medium in which the oil is entrapped commonly is quantified in terms of 'Darcy units', represented by the symbol D. For a given porous medium and fluid, Darcy's law states that the fluid flow rate is directly proportional to the pressure drop; the flow rate also is inversely proportional to the viscosity of the fluid.

In a porous medium of 1.0 D permeability, a volumetric flow rate of $1.0 \text{ cm}^3 \text{ s}^{-1}$ will be produced through a 1.0 cm^2 area (perpendicular to the flow direction) at a pressure drop per unit length of 1.0 atm cm^{-1} for a fluid of 1.0 centipoise viscosity ($1.0 \text{ cP} = 10^{-2} \text{ g cm}^{-1} \text{ s}^{-1} = 10^{-3} \text{ kg m}^{-1} \text{ s}^{-1}$). Hence, $1.0 \text{ D} = 1.0 \text{ cP cm}^2 \text{ atm}^{-1} \text{ s}^{-1}$. In a consistent system of units, k in equation (1) has units of area (cm^2, m^2, ft^2, *etc.*). For conversion to SI units, $1.0 \text{ D} = 0.987 \text{ } \mu\text{m}^2$.

$$v = \dot{V}/S = k(\Delta P/L)\eta^{-1} \tag{1}$$

where v = linear velocity (cm s^{-1}), $\dot{V}$ = volumetric flow rate (cm^3 s^{-1}), S = area normal to the flow direction (cm^2), ΔP = pressure drop (atm), L = length of porous medium (cm), η = viscosity (cP), k = permeability coefficient (Darcy).

While geological formations vary greatly in the nature of their permeability, porosity and mineralogy, all reservoirs present certain common problems to the microbiologist. The formations are very large, present monumental surface areas and can have excessive capacities to buffer, poise or otherwise resist attempts to adjust or improve the *in situ* environment. Reservoirs are anaerobic and increase in temperature and pressure with depth. Water which may be present with the oil can be highly saline or of a pH unsuitable for microbial growth. Perhaps most importantly the reservoir is inaccessible. Once injected, the biological system cannot be adjusted or even closely monitored.

Obviously the immediate challenge to the microbiologist is to determine from the known chemical and physical parameters of the target reservoir whether microbial activity is even feasible in the formation. Overall prospects are good. In a review of data bases for all oil reservoirs in nine oil producing states in the continental United States, Clark and colleagues (1981) found that a large number of reservoirs were potential targets for MEOR technology. The criteria for selection included a pH of from 4 to 8, temperature less than 75 °C and a salinity of less than 10%. Consideration of the relative rates of increase of temperature and pressure with depth suggests that generally microbial activity will be limited by high temperature in deep reservoirs before inhibitory pressures are reached.

A less obvious obstacle to the application of MEOR is the problem of injection and dispersion of even small bacteria underground. For example, very 'tight' rock formations of less than 75 mD permeability usually have a pore size distribution too small to allow easy passage of bacterial cells. Clark *et al.* (1981) recognized this as a serious restriction which significantly reduces the number of potential target reservoirs for MEOR (Table 1).

Table 1 Percent of Oil Reservoirs Potentially Suitable for a Microbially Enhanced Recovery Process[a]

State	Percent allowing growth	Plus easy injection
Oklahoma	50	16
Texas	43	22
Louisiana	39	36
Kansas	51	20
California	79	65
Colorado	66	43
Mississippi	6	5
New Mexico	73	28
Wyoming	60	25

[a] Clark *et al.*, 1981.

16.3 LIMITS TO MICROBIAL ACTION UNDERGROUND

Use of living organisms in the reservoir demands that a suitable carbon source be present. In an anaerobic environment, crude oil may not meet this demand.

Partial biodegradation of petroleum is a fast facile process in an aerobic environment where a biologically catalyzed reaction of oxygen with the pristine hydrocarbon can initiate the degradation process. Once oxygen atoms have been introduced into the hydrocarbon, numerous further biologically mediated reactions including anaerobic ones may ensue. In the absence of oxygen fast attack on pure hydrocarbons does not occur (Ward and Brock, 1978).

A number of research groups have demonstrated some biological activity on crude oil under anaerobic conditions (Moses and Springham, 1982; Davis, 1967; Jack *et al.*, 1985a). Radioisotopically labelled hexadecane has been metabolized by sulfate-reducing bacteria under laboratory conditions, albeit at a very slow rate (Davis, 1967). In our own research program, evolution of CO_2, H_2S and CH_4 from certain crude oils has been noted where less than 30 p.p.m. (v/v) of oxygen was present. This indicates that observable activity may arise even where very low levels of oxygen in the headspace gas over the culture exist. Use of radiolabelled alkanes confirmed hydrocarbons as the source of both the CH_4 and CO_2 observed (Jack *et al.*, 1985a). The rate of CH_4 production was enhanced when 700 p.p.m. headspace oxygen was introduced and shut down by 12 000 p.p.m. headspace oxygen (Figure 1). While the latter observation is consistent with the known sensitivity of methanogens to oxygen, the former observation suggests that oxygen at low levels improves the overall kinetics of gas production, presumably by initiating the degradation of susceptible hydrocarbons *via* the reaction sequence described above. Whether any attack occurs in the complete absence of oxygen is not clear. Certainly the overall yields of gas (extremely low) and low rates of gas production (months) indicate that oil is not a likely substrate for the support of rapid MEOR processes in the anaerobic oil reservoir.

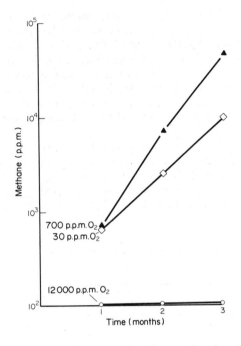

Figure 1   Effect of trace levels of headspace O_2 (p.p.m., v/v) on the rate of methane formation from paraffinic crude oil under anaerobic conditions

Our studies also indicate that the nature of the oil is critical to the observation of gas production under anaerobic conditions. Reproducible activity was limited to paraffinic and alkane rich crude oils. An aromatic oil lacking a high alkane content and two different heavy crude oils also deficient in simple alkanes showed almost no activity. This is consistent with previous studies on the biomethanation of crude oils (Muller, 1957) and with the general observation that the simple alkanes are one of the most susceptible oil fractions for aerobic degradation (Rubenstein *et al.*, 1977).

Anaerobic MEOR processes based on crude oil as the sole carbon source cannot be regarded as very promising. Yet the apparent advantage of utilizing the oil with the prospect of pervasive microbial action throughout the reservoir remains both tempting and perhaps a little intimidating since it seems probable that bacterial action would affect the most valuable portions of the oil. In some reservoirs, such as those in the North Sea, the long injection times required seem to preclude any other approach. Consequently, research on the direct use of petroleum is proceeding (Moses and Springham, 1982).

Overall MEOR processes have been and continue to be based on the injection of essential nutrients into the reservoir to support the desired microbial activity (Hitzman, 1983). Attack on the oil in the anaerobic reservoir is generally regarded as insignificant and bacterial activity as being localized about the injected nutrient package. Hence the problem of injecting and dispersing the MEOR system usually involves consideration of both bacteria and their required nutrients.

Figure 2 illustrates schematically the principal areas of concern in planning an MEOR application. Figure 3 expands the underground scenario. The detail in Figure 3 is particularly important since once the MEOR system is injected no alteration or intercession in the ensuing events is possible.

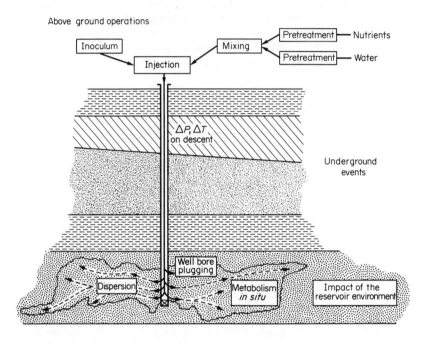

Figure 2 Schematic illustration of the use of bacteria *in situ* showing the operations on the surface and events underground

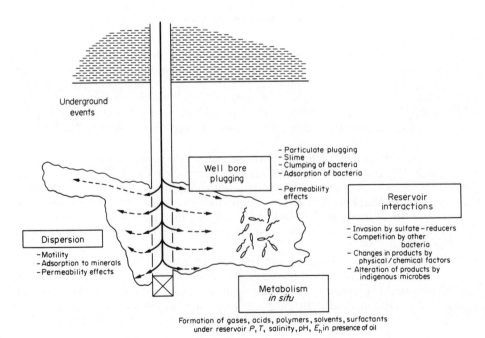

Figure 3 Events underground: details of factors and potential problems

16.4   WELL BORE PLUGGING

The critical step in injecting bacterial systems into the oil reservoir is the passage of bacteria from the well into the rock through the formation face. Where extended water injection for disposal or for secondary oil production has been carried out in the field, plugging of the formation face by a build up of biomass can take place to the extent that the 'injectivity' of the well is reduced sufficiently to jeopardize the injection process (Clementz *et al.*, 1982). Laboratory studies (Ionescu, 1984; Shaw, 1982), using the natural bacterial populations found in tap water or even distilled water, have shown that up to a 70% reduction in the permeability of a model core can take place during extended water flooding due to the biofilms which develop on the core face. The sensitivity of the formation face to this sort of plugging is of concern since large volumes must pass through the formation face during injection of an MEOR system. This problem has been considered in some detail elsewhere (Jack *et al.*, 1985b).

To minimize injectivity loss the following factors must be avoided: filamentous or clumping bacteria, active polymer production during injection, fermentation gas bubble production in the formation and tacophilic organisms. Even with these precautions the natural tendency for biofilm development on surfaces in flowing nutrient streams (Bitton and Marshall, 1980) may preclude very long periods of injection. Various strategies have been suggested to minimize these plugging problems. These include the use of spores, small singly dispersed bacteria and various chemical agents which condition rock and/or cells to reduce sticking (Jang *et al.*, 1983a). Short injection times, reduced cell concentrations in the bacterial cell suspension (Jang *et al.*, 1983a) and high rock permeability will also alleviate filter cake formation on the rock face.

16.5   DISPERSION THROUGH THE RESERVOIR

Once bacteria have entered the formation they must be dispersed through the rock matrix to the desired location for MEOR activity. This involves movement of the cell or spore suspensions through the host rock, presumably in a water phase. As was recognized earlier, the pore size distribution in the rock imposes absolute limitations on the dispersion process. Where the pore throat sizes are similar to the size of the injected particles, plugging and exclusion occur. For this reason reservoirs of less than 100 mD are usually cited as being unsuitable for MEOR (Springham, 1984).

Aqueous bacterial suspensions will preferentially move into zones having a high water permeability. Since both oil and water and even a gas phase can share the rock pore space, water permeabilities depend on the phase distribution present in the reservoir as well as on the absolute permeability of the host rock. To further complicate the underground picture reservoirs are often very inhomogeneous. Unconsolidated sand reservoirs may be bedded with considerable variation in the vertical distribution of permeabilities. Limestone reservoirs are often fractured and have interconnected systems of solution cavities called vugs. The problem of knowing where injected bacteria will end up is not trivial.

Adsorptive mineral surfaces within the reservoir can retain even soluble substances. Short term retentions lead to chromatographic effects which cause various components in an injected nutrient package to travel at different rates. Jenneman *et al.* have studied the rate of travel of phosphate, ammonia nitrogen, glucose and protein through 10 cm thick Berea sandstone cores of 400 to 500 mD permeability in the laboratory (Jenneman *et al.*, 1985). In these experiments, the rate of travel was expressed as the number of times the water in the pore space had to be replaced before a certain nutrient level appeared in the effluent. Both glucose and nitrogen appeared quickly, attaining their initial concentrations by the fifth and fourth pore volume, respectively. Phosphate had attained 80% of initial concentration by the tenth pore volume but stayed at this level indicating that a significant fraction of this ion was retained in the core permanently. Even after sixteen pore volumes the effluent protein level was less than 40% of its initial value. Such results indicate that a complete nutrient package may become considerably segregated in travel through sandstone due to chromatographic effects. Under these circumstances, a scheduled injection protocol might be needed to ensure that all essential nutrients ended up together at the site of desired bacterial activity underground. The affinity of protein for the core minerals is sufficient to preclude its use. Given the extremely large suface areas available for adsorption in the geological formation, the level of protein loss observed could well become prohibitively expensive for deep injections into a sandstone reservoir.

Similar studies in our laboratories using a 1.8 m long column packed with coproduced sand from a western Canadian oil well gave analogous data for a core of higher permeability (Figure

4). In this case a relatively fresh water solution of sugar beet molasses was injected into a 4.1 D sand pack saturated with a brine solution. From Figure 4, it is evident that the sucrose in the molasses travels with the injected fresh water while the protein is retarded. Comparison of these results with those above suggests that the extent of protein retention may be a function of the formation composition and permeability.

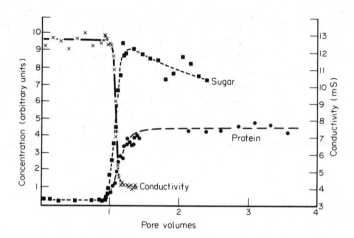

Figure 4   Relative rates of travel for fresh water and sugar beet molasses sugar and protein through a brine saturated sand packed column of length 1.8 m, permeability 3.5 D, and flow rate 0.1 ml min^{-1}

In any MEOR application which requires bacterial activity deep in the reservoir, the relative rates of travel and retention for all the essential components of the biological system must be known to ensure that everything needed reaches the site of application in sufficient quantity and at the same time.

Obviously the bacteria themselves are an essential component of the MEOR system. As discussed above, bacteria can be seriously retained at the formation face, but even deeper in the reservoir adsorption on mineral surfaces by, for example, charge/charge interactions may retain cells or spores from an injected suspension. Jang *et al.* have developed a deep filtration model to describe bacterial retention in laboratory cores (Jang *et al.*, 1983a, 1983b).

Viable cells of *Pseudomonas putida* have been injected through 4 cm thick sandstone cores at permeabilities as low as 196 mD. However, less than 1% of the injected cells were recovered in the effluent (Jenneman *et al.*, 1984). In packed sand cores of 4.1 D permeability the fraction of *Leuconostoc mesenteroides* cells retained was not negligible over a length of 1.8 m. Recent results in our laboratory using a sand-packed column indicate that about 90% retention of injected microbial cells may be achieved compared to the approximately 99% retention reported by Jenneman *et al.* (1984). It is probable that bacterial retention is dependent on the bacterial species and growth conditions as well as on rock parameters related to the permeability and mineralogy.

Where soluble nutrients have been previously dispersed, bacteria can follow through diffusive growth and active motility. Various research reports indicate significant rates for the penetration of bacteria through sandstone and sand packs saturated with stagnant nutrient solutions. The following rates are representative: *Bacillus*, 0.16 cm h^{-1} (Jang *et al.*, 1983a); *Pseudomonas*, 0.07 cm h^{-1} (Jang *et al.*, 1983a); sulfate-reducing bacteria, 0.008 to 0.15 cm h^{-1} (Zobell, 1958); and mixed anaerobes, 0.19 to 0.38 cm h^{-1} (Lee and Mueller, 1982). At very low permeabilities, this natural spreading rate is curtailed. Jenneman *et al.* found that for sandstone cores of less than 100 mD permeability, the rate of travel decreased logarithmically with the permeability while above 100 mD rates up to 0.47 cm h^{-1} were found for a *Bacillus* independent of the permeability (Jenneman *et al.*, 1984b). The chief mechanism for the penetration of the bacteria was their motility. Motile species moved up to ten times more quickly than non-motile ones (Jenneman *et al.*, 1984b).

Presuming that an MEOR system is injected with a suitable nutrient package, viable bacteria can be expected to grow through the reservoir at the sort of rates indicated. This could allow a more efficient and pervasive system than might be envisioned were the bacteria strictly limited to their injected location.

Based on data compiled from field tests over the last three decades (Hitzman, 1983), it is evident that dispersion of bacterial systems can occur over wide areas of an appropriate reservoir. Apparently retention effects do not present an insurmountable limitation in the field.

16.6 RESERVOIR INTERACTIONS

Whether bacteria are indigenous to oil reservoirs has been a point of discussion for many years (Davis, 1967). In effect the point is academic since drilling, completion and various injections normally carried out in the development of an oil field will introduce bacteria into the reservoir. Core samples taken on site from a new well being drilled adjacent to a field under active water flood contained viable bacteria in significant numbers (Figure 5). The distribution of organisms correlated with the permeability of the various sand layers in the reservoir. This is consistent with either an influx of organisms entering the area along the highest permeability streaks or an influx of nutrients selectively enhancing the bacterial population in these zones. In terms of MEOR, the essential consideration is that bacteria are probably already present underground in most locations and are most abundant in the very zones most likely to receive the injected MEOR system. This implies that injected organisms may face direct competition from previous arrivals for injected nutrients and that subsequent modification of agents produced by the MEOR system may occur.

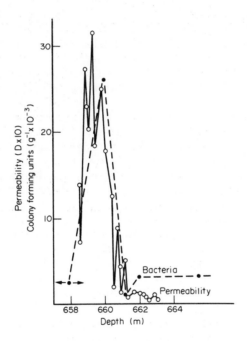

Figure 5 Correlation between bacterial numbers and permeability for a heavy oil reservoir in an unconsolidated sand formation

Sulfate-reducing bacteria are responsible for down hole corrosion problems and reservoir souring in many oil fields. Where the residual materials left by an MEOR application provide the nutrients for sulfate-reducing metabolism, a burst of growth of these organisms could eclipse the value of a successful MEOR operation through increased H_2S production and increased corrosion in well and production facilities.

In planning an MEOR operation the presence of competitors should be assumed and steps taken to ensure the selective advantage of the desired organism. Where sulfate-reducers are present, fermentations should be designed to minimize the opportunity for secondary growth by these bacteria.

16.7 METABOLISM

Bacteria are capable of producing a number of agents which can enhance oil recovery from a susceptible reservoir. The potentially useful products of MEOR metabolism can be categorized into five general classes: polymers, biomass, organic acids, gases, and biosurfactants and solvents. The selection of an MEOR system will depend on the production problems and promise of the target reservoir.

A major constraint to secondary oil production by waterflooding the reservoir to displace oil with injected water is the poorly matched mobilities of water and oil. Since water is more mobile underground than petroleum, water tends to poke pathways through the oil rather than displace it, a phenomenon called 'fingering'. A marked improvement can be made in the sweep efficiency of the process by adding viscosifying agents to the water phase. These include water soluble biological polymers such as xanthan and scleroglucan. While current recovery technology injects prepared solutions of these biopolymers from the surface, such agents could be made *in situ*. One patented process describes the use of bacterial cell suspensions of increasing viscosity as an appropriate drive fluid (Jack and Thompson, 1983).

Biomass and insoluble polymers have been claimed as selective plugging agents for high permeability zones (Jenneman *et al.*, 1984; Jack and DiBlasio, 1985). In this case increased sweep efficiency is obtained by improving the homogeneity of the reservoir. During injection, the MEOR system preferentially enters high permeability zones where subsequent biomass and polymer production causes disproportionate plugging. Since high permeability zones are more affected than low permeability zones, the final reservoir permeabilities become more closely matched. Such a scheme could be used to precondition the reservoir for some other production technology which would actually displace the oil. For example, selective plugging of high permeability water channels could improve water flooding yields. Selective plugging could even be used to choke off undesirable water flow underground. This latter application could be of use especially in heavy oil production (Jack and DiBlasio, 1985) where the high viscosity of the oil often results in wells drawing remote water instead of producing oil from the adjacent formation. 'Watering out' is a common fate for heavy oil wells in their primary production phase.

Organic acids produced through fermentation readily dissolve carbonates and could greatly increase permeability in local areas of a limestone reservoir (Bubela, 1983).

Fermentation gases can repressurize wells leading to displacement and production of oil through a revitalized gas drive mechanism. Well head pressure increases to 1.7 MPa have been reported for *in situ* molasses fermentations (Hitzman, 1983). Gas drives are most effective at displacing conventional or light crude oils. While CO_2 dissolution can cause heavy oil to swell and to decrease in viscosity (Miller and Jones, 1981), the practical limit to the amount of CO_2 which could be produced *in situ* from injected nutrients probably precludes this approach from large scale production in heavy oil reservoirs.

Biosurfactant and solvent production could facilitate the release and transport of hydrocarbons. Incremental oil release has been demonstrated using biosurfactants in the laboratory (Finnerty *et al.*, 1985). Given the large volumes and surface areas involved in the reservoir, the action of surface active agents may be limited to the region about the well bore. This zone is critical to well performance and even local effects which clean out the immediate formation or which affect the oil/water ratio in this region could greatly influence subsequent production.

16.8 MEOR FIELD APPLICATIONS

The idea of employing bacteria in oil recovery is not new. In the mid-1940s, Zobell observed that sulfate-reducing bacteria could displace oil from test cores (Davis, 1967). The idea of using bacteria in oil recovery was patented and sulfate-reducing bacteria were actually taken to field trial in eastern Europe. Unfortunately, sulfate-reducing bacteria were a poor choice. H_2S production and corrosion problems associated with these organisms, as well as their poor performance, resulted in the rejection of sulfate-reducers as MEOR agents.

In 1954 an alternate approach was field tested in Arkansas, USA. The anaerobe *Clostridium acetobutylicum* was injected with molasses as a fermentable substrate into an almost depleted oil reservoir. A well 400 m distant showed a marked increase in production but overall the increase amounted to only a few barrels of oil. Secondary fermentation reactions could be deduced from the products collected. These included the *in situ* conversion of microbial $H_2 + CO_2$ into methane and the production of extraordinary organic acids not typical of the injected culture. Carbonate dissolution by these organic acids has been cited as a possible mechanism of MEOR impact as has selective plugging (Jack, 1983).

In 1956 a Dutch patent reported the intentional use of bacteria able to form slime from an injected molasses solution. The stated aim was to equalize the permeability of the formation by selectively forming viscous plugs in regions of high permeability. Sparse details of a successful field test were included in the patent example (Hitzman, 1983).

These innovations set off a series of field tests in eastern Europe. In the next three decades field tests were undertaken in Czechoslovakia, Poland, Hungary and Romania. The most extensive tests were in Poland. In these applications, several hundred liters of inoculum were injected along with several tonnes of molasses into a single oil well. The well was then shut in for two months or so to allow the fermentation to proceed before it was returned to production. Stimulation of oil production, if it occurred, took two forms: a short lived effect which fell off with the decline in bacterial numbers and a longer lived production increase which has been ascribed to a well bore cleaning effect (Jack, 1983; Donaldson, 1985). This latter increase lasted in some instances over several years (Hitzman, 1983; Jack, 1983). One feature of special note was the observation that a stimulation was also seen on the arrival of viable bacteria at a well adjacent to the treated well (Jack, 1983). This strongly suggests that the short term stimulation effects observed were due to bacterial action and were not due to the incidental effects of MEOR injection and the shut in period always associated with the injection well in an MEOR experiment. Much of the useful data from these field trials as well as early work in the USSR have been compiled in English by Hitzman (1983).

The current resurgence of interest in MEOR is continuing with active research taking place in Europe, North America, Asia and Australia. New field applications are underway or are planned for the United States, Canada, Australia, Romania and possibly other countries. It is to be hoped that this round of field tests will provide a better definition of the commercial potential of MEOR processes. Where economically sensible MEOR systems can be identified, genetic engineering offers the opportunity for striking improvements in control and performance. Given the long term economic incentive for enhanced oil recovery, the outlook remains optimistic.

16.9 REFERENCES

Bitton, G. and K. C. Marshall (eds.) (1980). *Adsorption of Microorganisms to Surfaces*. Wiley, Toronto.

Bubela, B. (1983). Physical simulation of microbiologically enhanced oil recovery. In *Microbial Enhanced Oil Recovery*, ed. J. E. Zajic, D. G. Cooper, T. R. Jack and N. Kosaric, pp. 1–7. PennWell Books, Tulsa.

Clark, J. B., D. M. Munnecke and G. E. Jenneman (1981). *In situ* microbial enhancement of oil production. *Dev. Ind. Microbiol.*, **22**, 695–701.

Clementz, D. M., D. E. Patterson, R. J. Aseltine and R. E. Young (1982). Stimulation of water injection wells in the Los Angeles basin by using sodium hypochlorite and mineral acids. *J. Pet. Technol.*, **34**, 2087–2096.

Davis, J. B. (1967). *Petroleum Microbiology*. Elsevier, New York.

Donaldson, E. C. (1985). Effect of *Clostridium* on production well capillary pressure phenomena. In *Microbes and Oil Recovery, Proceedings of the International Conference on Microbial Enhancement of Oil Recovery, Fountainhead, Oklahoma*, May 20–25, 1984, ed. J. E. Zajic and E. C. Donaldson, vol. 1, pp. 181–189. Petroleum Bioresources, El Paso.

Finnerty, W. R., M. E. Singer, F. Ahene and H. Attaway (1985). The chemistry and application of a microbial biosurfactant for processing heavy oils. In *Microbes and Oil Recovery, Proceedings of the International Conference on Microbial Enhancement of Oil Recovery, Fountainhead, Oklahoma*, May 20–25, 1984, ed. J. E. Zajic and E. C. Donaldson. Petroleum Bioresources, El Paso.

Hitzman, D. O. (1983). Petroleum microbiology and the history of its role in enhanced oil recovery. In *Proceedings of the International Conference on Microbial Enhancement of Oil Recovery, Shanghri-la, Oklahoma*, May 16–21, 1982, ed. E. C. Donaldson and J. B. Clark, pp. 162–218. US Department of Energy, Bartlesville.

Ionescu, E. (1984). *The Effect of Water Quality on Laboratory Core Floods*. Petroleum Recovery Institute, Calgary.

Jack, T. R. (1983). Enhanced oil recovery by microbial action. In *Fossil Energy Extraction*, ed. T. F. Yen, F. K. Kawahara and R. Hertzburg, pp. 13–22. Ann Arbor Science Publishers, Ann Arbor.

Jack, T. R. and E. DiBlasio (1985). Selective plugging for heavy oil recovery. In *Microbes and Oil Recovery, Proceedings of the International Conference on Microbial Enhancement of Oil Recovery, Fountainhead, Oklahoma*, May 20–25, 1984, ed. J. E. Zajic and E. C. Donaldson, vol. 1, pp. 205–212. Petroleum Bioresources, El Paso.

Jack, T. R. and B. G. Thompson (1983). Patents employing microorganisms in oil production. In *Microbial Enhanced Oil Recovery*, ed. J. E. Zajic, D. G. Cooper, T. R. Jack and N. Kosaric, pp. 14–25. PennWell Books, Tulsa.

Jack, T. R., E. G.-H. Lee and J. C. Mueller (1985a). Anaerobic gas production: controlling factors. In *Microbes and Oil Recovery, Proceedings of the International Conference on Microbial Enhancement of Oil Recovery, Fountainhead, Oklahoma*, May 20–25, 1984, ed. J. E. Zajic and E. C. Donaldson, vol. 1, pp. 167–180. Petroleum Bioresources, El Paso.

Jack, T. R., J. C. Shaw, N. C. Wardlaw and J. W. Costerton (1985b). Microbial plugging in enhanced oil recovery. In *Microbial Enhancement of Oil Recovery*, ed. E. C. Donaldson, G. V. Chilingarian and Teh Fu Yen. Elsevier, Amsterdam.

Jang, L.-K., P. W. Chang, J. E. Findley and Teh Fu Yen (1983a). Selection of bacteria with favourable transport properties through porous rock for the application of microbial enhanced oil recovery. *Appl. Environ. Microbiol.*, **46**, 1066–1072.

Jang, L.-K., M. M. Sharma, J. E. Findley, P. W. Chang and Teh Fu Yen (1983b). An investigation of the transport of bacteria through porous media. In *Proceedings of the International Conference on Microbial Enhancement of Oil*

Recovery, Shanghri-la, Oklahoma, May 16–21, 1982, ed. E. C. Donaldson and J. B. Clark, pp. 60–70. US Department of Energy, Bartlesville.

Jenneman, G. U., R. M. Knapp, M. J. McInerny, D. E. Menzie and D. E. Revis (1984a). Experimental studies of *in situ* microbial enhanced oil recovery. *Soc. Pet. Eng. J.*, **24**, 33–37.

Jenneman, G. U., M. J. McInerny, R. M. Knapp and D. E. Menzies (1984b). The kinetics of microbial growth through porous media. Poster at the *International Conference on Microbial Enhancement of Oil Recovery, Fountainhead, Oklahoma*, May 20–25, 1984.

Lee, E. G.-H. and J. C. Mueller (1982). Acquisition of Microbial Cultures Project 1-03-501. B. C. Research, Vancouver, BC.

Miller, J. S. and R. A. Jones (1981). Physical characteristics of heavy oil after CO_2 saturation are studied in DOE tests. *Oil Gas J.*, July 6, 135–145.

Moses, V. and D. G. Springham (1982). *Bacteria and the Enhancement of Oil Recovery*, Applied Science Publishers, London.

Muller, F. M. (1957). On methane fermentation of higher alkanes. *Antonie van Leeuwenhoek*, **23**, 369–384.

Rubenstein, I., O. P. Strausz, C. Spyckerelle, R. J. Crawford and D. W. S. Westlake (1977). The origin of the oil sand bitumens of Alberta: a chemical and a microbiological study. *Geochim. Cosmochim. Acta*, **41**, 1341–1353.

Shaw, J. C. (1982). Bacterial Fouling of a Model Core System. M.Sc. thesis, Geology Department, University of Calgary, Calgary.

Springham, D. G. (1984). Microbiological methods for the enhancement of oil recovery. *Biotechnology and Genetic Engineering Reviews*, ed. G. E. Russel, vol. 1, pp. 187–221. Intercept, Newcastle-upon-Tyne.

Ward, D. M. and T. D. Brock (1978). Metabolism of hexadecane in sediments. *Geomicrobiol. J.*, **1**, 1–9.

Zobell, C. E. (1958). Ecology of sulfate-reducing bacteria. *Producers Monthly*, **22**, 12–29.

17

Microbial Growth Rate Measurement Techniques

D. W. AGAR
BASF, Ludwigshafen, Federal Republic of Germany

17.1 INTRODUCTION

17.1.1 Importance of Growth Rate Measurement

Knowledge of the amount of biomass in a system and its increase with time—the growth rate—is fundamental to any work in biotechnology, be it laboratory studies of microbial behavior or the monitoring and control of industrial fermentations. The spectrum of techniques available for growth rate measurement is correspondingly broad: from a simple brewer's hydrometer to the electronically complex Coulter Counter.

The growth rate depends on a variety of factors: the type of microorganism, the age of the culture, the growth medium and other physical and chemical environmental factors. It follows that a technique for measuring the growth rate may only be valid or suitable under a particular set of conditions. For this reason only rough guidelines for growth rate measurement techniques may be provided and the selected technique must be proven, and sometimes modified, in each case.

The technique for the measurement of growth rate also depends on the application. Industrial fermentations tend to use automated techniques and parameters indicating the desired product, *e.g.* an amino acid. Research work will often employ more specific and complicated techniques, *e.g.* radioactive tracers. Further, more detailed descriptions of microbial growth and its measurement may be found in the reviews of Cooney (1982), Koch (1981) and Mallette (1971).

17.1.2   Relationships Between Growth Rate Measurements

The growth rate may generally be defined as the increase of biomass with time. The basic problem of growth rate measurement lies in the selection of a suitable parameter to reflect accurately the amount of biomass under a variety of conditions. In all but exceptional circumstances, such as perfectly synchronous growth, the observed behavior and extensive properties of a microbial population, which might be used to measure the growth rate, represent mean values over the population. These properties are distributed differently among the individual microorganisms according to their stage in the cell cycle and their individual metabolic and growth processes (Figure 1).

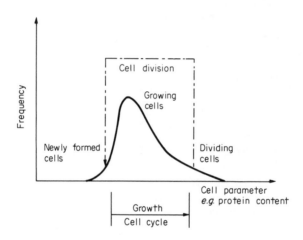

Figure 1   Distribution of a biomass component within a population of microorganisms

Only in a condition known as balanced growth, achieved in chemostats and dilute exponentially growing cultures, does this distribution (and thus the relationship between the parameters of the population) remain time invariant. In such instances any choice of an appropriate population-averaged parameter (Z_i) for biomass will give the same value for the growth rate as any other. The specific growth rate defined by $\mu_g = (1/Z_i)(dZ_i/dt)$ is therefore independent of the Z_i selected and can further be shown to be constant, *i.e.* balanced growth is exponential.

Generally this will not be the case, and the changing distribution within the population must be considered to describe growth and to take changing cell compositions and behavior into account (Figure 2). Traditionally growth has been measured using a single population-averaged parameter (*e.g.* turbidity), which correlates well with the biomass (taken to be the dry weight) under a variety of conditions. By measuring several population-averaged parameters (*e.g.* chemical compositions) changes in the structure of the population, as reflected in the mean values, may to some extent be allowed for. A more advanced and complicated approach is actually to measure the distribution of parameters within the population. Ideally this should permit a measurement of the biomass valid under any circumstances. A classification of growth measurement techniques according to this scheme is shown in Table 1.

The interaction between growth parameters during growth may be described by a population balance equation (Ramkrishna, 1978):

$$\frac{\partial f_1(\mathbf{z},\, t)}{\partial t} + \nabla \cdot \dot{\mathbf{Z}}(\mathbf{z},\, c) f_1(\mathbf{z},\, t) = -\, \sigma\,(\mathbf{z},\, c) f_1(\mathbf{z},\, t) + 2\int_v \sigma\,(\mathbf{z}',\, c) p(\mathbf{z},\, \mathbf{z}',\, c) f_1(\mathbf{z}',\, t) \mathrm{d} v' \qquad (1)$$

$$\text{(growth)} \qquad\qquad \text{(division)} \qquad\qquad \text{(birth)}$$

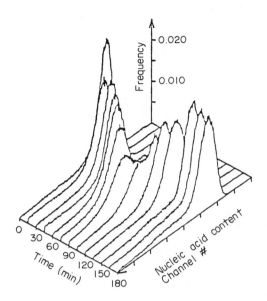

Figure 2 Shifts in the distribution of RNA within a synchronously growing population of the fission yeast *Schizosaccharomyces pombe* (from Agar and Bailey, 1982a. Reprinted by permission of John Wiley and Sons Inc.) A partial synchrony was achieved by a periodic heat-shock treatment. The flow microfluorometer measurements were taken over a period corresponding to a single cell cycle

Table 1 Scheme for Categorizing Growth Rate Measurement Techniques (Bailey *et al.*, 1979)

	Unsegregated (population averages)	*Segregated (individual cell)*
Unstructured (univariate)	Turbidimeter	Coulter Counter (size) (in some cases use of the flow microfluorometer may be applicable)
Structured (multivariate)	'Classical' analytical methods	Flow microfluorometer (composition, size structure)

where z ($= z_1, z_2, \ldots, z_n$) is the physiological state vector defining a single cell, c ($= c_1, c_2, \ldots, c_m$) is the environmental concentration vector, $f_1(z, t)$ is the number density function giving the distribution of z within the population, v is the region of admissible states for z, $\dot{Z}$ is the growth rate of a cell characterized by z in an environment defined by c, $\sigma(z, c)$ is the probability that a cell z in an environment c will divide between t and $t + dt$, and $p(z, z', c)$ is the probability that a cell z' will yield a cell z upon division in an environment c.

The interaction between growth and the environment may be expressed by:

$$\frac{\partial c}{\partial t} = \gamma \int_v R(z, c) f_1(z, t) dv' \tag{2}$$

where γ is the stoichiometric matrix, γ_{ij} is the stoichiometric coefficient of the ith component of c in the jth cellular reaction, and $R(z, c)$ is the intrinsic reaction rate vector of the j independent chemical reactions between the components of c and z.

All growth rate measurement techniques determine a component, or a function of several components, of the vectors z or c. Components of z which have been found to be good general indicators of biomass, such as dry weight, turbidity and cell number, are often referred to as direct methods, whereas the measurement of components of c or other components of z, such as DNA, are referred to as indirect methods.

The aim of growth rate measurement is to establish the biomass ($s = g(z)$) at successive times or directly to determine the biomass growth rate:

$$\bar{\mathbf{S}}(s, \mathbf{c}, t) = \int_V \frac{\nabla g(\mathbf{z}) \cdot \dot{\mathbf{Z}}(\mathbf{z}, \mathbf{c}) f_1(\mathbf{z}, t)}{\int_{\varrho_s} f_1(\mathbf{z}, t) d\varrho} \, d\mathbf{v}' \tag{3}$$

where ϱ_s is the hypersurface described by $g(\mathbf{z})$ for a given s.

In practice, of course, the biomass is usually taken to be proportional to a single population-averaged component of $\mathbf{z}$ or to the consumption/production of a component $\mathbf{c}$ and the equations then simplify to trivial forms:

$$X = K \int_{z_{i, \min}}^{z_{i, \max}} z_i \int_{\varrho_{z_i}} f_1(\mathbf{z}, t) \, d\varrho \, dz_i = KZ_i \tag{4}$$

where z_i is the component of $\mathbf{z}$ measured, ϱ_{z_i} is the hypersurface for a given z_i, X is the total amount of biomass present, K is a proportionality constant, and Z_i is the total amount of z_i present.

Another common simplified approach uses cellular carbon or nitrogen as an indicator of biomass and assumes the metabolism and growth to be described by simple chemical equations of constant stoichiometry:

$$a C_x H_y O_z + b O_2 + c N_u H_v O_w \rightarrow d C_\alpha H_\beta O_\gamma N_\delta + e H_2 O + f CO_2 + g C_n H_m O_p N_r + \Delta H \tag{5}$$
$$\text{(C source)} \qquad \text{(N source)} \qquad \text{(cell mass)} \qquad\qquad\qquad \text{(products)} \qquad \text{(heat)}$$

An element balance may thus be used to determine the amount of biomass present if the remaining components of $\mathbf{c}$ are known. Shifts in the metabolism or cell composition represented by this type of equation due to the stage of culture growth, the available and limiting substrate or physical factors, such as temperature and oxygen partial pressure, change the stoichiometry. It is often possible, however, to describe growth by several simple metabolic equations of known stoichiometry. In such cases only a few components of $\mathbf{c}$ need be monitored to select the pertinent equations and construct the balances.

Numerous studies have been carried out on the relationships between biomass yields and the uptake of specific nutrients or the release of specific products (Stouthamer, 1971; Pirt, 1975; Nagai, 1979). The basic concept involves the energy or ATP (adenosine triphosphate) yield from the prevailing metabolic pathway. The growth and maintenance functions of biomass have relatively constant energy requirements, thus the amount of substrate needed to provide this energy, and thus the yield factor, can be estimated.

This simple approach gives equations of the following forms:

Luedeking and Piret (1959) $\qquad \dfrac{dC_p}{dt} = \alpha \dfrac{dX}{dt} + \beta X$ $\hfill (6)$

Pirt (1975) $\qquad\qquad\qquad \dfrac{dC_S}{dt} = \dfrac{1}{Y_{X/S}} \dfrac{dX}{dt} + mX$ $\hfill (7)$

where X is the concentration of biomass present, C_S is the concentration of nutrient, C_P is the concentration of product, α and $Y_{X/S}$ are yield coefficients, and β and m are maintenance coefficients.

All the above methods are unsegregated and consider population-averaged quantities and behavior. Whilst they are tractable and in many cases, with simple metabolism and substrates, entirely adequate, they are always limited to specific situations by failing to consider the underlying growth processes of the population balance equation. With the advent of the flow microfluorometer it has become feasible to measure simplified versions of the distribution $f_1(\mathbf{z}, t)$ permitting a far more powerful and generalized approach to the determination of growth. Under conditions of balanced growth, the measurement of $f_1(\mathbf{z}, t)$ enables the growth rate function $\dot{\mathbf{Z}}(\mathbf{z}, \mathbf{c})$ to be evaluated by solution of the population balance equation (equation 1) (Agar and Bailey, 1981). Together with a suitable function, $g(\mathbf{z})$, to describe the biomass, this enables growth rates to be calculated from $f_1(\mathbf{z}, t)$ using equation (3). Knowledge of $g(\mathbf{z})$ can also be used to determine growth rates by means of equation (4), with $g(\mathbf{z})$ in place of z_i. In this case the distribution $f_1(\mathbf{z}, t)$ must be measured at several successive times.

Naturally, for most applications this technique is too involved. Furthermore, flow microfluorometry has only been used for two-dimensional *z* vectors (*e.g.* protein and DNA) due to experimental limitations. Nevertheless, this approach serves to illustrate the limitations inherent in the techniques normally used for growth measurement.

17.1.3 Comparative Microbial Growth Rate Measurement

The properties of the main groups of microorganisms of importance in the measurement of their growth rate are shown in Table 2. Algae and protozoa have not been included because the parameters for these microorganisms vary over such a wide range. The actinomycetes, although classified as bacteria, resemble molds with thin hyphae (1–5 μm instead of 10–20 μm thick) in their growth behavior.

It should be appreciated that the table presents only general guidelines. Exceptions, such as the binary fission of the yeast *Schizosaccharomyces pombe* and the yeast-like growth mode of the mold *Mucor* under anaerobic conditions, abound. As the concept of doubling time is not strictly valid for the non-exponential growth kinetics of viruses and molds, the numbers given should be considered only as indicators of the relative growth rate.

Table 2 Properties of Microorganisms Relevant to Growth Rate Measurement (data taken from Bailey and Ollis, 1977)

Property (units)	Virus	Bacteria	Yeast	Mold
Size (μm)	0.01–0.1	0.3–3.0	4–7	10–150
Mass ($\times 10^{-15}$ kg)	0.02–0.4	1	10	—
Doubling time (h)	0.2	0.5	2	5
Growth mode	Obligate parasite	Unicellular short chains and clumps	Unicelial pseudomycelia	Mycelial diffuse or as pellet
Reproduction mode	Lytic cycle Lysogeny	Transverse binary fission	Budding Ascospores	Hypha chain elongation
Composition (% dry weight) *Elements*:				
C	50	46–52	46–52	45–55
H		*ca.* 10	*ca.* 7	*ca.* 7
N		10–14	6–9	4–7
O		*ca.* 20	*ca.* 30	*ca.* 35
P	0.5–5	2–3	0.4–4.5	0.8–2.6
Minerals	—	4–10	4–10	4–10
Water (% total wt.)		*ca.* 70	70–75	*ca.* 70
Macromolecules: (% dry wt.)				
Protein	50–90	50–70	35–45	25–40
RNA	5–50	*ca.* 10	3–7.5	5–10
DNA		*ca.* 5	0.05–0.25	
Carbohydrate	—	6–15	30–45	40–55
Lipid	1–25	5–10	4–10	5–10
ATP ($\times 10^{-18}$ kg/cell)	—	*ca.* 2	*ca.* 200	—

Possibly the most significant factor in the choice of a growth rate measurement technique is whether or not growth is mycelial. Such growth, being non-repetitive, does not obey the population balance equation. For this reason the measurement of mycelial growth will be considered separately later (Section 17.3.1).

Other factors which must be taken into consideration are: the existence of sporulation or sexual reproduction phases (in microscopic and colony counts), the formation of storage compounds such as glycogen or β-hydroxybutyric acid (in chemical analysis), the shape of the cells (in nephelometry), motility of the cells (in microscopic or colony counts) and any distorting change in the relative rates of cell division and growth, as can occur in unbalanced media (in enumeration methods).

Such characteristics, together with the class of microorganism, permit an initial selection of suitable techniques, or modifications, for growth rate measurement. For example, enumeration techniques are not feasible with molds, but are virtually the only possibility in the case of viruses. The final choice of method should be checked for compatibility with the biochemistry, metabo-

lism and cell cycle of the microorganism concerned. If possible, the correlation with an independent direct method, such as dry weight determination or turbidimetry, should be established.

17.2 MEASUREMENT TECHNIQUES

The techniques available for growth rate measurement may best be subdivided into those
involving z, the physiological state vector of the cells, and those involving c, the environmental
concentration vector. Cell enumeration can be assigned to the former class. The number of cells
is given by:

$$N = \int_V f_1(z, t)\mathrm{d}v' \qquad (8)$$

17.2.1   Enumeration of Microorganisms

Enumeration techniques may be classified as microscopic, electronic or biological. The first
enable direct observation of individual microorganisms, thus permitting a simultaneous morphological examination and observation of clumping and contamination effects. The electronic or
automated methods permit large numbers of cells to be counted, thus reducing the statistical
uncertainty implicit in all enumeration methods. They can also yield the distribution of biomass
properties within the population, the importance of which has already been discussed (Section
17.1.2). The biological methods determine the number of actively growing or viable cells in the
population.
Enumeration techniques have been reviewed in detail by Collins and Lyne (1976), Gavin and
Cummings (1973) and Meynell and Meynell (1970).

17.2.1.1   *Microscopic techniques*

The normal procedure involves the counting by microscopic examination of a sample, appropriately diluted and stained, in a counting chamber of known volume. A grid within the counting
chamber is used to facilitate the counting process and define sample fields for enumeration. The
Petroff–Hausser counting chamber is the most commonly used variety. The chamber is 20 μm
deep and each field has an area of 2.5×10^{-9} m^2. The depth of the field should always be confirmed by microscopic measurement to ensure that a constant known volume is being observed.
Usually 600 organisms must be counted and concentrations of less than 3×10^{14} cells m^{-3} are
recommended (Koch, 1981; Mallette, 1971).
The methods for counting clumps of cells and defining the exact point of cell division vary, but
the convention devised should be used consistently.
Microscopic examination may be carried out under phase contrast (in which case staining artifacts are avoided) or using fluorescent, negative or viable stains. A diluent containing 0.5 M sucrose, 0.2% formalin and 0.1% anionic detergent neutralized with disodium hydrogen phosphate
prevents motility and growth interfering with the counting process and avoids aggregation, adhesion and osmotic damage of the cells (Gerhardt, 1981; Koch, 1981). A final dilution with 0.1 M
hydrochloric acid produces stationary adsorbed cells, without aggregation, for easy counting.
A less widely used technique is the so-called 'comparative count', in which a known volume of
the sample is mixed with a similar volume of a standard suspension containing microscopic particles (latex spheres or blood cells) at a similar, but known, concentration. The ratio of microorganisms to standard particles, observed in the subsequent microscopic examination, may then
be used to calculate the concentration of microorganisms in the original sample (Mallette, 1971).
Methods using optical microscopy are simple, relatively cheap and rapid techniques for the
measurement of growth. Electron microscopy may be used for the enumeration of virus particles
(Gavin and Cummings, 1973). Microscopic organelle counts, *e.g.* of nuclei, have been used to
enumerate non-unicells (Pirt, 1975).

17.2.1.2   *Automated techniques*

Two instruments are presently available for the automated counting of cells: the Coulter Counter and the flow microfluorometer. Count rates of 3–5×10^3 cells s^{-1} can be achieved. The

Coulter Counter may be used for particles in the size range 0.5–200 μm. The flow microfluorometer has been employed to measure both viruses and mammalian cells.

The Coulter Counter measures the changes in the resistance which occur when a suspension of microorganisms in saline is drawn through a small (30–100 μm diameter) glass orifice. A transient increase in the resistance occurs when a cell passes through this aperture, the magnitude of which is proportional to the cell volume. These pulses are electronically sorted and counted for the passage of a metered volume of saline to yield the concentration and size distribution function of the microorganisms. The principle and operation of the Coulter Counter have been described in detail by Kubitschek (1971).

The Coulter Counter has several sources of errors which must be taken into consideration. Firstly, inert material, such as solid particles or bubbles, can also give rise to pulses and must therefore be eliminated from sample and saline alike. Secondly, unlike in microscopic enumeration, clumps and unseparated daughter cells may not be distinguished, thus tending to give lower counts and distort the cell volume distributions. An ultrasonic pretreatment may be used to counteract these effects (Pringle and Mor, 1975). A related phenomenon is coincidence, in which the simultaneous passage of two independent cells through the orifice gives rise to a single pulse. This may be allowed for by carrying out counts at several different dilutions. Finally, the 30 μm diameter orifice required to detect bacteria is liable to block and must be carefully cleaned.

The flow microfluorometer (Hutter and Eipel, 1979a; Melamid *et al.*, 1979) measures the fluorescence resulting when individual stained cells flow past an exciting laser beam (Figure 3). The amount of emitted fluorescence depends on the quantity of stain, thus permitting the distribution of a specifically quantitatively stained component of the biomass to be determined as well as the cell count. The use of specific stains for proteins (*e.g.* fluorescein isothiocyanate) and nucleic acids (*e.g.* propidium iodide) enables non-biological material to be eliminated from the enumeration. Fluorescent antibodies and viability stains may be used to count only component subpopulations (Hutter, 1978, 1979). Furthermore, by use of multiple lasers and/or optical processing of the fluorescence signal, joint distribution functions for cells which have been stained for more than one component may be obtained (Agar and Bailey, 1982a; Fazel-Madjlessi *et al.*, 1980; Hutter and Eipel, 1979b). In addition to the fluorescence of stained cells, the diffraction of the laser beam at various angles may be used to give information on cell size and shape. The device may also be set up to sort individual cells on the basis of measurements made (Horan and Wheeless, 1977).

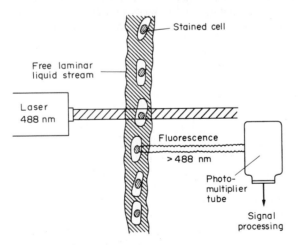

Figure 3 Schematic diagram of a flow microfluorometer (from Bailey *et al.*, 1979. Reprinted by permission of the New York Academy of Sciences)

From this impressive list of capabilities it can be seen that the flow microfluorometer is a powerful instrument with great potential for the measurement of microbial population dynamics (Bailey *et al.*, 1979). As with the Coulter Counter, clumping, delayed separation after cell division and coincidence can cause problems. The first two effects seem to be prevented by the staining procedures. More serious is the loss of biomass which can occur during the often multistage staining and washing processes. Simple, careful staining and the remedies given earlier can overcome these difficulties.

17.2.1.3 *Biological techniques*

These techniques rely on the macroscopic manifestations of growth observed in a suitably diluted sample. The dilution and diluent must be carefully chosen to give statistically meaningful results reflecting the number of viable cells in the original sample.

The most widely used technique is that of colony counting, in which samples of a tenfold serial dilution are spread on 0.1 m agar plates of a suitable medium with 1.5–2% agar. Following an incubation of 24–72 h, each cell originally present gives rise to a visible colony. At a suitably high dilution these colonies are distinct and may be counted. Plates with 30–300 colonies can be used for enumeration. Methods with subsurface colonies, such as pour or layered plates, permit work with motile organisms and, by inhibiting the overlap of adjacent colonies, permit higher counts (< 2000) (Koch, 1981). The counting and plating processes may be automated, but the method remains a very time consuming technique.

Two derivative methods should be mentioned. The slide culture technique of Postgate (1971) uses agar slides instead of plates and the number of microcolonies (> 4 cells) formed after a short incubation period is measured by microscopic examination. It may therefore be considered as a hybrid microscopic/biological technique. The membrane filter method (Collins and Lyne, 1976) is suitable for the enumeration of dilute samples (< 10^8 cells m^{-3}). A known volume of the fluid sample is passed through a porous membrane filter (usually 0.45 μm diameter pores) which retains the microorganisms present. The membrane is then allowed to absorb culture medium and incubated to enable colonies to develop in the usual way. If cell culture plates of an appropriate host are used instead of simple culture medium plates, the method may be extended to the enumeration of viruses. Each viral particle gives rise to a cell-free region (known as a plaque or focal lesion) in analogy to the development of a colony from a cell. The determination of so-called 'plaque forming units' is very sensitive to the conditions under which dilution and incubation are carried out.

The second class of biological enumeration techniques uses an 'all-or-nothing' response rather than a count. If multiple samples of a serial dilution are incubated in a suitable medium the number of positive responses, *i.e.* those showing growth after a given period, within the dilution series may be used to calculate the 'most probable number' of microorganisms with the aid of special tables (Gavin and Cummings, 1973). The technique is inaccurate, inefficient and time consuming. The advantages over the colony enumeration technique lie in the simplicity, the applicability to low concentrations of microorganisms and the possibility of using a product of growth, *e.g.* acid, instead of the microorganisms themselves, as a sensitive indicator for growth. Quantal measurement, as the technique for virus enumeration is known, uses the dilution at which 50% positive responses are obtained to measure viral concentrations.

17.2.2 Measurement of Biomass

17.2.2.1 *Dry weight*

Dry weight is the most direct technique available for determining biomass. Although being replaced by more advanced and specific techniques, it remains the yardstick against which alternatives are judged. The disadvantages of the technique lie in its slowness, poor sensitivity and inapplicability to situations in which non-microbial solids are present. The technique consists of three steps: removal of cells from the medium, drying and weighing. Cell removal is accomplished by centrifuging or filtration. Harvesting by centrifugation usually results in the loss of some cell material, either through incomplete removal or in the course of the decanting process. Filtration, on the other hand, requires careful examination of the behavior of the filter (0.22–0.45 μm diameter pores are usually used) during washing and drying.

Washing with an appropriate solution or solvent then removes residual traces of the medium. Non-microbial solids may sometimes be removed by special washing, *e.g.* calcium carbonate using mineral acids. The washing should be adequate to remove the medium present but not so severe as to result in leaching or lysis of the biomass.

Drying conditions must similarly represent a compromise between the need to remove water and the prevention of biomass loss through volatilization or oxidation. Typically 80 °C for 24 h or 110 °C for 8 h is used, although low-temperature vacuum drying or inert gas techniques may be used in certain circumstances. A discussion of dry weight procedures is given by Mallette (1971) and Pringle and Mor (1975).

The macroscopic, fibrous nature of mycelial forms makes them especially amenable to filtration, whilst the small size of viruses makes the technique unsuitable in their case.

17.2.2.2 Wet weight

This is a less accurate and faster technique than dry weight, in which the drying step is dispensed with. The variable amounts of water bound in and between the microorganisms make the procedure unreliable, even when the conditions for cell removal are standardized. The values obtained are usually about four times the dry weights. Wet weights and their determination are described by Gerhardt (1981).

The technique of packed-cell volume (PCV) differs only in that instead of the mass the volume of the cellular material is measured in specially calibrated centrifuge tubes. This volume is influenced by the conditions of centrifuging (which must therefore be carefully standardized), the tonicity of the medium (which can cause swelling of cells) and the morphology of the culture (which influences packing). Denser non-microbial solids can often be removed by an initial centrifuging stage, though the degree of separation is imperfect. Nestaas *et al.* (1981) describe a related technique, the filtration probe, for monitoring mycelial growth. Filtration instead of centrifugation is used to separate the biomass and the volume of the filter cake is relatively independent of mycelial morphology. Furthermore, the density of the filter cake may be predicted from the specific filtration resistance of the broth, enabling calculation of dry weight.

As with dry weight, these techniques are insensitive and can only be used for relatively concentrated microbial suspensions (> 1 kg m^{-3}).

17.2.2.3 Turbidimetry and nephelometry

In many situations the optical density of microbial suspensions exhibits a very close correlation with the amount of biomass present. As a result of this, combined with the speed, simplicity and convenience of the measurement technique (turbidimetry), this property is widely used to determine growth.

The attenuation of a light beam passing through a dilute microbial suspension is described by the empirical Beer–Lambert law:

$$I = I_0 e^{-\tau L} \tag{9}$$

$$\tau = \tau' X \tag{10}$$

$$\log\left(\frac{I_0}{I}\right) = \frac{\tau' L}{2.303} X \tag{11}$$

where I_0 is the intensity of the incident beam, I is the intensity of the attenuated beam, L is the length of the light path, τ is the turbidity or optical density of the suspension, and X is the concentration of the suspension.

This relationship applies as long as the left hand side of equation (11), known as the absorbance (A_λ), is less than 0.3. For higher concentrations the proportionality fails, attenuation being reduced by light rescattered in the original direction. The resultant curve is described by a parabola. Koch (1981) gives the following equation for the absorbance of a bacterial suspension at 420 nm wavelength:

$$A_{420} = (7.114 \times 10^{-3})X - (7.702 \times 10^{-6})X^2 \ (X \text{ in } 10^{-3} \text{ kg m}^{-3}) \tag{12}$$

For absorbances of greater than 0.3, calibration curves are employed or the sample is diluted with fresh medium to permit measurement in the linear region.

The instrumentation varies in complexity. The simple Klett–Summerson units, designed to measure optical densities directly in shake flask side arms, have a simple parallel beam arrangement with a filter and a CdS detector. Klett units employ a unique scale, in which a reading of 500 corresponds to an absorbance of unity. At the other end of the range are well-collimated spectrophotometers designed to work with special cuvettes and at an exact and adjustable wavelength. The wavelengths used are typically 420 or 660 nm. The former is more sensitive (the first constant in equation (12) is roughly proportional to the reciprocal of the wavelength squared), whereas the latter is less sensitive to the medium used, is not specifically absorbed by cell components and can be used at higher cell concentrations.

Instrument errors can arise from external stray light, variations in light source intensity, temperature changes due to heating effects and cuvette errors. The clarity and coloration of the medium as well as any non-microbial particles will influence the optical density. For this reason blanks should always be measured. If these factors are important and change with time during growth, turbidimetry is unsuitable. In some cases non-microbial solids may be removed by special pretreatments, *e.g.* hydrocarbon emulsions using propionic acid. The optical density is also adversely affected by cell clumping and settling. The effects are minimized by use of dispersing detergents and mixing.

The method is most suitable for cells in the $0.4–2 \times 10^{-18}$ m^3 size range; viruses are too small and the macroscopic nature of mycelia invalidates the Beer–Lambert law. Cell shape is usually not important, only becoming so when the cells are preferentially aligned, *e.g.* by convection currents.

Low temperatures, antibiotics or germicides, such as 0.5% formalin, prevent further growth after sampling. To avoid cell lysis and changes in the refractive index due to the tonicity, the culture medium is used as a diluent. For further information on turbidimetry the reviews of Mallette (1971) and Koch (1981) should be consulted.

Continuous turbidimetric systems have been devised for use in fermenters (Lee, 1981; Metz, 1981). These *in situ* systems often give problems due to the heterogeneity of the medium, *e.g.* air bubbles, or due to growth on the optical surfaces and have not yet found widespread application.

Nephelometry measures the light scattered at a 90° angle to the incident beam by the suspension. The angular dependence of the amount of light scattered is a characteristic feature of a given microorganism. The technique is more sensitive than turbidimetry: it can be used at low concentrations (10^{10} cells m^{-3}), with small particles (viruses) and can yield information on cell shape and internal structure. The disadvantages are a greater sensitivity to non-microbial particles and the need for a well-collimated light source (*e.g.* laser) and other expensive instrumentation.

17.2.3 Measurement of Biomass Components

The techniques for determining chemical components of the biomass may be subdivided into specific and non-specific methods. The latter include elemental and protein determinations and involve substances also found in the medium. The biomass must therefore be separated prior to the measurement. The former techniques, including DNA, NADH and ATP determinations, involve components found exclusively in the biomass. Separation of the biomass thus becomes less critical.

A comparison of the batch growth of the fission yeast *Schizosaccharomyces pombe*, as measured by several chemical determinations as well as by enumeration and direct methods, is presented in Figure 4.

Reviews of the chemical analysis of microbial cells have been presented by Herbert *et al.* (1971) and Hanson and Phillips (1981). Techniques specifically for yeasts are given by Stewart (1975).

17.2.3.1 Elemental analysis

Nitrogen determination is often used to indicate biomass. Nitrogen is a major component of cell dry weight (Table 2); the fraction, however, may vary considerably with nutritional and physiological conditions, especially in mycelia and when a storage compound is present.

The most common method, the Kjeldahl technique, converts all the nitrogen in a 1 mg sample into ammonia by digestion with sulfuric acid containing additives, such as selenium or permanganate. After neutralization this ammonia is determined by titration or colorimetrically using Nessler's reagent. 'Crude protein' values are obtained by multiplying the Kjeldahl nitrogen value by a factor of 6.25, though ideally non-protein nitrogen should also be accounted for.

The measurement of total organic carbon, although not widespread, has been used to determine biomass in the presence of non-microbial carbon-containing solids (Hashimoto *et al.*, 1982). These solids were first separated by a differential centrifugation. The carbon in the residual

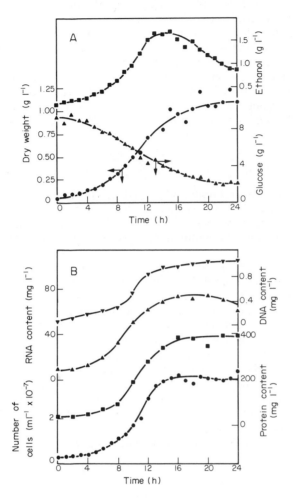

Figure 4 Determination of the batch growth curve of *Schizosaccharomyces pombe* using several measurement techniques. (A) Dry weight, g l⁻¹ (●); medium glucose content, mg l⁻¹ (▲); medium ethanol content, g l⁻¹ (■). (B) Cell numbers, m l⁻¹ × 10⁷ (●); culture protein content, mg l⁻¹ (■); culture RNA content, mg l⁻¹ (▲); culture DNA content, mg l⁻¹ (▼) (from Agar and Bailey, 1982b. Reprinted by permission of the Society for Analytical Cytology)

supernatant correlated well with DNA measurements. Carbon is determined by catalytic combustion to carbon dioxide at 900 °C followed by infrared measurement of this product. Alternatively the carbon dioxide can be converted to methane over a nickel catalyst at 350–450 °C, which is then determined using gas–liquid chromatography. Only small quantities (30×10^{-9} m³) of sample are required and the method is rapid and automatic. A colorimetric method, in which the absorbance of a sample digested in a sulfuric acid–dichromate mixture is measured, has also been proposed (Mallette, 1971).

Bennett and Williams (1957) have reported that phosphorus may be used to measure biomass during the growth of *E. coli*. The determination of phosphorus is carried out by converting all organic and inorganic forms to phosphate using acid digestion. The digest is then reacted with ammonium molybdate and reduction of the resultant phosphomolybdate yields a blue complex which is determined with a spectrophotometer.

17.2.3.2 Protein

As shown in Table 2, protein is the main macromolecular component of microorganisms. It is not always, however, confined to the biomass and the microorganisms themselves may secrete enzymes and amino acids into the medium. Furthermore, the term protein covers a multitude of different amino acid arrangements, each of which will respond differently to the specific reaction used for the determination. For these reasons the washing of the cells and the correct choice of standard (to resemble the proteins of the sample) are both important in protein determination.

A variety of techniques is available for protein measurement (Cooney, 1982), but only two are commonly used for microbial growth determination: the Biuret and the Folin–Ciocalteau (or Lowry) analyses. The Biuret method determines α-peptide linkages and may be applied to whole cells or alkali extracts. The non-specific reaction means that the type of protein is largely immaterial. In the Lowry method the aromatic amino acids tryptophan and tyrosine are measured, following an alkaline hydrolysis of the protein. The method is thus dependent on the amino acid composition of the protein and also responds to non-proteins such as glucose and urea. However, the Lowry analysis is about 60 times more sensitive than the Biuret analysis.

17.2.3.3 DNA

DNA (deoxyribonucleic acid) is found exclusively in cells. Thus, unlike protein, DNA determinations may be used to indicate biomass even in complex media containing proteins. Although the ratio of DNA to dry weight shows some variation during growth, the fluctuation is much less than that for other macromolecular components such as carbohydrates, lipids and RNA (Dean and Hinshelwood, 1966).

The problems with DNA determination are the extremely small amounts present (*e.g.* 2×10^{-14} g per cell in *E. coli*) and the presence of a large excess ($\times 100$ in *Schizosaccharomyces pombe*) of chemically similar RNA. Traditionally, DNA has been measured colorimetrically by the diphenylamine reaction of deoxyribose residues in a perchloric acid extract of the cells. The reagents for this reaction are unstable, and numerous substances, such as pentoses, interfere with the determination. A more sensitive fluorometric method for determining deoxyribose using diaminobenzoic acid, which may be applied to whole cells, has also been described (Hinegardner, 1971). A more recent technique employs the intercalating fluorescent dyes ethidium bromide and propidium iodide. These react quantitatively with double stranded nucleic acids to yield highly fluorescent, conjugated forms (Le Pecq and Paleotti, 1967). The extent of reaction depends on the physical structure (coiling) of the nucleic acids in a similar way to the dependence of the Lowry analysis on the chemical structure of proteins. A pretreatment with barium hydroxide or the enzyme RNAase is necessary to remove RNA from the cells. This technique is rapid and may be applied to whole cells.

17.2.3.4 Other

Like DNA, ATP (adenosine triphosphate) is confined to the biomass. Although the levels encountered can show dramatic variations ($0.3–3 \times 10^{-15}$ g per cell in bacteria), the amount in any given species is surprisingly independent of growth rate and only changes by a factor of two between starvation and exponential growth conditions.

The reaction for the estimation of ATP is as follows. First ATP is extracted from the cells using a special solvent designed to prevent enzymatic breakdown occurring. The extracts are then treated with the firefly enzyme luciferase to form a bioluminescent complex which, upon oxidation, yields an amount of light proportional to the ATP present. This bioluminescence is sensitive to quenching substances and heavy metals which inhibit luciferase. The reaction is very sensitive and can determine cell levels as low as 10^9 cells m^{-3}. This sensitivity, combined with the specificity for viable cells, makes the reaction suitable for sterility testing. Further applications are given in a publication from Lumac® (1980). The excellent correlation obtainable between dry weights and ATP measurements is illustrated in Figure 5.

A second component of the metabolic machinery of viable cells, NAD(P)H (reduced nicotinamide adenine dinucleotide), has also been shown to correlate with the biomass present (Zabriskie and Humphrey, 1978a). The culture fluorescence (excitation: 366 nm; emission: 460 nm) depends largely on NADH (> 50% in a culture of *Candida utilis*) and thus offers a simple measurement technique. Other fluorophores, such as chlorophyll and antibiotics, can distort the results. In addition, as shown in Figure 6, NADH responds to metabolic shifts. In the example shown, the shift from a glucose- to an ethanol-based metabolism yields a spurious fluorescence peak unrelated to the biomass (Beyeler *et al.*, 1981).

Although fluorescence is somewhat non-specific, the measurement technique permits on-line *in situ* determinations within bioreactors. An electrode system for the determination of biomass

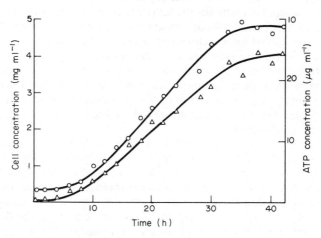

Figure 5 Correlation between ATP ($\triangle$) and biomass concentration ($\bigcirc$) during the batch growth of *Trichoderma viride* on glucose (from Hendy and Gray, 1979. Reprinted by permission of John Wiley and Sons Inc.)

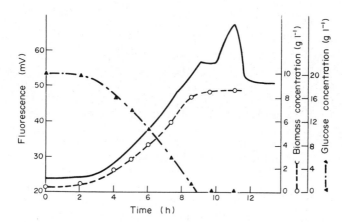

Figure 6 NADH fluorescence during the batch growth of *Candida tropicalis* (from Beyeler *et al.*, 1981. Reprinted by permission of Springer-Verlag)

based on the measurement of cell surface NADH groups has been described by Matsunaga *et al.* (1979).

17.2.4 Measurement of Biomass Environment

Growth rate measurement techniques involving *c*, the environmental concentration vector, are inherently indirect and require assumptions about metabolic behavior. Despite this they are often preferred on account of their ability to be used as continuous on-line measurements in bioreactors. At low concentrations of microorganisms, where the influence of growth on the environment is extremely small, they are often inadequate.

The distinction between the measurement of nutrients and products is artificial, as chemically and physically they are similar, as well as being governed by similar relationships (equations 6 and 7). Microorganisms implicitly acknowledge this fact sometimes by using the product of one fermentation stage as the substrate for the next. In light of this a more meaningful classification may be based on the type of component/measurement. Three classes can be identified: gas phase compositions, liquid phase compositions and liquid phase physical/physicochemical properties.

17.2.4.1 Gas composition

These techniques are almost exclusively concerned with oxygen and carbon dioxide. The physical nature of the measurements and the fact that they may be carried out without directly contacting or disturbing the growing microorganisms make them suitable for continuous on-line determinations. The older manometric and respirometric techniques are now seldom used and will not be considered.

Oxygen may be measured polarographically, paramagnetically or by using a mass spectrometer. These instruments may be used continuously for long periods, but care must be taken with the flow rate of the gas, its temperature, any sensitivity to the water content and the response times for changes (10–300 s).

The changes in the dissolved oxygen content of the medium are, in absolute terms, small because of the low oxygen solubility (*ca.* 10^{-2} kg m^{-3}). The oxygen uptake rate (OUR) can thus be calculated by a simple mass balance on the gas between the inlet and outlet (Cooney *et al.*, 1977). Mateles (1971) has shown that the yield factors of biomass with respect to oxygen and carbon source are related by the simple equations:

$$\frac{1}{Y_{X/O}} = \frac{1.5}{Y_{X/S}} - 1.33 \text{ (for yeast)} \tag{13}$$

$$\frac{1}{Y_{X/O}} = \frac{1.5}{Y_{X/S}} - 1.6 \text{ (for bacteria)} \tag{14}$$

where $Y_{X/O}$ is the yield factor with respect to oxygen, *i.e.* kg biomass dry weight per kg oxygen and $Y_{X/S}$ is the yield factor with respect to the carbon source, *i.e.* kg biomass dry weight per kg carbon source utilized.

The OUR alone can provide a reasonable estimate of the growth rate of microorganisms (Reuss *et al.*, 1976; Zabriskie and Humphrey, 1978b). The results are improved if carbon dioxide measurements are included in the calculation.

Carbon dioxide is measured continuously using infrared spectroscopy or mass spectrometry, discontinuously using gas–liquid chromatography or gravimetrically with barium hydroxide. The relative merits of the instrumentation for gas phase analysis are presented by Wang *et al.* (1979). The solubility of carbon dioxide is about 30 times greater than that of oxygen. The capacity of the medium is thus not negligible and changes in pH can affect the evolution of carbon dioxide.

The carbon dioxide production rate (CPR) can be used alone to measure growth (Boyles, 1977; Cooney, 1982). In combination with OUR, the CPR enables changes in yield factors to be allowed for. The ratio between the two measurements, the respiratory quotient, is used to determine the characteristic metabolic pathway, of which the stoichiometry is known. In this way the diauxic aerobic growth of *Saccharomyces cerevisiae* on glucose (Figure 7) may be accurately simulated (Zabriskie and Humphrey, 1978b). Further examples of this technique of material balancing, using gas phase composition measurements, are given by Cooney *et al.* (1977), Siebert and Hustede (1982) and Wang *et al.* (1977).

Mass spectrometry and gas–liquid chromatography have been employed to measure other components of the gas phase. The levels of hydrocarbons in the exit gas of a bioreactor enable the liquid phase compositions of these components to be calculated using Henry's law (Sukatsch and Nesemann, 1977). Ethanol contents in media have been measured using a similar principle (Bach *et al.*, 1978; Weaver *et al.*, 1978). Special semiconductor probes have been devised to measure hydrogen and ammonia in the gas phase (Winquist *et al.*, 1980). Gas phase analyses are discussed in greater detail in Volume 4, Chapter 21.

17.2.4.2 Liquid composition

The chemical complexity of the liquid phase, the presence of biomass and the need to maintain sterility during measurements mean liquid phase analysis is often quite complicated. The techniques may be subdivided into chemical, enzymic and physical methods.

The first class, chemical determinations, is described in great detail by Dawes *et al.* (1971). The techniques require prior separation of the biomass, are usually non-specific (*e.g.* total reducing sugar determined by the dinitrosalicylic acid analysis) and can only be adapted with difficulty for continuous on-line monitoring (*e.g.* ammonia determination; Cooney and Wang, 1976).

Examples of enzymic analyses are shown in Table 3; they have been reviewed by Guilbault (1976). The regular analysis technique is sensitive and specific, but requires separation of the

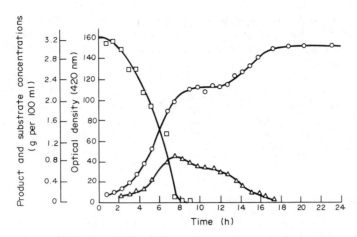

Figure 7 Batch growth curve for *Saccharomyces cerevisiae* (○) and medium glucose (□) and ethanol concentrations (△) (from Gilbert *et al.*, 1978. Reprinted by permission of the American Society for Microbiology)

biomass. The growth of *Saccharomyces cerevisiae* shown in Figure 7 may also be calculated from the accompanying enzymically measured glucose and ethanol profiles using the yield factors given by Bijkerk and Hall (1977). The enzyme electrode (Enfors and Molin, 1978; see also Volume 4, Chapter 20) employs a selective membrane containing an immobilized enzyme and may be used in the presence of biomass. The enzyme thermistor (Danielsson, 1982; Mosbach and Danielsson, 1981) uses the thermal effect (*ca.* 20 kJ mol^{-1}) of enzyme reactions to measure substrate levels. The temperature rise when a sample is passed over a fixed bed of immobilized enzyme is determined. *In situ* application of enzymic methods is prevented by the thermal lability of enzymes, which results in their destruction during sterilization procedures, and variations in the activity with time.

Table 3 Enzymic Analysis Systems (from Onken and Buchholz, 1982)

Substrate	Enzyme	Product
β-D-Glucose	Glucose oxidase	H_2O_2
	(+ catalase)	(O_2)
Saccharose	Invertase	(H_2O_2)
	(+ mutarotase + glucose oxidase)	
Lactase	β-Galactosidase	(H_2O_2)
	(+ glucose oxidase)	
Alcohols (short chain, primary)	Alcohol oxidase	H_2O_2
α-Amino acids	α-Amino acid oxidase	NH_4^+, H_2O_2
Urea	Urease	NH_4^+, CO_2
Penicillin	Penicillinase	H^+

On-line electrode measurement systems have been devised for antibiotics (Nilsson *et al.*, 1978) and vitamins (Mattiasson *et al.*, 1982). Although present at low levels, the high 'biological activity' of these substances facilitates their determination. The release of these products is, however, usually not closely related to the amount of biomass. Such substances, known as secondary metabolites, are thus unsuitable for the measurement of growth. Examples of primary metabolites which correlate with biomass levels include substances such as amino acids, urea, lactic acid and butanol. Enzyme production may or may not be growth dependent. Wood (1979) gives an example in which the enzyme laccase was used to follow the growth of *Agaricus bisporus*.

Physical methods used to ascertain the chemical composition of the liquid phase include oxygen (see Volume 4, Chapter 19) and specific ion electrodes (Kell, 1980) and methods using mass spectrometry to analyze the components diffusing through a separating membrane. The use of dissolved oxygen to measure growth rate is described by Hill and Robinson (1974) and by Boyles (1978). The low and variable uptake of minerals by microorganisms makes them unreliable indicators of growth. If ammonium ions are the sole source of nitrogen, the disappearance of this ion from the medium correlates well with growth. Sodium and potassium interfere with the determination of ammonium ion using specific ion electrodes. The dialysis approach in combination with mass spectrometry or gas–liquid chromatography has been used for the measurement of methanol (Yano *et al.*, 1978), ethanol (Dairaku and Yamané, 1979) and dissolved gases (Reuss *et al.*, 1975). Zabriskie and Humphrey (1978c) used dialysis followed by enzymic analysis for the continuous measurement of the glucose content of the liquid phase.

The use of radioactive tracers to monitor substrate uptake and elucidate metabolic pathways is a well-established laboratory procedure. Growth rates are measured by applying a pulse of radioactively labelled substrate and following its disappearance from the medium or accumulation of the biomass. Suitable radioisotopes and their application are given by Wood (1981) and Hanson and Phillips (1981). Generally the technique is only employed when the rate of synthesis of a specific component, *e.g.* DNA, or when a high sensitivity is desired (MacLeod *et al.*, 1966).

17.2.4.3 *Physical/physicochemical properties*

These correspond to complex functions of the environmental concentration vector and can reflect the influence of the biomass present. Their poor specificity and sometimes poor sensitivity is compensated for by their suitability for continuous on-line measurements.

By constructing an energy balance for a bioreactor the heat evolved by microbial metabolism may be estimated. This heat correlates well with the OUR and CPR of the culture (Cooney *et al.*, 1969) as well as with the biomass (Figure 8).

$$H_X = 0.519 \pm 0.013 \text{OUR} \tag{15}$$

$$H_X = 0.406 \pm 0.017 \text{CPR} \tag{16}$$

where H_X = rate of heat evolution (kW m^{-3}), OUR = oxygen uptake rate (mmol m^{-3} s^{-1}) and CPR = carbon dioxide production rate (mmol m^{-3} s^{-1}).

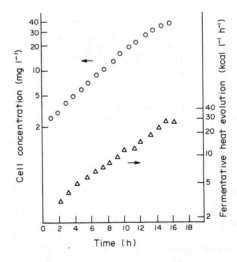

Figure 8 Cell concentration and heat of fermentation during the growth of *Saccharomyces cerevisiae* on molasses (from Cooney, 1982. Reprinted by permission of Verlag-Chemie)

For acceptable precision the 'heat of fermentation' must be comparable with the other energy inputs and outputs. This is only true for dense cell suspensions. Decoupling between heat evolution and growth, because of maintenance metabolism, can also occur (Mou and Cooney, 1976). In the laboratory, bioreactors may be specially designed as calorimeters to utilize this effect (Forrest, 1971; Monk, 1978). Such devices may also be employed for the identification of microorganisms (Russell *et al.*, 1975).

The control of pH has long been part of the bioreactor control system. The pH value represents an interaction of the consumption of ammonia, the production of organic acids and the buffering capacity of the medium and biomass. For this reason, measurements of the acid or base addition needed to maintain the desired pH are difficult to interpret in terms of growth, unless a single effect predominates. Changes in the nitrogen content of the microorganisms can also cause deviations of up to 27% in the correlation between growth and the alkali consumption needed to compensate for ammonia uptake (Concone *et al.*, 1978). Nevertheless, as shown in Figure 9, the correlation between ammonia addition for pH control and biomass can be remarkably good.

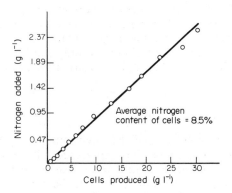

Figure 9 Correlation between the ammonia addition and cell growth during the pH-controlled batch growth of bakers' yeast (from Wang *et al.*, 1977. Reprinted by permission of John Wiley and Sons Inc.)

The viscosity of the liquid phase may reflect the medium (*e.g.* starch, carboxymethylcellulose, dextran), the biomass (for high concentrations) or a combination of both. Viscosities are in the range of $1–1500 \times 10^{-3}$ kg m^{-1} s^{-1}. For spherical unicells in an aqueous medium, the relationship between cell concentration and viscosity is given by Einstein's equation:

$$\frac{\mu_X}{\mu_R} = \frac{1 + 0.5\phi}{(1 - \phi)^2} \tag{17}$$

where ϕ is the volume fraction of the biomass, μ_X is the viscosity of the cell suspension and μ_R is the viscosity of the medium.

The morphology of the cells plays an important role in determining the viscosity. The viscosity is measured by determining the pressure drop for a known flow through a capillary, from which a value may be calculated using Poiseuille's equation (Onken and Buchholz, 1982). The apparatus is sterilizable and suitable for continuous on-line measurements. It has been used to monitor growth in antibiotic and polysaccharide fermentations. An example of the relationship between biomass and viscosity is presented in Figure 10.

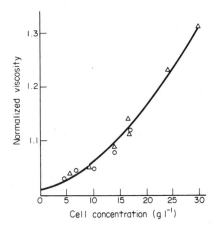

Figure 10 Viscosity of cell suspensions of *Hansula polymorpha* DL-1 grown in continuous culture on methanol. Measurements on cells harvested before ($\triangle$) and during ($\bigcirc$) steady state (from Perley *et al.*, 1979. Reprinted by permission of John Wiley and Sons Inc.)

The capacitance of the liquid phase has been used to measure microbial cell concentrations Hadley and Senyk, 1975; Gencer and Mutharasan, 1979; Richards *et al.*, 1978). A precondition for such measurements is a low medium conductance ($\leqslant 60\ \Omega^{-1}\ m^{-1}$) to avoid interference from a double layer effect due to the salts present. The proportionality between biomass and capacitance under such conditions is shown in Figure 11.

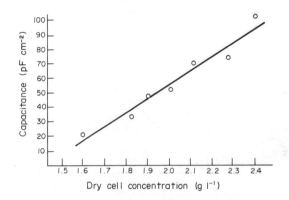

Figure 11   Capacity *versus* yeast concentration ($T = 29\ °C$, $i = 25\ \mu A$, $f = 10\ 000\ Hz$) (from Gencer and Mutharasan, 1979. Reprinted by permission of John Wiley and Sons Inc.)

The redox potential of the medium indicates the tendency of a solution to give up or receive electrons. In aerobic cultures the redox potential is a function of the dissolved oxygen and substrate levels (Onken and Buchholz, 1982). If oxygen levels are determined independently, the redox potential can be used to monitor glucose concentrations (Kjaergaard, 1977).

17.3 SELECTION OF TECHNIQUE

Pirt (1975) gives five criteria which decide the selection of a suitable growth rate measurement technique:
(1) properties of the biomass;
(2) properties of the medium;
(3) accuracy required;
(4) sensitivity required; and
(5) time demand of the method.
These criteria will be illustrated by considering special cases of growth measurement in the following four sections.

When taking samples for growth measurements, weight rather than volume should be the indicator of sample size. This avoids problems due to variable liquid densities, gas entrainment or large quantities of insoluble nutrients (*e.g.* cellulose). Samples should also be taken at a representative location and regions of stagnancy or film formation should be avoided.

Following sampling, further growth must be arrested as soon as possible, or, in the case of biological enumeration methods, allowed to proceed with as little disturbance as possible. The choice of inhibitor (*e.g.* sodium azide, formalin, low temperatures, antibiotics) should be made to minimize the distortion of the property subsequently measured. Similarly, to permit the growth of all viable cells present, the medium for colony counting is often made richer than that from which the sample was taken.

The size of the sample required is controlled by the sensitivity of the selected technique (Table 4).

17.3.1 Mycelial Growth

In the growth of molds and actinomycetes by the elongation of hyphal chains, the cells do not continually pass through a given life cycle and the growth process is localized at the tips of the hyphae. This leads to a much broader and more poorly defined age distribution of the component cells, which complicates the properties, behavior and metabolism of the resultant cell mass. In

Table 4 Comparison of the Sensitivities of some Methods of
Bacterial Biomass Estimation (from Pirt, 1975)

Method	Minimum dry mass of bacteria required for an estimation with an error of $< 2\%$ (mg)
Dry weight	50
Biuret protein	1.0
DNA	1.0
Folin–Ciocalteau protein	10^{-1}
Opacity	10^{-1}
Cell count	10^{-5}

addition, the macroscopic structures produced by mycelial growth can give rise to concentration gradients, with nutrients diffusing into the mass and products diffusing out. Again the net effect is to produce a sensitive, poorly defined growth process. Further, in response to high shear rates in bioreactors, mycelia may form a dense pellet instead of the normal fibrous form.

Calam (1971) describes growth measurement techniques in a specifically mycelial context. The poor definition of individual cells, their interconnection and tendency to form macroscopic structures make enumeration techniques and turbidimetry impossible. Non-Newtonian behavior can give problems in viscosity determination. To take variations in the state of the culture into account, morphological features such as the percentage of vacuolated, branching and autolyzed cells can be incorporated in the growth rate measurement.

The growth of surface mycelial cultures may be determined from the linear extent of the macroscopic colonies. The non-exponential growth kinetics so obtained are discussed by Pirt (1975).

17.3.2 Growth in Presence of Solids

Growth on solid substrates has assumed new importance with the interest in microbial utilization or transformation of cellulose. The removal of non-microbial solids using centrifugation and solvent treatments has already been mentioned. In the first case biomass may be lost with these solids, especially given the tendency of microorganisms to form films on solid surfaces. Solubilization procedures can extract material from the biomass.

To differentiate between biomass and non-microbial solids on the basis of chemical composition, a component found only in the former must be identified. For growth on cellulose, for example, protein measurements suffice to establish the amount of biomass present (Moreira *et al.*, 1978). With more complex substrates, such as sewage sludge, a more specific substance, such as DNA, is required (Hashimoto *et al.*, 1982). Interestingly, another biomass specific substance, ATP, is apparently unsuitable in this case (Thomanetz *et al.*, 1982).

Of the direct methods for growth measurement, only microscopic or biological enumeration is generally feasible in the presence of non-microbial solids without pretreatments. Even with these techniques, the adsorption of microorganisms on to the solids can cause problems. If components of the environmental concentration vector are to be determined, either the insoluble nutrients (*e.g.* cellulose by the anthrone reaction) or soluble products (*e.g.* reducing sugars) may be selected.

17.3.3 Growth in Mixed Populations

When only one component microorganism of a mixed population is to be measured, enumeration techniques are usually the only option. Microscopic and biological counts can utilize visible differences in cell and colony morphology to distinguish between different sub-populations. In the latter case, selective media or plates containing a specific indicator (*e.g.* 8-anilino-1-naphthalenesulfonic acid for Gram-negative microorganisms; Ramsey *et al.*, 1980) may be used. Fluorescent antibodies offer a very specific means of identifying a particular microorganism applicable to microscopic counts (Yoshioka *et al.*, 1982) and flow microfluorometry (Hutter, 1978). The Coulter Counter can only be used to resolve simple mixtures where an appreciable difference in the size of the component microorganisms exists (Tseng and Phillips, 1981).

The biological and metabolic similarity of many microorganisms prevents chemical compositions of the biomass or environment being used to measure the growth of a single sub-population. The exception is when the microorganism of interest possesses a unique characteristic, *e.g.* diaminopimelic acid, which can be used to follow the growth of Gram-negative organisms alone (El-Shazly and Hungate, 1966).

The determination of the fraction of viable biomass represents a special case of mixed population analysis. Biological enumeration implicitly determines only the viable sub-population. Using a viability stain such as methylene blue in microscopic enumeration and a corresponding fluorescent stain in flow microfluorometry (Hutter, 1979), these techniques may be adapted to determine only viable cells. Of the remaining techniques, measurements of the environmental concentration vector only indicate the metabolizing, *i.e.* viable, biomass present. The same applies for measurements of ATP and NADH (Miller *et al.*, 1978). All the other techniques described measure total biomass, *i.e.* including dead cells.

17.3.4 Microbial Growth in Industry

In industrial bioreactors the main purpose of growth rate measurement is to obtain data continuously and on-line, which may be used to monitor and control the fermentation. Unlike the laboratory, where time is not usually a factor and taking samples aseptically presents no problems, industrial applications require rapid response times and a large degree of automation. The requirements with regard to precision and sensitivity are sometimes less stringent.

Of the enumeration methods only electronic counters offer the speed needed for an on-line capability. Hatch *et al.* (1979) describe an on-line version of the flow microfluorometer. For similar reasons packed cell volumes and turbidity are the preferred parameters for direct biomass determination. Of the chemical components of the biomass only NADH lends itself to continuous *in situ* monitoring.

Despite the effort and research invested in the above techniques and other suitable possibilities, such as enzyme electrodes, only gas phase analyses and physical/physicochemical properties of the culture are presently in widespread industrial use. Growth measurement techniques for industrial fermenters have been reviewed by Cooney (1979), Onken and Buchholz (1982) and Sukatsch and Nesemann (1977).

The suitability of the techniques described earlier for these four types of growth measurement is summarized in Table 5.

Table 5 Suitability of growth rate measurement techniques in specific situations

Technique	Mycelial growth	Growth in the presence of solids	Growth in mixed populations	Viable cells only	Industrial microbial growth
Enumeration:					
Microscopic	−	+/−	+	+/−	−
Automatic	−	−	+	+/−	+/−
Biological	−	+/−	+	+	−
Direct methods:					
Dry weight	+	−	−	−	−
PCV	+	−	−	−	+
Turbidity	−	−	−	−	+
Analyses:					
Elemental	+	−	−	−	−
Macromolecules	+	+/−	+/−	−	−
ATP, NADH	+	+	−	+	+/−
Gas phase	+	+	−	+	+
Liquid phase	+	+	+/−	+	+/−
Physical properties	+/−	+/−	−	+/−	+

17.4 SUMMARY

A variety of different techniques is available for microbial growth rate measurement. As the properties and metabolism of the biomass may change during growth, the various methods will not necessarily give the same values. Whilst some parameters such as dry weight always correlate well with the amount of biomass, others such as NADH measurement may show important deviations. In addition to the limitations of the possible methods, the type of microorganism, the properties of the medium and the restrictions, such as precision, on the measurement must be taken into account in the selection of a suitable technique.

ACKNOWLEDGEMENT

The author is grateful to Professor J. E. Bailey of Caltech and Professor C. L. Cooney of MIT for reading through the manuscript and to Ms P. Brachmann and BASF for their help in its preparation.

17.5 REFERENCES

Agar, D. W. and J. E. Bailey (1981). Continuous cultivation of fission yeast: analysis of single-cell protein synthesis kinetics. *Biotechnol. Bioeng.*, **23**, 2315–2331.

Agar, D. W. and J. E. Bailey (1982a). Measurements and models of synchronous growth of fission yeast induced by temperature oscillations. *Biotechnol. Bioeng.*, **24**, 217–236.

Agar, D. W. and J. E. Bailey (1982b). Cell cycle operation during batch growth of fission yeast populations. *Cytometry*, **3**, 123–128.

Bach, H. P., W. Woehrer and M. Roehr (1978). Continuous determination of ethanol during aerobic cultivation of yeasts. *Biotechnol. Bioeng.*, **20**, 799–807.

Bailey, J. E. and D. F. Ollis (1977). *Biochemical Engineering Fundamentals*. McGraw-Hill, New York.

Bailey, J. E., J. Fazel-Madjlessi, D. N. McQuitty and M. F. Gilbert (1979). Measuring microbial population dynamics. *Ann. N. Y. Acad. Sci.*, **326**, 7–16.

Bennett, E. O. and R. P. Williams (1957). A comparison of methods for determining bacterial mass with particular emphasis upon the use of total phosphorus. *Appl. Microbiol.*, **5**, 14–16.

Beyeler, W., A. Einsele and A. Fiechter (1981). On-line measurements of culture fluorescence: method and application. *Eur. J. Appl. Microbiol. Biotechnol.*, **13**, 10–14.

Bijkerk, A. H. E. and R. J. Hall (1977). A mechanistic model of the aerobic growth of *Saccharomyces cerevisiae*. *Biotechnol. Bioeng.*, **19**, 267–296.

Boyles, D. T. (1977). A rapid method for measuring specific growth rate of microorganisms. *Biotechnol. Bioeng.*, **19**, 297–300.

Boyles, D. T. (1978). Specific growth rate measurement in an oxygen electrode chamber. *Biotechnol. Bioeng.*, **20**, 1101–1104.

Calam, C. T. (1971). The evaluation of mycelial growth. In *Methods in Microbiology*, ed. J. R. Norris and D. W. Ribbons, vol. 1, chap. 16, pp. 567–591. Academic, New York.

Collins, C. H. and P. M. Lyne (1976). *Microbiological Methods*, 4th edn., pp. 194–210. Butterworths, London.

Concone, B. R. V., P. A. Doin and A. G. Pinto (1978). Some major deviations for biomass determination by indirect method and estimation based on alkali consumption. *J. Ferment. Technol.*, **56**, 59–63.

Cooney, C. L. (1979). Computer application in fermentation technology. Perspective. *Biotechnol. Bioeng. Symp.*, **9**, 1–11.

Cooney, C. L. (1982). Growth of microorganisms. In *Biotechnology*, ed. H.-J. Rehm and G. Reed, chap. 2, pp. 73–97. Verlag-Chemie, Weinheim.

Cooney, C. L. and D. I. C. Wang (1976). Transient response of *Enterobacter aerogenes* under a dual nutrient limitation in a chemostat. *Biotechnol. Bioeng.*, **18**, 189–198.

Cooney, C. L., D. I. C. Wang and R. I. Mateles (1969). Measurement of heat evolution and correlation with oxygen consumption during microbial growth. *Biotechnol. Bioeng.*, **11**, 269–281.

Cooney, C. L., H. Y. Wang and D. I. C. Wang (1977). Computer-aided material balancing for prediction of fermentation parameters. *Biotechnol. Bioeng.*, **19**, 55–67.

Dairaku, K. and T. Yamané (1979). Use of the porous Teflon tubing method to measure gaseous or volatile substances dissolved in fermentation liquids. *Biotechnol. Bioeng.*, **21**, 1671–1676.

Danielsson, B. (1982). The enzyme thermistor. *Appl. Biochem. Biotechnol.*, **7**, 127–134.

Dawes, E. A., D. J. McGill and M. Midgley (1971). Analysis of fermentation products. In *Methods in Microbiology*, ed. J. R. Norris and D. W. Ribbons, vol. 6A, chap. 3, pp. 53–215. Academic, New York.

Dean, A. C. R. and C. N. Hinshelwood (1966). *Growth, Function and Regulation in Bacterial Cells*, pp. 87–89. Oxford University Press, London.

El-Shazly, K. and R. E. Hungate (1966). Method for measuring diaminopimelic acid in total rumen contents and its application to the estimation of bacterial growth. *Appl. Microbiol.*, **14**, 27–30.

Enfors, S. O. and N. Molin (1978). Enzyme electrodes for fermentation control. *Process Biochem.*, **13** (2), 9–11, 24.

Fazel-Madjlessi, J. E. Bailey and D. N. McQuitty (1980). Flow microfluorometry measurements of multicomponent cell composition during batch bacterial growth. *Biotechnol. Bioeng.*, **22**, 457–462.

Forrest, W. W. (1971). Microcalorimetry. In *Methods in Microbiology*, ed. J. R. Norris and D. W. Ribbons, vol. 6B, chap. 10, pp. 285–318. Academic, New York.

Gavin, J. J. and D. P. Cummings (1973). Enumeration of microorganisms. In *Handbook of Microbiology*, ed. A. I. Laskin and L. Lechevalier, vol. 1 (*Organismic Microbiology*), pp. 661–670. Chemical Rubber Company Press, Cleveland, Ohio.

Gencer, M. A. and R. Mutharasan (1979). Determination of biomass concentration by capacitance measurement. *Biotechnol. Bioeng.*, **21**, 1097–1103.

Gerhardt, P. (1981). Diluents and biomass measurement. In *Manual of Methods for General Bacteriology*, ed. P. Gerhardt, chap. 25, pp. 504–507. American Society of Microbiology, Washington, DC.

Gilbert, M. F., D. N. McQuitty and J. E. Bailey (1978). Flow microfluorometry study of diauxic batch growth of *Saccharomyces cerevisiae. Appl. Environ. Microbiol.*, **36**, 615–617.

Guilbault, G. G. (1976). *Handbook of Enzymatic Methods of Analysis.* Dekker, New York.

Hadley, W. K. and G. Senyk (1975). Early detection of microbial metabolism and growth by measurement of electrical impedance. In *Microbiology — 1975*, ed. D. Schlessinger, pp. 12–21. American Society of Microbiology, Washington, DC.

Hanson, R. S. and J. A. Phillips (1981). Chemical composition. In *Manual of Methods for General Bacteriology*, ed. P. Gerhardt, chap. 17, pp. 328–364. American Society of Microbiology, Washington, DC.

Hashimoto, S., M. Fujita and R. A. Baccay (1982). Biomass determination in the anaerobic digestion of night soil. *J. Ferment. Technol.*, **60**, 51–54.

Hatch, R. T., C. Wilder and T. W. Cadman (1979). Analysis and control of mixed cultures. *Biotechnol. Bioeng. Symp.*, **9**, 25–39.

Hendy, N. A. and P. P. Gray (1979). Use of ATP as an indicator of biomass concentration in the *Trichoderma viride* fermentation. *Biotechnol. Bioeng.*, **21**, 153–156.

Herbert, D., P. J. Phipps and R. E. Strange (1971). Chemical analysis of microbial cells. In *Methods in Microbiology*, ed. J. R. Norris and D. W. Ribbons, vol. 5B, chap. 3, pp. 209–344. Academic, New York.

Hill, G. A. and C. W. Robinson (1974). Measurement of aerobic batch culture maximum specific growth rate and respiration coefficient using a dissolved oxygen probe. *Biotechnol. Bioeng.*, **16**, 531–538.

Hinegardner, R. T. (1971). An improved fluorometric assay for DNA. *Anal. Biochem.*, **39**, 197–201.

Horan, P. K. and L. L. Wheeless, Jr. (1977). Quantitative single cell analysis and sorting. *Science*, **198**, 149–157.

Hutter, K.-J. (1978). Possible applications of flow cytometry in brewing-biological experiments. II. Immunofluorescence. *Brauwissenschaft*, **31**, 287–292.

Hutter, K.-J. (1979). Applications of flow cytometry in brewing-biological experiments. III. Viability test. *Brauwissenschaft*, **32**, 13–16.

Hutter, K.-J. and H. E. Eipel (1979a). Microbial determinations by flow cytometry. *J. Gen. Microbiol.*, **113**, 369–375.

Hutter, K.-J. and H. E. Eipel (1979b). Simultaneous measurements of DNA and protein content of microorganisms by flow cytometry. *Eur. J. Appl. Microbiol. Biotechnol.*, **6**, 223–231.

Kell, D. B. (1980). The role of ion-selective electrodes in improving fermentation yields. *Process Biochem.*, **15** (1), 18–23, 29.

Kjaergaard, L. (1977). The redox potential: its use and control in biotechnology. *Adv. Biochem. Eng.*, **7**, 131–150.

Koch, A. L. (1981). Growth measurement. In *Manual of Methods for General Bacteriology*, ed. P. Gerhardt, chap. 11, pp. 179–207. American Society of Microbiology, Washington, DC.

Kubitschek, H. E. (1971). Counting and sizing microorganisms with the Coulter Counter. In *Methods in Microbiology*, ed. J. R. Norris and D. W. Ribbons, vol. 1, chap. 17, pp. 593–610. Academic, New York.

Lee, Y. H. (1981). Pulsed light probe for cell density measurement. *Biotechnol. Bioeng.*, **23**, 1903–1906.

Le Pecq, J.-B. and C. Paole (1967). A fluorescent complex between ethidium bromide and nucleic acids. *J. Mol. Biol.*, **27**, 87–106.

Luedeking, R. and E. L. Piret (1959). A kinetic study of the lactic acid fermentation. Batch process at controlled pH. *J. Biochem. Microbiol. Technol. Eng.*, **1**, 393–412.

Lumac Systems Inc. (1980). *Lumac rapid microbial tests: Food industry.* Technical leaflet, Lumac Systems Inc., Titusville, Florida.

MacLeod, R. A., M. Light, L. A. White and J. F. Currie (1966). Sensitive rapid detection method for viable bacterial cells. *Appl. Microbiol.*, **14**, 979–984.

Mallette, M. F. (1971). Evaluation of growth by physical and chemical methods. In *Methods in Microbiology*, ed. J. R. Norris and D. W. Ribbons, vol. 1, chap. 15, pp. 521–566. Academic, New York.

Mateles, R. I. (1971). Calculation of the oxygen required for cell production. *Biotechnol. Bioeng.*, **13**, 581–582.

Matsunaga, T., I. Karube and S. Suzuki (1979). Electrode system for the determination of microbial populations. *Appl. Environ. Microbiol.*, **37**, 117–121.

Mattiasson, B., P.-O. Larsson, L. Lindahl and P. Sahlin (1982). Vitamin analysis with the use of a yeast electrode. *Enzyme Microb. Technol.*, **4** (5), 153–157.

Melamed, M. R., P. F. Mullaney and M. L. Mendelsohn (1979). *Flow Cytometry and Sorting.* Wiley, New York.

Metz, H. (1981). Continuous turbidity measurement in bioreactors. *Chem. Tech. (Heidelberg)*, **10**, 691–696.

Meynell, G. G. and E. Meynell (1970). *Theory and Practice in Experimental Bacteriology*, pp. 8–34. Cambridge University Press, Cambridge.

Miller, L. F., M. S. Mabee, H. S. Gress and N. O. Jangaard (1978). An ATP bioluminescence method for the quantification of viable yeast for fermenter pitching. *J. Am. Soc. Brew. Chem.*, **36** (2), 59–62.

Monk, P. R. (1978). Microbial calorimetry as an analytical method. *Process Biochem.*, **13** (12), 4–5, 8.

Moreira, A. R., J. A. Phillips and A. E. Humphrey (1978). Method for determining the concentration of adsorbed protein and cell biomass in cellulose fermentations. *Biotechnol. Bioeng.*, **20**, 1501–1505.

Mosbach, K. and B. Danielsson (1981). Thermal bioanalyzers in flow streams. Enzyme thermistor devices. *Anal. Chem.*, **53**, 83A–84A, 86A, 89A–91A, 94A.

Mou, D. G. and C. L. Cooney (1976). Application of dynamic calorimetry for monitoring fermentation process. *Biotechnol. Bioeng.*, **18**, 1371–1392.

Nagai, S. (1979). Mass and energy balances for microbial kinetics. *Adv. Biochem. Eng.*, **11**, 49–83.

Nestaas, E., D. I. C. Wang, H. Suzuki and L. B. Evans (1981). A new sensor, the filtration probe, for quantitative characterization of penicillin fermentation. II. The monitor of mycelial growth. *Biotechnol. Bioeng.*, **23**, 2815–2824.

Nilsson, H., K. Mosbach, S. O. Enfors and N. Molin (1978). An enzyme electrode for measurement of penicillin fermentation broth: a step toward the application of enzyme electrodes in fermentation control. *Biotechnol. Bioeng.*, **20**, 527–539.

Onken, U. and R. Buchholz (1982). Measuring methods for fermentations under sterile conditions. *Chem.-Ing.-Tech.*, **54**, 581–590.

Perley, C. R., J. R. Swartz and C. L. Cooney (1979). Measurement of cell mass concentration with a continuous-flow viscometer. *Biotechnol. Bioeng.*, **21**, 519–523.

Pirt, S. J. (1975). *Principles of Microbe and Cell Cultivation*. Blackwell Scientific Publications, Oxford.

Postgate, J. R. (1971). Viable counts and viability. In *Methods in Microbiology*, ed. J. R. Norris and D. W. Ribbons, vol. 1, chap. 18, pp. 611–628. Academic, New York.

Pringle, J. R. and J.-R. Mor (1975). Methods for monitoring the growth of yeast cultures and for dealing with the clumping problem. In *Methods in Cell Biology*, ed. D. M. Prescott, vol. 11, chap. 7, pp. 131–168. Academic, New York.

Ramkrishna, D. (1978). Statistical models of cell populations. *Adv. Biochem. Eng.*, **11**, 1–47.

Ramsey, W. S., E. D. Nowlan, L. B. Simpson, R. A. Misseng and M. M. Takeguchi (1980). Applications of fluorophore-containing microbial growth media. *Appl. Environ. Microbiol.*, **39**, 372–375.

Reuss, M., M. Piehl and F. Wagner (1975). Application of mass spectrometry to the measurement of dissolved gases and volatile substances in fermentation. *Eur. J. Appl. Microbiol. Biotechnol.*, **1**, 323–325.

Reuss, M., R. P. Jefferis, III and J. Lehmann (1976). Application of an on-line system of coupled computers to fermentation modelling. *GBF Monogr. Ser.*, No. 3 (Workshop Comput. Appl. Ferment. Technol., 1976), 107–124.

Richards, J. C. S., A. C. Jason, G. Hobbs, D. M. Gibson and R. H. Christie (1978). Electronic measurement of bacterial growth. *J. Phys. E.*, **11**, 560–568.

Russell, W. J., S. R. Farling, G. C. Blanchard and E. A. Boling (1975). Interim review of microbial identification by microcalorimetry. In *Microbiology — 1975*, ed. D. Schlessinger, pp. 22–31. American Society of Microbiology, Washington, DC.

Siebert, D. and H. Hustede (1982). Citric acid fermentation — biotechnological problems and possibilities of computer control. *Chem.-Ing.-Tech.*, **54**, 659–669.

Stewart, P. R. (1975). Analytical methods for yeasts. In *Methods in Cell Biology*, vol. 12, chap. 8, pp. 111–147. Academic, New York.

Stouthamer, A. H. (1971). Determination and significance of molar growth yields. In *Methods in Microbiology*, ed. J. R. Norris and D. W. Ribbons, vol. 1, chap. 21, pp. 629–663. Academic, New York.

Sukatsch, D. A. and G. Nesemann (1977). Automatic parameter determination in industrial fermentations. *Chem. Tech. (Heidelberg)*, **6**, 261–265.

Thomanetz, E., A. Sperandio and D. Bardtke (1982). Is ATP a suitable biomass parameter for active sludges? II. Working methods. *GWF, Wasser-Abwasser*, **123**, 147–151.

Tseng, M. M.-C. and C. R. Phillips (1981). Mixed cultures: Commensalism and competition with *Proteus vulgaris* and *Saccharomyces cerevisiae*. *Biotechnol. Bioeng.*, **23**, 1639–1651.

Wang, H. Y., C. L. Cooney and D. I. C. Wang (1977). Computer-aided bakers' yeast fermentation. *Biotechnol. Bioeng.*, **19**, 69–86.

Wang, H. Y., C. L. Cooney and D. I. C. Wang (1979). On-line gas analysis for material balances and control. *Biotechnol. Bioeng. Symp.*, **9**, 13–23.

Weaver, J. C., C. R. Perley and C. L. Cooney (1980). Mass spectrometer monitoring of a yeast fermentation. *Enzyme Eng.*, **5**, 85–88.

Winqvist, F., B. Danielsson, I. Lundstroem, and K. Mosbach (1982). Use of hydrogen sensitive palladium–MOS materials in biochemical analysis. *Appl. Biochem. Biotechnol.*, **7**, 135–139.

Wood, D. A. (1979). A method for estimating biomass of *Agaricus bisporus* in a solid substrate, composted wheat straw. *Biotechnol. Lett.*, **1**, 255–260.

Wood, W. A. (1981). Physical methods. In *Manual of Methods for General Microbiology*, ed. P. Gerhardt, chap. 18, pp. 309–318. American Society of Microbiology, Washington, DC.

Yano, T., T. Kobayashi and S. Shimizu (1978). Silicone tubing sensor for detection of methanol. *J. Ferment. Technol.*, **56**, 421–427.

Yoshioka, T., H. Terai and Y. Saijo (1982). Growth kinetic studies of nitrifying bacteria by the immunofluorescent counting method. *J. Gen. Appl. Microbiol.*, **28**, 169–180.

Zabriskie, D. W. and A. E. Humphrey (1978a). Estimation of fermentation biomass concentration by measuring culture fluorescence. *Appl. Environ. Microbiol.*, **35**, 337–343.

Zabriskie, D. W. and A. E. Humphrey (1978b). Real time estimation of aerobic batch fermentation biomass concentration by component balancing. *AIChE J.*, **24**, 138–146.

Zabriskie, D. W. and A. E. Humphrey (1978c). Continuous dialysis for the on-line analysis of diffusible components in fermentation broth. *Biotechnol. Bioeng.*, **20**, 1295–1301.

18

Assay of Industrial Microbial Enzymes

B. S. MONTENECOURT
Lehigh University, Bethlehem, PA, USA
and
J. O. CARROLL and R. P. LANZILOTTA
NOVO Laboratories Inc., Wilton, CT, USA

18.1 INTRODUCTION

The concept of enzymatic catalysis in the production of microbial primary and secondary products, especially alcoholic beverages, has been known since the late 1800s. Rapid strides have been made in enzyme chemistry during the first half of the present century, including crystallization of many enzymes and the elucidation of the primary, secondary, tertiary and quaternary structures of others. Commercial exploitation of microbial enzymes began long before scientists had reached a full understanding of the nature and properties of enzymes. An example was the common practice of the application of animal dung in the bating of hides in leather manufacture. Animals were generally the sources for production of the earliest enzymes (*e.g.* rennin), but often these sources were expensive and unreliable. In the latter part of this century, industrial enzymologists have turned to microbes as a source of their enzymes. Microbes have a number of advantages over animal sources. Growth of microorganisms is extremely rapid and can usually be achieved on inexpensive substrates. In many cases the enzymes are produced as extracellular proteins, which can easily be recovered from the fermentation broth. The specificity of microbial enzymes is equal to or in some cases better than those of animals. Microbial enzymes often have considerably higher temperature and a wider range of pH optima. In this chapter we will discuss the assay procedures which are recommended for a number of industrially important enzymes.

Assays of industrially useful enzymes pose certain unique problems to the enzymologist which are rarely encountered in the study of other enzymes. Industrial enzymes are infrequently purified to homogeneity before commercial application. Thus, the enzyme preparation generally consists of several different enzymatic activities in addition to the one of interest. In many cases, these problems have been somewhat overcome by the selection of microbial strains which produce low yields of the unwanted enzymes. However, the fact that the enzymes are in the crude state precludes definitive analysis of their kinetic parameters. Purified enzymes, on the other hand, exhibit strikingly different kinetic characteristics than those of the crude enzyme.

The substrates for industrial enzymes pose a second important problem. Although model substrates have been used in many studies, the results with model substrates are rarely useful in predicting the activity of the enzyme on the true substrate. Most industrial enzymes, especially the hydrolytic ones, will have application in the hydrolysis of extremely heterogeneous material. Not only is the substrate heterogeneous, but it is constantly changing during enzymatic hydrolysis. This situation again precludes the identification of definitive kinetic parameters. In many cases (*e.g.* pectinases, proteases, cellulases and amylases) the substrate is insoluble and may require the synergistic activity of several different enzymes to produce the desired industrial products.

In order to appreciate the complexity of industrial enzyme assays, it is necessary to understand the basic philosophy of enzymatic analysis. It is often stated that a biochemist would feel more comfortable using his colleague's toothbrush than his enzyme assay. It is this philosophy that has led to the myriad of variations in assay systems.

A wide variety of approaches has been taken to devise enzyme assays. These include direct and indirect spectrophotometric assays, coupled assays (with a second enzyme), and those based on solubility, viscosity, polarimetry, chromatography, pH and radioactivity. Within these assay parameters either product formation or disappearance of the substrate may be measured. The assays can be further divided into physical methods (solubility and viscosity) and direct chemical measurements. The most common methods used involve the measurement of product formation. Since the starting value is zero, it is easy to obtain large differences in raw data readings and thus minimize error. Measurement of loss of substrate is used in some assays because it is easier to detect than the products formed. However, the actual measurements made reflect small differences in large numbers and often introduce errors.

Major factors which will affect the measurement of enzyme activities are pH, temperature, degree of conversion, substrate concentration, cofactors and ionic strength. A given enzyme is generally active over a fairly narrow pH range. Not only does pH affect the enzyme but it often also affects the substrate, a fact that is often overlooked in enzyme assays. An example is the effect of pH on hemoglobin as a substrate for proteases. This effect will be discussed in further detail later. A second important effect of pH involves enzyme stability. pH activity profiles and pH stability profiles are different properties of the enzyme. With industrial enzymes, the pH stability profile may be more important than the activity profile.

Temperature affects all chemical reactions and enzyme activity is no exception. The general rule of thumb is that for every increase of 10 °C there is a doubling of enzyme activity. The temperature optimum of a given enzyme is condition dependent and not a basic property of the enzyme. Thus, temperature stability is very dependent upon pH, ionic strength, substrate concentration and total protein concentration. Generally speaking, the purer the enzyme the less stable it will be to heat inactivation. It should also be noted that assay time will affect the observed temperature optimum and the longer the assay, the lower the observed temperature optimum. As with pH, temperature stability may be a more important factor for industrial enzymes than temperature optima.

The degree of reaction is an important consideration when dealing with industrial enzymes. In kinetic models, such as Michaelis–Menten and Haldane, the assumption is made that there is no change in the substrate concentration during the reaction and further that there is no reversal of the reaction. The kinetic models require measurement of the initial reaction velocity and the extent of the reaction should be less than 1% of the substrate converted to the product. The rate of the reaction is dependent upon the substrate concentration. The general rule of thumb is to devise the assay so that the substrate concentration is 20–100 times the K_m value.

18.1.1 Kinetic Considerations

Michaelis–Menten/Briggs–Haldane Model

This theory assumes that the enzyme E first combines with the substrate S to form a complex ES, and that after conversion of substrate to product this complex breaks down to form free enzyme and product (Lehninger, 1983).

$$\text{E} + \text{S} \underset{k_{-1}}{\overset{k_1}{\rightleftharpoons}} \text{ES} \underset{k_{-2}}{\overset{k_2}{\rightleftharpoons}} \text{E} + \text{P} \tag{1}$$

From this concept Michaelis and Menten proposed that

$$V_0 = \frac{V_{max}[S]}{K_m + [S]} \qquad (2)$$

where V_{max} ($= k_2[E_{total}]$) represents the maximum velocity for conversion of substrate and K_m (the Michaelis–Menten constant) $= (k_{-1} + k_2)/k_1$.

Equation (2) is important to assay development since it predicts that at low substrate concentration (*i.e.* $[S] \ll K_m$) the reaction velocity V_0 will be dependent on enzyme [E] and substrate [S] concentration. At high substrate concentration (*i.e.* $[S] \gg K_m$), the enzyme is completely saturated with substrate and the equation for reaction velocity simplifies to

$$V_0 = V_{max} = k_2[E_{total}] \qquad (3)$$

since $[S]/([S] + K_m) \approx 1$ (White *et al.*, 1978; Fersht, 1977). This is convenient since the reaction rate under these conditions is a function of enzyme concentration only. Linear 'standard curves' of reaction progress *versus* enzyme dose can now be readily compiled.

Many enzymes require metal ion cofactors and these cofactors affect either the activity, the stability or both. It should be noted as indicated below that the same enzyme from different microbial sources may require a different cofactor. In addition, the metal ion concentration must also be considered in assaying enzyme activities.

Finally, a few comments are necessary with respect to the expression of units of enzyme activity. The true measure of the biocatalytic activity of an enzyme is the cleavage or degradation of a unit of substrate per unit of time. In accordance with the International Union of Biochemistry (Enzyme Nomenclature, 1978), one standard International Unit (SI unit), nanokatal, is the amount of enzyme required to hydrolyze one nanomole of substrate per second. This is a straightforward calculation where the number of bonds hydrolyzed can be measured. However, when dealing with heterogeneous substrates, a number of chemically defined substrates persist and new substrates are being generated throughout the hydrolysis. Thus it has been the custom in dealing with industrial enzymes to define the unit of activity in terms of units of product formed per unit of time (μmoles or nanokatals of product min^{-1} or s^{-1}). The μmol min^{-1} is the unit most prevalent in the literature. Users of industrial enzymes will find that often manufacturers will have their own ways of expressing activity units. In many cases, these units are arbitrary and are simply a comparative basis for price structure. Their relationship to absolute units (nmol s^{-1}) usually requires considerable interpretation.

The enzymes which will be considered in this chapter are either commercially in use today or show clear future industrial potential. These include β-(1,3; 1,4)-glucanases, glucose isomerase, glucose oxidase, invertase, lactase, lipases, pectinases, penicillin acylase, proteases and rennet. We have endeavored to include, where possible, assay values of the important kinetic parameters for each of the enzymes and to bring forward in the discussion any possible pitfalls and recommendations.

18.2 AMYLASES

A number of recent reviews on amylases have covered assay procedures (Fogarty and Kelly, 1979a), bond specificity (French, 1981) and overall characteristics and industrial application (Aunstrup, 1978, 1980; Fogarty and Kelly, 1979b, 1980). Only a brief summary of their biochemical properties will be included here. The amylase family of enzymes is a diverse group of starch-degrading enzymes ubiquitous in the microbial plant and animal kingdoms. They consist of three main groups having exosplitting, endosplitting and debranching activities (Figure 1). In general, amylases hydrolyze the bonds between adjacent glucose units in the starch molecule. The products formed are a function of the enzymatic activity, some retaining the configuration of the substrate and others undergoing inversion of the anomeric carbon to the β configuration. Upon reaction, there is a rapid interconversion between α and β forms which ultimately reach equilibrium.

Within the family of microbial amylases, three, namely α-amylase, β-amylase and glucoamylase (amyloglucosidase), have found wide scale industrial application. All three enzymes form the basis of the starch-processing industry in that they yield sugars, syrups and dextrins. The products of hydrolysis subsequently are used in the manufacture of food products and beverages. α-Amylase is also used in the baking and textile industry and has even found application in products

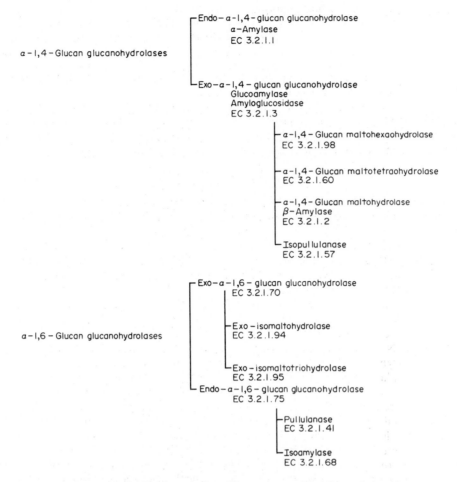

Figure 1 The family of starch-degrading enzymes and their interrelationship

involved with wallpaper removal. The debranching enzymes such as pullulanase and isoamylase have not to date found large scale industrial applications, but show great potential.

The substrate, starch, is a polysaccharide composed of two high molar mass components: amylose and amylopectin. These two polymers differ greatly in many physical properties. Amylose is a linear molecule of α-1,4-linked D-glucose units with a degree of polymerization of approximately 10^3 glucose units. Amylose, once solubilized in water, can retrograde, forming a highly insoluble substance due to aggregation and hydrogen bonding which is difficult to degrade. Amylopectin, on the other hand, is a highly branched molecule consisting of α-1,4-D-glucose chains with α-1,6 branchpoints arranged like a weeping willow. It has a degree of polymerization for the entire molecule on the order of 10^5 and an average chain length of 20 to 25 glucose moieties. Amylose molecules in solution exist primarily as random coils containing short, unstable helical segments. The helical form is stabilized by several complexing agents such as iodine. Amylose binds iodine to give a blue colored complex, with an iodine complex number (λ_{max} nm) of 650. Amylopectins and dextrins bind only a small amount of iodine (reddish colored complex) with a lower iodine complex number (λ_{max} nm) of 550. The iodine-binding capacity of a starch preparation can be used as an indication of its amylose to amylopectin ratio, which in turn is a function of the plant from which the starch granules are derived.

18.2.1 α-Amylases

18.2.1.1 Physicochemical properties

α-Amylase (endo-α-1,4-glucan 4-glucanohydrolase, EC 3.2.1.1) occurs widely in both bacteria and fungi. All α-amylases are endoacting enzymes which specifically cleave α-1,4 glucosidic linkages in amylose, amylopectin and glycogen, yielding sugars in the α-configuration. They are unable to hydrolyze the α-1,6 branchpoints in amylopectin, but are able to bypass this branchpoint. Hydrolysis of starch by α-amylases causes a rapid decrease in solution viscosity and iodine complex number. Two types of microbial α-amylases exist: saccharifying α-amylases and liquefying α-amylases. They are distinguished by the fact that saccharifying amylases produce an increase in reducing sugar about twice that of the liquefying α-amylase. The average molar mass of α-amylases is about 50 000 daltons. Larger values are reported for some bacterial enzymes containing subunits (*e.g. Bacillus* and *Bacteroides*). α-Amylases are generally stable within a pH range of pH 4.5–8.0, with an optimum activity between pH 4.8 and 6.5. The temperature range of these enzymes is between 35 and 90 °C. The presence of calcium stabilizes the enzyme to both pH and temperature. Many α-amylases are calcium metallo-enzymes containing at least one calcium atom per molecule of enzyme. Other important properties are summarized in Table 1.

Table 1 Physicochemical Properties of Important Microbial Amylases

	Endo-α-1,4-glucan glucanohydrolase, amylase (EC 3.2.1.1)	*Exo-α-1,4-glucan malthohydrolase, β-amylase (EC 3.2.1.2)*	*Exo-α-1,4-glucan glucanohydrolase, amyloglucosidase, glucoamylase (EC 3.2.1.3)*
Molar mass (dalton)	~50 000	35 000–160 000	50 000–112 000
pH optimum	3.5–7	5.5–7.5	4.0–6.0
Temperature optimum (°C)	35–90	37–55	40–70
Metal ion requirements	Ca^{2+}	—	—
Approximate K_m (mmol l^{-1})	1.0	0.2	18.5[a]
Specificity	α-1,4 linkages	α-1,4; cannot bypass α-1,6	α-1,4 linkages, α-1,6 linkages
Substrate	Amylose, amylopectin, glycogen	Amylose, amylopectin, glycogen (limited)	Amylose, amylopectin, glycogen, dextrins, maltose
Products	Maltodextrins	Maltose, β-limit dextrins	β-D-Glucose

[a] K_m for maltose as substrate.

The action pattern of α-amylases varies according to the microbial source of the enzyme and the experimental conditions, including chain length of the substrate, temperature, pH and the addition of stabilizing agents. A number of various models have been proposed for the interaction of α-amylase with the substrate, which include random-attack, preferred-attack, multiple- or repetitive-attack, and multiple-site (French, 1981). It has been proposed that the actual mechanism of attack may vary according to the experimental conditions and may even change during the course of hydrolysis under a standard set of conditions (Thoma, 1976).

The amylases of fungi and certain Gram-positive bacteria are reported to be glycoproteins with varying quantities of carbohydrate (Takahashi *et al.*, 1982). The presence of carbohydrate may enhance the stability of these enzymes to both pH and temperature.

18.2.1.2 Assays

α-Amylases are generally quantified by determination of the reducing groups formed during hydrolysis of certain types of commercially available soluble starch (*e.g.* Lintner, Merck Amylum and Zulkowsky). The simplest, but not always the most reliable, method is the dinitrosalicylic acid method as modified by Miller (1969). This assay, however, has been used with many variations. The following standard formulation is recommended: DNS reagent containing per liter: 10 g NaOH, 10 g dinitrosalicylic acid and 2 g phenol. This may be stored for up to 6 months in a brown bottle. Just before use, 50 mg of solid sodium metabisulfite per 100 ml of the DNS reagent is added. This reducing sugar assay is linear over a range from zero to about 9 mg of glucose provided that 3 ml of the aforementioned DNS reagent is used. It should be noted, however, that above 1 mg of reducing sugar the absorbance values (550 nm) may not fall within the reliable limits of the spectrophotometer and the sample will thus require dilution with water. This practice is frowned upon by biochemists, but is allowable with the DNS assay since the normal standard deviation is around $\pm 15\%$. The development of color in this spectrophotometric assay is achieved by boiling and the degree of color development is strongly dependent upon the length of time the samples are boiled. Different reducing sugars and reducing dextrins also produce variations in the color development (*e.g.* equimolar amounts of glucose and xylose will give different color intensities). Test results will vary depending upon the sugar solution used to prepare the standard curve. Moreover, the optimal boiling time will vary with the chain length of the dextrins.

Thus, it is advisable to determine experimentally the optimum boiling time and adhere closely to this time. After boiling, 1 ml of sodium potassium tartrate (40% w/v) is added to each tube to stabilize the color. The sequential addition of the components of DNS reagent yields more reproducible results than if a completely formulated reagent is employed. In addition, it has been shown that DNS does not give reducing values proportional to the reducing groups and that this variation may be due to differences in chain length of the oligosaccharides (Robyt and Whelan, 1972). A more suitable and definitive method is that of Somogyi–Nelson (Nelson, 1944; Somogyi, 1945) employing the alkaline copper reaction. This procedure is not dependent on chain length and gives a stoichiometric relationship between color development and the number of hemiacetal reducing groups present in the hydrolysate. It should be noted, however, that both of these methods are carried out under highly alkaline conditions, which result in some initial loss of reducing sugar (a total of about 40 μg for glucose analyzed by the DNS method). They are not recommended where a highly sensitive assay is required. The neocuproine reducing sugar assay (Dygert *et al.*, 1965) overcomes many of these problems since it is carried out under less alkaline conditions and is accurate in the range of 5–125 μg of reducing sugar.

The starch–iodine colorimetric method has been extensively employed for measurement of α-amylase activity (Street and Close, 1956; Wohlgemuth, 1908), especially in industrial situations. This method has undergone a number of modifications and the most commonly used modification is that of Sandstedt *et al.* (1939). The principle of this assay involves the interaction of the iodine with the amylose helix, as previously discussed. The development of color is dependent on chain length. Long chains give an intense blue color, oligosaccharides having a degree of polymerization (DP) of 10 to 20 yield a red color and smaller dextrins result in a colorless solution. The assay usually measures the time in which a starch substrate is degraded to give a specific color with iodine and this then is compared with a standard. One advantage of this assay is that it can specifically analyze for the liquefying enzyme activity and is relatively independent of the levels of exo-amylases.

A wide variety of viscometric assays also have been employed, by measuring either the loss of viscosity or the time required to reach a specified viscosity. The method described by Hagberg (1960) is the most commonly employed. These methods may be used to determine if an unknown enzymatic activity utilizes an exo- or an endo-splitting mechanism and are unaffected by the presence of low molar mass sugars and dextrins.

Several colorimetric starch substrates have been prepared and employed in assay procedures. A number of dyed amyloses, amylopectins and starches are utilized (Babson *et al.*, 1970; Ceska *et al.*, 1969; Frindhandler *et al.*, 1971; Kennedy, 1974; Klein *et al.*, 1969). The procedures are based on release of soluble dye, which can easily be separated from the undigested insoluble material and quantified spectrophotometrically. One disadvantage of these methods is that the bulky dyes may interfere with the natural crystallization and recrystallization of the substrate.

A powerful analytical tool for the measurement of α-amylases is the use of special HPLC columns which employ either acetonitrile/water or water solvent systems (Conrad and Fallick, 1974; Conrad and Palmer, 1976). These methods are capable of rapidly separating and quantifying starch hydrolysis products having a DP of between 1 and 15. These HPLC columns are now commercially available.

18.2.2 β-Amylases

18.2.2.1 *Physicochemical properties*

β-Amylase (α-1,4-glucan maltohydrolase; EC 3.2.1.2) is a saccharifying amylase that is widely distributed in plants and also has recently been found to be widespread in the microbial world. β-Amylases are exosplitting enzymes which attack amylose, amylopectin and glycogen from the nonreducing termini, resulting in the formation of maltose in the β-configuration. β-Amylase is specific for α-1,4 linkages and is unable to bypass α-1,6 branchpoints.

This enzyme is generally inhibited by sulfhydryl reagents such as *p*-chloromercuribenzoate (Rowe and Weill, 1962), *N*-ethylmaleimide (Thoma, 1974) and by Schardinger dextrins (Thoma and Koshland, 1960). Microbial β-amylases have a molar mass range between 35 000 and 50 000 daltons (Fogarty and Kelly, 1979a, 1979b), with the exception of the enzymes from several *Bacillus* species which have considerably higher molar masses (Shinke *et al.*, 1975). A pH optimum of 5–7 has been reported for most microbial β-amylases and a temperature optimum of 45–60 °C (Fogarty and Kelly, 1979a, 1979b). β-Amylases attack amylose, amylopectin, starch and glycogen with respectively decreasing efficiency. Other microbial β-amylases have been described which produce maltodextrins of various chain length. The most prevalent product is maltotetraose (*e.g.* enzymes from *Pseudomonas stutzeri* and *Enterobacter aerogenes*).

18.2.2.2 Assays

Since the main products of β-amylase hydrolysis are the disaccharides β-maltose and β-limit dextrin, any of the reducing sugar assays described in Section 18.2.1.2 for α-amylase may be employed. The HPLC method is probably the most sensitive and accurate, since disaccharides are easily separated from maltodextrins and the largest maltodextrins encountered (malto-tetraose) are still quite soluble.

18.2.3 Amyloglucosidases

18.2.3.1 Physicochemical properties

Amyloglucosidases (EC 3.2.1.3; glucoamylases) are exosplitting amylolytic enzymes which attack amylose, amylopectin and glycogen. They specifically hydrolyze α-1,4 and α-1,6 linkages from the non-reducing termini; however, the α-1,6 glucosidic linkages are cleaved less readily than the α-1,4 glucosidic linkages. The product of the reaction is β-D-glucose. The rates of hydrolysis are closely linked to the degree of polymerization and branching of the substrate. Although amyloglucosidases are widespread in higher eukaryotic organisms, in the microbial world they appear to be almost exclusively found in fungi. Amyloglucosidases vary in molar mass from 50 000 to 110 000 daltons. Smaller enzymes from *Aspergillus oryzae* and *Mucor rouxianus* have been reported (Miah and Yeda, 1977a, 1977b; Tsuboi *et al.*, 1974). In general, these enzymes are all glycoproteins with carbohydrate content varying between 15 and 22%. It is thought that the carbohydrate is linked through the *O*-glycosyl linkage of mannose to serine and threonine (Lineback and Aira, 1972).

Amyloglucosidases have an acidic pH optimum of 4.0 to 5.0, as one might expect for an extra-cellular fungal enzyme. The temperature optimum of the enzymes from mesophilic fungi is in the range 40 to 60 °C. Enzymes from the more thermophilic species (*e.g. Humicola*) have higher temperature optima. The reaction rates will vary widely with the DP of the substrate.

18.2.3.2 Assays

Since the major product of amyloglucosidase activity is glucose, any major spectrophotometric assay which is specific for glucose may be employed. The most commonly used method is the glucose oxidase/peroxidase assay (Hugget and Nixon, 1957). Other methods include the use of the dyes *o*-toluidine (Dobrick, 1958) and 2,6-dichlorophenolindophenol (Middleton, 1968). Glucose may, of course, be quantified by HPLC methods employing the appropriate standards or with a commercial glucose analyzer.

Substrates typically indicated for analysis of amyloglucosidases include soluble starches, such as those of Lintner and Zulkowsky, and more readily characterized substrates, such as maltose. Side activities in crude amyloglucosidase preparations (specifically α-amylase) can affect the hydrolysis of the starch substrates and perturb the assay results (unpublished observation).

Amyloglucosidases will also attack model substrates such as *p*-nitrophenyl-α-D-glucopyrano-side; however, as previously noted, the use of model substrates is not usually advisable.

18.2.4 Debranching Enzymes

18.2.4.1 Physicochemical properties

A number of debranching enzymes which exist in the microbial kingdom have been described. These include pullulanase, exopullulanase (both predominantly bacterial) and isoamylase (pre-dominantly fungal). These debranching enzymes are all specific for the α-1,6 linkage and are characterized by their ability to extensively degrade pullulan (*cf.* Section 18.2.4.2)

An additional class of debranching enzymes exists which can only attack a substrate which has been previously enzymatically modified. These are generally termed indirect debranching enzymes. The specificities of the various debranching enzymes have not as yet been precisely defined. Pullulanases have a pH range between 4.5 and 7.0 and a temperature optimum between 45 and 60 °C. A wide range of molar masses have been reported: 48 000 to 120 000 daltons (Fogarty and Kelly, 1979a). The main product of hydrolysis of pullulan is the trisaccharide, maltotriose.

Isoamylases, a second type of direct debranching enzyme, have been shown to hydrolyze α-1,6 branchpoints in glycogen and amylopectin, but are unable to hydrolyze pullulan (Lee and Whelan, 1971). Isoamylase, like pullulanase, has been most extensively studied in bacteria, although yeasts also appear to contain this enzyme. The pH optimum of isoamylases is between 4.0 and 6.5 and the temperature optimum between 40 and 55 °C (Fogarty and Kelly, 1979a). Molar masses between 50 000 and 100 000 daltons generally have been reported (Eisele *et al.*, 1972; Mercier *et al.*, 1972). The relative activity of the various debranching enzymes towards complex polysaccharide substrates is shown in Table 2.

Table 2 Action of Pullulanase and Isoamylase on Various Substrates

Substrate	Relative rate of hydrolysis	
	Isoamylase	Pullulanase
Pullulan	v. low	100
Amylopectin	100	15
Glycogen (oyster)	124	1
6^3-O-a-maltosyl-maltotriose	2.8	22
6^3-O-a-maltotriosyl-maltotriose	9.7	162
6^3-O-a-maltotriosyl-maltotetraose	33	146

18.2.4.2 *Assays*

Pullulanase can be analyzed by quantifying the reducing sugars released from pullulan (a polysaccharide of α-1,6-linked maltotriose units). Any reducing sugar assay is acceptable. The other debranching enzymes are unable to hydrolyze pullulan and are generally analyzed by measuring the increase in iodine staining number or increase in reducing sugar using amylopectin and glycogen as a substrate. Assay procedures for detecting indirect debranching enzymes will depend upon the specific pre-enzymatic modification required to activate the substrate.

18.2.5 α-Glucosidases

18.2.5.1 *Physicochemical properties*

α-Glucosidase (EC 3.2.1.20; α-D-glucosidase glucohydrolase; maltase) is ubiquitous in nature and catalyzes the specific release of α-D-glucose from the non-reducing end of α-1,4-linked glucose polymers. Substrates with a high degree of polymerization are rarely hydrolyzed and the enzyme shows a high degree of specificity for oligosaccharides with a degree of preference $G_2 > G_3 > G_4 > G_5$, *etc.* Thus, strictly speaking, maltase should not be considered a starch-degrading enzyme. Many α-glucosidases have a strict specificity for the glucose moiety and will attack substrates with an aglycone-substituted portion. α-Glucosidases may be either intracellular or extracellular and are capable of carrying out both hydrolytic and transglycosylating reactions. α-Glucosidases have a wide pH optimum of between 3.0 and 7.5 and an optimal temperature of between 40 and 60 °C (Fogarty and Kelly, 1979a). Fungal α-glucosidases tend to be proteins of higher molar mass (90 000–120 000 daltons) (Yamasaki *et al.*, 1977) than the bacterial enzymes (20 000–50 000 daltons) (Urlaub and Wober, 1978).

18.2.5.2 *Assays*

α-Glucosidases may be quantified by an assay procedure which will detect the end-product, glucose, following the hydrolysis of maltose. As in the case of cellobiase, an aryl substrate such as *p*-nitrophenyl-α-D-glucose also may be employed and the release of *p*-nitrophenol detected spectrophotometrically. The kinetics of hydrolysis of the various substrates will vary depending upon the source of the enzyme. Certain α-glucosidases show a greater specificity for one or the other substrates.

18.3 CATALASE

Catalase (hydrogen peroxide:hydrogen peroxide oxidoreductase; EC 1.11.1.6) catalyzes the splitting of hydrogen peroxide to molecular oxygen and water. Hydrogen peroxide addition is becoming an important method of cold sterilization of precursors of food products, in addition to its use as a bleach (textiles, paper pulp and hair products). The removal of excess H_2O_2 can be achieved by catalase treatment. Additional uses of the enzyme include the direct generation of O_2 for microbial growth, porous cement and foam rubber production. Catalase is also a common contaminant of microbial rennets. Catalases are widespread in aerobic microorganisms and generally function to protect the microorganism from the damaging effects of H_2O_2 formed when reduced flavoproteins or iron–sulfur proteins come together with O_2.

18.3.1 Physicochemical Properties of Catalase

Catalase has a molar mass of about 250 000 daltons (Herbert, 1955) and contains a heme prosthetic group; this prosthetic group is directly involved in activation of the substrate, H_2O_2. This interaction of enzyme and substrate forms the basis of the enzymatic analysis. The enzyme generally contains four subunits of equal size. It has a pH optimum around neutrality and a pI of 5.4.

18.3.2 Catalase Assays

The use of H_2O_2 as a substrate has been the long-standing method for catalase analysis. The disappearance of the peroxide is followed spectrophotometrically at 240 nm (Beers and Sizer, 1952, 1955). The reaction is first order and is fairly simple. However, this assay may not be used if the enzyme solution absorbs strongly between 230–250 nm.

Spectrophotometric assays employing H_2O_2 as a substrate have been reviewed (Aebi and Suter, 1969). Alternatives to the spectrophotometric measurement include the direct manometric determination of the O_2 evolved (Beers and Sizer, 1955), the indirect volumetric measure of the O_2 (Roeder, 1930), idiometric titration (Stern, 1932) and an O_2 electrode. An assay procedure employing perborate as a substrate has been described (Feinstein, 1949) and modified by Thomson and Klipfel (1957). However, these latter authors found the titrimetric and colorimetric methods for analysis of the residual substrate time-consuming and of doubtful accuracy. They consequently modified the perborate assay to allow spectrophotometric quantification by following the decrease in absorption at 220 nm (Thomson *et al.*, 1978). The reactions obey first order kinetics over the assay time.

18.4 CELLULASES

The assay of cellulases poses a unique enzymological nightmare. Not only is the natural substrate both insoluble and structurally variable, presenting the problem of an ill-defined concentration and chemical form, but the activity requires a multitude of endo- and exo-glucanases which must react synergistically in order to achieve the desired hydrolysis. The enzymes perform both hydrolysis and transglycosylating reactions and are subject to severe end product inhibition.

Microbial cellulases are ubiquitous in nature, active in every climate and temperature and responsible for recycling much of the carbon fixed annually into biomass. It is only recently that man has come to the realization that fossil fuels are in finite supply and that microbial cellulases could be harnessed as tools to unlock the carbon fixed in cellulose as a renewable energy resource. Thus, the major potential application of cellulases and hemicellulases is in their use in the hydrolysis of cellulosic and hemicellulosic biomass to simple sugars, both hexoses and pentoses, and in the food processing industry.

The major enzymes in the cellulase complex are outlined in Figure 2. Although many microorganisms are capable of degrading modified cellulosic substrates (*e.g.* carboxymethylcellulose), fewer organisms can effectively degrade natural crystalline cellulose. Those which contain a complete 'cellulase complex', including an efficient β-1,4-exoglucanase, generally are either of the genera Basidiomycetes or Deuteromycetes. However, in the last decade two thermophilic bacteria, *Thermomonospora fusca* and *Clostridium thermocellum*, have received considerable attention (Hägerdal *et al.*, 1978; Garcia-Martinez *et al.*, 1980). Both of these bacterial strains appear to possess a complete cellulase complex and are capable of rapid digestion of crystalline cellulose. In addition, as convenient methods of culturing rumen bacteria are developed, more effort will be placed on the study of these extremely efficient cellulolytic microbes. Since it is clear that rumen bacteria do degrade crystalline cellulose, a complete cellulase complex must be present in the rumen. Whether this complex is predominantly contributed by a single bacterial species or from a number of different bacteria is at present ill defined. Although the major emphasis to date has been placed on fungal cellulases, we will discuss assays for both fungal and bacterial cellulases since there is considerable potential for industrial use of the latter.

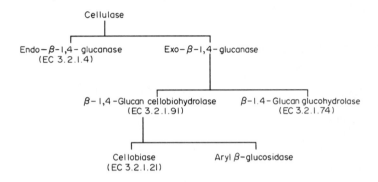

Figure 2 Enzymes of the cellulase complex

18.4.1 Physicochemical Properties of the Substrate

Whereas native cellulose is chemically homogeneous (containing only β-1,4-linked cellobiosyl moieties), it is extremely heterogeneous physically, since it contains areas of highly ordered hydrogen bonding (crystalline regions) and areas which are relatively disordered (amorphous regions). Generally speaking, the order in which the enzymes act synergistically are: initial attack by the endoglucanase in a random fashion on the amorphous regions, followed by sequential hydrolysis of the non-reducing end of the cellulose chain by cellobiohydrolase to yield cellodextrins and cellobiose. The synergistic action of these two enzymes serves to loosen the rigid hydrogen bonding between the parallel cellulose chains, resulting in a decrease in crystallinity of the cellulose. Since two enzymes must work in concert, the extent of hydrolysis is dependent upon the proportions of each of them in the cellulase preparations. A third exoglucanase has been reported in *Phanaerochaete chrysosporium* (Eriksson and Pettersson, 1975) and in *Penicillium funiculosum* (Wood and McCrae, 1982) which apparently is less specific and cleaves glycosidic linkages other than those broken by cellobiohydrolase.

18.4.2 Endo-β-1,4-glucanases

18.4.2.1 Physicochemical properties

The endoglucanases of fungal origin, which have been the most extensively studied, range in molar mass from 5300 to 55 000 daltons (Goksøyr and Eriksen, 1980; Lee and Fan, 1981; Ryu and Mandels, 1980) and in their neutral carbohydrate content from 0 to 44.7%. These enzymes catalyze the random splitting of internal β-1,4 glucosidic linkages and are highly active towards carboxymethylcellulose and phosphoric acid-swollen cellulose. Among the endoglucanases of fungi, a number of multiple forms have been described. An illustration of the multiplicity is the endoglucanases of *Trichoderma*, which have undergone extensive study (Shoemaker and Brown, 1978a, 1978b). Four electrophoretically distant glycoprotein endoglucanases have been purified and their substrate specificity determined. Endoglucanases I and II have a low affinity for cellulose and endoglucanases III and IV have a high affinity for cellulose. Endoglucanases II and IV effect a large change in specific fluidity in a fashion analogous to liquefying amylases, while endoglucanase III preferentially attacks short chain oligosaccharides (G_4–G_6) and is more like a saccharifying enzyme. The isoelectric points range from a pI of 3.4 to a pI of 5.0, depending upon the fungal origin. K_m values have been reported for only a few fungal enzymes. Individual enzymes from a commercial preparation of *T. viride* enzyme complex have been purified and their kinetic parameters determined (Shoemaker and Brown, 1978a, 1978b; Ladisch *et al.*, 1981). These are summarized in Table 3. Endoglucanase has been shown to hydrolyze cellobiose, cellotriose and cellotetraose as well as the soluble cellulose derivative carboxymethylcellulose and phosphoric acid-swollen cellulose. The hydrolysis of cellobiose is competitively inhibited by glucose and substrate inhibition has also been reported (Shoemaker and Brown, 1978a, 1978b). In addition to the hydrolytic properties of endoglucanase, synthesis of higher oligosaccharides from cellodextrins has been reported, suggesting a synthetic as well as a hydrolytic activity (Shoemaker and Brown, 1978a, 1978b).

Table 3   Kinetic Parameters for *Trichoderma* β-1,4-Glucanohydrolase

Substrate	V_{max}[a] (mmol min^{-1} mg^{-1} protein)	K_m[b] (mmol l^{-1})	K_i[a,b] (mmol l^{-1})
Cellobiose	0.498–0.58	1.0–1.6	0.98
Cellotriose	1.2	0.34	—
Cellotetraose	14–16	1–2	—

[a] Data from Shoemaker and Brown (1978a, 1978b) and Ladisch *et al.* (1981).   [b] K_i denotes the inhibition constant which quantitatively expresses the affinity of the inhibitor, I, for the enzyme. For competitive inhibition, the concentration of inhibitor [I] affects K_m and the rate expression is modified to:

$$v = \frac{V_{max}[S]}{[S] + K_m(1 + [I]/K_i)}$$

18.4.2.2   Assays

The most widely used assay for endoglucanase is that described by Mandels *et al.* (1976) which employs carboxymethylcellulose as a substrate. The major controversy over this assay surrounds the extent of hydrolysis of the substrate. Mandels *et al.* (1976) recommend that hydrolysis be allowed to proceed until 0.5 mg of reducing sugar has been released, which represents 5% hydrolysis of the substrate. On the other hand, other workers (*e.g.* Halliwell and Riaz, 1970) recommend that no more than 30 μg of reducing sugar be evolved from roughly the same initial substrate concentration; this is about 0.2% hydrolysis. One critical factor in this assay is the precise chemical composition of the substrate. Carboxymethylcellulose is available in a variety of forms of high, low or medium viscosity, which is a function of its degree of polymerization, and with a high, low or medium degree of substitution, which is an indication of the percentage of the glucose molecules which have been modified by carboxymethylation. Endoglucanases prefer long chain polymers (high DP). However, they only recognize β-1,4 glycosidic linkages between unsubstituted glucose moieties. The choice of the substrate will profoundly affect the final degree of hydrolysis and ultimately the number of activity units reported. If a highly substituted carboxymethylcellulose is employed, a range of enzyme concentrations should be tested to insure that the substrate is not limiting. As with the amylases, product formation is usually measured by a reducing sugar assay: dinitrosalicylic acid (Miller, 1969), Nelson–Somogyi (Somogyi, 1945) or neocuproine (Dygert *et al.*, 1965). The same precautions outlined for the amylase assay (*cf.* Section 18.2.1.2) should be followed with cellulases.

Other assay procedures have been described which measure the change in viscosity of carboxymethylcellulose (Shoemaker and Brown, 1978a, 1978b) or hydroxyethylcellulose (Nummi *et al.*, 1980). However, viscometric analyses are tedious and time consuming and in order to define the number of chemical bonds broken, detailed physicochemical methods involving light scattering and refractive index increments are required (Almin *et al.*, 1975).

18.4.3 β-1,4-Glucan Cellobiohydrolase

18.4.3.1 Specificity and kinetics

As is the case with endoglucanases, the cellobiohydrolases (EC 3.2.1.91) of *Trichoderma* have been the most extensively studied. The kinetic parameters for cellobiohydrolase are given in Table 4. Care should be taken in interpreting these data since recently two cellobiohydrolases, CBH I and CBH II, have been reported in *Trichoderma* (Fägerstam and Pettersson, 1979, 1980; Gritzali and Brown, 1979). These enzymes are immunologically distinct and CBH II appears to have three to four times higher specific activity. CBH I and CBH II have been shown to act synergistically with each other. It is not clear whether the kinetic data shown in Table 4 were obtained with a single CBH or a mixture of the two. Cellobiohydrolase is active on cellodextrins and fibrous cellulose, but not on cellobiose or CMC. Unlike endoglucanase, it does not appear to have a synthetic function. It is strongly end product inhibited by cellobiose and to a lesser extent by glucose. Crude preparations have been reported to exhibit substrate inhibition (Howell and Stuck, 1975).

Table 4   Kinetic Parameters for β-1,4-Glucan Cellobiohydrolase from *Trichoderma*[a]

Substrate	V_{max} (μmol min^{-1} mg^{-1} protein)	K_m (mmol l^{-1})	K_i glucose (mmol l^{-1})	K_i cellobiose (mmol l^{-1})
Cellobiose	No activity	—	—	—
Cellotriose	0.1	0.2	2.1	0.2
Cellotetraose	2.7	0.08	—	0.4
Fibrous a cellulose	0.0071	—	—	1.13

[a] Data from Hsu *et al.* (1980), Maguire (1977) and Ladisch *et al.* (1981).

18.4.3.2 Assays

Cellobiohydrolase (Figure 2) is a difficult enzyme to assay since its normal activity is dependent upon a synergism with endoglucanase. The most reliable assay is the quantification of cellobiose produced from Avicel (FMC Corporation, Princeton, NJ, USA) or filter paper (Whatman #1). If the enzyme is in its purified state, it will have very minimal activity against crystalline cellulose since its synergistic partner endoglucanase will be absent. A reducing sugar assay is usually not sensitive enough, with the possible exception of the neocuproine assay (Dygert *et al.*, 1965). The most reliable method for quantification of the cellobiose released is by high performance liquid chromatography, using either the method of Ladisch *et al.* (1978, 1979) or a commercial HPLC column (Aminex HPX-87, Bio-Rad Laboratories, Richmond, CA, USA). Both of these procedures use water as an eluant. Procedures employing acetonitrile/water mixtures are usually not sensitive enough due to the high refractive index background of the acetonitrile. An alternative substrate is phosphoric acid-swollen cellulose coupled with a reducing sugar assay or HPLC analysis. However, acid-swollen cellulose is not a reproducible substrate, and the enzyme must be absolutely pure as endoglucanases are also active on this substrate.

18.4.4 Cellobiase

18.4.4.1 Reactivity and kinetics

Cellobiase (EC 3.2.1.21), strictly speaking, is not part of the 'cellulase complex' since it does not attack cellulose. However, it is an important component if glucose is envisioned as the final product of cellulose hydrolysis. Values in the literature for the K_m for cellobiose vary between 1.5 and 2.7 mmol l^{-1} and the V_{max} between 33 and 185 mol min^{-1} mg^{-1} protein (Ladisch *et al.*, 1981). Cellobiase has a much higher apparent maximum velocity towards cellobiose than do the other cellulase enzymes. It is strongly non-competitively inhibited by glucose with a K_i value of 16.4 mmol l^{-1} (Ladisch *et al.*, 1977; Gong *et al.*, 1977). It also shows strong substrate inhibition (K_s = 31.5 mmol l^{-1}) (Emert *et al.*, 1974).

18.4.4.2 Assays

The assay procedures for cellobiase employ cellobiose as the substrate. Cellobiases will also attack model aryl substrates, such as salicin and *p*-nitrophenyl-β-D-glucoside. It should be noted, however, that many microorganisms possess a specific aryl β-glucosidase activity which will not act on cellobiose. Thus, if an assay procedure is to be used to predict the potential industrial usefulness of an enzyme, it should employ the appropriate substrate for this application. When cellobiose is used as a substrate, the glucose produced may be determined either colorimetrically with the glucose oxidase procedure (Huggett and Nixon, 1957) or by high performance liquid chromatography.

18.4.5 Assays for the Total Cellulase Complex

Since the cellulase complex represents a group of enzymes which must act in concert to achieve the desired hydrolysis of cellulose, a suitable assay system is needed which will allow the investigator to predict the efficiency of this complex to act on a true crystalline cellulosic material. A number of assays have been developed for this purpose, but the filter paper assay of Mandels and co-workers (1976) is the most widely employed. This assay, although far from ideal, represents a workable system which, in theory, employs a substrate containing at least some homologous regions of crystalline and amorphous cellulose compatible with an industrial substrate. Dyed Avicel substrates have also been used (Ng and Zeikus, 1980), where the extent of hydrolysis is quantified spectrophotometrically by the release of the dye.

18.4.6 Discussion

For insoluble cellulose, initial rates of hydrolysis are of little value because there is always some amorphous cellulose in the substrate and once this has been hydrolyzed the rates will fall off rapidly to zero if the cellulase complex being tested is incomplete. A cellulase assay suitable for industrial purposes must show the hydrolysis of both amorphous and crystalline areas of the cellulose. Since Whatman #1 filter paper is purported to be 95% α-cellulose, use of this as the assay substrate meets the criterion. Several modifications of the procedure have been described. A filter paper disc assay (Montenecourt *et al.*, 1978) has some promise because the antibiotic assay disks (740 E, Schleicher & Schuell, Keene, NH, USA) are easier to handle than the strips. They are also more difficult to hydrolyze, which alleviates problems with the cellulose fibers interfering with the dinitrosalicylic acid assay. This substrate meets the requirements of being a substrate in which there is a balance between amorphous and a crystalline cellulose. However, this method has not been universally adopted. Some laboratories have advocated the use of dyed Avicel as a substrate for assay of the cellulase complex (Leisola and Linko, 1976; Ng and Zeikus, 1980). However, dyed substrates represent a modification of the natural substrate and often will not predict accurately the ability of the cellulase complex to attack a crystalline cellulose.

In the late 1970s it was recognized by the Biotechnology Commission of IUPAC that some standardized procedures for the analysis of the cellulase enzymes were needed. As a result of the diversity of assay procedures being employed around the world, it was very difficult to compare and evaluate results from different laboratories. To address this problem, a document was prepared for IUPAC summarizing all of the major assay procedures (Ghose *et al.*, 1981). In addition, samples of a standard *Trichoderma* cellulase preparation, assay procedures and substrates were sent to about 25 international laboratories involved in cellulase research. The investigators were asked to carry out the analyses and report their results to Dr Mary Mandels at the US Army NRDC, Natick, MA, U.S.A. Roughly 75% of the laboratories responded and the assays described in detail here represent those which showed the least standard deviation among the different laboratories (M. Mandels, personal communication).

18.4.7 Bacterial Cellulase Assays

The assay procedures previously described were developed specifically for the enzymes of mesophilic fungi. Studies with cellulolytic bacteria suggest that prokaryotic cellulases may have somewhat different modes of action. Additionally, pH and temperature conditions for optimal cellulase activity will generally differ from most fungi. As a consequence, the assay procedures described will often need to be adapted to other cellulase systems by linearizing with respect to substrate, enzyme concentration and time and choosing the appropriate pH and temperature.

18.5 β-(1,3; 1,4)-GLUCANASES

18.5.1 Sources and Physicochemical Properties

β-Glucanase (1,3-(1,3; 1,4)-β-D-glucan 3(4)-glucanohydrolase; EC 3.2.1.6) enzymes are widely distributed in saprophytic bacteria and fungi. They are a heterogeneous group of enzymes with both exosplitting and endosplitting activity. In general, they attack β-glucans, lichenin and laminarin, although all microbial β-glucanases will not attack all of these substrates. β-Glucanases are important in the brewing and malting processes as they effect decreases in the viscosity of the barley β-glucan and facilitate filtration and extraction of the malt. They also may have potential application in the hydrolysis of yeast and fungal cell walls to increase the digestibility of this material for single cell protein.

Barley β-glucan contains four β-D-glucopyranosyl units linked by a 1,3 bond and two consecutive 1,4 bonds as the major structural unit (Parrish *et al.*, 1960). Some bacterial β-glucanases have been shown to have a binding affinity for the trimeric G4G3G position of the molecule (*i.e.* two β-1,4 linked glucose units joined by a β-1,3 bond to the next glucose moiety), hydrolyzing adjacent to the 1,4 linkages (Ducroo and Delecourt, 1972), and are unable to attack laminarin. Alternatively, some microbial laminarinases (β-1,3-glucanase) are unable to act against the mixed linkage (β-1,3; β-1,4)of barley β-glucan. Most commercial β-glucanases are from species of *A. niger* or *B. subtilis*.

The enzymes from bacteria have a molar mass of about 27 000 daltons, a pH optimum around neutrality (Suzuki and Kaneko, 1976) and a pH stability profile of pH 5–7. Their temperature optima are between 50 and 60 °C. Some bacterial enzymes are very specific and cleave only β-1,4 linkages adjacent to β-1,3 linkages in the polymers and are unable to attack polymers which contain exclusively β-1,4 linkages (cellulose) or β-1,3 linkages (laminarin) (Ducroo and Delecourt, 1972; Suzuki and Kaneko, 1976).

Both yeast and filamentous fungi have been shown to synthesize β-glucanases which are primarily exosplitting (Manners *et al.*, 1976). The fungal enzymes have a molar mass of about 43 000 daltons and split glucosyl residues from the non-reducing termini of β-1,3 glucan with inversion of the configuration of the anomeric carbon of glucose (Nagasaki *et al.*, 1977). They have a pH optimum between pH 5.0 and 6.0 and a temperature optimum around 50 to 60 °C. They are most active on the larger polymers with a K_m for laminarin of 66 mmol l^{-1} (assuming a molar mass of 4000 daltons for the laminarin). They show decreasing activity toward smaller polymers. In general, the yeast and fungal enzymes are glycoproteins (Lachance *et al.*, 1977; Notario *et al.*, 1976).

18.5.2 Assay Procedures

Until recently, barley β-glucan and laminarin were used exclusively for assay of β-glucanases. Activity was measured viscometrically or reductometrically (Enari and Markkanen, 1975; Fleet and Phaff, 1974; Manners and Wilson, 1974). Several problems are associated with these assays. Extraction of barley β-glucan is a tedious task which produces non-reproducible material. Laminarin is expensive, and both substrates tend to yield high reducing sugar backgrounds. On the other hand, viscometric assays with either substrate are non-reproducible primarily due to variations from batch to batch in the substrate.

Since lichenin has been shown to closely resemble barley β-glucan (Parrish *et al.*, 1960; Peat *et al.*, 1957) and is commercially available, methods have been developed to employ laminarin and lichenin as substrates with the neocuproine reducing sugar assay (Denault *et al.*, 1978). The high reducing sugar background of these substrates may be eliminated by sodium borohydride reduction. With the exo-β-1,3-glucanases, the products may be analyzed by HPLC. The use of sodium borohydride-reduced lichenin as a substrate now provides a sensitive assay for the endo-β-(1,3; 1,4)-glucanases (Denault *et al.*, 1978).

18.6 GLUCOSE ISOMERASE

18.6.1 Physicochemical Properties

Glucose isomerase (D-xylose ketol-isomerase; xylose isomerase; EC 5.3.1.5) has been reported in a wide variety of both Gram-negative and Gram-positive bacteria. This enzyme catalyzes the reversible isomerization of glucose to fructose. It also catalyzes the reversible isomerization of xylose to xylulose. Commercial enzymes are generally prepared from species of *Streptomyces* or *Bacillus*. Glucose isomerase has been extensively reviewed by Chen (1980a, 1980b). The molar mass of the enzyme varies from microorganism to microorganism, but is generally in the range of 120 000 to 175 000 daltons. The enzyme usually contains four subunits of approximately equal molar mass which are ionically associated (Hogue-Angeleti, 1975). No disulfide bonds have been detected. Most commercial enzymes contain Mn^{2+} or Co^{2+} in their active site and are activated by Mg^{2+}. Some enzymes require arsenate. The K_m for glucose using the soluble enzyme is in the range of 860–920 mmol l^{-1} and for xylose 5–9.3 mmol l^{-1} (Chen, 1980a, 1980b). Glucose isomerases have a fairly high temperature optimum (50–80 °C) and an optimal pH in the range of 7.0–9.0.

18.6.2 Assays with D-Glucose Substrate

A widely used assay for glucose isomerase is a polarimetric method. The optical activity of the product, fructose, is dramatically different from the optical activity of the substrate, glucose. The specific rotation for glucose is approximately +52.7° whereas that of fructose is −92° at 25 °C (Sober, 1973). As a consequence, fractional conversion curves can be established that will allow calculation of the percentage conversion of the substrate. The two sugars may also be determined by high performance liquid chromatography. A bench scale colorimetric assay can be used to measure fructose, *i.e.* the cysteine–carbazole reaction (Ashwell and Hickman, 1957). It should be noted that the cysteine–carbazole reaction is specific for keto sugars, but does not distinguish between the stereoisomers of the keto sugars.

The reader should note that all commercially available glucose isomerases are sold as immobilized enzymes since the isomerases are intracellular and therefore costly to produce. The analysis of these commercial products requires the use of column reactors (*cf.* Food Chemical Codex III, Food and Nutrition Board, 1981).

18.6.3 Assay with D-Xylose Substrate

In recent years, interest has been generated in the isomerization of xylose (a component produced by hydrolysis of the hemicellulose in biomass) to xylulose which then can be converted to ethanol by certain yeasts which cannot ferment D-xylose directly. Glucose isomerase is, in fact, xylose isomerase as indicated by the difference in K_m values (5.0–9.0 mmol l^{-1} for xylose and 860–920 mmol l^{-1} for glucose). The isomerization of xylose to xylulose can be measured polarimetrically, but the difference in optical rotation of the two isomers is not as great as for the glucose to fructose conversion ($[\alpha]_{25}^D$ values for xylose and xylulose are approximately 19° and 33°, respectively (Sober, 1973)). As a result, the polarimetric assay is less sensitive with pentose isomerization than with hexose isomerization. Xylose and xylulose may also be analyzed by HPLC, or spectrophotometrically with the ketose specific cysteine–carbazole assay (Ashwell and Hickman, 1957).

18.7 GLUCOSE OXIDASE

18.7.1 Sources and Physicochemical Properties

Glucose oxidase (EC 1.1.3.4; β-D-glucose: oxygen 1-oxidoreductase) catalyzes the oxidation of β-D-glucose to D-glucano-δ-lactone, which is immediately chemically converted to D-gluconic acid. The overall reaction is

$$Glucose + O_2 + H_2O \rightarrow gluconic\ acid + H_2O_2 \qquad (4)$$

The enzyme is commercially produced from species of *Aspergillus* and *Penicillium* and is widely used in clinical diagnosis as an indicator of carbohydrate metabolism. Immobilized glucose oxidase is used in commercial glucose analyzers for assay of glucose in whole blood, serum and plasma and has found recent application in analyzing syrups from starch and cellulose hydrolysis. The enzyme from *A. niger* contains 16% carbohydrate and also contains two flavin dinucleotide cofactors per molecule. The native enzyme has a molar mass of 160 000 daltons and consists of two polypeptide chains united by two disulfide bonds. Each subunit contains 1 atom of Fe^{2+} (O'Malley and Weaver, 1972). Glucose oxidase from fungal sources exhibits a maximal activity at pH 5.6 and at a temperature of 35–40 °C. The enzymes which have been studied in detail obey Michaelis–Menten kinetics with values for V_{max} of 16.95 mmol min^{-1} mg^{-1} protein and for K_m of 192.3 mmol l^{-1} (Gibson *et al.*, 1974). The only substrate for the enzyme is α-D-glucose in the pyranose ring form.

18.7.2 Assay Procedures

The assay system is based on the original observation of Keston (1956) that glucose oxidase can be coupled to peroxidase and chromogen. The reaction velocity is determined by an increase in absorbancy at 460 nm resulting from the oxidation of *o*-dianisidine through a peroxidase coupled system. The reaction is carried out in 0.1 mol l^{-1} potassium phosphate buffer, pH 6.0. *o*-Dianisidine (1%) stock solution is prepared in distilled water. Extreme care should be taken with *o*-dianisidine as the solid form has been reported to be carcinogenic (Toxic Substances Act of 1976, 45 Fed. Reg. 42, 854). Alternative chromogens are the ammonium salt of ABTS (2,2'-azinodi-3-ethylbenzothiazolinesulfonic acid) and a mixture of phenol and 4-aminophenazone (Trinder, 1969). Glucose solution (18% w/v) is allowed to stand overnight to allow mutarotation. A stock peroxidase solution (200 μg ml^{-1}) is prepared in distilled water. Just before use, an *o*-dianisidine buffer mixture is prepared by adding 0.1 ml *o*-dianisidine (1%) to 12 ml potassium phosphate buffer (0.1 mol l^{-1}, pH 6.0). This mixture is saturated with air. Activity is followed spectrophotometrically at 460 nm. This is a fairly straightforward assay since it involves a simple and homogeneous substrate, glucose. The only precautions are in the handling of the *o*-dianisidine and in some instances the possible need for a mutarotase to catalyze the conversion of glucose from the β to the α form.

18.8 INVERTASE

18.8.1 Sources and Physicochemical Properties

Invertase (β-D-fructofuranoside fructohydrolase; β-fructosidase; EC 3.2.1.26) is found as both an extracellular and an intracellular enzyme in a wide variety of fungi. Commercial sources of invertase are either yeast or *Aspergillus*. Invertase catalyzes the cleavage of the β-fructofuranosyl bond on the fructose side of sucrose, thereby liberating glucose in the α-D configuration. Raffinose (*O*-α-D-galactopyranoside-(1→6)-*O*-α-D-glucopyranosyl-(1→2)-β-D-fructofuranoside) and stachyose (*O*-α-D-galactopyranosyl-(1→2)-β-D-fructofuranoside) are also substrates for invertase. The enzyme shows specificity for β-fructofuranosyl moieties with unsubstituted terminal ends. It should be noted that α-D-glucosidase can hydrolyze sucrose, but is unable to hydrolyze raffinose. Care should be taken in analyzing crude enzyme extracts that the activity measured is in fact a true invertase and not an α-glucosidase. Invertase is used predominantly in the confectionery industry for the production of soft centered candies, fondant and chocolate coatings.

Yeast invertase is present in two forms, an intracellular enzyme which is carbohydrate-free and which has a molar mass of 135 000 daltons and an extracellular form with a molar mass of 270 000 daltons and which contains roughly 50% carbohydrate (Neumann and Lampen, 1967). The internal and external invertases have roughly the same K_m values (25–45 mmol l^{-1}, Gascon and Lampen, 1968). The internal and external invertases of *Neurospora* have also been shown to have similar K_m values (Tashiro and Trevithick, 1977). The pH optimum for the industrial enzyme (the external form) is pH 4.0 to 4.5, but the enzyme is both stable and active over a pH range of 3 to 7.5.

18.8.2 Assay Methods

Invertase is generally assayed by measuring the production of the products, glucose and fructose. The simplest method involves the colorimetric determination of glucose with glucose oxidase-peroxidase (Huggett and Nixon, 1957). The assays for glucose determination and invertase may be coupled since the temperature and pH optima of the two enzymes are similar. Glucose and fructose may also be analyzed by HPLC, or glucose may be determined by the use of a commercial glucose analyzer.

18.9 LACTASE (β-GALACTOSIDASE)

Lactase (galactohydrolase, β-galactosidase; EC 3.2.1.23), especially from *E. coli*, is one of the most studied enzymes to date. Molecular biologists and biochemists have even unlocked the key to control of expression of this enzyme and have utilized the lac-promoter to stimulate and control expression of activities cloned in recombinant plasmids. It is fortuitous that industrial enzymologists would ultimately find commercial application for this enzymatic activity. The enzyme is currently used to convert lactose in milk products to glucose and galactose. An increasing number of people are lactose intolerant and treatment of milk products with lactase allows them to digest the milk. An additional use is for the conversion of the lactose found in cheese whey to glucose and galactose, yielding a sweetened product which can be recycled as an additive to ice cream, egg nog, yogurt and other dairy products.

18.9.1 Physicochemical Properties

β-Galactosidase activity has been reported in both prokaryotes and eukaryotes. The molar mass range of bacterial enzymes is between 20 000 and 50 000 daltons. The bacterial lactases have a pH optimum around neutrality, whereas fungal lactases have a slightly more acidic pH optimum. The temperature optimum is between 50 and 70 °C. The K_m for the enzyme from *Bacilli* is on the order of 110 mmol l^{-1} (Goodman and Pederson, 1976), 20 mmol l^{-1} for the enzyme from *Streptomyces fragilis* (Uwajima *et al.*, 1972) and 64 mmol l^{-1} for *A. niger* (Hourigan, 1977). A new, more efficient lactase has been found in *Arthrobacter simplex*. This enzyme has a K_m value of 10.5 mmol l^{-1}. Mg^{2+}, K^+ and Mn^{2+} ions have been found to activate the enzyme and it appears sensitive to sulfhydryl reagents (Donelly *et al.*, 1977). *Kluyvermyces fragilis* is also a known source of industrially produced lactase (Novo Industri A/S). The enzyme from most microbial sources is inhibited by the product galactose with K_i values of 3 to 9 mmol l^{-1}. Lactase is an intracellular or periplasmic enzyme in bacteria, but is extracellular in many fungi.

18.9.2 Assays

The assay for lactase is fairly straightforward. Lactose is generally employed as the substrate and the products are detected colorimetrically with a glucose-specific assay (*e.g.* glucose oxidase-peroxidase assay, *cf.* Section 18.7.2), with a commercial glucose analyzer, or by HPLC. Chromogenic substrates may also be used, *e.g. o*-nitrophenyl-β-D-galactopyranoside. It should be noted, however, that the activity of the enzyme in milk may be quite different from its activity in dilute protein solutions under assay conditions. With the advent of commercial glucose analyzers which can detect glucose in the presence of turbid or colored solutions such as milk or blood, assay procedures employing the commercial substrate may become prevalent.

Furthermore, one complication which is noteworthy is that lactases characteristically produce reversion products, such as allolactose, which are in turn acted upon by the enzyme. The formation of these reversion products is concentration dependent.

18.10 LIPASES

18.10.1 Sources, Properties and Applications

Lipases (acylglycerol acylhydrolase; EC 3.1.1.3), unlike any of the enzymes discussed thus far, are not active in pure aqueous solutions. These enzymes hydrolyze esters of long-chain aliphatic acids from their glycerol backbone only at aqueous–lipid interfaces. Fat bodies, emulsion globules and lipoproteins serve as the stabilizers for lipases and have been termed the 'supersubstrate' (Brockerhoff and Jensen, 1974).

Lipases have current industrial use in the production of characteristic flavors of certain naturally ripened cheeses (*e.g.* blue cheese, cheddar, romano and mozzarella). As the biochemistry of lipases becomes better understood, the potential uses of lipases will grow. For example, the activation of milk lipoprotein lipase can result in the spoilage and loss of thousands of gallons of milk or milk products. On the other hand, addition of certain microbial lipases to processed milk products can result in a buttery flavor that greatly enhances the palatability of the product. Additional potential application of microbial lipases is in the field of antiatheromatic drugs and analysis of serum lipids.

The biochemistry of lipases is still in its infancy. Several recent reviews have contributed greatly to the basic knowledge of lipase specificity (Brockerhoff and Jensen, 1974; Jensen, 1983; Jensen *et al.*, 1983). In general, the specificity of a lipase is a function of the molecular properties of the enzyme, the nature of the substrate and the factors that affect binding of the enzyme and the substrate. Optimal association of the enzyme and the substrate presents an additional problem. For maximal activity, stable emulsions are required which provide maximum surface areas and adequate mixing. An additional problem arises from the belief that the free fatty acids which are the products of hydrolysis serve as end product inhibitors of the lipases. If emulsion conditions are not standardized, the free fatty acids are unable to reproducibly diffuse away from the enzyme–substrate complex.

Some of the major problems envisioned in the biochemistry of lipolysis are the vast variations in specificity. The same enzyme may show different rates of hydrolysis for monoglycerides, diglycerides and triglycerides. Moreover, there may be separate enzymes from the same source with independent specificities for the mono-, di- and tri-glycerides. The enzymes may show preference for the primary or secondary esters, or the action may be random. Additionally, certain enzymes may prefer certain fatty acid types and exhibit stereospecificity. In the light of these anomalies, it may be difficult to design an assay system which will cover the various specificities.

Kinetic constants for lipolytic enzymes are virtually meaningless. Although the enzymes are water soluble, the substrate is not. The problem is how to measure the concentration of an insoluble substrate consisting of an oil and water emulsion. It would seem that only those substrate molecules at the oil–water interface are available to the enzyme and the K_m value will vary as a function of the emulsion droplet size. The maximum velocity (V_{max}), therefore, is the only reasonably meaningful constant which can be applied to lipases and this value will vary widely depending on the substrate employed.

18.10.2 Assay Systems

It has been suggested that in the detection and assay of lipases the substrate and the reaction conditions be carefully chosen so as to meet the definition of a lipase (*i.e.* hydrolysis of long-chain acylglycerols at an oil–water interface) (Brockerhoff and Jensen, 1974). If the appropriate assay conditions are chosen, nonlipolytic esterases will not be detected.

The most universal substrate is trioleoglycerol (triolein) (18:1; 18:1; 18:1-triple esterified with oleic or 9-octadecanoic acid). This substrate meets the definition of being a long-chain acylglycerol. Olive oil may serve as a substitute for triolein since it contains 70% oleic acid esters and is considerably less expensive. Commercial olive oil, however, contains many impurities. These may be removed by activated charcoal, chromatography on magnesium silicate or by passage through neutral alumina in hexane/ethyl ether solvent (9:1) (Jensen, 1983). Tributyrylglycerol (tributyrin) (4:0, 4:0, 4:0) may also be used as a substrate, but it should be noted that this substrate does not meet the proposed criterion for the fundamental definition of the enzyme activity since the fatty acid chain is short. Although most lipases will hydrolyze this substrate, nonlipolytic esterase will also attack it. Emulsifiers such as sodium dodecylsulfate and bile salts are usually added to ensure maximum supersubstrate availability.

The pH optimum of most lipases is on the alkaline side, although acidic lipases have been reported for several fungi. Zwitterionic buffers of the Good series have been recommended (Good *et al.*, 1966; Brockerhoff and Jensen, 1974). Lipases may be active over a wide temperature range, 20–45 °C, as evidenced by the fact that they are active in large scale milk coolers.

The reaction of lipases is usually terminated by acidification of the reaction mixture which inactivates the lipases, converts soaps to free acids and prevents formation of emulsions during future extraction. Ca^{2+} and Mg^{2+} ions should be present during lipolysis to tie up the free fatty acids released and prevent fatty acid inhibition of the lipase. However, it is recommended that exposure to acid be brief and should be stopped at pH 4.0 as determined by an indicator: prolonged acid treatment will cause acyl migration of partial acylglycerols.

The release of free fatty acids from triolein or tributyrin can be detected quantitatively by a number of methods. Titration of the free fatty acids released is the simplest indication of lipase activity. The Ca^{2+} and Mg^{2+} ions must be sequestered by addition of a chelating agent such as EDTA before titration and the endpoint should be above pH 8.5. The Food Chemicals Codex (Food and Nutrition Board, 1981) provides a suitable industrial scale procedure. Colorimetry of Cu soaps of free fatty acids is a second method originally developed by Duncombe (1963) and later refined (Brunk and Swanson, 1981; Hron and Menahan, 1981; Shipe *et al.*, 1980). This assay is simple and sensitive and shows a good correlation coefficient with GLC analysis. Chromophoric substrates such as trinitrophenylaminolauric acid have been employed in lipase assays (Gatt *et al.*, 1981) and also provide a sensitive assay system.

The above assay procedures are adequate for rate studies and as an indication of the presence of lipases. However, they provide no knowledge of specificity. Gas liquid chromatography (Chapman, 1979) and high performance liquid chromatography (Payne-Wahl *et al.*, 1981) are the methods of choice for precise measurements of the products of lipolysis. (The reader is referred to Jensen (1983) for a comprehensive assessment of lipase assays.) Finally, it should be noted that lipases and esterases are two distinct classes of enzymes. It is beyond the scope of this overview to delineate the intricacies of differentiating between true lipases and true esterases.

18.11 PECTINASES

18.11.1 Uses and Properties

The pectinase family of enzymes includes both hydrolases and lyases (Figure 3). Pectin substances are structural polysaccharides found in the middle lamella in plant tissue. They consist of heteropolysaccharides which range in molar mass from 30 000 to 300 000 daltons. The main chain consists of partially methyl-esterified α-1,4-D-galacturonic acid. The demethylated compound is known as pectic acid or polygalacturonic acid. Side chains of rhamnose, galactose, xylose and arabinose may be present depending upon the plant source. The various microbial pectinases and their specificity are listed (Table 5). Several excellent reviews on pectinases have been published which cover the enzymology in greater detail (Fogarty and Ward, 1974; Kilara, 1982; MacMillan and Sheiman, 1974; Rombouts and Pilnik, 1980).

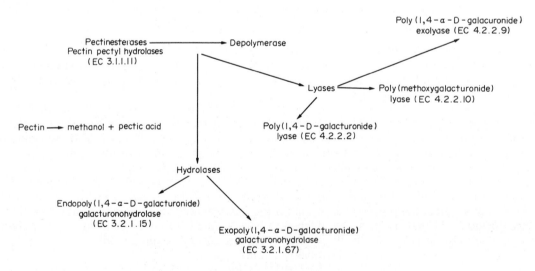

Figure 3 Pectin-degrading enzymes

Pectinases find application in the extraction and clarification of fruit juices and in the maceration and liquefaction of fruit and vegetable tissues in the preparation of nectar fruit juices. Pectinases are also used in the citrus industry to extract the maximum amount of soluble solids from the pulp and in the isolation of essential oils and carotenoid pigments from citrus peels. It is anticipated that the application of pectinases in the food industry will expand as they will allow use of a greater variety of raw materials and more complete recovery of the edible products. Assay procedures for pectinases have been reviewed in detail (Rexová-Benková and Marković, 1976).

Table 5 Microbial Pectinases and their Specificity[a,b]

Enzyme	Preferred substrate	Mode of attack	Products	Molar mass (dalton)	Optimum pH	K_m substrate (mg ml^{-1})
Hydrolases						
Endopolygalacturonase (EC 3.2.1.15)	Pectate	Random hydrolysis	G1, G2, G3	30 000–35 000	4.5–6.5	0.5–1.5
Exopolygalacturonase (EC 3.2.1.67)	Pectate	Exosplitting	G1			
Lyases						
Endopectate lyase (EC 4.2.2.2)	Pectate	Random, unsaturate bond between C-4 and C-5	Unsaturated oligogalacturonates	30 000–40 000	8.0–10.0	0.018–0.056
Exopectate lyase (EC 4.2.2.9)	Pectate	Exosplitting	Unsaturated dimers		9.00–10.0	0.9–5.0
Endopectin lyase	Pectin	Random	Low molar mass unsaturated oligogalacturonate methyl esters		6.0–8.0	0.9–5.0
Pectinesterase (EC 3.1.1.11)	Pectin	Deesterification	Methanol + Pectic acid	30 000–40 000	4.8–7.0	0.7

[a] Summarized from Rombouts and Pilnik (1980) and Atkinson and Mavituna (1983). [b] It should be noted that industrial pectinases contain many of the above enzymes, but not usually all.

18.11.2   Endopolygalacturonase Assays

Endopolygalacturonases may be measured either by quantification of the reducing groups liberated following enzyme attack or viscometrically. The preferred substrate for analysis is pectate. Reducing sugars may be determined by a number of colorimetric tests, but the method of Somogyi–Nelson (Somogyi, 1945) is recommended since methods which employ strong alkaline reaction conditions, such as the dinitrosalicylic acid method (Miller, 1969), result in high reducing sugar background due to β-elimination of the esterified substrate.

Viscometry is a very sensitive assay procedure for endopolygalacturonase. A 50% reduction in viscosity of a pectate solution corresponds to hydrolysis of only a few percent of the total glycosidic bonds (Rombouts and Pilnik, 1972). However viscometric analyses are tedious and large variations in reduction of specific viscosity may represent differences in the action patterns of the enzyme (random *versus* multichain attack) rather than the quantitative amount of the enzyme present.

18.11.3   Exopolygalacturonase

Exopolygalacturonases are specific for pectate. The most common assay systems quantify the reducing sugars formed as galacturonate monomers. A more specific determination has been described by Wagner and Hollmann (1976) which employs the enzymatic determination of galacturonate.

18.11.4   Endopectate Lyase

Endopectate lyases are also specific for pectate. They generally have a high pH optimum and show an absolute requirement for calcium. They catalyze the β-eliminative attack on the pectate substrate, which results in double bond formation between C-4 and C-5. Conjugation of the double bond with the carboxyl group at C-5 results in an absorption peak in the UV spectrum at 235 nm. A convenient spectrophotometric assay has been described by MacMillan and Phaff (1966).

18.11.5   Exopectate Lyase

Exopectate lyases show preferential attack of pectate over pectins and the polymethyl galacturonate–methylglucoside is not attacked by this enzyme. The exopectate lyases also have an unusually high pH optimum and in most cases show an absolute requirement for calcium. The enzyme attacks the pectate from the nonreducing end forming unsaturated dimers of galacturonate.

18.11.6   Endopectin Lyases

Endopectin lyases are predominantly produced by fungi. They show a wide pH optimum (5.2 to 8.7) (Rombouts and Pilnik, 1980) and are markedly stimulated by calcium. They preferentially attack highly esterified pectins and the products, which are unsaturated methylated oligogalacturonates, have a molar extinction coefficient of 5500 cm^2 $mmol^{-1}$ at 235 nm. The assay system of choice is the spectrophotometric method of Edstrom and Phaff (1964).

18.11.7   Pectinesterases

Pectinesterases preferentially act on pectin and specifically de-esterify pectin to form methanol and pectic acid. In general, pectinesterase prefers long chain pectin and the rate of hydrolysis decreases with decreasing chain length. The specificity of pectinesterase appears to be for the methyl ester group adjacent to free hydroxyl groups. Pectinesterases may be measured by titration of the free carboxyl groups which result from de-esterification (Rexová-Benková and Marković, 1976). A second method which is both sensitive and accurate measures the methanol released. This can be accomplished by converting the methanol into methyl nitrate and carrying out head space analysis using gas–liquid chromatography (Bartolome and Hoff, 1972). Alternatively, samples may be ether extracted and analyzed on commercially available GC columns. Analysis for the ethanol formed is the recommended method when dealing with acidic samples (*e.g.* fruit nectars), or when buffered samples are analyzed (Kilara, 1982).

18.11.8 The Pectinase Complex

Obviously all industrial preparations contain several of the above described enzymatic activities. The enzymes behave synergistically and indeed several activities are necessary to provide sufficient breakdown of the fruit pectin to facilitate juice recovery.

The activity determination of these crude preparations is complicated since the activity of interest, 'pectin degradation', is a composite of several enzymes. Significant variations in hydrolysis response can be expected with different substrates. The degree of esterification, for example, varies widely among the various botanical sources; apple pectin is on the high end with 70–85% esterification and grape pectin on the low end with about 45% (MacMillan and Sheiman, 1974; Pilnik and Voragen, 1970).

Although the monitoring of 'viscosity decrease with time' represents a reasonable approach to activity determination, this method suffers from the limitation that the results may not correlate well with juice clarification efficacy. In processes where the pectolytic enzyme mixture is being used for juice clarification, the extent of pectin degradation is generally followed by determining the quantity of 'isopropyl alcohol insoluble' solids. In the test, two parts acidified isopropyl alcohol are added to one part raw juice (unclarified) and the amount of sediment is noted. As the hydrolysis proceeds the amount of sediment decreases exponentially, following a time course similar to that of the viscosity reduction method. This value reaches 0% when the pectin has been degraded to a nominal chain length of roughly DP 10. This concept can be adapted as the basis for a convenient, application-oriented 'end point' assay (Cruess, 1958; Carroll, NOVO Lab., unpublished data).

18.12 PENICILLIN-MODIFYING ENZYMES

Enzymes which alter or degrade penicillin are produced by a wide variety of microorganisms, including fungi, yeast, actinomycetes and bacteria. Many of these enzymes irreversibly inactivate the penicillin moiety and are, therefore, not useful in the enzymatic modification of the antibiotic. However, two penicillin-modifying enzymes, penicillin acylase (Figure 4) and β-lactamase, do have industrial significance. Penicillin acylases are generally used to deacylate natural penicillins produced microbiologically to yield 6-aminopenicillanic acid (6-APA). This compound then serves as the starting nucleus for the synthesis of semisynthetic penicillins with superior clinical effectiveness. The second useful penicillin-modifying enzyme is β-lactamase. This enzyme cleaves the β-lactam ring to yield penicilloic acid. This enzyme is used clinically to determine penicillin levels in biological specimens and also in the detection of low levels of penicillin in consumer products such as milk which might otherwise cause allergic reactions. The latter enzyme is not employed industrially on a large scale and will not be considered here.

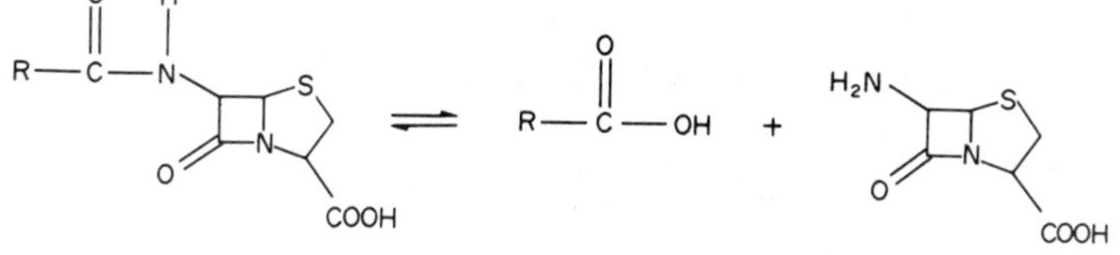

Figure 4 Penicillin hydrolysis (Lemke and Brannon, 1972)

18.12.1 Penicillin Acylases

Generally speaking there are three classes of penicillin acylases: phenoxymethyl penicillin acylase (penicillin V acylase), benzyl penicillin acylase (penicillin G acylase) and D-α-aminobenzylpenicillin acylase (ampicillin acylase).

Penicillin V acylase has been reported from a wide range of microbial sources. It is primarily an intracellular enzyme, although an extracellular enzyme has been isolated from certain *Streptomyces* (Vandamme, 1980). These enzymes are highly specific acylases with a pH optimum between pH 7 and 8 and a molar mass range from 65 000 to 80 000 daltons. In general, the temperature optimum is between 28 and 35 °C. However, some fungal enzymes have temperature optima around 50 °C (Schneider and Röhr, 1976; Waldschmidt-Leitz and Bretzel, 1964). The K_m varies between 2 and 17 mmol l^{-1}. This value varies widely depending upon whether the K_m value was determined with intact cells or with purified enzyme. The enzyme from *Erwinia* appears to have the highest K_m (35 mmol l^{-1}) (Vandamme and Voets, 1975). Inhibition kinetics (K_i) do not appear to have been determined for this activity.

Penicillin G acylase is also produced by a wide variety of bacteria and fungi. In contrast to the penicillin V acylase, penicillin G acylase has a broad substrate specificity and is considered to be an aspecific deacylating enzyme which will hydrolyze all phenylacetylated compounds. This enzyme is also predominantly an intracellular enzyme, although the enzymes from *Bacillus megaterium* are reported to be extracellular (Chaing and Bennett, 1967). The activity optimum is from pH 7 to 9 and at a temperature of 30 to 50 °C. The K_m shows a wide range (0.02 to 30 mmol l^{-1}), which illustrates the variations resulting from K_m measurements on intact cells, purified enzyme or immobilized enzyme. The molar mass of the enzyme is between 63 000 and 70 000 daltons for the intracellular enzyme and considerably larger for the extracellular enzyme from *Bacillus* (Acevedo and Cooney, 1973). Inhibition kinetics have been determined for penicillin G acylase. Phenylacetic acid acts as a competitive inhibitor ($K_i = 0.02$ mmol l^{-1}) and 6-APA as a noncompetitive inhibitor ($K_i = 1.5$ mmol l^{-1}). Many enzymes catalyze both the deacylating and the transacylase reaction, but the reverse reaction appears to proceed only under acidic pH conditions (pH 4.5–5.5).

Ampicillin acylase is reported to be produced primarily by pseudomonads, and many are absolutely specific for ampicillin. One strain has been shown to produce an enzyme which catalyzes the synthesis of ampicillin from hydrolysates of penicillin following addition of aminophenylacetic acid (Okachi *et al.*, 1973a, 1973b).

18.12.2   Assay Systems for Penicillin Acylase

All of the above penicillin acylases are amenable to the same assay systems since the product (6-APA) is the same for all of the enzymes. Assay systems for penicillin acylase suffer from the unstable nature of the 6-APA product. The most reliable methods depend on the separation of 6-APA by paper or thin layer chromatography of mixtures of penicillin and 6-APA. The 6-APA is then treated with phenylacetyl chloride and converted into penicillin G. The amount of penicillin G is then determined by bioauthographic methods (Cole *et al.*, 1975). Many qualitative thin layer chromatography methods have also been described, but they lack quantification. A colorimetric hydroxylamine method originally devised by Ford (1947) and Boxer and Everett (1949) for the detection of 6-APA may be employed, but it requires the extraction of residual penicillin at pH 2 with *n*-butyl acetate (Cole *et al.*, 1975).

Recently a new colorimetric method for 6-APA determination has been described (Kornfield, 1978). Penicillins do not interfere with this reaction. However, 6-aminopenicilloic acid does interfere. The method is based on the formation of a 2,4-pentanedione derivative of 6-APA followed by a second reaction with *p*-dimethylaminobenzaldehyde (Erlich's reagent), which results in a red product as a result of Schiff base formation which is measured at 538 nm. The absorbance response is linear between 0 and 350 g of 6-APA. Several other similar assay procedures have been described which were linear at much higher product ranges: 600 to 4000 μg ml^{-1} 6-APA (Bomstein and Evans, 1965; Saccani and Pitrolo, 1969). A rapid liquid chromatographic (HPLC) procedure is now being routinely used to monitor the products of penicillin acylase (Gestrelius, 1980). This method should overcome the problems of stability of 6-APA. Penicillin acylase activity is generally expressed as μmol of 6-APA min^{-1} under standard assay conditions.

18.13   PROTEASES

From an economical point of view, proteases are one of the most important industrial enzymes. A recent survey of the production figures for bacterial proteases alone indicated that 530 tons valued at US$66 million were produced and sold annually, worldwide (Eveleigh, 1981). These proteases are employed as biological detergents, as tanning agents and in the food industry (baking, brewing, cheese manufacturing, meat tenderizing). Proteases can be classified on a number of bases: the pH range over which they are active (acid, neutral or alkaline); the specificity of the substrate (keratinase, elastase, collagenase, *etc.*); or sometimes on their similarity to well-studied proteases, such as pepsin, trypsin or chymotrypsin. However, the classification of Hartley (1960) has been touted as the most satisfactory. This classification divides the endoproteases into four groups on the basis of their active site and sensitivity to various inhibitors: aspartic proteases (EC 3.4.23), metalloproteases (EC 3.4.24), serine proteases (EC 3.4.21) and cysteine proteases (EC 3.4.22). Some characteristics of the groups of proteases are *Bacillus licheniformis* (alkaline serine protease in detergents), *Mucor* (proteases which are utilized in the dairy industry) and *Aspergillus* (proteases which have application in the dairy industry and in baking). An additional use of the *Aspergillus* enzyme is in the solid substrate fermentation of soy sauce. However, in this case the enzyme is generally not purified, and is produced concurrently with the product (see Table 6).

Table 6 Major Microbial Proteolytic Enzymes

	Aspartic acid (EC 3.4.23)	Cysteine (EC 3.4.22)	Metallo (EC 3.4.24)	Serine (EC 3.4.21)
Major inhibitor	Pepstatin	Indoacetamide	Chelating agents	PMSF, DIFP[a]
Molar mass range (dalton)	30 000–45 000	34 000–35 000	19 000–37 000	18 000–35 000
pH optimum	3–5	2–3	5–7	6–11
Major source	*Aspergillus, Mucor, Endotria*	*Aspergillus*	*Bacillus, Aspergillus*	*Bacillus*
Metal ion requirement	Ca^{2+}	—	Zn^{2+}, Ca^{2+}	Ca^{2+}
Temperature optimum (°C)	40–55	40–55	65–85	50–70
Use	Dairy, baking, soy sauce	Brewing, soy sauce, baking	Brewing	Detergent

[a] DIFP, diisopropyl fluorophosphate; PMSF, phenylmethylsulfonyl fluoride.

18.13.1 Biochemical Properties

18.13.1.1 Serine proteases

Serine proteases represent the major enzymes employed in detergents. They are produced commercially from species of *Bacillus licheniformis* (Aunstrup, 1980), but have also been described for fungi of all major taxonomic groups (North, 1982). The active site of these alkaline proteases involves serine (221), histidine (64) and aspartic acid (32) residues. The amino acid sequences of two serine proteases from *Bacillus* have been determined (Markland and Smith, 1971; Aunstrup, 1980). The enzyme was found to be a single polypeptide chain of 274 amino acids. It lacked cysteine residues and carbohydrate and showed, in addition to its proteolytic activity, both transpeptidation and transesterification activity. The kinetic parameters of the esterase activity have been determined (Aunstrup, 1980). The physicochemical properties reported should be viewed with caution since this protease, as well as others discussed below, is subject to autodigestion and contamination with peptides which have adsorbed on to the enzyme.

Serine proteases will hydrolyze most peptide bonds and some ester bonds. Catalysis follows favorable collisions between the carbonyl carbon atom and the oxygen atom of the serine residue. A general base catalysis is followed by a general acid catalysis. The proton is then transferred from the imidazolium to the leaving group (Boyer, 1971). When synthetic amino acid esters are employed as substrates, those containing aromatic amino acids are preferentially attacked. Serine proteases are resistant to denaturing agents, 50% ethanol and anionic detergents. They are stable over a broad pH range (5.5 to 11.0), but are rapidly inactivated below pH 5 or above pH 11. Some serine proteases are stabilized by Ca^{2+}.

18.13.1.2 Metalloproteases

Metalloproteases are endopeptidases containing an essential metal ion, usually zinc and sometimes calcium. They have been most extensively studied in the genus *Bacillus* (Keay and Wildi,

1970; Keay *et al.*, 1972) and have also been found in *Penicillium* (Gripon *et al.*, 1980). The amino acid sequence of thermolysin, a metalloprotease of *Bacillus thermoproteolytius*, has been determined (Levy *et al.*, 1975) and found to contain 316 amino acids and to have a molar mass of about 35 000 daltons. Neutral metalloproteases preferentially hydrolyze the peptide bonds with hydrophobic side chains, such as phenylalanine and leucine. They have only weak esterase activity.

18.13.1.3 *Aspartic acid and cysteine proteases*

Aspartic acid and cysteine proteases are produced predominantly by fungi and are acid proteases with pH optima between pH 3 and 5. They have a molar mass between 30 000 to 45 000 daltons and contain either aspartic acid or cysteine in the active site (North, 1982). Some contain carbohydrate, for example the acid protease of *Mucor miehei* (Rickert and McBride-Warren, 1974), while others lack carbohydrate. Papain, ficin and bromelain are three fairly closely related nonmicrobial cysteine proteases which lack a glycoprotein portion. The most important acid proteases are those produced by species of *Aspergillus* and *Mucor* which are used in soy sauce production and in the milk coagulation step in cheese production.

18.13.2 Enzyme Analyses

A wide variety of assay procedures have been employed for the analysis of proteases. We have divided our discussion into procedures which employ complex substrates and those which utilize modified substrates. It should be noted that the various assays and substrates are interchangeable and that the assay conditions will depend upon the type of protease being analyzed.

18.13.2.1 *Complex substrate*

Proteases historically have been assayed using complex substrates such as casein and hemoglobin. Although model substrates such as amino acid methyl and ethyl esters have been employed for the assay of some proteases, these assays are most useful in studying proteases at the molecular level and are not informative for the industrial enzymologist.

The original casein assay was first described by Kunitz and later modified by Detmar and Vogels (1971). It was based on the photometric quantification of aromatic amino acids following trichloroacetic acid precipitation of the undigested substrate employing L-tyrosine as a standard. Generally speaking, Hammerstein casein is recommended (Nutritional Biochemicals, Cleveland, OH) as it represents the most standard casein preparation available. Both the temperature optimum and the pH optimum for maximal activity will vary according to the microbial source of the enzyme or the source being studied. It is thus advisable to carry out both pH and temperature curves to determine the individual requirements for the system. The reaction is terminated by addition of an equal volume of 10% (w/v) TCA which results in precipitation of the undigested casein. The soluble peptides released by action of the proteases may be determined in a number of ways. Many laboratories use a tyrosine standard curve since this amino acid absorbs strongly at 280 nm. Between 0 and 75 μg ml^{-1}, the absorbance of tyrosine follows Beer's Law. However, Keay and Wildi (1970) have shown that, if the soluble peptides are analyzed using the method of Lowry *et al.* (1951), higher protease values result. If extremely accurate measurements of protein hydrolysis are required, the amount of TCA-soluble nitrogen can be determined by carrying out Kjeldahl nitrogen determinations. It should be noted that all aromatic amino acids absorb at 275 nm. The content of these aromatic amino acids will vary from protein substrate to protein substrate and may even vary from batch to batch within the same substrate. A tyrosine standard curve may not adequately reflect the degree of proteolysis.

A similar method may be used employing denatured hemoglobin as a substrate. This is probably a better defined substrate since the complete amino acid sequence is known. A completely soluble hemoglobin is recommended (Worthington Biochemical Corp., Freehold, NJ). This method is sometimes called the HUT assay which stands for Hemoglobin Units Tyrosine. At the end of the incubation, the undigested hemoglobin is precipitated with an equal volume of 10% (w/v) TCA and filtered through Whatman #42 filter paper. The absorbance of the filtrate is determined at 275 nm and compared to that of a 0 to 75 μg ml^{-1} tyrosine standard curve. One HUT unit is defined as the amount of enzyme that produces per minute an absorbance at 275 nm equivalent to 1.10 μg ml^{-1} tyrosine (1.10 μg ml^{-1} tyrosine = 0.0084 O.D. units).

Of the two substrates discussed above, hemoglobin is the preferred one under acidic assay conditions; the HUT assay is typically carried out at pH 4.5. However, it can be used at higher pH values (>6.0), but a precipitate may form due to the low solubility of hemoglobin in the alkaline range. The pK of hemoglobin is 6.8 and the pK of casein 4.8. Both substrates are electrolytes and their solubility and secondary structure will vary as a function of pH.

18.13.2.2 Modified complex substrates

Casein and hemoglobin can be modified by reductive methylation and employed as substrates for protease assays. The advantage of the reductively methylated substrate is that the background reaction with the colorimetric reagent 2,4,6-trinitrobenzenesulfonic acid (TNBS) is substantially reduced. The dimethylated substrate may be purchased (Nutritional Biochemicals, Cleveland, OH) or prepared by the method as outlined below (Means and Feeney, 1968; Lin *et al.*, 1969). Hammerstein quality casein (Nutritional Biochemicals) (1.5 g) is dissolved in 0.1 M borate buffer (150 ml, pH 9.0) with warming and subsequently cooled to 0 °C. Sodium borohydride (300 mg) then is added with rapid stirring. A few drops of 2-octanol are added to reduce foaming. Formaldehyde then is added (3 ml) in increments of 0.1 ml over a period of 30 minutes. After addition of the formaldehyde, the solution is brought to pH 6 with 50% acetic acid and dialyzed against deionized water. The desalted proteins are lyophilized and stored at −20 °C. The reaction results in methylation of the ε-amino group of lysine. Reductive methylation of hemoglobin and gelatin may be carried out by the same procedure. At neutral and alkaline pH, the N,N-dimethylcasein may be used as a substrate. However, the casein is not soluble at extremely low pH values and thus dimethylhemoglobin is a better substrate.

The products of protease hydrolysis of reductively methylated protein substrates are generally determined by the method of Habeeb (1966) employing 2,4,6-trinitrobenzenesulfonic acid. The molar extinction coefficients of trinitrophenyl complexes vary from 1.1×10^4 to 1.5×10^4 depending upon the side chain, but most values range about 1.3×10^4. Using this latter value, results may be expressed as the number of primary amino groups present irrespective of the original peptide bond. The molar concentration of cleaved peptide bonds can be calculated using the following equation (Lin *et al.*, 1969):

$$[NH_2\text{-terminal}] = 7.69[10^{-5}][A_{340}] \tag{5}$$

Other types of modified substrates routinely used in protease assays are dyed proteins, azocoll, azocasein and dyed hide powder. Azocoll and dyed hide powder are insoluble substrates, while azocasein is soluble at alkaline pH and less soluble between pH 3.5 and 5.5. Those assays employing the insoluble substrates are sufficient for rough protease determinations, but unacceptable for precise kinetic analysis. Protease activity is monitored by analysis for the dye released after removal of the undigested substrate by filtrate or TCA precipitation. The absorbancy of the solubilized dye is determined at 520 nm. The unit is an arbitrary unit and to be meaningful must be standardized against some known purified protease.

More recently, a number of fluorescent assays for proteolytic activity have been described. The method formulated by Schwabe (1973) employs succinylated hemoglobin as a substrate. After reaction of the enzyme with the substrate for various periods of time, 1 ml of freshly prepared fluorescamine (0.1 mg ml^{-1} in acetone) is added to the assay tubes and the fluorescence emission measured at 475 nm. This assay system is reported to be 100 times more sensitive than the Lowry *et al.* (1951) method or the Anson (1938) method and does not involve a precipitation and separation step.

18.14 RENNET

18.14.1 Sources and Applications

Microbial rennet (EC 3.4.23.6) is an acid protease produced predominantly by fungi. It is considered here independently of the other acid proteases because the assay procedures for rennet are unique to this activity and not applicable to other proteases. Rennets are characterized by maximum activity and stability between pH 3.0 and 6.0. Their molar mass is around 35 000 daltons and the temperature optimum between 48 and 60 °C. The high temperature required to inactivate microbial rennets is a problem for the cheese industry. If the proteolytic activity is not inactivated, the increased hydrolytic activity of the rennet produces a bitter taste in the cheese and precludes the further use of the whey. On the other hand, complete inactivation would require raising the temperature of the milk above the normal temperatures for pasteurization. This problem has been largely overcome by *in vitro* modification of the enzyme to increase its heat lability. Microbial rennets have been reviewed in considerable detail (Sternberg, 1976). Microbial rennet is used in the manufacture of cheese to replace the dwindling supply of chymosin or rennin (EC 3.4.23.3) which historically was obtained from the fourth stomach of milk-fed calves.

The activity of interest is the ability to clot the $\varkappa$-casein in milk. Milk consists of three types of casein: α-, β- and $\varkappa$-casein. One of the roles of $\varkappa$-casein is to keep the casein micelles in solution and protect them against flocculation by calcium ions. The primary enzymatic reaction of microbial rennet is to break down the protective colloid ($\varkappa$-casein) and split off a glycomacropeptide. The micelle is now susceptible to coagulation by calcium ions. The 'ideal' rennet hydrolyzes only very specific bonds in the casein and is not a general hydrolytic enzyme. In general, rennets preferentially attack bonds between adjacent aromatic and hydrophobic amino acids producing macropeptides.

18.14.2 Assay of Microbial Rennet

Assay of clotting activity based on the time required to initiate clotting of milk has survived all trials to replace it. This assay suffers from lack of reproducibility, substrate standardization and variability of the end point observation. The assay seems to have survived because there is strict correlation to the actual performance of the enzyme in the cheese vat. The most common method is the visual observation of the formation of a clot or the sudden fracture of a film of milk on the wall of a bottle or a test tube (Berridge, 1952; Sommer and Matsen, 1935). A Formagraph instrument has recently been used to replace the visual observation with actual physical measurements based upon viscosity changes (McMahon and Brown, 1982) and an oscillatory deformation technique for measuring rigidity of the clots has also been described (Garnot and Olson, 1982).

Several other methods have been suggested. These include determination of the proteolytic breakdown of sodium caseinate under conditions similar to those of cheese-making (Clarke and Richards, 1973). A spectrophotometric assay has also been devised which measures the enzymatic activity on a synthetic hexapeptide which resembles a specific sequence in $\varkappa$-casein (Raymond *et al.*, 1972, 1973; Schatternkerk *et al.*, 1971). However, none of these assays provides information about the total activity of the rennet on $\varkappa$-casein. It should be noted that all acid proteases will effect coagulation of milk proteins. However, extensive nonspecific degradation of casein will result in the formation of bitter peptides and loss of the final yield of the cheese product. One method of determining whether the enzyme being assayed closely resembles a true rennet activity is to carry out a milk clotting assay and additionally a standard Anson (1939) assay which measures the proteolytic degradation of hemoglobin. If the ratio of rennet activity to Anson activity is high, then the enzyme is more like chymosin. If the ratio is low, nonspecific proteolysis is occurring and the enzyme would not be a good rennin substitute.

Several areas of caution concerning rennet assays are noteworthy. In the measurement of clotting time, the pH, Ca^{2+} concentrating and type of milk substrate will strongly affect rennet activity measurements. Nonfat dry skim milk is recommended as it has a fairly long shelf-life. Low heat skim milk should be used since heating will change the casein substrate. Whole milk has been used in the assay, but the presence of fat globules interferes with clotting observation. Whole milk also contains lipases which will act upon the fats to yield free fatty acids. These free fatty acids will in turn chelate the divalent cations (Ca^{2+}) and change the overall activity. The pH will also affect the activity. The assay is typically carried out at pH 6.3, but the pH of milk is pH 6.6 to 6.7. The pH will change depending upon how much $CaCl_2$ is added to the milk. Additionally, the mount of Ca^{2+} will affect the rennet activity. A high concentration of calcium can by itself clot milk.

18.15 NOMENCLATURE

DE   Dextrose equivalent, the total amount of reducing sugars expressed as dextrose as a
            percentage of total dry solids

DP	Degree of polymerization, the number of monomer units in an average polymer molecule
$[E]$	Enzyme concentration, mol l^{-1}
$[I]$	Inhibitor concentration, mol l^{-1}
K_i	Inhibition constant, mol l^{-1}
K_m	Michaelis–Menten constant, mol l^{-1}
K_s	Substrate inhibition constant, mol l^{-1}
k_1, k_2; k_{-1}, k_{-2}	Rate constant for forward and reverse reactions respectively
$[P]$	Product concentration, mol l^{-1}
pI	Isoelectric point
pK_a	Acid dissociation constant
$[S]$	Substrate concentration, mol l^{-1}
V_{max}	Maximal reaction rates, corresponding to the velocity when enzyme is completely saturated with substrate, mol min^{-1} mg^{-1} protein
V_0	Initial rate (assumes zero conversion of substrate to product), mol min^{-1} mg^{-1} protein

ACKNOWLEDGEMENT

The authors would like to thank R. G. Jensen, University of Connecticut, for making available his as yet unpublished reviews on lipases.

18.16 REFERENCES

Acevedo, F. and C. L. Cooney (1973). Penicillin amidase production by *Bacillus megaterium*. *Biotechnol. Bioeng.*, **15**, 493–503.

Aebi, H. and H. Suter (1969). Catalase (in blood). In *Biochemical Methods in Red Cell Genetics*, ed. J. J. Yuniss, pp. 255–288. Academic, New York.

Almin, K. E., K. E. Eriksson and L. G. Pettersson (1975). Extracellular enzyme system utilized by the fungus *Sporotrichum pulverulentum* (*Chrysosporium lignorum*) for the breakdown of cellulose. 2. Activities of the five endo-1,4-β-glucanases towards carboxymethylcellulose. *Eur. J. Biochem.*, **51**, 207–211.

Anson, M. L. (1938). Estimation of pepsin, trypsin, papain and cathespin with hemoglobin. *J. Gen. Physiol.*, **22**, 79–89.

Ashwell, G. and J. Hickman (1957). Enzymatic formation of xylulose 5-phosphate from ribose 5-phosphate in spleen. *J. Biol. Chem.*, **226**, 65–76.

Atkinson, B. and F. Mavituna (1983). *Biochemical Engineering and Biotechnology Handbook*, chap. 6. Nature Press, New York.

Aunstrup, K. (1978). Enzymes of industrial interest; traditional products. In *Annual Reports on Fermentation Processes*, ed. D. Perlman, vol. 2, pp. 125–154. Academic, New York.

Aunstrup, K. (1980). Proteinases. In *Microbial Enzymes and Bioconversions: Economic Microbiology*, ed. A. H. Rose, vol. 5, pp. 49–114. Academic, London.

Babson, A. L., S. A. Tenney and R. E. Megraw (1970). New amylase substrate and assay procedure. *Clin. Chem. Acta*, **16**, 39–43.

Bartolome, L. G. and J. E. Hoff (1972). Gas chromatographic methods for the assay of pectin methylesterase, free methanol and methoxy groups in plant tissues. *J. Agric. Food Chem.*, **20**, 262–266.

Beers, R. F., Jr. and I. W. Sizer (1952). A spectrophotometric method for measuring the breakdown of hydrogen peroxide by catalase. *J. Biol. Chem.*, **195**, 133–140.

Beers, R. F. and I. W. Sizer (1955). Catalase assay with special reference to manometric techniques. *Science*, **117**, 710–712.

Berghem, L. E. R., L. G. Pettersson and U. Axiö-Fredriksson (1975). The mechanism of enzymatic cellulose degradation. Characterization and enzymatic properties of a β-1,4 glucan cellobiohydrolase from *Trichoderma viride*. *Eur. J. Biochem.*, **53**, 55–62.

Berridge, N. J. (1952). Coagulation test apparatus. *Dairy Res.*, **19**, 328.

Bomstein, J. and W. G. Evans (1965). Automated colorimetric determination of 6-amino penicillanic acid in fermentation media. *Anal. Chem.*, **37**, 576–579.

Boxer, G. E. and P. M. Everett (1949). Colorimetric determination of benzyl penicillin. Colorimetric determination of total penicillins. *Anal. Chem.*, **21**, 670–673.

Boyer, P. D. (ed.) (1971). *The Enzymes*, Hydrolysis: Peptide Bonds, vol. 3. Academic, New York.

Brockerhoff, H. and R. G. Jensen (1974). *Lipolytic Enzymes*. Academic, London.

Brunk, S. D. and J. R. Swanson (1981). Colorimetric method for free fatty acids in serum validated by comparison with gas chromatography. *Clin. Chem.*, **27**, 924–926.

Ceska, M., K. Birath and B. Brown (1969). New and rapid method for the clinical determination of α-amylase activities in human serum and urine. Optimal conditions. *Clin. Chem. Acta*, **26**, 437–445.

Chapman, G. W., Jr. (1979). Gas chromatographic determination of free fatty acids in vegetable oils by a modified esterification procedure. *J. Am. Oil Chem. Soc.*, **56**, 77–79.

Chen, W.-P. (1980a). Glucose isomerase (a review). *Process Biochem.*, June/July, 30–36.
Chen, W.-P. (1980b). Glucose isomerase (a review). *Process Biochem.*, August/September, 36–41.
Chiang, C. and R. E. Bennett (1967). Purification and properties of penicillin amidase from *Bacillus megaterium*. *J. Bacteriol.*, **93**, 302–308.
Clarke, N. H. and E. L. Richards (1973). An assay for rennin. *N. Z. J. Dairy Sci. Technol.*, **8**, 152–155.
Cole, M., T. Savidge and H. Vanderhaeghe (1975). Penicillin acylase (assay). *Methods Enzymol.*, **43**, 698–705.
Conrad, E. C. and G. J. Fallick (1974). High pressure liquid chromatography (HPLC) and its application in the brewing laboratory. *Brewers Digest*, **49**, 72–74, 76, 78, 80.
Conrad, E. C. and J. K. Palmer (1976). Rapid analysis of carbohydrates by high pressure liquid chromatography. *Food Technol.*, **30** (10), 84, 86, 88–92.
Cruess, W. V. (1958). Pectin, jellies, and marmalades. In *Commercial Fruit and Vegetable Products*, ed. R. A. Brink, 4th edn., pp. 434–445. McGraw-Hill, New York.
Denault, L. J., W. G. Allen, E. W. Boyer, D. Collins, D. Kramme and J. E. Spradlin (1978). A simple reducing sugar assay for measuring β-glucanase activity in malt and various microbial enzyme preparations. *J. Am. Soc. Brew. Chem.*, **36**, 18–23.
Detmar, D. A. and R. J. Vogels (1971). Automated determination of protease for laundry compounds. *J. Am. Oil. Chem. Soc.*, **48**, 77–79.
Dobrick, L. J. (1958). Screening method for glucose of blood serum utilizing glucose oxidase and an indophenol indicator. *J. Biol. Chem.*, **231**, 403–409.
Donelly, W. J., I. N. Fhaolain and J. W. Patching (1977). Some properties of a multiple-form β-galactosidase from an *Arthrobacter* species. *Int. J. Biochem.*, **8**, 101–106.
Ducroo, P. and R. Delecourt (1972). Enzymic hydrolysis of barley β-glucans. *Wallerstein Lab. Commun.*, **35**, 219–228.
Duncombe, W. G. (1963). The colorimetric micro-determination of long-chain fatty acids. *Biochem. J.*, **88**, 7–10.
Dygert, S., L. H. Li, D. Florida and J. A. Thoma (1965). Determination of reducing sugar with improved precision. *Anal. Biochem.*, **13**, 367–374.
Edstrom, R. D. and H. J. Phaff (1964). Purification and properties of pectin trans-eliminase from *Aspergillus fonsecaeus*. *J. Biol. Chem.*, **239**, 2403–2408.
Eisele, B., I. R. Rasched and K. Wallenfels (1972). Molecular characterization of pullulanase from *Aerobacter aerogenes*. *Eur. J. Biochem.*, **26**, 62–67.
Emert, G. H., E. K. Gum, J. A. Lang, T. H. Liu and R. D. Brown, Jr. (1974). Cellulases. In *Food Related Enzymes*, ed. J. R. Whitaker, *Adv. Chem. Ser.*, **136**, pp. 79–100. American Chemical Society, Washington, DC.
Enari, T. M. and P. H. Markkanen (1975). Microbial β-glucanase in brewing. *Proc. Am. Soc. Brew. Chem.*, **33**, 13–17.
Enzyme Nomenclature (1978). Academic, London.
Eriksson, K.-E. and B. Pettersson (1975). Extracellular enzyme system utilized by the fungus *Sporotrichum pulverulentum (Chrysosporum lignorum)* for the breakdown of cellulose. 3. Purification and physico-chemical characterization of an exo-1,4-β-glucanase. *Eur. J. Biochem.*, **51**, 213–218.
Eveleigh, D. E. (1981). The microbial production of industrial chemicals. *Sci. Am.*, **245**, no. 3, 155–178.
Fägerstam, L. and L. G. Pettersson (1979). The cellulolytic complex of *Trichoderma reesei* QM9414. An immunochemical approach. *FEBS Lett.*, **98**, 363–367.
Fägerstam, L. G. and L. G. Pettersson (1980). The 1,4-β-glucan cellobiohydrolases of *Trichoderma reesei* QM9414, a new type of cellulolytic synergism. *FEBS Lett.*, **119**, 97–100.
Feinstein, R. N. (1949). Perborate as substrate in a new assay of catalase. *J. Biol. Chem.*, **180**, 1197–1202.
Fersht, A. (1977). *Enzyme Structure and Mechanism*, pp. 84–100. Freeman.
Fleet, G. H. and H. J. Phaff (1974). Lysis of yeast cell walls: glucanases from *Bacillus circulans* WL-12. *J. Bacteriol.*, **119**, 207–219.
Fogarty, W. M. and C. T. Kelly (1979a). Starch-degrading enzymes of microbial origin. *Prog. Ind. Microbiol.*, **15**, 87–150.
Fogarty, W. M. and C. T. Kelly (1979b). Developments in microbial extracellular enzymes. In *Topics in Enzyme and Fermentation Biotechnology*, ed. A. Wiseman, vol. 3, pp. 45–102. Halsted Press/Wiley, New York.
Fogarty, W. M. and C. T. Kelly (1980). Amylases, amyloglucosidases and related glucanases. In *Microbial Enzymes and Bioconversions: Economic Microbiology*, ed. A. H. Rose, vol. 5, pp. 115–170. Academic, London.
Fogarty, W. M. and O. P. Ward (1974). Pectinases and pectic polysaccharides. *Prog. Ind. Microbiol.*, **13**, 59–119.
Food and Nutrition Board (1981). *Food Chemical Codex*, 3rd edn., pp. 479–498.
Ford, J. H. (1947). Hydroxylamine method of determining penicillins. *Anal. Chem.*, **19**, 1004–1006.
French, D. (1981). Amylases: enzymatic mechanism. In *Trends in the Biology of Fermentations for Fuels and Chemicals*, ed. A. Hollaender, pp. 151–182. Plenum, New York.
Fridhandler, L., J. E. Berk and S. Take (1970). Applicability of a new synthetic dyed-labeled substrate for amylase assay. *Proc. Soc. Expl. Biol. Med.*, **133**, 1212–1216.
Garcia-Martinez, D. V., A. Shinmyo, A. Madia and A. L. Demain (1980). Studies on cellulase production by *Clostridium thermocellum*. *Eur. J. Appl. Microbiol. Biotechnol.*, **9**, 189–197.
Garnot, P. and N. F. Olson (1982). Use of oscillatory deformation technique to determine clotting times and rigidities of milk clotted with different concentrations of rennet. *J. Food Sci.*, **47**, 1912–1915.
Gascon, S. and J. O. Lampen (1968). Purification of the internal invertase of yeast. *J. Biol. Chem.*, **243**, 1567–1572.
Gatt, S., Y. Barenholz, R. Goldberg, T. Dinur, G. Besley, Z. Leibovitz-Ben-Gershaw, J. Rosenthal, R. G. Desnick, E. A. Devine, B. Shafit-Zagardo and F. Tsuruki (1981). Assay of enzymes of lipid metabolism with colored and fluorescent derivatives of natural lipids. *Methods Enzymol.*, **72**, 351–375.
Gestrelius, S. (1980). Immobilized penicillin acylase for production of 6-APA from penicillin-V. In *Enzyme Engineering*, vol. 5, ed. H. H. Weetal and G. P. Royer, pp. 439–442. Plenum, New York.
Ghose, T. K., B. S. Montenecourt and D. E. Eveleigh (1981). Measurement of cellulase activity. (Unpublished report of IUPAC Commission on Biotechnology.)
Gibson, Q. H., B. E. P. Swoboda and V. Massey (1974). Kinetics and mechanism of action of glucose oxidase. *J. Biol. Chem.*, **239**, 3927–3934.

Goksøyr, J. and J. Eriksen (1980). Cellulases. In *Microbial Enzymes and Bioconversions: Economic Microbiology*, ed. A. H. Rose, vol. 5, pp. 283–330. Academic, London.

Gong, C. S., M. R. Ladisch and G. T. Tsao (1977). Cellobiase from *Trichoderma viride*: purification, properties, kinetics, and mechanism. *Biotechnol. Bioeng.*, **19**, 959–981.

Good, N. E., G. D. Winget, W. Winter, T. N. Connolly, S. Izawa and R. R. N. Singh (1966). Hydrogen ion buffers for biological research. *Biochemistry*, **5**, 467–477.

Goodman, R. E. and D. M. Pederson (1976). β-Galactosidase from *Bacillus stearothermophilus*. *Can. J. Microbiol.*, **22**, 817–825.

Gripon, J.-C., B. Auberger and J. Lenoir (1980). Metalloproteases from *Penicillium caseicolum* and *P. roqueforti*: comparison of specificity and chemical characterization. *Int. J. Biochem.*, **12**, 451–455.

Gritzali, M. and R. D. Brown, Jr. (1979). The cellulase system of *Trichoderma*. *Adv. Chem. Ser.*, **181**, 237–260.

Habeeb, A. F. S. A. (1966). Determination of free amino groups in proteins by trinitrobenzenesulfonic acid. *Anal. Biochem.*, **14**, 328–336.

Hagberg, S. (1960). A rapid method for determining α-amylase activity. *Cereal Chem.*, **37**, 218–222.

Hägerdal, B. G. R., J. D. Ferchak and E. K. Pye (1978). Cellulolytic enzyme system of *Thermoactinomyces* sp. grown on microcrystalline cellulose. *Appl. Environ. Microbiol.*, **36**, 606–612.

Halliwell, G. and M. Riaz (1970). The formation of short fibres from native cellulose by components of *Trichoderma koningii* cellulase. *Biochem. J.*, **116**, 35–42.

Hartley, B. S. (1960). Proteolytic enzymes. *Annu. Rev. Biochem.*, **29**, 45–72.

Herbert, D. (1955). Catalase from bacteria. *Methods Enzymol.* **2**, 784–788.

Hogue-Angeletti, R. A. (1975). Subunit structure and amino acid composition of xylose isomerase from *Streptomyces albus*. *J. Biol. Chem.*, **250**, 7814–7818.

Hourigan, J. A. (1977). Kinetic studies of the enzymic hydrolysis of lactose. *Diss. Abstr.*, **37**, 6043.

Howell, J. A. and J. D. Stuck (1975). Kinetics of solka floc cellulose hydrolysis by *Trichoderma viride* cellulase. *Biotechnol. Bioeng.*, **17**, 873–893.

Hron, W. T. and L. A. Menahan (1981). A sensitive method for the determination of free fatty acids in plasma. *J. Lipid Res.*, **22**, 377–381.

Hsu, T. A., C. S. Gong and G. T. Tsao (1980). Kinetic studies of cellodextrins hydrolysis by exocellulase from *Trichoderma reesei*. *Biotechnol. Bioeng.*, **22**, 2305–2320.

Huggett, A. St. G. and D. A. Nixon (1957). Use of glucose oxidase, peroxidase, and *o*-dianisidine in determination of blood and urine glucose. *Lancet*, **273**, 368–370.

Jensen, R. G. (1983). Detection and determination of lipase (acylglycerol hydrolase) activity from various sources. *Lipids*, **18**, 650–657.

Jensen, R. G., F. A. Dejong and R. M. Clark (1983). Determination of lipase specificity. *Lipids*, **18**, 239–252.

Keay, L. and B. S. Wildi (1970). Proteases of the genus *Bacillus*, I. Neutral proteases. *Biotechnol. Bioeng.*, **12**, 179–212.

Keay, L., M. H. Mosley, R. G. Anderson, R. J. O'Connor and B. S. Wildi (1972). Production and isolation of microbial proteases. *Biotechnol. Bioeng. Symp.*, **3**, 63–92.

Kennedy, J. F. (1974). Chemically reactive derivatives of polysaccharides. *Adv. Carbohydr. Chem.*, **29**, 305–405.

Keston, A. S. (1956). Specific colorimetric enzyme reagents for glucose. Abstracts of Papers, 129th Meeting. American Chemical Society, Washington DC.

Kilara, A. (1982). Enzymes and their uses in the processed apple industry: a review. *Process Biochem.*, July/August., 35–41.

Klein, B., J. A. Foreman and R. L. Searcy (1969). The synthesis and utilization of cibachrom blue-amylose: a new chromogenic substrate for determination of amylase activity. *Anal. Biochem.*, **31**, 412–425.

Kornfield, J. M. (1978). A new colorimetric method for determination of 6-aminopenicillanic acid. *Anal. Biochem.*, **86**, 118–126.

Lachance, M. A., T. G. Villa and H. J. Phaff (1977). Purification and partial characterization of an exo-β-glucanase from the yeast *Kluyveromyces aestuarii*. *Can. J. Biochem.*, **55**, 1001–1006.

Ladisch, M. R., C. S. Gong and G. T. Tsao (1977). Corn crop residues as a potential source of single cell protein: kinetics of *Trichoderma viride* cellobiase action. *Dev. Ind. Microbiol.*, **18**, 157–168.

Ladisch, M. R., A. L. Huebner and G. T. Tsao (1978). High speed liquid chromatography of cellodextrins and other saccharide mixtures using water as the eluent. *J. Chromatogr.*, **147**, 185–194.

Ladisch, M. R., A. W. Anderson and G. T. Tsao (1979). Measurement of cellulolytic activity by low pressure liquid chromatography. *J. Liq. Chromatogr.*, **2**, 745–760.

Ladisch, M. R., J. Hong, M. Voloch and G. T. Tsao (1981). Cellulase kinetics. In *Trends in the Biology of Fermentations for Fuels and Chemicals*, ed. A. Hollaender, pp. 55–83. Plenum, New York.

Lee, E. Y. C. and W. J. Whelan (1971). Glycogen and starch debranching enzymes. In *The Enzymes*, 3rd edn., ed. P. D. Boyer, vol. 5. Academic, New York.

Lee, Y. H. and L. T. Fan (1981). Properties and mode of action of cellulase. *Adv. Biochem. Eng.*, **17**, 101–129.

Leisola, M. and M. Linko (1976). Determination of the solubilizing activity of a cellulase complex with dyed substrates. *Anal. Biochem.*, **70**, 592–599.

Lemke, P. M. and D. R. Brannon (1972). Microbial synthesis of cephalosporin and penicillin compounds. In *Cephalosporins and Penicillins, Chemistry and Biology*, ed. E. H. Flynn, pp. 370–437. Academic, New York.

Lehninger, A. L. (1983). *Principles of Biochemistry*, p. 214. Worth, New York.

Levy, P. L., H. K. Pangborn, Y. Bernstein, L. H. Ericsson, H. Neurath and K. A. Walsh (1975). Evidence of a homologous relationship between thermolysin and neutral protease A of *Bacillus subtilis*. *Proc. Natl. Acad. Sci. USA*, **72**, 352–361.

Lin, Y., G. E. Means and R. E. Feeney (1969). The action of proteolytic enzymes on *N*,*N*-dimethyl proteins. Basis for a microassay for proteolytic enzymes. *J. Biol. Chem.*, **244** (4), 789–793.

Lineback, D. R. and L. A. Aira (1972). Structural characterization of the two forms of glucoamylase from *Aspergillus niger*. *Cereal Chem.*, **49**, 283–298.

Lowry, O. H., N. J. Rosebrough, A. L. Farr and R. J. Randall (1951). Protein measurements with the folin phenol reagent. *J. Biol. Chem.*, **193**, 265–275.

MacMillan, J. D. and H. Phaff (1966). Exopolygalacturonate lyase from *Clostridium multifermentans. Methods Enzymol.*, **8**, 632–635.

MacMillan, J. D. and M. I. Sheiman (1974). Pectic enzymes. In *Food Related Enzymes*, ed. J. R. Whitaker, pp. 101–130. American Chemical Society, Washington, DC.

Maguire, R. J. (1977). Kinetics of the hydrolysis of cellulose by β-1,4-glucan cellobiohydrolase of *Trichoderma viride. Can. J. Biochem.*, **55**, 644–650.

Mandels, M., R. Andreotti and C. Roche (1976). Measurement of saccharifying cellulase. *Biotechnol. Bioeng. Symp.*, **6**, 21–34.

Manners, D. J., J. R. Stark, G. Wilson and J. Y. Brodie (1976). Some properties of a fungal β-D-glucanase preparation. *Carbohydr. Res.*, **49**, 383–388.

Manners, D. J. and G. Wilson (1974). Purification and properties of an endo-(1→3)-β-D-glucanase from malted barley. *Carbohydr. Res.*, **37**, 9–22.

Markland, F. S. and E. L. Smith (1971). Subtilisins: primary structure, chemical and physical properties. In *The Enzymes*, vol. 3, ed. P. D. Boyer, pp. 568–608. Academic, New York.

McMahon, D. J. and R. J. Brown (1982). Evaluation of formagraph for comparing rennet solutions. *J. Dairy Sci.*, **65**, 1639–1642.

Means, G. E. and R. E. Feeney (1968). Reductive alkylation of amino groups in proteins. *Biochemistry*, **7**, 2192–2201.

Mercier, C., B. M. Frantz and W. J. Whelan (1972). An improved purification of cell-bound pullulanase from *Aerobacter aerogenes. Eur. J. Biochem.*, **26**, 1–9.

Miah, M. N. N. and S. Ueda (1977a). Multiplicity of glucoamylase of *Aspergillus oryzae*. Part 1. Separation and purification of three forms of glucoamylase. *Die Stärke*, **29**, 191–196.

Miah, M. N. N. and S. Ueda (1977b). Multiplicity of glucoamylase of *Aspergillus oryzae*. Part 2. Enzymic and physicochemical properties of three forms of glucoamylase. *Die Stärke*, **29**, 235–239.

Middleton, J. E. (1968). Preparation and investigation of a stabilized glucose oxidase-peroxidase reagent for estimating glucose, using o-toluidine with an alkylaryl sulfonate and polyethylene glycol. *Clin. Chim. Acta*, **22**, 433–437.

Miller, G. L. (1969). Use of dinitrosalicylic acid reagent for determination of reducing sugar. *Anal. Chem.*, **31**, 426–428.

Montenecourt, B. S., D. E. Eveleigh, G. K. Elmund and J. Parcells (1978). Antibiotic discs — an improvement in the filter paper assay for cellulase. *Biotechnol. Bioeng.*, **20**, 297–300.

Nagasaki, S., K. Saito and S. Yamamoto (1977). Purification and characterization of an exo-β-1,3 glucanase from fungi imperfecti. *Agric. Biol. Chem.*, **41**, 493–502.

Nelson, N. (1944). A photometric adaption of the Somogyi method for the determination of glucose. *J. Biol. Chem.*, **153**, 375.

Neuman, N. P. and J. O. Lampen (1967). Purification and properties of yeast invertase. *Biochemistry*, **6**, 468–475.

Ng, T. K. and J. G. Zeikus (1980). A continuous spectrophotometric assay for the determination of cellulase solubilizing activity. *Anal. Biochem.*, **103**, 42–50.

Norman, B. E. (1980). New developments in starch syrup technology. In *Enzymes and Food Processing Symposia*, ed. G. G. Birch, N. Blakebrough and K. J. Parker, pp. 15–50.

North, M. J. (1982). Comparative biochemistry of the proteinases of eukaryotic microorganisms. *Microbiol. Rev.*, **46**, 308–340.

Notario, V., T. G. Villa and J. R. Villanueva (1976). Purification of an exo-β-D-glucanase from cell-free extracts of *Candida utilis. Biochem. J.*, **159**, 555–562.

Nummi, M., M. L. Niku-Paavola, T. M. Enari and V. Raunio (1980). Immunoelectrophoretic detection of cellulases. *FEBS Lett.*, **113**, 164–166.

Okachi, R., I. Kawamoto, M. Yamamoto, S. Takasawa and T. Nara (1973a). Isolation of penicillinase-deficient mutants of *Klyvera citrophilia* KY 3641. *Agric. Biol. Chem.*, **37**, 335–339.

Okachi, R., F. Kato, Y. Miyamura and T. Nara (1973b). Enzymic synthesis of D(−)-α-aminobenzyl-penicillin. VI. Selection of *Pseudomonas melanogenum* KY 3987 as a new ampicillin-producing bacteria. *Agric. Biol. Chem.*, **37**, 1953–1957.

O'Malley, J. J. and J. L. Weaver (1972). Subunit structure of glucose oxidase from *Aspergillus niger. Biochemistry*, **11**, 3527–3532.

Park, J. T. and M. J. Johnson (1949). A submicro determination of glucose. *J. Biol. Chem.*, **181**, 149–156.

Parrish, F. W., A. S. Perlin and E. T. Reese (1960). Selective enzymolysis of poly-β-D-glucans and the structure of the polymers. *Can. J. Chem.*, **38**, 2094–2104.

Payne-Wahl, P., G. F. Spencer, R. D. Plattner and R. O. Butterfield (1981). High performance liquid chromatography method for quantitation of free fatty acids, mono-, di-, and triglycerides using an infrared detector. *J. Chromatogr.*, **209**, 61–66.

Peat, S., W. J. Whelan and J. G. Roberts (1957). Structure of lichenin. *J. Chem. Soc.*, 3916–3924.

Pilnik, W. and A. G. J. Voragen (1970). Pectic substances and other uronides. In *The Biochemistry of Fruits and their Products*, ed. A. C. Hulme, vol. 1, pp. 53–87. Academic, New York.

Raymond, M. N., J. Garnier, E. Bricas, S. Cilianu, M. Blasnic, A. Chaix and P. Lefrancier (1972). Specificity of chymosin (rennin). I. Kinetic parameters of the hydrolysis of synthetic oligopeptide substrates. *Biochemie*, **54**, 145–154.

Raymond, M. N., E. Bricas, R. Salesse, J. Garnier, P. Garnot and B. Ribadeau-Dumas (1973). Proteolytic unit for chymosin (rennin) activity based on a reference synthetic peptide. *J. Dairy Sci.*, **56**, 419–422.

Rexová-Benková, L. and O. Marković (1976). Pectic enzymes. *Adv. Carbohydr. Chem.*, **33**, 323–385.

Rickert, W. S. and P. A. McBride-Warren (1974). Structural and functional determinants of *Mucor miehei* protease. III. Isolation and composition of the carbohydrate moiety. *Biochim. Biophys. Acta*, **336**, 437–444.

Robyt, J. F. and W. J. Whelan (1972). Reducing value methods for maltodextrins: I. Chain-length dependence of alkaline 3,5-dinitrosalicylate and chain-length independence of alkaline copper. *Anal. Biochem.*, **45**, 510–516.

Roeder, G. (1930). Catalase test [of milk]. *Milchwirtsch. Forsch.*, **9**, 516–590.

Rombouts, F. M. and W. Pilnik (1972). Research on pectin depolymerases in the sixties. A literature review. *CRC Crit. Rev. Food Technol.*, **3**, 1–26.

Rombouts, F. M. and W. Pilnik (1980). Pectic enzymes. In *Microbial Enzymes and Bioconversions*, ed. A. H. Rose. pp. 227–282. Academic, London.

Rowe, A. W. and C. E. Weill (1962). The inhibition of β-amylase by ascorbic acid. II. *Biochim. Biophys. Acta*, **65**, 245–251.

Ryu, D. D. Y. and M. Mandels (1980). Cellulases: biosynthesis and applications. *Enzyme Microb. Technol.*, **2**, 91–102.

Saccani, F. and G. Pitrolo (1969). Simultaneous determination of benzylpenicillin and 6-aminopenicillanic acid in fermentation broth. III. *Boll. Chim. Farm.*, **108**, 324–329.

Sandstedt, R. M., E. Kneen and M. J. Blish (1939). A standardized wohlgemuth procedure for alpha-amylase activity. *Cereal Chem.*, **16**, 712–723.

Schattenkerk, C., I. Holtkamp, J. G. M. Hessing, K. E. T. Kerling and E. Havinga (1971). Polypeptides. XIII. Relation between structure and capacity to function as rennin substrate. *Recl. Trav. Chim. Pays-Bas*, **90**, 1320–1322.

Schneider, W. J. and M. Roeher (1976). Purification and properties of penicillin acylase of *Bovista plumbea*. *Biochim. Biophys. Acta*, **452**, 177–185.

Schwabe, C. (1973). A fluorescent assay for proteolytic enzymes. *Anal. Biochem.*, **53**, 484–490.

Schinke, R., Y. Kunimi and H. Nishira (1975). Production and some properties of β-amylase of *Bacillus* sp. BQ10. *J. Ferment. Technol.*, **53**, 693–702.

Shipe, W. F., G. F. Senyk and K. B. Fountain (1980). Modified copper soap solvent extraction method for measuring free fatty acids in milk. *J. Dairy Sci.*, **63**, 193–198.

Shoemaker, S. P. and R. D. Brown, Jr. (1978a). Enzymatic activities of endo-1,4-β-D glucanases purified from *Trichoderma viride*. *Biochim. Biophys. Acta*, **523**, 133–146.

Shoemaker, S. P. and R. D. Brown, Jr. (1978b). Characterization of endo-1,4-β-D glucanases purified from *Trichoderma viride*. *Biochim. Biophys. Acta*, **523**, 147–161.

Sober, H. A. (ed.) (1973). *CRC Handbook of Biochemistry*, 2nd edn. CRC Press, Cleveland, Ohio.

Sommer, H. H. and H. Matsen (1935). The relation of mastitis to rennet coagulability and curd strength of milk. *J. Dairy Sci.*, **18**, 741–749.

Somogyi, M. (1945). A new reagent for the determination of sugars. *J. Biol. Chem.*, **160**, 61–68.

Stern, K. G. (1932). The catalase of colorless blood cells. *Hoppe-Seylers Z. Physiol. Chem.*, **204**, 259–282.

Sternberg, M. (1976). Microbial rennets. *Adv. Appl. Microbiol.*, **20**, 135–157.

Street, H. V. and J. R. Close (1956). Determination of amylase activity in biological fluids. *Clin. Chim. Acta*, **1**, 256–261.

Suzuki, H. and T. Kaneko (1976). Degradation of barley, glucan and lichenan by a *Bacillus pumilus* enzyme. *Agric. Biol. Chem.*, **40**, 577–586.

Takahashi, N., H. Toda, H. Nishibe and K. Yamamoto (1982). Isolation and characterization of taka-amylase A apoprotein deglycosylated by digestion with almond glycopeptidase immobilized on sepharose. *Biochim. Biophys. Acta*, **707**, 236–242.

Tashiro, Y. and J. R. Trevithick (1977). The role of carbohydrate in the glycoenzyme invertase of *Neurospora crassa*. *Can. J. Biochem.*, **55**, 249–256.

Thoma, J. A. and D. E. Koshland, Jr. (1960). Competitive inhibition by substrate during enzyme action. Evidence for the induced fit theory. *J. Am. Chem. Soc.*, **82**, 3329–3333.

Thoma, J. A. (1974). Interaction of β-amylase with substrates and inhibitors with comments on Koshland's induced-fit theory, a reply. *Eur. J. Biochem.*, **44**, 139–142.

Thoma, J. A. (1976). Models for depolymerizing enzyme: criteria for discrimination of models. *Carbohydr. Res.*, **48**, 85–103.

Thomson, J. F. and F. J. Klipfel (1957). Effect of X-irradiation on increase of liver tryptophan peroxidase produced by adrenal steroids. *Arch. Biochem. Biophys.*, **70**, 224–233.

Thomson, J. F., S. L. Nance and S. L. Tollaksen (1978). Spectrophotometric assay of catalase with perborate as substrate. *Proc. Soc. Expl. Biol. Med.*, **157**, 33–35.

Trinder, P. (1969). An improved colour reagent for the determination of blood glucose by the oxidase system. *Ann. Clin. Biochem.*, **6**, 24–29.

Tsuboi, A., V. Yamasaki and Y. Suzuki (1974). Two forms of glucoamylase from *Mucor rouxianus*. I. Purification and crystallization. *Agric. Biol. Chem.*, **38**, 543–550.

Urlaub, H. and G. Wober (1978). α-Glucosidase, a membrane-bound enzyme of α-glucan metabolism in *Bacillus amyloquefaciens*. *Biochim. Biophys. Acta*, **522**, 161–173.

Uwajima, T., H. Yagi and O. Terada (1972). Purification, crystallization, and some properties of β-galactosidase from *Saccharomyces fragilis*. *Agric. Biol. Chem.*, **36**, 570–577.

Vandamme, E. J. (1980). Penicillin acylases and β-lactamases. In *Economic Microbiology*, ed. A. H. Rose, vol. 5, pp. 467–522. Academic, London.

Vandamme, E. J. and J. P. Voets (1975). Properties of the purified penicillin V-acylase of *Erwinia aroideae*. *Experientia*, **31**, 140–143.

Wagner, G. and S. Hollman (1976). Uronic acid dehydrogenases from *Pseudomonas syringae*. Purification and properties. *Eur. J. Biochem.*, **61**, 589–596.

Waldschmidt-Leitz, E. and C. Bretzel (1964). Über Penicillinamidase, Reinigung und Eigenschaften. *Hoppe-Seylers Z. Physiol. Chem.*, **337**, 222–228.

White, A., P. Handler, E. Smith, R. Hill and I. R. Lehman (1978). *Principles of Biochemistry*. McGraw-Hill, New York.

Wohlgemuth, J. (1908). Über eine neue Methode zur quantitativem Bestimmung des diastratischen Ferments. *Biochim. Z.*, **9**, 1–8.

Wood, T. M. and S. I. McCrae (1982). Purification and some properties of a (1-4)-β-D-glucan glucohydrolase associated with the cellulase from the fungus *Penicillium funiculosum*. *Carbohydr. Res.*, **110**, 291–303.

Yamasaki, V., Y. Suzuki and J. Ozawa (1977). Three forms of α-glucosidase and a glucoamylase from *Aspergillus awamori*. *Agric. Biol. Chem.*, **41**, 2149–2161.

19

Dissolved Oxygen Probes

V. LINEK and J. SINKULE
Institute of Chemical Technology, Prague, Czechoslovakia
and
V. VACEK
Institute of Inorganic Chemistry, Prague, Czechoslovakia

19.1 INTRODUCTION

The increasing use of oxygen probes (OP) in various fields has stimulated a growing interest in their fundamental properties and operating principles. The diverse aspects of OP use have been discussed in a number of papers and reviews (see a short overview in Table 1).

Table 1   List of Reviews and Systematic Treatments of Oxygen Probe Theory and Practice

Hoare (1968)	Standard reference monograph on electrochemistry of oxygen electrodes
Hitchman (1978)	Monograph on electrochemical principles of oxygen probe and on its application
Fatt (1976)	Monograph: detailed account of construction and of application in biology, medicine and technology
Lee and Tsao (1979)	Review of probe construction, general design considerations, oxygen probe applications in various fields, sources of measurement errors due to probe characteristics
Kreuzer *et al.* (1980)	Comprehensive review of catheter type, skin and tissue oxygen probes
Beechey and Ribbons (1972)	Review of microbiological applications in short term experiments, including bare electrodes
Brown (1970)	Review of oxygen probes applications for long term experiments in microbiology
Davies (1962)	Review of electrochemical principles of oxygen probe prior to Hoare (1968) and Hitchman (1978) monographs, with summary of biological applications
Elsworth (1972)	Short review of galvanic oxygen probes applications in chemical engineering
Silver (1973)	Comprehensive review of microelectrodes and of their applications
Mancy *et al.* (1962), Kinsey and Bottomley (1963), Lucero (1968, 1969)	Papers containing systematic study of probe characteristics in relation to probe designs
Feder (1968), Kreuzer (1969), Kessler *et al.* (1973), Degn *et al.* (1976), Severinghaus *et al.* (1978), Gnaiger and Forstner (1983)	Books containing papers presented at meetings or conferences and/or chapters written on various problems with oxygen probe applications, mostly in the fields of medicine, physiology, biotechnology and environmental science

This chapter deals with the use of OP in chemical engineering and bioengineering, with emphasis on the state of the art of rational modelling of probe performance and the medium—probe interaction occurring within the external liquid film adjacent to the membrane.

Oxygen probe characteristics are determined by electrochemical processes occurring at the probe electrodes and in their surroundings, and by the kinetics of oxygen transport from the medium measured to the probe cathode (see Figure 1). Rational models of probe characteristics considering the diffusion mechanism of oxygen transport through layers E, M and L are preferred here to the still prevailing empirical descriptions of actual responses. Special attention is paid to measurements in viscous liquids, involving significant resistance of the external liquid film at the surface of the probe membrane against oxygen transfer, with or without a chemical reaction in the film.

19.2 PRINCIPLES OF OXYGEN MEASUREMENT BY POLAROGRAPHIC AND GALVANIC PROBES

19.2.1 Polarographic Probes

When a negative voltage sufficiently high with respect to a reference electrode (usually Ag/AgCl) is applied to a noble metal cathode, dissolved oxygen begins to reduce on the catalytically active centres of the cathode surface. The plot of electrode current, M, against voltage shows a plateau (see Figure 2). In this plateau region, called the limiting diffusion current (LDC), the current depends only on the rate of oxygen transport towards the cathode, and thus is proportional to the oxygen tension. In the kinetic (overvoltage) region (Figure 3), the reaction is not sufficiently rapid for the oxygen concentration at the cathode to be kept at zero; in the hydrogen wave region, water decomposition occurs. The position of the polarographic wave depends on the pH of the electrolyte near the cathode: as the pH is increased, the polarogram shifts to higher voltages, as shown in Figure 3. It is, therefore, preferable to choose a polarizing voltage below the onset of the hydrogen wave, thus minimizing the risk that a local increase in pH near the cath-

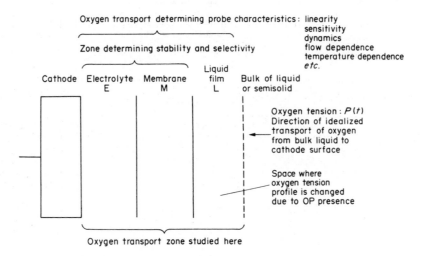

Figure 1 Oxygen transport zone

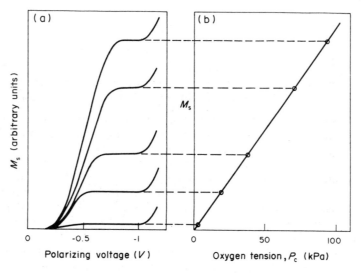

Figure 2 (a) Polarogram; (b) calibration curve

ode, caused by accumulation of OH⁻ ions formed in the cathodic reactions (Hitchman, 1978), will shift the cathode potential outside the LDC region. Silver cathodes are better than Pt in this respect, since their plateaux are longer and the kinetic (overvoltage) regions shorter (see Figure 3; Hoare, 1968). Small cathodes with high current densities exhibit poorly developed, shorter plateaux (Jensen *et al.*, 1978). Since the polarogram shape depends on the electrode material, electrolyte composition, size of the cathode, its purity and the condition of its surface, the optimum voltage should be found separately for each probe.

The overall electrode reactions are as follows:

at the cathode: $$O_2 + 2\,H_2O + 2\,e^- \rightarrow H_2O_2 + 2\,OH^- \tag{1}$$

$$H_2O_2 + 2\,e^- \rightarrow 2\,OH^- \tag{2}$$

at the anode: $$Ag + Cl^- \rightarrow AgCl + e^- \tag{3}$$

When reaction (2) is not sufficiently fast, H_2O_2 diffuses away from the cathode, changing the electrode reaction stoichiometry (Hoare, 1968). The production of hydrogen peroxide in oxygen probes has been studied by Lloyd and Seaton (1970), Tang *et al.* (1973), Hahn *et al.* (1975) and Hale and Hitchman (1980). Recently, Linek *et al.* (1985) have demonstrated that accumulation

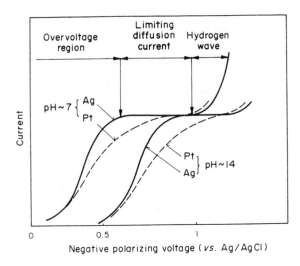

Figure 3   pH dependence of polarogram

of OH^- ions at the cathode retards reaction (2), thereby delimiting the range of conditions within which the probe provides reliable measurements.

19.2.2   Galvanic Probes

Galvanic types of probe use an electrode pair that generates a potential sufficient for the LDC to be attained, so that they do not require an external voltage source as is the case for polarographic types. A silver cathode and a lead anode are most frequently used and the reactions involved are:

at the cathode: $$O_2 + 2H_2O + 4e^- \rightarrow 4OH^- \tag{4}$$

at the anode: $$Pb \rightarrow Pb^{2+} + 2e^- \tag{5}$$

19.2.3   General Considerations

The principle of measurement is the same for both types of probe: the current is proportional to the oxygen flux towards the cathode. The electrode system is separated from the surrounding medium by a membrane permeable to O_2 (but also to other, undesired gases such as H_2S, SO_2, CO_2, Cl_2, *etc.*) and impermeable to ions.

It is advantageous to confine the main resistance to oxygen transfer within the membrane, where it is well-defined. Thus, the membrane not only prevents contamination of the electrode surfaces, but also ensures reproducible oxygen transport to the cathode. It is, however, unwise to increase the membrane resistance up to values at which the resistance of the liquid film becomes insignificant, since then the probe response is usually too slow. Therefore, the liquid film resistance should generally be taken into account.

A chemical reaction with dissolved oxygen occurring in the bulk of liquid, and thus also in the liquid film before the membrane, will affect oxygen transport through the film and, consequently, also the probe reading. When the chemical reaction is rapid enough, all of the transferred oxygen may be consumed within the film so that the probe reading is zero, even if the dissolved oxygen concentration in the bulk is quite high. Studies on situations involving a chemical reaction are, however, rather scanty (Lundsgaard *et al.*, 1978; Linek *et al.*, 1983a).

19.3   THEORY OF OXYGEN TRANSPORT TO THE PROBE CATHODE

The model situations solved so far are summarized in Table 2. Multilayer models have been developed primarily for the estimation of the sensitivity of the OP to the hydrodynamics of the external fluid phase (flow velocity), in which case the liquid film resistance to oxygen transport has to be considered in addition to that of the membrane layer.

When oxygen transport through the probe layers is described by a linear homogeneous differential diffusion equation, the principle of superposition can be used: the normalized probe response, $G(t)$, to an arbitrary oxygen tension, $P(t)$, starting from a steady state, P_{s1}, may be evaluated from the superposition integral (Carslaw and Jaeger, 1959):

$$G(t) = \frac{M(t) - M_{s1}}{M_{s2} - M_{s1}} = \int_0^t \frac{P(v) - P_{s1}}{P_{s2} - P_{s1}} \frac{dH(t-v)}{dt} dv \qquad (6)$$

where $H(t)$ is the normalized transient probe characteristic, and M_{s2} and M_{s1} are the steady-state readings at the oxygen tensions P_{s2} and P_{s1}, respectively. Thus, the transient characteristic $H(t)$ contains full information on the probe dynamics and is defined as a probe response to a step concentration change P_c before the oxygen transport zone (see Figure 1). Denoting z as p_L/P_c, the corresponding initial concentration profile for $z=0$ represents the steady state which is a necessary initial condition for the application of equation (6). This initial profile is shown in Figure 4. In practice, it is easy to realize an initial concentration profile where $z=1$; however, this case is not the steady-state concentration profile.

19.3.1 Principal Models of Oxygen Transport to the Cathode

The three-layer probe model (Figure 4) is defined by the relations:

$$\frac{\partial p_E}{\partial t} = D_E \frac{\partial^2 p_E}{\partial x^2} \qquad (7)$$

$$\frac{\partial p_M}{\partial t} = D_M \frac{\partial^2 p_M}{\partial x^2} \qquad (8)$$

$$\frac{\partial p_L}{\partial t} = D_L \frac{\partial^2 p_L}{\partial x^2} - k_j p_L^j m_L^{j-1} \qquad (9)$$

Table 2 Summary of Published Models for Oxygen Transfer through a Probe[a]

(A) Steady-state diffusion models

Ref.	Situation considered	Aim or range of applicability
Schuler and Kreuzer (1967, 1969), Aiba and Huang (1969b)	One-dimensional diffusion to planar cathode through layers E and M	Description of planar cathodes: dependence of flow sensitivity of probe on liquid flow velocity, viscosity of the liquid and cathode size
Grunewald (1970)	One-dimensional diffusion through membrane to half-spherical cathode from semi-infinite tissue	Interpretation of measurement of oxygen in tissue by microprobes
Albanese (1971)	One-dimensional diffusion to half-spherical cathode through membrane from an oxygen consuming tissue	Calibration of probe at discrete points of measurement of oxygen in tissue
Siu and Cobbold (1976)	Two-dimensional diffusion to planar disc cathode through membrane from semi-infinite film layer L	Multiwire cathode description: mutual influence of microcathodes by side diffusion; the analysis is incorrect, see Jensen *et al.* (1978)
Jensen *et al.* (1978)	One-dimensional diffusion to planar disc cathode through membrane, and side diffusion through E	Planar cathodes: dependence of probe sensitivity on cathode size and on thickness of electrolyte layer
Lundsgard *et al.* (1978)	One-dimensional diffusion through L and M layers with chemical reaction of zero, 1st order and Michaelis–Menten kinetics	Description of the effect of chemical reaction on probe sensitivity
Schneiderman and Goldstick (1978)	Two-dimensional diffusion to recessed cathode from infinite stagnant medium through membrane	Description of concentration profile round recessed microcathode: the effect of cathode size and depth of recession
Ultman *et al.* (1981)	One-dimensional diffusion through spherical E and M layers	Approximate description of the dependence of probe sensitivity on spherical cathode size and on thickness of layers E and M
Firouztalé *et al.* (1982)	One-dimensional diffusion through spherical E, M and L layers	Extension of previous paper Ultman *et al.* (1981) to include the layer L

(B) Non-steady-state diffusion models

Ref.	Type of model[b]	Layers considered	Cathode geometry	Film layer	Aim and/or range of applicability
Mancy et al. (1962), Benedek and Heideger (1970)	ODD	E,M	Planar	—	Solution for special case where $A_E = 1$
Aiba and Huang (1969a)	ODD	E,M	Planar	—	The effect of layer E on probe response slowdown
Lucero (1969)	Analogue	E,M /side diffusion is considered: slowdown described			
Aiba and Huang (1969a), Heineken (1970), Linek (1972)	ODD	M	Planar	—	Dependence of probe constant K_M on transport properties of membrane
Grunewald (1971), Lee *et al.* (1978)	ODD	M,L	Half spherical	Semi-infinite stagnant	Spherical microcathode; dependence of probe sensitivity on cathode size
Kok and Zajic (1975)	Lateral diffusion	M	Central well	—	Description of YSI probe response slowdown
Linek and Vacek (1976, 1977)	ODD	M,L	Planar	Noncapacitive	Liquid film resistance; for critical analysis, see Linek *et al.* (1983a)
Linek and Beneš (1977)	ODD	Two regions of membrane	Planar	—	General multiregion model description of slowdown
Linek *et al.* (1979a)	ODD	Regions of membrane and film	Planar	Noncapacitive	For criticism see Linek *et al.* (1983a)
Lee and Tsao (1979)	ODD	E,M,L	Planar	Noncapacitive	Formal extension of one-layer to three-layer model, see Linek *et al.* (1983a)
Linek *et al.* (1983a)	ODD	E,M,L	Planar	With chemical reaction	Dependence of probe characteristics on properties of individual layers E,M,L
Linek *et al.* (1983b)	ODD	M,L	Planar	With reaction of zero or 1st order	Analysis of the effect of initial conditions on dynamic response of probe
Vacek *et al.* (1985)	ODD	E,M,L	Spherical	Capacitive	Dependence of probe characteristics on individual layers

[a] Layers identified in Figure 1.   [b] ODD—one-dimensional diffusion type.

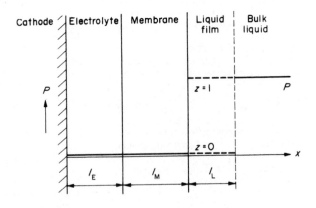

Figure 4 Coordinate system of three-layer diffusion model and initial concentration profiles of oxygen for transient probe characteristics, H ($z=0$), and for Dang's response (Dang *et al.*, 1977), D ($z=1$)

with the initial condition

$$t = 0: p_E = p_M = 0; p_L = zP_c; z = 0, 1 \qquad (10)$$

and the boundary conditions

$$x = 0: p_E(t > 0) = 0 \qquad (11)$$

$$x = l_E: p_E = p_M; k_E \frac{\partial p_E}{\partial x} = k_M \frac{\partial p_M}{\partial x} \qquad (12)$$

$$x = l_E + l_M: p_M = p_L; k_M \frac{\partial p_M}{\partial x} = k_L \frac{\partial p_L}{\partial x} \qquad (13)$$

$$x = l_E + l_M + l_L: p_L = P_c \qquad (14)$$

The above set of equations is written for zero initial condition. A non-zero stationary profile as the initial state can easily be converted to the condition described by equation (10) for $z=0$, but not for $z=1$. The above equations are written in terms of the equivalent partial pressure of dissolved oxygen, p (oxygen tension), whose profile is continuous at the boundaries between the layers. Equation (11) holds for LDC, as verified experimentally by Baumgärtl *et al.* (1974). The condition of equation (10) is chosen so as to include both the frequently used step concentration changes where $z=0$ and $z=1$. The current output of the probe, $M(t)$, is proportional to the oxygen flux towards the cathode, $N(t)$:

$$M(t) = -nFSN(t) = nFSk_E \frac{\partial p_E}{\partial x}\Big|_{x=0} \qquad (15)$$

The solution of the set of equations (7) to (15) is summarized in Table 3 which gives expressions for the steady-state flux, N_s, the transient characteristics $H(t) = M(t)/M_s$ for $z=0$, the difference, $W(t) = D(t) - H(t)$, between the normalized response $D(t) = M(t)/M_s$ for $z=1$ (Dang's response; Dang *et al.*, 1977) and the transient characteristics $H(t)$, and the corresponding zeroth moments

$$b_J = \int_0^\infty [1 - J(t)]dt, \text{ where } J = D, H \text{ or } (1 - W) \qquad (16)$$

In the prevailing probe design the effect of the electrolyte layer on the probe response may usually be neglected. Linek *et al.* (1983a) have shown when the general three-layer model can be reduced to the two-layer or single-layer variants. Expressions for two-layer and single-layer models are included in Table 3. The relationship between the above expressions and the actual responses will be discussed below.

Attempts to derive an expression for transient probe characteristics H and b_H for zero-order reaction failed because of a negative concentration in the film arising from the zero-order reaction term for small initial concentration in the film or for fast reaction rates. Sufficient conditions for this difficulty to disappear are $P_c > A_L^2 B_o/2$, and $z=1$. The expressions for $D(t)$ and b_D listed in Table 3 are, therefore, subject to these conditions.

A model of a spherical cathode has been introduced for microelectrodes (Grunewald, 1971; Lee *et al.*, 1978; Linek *et al.*, 1979a). The solution of the three-layer model for spherical geometry (Vacek *et al.*, 1985; see Figure 5) is also given in Table 3. For $l_L \rightarrow \infty$, *i.e.* for infinite stagnant medium, this solution reduces to an expression derived by Grunewald (1971).

19.3.2 Assumptions Underlying the Models

The derivation of the relations presented in Table 3 was based on the following assumptions: (i) only the normal component of the oxygen flux towards the cathode is significant; (ii) permeability, diffusivity, and thickness of each layer are constant; (iii) the electrode reactions proceed in the LDC region.

Assumption (i) is fulfilled only when the effect of side diffusion, whose magnitude depends on geometrical and transport factors, is negligible. The space within which side diffusion is effective may be approximated by a band of the thickness of all three layers, extending around the circumference of the cathode (Hitchman, 1978). The ratio, w, of the normal to non-normal oxygen diffusion towards the cathode can be estimated from the ratio of the band area to the cathode surface area:

$$w = Q_c(l_E + l_M + l_L)/S \qquad (17)$$

Table 3 Summary of Equations for One-dimensional Diffusion Transport of Oxygen to the Probe Cathode

Model	Layers considered[a]	Chemical reaction	Steady-state oxygen flux intensity, N_s	Transient characteristics for $z = 0$, $H(t)$	Dang's response or $W(t)$ function, $D(t)$ or $W(t)$	Zeroth moment of transient characteristics, b_H	Zeroth moment of Dang's response or of $W(t)$ function b_D or b_W
			(i) For planar cathode				
A	E, M, L	—	(A1)	(A2)	(A3)	(A4)	(A5)
B	M, L	—	(B1)	(B2)	(B3)	(B4)	(B5)
C	M, L	First order	(C1)	(C2)	(C3)	(C4)	(C5)
D	M, L	Zero order	(D1)	—	(D3)	—	(D5)
E	M	—	(E1)	(E2)	—	(E4)	—
F	M, L[b]	—	(F1)	(F2)	—	(F4)	—
			(ii) For spherical cathode				
P	E, M, L	—	(P1)	(P2)	—	(P4)	(P5)
S	M, L	—	(S1)	(S2)	—	(S4)	—

[a] Layers are identified in Figure 1.   [b] Noncapacitive liquid film is considered in this film model.

(A1) $N_s = -P_c(R_E + R_M + R_L)^{-1}$

(A2) $H(t) = 1 + 2(1 + L_E + L_L)\sum\limits_{n=1}^{\infty} \exp(-K_M\alpha_n^2 t)[Q(\alpha_n)]^{-1}$, where α_n are positive roots of the equation:

$\cos A_E\alpha(S_L\cos \alpha \sin A_L\alpha + \sin \alpha \cos A_L\alpha) + S_E\sin A_E\alpha(\cos \alpha \cos A_L\alpha - S_L\sin \alpha \sin A_L\alpha) = 0$

and $Q(\alpha) = L_E \cos A_E\alpha (\cos \alpha \cos A_L\alpha - S_L\sin \alpha \sin A_L\alpha) -$
$A_E \sin A_E\alpha (\sin \alpha \cos A_L\alpha + S_L \cos \alpha \sin A_L\alpha) +$
$\cos A_E\alpha [(1 + L_L) \cos \alpha \cos A_L\alpha - (A_L + S_L) \sin \alpha \sin A_L\alpha] -$
$S_E\sin A_E\alpha[(1 + L_L) \sin \alpha \cos A_L \alpha + (A_L + S_L)\cos \alpha \sin A_L\alpha]$

(A3) $W(t) = -2(1 + L_E + L_L)\sum\limits_{n=1}^{\infty} (1-\cos A_L\alpha_n)\exp(-K_M\alpha_n^2 t)[Q(\alpha_n)]^{-1}$

(A4) $b_H = \dfrac{b_M}{1 + L_E + L_L}[1 + 3 (L_E + L_L + 2L_EL_L) + 3A_E^2 (1 +L_L+L_E/3) +3A_L^2 (1 + L_E+L_L/3)]$

(A5) $b_W = -3b_MA_L^2$, where $b_M = (6K_M)^{-1}$

(B1), (B2), . . ., (B5) follow from (A1), (A2), . . ., (A5), respectively, for $R_E=L_E=A_E=S_E=0$

(C1) $N_s = - P_cA_L\sqrt{B_1}[R_MA_L\sqrt{B_1} \cosh A_L\sqrt{B_1} + R_L \sinh A_L\sqrt{B_1}]^{-1}$

(C2) $H(t) = 1 + \dfrac{2A_L}{\sqrt{B_1}} (L_L\sinh A_L\sqrt{B_1} + A_L\sqrt{B_1} \cosh A_L\sqrt{B_1})\sum\limits_{n=1}^{\infty} \dfrac{B_1 - \alpha_n^2}{Q(\alpha_n)} \exp(-K_M\alpha_n^2 t)$, where

$\alpha_n \neq 0$, $\alpha_n \neq \sqrt{B_1}$ and α_n are positive roots of the equations

$L_L\alpha \cos \alpha \sinh V + V\sin \alpha \cosh V = 0$ for $\alpha < \sqrt{B_1}$;

$L_L\alpha \cos \alpha \sin V + V\sin \alpha \cos V = 0$ for $\alpha > \sqrt{B_1}$; $V = A_L|B_1 -\alpha^2|^{0.5}$ and

$Q(\alpha) = V\sinh V [L_L\cos \alpha -(A_L^2+L_L)\alpha \sin \alpha]+\cosh V[(V^2-A_L^2L_L\alpha^2)\cos \alpha -A_L^2\alpha \sin \alpha]$ for $\alpha<\sqrt{B_1}$

$Q(\alpha) = -V\sin V [L_L\cos \alpha -(A_L^2+L_L)\alpha \sin \alpha] - \cos V[(V^2+A_L^2L_L\alpha^2)\cos \alpha+A_L^2\alpha \sin \alpha]$ for $\alpha>\sqrt{B_1}$

(C3) $W(t) = \dfrac{2A_L}{\sqrt{B_1}}(L_L\sinh A_L\sqrt{B_1} + A_L\sqrt{B_1} \cosh A_L\sqrt{B_1}) \sum\limits_{n=1}^{\infty} \alpha_n^2E(\alpha_n) \exp(-K_M\alpha_n^2 t)[Q(\alpha_n)]^{-1}$

where $E(\alpha) = (1 - \cosh V)$ for $\alpha<\sqrt{B_1}$ and $E(\alpha) = (1 - \cos V)$ for $\alpha>\sqrt{B_1}$, and
for $Q(\alpha)$ and α_n see (C2)

(C4) $b_H=b_M[3(A_L^2+L_L-L_L/B_1)\tanh A_L\sqrt{B_1} + 3A_LL_L/\sqrt{B_1}+ A_L\sqrt{B_1}][L_L \tanh A_L\sqrt{B_1}+ A_L\sqrt{B_1}]^{-1}$

(C5) $b_W = 6b_M(1 - \cosh A_L\sqrt{B_1})B_1^{-1}$

(D1) $N_s = (B_oA_L^2 - 2P_c) (2R_M + 2R_L)^{-1}$ for $P_c > B_oA_L^2 /2$

(D3) $D(t) = 1 - \dfrac{4(1 + L_L)}{B_oA_L^2 - 2P_c}\sum\limits_{n=1}^{\infty} [(B_o\alpha_n^{-2} + P_c)(\cos A_L\alpha_n - 1) +P_c] \exp(-K_M\alpha_n^2 t)[Q(\alpha_n)]^{-1}$

for $P_c > B_oA_L^2/2$; for α_n and $Q(\alpha)$ see (B2) and (A2)

(D5) $b_D=b_M\{[A_L^2L_L+3(A_L^2+L_L)+1](1+L_L)^{-1} - (12P_c-A_L^2B_o)A_L^2 [2(2P_c-A_L^2B_o)]^{-1}\}$ for $P_c > A_L^2B_o/2$

(E1) $N_s = - P_cR_M^{-1}$

(E2) $H(t) = 1 + 2\sum\limits_{n=1}^{\infty} (-1)^n\exp(-K_Mn^2\pi^2t)$

(E4) $b_H = b_M$

(F1) $N_s = -P_c(R_M+R_L)^{-1}$

(F2) $H(t) = 1+2(1+L_L)\sum\limits_{n=1}^{\infty} (-1)^n\exp(-K_M\alpha_n^2 t)[Q(\alpha_n)]^{-1}$

where $Q(\alpha)=(1+L_L+L_L^2\alpha^2)(1+L_L^2\alpha^2)^{-0.5}$ and α_n are positive roots of the equation $\alpha L_L \cos \alpha + \sin \alpha = 0$

(F4) $b_H = b_M(1+3L_L)(1 + L_L)^{-1}$

(P1) $\quad N_s = P_c\left\{R_M\dfrac{(1-d_E-d_M)^2}{1-d_M}\left[1 + L_E\dfrac{1}{1-d_E-d_M} + L_L\dfrac{1-d_M}{1+d_L}\right]\right\}^{-1}$

(P2) $\quad H(t) = 1 + 2\dfrac{L_E(1+d_L) + (1-d_E-d_M)\,[1+d_L+L_L(1-d_M)]}{1-d_M}\displaystyle\sum_{n=1}^{\infty}\dfrac{\alpha_n^2\exp(-K_M\alpha_n^2 t)}{Q(\alpha_n)}$

where α_n are positive roots of equation $(\alpha \cos A_L\alpha + \dfrac{d_L}{A_L}\sin A_L\alpha)$

$[(\alpha \cos \alpha + \dfrac{d_M}{1-d_M}\sin \alpha)S_E \sin A_E\alpha + (\alpha \cos A_E\alpha - \dfrac{d_E}{1-d_M}\dfrac{\sin A_E\alpha}{A_E})\sin \alpha] - S_L\sin A_L\alpha.$

$\{S_E \sin A_E\alpha[\alpha^2 \sin \alpha + \dfrac{d_M^2}{1-d_M}(\sin \alpha - \alpha \cos \alpha)] + (\alpha \cos A_E\alpha - \dfrac{d_E}{1-d_M}\dfrac{\sin A_E\alpha}{A_E})(d_M\sin \alpha - \alpha \cos \alpha)\} = 0$

$Q(\alpha) = (\alpha \cos A_L\alpha + \dfrac{d_L}{A_L}\sin \alpha A_L)[X(\alpha)L_E \cos A_E\alpha + (\dfrac{\cos \alpha}{1-d_M} - \alpha \sin \alpha)S_E \sin A_E\alpha + Z(\alpha)\cos \alpha +$

$Y(\alpha)\sin \alpha] + [(1+d_L)\cos A_L\alpha - \alpha A_L\sin A_L\alpha][X(\alpha)S_E \sin A_E\alpha + Z(\alpha)\sin \alpha] +$

$L_L\cos A_L\alpha\{S_E \sin A_E\alpha[\dfrac{d_M^2}{1-d_M}(\alpha \cos \alpha - \sin \alpha) - \alpha^2 \sin \alpha] + Z(\alpha)(\alpha \cos \alpha - d_M \sin \alpha)\} +$

$S_L\sin A_L\alpha\{L_E \cos A_E\alpha[\dfrac{d_M^2}{1-d_M}(\alpha \cos \alpha - \sin \alpha) - \alpha^2 \sin \alpha] + Y(\alpha)(\alpha \cos \alpha - d_M \sin \alpha) +$

$Z(\alpha)[(1-d_M)\cos \alpha - \alpha \sin \alpha] - S_E \sin A_E\alpha[2\alpha \sin \alpha + \alpha^2 \cos \alpha + \dfrac{d_M^2}{1-d_M}\alpha \sin \alpha]\}$

$X(\alpha) = \alpha \cos \alpha + \dfrac{d_M}{1-d_M}\sin \alpha;\ Y(\alpha) = \dfrac{1-d_E-d_M}{1-d_M}\cos A_E\alpha - \alpha A_E \sin A_E\alpha,\ Z(\alpha) = \alpha \cos A_E\alpha - \dfrac{d_E\sin A_E\alpha}{(1-d_M)A_E}$

(P4) $\quad b_H = b_M\{A_L^2(3+d_L)(1+L_E-d_E-d_M) + (1+d_L)\,[L_E(3-2d_M+A_E^2) + 1-d_E - $
$\qquad d_M + A_E^2(3-3d_M-d_E)] + L_L[2L_E(3-3d_M+d_M^2) + A_L^2(1-d_M-d_E)\times$
$\qquad (1-d_M) + A_E^2(1-d_M)(3-d_E-d_M) + (3-d_M)(1-d_E-d_M)]\}$
$\qquad \{L_E(1+d_L) + (1-d_E-d_M)[1+d_L+L_L(1-d_M)]\}^{-1}$

(P5) $\quad b_W = -A_L^2(3+d_L)/[6K_M(1+d_L)]$

(S1) $\quad N_s = P_c\left[R_M(1-d_M)\left(1 + L_L\dfrac{1-d_M}{1+d_L}\right)\right]^{-1}$

(S2) $\quad H(t) = 1 + 2\,[1+d_L+L_L(1-d_M)]\displaystyle\sum_{n=1}^{\infty}\dfrac{\alpha_n\exp(-K_M\alpha_n^2 t)}{Q(\alpha_n)}$

where α_n are positive roots of equation

$\sin \alpha\left[\alpha \cos A_L\alpha + (d_L-L_Ld_M)\dfrac{\sin A_L\alpha}{A_L}\right] + \alpha L_L \cos \alpha\dfrac{\sin A_L\alpha}{A_L} = 0$

$Q(\alpha) = \cos \alpha\left[\alpha(1+L_L)\cos A_L\alpha + (d_L+L_L-L_Ld_M)\dfrac{\sin A_L\alpha}{A_L}\right] +$

$\sin \alpha\left[\cos A_L\alpha(1+d_L-L_Ld_M) - \alpha\left(A_L + \dfrac{L_L}{A_L}\right)\sin A_L\alpha\right]$

(S4) $\quad b_H = b_M\dfrac{A_L^2(3+d_L) + 1+d_L+L_L[A_L^2(1-d_M)+3 - d_M]}{1 + d_L + L_L(1 - d_M)}$

The side diffusion effect increases with increasing value of w and, at constant w, with increasing ratio k_E/k_M. The side diffusion produces a slowdown of the transient probe characteristics, enhances the sensitivity of the OP and diminishes its dependency on the external fluid hydrodynamics, and may be responsible for response hysteresis; see Section 19.5.4.

Assumption (ii) may break down as a result of (a) non-uniform thickness of the electrolyte or membrane layer owing to mechanical stress which, in addition, may alter the structure of the polymer membrane material, and thus also its permeability; (b) local changes in the electrolyte concentration near the cathode due to accumulation of OH$^-$ ions, see Section 19.5.1; or (c) high pressures at which Henry's law is no longer valid and the permeability is pressure dependent.

Assumption (iii) may not be fulfilled for a number of reasons, *e.g.* the catalytic activity of the cathode may vanish after long-term use of the probe, or at large current densities (*i.e.* when using very thin membranes and high oxygen concentrations), local changes in the electrolyte pH may shift the cathode potential outside the LDC region or retard reaction (2).

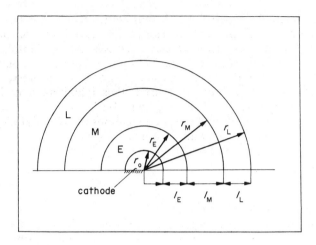

Figure 5   Spherical cathode and E, M, L layers

19.4   PROBE DESIGN

The diversity in the behaviour of real OP stems from the variety of combinations of several basic construction elements and design concepts. Comprehensive reviews of various probe designs have been given by Fatt (1976), Lee and Tsao (1979) and Kreuzer *et al.* (1980). Systematic considerations about OP design have also been presented by Kinsey and Bottomley (1963), Mancy *et al.* (1962), Lucero (1968, 1969), Schneiderman and Goldstick (1978), and Siu and Cobbold (1976).

19.4.1   Electrode Metals

Pt, Au and Ag are the most frequently used cathode metals. Pt cathodes stick well on glass but, unlike Au and Ag (Hitchman, 1978), show ageing effects (Aiba and Huang, 1969a), and like Ag are very susceptible to H_2S poisoning. Pt cathodes have to be polished (Lübbers *et al.*, 1969); their performances vary depending on their fabrication history, and in an acidic medium the reaction on the Pt surface may not proceed at a sufficient rate (Mancy *et al.*, 1962). Au cathodes are resistant to H_2S poisoning. Unlike Pt, the electrode reactions on Au and Ag are free from interfering effects of surface oxides. Gold purity governs the width of the current–voltage plateau, as well as the magnitude of residual current (Kinsey and Bottomley, 1963).

Ag/AgCl, Ag/Ag_2O and calomel electrodes are most frequently used as the anode. Ag_2O may flake off the anode and short-circuit the probe, a problem not encountered with AgCl (Phillips and Johnson, 1961). The anode should be as large as possible to avoid polarization (Lucero, 1969).

For galvanic probes, different pairs of electrode metals are used, of which Ag as the cathode and Pb as the anode are the most common combination. Ag is preferred for its good stability (Mancy *et al.*, 1962). Detailed accounts of the metal combination and their main properties have been given by Lee and Tsao (1979) and Kreuzer *et al.* (1980). A comprehensive review of electrochemical oxygen reduction on various metals has been given by Hoare (1968).

19.4.2   Electrolytes

An account of the electrolytes in use has been given by Lee and Tsao (1979). Efforts to suppress the effect of uncontrollable convection of the electrolyte layer on the oxygen flux have led to the development of a 'solid electrolyte' based on ion membranes (Niedrach and Stoddard, 1972), or to deposition of a thin porous cathode on the inner membrane surface (Bergman, 1970). Stabilization of the electrolyte layer by inserting a cellophane foil leads to a sluggish probe response (Lübbers *et al.*, 1969; Thunstrom and Severinghaus, 1978).

Aqueous solutions of salts of various concentrations are used in most probes. The saturated KCl solution used by Clark (1956) causes undesired deposition of Ag on the cathode, increasing the residual current. Phillips and Johnson (1961) have recommended 0.75 kmol m^{-3} KCl, in which this effect is suppressed. Another choice has been Cl^--free electrolytes (Severinghaus, 1978). KOH, in which substitution of Cl^- for OH^- does not take place, causes Ag_2O to flake off the Ag anode (Phillips and Johnson, 1961). However, the expected distortion of the reading due to replacement of the original Cl^- ions by OH^- has not been observed (Kinsey and Bottomley, 1963). Buffered solutions, which prevent changes in pH, have been recommended in the presence of CO_2 (Kimmich and Kreuzer, 1969; Severinghaus, 1968). Phosphate buffer, $KHCO_3$, $NaHCO_3$, K_2CO_3–$KHCO_3$ mixture (Lee and Tsao, 1979), or acetate buffer (Borkowski and Johnson, 1967) are used. Additions of $NaNO_3$ can reduce the vapour pressure of water above the electrolyte to the level of the medium measured, thus inhibiting undesired water diffusion through the membrane (Thunstrom and Severinghaus, 1978).

For galvanic probes, Mancy *et al.* (1962) tested solutions of KCl, $KHSO_4$, KOH and H_2SO_4; the highest probe sensitivity was found with 1 kmol m^{-3} KOH.

19.4.3 Membranes

The most important properties are the solubility and diffusivity of oxygen, water, and other gases in the membrane material (some of the data are listed in Table 4). Solubility and permeability data for a number of gases at various temperatures have been presented by Hwang *et al.* (1974). Other important membrane properties are the heat resistance and mechanical strength. Some membrane materials are hydrated to such an extent in aqueous media that their oxygen solubilities and diffusivities are significantly changed, *e.g.* the permeability of poly(vinyl chloride) (PVC) and poly(vinyl fluoride) (PVF) membranes increases several times in the presence of water. Permeabilities of polytetrafluoroethylene (PTFE), polyethylene (PE) and polypropylene (PP) are not affected by water, but are dependent on mechanical stress which apparently gives rise to structural changes in the polymers and an increase in permeability (Aiba and Huang, 1969a). Therefore, membranes have to be fixed to the cathode in such a way as to ensure uniform thickness and lack of mechanical stress. The particularly inelastic PTFE membrane cracks when stretched; on the other hand, PTFE has the advantage of high heat resistance and extremely low water permeability. The parameters of PE and PP are not stable over the long term, but these materials have conveniently low oxygen permeabilities. Natural and silicone rubber membranes squeeze out electrolyte from the space between the membrane and the cathode (Aiba *et al.*, 1968).

Table 4 Physical Properties of Membrane Materials[a]

Material	Oxygen $k_M \times 10^{16d}$ (g/g)[b]	(w/w)[c]	T (°C)	Oxygen $D \times 10^{11c}$ (g/g)[b]	(w/w)[c]	T (°C)	Carbon dioxide $k_M \times 10^{16d}$ (g/g)[b]	T (°C)	Water $k_M \times 10^{16d}$	T (°C)	Upper temperature limit (°C)
Polytetrafluoroethylene (PTFE)		29.2	20		1.25	20	57	26	28.4	38	230
	25.6	26.2	30								
Polyethylene (PE)											
high density	11.4	12.4	30	5.4	5.3	30	17.8	30	187	25	80
low density	19.2	21.4	30				44.4	30	190	25	
Polypropylene (PP)	13.4	12.4	30	2.4	2.6	30					
	5.8	5.0	20		1.4	20	22.8	25	159	25	130
Polystyrene (PS)	8.8		25								
	8.4	10.4	30				35	25	2180	25	80
Poly(vinyl chloride) (PVC)	1.6	2.86	20				10	25	940	25	50
	2.1		25								
Silicone rubber (SR)	284	286	30								
	1448		20	175	51.2	17	10 016	25	61 400	25	—

[a] Aiba *et al.* (1968), Aiba and Huang (1969a), Hwang *et al.* (1974). [b] g/g—measured in gas *vs.* gas phase. [c] w/w—measured in aqueous *vs.* aqueous phase. [d] $k_M = $ mol m^{-1} Pa^{-1} s^{-1}. [e] $D = $ m^2 s^{-1}.

19.4.4 Construction and Probe Types

In addition to the material and composition of the layers E, M and L, the factors controlling probe performance are the geometrical relations among the cathode and reference electrode (electrode shape and size), the electrolyte (size, location, shape and material of the electrolyte reservoir), and the membrane (thickness, size, method of fixing, areas in contact with electrolyte and the medium measured). Examples of construction of frequently used types of probes are shown in Figure 6. The cathode area ranges from a few μm^2 to several cm^2, and its magnitude determines the probe reading (see equation 15). The cathodes in use are planar [disc, multidisc, or multiwire (Siu and Cobbold, 1976); S-shaped and O-ring (Kimmich and Kreuzer, 1969); spiral (Kimmich *et al.*, 1981); wave shaped]; cylindrical [hollow cylinder (Mackereth, 1964), wire spiral (Borkowski and Johnson, 1967)] and in microelectrodes, conical and half-spherical (Lee *et al.*, 1978; Silver, 1973). The choice of cathode shape and electrolyte configuration may affect the role of side diffusion, which becomes more significant as the value of w (equation 17) is increased. The electrolyte may be spread into a thin film connecting the cathode and anode spaces, as in MN and HC probes and microelectrodes (this configuration is usually associated with short probe life); or the electrolyte film between the cathode and the membrane is connected with a larger reservoir placed at a distance from the cathode, and is exposed, to various degrees, to the contact with the surroundings through the membrane. The next section will show how various designs lead to different probe properties.

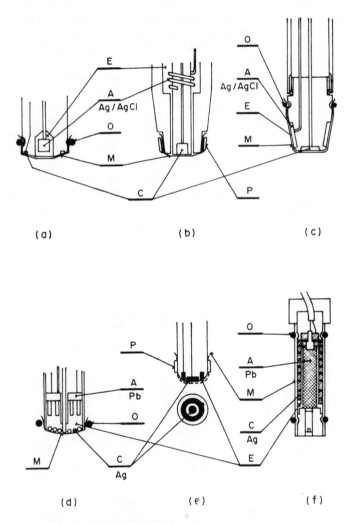

Figure 6 Construction of various oxygen probes (C, cathode; E, electrolyte; M, membrane; A, anode; O, O-ring; P, plastic collar). (a) YSI, (b) CV (c) HC (d) BJ (e) MN (f) MC

19.5 PROBE CHARACTERISTICS: RELATION TO PROBE DESIGN

The probe characteristics depend on the thicknesses, diffusivities and permeabilities of layers E, M and L (Figure 1), and on the cathode shape and size. These factors are considered below in connection with the individual characteristics. Probe characteristics have been discussed by a number of authors for various probe types: galvanic (Elsworth, 1972; Mancy *et al.*, 1962); polaro-graphic (Kinsey and Bottomley, 1963; Lucero, 1968, 1969); multicathode (Siu and Cobbold, 1976); catheter (Kreuzer *et al.*, 1980; Schuler and Kreuzer, 1967); recessed microelectrodes (Schneiderman and Goldstick, 1978); see also Table 1 for reviews.

19.5.1 Linearity

A probe is linear when its sensitivity (equation 19) does not depend on the oxygen tension. With the exception of steady-state measurements where empirically calibrated probes with repro-ducible nonlinear response can be applied, all OP applications require linear probes.

An initially linear probe may begin, for various reasons, to show nonlinear behaviour. For example, the oxygen reduction rate at a Pt cathode may decrease with time (Aiba and Huang, 1969a; Hitchman, 1978; Hoare, 1968): excessive oxygen chemisorption in the surface oxide layer may, like other adsorbed gases or impurities, prevent adsorption and thus also reduction of oxy-gen; or, conversely, the oxide layer may not be so developed as to catalyze peroxide decompo-sition at a sufficient rate. This situation does not arise at a sufficiently negative potential and high pH of the electrolyte in the cathode space (Hoare, 1968; Hitchman, 1978).

Nonlinearity of response may also be due to a high resistance of the Ag anode, arising from AgCl accumulation (Le Fevre; see Fatt, 1976, p. 44). Removing excess AgCl restores the linear-ity.

Linearity of probe behaviour may break down at higher oxygen tension, *e.g.* Maclennan and Pirt (1966) found their probe to be nonlinear at $P > 13$ kPa (100 torr). An analysis of probe linearity for Pt and Ag cathodes of various sizes and membranes of various thicknesses and materials has been given by Linek *et al.* (1984, 1985). Figure 7 shows probe responses to step changes in the gas-phase oxygen concentration. For low P_c, or with thick membranes of low per-meability, and small cathodes the response is monotonous and the probe linear. Increasing P_c and the cathode size, and reducing the thickness and increasing the permeability of membrane lead to a nonlinear steady-state reading, G_s, and, to a smaller extent, even to a nonlinear G_{max} defined in Figure 7a. Deviations from linearity, ΔG_s, defined in Figure 7b, are plotted in Figure 7d against $(P_c k_M d_K^2 / l_M l_E)$ describing local pH increase at the cathode surface. The upper limit of oxygen tension, P_{max}, up to which the probe is linear is given by equation (18) (see Figure 7d).

$$P_{max} = (6 \times 10^{-6}) l_E l_M k_M^{-1} d_K^{-2} \quad [Pa] \tag{18}$$

In contrast to N_2–O_2 interchange, no non-monotonous response profile has been observed for O_2–N_2 interchange. Equation (18) suggests a possibility of enhancing the linearity range by increasing l_E, l_M or by reducing the cathode size d_K or the membrane permeability k_M (*i.e.* by substituting PP for PTFE; see Table 4). Microcathodes are usually linear up to high oxygen con-centrations. Increasing the polarization voltage up to the onset of the hydrogen wave and thick-ening the electrolyte layer widens the linearity range of the probe. Even fast-response probes with large cathodes can be linear over the whole oxygen concentration range if a membrane of low permeability (PE, PP) is used.

In view of the character of the nonlinearity sources, it is advisable to repeat the test for linearity after each preparation.

19.5.2 Stability

Instability means a change in sensitivity with time, which may occur at various rates over the time scale of measurement. This has led to the introduction of the terms short-term and long-term instability.

Short-term instability may arise from:

(a) Reduction of surface oxides on the cathode. The initial stabilization period of a newly-pre-pared sensor ranges from several minutes to 10 h. This period can be shortened electrochemically (*e.g.* by increasing the current density) down to 7 min (Mindt and Eberhard, 1978).

(b) The presence of another component which passes through the membrane and (i) interferes with oxygen measurement (*e.g.* chlorine in tap water, Br_2, NO); (ii) alters the electrolyte compo-sition (CO_2); (iii) poisons the Pt cathode (H_2S); (iv) contaminates the cathode surface (reducible metal ions). The contamination may also proceed very slowly, leading to long-term instability.

(c) Movement of the membrane due to its poor fixing. The consequent movements of the elec-

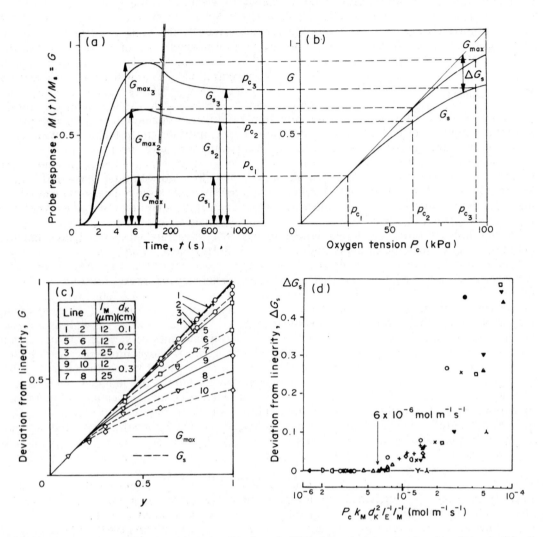

Figure 7 Linearity and stability of oxygen probes (Linek *et al.*, 1984): (a) probe responses at various P_c levels; (b) calibrating curves for G_{max} and G_s; (c) experimentally determined calibrating curves for PTFE membrane and Pt cathodes; (d) correlation of deviations from linearity ΔG_s and $P_c d_K^2 k_M (l_M l_E)^{-1}$. Electrolyte used: 2 kmol m^{-3} KCl.

Symbol	d_K (mm)		l_M (μm)		OP type
◁	0.5	Pt	25	PTFE	CV
◘	1	Pt	15	PP	CV
+	2	Pt	25	PTFE	CV
×	2	Pt	12	PTFE	CV
○	3	Pt	25	PTFE	CV
◇	3	Pt	12	PTFE	CV
●	3	Ag	25	PTFE	CV
▼	3	Ag	12	PTFE	CV
□	3	Pt	12	PTFE	CV
Y	7	Pt	15	PP	HC
人	7	Pt	12	PTFE	HC
■	1,9	Au	25	PTFE	YSI

trolyte layer, whose permeability is much higher than that of the membrane, change the proportion of side diffusion through the electrolyte, and thereby the instantaneous oxygen flux to the cathode (Severinghaus and Thunstrom, 1978).

(d) Local changes in pH near the cathode induced by high oxygen flux to the cathode and slow withdrawal of the OH$^-$ ions formed (Hitchman, 1978; Linek *et al.*, 1984, 1985).

Long-term instability may be due to:

(a) Slow side reactions (deposition of substances from the electrolyte layer). Specific cases of gaseous poisons have been discussed by Hitchman (1978). Severinghaus and Thunstrom (1978) have recommended cleaning or restoring the cathode surface by polishing with fine abrasives.

(b) Gradual depletion due to consumption of the anode or reduction of its area by anodic dissolution. With microcathodes, deposition of anodic Ag on the cathode and, consequently, an increase in sensitivity may occur at high Cl^- concentrations (Severinghaus and Thunstrom, 1978). Deposition of small particles of anode products on the cathode leading to an eventual build-up of a continuous bridge between the two electrodes is also a serious limitation on the detector stability (Hitchman, 1978). This problem does not arise, for example, in galvanic probes of the type described by Mancy *et al.* (1962), where the anodic reaction product, Pb^{2+}, is quite soluble in the KOH electrolyte and is not reduced on the cathode.

(c) Significant effect of the electrolyte layer diffusion resistance on the probe sensitivity. In this case, dilution or concentration of electrolyte due to the movement of water through the membrane leads to an enhancement or a reduction of sensitivity, respectively (Thunstrom and Severinghaus, 1978). PTFE is advantageous from this point of view, thanks to its extremely low water permeability.

(d) Membrane ageing associated with structural changes in the polymer material, leading to a gradual change in the transport characteristics of membrane (Aiba *et al.*, 1968; Kok and Zajic, 1975).

Probes of long-term stability have been designed even for measurements under extreme conditions, such as repeated steam sterilization, or measurements in sea depths over periods of months (Fatt, 1976; Lee and Tsao, 1979). Long-term stability is generally inherent in galvanic-type probes with a Ag cathode, and in probes with low intensities of oxygen flux to the cathode, *i.e.* with low sensitivity.

19.5.3 Sensitivity

The sensitivity, s, is defined as the ratio of the steady-state current output per unit surface area of cathode, M_s/S, to the partial pressure of oxygen, P, in the medium measured. From equations (15) and (A1, Table 3) it follows that

$$s \equiv M_s/(PS) = nF/(R_E + R_M + R_L) = nF/(l_E/k_E + l_M/k_M + l_L/k_L) \qquad (19)$$

So long as R_L is negligible with respect to (R_E+R_M), the sensitivity remains the same even in media of different oxygen solubilities; this means that the probe reading is the same in air and in air-saturated electrolyte solutions of oxygen concentrations differing by up to an order of magnitude. This has been verified experimentally by Kinsey and Bottomley (1963) and Mancy *et al.* (1962).

The electrolyte layer thickness is usually governed by the roughness of cathode surface, the latter being typically around 3 μm (Jensen *et al.*, 1978). The effect of electrolyte layer thickness on sensitivity has been investigated by Vogel *et al.* (1969), who found an optimum when the membrane was just in contact with the cathode: separating the membrane from, as well as pressing it onto, the cathode diminished the sensitivity (see also Fatt, 1976, p. 112). The electrolyte layer is usually sufficiently thin so that $R_E \ll R_M+R_L$. For example, for a 3 μm thick electrolyte layer of similar properties to those of water, $R_E = 1.14 \times 10^8$ mol^{-1} m^2 Pa s, while for a 25 μm PTFE membrane $R_M = 85.1 \times 10^8$ mol^{-1} m^2 Pa s. Exceptions are found in the cases where the electrolyte layer has been stabilized by a cellophane spacer soaked with the electrolyte, or where lens paper (Mancy *et al.*, 1962), or a very thin membrane has been employed; then the sensitivity depends on the electrolyte composition. By selecting electrolyte composition and concentration such that the oxygen flux to the cathode is reduced, one can increase the stability while preserving the response time (Thunstrom and Severinghaus, 1978).

As the cathode size is reduced, the sensitivity increases owing to the growing significance of side diffusion; see Section 19.3.2 (Jensen *et al.*, 1978; Kimmich and Kreuzer, 1969). With increasing side diffusion there is a decrease in the dependency on the external fluid flow hydrodynamics. However, the current load of the cathode may become so high that the current will be determined not only by diffusion, but also by the electrode kinetics (Jensen *et al.*, 1978).

The minimum measurable oxygen concentration is determined by the magnitude and stability of the residual current (RC; also called zero, background, dark or offset current), *i.e.* the current output observed when the sensor surroundings are free of oxygen. RC is not stable and so the lower sensitivity limit is determined primarily by its instability. Low RC and high stable RC may be compensated electrically and subtracted from the signal, respectively. Measurements with high unstable RC are useless.

The most frequent sources of RC are: reduction of electrochemically active impurities present in the electrolyte (*e.g.* metal ions) or on the cathode surface (noble metal oxides), or coming from the medium measured (*e.g.* chlorine from tap water); reduction of the water in the electrolyte solution, particularly when too high a polarizing voltage is used; electrical leakage through the insulating material separating the two electrodes; water-absorbing materials such as epoxy resin are unsuitable; an ideal insulating material is glass; oxygen dissolved in the insulating body of the sensor (when it is a plastic) or in the electrolyte reservoir. This effect can be suppressed by using suitable construction materials (Lucero, 1969).

19.5.4 Probe Dynamics

The probe dynamics are fully described by the transient characteristics. Sometimes the time required for a certain percentage (*e.g.* 95%) of the full scale response is used, but these characteristics do not allow one to evaluate the distortion of rapid processes due to the probe dynamics. The one-dimensional, single-layer diffusion model, (equation E2, Table 3), can be used (Mancy *et al.*, 1962; Linek *et al.*, 1978a) to describe a class of probes with planar disc cathodes of circular cross section over 0.4 mm, and with membranes sticking tightly to the cathode so that the electrolyte layer thickness is only given by the roughness of the cathode surface. The effect of the electrolyte layer on the probe dynamics may usually be neglected (Linek *et al.*, 1983a, 1984). As the electrolyte layer becomes thicker, the response time increases, but simultaneously the stability and the side diffusion are enhanced. Addition of $NaNO_3$ to the electrolyte can improve the probe stability (by reducing O_2 flux to the cathode and preventing penetration of water through the membrane) without increasing the response time; this is because oxygen permeability is lowered while oxygen diffusivity in the electrolyte remains unchanged (Thunstrom and Severinghaus, 1978). In an effort to eliminate the effect of the electrolyte layer on the oxygen transport, the layer is sometimes thinned down to a limit where OH^- ions cannot be drawn away from the cathode at a sufficient rate, and undesired changes in the electrolyte pH may occur (see Section 19.5.1).

Some probes show a slowdown in the last 20% of the response as compared with the rate predicted from one-dimensional models. This tailing of response may be due to:

(i) Side diffusion of oxygen through the electrolyte layer or membrane (Benedek and Heideger, 1970; Linek *et al.*, 1979a). Aiba and Huang (1969a) have demonstrated an increase in slowdown with increasing electrolyte layer thickness. The role of side diffusion may be enhanced when the membrane permeability is reduced, *e.g.* on replacing a PP membrane by PTFE (Hitchman 1978; Linek *et al.*, 1984).

(ii) Irregular thicknesses of electrolyte layer and membrane. In the BJ probe, for instance, the cathode consists of a wire in contact with the membrane, so that the electrolyte thickness is variable. An extensive slowdown for BJ type probes was found by Linek and Beneš (1977) and Linek *et al.* (1979a). Irregular stretching of the membrane may lead to non-uniform thickness.

(iii) Side diffusion of oxygen from the electrolyte reservoir if it is placed near the cathode and exposed to the surroundings, as is the case, for example, with the YSI probe (Kok and Zajic, 1975; Linek *et al.*, 1979a).

(iv) Electrochemical effects, *e.g.* insufficient rate of H_2O_2 decomposition, resulting in a slow establishment of a peroxide flux away from the cathode (Hitchman, 1978; Linek *et al.*, 1985). A slowdown may develop even in probes that did not show it initially. Aiba and Huang (1969a) ascribe this to electrochemical causes such as gradual changes in the catalytic properties of the cathode surface. The explanation advanced by Benedek and Heideger (1970) considers a decrease in the membrane stress resulting in a thickening of the electrolyte layer and an enhancement of side diffusion.

Probes with complicated dynamic behaviour (YSI, BJ, MC) are not suited to monitoring rapid processes; see Section 19.7.2. If probes show only a slight slowdown, their behavior can be described by the two-region, one-dimensional model (Table 3) with the parameters Z_1, K_{M1} and K_{M2}, which holds for negligible film resistance, *i.e.* for $R_L = 0$

$$H = 1 + 2Z_1 \sum_{n=1}^{\infty} (-1)^n \exp[-n^2\pi^2 K_{M1}t] + 2(1-Z_1)\sum_{n=1}^{\infty} (-1)^n \exp[-n^2\pi^2 K_{M2}t] \qquad (20)$$

The role of side diffusion is especially notable with microprobes, where the cathode size is comparable with, or less than, the membrane thickness. Transient characteristics of microprobes can therefore be expected to show slowdown. However, probe response need not be strongly influenced by the cathode size (Kinsey and Bottomley, 1963). When the slowdown is due to electrochemical effects or side diffusion of oxygen from the reservoir, the transient characteristics may show hysteresis, usually with the upstep response faster than the downstep response (YSI, BJ, MC probes). Diffusion models allowing for side diffusion can describe hysteresis only when it is due to the lack of symmetry of the initial and boundary conditions for the two steps. In probes with high-capacity reservoirs close to the cathode, this lack of symmetry is, in effect, due to different times of saturation of the probe with oxygen for the upstep and the downstep response (probes are usually stored in air, and the electrolyte reservoirs are air-saturated). Kok and Zajic (1975) studied the dependence of the transient characteristics on the time for which probes of this type had been kept in a medium of an initial specified concentration. When the electrolyte reservoir and the cathode are sufficiently distant from each other and connected by a thin film only, no hysteresis may occur (*e.g.* CV, MN probes) even when the probe exhibits a slowdown (Heineken, 1971; Linek *et al.*, 1984, 1985). However, Linek *et al.* (1984, 1985) showed that hysteresis in this type of probe frequently may be due to nonlinearity arising from a large current load at high oxygen concentrations.

19.5.5 Liquid Film Effect

19.5.5.1 *Effects of individual layers and of cathode geometry*

The continuous consumption of oxygen by a probe produces a depletion layer of decreasing oxygen content, not only in the electrolyte and membrane, but also in the liquid film adjacent to the membrane. Since the sensor measures oxygen tension in this region, such a depletion would lead to erroneously low values. Lilley *et al.* (1969) have developed a chronoamperometric technique of measurement: since during the short period of measurement the depletion layer does not reach the outer surface of the membrane, the probe signal is independent of stirring or flow rate and/or of fouling of the outer membrane surface. This technique is only suitable for very slow oxygen tension changes: oxygen tension in the electrolyte and the membrane must be equilibrated with that in the medium measured prior to initiating the measurement, which may take as long as 5 min (Lilley *et al.*, 1969). Stirring or increasing the rate of flow round the membrane reduces the extent of the depletion layer and suppresses its effect, *i.e.* a drop in the reading and delay in response time. In fact, the oxygen concentration which is to be measured is that of the liquid film layer where the actual interaction occurs between the probe and the medium. It is more difficult to gain an insight into the quantitative description of the liquid film effect than in the case of layers E and M. The effect of the liquid film on the probe characteristics has been described using the notion of a uniform stagnant film (Aiba and Huang, 1969b; Linek and Vacek, 1976; Linek *et al.*, 1983a; Schuler and Kreuzer, 1969) of effective thickness l_L, evaluated from the relation $R_L = l_L/k_L$ using a measured value of R_L.

From equations (15) and (A1, Table 3) for steady-state probe reading one can derive a gas/liquid ratio or stirring effect factor, f, for the dependency of sensitivity upon fluid flow

$$f = \frac{M_{sg}}{M_{sl}} = \frac{s_g}{s_l} = 1 + \frac{R_L}{R_E + R_M} = 1 + R_L s_g/(nF) \qquad (21)$$

The lowering of the flow dependency resulting from increasing the flow rate around the probe and reducing R_L can be characterized in terms of a critical velocity above which the stirring effect factor f remains below a chosen value, *e.g.* 1.03. Estimation of the critical velocity has been made by Aiba and Huang (1969b) and Linek and Vacek (1977) on the basis of experimental data on f as a function of fluid velocity and viscosity, and by Schuler and Kreuzer (1969) who applied the boundary layer theory.

In the one-dimensional models presented (Table 3), the liquid film effect is characterized by the parameter $L_L = R_L/R_M$. For probes with negligible electrolyte effect, the L_L value can be found directly from the measured stirring effect factor:

$$L_L = f - 1 \qquad (22)$$

The flow dependency factor, f, can be diminished by increasing the resistances R_E and R_M. In practice, decreasing R_L and increasing R_E and R_M is limited, for example, by requirements for probe response or by the impracticability of stirring in some cases. The stirring effect also decreases with increasing proportion of side diffusion. For oxygen side-diffusing from that part of the membrane which is not in direct contact with the cathode, the ratio $R_L/(R_E+R_M)$ is markedly lower on account of the longer diffusion path through the membrane and the electrolyte. This is utilized to suppress the stirring effect in O-ring, S- and wave-shaped cathodes having large values of w, and in multiwire cathodes. The effect of cathode size on the stirring effect has been studied by Aiba and Huang (1969b), and by Schuler and Kreuzer (1967). Their results are based on the use of dimensionless criteria, with the cathode diameter chosen, without an adequate justification, as the characteristic dimension. The correlation presented by Aiba and Huang (1969b) predicts, in contradiction to experimental findings, an enhancement of the stirring effect with reduction in cathode size. The effect of cathode size and recession on the stirring effect has been described by Schneiderman and Goldstick (1978). The effect of cathode radius on the stirring effect for spherical cathodes (Vacek *et al.*, 1985) can be assessed from the relation for the limiting case of stagnant medium. From equation (S1, Table 3) it follows that

$$f = 1 + k_M r_0/(k_L l_M) \tag{23}$$

When $k_M r_0/k_L l_M \ll 1$, the effect of the outer medium, and thus also the stirring effect, are negligible. Silver (1973) states that probes with cathodes of diameters under 1 μm become insensitive to the liquid flow even without a membrane.

The probe linearity may break down as a result of a change in hydrodynamic conditions during measurement. For example, inappropriate location of a probe with a small cathode may give rise to fluctuations of reading due to interaction with bubbles (Aiba and Huang, 1969b; Linek and Vacek, 1978, 1979; Merta and Dunn, 1976).

The film has much greater effect on the dynamic characteristics than on sensitivity. For example, with a catheter-type probe having a 6 μm thick membrane and an O-ring cathode 3 μm thick and ~1 mm in diameter, the liquid-phase measurement 95% response time is double that in the gas phase even when the stirring effect factor is only 1.003, and is still about 30% longer at high rates of flow round the membrane (100 cm s^{-1}) when the steady-state reading is no longer flow dependent (Kimmich and Kreuzer, 1969).

Unlike sensitivity, which depends only on the film resistance L_L, the dynamic characteristics are also a function of the film capacity expressed by the parameter A_L^2/L_L ($=A_L/S_L$). The effect of the film on the transient characteristics has only been quantified for probes with little or no slowdown, which can be described by one-dimensional single-region (see Table 3) or two-region models. For the two-region model, only the variant has been derived (equation 30) which disregards the film capacity component (model F, Table 3). Experimentally-determined slopes of the linear portions of transient characteristics ln $(1-H)$ *vs. t*, see Figure 8, indicate that the simplified model F is only adequate for low L_L and high S_L. The range where the liquid film capacity is insignificant is well documented in Figures 9a and 9b, which show the moments of transient characteristics in the form b_H/b_M, where b_M is the zeroth moment of the transient characteristics for the probe without a film. Figure 9b shows relative differences of zeroth moments, evaluated with (b_H from equation B4, Table 3) and without (b_H^∞ for $S_L \to \infty$ from equation F4, Table 3) considering the film capacity. This plot allows one to estimate the values of the stirring effect factor, f, or L_L (equation 22) up to which the above model can be used for a given S_L value.

19.5.5.2 Effect of chemical reaction

A chemical reaction in the bulk of liquid proceeds also within the liquid film before the membrane, changing, more or less, the concentration profile and the transport characteristics of the film (by increasing its capacity) and, ultimately, the probe response. Linear reactions, *i.e.* zero- and first-order reactions, lower the reading (Lundsgaard *et al.*, 1978) and shorten the response time (Linek *et al.*, 1983a) without disturbing the probe linearity. Reactions with nonlinear kinetics, such as enzymatic reactions with Michaelis–Menten kinetics (Lundsgaard *et al.*, 1978), or oxygen transport in blood (Siu and Cobbold, 1976) may be responsible for probe nonlinearity.

The analysis of the effect of zero- or first-order reaction on transient probe characteristics presented by Linek *et al.* (1983a) gives a guideline for deciding whether or not the effect of a reaction can be neglected. The extent to which the relevance of a first-order reaction depends on the parameters B_1, S_L and L_L is demonstrated in Figure 10. The way in which a significant effect of a chemical reaction is to be taken into consideration is presented in Section 19.6.2.2.

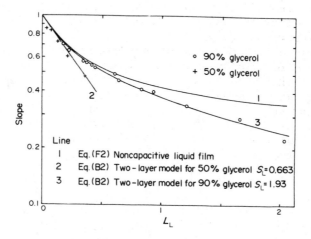

Figure 8 Comparison of experimental values (Linek and Vacek, 1977) of the slope of the linear part of $\ln(1-H)$ *vs. t* plot with the values calculated from the two-layer model (lines 2 and 3) for 50% and 90% glycerol solutions, respectively, and from one-layer model with noncapacitive liquid film (line 1) (Linek *et al.*, 1983b)

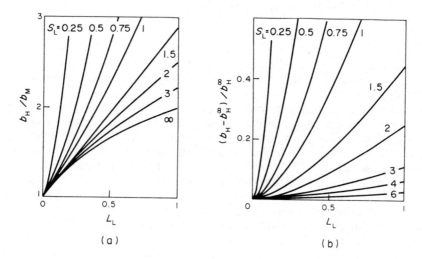

Figure 9 (a) The ratio b_H/b_M as a function of the parameters L_L and S_L (equation B4, Table 3); (b) the relative difference $(b_H - b_H^\infty)/b_H^\infty$ as a function of the parameters L_L and S_L

19.5.6 Effect of Temperature

Sensitivity (Kinsey and Bottomley, 1963; Lucero, 1969) and response time (Linek and Petrlík, 1979; Schuler and Kreuzer, 1969) increase with temperature by as much as 1 to 6% per 1 °C. For precise measurements it is therefore important to thermostat the system to within better than ± 0.1 °C. As the temperature is increased, both the diffusivities and permeabilities of the probe layers are enhanced. The effect of temperature can be incorporated into one-dimensional diffusion models in terms of the activation energies of the diffusivities and permeabilities of the respective layers:

$$D_i(T_2) = D_i(T_1) \exp\left[\frac{E_{Di}}{R}\left(\frac{1}{T_1} - \frac{1}{T_2}\right)\right] \tag{24}$$

$$k_i(T_2) = k_i(T_1) \exp\left[\frac{E_{ki}}{R}\left(\frac{1}{T_1} - \frac{1}{T_2}\right)\right] \tag{25}$$

where i = E, M, L. In the one-layer diffusion model the temperature dependence of sensitivity is

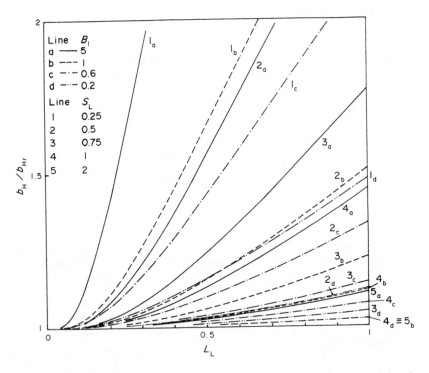

Figure 10 The effect of first-order reaction on zero-th moment of transient probe characteristics: dependence of the ratio b_H/b_{Hr}, given by equations (B4) and (C4), Table 3, on L_L, S_L and B_1 parameters

governed by the activation energy for the membrane permeability, E_{kM}, while the temperature dependence of the transient characteristics is controlled by the activation energy for the membrane diffusivity, E_{DM}, which determines the probe constant K_M. Agreement between published E_{DM} and E_{kM} data and values evaluated from the temperature dependences of probe characteristics was found to be better for PTFE than for PE and PP (Linek and Petrlík, 1979).

The effect of temperature on the sensitivity can be eliminated by electronic compensation with the use of a thermistor. Since the residual current and the diffusion flux of oxygen depend on temperature in different ways, the performance of the compensating circuits may not be equally satisfactory at low and at high O_2 concentrations (Kinsey and Bottomley, 1963). To achieve adequate compensation in situations encountered in common practice (measurements with and without film, using various kinds of membranes and electrolytes), it is necessary that the compensating circuits be sufficiently flexible. Automatic temperature compensation is suitable at temperature changes in the medium slow enough for the difference in the rates at which the thermistor and diffusion zone temperatures reach steady values to be insignificant. This problem has been analyzed in detail by Hitchman (1978) and by Lucero (1969). Electronic compensation of the effect of temperature on transient probe characteristics has not as yet been accomplished.

19.5.7 Effect of Pressure

The probe senses the partial pressure of oxygen, and its reading in a liquid of a given oxygen concentration should be independent of the total pressure (Mancy et al., 1962). In measurements in the gas phase, in contrast, a change in the total pressure alters the partial pressure of oxygen and, consequently, the probe reading. If the probe reading in a liquid does change with pressure, this may be a consequence of changes in the geometry determining oxygen transport towards the cathode (Lucero, 1968, 1969). Thinning the electrolyte layer by pressure, for example, may diminish or enhance the probe sensitivity depending on whether or not side diffusion plays a significant role. Moreover, Henry's law may no longer be valid at higher pressures (Hitchman, 1978; Lucero, 1968). When the electrode space is not completely filled with liquid, pressure changes may produce movements of the membrane and variations in the thickness and resistance of the electrolyte layer. This may be prevented by connecting the electrolyte reservoir with the surroundings by means of a freely movable diaphragm, equalizing the pressure differences and volume changes (YSI probes). Pressure changes may also be due to different thermal expansions of the individual parts of probe, especially when the electrolyte reservoir is large and tightly closed. Special probes have been designed with resistance to extreme conditions such as high pressures, external pressure shocks and vibrations, and liquid flow acceleration (Fatt, 1976; Lee and Tsao, 1979).

19.6   DETERMINATION OF PROBE CHARACTERISTICS

The essential characteristics for the interpretation of measured data are sensitivity and linearity, and, with nonsteady-state processes, the transient characteristics. In establishing these we distinguish two principal situations: (i) the characteristics can be measured directly under the same conditions, $\mathbb{C}_H$ (temperature, pressure, film, chemical reaction, level of interfering substances) at which the process under study occurs, $\mathbb{C} = \mathbb{C}_H$; (ii) the same conditions cannot be ensured, $\mathbb{C}_H \neq \mathbb{C}$, so that the measured characteristics must be corrected for the effect of the parameter which could not be adjusted to the desired level.

In what follows we assume the level of interfering substances to be constant. Their effect is then eliminated by taking the probe reading, M_g or M_l, as the difference between the true reading and the residual current.

19.6.1   Direct Determination of Probe Characteristics

19.6.1.1   Sensitivity and linearity

When the measurement is carried out in the gas phase, the calibration is based on partial pressure. Mixtures of oxygen and inert gases of various compositions are commercially available or can be made up with calibrated flow meters. Zero oxygen concentration is obtained by using a pure inert gas such as He, Ar, N_2O, cyclopropane, propane, butane or oxygen-free N_2. The use of sodium sulfite solution entails the risk of a rise in RC by introducing trace amounts of H_2S, which easily passes through the membrane. In liquids, the calibration can be based on (a) equivalent partial pressure of oxygen or degree of saturation, or (b) oxygen concentration in the liquid. At conditions of significant film resistance, the calibration has to be carried out with the same location of probe and in the same liquid as the measurement of the actual process studied. The liquid is saturated with gaseous mixtures of different levels of P_c.

Once the linearity of response has been verified (the check may conveniently be made in the gas phase), the calibration under conditions of significant film resistance can be done simply by fixing one point $M_{sl}(P)$ or $M_{sl}(c_l)$; then

$$s_l = M_{sl}/P \text{ for variant (a)} \tag{26}$$

$$s_l = M_{sl}/c_l = M_{sl}/(m_L P) \text{ for variant (b)} \tag{27}$$

Oxygen solubility, m_L, in a given liquid can be estimated from correlations, or determined by an independent analytical method. Another approach involves monitoring concentration changes produced by additions of known amounts of the oxygen-consuming coenzyme NADH (Robinson and Cooper, 1970).

For systems involving a chemical reaction, a direct method of probe calibration has not yet been devised.

19.6.1.2   Transient characteristics

The transient characteristic is obtained as a probe response to a step concentration change at the same values of the parameters L_L and A_L as in the process investigated. This is accomplished by measuring the transient characteristics at a liquid flow rate adjusted so as to provide the same L_L value as found by an independent experiment on the process under study. The same value of A_L is ensured by using the same liquid as in the process. If the transient characteristic is to be representative of the dynamics of all the three layers, the step concentration change must take place in front of the film, *i.e.* in the bulk of the liquid. This can be realized by placing the probe at the outlet of a three-way cock or a gate valve (adapted so as not to allow dead flow zones to be produced; see Figure 11). In simply turning the cock or opening the valve, liquid of initial oxygen concentration c_{11}, is replaced by the same liquid of an oxygen concentration $c_{12} \neq c_{11}$, flowing at the same rate. Luttman *et al.* (1978) have proposed an alternative photometric method allowing estimation the time course of c_1 independently of, and simultaneously with, the probe response.

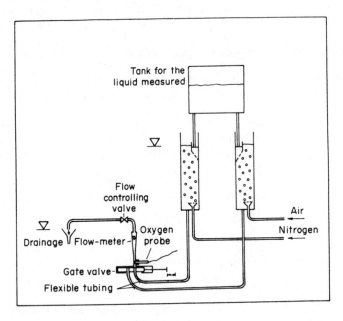

Figure 11 Experimental apparatus for determination of transient probe characteristics *H* with significant liquid film resistance

Dang *et al.* (1977) and Nakanoh and Yoshida (1980) have proposed a method for determining the transient characteristic at a significant liquid film resistance; the method consists essentially in creating a step concentration change by placing the probe at the measuring place after its equilibration in a nitrogen atmosphere. This method ensures that the film before the probe membrane will be identical with that existing in the measurement of the actual process. However, the initial conditions defining the transient probe characteristics are not fulfilled. Consequently, probe responses obtained by this method do not represent the transient characteristics (Linek *et al.*, 1983b), but can be converted to them (see Section 19.6.2.2).

For systems involving a chemical reaction, a direct method of establishing the transient characteristics has not yet been devised.

Special electronic analog circuits have been developed for compensating signal distortion due to probe dynamics (Kimmich *et al.*, 1981; Bourdaud and Lane, 1974; Luttman *et al.*, 1978).

19.6.2 Correction of Characteristics to Process Conditions

Correction for pressure effects can be made only for linear probes, in which case the sensitivity does not change with pressure, either in the gas phase (where probe reading, M_s, and oxygen partial pressure, P, are directly proportional to pressure), or in the liquid phase (where probe reading, M_s, and equivalent oxygen tension, P, in the bulk liquid are independent of pressure). The probe linearity may break down at higher oxygen concentrations and pressures, as a result of local increases in OH^- concentration near the cathode (see Section 19.5.1), or due to a pressure dependence of Henry's law constant (Hitchman, 1978; Lucero, 1969).

19.6.2.1 Sensitivity

The effect of temperature on sensitivity is usually well described by the Arrhenius equation, with the activation energy for sensitivity obtained empirically as the slope of the $\ln M_s$ *vs.* $(1/T)$ plot. The activation energy for probe sensitivity in a liquid may differ from the value in a gas when the film resistance is significant. In this case, one has to measure the activation energy for the probe reading, E_{M_s}, at the given L_L. Only in gas-phase measurements using a probe with a negligible electrolyte layer can the temperature effect be estimated from the activation energy for oxygen permeability in the membrane material (equation 25). Owing to a dependence of the permeability of a membrane on its condition and history (stretching, manufacturing procedure, age, *etc.*), this method cannot be regarded as reliable. When the probe calibration is based on the oxygen concentration in the liquid (equation 27), the temperature dependence of the probe reading, $M_s(T)$, is additionally related to the temperature dependence of solubility, $m_L(T)$.

From the above discussion, it follows that in automatic compensation of the temperature effect the temperature coefficient should be adjusted so that the probe reading (i) does not change with temperature (for calibration based on relative or absolute oxygen tension), or (ii) changes with temperature following the change in oxygen solubility in the medium being measured (for calibration based on concentration c_l).

The dependence of sensitivity on film resistance has to be measured directly. For probes with proven linearity, it suffices to fix a single point of the M_{sl} *vs.* P (or c_l) plot, see Section 19.6.1.1.

The effect of a reaction on sensitivity can be estimated from equations (B1), (C1) and (D1) in Table 3.

$$s_l/s_{lr} = (1+L_L)A_L\sqrt{B_1}/[L_L\sinh(A_L\sqrt{B_1}) + A_L\sqrt{B_1}\cosh(A_L\sqrt{B_1})] \tag{28}$$

$$s_{lr}/s_l = 1 - B_0A_L^2/(2P) \tag{29}$$

For a first-order reaction, the slope of the calibration M_s *vs.* P plot is altered according to equation (28); in the case of a zero-order reaction, the zero reading ($M_s = 0$) is shifted to a pressure value $P = B_0A_L^2/[2(1+L_L)]$, while the slope, $1/(L_L+1)$, remains unchanged (see equation 29). The method for estimating the parameters A_L, B_0 and B_1 is described in Section 19.6.3.

19.6.2.2 Transient characteristics

The effect of temperature on transient characteristics in the special case of probes with negligible electrolyte and film layers follows from equations (F2, Table 3) and (30), and is governed by the temperature dependence of K_M given by equation (24). In the two-region model (equation 30), the K_M constants for the two regions are assumed to depend on temperature in the same manner. The description of the temperature dependence for multilayer models would require establishing temperature dependences of more parameters, this being more laborious and time-consuming than direct determination of the transient characteristics at another temperature.

Effect of liquid film: in dynamic measurements, it is preferable to use a probe with a negligible electrolyte layer. An analysis of the role of electrolyte has been given by Linek *et al.* (1983a, 1984). In the following discussion we shall confine ourselves to the cases where the electrolyte layer can be neglected.

In the two-layer model (equation B2, Table 3), the effect of the film on the transient characteristics is incorporated through the parameters L_L and S_L. Provided L_L is not too high and the film capacity can be neglected (see Section 19.5.5.1), it is possible to apply the simpler model of non-capacitive electrolyte resistance, covering even the frequent case of transient characteristics showing a slight slowdown.

$$H = Z_1H_1 + (1-Z_1)H_2 \tag{30}$$

where

$$H_u = 1 + 2(1+L_{Lu})\sum_{n=1}^{\infty} \frac{(-1)^n(1+\beta_{nu}^2L_{Lu}^2)^{0.5}}{(1+L_{Lu}+L_{Lu}^2\beta_{nu}^2)} \exp(-\beta_{nu}^2 K_{Mu}t) \tag{31}$$
$$u=1, 2$$

and β_{nu} are the positive roots of the equation

$$L_{Lu}\beta_{nu}\cos\beta_{nu} + \sin\beta_{nu} = 0 \tag{32}$$

The procedure involves measurement of the transient characteristics in the gas phase, where $L_{Lu}=0$ and equation (30) reduces to equation (20), and estimation of the parameters $Z_1=Z_{1g}$,

K_{M1} and K_{M2} from equation (20). The parameters L_{L1} and L_{L2} can be calculated from equations (33) and (34), respectively, by substituting independently-found values of the stirring effect factor, $f = M_{sg}/M_{sl}$.

$$Z_{1g}\frac{1}{1+L_{L1}} + (1-Z_{1g})\frac{1}{1+L_{L2}} - \frac{1}{f} = 0 \tag{33}$$

$$L_{L2} = L_{L1}(K_{M2}/K_{M1})^{0.5} \tag{34}$$

The value of Z_1 for L_{L1} so obtained can be evaluated from:

$$Z_1(L_{L1}) = \{1+(1-Z_{1g})(1+L_{L1})[1+L_{L1}(K_{M2}/K_{M1})^{0.5}]^{-1}Z_{1g}^{-1}\}^{-1} \tag{35}$$

The forms of equations (33) and (35) previously derived by Linek *et al.* (1979a) are incorrect. The zeroth moment of transient characteristics reflecting the film resistance (Linek *et al.*, 1979a) is

$$b_H(L_L) = + \frac{1}{6}\sum_{u=1}^{2}\frac{Z_u}{K_{Mu}}\frac{3L_{Lu}+1}{L_{Lu}+1} \tag{36}$$

where $Z_2=1-Z_1$

Effect of initial conditions: in the method proposed by Dang *et al.* (1977) and Nakanoh and Yoshida (1980) for determining transient characteristics under conditions of film resistance, responses corresponding to equations (A3), (B3), (C3), (D3), and (A5), (B5), (C5), (D5) of Table 3, where $z=1$, are incorrectly considered as the transient characteristics, *i.e.* as if they were given by equation (A2), (B2), (C2), (D2), and (A4), (B4), (C4), (D4) where $z=0$, respectively. Zeroth moments of Dang's responses (Dang *et al.*, 1977), b_D, can be converted to the moments of transient characteristics, b_H, for example by using equations (B4) and (B5).

$$\frac{b_H}{b_D} = [L_L^3 + 3L_L^2 + S_L^2(1+3L_L)][S_L^2(1+3L_L) - 2L_L^3]^{-1} \tag{37}$$

The effect of chemical reaction on transient characteristics in the one-dimensional two-layer model is given by equations (B2), (C2) and (D2) of Table 3. A detailed analysis of this effect has been presented by Linek *et al.* (1983a, 1983b). If the reaction kinetic parameter B_1 is known, we can calculate the transient characteristics or their zero moment from transient characteristics for the system without a reaction, or from the response obtained by the method of Dang *et al.* (1977) for the system with an accompanying reaction, using relationships given in Table 3.

19.6.3   Estimation of Parameters L_L, S_L (or A_L), B_0, B_1, Z_1, K_{M1}, K_{M2} and b_M

The parameter L_L for a system without a chemical reaction is determined by the stirring effect factor (equation 22).

The parameters K_M, b_M, Z_1, K_{M1} and K_{M2} can conveniently be estimated from the transient characteristics in the gas phase, where $L_L \to 0$, using either the moment method (b_M from equation F4, Table 3), or the regression method with K_M calculated from equation (F2), and K_{M1}, K_{M2} and Z_1 from equation (30).

The parameter $S_L=[k_M\sqrt{D_L}/(k_L\sqrt{D_M})]=(L_L/A_L)$ can be evaluated from known membrane and film permeabilities and diffusivities; alternatively, one can determine experimentally the dependence $b_D(L_L)$ or $b_H(L_L)$ and evaluate S_L by regression from equations (B4) or (B5) of Table 3.

$$\frac{b_D}{b_M} = (3L_L + 1 - 2L_L^3/S_L^2)(1 + L_L)^{-1} \tag{38}$$

$$\frac{b_H}{b_M} = [3L_L + 1 + L_L^2(L_L+3)/S_L^2](1 + L_L)^{-1} \tag{39}$$

The parameters B_0 and B_1 can be estimated by the method described by Linek *et al.* (1983a).

When B_0 or B_1 is the parameter to be evaluated, one must solve the equations describing the process investigated together with the relationship between the probe reading and the actual oxygen concentration, as discussed in Section 19.7.1.

19.7 APPLICATION OF OXYGEN PROBES

19.7.1 Interpretation of Probe Signals

The problem of OP application can be formulated generally as follows. A probe of known characteristics, H, is used to monitor a process characterized by the parameters sought, q_j. The probe yields a signal, M_s or $M(t)$, which is more or less distorted as a result of the probe dynamics or of a chemical reaction occurring within the film. If the probe is so chosen that the distortion is negligible (see Section 19.7.2), we can simply fit the model concentration profile to the probe signal, which in this case corresponds to the oxygen concentration. When this is not the case, the choice lies between two methods: (i) model signal, M_m, obtained from the concentration profile c_m calculated from the model of the process is modified so as to include the distortion introduced by the probe used; this corrected model signal is then fitted to the experimentally found signal M; (ii) the experimental signal, M, and the transient probe characteristics, H, are used to calculate the experimental concentration profile, c, which is fitted to concentration profile c_m following from the process model. Approach (ii) is used, for example, in automatic compensation of probe dynamics, and is limited to the cases where the transient probe characteristics can be determined (and the compensation parameters adjusted) at the same conditions under which the process under study occurs.

Whereas in case (i) the allowance for distortion due to the probe consists of a simple integration of the convolution integral equation (6) (Leeuwen, 1979; Linek *et al.*, 1979a; Mueller *et al.*, 1967), the method of case (ii) involves a more laborious solution of the Volterra integral equation, equation (6), of the second kind (Wodick, 1972).

The experimental procedures and the individual steps of probe signal evaluation are shown in Figure 12. Block 1: laboratory or full-scale performance of the process under conditions ₵, with a concentration change $c(t)$. The probe responds to the concentration c by signal output M incorporating the effect of the film. Block 2: independent experimental determination of the parameters L_L and A_L under the conditions of the process, ₵. Block 3: experimental determination of probe characteristics under conditions ₵$_H$. It is desirable that ₵$_H$=₵. Block 4: conversion of the transient characteristics, $H(t)$, under conditions ₵$_H$ to those corresponding to conditions ₵, using a suitable probe model. The consequences of an improper choice of the model have been analyzed by Linek *et al.* (1978a). Block 5: the model of the process may express the dependence of oxygen concentration in the bulk fluid on time and the process parameters, $c_m(t,q_i)$ in numerical or analytical form. Block 6: evaluation of model probe signal, $M_m(t)$ or M_{sm}, consisting of integrating equation (6). Along with block 4, this step may be considered as a theoretical counterpart of block 1. The function $c_m(t,q_j)$ may be of analytical form (for simpler systems, *e.g.* exponential profile of supersaturation concentration in a fermentation batch; analytical integration of Michaelis–Menten kinetic equation), or may be expressed only numerically. Linek *et al.* (1978b), for example, used a numerical solution of partial differential equations describing nonsteady-state adsorption in a packed column, with longitudinal mixing considered. The transient characteristics may likewise be given only numerically. Analytical solutions of the convolution integral, equation (6), for special cases of c and H functions have been presented by Linek *et al.* (1978a).

If the zeroth moment of the signal, b_G, has only been found, we can take advantage of the reduction of the convolution integral in equation (6) to a sum of zeroth moments of the actual process, $b_m(q_j)$, and of the transient probe characteristics, b_H, obtained at ₵$_H$=₵ (Van Leeuwen, 1979). Then

$$b_G = b_m(q_j) + b_H \qquad (40)$$

The moment of Dang's response (Dang *et al.*, 1977), b_D, cannot be substituted for b_H, as is usually done; it has to be converted to b_H using equation (37).

The above method can only be used for a small number of parameters q_j (1 or 2 at the most, see Linek and Beneš, 1978). The moment methods are substantially simpler and yield certain values of the required parameter even when an inadequate model of the process has been used, whereas the regression methods usually do not converge in such a case so that the risk of evaluating absurd results is less. Some authors erroneously regard this property of the regression approaches as a disadvantage. The regression method allows one to evaluate the zero time as a parameter, whereas in the moment method this time must be known beforehand. Unfortunately, experimental determination of the zero time is generally subject to large error which propagates into the resulting q_j values (Linek *et al.*, 1979b).

Block 7: an inverse operation to block 6. The relations for the transient characteristics of the

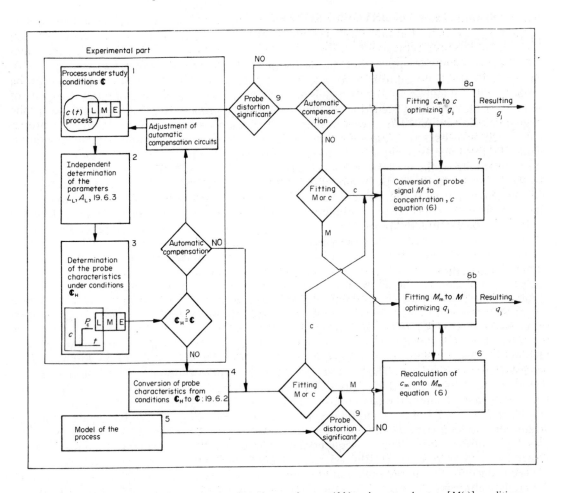

Figure 12   Evaluation of oxygen probe signals under steady-state (M_s) and nonsteady-state [$M(t)$] conditions

first and second orders have been presented by Wodick (1972). Block 8: common optimization methods of parameter estimation by fitting the model probe response (M_m,c_m) to the experimental probe response (M,c), using the criterion of least squares. For the moment methods the fitting reduces to trivial algebraic operations. Block 9: analysis of the significance of signal distortion. The analyses in conjunction with specific problems have been presented, *e.g.* by Nakanoh and Yoshida (1980), Linek (1972), and Benedek and Heideger (1970).

19.7.2   Probe Selection

The requirements for probes depend on the rate of the process studied and on the time scale of measurement. If recalibration *in situ* is possible, even probes of short-term stability may be used for long-term measurements. A device for probe recalibration *in situ* has been described by Kok (1976). Short-term stability is also a minimum requirement for probes used in short-term measurements of steady-state or slow (with respect to the probe response time) nonsteady-state processes. When the probe characteristics measured under conditions C_H are converted to those at the process conditions C, involving, for example, a chemical reaction, at least the steady-state probe reading has to be described by a one-dimensional or other suitable model.

In short-term nonsteady-state measurements of rapid processes, there is an additional requirement of constant temperature (for the above given reasons automatic temperature compensation cannot be used). According to the method of evaluation, the requirements for the dynamic probe behavior are as follows: (a) the shape of the transient characteristics of linear probe is immaterial if the characteristics are measured so that $C_H = C$; (b) if the transient characteristics are measured at $C_H \neq C$, and thus have to be converted to those for conditions C, the dynamic probe behavior must be described by a suitable model. It is usually one of the following models:

(i) A one-dimensional model of the probe not exhibiting a slowdown, *i.e.* not subject to decreasing response time: the film resistance and capacity can be allowed for (equations A2, B2; Table 3), but it is preferable to use a probe with negligible film capacity. Application of non-capacitive models, equation (F2), requires a sufficiently low value of parameter L_L (see Figure 9b), achievable by using sufficiently thick or poorly permeable membranes. The use of a fast probe with a more permeable membrane enhances the effect of the liquid film and also the sensitivity of the parameters q_j to experimental errors in L_L and A_L. The effect of reaction on the probe characteristics diminishes with decreasing L_L, and may, therefore, be suppressed by a proper choice of membrane. (ii) A non-capacitive, two-region model, equation (30), for probes exhibiting not too pronounced a slowdown.

Benedek and Heideger (1970) recommended an empirical check for probe signal distortion by repeated measurements using membranes of different thicknesses; the evaluated parameters q_j (*e.g.* k_la) should be independent of the membrane thickness. The parameter sensitivity of steady-state and dynamic methods of k_la determination using the first order probe response has been analyzed by Keitel and Onken (1981).

19.7.3 Comments on the Dynamic Method of k_la Measurement in Fermenters

The volumetric mass transfer coefficient, k_la, determined by dynamic methods in fermenters is mostly evaluated by the moment method, with the probe response approximated by the first order response (Fuchs *et al.*, 1971; Keitel and Onken, 1981) or, less frequently, by the second order response (Dang *et al.*, 1977; Ruchti *et al.*, 1981). We should like to point out the neglected fact that this approach is only justified when (i) the transient characteristic of the probe used (or its moment) is established at the same conditions under which the process studied occurs; (ii) the model of the process has been verified by an independent experimental method. Unfortunately, these conditions are not fulfilled in most of the methods in use.

The method proposed by Dang *et al.* (1977) and Nakanoh and Yoshida (1980) involves identical conditions in the process and in the determination of probe characteristics, but the response obtained does not represent the transient characteristics because the initial condition $z=0$ in the film is not fulfilled (see Section 19.6.2.2). Empirical models such as first or second-order models of time lag do not allow Dang's response (Dang *et al.*, 1977) to be converted to the transient characteristics. Conversion of the moment of Dang's response, b_D, to the moment of transient characteristics, b_H, can be done using models summarized in Table 3. However, these models are only applicable to probes with no or slight slowdown, so that the main advantage of the moment methods, *i.e.* the option to use a probe with transient characteristics of arbitrary shape, is lost. In comparing and assessing the regression and moment methods, Lee and Tsao (1979) and Ruchti *et al.* (1981) overlooked these circumstances, giving preference to the moment methods (see Section 19.7.1, Figure 12, block 6). Taking Dang's response (Dang *et al.*, 1977) as the transient characteristics always leads to overestimation of the probe response time, the more so the more significant the film effect (Linek *et al.*, 1983b). The resulting k_la value is then underestimated, the error being greater at higher values (Keitel and Onken, 1981). This is one of the reasons for the differences found in k_la values determined by steady-state and dynamic methods in one and the same batch.

It is a common practice that the suitability of models adopted for the process is not tested. It has been shown that the models assuming ideal mixing in the liquid and gas phases fail to describe the situations where simultaneous transport of more than one component, *e.g.* N_2 or He in addition to O_2, occurs across the phase boundary. Neglecting the transport of the additional components leads to underestimating the resulting k_la value; the discrepancy increases with lowering of the rate of bubble coalescence in the batch, and with higher k_la values (Linek *et al.*, 1981). When the experiment is conducted so that simultaneous transport of other components is eliminated, the k_la values obtained agree well with those found by the steady-state method even at high rates of aeration (Linek and Vacek, 1982; Linek *et al.*, 1982). Such experiments can be carried out, for instance, by introducing pure oxygen into a batch from which dissolved gases have been removed by applying a vacuum.

The probe signal, and hence the k_la evaluated from it, varies depending on the probe location in the fermenter. One must, therefore, avoid (i) placing the probe in the stagnant region, (ii) direct contact of bubbles with the probe membrane, since the signal may be distorted unless the oxygen concentration in the bubbles is in equilibrium with that in the liquid. If impingement of bubbles on the membrane is precluded, then the probe signal reflects the oxygen concentration in the bulk of liquid. Such an optimum location of the probe has to be determined case by case. This may be done using a model system of oxygen absorption into a sodium sulfite solution containing a sufficient amount of cobalt catalyst to ensure zero oxygen concentration in the bulk of the liquid even under conditions of vigorous aeration (Linek and Vacek, 1978). At the optimum location, the probe signal remains zero even if the batch is vigorously aerated (Linek *et al.*, 1979b).

19.8 NOMENCLATURE

$A_i = l_i \sqrt{D_M}/(l_M \sqrt{D_i}) = \sqrt{(K_M/K_i)}$

b_G zeroth moment of probe signal

b_J zeroth moment of function J; $J = D$, W, or H; s

$b_M = 1/(6K_M)$; s

b_m zeroth moment of actual process

$B_0 = \bar{k}_0/(m_L K_M)$

$B_1 = \bar{k}_1/K_M$

 c molar oxygen concentration; mol m^{-3}

 $\mathbb{C}$ vector of conditions P, T, L_L, A_L, B_0, B_1

 $d_i = l_i/(r_0 + l_E + l_M)$

d_K diameter of cathode; m

 D Dang's response (Dang *et al.*, 1977); response to initial concentration profile where $z=1$ (Figure 4)

D_i oxygen diffusion coefficient in layer i; m^2 s^{-1}

E_{Di} activation energy of oxygen diffusivity in layer i; J mol^{-1}

E_{ki} activation energy of oxygen permeability in layer i; J mol^{-1}

 $f = M_{sg}/M_{sl} = s_g/s_l$; gas/liquid ratio or stirring effect factor

 F Faraday constant; C equivalent^{-1}

 G normalized response $= M(t)/M_s$

 H transient characteristics of probe

$k_i = m_i D_i$; permeability of oxygen in layer i; mol m^{-1} Pa^{-1} s^{-1}

$\bar{k}_j$ kinetic constant of jth order reaction; mol^{1-j} m^{3j-3} s^{-1}

k_l mass transfer coefficient in liquid; m s^{-1}

$K_i = D_i/l_i^2$; s^{-1}

l_i thickness of layer $_i$; m

$L_i = R_i/R_M = l_i k_M/(l_M k_i)$

m_i oxygen solubility ($m_i = c_i/p_i$); mol m^{-3} Pa^{-1}

 $M = nFSN$; probe signal: M_s or $M(t)$; C s^{-1}

 n stoichiometric coefficient, equation (4)

 N intensity of oxygen flux; mol m^{-2} s^{-1}

$N_s = \lim_{t \to \infty} N(t)$; mol m^{-2} s^{-1}

p_i equivalent partial pressure of oxygen in layer i; Pa

P_c step change; Pa

P_{max} maximum oxygen tension for probe linearity, equation (18); Pa

$P(t)$ equivalent oxygen tension in bulk liquid; Pa

 q_j parameters of process

Q_c cathode circumference; m

 r radial coordinate (see Figure 5); m

r_o radius of spherical cathode; m

 R universal gas constant; J mol^{-1} K^{-1}

$R_i = l_i/k_i$; diffusional resistance of layer i; m^2 Pa s mol^{-1}

 s sensitivity $= M_s/(SP)$; C m^{-2} Pa^{-1} s^{-1}

 S surface area of cathode; m^2

$S_i = L_i/A_i$

 t time; s

 T absolute temperature; K

 $w = Q_c(l_E + l_M + l_L)/S$

$W(t) = D(t) - H(t)$

 x coordinate (see Figure 4); m

 y volume fraction of oxygen in dry gas phase

z parameter of initial condition in liquid film; equation (10)
Z_1 parameter of two-region model

Subscripts
  D Dang's response
  E electrolyte
  g gas phase
  G corresponding to process G
  H corresponding to transient characteristics H
  i E, M, L
  l liquid phase
  L liquid film
  m corresponding to model
  M membrane
  r in the presence of reaction
  s steady-state value
  u region of probe
  W corresponding to function (D–H)

Superscript
  ∞ corresponding to $S_L \to \infty$ (see Figure 9)

Abbreviations
   BJ Borkowski–Johnson galvanic probe (Borkowski and Johnson, 1967)
   CV Čerkasov polarographic probe (Linek *et al.*, 1984)
   HC Hospodka–Čáslavský polarographic probe (Hospodka and Čáslavský, 1965)
  LDC limiting diffusion current
   MC Mackereth galvanic probe (Mackereth, 1964)
   MN Mancy galvanic probe (Mancy *et al.*, 1962)
   OP oxygen probe
   PE polyethylene
   PP polypropylene
 PTFE polytetrafluorethylene
   RC residual current
  YSI Yellow Springs Instrument polarographic probe, Model G-1678-5 (Yellow Springs Inst.
       Co., Yellow Springs, OH, USA)

19.9 REFERENCES

Aiba, S. and S. Y. Huang (1969a). Oxygen permeability and diffusivity in polymer membranes immersed in liquids. *Chem. Eng. Sci.*, **24**, 1149–1159.
Aiba, S. and S. Y. Huang (1969b). Some consideration on the membrane-covered-electrode. *J. Ferment. Technol.*, **47**, 372–381.
Aiba, S., M. Ohashi and S. Y. Huang (1968). Rapid determination of oxygen permeability of polymer membranes. *Ind. Eng. Chem., Fundam.*, **7**, 497–502.
Albanese, R. A. (1971). Use of membrane-covered oxygen cathodes in tissue. *J. Theor. Biol.*, **33**, 91–103.
Baumgärtl, H., W. Grunewald and D. W. Lübbers (1974). Polarographic determination of the oxygen partial pressure field by Pt microelectrodes using the O_2 field in front of a Pt macroelectrode as a model. *Pflügers Arch.*, **347**, 49–61.
Beechey, R. B. and D. W. Ribbons (1972). Oxygen electrode measurements. In *Methods in Microbiology*, ed. J. R. Norris and D. W. Ribbons, vol. 6B, chap. 2, pp. 25–53. Academic, London.
Benedek, A. A. and W. J. Heideger (1970). Polarographic oxygen analyzer response: the effect of instrument lag in the nonsteady-state reaeration test. *Water Res.*, **4**, 627–640.
Bergman, I. (1970). Improvements in or relating to membrane electrodes and cells. *US Pat.* 1 200 595.
Borkowski, J. D. and M. J. Johnson (1967). Long-lived steam-sterilizable membrane probes for dissolved oxygen measurement. *Biotechnol. Bioeng.*, **9**, 635–639.
Bourdaud, D. and A. G. Lane (1974). An electronic dynamic compensator for response time of oxygen electrodes. *Biotechnol. Bioeng.*, **16**, 279–284.
Brown, D. E. (1970). Aeration in the submerged culture of microorganisms. In *Methods in Microbiology*, ed. J. R. Norris and D. W. Ribbons, vol. 2, chap. 5, pp. 124–174. Academic, London.
Carslaw, H. S. and J. C. Jaeger (1959). *Conduction of Heat in Solids*. Clarendon Press, Oxford.
Clark, L. C., Jr. (1956). Monitor and control of blood and tissue oxygen tension. *Trans. Am. Soc. Artif. Intern. Organs*, **2**, 41–45.

Dang, N. D. P., D. A. Karrer and I. J. Dunn (1977). Oxygen transfer coefficients by dynamic model moment analysis. *Biotechnol. Bioeng.*, **19**, 853–865.

Davies, P. W. (1962). The oxygen cathode. In *Physical Techniques in Biological Research*, ed. W. L. Nastuk, vol. 4, chap. 3, pp. 137–179. Academic, New York.

Degn, H., I. Balslev and R. Brook (1976). *Measurement of Oxygen*. Elsevier, New York.

Elsworth, R. (1972). The value and use of dissolved oxygen measurement in deep culture. *Chem. Eng. (London)*, February, No. 258, 63–71.

Fatt, I. (1976). *Polarographic Oxygen Sensors*. CRC Press, Cleveland, OH.

Feder, W. (1968). Bioelectrodes. *Ann. N. Y. Acad. Sci.*, **148**, 1–287.

Firouztalé, E., M. J. Skerpon and J. S. Ultman (1982). A spherical model of the Clark electrode in a flowing liquid medium. *J. Electroanal. Chem.*, **134**, 1–10.

Fuchs, R., D. D. Y. Ryu and A. E. Humphrey (1971). Effect of surface aeration on scale-up procedures for fermentation processes. *Ind. Eng. Chem., Process Des. Dev.*, **10**, 190–196.

Gnaigner, E. and H. Forstner (1983). *Polarographic Oxygen Sensors. Aquatic and Physiological Applications*. Springer Verlag, Berlin.

Grunewald, W. (1970). Diffusion error and O_2 consumption of the Pt electrode during P_{O_2} measurement in the steady state. *Pflügers Arch.*, **320**, 24–44.

Grunewald, W. (1971). Einstellzeit der Pt-Elektrode bei Messungen nicht stationärer O_2-Partialdrucke. *Pflügers Arch.*, **322**, 109–130.

Hahn, C. E. W., A. H. Davis and W. J. Albery (1975). Electrochemical improvement of the performance of P_{O_2} electrodes. *Resp. Physiol.*, **25**, 109–133.

Hale, J. M. and M. L. Hitchman (1980). Some considerations of the steady-state and transient behaviour of membrane-covered dissolved oxygen detectors. *J. Electroanal. Chem.*, **107**, 281–294.

Heineken, F. G. (1970). On the use of fast-response dissolved oxygen probes for oxygen transfer studies. *Biotechnol. Bioeng.*, **12**, 145–154.

Heineken, F. G. (1971). Oxygen mass transfer and oxygen respiration rate measurements utilizing fast response oxygen electrodes. *Biotechnol. Bioeng.*, **13**, 599–618.

Hitchman, M. L. (1978). *Measurement of Dissolved Oxygen*. Wiley, New York.

Hoare, J. P. (1968). *The Electrochemistry of Oxygen*. Interscience, New York.

Hospodka, J. and Z. Čáslavský (1965). Design and application of electrodes for the determination of dissolved oxygen, *Folia Microbiol. (Prague)*, **10**, 186–195.

Hwang, S. T., C. K. Choi and K. Kammermayer (1974). Gaseous transfer coefficients in membranes. *Sep. Sci.*, **9**, 461–478.

Jensen, O. J., T. Jacobsen and K. Thomsen (1978). Membrane-covered oxygen electrodes. I. Electrode dimensions and electrode sensitivity. *J. Electroanal. Chem.*, **87**, 203–211.

Keitel, G. and U. Onken (1981). Errors in the determination of mass transfer in gas–liquid dispersions. *Chem. Eng. Sci.*, **36**, 1927–1932.

Kessler, M., D. F. Bruley, L. C. Clark, Jr., D. W. Lübbers, I. A. Silver and J. Strauss (1973). *Oxygen Supply*. University Park Press, London.

Kimmich, H. P. and F. Kreuzer (1969). Catheter P_{O_2} electrode with low flow dependency and fast response. *Prog. Respir. Res.*, **3**, 100–110.

Kimmich, H. P., M. N. Kuypers, J. M. L. Engels and H. G. R. Maas (1981). Disposable solid state oxygen sensor. *Proceedings of the International Symposium on Oxygen Transport to Tissue* (preprint).

Kinsey, D. W. and R. A. Bottomley (1963). Improved electrode system for determination of oxygen tension in industrial applications. *J. Inst. Brew.*, **69**, 164–176 (*Chem. Abstr.*, **59**, 8114).

Kok, R. (1976). An *in situ* dissolved oxygen probe calibrator. *Biotechnol. Bioeng.*, **18**, 729–735.

Kok, R. and J. E. Zajic (1975). Dynamic response of a polarographic oxygen probe. *Biotechnol. Bioeng.*, **17**, 527–539.

Kreuzer, F. (1969). Oxygen pressure recording in gases, fluids and tissues. *Prog. Respir. Res.*, **3**, 1–164.

Kreuzer, F., H. P. Kimmich and Březina (1980). Polarographic determination of oxygen in biological materials. In *Medical and Biological Application of Electrochemical Devices*, ed. J. Koryta, pp. 173–261. Wiley, New York.

Lee, Y. H. and G. T. Tsao (1979). Dissolved oxygen electrodes. *Adv. Biochem. Eng.*, **13**, 36–86.

Lee, Y. H., G. T. Tsao and P. C. Wankat (1978). Ultramicroprobe method for investigating mass transfer through gas–liquid interface. *Ind. Eng. Chem., Fundam.*, **17**, 59–66.

Lilley, M. D., J. B. Story and R. W. Raible (1969). The chronoamperometric determination of dissolved oxygen using membrane electrodes. *J. Electroanal. Chem.*, **23**, 425–429.

Linek, V. (1972). Determination of aeration capacity of mechanically agitated vessels by a fast response oxygen probe. *Biotechnol. Bioeng.*, **14**, 285–289.

Linek, V. and B. Beneš (1977). Multiregion, multilayer, nonuniform diffusion model of an oxygen electrode. *Biotechnol. Bioeng.*, **19**, 741–748.

Linek, V. and P. Beneš (1978). Comparison of regression and moment methods of evaluation of oxygen probe responses. *Biotechnol. Bioeng.*, **20**, 903–912.

Linek, V. and J. Petrlík (1979). Temperature characteristics of oxygen probes. *Sci. Pap. Prague. Inst. Chem. Technol.*, **K14**, 107–113.

Linek, V. and V. Vacek (1976). Oxygen electrode response lag induced by liquid film resistance against oxygen transfer. *Biotechnol. Bioeng.*, **18**, 1537–1555.

Linek, V. and V. Vacek (1977). Dynamic measurement of the volumetric mass transfer coefficient in agitated vessels: effect of the start-up period on the response of an oxygen electrode. *Biotechnol. Bioeng.*, **19**, 983–1008.

Linek, V. and V. Vacek (1978). Measurement of fermentor aeration capacity by a fast-response oxygen electrode in medium-air dispersions. *Biotechnol. Bioeng.*, **20**, 305–308.

Linek, V. and V. Vacek (1979). Measurement of fermentor aeration capacity by a fast-response oxygen electrode in medium-air dispersions. *Biotechnol. Bioeng.*, **21**, 907–908.

Linek, V. and V. Vacek (1982). Consistency of steady-state and dynamic methods of $k_l a$ determination in gas–liquid dispersions. *Chem. Eng. Sci.*, **37**, 1425–1429.

Linek, V., P. Beneš, F. Hovorka and V. Vacek (1978a). Correction of the dynamic response of oxygen probe for measurement of kinetic parameters of fast processes by dynamic methods. Comparative study. *Collect. Czech. Chem. Commun.*, **43**, 2980–3001.

Linek, V., P. Beneš, J. Sinkule and Z. Křivský (1978b). Simultaneous determination of mass transfer coefficient and of gas and liquid axial dispersions and holdups in a packed absorption column by dynamic response method. *Ind. Eng. Chem., Fundam.*, **17**, 298–305.

Linek, V., P. Beneš and V. Vacek (1979a). Oxygen probe dynamics in flowing fluids. *Ind. Eng. Chem., Fundam.*, **18**, 240–245.

Linek, V., J. Stejskal, J. Sinkule and V. Vacek (1979b). Determination of the aeration capacity of bubble columns by dynamic method. The influence of axial dispersion and of the start-up period. *Collect. Czech. Chem. Commun.*, **44**, 2583–2597.

Linek, V., P. Beneš and F. Hovorka (1981). The role of interphase nitrogen transport in the dynamic measurement of the overall volumetric mass transfer coefficient in air-sparged systems. *Biotechnol. Bioeng.*, **23**, 301–319.

Linek, V., J. Sinkule and V. Vacek (1983a). Oxygen electrode dynamics: three layer model: chemical reaction in the liquid film. *Biotechnol. Bioeng.*, **25**, 1401–1418.

Linek, V., V. Vacek and J. Sinkule (1983b). Check for consistency of experimental determination of transient characteristics of oxygen probes with significant liquid film resistance. *Biotechnol. Bioeng.*, **25**, 1195–1200.

Linek, V., P. Beneš, V. Vacek and F. Hovorka (1983c). Analysis of differences in $k_l a$ values determined by steady-state and dynamic methods in stirred tanks. *Chem. Eng. J.*, **25**, 77–88.

Linek, V., P. Beneš and V. Vacek (1984). An experimental study of oxygen probe linearity and transient characteristics in the high oxygen concentration range. *J. Electroanal. Chem.*, **169**, 233–257.

Linek, V., J. Sinkule and V. Vacek (1985). Effect of accumulation of cathodic reaction products in the Clark oxygen probe on the probe linearity, signal overshooting, tailing and hysteresis of transient characteristics for step changes in the concentration of oxygen. *J. Electroanal. Chem.*, **187**, 1–30.

Lloyd, B. B. and B. Seaton (1970). Storage effects in an oxygen electrode. *J. Physiol. (London)*, **207**, 29P–30P.

Lucero, D. P. (1968). Performance characteristics of membrane-covered polarographic gas detectors. *Anal. Chem.*, **40**, 707–711.

Lucero, D. P. (1969). Design of membrane-covered polarographic gas detectors. *Anal. Chem.*, **41**, 613–622.

Lundsgaard, J. S., J. Grønlund and H. Degn (1978). Error in oxygen measurement in open systems owing to oxygen consumption in unstirred layer. *Biotechnol. Bioeng.*, **20**, 809–819.

Luttmann, A., K. Mückenhoff and H. H. Loeschcke (1978). Fast measurement of the CO_2 partial pressure in gases and fluids. *Pflügers Arch.*, **375**, 279–288.

Lübbers, D. W., H. Baumgärtl, H. Fabel, A. Huch, M. Kessler, K. Kunze, H. Riemann, D. Seiler and S. Schuchhardt (1969). Principle of construction and application of various platinum electrodes. *Prog. Respir. Res.*, **3**, 136–146.

Mackereth, F. J. H. (1964). An improved galvanic cell for determination of oxygen concentrations in fluids. *J. Sci. Instrum.*, **41**, 38–41.

Maclennan, B. G. and S. J. Pirt (1966). Automatic control of dissolved oxygen concentration in stirred microbial cultures. *J. Gen. Microbiol.*, **45**, 289–302.

Mancy, H. K., D. A. Okun and C. N. Reilley (1962). A galvanic cell oxygen analyzer. *J. Electroanal. Chem.*, **4**, 65–92.

Merta, K. and I. J. Dunn (1976). Oxygen electrode characteristics. *Biotechnol. Bioeng.*, **18**, 591–593.

Mindt, W. F. H. and P. Eberhard (1978). A fast stabilization technique for oxygen sensors with low cathode current densities. *Acta Anaesthiol. Scand., Suppl.* **68**, 63–67.

Mueller, J. A., W. C. Boyle and E. N. Lightfoot (1967). Effect of the response time of a dissolved oxygen probe on the oxygen uptake rate. *Appl. Microbiol.*, **15**, 674–676.

Nakanoh, M. and F. Yoshida (1980). Gas absorption by Newtonian and non-Newtonian liquids in a bubble column. *Ind. Eng. Chem., Process Des. Dev.*, **19**, 190–195.

Niedrach, L. W. and W. H. Stoddard (1972). Sensor with anion exchange resin electrolyte. *US Pat.* 3 703 457.

Phillips, D. H. and M. J. Johnson (1961). Measurement of dissolved oxygen in fermentations. *J. Biochem. Microbiol. Technol. Eng.*, **3**, 261–275.

Robinson, J. and J. M. Cooper (1970). Method of determining oxygen concentrations in biological media, suitable for calibration of the oxygen electrode. *Anal. Biochem.*, **33**, 390–399.

Ruchti, G., I. J. Dunn and J. R. Bourne (1981). Comparison of dynamic oxygen electrode methods for the measurement of $k_l a$. *Biotechnol. Bioeng.*, **23**, 277–290.

Schneiderman, G. and T. K. Goldstick (1978). Oxygen electrode design criteria and performance characteristics: Recessed cathode. *J. Appl. Physiol.*, **45**, 145–154.

Schuler, R. and F. Kreuzer (1967). Rapid polarographic *in vivo* oxygen catheter electrodes. *Respir. Physiol.*, **3**, 90–110.

Schuler, R. and F. Kreuzer (1969). Properties and performance of membrane-covered rapid polarographic oxygen catheter electrodes for continuous oxygen recording *in vivo*. *Prog. Respir. Res.*, **3**, 64–78.

Severinghaus, J. W. (1968). Measurement of blood gases: P_{O_2} and P_{CO_2}. *Ann. N. Y. Acad. Sci.*, **148**, 115–132.

Severinghaus, J. W. (1978). Alkaline chloride-free oxygen electrolytes. *Acta Anaesthiol. Scand., Suppl.*, **68**, 73–75.

Severinghaus, J. W. and A. Thunstrom (1978). Problems of calibration and stabilization of tcP_{O_2} electrodes. *Acta Anaesthiol. Scand., Suppl.*, **68**, 68–72.

Severinghaus, J. W., J. Peabody, A. Thunstrom, P. Eberhard and E. Zappia (1978). Workshop on methodologic aspects of transcutaneous blood gas analysis. *Acta Anaesthiol. Scand., Suppl.*, **68**, 1–144.

Silver, I. A. (1973). The oxygen micro-electrode. Oxygen transport to tissue, instrumentation, methods, and physiology, *Adv. Exp. Med. Biol.*, **37A**, 7–15.

Siu, W. and R. S. C. Cobbold (1976). Characteristics of a multicathode polarographic oxygen electrode. *Med. Biol. Eng.*, **14**, 109–121.

Tang, T. E., R. E. Barr, V. G. Murphy and A. W. Hahn (1973). A working equation for oxygen sensing disc electrodes. Oxygen transport to tissue, instrumentation, methods, and physiology. *Adv. Exp. Med. Biol.*, **37A**, 9–25.

Thunstrom, A. M. and J. W. Severinghaus (1978). The use of concentrated $NaNO_3$ in tcP_{O_2} electrodes to diminish O_2 consumption and dehydration drift. *Acta Anaesthiol. Scand., Suppl.*, **68**, 55–62.

Ultman, J. S., E. Firouztalé and M. J. Skerpon (1981). A spherical model of the Clark electrode. *J. Electroanal. Chem.*, **127**, 59–66.

Vacek, V., V. Linek and J. Sinkule (1985). Transient diffusion of oxygen through three spherical layers of Clark oxygen sensor. *J. Electrochem. Soc.*, in press.

Van Leeuwen, C. (1979). Dynamic measurement of the overall volumetric mass-transfer coefficient in air-sparged systems. *Biotechnol. Bioeng.*, **21**, 2125–2131.

Vogel, H. R., O. Harth and G. Thews (1969). Continuous recording of P_{O_2} in respiratory air by a rapid platinum electrode. *Prog. Respir. Res.*, **3**, 42–46.

Wodick, R. (1972). Compensation of measuring errors produced by finite response time in polarographic measurements with electrodes sensitive to oxygen and hydrogen. *Pflügers Arch.*, **336**, 327–344.

20

Enzyme Probes

B. DANIELSSON
University of Lund, Lund, Sweden

20.1 INTRODUCTION

In no other field has the use of immobilized enzymes become as widespread as in analysis. There are several reasons for this situation: immobilized enzymes can easily be combined with various transducers; the operational stability and the reproducibility of the analytical procedure is increased, which is important in routine analysis; the enzyme cost and presumably also the total reagent cost per sample is reduced. This fact, furthermore, makes the use of more sophisticated enzymes affordable, since only small amounts will be required. The use of an immobilized enzyme also facilitates the construction of enzyme probes, *i.e.* the combination of an enzyme and a transducer converting, for example, concentration, temperature or pH to another quantity, such as a voltage.

Various concepts for applying the enzyme in proximity to the transducer have been developed, such as coating or entrapping the enzyme into or around the transducer which, for instance, could be a pH electrode in the case of an enzymic reaction involving protons giving a pH shift. We then have a probe that can simply be dipped into the sample solution to sense the actual metabolite concentration. If the analysis is to be performed in a flow path it is usually advantageous to apply the enzyme immobilized in a support material in a small column. The transducer can then be inserted in the enzyme bed or in a flow cell after the column. This arrangement offers high flexibility, both in the choice of enzyme preparation and transducer and in particular has been used with temperature transducers in so-called 'enzyme thermistors'.

The inherently high specificity of enzymes allows them to be combined with less specific transducers, yet still produce enzyme probes with sufficient specificity. A heat (temperature) sensor such as a thermistor is a most general transducer, since it can detect almost any chemical reaction provided the enthalpy change is not zero. Enzymic reactions as a rule are rather exothermic and therefore thermal enzyme probes will be highly versatile. The heat transducer has the advantage of being capable of directly following the enzymic reaction with no need for coupled reactions to give, for instance, a color change as in colorimetry. It can be used on crude or colored samples without problems. The sensitivity to all temperature changes is, however, at the same time a considerable drawback and special care has to be taken to eliminate interferences by extraneous heat sources or sinks.

On the other hand the transducer can be more specific as in the cases of a mass filter or a spectrophotometer tuned to special mass or wavelength. The overall specificity of the analysis is thereby increased, although the applicability of the transducer is less general.

General aspects of transducers in enzyme based analysis and the impact of newer physical and electronic achievements have recently been discussed by Weaver and Burns (1981).

This chapter will be mainly concerned with enzyme electrodes and enzyme thermistors. Enzyme electrodes are widely used today and are relatively simple to construct. There is also a wide variety of electrodes available that can be used in enzyme probes. Thermal enzyme probes such as enzyme thermistors have a great potential in biochemical analysis due to their versatility and high operational stability. They appear especially attractive in continuous monitoring.

The combination of microelectronics with immobilized enzymes opens a new, interesting field (Weaver and Burns, 1981). Some work on ChemFETs or enzyme transistors, *i.e.* the combination of enzymes and semiconductor structures, will be described here.

A few other examples of enzyme–transducer combinations will also be given, merely to indicate the multitude of possibilities that exists. Finally, the concept of microbe and organelle sensors is briefly introduced. The importance of this type of probe is steadily increasing, especially since techniques for immobilized microorganisms and organelles have improved.

Further reading on enzyme probes can be found in works by Carr and Bowers (1980), Guilbault (1982, 1984) and in Mosbach (1976, 1986).

20.2 ENZYME ELECTRODES

The enzyme electrode concept was introduced by Clark and Lyons (1962) with a device in which the enzyme glucose oxidase was retained between two Cuprophan membranes over an amperometric oxygen electrode. In the presence of glucose, oxygen was consumed:

$$\text{D-glucose} + O_2 \xrightarrow{\text{glucose oxidase}} \text{D-glucono-1,4-lactone} + H_2O_2 \qquad (1)$$

Since then a great variety of enzyme electrodes has been developed and these can be systematized in many different ways, physically as well as chemically. In principle, the compound of interest reacts in an enzyme-catalyzed reaction in which one reactant can be measured electrochemically. The enzyme is usually applied closely to the sensing part of the electrode by physical or chemical entrapment, or by immobilization techniques, but it could also be used in a flow system immobilized on a suitable support and packed in a small column, the elecrode being mounted in the flow stream from the enzyme bed. The latter arrangement has some definite advantages, especially in flow analysis, and it will therefore also be treated here.

For more detailed texts on enzyme electrodes the reader is referred to recent reviews by Guilbault (1982), Carr and Bowers (1980), and Barker and Somers (1978).

20.2.1 Enzyme Electrodes Based on Oxygen Electrodes

The first combination of an enzyme and an electrode described was based on an oxygen electrode (Clark and Lyons, 1962). Updike and Hicks (1967), who introduced the term 'enzyme electrode', used a dual cathode Clark oxygen electrode covered with glucose oxidase immobilized in polyacrylamide gel (Figure 1). Oxygen electrodes have continued to be very attractive due to the large number of enzymic and microbial processes to which they can be applied. There are today well over 50 different enzymes classified as oxidoreductases with oxygen as acceptor (oxidases). The oxidases are generally very stable and can be found for a wide variety of substrates: aldehydes, alcohols, carbohydrates, amino acids, phenols, thiols, steroids, *etc.* Many of them are commercially available, *e.g.* glucose oxidase, alcohol oxidase, pyruvate oxidase and oxalate oxidase. Another rather large group of oxidoreductases acting on a variety of substrates is formed by the oxygenases: monooxygenases and dioxygenases incorporating one or two atoms of molecular oxygen, respectively. Not so many of them are as yet to be found on the market, but lactate oxidase, for example, is a well-known monooxygenase that has been on the market for some years.

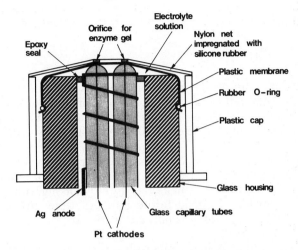

Figure 1 Dual cathode enzyme electrode according to Updike and Hicks (1967). (Reprinted with permission of Macmillan Journals Ltd)

The Clark pO_2-electrode normally used in oxygen based enzyme electrodes has an oxygen permeable membrane (Teflon[TM], polypropylene, silicone rubber) covering a platinum electrode. Oxygen diffusing through the membrane is reduced at the Pt cathode that is kept at a fixed potential with respect to a reference electrode, *e.g.* Ag/AgCl. The current is proportional to the oxygen concentration (amperometry). Depending on design and membrane properties, the response time for a polarographic pO_2-electrode is from seconds to 1–2 min. Detailed descriptions of oxygen-based enzyme electrodes can be found in handbooks (Guilbault, 1984; Mosbach, 1976), and only brief mention of some significant probe types will be made here.

Naturally, a large number of reports deal with glucose oxidase electrodes for glucose determination, both because of the demand for this important assay in clinical chemistry, biotechnology, food chemistry and many other areas and because of the favorable properties of the enzyme glucose oxidase: stability, specificity and availability in highly active preparations. In addition to the two pioneering glucose electrodes already mentioned, the electrode described by Nanjo and Guilbault (1974) can be mentioned as a successful development of this concept. Again a polarographic measurement of oxygen depletion was made by amperometry at −0.6 V. Glucose oxidase was immobilized by crosslinking to albumin with the bifunctional reagent glutaraldehyde. The cathode was a fairly large (0.5 cm in diameter) Pt-disc attached to the end of a 1 cm diameter cylindrical plastic rod. The enzyme–albumin preparation was held against this Pt inlay with a nylon net. This electrode, which could be used for over a year, had a rather quick response time of about 1 min, which is adequate for most applications. The sensitivity was 0.005 mmol l^{-1} and the linear range extended up to 1.28 mmol l^{-1}.

Considerably faster response was noted, however, for a design by Koyama *et al.* (1980) involving an ultrafilter membrane. Their enzyme sensor consisted of the following parts: over the Pt cathode was a 12.5 μm Teflon[TM] membrane followed by the enzyme membrane, which was made from cellulose triacetate in a way that produced a very porous and hydrophilic structure. This fragile membrane had to be protected by a specially made ultrafiltration membrane cutting off molecules larger than 10^4 Daltons. Besides the protective role of the membrane, the authors found that it prevented contamination of the enzyme by blood constituents and that it acted as a noise filter attenuating mechanical vibrations and thereby reducing the noise current from the electrode. The sensor had a very rapid response of only 10 s and a linear operating range of 0.003 to 2 mmol l^{-1}.

By coupling to other enzymic reactions, the applicability of an enzyme probe can be extended. Thus, a glucose electrode can be used for determination of a variety of sugars by incorporating a preceding enzymic step, such as hydrolysis of a disaccharide or an isomerization. The same situation applies for other glucose probes, such as glucose oxidase charged enzyme thermistors. Some additional examples of the use of sequentially working enzymes will be given in Section 20.3.3. Satoh *et al.* (1976) developed a sucrose electrode using invertase, mutarotase and glucose oxidase immobilized in a collagen membrane:

$$\text{sucrose} + H_2O \xrightarrow{\text{invertase}} \alpha\text{-D-glucose} + \text{D-fructose} \qquad (2)$$

$$\downarrow \text{mutarotase} \qquad (3)$$

$$\begin{array}{c}\text{D-gluconolactone} \\ + H_2O_2 + O_2\end{array} \xleftarrow{\text{glucose oxidase}} \beta\text{-D-glucose} \qquad (1)$$

This electrode consisted of a double membrane in contact with the Pt cathode. The outer layer was a collagen membrane (50 μm thick) and the inner membrane was a 27 μm thick Teflon™ film. The electrode further contained an alkaline electrolyte and a lead anode. A steady state reading was attained within 3 min and the electrode was linear with sucrose concentrations of 1 to 10 mmol l^{-1}.

An interesting design for a multienzyme electrode with specificity for sucrose, maltose or lactose, depending on the enzymes attached, was presented by Cordonnier *et al.* (1975). The enzymes were immobilized in a magnetic membrane by crosslinking with albumin and the bifunctional reagent, glutaraldehyde. Magnetic ferrite particles were physically entrapped in the enzyme layer and the membrane could be simply attached to the surface of a Clark pO$_2$-electrode with the aid of a cylindrical magnet.

The problem of measuring glucose in samples with variable oxygen content has been addressed by Romette *et al.* (1979). Normally the pO$_2$ of the sample is stabilized before or during the measurements. The electrode described by Romette and coworkers (1979), however, contained enough oxygen itself to compensate for variations in sample oxygen content. The enzyme, glucose oxidase, was crosslinked with gelatin using glutaraldehyde on a 6 μm thick polypropylene membrane covering the pO$_2$-electrode. The concentration of oxygen was about 20 times higher in the membrane than in the surrounding solution. To allow quick recovery of the oxygen consumed, the electrode was equilibrated with air between measurements.

A new principle for the construction of oxygen dependent electrodes was presented by Enfors (1981). To permit analysis in a fermentation broth even at low or zero oxygen concentration, the electrode was designed to be inserted in a fermenter and to be self-supplementing with oxygen. To reduce the demand for oxygen, the glucose oxidase was coimmobilized with catalase. The total oxygen demand was thereby reduced by 50% by the reaction:

$$H_2O_2 \xrightarrow{\text{catalase}} H_2O + 0.5\,O_2 \qquad (4)$$

Moreover, removal of the hydrogen peroxide protects the glucose oxidase from deactivation.

The enzymes were mixed with albumin and crosslinked with glutaraldehyde on a Pt gauze, which was then placed over a galvanic electrode (Figure 2). This Pt gauze also was utilized as the electrolysis anode with a Pt coil wired around the electrode body as cathode to produce oxygen at the site of the enzymic reaction by electrolytic decomposition of water. Another oxygen electrode served as a reference electrode and the current through the electrolysis cell was regulated by a PI-controller to keep the differential signal from the two oxygen electrodes at zero potential. The electrolytic current as measured by a microammeter was linearly correlated with glucose up to 2 g l^{-1}, regardless of the oxygen tension of the sample.

A considerable part of the literature on oxygen based enzyme electrodes has been devoted to glucose oxidase probes, but there is a rapidly growing number of applications of other enzymes as well. Of these, this chapter will deal with alcohol oxidase, amino acid oxidases, monoamine oxidase, choline oxidase, phenol hydroxylase and pyruvate oxidase. Some oxidases will also be discussed in the following section on hydrogen peroxide based electrodes.

Guilbault and Lubrano (1974) used alcohol oxidase obtained from the mycelium of a Basidiomycetes species in combination with amperometric monitoring of the hydrogen peroxide formed. This enzyme catalyzes the oxidation of lower primary aliphatic alcohols according to the reaction:

$$RCH_2OH + O_2 \rightarrow RCHO + H_2O_2 \qquad (5)$$

The results indicated that the method should be adequate for clinical determination of blood ethanol, although methanol gave serious interference.

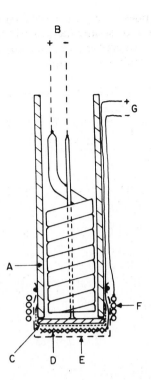

Figure 2 Oxygen-stabilized enzyme electrode. (A) electrode glass body; (B) oxygen electrode output; (C) oxygen electrode Teflon^{TM} membrane; (D) Pt-gauze with immobilized enzyme (electrolysis anode); (E) dialysis membrane; (F) Pt coil wired around the electrode (electrolysis cathode); and (G) electrolysis electrode terminals. (Reproduced with permission of Butterworth Scientific Ltd. from Enfors, 1981)

A vastly improved alcohol electrode, selective for ethanol, was presented by Nanjo and Guilbault (1975). Instead of measuring hydrogen peroxide, the decrease in current as oxygen was depleted from the solution was measured at an applied potential of -0.6 V *vs.* the standard calomel electrode. The apparent substrate selectivity with this system was very different from earlier reported values for the soluble enzyme, showing much higher reactivity for ethanol than for methanol. Also, the reactivity towards *n*-propanol and *n*-butanol was much higher with the dissolved method than for a spectrophotometric determination of the hydrogen peroxide formed by the soluble enzyme. This alcohol electrode was useful for the assay of blood ethanol in concentrations as low as 100 mg l^{-1}. Additionally, it could be used for the determination of aldehydes and carboxylic acids. One determination took 5 min.

Alcohol oxidase from another source (*Candida boidinii*) was used by Gulberg and Christian (1981) in a continuous flow determination of blood ethanol with an oxygen electrode. The electrode was placed in a specially designed Plexiglas flow-cell and the oxygen concentration was amperometrically determined by monitoring the current at a potential of -0.6 V *vs.* Ag–AgCl. Various types of porous glass beads packed in long columns, as well as nylon tubing (1 mm i.d.) were tested as enzyme supports in their continuous flow system. Silanized glass beads were found to give the best results, although the column had to be changed periodically because of pressure build-up when blood samples were processed. A 1:100 dilution resulted in a linear range of 2 to 30 mg l^{-1}.

Enzyme electrodes for determining amino acids can be constructed in different ways using amino acid oxidases. From the reaction:

$$RCH(NH_3^+)CO_2^- + H_2O + O_2 \xrightarrow{\text{amino acid oxidase}} RCOCO_2^- + NH_4^+ + H_2O_2 \qquad (6)$$

it can be seen that use of the following electrodes should be possible: pH, NH$_3$ or NH$_4^+$, O$_2$ and H$_2$O$_2$.

Hydrogen peroxide, however, reacts further:

$$RCOCO_2^- + H_2O_2 \rightarrow RCO_2^- + CO_2 + H_2O \qquad (7)$$

which makes amperometric determination of hydrogen peroxide less attractive. Nanjo and Guilbault (1974) extended the measurements based on O$_2$ depletion of glucose to L-amino acids by use of immobilized L-amino acid oxidase. Their electrode was specific for L-amino acids, but not very selective for the side chain. The response for various amino acids was established by rate and steady state measurements of both dissolved oxygen and hydrogen peroxide. Relative reactivities for L-amino acids were also studied by Tran-Minh and Broun (1975) using an L-amino acid oxidase.

Matsumoto and coworkers (1980) employed a polarographic oxygen probe to determine choline with use of immobilized choline oxidase on a partially aminated polyacrylonitrile membrane. The calibration curve for choline was linear from 0 to 0.1 mmol l^{-1}. One assay took only 1 min, with a response time of 7 s for choline. The probe could be used in fermentation media and, furthermore, phospholipids were determined in serum with native phospholipase D and the choline electrode.

The use of a monoamine oxidase electrode for testing the freshness of meat was proposed by Karube *et al.* (1980).They used monoamine oxidase immobilized in a collagen membrane in combination with amperometric determination of oxygen consumption to establish the response for various amines known to be produced at putrefaction. The response of the electrode was linear in the range of 0.05–0.2 mmol l^{-1} of tyramine.

An enzyme electrode for quantitative determination of phenol at micromolar concentrations (0.5–50 μmol l^{-1}) has been described by Kjellén and Neujahr (1980). Phenol hydroxylase from *Trichosporon cutaneum* was immobilized by different methods and attached to the surface of a Clark oxygen electrode. The probe had to be preconditioned with NADPH for optimal performance.

Finally, the possibility of determining enzymic activities with an enzyme electrode should be noted. The assay described by Mizutani *et al.* (1980) for glutamate pyruvate transaminase (alanine aminotransferase) with use of a pyruvate oxidase sensor can be used as an example. Pyruvate oxidase was immobilized by adsorption on a porous acetylcellulose membrane and combined with an oxygen electrode to provide a pyruvate sensor (0.1 to 0.8 mmol l^{-1}). Glutamate pyruvate transaminase could be determined in the range 0.5 to 180 I.U. l^{-1} by its effect on the pyruvate production in the alanine-α-ketoglutarate reaction.

20.2.2　Enzyme Electrodes Based on Hydrogen Peroxide Electrodes

As an alternative to measurement of oxygen depletion in oxidase reactions, the hydrogen peroxide produced by the reaction could be measured by a polarographic electrode. A limitation is the disappearance of hydrogen peroxide by enzymic or nonenzymic reactions which affects the sensitivity and reproducibility of the method. Catalase, which is a common contaminant in oxidase preparations, is a source of error even in small amounts due to its extremely high turn-over rate. Nevertheless, enzyme electrodes based on hydrogen peroxide determination are quite common and all commercial glucose analyzers on the American market utilize such electrodes (Keyes *et al.*, 1979).

As with oxygen electrodes, most work has been done on glucose probes. Nanjo and Guilbault (1974) compared both techniques for glucose (with glucose oxidase) and for L-amino acids (with L-amino acid oxidase) by determining oxygen depletion at −0.6 V *vs.* the standard calomel electrode (SCE) and hydrogen peroxide production at +0.6 V *vs.* SCE, both by initial rate and by steady state techniques. In all cases they found the sensitivity of the oxygen measurements to be superior. In the Biostator® system (Miles Laboratories, Elkhart, IN, USA), which is a computerized, feedback control system for dynamic control of blood glucose concentrations in diabetics, the glucose sensor is a combination of a glucose oxidase membrane with a Clark type electrode for hydrogen peroxide measurement. The enzyme is immobilized in a dual function membrane (Fogt *et al.*, 1978). The primary (outer) membrane serves as a prefilter for macromolecules and cells, whereas glucose can readily pass into the enzyme layer. The secondary (inner) membrane is a fine filter and allows only the hydrogen peroxide generated by the enzymic reaction and molecules smaller than 250 Daltons to reach the electrode. The membranes permit continuous on-line glucose analysis on diluted blood with both short- and long-term stability. The response is precise, accurate and fast, permitting determination of blood glucose within 2 min from the moment the blood leaves the patient. Furthermore, the response is linear over the range of clinical interest, virtually without interferences. Obviously, this instrument should also be useful for continuous monitoring of, for instance, fermentation processes.

A dual function membrane was also employed by Tsuchida and Yoda (1981) in their glucose electrode (Figure 3). An asymmetric membrane with a skin layer was prepared by casting acetylcellulose dissolved in acetone and cyclohexanone on a glass plate in a 300 μm thick film. After drying, a membrane with a thickness of 15.3 μm was obtained. Glucose oxidase, albumin and glutaraldehyde were added to the porous side of the membrane, and after immobilization and washing, the glucose oxidase membrane was placed over the hydrogen peroxide electrode. The asymmetric membrane was found to be permeable to hydrogen peroxide, but excluded efficiently other electrically oxidizable substances such as uric acid and ascorbic acid. The porous side of the membrane protected the enzyme layer from contamination by sample constituents and the membrane could be used for 2 to 4 weeks for 1000 assays of D-glucose in human whole blood. The electrode responded linearly to D-glucose over the concentration range 0 to 10 g l^{-1} (0 to 55 mmol l^{-1}) within 10 s.

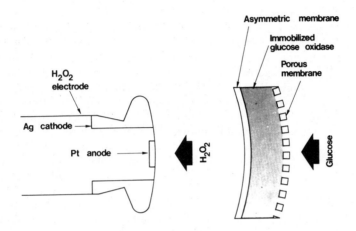

Figure 3 Glucose electrode with dual function membrane developed by Tsuchida and Yoda (1981). (Reprinted with permission of Butterworth Scientific Ltd)

Keyes and coworkers (1979) have evaluated three commercial glucose analyzers, all utilizing glucose oxidase and amperometric determination of the H_2O_2 formed. The instrument from Yellow Springs Instrument Co. (Yellow Springs, OH, USA) is the only one to employ a 'real' enzyme electrode. The probe consists of a Pt anode polarized at +0.7 V *vs.* the Ag cathode and is covered by three membranes. The inner cellulose acetate membrane has such a small pore size that only H_2O_2 and water can reach the probe. The intermediary layer contains immobilized glucose oxidase and the outer polycarbonate layer prevents glucose oxidase from passing out to the solution and also prevents catalase as well as other proteins and macromolecules from penetrating into the enzyme layer. The instrument operates in a kinetic mode. About 300 samples (25 μl) can be injected ($\leqslant$ 40 samples h^{-1}) before the membrane has to be replaced.

The other two instruments, the Kimble glucose analyzer (Owens-Illinois, Toledo, OH, USA) and the Leeds and Northrup (L & N) glucose analyzer (Sumneytown Pike, PA, USA) employ separate enzyme columns. The H_2O_2 formed by the enzymic reaction is measured by a three electrode potentiostat. To prevent oxidizable substances other than H_2O_2 from reaching the electrode, a membrane is placed between the electrode and the flowing stream. The porous alumina support is loaded with enough glucose oxidase for total conversion of all β-D-glucose present in the sample. The column can be used for 700 to 2000 samples at a sampling rate of up to 45 per hour, using 2 μl samples.

The L & N analyzer utilizes a column of glucose oxidase immobilized on a porous inorganic support placed directly in the flow stream. In a dual channel version, one channel can be used without active enzyme as a reference channel to improve accuracy. The second channel can also be used for the simultaneous determination of sucrose or lactose by adding the appropriate enzymes: invertase and mutarotase for sucrose and β-galactosidase for lactose. In all cases, the H_2O_2 generated is detected by a potentiometric three electrode system.

20.2.3 Ammonium and Ammonia Based Enzyme Electrodes

Much of the knowledge of enzyme electrode work and design emanates from studies of ammonia electrodes, especially in urea analysis. The various steps taken in the continuous improvement of urea electrodes have been elucidated by Guilbault (1982). Guilbault and Montalvo (1970) proposed several designs of enzyme electrodes for urea based on the reaction:

$$(NH_2)_2CO + 2 H_2O \xrightarrow{\text{urease}} NH_4^+ + NH_3 + HCO_3^- \qquad (8)$$

The generation of ammonium ions was measured with an NH_4^+-selective cationic electrode. The enzyme, urease, was immobilized and applied in a 60 to 350 μm thick layer of polyacrylamide on the surface of the electrode. Later a thin film of cellophane was wrapped around the enzyme layer to prevent leaching of the urease. These electrodes were useful in the range 0.05 to 1 mmol l^{-1} of urea and had a lifetime of 2 to 3 weeks. Their most serious drawback was the interference by Na^+ and K^+ with the NH_4^+ measurements. Guilbault and Nagy (1973) improved the selectivity by using a silicone rubber-based nonactin (an antibiotic) ammonium ion-selective electrode, but the most significant step towards a urease electrode free from interferences was taken by the introduction of ammonia sensitive gas probes. Anfält *et al.* (1973) crosslinked urease directly onto the surface of an ammonia gas electrode by means of glutaraldehyde. Even at the rather low pH (7–8) suitable for the enzymic reaction, sufficient NH_3 was released to allow direct measurement of urea in the presence of large amounts of Na^+ and K^+. The response time was 2 to 4 min.

Utilizing the air-gap electrode for ammonia invented by Růžička and Hansen (1974), Guilbault and Tarp (1974) made an improved, totally interference-free, direct-reading electrode for urea. A thin layer of urease chemically bound to polyacrylamide was placed at the bottom of the measuring chamber. The reaction was run at pH 8.5, giving a linear range of 0.05 to 30 mmol l^{-1}.

A different approach was taken by Pastathopoulos and Rechnitz (1975) in their design of a double membrane electrode for urea. A thin layer of EDTA-stabilized urease was applied between the membranes covering the electrode. The lifetime, precision and accuracy of the electrode, which developed a linear response over the range 0.5 to 70 mmol l^{-1} of urea, was found appropriate for clinical analysis and the probe functioned reliably with blood samples.

Finally the enzyme reactor–electrode concept proposed by Ögren and Johansson (1978) should be mentioned. By using a separate enzyme reactor (0.4 ml), in which the urease was immobilized by glutaraldehyde to porous glass beads, and by mixing the effluent with sodium hydroxide, urea could be quantitatively converted to NH_3 for urea concentrations in the range of 0.05 to 30 mmol l^{-1} and a flow rate of 40 ml h^{-1}. Their urease reactor could furthermore be used for the determination of trace amounts of Hg(II). A commercial instrument designed on the same concept was described by Gray *et al.* (1977; Kimble BUN analyzer, Owens-Illinois, Inc.). The ammonia produced was detected by a gas sensor and the instrument was intended for clinical use. A flow injection system for urea has been developed by Gorton and Ögren (1981). Here too urea is converted to ammonia in a reactor with immobilized urease and detected by an NH_3 gas membrane electrode. A linear range of 10^{-1} to 10^{-4} mol l^{-1} was observed. The samples (0.1 ml) are passed over an on-line dialyzer, at a flow rate of 1 ml min^{-1}.

Oxidative determination of amino acids by amino acid oxidases (see equation 6) produces NH_4^+ that can be detected with NH_4^+- or NH_3-sensitive electrodes. Enzyme electrodes specific for several D-amino acids have been developed by Guilbault and Hrabankova (1971). As can be seen from reviews of this field (Guilbault, 1982), there are many different electrodes of this type, for both L- and D-amino acids. The selectivity of these probes varies with the enzyme employed and can sometimes be rather broad. In some cases, however, highly specific enzymes are available for amino acid analysis. Such is the case with the potentiometric sensor for L-histidine developed by Walters *et al.* (1980). This probe was made by immobilizing histidine ammonia-lyase (EC 4.3.1.3) on an ammonia gas sensing electrode. The response was linear for 0.03 to 10 mmol l^{-1} of histidine, and the electrode responded only to histidine and ammonia.

20.2.4 Enzyme Probes Based on pH Electrodes and Some Other Probe Configurations

Many enzymic reactions are associated with the uptake or generation of protons. In principle, such reactions could be followed by measuring the pH of the reaction mixture. Thus, simple, inexpensive enzyme probes were designed for urea, penicillin and glucose (Nilsson *et al.*, 1973) based on this concept. The enzymes were either encased with a dialysis membrane tightly placed over the glass bulb or entrapped in polyacrylamide gel polymerized directly onto the pH electrode in a 400 μm thick layer. In the urea electrode, the pH change associated with the urease reaction (equation 8) produced a linear response over the range 0.05 to 20 mmol l^{-1} with a sensitivity of 0.8 pH unit per 10 mmol l^{-1}. In the glucose sensor, the gluconic acid formed in the reaction (equation 1) produced a linear response in pH (0.85 units per decade) in the range of 0.1 to 1 mmol l^{-1}.

The determination of antibiotics using β-lactamases is of considerable biotechnological interest. A pH-based penicillin electrode was also prepared by Papariello and coworkers (1973) by immobilizing penicillinase (β-lactamase) in a thin membrane of polyacrylamide gel cast around a glass pH electrode. This electrode had a fast response, 30 s, in the range 0.1 to 50 mmol l^{-1} of sodium ampicillin.

The response of a pH-based enzyme probe is directly dependent on the buffering capacity of the medium, which has to be controlled by a very weak (10^{-3} mmol l^{-1}) buffer to permit good sensitivity. This is, however, a serious limitation, especially in work with crude samples, and has been the subject for detailed studies by Enfors (see Enfors and Nilsson, 1979). In this study they also described a new electrode configuration of considerable practical interest: an autoclavable enzyme electrode intended for measurement of penicillin in fermentation (Figure 4). The electrode can be sterilized *in situ* and then the enzyme solution is introduced *via* the filling port between the cellophane membrane and the pH sensor.

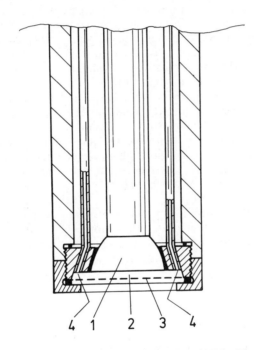

Figure 4 Sterilizable enzyme electrode according to Enfors and Nilsson (1979). (1) pH glass electrode; (2) enzyme chamber; (3) cellophane membrane; and (4) transport tubes for enzyme solution. (Reprinted with permission of Butterworth Scientific Ltd)

pCO$_2$ electrodes have in some cases been found to make very specific enzyme electrodes. Several decarboxylases are commercially available and, for example, those acting on amino acids are often more specific than other amino acid metabolizing enzymes (oxidases). Thus, a totally specific L-lysine electrode, useful for the assay of grains and foodstuffs, was described by White and Guilbault (1978). L-Lysine decarboxylase and albumin were crosslinked with glutaraldehyde on a gas-permeable Teflon^{TM} membrane of a p$_{CO_2}$ sensor yielding a quite stable enzyme probe with a linear L-lysine concentration range of 0.05 to 100 mmol l^{-1}. The only limitation observed was the long response time (5 to 10 min).

The urease reaction produces bicarbonate (equation 8), which fact has been used by Tran-Minh and Broun (1975), among others, for creating a urease electrode based on a p$_{CO_2}$ probe. In their design, the enzyme preparation of urease, crosslinked with albumin using glutaraldehyde, included carbonic anhydrase, which was found to improve the performance of the enzyme electrode by accelerating the conversion of HCO_3^- to CO_2. The enzyme film prepared was attached to the gas-sensing probe with a Teflon^{TM} membrane, resulting in a specific urea sensitive electrode with linear response in the range 0.3 to 10 mmol l^{-1}.

Electrodes specific for amygdalin based on solid-state cyanide electrodes have been described. The underlying reaction is catalyzed by β-glucosidase:

$$\text{amygdalin} \rightarrow \text{HCN} + 2\,\text{glucose} + \text{benzaldehyde} \qquad (9)$$

In the amygdalin electrode described by Mascini and Liberti (1974), the enzyme was physically retained on the surface of the cyanide sensor by a thin dialysis membrane. A linear response curve was obtained from 0.1 to 100 mmol l^{-1}.

I^- electrodes have been frequently used for measuring hydrogen peroxide. The enzyme peroxidase catalyzes the reaction:

$$H_2O_2 + 2\,I^- + 2\,H^+ \rightarrow I_2 + 2\,H_2O \qquad (10)$$

In the glucose electrode developed by Nagy *et al.* (1973), the hydrogen peroxide produced by the oxidation of glucose by glucose oxidase (equation 1) was quantified by monitoring the depletion of I^- at the surface of an I^- sensor. Glucose oxidase was applied to the sensor by three different procedures. Measurements were performed in flow streams as well as in stationary solutions. After removing uric acid and other possible reducing agents from the blood samples tested, flow-stream experiments with two I^- sensors provided accurate and convenient glucose determinations.

Finally, the possibility of extending the field of application and increasing the sensitivity of enzyme electrodes by their adoption in enzyme immunoassays should be recognized. Catalase and peroxidase are common marker enzymes, both of which can be conveniently assayed by oxygen probes. Aizawa *et al.* (1980a) have constructed such an immunosensor for determination of α-fetoprotein (AFP). Anti-AFP antibodies were covalently bound to a derivatized cellulose triacetate membrane and mounted on an oxygen probe. After completion of binding of free and catalase-labelled AFP, hydrogen peroxide was added and the sensor examined for catalase activity by amperometric measurement. AFP could be determined in the range 10^{-11} to 10^{-8} g ml^{-1}. Potentiometric rate measurements with an ammonia gas-sensing electrode were used by Gebauer and Rechnitz (1981) in their adenosine deaminase labelled immunoassay. As a model, the haptenic group dinitrophenyl (DNP) was studied. Detection limits of 50 ng of anti-DNP antibody and 10^{-10} mol l^{-1} of DNP were observed when incubation was carried out at pH 7.5 and assays made with the NH_3 electrode at pH 9.0. Further examples can be found in a recent review by Mattiasson and Borrebaeck (1980).

Another, related form of biospecific interaction was utilized by Borrebaeck and Mattiasson (1980) in their study of lectin–carbohydrate interactions. The immobilized lectin, concanavalin A, placed in close proximity to an oxygen sensor enabled quantification and determination of association constants for lectin–carbohydrate interactions. Methyl-α-D-mannose was determined down to 0.5 μg ml^{-1}.

20.2.5 Applications in Biotechnology

Enzyme electrodes have now been in common use for 10 to 15 years and reliable methodology exists in many areas of biotechnology, such as for specific amino acid and carbohydrate analysis. The market for enzyme electrode equipment and accessories is continuously growing and the importance of enzyme electrode based analysis can be expected to increase. A selection of enzyme electrode determinations of interest in biotechnology has been compiled in Table 1.

20.3 THERMAL ENZYME PROBES — ENZYME THERMISTORS

20.3.1 Introduction

The first practical, simple enzyme calorimeters combining immobilized enzymes with calorimetry appeared around 1975. Before that the potential of calorimetry as a bioanalytical tool had long been recognized and there were several good instruments for biochemical studies on the market (Spink and Wadsö, 1976; Martin and Marini, 1979). They were, however, too expensive, complicated and time-consuming to be useful for routine analysis. Almost simultaneously several groups then introduced comparatively simple devices based on immobilized enzymes taking advantage of their properties as specific, reusable, stable and 'concentrated' sensors combined with temperature transducers, usually thermistors. Consequently, we called our own devices

Table 1 Enzyme Electrodes of Interest for Biotechnology

Type	Enzyme	Sensor	Range (mol l^{-1})	Ref.
Acetic, formic acids	Alcohol oxidase	Pt (O_2)	10^{-1}–10^{-4}	Nanjo and Guilbault (1975)
Alcohols	Alcohol oxidase	Pt (O_2)	2×10^{-2}–10^{-4}	Guilbault et al. (1983)
L-Amino acids	L-Amino-acid oxidase	Pt (H_2O_2)	10^{-3}–10^{-5}	Nanjo and Guilbault (1974)
D-Amino acids	D-Amino-acid oxidase	NH_4^+	10^{-2}–5×10^{-5}	Guilbault and Hrabankova (1971)
Amygdalin	β-Glucosidase	CN^-	10^{-1}–10^{-4}	Mascini and Liberti (1974)
Choline	Choline oxidase	Pt (O_2)	2×10^{-4}–10^{-5}	Matsumoto et al. (1980)
Glucose	Glucose oxidase	Pt (O_2)	2×10^{-3}–3×10^{-6}	Koyama et al. (1980)
Glucose	Glucose oxidase	Pt, O_2-generating	10^{-2}	Enfors (1981)
L-Histidine	Histidine ammonialyase	NH_3	10^{-2}–3×10^{-5}	Walters et al. (1980)
Lactose, maltose	Lactase or maltase + glucose oxidase	Gas (O_2)	4×10^{-3}	Cordonnier et al. (1975)
L-Lysine	L-Lysine decarboxylase	CO_2	10^{-1}–5×10^{-5}	White and Guilbault (1978)
Penicillin	β-Lactamase	pH	3×10^{-2}–10^{-4}	Enfors and Nilsson (1979)
Phenol	Phenol hydroxylase	Gas (O_2)	5×10^{-5}–5×10^{-7}	Kjellén and Neujahr (1980)
Sucrose	Invertase + mutarotase + glucose oxidase	Pt (O_2)	10^{-2}–10^{-3}	Satoh et al. (1976)

'enzyme thermistors' (Mosbach and Danielsson, 1974) by analogy with the term 'enzyme electrode' which was then a well-established designation. Although somewhat misleading, both terms are still in common usage as they indicate the simplicity of the concepts made possible by the proximal arrangement of the enzyme around or close to the transducer (Figure 5). The apparatus presented by Canning and Carr (1975) was in a later publication called an (immobilized enzyme) flow enthalpimetric analyzer (Bowers and Carr, 1976). In a review Grime (1980) also suggested the designation flow enthalpimeter for simple calorimeters of this kind. The simplest concept of all was that of the thermal enzyme probe (TEP) as presented by Cooney *et al.* (1974). Here the enzyme was attached directly on a thermistor probe, while a second thermistor was used as a reference. The pair of thermistors was then dipped into the sample solution which was kept in a carefully thermostatted waterbath, and the temperature difference measured by the two thermistors could be related to the substrate concentration. The sensitivity was, however, poor since only a minor part of the heat evolved could actually be detected as a temperature change. This design was later improved to some extent by Tran-Minh and Vallin (1978) who surrounded the thermistor probe by a glass jacket, thereby reducing the escape of heat from the probe. A related, simple apparatus, a small volume microcalorimeter in which the enzyme was immobilized on the surface of a Peltier element, was presented by Pennington (1974).

All these devices and some others not mentioned here have been more thoroughly discussed in a recent review (Danielsson *et al.*, 1981a). Here it suffices to point out that the TEPs generally had one order of magnitude or lower sensitivity than had constructions in which the temperature increase was measured over a small, well-insulated and thermostatted column containing the immobilized enzyme (see Figure 5). Such flow-through arrangements, which offer the additional advantage of permitting continuous analysis, have been exploited by Mosbach *et al.* (1975), Canning and Carr (1975), Schmidt *et al.* (1976) and Aizawa *et al.* (1979). The original TEP concept was modified by Fulton *et al.* (1980) to utilize a laminar flow.

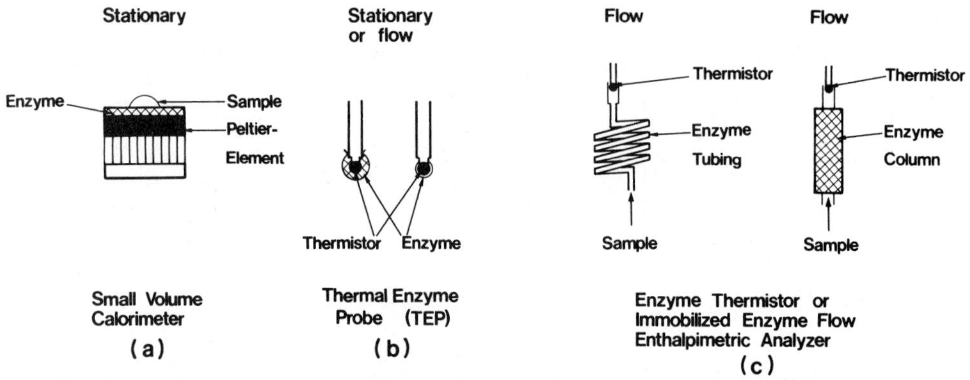

Figure 5 Schematic representations of (a) small volume calorimeter; (b) thermal enzyme probes (TEP); and (c) enzyme thermistor or immobilized enzyme flow enthalpimetric analyzer. (Reprinted from Mosbach and Danielsson, 1981, with permission of American Chemical Society)

Most of these closely related devices have a rather similar performance and, as can be established from reactions of known enthalpy (Table 2), or from electrical calibration (Danielsson *et al.*, 1979a), as much as 80% of the heat evolved in such 'semiadiabatic' instruments can be registered as a temperature change. This implies that for a given substrate present at a concentration of 1 mmol l^{-1} and with a molar enthalpy change of 80 kJ mol^{-1}, a peak height corresponding to 10^{-2} °C or higher will be obtained and requires a temperature resolution of 10^{-4} °C to give 1% accuracy in the measurement. Most enzymic reactions are accomplished by considerable heat production in the range 5 to 100 kJ mol^{-1} (Table 2), so measurements down to a concentration of 0.1 mmol l^{-1} should not present any problems.

20.3.2 Enzyme Thermistors

20.3.2.1 Apparatus and procedures

Enzyme thermistors and similar thermal flow analyzers now appear to be the most successful type of instrument amongst existing enzyme calorimeters. The enzyme thermistor developed in

Table 2 Molar Enthalpies of Reaction of some Enzyme-catalyzed Reactions

Enzyme	Substrate	$-\Delta H$(kJ mol^{-1})	Ref.
Catalase	Hydrogen peroxide	100.4	Nelson and Kiesow (1972)
Cholesterol oxidase	Cholesterol	52.9	Rehak and Young (1978)
Glucose oxidase	Glucose	80	Schmidt *et al.* (1976)
Hexokinase	Glucose	27.6 (74.9)[a]	McGlothlin and Jordan (1975)
Lactate dehydrogenase	Sodium pyruvate	62.1	Brown (1969)
Penicillinase	Penicillin G	67.0 (114.7)[a]	Grime and Tan (1979)
Trypsin	Benzoyl-L-arginineamide	27.8	Brown (1969)
Urease	Urea	6.6	Brown (1969)
Uricase	Urate	49.1	Rehak and Young (1978)

[a] The ΔH values in parentheses include protonation of Tris (-47.7 kJ mol^{-1}; Grime and Tan, 1979).

our laboratory has previously been described in some detail by Danielsson *et al.* (1981a). A recent design of our apparatus is schematically shown in Figure 6. Inside a temperature controlled (25, 30 or 37 °C) aluminum cylinder is another aluminum cylinder separated by a small airspace providing a certain degree of thermal insulation. Heat is transported between the two blocks by convection and by the fluid pumped from the main heat exchanger in the outer cylinder to the short, secondary heat exchanger in the inner cylinder. Furthermore, after passing the columns, the liquid pumped through the system transports heat from it, including the majority of the heat produced by the enzymic reaction. The columns will, therefore, be surrounded by a very stable temperature.

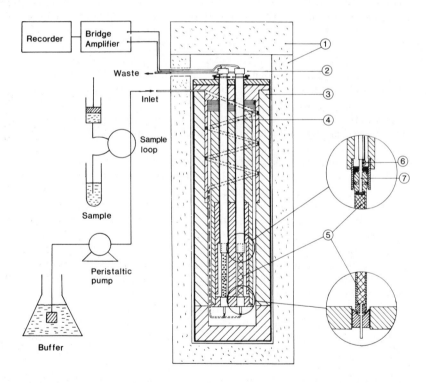

Figure 6 Enzyme thermistor set-up. Enzyme thermistor with aluminum constant temperature jacket: (1) polyurethane insulation; (2) Plexiglass tube with bayonet lock for column insertion; (3) thermostatted aluminum cylinder, height 250 mm; (4) heat exchangers; (5) plastic column; (6) thermistor attached to a gold capillary; and (7) column outlet

The temperature at the outlet of the column is measured with a small thermistor mounted on a short gold capillary tube. We have commonly used a 1 mm bead thermistor from Victory Engineering Corporation (Veco, Springfield, NJ) type 51A70 with a resistance of 100 kΩ at 25 °C and a temperature coefficient of -4.6% °C^{-1}. In many cases we also use a reference thermistor mounted in the inner block for differential measurements with increased baseline stability.

Temperature registration is made by a DC-coupled Wheatstone bridge equipped with a chopper-stabilized amplifier (MP 221, Analogic Corporation, Wakefield, MA) and wire-wound precision resistors with low temperature coefficient (Econistor, type 8E16 from General Resistance,

Bronx, NJ). At maximum sensitivity this bridge produces a 100 mV change in the recorder signal for a temperature change of 10^{-3} °C. The lowest practically useful range mainly dictated by noise created by friction and turbulence in the column, is, however, usually higher, typically 10^{-2} °C. In this context, it can be mentioned that a substrate concentration of 0.5 to 1 mmol l^{-1} in many reactions gives a temperature change of 10^{-2} °C.

A change of columns is very easy to perform as the column is simply inserted into the end of a Plexiglas tube, containing the outlet tubing and the temperature sensor, by which it is fitted into the apparatus. Columns of different diameters (max. i.d. 8 mm) and bed heights (max. 30 mm) can be accommodated. A commonly used column is 4 mm i.d. and has a height of 20 mm (0.25 ml). In addition, nylon tubing can also be used as enzyme support. The nylon tubing is wound around a special adaptor which fits into the column holder and connects the tubing to the flow system. Nylon tubing has definite advantages when crude samples containing particulates are tested (Mattiasson *et al.*, 1981), but suffers from low enzyme loading capacity. Thus, the normally employed carrier material is CPG (controlled pore glass), which offers high enzyme coupling capacity, good mechanical and chemical and microbial stability as well as relatively simple coupling procedures. Other materials have also been used (Danielsson *et al.*, 1981a) and one important factor to consider is that different carrier materials can show considerable variations in their adsorption pattern of sample constituents depending on the actual surface distribution of ionic, hydrophobic, *etc.* groups. Since such adsorption almost certainly will give rise to nonspecific heat effects and may even affect the enzymic reaction, the choice of enzyme support material could be rather crucial. We have, however, good experience with CPG, which we have used with pore sizes in the range of 50 to 200 nm and with particle size usually around 80 mesh. Regardless of whether we have started with untreated CPG or with alkyl(propyl)amino-derivatized CPG from different suppliers, we have almost invariably experienced the same good results employing glutaraldehyde for activating the glass and binding of the enzyme (Weetall, 1976). One point worth special mention is that we generally apply rather a large excess of enzyme to an enzyme thermistor column — typically 100 units, if affordable. This ensures good operational stability, *i.e.* unchanged performance of the system over a long series of samples or periods of continuous monitoring. A column life of several weeks or even months is expected.

An instrumental arrangement for enzyme thermistor analysis, including an inactive reference column for compensating for nonspecific heat, was devised by Mattiasson *et al.* (1976). Such a split-flow arrangement has proven valuable in many situations involving crude samples, *e.g.* for the triglyceride determination described by Satoh *et al.* (1981), but in most cases it is possible to avoid nonspecific effects by careful design of the procedure.

In all cases buffer is continuously pumped through each channel in use at a flow rate of 0.5 to 3 ml min^{-1}. Samples are generally introduced as short pulses with a chromatography valve equipped with a 0.25 to 1 ml loop. Thermal steady state will not be obtained for short sample pulses, but the enzymic reaction will generate a temperature peak, the height of which is normally taken as a measure of the substrate concentration. The linear range of the relationship between temperature peak height and substrate concentration is usually at least 10^{-5} to 10^{-1} mol l^{-1}, when not limited by reactant concentrations (as oxidases are by the supply of oxygen) or by other factors. As has been discussed previously by Danielsson *et al.* (1981a), the area under the temperature peak or the slope of the ascending peak are also useful measures of substrate concentration. The maximum number of samples that can be analyzed by the present system is 15 to 30 per hour with the above mentioned sample volumes. Decreasing the sample volume to 0.1 ml and increasing the flow rate to 2 to 3 ml min^{-1} will permit a throughput of 1 sample per minute. In order to demonstrate simplified sample handling we have modified a sample changer for a Technicon AutoAnalyzer system and interfaced it with a small desktop computer that also controls the sampling valve. This system can be left unattended over night, and if the sample series has also included a set of standards, the computer is programmed to calculate the concentration of each sample based on peak heights or peak areas.

It should be pointed out that when the sample pulse length is increased sufficiently (more than 1 to 5 min), the temperature response will eventually reach a constant value proportional to the concentration. This value may be twice as high as the peak height normally obtained.

20.3.2.2 *Clinical results*

All of the simple enzyme calorimetric devices so far introduced have first been applied to metabolites primarily of clinical concern. In clinical chemistry there is a wide-spread need for simple, dedicated instruments for determination of blood glucose, urea and cholesterol, to mention some of the most frequent ones. These instruments should be used as back-up instruments, for acute analysis, and as routine instruments in small laboratories. A number of apparatuses based on various measuring principles, including spectrophotometry and ion selective electrode methodology, using chemical as well as enzymic procedures, have found their niches in this market. The versatility of enzyme calorimetry is an obvious advantage for this kind of instrument. It can easily be adapted for different assays by simply changing the enzyme column.

Useful routine methods for the determination of urea serum based on immobilized urease have been described by Bowers *et al.* (1976), Danielsson *et al.* (1976) and Krisam and Schmidt (1977). It is quite clear that the demands for sensitivity, reproducibility, accuracy, analysis speed and simplicity all are adequately met by these methods. This also holds true for the different glucose assays developed. For determination of glucose, either hexokinase or glucose oxidase has been used, the latter usually together with catalase. Hexokinase was used by Bowers and Carr (1976) in their flow enthalpimetric system in the range of 0.5 to 50 mmol l^{-1} (40 samples per hour). We preferred glucose oxidase because, first, we found it more stable, and second, it does not need any cofactor (Danielsson *et al.*, 1977). A disadvantage is, however, that linearity is obtained only to about 0.7 mmol l^{-1} (with catalase included), so samples often have to be diluted at least 100-fold; however, small serum volumes (5 to 20 μl) also can be injected directly into the buffer stream to the enzyme thermistor.

Similar results were obtained by Schmidt *et al.* (1976), who also used glucose oxidase coimmobilized with catalase, and by Marconi (1978), who used these enzymes fiber entrapped and thereby reported a wider linear range (up to 5 mmol l^{-1} of glucose) although at the expense of sensitivity and probably also of operational stability.

Determination by enzyme calorimetry of other substances of clinical interest was recently surveyed by Mosbach and Danielsson (1981) and by Danielsson *et al.* (1981a) (see also Table 3). Triglyceride determination can be mentioned as an example of later studies in this field (Satoh *et al.*, 1981). Lipoprotein lipase (EC 3.1.1.34) was immobilized on CPG (controlled pore glass) with a pore size of 200 nm and the assay was performed with a split-flow apparatus in Tris buffer containing 0.5% Triton X-100. A linear temperature response was obtained for 0.05 to 10 mmol l^{-1} of tributyrin and for 0.1 to 5 mmol l^{-1} of triolein. The triglyceride concentration in serum samples could be directly determined after twofold dilution up to a concentration of 3 mmol l^{-1}, and the results agreed well with those obtained with conventional, spectrophotometric, enzymic methods.

Table 3 Substances Analyzed with Enzyme Thermistors and Related Thermal Probes

Substance	Enzyme	Concentration range (mmol l^{-1})	Ref.
Ascorbic acid	Ascorbic acid oxidase	0.05–0.6	Mattiasson and Danielsson (1982)
Cellobiose	β-Glucosidase + glucose oxidase + catalase	0.05–5	Danielsson *et al.* (1981b)
Cephalosporin	Cephalosporinase	0.005–10	Danielsson *et al.* (1979b)
Ethanol	Alcohol oxidase	0.01–1	Guilbault *et al.* (1983)
Galactose	Galactose oxidase	0.01–1	Mattiasson and Danielsson (1982)
Glucose	Glucose oxidase + catalase	0.002–0.8	Danielsson *et al.* (1977); Schmidt *et al.* (1976)
Glucose	Hexokinase	0.5–25	Bowers and Carr (1976)
Lactate	Lactate oxidase	0.01–1	Danielsson *et al.* (1981a)
Lactose	Lactase + glucose oxidase + catalase	0.05–10	Mattiasson and Danielsson (1982)
Oxalic acid	Oxalate oxidase	0.005–0.5	Winquist *et al.* (1985)
Penicillin G	Penicillinase (β-lactamase)	0.01–500	Mattiasson *et al.* (1981)
Sucrose	Invertase	0.05–100	Mattiasson and Danielsson (1982)
Triglycerides	Lipoprotein lipase	0.1–5	Satoh *et al.* (1981)
Urea	Urease	0.01–500	Bowers *et al.* (1976) Danielsson *et al.* (1976)

Recently, a study of oxalate determination using oxalate oxidase was finished (Winquist *et al.*, 1985). The temperature response was linear in the range of 0.01 to 0.5 mmol l^{-1} with a rather high sensitivity (10^{-2} °C mmol^{-1}). The oxalic acid concentration could be determined directly in 5- to 10-fold diluted serum or urine. In addition, various fruit juice and vegetable extracts were tested with good results.

20.3.2.3 *TELISA*

For compounds present at levels below the detection range for direct calorimetric determination, the enzyme thermistor can be applied in an enzyme immunochemical assay (EIA) based on the interaction between an antigen and a specific antibody in which one of the moieties involved in the interaction is labelled with an enzyme. Figure 7 shows the reaction scheme of the TELISA (thermometric enzyme linked immunosorbent assay) technique introduced by Mattiasson *et al.* (1977). In principle, the enzyme thermistor column is filled with immunosorbent, such as antibodies immobilized on Sepharose CL 4B. The antigen to be determined and an enzyme-labelled (*e.g.* catalase) antigen are introduced into the flow, whence the amount of catalase-bound antigen remaining bound to the column is a function of the content of antigen. The less antigen that is present in the sample, the more catalase-bound labelled antigen will be found in the column and thus evolve more heat after subsequent introduction of H_2O_2, the substrate of catalase. With the TELISA technique analysis is possible down to 10^{-13} mol l^{-1}. In addition, the same immobilized antibody preparation can be used a few hundred times. The lower sensitivity inherent to this nonequilibrium assay is outweighed by quicker analysis (one assay, including regeneration takes only 12 min), reproducibility and simplicity (Borrebaeck *et al.*, 1978).

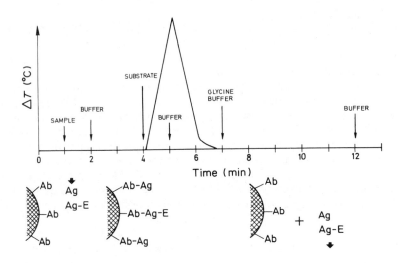

Figure 7 Schematic presentation of a reaction cycle in the TELISA procedure. The arrows indicate changes in the perfusing medium (flow rate 0.8 ml h^{-1}). The cycle starts with buffer. The enzyme thermistor column now contains only immobilized antibodies. At the arrow, the 'sample', a mixture of antigen and catalase-bound antigen, is introduced. The system is then washed with buffer for 2 min. The sites of the antibodies of the column are now occupied by antigen, as well as by catalase-labelled antigen. The amount of catalase bound is measured by registering the heat produced during a 1 min pulse of the substrate, 1 mmol l^{-1} H_2O_2. After a 5 min wash with 0.2 mol l^{-1} glycine–HCl, pH 2.2, the system is ready for another assay. (Reprinted from Mattiasson *et al.*, 1977, with permission of Elsevier Publishing Co)

20.3.3 Applications in Biotechnology

Although most thermal enzyme probe or enzyme calorimeter applications initially were in clinical chemistry, analysis in biotechnology has emerged as an even more promising field. Obviously, several of the calorimetric procedures proposed for clinical chemistry are of direct interest in biotechnology, for instance for glucose and many other sugars, lactate, oxalate and ethanol. There are many reasons why biotechnology should be an interesting field for enzyme calorimetric analysis. First, concentrations of many metabolites of primary interest can be found in a range one or two orders of magnitude higher than in clinical chemistry. In fact, reproducible dilution can sometimes be a bit of a problem when concentrations are very high. Secondly, the high specificity inherent to an enzymic reaction is a definite advantage, since samples are usually quite complex. Inhibition or interference with the enzymic reaction, which sometimes can be a problem in clinical chemistry, is very much less so in biotechnology since the dilution required also brings down the concentration of any interfering substances. Thirdly, the sample solutions are often turbid or colored, which may affect a colorimetric assay. As a fourth point it can be mentioned that there actually is a lack of reliable, accurate and simple one-line assays adapted for biotechnology, while for many years the needs of clinical chemistry have been much better accommodated.

As has already been discussed, glucose can be measured with high sensitivity using glucose oxidase coimmobilized with catalase. The fact that the measuring range is limited to upwards of 0.7 to 0.8 mmol l^{-1} by the available supply of oxygen is a drawback in biotechnological applications with high glucose concentrations, since it necessitates about a 100-fold dilution of the sample. This can be difficult to accomplish reproducibly in continuous flow systems. A calorimetric glucose assay based on hexokinase, which gives a linear response up to at least 10 mmol l^{-1}, on the other hand is hampered by the need for a cofactor (ATP) and by a less stable enzyme, both of which are important factors in continuous assays. We have, however, obtained very satisfying results with glucose oxidase/catalase with discrete samples (Danielsson *et al.*, 1981b) as well as in continuous monitoring (Danielsson *et al.*, 1979a).

Combining a glucose oxidase/catalase thermistor with a disaccharide hydrolyzing enzyme results in a more general approach. The immobilized disaccharidase can either be placed in the enzyme thermistor column or applied in a column outside the system. The latter example is usually favored as the enthalpy of the hydrolysis is low and does not contribute much to the total heat production. This way the column volumes available are better utilized (Danielsson *et al.*, 1981b). In that study we found that cellobiose could be directly measured only at concentrations above 10 mmol l^{-1} using β-glucosidase as the sole enzyme. We also found that when salicin was used as substrate, the reaction enthalpy was considerably larger, permitting readings of salicin concentrations down to 0.25 mmol l^{-1}. When the β-glucosidase column was combined with a glucose oxidase/catalase enzyme thermistor more useful results were obtained with cellobiose as well. The linear range was 0.1 to 10 mmol l^{-1} and the operational stability was satisfactory for practical measurements on broths containing cellulose degradation products. If desired, a dialyzer unit could be incorporated (Technicon, 12 in) although the sensitivity then was reduced to 0.5 mmol l^{-1} of cellobiose. If a more active β-glucosidase column had been available, the sensitivity would have been even better; only about 20% of the cellobiose was converted.

More complete hydrolysis is easily obtained in a similar system for lactose determination using β-galactosidase that can be obtained in highly active preparations. In the study reported by Mattiasson and Danielsson (1982), a working range of 0.25 to 5 mmol l^{-1} was studied for the three enzyme system shown below:

$$\text{lactose} + H_2O \xrightarrow{\text{lactase}} \text{D-glucose} + \text{D-galactose} \tag{11}$$

$$\text{D-glucose} + O_2 \xrightarrow{\text{D-glucose oxidase}} \text{D-gluconon-1,4-lactone} + H_2O_2 \tag{1}$$

$$H_2O_2 \xrightarrow{\text{catalase}} H_2O + 0.5\, O_2 \tag{4}$$

in which the lactase column was placed outside before the enzyme thermistor.

A similar procedure could be applied to sucrose determination. In this case, however, it is interesting to note that the disaccharidase invertase produces much more reaction heat. Thus, sucrose can be directly determined in the range 1 to 200 mmol l^{-1} using invertase as single enzyme in the enzyme thermistor (Mattiasson and Danielsson, 1982). This assay compares very favorably with other procedures for sucrose determination. Since only one enzyme is required, the specificity is high and the sucrose concentration can be directly determined, regardless of any glucose present in the sample. Even if the sensitivity after all is rather low, it is sufficient for most biotechnological applications. Furthermore, the enzyme is highly stable, which makes continuous monitoring based on this system very reliable (Mandenius *et al.*, 1981).

The report by Mattiasson and Danielsson (1982) also included assays for galactose and ascorbic acid. Galactose was determined in the range 0.5 to 10 mmol l^{-1} with the use of galactose oxidase, which is a high concentration for an oxidase, thus indicating low catalytic activity of the enzyme column. The single-day performance of the system was good, but the enzyme displayed a rather poor long-term stability due to loss of Cu^{2+}, although the activity could be repeatedly restored by flushing the column overnight with buffer containing Cu^{2+}.

Ascorbic acid was assayed in the range of 0.5 to 0.6 mmol l^{-1} by the enzyme ascorbic acid oxidase, which was bound to the enzyme thermistor column by reversible, biospecific immobilization. A 0.5 ml column containing concanavalin A–Sepharose was placed in the enzyme thermistor and 4.5 units of ascorbate oxidase in buffer solution were introduced by a peristaltic pump. The enzyme (which has to be a glucoprotein) readily binds to the concavalin A and the column is immediately ready for use. When it is time to recharge the column, the old enzyme is first washed out by a pulse of 0.1 mol l^{-1} glycine–HCl, pH 2.2 and, after reconditioning to the perfusion buffer, fresh enzyme can be introduced. This is a convenient procedure when using a labile enzyme that has to be replaced regularly. One advantage is that the recharging can be done in a short time without removing the column from the instrument.

Excellent results have been obtained from enzyme thermistor analysis of penicillins and other antibiotics in fermentation solution. The normal concentration range is suitable for this type of analysis and sample preparation procedure is simple; these are further advantages. One company runs this assay system for penicillin V determination in 250 samples per day (30 per hour) with only weekly change of enzyme columns. The β-lactamases used for these analyses can be obtained with different specificities, including those having a more pronounced specificity for cephalosporins (Danielsson *et al.*, 1979b).

Various possibilities for sample treatment and enzyme immobilization procedures for penicillin determination in fermentation broths were recently discussed by Mattiasson *et al.* (1981). Figure 8 illustrates the difference in operating range obtained with CPG- and nylon tubing-bound penicillinase. While the nylon tubing better withstands fouling, its enzyme binding capacity is low and the CPG-column gives higher sensitivity and a wider measuring range. Determination of penicillins by enzymatic enthalpimetry was studied by Grime and Tan (1979). They also used a simple calorimeter construction, but with soluble enzyme kept in a small Dewar vessel into which the samples were injected; an alternative name for this technique is direct injection enthalpimetry (DIE). Their report gives interesting data for enthalpies and reaction rates for various penicillins.

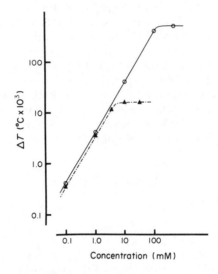

Figure 8 Calibration curves for penicillin G. The linear ranges obtained for CPG-bound (○) and for nylon tubing-coupled penicillinase (▲) are compared

In a recent investigation, Guilbault *et al.* (1983) compared enzyme electrode and enzyme thermistor assays for alcohols based on the alcohol oxidase from *Candida boidinii*. The stability of the alcohol oxidase in the enzyme thermistor method was very dramatically changed by coimmobilizing with catalase. The catalase continuously removes the hydrogen peroxide formed and this is the most probable reason for the increased stability. Besides ethanol, propanol, butanol and (even better) methanol were found to give good responses with both techniques. In many cases, however, ethanol is the predominant component, and then both methods are quite useful alternatives in biotechnological analysis. The good long-term stability of the enzyme thermistor method makes it particularly suitable for continuous monitoring. A typical calibration curve for different substrates is shown in Figure 9.

20.3.4 Continuous Monitoring

As already pointed out, temperature measurements with a thermal probe such as a thermistor preferably are made in a continuous fashion, and the enzyme thermistor signal is linearly related to the substrate level generally over a very wide range. By applying a large excess of enzyme on

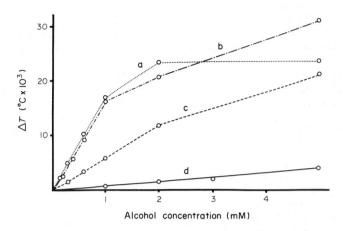

Figure 9 Calibration plots for (a) methanol; (b) ethanol; (c) butanol; and (d) propanol obtained with an alcohol oxidase/catalase column measuring the total heat change

columns with good mechnical durability, the system is designed for good operational stability. Constant response for a given substrate concentration is obtained over long periods of time. Only when very small concentration changes are to be detected at sensitive measuring ranges (*e.g.* 10^{-2} °C) does the baseline need rechecking, maybe every 4 to 6 h. As one objective in the design of the enzyme thermistor and similar systems has been to make simple instruments, they cannot accommodate the rather big variations in ambient temperature (maybe ± 5 °C or more) in the places where these instruments are used. This is to be compared with conventional calorimeters normally being run in thermostatted rooms. However, this drawback is easily eliminated in an automated system in which a valve can be programmed to introduce blank solution for baseline control at suitable intervals. If a second valve is included for standard samples, the system can also be automatically calibrated whenever required. In our own system, the computer used for process control based on the enzyme thermistor signal, can be used for supervising the calibration procedures as well.

When the measuring system is operated in connection with a computer, it is easy to sample at any selected interval and with any volume. In many cases, continuous monitoring may not require a reading every second; perhaps only once or a couple of times per hour would be sufficient. In such a case, intermittent sampling could give higher accuracy and longer column life. The examples given below are, however, all from genuine continuous monitoring.

One of the first studies in which the enzyme thermistor was used in a model system for process control was published by Danielsson *et al.* (1979a). In that work, the flow rate through a plug-flow reactor containing β-galactosidase immobilized on Sepharose was regulated to give a constant glucose level in the effluent, as measured by a glucose oxidase/catalase enzyme thermistor. The enzyme thermistor signal was converted by a PID-controller to a control signal for the pump feeding the enzyme reactor. Thus, variations in the concentration of the incoming lactose solution (whey), as well as changes in the catalytic capacity due to fouling, could be continuously compensated.

Our current studies in process control are focussed on control of pilot plant reactors for ethanol fermentation. The principal design of our model system is shown in Figure 10. The ultimate goal is simultaneous control of the substrate supply to the fermenter and control of the ethanol recovery in order to maintain the substrate concentration at an optimal level. A third variable to control is the final utilization of the remaining substrate. This may be achieved in a plug flow reactor containing immobilized yeast of a type resistant to higher ethanol concentrations. The main fermenter is a continuous stirred tank reactor which, for practical reasons, usually is loaded with yeast immobilized in alginate beads. The integrated system is operated with a PDP-11 computer (Digital Equipment), while subsystems have been studied using simpler computers, such as a hobby type computer or even analog controllers. Thus, Mandenius *et al.* (1980) showed that the sucrose feed to a fermenter could be adequately controlled by an invertase thermistor in connection with a PI-controller. The control task was facilitated by the fact that no nutrients were added during the fermentation, resulting in a rather low activity. In a subsequent study on a more efficient fermenter, Mandenius *et al.* (1981) found that the simple analog controller sometimes ended up in oscillation. However, decreasing the proportional band to 1% resulted in an on–off control that reduced the oscillation to an acceptable value, 5% of the set level. More sophisticated control, however, requires a computer, especially when control of recovery of ethanol is taken into account and when control of the final utilization of substrate in a column reactor is included. By these studies it has also been demonstrated that the enzyme thermistor is a powerful and reliable tool for process monitoring.

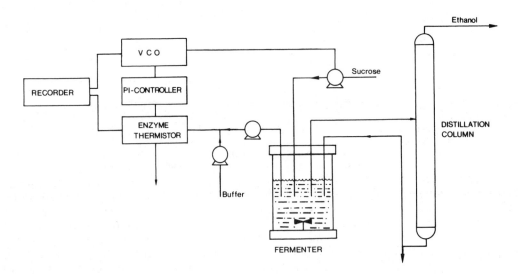

Figure 10   Model system for studying the enzyme thermistor as a monitor and control instrument for fermentation processes

Danielsson *et al*. (1981a) have previously discussed different possibilities of measuring enzyme activity in solution with an enzyme thermistor unit. Simply by replacing the enzyme column with an inert column or a reaction coil, the apparatus can be directly used for this purpose. Substrate solution and the enzyme solution (the sample) are mixed and after temperature equilibration to eliminate mixing heats, the mixture is pumped through the reaction coil. The temperature increase during this passage is a measure of the enzymic activity (Danielsson and Mosbach, 1979). The sensitivity is about 0.1 units ml^{-1}. At least the same sensitivity can be obtained with an alternative technique involving biospecific absorption. If an enzyme solution is pumped through an enzyme thermistor associated, for example, with a concanavalin A–Sepharose column, the enzyme will bind to the column and subsequently can be quantified by introduction of substrate to the enzyme in excess. The resulting temperature response will be proportional to the enzyme amount. The sensitivity is determined by the sample volume applied. Finally, the column is washed with glycine–HCl and is then ready for a new sample.

The enzyme thermistor recently has been successfully applied as an instrument for the specific monitoring of some common chromatographic procedures: gel filtration, ion exchange chromatography, and affinity chromatography (Danielsson *et al*., 1981c). Since a calorimeter can be used for continuous monitoring of enzymic activities directly on crude samples, it provides possibilities for direct localization of a specific component in a complex chromatogram as, for instance in the initial steps of an enzyme purification scheme (Figure 11). Furthermore, elution in affinity chromatography is frequently accomplished with coenzymes having a strong UV-absorbance thereby often precluding direct on-line assay of a particular enzyme, either by UV monitoring or by spectrophotometric monitoring of changes in NAD(P)H concentration. Calorimetric detection of the eluted enzyme activity should thus have definite advantages in this area as well.

Finally, with the enzyme thermistor unit it is possible not only to monitor continuously a product, *e.g.* penicillin, as it is formed in a fermentation process, as well as substrates, *e.g.* lactose, being consumed, or the activity of a particular enzyme, but also to register simultaneously the overall thermal behavior of such a microbial process, yielding a thermogram (power–time curve), which provides additional valuable information. Again different procedures are possible. The microorganism can be immobilized and placed in a column in the same way as an enzyme, resulting in a 'microbe thermistor', as recently used by Mandenius *et al*. (1981) to study the metabolic response of yeast to different ethanol concentrations. Alternatively, a microorganism suspension can be treated the same way as an enzyme solution in the enzyme activity assays

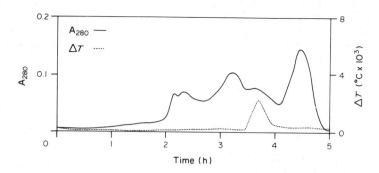

Figure 11 Gel filtration of 1 ml crude yeast extract on an Ultrogel AcA 44 column (53 × 2.5 cm) eluted with 0.2 mol l^{-1} Tris–HCl, 0.0133 mol l^{-1} MgCl$_2$, pH 7.8, at a flow rate of 0.75 ml l^{-1}. For thermal analysis, the effluent was mixed with a substrate solution containing 0.54 mol l^{-1} glucose, 0.011 mol l^{-1} ATP, flow rate 0.2 ml l^{-1}. The UV absorbance at 280 nm was registered with a Uvicord S (LKB). (Reprinted from Danielsson *et al.*, 1981c, with permission of Academic Press)

described above. Thus, a sample can be drawn continuously from a fermenter to a calorimetric unit to provide information on the metabolic condition in the fermenter (Danielsson *et al.*, 1981a).

20.4 ENZYME TRANSISTORS — ChemFETs

There is a rapidly growing interest in devices combining semiconductor technology and enzymes or other biological systems. Semiconductor components used in most studies are field effect transistors (FET). Such components made sensitive to ionic species or chemical agents are thus called ISFETs or ChemFETs. For a combination with an enzyme, the term ENFET has been used. Danielsson *et al.* (1979c) combined immobilized enzymes with a hydrogen- and ammonia-sensitive FET and called their device an 'enzyme transistor'. In conclusion, the designation 'bio-chip' could be used to sum up the activities incorporating biology in microelectronics. Reports from this area have so far been rather preliminary, but the expectations are high as can be seen from a short survey by Yanchinsky (1982) of current studies in some laboratories, industrial as well as university institutes. What makes the concept of the biochip so attractive is above all its potential for miniaturization. The active area of the semiconductor being used is very small and it should be possible to introduce a biochip into a blood vessel of a patient to measure a metabolite by a suitable enzyme attached to the semiconductor. Perhaps even different spots of the silicon chip could be covered by different enzymes to allow simultaneous determination of several metabolites.

The circuit diagram developed by Caras and Janata (1980) for determination of penicillin by an ENFET probe is shown in Figure 12. The transistor chip contained two pH-sensitive ISFETs, but could nevertheless be mounted in a 2 mm diameter dual lumen PVC catheter. The gate of one of the ISFETs had a glutaraldehyde crosslinked albumin–penicillinase (β-lactamase) membrane, while the other gate had only a crosslinked albumin membrane. Although only 2.5×10^{-4} I.U. of enzyme were applied on its small sensing area (0.5 mm^2), this ENFET had a lifetime of 2 months with only a slight loss of activity. Penicillin measurements were made in differential current mode, measuring the drain current difference between the penicillin-sensitive gate and the reference gate. The detection limit was approximately 0.1 mmol l^{-1} of penicillin and the calibration curve was linear up to 25 mmol l^{-1} at a buffer capacity of 0.02 mol l^{-1}. The differential mode of operation provided automatic compensation for changes of ambient pH and temperature.

The application of a hydrogen- and ammonia-sensitive palladium coated semiconductor device (Pd-MOSFET) in enzymic analysis was studied by Danielsson *et al.* (1979c). The response of this gas-sensitive device to NH$_3$, liberated both nonenzymically as well as by the enzymes urease and creatininase, was investigated. The p-channel Pd-MOSFET used resembled an ordinary field effect transistor of MOS-type (metal oxide semiconductor), but had a catalytically active gate metal, Pd, instead of, for example, Al. The transistor chip also contained a heater and a control circuit to give a constant temperature of 100 to 150 °C, to eliminate water and to speed up the reactions at the Pd surface. Hydrogen or gases containing hydrogen, such as NH$_3$ or H$_2$S, are dissociated on the catalytically active surface to atomic hydrogen, and a fraction of the hydrogen

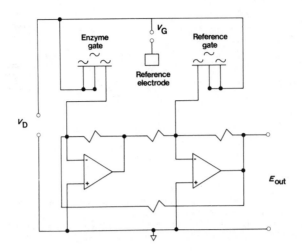

Figure 12 Circuit diagram for measuring differential drain current of an ENFET probe. (Reprinted from Caras and Janata, 1980, with permission of the American Chemical Society)

atoms rapidly diffuses through the Pd-film to form a dipole layer at the Pd–SiO$_2$ interface. The characteristics of the component are thereby altered in such a way that if the current through it is kept constant, the threshold voltage, which can be registered by a potentiometric recorder will be a function of the H$_2$ (or NH$_3$) gas concentration. H$_2$ could be determined with a sensitivity of 10^{-6} mass fraction and NH$_3$ with a sensitivity of 10^{-5} mass fraction.

The response of the Pd-MOSFET to NH$_3$ released from NH$_4$Cl solutions by adding NaOH was studied in an arrangement resembling the air-gap electrode described above. Further, the response to NH$_3$ formed upon enzymic hydrolysis of urea by urease was investigated. Urea concentrations of 0.1 mmol l^{-1} could easily be detected. Finally, the Pd-MOSFET was applied in a flow-cell for monitoring the ammonia present in the effluent from an enzyme thermistor containing immobilized creatininase (EC 3.5.4.21). The response of the Pd-MOSFET when samples with different creatinine concentrations were introduced was in good agreement with the enzyme thermistor signal.

In other studies, Pd-MOS components with a more pronounced hydrogen sensitivity were investigated. Besides Pd-MOSFETs, Pd-MOS capacitors can be used (Winquist *et al.*, 1982). With such a component mounted in a small heated chamber (105 °C) applied in the experimental set-up shown in Figure 13, the hydrogen production from, for example, a small column containing *Clostridium acetobutylicum* immobilized in alginate beads could be followed during several days. The applicability of such H$_2$-sensitive devices is vastly increased by exploitation of the enzyme hydrogen dehydrogenase (hydrogenase, EC 1.12.1.2), which is capable of reducing coenzymes such as NAD$^+$ or NADP$^+$ by heterolytic cleavage of molecular hydrogen:

$$NAD^+ + H_2 \rightleftharpoons NADH + H^+ \tag{12}$$

Through this reversible reaction, other dehydrogenase reactions can thus be followed by a H$_2$-sensitive Pd-MOS component. Using a small column of hydrogenase bound to controlled pore glass in the set-up shown in Figure 13, Winquist *et al.* (1982) could demonstrate a linear relationship between hydrogen formation and incoming NADH concentration. NAD$^+$ could be quantified through the hydrogen consumption when introduced in a solution supplied with a known concentration of hydrogen. As for NADH, a linear calibration curve was obtained for NAD$^+$ over the range tested, 0.1 to 0.8 mmol l^{-1}. The sensitivity was adequate for practical determinations as was demonstrated in a coupled assay for ethanol using alcohol dehydrogenase.

20.5 OTHER ENZYME–TRANSDUCER COMBINATIONS

This chapter on enzyme probes has focussed on enzyme electrodes and thermal enzyme probes or enzyme thermistors. Currently these are the most important enzyme probes and they also can be expected to keep that position in the near future. There are, however, many other promising enzyme–transducer combinations that may become very important in a longer perspective. (*cf.* Weaver and Burns, 1981). A number of optical techniques could be envisaged to give interesting

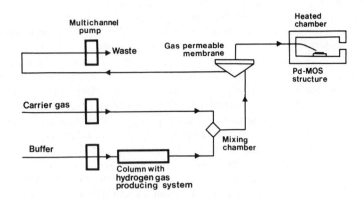

Figure 13 Experimental set-up for measuring gaseous hydrogen produced by biological systems. (Reprinted from Winquist *et al.*, 1982, with permission of The Humana Press Inc.)

combinations with immobilized enzymes, especially in connection with fiber optics or light guides. Developments of light sources and detectors have laid the ground for smaller and simpler instruments that could be combined with enzymes, *e.g.* attached to the end of a 'light pipe', resulting in colorimetric or fluorimetric enzyme probes. Photoacoustic spectroscopy appears to have special advantages for studies of opaque preparations of immobilized enzymes, but at present involves comparatively expensive instrumentation.

In recent years, bioluminescence has found important applications in many branches of analysis. The most striking feature of these assays is their high sensitivity. Some studies have involved immobilized biological systems, as will be exemplified here by the concept devised by Kobayashi *et al.* (1981), which can be regarded as a sort of enzyme probe. The enzymes studied, xanthine oxidase, peroxidase, glucose oxidase, uricase and cholesterol oxidase, were immobilized in a photocrosslinked resin (xanthine oxidase and peroxidase), or on anion-binding cellulose beads. In the cases of xanthine oxidase and peroxidase, an enzyme film 0.5 mm thick was formed that was attached to the inside of a test tube placed close to a photomultiplier tube. For determining the concentration of xanthine or hydrogen peroxide, the luminescence rate was determined by photon counting after adding a small amount of a synthetic *Cypridina* analog luciferin. The detection limits were 0.02 μmol l^{-1} for both xanthine and hypoxanthine, while the detection limit for hydrogen peroxide was 0.2 μmol l^{-1}. For the measurement of glucose, cholesterol or uric acid, the appropriate enzyme was added to a test tube with a peroxidase membrane attached to the inner wall. Luminescence rate determination following the peroxidase reaction on the hydrogen peroxide formed allowed glucose, cholesterol and uric acid to be detected at concentrations down to 3, 8 and 8 μmol l^{-1}, respectively.

Another new optical method is the dry film technology developed by Eastman Kodak Co. (Curme *et al.*, 1978). Chemical determinations have been described for serum components such as glucose, urea, bilirubin, triglycerides and α-amylase. Here, the dry film technique for the enzymic determination of total cholesterol developed by Dappen and coworkers (1982) will be described. The dry film consists of a transparent film base coated with a gelatin layer and a porous spreading layer containing a white pigment, $BaSO_4$. The gelatin layer enhances reaction kinetics and improves sample distribution. The spreading layer has the function of uniformly spreading and metering the serum sample and providing a white reflective background. Furthermore, the spreading layer contains all the necessary components for the detection of cholesterol: enzymes, dye precursor, surfactant (Triton X-100) and buffer. Cholesterol esters are hydrolyzed to cholesterol by cholesterol ester hydrolase. Cholesterol oxidase then oxidizes all cholesterol to cholest-4-en-3-one and hydrogen peroxide, which finally in the presence of peroxidase oxidizes a triarylimidazole leuco dye to a colored product. A cartridge containing 50 test films is inserted in the analyzer and 10 μl of sample is applied to the dry film. After 5 min of incubation at 37 °C, the intensity of the dye formed is measured by reflectance densitometry through the film base. Results obtained by this technique correlated well with those obtained by a conventional method and the assay was found to be precise and relatively free of interferences.

A quite different combination of a physical technique and immobilized enzymes was proposed by Weaver *et al.* (1976). The volatile compounds produced by an immobilized enzyme catalyzed reaction were detected by a mass spectrometer which was interfaced *via* a semipermeable membrane. This technique assumes that at least one of the reactants is reasonably volatile and can penetrate the membrane and evaporate into the vacuum of the mass spectrometer. Since the enzyme provides the substrate specificity, the mass spectrometer can be fairly simple and can often be operated at only one mass peak. Weaver *et al.* (1976) used this technique for rapid determination of urea and NADH. For urea determination, immobilized urease was utilized (equation 8). While both the ammonia and the carbon dioxide formed were volatile and permeated the membrane, CO_2 was best suited for measurement, since the NH_3 peak was disturbed by OH fragments produced from H_2O. NADH was quantified by immobilized alcohol dehydrogenase supplying acetaldehyde at excess concentrations. The versatility of simple mass spectrometers in direct monitoring of biological reactions is underlined by a later study (Pungor *et al.*, 1980), in which a computer controlled mass spectrometer was used to monitor N_2, O_2, CO_2 and ethanol in the gas phase and N_2, O_2, CO_2 and ethanol in the liquid phase during the fermentation of *Saccharomyces italicus*. The relatively high cost of the mass spectrometer, even that of a simple quadrupole instrument, can thus be outweighed by its multichannel capacity.

In conclusion, this somewhat disparate collection of examples of enzyme–transducer combinations together with the preceding section serves to indicate the span of possibilities for such combinations, rather than to be a complete account.

20.6 MICROBE AND ORGANELLE PROBES

In recent years, microbial sensors have received increasing attention. The transducers have usually been electrochemical in nature, but also thermal probes (Danielsson *et al.*, 1981a) and gas-sensing semiconductor devices (Winquist *et al.*, 1982) can be used. Instead of pure enzyme, the bioselective membrane or reactor could contain organelles such as mitochondria, whole cells, or even intact tissue slices. Such biocatalysts extend the range of biospecific probes beyond the possibilities of conventional enzyme probes. When multistep reaction paths are required, or in situations where the appropriate enzyme is difficult to isolate or unstable in a purified state, organelles or whole cells are advantageous. They also reduce the labor and cost of preparation of biospecific probes. Possible disadvantages are loss of specificity and low sensitivity.

Aizawa and coworkers (1980b) used nonphosphorylating electron transport particles (ETP) from beef heart mitochondria in an electrochemical device for the determination of NADH or succinate. The choice of method for immobilization was critical and of several methods evaluated only entrapment in agar was found to give acceptable stability results. A 400 μm thick agar membrane containing ETP was placed over a galvanic oxygen probe. Succinate and NADH were oxidized by the ETP with consumption of oxygen and could be measured over a concentration range of 0.05 to 0.4 mmol l^{-1}.

Karube *et al.* (1981) used immobilized nitrifying bacteria for the amperometric determination of ammonia gas. Their microbial electrode consisted of a galvanic oxygen electrode covered by a Teflon^{TM} membrane. The bacteria were retained on an acetylcellulose membrane held against the oxygen probe by a gas permeable Teflon^{TM} membrane.

The nitrification was accomplished by two kinds of bacteria: a *Nitrosomonas* species that utilized ammonia as the sole source of energy

$$NH_3 + 1.5\,O_2 \xrightarrow{\textit{Nitrosomonas sp.}} NO_2^- + H_2O + H^+ \tag{13}$$

and a *Nitrobacter* species that oxidized nitrite

$$NO_2^- + 0.5\,O_2 \xrightarrow{\textit{Nitrobacter sp.}} NO_3^- \tag{14}$$

This microbial electrode could be used for the determination of NH_3 in the range of 0.1 to 42 mg l^{-1}. The selectivity was acceptable with no interference of amines. The sensor was stable for more than 10 days. Other microbial sensors, including those for automatic measurement of BOD and ammonia in waste-waters, a microbial sensor for on-line measurement of acetic acid in fermentation broth and a microprocessor equipped glutamate sensor, have been described by Hikuma *et al.* (1981).

The concept of tissue-based membrane electrodes was recently discussed by Rechnitz (1981). Such electrodes have been constructed for the determination of, for example, glutamine using porcine kidney tissue in combination with an ammonia gas-sensing electrode, and for glutamic acid using a plant tissue electrode devised from the mesocarp skin layer of the growing portion of a yellow squash with a CO_2 gas-sensing probe as transducer. An optimization strategy was proposed by Arnold and Rechnitz (1982) for the development of tissue-based electrodes and was demonstrated through the construction of a rabbit liver based membrane electrode for guanine.

20.7 CONCLUSIONS

Progress in the technology of enzyme immobilization has been of great importance for the evolution of enzyme probes. A stable enzyme membrane or reactor with high specific activity is one prerequisite for a functional and reliable enzyme probe. The other one is a suitable transducer. Today, a wide variety of ion-selective electrodes and gas-sensing probes are available that could be used for the construction of enzyme electrodes. Hundreds of such electrodes have been described and many of them have been put into practical use in various analytical tasks in clinical chemistry and biotechnology. There are several commercially available instruments built in the USA and Japan which are based on enzyme electrodes, and on self-contained enzyme probes to be connected to an oxygen meter or a pH meter. Such probes permit simple, inexpensive and rapid determinations of metabolites to be made with the simplicity of a pH measurement. In medicine, enzyme probes could even be implanted to follow the progress of therapy. Much of the experiences from clinical chemistry, which so far has been the most innovative field of enzyme electrode analysis, can be directly transferred to biotechnology. Continuous, on-line analysis of specific components during fermentation is one possibility provided for by enzyme probes inserted in the fermenter, or by continuous-flow analysis of a sample stream from the fermenter. Such direct information could be a very valuable complement to the indirect information obtained through physical variables, such as pH, pCO_2 and viscosity.

The potential of continuous monitoring has also been demonstrated using enzyme thermistors, another important group of enzyme probes. Like enzyme electrodes, they are insensitive to the optical properties of the sample. This is a factor of importance in work with colored or turbid samples, which are common in biotechnology.

Of all transducers, temperature sensors offer the greatest flexibility in the choice of enzyme for a particular application. Since most enzymic reactions are associated with considerable heat production, only one enzymic step is usually required, which minimizes the risk for interferences. Consequently, thermal analyzers have a wider spectrum of applications than any other instrument. The analytical procedure is simply changed just by changing the enzymic reactor. The sensitivity range and the operational stability of enzyme thermistors make them well suited for biotechnological applications. What kind of enzyme probes the future will bring can be difficult to assess. Enzyme transistors or ChemFETs are likely to gain widespread use due to their potential sensitivity and the possibilities of miniaturization and direct integration with microelectronics. Progress in applied optics also will probably have a great impact on the development of new types of enzyme probes.

Finally, the use of immobilized cells or organelles will open new analytical possibilities when pure enzymes are not available, or when multistep enzymic reactions are required. In addition to such microbial sensors, probes based on intact tissue of animal or even plant origin have now come into use.

20.8 REFERENCES

Aizawa, M., Y. Watanabe and S. Suzuki (1979). Biospecific thermal analyzer coupled with a flowthrough immobilized enzyme column. *J. Solid-Phase Biochem.*, **4**, 131–141.
Aizawa, M., A. Morioka and S. Suzuki (1980a). An enzyme immunosensor for the electrochemical determination of the tumor antigen α-fetoprotein. *Anal. Chim. Acta*, **115**, 61–67.
Aizawa, M., M. Wada, S. Kato and S. Suzuki (1980b). Immobilized mitochondrial electron transport particle for NADH determination. *Biotechnol. Bioeng.*, **22**, 1769–1783.
Anfält, T., A. Graneli and D. Jagner (1973). A urea electrode based on the ammonia probe. *Anal. Lett.*, **6**, 969–975.
Arnold, M. A. and G. A. Rechnitz (1982). Optimization of a tissue-based membrane electrode for guanine. *Anal. Chem.*, **54**, 777–782.
Barker, A. S. and P. J. Somers (1978). Enzyme electrodes and enzyme based sensors. In *Topics in Enzyme and Fermentation Biotechnology*, ed. A Wiseman, vol. 2, pp. 120–151. Ellis Horwood Ltd., Chichester.
Borrebaeck, C. and B. Mattiasson (1980). Lectin–carbohydrate interactions studied by a competitive enzyme inhibition assay. *Anal. Biochem.*, **107**, 446–450.
Borrebaeck, C., B. Mattiasson and K. Svensson (1978). In *Enzyme Labelled Immunoassay of Hormones and Drugs*, ed. S. B. Pal, pp. 15–27. de Gruyter, Berlin.
Bowers, L. D. and P. W. Carr (1976). An immobilized-enzyme flow-enthalpimetric analyzer: application to glucose determination by direct phosphorylation catalyzed by hexokinase. *Clin. Chem.*, **22**, 1427–1433.

Bowers, L. D., L. M. Canning, Jr., C. N. Sayers and P. W. Carr (1976). Rapid-flow enthalpimetric determination of urea in serum, with use of an immobilized urease reactor. *Clin. Chem.*, **22**, 1314–1318.

Brown, H. D. (1969). Calorimetry of enzyme-catalyzed reactions. In *Biochemical Microcalorimetry*, ed. H. D. Brown, pp. 149–164. Academic, New York.

Canning, L. M., Jr. and P. W. Carr (1975). Rapid thermochemical analysis *via* immobilized enzyme reactors. *Anal. Lett.*, **8**, 359–367.

Caras, S. and J. Janata (1980). Field effect transistor sensitive to penicillin. *Anal. Chem.*, **52**, 1935–1937.

Carr, P. W. and L. D. Bowers (1980). Immobilized enzymes in analytical and clinical chemistry: Fundamentals and applications. *Chem. Anal. (N.Y.)*, **56**, 197–310.

Clark, L. C., Jr., and C. Lyons (1962). Electrode systems for continuous monitoring in cardiovascular surgery. *Ann. N.Y. Acad. Sci.*, **102**, 29–45.

Cooney, C. L., J. C. Weaver, S. R. Tannenbaum, D. V. Faller, A. Shields and M. Jahnke (1974). The thermal enzyme probe — a novel approach to chemical analysis. In *Enzyme Engineering*, ed. E. K. Pye and L. B. Wingard, Jr., vol. 2, pp. 411–417. Plenum, New York.

Cordonnier, M., F. Lawny, D. Chapot and D. Thomas (1975). Magnetic enzyme membranes as active elements of electrochemical sensors. Lactose, saccharose, maltose bienzyme electrodes. *FEBS Lett.*, **59**, 263–267.

Curme, H. G., R. L. Columbus, G. M. Dappen, T. W. Eder, W. D. Fellows, J. Figueras, C. P. Glover, C. A. Goffe, D. E. Hill, W. H. Lawton, E. J. Muka, J. E. Pinney, R. N. Rand, K. J. Sanford and T. W. Wu (1978). Multilayer film elements for clinical analysis: general concepts. *Clin. Chem.*, **24**, 1335–1342.

Danielsson, B. and K. Mosbach (1979). Determination of enzyme activities with the enzyme thermistor unit. *FEBS Lett.*, **101**, 47–50.

Danielsson, B., K. Gadd, B. Mattiasson and K. Mosbach (1976). Determination of serum urea with an enzyme thermistor using immobilized urease. *Anal. Lett.*, **9**, 987–1001.

Danielsson, B., K. Gadd, B. Mattiasson and K. Mosbach (1977). Enzyme thermistor determination of glucose in serum using immobilized glucose oxidase. *Clin. Chim. Acta*, **81**, 163–175.

Danielsson, B., B. Mattiasson, R. Karlsson and F. Winquist (1979a). Use of an enzyme thermistor in continuous measurements and enzyme reactor control. *Biotechnol. Bioeng.*, **21**, 1749–1766.

Danielsson, B., B. Mattiasson and K. Mosbach (1979b). Enzyme thermistor analysis in clinical chemistry and biotechnology. *Pure Appl. Chem.*, **51**, 1443–1457.

Danielsson, B., I. Lundström, K. Mosbach and L. Stiblert (1979c). On a new enzyme transducer combination: the enzyme transistor. *Anal. Lett.*, **12**, 1189–1199.

Danielsson, B., B. Mattiasson and K. Mosbach (1981a). Enzyme thermistor devices and their analytical applications. *Appl. Biochem. Bioeng.*, **3**, 97–143.

Danielsson, B., E. Rieke, B. Mattiasson, F. Winquist and K. Mosbach (1981b). Determination by the enzyme thermistor of cellobiose formed on degradation of cellulose. *Appl. Biochem. Biotechnol.*, **6**, 207–222.

Danielsson, B., L. Buelow, C. R. Lowe, I. Satoh and K. Mosbach (1981c). Evaluation of the enzyme thermistor as a specific detector for chromatographic procedures. *Anal. Biochem.*, **117**, 84–93.

Dappen, G. M., P. E. Cumbo, C. T. Goodhue, S. Y. Lynn, C. C. Morganson, B. F. Nellis, D. M. Sablauskas, J. R. Schaeffer, R. M. Schubert, R. E. Snoke, G. M. Underwood, C. D. Warburton and T.-W. Wu (1982). Dry film for the enzymic determination of total cholesterol in serum. *Clin. Chem.*, **28**, 1159–1162.

Enfors, S.-O. (1981). Oxygen-stabilized enzyme electrode for D-glucose analysis in fermentation broths. *Enzyme Microb. Technol.*, **3**, 29–32.

Enfors, S.-O. and H. Nilsson (1979). Design and response characteristics of an enzyme electrode for measurement of penicillin in fermentation broth. *Enzyme Microb. Technol.*, **1**, 260–264.

Fogt, E. J., L. M. Dodd, E. M. Jenning and A. H. Clemens (1978). Development and evaluation of a glucose analyzer for a glucose-controlled insulin infusion system (Biostator®). *Clin. Chem.*, **24**, 1366–1372.

Fulton, S. P., C. L. Cooney and J. C. Weaver (1980). Thermal enzyme probe with differential temperature measurements in a laminar flow-through cell. *Anal. Chem.*, **52**, 505–508.

Gebauer, C. R. and G. A. Rechnitz (1981). Immunoassay studies using adenosine deaminase enzyme with potentiometric rate measurement. *Anal. Lett.*, **14**, 97–109.

Gorton, L. and L. Ögren (1981). Flow injection analysis for glucose and urea with enzyme reactors and on-line dialysis. *Anal. Chim. Acta*, **130**, 45–53.

Gray, D. N., M. H. Keyes and B. Watson (1977). Immobilized enzymes in analytical chemistry. *Anal. Chem.*, **49**, 1067A–1078A.

Grime, J. K. (1980). Biochemical and clinical analysis by enthalpimetric measurements — a realistic alternative approach? *Anal. Chim. Acta*, **118**, 191–225.

Grime, J. K. and B. Tan (1979). The determination of some selected penicillins by enzymatic enthalpimetry. *Anal. Chim. Acta*, **107**, 319–326.

Guilbault, G. G. (1982). Ion selective electrodes applied to enzyme systems. *Ion Sel. Electrode Rev.*, **4**, 187–231.

Guilbault, G. G. (1984). *Analytical Uses of Immobilized Enzymes*. Dekker, New York.

Guilbault, G. G. and E. Hrabankova (1971). New enzyme electrode probes for D-amino acids and asparagine. *Anal. Chim. Acta*, **56**, 285–290.

Guilbault, G. G. and G. J. Lubrano (1974). Amperometric enzyme electrodes. Part III. Alcohol oxidase. *Anal. Chim. Acta*, **69**, 189–194.

Guilbault, G. G. and J. G. Montalvo (1970). An enzyme electrode for the substrate urea. *J. Am. Chem. Soc.*, **92**, 2533–2538.

Guilbault, G. G. and G. Nagy (1973). Improved urea electrode. *Anal. Chem.*, **45**, 417–420.

Guilbault, G. G. and M. Tarp (1974). A specific enzyme electrode for urea. *Anal. Chim. Acta*, **73**, 353–365.

Guilbault, G. G., B. Danielsson, C. F. Mandenius and K. Mosbach (1983). A comparison of enzyme electrode and thermistor probes for assay of alcohols using alcohol oxidase. *Anal. Chem.*, **55**, 1582–1585.

Gulberg, E. L. and G. D. Christian (1981). The use of immobilized alcohol oxidase in the continuous flow determination of ethanol with an oxygen electrode. *Anal. Chim. Acta*, **123**, 125–133.

Hikuma, M., T. Yasuda, I. Karube and S. Suzuki (1981). Application of microbial sensors to the fermentation process. *Ann. N. Y. Acad. Sci.*, **369**, 307–320.

Karube, I., I. Satoh, Y. Araki, S. Suzuki and H. Yamada (1980). Monoamine oxidase electrode in freshness testing of meat. *Enzyme Microb. Technol.*, **2**, 117–120.

Karube, I., T. Okada and S. Suzuki (1981). Amperometric determination of ammonia gas with immobilized nitrifying bacteria. *Anal. Chem.*, **53**, 1852–1854.

Keyes, M. H., F. E. Semersky and D. N. Gray (1979). Glucose analysis utilizing immobilized enzymes. *Enzyme Microb. Technol.*, **1**, 91–94.

Kjellén, K. G. and H. Y. Neujahr (1980). Enzyme electrode for phenol. *Biotechnol. Bioeng.*, **22**, 299–310.

Kobayashi, T., K. Saga, S. Shimizu and T. Goto (1981). The application of chemiluminescence of a *Cypridina* Luciferin analogue to immobilized enzyme sensors. *Agric. Biol. Chem.*, **45**, 1403–1408.

Koyama, M., Y. Sato, M. Aizawa and S. Suzuki (1980). Improved enzyme sensor for glucose with an ultrafiltration membrane and immobilized glucose oxidase. *Anal. Chim. Acta*, **116**, 307–314.

Krisam, G. and H.-L. Schmidt (1977). Development and properties of caloric systems for substrate determinations with immobilized enzymes. In *Application of Calorimetry in Life Sciences*, ed. I. Lamprecht and B. Schaarschmidt, p. 39. de Gruyter, Berlin.

Mandenius, C. F., B. Danielsson and B. Mattiasson (1980). Enzyme thermistor control of the sucrose concentration at a fermentation with immobilized yeast. *Acta Chem. Scand.*, *Ser. B*, **34**, 463–465.

Mandenius, C. F., B. Danielsson and B. Mattiasson (1981). Process control of an ethanol fermentation with an enzyme thermistor as a sucrose sensor. *Biotechnol. Lett.*, **3**, 629–634.

Marconi, W. (1978). Biomedical applications of enzymatic fibres. In *Enzyme Engineering*, ed. G. Broun, G. Manecke and L. B. Wingard, Jr., vol. 4, pp. 179–186. Plenum, New York.

Martin, C. J. and M. A. Marini (1979). Microcalorimetry in biochemical analysis. *Crit. Rev. Anal. Chem.*, **8**, 221–286.

Mascini, M. and A. Liberti (1974). An enzyme-coupled cyanide solid-state electrode. *Anal. Chim. Acta*, **68**, 177–184.

Matsumoto, K., H. Seijo, I. Karube and S. Suzuki (1980). Amperometric determination of choline with use of immobilized choline oxidase. *Biotechnol. Bioeng.*, **22**, 1071–1086.

Mattiasson, B. and C. Borrebaeck (1980). Novel approaches to enzyme immunoassay. In *Enzyme Immunoassay*, ed. E. T. Maggio, pp. 213–248. CRC Press, Boca Raton FL.

Mattiasson, B. and B. Danielsson (1982). Calorimetric analysis of sugars and sugar derivatives with aid of an enzyme thermistor. *Carbohydr. Res.*, **102**, 273–282.

Mattiasson, B., B. Danielsson and K. Mosbach (1976). A split-flow enzyme thermistor. *Anal. Lett.*, **9**, 867–889.

Mattiasson, B., C. Borrebaeck, B. Sanfridson and K. Mosbach (1977). Thermometric enzyme linked immunosorbent assay: TELISA. *Biochim. Biophys. Acta*, **483**, 221–227.

Mattiasson, B., B. Danielsson, F. Winquist, H. Nilsson and K. Mosbach (1981). Enzyme thermistor analysis of penicillin in standard solutions and in fermentation broth. *Appl. Environ. Microbiol.*, **41**, 903–908.

McGlothlin C. D. and J. Jordan (1975). Enzymatic enthalpimetry, a new approach to clinical analysis: glucose determination by hexokinase catalyzed phosphorylation. *Anal. Chem.*, **47**, 786–790.

Mizutani, F., K. Tsuda, I. Karube, S. Suzuki and K. Matsumoto (1980). Determination of glutamate pyruvate transaminase and pyruvate with an amperometric oxidase sensor. *Anal. Chim. Acta*, **118**, 65–71.

Mosbach, K. (1976). Immobilized enzymes. *Methods Enzymol.*, **44**.

Mosbach, K. (1986). *Methods Enzymol.*

Mosbach, K. and B. Danielsson (1974). An enzyme thermistor. *Biochem. Biophys. Acta*, **364**, 140–145.

Mosbach, K. and B. Danielsson (1981). Thermal bioanalyzers in flow streams — Enzyme thermistor devices. *Anal. Chem.*, **53**, 83A–94A.

Mosbach, K., B. Danielsson, A. Borgerud and M. Scott (1975). Determination of heat changes in the proximity of immobilized enzymes with an enzyme thermistor and its use for the assay of metabolites. *Biochim. Biophys. Acta*, **403**, 256–265.

Nagy, G., L. H. Von Storp and G. G. Guilbault (1973). Enzyme electrode for glucose based on an iodide membrane sensor. *Anal. Chim. Acta*, **66**, 443–455.

Nanjo, M. and G. G. Guilbault (1974). Enzyme electrode for L-amino acids and glucose. *Anal. Chim. Acta*, **73**, 367–373.

Nanjo, M. and G. G. Guilbault (1975). Amperometric determination of alcohols, aldehydes and carboxylic acids with an immobilized alcohol oxidase enzyme electrode. *Anal. Chim. Acta*, **75**, 169–180.

Nelson, D. P. and L. A. Kiesow (1972). Enthalpy of decomposition of hydrogen peroxide by catalase at 25 °C (with molar extinction coefficients of H_2O_2 solutions in the UV). *Anal. Biochem.*, **49**, 474–478.

Nilsson, H., A.-C. Åkerlund and K. Mosbach (1973). Determination of glucose, urea and penicillin using enzyme–pH electrodes. *Biochim. Biophys. Acta*, **320**, 529–534.

Ögren, L. and G. Johansson (1978). Determination of traces of mercury(II) by inhibition of an enzyme reactor electrode loaded with immobilized urease. *Anal. Chim. Acta*, **96**, 1–11.

Papariello, G. J., A. K. Mukherji and C. M. Shearer (1973). A penicillin selective enzyme electrode. *Anal. Chem.*, **45**, 790–792.

Papastathopoulos, D. S. and G. A. Rechnitz (1975). A urea-sensing membrane electrode for whole blood measurements. *Anal. Chim. Acta*, **79**, 17–26.

Pennington, S. N. (1974). Use of enzymes in thermal chemical (calorimetric) analysis. *Enzyme Technol. Digest.*, **3**, 105–114.

Pungor, Jr, E., E. Schaefer, J. C. Weaver and C. L. Cooney (1980). Direct monitoring of a fermentation in a computer–mass spectrometer–fermentor system. In *Advances in Biotechnology*, ed. M. Moo-Young, C. W. Robinson and C. Vezina, vol. 1, pp. 393–398. Pergamon, Toronto.

Rechnitz, G. A. (1981). Bioselective membrane electrode probes. *Science.*, **214**, 287–291.

Rehak, N. N. and D. S. Young (1978). Prospective applications of calorimetry in the clinical laboratory. *Clin. Chem.*, **24**, 1414–1419.

Romette, J.-L., B. Froment and D. Thomas (1979). Glucose-oxidase electrode. Measurements of glucose in samples exhibiting high variability in oxygen content. *Clin. Chim. Acta*, **95**, 249–253.

Růžička, J. and E. H. Hansen (1974). A new potentiometric gas sensor — the air-gap electrode. *Anal. Chim. Acta*, **69**, 129–141.

Satoh, I., I. Karube and S. Suzuki (1976). Enzyme electrode for sucrose. *Biotechnol. Bioeng.*, **18**, 269–272.

Satoh, I., B. Danielsson and K. Mosbach (1981). Triglyceride determination with use of an enzyme thermistor. *Anal. Chim. Acta*, **131**, 255–262.

Schmidt, H.-L., G. Krisam and G. Grenner (1976). Microcalorimetric methods for substrate determination in flow systems with immobilized enzymes. *Biochim. Biophys. Acta*, **429**, 283–290.

Spink, C. and I. Wadsö (1976). Calorimetry as an analytical tool in biochemistry and biology. *Methods Biochem. Anal.*, **23**, 1–159.

Tran-Minh, C. and G. Broun (1975). Construction and study of electrodes using cross-linked enzymes. *Anal. Chem.*, **47**, 1359–1364.

Tran-Minh, C. and D. Vallin (1978). Enzyme-bound thermistor as an enthalpimetric sensor. *Anal. Chem.*, **50**, 1874–1878.

Tsuchida, T. and K. Yoda (1981). Immobilization of D-glucose oxidase onto a hydrogen peroxide permselective membrane and application for an enzyme electrode. *Enzyme Microb. Technol.*, **3**, 326–330.

Updike, S. J. and G. P. Hicks (1967). The enzyme electrode. *Nature (London)*, **214**, 986–988.

Walters, R. R., P. A. Johnson and R. P. Buck (1980). Histidine ammonia-lyase enzyme electrode for determination of L-histidine. *Anal. Chem.*, **52**, 1684–1690.

Weaver, J. C. and S. K. Burns (1981). Potential impact of physics and electronics on enzyme-based analysis. *Appl. Biochem. Bioeng.*, **3**, 271–308.

Weaver, J. C., M. K. Mason, J. A. Jarrell and J. W. Peterson (1976). Biochemical assay by immobilized enzymes and a mass spectrometer. *Biochim. Biophys. Acta*, **438**, 296–303.

Weetall, H. H. (1976). Covalent coupling methods for inorganic support materials. *Methods Enzymol.*, **44**, 134–148.

White, W. C. and G. G. Guilbault (1978). Lysine specific enzyme electrode for determination of lysine in grains and foodstuffs. *Anal. Chem.*, **50**, 1481–1486.

Winquist, F., B. Danielsson, I. Lundström and K. Mosbach (1982). Use of hydrogen-sensitive Pd-MOS materials in biochemical analysis. *Appl. Biochem. Biotechnol.*, **7**, 135–139.

Winquist, F., B. Danielsson, J.-Y. Malpote, M.-B. Larsson and L. Persson (1985). Enzyme thermistor determination of oxalate with immobilized oxalate oxidase. *Anal. Lett.*, **18**, 573–588.

Yanchinsky, S. (1982). Biochips speed up chemical analysis. *New Sci.*, **93**, 236.

21

Analysis of Fermentation Gases

H. Y. WANG
University of Michigan, Ann Arbor, MI, USA

21.1 INTRODUCTION

Gas exchange rates of a fermentation process have long been regarded as useful parameters to measure because they often provide an indication of growth, which otherwise may not be readily available, and the signals are not directly influenced by the culture medium constituents. Both oxygen uptake and carbon dioxide production rates have been used to study aerobic fermentation. The ratio of carbon dioxide production to oxygen uptake, called the respiratory quotient (RQ), provides additional information such as indication of the kind of substrate being utilized by the microorganism.

During microbial growth, the carbon energy source is aerobically converted to biomass, carbon dioxide and water. Growth is expressed by the general stoichiometric relationship:

$$a\,C_iH_jO_k + b\,O_2 + c\,N_mH_nO_p \rightarrow d\,C_gH_rO_sN_t + e\,H_2O + g\,CO_2 \tag{1}$$

where a, b, c, d, e, and g are moles of substrates and products, and $C_iH_jO_k$ and $N_mH_nO_p$, represent the carbon-energy and nitrogen sources respectively. Biomass is represented by $C_gH_rO_sN_t$, where g, r, s and t correspond to the number of atoms in an empirical unit of cell mass. For example, the yeast, *Saccharomyces cerevisiae* can be represented with an empirical formula for cell composition $C_6H_{10}NO_3$ (Wang *et al.*, 1977). Obviously, the cellular composition changes with various environmental conditions, as demonstrated by Herbert (1976).

Both O_2 and CO_2 concentrations in equation (1) can be measured since they are gases. The carbon and nitrogen sources can be gases too and, in this case, be measured with suitable instruments. For example, methane has been used as a carbon-energy source in SCP production and gaseous NH_3 has been used for both nitrogen source and pH control purposes. Both gases can be measured with corresponding instruments. In many fermentations, there is significant product formation in addition to cell mass, carbon dioxide and water. Product formation can be expressed by the general stoichiometric relationship:

$$\alpha\,C_iH_jO_k + \beta\,O_2 + \gamma\,NH_3 \rightarrow \delta\,C_wH_xO_yN_z + \varepsilon\,H_2O + \eta\,CO_2 \tag{2}$$

the product ($C_wH_xO_yN_z$) formed can be quite volatile and be measured by gas analyses. The extent of product formation can be observed on the gaseous measurement if β/η (2) is not equal to b/g (1). This ratio is called the respiratory quotient (RQ), the value of which depends on the product yield and the nature of the product and substrate compositions. For instance, when *S. cerevisiae* is grown on glucose, ethanol production occurs as a frequent by-product, the amount of which is dependent on the environmental conditions. For ethanol formation, equation (2) becomes:

$$C_6H_{12}O_6 \rightarrow 2\,C_2H_5OH + 2\,CO_2 \tag{3}$$

Since the synthesis of ethanol involves only the production of CO_2 without corresponding uptake of O_2, it would be expected that gas measurement of O_2 and CO_2 becomes an indicator of ethanol production. Wang *et al.* (1977, 1979a) showed that ethanol production rate is indeed related to O_2 and CO_2 measurements, especially the ratio RQ.

It is well established that the monitoring and control of the gaseous environment in a fermenter is often of major importance for the efficient production of microbial products. Both oxygen and carbon dioxide have been shown to be the major variables and play a major role in many fermentations. Various instruments have been used to measure oxygen uptake and carbon dioxide production such as orsat analyses, process gas chromatographs, paramagnetic and infra-red analyzers. Recently, increasing interest in using mass spectrometer-based gas analyzers for research and plant control has been developed. A comparison between various gas analysis techniques is shown in Table 1. Some of the instruments, such as the paramagnetic based oxygen analyzer or the infrared CO_2 analyzers, are specific for a single gas component while others, such as process gas chromatography or mass spectrometer gas analysis, can be used for multiple gas measurements. Criteria for a good fermentation gas monitoring device are shown in Table 2. These include fast response time, stability and specificity.

Table 1 Comparison of Existing Sensors for Gas Analyses

Gas measurements	Advantages	Disadvantages
Infrared analyzer (CO_2)	Very specific	Drift; slow response time
Thermal conductivity (CO_2)	Less expensive	Interference by other gases
Mass spectrometry (CO_2, O_2, others)	Versatile; can also be used for other purposes; fast response time	Expensive
Paramagnetic analyzer (O_2)	Very specific	Drift; interference by water vapor; slow response time
Electrolyte fuel cell (O_2)	No interference by water vapor	Interference by combustibles
Process Gas chromatograph (CO_2, O_2, others)	Can be used for other purposes	Discrete analyses; accuracy reduced

Table 2 Properties Desired in a Gas Analysis System

Speed of response
Insensitivity to humidity
Insensitivity to changes in atmospheric pressure
Insensitivity to changes in stream volume due to low respiratory quotient
Stability
Accuracy
Specificity to the gas

21.2 MANOMETRIC METHODS AND GAS ELECTRODES

Manometric methods have long been used to measure oxygen consumption and carbon dioxide evolution by biological cultures. In fact, most of the earlier information on cell respiration and metabolic studies has been obtained by the use of manometric methods and devices. Several automated manometric techniques were described and discussed by Umbreit *et al.* (1959). The major difference between the described devices is in the method of sensing the change in partial pressure of oxygen and carbon dioxide. All devices use some form of transducer to convert the change in pressure to an electrical signal which can be fed into a recorder. A typical respirometer is described by Arthur (1965). It operates in the following manner. As the bacterial culture in the enclosed chamber utilizes oxygen and as the generated CO_2 is absorbed by alkali, the partial pressure of the oxygen in the atmosphere decreases, causing the manometer fluid in the closed leg of an oil manometer to rise. The fluid level changes in the manometer can be measured and converted to an electrical signal. If the amount of carbon dioxide evolved needs to be measured, the respiratory carbon dioxide is trapped by alkali in wash bottles. The wash bottles are replaced at regular time intervals, and the carbonate precipitated as barium carbonate, which is centrifuged, dried and weighed. This method of measuring oxygen uptake and carbon dioxide is not only quite time consuming but also has the drawback of limited applicability and sensitivity.

In the early 1960s, several forms of gas-sensing electrodes were developed (Fatt, 1964). The most commonly used pO_2 electrode is the membrane-covered Clark type platinum polarographic electrode (Clark, 1956). The usual pCO_2 electrode operates essentially as a membrane-covered pH electrode as described by Severinghaus and Bradley (1958). These electrodes can be used to measure oxygen or carbon dioxide partial pressure in the gas streams, and are still the cheapest way to monitor these gases. Unfortunately, maintenance of these electrodes is difficult, their sensitivity to small gas composition changes is very low and the temperature coefficient of these gas electrodes is high. The electrodes are also sensitive to variations in gas flow rates, therefore standard operating conditions must be maintained throughout the monitoring period. They have been used more successfully for dissolved oxygen and carbon dioxide measurements in the liquid culture.

21.3 CONTINUOUS FLOW OXYGEN AND CARBON DIOXIDE ANALYZERS

Oxygen uptake and carbon dioxide production are readily measured by means of a paramagnetic gaseous analyzer and an infrared carbon dioxide analyzer. These instruments are readily available at moderate cost and can be interfaced with a computer. They can be multiplexed for use with many fermenters, or used only for an individual fermentation. When the gas flow rate is measured concurrently, oxygen and carbon balances on the gas flow in and out of the fermenter can be used to calculate the oxygen uptake rate (OUR) and carbon dioxide evolution rate (CER):

$$\text{OUR} = \frac{F_N}{V} \left[\frac{p_{o,in}}{P_t - p_{o,in} - p_{w,in} - p_{c,in}} \right] - \left[\frac{p_{o,out}}{P_t - p_{o,out} - p_{w,out} - p_{c,out}} \right] \tag{4}$$

$$\text{CER} = \frac{F_N}{V} \left[\frac{p_{c,out}}{P_t - p_{c,out} - p_{w,out} - p_{o,out}} \right] - \left[\frac{p_{o,in}}{P_t - p_{c,in} - p_{w,in} - p_{o,in}} \right] \tag{5}$$

where F_N is the flow rate of inert gas (N_2; $1 \, m^{-3}$); p_c, p_o and p_w are the partial pressures of carbon dioxide, oxygen and water, respectively (kPa); P_t is the total pressure (kPa); and V is the liquid volume of the fermenter (m^3). The suffixes in and out indicate whether the value is from the inlet or outlet gas stream of the fermenter.

Oxygen consumption and carbon dioxide evolution have been used to correlate cell growth and metabolic activity (Humphrey, 1973; Cooney *et al.*, 1977). Since no liquid sampling is required and analysis is continuous, OUR and CER determinations have become powerful indirect measurement techniques to follow cell growth and metabolic activity.

21.3.1 Oxygen Analysis

There are two methods for the continuous analysis of oxygen in the gas streams. One depends on the paramagnetic properties of oxygen (Elsworth, 1970) and the other is based on an electrochemical fuel cell using zirconium oxide or others as a gas-sensing electrode. For the paramagnetic oxygen analyzer or the so called 'Pauling' type analyzer (Elsworth, 1970), a dumb-bell formed by two hollow glass spheres is suspended on a taut fiber in a magnetic field. When a gas containing O_2 is introduced, the magnetic force is changed. The spheres are repulsed and rotate to a new position until the torque on the fiber again balances the magnetic field. The degree of rotation is a measure of the oxygen concentration. Recent advances in solid state circuitry have helped to stabilize the machine tremendously. Special precautions need to be taken in treating the sampling gas. It should be freed from dust by filtration and the water saturated gas stream must be dewatered because water molecules are paramagnetic and introduce error. Magnetic properties of the gas molecules are a function of temperature, hence the temperature of the analyzer must be controlled. The electrochemical fuel cell type oxygen sensor (Preier, 1983) has been used to measure and control oxygen content in a furnace. It has a limited life span and the accuracy of the sensor is much worse than that of the paramagnetic analyzer.

The continuous measurement of a small amount of oxygen consumption using the analyzers tends to be very difficult since the analysis is affected by very slight changes in atmospheric pressure and the relative humidity. In this case, the instruments also are measuring the difference between two fairly high concentrations of oxygen, which makes the exact measurement of fermentation oxygen uptake very difficult.

21.3.2 Carbon Dioxide Analysis

Monitoring carbon dioxide production is another means by which to follow growth activity, either by itself or as a means of complementing oxygen uptake rate measurements (Humphrey, 1971). The method for calculating carbon dioxide evolution rate has already been described in equation (5). Exact calculation requires knowledge of the oxygen partial pressure and gas flow rate (Humphrey, 1971).

There are two methods of analyzing carbon dioxide in a fermentation gas stream. First, there is the infrared analysis which is specific for CO_2 monitoring (Elsworth, 1970). The other is the thermal conductivity method, which is less specific to CO_2 monitoring because other gaseous components, such as H_2O and H_2, affect its measurement (Elsworth 1970). The principle of the infrared analysis is as follows. A null-balance type non-dispersive analyzer is used to examine absorption in the wavelength range 3–15 μm. The light beam, after passing through the absorption cell through which the sampling gas is flowing, is scanned by a detector that is selective for the wavelength at which CO_2 absorbs. The intensity of the transmitted light is compared with a reference cell filled with CO_2-free gas. The difference between the two signals is a measure of the amount of CO_2 present in the gas and may be displayed directly on an indicator–recorder (Elsworth, 1970). This principle is also used in infrared analyzers to measure hydrocarbons and other infrared sensitive gas components (Elsworth, 1970). The solid state electronic circuitry developed recently has helped to stabilize this type of machine and thus made them practically maintenance free.

The thermal conductivity analyzer is cheaper and less specific. It is not suitable for monitoring anaerobic cultures or if the measurement has to be accurate to within ±5–10%. In practice continuous measurement of small amounts of CO_2 is preferred to detection of small amounts of O_2, because CO_2 measurement is practically unaffected by small changes in atmospheric pressure and relative humidity. In many aerobic and anaerobic fermentations, CO_2 is a good growth indicator during microbial cultivation. Since the solubility of CO_2 in aqueous solution is heavily dependent on the pH and the buffering agents in the fermentation broth, changes in CO_2 values should be compared with pH changes to separate the influence of pH on this measurement.

The first exact description of using these continuous flow gas analyzers for the monitoring and control of microbial fermentation was given by Fiechter and von Meyenburg (1968). Nyiri *et al.* (1975) have examined the gas exchange conditions of a *Candida utilis* culture. Wang *et al.* (1979a) have demonstrated that the information obtained by these analyzers can be used to control an aerobic yeast fermentation.

21.4 PROCESS GAS CHROMATOGRAPHY

Gas chromatography has become one of the most important off-line analytical techniques presently used in laboratories, and will probably continue to be so for the foreseeable future. Various types of GC columns such as packed, capillary and preparative-scale columns are being used to separate gas mixtures. Various types of detectors, such as thermal conductivity, flame ionization, electron capture, *etc.*, are used to detect and measure the different components as they emerge from the column (Figure 1). The analysis includes calculating the area of a chromatography peak which is proportional to the number of gas molecules of the component reaching the detector. Unfortunately, none of the detectors available responds to the same extent for an equal number of molecules of different compounds. For this reason, a response factor must be ascertained experimentally for each compound. This can be obtained by injecting a known amount of the compound into the gas chromatograph and then measuring the area of the relevant response peak.

A gas chromatographic system was developed at the Shell Laboratories, UK for the monitoring

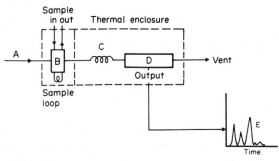

A Carrier gas – helium
B Sample valve – injects sample loop volume into carrier gas stream
C Column – separates components of sample
D Detector – detects components as they emerge from column
E Chromatogram – detector signal plotted *vs.* time

Figure 1 The basic layout of a gas chromotograph

of fermentation gases from a bank of bench-scale microbial cultures grown on methane gas (Harrison, 1978). It was necessary to base the calibration on the integration of peaks and to correct continuously for base-line changes in order to obtain the type of precision of ±1% required. The calibration was checked once per cycle together with the peak position. Gas chromatography also offers the possibility of measuring dissolved gases by using a sampling tube comprised of semipermeable polymeric material which is purged with a carrier gas.

The major drawback of gas chromatography is the complexity of the gas sampling technique and the discrete nature of the analyses. Each analysis may take 10–20 minutes and caution must be taken to avoid over-lapping of the peaks, thus creating interference with the specific analyses. Other than the mass spectrometer, gas chromotography is the only other means available of measuring multiple gas components simultaneously with one machine.

21.5 MASS SPECTROMETER-BASED GAS ANALYZERS

Problems which have been encountered with the traditional methods described in the previous sections, such as low specificity, slow response, unstable calibration, sensitivity to humidity, sensitivity to changes in gas volume caused by respiration and sensitivity to changes in atmospheric pressure, need to be overcome by finding a better measuring technique. Speed of response is particularly important if a single analyzer is to be used to sample a large number of fermenters. Typical paramagnetic or infrared analyzers may take up to 10–15 minutes to reach a new equilibrium after the gas sample has been changed. If six fermenters are being monitored sequentially, then each gas stream can only be measured once an hour.

There has been an upsurge of intensive research in applying an on-line process mass spectrometer in the petro-chemical and fermentation industries (Ahlstron and Shaver, 1976; Tonge, 1980; Buckland and Fastert, 1982). Some of the problems encountered with the traditional methods of gas analysis can be largely eliminated by using an on-line mass spectrometer. The potential of an on-line mass spectrometer gas analyzer goes much further than measuring oxygen and carbon dioxide in the exit gas stream from a fermenter. The most prominent application is its use in the identification and measurement of other volatile substrates or products. It is also practical to use this type of mass spectrometer to measure dissolved gases by passing a purge stream of carrier gas through gas permeable tubing in contact with the fermenting broth.

In 1910, J. J. Thompson invented and used an instrument that was to evolve into the present day mass spectrometer. During the next 30 years, various investigators used self-built mass spectrometers to elucidate the atomic weight of the elements and to separate isotopes. The first use of the instrument was as an analytical tool for determining the molecular structure of unknown compounds, as well as for the quantification of these compounds. With the introduction of heated sample systems, the mass spectrometer was soon applied to many areas of chemistry. Today, a quick glance at the literature reveals that it has been a widely used analytical technique in atmospheric research, medicine and environmental monitoring (Ahlstron and Shaver, 1976).

Reuss *et al.* (1975) published the first report on the use of mass spectrometry to monitor fermentation. These authors used a Finnegan model 3000 quadrupole mass spectrometer and a specially constructed membrane-covered inlet system, which was immersed directly into the fermentation broth to follow the concentrations of methanol, dissolved CO_2 and dissolved O_2 during the course of a yeast fermentation on methanol. A number of industrial fermentation companies are employing mass spectrometry (for example, Perkin-Elmer model 1200) in pilot plants for the analysis of off-gases, primarily O_2 and CO_2 (Buckland and Fastert, 1982). Pungor *et al.* (1980) constructed a low resolution quadrupole mass spectrometer (QMS) system from basic components at a cost of US $20 000. The instrument was interfaced to a PDP 11/10 computer for control, data acquisition and data analysis. The 95% response time of this self-made instrument is about one minute and dependent on the time required to establish a steady-state flux of molecules across the membrane from the sample stream to the vacuum chamber. However, the QMS measurements showed errors as large as 40% as a consequence of decay in electron multiplier gain and drifts in vacuum chamber pressure which occurred during fermentation monitoring. This is not observed in the commercially available mass spectrometer-based gas analyzers, such as the Perkin-Elmer model 1200.

There are two other types of commercial mass spectrometer-based gas analyzers besides the QMS system. One type measures fixed, simultaneously detectable mass peaks. It is called the multiple collector mass spectrometer. The Perkin-Elmer MGA-1200 is a typical example of this type of machine (Figure 2). The other type is based on monitoring adjustable mass peaks and utilizes the time of flight principle. A typical example of this latter type of machine is the CVC-Superspec 600. Species of different mass numbers are separated by ionizing the sample, subjecting the ions to a negative polarity accelerating field, and allowing the accelerated ions to pass through a drift region. The lighter ions travel faster than the heavier ions, producing mass separation.

The Perkin-Elmer machine is primarily suitable for a production plant where the analyzer is dedicated to one or a few specific monitoring and control purposes. The instrument is very reliable and almost maintenance free. One major drawback of this type of instrument is that it is not versatile enough for research purposes. If a new gas is introduced and needs to be monitored, a major overhaul of the machine is required in order to measure that particular new gas. This problem can be solved by using the second type of machine.

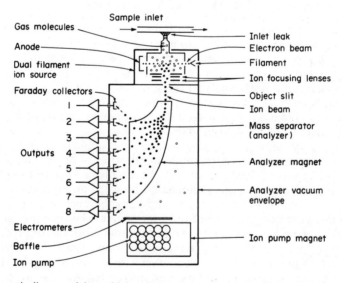

Figure 2   Schematic diagram of the multiple collector mass spectrometer (Courtesy of Perkin-Elmer)

The most critical step in continuous gas analysis by a mass spectrometer is the introduction of the gas sample into the high vacuum of the mass spectrometer. The capillary inlet system used for the commercial gas analyzer is adequate for this purpose. The gas stream should be clean of particulate matter and condensed water to prevent plugging of the capillary system. The Perkin-Elmer system can be stable over a month without calibration and the response time is as little as 10 s (Buckland and Fastert, 1982). Various applications of the on-line mass spectrometer and existing literature on this subject recently have been reviewed by Heinzle *et al.* (1983).

21.6 THE FERMENTATION GAS SAMPLING SYSTEM

Before the stream of vent gas is carried into the gas analyzer from the fermenter, additional pretreatment must be done to ensure proper reading from the gas analyzers. At least one liquid separator must be used to separate condensed water from the gas stream. Each stream passes through a needle valve and flowmeter where the flow is set to about 0.5–1.0 l min^{-1} (0 °C, 101.3 kPa; Figure 3). If the gas analyzers are used to monitor more than one fermenter, a sampling manifold has to be installed with manual and automatic operation. The gas streams are selected sequentially for analysis with the others being vented. Obviously, there should be no appreciable mixing of the streams. The system should have a low holdup volume so as to have a fast response time so that the limiting factor will be the response time of the analyzer. The sampling valves must be durable enough to withstand constant switching every few seconds. The analyzer must be easily interfaced with a computer for automatic operation and should also send an indication back to the computer of which fermenter is being sampled.

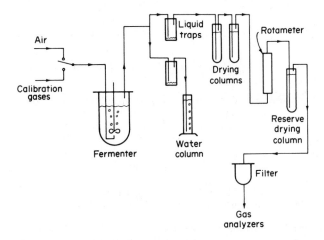

Figure 3 Pretreatment of the fermentation gas stream before entering the gas analyzer

Finally, any sampling stream not being analyzed should be vented so that the long exhaust tubings running from the fermenters into the manifold will be kept swept with fresh sample gas. Figure 4 shows a schematic drawing of a typical sampling manifold which has been developed to meet these requirements (Dobry, 1978). The sampling manifold consists of a network of four-way solenoid valves controlled by digital relays. Each of the relays is controlled by a digital output from the computer (or other types of digital controllers). Each tier of valves is driven by single command. Hence, the three tiers of valves shown in Figure 4 represent a three-bit binary number whose value can be zero to seven. In this way any one of the eight fermenters connected to the manifold can be sampled. More fermenters (up to 32) can be sampled using an additional tier of sampling valves. It is important to use heavy duty sampling valves because of the frequently cyclic nature of the operation. Similar sampling systems have been used in the fermentation industry with success.

21.7 CONCLUSIONS

One of the major limitations in the use of liquid culture analysis in fermentation is the need for, and difficulty of, continuous sterile sampling. Fermentation gas analyses can be done continuously without interfering with the liquid culture. Fermentation gas analyses have also been shown to be valuable metabolic indicators, both in research and plant environments. The initial high capital costs of these instruments are worthwhile if the fermentations can be monitored and controlled accordingly. However, to use these measured values, appropriate and accurate measurements and data reduction must be provided. Factors such as gas analyzer drift and time response phenomena can produce significant errors in the measurement of the fermentation gas concentrations (Swartz and Cooney, 1978). Careful selection and analysis of the measurement system will reduce these errors.

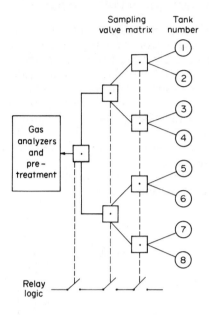

Figure 4 Arrangement of a gas sampling-manifold system

21.8 REFERENCES

Ahlstron, R. C. and F. A. Shaver (1976). The development and application of an on-line process mass spectrometer. Presented at the 31st Annual Instrumentation Symposium for the Process Industries, Texas A & M Univ., College Station, Texas. Jan. 21–22.

Arthur, R. M. (1965). Automatically recording respirometer. *Appl. Microbiol.* **13**, 125–127.

Buckland, B. C. and H. Fastert (1982). Analysis of fermentation exhaust gas using a mass spectrometer. *Computer Applications in Fermentation Technology, Third International Conference*, pp. 119–126. Society of Chemical Industry, London.

Clark, L. C. (1956). Oxygen sensing electrode. *Trans. Am. Soc. Artif. Intern. Organs*, **21**, 41–44.

Cooney, C. L., H. Y. Wang and D. I. C. Wang (1977). Computer aided material balancing for prediction of fermentation parameters. *Biotechnol. Bioeng.*, **19**, 55–67.

Dobry, D. D. (1978). Computer-coupled mass spectrometer for respiratory gas analysis in the fermentation plant. Presented at the 78th ASM Meeting, Las Vegas, Nevada, May 14–19.

Elsworth, R. (1970). The measurement of oxygen absorption and carbon dioxide evaluation in stirred deep cultures. *Methods Microbiol.*, **2**, 213–228.

Fatt, I. (1964). Rapid-responding carbon dioxide and oxygen electrodes. *J. Appl. Physiol.*, **19**, 550–553.

Fiechter, A. and K. von Meyenburg (1968). Automatic analysis of gas exchange in microbial systems. *Biotechnol. Bioeng.*, **10**, 535–549.

Harrison, D. E. F. (1978). Automated gas analysis. Presented at the 2nd International Conference on Computer Applications in Fermentation Technology, Univ. of Pennsylvania, Philadelphia, PA. August 28–30.

Heinzle, E., K. Furukawa, I. J. Dunn and J. R. Bourne (1983). Experimental methods for on-line mass spectrometry in fermentation technology. *Biotechnology*, **25**, 181–188.

Herbert, D. (1976). Application and New Fields. In *Continuous Culture*, eds. A. C. R. Dean, D. C. Ellwood, C. G. T. Evans and J. Melling, pp. 1–30. Ellis Horwood, London.

Humphrey, A. E. (1971). Computer control of fermentation processes. Proc. Labex. Symp., Earls Court, London.

Humphrey, A. E. (1973). Rationale for and principles of computer-coupled fermentation system. *The First European Conference on Computer Process Control Fermentation*. Institut National de la Récherche Agronomique, Dijon, France.

Nyiri, L. K., G. Toth and M. Charles (1975). On-line measurement of gas-exchange conditions in fermentation processes. *Biotechnol. Bioeng.*, **17**, 1663–1678.

Preier, H. (1983). Solid state sensors for gas analysis. *Process Automation*, **1**, 29–33.

Pungor, E., C. R. Perley, C. L. Cooney and J. C. Weaver (1980). Continuous monitoring of fermentation outlet gas using a computer coupled MS. *Biotechnol. Lett.*, **2**, 409–414.

Reuss, M., H. Piehl and F. Wagner (1975). Application of mass spectrometry to the measurement of dissolved gases and volatile substances in fermentation. *Eur. J. Appl. Microbiol. Biotechnol.*, **1**, 323–325.

Severinghaus, J. W. and A. F. Bradley (1958). Carbon dioxide electrode. *J. Appl Physiol.*, **13**, 515–517.

Swartz, J. R. and C. L. Cooney (1978). Instrumentation in computer-aided fermentation. *Process Biochem.*, Feb. 3–7.

Thompson, J. J. (1910). Rays of positive electricity. *Philos. Mag.* **20**, 752–767.

Tonge, G. M. (1980). Instrumentation and control in fermentation: the application of computer controlled mass spectrometry. Abstracts of VIth International Fermentation Symposium, London, Canada. July 20–25.

Umbreit, W. W., R. H. Burris and V. F. Stauffer (1959). *Manometric Techniques*, 3rd edn. Burgess Publishing Co., Minneapolis, MN.

Wang, H. Y., C. L. Cooney and D. I. C.Wang (1977). Computer aided baker's yeast fermentation. *Biotechnol. Bioeng.*, **19**, 69–68.

Wang, H. Y., C. L. Cooney and D. I. C. Wang (1979a). Computer control of baker's yeast production. *Biotechnol. Bioeng.*, **21**, 975–995.

Wang, H. Y., C. L. Cooney and D. I. C. Wang (1979b). On-line gas analysis for material balances and control. *Biotechnol. Bioeng. Symp. Ser.*, **9**, 13–23.

22

Surface Thermodynamics of Cellular and Protein Interactions

D. R. ABSOLOM

The Hospital for Sick Children, Toronto, Ontario, Canada and University of Toronto, Toronto, Ontario, Canada and State University of New York, Buffalo, NY, USA

and

W. ZINGG and A. W. NEUMANN

The Hospital for Sick Children, Toronto, Ontario, Canada and University of Toronto, Toronto, Ontario, Canada

22.1   INTRODUCTION

Adhesion of cells and adsorption of proteins, as well as the reverse processes of detachment, play an important role in several areas and disciplines. Consequently, there are many diverse approaches to these problems so that the whole field may appear to be extremely complex, if not chaotic. In an attempt to bring some clarity into this state of affairs, we have introduced a simple thermodynamic model to try to elucidate the question of whether there are any general patterns

433

of cellular and protein interactions. In this chapter, we shall outline the surface thermodynamic modelling of processes such as cell and protein adhesion, as well as ingestion of small particles, *e.g.* bacteria, by phagocytic cells. The model develops expressions for such quantities as the free energy of adhesion and the free energy of engulfment. These free energies are expressions for the driving force for these processes. Negative free energies imply that the process being considered is favoured and positive values indicate that the process is unlikely to occur. Thus the model provides the possibility to compare the experimentally obtained extent of the reaction with the thermodynamic prediction, *i.e.* the model is a predictive one.

In view of the remarkable agreement between the experimental results and the theoretical predictions, it is possible to invert the procedures and the aims of the investigation. However, working out the above thermodynamic model involves the prior determination of the surface tensions of cells and proteins by independent means. Having established that ensuing predictions are borne out by experiment, it becomes possible to deduce from the experiments the surface tensions of the cells and/or proteins investigated, resulting in novel analytical procedures to characterize the surface properties of such systems. These approaches are particularly useful in distinguishing between two only slightly different samples, such as normal and pathological cells of the same type.

Finally, the understanding of cellular and protein interactions on the level of macroscopic thermodynamics will offer new strategies to promote or hinder processes such as cell adhesion and protein adsorption to various substrates. Some of these strategies have given rise to the development and/or improvement of several preparative procedures, *e.g.* the selective fractionation of large quantities of human polymorphonuclear leukocytes by desorption from nylon fibres, the desorption of protein in processes such as hydrophobic chromatography, and the dissociation of antigen–antibody complexes, crucial to affinity chromatography.

22.2   BASIC THERMODYNAMIC CONSIDERATIONS

One of the most important aspects of thermodynamics is that it provides an answer to the question of the directionality of a natural process. Thermodynamics tells us that a properly identified thermodynamic potential, in our case the free energy function, will be minimized at equilibrium. This implies that the process under consideration, *e.g.* protein adsorption, will be favoured if it causes this function, pragmatically called the free energy F, to decrease. If F increases, then the process is not possible.

22.3   THE THERMODYNAMIC MODEL

In order to make a quantitative prediction, we have to 'model' the process. Consider a particle (P), *e.g.* a cell, initially suspended in a liquid (L), attaching to a solid (S) which is also immersed in this liquid (Figure 1). In the absence of external forces, as well as in the absence of any specific biochemical interactions, *i.e.* when we consider the cell simply as a deformable polymer particle, the change in the free energy ΔF^{adh} due to the process of cell adhesion is

$$\Delta F^{adh} = \gamma_{PS} - \gamma_{PL} - \gamma_{SL} \tag{1}$$

where γ_{PS} and γ_{PL} are the particle–solid and particle–liquid interfacial tensions, respectively, and γ_{SL} is the solid–liquid interfacial tension. The change in free energy is the maximum work obtainable from a process, here the adhesion of a cell, and the interfacial tensions are, by definition, the work required to generate the interfacial areas under consideration.

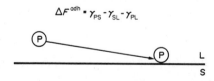

Figure 1   Schematic representation of the process of particle adhesion. P, particle; L, suspending liquid; S, substrate

If we pursue this type of approach and find patterns of agreement between model predictions and experimental observations which are broad enough to exclude statistical coincidence, we gain insight into the processes which might allow us to select materials and design procedures which would help to minimize or reduce undesired processes. Partial agreement and particularly the pattern in which the model breaks down also may provide valuable information, *e.g.* on specific biochemical interactions in the system which are not contained in the model.

22.4 THE EQUATION OF STATE APPROACH

The crucial point in the actual application of the model is the obtaining of values for the interfacial tensions γ_{PS}, γ_{SL} and γ_{PL} in equation (1). It is possible to obtain these data from an equation of state approach (Neumann *et al.*, 1974), which allows the calculation of a surface tension γ_{SV} of a solid (S) in contact with its vapour (V), from a measured liquid surface tension γ_{LV} and the contact angle θ which a drop of the liquid makes with the solid. Briefly, the technique utilizes Young's equation

$$\gamma_{SV} - \gamma_{SL} = \gamma_{LV}\cos\theta \tag{2}$$

(Young, 1805) where γ_{SV}, γ_{SL} and γ_{LV} are, respectively, the interfacial tension between a solid substrate S and the vapour phase V, γ_{SL} between S and the liquid L, and γ_{LV} between L and V; θ represents the contact angle. Of the four quantities in equation (2), only γ_{LV} and θ are readily determined experimentally; determination of a further relationship between these and the other two quantities is required. It has been shown by thermodynamic considerations that an equation of state of the form

$$\gamma_{SL} = f(\gamma_{SV}, \gamma_{LV}) \tag{3}$$

must exist (Ward and Neumann, 1974). Using experimental contact angles and liquid–vapour interfacial tensions, equation (3) has been formulated explicitly (Neumann *et al.*, 1974). In conjunction with experimental values for γ_{LV} and θ, γ_{SV} and γ_{SL} may thus be determined from equations (2) and (3). Considering equation (3) as a generic equation to calculate any interfacial tensions γ_{12} from given interfacial tensions γ_{13} and γ_{23}, where the subscripts 1, 2 and 3 refer to different phases, all the relevant interfacial tensions in equation (1) can be computed from the surface tension of the liquid and contact angles of the solid phases involved in the process under consideration. Thus, explicit thermodynamic predictions for the process, *e.g.* cell adhesion, may be obtained. Computer programs which facilitate such computations are available from the authors upon request (Neumann *et al.*, 1980b). Alternatively, recently published tables (Neumann *et al.*, 1980a) may be used.

22.5 IMPLICATIONS OF THE MODEL

Our simple thermodynamic model in conjunction with the equation of state relations predicts patterns of behaviour considerably more complex than might be expected. As an illustration, we show in Figure 2 the calculation of the free energy of adhesion (ΔF^{adh}) for the attachment of a single cellular species as a function of varying substrate surface tension γ_{SV}, for the two cases where the liquid surface tension γ_{LV} is either smaller or larger than the surface tension of the cellular species itself. For

$$\gamma_{LV} > \gamma_{PV} \tag{4}$$

where γ_{PV} is the interfacial tension between the cell and vapour, ΔF^{adh} increases with increasing γ_{SV} (increasing substrate hydrophilicity), thus predicting decreasing cell adhesion with increasing substrate surface tension γ_{SV} over a wide range of γ_{SV} values. On the other hand, when

$$\gamma_{LV} < \gamma_{PV} \tag{5}$$

the opposite pattern of behaviour is predicted, *i.e.* increasing cell adhesion with increasing γ_{SV}. For the case of the equality

$$\gamma_{LV} = \gamma_{PV} \tag{6}$$

ΔF^{adh} becomes equal to zero independent of the value of γ_{SV}, *i.e.* in this special case the extent of cell adhesion is independent of substrate surface tension. It is thus clear that the often asked important question whether the extent of cellular adhesion is expected to increase or decrease with increasing surface tension (increasing hydrophilicity) of the substrate is an incomplete question: the pattern of adhesion depends on the relative surface tensions of the particle, substrate and suspending liquid.

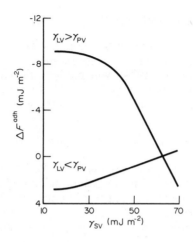

Figure 2   Dependence of the change in free energy on substrate surface tension (γ_{SV})

22.6   RESULTS

We have conducted several experiments in which a comparison is made between the observed extent of the reaction and the thermodynamic driving force or, in other words, the extent of cellular adhesion (or protein adsorption) in comparison to the sign and magnitude of the free energy, ΔF^{adh}, for the process. The experimental details of these investigations have been described elsewhere (Absolom *et al.*, 1979, 1980; Neumann *et al.*, 1979b; van Oss *et al.*, 1981a). For these studies six well characterized polymeric substrates were employed. These had a wide range of surface properties. They were poly(tetrafluoroethylene) (surface tension $\gamma_{SV} = 17.6$ mJ m^{-2}) polystyrene ($\gamma_{SV} = 25.6$), low density polyethylene ($\gamma_{SV} = 32.5$), acetal ($\gamma_{SV} = 44.6$), poly(ethyleneterephthalate) ($\gamma_{SV} = 47.0$) and sulfonated polystyrene ($\gamma_{SV} = 66.7$).

22.6.1   Cell Adhesion

22.6.1.1   Granulocytes

We have investigated in detail the adhesion of human granulocytes to these polymer surfaces (Absolom *et al.*, 1979, 1980; Neumann *et al.*, 1979b). Two important points about these experiments should be noted: (1) the cells were isolated and then resuspended, after several centrifugation washes, in a protein-free liquid medium, thus ensuring that the polymer surfaces were not precoated with a protein layer before cell adhesion occurred; and (2) the isolated, washed granulocytes then were resuspended in protein-free liquids having different surface tensions (γ_{LV}). The liquid surface tensions γ_{LV} ranged from approximately 73 to 65 mJ m^{-2}. The variation in surface tension was achieved through the incorporation of varying amounts of a water soluble, low molecular weight agent such as dimethyl sulfoxide (DMSO). Thereafter the cells were exposed to the various substrates under standard conditions. Typical results are given in Figure 3. The theoretical predictions, *i.e.* whether or not the extent of adhesion will increase or decrease with increasing substrate surface tension, developed from free energy calculations (*cf.* Figure 2) using the relevant interfacial tensions described in equation (1), are borne out completely by the experimental observations (Figure 3). At the lowest DMSO concentration, *i.e.* at the highest surface tension γ_{LV} of the suspending medium, cell adhesion decreases with increasing γ_{SV} of the substrate. As the DMSO concentration is increased and the liquid surface tension γ_{LV} correspondingly decreased, the difference in the extent of cell adhesion to the various substrates becomes less pronounced. At certain intermediate liquid surface tensions γ_{LV}, adhesion becomes virtually independent of γ_{SV} and finally, at even lower values of the surface tension γ_{LV} (produced through the admixture of increased volumes of DMSO into the buffer), cell adhesion increases with increasing substrate surface tension γ_{SV}.

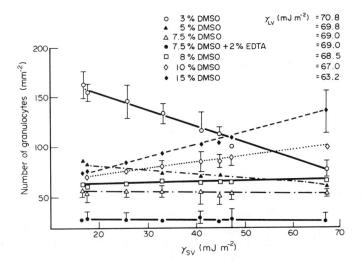

Figure 3 Granulocyte adhesion for the various liquid surface tensions, error limits are 95% confidence. (For graphical reasons error limits are given only for some cases; the errors are similar in all cases.) Reproduced from Absolom *et al.* (1979), with permission of The American Society for Artificial Internal Organs, Inc.

It was also shown that the extent of cell adhesion depends on the surface tension of the adhering particles. For this purpose blood cells of the same type, *viz.* granulocytes, which had modified surface properties, were used (Absolom *et al.*, 1980). These physiologically defective cells were obtained from patients with juvenile periodontal disease (Cianciola *et al.*, 1977). In these cases, the cell adhesion experiments resulted in plots with similar trends to those shown in Figure 3. In the case of the pathological granulocytes, however, the slopes of the straight lines were markedly different from those shown in Figure 3 for normal human granulocytes.

In summary, it is apparent from Figure 3 that the pattern of cell adhesion is different in three regions: (1) $\gamma_{LV} > \gamma_{PV}$, where adhesion decreases with increasing γ_{SV}; (2) $\gamma_{LV} < \gamma_{PV}$, where adhesion increases with increasing γ_{SV}; and (3) $\gamma_{LV} = \gamma_{PV}$, where the extent of adhesion is independent of γ_{SV}.

These observations are in accord with the theoretical predictions of the model as depicted in Figure 2.

22.6.1.2 Platelets

Similar adhesion studies have been performed with platelets. In these experiments, in spite of careful washing of the platelets, it has never been possible to obtain a surface tension of the suspending medium γ_{LV} greater than the surface tension of the platelets ($\gamma_{PV} \simeq 68.0$ mJ m^{-2}) (Hum *et al.*, 1975; Neumann *et al.*, 1975, 1979c). This lowered surface tension of the medium is due to the release of surface active materials (such as proteins, amines and/or lipids) by the platelets. Thus when $\gamma_{LV} < \gamma_{PV}$ the thermodynamic model predicts that the extent of platelet adhesion should increase with increasing γ_{SV} of the polymer substrate. As illustrated in Figure 4, this did in fact occur.

Thus the only real difference between granulocyte and platelet adhesion is that carefully washed granulocytes when resuspended in a protein-free buffer show a decreasing adhesion with increasing γ_{SV} when $\gamma_{LV} > \gamma_{PV}$, whereas platelets do not (because the latter exude a surfactant which makes it impossible for γ_{LV} to be greater than γ_{PV}).

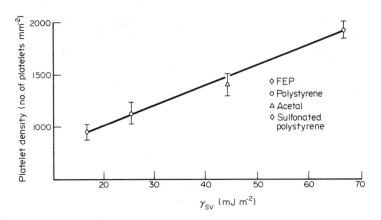

Figure 4 Platelet adhesion to various substrates as a function of γ_{SV}. Errors are 95% confidence limits. Reproduced from Absolom *et al.* (1979), with permission of The American Society for Artificial Internal Organs, Inc.

22.6.2 Protein Adsorption

During the course of these adhesion studies (Absolom *et al.*, 1979, 1980; Hum *et al.*, 1975; Neumann *et al.*, 1975, 1979b, 1979c; Zingg *et al.*, 1981), we noted the effect of various proteins in solution and pre-adsorbed onto the polymer surface on cell adhesion. Fibrinogen and immunoglobulin G promoted cell adhesion, whereas albumin resulted in a decreased adhesion to the various polymer surfaces. These effects were presumably due to the prior adsorption of the proteins onto the polymers, thereby modifying the surface properties of the polymer substrates. These observations, in conjunction with the generally accepted opinion that protein adsorption precedes cell adhesion, led us to investigate the role of interfacial tensions in determining the extent of adsorption of serum proteins onto various polymers. In particular, we were interested in determining to what degree protein adsorption followed the pattern of cell adhesion and whether it could be described by the same thermodynamic principles.

For this purpose, a strategy similar to that described for the granulocyte adhesion studies was adopted. We investigated the role of substrate (γ_{SV}), suspending liquid (γ_{LV}) and protein (γ_{PV}) surface tensions on the extent of protein adsorption. Human serum albumin, α_2-macroglobulin, immunoglobulin G and immunoglobulin M were studied (van Oss *et al.*, 1981a). The results of these experiments confirmed, in all cases, the predictions of the thermodynamic model. Typical results of a protein adsorption experiment for human immunoglobulin IgM from various aqueous DMSO solutions are given in Figure 5. The protein adsorption curves in Figure 5 have a striking similarity with the curves for granulocyte adhesion in Figure 3. It is apparent from Figure 5 that for $\gamma_{LV} > \gamma_{PV}$ adsorption of the protein decreases with increasing substrate surface tension, γ_{SV}. When $\gamma_{LV} < \gamma_{PV}$, protein adsorption increases with increasing γ_{SV}. For the limiting case, *i.e.* when $\gamma_{LV} = \gamma_{PV}$, protein adsorption is independent of γ_{SV}. The extent of adsorption for a single protein species is dependent on both γ_{LV} and γ_{SV}, in accord with the thermodynamic model.

Therefore it would seem that the simple thermodynamic model depicted in Figure 1, together with an equation of state approach for obtaining interfacial tensions, describes the qualitative features of protein adsorption and cell adhesion to a wide range of polymer surfaces. These results suggest that the initial protein–substrate or granulocyte–substrate interaction is non-specific in a biochemical sense, *i.e.* it is not receptor mediated.

22.6.3 Other Factors

Although the thermodynamic predictions, as discussed above, do indeed describe the qualitative experimental adhesion/adsorption data of cells and proteins to a range of polymers under varying conditions of γ_{LV}, it is also clear that the model does not describe the phenomena completely. At the particular value of γ_{LV} (characteristic of the cell or protein of interest) where the adhesion or adsorption is identical to all the surfaces,

$$\Delta F^{\text{udh}} = 0 \tag{7}$$

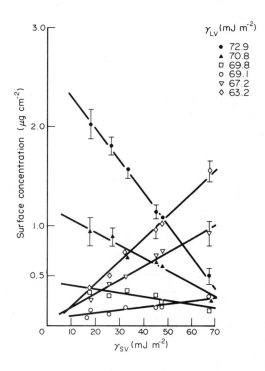

Figure 5 Adsorption of immunoglobulin M for various liquid surface tensions; error limits, 95% confidence. For graphical reasons error limits are given only for some cases; the errors are similar in all cases. Reproduced from van Oss *et al.* (1981a), with permission of Elsevier/North Holland Biomedical Press

Equation (7) predicts that there is no thermodynamic driving force to promote the adhesion or adsorption of cells/proteins to the polymer surfaces. Nevertheless, in both the protein adsorption and cell adhesion studies, even under those conditions where $\Delta F^{adh} = 0$, a small, but nevertheless positive, level of adhesion is observed. This residual adhesion must be due to interactions other than van der Waals attraction, *e.g.* electrostatic interactions and plurivalent cationic bridging (Absolom *et al.*, 1980).

In addition, under physiologically relevant conditions the adhesion or adsorption process is considerably more complex than the 'ideal' systems discussed above. Consequently, several additional factors need to be taken into consideration in such studies *e.g.* the effect of flow rate on cell adhesion, protein–protein or protein–cell interactions, surface topography and the kinetics of adhesion, *etc.* We have discussed such factors elsewhere (Absolom *et al.*, 1979; Neumann *et al.*, 1983a, 1983b; Zingg *et al.*, 1981).

22.7 PROCEDURES FOR DETERMINING THE RELEVANT INTERFACIAL TENSIONS

The crucial point in the application of the thermodynamic model is the problem of obtaining values for the interfacial tensions in equation (1). Over the past decade, several independent techniques have been developed which allow the determination of the surface tension of biological entities. This methodology is the subject of a review article to be presented elsewhere (Neumann *et al.*, 1983a) and will only be briefly dealt with here.

22.7.1 Contact Angle Measurements

At this time, the equation of state approach for determining interfacial tensions (Neumann *et al.*, 1974) from contact angle measurements is the simplest means for obtaining the surface tension of biological materials. For obvious physiological and osmotic reasons, the liquid of choice for these measurements is physiological saline (0.15 M saline, pH 7.2). This method has been used to determine the surface tension of human lymphocytes (van Oss *et al.*, 1975), granulocytes (Absolom *et al.*, 1979, 1981; van Oss *et al.*, 1975, 1982a), macrophages (Thrasher *et al.*, 1973), erythrocytes (van Oss *et al.*, 1975), porcine platelets (Absolom *et al.*, 1982b), serum proteins (van Oss *et al.*, 1981a) and several strains of bacteria (Absolom *et al.*, 1982a, 1982b, 1983; van Oss *et al.*, 1975, 1982a). The contact angle which a drop of saline makes with a layer of the protein or cellular material is determined at regular time intervals (Absolom *et al.*, 1981, 1982a, 1982b; van Oss *et al.*, 1975, 1981a, 1982a). Initially this value is close to zero degrees, but due to water evaporation rapidly increases to reach a plateau value that is characteristic of the species being examined. The plateau value is usually stable for approximately 30 minutes (Figure 6). The surface tension of the biological substrate can then be determined as explained above, using the contact angle corresponding to the plateau.

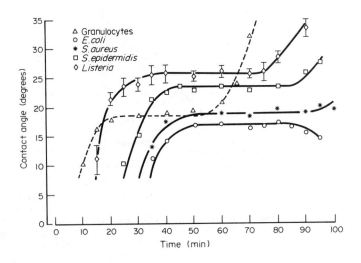

Figure 6   Contact angles of saline on non-opsonized bacteria and human granulocytes as a function of water evaporation from the wet biological substrate, measured in terms of time. Each symbol is the average of ten individual contact angle readings on each of four different drops, at approximately the same time, on one and the same substrate. Reproduced from Absolom *et al.* (1982a), with permission of Alan R. Liss, Inc.

22.7.2   Adhesion/Adsorption Technique

If the slopes of the straight lines in Figures 3 and 5 are calculated and plotted against the various liquid surface tensions (γ_{LV}) employed in these studies, an independent method is developed for the determination of the effective surface tension of the adhering particles (Figure 7). When the slope is zero, $\gamma_{LV} = \gamma_{PV}$, the surface tension of the adhering particles or protein molecules (Absolom *et al.*, 1979; Neumann *et al.*, 1979b; van Oss *et al.*, 1981a) as described earlier in this chapter (see Section 22.5). The cellular and protein surface tension results obtained with this method are in excellent agreement with the values obtained from contact angle measurements. Although the adsorption technique is more laborious, its importance lies in the fact that the methodology is fundamentally independent of contact angle measurements; it therefore may be used to confirm the results obtained by the contact angle method.

22.7.3   Freezing Front Technique

Particles or cells immersed in the liquid phase of a solidifying melt will be pushed by the advancing solidification front if the free energy for engulfment is positive (Spelt *et al.*, 1982). However, as the solidification front velocity is increased, a limiting or critical velocity V_c will be reached at which viscous drag becomes greater than the thermodynamic repulsion, so that engulfment then occurs (Neumann *et al.*, 1979a). From the experimental data an empirical relation could be established between ΔF^{adh} of the particles or cells, their diameter D, V_c and other system properties (Omenyi *et al.*, 1980). Recently we have used this technique to obtain quantitative surface tension data on polymer particles which agree extremely well with the values obtained by other methods (Omenyi *et al.*, 1981a). This method has been extended for use with biological

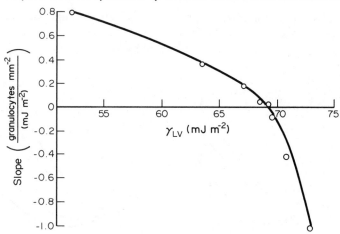

Figure 7 Slopes of the straight line data of Figure 3 *versus* γ_{LV}. Slope is zero for $\gamma_{LV} = \gamma_{PV} = 69.0$ mJ m^{-2}. The points are computer curve fitted to a second order polynomial and the intercepts taken from that curve fit. Reproduced from Absolom *et al.* (1979), with permission of The American Society for Artifical Internal Organs

materials suspended in water; the advancing solidification front then consists of ice. To date we have used this technique to determine the surface tension of glutaraldehyde-fixed erythrocytes of several animal species: human, horse, canine, turkey and chicken (Absolom *et al.*, 1985), as well as fresh human granulocytes and lymphocytes (Spelt *et al.*, 1982). This technique has the advantage that the cells of interest are maintained in an aqueous medium and thus are not exposed to an air interface (as in the contact angle approach) nor to low molecular weight solutes such as dimethyl sulfoxide (as in the adhesion approach).

22.7.4 Droplet Sedimentation

The maximum stability of suspensions occurs when the van der Waals attraction between particles or cells is reduced to zero, which happens when the surface tension of the liquid suspending medium (γ_{LV}) equals that of the cells (γ_{PV}). Thus, by determining the γ_{LV} at which the highest concentration of cells can be supported by a density gradient (*e.g.* of D_2O) without giving rise to instability (as judged by droplet sedimentation), the value for γ_{PV} of the cell is easily found (Omenyi *et al.*, 1981b, 1982). For glutaraldehyde-fixed human erythrocytes $\gamma_{PV} \simeq 65$ mJ m^{-2} (at 26 °C) was found, which conforms well with the value of 64.5 mJ m^{-2} found for the same cells with the solidification front method.

22.7.5 Biological Methods

The process of phagocytic ingestion of small particles (< 2 μm) can also be used to determine the surface tension of these particles. By direct analogy to the cell adhesion/protein adsorption studies, if the phagocytes and particles are suspended in liquid media of different surface tensions γ_{LV} (produced through the admixture of varying small amounts of DMSO), then the extent of particle engulfment will also be different. When $\gamma_{LV} > \gamma_{CV}$ (where γ_{CV} is the surface tension of the phagocytic cells and γ_{PV} is the surface tension of the ingested particles), then the most hydrophobic particles are most avidly ingested, whereas when $\gamma_{LV} < \gamma_{CV}$ then the more hydrophilic particles are more readily engulfed. If, however, we consider ingestion of only one type of particle as a function of varying γ_{LV}, it is clear from a theoretical aspect that particle engulfment should proceed from a maximum level at high γ_{LV} through a minimum to an increased level of ingestion at comparatively low γ_{LV} values. The minimum ingestion should occur where $\gamma_{PV} = \gamma_{LV}$. We have experimentally tested this hypothesis using human granulocytes and five strains of non-opsonized bacteria (Absolom *et al.*, 1982a). In all cases minimum bacterial ingestion occurred at that value of γ_{LV} that corresponded closely to the γ_{PV} value determined by other, independent methods (Absolom *et al.*, 1983; van Oss *et al.*, 1975). Such methodology requires that the phagocytic ingestion process is non-specific in character and is not controlled by specific membrane–ligand interactions (*e.g.* Fc$_\gamma$ receptors). Such conditions can be achieved by using non-opsonized bacteria in an essentially protein-free buffer medium (Absolom *et al.*,

1982a). Similar results were obtained in phagocytosis experiments with porcine platelets as phagocytes (Figure 8) (Absolom *et al.*, 1982b).

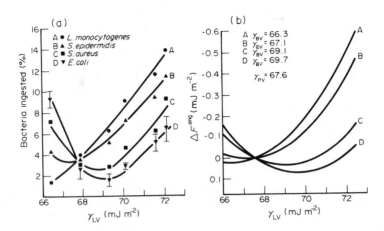

Figure 8   Free energy of engulfment as a function of the surface tension of the suspending aqueous media for the four species of non-opsonized bacteria (B) and the experimentally determined extent of the same bacteria ingested by pig platelets in the same aqueous media (A). Errors are 95% confidence limits; for graphical reasons error limits are given only for some cases; the errors are similar in all cases. Reproduced from Absolom *et al.* (1982b), with permission of Academic Press, Inc.

22.8   APPLICATIONS OF SURFACE THERMODYNAMICS TO BIOTECHNOLOGY

22.8.1   Hydrophobic Chromatography

The thermodynamic model may be applied to develop preparative procedures for the isolation or purification of large quantities of granulocytes and serum proteins. These procedures require that one initially maximizes cell or protein adhesion to a given solid substrate, prior to achieving selective fractionation of the particles or molecules of interest by desorption from the matrix material. We have previously described in detail the conditions under which such a desorption can be achieved (van Oss *et al.*, 1978, 1980a). Briefly, what is required is that the surface tension of the suspending buffer is lowered to a surface tension value intermediate between that of the protein or cell of interest and the solid carrier. Under such conditions the free energy of adhesion is positive, and the van der Waals interactions between the cells/protein and the polymer become repulsive (van Oss *et al.*, 1980b). If at the same time a chelating agent is added to the elution medium so as to obviate the effect of electrostatic interactions due to plurivalent cationic bridging, complete desorption of large quantities of cells or proteins from the polymer material ensues. This procedure has been applied to protein purification by means of hydrophobic chromatography (van Oss *et al.*, 1979b) and to granulocyte elution from nylon fibres (Absolom *et al.*, 1981) as illustrated in Figures 9 and 10, respectively. Figure 10 indicates the role of both electrostatic (specifically cationic bridging) and van der Waals interactions in determining the overall cell adhesion to a polymer surface. When the electrical interactions between the cell and nylon matrix are overcome (through the incorporation of a chelating agent into the buffer), a cell yield of approximately 30% is achieved. When, however, the attractive van der Waals interaction is rendered repulsive, through the lowering of the surface tension of the buffer to a value intermediate between that of the adhering cells and the nylon, a three fold increase in cell yield is achieved. This enhanced yield occurs at the point where $\gamma_{LV} = \gamma_{PV}$; see Figure 10.

22.8.2   Antigen–Antibody Dissociation and Affinity Chromatography

There were hitherto a number of methods for dissociating antigen–antibody (AG–AB) complexes, usually employing low pH and/or 'chaotropic salts'. Such methods sometimes work, at least partially, but they generally cause some degree of protein denaturation, which may or may not be partly reversible. The reason for the unreliability of the older empirical methods for disso-

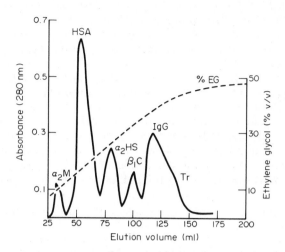

Figure 9 Elution of human serum proteins from a phenyl-sepharose column, after adsorption of whole human serum, with an elution medium of increasing concentration of ethylene glycol. The following fractions were isolated: α_2-macro-globulin (α_2M); serum albumin (HSA); α_2 glycoprotein HS (α_2HS); β_{1c} complement component (β_{1c}); immunoglobulin G (IgG); transferrin (Tr). Reproduced from van Oss *et al.* (1979a), with permission of Marcel Dekker, Inc.

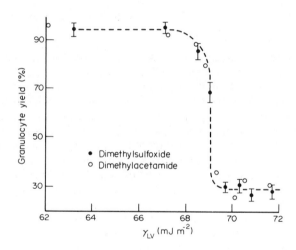

Figure 10 Granulocyte elution from nylon fibres as a function of the surface tension of the eluting buffer. When a chelating agent (*e.g.* sodium ethylenediaminetetraacetic acid) is incorporated into the buffer a cell yield of approximately 30% is achieved. When in addition to the above the liquid surface tension is lowered the cell yield increases to approximately 95%. Reproduced from Absolom *et al* (1981), with permission of J. B. Lippincott Co.

ciating AG–AB complexes lies, as with cell adhesion (see above), in the fact that it usually is not realized that in most cases AG is bound to AB through electrostatic and van der Waals bonds, so that to effect complete dissociation it is necessary to break both types of bonds simultaneously, by different methods. Extremes of pH affect electrostatic, but not van der Waals, bonds. To dissociate the latter type of bond, the free energy for the system must be made positive. This can be achieved by lowering the surface tension γ_{LV} of the liquid medium. Thus, for dissociating most types of AG–AB bonds one must use high (or low) pH and a decrease in γ_{LV}. In practice it is preferable to neutralize the electrostatic bonds by means of high rather than low pH, as an alkaline pH (of the order of 9.5 to 10) tends to be less denaturing to most proteins than an acidic pH (van Oss *et al.*, 1979a). Lowering of the liquid surface tension γ_{LV} can be achieved by admixture of low molecular weight agents such as ethylene glycol or dimethyl sulfoxide to the buffer. By this procedure we have completely dissociated specific protein–AB immune complexes (van Oss *et al.*, 1979a), bacteria–AB complexes (Hovanec *et al.*, 1980) and several types of blood group–AB complexes (de Groot *et al.*, 1980; van Oss *et al.*, 1981b).

In affinity chromatography, an AG (or its corresponding specific AB) is covalently attached to a solid carrier and used in that form for the isolation of its AB (or AG, as the case may be).

Therefore, if one can elicit the formation of specific ABs to a given compound, a column made with those ABs as the ligand may then serve for the specific binding of the compound (*i.e.* AG) in question. The attachment or binding of AG to AB, in an aqueous medium, is easy; the subsequent quantitative elution of active counterligand often is difficult. However, if one observes the same principles as discussed above for the dissociation of most AG–AB systems, *i.e.* a combination of an increased pH (9.5) for neutralization of the electrostatic attraction and a decreased liquid surface tension ($\gamma_{LV} \simeq 50$ mJ m^{-2}) for achieving a positive free energy, total elution is possible (Absolom, 1981; van Oss *et al.*, 1982b). In cases of exclusively van der Waals type AG–AB interaction, just rendering ΔF^{adh} positive, through the reduction of the liquid surface tension γ_{LV} (to a value intermediate between the surface tension of the antigenic determinant and the corresponding antibody binding site), will suffice to ensure complete elution. For example, when dextran (Sephadex® beads) is used as the carrier, antidextran antibodies can be completely eluted at $\gamma_{LV} \simeq 47$ mJ m^{-2}, at neutral pH (Absolom *et al.*, in preparation).

22.8.3 Bacterial Adhesion

Bacterial adhesion to various surfaces plays an important role in various situations such as tooth decay, ship fouling, waste water treatment, *etc.* We have recently shown that it is possible to describe the process of bacterial adhesion by surface thermodynamic considerations (Absolom *et al.*, 1983). Indeed, bacterial adhesion appears to follow identical patterns of behaviour to those described above for granulocyte adhesion and protein adsorption.

22.8.4 Particle and Suspension Stability

In several areas it is often important to be able to use suspensions having the maximum possible particle concentration, *e.g.* in electrophoresis or density gradient centrifugation. In addition, such suspensions should possess a high degree of stability. On the termination of these procedures it is necessary to maintain the separation of the various fractions in order to achieve the greatest resolution of the applied sample. If these zones are not stable, then the particles contained therein will disperse and mix with adjacent fractions. Omenyi *et al.* (1981b, 1982) have shown that it is possible to achieve a four-fold increase in particle concentration and yet maintain suspension stability. This was achieved by controlling the surface thermodynamics of the system. The useful increase in particle concentration was obtained by utilizing conditions under which there was maximum electrostatic repulsion and a negligible van der Waals attraction between the particles. Such a situation was achieved by incorporating increasing amounts of dimethyl sulfoxide into the suspending medium, which resulted in a gradual lowering of the surface tension of the liquid, γ_{LV}. Under otherwise identical conditions, the particle concentration (N_o) that could be supported on a D$_2$O cushion was seen to vary with γ_{LV} (Figure 11). The maximum particle concentration was achieved when $\gamma_{LV} = \gamma_{PV}$.

22.8.5 Other Applications

There are many applications in which, knowingly or often unknowingly, one uses a positive free energy to achieve separation.

Compatibility and phase separation of two polymers when dissolved in one and the same solvent is one example (van Oss *et al.*, 1979c). The detachment of soil particles from fabrics in processes of detergency and dry cleaning may also be understood in terms of positive free energy of adhesion and thus repulsive van der Waals interactions. Enhanced oil recovery and bitumen extraction from oil sands, possibly using natural biosurfactants (van Oss *et al.*, 1980a), might also be potential areas of application of the approach.

22.9 CONCLUSIONS

A thermodynamic model providing qualitative and quantitative insight into the attachment and detachment of cells and proteins with respect to solid surfaces is proposed. Crucial phenomena, such as the reversal of the trend of cell adhesion when γ_{LV} changes from $\gamma_{LV} > \gamma_{PV}$ to $\gamma_{LV} < \gamma_{PV}$,

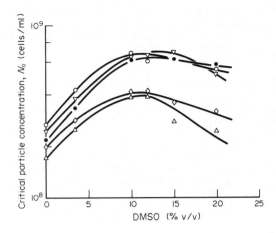

Figure 11 Suspension stability as a function of liquid surface tension. Critical particle concentration (no. of erythrocytes ml^{-1}) is plotted against the percentage (v/v) of added dimethyl sulfoxide (DMSO). Maximum stability occurs when $\gamma_{PV} = \gamma_{LV}$. $\triangle$, Turkey; $\lozenge$, Chicken; $\bigcirc$, Horse; $\triangledown$, Canine; $\bullet$, Human. Reproduced from Omenyi *et al.* (1982), with permission of Marcel Dekker, Inc.

and desorption of cells from a hydrophobic matrix when $\gamma_{LV} = \gamma_{PV}$, are convincing evidence that the proposed thermodynamic model is a good starting point for the study of such phenomena. This point is further reinforced by the fact that protein adsorption and bacterial adhesion follow the same patterns as eukaryotic cell adhesion.

Consideration of the surface thermodynamic parameters of a system permits us to explain and quantify the methodology used in preparative procedures for the isolation of large quantities of cells or proteins by means of hydrophobic chromatography.

Use of the concepts outlined in this chapter permit complete dissociation of antigen–antibody complexes under non-denaturing conditions. It is anticipated that the approach outlined will be of use in understanding and developing various other separation procedures.

ACKNOWLEDGEMENTS

This work was supported in part by the Medical Research Council of Canada ($\sharp$ MT 5462) and the Natural Sciences and Engineering Research Council of Canada ($\sharp$ A 8278). One of the authors (DRA) acknowledges support of the Ontario Heart Foundation through a Senior Fellowship.

22.10 REFERENCES

Absolom, D. R. (1981). Affinity chromatography. *Sep. Purif. Methods*, **10**, 239–286.
Absolom, D. R., A. W. Neumann, W. Zingg and C. J. van Oss (1979). Thermodynamic studies of cellular adhesion. *Trans. Am. Soc. Artif. Intern. Organs*, **25**, 152–156.
Absolom, D. R., C. J. van Oss, R. J. Genco, D. W. Francis and A. W. Neumann (1980). Surface thermodynamics of normal and pathological human granulocytes. *Cell Biophysics*, **2**, 113–126.
Absolom, D. R., C. J. van Oss and A. W. Neumann (1981). Elution of human granulocytes from nylon fibres by means of repulsive van der Waals forces. *Transfusion*, **21**, 642–657.
Absolom, D. R., C. J. van Oss, W. Zingg and A. W. Neumann (1982a). Phagocytosis as a surface phenomenon: Opsonization by aspecific adsorption of IgG as a function of bacterial hydrophobicity. *J. Reticuloendothel. Soc.*, **31**, 59–70.
Absolom, D. R., D. W. Francis, W. Zingg, C. J. van Oss and A. W. Neumann (1982b). Phagocytosis of bacteria by platelets: Surface thermodynamic aspects. *J. Colloid Interface Sci.*, **85**, 168–177.
Absolom, D. R., F. V. Lamberti, Z. Policova, W. Zingg, C. J. van Oss and A. W. Neumann (1983). Surface thermodynamics of bacterial adhesion. *J. Appl. Environ. Microbiol.*, **46**, 90–97.
Absolom, D. R., Z. Policova, E. Moy, W. Zingg and A. W. Neumann (1985). Determination of the surface tension of various species of erythrocytes by means of the solidification front technique. *Cell Biophysics*, in press.
Cianciola, L. J., R. J. Genco, H. R. Patters, J. McKenna and C. J. van Oss (1977). Defective polymorphonuclear leukocyte function in a human periodontal disease. *Nature (London)*, **265**, 445–446.
de Groot, E. R., M. C. Lamers, L. A. Aarden, R. J. T. Smeenk and C. J. van Oss (1980). Dissociation of DNA/anti-DNA complexes at high pH. *Immunol. Commun.*, **9**, 515–528.
Hovanec, D. L., D. R. Absolom, C. J. van Oss and E. A. Gorzynski (1980). Relationship to coagglutination of immunoglobulin class dissociated from *Escherichia coli*–antibody complexes. *J. Clin. Microbiol.*, **12**, 608–609.
Hum, O. S., A. W. Neumann and W. Zingg (1975). Platelet interaction with smooth solid substrates determined by an open static method. *Thromb. Res.*, **7**, 461–470.

Neumann, A. W., R. J. Good, C. J. Hope and M. Sejpal (1974). An equation-of-state approach to determine surface tensions of low-energy solids from contact angles. *J. Colloid Interface Sci.*, **49**, 291–304.

Neumann, A. W., C. J. Hope, C. A. Ward, M. A. Herbert, G. W. Dunn and W. Zingg (1975). The role of surface thermodynamics in thromboresistance of biomaterials. *J. Biomed. Mater. Res.*, **9**, 127–142.

Neumann, A. W., S. N. Omenyi and C. J. van Oss (1979a). Negative Hamaker coefficients. I. Particle engulfment or rejection at solidification fronts. *Colloid Polymer Sci.*, **257**, 413–419.

Neumann, A. W., D. R. Absolom, W. Zingg and C. J. van Oss (1979b). Surface thermodynamics of leukocyte and platelet adhesion to polymer surfaces. *Cell Biophysics*, **1**, 79–92.

Neumann, A. W., M. A. Moscarello, W. Zingg, O. S. Hum and S. K. Chang (1979c). Platelet adhesion from human blood to bare and protein coated polymer surfaces. *J. Polymer Sci. Polymer Symp.*, **66**, 391–398.

Neumann, A. W., D. R. Absolom, D. W. Francis and C. J. van Oss (1980a). Conversion tables of contact angles to surface tensions. *Sep. Purif. Methods*, **9**, 69–163.

Neumann, A. W., O. S. Hum, D. W. Francis, W. Zingg and C. J. van Oss (1980b). Kinetic and thermodynamic aspects of platelet adhesion from suspension to various substrates. *J. Biomed. Mater. Res.*, **14**, 499–509.

Neumann, A. W., D. R. Absolom, D. W. Francis, S. N. Omenyi, J. K. Spelt, Z. Policova, W. Zingg and C. J. van Oss (1983a). Measurement of the surface tensions of blood cells and other particles. *Ann. N.Y. Acad. Sci.*, **416**, 276–298.

Neumann, A. W., D. R. Absolom, W. Zingg, C. J. van Oss and D. W. Francis (1983b). Surface thermodynamics of protein adsorption, cell adhesion and cell engulfment. In *Biocompatible Polymers*, ed. M. Syzcher, pp. 89–102. Thermo Electron, Walton, MA.

Omenyi, S. N., R. P. Smith and A. W. Neumann (1980). Determination of solid/melt interfacial tensions and of contact angles of small particles from the critical velocity of engulfing. *J. Colloid Interface Sci.*, **75**, 117–125.

Omenyi, S. N., A. W. Neumann and C. J. van Oss (1981a). Attraction and repulsion of solid particles by solidification fronts. I. Thermodynamic effects. *J. Appl. Physics*, **52**, 789–795.

Omenyi, S. N., R. S. Snyder, C. J. van Oss, D. R. Absolom and A. W. Neumann (1981b). Effects of zero van der Waals and zero electrostatic forces on droplet sedimentation. *J. Colloid Interface Sci.*, **81**, 402–409.

Omenyi, S. N., R. S. Snyder, D. R. Absolom, C. J. van Oss and A. W. Neumann (1982). Enhanced erythrocyte suspension layer stability achieved by surface tension lowering additives. *J. Dispersion Sci. Technol.*, in press.

Spelt, J. K., D. R. Absolom, W. Zingg, C. J. van Oss and A. W. Neumann (1982). Surface tension determination of biological cells by means of the freezing front technique. *Cell Biophysics*, **4**, 117–132.

Thrasher, S. G., T. Yoshida, C. J. van Oss, S. Cohen and N. R. Rose (1973). Alteration of macrophage interfacial tension by supernatants of antigen-activated lymphocyte cultures. *J. Immunol.*, **110**, 321–326.

van Oss, C. J., C. F. Gillman and A. W. Neumann (1975). *Phagocytic Engulfment and Cell Adhesiveness as Surface Phenomena.* Dekker, New York.

van Oss, C. J., A. W. Neumann, S. N. Omenyi and D. R. Absolom (1978). Repulsive van der Waals interactions: Their role in various separation methods. *Sep. Purif. Methods*, **7**, 245–271.

van Oss, C. J., D. R. Absolom and A. W. Neumann (1979a). Repulsive van der Waals forces. II. Mechanism of hydrophobic chromatography. *Sep. Sci. Technol.*, **14**, 305–317.

van Oss, C. J., S. N. Omenyi and A. W. Neumann (1979b). Negative Hamaker coefficients. II. Phase separation of polymer solutions. *Colloid Polymer Sci.*, **257**, 737–744.

van Oss, C. J., D. R. Absolom, A. L. Grossberg and A. W. Neumann (1979c). Repulsive van der Waals forces. 1. Complete dissociation of antigen–antibody complexes by means of negative van der Waals forces. *Immunol. Commun.*, **8**, 11–29.

van Oss, C. J., D. R. Absolom and A. W. Neumann (1980a). Application of net repulsive van der Waals forces between different particles, macromolecules or biological cells in liquids. *Colloids and Surfaces*, **1**, 45–56.

van Oss, C. J., A. W. Neumann, R. J. Good and D. R. Absolom (1980b). Influence of extremely small attractive as well as repulsive van der Waals–London forces on cell interactions. *Adv. Chem. Ser.*, **188**, 107–114.

van Oss, C. J., D. R. Absolom, A. W. Neumann and W. Zingg (1981a). Determination of the surface tension of proteins. I. Surface tension of native serum proteins in aqueous media. *Biochim. Biophys. Acta*, **670**, 64–73.

van Oss, C. J., D. Beckers, C. P. Engelfriet, D. R. Absolom and A. W. Neumann (1981b). Elution of blood group antibodies. I. Red cell antibodies. *Vox Sanguis*, **40**, 367–371.

van Oss, C. J., D. R. Absolom and A. W. Neumann (1982a). Phagocytosis as a surface phenomenon. In *The Reticuloendothelial System*, eds. H. Friedman, M. R. Escobar and S. M. Reichert, vol. 4, pp. 187–236, Plenum, New York.

van Oss, C. J., D. R. Absolom and A. W. Neumann (1982b). Role of attractive and repulsive van der Waals forces in affinity and hydrophobic chromatography. In *Affinity Chromatography and Related Techniques*, ed. T. C. J. Gribnau, J. Visser and R. J. F. Nivard, pp. 29–37. Elsevier, Amsterdam.

Ward, C. A. and A. W. Neumann (1974). On the surface thermodynamics of a two-component liquid-vapor-ideal solid system. *J. Colloid Interface Sci.*, **49**, 286–290.

Young, T. (1805). An essay on the cohesion of fluids. *Philos. Trans. R. Soc. London, Ser. B*, **95**, 65–71.

Zingg, W., A. W. Neumann, A. B. Strong, O. S. Hum and D. R. Absolom (1981). Platelet adhesion to smooth and rough hydrophobic and hydrophilic surfaces under conditions of static exposure and laminar flow. *Biomaterials*, **2**, 156–158.

23

Biological Methods of Detecting Hazardous Substances

E. R. NESTMANN

Department of National Health and Welfare, Ottawa, Ontario, Canada

23.1 INTRODUCTION

The purpose of this chapter is to examine the field of genetic toxicology or environmental mutagenesis. This discussion will involve mutagenicity testing with microbial systems. Methods are described in general, but the emphasis is placed upon their use and the interpretation of results. Test modifications are described that enable the study of different types of samples (*e.g.* volatiles, mixtures), and several examples from the author's laboratory are presented.

Toxicology has been described as 'the basic science of poisons' (Doull *et al.*, 1980) and as such has been studied for many centuries, as noted by references in ancient writings. Genetic toxicology as a discipline, on the other hand, has been a recent development, growing from basic interests in mutagenesis which have evolved into concerns about environmental agents in foods and drugs and about pollution. The apparent relationship between mutagenesis and carcinogenesis has been largely responsible for the tremendous growth of interest and research in this field, as many investigators attempt to find short-term testing methods to identify carcinogens as mutagens (*e.g.* Ames *et al.*, 1972). This correlation between mutation and cancer, however, has actually impeded the acceptance, by regulators as well as by certain workers in the field, of mutagenicity as a toxicological entity in its own right. In spite of this tendency, the discipline of genetic toxicology has begun to gain recognition in the overall toxicological evaluation process through the efforts and recommendations of concerned geneticists (Information Letter, 1981). In contrast to cancer, which generally affects people in their later years, many genetic defects are crippling from birth, resulting in a larger loss of potential productivity and a greater financial burden upon the family and/or society. Heredity, of course, also contributes to leading causes of adult death, such as heart disease and stroke, and germinal mutations are passed on to future generations. Efforts to identify environmental mutagens which could cause increases of this genetic load should be a high priority.

23.2 METHODS IN ENVIRONMENTAL MUTAGENESIS

The overall purpose of different methods used in genetic toxicology is to detect either genetic changes or indications of DNA damage. The basic principle is to use an organism (or *in vitro* system) that can show phenotypic (or other detectable) changes resulting from insult to DNA. It is not within the scope of this chapter to list these tests, much less to discuss them in detail. Over 100 techniques have been described in the literature, ranging from treatment of isolated DNA plasmids *in vitro* (Yoon, 1982) to various methods of monitoring genetic changes in man (Bora *et al.*, 1982). However, they can be categorized on the basis of organism (*i.e.* microorganisms, insects, plants, mammals) or genetic endpoint (*e.g.* point or chromosome mutation, DNA damage or its repair). These assays are introduced in a comprehensive text by Brusick (1980) which also contains detailed test protocols.

The higher the proportions of known mutagens and carcinogens identified by a test, the greater its 'sensitivity'. On the other hand, the 'specificity' of tests in identifying mammalian mutagens and carcinogens may be reduced if more 'positives' are found in the sensitive non-mammalian or *in vitro* assays. The terms sensitivity and specificity are usually used to compare results of short-term tests with results of cancer bioassays. Thus, a test with high sensitivity shows few 'false negatives' and one with high specificity has few 'false positives' (Cooper *et al.*, 1979; Hoffmann, 1982; Purchase, 1982). This terminology is objectionable for several reasons. At times, cancer bioassay results themselves could be incorrect due to improper design or interpretation, insensitive test species, insufficient sample size *etc.* In such cases, comparisons could imply that the short-term test results are incorrect, spurious or artifactual. In other cases, there are metabolic differences which are responsible for disparate responses, *e.g.* the carcinogen *o*-toluidine which is non-mutagenic in bacteria (Ashby and Styles, 1978) and the non-carcinogen caffeine which is a bacterial mutagen (Demerec *et al.*, 1948). Thus, it is more appropriate to call such agents 'bacterial mutagens' or 'bacterial non-mutagens', rather than 'false positives' or 'false negatives'—terms which should be reserved for reponses due to artifacts (*e.g.* Harrington *et al.*, 1983).

Since no individual mutagenicity test is sufficiently sensitive to detect all mammalian mutagens and carcinogens, the recommended strategy is to combine two or more complementary assays in a test battery (Bridges, 1976). Such a combination usually includes: a microbial test, using bacteria (Ames *et al.*, 1975) or yeast (Nestmann and Lee, 1983); an assay using mammalian cells in culture (Bradley *et al.*, 1981; Hsie *et al.*, 1981; Preston *et al.*, 1981); and a short-term mammalian *in vivo* assay (Preston *et al.*, 1981). Given the wide variety of assays to choose from, many different batteries of tests can be used. The choices may depend upon the chemical class or physical state of the compound, the cost, the expertise available, and the intent of the study. For investigators who do not wish or cannot work with mammalian systems, a battery of microbial tests has been proposed (Nestmann *et al.*, 1979c).

Certain compounds have been found to require chemical modification (activation) by metabolic enzymes before exhibiting their carcinogenic effects (Miller and Miller, 1969). To accomplish this, a mammalian activating system in the form of a tissue homogenate has been incorporated into non-mammalian and *in vitro* testing procedures (Malling, 1971; Ames *et al.*, 1973). Increased levels of activating enzymes in tissue homogenates (usually of liver) are obtained when the animal used as the source is treated with inducing chemicals such as Aroclor 1254 (or phenobarbital) (Ames *et al.*, 1975). Recent work has focused on finding appropriate substitutes for enzyme-inducing polychlorinated biphenyls (PCB's, *e.g.* Aroclor) (Yoshikawa *et al.*, 1982), since they are an environmental concern and even banned in some countries. The incorporation of enzymes (S9 or microsome preparations) in mutagenicity tests allows a degree of mammalian metabolism *in vitro*; however, these preparations cannot reproduce accurate *in vivo* conditions for a variety of reasons, such as assimilation, transport, deactivation, tissue specificity *etc.* Thus, genetic activity may be detected in short-term mutagenicity tests, but mutagenic potential in man must be extrapolated at present from *in vivo* mammalian experiments that will allow quantitative estimates of risk (Nestmann and Douglas, 1981).

23.3 SELECTION OF CHEMICALS FOR TESTING

The reasons for selecting certain chemicals for testing are as varied as the types of investigators and agencies engaged in toxicological research. In general, several criteria serve as important guidelines: amount of production of the compound; demonstrated presence and persistence in the environment; potential for exposure to the compound and degree of chemical relatedness of the compound to known mutagens or carcinogens (Criteria for Selection, 1982). The primary concern is the protection of man or danger to other susceptible organisms in the environment. If human health is involved, potential mutagen sources to examine are foods (Powrie *et al.*, 1982), air (Chrisp *et al.*, 1978), water (Nestmann *et al.*, 1979b; Kool *et al.*, 1982), the workplace (Bryant and McCalla, 1982) and consumer products (Nestmann *et al.*, 1981c; 1984). Potential mutagenicity in exposed species in addition to man is also a concern with certain pollutants such as effluents from pulp and paper mills (Douglas *et al.*, 1982). The many compounds that are included in all the categories above provide the subject matter for different types of studies, such as the screening of many chemicals with a single test (Nestmann *et al.*, 1980c), or more comprehensive studies of one or more compounds with a battery of assays (Douglas *et al.*, 1981; Ellenton *et al.*, 1981).

Effects of genotoxic agents can be studied with a variety of strategies. One example is population monitoring by assaying for genetic changes in cells of people who have been subjected to therapeutic, accidental or work-related exposures, or to chemical, physical or biological agents (Bora *et al.*, 1982; Heddle, 1982). One such method involves culturing human lymphocytes from human subjects and assaying for chromosome damage manifested by gene mutation (Albertini *et al.*, 1982), sister chromatid exchange (Carrano and Moore, 1982), or chromosome aberrations (Kǔcerová, 1982). Another method is to examine sperm cells of exposed males for abnormalities (Wyrobek, 1982). The absence of genetic damage in the somatic or germinal cells does not mean that exposure has not occurred. However, the presence of such damage provides sufficient evidence that there have been biologically relevant exposures, providing that proper controls have been considered. In the latter case, target tissue specificity and lifestyle may determine whether or not mutagenic or carcinogenic effects may ultimately be expected.

Another strategy is to monitor occupational exposures by assaying concentrates of urine (Yamasaki and Ames, 1977), breast fluid (Petrakis *et al.*, 1980) or extracts of feces (Bruce *et al.*, 1977), using an *in vitro* test, to see if genotoxic agents are present in the donors. For example, an investigator might choose to monitor occupational exposures by assaying urine from groups of workers who have been subject to varying degrees of potential exposure (Falck *et al.*, 1980), while researchers pursuing possible effects of diet on development of colon cancers might, on the other hand, be more interested in fecal mutagenicity (Dion *et al.*, 1982). These studies provide evidence of the subjects' exposure to genotoxic agents, but in the absence of demonstrable effects on human cells, the relevance to the subject is not known.

23.4 MUTAGENICITY TESTING WITH MICROBIAL SYSTEMS

23.4.1 Screening with the *Salmonella*/Mammalian-Microsome Assay

23.4.1.1 The method

The *Salmonella* test for the detection of chemical mutagens was originally described by Ames (1971). The test has been the subject of considerable development in Ames' laboratory in an effort to isolate improved bacterial strains to detect a wider variety of mutagenic compounds. Although the test is widely referred to as the 'Ames Test', its proper name was established as the '*Salmonella*/mammalian-microsome assay', and shortened to the *Salmonella*/microsome assay by Ames and his co-workers in their highly cited 'methods' paper (Ames *et al.*, 1975). By that time, use of mammalian liver homogenate (S9) for metabolic activation of promutagens (Ames *et al.*, 1973) and of more sensitive tester strains containing R factor plasmids (McCann *et al.*, 1975b) had increased the sensitivity of the original test dramatically. Although the test has been modified for specific uses by many investigators (see Section 23.4.1.3), leading to published recommendations concerning its use (de Serres and Shelby, 1979a,b), the test has remained relatively unchanged.

It was estimated in 1977 (McCann and Ames, 1977) that the *Salmonella* assay was in routine use in over 1000 laboratories throughout the world. Due to the relative simplicity of the test, its low cost, the availability of the strains, and the standard protocol (Ames *et al.*, 1975; Maron and Ames, 1983), a very large data base has been established with this test, and it continues to grow with hundreds of publications appearing annually in a variety of specialized and multidisciplinary journals.

Until recently, the five tester strains recommended for use in the *Salmonella* assay by Ames *et al.* (1975) have been TA1535, TA100, TA1537, TA1538 and TA98. Various strains are used because they contain different genetic deficiencies (mutations) in the genes that control the bio-synthesis of the amino acid histidine. These His⁻ mutations revert to His⁺ by different types of mutations induced by different classes of mutagens. The advantage of a reversion assay (His⁻ to His⁺) is the ability to visualize induced His⁺ mutant colonies on a selective medium which does not allow the His⁻ 'parental' bacteria to grow into colonies. Strains TA1535 and TA100 contain the *hisG46* mutation caused by the substitution of an improper base in the *hisG* gene. Two strains with the same mutant gene are used for detection of agents causing base substitution mutations because TA100 also contains plasmid pKM101 which enhances mutability. An analogous pair of strains, TA1538 and TA98, contains the frameshift *hisD3052* mutation with a repetitive (–C–G–C–G–C–G–C–G–) sequence nearby (Ames *et al.*, 1975). Histidine independence arises by either an addition or deletion of the proper numbers of bases to restore the reading frame. These frameshift mutations are promoted seemingly by shifted base pairing in the repetitive sequences. Strain TA1537 contains an added C in a run of six cytosines resulting in the *hisC3076* mutation, also reverted by frameshift mutagens, but not necessarily the same compounds that revert the *hisD3052* mutation in strains TA1538 and TA98. Ames *et al.* (1975) recommended that all five of these strains be used, but TA1538 could be deleted for screening purposes since TA98 was more sensitive for all mutagens they had tested. However, it has been found since that some compounds are more mutagenic in strain TA1538 compared to TA98 (Nestmann *et al.*, 1979a; Nestmann and Kowbel, 1979; Nestmann *et al.*, 1980b, 1981a), as shown in Figures 1 and 2.

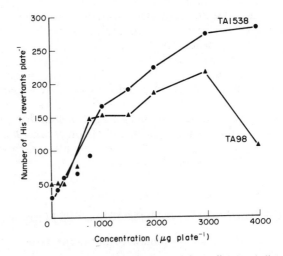

Figure 1 Dose-related mutagenicity of the dye rhodamine B in the *Salmonella*/mammalian-microsome assay. Materials and methods are the same as reported previously (Nestmann *et al.*, 1979a). Mutagenic activity of rhodamine B was dependent upon the presence of S9 mix

Ames *et al.* (1975) suggested that a compound is non-mutagenic in the test if at least 0.5 mg was tested (solubility and cell viability permitting) and if the maximal number of revertants was less than two times the number in the control. Conversely, a compound is positive when the largest response was two times the spontaneous level with evidence of dose-related increases in mutation induction. An extensive comparison of this 'modified two-fold rule' with other statistical analyses has shown that the former method of interpreting results is preferable (Chu *et al.*, 1981). The results for strain TA1538 and TA98 are clearly positive in Figures 1 and 2, but the positive response of strain TA1537 in Figure 2 is not as pronounced. Figures 1 and 2 also serve as

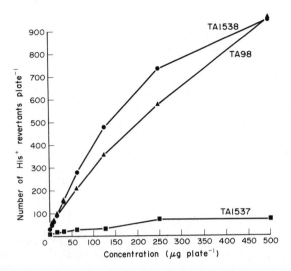

Figure 2 Dose-related mutagenicity of the dye rhodamine 6G in *Salmonella*. Materials and methods were described previously (Nestmann *et al.*, 1979a). S9 mix was required for mutagenic activity

examples of how mutagenicity data are presented; in this case, responses of two rhodamine dyes (B and 6G) in frameshift detectable strains of *Salmonella*. It is evident from these figures that numbers of mutants, or His$^+$ revertants, per Petri dish (or plate) increase as the doses of mutagen added to the plates increase. Figure 1 shows that toxic doses of the test chemical cause the numbers of mutants to decrease as the bacteria (parental and revertant) are killed. On the other hand, a toxic dose was not found for the test compound in Figure 2, and the maximal dose tested was at the limit of its solubility.

The most recent developments in the *Salmonella* test involve the construction and validation of two new tester strains; TA97 (Levin *et al.*, 1982b) and TA102 (Levin *et al.*, 1982a). Strain TA97 detects the same array of frameshift mutagens as TA1537, since it also contains an extra C in a run of six cytosines at a different mutational site (locus) *hisD6610*. Strain TA97 contains plasmid pKM101, is more sensitive, detects additional mutagens and has, therefore, been recommended as the replacement for TA1537 in the standard battery of *Salmonella* tester strains (Levin *et al.*, 1982b). Strain TA102, containing pKM101 and a second multi-copy plasmid with the His$^-$ mutation, enables detection of a range of oxidative mutagens that were not detected previously in the *Salmonella* assay. Therefore, Levin *et al.* (1982a) recommended its addition to the set of tester strains. In a recent revision of the 1975 'methods' paper (Ames *et al.*, 1975), Maron and Ames (1983) have recommended the use of tester strains TA97, TA98, TA100 and TA102 in the standard *Salmonella* assay.

23.4.1.2 Interpretation of results

Ames' original intention in designing the *Salmonella* test system was to provide an effective and inexpensive method, using bacteria, to identify chemicals that could be mutagenic or carcinogenic in man (Ames, 1971). Many bacterial mutagens are carcinogenic and one of Ames' endeavours in 'validating' the test has been to show high correlations between *Salmonella* mutagenicity and animal carcinogenicity (McCann *et al.*, 1975a; McCann and Ames, 1976). The percent correlations vary depending upon choice of compounds to be tested (Purchase, 1982). However, as the test has become more accepted and popular, the misconception continually appears, mostly but not exclusively in the popular press, that *Salmonella* 'positive' compounds are (necessarily) carcinogenic. It is imperative to recognize the meaning of test results, and the *Salmonella* assay is just one example. As mentioned previously (Section 23.2), there are many reasons for different results in different species and test systems. It is usually safe to say that a given chemical is positive or negative in a certain assay performed with a specific protocol. Such results indicate that these responses might be found in man, but some form of acceptable whole mammal test is required (Nestmann and Douglas, 1981).

23.4.1.3 *Modification of the assay for detection of selected chemicals*

(i) *Pre-incubation*

Perhaps the most useful assay modification that was not developed by Ames' group itself is the widely used pre-incubation method (Yahagi *et al.*, 1975), which enables detection of several mutagens (*e.g.* nitrosamines) otherwise undetectable in the standard plate incorporation assay (Yahagi *et al.*, 1977). This modification involves a preliminary incubation step in liquid suspension (bacterial cells, test chemical and S9 mix), addition of soft agar after 20 to 30 min, and finally pouring this mixture as an overlay on an agar plate as in the standard assay. The increased sensitivity of this protocol may be due to better interaction between active metabolites and cells in the absence of the agar matrix. This type of experiment does not involve a net increase in time to perform, but the logistics of timing and coordinating the pre-incubation step adds a step of slightly more complexity. Not only has the 'pre-incubation' assay been used routinely for evaluation of individual chemicals (Prival *et al.*, 1982), but it has been used for screening of large numbers of compounds (Haworth *et al.*, 1983).

(ii) *Exposure chambers for volatile liquids and gases*

Recognizing that a plate incorporation procedure was not suitable for the testing of liquids with low boiling points or of gases, Ames *et al.* (1975) suggested the use of sealed desiccators as exposure chambers, as performed earlier by investigators testing vinyl chloride (Rannug *et al.*, 1974; Bartsch *et al.*, 1975). A more elaborate container for such mutagenicity studies was described in a more recent report (Barber *et al.*, 1981). One of the most extensive surveys using the desiccator modification of the *Salmonella* assay involved the screening of 71 contaminants found in drinking water (Simmon *et al.*, 1977). Many of these compounds were volatile liquids or gases and, hence, were tested in desiccators adapted for evacuation of air as the sample was introduced into the chamber's atmosphere. This procedure was adapted further by adding a sampling port in the desiccator to allow for periodic measurements of the atmosphere inside as part of studies of the mutagenicity of paint removing products containing dichloromethane (Nestmann *et al.*, 1981c). If sampling is not performed, only the dose delivered is known and the precise exposure level is unknown; data are presented as shown in Figure 3. This example shows that different amounts of a liquid paint remover caused dose-related increases in mutant production in three *Salmonella* tester strains exposed within desiccators. Since dichloromethane (methylene chloride) is the principal component of paint removers and its mutagenic specificity is the same as the paint removers (Table 1), it was concluded that dichloromethane is responsible for the mutagenicity of paint removing products containing that chemical (Nestmann *et al.*, 1981c). For the sake of comparing results with other findings or to determine the relevance of these doses with respect to measured amounts in the air of work-places or homes where such products are used, it is preferable to know the actual exposure levels within the desiccators. Figure 4 shows additional mutagenicity data as an example of how they can be expressed as a function of the levels of dichloromethane to which the bacteria were exposed, as measured by gas chromatography. Exposures have been determined for some chemicals in an even more rigorous fashion as the dose which actually interacts with the genetic material of the target call, so-called molecular dosimetry (Aaron, 1976; Lee, 1976).

(iii) *Co-mutagens*

Co-mutagens are non-mutagens which, in combination with other compounds with no (or low) mutagenicity, contribute to (increased) mutagenic effects. Much of the work with co-mutagens has been performed in the laboratory of T. Sugimura. In the testing of certain compounds (aniline, *o*-toluidine, yellow OB, *N,N*-diphenylnitrosamine), addition of norharman (*β*-carboline), a non-mutagenic product of tryptophan or protein pyrolysis, was responsible for changing negative to mutagenic responses (Nagao *et al.*, 1977; Sugimura *et al.*, 1980). Subsequently, Ashby and Styles (1978) offered an explanation for this observed co-mutagenicity as the provision of an alternate substrate (norharman) for enzymes that are required for the metabolic conversion of these promutagens into active mutagens. Other compounds (azo dyes, trypan blue and ponceau R) were mutagenic only when riboflavin was included in the S9 mix (Matsushima *et al.*, 1980). For certain azo compounds, riboflavin enhances the reduction of azo bonds in the formation of free aromatic amines which are mutagenic in the presence of S9 for further activation. However, riboflavin-enhanced azo reduction can lead to a loss of mutagenicity for other compounds (Matsushima *et al.*, 1980; Ashby *et al.*, 1983). Prival and Mitchell (1982) found that riboflavin 5′-phosphate was more satisfactory than riboflavin for azo reduction due to its higher solubility in water.

Table 1 Mutagenicity of Dichloromethane in
Salmonella Bacteria in Desiccators as Exposure
Chambers[a]

Dose	*His+ revertants plate*[-1b]		
ml desiccator[-1]	TA1535	TA100	TA98
Control[c]	13	154	32
0.15	15	268	43
0.30	24	(401)	(73)
0.60	25	(789)	(138)
1.20	(34)	(1084)	(164)
Control[d]	(33)	(823)	(159)

[a] The materials and methods for this experiment were the same as described previously (Nestmann *et al.*, 1981c). [b] Results are averages of triplicate plates; values in parentheses are at least twice the negative contol. [c] The negative contol plates were enclosed in a sealed desiccator containing air. [d] The positive control plates contained 0.001 mg 4-nitroquinoline oxide.

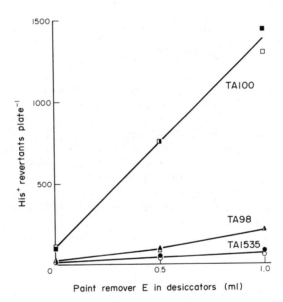

Figure 3 Dose-related mutagenicity of a liquid paint remover (identified only by code 'E') in the *Salmonella*/microsome assay. The experimental procedures using desiccators as exposure chambers are described in a previous report also dealing with these paint removing products (Nestmann *et al.*, 1981c). The open symbols are results without S9 mix and closed symbols with S9 mix

Other studies have shown that cigarette smoke condensate or urine concentrate from a smoker or a non-smoker acts comutagenically in enhancing the mutagenicity of 2-aminoanthracene (2–AA) but not benzo(a)pyrene (Hannan *et al.*, 1981; de Raat, 1982). This response was explained by: (1) the observation that maximal activation of 2–AA occurs at a much lower dose of S9 than is normally used in the *Salmonella* assay; and (2) the supposition that the co-mutagens in these cases were competing for S9, thereby reducing the amount available for activation of 2–AA and, thus, increasing its mutagenicity (de Raat, 1982).

(iv) Activation of glycosides

People are exposed to compounds that are potentially mutagenic through ingestion, inhalation, absorption, or injection. These chemicals may remain unchanged in the body, they may be metabolized into other active or inactive forms, and certain mutagens may react with carbohydrates to form non-mutagenic, conjugated glucuronides. These compounds, metabolites and conjugates are often excreted in the urine, which may be monitored for mutagenic activity (Durston and Ames, 1974; Yamasaki and Ames, 1977). The decision to treat urine concentrates with glucuroni-

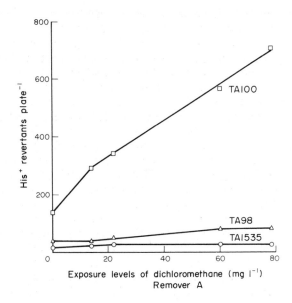

Figure 4 Dose-related mutagenicity of a liquid paint remover (identified by code 'A') in the *Salmonella*/microsome assay. The experimental procedures for the determination of mutagenicity and for the measurements of dichloromethane by gas chromatography are described fully in a previous report (Nestmann *et al.*, 1981c)

dase to release potential mutagens from non-mutagenic glucuronides appears to be somewhat controversial (Anderson, 1981). If the purpose of a study is to determine the relationship between urine mutagenicity and bladder cancer, the urine (or concentrate) itself should be tested as it is, with only the appropriate metabolic activating system (S9), which does not include glucuronidases. This is the material to which the bladder is exposed and, therefore, is the logical choice for testing. However, if the study involves monitoring urine mutagenicity to determine whether or not the subject has been exposed to mutagens, it is important to use glucuronidase or a similar enzyme to maximize the chance of mutagen detection (*e.g.* Nguyen *et al.*, 1982). In the case of smokers' urine, Yamasaki and Ames (1977) reported that no additional activity was found in the presence of a glucuronidase.

On the other hand, glycosides in the intestine may be activated to mutagenic (or carcinogenic) forms by hydrolysis, that is, be catalyzed by enzymes released by natural flora of the gut. Since the usual enzymatic preparations employed in mutagenicity testing do not contain enzymes that cleave glycosides, Tamura *et al.* (1980) have suggested the addition of fecalase, a fecal extract which contains various glycosidases, when testing dietary items for mutagenicity. Such an experimental approach would provide a more realistic estimate of actual exposure of the intestinal tract to ingested mutagens, especially of plant origin, since an enormous variety of substances are glycosides in plants (Tamura *et al.*, 1980).

(v)   Derivatives of Ames' tester strains

The strains developed through the years by Ames and his coworkers have undergone certain modifications in others' laboratories to facilitate their usefulness in testing or in the interpretation of test data. For example, from pKM101-containing strains (ampicillin and cycloheximide resistant), Whong *et al.* (1982) have constructed multi-drug resistant strains (TA98W and TA100W) which are also resistant to streptomycin and 8-azaguanine. These strains may be used in an apparatus designed to collect and trap airborne mutagens in the workplace. Contamination by other microorganisms is eliminated since the growth medium contains these four antibiotics.

Rosenkranz and coworkers have isolated and characterized several derivatives of Ames' tester strains which are deficient in bacterial nitroreductases, enzymes capable of metabolizing (reducing) nitro groups of certain compounds to proximate mutagens (McCoy *et al.*, 1981). Originally, these mutants were isolated as resistant derivatives of the Ames' tester strains which were exposed to nitrofurans or niridazole (Rosenkranz and Poirier, 1979; Rosenkranz and Speck, 1975; 1976). Comparisons of mutagenic responses of nitro compounds in the normal and nitroreductase-deficient strains have shown that the latter are still mutable by nitropyrene. Selection of another nitroreductase mutant unable to reduce nitrated pyrenes has led to the conclusion that nitroreduction that takes place in bacteria is due to a 'family' of very active enzymes. Nitroreductase activity in mammals is lower by comparison (McCoy *et al.*, 1981).

Another set of derivatives of the Ames' strains was developed to enable investigators to determine accurately what degree of killing occurs when compounds are tested in the *Salmonella* assay (Waleh *et al.*, 1982). A major drawback with the *Salmonella* test has been the lack of a suitably accurate method to make meaningful comparisons of mutagen potency on the basis of survival, essential in the estimation of the 'relative mutagenic efficiencies' of various compounds (Haynes and Eckardt, 1979). Various estimates of bacterial killing in the *Salmonella* assay have been described, by microscopic examination of the background lawn of His$^-$ bacteria (Ames *et al.*, 1975; McGregor *et al.*, 1984), or by dilution and plating after an initial preincubation in suspension (Plewa and Johnson, 1983). The method devised by Waleh *et al.* (1982) has the advantage of a nearly exact reconstruction of the procedures and conditions involved in the *Salmonella* test using two types of bacterial strains derived from Ames' testers. For each of the five formerly standard tester strains, TA1535, TA100, TA1537, TA1538 and TA98 (Ames *et al.*, 1975), there is a pair of His$^+$ and His$^-$ derivatives. The former is used to measure survival and the latter 'filler' cells to simulate the background lawn of growing cells. The assay is performed using exactly the same procedure as the standard plate incorporation test, with the exception that the series of survival plates are initially seeded with approximately 10^8 filler cells and 500 *His$^+$* cells. Although not specified by Waleh *et al.* (1982), a reasonable approach to using the accompanying toxicity assay would be to determine which strain(s) is reverted (and/or killed) by the test chemical and limit the additional work to the repeat experiment and determine survival only in the pair(s) of strains corresponding to the sensitive strain(s) identified in the initial screen.

Recently, a new plasmid (pAA95) was derived from pKM101, the plasmid present in the *Salmonella* tester strains. Plasmid pAA95 retains the properties which make pKM101 useful in testing, but it is smaller and lacks certain undesirable genes. The *Salmonella* strain (AAP311) carrying pAA95 grows faster than its analog (TA100) containing pKM101 (Potter *et al.*, 1983), and it is possible that pAA95 will be more stably maintained than pKM101. Work is in progress to replace pKM101 in the other tester strains (TA97, TA98 and TA102) with pAA95.

23.4.1.4 *Precautions for testing*

(i) *Purity of compounds*

An important consideration that is often overlooked in mutagenicity testing is the purity of the test compounds. It is possible that a positive response reported for a compound is due to contamination by impurities formed during synthesis, processing or storage. Examples of such compounds include rhodamine dyes B and 6G. An initial investigation into the mutagenicity of these dyes revealed that the commercial (not pure) dyes were mutagenic in *Salmonella* and induced DNA damage in Chinese hamster ovary (CHO) cells (Nestmann *et al.*, 1979a). That study also showed that rhodamine B, purified by thin layer chromatography (TLC), did not show the mutagenicity found with the commercial product, but the pooled impurities showed the same mutagenic response as the commercial dye. The preliminary evidence with rhodamine 6G suggested that the purified dye was just as mutagenic as the product (Nestmann *et al.*, 1979a). Subsequently, it was shown using a different solvent system in TLC analysis, that purer samples and purified rhodamine 6G were non-mutagenic and that the original product investigated did contain a mutagenic impurity (Matula *et al.*, 1982). Moreover, it was shown that the purified samples of rhodamine B that were non-mutagenic in *Salmonella*, induced gene conversion and reverse mutation in yeast (Matula *et al.*, 1982) and DNA damage, chromosome aberrations and SCE in CHO cells (Douglas *et al.*, 1980). These studies provide evidence not only of the importance of a compound's purity in evaluating its mutagenicity, but also of the advisability of using a battery of tests as discussed in a previous section.

Another example of a compound with known impurities is 1,1,1-trichloroethane (TCE). TCE has been shown to be mutagenic in *Salmonella* using desiccators as exposure chambers (Nestmann *et al.*, 1984). Because different samples of TCE showed different effects in *Salmonella*, it is possible that an impurity, not TCE, is responsible for the mutagenicity observed. Table 2 compares the composition of TCE from two suppliers, one sample (Fisher) showing higher mutagenicity than the other (Nestmann *et al.*, 1984). Both sources show traces of mutagenic compounds. The difficulty in testing a purified solvent such as TCE is the near impossibility of obtaining a pure sample without codistillation or coelution of other compounds with almost identical properties (R. Otson, Department of National Health and Welfare, Ottawa, personal communication).

Table 2 Composition of Technical Grades of 1,1,1-Trichloroethane

| Source[a] | Unique components[b] | Mutagenicity | | Reference |
		Result	Test	
Fisher	tetrachloroethylene	−	*E. coli*	Greim *et al.* (1975)
		+	*Salmonella* spot test	Černá and Kypěnová (1977)
		+	Host-mediated assay	Černá and Kypěnová (1977)
	1,1,2-trichloroethane	−[c]	*Salmonella*	Simmon *et al.* (1977)
		−	*Salmonella*	Barber *et al.* (1981)
	p-dioxane	−	*Salmonella*	Nestmann *et al.* (1984)
Aldrich	1,1-dichloroethylene	+	*E. coli*, −S9	Greim *et al.* (1975)
		+[c]	*Salmonella*, −S9	Simmon *et al.* (1977)
		+	*Salmonella* spot test	Černá and Kypěnová (1977)
		+	Host-mediated assay	Černá and Kypěnová (1977)
	1,2-dichloroethylene	−[c]	*Salmonella*	Simmon *et al.* (1977)

[a] Both sources are described in more detail in a previous report (Nestmann *et al.*, 1984)   [b] Both sources contained trichloroethylene which is mutagenic (Greim *et al.*, 1975; Simmon *et al.*, 1977)   [c] Using a desiccator assay

The question remains, however, concerning the relevance of testing a pure substance when the public is exposed to commercial products that may contain mutagenic impurities. It is important, for different reasons, to know what effects both types of preparations possess and, when describing the activity of a certain compound, to be assured that the properties described are due to that compound and not to contaminating impurities. On the one hand, hazard evaluations must be conducted with substances to which the public is actually exposed, but information on the pure chemicals is also important. If impurities are present which contribute to mutagenicity or, perhaps more interestingly, which eliminate genetic effects of the pure chemical, further study is warranted to determine the actual risks involved with exposure. It is also essential to have data on pure chemicals if structure–function relationships among closely-related chemicals are to be at all meaningful.

(ii) Solvent interactions

One of the most potentially time-consuming steps in mutagenicity testing is choosing an appropriate solvent, since most test chemicals are water insoluble. Most commercial testing laboratories deal only with coded samples or with unidentified mixtures which do not permit consultation of the literature for appropriate solvents. Rather than taking time either to consult an index, when the compound is known, or to experiment with different solvents, many investigators rely on dimethyl sulfoxide (DMSO) for dissolving and diluting test agents (Ames *et al.*, 1975). Yahagi *et al.* (1977) showed that dissolution of dimethyl nitrosamine in DMSO inactivates this mutagen, leading to a falsely negative result. Work in our laboratory has shown that both water and DMSO are inappropriate for certain chemicals, as summarized in Table 3. For example, in a screening study with coded samples, trichloroacetic acid (TCA) was reported to be positive, but several attempts in another laboratory to confirm this finding were unsuccessful. The difficulty was identified as the choice of solvent. The second laboratory, knowing the identity of the compound, used water, whereas the initial laboratory dissolved the unidentified chemical in DMSO. TCA and DMSO react to produce an unstable mutagen that subsequently disappears (Nestmann *et al.*, 1980a). Additional examples of solvent effects and influences have led to the following recommendations for a two-fold strategy to help prevent reporting of artifactual results: (1) results from initial screening experiments should be confirmed in repeat experiments using a different solvent; and (2) a different solvent should be used in confirming the findings of other investigators. If the results are similar using both solvents, the investigator will be reassured that the negative or positive effects found are due to the test agent, not to an artifactual interaction involving the solvent. Differences in results will indicate a need for further study (Nestmann *et al.*, 1985). Although some extra effort is involved in identifying a suitable second solvent, Ames' group has shown that eleven organic solvents, in addition to DMSO, are compatible with the *Salmonella* assay (Maron *et al.*, 1981). Nevertheless, the usual procedure in testing involves repeating each experiment to demonstrate reproducibility, and results with two solvents means little extra effort and will lead to much greater confidence in the veracity of the findings.

Table 3 Examples of Solvent Effects

Compound	Solvent	Result	Reference
Dimethylnitrosamine	DMSO	−	Yahagi *et al.* (1977)
	Water	+	
Dimethylcarbamyl chloride	DMSO	+	Nestmann *et al.* (1980a)
	Water	−	
	None	+	
Trichloroacetic acid	DMSO (0h)	−	Nestmann *et al.* (1980a)
	DMSO (0.2h)	+	
	DMSO (1h)	−	
	Water	−	
p-Phenylenediamine	DMSO (0h)	−	Burnett *et al.* (1982)
	DMSO (4h)	+	
	Water	−	

23.4.2 Screening for Mutagens not Detected in the *Salmonella* Test — the Microbial Battery

The *Salmonella* assay is unable to detect all mammalian mutagens and carcinogens as bacterial mutagens for various reasons, including some that have been referred to in Section 23.2. Therefore, the approach to testing using a battery of assays is logical. Test batteries often involve tissue culture or short-term animal studies—assays that will not be discussed in the present review. However, limiting the scope of assays to microbial tests does not preclude a discussion of screening chemicals with a battery approach. This section will describe certain assays using bacteria and yeast and their applications in complementing each other.

23.4.2.1 The fluctuation test for weak mutagens

Making use of the now classic fluctuation assay of Luria and Delbrück (1943) to prove the origin of spontaneous mutants, a simple fluctuation test was devised by Green *et al.* (1976) to enable detection of mutagens at lower doses than in plate tests. Another important application of this methodology is to provide more definitive results on compounds which give a questionable, or weakly positive effect in the *Salmonella* plate assay. For example, Green *et al.* (1976), working with the mutagens methylmethane sulfonate and dichlorvos, detected mutagenic effects at lower doses using the fluctuation test than with the *Escherichia coli* plate assay. They also found mitomycin C to be positive only with the fluctuation method. Nestmann *et al.* (1979c) found that lead chromate was either negative or only weakly positive (2.2 times the spontaneous control) in the *Salmonella* assay with strains TA1535 and TA100, respectively, which revert through base substitution mutation. Although this compound was also negative in the *E. coli* plate test, it was clearly mutagenic in the *E. coli* fluctuation test (Nestmann *et al.*, 1979c), using a strain which also reverts by base substitution. Due to the extra work involved with the fluctuation assay, Nestmann *et al.* (1979c) proposed that it is best used as a supplementary assay in cases of questionable results.

Although originally the fluctuation test was an *E. coli* assay, it is amenable and perhaps preferable to use with Ames' *Salmonella* strains due to the extensive engineering and validation that have taken place with these strains compared to *E. coli*. In a direct comparison between the *Salmonella* plate assay and a *Salmonella* fluctuation test, a dye (thiodiphenyl-4,4'-diazo-bissalicylic acid) was found to be mutagenic in the fluctuation test at 1/40 the dose required for mutagenicity in the plate test (at the same level of statistical significance, $p<0.01$) using strain TA98 (Nestmann *et al.*, 1981a).

Gatehouse (1978) showed that the fluctuation test could be performed using multi-well microtitre plates rather than test tube racks of tubes as originally described by Green *et al.* (1976). We have found that use of the multi-well plates generally results in a reduction of sensitivity (Kowbel, Harrington and Nestmann, unpublished), as predicted and discussed by Green *et al.* (1976).

For certain compounds, the sensitivity of the fluctuation test can be enhanced even more by using a modified incubation procedure. Kowbel and Nestmann (1981), by fortuitously leaving an experiment in an unheated incubator for 24h of the standard 72h incubation period, found that such a temperature regimen enabled the detection of mutagenicity by *N*-methyl-*N*'-nitro-*N*-nitrosoguanidine (NTG) at a concentration 1/33 of the amount required using the standard incubation protocol. It is not clear whether this temperature effect is due to reduced volatility of NTG at the lower temperature, reduced capability of DNA repair at a less optimal temperature, or some other factor(s). Lead chromate, a non-volatile mutagen, was found also to be mutagenic at a lower dose with the modified procedure, indicating that volatility may be a contributing factor but it is not solely responsible for the enhanced sensitivity of the assay.

23.4.2.2 *Detection of genetic endpoints other than bacterial mutation*

The *Salmonella* reverse mutation assay of Ames *et al.* (1975; Maron and Ames, 1983) is the most commonly used test in the field. Bacterial assays also have been developed using other organisms such as *E. coli* (Green and Muriel, 1976) or *Klebsiella pneumoniae* (Voogd *et al.*, 1974), using other methods such as the fluctuation test (Green *et al.*, 1976) and other endpoints such as forward mutation (Ames, 1971; Skopek *et al.*, 1978), recombination (Norin and Goldschmidt, 1979), DNA damage (Slater *et al.*, 1971), gene duplication (Hoffmann *et al.*, 1978) or lysogenic or SOS induction (Levine *et al.*, 1978; Ben-Gurion, 1978; Elespuru and Yarmolinsky, 1979; Quillardet *et al.*, 1982). Many of these techniques are perceived to have arisen as research curiosities unintended for widespread use as screening tests. Most investgators feel that only one bacterial assay should be included in a test battery and that the *Salmonella* assay is the method of choice due to its common usage and continual development and validation. For these and other reasons, attention has turned to the use of fungi as eukaryotic microorganisms with certain advantages for testing. These organisms feature a true nucleus and a life cycle which includes mitosis, meiosis and sporulation. Genetic endpoints that can be assayed include gene conversion, mitotic recombination, aneuploidy, and non-specific chromosome damage in addition to forward and reverse mutation *etc.* (von Borstel, 1981).

Although the filamentous fungi *Neurospora crassa* (de Serres and Malling, 1971) and *Aspergillus nidulans* (Roper, 1971) have been used in studies of environmental mutagenesis by certain investigators for a number of years, single-celled yeasts are more commonly employed as fungal assays. Zimmermann is widely regarded as a primary and sustaining influence in the conception of yeast testing and in the development of tester strains useful for screening chemicals for non-mutational genetic effects, such as gene conversion, recombination and aneuploidy (Zimmermann *et al.*, 1966; 1975). Zimmermann's strains of the budding yeast *Saccharomyces cerevisiae* and additional strains developed in other laboratories have been described and compared by von Borstel (1981). Mutation induction has also been studied using the fission yeast *Schizosaccharomyces pombe* (Nasim, 1968; Loprieno, 1978). In addition, DNA damage, as a function of differential survival between DNA repair-proficient and DNA repair-deficient strains, has been studied using an assay with both budding and fission yeasts (Nestmann *et al.*, 1982).

The utility of inclusion of a yeast assay in a battery of microbial tests has been demonstrated repeatedly with various agents. One example is actinomycin D which is non-mutagenic in the *Salmonella* assay (Benedict *et al.*, 1977) and in the *E. coli* fluctuation test (Nestmann *et al.*, 1981b), but it does induce mitotic recombination in yeast (Nestmann *et al.*, 1981b). Another study of 42 chemical constituents of pulp and paper mill effluent (Nestmann and Lee, 1983) compared two yeast tests which detected either gene conversion (Zimmermann *et al.*, 1975) or reverse mutation (von Borstel, 1981). Although none induced gene conversion in yeast, eight compounds were mutagenic, whereas only three had been positive in *Salmonella* (Nestmann *et al.*, 1980c), providing additional support of the value of yeast as an organism for mutagenicity testing.

23.4.2.3 *Relevance of data from microbial tests*

Although the relationship between carcinogenicity and short-term mutagenicity tests remains an unanswered question to experts considering the problem (ICPEMC Committee 2, 1982), considerable effort has gone into classifying tests on the basis of their predictability of carcinogenicity. Purchase (1982) has proposed criteria that can be used in the validating process for the purpose of determining the predictive values of short-term tests. Tests were classified as 'developing', 'developed', or 'established predictive' tests. Performance characteristics (sensitivity, specificity and predictive value) were calculated for developed tests in the last two categories. Only two bacterial tests, the *Salmonella* and *E. coli* WP2 assays, were considered to be established, and their predictive values ranged from 65 to 95% for the 16 different validating studies referenced (Purchase, 1982). An important factor in evaluating test performance is the prevalence of carcinogens in the number of chemicals tested—the accuracy of a positive result predicting a carcinogen, even for tests with high sensitivity and specificity, decreases dramatically as prevalence drops below 50%. This relationship is especially disturbing in the case of random screening of environmental contaminants, since most chemicals are non-carcinogens. The reason for high success rates of tests, of course, is due to bias in the selection of test chemicals with high proportions of carcinogens. In practice, the investigator hopefully will screen chemicals perceived with reasonable suspicion to have genetic activity. With limited resources, careful consideration of the properties of potential test chemicals can reduce the number of non-carcinogens (non-mutagens) tested. This results not only in a saving of time and resources but a higher predictive value of the positive results found.

23.5 TESTING COMPLEX MIXTURES

23.5.1 Studying the 'Whole'

A logical approach to testing mixtures, *i.e.* material composed of more than one compound, would appear to be an assay of its net mutagenic effect using a standard experimental protocol. This approach is useful, but it also presents certain difficulties both in test performance and in interpretation. Testing mixtures of unknown composition leads to the practical problem of selecting a suitable solvent to deliver the chemicals to the target cells. Not only is it unlikely that all compounds will be equally soluble in any solvent, but there is the potential problem of artifacts due to solvent–solute interactions and reactions (Nestmann *et al.*, 1980a; 1985). Solvent effects on mixtures are demonstrated by observations that test solutions of extracts of drinking water in dimethyl sulfoxide (DMSO) lose mutagenic activity after storage for 2 to 3 days at 4 °C (Grabow *et al.*, 1981). In addition, the same mixture, dissolved in different solvents, can have similar dose-effect curves comparing initial linear portions (Nestmann, 1983), as shown in Figure 5, or quite different responses, as shown in Figure 6, for extracts of drinking water.

Certain mixtures are sufficiently active to induce genetic effects in their unadulterated form, for example, effluent from a Canadian pulp and paper mill (Douglas *et al.*, 1982). On the other hand, the levels of chemical contaminants in drinking water are very low, leading to either negative responses or to artifactually positive results in bacterial fluctuation assays (Forster *et al.*, 1983; Harrington *et al.*, 1983).

Some environmental samples (*e.g.* urine) and most foods contain certain nutrients (*e.g.* the amino acid histidine) that may influence the growth and/or the metabolism of test organisms (*e.g.* His⁻ strains of *Salmonella* bacteria). Thereafter, in many cases, it is desirable to adsorb the traces of organic compounds onto non-polar macroreticular resins, producing a concentrated sample suitable for chemical analysis (LeBel *et al.*, 1979) or mutagenicity testing (Nestmann *et al*, 1979b). On the other hand, certain samples such as filtrates of air are largely composed of particulate materials with adsorbed organic compounds which must be extracted before testing (Pellizzari *et al.*, 1978). Typically, liquid samples such as shale oil are fractionated by liquid–liquid partitioning into basic, acidic and neutral components for mutagenic evaluation (Pelroy and Petersen, 1979). Alternatively, the liquid mixture can be split into successive fractions of polar, intermediate and non-polar constituents by use of preparative reverse-phase high-performance liquid chromatography or HPLC (Bryant and McCalla, 1982). Comparisons of these extraction methods have been made on certain types of mixtures, as reviewed by Nestmann (1983).

Whatever extraction procedure is used, the amount of resulting material usually is quite limited, leading to the adoption of certain criteria and priorities in testing, exploiting what may be known about the samples — thus eliminating non-informative aspects of the test. For example, since mutagens in extracts of Canadian drinking water are direct-acting, and the addition of S9 for metabolic activation leads to reductions in mutagenic responses (Figure 5), experiments with S9 are informative but not required (Nestmann *et al.*, 1983). Conversely, extracts of drinking water from surface sources in the Netherlands require S9 for maximal mutagenic responses (Kool *et al.*, 1981). *Salmonella* strain TA100 appears to be the most sensitive strain for detecting mutagenic activity of extracts of Canadian drinking water (Nestmann *et al.*, 1983), as shown in Figure 5, whereas strain TA1538 has been suggested for testing of concentrates of urine from cigarette smokers (Yamasaki and Ames, 1977). Therefore, for these tests the available material can be studied more comprehensively using multiple doses in one strain rather than one dose in 4 to 5 strains.

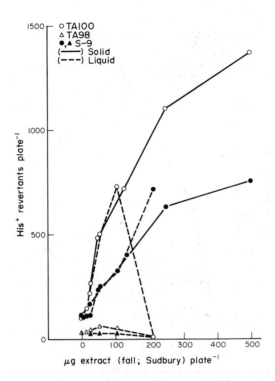

Figure 5 Dose-related mutagenicity of an extract of a drinking water sample from the treatment plant of Sudbury, Ontario, Canada. *Salmonella* strains TA100 and TA98 were used, with and without S9 for metabolic activation. The extract was prepared by running 200 l of water through an XAD–2 column and eluting the non-polar organics adsorbed to the resin with 300 ml acetone:hexane (85:15). This volume was reduced either to dryness (solid) or to 10 ml (liquid). Mutagenicity testing was performed with the liquid material and with the dry material dissolved in DMSO (Nestmann *et al.*, 1979b)

Regardless of the preparatory steps taken in the testing of mixtures, the investigator is left with the problem of interpreting the results. The net mutagenic response from a mixture is not usually a simple sum of the amounts of various mutagens present. This conclusion has been demonstrated in two different ways. Pelroy and Petersen (1978) found that the mutagenicity of a known quantity of a promutagen decreased in the presence of increasing amounts of shale oil fractions, and a similar observation was reported by Shahin and Fournier (1978). Salamone *et al.* (1979) examined the problem by studying model mixtures, in this case a pair of promutagens. They showed that the individual mutagenic effect of one compound could be reduced by adding the other. The explanation for both types of observation is concerned with availability of enzymes in the S9 for activation, as previously mentioned in the discussion of comutagenicity in Section 23.4.1.3(*iii*) and studied in detail by Haugen and Peak (1983). These studies emphasize the danger of relying exclusively on results of tests with mixtures (Shahin and Fournier, 1978). In practice, the individual components may be fractionated naturally in the environment, due to differential solubility, biodegradation *etc.*, or in the body, due to differential absorption, metabolism, bioaccumulation *etc.*

23.5.2   Studying Constituents of a Mixture

When studying the toxicity of a mixture, it is important to determine what constituent(s) is responsible. The steps in characterizing the chemical composition of a mixture involve an initial appreciation of what type of material it is (*e.g.* oil, air filtrate, effluent *etc.*) and what physical, chemical or biological treatments the material has received (*e.g.* exposure to sunlight, chlorination, microbiological degradation *etc.*). As individual chemicals are identified and their levels in the mixture are determined, literature searches often will reveal which of these compounds have been studied previously. For mutagenicity, the Environmental Mutagen Information Centre (EMIC, Oak Ridge, Tennessee, US) continues to provide valuable searches free of charge for

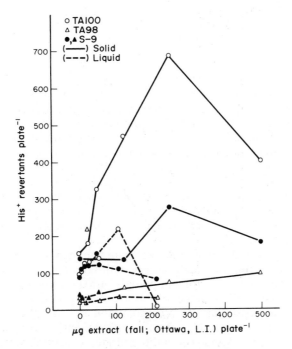

Figure 6 Dose-related mutagenicity of an extract of a drinking water sample from a treatment plant in Ottawa, Ontario, Canada. Details are given in the legend to Figure 5

clients who make such requests. For compounds with no published data, testing may be warranted, following the criteria mentioned previously in Section 23.3.

If known mutagens or carcinogens are present, regardless of whether or not the mixture's net mutagenic effect is positive, efforts should be made either to protect workers, the public and the environment from exposure or contamination, or to remove, or at least to reduce the levels of, these hazardous chemicals from the mixtures. It is practical to monitor the potential hazard of a mixture on the basis of one or a few toxic compounds. The success of alternate treatments or of clean-up procedures can be followed by such monitoring. However, caution is required since reduction in levels of certain chemicals may be accompanied by increases of unmonitored, equally-dangerous compounds.

23.5.3 Coupled Bioassay/Chemical Fractionation of Mixtures

A third approach to the study of mixtures was proposed by Loper and Lang (1978) whereby initial sample fractionation by liquid–liquid fractionation would be followed by HPLC separation of components into subfractions. Only subfractions showing mutagenic activity (in this case in the *Salmonella* assay) would be analyzed further for chemical identification. Successful applications of this approach have been reported for extracts of drinking water (Loper and Tabor, 1980) and of pulp and paper mill effluent (Douglas *et al.*, 1982). The advantage of this system is to enable the investigator to focus attention on and devote analytical capability to the mutagenic compounds. With completion of such work, one knows with greater certainty that the compounds identified do contribute to the net mutagenicity of the mixture. One disadvantage is that, by separating components, antagonistic and synergistic interactions may be overlooked since the compounds responsible for these effects could be present in subfractions not chosen for further study. Another difficulty with this approach, given that the effects of mutagens may be masked by other constituents, is that subsequent emphasis could be placed on only certain of the mutagens, and these might not be the most potent.

It must be reemphasized that the components of mixtures may or may not remain together when people or other organisms are exposed. Therefore, predicted effects of mixtures due to synergisms would be exaggerated, but more importantly, effects due to antagonisms could be severely underestimated. These difficulties in interpretation support a combined strategy in the testing of mixtures, using all three approaches discussed in this section, as demonstrated in a recent report by Douglas *et al.* (1982).

23.6 SUMMARY

The field of genetic toxicology has evolved from concern about the mutagenic properties of environmental agents found in foods, drugs and our surroundings. Growth of genetic toxicology has been stimulated by the quest to find short-term tests for identification of potential carcinogens as mutagens. This chapter has concentrated on the use of microbial test systems, emphasizing examples and problems within the author's range of interests and experience.

The *Salmonella*/mammalian-microsome assay, commonly referred to as the 'Ames Test', is the best known and most widely used mutagenicity test. Recently, Ames and his colleagues have developed new strains and published revised guidelines for this test. These and other developments have been described in considerable detail, emphasizing the importance of recognizing different properties of various compounds and mixtures. The importance of modifying test methods to detect the possible genetic activity of diverse substances, rather than conforming to rigid or standardized testing procedures, is emphasized.

No single mutagenicity test is capable of detecting all mutagens and potential carcinogens, and the accepted approach to testing involves a battery of different assays. For microbial testing, the fluctuation test has proven to be useful for compounds with questionable results in plate incorporation assays. The yeast assays allow detection, in a eukaryote, of genetic endpoints other than mutation, such as gene conversion, mitotic recombination and aneuploidy.

The analysis of complex mixtures for potential mutagenicity involves an integrated strategy using three approaches: chemical analysis can identify the constituents which may be tested for mutagenicity; testing of the mixture can reveal the net mutagenic effect of the material; and mutagenicity-directed fractionation can identify which fraction, if not which compound(s) is (are) responsible for the mutagenicity of the mixture. Any or all of these approaches should be used with the realization that various additive, synergistic, and/or antagonistic interactions among constituents are likely to occur.

Finally, it is important to recognize that short-term tests are useful for research, testing and monitoring, but they are not definitive in determining risk to man from various mutagenic agents. Human risk of mutation or cancer can be extrapolated only from appropriate *in vivo* mammalian bioassays.

ACKNOWLEDGEMENTS

I wish to thank my collaborators on certain aspects of the work described here, including D. J. Kowbel, E. G.-H. Lee, G. L. LeBel, R. Otson, P. Bothwell and T. Harrington; Drs. D. H. Blakey, G. R. Douglas, A. F. Hanham and A. P. Hugenholtz for valuable suggestions concerning the manuscript; and Kathy Nesbitt for typing the manuscript.

REFERENCES

Aaron, C. S. (1976). Molecular dosimetry of chemical mutagens: selection of appropriate target molecules for determining molecular dose to the germ line. *Mutat. Res.*, **38**, 303–310.

Albertini, R. J., D. L. Sylwester and E. F. Allen (1982). The 6-thioguanine-resistant peripheral blood lymphocyte assay for direct mutagenicity testing in humans. In *Mutagenicity, New Horizons in Genetic Toxicology, Cell Biology: A Series of Monographs*, ed. J. A. Heddle, pp. 305–336. Academic, New York.

Ames, B. N. (1971). The detection of chemical mutagens with enteric bacteria. In *Chemical Mutagens: Principles and Methods for Their Detection*, ed. A. Hollaender, vol. 1, pp. 267–282. Plenum, New York.

Ames, B. N., W. E. Durston, E. Yamasaki and F. D. Lee (1973). Carcinogens are mutagens: a simple test system combining liver homogenates for activation and bacteria for detection. *Proc. Natl. Acad. Sci. USA*, **70**, 2281–2285.

Ames, B. N., E. G. Gurney, J. A. Miller and H. Bartsch (1972). Carcinogens as frameshift mutagens: metabolites and derivatives of 2-acetylaminofluorene and other aromatic amine carcinogens. *Proc. Natl. Acad. Sci. USA*, **69**, 3128–3132.

Ames, B. N., J. McCann and E. Yamasaki (1975). Methods for detecting carcinogens and mutagens with the *Salmonella*/mammalian-microsome mutagenicity test. *Mutat. Res.*, **31**, 347–363.

Anderson, D. (1981). Mutagenicity testing. *Rev. Environ. Health*, **3**, 369–433.

Ashby, J. and J. A. Styles (1978). Comutagenicity, competitive enzyme substrates, and *in vitro* carcinogenicity assay. *Mutat. Res.*, **54**, 105–112.

Ashby, J., P. A. Lefevre and R. D. Callander (1983). The possible role of azoreduction in the bacterial mutagenicity of 4-dimethylaminoazobenzene (DAB) and 2 of its analogs (6BT and 5I). *Mutat. Res.*, **116**, 271–279.

Barber, E. D., W. H. Donish and K. R. Mueller (1981). A procedure for the quantitative measurement of the mutagenicity of volatile liquids in the Ames *Salmonella*/microsome assay. *Mutat Res.*, **90**, 31–48.

Bartsch, H., C. Malaveille and R. Montesano (1975). Human, rat and mouse liver-mediated mutagenicity of vinyl chloride in *S. typhimurium* strains. *Int. J. Cancer*, **15**, 429–437.

Benedict. W. F., M. S. Baker, L. Haroun, E. Choi and B. N. Ames (1977). Mutagenicity of cancer chemotherapeutic agents in the *Salmonella*/microsome test. *Cancer Res.*, **37**, 2209–2213.

Ben-Gurion, R. (1978). A simple plate test for screening colicine-inducing substances as a tool for the detection of potential carcinogens. *Mutat. Res.*, **54**, 289–295.

Bora, K. C., G. R. Douglas and E. R. Nestmann (eds.) (1982). *Progress in Mutation Research*, vol. 3. *Proceedings of the International Symposium on Chemical Mutagenesis, Human Population Monitoring and Genetic Risk Assessment*, Elsevier/North-Holland, Amsterdam.

Bradley, M. D., B. Bhuygan, M. C. Francis, R. Langenbach, A. Peterson and E. Huberman (1981). Mutagenesis by chemical agents in V79 Chinese hamster cells: A review and analysis of the literature. A report of the Gene–Tox Program. *Mutat. Res.*, **87**, 81–142.

Bridges, B. A. (1976). Short term screening tests for carcinogens. *Nature (London)*, **261**, 195–200.

Bruce, W. R., A. J. Varghese, R. Furrer and P. C. Land (1977). A mutagen in human feces. In *Origins of Human Cancer*, eds. H. H. Hiatt, J. D. Watson and J. A. Winsten, pp. 1641–1644. Cold Spring Harbor, New York.

Brusick, D. (1980). *Principles of Genetic Toxicology*. Plenum, New York.

Bryant, D. W. and D. R. McCalla (1982). Mutagenicity and lung cancer in a steel foundry environment. In *Mutagenicity, New Horizons in Genetic Toxicology*, ed. J. A. Heddle, pp. 89–115. Academic, New York.

Burnett, C., C. Fuchs, J. Corbett and J. Menkart (1982). The effect of dimethylsulfoxide on the mutagenicity of the hair dye *p*-phenylenediamine. *Mutat. Res.*, **103**, 1–4.

Carrano, A. V. and D. H. Moore (1982). The rationale and methodology for quantifying sister chromatid exchange in humans. In *Mutagenicity, New Horizons in Genetic Toxicology, Cell Biology: A Series of Monographs*, ed. J. A. Heddle, pp. 267–304. Academic, New York.

Černá, M. and H. Kypěnová (1977). Mutagenic activity of chloroethylenes analysed by screening system tests. *Mutat. Res.*, **46**, 214–215.

Chrisp, C. E., G. L. Fisher and J. E. Lammert (1978). Mutagenicity of filtrates from respirable coal fly ash. *Science*, **199**, 73–75.

Chu, K. L., K. M. Patel, A. H. Lin, R. E. Tarone, M. S. Linhart and V. C. Dunkel (1981). Evaluating statistical analyses and reproducibility of microbial mutagenicity assays. *Mutat. Res.*, **85**, 119–132.

Cooper II, J. A., R. Saracci and P. Cole (1979). Describing the validity of carcinogen screening tests. *Br. J. Cancer*, **39**, 87–89.

Criteria for Selection of Chemical Substances for Mutagenicity Testing (1982). A report of the Department of National Health and Welfare/Department of Environment, Environmental Contaminants Advisory Committee on Mutagenesis. Information Directorate, Department of National Health and Welfare, Canada.

Demerec, M., B. Wallace and E. M. Witkin (1948). The gene. *Carnegie Inst. Wash. Year Book*, **47**, 169–176.

deRaat, W. K. (1982). The synergistic effect of urine concentration on 2-aminoanthracene in the Ames test. *Cancer Lett.*, **16**, 347–351.

deSerres, F. J. and H. V. Malling (1971). In *Chemical Mutagens, Principles and Methods for Their Detection*, ed. A. Hollaender, vol. 2, pp. 311–342. Plenum, New York.

deSerres, F. J. and M. D. Shelby (1979a). The *Salmonella* mutagenicity assay; recommendations. *Science*, **203**, 563–565.

deSerres, F. J. and M. D. Shelby (1979b). Recommendations on data production and analysis using the *Salmonella*/microsome mutagenicity assay. *Mutat. Res.*, **64**, 159–165.

Dion, P. W., E. B. Bright-See, C. C. Smith and W. R. Bruce (1982). The effect of dietary ascorbic acid and α-tocopherol on fecal mutagenicity. *Mutat. Res.*, **102**, 27–37.

Douglas, G. R., C. E. Grant, R. D. L. Bell, J. M. Wytsma, E. R. Nestmann and D. J. Kowbel (1980). Effect of impurities and metabolic activation on the mutagenic activity of rhodamine B in bacterial and mammalian cells. *Environ. Mutag.*, **2**, 289–290 (abstract).

Douglas, G. R., E. R. Nestmann, C. E. Grant, R. D. L. Bell, J. M. Wytsma and D. J. Kowbel (1981). Mutagenic activity of Diallate and Triallate determined by a battery of *in vitro* mammalian and microbial tests. *Mutat. Res.*, **85**, 45–56.

Douglas, G. R., E. R. Nestmann, A. B. McKague, O. P. Kamra, E. G.-H. Lee, J. A. Ellenton, R. Bell, D. Kowbel, V. Liu and J. Pooley (1982). Mutagenicity of pulp and paper mill effluent; a comprehensive study of complex mixtures. In *Applications of Short-Term Bioassays in the Analysis of Complex Environmental Mixtures*, ed. M. Waters, S. Sandhu, J. Lewtas, L. Claxton, N. Chernoff and S. Nesnow, vol. III, pp. 431–459. Plenum, New York.

Doull, J., C. D. Klassen and M. O. Amdur (eds.) (1980). *Casarett and Doull's Toxicology: The Basic Science of Poisons*, 2nd edn. Macmillan, New York.

Durston, W. E. and B. N. Ames (1974). A simple method for the detection of mutagens in urine: studies with the carcinogen 2-acetylaminofluorene. *Proc. Natl. Acad. Sci. USA*, **71**, 737–741.

Elespuru, R. K. and M. B. Yarmolinsky (1979). A colorimetric assay of lysogenic induction designed for screening potential carcinogenic and carcinostatic agents. *Environ. Mutag.*, **1**, 65–78.

Ellenton, J. A., G. R. Douglas and E. R. Nestmann (1981). Mutagenic evaluation of 1,1,2,3-tetrachloropropene, a contaminant in pulp mill effluents, using a battery of *in vitro* mammalian and microbial tests. *Can. J. Genet. Cytol.*, **23**, 17–25.

Falck, K., M. Sorsa, H. Vainio and I. Kilpikari (1980). Mutagenicity of urine of workers in rubber industry. *Mutat. Res.*, **79**, 45–52.

Forster, R., M. H. L. Green, R. D. Gwilliam, A. Priestly and B. A. Bridges (1983). The use of the fluctuation test to detect mutagenic activity in unconcentrated samples of UK drinking waters. In *Water Chlorination, Environmental Impact and Health Effects*, ed. R. L. Jolley, W. A. Brungs, J. A. Cotruvo, R. B. Cumming, J. S. Mattice and V. A. Jacobs, vol. 4, book 2, *Environment, Health and Risk*, pp. 1189–1198. Ann Arbor Science, Ann Arbor, MI.

Gatehouse, D. (1978). Detection of mutagenic derivatives of cyclophosphamide and a variety of other mutagens in a 'microtitre'[R] fluctuation test, without microsomal activation. *Mutat. Res.*, **53**, 289–296.

Grabow, W. O. K., P. G. van Rossum, N. A. Grabow and R. Denkhaus (1981). Relationship of the raw water quality to mutagens detectable by the Ames *Salmonella*/microsome assay in a drinking water supply. *Water Res.*, **15**, 1037–1043.

Green, M. H. L. and W. J. Muriel (1976). Mutagen testing using Trp[+] reversion in *Escherichia coli*. *Mutat. Res.*, **38**, 3–32.

Green, M. H. L., W. J. Muriel and B. A. Bridges (1976). Use of a simplified fluctuation test to detect low levels of mutagens. *Mutat. Res.*, **38**, 33–44.

Greim, H., G. Bonse, Z. Radwan, D. Reichert and D. Henschler (1975). Mutagenicity *in vitro* and potential carcinogenicity of chlorinated ethylenes as a function of metabolic oxirane formation. *Biochem. Pharmacol.*, **24**, 2013–2017.

Hannan, M. A., L. Recio, P. P. Deluca and H. Enoch (1981). Comutagenic effects of 2-aminoanthracene and cigarette smoke condensate on smokers' urine in the Ames *Salmonella* assay system. *Cancer Lett.*, **13**, 203–212.

Harrington, T. R., E. R. Nestmann and D. J. Kowbel (1983). Suitability of the modified fluctuation assay for evaluating the mutagenicity of unconcentrated drinking water. *Mutat. Res.*, **120**, 97–103.

Haugen, D. A. and M. J. Peak (1983). Mixtures of polycyclic aromatic compounds inhibit mutagenesis in the *Salmonella*/microsome assay by inhibition of metabolic activation. *Mutat. Res.*, **116**, 257–269.

Haworth, S. R., T. E. Lawlor, K. Mortelmans, W. Speck and E. Zeiger (1983). *Salmonella* mutagenicity test results 250 chemicals using both Aroclor 1254 induced rate and hamster S–9. *Environ. Mutag.*, **5**, Suppl. 1, 3–142.

Haynes, R. H. and F. Eckardt (1979). Analysis of dose-response patterns in mutation research. *Can. J. Genet. Cytol.*, 277–302.

Heddle, J. A. (ed.) (1982). *Mutagenicity, New Horizons in Genetic Toxicology. Cell Biology: A Series of Monograṭ* Academic, New York.

Hoffmann, G. R. (1982). Mutagenicity testing in environmental toxicology. *Environ. Sci. Technol.*, **16**, 560A–574A.

Hoffmann, G. R., R. W. Morgan and R. C. Harvey (1978). Effects of chemical and physical mutagens on the freque of a large genetic duplication in *Salmonella typhimurium*. I. Introduction of duplications. *Mutat. Res.*, **52**, 73–80.

Hsie, A. W., D. A. Casciano, D. B. Couch, D. F. Krahn, J. P. O'Neill and B. L. Whitfield (1981). The use of Chinese hamster ovary cells to quantify specific locus mutation and to determine mutagenicity of chemicals. A report of the GENE-TOX program. *Mutat. Res.*, **86**, 193–214.

ICPEMC Committee 2 Final Report (1982). Mutagenesis testing as an approach to carcinogenesis. *Mutat. Res.*, **99**, 73–91.

Information Letter No. 599 (1981). Health Protection Branch, Health and Welfare Canada, Ottawa.

Kool, H. J., C. F. van Kreijl, J. H. van Kranen and E. de Greef (1981). The use of XAD-resins for the detection of mutagenic activity in water, I. Studies with surface water. *Chemosphere*, **10**, 85–98.

Kool. H. J., C. F. van Kreijl and B. C. J. Zoeteman (1982). Toxicology assessment of organic compounds in drinking water. *CRC Crit. Rev. Environ. Control*, **12**, 307–357.

Kowbel, D. J. and E. R. Nestmann (1981). Enhanced sensitivity of the fluctuation assay using a modified incubation procedure. *Mutat. Res.*, **91**, 33–36.

Kûcerová, M. (1982). Chromosomal aberrations induced in occupationally exposed persons. In *Mutagenicity, New Horizons in Genetic Toxicology. Cell Biology: A Series of Monographs*, ed. J. A. Heddle, pp. 241–266. Academic, New York.

LeBel, G. L., D. T. Williams, G. Griffith and F. M. Benoit (1979). Isolation and concentration of organophosphorus pesticides from drinking water at the ng/L level, using macroreticular resin. *J. Assoc. Off. Anal. Chem.*, **62**, 241–249.

Lee, W. R. (1976). Molecular dosimetry of chemical mutagens. Determination of molecular dose to the germ line. *Mutat. Res.*, **38**, 311–316.

Levin, D. E., E. Yamasaki and B. N. Ames (1982a). A new *Salmonella* tester strain, TA97, for the detection of frameshift mutagens. A run of cytosines as a mutational hot spot. *Mutat. Res.*, **94**, 315–330.

Levin, D. E., M. Hollstein, M. F. Christman, E. A. Schwiers and B. N. Ames (1982b). A new *Salmonella* tester strain (TA102) with A·T base pairs at the site of mutation detects oxidative mutagens. *Proc. Natl. Acad. Sci. USA*, **79**, 7445–7449.

Levine, A., P. L. Moreau, S. G. Sedgwick and R. Devoret (1978). Expression of a bacterial gene turned on by a potent carcinogen. *Mutat. Res.*, **50**, 29–35.

Loper, J. C. and D. R. Lang (1978). Mutagenic, carcinogenic and toxic effects of residual organics in drinking water. In *Application of Short-Term Bioassays in the Fractionation and Analysis of Complex Environmental Mixtures*, ed. M. D. Waters, S. Nesnow, J. L. Huisingh, J. S. Sandhu and L. Claxton, pp. 512–528. Plenum, New York.

Loper, J. C. and M. W. Tabor (1980). Detection of organic mutagens in water residues. In *Short-Term Bioassays in the Analysis of Complex Mixtures*, ed. M. D. Waters, S. S. Sandhu, J. L. Huisingh, L. Claxton and S. Nesnow, vol. II, pp. 155–165. Plenum, New York.

Loprieno, N. (1978). Use of yeast as an assay system for industrial mutagens. In *Chemical Mutagens, Principles and Methods for Their Detection*, ed. A. Hollaender and F. J. deSerres, vol. 5, pp. 25–53. Plenum, New York.

Luria, S. E. and M. Delbrück (1943). Mutations of bacteria from virus sensitivity to virus resistance. *Genetics*, **28**, 491–511.

Malling, H. V. (1971). Dimethylnitrosamine: formation of mutagenic compounds by interaction with mouse liver microsomes. *Mutat. Res.*, **13**, 425–429.

Maron, D. M. and B. N. Ames (1983). Revised methods for the *Salmonella* mutagenicity test. *Mutat. Res.*, **113**, 173–215.

Maron, D., J. Katzenellenbogen and B. N. Ames (1981). Compatibility of organic solvents with the *Salmonella*/microsome test. *Mutat. Res.*, **88**, 343–350.

Matsushima T., T. Sugimura, M. Nagao, T. Yahagi, A. Shirai and M. Sawamura (1980). Factors modulating mutagenicity in microbial tests. In *Short-Term Mutagenicity Test Systems for Detecting Carcinogens*, ed. K. Norpoth and R. C. Garner, pp. 271–285. Springer-Verlag, Berlin.

Matula, T. I., R. Downie, A. G. Butterfield and E. R. Nestmann (1982). Mutagenicity of rhodamine dyes B and 6G and their impurities in *Salmonella* and *Saccharomyces cerevisiae*. *Environ. Mutag.*, **4**, 378–379 (abstract).

McCann, J., E. Choi, E. Yamasaki and B. N. Ames (1975a). Detection of carcinogens as mutagens in the *Salmonella*/microsome test: assay of 300 chemicals. *Proc. Natl. Acad. Sci. USA*, **72**, 5135–5139.

McCann, J., N. E. Spingarn, J. Kobori and B. N. Ames (1975b). Detection of carcinogens as mutagens: bacterial tester strains with R factor plasmids. *Proc. Natl. Acad. Sci. USA*, **72**, 979–983.

McCann, J. and B. N. Ames (1976). Detection of carcinogens as mutagens in the *Salmonella*/microsome test: assay of 300 chemicals: Discussion. *Proc. Natl. Acad. Sci. USA*, **73**, 950–954.

McCann, J. and B. N. Ames (1977). The *Salmonella*/microsome mutagenicity test: predictive value for animal carcinogenicity. In *Origins of Human Cancer*, ed. H. H. Hiatt, J. D. Watson and J. A. Winsten, pp. 1431–1450. Cold Spring Harbor, New York.

McCoy, E. C., H. S. Rosenkranz and R. Mermelstein (1981). Evidence for the existence of a family of bacterial nitroreductases capable of activating nitrated polycyclics to mutagens. *Environ. Mutag.*, **3**, 421–427.

McGregor, D., R. D. Prentice, M. McConville, Y. J. Lee and W. J. Caspary (1984). Reduced mutant yield at high doses in the *Salmonella*/activation assay: The cause is not always toxicity. *Environ. Mutag.*, **6**, 545–557.

Miller, J. A. and E. C. Miller (1969). The metabolic activation of carcinogenic amines and amides. *Prog. Exp. Tumor Res.*, **11**, 273–301.

Nagao, M., T. Yahagi, M. Honda, Y. Seino, T. Matsushima and T. Sugimura (1977). Demonstration of mutagenicity of aniline and *o*-toluidine by norharman. *Proc. Jpn. Acad.*, **53B**, 34–37.

Nasim, A. (1968). Repair-mechanisms and radiation-induced mutations in fission yeast. *Genetics*, **59**, 327–333.

Nestmann, E. R. and D. J. Kowbel (1979). Mutagenicity in *Salmonella* of rhodamine WT, a dye used in environmental water-tracing studies. *Mutat. Res.*, **68**, 389–392.

Nestmann, E. R. (1983). Mutagenic activity of drinking water. In *Carcinogens and Mutagens in the Environment*, ed. H. F. Stich, vol. III, *Naturally Occurring Compounds: Epidemiology and Distribution*, pp. 137–147. CRC Press, Boca Raton, FL.

Nestmann, E. R. and E. G.-H. Lee (1983). Mutagenicity of constituents of pulp and paper mill effluent in growing cells of *Saccharomyces cerevisiae*. *Mutat. Res.*, **119**, 273–280.

Nestmann, E. R., G. R. Douglas, T. I. Matula, C. E. Grant and D. J. Kowbel (1979a). Mutagenic activity of rhodamine dyes and their impurities as detected by mutation induction in *Salmonella* and DNA damage in Chinese hamster ovary cells. *Cancer Res.*, **39**, 4412–4417.

Nestmann, E. R., G. L. LeBel. D. T. Williams and D. J. Kowbel (1979b). Mutagenicity of organic extracts from Canadian drinking water in the *Salmonella*/mammalian-microsome assay. *Environ. Mutag.*, **1**, 337–345.

Nestmann, E. R., T. Matula, G. Douglas, K. Bora and D. Kowbel (1979c). Detection of the mutagenic activity of lead chromate using a battery of microbial tests. *Mutat. Res.*, **66**, 357–365.

Nestmann, E. R., I. Chu, D. J. Kowbel and T. I. Matula (1980a). Short-lived mutagen in *Salmonella* produced by reaction of trichloroacetic acid and dimethyl sulphoxide. *Can. J. Genet. Cytol.*, **22**, 35–40.

Nestmann, E. R., D. J. Kowbel and J. Ellenton (1980b). Mutagenicity in *Salmonella* of fluorescent dye tablets used in water tracing. *Water Res.*, **14**, 901–902.

Nestmann, E. R., E. G.-H. Lee, T. I. Matula, G. R. Douglas and J. C. Mueller (1980c). Mutagenicity of constituents identified in pulp and paper mill effluents using the *Salmonella*/mammalian-microsome assay. *Mutat. Res.*, **79**, 203–212.

Nestmann, E. R. and G. R. Douglas (1981). New tools for assessing mutagenic and carcinogenic risk of municipal and industrial wastewaters. In *Advances in Biotechnology*, ed. M. Moo-Young and C. W. Robinson, vol. 2, pp. 649–658. Pergamon, Toronto.

Nestmann, E. R., D. J. Kowbel and J. A. Wheat (1981a). Mutagenicity in *Salmonella* of dye used by defence personnel for the detection of liquid chemical warfare agents. *Carcinogens.*, **2**, 879–883.

Nestmann, E. R., A. Nasim, R. Haynes and D. J. Kowbel (1981b). Genetic activity of actinomycin D in *Saccharomyces cerevisiae* but not in *Escherichia coli*. *Mutat. Res.*, **89**, 229–236.

Nestmann, E. R., R. Otson, D. Williams and D. J. Kowbel (1981c). Mutagenicity of paint removers containing dichloromethane. *Cancer Lett.*, **11**, 295–302.

Nestmann, E. R., E. R. Stephen, D. J. Kowbel and A. Nasim (1982). Differential survival as an indicator of potential mutagenicity using repair deficient strains of *Saccharomyces cerevisiae* and *Schizosaccharomyces pombe*. *Can. J. Genet. Cytol.*, **24**, 771–775.

Nestmann, E. R., R. Otson, G. L. LeBel. D. T. Williams, E. G.-H. Lee and D. C. Biggs (1983). Correlation of water quality parameters with mutagenicity of drinking water extracts. In *Water Chlorination, Environmental Impact and Health Effects*, ed. R. L. Jolley, W. A. Brungs, J. A. Cotruvo, R. B. Cumming, J. S. Mattice and V. A. Jacobs, vol. 4, book 2, *Environment, Health and Risk*, pp. 1151–1164. Ann Arbor Science, Ann Arbor, MI.

Nestmann, E. R., R. Otson, D. J. Kowbel, P. D. Bothwell and T. R. Harrington (1984). Mutagenicity in a modified *Salmonella* assay of fabric-protecting products containing 1,1,1-trichloroethane. *Environ. Mutag.*, **6**, 71–80.

Nestmann, E. R., G. R. Douglas, D. J. Kowbel and T. R. Harrington (1985). Solvent interactions with test compounds and recommendations for testing to avoid artifacts. *Environ. Mutag.*, **7**, 163–170.

Nguyen, T. V., J. C. Theiss and T. S. Matney (1982). Exposure of pharmacy personnel to mutagenic antineoplastic drugs. *Cancer Res.*, **42**, 4792–4796.

Norin, A. J. and E. P. Goldschmidt (1979). Effect of mutagens, chemotherapeutic agents and defects in DNA repair genes on recombination in F′ partial diploid *Escherichia coli*. *Mutat. Res.*, **59**, 15–26.

Pellizzari, E. D., L. W. Little, C. Sparacino, T. J. Hughes, L. Claxton and M. D. Waters (1978). Integrating microbiological and chemical testing into the screening of air samples for potential mutagenicity. In *Application of Short-Term Bioassays in the Fractionation and Analysis of Complex Environmental Mixtures*, ed. M. D. Waters, S. Nesnow, J. Huisingh, S. S. Sandhu and L. Claxton, pp. 331–351. Plenum, New York.

Pelroy, R. A. and M. R. Petersen (1978). Mutagenicity of shale oil compounds. In *Application of Short-Term Bioassays in the Fractionation and Analysis of Complex Environmental Mixtures*, ed. M. D. Waters, S. Nesnow, J. L. Huisingh, S. S. Sandhu and L. Claxton, pp. 464–475. Plenum, New York.

Pelroy, R. A. and M. R. Petersen (1979). Use of Ames test in evaluation of shale oil fractions. *Environ. Health Perspect.*, **30**, 191–203.

Petrakis, N. L., C. A. Maack, R. E. Lee and M. Lyon (1980). Mutagenic activity in nipple aspirates of human breast fluid. *Cancer Res.*, **40**, 188–189.

Plewa, J. J. and G. V. Johnson (1983). The effects of estimated dose and toxicity on statistical analysis of dose-response curves for the *Salmonella* assay. *Environ. Mutag.*, **5**, 394 (abstract).

Potter, A. A., G. Selvaraj, E. R. Nestmann and V. N. Iyer (1983). Improved Salmonella tester strain carrying a pKM101-derived plasmid. *Mutat. Res.*, **122**, 7–12.

Powrie, W. D., C. H. Wu, M. P. Rosin and H. F. Stich (1982). Mutagens and carcinogens in food. In *Progress in Mutation Research*, vol. III, *Chemical Mutagenesis, Human Population Monitoring and Genetic Risk Assessment*, ed. K. C. Bora, G. R. Douglas and E. R. Nestmann, pp. 187–199. Elsevier Biomedical, Amsterdam.

Preston, R. J., W. Au, M. A. Bender, J. G. Brewen, A. V. Carrano, J. A. Heddle, A. F. McFee, S. Wolff and J. S. Wassom (1981). Mammalian *in vivo* and *in vitro* cytogenetic assays: A report of the U.S. EPA's Gene-Tox Program. *Mutat. Res.*, **87**, 143–188.

Prival, M. J., S. J. Bell, V. D. Mitchell, M. D. Peiperl and V. L. Vaughn (1984). Mutagenicity of benzidine and benzidine-congener dyes and selected monoazo dyes in a modified *Salmonella* assay. *Mutat. Res.*, **136**, 33–47.

Purchase, I. F. H. (1982). ICPEMC Working Paper 2/6. An appraisal of predictive tests for carcinogenicity. *Mutat. Res.*, **99**, 53–71.

Quillardet, P., O. Huisman, R. D'ari and M. Hofnung (1982). SOS chromotest, a direct assay of induction of an SOS function in *Escherichia coli* K-12 to measure genotoxicity. *Proc. Natl. Acad. Sci. USA*, **79**, 5971–5975.

Rannug, U., A. Johansson, C. Ramel and C. A. Wachtmeister (1974). The mutagenicity of vinyl chloride after metabolic activation. *Ambio*, **3**, 194–197.

Roper, J. A. (1971). Aspergillus. In *Chemical Mutagens, Principles and Methods for Their Detection*, ed. A. Hollaender, vol. 2, pp. 343–363. Plenum, New York.

Rosenkranz, H. S. and L. A. Poirier (1979). An evaluation of the mutagenicity and DNA modifying activity in microbial systems of carcinogens and non-carcinogens. *J. Natl. Cancer Inst.*, **62**, 873–892.

Rosenkranz, H. S. and W. T. Speck (1975). Mutagenicity of metronidazole: activation by mammalian liver microsomes, *Biochem. Biophys. Res. Commun.*, **66**, 520–525.

Rosenkranz, H. S. and W. T. Speck (1976). Activation of nitrofurantoin to a mutagen by rat liver nitroreductase. *Biochem. Pharmacol.*, **25**, 1555–1556.

Salamone, M. F., J. A. Heddle and M. Katz (1979). The use of the *Salmonella*/microsome assay to determine mutagenicity in paired chemical mixtures. *Can. J. Genet. Cytol.*, **21**, 101–107.

Shahin, M. M. and F. Fournier (1978). Suppression of mutation induction and failure to detect mutagenic activity with Athabasca tar sand fractions. *Mutat. Res.*, **58**, 29–34.

Simmon, V. F., K. Kauhanen and R. G. Tardiff (1977). Mutagenic activity of chemicals identified in drinking water. In *Progress in Genetic Toxicology, Developments in Toxicology and Environmental Science*, ed. D. Scott, B. A. Bridges and F. H. Sobels, vol. 2, pp. 249–258. Elsevier/North-Holland, Amsterdam.

Skopek, T. R., H. L. Liber, J. J. Krolewski and W. G. Thilly (1978). Quantitative forward mutation assay in *Salmonella typhimurium* using 8-azaguanine resistance as a genetic marker. *Proc. Natl. Acad. Sci. USA*, **75**, 410–414.

Slater, E. E., M. D. Anderson and H. S. Rozenkranz (1971). Rapid detection of mutagens and carcinogens. *Cancer Res.*, **31**, 970–973.

Sugimura, T. (1982). Mutagens, carcinogens, and tumor promoters in our daily food. *Cancer*, **49**, 1970–1984.

Sugimura, T., K. Wakabayashi, M. Yamada, M. Nagao and T. Fujino (1980). In *Mechanisms of Toxicity and Hazard Evaluation*, ed. B. Holmstedt, R. Lauwerys, M. Mercier and M. Roberfroid, pp. 205–217. Elsevier/North-Holland Biomedical, Amsterdam.

Tamura, A., C. Gold, A. Ferro-Luzzi and B. N. Ames (1980). Fecalase: a model for activation of dietary glycosides to mutagens by intestinal flora. *Proc. Natl. Acad. Sci. USA*, **77**, 4961–4965.

von Borstel, R. C. (1981). The yeast *Saccharomyces cerevisiae*: an assay organism for environmental mutagens. In *Short-Term Tests for Chemical Mutagens*, ed. H. F. Stich and R. H. C. San, pp. 161–174. Springer, New York.

Voogd, C. E., J. J. van der Stel and J. J. J. A. A. Jacobs (1974). The mutagenic action of nitroimidazoles. I. Metronidazole, nimorazole, dimetridazole and ronidazole. *Mutat. Res.*, **26**, 483–490.

Waleh, N. S., S. J. Rapport and K. Mortelmans (1982). Development of a toxicity test to be coupled to the Ames *Salmonella* assay and the method of construction of the required strains. *Mutat. Res.*, **97**, 247–256.

Whong, W.-Z., W. G. Sorenson, J. Stewart and T. Ong (1982). Development of an *in situ* microbial test system for airborne mutagens. *Environ. Mutag.*, **4**, 416 (abstract).

Wyrobek, A. J. (1982). Sperm assays as indicators of chemically induced germ-cell damage in man. In *Mutagenicity, New Horizons in Genetic Toxicology. Cell Biology: A Series of Monographs*, ed. J. A. Heddle, pp. 337–350. Academic, New York.

Yahagi, T., M. Degawa, Y. Seino, T. Matsushima, M. Nagao, T. Sugimura and Y. Hashimoto (1975). Mutagenicity of carcinogenic azo dyes and their derivatives. *Cancer Lett.*, **1**, 91–96.

Yahagi, T., M. Nagao, Y. Seino, T. Matsushima, T. Sugimura and M. Okada (1977). Mutagenicities of *N*-nitrosamines in *Salmonella*. *Mutat. Res.*, **48**, 121–130.

Yamasaki, E. and B. N. Ames (1977). Concentration of mutagens from urine by adsorption with the nonpolar resin XAD–2: cigarette smokers have mutagenic urine. *Proc. Natl. Acad. Sci. USA*, **74**, 3555–3559.

Yoon, K. (1982). Localized mutagenesis of the tetracycline promotor region in pBR322 by 4,5′,8–trimethylpsoralen. *Mutat. Res.*, **93**, 253–262.

Yoshikawa, K., T. Nohimi, R. Miyata, M. Ishidate Jr., N. Ozawa, M. Isobe, T. Watabe, T. Kada and T. Kawachi (1982). Differences in liver homogenates from Donryu, Fisher, Sprague-Dawley and Wistar strains of rat in the drug-metabolizing enzyme assay and the *Salmonella*/hepatic S9 activation test. *Mutat. Res.*, **96**, 167–186.

Zimmermann, F. K., R. Schwaier and U. von Laer (1966). Mitotic recombination induced in *Saccharomyces cerevisiae* with nitrous acid, diethylsulfate and carcinogenic, alkylating nitrosamide. *Z. Vererbungsl.*, **98**, 230–246.

Zimmermann, F. K., R. Kern and H. Rosenberger (1975). A yeast strain for simultaneous detection of induced mitotic crossing over, mitotic gene conversion and reverse mutation. *Mutat. Res.*, **28**, 381–388.

24

Laboratory and Equipment Design for Containment of Biohazards

K. ALLNER

Public Health Laboratory Service, Porton Down, Salisbury, Wiltshire, UK

24.1 INTRODUCTION

During the last decade an upsurge of interest in biotechnology has been matched with equal attention and concern as to how activities in this field can be performed safely. In an attempt to secure the health, safety and welfare of people at work, legislation was passed in the United States in the form of the Occupational Safety and Health Act (1970) and in Great Britain with the Health and Safety at Work Etc. Act (1974). Basic criteria for the design of laboratory facilities and equipment for work involving hazardous biological materials must take the requirements of such legislation into consideration. Laboratory personnel, the community and environment must be protected against the hazards which may be associated with the use of microbial cells and their products. Although it has been stated that it is unlikely that dangerous pathogens will be used on a large industrial scale (Sargeant and Evans, 1979), the potential for microorganisms to infect personnel or contaminate the environment must not be underestimated. The best form of protection of personnel is effected by physically separating them from the microorganisms with which they work. This is known as containment (Melling and Allner, 1981). Appropriate physical containment measures for work involving pathogens or genetically manipulated microorganisms will continue to be the primary defence against infection of laboratory personnel (National Institutes of Health, 1978; Centers of Disease Control 1981) because it is generally accepted that the health hazards associated with the applications of biotechnology are still unknown.

Various techniques are summarised by Zaugg and Swarz (1982): they include recombinant DNA research and its application to industrial processes in industries such as energy, mining, pollution control and the pharmaceutical industry (interferon, insulin, human Growth Hormone).

As scale-up operations are becoming technically feasible, regulatory bodies on both sides of the Atlantic are making their position clear. In September 1980 in the USA, the Recombinant DNA Advisory Committee (RAC) of the National Institutes of Health (NIH) recommended that an *ad hoc* working group should be established to 'advise on procedures pertaining to large scale operations' and the Large Scale Review Working Group was set up. As from April 1982, procedures involving manipulation and growth of recombinant DNA organisms in the USA in greater volumes than 10 l are covered by NIH guidelines (Milewski, 1982). Responsibility for reviewing protocols, evaluation of host–vector systems, inspection of physical facilities and setting containment levels will rest with the local Institutional Biosafety Committees. In the UK, the Genetic Manipulation Advisory Group in GMAG Note 12 (revised) (Working Party, 1976) makes it clear that similar responsibilities will be placed on local biological safety committees. Statutory provision is made for the inspection of premises by the Health and Safety Executive.

24.2 BUILDING DESIGN

24.2.1 General Building Design and Layout

A critical appraisal of the specific needs of the laboratory will determine certain design features. The requirements of primary and secondary containment must be clearly defined. Usually, there is a need for higher containment in the early stages of processing, in laboratories operating under negative pressure. The finishing stages of the product may require clean room standards under positive pressure as illustrated by the installation depicted in Figure 1. A simple layout is shown in Figure 2.

A more sophisticated arrangement may comprise at least two floors with a fermenter hall forming the focal point of the laboratory complex, surrounded by satellite laboratories where downstream processing would be carried out. Figure 3 shows a modern fermenter hall in use at G. D. Searle and Co. Ltd. Undoubtedly, there would be a requirement for a general microbiology laboratory, another where genetic manipulation work might be conducted, a biochemical laboratory, a dark room, and a cold room with storage facilities for the final products awaiting despatch. A schematic arrangement is shown in Figure 4. It may be desirable for the fermentation procedures and subsequent processing to be, as far as possible, computer controlled and a control room should be incorporated in the design plan.

24.2.2 Access

A reception area, with perhaps an outer office, would be necessary for receiving visitors and to provide a means of access for laboratory staff. Some form of security arrangement should be

Figure 1 View of the sterile vaccine blend area maintained at a slight positive pressure above ambient, showing HEPA filters in the ceiling, double ended autoclave on the left and vessels designed for the total exclusion of contaminating microorganisms. (Photograph courtesy of Hoechst Pharmaceuticals Manufacturers, Milton Keynes, Bedfordshire, UK)

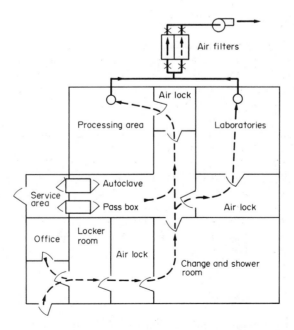

Figure 2 Simplified diagram of general building layout. The arrows show direction of air flow

Figure 3 Technicians adjusting a heat exchanger (ground floor) and checking the fermentation vessels (first floor) in the fermenter hall at Searle Research and Development, Biological Development Pilot Plant. (Photograph by kind permission of Searle Research and Development Division of G. D. Searle and Co. Ltd., High Wycombe, Buckinghamshire, UK)

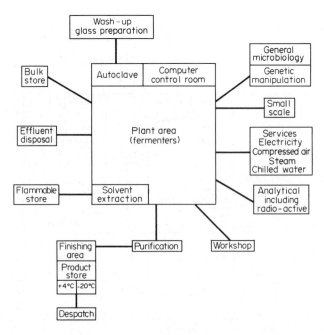

Figure 4 A sophisticated laboratory complex with the fermenter hall as the focal point showing connecting satellite laboratories and support areas

employed so that only authorized personnel may enter the laboratory complex. A card key system, combination security lock or entry by a computer controlled system might be considered. A service area should be included for the delivery of supplies and despatch of the finished products. A locker room may be situated adjacent to the office where outer clothing may be removed, and from here staff may proceed through an airlock, with interlocking doors, into the change and shower area. Depending upon the nature of the work to be undertaken, staff should then don suitable protective clothing and proceed through another airlock into the laboratories. A pass box situated near the service area enables media, chemicals and laboratory apparatus to be brought into the laboratory. The pass box is equipped for surface decontamination, and waste materials may be passed out this way or through a double ended autoclave with interlocking doors.

24.2.3 Laboratories

Detailed design of laboratories cannot be listed here because of the many and diverse activities to be conducted therein. Apart from the general microbiology laboratory, it is likely that a laboratory specifically designed to handle genetically-manipulated organisms would be required. In the UK, the report of the Working Party on the Practices of Genetic Manipulation lists the standards which should be met by such a laboratory. Other countries have their own standards. In all but the lowest containment laboratory (P1), the air from the laboratory must be extracted through an independent high efficiency particulate air (HEPA) filtration unit. The laboratories should meet the following specifications: (1) they should not be sited adjacent to, or open from, corridors used by the general public; (2) they must be capable of being sealed to facilitate fumigation; (3) they must be rodent proof; (4) they should not be subject to flooding; (5) they should not be sited near a known fire hazard; and (6) they must be locked at all times when not in use. It is desirable that a capability for hosing down the fermenter hall should be provided. Water may be permitted to run to waste *via* a 15 cm water trap or to a disposal tank prior to sterilization.

If finishing processes are envisaged within the complex, this area should be regarded as a clean room and should be designed to meet clean room technology standards as set out in the US Federal Standard 209B (1973), the British Standard 5295 (British Standards Institution, 1976) or other relevant national standards.

24.2.4 Animal Holding Facility

In a number of circumstances, for example vaccine production, it might be desirable to possess an animal holding facility where prepared vaccines may be potency tested. Detailed design of such a facility and safety in the animal house is a study in itself and, because of the wide variation in the type of animal rooms encountered, is well beyond the scope of this chapter. Suffice it to say that at least a clean area, including perhaps a separate space for breeding colonies, should be physically separated from experimental animal holding rooms. The latter should comprise an outer lobby, changing and shower areas, airlock and laboratory, the whole complex providing an environment of increasing negative pressure terminating in the laboratory where air is exhausted through HEPA filters. The fitting of prefilters prior to the HEPA filters is most important here because of the amount of particulate matter in the air that is presented to the filters in the form of animal fur and feathers, bedding, foodstuffs and dust. Interior building services, fittings and finishes should be much the same as described for the remainder of the laboratory complex. Apart from the provision of a double ended autoclave to sterilize waste products from the animals, bedding, foodstuffs and animal carcases, an incinerator will be needed for total destruction of such materials. Sterilizable modern materials for the construction of animal cages are now available avoiding the use of older, heavier and more cumbersome models.

Primary containment of experimental laboratory animals is not always easy but smaller mammals, *e.g.* mice, hamsters, guinea pigs and rats, have been successfully kept in flexible isolators maintained under negative pressure. Keeping primates under such conditions is virtually impossible. However, regulatory bodies responsible for the preparation of safe codes of practice appear convinced that this should be the objective. Scientists and those involved in animal husbandry remain sceptical and argue that primary containment in this instance poses a more serious threat to the operator than providing the experimentalist with some form of protective clothing (secondary containment).

24.3 DESIGN FEATURES

24.3.1 Ventilation

Good ventilation is probably the most important single safety feature to be incorporated in the entire building design. The ventilation system should prevent the escape of hazardous biological materials into uncontrolled areas, including the environment. The building itself should be a completely enclosed structure having a total loss ventilation system so that air flows from areas of lesser contamination into areas of greater contamination as indicated by the arrows in Figure 2, thus creating pressure zones as indicated in Figure 5. A supply of clean air is desirable, if not essential, to certain areas. Pressure gradients are maintained in the building until the air is finally expelled to the atmosphere through HEPA filters which may be wall mounted as flush fittings. The life of a HEPA filter will be greatly prolonged if a roughing and/or a medium efficiency prefilter is interposed in the airstream prior to the absolute filter. Prefilters should be so arranged as to facilitate easy removal from within the laboratory when the resistance to the airflow becomes excessive. Care must be taken to ensure the sealing of filters into ducting to prevent unfiltered air by-passing the system. They should also be capable of being decontaminated *in situ*. A wide range of such filters are commercially available to meet all requirements. Each filter is quality controlled by the manufacturer, and should have an NaCl penetration of <0.003% and be tested in accordance with British Standard 3928 (British Standards Institution, 1965) or a similar standard. A typical arrangement avoiding long lengths of potentially contaminated ducting is shown in Figure 6.

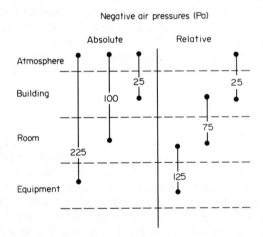

Figure 5   Air pressure gradients within the laboratory complex showing negative air pressures (Pa) relative to ambient and between pressure zones

In case of mains electricity failure, the supply and extract fans should be electrically interlocked. A standby generator or batteries might afford an emergency supply until normal power can be restored. The fans must be mechanically interlocked, and air pressure switches in the ducting would be activated if the airflow dropped below design intent due to belt failure, fan impeller slippage, or any other mechanical failure, thus ensuring a still air situation in the laboratory. Consequently, it is not possible for a negatively-pressurized biohazard area to become positive under any circumstances. Local and central audible and visual alarms would alert laboratory and administrative staff of ventilation failure. Ducting should be constructed of welded galvanized mild steel which is seamed and sealed.

It is arguable whether air from fermentation halls needs to be HEPA filtered if primary containment of fermenters is adequate. This is very much a matter of opinion and depends largely on the microbiological agent being handled. If work with pathogens is envisaged, it would seem sensible to extract air through HEPA filters as shown in Figure 6.

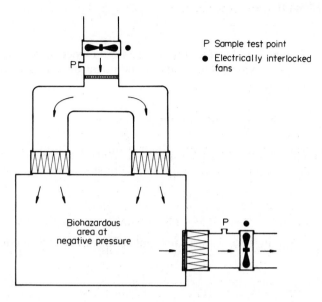

P Sample test point

● Electrically interlocked fans

Biohazardous area at negative pressure

Figure 6 Elevation showing room ventilation system and HEPA filter assembly for the removal of biohazardous particles. The arrows indicate direction of air flow. Two HEPA filters are shown at ceiling level to ensure good air distribution. The rate of air flow will be adjusted to create the required negative pressure within the laboratory

24.3.2 Lighting

Daylight only may be required in some areas, but it may be supplemented by artificial means in others. Batten type fluorescent fittings are probably the most acceptable. They should be flush mounted in the ceiling so that tubes may be changed from a service area above. The light housings should be sealed and become an integral part of the room. A recommended light intensity may be 800 to 1000 lx. Provision should be made for some form of emergency lighting.

24.3.3 Temperature Control

A comfortable working temperature should be provided. Excessive heat generation from apparatus could necessitate the installation of air conditioning. Such units are best mounted on the roof remote from the laboratory. With total loss ventilation it is obvious that this sort of air conditioning is expensive and not conducive to conservation of energy.

24.3.4 Interior Finishes: Walls/Ceilings

An impervious finish must be provided which is resistant to chemicals, fumigants and disinfectants. Welded sheets of poly(vinyl chloride) (PVC) of at least 2 mm thickness are commonly used. Alternatively, several applications of a good quality gloss paint may be acceptable. However, it is imperative that whatever surface is chosen it should withstand being washed down with water or suitable disinfectant when the presence of biohazardous material is suspected.

24.3.5 Interior Finishes: Floors

A highly suitable material to use is heavy duty PVC. A commercially available product incorporates carborundum as an essential non-slip safety feature. Flooring material should continue up the walls for several centimetres and be bonded to the wall surface in some manner. Hosing down such areas as fermenter halls will sometimes be necessary and a drain should be sited in the floor for water to run to waste through a 15 cm water trap to a sump or sterilizable waste disposal unit.

24.4 ESSENTIAL SERVICES

24.4.1 Water

In addition to the normal supply of water there may be special requirements for distilled, demineralized or pyrogen-free water. Such units should be installed in a service area, outside the laboratory.

24.4.2 Steam

High pressure steam at 689 kPa (6.89 bar) should be made available to the building. Reducing valves should be installed to supply low pressure steam for sterilizing apparatus at 175 kPa.

24.4.3 Electricity

It is advisable to incorporate only waterproof connections to electrical equipment and apparatus.

24.4.4 Gas

The usual supply of town or natural gas will be required.

24.4.5 Compressed Gases

A remote store for compressed gases should be provided to facilitate the separation of flammable and non-flammable gases to comply with local regulations. Compressed gas should be piped from cylinders sited outside the building to the equipment for which it is required. Duplicate cylinders with an automatic changeover device will ensure continuity of supply to essential equipment.

24.4.6 Communications

Apart from the traditional telephone communications, there would be a need for alarms to alert personnel to leave the area in case of fire or for any other emergency. It is strongly suggested that there should be a separate fire alarm signal easily distinguishable from all others and audible from all parts of the building. It is also imperative that all alarms are tested regularly and that the evacuation routine is written into a code of practice.

24.5 MONITORING AND COMPUTER SYSTEMS

Programmable electronic systems (PES) are likely to be employed in the control of industrial processes (HSE OP2 Health and Safety Executive, 1981), since such equipment is becoming less expensive and more readily available. A great number of problems remain to be solved before PES becomes commonplace, but developments in this field are rapid and techniques are likely to be refined in the future. Three modes of operation may be envisaged.

Mode I: the computer receives signals from the plant or machine to which it is linked, processes the information and displays it on a screen. The operator retains decision and control of the plant.

Mode II: the computer acts as a link between operator and plant and its control is limited to instructions given by the operator. Decisions are retained by the operator and control is exercised by the computer.

Mode III: decisions and control are exercised by the computer, but with limited manual intervention.

There are a number of basic safety implications in installing such a system. New and proposed biofermentation plants demand computer controlled operation and it is essential that instructions transmitted to a valve/pump have been obeyed. An important feature of the 'Zephyr' type (APV, Crawley, UK) valve control unit, Figure 7, is the provision of the two reed switches. These switches, operated by movements of the valve spindle, feed back actual position information (either open or closed) for electrical interlocking with other valves and equipment, and for remote indication on the control panel. This feature ensures that panel indications are always correct and, by the application of control logic to the pair of switches, any feedback modification can easily be detected. The reed switches may be manually adjusted. Any excessive delay between the operation of the two switches indicates early warning of possible valve failure. This delay is analysed by the computer and an alarm condition is indicated and recorded. Maintenance procedures may then be implemented.

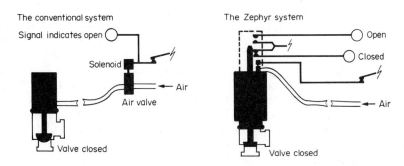

Figure 7 Computer control: electrical signal feedback from valves, comparing the conventional system and the 'Zephyr' system (APV, Crawley, Sussex, UK)

24.6 CONTAINMENT: EQUIPMENT AND APPARATUS

24.6.1 Safety Cabinets

Conflicting needs to protect the operator on the one hand, and the product on the other, have led to the development of different types of safety cabinets. Debate continues about the suitability of one type of cabinet over another for specific purposes and manufacturers compete over design and cost implications. The advantages and shortcomings of a particular safety cabinet must be carefully assessed by the customer before purchase. A number of manufacturers will supply standard models, or custom built safety cabinets on request at extra cost. The three basic models are shown schematically in Figure 8, and are the subject of British Standard 5726 (British Standards Institution, 1979) and WHO guidelines (World Health Organization, 1981). The Class I microbiological safety cabinet is an open fronted exhaust protective cabinet which offers adequate protection to the operator from inhalation of aerosols generated by low to moderate pathogens being handled inside the cabinet.

The Class II microbiological safety cabinet is an open fronted cabinet in which the work space is flushed with an undirectional (so called laminar flow) downward airflow of filtered air to protect the product from airborne contaminants. Simultaneously, an inward flow of air through the working aperture offers a degree of protection to safeguard the operator from inhalation of aerosols generated in the cabinet.

The Class III microbiological safety cabinet is a totally enclosed cabinet designed for work with very dangerous pathogens where complete isolation of the organism from the operator is required. Air is drawn into the cabinet through a prefilter and HEPA filter of 99.997% efficiency, and is exhausted to atmosphere through a prefilter and HEPA filter of 99.997% efficiency at 0.3 μm particle size. Access is gained to the work space through ports, fitted with rubber gauntlets, in the front of the cabinet.

Special attention should be given to the siting of microbiological safety cabinets within the laboratory (Collins *et al.*, 1977). If the building is furnished with a ventilation system, it is important to make sure that switching on the cabinet does not upset the airflow within the room. Exhausted air is best discharged to atmosphere through individual ducting, but on occasion a thimble system connected to the room ventilation may be adequate.

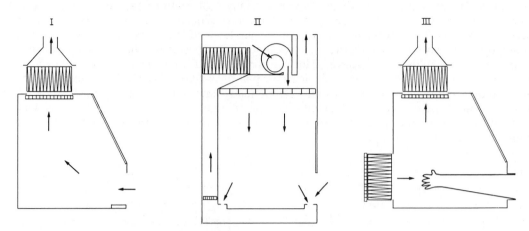

Figure 8   The three basic models of microbiological safety cabinets I II and III are shown. The direction of air flow is indicated by the arrows. Air is extracted through HEPA filters in each case. Supply air is also HEPA filtered into the work area of the class III safety cabinet

24.6.2   Flexible Isolators

Plastic film isolators were originally developed for the maintenance of gnotobiotic animals (Trexler, 1976). In recent years they have gained application in other areas, such as patient management, and as an alternative to facilities offered by the conventional type of microbiological safety cabinet (Smallridge, 1981). Applications in industrial processes have so far been limited, but a flexible isolator has been found satisfactory to provide primary containment of a high-pressure homogenizer (Dunnill, 1982) used for cell disruption in the isolation of intracellular microbial enzymes. A PVC halfsuit supplied with HEPA filtered air provides the operator with safe access to heavy equipment not possible in a conventional safety cabinet.

24.6.3   Special Purpose Cabinets for the Safe Handling of Biohazardous Materials

Conventional Class III microbiological safety cabinets permit only relatively small volumes of biohazardous material to be handled. Invariably the need arises to grow larger volumes of organisms in order to seed still larger vessels (fermenters). The Porton Mobile Enclosed Chemostat (Harris-Smith and Evans, 1968) has been developed, incorporating all the safety features afforded by a conventional Class III safety cabinet. The controls governing stirrer speeds, aeration and pH may be panel mounted on the exterior of the cabinet. A dunk tank mounted on the side of the cabinet and filled with suitable disinfectant provides access for taking materials in and out of the process chamber whilst the cabinet is in operation. Inlet and extract HEPA filters may be so arranged as to create an airflow of $3 \text{ m}^3 \text{ min}^{-1}$ and a negative pressure of about 200 Pa within the cabinet. It is a simple matter to connect two or more cabinets together *via* the dunk tank system in order that further downstream processing such as cell disruption, centrifugation or filtration may be carried out under similar containment conditions as those applied to the growth of the organism.

24.6.4   Autoclaves

The provision of a double ended autoclave will be essential, as was indicated in Section 24.2.2. It will provide a means of sterilizing media, glassware and other apparatus entering the laboratory (although media are best sterilized by steam in the fermenter prior to use) and the sterilization of waste materials from the laboratory complex. Design features should comply with British Standard 3970 (British Standards Institution, 1976) or the appropriate national standard, and include temperature-activated door locking mechanisms which prevent the opening of either door until the load temperature drops below 80 °C. The door interlocking device should not permit the opening of the door on the clean side until a successful autoclave run has been completed, if the door on the dirty side has been opened. The stainless steel chamber should be jacketed and contain steam under pressure during the sterilizing process and may be utilized as a cold jacket for rapid cooling after sterilizing. The provision of one or more thermocouple probes, with a direct read out facility, should be included in the specification for strategic siting within the load to ensure that the selected sterilizing temperature is reached. Exhausted air from the chamber should pass through a 0.2 μm microbial rated filter to ensure total retention of microorganisms.

24.6.5 Fermenters

The principles of fermentation have been well established for a good many years, but it is only comparatively recently that equipment originally designed for the brewing industry has found wider applications in the field of biotechnology. The concept of primary containment has been applied to fermenters with a great deal of success, irrespective of whether the fermentation process is conducted under anaerobic or aerobic conditions.

Fermenters will usually be cylindrical in shape, constructed of stainless steel and incorporating a cooling jacket or coil. A means of aeration and agitation of the tank contents must be provided. Figure 9 shows the manner in which air may be introduced into the fermenter through a filter to remove oil and water (<2%), filtering to about 98% efficiency at 0.2 μm rating, and finally through a HEPA filter immediately before entry into the fermenter. Exhaust air from fermenters of a capacity larger than 200 l must pass through a cyclone separator, a coalescing filter and finally through a cartridge filter (0.2 μm rating), ensuring total retention of microorganisms and effective sterile filtration in wet or dry conditions, before venting to atmosphere. Bleed-off points may be sited as shown (Figure 9) so that air passes through a condenser and a 0.2 μm rated filter to a mass spectrometer for gas analysis. In systems where fermenters of <200 l are employed, the cyclone separator may be omitted.

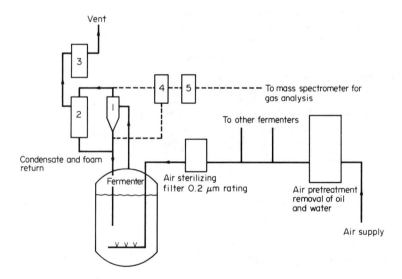

Figure 9 Schematic illustration of the method of supplying sterile air to a large fermenter (> 100 l) or a group of fermenters, and the means of preventing the escape of biohazardous particles from the fermenter into the environment. 1, cyclone separator; 2, coalescing filter; 3, 0.2 μm rating filter; 4, condenser; 5, 0.2 μm rating filter

The agitator drive assembly, situated at the top or bottom of the fermenter, must be provided with an efficient seal such as a 'Crane' (Crane Packing Ltd.) seal, which may be lubricated with condensate when the fermenter is in use. These are elastomeric bellows seals ensuring perfect sealing contact for use in a wide range of applications. The bellows are clear of the shaft being forward of the point of drive. They also offer a large static sealing area and all parts can be easily replaced. An alternative arrangement involving a steam barrier is shown in Figure 10 (White, 1983). A flange fitted with a double 'O' ring seal and steam barrier (Figure 11a) encircles the fermenter at the junction between the body and top dishing of the vessel, ensuring exclusion of contaminating organisms during fermentation and preventing the escape of biohazardous material into the atmosphere. It may be possible to rely entirely on the double 'O' ring seal without a steam barrier, as shown in Figure 11b. During fermentation it may be necessary to make additions to or take samples from the contents of the fermenter. A sampling adaptor (Figure 12) may be used for introducing the inoculum and other agents, such as antifoam, into the fermenter. It may serve also as a sampling point for the removal of samples. The adaptor is welded onto the side of the fermenter and consists of a stainless steel body with an internal thread, a rubber diaphragm, a metal washer, a threaded retaining ring and a removable threaded cap. Materials are introduced or withdrawn by puncturing the self-sealing rubber diaphragm with a hypodermic needle housed in a specially designed stainless steel holder (shown in Figure 13). The holder protects the operator from accidental injury and will prevent escape of biohazardous materials on removal of the needle from the adaptor. The sterilized assembly, with the needle retracted between the two diaphragms, is screwed into the sampling adaptor with a pledget of cotton wool soaked in a suitable disinfectant between the assembly diaphragm and the sampling adaptor diaphragm. The needle is thrust through the forward assembly diaphragm and the adaptor diaphragm. Materials may be inoculated into the fermenter or samples withdrawn from it into suitable containers for further assessment. The needle is withdrawn through the two diaphragms and the assembly removed from the adaptor, the cap of which is now replaced. Thus, the contaminated needle is contained between the two diaphragms of the assembly which can be autoclaved. The introduction of larger volumes of materials to the fermenter may be made *via* stainless steel piping, welded directly into the body of the vessel and protected by steam valves. Similarly, a less sophisticated method of sampling often used on smaller equipment is to weld a stainless steel pipe from the body of the fermenter *via* a steam sterilizable valve, directly to a hooded sampling port terminating inside a Class III safety cabinet operating under negative pressure. To sample the contents of the fermenter, the valve is opened and a sample is collected in a suitable container. On completion of the sampling exercise, the valve is closed and the assembly is steam sterilized to this point.

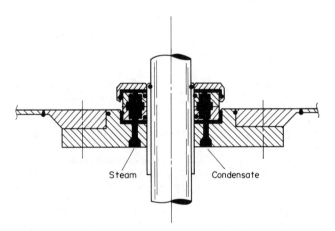

Steam   Condensate

Figure 10   An assembly on the agitator shaft of a fermenter with double 'O' ring seals and extra security afforded by a steam purge (Diagram by kind permission of R. S. White, APV, Crawley, Sussex, UK)

Other probes for the monitoring of pH, temperature control and oxygen tension will be required. Double 'O' ring seals may be utilized to maintain primary containment (Figure 14a). Extra security is assured by introducing a steam seal (White, 1983), as shown in Figure 14b. It must be emphasized that the effectiveness of 'O' ring seals will depend upon good servicing and maintenance.

All pipework should be of welded stainless steel. An orbital welding system consistently produces welded joints to the high standards required. It ensures that the weld bead on the inside of the pipe has a smooth, clean surface and the entire joint is free from crevices which could trap contaminated liquids and harbour microorganisms. To achieve these necessary qualities, robust, commercially available welding heads are coupled to highly specialized control equipment which has been specifically developed by research engineers. These electronic welding power con-

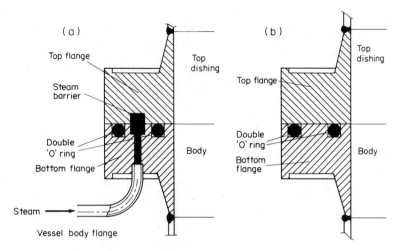

Figure 11 Sections through the vessel body flange at the junction between the body and top dishing of a fermenter: (b) shows the conventional double 'O' ring seal; (a) demonstrates the extra protection afforded by a steam barrier (R. S. White, APV, Crawley, Sussex, UK)

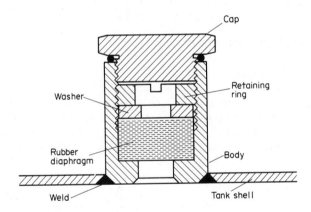

Figure 12 Sampling adaptor which is welded to the body of a fermenter and which permits addition or withdrawal of materials without comprising the microbial contents of the vessel or contaminating the environment. (Diagram by Glenwood B. Achorn)

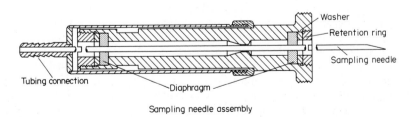

Sampling needle assembly

Figure 13 Design of a stainless steel holder to protect the operator from accidental injury and to prevent escape of biohazardous materials on removal of the needle from the sampling adaptor

trollers ensure consistency of welding current under all operating conditions. The tungsten inert gas (tig) method of operation uses argon as the shielding for both the arc and the molten weld inside the pipe.

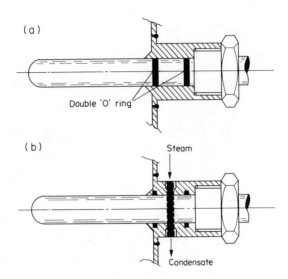

Figure 14   Diagrams to emphasize the use of 'O' ring seals protecting probes entering the fermenter: (a) shows double 'O' ring seals on the probe itself; (b) illustrates further protection by means of a steam seal (R. S. White, APV, Crawley, Sussex, UK) and the 'O' ring seals are recessed into the sectioned welded adaptor

24.6.6 Downstream Processing Apparatus

24.6.6.1 *Separators*

The optimum harvest time for a particular fermentation process will depend upon a number of factors. Having terminated the fermentation, the next phase will be that of separating the solids from the liquor. The cells will be required for further treatment if intracellular proteins are being sought, or the liquor will be recovered if extracellular proteins are to be extracted. Figure 15 illustrates the principle of a clarifier system (Westfalia Separator Ag, West Germany) which incorporates a steam sterilizing facility for aseptic processing, for example in the production of vaccines. The fermenter may be connected directly to the separator by a stainless steel transfer line which is capable of being sterilized by steam at 121 °C for 60 minutes. The material may be gravity fed or pumped *via* the transfer line to the separator. The ever increasing demand for larger volumes of material for production purposes has in turn brought about the development of apparatus for the continuous separation of cells and liquor.

24.6.6.2 *Cell rupture*

The disintegration of microorganisms follows that of concentration of material in the downstream processing for the production of intracellular proteins. Hughes *et al.* (1971) reviewed the physical and chemical methods of cell disruption available at that time. Since then, designs of high speed homogenizers and high speed agitator mills have been developed and improved upon, and these mechanical methods are used today in enterprises involving large scale production. Alternative chemical techniques are limited to small scale and pilot studies. Limon-Lason and colleagues (1979) described the properties of a high speed bead mill where disruption occurred beneath a liquid surface in a totally contained chamber. Woodrow and Quirk (1982) investigated the suitability of a bead mill, the Dyno-mill model KDL, for the release of intracellular bacterial enzymes. Their results suggest that the larger KD5 and KD15 models should be able to produce cell breakage at flow rates of the order of 125 and 379 l h^{-1}, equivalent to about 35 and 100 kg cells h^{-1}. The need for secondary containment whilst using a high speed homogenizer, and subjecting microbial cells to a pressure of up to 5×10^4 kN m^{-2} and subsequent return to atmospheric pressure, was discussed by Dunnill (1982). Having successfully disrupted viable microbial cells, the chances of infection from subsequent stages in processing is greatly reduced, if not totally eliminated, but there is still the possibility of contamination of personnel with allergenic material.

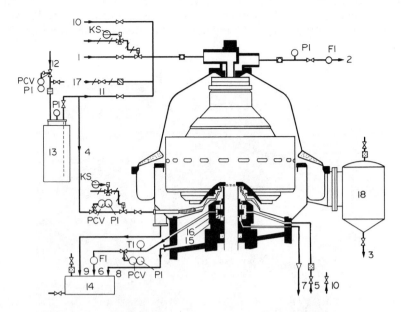

Figure 15 Westfalia clarifier system with steam sterilizing facility for aseptic processing. 1, feed; 2, discharge; 3, bacterial concentrate discharge; 4, operating water; 5, sealing liquid for slide-ring seal; 6, used sealing liquid; 7, leakage, lower slide-ring seal; 8, frame drain and leakage, upper slide-ring seal; 9, operating water discharge; 10, steam feed; 11, flushing water; 12, sterile air; 13, storage tank for operating and flushing water; 14, sterilizing tank; 15, double action slide-ring seal; 16, sealing chamber; 17, sterile air; 18, concentrate catcher. (Courtesy of Westfalia Separator AG, Oelde, West Germany)

24.6.6.3 Ultrafiltration

Modern techniques involving the use of a tangential flow system of filtration can now be offered as an alternative to centrifugation for the concentration of microbiological particles. Moreover, this system eliminates damage to shear-sensitive material and therefore prevents the loss of biological activity. The method has found application in the purification of interferon (Zoon *et al.*, 1979) and the concentration of arenaviruses (Gangemi *et al.*, 1977). Using the Pellicon Cassette System (Millipore Corporation, Bedford, MA, USA), 30 to 100-fold concentrations were obtained in less than 2 h without significant loss of virus infectivity, suggesting that the system might be used to concentrate viral antigens for vaccine production. The Pellicon Cassette System operates in a completely enclosed system preventing aerosol release of organisms from the apparatus. The manufacturers claim that several pharmaceutical companies currently use cassette systems in parallel to process up to 1000 US gallon batches of fluid. Walker and Foster (1981) report that ultrafiltration has been applied in the production of *Clostridium tetani* toxoid, following centrifugation, as part of a further purification procedure.

24.7 PROTECTIVE CLOTHING

The design of equipment to contain biohazardous material (primary containment) under all possible conditions is most difficult. If one accepts the possibility of the liberation of microorganisms from the apparatus into the environment due to poor design or through a fault developing in the equipment, some form of secondary containment must be provided.

Respiratory protection is best afforded by so-called positive pressure suits. Two which have proved reliable and efficient are the Martindale Positive Pressure Powered Respirator and the Vickers Medical High Efficiency Respirator. The Martindale Positive Pressure suit is completely self-contained and powered by batteries which drive air through HEPA filters and deliver it to the face piece, hood and blouse at the rate of 6 ft^3 min^{-1}. The sealed nickel cadmium fully charged battery carries sufficient power for seven hours' continuous use, and may be clipped onto the waistband of the suit. The model known as the 'Harwell' (registered trade mark of UK Atomic

Energy Authority for which Martindale are registered users) for personnel protection from high toxic dusts, which includes aerosols of microorganisms, is shown in Figure 16. Waste air is dispelled to the atmosphere at the waistband and cuffs (Martindale Protection Ltd., Neasden Lane, London, UK).

Figure 16 Demonstrates the use of protective clothing. The operator is wearing a Martindale positive pressure suit ('Harwell' type) providing HEPA filtered air at a positive pressure within the suit to that outside. The operator is protected against infection from microbiological hazards which may be present in the laboratory atmosphere

The Vickers Medical High Efficiency Respirator consists of a self-supporting PVC hood with a large clear vizor. HEPA filtered air reaches the hood through a flexible hose. An optional extra warns the operator of low battery voltage. The respirator incorporates a PVC 'cape' which finishes just below shoulder level so that disposable protective clothing must be worn underneath.

24.8 WASTE DISPOSAL AND TREATMENT OF BIOHAZARDOUS MATERIALS

24.8.1 General

Waste management is often regarded as the 'poor relation' in a research context, because it is the termination of a chain of events when most of the 'exciting exercises' have been completed. The 'blind eye' approach to the subject cannot be perpetuated and experimentalists must ensure that the waste generated by their activities in the laboratory is disposed of in a safe and responsible manner (US EPA 1982).

The types of waste may be summarized as: air and gaseous waste, solid waste, liquid waste and radioactive waste. The manner of disposal of such waste falls into the following categories: steam sterilization or autoclaving, incineration, dry heat sterilization, gas or vapour sterilization, chemical disinfection, irradiation or filtration. With this array of options, is it any wonder that in the absence of any clear standards, confusion and uncertainties persist in the minds of operators as to the best method of disposal for a particular type of waste?

24.8.2 Decontamination of Air from the Work Environment

Should an accident occur to equipment during processing, causing the uncontrollable release of microorganisms into the atmosphere, if the laboratory is furnished with efficient HEPA filtered exhaust ventilation, then any aerosol will be trapped within the filter material, thus protecting the environment beyond the building itself. Given time, the aerosol within the laboratory will become so diluted as to be non-existent if the ventilation system is maintained.

Depending upon the nature of the microorganisms being handled, it may be sufficient to spray contaminated surfaces with a suitable disinfectant which is left for a specified time before being hosed down with water. If a pathogen is being handled, however, the plenum and extract air supplies should be sealed off, producing a 'still air' situation and a suitable fumigant should be released into the atmosphere under controlled conditions for a specified time. On re-establishing the ventilation, the area will gradually clear of the fumigant. Formaldehyde vapour is commonly used for this purpose (Darlow, 1958; Tulis, 1973), despite allegations that formaldehyde vapour is potentially carcinogenic (Swenberg *et al.*, 1980). The Health and Safety Executive has produced a toxicity review of formaldehyde (HSE, 1981).

24.8.3 Air/Gases from Apparatus

Cartridge filters are designed to operate efficiently in a variety of vent applications, particularly where repeated *in situ* steam sterilization of the filter is necessary. They are suitable for the 100% retention of microorganisms and filtering air vented from autoclaves, fermenters or other apparatus. Mounted externally to the apparatus, the filters should withstand a normal and reverse flow of steam at temperatures up to 140 °C. Penetration is less than 0.001% when tested to British Standard 3928 (British Standards Institution, 1965) with particles in the range of 0.02 to 2.0 μm diameter. Modern materials of polypropylene and poly(tetrafluoroethylene) (PTFE) melt-sealed together are used by some manufacturers to produce a lightweight filter cartridge with no metal parts incorporated to rust or corrode.

24.8.4 Decontamination of Apparatus

24.8.4.1 *Fermenters and downstream processing apparatus*

In the future, it is likely that a great deal of automation will be applied to fermentation and downstream processing techniques. Where possible the method of sterilization is steam under pressure, *e.g.* steam at 121 °C for 30 minutes. Lengths of stainless steel pipework protected by steam valves may also be sterilized by this method.

24.8.4.2 *Heat sensitive apparatus*

If the apparatus consists of a number of electronic components, steam sterilization is not possible and gaseous sterilization is the preferred method of decontamination. Death *et al.* (1982) showed that a piece of apparatus can be exposed to formaldehyde vapour many times without any deleterious effect. Ethylene oxide gas or paraformaldehyde vapour (Taylor *et al.*, 1969) provide alternative methods of decontamination of apparatus. However, ethylene oxide is rarely used (unless no other method exists) because of its explosive properties when mixed with air. If gaseous methods of sterilization cannot be used, consideration should be given to irradiation techniques involving the emission of deep penetrating γ-rays from ^{60}Co (Eymery, 1973), but this requires considerable capital outlay.

24.8.5 Solid Waste

This term embraces discarded microbiological culture, cell sludge or debris, glassware and disposable plasticware, animal carcasses, bedding, foodstuffs and excrement, sharps (hypodermic needles, scalpels, *etc.*) and protective clothing. There is no doubt that these materials are best autoclaved, followed in some instances by incineration. A pre-vacuum type autoclave should be selected and the criterion for time/temperature exposure for destruction is usually calculated as double the exposure time to kill a standard suspension of *Bacillus stearothermophilus*. Contaminated clothing will require a porous run. The siting of one or more thermocouples, strategically placed in the centre, top and bottom of the load, will indicate that the correct temperature has been reached and maintained.

24.8.6 Liquid Waste

This will include water, blood and blood products, media and effluent from fermenters, vaccine plants and animal houses. Where small volumes are concerned, chemical treatment may be considered, but if protein is present in the waste, care should be taken to ensure that the disinfectant is not inactivated by its presence. For larger volumes of water or effluents from fermenters, heat treatment is the only sure way of effecting sterilization. This is usually achieved by providing some form of holding tank, which may also serve as the sterilizing vessel, providing steam at 128 °C (253.9 kPa steam pressure) for 30 minutes. In exceptional circumstances a holding tank may be permitted, containing effluent to which formalin may be added and which is held for a specified time and continuously mixed. The required concentrations of formalin and holding period must be carefully assessed to ensure 'double' killing time (under the conditions proposed) for the contaminating organisms being handled. Care must be taken to ensure that concentrated formaldehyde does not come into contact with any source of free chlorine, such as is contained in some disinfectants, for fear of producing bis(chloromethyl) ether, a potent lung carcinogen.

24.9 SUMMARY

Design criteria for laboratories and equipment with regard to maintaining a safe working environment and to prevent the escape of biohazardous materials into the atmosphere have been discussed. Emphasis and priorities must be given to providing primary containment where possible. Not only must personnel be protected from infectious and allergenic hazards, but the product must be kept free of contaminants, the presence of which might render valuable materials commercially useless. Where primary containment is not possible, consideration should be given to secondary containment.

It should not be presumed that the level of containment described in this chapter is always necessary, but rather that these indicate measures which may be taken and the techniques available. The degree of containment will necessarily vary depending upon the nature of the work to be carried out. In the absence of standards in this difficult field it is mostly a question of fine judgement on the part of the scientists regarding the amount of containment which should be applied to a specific project; hopefully common sense will prevail.

The equipment referred to in this chapter is not an exhaustive list of the apparatus available on the market, nor is it implied that the pieces of equipment mentioned are necessarily the best of their kind. Rather the aim has been to provide the reader with an illustration of the types of equipment available.

ACKNOWLEDGEMENTS

I am indebted to the widow of the late Glenwood B. Achorn, Jr. for permission to use some of the material originally prepared by her husband for inclusion in this chapter, and to colleagues at the Fermentation Pilot Plant, Public Health Laboratory Service, Centre for Applied Microbiology and Research, Porton Down, Salisbury, Wiltshire, UK for their advice.

24.10 REFERENCES

British Standards Institution, London (1965). British Standard 3928. Method of test for low-penetration air filters.
British Standards Institution, London (1966). British Standard 3970. Specification for steam sterilisers. Part I: Sterilisers for porous loads; Part II: Sterilisers for bottle fluids.
British Standards Institution, London (1976). British Standard 5295. Environmental cleanliness in enclosed spaces. Parts I, II and III.

British Standards Institution, London (1979). British Standard 5726. Specification for microbiological safety cabinets. (Amendments April 1982 British Standard Institute Number 3693).

Centers of Disease Control (1981). *Proposed Biosafety Guidelines for Microbiological and Biomedical Laboratories.* Centers of Disease Control, Atlanta, GA.

Collins, C. H., E. G. Hartley and R. Pilsworth (1974, revised 1977). *The Prevention of Laboratory Acquired Infection.* Public Health Laboratory Service, Monograph Series No. 6. HMSO, London.

Darlow, H. M. (1958). The practical aspects of formaldehyde fumigation. *Monthly Bull. Minst. Health,* **17**, 270–273.

Death, J. E., B. E. Hallin and G. J. Harper (1982). Decontamination of automated laboratory equipment. *J. Clin. Pathol.,* **35**, 580–581.

Dunnill, P. (1982). Biosafety in the large-scale isolation of intracellular microbial enzymes. *Chem. Ind.,* **22**, 877–879.

Eymery, R. (1973). Design of radiation sterilization facilities. In *Industrial Sterilization* (B-D Technology Series), ed. G. B. Phillips and W. S. Miller, chap. 11, pp. 153–179. Duke University Press, Durham, NC.

Gangemi, J. D., E. V. Connell, B. G. Mahlandt and G. A. Eddy (1977). Arenavirus concentration by molecular filtration. *Appl. Environ. Microbiol.,* **34**, 330–332.

Genetic Manipulation Advisory Group (1982). Large-scale use of genetically manipulated organisms. *GMAG Note 12 (revised).* Medical Research Council, London.

Harris-Smith, R. and C. G. T. Evans (1968). The Porton mobile enclosed chemostat (The POMEC). In *Continuous Cultivation of Micro-organisms. Proceedings of the 4th Symposium,* pp. 391–410. Academia, Prague.

Health and Safety Executive (1981). *Formaldehyde Toxicity Review 2.* HMSO, London.

Health and Safety Executive (1981). Safety implications of the uses of programmable electronic systems in factories. In *Microprocessors in Industry.* Health and Safety Executive Occasional Paper Series: OP2, HMSO, London.

Hughes, D. E., J. W. T. Wimpenny and D. Lloyd (1971). The disintegration of micro-organisms. *Methods Microbiol.,* **5B**, 1–54.

Limon-Lason, J., M. Hoare, C. B. Osborn, D. J. Doyle and P. Dunnill (1979). Reactor properties of a high-speed bead mill for microbial cell rupture. *Biotechnol. Bioeng.,* **21**, 745–774.

Melling, J. and K. Allner (1981). The containment of micro-organisms. In *Essays in Applied Microbiology,* ed. J. R. Norris and M. H. Richmond, chap. 11. Wiley, New York.

Milewski, E. (1982). Large-scale procedures under the NIH Guidelines. *Recombinant DNA Tech. Bull.,* **5**, 88–91.

National Institutes of Health (1978). *Laboratory Safety Monograph. A Supplement to the NIH Guidelines for Recombinant DNA Research.* National Institutes of Health, Bethesda, MD.

Sargeant, K. and C. G. T. Evans (1979). Hazards involved in the industrial use of micro-organisms. A study of the necessity, content and management principles of a possible community action. *Biological Sciences.* Commission of the European Communities. EUR 6349 EN.

Smallridge, G. J. (1981). Flexible film isolators. *Int. Labmate,* **5**, Issue 1.

Swenberg, J. A., W. D. Kerns, R. I. Mitchell, E. J. Gralla and K. L. Pavkov (1980). Induction of squamous cell carcinomas of the rat nasal cavity by inhalation exposure to formaldehyde vapour. *Cancer Res.,* **40**, 3398–3402.

Taylor, L. A., M. S. Barbeito and G. G. Gremillion (1969). Paraformaldehyde for surface sterilization and detoxification. *Appl. Microbiol.,* **17**, 614–618.

The Occupational Safety and Health Act of 1970. United States Occupational Safety and Health Administration. Public Law 91-596, 91st Congress S.2193. USA.

The Health and Safety at Work Etc. Act 1974. HMSO, London.

Trexler, P. C. (1976). The development of isolators. *Post-Grad. Med. J.,* **52**, 545–549.

Tulis, J. J. (1973). Formaldehyde as a sterilant. In *Industrial Sterilization* (B-D Technology Series) ed. G. B. Phillips and W. S. Miller, chap. 13, pp. 209–238. Duke University Press, Durham, NC.

US Environmental Protection Agency (1982). *Draft Manual for Infectious Waste Management.* Office of Solid Waste and Emergency Response, Washington, DC.

US Federal Standard 209B (1973). (Amendment No. 1 1976) *Clean Room and Work Station Requirements. Controlled Environment.*

Walker, P. H. and W. H. Foster (1981). Bacterial vaccine production. In *Essays in Applied Microbiology,* ed. J. R. Norris and M. H. Richmond, chap. 9. Wiley, New York.

White, R. S. (1983). APV, Crawley, UK. Personal communication.

Woodrow, J. R. and A. V. Quirk (1982). Evaluation of the potential of a bead mill for the release of intracellular bacterial enzymes. *Enzyme Microb. Technol.,* **4**, 385–389.

Working Party on the Practice of Genetic Manipulation (1976). Report. Cmnd. 6600. HMSO, London.

World Health Organisation (1981). *Guidelines for Biological Safety Cabinets.* WHO COS/SMM/81.82. World Health Organisation, Geneva.

Zaugg, R. H. and J. R. Swarz (1982). Industrial use of applied genetics and biotechnologies. *Recombinant DNA Tech. Bull.,* **5**, (1), 7–13.

Zoon, K. C., M. E. Smith, P. J. Brigden, D. zur Nedden and C. B. Anfinsen (1979). Purification and partial characterisation of human lymphoblastoid interferon. *Proc. Natl. Acad. Sci. USA,* **76**, 5601–5605.

GOVERNMENTAL REGULATIONS AND CONCERNS

25
Introduction

C. W. ROBINSON
University of Waterloo, Ontario, Canada

Throughout the overall development stages of any new technology, including a biotechnology, adequate steps must be taken either to maintain the confidentiality of proprietary information ('trade secret') or to protect against its unauthorized use by others ('patent'). Otherwise, the potential economic return to the developer/investor may not be fully achieved. Therefore, we begin **Section 2** with an extensive chapter on *Patenting* requirements as established by the governments of numerous countries in which extensive research activities on and/or commercializations of various biotechnologies already exist. Chapter 26 also discusses the trade secret approach as an alternative to patenting; in addition, it emphasizes the necessity of keeping proper records of research results and offers details of how best to establish and validate such records.

Clearly, the general public and their representatives in government generally are well aware of and have well received the various socio-economic benefits that have been conferred by the applications of 'conventional' biotechnologies and their products, *e.g.* antibiotics, insulin, vaccines, biotreatment of municipal and industrial wastewaters, biofuels, *etc.* Processes based on and products derived from the 'new' biotechnologies show great promise to meet existing societal needs in a more efficient manner (*e.g.* insulin from recombinant microorganisms) or to meet newly-recognized needs based on economic and/or environmental concerns (*e.g.* improved agricultural or forestry yields by acquired or enhanced nitrogen-fixing capabilities). Many of these newly identified or increasing prominent needs have been discussed from the perspective of biotechology in **Section 1**, particularly in Chapters 3–16.

Particularly in technically developed countries and in so-called 'information-based' societies, there is growing public awareness (albeit sometimes based on incomplete information) and concern, reflected in the regulatory proposals or actions of governments, about the potential impact of new chemical substances and by-products of their manufacture on human health and either local or global ecology. Indeed, some of the impetus for the development of biotechnology-based products has arisen from a desire to provide alternatives or replacements for certain chemical-process-based products now in widespread use (*e.g.* narrow-spectrum microbial insectides to replace broader-spectrum chemical pesticides). However, there has been public apprehension about potential hazards that may be associated with research on and eventual production scale application of the 'new' biotechnologies, especially those based on recombinant DNA technologies.

It is difficult to assess accurately the depth and scope of such concerns as held by the general public and to assess the probable response to those concerns and others by governmental regulatory agencies, or, indeed, by the courts. Nonetheless, some of these concerns already have been manifested in the form of a legal action seeking an injunction to prevent the field testing in the United States of a recombinant form of *Pseudomonas syringae* designed to reduce frost damage to fruit and vegetable plants (Anon., 1984a). The District Court of the District of Columbia granted the requested injunction (Anon., 1984b). Subsequently, the US National Institutes of Health prepared an environmental assessment of the proposed project, and on February 7, 1985 the US Department of Justice and the Regents of the University of California filed a court motion requesting a partial stay of the injunction (Anon., 1985a). The outcome of this motion was not known at the time Volume 4 went to press.

Although it is beyond the scope of this introduction to discuss the scientific potential and socio-economic merits of this particular rDNA product, or the basis of the claims against its introduction into the environment, the message this action and counter-action brings seems quite clear. If biotechnology-based industries and applications are to reach their full potential in a time span controlled solely or largely by only the need to overcome technico-scientific barriers, the advocates and practitioners of biotechnology must fully address (*i.e.* up to the practical limits of existing risk assessment methodologies) the scientific and technological aspects of biohazard assessment and control in order to alleviate public concerns and to temper the regulatory environment. Thoughts similar to these recently were expressed by the President of the US National Academy of Sciences (Press, 1984).

Accordingly, **Section 2** is designed, in part, to provide an overview of known or potential biohazards associated with the methodologies, processes or products of biotechnology. Chapters 27–30 deal with some *Biological Substances of a Hazardous Nature* (the reader is reminded that Chapters 23 and 24 in **Section 1** dealt with substance hazard detection by biological methods and with containment methods, respectively). Chapters 31–33 immediately following discuss *Regulations on Hazardous Materials from Bioprocesses*, including the biocatalysts/bioproducts themselves (especially recombinant organisms), products and effluents in principally the United States and Japan. Special emphasis has been placed on the evolving situation in the United States as it is there that general governmental and judicial procedures for developing and implementing regulatory statutes and standards have been most extensively established.

As previously noted in Chapter 1, modern biotechnology is a rapidly developing field with respect to its scientific and technological aspects. Therefore, it is only to be expected that self-regulatory and government-regulatory policies and procedures for the many facets of biotechnology and their applications also will be in a state of continual evolution and modification. While Chapters 31–33 provide excellent overviews and summaries of their respective topics at the time they were written, readers are advised to check directly with the appropriate regulatory agency in order to apprise themselves of the current requirements for compliance. In this connection, it is interesting to note that several United States government agencies, all of which have claimed jurisdiction to regulate biotechnological processes and products, recently issued a joint proposal for rationalizing the regulatory process in that country (Anon., 1985b). The proposal advocates the formation of a two-level review and approval mechanism. Each of the five agencies (Food and Drug Administration, Environmental Protection Agency, Department of Agriculture, National Institutes of Health—in which is located the established Recombinant-DNA Advisory Committee—and the National Science Foundation) would establish a scientific advisory board to review applications involving advanced biotechnological techniques. Two members from each of the agency boards would comprise the Biotechnology Science Board (BSB), chaired by the Assistant Secretary of Health, Department of Health and Human Services. The BSB would function as a review agency for all applications involving the use of recombinant-DNA and -RNA and cell fusion techniques, to develop guidelines for similar types of applications. In addition, the BSB is conceived as being the appropriate place for public discussions of biotechnology.

Of course, there are a multitudinous number of biotechnology products which, in terms of their targeted function, normally are not considered hazardous to humans or animals when administered or ingested at safe dosages. Indeed, their role in the receptor organism is viewed as being highly beneficial with respect to medical or nutritional aspects. Nonetheless, the use of such a product in the manner recommended must be demonstrated as being not only efficacious with respect to its primary function, but also lacking in detrimental side-effects. Pharmaceuticals and microbial proteins (historically given the generic name 'single-cell protein', regardless of the morphology of the microbial source) are two such types of products. Acceptance (*i.e.* approval to market the product) of these products is subject to government regulation.

These regulations and the documentation required to be submitted are described in *Regulations on Pharmaceuticals and Single-cell Proteins*. Chapters 34 and 35 deal with pharmaceuticals' acceptance procedures in the United States and Canada and in Japan, respectively. Nituch in the chapter on *Acceptance of New Drug Products in the United States and Canada* describes the complexity of data collection, documentation and submission of the application and the generally lengthy time taken by the review process under the present operating procedures. However, readers with a particular interest in this area will be heartened by a report (Hanson, 1985) that the US Food and Drug Administration (FDA) plans to streamline its new drug application requirements, effective May 23, 1985. It is reported (Hanson, 1985) that 'the new procedures are expected to shorten the review time up to six months and lessen the paperwork burden as much as 70%'. Concurrently, the FDA plans to intensify its monitoring of marketed drugs to detect harmful effects. In addition, the FDA may approve new drugs solely on the basis of foreign clinical studies that meet US testing criteria, as long as those studies are applicable to the US population (Hanson, 1985). Formerly, at least some data from US-based clinical trials were required to be included in the application.

Single-cell protein for applications in human foods and animal feeds is discussed in Chapters 36 and 37, respectively.

Undoubtedly, many governments and the societies they represent have ambivalent feelings about the development and large-scale implementation of any technology, including biotechnology. On a perhaps 'negative' note, they may feel the need (either self-initiated or in response to publicly expressed concerns) to regulate the introduction of the technology and its products. This regulation is meant to reduce to an acceptable level the probability of detrimental impacts on the population or on the environment (whether the means applied—the regulations and procedures—are commensurate with the potential risks involved with respect to a particular technology and its products is, of course, a subject of considerable continuing debate in many jurisdictions).

Regulation, however benign it may be, results in slowing the pace of development and commercialization. On the other hand, and in a 'positive' sense, governments are eager to promote the use of new technologies which, in addition to the primary products *per se*, often are viewed as the source of other tangible benefits: increased employment, increased national pool of highly trained professionals and technicians, increased national revenue from exports (or decreased expenditures on imports), *etc*. The promotion of biotechnology R & D worldwide, by means of direct or indirect financial assistance from national governments and international development agencies, is dealt with in Chapters 38 and 39 under the general heading *Development and Assistance Programmes in Biotechnology*, with which **Section 2** concludes. Chapter 39 also presents and discusses the considerable cooperative efforts that have been and continue to be made by international organizations (particularly those of the United Nations systems) and non-governmental bodies to promote increased quantity and significance of R & D and increased training of scientists and engineers in biotechnology in developing countries.

REFERENCES

Anon. (1984a). Rifkin seeks injunction against recombinant test. *Genet. Eng. News*, **4** (4), 1, 36 (Jan. 23).
Anon. (1984b). District judge quashes biotechnology project. *Chem. Eng. News*, **62** (21), 23 (May 21).
Anon. (1985a). Genetic engineering: court asked to allow field testing. *Chem. Eng. News*, **63** (7), 6 (Feb. 18).
Anon. (1985b). Biotechnology products: Federal agencies issue control policy. *Chem. Eng. News*, **63** (1), 8 (Jan. 7).
Hanson, D. (1985). New drug application rules shorten review time. *Chem. Eng. News*, **63** (10), 16–17 (Mar. 11).
Press, F. (1984). The challenge for biotechnology. *Chem. Eng. News*, **62** (49), 5 (Dec. 3; editorial).

26

Patenting Biotechnological Processes and Products

R. SALIWANCHIK
Richland, MI, USA

26.1 PREPARATION AND PRESERVATION OF INVENTION RECORDS

The legal protection of biotechnological processes and products is critical to both the researcher and the businessman. Each has a stake in obtaining effective legal protection for these innovations. The researcher, in addition to receiving recognition and reward for innovations, will often find that proper legal protection will afford needed monetary support for further investigations. The businessman can remain in business only if the business returns a profit. Effective legal protection for innovations of technical employees is a must if the businessman is to profit from such developments. Acquiring legal protection for biotechnological processes and products is not fraught with uncertainty if those charged with obtaining such protection understand the intracacies of biotechnology and the available legal systems that can be used to protect such innovations. An initial requirement in this legal process is the preparation and preservation of invention (innovation) records.

Experience has shown that there is a wide diversity among researchers in the methods of preparation and preservation of scientific (invention) records. The reason for this is a possible lack of appreciation by the researcher of the manner in which experimental data are best kept. Perhaps this is no different from the general lack of knowledge by members of the public of the best means for maintaining tax records to support their income tax returns. The knowledgeable researcher, like the knowlegeable taxpayer, understands the legal requirements for records and learns the best methods for satisfying these requirements.

Proper procedures for recording experimental work, and the preservation of such records, should be a subject within the curriculum of all researchers, but many years of working with them have shown that little if any formal training is devoted to these critical aspects of professional life. It is only when a researcher is exposed to the services of a lawyer that this entire spectrum of proper record keeping comes into focus. Many of these researcher–lawyer associations are initially formed when a researcher hires into a corporation and meets corporate patent counsel. Researchers in academe are, in some cases, left uninformed much of their productive lives, at least until they have invented something that calls for the services of a patent lawyer.

Whereas a corporate researcher is informed early on by the corporate patent attorney about proper record keeping, the researcher in academe may well find after a valuable invention has been made and then exposed to a patent attorney that proper record keeping has not been practiced. The practical effect of these two diverse situations is that the corporate researcher, having been informed of proper record keeping, may be in a better position to legally protect an invention than his colleague in academe, who may suffer ultimate loss of legal rights because of improper recording of the experimental details leading to the invention.

Addressing this subject of record keeping during the formal training of the researcher is desirable; permitting subsequent events (such as the creation of an invention) to initiate this vital area of instruction deprives the researcher of an ingredient basic to the practice of a successful career.

Before digging into the details of proper record keeping procedures, let us briefly review instances where good scientific records are helpful, if not indispensable.

Recording experimental work in a systematic, relatively permanent, and legible manner serves in all cases to give the investigator a foundation for future experimentation, for only in knowing what has been done and how it has been done does it become scientifically effective to plan and conduct future experiments. Incomplete or haphazard records of experiments can lead only to uncertainty in the evaluation of data and the planning of future experiments. So, strictly from the researcher's own point of view in pursuing a meaningful, non-repetitive line of experimentation, proper recording is essential.

As the experiment moves toward a completed work that appears to be inventive, thoughts of the future in terms of what to do with the invention inevitably spring forth. Suddenly an image of peer recognition coupled with reward excites the inventor. Almost simultaneously there arises a feeling of insecurity and uncertainty as to what to do with the invention to protect it from those who would exploit it and not pay proper tribute to the innovator. These are true feelings — they are not science fiction concoctions. It is an absolute truism that those who sleep on their rights may wake to find them irretrievably lost, as if everything were a dream. The rewards of proper record keeping inure to the researcher, but he or she does not stand alone at the time an invention is made.

Since the vast majority of scientists and engineers engaged in research are employed by someone, this someone also has a vital stake in realizing a benefit from the researcher's innovations. Thus, if steps have been taken to record experiments faithfully in a relatively permanent and legible manner, both the researcher and the employer can rest easier in knowing that a firm foundation for their expectations has indeed been laid. This foundation will support the inventor's position of being the first inventor when others would claim the right. This determination of inventorship rights *via* an interference proceedings before the United States Patent and Trademark Office (PTO) often turns on the quality of the records of the experimental work leading to the invention. If the records are good, then the inventorship determination is a more certain event. In the absence of good records, even the true first inventor can, at times, expect to come out second best. Such is the nature of our judicial system that it must rely on the best evidence to make the determination of first inventorship. This best evidence must be developed by the researcher from the beginning of the experiment; it cannot be manufactured when the experiment is essentially completed and the value of the experiment becomes apparent.

As indicated previously, the procedures for obtaining legal protection for biotechnology are not fraught with uncertainty if those charged with this responsibility understand the methods available to accomplish this goal. So too, proper record keeping is not an uncertain journey; rather, it can be a journey on a well-marked trail. Some of the basic characteristics of a good record of experimental work are as follows:

(1) a bound laboratory notebook should be used to record all experiments;

(2) the recording of an experiment should begin by placing the date in the upper right corner of the page;

(3) all notebook entries should be in ink of a single color;

(4) the planned details of the experiment should then be written out with an indication as to why the experiment is being done and of the result desired;

(5) if some of the experimental work is done by another person(s), for example an assayist, then the data obtained from the assayist should be recorded in the bound laboratory notebook as soon as they are received by the submitter of the material that was assayed;

(6) the recording of assay data should be made on a notebook page dated the date the assay data are received by the submitter, and the assay data should be identified by reference to a prior notebook page that contains a description of the material upon which an assay was requested;

(7) as the experiment is conducted and results obtained, entries should be made directly into the notebook;

(8) if an incorrect entry is made, a single line should be drawn through the incorrect entry; there should be no attempt to obliterate or erase the incorrect entry;

(9) conclusions and observations concerning the experiment should be entered into the notebook; remember, even negative experimental results are important and, therefore, should be recognized in writing in the notebook;

(10) each page of the notebook should be initialed and dated by the person doing the experiment; this person may very well be a laboratory assistant;

(11) each page of the notebook should be initialed and dated by the person who may have supervised the person who actually conducted the experiment;

(12) the experimenter should initial and date each notebook page immediately after a page is completed;

(13) the recording of the conclusion of an experiment should be followed by the signature of the experimenter and the signature of a witness, *i.e.* a person who observed and understood the experiment or a person to whom the experimenter disclosed the experimental details and who understood the experiment;

(14) the experimenter should sign the notebook immediately upon completion of the experiment, and the aforementioned witness should sign soon thereafter; and lastly,

(15) when a notebook page is not filled out completely, a diagonal line should be drawn through the unwritten portion.

Data entered into a notebook should be accurate and legibly written. Perhaps a course in handwriting for all researchers is a good idea. Perusing a particularly illegible notebook often brings this idea to mind. The notebook entry should be done by the experimenter, or by the laboratory assistant, if that person's handwriting is better. In-house jargon that is not known in the particular art of the experimenter should be avoided or duly identified. For example, reference to an in-house assay procedure by 'Joe's assay' is without probative value standing alone. Either extraneous evidence will have to identify 'Joe's assay' or, better still, the details of the assay should be set forth in the experimenter's notebook. If the assay is used frequently, then a simple entry of the assay details can thenceforth serve as a reference point for uses of the assay. Likewise, the use of trademark materials should be accompanied either by the source of the trademark materials or the composition of the materials. In summation, the entry of notebook data should be done with an eye to what a person skilled in the particular art of the experiment would understand.

Recording, and thus proof, of experiments also can be accomplished by use of in-house memos and reports to superiors and associates. Again, such reports are best when the disclosed data can be understood by an 'outside' person in the same manner as described for the understanding of notebook entries.

Recorded experimental work should be preserved in a way to maintain confidentiality and permanence of the record. Laboratory notebooks are valuable entities that require the maintenance of confidentiality in their access and dissemination. Permitting notebooks to be shelved in an open area calls for loss and possible theft of this record. Many companies make photographic copies of laboratory notebooks to prevent loss. This does not solve the theft aspect, which can be a very real possibility in a dynamic research area such as genetic engineering.

26.2 MICROBIOLOGICAL CULTURE DEPOSITS AND PATENT PROTECTION IN GENERAL

The recording and preservation of invention records are not, as indicated previously, idle acts. Assuming that such invention records evidence patentable inventions, much must still be done before patents are actually obtained. Basic to the obtention of patent rights for biotechnological inventions is the preliminary step of making a culture deposit if a particular microbe is involved in the invention.

Culture deposits are essential in conjunction with an application for patent rights in all of the industrially important countries granting patents for biotechnological inventions. There is a common denominator among all these countries in the type of culture deposit that is acceptable for patent purposes. Basically, a pertinent culture must be deposited with a recognized culture repository before a patent application is filed in the particular country (Argoudelis *et al.*, 1970).* Recognized culture repositories are those which have the facilities and staff to store and maintain culture deposits and disseminate them at the proper time. The time of distribution of a culture deposited in conjunction with a patent application is determined by the patent law of the particular country wherein the patent application is filed. For example, the United States patent law requires the availability of a culture deposit at the time a US patent, disclosing the culture deposit, is issued (Argoudelis *et al.*, 1970). On the other hand, a culture deposited in conjunction with a patent application in the Federal Republic of Germany must be made available to the public 18 months from the effective filing date of the patent application. The effective filing date is the initial date on which a patent application was filed. For example, if the initial filing was in the German patent office, then the culture deposit must be available to the German public 18 months from this date. If the initial filing was in the United States PTO, and a corresponding West German application was filed (in the Federal Republic of Germany) within one year of the United States filing date, then the culture deposit would have to be available to the West German public 18 months from the United States filing date. At such time no such Federal Republic of Germany patent has as yet issued. Thus there is a significant difference among some countries concerning the release of cultures deposited for patent purposes. These differences in release will be discussed further as individual country patent systems are analyzed.

As stated previously, cultures are deposited if involved in an invention for which a patent is sought. This means that if the invention cannot be practiced without possession of a particular culture, and the culture is not already known and available to the public, then the patent applicant must deposit the culture in a recognized culture repository before a patent application is filed in the patent office of the country from which a patent is sought. The culture can be deposited in a recognized culture repository outside the country from which a patent is sought.†

Depositing the culture is not enough in and of itself; the specific accession number given to the culture deposit by the repository must be disclosed in the patent application when it is filed in the

---

* The United States Court of Customs and Patent Appeals (CCPA) in this case held that a culture deposited prior to filing a United States patent application was sufficient under the United States patent law. The culture deposit in the Argoudelis case was available to (1) persons having legal access to the patent application during the pendency of the application in the PTO, and (2) the public upon the issuance of a patent disclosing the culture. The PTO subsequently adopted the Argoudelis holding and revised their *Manual of Patent Examining Procedure* (MPEP) in 1975. This manual may be purchased from the US Government Printing Office, Washington, DC, 20402.

† This issue was resolved in the United States by the CCPA in the case Feldman *v.* Aunstrup (1975). In this case the repository was the Centraalbureau voor Schimmelcultures (CBS) in the Netherlands.

patent office. There is a variation of this rule in the United Kingdom and European patent systems, which are discussed later, but the safest procedure to accommodate all countries is to disclose the accession number when a patent application is filed in the patent office. Another safe procedure is to also disclose the date the culture was deposited in the repository. Though these data are not required in most countries, the United Kingdom and France do require it.

The culture that is deposited should be the best culture known to the patent applicant to carry on the invention at the time the patent application is filed in the patent office. This requirement is necessary in the United States, where the Patent Statute requires a disclosure of the best mode for practicing the invention. Though patent acts in other countries may not contain this express requirement, it is likely that most, if not all, patent laws would be construed by a court to require such a disclosure. Realistically, it would be very hazardous to argue before any patent court in any country that their patent law did not require a patent applicant to disclose the best means for practicing the invention in return for a patent grant.

Culture deposits have been well known in many of the long-standing biotechnological arts, for example, in the antibiotic art. Numerous patents have been issued throughout the world wherein processes for making antibiotics with novel microbes are disclosed in the patent.* These patents have proven to be valuable assets not only to the patentees and their assignees, but also to the scientific public. The benefits to the patentees and their assignees are obvious. The benefits to the scientific public come in access to new cultures that can be ordered from the designated repository. The scientific public can experiment with these cultures and, hopefully, make new inventions or improvements on the 'old' invention. In other words, the disclosure of the patented invention and accessibility to a new culture add to the scientific armada available to the researcher.

In recent times the field of genetic engineering has exploded on the biotechnological scene. Much of the genetic engineering research uses cultures to achieve certain goals. For example, a culture can be the host of a desired vector such as a plasmid; it may also be the host of a cloned gene.† Such cultures most likely must be deposited in a culture repository if a patent is sought. The test is whether the invention can be carried on only by physical possession of the culture; if so, then the culture must be deposited. Another consideration, as discussed previously, is whether the best mode of practicing the invention is with the culture; if so, then the culture must be deposited. Choosing a different course in conjunction with such genetic engineering inventions is an invitation to future litigation even if a patent is granted.

No discussion of culture deposits would be complete without reviewing the Budapest Treaty on the International Recognition of the Deposit of Microorganisms for the Purposes of Patent Procedure.‡ For brevity, this treaty is normally referred to as the Budapest Treaty for Depositing Cultures. Basically, the treaty allows a patent applicant to make a single culture deposit with a recognized repository. Thereafter, this single deposit will satisfy the patent law disclosure requirements of all the countries signatory to the treaty. At the present time, all the industrially important countries have signed the treaty. The purpose of the treaty is to reduce the cost of depositing patent cultures in multiple repositories in conjunction with multiple patent filings. This purpose has been accomplished. The treaty does not deal with the critical aspect of culture release to the public. Thus the patent applicant must handle this apart from the depositing procedure, as discussed previously. The importance of this release aspect cannot be overemphasized. Any neglect in making a culture deposit available at the required time in a particular country can result in an irretrievable loss of patent rights. As indicated, there is a lack of uniformity among various countries' patent systems on when a culture deposit must be made available to the public. The laws of each individual country must be consulted on this point. The Budapest Treaty has no bearing on release of the culture deposit.

Appendix C contains a form used by the NRRL repository in Peoria, Illinois, for depositing under the Budapest Treaty. Forms of a similar nature can be obtained by contacting any reposit-

---

* A representative US patent is US 3 988 441, for 'Antibiotic U-43,120 and Process for Preparing Same.' It was issued on Oct. 26, 1976, to the inventors Ladislav J. Hanka and Paul F. Wiley. The disclosure section of this patent represents a type of antibiotic patent disclosure that is considered sufficient by the patent offices of the industrially important countries of the world. A copy of this patent is found in Appendix A.

† A plasmid designated as pUC6 is disclosed and claimed in US patent 4 273 875, in which Jack J. Manis is the patentee. pUC6 is harbored in the host microbe *Streptomyces espinosus* biotype 23724a, NRRL 11439. 'NRRL' refers to the culture repository located at the Northern Regional Research Center, US Department of Agriculture, Peoria, Illinois, 61604. US Patent 4 332 898, with Fritz Reusser as the patentee, relates to a novel hybrid plasmid named pUC1021. US patent 4 332 900, Jack J. Manis and Sarah K. Highlander as patentees, covers the construction of co-integrate plasmids from plasmids of *Streptomyces* and *Escherichia*, and the novel plasmids themselves: see Appendix B.

‡ The treaty in its entirety can be found in the *Official Gazette of the United States Patent and Trademark Office* (Aug. 23, 1977), **961**, 21–36.

ory recognized under the Budapest Treaty. Most of the well-known culture repositories throughout the world have been accepted or are expected to be accepted under the treaty. In the United States, both the Agricultural Research Culture Collection (NRRL) and the American Type Culture Collection (ATTC), in Rockville, Maryland, are accepted repositories.

26.3 PATENTING BIOTECHNOLOGICAL PROCESSES AND PRODUCTS IN THE UNITED STATES

Biotechnological processes and products comprise a broad range of subject matters. Included are antibacterials, vaccines, hybridomas, monoclonal antibodies, antivirals, antineoplastic agents, coccidiostats, biological herbicides, fermentation and chemical processes for making all these and other biologically-active entities, equipment used to make these entities, such as fermenters, filters and the like, and laboratory equipment used to develop processes for making biological entities—this would include assay instruments and reagents. In the fast-moving genetic engineering field, all the processes used to clone genes or to purify and assemble nucleotide sequences, enzymes, cloned genes, hybrid plasmids, new microorganism hosts, and processes for making useful chemical compounds by employing cloned genes are included. This list is not meant to be exhaustive; it merely illustrates the magnitude of the art we conveniently refer to as 'biotechnological processes and products'. The other chapters in this series bring specific information on these varied areas into full focus. The consideration here is to ferret out the significant biotechnological developments that can be made the subject of patents. As will be related, patent systems in the various countries are not uniform in the recognition of subject matter that may be patented under each country's patent law. The patent law of the United States is a rather broad law insofar as it allows virtually any biotechnological process and product to be patented if other statutory (legal) requirements are met.

26.3.1 Patentable Subject Matter Under the US Patent Act

The US Patent Act defines subject matter that can be patented as follows (United States Code, Title 35, 1952):

> *§101. Inventions patentable*
>
> *Whoever invents or discovers any new and useful process, machine, manufacture, or composition of matter, or any new and useful improvement thereof, may obtain a patent therefor, subject to the conditions and requirements of this title.*

The language of this statute has been interpreted by the United States courts as being expansive in nature. This means that words of the statute such as 'manufacture' or 'composition of matter' include anything made by man. The United States Court of Customs and Patent Appeals (CCPA) in the case of *In re* Bergy, Coats, and Malik; *In re* Chakrabarty, decided on March 29, 1979, quoted a statement from the United States House and Senate reports that accompanied the 1952 reenactment of the United States Patent Law as follows:

> *A machine, or a manufacture* * * * *may include anything under the sun that is made by man.*
> *201 USPQ 383.*

This statement by the CCPA was made in support of the Court's position that living microbes could be patented under the United States Patent Law.

Merely because a particular subject matters falls within a statutory class that can be patented does not mean that a patent is automatically granted. There are other requirements that also must be met.

26.3.2 Statutory Requirements for Obtaining a Patent

The main substantive requirements contained in the United States Patent Law are as follows: (1) the invention must be new or novel; (2) the invention must be useful; and (3) the invention must not be obvious from what is known in the prior art.

Before we elaborate on the above, it should be recognized that 'invention' is not synonymous with 'patentable invention'. A particular innovation may be new, unobvious and useful to the person(s) responsible for the innovation, but it still may not be patentable under the laws of the country from which a patent is sought. The novelty, usefulness and unobviousness of an invention under the United States Patent Law are tested by reference to the hypothetical person of ordinary skill in the art to which the invention pertains. If the United States Patent and Trademark Office (PTO) examiner determines that such a hypothetical person would find the invention new, useful and unobvious, then the major barriers to granting a patent will have been overcome. Thus the innovator's personal belief regarding the novelty, usefulness and unobviousness of an invention is not the deciding factor for patentability. However when an innovator is knowledgeable in the prior art in the invention area, it is likely that the innovator's belief is a true reflection of what would be found by the hypothetical person of ordinary skill in the art. Because of this, it is incumbent upon an innovator to make a personal appraisal of the invention with regard to its novelty, usefulness and unobviousness. This assessment should be made prior to any consideration of taking patent action, such as filing a patent application, and would be made most advantageously with the aid of a patent lawyer. A patent lawyer is a necessary aid because there are specific patent law rules that must be applied to the prior art. For example, the prior art, in order to be anticipating prior art, must enable a person skilled in the art to successfully carry on a disclosed invention. If the prior art is deficient in this or other factors, the filing of a patent application may be the prudent thing to do. The patent lawyer's expertise in these matters is absolutely necessary and should be sought early in the inventing process.

For those desiring a more detailed analysis of the United States statutory requirements of novelty, usefulness and unobviousness with regard to biotechnological inventions the author suggests an article written for *Developments in Industrial Microbiology* (Saliwanchik, 1978a) or the book entitled *Legal Protection for Microbiological and Genetic Engineering Inventions* (Saliwanchik, 1982).

Additionally, students of patent law can profit from a reading of the CCPA opinion in the landmark Bergy case (1979), referred to above. In its opinion, the CCPA gives a detailed analysis of the United States Patent Law and its applicability to an invention comprising a living microbe.

The obtention of a United States patent for a biotechnological process or product invention is, indeed, a valuable asset for the patentee.

26.3.3 Protective Features of a United States Patent

A United States patent affords the patentee the legal right to prevent anyone else from practicing the claimed invention in the United States and its possessions for a term of 17 years from the date the patent is issued. Though 17 years may seem like a long time to protect a patentee's rights, there are some instances in which this time period is painfully short in terms of recouping costs of research incurred by the patentee or employer of the patentee. These instances are found primarily in areas where invention subject matter must pass government regulatory review before such subject matter can be marketed. Inventions involving antibiotics are illustrative of those that may require many years of government review in order to reach the market-place. Should a patent be obtained years prior to the completion of the necessary government review, the effective patent term can be much less than the 17-year period. Because of this inequitable situation, which is argued as having a negative effect in stimulating costly research programs, there is now a determined attempt in the United States Congress to enact legislation that will allow for government regulatory review and restore the proper patent term to these much-needed innovators.*

A United States patent is valuable not only because of the 17-year term of protection which it grants, but also because both product and process claims are available. The significance of this can be illustrated by the following. If a product claim is obtained for a new antibiotic, then the claim covers the product made by any process, whether practiced in the United States or elsewhere, if the product is used or sold in the United States. If a process claim for making the antibiotic is obtained, then the claim is effective only to prevent practice of the claimed process in the United States and its possessions. Variations of the claimed process that cannot be termed equivalents can be practiced without infringing the claimed process. This illustrates the narrow protection which may result when only a process claim is obtained. Of course, if the process claimed is the only way to make the antibiotic, then its value is enhanced. Still, in order to provide meaningful patent protection for such a process in this day and age, there must be a suf-

---

* The 'Patent Term Restoration Act' (H.R. 1937 and S. 255) was enacted into law in 1984 (P.L. 98–417).

ficient number of foreign patents also covering the key process. This aspect of obtaining patents in multiple countries can be a costly procedure. However, if the invention is economically significant, the cost factor of obtaining world-wide patent coverage is not a negating reason.

26.4 CANADIAN PATENTS FOR BIOTECHNOLOGICAL PROCESSES AND PRODUCTS

For several reasons, which shall be detailed, Canadian patents are not high on the list of those seeking patent protection for their inventions. The Canadian government has severely limited the value of their patent system by the use of compulsory licensing provisions in the Canadian Patent Act (1970). It is told that this undesirable patent feature is a trade-off to appease Canadian consumer activists desiring lower prices for pharmaceuticals. Whatever the reason for the provisions, many acknowledge that the expense of Canadian patent filing is not justified by the worth of the grant. As a result, Canadian patent filing for biotechnological inventions presently is not as extensive as it could be.

26.4.1 Patentable Subject Matter Under the Canadian Patent Act

The Canadian Patent Act defines 'invention' as

> . . . *any new and useful art, process, machine, manufacture or composition of matter, or any new and useful improvement in any art, process, machine, manufacture or composition of matter.*

This statutory list of patentable subject matter is not unlike that found in the United States Patent Act. Also in agreement with the United States Patent Act is Chapter P-4, Section 28. (3) which states,

> *No patent shall issue for an invention that has an illicit object in view, or for any mere scientific principle or abstract theorem.*

A recent decision by the Canadian Patent Appeal Board and the Commissioner of Patents, dated March 18, 1982, interprets the Canadian Patent Act to allow patent claims for life forms.* The application in issue concerned a 'Foam Flotation Activated Sludge Process',† with the Abitibi Company of Toronto as the assignee. The claims rejected by the Canadian patent examiner were directed to various microbes. In deciding the appeal in favor of the appellants, the Canadian Board relied heavily on decisions throughout the world that already had recognized life forms as patentable subject matter. Still the Canadian Board and Commissioner of Patents go further than decisions in other countries by recognizing, rightly it is believed, that

> *[i]f an inventor creates a new and unobvious insect which did not exist before (and thus is not a product of nature), and can recreate it uniformly and at will, and it is useful (for example to destroy the spruce bud worm), then it is every bit as much a new tool of man as a micro-organism. With still higher life forms it is of course less likely that the inventor will be able to reproduce it at will and consistently, as more complex life forms tend to vary more from individual to individual. But if it eventually becomes possible to achieve such a result, and the other requirements of patentability are met, we do not see why it should be treated differently.*

Another desirable aspect in the Canadian Abitibi decision is the recognition of the fact that a culture deposit is the key to the practicing of a biotechnological invention. So long as the pertinent culture deposit is available to a person skilled in the art, the invention disclosure is sufficient. This point is emphasized here because of a contrary position taken by the German courts which is discussed later. The Canadian Board stated on this issue that

> *Section 36 requires that the application should set forth the steps of making the invention, in this case the new microorganism. Now the creation of a new microorganism by mutation, or by other means, is fraught with considerable difficulty, and it is by no means certain that the inventor, or others following his directions, will be able to produce it again using the original method of manufacture. However a microorganism, being living matter, will reproduce itself on the proper culture medium, so that the inventor can maintain his supply indefinitely. If he places samples of the organism in a culture collection to which others have access, they too will*

---

* A copy of the unpublished Board decision was obtained from W. John McClenahan, of the Canadian firm McFadden, Fincham & Co., Patent & Trade Mark Agents, 251 Bank Street, Ottawa, Ontario, Canada K2P 1X3. The agent for the Abitibi appellants was Gowling & Henderson, Box 466, Terminal A, Ottawa, Ontario, K1N 853.

† The inventors were James E. Zajic, Martha A. Hill, Donald F. Manchester and Karel Muzika, of the University of Western Ontario in London (Canadian Patent Application 2 57 177, unpublished).

be able to reproduce the organism, and thus have access to his invention, and use it once the patent expires. The question will consequently arise: is the deposition of the invention in the culture collection sufficient to satisfy the requirements of Section 36?

We do not see why it would not be. It would certainly permit others to make the invention, i.e. the microorganism. It will enable the public to do what the patentee has invented, as called for by Sec. 36, i.e. to make the microorganism, and in most instances by the easiest, most certain, most efficient, and best mode. This, we think, satisfies the requirements of the Act.

This is indeed the solution accepted by the House of Lords in the case of the American Cyanamid Company (Dann's) Patent, 1971 RPC 42, which recognized that deposition of a microorganism in a national culture collection would be an adequate description of the invention in question because a person of ordinary skill could put the invention into practice.

If deposition of a microorganism in a culture collection is sufficient disclosure of it when an applicant claims a process utilizing that organism, it seems strange indeed to hold it is inadequate when the organism itself is claimed. In both instances the public needs the organism to work the invention, and in both instances it has it, through the culture collection. What we should be concerned with is making the invention available.

Certainly the inventor should describe his original method of production, and with such clarity that if it can be repeated others could do so. But if the organism can subsequently only be reproduced from itself, we do not see why the inventor should be deprived of his reward provided, by deposition, he makes it available to others. Indeed where it is possible we believe he should make use of both methods of disclosure to reduce the danger of his invention being lost to mankind.

The Canadian Abitibi decision is certainly a favorable one for biotechnological inventors.

26.4.2 Statutory Requirements for Obtaining a Patent

Chapter P-4, Section 41.(1) of the Canadian Patent Act states:

In the case of inventions relating to substances prepared or produced by chemical processes and intended for food or medicine, the specification shall not include claims for the substance itself, except when prepared or produced by the methods or processes of manufacture particularly described and claimed or by their obvious chemical equivalents.

Also, Chapter P-4, Section 41. (3) states:

In the case of any patent for an invention intended or capable of being used for the preparation or production of food, the Commissioner shall, unless he sees good reason to the contrary, grant to any person applying for the same a licence limited to the use of the invention for the purposes of the preparation or production of food but not otherwise; and, in settling the terms of the licence and fixing the amount of royalty or other consideration payable, the Commissioner shall have regard to the desirability of making the food available to the public at the lowest possible price consistent with giving to the inventor due reward for the research leading to the invention.

This is followed by Chapter P-4, Section 41. (4) which says:

Where, in the case of any patent for an invention intended or capable of being used for medicine or for the preparation or production of medicine, an application is made by any person for a licence to do one or more of the following things as specified in the application, namely:

(a) where the invention is a process, to use the invention for the preparation or production of medicine, import any medicine in the preparation or production of which the invention has been used or sell any medicine in the preparation or production of which the invention has been used, or

(b) where the invention is other than a process, to import, make, use or sell the invention for medicine or for the preparation or production of medicine,

the Commissioner shall grant to the applicant a licence to do the things specified in the application except such, if any, of those things in respect of which he sees good reason not to grant such a licence; and, in settling the terms of the licence and fixing the amount of royalty or other consideration payable, the Commissioner shall have regard to the desirability of making the medicine available to the public at the lowest possible price consistent with giving to the patentee due reward for the research leading to the invention and for such other factors as may be prescribed.

These statutory provisions are set out verbatim to show the severe limitations of a Canadian patent in the medicinal and food sectors, the current bastions of the biotechnology industry.

Aside from these undesirable features of the Canadian Patent Act, the procedure for filing patent applications remains essentially as found in the United States.

26.4.3 Protective Features of a Canadian Patent

A Canadian patent has a term of 17 years from the date of grant. Again, this is identical to a United States patent. This Canadian patent gives the patentee

> . . . *the exclusive right, privilege and liberty of making, constructing, using and vending to others to be used the said invention* . . .

(Chapter P-4, Section 46)

The Canadian Patent Act goes further than the United States Patent Act in establishing a presumption in favor of a patentee, to wit:

> *In an action for infringement of a patent where the invention relates to the production of a new substance, any substance of the same chemical composition and constitution shall, in the absence of proof to the contrary, be deemed to have been produced by the patented process.*
> [Chapter P-4, Section 41. (2)]

Thus the burden is on the alleged infringer to show non-infringement. The United States patent law requires the patentee to present the initial proof of infringement. Oftentimes this is a difficult assignment that requires the use of sophisticated legal procedures.

On balance, the Canadian Patent Act is not a sought-after protective mechanism for biotechnological inventions. It is hoped that the Canadian Patent Office Board of Appeals decision in the Abitibi case, discussed above, will begin to move Canadian law into a more favorable position to encourage biotechnological inventors to file patent applications in Canada. A current concise review of Canadian patents and their effect on United States drugmakers can be found in *Business Week* (1982).

26.5 PATENT PROTECTION IN EUROPE FOR BIOTECHNOLOGICAL PROCESSES AND PRODUCTS

Generally speaking, patent protection for biotechnological processes and products in Europe is comparable to that obtainable in the United States. European countries issue patents for the older biotechnology, such as various fermentation processes and products, and also recognize the innovations of genetic engineering as patentable subject matter. As a matter of fact, countries such as Great Britain and the Federal Republic of Germany were ahead of the United States in recognizing the patentability of living microorganisms. Despite this forward approach of the European countries, they have an unfortunate procedural practice that is difficult to justify. This detrimental requirement for biotechnological innovators is that a pertinent microbe deposit must be made available to the public upon the initial publication of a patent application disclosing the same. Such initial publication, which occurs approximately 18 months from the patent application effective filing date, does not guarantee to any extent that patent rights will ultimately be granted. Yet the required release of the pertinent microbe to the public allows the public to practice the invention without paying homage to the patent applicant. The uniqueness of microbe deposits would seem to be reason enough to allow the patent applicant to maintain the culture deposit in absolute secrecy until a patent is granted. This is the procedure in the United States. It is eminently fair to the biotechnological innovator and gives the public access to a valuable entity in return for a patent grant. Japanese culture release procedure approaches that of the United States. Although Japan publishes patent applications as do European countries, Japan does not require release of a deposited culture at the initial publication of the patent application. Upon publication of a Japanese patent after examination of the patent application on the merits, the culture must be made available to the public. This second publication of a Japanese patent occurs merely months (approximately three to four) from the date a Japanese patent will issue. The adoption of the Japanese practice by European countries would result in a great service having been done for biotechnological innovators without harming the public.

26.5.1 The United Kingdom

The British Patents Act 1977 made several important changes in the existing law to harmonize British law more closely with the European Patent Convention. These changes were primarily procedural in nature and did not disturb the scope of protection afforded biotechnological products and processes by the existing patent law. As in the United States, claims in a patent application to microbiological processes and the microbes *per se* are patentable subject matter. The ability of the British to move with new technology is illustrated by the fact that they recognized the patentability of new microbes *per se* several years before the issue was settled in the United States. The British recognition of the patentability of this technology runs parallel to the practice of the Federal Republic of Germany, although the latter practice appears to be more artificially restricted.

The British law will allow the claiming in a patent application of a microbe by its genus–species name (binomial) with the pertinent repository accession number, *e.g.* NRRL 0000. Of course, the patent application disclosure must contain a description (taxonomy) of the microbe and details concerning the repository address. Also, it is better practice to disclose the date the pertinent microbe was deposited with the culture repository. Though the repository accession number and date of deposit can be supplied to the British Patent Office within two months of the British patent application filing date, it is desirable to have these data in the patent application when the application is filed to prevent the human act of forgetting to submit these additional data within the stated period. A check list of necessary things to do when filing a patent application in Britain can be of immense value.

The United States patent law does not require the disclosure of the date of deposit of a pertinent culture. It is difficult to understand the need for such information in a patent application. The disclosure of the repository in which the culture is deposited, the taxonomy of the microorganism and the accession number are sufficient to enable a person skilled in the microbiological art to practice the disclosed and claimed invention. The inclusion of the date of deposit is surplus and thus should not be required. Perhaps those who set up the rule, or requirement, did not realize that a culture deposit accession number is not assigned until a culture deposit has been received and accepted by the culture repository. Further, an accession number is specific for the one culture received. Thus the requirement of deposit date in a patent application is an idle act that would best be eliminated.

British patent law allows for claiming genetic engineering processes and products in a patent application. In this array of innovations would be found plasmids, cloned genes, processes for sequencing and/or purifying nucleotide sequences, restriction enzymes, and the like.

Whatever the biotechnological invention, the British law requires that it be new and that it involve an inventive step. Newness or novelty means that any written or oral disclosure of the pertinent invention in any country before the priority date is relevant. Further, public use of the claimed invention in any country before the priority date negates novelty.

The priority date is the date of the initial patent application filing date, generally referred to as the effective filing date. This term was discussed previously with reference to culture deposit release (see Section 26.2). Let me illustrate this point again to foster a better understanding. If a United States patent applicant files a British patent application within 12 months of the corresponding United States patent filing date, then the effective or priority date is the United States filing date. Thus the disclosure or use of the invention before the British filing date does not negate novelty of the invention for British patent purposes. However, if the British patent application referred to above was filed outside the 12-month period, then the publication or use of the invention before the British patent application filing date would negate the novelty of the invention under British law.

The inventive step required by British law is similar to that used in the United States and under the European Patent Convention. This means that if the claimed invention is considered obvious in view of the prior art by a person of ordinary skill in the art, then there would be no inventive step. The British patent examiner makes this determination of obviousness, since the examiner is presumed to be a person of ordinary skill in the art. The burden then falls upon the patent applicant to rebut this position of obviousness by use of competent evidence. There are many ways to do this sort of thing and only an experienced patent attorney should be allowed to proffer the evidence to the British Patent Office. Procedures which have been successfully employed in the United States, and which should be applicable under the new British law, are set out in the book *Legal Protection for Microbiological and Genetic Engineering Inventions* (Saliwanchik, 1982).

British patents afford good protection to many biotechnological inventions other than microbes and processes using the same. This scope of protection can be summarized as follows.

(1) New chemical compounds and compositions and methods of making compounds are patentable, irrespective of the intended use of the compounds.

(2) The discovery of a pharmaceutical or veterinary activity in old products not previously

known to have any pharmaceutical or veterinary activity is patentable by a form of product claim.

(3) The discovery of a new pharmaceutical or veterinary activity in old products known to have a different pharmaceutical or veterinary activity is not patentable.

(4) Methods of treating humans by surgery or therapy are not patentable.

(5) Methods of treating animals by surgery or therapy are not patentable.

(6) Other agricultural methods, for example methods using herbicides, are patentable.

The disclosure of an invention in a British patent application should be an enabling disclosure. This enablement requirement is not unlike that found in the patent laws of all the other industrially important countries. If the patent application disclosure would not allow a person of ordinary skill in the art to practice the claimed invention without undue experimentation, then the disclosure is not enabling.

British patents offer protection for the claimed invention for 20 years from the application filing date in Britain. This protection covers British territorial waters and British-designated areas of the European continental shelf.

26.5.2 German Federal Republic (GFR)

Microbiological processes and the products thereof are patentable subject matter under the German patent law (1976). Additionally, microorganisms *per se* are patentable. Thus there is no prohibition in the German patent law for patenting living entities so long as the other requirements of the German patent law are met. One of these requirements limits the scope of living entities that can be patented. This requirement is that a disclosure of a living entity, *i.e.* a microorganism, must be such that the microorganism can be reproduced by a person skilled in the art. The GFR Patent Office will accept a claim for a mutant microbe if the application disclosure enables a person skilled in the art to make the mutant microbe. This written disclosure must be sufficient in and of itself, even though a subculture of the claimed mutant microbe has been deposited with a recognized culture repository and the repository accession number is disclosed in the patent application. Why anyone skilled in the art would want to make the mutant microbe if a subculture is readily available is not understood. Still, this requirement of reproducibility is adhered to by the GFR Patent Office. Though the effect of this requirement on claiming a mutant microbe can be nil, since the mutation process can be described in writing, the effect on claiming a biologically pure culture of a novel microbe is devastating. In the case of a biologically pure culture, again the best method of enabling a person skilled in the art to make the microbe is to make it possible to obtain a subculture from a disclosed repository so that new microbes can be cultured. Yet the GFR Patent Office will not recognize this and instead insists on a written disclosure of how to make the biologically pure culture. Fortunately for the microbiological inventor the German Federal Republic requirement is not followed in other countries.*

Like the British patent law discussed previously, GFR patent law requires the release of a pertinent microbe deposit upon the initial publication of a West German patent application. This publication occurs 18 months from the effective filing date of the patent application.

The term of a West German patent is 20 years from the date of filing a West German patent application, not from the priority date. This term is favorable, but should be viewed in conjunction with the fact that often a patent is not obtained for three to five years after a West German filing. Thus, the real effective term of a West German patent is much less than the 20 years. This, coupled with the yearly maintenance fees and taxes which must be paid, makes a West German patent filing an expensive venture. For that matter, most patent filings throughout the world are becoming quite expensive. This fact will no doubt cause a more careful review of inventions before decisions are made to pursue the patent route. A favorable aspect of the West German law is the provision allowing for a delayed examination of the GFR patent application. An applicant can request examination within seven years after the application is filed in West Germany. If this request is not made, then the application automatically lapses. The apparent advantage of a delayed examination system is that it allows the patent applicant a period of time to evaluate the value of the invention and the desirability of obtaining a patent and paying the added costs of procuring a patent.

---

* See the discussion of Canadian patent law and the recent Canadian Patent Office Board of Appeals decision in the Abitibi case (Section 26.4.1). The Canadian Board took a position opposite that of the German Patent Law.

26.5.3 Belgium

Belgian patent law (1854) allows patents for microbiological inventions, including microbes *per se*. When reference is made to microbiological inventions, it should be understood that genetic engineering inventions of a microbiological nature are also included. Inventions which encompass microbial hosts, cloned genes, and the like are patentable subject matter. Also, genetic chemical-type inventions, such as synthesizing and purifying nucleotide sequences, are patentable subject matter.

Subject matter that is not patentable in Belgium includes natural products *per se* and the discovery of natural phenomena. These limitations are similar to those found in other major patent systems.

At the present time, Belgium is rather loose with regard to culture deposits in conjunction with the filing of a patent application. Such a requirement is not specified in the Belgian patent law, but the better practice would be to deposit the culture as is done under the United States patent law, for example, and avoid the possibility of future litigation.

Belgian patent law allows for publication of patent applications 18 months from the priority date. Whether a deposited culture should be released at this time is a decision that the patent applicant must make since there is no statutory requirement in Belgium to this effect. If the patent applicant has also filed in a country, such as Great Britain or the German Federal Republic, which requires release of the culture, then there is no problem because the culture will be available. It is unlikely that a patent applicant will file in Belgium without also filing in Great Britain and the German Federal Republic. Thus, so long as the laws of Great Britain or West Germany require release of the culture at first publication of the patent application, the vagary of Belgium's or other countries' laws on the point is *de minimis*.

The Belgian patent system is not much more than a registration system since there is no preliminary novelty examination. A patent application is filed, it is published 18 months from the priority date, and subsequently a patent is granted.

26.5.4 The Netherlands

Patentable subject matter under the Dutch patent law (Netherlands, Patent Acts of the Kingdom 1910/78) includes microbiological processes and microbes *per se*. Again, this would encompass genetic engineering inventions concerning such entities. Claims to the microbes *per se* can recite the binomial name and the repository-assigned accession number, *e.g. Mycobacterium phlei* NRRL B-8154.

The examination of Dutch patent applications, as contrasted to, for example, that of Belgian applications, is extensive. Dutch patent examiners are very thorough in their examination for novelty and unobviousness of the claimed invention. Rejections on obviousness over the prior art are especially difficult to overcome by the patent applicant, as it appears that Dutch patent examiners readily form conclusions of obviousness. Further, these conclusions are not easily reversed, even with use of probative affidavit evidence. In short, Dutch patent rejections based on obviousness are a real problem for microbiological patent applicants.

The Dutch patent system is rather lenient in its culture deposit requirements. It is not necessary to deposit a pertinent culture before the priority filing date nor before the filing date of the Dutch patent application. During the examination procedure of the Dutch patent application, which examination can be delayed up to seven years from the filing date in the Netherlands, the Dutch patent examiner will request that the pertinent microbe be deposited. This seems to be a very favorable feature for microbiological inventors, one that could well be used to improve patent systems in other European countries.

The Dutch system requires the release of the culture deposit to the public when the patent application is published for opposition purposes. This period of opposition occurs after the Dutch patent examiner has found the claimed invention to be patentable. The culture release feature is identical to the Japanese procedure discussed below.

The Dutch patent system is a strict novelty system, *i.e.* publication of the claimed invention before the priority filing date will bar patent rights except under certain specified situations. The term of a Dutch patent is 20 years from the Dutch filing date. The Dutch system has a delayed examination feature like that discussed previously for West Germany.

26.5.5 France

French patent law (1978) provides a broad range of protection for biotechnological inventions. Chemical compounds *per se* are patentable, as are chemical and microbiological processes. Thus the areas of microbiological and genetic engineering inventions are well covered. Included within this coverage are microbes *per se*.

The culture deposit requirements under the French patent law are similar to the British requirements discussed previously. This includes the necessity to disclose the culture deposit date of the pertinent microbe in the patent application when it is filed with the French Patent Office, if the culture is new and not available to the public. This deposited culture must be made available to the public upon the initial publication of the French patent application, which occurs 18 months from the application priority filing date. A similar provision is found in the British and West German systems discussed previously.

Recent prosecution of French patent applications for biotechnological inventions indicates that French patent examiners are applying a more demanding test of unobviousness to the claimed invention than they have in the past.

26.5.6 European Patents

The European Patent refers to the patent issued under the European Patent Convention (EPC) (1977). This is a single-application patent system that services 11 countries: Austria, Belgium, France, Federal Republic of Germany, Italy, Liechtenstein, Luxemburg, The Netherlands, Sweden, Switzerland and the United Kingdom. This system is an attempt by the member countries to provide a unitary patent system and reduce the overall cost of patent protection. Under the system a single patent application is filed and processed under a single set of rules. Upon a finding of patentable subject matter, national patents of the member countries can be issued.

An advantageous feature of this patent system is that applications can be filed in English, French or German. However, nationals of member states that have other official languages have leeway to file certain documents before translations to one of the above three languages have to be provided.

When a European Patent application is filed, the patent applicant must designate EPC countries where patents are desired. This designation can be reduced in number but not increased during the subsequent patent application processing. Thus the normal practice is to designate a rather full complement of countries at the outset. Designation of countries will be influenced by the nature of the invention and a consideration of where it might be commercially practiced. For example, microbiological process inventions should be considered for filing in EPC countries known for their fermentation capabilities, *e.g.* the German Federal Republic, Italy, The Netherlands and the United Kingdom.

The EPC system provides broad patent protection for biotechnological inventions. This protection encompasses microbiological processes and microbes *per se*. Within this scope would come genetic engineering inventions such as host microbes, cloned genes, plasmids, and the like.

Where a microbe is new and unavailable to the public, and such microbe is necessary to practice the disclosed and claimed invention, then the microbe culture must be deposited in a recognized culture repository; this act must precede the filing of the EPC patent application. In addition, the deposit accession number assigned by the culture repository eventually must be disclosed in the EPC patent application. This culture accession number can be provided within a period of 16 months after the date of filing the EPC patent application, or, if priority is claimed for a previous patent application filing, *e.g.* a United States application, 16 months from such filing. Thus, if a United States patent application is filed initially, an EPC application will be filed within 12 months in order to obtain the benefit of the United States filing date as the priority date. In such a case, the disclosure of the culture deposit accession number will not be a problem because the United States patent law requires that disclosure when the United States application is filed.

As in the British and German Federal Republic patent systems, the EPC patent application is published 18 months after the priority date. At this time, deposited and disclosed culture deposits must be made available to the public. This is clearly a disadvantageous feature for biotechnological inventors. The EPC system provides for conditioning the culture release by a contractual agreement between the patent applicant and the person(s) desiring the culture deposit. Such an agreement falls far short of protecting the biotechnological inventor's real interests since, as any microbiologist knows, once a culture is in hand, it is a relatively easy task to subculture and transfer samples thereof surreptitiously. The best situation from a biotechnological inventor's position is that there be no required culture deposit release until patent rights have been granted.

When we are discussing culture repositories, the aspect of recognized repositories under the EPC system must be identified. Such repositories are those that are known in the microbiological art as being competent to maintain, store and disseminate various microbes. Some recognized repositories are more adept at handling viruses, others at handling *Streptomyces*, bacteria and the like. Depending on the nature of the biotechnological invention, certain culture repositories will be more appropriate for a particular invention than other repositories. Presently, there are a total of nine culture repositories that have been recognized as competent by the EPC system (Table 1). Each of these repositories is quite competent, and though there is some duplication in microbes that can be deposited in them, there is also a degree of specialization. The biotechnological inventor can readily obtain details of the depositing operations of the recognized repositories by corresponding with them.

Having discussed the uniqueness of the EPC in terms of a single patent application that blossoms forth into a number of national patents, it should be restated that the EPC system provides good patent coverage for biotechnological inventions. An EPC patent can cover chemical products *per se* and agricultural and horticultural processes. The EPC recognizes microbiological processes and products produced thereby as patentable subject matter. Also, genetic engineering inventions such as vectors, hosts, cloned genes, nucleotide sequences, and the like are patentable subject matter. In short, the EPC scope of patentable subject matter is in line with that of the major industrialized patent systems such as the United States, the United Kingdom, the Federal Republic of Germany, Japan and Canada.

Examination of an EPC patent application is somewhat similar to that found in the United States. There is a test for novelty and one for obviousness. The novelty test is unlike the United States test for novelty since the EPC test requires absolute novelty. This means that any publication of the claimed invention before the priority filing date is a statutory bar to the obtention of an EPC patent. In the United States an invention can be the subject of a United States patent application so long as the United States application is filed within one year of the publication of the invention.

The obviousness test in the EPC system is very reasonable and very similar to that of the United States system. Thus, evidence used in the United States to overcome a rejection of obviousness is acceptable type of evidence in the EPC system.

Except for the required release of a pertinent culture deposit at the initial publication of an EPC patent application, the EPC system is a sound and beneficial patent system for biotechnological inventors. Since the system has only been in force since October 7, 1977, it is still too early to report on court cases dealing with EPC patents. However, it is expected that such cases when they occur will, on balance, support the EPC system.

The term of a European patent is 20 years from the date of filing in all designated contracting countries.

26.6 PATENT PROTECTION IN SWITZERLAND FOR BIOTECHNOLOGICAL PROCESSES AND PRODUCTS

The Swiss patent law gives patent protection for inventions which are applicable in industry. Accordingly, much of what is invented in the nature of biotechnological processes and products can be patented under the Swiss Patent Act. The new Swiss patent law came into force on January 1, 1978.

26.6.1 Patentable Subject Matter Under the Swiss Patent Act

Microbiological processes and products obtained by such processes are patentable under the Swiss Patent Act. Within this scope of patentable subject matter would fall antibiotics, amino acids, and the like, and the microbiological processes for preparing such products. In the genetic

Table 1   Culture Repositories Recognized by the EPC

Agricultural Research Culture Collection (NRRL)
1815 North University Street
Peoria, IL 61604, USA

American Type Culture Collection (ATCC)
12301 Parklawn Drive
Rockville, MD 20852, USA

Centraalbureau voor Schimmelcultures (CBS)
Oosterstraat 1
P.O. Box 273
3740 AG
Baarn
The Netherlands

Culture Centre of Algae and Protozoa
36 Storey's Way
Cambridge, CB3 0DT
United Kingdom

Deutsche Sammlung von Mikroorganismen (DSM)
Grisebachstrasse 8
D–3400 Göttingen
Federal Republic of Germany

Fermentation Research Institute (FRI)
Agency of Industrial Science and Technology
Ministry of International Trade and Industry
1–3 Higashi 1-chome, Yatabe-machi
Tsukuba-gun, Ibaraki-ken 305
Japan

National Collection of Industrial Bacteria (NCIB)
Torry Research Station
P.O. Box 31
135 Abbey Road
Aberdeen AB9 8DG
United Kingdom

National Collection of Type Cultures (NCTC)
Central Public Health Laboratory
Colindale Avenue
London NW9 5HT
United Kingdom

National Collection of Yeast Cultures (NCYC)
Food Research Institute
Colney Lane
Norwich
Norfolk NR4 7UA
United Kingdom

engineering field, subject matter such as novel polypeptides and the microbiological processes for preparing such products would be included. Such microbiological processes could utilize microbes carrying cloned genes that express the desired polypeptide. It is also feasible in the present genetic engineering art for antibiotics to be produced by using microbes that have been genetically engineered to express higher titers of a desired antibiotic.

The Swiss patent law does not allow for the grant of patents for new varieties of plants or animal breeds, or for essentially biological processes for producing plants or breeding animals. This is a severe limitation in the Swiss patent law for the many inventions which are expected to emanate from genetic engineering research.

As with other patent systems throughout the world, the Swiss system also excludes a standard list of inventions that might be contrary to public order or morality. Offhand it is not seen how such an exclusion would affect adversely the affairs of biotechnological inventors.

Other exclusions from patent protection in Switzerland are

methods of surgical or therapeutic treatment and of diagnosis applied to the human body or to the bodies of animals. (Sect. 2[b])

This exclusion can adversely affect biotechnological inventors in the rapidly emerging field of monoclonal antibodies. The use of such antibodies as diagnostic tools is already a reality. It is believed that in short order monoclonal antibodies may occupy a significant position in the therapeutic treatment of humans and animals.

26.6.2 Statutory Requirements for Obtaining a Patent in Switzerland

Switzerland is an absolute novelty country. Therefore, publication of an invention prior to the priority filing date of a patent application for the published invention will in most instances bar the obtention of patent rights in Switzerland. The Swiss Patent Act states in Sections 7 and 7b that

> 7.-(1) *An invention that is not included in the state of the art shall be deemed to be new.*
>
> 7.-(2) *The state of the art shall consist of everything that has been made accessible to the public prior to the filing or priority date, by means of a written or oral description or a practice, or by any other means.*

However,

> 7b. *Where the invention has been made accessible to the public during the six months prior to the application date or priority date, this disclosure shall not be included in the state of the art when it is the direct or indirect result of:*
>
> (a) *a manifest abuse in relation to the applicant or his legal predecessor; or*
>
> (b) *the fact that the applicant or his legal predecessor has displayed the invention at an official or officially recognized exhibition within the meaning of the Convention of November 22, 1928, concerning international exhibitions, where he has declared the fact at the time of filing and has produced sufficient supporting evidence in good time.*

Section (a) protects the inventor–applicant from losing Swiss patent rights because of publication by another without the applicant's permission. The applicant has six months from such a publication to file an application.

Section (b) also allows publication of the invention at a recognized exhibition, perhaps as a part of a poster session, without loss of patent rights as long as an application is filed within six months of the publication date.

It should be recognized that these kinds of exemptions are not available in all the other strict novelty patent systems. Thus the better practice is to 'patent first, publish second' (Saliwanchik, 1982).

In obtaining a Swiss patent for a biotechnological invention that uses a new and previously unavailable microbe, Section 27 of the Swiss Patent Rules must be recognized. This section is as follows:

> 27.—(1) *Where an invention concerning a microbiological process or a product obtained by such a process entails the use or production of a microorganism to which the public does not have access and which cannot be described in the technical documents in such a way as to enable a person skilled in the art to carry out the invention, the applicant shall supplement the description of the invention by the reference in it to the deposit of a microorganism culture.*
>
> (2) *The culture shall be deposited, not later than on the filing date of the application, with an institution recognized by the Bureau possessing a culture collection.*
>
> (3) *The deposited culture shall form an integral part of the description by virtue of the fact that the description refers to it.*
>
> (4) *The reference shall contain the following information:*
>
> (a) *the designation of the institution possessing the culture collection;*
>
> (b) *the date of deposit;*
>
> (c) *the serial number under which the culture is registered with the institution possessing the culture collection.*
>
> (5) *When the date and serial number of the deposit are indicated within two months following the date of deposit, they shall be deemed to have been deposited on that date.*
>
> (6) *The release of samples of the culture to third parties may be made subject to the condition that those third parties communicate to the institution possessing the culture collection their names and addresses for the information of the party having made the deposit, and that they undertake not to make the culture available to other parties.*

Section 27 can be compared to British patent practice, which also requires the disclosure of the date the culture is deposited.

26.6.3 Protective Features of a Swiss Patent

As with patents in other countries, the measure of protection in a Swiss patent are the claims. The Swiss law allows for 10 claims tax-free. A tax is assessed for claims in excess of 10.

Swiss law has a rather elaborate set of rules on the structuring of claims. Anyone venturing into Switzerland with an intent to patent an invention must engage the services of a Swiss patent person to ensure compliance with Swiss rules for presenting claims. For one thing, microbes *per se* can be claimed only in dependent form to a process claim, for example: means for implementing the process of claim 1, characterized by use of the microbe *E. coli*, NRRL 0010. Interestingly, where a process claim discloses the use of a mutant microbe and the patent application discloses how the microbe was made, then the parent microbe must be deposited with a culture repository and subsequently made available to the public. The mutant, apparently, does not need to be deposited under such a condition. As a practical matter, it is recommended that the mutant also be deposited and made available. Clearly, in practicing the disclosed and claimed invention a person skilled in the art would find access to a subculture of the mutant more enabling than any written disclosure of how the mutant was made.

The maximum term of a Swiss patent is 20 years from the Swiss filing date of the application. However, a Swiss patent can expire prematurely (a) if the owner renounces it by a written declaration to the Federal Bureau of Intellectual Property; (b) if an annual fee due is not paid within the prescribed time (Federal Law on Patents for Inventions, Section 15).

There are a series of legal restrictions on Swiss patent rights detailed in Sections 35 to 40 of the Federal Law on Patents for Inventions. These restrictions severely limit the value of Swiss patents. There is a right of prior use which is stated in Section 35 as follows:

> 35.—(1) A patent shall have no effect against persons who, prior to the date of filing of the patent application, were in good faith professionally using the invention in Switzerland or had made special preparations for that purpose.
>
> (2) Any such person may use the invention for the purpose of their trade or business; this right may be transferred only by assignment or inheritance, together with the trade or business.

Section 36 gives a subsequent patentee the right to obtain a license from a dominating patent holder.

If a patentee has not worked the patented invention in Switzerland 'after a period of three years from the date of grant of a patent, but not sooner than four years after the filing of the application, then any person may be granted a license to use the invention' (Section 37).

Licenses deemed to be in the public interest are covered by Section 40 as follows:

> 40.—(1) Where the public interest so dictates, the person to whom the owner of the patent has, without sufficient reason, refused to grant the license requested may apply to the court for the grant of a license to use the invention.
>
> (2) The judge shall decide upon the extent and duration of the license and upon the compensation to be paid.

26.7 PATENT PROTECTION IN JAPAN FOR BIOTECHNOLOGICAL PROCESSES AND PRODUCTS

The importance of the patent system to the biotechnological inventor in Japan cannot be overstated. Biotechnology is big research and business in Japan, as everyone knowledgeable in the field recognizes. Legally protecting the fruits of this research is the basic purpose of the Japanese patent system.* On the whole, this patent system does a good job in achieving its designated goal. Evidence of this might be found in the fact that when patent applications are filed in countries outside the origin of the inventor, Japan is generally a country where an application is filed. This indicates the stature of the Japanese patent system in the eyes of those seeking world-wide legal protection for their biotechnological inventions. Of course, the fact that Japanese industry is highly capable in the biotechnological field also bears on the decision to file patent applications in Japan. One of the most difficult and costly aspects of filing in Japan is the need for translating the patent application into the Japanese language. A seemingly desirable feature of the Japanese patent law is that upon filing a Japanese patent application, the patent applicant can allow the application to lie dormant in the Japanese Patent Office for up to seven years. Within this seven-year period, the applicant must request examination of the application for patentability. If this

---

* Japanese Patent Law of 1978, Chapter I, Section 1 states, 'The purpose of this Law shall be to encourage inventions by promoting their protection and utilization so as to contribute to the development of industry.'

request is not made, then the application automatically lapses. A similar delayed examination system is also available under the German Federal Republic and The Netherlands patent laws, as discussed previously.

26.7.1 Patentable Subject Matter Under the Japanese Patent Act

Microbiological processes as well as microbes *per se* are patentable. In this category are found microbiological processes for preparing antibiotics and nutrients such as amino acids. Antibiotic compounds as well as other chemical compounds are patentable in Japan. In the same view, the many inventions emanating from genetic engineering research concerning microbial hosts, vectors, cloned genes, nucleotide (DNA) sequences, enzymes and the like are patentable subject matter. Table 2 lists genetic engineering patent applications which were filed with the Japanese Patent Office during the period January, 1980 to March, 1981 (*Patents and Licensing*, 1981).

Table 2 Gene Engineering Patent Applications Filed in
Japan (1980–1981)[a]

Applicant	Cases
US	
Upjohn	9
Cetus	1
G. D. Searle	1
Harvard University	2
University California	1
Stanford University	1
France	
Pasteur Institute	3
French Research Evaluation Bureau	1
West Germany	
Life Engineering Institute	1
Antilles, the Netherlands	
Biogen	1
Japan	
Ajinomoto	3
Mitsubishi Chemical Industries	1
Sumitomo Chemical	1
Noda Industrial Science Institute	2
Agency of Industrial Science and Technology, MITI	1

[a] Adapted from *Patents and Licensing* (1981).

Since research in the area of recombination of the genes of 'higher animals' is not permitted in Japan at present, inventions in this area are not patentable.

Unpatentable inventions are dealt with in Chapter I, Section 32 of the Japan Patent Law (1978) as follows:

> 32.—*The following inventions shall not be patented, notwithstanding Section 29:*
> (i) *inventions of substances manufactured by the transformation of the atom;*
> (ii) *inventions liable to contravene public order, morality or public health.*

On balance, the Japan Patent Law recognizes as patentable subject matter a rather complete scope of inventions that can be expected to emanate from the total biotechnological field. Perhaps as time passes there will be a recognition that genetic engineering research in animals is a valid endeavor and, thus, should be supported by the patent system. The expected developments in genetic engineering research relating to farm animals could certainly go far to benefit the human race in terms of better food production. Such innovation is of interest to all countries in view of projected populations and food problems. The previously discussed Canadian Abitibi decision (see Section 26.4.1) is an encouraging sign that genetic engineering inventions encompassing higher living forms than microbes will be patentable subject matter in more countries as time passes.

26.7.2 Statutory Requirements for Obtaining a Patent

The filing of a Japanese patent application should be done with care commensurate with that used in filing a patent application under the patent system of most countries. The disclosure of the invention in the application must be of such a caliber that a person skilled in the art can practice the invention from the disclosure without undue experimentation. In the field of biotechnological inventions, a written disclosure of the invention may not be enough to accomplish this goal. For example, if a new and publicly-unavailable microbe is used in the invention, then a written characterization of the microbe is not a sufficient disclosure. Additionally, it is essential, as it is in other countries, that a subculture of the pertinent microbe be deposited with a recognized culture repository before the Japanese patent application is filed. The culture accession number assigned by the culture repository must then be disclosed in the Japanese patent application when it is filed. Even though a culture is deposited it is still necessary for the written patent application disclosure to contain a full taxonomic description of the new microbe. For example, where the new microbe is an actinomycete, the following taxonomic features, as delineated by the Japanese Patent Office, must be disclosed.

(a) Morphological characteristics—describe branching (whether simple branching or axle branching) and form (whether straight, curve, loop or spiral shape) of spore-forming hypha; the number of spores (whether monospore or more, dispore or more, or 10 spores or more); surface structure and size of spore; existence of flagellated spores; existence of sporangium; growth location of sporophore (whether basal hypha or aerial hypha); and, if necessary, fission condition of hypha (mode of fission and time) and existence of formability of sclerotium.

(b) Growth condition on various media—(1) sucrose–nitrate–agar medium, (2) glucose–asparagine–agar medium, (3) glycerine–asparagine–agar medium, (4) starch–agar medium, (5) tyrosine–agar medium, (6) nutrient agar medium, (7) yeast–malt–agar medium and (8) oatmeal–agar medium.

With regard to growth condition on these media, color of the surface of colony, color of the surface and reverse of basal hypha, diffusible pigment into medium, and so on should be mentioned.

(c) Physiological properties—(1) range of growth temperature conditions, (2) liquefaction of gelatine (on glucose–peptone–gelatine medium), (3) starch hydrolysis (on starch–agar medium), (4) coagulation and peptonization of skim milk, (5) melanin-like pigment production (on tyrosine–agar medium and peptone–yeast–iron agar medium).

(d) Utilization of the following carbon sources (on Pridham–Gottlieb's agar medium)—(1) L-arabinose, (2) D-xylose, (3) D-glucose, (4) D-fructose, (5) sucrose, (6) inositol, (7) L-rhamnose, (8) raffinose and (9) D-mannitol.

In general, it can be stated that when a taxonomic description is provided it should be in accord with what is recognized in the art as being sufficient to describe the particular type of microbe involved. Experience with prosecuting numerous biotechnological patent applications throughout the world, including in the United States and Japan, has shown that the Japanese patent examiner is more demanding in the area of taxonomic data than patent examiners in any other country.

With regard to depositing the pertinent microbe, it is now possible to use a recognized culture repository outside Japan. This has resulted from the Japanese signing of the Budapest Treaty (see Section 26.2 for a discussion of the Budapest Treaty).

26.7.3 Protective Features of a Japanese Patent

Japanese patents can contain both product and process claims. Product claims are acknowledged to be better since they cover the claimed product by whatever method it is produced. On the other hand a process patent covers only the process claimed. A process claim in Japan is broader in scope than in the United States. Under Japanese patent law, if a product is imported into Japan, where there exists only a process type claim covering the preparation of the product, there is a presumption in favor of the patentee that the imported product was made by the process patented in Japan. In such a case, the patentee can obtain an injunction barring entry of the product into Japan. The person desiring to import the product into Japan can rebut the presumption by showing that a process different from the patented process was used to make the product. Thus, the Japanese patent law has extraterritorial effect in the sense that a patentee of a process can prevent the importation of a product into Japan if the process patented in Japan was used to make the product.

In contrast, a United States patent with process claims does not have extraterritorial effects as found in the Japanese system. This means that even though a patented United States process was used to make a product that is then imported into the United States, the United States patent system is ineffective to bar entry of the product. The only recourse in the United States is under the Tariff Act, which is not effective unless the imported product can be shown to have a detrimental effect on United States industry. This places a burden upon the United States patentee to make the case of unfair competition.

A Japanese patent can exist for 15 years from publication on the merits, but not exceeding 20 years from the Japanese filing date.

The biotechnological inventor claiming a mutant microbe *per se* must structure the claim so as to include more than the mutant microbe name and deposit number. Presently, the Japanese patent examiners require that such claims also contain distinguishing differences between the mutant and the parent microbe. Critical differences in, for example, antibiotic or other metabolite production is not being readily accepted by the Japanese patent examiners. It seems that the Japanese examiner considers morphological differences more definitive and desirable in the claims. This position does not seem to be scientifically valid, since many mutants have essentially the same morphological properties as the parent microbe, but differ significantly in ability to produce a desirable product, *e.g.* an antibiotic or a steroid intermediate. The mutation giving rise to the mutant may affect only a particular gene(s) concerning product formation without affecting anything governing the morphological characteristics of the parent microbe. Therefore, distinguishing differences should be those differences which make the mutant a useful entity (Saliwanchik, 1977). It is the utility of any invention, microbe or otherwise, which gives rise to the interest in it in Japan or elsewhere. Because of what useful inventions can do for the public good, there is a social justification for the granting of patent monopoly rights when the inventions are fully disclosed to the public.

26.8 TRADE SECRET PROTECTION AS AN ALTERNATIVE TO PATENTS

Patent protection is a good form of legal protection for a large number of biotechnological inventions. Some of the inventions that can be adequately protected by patents are listed in Table 3. This list only illustrates the types of inventions that can be protected effectively by the use of patents. Inventions that are not necessarily patentable under the laws of a particular country, or are arguably protectable at best, would be better held as trade secrets.

Trade secret protection does not emanate from an examination procedure by a government agency as does patent protection. While patents are referred to as being available because of statutory authority, trade secret protection arises from the common law recognition that a person is not obligated to disclose to anyone that which comes under the name of 'idea' or 'invention'. This means that if a person thinks of a new way of doing something, for example of making an old antibiotic, this person is not obligated by any law to disclose the process to the public. The process can be maintained in secrecy as a trade secret.

There are advantages and disadvantages in maintaining an invention as a trade secret, as opposed to disclosing the invention to a governmental agency (the patent office) to obtain a patent right. These factors should be recognized before any conscious choice is made between trade secret and patent as the form of legal protection for a particular invention.

26.8.1 Advantages of a Trade Secret

(1) Trade secret protection is not limited in time so long as the information or invention that is the *res* (*i.e.* thing (L.), or basis) of the trade secret is kept secret: secrecy as well as novelty are the touchstones of a trade secret.

(2) Where a patent will only be granted for an invention that meets the statutory requirements of patentability in the country where a patent is sought, a trade secret can be maintained without meeting any statutory tests so long as the trade secret is unique and secret.

(3) When an invention is maintained as a trade secret, competitors are not apprised of the secret as they are when patents are the legal form of protection and a full disclosure of the invention to the governmental agency granting the patent is the *quid pro quo* for the patent.

(4) It is true that the enforcement of a patent right can be a very expensive undertaking. On the other hand, the maintenance of a trade secret can be accomplished with a fraction of the cost, since trade secret costs are primarily internal ones, to maintain control over the dissemination of the trade secret.

(5) Trade secret as a form of legal protection appears to be especially appealing for those bio-

Table 3 US Patents Exemplifying Biotechnological Inventions

Antibiotics

US Pat. 3 814 796	Antibiotics zorbonomycin and zorbonomycin B and process for the preparation thereof
US Pat. 4 209 611	Dihydrosteffimycin compounds
US Pat. 4 263 404	Rifamycin R and microorganisms useful in the production of rifamycins
US Pat. 4 267 112	Antibiotic U-58 431
US Pat. 4 275 155	Manufacture of aminoglycoside antibiotics
US Pat. 4 276 412	3-Trehalosamine compounds
US Pat. 4 277 466	Complexes of DNA and esters derived from daunorubicine, their preparation and use
US Pat. 4 310 627	Process of producing 3-trehalosamine by fermentation

Genetic Engineering

US Pat. 4 237 224	Process for producing biologically functional molecular chimeras
US Pat. 4 264 731	DNA joining method
US Pat. 4 273 875	Plasmid and process of isolating same
US Pat. 4 293 652	Method for synthesizing DNA sequentially
US Pat. 4 332 898	Hybrid plasmid and process of making same
US Pat. 4 332 900	Construction of co-integrate plasmids from plasmids of *Streptomyces* and *Escherichia*

Transformation of Steroids

US Pat. 4 211 841	Process for microbial transformation of steroids
US Pat. 4 273 872	Method for the microbiological conversion of steroids

technological inventions that cannot be readily ascertained by the public. For example, the identity of high-producing antibiotic microbes in a production unit cannot be ascertained from the marketing of the antibiotic product. Accordingly, such high-producing microbes are best kept as a trade secret. Should such inventions be protected by patents, as they could be, the valuable microbes would have to be made available to the public in return for the patent right. True, the public would not be able to practice the patented invention using the microbe during the valid term of the patent, but the public could experiment with the microbe and, by mutation, perhaps make a better microbe. Theoretically, the patentee still has a dominant position on any newly developed mutants, but in practice, information about the mutant would at best be difficult to obtain.

26.8.2 Disadvantages of a Trade Secret

There are disadvantages in maintaining an invention as a trade secret. This is particularly true when the invention is ostensibly patentable under the laws where a patent can be obtained.

(1) Trade secret protection is viable only so long as the secret remains.

(2) It is perfectly proper to try to discover a trade secret by legitimate means, *e.g.* reverse engineering of a product presented to the public. As indicated above, reverse engineering is not always a feasible act for many biotechnological inventions. Therefore, the biotechnological field may occupy a unique position in terms of inventions that lend themselves to trade secret protection.

(3) Another undesirable aspect of trade secret protection is that there can be no enabling publication if the trade secret protection is to continue. This means that researchers in an industrial or academic environment cannot speak or write in an enabling manner about the trade secret. Needless to say, this may not settle well with many investigators (Saliwanchik, 1978b). When a decision must be made between patents or trade secrets as the legal protection for an invention, the above discussion of advantages and disadvantages of a trade secret should be reconciled with

previously disclosed features and requirements of obtaining patents under various patent systems. Another factor to be taken into account, at least in countries outside the United States, is that a person who maintains an invention as a trade secret is not precluded from practicing the trade secret even though a patent subsequently issues on the invention. There are attempts being made to add this feature to the United States law but success has not been achieved. If such a system is available in the United States ultimately, then the importance of the patent system in the United States, as well as in other countries, for biotechnological inventions will have diminished significantly. At the same time, publication of innovations in the field of biotechnology will dry up. Should we reach this point, the purpose of the patent system, to promote the progress of science and the useful arts, will have been dealt a serious blow. Science and technology will revert to a system of no disclosure, as practiced in early times. Efforts should be directed to achieve a balance between patent and trade secret protection, a balance that will promote the progress of science and engineering and protect the legitimate legal interest of the biotechnological innovator.

26.9 REFERENCES

Argoudelis, A. D., C. DeBoer and T. E. Eble, *In re* (1970). *U.S. Court of Customs and Patent Appeals, U.S. Patents Quarterly*, **168**, 99–103.
Belgian Patent Law (1854). WIPO Publications Section, 34, chemin des Colombettes, 1211 Geneva 20, Switzerland.
Bergy, M. E., J. H. Coats and V. S. Malik, *In re*; Chakrabarty, A. M., *In re* (1979). *U.S. Court of Customs and Patent Appeals, U.S. Patents Quarterly*, **201**, 352–394.
British Patents Act (1977). Her Majesty's Stationery Office, London, England.
Business Week (March 1, 1982), p. 34.
Canada Patent Act, R. S., c. 203, s.1 (1970). Canadian Government Publishing Centre, Supply & Services Canada, Hull, Quebec, Canada KIA OS9.
European Patent Convention (1977). WIPO Publications Section, 34, chemin des Colombettes, 1211 Geneva 20, Switzerland.
Feldman, L. I. *v.* K. Aunstrup (1975). *U.S. Court of Customs and Patent Appeals, U.S. Patents Quarterly*, **186**, 108–113.
French Patent Law (1978). Association des Conseils en Propriété Industrielle, 21 Rue de Léningrad, 75008 Paris, France.
German Patent Law (1976). *The New German Patent Law (1981) in English and German*, H. Kinkeldey and W. Stockmair-Weinheim (translators). Verlag Chemie, Basel, Switzerland.
Japanese Patent Law (1978). WIPO Publications Section, 34, chemin des Colombettes, 1211 Geneva 20, Switzerland.
Netherlands, Patents Act of the Kingdom (1910/1978). WIPO Publications Section, 34, chemin des Colombettes, 1211 Geneva 20, Switzerland.
Patents and Licensing (April, 1981) **XI**, 2. Japan Engineering News, Inc., Tokyo.
Saliwanchik, R. (1977). Patentable distinctions among microorganisms. *Dev. Ind. Microbiol.*, **18**, 327–331.
Saliwanchik, R. (1978a). Patenting microbiological inventions. *Dev. Ind. Microbiol.*, **19**, 135–139.
Saliwanchik, R. (1978b). Microbiological inventions: Protect by patenting or maintain as a trade secret? *Dev. Ind. Microbiol.*, **19**, 273–277.
Saliwanchik, R. (1982). *Legal Protection for Microbiological and Genetic Engineering Inventions*. Addison-Wesley, Reading, MA.
Switzerland Federal Law on Patents for Invention (1954/1976); Ordinance on Patents for Inventions (Patent Rules) (1977). WIPO Publications Section, 34, chemin des Colombettes, 1211 Geneva 20, Switzerland.
United States Code, Title 35—Patents, Section 101 (1952). US Government Printing Office, Washington, DC, 20402.

APPENDIX A: US PATENT 3 988 441 — ANTIBIOTIC DISCLOSURE

United States Patent [19]

Hanka et al.

[11] **3,988,441**

[45] **Oct. 26, 1976**

[54] **ANTIBIOTIC U-43,120 AND PROCESS FOR PREPARING SAME**

[75] Inventors: **Ladislav J. Hanka; Paul F. Wiley,** both of Kalamazoo, Mich.

[73] Assignee: **The Upjohn Company,** Kalamazoo, Mich.

[22] Filed: **Oct. 2, 1975**

[21] Appl. No.: **619,074**

[52] **U.S. Cl.**............................... **424/117**; 195/80 R
[51] **Int. Cl.²**.. **A61K 35/74**
[58] **Field of Search**.................... 424/117; 195/80 R

[56]   **References Cited**
          UNITED STATES PATENTS

3,655,876   4/1972   Sciavolino et al. 424/117

3,857,936   12/1974   Avgoudelis 424/117

Primary Examiner—Jerome D. Goldberg
Attorney, Agent, or Firm—Roman Saliwanchik

[57]   **ABSTRACT**

New antibiotic U-43,120 producible by the controlled fermentation of the new microorganism *Streptomyces paulus*, NRRL 8115.

This antibiotic is active against Gram-positive and Gram-negative bacteria. Accordingly, it can be used in various environments to eradicate or control such bacteria.

2 Claims, 5 Drawing Figures

FIGURE 1.

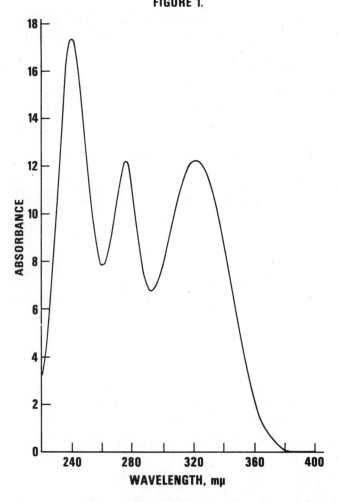

U.S. Patent   Oct. 26, 1976   **3,988,441**

FIGURE 2.

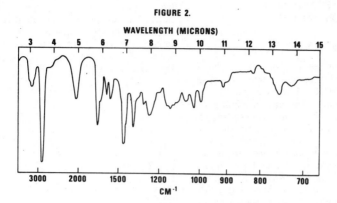

FIGURE 3.

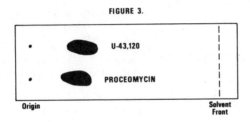

FIGURE 4.

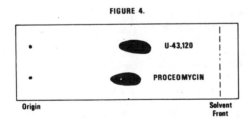

FIGURE 5.

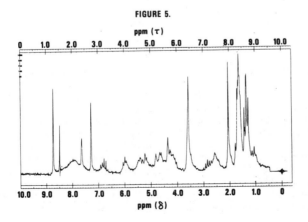

3,988,441

1

BRIEF SUMMARY OF THE INVENTION

The novel antibiotic of the invention, U-43,120 is obtained by culturing a new microorganism *Streptomyces paulus*, NRRL 8115, in an aqueous nutrient medium under aerobic conditions. Antibiotic U-43,120 and its metal salts have the property of adversely affecting the growth of Gram-positive bacteria, for example, *Staphylococcus aureus*, *Streptococcus hemolyticus*, *Diplococcus pneumoniae*, and *Streptococcus faecalis;* and Gram-negative bacteria, for example, *Escherichia coli*, *Proteus vulgaris*, *Klebsiella pneumoniae*, and *Pseudomonas aeruginosa*. Accordingly, U-43,120 and its metal addition salts can be used alone or in combination with other antibiotic agents to prevent the growth of or reduce the number of bacteria, as disclosed above, in various environments.

DETAILED DESCRIPTION OF THE INVENTION

Chemical and Physical Properties of U-43,120

Molecular Formula: $C_{34}H_{44}N_2SO_{18}$ (Calculated on the basis of analysis).

Elemental Analysis: Calcd.: C, 51.00; H, 5.53; N, 3.50; S, 4.01; O, 35.96. Found: C, 50.64, 51.34; H, 5.88, 6.00; N, 3.60, 3.65; S, 4.07, 4.06; O, 34.41.

Molecular Weight: 800.6

Melting Point: 119°–122° C.

Specific Rotation: $[\alpha]_D^{25} = +9.3°(c, 1, CHCl_3)$.

Ultraviolet Absorption Spectrum:

The ultraviolet absorption spectrum of U-43,120 is reproduced in FIG. 1 of the drawings.

Infrared Absorption Spectrum:

U-43,120 has a characteristic infrared absorption spectrum in a mineral oil mull as shown in FIG. 2 of the drawings. Peaks are observed at the following wave lengths expressed in reciprocal centimeters: 3560, 2070, 1730, 1690 (shoulder), 1620, 1580, 1295, 1260, 1140, 1060, 1025, 993, 913, 820, 753 and 725.

Thin layer chromatography comparison with proceomycin:

FIG. 3 of the drawings shows the movement of antibiotic U-43,120 on a silica plate (silica gel on microscope slide) using a solvent system consisting of ethyl acetate:ethanol:water (92:5:3).

FIG. 4 of the drawings shows a comparison of the movement of antibiotic U-43,120 and proceomycin on a silica plate, as described above. This plate was developed with the solvent system ethanol:methanol:water (10:9:1).

In the tlc of both FIGS. 3 and 4, the starting material is dissolved in methylene chloride at a concentration of 5 mg/ml. Two γ of solution is then applied to the tlc plate.

In addition to the above tlc of antibiotic U-43,120, it has also been determined that U-43,120 has an R_f of 0.62 inches on a silica plate using the solvent system methyl ethyl ketone:acetone:water (70:20:11).

Solubilities:

Soluble in chlorinated hydrocarbons, ethyl acetate, lower alcohols, pyridine, and acetone.

Nuclear Magnetic Resonance (NMR):

The NMR spectrum of antibiotic U-43,120 at 60 megacycles is shown in FIG. 5 of the drawings. The

2

NMR spectrum was observed on a Varian A-60 spectrometer on a solution (ca. 0.5 ml., ca. 15% concentration) of the sample of U-43,120 in deuterated pyridine (D_5C_5N). The spectrum was calibrated against internal tetramethylsilane and the precision of the $\Delta\gamma$ was $> \pm$ 1 c.p.s. Frequencies were recorded in c.p.s. downfield from tetramethylsilane.

Antibacterial Activity of U-43,120

Organism	No. of Strain	Minimal Inhibitory Concentration ($\mu g/ml$)
Staphylococcus aureus	UC 80	15.6
Streptococcus hemolyticus	UC 152	62.5
Streptococcus faecalis	UC 3235	500
Escherichia coli	UC 51	500
Proteus vulgaris	UC 93	500
Klebsiella pneumoniae	UC 57	500
Pseudomonas aeruginosa	UC 95	250
Diplococcus pneumoniae	UC 41	31.2

NOTE: UC refers to The Upjohn Company Culture Collection.

The tube dilution assay procedure was conducted with the medium BHI (Brain Heart Infusion Broth, Difco, Detroit, Michigan). Assay tubes (13 mm × 100 mm) were prepared in the customary manner as set out in Snell, E. E., Vitamin Methods, Volume 1, Academic Press, Inc., New York 1950, p. 327. Test organisms grown for 18 hours at 37° C. were used to inoculate the test medium. The assays were read at 17 hours.

The UC cultures referred to above can be obtained from The Upjohn Company of Kalamazoo, Michigan, upon request.

THE MICROORGANISM

The microorganism used for the production of U-43,120 is *Streptomyces paulus*, NRRL 8115.

A subculture of this microorganism can be obtained from the permanent collection of the Northern Regional Research Laboratory, U.S. Department of Agriculture, Peoria, Illinois, U.S.A. Its accession number in this depository is NRRL 8115. It should be understood that the availability of the culture does not constitute a license to practice the subject invention in derogation of patent rights granted with the subject instrument by governmental action.

The microorganism of this invention was studied and characterized by Alma Dietz of The Upjohn Research Laboratories.

A *Streptomyces sp.* isolated from soil has been characterized. This culture may be placed in the Helvolus series of Gauze [Gauze, G. F., T. P. Preobrazhenskaya, E. S. Kudrina, N. O. Blinov, I. D. Ryabova, and M. A. Iveshnikova. 1957. Problems in the classification of antagonistic actinomycetes. State Publishing House for Medical Literature, Moscow. English edition translated by Fritz Danga; David Gottlieb (ed.). The American Institute of Biological Sciences, Washington, D.C.], the Streptomyceten mit griseus-Luftmycel of Hütter [Hütter, R. 1967. Systematik der Streptomyceten. S. Karger, Basel. 382p.], the variants of *Actinomyces albus* of Krasil'nikov [Krasil'nikov, N. A. 1941. Keys to Actinomycetales. Academy Sciences of the U.S.S.R., Institute of Microbiology. English edition published for the U.S. Department of Agriculture and the National Science Foundation, Washington, D.C., by the Israel Program for Scientific Translations. Jerusalem. 1966.], or the "Yellow series (17.43b)" of Pridham and Tresner in Bergey's Manual, 8th Edition [Buchanan, R. E., and

3,988,441

3

N. E. Gibbons. 1974. Bergey's Manual of Determinative Bacteriology, Eight Edition, The Williams and Wilkins Co., Baltimore]. The soil isolate is differentiated from species in the references cited by the characteristics noted in the Tables. In Table 4 the culture is differentiated from *Streptomyces albidoflavus* [Buchanan, R. E., and N. E. Gibbons. 1974. Bergey's Manual of Determinative Bacteriology, Eight Edition, The Williams and Wilkins Co., Baltimore] [Shirling, E. B., and D. Gottlieb. 1969. Cooperative description of type cultures of *Streptomyces*. IV. Species descriptions from the second, third and fourth studies. Int. J. Syst. Bacteriol. 19:391–512] and *Streptomyces globisporus* [Buchanan, R. E., and N. E. Gibbons. 1974. Bergey's Manual of Determinative Bacteriology, Eight Edition, The Williams and Wilkins Co., Baltimore] [Shirling, E. B., and D. Gottlieb. 1968. Cooperative description of type cultures of *Streptomyces*. III. Additional species descriptions from first and second studies. Int. J. Syst. Bacteriol. 18:280–399]. [Sugimoto, Hiroshi. 1967. Lysis of yeast cell wall by enzymes from streptomycetes. Agr. Biol. Chem. 31:111–1123] the cultures to which it appeared similar.

S. albidoflavus is reported to have spiral spore chains by Gauze [Gauze, G. F., T. P. Preobrazhenskaya, E. S. Kudrina, N. O. Blinov, I. D. Ryabova, and M. A. Iveshnikova. 1957. Problems in the classification of antagonistic actinomycetes. State Publishing House for Medical Literature, Moscow. English edition translated by Fritz Danga; David Gottlieb (ed.). The American Institute of Biological Sciences, Washington, D.C.] and is placed in his "Albus Series". *Streptomyces globisporus* of Krasil'nikov [Krasil'nikov, N. A. 1941. Keys to Actinomycetales. Academy Sciences of the U.S.S.R., Institute of Microbiology. English edition published for the U.S. Department of Agriculture and the National Science Foundation, Washington, D.C., by the Israel Program for Scientific Translations. Jerusalem. 1966] belongs to the variants of *Actinomyces globisporus*.

The culture characterized is considered to be a new species of Streptomyces. The consideration is justified by the differences noted in Table 4 and in the references cited for the cultures with which it was compared. The new soil isolate is designated *Streptomyces paulus sp*. n. It is understood that this culture is the type species and that it will become the type variety should cultures with similar properties be isolated.

Color characteristics.

4

Aerial growth cream to olive. Melanin negative. Appearance on Ektachrome is given in Table 1. Reference color characteristics are given in Table 2. The culture may be placed in the yellow color group of Tresner and Backus [Tresner, H. D., and E. J. Backus. 1963. System of color wheels for streptomycete taxonomy. Appl. Microbiol. 11:335–338].

Microscopic characteristics.

Spore chains long, flexuous (RF) in the sense of Pridham et al. [Pridham, T. G., C. W. Hesseltine, and R. G. Benedict. 1958. A guide for the classification of streptomycetes according to selected groups. Placement of strains in morphological sections. Appl. Microbiol. 6:52–79]. Spore chains may be in tufts. Spores, examined with the scanning electron microscope, appear spherical with a smooth surface. The procedure was that cited by Dietz and Mathews [Dietz, A., and J. Mathews. 1960. Classification of Streptomyces spore surfaces into five groups. Appl. Microbiol. 21:527–533].

Cultural and biochemical characteristics.

Cultural and biochemical characteristics are cited in Table 3.

Carbon utilization.

In the synthetic medium of Pridham and Gottlieb [Pridham, T. G., and D. Gottlieb. 1948. The utilization of carbon compounds by some Actinomycetales as an aid for species determination. J. Bacteriol. 56:107–114], *S. paulus* growth on the control (basal medium without added compound) was moderate. Growth was good on D-xylose, L-arabinose, D-fructose, D-galactose, D-glucose, D-mannose, maltose, cellobiose, dextrin, soluble starch, glycerol, D-mannitol, salicin, sodium citrate, and sodium succinate; moderate on rhamnose, sucrose, lactose, raffinose, inulin, dulcitol, D-sorbitol, and inositol.

Table 1

Appearance of *Streptomyces paulus* on Ektachrome*		
Agar Medium	Surface	Reverse
Bennett's	Cream-white	Tan
Czapek's sucrose	Cream-white	Colorless
Maltose-tryptone	Cream-white	Tan
Peptone-iron	Cream-white	Tan
0.1 % Tyrosine	Trace cream-white	Tan
Casein starch	Trace cream-white	Very pale tan

*Dietz, A. 1954. Ektachrome transparencies as aids in actinomycete classification. Ann. N. Y. Acad. Sci. 60:152–154.

Table 2

			Reference Color Characteristics of *Streptomyces paulus* ISCC-NBS Color Charts Illustrated with Centroid Colors*	
Agar Medium	Determination	Chip No.	Color	
Bennett's	S	90	gy.y.	Grayish yellow
		92	y.white	Yellowish white
	R	95	m.Ol.Br.	Deep olive
	P	72	d.OY	Dark orange yellow
Czapek's sucrose	S	92	y.white	Yellowish white
	R	92	y.white	Yellowish white
	P	—	—	—
Maltose-tryptone	S	92	y.white	Yellowish white
		93	y.Gray	Yellowish gray
	R	95	m.Ol.Br.	Deep olive
	P	94	l.Ol.Br.	Olive
Hickey-Tresner	S	92	y.white	Yellowish white
		90	gy.y.	Grayish yellow
	R	77	m.y.Br.	Yellowish brown
	P	76	l.y.Br.	Dull yellowish brown
Yeast extract-malt extract (ISP-2)	S	89	p.y.	Pale yellow
		90	gy.y.	Pale yellow
	R	96	d.Ol.Br.	Dark brown
	P	95	m.Ol.Br.	Dark brown

3,988,441

Table 2-continued

Agar Medium	Determi-nation	Chip No.	Color	Reference Color Characteristics of *Streptomyces paulus* ISCC-NBS Color Charts Illustrated with Centroid Colors*
Oatmeal	S	92	y.white	Yellowish white
(ISP-3)	R	87	m.Y.	Yellow Brazil Wood (yellow)
		88	d.Y.	Yellow
	P	87	m.Y.	Yellow Brazil Wood (yellow)
Inorganic-	S	90	gy.y.	Pale yellow
salts-starch		92	y.white	Yellowish white
(ISP-4)		to		
		105 (edge)	gy.g.Y	Grayish greenish yellow
	R	95	m.Ol.Br.	Dark brown
	P	91	d.gy.Y.	Dark grayish yellow
Glycerol-	S	92	y.white	Yellowish white
asparagine	R	72	d.O.Y.	Dark orange yellow
(ISP-5)	P	72	d.O.Y.	Dark orange yellow

S = Surface
R = Reverse
P = Pigment
*Kelly, K.L., and D.B. Judd. 1955. The ISCC-NBS method of designating colors and a dictionary of color names. U.S. Dept. of Comm. Circ. 553, Washington, D.C.

Table 3

Medium	Cultural and Biochemical Characteristics of *Streptomyces paulus* Surface	Reverse	Other Characteristics
Agar			
Peptone-iron	Pale pink	Yellow-tan with red edge	No pigment Melanin negative
Calcium malate	Trace pale yellow	Pale yellow	No pigment Malate slightly solubilized undergrowth
Glucose asparagin	Cream	Yellow	Pale yellow pigment
Skim milk	—	Tan	Tan pigment Casein solubilized
Tyrosine	Cream	Tan	Tan pigment Tyrosine solubilized
Xanthine	Cream	Cream olive	Pale yellow-tan pigment Xanthine solubilized
Nutrient starch	Cream	Cream olive	Pale yellow-tan pigment Starch solubilized
Yeast extract-malt extract	Pale olive-cream	Tan-brown	Pale tan pigment
Peptone-yeast extract-iron (ISP-6)	Very slight trace white	Pale yellow tan	No pigment
Tyrosine (ISP-7)	Cream	Yellow tan	No pigment
Gelatin			
Plain	Cream white	—	Yellow pigment ¼ Liquefaction complete
Nutrient	White	—	Yellow pigment Liquefaction complete
Broth			
Synthetic nitrate	Trace white aerial growth on thin surface pellicle	—	Poor compact bottom growth Nitrate not reduced to nitrite
Nutrient nitrate	Cream aerial growth on surface pellicle	—	Yellow pigment No bottom growth Nitrate reduced to nitrite
Litmus milk	Blue-gray aerial growth on surface pellicle Blue-gray-green surface ring	—	Peptonization-partial to complete Litmus reduced pH 7.4–7.6

Table 4

Test Condition	Comparison of *Streptomyces paulus*, *Streptomyces albidoflavus*, and *Streptomyces globisporus* *Streptomyces paulus* NRRL 8115	UC 2190 (CBS)*	*Streptomyces albidoflavus* CBS 416.34 (ISP 5455)**
Spore chain morphology	Section Rectiflexibilis (RF long)	Section Rectiflexibilis (RF long)	Section Rectiflexibilis (RF long)
Spore surface	Smooth	Smooth	Smooth
Spore chains	Abundant	Abundant	Sparse
Aerial Mass Color	Yellow	Yellow	White or gray
Carbon utilization			
D-glucose	Good	Good	Good
L-arabinose	Poor (doubtful)	Poor	Good
Sucrose	Poor (doubtful)	Negative	Poor
D-xylose	very good	Very good	Doubtful
Inositol	Poor	Negative	Doubtful
D-fructose	Very good	Good	Doubtful
D-mannitol	Very good	Very good	Negative

3,988,441

Table 4-continued

Comparison of *Streptomyces paulus*, *Streptomyces albidoflavus*, and *Streptomyces globisporus*

Rhamnose	Good	Negative	Negative
Raffinose	Negative	Negative	Negative
Calcium malate agar	Malate slightly solubilized	Malate solubilized	—
Peptone-iron agar	Pale pink aerial growth	No aerial growth	—
Plain gelatin	Complete liquefaction	No liquefaction	—
Nutrient gelatin	complete liquefaction	Trace liquefaction	—
Litmus milk	Litmus reduced pH 7.4–7.6	Litmus reduced pH 6.9	—
Czapek's sucrose agar	Good aerial growth	No aerial growth	—
Antibiotic Produced	U-43120	None cited	None cited

	Streptomyces globisporus	
Test Condition	UC 5398 (NRRL B-2872)[***]	INMI 2302 (ISP-5199)[****]
Spore chain morphology	Section Rectiflexibilis (RF long)	Section Rectiflexibilis (RF long)
Spore surface	Smooth	Smooth
Spore chains	Abundant	Good
Aerial Mass Color	Yellow	Yellow
Carbon utilization		
D-glucose	Good	Good
L-arabinose	Good	Good
Sucrose	Good	Good
D-xylose	Very good	Negative
Inositol	Very good	Negative
D-fructose	Very good	Good
D-mannitol	Negative	Good
Rhamnose	Very good	Good
Raffinose	Negative	Negative
Calcium malate agar	Malate not solubilized	—
Peptone-iron agar	Cream white aerial growth	—
Plain gelatin	Trace liquefaction	—
Nutrient gelatin	Complete liquefaction	—
Litmus milk	Litmus reduced pH 7.7	—
Czapek's sucrose agar	Fair aerial growth	Excellent
Antibiotic Produced	None cited	None cited

NOTE: UC refers to The Upjohn Company Culture Collection.
[*]Received in 1954 before CBS cultures were numbered. (probably CBS 416.34 = type strain)
[**]Shirling, E.B., and D. Gottlieb. 1969. Cooperative description of type cultures of Streptomyces. IV. Species descriptions from the second, third and fourth studies. Int. J. Syst. Bacteriol. 19:391–512.
[***]Tresner, H.D., and E.J. Backus. 1963. System of color wheels for streptomycete taxonomy. Appl. Microbiol. 11:335–338.
[****]Shirling, E.B., and D. Gottlieb. 1968. Cooperative description of type cultures of Streptomyces. III. Additional species descriptions from first and second studies. Int. J. Syst. Bacteriol. 18:280–399.

The new compound of the invention is produced when the elaborating organism is grown in an aqueous nutrient medium under submerged aerobic conditions. It is to be understood, also, that for the preparation of limited amounts surface cultures and bottles can be employed. The organism is grown in a nutrient medium containing a carbon source, for example, an assimilable carbohydrate, and a nitrogen source, for example, an assimilable nitrogen compound or proteinaceous material. Preferred carbon sources include glucose, brown sugar, sucrose, glycerol, starch, cornstarch, lactose, dextrin, molasses, and the like. Preferred nitrogen sources include cornsteep liquor, yeast, autolyzed brewer's yeast with milk solids, soybean meal, cottonseed meal, cornmeal, milk solids, pancreatic digest of casein, fish meal, distillers' solids, animal peptone liquors, meat and bone scraps, and the like. Combinations of these carbon and nitrogen sources can be used advantageously. Trace metals, for example, zinc, magnesium, manganese, cobalt, iron, and the like, need not be added to the fermentation media since tap water and unpurified ingredients are used as components of the medium prior to sterilization of the medium.

Production of the compound of the invention can be effected at any temperature conducive to satisfactory growth of the microorganism, for example, between about 18° and 40° C., and preferably between about 20° and 28° C. Ordinarily, optimum production of the compound is obtained in about 3 to 15 days. The medium normally remains acidic during the fermentation. The final pH is dependent, in part, on the buffers present, if any, and in part on the initial pH of the culture medium.

When growth is carried out in large vessels and tanks, it is preferable to use the vegetative form, rather than the spore form, of the microorganism for inoculation to avoid a pronounced lag in the production of the new compound and the attendant inefficient utilization of the equipment. Accordingly, it is desirable to produce a vegetative inoculum in a nutrient broth culture by inoculating this broth culture with an aliquot from a soil, liquid N₂ agar plug, or a slant culture. When a young, active vegetative inoculum has thus been secured, it is transferred aseptically to large vessels or tanks. The medium in which the vegetative inoculum is produced can be the same as, or different from, that utilized for the production of the new compound, so long as a good growth of the microorganism is obtained.

A variety of procedures can be employed in the isolation and purification of the compound of the subject invention from fermentation beers, for example, solvent extraction, partition chromatography, silica gel chromatography, liquid-liquid distribution in a Craig apparatus, adsorption on resins, and crystallization from solvents.

In a preferred recovery process the compound of the subject invention is recovered from its culture medium by separation of the mycelia and undissolved solids by conventional means, such as by filtration or centrifugation. The antibiotic is then recovered from the filtered or centrifuged broth by extraction. For the extraction of U-43,120 from the filtered broth, water-immiscible organic solvents in which it is soluble, for example, chloroform, ethylene dichloride, ethyl acetate, and methylene chloride (preferred) can be used. Advantageously, the extraction is carried on after the filtered beer is adjusted to a pH of about 2 to 7 with a mineral acid. The methylene chloride extracts are combined and evaporated to dryness under vacuum.

The first step in purification of the methylene chloride extract, as described above, is the use of silica gel chromatography using as solvents ethyl acetate, ethyl acetate-ethanol-water, ethyl acetate-water, and chloroform-ethyl acetate. The active fractions from the silica gel column can be further purified by countercurrent distribution using cyclohexane-ethyl acetate- 95% ethanol-water (1:1:1:1) as the solvent system. Fractions from the countercurrent distribution can be subjected to further silica gel chromatography to obtain a more active preparation which is thus more pure. A final purification step to yield a pure crystalline product is conducted by using chloroform to crystallize the product from the silica gel chromatography.

Preparations of U-43,120 can be purified by repeated silica gel chromatography, as described above, without resort to countercurrent distribution.

Hereinafter are described non-limiting examples of the process and products of the subject invention. All percentages are by weight and all solvent mixture proportions are by volume unless otherwise noted.

EXAMPLE 1

Part A. Fermentation

A frozen vegetative seed of *Streptomyces paulus*, NRRL 8115, is used to inoculate a series of 500-ml Erlenmyer flasks each containing 100 ml of sterile seed medium consisting of the following ingredients:

Glucose monohydrate	25 g/l
Pharmamedia*	25 g/l
Tap water q.s.	1 liter

*Pharmamedia is an industrial grade of cottonseed flour produced by Traders Oil Mill Company, Fort Worth, Texas.

The seed medium presterilization pH is 7.2. The seed inoculum is grown for 2 days at 25° C. on a Gump rotary shaker operating at 250 r.p.m. and having a 2½ inch stroke.

Seed inoculum (5%), prepared as described above, is used to inoculate a series of 500 ml fermentation flasks containing 100 ml of sterile fermentation medium consisting of the following ingredients:

Glucose monohydrate	10 g/l
Malt extract	30 g/l
Wilson's Peptone Liquor No. 159*	20 g/l
Corn steep liquor	5 g/l
Tap water q.s.	1 liter

*Wilson's Peptone Liquor No. 159 is a preparation of hydrolyzed proteins of animal origin.

The inoculated fermentation flasks are incubated at a temperature of 25° C. for 5 days while being shaken on a Gump rotary shaker operating at 250 r.p.m. and having a 2½ inch stroke. Foaming in the fermentation flasks is controlled by the antifoam agent UCON (a synthetic defoamer supplied by Union Carbide, N.Y., N.Y.). A representative 5-day fermentation has the following titers of antibiotic in the fermentation broth:

Day	Assay, BU/ml
2	0
3	38
4	89
5	101

The assay is a disc plate biounit assay using the bacterium *Bacillus subtilis* grown on a synthetic medium. The synthetic medium has the following composition:

$Na_2HPO_4 \cdot 7H_2O$	1.7 g
KH_2PO_4	2.0 g
$(NH_4)_2SO_4$	1.0 g
$MgSO_4$	0.1 g
Glucose	2.0 g
Bacto Agar*	15.0 g
Distilled water	1 liter
Metallic ion stock solution**	1 ml

*Bacto Agar provided by Difco Laboratories, Detroit, Michigan.
**Metallic ion stock solution consists of the following:

$NaMoO_4 \cdot 2H_2O$	200 µg/ml
$CoCl_2$	100 µg/ml
$CuSO_4$	100 µg/ml
$MnSO_4$	2 mg/ml
$CaCl_2$	25 mg/ml
$FeCl_2 \cdot 4H_2O$	5 mg/ml
$ZnCl_2$*	5 mg/ml

*$ZnCl_2$ has to be dissolved separately using a drop of 0.1N HCl for 10 ml of water. The stock solution is heated to bring all the compounds in solution, kept standing for 24 hours, and sterile filtered.

This medium is inoculated with a spore suspension of *B. subtilis* (1.5×10^{10} cells/ml) at a rate of 0.5 ml/liter. The beer samples are applied to 12.5 mm diameter adsorbent paper discs (0.08 ml/disc), the assay system is incubated overnight at 37° C., and the zones of inhibition are measured. The potency of the sample is related to the diameter of the inhibition zone by means of the usual standard curve.

A biounit (BU) is defined as the concentration of the antibiotic which gives a 20 mm zone of inhibition under the above assay conditions. Thus, if for example a fermentation beer, or other solution containing the antibiotic, needs to be diluted 1/100 to give a 20 mm zone of inhibition, the potency of such beer or solution is 100 BU per ml.

Part B. Recovery

Antibiotic U-43,120 in beers is detected and assayed by the use of tlc and antibacterial assays. Thin layer chromatograms are run on silica gel plates using 95% aqueous ethanol-water (75:25 v/v) as the solvent system. Bioactivity is detected by bioautography using *B. subtilis*-seeded agar trays as disclosed in Example 1, Part A.

Whole fermentation beer (720 ml), obtained as described above, is filtered with the aid of diatomaceous earth (36 grams) as a filter aid. The filter cake is washed with 180 ml of water. The filtrate (clear beer) has a pH of 6.1 and is adjusted to pH 2.0 with 6 N sulfuric acid. This acidic solution is extracted with four 160-ml portions of methylene chloride. The combined methylene chloride extracts are evaporated to dryness under reduced pressure. The clear beer, the extracted aqueous and the residue are assayed by the disc-plate procedure using *B. subtilis* in synthetic medium, as described above. The results are 18.3 Bu/ml, 4.2 Bu/ml

3,988,441

11

and 56.5 Bu/mg respectively, with 60% of the original antibiotic activity in the solid residue.

Part C. Purification

(No. 1) Chromatography on Silica Gel

Three hundred grams of silica gel (E. Merck, Darmstadt) is packed in a 29-mm diameter column using a solvent system consisting of ethyl acetate-ethanol-water (92:5:3). Five grams of a crude preparation of antibiotic U-43,120, prepared as described above, is dissolved in 20 ml of the above solvent system and applied to the silica gel column while washing the column down with two 20-ml portions of solvent. The solvent is eluted with the same solvent system and 240 twenty-ml fractions are collected. Fractions 57–95 are combined on the basis of high activity against *B. subtilis* in a disc-plate asay as described above. Evaporation of the above combined fractions under reduced pressure gives 500 mg of a preparation of antibiotic U-43,120 assaying 118 Bu/ml against *B. subtilis*. This preparation has four to seven times the activity of the starting material, thus giving an activity yield of 50 to 60%.

(No. 2) Countercurrent Distribution

Fifty grams of the preparation of antibiotic U-43,120 prepared as described in Part C, No. 1, is purified further by countercurrent distribution in a 200-tube, 50-ml per phase machine using cyclohexane-ethyl acetate-95% ethanol-water (1:1:1:1) as the solvent system and running 200 transfers. The contents of the tubes are analyzed by weight and by activity against *B. subtilis*. The peak activity is at fraction 150, and the peak weight is at fraction 190. Fractions 116–175 are combined, and the pool is evaporated to dryness under reduced pressure. The resulting residue is assayed using *B. subtilis* and is found to be about 5 times as active as the starting material, and about 30% of the activity is recovered.

(No. 3) Second Chromatography on Silica Gel Column

Fourteen and four-tenths grams of a preparation of antibiotic U-43,120 which has been purified by countercurrent distribution, as described above, is chromatographed on 1080 grams of silica gel using ethyl acetate-ethanol-water (92:5:3) as the solvent system. The antibiotic material is dissolved in 70 ml of solvent and put on the column and further washed down with two more 70-ml portions of solvent. The column is eluted with the same solvent system and 686 twenty-ml fractions are collected. The fractions are analyzed by weight and activity against *B. subtilis*. Fractions 204–395, which have the highest antibiotic activity, are combined and evaporated under reduced pressure. The resulting residue weighs 3.3 g and contains about 90% of the antibiotic activity put on the column. Crystallization of 2.7 g of this U-43,120 product from chloroform gives 0.74 g of a crude crystalline preparation of U-43,120. Five recrystallizations from chloroform gives an essentially pure colorless preparation of antibiotic U-43,120.

The antibiotic of the subject invention has similar properties to those reported for senfolomycins A and B (L. A. Mitscher et al., Antimicrobial Agents and Chemotherapy p. 828 (1965). Senfolomycin A differs from antibiotic U-43,120 because of different R_f values in methyl ethyl ketone-acetone-water (70:20:11) on silica gel plates. Antibiotic U-43,120 moves faster than senfolomycin A on these plates. The optical rotation of senfolomycin B in CH_3OH is −60°, whereas the optical

12

rotation of antibiotic U-43,120 in CH_3OH is −34.9°. Thus, antibiotic U-43,120 is clearly differentiated from senfolomycin A and B.

Antibiotic U-43,120 also has properties similar to those for preceomycin (H. Tsukiura et al., *J. Antibiotics*, [Tokyo] Ser. A.17, 225, (1964). Antibiotic U-43,120 is differentiated from proceomycin by the following criteria:

	Proceomycin	U-43,120
Color	Yellowish-orange	White
$[\alpha]_D$	−2.2° (ethanol)	−34.9 (CH_3OH)
R_f (ethyl acetate-ethanol-water; 92:5:3)	0.25	0.31
R_f (ethanol-methanol-water; 10:9:1)	0.49	0.58

Thus, antibiotic U-43,120 is clearly differentiated from proceomycin. See also FIG. **4** of the drawings.

Salts of antibiotic U-43,120 can be formed with inorganic cations, for example, sodium, potassium, lithium, and calcium since antibiotic U-43,120 is weakly acidic. Such salts can be prepared, as for example, by suspending antibiotic U-43,120 in water, adding a dilute base until the pH of the solution is about 10.0 to 11.0, and freeze-drying the solution to provide a dried residue consisting of the U-43,120 salt. Salts of antibiotic U-43,120 can be used for the same biological purposes as the parent compound.

Antibiotic U-43,120, or its salts, can be used to disinfect washed and stacked food utensils contaminated with *S. aureus;* they also can be used as disinfectants on various dental and medical equipment contaminated with *S. aureus* or *Streptococcus hemolyticus*. Further, antibiotic U-43,120, or its salts, can be used as an industrial preservative, for example, as a bacteriostatic rinse for laundered clothes and for impregnating paper and fabrics; and, it is useful for suppressing the growth of sensitive organisms in plate assays and other microbiological media.

The invention described herein was made in the course of, or under, contract PH 43-NCl-69-1023 with the National Cancer Institute, National Institutes of Health, Bethesda, Maryland 20014.

We claim:

1. Antibiotic U-43,120, which is active against various Gram-positive and Gram-negative bacteria and which in its essentially pure form:
   a. has the molecular formula $C_{34}H_{44}N_2SO_{18}$ (calculated on the basis of analysis);
   b. has the following elemental analysis: C, 50.64, 51.34; H, 5.88, 6.00; N, 3.60, 3.65; S, 4.07, 4.06; O, 34.41;
   c. has a specific rotation of $[\alpha]_D^{25} = +9.3°$ (c, 1, $CHCl_3$);
   d. is soluble in chlorinated hydrocarbons, ethyl acetate, lower alcohols, pyridine and acetone;
   e. has a characteristic UV spectrum as shown in FIG. **1** of the drawings;
   f. has a characteristic infrared absorption spectrum when dissolved in a mineral oil mull as shown in FIG. **2** of the drawings;
   g. has a characteristic and differentiating thin layer chromatography spectrum as shown in FIGS. **3** and **4** of the drawings; and

3,988,441

13

h. has a characteristic NMR spectrum as shown in FIG. 5 of the drawings, or inorganic cationic salts thereof.

2. A process for preparing antibiotic U-43,120, as defined in claim 1, which comprises cultivating *Streptomyces paulus*,

14

having the identifying characteristics of NRRL 8115, in an aqueous nutrient medium containing a source of assimilable carbohydrate and assimilable nitrogen under aerobic conditions until substantial antibiotic activity is imparted to said medium.

* * * * *

5

10

15

20

25

30

35

40

45

50

55

60

65

APPENDIX B: US PATENT 4 332 900 — CONSTRUCTION OF CO-INTEGRATE PLASMIDS FROM PLASMIDS OF STREPTOMYCES AND ESCHERICHIA

United States Patent [19]

Manis et al.

[11] **4,332,900**

[45] **Jun. 1, 1982**

[54] **CONSTRUCTION OF CO-INTEGRATE PLASMIDS FROM PLASMIDS OF STREPTOMYCES AND ESCHERICHIA**

[75] Inventors: **Jack J. Manis**, Portage; **Sarah K. Highlander**, Oshtemo Township, Kalamazoo County, both of Mich.

[73] Assignee: **The Upjohn Company**, Kalamazoo, Mich.

[21] Appl. No.: **192,727**

[22] Filed: **Oct. 1, 1980**

[51] Int. Cl.³ **C12N 15/00**; C12N 1/20; C12N 1/00; C12P 19/34
[52] U.S. Cl. **435/172**; 435/91; 435/253; 435/317; 435/820; 435/849
[58] Field of Search 435/172, 253, 68, 317

[56] **References Cited**
U.S. PATENT DOCUMENTS

4,237,224 12/1980 Cohen et al. 435/68
4,273,875 6/1981 Manis 435/253

OTHER PUBLICATIONS

Muray et al. in *Recombinant DNA and Gentic Experimentation.*
Morgan et al., (ed.) Pergamon Press, (1979), pp. 53–64.
Blattner et al., Science 196, 161, (1977).
Armstrong et al., Science 196, 172, (1977).
Bolivar, Gene 4, 121, (1978).

Primary Examiner—Alvin E. Tanenholtz
Assistant Examiner—James Martinell
Attorney, Agent, or Firm—Roman Saliwanchik

[57] **ABSTRACT**

Novel chemical compounds, recombinant plasmids pUC1019 and pUC-1020, which are obtained by covalent linkage of ca. 4.2 kb BclI restriction endonuclease fragment of the *Streptomyces espinosus* plasmid pUC6 into the BamHI endonuclease site of the E. coli plasmid pBR322. Plasmid pUC1024 is obtained by restructuring plasmid pUC1019. These plasmids are useful as cloning vehicles in recombinant DNA work. For example, using DNA methodology, a desired gene, for example, the insulin gene, can be inserted into the plasmids and the resulting plasmids can then be transformed into a suitable host microbe which, upon culturing, produces the desired insulin.

14 Claims, 5 Drawing Figures

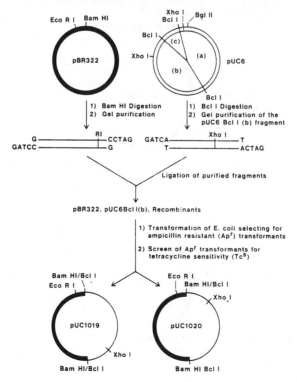

Figure 1

CONSTRUCTION OF pBR322:pUC6 RECOMBINANT PLASMIDS

U.S. Patent   Jun. 1, 1982   4,332,900

Figure 2

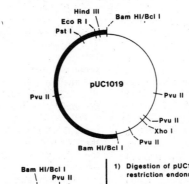

Figure 3

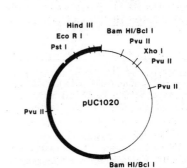

Figure 4

CONSTRUCTION OF RECOMBINANT PLASMID pUC1024

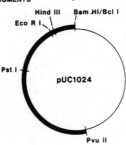

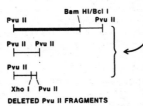

1) Digestion of pUC1019 DNA With restriction endonuclease Pvu II

2) Isolation of the appropriate Pvu II fragment by gel electrophoresis

3) Ligation of the purified fragment with T4 DNA Ligase

4) Transformation of E. coli and selection for ampicillin resistance

DELETED Pvu II FRAGMENTS

Figure 5

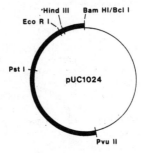

4,332,900

<table>
<tr><td>1</td><td>2</td></tr>
</table>

CONSTRUCTION OF CO-INTEGRATE PLASMIDS
FROM PLASMIDS OF STREPTOMYCES AND
ESCHERICHIA

BACKGROUND OF THE INVENTION

The development of plasmid vectors useful for recombinant DNA genetics among microorganisms is well known. The editorial in Science, Vol. 196, April, 1977, gives a good summary of DNA research. This editorial is accompanied by a number of supporting papers in the same issue of Science.

Similar DNA work is currently being done on industrially important microorganisms of the genus Streptomyces. [Bibb, M. J., Ward, J. M., and Hopwood, D. A. 1978. "Transformation of plasmid DNA into Streptomyces at high frequency." Nature 274, 398–400.] Though plasmid DNA's have been detected in several streptomycetes [Huber, M. L. B. and Godfrey, O. 1978. "A general method for lysis of Streptomyces species." Can. J. Microbiol. 24, 631–632.] [Schrempf, H., Bujard, H., Hopwood, D. A. and Goebel, W. 1975. "Isolation of covalently closed circular deoxyribonucleic acid from *Streptomyces coelicolor* A3(2)." J. Bacteriol. 121, 416–421.] [Umezawa, H. 1977. "Microbial secondary metabolites with potential use in cancer treatment (Plasmid involvement in biosynthesis and compounds)." Biomedicine 26, 236–249.], [Malik, V. S. 1977. Preparative Method for the isolation of super-coiled DNA from a chloramphenicol producing streptomycete. J. Antibiotics 30, 897899], only a few streptomycete plasmids have been physically isolated and extensively characterized [Schrempf, supra]. See also [Bibb, M., Schottel, J. L., and Cohen, S. N. 1980. A DNA cloning system for interspecies gene transfer in antibiotic-producing Streptomyces. Nature 284, 526–531.] and [Thompson, C. J., Ward, J. M. and Hopwood, D. A. 1980. DNA cloning in Streptomyces:resistance genes from antibiotic-producing species. Nature 286, 525–529.] The existence of other plasmids in the genus Streptomyces has been inferred from reported genetic data as follows:

(1) Akagawa, H., Okanishi, M. and Umezawa, H. 1975. "A plasmid involved in chloramphenicol production in *Streptomyces venezuelae:* Evidence from genetic mapping." J. Gen. Microbiol. 90, 336–346.

(2) Freeman, R. F. and Hopwood, D. A. 1978. "Unstable naturally occurring resistance to antibiotics in Streptomyces." J. Gen. Microbiol. 106, 377–381.

(3) Friend, E. J., Warren, M. and Hopwood, D. A. 1978. "Genetic evidence for a plasmid controlling fertility in an industrial strain of *Streptomyces rimosus.*" J. Gen. Microbiol. 106, 201–206.

(4) Hopwood, D. A. and Wright, H. M. 1973. "A plasmid of *Streptomyces coelicolor* carrying a chromosomal locus and its inter-specific transfer." J. Gen. Microbiol. 79, 331–342.

(5) Hotta, K., Okami, Y. and Umezawa, H. 1977. "Elimination of the ability of a kanamycin-producing strain to biosynthesize deoxystreptamine moiety by acriflavine." J. Antibiotics 30, 1146–1149.

(6) Kirby, R., Wright, L. F. and Hopwood, D. A. 1975. "Plasmid-determined antibiotic synthesis and resistance in *Streptomyces coelicolor.*" Nature 254, 265–267. ·

(7) Kirby, R. and Hopwood, D. A. 1977. "Genetic determination of methylenomycin synthesis by the

SCPI plasmid of *Streptomyces coelicolor* A3(2)." J. Gen. Microbiol. 98, 239–252.

(8) Okanishi, M., Ohta, T. and Umezawa, H. 1969. "Possible control of formation of aerial mycelium and antibiotic production in Streptomyces by episomic factors." J. Antibiotics 33, 45–47.

Plasmid pUC6 was isolated from *Streptomyces espinosus* biotype 23724a, NRRL 11439.

Plasmid pBR322 is a well known plasmid which can be obtained from *E. coli* RR1, NRRL B-12014. The restriction endonuclease map for pBR322 is published; Sutcliff, J. G. "pBR322 restricting map derived from the DNA sequence: accurate DNA size markers up to 4361 nucleotide pairs long." Nucleic Acids Research 5, 2721–2728, 1978. This map is incorporated herein by reference to the above publication.

BRIEF SUMMARY OF THE INVENTION

Plasmids pUC1019 and pUC1020 are obtained by the in vitro linkage of ~4.2 kb BclI restriction endonuclease fragment of the *S. espinosus* plasmid pUC6 into the BamH1 endonuclease site of the *E. coli* plasmid pBR322.

Plasmids pUC1019 and pUC1020 constitute the insertion of this BclI fragment in the two possible orientations in pBR322. In a like manner, the 4.1 and 0.9 kb BclI restriction fragments of pUC6 can be recombined with pBR322.

The invention further includes the construction of plasmid pUC-1024, which is derived from pUC1019 by in vitro deletion of DNA sequences between endonuclease PvuII sites. Plasmid pUC1024 is ~3.5 kilobases (kb) smaller than pUC1019.

The plasmids, advantageously, are transformed into a suitable host, for example, *E. coli.*

Plasmids pUC1019 and pUC1020 are recombinant DNA molecules consisting of the entire genome of the small ($\sim2.6\times10^6$ dalton) high copy number ($\sim$ 30/chromosome) *E. coli* plasmid pBR322 and ~46% (4.2 kb) of the genome of the small ($\sim6.0\times10^6$ daltons) high copy number ($\sim$30/chromosome) *S. espinosus* plasmid pUC6. Plasmids pUC1019 and pUC1020 contain single sites for the restriction endonucleases EcoRI, PstI, HindIII and XhoI. The XhoI site will also allow the cloning of SalI restriction fragments. Hence plasmids pUC1019 and pUC1020 represent DNA molecules which may function as vectors in both *E. coli* and Streptomyces and represent valuable intermediates for the development of better host-vector systems.

Plasmids pUC1019, pUC1020, and pUC1024 are characterized by the restriction maps shown in FIGS. 2, 3, and 5, respectively, of the drawings. The restriction endonuclease abbreviations shown in the drawings are standard and well known in the art.

Plasmid pUC6 is obtainable from the novel microorganism *Streptomyces espinosus* biotype 23724a, NRRL 11439. This plasmid can be obtained from NRRL 11439 by growing the culture on a suitable medium, fragmenting the mycelia, incubating the fragmented mycelia, harvesting the culture after a suitable time, and then lysing the mycelia. From this lysate it is possible to isolate essentially pure pUC6. pUC6 is characterized by standard characterization tests which include its molecular weight, approximately 6.0 megadaltons, sensitivity to restriction endonucleases, infra, and presence at 20–40 copies per *S. espinosus* NRRL 11439 cell.

4,332,900

3

REFERENCE TO THE DRAWINGS

FIG. 1—This shows the isolation scheme of the plasmids.

FIG. 2—Restriction endonuclease cleavage map for pUC1019.

FIG. 3—Restriction endonuclease cleavage mpa for pUC1020.

FIG. 4—Construction of recombinant plasmid pUC1024.

FIG. 5—Restriction endonuclease cleavage map for pUC1024.

The maps are constructed on the basis of plasmids pUC1019 and pUC1020 having a molecular weight of ca. 5.7 megadaltons or a molecular length of ca. 8.6 kilobases (kb). Plasmid pUC1024 has a molecular length of ca. 5.1 kb. It retains the locus conferring genetic instability to plasmid pUC1019 in *E. coli* hosts. Plasmid pUC1024 has single restriction sites for the endonucleases EcoRI, HindIII, PstI and PvuII. The PvuII site will allow the cloning of blunt ended DNA fragments from a wide range of restriction enzyme digests or from other properly prepared DNA fragments. The restriction endonuclease abbreviations are as follows: (1) BglII is an enzyme from *Bacillus globigii;* (2) BclI is an enzyme from *Bacillus caldolyticus;* (3) PvuII is an enzyme from *Proteus vulgaris;* and (4) XhoI is an enzyme from *Xanthomonas holicola.*

pUC1019, pUC1020, and pUC1024, can be used to create recombinant plasmids which can be introduced into host microbes by transformation. The process of creating recombinant plasmids is well known in the art. Such a process comprises cleaving the isolated vector plasmid at a specific site(s) by means of a restriction endonuclease, for example, BglII, XhoI, and the like. The plasmid, which is a circular DNA molecule, is thus converted into a linear DNA molecule by the enzyme which cuts the two DNA strands at a specific site. Other non-vector DNA is similarly cleaved with the same enzyme. Upon mixing the linear vector or portions thereof and non-vector DNA's, their single-stranded or blunt ends can pair with each other and in the presence of a second enzyme known as polynucleotide ligase can be covalently joined to form a single circle of DNA.

The above procedure also can be used to insert a length of DNA from a higher animal into pUC1019, pUC1020, or pUC1024. For example, the DNA which codes for ribosomal RNA in the frog can be recombined with pUC1019 DNA. The resulting circular DNA molecules consist of plasmid pUC1019 with an inserted length of frog rDNA.

The recombinant plasmids containing a desired genetic element, prepared by using pUC1019, pUC1020, or pUC1024, can be introduced into a host organism for expression. Examples of valuable genes which can be inserted into host organisms by the above described process are genes coding for somatostatin, rat proinsulin, interferon, and proteases.

The usefulness of plasmids pUC1019, pUC1020, and pUC1024 is derived from their capacity to function as plasmid vectors in industrially important microorganisms, e.g., Streptomyces. Also, pUC1019, pUC1020, and pUC1024 are especially useful because of their single restriction sites. Hence, cloning of genetic information from Streptomyces into pUC1019, pUC1020, or pUC1024 provides a means of increasing the production

4

of commercially important products from these organisms, e.g., antibiotics.

This approach is compared to the concept of cloning genes for antibiotic production into the well characterized *Escherichia coli* K-12 host-vector system. The *E. coli* system has the disadvantage that it has been found that genes from some Gram-positive organisms, e.g., Bacillus, do not express well in the Gram-negative *E. coli* host. Likewise, plasmids from Gram-negative organisms are not maintained in Gram-positive hosts, and Gram-negative genetic information is either expressed poorly or not at all in Gram-positive hosts. This clearly argues for the advantage of a Gram-positive host-vector system and argues the usefulness of plasmid pUC1019, pUC1020, or pUC1024, in such a system.

In general, the use of a host-vector system to produce a product foreign to that host requires the introduction of the genes for the entire biosynthetic pathway of the product to the new host. As discussed above, this may lead to problems of genetic expression, but may also generate new and/or increased problems in the fermentation of the microorganisms and in the extraction and purification of the product. A perhaps more useful approach is to introduce a plasmid vector into a host which normally produces the product and clone onto that plasmid the genes for biosynthesis of the product. At the very least, problems of fermentation and product extraction and purification should be minimized. Additionally, in this cloning system it may not be necessary to clone and amplify all the genes of the biosynthetic pathway, but rather it may be necessary only to clone regulatory genes or genes coding for the enzymes that are rate limiting in product biosynthesis. Since pUC1019 and pUC1020 are cointegrate plasmids, they can be used to clone DNA sequences in *E. coli* or within the genera of Streptomyces and Micromonospora, as well as within other microbes.

DETAILED DESCRIPTION OF THE INVENTION

The Microorganisms and Plasmids

The following microorganisms are available from the permanent collection of the Northern Regional Research Laboratory, U.S. Department of Agriculture, Peoria, Ill., U.S.A.

NRRL B-12110—*E. coli* CSH50
NRRL B-11439—*S. espinosus* biotype 23724a
NRRL B-12014—*E. coli* RR1 (pBR322)
NRRL B-12252—*E. coli* CSH50 (pUC1019)
NRRL B-12253—*E. coli* CSH50 (pUC1020)
NRRL B-12254—*E. coli* RR1 (pUC1024)
NRRL B-12186—*E. coli* RR1

These deposits are available to the public upon the grant of a patent to the assignee, The Upjohn Company, disclosing them. The deposits are also available as required by foreign patent laws in countries wherein counterparts of the subject application, or its progeny, are filed. However, it should be understood that the availability of a deposit does not constitute a license to practice the subject invention in derogation of patent rights granted by governmental action.

The following examples are illustrative of the process and products of the subject invention but are not to be construed as limiting. All percentages are by weight

4,332,900

5

and all solvent mixture proportions are by volume unless otherwise noted.

EXAMPLE 1

Isolation of Vector pBR322 DNA from *E. coli* NRRL B-12014

A 100 ml. culture of *E. coli* RR1 (pBR322) is grown overnight in L-broth which consists of the following ingredients:

Bacto tryptone (Difco)	10 g./liter
Bacto yeast extract (Difco)	5 g./liter
NaCl	5 g./liter
Ampicillin	50 mg./liter

The cells are recovered by centrifugation at $17,000 \times g$. for 10 minutes in a refrigerated centrifuge. The pellet is suspended in 2.5 ml. 50 mM tris buffer (pH 8) containing 25% sucrose. One-half ml. of lysozyme stock solution is added (5 mg./ml. in TES buffer). The mixture is allowed to stand in ice for 5 minutes. At this point 1 ml. EDTA (0.25 M, pH 8) is added and the mixture is again allowed to stand in ice for 5 minutes. One and a quarter ml. of 5 N NaCl and 1 ml. 10% SDS (sodium dodecyl sulfate) are added. The mixture is shaken on a Vortex and incubated at 37° C. for 20 minutes. Then 10 μl of ribonuclease (20 mg./ml.) is added and the sample is again incubated at 37° C. for 20 minutes. The mixture is then kept in ice overnight and then centrifuged at $35,000 \times g$. for 30 minutes in a refrigerated centrifuge. 2 ml. of the supernatant solution (lysate) are carefully removed with a pipette. Four and one-half ml. of TES buffer (30 mM tris.HCl, pH 8, 5 mM EDTA.Na$_2$, 50 mM NaCl) are mixed with 1.5 ml. EtBr (ethidium bromide) stock (1 mg./ml. in TES buffer) and 7.5 g. solid CsCl. After the salt has dissolved, 2 ml. of the lysate, described above, is added and the mixture is transferred into a polyallomer tube fitting a titanium 50 (50 Ti) head (Beckman ultracentrifuge). The tubes are filled to the top with mineral oil and centrifuged in a Beckman ultracentrifuge at 40,000 rpm in a 50 Ti head at 15° C. for at least 2 days. The DNA is located under a long wave UV-lamp and the heavier band containing the plasmid DNA is removed with a syringe by puncturing the tube wall from the side. The samples are extensively dialysed against 200 volumes of TES buffer at 4° C. Following dialysis, 1/10 sample volume of a 3 M Na acetate stock solution is added and the plasmid DNA is precipitated by the addition of 2 volumes of cold ethanol. The resulting pellet is then lyophilized and redissolved in 200 μl 10 mM tris buffer, pH 7.8 containing 1 mM EDTA.Na$_2$ and frozen for storage.

EXAMPLE 2

Isolation of Plasmid pUC6 from a Biologically Pure Culture of *Streptomyces espinosus*, biotype 23724a, NRRL 11439

The spores from a biologically pure culture of *Streptomyces espinosus* biotype 23724a, NRRL 11,439, are inoculated into 10 ml. of the following Difco Antibiotic Medium No. 3 Broth (Difco Labs., Detroit, Mich.): 0.15% Beef extract; 0.15% yeast extract; 0.5% peptone; 0.1% glucose; 0.35% NaCl; 0.368% K$_2$HPO$_4$; 0.132% KH$_2$PO$_4$.

The medium has previously been sterilized in a 50 ml. Erlenmeyer flask. After inoculation, the flask is incu-

6

bated at 37° C. for about 36 to 48 hours on a Gump or New Brunswick rotary shaker operating at 100–250 rpm. Upon completion of the incubation, the mycelia-broth suspension in the flasks is homogenized under sterile conditions and is then mixed in a sterile 125 ml. Erlenmeyer flask containing 10 ml. of the above medium and also, advantageously 68% (w/v) sucrose and 1% (w/v) glycine. The addition of sucrose and glycine facilitates the subsequent lysing of the cells. The amounts of sucrose and glycine in the medium can be varied by routine adjustments with the goal being to facilitate the subsequent lysing of the cells. The flask is then incubated further for another 36 to 48 hours at 37° C. on a Gump rotary shaker, as above. After this incubation, the mycelia are separated from the broth by low speed centrifugation, for example, at $6000 \times g$. for 15 minutes at 4° C. and decantation of the supernatant from the mycelial pellet.

The supernatant is discarded and the pellet is resuspended in 1.5 ml. of an isotonic buffer containing ethylenediaminotetraacetic acid (EDTA) and sucrose, e.g., TES buffer [0.03 M tris(hydroxymethyl)aminomethane (Tris), 0.005 M EDTA and 0.05 M NaCl; pH=8.0] containing 20% (w/v) sucrose. Next, 1.5 ml. of a 5 mg./ml. solution of lysozyme in the same buffer is added and the mixture is incubated at 37° C. for 30 minutes with occasional mixing. Then, 1.5 ml. of 0.25 M EDTA (pH=8.0) is added and this mixture is incubated 15 minutes at 37° C. Subsequently, the cell suspension is lysed by the addition of 2.5 ml. of a lytic mixture, e.g. [1.0% (w/v) Brij-58 (a detergent sold by Pierce Chem. Co., Rockford, Ill.), 0.4% (w/v) deoxycholic acid, 0.05 M Tris (pH=8.0) and 0.06 M EDTA] and incubation of this mixture at 37° C. for 20 minutes. The lysate is then sheared by passing it 5–10 times through a 10 ml. pipette. The sheared lysate is then digested with ribonuclease (140 μg/ml.) and pronase (300 μg/ml.) for an additional 20 minutes at 37° C. Alternatively, the cell-lysozyme-EDTA mixture can be digested with ribonuclease and pronase before lysis with a lytic agent such as 2% sodium dodecyl sulfate in water.

This crude lysate material is then mixed with a salt, for example, cesium chloride (preferred), and cesium sulfate, and the intercalating dye ethidium bromide to give a solution of density $\rho = 1.550$. This solution is centrifuged to equilibrium at $145,000 \times g$. (isopycnic density gradient centrifugation). The covalently closed circular plasmid DNA is then visible in the centrifuge tube under long wave ultraviolet (320 nm) illumination as a faint fluorescent band below the intensely fluorescent band of linear chromosomal and plasmid DNA's.

Covalently closed circular plasmid DNA is prepared for characterization by removing it from the isopycnic gradients, extracting the ethidium bromide by two treatments with one-third volume of isopropyl alcohol and then dialyzing the aqueous phase against an appropriate buffer, e.g., 0.1×SSC buffer (0.015 M NaCl, 0.0015 M sodium citrate; pH=7.4) to yield essentially pure pUC6.

Characteristics of pUC6

Molecular Weight: ca. 6.0 megadaltons
Copies per Cell: 20–40

Restriction Endonuclease Sensitivities: pUC6 has the following sensitivies to restriction endonucleases.

7

Plasmid Sensitivities to Restriction Endonucleases

# Cleavage Sites		# Cleavage Sites	
Enzyme	pUC6	Enzyme	pUC6
BglI	>7	BglII	1
BamHI	0	HpaI	0
HpaII	Many	HindIII	0
EcoRI	0	KpnI	0
PstI	0	PvuII	4
MboII	>5	AvaI	>7
XbaI	0	XhoI	2
SalI	5–6	SmaI	>5
HincII	>7	BclI	3

These results were obtained by digestion of pUC6 DNA in the presence of an excess of restriction endonuclease. The number of restriction sites were determined from the number of resolvable fragments in either 0.7 or 1.0% agarose gels.

EXAMPLE 3

Preparation of Plasmids pUC1019 and pUC1020

Plasmids pUC6 and pBR322, prepared as described above, are linearized by digestion with restriction endonucleases BclI and BamHI respectively. Plasmid pBR322 DNA is digested with BamHI restriction enzyme by mixing ~50 µl. of DNA (~0.5 µg) solution in TE buffer (0.01 M Tris .HCl, 0.001 M Na$_2$ EDTA; pH 8.0) with 50 µl of 2X restriction buffer (Post et al., Cell 15, 215–229, 1978) and 4 units of BamHI enzyme preparation. This mixture is incubated at 37° C. for 1 hour. The digest is then applied to a 1% preparative low melting point agarose gel and electrophoresed for ~3 hours at 50 volts and 4° C. The resolved DNA fragments are visualized by ethidium bromide staining and long wave ultraviolet light illumination. The region of the gel containing the DNA is excised from gel and heated to 65° C. in the presence of 1.5 ml. of TE buffer to melt the gel and release the DNA from the gel matrix. This suspension is chilled and centrifuged at 37,000×g to pellet the agarose. The supernatant is decanted and saved. The agarose pellet is extracted a second time by heating to 65° C. with TE buffer. The two supernatants are pooled and ethanol precipitated by the addition of 0.1 volume of Na Acetate and 2 volumes 95% ethanol at −20° C. The DNA precipitate is collected by centrifugation at 85,000×g at 4° C. for 60 minutes. The precipitate is redissolved in 100 µl of TE buffer. This sample is used for ligation as described below.

Plasmid pUC6 is subjected to BclI digestion in a reaction mixture containing 50 µl of pUC6 DNA (~0.5 µg) solution in TE buffer and 50 µl of 2X BclI restriction buffer and 4 units of BclI enzyme. This mixture is incubated at 37° C. for one hour and the digest placed in and isolated from a preparative agarose gel electrophoresis system as described above.

For ligation, 25 µl of BamHI digested pBR322 DNA, 25 µl of BclI digested pUC6 DNA and 20 µl DD H$_2$O are combined. Ten µl 100 mM DDT, 10 µl 50 mM MgCl$_2$ and 10 µl of 0.5 mM ATP are combined with the restricted DNA mixture. Finally, 1.0 unit of T$_4$ DNA ligase is added and the sample is kept in ice for 1–2 days.

For transformation into *E. coli* CSH50, inoculum is grown overnight in L-broth and diluted 1:100 into fresh L-broth the next day. The cells are incubated at 37° C. and allowed to grow to an OD$_{650}$ of 0.2. At this point 50 ml. of culture is centrifuged in the cold, the pellet resus-

8

pended in 20 ml. cold 100 mM CaCl$_2$, incubated at 0° C. for 20–25 minutes and centrifuged again. The pellet is then resuspended in 0.5 ml. cold 100 mM CaCl$_2$ solution and kept at 0°–4° C. for 24 hours. (Dagert, M. and Ehrlich, S. D. 1979, Gene 6: 23–28) One hundred µl of pUC1019 and pUC1020 ligase mixture (see above) is mixed with 500 µl cell suspension. This mixture is kept in ice for 10 minutes, and then at 37° C. for 5 minutes. Ten to 20 ml. of L-broth is added and the cell suspension is incubated 1–2 hours at 37° C. Next, 100 µl aliquots are plated on freshly prepared agar plates containing 25 ml. of L-broth, 1.5% agar, and 50 µg of ampicillin/ml. Colonies are selected and scored for tetracycline sensitivity.

Suspected recombinant DNA containing transformants are then grown in 25 ml. cultures. Cleared lysates are prepared by pelleting the cells from the culture medium at ~10,000×g. The pellet is resuspended in 10 ml. of cold TES buffer (30 mM Tirs.HCl, 5 mM Na$_2$ EDTA and 50 mM NaCl, pH 8.0) and pelleted again. This pellet is resuspended in 1 ml. of TES buffer containing 20% sucrose. 0.2 ml. of lysozyme solution (5 mg./ml. in TES) is added and incubated on ice 15 minutes at which time 0.4 ml. of 0.25 M Na$_2$ EDTA (pH 8.0) is added and the incubation continued 15 minutes. 1.6 ml. of a lytic mix (1% Brij 58, 0.4% Na deoxycholate, 0.05 M Tris.HCl, 62.5 mM Na$_2$ EDTA; pH 8.0) is added and the lysate incubated an additional 15 minutes at 4° C. The lysate is sheared by passage 5 times through a 10 ml. pipette. The bulk of the cellular DNA and debris are removed by centrifugation at 48,000×g for 30 minutes. The cleared lysate is digested successively for 15-minute intervals with pancreatic RNAse A (100 mcg./ml.) and Pronase (200 mcg./ml.) at 37° C. These lysate are then centrifuged in a CsCl-ethidium bromide isopycnic density gradient. Plasmid DNA isolated from these gradients is characterized by digestion with restriction endonuclease.

EXAMPLE 4

Preparation of Plasmid pUC1024

Plasmid pUC1019, prepared as described in Example 3, is linearized by digestion with restriction endonuclease PvuII as follows:

Approximately 0.5 µg of pUC1019 DNA in 25 µl of TE buffer is mixed with an equal volume of 2X PvuII restriction enzyme buffer (0.3 M NaCl, 12 mM Tris.HCl [pH 7.4], 12 mM MgCl$_2$, 12 mM 2-mercaptoethanol) and two units of PvuII restriction enzyme. This sample is digested for one hour at 37° C.

The resulting digest is then applied to a 1% preparative low melting point agarose gel and electrophoresed for ~3 hours at 50 volts and 4° C. The remainder of the agarose gel procedure is as described above in Example 3. The appropriate PvuII fragment is isolated and ligated, as described above in Example 3.

Transformation of the ligated plasmid, which is now plasmid pUC1024, into *E. coli* RR1 is as described in Example 3.

Recombinant plasmids were characterized by cleavage with restriction endonucleases.

Restriction endonucleases were obtained as commercial preparations from Miles Laboratoris, Bethesda Research Laboratories, and New England Biolabs. Enzyme diges-tions were prepared in accordance with the conditions specified by the suppliers using at least a two-fold excess of endonuclease.

4,332,900

9

The digested samples were applied to 0.7–1% agarose gels and were electrophoresed for 2 hours at a constant applied voltage of 10–15 v/cm of gel height. [Sharp, P. A., Sugden, J. and Sambrook, J. 1973. Detection of two restriction endonuclease activities in *Haemophilus parainfluenzae* using analytical agarose-ethidium bromide electrophoresis. Biochemistry 12, 3055–3063]. The molecular weights of restriction fragments were determined relative to the standard migration patterns of bacteriophage lambda DNA digested with enzyme HindIII [Murray, K. and Murray, N. E. 1975. "Phage lambda receptor chromosomes for DNA fragments made with restriction endonuclease III of *Haemophilus influenzae* and restriction enconuclease I of *Escherichia coli*." J. Mol. Biol. 98, 551–564] or EcoRI [Helling, R. B., Goodman, H. M. and Boyer, H. W. 1974. Analysis of endonuclease R.EcoRI fragments of DNA from lambdoid bacteriophages and other viruses by agarose-gel electrophoresis. J. Virology 14, 1235–1244].

Cointegrate plasmids pUC1019 and pUC1020, and restructured plasmid pUC1024, can be isolated from their *E. coli* hosts by well known procedures, e.g., using the cleared lysate-isopycnic density gradient procedures described above. Once transformants containing pUC1019, pUC1020, or pUC1024 are identified, they are separated as pure entities in a pure culture. These plasmids can be differentiated as distinct entities by their unique restriction patterns as would be predicted by their restriction maps.

As shown herein, plasmid pUC1019 can be used as a substrate for in vitro deletion restructuring, e.g., pUC1019 can be digested with PvuII to form a derivative lacking some of the PvuII restriction sites. This advantageously also gives a smaller plasmid.

Examples of other vectors which can be used in the invention as a substitute for pBR322 are pBR313, which codes for ampicillin and tetracycline resistance, pSC101, which codes for tetracycline resistance, pCR11, which codes for kanamycin resistance, λ bacteriophage vectors, for example, charon phages, and yeast 2μ plasmid DNA.

Examples of other hosts for the vectors are any *E. coli* K-12 derivative [Bacteriological Reviews, December 1972, pages 525–557] (these have been approved by the NIH Guidelines) and yeasts, other fungi, or other bacteria. It is recognized that these latter hosts would also have to be approved by the NIH Guidelines.

The work described herein was all done in conformity with physical and biological containment requirements specified in the NIH Guidelines.

We claim:

1. *E. coli* CSH50 (pUC1019) having the deposit accession number NRRL B-12252.

10

2. *E. coli* CSH50 (pUC1020) having the deposit accession number NRRL B-12253.

3. Cointegrate plasmid pUC1019, characterized as shown by the restriction map in FIG. 2 of the drawings.

4. Cointegrate plasmid pUC1020 characterized as shown by the restriction map in FIG. 3 of the drawings.

5. A process for preparing cointegrate plasmids pUC1019 and pUC1020 which comprises:
  (a) linearizing plasmid pBR322 by BamHI endonuclease to obtain linear plasmid DNA;
  (b) linearizing plasmid pUC6 by BclI endonuclease to obtain linear plasmid DNA; and
  (c) ligating said linear plasmid DNA from pBR322 and pUC6 to obtain cointegrate plasmids pUC1019 and pUC1020.

6. A process for cloning ca. 4.2 kb BclI restriction endonuclease fragment of plasmid pUC6 into a suitable host which comprises:
  (a) linearizing plasmid pBR322 by BamHI enconuclease to obtain linear plasmid DNA;
  (b) linearizing plasmid pUC6 by BclI endonuclease to obtain linear plasmid DNA;
  (c) ligating said linear plasmid DNA from pBR322 and pUC6 to obtain cointegrate plasmids pUC1019 and pUC1020; and,
  (d) transforming said cointegrate plasmids into said suitable host.

7. A process, according to claim 6, wherein said suitable host is a bacterium.

8. A process, according to claim 7, wherein said bacterium is *E. coli* CSH50.

9. *E. coli* RR1 (pUC1024) having the deposit accession number NRRL B-12254.

10. Plasmid pUC1024, characterized as shown by the restriction map in FIG. 5 of the drawings.

11. A process for preparing plasmid pUC1024 which comprises:
  (a) digestion of pUC1019 DNA with restriction endonuclease PvuII to obtain fragmented linear plasmid DNA; and,
  (b) ligating the PvuII fragment of said plasma DNA to obtain plasmid pUC1024.

12. A process for cloning plasmid pUC1024 into a suitable host which comprises:
  (a) digestion of pUC1019 DNA with restriction endonuclease PvuII to obtain fragmented linear plasmid DNA;
  (b) ligating the PvuII fragment of said plasmid DNA to obtain plasmid pUC1024; and,
  (c) transforming said plasmid into said suitable host.

13. A process, according to claim 12, wherein said suitable host is a bacterium.

14. A process, according to claim 13, wherein said bacterium is *E. coli* RR1.

* * * * *

APPENDIX C: AGRICULTURAL RESEARCH CULTURE COLLECTION (NRRL) FORM FOR BUDAPEST TREATY DEPOSITS

AGRICULTURAL RESEARCH CULTURE COLLECTION (NRRL)
<u>(International Depositary Authority)</u>

Fermentation Laboratory, Northern Regional Research Center
Science and Education Administration, U.S. Department of Agriculture
1815 North University Street, Peoria, Illinois 61604 U.S.A.

<u>ADDRESS SHIPMENTS AND FORMS TO:</u> Mr. A. J. Lyons at above address

MICROORGANISM DEPOSIT UNDER THE BUDAPEST TREATY ON THE INTERNATIONAL RECOGNITION OF THE DEPOSIT OF MICROORGANISMS FOR THE PURPOSES OF PATENT PROCEDURE (Budapest Treaty)

> THIS BOX FOR NRRL USE ONLY
> Date and Nature of Material Received: _____
>
> Date Deposit Accessioned: _____
>
> as NRRL No. _____
>
> *(Contingent on depositor's verifying acceptability of Agricultural Research Culture Collection (NRRL)'s processed material as equivalent to original material deposited)

DEPOSIT STATEMENT
<u>THIS FORM MUST BE COMPLETED IN ENGLISH</u>
[Our present policies are stated in Industrial Property No. 1, pp. 22–24 (1981)]

1. Name of Microorganism and Depositor's Strain Designation (acronym, sigla, abbreviation + number, symbols)
   _____

   This microorganism is: a bacterium ☐, an <u>Actinomycetales</u> ☐, a mold ☐,

   a yeast ☐, an alga ☐, a strain containing recombinant DNA molecules ☐,

   a strain containing its own naturally occurring plasmid(s) ☐, a strain

   containing inserted naturally occurring plasmid(s) from another host ☐,

   a strain containing inserted constructed plasmid(s) ☐, a strain containing

   a virus of any kind ☐ .

2. Is this strain being deposited under the Budapest Treaty? _____

3. Name and Address of Depositor:* _____

   _____

   _____

   _____

   _____

*Viability test reports and other correspondence will be sent to above party. Our processed material from the original deposit will be sent to above party for checking. If there is no response within 3 months from date of shipment, our processed material will be considered equivalent to the original deposit in terms of viability and performance attributed to the strain.

4. Indicate the properties of the microorganism which are or may be dangerous to health or the environment, or indicate that the depositor is not aware of such properties.

   _____

   _____

5. Is this strain zoopathogenic? _____ phytopathogenic? _____

6. Recommended Conditions for Optimal Cultivation of the Microorganism and for Testing Its Viability:* _____

   _____

   _____

   _____

   *Progeny of strains will be preserved at the Agricultural Research Culture Collection (NRRL) as lyophilized preparations or, in some cases, as agar slant cultures overlayed with mineral oil. All materials will be stored at 3-5° C.

7.* Lowest permissible NIH** (U.S.A.) Physical Containment level for processing and viability testing (P1, P2, P3, P4):
   Lowest permissible CDC*** (U.S.A.) Biosafety level for processing and viability testing (1, 2, 3, 4): _____

   * The Agricultural Research Culture Collection (NRRL) can, at present, process progeny of strains only at Physical Containment level P1 or Biosafety Level 1 or less.

   ** U.S. Department of Health and Human Services, National Institutes of Health, Bethesda, Maryland 20205, U.S.A., November 1980. Guidelines for Research Involving Recombinant DNA Molecules.

   *** U.S. Department of Health and Human Services, Public Health Service, Centers for Disease Control, Office of Biosafety, Atlanta, Georgia 30333, U.S.A. 1980. Proposed Biosafety Guidelines for Microbiological and Biomedical Laboratories.

8. Are you willing to waive the right to be informed of all requests for progeny of this strain? (This is allowed under the Budapest Treaty but will require additional correspondence.) _____

9. I understand and agree that the deposit may not be withdrawn by me or any representative of my organization for the period specified in Rule 9.1 of the Budapest Treaty (at least 30 years after the date of accessioning).

   Date _____ Signature of Depositor _____

   _____   _____
   (on behalf of)   (Typed name of depositor)

27

Control of Toxic and Inhibitory Contaminants in Biological Wastewater Treatment Systems

H. MELCER
Environment Canada, Wastewater Technology Centre, Burlington, Ontario, Canada

27.1 INTRODUCTION

The role of biological wastewater treatment in the control of trace contaminants discharged to natural receiving waters is under close scrutiny in view of the potentially high cost of minimizing the environmental impact of such compounds. Many researchers in North America are engaged in evaluating the degree to which man-made contaminants can be removed by conventional biological treatment, encouraged by the knowledge that biological systems have proven to be generally more cost effective than physical/chemical systems. Considerable data have been published to demonstrate the effectiveness of biological wastewater treatment systems in removing trace contaminants from the aqueous phase.

Lists of trace contaminants of concern have been prepared by different regulatory/advisory agencies. The significance of each is related to the role of the originating agency. The most actively researched is the list of 129 compounds and heavy metals designated as priority pollutants by the US Environmental Protection Agency (EPA) (NRDC, 1975). This is not an exhaustive list since there are other compounds that have been shown to be toxic or inhibitory to biological wastewater treatment systems. Long-term human health effects are very complex and difficult to evaluate. To date they have not been determined for many of the compounds that can be present in aqueous discharges to receiving waters. At what concentration their presence is significant is not known at this time. Thus, current philosophy in wastewater treatment is to minimize the discharge of all trace contaminants of concern.

The analysis of trace contaminants in wastewaters is complex, time-consuming and very expensive. The development of improved analytical procedures is the subject of considerable research. Until improved procedures are available, evaluation of the fate of trace contaminants in treatment systems will remain semi-quantitative. While the high cost of chemical analysis precludes a rigorous examination of wastewaters, a list of those contaminants of greatest concern may be used as a measure by which the performance of a wastewater treatment plant is judged. This paper focusses upon the list of US EPA priority pollutants for which there is the most comprehensive data base published to date.

The major studies on the presence and fate of trace contaminants in full scale treatment plants have been sponsored by the US EPA. They have addressed the municipal sector (Burns and Roe Industrial Services Corporation, 1982; Cohen and Hannah, 1981) and the chemical industry (Tischler and Kocurek, 1982). They demonstrated that biological wastewater treatment processes (activated sludge systems mainly were assessed) were effective for removal of the majority of trace contaminants. In summary, of the US EPA organic priority pollutants, 10 to 15 were found in municipal or industrial effluents from biological wastewater treatment plants at concentrations greater than detection levels, usually in the range 10 to 30 (mg m^{-3}). More rigorous investigations carried out at pilot scale at the US EPA Testing and Evaluation Center confirmed these findings for both volatile contaminants (Petrasek *et al.*, 1983) and semi-volatile acid and base/neutral extractable priority pollutants (Petrasek *et al.*, 1982).

27.2 REACTOR DESIGN CONSIDERATIONS

The biodegradation of anthropogenic contaminants is reported to be relatively difficult, possibly because indigenous microorganisms are unable to produce the enzymes required for their metabolism to intermediates that can enter established biochemical pathways (Alexander, 1973; Kobayashi and Rittmann, 1982). Simple, bench-scale biological tests are commonly used to assess the biodegradability of candidate contaminants. However, the utility of such tests in predicting the biodegradability of these contaminants in biological wastewater treatment systems is limited. Many factors relating to environmental conditions and the physicochemical characteristics of a particular contaminant can create antagonistic or synergistic effects that can distort predictions.

There are special considerations to be made when dealing with low contaminant concentrations. Rittmann and McCarty (1980) demonstrated that steady-state bacterial mass and substrate utilization declined to negligible quantities when the substrate concentration in a biofilm reactor approached a minimum threshold value, typically in the range 100 to 10 000 (mg m^{-3}) for aerobic systems whereas desired effluent concentrations are of the order of 1 mg m^{-3}. Not only is substrate concentration so low that it may be insufficient to induce the production of necessary enzymes, but also the slow substrate utilization kinetics that occur with very low concentrations provide too little energy flux to sustain the microorganisms (Kobayashi and Rittmann, 1982). They suggested utilization of oligotrophic bacteria for the removal of contaminants present at very low concentrations. These microorganisms are adapted to life under low nutrient conditions and generally have a high surface area/volume ratio and a high affinity for substrate. The minimum substrate concentration needed for measurable growth is lower than for eutrophic (high nutrient) organisms but the maximum growth rate is also lower. An important characteristic of these organisms is that they often appear to have multiple inducible enzymes, can shift metabolic pathways and can often use mixed substrates.

There are an increasing number of techniques other than biodegradability tests that are being used to predict trace contaminant removal from wastewaters. They are based upon the recognition of at least three mechanisms, other than biological oxidation, that are responsible for the removal of trace contaminants in biological treatment plants. They are volatilization, adsorption and chemical oxidation (Matter-Muller *et al.*, 1980; Heck *et al.*, 1981). The techniques adopt a molecular approach to treatability using physical and chemical properties and the chemical structures of specific organic contaminants to characterize their behaviour in a particular type of treatment process (Mackay and Paterson, 1981; Mackay, 1982). Some researchers, for example Strier and Gallup (1983), advocate that compounds having similar physical and chemical properties will be treated similarly and removed to the same extent. This approach, known as structure–activity analysis, predicts grouping of molecular analogues with common mechanisms of removal and may be used as a first approximation in determining the potential for removal of specific contaminants.

The increasingly complex microbiological problems associated with the degradation of anthropogenic contaminants has been recognized by changes in biological reactor design. It is well recognized that increasing the hydraulic retention time (HRT) of a completely mixed reactor will dilute, to more tolerable levels, the concentration of raw wastewater contaminants that are biodegradable but inhibitory to reactor biomass. There is a trade-off between the cost of increasing HRT *versus* the degree of treatability that can be achieved.

Powdered activated carbon (PAC) is often added in the case of industrial wastewater treatment where biomass inhibition is evident. The role that PAC plays in controlling inhibition is not well understood. However, it is thought that the PAC adsorbs a proportion of the inhibitory compounds and minimizes concentration effects, allowing biological metabolism to proceed. The addition of PAC is not, however, a panacea for poor operation of biological treatment plants.

Control of solids retention time (SRT) is probably the most significant change in reactor design that has occurred. This has led to changes in the operation of suspended growth systems and the advent of new reactor configurations that promote SRT control by establishing a fixed-film biomass. The retention of microbial populations by SRT control is important since the microorganisms of interest, such as the oligotrophs, may grow slowly resulting in either washout (total loss from a conventionally designed system), or domination by other microbial systems. The relationship between microbial growth rate and SRT is given by

$$SRT = 1/\mu$$

where μ = specific growth rate (d^{-1}).

To prevent washout, a minimum SRT must be maintained such that microbial growth can proceed at or less than its maximum possible rate, μ_{max},

$$SRT_{min} > 1/\mu_{max}$$

To minimize construction costs, it is desirable that HRT and SRT be adjustable independently, that is by implementing sludge recycle and a sludge wasting procedure. With this flexibility, HRT may be minimized to reduce reactor volume while SRT is maximized to improve contaminant removal rates and prevent washout.

A study of the biodegradation of pentachlorophenol by Wukasch *et al.* (1981) well illustrates the concept of utilizing SRT control for removal of trace contaminants. Four bench-scale activated sludge systems were each operated at a different SRT in the range 3 to 15 days. Pentachlorophenol was spiked into the feed of each reactor at a concentration of 350 mg m^{-3}. The resulting effluent pentachlorophenol concentration was reduced with increasing SRT.

Similar trends have been reported by Kincannon and Stover (1981) for the removal of a range of contaminants including benzene, nitrobenzene, 1,2-dichlorobenzene, 2,4-dichlorophenol, 1,2-dichloroethane and tetrachloroethane.

A bench-scale evaluation of the predenitrification–nitrification technology operated in the activated sludge mode for the treatment of coke plant wastewater illustrates the optimization of reactor design for contaminant control by manipulating HRT and SRT. Until recently, the state-of-the-art biological treatment of coke plant wastewater was activated sludge oxidation of phenol and thiocyanate followed by alkaline chlorination of cyanide and ammonia. This technology did not address the control of polynuclear aromatics (PNAs) that are normally present in coke plant wastewater. Bridle *et al.* (1980) identified that a two-stage anoxic–aerobic reactor design could achieve the simultaneous removal of all the above inorganic contaminants and most of the organic contaminants. Optimum conditions were defined as 30 to 40 days system SRT, 1 day anoxic HRT and 3 day aerobic HRT and a low level of 0.5 kg m^{-3} of PAC (as reactor concentration). Removals of 99.9% total phenols, 99.5% thiocyanate, 99% ammonia and 90% cyanide (amenable to chlorination) were achieved under these conditions. Most of the trace organic contaminants present in the raw wastewater were also removed under these operating conditions. Selected data are shown in Table 1. Also, analysis of the biological sludges indicated the accumulation of some PNAs, notably, indenopyrene, naphthalene, pyrene and benz[*a*]anthracene. Mass balances showed more than 90% of the phthalates and naphthalene were removed but it could not be deduced from the data what proportion was removed by biodegradation or by adsorption. In contrast, the indenopyrene and pyrene remained adsorbed on the sludge, that is they were not degraded and reached high concentrations.

A substantial accumulation of organic contaminants can occur when SRT control is practised and illustrates the bioconcentration role of the sludge. For example, in Table 1, benz[*a*]anthracene was not detected in the feed but was present at 0.36 mg kg^{-1} on a dry weight basis in the sludge. Mass balances showed that, assuming no losses to other mechanisms, a concentration factor of approximately 4000 was in effect. The feed concentration of benz[*a*]anthracene would then have been 0.09 mg m^{-3} which was below the detection limit of 0.6 mg m^{-3}. It could be concluded that the trace contaminants of concern may be present in raw wastewater at concentrations less than detectable level but, with SRT control, some contaminants can be adsorbed to significant levels in the sludge.

Fixed-film reactors are rapidly gaining acceptance as an improved design alternative to suspended growth systems where population retention is important. Fixed-film processes will foster

Table 1 Selected Trace Organic Contaminant Analysis—Suspended Growth System[a]

	Raw wastewater (mg m^{-3})	System effluent (mg m^{-3})	Reactor sludge[b] (mg kg^{-1})	Detection limit (mg m^{-3})
Anthracene	1	0.3	ND	0.25
Benz[a]anthracene	ND	ND	0.36	0.6
Benzo[a]pyrene	5	0.4	ND	0.1
Chrysene	4	0.17	ND	0.4
Fluoranthene	2.6	0.1	ND	0.1
Fluorene	4	0.15	ND	1.0
Indenopyrene	7	0.2	9.50	2.0
Naphthalene	760	ND	7.90	10
Phenanthrene	3.5	0.7	ND	0.4
Pyrene	2.2	0.7	0.74	0.6
Diethyl phthalate	300	100	9.10	10
Bis(2-ethylhexyl) phthalate	T[c]	T[c]	11.0	10

[a] After Bridle *et al.* (1980). [b] Sludge values reported on a dry weight basis. [c] T < 10 (mg m^{-3}); ND—not detected.

long cell retention times and enhance growth of slow growing microorganisms. Furthermore, fixed-film processes can provide cell concentrations of an order of magnitude higher than those found in suspended growth systems, an effect which is thought to enhance the removal of organic contaminants (Matter-Muller *et al.*, 1980). This could be attributed to an increase in substrate utilization rate as substrate concentration decreases within the film, an effect observed by Howell and Atkinson (1976). Physical attachment of microorganisms improves cell retention during shock loading conditions which in an activated sludge system could lead to a poor-settling sludge.

An example of the application of a fluidized bed fixed-film reactor for the control of trace contaminants in an industrial wastewater is given below.

27.3 CASE HISTORY

Biological fluidized bed technology development has been a major activity at Environment Canada's Wastewater Technology Centre. Most recently, a pilot-scale, two-stage coupled anoxic–aerobic fluidized bed system was operated over a two year period to evaluate the feasibility of removing a range of contaminants from coke plant wastewaters. A flow schematic of the system is shown in Figure 1. The system was operated in the predenitrification–nitrification mode to minimize energy and alkalinity consumption. The pilot plant consisted of an anoxic denitrification fluidized bed reactor, 0.15 m in diameter, and an oxygenic nitrification fluidized bed reactor, 0.29 m in diameter. The initial empty reactor volumes were 0.059 m^3 and 0.21 m^3 for the anoxic volumes and oxygenic beds respectively; however, the bed heights and reactor volumes were adjustable by relocation of the position of the sand–biomass separation systems. High internal recycle rates were necessary for both reactors to maintain the required fluidization fluxes, to ensure an adequate supply of oxygen to the nitrification process and to return nitrate and nitrite to the denitrification reactor. The support medium in both reactors was quartzite sand with an effective size 0.048 mm. The raw wastewater was ammonia still effluent drawn from a Canadian integrated steel mill. Details of the system, startup and performance data are reported elsewhere (Nutt *et al.*, 1981, 1983).

The coupled fluidized bed system was shown to be capable of achieving complete nitrification and denitrification of undiluted coke plant wastewater without the addition of PAC to the system. Total nitrogen removal efficiencies in excess of 90% were achieved at a total system HRT of 0.67 day and SRTs in the range 20 to 40 days. Also, greater than 90% removal of total organic carbon, thiocyanate and phenolic compounds was measured.

Seven grab samples of raw wastewater were collected on a non-routine basis for trace contaminant analysis. Eighteen US EPA base/neutral extractable priority pollutants were identified at greater than trace concentrations (1 mg m^{-3}) in more than one sample, including PNAs such as acenaphthylene, acenaphthene, anthracene/phenanthrene, fluoranthene, naphthalene and pyrene (Melcer and Nutt, 1983). In addition, 15 base/neutral extractable contaminants, not included in the US EPA priority pollutant classification, were identified at greater than trace levels in more than one untreated sample. Heterocyclic nitrogenous compounds such as quinoline, isoquinoline,

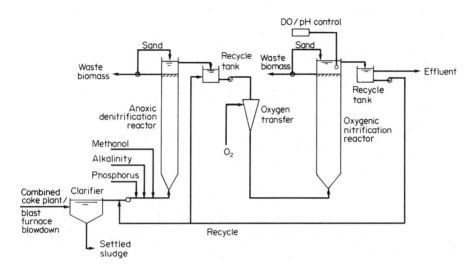

Figure 1 Process flow sheet of two-stage coupled biological fluidized bed system

indole and carbazole were consistently present at concentrations approaching or exceeding 10 000 mg m^{-3} (Melcer and Nutt, 1983).

The fluidized bed system effluent was typically sampled at the completion of each pseudo-steady-state run to determine the ability of the process to effect trace contaminant control. (Although the system had stabilized in terms of effluent quality parameters, the biomass was not considered to have reached equilibrium in each run because of the high system SRTs at which the fluidized beds were operated. These, then, were considered to be pseudo-steady-state conditions.) The treated effluent quality from the fluidized bed system is presented in Table 2 in terms of US EPA base/neutral extractable priority pollutants and non-priority base/neutral extractable trace organic contaminants. Only four priority pollutants were identified at greater than trace concentrations in more than one treated effluent sample. Two of these were phthalate esters. The others were anthracene/phenanthrene and nitrobenzene. Anthracene was also identified by Bridle *et al.* (1980) in the treated effluent from suspended growth systems treating coke plant wastewater. Nitrobenzene was not identified in any of the feed samples but could be a metabolic product of the oxidation of aniline, which was identified in the raw wastewater. Ten base/neutral extractable non-priority contaminants were found at greater than trace concentrations in more than one treated effluent sample. None of these compounds were detected at concentrations in excess of 30 mg m^{-3}.

Samples of fluidized bed reactor sludges were also examined for base/neutral extractable contaminants. There were insufficient data to identify differences in the trace organic concentrations between sludges taken from the nitrification and denitrification reactors. On an overall basis, however, five base/neutral extractable priority pollutants were measured at greater than trace concentrations (0.1 mg kg^{-1} on a dry weight basis) in more than one sample. Three of these compounds were phthalate esters which accumulated in the sludge to concentrations of 0.2 to 30 mg kg^{-1}. Bridle *et al.* (1980) reported similar concentrations of these phthalate esters in biomass from suspended growth reactors treating coke plant wastewater. Other base/neutral priority pollutants consistently identified in the sludges were acenaphthylene and pyrene, found at concentrations up to 2 mg kg^{-1}. None of the non-priority pollutant base/neutral extractable contaminants were consistently identified at greater than trace levels.

In spite of the relatively high system SRTs employed, minimal accumulation of trace contaminants was observed indicating that the majority of these contaminants were biologically degraded. Although a direct comparison of influent and effluent concentrations is difficult in view of the variability in analytical data, it appears that the biological fluidized bed process is capable of effecting a high degree of trace contaminant control based on the order of magnitude differences in influent and effluent trace organic contaminant concentrations.

Table 2 Selected Trace Organic Contaminant Analysis—Fluidized Bed System[a]

	Raw wastewater (mg m^{-3})[b]	System effluent (mg m^{-3})[c,e]	Reactor sludge (mg kg^{-1})[d,e]
B/N Priority Pollutants			
Acenaphthene	2–15	T	T
Acenaphthylene	7–33.3	T	0.2–0.8
Anthracene/phenanthrene	30–227	T–21	T
Fluoranthene	8–54	T	T
Fluorene	4–33	T	T
Naphthalene	3–81	T	ND
Pyrene	3–49	T	T–2.0
Di-*n*-butyl phthalate	3–173	T–90	0.2–34.0
Bis(2-ethylhexyl) phthalate	3–8	1–4	T–13.0
Nitrobenzene	ND[e]	T–10	ND
Non-B/N Priority Pollutants			
Isoquinoline	1000–7000	1–10	ND
2-Methylnaphthalene	2–111	10	T
Indole	23–26 000	20	T
Carbazole	330–2200	10	T–0.2
Quinoline	805–17 000	20	T
4-Methylquinoline	3–274	T	ND

[a] After Melcer and Nutt (1983). [b] Based on 7 samples. [c] Based on 8 samples. [d] Based on 9 samples, reported on a dry weight basis. [e] T < 1 mg m^{-3} for liquids; T < 0.1 mg kg^{-1} for sludges; ND—not detected.

27.4 SUMMARY

The problems associated with biological degradation of trace contaminants are reviewed. Biological reactor design has changed to accommodate these problems. In particular, the advent of fixed-film systems and the use of SRT control in both suspended growth and fixed-film systems are probably the most important changes in reactor design to influence biological degradation of trace contaminants. Examination of the performance of well-operated biological treatment plants has demonstrated at bench-, pilot- and full-scale the utility of SRT control for the control of trace contaminants.

27.5 REFERENCES

Alexander, M. (1973). Biotechnology report. Nonbiodegradable and other recalcitrant molecules. *Biotechnol. Bioeng.*, **15**, 611–647.
Bridle, T. R., W. K. Bedford and B. E. Jank (1980). Biological treatment of coke plant wastewater for control of nitrogen and trace organics. Presented at the 53rd Annual WPCF Conference, Las Vegas, NV.
Burns and Roe Industrial Services Corporation (1982). Fate of priority toxic pollutants in publicly-owned treatment works—Final Report, vol. 1, EPA-440/1-82/303.
Cohen, J. M. and S. A. Hannah (1981). National survey of municipal wastewaters for toxic chemicals. Presented at the 8th US–Japan Conference on Sewage Treatment Technology, Cincinnati, OH.
Heck, H., D. G. Parker, L. J. Thibodeaux and R. L. Dickerson (1981). Measurement of volatile chemical emissions from aerated stabilization basins. Presented at the 1981 ASCE Conference on Environmental Engineering, Atlanta, GA.
Howell, J. A. and B. Atkinson (1976). Influence of oxygen and substrate concentrations on the ideal film thickness and the maximum overall substrate uptake rate in microbial film fermenters. *Biotechnol. Bioeng.*, **18**, 15–35.
Kincannon, D. F. and E. L. Stover (1981). Fate of organic compounds during biological treatment. Presented at the 1981 ASCE Conference on Environmental Engineering, Atlanta, GA.
Kobayashi, H. and B. E. Rittmann (1982). Microbial removal of hazardous organic compounds. *Environ. Sci. Technol.*, **16**, 170A–183A.
Mackay, D. (1982). Correlation of bioconcentration factors. *Environ. Sci. Technol.*, **16**, 274–278.
Mackay, D. and S. Paterson (1981). Calculating fugacity. *Environ. Sci. Technol.*, **15**, 1006–1014.
Matter-Muller, C., W. Gujer, W. Griger and W. Stumm (1980). Non-biological elimination mechanisms in a biological sewage treatment plant. *Prog. Water Technol.*, **12**, 299–314.
Melcer, H. and S. G. Nutt (1983). Removal of trace organic contaminants from coke plant wastewater and blast furnace blowdown in a biological fluidized bed system. Presented at the US EPA Symposium on Iron and Steel Pollution Abatement Technology for 1983, Chicago, IL.
Natural Resources Defense Council (1975). M. C. *vs.* Train, 1975, 8 ERC, 2120, 2122-29, D.D.C.
Nutt, S. G., H. Melcer, I. J. Marvan and P. M. Sutton (1983). Treatment of coke plant wastewater in the coupled prede-

nitrification nitrification fluidized bed process. *Proc. 37th Ind. Waste Conf., Purdue Univ., W. Lafayette, Indiana*, 527–536. Ann Arbor Science, Ann Arbor, MI.

Nutt, S. G., H. Melcer, I. J. Marvan and P. M. Sutton (1984). Two-stage biological fluidized bed treatment of coke plant wastewater for nitrogen control. *J. Water Pollut. Control Fed.*, **56** (7), 851–7.

Petrasek, A. C., B. M. Austern and T. W. Neiheisel (1983). Removal and partitioning of volatile organic priority pollutants in wastewater treatment. Presented at the 9th US–Japan Conference on Sewage Treatment Technology, Tokyo, Japan.

Petrasek, A. C., I. J. Kugelman, B. M. Austern, T. A. Pressley, L. A. Winslow and R. H. Wise (1983). Fate of toxic organic compounds in wastewater treatment plants. *J. Water Pollut. Control Fed.*, **55** (10), 1286–96.

Rittmann, B. E. and P. L. McCarty (1980). Model of steady-state biofilm kinetics. *Biotechnol. Bioeng.*, **12**, 2343–2357.

Strier, M. P. and J. D. Gallup (1983). Removal pathways and fate of organic priority pollutants in treatment systems: chemical considerations. *Proc. 37th Ind. Waste Conf., Purdue Univ., W. Lafayette, Indiana*, 813–818. Ann Arbor Science, Ann Arbor, MI.

Tischler, L. F. and D. Kocurek (1982). The CMA/EPA five plant study—biological treatment of toxic organic pollutants. Presented at the 55th Annual WPCF Conference, St. Louis, MO.

Wukasch, R. F., C. P. L. Grady and E. J. Kirsch (1981). Prediction of the fate of organic compounds in biological wastewater treatment systems. *AIChE Symp. Ser. (Water 1980)*, **77** (209), 137–143.

28

Mycotoxin Hazards in the Production of Fungal Products and Byproducts

J. E. SMITH
University of Strathclyde, Glasgow, UK

28.1 INTRODUCTION

Filamentous fungi are being explored in established and new biotechnological processes for the production of many important products including enzymes, organic acids, SCP, fungal insecticides and many oriental fermentations (Smith and Berry, 1975; Smith *et al.*, 1980, 1982). In recognizing the undoubted beneficial aspects of fungi their less beneficial attributes must always be determined for each process. Fungi used in industrial processes are seldom realistically considered from a health aspect and although there are few examples of mortality to man by fungi there are enumerable examples of serious illness arising from occupational exposure to certain fungi and to the consumption of food contaminated with toxigenic fungi.

28.1.1 Fungal Infections

The infectious nature of fungi to animals and man is well documented (Emmons *et al.*, 1977). The infections or mycoses resulting from the growth of certain pathogenic fungi in animals and man can lead to inflammation, sickness and even, on occasions, death. Truly pathogenic fungi have so far not been used in biotechnological processes. However, there is increasing use of *Aspergillus fumigatus*, a thermotolerant cellulose degrader, in laboratory studies related to ligno-cellulose degradation. *A. fumigatus* is a well-known opportunistic pathogen causing respiratory infections and death in man and animals as well as mycotic infections in cattle. Other *Aspergillus* spp. have also been occasionally reported as weak pathogens, especially *A. flavus*, *A. niger* and *A. nidulans* while *Absidia* spp. and *Mucor pusillus* may cause phycomycosis and gastric ulceration (Emmons *et al.*, 1977).

28.1.2 Fungal Allergies

Chronic exposure to fungal spores can occur in many occupations, in particular in agriculture as well as in many biotechnological industries. Airborne spores can cause two types of allergy, immediate allergy and allergic alveolitis (Lacey, 1975; Emmons *et al.*, 1977). In immediate allergy the symptoms resemble hay fever and can be accompanied by asthma within minutes of exposure. It is mostly caused by field fungi such as *Cladosporium* and *Alternaria*, with spores larger than 5 μm, which are deposited in the nose and the alveoli. Allergic alveolitis, *e.g.* farmers lung and bagassosis, results in fever and breathlessness within several hours of exposure. Although most often caused by thermophilic actinomycetes several fungi such as *A. fumigatus*, *A. clavatus* and *Penicillium* spp. can be the causal agents. In these cases the spores are smaller than 5 μm and can penetrate to the deepest parts of the lung. In farm animals chronic exposure to fungal spores can lead to reduced productivity.

28.1.3 Fungal Poisons

Many fungi can be directly poisonous to man and less occasionally to animals. Poisoning by macrofungi or mushrooms is termed mycetismus and by micro or mould fungi as mycotoxicosis.

28.1.3.1 *Mycetismus*

Mushroom-type fungi have had a long historical involvement with man as a source of food and flavour additives, *e.g. Agaricus bisporus* (the white commercial mushroom), *Lentinus edodes* (the Japanese Shii-ta-ke mushroom), *Volvariella volvaceae* (the Asiatic Padi-straw mushroom) and *Tuber melanogaster* (the European truffle). *A. bisporus* and *L. edodes* cultivations are, in particular, major biotechnological processes practised throughout the world and involve complicated solid-state fermentation practices (Hayes and Nair, 1974; Smith, 1969, 1972). There is no evidence of major toxicological syndromes arising from these fungal products other than infrequent allergic responses, *e.g.* headaches or slight gastroenteric upsets.

Since other groups of the macrofungi, in particular wood-rotting Basidiomycetes, may well find applications in future biotechnological processes (lignocellulose degradation, enzymes, food flavours, *etc.*) a cautionary statement must be made to indicate the ability of many of these fungi to produce toxic symptoms in man. The most dangerous fungi include those responsible for cytolytic poisoning since the initial gastroenteric symptoms caused by the fungi may obscure the liver and kidney attack which may not become apparent for several days. Only small amounts of the fungus need to be eaten to cause extremely dangerous, even fatal, poisoning. The range of toxic types are shown in Table 1 and fuller details can be obtained from Lincoff and Mitchel (1977), Rumack and Salzman (1978) and Pegler and Watling (1982). Such fungi that are toxic to man seem to have little effect on most grazing animals. Cattle are indeed renowned mycophagists.

28.1.3.2 *Mycotoxicosis*

In contrast, mycotoxicosis is poisoning by the ingestion of toxins of fungal origin in foods and feeds which have been altered or damaged by the growth of certain toxin-producing mould fungi.

Table 1 Classification of Toxic Mushrooms[a]

(I)	*Cytolytic poisoning*
	(a) Cyclopeptides, *viz.* amatoxins (cyclic octapeptides) and phallotoxins (cyclic heptapeptides): *Amanita phalloides*, *A. virosa*, *Conocybe filaris*, *Lepiota* spp. and *Galerina unicolor*
	(b) Orellannis, *viz.* orellamine, grzymaline, cortinarine and two benzonines: *Cortinarius speciosissimus*, *C. gentilis*, *C. cinnamomeus*
	(c) Gyromitrin (monomethyl hydrazine): *Gyromitra esculenta* and some over-mature *Morchella* spp.
(II)	*Haemolytic compounds* (possibly phallolysin) *Amanita rubescens*, *A. vaginata*
(III)	*Coprine poisoning* (alcohol related) *Coprinus atramentarius*, *Pholiota squarrosa*
(IV)	*Psychotropic poisoning*
	(a) Muscinnol, ibotenic acid, muscazone: *Amanita muscaria*, *A. pantherina*
	(b) Indole group (psilocin, psilocybin): *Psilocybe semilanceata*, *P. cubensis*, *Panaeolus* spp.
	(c) Hordenine, *m*-methyltyramine, tyramine: *Lactiporus sulphureus*, *Meripilus giganteus*
(V)	*Muscarine poisoning* *Amanita muscaria*, *A. pantherina*, *Clitocybe* and *Inocybe* spp.
(VI)	*Gastroenteric irritants* *Entoloma sinuatum*, *Paxillus involutus*, *Boletus satanus*, *Hebeloma* spp., *Stropharia*, *Trocholoma*, *Russula* and *Lactarus* spp.

[a] Adapted from Pegler and Watling (1982).

Mycotoxins are, in general, low molecular weight non-antigenic fungal metabolites capable of causing a toxic response in man and animals. Mycotoxins are generally synthesized typically, but not always, on grains and other low moisture plant products but may also be found in many fungal fermentations, in milk and milk products and in certain animal tissues (Table 2). They can enter the animal or human system by ingestion, inhalation or direct skin contact and very small amounts of these toxins can cause significant health problems. In practice, mycotoxins have been shown to have major significance in animal production, *e.g.* poultry, pigs and cattle. However, there is now much evidence to relate specific mycotoxins as the causal agents in certain diseases of man (Shank, 1978; WHO, 1979). At high concentrations many mycotoxins can produce acute disease syndromes while at lower levels may be carcinogenic, mutagenic, teratogenic or oestrogenic, may reduce the growth rate of young animals and can even interfere with natural mechanisms of resistance and impair immunologic responsiveness making animals more susceptible to infections (Pier *et al.*, 1980).

Table 2 Possible Routes for Mycotoxin Contamination of Human and Animal Foods[a]

(1)	Mould-damaged foodstuffs:
	(a) Agricultural produce, *e.g.* cereals, oilseeds (groundnuts), fruits, vegetables
	(b) Consumer foods (secondary infections)
	Compounded animal feeds (secondary infections)
(2)	Residues in animal tissues and animal products, *e.g.* milk, dairy produce, meat
(3)	Mould-ripened foods, *e.g.* cheeses, fermented meat products, oriental fermentations
(4)	Fermented products, *e.g.* microbial proteins, enzymes, food additives, *e.g.* vitamins

[a] Adapted from Jarvis (1976).

28.2 CHEMISTRY AND BIOLOGY OF MYCOTOXINS

Mycotoxins are in general secondary metabolites of relatively wide chemical composition and are produced by a limited range of fungal species (Turner, 1971). Specific names of mycotoxins can be derived from several routes, *viz.* (a) *aflatoxin* formed from the *A* from *Aspergillus fla* from the species *flavus* and the appended noun toxin; (b) by a descriptive chemical name, *diacetoxyscirpenol*; and (c) by the toxic manifestation, *vomitoxin*.

The mycology of mycotoxin production has been well examined and although many fungal genera have been shown to produce toxins, most of the common mycotoxins are produced within three genera, *viz.* aspergilli, penicilli and fusaria. Although there is some advantage in identifying the contaminating fungi present during outbreaks of mycotoxicosis this in itself is not sufficient and it remains absolutely essential to identify the specific toxins since: (a) the presence of the fungus is no assurance that it was producing the toxin; (b) a specific fungus may be able to produce more than one toxin; (c) a given toxin may be produced by different genera of fungi; and (d) a given toxin may persist in a product when the fungus producing it is no longer present.

Mycotoxins generally occur in natural products as single entities often associated with near monoculture of the producer toxigenic mould. However, the multiple presence of mycotoxins can occur particularly in compounded animal feeds where the raw materials have been derived from various sources. Actual multiple organism cultures generally do not show evidence of mycotoxins (Mislivic, 1977).

Mycotoxin production by a toxigenic mould will depend on the interrelationship between three conditions: (1) the actual presence of the toxigenic mould; (2) a suitable substrate for the growth of the mould; and (3) an environment suitable for the growth of the mould.

Analytical methods for the detection, identification and quantification of all of the main mycotoxins are readily available. The procedures involved in this methodology include extraction of the toxins from organic samples, a clean-up procedure (thin layer chromatography, column chromatography, gel filtration chromatography, dialysis or solvent partition), analysis (silica gel thin layer chromatography, high performance liquid chromatography or field absorption mass spectrometry), confirmation of identity and quantification (Pohland *et al.*, 1979; Steyn, 1981).

Although the chemical identification of most mycotoxins is now routine the main problem with the analysis of mycotoxins from natural products is that of obtaining a suitable representative sample for analysis. Sampling techniques should be designed to achieve statistically sound information concerning the presence or absence of mycotoxins in a defined population or organic sample, *e.g.* grain, animal feed, mould products, fermented foods, *etc.* A set of guidelines for mycotoxin evaluation in cereals and cereal products can also be used for other possible mycotoxin sources (Davis *et al.*, 1980).

Analysis for individual mycotoxins in raw materials for food, feed or biotechnological processes can be very expensive and for this reason analytical procedures should seek to achieve simultaneous analysis of several mycotoxins. Many such methods now exist but the search continues for analytical methods that require less time, less solvent and less toxic solvents without sacrificing reliability (Steyn, 1981).

The best known of the mycotoxins are the family of nonaketides, the aflatoxins, produced by *Aspergillus flavus* and *A. parasiticus*. There are many chemical forms of the aflatoxins, B_1 (**1a**) and G_1 (**2**) are the dominant forms in nature while lactating animals consuming B_1 will convert it to M_1 (**1b**) in milk.

(**1a**) R = H
(**1b**) R = OH

(**2**)

Ochratoxin A (**3**) is a chlorine-containing pentaketide dihydroisocoumarin linked to L-β-phenylalanine, which was originally identified from *A. ochraceus* but can also be produced by many *Penicillium* spp., in particular *P. viridicatum*.

Sterigmatocystin (**4**) is a mycotoxin produced by several species of *Aspergillus*, in particular *A. versicolor* and one species of *Bipolaris*. Structurally sterigmatocystin resembles aflatoxin and is basically a xanthone nucleus attached to a bifuran ring. It is considered to be a precursor in the biosynthesis of aflatoxin.

Citrinin (**5**) is a yellow cyclic compound with a free carboxylic acid group and is produced by

(3)   **(4)**

several *Penicillium* spp., in particular *P. citrinum*, *P. viridicatum* and *P. expansum*, as well as some *Aspergillus* spp. Patulin (**6**) is an unsaturated lactone and is produced by several *Penicillium* and *Aspergillus* spp. Penicillic acid (**7**) is an unsaturated lactone-type compound of low potency and is produced by several *Aspergillus* and *Penicillium* spp.

(5)   **(6)**   **(7)**

The genus *Fusarium* is of wide occurrence and is found most often in high moisture grain and other products (WHO, 1979). Zearalenone (**8**) is the most recognized *Fusarium* toxin and is the generic name for a complex macrocylic molecule. It can be produced by many species, in particular *F. roseum*, *F. tricinctum*, *F. culmorum* and *F. moniliforme*. T-2 toxin (**9**), diacetoxyscirpenol and vomitoxin (deoxynivalenol) (**10**) are members of a chemical group of metabolites termed trichothecenes, a family of epoxide-containing sesquiterpenoid mycotoxins. They are produced by many *Fusarium* spp. but also by other genera such as *Myrothecium*, *Trichoderma*, *Cephalosporium* and *Stachybotrys*. In general trichothecenes are found at low temperatures (0–20 °C).

(8)   **(9)**

(10)

28.3   THE FORMATION OF MYCOTOXINS

28.3.1   The Role of the Environment

Mycotoxins can be produced by fungi growing in different substrate relationships, *viz.* (1) on the living plant (*field fungi*); (2) on high moisture organic material (*advanced decay fungi*); and (3) on low moisture organic material (*storage fungi*) (Table 3).

Table 3   Fungi which Produce Mycotoxins Classified by Habitat[a]

Fungi growing in the living plant

Claviceps purpurea	*Aspergillus flavus*
Sclerotinia sclerotiorum	*Rhizoctonia leguminicola*
Fusarium graminearum	*Helminthosporium biseptatum*
(*Gibberella zeae*)	

Fungi growing in decaying plant material

Pithomyces chartarum	*Fusarium graminearum*
Stachybotrys atra	*Chaetomium globosum*
Periconia minutissima	*Dendrodochium toxicum*
Fusarium sporotrichoides	*Myrothecium verrucaria*
Cladosporium spp.	*Trichothecium roseum*
Alternaria longipes	*Trichoderma viride*

Fungi growing on stored plant material

Aspergillus flavus	*Penicillium islandicum*	*Chaetomium globosum*
A. parasiticus	*P. citrinum*	*Fusarium graminearum*
A. versicolor	*P. rubrum*	*F. tricinctum*
A. ochraceus	*P. citreoviride*	*F. nivale*
A. clavatus	*P. cyclopium*	*F. moniliforme*
A. fumigatus	*P. viridicatum*	
A. rubrum	*P. urticae*	
A. chevalieri	*P. verruculosum*	
	P. puberulum	
	P. expansum	
	P. rugulosum	
	P. palitans	
	P. roqueforti	

[a] Hesseltine (1976).

The moisture level of the substrate has a major determining influence on the type of toxigenic fungus that will grow and produce toxins. Thus, high moisture levels (20–25% wet weight) are required for growth of some field fungi and most advanced decay fungi (in particular the fusaria) whereas storage fungi are able to colonize substrates with moisture levels as low as 12–13% (mostly penicilli and aspergilli) (Hesseltine, 1976).

There has been a widely held misconception that most mycotoxin production occurs in dry or stored products. This can no longer be considered true. Detailed studies have shown that several typical storage moulds such as *Aspergillus flavus* can be found growing on many crops before harvest, *e.g.* peanuts, maize and cotton seed (Stoloff, 1976). While insect damage is one of the main factors allowing mycotoxin production on living plants many other predisposing conditions are now recognized, *viz.* stress factors on the growing plant, mechanical damage, mineral nutrition deficiencies and unseasonal temperatures. After harvest and during shipment, storage and utilization of organic commodities (including many biotechnological processes), mycotoxin production can be influenced by many factors including moisture level, temperature, aeration, insects and other microorganisms, duration of storage, types of chemical treatments, mould spore infection density, storage conditions such as leakage of water or condensation, biological heating and the chemical nature of the raw material (Hesseltine, 1976; Stoloff, 1976).

The most relevant and critical factors in the environment controlling mycotoxin formation are moisture content or water activity of the product and temperature together with the duration of incubation. Water activity (a_w) is the ratio of the vapour pressure of water in a product or solution to that of pure water at the same temperature. Water activity and equilibrium relative humidity (e.r.h.) are numerically equivalent but e.r.h. is expressed as a percentage; thus a_w 0.8 = 80% e.r.h. Water activity is a more useful parameter than water content since it reflects the availability of water for metabolic processes. Different products with the same water activity may have very different water contents; thus oilseeds have a higher water activity at a given water content than starchy cereal seeds (Corry, 1978).

Microorganisms including toxigenic moulds have characteristic optimum and limiting water activities and temperatures, and these are interrelated. For a given microorganism the lowest minimum water activity for growth occurs at or slightly above the optimum temperature while the maximum temperature for growth may be highest at rather low water activity (Ayerst, 1966).

In this way organic substrates that have water activities of less than 0.65 are inherently immune from toxigenic mould growth and subsequent toxin production. Thus if storage can be maintained at or below this level there will be no new mycotoxin production. When the water activity factor exceeds this arbitrary level in organic raw materials or finished products then in the presence of toxigenic moulds such materials will become increasingly susceptible to colonization and possible toxin formation. Thus in biotechnological industries the need to control moisture levels in raw materials and products becomes of cardinal importance.

28.3.2 Techniques of Storage

Many biotechnological products such as antibiotics are produced under aseptic conditions and should be free of contaminating toxigenic moulds. Provided secondary contamination is avoided there should be no mycotoxin hazard. However, many of the raw materials for biotechnological processes will by necessity require to be stored for variable periods of time before being used.

In practice, agricultural products such as cereals and oil seeds, finished animal feeds and many human foods can be held under storage for variable periods of time. Mould growth in stored grain can (a) cause a decrease in future germinability (*e.g.* malting); (b) change the nutrient content or quality (feed and food production); (c) cause production of mycotoxins; and (d) support the growth of microflora that can become a health hazard not only to farm animals but also to associated workers (respiratory diseases in operators) (Tuite and Foster, 1979). Prevention or inhibition of mould growth in storage conditions is usually accomplished by modifying the interseed or interparticle environment, *e.g.* moisture, temperature or gaseous environment.

Proper long term storage has been both a microbiological and an engineering problem. In most cases a drying regime will reduce the moisture level below the value suitable for mould spore germination. Following on, the material will be placed under storage conditions where environmental control by engineering practice will attempt to hold the moisture level at the correct value. Within large volumes of stored materials changing environmental conditions or physical damage to the containment vessel, allowing the entry of water, can lead to migration of moisture, condensation and the formation of moist pockets. Under such conditions mould spore germination, mycelial proliferation and possible mycotoxin production can occur.

There exists an extensive range of drying techniques depending on the range and efficiency of instrumentation that is available. Solar drying systems may soon be realized.

Storage capabilities of grains will be further improved by the production of improved varieties that dry more rapidly, damage less when handled and resist insect and mould attack in field and storage conditions.

More recently there has been an increasing use of chemical preservatives used in conjunction with drying procedures or directly with high moisture materials (Smith, 1982). The use of chemical preservation for agricultural crops and organic materials allows greater flexibility during storage, decreased capital outlay and greater speed of treatment (Tuite and Foster, 1979). Chemicals used must demonstrate low mammalian toxicity with wide and long lasting fungal inhibitory ability. By far the most successful compounds have been organic acids, in particular propionic acid, while acetic and formic acids, although much less efficient, are often used synergistically in combination with propionic acid (Huitson, 1968; Stewart *et al.*, 1977; Lacey *et al.*, 1980).

28.4 THE INFLUENCE OF MYCOTOXINS ON ANIMAL AND HUMAN HEALTH

28.4.1 Animal Mycotoxicoses

Mycotoxins have their main influence on assembled groups of animals such as poultry, pigs and dairy and feedlot cattle because of the high intake of stored concentrated plant-derived foods (the major natural source of most mycotoxins). From a veterinary point of view six criteria can normally be associated with mycotoxicoses of animals, *viz.* (1) they frequently arise as veterinary problems whose true cause is not immediately obvious; (2) the mycotoxicoses or disorders are not transmissable from one animal to another being neither infectious nor contagious; (3) treatment with drugs or antibiotics usually has little effect on the course of the disease; (4) in field outbreaks the trouble is often seasonal as particular climatic sequences may favour toxin production by the mould fungus; (5) careful study indicates association with specific foodstuffs, *e.g.* peanuts and groundnut meal, corn and other cereals; and (6) examination of suspected food stuffs may or may not reveal signs of fungal activity.

A further confusing factor concerning diagnosis is that animals will display variable susceptibility to mycotoxins according to sex, age and species of animal and to the specific toxin(s) involved.

Mycotoxicoses in animals will normally assume one of three identifiable forms (Pier *et al.*, 1977; Pier *et al.*, 1982).

28.4.1.1 Acute primary mycotoxicoses

An acute disease syndrome such as hepatitis, haemorrhage, nephritis, necrosis of oral and enteric epithelia or death can arise when high to moderate concentrations of mycotoxins are consumed by susceptible animals. Studies on the acute effects of mycotoxins have shown that virtually every system of the animal body can be influenced by one or a combination of mycotoxins (Morehouse, 1979), *viz.* (1) digestive system: (diarrhoea, intestinal haemorrhage or hepatotoxic effects manifested in liver necrosis, bile duct proliferation and fibrosis, *aflatoxin*; caustic effects on mucous membranes, *T-2 toxin*; bile duct occlusion, *sporidesmin*; salivation, *slaframine*; feed rejection, *vomitoxin*); (2) cutaneous system: photosensitization, *sporidesmin*; necrosis and sloughing of the extremities, *ergot*; (3) vascular system: increased vascular fragility, haemorrhage into body tissues, *aflatoxin, dicoumarin*; (4) respiratory system: adenomatosis, *4-ipomeanon*; (5) nervous system: tremors, incoordination, mania, ataxia, coma, *tremorgens, ergotamine*; (6) reproductive system: infertility, prolonged oestrogens, *zearalenone*; and (7) urinary system: nephrosis, uraemia, *ochratoxin A, citrinin*.

28.4.1.2 Chronic primary mycotoxicoses

This is now recognized as the most common form of mycotoxicoses and occurs with moderate to low levels of mycotoxins, resulting in reduced productivity of animals in the form of slower growth rates, reduced reproductive efficiency and inferior market quality of animals such as poultry. In most cases the effects will occur without any specific primary disease syndrome. In practice this phenomenon can only be properly diagnosed in carefully monitored systems and most often would be mistaken for other microbial diseases or dietary deficiencies.

28.4.1.3 Secondary mycotoxin diseases

This disease response arises from the intake of very low concentrations of certain mycotoxins (p.p.b.) and the affected animals become predisposed to infectious diseases by the impairment of immunogenesis and native mechanisms of resistance. It has been demonstrated that several mycotoxins can suppress the cellular immune system. In this way secondary mycotoxin diseases will simply enhance the infection processes to which the host is predisposed (Pier *et al.*, 1978; Giambrone *et al.*, 1978). Aflatoxin consumption has been associated with increased susceptibility to salmonellosis, candidosis and coccidiosis of poultry, fascidiosis in calves and salmonellosis and dysentery of pigs (Pier *et al.*, 1978; Hamilton and Harris, 1971).

Although most mycotoxins have specific target organs within the body they can, in general, still affect other parts, resulting in many clinical signs that are difficult to differentiate from various well-known diseases. A set of criteria or guidelines has been set down by Morehouse (1979) which could help to rationalize these difficulties, *viz.* (1) recognizing the clinical signs consistent with a given mycotoxicosis; (2) failure to isolate infectious agents or observation of failure to respond to antibiotic therapy; (3) probability of exposure to fungal toxins, *i.e.* careful examination of feed (often gone by the time the disease problems are presented), storage conditions, feeding mechanisms and consideration of the environmental factors favourable to fungus development and toxin production; (4) evidence of feed refusal or their lack of palatability; and (5) test feeding of small groups of animals with suspect feed observing that the young of most species are usually more susceptible than the old.

28.4.2 Human Mycotoxicoses

Throughout the world, in the course of history, mycotoxins have played a significant role in human health. Cardiac beriberi in Japan, ergotism in Europe and alimentary toxic aleukia and stachybotryotoxicosis in Russia have been well authenticated as true mycotoxicoses of humans (Shank, 1978; Table 4). The potential involvement of the causal organisms of these diseases in modern biotechnological processes is most unlikely and will not be further pursued.

Table 4 Some Mycotoxins and Associated Mycotoxicoses of Man and Animals[a]

Mycotoxin	Fungi responsible	Mycotoxicosis or symptoms	Animals affected
Aflatoxin	*Aspergillus flavus, A. parasiticus*	Hepatitis, cirrhosis, hepatoma	Man, poultry, pigs, cattle
Citrinin	*Aspergillus ochraceus, Penicillium viridicatum, P. citrinum*	Nephrotoxicosis	Pigs
Citreoviridin	*Penicillium citreoviride*	Cardiac beriberi	Man
Diacetoxyscirpenol, T-2 toxin and other trichothenes, vomitoxin	*Fusarium tricinctum, F. poae, F. graminearum, F. culmorum*	Alimentary toxic aleukia, haemorrhage, vomiting, rejection of food, dermatitis	Man, cattle, sheep, horses, pigs, poultry
Luteoskyrin	*Penicillium islandicum*	Hepatitis, hepatoma	Man, chickens
Moniliformin	*Fusarium moniliforme*	Leucoencephalomalacia	Horses, donkeys
Ochratoxin	*Aspergillus ochraceus, Penicillium viridicatum*	Nephrotoxicosis	Pigs, cattle, man?
Patulin	*Aspergillus clavatus, Byssochlamys nivea, Penicillium urticae, P. patulum, P. expansum*	Neurotoxicosis, pulmonary oedema, vomiting, dermatitis, sarcoma	Cattle, man
Penicillic acid	*Aspergillus ochraceus, Penicillium viridicatum*	Carcinoma	Rats
Penitrem A	*Penicillium crustosum*	Tremors, convulsions	Sheep
Rubratoxin B	*Penicillium rubrum*	Hepatitis, haemorrhage	Cattle, pigs, dogs
Satratoxin	*Stachybotrys atra*	Stachybotryotoxicosis, dermatitis	Man, horses, cattle, sheep, pigs, buffalo
Sterigmatocystin	*Aspergillus versicolor, A. nidulans*	Hepatoma	Man
Zearalenone	*Fusarium graminearum, F. culmorum*	Vulvovaginitis, abortion, infertility, splay leg in piglets	Pigs
Verruculogen	*Penicillium simplicissimum*	Tremors	Sheep

[a] After Austwick (1975).

It is difficult to make a true assessment of human susceptibility to mycotoxins since it is not possible to submit human subjects to the types of mycotoxin experiments that have been successfully performed with animals. Some attempts have been made to extrapolate animal data to man but the phenomenon of species variation renders such data of limited value. However, by analyzing the contamination level of human foods with mycotoxins and by observing the type and severity of diseases associated with such consumption it may be possible to estimate to a limited degree man's susceptibility to mycotoxins (WHO, 1979).

Health assessment of the aflatoxins has been widely studied throughout the world while chronic studies involving zearalenone and patulin, and subacute studies of penicillic acid are being completed in the USA (CAST, 1979). Wide ranging studies with other mycotoxins will be dependent on the level of regular contamination in foods and feeds.

The impact of aflatoxins on human health is, however, difficult to evaluate (Hayes, 1980). Epidemiological studies in Thailand, Kenya and Mozambique have clearly implied a significant correlation between the estimated ingestion of aflatoxin-contaminated food and primary liver cancer (Heathcote and Hibberd, 1978).

Aflatoxin B_1 has also been detected in tissues of children from South East Asia, Czechoslovakia and the USA who have died from Reyes syndrome (Shank, 1978). Reyes syndrome, defined as encephalopathy and fatty degeneration of the viscera, is a severe paediatric disease of high mortality and unknown aetiology.

It is now well recognized that mycotoxins, in particular aflatoxins, ochratoxins and zearalenone-type toxins, are difficult to eliminate completely from food products. An administrative guideline of 20 p.p.b. upper limit for aflatoxins in all foods is widely accepted throughout the world. But the direct risks to humans from ingestion of the aflatoxins cannot even yet be precisely measured. For this reason, risk assessments must rely on mathematical treatment of data from animal toxicological and human epidemiological studies. Epidemiological studies relating aflatoxin to primary liver cancer in humans conducted in Uganda, Kenya, Thailand, India and Taiwan strongly suggested that aflatoxin causes human cancer. Aflatoxin intake in certain diets was shown to range from 3.5–222 μg/kg body weight per day (corresponding to 0.1–7.4 p.p.b. in the diet) with the cancer rate ranging from 0.7–13.00 per 100 000 population/year.

In the USA only corn and peanuts have been considered in the risk estimate made by the FDA since these products constitute the major sources of exposure of humans to aflatoxins (WHO, 1979). However, trace levels of aflatoxins are continuously present in most world food supplies and under some conditions the concentration can be high enough to be hazardous. For example if each day a child weighing 35 kg consumed 2 oz of peanut butter that contained the maximum allowable limit of 20 p.p.b of aflatoxin, the intake would equal about 1 μg. This would be equivalent to about 1/8 of the exposure of humans to aflatoxin in the Inhambane area of Mozambique which has the highest known incidence of aflatoxin-associated liver cancer in the world (CAST, 1979). Certainly continuous caution must be exercised (WHO, 1979).

Yet perhaps too much attention has been given to the cancer-inducing potential of aflatoxins and other mycotoxins. The possibility that low levels of mycotoxins could cause secondary mycotoxin disease syndromes in man, in a way similar to those now being identified in animal populations, cannot be overlooked. If this were in fact true then the real implications of mycotoxins in human foods would be much more threatening. Until this is proved otherwise every effort must be made to exclude or eliminate mycotoxins from the human dietary chain.

Finally, in scientific practice laboratory workers dealing with toxigenic fungi or mycotoxins are prime subjects for accidental or careless exposure which may lead to skin contact, ingestion or inhalation of toxins. Aerosols are easily generated from mould dusts (agricultural and brewery hazard), spillage or pipetting from toxin solutions, and scraping off of thin layer plates during diagnostic studies. Rigorous controls must always be practised under these conditions and long term health assessment should be considered for operators regularly handling mycotoxins or mycotoxin-contaminated products (WHO, 1979).

28.5   MYCOTOXIN ENTRY INTO FOOD CHAINS

The main hazards of mycotoxins to man and animals will come by way of the food chain (Table 2) (Jarvis, 1976). Mycotoxin adulteration to food can occur either by direct or indirect contamination (WHO, 1979).

Direct contamination will be caused by the growth of toxigenic fungi in food and related types of products during some stage of production, processing, storage or transport. Contamination of finished foods with moulds invariably leads to exclusion of that product from the food chain in developed societies. However, in many developing countries, such luxuries are not always considered and moulded foods, such as cereals, are often a regular part of the diet. In certain parts of Africa where liver cancer is extremely prevalent the regular presence of aflatoxins in the cereal diet has been recognized (CAST, 1979). Stricter adherence to good food hygiene would greatly reduce mycotoxin presence in the human food chain (WHO, 1979). Where mould fungi are used in food and related product formation, *e.g.* enzymes, organic acids and many oriental foods, particular care must be taken to avoid mycotoxin production by the producer organisms, or later contamination of the finished products.

In contrast, indirect contamination of foods can occur when an ingredient of a food process has previously been contaminated with toxigenic moulds and although the mould may have been removed during processing, mycotoxins may well remain unseen in the product. Indirect contamination of cereals could well represent an important point of entry into several biotechnological processes, *e.g.* brewing, baking, *etc.* Mycotoxin residues in animal tissues and animal products can occur by consumption of mycotoxin-contaminated foods or cereals by animals (WHO, 1979).

28.5.1   Indirect Contamination

28.5.1.1   Cereal products

Mycotoxin contamination of cereal products represents the main dietary intake of mycotoxins to most farm animals, *e.g.* poultry, pigs and dairy cattle (Table 5). Mycotoxin contamination is more usually associated with low-grade cereals which normally do not enter the human food chain (Shotwell *et al.*, 1969). The main mycotoxins detected in cereals in regular occurrence include the aflatoxins, sterigmatocystin, ochratoxin A, zearalenone, T-2 toxin and vomitoxin

(Scott, 1973; Stoloff, 1976; WHO, 1979; Anon., 1980; Vesonder and Hesseltine, 1981; Buckle, 1981). Maize or corn is now considered to be one of the main dietary sources of mycotoxins not only to animals but also to man. This is due in large part to the size of the seeds and also to the increasing practice of harvesting grains at high moisture levels (20–25%).

The groundnut has become a major source of dietary protein for animals. However, there can now be little doubt that a high proportion of imported groundnut meal is contaminated with variable levels of aflatoxins (Anon., 1980; WHO, 1979). Restriction of groundnut importation is now coming into practice in many countries and it is hoped by doing so that the producer countries will set about and remedy the inefficient drying and product handling procedures currently in practice.

The current good standards of crop handling and storage in most developed economies does mean that mycotoxin levels in cereal raw materials is generally at the lower levels of detection limits (Anon., 1980; Kiermeier, 1978; Bullerman, 1979). However, isolated examples do occur where much higher toxic levels can occur. Constant vigilance is required (WHO, 1979).

28.5.1.2 Animal products

Some indirect exposure to mycotoxins could occur from the consumption of products of animal origin where animals have been fed with mycotoxin-contaminated feed. Of particular relevance are the aflatoxins and ochratoxin A.

The ability of certain lactating animals such as cows, sheep and goats to convert aflatoxin B_1 into aflatoxin M_1 and to excrete this carcinogenic metabolite into their milk is undoubtedly of some practical concern. The transfer ratio (*i.e.* consumed B_1 to excreted M_1) is approximately 200:1 and constitutes a potential hazard in particular to babies but also in many products derived from milk. Although the effect of such low concentrations of M_1 in milk are not known, the mere identified presence of a potential carcinogen cannot readily be overlooked. Aflatoxin M_1 has been detected in commercial fluid milk in the USA, Germany, Britain and South Africa, and in non-fat dried milk and cottage cheese in the USA and in various cheeses (Stoloff, 1976; Anon., 1980; van Walbeek, 1973; Bullerman, 1979).

There is little or no danger in the carry-over of mycotoxins in the muscle tissue of most animals consuming mycotoxin-contaminated feeds, because the transfer ratio is very high, *ca.* 5000–14 000:1 (Simpson, 1977) (Table 6). However, aflatoxins have been found in livers and kidneys of pigs in feeding trials with aflatoxin-contaminated diets (Jarvis, 1976) while somewhat similar relationships were found for ochratoxin A (Jarvis, 1976; Krogh, 1977; Madsen *et al.*, 1982).

28.5.2 Direct Contamination

28.5.2.1 Oriental fermentations

Mould fungi are widely used in the Orient and to a limited extent in the West for the production of such foods as soy sauce, tempeh, miso and many other products. Such products are produced both under commercial conditions and by home or cottage industry.

The main moulds used in oriental fermentations are strains of *Aspergillus oryzae* and *A. sojae*. These fungi are close relatives to *Aspergillus flavus* the main aflatoxin-producing *Aspergillus* (Raper and Fennell, 1965). It is because of this taxonomic proximity that much care must be given to correct identification of the producer strains. It is not surprising, therefore, that careful mycotoxin analyses have been carried out by, in particular, Japanese food toxicologists. Investigations have failed to find aflatoxin-producing fungi among the *Aspergillus* moulds used by the main manufacturers of food fermentations (Yokotsuka and Sasaki, 1981). Although some cultures produced fluorescent compounds having *Rf* values resembling those of aflatoxins, further investigations have indicated that these compounds were seven kinds of non-toxic pyrazine compounds including flavacol, isocoumarin compounds, lumichrome and compounds other than aflatoxins. A wide range of industrial moulds were also examined for the production of aspergillic acid, kojic acid, 8-nitropropionic acid and oxalic acid. Although these compounds were found to be produced by some of the strains it was considered that they did not constitute a toxic hazard to humans (Yokotsuka *et al.*, 1969). Yokotsuka (1977) was unable to detect aflatoxins, patulin, ochratoxin A or sterigmatocystin in commercial *Aspergillus sojae* cultures used for soy sauce production. Similarly, Sasaki (1980) has checked 33 strains of commercial *Aspergillus* spp. for

Table 5 Representative Fungi and Mycotoxins Associated with Various Agricultural Communities[a]

Mycotoxin (disease)	Fungus		Natural occurrence
Aflatoxins	*Aspergillus flavus*		Peanuts, corn, cottonseed,
	A. parasiticus		Brazil nuts, pecans, rice, sorghum
Alimentary toxic aleukia (ATA)	*Fusarium* spp.		
	Cladosporium spp.		Grains
Citrinin	*Penicillium citrinum*	*P. notatum*	Wheat, oats, rye, rice
	P. viridicatum	*P. palitans*	
	P. citreo-viride	*P. steckii*	
	P. expansum	*P. corylophilum*	
	P. fellutanum	*Aspergillus candidus*	
	P. implicatum	*A. niveus*	
	P. jensenii	*A. terrreus*	
	P. lividum		
Ergot alkaloids	*Claviceps purpurea*		Rye and other grains, dallis grass and other pasture grasses
Ochratoxins	*Aspergillus ochraceus*		Corn, oats, barley, rye, wheat,
	A. melleus		white beans, peanuts,
	A. sulphureus		cottonseed meal, Braxil nuts,
	Penicillium viridicatum		citrus fruits, tobacco, coffee
	P. cyclopium		
Patulin	*Penicillium expansum*	*P. leucopus*	Apple juice, various processed
	P. cyclopium	*P. melinii*	fruits
	P. claviforme	*P. novae-zeelandiae*	
	P. divergens	*P. urticae (P. patulum)*	
	P. equinum	*Aspergillus clavatus*	
	P. granulatum	*A. giganteus*	
	P. griseofulvum	*A. terreus*	
	P. lanosum	*Byssochlamys nivea*	
	P. lapidosum		
Penicillic acid	*Penicillium cyclopium*	*P. roqueforti*	Corns, beans, tobacco
	P. aurantio-virens	*P. simplicissimum*	
	P. baarnense	*P. stoloniferum*	
	P. fennelliae	*P. viridicatum*	
	P. janthinellum	*Aspergillus ochraceus*	
	P. lividum	*A. alliaceus*	
	P. martensii	*A. melleus*	
	P. palitans	*A. sclerotiorum*	
	P. puberulum	*A. sulphureus*	
Rubratoxin	*Penicillium rubrum*		Corn, various grains
	P. purpurogenum		
Stachybotryo-toxicosis	*Stachybotrys chartarum* (*S. alternans*)		Hay
Sterigmatocystin and derivatives	*Aspergillus amstellodami*	*A. versicolor*	Grains, green coffee,
	A. chevalieri	*Bipolaris sorokiniana*	miscellaneous foodstuffs
	A. flavus	*Penicillium luteum*	
	A. nidulans		
Tremorgens	*Aspergillus clavatus*	*P. palitans*	Peanuts, various mouldy
	A. flavus	*P. paxilli*	commercial feeds, rice
	A. fumigatus	*P. puberulum*	
	Penicillium crustosum	*P. verruculosum*	
	P. cyclopium	*Rosellinia necatrix*	
	P. martensii		
Tricothecenes	*Fusarium avenaceum*	*F. monoiliforme*	Corn and various grains,
	F. culmorum	*F. oxysporum*	contaminated feed
	F. equiseti	*F. poae*	
	F. graminearum (*F. roseum, Gibberella zeae*)	*F. solani* *F. sporotrichioides*	
	F. lateritium	*F. tricinctum*	
Yellowed rice disease	*Pencillium citreo-viride*		Rice
	P. islandicum		
	P. rugulosum		
Zearalenone	*Fusarium graminearum* (*F. roseum*)	*F. oxysporum* *F. sporotrichioides*	Corn and various grains
	F. moniliforme	*F. tricinctum*	

[a] Adapted from Davis and Diener (1978).

Table 6 The Relationship between Quantity of
Toxins Present in Feed to Final Tissue Level[a]

Product	Aflatoxin	Feed : tissue ratio
Milk	M_1	300
		330–700
		1400
		35–200
		100
		66
	M_2	35
Cheese	M_1	0 detected
Eggs	B_1	2200
Poultry meat	B_1	0 detected
liver		330
muscle		1000
Beef liver		14 000
Meat general		300–1000
Pig liver	B_1	480
	M_1	2100
kidney	B_1	650
	M_1	1300
blood	B_1	1100
	M_1	5000
ham	B_1	1500
	M_1	6700

[a] Simpson (1977).

aflatoxin, sterigmatocystin, ochratoxin A, patulin, cyclopiazonic acid and penicillic acid. With the exception of a few strains producing cyclopiazonic acid all proved negative.

28.5.2.2 Cheeses and fermented meat products

Many recognized cheeses owe some part of their distinctive flavour to the involvement of specific *Penicillium* spp. *e.g.* Stilton, Roquefort, Camembert and Danish Blue. Similarly *Penicillium* spp. are widely used in certain fermented meat products. The possible hazards of mycotoxin presence in these products has been extensively studied and reviewed by Bullerman (1979) and Kiermeier (1978). Although most of the fungi involved in these processes do produce secondary metabolites they are not, in general, considered to cause any health hazard. Secondary contamination of these products by toxigenic fungi cannot be excluded and must be guarded against.

The carry-over of aflatoxin M_1 from contaminated milk is causing much concern in the dairy industries including cheese and yoghurt manufacture (WHO, 1979; Anon., 1980).

28.5.2.3 Pure culture mould fermentations

A wide spectrum of industrial processes involve the use of mould fungi for the production of fungal biomass, organic acids, enzymes, antibiotics, flavour compounds, *etc.* (Smith and Berry, 1974; Smith *et al.*, 1980, 1982). In all cases the producer organisms have been carefully screened and analyzed for potential mycotoxin production. In many cases organisms have been used for several decades (*e.g. A. niger* for citric acid production) without any observable toxic manifestations.

Stringent tests must be carried out on enzymes produced by mould fermentations and destined to be used in human food or beverage processes. No main type of mycotoxin has been identified in such preparations but care must always be exercised.

Of particular significance is the use of certain moulds for SCP processes where the product is destined for human or animal consumption. *Fusarium* spp. have been used for the production of mycoprotein by Rank Hovis and McDougall in England, while *Paecilomyces variotii* has been the fungus used in the Finnish Pekilo Process for SCP production. This last process appears to have been discontinued. However, in each case it has been required that the final product must undergo long, exacting and costly feeding tests before final acceptance. Mycoprotein would appear to have achieved this accolade and will soon become an acceptable food addition.

SCP processes invariably use continuous operations, and Sargeant and Evans (1979) have implied some degree of caution when using fungi. In continuous fermentations, fermenter runs may need to be maintained genetically homogeneous for many hundreds of days and thousands of generations. Phenotypic changes in response to sudden and unplanned environmental changes could have marked influence on product formation. There is almost no information on how fungi will respond to these situations since almost all studies on toxin formation have been in batch cultures. However, it is encouraging to note that studies with an aflatoxin producing *A. flavus* strain have demonstrated that toxin production decreases rapidly in chemostat culture (S. Habeeb, J. E. Smith and B. Kristiansen, unpublished material).

Thus to ensure a mycotoxin-free product from mould fermentations the following conditions should be closely followed: (1) the use of mycotoxin-free raw materials; (2) the use of pure starter cultures which have no known ability to produce mycotoxins; and (3) the avoidance of contamination by mycotoxin-producing moulds from the air and apparatus during processing, transportation and storage of products.

28.6 REGULATIONS AND CONTROL OF MYCOTOXINS

Many factors are involved in the control of mycotoxins in foods and feeds. Undoubtedly the most effective way is to exclude toxigenic mould growth totally. However, it becomes extremely difficult to exert this absolute control over all aspects of raw material handling, *i.e.* at the growing stage, harvesting, transporting, storage and ultimate processing. For specific mould processes, excluding prior contamination of the raw materials, it should be possible to exert good control over toxigenic mould involvement. At product level, control of mycotoxins will utilize quality control procedures to detect and remove contaminated products from commercial outlets before they reach the consumer. However, this is a very difficult and complex operation.

The stability of mycotoxins (in particular aflatoxins) during food processing and the ability to remove mycotoxins from foods by physical and chemical means has been extensively reviewed by Bullerman (1979), WHO (1979), Rodericks *et al.* (1976) and FAO (1977). Physical methods of separating out moulded seeds are of more practical value with large seeded crops *e.g.* nuts. Chemical methods of treatment, although of value for animal feeding, do tend to exclude the product from direct human utilization.

The major problem lies with the lack of information for most mycotoxins on toxicity, carcinogenicity and teratogenicity to humans, the extent of contamination in foods and how stable the mycotoxins are during processing, *etc.*

Aflatoxins have become generally accepted to be poisonous and deleterious and are now widely regulated against in foods. Several countries have established regulatory programmes towards aflatoxins in foods and feeds and include Canada, Brazil, Denmark, Britain, France, Hungary, India, Japan, the Netherlands, South Africa, the Federal Republic of Germany and the USA. These regulations contain guidelines for aflatoxin contamination of human foods in the range of 5–30 p.p.b. (Rodericks, 1976). In animal feedstuffs the regulatory amounts of aflatoxin vary somewhat with the type of animal (Table 7; Fertilisers and Feeding Stuffs (Amendment) Regulations, 1976).

Table 7   Prescribed Limits for Undesirable Substances in Feedstuffs

Substance	Feedstuffs	Maximum content in the feedstuff as found (mg kg^{-1})
Aflatoxin B$_1$	Straight feedstuffs	0.05
	Whole feedstuffs for cattle, sheep and goats (except dairy cattle, calves, lambs and kids)	0.05
	Whole feedstuffs for pigs and poultry (except piglets and chicks)	0.02
	Other whole feedstuffs	0.01
	Complementary feedstuffs	0.02

It is not yet known if a safe tolerance level of mycotoxins can be realistically achieved. In the absence of such tolerance values practical limits, in particular aflatoxins in foods and feeds, have been based primarily on the limits of detection and measurement of analytical methods. The current acceptable levels may well be reduced since refinement in methodology has led to even lower detectable limits.

Levels of aflatoxin in whole milk, skim milk and low fat milk must not exceed 0.5 p.p.b. The setting of lower values of aflatoxin levels in milk was related to the higher consumption of milk by infants and young children (WHO, 1979).

All established standards or guidelines have only been determined for aflatoxin. Currently the FDA is examining toxicological data and contamination incidences of several other mycotoxins including ochratoxin A, patulin, penicillic acid, sterigmatocystin, zearalenone and T-2 toxin. In particular, the known accumulation of ochratoxin in specific animal tissues and the presence of zearalenone in ruminal milk may well hasten legislation for these mycotoxins (see Appendix 1 for additional references on regulations).

28.7 CONCLUSIONS

The best way to control the presence of mycotoxins is to prevent their production. Correct application of food and feed technology necessary to prevent proliferation of potentially hazardous fungi should be achieved, including maintenance of product quality during growth, harvesting, transportation, processing and storage of organic raw materials. Probably the most important factors for controlling the presence of mycotoxins in the initial raw materials to be used in certain biotechnological processes are the prevention of damage to crops during harvest and rapid post-harvest reduction of moisture to levels below those required for fungal growth. In biotechnological processes using fungi the culture purity must always be of paramount importance. Not withstanding, all products from fungal processes should be routinely screened for any possible mycotoxin presence. The costs for such testing is small compared with the possible consequences of human toxicology. WHO (1979) have set out a list of recommendations for further study to improve the understanding of the human implications of mycotoxins (Table 8).

Table 8 Recommendations for Further Mycotoxin Studies[a]

(a)	There is a need for more information concerning the occurrence of mycotoxins in various parts of the world and the possible daily intake of mycotoxins by man.
(b)	Further studies should be undertaken on factors affecting fungal growth and mycotoxin formation in foodstuffs, under pre-harvest, post-harvest and storage conditions.
(c)	The effects of various cooking processes on the levels of mycotoxins in foods should be elucidated.
(d)	Better methods should be developed for the rapid detection and measurement of mycotoxin levels in foodcrops.
(e)	Sampling has proved to be the most difficult step in the surveillance of food commodities. The development of reliable, internationally accepted sampling procedures is strongly recommended.
(f)	Better methods should be developed for the identification and measurement of mycotoxins in human tissues, body fluids and excreta.
(g)	A network of reference centres should be established to assist member States in confirming the identity of individual mycotoxins found in human foods and tissues. These reference centres should also provide mycotoxin reference samples, upon request, to reinforce the intercomparability of analytical results obtained in different parts of the world.
(h)	Better understanding is needed of the role of mycotoxins in human diseases. Where association between exposure to mycotoxins and the incidence of certain diseases is suspected, detailed epidemiological studies should be carried out.
(i)	Improved diagnostic methods for the effects on health of mycotoxins are needed, particularly methods for the detection of early changes that occur before the development of irreversible effects.
(j)	Attempts should be made to monitor exposure levels and to search for effects in workers handling pure mycotoxins or contaminated materials. This could provide important information on the effects of chronic exposure to mycotoxins and also indicate the need for safety measures.

[a] WHO (1979).

In conclusion, it is appropriate to end with the following paragraph from Shank (1978): 'The most relevant question, and the most difficult to answer, is what is to be done about mycotoxins in the human environment? Exposure must be minimised, of course, but to what level? At what level of mycotoxin should food be condemned or destroyed? The logical approach to these questions is to find a level at which the risk to human health is tolerable. Unfortunately, a scientific means of determining these levels is not available, and who is to bear the burdensome responsibility of determining what risks to human health are tolerable? One must not quit in despair, but strive even more diligently, in the laboratory with animals and in the field with humans, to seek better answers to the problems of mycotoxins and human health.'.

28.8 REFERENCES

Anon. (1980). *Survey of Mycotoxins in the United Kingdom*. Food Surveillance Paper No. 4. Ministry of Agriculture, Fisheries and Food, London, 40150.
Austwick, P. K. C. (1975). Mycotoxins. *Br. Med. Bull.*, **31**, 222–229.
Ayerst, G. (1966). The influence of physical factors on deterioration by moulds. *Microbiological Deterioration in the Tropics*. Society of Chemical Industry Monograph 23, pp. 14–20. Society of Chemical Industry, London.
Buckle, A. E. (1981). The occurrence of mycotoxins in barley stored on farms in England and Wales. *Mycotoxins in Animal Feeding-stuffs*, pp. 21–25. Llandrindod Wells, ADAS.
Bullerman, L. B. C. (1979). Significance of mycotoxins to food safety and human health. *J. Food Protect.*, **42**, 65–86.
CAST (1979). Aflatoxin and other mycotoxins: an agricultural perspective. *Council of Agriculture and Technology Report No. 80*.
Corry, J. E. L. (1978). Relationships of water activity to fungal growth. In *Food and Beverage Mycology*, ed. L. R. Beuchat, pp. 45–82. Avi Publishing Co., CT.
Davis, N. D. and U. L. Diener (1978). Mycotoxins. In *Food and Beverage Mycology*, ed. L. R. Beuchat, pp. 397–444. Avi Publishing Co., CT.
Davis, N. D., J. W. Dickens, R. L. Freil, P. B. Hamilton, O. L. Shotwell, T. D. Wyllie and J. F. Fulkerson (1980). Protocols for surveys, sampling, post collection handling, and analysis of grain samples involved in mycotoxin problems. *J. Assoc. Anal. Chem.*, **63**, 95–102.
Emmons, C. W., C. H. Binford, J. P. Utz and K. J. Kwan-Chung (1977). *Medical Mycology*, 3rd edn. Lea and Febiger, PA.
FAO (1977). Mycotoxins. *Report of the Joint FAO/WHO/UNEP Conference on Mycotoxins, Nairobi*, 19–27 September 1977. FAO Food and Nutrition Paper 2.
Giambrone, J. J., D. L. Ewert and R. D. Wyatt (1978). Effect of aflatoxin on the humoral and cell-mediated immune systems of the chicken. *Am. J. Vet. Res.*, **39**, 304–308.
Hamilton, P. B. and J. R. Harris (1971). Interactions of aflatoxins with *Candida albicans* infections and other stresses in chickens. *Poultry Sci.*, **50**, 906–912.
Hayes, W. A. (1980). Mycotoxins: a review of biological effects and their role in human diseases. *Clin. Toxicol.*, **17**, 45–83.
Hayes, W. A. and N. G. Nair (1974). The cultivation of *Agaricus bisporus* and other edible mushrooms. In *The Filamentous Fungi*, ed. J. E. Smith and D. R. Berry, vol. 1, pp. 212–248. Edward Arnold, London.
Heathcote, J. G. and J. R. Hibberd (1978). *Aflatoxins: Chemical and Biological Aspects*. Elsevier, Amsterdam.
Hesseltine, C. W. (1976). Conditions leading to mycotoxin contamination of foods and feeds. In *Mycotoxins and other Fungal Related Food Problems*, ed. J. V. Rodericks, Advances in Chemistry Series No. 5. 149, Washington DC.
Huitson, J. J. (1968). Cereals preservation with propionic acid. *Process Biochem.*, **3**, 31–32.
Jarvis, B. (1976). Mycotoxins in food. In *Microbiology in Agriculture, Fisheries and Food*, ed. F. A. Skinner and J. G. Car, pp. 251–267. Academic, London.
Kiermeier, F. (1978). Einfichrung in die Mykotoxin-Problematik. *Zeitsch. Leben. Untersuch. Forsch.*, **167**, 117–127.
Krogh, P. (1977). Ochratoxins. In *Mycotoxins in Human and Animal Health*, ed. J. V. Rodericks, C. W. Hesseltine and M. A. Mehlman, pp. 489–498. Pathotox Publishers Inc., IL.
Lacey, J. (1975). Potential hazards to animals and man from microorganisms in fodder and grain. *Trans. Br. Mycol. Soc.*, **65**, 171–184.
Lacey, J., S. T. Hill and M. A. Edwards (1980). Microorganisms in stored grain: their enumeration and significance. *Trop. Stored Prod. Inform.*, **39**, 19–33.
Lincoff, G. and D. H. Mitchel (1977). *Toxic and Hallucinogenic Mushroom Poisoning: A Handbook for Physicians and Mushroom Hunters*. Van Nostrand Reinhold Co., Amsterdam.
Madsen, A., H. P. Martensen and B. Hald (1982). Feeding experiments with ochratoxin A contaminated barley for bacon pigs. *Acta Agric. Scand.*, **32**, 225–239.
Mislivec, P. B. (1977). Toxigenic fungi in foods. In *Mycotoxins in Human and Animal Health*, ed. J. V. Rodericks, C. W. Hesseltine and M. A. Mehlman, pp. 469–477. Pathotox Publishers Inc., IL.
Morehouse, L. G. (1979). Mycotoxicoses of the bovine with reference to fungi and toxins associated with disease. *The Bovine Practitioner*, **14**, 175–180.
Pegler, D. N. and R. Watling (1982). British toxic fungi. *Bull. Br. Mycol. Soc.*, **16**, 66–75.
Pier, A. C., S. J. Cysewiski and J. L. Richard (1977). Mycotoxins as a veterinary problem. In *Mycotoxins in Human and Animal Health*, ed. J. V. Rodericks, C. W. Hesseltine and M. A. Mehlman, pp. 745–750. Pathotox Publishers Inc., IL.
Pier, A. C., J. L. Richard and S. J. Cysewiski (1980). Implications of mycotoxins in animal disease. *J. Am. Vet. Med. Assoc.*, **176**, 719–724.
Pier, A. C., J. L. Richard and J. R. Thurston (1978). The influence of mycotoxins on resistance and immunity. In *Interaction of Mycotoxins in Animal Production*, pp. 56–66. National Academy of Science, Washington, DC.
Pohland, A. E., C. W. Thorpe and S. Nesheim (1979). Newer developments in mycotoxin methodology. *Pure Appl. Chem.*, **52**, 213–223.
Raper, K. B. and D. I. Fennell (1965). *The Genus Aspergillus*. Williams and Wilkins, Baltimore, MD.

Rodericks, J. V., C. W. Hesseltine and M. A. Mehlman (eds.) (1976). *Mycotoxins in Human and Animal Health*. Pathotox Publishers Inc., IL.

Rumack, B. H. and E. Salzman (1978). *Mushroom Poisoning: Diagnosis and Treatment*. CRC Press Inc., Boca Raton, FL.

Sargeant, K. and C. G. T. Evans (1979). Hazards involved in the industrial use of microorganisms. Commission of the European Communities, Luxembourg.

Sasaki, M. (1980). Quoted in T. Yokotsuka and M. Sasaki (1981). *Advances in Biotechnology*, ed. M. Moo-Young and C. W. Robinson, vol. 2, p. 466. Pergamon, Toronto.

Shank, R. C. (1978). Mycotoxicosis of man: dietary and epidemiological considerations. In *Mycotoxic Fungi, Mycotoxins, Mycotoxicoses*, ed. T. D. Wyllie and L. G. Morehouse, vol. 3, pp. 1–19. Dekker, New York.

Shotwell, O. L., C. W. Hesseltine, H. R. Burmeister, W. F. Kwolek, G. M. Shannon and H. H. Ball (1969). Survey of cereal grains and soy beans for the presence of aflatoxin. II. Corn and soy beans. *Cereal Chem.*, **46**, 454–469.

Simpson, A. D. F. (1977). *Mycotoxins and the Food Chain*. R. H. M. Private publication.

Smith, J. E. (1969). Commercial mushroom production, 1. *Process Biochem.*, May, 43–46.

Smith, J. E. (1972). Commercial mushroom production, 2. *Process Biochem.*, May, 24–26.

Smith, J. E. (1982). Mycotoxins and animal feeds: Occurrence and control. *The Feed Compounder*, **2**, 13–17.

Smith, J. E. and D. R. Berry (eds.) (1974). *The Filamentous Fungi*, vol. 1. Edward Arnold, London.

Smith, J. E., D. R. Berry and B. Kristiansen (eds.) (1980). *Fungal Biotechnology*. Academic, London.

Smith, J. E., D. R. Berry and B. Kristiansen (eds.) (1982). *The Filamentous Fungi*, vol. 4. Edward Arnold, London.

Stewart, R. G., R. D. Wyatt and M. D. Ashmore (1977). The effects of various antifungal agents on aflatoxin production and growth characteristics of *Aspergillus flavus* in liquid media. *Poultry Sci.*, **56**, 1630–1635.

Steyn, P. (1981). Multimycotoxin analysis. *Pure Appl. Chem.*, **53**, 891–901.

Stoloff, L. (1976). Occurrence of mycotoxins in foods and feeds. In *Mycotoxins and other Fungal Related Food Problems*, ed. J. V. Rodericks, Advances in Chemistry Series, No. 149, pp. 23–50. American Chemical Society, Washington, DC.

Tuite, J. and G. H. Foster (1979). Control of storage diseases of grain. *Annu. Rev. Phytopathol.*, **17**, 343–366.

Turner, W. B. (1971). *Fungal Metabolites*. Academic, London.

van Walbeek, W. (1973). Fungal toxins in foods. *Can. J. Food Sci. Technol.*, **6**, 96–110.

Vesonder, R. F. and C. W. Hesseltine (1981). Vomitoxin: natural occurrence on cereal grains and significance as a refusal and emetic factor to swine. *Process Biochem.*, June, 12–14.

WHO (1979). *Environmental Health Criteria 2 Mycotoxins*. World Health Organisation, Geneva.

Yokotsuka, T. (1977). Presentation in SCOGS Hearing, at Hilton Hotel, Bethesda, Maryland, USA (July 26).

Yokotsuka, T. and M. Sasaki (1981). Risks of mycotoxin in fermented foods. In *Advances in Biotechnology*, ed. M. Moo-Yong and C. W. Robinson, vol. 2, pp. 461–466. Pergamon, Toronto.

Yokotsuka, T., K. Oshita, T. Kikuchi, M. Sasaki and Y. Asao (1969). Studies on the compounds produced by moulds. VI. Aspergillic acid, kojic acid, β-nitropropionic acid and oxalic acid in solid koji. *J. Agric. Chem. Soc. Jpn.*, **43**, 189–196.

APPENDIX 1

A list of further references to be consulted for further information on international regulations on aflatoxin

Health Laws and Regulations—United States of America
World Health Organisation,
Geneva, Switzerland.
International Digest of Health Legislation,
vol. 29, no. 3, pp. 648–686, 1978, in English.

Problems with Legislation covering Tolerances for Aflatoxins in Foods
Zur Problematik von Höchstmengenregelungen für Aflatoxine in Lebensmitteln,
Muecke, W.
Kleiststrasse 20, 8012 Ottobrunn, Federal Republic of Germany.
Lebensmittelchemie und Gerichtliche Chemie,
vol. 31, no. 6, pp. 109–110, 1977, in German.

Significance of Mycotoxins from the Health Point of View
Zur Gesundheitlichen Bedeutung der Mykotoxine,
Roll, R.
Zeitschrift für das Gesamte Lebensmittelrecht (ZLR),
vol. 40, ref. 4 (2), pp. 187–198, 1977, in German.

Brazil Nuts, Pistachio Nuts, and Other Foods and Feeds. Availability of Guidelines
United States of America, Food, Drug Administration,
Washington, DC, USA.
Federal Register,
vol. 43, no. 11, Jan. 17, pp. 2444–2445, 1978, in English.

Aflatoxin Contamination of Milk. Establishment of Action Level
United States of America, Food, Drug Administration,
Washington, DC, USA.
Federal Register,
vol. 42, no. 234, Dec. 6, p. 61630, 1977, in English.

National Food Administration Ordinance on Foreign Matter in Foods

Sweden, Statens Livsmedelsverk.
Statens Livsmedelsverks Forfattningssamling,
SLV FS 1978:34 12 pp. 1978, in Swedish.

National Food Administration Ordinance on Foreign Substances in Foods
Sweden, Statens Livsmedelsverk.
Statens Livsmedelsverks Foerfattningssamling,
Publishing Details: Box 622, S–751 25 Uppsala, Sweden.
SLV FS 1977:7, 6 pp. 1977, in Swedish.

Health Laws and Regulations—Benelux Economic Union
World Health Organisation,
Geneva, Switzerland.
International Digest of Health Legislation,
vol. 29, no. 1, p. 5, 1978, in English.

Practical Standards for Preparation and Checking of Standard Aflatoxin Solutions
Cantafora, A.; Cirilli, G.
Italy, Associazione Analisti Chimici Mycotoxin Symposium.
Technica Molitoria,
vol. 30 (2), pp. 126–127, 1979; in Italian.

Health Laws and Regulations—Belgium
World Health Organisation,
Geneva, Switzerland.
International Digest of Health Legislation,
vol. 29, no. 4, pp. 721–726, 1978, in English.

Health Laws and Regulations—Federal Republic of Germany
World Health Organisation,
Geneva, Switzerland.
International Digest of Health Legislation,
vol. 29, no. 1, pp. 84–104, 1978, in English.

29

Health Hazards During Microbial Spoilage

E. C. HILL
E. C. Hill and Associates, Cardiff, UK

29.1 INTRODUCTION

The culture and processing of microorganisms may involve health risks which are largely well recognized and governed by legislation or codes of practice (see Volume 4, Chapters 31–37). In other areas of biotechnology, particularly microbial fouling, spoilage and corrosion, potential health hazards are less well defined and are more likely to be insidious. These are the health risks which will be described in this chapter. Holdom (1976) has given an appraisal of theoretical risks in the metal-working industries. Hill and Al-Zubaidy (1979) have categorized these, and these categories can be modified to be of more general application as follows: (1) inadvertant contamination by pathogenic microorganisms; (2) infections due to opportunistic pathogens; (3) inhalation of living or dead microbes invoking a serological response; (4) microbial enzymes and toxins affecting skin or lung directly; (5) microbial enzymes aggravating chemical irritancy; and (6) products of primary metabolism constituting a health risk.

Of these only (1) and (2) are likely to be obviously related to the working environment. The other categories may be suspected but are unlikely to be the subject of a serious investigation.

29.2 HAZARD ROUTES

29.2.1 Air Borne

For an air-borne hazard to exist it is not sufficient for the organism or its products to exist; there must also be access to the lung. The hazardous material must be present as a gas or a finely dispersed aerosol. It is generally held that only particles below 5 μm are capable of reaching the lung alveoli (Brown *et al.*, 1950), although larger particles may gain access where there is some

562   Governmental Regulations and Concerns

lung disability. Such small particles have an extremely low free fall velocity, which can be calculated from Stoke's law, as indicated in Figure 1, and they will remain suspended for many hours. During this time desiccation of aqueous droplets will occur, although hygroscopic materials present may modify this, and thus the physical properties of the droplet will change. At the same time the viability of organisms present may be lost, particularly that of Gram-negative bacteria. The Andersen sampler (Andersen, 1958) can be employed to quantify the numbers of viable particles present in air according to their size.

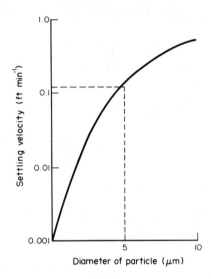

Figure 1 Settling velocity of spherical particles of unity specific gravity calculated from Stoke's law. Inhalable particles penetrating to the alveoli are enclosed within the dotted area

From a consideration of the free fall velocity and known airflows it is possible to make relevant deductions. For example in the cooling tower shown in Figure 2 it is possible to say that all water droplets below *ca.* 300 μm in size which pass the drift eliminator will escape through the top of the tower. A domestic shower is in effect a mini-cooling tower and measurements indicate induced upward air-flows of *ca.* 0.4 m s^{-1}. Thus spray droplets below *ca.* 90 μm in size, aerosolized from the drain, will move upwards and could be inhaled.

29.2.2 Skin Contact

Significant microbial fouling or spoilage necessitates an aqueous environment and the nature of the job may require hand immersion in the aqueous phase or proximity within the spray range. Microbial hazards may be promoted by skin damage due to other causes, such as chemical irritation or mechanical abrasion.

29.2.3 Ingestion

There is unlikely to be a hazard due to the ingestion of spoilage microorganisms or their products. Even if enteric pathogens are present they are unlikely to be ingested in quantities approaching the threshold infective dose unless the spoilage is of a food product.

29.3 CONTAMINATION BY PATHOGENIC ORGANISMS

True pathogens tend to have restricted powers of degradation and synthesis and are unlikely to have a spoilage role. Thus their numbers will not increase and any hazard is determined by their propensity to survive. There has been considerable concern that metal-working fluids could be contaminated with pathogenic microbes from some of the work-force and that these organisms could be disseminated to other staff. Workers have considerable contact with and exposure to these fluids and thus we might expect to find organisms of wound sepsis and also dermatophytes.

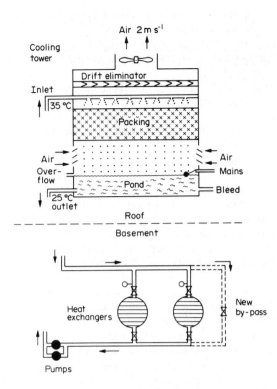

Figure 2 Layout of typical cooling tower, not to scale. The position of a by-pass, constructed to facilitate cleaning operations, is indicated

The open flow of metal-working coolant encourages its use as a waste depository not only for half-eaten food, cigarettes and sputum but also for urine. Shop floor pets, rodents and birds will contribute their faeces. Thus there is ample opportunity of access for respiratory and intestinal tract pathogens, and many authors have reported the presence of *Klebsiella*, *Staphylococcus*, *Streptococcus*, *Shigella* and *Salmonella* and some recent investigations for dermatophytes are indicating that they are present. These results are as expected, but the bulk of evidence suggests that pathogens do not survive for long, particularly the Gram-positive pathogens. Bennett and Wheeler (1954) added 18 pathogenic Gram-negative bacteria and 12 Gram-positive pathogens to sterile cutting oil emulsions and found that the latter generally survived for less than a day. Gram-negative bacteria fared better, *Shigella* surviving for 76 days. Metal-working fluids vary in their chemistry and have a normal burden of spoilage organisms but in general these findings have been amply confirmed (Pivnick *et al.*, 1954; Lloyd *et al.*, 1975 and others). *Staphylococcus* is rarely isolated from cutting fluids although it must be constantly introduced, but its survival is known to be a matter of hours, particularly when an antagonistic spoilage flora is present (Rossmore and Williams, 1967). It would in fact hardly survive long enough for a field sample to be transported to the laboratory for testing. *Klebsiella* survival is rather better and it is frequently isolated from machine shop samples (Tant and Bennett, 1954). Hill (1983) has reviewed all possible microbial factors which could affect the health of metal-workers using infected fluids.

Other industrial situations undoubedly suffer similar contamination by pathogens. There is no good evidence that industrial fluids are implicated epidemiologically in disease and it is reasonable to assume that they are no more hazardous as vectors than the lavatory seat or cloakroom towel.

29.4 OPPORTUNISTIC PATHOGENS

One of the common opportunistic pathogens, *Pseudomonas aeruginosa*, poses the greatest potential hazard. It is prevalent as a serious pathogen in the hospital environment where the host resistance is low and is commonly found as a spoilage organism in petroleum products (Al-Zubaidy, 1978; Hill, 1979). It can be isolated from about a quarter of field samples and where it does occur it is often the dominant spoilage organism. Isolates tend to have the same pattern of antibiotic resistance and are of similar pyocin types as strains isolated from clinical material, yet there is no obvious pattern of infection in exposed staff. Brown, Glover and Hill (unpublished work) failed to detect respiratory tract colonization in 38 machine tool operators although it was demonstrated that six of the machine tool coolants contained *Pseudomonas aeruginosa* and that inhalable aerosols were being created.

Aspergillus fumigatus strains are commonly isolated from fuels and lubricating oils as spoilage organisms in warm systems. They attack hydrocarbons avidly but although Thomas (1973) has demonstrated serological similarities to known pathogenic strains, the hydrocarbon-oxidizing strains have some unusual characteristics, particularly pigmentation, and their clinical significance is not known.

Thus the actual risk from these pathogens is probably small, but we should bear in mind that opportunism is as much a property of the host as the parasite and it would not seem prudent to expose compromised hosts to spoilage situations generating these organisms. For example it is believed that there have been some incidents of systemic infection by *Pseudomonas* of industrial origin of workers being treated with immunosuppressive drugs, and there is often concern expressed about eye infections in metal-working shops.

Since 1976 we have known of a new opportunistic pathogen, *Legionella pneumophila*. It is obviously a common environmental organism and has been isolated from mud, rivers and lakes (Morris *et al.*, 1979; Fliermans *et al.*, 1979; Fliermans *et al.*, 1981) and unfortunately has found a convenient niche in man-made water systems. The inhalation of aerosols containing the organism can cause the type of pneumonia dramatically named Legionnaire's disease. The successive steps leading to its recognition and culture have been described by Fraser and McDade (1979). One disturbing aspect of this event is the exposure of our inability to grow organisms with an unusual metabolism on conventional laboratory media and the somewhat chilling realization that many more microbes must commonly exist which we fail to isolate by conventional methods. *Legionella* does not ferment carbohydrates or use organic acids and it obtains its carbon and energy from amino acids. Cysteine and iron are obligatory, the optimum temperature is *ca.* 35 °C and it prefers an enriched CO_2 atmosphere. We can surmise that its growth in nature is dependent upon a close association with other fouling organisms. It has the ability of very prolonged survival in water (up to 415 days has been recorded) and hence its isolation from the environment does not necessarily imply proliferation at that point.

One man-made niche which seems particularly appropriate is a cooling tower. There is normally substantial microbial fouling and reference to Figure 2 will suggest that the return pipe line from the heat exchangers will be an almost ideal location. Several outbreaks of the disease have implicated cooling towers but many other situations must exist which are rich in organic matter and warm and are conducive to growth. The nutritional aspects of cooling towers are favourable as dust and vapours are scrubbed from the ascending air by the descending water and become concentrated by evaporation. Our exposure to the organism is probably high. Kurtz *et al.* (1982) monitored 14 cooling towers and found *Legionella* in six at the beginning of the summer and 11 at the end. As with other opportunistic pathogens we still need an inhalable aerosol in sufficient quantity to constitute an infective dose and it must reach a susceptible host, in this case usually an ageing male with some respiratory distress.

Preliminary studies have suggested that proliferation is best in an aqueous system high in dissolved solids and iron and it is therefore a sensible alleviatory measure to counteract evaporation effects by regularly partially dumping and refilling cooling towers. Kurtz *et al.* (1982) achieved little success in controlling *Legionella* in cooling towers by manual additions of tributyltin oxide, a quaternary ammonium compound and chlorine, but some transient success was achieved with a chlorinated phenol. This was in contradiction to favourable laboratory tests for all these biocides.

Automated additions of biocide and corrosion inhibitor, coupled to partial dumping controlled by conductivity have been achieved in the SM Phobro-Mat (Schulke and Mayr GmbH, Hamburg) and this has overcome the problems of manual operations.

Very many other man-made systems must encourage proliferation of this and similar organisms and it is perhaps surprising that the incidence of pneumonias due to these agents is so low.

A rather different situation exists in food spoilage where the opportunistic enteric pathogen, *Yersinia enterocolitica*, has the unusual ability to produce spoilage at low temperature but can colonize the gut and flourish at body temperature when ingested, but this is beyond the scope of this review.

In many spoilage situations, low grade opportunistic pathogens, such as *Proteus* and *Pseudomonas fluorescens*, are present and are proliferating, but little is known of their significance in infective processes.

29.5 INHALATION OF MICROORGANISMS

It has been noted already that inhalable particles must in general be sub 5 μm in size. There is ample opportunity to produce such aerosols, particularly in metal-working operations. Table 1 is derived from Hill and Al-Zubaidy (1979) and gives the numbers and sizes of particles containing viable bacteria. A number of determinations were made (not at the same time) and it can be seen that large numbers of inhalable, viable particles can be present. Viable numbers decrease sharply further away from the source, either by dilution due to ventilation or actual loss of viability. Both viable and non-viable microbes can invoke a serological response. Al-Zubaidy (1978) noted that the presence of oil emulsion stimulated antibody response to *Pseudomonas aeruginosa* when laboratory animals were exposed to aerosols; this could be anticipated from the long established use of oil in injection material to promote antibody production.

Table 1 Viable Bacteria in Air near a Grinding Machine as Determined by an Andersen Sampler[a]

	Number of particles per m³	
Particle size (μm)	*0.6 m distant*	*1.2 m distant*
> 9.2	18 000	300
5.5–9.2	18 000	70
3.3–5.5	10 000	260
2.0–3.3	18 000	35
1.0–2.0	4 000	7
< 1.0	1 000	7
Total	69 000	679

[a] Data derived from Hill and Al-Zubaidy (1979).

Jarvholm (1981) noted that the immunoglobulin IgM titres in three groups of workers, namely turners, grinders and hardeners, were raised in comparison to a control group and speculated that this could possibly be because of exposure to bacteria in oil aerosols. Hill (1983) determined serum antibody titres to *Pseudomonas aeruginosa* of various groups of workers. Figure 3 gives the overall results for workers in an automotive factory and a non-metal working factory and they suggest a greater exposure and response of the former group, although the actual occupation of individual workers was not in fact known. However, although there seems good reason to believe that there is a serological response to exposure to infected oil aerosols there is at present no attribution of clinical significance.

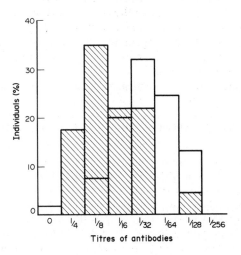

Figure 3 Antibody titres to *Pseudomonas aeruginosa* in serum of work-force at an automotive factory (open blocks) and a synthetic fibre factory (hatched blocks) as a control (Jarvis and Hill, unpublished work)

Much more is known of the clinical significance of exposure to aerosols from infected water in humidifiers and to aerosols from 'mouldy' hay. In both cases there is considerable sensitization and respiratory distress. In the former, the condition, aptly named 'humidifer fever', is characterized by the production of antibodies to the flora and fauna in the humidifier pond. These antibodies appear to be specific to amoebae which have ingested Gram-negative bacteria. A recent summary of current beliefs has been given by Edwards (1981).

The spoilage of moist hay is accompanied by a rise in temperature; this is selective for thermophilic cellulolytic organisms such as *Micropolyspora faeni* and *Thermoactinomyces vulgaris*. The clinical condition called farmer's lung is characterized by the production of antibodies to the first named organism and sometimes to the second also.

Neither of these clinical conditions incur infection of the host.

29.6 MICROBIAL ENZYMES AND TOXINS

Gram-negative bacteria are common spoilage organisms, particularly in petroleum products, and they generally produce endotoxins. These will be most obvious when cells are killed. The introduction of the *Limulus* amoebocyte lysate test (Levin and Bang, 1964) has provided a convenient tool by which they can be measured. Johansen (unpublished work) has related endotoxin levels in metal-working fluids to bacterial infection and his results are reproduced in Figure 4. Such high levels are likely to have clinical significance, possibly by increasing inflammation at minor skin lesions.

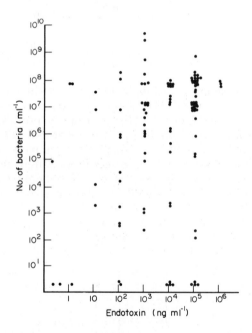

Figure 4   Relation between viable count of bacteria and endotoxin level (as determined by the *Limulus* amoebocyte lysate test) in field samples of metal-working fluids (Johansen, unpublished work)

Pseudomonas aeruginosa is an exotoxin producer, a rather unusual property for a Gram-negative bacterium (Born *et al.*, 1977; Pollock *et al.*, 1977). Hodgson (1970) has described a keratolysin and it could be postulated that this would degrade protective skin keratin and thus increase the irritancy of oil and chemicals. Singh (1976) has implicated *Pseudomonas aeruginosa* in eczema and Jacobson *et al.* (1976) has implicated it in skin rash. An involvement in dermatitis in the metal-working industries of endotoxins and exotoxins is more in keeping with field observations than any infective process.

Corneal damage due to *Pseudomonas* exotoxin has been demonstrated by Iglewski *et al.* (1977).

Responses to inhalation of living and dead bacteria have already been mentioned. More specifically there are known serological responses, usually with clinical symptoms such as malaise, fevers and respiratory distress, when endotoxins are inhaled. In many cases the active stimulatory agent is not known but occupational diseases are recognized due to inhalation of aerosols or dusts from cotton mills, grain elevators, animal processing plants and sewage works. In each case there is a demonstrable presence of Gram-negative bacteria.

Maria and Burrell (1980) have described pulmonary responses in rabbits to *Escherichia coli* endotoxin and they conclude that much more investigation is called for in this field to elucidate the role of endotoxins in occupational lung diseases.

29.7 MICROBIAL METABOLIC PRODUCTS

Microbial spoilage can produce gross changes, such as extremes of pH, which could have a direct irritant consequence or promote irritancy due to other substances. Some organic acids are known to be specific skin irritants. Of most concern is the metabolic product hydrogen sulfide, which has a threshold limit value in the UK of 10 mg m^{-3} of air and is also a skin irritant. Increasing concentrations in air cause eye irritation, dizziness and eventually death. Hydrogen sulfide kills in 30 min at exposure levels of 800–1000 mg m^{-3} and it is more toxic than hydrogen cyanide. There are recorded fatalities during tank cleaning operations after spoilage of ship cargoes. The gas progressively anaesthetizes the sensory organs and hence leads to a false impression that it has disappeared. Hydrogen sulfide is the objectionable ingredient of 'Monday morning odour' in machine shops and it seems prudent to consider excessive generation of this gas as a potential health hazard.

There have been some misgivings that prolonged use of metal-working oils could lead to the accumulation of carcinogenic polynuclear aromatics. Eyres (1981) has measured levels of these and detected increases but he concludes that they are within acceptable limits.

29.8 OTHER HEALTH FACTORS

The control of spoilage frequently involves the application of biocides. These are all irritant or toxic to a degree and their use should be under close supervision. Those biocides which are used in liquids may be aerosolized in appreciable quantities. For example a biocide added to a 2% cutting oil emulsion at 1000 p.p.m. could be aerosolized when the fluid is used. The only legal restriction on oil aerosolization is that the oil mist should not exceed 5 mg m^{-3} (UK threshold limit value). At this point the concentration of biocide in air would be 0.25 mg m^{-3}. Threshold limit values (UK) for chlorine and formaldehyde in air are 1 and 2 p.p.m. respectively. In most individuals formaldehyde vapour causes lachrymation at *ca.* 4 p.p.m. and irritant effects on the nose and throat at *ca.* 10 p.p.m. However recent fears of a carcinogenic role have not been substantiated for humans; formaldehyde is a known skin sensitizer. Toxicity data on other biocides are well documented and should be given due regard when planning antimicrobial regimes. Hill (1982) has given practical guidance on the use of biocides in petroleum products.

29.9 CONCLUSIONS

There is acceptable evidence that microbial fouling and spoilage can create a health hazard in water systems and certain natural products, such as hay, cotton and food. In other situations, similar organisms exist and presumably have a potential for adversely affecting health in a variety of ways. Indeed in some systems it is difficult to believe that there is no potential health hazard, for example when the spoilage organism is *Pseudomonas aeruginosa*. However the health risks from such microbial spoilage would tend to be insidious and to be masked by non-microbiological factors such a chemical irritation or abrasion.

There are no standards for acceptable levels of microorganisms although there may be rule of thumb guides such as less than 10^5 bacteria ml^{-1} in recirculated water or metal-working fluids. In the light of current knowledge it would be prudent to minimize exposure to endotoxins, exotoxins and opportunistic pathogens and to avoid stagnant situations which encourage hydrogen sulfide generation.

The reduction of the microbial burden can be achieved by chemicals, which may introduce their own hazards, but there is almost always scope for improving plant design and house-keeping to minimize infection levels. There is usually economic justification for antimicrobial procedures and it seems reasonable to consider that effective measures would have a bonus in improved health.

29.10 REFERENCES

Al-Zubaidy, T. S. (1978). Pulmonary and immune responses to infected and non-infected oil emulsions. Ph.D. Thesis, University of Wales, Cardiff.

Andersen, A. A. (1958). Sampler for the collection, sizing and enumeration of viable air-borne particles. *J. Bacteriol.*, **76**, 471–484.

Bennett, E. O. and H. O. Wheeler (1954). Survival of bacteria in cutting oil. *Appl. Microbiol.*, **2**, 368–371.

Born, M. J., M. L. Vasil, J. C. Sadiff and B. H. Iglewski (1977). Incidence of exotoxin production by *Pseudomonas* spp. *Infect. Immunol.*, **16** (1), 326–366.

Brown, J. H., K. M. Cook, F. G. Ney and T. Hatch (1950). Influence of particle size upon the retention of particulate matter in the human lung. *Am. J. Public Health*, **40**, 450–459.

Edwards, J. H. (1981). Humidifier fever. *Proc. R. Soc. Health 88th Health Congress, London*, 13–16th April, 11–12.

Eyres, A. R. (1981). Polycyclic aromatic hydrocarbon contents of used metal-working oils. Paper IP 81–002. Inst. Petrol, London.

Fliermans, C. B., W. B. Cherry, L. H. Orrison and L. Thacker (1979). Isolation of *Legionella pneumophilia* from non-epidemic related aquatic habitats. *Appl. Environ. Microbiol.*, **37**, 1239–1242.

Fliermans, C. B., W. B. Cherry and L. H. Orrison (1981). Ecological distribution of *Legionella pneumophila. Appl. Environ. Microbiol.*, **41**, 9–16.

Fraser, D. W. and J. E. McDade (1979). Legionellosis. *Sci. Am.*, **241**, 82–96.

Hill, E. C. (1979). Microbial infection of lubricating and other oils and its significance. Technical Paper MF 79-391. Soc. Man. Eng. (USA).

Hill, E. C. (1982). Preservation in specialised areas. A. Cutting oil emulsions. B. Fuels and lubricants. In *Principles and Practice of Disinfection, Preservation and Sterilisation*, ed. A. D. Russell, W. B. Hugo and G. A. Aycliffe, pp. 343–357. Blackwell, Oxford.

Hill, E. C. (1983). Microbial aspects of health hazards from water-based metal working fluids. *Tribol. Int.*, **16**, 136–140.

Hill, E. C. and T. S. Al-Zubaidy (1979). Some health aspects of infections in oil and emulsions. *Tribol. Int.*, **12**, 161–164.

Hodgson, G. (1970). Cutaneous hazards of lubricants. *Ind. Med. Surg.*, **39** (2), 41–46.

Holdom, R. S. (1976). Microbial spoilage of engineering materials 3. Are infected oil emulsions a health hazard to workers and to the public? *Tribol. Int.*, **9**, 271–281.

Iglewski, B. H., R. P. Burns and I. K. Gipson (1977). Pathogenesis of corneal damage from *Pseudomonas* exotoxin A. *Invest. Ophthalmol. Visual Sci.*, **16** (1), 73–76.

Jacobson, J. A., A. W. Hoadley and J. J. Farmer (1976). *Pseudomonas aeruginosa* serogroup II and pool associated skin rash. *Am. J. Public Health*, **66**, 1092–1093.

Jarvholm, B. (1981). Mineral oils. Studies on health effects in the engineering industries. Collection of papers published by Department of Occupational Medicine, Sahlgrenska Sjukhuset, Gothenburg, Sweden.

Kurtz, J. B., C. L. R. Bartlett, U. A. Newton, R. A. White and N. L. Jones (1982). *Legionella pneumophila* in cooling water systems. *J. Hyg.*, **88**, 369–381.

Levin, J. and F. B. Bang (1964). A description of cellular coagulation in the *Limulus. Bull. Johns Hopkins Hospital*, **115**, 337.

Lloyd, G., G. I. Lloyd and J. Schofield (1975). Enteric bacteria in cutting oil emulsion. *Tribol. Int.*, **8**, 27–29.

Maria, T. F. and R. Burrell (1980). Effects of inhaled endotoxin-containing bacteria. *Environ. Res.*, **23**, 87–97.

Morris, G. K., C. M. Patton, J. C. Feeley, S. E. Johnson, G. Gorman, W. T. Martin, P. Skaliy, G. F. Mallison, B. D. Politi and D. C. Mackel (1979). Isolation of the Legionnaire's Disease bacterium from environmental samples. *Ann. Intern. Med.*, **90**, 664–666.

Pivnick, H., W. E. Engelhard and T. L. Thompson (1954). The growth of pathogenic bacteria in soluble oil emulsions. *Appl. Microbiol.*, **2**, 140–142.

Pollock, M., N. S. Taylor and L. T. Callahan (1977). Exotoxin production by clinical isolates of *Pseudomonas aeruginosa. Infect. Immunol.*, **15**, 776–780.

Rossmore, H. W. and B. W. Williams (1967). Survival of coagulase-positive staphylococci in soluble cutting oils. *Health Lab. Sci.*, **4**, 160–165.

Singh, V. (1976). Eczema: a common skin disorder. *Curr. Med. Pract.*, **20**, 214–218.

Tant, C. O. and E. O. Bennett (1954). The growth of aerobic bacteria in metal-cutting fluids. *Appl. Microbiol.*, **6**, 388–391.

Thomas, A. R. (1973). Fungal growth in petroleum products. Ph. D. Thesis, University of Wales, Cardiff.

30

Carcinogenic, Mutagenic and Teratogenic Biologicals

J. M. PARRY and E. M. WATERS

University College of Swansea, UK

30.1 INTRODUCTION

An extensive range of biologicals are capable of producing toxic responses in animals exposed to their influence. The range of toxic effects produced is extensive, varying from lethality to allergic responses such as those of pollen-induced hay fever. This chapter will be concerned with three of the long-term effects of toxic exposure and the techniques which may be used to measure the potential of biological agents to produce such effects in mammals.

The life cycle of a mammal can, for convenience, be divided into three interrelated stages, the embryo, the somatic stage and the germ cell stage, the products of which, after gamete fusion, give rise to new embryos. Each of these stages is sensitive to modifications which result in characteristic forms of long-term damage. In the embryo, modifications of development are defined as teratogenic and lead to lethality or the production of abnormal embryos or foetuses. In somatic cells, carcinogenic effects result in the production of tumours, which are abnormal tissue masses which exhibit various degrees of uncontrolled and uncoordinated growth. Mutagenic changes involve modifications of the content of the informational molecule which, if transmitted by the germ cell, may lead to defects in the embryonic and somatic cells.

Each somatic cell of an organism carries the complete 'gene or informational sequence' for the species. During embryo development in higher organisms, such as mammals, the individual cells become specialized into tissues and organs in a manner whereby only selected groups and sequences of information are active and transcribe gene products. In the adult organism these differentiated tissues remain under coordinated control throughout the life span. For the purpose of this chapter, the three endpoints can be generalized as follows:

(1) Mutagenesis produces changes in the information content of the units of inheritance, the genes.

(2) Teratogenesis produces changes in the differentiation of cell types and organs during embryogenesis.

(3) Carcinogenesis produces changes in the status of the somatic cells leading to de-differentiation and uncontrolled proliferation.

There are numerous examples in the scientific literature of biologicals, both whole organisms and the products of organisms (such as fungal toxins), which are capable of producing damage to the three stages of the mammalian life cycle and which may result in modifications of 'normal' development.

30.2 MUTAGENIC EFFECTS

The units of inherited information, the genes, are composed of the molecule deoxyribonucleic acid (DNA) and are arranged in linear sequences in the chromosomes. Genes are capable of undergoing changes in their content to produce new forms, or mutants. This process of mutation occurs spontaneously, but may be dramatically increased by a variety of agents. The products of such mutant genes may lead to a range of inherited diseases and have also been implicated in the production of cancer (Ames *et al.*, 1973; Miller, 1978).

For a cell exposed to a DNA-damaging agent, there is a range of potential consequences ranging from lethality to a variety of genetic changes as illustrated in Figure 1. The changes in the informational content of the cell may occur at a number of levels: at the level of the gene itself (point mutations), by rearrangement of the sequence of genes in the chromosome (chromosome aberrations) or by changes in the number of chromosomes which lead to deviations from the normal characteristic karyotype of the cell (aneuploidy and polyploidy). Point mutational changes at the level of the gene may be subdivided into two groups: base substitution changes and frameshift changes. Base substitution changes involve the replacement of specific nucleotide bases by other bases, which lead to the production of a messenger RNA molecule which contains an informational sequence which deviates from the 'normal' or 'wild type'. Frameshift changes involve the removal or insertion of bases in the DNA which results in changes in the 'reading frame' of the genetic information at the level of protein synthesis. The types of changes are illustrated in Figure 2. Such changes in nucleotide sequence may lead to the production of proteins containing an abnormal sequence of amino acids (mis-sense) or to the production of a non-coding triplet which leads to chain termination and the production of a shortened protein (non-sense).

Changes in the organization of information within a chromosome may lead to a number of consequences which are illustrated in Figure 3. Breakage of chromosomes may lead to the production of chromosomes with missing sequences of genes (deletions). Multiple damaging events occurring between different chromosomes may lead to the illicit rejoining of chromosomal fragments producing atypical chromosomes with new arrangements of genes (translocations). Misdivision events with unbalanced segregation of chromosomes (non-disjunction) may result in the production of cells with abnormal chromosome numbers.

At least a fraction of the agents of biological origin to which man is exposed may be capable of producing mutagenic damage in exposed individuals. Such damage may give rise to increases in the frequency of congenital abnormalities in future generations which inherit mutant genes. Thus, the assay of such potential mutagenic activity represents an important part of toxicological evaluation.

30.3 CARCINOGENIC EFFECTS

In mammals, tumours arise most frequently in those tissues in which cells are continuously being replaced and which are exposed to various environmental exposures, such as those of the skin and the epithelium of the alimentary canal and respiratory tracts. However, many tumours

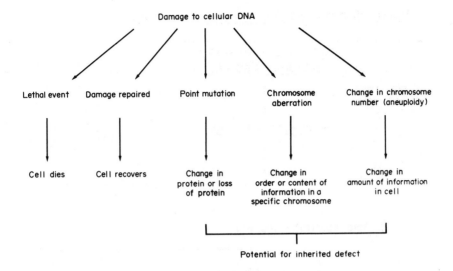

Figure 1 Consequences of damage to DNA of living cell

also originate from cells of more stable origins such as those of the liver and thyroid. Nearly all tumours may be classified on the basis of their tissues of origin into epithelial or connective tissue neoplasms.

The most important functional classification of tumours is into the benign and malignant forms. The benign tumours remain at the site of origin as a cell mass whereas malignant tumours invade local tissues and can also be spread by the lymphatics, blood stream and body cavities to other sites where secondary tumours arise (metastasis). The term cancer is used for all malignant tumours regardless of origins and carcinogenesis for the changes involved in the development of malignant tumours (Anderson, 1980).

The cellular events leading to the production of cancer are complex. However, it is generally accepted that a variety of steps are involved in the process ranging from an 'initiation event' which involves the production of a cellular lesion to secondary events which lead to the conversion of the potentially carcinogenic lesions into a malignant tumour. There is now a considerable amount of information available, which is beyond the scope of this chapter, which implicates environmental agents in the production of a high proportion of human tumours.

There are examples of biologicals which have been implicated in the various events of cancer formation, and this involvement provides a convenient method of classification of activity. For example, some materials such as some aflatoxins are capable of producing the whole range of carcinogenic events and are classified as complete carcinogens, whereas others such as croton oil are capable of stimulating secondary events only after exposure to some initiating agent, and are classified as cocarcinogens or tumour promoters.

A considerable proportion of the initiating agents and complete carcinogens have been shown to be capable of inducing mutagenic events in experimental systems, whereas less biological information is available on the activity of tumour promoters. The assay of a biological agent for its potential carcinogenic response thus requires the determination of the ability of the test substance to stimulate a variety of cellular events.

30.4 TERATOGENIC EFFECTS

Exposure of the mammalian embryo and foetus to a variety of agents may result in a range of deleterious effects with a wide spectrum of severity. Some are so profound as to be incompatible with life *in utero* and lead to abortion and reabsorption of the embryo. Others can be recognized at birth as gross congenital abnormalities. In contrast, others are so mild that they can only be recognized at a later stage as various functional defects such as mental or growth retardation. These latter effects can generally not be observed in animal experiments by the very nature of their design (see later).

The three classic teratogenic effects observable in the mammalian embryo are: (1) embryonic,

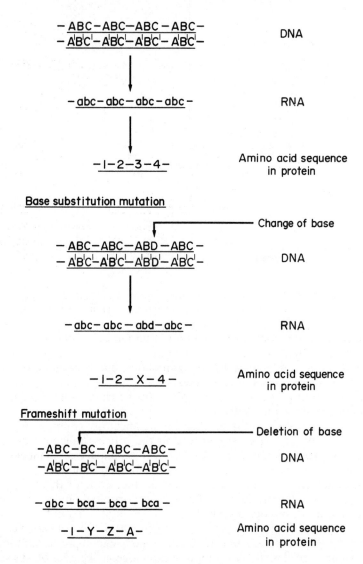

Figure 2 The induction of point mutation

foetal or neonatal death; (2) intrauterine and/or extrauterine growth retardation; and (3) congenital malformation at birth.

In rats and mice all the above effects may be observed in a single litter of teratogen-exposed animals after fertilization. The probability of finding one or more of these effects in an exposed embryo depends upon a number of factors such as the total absorbed quantity of teratogen, the rate of delivery and the stage of gestation (embryo development) at which exposure occurs.

Of these factors the stage of gestation at which exposure occurs is of particular importance as not all the stages of development have equal sensitivity to teratogenic damage. This factor must be considered in experimental design for the detection of teratogenic potential.

Gestation may be divided into three basic stages: (1) pre-implantation, which extends from fertilization until the time at which the embryo attaches to the wall of the uterus; (2) organogenesis, the period during which the major organs differentiate; and (3) the foetal stage, where growth occurs.

The effects of exposure of the embryo to a potential teratogen show a relationship to the stage of embryo development at which exposure occurs. Lethal effects are generally at their highest for exposure just after fertilization and are reduced near term. Growth retardation can result from

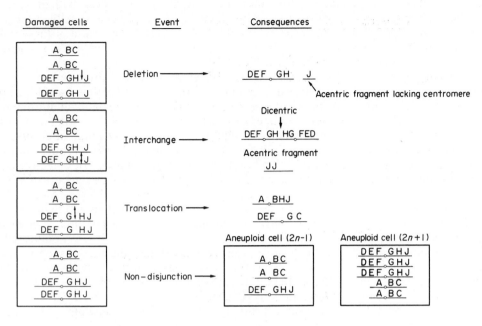

NB. For the sake of simplicity the cell is shown with only two chromosomes
| Indicates an exchange of chromosomal material

Figure 3 Production of chromosome damage

exposure at any time during gestation after the period of implantation. However, in general, retardation is at its greatest if exposure occurs during the period of organogenesis. The induction of congenital malformations is also generally at a maximum when exposure occurs during the period of organogenesis (approximately days 14 to 50 in human pregnancy). We should however point out that the periods of sensitivity are generalities and that the vast range of potentially active biologicals may produce different sensitivities. Such variations may be dependent upon a whole complex of factors such as tissue concentrations, detoxification mechanisms and the role of the placental barrier. These represent some of the unknowns in the study of teratogens.

30.5 MUTAGENICITY TESTING

Long-term genetic damage which may give rise to congenital abnormalities may be assayed by the use of specialized strains of mice exposed to test chemicals and the damage produced determined in their progeny. Such assays are designed to provide information which is of direct relevance to potential human exposure. Examples of such assays are the Specific Locus Test, which involves the measurement of point mutations, and the Dominant Lethal Assay, which involves the measurement of embryo toxicity.

The Specific Locus Test is an assay for the induction of mutations in the mouse and has been extensively used for over 30 years. The test involves the mating of wild-type male mice, which have been exposed to a test substance, with untreated female mice homozygous for (*i.e.* carrying two identical copies of) a series of recessive genes (must be present in two copies to be observable) affecting a range of phenotypes (generally coat colour and eye markers). The resulting heterozygous (carrying two different copies) offspring show the wild-type phenotypes of the genes unless a mutation has occurred in one of the wild-type genes of the treated male mouse, leading to the presence of two defective copies of the gene concerned. The test involves the scoring of large number of progeny for the presence of new chemical- or radiation-induced mutations. The great advantage of the test is that it assays the induction of mutations transmitted to offspring *via* the gametes. However, because of the large number of mice utilized the test has been used with only a relatively small number of potentially mutagenic agents.

The Dominant Lethal Assay in the mouse involves the measurement of dominant mutations (*i.e.* those that are observed in the first generation after treatment even when only one copy is present) produced in sperm. Such mutations are expressed in embryos to produce lethality and this is measured as a reduction in litter size after fertilization of treated sperm with untreated ova. Reduction in litter size may result from either the failure of fertilized eggs to implant or to develop into a normal embryo after the process of implantation. The prime cause of these defective sperm has been shown by Brewen *et al.* (1975) to be primarily the result of chromosome damage and aberration formation in the treated male sperm.

The procedure for the measurement of dominant lethal mutations produced by a potential chemical mutagen involves the treatment of sexually mature mice (8 to 10 weeks old) of proven fertility with the test agent by a variety of routes of administration. Treated and untreated males are mated with virgin females over a number of weeks to allow for the sampling of a variety of stages in sperm development. The females are killed 17 days after mating and, after dissection of the contents of the uterus, are examined for the presence of normal and abnormal foetuses and defective implantations. A statistically significant increase in the number of defective embryos provides convincing evidence that the agent under test produces genetic damage in male mice which can be transmitted *via* the sperm.

A number of other *in vivo* mammalian assays, such as the Micronucleus and Spot Test, which are both performed in the mouse, are in use for studying the mutagenic potential of environmental agents. Detailed descriptions of these and other assays have recently been published in the results of a number of collaborative studies designed to validate the use of such systems (de Serres and Ashby, 1981; Parry, 1982). However, animal tests, particularly those involving the long-term husbandry of the animals, are extremely expensive and the results of such studies are often difficult to relate to human conditions. Over the past 15 years considerable efforts have been made to develop short-term tests which may act as partial or complete replacements for animal tests.

A problem posed by the replacement of *in vivo* animal tests with *in vitro* tests is the reproduction in the *in vitro* situation of the metabolic capacity of the whole animal. In the whole animal there is the capacity to convert innocuous compounds into mutagenic chemical species, the electrophiles, which are capable of interacting with DNA. Such features are lacking in the metabolism of bacterial cells and also because of tissue specificity may not exist in the types of cells capable of growth under tissue culture conditions. These problems have been partially solved by the use of rat liver (and other tissues) microsome preparations, together with the appropriate cofactors, which reproduce *in vitro* some of the characteristics of the metabolism of the whole animal (Malling, 1971; Garner *et al.*, 1972).

Short-term assays for the prescreening of potentially mutagenic agents can be divided into three basic groups based upon the nature of the cellular endpoints which are being assessed. These may be classified as follows: (1) tests for primary DNA damage; (2) point mutation tests; (3) assays for chromosomal damage.

Tests for primary DNA damage are based upon the concept that mutagens are capable of inducing lesions or structural changes in the DNA of an exposed organism. In some cases the specific DNA lesions induced may be identified, but this is not generally the case with the vast majority of potentially hazardous environmental agents. However, a fundamental property of those DNA lesions that initiate genetic damage is that they are susceptible to the action of cellular enzymes which are capable of removing the lesions and repairing the resultant gaps in the DNA by the replacement of the missing nucleotides with new ones. A detailed description of such processes is beyond the scope of this chapter but comprehensive reviews are to be found elsewhere (*e.g.* Hanawalt *et al.*, 1979).

This process of DNA repair activity can be assayed as an indirect measure of DNA lesion formation and exposure to potentially genotoxic agents. DNA repair activity can be assayed directly by the measurement of the incorporation of radiolabelled nucleotides into the region of DNA repair, a process which is called unscheduled DNA synthesis because it occurs outside the normal DNA synthetic period (Rasmussen and Painter, 1966).

Indirect measurements of DNA repair may be performed by the use of pairs of strains of suitable microbial cultures such as bacteria, one of which is repair-proficient and the other of which is defective in one of the enzymes of DNA repair. In the presence of a mutagenic agent the repair-deficient strain will show increased sensitivity compared to the repair-proficient strain, as unrepaired lesions will generally be lethal. Such an assay has been extensively utilized and details of its use have been described in detail by Rosenkranz *et al.* (1981).

By far the most widely used short-term genetic assay is the point mutation assay involving the use of cultures of bacteria. Its most popular form is the agar-overlay test developed by Ames and his coworkers (Ames *et al.*, 1977). This assay involves the use of cultures of *Salmonella typhimurium* auxotrophic for the amino acid histidine (*i.e.* they require histidine for growth in minimal medium). In the presence of a mutagenic agent these auxotrophs undergo mutation to prototrophy (*i.e.* they no longer require histidine for growth) and are capable of growth on a selective medium which lacks histidine. The number of prototrophic colonies which grow on selective agar plates provides a quantitative measure related to the activity of the compound under test. The agar-overlay test has undergone a range of modifications to increase its sensitivity and suitability for the screening of a wide range of environmental agents, and the general principles of the test are shown in Figure 4.

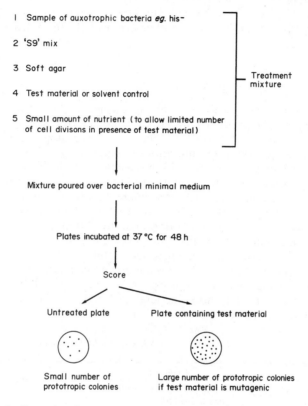

Figure 4 General outline of the protocol used in microbial mutagenicity assay

The general principles of the *in vitro* point mutation assay can be extended to higher organisms such as the fungi and to mammalian cells under tissue culture conditions. In the case of mammalian cells, cultures of a variety of different cell types have been utilized which are sensitive to the action of various growth-inhibiting drugs such as ouabain and 8-azaguanine. After exposure to test chemicals, mutations may be produced which may give rise to cells which are capable of producing observable clones when grown on a selective medium containing the drug. The enumeration of these clones provides an estimate of the mutagenic potential of the test chemical (Arlett, 1982).

Cultured mammalian cells may also be used to assay the induction of chromosome damage after mutagen exposure (Nichols *et al.*, 1977). Such cultures are exposed to the test agent for a time period which will permit the induction of chromosome damage. Cells are then allowed to undergo cell division, during which the primary damage is converted into chromosome aberrations, and are then blocked at the metaphase stage of division by the use of the spindle inhibitor such as colcemid. This inhibition allows for the accumulation of chromosome aberrations at a readily observable stage of cell division. After fixation and staining, the cells are then scored for the presence of chromosome aberrations of various types and the relationship between the dosage of the test compound and the quantitative response of aberration yield determined. This type of assay may also be used with samples of human blood in which cell division is stimulated by the addition of the drug phytohaemoglutinin. Assays for the production of chromosome aberrations represent a vital part of a genetic toxicology testing programme as this type of genetic event cannot be observed in the bacteria because of their size and chromosomal organization.

As with other *in vitro* assays, cultured mammalian cells also require the presence of a mammalian metabolic activation preparation to reproduce some of the aspects of whole animal metabolism of foreign compounds. At the present time the procedures for the use of such activation mixes with mammalian cells are far from optimal and their presence often leads to cellular toxicity (Scott, 1982). To overcome some of the problems of the requirement for metabolic activation preparation with mammalian cells, attempts have been made to culture cell types which retain some aspects of whole animal metabolism. One such attempt is the use of the rat liver epithelial cell line RL4 developed by Dean and coworkers (Dean and Hodson-Walker, 1979). This cell line is capable of activating a number of indirect mutagens into mutagenic chemical species and shows high sensitivity to chromosomal damage (Scott, 1982).

The use of a series of assay systems to evaluate the effect of biologicals upon the three types of genetic endpoint described provides a comprehensive package of information and the potential mutagenesis of a test agent.

30.5.1 Naturally Occurring Mutagens

Mutagenicity assays are extensively used to prescreen possible carcinogenic agents prior to long-term animal studies. The predictive nature of mutagenicity assay is well illustrated by reference to the potent carcinogen, aflatoxin B1. Aflatoxin B1 is mutagenic in a variety of assay systems and endpoints ranging from point mutations in bacteria to *in vivo* chromosome damage in mice (IARC, 1982). A summary of aflatoxin mutagenicity is given in Table 1. Aflatoxin B1 is metabolized primarily by the liver microsomal mixed function oxidase system. During its metabolism, highly reactive electrophilic species are generated which may interact with nucleophilic centres in DNA, leading to mutations. The metabolite, aflatoxin B1 2,3-epoxide, has been demonstrated in human, rat and hamster liver preparations (Garner, 1973; Swenson *et al.*, 1973; Swenson *et al.*, 1974) and is likely to be the ultimate carcinogen and mutagen for this compound. Attempts to isolate this epoxide have been unsuccessful due to its high reactivity. However, Swenson and coworkers (1975) have utilized the more stable aflatoxin 2,3-dichloride as a model compound. This electrophilic epoxide has proved to be a more potent bacterial mutagen and animal carcinogen than the parent compound, lending weight to the theory that the epoxide is indeed the ultimate mutagen (Swenson *et al.*, 1975).

Table 1 Summary of Aflatoxin B1 Mutagenicity

Genetic change	Organism
Point mutation	*Salmonella typhimurium*
Point mutation	*Bacillus subtilis*
Point mutation	*Neurospora crassa*
Point mutation	*Drosophila melanogaster*
Point mutation	Chinese hamster V79 cells
Point mutation	Rat liver cells
Chromosome aberrations	Human lymphocytes (*in vitro*)
Sister chromatid exchanges	Human lymphocytes (*in vitro*)
	Chinese hamster cells
	C57BL/6J mice (*in vivo*)
Cell transformation	Syrian hamster SHE cells
	Guinea pig foetal cells
	Mouse 3C3 cells

Source: IARC (1982).

30.6 SCREENING FOR CARCINOGENIC POTENTIAL

The ability of an agent to induce cancer in experimental animals is most comprehensively determined in a long-term animal study. The basic principles of such a study are that animals are exposed on a regular basis to a test compound over a significant period of its life cycle and at the end of the exposure period the animals are killed and examined for the presence of tumours in all organs. Clearly, for the assay to be reliable, the test requires high standards of animal husbandry over the test period, adequate numbers of test animals, skilled personnel for the evaluation of the presence and frequency of tumours and the use of strains of known performance (Ministry of Health and Welfare, Canada, 1975; Sontag *et al.*, 1975).

For a new substance with a potential human exposure, testing is usually performed in at least two mammalian species, normally the rat and the mouse. However, there are some circumstances where other species may be appropriate, for example, if the test agent is suspected of being a potential bladder carcinogen other species may be chosen. The test is generally started just after weaning and acclimatization of the animals, *i.e.* before six weeks in the rat and mouse. It may however be appropriate for the test to be started earlier and in some cases *in utero* exposure may be required. If exposure to the test agent is to cover a significant portion of the animal lifespan then it must continue over a period of 18–24 months in mice and 24–30 months in rats.

The most common method of exposure of the animal to the test agent is to incorporate it into the diet over the assay period. However, a more satisfactory approach is for the method of exposure to be related to potential human exposure (which may or may not be in the food supply). An adequate study requires exposure to at least three dose groups plus an appropriate control with at least 50 animals of each sex in the treatment groups. Dose levels are generally selected on the basis of the maximum tolerated dose, which is the highest dose which is predicted (from a previous toxicity study) not to significantly reduce life span due to some effect other than the induction of tumours. The doses selected are then the maximum tolerated dose and some fraction of this value such as a half and a quarter.

During the course of a long-term study animals are carefully maintained and observed. Deaths occurring during the study must be kept to a minimum to avoid the excessive loss of animals which would render the study invalid and each dead animal examined in detail. At the end of the study period the animals are killed and subjected to detailed pathological examination. All tissues are examined and any tumours present classified into malignant and non-malignant lesions.

It should be clear from the above general outline that a long-term animal study is an expensive and time-consuming assay to undertake, and for optimum performance there is a requirement for considerable pre-existing information, *e.g.* toxicity levels, without which the assay cannot be expected to provide optimal data. Considerable efforts have therefore been made to develop short-term assays which have a predictive value in the determination of carcinogenic potential.

As was pointed out earlier, there is a high degree of correlation between the ability of an agent to induce mutagenic damage and its carcinogenic potential. The use of the genetic techniques described earlier thus provides valuable complementary assays in the estimation of carcinogenic potential.

Another valuable short-term test is the so-called Cell Transformation Assay. This assay is based upon the observations that cultured cells which have been exposed to an initiating agent (presumably causing a DNA change) are not thought to be capable of undergoing tumour formation until a further biological event takes place. This process of transformation presumably involves the conversion of the initiated cell into a neoplastic state where the cells proliferate and produce clones with tumour-forming potential. This transformation step is associated with the acquisition of certain morphological and growth characteristics which may themselves be used as criteria for prior exposure to a carcinogenic agent (*e.g.* Pienta *et al.*, 1977).

Under tissue culture conditions normal untransformed fibroblast cultures form monolayers over growing surfaces until the plate is completely covered, at which point contact inhibition occurs and cell division ceases. In contrast, transformed cells may continue to grow and multilayered regions may be observed in growth plates. Such changes in growth patterns may also be correlated with the ability of the transformed cell to form distinct colonies in soft agar medium. The final test that neoplastic transformation has taken place is the transfer of some of these cells to a host animal where tumour proliferation can be observed. Using the criteria of the appearance of the morphological characteristics of transformation, a variety of cultured cell types may be used as a prescreen for carcinogenic materials.

30.6.1 Carcinogenic Biologicals

While most attention is focused on the carcinogenic potential of man-made chemicals, it is often forgotten that some of the most potent carcinogens are of natural origin. Examples of carcinogenic biologicals are to be found among the viruses, plant and bacterial products and, in particular, among fungal metabolites (see Table 2).

30.6.1.1 *Viruses*

The association between viral infection and certain human cancers (*e.g.* some leukaemias) has long been suspected though little conclusive scientific evidence is available. The most convincing

Table 2 A Selection of Naturally Occurring Carcinogens

Compound	Source	Target species	Ref.
1. MYCOTOXINS			
Aflatoxin B1	*Aspergillus flavus*	Mouse, hamster, rat	a
Cycloclorotine	*Pencillium* sp.	Mouse	b
Griseofulvin	*Penicillium griseofulvum*	Rat	b
Luteoskyrin	*Penicillium* and *Mycelia sp.*	*Mouse*	b
Ochratoxin A	*Aspergillus/Penicillium* sp.	Mouse, rat	b
Patulin	*Penicillium* sp.	Rat	b
Sterigmatocystin	*Aspergillus* sp.	Mouse, rat	b
2. ANTIBIOTICS			
Actinomycin D	*Streptomyces* sp.	Rat	c
Azaserine	*Streptomyces fragilis*	Rat	b
Daunomycin	*Streptomyces* sp.	Rat	b
Mitomycin C	*Streptomyces caespitosus*	Rat	b
3. ALKALOIDS and EXTRACTS			
Cycasin	*Cycas revoluta, Cycas*		
	circinalis seeds	Mouse, rat, guinea pig, rabbit	b
Safrole	Sassafras oil	Mouse, rat	b
Vinblastine	*Vinca rosea*	Mouse, rat, hamster	d
Vincristine	*Vinca rosea*	Mouse, rat, monkey	d
4. PLANTS			
Bracken fern		Cattle, rat, mouse	e
Laburnum seeds		Rat	e
5. PYRROLIDIZINE ALKALOIDS			
Lasiocarpine	*Heliotropium* sp.	Rat	b
Monocrotaline	*Crotalaria* sp.	Rat	b
Retrorsine	*Senecio* sp.	Rat	b
6. MISCELLANEOUS			
Ethionine	*Escherichia coli*	Rat	e

a, Wogan and Busby (1980); b, IARC (1976); c, IARC (1982); d, IARC (1981); e, Miller and Miller (1976).

evidence for viral involvement in human cancer is of the association between Epstein–Barr virus (EBV) and Burkitt's lymphoma (BL). Burkitt's lymphoma is a malignant lymphoma which occurs at high frequency in certain areas of Africa and New Guinea. The most commonly affected group are 6 to 10 year old children. Burkitt's suggestion that the climate-related geographical distribution and time–space clustering pattern of the disease indicated an insect-vectored viral aetiology stimulated the search for a causative agent. Examination of lymphoma tissues revealed the presence of a herpes-like virus which produced an immunological reaction when exposed to the serum of BL patients. The newly identified virus was named Epstein–Barr virus after its discoverers (Klein, 1973).

EBV is now linked with a number of acute (heterophile positive infectious nucleosis) and chronic (African or EBV positive BL and undifferentiated nasopharyngeal carcinoma) proliferative diseases. Studies of BL tissue have revealed some interesting facts. All EBV positive tumours are monoclonal in origin, *i.e.* they have arisen from a single transformed cell and show a specific chromosomal change, a portion of chromosome 8 having been transferred to chromosome 14. These observations have led to the following speculations on the development of the disease: (1) EBV infects the B-lymphocytes (short-lived lymphocytes responsible for immunoglobulin production) leading to their immortalization. (2) An environmental factor (such as malaria) weakens the host's immune system. (3) A specific chromosomal translocation occurs. Thus, the viral infection can be considered to initiate the multistage process culminating in the production of neoplasia (Gilden and Rabin, 1982).

30.6.1.2 Aflatoxin

Aflatoxins in general and aflatoxin B1 in particular are among the most potent known carcinogens. Toxic and carcinogenic effects have been demonstrated in a wide variety of animal species and circumstantial evidence implicates them as environmental carcinogens for man (Wogan and Busby, 1980). Aflatoxin is a secondary metabolite of the fungus *Aspergillus flavus* and is commonly found on grains and nuts which have been improperly dried at harvest and subsequently stored at high temperature. Use of contaminated grain as food and also drinking the milk of cattle fed on aflatoxin-contaminated fodder are the principal routes of human exposure. Surveys in aflatoxin-prevalent areas such as Uganda, Thailand and Swaziland have demonstrated levels of up to 1 mg of aflatoxin per kg of food (IARC, 1976).

Aflatoxin poisoning was first described following an outbreak of hepatic necrosis on English poultry farms in 1960/1961. Fungal-contaminated peanut meal was shown to be the cause and subsequent feeding of this meal to rats produced the first reports of hepatoma induction (Lancaster *et al.*, 1961). Since then the carcinogenic potency of aflatoxin B1 has been amply demonstrated in many species of rodent, non-human primates, birds and fish. In all species the liver is the target organ, although tumours at other sites have been reported (Wogan and Busby, 1980). A summary of the hepatotoxicity of aflatoxin B1 for a variety of species is given in Table 3.

Table 3 Summary of Aflatoxin Hepatocarcinogenicity[a]

Species	Treatment	Tumour incidence
Monkey, rhesus (male)	1.655 g total over 5.5 y	1/1
Monkey, rhesus (female)	1.655 g total over 5.5 y	1/1
Monkey, rhesus (female)	0.504 g total over 6 y	1/1
Monkey, rhesus (male and female)	99–842 mg total over 3.8–6 y	3/3
Marmoset	3 mg total over 50–55 wk	1/3
	5.04–8.54 mg total over 87–94 wk	2/3
Tree shrew (male and female)	24–66 mg total over 74–172 wk	9/12
Ferret	0.3–2 p.p.m. in diet for 28–37 mo	7/9
Rat, Fischer	1 p.p.m. in diet for 33 wk	3/6
Rat, Fischer	1 p.p.m. in diet for 41–64 wk	18/21
Rat, Fischer	55 p.p.b. in diet for 71–97 wk	20/25
Rat, Porton	1 p.p.m. in diet for 20 wk	19/30
Rat, Wistar	1 p.p.m. in diet for 21 wk	12/14
Mouse, Swiss	150 p.p.m. in diet for 20 mo	0/60
Mouse, C57Bl/6NB	1 p.p.m. in diet for 20 mo	0/30
Mouse, C3HfB/HEN	1 p.p.m. in diet for 20 mo	0/30
Mouse	60 μg g^{-1} bw, 3 ip doses	16/16
Duck	30 p.b.b. in diet for 14 mo	8/11
Rainbow trout	4 p.p.b. in diet for 1 y	15%
	8 p.p.b. in diet for 1 y	40%
	20 p.b.b. in diet for 1 y	65%
Salmon	12 p.b.b. in diet for 20 mo	50%
Guppy	6 p.p.m. in diet for 11 mo	7/113

[a] Modified from Wogan and Busby (1980).

30.7 TERATOLOGY ASSAY

The aim of a teratology assay is to determine whether exposure of a pregnant animal to a test substance is capable of inducing embryonic and foetal damage. The species most commonly used for such studies are rat, mouse, hamster and rabbit, with the rat and rabbit being the species of choice. However, under some circumstances, primate studies may also be necessary for adequate evaluation of the test agent (Collins and Collins, 1976; Shepard *et al.*, 1975).

Young, mature pregnant females of uniform age and size are divided into four test groups of which three are exposed to the test material and the fourth acts as a vehicle control. An adequate study requires the use of at least 20 pregnant rats or 12 pregnant rabbits, the aim being to produce sufficient offspring for adequate examination of the foetuses and to establish the statistical significance of the test.

The highest test dose should produce some maternal effect (such as weight loss) but not produce excess foetal losses. The other two doses are some fraction of this such as one half or one quarter. Animals are exposed to the test substance on a daily basis by some appropriate route (commonly in the diet) soon after implantation and continuing well into the period of foetal development. For the rat this is day 6 through day 15 and for the rabbit day 7 through day 18 after fertilization. Alternatively, the animal may be dosed from early implantation to just before the expected delivery date. The overall experimental design requires some prior knowledge of the properties of the test material. For example, with highly toxic materials shorter dosage periods are preferable.

Foetuses are not allowed to reach term but are delivered by hysterectomy about one day prior to term. At sacrifice, the dam is examined in detail for gross abnormalities and pathological damage which may have influenced the pregnancy. After removal of the uterus the contents are examined for embryonic or foetal death and live foetuses. Each foetus is carefully examined for the presence of abnormalities. Increases in the frequency of abnormalities in the treated groups over the control are indicative of the action of a teratogenic agent.

30.7.1 Teratogenic Biologicals

The ability of naturally occurring agents to produce foetal death and malformation is well documented. Plants, specific plant alkaloids, fungal products and viral infections are all capable of inducing a broad spectrum of malformation (see Tables 4 and 5). As already stated, the type of abnormality produced by any teratogenic agent is dependent on a number of factors: (1) the genotype of the conceptus, (2) developmental stage at which exposure occurs, and (3) maternal factors. These principles are clearly apparent when considering the effect of the various biological agents on developing embryos.

Table 4 Teratogenic Plants

Species	Target species	Ref.
1. PLANTS		
Locoweed (*Astragalus* sp.)	Sheep, cattle, horse	a
Lupins (*Lupinus* sp.)	Cattle	a
Jimsonweed (*Datura stramonium*)	Pig	a
Wild black cherry (*Prunus serotina*)	Pig	a
Chick pea (*Lathyrus cicera*)	Cattle, sheep	a
Sweat pea (*Lathyrus odoratus*)	Baboon	a
False hellebore (*Verratrum californicum*)	Sheep	a
Tobacco (*Nicotiana tabacum*)	Pig	a
Poison hemlock (*Conium maculatum*)	Cattle	a
Sudan grass (*Sorghum vulgare*)	Horse	b
Pine (*Pinus ponderosa*)	Sheep	b
Mimosa (*Leucaena* sp.)	Sheep, pigs	b
2. MYCOTOXINS		
Aflatoxin B1 (*Aspergillus flavus*)	Mouse, rat, hamster	a
Ergotamine (*Claviceps purpurea*)	Mouse, rat, rabbit	a
Ochratoxin A (*Aspergillus flavus*)	Mouse	a
Rubratoxin B (*Penicillium*)	Mouse	a
3. ANTIBIOTICS		
Actinomycin D (*Streptomyces* sp.)	Rat, rabbit	c
Azaserine (*Streptomyces fragilis*)	Rat	c
Chloramphenicol (*Streptomyces venezuelae*)	Rat	c
4. ALKALOIDS		
Cycasin (*Cycas revoluta* and *Cycas circinalis* seeds)	Hamster	c
Vinblastine (*Vinca rosea*)	Mouse, rat, hamster	d
Vincristine (*Vinca rosea*)	Mouse, hamster, monkey	d
5. MISCELLANEOUS		
Snake venoms (*Vispera* and *Naja* spp.)	Mouse, chicken	a
Tetanus toxin (*Clostridium tetani*)	Chicken	a
Endotoxin (*Escherichia coli*)	Rat	a

a, Wilson (1977); b, Selby (1978); c, IARC (1976); d, IARC (1981).

Teratogenic plants can have devastating consequences in the farming community and much interest has been aroused by such plant species.

30.7.2 Lupins (*Lupinus* sp.)

The incidence of epidemics of 'crooked calf disease' in range-grazing cattle has been linked with maternal ingestion of lupins. The severity of the disease varies both as a consequence of the

Table 5 Selected Teratogenic Infectious Agents[a]

Viral group	Species	Effect
1. ARENAVIRIDAE		
Lymphocytic choriomeningitis	Mouse, rat	Foetal death, ataxia, cerebellar hypoplasia
2. HERPETOVIRIDAE		
Cytomegalovirus	Man,	Multiple CNS effects
	Pig, guinea pig	Foetal death
Varicella-Zoster	Man	
Feline herpes virus	Cat	Foetal death
Herpes simplex virus	Man	Multiple CNS effects
3. ORTHOMYXIVIRIDAE		
Influenza	Monkey, hamster, rhesus monkey	Hydrocephalus
4. PARVOVIRIDAE		
Feline panleukopenia virus	Cat	Cerebellar hypoplasia
	Ferret	Ataxia
Rat virus	Rat, hamster, cat, ferret	Cerebellar hypoplasia
Minute virus	Mouse	Cerebellar hypoplasia
5. PARAMYXOVIRIDAE		
Mumps virus	Hamster	Hydrocephalus
Newcastle disease virus	Chicken	Micrencephaly, neural tube defects
Parainfluenza type 2	Hamster	Hydrocephalus
6. POXVIRIDAE		
Variola, vaccinia	Man	Foetal death
7. REOVIRIDAE		
Reovirus type 1	Hamster, mouse, rat, ferret	Hydrocephalus
Bluetongue virus	Sheep, mouse, hamster	Hydancephaly, parencephaly
8. TOGAVIRIDAE		
Rubella	Man	Multiple systemic effects
Hog cholera virus	Pig	Foetal death, multiple CNS effects
Bovine diarrhoea–mucosal disease virus	Cattle, sheep	Cerebellar hypoplasia, hydrocephalus, blindness
Venezuelan equine encephalitis virus	Man, rhesus monkey	Multiple CNS effects
9. NON-VIRAL AGENTS		
Toxoplasma gondii (toxoplasmosis)	Man	Foetal death, multiple CNS effects
Treponema pallidum (syphilis)	Man	Multiple CNS and skeletal effects, abortion

[a] Adapted from Mims (1981); Kurent and Sever (1977)

time of exposure and the maturity of the lupin plant consumed. Maternal ingestion of lupins between days 40 and 70 of gestation produces the most severe response and up to a third of the herd may be affected. The syndrome is characterized by arthrogrypotic malformation—permanent flexure or contracture of the joint. Spinal curvature, twisted neck and cleft palate also occur. These osseus changes are permanent and worsen as the animal ages. The lupin alkaloid, anagrine, has been identified as the teratogenic agent (Keeler, 1978).

30.7.3 Tobacco Stalks (*Nicotiana tabacum*)

Epidemics of arthrogrypotic pigs have been linked with the ingestion of tobacco stalks by pregnant sows. Ingestion of stalks between days 10 and 35 of pregnancy produces maximum damage. Initially, nicotine was suspected of being the causative agent, but experimental application of nicotine to pregnant sows failed to induce the syndrome (Crowe, 1978). On the basis of studies with other arthrogrypotic-producing plants, Keeler (1978) has hypothesized that the tobacco alkaloid, anabasine, may be responsible.

30.7.4 False Hellebore (*Veratrum californicum*)

Cyclopia and associated cephalic disorders such as shortened upper jaw, protruding upper jaw and a skin-covered proboscis above the single eye (commonly known as monkey-faced lambs) are observed following maternal ingestion of false hellebore. Another alkaloid, aptly named cyclopamine, is thought to be responsible (Keeler, 1978).

30.7.5 Locoweed (*Astragalus lentiginosus*)

In contrast to the above examples where no maternal effects are apparent, animals ingesting locoweed show well-defined toxic lesions. In pregnant animals, general vacuolization of the maternal tissues leads to decreased foetal blood supply and abortion. Foetal tissues become distended and oedematous with musculo-skeletal deformities present in some foetuses (Van Kampen *et al.*, 1978).

30.7.6 Viruses

Since the first observations by Gregg in 1941 of the association between maternal rubella infection and the incidence of cataracts in the offspring, a large body of scientific evidence has accumulated associating maternal viral infection with subsequent malformations in the offspring. A selection of virally induced malformations are listed in Table 5.

Virus infections are aetiologically related to a variety of malformations, many of which resemble the classic defects of embryogenesis. The majority of viral infections result in central nervous system (CNS) lesions. In human populations, the exact proportion of CNS lesions attributable to viruses is difficult to gauge. However, whereas the infant mortality figures in general have continued to decline, the incidence of congenital malformations has remained constant and the involvement of an environmental agent (such as a virus) is indicated. As with all teratogenic agents, the results of exposure are dependent on a number of factors. The possible outcome of maternal viral infection during pregnancy is summarized in Figure 5.

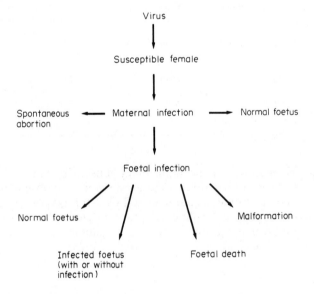

Figure 5 Possible consequences of maternal viral infection during pregnancy

30.7.6.1 Rubella

Rubella is the classic model of virally induced teratogenesis in humans. Maternal rubella infection was initially linked to an increased incidence of cataracts in the offspring (Gregg, 1941). Following rubella epidemics in the 1960s, an expanded rubella syndrome was characterized in which a wide spectrum of associated defects including congenital heart disease, cataract, deafness, liver and blood abnormalities, radiolucency of the long bones and delayed dentition are apparent (Catalano and Sever, 1971).

Rubella demonstrates clearly the classic association between the type and frequency of abnormality produced and gestational age at infection. Infection in the first trimester of pregnancy results in the highest incidence of abnormal offspring. During the first month, infection results in 50% abnormal; this drops to 22% in the second month and 6–8% in the third to the fifth month. Second trimester infection usually results in more subtle effects manifest as hearing loss, eye defects and motor mental retardation.

30.7.6.2 Cytomegalovirus

Cytomegalovirus is carried by approximately 3–5% of all pregnant females, thus the potential for intrauterine infection of the foetus is considerable leading to increased abortion rate and a range of central nervous system malformations characterized by microcephaly, hydrocephaly, encephalitus, seizures and mental retardation (Sever, 1980).

30.7.6.3 Herpes simplex virus

Herpes simplex virus of man occurs in two distinct forms. Herpes simplex I causes the common cold sore infections while simplex II is primarily associated with genital infections. Intrauterine infection during the first trimester of pregnancy leads to increased abortion rates. Congenital abnormalities in surviving foetuses resemble those seen following cytomegalovirus infection, namely central nervous system abnormalities, seizures and mental retardation. Maternal herpes simplex II infections at term are correlate with high foetal risk: infection occurs *via* contact with the infected birth canal. Effects range in severity from a few skin lesions to total generalized disease with jaundice, fever, encephalitis, circulatory collapse and death (Catalano and Sever, 1971; Sever, 1980).

30.8 APPLICATION OF THE ASSAY SYSTEMS

The assays described here may be utilized to determine the potential long-term effects of biological materials. The value of their use is that they allow for predictions to be made of the consequences of human exposure to such materials. Each assay described involves the exposure of test organisms to concentrations of test material at maximum tolerable concentrations far in excess of potential human exposure to such materials.

In general, human exposure involves exposure at comparatively low concentrations and possibly for long periods of time for products such as novel foods. Such long-term exposure presents the problem that some potential human effects may not be observable in the animal studies. Under such conditions human monitoring of exposed persons and epidemiological investigation are appropriate. Such studies require extensive investment in time and labour, and cooperation between suppliers, exposed individuals and their medical advisers.

When a new biological material is produced where usage involves potential human exposure then there is considerable advantage to be gained from extensive pretesting. The tests described here are recommended by an extensive range of national bodies responsible for the regulation of a variety of products. Most such authorities offer an alternative range of assays for the testing of particular types of products and patterns of usage.

For maximum efficiency of use, the decision as to the most appropriate tests to use is assisted by the acquisition of the available data on related materials. The collection of such information requires extensive study of the published literature. This study is facilitated by the use of specialized information centres and data bases now available in a computerized form.

Such information will ideally allow for the comparison of the response of related materials in a variety of test systems and most important the selection of the appropriate assay for the particular circumstances of the product usage.

30.9 SOURCES OF NEW BIOLOGICAL MATERIALS

The increasing availability of a new range of biological techniques such as that of the large-scale production of the products of pre-existing organisms and the production of new material by the use of gene cloning techniques provides a considerable challenge to the toxicologist. Such techniques may result n the synthesis of biological materials for which little information may be available as to their long-term consequences on the exposed human population.

We have illustrated in this chapter the potential long-term hazards of a variety of biologicals. In view of potential usage of these materials it will therefore be an important priority for new material such as pharmaceuticals and novel foods to be thoroughly tested for their potential long-term effects prior to extensive human exposure.

ACKNOWLEDGEMENT

We thank Miss K. Goode for the preparation of the figures.

30.10 REFERENCES

Ames, B. N., W. E. Durston, E. Yamasaki and F. D. Lee (1973). Carcinogens are mutagens: a simple test system combining liver homogenate for activation and bacteria. *Proc. Natl. Acad. Sci. USA*, **69**, 2281–2285.
Ames, B. N., J. McCann and E. Yamaski (1977). Methods for detecting carcinogens and mutagens with the Salmonella/mammalian-microsome mutagenicity test. In *Handbook of Mutagenicity Test Procedures*, ed. B. J. Kilbey *et al.*, pp. 1–37. Elsevier, Amsterdam.
Anderson, J. R. (1980). *Muir's Textbook of Pathology*. Arnold, London.
Arlett, C. F. (1982). The use of cultured mammalian cells in the UKEMS collaborative genotoxicity trial (1981). *Mutation Res.*, **100**, 271–276.
Brewen, J. G., H. S. Payne, K. P. Jones and R. J. Preston (1975). Studies on chemically induced dominant lethality. I. The cytogenetic basis of MMS-induced dominant lethality in post-meiotic germ cells. *Mutation Res.*, **33**, 239–250.
Catalano, L. W., Jr. and J. L. Sever (1971). The role of viruses as causes of congenital defects. *Annu. Rev. Microbiol.*, **25**, 255–282.
Collins, T. F. X. and E. V. Collins (1976). Current methodology in teratogenicity research. In *Advances in Modern Toxicology*, ed. M. A. Mehlman, pp. 155–175. Hemisphere, Washington, D.C.
Crowe, M. W. (1978). Tobacco—a cause of congenital arthrogryposis. In *Effects of Poisonous Plants on Livestock*, ed. R. F. Keeler, K. R. Van Kampen and L. F. Jones, pp. 419–427. Academic, New York.
Dean, B. J. and G. Hodson-Walker (1979). An *in vitro* chromosome assay using cultured rat liver cells. *Mutation Res.*, **64**, 329–337.
de Serres, F. J. and J. Ashby (1981). *Evaluation of Short-term Tests for Carcinogens*. Elsevier/North Holland, Amsterdam.
Garner, R. C. (1973). Chemical evidence for the formation of a reactive aflatoxin B1 metabolite by hamster liver microsomes. *Fed. Eur. Biochem. Soc. Lett.*, **36**, 261–264.
Garner, R. C., E. C. Miller and J. A. Miller (1972). Liver microsomal metabolism of aflatoxin B1 to a reactive derivative toxic to *Salmonella typhimurium* TA 1530. *Cancer Res.*, **32**, 2058–2066.
Gilden, R. V. and H. Rabin (1982). Mechanisms of viral tumorigenesis. *Adv. Virus Res.*, **27**, 281–334.
Gregg, N. (1941). Congenital cataracts following German measles in the mother. *Trans. Ophthalmol. Soc. Aust.*, **3**, 35–46.
Hanawalt, P. C., P. K. Cooper, A. K. Ganesan and C. A. Smith (1979). DNA repair in bacteria and mammalian cells. *Annu. Rev. Biochem.*, **48**, 783–836.
IARC (1976). Some naturally occurring substances. *Monographs on the Evaluation of Carcinogenic Risk of Chemicals to Man*, vol. 10. IARC, Lyon.
IARC (1981) Some antineoplastic and immunosuppressive agents. *Monographs on the Evaluation of Carcinogenic Risk of Chemicals to Humans*, vol. 26, IARC, Lyon.
IARC (1982). Chemicals, industrial processes and industries associated with cancer in humans. *Monographs on the Evaluation of Carcinogenic Risks of Chemicals to Humans*, vol. 1–29, suppl. 4. IARC, Lyon.
Keeler, R. F. (1978). Alkaloid teratogens from *Lupinus, Conium, Veratrum* and related genera. In *Effects of Poisonous Plants on Livestock*, ed. R. F. Keller, K. R. Van Kampen and L. F. James, pp. 397–408. Academic, New York.
Klein, G. (1973). The Epstein–Barr virus. In *The Herpesviruses*, ed. A. S. Kaplan, chap. 16, pp. 521–555. Academic, New York.
Kurent, J. E. and J. L. Sever (1977). Infectious diseases. In *Handbook of Teratology. 1. General Principles and Etiology*, ed. J. G. Wilson and F. C. Fraser, chap. 6, pp. 225–259. Plenum, New York.
Lancaster, M. C., F. P. Jenkins and J. McL. Philip (1961). Toxicity associated with certain samples of groundnuts. *Nature (London)*, **192**, 1095–1096.
Malling, H. V. (1971). Dimethyl-nitrosamine; formation of mutagenic compounds by interaction with mouse liver microsomes. *Mutation Res.*, **13**, 425–429.
Miller, E. C. (1978). Some current perspectives on chemical carcinogenesis in humans and experimental animals. *Cancer Res.*, **38**, 1479–1496.
Miller, J. A. and E. C. Miller (1976). Carcinogens occurring naturally in foods. *Fed. Proc.*, **35** (6), 1316–1321.
Mims, C. A. (1981). Vertical transmission of viruses. *Microbiol. Rev.*, **45** (2), 267–286.

Ministry of Health and Welfare, Canada (1975). The testing of chemicals for carcinogenicity, mutagenicity and teratogenicity. *Health Protection Branch*, pp. 1–68.

Nichols, W. W., R. C. Miller and C. Bradt (1977). *In vitro* anaphase and metaphase preparations in mutation testing. In *Handbook of Mutagenicity Test Procedures*, ed. B. J. Kilbey *et al.*, pp. 225–233. Elsevier, Amsterdam.

Parry, J. M. (1982). The genetic toxicology trial of 4CMB, 4HMB and BC as assessed by the UKEMS genotoxicity trial 1981. *Mutation Res.*, **100**, 411–416.

Pienta, R. J., J. A. Poley and W. B. Lebherz (1977). Morphological transformation of early passage golden Syrian hamster embryo cells derived from cryo-preserved primary cultures as a reliable *in vitro* bioassay for identifying diverse carcinogens. *Int. J. Cancer.*, **19**, 642–655.

Rasmussen, R. E. and R. B. Painter (1966). Radiation-stimulated DNA synthesis in cultured mammalian cells. *J. Cell Biol.*, **29**, 11–19.

Rosenkranz, H. S., J. Hyman and Z. Leiser (1981). DNA polymerase deficient assay. In *Progress in Mutation Research*, ed. F. J. de Serres and J. Ashby, vol. 1, pp. 210–218. Elsevier/North Holland, Amsterdam.

Scott, D. (1982). UKEMS collaborative genotoxicity trial, cytogenetic tests of 4CMB, BC and 4HMB: summary and appraisal. *Mutation Res.*, **100**, 313–331.

Selby, L. A. (1978). Epidemiological investigations of poisonous plant-induced birth defects. In *Effects of Poisonous Plants on Livestock*, ed. R. F. Keeler, K. R. Van Kampen and L. F. James, pp. 441–449. Academic, New York.

Sever, J. L. (1980). Infectious causes of human reproductive loss. In *Embryonic and Fetal Death*, ed. I. H. Porter and E. B. Hook, pp. 169–175. Academic, New York.

Shepard, T. H., J. R. Miller and M. Morris (1975). Methods for detecting environmental agents that produce congenital defects. *Proceedings of the Gaudeloupe Conference sponsored by L'Institut de la Vie*. North-Holland, Amsterdam.

Sontag, J. M., N. P. Page and Saffriotti (1975). *Guidelines for Carcinogen Bioassay in Small Rodents*, pp. 1–65. National Cancer Institute, Bethesda, MD.

Swenson, D. H., J. A. Miller and E. C. Miller (1973). 2,3-Dihydro-2, 3-dihydroxyaflatoxin B1: an acid hydrolysis product of an RNA–Aflatoxin B1 adduct formed by hamster and rat liver microsomes *in vitro*. *Biochem. Biophys. Res. Commun.*, **53**, 1260–1267.

Swenson, D. H., E. C. Miller and J. A. Miller (1974). Aflatoxin B1–2,3-oxide: evidence for its formation in rat liver *in vivo* and by human liver microsomes *in vitro*. *Biochem. Biophys. Res. Commun.*, **60**, 1036–1043.

Swenson, D. H., E. C. Miller and J. A. Miller (1975). Reactivity and carcinogenicity of Aflatoxin B1–2,3-dichloride, a model for the putative 2,3-oxide metabolite of Aflatoxin B1. *Cancer Res.*, **35**, 3811–3823.

Van Kampen, K. R., R. W. Rhees and L. F. James (1978). Locoweed poisoning in the United States. In *Effects of Poisonous Plants on Livestock*. ed. R. F. Keeler, K. R. Van Kampen and L. F. James, pp. 465–471. Academic, New York.

Wilson, J. G. (1977). Environmental chemicals. In *Handbook of Teratology. 1. General Principles and Etiology*, ed. J. G. Wilson and F. C. Fraser, chap. 9, pp. 357–385. Plenum, New York.

Wogan, G. N. and W. F. Busby, Jr. (1980). Naturally occurring carcinogens. In *Toxic Constituents of Plant Foodstuffs*, ed. I. E. Liener, 2nd edn., chap. 11, pp. 329–369. Academic, New York.

Wright, H. T., Jr. (1973). Cytomegaloviruses. In *The Herpesviruses*, ed. A. S. Kaplan, chap. 12, pp. 353–388. Academic, New York.

31

Governmental Regulation of Recombinant DNA Research and Manufacturing Processes

T. O. McGARITY
University of Texas School of Law, Austin, TX, USA
and
K. O. BAYER
Law Offices of Pluymen & Bayer, Austin, TX, USA

31.1 INTRODUCTION

From their inception, recombinant DNA (rDNA) technologies have received very careful scrutiny from a concerned public.[1] A well-publicized letter from four prominent rDNA researchers to their fellow scientists pointed out some of the possible hazards of their research and suggested a moratorium on specified types of research. This suggestion precipitated an intense political debate that resulted in the promulgation in 1976 of guidelines which became binding upon federal agencies and recipients of federal grants (NIH Guidelines; National Institutes of Health, 1981a). In the intervening years, many of the early fears that rDNA research might endanger human beings have been laid to rest, and scientists have developed special fragile strains of microorganisms, such as *E. coli* Chi-1776, that ensure relatively effective 'biological containment' of rDNA molecules.

At the same time, scientists and engineers in university and corporate laboratories have been designing large-scale facilities for the production of useful products from microorganisms containing rDNA molecules. The last two years have witnessed a geometric growth in the efforts of United States' firms to commercialize the outputs of rDNA research (OTA Report; Office of

Technological Assessment, 1981). As the focus of attention accordingly shifts from laboratory research to applied research and application, it is appropriate to examine the legal mechanisms that exist in the United States and elsewhere for assessing the risks to health and safety that large-scale application of rDNA research pose to man and the environment and for reducing those risks, if necessary, to an appropriate level.

This chapter will examine several federal, state and local statutes and regulations that arguably give federal or state agencies authority to regulate large-scale industrial applications of rDNA research. The chapter will begin with a very brief discussion of the kinds of regulatory mechanisms that might be desirable in an adequate regulatory regime. It will then examine each of the relevant statutes to ascertain the extent to which those statutes contain these desirable elements. This chapter will not probe the potential risks and benefits of particular large-scale technologies; it will merely describe legal mechanisms available for reducing risks should they be discovered.

31.2   ELEMENTS OF A DESIRABLE REGULATORY REGIME

It is entirely possible that large-scale applications of rDNA research will be so risk-free that little, if any, regulatory controls will be called for. However, at the present stage of development of the technology, it is probably too early to draw this conclusion. Most certainly, we can expect one important use for rDNA technologies will be to produce biochemical compounds, such as the brain hormone somatostatin, which have previously been very rare in nature. These new chemicals may have unanticipated effects on exposed organisms, especially workers in large-scale facilities. In addition, some rDNA technologies may produce hazardous by-products that pose risks to workers or to the environment. Moreover, some uses of rDNA technologies call for broad-scale application of organisms containing rDNA to the environment.[2] Consequently, it would probably be unwise to move forward with this new technology until adequate risk assessments are performed. It would, likewise, be prudent to have regulatory mechanisms available to prevent the introduction of unacceptably dangerous technologies, to monitor the technologies that are introduced in actual operation, and to intervene to eliminate or reduce risks once they have become manifest in existing operations.

A desirable regulatory regime might therefore include some mechanism for acquiring data concerning the risks posed by technologies before they are introduced. It might require the proponent of the technology to analyze and test the technology for possible risks to society before using the technology, or it at the very least could require the proponent to give the regulatory entity access to the technology for the purpose of performing its own tests and data gathering activities. In addition, it would be desirable to provide for monitoring the technology in operation to detect any previously unanticipated risks and to observe the efficacy of any previously imposed regulatory constraints.

The regulatory entity should also possess some mechanism for assessing risks and comparing those risks to the potential benefits of the technology. Again, this function could be performed by either the proponent of the technology or by the regulatory entity, but the inherent susceptibility of this subjective function to (perhaps unintended) bias suggests that this function be performed by the regulatory entity or at least under its careful and skeptical scrutiny.[3]

Should the regulatory entity determine the risks posed by a particular technology are too great, it must have available regulatory tools to require the users of the technology to reduce or eliminate such risks. This would necessitate some standard-setting or licensing mechanism capable of requiring the users of the technology to employ it only in specified ways or only with the addition of certain prescribed risk-reduction technologies. Finally, the regulatory entity must have available to it methods for providing positive or negative incentives capable of ensuring that the prescribed regulatory controls are in fact observed.

31.3   CURRENTLY EXISTING REGULATORY REGIMES

31.3.1   The National Institutes of Health Guidlines

Following a vigorous public debate on the risks and benefits of conducting laboratory research with rDNA, the National Institutes of Health (NIH), on July 7 1976, issued its 'Guidelines for Research Involving Recombinant DNA Molecules' (National Institutes of Health, 1981a). Although they have been amended several times in the intervening years, the Guidelines remain

binding on all research conducted by federal agencies and all research sponsored by federal granting agencies.[4] These Guidelines require the establishment of Institutional Biosafety Committees,
detail the responsibilities of those Boards, categorize rDNA research, and prescribe varying
levels of biological and physical containment for various experiments.[5] Other countries have
similar regulations for rDNA research, although they have historically been less detailed and less
stringent than the United States' Guidelines.[6]

As their name implies, the NIH Guidelines are largely concerned with research involving
rDNA molecules, and not with practical applications of rDNA technologies. The Guidelines do,
however, briefly address large-scale projects involving rDNA in laboratories receiving federal
support. They prohibit large-scale experiments using more than 10 l of culture, unless it can be
shown to the satisfaction of the Director of NIH that the rDNAs used in the experiment are rigorously characterized and the absence of harmful sequences has been established (NIH Guidelines,
1981a, at § I-D-6).

This prohibition would, of course, constitute a significant impediment to industrial use of
rDNA technologies were it binding upon private entities. The NIH Guidelines are not binding,
however, on private entities that do not receive federal funding for rDNA research. The only
sanction that NIH has available to it for a violation of the Guidelines is to cut off NIH funding for
a laboratory. Therefore, the only arguably relevant requirement for large-scale projects using
rDNA technologies is not at all binding upon the corporate entities that are likely to be developing these large-scale projects. Although many corporations in the United States have voluntarily
agreed to abide by the NIH Guidelines,[7] they, not surprisingly, have declined to adhere to the
prohibition on large-scale use of rDNA organisms.

Recognizing its inability to require companies to comply with the prohibition, NIH in 1980 promulgated 'Physical Containment Recommendations for Large-Scale Uses of Organisms Containing Recombinant DNA Molecules' (45 *Fed. Regist.* 24 968). These 'recommendations' are
intended to serve merely as a guidance function. Like the NIH Guidelines, the recommendations
attempt to categorize large-scale projects according to levels of expected risk. The recommendations then prescribe three increasingly stringent levels of physical containment for large-scale
projects. In addition, the NIH Recombinant DNA Advisory Committee stands ready to consult
with companies attempting large-scale efforts and to review suggested safety precautions. Nevertheless, none of the NIH requirements are legally binding upon private companies involved in
large-scale applications of rDNA technologies, and those companies are free to ignore the Guidelines and recommendations.

31.3.2 The Occupational Safety and Health Act

The Occupational Safety and Health Act of 1970 (OSHAct) (29 U.S.C. §§ 651–678, 1976) was
intended by Congress to 'assure so far as possible every working man and woman in the [United
States] safe and healthful working conditions . . . . ' (29 U.S.C. § 651). The statute creates an
Occupational Safety and Health Administration (OSHA) within the Department of Labor to
establish and enforce occupational safety and health standards (29 U.S.C. § 553), an independent
Occupational Safety and Health Review Commission (OSHRev) to adjudicate alleged violations
of OSHA standards and of the statute's general duty clause (29 U.S.C. § 661), and a National
Institute for Occupational Safety and Health (NIOSH) in the Department of Health and Human
Services (HHS) to gather data, assess risks and recommend occupational safety and health sta-
dards to OSHA (29 U.S.C. § 671). The Act thus establishes a comprehensive scheme for protecting workers from workplace hazards.

The OSHAct gives OSHA and NIOSH extensive authority to acquire and analyze data on
possible diseases attributable to rDNA technologies. Section 20 of the OSHAct (29 U.S.C. § 669)
empowers NIOSH to gather information on workplace safety and 'conduct (directly or by grants
or contracts) research, experiments, and demonstrations relating to occupational safety and
health . . . .' (29 U.S.C. § 669). Of particular importance to rDNA activities is the command in
Section 20(a)(4) that NIOSH 'conduct special research, experiments and demonstrations relating
to occupational safety and health as are necessary to explore new problems, including those
created by new technology in occupational safety and health, which may require ameliorative
action beyond that which is otherwise provided for in the operating provisions of this Act.' (29
U.S.C. § 669(a) (4)). Pursuant to this authority, NIOSH is currently conducting a study of risks to
workers in the fermentation industry with the ultimate goal of performing a general assessment of
those risks (McKinney, 1981).

Under Section 8(a) of the OSHAct, both NIOSH and OSHA have authority to enter workplaces at reasonable hours to investigate in a reasonable manner conditions in those workplaces (29 U.S.C. § 657).[8] In addition, Section 8(c)(1) of the OSHAct requires employers to 'make, keep and preserve' such records as the Secretary of Labor (in cooperation with the Secretary of Health and Human Services) prescribes by regulation as 'necessary and appropriate for the enforcement of [the OSHAct] or developing information regarding the causes and preventing of occupational accidents and illnesses' (29 U.S.C. § 657). Such regulations may 'include provisions requiring employers to conduct periodic inspections' (29 U.S.C. § 657). These provisions would seem to provide adequate authority for OSHA to require employers to monitor periodically the workplace environment around potential sources of rDNA-containing organisms (such as fermenters) for microorganisms that might escape from physical containment. Arguably, the section also gives OSHA authority to require employers to convey to the agency the identities of host, vectors, genetic sequences, products and by-products of particular rDNA technologies. The crucial question would be whether such reporting requirements would be 'necessary and appropriate' to 'developing information regarding the cause and preventions of occupational accidents and illnesses'.

Section 8(c)(2) of the OSHAct requires the Secretary of Labor (in cooperation with the Secretary of HHS) to prescribe regulations 'requiring employers to maintain accurate records of, and to make periodic reports on, work related deaths, injuries and illnesses [other than minor injuries]' (29 U.S.C. § 657). In addition, Section 8(c)(3) makes it mandatory for OSHA to issue regulations requiring employers to maintain accurate records of employee exposures to potentially toxic materials or harmful physical agents which are required to be monitored under standards issued pursuant to OSHA's standard-setting authority (29 U.S.C. § 657). These provisions would appear to give OSHA the authority, if it so desired, to establish a registry of worker diseases that result in injuries that are not minor. Whether or not OSHA has authority to require reporting of worker exposures to microorganisms containing rDNA in the absence of clearly demonstrated serious injury depends upon whether OSHA has prescribed a standard under Section 6.

Section 6(b) (1) gives OSHA authority to promulgate occupational safety and health standards 'in order to serve the objectives of this Act . . . .' (29 U.S.C. § 655).[9] The term 'occupational safety and health standard' is defined in § 3(8) to mean 'a standard which requires conditions, or the adoption of one or more practices, means, methods, operations, or processes, reasonably necessary and appropriate to provide safe or healthful employment and places of employment' (29 U.S.C. § 652). Section 6(b)(5) further elaborates on occupational safety and health standards for 'toxic materials or harmful physical agents'. For those substances, OSHA must set the standard 'which most adequately assures, to the extent feasible, on the basis of the best available evidence that no employee will suffer material impairment of health or functional capacity even if such employee has regular exposure to the hazard dealt with by such standard for the period of his working life' (29 U.S.C. § 655).

The Supreme Court of the United States has recently wrestled with the meaning of these three interrelated phrases in cases involving OSHA's attempts to set standards for workplace exposure to benzene and cotton dust. In the *Benzene* case, *Industrial Union, AFL-CIO v. American Petroleum Institute*, 100 S. Ct. 2844 (1980), a plurality of the Supreme Court held that OSHA may not promulgate a standard requiring reduced exposure to a toxic substance or harmful physical agent unless such reduced exposure is 'reasonably necessary and appropriate to remedy a significant risk of material health impairment' (*Id.* at 2863). OSHA was not at liberty merely to assume that a substance which causes leukemia in workers at 100 p.p.m. poses a significant risk of workers contracting leukemia at current exposure levels of 10 p.p.m. or less. However, in the *Cotton Dust* case, *American Textile Manufacturers Institute v. Donovan*, 101 S. Ct. 2478 (1981), the Court held that OSHA was not required to weigh a standard's costs against its benefits prior to imposing it upon employers. So long as a standard is necessary to reduce or eliminate a significant risk, OSHA may require employers to implement all feasible control devices, even over the employers' objections that the costs of implementing such controls far exceed the projected benefits.

Although one might plausibly argue that microorganisms containing rDNA are not 'toxic materials or harmful physical agents', it would seem that Congress intended those terms to be sufficiently inclusive to reach rDNA-containing organisms which pose a 'significant risk' to workers. Even if genetically-engineered industrial microorganisms do not fit this definition, OSHA could still promulgate standards under § 6(b) (2) (29 U.S.C. § 655) so long as such a standard was 'reasonably necessary to provide safe or healthful employment or places of employment' (29 U.S.C. § 652).

Section 6(b) (5) thus appears to provide adequate authority for setting standards for physical containment. OSHA might specify an ambient exposure level of the relevant microorganism, products and by-products; it might specify requirements for containment vessels; or, in the extreme, it might specify standards for personal protective gear.[10] The authority in Section 6 to prescribe standards for biological containment is more speculative. A requirement that the host microorganism should not be capable of surviving in the natural environment could be called a standard 'dealing with' toxic materials or harmful physical agents. However, if the industrial process could not be carried out with host microorganisms capable of such biological containment, a biological containment requirement would presumably not be feasible and therefore beyond OSHA's authority.

Any occupational health standard promulgated under Section 6 must 'prescribe the use of labels or other appropriate forms of warning as are necessary to insure that employees are apprised of all hazards to which they are exposed, relevant symptoms and appropriate emergency treatment, and proper conditions and precautions of safe use or exposure' (29 U.S.C. § 655). This section would allow OSHA to require manufacturers to educate employees about the risks of substances for which the agency prescribes Section 6 standards. In addition, OSHA has proposed pursuant to Section 6(b) (7) of the OSHAct general labeling regulations that require employers to inform employees of risks of hazardous materials in the workplace [47 *Fed. Regist.* 187 (1982) (advance notice of proposed rule)]. Finally, a Section 6 occupational health standard must, where appropriate, 'provide for monitoring or measuring employee exposure at such locations and intervals, and in such manner as may be necessary for the protection of employees'. Moreover, such standards may prescribe, where appropriate, the type and frequency of medical examinations or other tests which shall be made available by the employer (or at his cost) to exposed employees. This section gives adequate authority to require monitoring of employee exposure and health, but only after the Secretary has determined that sufficient hazard exists to warrant the promulgation of an occupational health standard. Such monitoring, however, should be valuable in assessing the efficacy of the standard in operation.

Employers who violate OSHA standards may be issued citations which carry civil penalties of up to $1000 per violation. Willful or repeated violations can result in civil penalties of up to $10 000 per violation, and the willful violation of any Section 6 standard can also result in a felony conviction punishable by a fine of not more than $10 000, imprisonment for up to six months, or both.

31.3.3 The Federal Food, Drug and Cosmetic Act, Fair Packaging and Labeling Act and Public Health Service Act

Although perhaps not as clearly defined as that of OSHA and NIOSH, the authority to regulate rDNA products and technology may be shared with the Food and Drug Administration (FDA) under such statutes as the Federal Food, Drug and Cosmetic Act (FFDCA) (21 U.S.C. § 301, *et seq.*), Fair Packaging and Labeling Act (FPLA) (15 U.S.C. § 1415 *et seq.*), and the Public Health Service Act (PHSA 42 U.S.C. § 262). Though none of these statutes expressly covers emerging rDNA technologies[11], through each the FDA has a broad grant of authority and discretion to protect the public health by regulating the manufacturing process and resulting product, public representations made about that product and potential exposure to a product that may threaten public welfare.[12]

The FDA is divided into bureaus which take primary responsibility for particular products, for example, drugs, biologics, foods, medical devices and veterinary medicines. These divisions have authority under the FFDCA to regulate areas such as antibodies and insulin production (21 U.S.C. §§ 356, 357), pesticide tolerance levels (21 U.S.C. § 346a), drug production and marketing (21 U.S.C. § 360) and cosmetics (21 U.S.C. §§ 361–363). The Biologic Division, renamed the National Center for Drugs and Biologics in 1982, regulates products such as viruses, therapeutic sera, toxins and antitoxins, blood and blood products such as plasma, and other related products under the authority of the PHSA as well as the FFDCA. That division will become increasingly important as a regulator as recombinant processes are used to duplicate biologic drugs.[13]

With this outline of the organization of FDA as background, the remainder of this section will focus primarily on FDA regulation (potential and actual) of the microbiological production of pharmaceuticals. Although this approach will leave some questions unique to food and cosmetics unanswered, as a practical matter the immediately important regulatory issue will be discussed in the context of products that will likely be the first commercially significant products of the new technology. These products include hormones, vaccines, interferons and antibody-based diagnostic tests.

Under the FFDCA, before a new drug for human consumption is allowed on the market, experts in the field must find it 'generally recognized as safe and effective' (GRASE) for uses recommended by its labeling (21 U.S.C. §§ 321, 351). Otherwise, the new product must be proven safe and effective through a premarketing approval process (21 U.S.C. § 355) which follows a three-step test: (1) whether the product is a 'drug' within the meaning of the FFDCA; (2) if the product is a drug, whether it is a 'new drug'; and (3) if deemed new, whether the product qualifies for an exemption.[14] If the product is a non-exempt new drug, a New Drug Application (NDA) and premarket approval are required.

The submission of an NDA is the key to premarket approval. The NDA must include: clinical tests that demonstrate the safety and efficacy of the new drug; a list of all active and inert ingredients used in manufacturing the new drug; a full description of the manufacturing process, including packaging and labeling; and if requested, samples of the product (21 C.F.R. § 314.1). See also Chapter 34 of this volume. For a drug to be considered 'safe and effective', the NDA must also contain evidence of clinical tests on human subjects. Before tests on humans may begin, however, manufacturing applicants must file a Notice of Claimed Investigational Exemption (IND) which serves as a temporary approval pending human experimentation. The IND must also contain evidence of thorough and satisfactory animal testing (21 C.F.R. § 312.1). Known as a 'full NDA' procedure, the compliance process is estimated to require 5–17 years and cost millions of dollars (Korwek and Trinker, 1981).

In some cases, the product, while technically a 'new drug', might be a generic duplicate, or it may contain an active ingredient identical to an approved NDA drug. To prevent unnecessary, corroborative and duplicative testing, FDA regulations provide for Abbreviated New Drug Applications (ANDAs). ANDAs for so-called 'me-too' drugs may gain approval with reliance on information already on file with FDA (21 C.F.R. § 314.1(f)). An ANDA may only be used, however, to rely on information approved prior to 1962.[15] The ANDA must contain sufficient scientific evidence to demonstrate that the drug performs in the same manner as the pre-1962 drug on which it relies, and that the producer is complying with FDA's Good Manufacturing Practice Regulations (GMPs) for drugs (21 C.F.R. § 314.2(f)(1)).

It is also possible for a manufacturer to petition FDA to change or modify the conditions approved for that product in an existing NDA by submitting a Supplemental New Drug Application (SNDA) that explains the proposed modification (21 C.F.R. § 314.8). FDA then reviews the request, and determines whether the modifications include changes which 'may alter conditions of use, labeling, the safety, effectiveness, identity, strength, quality, purity or adequacy of the manufacturing methods, facilities or controls to preserve them' (21 C.F.R. § 314.8(a)(2) and (c)). 'Immaterial' changes are merely changes in personnel, not affecting facilities or production, or changes in only the inert ingredients of the drug (21 C.F.R. § 314.8(a)(5)). Manufacturers switching from traditional methods of production to the use of rDNA technology may attempt to use the SNDA process to take advantage of that streamlined and significantly less-costly application (Korwek and Trinker, 1981). FDA, however, must still determine that such a transition is 'immaterial' and given the current controversy and interest surrounding all rDNA products, such determination seems unlikely.

For those 'me-too' applicants relying on drugs approved after 1962, FDA began an unofficial 'Paper NDA' policy which substantially tracks the ANDA procedures. It is possible that FDA will soon incorporate these two application procedures, thus doing away with the artificially created chronological division (46 *Fed. Regist.* 27 396 (1981)).

As mentioned above, new products which are 'generally recognized as safe and effective' (GRASE) are categorized as 'old drugs' and, therefore, do not require premarket review. FDA recognized that products of the developing and still controversial recombinant technologies would not meet the GRASE criteria, and in 1981 began shaping a policy which would consider all rDNA drugs 'new drugs' (Miller, 1981). Apparently, that policy will extend to those manufacturers, with approved NDAs, which are switching to recombinant technologies and to generic 'me-too' recombinant versions of existing drugs. This excludes the possibility of an abbreviated application such as an SNDA or a 'Paper NDA' and would require each new recombinant product to undergo complete testing[16] (Korwek and Trinker, 1981). An NDA requirement will also provide some agency review of the rDNA production process, a guard against the popular fear that unmonitored recombinant techniques could result in unpredicted and uncontrollable mutant strains.

FDA's delay in developing a formal policy for dealing with recombinant technologies is due in large part to questions raised regarding the scope of the Agency's authority and jurisdiction. At issue is whether FDA's powers to protect are limited to the drug consumer, or extend to considerations of the environment, workers or the non-consuming public. Because its authority may extend further than the drug-consuming public, FDA is considering a proposal to amend the FFDCA regulations to include the NIH Guidelines as part of the 'Good Manufacturing Practices' (GMP), which must be demonstrated by 'new drug' applicants (21 U.S.C. § 351; 43 *Fed. Regist.* 60 134 (1978); Korwek, 1981).

FDA is basing this proposed extension of the Guidelines to the presently unregulated private firms primarily on the adulteration provisions of the FFDCA. Section 501 of the FFDCA makes the marketing of an 'adulterated' drug (21 U.S.C. § 331) unlawful. Under Section 501(a)(2)(B), a drug may be adulterated if, among other things, it was not produced in a facility which can demonstrate conformity with 'GMPs'. The GMP regulations set out requirements for virtually every segment of the manufacturing process (21 C.F.R. § 211 *et seq.* (drugs) and §§ 600–680 (biologics)) and have withstood several legal challenges.[17] Incorporating existing NIH guidelines into the GMP regulations would require all rDNA testing laboratories and plants, regardless of whether they receive NIH funds, to comply with the only rDNA production technique standards developed to date.

Because compliance with NIH Guidelines is costly for private firms mass-producing rDNA products, a legal challenge to the incorporation of the Guidelines into the GMPs seems inevitable. Opponents argue that it is beyond FDA's statutory authority to incorporate the Guidelines, given the 'different primary objectives of the GMP regulations, the FFDCA and the Guidelines' (Korwek, 1981; Tooze, 1981). The purpose of the FFDCA and GMP regulations is to assure the quality, purity and safety of products to which the drug-consuming public is exposed. The NIH Guidelines, on the other hand, are designed to specifically monitor rDNA experiments and mandate safety precautions and containment levels (Korwek, 1981; Rosenblatt, 1982). Although there is some overlap between the Guidelines and the GMP regulations, and compliance with either may produce similar results, the extent of FDA's authority to incorporate the Guidelines into GMPs remains unclear.

FDA could find support for extending its regulatory efforts beyond its protection of the drug-consuming public in the National Environmental Policy Act (NEPA, 42 U.S.C. § 4321 *et seq.*). NEPA admonishes all federal agencies to consider the environmental effects of significant government action and requires the submission of an Environmental Impact Statement (EIS) to discuss the effect of such proposed action on the environment (42 U.S.C. § 4332).

FDA responded to its duties under NEPA by including the EIS requirement in FDA regulations (21 C.F.R. Part 25). These regulations temper the burdens of the application process by simultaneously exempting large categories of regulated products, such as some food and drug additives (21 C.F.R. § 25.1(f)). Exempted drugs and additives thus are assumed to have no significant adverse effect on the quality of the human environment.

To encourage private industry compliance with NIH Guidelines, FDA suggested a policy to allow compliance with the Guidelines to obviate the need for an EIS (43 *Fed. Regist.* 60 134; Korwek, 1981). Such a policy is vulnerable to attack, however, since it is unclear whether FDA could impose any restrictions or criteria on the content of the EIS in the absence of a statutory environmental duty on the applicant. Courts generally will not require more from an agency or applicant than a mere demonstration that environmental effects have been considered and will not require the conclusion that there can be no adverse environmental effects from the proposed action.[18]

Another potential source of FDA authority to regulate the products of rDNA technology is the Public Health Service Act (PHSA) (42 U.S.C. § 262 *et seq.*). Under the PHSA, the Department of Health and Human Services, with FDA, is authorized to pass such regulations as are necessary to prevent the possible spread of interstate or foreign communicable diseases. The FDA uses its authority under the PHSA to regulate the sale of shellfish and small turtles, prohibit travel of individuals believed to have been exposed to infectious diseases, and to quarantine areas of suspected infection.[19] Opponents argue, however, that the nature of the provision, which focuses on the quarantine of individuals, seems to limit FDA to that type of restriction. Moreover, say the opponents, the PHSA itself was only designed to prevent human exposure to communicable disease and should not be extended to cover danger to animals or plants (Korwek, 1981).

Despite the lack of an official FDA position and the heated theoretical debate about what FDA authority might be, FDA is regulating rDNA technology. FDA's own assessment (Miller, 1981) of the pharmaceutical products produced using rDNA techniques indicates that among those falling under the regulatory purview of the various Bureaus are: (1) *drugs*—human insulin, human growth hormone (or analogs), thymosin, ACTH, endorphins, DNA; (2) *Biologics*—interferons, vaccines (including hepatitis B and influenza), serum albumin, urokinase, DNA; (3) *Foods*—certain enzymes used in food processing; and (4) *medical devices—in vitro* diagnostic tests (*e.g.* glucose oxidase); specific *in utero* diagnostic tests (*e.g.* thalassemia, sickle-cell anemia).

Anticipating INDs and NDAs for substances produced using rDNA technology, the FDA added to its staff several M.D.s and Ph.D.s with recent laboratory experience in both the science and technology of rDNA (Miller, 1981). With this new expertise, combined with their long experience in premarket approval procedures and the GMP regulations, FDA believes it can accommodate the new manufacturing technology. However, with different statutory requirements for different product classes (drugs, food additives, biologics and devices), FDA intends to regulate on a product-by-product basis through its Bureaus.

Recently, the FDA Commissioner established the rDNA Coordinating Committee within FDA (Miller, 1981). The Committee is chaired by the Deputy Associate Commissioner for Health Affairs and Science and has representatives from the Bureaus, the Office of Regulatory Affairs, Executive Director for Regional Operations (EDRO) and General Council. The Committee will serve as the focus of rDNA policy issues, coordinating, advising and interacting with FDA's various organizational units.

Enforcement of FDA regulations regarding rDNA research and manufacturing would be achieved by the same methods that FDA enforces all its other regulations (21 U.S.C. §§ 332–337). Within the United States, compliance with the FFDCA is secured through periodic inspections of facilities and products, analysis of samples, educational activities, as well as administrative and judicial proceedings. Adulterated or misbranded products may be voluntarily destroyed or recalled from the market by the shipper, or may be seized by U. S. marshals on orders obtained by the FDA from Federal district courts. Persons or firms responsible for illegal products may be fined and/or imprisoned. Continued violations may be prohibited by Federal court injunctions, punishable by contempt of court.

Product recalls have become a major means of consumer protection under the law. Recall of a product from the market, by the manufacturer, is generally the fastest and most effective way to protect the public. A recall may be initiated by the manufacturer or shipper of the product, or requested by FDA. While cooperation in a recall may make court proceedings unnecessary to remove the product from the market, it does not relieve a person or firm from liability for violations. An FDA statement of Enforcement of Policy published in the Federal Register of June 16, 1978, provides guidelines on recall procedure and industry responsibilities.

Seizure is a civil action against a specific lot of goods to remove them from the channels of commerce. After seizure, the goods may not be tampered with, except by permission of the court. The owner or claimant of the seized merchandise is usually given about 30 days by the court to decide on his course of action. If the owner does nothing, the goods will be disposed of by the court. The owner may decide to contest the Government's charges and the case will be scheduled for trial; or the owner may request permission of the court to bring the goods into compliance with the law. The owner of the goods is required to provide a bond to assure that the orders of the court will be carried out, and must pay for FDA supervision of any compliance procedure.

31.3.4 The Toxic Substances Control Act

The Toxic Substances Control Act (TSCA) (15 U.S.C. § 2601 *et seq.* (Supp. 1981)) was enacted in 1976 with the purpose of providing a comprehensive mechanism for gathering data on the health and environmental effects of chemical substances, assessing their risks, and ensuring that manufacture, distribution, use and disposal of toxic substances do not pose unreasonable risks to man and the environment. The U. S. Environmental Protection Agency (EPA) is charged with administering and enforcing the statute (15 U.S.C. § 2602(1)). To date, EPA's implementation efforts have been slow and halting, and recent budgetary cuts will ensure that TSCA remains a relatively ineffective regulatory tool. Nevertheless, TSCA does give EPA broad authority to regulate toxic substances should EPA ever decide to exercise that authority.

Under Section 5 of TSCA (15 U.S.C. § 2604), the manufacturer of a new 'chemical substance' must submit to EPA a notice of its intention to manufacture or process such substance.[20] The notice must include, insofar as is reasonably ascertainable: (1) name, chemical identity and chemical structure; (2) categories or proposed categories of use; (3) estimates of total amounts to be manufactured, processed or used; (4) description of by-products; (5) existing health and safety data; (6) estimates of the number of people who will be exposed to the substance; (7) manner of disposal (15 U.S.C. § 2604(d)).

Section 5 would appear to give EPA sufficient authority to require companies to notify the agency of any new chemical products or by-products that result from fermentation or other large-scale use of industrial microorganisms containing rDNA. It would not necessarily give EPA authority to require companies to disclose to EPA solvents or other existing chemicals used in the fermentation extraction process.[21] Section 5 does not give EPA authority to require that companies give it premanufacture notification of hosts and vectors, because those would generally not come within the definitions of 'chemical substance' or 'mixture'.[22] On the other hand, Section 5 may give EPA authority to require that companies notify it of the identity of the genetically manipulated DNA that is within the host cell, because the DNA may be a 'chemical substance', or a 'mixture'.

The term 'chemical substance' is defined in Section 3(2) broadly to include 'any organic or inorganic substance of a particular molecular identity, including . . . any combination of such substances occurring in whole or in part as a result of a chemical reaction occurring in nature, and any uncombined radical' (15 U.S.C. § 2602(2)).[23] The word 'mixture' is also defined broadly to include 'any combination of two or more chemical substances if the combination does not occur in nature and is not, in whole or in part, the result of a chemical reaction (15 U.S.C. § 2602(8)). The rDNA within a genetically-engineered microorganism would seem to fit the definition of 'chemical substance'. It has a particular molecular identity, although it is not always ascertainable. It is the presence of that substance within the microorganism that allows the 'microbial factory' to be industrially useful. It might be argued that the combination of genes does not 'occur in nature'. Yet if this is the case, then the DNA would appear to come within the definition of 'mixture'. Clearly, this difficult question is ripe for litigation, should EPA decide to require premanufacture notification for the DNA of genetically-engineered industrial microorganisms. If the courts refuse to find that the rDNA within a microorganism is a 'chemical substance', then TSCA will be largely unavailable to regulate the large-scale use of rDNA technologies.

Even if rDNA is a 'chemical substance', Section 5(h)(4) allows the EPA to exempt the manufacture of any new chemical substance from all or part of Section 5's requirements if EPA determines that the manufacture, processing, distribution or use of the chemical will not present an unreasonable risk of injury to health or the environment' (15 U.S.C. § 2604(h)(4)).[24] In addition, Section 5(h)(5) allows the EPA to exempt from Section 5's notification and test data submission requirements 'any chemical substance (A) which exists temporarily as a result of a chemical reaction in the manufacturing or processing of another chemical substance and (B) to which there is not, and will not be, human or environmental exposure'. The DNA within organisms used in the fermentation process, for example, would appear to come within the spirit of Section 5(h)(5), if the organisms are destroyed when the fermentation is complete. In any event, exemptions may be granted under Section 5(h)(4) for fermentation processes and perhaps unenclosed processes that are determined to pose no unreasonable risks. This may be one way to deal with the expected flood of premanufacture notification filings that would otherwise result as organisms are 'fine tuned' over time to achieve maximum output.

Section 8(a) of TSCA gives EPA authority to require companies (other than small manufacturers)[25] to 'maintain such records and . . . submit . . . such reports as [EPA] may reasonably require' (15 U.S.C. § 2607(a)). Such records and reports may include, insofar as is known to the company or insofar as is reasonably ascertainable: (1) the common or trade name, the chemical identity and the molecular structure of each chemical substance or mixture for which such a report is required; (2) the categories or proposed categories of use of each such substance or mixture; (3) the total amount of each such substance and mixture manufactured or processed, reasonable estimates of the total amount manufactured or processed, the amount manufactured or processed for each of its categories of use, and reasonable estimates of the amount to be manufactured or processed for each of its categories of use or proposed categories of use; (4) a description of the by-products resulting from the manufacture, processing, use or disposal of each such substance or mixture; (5) all existing data concerning the environmental and health effects of such substance or mixture; (6) the number of individuals exposed and reasonable estimates of the number who will be exposed to such substance or mixture in their places of employment and the duration of such exposure; and (7) in the initial report under paragraph (1) on such substance or mixture, the manner or method of its disposal, and in any subsequent report on such substance or mixture, any change in such manner or method (15 U.S.C. § 2607(a)(2)).

This would appear to give EPA adequate authority to require firms to make EPA aware of potential systematic employee and environmental exposure to rDNA and its products and by-products in the fermentation industry and in other large scale applications. In addition to EPA's recordkeeping requirements, TSCA Section 8(a)(2)(E) gives EPA authority to require manufacturers to submit 'all existing data concerning the environmental and health effects of an existing chemical substance', insofar as they are known or reasonably ascertainable (15 U.S.C., § 2607(a)(2)(E)). This requirement could provide EPA with existing studies on the survivability of host organisms in human and other environmental systems. In addition, EPA could require that companies make available to EPA existing studies characterizing genetically-engineered organisms and any existing risk assessment performed on such organisms that are in their files.

Further, Section 8(c) of TSCA independently requires the manufacturer, processor, and distributor of a chemical substance to 'maintain records of significant adverse reactions to health or the environment alleged to have been caused by the substance' (15 U.S.C. § 2607(c)), and Section 8(e) requires manufacturers, processors and distributors who obtain information 'which reasonably supports the conclusion that such substance and mixture presents a substantial risk of injury to health or the environment' to inform the EPA immediately of such information (15 U.S.C. § 2607(e)). These sections would appear to provide sufficient authority to require the reporting of diseases caused by exposure to products and by-products of an rDNA technology. Whether they give EPA authority to require reporting of diseases caused by genetically-engineered pathogenic organisms depends upon whether it can be said that a particular disease was 'caused by' the DNA within the organisms (assuming that the DNA is a 'chemical substance' or 'mixture'). Similarly, whether the 'substantial risk' notice requirement applies to pathogenic organisms depends upon whether it can be said that the DNA (the 'chemical substance' or 'mixture') presents a substantial risk of injury.

Section 11 of TSCA (15 U.S.C. § 2610) allows a duly designated representative of EPA to inspect any establishment in which chemical substances or mixtures are manufactured or processed. This would appear to grant EPA sufficient authority to require employers to allow EPA employees or designated representatives to monitor facilities for human and environmental exposure to microorganisms and for possible diseases caused by such organisms. It would not appear to grant authority for EPA to require firms to self-monitor. However, Section 6(a)(4) of TSCA (15 U.S.C. § 2605 (a)(4)) provides that as part of a Section 6 rule, EPA may require manufacturers and processors of a chemical substance to 'monitor or conduct tests which are reasonable and necessary to assure compliance with the rule'. But before the agency may promulgate a Section 6 rule, it must first make the threshold determination that the manufacture or processing of the chemical substance 'presents or will present an unreasonable risk of injury to health or the environment' (15 U.S.C. § 2605(a)). Hence, Section 6(a)(4) will not be useful as a mechanism for acquiring information for initial assessments of risks posed by rDNA technologies.

EPA's extensive data gathering authority is supplemented further by Section 4 of TSCA, which allows EPA to require that testing be conducted, in accordance with specified standards, on a substance or mixture to develop data with respect to its health and environmental effects for which there is currently an insufficiency of data to determine whether or not the manufacture or processing of the substance will present an unreasonable risk. Before testing may be ordered, the agency must first determine that: (1) the manufacture, processing, distribution, use or disposal of the substance may present an unreasonable risk of injury to health or the environment or the substance will be produced in substantial quantities and will enter the environment in substantial quantities or result in substantial human exposure, *and* (2) there are insufficient data and experience to predict the effects of the substance on health or the environment, *and* (3) testing is necessary to develop adequate data (15 U.S.C. § 2603(a)).[26]

These determinations will be very difficult to make in the abstract. It is possible to imagine that some forms of genetically altered DNA in some kinds of microorganisms could produce substances that are harmful to humans and the environment. Existing risk assessments on a few industrially useful microorganisms, however, indicate that human or environmental exposure to such organisms is likely to pose negligible risks no matter how they are genetically altered. Hence, EPA could perhaps justifiably decline to require testing for whole strains of genetically-altered bacteria once the safety of the bacterium itself has been established (*e.g.* E. coli K_{12} and X_{1776}). On the other hand, EPA might require testing on the effects of DNA in microorganisms which are not yet fully characterized and determined to be adequately safe.

Even if EPA cannot determine that genetically-altered DNA within a microorganism may pose unreasonable risks, it can require testing if the substance will be produced in substantial quantities and will enter the environment in substantial quantities or result in substantial human exposure. Enclosed fermentation processes probably do not cross this threshold. However, large scale industrial use of genetically altered microorganisms in the environment (*e.g.* for leaching minerals or digesting oil spills) could easily meet this threshold requirement.

After gathering data, requiring testing, and assessing the potential risks of a chemical substance, EPA may take a variety of regulatory actions. If EPA determines that there is a 'reasonable basis to conclude' that the manufacture, processing, distribution, use or disposal of a chemical substance will present an 'unreasonable risk of injury to health or the environment', it may apply one or more of the following requirements (using the least burdensome): (1) a requirement *prohibiting* the manufacture, processing and distribution of the substance entirely or for a particular use; (2) a requirement *limiting* the amount of the substance which may be manufactured, processed and distributed; (3) a *labeling* or *warning* requirement; (4) a *recordkeeping* requirement; (5) a *monitoring* requirement; (6) a requirement *prohibiting* or otherwise *regulating* any manner or method of *commercial use* of the substance; (7) a requirement *prohibiting* or otherwise *regulating* any manner or method of disposal of the substance; (8) a requirement directing manufacturers or processors to give *notice* to the public and to distributors of such unreasonable risks and to *replace* or *repurchase* such substances as elected by the recipient of the notice (15 U.S.C. §§ 2605(a)(1)–(7) (emphases added)).

In addition, if EPA has a 'reasonable basis to conclude' that a particular manufacturer or processor is manufacturing or processing a chemical substance or mixture in a manner which unintentionally causes it to present an unreasonable risk, EPA may require the manufacturer or processor to submit a description of its quality control procedures (15 U.S.C. § 2605(b)(1)). If EPA determines that the quality control procedures are inadequate to prevent the substance from posing an unreasonable risk, it may order the manufacturer to revise such quality control procedures to the extent necessary to remedy such inadequacy (15 U.S.C. § 2605(b)(2)).

Assuming that the 'unreasonable risk' threshold is met and that the DNA in microorganisms is a 'chemical substance' or 'mixture', there is clearly sufficient authority in Section 6 for EPA to require physical containment of those organisms. Moreover, Section 6(a)(5) may provide adequate authority to require biological containment. Specifying appropriate hosts is arguably regulating the commercial use of the DNA. In addition, EPA may have authority to require biological containment under its Section 6(b) authority to revise quality control procedures. However, Section 6(b) would presumably require a showing that a less risky host could produce the same end-product. Finally, the labeling or warning provisions of Sections 6(a)(3) and 6(a)(7) provide adequate authority to educate and inform workers and others who might be exposed to genetically engineered microorganisms or their products or by-products.[27]

Violations of rules promulgated under Sections 4, 5 and 6 of TSCA can result in civil penalties of up to $25 000 per violation (15 U.S.C. § 2615(a)(1)). Such penalties may be assessed administratively with judicial review in a court of appeals (15 U.S.C. § 2615(a)(2)). Knowing or willful violations can result in an additional fine of $25 000 per violation, or imprisonment for one year or both (15 U.S.C. § 2615 (b)).

31.3.5   The Clean Air Act, Clean Water Act and Other Miscellaneous Federal Acts

31.3.5.1   The Clean Air Act

If the industrial process using rDNA technologies results in the release of 'criteria' pollutants or currently listed hazardous air pollutants, the firm must comply with Parts A, C and D of Title I of the Clean Air Act (CAA) (42 U.S.C. §§ 7401–7508, 7521–7574, 7601–7626 (Supp. 1981)).[28] It is, however, unlikely that the normal operation of a fermentation plant or large-scale release process will result in emissions of these pollutants, unless the firm decides to dry liquid wastes and incinerate them (*cf.* USEPA, 1976). Since it is unlikely that organisms containing rDNA molecules will be released from 'numerous or diverse mobile or stationary sources', they will probably never be regulated as criteria pollutants.[29] Therefore, if such organisms are regulated at all under the CAA, they will probably be regulated as hazardous pollutants under Section 112 of the Act (42 U.S.C. § 7412).

Section 112 of the CAA gives EPA authority to promulgate emission standards for 'hazardous air pollutants'.[30] 'Hazardous air pollutants' are those non-criteria pollutants that EPA determines 'may reasonably be anticipated to result in an increase in mortality or an increase in serious irreversible or incapacitating reversible illness' (42 U.S.C. § 7412(a)(1)). Before EPA can regulate rDNA-containing microorganisms under this section, it must first determine that a microorganism or its products or by-products crosses this threshold. Although the burden on EPA is not a high one, it still requres the EPA to come forward with some evidence of risk. The procedures for promulgating a hazardous air pollutant standard are, however, tedious and time consuming, and EPA is reluctant to employ them. It is, therefore, more likely that EPA will address potentially hazardous genetically engineered microorganisms under TSCA.

If EPA discovered that a firm was using a microorganism that could be characterized as a hazardous pollutant, it could probably require adequate physical containment through an ambient air standard or a design equipment or operational standard because such standards must provide for an 'ample margin of safety' (42 U.S.C. § 7412(b)(1)(B)). Although it may be unlikely that an organism which escaped a fermentation vessel would travel very far in the air, the term 'air pollutant' includes 'any air pollution agent', including any 'biological . . . substance or matter which is emitted into or otherwise enters the ambient air' (42 U.S.C. § 7602(g)). Since airborne exposure to hazardous microorganisms would probably affect workers in the first instance, the OSHAct may be the more appropriate vehicle for this regulation.[31]

31.3.5.2 *The Clean Water Act*

Under Sections 301, 304 and 306 of the Clean Water Act, EPA may promulgate technology-based effluent limitations and guidelines for categories and subcategories of new and existing dischargers of conventional pollutants into the navigable waters of the United States (33 U.S.C. §§ 1311, 1314, 1316 (Supp. 1981)).[32] 'Conventional pollutants' include biological oxygen demand (BOD), chemical oxygen demand (COD), pH, fecal coliform and suspended solids. Effluent limitations and guidelines for various categories and subcategories of industries specify 'best practical control technology' for these conventional pollutants and future limitations and guidelines will specify 'best conventional control technology' (33 U.S.C. §§ 1311(b), 1314(b)). If a source does not belong to a category for which effluent limitations and guidelines have been promulgated, it must still obtain a National Pollution Discharge Elimination System (NPDES) permit under Section 402 (33 U.S.C. § 1342). 'Best conventional control technology' will be presented on a case-by-case basis during the permitting process.

EPA has promulgated effluent limitations and new source performance standards for the pharmaceutical industry which specify the amounts of conventional pollutants (BOD, COD and pH) that can be emitted into the navigable waters by new and existing fermentation processes. These limitations would apply to new sources and modified existing sources employing genetically-engineered microorganisms to produce pharmaceutical products. They do not apply to firms that utilize genetically-engineered microorganisms to produce other products, although EPA could promulgate limitations for such processes in the future. However, any source that discharges any pollutants into the navigable waters must have a permit under Section 402 of the Clean Water Act, and state (or in some cases federal) permitting authorities must prescribe technology based effluent limitations on a source-by-source basis for all pollutants discharged into the navigable waters.

While it is quite likely that industrial use of rDNA technologies will produce conventional water pollutants (especially BOD), the regulation of these pollutants is not particularly directed at the toxicity of the organisms. Under Section 301(b)(2), however, EPA must prescribe technology-based effluent limitations and guidelines for designated toxic substances reflecting the 'best available technology economically achievable' (33 U.S.C. § 1311 (b)(2)). 'Best available technology' can include zero discharge. At present, the list of designated chemicals does not contain any organisms containing rDNA molecules, but EPA may add to that list if it so desires. In determining whether to list a substance, EPA must take into account the toxicity of the pollutant, its persistence and its degradability (33 U.S.C. § 1317(a)(1)).

Section 307 also empowers EPA to promulgate toxic effluent standards, which are more stringent than toxic effluent limitations (33 U.S.C. § 1317(a) (2)). Toxic effluent standards are water quality-based and must be set at a level that provides an ample margin of safety.[33] This can mean zero discharge, if any pollutant in receiving waters will prevent the attainment of an ample margin of safety. The essential difference between toxic effluent limitations and toxic effluent standards is that limitations are prescribed on the basis of pollutant removal technology, whereas standards are prescribed according to the effect of the discharge on the quality of the receiving waters. Either approach could effectively regulate the discharge of microorganisms containing rDNA molecules should EPA determine that such organisms constitute toxic pollutants.

31.3.5.3 *The Solid Waste Disposal Act, the Resources Conservation and Recovery Act, and the Marine Protection, Research and Sanctuaries Act*

Regulation of the disposal of solid wastes generated by rDNA technology will essentially be accomplished by EPA under the same statutory authority that governs disposal of solid waste from any laboratory or manufacturing process. Like the Clean Air Act and the Clean Water Act, the Solid Waste Disposal Act (SWDA) (42 U.S.C. § 3251) as amended by the Resource Conservation and Recovery Act (RCRA) (42 U.S.C. § 6901) is used to regulate pollutants based on what they are rather than what process generated them. The RCRA amendments to the SWDA have advanced considerably federal involvement in the regulation of solid waste disposal and management, particularly in the area of hazardous waste management. In spite of these substantial advances, however, federal regulation of solid waste is not as far advanced as is federal involvement in air and water pollution control.

The solid wastes generated, stored, transported, treated and disposed as a result of the application of rDNA technology could come under stringent regulation if they are hazardous wastes. Hazardous waste is defined in Section 1004(5) of subtitle C of RCRA as follows: 'a solid waste or combination of solid wastes, which because of its quantity, or physical, chemical, or infectious characteristics may—(A) cause, or significantly contribute to an increase in mortality or an increase in serious irreversible, or incapacitating reversible illness; or (B) pose a substantial present or potential hazard to human health or the environment when improperly treated, stored, transported, or disposed of, or otherwise managed' (42 U.S.C. § 6903 (5)).

To regulate all the wastes that could fit within this broad definition, EPA has promulgated detailed hazardous waste regulations which generally became effective beginning in November 1980 (40 C.F.R. Pt. 260, 45 *Fed. Regist.* 12 722–12 746, 33 063–33 588, 34 560–34 708 (February, May 1980)). Although the regulations do not specifically deal with wastes from laboratories or industry engaged in rDNA activity, it is easy to understand how these rules will apply to the disposal of much of the solid wastes generated by both research and applications involving the new technology.

The regulations are intended to track hazardous wastes from the point of generation to ultimate storage, treatment or disposal and to make accountable all parties that have any significant role in waste generation or management. The characteristics of hazardous wastes are identified, and particular wastes are listed as hazardous (40 C.F.R. §§ 261.3 and 261.20 *et seq.*). The most important characteristics of a hazardous waste are ignitability, corrosivity, reactivity and 'extraction procedure' (EP) toxicity (a procedure in which components are extracted from waste in a manner designed to simulate the leaching action that occurs in landfills) (42 C.F.R. §§ 261.21, 261.22, 261.23 and 261.24). Standards are established for generators, transporters and disposers of hazardous wastes (40 C.F.R. Pt. 262). Persons who generate, store, treat, transport, dispose of or otherwise handle hazardous wastes must allow inspection of their facilities and records. Hazardous waste handlers must comply with an extensive manifest system and must file annual reports (40 C.F.R. Pt. 263). All storage, treatment and disposal facilities must be approved by EPA (40 C.F.R. Pts. 264 and 265). The waste itself must be packaged, labeled, marked and placarded in accordance with the requirements of 49 C.F.R. Pts. 172, 173, 178 and 179. Subtitle C also provides for coordination with the regulations of the Secretary of Transportation issued under the Hazardous Materials Transportation Act (42 U.S.C. § 6923(b)). The provisions of federal enforcement of the hazardous waste management title (Section 3008) in general follow the enforcement provisions of the Federal Water Pollution Control Act discussed above.

Section 3013 (42 U.S.C. § 6934) authorizes the EPA to require the owner or operator of a facility to conduct monitoring, testing or analysis, when it determines that the presence of any hazardous waste at its facility, or the release of any such waste from the facility, may present a substantial hazard to human health or the environment. The significance of Section 3013 can hardly be overstated, because it provides the Administrator with the necessary power to order continuous monitoring and to obtain information from inactive (as well as active) disposal sites.

A number of lawsuits were brought to challenge many aspects of the hazardous waste regulations. The litigation was consolidated in the U.S. Court of Appeals for the District of Columbia Circuit in *Shell Oil Co. v. EPA*, No. 80–1532 (D.C. Cir.). In the course of consolidation, EPA agreed that a number of issues raised legitimate concerns and on August 19 1980 published its list which contained matters it considered appropriate for reevaluation and remodification. Since then, many of these issues have been resolved by amendment of the regulations and some others by the 1980 amendment of RCRA (45 *Fed. Regist.* 55 386 (1980)).

Despite the progress in the *Shell* case, several issues important to those involved in rDNA activity still remain. These issues include the small quantity generator exemption, on-site accumulation and spill clean-up. Because of the very large number of facilities and persons that could be subject to RCRA immediately, EPA exempts a category of generators handling hazardous wastes in small quantities (40 C.F.R. Pt. 261.5). The small quantity generator, namely a generator who produces less than 1000 kg of hazardous substances per month, is therefore temporarily exempted, although those generating wastes with greater toxicity lose their exemption at substantially lesser amounts. Small quantity generators are clearly subject to some of RCRA's disposal controls, but they are not subject to the full requirements of Subtitle C. This exemption undoubtedly will be important for almost all rDNA research facilities and some manufacturers if their solid wastes possess one of the four hazardous waste characteristics or are listed as hazardous wastes.

The regulations permit the accumulation of a small amount of hazardous wastes around a plant without requiring an RCRA storage permit, if such wastes are shipped off site within 90 days, and if certain container standards are met (40 C.F.R. § 262.34). The rule allowing such accumulations in short-term storage is also applicable to wastes stored for 30 days or less, if such wastes are then treated, stored and disposed of on-site in a permitted facility.

The reach of the RCRA regulations to a common laboratory problem might be surprising to many rDNA researchers. Spill is defined in 40 C.F.R. § 260.10(a) so that spills of some commercial chemical products, such as those that might be used in rDNA research or manufacturing, would constitute the creation of hazardous waste. It follows that many efforts to clean up such a spill would fall under the definition of treatment or storage, as defined in Section 260.10. To carry on such treatment or storage, a person needs an appropriate permit under RCRA. In the nature of such a clean-up operation, it is impossible to obtain a permit before the clean-up is completed. Actions taken in response to a spill do not require an RCRA permit until the immediate emergency is over, at which time the person is to be treated as a generator of hazardous waste and must then follow the appropriate provisions of the regulations. These include the requirement of proper packaging and labeling of the recovered wastes, and of filling out appropriate manifests. A generator requires a generator identification number, and, in any operation in which spills may be expected, a generator may be well advised to file for a generator permit ahead of time.

Although EPA has failed to propose any RCRA regulations dealing specifically with the solid waste products of rDNA research or manufacturing, the regulations address several other issues that might be important in the disposal of rDNA-associated solid wastes. As part of its regulations dealing with open dumping, which is prohibited by Section 4005(c) of RCRA, EPA requires disposal facilities to minimize 'through the periodic application of cover material or other techniques as appropriate so as to protect public health' the 'on-site population of disease vectors' (40 C.F.R. § 257.3–6). Because genetically-engineered microorganisms used in research, manufacturing or wastewater treatment could be 'disease vectors' and might be disposed of at a landfill, this regulation seems to apply. EPA, however, limits the definition of disease vector to 'rodents, flies and mosquitos capable of transmitting disease to humans'. EPA could, of course, extend this definition to include microorganisms, but does not presently intend to do so. Regardless of the possible disease vectors, sewage sludges (including those from biological treatment plants) must be treated either by a 'Process to Significantly Reduce Pathogens' or to 'Further Reduce Pathogens' and the disposal site's availability to crops, animals or people is limited for at least a year.

EPA has proposed listing as hazardous wastes infectious wastes generated in health care facilities, laboratories handling etiologic agents and sewage treatment facilities unless the wastes are sterilized or incinerated as required by EPA (45 *Fed. Regist.* 33 087 (1980)). When this proposal was made, EPA did not have an infectious waste-listing criterion. EPA claims it has now developed this criterion but is deferring action on infectious wastes until it can identify the treatment methods that could be used to exempt wastes from the regulation. EPA could attempt to include wastes from rDNA research laboratories in the definition of infectious wastes and thus these wastes could become 'listed' hazardous wastes. EPA, however, has not given any indication that they intend to regulate rDNA waste disposal activities in this manner.

EPA has recognized that a rule designating all waste mixtures containing listed hazardous wastes as hazardous could create some unintended results. Of particular importance to manufacturing plants with on-site research and development laboratories, is EPA's exemption from the mixture rules for laboratory wastewater containing listed hazardous wastes (46 *Fed. Regist.* 56 582 (1981)). The exemption is available if annual average concentration of laboratory wastewaters in the generator's total wastewater flow is less than 1%. EPA believes that this limitation will typically assure laboratory-derived listed hazardous waste concentrations in wastewaters of less than 1 p.p.m., which EPA 'deems will not pose a substantial hazard to human health or the environment'.

Ocean disposal has thus far constituted a relatively minor aspect of the disposal of solid wastes. Congress, however, considered the prospects for the future to be potentially dangerous. In addition to the Water Pollution Control Act Amendments of 1972, it enacted the Marine Protection, Research and Sanctuaries Act of 1972 (33 U.S.C. § 1401), generally referred to as the

'Federal Ocean Dumping Act'. The Act was subsequently amended in 1974 to make the law conform with treaty responsibilities under the Convention of the Prevention of Marine Pollution by Dumping of Wastes and Other Matter.

Although the prohibition of ocean dumping is more relevant to a discussion of international protection of the waters of the sea, it is also relevant to the subject of domestic waste disposal. The purpose of the law is to regulate the transportation of materials from the United States for dumping into ocean waters, and the dumping of material transported from outside the United States if the dumping occurs in ocean waters over which the United States has jurisdiction or may exercise control.

Ocean dumping generally refers to dumping in the open sea. The law prohibits dumping of 'matter of any kind and description, including but not limited to, *solid waste*, incinerator residue, garbage, sewage, *sewage sludge*, munitions; radiological, *chemical and biological warfare agents*, radioactive materials; *chemical, biological and laboratory wastes*, wrecked or discarded equipment; rock, sand, excavation debris; industrial, municipal, agricultural and other wastes' (emphasis added) (33 U.S.C. § 1402(c)). The ocean dumping law closely parallels the Federal Water Pollution Control Act in its approach, and in its reliance on a permit system. Other parallels, in the prohibition of toxic discharges and in administrative provisions, can also be discerned.

The ocean dumping law prohibits unconditionally the dumping of any radiological, chemical or biological warfare agents, or of any high-level radioactive wastes (33 U.S.C. § 1411(b)). The dumping of any other kinds of material is also prohibited, but it may be authorized by permits issued under the law. EPA may issue permits to dump materials, including materials commonly defined as solid waste, after notice and opportunity for public hearing, if it determines that such dumping will not 'unreasonably degrade or endanger human health, welfare or amenities, or the marine environment, ecological systems, or economic potentialities' (33 U.S.C. § 1412(a)). EPA must consider the need for the proposed dumping, the effects of such dumping on health and on economic, esthetic and recreational values, the effect of dumping on fisheries, fishing resources, as well as on plankton, fish, shellfish, wildlife, shorelines and beaches, and on marine ecosystems generally. The Administrator must also consider the persistence of the effects, and the effect of dumping particular volumes and concentrations of materials.

Violations of the law or of the regulations promulgated under it are punishable by a civil penalty of not more than $50 000 for each violation, to be assessed by the Administrator (33 U.S.C. § 1415(a)). The Administrator may also revoke or suspend any dumping permit if the law has been violated, after notice and opportunity for a hearing.

31.3.5.4 *The Federal Insecticide, Fungicide and Rodenticide Act*

It is conceivable that rDNA technologies might be used to create organisms (such as insect or plant pathogens) that could be used as pesticides. If so, the Federal Insecticide, Fungicide and Rodenticide Act (FIFRA) (7 U.S.C. § 135 *et seq.* (Supp. 1981)) will probably be relevant to the manufacturers' research and development plans. That statute states that no pesticide may be sold, distributed or used within the United States unless it is registered with the U.S. Environmental Protection Agency. The word 'pesticide' is defined very broadly to include 'any substance or mixture of substances intended for preventing, destroying, repelling, or mitigating any pest' (7 U.S.C. § 136).

Before a pesticide may be registered with EPA, the applicant for registration must demonstrate that: (1) its composition is such as to warrant the proposed claims for it; (2) its labeling and other material required to be submitted comply with the requirements of (the Act); (3) it will perform its intended function without unreasonable adverse effects on the environment; (4) when used in accordance with widespread and commonly recognized practice it will not generally cause unreasonable adverse effects on the environment (7 U.S.C. § 136(c) (5)). The applicant must make these showings on the basis of scientific studies performed by the registrant on the efficacy and safety of the pesticide. EPA has promulgated extensive registration guidelines that describe the kinds of studies EPA will accept for purposes of making the above showings (40 C.F.R. § 162.5 (1982)). The guidelines, however, do not apply to biological pesticides, which are regulated on a case-by-case basis (43 *Fed. Regist.* 29 696 (1978)).[34] To aid applicants in conducting efficacy studies, Section 5 of the Act provides for the issuance of experimental use permits that allow limited quantities of unregistered pesticide to be used in specifically defined areas (7 U.S.C. § 136c).

Although EPA has no special guidelines addressing pesticidal microorganisms that contain rDNA, new genetically-engineered organisms will have to be approved by EPA before they may be marketed. It would, therefore, be sensible to follow EPA's guidelines in testing the product. The testing process itself can be very expensive; some cost estimates range as high as $5 million dollars (*see* S. Rep. No. 334, 95th Cong., 1st Sess. 30 (1977)). A manufacturer must, therefore, have a very promising pesticide before it would be profitable to attempt to obtain a license from EPA.

Civil penalties for unlawful sale or distribution of an unregistered pesticide can range up to $5000 for each offense (7 U.S.C. § 136b(a) (1)). Knowing violations of the Act are punishable by fines of up to $25 000 per violation or imprisonment for up to one year or both (§ 14(b) (1) or 7 U.S.C. § 136b(b) (1)).

31.4 INTERNATIONAL REGULATION OF rDNA RESEARCH

Attempts to regulate rDNA have truly been those of the international scientific community. For example, in the early 1970s, some American molecular biologists called for a world-wide moratorium on genetic experiments until guidelines and safeguards could be developed.[35] In 1974, a group of scientists from the United Kingdom, under the direction of Lord Ashby, took a serious look at current genetic engineering experiments. They published a report that echoed earlier American opinion: genetic engineering research should continue, but not without careful monitoring. In early 1975, molecular biologists from around the world met at Pacific Groves, California (the 'Asilomar Conference'), to discuss the future of genetic engineering, including the impact of regulation on the development of rDNA technology. The conferees lifted their self-imposed moratorium on genetic engineering research and proposed guidelines for future studies.

A few months after Asilomar, a working party in the United Kingdom was assigned the task of developing genetic engineering guidelines. Sir Robert Williams led the task force in its efforts to establish a laboratory practice code. The Williams report recommended that a new agency regulate research *via* development of a body of case law and precedent (Tooze, 1981). The suggestions of the Williams report, submitted by a committee of the Department of Science and Education, were adopted.

In 1976 the Genetic Manipulation Advisory Group (GMAG) was formed, pursuant to Williams report recommendations, to oversee rDNA research performed in the United Kingdom. The GMAG's responsibility was to review every rDNA research proposal and develop unique guidelines for each project. Drafters of the Williams report anticipated the precedential value of GMAG decisions, which were to form a generally-applicable scheme of regulation. Merely precatory until 1978, the GMAG's proposed guidelines have the force of law after an act of Parliament amended regulations pursuant to the Health and Safety at Work Act, to cover any research involving rDNA molecules (Zilinskas, 1978). The United Kingdom remains one of the few countries that have legally-binding regulations governing genetic engineering research.

European scientists watched the developments at Asilomar intently, and in 1976, when the NIH Guidelines were finally released, worked seriously to adapt those guidelines to their own use. Countries following those developments had their choice of two approaches once the NIH Guidelines were released. First, the Williams committee guidelines generally outlined the categories of physical containment and depended on the GMAG, with its regulatory decisions, to supply the special considerations of individual projects. Second, the NIH Guidelines, on the other hand, relied on the 'encyclopaedic approach', which provided examples of containment for almost every conceivable experiment (Tooze, 1981). Dialogue between countries continued, but no international uniform system of regulation developed as countries chose one method or the other to fit particular needs. At that time, scientists opined that the Asilomar conferees' fears were overstated, with resultant NIH Guidelines that were unnecessarily stringent (Tooze, 1981).

France responded immediately to the Asilomar report (published one year before the NIH Guidelines) and implemented those suggestions until 1976. The Commission de Control shared the European scientific community's antipathy towards the NIH Guidelines and chose not to adopt them. By 1977, the Commission produced a less stringent set of its own. That move placed U.S. agencies in the not unfamiliar position of promulgating regulations more onerous than their international counterparts. For the first time, the NIH began watching Europe for guidance (Tooze, 1981).

Sweden followed the United Kingdom's example, and in 1979 passed a similar ordinance pursuant to its Work Environment Act. The Swedish ordinance regulates 'new or untried' uses of rDNA research. That terminology is ambiguous and may create constructional problems leading to hindrance of enforcement (Tooze, 1981). Interestingly, the Federal Inter-Agency Committee on Recombinant DNA Research (RAC), formed in the United States in 1976, determined that no U.S. agency had the power to promulgate a similar regulation without the passage of a new enabling act (Tooze, 1981).

The Federal Republic of Germany has also had problems enacting legislation to enforce its research guidelines. The Central Commission for Biological Safety (ZKBS) supervises compliance with the West German guidelines, which closely track the NIH Guidelines. In the absence of legislation, however, the guidelines merely require voluntary compliance.

In 1979, the NIH succumbed to generally-accepted scientific opinion and significantly relaxed the existing Guidelines. Predictably, European countries closely watched American decision-making and began to consider amendment of their own guidelines. In Britain, where regulation had taken the form of a set of GMAG precedents set on a case-by-case basis, the adoption of less stringent standards was difficult, though not impossible. In 1978 GMAG developed, and adopted on a trial basis, a mathematical model to quantify the biohazards of genetic experimentation. GMAG's formula achieved the desired result of less stringent regulation. By 1980 United Kingdom molecular biologists expected to be competitive again with their counterparts (Tooze, 1981).

In 1980, the NIH again relaxed its guidelines, but these changes were not equal to a major innovation by the GMAG. Though it meant relinquishing substantial control over rDNA experiments, the GMAG decided to limit its review to only those projects involving known pathogens or organisms producing toxins. Simpler projects would receive cursory review by 'institutional safety committees' (Tooze, 1981). Since most European countries follow the regulatory actions of the United States or the United Kingdom, and since Europe typically strives for uniformity, it can be expected that European regulations will become less stringent.

Not surprisingly, the private industries of other developed countries were watching these developments closely. In 1980, the Japanese Ministry of International Trade and Industry enthusiastically encouraged the development of genetic engineering, a challenge that commercial manufacturers enthusiastically accepted. Japanese guidelines, simpler than (though modeled after) the 1980 version of the NIH Guidelines, are theoretically unenforceable except for government-sponsored research. Practically, however, the symbiotic relationship between Japanese industry and government compels compliance (Henderson, 1981).

In addition to the activities of national governments discussed above, the rapid development of genetic engineering has led to a heightened interest by international public health organizations as well as the formation of international committees to monitor and guide the development of recombinant technologies (Zilinskas, 1978). The European Molecular Biology Organization (EMBO) is perhaps the oldest group currently involved in the development of regulatory guidelines. The European Molecular Biology Conference (EMBC) formed EMBO in 1964 and has funded it since 1970. EMBO's goal is to encourage development of molecular biology in Europe by sponsoring research and facilitating collaboration. Sixteen European countries and Israel comprise the EMBC, and each receives regular reports from the EMBO's Standing Advisory Committee on rDNA Research. The EMBO has always taken an active role in the development of genetic engineering guidelines. The organization appeared at Asilomar and later drafted regulation proposals, which were frequently relied upon in NIH guideline drafting and subsequent revision (Tooze, 1981).

Other international groups became involved with rDNA regulation with various levels of participation. Although the World Health Organization (WHO) has contributed somewhat to the international debate, it is more concerned with safety measures, transportation controls and organism classification (Cripps, 1980). A possibility exists, however, of promulgating international regulations under the authority of Article 21 of the WHO Constitution permitting WHO to issue sanitary and quarantine regulations and to implement other procedures designed to prevent the international spread of disease. Article 21, like Section 361 of the U.S. Public Health Service Act (discussed above) does not mention occupational or environmental controls (Shubber, 1976). Accordingly, WHO's authority to promulgate such standards is necessarily based on the implied powers theory and a liberal interpretation of Article 21. That authority is most likely too weak for the promulgation of binding international genetic engineering standards.

Another organization, the International Council of Scientific Unions (ICSU), adopted as one of its primary goals the facilitation of information exchange among molecular biologists. Through its Committee on Genetic Engineering (COGENE), the ICSU sponsored workshops and conferences and taught genetic engineering techniques in Brazil, India and Yugoslavia (Tooze, 1981). COGENE has also conducted risk assessment experiments and examined the large-scale industrial use of the technology. By way of contrast with some of the other interested international organizations, it has been extremely systematic in obtaining the information on which its recommendations are based and, accordingly, it has made a useful contribution to the literature on the status of the technology and the diverse national attempts to regulate it (Cripps, 1980). For example, the committee has distributed detailed questionnaires that by July 1 1979 had been returned by more than 50 countries. The interim report (Working Group, 1980) based on these statistics indicated that at the end of July, 1979, hundreds of rDNA projects were under way in approximately 250 laboratories in 26 countries. Twenty-eight nations had developed guidelines for the conduct of the research, but it was significant that only eight of these had prepared their own guidelines; the rest had adopted the NIH Guidelines (National Institutes of Health, 1981a), either *in toto* or in modified form.

The COGENE working group also noted that supervisory procedures and methods of enforcement are substantially different between countries and concluded that 'identical experiments require strikingly different levels of precaution in different nations. Moreover, although the general approach to biological and physical containment first proposed in the U.S. guidelines has been adopted by most other nations, the specific characteristics of each containment level differ markedly among different nations' (Cripps, 1980).

Scientists and researchers from 18 countries also formed the European Science Foundation (ESF) to provide support and funding for general research. Naturally concerned with the future of rDNA research in Europe, the ESF hosted a conference in 1976 attended by representatives of the RAC, NIH and the Canadian Medical Research Council (Tooze, 1981). Armed with information exchanged at that conference and with the NIH guidelines as a model, the ESF issued a series of reports advocating a general relaxation and uniformity of existing guidelines in Europe. Satisfied that this goal of uniformity had been achieved, the ESF disbanded itself in January, 1981 (Tooze, 1981).

Europe's perceived need for a uniform system of regulation even reached the attention of the European Economic Community (EEC), better known as the Common Market. The EEC's governing body interpreted the Common Market's formative treaty to allow EEC involvement in and support of scientific research with conceivable impact on Market members. Though plagued with cumbersome procedural delays, the EEC's Commission would like to assume a strong role in the promulgation and enforcement of uniform rDNA guidelines for member nations (Tooze, 1981).

As is clear from the above discussion, scientists in many countries have been at the forefront of the debate surrounding the regulation of rDNA research. Undoubtedly, this active participation will continue.

31.5 NOTES

1. For accounts of the political debate over rDNA research and the development of the NIH Guidelines, *infra*, see, for example, Goodfield (1977); Rogers (1977); Wade (1977); Subcommittee, Science, Research and Technology (1977).

2. One obvious use for organisms containing rDNA that would involve broad exposure to the environment would be the application of such organisms to spent ore deposits for purposes of leaching valuable metals from the ore. See, for example, OTA Report (1981); Demain and Soloman (1981).

3. For analyses of the inherent susceptibility of risk and benefit assessments to manipulation and bias, see Kennedy (1981); Rodgers (1980).

4. Technically, the NIH Guidelines are binding only upon research sponsored by the National Institutes of Health (NIH Guidelines at Section IV-B). However, virtually all other federal agencies have agreed to make the NIH Guidelines binding on research that they support. See Subcommittee, Health and Environment (1977), testimony of Dr. Donald B. Fredrickson, Director, National Institutes of Health, at 320, 327.

5. See NIH Guidelines (1981a). Interestingly, the most recent version of the NIH Guidelines exempts research conducted with *E. coli* K12, the bacterium that will probably be the predominant organism used in early industrial applications of rDNA technologies.

6. For excellent descriptions of the evolution of the regulation of rDNA research in countries other than the United States, see Henderson (1981); Tooze (1981).

7. See, for example, Subcommittee, Science, Research and Technology, testimony of Dr. Ronald E. Cape, President Cetus Corp, at 91 and testimony of Pharmaceutical Manufacturers Association, at 374. Private companies in Japan and Europe have also cooperated voluntarily

with the applicable guidelines in their respective countries. See Henderson (1981), at 897–98; Tooze (1981), at 879–80.

8. NIOSH acquires the power to enter workplaces and investigate through Section 20(b) (20 U.S.C. § 669).

9. 29 U.S.C. § 655 (1976). In addition, every employer under Section 5(a) (1) (29 U.S.C. § 264) has a 'general duty' to furnish each of his employees 'employment and a place of employment which are free from recognized hazards that are causing or are likely to cause death or serious physical harm to his employees'. OSHA can enforce this general duty clause on a case-by-case basis.

10. OSHA has in the past preferred engineering and design controls over personal protective devices, although it may be changing its position on this issue.

11. FDA has defined recombinant DNA molecules as either (1) molecules that are constructed outside living cells by joining natural or synthetic DNA segments to DNA molecules that can replicate in a single cell, or (2) DNA molecules that result from the replication of such molecules. 43 *Fed. Regist.* 33 042 at 33 069 (1978).

12. Despite FDA's broad authority, in the 1977–78 session, House and Senate committees drafted 12 and 13 bills, respectively, to regulate rDNA research in both public and private laboratories. None of the bills passed (Tooze, 1981).

13. 'Interferon Plus', a combination of a biologic, interferon and vitamin, is a biologic drug and the marketing of the drug without FDA approval violated PHSA 42 U.S.C. § 262(a) and FFDCA 21 U.S.C. § 351(a) and 352. *Biotics Research Corp. v. Heckler*, FOOD DRUG COS. L. REP. (CCH) § 38 226, _____ F. Supp. _____ aff'd 710 F.2d 1375 (9th Cir. 1983).

14. Courts and FDA share concurrent jurisdiction to determine whether the product is a 'new drug', but FDA retains primary authority because the agency has technical expertise in the area. See *Weinberger v. Hynson, Westcott and Dunning, Inc.*, 412 U.S. 609 (1973); *Ciba Corp. v. Weinberger*, 412 U.S. 640, 643–44 (1973); *Weinberger v. Bentex Pharmaceuticals, Inc.* 412 U.S. 645, 653–54 (1973).

15. In 1962, when Congress amended the FFDCA to require proof of efficacy as well as safety, FDA was forced to develop a program to review the large number of pre-1962 approvals. The National Academy of Sciences–National Research Council (NAS–NRC) was hired to review the existing approvals and issue Drug Efficacy Study Implementation (DESI) notices for each active ingredient that complied with the efficacy requirement. In an effort to keep existing 'me-too' approvals based on pre-1962 drugs from repeating work done by NAS–NRC, FDA issued the ANDA regulations.

16. For example, FDA recently required an NDA before a biosynthetic insulin could be marketed. This NDA was approved in November, 1982. FOOD DRUG COS. L. REP. ¶ 42 286.

17. See *U.S. v. Articles of Drug Labeled Colchine*, 442 F. Supp. 1236, 1241 (S. D. N. Y. 1978); *Nat'l. Assoc. of Pharmaceutical Manufacturers v. FDA*, 487 F. Supp. 412, *aff'd* 637 F.2d 877 (2d Cir. 1978) *reh and reh en banc den'd*; *Nat'l Nutritional Foods Ass'n v. Weinberger*, 512 F.2d 688, 696, *cert. den.* 423 U.S. 829, 96 S.Ct. 44 (1975).

18. In *Mack v. Califano*, petitioners sought a preliminary injunction to keep the Federal Cancer Research Center from testing the biological properties of polyoma DNA cloned in bacterial cells. Petitioners challenged the adequacy of the respondent's EIS, a challenge the Court rejected, believing the EIS demonstrated that a 'hard look' had been taken. 447 F. Supp. 668 (D.D.C. 1978).

19. *Louisiana v. Mathews*, 427 F. Supp. 174 (D.C. La. 1977); *U.S. ex. rel. Siegal v. Shinnick*, 219 F. Supp. 789 (D.C. N.Y. 1963).

20. The term 'manufacture' is defined somewhat circularly to mean 'to import . . . produce, or manufacture.' (§ 3(7), 15 U.S.C. § 2602(7)).

21. Such chemicals should, however, be listed in EPA's inventory of existing chemical substances (15 U.S.C. § 2607(b)).

22. See text accompanying note 23, *infra*. Arguably, a small viral or plasmid vector would be a 'chemical substance' within the meaning of Section 3(2) of TSCA (15 U.S.C. § 2602(2)).

23. The definition, however, explicitly excludes pesticides, tobacco, food, food additives, drugs and cosmetics, and source material, special nuclear material and by-product material (15 U.S.C. § 2602(2) (B)).

24. EPA has recently exempted certain chemicals used in enclosed photographic processes. See 40 C.F.R. § 710.4(d) (5) (1982).

25. Small manufacturers are defined according to standards prescribed by EPA, after consultation with the Small Business Administration (40 C.F.R. 710–2 (1981)).

26. TSCA establishes a committee to make recommendations to the Administrator concerning the priority with which he should require testing (15 U.S.C. § 2603(e)).

27. Note, however, that Congress in Section 6(c) (1) of TSCA has indicated an intention that EPA use regulatory tools other than Section 6(a) or TSCA when feasible. Section 6(c) (1) provides that 'If the Administrator determines that a risk of injury to health or the environment could be eliminated or reduced to a sufficient extent by actions taken under another Federal law (or laws) administered in whole or in part by the Administrator, the Administrator may not promulgate a rule under subsection (a) to protect against such risk of injury unless the Administrator finds, in the Administrator's discretion, that it is in the public interest to protect against such risk under this Act. In making such a finding the Administrator shall consider (i) all relevant aspects of the risk, as determined by the Administrator in the Administrator's discretion, (ii) a comparison of the estimated costs of complying with actions taken under this Act and under such law (or laws), and (iii) the relative efficiency of actions under this Act and under such law (or laws) to protect against such risk of injury (15 U.S.C. § 2605(c) (1)).'

28. The currently listed criteria pollutants are sulfur dioxide, particulate matter, carbon monoxide, ozone, hydrocarbons, nitrogen dioxide and lead (40 C.F.R. Pt. 50 (1982)). The currently listed hazardous air pollutants are asbestos, beryllium, mercury and vinyl chloride (40 C.F.R. § 61 (1982)).

29. One of the preconditions to listing a pollutant as a priority pollutant is that it should result from 'numerous or diverse mobile or stationary sources' (42 U.S.C. § 7408(a) (1)).

30. If it is not 'feasible' to prescribe or enforce an emissions standard for a particular pollutant, EPA may instead promulgate a design, equipment, work practice or operational standard, or combination thereof (42 U.S.C. § 7412(e)).

31. Note, however, that if EPA decides to regulate emissions of microorganisms under any of its statutory authorities with the object of protecting workers, then OSHA would have no authority to deal with that subject matter (29 U.S.C. § 653(b) (1976)).

32. The term 'navigable waters' is defined very broadly to include virtually every stream in the United States (33 U.S.C. § 1362). See *United States v. Phelps Dodge*, 381 F. Supp. 1181 (D. Ariz. 1975); *United States v. Holland*, 373 F. Supp. 665 (M. D. Fla. 1974).

33. See generally *Environmental Defense Fund v. EPA*, 12 ERC 1353 (D. C. Cir. 1978); *Hercules, Inc. v. EPA*, 12 ERC 1376 (D.C. Cir. 1978). However, EPA must consider the extent to which effective control is being or may be achieved under other regulatory authority (33 U.S.C. § 1317(a) (2)).

34. In addition, EPA has proposed to exempt certain biorational pesticides that are currently being effectively regulated by other federal agencies. (See 46 *Fed. Regist.* 18 322 (1981).)

35. See letter from Maxine Singer and Dieter Soll, reprinted in Guidelines for DNA Hybrid Molecules, *Science*, **181**, 1114 (1973), and letter from Paul Berg, reprinted in Potential Biohazards of Recombinant DNA Molecules, *Science*, **185**, 303 (1974).

31.6 ABBREVIATIONS

ANDA	Abbreviated New Drug Application (FDA, USA)
BOD	Biological oxygen demand (EPA, USA)
CAA	Clean Air Act (EPA, USA)
C.F.R.	Code of Federal Regulations
COD	Chemical oxygen demand (EPA, USA)
COGENE	Committee on Genetic Experimentation (International Council of Scientific Unions, USA)
CWA	Clean Water Act (EPA, USA)
DNA	Deoxyribonucleic acid
EDF	Environmental Defense Fund
EDNY	Eastern District Court of New York
EDRO	Executive Director for Regional Operations (FDA, USA)
EEC	European Economic Community (Britain)
EIAR	Environmental Impact Analysis Report (FDA, USA)
EIS	Environmental Impact Statement (EPA, USA)
EMBC	European Molecular Biology Conference
EMBL	European Molecular Biology Laboratory

EMBO	European Molecular Biology Organization
EP	Extraction procedure
EPA	Environmental Protection Agency
ERC	Environmental Reporter Cases
ESF	European Science Foundation
FAO	Food and Agriculture Organization
FDA	Food and Drug Administration
Fed. Regist.	Federal Register
FFDCA	Federal Food, Drug and Cosmetic Act (FDA, USA)
FIFRA	Federal Insecticide, Fungicide and Rodenticide Act (USA, EPA)
FPLA	Fair Packaging and Labeling Act (FDA, USA)
FPP Act	Federal Plant Pesticide Act
F. Supp.	Federal Supplement
GAO	General Accounting Office (USA)
GMAG	Genetic Manipulation Advisory Group (UK)
GMP	Good Manufacturing Practice Regulation (FDA, USA)
HGH	Human growth hormone
HHS	Department of Health and Human Services (USA)
HMSO	Her Majesty's Stationery Office (UK)
HSE	Health and Safety Executive (UK)
ICSU	International Council of Scientific Unions
IND	Claimed Investigational Exemption for a New Drug (FDA, USA)
MITI	Ministry of International Trade and Industry (Japan)
NDA	New Drug Application (FDA, USA)
NEPA	National Environmental Protection Act (EPA, USA)
NIH	National Institutes of Health (USA)
NIOSH	National Institute for Occupational Safety and Health (OSHA, USA)
NPDES	National Pollution Discharge Elimination System (EPA, USA)
OSHA	Occupational Safety and Health Administration (USA)
OSH Act	Occupational Safety and Health Act of 1970 (OSHA, USA)
OSH Rev	Occupational Safety and Health Review Commission (OSHA, USA)
OTA	Office of Technological Assessment (USA)
PHSA	Public Health Service Act (FDA, USA)
RCRA	Resource Conservation and Recovery Act (EPA, USA)
rDNA	Recombinant deoxyribonucleic acid
SNDA	Supplemental New Drug Application (FDA, USA)
S. Rep.	Southern Reporter (USA)
SWDA	Solid Waste Disposal Act (EPA, USA)
TSCA	Toxic Substances Control Act (EPA, USA)
UNEP	United Nations Environmental Program
UNESCO	United Nations Educational, Scientific and Cultural Organization
U.S.C.	United States Code
WHO	World Health Organization
ZKBS	Central Commission for Biological Safety (West Germany)

31.7 REFERENCES

Aharonowitz, Y. and G. Cohen (1981). The microbiological production of pharmaceuticals. *Sci. Am.*, **245** (3), 140–153.

Birnbaum, J. (1982). FDA proposes sweeping rule changes to hasten process of approving drugs. *Wall Street Journal*, June 24, 4.

Commission of the European Communities Document (1980). COM (80), vol. 467.

Cripps, Y. (1980). *Controlling Technology; Genetic Engineering and the Law.* Prager, New York.

Demain, A. and N. Solomon (1981). Industrial microbiology. *Sci. Am.*, **245** (3), 66–75.

Goldberg, M. and H. Miller (1981). The role of FDA in the regulation of the products of recombinant DNA technology. *Recombinant DNA Tech. Bull.*, **4**, 15.

Goodfield, J. (1977). *Playing God.* Random House, New York.

Henderson, D. F. (1981). Japanese regulation of recombinant DNA activities. *U. Toledo L. Rev.*, **12**, 891–901.

Her Majesty's Stationery Office (1978). *Health and Safety at Work: Genetic Manipulation.* HMSO, London.

Kennedy, D. (1981). Cost–benefit analysis of entitlement problems: A critique. *Stanford L. Rev.*, **33**, 387–445.

Korwek, E. (1981). The NIH guidelines for recombinant DNA research and the authority of FDA to require compliance with the guidelines. *Jurimetrics J.*, **21**, 264–283.

Korwek, E. L. and D. S. Trinker (1981). Perspectives on the FDA status of drug products manufactured by the recombinant DNA technique. *Food, Drug, Cosmetic L. J.*, **36**, 517–535.

McKinney, R. (1981). Telephone interview of September 18.

Medical Research Council (1980). *Revised Guidelines for the Categorisation of Recombinant DNA Experiments*, GMAG Note 14.

Miller, H. I. (1981). The impact of new technology on regulation by the FDA: recombinant DNA technology. *Food, Drug, Cosmetic L. J.*, **36**, 348–353.

National Institutes of Health (1981a). Guidelines for research involving recombinant DNA molecules (also cited as NIH Guidelines). Promulgated, 41 *Fed. Regist.* 27 902 (1976), subsequently amended several times. National Institutes of Health, Bethesda, MD.

National Institutes of Health (1981b). *Recombinant DNA Research*, vol. 6.

Office of Technology Assessment (1981). Impacts of applied genetics (hereinafter cited as OTA Report). Office of Technology Assessment, US Congress, Washington, DC.

Rodgers, W. H., Jr. (1980). Benefits, costs, and risks: oversight of health and environmental decisionmaking, *Harvard Envl. L. Rev.*, **4**, 191–226.

Rogers, M. (1977). *Biohazard*. Alfred A. Knopf, New York.

Rosenblatt, D. (1982). The regulation of recombinant DNA research: the alternative of local control. *Environ. Affairs*, **10**, 37–78.

Shubber, S. (1976). The role of WHO in environmental pollution control. *Earth L. J.*, **2**, 363–392.

Subcommittee on Health and the Environment, U.S. House of Representatives Committee on Interstate and Foreign Commerce (1977). Hearings, 95th Cong., 1st Sess.

Subcommittee on Science, Research and Technology, U.S. House of Representatives Committee on Science and Technology (1977). *Hearings on Science Policy Implications of DNA Recombinant Molecule Research*, 95th Cong., 1st Sess.

Subcommittee on Science, Research and Technology, U.S. House of Representatives Committee on Science and Technology (1978). *Report on Science Policy Implications of DNA Recombinant Molecule Research*, 95th Cong., 2nd Sess.

Subcommittee on Science, Research and Technology, U.S. House of Representatives Committee on Science and Technology (1980). Report on the Food and Drug Administration's *Process for Approving New Drugs*, 96th Cong., 2nd Sess.

Subcommittee on Science, Technology, and Space, U.S. Senate Committee on Commerce, Science and Transportation (1978). *Oversight Report: Recombinant DNA Research and Its Applications*, 95th Cong., 2nd Sess.

Subcommittee on Science, Technology, and Space, U.S. Senate Committee on Commerce, Science and Transportation (1980). *Hearing on Industrial Applications of Recombinant DNA Techniques*, 96th Cong., 2nd Sess.

Tooze, J. (1981). International and European regulation of recombinant DNA research. *U. Toledo L. Rev.*, **12**, 869–890.

United States Environmental Protection Agency (1976). *Development Document for Interim Final Effluent Limitations Guidelines and Proposed New Source Performance Standards for the Pharmaceutical Manufacturing Point Source Category*, pp. 91–94.

Wade, N. (1977). *The Ultimate Experiment*. Walker, New York.

Working Group on Recombinant DNA Guidelines (1980). *Report to COGENE*.

Zilinskas, R. A. (1978). Recombinant DNA research and the international system. *S. Cal. L. Rev.*, **51**, 1483–1501.

32

United States and Canadian Governmental Regulations Concerning Biohazardous Effluents

S. KRIMSKY
Tufts University, Medford, MA, USA
and
A. H. FRAENKEL*
Harvard School of Public Health, Boston, MA, USA

*Present address: Ohio State University.

32.1 INTRODUCTION

32.1.1 Scope of the Chapter

The discovery of plasmid-mediated *in vitro* DNA recombination in 1972, resulted in a renaissance for commercial fermentation technology. The prospects of applying techniques in molecular biology to the development of a wide range of microbial products has given rise to several hundred new firms worldwide within a few years. While applications for genetically engineered microorganisms were being sought and commercial markets assessed, scientists and the public debated the safety of rearranging genetic material across disparate species. Some skeptics called for new laws, while others were satisfied to establish regulations under the authority of existing laws. Throughout a decade of intense controversy, primarily centered in the United States and Britain, many scientists emphasized that the risks associated with the manufacture of novel organisms were purely conjectural.

Much of the early discussion over biohazards was almost exclusively directed to gene-splicing experiments. As the debate over laboratory-scale recombinant DNA manipulations subsided, the focus shifted to industrial scale activities where environmental release and occupational health are highlighted. The question at the heart of the issue is: 'Will commercial biotechnology be clean in comparison to the chemical industry which is plagued with problems of treatment, transportation, storage and disposal of hazardous wastes?'

The decades of the 1960s and 1970s brought an unprecedented public response to the chemical hazards facing man and the environment. Throughout the industrial world, major pieces of legislation were enacted to protect land, water, air resources and the working environment from chemical contaminants. The commercial developments that were taking place to exploit the use of gene-splicing brought a new air of apprehension to some environmentally-minded individuals, government officials and public advocacy groups. Central to their concerns were the following possibilities: (1) a novel organism might be released into the environment with unpredictable and possibly irreversible effects on the environment; (2) a new microbial agent, infectious to humans or animals, might be released, or a conventional pathogen might have its host range broadened; (3) a rapid rise in large-scale biotechnology could result in bioeffluents that would place additional stresses on the quality of land and water resources; (4) organisms engineered to perform useful functions in the environment might produce adverse secondary effects of an unanticipated nature.

Initiated first by Britain (Working Party, 1976), followed soon after by the United States (National Institutes of Health, 1976), guidelines regulating experiments involving the production of recombinant DNA molecules were adopted by many countries actively pursuing research in this area. Scientists questioned the rationale that selects out genetically engineered organisms for regulation. The laboratory use of pathogens and infectious viruses has been an area of concern for the public health community. Infectious disease experts can point to studies of hospital-induced (nosocomial) infections traced in part to clinical laboratories and the lack of adequate attention to the disposal of infectious waste products. By 1980 many states had adopted regulations for the management of infectious wastes. Special treatment of infectious waste materials is required by some Canadian provinces. But in the laboratory uses of infectious agents, even in large clinical laboratories, the volumes have not been large relative to an industrial scale, and thus federal environmental initiatives which respond to the scale of pollution have been minimal.

In the food and pharmaceutical industries, the fermentation technology has utilized mainly non-pathogenic microorganisms cultured from natural environments, or their laboratory-induced hybrids. As a consequence, little attention has been given to regulating industrial effluent consisting of biological entities that are plentiful in nature. But as genetically engineered bacterial strains constructed in the laboratory are introduced into the industrial fermentation process, or released into the environment as pesticides or pollution degraders, new expressions of concern can be heard among those who are inclined to draw connections between the developments in synthetic chemistry and those in synthetic biology.

The goals of this chapter are threefold. First, it examines six areas where the use of microorganisms in an industrial setting is expected to grow as a consequence of the revolution in applied molecular genetics that is currently taking place. Second, it reviews the laws in the United States and Canada which are or may be applicable to the regulation of biological agents released into the environment. Third, it summarizes statutory regulations and guidelines that have been issued in the United States and Canada pertaining to the release of biological agents. The inventory of environmental laws and regulations includes those in the area of occupational safety and health.

In contrast to the use of chemicals in industrial processes, the application of microbial agents in production or as pesticides has been small. This is reflected in the fact that few regulations have been established for contaminants from bioprocess sources. This study distinguishes between the authority to regulate and the actual regulations. For example, the United States Environmental Protection Agency may have a legal authority to regulate biological waste products, but it has not chosen to exercise this authority thus far.

32.1.2 Bioprocess Sources and Pathways for Environmental Release

The term bioprocess in this study is used in a general sense to mean any human activity that produces or transports microbial agents, disperses them into the environment, or uses them as part of a system of production. The laws and regulations reviewed were chosen with consideration of the following pathways through which biological agents are released into the environment: human to human, animal to animal, or plant to plant contact; transport of etiologic agents; laboratory and clinical wastes; manufacturing processes involving large cultures of bacteria or viruses; large-scale release of biological agents into the environment; release of biological agents from experimental laboratories, through drains, human, insect or rodent vectors.

In the pharmaceutical and food industries that employ fermentation technologies, biological agents may be released inadvertently during the venting of gases, through the wastewater effluent stream or by the disposal of the sludge resulting from the fermentation process. Some technologies, such as wastewater treatment, use the indigenous organisms within the effluent stream by enhancing their growth and thereby helping them break down organic contaminants. Some efforts are underway to improve the efficiency of secondary treatment plants by adding specially cultured or genetically engineered bacteria. In these uses, the microorganisms have a natural pathway into the environment unless these biodegraders of organic contaminants are themselves disinfected after their work is completed.

Relatively little legislation has been conceived with the release of biological agents in mind. It has been a small effort compared to the regulatory activities for chemical agents and radioactive materials. The legislation which has been enacted is concerned primarily with pathogenic agents infectious to humans, plants and animals. As an example, under the quarantine acts certain organisms or their hosts are prohibited from being imported into North America. Beyond direct infectivity, there are clearly other ways that biological agents can be ecologically troublesome. They may interfere with biochemical pathways, increase the coliform bacteria levels in drinking water or affect the biological oxygen demand in lakes, rivers and streams. The guidelines that were issued in the United States, Canada and elsewhere for genetically modified organisms are based upon a broader interpretation of biohazardous materials than what is understood by a pathogenic or infectious agent. That is also true with regulations for the use of biological pesticides, where studies are required that include host specificity, allergenic effects and toxicity prior to the granting of a permit for their use.

32.1.3 Overview of Environmental Laws

32.1.3.1 United States

Regulation has been the preferred method of dealing with environmental problems in the United States. American governments at all levels have chosen to regulate sources of environmental pollution, although alternative forms of governmental intervention have been suggested (Stewart and Krier, 1978).

Several major pieces of legislation have been enacted by Congress to deal with the problems of air pollution, water pollution and hazardous substances. The federal agency given primary responsibility for implementation and enforcement of government policy as directed by these laws is the Environmental Protection Agency (EPA). EPA's enforcement role covers a number of areas including stationary sources of air pollution, discharges into the nation's waters, mobile sources of air pollution, hazardous waste sites, toxic substances, solid wastes, drinking water and pesticides.

In many instances the federal and state governments share the burden of implementation and enforcement. State agencies analogous to the federal EPA regulate and enforce state environmental laws and state components of federal programs. Some federal programs are predicated on direct state regulation with varying degrees of federal overview, others on direct federal regulation only until states develop programs which are in compliance with federal guidelines. When compliance is achieved, the state directs the regulatory program. A third category of regulation involves direct federal regulation with little or no state involvement.

The Clean Air Act (33 U.S.C. § 7401 *et seq.*) established federal controls on air pollution. It was last amended in 1977 and it is that version of the Act which is now in effect. The requirements of the Act are expressed in terms of ambient air standards which are created by statute, federal or state regulations, or permits. It was intended by Congress that the states would be primarily responsible for administering the Clean Air Act requirements through state implementation plans (SIPs).

The Federal Water Pollution Control Act (FWPCA) (42 U.S.C. § 1251 *et seq.*) was shaped in its present form by the Amendments of 1972. The goals of the Act are: (1) the achievement of swimmable and fishable waters by 1983, and (2) elimination of pollution discharge by 1985. Under the law, effluent limitations were established for industrial and municipal sources of water pollution. The Clean Water Act went through certain revisions in 1982–83, including re-authorization of some programs and clarification of particular sections.

In response to the problems posed by hazardous waste and toxic substances which may enter the environment, Congress passed three major pieces of legislation. The Toxic Substances Control Act of 1976 (TSCA) (15 U.S.C. § 2601 *et seq.*) is a product control law in contrast to a pollution control law. TSCA regulates the manufacture, distribution and sale of chemicals which 'may present an unreasonable risk of injury to health or the environment'. The Resource Conservation and Recovery Act (RCRA) (42 U.S.C. § 6901 *et seq.*) gives authority to the federal government to provide technical and financial assistance to state and local governments and interstate agencies to promote improved solid waste management techniques, and to issue regulations for the treatment, storage, transportation and disposal of hazardous wastes. The Comprehensive Environmental Response, Compensation and Liability Act of 1980 (42 U.S.C. § 9601 *et seq.*) provides the EPA with authority to require generators, transporters, treaters and disposers of hazardous wastes to remedy actual or potentially endangering hazardous waste sites and associated damage to natural resources. The legislation established a 'Superfund' and authorizes use of the fund's resources to perform the remedy if the responsible party fails to do so.

Further examination of these basic elements in the United States regulatory arsenal is necessary to understand their applicability to the industrial release of biological agents. While EPA has principal jurisidiction over industrial waste streams, other federal and state agencies, under public health statutes, can regulate the release of pathogenic agents in the environment from non-industrial sources.

32.1.3.2 Canada

In Canada, the demarcation between federal and provincial jurisdictions is framed in the British North America (B.N.A.) Act of 1867, a principal document of the Constitution of Canada along with its amendments in the Canada Act of 1982. Most of the federal powers are provided through Section 91 of the B.N.A. Act while Section 92 delineates the powers of the provinces. Amendments to Section 92 are found in Part VI of the Canada Act of 1982.

The B.N.A. Act does not address the environment specifically. Legal authority in this area has evolved through judicial interpretation. Some environmental problems lie within the jurisdiction of both the provincial and federal governments. Water pollution is a case in point. Whereas federal laws are designed to protect the fisheries, provincial laws are enacted to protect the public health.

The full impact of the new Canadian Constitution on the B.N.A. Act and its delineation of powers between federal and provincial jurisdictions is not fully understood at this time. Therefore, this chapter emphasizes historically-developed environmental jurisdictions.

Both federal and provincial governments have powers with respect to agriculture. Each can pass laws regulating fertilizers, feed products and pesticides. In the case of pesticides, the provinces have promulgated additional regulations beyond those of the Parliament. Where there are areas of overlapping jurisdictions, the federal government will at times resort to measures involving negotiation and coordination with provincial authorities. And when federal and provincial statutes are in conflict, federal laws generally prevail.

Chemical contaminants in the environment are regulated by the federal government under four main acts: The Fisheries Act, the Clean Air Act, the Canada Water Act and the Environmental Contaminants Act. In most instances these Acts define an environmental contaminant broadly enough to include biological agents. The exception is the Environmental Contaminants Act which applies exclusively to 'inanimate matter'.

According to Section 92 of the B.N.A. Act, the provinces are given authority over the working environment and waste disposal. The federal government can enact occupational health and safety statutes for a select number of industries that fall under its jurisdiction, subject to Part IV of the Canada Labour Code. Generally, federal legislation has been upheld where problems have taken on 'national dimensions' or become a matter of 'national concern'.

Ince (1976) makes the following observations about the genesis of environmental law in Canada: 'Because provincial powers of legislation are framed in such general terms, it is very difficult to limit provincial powers. Over the past century, the courts have, quite understandably, interpreted these powers very broadly which has enabled the provincial legislatures to deal with a vast number of areas. On the basis of the provinces' powers to control property, civil rights and local matters, a great deal of environmental legislation is authorized. We can safely say that these powers allow the provinces to legislate on land, air, water and noise pollution, land use control, parks and industrial regulation'.

Indeed, many provinces have enacted laws and issued regulations for chemical contaminants, radiation, clean air and water, waste management, pesticides and the working environment. These statutes were not developed with biological agents in mind, although in some cases the language is sufficiently broad to justify their application to microbial forms, if certain conditions are satisfied. In contrast to the strictly formal and mandatory framework for regulatory policy in the United States, Canada is known for a more informal and discretionary system of lawmaking. Within this framework there is considerable federal–provincial negotiation, consultation and cooperation, as well as industry–government consultations.

Before U.S. and Canadian laws and regulations are examined more fully with respect to the release of biological agents, a brief review is given of the major applications of microorganisms in industry.

32.2 POTENTIAL HAZARDS OF BIOPROCESSES

32.2.1 Pharmaceutical Industries

The conventional organisms that are used to manufacture antibiotics consist of a relatively narrow taxonomic range. Nearly a thousand distinct antibiotics are derived from six genera of filamentous fungi including molds of the genus *Cephalosporium* and the genus *Penicillium*. By inducing mutations through radiation and chemical substances, these two molds were the main source of antibiotics for 30 years. Other drugs manufactured by microbial fermentation techniques are viral and bacterial antigens, antifungal agents, antitumor drugs, alkaloids and vitamins.

Fermentation technology has been the principal process for manufacturing pharmaceuticals. Volumes produced in the fermentation vats may be as high as 100 000 l. A brief description of the process is given by Aharonowitz and Cohen (1981).

Recombinant DNA technology opens up opportunities for the biosynthesis of drugs, hormones and other biologically active substances by microorganisms containing the inserted relevant information. Somatostatin (a hormone made in the hypothalamus), insulin, growth hormone and interferons (antiviral agents) are currently being synthesized by *E. coli* K12 which have the requisite gene inserts. Gene-splicing techniques are also beginning to revolutionize vaccine production. Molecular cloning makes it possible to manufacture in large quantities non-virulent, nonselfreplicable segments of a virus that can be used to immunize a host.

32.2.2　Food Industries

The food processing industry uses microbial activity in two ways (Office of Technology Assessment, 1981): (1) inedible biomass is transformed by microorganisms into food for human consumption or animal feed; (2) organisms are used in food processing either by acting directly on food or by providing materials that can be added. Enzymes and vitamins are examples of the latter use. Food processing has utilized enzymes extracted from plants and animals. Microbial production of them has become economically competitive in some cases. Bacteria and molds also are used to make vitamins, such as riboflavin (*Ashbya gossypii*) and vitamin B12 (*Propionibacterium shermanii* and *Pseudomonas denitrificans*).

New developments in genetic engineering, *e.g.* protoplast fusion, are expected to broaden the use of fermentation in the food industry (Demain and Solomon, 1981). The genomes of two distinct species can be brought into a single cell. Recombinant DNA technology establishes genes as interchangeable elements capable of being transplanted between diverse organisms. Food processing need not depend exclusively on the enzymes found naturally in microorganisms; even human enzymes can be produced in large quantities by cloning their DNA in bacteria.

32.2.3　Energy Production

The production of liquid fuels through fermentation can be improved upon in two ways through genetic engineering (Office of Technology Assessment, 1981). First the genetic manipulation of plant seeds may yield better quality and greater quantity of biomass. Second, microbial mutants are being sought to improve the efficiency of converting agricultural and forest biomass into liquid fuel. Ethanol, among the most important organic substances in the chemical industry, has attracted significant attention in the biotechnology field. The genetic programming of conventional organisms used in the fermentation of ethanol, which include yeasts, *Zymomonas mobilis*, *Clostridium thermocellum* and *Trichoderma reesei*, has been proposed 'to increase the amount of certain enzymes in the cell or to replace one enzyme with another that has a higher specific activity' (Eveleigh, 1981, p. 168).

In a report entitled *Biotechnology: A Development Plan for Canada* (Task Force on Biotechnology, 1981), it was noted that the production of methane from the fermentation of agricultural, industrial and domestic wastes is another fuel prospect in addition to ethanol. Moreover, the report states, the bioproduction of substitute fuels to replace hydrocarbon-based conventional crude oil derivatives, 'could be of significance in determining alternate energy strategies for Canada'.

Some concern has been raised about the ecological impacts of genetically modified organisms released into the environment either purposefully or inadvertently through the waste stream. Potential hazards of disturbing biochemical pathways or altering ecological balances have been cited (Krimsky, 1982, p. 122; Wright, 1982). In Massachusetts, several communities have passed ordinances that regulate fermentation with recombinant organisms (Krimsky *et al.*, 1982) to insure that novel organisms are adequately contained or appropriately treated prior to being released in the waste stream.

32.2.4　Waste Treatment

Microbial activity is used in the detoxification and degradation of sewage and industrial wastes. Secondary sewage treatment facilities use some form of biological activity to degrade organic materials. The activated sludge process developed early in this century has depended upon the indigenous microorganisms in the waste stream. The activity of the microorganisms may be enhanced by additives or environmental controls, such as pH and temperature.

More recently, sludges have been inoculated with mixtures of microorganisms which are designed to accelerate the degrading process or broaden the types of chemicals broken down by bacteria. The next important breakthrough in this area will be made through genetic engineering. Microorganisms are being genetically constructed to degrade specific compounds found in industrial wastes. Organisms have been developed which are successful at the laboratory scale in degrading industrial organic compounds such as polychlorinated biphenyls (PCBs) and the herbicide 2,4,5-T. The Canadian Task Force on Biotechnology (Task Force on Biotechnology, 1981) cited two advantages offered by biological processes over other methods of detoxification of effluents. First, biological processes are readily adaptable to the varied composition of wastes and conditions of degradation. Second, a large selection of microorganisms can degrade a wide variety of substances. The Task Force reported that 'increased pressure upon new and existing industries to invest more heavily in waste treatment and pollution control could spur more development in this application of biotechnology'.

Some policy experts are concerned that novel organisms developed for sewage treatment may constitute new forms of pollution. Johnston (1981) discussed some environmental outcomes of enhanced sewage treatment through new microbial systems. '[T]he use of microbes for pollution treatment and control must be viewed as an irreversible release of the organism and its DNA to the environment . . . . Even if consideration were restricted to a contained system such as a sewage treatment plant, where disinfection of sludge and effluents could be undertaken, it would be unrealistic to guarantee that no viable organisms would be released . . . . Unlike biological wastes from pharmaceutical manufacturers . . . the sheer volume of effluent from waste treatment facilities makes the disinfection impractical. Economic constraints alone negate the effective sterilization of such waste streams . . . . It is imperative therefore, that any use of engineered organisms for waste treatment be considered a deliberate, irreversible release of the organism and its DNA to the environment.'

32.2.5 Bioorganic Chemistry

Industrial organic chemicals including enzymes, acids and solvents have been manufactured by microbial fermentation. There are many organic chemicals which can be made either by chemical synthesis or by fermentation. The deciding factor has been economics. According to Eveleigh (1981), microorganisms yield products for some 200 substances of commercial value; only a few of those are currently manufactured by biological methods. This group includes ethanol, *n*-butanol, acetone, acetic acid, citric acid, lactic acid, amino acid and various enzymes. Solvents like *n*-butanol, acetone and glycerol, once manufactured by biological methods, are currently produced mainly by chemical synthesis. Efforts are underway to exploit the use of thermophilic bacteria and renewable-resource-based feedstocks to again manufacture solvents by fermentation.

Zaugg and Swarz (1982) cite three genera of organisms that will probably be used in chemical production: *Pseudomonas*, *Acinetobacter* and *Flavobacterium*. The authors claim little is known about the effects these organisms have on man and express some apprehension about the introduction of genetically engineered organisms in production. 'The chemical industry has a poorer record than the pharmaceutical industry in areas related to worker safety and environmental protection, causing one to be apprehensive of the new technologies for which industry-wide experience is limited.'

32.2.6 Biological Agents Used in the Environment

Both current and contemplated uses of microorganisms in non-contained environments are designed to improve operations in the fields of metallurgy, agriculture and *in situ* pollution control. Bacteria have been used to extract metals from low grade ores. Copper and uranium are mined and commercially purified by a process that includes leaching by bacteria of the genus *Thiobacillus*. The bacterial leaching process takes two principal forms: organisms act directly on the ore to extract the metal, or they produce other chemicals which make separation of the metal possible.

Zaugg and Swarz (1982) cite some potential hazards in the use of organisms in the mining industry. '[A]ll foreseeable applications of biological processes . . . involve microbial systems operating in relatively open environments, such as slag heaps or tailings ponds. Consequently, there are risks that microorganisms or their metabolic products will inadvertently contaminate the local ecology'. The authors offer three examples: (1) bacterial leaching operations that generate large quantities of sulfuric acid could acidify water supplies; (2) *Thiobacilli* and related species may acquire the traits to infect humans; (3) metals concentrated by bacteria from dilute mine waters can accumulate in the food chain.

The organism *Xanthomonas campestris* produces xanthan gum which is used for oil recovery by increasing the viscosity of water used to displace the oil. And while some organisms are used in oil recovery operations, another has been developed to degrade crude oil. The General Electric Corporation was awarded the first patent on a microorganism *sui generis* in the United States after the Supreme Court upheld a lower court decision to grant a patent for a strain of *Pseudomonas putida* (*Diamond* v. *Chakrabarty*, 1980). At the present time the organism has not been marketed by the company. According to the Office of Technology Assessment (OTA), bacteria have been used to degrade gasoline in underground spills and strains are available for breaking down highly toxic substances such as pentachlorophenol. Alexander (1981) reports on a growing interest in biodegradation of toxic chemicals.

The Office of Technology Assessment (OTA) cites some ecological risks over the release of such organisms into the environment (OTA, 1981, p. 118). 'Introducing large numbers of genetically engineered microorganisms into the environment raises questions of possible ecological disruption, and liability if damage occurs to the environment or human health . . . the present lack of sufficient scientific knowledge, scientists, and interdisciplinary teams, and the concerns for ecological safety present the major obstacles to the use of genetic engineering in microbial leaching.'

OTA raises some environmental concerns about the use of organisms in oil recovery also. 'All strains of *Xanthomonas*, which produce xanthan gum polymer, are plant pathogens. Other microorganisms with potential, such as *Sclerotium rolfii* and various species of *Aureobasidium*, have been associated with lung disease and wound infections, respectively' (OTA, 1981).

The OTA study (OTA, 1981) expresses a cautious optimism about the large scale use of microorganisms in the environment. 'Immediate environmental and legal concerns, therefore, arise from the potential risks associated with the release of microorganisms into the environment. When they naturally cause disease or environmental disease or environmental disruption, their use is clearly limited. And when they do not, genetic engineering raises the possibility that they might.'

An agricultural application which requires spreading organisms in the environment is the use of biological insecticides, fungicides and herbicides. It is estimated that 100 known species of bacteria are pathogenic to insects, but only a few species have been developed into commercial insecticides (OTA, 1981). The bacterial agents are *Bacillus popilliae*, *Bacillus thuringiensis* and *Bacillus moritai*. Here, too, genetic engineering may make it possible to increase the potency of bacterial strains by doubling the genes that code for the toxins which destroy the insect pests.

Supporters of biorational pesticides as a substitute for synthetic chemical agents believe that these organisms occupy a narrow ecological niche for which there is no evidence of departure. The viral pesticide agents have shown no potential for adverse human effects. Moreover, viruses may occur naturally at much higher levels than result from their intentional use as pest control agents. On the other side are those who support extensive testing. They cite the potential for the latency of viral agents in non-target species, including man, where the virus may interact with the host's genetic material and could be involved in a disease etiology. Second, they cite the capacity of the viruses to mutate which may change their specificity and selectivity (EPA, 1979b).

32.3 LAWS APPLICABLE TO BIOLOGICAL AGENTS

32.3.1 United States

32.3.1.1 *Authority under the Federal Food, Drug and Cosmetic Act (FFDCA)*

The Federal Food, Drug and Cosmetic Act (21 U.S.C. § 301 *et seq.*) and Section 351 of the Public Health Service Act (42 U.S.C. § 201 *et seq.*) give the Food and Drug Administration (FDA) authority to regulate food, drugs, biologics, medical devices and veterinary medicines. The statutes are broad enough to cover recombinant DNA (rDNA) techniques, and the agency's authority will extend to products it ordinarily regulates when they are manufactured by newly developed genetic engineering methods.

University and industrial laboratories are developing sophisticated techniques which will be used to create products for which FDA approval must be obtained before they can be marketed. Several examples of these products, some in the experimental stage, others closer to large scale production and marketing are given by Miller (1981): (1) drugs—human insulin, human growth hormone, thymosin, ACTH, endorphins; (2) biologics—interferons, vaccines, serum albumin, urokinase; (3) enzymes used in food processing; (4) medical devices—*in vitro* diagnostic tests for thalassemia or sickle cell anemia; (5) veterinary medicine—interferons, animal growth hormones.

Although the FFDCA focusses on products, the statute can be used to control production processes as well. FDA can regulate chemicals being produced for use in a product regulated by the agency. This authority extends to chemicals used in experiments designed to develop products which are within the purview of FDA. In theory, FDA can regulate the development of genetically engineered host organisms, as well as plasmids and vectors used in rDNA processes, if the procedure involves making a product for which FDA approval is necessary. The agency has not determined how or whether it will exercise this authority, the extent of which is unclear.

32.3.1.2 Public Health Services Act (PHSA)

Section 361 of the Public Health Services Act (42 U.S.C. § 264) authorizes the Department of Health and Human Services (DHHS) 'to prevent the introduction, transmission, or spread of communicable diseases'. This might provide authority to require adherence to Section I-D-2 of the National Institutes of Health rDNA Guidelines (NIH, 1982), which currently restricts experiments that deliberately form rDNA molecules containing genes for the biosynthesis of certain toxins lethal to vertebrates. The Recombinant DNA Advisory Committee (RAC) of the NIH voted in February, 1982 to drop the prohibition for this type of experiment and requires instead prior approval by the committee, NIH and the local institutional biosafety committee. The recommendation has been approved by the acting director of the National Institute of Allergy and Infectious Diseases of NIH, who currently has authority to promulgate rDNA guidelines for institutions receiving funding from NIH.

Aside from the Guidelines, Section 361 of the PHSA could potentially be used if FDA concluded that an rDNA product could be harmful to humans or the environment, but no regulations have yet been promulgated.

32.3.1.3 Resource Conservation and Recovery Act of 1976 (RCRA)

One of the objectives of RCRA (42 U.S.C. § 6901 *et seq.*) stated in § 1003(4), is to promote the protection of health and the environment by 'regulating the treatment, storage, transportation, and disposal of hazardous wastes'. A 'hazardous waste' is defined in § 1004(5) as a solid waste which, because of its quantity, concentration, or physical, chemical or infectious characteristics may (a) cause, or significantly contribute, to an increase in mortality or an increase in serious irreversible, or incapacitating illness; or (b) pose a substantial or potential hazard to human health or the environment.

RCRA is administered by the Environmental Protection Agency (EPA), which has promulgated regulations for the generation, transportation, storage and disposal of hazardous wastes (40 C.F.R. Parts 261–264, 1980). Numerous provisions of these regulations are being litigated.

Solid wastes, considered hazardous wastes, and thus regulated under RCRA, are identified and listed in 40 C.F.R. Part 261, 1980. In addition to listing hazardous wastes, EPA has prescribed an 'extraction procedure' (EP) (40 C.F.R. Part 261, Appendix II). Solid wastes which exhibit EP toxicity as well as those listed as hazardous wastes in Subpart D will be regulated as hazardous wastes. EPA has also prescribed processes to reduce pathogens in solid waste (*Fed. Regist.*, 1979).

If pathogenic organisms remain in the waste after treatment or it contains substantial amounts of heavy metals or toxic chemicals, it is probably hazardous waste and the generator, transporter and disposer of the waste must comply with EPA requirements. EPA has issued standards applicable to generators (40 C.F.R. Part 262) and transporters (40 C.F.R. Part 263) of hazardous waste and owners and operators of hazardous waste treatment, storage and disposal facilities (40 C.F.R. Part 264). Section 3002 of RCRA authorizes the use of a 'manifest system' to track shipments of hazardous waste from the time it is generated until final disposal. If a manufacturer determines that his waste is hazardous, he must comply with EPA regulations establishing standards for the manifest system (40 C.F.R. § 261.5(c)(5)).

Manufacturers whose solid waste includes microorganisms will be subject to EPA hazardous waste regulations issued under RCRA if the waste after treatment contains pathogenic organisms, EP toxic substances, or is a listed hazardous waste in nonexempt quantities.

32.3.1.4 Clean Air Act (CAA) (42 U.S.C. § 7401 et seq.)

If an industrial process results in the release of 'criteria pollutants', *i.e.* hazardous air pollutants for which standards have been set by EPA, the company must comply with Parts A, C and D of

Title I of the Clean Air Act (40 C.F.R. Part 50). It is not likely, however, that the operation of a fermentation plant will result in emission of criteria pollutants unless the liquid wastes of the fermentation process were to be dried and incinerated (EPA, 1976).

Section 112 (42 U.S.C. § 7412) authorizes EPA to set emission standards for non-criteria hazardous air pollutants which 'may reasonably be anticipated to result in an increase in mortality or an increase in serious irreversible or incapacitating reversible illness'. If EPA can demonstrate that a microorganism or a product synthesized by a microorganism which is emitted into the air poses such a risk it may promulgate hazardous air pollution regulations establishing emission standards. However, since the procedure for promulgating a hazardous air pollutant standard is time consuming, it has been suggested that EPA will be more likely to regulate airborne microorganisms containing recombinant DNA molecules or their hazardous by-products under the Toxic Substances Control Act (McGarity, 1981).

The term 'air pollutant' as defined in § 302(g) includes a 'biological . . . substance . . . which is emitted into or otherwise enters the ambient air'. Microorganisms which escape from a fermentation vessel would fit within this definition, however, if it is unlikely that they travel beyond the workplace, they would more likely be regulated by the Occupational Safety and Health Administration (OSHA).

As the biotechnology industry develops and microorganisms produce potentially hazardous products on a large scale, the danger of emission of hazardous air pollutants may increase. The CAA may play a larger role in regulation of the industry at that time.

32.3.1.5 Federal Water Pollution Control Act (FWPCA) as amended by the Clean Water Act of 1977 (CWA) (33 U.S.C. § 1251 et seq.)

The FWPCA regulates the discharge of conventional pollutants, which include biological materials, into navigable waters of the U.S. Under Sections 301, 304(b), (c) and 306(b), EPA may promulgate technology-based controls for new and existing dischargers of conventional pollutants. Conventional pollutants are defined in Section 304(a)(4) to include biological oxygen demanding pollutants (BOD), suspended solids, fecal coliform and pH. Dischargers of conventional pollutants are to be regulated assuming application of 'best conventional pollution control technology' (BCT) (Section 301). Even if a source does not belong to a category for which effluent limitations and guidelines have been promulgated, it must still obtain a National Pollutant Discharge Elimination System (NPDES) permit under Section 402. EPA has promulgated effluent limitations and new source performance standards for the pharmaceutical industry which specify the amounts of conventional pollutants that can be discharged into navigable waters by new and existing fermentation processes (40 C.F.R. Part 439).

The category of toxic water pollutants is regulated under a different scheme from conventional pollutants. Section 307 adopts the list of 65 toxic water pollutants (40 C.F.R. § 401.15) and empowers EPA to add or withdraw substances from this list.

EPA can prescribe, under Section 307(a)(2), toxic effluent limitations, which require the use of 'best available technology' (BAT) for treating toxic pollutants. In compiling a list of toxic pollutants, EPA must take into account the toxicity, persistence and degradability of the pollutant. For toxic water pollutants, there are no water quality based modifications of BAT. Under Sections 307(a)(2) and (a)(4), EPA is empowered to promulgate toxic effluent standards which are more stringent than toxic effluent limitations. They are ambient standards which must be set at a level that provides an ample margin of safety.

Depending upon the category into which waste products of the biotechnology industry fall, EPA can regulate discharges into navigable waters under the scheme provided by FWPCA for toxic, conventional or residual pollutants. The most stringent regulations apply to toxic pollutants which must be treated using BAT.

32.3.1.6 Toxic Substances Control Act (TSCA) (15 U.S.C. § 2601 et seq.)

The purpose of TSCA is to develop data concerning potentially hazardous chemicals and to authorize the EPA to regulate chemicals used both in research and commerce so as to minimize risk to human health and the environment (15 U.S.C. § 2601(b)). It was intended by Congress to be used to fill in regulatory gaps left by other environmental statutes. If EPA can regulate under another statute, it is obligated to do so before using TSCA. The provisions of TSCA apply to research as well as manufacturing and only Section 5 which requires premarket notification for a 'significant new use' of an existing chemical substance is not applicable to university research.

A key legal issue with respect to the applicability of TSCA to genetically engineered micro-organisms is whether the recombinant DNA within the microorganism will be considered a 'chemical substance' or a 'mixture'. A 'chemical substance' is defined in Section 3(2) to include 'any organic or inorganic substance of a particular molecular identity' including 'any combination of such substances occurring in whole or in part as a result of a chemical reaction or occurring in nature'. The statute excludes pesticides, tobacco and substances subject to FFDCA or the Atomic Energy Act. It is unclear whether the definition would include microorganisms containing recombinant molecules and the issue will have to be litigated if EPA chooses to use its authority under TSCA to regulate microorganisms.

In promulgating regulations under TSCA, EPA took the position that 'the definition of chemical substances does not exclude life forms which may be manufactured for commercial purposes . . .' (*Fed. Regist.*, 1977). However, in a letter responding to a Senate inquiry, former EPA administrator Douglas M. Costle (1978) stated that although EPA agreed that rDNA molecules were 'chemical substances' within the meaning of Section 3, it wasn't clear that the host organisms containing these molecules were subject to TSCA. Some legal experts (McGarity, 1981) believe that if EPA were to assert such authority and its initiative survived legal challenge, TSCA could potentially be used to regulate the biotechnology industry.

Section 4 authorizes EPA to require testing of chemicals which 'may present an unreasonable risk of injury to health or the environment' and for which there exist insufficient data and experience to predict the risk, or chemicals which will be produced in substantial quantities and which may reasonably be expected to enter the environment in substantial quantities.

Assuming that spliced DNA is held to be a 'chemical substance' or a 'mixture', EPA must first find that it may present an 'unreasonable risk' before it could require testing. Although it is possible that in the future harmful strains will be developed, present scientific evidence does not support a finding that genetically altered microorganisms now being utilized for rDNA work present an unreasonable risk. However, the testing requirement in Section 4 could be imposed by EPA if new or less well-characterized host strains are used.

The second requirement of Section 4, that the chemical will be entering the environment in substantial quantities, also serves to reduce the probability that testing would be required of the genetic engineering industry. Since the fermentation process is enclosed, it is unlikely that escaping microorganisms or chemicals would cross the 'substantial quantities' threshhold. However, this may not be the case for non-pharmaceutical use of genetically engineered microorganisms, such as those used for the digestion of spills.

TSCA authorizes EPA to compel notification of new chemicals or mixtures before they are manufactured, to require submission of relevant records, to inspect manufacturing facilities and to require testing in order to generate adequate risk data on proposed new chemicals.

In 1983 EPA began exploring the agency's role in biotechnology. The EPA Administrator's Toxic Substances Advisory Committee recommended that the agency apply TSCA to regulate the release of genetically modified organisms into the environment. EPA will use its experience with regulating microbial pesticides as a basis for addressing the problem (see Section 32.4.1.1).

In conclusion, the authority of EPA to regulate the biotechnology industry and university researchers under TSCA will be established only if the courts hold that recombinant DNA molecules, vectors and host organisms containing altered DNA, or even natural biological entities that might present a hazard, are 'chemical substances' or 'mixtures' as these are defined in Section 3.

32.3.1.7 *Marine Protection, Research and Sanctuaries Act of 1972*

The Marine Protection, Research and Sanctuaries Act of 1972 (33 U.S.C. § 1401 *et seq.*) was passed by Congress in order to 'prevent or strictly limit' the use of the oceans for disposal of wastes of almost any type, including those generated by the pharmaceutical industry. Title I regulates ocean dumping and the transportation for dumping of waste materials through a permit program operated by the EPA and the Army Corps of Engineers. Title II directs the Secretary of Commerce to conduct research on the effects of ocean dumping through the Office of Marine Pollution Assessment.

The Act defined 'unreasonable degradation' of the environment in terms of a balance between environmental and public health factors and the availability of alternatives to dumping (33 U.S.C. § 1412(a)). Current regulations ban the dumping of sludge in the ocean unless it has been determined that there is not a better alternative method of disposal. EPA makes this determination on a case-by-case basis after balancing the statutorily mandated factors.

Under 1982 draft proposals, EPA would make it easier to obtain permits to dispose of dredged material and low level radioactive wastes. EPA has identified the pharmaceutical industry as one which would benefit from the revised rules. There is uncertainty at this time as to whether these proposals will be promulgated as regulations.

32.3.1.8   *Federal Insecticide, Fungicide and Rodenticide Act (FIFRA)*

Chemical and biological pest control agents are regulated by EPA under the Federal Insecticide, Fungicide and Rodenticide Act (7 U.S.C. § 136 *et seq.*) as amended in 1972, 1975 and 1978. Section 2(u) of FIFRA defines a pesticide as 'any substance intended for preventing, destroying, repelling, or mitigating any pest, and . . . any substance or mixture of substances intended for use as a plant regulator, defoliant, or desiccant'. The language of FIFRA provides EPA with very broad regulatory authority. Each new pesticide must be registered by the agency before it can be marketed. Prior to registration EPA evaluates the safety and efficacy of the agents. The registration must be approved if the pest control agent performs its intended function without unreasonable adverse effects on the environment, and when used in accordance with accepted practice it will not cause unreasonable adverse effects on the environment. The term 'unreasonable adverse effects on the environment' is defined in the Act as 'any unreasonable risk to man or the environment, taking into account the economic, social and environmental costs and benefits of the use of any pesticide' (§ 136(a)(c)(5)). A pesticide which has been registered may be cancelled if its continued use poses a 'substantial question of safety' to man or the environment. EPA may also ban the production and distribution of a pesticide if it presents an imminent hazard.

Responding to the differences in biochemical activity between chemical and biological pesticides, EPA issued two sets of guidelines for registering pest control agents. (EPA, 1978a; EPA, 1980). The justification for the separate registration requirements is provided by the agency: 'The vast majority of the more than one thousand pesticide active ingredients regulated by EPA . . . are man-made organic and inorganic chemicals and are innately toxic. Less than one percent of the pesticide active ingredients registered by the Agency are inherently different in their mode of action from most organic and inorganic compounds. This small group is exemplified by the living or replicable biological entities, such as viruses, bacteria, fungi and protozoans. Naturally occurring biochemicals, such as plant growth regulators and insect pheromones, also function by modes of action other than innate toxicity'. EPA defines this group of pest control agents as 'biorational pesticides'. A discussion of regulations for 'biorational pesticides' is given in Section 32.4.

32.3.2   Canada

In a pattern similar to that of the United States, during the 1970s the Canadian Parliament responded to societal concerns over chemical contamination of the environment with the passage of new laws followed by a host of regulations. However, in contrast with the United States, the federal government in Canada does not have principal jurisdiction over the release of hazardous agents into the environment. The authority is divided between the federal and provincial sectors. In effect, Canada has two tiers of pollution control statutes and occupational health and safety laws, those passed by Parliament with application to federal jurisdictions and those passed by the provinces. In cases where jurisdictions overlap, the federal government has preemptive authority. Many of the laws reviewed in this section were not explicitly enacted for controlling biological agents but where the statutory language is framed broadly enough, their applicability to such agents becomes manifest.

32.3.2.1   *Environmental Contaminants Act (S.C. 1974–75, C.72)*

The goal of the Environmental Contaminants Act (ECA) is 'to protect human health and the environment from substances that contaminate the environment'. The Act defines 'substance' quite explicitly in a way that excludes biological agents. 'A "substance" means any distinguishable kinds of inanimate matter' capable of becoming dispersed or transformed in the environment, into something of the same chemical moiety or having similar chemical properties (Sec. 2).

The Act is under the joint jurisdiction of the Minister of the Department of Environment and the Minister of the Department of National Health and Welfare. Under the law, the Governor in Council may establish the maximum quantity or concentration of a substance released into the environment in the course of any commercial, manufacturing or processing activity, and to establish a schedule of substances which may not be imported, manufactured, processed, offered for sale or used.

If the Ministers have reason to conclude that a substance may endanger human health or be a threat to environmental quality they may (1) require commercial producers of the substance or class of substances to notify the government of such activities and provide information about the substances; (2) require producers and importers of the substance or any product containing it to conduct tests which the Ministers may reasonably require. Like the Toxic Substances Control Act in the United States, the ECA is designed to regulate chemicals before they are put into the market place. The Act requires the Canadian federal government to consult with the provinces and industry prior to taking any regulatory action (Sec. 5(1)(b)). Like other federal environmental control statutes, the ECA is an enabling law which becomes realized when specific regulations are made under it.

While the ECA is directed at the regulation and control of inert chemical substances and not organisms, it could have jurisdiction over chemicals produced by microorganisms. If chemical-producing bacteria are used in an industrial process, and the release of the organisms into the environment is tantamount to a release of the chemicals, then it is plausible that the ECA might be applied, but as a control measure for the chemical substances. The applicability of the law to microbially-produced chemicals will depend upon the large release of organisms that synthesize chemicals which endanger humans or the environment. A distinct area of ambiguity is whether toxin-producing bacteria fall under the Act since the toxins themselves are chemical products with harmful side effects. In whichever manner the chemicals are produced, there is a 500 kg cut-off for manufacture or import below which the law does not apply, making its application to biological agents even a more remote possibility.

32.3.2.2   *Pest Control Products Act, 1968–69 (R.S.C. 1970, C. P-10)*

The Pest Control Products Act (PCPA) regulates products used for the control of pests and the organic functions of plants and animals. It applies to any 'product, device, organism, substance or thing that is manufactured, represented, sold or used as a means for directly or indirectly controlling, preventing, destroying, mitigating, attracting or repelling any pest' (Sec. 2). The Minister of Agriculture administers the Act and is responsible for 'prescribing the form, composition and other standards for (pest) control products' (Sec. 5).

Under the PCPA, any pest control product sold or imported into Canada must be registered, conform to prescribed standards, and be labeled and packaged as determined by the Minister (Sec. 4). To satisfy the registration requirement, a producer or importer must provide data that demonstrate the product is both safe and effective under acceptable standards of use.

The use and manufacture of pesticides is controlled not only by federal laws and regulations in Canada, but also by a variety of provincial legislation and regulations.

The regulations issued under the Act provide the registration requirements for pest control products. Despite the broad statutory language, the regulations issued are targeted toward chemical agents, *i.e.* hazard information refers to toxicological data, but not information about infectious agents.

32.3.2.3   *Clean Air Act (S.C. 1970–71–72, C.47)*

This Act gives the Governor in Council authority to establish national ambient air quality standards for contaminants emitted into the air by stationary sources. To be regulated, the emissions must impose risks to health or be in violation of an international agreement. For the purpose of the Act, an air contaminant is defined as 'a solid, liquid, gas or odour, or a combination of any of them that, if emitted into the ambient air, would create or contribute to the creation of air pollution' (C.47, Sec. 2). Air pollution is defined as a condition of the ambient air that, from the presence of contaminants, 'endangers the health, safety or welfare of persons, interferes with the normal enjoyment of life or property, endangers the health of animal life, or causes damage to plant life or property' (C.47, Sec. 2).

Three ambient air quality objectives are specified: maximum tolerable limit; maximum acceptable limit; and maximum desirable limit. The government's standards are only guidelines and not legally enforceable. The objectives provide a signal to industry about potential federal regulations, and to the provinces that they might adopt the emission standards into their own regulations. The Minister of Fisheries and Forestry may enter into agreements with the provincial governments for the purpose of 'facilitating the formulation, coordination and implementation of policies and programs designed for the control and abatement of air pollution . . . ' (C.47, Sec. 19).

The description in the Act of an air contaminant would seem to include biological organisms that might be released in large-scale fermentation processes as part of the gaseous effluent. For regulation, the burden is on the government to demonstrate that the gaseous effluent, if it contains viable organisms, represents a hazard to human health, endangers the environment or interferes with the normal enjoyment of life.

32.3.2.4 Canada Water Act (R.S.C. 1970, 1st Supplement, C.5)

This Act authorizes the establishment of water quality management areas and comprehensive water resource management programs by the federal government in collaboration with provincial governments. Under the management of water quality are included the establishment of waste treatment facilities and water quality management plans. These plans make recommendations about the types of wastes and the quantities that may be discharged into the water as well as the conditions for such discharge. Waste is defined broadly as 'any substance that, if added to any waters, would degrade or alter . . . the quality of those waters to an extent that is detrimental to their use by man or any animal, fish or plant that is useful to man . . . ' (Sec. 2). This Act is an appropriate vehicle for two types of biological releases into the environment. It would appear to cover the processing of sludges from wastewater treatment plants that may contain viable organisms. The Act may also have jurisdiction over the use of novel organisms in the waste treatment process, since it includes recommendations for the types of treatment facilities necessary to achieve prescribed standards.

32.3.2.5 Fisheries Act (R.S.C. 1970, and amendments S.C. 1977, C.35)

Authority for the Federal Fisheries Act (FFA) arises out of federal jurisdiction over the seacoast and inland fisheries. The Act has broad powers making almost any discharge to waters inhabited by fish under its purview. It prohibits the disposal of deleterious substances into such waters. The substances referred are considered to be those which, when found in water, make it 'deleterious to fish or to the use by man of fish that frequent that water' (Sec. 7). The federal government may specify substances it classifies as deleterious under the Act and establish conditions under which substances can be legally disposed.

The Water Pollution Control Directorate of Environment Canada has issued regulations and guidelines under FFA for certain industrial sectors pertaining to the control of the discharge of conventional pollutants. According to Cornwall (1982), 'none of these regulations and guidelines have included biological agents as parameters; nor do they involve the fermentation industry directly'.

The Water Pollution Control Directorate is actively addressing the control of the discharge of toxic chemicals; the regulation of biological agents in wastewaters is not currently under consideration. However, the Directorate has issued a report on the lack of effective treatment processes for the reduction of *Salmonella* populations in the liquid effluent from meat and poultry plants (WPCD, 1976). The report recommended disinfection of the treated effluents before the *Salmonella* organisms are discharged into the environment.

32.3.2.6 Quarantine Act (R.S.C. 1970, C.33, 1st Supplement)

This Act gives the Minister of National Health and Welfare authority to prevent the introduction into Canada of infectious agents or contagious diseases. The Minister publishes a schedule of such agents or diseases that fall under the regulations. A quarantine officer may order the quarantine of any person arriving in or departing from Canada who may be in possession of such agents, be a carrier of infections, be infected with insects that may be carriers of such agents, or may have been in close proximity with a person satisfying these conditions. The officer may also require the disinfection of any conveyance found to be carrying agents of an infectious or contagious disease. The Act has no jurisdiction over industrial process releases of microorganisms, the disposal of infectious materials, or the inter- or intra-province transport of biological agents.

32.3.2.7 Canada Shipping Act (R.S.C. 1970, C.S-9 as amended in S.C. 1971, C.27)

The Governor in Council may issue regulations prohibiting the discharge from ships of pollutants to all Canadian waters and fishing zones under federal jurisdiction. A pollutant is defined (C.27, Part 19, Sec. 736) as 'any substance that, if added to any waters, would degrade or alter or form part of a process of degradation or alteration of the quality of those waters to an extent that is detrimental to their use by man or by any animal, fish, or plant that is useful to man . . . '. The Act would apply to the discharge from ships of fermentation sludges containing biological materials if they threaten to alter the ecological balance of nature. However, for the purposes of implementation, the Governor in Council classifies substances as pollutants.

32.3.2.8 The provinces

Under Section 92 of the British North America (B.N.A.) Act, provincial legislatures have authority within their provinces over matters of manufacturing, municipal institutions, property and civil rights, the working environment, and waste disposal. Although the B.N.A. Act provides no federal–provincial division of authority on environmental affairs, the provinces have developed considerable responsibility and autonomy for the regulation of environmental contaminants. Most of the provinces have a variety of pollution control statutes in conjunction with more established public health laws that regulate the discharge of effluent into the air, water or land. Ilgen (1981) finds that 'while provincial governments have found their own department of environment, there seems thus far to be a willingness to cede these residual powers in the control of toxic substances to the federal government'.

The regulatory instrument used most frequently in provincial regulation for the discharge of pollutants is a system of licensing, permitting, or issuance of guidelines. In cases where the federal government establishes minimal environmental quality standards or performance criteria for air or water, the provinces build upon these standards in their regulations. For those situations where there is overlapping jurisdictions, coordinated regulatory agreements are established between federal and provincial governments.

For this section it will suffice to illustrate a few examples of provincial statutes that pertain to biological effluent.

In Alberta, air pollutants are regulated by the Clean Air Act (1971, amended 1972, 1974) and Standards for Incineration. Bacteriological agents and viruses may fall under the incineration classification 'animal solids and organic wastes'. There are no standards for incineration of pesticide materials and other hazardous wastes, and disposal of these materials in this way is not allowed. No standards have been issued which regulate airborne microorganisms.

The disposal of biological agents into the waters of Alberta fall under the Clean Water Act of Alberta (1971, amended 1972, 1974). Industrial plants discharging their wastewater to municipal sewage systems are exempted from the Clean Water Act, but municipalities are responsible for meeting effluent standards. A water contaminant according to the Alberta statutes means 'any solid, liquid, gas or combination of any of them . . . '.

The land disposal of sludges from municipal and industrial wastewater treatment are regulated by the province, as is the burial of biological waste which is governed by the Public Health Act.

For Ontario, the principal law for controlling emissions is the Environmental Protection Act (1971). The Act prohibits discharges into the natural environment of any contaminant that exceeds what is prescribed in regulations. Discharge of contaminants must be approved by the Ministry according to waste management criteria and approved disposal sites. The Water Resources Act (1970, amendments, 1972) prohibits the release into the water of any substance which may impair its quality. Impairment may be said to take place if the material discharged causes or may cause injury to any person, animal or other living thing.The Act established authority over sewage treatment systems and accidental discharges of contaminants.

In British Columbia, the chief legislation for protecting water quality is the Pollution Control Act (1967). This law allows the province to set objectives for the discharge of effluent from a variety of industrial sources (including those that use fermentation processes, such as breweries and distilleries) to marine and fresh waters. Water quality objectives are in terms of BOD levels, suspended solids and pH range. A Pollution Control Board designated by the provincial government sets environmental quality objectives for specific industries. These objectives function as unenforceable guidelines. The Director of the Pollution Control Branch exercises considerable discretion in issuing permits for specific emission sources, in holding hearings and in response to permit objections. The discretionary powers of pollution control agencies exists in all the provinces and has come to be understood as Canada's flexible approach to regulation.

32.4 GUIDELINES AND REGULATORY ACTIONS FOR THE RELEASE OF BIOLOGICAL AGENTS

32.4.1   United States

32.4.1.1   Environmental Protection Agency

Under the Resource Conservation Recovery Act, the EPA acquired authority to promulgate regulations for any waste, chemical or biological, that may be harmful to man or the environment. On 18 December 1978, EPA published a proposed plan for the disposal and treatment of hazardous wastes (EPA, 1978b). The plan included infectious wastes as a special category of hazardous wastes. Listed in the category of infectious wastes was the classification of etiologic agents issued by the Centers for Disease Control (1974). Subsequent to the publication of the proposed rule, EPA decided not to adopt regulations for infectious waste materials. It chose instead to develop an infectious waste management plan. This is purely an information report on state of the art procedures for handling infectious wastes. Although the plan has not yet been released by the agency it indicates that EPA will not be regulating infectious wastes, but merely providing industry with a range of options for managing their handling and disposal.

Most states have laws and regulations for the treatment of infectious wastes. These laws generally require autoclaving, incineration or processing by an approved treatment method before disposal in a sanitary landfill or discharge into the sewers. Some states treat infectious wastes as a subcategory of hazardous wastes, but many in this group have not yet promulgated regulations.

EPA's regulatory umbrella extends to pesticides through the Federal Insecticide, Fungicide and Rodenticide Act. As previously discussed (Section 32.3.1.8), before distribution or sale, a new pesticide must be registered with EPA. Under authority of FIFRA, the agency evaluates the safety and efficacy of the product. The registration must be approved if the pesticide performs its intended function without unreasonable adverse effects on man or the environment.

Prior to 1979, EPA had no formal policy on the registration of biological pesticides. Organisms were registered on an *ad hoc* basis. Only a few microorganisms were accepted as pest control agents over a period of 30 years. *Bacillus popilleae* was registered by the Department of Agriculture in 1948; *Bacillus thuringiensis* received its registration by that agency in 1960.

In May 1979, EPA issued a policy statement on the regulation of biorational pesticides, a subcategory of which includes microbial and viral agents. Biorational pesticides are defined as 'biological pest control agents and certain naturally occurring biochemicals (*i.e.* pheromones) which are inherently different in their mode of action from most organic and inorganic pesticide compounds currently registered with EPA. Among the organisms covered in the policy are viruses, bacteria, fungi, protozoa and algae' (Environmental Protection Agency, 1979a).

On 24 March 1981, EPA published a proposed rule exempting from regulation under FIFRA certain classes of biological control agents (defined as 'any living organism applied to or introduced into the environment to control the population or biological activities of another life form which is considered a pest under Section 2(t) of FIFRA' (*Fed. Regist.*, 1981).

Section 25(b)(1) of FIFRA states that EPA can exempt certain biological control organisms from regulation, if they are adequately regulated by another federal agency or are of a character which is unnecessary to be subject to the Act. Under FIFRA, the definition of pesticide includes many diverse forms of macroscopic life such as birds, insects and aquatic mammals. EPA has not attempted to regulate the macroscopic control agents.

Everything except the following biological control agents are exempted from EPA pesticide registration: (1) viruses; (2) bacteria, including actinomycetes, rickettsia, mycoplasmas; (3) protozoa; (4) fungi of lower taxonomic order than the subdivision Basidiomycotina, or as members of the class Teliomycetes, or the sub-class Phragmobasidiomycetidae of the Basidiomycotina; (5) organisms classified as members of Class I, Schizophyceae, of Division I of the Plant Kingdom, Protophyta, including blue-green algae.

The EPA has drafted a 433 page set of working guidelines for registration of biological control agents (EPA, 1980). Under these guidelines, EPA sets requirements for the performance, toxicology and registration procedures of biorational pesticides. It also establishes criteria for the assessment of the environmental survival of released agents, their host range and their potential effects on non-target organisms. The draft guidelines stipulate that data requirements for genetically engineered bacteria will be determined on a case by case review.

The following biological agents have received registration for use in the United States: (1) *Bacillus popilliae*, reg. 1948; (2) *Bacillus lentimorbus*, reg. 1948; (3) *Bacillus thuringiensis*, reg. 1969; (4) *Heliothis*, nuclear polyhedrosis virus (NPV), reg. 1975; (5) *Agrobacterium radiobacter*, reg. 1979; (6) *Nosema locustae*, reg. 1980; (7) *Hersitella thompsoni*, reg. 1981; (8) *Bacillus thuringiensis, israliensis* variety, reg. 1981; (9) *Bacillus thuringiensis, aiswaga* variety, reg. 1981; (10) *Phytophthora palmivora*, reg. 1981

32.4.1.2 *Department of Health and Human Services*

The Public Health Service of HHS has played a role in establishing guidelines for the handling of biological agents primarily in laboratory settings. Two agencies under the PHS relevant in this regard are the Centers for Disease Control (CDC) and the National Institutes of Health (NIH).

The CDC publishes the voluntary guidelines entitled *Classification of Etiologic Agents on the Basis of Hazard* (CDC, 1974). This document provides a standard for assessing the hazards of numerous infectious agents. It also defines a set of minimal safety conditions for four classes of etiologic agents. Physical containment recommendations for handling the agents are made according to the degree of hazard.

Since 1976, the National Institutes of Health has issued guidelines (National Institutes of Health, 1976) for experiments that involve the use of recombinant DNA technology. Since NIH is not a regulatory agency, the guidelines apply exclusively to institutions receiving funding from HHS. However, other agencies of government have also made the NIH Guidelines a requirement for projects that they fund. NIH has no statutory authority to mandate the use of its guidelines by private companies, but it has instituted a program of voluntary compliance that allows industry to demonstrate adherence to the rules.

Until recently, large-scale work with recombinant DNA technology was prohibited by the NIH Guidelines and could be undertaken only with approval of the Director of NIH after review by the Recombinant DNA Advisory Committee. In the latest revisions of the Guidelines (NIH, 1982) the prohibitory language has been eliminated.

For large-scale experiments, defined as experiments with greater than 10 l of culture, investigators are required to obtain approval from their local Institutional Biosafety Committee (IBC) before initiation. The containment for such experiments is decided by the IBC with guidance from the large-scale containment recommendations published by the NIH. An experiment involving the deliberate release into the environment of any organism containing recombinant DNA can only be initiated when it has been reviewed by the RAC and NIH and approval has been obtained by the IBC (NIH, 1982).

Private firms that register their IBCs with the NIH and participate in the Voluntary Compliance Program are making a non-binding commitment to follow the same procedures for the review of experiments and containment requirements as any institution funded by NIH.

The interstate shipment and packaging of etiologic agents in transit is regulated by the Public Health Service of HHS through the CDC. According to the regulations, no person may transport an etiologic agent through interstate traffic unless the material is packaged to withstand leakage of contents, shocks and pressure changes experienced in ordinary handling. The list of etiologic agents covered by the regulations include bacterial, fungal, viral and rickettsial agents. The shipping regulations specify packaging for volumes less than and exceeding 50 ml, labeling, and responsibility when leakage is discovered in a package bearing an etiologic agents/biomedical materials label (42 C.F.R. § 72.1–72.5).

Under the regulations an 'etiologic agent' means 'a viable microorganism or its toxin which causes, or may cause, human disease' (42 C.F.R. § 72.25). The regulations of etiologic agents derive from Section 361 of the Public Health Service Act, which authorizes the Secretary of HHS to 'make and enforce . . . such regulations as in his judgement are necessary to prevent the introduction, transmission, or spread of communicable diseases . . . from one state . . . into any other state . . . '. This section, although broad in scope, is limited to those agents or bioprocesses that involve the transmission of communicable diseases.

Section 353 of the PHS Act gives CDC the general authority to license and control the operation of clinical laboratories.

32.4.1.3 Food and Drug Administration (DHHS)

The Federal Food, Drug and Cosmetic Act authorizes the Food and Drug Administration to regulate the manufacture of drugs and biologics. The rules promulgated for this purpose are designed primarily with the quality of the drug in mind. FDA has no statutory powers to establish regulations pertaining to the release of organisms into the environment or to the exposure of workers to biological agents. However, environmental safeguards are a by-product of regulating the manufacture of pharmaceuticals or food products. The conditions required for preventing contamination of drugs are similar to those necessary for containing organisms during the manufacturing process. Under Part 211 of the FDA regulations for the manufacture of biologics (21 C.F.R.), standards are established for the construction and maintenance of facilities and equipment. By minimizing the contamination of drugs during the manufacturing process, the regulations also serve to inhibit the release of organisms into the environment. The regulations state in part: '(1) Minimize contamination of products of extraneous adulterants, including cross-contamination of one product by dust or particles of ingredients, arising from the manufacture, storage, or handling of another product; (2) Minimize dissemination of microorganisms from one area to another'.

In 1978 FDA published a Notice of Intent to Propose Regulations which would require all experimental work using rDNA techniques to be done according to the NIH Guidelines (*Fed. Regist.*, 1978), or the resulting product would be rejected when submitted to FDA for approval. No further action has been taken pursuant to this notice; however, FDA strongly encourages industry to comply with the Guidelines and to inform FDA that they have done so (Miller, 1981).

In the 1978 Notice of Intent, FDA also proposed that the Guidelines be incorporated into the Good Manufacturing Practice (GMP) Regulations for those products produced by rDNA technology. However, it is unlikely that FDA has the authority to require adherence to all the provisions of the Guidelines (Korwek, 1981).

The purpose of the food and drug GMP Regulations is to assure that products affected are safe for human consumption. A very different focus is given in the Guidelines, which specify practices for conducting basic research in order to minimize risks to workers, the public and the environment. The legal authority to incorporate the Guidelines into GMP Regulations remains unclear since the aims of each set of rules are different.

The Good Laboratory Practices (GLP) Regulations (21 C.F.R. § 58.1 (1980)) are intended to insure the quality and integrity of safety data submitted in support of work on which FDA regulations are based. They are not directed to the safety of laboratory workers, the public or the environment and do not apply to basic research. They cover personnel, animal care and supply facilities, laboratory areas, testing facilities, operations, records and reports but are designed only to enhance the reliability of submitted data. It is unclear whether FDA has the authority to issue similar regulations to protect laboratory workers and the public. The primary focus of FDA is on products, ensuring their safety, effectiveness and purity, not environmental protection.

32.4.1.4 Department of Agriculture

The Department of Agriculture (USDA) has two important functions in the regulation of biological agents. Under the federal Quarantine Laws it has jurisdiction over the importation and interstate shipment of plants and animals and host pathogens. The Department also regulates vaccines, serums and related products used in the treatment of domestic animals under the auspices of the Virus, Serum, Toxin Act. This Act provides for premarket clearance and licenses for both product and factory. Similarly, under the Quarantine Laws, permits are required for the interstate or internation transport of plant and animal pathogens.

A plant pathogen is defined as any living organism that is injurious to plants including insects, nematodes, fungi, bacteria and viruses. Before plant pathogens or any vector for such can be imported from any country or transferred between states, permits must be obtained from USDA's Animal and Health Inspection Service (APHIS). Vectors are defined by APHIS as all animals which have been treated or inoculated with organisms which are diseased or infected with any contagious, infectious, or communicable diseases of animals or poultry, or which have been exposed to any such diseases. USDA maintains a similar, but voluntary quarantine and permit program for importation and release of non-pest biological control organisms.

32.4.2   Canada

32.4.2.1   Medical Research Council

In February 1977, seven months after the United States published gene-splicing rules, the Medical Research Council of Canada issued its *Guidelines for the Handling of Recombinant DNA Molecules and Animal Viruses and Cells* (Medical Research Council of Canada, 1980). The Guidelines have been revised on several occasions. The MRC requires all research it funds to conform to the rules established in this document. But it has no jurisdiction over privately funded work. The Guidelines contain standards of practice, safety procedures and physical containment levels for different classes of organisms. Where large volume work is underway, the Guidelines state: 'It appears evident that, as the volume or concentration of agent used in an experiment increases, the problems encountered in minimizing risk to investigators or the environment also increase. The procedure described in these Guidelines relate mainly to the scales of operation normally encountered in a research laboratory'.

It is notable that the MRC Guidelines require a case by case review of two types of activities that are especially pertinent to commercial operations: (1) deliberate release into the environment of any organism containing recombinant DNA; (2) large-scale experiments (more than 10 l) unless the recombinant DNA molecules are rigorously characterized and shown to be free of harmful genes. If such work is permitted, the MRC may set the physical and biological conditions for executing the experiments.

32.4.2.2   Department of Agriculture

Under authority of the Pest Control Products Act (see Section 32.3.2.2), the Minister of Agriculture has issued regulations for the registration of pest control products (C.R.C., Vol. XIII, C.1253). The regulations also apply to microorganisms that are imported for personal use and in small quantities. Under Section 4 of the regulations, control products other than live organisms are given an exemption if they are 'imported into Canada for the importer's own use', and 'if the total quantity of the control product being imported does not exceed 500 grams by mass and 500 millilitres by volume and does not have a monetary value exceeding ten dollars'.

Three categories of exemptions are written into the regulations for agents covered by the Food and Drugs Act: (1) viruses, bacteria or other microorganisms used in humans or domestic animals; (2) control of microorganisms on articles intended to come directly into contact with humans or animals for the purpose of treating disease; (3) the control of microorganisms in premises in which food is manufactured, prepared or kept. Currently the Minister of Agriculture has registered *Bacillus thuringiensis* and will soon extend limited registration to the *israliensis* variety. The Department of Agriculture is in the process of issuing guidelines for the registration of biological pesticides.

In addition to falling under the federal Pest Control Products Regulations, a biological control agent also falls under the authority of provincial pesticide registration requirements.

32.4.2.3   Provincial regulations

It is not within the scope of this chapter to undertake a review of provincial regulations that apply to biological agents. But we can draw some conclusions from a preliminary survey. The primary body of Canadian provincial regulations that restrict the release of biological agents into the environment are in the area of hazardous waste and pesticide controls. Some provinces have adopted sludge management regulations and guidelines for disposing of infectious wastes. For example, Alberta has standards for the incineration of animal solids and organic wastes including carcases, organs, and hospital and laboratory wastes. A permit and licensing system specifies design and operational requirements for facilities. Alberta also issues *Guidelines for the Application of Municipal Wastewater Sludges to Agricultural Lands* (Alberta Environment, undated). Four categories of potential human pathogens found in domestic wastewater cited in the Guidelines are bacteria, protozoan parasites, Helminth parasites and viruses. The land application of sludge is cited as a public health risk because: (1) survival of various infectious agents in soil will result in the contamination of forage and edible crops; and (2) runoff and percolation will result in the contamination of water. The Guidelines emphasize proper site selection, proper sludge handling and application, proper crop selection and appropriate land use as a way to reduce health risks. A 'letter of permission' from Alberta Environment must be obtained by the owner of a sludge-producing facility before sludge can be applied to land.

Provinces have issued guidelines respecting the discharge of industrial wastes into municipal sewerage systems. These guidelines are not enforced by statute but rather provide suggested procedures for a municipality to use when dealing with industry. According to guidelines published by Alberta (Alberta Environment, 1978), special care should be taken in treatment of pharmaceutical wastes. 'Wastes from pharmaceutical plants producing penicillin and similar antibiotics are strong (high BOD) and generally should not be treated with domestic sewage, unless the extra load is considered in the design and operation of the treatment plants.'

No special regulations or guidelines have been issued in the provinces as a response to the developments in recombinant DNA technology.

Clean Air Regulations at the provincial level are framed sufficiently broadly to include airborne infectious wastes. Alberta's Clean Air Regulations (216/75 Part I, Sec. 5), which prohibit the release of toxic contaminants into the atmosphere, include debris from animal cadavers, animal manure and pathological wastes.

British Columbia publishes *Pollution Control Objectives for Food-processing, Agriculturally-oriented, and Other Miscellaneous Industries* (Water Resources Service, 1980). These are recommended minimum objectives in issuing orders to industry for waste discharges to air and water. They reflect the more flexible and less formal regulatory orientation that has come to be characteristic of Canadian environmental policy.

32.5 OCCUPATIONAL HEALTH AND SAFETY

32.5.1 United States: Occupational Safety and Health Act

The Occupational Safety and Health Act (29 U.S.C. § 651 *et seq.*) (OSH Act), gives the Secretary of Labor power to require employers to provide a safe workplace for their employees. Every employer, under Section 5(a)(1), has a general duty to furnish each of his employees with a place of employment 'free from recognized hazards that are causing or are likely to cause death or serious physical harm'.

OSHA's general authority to promulgate occupational health or safety standards is found in Section 6(b)(2). A standard as defined in Section 3(8) must be 'reasonably necessary or appropriate to provide safe or healthful employment and places of employment'. Section 6(b)(5) of OSH Act gives the Secretary of Labor authority to promulgate 'standards dealing with toxic materials or harmful physical agents'. OSHA 'shall set the standard which most adequately assures, to the extent feasible, on the basis of the best available evidence, that no employee will suffer material impairment of health or functional capacity even if such employee has regular exposure to the hazard . . . for the period of his working life'.

The U.S. Supreme Court has twice interpreted this section. The court has held that before invoking Section 6(b)(5), OSHA must make the threshold determination that there exists a significant risk (*Industrial Union Dept.* v. *Am. Pet. Inst.*, 1980). Once the threshold is crossed, however, OSHA can require an employer to do whatever is feasible to limit the risk (*Am. Text. Manu. Inst.* v. *Donovan*, 1981).

OSHA's authority under Sections 6(b)(2) and 6(b)(5) appears adequate to set standards regulating the physical environment in research areas and in the fermentation plant (McGarity, 1981). It could set standards for ambient air levels or enhanced physical containment for the fermentation process, and in general has statutory authority for regulating the working conditions of employees involved in all phases of the biotechnology industry. But before OSHA can initiate regulations, it is generally conceded that the existence of a hazard will have to be demonstrated;

mere speculations will not suffice to meet the standards of 'recognized hazards'. Reviewing OSHA's role in industrial bioengineering, Korwek (1982) concludes: ' . . . hazards associated with commercialization of biotechnology could not easily be made the subject of a permanent standard until definitive data become available demonstrating that a particular type of danger exists, that it is significant and that it is likely to cause material impairment as well'.

The National Institutes of Occupational Safety and Health (NIOSH), the research agency created by the OSH Act, evaluates new technologies to determine whether they may have associated occupational health and safety risks. NIOSH conducted a walk-through industrial hygiene survey of four companies involved in research and development of rDNA techniques applied to fermentation processes. For each company, NIOSH assessed the following areas: (1) potential for worker exposure; (2) engineering controls of physical containment design; (3) work practices; (4) medical surveillance; (5) validation procedures; (6) emergency and accident procedures; (7) environmental monitoring; (8) number of exposed workers; (9) employee training and education. In a preliminary report which has not been formally released by NIOSH and does not represent NIOSH policy the following recommendations were made: (1) a medical surveillance program should consist of pre-employment and periodic follow-up examinations; (2) systematic preventive maintenance should be established; (3) a registry of all workers using rDNA processes should be established; (4) environmental monitoring programs should be implemented; (5) physical containment of the fermentation process should be extended through the cell lysis and product extraction stages; and (6) the P1-LS production area should be segregated from other production or research operations.

Currently both OSHA and NIOSH are monitoring the industrial developments in biotechnology. Both agencies have representatives on a large-scale working group of the Recombinant DNA Advisory Committee (RAC). A report of the CDC/NIOSH Ad Hoc Working Group on Medical Surveillance for Industrial Applications of Recombinant DNA (CDC/NIOSH Ad Hoc Working Group, 1981) concluded that 'medical surveillance of industrial workers engaged in commercial applications of recombinant DNA techniques can play a valuable auxiliary role in protecting worker health'. The CDC/NIOSH study also emphasized the importance of physical containment of recombinant DNA organisms and their products, as a first line of defense against occupational exposures; a second level of protection cited was the use of attenuated or debilitated organisms, commonly known as biological containment.

32.5.2 Canada Labour Code

Jurisdiction over occupational health and safety is granted to the provinces by Section 92 of the British North America Act. Although the provinces have been primarily responsible for occupational environments, the Canadian Parliament can regulate the workplaces of a small number of industries that fall under federal jurisdiction. The provinces are also restricted from enacting worker safety legislation which pertains to issues of interprovincial trade or commerce.

Federal occupational health and safety regulations are authorized through the Canada Labour Code and administered by Labour Canada. Chapter 997 of the Labour Code entitled the Canada Dangerous Substances Regulations spells out the responsibilities of employers for the use of hazardous substances in the workplace. A dangerous substance is defined (Sec. 2) as 'any substance that, because of a property it possesses, is dangerous to the safety or health of any person who is exposed to it'. The Act was written for chemical substances and radiation emitting devices, although the statutory language clearly is broad enough to include biological agents should they be designated a dangerous substance by the Minister of Labour. The regulations consist of a set of general performance standards where there is considerable room for discretionary interpretation.

The general duties of the employer are stipulated in Section 4 of the regulations. 'No employer shall use in his operations a dangerous substance or radiation emitting device, if it is reasonably practicable to use a substance or device that is not dangerous. Where it is necessary for an employer to use a dangerous substance or a radiation emitting device in his operations and more than one kind of such substance or device is available, he shall to the extent that is reasonably practicable use the one that is least dangerous to his employees.'

The Act also requires that every employer 'ensure that any dangerous substance that may be carried by the air is confined as closely as is reasonably practicable to its source' (Sec. 9). The airborne contaminants in question refer to those on a published list where recommended threshold values are given.

The Minister of Labour is also given authority to require employers within federal jurisdiction to institute a medical examination for an employee who might be endangered by working with a dangerous substance.

When federal and provincial statutes are in conflict, the federal legislation preempts the provincial statute. And as a general rule, the legislation in the jurisdiction with the more stringent rules prevail.

There is nothing in the Act which requires pretesting of new substances to determine their effect on workers. In general, the evaluation of substances occurs after they have been introduced into the workplace. The Ministry of Labour utilizes toxicity information compiled by the Departments of the Environment and National Health and Welfare. No regulations at the federal level are specifically concerned with the exposure of workers to biological agents.

By the early 1970s new occupational health and safety legislation was passed in most of the provinces. Like their federal counterpart, the provincial regulations concentrated on chemical contaminants, radiation and pesticides.

32.6 CONCLUSIONS

Microorganisms have been put to the service of society in a variety of industrial processes, most extensively in the food and pharmaceutical sectors. Until recently, organisms employed in the fermentation technology of these industries originated from natural sources. This is one of the reasons why the use of biological agents and their release into the environment through the waste stream has not been viewed as a public health problem. Other factors affecting past regulatory responses to bioprocesses include their low volume of use compared to chemical processes, and the fact that microbial agents can be rendered harmless by disinfection or steam sterilization, unlike many toxic chemicals which require high temperature incineration.

As a result of the revolution that has taken place in applied genetics, especially with the recombination of genes *in vitro*, microorganisms can be engineered with the capacity to synthesize products which are coded for by exogenous DNA. These developments have brought about a renewed interest in microbial fermentation. In the chemical industry it is being pursued as an alternative to the costly synthesis of industrial chemicals from petroleum stocks.

The application of molecular genetics to industrial processes has raised concerns about the dissemination of novel organisms into the environment. Except for guidelines that regulate government-funded research, there have been no new statutes enacted in the U.S. or Canada responding to the developments in biotechnology. The federal governments of these countries actively regulate the release of biological agents in three general areas. First, quarantine laws restrict the movement of humans, plants and animals harboring infectious pathogens. Second, biological pesticides must meet registration criteria before they can be imported, marketed or disseminated. Third, as previously noted, federal agencies in both countries have issued regulations for laboratory experiments involving the use of hazardous or potentially hazardous biological agents and recombinant DNA molecules. The guidelines for the experimental use of microorganisms are not mandatory for the private sector.

The three areas of government regulation of the release of biohazardous materials are not applicable to industrial processes, such as those where genetically engineered organisms are employed in large-scale fermentation vessels.

Many of the laws enacted in Canada and the U.S. to protect the environment are broad enough in their statutory language to include the discharge of biohazardous materials. However these same laws that protect the air, water and land from contaminants require for their implementation the establishment of a performance or environmental quality standard. Unless a clear case can be made that biological agents are a viable part of the effluent and that they represent an endangerment to human health or the environment, existing laws such as the Clean Air and Clean Water Acts are not likely to be applied.

In the U.S. and Canada, federal, provincial and/or state laws regulate the disposal of hazardous wastes. It is generally accepted that pathogenic and infectious materials fall under this category. The U.S. Environmental Protection Agency has decided against taking action under RCRA to regulate the disposal of biological waste products. Canada has no comprehensive federal hazardous wastes regulations similar to RCRA within which infectious agents could be included. At a subindustrial scale, states and provinces concerned about the release into the environment of pathogenic wastes from clinical laboratories regulate the disposal of such materials under the public health laws.

Both Canada and the U.S. have enacted laws regulating the introduction of toxic agents into the industrial process with passage of the Environmental Contaminants Act and the Toxic Substances Control Act, respectively. But these Acts may not be applicable to life forms, since their statutory language refers explicitly to chemical substances.

In the food and drug industries, U.S. and Canadian federal roles have been principally to ensure the safety, efficacy and purity of the products. The protection against the release of organisms into the environment or against worker exposure is only a secondary effect of the governments' responsibilities to certify good manufacturing practices designed to protect the public from poor quality or contaminated drugs.

Certain uses of microorganisms in the environment (except for pesticides) such as oil-eating bacteria, microbial leachates, or organisms that degrade toxic chemicals, do not fall unambiguously within the jurisdiction of current statutes in the U.S. or Canada. No regulations exist for the private sector that cover the dissemination of novel biotypes into the environment. The NIH Recombinant DNA Guidelines have incorporated a voluntary compliance program for industry which covers such releases. But these Guidelines cannot be enforced on industry. Furthermore, if present trends continue in the U.S., the rDNA Guidelines will be non-existent in a few years. The Canadian rDNA Guidelines have followed the general trend toward relaxation.

Both U.S. and Canadian laws can be used to control occupational exposure to hazardous biological agents. However, no special regulations have been promulgated in this area. Occupational health and environmental agencies in the U.S. have taken a wait and see posture to the burgeoning biotechnology industry. Provincial authorities which have principal jurisdiction over worker health and safety have not initiated any rule-making or guidelines for the genetic engineering industry.

32.7 REFERENCES

Aharonowitz, Y. and G. Cohen (1981). The microbial production of pharmaceuticals. *Sci. Am.*, **245**, 141–152.
Alberta Environment (undated). *Guidelines for the Application of Municipal Wastewater Sludges to Agriculture Lands in Alberta*. Alberta Ministry of the Environment, Edmonton, Alberta.
Alberta Environment (1978). *Guidelines for the Control of Industrial Wastes Discharging to Municipal Sewerage Systems*, July. Alberta Ministry of the Environment, Edmonton, Alberta.
Alexander, M. (1981). Biodegradation of chemicals of environmental concern. *Science*, **211**, 132–138.
American Textile Manufacturers Inst. v. *Donovan* (1981). 452 U.S. 488.
Centers for Disease Control (1974). *Classification of Etiologic Agents on the Basis of Hazard*, 4th edn. July.
Centers for Disease Control—National Institute for Occupational Safety and Health (1981). *Report of the CDC/NIOSH Ad Hoc Working Group on Medical Surveillance for Industrial Applications of Recombinant DNA*.
Cornwall, G. M. (1982). Director general, Water Pollution Control Directorate, Environment Canada, Ottawa, Ontario, personal correspondence, 25 May.
Costle, D. M. (1978). Letter to Adlai E. Stevenson Jr., Chairman, Subcommittee on Science, Technology and Space, U.S. Senate Committee on Commerce, Science and Transportation, in *Oversight Report, Recombinant DNA and Its Applications*, 95th Congress, August, p. 88.
Demain, A. L. and N. A. Solomon (1981). Industrial microbiology. *Sci. Am.*, **245**, 66–75.
Diamond v. *Chakrabarty* (1980). 447 U.S. 303.
Environmental Protection Agency (1976). *Development Document for Interim Final Effluent Limitations Guidelines and Proposed New Source Performance Standards for the Pharmaceutical Point Source Category*, pp. 91–94.
Environmental Protection Agency (1978a). Registration of pesticides in the United States: Proposed guidelines. *Fed. Regist.*, **43**, 29 696–29 741.
Environmental Protection Agency (1978b). Hazardous waste: Proposed guidelines and regulations and proposal on identification and listing. *Fed. Regist.*, **43**, 58 946–59 028.
Environmental Protection Agency (1979a). Regulation of 'biorational' pesticides; policy statement and notice of availability of background document. *Fed. Regist.*, **44**, 28 093.
Environmental Protection Agency, U.S., Office of Pesticide Programs (1979b). Background paper: regulation of biorational pesticides.
Environmental Protection Agency (1980). Proposed guidelines for regulating pesticides in the United States, subpart M. Data requirements for biorational pesticides, September 29.
Eveleigh, D. E. (1981). The microbiological production of industrial chemicals. *Sci. Am.*, **245** (3), 154–178.
Fed. Regist. (1977). Toxic substances control. Inventory reporting requirements. **42**, 64 572.
Fed. Regist. (1978). Recombinant DNA. Intent to propose regulations. **43**, 60 134–60 135.
Fed. Regist. (1979). Criteria for classification of solid waste disposal facilities and practices. Final, interim final, and proposed regulations. **44**, 53 438–53 464. *Ibid*. Amendment. **44**, 53 465–53 468.
Fed. Regist. (1981) Certain biological control agents; proposed exemption from regulations. **46**, 18 322–18 325.
Ilgen, T. L. (1981). Regulating toxic substances: The case of Canada. Unpublished paper delivered at the annual meeting of the American Political Science Association, 3–6 September, p. 31.
Ince, J. G. (1976). *Environmental law: A study of legislation affecting the environment of British Columbia, Vancouver*, p. 23. Centre for Continuing Education, University of British Columbia, Vancouver, BC.
Industrial Union Dept., *AFL-CIO* v. *Am. Pet. Inst.* (1980). 448 U.S. 607.

Johnston, J. B. (1981). The development of new pollution control technologies using genetic engineering methods—an assessment of problems and opportunities. *Proceedings, Genetic Engineering International Conference*, 6–10 April 1981, Reston, VA, vol. 4, pp. 184–191. Battelle Memorial Institute, Seattle, WA.

Korwek, E. L. (1981). The NIH Guidelines for recombinant DNA research and the authority of FDA to require compliance with the Guidelines. *Jurimetrics J.*, **21**, 264–283.

Korwek, E. L (1982). OSHA regulations of industrial applications of recombinant DNA technology. *Recombinant DNA Technical Bulletin*, **5**, 19–29, March.

Krimsky, S. (1982). *Genetic Alchemy: The Social History of the Recombinant DNA Controversy*, p. 122. MIT Press, Cambridge, MA.

Krimsky, S., A. Baeck and J. Bolduc (1982). *Municipal and State Recombinant DNA Laws*. Tufts University, Medford, MA.

McGarity, T. O. (1981). Unpublished materials prepared for ALI-ABA course of study on industrial applications of genetic engineering, 29–31, October.

Medical Research Council of Canada (1980). *Guidelines for the Handling of Recombinant DNA Molecules and Animal Viruses and Cells*. Ministry of Supply and Services, Ottawa, Ontario.

Miller, H. I. (1981). The impact of new technology on regulation by the FDA recombinant DNA technology. *Food, Drug Cosmetic Law J.*, **36**, 348–353.

National Institutes of Health (NIH) (1976). Guidelines for research involving recombinant DNA molecules (NIH Guidelines). *Fed. Regist.*, **41**, 27 902.

National Institutes of Health (1982). Guidelines for research involving recombinant DNA molecules. *Fed. Regist.* **47**, 17 180–17 198.

Office of Technology Assessment (1981). *Impacts of Applied Genetics*. U.S. Gov't. Print. Off., Washington, DC.

Stewart, R. B. and J. E. Krier (1978). *Environmental Law and Policy*, pp. 326–332. Bobbs Merrill, New York.

Task Force on Biotechnology (1981). *Biotechnology: A Development Plan for Canada*. Report to the Minister of State for Science and Technology. Ministry of Supply and Services Canada, Ottawa, Ontario.

Water Pollution Control Directorate (1976). *A Bacteriological Investigation of Meat and Poultry Packing Plant Effluents with Particular Emphasis on* Salmonella. Report EPS 3-WP-76-9, December. Environment Canada, Ottawa, Ontario.

Water Resources Service, Victoria, British Columbia (1980). *Pollution Control Objectives for Food-processing, Agriculturally Oriented, and Other Miscellaneous Industries of British Columbia*. Ministry of Water Resources, Victoria, British Columbia, Queen's Printer for British Columbia.

Working Party (1976). *Report of the Working Party on the Practice of Genetic Manipulation*. HMSO, London, Cmnd 6600 (Working Party).

Wright, S. (1982). The status of hazards and controls. *Environment*, **20**, 51–53; **24**, 12–17.

Zaugg, R. H. and J. R. Swarz (1982). Industrial use of applied genetics and applied biotechnology. *Recombinant DNA Technical Bulletin*, **5**, 7–13, March.

33

Regulations for Ultimate Disposal of Biohazardous Materials in Japan

R. SUDO and M. OKADA
National Institute for Environmental Studies, Ibaraki, Japan

33.1 INTRODUCTION

Wastes are finally discharged to the environment in the form of landfill or ocean dumping, except for those which are reused as useful resources. It is basically required of these ultimate disposal methods that the disposed wastes should not disturb the living environment and natural ecosystems. It is necessary, therefore, to isolate those materials which have the potential of damaging human health by means of assimilation through the food chain, water use, *etc.* when they are released to the environment.

The Japanese standards related to final disposal of biohazardous materials are provided for in the Waste Disposal and Public Cleaning Law (Law No. 137 of 1970) and the Marine Pollution Prevention Law (Law No. 136 of 1970). These laws are designed to prevent environmental pollution associated with the ultimate disposal of wastes. Final disposal is being made by landfill and ocean dumping, and the standards are allotted in the above mentioned laws as shown in Figure 1 (Environmental Agency Japan, 1981).

The basic premises used to establish these standards related to final disposal of wastes are as follows.

(1) Ultimate disposal of wastes should be conducted by landfill as a rule, and disposal by ocean dumping is permitted in exceptional cases only when particular problems are associated with the disposal by landfill.

(2) As a rule, disposal of wastes by landfill may be conducted for all wastes. However, it is a requirement that specific wastes be subjected to certain pretreatments, such as crushing, incineration and cutting to prevent environmental pollution due to the landfill.

(3) Disposal of waste by ocean dumping is permitted only in exceptional cases, and the scope of the wastes which are permitted to be disposed of by ocean dumping should be limited as much as possible. The disposal of specific wastes by ocean dumping, however, is permitted only after they are subjected to pretreatments such as crushing, grinding and incineration.

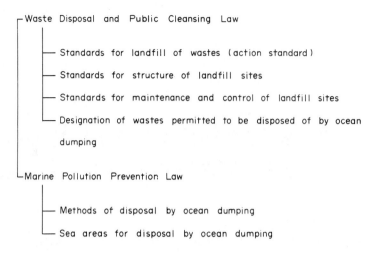

Figure 1

(4) The standards for final disposal of the wastes containing toxic substances such as mercury, cadmium and chromium(VI) should be severer than those for other wastes.

(5) Wastes to be dumped into the ocean are classified into two categories. One is degradable wastes, which shall be returned to the ocean by dispersion. The other is degradation-resistant wastes; these are intended to be precipitated and accumulated on the sea bottom as quickly as possible. For these latter wastes, the sea dumping areas are restricted and the sea bottom space is used as the place for disposal only under unavoidable circumstances.

33.2   CLASSIFICATION OF BIOHAZARDOUS MATERIALS

The Waste Disposal and Public Cleansing Law classifies wastes into domestic wastes and industrial wastes; in addition, it classifies industrial wastes into 19 kinds and provides a standard for disposal of each one of them (see Figure 2).

If the concentration of hazardous substances in an industrial waste exceeds the criteria as a result of the specific test, the waste is defined as hazardous and subject to severer standards for disposal than those applicable to non-hazardous wastes. Mercury and its compounds, cadmium and its compounds, lead and its compounds, chromium(VI) compounds, arsenic and its compounds, cyanides, PCBs and organic chlorides (applicable only in the case of ocean dumping) are designated as hazardous substances.

The standards for verification of hazardous wastes and an outline for testing methods are shown in Table 1.

33.2.1   Wastes to be Disposed of by Ocean Dumping

Organic sludge, water-soluble inorganic sludge, waste acid and waste alkali should be discharged into the ocean by a dispersion-type discharge method (discharged under the sea surface during navigation). It is most likely that the majority of the hazardous substances contained in the wastes would be diffused into the surrounding water immediately after discharge. The contents of the hazardous substances contained in these wastes, therefore, are restricted.

The criteria for judgement are established based on a rule where the critical threshold value for each hazardous substance is basically 500 times the environment standard of the substance related to water pollution, assuming that the concentration of the hazardous substance will be less than the environmental standard at a radius of 10 m from the point of discharge by the dilution effect. In addition, it was decided to provide a difference in the standard between sludge and waste acid/waste alkali because the extent of dissolution into the surrounding water is different between them; 100 times the environmental standard (or water quality standard for fisheries) for waste acid and waste alkali and 500 times for sludges.

Water-insoluble sludges are discharged by concentration-type discharge methods (discharge is

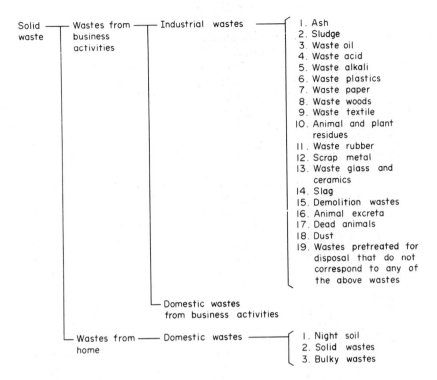

Figure 2 Classification of wastes

made in a form having a specific gravity of 1.2 or higher; no discharge is made in the form of powder; discharge is made under the sea). These sludges will be accumulated in fixed and limited sea areas and, therefore, the concentration of the hazardous substances which are dissolved into the surrounding water during the sinking process must be taken into consideration. The assessment is based on the estimated extent of the hazard by means of conducting elutriate tests of hazardous substances in the wastes. The standards are based on the concentration of the hazardous substances which are dissolved in the solvents, using a solvent volume ten times that of the waste. The standard values were set to be equal to those of the wastewater discharge standard of the Water Pollution Control Law (Environmental Agency Japan, 1981). Disposal of sludge containing organic chlorides (excluding PCBs) is restricted by the content in all cases.

The government of Japan ratified THE CONVENTION ON THE PREVENTION OF MARINE POLLUTION BY DUMPING OF WASTES AND OTHER MATTER in October 1980. In order to comply with the conditions of the Convention, the wastes disposal standards related to disposal by ocean dumping (Waste Disposal and Public Cleansing Law) were amended by Cabinet Order. After this amendment, subhazardous wastes have not been dumped at sea. Subhazardous wastes are those such as sludge, cinder, dust, waste acid, waste alkali and slags that contain significant amounts of copper and its compounds, zinc and its compounds or fluorides. Although subhazardous wastes are regarded as not particularly toxic, they may deteriorate the oceanic ecosystem. If the concentration of above-mentioned subhazardous substances in a waste, as determined by the specific tests (similar to those in Table 1), exceeds the criteria listed in Table 2, then the waste is defined as a subhazardous waste.

33.2.2 Wastes to be Disposed of by Landfill

The extent of dissolution of hazardous substances contained in wastes into the water which leaches from landfill sites to subsurface water and public water bodies is not well known. Therefore, it was decided to set standards on hazardous substances as determined by the elutriate tests as shown in Table 1. The criteria for judgement are eased slightly compared to the case of ocean dumping, because seizure and control of artificially polluted areas are easy with landfill sites compared to ocean dumping and also the fact that hazardous substances may be adsorbed by the soil

Table 1 Standards for Verification of Hazardous Wastes (summary of the 'Ordinance of Prime Minister's Office on Verification of Hazardous Industrial Waste', latest revision, Ordinance No. 3, 1977)

	Landfill			Ocean dumping		
	Sludge	(1) Slag (2) Dust (3) Cinder	Water-insoluble inorganic sludge	(1) Slag (2) Dust (3) Cinder	(1) Organic sludge (2) Water-soluble inorganic sludge	(1) Waste acid (2) Waste alkali
Testing methods	a	a	b	b	c	c
Alkylmercury compounds	N.D.[d]	N.D.	N.D.	N.D.	N.D.	N.D.
Mercury and its compounds	Not more than 0.005 mg l^{-1}	Not more than 0.005 mg l^{-1}	Not more than 0.005 mg l^{-1}	Not more than 0.005 mg l^{-1}	Not more than 2 mg l^{-1}	Not more than 0.05 mg l^{-1}
Cadmium and its compounds	0.3	0.3	0.1	0.1	5	1
Lead and its compounds	3	3	1	1	50	10
Organic phosphorus compounds	1	—	1	—	5	1
Chromium (VI) compounds	1.5	1.5	0.5	0.5	25	5
Arsenic and its compounds	1.5	1.5	0.5	0.5	25	5
Cyanide compounds	1	—	1	—	5	1
PCBs	0.003	0.003 (not applied to slag)	0.003	0.003 (not applied to slag)	0.15	0.03
Organic chlorine compounds	—	—	40 mg l^{-1} content in sample	—	40	8

[a] Dilute 100 g (wet-base) of wastes with 1 l of water, shake the slurry for 6 h, separate by filtration, and measure the concentrations in the resulting solution. [b] As in a, but taking 30 g (dry-base) of wastes other than slag, or 100 g (wet-base) of slag. [c] Measure the content of the substances in the wastes. [d] Not detectable.

Table 2 Standards for Verification of Subhazardous Wastes for Disposal by Ocean Dumping

	Slag, water-insoluble inorganic sludge, cinder, dust	Water-soluble inorganic sludge, organic sludge	Waste acids, waste alkalis
Copper and its compounds	Not more than 3 mg l^{-1}	Not more than 70 mg l^{-1}	Not more than 15 mg l^{-1}
Zinc and its compounds	5	450	90
Fluorides	15	1000	200

shall be considered (three times the wastewater discharge standard for cadmium, lead, chromium(VI) and arsenic; the same as the wastewater discharge standard for others).

33.3 REGULATIONS FOR LAND DISPOSAL

The standards for the disposal of wastes by landfill (standards for the actions to be taken for the disposal of wastes by landfill), standards for structure, maintenance and control of landfill sites are provided for in the Waste Disposal and Public Cleansing Law.

33.3.1 Standards for Wastes to be Disposed of by Landfill

Standards for each kind of waste for disposal by landfill are provided to minimize environmental pollution associated with the landfill. A part of the standards considers reuse of filled lands. These standards are summarized in Tables 3 and 4.

With respect to industrial wastes which are judged as hazardous wastes, they are considered to be satisfactory for landfill if the standards indicated in Table 4 are observed and the wastes have been changed into non-hazardous wastes by treatment with certain methods. In the case where they are not changed into non-hazardous wastes, disposal by landfill should be made in places isolated from public water bodies and subsurface water.

Wastes which are judged to be hazardous by the presence of contained mercury and/or cyanide should be solidified in accordance with the following standards before they are landfilled: (1) hydraulic cement shall be used as a binder; the content of cement shall not be less than 150 kg m^{-3}, (2) the strength of solidified wastes shall not be less than 10 kg cm^{-2} in terms of an unconfined compression test, and (3) dimensions of solidified wastes shall comply with the following requirements: (a) volume/surface areas $\geq$ 1 cm, (b) maximum dimension/minimum dimension $\leq$ 2, and (c) minimum dimension $\geq$ 5 cm.

33.3.2 Standards for Structure, Maintenance and Control of Landfill Sites

Landfill sites were characterized as a type of waste disposal facility in accordance with the amendment of the Waste Disposal and Public Cleansing Law of 1976, and the standards for structure, maintenance and control were established.

The landfill sites are classified into four types in accordance with the kind of discharged wastes (Table 5), and standards for structure, maintenance and control are provided for each one of them. Tables 6 and 7 summarize these standards. Very severe restrictions are applied to landfill sites for hazardous industrial wastes, whereas moderate restrictions are applied to those used for wastes that are regarded to cause almost no human health problems. It is also considered that landfill activities should not deteriorate the qualities of public and subsurface waters.

33.4 REGULATIONS FOR OCEAN DUMPING

Standards for the wastes which are permitted to be disposed of by ocean dumping are provided for in the Waste Disposal and Public Cleansing Law, and the methods of discharge of wastes and the permitted sea areas for ocean dumping are provided for in the Marine Pollution Prevention Law.

Table 3 Standards for Disposal of Domestic Wastes

| | Standards for ocean dumping | |
Standards for landfill	Standards for wastes	The sea areas and the methods of discharge
Common standards 1. Landfill shall be made in such a way that prevents wastes from scattering or flowing out 2. Landfilling sites shall be constructed in such a way that preservation of the living environment may not be hindered 3. Landfill shall be made in sites which are enclosed by enclosures and which can be identified as landfilling sites by billboards 4. Necessary measures shall be taken to prevent leachate from landfill sites from polluting public waters and underground water 5. Necessary measures shall be taken to prevent offensive odor from leaking out of the landfilling sites 6. The thickness of a layer of domestic wastes (excluding those of which ignition loss was reduced to 15% or less) shall be covered with approximately 50 cm of soil and sand, except when the area of the landfilling site is 10 000 m^2 or less, or when the volume of it is 50,000 m^3 or less, or when landfill is conducted using an underground space 7. Necessary measures shall be taken to prevent the growth of rats, mosquitoes, flies or other vermin at landfilling sites	Ocean dumping shall be avoided when it is considered that the said wastes can be disposed of by landfill	1. Concentration-type discharge method Necessary measures shall be taken to enable the said wastes to precipitate and accumulate on the sea bottom as soon as possible 2. Dispersion-type discharge method Necessary measures shall be taken to discharge the said wastes little by little, and to enable them to disperse in the sea as soon as possible 3. Efforts shall be made to avoid places where the growth of marine animals and plants may be hindered by the disposal of wastes
Sludge from septic tank 1. Landfill shall be conducted after treatment in one of following ways: (a) treatment in human excreta treatment facilities, (b) incineration in incineration facilities, (c) mixture of slaked lime at 0.5% or more 2. Reclamation on surface water shall be conducted after the treatment in one of the following ways: (a) treatment in human excreta treatment facilities, (b) mixture of slaked lime at 0.5% or more	Only the following wastes are allowed to be dumped into sea: (a) those in which 0.1% or more ferrous sulfate or ferric chloride is mixed, (b) those which are crushed	**C sea area** Dispersion-type discharge method
Human excreta 1. Landfill shall be conducted after treatment in either of the following ways: (a) treatment in human excreta treatment facilities, (b) mixture of slaked lime at 0.5% or more 2. Reclamation on surface water shall be conducted after treatment in human excreta treatment facilities	As for sludge from septic tank	As for sludge from septic tank

Table 4 Standards for Disposal of Industrial Wastes (Bioprocess Sources Only)

	Standards for ocean dumping	
Standards for landfill	*Standards for wastes*	*The sea areas and the methods of discharge*
Common standards 1. Landfill shall be conducted in such a way that prevents wastes from scattering or flowing out	Ocean dumping shall be avoided when it is considered that the said wastes can be disposed of by landfill	1. Concentration-type discharge method Necessary measures shall be taken to enable the said wastes to precipitate and accumulate on the sea bottom as soon as possible
2. Landfilling sites shall be constructed in such a way that preservation of the living environment may not be hindered		2. Dispersion-type discharge method Necessary measures shall be taken to discharge the said wastes little by little, and to enable them to disperse in the sea as soon as possible
3. Landfill shall be conducted in sites which are enclosed by enclosures and which can be identified as landfilling sites by billboards 4. Necessary measures shall be taken to prevent leachate from landfill sites from polluting public waters and underground water 5. Necessary measures shall be taken to prevent offensive odor from leaking out of the landfilling sites 6. Necessary measures shall be taken to prevent the growth of rats, mosquitoes, flies or other vermin at landfilling sites		3. Efforts shall be made to avoid places where the growth of marine animals and plants may be hindered by the disposal of wastes
Sludge 1. Landfill shall be conducted after treatment in one of the following ways: (a) incineration by incineration facilities, (b) reduction of water content to 85% or less. In the case of organic sludge, standards for putrescible wastes are applied 2. Reclamation on surface water shall be conducted after the following treatment: (a) no pretreatment needed for inorganic sludge, (b) incineration in incinerator for organic sludge	It is banned to dump the following wastes into the sea: (a) those containing oil, (b) those containing phenols	1. Organic sludge and water-soluble inorganic sludge **C sea area** Dispersion-type discharge method 2. Water-soluble sludge **B sea area** Concentration-type discharge method

Table 5 Classification of Final Disposal of Wastes

Type of final disposal site		*Scale requiring reporting*	*Wastes to be discharged*
Final disposal site of general wastes		1 m² or more	General wastes
Final disposal site of industrial wastes	Isolation type (Type 1)	All	Hazardous industrial wastes
	Control type (Type 2)	1000 m² or more	Industrial wastes other than those discharged to final disposal sites of isolation type and stability type
	Stability type (Type 3)	3000 m² or more	Waste plastics, rubber scrap, metal scrap, waste rubber, scrap metal, waste glass and ceramics, demolition wastes

Table 6 Standards for the Structure of Final Disposal Sites

Structure, facility, etc.	Final disposal site of general wastes	Final disposal site of industrial wastes		
		Type 1	Type 2	Type 3
1. Enclosure around the landfill site for preventing entry of people	●	●	●	●
2. Notice boards and other facilities indicating that the subject site is a final disposal site of wastes	●	●	●	●
3. Landslide preventative measures (when required) for preventing landslide of the ground	●	●	●	●
4. Settlement preventive measures (when required) for preventing settlement of the facilities provided in the final disposal site	●	●	●	●
5. Retaining wall, dams and other facilities for preventing outflow of wastes	●		●	●
6. Water-interrupting measures capable of preventing leaching of water-containing wastes, rainwater, *etc.* out of the landfill site	●		●	
7. Conduits and other water-gathering facilities capable of selectively gathering water-containing wastes	●		●	
8. Water-treating facilities capable of causing water-containing wastes and so forth gathered with the facilities of item 7 above to conform to the water discharge standards of the Water Pollution Prevention Act	●		●	
9. Open pit and other facilities capable of preventing inflow of surface water into landfill site	●		●	
10. External periphery separating facilities (of concrete; minimum thickness 15 cm, minimum unconfined compression strength 250 kg cm^{-2})		●		
11. Internal separating facilities (of concrete; minimum thickness 10 cm, minimum unconfined compression strength 250 kg cm^{-2})		●		

33.4.1 Standards for Wastes to be Disposed of by Ocean Dumping

The wastes which are permitted to be disposed of by ocean dumping, the sea areas where said wastes may be discharged, and the methods of discharge are indicated in Tables 3 and 4.

In relation to bioprocess sources, the following wastes are prohibited for ocean dumping: (1) sludges that are judged to be hazardous with respect to mercury and its compounds, cadmium and its compounds, PCBs and organic chlorides, (2) dust and cinder which are judged to be hazardous with respect to mercury and its compounds, cadmium and its compounds, and PCBs, and (3) hazardous waste acid and waste alkali.

Even if disposal by ocean dumping is permitted, hazardous or subhazardous industrial wastes should be solidified in accordance with the following standards before they are dumped into the ocean: (1) hydraulic cement shall be used as a binder, (2) the strength of solidified wastes shall be not less than 100 kg cm^{-2} in terms of an unconfined compression test, (3) dimensions of solidified wastes shall comply with the following requirements: (a) volume/surface area $\geq$ 5 cm, (b) maximum dimension/minimum dimension $\leq$ 3 and (c) minimum dimension $\geq$ 30 cm, (4) waste and hydraulic cement shall be homogeneously mixed.

33.4.2 Standards for Methods of Ocean Dumping

Two methods for ocean dumping are indicated in Tables 3 and 4, *i.e.* concentration- and dispersion-type discharges. Wastes that cannot be degraded easily in the ocean shall be discharged by the concentration-type discharge method to be accumulated in the fixed sea areas. The following requirements should be satisfied for this discharge: (a) discharge shall be made in a state of specific gravity of 1.2 or over, (b) discharge shall not be made in a state of powder, and (c) discharge shall not be made when a ship is navigating.

The dispersion-type discharge method shall be applied to the wastes that can be easily

Table 7 Standards for the Maintenance and Control of Final Disposal Sites

Structure, facility, etc.	Final disposal site of general wastes	Final disposal site of industrial wastes		
		Type 1	Type 2	Type 3
1. Necessary measures shall be taken so as not to allow the wastes to scatter or flow out	●	●	●	●
2. Necessary measures shall be taken so as not to allow offensive odor to be disperted to the exterior of the disposal site	●	●	●	●
3. Necessary measures shall be taken for preventing occurrence of fire, and the site shall be equipped with fire fighting facilities including fire extinguishers	●	●	●	●
4. Sprinkling of chemicals and other necessary measures shall be taken so as not to allow rats to live and not to allow generation of mosquitoes, flies and other harmful insects	●	●	●	●
5. Necessary measures shall be taken so as not to allow entry of people through the enclosure	●	●	●	●
6. Notice boards and other facilities shall be easily observable, and the description shall be rewritten if necessary	●	●	●	●
7. Retaining walls, dams and other facilities shall be periodically inspected, and necessary measures shall be taken if there is a fear of occurrence of damage to them	●		●	●
8. Water-interrupting measures shall be periodically inspected, and necessary measures shall be taken if there is a fear of occurrence of reduction of water-interrupting effect	●		●	
9. Water examination of subsurface water in the peripheral area shall be carried out	●	●	●	
10. Measures for preventing inflow of rainwater shall be taken. (Water-gathering facilities are not required at this disposal site.)	●	●	●	
11. Dewatering facilities shall be maintained and controlled, periodically inspected and water examination shall be carried out so that the discharged water conforms to the water discharge standards	●		●	
12. Removal of earth and sand and other necessary measures shall be taken for the open pits and other facilities for maintaining their functions	●	●	●	●
13. Ventilating equipment shall be provided for discharging gases	●			
14. Covering with earth shall be made to the thickness of about 50 cm and openings shall be closed on completion of landfill	●		●	
15. Records of maintenance and control shall be kept for five years	●	●	●	●
16. Closing of the disposal site shall be made with measures for preventing scattering and outflow of wastes, contamination by decoction and occurrence of fire suitably taken	●	●	●	●
17. Stagnant water in the landfill site shall be discharged before commencement of landfill		●		
18. External periphery separating facilities and internal separating facilities shall be periodically inspected, and necessary measures shall be taken if there is a fear of occurrence of damage and outflow of water containing wastes		●		
19. The landfill site shall be closed with the covering which satisfies the requirements equivalent to those for external periphery separating facilities on completion of landfill		●		
20. The covering stated in item 19 above shall be periodically inspected, and necessary measures shall be taken if there is a fear of occurrence of damage to them		●		

degraded in the ocean after dispersion. The following requirements should be satisfied for the dispersion-type discharge: (a) discharge shall be made below the sea surface, and (b) discharge shall be made when a ship is navigating.

33.4.3 Sea Areas for Discharge

The sea areas for discharge of wastes are divided into three, *i.e.* sea areas for discharge of hazardous industrial wastes (sea area A, Figure 3), sea areas for discharge of wastes which are not degradable in the ocean (sea area B, Figure 3), and sea areas for discharge of wastes which are degradable in the ocean (sea area C, Figure 3).

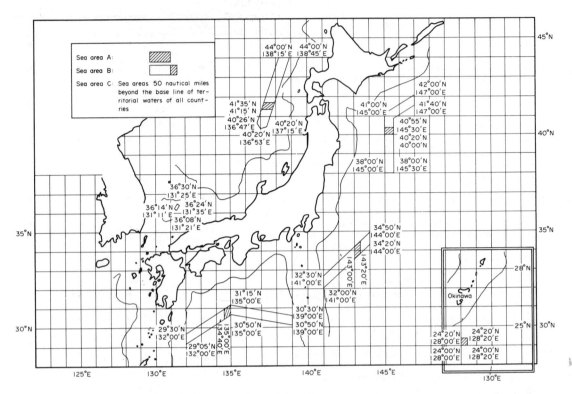

Figure 3   Sea areas for dumping of wastes around Japan

Sea area A was designated by taking the following criteria into consideration: (a) the bottom of the sea area should be flat so as not to allow solidified wastes to be broken up due to rolling down inclines, (b) the sea area should be located in an area where the water depth is 3500 m or deeper so that the surface water will not be affected even if a hazardous substance is dissolved from the solidified wastes. Also, the maximum water depth should be 6000 m so that breakage of solidified waste will not occur due to hydraulic pressure, (c) the sea area should be approximately equidistant from the main ports, (d) sea area A should be smaller than sea area B for ease of monitoring which must be severer than for sea area B.

Sea area B meets the following requirements: (a) the sea area should be located outside of ocean currents, such as the Kurile and Kuroshio currents, to prevent dispersion of wastes and to enhance precipitation and accumulation of wastes on the sea bottom, (b) the minimum water depth of the sea area should be 1500 m so as not to disturb trawl fishing, (c) the sea area should not be located in fish spawning areas nor in main fishing areas, (d) the sea area should be located so as not to damage any submarine cable, and (e) the sea area should be approximately equidistant from the main ports. Furthermore, its width should be 20 nautical miles, taking into consideration the ease for position finding of the waste discharge ships and monitoring efforts.

Sea area C is specified from the following criteria: (a) the sea area should be located in a place such that discharged wastes can be readily diffused into the ocean currents, such as the Kurile and Kuroshio currents, to enhance dilution and degradation of the wastes, and (b) the sea area should be located in a place where the dumped wastes will not be returned to coastal areas such as bays and inland seas.

33.5   SUMMARY

Japanese regulations for ultimate disposal of biohazardous materials are briefly introduced. These disposal activities are regulated by two laws: the Waste Disposal and Public Cleansing Law and the Marine Pollution Prevention Law. The first one is the major law and provides for standards for verification of wastes to be disposed of and for landfill activities and sites. Ocean dumping is not recommended in Japan as a rule, whereas wastes which are permitted and/or ordered to be discharged by ocean dumping based on the first law are also regulated by the second law which governs the methods and areas of ocean dumping.

33.6 REFERENCES

Environmental Agency Japan (1981). *Environmental Laws and Regulations in Japan. III. Water.* Environment Agency.
Minister's Office, Environmental Agency Japan (1982). *Kankyo Roppo 1982 edn.* (in Japanese: Environmental Laws and Regulations). Chuo Hoki, Tokyo.

34

Acceptance of New Drug Products in the United States and Canada

N. E. NITUCH
Ayerst Laboratories, Saint-Laurent, Quebec, Canada

34.1 INTRODUCTION

'We can lick gravity, but sometimes the paperwork is overwhelming.'

Werner von Braun

34.1.1 Background

In both the United States and Canada information on 'new drugs' must be submitted to, reviewed and, prior to marketing, approved by the respective regulatory agencies, namely the Food and Drug Administration (FDA) and the Health Protection Branch (HPB). While legislation concerning drugs in general has a much longer history, this requirement came into effect in the United States in 1938 and in Canada in 1951 (Temin, 1980; Pugsley, 1967). The extent and focus of this governmental scrutiny of supportive documentation has changed and continues to change with developments in science and specific agency philosophy. Today, a drug not recognized (by different criteria in the two countries) as both safe and effective for the uses recommended by the manufacturer is deemed a 'new drug' and is subject to the 'approval' of a New Drug Application (NDA, US) or Submission (NDS, Canada) before it can be sold. This chapter will present an overview of the requirements for such submissions. Potential sponsors of submissions should consult the full regulatory requirements presented in detail in the appropriate governmental acts, regulations and interpretive documents identified in the text.

34.1.2 Applicable Legislation and Regulations

Copies of the US Federal Food, Drug and Cosmetic Act and the regulations for its enforcement, Title 21 of the Code of Federal Regulations (21 CFR), are available from the US Government Printing Office (Table 1). The *Federal Register*, the official publication for notification of proposed and final regulations, can also be obtained from this source.

Copies of the Canadian Food and Drugs Act, the Food and Drug Regulations and the *Canada*

Table 1   Sources of Referenced Acts, Regulations, Forms, Guidelines and Governmental Publications

Source	Item(s)
Superintendent of Documents US Government Printing Office Washington, DC 20402	Federal Food, Drug and Cosmetic Act Title 21 Code of Federal Regulations The *Federal Register* FDA Bureau of Drugs Clinical Guidelines (see Table 2)
Canadian Government Publishing Center Ottawa, Ontario K1A 0S9	Office Consolidation of the Food and Drugs Act and Regulations *Canada Gazette Part II* Good Manufacturing Practices for Drug Manufacturers and Importers
Clerical Services Health Protection Branch Banting Building Tunney's Pasture Ottawa, Ontario K1A 0L2	HPB Information Letters (Drugs)
FDA Bureau of Drugs 5600 Fishers Lane Rockville, MD 20857	Form FD-356H NDA binders
Drug Regulatory Affairs Division Drugs Directorate (HPB) Tunney's Pasture Ottawa, Ontario K1A 0L2	*Guidelines for Preparing and Filing New Drug Submissions*
Bureau of Human Prescriptions Drugs Health Protection Branch Place Vanier, Tower 'B' 355 River Road Vanier, Ontario K1A 1B8	HPB Form 3011 *Guide for the Preparation of the Chemistry and Manufacturing Portions of New Drug Submissions* Preclinical Toxicologic Guidelines *Guidelines for Product Monographs: Drugs for Use in Humans (1979)* *Draft Guide for the Labelling of Drugs for Human Use (1981)*
Freedom of Information Staff (HFI-35) Food and Drug Administration Room 12A-16 5600 Fishers Lane Rockville, Maryland 20857	*Guidelines: Manufacturing and Controls for IND's and NDA's* FDA Office of Planning and Evaluation Report: *Approvals and Non- approvals of New Drug Applications During the 1970s* Guidelines for Toxicology Requirements for NDA's GLP's: Questions and Answers Summary Basis of Approval for (specify drug) Bureau of Drugs Statistical Evaluation Branch: Draft General Statistical Documentation Guide for Protocol Development and NDA Submissions (April, 1980) 'Goals for Submittal of Data to NDA's'—Presentation by Dr. Marion J. Finkel, 14th Annual Meeting of the Drug Information Association, Philadelphia, June 19, 1978
Pharmaceutical Manufacturers' Association 1155, Fifteenth Street N.W. Washington, D.C. 20005	*PMA Guidelines for the Production, Packing or Holding of Bulk Pharmaceutical Chemicals—Second Edition*

Gazette, wherein new regulations or amendments are published, are available from the Canadian Government Publishing Centre (Table 1). Proposed regulatory changes accompanied by an invitation for comment are published in HPB Information Letters. Subsequent publication of summaries of the comments received and the government's response to them, as is done in the United States through *Federal Register* publication, is a relatively new feature of this HPB document. Potential sponsors of submissions can request inclusion on the mailing list or obtain back issues by writing to the Health Protection Branch (Table 1).

34.1.3 Required Form and Content

The required form and content of the NDA is described in 21 CFR regulation 314.1. These are also outlined in Form FD-356H which must accompany the submission. Copies of this form and the required special binders for the submission are available from the FDA's Bureau of Drugs (Table 1).

The sources for Canadian *Guidelines for Preparing and Filing New Drug Submissions* and HPB form 3011 (which must accompany a New Drug Submission) are also listed in Table 1.

In compiling a submission, every effort should be made to present the required information in a clear, logical and organized manner to facilitate its review. Tables of contents should be detailed. Whenever possible, schematics, tabulations and/or graphical displays should accompany text. Related subject areas and conclusions based on several areas of work should be adequately cross-referenced. Within sections, individual reports should be presented in descending order of significance; substantive data should be clearly identified as such together with the reasoning for being so designated and should precede supportive data.

Both countries require summaries of certain specified areas of the submission. It is felt that the time and effort expended in their preparation pays direct dividends in rapid government review. Well-prepared and cross-referenced summaries greatly enhance the reviewer's grasp of the totality of the evidence being presented and the significance and place of individual reports therein.

While the form and content requirements are to be strictly adhered to, both governments would probably be open to discussions of modifications for specific submissions where either special characteristics of the drug so warrant, or where an alternative organization which may enhance review is proposed.

34.2 DRUG SUBSTANCE

34.2.1 Preparation and Quality Control

This portion of the new drug submission will require special attention by sponsors of applications based on biotechnology. The use of novel manufacturing technology will undoubtedly raise new concerns at the interface of science and regulation. Moreover, an informational and experiential gap may exist between regulatory scientists and those involved in proposing the use of novel, esoteric and often proprietary processes which lie at the current frontiers of science and technology. The descriptions of such processes in a submission should therefore be as well organized as possible, and should be presented in a lucid and detailed manner. It is to be noted that even with existing technologies, regulators often find this section of submissions deficient in non-substantive aspects due to inadequate, incomplete or unclear descriptions of procedures, equipment and components.

An excellent reference entitled *Guide for the Preparation of the Chemistry and Manufacturing Portions of New Drug Submissions* is available from the Pharmaceutical Evaluation Division of HPB's Bureau of Human Prescription Drugs (Table 1). A US document of much earlier vintage, entitled *Guidelines: Manufacturing and Controls for IND's and NDA's*, is available from the FDA. Additional guidance for sponsors of US submissions can be found in the analysis of deficiencies in the chemistry and manufacturing sections of NDAs in an FDA Office of Planning and Evaluation report, entitled *Approvals and Non-approvals of New Drug Applications During the 1970s*. Copies of these two documents can be obtained from FDA's Freedom of Information Staff (Table 1).

A full list of the articles used in the manufacture of the drug substance is required in the NDA and NDS. The listing should include all of the ingredients used in the preparation of the drug substance, including its extraction, isolation and purification, regardless of whether they undergo any change or are removed during the process. The ingredients should be identified by their established names, where these exist, or by their complete chemical name, using structural formulae, when necessary, for specific identification. If proprietary preparations are used, their complete quantitative composition should be given.

Specifications, including identity, purity and potency, where applicable, should be provided for all materials used in the production of the drug substance. The methods of testing used to assure conformance to these specifications should also be provided. Microorganisms, enzymes or hybridomas should be fully characterized, including a description of their preparation and evidence of their identity and purity. The procedures and standards used to maintain their characteristics should be described in detail. Exceptionally good quality control systems should exist to assure no contamination of viable components.

The method of production, isolation and purification on a commercial scale should be fully described. Flow charts of the production sequence are recommended. Each step of the process should describe the name and amounts of materials used, the equipment being utilized, the synthetic step being effected, the conditions being controlled (time, temperature, pH, *etc.*), and the yield obtained. In-process quality control checks and associated testing methods should be supplied. Specifications for intermediates or precursors should be supplied in cases where these are isolated and controlled with respect to purity or other characteristics prior to being used in a subsequent step in the process. If the drug substance is claimed sterile, this section should include a complete description of the method(s) used to achieve sterility and the controls used to maintain sterility during storage and transportation.

Specifications for the finished drug substance which will ensure adequate batch-to-batch control of identity, potency and purity should be provided, together with fully described analytical methods used to test these characteristics. The use of new production technologies may require newer, more sensitive and specific methods based on techniques developed in molecular biology or analytical chemistry. It is essential that identity can be unequivocally determined and that impurities, especially those structurally similar to the active moiety, are stringently controlled. Using these specifications, a comparison should be made between material produced on a commercial scale and that obtained from pilot plant or laboratory scale production which may have been used in early preclinical and clinical trials. Such a comparison will assure that no physical, chemical or biological changes have resulted from production scale-up. Complete analytical results for several batches of the drug substance, including those used in toxicological and clinical testing should be supplied.

In the United States, approval of a product application also constitutes approval of the sponsor's specific and general processing techniques. The NDA must contain a full description of the facilities where the drug substance is produced and tested, the equipment available, the qualifications and educational background and experience of personnel responsible for the various aspects of production and quality control, and of the written general and specific operating procedures being utilized. The adequacy of the facilities, processing and quality control techniques and procedures are reviewed by on-site inspections. Although the Current Good Manufacturing Practice Regulations (21 CFR 210–211) are not directly applicable, the standards of Part 211 are utilized by FDA as guidelines during inspections of manufacturers of bulk drug components. For additional guidance in this area, potential sponsors of submissions should consult the *PMA Guidelines for the Production, Packing, or Holding of Bulk Pharmaceutical Chemicals—Second Edition*, available from the Pharmaceutical Manufacturers Association (Table 1).

In Canada, information along the lines outlined above should be submitted for material of foreign origin, especially when other than conventional technology is applied. Inspection of drug substance manufacturing facilities prior to NDS approval is discretionary, and for material coming from outside Canada the resident drug regulatory agency may be consulted by HPB. Inspection by HPB is, however, required for the various preparations of biologic origins which are listed in Schedule D of the Food and Drugs Act. The listed products, which include insulin, gonadotrophins, drugs (other than antibiotics) prepared from microorganisms, and all drugs obtained by recombinant DNA procedures, must be licensed in accordance with Division 4 of the Regulations in addition to meeting the new drug requirements. The requisite licence is issued following inspection of the manufacturing facilities and testing of samples by the Bureau of Biologics of HPB's Drugs Directorate. In addition, copies of those portions of the NDS which cover manufacturing and quality control must also be provided to this bureau for review. The required information is listed in the guideline for NDS format identified in Section 34.1.3.

34.2.2 Proof of Structure

The results of investigations performed to elucidate and/or confirm the structure of the drug substance should be submitted. Elemental analysis, UV, IR and NMR resonance spectroscopy, and mass spectrometry are among the techniques considered acceptable. Copies of the actual spectra with assignments should be supplied. Alternative methods, especially for products which present difficulties with respect to structural investigation, would undoubtedly be accepted provided they are based on sound science and are adequately described.

34.2.3 Characterization

Comprehensive information on the chemical and physical properties of the new drug substance should be submitted. These data will influence the development of quality control specifications and analytical procedures for the drug substance and its dosage form(s), and the establishment of dosage form formulations, production techniques, stability protocols, packaging systems and storage recommendations.

34.2.3.1 Physicochemical properties

The chemical structure, molecular formula and molecular weight should be given together with details of the substance's physicochemical properties (*e.g.* physical form, melting point, solubility in common solvents, pH of aqueous solutions, particle size distribution, *etc.*). Information should also be provided on the possible existence of isomers and polymorphs of the drug substance. Where applicable, details of specifications and methods for their control should be provided.

34.2.3.2 Impurities

With the advent of new, specific and sensitive analytical methodologies, the identification and control of impurities in drugs has become a major regulatory concern. Sponsors of submissions should provide a detailed discussion of known and theoretical impurities which may arise during the preparation of the drug substance, including their structures. This should cover impurities which are likely to be present from the process itself, the ingredients, intermediates or the by-products from various steps. The results of studies carried out to determine the impurities actually present should also be provided. Where possible, impurities should be isolated and identified. Some current analytical methods can identify extremely low levels of impurities for which reference standards cannot be utilized; however, efforts should be made to prepare such standards for major contaminants. Major impurities and those of known toxicity should be individually controlled, while those found in low quantities can be limited as a group. A purity specification should be developed on the basis of the bioactivity and toxicity of the impurities, batch-to-batch production experience, and on the levels seen in material which has been used in preclinical and clinical testing. A full description of the methods used to assess impurities together with the results of their validation should be presented.

34.2.3.3 Stability

The stability of the drug substance should be studied under both normal and exaggerated conditions to determine its susceptibility to factors such as heat, light, moisture and oxidation. The results of stability studies on at least two batches, specifying the container/closure system used, should be submitted. Information should be provided on degradation pathways and the actual degradation products found. Full descriptions of analytical methods employed, including data to demonstrate their specificity, accuracy and precision, should also be submitted. Depending on the results obtained and based on criteria such as those outlined earlier for impurities, limits and/ or special storage requirements may be necessary to control degradation products in the drug substance.

34.2.3.4 Reference standard

The method of preparation of the reference standard should be described if different from that provided for the drug substance. The reference standard should be of the highest purity that can be reasonably obtained and any additional purification procedures should be provided in detail. The results of analytical testing performed to establish the identity, purity and potency of the reference standard should also be submitted.

34.3 PRECLINICAL STUDIES

The results of preclinical studies establish the basic biological characterization of the potential new drug. Although *in vitro* systems may be useful for certain drugs such as antibiotics, most drugs will require *in vivo* studies. The division of preclinical studies into various categories is one of convenience since, for example, pharmacologic and toxicologic studies employ similar scientific principles and are interdependent. Studies in the different disciplines must be integrated. Acute toxicity studies will provide needed information for the pharmacologist, whose work may then influence the toxicologist's design of long-term studies. Although the predictive value of observations in animals to man is far from absolute, these studies can estimate the safety, efficacy and mechanism of action of the drug substance and should form the basis for the design of initial human studies. These studies may, in turn, uncover unexpected results, therapeutically beneficial or adverse, which may call for further preclinical work in animals.

It is desirable that the animal species used in preclinical studies metabolize the drug in a manner similar to man, but this is not always possible. Marked differences between species should alert the developers of the drug to the importance of early investigations of the metabolic pattern in man so that more meaningful animal studies, if necessary, can be carried out. Different interspecies reactions to a drug can often be explained by important differences in its biotransformation or pharmacokinetic properties; however, other factors such as differences in receptors, reactions following receptor binding, or endogenous intermediary metabolism may be involved.

34.3.1 Pharmacology

The purpose of defining the pharmacological profile of a potential new drug is to justify its study in man with respect to both desired and undesired effects. The kind of studies carried out will depend on the nature, actions and possible uses of the substance and consequently cannot be subject to strict guidelines. Submissions should present the details of all experiments carried out to define the pharmacology of the drug including descriptions of methodology, interpretation of results and conclusions. The minimum effective dose should be identified, where relevant, and descriptions of dose–effect relationships in each species studied should be provided. Summaries, cross-referenced to the location of the basic scientific data in the submission, should be prepared to outline the following: (a) primary action: results of studies of the pharmacological activity related to the primary therapeutic effects in man; and (b) secondary action: results of studies related to the secondary pharmacological effects of the substance which may be relevant to its expected use or its adverse effects. The results of other miscellaneous pharmacologic studies which are not directly related to the drug's proposed use in man should also be summarized and submitted. These may be considered pertinent to safety and might include other actions demonstrated in the sponsor's pharmacology screening program.

34.3.2 Microbiology

Submissions for proposed anti-infectives should provide the results of investigations carried out to delineate the substance's antimicrobial spectrum. Methodology should be fully described and minimum inhibitory concentrations (MICs) should be provided on relevant clinical isolates and standard laboratory strains. Data should be supplied on the effect of inoculation size on the determination of MICs. The results of studies in experimentally-infected animals to evaluate potential for therapeutic efficacy may also be required.

Information should be submitted on the mode of action of the drug and to support claims of bactericidal and/or bacteriostatic activity, specifying the concentration at which this occurs.

Studies designed to measure the incidence, type and rate of development of resistance by target organisms should be carried out at various drug concentrations. The results of both *in vitro* and *in vivo* experiments should be submitted and should include information on the possible effects of the drug's metabolite(s).

Information should also be provided on the drug's effects on natural body flora and the potential for overgrowth and superinfection.

34.3.3 Metabolism Studies

These studies should elucidate the absorption, distribution, biotransformation and elimination of the drug. The data should be generated in the species and by the route(s) of administration used in toxicology studies and in studies of primary pharmacologic activity. Where possible, the relationship between blood levels and pharmacologic or toxic effects should be determined.

The methods used to measure the levels of the drug or its metabolites in biological fluids and tissues should be fully described and should include data to demonstrate their specificity and sensitivity.

34.3.4 Toxicology Studies

Toxicological studies seek to estimate the risks which may be associated with the use of a drug in man. This type of animal testing is routine, yet despite its advantage of allowing exaggeration of drug insult the accuracy of the estimate cannot be assured. While positive findings may be predictive of potential toxicity in man, negative results do not guarantee safety. Examples in which animal data have been suggestive of toxic effects in man are balanced by experiences such as the practolol oculomucocutaneous syndrome and the renal and cutaneous problems associated with captopril which were not seen in animals.

Guidelines for toxicology requirements for new drug applications are available from the FDA through their Freedom of Information Office (Table 1). Canadian guidelines can be obtained from the HPB Bureau of Human Prescription Drugs (Table 1). The extent, type and duration of testing required depends on the proposed duration of treatment in man, the condition(s) for which the drug is to be recommended, and on the particular nature of the drug involved. The guidelines cited above are not strict protocols and should be used for general guidance only. Sponsors of submissions should communicate directly with the appropriate regulatory agency involved to determine the specific requirements for their particular product.

In addition, sponsors of US submissions must ensure compliance to the regulations governing the conduct of nonclinical laboratory studies (21 CFR Part 58). These 'Good Laboratory Practice' (GLP) regulations have been put into effect to assure the quality and integrity of safety data submitted to the US government by establishing standards for animal facilities, animal care practices, training and qualification of personnel, recording and handling of data, administration of substances, maintenance of records and reporting of results. Interpretive guidelines for these regulations are also available from FDA's Freedom of Information Office (Table 1).

The Canadian Health Protection Branch has issued a similar standard for these types of studies which is based on voluntary compliance.

34.3.4.1 Acute toxicity

These studies define the lethal dosage range of a product following the administration of a single dose or several closely spaced doses. Several species, including those in which the pharmacological and subsequent long-term toxicological studies have been carried out, should be used. The product is usually administered to animals of both sexes, unless the proposed human use will be restricted to only one sex. The product is administered by various routes, including those to be recommended in man. Observations such as the onset and duration of signs of toxicity and time of death are recorded.

Submissions should present the results of these studies in summary form organized by species and route of administration. Dose levels, vehicle(s) used, numbers of animals per dose level, and weight, sex and age of the animals should be identified. LD_{50} values, with confidence limits, should be provided together with a description of the methods used for their calculation. In those species where the use of large numbers of animals is neither practical not feasible, an approximation of the LD_{50} value is acceptable. Signs of toxicity, times of death and other pertinent observations such as species, sex or age differences should be described. The location of detailed reports and basic data in the submission should be identified.

34.3.4.2 *Chronic toxicity*

These studies define the behavioral, functional and histopathologic effects of prolonged administration of the drug. Their duration is related to the proposed duration of therapy in man and general guidance in this respect can be found in the guidelines identified earlier in this section.

Chronic toxicity studies are performed in at least two species of animals, one rodent and one nonrodent. Again, species selection must be based on similarity to man with respect to absorption, distribution, metabolism and excretion of the drug being studied.

At least three different dose levels are generally used. Their selection should be based on the pharmacology of the drug and on preliminary dose-range finding toxicity studies. The highest dose chosen should be one that produces toxic effects, but which allows a majority of the animals to survive. The lowest dose selected should elicit appropriate pharmacological effects in the test species. In addition, at least one intermediate dose should be included to define incipient effects and dose–response relationships. A control group is mandatory for comparative purposes.

The drug is administered by the route(s) to be recommended in human therapy. The dosage regimen should also be related to human use; if the drug is intended to be given to man on a daily basis it should be administered to animals for seven days per week.

Observations to be recorded during the conduct of long-term studies will vary depending on study objective(s), the species used and the nature of the drug. Changes in appearance, general conditions, behavior, rate of body weight gain, food and water consumption, hematology, clinical chemistry and urinalysis parameters, and where possible physiologic and pharmacologic manifestations should be recorded. Comprehensive pathological studies describing both macroscopic and microscopic findings are also carried out.

Submissions should present the results of these studies in summary form, organized by species, duration and route(s) of administration, according to the format prescribed by the applicable regulatory agency (see Section 34.1.3). These summaries must give full details of methodology: dose levels, species, weight, age, sex, number of animals/dose, treatment periodicity, method and route of administration, duration and an outline of the parameters measured, including a list of organs and tissues submitted to pathological examination. Results should list all pertinent observations, including identification of those parameters unaffected by treatment. The incidence, severity and dose–response relationships of observed alterations should be described. Where possible, those alterations associated with the pharmacological actions of the drug are to be distinguished from those considered to be related to its toxicity. The use of tabular presentations can often enhance the comprehension of these large and detailed studies. These summaries should also be adequately cross-referenced to the individual reports and raw data presented elsewhere in the submission.

34.3.4.3 *Reproduction and teratology studies*

Studies of the effects of the drug on the reproductive processes in at least two animal species are usually required for new drugs. These studies, based on a generalized experimental design developed by the FDA in 1966, evaluate possible effects on male and female fertility, conception, transport, implantation and development of the zygote, parturition, and the viability and development of offspring. They are divided into three segments: (i) general study of fertility and reproductive performance, (ii) teratology, and (iii) perinatal and postnatal studies. The results should be presented by species and route of administration, giving full details of methodology including dose levels, time of drug administration relative to stage of pregnancy, parameters studied and methods used to examine offspring. The effects of the drug on the mother, pregnancy and the fetus should be fully described and a discussion presented on the relationship of the doses used in these studies to those known to cause toxic effects in the same species and to the proposed human dosage recommendations.

34.3.4.4 *Special studies*

Carcinogenicity testing is required for some compounds prior to marketing. These studies are mandatory for drugs which are chemically related to or which produce metabolites similar to known carcinogens. For other drugs, the criteria for this requirement are less well defined and are based on individual evaluation. Carcinogenicity studies are complex, of long duration and are influenced by many factors. Potential sponsors of submissions should discuss the requirements for their drug with the appropriate agency during the course of its development and, where such studies are required, consult with the regulatory body on the design of experimental protocols prior to their initiation.

In addition to the testing outlined above, certain drugs will require further toxicological evaluation which is related to the particular nature of the drug itself or its intended use. Studies on tissue irritation, antigenicity, dependence liability and skin sensitization are examples of such special tests.

The submission should present a global discussion of the results of preclinical testing, wherein the safety and efficacy of the drug based on animal studies alone is evaluated. Important or unusual findings should be highlighted and interpreted.

34.4 DOSAGE FORMS

The information required in submissions concerning dosage from manufacturing and testing is essentially the same in type, extent and manner of presentation as that outlined earlier for the drug substance. The guidelines identified in Section 34.2.1 address themselves to both the drug substance and pharmaceutical dosage form.

34.4.1 Manufacture

A comprehensive description of the manufacturing process should be presented in sufficient detail to permit an evaluation of the adequacy of the methods used to assure the identity, strength, quality and purity of the drug.

The quantitative formula should list the name and quantity of each ingredient in the unit to be sold (*i.e.* mg/tablet or mg/ml for liquids). A batch formula representative of that to be used commercially should also be submitted. This must include all components used, regardless of whether they appear in the final product or are removed during the manufacturing process. Where excess quantities ('overages') are added for stability and/or manufacturing loss, this should be clearly identified and justified with supporting data. It should be noted that the use of ranges for ingredients and the listing of alternates for excipients is allowed by the FDA, but is not normally accepted by the HPB.

The manufacturing processes, including filling and packaging operations, should be fully described including such details as actual operating conditions, equipment used and in-process testing carried out.

US submissions must include descriptions of the physical facilities where operations are conducted and the qualifications and training of the personnel involved. In general, Canadian submissions should provide this type of information for parenteral drugs and for those being imported from foreign manufacturers. US submissions must also encompass other requirements related to the good manufacturing practice (GMP) regulations. Similar regulations have recently been promulgated in Canada; however, assessment of compliance is not tied to the New Drug Submission approval process, but is carried out by subsequent on-site inspection by the HPB's Bureau of Drug Quality. An interpretive document for these new GMP regulations can be purchased from the Canadian Government Publishing Center (Table 1).

34.4.2 Quality Control

Specifications and test methods for the control of all components of the dosage form should be submitted. These include active ingredient(s), excipients, including those such as solvents which do not appear in the finished product, and packaging materials which come in direct contact with the drug. If an ingredient is recognized in an official pharmacopial compendium, it is sufficient to state that it will be tested and accepted according to the written standard. Acceptance criteria and methods should also be submitted for intermediates such as tablet cores, granulations, blends or bulk solutions.

The final dosage form specifications should provide for batch-to-batch assurance of the identity, strength, quality and purity declared in the product's labelling. In addition, the specifications should include other tests and acceptance criteria related to the particular dosage form. Examples of these are weight variation and/or content uniformity for capsules, tablets and other solid dosage forms, tests for homogeneity, disintegration and/or dissolution for capsules and tablets, and sterility, pyrogenicity, safety, clarity, particulate matter and pH for injectables.

The analytical methods proposed for testing the final dosage form should be described in full detail. These methods must be accurate, precise, reproducible, sensitive, specific and sufficiently rugged to allow broad utilization. Data supporting these characteristics must also be submitted in both the US and Canada.

In the United States, samples must be submitted with the NDA which are then analyzed in the FDA's own laboratories using the proposed methods. If the methods meet the criteria listed above, they are deemed validated and become 'official' methods. Method validation by the FDA can delay submission approval. Such delays can be largely avoided by clear and complete descriptions of analytical procedures which facilitate their use by other than the sponsor's personnel. While the Canadian regulations have provisions which would allow similar verification of analytical techniques, it is not normally a requisite component of NDS clearance.

34.4.3 Stability Studies

Submissions must present the results of studies carried out to determine the appropriate storage conditions for the drug and the time period during which the product will remain within specifications when so stored. These data must be generated on the finished dosage form in the container/closure system(s) in which it will be marketed. If the product is to be reconstituted at the time of dispensing, data should also be submitted on the solution so prepared according to the proposed label directions. Similarly, compatibility and stability data are required to identify expiry times for any mixtures (*e.g.* with sodium chloride, dextrose, or other agents in intravenous administration) recommended in the labelling.

Stability studies should be based on at least two or three batches and should address the chemical, physical and, where applicable, the microbiological integrity of the dosage form. The primary data base should be the results of frequent assays (enough to allow statistical analysis) carried out on the product maintained under the recommended storage conditions. The results of accelerated testing using elevated temperatures and other stress conditions generate important information, but are usually considered only supporting evidence in the establishment of the expiry date.

The method for stability analysis should be capable of distinguishing the intact drug from its degradation products. A fully detailed description of the method should be provided, together with experimental evidence to support its specificity, sensitivity, accuracy and precision. Where the primary assay method lacks specificity, other tests capable of detecting known breakdown products must be used to measure and, with appropriate specifications, limit degradation products.

Physical changes in the dosage form can be as critical as chemical instability of the active ingredient. Physical changes such as crystal growth, change in crystal form, increase or decrease in dissolution and/or disintegration time can have important therapeutic consequences. Stability studies should, therefore, incorporate visual and organoleptic methods to detect gross physical changes and appropriate instrumental methods that are capable of detecting changes in the physical state of the dosage form.

If a preservative is used to prevent microbial growth in a product, stability studies should incorporate tests (both assay and microbiological challenge) to ensure that adequate levels are present up to the expiry date.

Submissions should present the results of initial and subsequent chemical, physical and biological testing and include information on the formulation, batch numbers, batch sizes, manufacture dates, type of container/closure system, storage conditions, test dates and proposed labelling statements relating to stability.

A discussion should be presented on theoretical degradation pathways accompanied by any available data on breakdown products actually found. Recommendations for storage conditions and an expiry date should be proposed and supported by an evaluation and statistical analysis of the data submitted.

34.5 CLINICAL STUDIES

The requirement for the provision of 'substantial evidence' of efficacy in addition to evidence on safety was incorporated into US and Canadian drug legislation in 1962 and 1963, respectively

(Temin, 1980; Pugsley, 1967). Both the FDA and HPB recognize clinical studies as the primary source of this evidence. The FDA further stipulates that these studies must be 'adequate and well controlled' and defines the attributes that a study must possess in order to be so described (21 CFR Part 314.111 (5) ii). Neither agency, however, provides a clear definition of what constitutes 'substantial evidence'.

The objective of clinical studies is to assess the utility of a drug in the treatment or prophylaxis of a disease or condition and to measure its risks or undesirable effects. The results of these studies in man form the basis of the difficult and complex benefit–risk decisions that must be made by the regulatory agencies in their acceptance or rejection of a new product. However, the extrapolation of results generated in a controlled clinical trial setting with a selected patient population to those expected when the drug is broadly used in an unmonitored fashion in less well-defined populations is tenuous. Feinstein (1971) contends that the current procedures for assessing safety and efficacy, while generally accepted, are 'oversimplified, naïve, and grossly inadequate for the needs of clinical medicine'.

Given the above and the complexity of both the conduct and evaluation of clinical trials, it is not surprising that they represent a major portion of drug development costs.

The design and extent of clinical investigations will depend on the nature of the drug and the disease being studied. They should be of a standard consistent with the current North American 'state of the art' in clinical science for the therapeutic field involved. A series of guidelines for the clinical evaluation of new drugs in over 20 therapeutic categories has been prepared by the FDA in collaboration with other organizations and individuals (Table 2). These guidelines can be purchased from the US Government Printing Office (Table 1). Additional guidance can be obtained from the Summary Basis of Approval (SBA) document issued by the FDA for other new drug products approved in the therapeutic category of interest. This document consolidates the data on which FDA approval was based and includes information on preclinical and clinical studies and labelling. These SBAs for specific drugs can be obtained through FDA's Freedom of Information Staff (Table 1). The basic information from these sources is generally applicable in Canada.

Table 2   FDA Bureau of Drugs Clinical Guidelines

Title	Order number
General Considerations for the Clinical Evaluation of Drugs	GPO 017–012–00245–5
General Considerations for the Clinical Evaluation of Drugs in Infants and Children	GPO 017–012–00246–3
Guidelines for the Clinical Evaluation of Antidepressant Drugs	GPO 017–012–00247–1
Guidelines for the Clinical Evaluation of Antianxiety Drugs	GPO 017–012–00248–0
Guidelines for the Clinical Evaluation of Anti-Infective Drugs (Systemic) (Adults and Children)	GPO 017–012–00250–1
Guidelines for the Clinical Evaluation of Anti-Anginal Drugs	GPO 017–012–00259–5
Guidelines for the Clinical Evaluation of Anti-Arrhythmic Drugs	GPO 017–012–00256–1
Guidelines for the Clinical Evaluation of Antidiarrheal Drugs	GPO 017–012–00257–9
Guidelines for the Clinical Evaluation of Gastric Secretory Depressant (GSD) Drugs	GPO 017–012–00252–8
Guidelines for the Clinical Evaluation of Hypnotic Drugs	GPO 017–012–00253–6
Guidelines for the Clinical Evaluation of Local Anesthetics	GPO 017–012–00255–2
Guidelines for the Clinical Evaluation of Anti-Inflammatory Drugs (Adults and Children)	GPO 017–012–00258–7
Guidelines for the Clinical Evaluation of Antacid Drugs	GPO 017–012–00261–7
Guidelines for the Clinical Evaluation of G.I. Motility-Modifying Drugs	GPO 017–012–00262–5
Guidelines for the Clinical Evaluation of Laxative Drugs	GPO 017–012–00263–3
Guidelines for the Clinical Evaluation of Psychoactive Drugs in Infants and Children	GPO 017–012–00281–1
Guidelines for the Clinical Evaluation of Bronchodilator Drugs	GPO 017–012–00271–4
Guidelines for the Clinical Evaluation of Drugs to Prevent, Control and/or Treat Peridontal Disease	GPO 017–012–00272–2
Guidelines for the Clinical Evaluation of Drugs to Prevent Dental Caries	GPO 017–012–00273–1
Guidelines for the Clinical Evaluation of Analgesic Drugs	GPO 017–012–00283–8
Guidelines for the Clinical Evaluation of Drugs Used in the Treatment of Osteoporosis	GPO 017–012–00284–6
Guidelines for the Clinical Evaluation of Lipid-Altering Agents in Adults and Children	GPO 017–012–00288–9
Guidelines for the Clinical Evaluation of Antiepileptic Drugs (Adults and Children)	GPO 017–012–00292–7
Guidelines for the Clinical Evaluation of Antineoplastic Drugs	GPO 017–012–00294–3
Guidelines for the Clinical Evaluation of Radiopharmaceutical Drugs	GPO 017–012–00301–0
Guidelines for the Clinical Evaluation of General Anesthetics	GPO 017–012–00303–6

Submissions should present the results of clinical trials in the format required by the appropriate agency (see Section 34.1.3). Additional guidance, primarily for NDAs, can be found in the following documents which are available through FDA's Freedom of Information Staff (Table 1):

(i) Draft General Statistical Documentation Guide for Protocol Development and NDA Sub-missions (April, 1980), prepared by FDA's Bureau of Drugs' Statistical Evaluation Branch; and (ii) the text of a presentation delivered at the 14th Annual Meeting of the Drug Information Association in Philadelphia (June 19, 1978) by the then Associate Director for New Drug Evalu-ation, Dr. Marion J. Finkel. The latter, entitled 'Goals for Submittal of Data to NDA's', addresses the presentation of clinical data in the 'expanded summary' of the NDA.

The presentation of results should be divided into several categories. Controlled studies which are considered primary or pivotal evidence should be grouped by indication and presented in des-cending order of the 'weight' or importance attributed to them. Uncontrolled or incompletely controlled studies and clinical pharmacology studies should be presented separately. Reports of published clinical studies should be reviewed and analyzed and also presented in the submission.

An overall summary should discuss the combined results from all study categories and present composite, balanced conclusions regarding the therapeutic benefits and potential risks associated with the use of the drug as well as recommendations for its proposed terms of use once marketed.

It should be noted that the FDA requires that some of the clinical investigations upon which approval is to be based must have been performed by investigators within the United States. The HPB does not have a similar requirement in Canada.

34.6 LABELLING

The term 'labelling' in both the US and Canadian context covers not only the labels affixed to the drug product's container, but also any accompanying matter which transmits information on the drug's value or its terms of use. This includes the package insert and, in Canada, a document entitled the Product Monograph. A product's 'labelling' is, in essence, the distillation of the information gathered during a drug's development and presented to government as the basis of approval to the practical recommendations for its medical use. The product's approved 'labelling' also functions to define and limit the promotional statements that can be made regarding its use in subsequent advertising to the medical profession.

Submissions should present draft labels and annotated proposed 'labelling' which identifies the location of evidence in the submission supporting the recommendations made therein.

United States regulations concerning labelling (21 CFR part 201) are quite detailed. Some additional guidance with respect to the intent and philosophy behind recent new content and for-mat requirements for human prescription drugs can be found in the preamble to their publication (Food and Drug Administration, 1979). In Canada one should adhere to the following guidelines, available from HPB's Bureau of Human Prescription Drugs (Table 1): *Guidelines for Product Monographs, Drugs for Use in Humans* (1979) and the 1981 *Draft Guide for the Labelling of Drugs for Human Use*.

34.7 ACCEPTANCE

The regulations in the US and Canada indicate that a judgment on the acceptance or rejection of a submission will be rendered in 180 and 120 days, respectively, after filing. In the US, how-ever, this '180-day clock' only starts when the submission is deemed complete. The criteria for completeness, on the basis of which FDA may refuse to file an NDA, are outlined in 21 CFR part 314.110. In either country this determination of acceptance or rejection cannot normally be expected within 180 or 120 days following submission.

In the US, the mean review time for all 96 NDAs approved in 1981 was 24.4 months. For the 27 NDAs for new chemical entities approved during that year, the mean time was 30.7 months. The FDA also uses a system of classification for new chemical entities which includes a rating of their therapeutic potential. The review of submissions on drugs with a high rating is given priority and approval is often granted in one year.

In Canada, it is rare, but not unusual, to have some feedback from the HPB on an NDS in about 120 days following its submission. The average time for clearance, based on surveys from 1978 to 1981 by the Pharmaceutical Manufacturers' Association of Canada, is about one year. The surveys covered 115 new drug submissions, of which 75 were cleared during the period exam-ined.

The time taken by both regulatory agencies in accepting or rejecting submissions is undoubt-edly influenced by the nature of the drugs and diseases involved and governmental administrative practices. There is no question, however, that the quality of the scientific work carried out and, more importantly, its presentation in well-organized submissions are major determinants of timely consideration of new drug submissions.

Following acceptance, the holder of an 'approved' NDA or 'cleared' NDS must comply with a number of other regulations regarding the continued sale of the new drug. In addition, changes to either the production, testing or packaging of the product and its components or its recommendations for use, among other things, must be covered by supplements to the original submission. These changes cannot be effected until these additional submissions are reviewed and accepted.

34.8 CONCLUSIONS

North American drug regulatory requirements continue to evolve. Changes in submission format are currently being discussed in both the US and Canada and other mechanisms are being examined to reduce the regulatory burden for both industry and government while ensuring optimum therapeutic benefit with a minimum of risk to patients.

Acceptance of products by regulatory agencies is the culmination of significant investments in time and expense in research and development. Future return on these investments depends not only on the quality of the scientific work carried out and the eventual strength of marketing efforts, but also to a great extent on regulatory expertise and a professional approach to the presentation of data to governments in order to gain that all important 'approval to market'.

34.9 REFERENCES

Pugsley, L. I. (1967). The administration and development of federal statutes on foods and drugs in Canada. *Med. Services J.*, **23**, 387–449.

Temin, P. (1980). *Taking Your Medicine. Drug Regulation in the United States*. Harvard University Press, Cambridge, MA.

Feinstein, A. R. (1971). Clinical biostatistics IX. How do we measure 'safety and efficacy'? *Clin. Pharmacol. Ther.*, **12**, 544–558.

Food and Drug Administration (1979). Final rule: labelling and prescription drug advertising; content and format for labelling for human prescription drugs. *Fed. Regist.*, **44**, 37 434–37 461.

35

Japanese Governmental Procedures for Approving the Manufacture or Importation of Pharmaceuticals

S. ESUMI
Kaken Pharmaceutical Co. Ltd., Tokyo, Japan

35.1 INTRODUCTION

The Pharmaceutical Affairs Law (Law No. 145, 1960) defined the term 'drug' as follows: (1) articles recognized in the Japanese Pharmacopoeia; (2) articles (other than quasi-drugs) which are intended for use in the diagnosis, cure or prevention of disease in humans or animals, and which are not instruments; and (3) articles (other than quasi-drugs and cosmetics) which are intended to affect the structure or any function of the bodies of humans or animals, and which are not instruments.

When manufacturing any of the articles as defined above, *i.e.* drugs, it is necessary to obtain a license for manufacture of the drug for each manufacturing establishment (Articles 12 and 13 of the Pharmaceutical Affairs Law). In the case of licenses for the manufacture of drugs, the major concern is whether or not the applicant possesses sufficient manpower and facilities to manufacture the drug. Therefore, licenses must be obtained for each manufacturing establishment (factory).

When applying to manufacture a drug which is not exempt from the approval process and which already is recognized in the Japanese Pharmacopoeia, approval must be obtained for each item individually before the license is granted (Article 14 of the Pharmaceutical Affairs Law). In the case of approvals, the problem is whether or not the article to be manufactured is suitable as a drug from the standpoints of safety and efficacy. Therefore, it is necessary for each party (enterprise) which is to manufacture the drug to obtain approval.

Drugs recognized in the Japanese Pharmacopoeia formerly did not require approvals and could be manufactured with only a manufacturing license, but because of the revision of the Pharmaceutical Affairs Law, drugs recognized in the Japanese Pharmacopoeia have required approvals since September 30, 1980. However, among the drugs recognized in the Japanese Pharmacopoeia those which are specified by the Minister of Health and Welfare as not requiring approval on the basis of the specifications in Article 14, Paragraph 1 of the Law shall not require approval. This relation is shown in Figure 1.

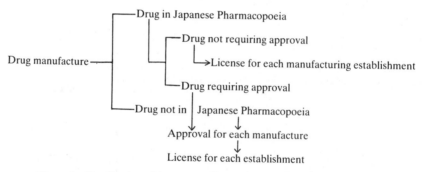

Figure 1   Classification of drugs according to the manufacturing procedures

35.2   RANGE OF DRUGS

As was described previously, drugs are as defined in the Pharmaceutical Affairs Law. However, there are certain items which are very difficult to classify precisely.

When vitamin B_1 is added to foods, it is handled as a food additive. In such cases, the same material can be handled both as a drug and not as a drug, and the evaluation of whether or not it is a drug depends mainly on the intended use rather than on the properties of the material.

Except for substances such as starches and lactose recognized in the Japanese Pharmacopoeia, items such as vehicles, lubricants, coloring agents, flavoring and odorizing agents, stabilizers and preservatives cannot, in principle, be considered as drugs on their own. Therefore, it is necessary to investigate carefully whether or not the item to be manufactured can be classified as a drug. There are also cases where the manner of describing the effects and indications of the item in the application form decides whether the item is a drug or not. Table 1 shows the jurisdiction of the various divisions of the Ministry of Health and Welfare and other Ministries concerning the licensing of drugs and related items.

35.3   APPROVAL FOR MANUFACTURING DRUGS

The investigations will differ in accordance with the items applied for. Figure 2 shows the classification with respect to the way of handling standards and test methods.

Table 1 The Jurisdiction of the Various Divisions of the Ministry concerning the Licensing of Drugs

Drug	*Ministerial division*
Manufacture and import of drugs and quasi-drugs	Evaluation and Registration Division, Pharmaceutical Affairs Bureau, Ministry of Health and Welfare
Manufacture and import of medical devices Manufacture and import of cosmetics	Medical Devices and Cosmetics Section, Evaluation and Registration Division
Manufacture and import of antibiotics Manufacture and import of biological preparations	Biologics and Antibiotics Division
Manufacture and import of drugs for use in both humans and animals	Evaluation and Registration Division (in cooperation with the Ministry of Agriculture, Forestry and Fisheries)
Manufacture and import of veterinary drugs and agricultural chemicals	Hygiene Section, Livestock Bureau, Ministry of Agriculture, Forestry and Fisheries

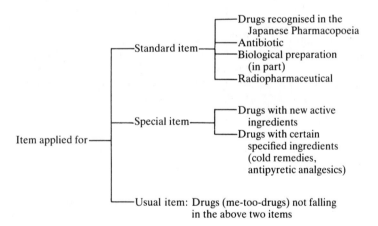

Figure 2 Classification of drugs according to the method of handling standards and test methods

As shown in Figure 2, there are three types of drugs among those drugs not recognized in the Japanese Pharmacopoeia for which stipulations have been enacted for standards and test methods. When a new drug application is made, the item category is stipulated by the Ministry at the same time as the approval is given.

35.3.1 Drugs Recognized in the Japanese Pharmacopoeia

Among the drugs requiring approval, approval procedures for biological preparations and antibiotics will be handled by the Biologics and Antibiotics Division and those of other drugs by the Evaluation and Registration Division as long as this does not run counter to any of the standards, *etc*. in the Japanese Pharmacopoeia.

35.3.2 Antibiotics

Approvals to manufacture antibiotics are granted by the Biologics and Antibiotics Division of the Pharmaceutical Affairs Bureau on the basis of the Japan Antibiotics Standards.

35.3.3 Biological Preparations

Examples of these are vaccines, toxoids, antitoxins and whole blood preparations, *etc.* In the case of biological preparations, the application form received from the prefectural governments is investigated by the Biologics and Antibiotics Division. If the results of these investigations are satisfactory according to the criteria for biological preparations set by the Government, approval is then granted by the Biologics and Antibiotics Division. For some applications, investigations also include the results of tests entrusted to the National Institute of Health.

35.3.4 Radiopharmaceuticals

The Evaluation and Registration Division of the Pharmaceutical Affairs Bureau handles the business of the approvals for the manufacture of radiopharmaceuticals, and if there are no problems regarding the Minimum Requirements of Radiopharmaceuticals, approval procedures are undertaken.

Standards are established for the new drugs described in Sections 35.3.2 to 35.3.4 through deliberations of the Central Pharmaceutical Affairs Council (set by government officials).

35.3.5 Drugs Requiring Special Investigations

Drugs requiring special investigations must undergo investigations other than those to determine their suitability given in the 'Standards and Test Methods' of the application form. Such drugs are those with new active ingredients not yet approved, or within two years after approval, and drugs containing certain specified ingredients with antipyretic analgesic action, such as aspirin.

Since this type of drug must undergo special investigation for which a fee must be paid, the application form is sent to the National Institute of Hygienic Sciences or the National Institute of Health after being investigated by the appropriate Division. The Institute not only investigates the contents of the application form but also performs actual analytical investigations using the standards and test methods given in the form. The results are reported back to the appropriate Division which takes the necessary steps on the basis of this report (see Figure 5).

In the case of drugs with new active ingredients, the application form is sent to the National Institute of Hygienic Sciences and National Institute of Health and is simultaneously investigated by the New Drug Investigation Subcommittee.

35.3.6 Narcotics, Exempt Narcotic Preparations, Stimulant Drugs and Stimulant Drug Raw Materials

After an investigation by the Evaluation and Registration Division, application forms for approval to manufacture narcotics or exempt narcotic preparations are forwarded to the Narcotics Section.

Investigations are conducted in the Narcotics Section to determine if the applicant is suitable to handle narcotics and to study the referred items for approval of the narcotics quota.

35.3.7 Drugs for Both Animals and Humans

In the case of applications for approval to manufacture drugs which can be used for the diagnosis, treatment or prevention of diseases in animals as well as humans, the applications are first investigated with respect to application to humans by the Pharmaceutical Affairs Bureau in the Ministry of Health and Welfare. If this investigation proves satisfactory, the application is then forwarded to the Hygiene Section, Livestock Bureau of the Ministry of Agriculture, Forestry and Fisheries. After this Bureau has conducted an investigation into use of the drug on animals, the application is returned to the Pharmaceutical and Supply Bureau where the necessary steps are taken on the basis of the results of the investigation by the Livestock Bureau (see Figure 5).

Figure 3 shows the handling of investigations of applications for manufacturing approval.

35.3.8 Ethical and Proprietary Drugs

Ethical drugs are those to be used by a physician or a dentist, or drugs used under the direction or through the prescription of a physician or dentist. Proprietary drugs are those drugs which

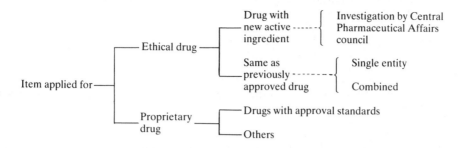

Figure 3   Handling of investigations of applications for manufacturing approval

have a mild action if used correctly within fixed directions and dose range and are safe enough to be sold directly to the public. However, the classification of a drug as ethical or proprietary is often not clear just from the nature of the drug itself and the applicant must make a clear distinction in the application form.

The range of drugs which will not be approved as proprietary drugs has been specified by a notification of the Pharmaceutical Affairs Bureau. In principle, drugs with new active ingredients will not be approved as proprietary drugs unless they are shown to be sufficiently safe through application by a physician.

The investigations are carried out according to this classification, but naturally there are differences between ethical and proprietary drugs in the aspects to be investigated.

35.3.9   New Drugs

New drugs are defined as those drugs for which the chemical structure, composition, or indications are not generally known. In Article 14, 2 in the Pharmaceutical Affairs Law, it is specified that new drugs are 'drugs which are designated by the Minister of Health and Welfare at the time of granting of manufacturing approval as drugs for which the active ingredients and quantities, directions and dosage, indications and effects, *etc.* are different from those of drugs which have already been approved for manufacture or import', and 'drugs with new characteristics which require investigations of the results of use after approval'. The range is decided together with the materials which must be submitted at the time of application in accordance with Notification No. 698 of the Pharmaceutical Affairs Bureau, dated 30 May, 1980.

The granting of approval for the manufacture of new drugs is based on the report of the Central Pharmaceutical Affairs Council after thorough investigation of the drug in question. However, within the Central Pharmaceutical Affairs Council, investigations are first conducted by the investigation Subcommittee concerned, as shown in Figure 4.

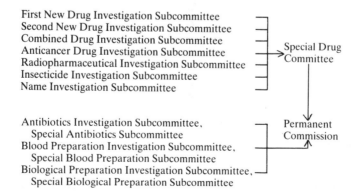

Figure 4   Investigations by the subcommittee within the Central Pharmaceutical Affairs Council

The Combined Drug Investigation Subcommittee investigates combined ethical drugs. When the combined drug has ingredients which are new drugs, it is investigated by both the New Drug Investigation Subcommittee and by the Combined Drug Investigation Subcommittee.

Each of the investigation subcommittees and the commission are made up of specialists such as physicians and pharmacists. The New Drug and Combined Drug Investigation Subcommittees generally meet twice a month and the other investigation subcommittees, the special committee and the commission meet once every three or four months.

35.3.10 Drugs for which Revisions of Cabinet Orders, Ministerial Ordinances or Notifications are Necessary

In the case of approval of new drugs, there are various regulations in accordance with the pharmacological action, *etc.* of the drug. For example, drugs with strong toxicity are designated as deadly poisons or powerful drugs, and drugs which can only be sold by prescription or under the direction of a physician are designated as prescription drugs. Other designations include narcotics and habit-forming drugs and special investigation drugs.

In principle, approval to manufacture such drugs is granted only after the revision of the cabinet order, ministerial ordinance or notification in question has been recorded in the Official Gazette.

Figure 5 gives a detailed account of the handling of applications for approval to manufacture drugs.

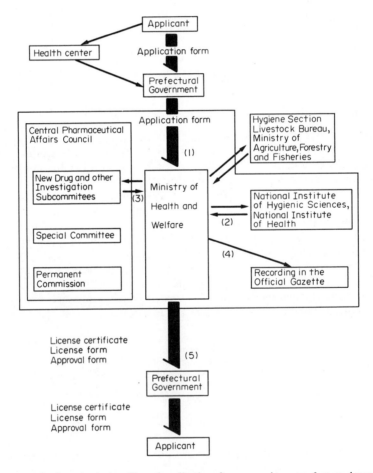

Figure 5 Steps in the handling of applications for approval to manufacture drugs

35.4 CHANGES IN FORMULATION OR MANUFACTURING PROCEDURES

When partial changes are to be made in the approved ingredients of drugs which have already been approved, an application must be made for a partial change in the approved ingredients on the basis of Article 14, Paragraph 4, of the Pharmaceutical Affairs Law.

When manufactured items of a licensed drug manufacturing establishment are to be increased or changed, an application for a license to change or add to the manufactured items must be on the basis of Article 18, Paragraph 1 of the Pharmaceutical Affairs Law.

There is also an application form for renewal of the license for a drug manufacturing establishment. However, these are all handled in the same way as the application for the license of a drug manufacturing establishment.

Approvals to import drugs are handled in almost the same way as the manufacturing approvals (Article 23, Pharmaceutical Affairs Law).

35.5 APPLICATION PROCEDURES FOR LICENSES TO MANUFACTURE DRUGS AND QUASI-DRUGS

Persons wishing to manufacture (or import) drugs must obtain licenses from the Minister of Health and Welfare in accordance with Article 12, Paragraph 1 of the Pharmaceutical Affairs Law.

The two prerequisites for this licensing are first that the drug (excluding drugs recognized in the Japanese Pharmacopoeia and specified by the Minister of Health and Welfare) or quasi-drug to be manufactured or imported must have been approved by the Minister of Health and Welfare in accordance with the stipulations in Article 14 of the Pharmaceutical Affairs Law, and second that the equipment in the manufacturing (business) establishment is in conformity with the stipulations in 'The Construction and Equipment Regulations for Pharmacy', Ordinance No. 2 of the Ministry of Health and Welfare, dated 1 February, 1961, and that the manufacturing establishment shall meet the personnel requirements in Article 6, Item 2(a) to (e) of the Pharmaceutical Affairs Law, and have technicians in direct control of manufacture (import). The Good Manufacturing Practices (GMP) were specified by an administrative guidance entitled 'Handling of drug manufacturing license applications in accordance with conformity to GMP standards' (Notification No. 179 of the Pharmaceutical Affairs Bureau, dated 27 February, 1976) and these GMP standards had to be met by all factories except those engaged in special manufacturing.

However, with the partial revision of the Structural and Equipment Regulations for Pharmacies, *etc.* (Ministerial Ordinance No. 32 of the Ministry of Health and Welfare, gazetted on 16 August, 1980) and the enactment of the Japanese GMP (Ministerial Ordinance No. 31 of the Ministry of Health and Welfare, gazetted on 16 August, 1980), the GMP became law and from 30 September 1980 there must be conformity with the GMP.

However, with respect to structures and equipment (except equipment and utensils required for tests and inspections), the factories of those manufacturing drugs on the basis of a license obtained in accordance with Article 12, Paragraph 1 of the Pharmaceutical Affairs Law prior to enactment of the above-mentioned ordinances shall be permitted, as an interim measure, to conform to the revised standards by 31 March, 1981 (31 March, 1983 for factories manufacturing only blood preparations for which there are no lots). If these conditions are met, any person or organization may apply (cited from *Drug Approval and Licensing Procedures in Japan*, published in 1981 by Yakugyo Jiho, Tokyo, Japan).

35.6 DATA AND INFORMATION REQUIRED WHEN FILING THE APPLICATION FOR APPROVAL

The following explanations are given in the order of the columns in the drug manufacturing application form.

35.6.1 Non-proprietary Names

As specified in Article 50 of the Pharmaceutical Affairs Law, the non-proprietary names of drugs not recognized in the Japanese Pharmacopoeia, if they have such names, must be indicated.

35.6.2 Proprietary Names

In principle, the proprietary name of a drug can be devised optionally by the manufacturer, but in order to maintain the 'dignity' of drugs and protect public health and hygiene, approvals will not be granted for false or exaggerated names, *etc.*

35.6.3 Ingredients and Quantities

The items to be entered in this column have been specified in various notifications (No. 76 of 3 April, 1964 and No. 718 of 30 May, 1980) and the stipulations in these notifications must be considered when filing. The following are the main items which must be entered in this column: (a) the names of the ingredients, (b) the quantities of the ingredients used, (c) standards of the ingredients used, (d) the form of the product (there are cases where this is made clear in the manufacturing method column), and (e) the purpose of including ingredients other than the active ingredients.

35.6.4 Manufacturing Method

The entire manufacturing process from the collection and mixing of the raw materials to the final packaging shall be entered simply for each process. In such cases, auxiliary agents used in the manufacture or processing, such as solvents for each process, and the materials of the containers or wrappers, *etc.* shall be clearly indicated (notification No. 718 of the Pharmaceutical Affairs Bureau, dated 30 May 1980). The manufacturing method column must contain information which makes clear the type of product to be manufactured in accordance with the information given in the Ingredients and Quantities column.

35.6.5 Directions and Dose

The directions and dose of a drug are among the most important factors in obtaining the desired results of the treatment. As was described previously, the directions and dose must be decided in accordance with the ingredients and quantities of the drug.

35.6.6 Indications and Efficacy

35.6.6.1 *Indications of ethical drugs*

Indications and effects of ethical drugs should be entered in the form of disease and symptom names using medical terminology whenever possible.

35.6.6.2 *Indications of proprietary drugs*

(1) Entries must be easy for the consumer to understand. (2) Indications related to infectious diseases designated by law and other serious diseases will not be approved. (3) Vague and wide-ranging indications are to be avoided. (4) Indications which could lead to erroneous use will not be approved. (5) Repetition of indications are to be avoided. (6) Indications of combined drugs are to be investigated sufficiently. (7) Indications are limited due to dose form (drinks and ampoules). (8) Indications specified in administrative regulations must be strictly adhered to.

35.6.7 Storage Conditions and Expiry Date

It is necessary to consider carefully the composition and form of the drug, and determine whether or not the quality of the drug will be maintained by ordinary storage conditions. If it is considered that the quality cannot be maintained by such conditions, the most suitable conditions must be found on the basis of stability test results. It is essential that suitable storage conditions and expiry date be entered in this column on the basis of careful studies when required for quality assurance. The column can be left blank for drugs which do not require any particular storage conditions and expiry date.

35.6.8 Standards and Test Methods

35.6.8.1 *Items required in standards and test methods entries*

In principle the 16 items shown in Table 2 must be included in the standards and test methods for each of the components entered in the Ingredients and Quantities column, and these standards and test methods must be entered in the Standards and Test Methods column. The test items which must be entered for each type of pharmaceutical preparation are as shown in Table 3, in principle.

Table 2 Items Required in Standards and Test Method Entries

No.	Entry item	No.	Entry item
1	Name	9	Rational values
2	Structural or empirical formula	10	Purity
3	Molecular formula and weight	11	Loss on drying or water content
4	Base material	12	Loss on ignition
5	Contents standards	13	Residue on ignition, total ash, or acid-insoluble ash
6	Manufacturing method	14	Special tests
7	Description	15	Pharmaceutical preparation test
8	Identification	16	Assay

Table 3 Test Items which Must be Entered for Each Type of Pharmaceutical Preparation[a,b,c]

Preparation			Item no. 1	2	3	4	5	6	7	8	9	10	11	12	13	14	15	16
Base material	Drugs with determined structural composition	Organic drugs	O	O	O	△	O	×	O	O	O	O	O	×	O	△	×	O
		Inorganic drugs	O	△	O	△	O	×	O	O	O	O	O	△	△	△	×	O
	Drugs with no determined structural composition		O	×	×	O	△	△	O	O	△	O	△	△	△	△	×	O
Preparation	Oral administration	Tablets, capsules pills	O	×	×	×	O	△	O	O	×	×	×	×	×	△	O	O
		Powders, granules	O	×	×	×	O	△	O	O	×	×	×	×	×	△	O	O
		Liquids	O	×	×	×	O	△	O	O	△	△	×	×	×	△	△	O
	Natural drugs		O	×	×	△	O	△	O	O	×	△	O	×	△	△	O	△
	Injection	Parenteral injections (in ampoules)	O	×	×	×	O	△	O	O	△	△	×	×	△	△	O	O
		Drugs for injection (in vials or ampoules)	O	△	△	△	O	△	O	O	△	△	△	×	△	△	O	O
	External use		O	×	×	×	O	△	O	O	×	×	×	×	×	△	O	O
Drugs such as natural drugs or animal or plant extracts for which effective ingredients are not clear			O	×	×	O	×	O	O	O	×	×	O	×	O	△	O	×

[a] O indicates the item is required, △ indicates the item is required if the test method is possible, and × indicates the item is not required.
[b] The manufacturing method (item 6) is to be entered in the case of preparations for which the active ingredient(s) is not clear when the content standards and assay are technically difficult. [c] Items 1–16 are identified in Table 2.

Drugs which in principle must undergo item 15, the pharmaceutical preparation test, are shown in Table 4. The test items are shown on the right of the table.

35.7 REGULATIONS AFTER APPROVAL

35.7.1 Side-effect Reports

35.7.1.1 *Outline*

Companies manufacturing or importing drugs have been instructed by notifications of the Pharmaceutical Affairs Bureau to make the necessary reports on side-effects, *etc.*, but in accordance with the revision of the Pharmaceutical Affairs Law in 1979, the submission of reports on side-effects, *etc.* has become an obligation on the basis of Article 69 of the Law. Details concerning the handling of these reports are specified in Article 62, 2 of the Enforcement Regulations of the Pharmaceutical Affairs Law and Notification No. 483 of the Pharmaceutical Affairs Bureau, dated 10 April, 1980 (No. 4, 2). These are outlined in the following sections.

Table 4 Drugs which must Undergo Item 15 (Table 2), the Pharmaceutical Preparation Test

Preparation	Test
Folded powders	Weight discrepancy test
Tablets, pills, capsules, folded granules, lozenges	Weight discrepancy test and disintegration test
Granules, tablets coated with sugar, *etc.*	Disintegration test
Elixirs, medicated spirits, tinctures, liquid extracts	Alcohol number measurement, ketone, isopropanol and *t*-butyl alcohol test, methanol test
Aqueous injections	Non-soluble impurities test, actual volume or actual weight discrepancy test, glass container test, sterility test
Non-aqueous injections using plant oils as solvents	Non-soluble impurities test, actual amount or weight variation test, mineral oil test, glass container test, sterility test
Intravenous injections with single doses exceeding 10 ml	Non-soluble impurities test, actual amount or weight variation test, glass container test, pyrogen test, sterility test
Suppositories	Melting point test
Adhesive plasters	Adhesion test
Eye drops	Non-soluble impurities test, sterility test
Non-water soluble eye drops using plant oils as solvents	Non-soluble impurities test, mineral oil test, sterility test
Aerosols	Ignition test, explosion test, heating test, relation between spray time and spray amount[a]

[a] For the heating test and the relation between the spray time and amount, only the data need be attached.

35.7.1.2 *Side-effects to be reported*

When a company receives information from a hospital, pharmacy or sales agent or obtains case or study reports thought to correspond to (i)–(iii) below from the company's own investigations concerning a drug which the company manufactures or imports, it must report the details, its opinions, *etc.* to the Ministry of Health and Welfare within 30 days. (i) Serious side-effects (those which might result in death, disability, or disability and death, and those which are found to be severe by the physician in charge in cases where recovery is difficult, *etc.*); (ii) unknown side-effects (all side-effects except those included in the Precautions for Use as specified in Article 52, Item 1 of the Law or which are clearly included from a medical standpoint even though the expression is different); and (iii) research reports which show that major side-effects such as cancer might appear, that there are major changes in the side-effect tendencies such as the number of cases, incidence and onset conditions, or that the approved indications and effects are not obtained.

35.7.1.3 *Precautions concerning drug side-effect investigations*

The necessity of side-effect investigations after marketing has already been explained for new drugs. Since experience of using drugs other than new drugs is also accumulated after marketing, the appearance of side-effects can be understood more clearly than in the case of new drugs. However, there have been reports of new side-effects which have never been seen during the previous 5 to 6 years or which had appeared more than 10 years after the drug went on the market. With the progress made in medicine and pharmacology, new safety-related problems which were not foreseen previously have appeared. New side-effects also occur because of changes in the social environment, *etc.* Even with known side-effects, there are sometimes high incidences of severe side-effects or major changes in the incidence, degree and symptoms, and it is necessary to take some sort of safety measures.

In cases where it is clear from the results of an investigation of the accumulated side-effect information that these are cases corresponding to (i)–(iii) in Section 35.7.1.2, the company shall investigate the following measures in accordance with requirements in addition to making a report to the Ministry of Health and Welfare: (a) measures such as changing the directions and dose, revising the Precautions for Use, temporarily suspending sales, suspending sales and recalls; and (b) transfer of information to related persons such as physicians and pharmacists.

The report submitted to the Ministry of Health and Welfare should only be submitted after careful surveys and investigations, but in cases where major side-effects must be handled urgently or where surveys and investigations require a lot of time, the information concerned should be reported to the Ministry of Health and Welfare as soon as possible with the reason attached.

35.7.2 Re-examination of New Drugs

When approvals are given for the manufacture or import of new drugs, detailed data have to be submitted and careful deliberations are carried out, but since the number of clinical trial cases, *etc.* up to the time of approval is limited, a new drug re-examination system has been established to investigate the results of use, *etc.* of new drugs by means of PMS (post marketing surveillance) or fourth-phase trials performed after approval, and to carry out a re-examination of safety, *etc.* after six years, in principle (Article 14, 2 of the Pharmaceutical Affairs Law).

35.7.2.1 Drugs to be re-examined

Whether or not a drug applied for is subject to re-examination is based on whether or not the active ingredients and their quantities, the directions, indications and effects, *etc.* clearly differ from those of drugs already approved, *i.e.* whether or not the drug has 'newness' requiring an investigation of the results of use, *etc.* after approval as a new drug. The decision is made by the Minister of Health and Welfare when the approval to manufacture or import each drug is granted.

The time when the re-examination is to be performed is specified or indicated by the Minister of Health and Welfare at the time of approval of the new drug and shall be as follows, in principle.

(i) Six years after the date of approval for the following new drugs: (a) drugs which have active ingredients which have not been used as active ingredients in drugs recognized in the Japanese Pharmacopoeia or in drugs already approved for manufacture or import (drugs with new active ingredients); (b) combined ethical drugs which have active ingredients or combination ratios which differ from those of combined drugs recognized in the Japanese Pharmacopoeia or combined drugs which have already been approved for manufacture or import as ethical drugs (new combined ethical drugs); (c) drugs which have the same active ingredients as drugs which have already been approved for manufacture or import, but which have different routes of administration (oral, subcutaneous, intramuscular, intravenous, percutaneous, per rectum, per vagina, eye or ear drops, inhalation, *etc.*; drugs with new administration routes).

(ii) Four years after the date of approval for the following new drugs: (a) drugs which have the same active ingredients and administration routes as drugs which have already been approved for manufacture or import, but for which clearly different indications and effects have been added (some drugs with new indications); (b) drugs which have the same active ingredients and administration routes as drugs which have already been approved for manufacture or import but which have clearly different doses and directions such as in cases where a different mechanism of action can be expected because of the large increase in dose (some drugs with new doses).

(iii) At the end of the remaining re-examination period for the new drug concerned in the case of drugs which have the same active ingredients, quantities, directions and dose, indications and dose, indications and effects, *etc.* as new drugs and which are approved during the period of re-examination of the new drug concerned.

However, even though they correspond to (i)–(iii) above, the following drugs shall not be considered as new drugs subject to re-examination in accordance with the essence and purpose of the re-examination system.

(1) Drugs such as diagnostic agents used outside the body and insecticides which are not used in direct contact with the human body.

(2) Combined ethical drugs such as compound digestive enzymes, mild acting poultices, artifi-

cial tears, nutrients for oral use or incubation and Chinese medicines, which are judged in an overall evaluation not to have 'newness' as drugs.

35.7.2.2　Evaluation standards

The re-examination by the Minister of Health and Welfare shall be performed by confirming that the drug for which the re-examination application has been made does not correspond to any of the items in Paragraph 2 of Article 14 of the Pharmaceutical Affairs Law on the basis of knowledge obtained during the re-examination.

35.7.3　Drug Efficacy Evaluation

The first demands for the evaluation of drug efficacies in Japan were from the field of public health.

The progress of medicine and the pharmaceutical sciences in recent years has resulted in the development of excellent new drugs through the introduction of comparative test methods such as the double blind test, but at the same time, it also has caused many doubts concerning the efficacy of drugs currently in use. In August, 1970, the Minister of Health and Welfare established a Round-table Conference on the Problem of Drug Efficacy consisting of 11 experts in the medical and pharmaceutical fields. This conference was entrusted with investigating drug efficacy re-evaluation methods. The conference submitted its report to the Minister of Health and Welfare on 7 July, 1971.

This long report described in detail the past status of approval and licensing in Japan, the way things should be done in the future, the necessity for re-evaluation, *etc.* It also indicated concrete measures for the re-evaluation of drugs presently on the market. The main points are as follows.

(a) The range of drugs for re-evaluation shall be, in principle, all those which the related enterprises intend to manufacture and market in the future, except new drugs or combined ethical drugs approved on or after 1 October, 1967.

(b) The re-evaluation shall be handled by the special subcommittee newly established for this purpose within the Central Pharmaceutical Affairs Council. Specialized investigation committees under this subcommittee shall be established for the various types of drug indications.

(c) Re-evaluation priority shall be given to drugs which are relatively easy to re-evaluate, such as psychotropic drugs, and to drugs for which there is considerable public demand for re-evaluation, such as vitamins.

(d) The basic method of re-evaluation shall be based on the submission within a set period of the required data collected and compiled by the manufacturers of the items concerned.

(e) Since proprietary drugs can not be re-evaluated in the same manner as ethical drugs because of their different aims, further investigations shall be performed after sufficient studies of the role of proprietary drugs in public health, *etc.*

In accordance with this report, the re-evaluation was undertaken by means of administrative guidance from December, 1971, but the re-evaluation became law with the partial revision of the Pharmaceutical Affairs Law in October, 1979. The provisions in the revised law were as follows (Re-evaluation of Drugs):

Article 14, 3: When the Minister of Health and Welfare designates ranges of drugs to be re-evaluated on hearing the opinion of the Central Pharmaceutical Affairs Council and this is made public, persons who have received approval for the manufacture of drugs shall receive re-evaluations of designated drugs by the Minister of Health and Welfare.

2: The re-evaluations of the Minister of Health and Welfare shall be performed by confirming that the drugs specified in the previous Paragraph do not conform to any of the items in Paragraph 2, Article 14 on the basis of findings obtained in the re-evaluation.

3: The publication specified in Paragraph 1 shall be accompanied by indications of data to be submitted by the person who is subject to the re-evaluation, and the deadline for the submission of such data.

As is clear from these provisions, the re-evaluations as specified by law are carried out whenever required and therefore, after completion of the re-evaluations by means of administrative guidance, re-evaluations based on the law will be performed constantly.

As of August, 1980, no items have been specified for re-evaluation based on the law, but specifications in ordinances such as those for cancellation of approvals and the partial revision of approval items based on the revised Pharmaceutical Affairs Law have been applied as measures after completion of re-evaluation for items which have already been re-evaluated in accordance with administrative guidance (cited from *Drug Approval and Licensing Procedures in Japan*, published in 1981 by Yakugyo Jiho, Tokyo, Japan).

36

Acceptance of Single-cell Protein for Human Food Applications

N. S. SCRIMSHAW
Massachusetts Institute of Technology, Cambridge, MA, USA

36.1 INTRODUCTION

The safety and nutritional value of single-cell protein (SCP) sources must be established by rigorous testing. When the first international conference on SCP was held at MIT in 1967, there was scant information available and no internationally recognized guidelines for obtaining it. Between 1970 and 1974 the Protein Advisory Group of the UN System (PAG) developed a series of guidelines for testing novel sources of protein, three of which were specifically directed at the production of SCP for human consumption (PAG Guidelines, 6, 7 and 12, Anon., 1983a, b, c), and the fourth (No. 15, Anon., 1983d) with the safety of the products from animals fed SCP. These have each been recently reviewed, revised by a task force convened by the United Nations University and reissued as UNU–PAG guidelines retaining the same numbers. This series was completed by the report of the meeting at WHO headquarters in Geneva in 1975 on the evaluation and acceptable nucleic acid levels of single-cell protein for human consumption (Anon., 1975a).

Throughout this period experience was accumulating with the feeding of various SCP preparations to human subjects on the basis of the above guidelines. In 1972 in Aix-en-Provence experimental evidence was presented that yeast grown on *n*-alkanes in Grangemouth, Scotland, and on gas oil in Lavera, France, by British Petroleum Ltd. and designated TOPRINA met all requirements for the initiation of clinical trials (de Pontanel, 1972). Soon afterwards a conference in Rome in 1973 presented similar information for the yeast produced by Liquichimica S.p.A. on *n*-alkanes and designated LIQUIPRON (Davis, 1974).

By the time of a PAG organized conference on SCP in Brussels in 1976 the bacteria grown on methanol by Imperial Chemical Industries Ltd. and identified as PRUTEEN was also reported to have undergone extensive trials in experimental and farm animals and to be ready for clinical testing. At the meeting in Aix-en-Provence, clinical studies done at MIT were reported (de Pontanel, 1972) for a yeast and a bacterium grown on ethanol derived from petroleum that had also fulfilled all of the requirements of the PAG preclinical guidelines.

At the Brussels conference, three issues were singled out for special examination and evaluation of the already extensive data that were available. These were (i) the utilization and safety of fatty acids with uneven numbers of carbon atoms in some hydrocarbon-grown yeasts; (ii) the significance and fate of residual paraffins found in some of these yeasts; and (iii) the strength of the evidence for lack of toxicity or carcinogenicity of the single-cell products being proposed for animal feeding.

The PAG working group on SCP meeting at this time concluded that the data 'demonstrated convincingly' the safety of the specific SCP products reviewed; Toprina, Liquipron and Pruteen. However, doubts continued to be expressed, mainly at the political level in Italy and Japan, therefore another international symposium was held in Milan, Italy that reviewed additional data and reinforced the above conclusions (Scrimshaw, 1979). By the time of the next international conference held in Paris in 1981, information was available on the clinical trials of a yeast grown on methanol by Phillips Petroleum Co. (Scrimshaw and Udall, 1981).

In this chapter we will add unpublished information on the clinical evaluation of filamentous microfungi produced by Rank Hovis McDougall on a starch substrate, and by Tampella Inc. on sulfite liquor, as well as some other relevant experience gained in our studies at MIT.

36.2 GUIDELINES FOR CLINICAL TRIALS

While the revised PAG/UNU Guideline No. 6, *Preclinical Testing of Novel Sources of Food* (Anon., 1983a), applies to all novel protein sources, it has been most used as a standard for single-cell protein studies. It stresses that no new protein source should be submitted for clinical trials until it has been successfully fed to one rodent and one non-rodent species in short-term toxicological tests in which the experimental protein source is fed at the highest practical level. A full battery of biochemical tests is required, and when the animals die or are sacrificed, they should be examined for gross pathology and organ weight, and histopathological examination of the principal organs and tissues should be carried out.

These short-term studies should be supplemented by examination of the new material using a battery of mutagenicity tests in both prokaryotic and eukaryotic systems. The short-term studies can then be extended to a multigeneration study that will assure satisfactory reproduction and lactation. The F_2 generation can be used for chronic toxicity and carcinogenicity studies and will provide information on possible teratological effects. All of the published clinical studies of new single-cell proteins have been on materials that have passed through the preclinical testing requirements of PAG Guideline No. 7 (Anon., 1983b).

There is an additional guideline, No. 12, *Production of Single-Cell Protein for Human Consumption* (Anon., 1983c), that lays down a series of additional requirements. To begin with, the final product should contain no living cells derived from the fermentation process. In view of the variety of carbon substrates, attention must be directed to the composition of the medium to ensure that it does not contain chemical compounds that are regarded as health hazards. If antifoam agents, detergents and flocculants are introduced, they must either meet minimum safety requirements or be capable of complete subsequent removal.

In order to ensure product quality and uniformity, process variables of the fermentation process such as temperature, air or oxygen, pH, cell growth rate and cell concentration must be carefully controlled. The guideline points out that this will require continuous careful monitoring. Attention must also be paid to maintaining the integrity of the original strain of organism. This will require an appropriate series of microbiological and biochemical tests to demonstrate the stability of the organisms and to ensure the absence of undesirable contaminants.

Constituents to be analyzed in new single-cell protein products include triglycerides, phospholipids, steroids, ash, minerals of biological importance such as iron, iodine and zinc, alkali, alkaline earth and heavy metals, as well as dietary fiber. The limits for viable and contaminating microorganisms are also given.

The original PAG Guideline No. 7 (Anon., 1983b) placed considerable stress on studies of nutritional value and described in detail both nitrogen balance in children and adults, and growth studies in children for this purpose. In practice the protein value of the various SCPs for human subjects can be predicted with sufficient accuracy by the use of amino acid score or experimental animals. In fact, it can be assumed that any food that supplies sufficient protein with an acceptable amino acid pattern and a useful content of other essential nutrients will be of value to the consumer if it is well tolerated and acceptable. Accordingly, the revised guideline issued in 1983 places primary emphasis on tolerance studies in relatively large numbers of subjects to determine acceptability and the frequency of allergic and other undesirable reactions.

No amount of testing in experimental animals, including in the higher apes, can serve to predict the kinds of intolerance to SCPs that have been encountered in human subjects. In fact, no problems have been encountered with well processed SCPs prepared under carefully defined and maintained conditions. This is probably because the reactions observed appear to have been allergic in nature and idiosyncratic in humans (Scrimshaw, 1975).

The most common symptoms to be looked for are gastrointestinal intolerance and skin rashes. Since these generally appear within two weeks, a 30-day trial is sufficient to detect reactions that occur with enough frequency to be of concern. In general, 25 to 50 are sufficient to study initially. All subjects should be in good health, as determined by medical history, physical examination, and routine blood and urine tests. Subjects should be chosen without reference to their personal and family history of allergies since it is essential to determine the prevalence of symptoms in the general population.

Various allergic reactions may occur in some individuals after the consumption of almost any common food, especially if it is a protein source. It is the frequency of such reactions that must be evaluated rather than their expected absence. Moreover, when persons know they are, or believe they are, consuming a novel food, minor symptoms unrelated to the material ingested may be exaggerated or even imagined. For this reason it is essential to feed a control group simultaneously.

Ideally, the trial is conducted so that individuals are randomly assigned to experimental and control groups, stratified by sex, in a double-blind cross-over design. Only after a low frequency of adverse symptoms is assured can additional trials be conducted without the necessity of a control group. It is inadvisable to test at a level markedly in excess of that at which the product is likely to be consumed. Even traditional foods tested at excessive levels might well give negative results.

In general, 25, and preferably 50 human subjects should be studied initially under carefully controlled conditions. The subjects should be in good health as determined by medical history, physical examination and routine blood and urine tests. The test material is to be consumed in addition to an *ad libitum* diet. Since this will be highly variable and may include meals or foods that in themselves cause symptoms, a control group of adequate size is essential. In addition each subject should keep a diary of all significant departures from the usual in diet, symptoms or activities. Even changes in mood, appetite, sleep patterns, libido and other subjective reactions should be recorded. These diaries often prove of great value in the retrospective interpretation of the results of a tolerance trial by making possible the detection of results that are unrelated to the test material or, conversely, identifying responses that were not obvious during the trial.

The guideline suggests that an appropriate study design is two four-week periods separated by a few days interval. In the first period, half of the subjects ingest daily a fixed amount of the experimental material and the remainder a control material that cannot be identified as different from that given to the experimental group. If this is not possible, then it must still be impossible for either group to know whether or not they are receiving the experimental material. After the initial four-week phase and an interval, the groups can be reversed if an undue number of reactions have not occurred. If this is the case, then the trial must be terminated and the code broken. Not only must the trial be terminated at any time it is apparent that an undue number of adverse reactions are occurring, but also for ethical reasons subjects must be allowed to leave the study at any time if they desire. Administration of the material must be done for a minimum of five and preferably six or even seven days a week. Experience indicates that individuals not developing symptoms in the first 20 days are not likely to do so thereafter.

Any method of oral administration will suffice that facilitates a double-blind trial in which neither the subject nor the supervising physician knows which group is receiving the experimental material. Among the possibilities are cookies, cakes and puddings. SCPs that are bland powders can often be tested by allowing the subjects to mix them into bouillon or any of a variety of fruit juices according to their individual preference, and they can vary their choice from day to day. For this to be acceptable, however, there must be a similar control material for which tolerance is already well established.

The level of feeding should be based on the intended level of use. Yeast as a vitamin supplement may be consumed at less than 5 g daily, as a food additive for its functional properties rarely more than 10 g, but as a significant protein supplement the amount might be 30 g or more. The upper limit is determined by the nucleic acid content. As discussed in a later section, it should not add more than 2 g of nucleic acid per day to the usual diet. If it is intended eventually to be incorporated into staple foods, it should be given daily in amounts approximating the upper limits of possible use. For a totally new food source it may be prudent to conduct a pilot trial with a small number of subjects at a lower level.

When evaluating the data, two contrary effects may be noted: intolerance, to which the subject soon becomes adapted, and boredom from eating the same food over a prolonged period. Guideline No. 7 (Anon., 1983b) advises a battery of laboratory tests at the beginning and end of each period. These are wise precautions and helpful in convincing regulatory agencies of the safety of a novel protein source, but they have never demonstrated changes in our studies except for serum and urinary uric acid levels that are proportional to the nucleic acid content of the material fed. Once an allergic reaction is observed in a group of subjects, there are some useful specific laboratory procedures not mentioned in the guideline that are described in a later section.

If the preliminary testing goes well, an additional 100 individuals should be fed the material to be statistically confident that the level of intolerance is acceptable for general use. It should also be noted that once production and commercial use of any product have commenced, it is advisable to have a mechanism by which any adverse reactions may be monitored in order that corrective changes may be introduced so that only the kind of product originally tested reaches the consumer.

36.3 TOLERANCE STUDIES

36.3.1 Yeasts and Bacteria

Experience with the human feeding of SCP long antedates the development of SCP based on petroleum hydrocarbons. It comes from more than 70 years of use of *Candida utilis* grown on a variety of substrates, most commonly molasses, starch or sulfite liquor, as a food additive or vitamin source. As commonly used, it is on the GRAS (Generally Recognized as Safe) list of the US Food and Drug Administration and is used in small quantities without restriction in processed foods throughout the world. In addition, the origins of the use of brewers' yeast and bakers' yeast are lost in antiquity. Bacteria are, of course, necessary for the production of many traditional foods, for example various species of *Lactobacillus* are essential to fermented milks, some cheeses, sauerkraut, pickles and soy sauce, *Propionibacterium* makes the holes in Swiss and Emmentaler cheeses, and *Bacterium linens* gives the flavor to Limburger cheese.

As noted earlier, the most extensive use of yeast as food occurred in Germany and Eastern Europe during World Wars I and II when it actually became a meat extender and meat substitute (Prescott and Dunn, 1959). There have also been a number of formal nutritional value and tolerance trials. While most of these have been free of any adverse gastrointestinal effects (Bressani, 1968; Fukui *et al.*, 1960; Goyco, 1959) this is not true of all. At levels over 15 g daily poor acceptability and loose stools were encountered in some investigations (Bressani, 1968; von Loesecke, 1946; Klapka *et al.*, 1958), but other workers have given far higher quantities with no adverse symptoms (Bressani, 1968).

During unpublished feeding trials held during World War II, severe gastrointestinal reactions developed quite suddenly after 30 days of uneventful consumption. The cause was not determined, but contamination would seem to be one possibility. Our MIT experiences with feeding commercially available sulfite-grown *C. utilis* from Lakes States Yeast Co. have been reported previously (de Pontanel, 1972; Scrimshaw and Udall, 1981; Scrimshaw, 1972). When 12 subjects were fed 90 g of yeast per day for 11 weeks, we encountered no adverse gastrointestinal reactions. However, a mild papular rash appeared on the palms and soles of eight of these subjects. It was painless and disappeared spontaneously when the yeast feeding was discontinued. Similar cutaneous reactions were observed in three of five subjects consuming 135 g per day and one out of 11 subjects consuming 46 g daily. No such lesions were observed with a glucose-grown *C. utilis* fed at 90 g per day to 21 subjects for 90 days.

In 1977 we began a series of double-blind studies with experimental yeasts and utilized the Lakes States *C. utilis* as the control. For several years we encountered no problems of any kind in feeding 20 g per day of this material, in addition to the usual diet, to a total of 90 individuals for 30 days despite reactions in the experimental group as described below. However, in the fall of 1979 five of 28 control subjects receiving this yeast developed gastrointestinal symptoms that included some degree of nausea, vomiting and diarrhea, a syndrome that we had previously identified as NVD, and one individual developed a papular rash on his palms and soles (Scrimshaw and Udall, 1981). It seems clear that in this case some difference in processing or quality control must have occurred. This reinforces experience with other SCPs that apparently minor differences in processing conditions can cause problems even when the yeast is fed at low levels comparable to food additive use.

In 1970 we approached, with great circumspection, the evaluation of the first SCP products grown on a petroleum-derived hydrocarbon substrate, even though the substrate itself, pure ethanol, could not be an issue (Scrimshaw and Udall, 1981). One was a bacterium (*Acinetobacter calcoaceticus*) and the other a yeast identified only as WO-100, both treated by alkaline hydrolysis and produced by the Nestlé Company, in full compliance with PAG Guideline No. 6 (Anon., 1983a). Nine of the 50 adult volunteers fed the yeast and eight of 50 fed the bacterium at levels of 20 g per day developed a moderate to severe NVD syndrome. None had significant changes on a battery of biochemical tests and all recovered rapidly. It was subsequently demonstated that acid-washing rendered the bacterial product innocuous to the 10 subjects tested who had previously shown the NVD reaction to this material.

The next material that we tested in human subjects was a strain of *C. lipolytica* grown by the British Petroleum Co. (BP) in Grangemouth on *n*-alkanes and which was RNA-reduced by alkaline hydrolysis (Scrimshaw, 1972). Seven of the 106 individuals consuming this material developed mild forms of the NVD syndrome. No comparable symptoms appeared in the control subjects. A characteristic of the NVD reaction in all of these trials was its appearance between nine and 18 days and the ability of most subjects to consume the materials for up to 90 days with no symptoms.

A new lot of the BP yeast that was acid-washed caused no symptoms in 56 subjects at first, but a second lot of the material that was supposed to have been processed in the same way began to cause mild NVD reactions that required discontinuing the trial. We later learned that one of 20 adults fed the same material in Sweden showed a similar reaction (Abrahamsson *et al.*, 1971). As mentioned earlier, small variations in processing can evidently have significant effects on human tolerance to SCP products.

In 1975 we completed an uneventful tolerance study with 106 subjects fed 20 g of *C. lipolytica* grown by BP in Lavera, France on a gas oil substrate and solvent-extracted with hexane (Scrimshaw, 1979). We were not so fortunate with a processed *C. utilis* grown on beet molasses and corn steep liquor. This was the type of yeast sold commercially for use as a food additive in Europe, except that it had been subjected to a short period of alkaline hydrolysis at 80 °C to reduce RNA. This material, fed at 35 g per day, caused a high frequency of skin rashes, mainly on the palms and soles, but in one case extending over the entire body. This was remarkable because this yeast, RNA-reduced in a similar way but at 130 °C, had been fed to eight subjects at a level of 50–60 g per day in a 15-day metabolic balance study with no detectable adverse reactions. Once again, a processing difference determined whether or not the material would cause adverse reactions in human subjects.

Extensive tolerance studies have also been completed on a *Pichia* sp. yeast. When adult subjects were fed cookies containing 20 g of the whole cells per day or control cookies in a double-blind, cross-over design for 30 days each, adverse cutaneous and gastrointestinal reactions required that the study be terminated 18 days after it was started (Scrimshaw and Dillon, 1979). At this time 25 subjects each had received the material for 18, 14 and 11 days, respectively. When material that had been subjected to RNA reduction to 2% by alkaline hydrolysis was used, six of 17 individuals developed a characteristic rash on hands and feet on the second to ninth day of the study. Use of a different medium but the same method of RNA reduction resulted in a material that was much more acceptable, although nine of 41 subjects developed very mild cutaneous rashes on their hands and in some cases also on their feet during the fifth to fourteenth day of the 30-day test period. A further modification of the processing resulted in a material that caused no adverse reactions in eight subjects tested in a 30-day study at the same level of intake.

Administration of a few grams of thoroughly washed and boiled cells of *Hydrogenomonas eutropha* caused severe NVD, and in some subjects vertigo and asthenia (Calloway and Kumar, 1969; Waslien *et al.*, 1969). A commercial lot of *Aerobacter aerogenes* (marketed as *E. coli*) produced comparable symptoms. It should be re-emphasized that none of the rats, mice, dogs, swine, monkeys and chimpanzees tested with proportionately much larger doses of these materials than given to the human subjects showed any detectable adverse reactions.

36.3.2 Filamentous Microfungi

In 1977 we fed 20 g per day of RNA-reduced *Fusarium graminearum* A3/5, produced by Rank

Hovis McDougall Ltd. on starch, to 100 subjects daily for 30 days in a double-blind tolerance trial that was completed wholly uneventfully. There were no subjective reactions and no changes in the results of a battery of biochemical tests performed at the start and end of each period. More recently we have fed 10 g daily of 'Pekilo Protein', a filamentous microfungus *Paecilomyces varioti* grown on sulfite liquor by Tampella Inc. (Scrimshaw, 1983), to 50 subjects for 30 days. This trial was also completed uneventfully.

There is no lack of practical experience with human consumption of a variety of mushrooms, but no systematic tolerance studies. There is, however, evidence from Bulgaria that an edible mushroom, grown in submerged culture so that it remains in mycelial form, continues to be well tolerated in experimental and farm animals and in human subjects (Torev, 1983). The organism is *Polyporus squamous*-64. Experimentally, a variety of energy substrates have been successfully used, including starch, sucrose and sulfite liquor, although the clinical trials were done with material grown using starch as the energy source.

36.3.3 Algae

For the single-cell algae that have been fed experimentally to human subjects, *Chlorella* (Dam *et al.*, 1965; Waslien *et al.*, 1970; Lee *et al.*, 1967), *Spirulina* (Bourges *et al.*, 1971; Anon., 1975b) and *Scenedesmus* (Dam *et al.*, 1965; Müller-Wecker and Kofranyi, 1973; Soeder and Pabst, 1975), the principal problems encountered have been strong odor and taste. A homogeneous mixture of *Chlorella* and *Scenedesmus* that had been autoclaved at 160 °C was fed by Powell *et al.* (1961) to five healthy young men. Although the spinach-like flavor predominated in all foods supplemented with the algae, it was possible to feed amounts varying from 10 to 500 g per person per day. Amounts up to 100 g were tolerated by all, but when larger amounts were added, gastrointestinal symptoms developed. These included nausea, vomiting, abdominal distention, flatulence, lower abdominal cramping pains and bulky hard stools. No other evidence of toxicity was observed, and the gastrointestinal symptoms cleared up promptly when the algae feeding was discontinued.

Spirulina was traditionally eaten by both the natives on the shores of Lake Chad in Central Africa and the Aztecs in Mexico, but apparently in relatively small quantities as a green and not as a major food (Bourges *et al.*, 1971). *Spirulina* produced in the alkaline Lake Texcoco in Mexico has been incorporated at low levels into the diets of hospitalized children in Mexico City and found acceptable and well tolerated (Durand-Chastel and Clement, 1975).

Scenedesmus produced experimentally in Dortmund, West Germany has been incorporated at five successive intake levels of four weeks duration, each up to 80 g per person per day, at which level it became the sole protein source. There were no negative effects observed and the acceptability was stated to be satisfactory (Soeder and Pabst, 1975). In addition, from 30 to 50 g of *Scenedesmus* powder per day, mainly through a gastric tube, was given in the treatment of more than 50 hospital patients without negative side effects.

36.3.4 Nucleic Acid Limitations

The 1975 PAG working group meeting on the clinical evaluation and acceptable nucleic acid levels of SCP for human consumption thoroughly and authoritatively explored this topic (Anon., 1975). Any rapidly growing cell contains relatively large amounts of nucleic acids. Ribonucleic acid (RNA), the predominant source of purines in SCP products, produces an increase of urinary uric acid of 100 to 150 mg per g RNA, which means that only part of the purines in this nucleic acid is absorbed and eventually excreted as uric acid through the kidneys. The ratio between RNA administered and uric acid excreted appears to be constant over a broad range, up to excretion values of at least 1750 mg per day. The effect of deoxyribonucleic acid (DNA) is about half that of RNA. In SCP products, RNA is the predominant uricogenic agent followed by DNA, with other nucleotides constituting only a negligible fraction.

Since a substantial part of the total nitrogen in SCP comes from nucleic acid unless specially processed to reduce it, the conventional calculation of 'crude protein' by multiplying total nitrogen by 6.25 is a serious overestimate of protein and results in an underestimate of the proportion of nucleic acid relative to protein. Also to be taken into account is the pyrimidine nitrogen of nucleic acid. Since the nitrogen content of pyrimidines is about 40% that of purines and both are present in equimolecular amounts in most nucleic acids, the purine nitrogen can be multiplied by the factor 1.4 to obtain nucleic acid nitrogen. A corrected protein nitrogen (still containing nitrogen from lipids and polysaccharides) is obtained by subtracting 1.4 times purine nitrogen from total nitrogen. Multiplying by the proper factors, 6.25 for protein and 9.0 for nucleic acids from purine nitrogen, gives the corrected values for protein and nucleic acids, respectively. Nevertheless, for most purposes 'crude protein', $N \times 6.25$, will continue to be used to describe the 'protein' content of SCPs as for other food proteins. The result, of course, is that the protein value in biological tests will be less than if the 'true' protein value were the basis for the calculation.

The relationship between dietary nucleic acid consumption and blood and urine levels of uric acid is critical to ascertaining how much nucleic acid from SCPs can be safely incorporated into human diets. In order to determine this more precisely, we fed known levels of yeast for nine days to individuals on a constant formula diet (Edozien *et al.*, 1970). The results were a linear increase from 4.5 to 8.9 mg dl^{-1} of uric acid at levels of intake from 0 to 90 g of yeast per day, which resulted in plasma urate levels rising from 4.5 to 8.9 g dl^{-1} and urinary urate from 510 to 1850 mg per day. According to the observations of Waslien *et al.* (1970), 1 g of yeast RNA per day in purine-free diets results in plasma urate levels of 6.5 g dl^{-1} compared with 8.4 g for those ingesting 4 g of yeast RNA. We proposed a 2 g limit to the amount of SCP nucleic acid acceptable as an addition to the usual diets of adults. We subsequently fed 136 male adults 1.5 to 2 g of SCP RNA daily for three to six months as additions to their usual diets and concluded that the resulting plasma and urinary uric acid values remained within acceptable limits for most subjects. Individuals with gout or a tendency to gout must still avoid significant intake of SCP-containing foods just as they are advised to limit their intakes of organ meats and other common foods with a high nucleic acid content.

The above information on uric acid levels in blood and urine in relation to diet has been derived from studies on populations consuming Western European or North American diets. It can be predicted that the predominantly vegetable diets of the majority of persons in developing countries would contain less nucleic acid and therefore these populations would tolerate somewhat higher levels of SCP intake without risk (Anon., 1975a). Conversely, the diets high in organ meats of some European populations may require lower limits on the amount of permissible additional nucleic acid from SCP. This is of particular interest, since one potential use for SCP is as a meat extender in sausages.

In reaffirming 2 g of nucleic acid per day from SCP as a safe practical limit for most adult populations, the 1975 PAG working group asserted that, even by making the unfavorable assumption that this is all RNA, the serum uric acid would rise by only 1.8 mg dl^{-1}. This would not involve an appreciable risk of gouty arthritis. The question of urinary lithiasis was addressed separately.

Based on the studies cited above, the working group recommended that the urine uric acid excretion should not exceed approximately 1000 mg per 24 hours. At this level of uric acid excretion, and under favorable circumstances of urine output (not less than 1500 ml per 24 hours) and urine pH (not less than 5.6 in the absence of infection), uric acid urolithiasis should not constitute an unacceptable risk.

Some urinary tract infections increase the urine pH and might therefore increase uric acid solubility. Nonetheless, such infections are not normal and should be treated vigorously. Every effort must be made to ensure rates of urine flow higher than 1500 ml per 24 hours in all types of urinary tract infection and in urolithiasis. The regular ingestion of a predominantly vegetable, and therefore an alkaline-ash, diet in developing countries would be an additional favorable factor that would tend to increase urine pH and hence uric acid solubility.

The limit of 2 g of nucleic acid from SCP added to the daily diet assumes that the total nucleic acid from all sources will not exceed 4 g per day. This limit would be sufficient to protect 'nearly all' of the normal population, or in practical terms, the mean plus two standard deviations, *i.e.* all but 2.5% of the normal population. Thus, when the nucleic acid content of the usual diet varies within the limits of 1–2 g, an upper limit of 2 g of additional nucleic acid per day would mean that relatively few individuals would be expected to approach this limit for practical reasons. The further chance of maximum nucleic acid SCP intake coinciding with maximum nucleic acid intake from other dietary sources was judged small, since any large amount of SCP-containing foods would normally be replacing other protein in the diet.

It should be noted that the higher the proportion of total protein requirements proposed to be met from SCP, the lower the ratio of nucleic acid to total protein will need to be. In theory, a non-RNA-reduced yeast with 8% nucleic acid would present no problem when used at 3 to 5% levels as a food additive for its functional properties or its vitamin content. A reduction of the RNA content to 3% would permit its use as a significant protein supplement. Reduction of it below 1% would facilitate its use as a major dietary protein source even in weaning foods and infant formulas, providing that other criteria were met.

The amount of dietary nucleic acid safely tolerated is related to body weight, so that the safe levels for children would be proportionately less than for adults. Table 1 represents such a calculation taken from the working group report. However, epidemiological studies indicate that boys up to the age of puberty and women until menopause have consistently lower serum uric acid than adult males in the same population (Mikkelsen *et al.*, 1965; Hall *et al.*, 1967). This should give a further margin of safety to these groups if the above recommendations are followed.

Table 1 Permissible Level of Intake of Nucleic Acid from SCP for Different Age Groups and Sex[a]

Age group (years)	Sex	Body weight (kg)	Intake of nucleic acid (purine N×9; g)
Adult	M	65	2.0
Adult	F	55	1.7
16–19	M	63	1.9
16–19	F	54	1.7
13–15	M	51	1.6
13–15	F	50	1.5
10–12	M	37	1.1
10–12	F	38	1.2
7–9	M, F	28	0.9
4–6	M, F	20	0.6
1–3	M, F	13	0.4

[a] From Anon. (1975a).

36.4 NUTRITIONAL VALUE

Depending on the organism, SCP protein content ranges from approximately 50% to over 70% 'crude protein' calculated as N × 6.25. If it has been well processed, its protein quality is comparable to that of other good protein sources of vegetable origin. However, processing conditions can make a great deal of difference. When in 1916 Funk *et al.* fed *C. utilis* as the sole source of protein in a diet, they found the protein to be poorly assimilated and of limited food value. On the other hand, most authors since have reported quite acceptable digestibility and protein values. In conventional studies in experimental animals and in human subjects fed at the deficient levels of protein intake required to obtain values comparable to those in animal studies, methionine is the limiting amino acid in SCP as it is in legumes. When legume proteins are fed to human subjects at requirement levels, however, a deficiency of methionine can no longer be detected, and adding this amino acid does not improve nitrogen retention (Scrimshaw and Young, 1979). There is little doubt that the same is true for SCPs, although the experimental evidence for this from studies based on currently recommended methods (Pellett and Young, 1980) is not available. What this would mean is that digestibility and the amount of amino acid nitrogen are the main limiting factors in the utilization of SCP protein rather than amino acid pattern.

Protein digestibility is not markedly affected by the level of protein intake, but biological value and net protein utilization are higher when the protein intake is adequate (Pellett and Young, 1980). Net protein utilization is in the same range as that of legumes. The reported digestibility of yeast varies from 52–87%, no different from those of conventional sources of vegetable protein. Some of the available information on the protein value of yeast, bacteria, microfungi and algae is given in Table 2.

Animal studies indicate that algae, including *Chlorella*, *Scenedesmus* and *Spirulina* grown under proper conditions, although somewhat deficient in methionine, are equally good protein sources. There is no problem in incorporating these materials into human diets as minor ingredients that add texture, color and flavor. The Japanese diet makes extensive use of algae in this way. It is quite another matter, however, to feed them as significant protein sources because of their strong, bitter flavor. It is quite clear that for their widespread acceptance, methods to improve both acceptability and digestibility would be needed, and the result would be a different product. While it should be possible to maintain and even improve protein value as well as flavor through such processing, the economic considerations are so unfavorable that the nutritional value of such materials has not been studied in human subjects.

Table 2 Digestibility and Protein Value of Various 'Single-cell Proteins'

Genus and species	No. of subjects	Protein level	Digestibility	Biological value	Reference
Yeasts					
Candida utilis	7	28 g	83 ± 7[b]	70 ± 5[b]	Waslien *et al.* (1970)
Candida utilis	7	51 g	87 ± 3[b]	58 ± 6[b]	Waslien *et al.* (1970)
Candida utilis	—	—	70–90	52–87	Bressani (1968)
Candida utilis	52	20 g	64		c
Candida utilis[a]	8	0.35 g/kg	84	58	c
Pichia spp.	6	—	84[b]	54[b]	Scrimshaw and Udall (1981)
Pichia spp.[a]	3	—	83[b]	51[b]	Scrimshaw (1982)
Bacteria and fungi					
Fusarium graminearum	13	0.30 g/kg	78[b]	65[b]	Udall *et al.* (1984)
Fusarium graminearum	6	0.30 g/kg + methionine	79[b]	73[b]	Udall *et al.* (1984)
Paecilomyces varioti	6	0.35 g/kg	81[b]	54[b]	Udall *et al.* (1984)
Algae					
Scenedesmus obliquus	5	46 g	68	—	Dam *et al.* (1965)
Chlorella pyrenoidosa	5	38 g	57	—	Dam *et al.* (1965)
Chlorella pyrenoidosa	6	62 g	59	—	Dam *et al.* (1965)
Chlorella pyrenoidosa	6	38 g	66[b]	—	Lee *et al.* (1967)
Chlorella sorokiniana	?	30 g	89 ± 4[b]	79 ± 12[b]	Waslien *et al.* (1970)
	?	51 g	82 ± 6[b]	60 ± 6[b]	Waslien *et al.* (1970)

[a] RNA-reduced. [b] Corrected for endogenous N loss. [c] Unpublished MIT data.

SCPs are also valuable sources of B vitamins, as indicated in Table 3.

36.5 OTHER CHARACTERISTICS AFFECTING USE AS FOOD

Assuming satisfactory human tolerance, good nutritional value and even favorable economic considerations, there remains the problem of finding acceptable food uses. Since the primary cellular product of yeast and bacteria production is a dry powder with flavors ranging from bland to bitter, its ready incorporation into human diets is not assured. The addition of such materials to cookies, cakes, puddings and other processed foods in small amounts is quite feasible if there is some incentive to do so. Vitamin enrichment may be one such incentive, but the amount of protein added in this way is usually not nutritionally significant enough to have market value. Until now, except for war-time use, the principal reason for adding torula yeast to processed foods has been some functional advantage.

Autolyzed or hydrolyzed yeast extract has been used as a flavoring agent to give a meaty or nutty flavor to foods. Often it has a dark brown color that makes it useful as a coloring agent. As a functional ingredient, SCP, like many other proteins, can serve as a surface active agent in emulsions, foams, gels and suspensions to maintain structure. The use of yeast as a carrying agent for flavoring packaged almonds is another example. Unfortunately, all of these uses involve such low levels of intake that the nutritional potential of the SCP is not involved.

The use of SCP from yeast and bacteria as major protein sources will require some formulation into a food with acceptable flavor, texture and appearance. The process of texturizing, coloring and flavoring protein isolates through spinning or extrusion has been highly developed for soy protein and can be applied equally well to SCP isolates. Thus far, the cost has limited the use of such products from soy protein, and they would be prohibitive in cost if made from SCP at the present time. The best immediate hope for the use of yeast and bacteria would seem to be an increase in their use as food additives in ways that will lead to a large volume at relatively low percentages of the diet. This would not require the increased cost of RNA-reduction or production of protein isolates or concentrates. In the long run, however, there is no doubt that palatable protein-rich foods can be produced from these materials when economic considerations and shortage of protein from conventional sources justify them.

Also of immediate interest is the fact that filamentous microfungi already possess a meat-like texture and can be readily formed into analogs of conventional foods or used to develop attractive new foods. They can also contribute the functional property of texture to processed foods to which they are added. While the economics of doing this are still uncertain, Rank Hovis McDougall Ltd. of the UK began marketing such materials on a trial basis in 1982. A similar product produced by Tampella Inc. has attractive properties in the moist cake remaining after fil-

Table 3 The Vitamin Contents of Some Food Yeasts[a, b]

Food yeast	Thiamine	Riboflavin	Nicotinic acid	Pantothenic acid	Pyridoxin	Folic acid	Biotin	p-Aminobenzoic acid	Medium
Bakers' compressed	20–40	60–85	200–700	180–330	—	—	0.6–1.8	—	—
Bakers' foil	80–150	50–65	180–400	120–160	—	—	0.5–1.5	—	—
Brewers'	104–250	25–80	300–627	72–86	23–40	19–30	1.1	15–40	—
Candida arborea	31.3–33.1	52.3–69.5	492.3–580.2	—	—	14.8–16.0	—	—	Molasses, beet, cane
Candida arborea	13–33	46–70	301–580	—	—	12–26	0.24–3.2	11–21	—
Hansenula strain	8.5	54	590	180	—	1.7	1.7	16	Wood sugar stillage
Mycotorula strain	5.3	59	600	—	—	3.1	1.8	31	Wood sugar stillage
Oidium lactis	12–29	40–55	186–248	—	—	6–15	1.30–2.1	—	Molasses, beet, cane
Oidium lactis	20.1–29.0	39.9–55.0	192.8–247.5	—	—	5.6–7.8	—	—	—
Saccharomyces cerevisiae	28–41	39–62	277–568	—	—	19–36	0.45–3.6	11–62	Spent sulfite liquor
Candida utilis	5.3	42.7	417.3	39.2	33.4	21.5	2.3	—	Molasses, beet, cane
Candida utilis	35.4–38.1	54.2–62.0	511.3–600.00	—	—	10.6–15.2	—	—	Molasses, beet, cane
Candida utilis	22	54	440–490	—	—	—	—	—	Molasses, beet, cane

[a] μg per g dried yeast.   [b] Dunn, 1975.

tration. With shelf life extended by freezing or refrigeration, the material can be a substantial component of sausages and other meat products. Because its texture and flavor are those of fresh mushrooms, it could also find use as an ingredient in soup and other foods in which mushrooms are now incorporated. As noted above, there does not appear to be much prospect for the use of algal SCP as a significant protein source in human diets.

36.6 REFERENCES

Abrahamsson, L., L. Hambraeus, Y. Hofvander and B. Vahlquist (1971). Single-cell protein in clinical testing. A tolerance test in healthy adult subjects comprising biochemical, clinical and dietary evaluation. *Nutr. Metab.*, **13**, 186.

Anon. (1975a). PAG *ad hoc* working group meeting on clinical evaluation and acceptable nucleic acid levels of SCP for human consumption. *PAG Bull.*, **54** (3), 17.

Anon. (1975b). Report of the 3rd meeting of the PAG *ad hoc* working group on single-cell protein, Cambridge, Massachusetts, 1973. In *PAG Compendium*, vol. C2, p. 2211. Wiley, New York.

Anon. (1983a). PAG/UNU Guideline No. 6: Preclinical testing of novel sources of food. *Food Nutr. Bull. (United Nations University)*, **5** (1), 60.

Anon. (1983b).PAG/UNU Guideline No. 7: Human testing of novel foods. *Food Nutr. Bull. (United Nations University)*, **5** (2), 77.

Anon. (1983c). PAG/UNU Guideline No. 12: The production of single-cell protein for human consumption. *Food Nutr. Bull. (United Nations University)*, **5** (1), 64.

Anon. (1983d). PAG/UNU Guideline No. 15: Nutritional and safety aspects of protein sources for animal feeding. *Food Nutr. Bull. (United Nations University)*, **5** (1), 67.

Bourges, H., A. Sotomayor, E. Mendoza and A. Chavez (1971). Utilization of the alga *Spirulina* as a protein source. *Nutr. Rep. Int.*, **4** (1), 31.

Bressani, R. (1968). The use of yeast in human foods. In *Single-Cell Protein*, ed. R. I. Mateles and S. R. Tannenbaum, p. 90. MIT Press, Cambridge, MA.

Calloway, D. H. and A. M. Kumar (1969). Protein quality of the bacterium, *Hydrogenomonas eutropha*. *Appl. Microbiol.*, **17**, 176.

Dam, R., S. Lee, P. C. Fry and H. Fox (1965). Utilization of algae as a protein source for humans. *J. Nutr.*, **86**, 376.

Davis P. (ed.) (1974). *Single Cell Protein. Proceedings of the International Symposium, Rome, Italy, 1973.* Academic, New York.

de Pontanel, G. (ed.) (1972). *Proteins from Hydrocarbons. Proceedings of the 1972 Symposium, Aix-en-Provence.* Academic, New York.

Dunn, C. G. (1975). Uses of: A. Yeast and yeast-like microorganisms in human nutrition. B. Bacteria and bacteria-like microorganisms in human nutrition. In *PAG Compendium*, vol. C2, p. 2051. Wiley, New York.

Durand-Chastel, H. and G. Clement (1975). *Spirulina* algae: food for tomorrow. In *Proceedings of the IXth International Congress on Nutrition*, ed. A. Chavez, H. Bourges and S. Basta, vol. 3, p. 85. S. Karger, Basel, Switzerland.

Edozien, J. C., U. U. Udo, V. R. Young and N. S. Scrimshaw (1970). Effects of high levels of yeast feeding on uric acid metabolism of young men. *Nature (London)*, **228**, 180.

Fukui, T., A. Tokas, T. Miyoski, M. Machiguchi and T. Sasahi (1960). Studies on the digestion and absorption of torula yeast (*Torula utilis, Mycotorula japonica*). 2. Digestion and absorption in the human body. *Shikoku Acta Med.*, **16**, 706.

Funk, C., W. G. Lyle and D. McCoskey (1916). The nutritive value of yeast, polished rice and white bread as determined by experiments on man. *J. Biol. Chem.*, **17**, 173.

Goyco, J. (1959). Nitrogen balance of young adults consuming a deficient diet supplemented with torula yeast and other nitrogenous products. *J. Nutr.*, **69**, 49.

Hall, A. P., P. E. Barry, T. R. Dawber *et al.* (1967). Epidemiology of gout and hyperuricemia. *Am. J. Med.*, **42**, 27.

Klapka, M. R., G. A. Duby and P. L. Pavcek (1958). Torula yeast as a dietary supplement. *J. Am. Diet. Assoc.*, **34**, 1317.

Lee, S. K., H. M. Fox, C. Kies and R. Dam (1967). The supplementary value of algae in human diets. *J. Nutr.*, **92**, 281.

Mikkelsen, W. M., H. J. Dodge and H. Valkenburg (1965). The distribution of serum uric acid levels in a population unselected as to gout or hyperuricemia: Tecumseh, Michigan. *Am. J. Med.*, **39**, 242.

Müller-Wecker, H. and E. Kofranyi (1973). Einzeller als ausätzliche Nahrungsquelle. *Hoppe-Seyler's Z. Physiol. Chem.*, **354**, Suppl., 1034.

Pellett, P. L. and V. R. Young (eds.) (1980). *Nutritional Evaluation of Protein Foods, The United Nations University World Hunger Programme Food and Nutrition Bulletin Supplement 4.* United Nations University, Tokyo, Japan (WHTR-3/UNUP-29).

Powell, R. C., E. M. Nevels and M. E. McDowell (1961). Algae feeding in humans. *J. Nutr.*, **75**, 7.

Prescott, S. C. and C. G. Dunn (1959). *Industrial Microbiology*, 3rd edn. McGraw-Hill, New York.

Scrimshaw, N. S. (1972). The future outlook for feeding the human race. The PAG's recommendations Nos. 6 and 7. In *Proteins from Hydrocarbons*, ed. G. de Pontanel, p. 189. Academic, New York.

Scrimshaw, N. S. (1975). Single-cell protein for human consumption—an overview. In *Single-Cell Protein II*, ed. S. T. Tannenbaum and D. I. C. Wang, p. 24. MIT Press, Cambridge, MA.

Scrimshaw, N. S. (1979). Summary of the PAG symposium: investigations on single-cell protein. In *Single-Cell Protein— Safety for Animal and Human Feeding*, ed. S. Garattini, S. Paglialunga and N. S. Scrimshaw, p. 189. Pergamon, Oxford.

Scrimshaw, N. S. (1983). Non-photosynthetic sources of single-cell protein—their safety and nutritional value for human consumption. In *A Systems Analysis Approach to the Assessment of Non-Conventional Protein Production Technologies. Proceedings of a Task Force Meeting, International Institute of Applied Systems, Sofia, Bulgaria, 1982*, ed. J. T. Worgan, IIASA, Laxenburg, Austria (CP-83-30).

Scrimshaw, N. S. and J.-C. Dillon (1979). Allergic responses to some single-cell proteins in human subjects. In *Single-Cell Protein—Safety for Animal and Human Feeding*, ed. S. Garattini, S. Paglialunga and N. S. Scrimshaw, p. 171. Pergamon, Oxford.

Scrimshaw, N. S. and J. Udall (1981). The nutritional value and safety of single-cell protein for human consumption. In *Proceedings of the International Colloquium on Proteins from Single-Cell Organisms, Association pour la Promotion Industrie-Agriculture (APRIA), Paris, France, 1981*.

Scrimshaw, N. S. and V. R. Young (1979). Soy protein in adult human nutrition: a review with new data. In *Soy Protein and Human Nutrition*, ed. H. K. Wilcke, D. T. Hopkins and D. H. Waggle, p. 121. Academic, New York.

Soeder, C. J. and W. Pabst (1975). Production, properties, preclinical and clinical testing of *Scenedesmus* 276 3a. In *PAG Compendium*, vol. C2, p. 2114. Wiley, New York.

Torev, A. (1983). Feed and food protein from higher fungi mycelia. *Food Nutr. Bull. (United Nations University)*, **5** (3), 65.

Udall, J. N., C. Lo, V. R. Young and N. S. Scrimshaw (1984). The tolerance and nutritional value of two microfungal foods in human subjects. *Am. J. Clin. Nutr.*, **40**, 285.

von Loesecke, H. W. (1946). Controversial aspects: yeast in human nutrition. *J. Am. Diet. Assoc.*, **22**, 485.

Waslien, C. I., D. H. Calloway and S. Margen (1969). Human intolerance to bacteria as food. *Nature (London)*, **221**, 84.

Waslien, C. I., D. H. Calloway, S. Margen and F. Costa (1970). Uric acid levels in men fed algae and yeast as protein sources. *J. Food Sci.*, **35**, 294.

37

Acceptance of Single-cell Protein for Animal Feeds

D. A. STRINGER
UNICELPE, Brussels, Belgium

37.1 INTRODUCTION

Single-cell protein (SCP) in this context relates to unicellular organisms grown specifically for their protein content under controlled fermentation conditions on an industrial scale. The organisms employed may be fungi, yeasts or bacteria. The substrates for cell growth include conventional materials such as molasses, sugar or starch and also unconventional materials like methanol, ethanol, ethane and petroleum hydrocarbons. Research into the production of single-cell protein on an industrial scale has been undertaken by a range of companies whose main financial strengths vary from food to petrochemicals (Table 1). The products that have originated from these organizations have not automatically followed the market sector that convention may have expected by company tradition, for example Amoco developed a yeast for human consumption based on ethanol fermentation. Exxon in conjunction with Nestlé examined food grade protein from both yeasts and bacteria. Nevertheless the greatest potential interest in single-cell protein has been in the animal feed sector where safety, acceptability of the feed and the ability to sustain animal growth are the main criterea of assessment, given of course suitable technical processes and economic environment. The universal view adopted by the producers of SCP for animal feed is one of achieving acceptance by the animal feed market and regulatory authorities, and then further developing the materials for human food taking into account nucleic acid content, functionability and human tolerance, once product safety has been established on a broad scale. This chapter deals only with the first stage, that is to say acceptance as an animal feed.

37.2 SAFETY ACCEPTANCE OF SCP

The introduction of the concept of food or feed grade protein from non-conventional sources initially posed problems for regulatory authorities. Single-cell proteins evolved during a period

Table 1 Organizations Involved in SCP Development

Type of substrate	Organism	Organization/location	Production	Use[a] A	Use[a] H	Commercial status
Starches, sugars/yeast						
Molasses	*C. utilis*	Cuban Sugar	70 000	+	?	Product being developed for H application
Lactose/whey	*Klugveromyces fragilis*	Bellyeast/France	6 000	+	+	High value application sought
Sulfite liquor	*C. utilis*	Pekilo/Finland	3 000	+		
Starches, sugars/fungi						
Hydrolysed starch	*Fusarium graminarium*	RHM/UK	Pilot plant		+	Awaiting scale up
Alcohols/yeasts						
Ethanol	*C. utilis*	Amoco/USA	5 000		+	High value markets
Methanol	*Pichia pastoris*	IFP/France	Pilot plant	+		
Methanol	*Pichia* spp.	Phillips/USA	Pilot plant		+	Research phase
Alcohols/bacteria						
Methanol	*Methylophilus methylotrophus*	ICI/UK	55 000	+	?	Successful large scale development
Methanol	*Methylomonas clara*	Hoechst/Germany	Pilot plant	?	+	Research phase
Hydrocarbons/yeast						
Gas oil	*Candida tropicalis*	BP/France	15 000	+		Project closed
	Candida spp.	VEB/DDR	100 000	+		Commercial development
n-Alkanes	*Candida lipolytica*	BF/UK	4 000	+	?	Project closed
	Candida lipolytica	ANIC/Italy	100 000	+		Project closed
	Candida maltosa	Liquichemica/Italy	100 000	+		Project closed
	Candida tropicalis	IFP/France	Pilot plant	+		Project closed

[a] A, animal; H, human.

when the need for toxicological testing of additives used in human and animal nutrition was a well accepted fact. Toxicological testing of food products was a novel concept in the early 1970s. Problems arose as to how SCP materials should be tested and the extent of that testing procedure. Ten years' experience has largely resolved that situation.

37.2.1 Guidelines for Testing SCP

Various guidelines have been compiled by national and international groups from regulatory and academic backgrounds with the intention of assisting the manufacturers of these products, and also other regulatory bodies, on the type and degree of testing required. The initial attempts to produce guidelines resulted in documents which reflected the specific concerns, expertise and background of the compilers. The later documents, such as those compiled by the Protein Advisory Group (PAG) of the United Nations System (Anon., 1974a, 1974b, 1983) and IUPAC (Anon., 1974c; Hoogerheide *et al.*, 1979), presented a more balanced view of the requirements. Recently the European Economic Community has published extensive guidelines to meet its own regulatory problems for materials derived from hydrocarbon substrates (Anon., 1983).

The concepts followed by all these latter guidelines follow well established toxicological experience. A comparison of the concepts developed by the PAG, IUPAC and the EEC show that they are very similar to those compiled by the United States Scientific Committee of the Food Safety Council (Anon., 1978) where evaluation programmes were defined specifically for food and feed additives (see Table 2). The feature that is immediately apparent when toxicological requirements are examined is the similarity in the extent of work required, irrespective of whether the product is intended for animal or human feeding. PAG Guideline No. 6 (Anon., 1974a), which is directed towards preclinical testing prior to human tolerance and nutrition studies, has, in general, set the trend for the type of work required. It was originally written with SCP in mind, but its latest revision is aimed at any novel food source irrespective of source and nature and as a result the degree of obligation for the various toxicological tests is somewhat alleviated, based on degrees of novelty and the safety understanding existing for products which could be deemed to fall within its remit. This concept is further developed by the EEC Guidelines which exclude the need for testing materials originating from conventional substrates. It appears to be their opinion, taken from a regulatory standpoint, that it is only the SCPs derived from hydrocarbons and analogous products that need a rigorous toxicological evaluation.

Table 2 Comparison of Toxicological Requirements Defined by Major Guidelines for Testing Novel Foods and Feeds

	Human food tests			Animal feed tests	
	Food Safety Council	Protein Advisory Group No. 6	IUPAC	Protein Advisory Group No. 15	EEC
Classical animal toxicology					
(a) Short term tests					
Rat—acute	+	+	+		+
—90 day study	+	+	+	+	+
—teratology	+		+	+	Optional
(b) Long term tests					
Rat—2 year or lifespan	+	+	+	+	+
—multigeneration	+	+	+	Optional	Optional
Features to be examined in (a) and (b)					
Live weight/feed intake	+	+	+	+	+
Haematology	+	+	+	+	+
Biochemistry, blood and urine	+	+	+	+	+
Pathology, gross and microscopic	+	+	+	+	+
Reproductive indexes	+	+	+	+	Optional
Need for a 2nd species					
Dog—3–6 months		+			
2nd rodent, *e.g.* mouse lifespan	+				+
Target species					
Short term tests					
e.g. Broiler chicken—56 days			+	+	Optional
Laying hen—1 year			+	+	Optional
Growing pig—120 days			+	+	Optional
Reproducing pig—1 year			+	+	Optional
Features to be examined as classical animal toxicology					
Metabolic studies	+	+	+	+	+
Mutagenic studies					
in vivo e.g. dominant lethal, cytogenetic tests		+	+	+	+
in vitro e.g. Ames	+		+	+	+

37.2.2 Nature of SCP Products

Prior to discussing the nature of the tests as defined in the above guidelines, it is necessary to examine the features of the SCP product as this will define, to a large extent, the manner and extent of the toxicological tests, both in the way that they are carried out and with regard to interpretation of the data eventually produced.

It must be appreciated first of all that SCP is a catch-all phrase. It is a term that has developed to describe proteinaceous materials, produced by diverse varieties of unicellular organisms, which can differ substantially in their chemical composition both in their own right and also as a result of the substrate used for cell growth. An understanding of the type of cell, mode of production and chemical analysis of the cellular material is, therefore, the first step in defining the risks that may accrue from the material.

Thus when comparing fermentation substrates based on methane or gas oil (taken to represent hydrocarbon extremes), it is likely that residues will persist from the gas oil process which, in turn, raises questions as to how they are to be reduced to acceptable residual levels, what solvents are to be employed, and what residual concentrations of solvents are acceptable.

Generally speaking any residual sugar substrate would be thought of as harmless, and methanol and ethanol residues are difficult to detect because of their low concentration; for methanol it is less than 4 p.p.m., which is the detection limit. Considerable discussion of the potential biological effects of residual *n*-alkanes has taken place; this will be discussed later.

An understanding of the substrate also poses questions related to the contaminants that may be present therein. Thus for the gas oil processes in particular, a rigorous chemical analysis is required for polycyclic aromatic hydrocarbons and in particular those with known carcinogenic potential.

The aqueous medium, the quality of the water used for fermentation and the quantity of chemicals used within the process, all have an effect upon the final product. When complete recycle of water occurs within the process then the only route for escape of any non-volatile material is within the dried fermentation product. Thus constraints need to be placed upon the heavy metal content of the final product. Typical values are Hg 0.1 p.p.m., As 2 p.p.m., Pb 5 p.p.m. and F 150 p.p.m., although experience shows that values much lower than these are obtained. Put simply, what goes into the fermenter system must come out, and hence if crude chemicals are used as either process aids (pH, flocculation, *etc.*) or as trace ingredients for cell growth, then their contaminants will be found in the final product. In practice the specification most difficult to meet is the mercury level, reflecting the conventional mode of production of caustic soda using the Caster–Kellner process.

Chemical analysis for mycotoxins must be regarded as essential for those products of fungal origin. For other SCP products the presence of mycotoxins is more likely to indicate bad storage conditions and/or insufficient water removal at the drying stage of product preparation and consequential mould growth.

Other features of the chemical analyses that are easily performed relate more to the nutritional potential of the product. Of particular importance in this context are the protein, fat and carbohydrate contents, the amino acid and fatty acid profiles, as well as the vitamin and mineral contents. Knowledge of the above is vitally important to the formulation of animal diets both in a commercial sense and, more importantly at this stage, to the correct composition and balance of the test diets to be used for toxicological evaluation.

The peculiarities of the materials also need to be demonstrated. For example, the cell walls of yeast consist of chitin, bacterial cell walls contain small quantities of non-utilizable D-amino acids, specifically D-analine and D-glutamic acid both of which are intimate components of the cell wall molecular structure. Non-protein nitrogen components also need to be quantified. Ammoniacal nitrogen, amides, nitrates and nitrites, and nucleic acids all contribute to the total nitrogen of the material, but not to the protein nitrogen which is best expressed as the sum of the nitrogen contained within the amino acids. Nucleic acids are not harmful to animals or man, although with the primate species there is a limitation placed on their catabolism, due to the absence of uricase (Prosser, 1952), which has resulted in the concept that ingestion of these materials from single-cell protein sources should not exceed 2 g d^{-1} (Anon., 1975). These constraints, however, do not apply to non-primate mammalian species or other vertebrates.

When evaluating single-cell proteins it is important to remember that one is dealing with dried dead cells, and that the products are not materially different to other materials used in animal feed. Nevertheless the non-conventional nature of their origin, which can include both types of organism as well as substrates, led to the concept that the products should be evaluated and demonstrate freedom from toxicity.

37.3 TOXICOLOGICAL TESTING OF SCP

Single-cell proteins are intended for consumption to provide dietary nitrogen. This implies that the main thrust of the toxicological testing is *via* the oral route. Such a statement may appear obvious, but it certainly infers that dermal or respiratory toxicology are not major sectors for investigation.

The fact that single-cell proteins are also intended for animal feeding widens the potential study profile for the safety evaluation, as in these circumstances it is possible to conduct the toxicological study in the final target species. In the main, toxicological studies are made to define the risk to humans; in this context it is conventional to use inbred laboratory rodents and dogs because of their consistency in response to diet. The variations induced by the test material are then extrapolated to humans and the risk, if any, assessed prior to use. Clearly this kind of extrapolation is not needed when studies can be made on the animals that will actually consume the product. However, as toxicological studies are not conventionally made on pigs and chickens, there is little understanding of the variability in the biological profiles normally examined in rats and dogs. The result has been that tests have been repeated on the conventional laboratory animal species as well as the target species, and this quite justifiably leads to the claim that more toxicological evaluations have been made on SCP than any other food material.

Toxicological testing of foods as opposed to food additives presents substantial problems associated with the experimental diets. For food additives, pesticides or other chemicals of potentially doubtful safety, the test material is incorporated in an animal diet at the normal usage level and also at 100 or even 1000 times its potential usage. It automatically follows that these factors cannot be applied to food (Newberne and McConnell, 1978). Instead the concept of partial or total replacement of the normal dietary protein components (*e.g.* fish or soya) is employed. Synthetic diets may be employed for short term tests where the SCP can replace casein and supply all the dietary protein, but the use of these diets is not recommended for long term tests (*e.g.* 2 year) in laboratory animals. The incorporation rate in conventional cereal-based diets will depend upon the crude protein level normally required by the animal. The point is illustrated in Table 3 using a simplified situation. The dietary contribution from SCP, assuming total substitution of all fish and soya, is greater for broilers than rats, the crude protein level of the diet being the critical factor. If higher incorporation rates are used than those indicated in Table 3, the crude protein content of the diet rises and it is necessary to make the appropriate compensatory changes in the control diet which in turn leads to the use of two control diets, one at the normal level and one with an elevated concentration. In this way the biological effects of high dietary protein and high inclusion rates of single-cell protein may be distinguished.

Table 3 Effect of Substituting SCP[a] into Simplified (i) Rat and (ii) Broiler Diets[b]

(i) Rat (normal dietary crude protein = 18%)

			Experimental diets					
			Replacement of supplementary protein (%)					
Dietary ingredient	*Control diet*		*33*		*66*		*100*	
	%	%CP	%	%CP	%	%CP	%	%CP
Cereal	74	7.4	74	7.4	78	7.8	81	8.1
Soya	16	7.0	16	7.0	7	3.1	—	—
Fish	5	3.5	—	—	—	—	—	—
SCP	—	—	5	3.5	10	7.0	14	9.8
Minerals and vitamins	5	—	5	—	5	—	5	—
Total	100	17.9	100	17.9	100	17.9	100	17.9

(ii) Broiler (normal dietary crude protein = 22%)

			Experimental diets					
			Replacement of supplementary protein (%)					
Dietary ingredient	*Control diet*		*22*		*45*		*100*	
	%	%CP	%	%CP	%	%CP	%	%CP
Cereal	62	6.2	62	6.2	66	6.6	74	7.4
Soya	28	12.3	28	12.3	19	8.4	—	—
Fish	5	3.5	—	—	—	—	—	—
SCP	—	—	5	3.5	10	7.0	21	14.7
Minerals and vitamins	5	—	5	—	5	—	5	—
Total	100	22.0	100	22.0	100	22.0	100	22.1

[a] Based on 70% crude protein.   [b] Dietary formulations and crude protein (CP) contributions to final diet, based on cereals (10% CP), soya (44% CP), fish and SCP (70% CP).

Publication, in a conventional sense, of routine toxicological studies, although supported in principle by the regulatory bodies, finds few supporters in those conducting the experiments, or the editors of journals, unless some abnormality is detected or totally novel materials have been tested. Few references will, therefore, be found relating to the toxicological studies on single-cell proteins. The exception refers to the early work conducted on behalf of BP for the then novel product TOPRINA (de Groot, 1970a, 1970b, 1971) in which no abnormal effects were detected for either the gas oil (TOPRINA L) or the purified alkane produced (TOPRINA G) materials. The absence of data from the formal scientific literature does not imply that the work, as defined in the guidelines mentioned earlier, has not been carried out for if this were the case, no regulatory approval would have been granted. The fact that TOPRINA (BP) and PRUTEEN (ICI) have been accepted widely by the governmental regulatory bodies gives an indication that the work has indeed been done to the required standards, and that there is no indication of toxicological risk.

37.4 SPECIFIC PROBLEMS RELATING TO THE ACCEPTANCE OF SCP

Although TOPRINA was accepted by a majority of governments as a safe and acceptable SCP, it still promoted considerable discussion which has been reviewed by Bellucci (1980). The principal points of discussion have been residual alkanes, polycyclic aromatic hydrocarbons, fatty acids with an uneven chain length and the pathogenicity of the microorganism.

Residual alkanes found within TOPRINA G or LIQUIPRON (Liqui Chemica, Spa) amounts to 0.25% with occasional maximum values of 0.4% (Bizzi *et al.*, 1979b; Di Muccio *et al.*, 1979). Considerably higher levels would exist in TOPRINA L if solvent extraction was not employed to remove these materials (Champagnat and Adrian, 1974). Animals fed SCP containing *n*-alkanes tend to accumulate these materials within the adipose tissue, which posed questions about their suitability for human feeding. Subsequent analysis of conventional foods and feeds revealed the presence of *n*-alkanes (Di Muccio *et al.*, 1979; Lester, 1979, Valfre *et al.*, 1979). The feeding of TOPRINA or LIQUIPRON brought about an elevation of paraffin levels in the adipose tissue levels of pigs, rats, monkeys and trout but not chickens. Deposition was time and dose related (Bizzi *et al.*, 1979b; Shacklady, 1979). Metabolic studies showed that 65% of the ingested alkanes were metabolized to CO_2, 20% were converted to fatty acids and stored within the adipose tissue, 7% were stored unchanged with the adipose tissue, and 8% were converted into proteins and carbohydrates (Tulliez and Bories, 1979). The interesting feature that emerged from this type of investigation was not the *n*-alkane content of meat derived from SCPs but the widespread distribution of paraffins in food which, with hindsight, was predictable as they are permitted as food additives.

The gas oil processes, because of the nature of crude petroleum, bring with them concerns about the levels of polycyclic hydrocarbons and particularly 1,2,5,6-dibenzanthracene, 3,4-benzopyrene and 1,12-benzoperylene. Solvent extraction was therefore used in the BP Lavera process to remove any such compounds, along with residual paraffins (Champagnat and Adrian, 1974). The molecular sieving process subsequently developed resulted in a very pure *n*-alkane stream in which polycyclics could be detected. Analysis of yeasts grown on alkanes shows the content of the various polycyclic aromatics to be similar to, and, in certain cases, less than various samples of bakers' yeast derived from various sources (Grimmer and Wihelm, 1969).

Fatty acids with both odd and even chain lengths are synthesized by cells that use paraffins as a substrate. The odd chain fatty acids account for 56% and 41% of the total fatty acids in TOPRINA and LIQUIPRON respectively (Tomassi and Serlupi-Crescenzi, 1979; Bizzi *et al.*, 1979a, 1980). The fact that odd chain fatty acids accumulate within the adipose tissue following consumption of alkane-yeast is not surprising. This situation is somewhat different to that described earlier for alkanes in that odd chain fatty acids, although occurring naturally in fish and milk, do so at low concentrations (between 1 and 3% of total fatty acids), whilst values up to 10% have been seen in pigs fed 30% TOPRINA (Alimenti *et al.*, 1979). Using rats (Bizzi *et al.*, 1979a, 1980) showed that a plateau occurred after two months' feeding and that no further accumulation occurred over a total of 15 months. The situation is widespread as it is found in all animal species examined (mice, rats, monkeys, chickens, pigs, cows). All the fatty tissues of the animal are affected including plasma lipids, platelet lipids, liver microsomes, liver phospholipids, triglycerides and cholesterol, and, at high dietary levels, the heart phospholipids. The effect of dietary treatment was least on the brain, intermediate on kidney, liver and heart, and most marked on fat deposits. It should be noted that odd chain fatty acids were also found in control diet fed animals but the concentration was not so high as those fed alkane-yeasts. A relay toxicity study with rats failed to show any differences in the level of odd chain fatty acids when the rats were fed a diet containing 10% egg material derived from hens that had been fed LIQUIPRON for 15 months and where the fatty acid profile of the egg contained 3.3% odd chain fatty acids relative to 0.4% in the control. This situation is reassuring as far as humans are concerned. In a rat reproduction experiment Tomassi and Serlupi-Crescenzi (1979) showed that the odd chain fatty acids crossed the placental barrier, but no other effects of significance were noted. The underlying problem, however, is whether the accumulation of the odd chain fatty acids influences physiological of biochemical functions and specifically those associated with membrane function. Metabolism apparently presents no problems; oxidation takes place as far as the terminal three carbon atom fragment which is bound to coenzyme A and subsequently metabolized to succinyl-CoA where it enters the Krebs cycle. Tomassi and Serlupi-Crescenzi (1979) showed that mitochondrial oxidative phosphorylation was not impaired in the rat using either liver or heart mitochondria. Behaviour patterns as shown by swimming and rotating rod tests with mice showed no differences attributable to diet. The overall conclusion that is reached from these investigations is that despite the differences in fatty acid profile between the control and test animals, especially in the structural lipids, the normal functions are not impaired and the changes described have no toxicological significance.

Cyclopropane fatty acids have been detected in the lipid component of bacterial SCPs. In the case of PRUTEEN 9,10-methylenehexadecanoic acid accounts for 1.8% of the total fatty acids or 0.24% of the final product. Traces of this acid (0.15%) have been found in the perirenal fat of pigs following the use of 30% PRUTEEN in the diet over a three month period. The fatty acid is labile and concentrations fall 50% over a six week withdrawal period (Waterworth and Stringer, unpublished data). Tracer studies with 9,10-methylenehexadecanoic acid given to pigs show that the material is catabolized to 9,10-methylenehexanedioic acid, which is excreted in the urine (Woollen and Carter, private communication). The data obtained are very similar to those generated by Eiselle *et al.* (1977) who studied the metabolism of a cyclopropene fatty acid (sterculic acid) in the rat. In the latter case the cyclopropene fatty acid is reduced to a cyclopropane which is followed by co-oxidation of the terminal methyl group to an acid and when accompanied by both α- and β-oxidation is degraded to short chain dicarboxylic acids which are found in rat urine. Similar excretory products have been found in human urine (Lindstedt *et al.*, 1974) but no explanation was given for the metabolic origin. Our own observations show that the presence of cyclopropane fatty acids in human urine is widespread and although no firm evidence as to the food components responsible exists, it seems likely that sterculic acid, as a component of cottonseed oil and hence vegetable oil products, or lactobacillic acid as a component of yoghurt, are the likely origins.

The safety of the process worker and the human population surrounding the SCP factory has prompted discussions on the potential pathogenicity of the microorganism used in the SCP process. Pathogenicity is evaluated by the injection of the viable organism into the body cavity or body fluids of a mammalian species, and evaluating the mortality induced or the survivability and degree of invasion and damage caused by the organism within the host assuming no mortality. The situation is further complicated by inadequacies associated with organism nomenclature as a substantial number of the yeasts and bacteria used in single-cell protein processes have been discovered from a wide screening programme and are not easily related to existing type species. Generic and specific names have been assigned to the organisms by microbiologists with little discussion of the problem with medical microbiologists. To be specific, *Candida albicans* is taken as the type species for all yeasts belonging to the genus *Candida* and this stems from the time when medical microbiology dominated the science. *C. albicans* is a pathogenic organism, that is to say in an experimental situation it kills mice when injected into the body. However, mortality is not necessarily the result when other *Candida* species are similarly administered, some organisms may be parasitic to greater or lesser degrees and, indeed, there are species of this organism which fail even to meet the criterion of survivability within the host, never mind the intermediate parasitic or final mortality conditions. It is unfortunate for single-cell protein that many of the yeasts used appear to be classified as *Candida* which is automatically accompanied by the concept of pathogenicity. Non-pathogenicity has, therefore, to be proven against a background of uncertainty related to organism classification and also what pathogenicity really implies. Fortunately the situation is now largely resolved. The non-pathogenicity of a large number of potential SCP organisms has been established. Organisms can in effect be placed in groups; thus for *Torulopsis*, *Hansenula polymorpha* and *C. lipolytica* there is no mortality and viable cells are not detectable from injected mice and hence must be thought of as non-pathogenic, for *C. maltosa*, *C. utilis* and *Saccharomyces cerevisiae* (bakers' yeast) there is no mortality but viable cells may be recovered from the host animal for a short period and these may be regarded as relatively innocuous, whilst *C. albicans* and *C. tropicalis* induce renal damage and mortality (Holzschu *et al.*, 1979) and must be thought of as pathogenic. Using similar test methods to those employed above, that is to say with normal or cortisone compromised mice, the bacterial organism used for the production of PRUTEEN, *Methylophilus methylotrophus*, is also non-pathogenic, a feature that would be expected from a methanol obligate.

37.5 NUTRITIONAL TESTING OF SCP

For a product to gain acceptance as a feed, the animals fed the product must consume the material and grow in both a physical and economic sense. The first criterion is palatability and the second is satisfactory growth rate and feed conversion.

Animal nutrition has evolved from an art, that is to say individual knowledge and experience based on a feeding livestock, to a relatively exact science in which the major nutrients that contribute to the animal's well-being are expressed numerically. There is a fairly good understanding of the relationship between animal growth and feed conversion response to nutrient density (the amount of energy and of the various amino acids that are in the diet). Regressional analysis of these features is practised by the more advanced animal feed manufacturing companies applying economic parameters to the benefits occurring (live weight gain, *etc.*), against cost (gross ingredient cost and relative value of each nutrient within each raw material). Linear programming of the dietary ingredients to produce the optimal diet at least cost is the first stage in this process, which in turn implies a thorough understanding of the numerical values to be ascribed to the raw material and their variability.

Against this background it naturally follows that single-cell proteins have to be nutritionally defined. The major economic features are: gross energy; available energy, expressed as metabolizable energy, net energy, *etc.*, depending upon the class of livestock; crude protein content; amino acid content, particularly lysine, methionine, cystine, tryptophan and threonine; mineral content, particularly Ca, P, Na, K and Cl which are all major nutrients; and fatty acids, particularly octadecenoic acid (C18:1). Other nutrients which are of benefit, but which will not contribute to the cost of the diet, include vitamins and trace mineral ions. Deficiencies of both of these will adversely affect the growth rate, but inadequacies are easily corrected by addition of trace quantities to the diet and hence the nutritional characters of a single raw material should not be equated with being equivalent to a complete diet.

Experiments need to be carried out using the species that will consume the SCP in its diet. There is little merit from an applied nutritional sense in conducting assays for biological value and net protein utilization with the rat. The aforementioned tests may have value in assessing the nutritional effects of changing the process parameters used in the manufacture of the SCP, but have no commercial value in ascribing the economic benefits that are given by the nutrients within the protein.

37.5.1 Specific Nutritional Problems Relating to SCP

The high rate of cellular division which operates within the fermentation system leads to a relatively high concentration of nucleic acid in the final product. Questions are frequently posed about the nutritional value of this component. As far as the animal is concerned, nucleic acids are composed of purine and pyrimidine bases, ribose and deoxyribose and phosphorus. For the single-cell protein PRUTEEN, which contains 70% crude protein, including 12.5% nucleic acid, the contributions to the metabolic pool from nucleic acids are 2.04% N, 1.22% P and 5.04% ribose and deoxyribose. The fact that these nutrients are potentially available is shown by the high nitrogen digestibility values of 92% obtained in both pigs and poultry, and this will include the nucleic acid component. It is likely that the nitrogen, phosphorus and sugar components contribute positively to the nutrient pool. Ribose and deoxyribose are readily incorporated into the pentose and hexose phosphate cycles and, therefore, contribute to the available energy. The phosphorus within PRUTEEN is 100% available; within the 2.4% total content 50% is derived from nucleic acid indicating thereby the positive contribution from the nucleotide source. In the degradation of both pyrimidine and purine bases there is release of ammonium ions that are capable of contributing to the non-specific nitrogen pool and which may be used for the synthesis of non-essential amino acids in periods of nitrogen limitation. The overall contribution of nucleic acids to the nutrient pool is shown in Figure 1.

Mineral balance was shown to be a problem for evaluation of SCP in both nutritional and toxicological studies. In general potassium levels within SCPs are low (*ca.* 0.1%) in contrast to conventional feed ingredients (0.3–0.6% in cereals, 0.6–0.8% in fish meal and 2.0% in soya). Formulation of diets containing high SCP levels can lead to diets that are borderline relative to the animal's potassium requirement. Addition of potassium salts to the diet overcomes the SCP induced growth rate reduction and kidney hypertrophy (Gatumel and Shacklady, 1977) or mortality and visceral gout observed in chickens (Talbot, 1978). The latter study also indicated the importance of distinguishing between sodium and chloride ions, which in conventional applied nutrition are considered together as 'salt'.

Apart from the above specific examples indicating the peculiarities of SCP, these materials may be used as nutrient sources in the same manner as soyabean meal, fish meal or even skimmed milk powder. All that is required is to formulate diets which are balanced and do not suffer from any nutrient deficiencies. Under these circumstances single-cell proteins behave in the same manner as conventional feed ingredients and present no greater problems than products conventionally utilized by the animal feed industry.

37.6 SPECIFIC PROBLEMS RELATING TO THE HEALTH OF WORKERS INVOLVED WITH SCP MANUFACTURING

Single-cell proteins, when in a fine particulate form, cause adverse clinical reactions in humans exposed to a dust cloud. In this respect single-cell proteins are no different to other proteinaceous

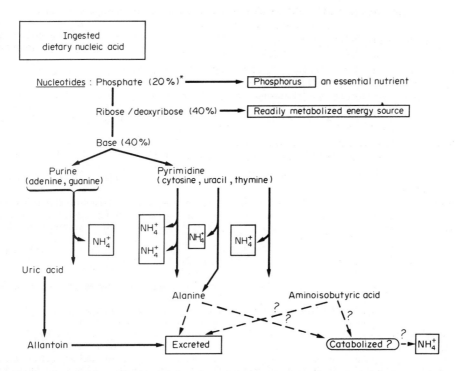

Figure 1 Nutrient contribution to body pools from ingested nucleic acid. NH_4^+ enters the active nitrogen pool available for synthesis of non-essential amino acids and other nitrogenous compounds, the excess is excreted. *Approximate proportion by weight of intact nucleic acids

materials. The clinical symptoms caused by an SCP derived from *Methylomonas methanolica* have been described by Ekenvall *et al.* (1983). The features described are likely to apply in the majority of situations to all SCP products. However, simple physical control of absolute dust levels within the factory and/or treatment of the material with an edible oil to completely dampen the dust yields a situation in which allergic reactions may be completely obviated (Mayes, 1982). It is clear that in this situation one group's problem has become another's opportunity, that being a fairly common situation!

37.7 SUMMARY

The acceptance of single-cell proteins for animal feeding depends upon exhaustive nutritional and toxicological examination. Those SCPs that are produced from substrates of natural origin are generally regarded as safe and need to be evaluated simply for their ability to support efficient animal growth. In contrast those derived from hydrocarbon feedstocks need to be evaluated for safety by a series of toxicological tests. The major areas of concern relate to organism pathogenicity, residual substrates, polycyclic hydrocarbons, unusual cellular metabolites and allergic reactions by humans. The evidence provided to support SCPs produced on an industrial scale indicates no major problems that would preclude their use. The experience generated during the last 15 years shows that these products have nutritional value and are capable of substituting other more conventional animal feed ingredients. The main feature inhibiting development of single-cell products is the capital cost of the equipment required for their production; in the case of those grown on hydrocarbon related substrates the escalating cost of raw materials needed for their growth and the relative value of conventional feed ingredients, which over the last decade have not inflated in price to the same extent as industrially related products, are important factors.

37.8 REFERENCES

Alimenti. R. and 37 co-authors (1979). Indagini nutrizionali sulla Toprina (lieviti coltivati su *n*-paraffine). *Ann. Inst. Super. Sanite*, **15**, 649–690.
Anon. (1974a). PAG Guideline No. 6 for preclinical testing of novel sources of protein. *PAG Bull.*, **4** (3), 17–31.

Anon. (1974b). PAG guideline No. 15 on nutritional and safety aspects of novel protein sources for animal feeding. *PAG Bull.*, **4** (3), 11–17.

Anon. (1974c). Proposed guidelines for testing of single cell protein destined as major source of animal feed. General standards of identity for SCP to be used in animal feed. Specific standards of identity of SCP prepared from hydrocarbons. *IUPAC Technical Report No 12.*

Anon. (1975). 5th meeting PAG *ad hoc* working group on clinical evaluation and acceptable nucleic acid level of single cell protein for human consumption. *PAG Bull.*, **5** (3), 17–26.

Anon. (1978). Proposed system for food safety assessment. *Food Cosmetic Toxicol.*, **16**, Suppl. 2, 1–136.

Anon. (1983a). PAG/UNU Guideline No. 6: preclinical testing of novel sources of food. *Food Nutr. Bull.*, **5** (1), 60–66.

Anon. (1983b). PAG/UNU Guideline No. 15: nutritional and safety aspects of protein sources for animal feeding. *Food Nutr. Bull.*, **5** (1), 67–70.

Belluci, P. (1980). *La Bioproteine*. Feltrinelli, Milano.

Bizzi, A., M. T. Tacconi, E. Veneroni, A. Jora, M. Salmona, G. de Gaetano, S. Paglialunga and S. Garattini (1979a). Biological significance of uneven fatty acids in various animal species fed diets containing single cell proteins. In *Single Cell Protein—Safety for Animal and Human Feeding*, ed. S. Garattini, S. Paglialunga and N. S. Scrimshaw, pp. 102–125. Pergamon, Oxford.

Bizzi, A., E. Veneroni, M. T. Tacconi, M. Cini, A. Guatani, I. Bartosek, R. Modica, N. Santoro, S. Paglialunga and S. Garattini (1979b). Biochemical and toxicological studies of *n*-hydrocarbons present in single cell proteins. In *Single Cell Protein—Safety for Animal and Human Feeding*, ed. S. Garattini, S. Paglialunga and N. S. Scrimshaw, pp. 47–58. Pergamon, Oxford.

Bizzi, A., E. Veneroni, M. T. Tacconi, A. M. Codegoni, R. Pagani, M. Cini and S. Garattini (1980). Accumulation and metabolism of uneven fatty acids present in single cell protein. *Toxicol. Lett.*, **5**, 227–240.

Champagnat, A. and J. Adrian (1974). *Petrole et Proteines*. Doin, Paris.

de Groot, A. P., H. P. Til and V. J. Feron (1970a). Safety evaluation of yeast grown on hydrocarbons. I. One-year feeding study in rats with yeast grown on gas-oil. *Food Cosmetic Toxicol.*, **8**, 267–276.

de Groot, A. P., H. P. Til and V. J. Feron (1970b). Safety evaluation of yeast grown on hydrocarbons. II. One-year feeding study in rats with yeast grown on pure *n*-paraffins. *Food Cosmetic Toxicol.*, **8**, 499–507.

de Groot, A. P., H. P. Til and V. J. Feron (1971). Safety evaluation of yeast grown on hydrocarbons. III. Two-year feeding and multigeneration study in rats with yeast grown on gas-oil. *Food, Cosmetic Toxicol.*, **9**, 787–800.

Di Muccio, A., L. Boniforti, A. Palomba, M. P. Bernardini and M. Delise (1979). Idrocarburi saturi negli alimenti. Metodo d'analisi e valori riscontrati in alcuni alimenti per uso umano e in campioni da organismi unicellulari. *Ann. Inst. Super. Sanite*, **15**, 525–540.

Eisele, T. A., J. K. Yoss, J. E. Nixon, N. E. Pawlowski, L. M. Libbey and R. O. Sinnhuber (1977). Rat urinary metabolites of [9,10-methylene-^{14}C]sterculic acid. *Biochim. Biophys. Acta*, **488**, 76–87.

Ekenvall, L., B. Bolling, C.-J. Gothe, L. Ebbinghaus, L.-V. von Stedingk and J. Wasserman (1983). Single cell protein as an occupational hazard. *Br. J. Ind. Med.*, **40**, 212–215.

Gatumel, E. and C. A. Shacklady (1977). Problemes lies a l'etude toxicologique des proteines d'organismes unicellulaires (S.C.P.). *Med. Nutr.*, **13**, 387–394.

Grimmer G. and G. Wihelm (1969). Der Gehalt Polycyclisher Kohlenwasserstoffe in europaischen Hefen. *Deutsche Lebensmittel-Rundschau*, **8**, 229–231.

Holzschu, D. L., F. W. Chandler, L. Ajello and D. G. Ahearn (1979). Studies on potential pathogenicity of industrial yeasts for normal and cortisone-treated mice. In *Single Cell Protein—Safety for Animal and Human Feeding*, ed. S. Garattini, S. Paglialunga and N. S. Scrimshaw, pp. 20–29. Pergamon, Oxford.

Hoogerheide, J. C., K. Yamada, J. D. Littlehiles and K. Ohno (1979). Guidelines for testing of single cell protein destined as protein source for animal feed II. *Pure Appl. Chem.*, **51**, 2537–2560.

Lester, D. E. (1979). Normal paraffins in living matter—occurrence metabolism and pathology. *Prog. Food Nutr. Sci.*, **3**, 1–66.

Lindstedt, S., G. Steen and E. Wahl (1974). 3,4-Methylenehexanedioic acid—a previously unknown compound in human urine. *Clin. Chim. Acta*, **53**, 143–144.

Mayes, R. W. (1982). Lack of allergic reactions in workers exposed to 'PRUTEEN' (bacterial single cell protein). *Br. J. Ind. Med.*, **39**, 183–186.

Newberne, P. M. and R. G. McConnell (1978). Special problems in assessing the toxicity of major dietary components. In *Chemical Toxicology of Food*, ed. C. L. Galli, R. Paoletti and G. Vettorazzi, pp. 185–232. Elsevier, Amsterdam.

Prosser, C. L. (1952). *Comparative Animal Physiology*. Saunders, Philadelphia.

Shacklady, C. A. (1979). *n*-Paraffins in tissues of animals fed on alkane-grown yeasts. In *Single Cell Protein—Safety for Animal and Human Feeding*, ed. S. Garattini, S. Paglialunga and N. S. Scrimshaw, pp. 126–132. Pergamon, Oxford.

Talbot, C. J. (1978), Sodium, potassium and chloride imbalance in broiler chicks. *Proc. Nutr. Soc.*, **37**, 53A.

Tomassi, G. and G. Serlupi-Crescenzi (1979). Effect of feeding odd-chain fatty acids on litter size and tissue composition of rats. In *Single Cell Protein—Safety for Animal and Human Feeding*, ed. S. Garattini, S. Paglialunga and N. S. Scrimshaw, pp. 59–80. Pergamon, Oxford.

Tulliez, J. E. and G. F. Bories (1979). Metabolic pathway of *n*-paraffins in mammals. In *Single Cell Protein—Safety for Animal and Human Feeding*, ed. S. Garattini, S. Paglialunga and N. S. Scrimshaw, pp. 160–166. Pergamon, Oxford.

Valfre, F., G. Bosi, P. Bellezza, O. Olivieri and S. Moca (1979). Effect of feeding *n*-paraffins on animal tissue levels. In *Single Cell Protein—Safety for Animal and Human Feeding*, ed. S. Garattini, S. Paglialunga and N. S. Scrimshaw, pp. 133–147. Pergamon, Oxford.

38

Government Programs for Biotechnology Development and Applications

L. A. SLOTIN
Ministry of State for Science and Technology, Ottawa, Ontario, Canada

38.1   INTRODUCTION

At the beginning of the 1970s, there was an underlying assumption throughout the developed world that the rate of economic growth experienced during the 1960s would simply carry forward. Although it was also assumed that science and technology would be important driving forces in this growth, attention focussed primarily on the fact that new technologies are not socially and environmentally benign. Thus the full implication of developments in science and technology, not only their economic ones, needed to be examined before adoption.

The assumptions about continued economic growth turned out to be too optimistic. Rapidly escalating energy and petroleum feedstock costs, inflation, instabilities in international exchange and shortages of skilled people led to declining productivity and reduced efficiencies. As a result, views on the role of science and technology had to be adjusted to reflect this more modest economic scenario. In this context, social and environmental considerations remained important, but the trade-offs between economic and non-economic factors shifted in favor of the economic. The interest of governments towards science and technology began to focus more sharply on economic potential than on any other objectives.

This shift in emphasis was common to all industrialized countries. Science and technology came to be viewed as an important means through which a country's competitive position could be improved and higher rates of growth achieved. Some countries (*e.g.* Germany, Sweden, Japan and France) undertook massive efforts to capture the economic benefits of science and technology. Many countries began to analyze technologies in a systematic way to assess their importance to economic development. The integration of science policy with economic policy is now an ongoing activity in almost all developed countries.

It is against this backdrop that the area of technology embodied by the term *biotechnology* finds itself. Many people would argue that biotechnology is merely an up-dated term for an area of technology which has been exploited for centuries, namely fermentation. There is, of course, no validity to such claims. Biotechnology today is characterized by a highly sophisticated, research-intensive exploitation of biological processes in ways which were largely unknown a decade ago. Techniques of recombinant DNA, fused cells, plant-cell culture, enzymes and enzyme systems, coupled with the developments in systems and process engineering are the elements which have attracted, globally, the excitement of scientists, industrialists and government planners to the field of biotechnology. The massive financial investments being made by industries and governments towards building up the manpower and technical expertise to be able to respond to the market opportunities as they arise are clear testimonials to the recognition of the newness and vast potential of biotechnology.

Changes to existing manufacturing processes together with the development of new products and processes in areas as diverse as energy, mining, agriculture, health care, forestry, food, chemicals and waste management represent the opportunities presented by biotechnology. Many nations have begun to assess these opportunities and set in place mechanisms to ensure the realization of the potentials. This article describes some of the approaches and initiatives being taken by the governments of several countries as their response to the economic prospects of biotechnology.

38.2 OVERVIEW

Biotechnology, from a world-wide government perspective, is in varying stages of evolution. As a result, when one tries to examine the approaches being taken by different governments towards the development of biotechnology, one is immediately confronted by a seemingly endless spectrum of activities. This spectrum ranges from specifically-designed studies to large infusions of financial resources, from new program initiatives to the development of policy frameworks. Each activity is important and is a reflection of the approach to strategic technologies taken by nation governments with regard to their own indigenous needs. Comparisons, except extremely superficial ones, are not instructive at this stage and therefore will not be attempted.

The speed with which biotechnology developments are occuring within various governments makes an article of this type largely out-of-date before it is published. This difficulty cannot be avoided entirely, but an attempt will be made to identify future directions, wherever possible.

Finally, the countries examined in this article are not the only ones in which government activity in biotechnology is apparent. Rather, they are considered by the author to be a reasonable sampling of the different ways in which governments are approaching biotechnology. Also, the activities of individual countries which are described herein do not represent an exhaustive list. Rather they are meant as a sampling of initiatives which will provide the reader with an introduction to the government approaches being pursued in individual nations.

38.3 AUSTRALIA

38.3.1 Commonwealth Scientific and Industrial Research Organization

Australia's principal government research establishment is the Commonwealth Scientific and Industrial Research Organization (CSIRO). It is within the CSIRO that the development of biotechnology initiatives by the Australian government began.

In 1978 the Executive of the CSIRO appointed a Review Committee to carry out an assessment of the possible applications of recombinant DNA and associated techniques in CSIRO programs, especially with respect to the strategic research obligations of the Organization. The report of the Committee, *Recombinant DNA Techniques in Research* (Anon., 1979c), revealed that major opportunities for Australia will be found in the application of biotechnology in plant breeding, principally because technology is already available for the culturing of plant cells and their subsequent induction to form propagating materials.

When it came to specific program proposals, however, the Committee was faced with a serious problem: the lack of a significant microbiological research capability within the CSIRO to capitalize upon the application of the new techniques both within and outside the plant area. Rather than propose specific initiatives, therefore, the Committee recommended that another review relating to industrial microbiology be carried out by the CSIRO. The Executive agreed, but broadened the terms of reference of the new review to cover what is now commonly held to be biotechnology. The review was commissioned in 1980 and the biotechnology report (Anon., 1981d) entitled *Biotechnology Research and Development* was published in November 1981. The report also contained an up-dated version of the previous recombinant DNA report.

As a result of the recommendations contained within the report *Biotechnology Research and Development*, the CSIRO has adopted biotechnology as a priority area and will pursue biotechnology within and across a number of its units. Although the financial details are not yet available, the CSIRO spent approximately $4.7 million (US) in 1981 on biotechnology involving 110 people spread across 10 Divisions and Units within the CSIRO Institutes of Animal and Food Sciences, Biological Resources, and Energy and Earth Resources. Moreover, in 1982 an Industrial Microbiology Unit was established in Melbourne to undertake research in microbial genetics, gene technology and industrial fermentations associated with the utilization of renewable materials and the treatment of wastes.

The establishment of an Industrial Microbiology Unit can be regarded as a positive step, albeit within only one government department (CSIRO). However, other recommendations contained in the report concerning manpower, linkages with universities and industry, and international collaborations, have not yet been acted upon. Moreover, recommendations concerning market assessment and intersectorial linkages, post-graduate training requirements and the establishment of pilot-scale facilities, which fall within the jurisdictions of other government departments, *i.e.* Ministry of Industry, Ministry of Education, Ministry of Science and Technology (MOST), must await decisions by these institutions.

38.3.2 Genetic Engineering Symposium

In early 1981, the Prime Minister requested the Minister of Science and Technology to hold a symposium on the likely industrial applications of recombinant DNA research and the implications for primary and secondary industry. The symposium, *Genetic Engineering: Commercial Opportunities in Australia*, was held on November 18–20, 1981 and jointly sponsored by the Department of Science and Technology, the Bank of New South Wales and the Australian Institute of Management. By bringing together industrial, research, marketing, finance, legal and government sectors, the symposium organizers sought to alert these sectors to the opportunities presented by genetic engineering, to examine the strengths and weaknesses in Australian genetic engineering activities and to present an overview of international developments in the area. The symposium was also designed to provide a basis for advice to the Government on whether or not there is a need for Government assistance in the commercial development of biotechnology and applied genetics in Australia (Pearce and Preston, 1981).

It is too early to tell what effect the symposium might have in terms of government action. Nevertheless, it did represent a laudable attempt to sensitize all sectors to a rapidly evolving field.

38.3.3 General Incentives

There are a number of government incentives which, although not specifically targetted for biotechnology, will undoubtedly play a role in the promotion of this area. The first is the Industrial Research and Development Incentives Scheme. Under the Scheme, which is the principal government program offering support to companies for industrial research and development (R and D), the maximum eligible commencement grant, aimed to develop R and D experience or expertise, is $42 000 (US). Project grants, on the other hand, are considered to be the main focus of the scheme. The maximum project grant is 50% of project costs, with an annual limit of $786 000 (US). Assistance is also provided, under the Scheme, to R and D projects being undertaken by universities or non-profit organizations if the work cannot be undertaken in industry.

In addition to these programs, there are a number of other measures which have been undertaken or are under consideration to improve the interaction between research and industry. The Government has, for instance, stated its intention to increase the contracting to industry of applied research as a mechanism for enhancing industrial R and D. Furthermore, as part of the Government's role in enhancing the flow of information to industry and between sectors, seminars and workshops such as the Technology Assessment Workshops held in 1978 and 1979 and the forums on creating high technology enterprises held in 1981 were important initiatives.

Finally, of great importance in the general infrastructure support areas, are the funding programs of the Tertiary Education Commission, a good deal of which support research activities. A particularly relevant and recent initiative in the education area is the center of excellence program. This program is designed to support the creation of particular centers throughout the country which are recognized as unique in terms of their scientific and educational abilities (Anon., 1981m).

38.3.4 Conclusion

There can be little doubt that biotechnology has been recognized within Australia as an area of technology strategic to national economic development. At least one agency, the CSIRO, has taken definitive action with the establishment of an Industrial Microbiology Unit and the overall treatment of biotechnology as a priority within its various activities. However, no Government policy on biotechnology is yet forthcoming and, therefore, a comprehensive package of measures to stimulate its development are not yet evident. One could best describe the situation as one in which Government is attempting to create an awareness among industries, universities, research organizations, state governments and various Commonwealth Government departments, of the potentials of biotechnology, as a prelude to the possible preparation of a national policy.

38.4 BRAZIL

In 1975, the Brazilian government, in an attempt to reduce its almost total dependence on imported oil, launched its PROALCOOL program. With the goal to reduce gasoline consumption by 45%, over $5 billion (US) are being invested by the Government to attain production figures of 10.5 billion l of ethanol per annum by 1985 (Bazin, 1979). Coupled to the commitment to alcohol production is the production of ethanol-burning vehicles. In mid-1981 over 400 000 ethanol-utilizing vehicles were in service in Brazil.

Cane sugar is the substrate of choice for the moment, but there are indications that manioc (cassava plants) will prove more advantageous. For example, PROALCOOL's biggest single project under construction is the 2 million l per day Jaiba distillery, owned by the Ometo Group, the world's largest producer of sugar and alcohol. However another firm, Sinop Agropecuario, will use manioc to produce 1 million l per day.

Some economic experts are predicting that the PROALCOOL program will lose impetus if world sugar prices start rising again. It is argued that to make the technological option irreversible and independent of the fluctuations of the international commodity market, new distilleries should be built which operate directly from raw cane, by-passing sugar production and storage. The Institute of Sugar and Alcohol, which administers the PROALCOOL program, approved the construction of 87 such new autonomous distilleries. But it also approved the construction of 139 standard distilleries which will double the number of distilleries operating from sugar as an intermediate. This decision leaves some doubts as to the earnestness of the Government's commitment to a fully independent alcohol program (Bazin, 1979).

There are other concerns on the horizon. In 1980, 400 million l of alcohol were exported. Whether this was a one-time aberration due to over-supply or a trend reflecting the instabilities in the price of oil and alcohol remain to be seen (Anon., 1981e). Other criticisms point to the destruction of unique ecosystems, the exploitation of working people, the misuse of the land and the encouragement of massive spraying programs with potentially harmful pesticides (Anon., 1981f).

Nevertheless Brazil is proceeding with its program. Although the technology for alcohol production is antiquated relative to the developments of modern-day biotechnology, there are indications that the Government is also financing research into bioreactor design, genetic engineering and high-density plantation development.

38.4.1 Conclusion

Driven by the absolute necessity to reduce its dependence on imported petroleum, Brazil has embarked upon an ambitious national program in alcohol production. There can be little doubt that considerable environmental, social and economic upheavals will accompany this initiative. The key to Brazil's success will be their ability to recognize and be sensitive to all of the changes which will occur.

38.5 THE UNITED KINGDOM

38.5.1 Joint Working Party Report

In early 1979, the Advisory Council for Applied Research and Development (ACARD), together with the Royal Society and the Advisory Board for the Research Councils (ABRC), set up a Joint Working Party to study biotechnology. The Working Party was instructed to review the prospects offered by biotechnology and to recommend action by Government or other bodies to facilitate British industrial development in biotechnology.

The Working Party Report (Anon., 1980c) stressed that 'the present structure of public and private support for R and D is not well suited to the development of a subject like biotechnology which, at the moment, straddles the divisions of responsibility both among government departments and among research councils and the arbitrarily defined fields of applied and fundamental research'.

The Working Party recommended the establishment of a joint committee for biotechnology to coordinate the biotechnological work of the five research councils, and that a parallel interdepartmental steering committee be set up for government departments. The Working Party also recommended that the research councils should double their commitments to biotechnology to a level of at least $5.3 million (US) per year and that government departments should commit $4.4 million (US).

With respect to industrial development, one specific recommendation was that the National Enterprise Board and the National Research Development Corporation jointly set up a biotechnology company with an allocation of $3.5 million (US) per year for five years. Finally, the expansion of university centers of excellence in biotechnology was recommended, with the creation of a minimum of 20 new teaching and research posts over a five year period accompanied by capital investment of approximately $3.5 million (US).

It fell to the Secretary of State for Industry (SSI) to respond to the Working Party report and the answer was forthcoming in March 1981. In his response, the Secretary argued that while biotechnology will undoubtedly play a large role in future industrial development, it is held that 'the private sector route offers the best prospect of building a vigorous efficient industry in those areas of the technology most appropriate to the United Kingdom' (Anon., 1981c). In the response, the need for special government intervention is down-played, particularly with respect to the Working Party's recommendations for increased resources. It was felt that the myriad of existing government research and commercial support measures would be sufficient to provide the appropriate climate for the continued growth of a biotechnology-based industry. Specific thrusts with respect to training were also felt to be unnecessary, since clear signals from the employment market were not yet forthcoming.

While considerable derision in the British scientific community greeted the Government's seemingly obvious non-interventionist stance with respect to biotechnology, there was at least one hopeful sign. In the response it was stated that the Government recognized the need to coordinate the activities of Government Departments in this area. As a result 'the Government is establishing machinery to do this so as to ensure the coherence of Government policies, and to review national and international developments. Close contacts will be established in this area with the Research Councils, the National Research Development Corporation (NRDC), the National Economic Development Council (NEDC) and the National Enterprise Board (NEB)' (Anon., 1981c).

Although one might conclude upon reading the above that the British Government was content to allow biotechnology to evolve naturally in Britain, it is certainly clear that subsequent events described below lead to a different conclusion.

38.5.2 British Technology Group

The British Technology Group (BTG), the marriage between the NEB and the NRDC, is developing a strategy for the long-term support of biotechnology. Provided the private sector is willing to match the money, BTG expects to invest about $8.8 million (US) in the next five years, in addition to a current (1982) commitment of $23 million (US) in over 40 British projects (Fishlock, 1982).

The 40 projects presently in the portfolio span the entire range of commercial activities, from the funding of academic research and help to set up production lines, to strategic investments with such partners as Rank Hovis McDougall and the Prudential Assurance Company. The involvement with Celltech and the Agricultural Research Council and Speywood Laboratories Ltd. are discussed below as examples of the type of ventures in which BTG is presently involved.

38.5.3 Celltech

Set up following the specific recommendation by the Joint Working Party Report (Anon., 1980c), Celltech is a biotechnology company intent upon becoming a knowledge-intensive business supplying expertise to other firms or establishing small-scale, sophisticated manufacturing plants (Fishlock, 1981). Celltech's backing capital [initially $21 million (US)] is made up of 14% from the Prudential Assurance Company, 14% from the Midland Bank, 14% from the British and Commonwealth Shipping Company and 14% from Technical Development Capital Ltd. The remaining 44% is put forward by the Government through the National Enterprise Board (now part of the British Technology Group). To further demonstrate the role of Government in Celltech, it is important to note that the company has first right of refusal to all developments arising out of research supported by the Medical Research Council (MRC). Celltech can also act as a broker for the licensing of MRC developments not of direct interest to itself. In this way the linkage between British medical researchers and the commercial world is expected to grow.

Celltech will also have an influence over universities and polytechnic departments. The company intends to establish working relationships with these sectors and will, therefore, be a significant influence over the future directions of some segments of British research.

The corporate plan of Celltech calls for a total staff of about 150 by mid-1983, with approximately 40 engaged in recombinant DNA research and 50–60 on cell hybridization.

38.5.4 Agricultural Research Council

A biotechnology company which will tap the resources of the Agricultural Research Council (ARC) is being established by the British Technology Group (formerly the National Enterprise Board and the National Research Development Corporation) at a cost of $26–35 million (US). The company will exploit the genetic engineering and related experience of the ARC's various institutes and will market products arising from plant and animal research. This venture will be modelled on similar lines to Celltech, with the British Technology Group prepared to pay up to one-half the cost.

The new company will receive all the patents covering plant and animal genetics formerly held by the National Research Development Corporation. The company will attempt to fill the predevelopment gap between research discoveries and their commercial exploitation.

It is anticipated that close relationships will be fostered between the new company and the ARC institutes. The institutes will provide the basic discoveries while the company will support post-doctoral researchers at the institutes.

In what must be viewed as a related development, the ARC will establish a new research centre in mid-1982. This research centre will house 12 researchers and provide a focal point for ARC work on monoclonal antibodies. Costing approximately $0.17 million (US) in capital and $0.1 million (US) per year in operating funds, the centre will pursue animal vaccine development and will have production facilities large enough to conduct commercial trials. The strength of the relationship between this research centre and the company mentioned above will be a major determinant in any future progress in Britain's attempts to capitalize upon the agricultural applications of biotechnology.

38.5.5 Speywood Laboratories Ltd.

In a third initiative, the Government, through the British Technology Group (BTG), is investing $3.4 million (US) in Speywood Laboratories Ltd. The company's chief concern is the development of new techniques for blood protein manufacture. Prutec Limited, a subsidiary of the Prudential Assurance Company, will be matching the BTG's investment.

Since BTG acts as a 'broker' between industries and universities, the investment in Speywood is important insofar as it gives the company access to expertise on recombinant DNA technologies in universities.

38.5.6 Science and Engineering Research Council

In what must be interpreted as a significant sign of direct Government involvement in the development of biotechnology, the British Science and Engineering Research Council (SERC) in November 1981 launched a new directorate to foster collaboration in biotechnology between universities and industry and to entice British biotechnologists to remain in Britain.

The biotechnology directorate reports to both the science and engineering boards of the Council, reflecting the broad spectrum of research that biotechnology encompasses. While the budget of the SERC was approximately $350 million (US) in 1981, the biotechnology directorate's share was only $1.7 million (US), but it is expected to rise to $4.2 million (US) by 1984–85.

Most of the resources will be spent on fostering collaboration through schemes already used by the SERC to encourage industrial involvement in university research. These include the teaching company scheme and Cooperative Awards in Science and Engineering (CASE), both of which support post-graduate students on research projects relevant to collaborating companies, and the cooperative grant scheme whereby SERC and collaborating companies share in the cost of research projects in university laboratories.

The SERC directorate is stressing collaboration on fermentation, enzyme and immobilized cell technology, separation and concentration technology, product processing and recombinant DNA research. The directorate works closely with the Department of Industry, which may take over the support of projects approaching the development stage, as well as closely with the Agricultural Research Council and the Medical Research Council.

38.5.7   Conclusion

While the British Government continues to subscribe to the theory that the growth of biotechnology in Britain should depend on market forces rather than government intervention, it is clear that through a number of government departments and agencies biotechnology is being vigorously promoted. The government intervention, however, is directly aligned with the commercial world. The sentiment seems to be that through this approach a pull will be generated upon nonspecific support programs, especially training and research, thus reorienting resources through reactive as opposed to interventionist mechanisms. The question one raises is whether or not the market approach can provide enough of an early warning to ensure that reallocated Government resources will be utilized effectively. Another question which is also unanswered is whether or not the fiscal and regulatory climate in Britain is commensurate with the demands of biotechnology for strong industrial involvement.

38.6   CANADA

From a government-wide perspective, the first overall examination of biotechnology in Canada was initiated by the Ministry of State for Science and Technology (MOSST). In June 1980, MOSST produced a Background Paper entitled *Biotechnology in Canada* (Anon., 1980b). The purpose of this paper was two-fold: first to describe briefly the nature of biotechnology and some of the opportunities it presents, and second to identify existing Canadian biotechnological activities in governments, industries and universities.

Following the release of the Background Paper, the Minister of Science and Technology established a Task Force of experts to advise him on the approach which Canada should pursue in developing and capitalizing upon biotechnology. The Task Force, made up of representatives from industry and universities, submitted a report entitled *Biotechnology: a Development Plan for Canada* (Anon., 1981b).

38.6.1   Report of the Task Force on Biotechnology

In its report, the Task Force on Biotechnology stated that 'a practically non-existent biotechnological industrial base, a rapidly shrinking Federal Government research capability and a highly fragmented and unfocussed university effort are the major features of Canada's current biotechnological activities' (Anon., 1981b). To rectify this situation, the Task Force recommended that a 10 year National Biotechnology Development Plan be implemented. The objective of this Plan was to create in Canada the climate which would encourage the establishment and growth of a variety of industries based upon biotechnology. To achieve this objective, the Task Force felt that it was essential to create a strong, viable research and development infrastructure with which to support an emerging biotechnological industry.

The Plan proposed by the Task Force had eight elements. These elements were: (1) commitment; (2) industrial stimulation; (3) interdisciplinary science base; (4) manpower; (5) directions; (6) regulation; (7) international collaboration; and (8) organization.

To support these elements, the Task Force recommended that a minimum average annual direct expenditure of $41 million (US) was necessary from the Federal Government. The MOSST, as an advisory body to the Government, placed these recommendations before the Government.

In May, 1983, the Government announced that $61 million (Can.) would be spent to establish a Biotechnology Research Institute in Montreal, to be operated by the National Research Council (NRC); also, $6 million (Can.) would be provided to expand the NRC's laboratory in Saskatoon, which would be renamed The Institute for Plant Biotechnology. In addition to these expenditures, a further $22 million (Can.) was allocated for other Canadian biotechnology acitivities.

38.6.2 Natural Sciences and Engineering Research Council (NSERC)

The NSERC is the largest single agency in the Federal Government which finances research in Canadian universities. Through an array of research grants, scholarships, equipment support and technology transfer programs the Council awarded over $184 million (US) to universities in 1982–83. In 1981–82 $5.2 million (US) involving 226 separate awards were provided for biotechnology-related research. Beginning in 1982–83 the Council initiated a Strategic Grants Program in Biotechnology. This program, which could result in expenditures as high as $3.3 million (US) in that year, will encourage interdisciplinary, group research and training in various aspects of biotechnology.

Another program of the NSERC which could have implications for biotechnology is the Project Research Applicable in Industry (PRAI) program. Under this program, grants are provided for further development of university research to the point where it can be transferred to industry. The total budget for the PRAI program in 1980–81 was $1.3 million (US).

38.6.3 National Research Council (NRC)

The NRC is the principal government research establishment, and among other areas of interest is also involved in biotechnological research and development. In 1980–81 some 227 people and nearly $7.4 million (US) were being utilized in areas such as fermentation technology, microbial cell processes, molecular genetics, plant cell culture, enzymology and cell proliferation. One program of the NRC which has been designed to support the development of industries based upon the research accomplishments of the NRC is known as the Program for Industry/Laboratory Projects (PILP). Another program, the Industrial Research Assistance Program (IRAP) provides salary support for research performed in Canadian industries. It is anticipated that over the next five years demands for biotechnology projects in industry from both the PILP and IRAP programs will amount to $5.7 million (US). Recent additions to NRC's biotechnology development responsibilities are discussed in Section 38.6.1.

38.6.4 Agriculture Canada

Agriculture Canada's activities in biotechnology, although more narrowly focussed than those of the NRC, nevertheless contribute significantly to Canada's biotechnological efforts. In 1981–82 over $6.7 million (US) were spent in areas such as plant strain improvement, nitrogen fixation, cellulose and waste utilization, integrated pest management and food processing technology.

38.6.5 Provincial Government Activities

Several of the Canadian provinces have recently taken steps to build a biotechnological research and development capability. In Alberta, the Alberta Research Council established a Frontier Sciences Division in 1979 with an initial thrust into biotechnology. The Division is expected to grow from two to ten scientists by 1984; its orientation will be towards the agricultural and energy resource sectors of the province (Anon., 1979b). In a related development, the Alberta Research Council, in late 1981, signed a $6.6 million (US), four year, joint agreement with a Canadian biotechnology firm (BIOLOGICALS) to develop new processes based on the design of enzymes for industrial applications.

In February 1981, the Province of Ontario, together with two Canadian corporations, John Labatt Ltd. and the Canada Development Corporation, announced their intention to establish a biotechnology company. The venture, known as Allelix Inc., will receive $86 million (US) over a 10 year period from its backers. The areas of interest to the firm, which became fully operational in 1983, are those applications of biotechnology related to the natural resource sectors. Up to 100 professionals and technicians will be employed and the company will pursue contractual arrangements with both academic institutions and private sector companies.

In late 1981 the Government of Quebec issued its Programme d'Interventions pour le Développement des Biotechnologies au Québec. In forecasting the expenditure of nearly $50 million (US) over five years, the Quebec Government will pursue such elements as the identification of centres of expertise, fellowships in both industry and universities as well as the enhancement of communication amongst all sectors. Implementation of this program is expected in 1982 (Anon., 1981k).

38.6.6 Conclusion

Although no definitive Federal response has yet been forthcoming to the report of the Task Force on Biotechnology, nevertheless the report has had a marked stimulative effect upon all sectors. Unless some Federal leadership is generated however, it is feared that the various activities in Provincial Governments will further balkanize an already fragmented effort.

The Task Force report was the only report which this author has seen to recognize that such elements as fiscal policy, procurement policy and regulation are an integral part of developing a comprehensive national approach towards biotechnology within the broader framework of a country's industrial development strategy. Whether or not the Canadian Government is prepared to pursue this approach remains to be seen.

38.7 EUROPEAN ECONOMIC COMMUNITY

38.7.1 Biomolecular Engineering Program

Biotechnology, or biomolecular engineering as it is referred to within the European Economic Community (EEC), is regarded as a key sector in the Community's effort to develop a coherent industrial strategy. As such it has become part of the overall Community policy activity on innovation.

On April 1, 1982, a biomolecular research and training program for the EEC was launched. Although a 25 000 ECU (European Currency Unit) package covering four years was originally proposed, the Community decided that 8000 ECU will be provided for a first phase (April 1982–March 1984) with the amount for the second phase to be reconsidered after two years (Anon., 1981i). The program stems from Europe's increasing concern over: (i) food supply, industrial crops, land use and the Common Agricultural Policy; (ii) the future basis of the European chemical industry; and (iii) Europe and the Third World: trade and technology transfer. The specific elements of the program will be: (a) the development of new reactors using immobilized multienzyme systems, including those requiring multiphase environment and cofactor regeneration; (b) the development of bioreactors for the production of human detoxification substances; (c) the transfer of genes from diverse sources to the bacterium *Escherichia coli*, the yeast *Saccharomyces cerevisiae* and other suitable organisms; (d) the development of cloning systems; and (e) the improvement of methods for detecting contamination and for the assessment of possible risks associated with applications in agriculture and industry of biomolecular engineering.

The program will provide up to 50% of the cost of research, most of which will be performed in national laboratories. Particular stress is to be placed upon training and promoting the mobility of personnel between the collaborating organizations. Training contracts last one to two years and both 'employed scientists' (including those employed by industry) and 'unemployed scientists' (who recently completed a doctoral thesis or post-doctoral assignment) are eligible (Wenz, 1982).

38.7.2 Biosociety Program

The research program of the EEC which undertakes forecasting and assessment in the field of science and technology (FAST), was established in 1978. The main aim of the research program is to contribute to the definition of long-term community research and development objectives and priorities, and thus to the development of a coherent science and technology policy in the long-term.

Coordination of the European effort in biotechnology is carried out primarily under the auspices of the FAST program and its subprogram on Biosociety. Studies are the principal activities of the subprogram with (i) the systematic gathering of information describing current European biotechnology capability, (ii) the identification of opportunities for the development of biotechnology within the Community, and (iii) the identification of significant research and development activities, as present examples. Through a network including study contractors and their associates, interdirectorate contacts throughout the European Community, the European Federation of Biotechnology, *etc.*, the Biosociety program represents an important communicative linkage amongst member countries (Anon., 1979a).

In a related exercise, FAST is presently undertaking an analysis and comparison of national perceptions and responses to biotechnology. Based on key national reports and proceedings of international conferences (1974–1982), the purpose of this examination is to relate this comparative assessment to strategic needs and operational policies at the Community level (Cantley and Sargeant, 1981).

38.7.3 Conclusion

The multi-year funding program for biotechnology within the EEC reflects that Community's concern over its future ability to compete with other nations in this area. The program is a noble attempt to strengthen the Community's research capability in biotechnology, but is not without problems. For example, while the need to develop an overall complementary plan is recognized, some member countries worry that through the program which has been articulated, they will have to make available information which is potentially proprietary. As difficult as it is for a specific country to develop a rational, concerted, integrated approach towards biotechnology, the difficulties are compounded many-fold when several countries, such as those in the EEC, with its varying cultural, economic, scientific, industrial and social approaches, attempt similar action.

As more and more nations turn their national (government) laboratories towards the service of industry, it will become increasingly difficult to pursue the collaborative approaches envisaged by the EEC. Perhaps the only hope for the success of such a program is that within the next few years the directions and true potentials of biotechnology will become much clearer, enabling countries to better separate proprietary from non-proprietary areas of research.

38.8 FRANCE

38.8.1 Advice to the Government

In 1979, the role of the biological sciences in France's future industrial development began to be recognized. This recognition was initiated by the report *Sciences de la Vie et Société*, originally commissioned by the Prime Minister (Gros *et al.*, 1979). Although the main body of the report dealt with the state of the art across all of the life sciences, an annex, *Biotechnologies et Bio-Industrie*, published separately, concentrated upon all aspects of biotechnology and its implications for French society, industry and the economy (de Rosnay, 1979). Both documents, in describing the strengths and weaknesses of French biotechnology, failed to come up with specific recommendations which could be translated into Government initiatives. This task was left to the Délégation Générale à la Recherche Scientifique et Technique (DGRST).

In December 1980, *La Biotechnologie Demain* (Pelissolo, 1980) was submitted to the French Prime Minister. Presented by the DGRST, this report laid the groundwork for the initiation of a biotechnology program. The report stressed the need for a central coordinating body to ensure that strong, communicative linkages are developed within government as well as between government, universities and the private sector. The French strategy, the report recommended, should be one which builds upon existing excellence, as opposed to establishing new institutions. In this connection the report cites L'Institut Pasteur, Institut National de la Recherche Agronomique (INRA), Centre National de la Recherche Scientifique (CNRS), Institut National de la Santé et de la Recherche Médicale (INSERM), le Commissariat a l'Energie Atomique, l'Institut Français du Petrole and several universities as institutions upon which French biotechnology should be built. In this regard, in February 1980 Le Groupement Génie Génétique, G-3, a public sector organization, was established to coordinate and support biotechnological developments. G-3 is financed and run chiefly by the Institut Pasteur and CNRS, with minor contributions from INSERM and INRA. It has a laboratory and a scientific staff of about eight at the Institut Pasteur and spent about $0.75 million (US) in 1980–81, mainly on the development of a vaccine against hepatitis B. G-3 also does contract research for industry.

The development of the appropriate human resources was felt to be more important than adequate financial resources. Recommendations to strengthen existing curricula, to permit the interchange of researchers both nationally and internationally, and to promote more workshops, seminars and training programs, reflect this conclusion. The need to have access to developments and expertise world-wide gave rise to the emphasis placed upon pursuing international collaborations.

The need for industrial awareness of biotechnology and the opportunity to capitalize upon existing expertise in France was also noted in the report, with the recommendation to establish specific centers which could effect technology transfer. Although not the result of this particular recommendation, the creation of a commercial operation called Transgène is consistent with this thought. Transgène was set up in 1980 by a consortium of companies which includes Elf Acquitaine, Air Liquide, Moet-Hennessey, Roussel-Uclaf, BSW-Gervais-Danone and Rhône Poulenc. Organized by the Compagnie Financière de Paris et des Pays Bas (Paribas), Transgène received initial capitalization of $70 million (US). Initial plans called for an establishment of 50–80 people with a close working relationship with CNRS, INRA, INSERM and the Institut Pasteur. Transgène will also have close ties with the university community, principally *via* contract research. Thus Transgène could play an important role for France in the transfer and development of technology based upon biological processes. Transgène is to be followed by Immunotech (to make diagnostic kits) and Hybridolab (which will make vaccines).

It is interesting to note that Elf Acquitaine, one of Transgène's shareholders, is presently building a $17.5 million (US) research centre for biotechnology. Elf, which is one of Europe's largest oil companies, is over 60% owned by the Government. Over 400 researchers will be involved in applications of genetic engineering and enzyme technology to agriculture and chemicals manufacture. Elf also runs a program with the University of Compiègne to develop enzyme engineering.

In another indirect example of French Government involvement in biotechnology, the pharmaceutical company Sanofi, owned by Elf, has the first rights to any developments at the Institut Pasteur that could be turned into saleable health products.

38.8.2 Government Response

The response of the French Government was rapid and forceful. The first development was the announcement that biotechnology would become one of five priority technical areas for France's future industrial development. A joint Government–industry plan totalling some $28 billion (US) over five years was established to promote these technical areas.

The second development was the establishment of a coordinating group for biotechnology (Mission des Biotechnologies) composed of representatives from government, industries and universities. This coordinating group has the responsibility for all government resources to be allocated to biotechnology, as well as the mandate to develop the appropriate mechanisms or programs. The coordinating group also has an advisory function to the Government, so that the Government is kept informed of the various issues which may arise throughout the evolution of biotechnology towards the commercial stage.

Because of the coordinating group's mandate and prominence, a strong iterative relationship has emerged amongst 19 government departments and agencies. This relationship has permitted the group to develop four central themes for the allocation of resources. These themes are: (i) training and information, (ii) basic research; (iii) transfer centers and logistics; and (iv) industrial research contracts. In addition to resources already allocated for biotechnology through different government programs, new resources of $10.7 million (US) are at the disposal of the coordinating group for 1982.

To date a system of 60–80 training and retraining bursaries has been put in place, together with generous travel support. Particular emphasis is being placed upon the movement of researchers, even for short terms, into industry. Also being pursued is the continuous delivery of information to the industrial milieu concerning developments in the basic sciences. The coordinating group will become the focal point where industry can go to familiarize itself with any aspect of biotechnology. Much more controversial, but nevertheless interesting, is the intention to offer financial incentives (such as removal allowances, better salaries and accelerated promotion), to entice eminent researchers to lesser known universities in rural areas whose local economies could benefit from the application of biotechnology (Yanchinski, 1982). Finally, medium-term contracts with companies of all sizes are being supported, particularly in the areas of chemicals production, health care products, bioreactors, agriculture and food.

There is the recognition in France that the approach being pursued in biotechnology must be part of a broader industrial policy which contains both research and production considerations. It is not surprising, therefore, that a strong working relationship has already emerged between the coordinating group and the Government Comité d'Orientation pour le Développement des Industries Strategiques.

38.8.3 Conclusion

While it is clear that most of the programming detail of the French Government's biotechnology initiative remains to be developed, a major commitment has been made with the backing of all sectors. The emphasis on training, basic research and technology transfer is laudable and picks up on the recommendations made by the DGRST. One wonders, however, if the appropriate industrial climate, *i.e.* tax policies, regulations, procurement policies, trade policies *etc.*, will accompany the present thrusts. The DGRST report, for example, made special note of the importance of a clearly enunciated patent policy so that no ambiguities might arise which weaken the protection for firms considering high-risk long-range development plans. Equal attention must be paid to this and other issues if France is to realize the economic returns from biotechnology which it covets.

38.9 FEDERAL REPUBLIC OF GERMANY

The Federal Republic of Germany regards biotechnology as being a future-oriented key technology, whose application could make important contributions to the improvement of the human condition as well as the strengthening of economic growth.

In 1972, the Government began providing targetted support for biotechnology to institutions and individual projects in order to create an efficient research capacity in both the public and industrial sectors, and to catch up with international developments. Of particular note was the establishment of the Gesellschaft für Biotechnologische Forschung MbH (GBF) (Society for Biotechnological Research) as a major scientific institution. Although the GBF will be described in more detail elsewhere, it is important to note that its project support was concentrated on the fields of bioengineering, product-oriented microbiology, cell culture techniques, enzyme technology and the development of bioreactors.

At the same time, measures introduced by the Länder and research promotion organizations, as well as additional supporting measures on the part of the Federal Government, have considerably widened the scope of fundamental biological research performed at universities and at establishments which are either completely or partly financed by the Federal Government. Between 1972 and 1978, the Federal Government invested a total of $62 million (US) in biotechnology research and development.

38.9.1 BMFT-Leistungsplan—Biotechnology

The second phase of Government initiatives into biotechnology began in 1979 with the publication of a plan and financing for the period 1979–1983 (Anon., 1980d). With a budget in excess of $74 million (US), the West German Government is pursuing six goals in the development and application of biotechnology. These are as follows:

(i) Securing of adequate food supplies as well as providing protein supplies to the Third World: implementation of projects in the field of the production of unconventional basic food substances and additives is mainly the task of industrial firms, while investigation concerning nutritional and medical evaluation is the responsibility of public research institutions. As a rule, the project management for biotechnology at the Kernforschungsanlage Jülich GmbH (KFA) (Nuclear Research Establishment) supervises such projects.

The Bundesministerium für Forschung und Technologie (BMFT) (Ministry for Research and Technology) has also played a pivotal role in this area over the last few years. Over $17 million (US) have been provided for the promotion of techniques for the industrial manufacture of single-cell proteins. Suitable microorganisms capable of metabolizing paraffins, alcohols or cellulosics have been examined and isolated, and biotechnological processes for the mass production of microorganisms have been developed. At a pilot plant in Frankfurt/Main, 3×10^6 kg per annum of protein-rich product have been produced from methanol. In addition to animal feed producers and firms in the capital goods sector, 18 university institutes and national research centers participated in the activities aimed at developing and testing the safety of the protein-rich product.

(ii) Preservation or restoration of the environment: both the development of biological pest control agents and the degradation of toxic substances are being pursued. The main accent of the activities for the development of biological pesticides is on how to select suitable production strains, how to form products and how to formulate and produce the developed preparations. Research on potential hazards is carried out selectively in publicly supported institutions. Research into the biological treatment of wastewater and waste air is conducted in close cooperation between industry and public institutions. The KFA is the project manager for these activities with $17.6 million (US) allocated under the Leistungsplan.

(iii) Improvement of medical diagnostics and treatment: most of the activities supported under this goal can be classed as fundamental research. Therefore, it is primarily the universities which participate in their implementation. A particularly important role is played by the GBF, with its sections for animal and plant cell cultures, in the scaling-up of the new processes of potential interest to industry. Increased participation by industrial firms is foreseen in the near future with $16.2 million (US) available through the Leistungsplan.

(iv) Production and recovery of raw materials: the conversion of biomass and organic wastes into chemical raw materials as well as the recovery of metals and minerals are the major foci of this activity and will receive $9.7 million (US) over the period of the Leistungsplan.

(v) Development of biotechnological processes: the most important piece of the research work is the development of bioreactors. The issues to be addressed are mainly thermodynamic and physical parameters as well as cell suspension parameters and data on the influence of various physicochemical phenomena on cells. On the basis of developments, new bioreactors of various sizes are to be designed, tested and developed for operation. The program will be carried out in close cooperation with the Deutsche Gesellschaft für chemisches Apparatewesen e.V. (DECHEMA) and the expert committee of Bioverfahrenstechnik (biological process engineering) of the Association of German Engineers (VDI), including some university institutes and the GBF. Over $15.5 million (US) will be supervised by the KFA.

(vi) Fundamentals of biotechnological research: the most important organization involved in pursuing the fundamentals of biotechnology is the Gesellschaft für Strahlen-und Umwelt-forschung (GSF) (Society for Radiation and the Environment), which combines application-oriented research and development with long-term basic research. Basic research is performed in the GSF's departments of genetics, metabolic regulation, material transport, molecular biology and microbiology. Their role is to develop close relationships with university institutes and the institutes of the Max-Planck-Gesellschaft (MPG) and the Fraunhofer Gesellschaft (FLG) which do research in the relevant fields.

Unlike the general support accorded the GBF, the Institute of Biotechnology of the KFA-Jülich and the participating institutes of the MPG and FLG, project support for fundamental research under the Leistungsplan [$4.3 million (US)] is mainly directed to the bioscience institutes of the universities. These projects are supervised by the project management for biotechnology at the KFA-Jülich.

38.9.2 Gesellschaft für Biotechnologische Forschung MbH

The Gesellschaft für Biotechnologische Forschung MbH (GBF) or Society for Biotechnological Research grew out of the former Institute for Research in Molecular Biology that was founded by the Volkswagenwerk Foundation in 1968. Presently financed almost entirely by the Government [$11.5 million (US) in 1982], the scientific staff numbers 163, although this number increases three- to four-fold if post-doctoral and Ph.D. students are included.

The size and budget of the GBF allows it to carry out far bigger projects and work on a far larger scale than individual universities could manage. It can also assemble a staff of experts in all the various forms of biotechnology to provide broad solutions to a variety of problems. Located at Braunschweig-Stockheim, there are six 660 l fermenters, two with 900 l capacity and one at 4500 l. The scale of the enterprise allows them to take a promising discovery, develop it, show that it can be scaled-up to commercial volumes and then take its production to a semi-industrial scale (Anon., 1981g). The simple philosophy behind GBF is to encourage industry to take up their ideas, products and processes. Some of the two-phase solubility systems and microprocessor-aided fermenters developed at GBF are beginning to be taken up by West German industry.

While GBF has been accused by some of not producing top class research, it certainly cannot be labelled an idle institution. While some research groups might sit back and watch their research results go unexamined at higher levels of production, the GBF will certainly be giving each of its developments a long commercial look.

38.9.3 Conclusion

The impression one has from examining the West German Government's approach to biotechnology development is one of careful orchestration and clear understanding of the objectives to be pursued. A new *Leistungsplan Biotechnologie* to cover 1983–1987 is anticipated soon and will undoubtedly place increased emphasis on particular developments.

Although one might argue that the large concentration of biotechnological research and development activity within government agencies could result in advances insensitive to the commercial sector, it should be pointed out that West German industry has a long standing tradition of close collaboration with government. Biotechnology will undoubtedly be no exception in this regard.

38.10 JAPAN

In a country in which 3–5% of its Gross National Product is derived from its fermentation industries, it is hardly surprising to find a complex array of government incentives designed to maintain the strength of this industry, as well as take advantage of the new techniques of modern-day biotechnology.

At the outset it is important to realize that initiatives seemingly taken individually by companies, government institutions or universities in Japan are actually the result of a complex iterative process involving all sectors. Moreover, the support given by each of these sectors to the others is such that it is virtually impossible to attribute any initiative to only one of these sectors. It will be important to keep these points in mind in appreciating the following.

38.10.1 Biotechnology Beginnings

The importance of the promotion of the life sciences in Japan was first identified in the fifth report of the Science and Technology Council, a Council which is chaired by the Prime Minister of Japan. The report, entitled *Fundamentals of Comprehensive Science and Technology Policy for the 1970s* (Anon., 1971) led to the establishment of a Life Sciences Panel; in their final report (Anon., 1980a), nine fields were singled out for special attention. These were: (i) research on life in general as well as various functions of organs of living things; (ii) ecology surrounding man; (iii) scientific study of the human mind; (iv) promotion of health and medical science; (v) energy-related life sciences; (vi) industrial utilization of biology; (vii) security of food resources; (viii) applications to problems of human populations; and (ix) recombinant DNA.

As a further part of the background to the Japanese biotechnology story the report noted that spending by the Government on life science research including health sciences was $101 million (US) in 1980, and within this $6.5 million (US) were for biotechnology. By comparison, private investment on life science research and development among five major companies was reported to have been $40 million (US) in 1979.

As of 1980 approximately 12 000 scientists were engaged in some form of microbiological work in Japan. Over 70% of these were employed in industry, 20–25% in universities and the remainder in government institutions. Each year over 300 students receive their doctorates in microbiology. Because of the extensive interplay between the three sectors, very rapid commercialization of fundamental discoveries occurs.

38.10.2 Biotechnology Research in Universities

The Ministry of Education, Culture and Science is the major support vehicle for government-sponsored research in universities. The Ministry has a special program to support DNA research in universities through a series of scientific research subsidies. The program, designated Recombinant DNA Experimental Technology, supported 18 separate research themes in 1980 for a total of $1 million (US) (Anon., 1980a). In 1981 seventeen research themes were supported for $1 million (US). In addition to this program, the Ministry has also recently financed the establishment of two DNA research facilities: one within the Institute of Medical Science at the University of Tokyo and the other at the Research Institute for Microbial Diseases, Osaka.

38.10.3 Science and Technology Agency

A crown corporation of the Science and Technology Agency, which is called the Institute for Physical and Chemical Research, has a division responsible for life science promotion. This division has five research projects which it is presently financing, two of which are of direct relevance to biotechnology. The first is the study and development of novel bioreactors while the second is the development of novel microbial technology. In 1980, $3.9 million (US) were provided for all five projects.

Beginning in 1981, the Science and Technology Agency implemented a new program to initiate and coordinate multiministerial research. The budget allocated to this program, called the Science and Technology Promotion and Coordination budget, is generally appropriated along the current interests of the Science and Technology Council. For the case of biotechnology, the most noteworthy project being supported is one concerning the 'extraction, analysis and synthesis of DNA'. In 1981 10 intragovernmental groupings shared a budget of $1.2 million (US) under this title.

38.10.4 Ministry of International Trade and Industry

As is already apparent, Japan's push into knowledge-intensive industries in the 1980s has targetted biotechnology as a key area. As a result of biotechnology, the Japanese hope to wean their chemical industry from imported oil and develop new products for industries ranging from pharmaceuticals and food to textiles.

Central to Japan's concerted push into the industrialization of biotechnology is the Ministry of International Trade and Industry (MITI). Within the Fermentation Research Institute of MITI, research has been conducted for several years on the genetic and biochemical breeding of microorganisms. The intramural budget of the institute was $26 million (US) in 1980, but less than $13 million (US) in 1981 as MITI shifted over 50% of the intramural budget into extramural support.

In 1981 MITI also began a new 10 year program titled 'Basic Technology for Next Generation Industries'. Biotechnology is one of three main themes to be pursued within this program.

The new biotechnology thrust will feature $2.8 million (US) to finance biotechnological research within a newly-formed biotechnology technical association. The new association consists of 14 companies undertaking research in cooperation with three MITI organizations: The National Chemical Laboratory for Industry, the Polymer and Textile Research Institute and the Fermentation Research Institute (Anon., 1981a). The research undertaken will centre on three areas of biotechnology: recombinant DNA, bioreactor development and large-scale cell cultures. Such research will benefit many companies pursuing biotechnology in Japan, but the primary aim is to aid Japan's chemical industry. This objective is evidenced by the names of the companies involved: Sumitomo Chemical, Mitsui Toatsu Chemicals, Mitsubishi-Kasei Institute of Life Science, Kao Soap, Daicel Chemical Industries, Denki Kagaku Kogyo, Mitsui Petrochemical Industries, Mitsubishi Gas Chemical, Mitsubishi Chemical Industries, Asahi Chemical Industry, Ajinomoto, Kyowa Hakko Kogyo, Takeda Chemical Industries and Toyo Jozo.

38.10.5 Conclusion

It is in the high-volume applications of biotechnology that Japan's expertise in biologically-based production could give the nation a decided advantage over other countries. Besides leading in fermentation technology, Japan has also pioneered the use of immobilized enzymes to create continuous production systems.

Some 200 Japanese companies and institutes are already actively involved in biotechnology. In addition to an approximate $220 million (US) per year being spent, many of the firms are establishing a variety of linkages with foreign concerns. With the support of national policies on biotechnology, strong, creative programming through MITI and the Science and Technology Agency and a well-developed infrastructure which integrates the strengths of government, industrial and academic institutions, the Japanese represent a formidable force in the international pursuit of biotechnology.

38.11 THE NETHERLANDS

38.11.1 Biotechnology Coordinating Committee

In May 1980, the Dutch Government set up a committee to coordinate biotechnology research in the Netherlands, to define areas requiring future investment and to increase the emphasis on commercial applications. The committee, which brought together experts from industry, the Governmental applied research organization (TNO) and universities, is a by-product of the Dutch Government's drive to stimulate innovation, which was launched in 1979.

The committee, in its report released in April 1982, opted for the rapid strengthening of intersectorial linkages, rather than setting up specific centers. The committee urged that the Government should provide extra support to stimulate biotechnology, specifically $30 million (US) between then and 1988. This money would come from the Government's fund for industrial innovation and be in addition to the planned biotechnology budget, which is much larger (Schuuring, 1982).

The committee concluded that efforts should be concentrated in sectors where Dutch companies are traditionally strong, *i.e.* agriculture, the dairy industry, fermentation and antibiotics production. Areas highlighted for future research include the development of host-vector systems for use in applied research, somatic cell hybridization, second-generation biotechnological reactors for enzyme production and the isolation of useful products from process liquids (Becker, 1982).

Characteristic of the genetic engineering scene in the Netherlands is serious concern about the risks of recombinant DNA experiments with cells of higher organisms. The strictest regulations in the world apply in the Netherlands. This was the reason why two commercial firms transferred their activities in the field of genetic engineering to other places: Unilever now carries out genetic engineering experiments in their Belgian laboratories and Gist-Brocades continues its genetic engineering experiments in Britain. However, restrictions on recombinant DNA work recently have been relaxed. P3 genetic engineering experiments, which previously were prohibited, now are permitted. In addition, the Dutch *Ad Hoc* Committee on Recombinant-DNA Activities is circulating for industrial comment draft guidelines for industrial output in batches greater than 10 l (Anon., 1983).

Although the Government tries to encourage and promote promising biotechnological development, its attention seems fixed on the risks associated with: (i) possible contamination by pathogenic microorganisms; (ii) the harmful side-effects of biotechnological products; and (iii) unforeseen changes in the quality of biotechnological products.

With these risk elements in mind, there is little incentive for the Government to encourage commercial enterprises, especially as the Government is under constant pressure by the media and public opinion.

A final important factor in Dutch biotechnology is that Dutch industries cannot be sure that contract research with university institutions can be kept confidential, an important aspect of market strategy and competition. The Dutch universities are 'democratized'. This means that on all levels elected councils are the final authority in terms of deciding upon collaborative projects and have access to all information. At all levels one-third of the membership of these councils are scientific personnel, one-third technical and administrative personnel, and one-third students. Industry has no guarantee that all the details of work carried out under contract in the field of genetic engineering or any other aspects of biotechnology will not be made public. As a result, no contract research between industries and university laboratories is occurring. Yet it is within the uiniversities that most of the Dutch biotechnology expertise and experience resides.

38.11.2 Conclusion

Although the Dutch Government appears to be trying to stimulate biotechnology within the Netherlands, the prospects, at least in the short-term, for the commercialization of developments from the application of recombinant DNA techniques, are dim.

38.12   THE UNITED STATES OF AMERICA

38.12.1   Report of the Office of Technology Assessment

When one examines biotechnology in the United States, one is immediately struck by the vitality of the enterprise. Small, innovative companies have sprung up like mushrooms throughout the country, pursuing everything from pharmaceutical development to waste treatment. Genetic engineering is being pursued in many sectors, building upon a molecular biology research establishment replete with Nobel laureates. Large companies have begun investing millions of dollars in both in-house research capacity, buildings, acquisitions and partnership arrangements. Given this activity, the question one must legitimately ask is whether or not there is a need for direct US Government intervention in the promotion of biotechnology. This question was one of the major issues brought forward in the report *Impact of Applied Genetics: Micro-Organisms*, *Plants and Animals* prepared by the Office of Technology Assessment (OTA) and submitted to the Congressional Board of the 97th Congress in April, 1981 (Anon., 1981h).

The chief argument for US Government subsidization of R and D in biotechnology is that Federal help is needed in areas such as general (generic) research in the form of highly speculative areas not being developed by industry. The argument against this kind of support is that industry will develop everything of commercial value on its own. In its report, the OTA offers a number of options for addressing this issue. Congress could, it suggests, specifically ear-mark funds for biotechnology within the budgets of appropriate agencies. On the other hand, a separate institute for funding biotechnology could be established. The latter option was not felt to be reasonable, given that the mandate of such an institute would undoubtedly overlap the mandates of other agencies, resulting in increased bureaucracy and competition for funds at the agency level.

In another vein, the OTA report (Anon., 1981h) suggested that Congress consider establishing biotechnology research centers in universities, as well as specialized grants for training in biological engineering to redress the problems in manpower supply. Also, selective use of the tax system was suggested as a means to increase the supply of venture capital to biotechnology firms and to ensure a greater flow of contributions to universities for research.

While Congress has not yet acted upon the suggestions contained within the OTA report, it is interesting to note that at least certain elements within Congress feel that a new kind of national guideline may be needed to regulate the biotechnology business. With special reference to the agreement between Hoechst AG, the Frankfurt-based chemicals company, and Massachusetts General Hospital, concern has been raised about the ethics of publicly-supported institutions assisting foreign and not American corporations. Underway is an intensive scrutiny of all agreements being struck between universities and industry and the possibility looms for tighter control over the financing and exploitation of research activities in the country, particularly those supported by the National Science Foundation and the National Institutes of Health.

38.12.2   General Climate

While there is some concern in the US that American business may lose its lead in international competition because there is no explicit national policy for integrating cooperative activity between government and industry, it is certainly clear that the Government, either consciously or subconsciously, has created a general climate in which the development of biotechnology as well as other areas of technology can flourish. In fact there are a number of improvised approaches that are designed to foster the industrial application of new technology. The goals of these approaches are: (i) to encourage venture capital investments in new high technology companies; (ii) to encourage established medium and large-sized companies to increase their research activity, particularly in the development of new technologies; and (iii) to enhance the transfer of technological innovations by promoting linkages between universities and industry.

Four incentives used to achieve these goals are favorable tax policies, favorable patent policies, a favorable regulatory environment and direct Federal subsidy of some research and development programs.

38.12.2.1   Tax incentives

First, to encourage established companies to increase their research activity, a 25% tax credit is allowed for incremental research expenditures. This includes both labor and capital components of the research. A second new, major tax incentive encourages investment in research and development by allowing an accelerated depreciation allowance for equipment and buildings used for research and experimentation. A third recently implemented tax incentive is a reduction in the rate of capital gains taxes. This latter incentive is particularly attractive to investors in new high technology companies that normally put all earnings back into the company and pay no dividends. These three tax incentives—tax credit, accelerated depreciation allowance and reduced capital gains tax rate—have generated a new type of financing of industrial research and development using private capital. It is predicted that the effect of these incentives will soon dwarf corporate funding. More specifically, limited partnership groups of individual investors are being established to finance industrial research. Income from the investment is highly tax sheltered and attractive to large investors. In many cases the partnership is started by the companies themselves and there is no public offering. Raising capital through limited partnerships is a particularly useful mechanism for middle size companies to finance their R and D activities. Another new incentive is the treatment of stock options as capital gains rather than ordinary income. This makes it easier for companies to give stock options to employees. It is a particularly attractive incentive for the use of new companies in recruiting scientists.

38.12.2.2 Patent policies

Another set of incentives contributing to a healthy environment for technology industries is favorable patent policies. In the United States a major new patent law is being implemented for inventions derived from government-supported research. The new act established a single government-wide policy on the ownership of patent rights resulting from Federally-supported research. It replaces more than 20 inconsistent statutes, each relating to a different government agency. Prior to the Act, Government determined the ownership of patent rights resulting from government-supported research. Under the new act, the research institute has the right to keep its inventions, except in specified exceptional circumstances. The law provides that the Government receives a non-exclusive, royalty-free license to use the invention. It covers government-supported research performed by small businesses and non-profit organizations, including universities. It is intended to encourage patent applications by small businesses, which are defined as companies with up to 500 employees. The new patent law is also intended to encourage university–industry cooperative research. Assured of ownership of inventions even if the university's research involves government funding, the universities are able to offer future licenses in exchange for private sector research support.

Another patent bill, which has passed in the Senate and is pending in the House, is the 'Patent Term Restoration Act' (Anon., 1981j). This bill is designed to encourage innovation, particularly in the pharmaceutical industry. It assures a full 17 years of patent protection to the inventor by adding back the amount of time that may be required for government testing after the patent is issued.

38.12.2.3 Favorable regulatory environment

Regulations can have a stifling effect on the environment in which high technology businesses can thrive if disincentives are not screened out. The removal of disincentives can create a favorable regulatory environment. Several examples can serve to illustrate this point. The primary example is the avoidance in the United States of national legislation regulating the use of recombinant DNA molecules in research or in industrial processing. During the period 1976–1978, no less than 16 bills were introduced into the Congress designed to regulate this technology. Had such inhibitory legislation been enacted, it is certainly debatable whether or not the US genetic engineering industry would have grown to its present state, whether or not much of the basic research that was important for the recent further development of the technology would have been done, whether or not many scientists would have done their research outside the US, and whether or not a much lower level of research activity would have been much more expensive to perform because of the containment, monitoring and surveillance requirements. A follow-up example is the evolution of guidelines controlling recombinant DNA use. During the first two years, the guidelines have evolved in a direction of less stringency and this direction has been a tremendous stimulus to enhance research in both academic and industrial laboratories.

Another example under the heading of a favorable regulatory environment is the Food and

Drug Administration's (FDA) handling of the evaluation of the safety and efficacy of human insulin made in bacterial cells. The drug must be evaluated as a new drug, but the FDA is following a procedure known as the 'fast track'. This means that at all steps, this drug gets priority scheduling.

A final example illustrating a favorable regulatory environment intended to promote a vibrant climate for industrial development is the interpretation of existing regulation in a less restrictive framework. In the US, anti-trust laws determine to what degree companies may engage in joint ventures for R and D or marketing. These laws can deter companies that may want to develop in certain areas from doing so, or they may inhibit two or more companies in the same industry from cooperative research. To dispel the perception that anti-trust policy inhibits innovation and to clarify the Government's position on collaboration among firms in research and development, the Justice Department was asked to prepare guidelines concerning these matters. Their report (unpublished, personal communication), in essence, gives *carte blanche* for collaborative research and a relatively free hand in collaborative development programs.

38.12.2.4   *Government support of industrial research*

In addition to a favorable tax policy a favorable patent policy and a favorable regulatory environment, the Government also supports research in industry directly, particularly in small businesses to promote technological innovation. The Small Business Innovation Research Program developed by the National Science Foundation operates on a three-phase concept for each project. Phase I awards are limited to $30 000 for a six month period to determine the technical feasibility of the proposed research project. Phase II awards, which average about $200 000 and can be carried out for as long as 24 months, are to support in-depth research. It is intended that this research, if successful, will lead to technology that is ready for commercial development. Phase III is carried out with private funds and leads, if successful, to a commercially marketable product. The underlying principle of the three-phase concept is that government should support the higher risk parts of the project and that venture capital or a larger manufacturer should be willing to support the project after the risk of technological failure is reduced. This program was initiated in 1977 and has shown substantial success in terms of the response of the small business community and in terms of the availability of private support for phase III. It is also important to note that of the 24 research topics selected by the National Science Foundation to be pursued under the Small Business Innovation Research program in 1982, at least two, alternative biological resources and biochemical engineering, could enhance the development of biotechnology (Anon., 1982).

A program within the National Science Foundation designed to promote technology transfer is the industry–university cooperative research program. It encourages collaborative research, with the Government supporting the academic partners and the company paying for its share of the research costs. The program is designed to explore the scientific basis for new technological possibilities, or to address important problems of current technology. The underlying principle of this program is to effectively marshal the use of suitable scientific talent and facilities to attack problems of importance to industry. The collaboration does not have to be a one-to-one arrangement, but can involve scientists of several universities and one company, or vice versa.

There is a cloud looming on the horizon, however, in the area of availability of appropriate human resources. Tight budgets and moves to consolidate existing programs have almost entirely eliminated training support as a specific program within either the National Science Foundation or the National Institutes of Health. As a consequence, support for trainees is derived from operating grants, with the individual grantee responsible for human resource selection. There is a growing fear that without a national scholarship competition, mediocrity will set in as academics scramble for needed graduate students. With graduation levels from secondary schools expected to drop by over 30% by the end of the 1980s and the quality of the graduates declining as well, human resources will loom larger and larger as a rate-limiting step in the development of the United States' scientific and technical capacity. The short-sightedness of education support reductions may come back to haunt the United States in the late 1980s and 1990s.

38.12.3   Conclusion

It is probably safe to say that a distinct national policy for biotechnology is an unlikely scenario in the US. While the OTA report (Anon., 1981h) suggested that more might be done by government in supporting research, in creating a favorable climate for industrial investment and in promoting innovation through increased interactions, many existing measures are already having a positive effect, even though their influence is not restricted to biotechnology. It will be interesting to watch, in coming years, whether or not closer relationships between government and industry will have to be forged in order to strengthen the ability of the US to compete internationally against other nation-backed industries. Of equal importance will be the approaches developed to ensure a continuing base of appropriately-trained human resources.

38.13 CONCLUDING REMARKS

Upon digesting the preceding sections in this chapter one sees that a nation's approach to the promotion and development of biotechnology will be the result of a myriad of interrelated social, economic, industrial, scientific and political factors. For example, the *laissez-faire* approach of the United States would appear to be consistent with a free enterprise, non-interventionist role for government. On the other hand we find France, for example, exhibiting a highly centralist, government-directed approach, an approach which, one observes, has been strengthened by the election of a socialist regime. The other countries fall somewhere between these two poles.

The immediate conclusion one may be drawn to, therefore, is that there is no single means whereby a government must pursue biotechnology. In fact, there is. Whether or not it is expressly stated by a national government, the approach to biotechnology must include the following elements: (1) a solid commitment to building up the requisite scientific and technical infrastructure within a country so that it can develop indigenous applications of biotechnology, as well as adapt foreign developed technology; (2) the creation of an industrial climate which reduces risk, encourages innovation, supports export penetration and creates market opportunities; (3) the encouragement of strong linkages between the 'three solitudes': government, industry and university; and (4) the promotion of a social awareness of the overall impact of biotechnology. To ignore any of these elements is to invite serious weaknesses in international competitiveness as well as inefficient domestic manufacturing capabilities.

Biotechnology, therefore, is not simply doing good science, is not simply a series of tax policies aimed at encouraging investment, is not simply a reduction in the barriers between the sectors and is not simply social awareness. Rather it is the integration of these elements which is important, and it is the role of government to see that this integration is pursued. The ways and mechanisms which governments use to effect this integration will vary depending upon existing situations. Where there can be no variability is in the four basic elements. France, Germany, Japan and the United States are well on their way towards achieving the necessary integration. Other countries such as Canada, Great Britain and Australia are still behind.

Towards the end of the twentieth century, technology will assume a dominant role in a world seeking greater efficiencies, more personalized service and minimal disruption to the biosphere. Those governments which understand the environment within which technology moves will be those in the best positions to take advantage of the opportunities presented. Biotechnology presents some of the most important opportunities and, therefore, challenges the world has faced in modern times. The question which remains to be answered is whether or not world governments are equal to the challenges biotechnology poses.

38.14 REFERENCES

Anon. (1971). *Fundamentals of Comprehensive Science and Technology Policy*. Science and Technology Council, Tokyo, Japan (April 21).
Anon. (1979a). *FAST Work Programme, Annex 3*. Biosociety, Bruxelles, Belgium (December 18).
Anon. (1979b). *Long Range Plan*. Alberta Research Council, Edmonton, Canada (December).
Anon. (1979c). *Recombinant DNA Techniques in Research*. Commonwealth Scientific and Industrial Research Organization, Canberra, Australia (May).
Anon. (1980a). *Basis for Policy for Promotion of Recombinant DNA Research*. Science and Technology Council, Tokyo, Japan (April).
Anon. (1980b). *Biotechnology in Canada*, MOSST Background Paper Number 11. Ministry of State for Science and Technology, Ottawa, Canada (June).
Anon. (1980c). *Biotechnology*. Report of a Joint Working Party, Advisory Council for Applied Research and Development, Advisory Board for the Research Councils, The Royal Society, London, England (March).
Anon. (1980d). *BMFT-Leistungsplan 04, Biotechnologie, Plan periode: 1979–1983*, Bonn, West Germany (January).
Anon. (1981a). A Government boost for the biotechnology industry. *Business Week*, December 14, 110.

Anon. (1981b). *Biotechnology: A Development Plan for Canada*. A report of the Task Force on Biotechnology to the Minister of State for Science and Technology, Ottawa, Canada (February).

Anon. (1981c). *Biotechnology*. Presented to Parliament by the Secretary of State for Industry by Command of Her Majesty, London, England (March).

Anon. (1981d). *Biotechnology Research and Development*. Commonwealth Scientific and Industrial Research Organization, Canberra, Australia (November).

Anon. (1981e). Brazil's ethanol fuel: A victim of success? *Business Week*, August 24, 42H.

Anon. (1981f). How Brazil's gasohol scheme backfired. *New Scientist*, **91**, 132.

Anon. (1981g). GBF—Gesellschaft für Biotechnologische Forschung mbH. Programme budget. Braunschwiegh-Stockheim, West Germany.

Anon. (1981h). *Impacts of Applied Genetics: Micro-Organisms, Plants and Animals*. Congress of the United States, Office of Technology Assessment, Washington, DC (April).

Anon. (1981i). *Official Journal of the European Communities*, No. L375. Bruxelles, Belgium (December 30).

Anon. (1981j). *Patent Term Restoration Act, HR 6444*. US Congress, Washington, DC (November).

Anon. (1981k). *Programme d'Inverventions Pour le Développement des Biotechnologies au Québec, Phase I: 1982–1986*. Secrétariat au Développement Scientifique, Québec, Canada (October).

Anon. (1981m). *Science and Technology Statement 1980–81*, Minister for Science and Technology, Canberra, Australia (April).

Anon. (1982). *Bulletin: National Science Foundation*, **9** (7), March, Washington, DC.

Anon. (1983). Japan, Netherlands ease r-DNA containment rules. *McGraw-Hill's Biotechnol. Newswatch*, **3** (4), 5.

Bazin, M. (1979). Brazil: running on alcohol. *Nature (London)*, **282**, 550–551.

Becker, J. (1982). Going Dutch. *Nature (London)*, **295**, 91.

Cantley, M. and K. Sargeant (1981). *FAST Sub-Programme Biosociety: Towards Conclusions*. An interim report to the ACPM (Associate Committee on Plant Materials), Bruxelles, Belgium (22/23 June).

de Rosnay, J. (1979). *Biotechnologies et Bio-Industrie*. Document complementaire au Rapport Science de la Vie et Société. La Documentation Française, Paris, France.

Fishlock, D. (1981). Celltech applied British genius to British genes. *Financial Times*, October 9.

Fishlock, D. (1982). Big plans for small biotechnology. *Financial Times*, March 31, 23.

Gros, F., F. Jacob and P. Royer (1979). *Science de la Vie et Société*. Rapport au Président de la République. La Documentation Française, Paris, France.

Pearce, S. M. and H. S. Preston (1981). *Genetic Engineering—Commercial Opportunities in Australia*. Secretariat Background Paper, Genetic Engineering Symposium, November 18–20. Department of Science and Technology, Australia.

Pelissolo, J. -C. (1980). *La Biotechnologie Demain*. Rapport à Monsieur le Premier Ministre, Délégation Générale à la Recherche Scientifique et Technique (Paris). La Documentation Française-Paris 1981-ISBN: 2-11-000670-6. France (December).

Schuuring, C. (1982). Dutch go-ahead. *Nature (London)*, **296**, 698.

Wenz, C. (1982). Bio-fund open. *Nature (London)*, **296**, 190.

Yanchinski, S. (1982). France entices its biotechnologists into industry. *New Scientist*, **93**, 767.

39

The Role of International Organizations in Biotechnology: Cooperative Efforts

E. J. DASILVA
UNESCO, Paris, France
and
C. G. HEDEN
Karolinska Institute, Stockholm, Sweden

39.1 INTRODUCTION

Microorganisms have always posed threats and offered opportunities that have transcended national borders. Consequently, microbiologists traditionally have had an international outlook, and there are in fact several examples of instances where the fight against endemic diseases, *e.g.*

717

the eradication of smallpox under the aegis of the WHO, or joint efforts in the practical use of beneficial microorganisms, have progressed unimpeded by political and military conflicts. One can even cite cases where a professional verdict over the misuse of microorganisms for military purposes (Heden, 1970) has played a significant role in paving the way towards disarmament efforts. In fact, such efforts have managed to keep the spectre of chemical and biological warfare at bay until fairly recently.

Needless to say, such initiatives are much easier to be undertaken by non-governmental organizations (NGOs) such as the International Union of Microbiological Societies (IUMS), the Pugwash Movement and the Stockholm International Peace Research Institute (SIPRI) rather than by inter-governmental organizations (IGOs) which, by necessity, tend to focus on national interests *vis-à-vis* technical considerations.

However, in fields where microbiology provides an important technical dimension, this discipline often has had a significant impact on the decision-making process, probably as a consequence of the common outlook of involved expertise and excellence. The World Health Organization, for instance, prepared a background document (WHO, 1970) which served as a very important guideline for the negotiations that led to a treaty that prohibited biological warfare (SIPRI, 1973). Similarly, the microbiological aspects of gene pool management were given prominence when the United Nations Environment Programme (UNEP) was launched in the early 1970s.

Actually, the interaction between inter-governmental (IGOs) and non-governmental organizations (NGOs) is healthier in the field of microbiology than it is with regard to other bodies that are concerned with most other disciplines. For instance, it is quite common for UN agencies to call on NGOs for expert advice in microbiology (Porter, 1979). In one instance this even made it necessary for the International Cell Research Organization (ICRO) to function in an advisory capacity to UNEP and the United Nations Educational, Scientific and Cultural Organization (UNESCO). Normally the relation is more *ad hoc*, exemplified, for instance, by the recent request of the United Nations Industrial Development Organization (UNIDO) to the International Federation of Institutes for Advanced Study (IFIAS) to provide expertise on the establishment of an International Center for Genetic Engineering and Biotechnology (ICGEB).

The interaction between IGO and NGO concerns is also broad when microbiology plays a part in the programmes of aid organizations such as the International Development Research Center (IDRC) and the International Foundation for Science (IFS) which support individual and national projects related to biological nitrogen fixation, microbiological upgrading of starchy crops or agricultural residues as well as several other areas in which biotechnology is relevant (IFS, 1980). Both organizations receive governmental support; IDRC is directly funded by the Canadian Government (IDRC, 1980), whereas IFS receives its support from a large number of national academies as well as from UNESCO (Barnaby, 1981; Revelle, 1981). Nevertheless, in both instances the operations are guided by executive organs that have a large component of 'foreign' experts selected in their personal capacity only.

Finally, there are instances where technical problems involving biotechnology concern only a few nations, but yet require broad-based international agreements. An example is space quarantine (Werber, 1973) which had to be given consideration as a result of the handling of returned lunar samples, and for the formulation of standards for space probe sterilization in preparation for the first physical contact with the planet Mars. The stringency of the containment and sterilization requirements in these cases was a great challenge for many bioengineers both in the US and in the USSR. Microbial technologies which are only used by a few of the technically-advanced countries might obviously be of considerable international interest. In retrospect, the measures taken in the form of a 'Lunar Receiving Laboratory' (Duke and Reynold, 1973) and the implementation of a sterilization programme costing millions of dollars annually may have seemed excessive. However, if extraterrestrial life forms had been detected, one needs little imagination to visualize the public outcry against the International Space Research Organization (COSPAR)—had it not initiated appropriate international discussions. A strong negative reaction no doubt would also have developed if the early fears with regard to the recombinant DNA technology (Grobstein, 1976) had been reinforced by experimental proof of a potential global threat.

At the present time the exotic areas mentioned above attract less attention than the potential impact of applied microbiology and bioengineering on developing countries and the need of standards for industrial products based on various biotechnologies. Reference to several standards is made in the following review of a number of UN inter-governmental organizations that are involved in applied microbiological and biotechnological activities.

39.2 THE UNITED NATIONS SYSTEM AND ITS PROGRAMMES IN BIOTECHNOLOGY

Several organs of the United Nations System and its Specialized Agencies are involved in the promotion of activities in biotechnology. In the past decade, several UN Conferences have made specific reference to the potential of applied microbiology and biotechnology for socio-economic growth in the industrialized countries and overall technological development in particularly the developing countries. In the following pages, a review has been made of the various UN-sponsored activities such as the promotion of training, the organization of specialized conferences, the seeding of industrial growth, and the catalyzing of technological progress through collaboration between the developed and developing countries (Pramer, 1984).

39.2.1 The Economic and Social Council (ECOSOC) — United Nations (UN)

The UN was established on 26 June 1945 on the basis of proposals worked out at the United Nations Conference on International Organization. The UN officially came into existence on 24 October 1945. The Economic and Social Council (ECOSOC) of the UN is responsible, amongst other issues, for studies, reports and recommendations on international economic, social, cultural, educational, health and other related matters. ECOSOC functions also on a regional basis through its regional commissions in Africa (ECA), Asia and the Pacific (ESCAP), Europe (ECE), Latin America (ECLA)* and Western Asia (ECWA).

In 1955, the Protein–Calorie Advisory Group of the United Nations (PAG) was established under the sponsorship of the Food and Agricultural Organization of the United Nations (FAO), the World Health Organization (WHO), the United Nations Children's Fund (UNICEF), the International Bank for Reconstruction and Development (IBRD), and the United Nations. PAG advises on technical and policy aspects of nutrition in relation to development, *e.g.* single-cell-protein (SCP) production, as well as on problems associated with protein–calorie malnutrition and protein resources. For several years it released monographs and PAG statements and guidelines in the *PAG Bulletin* which has now been incorporated into and continued in the *Food and Nutrition Bulletin* of the United Nations University (UNU) (see also Section 39.2.11).

The UN, either through the regional commissions of ECOSOC or through its then office for Science and Technology and its successor organ, the UN Centre for Science and Technology for Development (CSTD), has promoted microbial technology, *e.g.* through meetings on:
 (1) Biogas Technology and Utilization, Fiji, November 1976.
 (2) Rural Energy Development (Roving Seminar), Thailand, Philippines, Iran and Indonesia, July–October 1977.
 (3) Waste Recycling Technology for Development, UN, New York, December 1978.
 (4) Non-food Agricultural Materials for Energy and Development, UN, New York, November 1979.
 (5) The Importance of Biotechnology for Future Economic Development, Szeged, Hungary, June 1985.

The above is not an exhaustive list, but only an indication of the attention given by the United Nations itself (Table 1) to the potential of microbial technology and its allied biotechnologies for economic development and social progress.

39.2.2 Food and Agricultural Organization of the United Nations (FAO)

FAO, established in 1945 at Quebec, aims at raising the levels of nutrition and standards of living of the peoples of its 146 member nations; securing improvements in the efficiency of production and distribution of all food and agricultural products; bettering rural development; contributing towards an expanding world economy, and ensuring humanity's freedom from hunger.

FAO's motto '*Fiat Panis*' implies the use of a microorganism in the production of an essential food for man. Similarly, the applications of microbiology play a role in many of FAO's overall activities in the production of food, fiber and shelter, through agriculture, fishery and forestry. As a result the activities related to the use and control of microorganisms are dispersed in the following nine technical divisions:
 (1) Animal Protection and Health Division
 (2) Joint FAO/IAEA Division

* Known as ECLAC (Economic Commission for Latin America and the Caribbean) since 27 July 1984.

Table 1 Some Biotechnological Activities of the UN System and its Specialized Agencies[a,b]

Agency	Activity
ECOSOC–UN	Production of PAG Guidelines; New York, USA/Tokyo, Japan
FAO	Residue Utilization — Compendium of Technologies; quantitative survey 1977–1981; see also under FAO (Section 39.2.3); Rome, Italy
IAEA	Radiation Microbiology; Bangkok, Thailand, 1976
	Improvement of Energy Production for Agriculture with emphasis on methane and alcohol aided by Nuclear Techniques, Vienna, Austria, 1980
UNESCO[c]	Annual course on microbial engineering at Osaka University, Japan; organization of GIAM conferences; support to training in fermentation technology, culture collections, bio-energy; award of Carlos Finlay Prize for outstanding achievements in microbial technology; Paris, France
UNEP	Promotion of the environmental aspects of microbial technologies; Nairobi, Kenya
UN/UNDP	Demonstration of biogas and allied technologies in the People's Republic of Yemen, Tanzania, Philippines (1974–1982)
UNICEF	Provision of basic services to children *via* village and rural biogas systems; New York, USA
UNIDO	The Case of *Spirulina*; Mexico, 1980/Vienna, Austria, 1980
	Fermentation Technology (FERMITECH); Bangkok, Thailand, 1982
UNITAR	Microbial Energy Conversion; Gottingen, F.R. Germany, 1976
UNU	Rural Energy Systems with emphasis on biogas; Guangzhou, China, 1982
WHO	Biological Control of Insects and Pests; Geneva, Switzerland, 1980

[a]This is not an exhaustive list. [b] See also Tables 3 and 4. [c] See also sections on IFS, IOBB, IUMS and WFCC.

(3) Land and Water Development Division
(4) Plant Production and Protection Division
(5) Agricultural Services Division
(6) Commodities and Trade Division
(7) Food Policy and Nutrition Division
(8) Fishery Industries Division
(9) Forestry Industries and Trade Division

Considerations of use and control of microorganisms in food are of special interest to the fishing industry. Control of microbial growth in fish and fish products is essential to prevent economic losses, wastage of urgently required protein and public health hazards from pathogenic microorganisms. On the other hand, some fish products, for instance the fish sauces of Southeast Asia, are produced by fermentation. Again, the possibility of ensiling waste fish by microbiological production of lactic acid has been investigated and has great promise. The activities of the Fisheries Department of FAO incorporate the above aspects. Field projects include the training of quality control and inspection staff, setting up quality control and fish technology research laboratories and advising on production methods to make better and more fish products.

Similar activities are undertaken by the FAO Technical Divisions engaged in the processing of fruits and vegetables, meat and dairy products. Mycotoxins in foods have been the subject of increasing concern. Recent research in monitoring activities has shown the presence of mycotoxins in a wide variety of foods. FAO has instituted a number of steps to assist member governments in assessing the existing situation, and in developing methods to combat problems caused by mycotoxins.

Since 1973 the agenda on the annual session of the Intergovernmental Group on Oilseeds, Oils and Fats has included an item on 'Problems of Aflatoxins in Oilseeds and Oilcakes'. The Group's basic interest is to help countries to understand the problems raised by aflatoxins and to reduce the likelihood of aflatoxin contamination. The food preservation section, *i.e.* the joint FAO/IAEA Division, is engaged in the control of saprophytic and pathogenic microorganisms in food by the use of ionizing radiations. For example, the problem of combatting contamination with *Clostridium botulinum* has been intensively studied with special reference to the radiation preservation of fresh fish.

Through its field programme, FAO is engaged in the production of food products derived from microbial processes. Examples are bread, pickles, cheese and fermented foods such as gari, baker's yeast and alcohol.

FAO's activities in the field of novel protein production are maintained through the Protein/Calorie Advisory Group of the United Nations that was formed in 1955. Guidelines laid down by PAG have been used in the extensive programmes of nutritional and toxicological testing of hydrocarbon-derived single-cell protein for use in animal feeding.

FAO is promoting international efforts to control major diseases of livestock and poultry by assisting in the production of vaccines, therapeutic antisera, diagnostic reagents and cell cultures. FAO gives assistance to developing countries in producing vaccines against diseases of economic importance such as Newcastle disease, foot-and-mouth disease, rinderpest, hemorrhagic septice-mia, contagious bovine pleuropneumonia and theileriasis. Most of FAO's animal health projects are involved in diagnostic activities and several projects are in the operational phase.

For the control of livestock and poultry diseases, FAO cooperates with OIE (Office International des Epizooties), UNDP, WHO, EEC, PAHO (Pan American Health Organization) and OAU (Organization for African Unity). For instance, FAO is involved in an internationally coordinated programme for the control of African animal trypanosomiasis in close cooperation with WHO, UNDP, OAU and the World Bank. This programme involves not only the control of the organisms in the host animals, but also the control of the vectors, tsetse flies.

Within the integrated pest control concept, plant protection measures are based on the combined use of modern pesticides and natural mortality factors. In these, microbial agents have always been the main point of discussion within the FAO Panel of Experts on Integrated Pest Control, established in 1966. FAO is closely collaborating with WHO on developing guidelines for the safety evaluation of microbial agents. The FAO publication *The Use of Viruses for the Control of Insect Pests and Disease Vectors* refers to this activity and safety testing. In the past, the use of microbial agents for pest control has been a working tool in several of the field projects of FAO's Plant Protection Service. Special mention should be made of the successful project 'Research on the control of the coconut palm rhinoceros beetle', that made use of viruses, a fungus and entomophagous insects. In collaboration with UNEP, a global programme on the development and application of integrated pest control, just begun, will be of considerable help in obtaining satisfactory results.

The Plant Production and Protection Division and the Land and Water Division of FAO have developed a joint programme on Biological Nitrogen Fixation, with special emphasis on the use of *Rhizobium*. In December 1974 a panel of consultants in legume bacteriology met in Rome to formulate an action programme promoting the use of symbiotic nitrogen fixation in developing countries. Since then, a *Handbook of Tropical Forage Legumes*, which deals with various aspects of symbiotic nitrogen fixation by *Rhizobium*, has been prepared. In addition, support has been made available to the MIRCEN networks in South America and East Africa. The activities of the Land and Water Development Division have concentrated on exchange of information and promotion of field projects. Examples of these activities are a soils bulletin entitled *A Practical Manual of Soil Microbiology Laboratory Methods*; field projects in Asia, Latin America, Africa and Europe dealing with soil microbiology; soil organic matter and improvement of soil physical properties; inoculation with *Rhizobium*; and field experimentation with and demonstration of the combined use of organic matter and mineral fertilizers in the framework of the FAO Fertilizer Programme.

The FAO/WHO Codex Alimentarius Commission (CAC) was established in 1962 jointly by WHO and FAO to implement the joint FAO/WHO Food Standards Programme. The aims of the programme are to protect the health of the consumers and ensure fair practices in the food trade and to promote coordination of all food standards work undertaken by international governmental and non-governmental organizations. The Commission has responsibility for the publication of international product standards, the series of international maximum limits for pesticide residues, and the international codes of hygenic practice for different commodities. The Codex Committee on Food Hygiene has elaborated a code of General Principles of Food Hygiene which has been adopted by the Codex Alimentarius Commission.

The Commission has 27 subcommittees. Six of these 'Codex Committees' work on general matters (principles, labelling, hygiene, additives, biocide residues and analytical practices), and 17 are concerned with the elaboration of global or regional standards for a variety of products (*e.g.* vegetables and fruits, cereals, dairy products, fats and oils, meat, juices, frozen foods). In addition there are four coordinating committees (for Africa, Asia, Europe and Latin America) and an Executive Committee that is responsible for the activities between regular Plenary Sessions.

So far the Codex work has resulted in 174 product standards and more than 1500 borderline values for biocide residues. In addition, 25 *Codes of Practice* have been published as well as some 70 *Codes of Hygienic Practice* and some 70 guidelines for sampling and analysis. In view of the large international trade in cereals, much of the current interest is focused on the handling methods employed in this field. Also, the concern about biocide residues has been widened to include industrial pollutants.

As a result of the recommendations made by the UN Conference on Human Environment (Stockholm, 1972), FAO and WHO have been given responsibility for the institution and coordination of international programmes on research and monitoring of food contamination by chemical and biological agents (*Recommendation No. 78*). It was also recommended (*Recommendation No. 82*) that the Codex Alimentarius Commission be given increased support to develop international standards for pollutants in food and a Code of Ethics for the international food trade. A further part of this recommendation was that the FAO/WHO activities in food control be strengthened. UNEP has allocated funds for the purpose.

39.2.3 International Atomic Energy Agency (IAEA)

Established in 1956, IAEA fosters, encourages and advises on the development of the peaceful uses of atomic energy throughout the world. It promotes the wider use of radioisotopes and radiation sources in research, industry, agriculture, medicine and hydrology. Its work is carried out in close cooperation with a number of international and national organizations. In keeping with its global objectives, the Agency's radiation biology programme of the Life Sciences Division has been designed to promote the practical applications of radiobiological effects in the fields of nutrition, health and medical care as well as for the production of some essential substances of welfare and economic interest.

One of the major contributions of the Agency's coordinated research programme on applied radiation genetics is the initiation and the establishment of appropriate research trends. In this programme participating institutes in Nigeria, Greece, Malaysia, Singapore and the Philippines carried out relevant research on their respective traditional fermentation practices. Fermentation of cassava starch in Nigeria was attempted using a radiation-induced mutant of *Rhizopus*. The mutant strain of *Rhizopus* combined the characteristics of higher contents of protein and the amylolytic enzyme which facilitated the use of the starchy substrate of cassava. Fermentation studies and practices involving the use of improved mutant strains for protein production were supported. Traditional substrates were carob bean extracts (Greece), peanut cakes (for Ontjom) in Malaysia and coconut waters (Philippines). The Agency has also contributed to the development of radiation sterilization practices in the developing and developed countries. Radiobiological details relevant to the application of radiation as a sterilization method — such as the radiosensitivity of different contaminant microorganisms, the influence of environmental conditions prevailing during and after irradiation on the radiation responses of microbes, as well as the control and routine monitoring of the microbial contaminants on production sites — have been disseminated through internationally coordinated research programmes, training courses and the publication of technical manuals. The Agency's programme activities on radiation sterilization of medical supplies have increased interest in such practices in a number of countries, *e.g.* Czechoslovakia, Egypt, Greece, Hungary, India and the Republic of Korea. The work of the Agency is also carried out in close cooperation with a number of national and international organizations (IAEA, 1981a, 1981b). In cooperation with the WHO, the IAEA/WHO secondary standards dosimetry laboratories (SSDLs) have helped reduce problems and uncertainties concerning the administration of medical doses. The Agency, in 1981, assisted in the setting up of SSDLs in Ecuador, Indonesia, Malaysia, Pakistan, Thailand and Venezuela.

In 1964, FAO and IAEA decided to establish a joint programme for the specific purpose of assisting member states applying nuclear techniques to develop food and agriculture. A joint FAO/IAEA Division of Isotope and Radiation Applications of Atomic Energy for Food and Agricultural Development was established at the IAEA Headquarters in Vienna. Connected with this Division is the Agricultural Section of the IAEA Laboratory at Seibersdorf, outside Vienna.

The objectives of this joint FAO/IAEA programme are to exploit the potential of isotopes and radiation applications in research and development to increase and stabilize agricultural production, to reduce production costs, to improve the quality of food, to protect agricultural products from spoilage and losses, and to minimize pollution of food and the agricultural environment. The programme supplements priority areas in FAO and IAEA activities where isotope and radiation methods are particularly promising.

The Joint Division is organized by subject into six sections (soil fertility, irrigation, and crop production; plant breeding and genetics; animal production and health; insect and pest control; agrochemicals and residues; and food preservation) and executes a joint programme that focuses on: coordination and support of research; technical assistance, including training; and dissemination of information. Over 300 research institutions in member states are currently cooperating in some 25 coordinated research programmes. Each programme attempts to solve a practical problem of economic significance for developing countries. Institutes in developing countries are normally given nominal financial support, whereas institutes in the developed countries participate without payment. Each programme usually lasts up to five years. The joint FAO/IAEA programme is also responsible for some 80 technical assistance projects in 46 developing countries, as well as for the provision of training, expertise and specialized equipment. Among these there are four large-scale projects in Brazil and Peru in cooperation with UNDP and in Bangladesh and India (Swedish International Development Authority — SIDA). Large-scale projects in Indonesia, Sudan and Thailand are being planned. Since its inception, the Joint Division has helped organize, supervise or has participated in 87 agricultural training courses and study tours in 33 member states as well as at the IAEA Laboratory. Since 1969 some 800 fellowships in agriculture (average length of six to eight months) have been awarded. Scientific meetings are important in giving scientists from member states the opportunity to exchange views and to keep up-to-date in specialized fields. Approximately two symposia and two seminars annually are organized as part of the joint FAO/IAEA programme. Some 100 publications have been issued since 1964 in fields relevant to the joint FAO/IAEA programme.

Finally, the food irradiation activities of the Joint Division should be mentioned as an effective means of preservation of food, because it is an environmentally clean process which is low in energy requirement and does not leave any residue in the treated product.

The primary microbiological considerations in the development of food-irradiation operations include:

(1) Effective destruction by radiation of the microorganisms harmful to human health (radicidation).
(2) Extension of the shelf-life of perishable foods by reducing the spoilage of microorganisms in it (radurization).
(3) Elimination of toxigenic microorganisms and prevention of toxin production.

In some of the developed countries, nuclear techniques in food and agriculture have more or less been assimilated into standard laboratory techniques as part of the normal university curriculum. This is rarely so in developing member states. Moreover, agricultural information is not transferable since the agricultural ecology of developed and developing countries differs in soils, climate and crop yield. The joint FAO/IAEA programme is designed essentially to bridge this gap and to assist efforts to solve, particularly, agricultural economic problems under the prevailing socio-economic and ecological conditions.

39.2.4 United Nations Children's Fund (UNICEF)

UNICEF was established in 1946 by resolution 57 (1) of the UN General Assembly as the United Nations Children's Emergency Fund (UNICEF). In 1953 it was unaminously voted to continue for an indefinite period, enabling its programmes to be carried out in Asia, Africa, Europe and Latin America with the official name being shortened to UN Children's Fund with the retention of the well-established acronym UNICEF.

In working with the governments of several developing countries, the main areas of UNICEF aid are health, education, training, nutrition and family and child welfare. In the pursuit of the fundamental goals of helping developing countries meet some of the immediate needs for their young and of strengthening long-range services for children as essential parts of overall developmental efforts, UNICEF has undertaken a key role. Permanent health services in controlling diseases (malaria, tuberculosis, yaws, leprosy, trachoma, water-borne diseases like salmonellosis and measles) and the organization of training programmes for personnel in developing countries at all levels of work are some of the activities of UNICEF that have some microbial biotechnological components.

UNICEF has collaborated with WHO in combatting the spread of communicable diseases. Joint UNICEF/WHO-assisted mass penicillin treatment campaigns were carried out some two decades ago against endemic treponematoses (yaws, bejel and pinta), resulting in an interruption of the transmission of such diseases. The resurgence of endemic foci in West and Central Africa, Southeast Asia, the Western Pacific, the Caribbean countries and South Africa, indicates the development of appropriate surveillance techniques coupled with integrated eradication programmes. With regard to leprosy, UNICEF cooperates with WHO with respect to the provision of standardized chemotherapeutic drugs as well as training facilities in leprosy health-care systems.

UNICEF was awarded the Nobel Peace Prize in 1965.

39.2.5 United Nations Development Programme (UNDP)

UNDP, established in 1965 through the merger of the then Expanded Programme of Technical Assistance (EPTA) and the United Nations Special Fund, helps developing and low-income countries build more productive, dynamic societies and economies on the basis of the most effective use of their own natural resources and human talents. At the request of governments, UNDP supports projects that are designed to help these countries attract development capital, build trained cadres of skilled manpower, improve educational systems and apply modern technologies to expand agriculture and industry, and to improve medical, social and administrative systems. Projects are normally carried out in the field by the UN specialized agencies as well as other international institutions.

Given below are some UNDP-supported projects in different fields of biotechnology:

(1) *Latin American Network of Biotechnology Centres.* This project, to be executed by UNESCO and UNIDO, provides a network mechanism to coordinate efforts and to stimulate the collaboration of research groups in Latin America. Participating countries in the network are: Argentina, Brazil, Chile, Costa Rica, Cuba, Mexico and Venezuela.

The primary goals of the project are to: (a) use the techniques of biotechnology to attack problems of health, agriculture and industry facing the Latin American region; (b) strengthen existing research groups that are using biotechniques such as genetic engineering and fermentation technology to solve problems of special relevance to Latin America in the health, agricultural and industrial sectors; (c) strengthen the dialogue between scientists, industrialists and governments; (d) stimulate the creation of new bio-industries and investments in biotechnological research.

(2) *Maximizing Crop Production through Biological Nitrogen Fixation (BNF).* This project, to be executed directly by UNDP, is geared to the achievement of inter-institutional cooperation in the transfer of technology across continents. The main objectives of the project are to: (a) train developing country personnel in BNF technology associated with cowpeas and soybeans in Africa and beans and pasture legumes in Latin America and the Caribbean with a view to increasing the yields of these crops; (b) carry out limited research in Latin America on pertinent aspects of BNF.

39.2.6 United Nations Educational, Scientific and Cultural Organization (UNESCO)

The UNESCO programme in environmental and applied microbiology and biotechnological research traces its origins back to 1946, when UNESCO supported research that was geared to the conservation and applied use of microorganisms.

Twenty-one years ago the International Cell Research Organization (ICRO) came into being with the support of UNESCO and with the objective of advancing cell biology, particularly in developing countries. Within a few years of its foundation, ICRO was specifically invited to promote research on microorganisms and a Panel on Microbiology was formed. ICRO, through the Panel, was charged with various tasks, but principally: to establish an international network for the preservation and exchange of cultures; to promote the use of microorganisms as a natural resource; and to assist in the training of microbiologists (Porter, 1977). Since that time, UNESCO activity in the microbiological field has been done in cooperation with ICRO and with the International Organization for Biotechnology and Bioengineering (IOBB) and the World Federation for Culture Collections (WFCC), both of which were founded in the early 1970s with UNESCO support and encouragement.

In 1972 the UN Environment Programme (UNEP) joined forces with UNESCO and ICRO for the express purpose of preserving microbial gene pools and making them accessible to developing countries. Thus the Panel became the UNEP/UNESCO/ICRO Panel on Microbiology and helped set up a worldwide programme on environmental, applied microbiological and biotechnological research. Additional support has been given by such United Nations agencies as FAO, WHO, the United Nations Industrial Development Organization (UNIDO) and United Nations University (UNU).

39.2.6.1 MIRCEN network

A major development of the UNEP–UNESCO joint venture (1975–1984) was the establishment of a world network of microbiological resource centers — MIRCENs (Table 2). Among their other responsibilities, the MIRCENs are designed:

(1) to provide the infrastructure for the building of a world network which would incorporate regional and inter-regional functional units geared to the management, distribution and utilization of the microbial gene pool;

(2) to strengthen efforts relating to the conservation of microorganisms with emphasis on *Rhizobium* gene pools in developing countries with an agrarian base;

(3) to foster the development of new inexpensive technologies that are native to the region;

(4) to promote the applications of microbiology in the strengthening of rural economies;

(5) and to serve as focal centres for the training of manpower and the imparting of microbiological knowledge.

Table 2 Worldwide MIRCEN Network in Applied, Environmental and Biotechnological Research

Location	Speciality	Contact Address
Argentina	Alcohol production, kinetics of biomethanation, SCP production	F. Sineriz, Planta Piloto de Procesos Industriales Microbiologicos (PROIMI), Av. Belgrano y Pasaje Caseros, 4000 S.M. de Tucuman, Argentina
Australia	Registration of data in culture collections in developed and developing countries	V. B. D. Skerman, World Data Centre MIRCEN, University of Queensland, St. Lucia 4067, Brisbane, Australia
Brazil	*Rhizobium* technology	J. Freire, *Rhizobium* MIRCEN, IPAGRO, Caixa Postal 776, Porto Alegre, Brasil RS[a]
Canada	Fermentation technology	M. Moo-Young and D. Howell, University of Waterloo, Ontario, Canada N2L 3G1[b]
Egypt, Arab Republic of	Biotechnology (enzyme technology, general)	M. I. N. Magdoub, Ain-Shams University, Cairo, A.R.E.
Guatemala	Biotechnology (alcohol and SCP production)	C. Rolz, Central American Research Institute for Industry, Guatemala
Japan	Biotechnology (immobilized enzymes, fermentation technology, waste recycling)	H. Taguchi, International Centre for Microbial Engineering, University of Osaka, Osaka 565, Japan
Kenya	*Rhizobium* technology	S. Keya and S. Imbamba, Department of Soil Science, University of Nairobi, P.B. 30197, Kenya
Senegal	*Rhizobium* technology	M. Gueye, Centre Nationale de Recherche Agricole, Institut Senegalais de Recherche Agricoles, Bambaey, B.P.51, Senegal
Sweden	Biotechnology (enzyme engineering)	C. G. Heden, Department of Bioengineering, Karolinska Institute, S-10401 Stockholm, Sweden
Thailand	Biotechnology (food fermentations, waste recycling)	P. Attasampunna, Thailand Institute for Scientific and Technological Research, Bangkhen, Bangkok, Thailand
UK	Biotechnology, bioengineering, enzyme engineering, biodeterioration	A. T. Bull, Biotechnology MIRCEN, Institute of Biotechnological Studies, London, UK[c]
USA	*Rhizobium* technology	D. Weber and H. Keyser, Cell Culture and Nitrogen-Fixation Laboratory, U.S.D.A., Beltsville, MD 20705, USA
	Rhizobium technology	J. P. Roskoski, NIFTAL MIRCEN and Project, College of Tropical Agriculture and Human Resources, University of Hawaii, Paia, Hawaii 96779, USA
USA	Marine and aquatic microbiology	R. Colwell, Department of Microbiology, University of Maryland, College Park, MD 20742, USA

[a] In cooperation with the Universidade Federale do Rio Grande do Sul, Porto Alegre, Brasil RS. [b] In cooperation with the University of Guelph, Guelph, Ontario N2G 2W1, Canada. [c] In cooperation with the Biotechnology Institute and Studies Centre Trust, c/o The Polytechnic of Central London, School of Engineering and Science, London W1M 8JS, UK and the Commonwealth Mycological Institute, MIRCEN, Ferry Lane, Kew, Surrey, UK.

The first development of the MIRCEN world network was to establish the World Data Center (WDC) for microorganisms at the University of Queensland, Brisbane, Australia. The WDC houses a master copy of the *World Directory of Collections of Cultures of Microorganisms* and serves as a pivotal point for fostering development of culture collections in developing countries. It also strengthens interactions with activities concerning culture collections in developing countries and developed areas.

The work at the MIRCEN at the Karolinksa Institute in Stockholm, Sweden has centered on development of microbiological techniques for applying pattern recognition methods for identification of microorganisms, as well as other rapid methods for identification, including microtiter plate methods. Environmental studies are also under way, as well as production of ethanol in liquid two-phase systems. Such studies contribute to the WDC MIRCEN which is of obvious benefit to world biotechnologists.

The MIRCEN network has pioneered the organization of a series of MIRCENET computer conferences on biogas production, biconversion of lignocellulose for fuel, food and fodder for rural development in developing countries and anaerobic digestion. The plans of MIRCENET are:

(A) to help initiate closed computer conferences on defined topics such as:
   (1) microbiology;
   (2) biological nitrogen fixation;
   (3) biotechnology and integrated biomass systems;
   (4) biogas;
   (5) mushroom cultivation;
   (6) microbiology for self-sustaining communities;
   (7) networking for culture collections.
(B) To act as an information source for:
   (1) listings of important upcoming international and regional events in applied microbiology and biotechnology;
   (2) reviews of the activities in other relevant computer conferences;
   (3) identification of microbial resources;
   (4) listings of relevant bibliographic material.
(C) To provide a platform for discussions of brief presentations that are relevant to MIRCEN network activities (*e.g.* microbiology for rural development, socio-economic impact of applied microbiology).
(D) To distribute printouts of MIRCENET entries at regular intervals.

As a sequel to a request from the Japanese Federation of Culture Collections, a group of specialists met in July 1966, under the auspices of UNESCO, to consider problems relating to culture collections. At that time it was recommended that a survey of culture collections be carried out. The International Association of Microbiological Societies' (now the International Union of Microbiological Societies') section on culture collections agreed to survey the world culture collections, with the resulting publication in 1972 of the *World Directory of Collections of Cultures of Microorganisms*. A second edition of the directory is funded by UNESCO, FAO, WHO, UNU, UNIDO, UNEP and the European Economic Commission. The MIRCEN has released the second edition of the *World* Rhizobium *Catalogue* (Table 3).

In the region of East Africa, the Nairobi MIRCEN focuses on *Rhizobium* technology, playing a pivotal role in the conduct of research and training concerning *Rhizobium* holdings in Ethiopia, Kenya, Malawi, Rwanda, Sudan, Tanzania, Uganda, Zambia and Mauritius. Training courses have been organized and symposia have been held on agronomy, plant breeding, physiology, crop protection for farming systems, and nitrogen fixation.

In attempting to bring into contact scientists in Africa working on all aspects of biological nitrogen fixation, and with a view to sharing experiences, information, evaluating current local progress and, ultimately, charting a course for the efficient exploitation of BNF in the farming systems of Africa, the MIRCEN has pioneered the institutionalization of the African Association for Biological Nitrogen-Fixation.

Similarly, the MIRCEN at the Instituto de Pesquisas Agronomicas, Brazil, in collaboration with the Universidade Federale do Rio Grande do Sul, has emphasized nitrogen fixation in Latin America, with the objective of promoting *Rhizobium* technology. A large culture collection is maintained, with cultures distributed to research laboratories and inoculant factories. Training of researchers, extension workers and industrial technical staff is also carried out. The MIRCEN has produced a *Rhizobium* catalogue based on a survey of 56 institutions in Argentina, Chile, Uruguay, Bolivia, Brazil, Peru, Venezuela, Columbia, Panama, the Dominican Republic, El Salvador and Mexico. In addition, the MIRCEN has strengthened the role and activities of the Latin American Association of Rhizobiologists in promoting the use of biofertilizers in the region.

Research projects are under way in the area of biological nitrogen fixation at the MIRCEN located at the University of Hawaii (NIFTAL Project). The focus of the research is nitrogen fixation by tropical agricultural legumes, with core budget support from the US Agency for International Development and special funds from several organizations, including UNESCO. In conjunction with the NIFTAL Project, a MIRCEN at the Cell Culture and Nitrogen Fixation Laboratory at Beltsville, MD, USA also carries out studies on collection, characterization, documentation and preservation of *Rhizobium*, on distribution of cultures of *Rhizobium* for research and inoculum production in developed and developing countries, and on microbial germ plasm of useful nitrogen-fixing organisms. Recently, for the region of French-speaking Africa, a MIRCEN at Bambey, Senegal, has been established.

The MIRCEN at Ain Shams University, Cairo, Egypt, promotes activities in the fields of biotechnology. More than 1000 cultures are available in the various laboratories in that region with formal links to the MIRCEN. Training courses on conservation of microbial cultures and biotechnologies of interest to the region have been held at the MIRCEN.

The MIRCEN work in the Caribbean region is coordinated through the Central American Research Institute for Industry in Guatemala, which serves Central America in the field of biotechnology. Subsequently, training courses in bioengineering have been held, with participants from Costa Rica, Nicaragua, Honduras, Ecuador, Guatemala, the United States, Uruguay, Peru, Venezuela, El Salvador, Paraquay and the Dominican Republic.

The MIRCEN in Bangkok, Thailand, at the Thailand Institute of Scientific and Technological Research, serves the microbiological community of Southeast Asia *via* exchange of economically important microbial strains, training and fellowship programmes, and promotion of research on organisms in areas of microbiology appropriate to Southeast Asia. Active in culture collection activities and responsible for the microbial culture collection development in that region, it serves Thailand, Indonesia, Malaysia, the Philippines, the Republic of Korea and Singapore.

A significant activity of the MIRCENs are the training courses. About 60 training courses, based on a traditional ICRO pattern, have been held in developing countries on nitrogen fixation; fermentation technology; waste treatment and recycling; fermented foods; biological pest control; veterinary microbiology; environmental microbiology, including biomass and biofuel production; culture collection maintenance; and related subjects. The courses last about three weeks and include 15–30 participants, with not more than one-third originating in the host country. At least half of the conference is spent in bench work, with the faculty consisting of experts from the region supplemented with professors from abroad selected in consultation with the panel. In conjunction with such training courses, and other international meetings, panel members may be asked to stop off in several countries to undertake missions or give seminars. The latter are short, inexpensive and have many objectives: technical and public lecturing; promoting indigenous development; providing information on availability of financial resources;

Table 3 Some Details on the Contents of the *World Directory on Collections of Cultures of Microorganisms* and the *World* Rhizobium *Catalogue*

(A) *World Directory on Collections of Cultures of Microorganisms*

Country	Number of collections	Country	Number of collections
Argentina	6	Mexico	2
Australia	41	Netherlands	4
Austria	2	New Zealand	9
Belgium	3	Nigeria	2
Brazil	8	Norway	3
Bulgaria	2	Papua New Guinea	1
Canada	30	Philippines	7
Chile	2	Poland	6
China	1	Portugal	1
Colombia	1	South Africa	3
Czechoslovakia	13	Senegal	1
Denmark	3	Roumania	2
Ecuador	1	Singapore	1
Finland	2	Spain	2
France	15	Sri Lanka	1
FGR	14	Sweden	1
GDR	3	Switzerland	3
Greece	2	Thailand	9
Hungary	9	Turkey	2
India	26	Uganda	2
Indonesia	2	UK	36
Ireland	2	USA	30
Israel	2	USSR	6
Italy	11	Venezuela	3
Japan	11	Yugoslavia	1
Jordan	1	Zimbabwe	2
Malaysia	2		

Total number of countries: 53
Total number of collections: 335

Table 3—*continued*

(B) *World* Rhizobium *Catalogue*[a]

Country	Number of collections 1971	1981	Country	Number of collections 1971	1981
Argentina	—	4	Nigeria	—	1
Australia	4	10	Papua New Guinea	1	1
Belgium	1	1	Peru	—	1
Brazil	2	3	Philippines	—	1
Bulgaria	1	1	Poland	1	3
Canada	3	3	Romania	1	1
Chile	1	1	Senegal	—	1
Colombia	—	1	Singapore	—	1
Czechoslovakia	1	2	South Africa	1	—
Denmark	1	1	Spain	—	1
Finland	1	1	Sweden	1	1
France	1	1	Switzerland	—	1
GDR	2	2	Tanzania	1	1
Hungary	2	3	Thailand	—	2
India	4	10	Turkey	—	1
Indonesia	—	1	United Arab	2	2
Ireland	1	2	Republic of Egypt		
Israel	—	2	UK	3	5
Japan	3	6	Uruguay	—	1
Kenya	—	1	USA	9	45
Malawi	—	1	USSR	4	6
Malaysia	—	2	Venezuela	—	1
Mexico	—	1	Yugoslavia	2	3
Netherlands	3	—	Zimbabwe	1	1
New Zealand	1	3			

[a] *IBP World Catalogue of* Rhizobium *Collections*, compiled by O.N. Allen and E. Hamatova, ed. F.A. Skinner, 1971, Knapp, Drewett & Sons Ltd., Kingston-upon-Thames, UK.

scouting for potential sites for training courses; improving knowledge of native techniques and technologies and appreciation of local problems.

UNESCO has also been instrumental in supporting long-term postgraduate courses in microbial engineering and biotechnology. The first of these was established at Osaka University in 1973 with a particular focus on Southeast Asia, and a decade later a second course is provided by the Institute for Biotechnological Studies (University College London, Polytechnic of Central London and the University of Kent). Similar courses are being implemented in Argentina, Chile, Czechoslovakia and Portugal.

UNESCO has built upon its pioneering role since 1946 through its MIRCEN networks that are guided by the UNEP/UNESCO/ICRO Panel on Microbiology (DaSilva *et al.*, 1982; Colwell, 1983; Bull and DaSilva, 1983). Clearly, the networks and MIRCENs have made a significant difference in the way microbiology is practiced in developing countries. The programme has served to integrate microbiology infrastructures of developed and developing countries by promoting the holding of international conferences in developing countries and helping to introduce problems of developing countries into the programmes of conferences held in developed countries. The success of these activities is due largely to a policy of cooperation at the working level, with many governmental, intergovernmental and nongovernmental organizations participating, all of which provide a constellation of activities, with core funding from UNEP and UNESCO and with additional funding from FAO and UNIDO providing a high multiplier factor.

The MIRCENs network publishes newsletters, including *MIRCEN NEWS*. In addition, Oxford University Press in association with UNESCO copublishes the *MIRCEN Journal of Applied Microbiology and Biotechnology* (Vol. 1, 1985).

39.2.6.2   *GIAM conferences*

Microbial technology, with its branches in agriculture, cellular, molecular, genetic, veterinary and radiation biology research, industry, engineering and medicine, is a discipline *par excellence* for inter-agency cooperation, provision of close contacts, avoidance of duplication of projects and waste of funds. Examples of such inter-agency cooperation are International Conferences on the 'Global Impacts of Applied Microbiology' (GIAMs) which in previous years were organized by the Panel on Microbiology and cosponsored by several UN agencies and nongovernmental bodies (DaSilva, 1975). These conferences are aimed at appraising and presenting to high government officials, administrators, research workers and students, the latest developments that have scientific, economic and social applications. GIAM I (Sweden, 1963), GIAM II (Ethiopia, 1967), GIAM III (India, 1969), GIAM IV (Brazil, 1973), GIAM V (Thailand, 1977), GIAM VI (Nigeria, 1980) and GIAM VII (Finland, 1985) have led to a cognisance of microbial technology as a means towards technological development by governmental and policy decision-makers, to the emergence of microbiology in Africa, the establishment of university departments in Asia, Latin America and Africa, to the updating of microbiological curricula, the organization of national courses and fellowships, and close collaboration between scientists from the developed and developing countries.

GIAM I helped to catalyze research in microbial technology in Europe. GIAM II led to a five-year bilateral programme between the then Haile Selassie I University and the Hadassah Medical School in an attempt to strengthen microbiology in Ethiopia. GIAM III led to the establishment of University Departments at Bombay and Goa and the institution of a biennial GIAM III Fellowship at the University of Bombay. In GIAMs II and III there was strong participation in the UNESCO/IBP Symposium on SCP (single-cell protein), a field in which UNESCO, through the UNESCO/ICRO Panel on Microbiology, has closely cooperated with the working group on single-cell proteins of the United Nations system.

GIAM IV and V consolidated the established base-line in the developing and shaping of ideas and plans for the future of microbiology in the developing countries (Gyllenberg and Pietinen, 1973). The conferences have also served to 'raise the windows' on the research programmes of microbiology in the developed countries in tackling problems that 'hurt humanity and the environment', as well as to reveal motivated concern and social responsibility in the execution of research projects focusing on overall development and recent progress resulting from the impact of enzyme engineering. GIAM VI resulted in the establishment of the African network for microbiology in which the MIRCENs for East and West Africa play pivotal roles. GIAM VII at Helsinki, Finland, focused on bioinformatics (patents) and the socio-cultural aspects of biotechnology.

The importance of applied microbiology and biotechnology for the developing nations is significant, with benefits to be derived in fields as diverse as agriculture, the fermentation industry, public health, water supply and sanitation, environmental conservation and resource management, and production of food fodder and energy. The applications of microbiology are strongly interdisciplinary, interfacing engineering, applied mathematics, medicine, agriculture, the veterinary sciences, food science, toxicology and other related areas.

39.2.7 United Nations Environment Programme (UNEP)

UNEP, established in 1972 by the General Assembly in response to the report of the Secretary-General on the UN Conference on the Human Environment (5–16 June, 1972, Stockholm), is designed to promote international cooperation in the field of the environment. To this end, UNEP recommends policies and provides general policy guidance for the direction and coordination of environmental programmes within the UN system. It plays an important role in identifying significant environmental issues, framing objectives and priorities for action in dealing with these issues and in developing a concerted approach to such action by the UN system, other intergovernmental organizations, national institutions and nongovernmental organizations. The discipline of microbiology in this context is significant, especially when one considers the interactions of microbes in the micro- and macro-environments of all living forms. Creative thought and generations of skilled manpower have brought forward key microbial strains that have a direct relevance to economic development, environmental industry and judicious management of our natural resources, some of which are renewable.

UNEP, constantly aware of the potential of microorganisms in environmental management and development and as a follow-up of resolutions 39–45 of the Stockholm Conference, embarked together with UNESCO and the international scientific community through the International Cell Research Organization (ICRO) on a worldwide programme for preserving and increasing the accessibility of microbial gene pools to developing countries. Consequently, at a Panel working group meeting on Microbiology at UNEP headquarters in Nairobi early in 1974, the concept of the Microbiological Resources Centres (MIRCENs) was formulated (see Section 39.2.6).

Cooperation and coordination within the United Nations agencies and organizations are carried out at various levels and through several mechanisms. One such mechanism is the Administrative Committee on Coordination (ACC), which is chaired by the Secretary-General of the

United Nations and whose members include the executive heads of the specialized agencies such as FAO and UNESCO, as well as a number of senior officials of the UN Secretariat. Another mechanism is thematic joint programming.

Joint programming exercises are organized by UNEP. In this manner, possible duplication or overlapping of activities or gaps between the programmes of the various agencies are avoided. Thematic joint programming meetings on microbial genetic resources were held in September 1981 at UNESCO headquarters in Paris, immediately following a meeting of the Ecosystem Coordination Group, a programming and coordinating group established by UNEP whose members include FAO, IUCN (International Union for the Conservation of Nature) and UNESCO. In addition to cooperation with the UN agencies, UNEP cooperates with other international governmental and non-governmental organizations that have specialized interests or expertise. For example, in 1982, in cooperation with the Scientific Committee on the Problems of the Environment (SCOPE) of the International Council of Scientific Unions (ICSU), the SCOPE/UNEP Nitrogen Unit, together with several other sponsors (including the UNEP/UNESCO/ICRO Panel on Microbiology) organized a three-week training course on the microbiology of nitrogen cycling at the University of Dacca in Bangladesh. The Unit is also completing a new version of the review of the nitrogen cycle to update the publication in this area last issued in 1975.

Within the framework of the joint UNEP/USSR Commission for UNEP project, several activities on microbiological leaching of metals from ores were developed. A training course and seminar was held in Moscow in May/June 1982, attended by 17 participants from developing countries. An international seminar on modern aspects of microbiological hydrometallurgy immediately followed, with 15 leading international experts presenting reports. A state-of-the-art report was produced by Soviet experts, and 34 technical papers intended for teaching purposes were prepared by the international teaching staff. A manual on microbiological leaching of metals from ores for scientists and practitioners — based on the teaching material — is being prepared by an international board of editors for publication early in 1985.

In managing water weeds, UNEP promoted activities in various areas, specifically with regard to biogas fertilizer, stock feed and purification of water wastes in the leather, sugar and other industries. In the period 1976–1982, UNEP has funded a number of projects relating to the establishment of experimental rural energy centres in Indonesia, the Philippines, Senegal, Somalia and Sri Lanka. Some of these have biotechnological components. The aim of the centre in Sri Lanka was to demonstrate the feasibility of harnessing locally available renewable sources of energy and providing cheap energy for domestic and community use and for agro-based industries and cottage crafts. At the village of Pattiyapola, four windmills, a photovoltaic panel and a biogas plant were set up. The project, the first of its type, demonstrated the technical viability of producing electricity, on a small scale, from renewable sources of energy (the sun, wind and biogas). However, the technologies and systems used in the project were not economically competitive with conventional sources. Cost analysis yielded a figure of US $810 per megawatt-hour of electricity, 3–5 times the cost of electricity from diesel generators, which is primarily due to the high costs of the components. The possibility of producing these components locally is being reconsidered with a view to lowering the cost of electricity production.

UNEP has also collaborated with FAO and WHO in the area of surveillance of human taeniasis and cattle cystericosis. Draft guidelines for the prevention, surveillance and control of taeniasis/cystericosis have jointly been drawn up and are being reviewed before release. The FAO/UNEP/WHO *Guidelines for Surveillance, Prevention and Control of Echinococcus/Hydatidosis*, first issued by WHO in 1981, have been reissued by FAO in 1982.

Listed below are some UNEP projects concerning the environmental aspects of biotechnology.

(1) Contribution to the state of knowledge on field application of microbial technologies for environmental betterment.
(2) Promotion of the field application of microbial technologies for environmental betterment in developing countries.
(3) Development of a coordinated international programme in biological nitrogen fixation.
(4) Microbial technologies to overcome environmental problems from persistent pollutants.

39.2.8 United Nations Financing System for Science and Technology for Development (UNFSSTD)

As a consequence of the UN Conference on Science and Technology for Development (UNCSTD) at Vienna, Austria in 1979, an Interim Fund for Science and Technology for Development was set up specifically to aid the developing countries benefit from the applications of science and technology for development. In 1982, the General Assembly confirmed the long-term arrangements for the UN Financing System for Science and Technology for Development on a permanent basis. Based in New York, USA, the overall supervision of the UNFSSTD is entrusted to UNDP.

In 1982, UNFSSTD awarded the African Regional Centre for Technology (ARCT) a grant to define the steps for a full-scale project, especially since ARCT is expected to play a crucial role in the implementation of the Lagos Plan of Action for African economic growth. ARCT's activities concentrate on technologies for the reduction of post-harvest losses and for the use of biomass in energy production. Furthermore, a number of UNFSSTD projects have experimented with and used biotechnologies for converting agricultural and unutilized raw materials into useful products such as alcohol or proteins (Table 4). The potential impact of such work, if it were systematically applied to the developing world, is acknowledged to be considerable. As a consequence, the Government of Yugoslavia hosted in June 1982 a seminar on microbiological conversion for productive cross-fertilization of ideas and research results arising from efforts underwritten by UNFSSTD and others. Representatives of UNFSSTD-financed projects in the Philippines, Guatemala and Sudan, along with front-line researchers from biotechnology institutions in Mexico, Canada and Hungary, forged direct linkages for the continuous exchange of information among UNFSSTD-financed projects. The meeting produced detailed recommendations for establishing a network of microbiology institutions in the region for follow-up by Yugoslav authorities.

Table 4 Some Biotechnological Projects Supported by UNFSSTD

Country	Project	Executive agency
Bangladesh	Strengthening the Institute of Natural Drugs Research and Development	WHO
Ivory Coast	Medical research of haemotosis in black Africans	Government
Korea, Republic of	Pilot plant for recycling waste by-products	Government
Lesotho	Development of solar energy and biogas production	UNESCO
Philippines	Industrial chemicals from indigenous carbohydrate raw materials	UNIDO
Sudan	Cellulose technology research unit	UNIDO
Uruguay	Industrial methods to inactivate FMD virus in meat products for export markets	Government
Papua New Guinea	Sago starch hydrolysis and fermentation	Government
Yugoslavia	Microbiological conversion of raw materials and by-products of agriculture into proteins, alcohol and other products (seminar coupled to preliminary study)	Government
Regional	Ex-FERM process of fermentation ethanol (at ICAITI MIRCEN, Guatemala)	Government
Regional	Establishment of regional fermentation programming for antibiotics and other pharmaceuticals in Latin America	UNIDO
Interregional	Strengthening of capabilities in the use of agricultural systems	FAO

39.2.9 United Nations Industrial Development Organization (UNIDO)

39.2.9.1 Historical background

Following the United Nations General Assembly resolution 2152 (XXI) of 17 November 1966, UNIDO came into being in 1967, with the aim of promoting and accelerating the industrialization of the developing countries within the framework of the New International Economic Order and the activities of the agencies of the UN system. Since 1975, when the Lima Declaration and Plan of Action on Industrial Development and Co-operation was adopted by the Second General Conference in Peru, special attention has been given to efforts aimed at raising the world share of developing countries in manufacturing production from an estimated 11% to 25% by 2000 AD.

Amongst its activities concerning the promotion of industrial research and technological development, UNIDO has organized several activities in the field of applied microbiology and biotechnology. At the United Nations Conference on Science and Technology for Development (Vienna, August 1979), many speakers referred to the impact of various recent advances. As a consequence, and in order to promote meaningful technology assessment, UNIDO embarked upon a round of symposia, and commissioned a series of working papers on various high-key technologies that dealt with both threats and opportunities to developing countries. Micro-electronics and lighter-than-air transport systems were, for instance, studied within the framework of UNIDO's Technology Programme, but the biggest effort was directed towards genetic engineering and biotechnology.

39.2.9.2 International Center for Genetic Engineering and Biotechnology (ICGEB)

The study referred to above started with a conference on genetic engineering and involved several commissioned papers, some technical and others with an emphasis on the socio-economic implications of genetic engineering and biotechnology (Heden, 1981). This preparatory phase crystallized into a report (UNIDO, 1981) that called for the establishment of an International Center for Genetic Engineering and Biotechnology (ICGEB). Recommendations submitted for consideration by governments and to serve as a basis for a future work plan were:
(1) An International Center for Genetic Engineering and Biotechnology (ICGEB) should be established on the lines suggested in the report.
(2) UNIDO should follow up its initiative, pursue the question of establishment of the Center vigorously and continue to associate itself fully and actively in this activity.
(3) It should continue to associate the leading experts in the field in the setting up of the Center.
(4) It should initiate further consultations with interested United Nations agencies such as FAO, UNESCO, UNU and WHO and other international organizations such as AMBO (Asian Molecular Biology Organization), EMBO (European Molecular Biology Organization), ICRO and IFIAS.
(5) It should mobilize resources to create a small unit with a full-time project coordinator who would pursue the several activities leading to the establishment of the Center.
(6) It should carry out negotiations with interested governments and convene a meeting of participating governments where they could announce their participation and financial contribution and formally establish the Center.

Six countries offered to be the location of ICGEB at the end of a high-level meeting of eminent scientists and government officials from 35 countries, held at Belgrade, Yugoslavia, December 1982. The meeting, organized by UNIDO and the Government of Yugoslavia, affirmed that such a Center be established as soon as possible, with activities including training, research, application and information. A sub-committee comprised of an expert from either Argentina or Mexico, and one each from China, Hungary, Indonesia, Nigeria, Sweden and Yugoslavia, was nominated to visit and evaluate potential sites in Belgium, Cuba, India, Italy, Pakistan and Thailand. The committee named the cities of Bangkok, Brussels and Trieste as likely sites for ICGEB since they offered the best chances for attracting support from the international scientific community as well as financial and political assistance.

Statutes establishing the International Center for Genetic Engineering and Biotechnology were adopted at Madrid on 14 September 1983, at the close of a one-week meeting of 49 developed and developing countries. No decision on a location was taken. According to the statutes, the Center will promote international cooperation in developing and applying peaceful uses of genetic engineering and biotechnology, especially for developing countries. It will assist developing countries in strengthening their scientific and technological capabilities in the field, and will help with activities at regional and national levels. In this regard, existing national, subregional, regional and international networks are expected to participate as 'affiliated networks'.

Aside from research and development, the training of scientific and technological personnel from the Third World, both at the Center and elsewhere, will be one of the main functions of the Center. Advisory services will be provided to members to develop national technological capabilities. Among the Center's other functions are a programme of bioinformatics and collection and dissemination of information.

The statutes will enter into force when at least 24 states have deposited instruments of ratification and sufficient financial resources are ensured. At a meeting on 27 January 1984, at UNIDO, the Committee Headquarters, it was decided that the International Center consist of two component centers, to be located at Trieste, Italy, and New Delhi, India. Italy has offered funds totalling US $37 million and India US $19 million for the Center.

A draft program for the two-component international center promotes research on agriculture and animal and human health at New Delhi, and on energy, industrial microbiology, fermentation technology and protein engineering at Trieste. Attention at the latter facility will be given to the bioconversion of biomass, industrial scale fermentations and the physiology of micro-organisms deployed in oil recovery. Biological nitrogen fixation, stress tolerance in plants, vaccines, immunogenetics and tropical diseases will be pursued at the New Delhi site.

Simultaneously, UNIDO's Pharmaceuticals Division is also promoting the establishment of sub-regional centers in the field of biotechnology. An evaluation mission[1] has already visited Mexico, Costa Rica, Cuba, Guatemala, Guyana, Honduras, Nicaragua and Panama and recommended the establishment of a sub-regional centre for research and industrial development in the field of antibiotics. Another evaluation mission[2] visited Argentina, Bolivia, Brazil, Columbia, Chile, Ecuador, Paraguay, Peru, Uruguay and Venezuela and recommended the establishment of a sub-regional centre for genetic engineering and biotechnology in South America. Plans for both the sub-regional centres are being finalized and expectations are that they will interact closely with ICGEB.

In monitoring the latest developments in genetic engineering and biotechnology, UNIDO publishes the *Genetic Engineering and Biotechnology Monitor*.

39.2.10 United Nations Institute for Training and Research (UNITAR)

The UN General Assembly established UNITAR as an autonomous institution within the framework of the UN system in 1953. UNITAR was designed to catalyze the effectiveness of the United Nations in achieving its objectives (world peace and security, and economic and social progress) by providing facilities for high priority training and research projects. UNITAR's activities are comprised of (i) Project on the Future, (ii) research and (iii) training programmes. The Project on the Future concerns studies and international conferences on problems of the future of concern and interest to the UN system and with particular emphasis to the New International Economic Order. Current areas of study are petroleum and natural gas, development planning in Africa and Latin America, development strategies for arid zones, microbiological sources of energy, and geomicrobiology (Chopra and Morehouse, 1981).

39.2.11 United Nations University (UNU)

Established in 1973, the UNU functions as an international community of scholars engaged in research, postgraduate training and dissemination of knowledge in accordance with the purposes and principles of the Charter of the United Nations. Special attention is given to the continuing growth of vigorous academic and scientific communities in the developed and particularly in the developing countries. In operation, the UNU consists of a global network of institutions and scholars that are involved in a number of projects concerning the social sciences, the humanities and the natural and applied sciences. By 1981 the UNU comprised 28 associated institutions and some 120 research and training units in more than 60 countries. Several of these came into being within the framework of UNU's former programme on the Use and Management of Natural Resources that was phased out in 1981.

Major activities are: 'Regional and Global Studies', 'Global Learning', 'Development Studies' and 'Institutional Planning and Resource Development'. Five major themes cut across the programmes and many of them obviously have to take biotechnology into consideration. This is particularly true for 'Science and Technology and their Social and Ethical Implications', but certainly also for 'Hunger, Poverty, Resources and the Environment' as well as the themes related to conflict resolution and the global economy that take many developments in genetic engineering and biotechnology into consideration. The importance for developing countries to build up their own basic and applied science capabilities is emphasized, and the need to meet the threat of increased dependency posed by such fields as biotechnology and micro-electronics is also expressed.

The implications of biotechnology for meeting the Third World's needs for food and energy are, of course, extremely important and certainly attract attention within ongoing sub-programmes in areas concerned with the ecological basis for rural development, the interlinkages between food and energy, as well as a number of local developments in biotechnology and genetic engineering.

In collaboration with the Institute of Applied Science and Technology, University of Guyana,

---

1. Document/IO.489, 4 January 1982, available from UNIDO, Vienna, Austria.
2. Document/IO.519, 23 December 1982, available from UNIDO, Vienna, Austria.

UNU in 1985 outlined a project 'Dissemination of Biogas Technology to Rural Areas of Guyana' with the aim of stimulating the use of biogas systems for methane-rich gas generation, high grade fertilizer production and waste disposal in Guyanan rural areas.

The medium-term perspective notes the 'urgent need for careful scientific assessment and development of traditional biotechnologies into socially appropriate capital-saving, energy-conserving and employment-generating technologies' and intends to make full use of the strength that the United Nations University has already built up in the areas of world hunger, energy and eco-development (UNU, 1982).

For an even greater impact, the establishment of some new incorporated institutions is now being considered as part of UNU's institutional development. Programmes and allied projects concern the Institute for Natural Resources in Africa and a Regional Nutritional Center for the Gulf Countries that, for instance, may in time come to involve some bioengineers. However, two current feasibility studies on an International Institute for Biofuels Research at Guangzhou in the People's Republic of China and a national biotechnological institute at Caracas, Venezuela, are of more direct interest. They might well come to function as powerful catalysts, particularly in the fields of energy and genetic engineering.

One of the major functions of the United Nations University is to improve the exchange of research knowledge and its dissemination between scientists, scholars, researchers and decision-makers in the areas of solar, bioconversion and energy technology applicable to developing countries. The monthly publication *Abstracts of Selected Solar Energy Technology* (*ASSET*), in this regard, links scientists and engineers active in these fields, particularly in the developing countries. An activity of the Energy Systems and Policy Sub-programme of the UNU, *ASSET* in 1981 was the official newsletter of the UN Conference on New and Renewable Sources of Energy. In 1982, UNESCO and UNU, in view of the abundance of information on new and renewable sources of energy coming from the Third World and often delayed in publication in international journals, launched a subsidiary activity of *ASSET*, *i.e.* a clearing house for unpublished documents. *ASSET*, to date, has carried several important articles on biotechnologies relating to Third World development.

A history of the PAG formation and some of its activities has been referred to earlier (Section 39.2.1). Probably no activity of the United Nations system has been more useful or influential than the issuance of Guidelines for the development of new protein sources for animal and human feeding. In view of the lapse of over 10 years when these Guidelines were originally issued, it was deemed appropriate to re-examine them and make revisions as necessary. To accomplish this, UNU consulted with each of the three scientific unions principally concerned, *i.e.* the International Union of Nutritional Sciences (IUNS), the International Union of Food Scientists and Technologists (IUFoST), and the International Union of Pure and Applied Chemistry (IUPAC).

The revisions have now been completed and republished as PAG–UNU Guidelines in the UNU *Food and Nutrition Bulletin*. Guidelines 6, 12 and 15 have appeared in Volume 5, No. 1 and Guideline 7 in Volume 5, No. 2 of the *Bulletin* of 1983. Extensive revisions were made only in Guideline 7.

PAG Guideline 6, on the Pre-Clinical Testing of Novel Sources of Food, details the categories of information required on the product, including chemical and physical properties, nutrient composition, sanitary properties, and identification of naturally occurring toxins. The bulk of this Guideline, which has evolved into an international standard for judging when new protein sources can be safely subjected to trials in human subjects, is devoted to a description of testing procedures in experimental and farm animals and to the statistical interpretation of test results.

Guideline 7, dealing with the Human Testing of Novel Foods, was developed to describe the procedures for determining acceptability and nutritional value of new protein sources in human subjects. The revised Guideline reflects subsequent experience with both gastrointestinal and cutaneous allergies associated with some proposed single-cell protein sources. It stresses the importance of tolerance studies with 25–50 subjects fed either the experimental material or a conventional one in a double-blind crossover design. It proposes that the level fed be well above that of intended use, but excessive intakes are not necessary to reveal intolerance. No emphasis is given to feeding studies to determine protein value since this can be sufficiently predicted from the amino acid composition and pre-clinical studies in experimental animals.

Guideline 12, a supplementary pre-clinical guideline, was developed in response to the growing interest in the feasibility of incorporating into human diets yeasts and bacteria grown on hydrocarbon as well as carbohydrate substrates. It is equally applicable to filamentous microfungi. It deals with specifications for identity standards, substrate purity, control over process variables, and detailed data on composition.

The last of the Guidelines to be developed, No. 15, was required in relation to the safety of using animal products, produced by feeding single-cell proteins grown on substrates derived from petroleum to animals, for human consumption. It points out the impracticality of identifying traces of contaminants by feeding such products to experimental animals. Instead, reliance must be on sensitive analytical procedures to determine the levels of any undesirable residues that could arise from either the source or the media from which the proteins are derived, or from solvents or other agents employed in any stage of processing. Nutritional value should be determined in the species for which use is intended, and, as far as possible, under practical conditions. The types of studies that are relevant to establish pre-clinical safety are equally relevant in determining suitability for feeding to farm animals.

The United Nations University is setting up a network of institutes to carry out modern biotechnological research on animal and human vaccines. The network is to include laboratories in Latin American and Caribbean countries and industrialized countries.

The objective of the network is to facilitate cooperation among institutes to carry out research with clear and common goals and methodologies and to encourage the use of newer techniques of biotechnology (such as genetic engineering) and of institutions in developing countries to use these techniques in the furtherance of advanced science. This network of institutes will eventually operate in close cooperation with the proposed UNU Institute of Biotechnology in Venezuela. In the meantime it will function within the framework of the United Nations University.

Mechanisms available for establishing this network of collaborating research laboratories include:

(1) Meetings to organize the project and to present research findings.
(2) Workshops in collaborating laboratories to make junior personnel aware of the newer techniques and to train them therein.
(3) Post-doctorate fellowships (about 1 year) for well-qualified young scientists from developing countries to be tenable in laboratories in industrialized countries.
(4) Scientific exchange visits (short-term) of key senior personnel.
(5) Financial support of individual research projects.

39.2.12 World Bank Group

The term World Bank has been used to group three international financial institutions:
International Bank of Reconstruction and Development (IBRD)
International Development Association (IDA)
International Finance Corporation (IFC)
In common parlance, the name World Bank is used also as an acronym for IBRD which was established in 1945. The World Bank, conceived at the Bretton Woods Conference in 1944, aims at promoting economic development of its members states and extending technical assistance on matters relating to economic development. Until the early 1970s, nearly two-thirds of all economic assistance had been in the electrical power and transportation sectors, the remainder being devoted to water supply, agriculture, industry, tourism and communications. In recent years, agriculture and rural development has been the most important sector of lending and the energy sector has been enlarged to include the fuel energies. Both these sectors have strong biotechnological components which are referred to in some of the World Bank reports and publications:

(1) Energy in the Developing Countries, August 1980.
(2) Alcohol Production from Biomass in Developing Countries, September 1980.
(3) Renewable Energy Resources in the Developing Countries, November 1980.
(4) Agricultural Research (Sector Policy Paper), June 1981.
(5) Mobilizing Renewable Energy Technology in Developing Countries: Strengthening Local Capabilities and Research, July 1981.
(6) Evaluation of Microbial Technologies involved in fuel production, agriculture and forestry (S & T Report, No. 36), August 1981.

39.2.13 World Health Organization (WHO)

The main responsibility for intergovernmental coordination in the fight against infectious diseases rests on the World Health Organization, which was created in 1946. WHO is located in Geneva with regional offices and a network of reference laboratories that span the whole world. Even if infectious diseases attract much public attention, a quick look at the Organization's definition of health indicates a wide range of other activities where applied microbiology and biotechnology have to play a role: '*Health is a state of complete physical, mental and social well-being, and not merely the absence of disease or infirmity*' (preamble to the WHO Constitution). This statement underlines the importance not only of such areas as family planning, nutrition and health education, but also of the need for diagnostic aids and monitoring techniques that can support the surveillance activities which provide the foundation for effective health delivery systems.

WHO has several programmes that have strong microbial technological components. For example, the objectives of the programme on prophylactic, therapeutic and diagnostic substances are to promote drug quality assurance in developing countries; to recommend standard methods for evaluation of pharmaceutical preparations; to ensure regular exchange of information on the safety and efficacy aspects of pharmaceuticals and biologicals; and to assist in the standardization and quality control of biological substances, particularly in the developing countries (WHO, 1980).

For countries that depend essentially on the supply of imported drugs, an approach designed to help in assessment of such products is the WHO certification scheme on the quality of pharmaceutical products moving in international commerce. Concerning drug quality control, 58 countries had agreed to participate in the WHO certification scheme by mid-1980. Two specialized institutions, *i.e.* the Drug Quality Institute in Brazil and the Caribbean Drug Testing Laboratory in Jamaica, created in collaboration with WHO, will further enhance drug quality assurance in the region of the Americas.

International chemical reference standards for the quality control of pharmaceuticals are established and distributed by the WHO Collaborating Centre for Chemical Reference Substances in Solna, Sweden. During 1982–1983 the use of basic tests for pharmaceuticals was reviewed with due consideration to the constraints encountered when applying them in national distribution services. WHO activities in the selection of international non-proprietary (generic) names for pharmaceuticals continue: at the end of 1980 the number of these names had reached almost 4400. Attention is now being given to the production of international specifications for simple dosage forms for essential drugs. Volumes 1 and 2 of the *International Pharmacopoeia* have been produced in 1979 and 1981, respectively, and a third volume in the series is soon to be completed. Nearly 100 international chemical reference substances have now been established. The establishment of secondary reference materials for pharmaceuticals at the regional level is currently being promoted, especially among the countries of Latin America and of the Western Pacific Region. WHO has a special Panel on Biological Standardization and International Pharmacopoeia.

With relation to biological substances, the demand for international standards and reference preparations is continuously felt on account of the urgency of quality control. There are now 90 standard, 98 reference preparations and 150 reagents, established over the last 46 years, covering biological substances (antibiotics, antibodies, antigens, blood products and endocrinological substances). Several of the distributed standards go to the developing countries for the quality control of imported or locally produced biologicals, since national control authorities in these countries need the international standards in order to develop national or regional standards.

Worldwide interest in biosafety programmes has been aroused by the increasing risks for investigators and laboratory workers in biomedical research institutions and by illness or fatalities following exposure to the pathogenic agent. Airline and postal workers are concerned over the transport of biological materials. Many countries are developing special containment laboratories and hospital facilities for the isolation and treatment of persons exposed to or infected with dangerous pathogens. These problems are the subject of collaborative activities on safety measures in microbiology. In this regard, the role and functions of the large network of WHO collaborating centers for communicable diseases are being strengthened[1]. Their previous primarily reference and research functions are being expanded to include advisory and training responsibilities as part of technical cooperation with and among countries. For example, the WHO Regional Office for Europe convened a working group in 1982 at Dublin, Republic of Ireland 'to assess the likelihood of any potential hazards being associated with widespread implementation of biotechnological methods and to suggest steps which could be implemented, where necessary, to reduce possible risks to an acceptable level.

The use of chemical and biological pesticides will probably be the principal communicable disease control method of choice during the next two decades. Current WHO activities at the regional and global levels focus *inter alia* on:

(1) intensifying research on the ecology of the most important vectors and pests;
(2) re-evaluating now available chemical pesticides as well as the development of formula-

---

1. Organizational study on the role of WHO expert advisory panels and committees and collaborating centres in meeting the needs of WHO regarding expert advice and in carrying out technical activities of WHO (document EB65/1980/REC/1).

tions, specifications and guidelines for their safe and effective use in both wide-scale and village vector control operations;

(3) accelerating the development of biological control agents, with particular emphasis on those that could control malaria and filariasis vectors;

(4) intensifying studies on potential human hazards and environmental risks that may result from the use of biological and chemical pesticides in public health, and producing guidelines for their safe use;

(5) developing more useful methods for field evaluation;

(6) cooperating with countries in dealing with or preventing outbreaks of vector- or rodent-borne diseases.

WHO promotes the evaluation and testing of pesticides and bioinsecticides through field trials at its collaborating centres, *e.g.* the Interregional Vector Biology and Control at Maracay, Venezuela. The African Region benefits from the Onchocerciasis Control Programme in the Volta River Basin and the *Simulium* Control Programme in the Senegal River Basin, Guinea, Nigeria, Uganda and Zaire. Such activities are closely coordinated with those of FAO concerning the specification for agricultural pesticides.

Against this background it is natural that early on the WHO looked into the potential of various advanced technologies, including those made possible through the advances in enzyme engineering. In the last few decades the latter has opened up a range of new possibilities to increase the sensitivity and specificity of many laboratory procedures. It goes without saying that the biotechnologies related to tissue culture for the production of monoclonal antibodies and drugs like interferon have also been given close attention at WHO. This is particularly true in the venereal, diarrheal and tropical disease programmes, where also the potential of the DNA-hybrid technologies is now being actively explored. In this regard, it is worthwhile to note that a 'Working Group on Safety of Microbial Control Agents' was established in 1971, in France, at a meeting of the Society of Invertebrate Pathology (SIP) concerning safety regarding health and environment of candidate biological control agents. The Repository for Data on the Safety of Insect Pathogens has been located at the Research Unit on Vector Pathology of the Memorial University of Newfoundland, Canada, since 1972 (Laird, 1981).

The tropical disease programme of WHO, sponsored by UNDP and the World Bank, can serve as an example of the fact that transdisciplinary research efforts are often required (WHO, 1975). The programme is aimed at malaria, schistosomiasis, filariasis, trypanosomiasis, leishmaniasis and leprosy. These diseases, which threaten a thousand million people, sap the vitality and diminish the economic capacity of many developing countries. Consequently they constitute a major challenge to the 156 nations that make up the WHO. Through its Advisory Committee on Medical Research (ACMR) and a range of Expert Panels, WHO adopted a transdisciplinary outlook on malnutrition and infectious diseases that is necessary for a comprehensive, long-term goal-oriented medical research programme in the field of tropical diseases.

In starting the Special Programme on Tropical Diseases in 1975, WHO could readily draw on its research and training centers for immunology in developing countries (Nigeria, Kenya, Brazil, Mexico, India, Lebanon and Singapore) and on their parent centers in Switzerland.

Applied immunology has always played an important role for WHO, which can take credit for the widespread use of many vaccines where advances in biotechnology have influenced both quality and price. However, the production of vaccines against such single agents as diphtheria, measles and poliomyelitis is much simpler than the prophylaxis against tropical disease agents which are difficult or impossible to cultivate. In the case of malaria, there are also at least 50 and probably more than 100 different varieties of malaria 'so that persons living in hyperendemic areas, where they are constantly exposed to reinfection, may spend the first seven or eight years of their lives suffering from infection by one or other variety before they become immune to them all — if they have not died meanwhile' (Humphrey, 1976). Or take the case of schistosomiasis, where the adult worm disguises itself with a coating that makes it invisible to the immune system of the host. This results in failure to detect the protozoa hidden in the cells of people who suffer from leishmaniasis.

Obviously the advances in tissue culture, recombinant DNA techniques and the availability of susceptible experimental animals, like the four-banded armadillo for leprosy, are important for the production of antigens, but equally obviously many additional research approaches must be explored besides vaccination. Biotechnology is a component of several of these, for instance in the production of novel antibiotics and other potent drugs, and in the development of improved methods for vector control (biological control agents, pheromones, *etc.*). The facts that the malaria parasites, the trypanosomes, the *Leishmanias* and the filariae are all insect-borne, and that the schistosomes are transmitted by snails, underline the importance of vector control. At present, this approach is hampered by the increasing cost of pesticides, by insecticide resistance and by the risk of environmental pollution.

Finally, the need for simple tests which can indicate if a person is, or has been, infected by a particular parasite requires research on labelling techniques often based on developments in the field of enzyme engineering.

39.2.14 World Intellectual Property Organization (WIPO)

WIPO, a specialized agency of the United Nations System since 1974, came into existence on 26 April 1970. The aims of WIPO, based in Switzerland, are to promote the protection of intellectual property throughout the world through cooperation among member states in collaboration with other international organizations; and to ensure administrative cooperation among the intellectual property unions (Table 5). Some of these may have biotechnological implications.

Table 5   Some WIPO Activities with likely Biotechnological Implications

Activity/Instrument	Purpose
Union	
Paris	Protection of industrial property
Madrid	International registration of marks
The Hague	International deposit of industrial designs
Nice	International classification of goods and services relating to the registration of marks
Lisbon	Protection of the appellations of origin and their international registration
Locarno	International classification of industrial designs
IPC (International Patent Classification Union)	Establishment of worldwide uniformity in patent classification
PCT (details available from WIPO)	Protection of inventions through the examination of international applications
Berne	Protection of literary works
Convention and Treaties	
Rome	Protection of performers, broadcasting organizations
Geneva	Protection against unauthorized duplication of phonograms
Brussels	Distribution of satellite transmitted programme-carrying signals; protection of new varieties of plants
Budapest	Provision of international recognition of the deposit of microorganisms for the purposes of patent procedure
Geneva	International recording of scientific discoveries
Madrid	Double taxation of copyright royalties

Intellectual property is comprised of (i) industrial property in inventions, trademarks and industrial designs and (ii) copyright in literary, artistic, photographic and cinematographic works. WIPO activities help facilitate the conclusion of new international treaties and harmonization of national legislations; the extension of legal–technical assistance to developing countries; the maintenance of services for international registration as well as cooperation among member states; and the dissemination of information.

To aid the transfer of technology from highly industrialized to developing countries, a permanent programme concerning the development of cooperation in industrial property has been formulated. The programme is designed to promote and encourage inventive and innovative activity in developing countries with a view to strengthening their technological capacities and national institutions in the field of industrial property. The acquisition of technology related to such property under fair and reasonable terms and conditions is also supported.

39.3 INTERNATIONAL NON-GOVERNMENTAL ORGANIZATIONS

39.3.1 International Cell Research Organization (ICRO)

On 22 June 1962, ICRO was founded in UNESCO House as a non-governmental organization specifically designed to assist UNESCO in the implementation of its cell biology programme. The rationale for such a programme is constituted by the fact that scientific expertise is concentrated in the developed regions of the world rather than in the developing countries which are in need of a strong scientific infrastructure for helping solve their vital problems. ICRO fulfills the essential role of a clearing house, mediating the flow of science and scientists in cell biology from where these can be found towards areas of greatest need. For carrying out this task, ICRO has been building up networks of selected individual scientists in both the developed and the developing countries of the world. ICRO members are recruited from the scientific community not only for the level of their expertise, but also for their unreserved commitment to the ideals of international scientific cooperation across the ephemerous political, social, cultural and economical borderlines in the world. They also put at ICRO's disposal their educational experience, their pluriform connections in the scientific world and, not seldom, their own institutes as an operational base for programme activities like a training course.

ICRO's governing body is the Council which has about 25 members. These are elected from nominees of the panels and from representatives of the following international scientific organizations: International Council of Scientific Unions (ICSU), Council for International Organizations of Medical Sciences (CIOMS), International Union against Cancer (IUCC), International Association of Microbiological Societies (IAMS), International Union of Pure and Applied Biophysics (IUPAB), International Union of Biological Sciences (IUBS), International Union of Biochemistry (IUB), International Union of Physiology Sciences (IUPS), International Federation of Cell Biology (IFCB), and International Society of Developmental Biology (ISDB).

The Council nominates from among its members the Executive Committee which meets at least once a year and carries out the programme according to the decisions of the Council. ICRO members are individual scientists known internationally for their competence in their respective field of cell research. The main source for the enlargement of ICRO membership comes from members of the teaching staff of successive ICRO training courses, but the nomination of scientists from hitherto ill represented regions of the world is encouraged. The main instrument of ICRO is the short-term training course. In addition, a modest fellowship programme is sometimes operated. Both serve the purpose of fostering research capability in cell biology at the level of fundamental knowledge and that of practical application. Special emphasis is being placed on strengthening manpower and scientific infrastructure in the developing regions of the world.

ICRO is, of course, subject to the dynamic changes in the world of science, in which it is rooted, and at the same time to the equally dynamic changes in UNESCO which it tries to serve. In the scientific world, where the promises of future gains from research are as high as ever before, an almost universal attempt to reach an elusive balance between applied and fundamental research characterizes our decade in which the shortage of funds and diverse vicissitudes create obstacles to the development of science, both nationally and internationally. Thus ICRO has adopted, over the 15 years of its existence, a flexible, pragmatic and pluriform approach without sacrificing its ideals of truly universal, international, scientific endeavour. Some of its activities are directed towards advanced fundamental research, others to problem-solving applied training, but in all cases the interests of the developing countries take priority. ICRO further seeks coordination and cooperation, and orients its activities in such a way as to promote the benefits that applied cell biology can offer underprivileged populations in rural areas.

Such an approach leads to an increase of membership from developing countries in working panels and executive bodies of ICRO, to a strengthening of scientific infrastructure in developing regions through focus of ICRO activities on regional centers and to the identification and study of basic research questions underlying practical problems of developing countries. In this way, ICRO contributes towards a balanced participation of cell biologists from all over the globe in the work required to gain the utmost benefit from modern cell biology for the solution of major world problems. Thus, not only existing knowledge is more fully exploited, but also the time span between scientific breakthroughs and their application is reduced by anticipation through advanced training.

Early in its history, ICRO made the decision to invest a large part of its resources in the advanced training of young cell biologists. The means chosen was the training course, which offers to the young worker intensive and advanced experience in the laboratory under personal guidance by distinguished researchers. This programme has prospered; the ICRO courses are well-recognized, and their influence felt throughout the world community of life-scientists.

The typical UNESCO/ICRO course is designed for young scientists who have completed their academic training and are at the beginning of their careers in research. Some courses serve more mature scientists who need a command of the very newest techniques and concepts which are not widely accessible. The range of subjects of the courses is a broad one, unified by the principle that the properties and activities of cells are the foundation of all life processes. They include molecular biology, recent methodological progress, microbiology, membrane biology, development biology, *in vivo* and *in vitro* cell differentiation, immunology, cellular genetics, regulatory mechanisms, virology and neurobiology at the cellular level. The scientific value of these fields as well as the potential value of their applications are taken into account in the setting of priorities and the planning of the courses. The ideal has been to offer courses in all parts of the world to students from all parts of the world, with special concern for the developing countries. The record shows the approach to the ideal in the increasing geographical spread of the locations of the courses and of the participants, both students and instructors. On occasion, ICRO may join with other international organizations in the sponsorship of a course.

The training courses provide two to three weeks of the most intensive work; typically, work occupies six days a week and very often extends into the evening. Of this full-time work, 60% or more is devoted to laboratory investigation, the rest to lectures, demonstrations and evaluation of data. The laboratory work is often based on manuals written by the instructors expressly for the courses. The close and continuous interaction among students and teachers is enhanced by the fact that they usually live in the same quarters and take their meals together.

The scientific work of ICRO is overseen by five working Panels, which are:
(1) Panel on Structure, Function and Regulation.
(2) Panel on Genetics, Reproduction and Development.
(3) Panel on Applied Microbiology and Biotechnology.
(4) Panel on Cell Biology of Higher Plants and Plant Cell Biotechnology.
(5) Panel on Animal Cell Biotechnology.

As a result of official action at the 1965 General Conference of UNESCO, the Director General invited ICRO to establish a panel for the promotion of applied microbiology. This panel was appointed as an advisory body for a separate programme of UNESCO. In 1975 the Executive Director of the United Nations Environment Programme (UNEP) asked the panel to serve in an advisory capacity also to this UN body. This double appointment, and the many direct applications of microbiology, imposed on the panel an autonomous and more diversified activity. The Panel, previously referred to as the UNEP/UNESCO/ICRO Panel on Microbiology, is now known as the ICRO Panel on Applied Microbiology and Biotechnology. The Panel's activities with respect to MIRCENs and the GIAM conferences have been discussed in Section 39.2.6.

39.3.2   The International Federation of Institutes for Advanced Study (IFIAS)

The International Federation of Institutes for Advanced Study (IFIAS) was founded in 1972 at the initiative of the late Swedish Nobel Laureate Arne Tiselius, under the auspices of the Nobel Foundation of Sweden and the Rockefeller Foundation of the USA. It emerged from a commonly expressed need among scientists and intellectuals in many parts of the world for anticipatory analysis of world problems, and for an increased social responsibility on the part of scientists. The organization makes special efforts to involve young scholars in its work.

IFIAS is located in Stockholm and is financed by membership fees from 29 member institutes spread over the globe; by grants from individuals, foundations, UN bodies, the European Community and private corporations; and by overheads on projects. The latter are normally funded by special research grants from private and governmental sources.

When one looks back over the first decade of operations, it becomes obvious that the thrust of IFIAS programmes is centered around two main issues, *i.e.* (i) What is the 'carrying capacity' of the Earth, and which are the natural or man-made constraints? and (ii) What is the potential of science and technology to offer solutions in time, in view of various socio-cultural and institutional constraints? Special attention has been given to renewable resources and the requirements of a sustainable biomass production. For example, within the framework of the UNEP/ IFIAS 'Save our Soils' project, 15 case studies focusing on the technical, economical, legislative and social constraints to proper soil management were undertaken. This has led to many projects where the potential of biotechnology has been considered. This also has been the case in studies on the impact of climate change on the character and quality of human life and in the evaluation of the possibility of a self-reliant development. The latter programme, which emerged as a consequence of a comprehensive study of the developments in enzyme engineering (IFIAS, 1975), asks the question 'Is food and fuel self-reliance possible?' (Anderson, 1979), and involves both projects that have a socio-economic emphasis ('enhancing bioproductivity of coastal zones in developing countries' and 'nutrition, health and education') and a technological thrust ('biological fuel cell and bioelectric systems'). The possibilities offered by biotechnology and genetic engineering have been prominent in many of these projects.

39.3.3   International Foundation for Science (IFS)

The International Foundation for Science is an organization that provides promising young scientists and technologists with support for their research work in developing countries. IFS was founded in 1972 and is supported by scientific academies and research councils in 58 countries, of which two-thirds are in the developing part of the world, as well as by UNESCO. The Foundation is governed by an International Board of Trustees and is located in Stockholm. Traditional IFS support is in the form of fairly small grants (average about US $7000) and scientific advice. The criteria for a research grant are the scientific quality and promise of the researcher and the research project, and the relevance of the project to the needs of the country concerned. Applicants must be from the developing world and carry out research in their own respective countries.

IFS support is currently limited to seven research subjects, all of which have a biotechnological dimension: aquaculture, animal production, vegetables (including oil seeds and fruits), mycorrhizae and afforestation, fermentation and applied microbiology, natural products, and rural technology. Both the aquaculture and animal production themes, for instance, consider new feed resources such as agricultural waste; the plant theme includes microbial symbiosis; and the forestry theme has a substantial fungal component. Although the extraction of natural products and the development of rural technologies could benefit from advances in biotechnology, the field has its center of gravity in the theme 'fermentation and applied microbiology'. Between 1974 and 1980, IFS support to this theme was of the order of 44 research projects for a total cost of US $427 000 (IFS, 1980). Research areas covered were: fermented foods, single-cell protein, mushroom cultivation, mycotoxins, production of chemicals, lactic acid fermentations and energy-related fermentations.

39.3.4 International Organization for Biotechnology and Bioengineering (IOBB)

This non-profit organization was formally established in 1968 by Professors C. G. Heden, I. Malek, G. Terui, H. G. Gyllenberg and E. C. Gaden, Jr. on the occasion of the 3rd International Fermentation Symposium at New Brunswick, New Jersey, USA. Supported morally and financially by UNESCO, IOBB has been responsible for various training courses in biotechnology and bioengineering, and promotes exchange of information through a newsletter *Biotechnologia*. The objectives of IOBB are:

(1) To maintain contact with laboratories and institutes working in the field throughout the world and especially in the developing countries.
(2) To arrange and apply fellowships for educational visits in the developing countries and for people in the developing countries to attend courses, symposia, congresses, universities, laboratories and institutes in other countries.
(3) To maintain committees as may be appropriate for international cooperation and agreement.
(4) To arrange meetings, symposia and seminars to ensure the dissemination of new or pertinent knowledge in the field and to maintain contact between people and organizations in the field.
(5) To maintain contact with relevant UN organizations.
(6) To encourage research in biotechnology and bioengineering, and participate to realize the results in practice.
(7) To organize training programmes for biotechnologists and bioengineers of all nations (Lilly, 1975).

Membership of IOBB is open to laboratories and institutes concerned with biotechnology and bioengineering. A Member Laboratory or Institute consists of a laboratory or laboratories possessing facilities for and conducting an active programme of research and/or teaching in biotechnology and/or bioengineering. A group of laboratories in close physical proximity whose interests are complementary may constitute a single center. (Lilly, 1975). IOBB Member Laboratories exist in Australia, Argentina, Belgium, Brazil, Canada, Chile, Czechoslovakia, Finland, France, Federal Republic of Germany, Guatemala, India, Israel, Japan, Kuwait, Malaysia, Mexico, New Zealand, Nigeria, Philippines, Poland, Singapore, Spain, Sweden, Turkey, UK, Thailand, USA and Venezuela.

IOBB has established the following working committees: Single Cell Protein; Bioenergy and Biofuels; Education and Training; Bioreactor Design and Operation; Microbial Process Development; Biotechnology in Developing Countries; Immobilized Enzymes and Cells; Biological Wastetreatment and Pollution Control.

IOBB has international relationships with the UNEP/UNESCO/ICRO Panel on Microbiology, IUMS, IFIAS, and the International Unions of Pure and Applied Chemistry (IUPAC) and of Pure and Applied Biophysics (IUPAB).

39.3.5 International Organization for Standards (ISO)

Following a meeting in London in 1946, delegates from 25 countries decided to create a new organization to follow up on the work of the International Federation of National Standardizing Associations (1926–1942) and of the United Nations Standards Coordinating Committee (1944–1946) with the specific purpose of facilitating 'the international coordination and unification of industrial standards'. The International Organization of Standards (ISO) works through some 2000 technical bodies involving more than 100 000 experts. It has published 4580 standards in all fields except electrical and electronic engineering, which is the responsibility of a separate organization (ISO, 1982).

ISO has 72 member bodies, 70% of which are governmental organizations or organizations incorporated by public law. There are also 17 correspondent members, which normally are organizations in developing countries which do not yet have their own national standards body.

The work of ISO is carried out through technical committees that may establish subcommittees and working groups. By the end of 1981 there were 162 technical committees, 599 subcommittees and 1430 working groups.

Almost 400 international organizations have liaison status with ISO, including all UN specialized agencies working in similar fields. The reverse is also true, *i.e.* ISO has consultative status with the UN Economic and Social Council (ECOSOC) and with nearly all other bodies in the UN family. Often this relation paves the way for a certain division of labor. This can be illustrated by the ISO Technical Committee on Agricultural Food Products (TC 34), which concentrates on methods for sampling and analysis whereas the Codex Alimentaricus Commission (see below) is more concerned with the establishment of quality standards.

From the point of view of biotechnology it might be of interest to note that TC 34 has 15 subcommittees (for instance on animal and vegetable propagation materials, animal feedstuffs, milk and milk products, meat and meat products and microbiology) as well as a large number of working groups concerned with matters such as storage, non-meat proteins, analytical methods for pet foods, antibiotics, *Clostridium perfringens*, enterococci, Lactobacteriaceae, mesophilic clostridial spores, psychrotrophs, yeasts and moulds.

For a bioengineer it might also be of interest to know that there exist Technical Committees on Starch (including derivatives and by-products) — TC 93, and on much of the equipment he uses (compressors — TC 118; computers — TC 97; laboratory glassware — TC 48; pumps — TC 115; refrigeration — TC 86; sieves — TC 24; vacuum technology — TC 112; and valves — TC 153). Also air quality (TC 146) and water quality (TC 147) are the concern of special technical committees. However, with regard to laboratory instrumentation, most of the standardization work is actually done by the World Association of Societies of Pathology, which issues a regular newsletter including reports from its Commission on World Standards (CP). Even though this concentrates its attention on instrumentation for clinical laboratories, many of its studies on instrument repair and maintenance are of wider relevance — as is also the focus on biotechnology in clinical chemistry: monoclonal antibodies and microbial enzymes. With regard to safety standards for DNA-hybrid work, special reference should be made to the NIH guidelines. The handling of pathogens is covered in a safety manual, 'Safety Measures in Microbiology,' recently issued by WHO; and the special hazards involved in the industrial use of microorganisms have been studied as a basis for guidelines by the Commission of the European Communities (Sargeant and Evans, 1979).

The need for quality standards and internationally agreed methods of examination of biological products was felt very early in the food and drug field. However, it was not until 1962, after a UN conference on food standards that the 'Codex Alimentaricus Commission' (CAC) was created. International activities concerning microbiological specifications for foods are covered in depth in a review by Olson (1978). See also Section 39.2.2.

The Commission is located at the headquarters of the United Nations Food and Agriculture Organization (FAO) in Rome, and it is responsible for the implementation of a joint FAO/WHO programme for the standardization of foods. Its work is reflected in regular publications that describe the standards and their application in the member states (121 countries as of July 1981). The basic aims of the programme are to protect the health of consumers, to foster sound commercial practices and to harmonize legislation in order to make international trade easy.

The availability of well-characterized microbial strains is, of course, a prerequisite for most activities in biotechnology. This is the concern of the Commission for Culture Collections of the International Union of Biological Sciences (IUBS). Every four or five years this body arranges international conferences, normally associated with training courses.

39.3.6 International Union of Microbiological Societies (IUMS)

Founded in 1930 as the International Association of Microbiological Societies (IAMS) and recognized as the Division of Microbiology of the International Union of Biological Sciences (IUBS), IUMS adopted its present title in 1980. The aims of IUMS are to encourage the continuity of international microbiological congresses, organize meetings and symposia, encourage research and the establishment of scholarships, and maintain contact with the UN system and other professional microbiological societies. Several of its activities are carried out in close cooperation with UNESCO and the UNEP/UNESCO/ICRO Panel on Microbiology.

Following reorganization at the 10th International Congress, IUMS is comprised of the following:

Sections:	Bacteriology, Virology and Mycology
International Committees:	Economic and Applied Microbiology
	Microbiological and Immunological Documentation
	Microbial Ecology
	Systematic Bacteriology (with 24 sub-committees)
	Nomenclature of Viruses
	Food Microbiology and Hygiene
Commissions:	Microbiological Specifications
	Microbiological Standardization
	Microbial Genetics
	Yeasts and Yeast-like Organisms
Federations:	World Federation for Culture Collections
	International Federation for Enteric Phage Typing

IUMS publishes a monthly *International Journal of Systematic Bacteriology* and a bimonthly titled *Intervirology*, in addition to 'The International Code of Nomenclature of Bacteria'. Comprised of about 60 000 members in 56 countries, through their national societies, membership in IUMS on a geographical basis is six countries in Africa, nine in North and South America, 13 in Asia, two in Oceania and 24 in Europe.

With regard to standards and testing of biological products other than foods, an area of particular interest to bioengineers is represented by the vaccines. In this field WHO issues the official standards, but much of the groundwork is done by the 'Commission on Microbiological Standardization' of IUMS. This Commission was initiated in 1955 when manufacturers, state controllers and European research workers met in Lyons in order to try to coordinate their activities to reach agreement on terminology and requirements. It first became a Permanent Section of IAMS (which later became IUMS) and in 1970 took the name 'International Association of Biological Standardization'. This has 550 members in 54 countries, organizes congresses and symposia, and issues a journal on Biological Standardization.

The International Congresses of the Association normally take place every two years and cover problems related to research, production and control of immunological products in the fields of human and veterinary medicine. They were earlier published as a series, *Progress in Immunobiological Standardization*, that supplemented the *Symposia Series in Immunobiological Standardization*. However, the two series have now been combined under the name *Developments in Biological Standardization*.

39.3.7 World Federation for Culture Collections (WFCC)

In 1962 the Canadian Committee on Culture Collections (an associate committee of the National Research Council of Canada) convened a Conference on Culture Collections. In recognition of the world-wide importance of culture collections, the Conference, attended by 266 scientists from 28 countries, recommended that the then International Association of Microbiological Societies (IAMS) establish a Section on Culture Collections (Martin, 1963). At its meeting in Tokyo in 1968, the Section recommended that consideration be given to the establishment of a World Federation for Culture Collections with a constitution in keeping with the proposed new Statutes of IAMS. In 1970, at the Xth International Congress at Mexico City, the Section on Culture Collections was reconstituted as the World Federation for Culture Collections (Lapage, 1972).

In parallel with these developments, the Japanese Federation of Culture Collections, through the Japanese Delegation to UNESCO, in 1963 submitted a proposal for a 10-year programme for research in microbiology. This proposal, emphasizing the need for an international federation of culture collections and the establishment of specialized culture collections throughout the world, was referred to IAMS. In 1964, IAMS presented its brief to UNESCO, which initiated action on the programme by convening a meeting of culture collection specialists in Paris in 1966. Consequently, in 1968, an International Conference on Culture Collection was held in Tokyo, Japan, and it was resolved that an international federation for culture collections should be established. A draft constitution for the Federation was submitted to the Section on Culture Collections at the Congress in Mexico City in 1970, and with the amendments agreed by the Section on Culture Collections forms the Statutes of the World Federation for Culture Collections (WFCC, 1972).

The WFCC has approached the problem of cataloguing the known and available microbial genetic resources through the establishment of a World Data Centre which is located in the University of Queensland, Brisbane, Australia (Skerman, 1973). The early stages of the development of the World Data Centre were supported by funds from UNESCO, WHO and the Commonwealth Scientific and Industrial Research Organization (Australia). The National Research Council of Canada and the University of Queensland contributed significantly by way of staff and material assistance. Since 1975, through agreement with UNEP, UNESCO and the University of Queensland, the World Data Centre gained formal international status and was assured of continuity.

The WFCC World Data Centre maintains close liaison with international bodies such as the WHO Reference Centres, the International Centre of Information on Antibiotics (WHO), Mutant Registry of ICRO, the *Rhizobium* Collection of IBP (FAO), Codata (ICSU) and the Information Retrieval Service (UNESCO).

The Data Centre has as its prime role the collection, storage, retrieval and distribution of information about culture of microorganisms (Table 6). The necessary software has been developed to allow computer storage and manipulation of all of this information. An extensive second edition of the *World Directory of Collections of Cultures of Microorganisms* has been produced. The book consists of a directory to 335 collections from about 50 countries and a list of species maintained therein. The World Data Centre is developing the use of systems of recording culture collection data and a WFCC committee on Data Processing has been formed to correlate and devise methods of data processing for various types of microorganisms.

Table 6   Research Activities and Facilities of the World Data Centre MIRCEN

(A) *Facilities available to culture collection curators and researchers*
   1. The acceptance of data from research workers, matching it against data in the centre, and supplying the output, either classified or unclassified, to the research worker
   2. The location of cultures in various culture collections
   3. The provision of information on the geographical distribution of organisms, host ranges, or the occurrence of organisms with specific properties
   4. The provision of lists of all collections maintaining particular organisms
   5. The provision of lists of content of any collection
   6. Supplying culture collections with their own collection data in punch-card form if they wish to indulge in some local operations using the data

(B) *Research activities*
   1. Collection of strain data
   2. Compilation of data on economically important strains (*e.g. Rhizobium*)
   3. Updating of data of the *World Directory of Collections of Cultures of Microorganisms*
   4. Interactions with other MIRCENS in culture collection work
   5. Provision of training facilities within the MIRCEN context

---

The WDC with support from FAO and UNESCO has produced the World Catalogue of *Rhizobium* Collections. Valuable information on 144 collections in 48 countries is contained in the book. In this connection, an initial compilation of standardization methods has been published and produced (Skerman, 1969). It provides a starting point from which a critical assessment of such methods with comparative trials between each collection is planned.

In the area of training, the WFCC has organized several culture collections conferences (Tokyo, 1968; Sao Paulo, 1973; Bombay 1976; and Brno, 1981) as well as several specialized culture collection courses at Osaka (1972), Sao Paulo (1973), Bombay (1976), Brno (1981) and Istanbul (1983).

The WFCC is continuing and developing liaison with other international bodies, such as the Reference Centres of WHO, the International Centre of Information on Antibiotics (WHO), the

Mutant Registry of ICRO and the *Rhizobium* Collection of IPB (FAO). Over the past years there has been close liaison between the WFCC Committee on Patent Organisms and the World Intellectual Property Organization. As a Federation within IAMS, the WFCC is a Commission within the Division of Microbiology of IUBS. At the IUBS General Assembly held in Norway in 1973, the WFCC was also admitted as a Commission to the Divisions of Botany and Zoology. The WFCC has an international representation for the role of culture collections in the maintenance of microorganisms and tissue cell lines from all three major divisions of living organisms.

39.3.8 Committee on Genetic Experimentation (COGENE)

This Committee was founded in 1976 by the 16th General Assembly of the International Council of Scientific Unions (ICSU). The aims of the Committee are to review, evaluate and make available information on practical and scientific benefits, safeguards and containment facilities. In addition, consideration is given to environmental health-related and other consequences arising from the disposal of biological agents constructed by recombinant DNA techniques.

COGENE provides a unifying forum in the area of genetic experimentation for ICSU and its Unions (chemistry, biological sciences, biochemistry, biophysics, physiological sciences, pharmacology, nutrition, immunology). The value of COGENE to biologists has been stressed as that of providing an international, independent body that can address questions that transcend national interests or to which reference can be made when particular problems arise.

COGENE came into existence at a time in which potential hazards of recombinant DNA research were a matter of concern in certain circles. The two first working groups were on guidelines and risk assessment. As long as it seems necessary, COGENE will continue to evaluate national guidelines for the regulation of recombinant DNA research and the existing efforts to assess conjectural risks with the help of the present working groups. In the area of training, COGENE has organized training courses in recombinant DNA techniques in Sao Paulo (Brazil), Varanasi (India), Ljubljana (Yugoslavia) and Hong Kong. Future such activities are planned for Latin America and Africa. Following the example of the 1979 symposium sponsored with the Royal Society and the 1981 conference co-sponsored with the Italian National Research Council, COGENE seeks to stimulate open and timely discussion of issues of concern to basic scientists engaged in genetic experimentation.

39.4 LIST OF ABBREVIATIONS AND ADDRESSES OF INTERNATIONAL ORGANIZATIONS

AMBO   Asian Molecular Biology Organization

ARCT   African Regional Centre for Technology
Km4, 5 Route de Rufisque
P.P. 2435, Dakar
Senegal

CAC   Codex Alimentarius Commission
(see FAO)

COGENE   Committee on Genetic Experimentation
COGENE Secretariat
Institut Jacques Monod
2 Place Jussieu, 75251 Paris Cedex 05
France

COSPAR   Committee on Space Research
COSPAR Secretariat
51 Boulevard de Montmorency
75016 Paris
France

ECOSOC   Economic and Social Council of the United Nations
United Nations Headquarters
New York, NY 10017
USA

EMBO   European Molecular Biology Organization
              EMBO Secretariat
              Postfach 1022 40
              D-6900 Heidelberg 1
              Federal Republic of Germany

FAO   Food and Agricultural Organization of the United Nations
              Via delle Terme di Caracalla
              00100 Rome
              Italy

GIAM   Global Impacts of Applied Microbiology
              (International Conferences)

IAEA   International Atomic Energy Agency
              Vienna International Centre
              PO Box 100
              A-1400 Vienna
              Austria

IBRD   International Bank for Reconstruction and Development
              1818 H Street NW
              Washington, DC 20433
              USA

ICGEB   International Centre for Genetic Engineering and Biotechnology
              c/o UNIDO
              Vienna International Centre
              PO Box 300
              A-1400 Vienna
              Austria

ICRO   International Cell Research Organization
              c/o UNESCO
              7 Place de Fontenoy
              75700 Paris
              France

IDA   International Development Association
              1818 H Street NW
              Washington, DC 20433
              USA

IDRC   International Development Research Centre
              Box 8500
              Ottawa K1G 3H9
              Canada

IFC   International Finance Corporation
              1818 H Street NW
              Washington, DC 20433
              USA

IFIAS   International Federation of Institutes for Advanced Study
              Ulriksdal Slott
              S-17171 Solna
              Sweden

IFS   International Foundation for Science
              Grev Turegatan 19
              S-11438 Stockholm
              Sweden

IGO   Intergovernmental organization

IOBB	International Organization for Biotechnology and Bioengineering IOBB Secretariat, ICAITI PO Box 1552 Guatemala City Guatemala
ISO	International Organization for Standardization 1 rue de Varembé CH-1211 Geneva 20 Switzerland
IUCN	International Union for Conservation of Nature and Natural Resources Avenue du Mont Blanc 1196 Gland Switzerland
IUFoST	International Union of Food Science and Technology c/o CSIRO, Division of Food Research PO Box 52 North Ryde, NSW 2113 Australia
IUMS	International Union of Microbiological Societies IUMS Secretariat, Department of Genetics The University of Newcastle Ridley Building Claremont Place Newcastle-upon-Tyne NE1 7RU UK
IUNS	International Union of Nutritional Sciences IUNS Secretariat, Institute of Biology 20 Queensberry Place London SW7 2DZ UK
IUPAC	International Union of Pure and Applied Chemistry IUPAC Secretariat 2–3 Pound Way Cowley Centre Oxford OX4 3YF UK
MIRCEN	Microbiological Resource Centers (See Table 2)
NGO	Nongovernmental organization
OAU	Organization for African Unity PO Box 3243 Addis Abbaba Ethiopia
OIE	Office International des Epizooties 12 rue de Prony 75017 Paris France
PAG	Protein–Calorie Advisory Group of the United Nations Room A-606, United Nations New York, NY 10017 USA
PAHO	Pan-American Health Organization 525 23rd Street NW Washington, DC 20433 USA

SCOPE	Scientific Committee on Problems of the Environment
	SCOPE Secretariat
	51 Boulevard de Montmorency
	75016 Paris
	France

SIDA	Swedish International Development Authority
	S-105 25 Stockholm
	Sweden

SIPRI	Stockholm International Peace Research Institute
	Sveavägen 166
	S-113 46 Stockholm
	Sweden

UNEP	United Nations Environment Programme
	PO Box 30552
	Nairobi
	Kenya

UNESCO	United Nations Educational, Scientific and Cultural Organization
	7 Place de Fontenoy
	75700 Paris
	France

UNFSSTD	United Nations Financing System for Science and Technology for Development
	1 United Nations Plaza
	New York, NY 10017
	USA

UNICEF	United Nations Children's Fund
	866 United Nations Plaza
	New York, NY 10017
	USA

UNIDO	United Nations Industrial Development Organization
	Vienna International Centre
	PO Box 300
	A-1400 Vienna
	Austria

UNITAR	United Nations Institute for Training and Research
	801 United Nations Plaza
	New York, NY 10017
	USA

UNU	United Nations University
	Toho Seimei Building
	15-1 Shibuya 2-chome
	Shibuya-ku, Tokyo 150
	Japan

WFCC	World Federation for Culture Collections
	c/o Food Research Institute
	Colney Lane
	Norwich NR4 7UA
	UK

WHO	World Health Organization
	Ave Appia
	1211 Geneva 27
	Switzerland

WIPO   World Intellectual Property Organization
        34 Chemin des Colombettes
        1211 Geneva 20
        Switzerland

—   World Bank
        (see IBRD)

39.5 REFERENCES

Anderson, R. (1979). *Biological Path to Self-Reliance*. Van Nostrand, New York.
Barnaby, W. (1981). International foundation for science: aid for young scientists. *Bull. At. Sci.*, November, 26
Bull, A. T. and E. J. DaSilva (1983). World networks for microbial technology. *Soc. Gen. Microbiol. Q.*, **10**, 6–7.
Chopra, R. and W. Morehouse (1981). *Frontier Technologies, Developing Countries, and the United Nations System after Vienna*. UNITAR Science and Technology Working Series, Paper No. 12, New York.
Colwell, R. R. (1983). A world network for environmental, applied and biotechnological research. *ASM News*, **49**, 72–73.
DaSilva, E. J. (1975). *Life Sciences Programme: Microbiology, Cell and Molecular Biology and Interdisciplinary Brain Research: A Survey and Analysis: 1961–1974*. UNESCO Document SC/WS/70, Paris.
DaSilva, E. J., P. Attasampunna, M. El-Nawawy, J. Freire, J. Halliday, S. Keya and C. Rolz (1982). Les banques de souches. *Biofutur*, October, 35–40.
Duke, M. B. and M. A. Reynold (1973). Lunar sample quarantine procedures: interaction with non-quarantine experiments. Paper L.7.5, COSPAR Conference, Konstanz, West Germany.
Grobstein, C. (1976). Recombinant DNA research: beyond the NIH Guidelines. *Science*, **194**, 1133–1135.
Gyllenberg, H. G. and P. Pietinen (1973). *Future of Microbiology in Developing Countries*. Report Series, Department of Microbiology, University of Helsinki.
Heden, C. G. (1970). A professional verdict over BW. *New Sci.*, 10 September, 518–520.
Heden, C. G. (1981). *The Potential Impact of Microbiology on Developing Countries—A Fountainhead of Hope or a List of Lost Opportunities?* Report to UNIDO, April 15, Vienna.
Humphrey, J. M. (1976). The social and scientific challenge of parasitic diseases in developing countries. *Sci. World*, **20**, 11.
IAEA (1981a). *The Agency's Technical Co-operation Activities in 1981*. Document GC (xxvi)/INF/206, Vienna.
IAEA (1981b). *The Annual Report for 1981*. Document GC (xxvi)/664, Vienna.
IDRC Reports (1980). Communications Division, IDRC, P.O. Box 8500, Ottawa, Canada K1G 3H9.
IFIAS (1975). *Socio-Economic and Ethical Implications of Enzyme Engineering*. IFIAS Report No. 10, Solna, Sweden.
IFS (1980). *Directory of Grantees and their Projects 1974–1980*. The International Foundation for Science, Sibyllegatan 47, Stockholm, S-11442, Sweden.
ISO (1982). *ISO Memento*. Geneva.
Laird, M. (1981). Repository for data on the safety of insect pathogens. In *Microbial Control of Pests and Plant Diseases 1970–1980*, ed. H. D. Burges. Academic, London.
Lapage, S. P. (1972). World Federation for Culture Collections: Xth International Congress for Microbiology: minutes of the Extraordinary Meeting of the Provisional Board. *Int. J. Syst. Bacteriol.*, **22**, 404–405.
Lilly, M. D. (1975). International Organization for Biotechnology and Bioengineering. *Biotechnol. Bioeng.*, **17**, ii–iii.
Martin, S. (1963). Culture collections: perspectives and problems. *Proc. Specialists' Conf. Culture Collections, Ottawa, 1962*. University of Toronto Press.
Olson, J. C., Jr. (1978). Microbiological specifications for foods: international activities. *Food Technol.*, **32**, 55–57.
Porter, J. R. (1977). UNEP/UNESCO/ICRO Program in Microbiology. *ASM News*, **43**, 9–14.
Porter, J. R. (1979). Recognition of microbiology by international organizations. *ASM News*, **45**, 144–145.
Pramer, D. (1984). Microbiology needs of developing countries: environmental and applied problems. *ASM News*, **50**, 207–209.
Revelle, R. (1981). A good start, but miles to go. *Bull At. Sci.*, Nov., 27.
Sargeant, K. and C. G. T. Evans (1979). *Hazards Involved in the Industrial Use of Microorganisms*, Document EUR6349EN. Commission of the European Communities, Brussels, Belgium.
SIPRI (1973). Convention on the Prohibition of the Development, Production and Stockpiling of Bacteriological (Biological) and Toxin Weapons and their Destruction. *SIPRI Series*, Vol. III. The Problem of Chemical and Biological Warfare. Almqvist and Wiksell, Stockholm.
Skerman, V. B. D. (1969). *Abstr. Microbiol. Methods*. Wiley, New York.
Skerman, V. B. D. (1973). Statement on the WFCC Center for storage, retrieval and classification of data on microorganisms. *Int. J. Syst. Bacteriol.*, **23**, 477.
UNIDO (1981). *The Establishment of an International Center for Genetic Engineering and Biotechnology (ICGEB)*. Report of a Group of Experts. Document 1./S/254.
UNU (1982). Medium-term perspective of the United Nations University 1982–1987. *United Nations University Newsletter*, January.
Werber, M. (1973). *Objectives and Models of the Planetary Quarantine Program*. NASA Document SP-344, Washington, DC.
World Federation for Culture Collections (1972). Statutes. *Int. J. Syst. Bacteriol.*, **22**, 407.
WHO (1970). *Health Aspects of Chemical and Biological Weapons*. World Health Organization, Geneva.
WHO (1975). *WHO's Role in the Development and Co-ordination of Research in Tropical Diseases*. Document WHA 28.71, World Health Organization, Geneva.
WHO (1980). *Proposed Programme Budget for the Financial Period 1982–1983*. Document PB/82–83, Geneva.

WASTE MANAGEMENT AND POLLUTION CONTROL

40
Introduction

J. A. HOWELL
University College of Swansea, UK

The waste treatment chapters in Section 3 are intended to complement those chapters in **Volume 1** which describe the fundamental characteristics of biochemical, microbiological and engineering processes. In spite of this, some attempt has been made to give the particular fundamental aspects of microbiology, biochemistry and engineering which apply particularly to the problems of waste treatment, and the early chapters describe several systems which differ from those generally found in fermenters.

One obvious difference is in the raw material itself, which is under far less control than in the synthetic processes. The waste treatment organisms must grow on whatever comes down the pipe rather than on a specially prepared medium. Painter discusses these points. As a result of this, control of waste treatment plants can be quite tricky, and a substantial section on control has been provided. This is divided into a number of sections including those on the difficulty of measurement of the important parameters; one on successful strategies adopted in waste treatment, and for the iconoclastic view, another chapter which suggests that maybe too much emphasis has been placed on the control of wastewater treatment plants which would be better left to the microorganisms to get on with.

The microbiology of waste treatment plants is usually more complex, there being many organisms present interacting with each other. When the incoming stream is subject to disturbances and the plant suffers shock loads, the population of microorganisms changes, and may result in a stable or unstable process, dependent on the ability of the organisms to adapt and adjust to these changing concentrations. Hamer's chapter on microbiology introduces the reader to the variety of organisms present and discusses their interactions.

The stoichiometry of waste treatment is complicated by the vast array of compounds that enter into the process. It is therefore often futile to discuss in detail the decomposition of every single compound that is being treated, and so stoichiometry, instead of being based on elemental analysis of the single component, is often based on chemical oxygen demand, which is a conserved quality and can be related to electron acceptors and used in electron balances. It makes a useful base for analysis.

It is impossible to work on a convenient molar basis for a waste stream. There is difficulty in identifying exactly which molecules are present, and so a mass balance approach is adopted for stoichiometry. In particular cases, this may appear more complex, but in fact it considerably simplifies the overall approach. Irvine and Bryers show how traditional molecular-based stoichiometry arguments can be adapted to mass balances.

There are, however, some molecules or entities which are so widespread in waste streams that they need to be discussed more extensively, notably the lignocellulosics. Their degradation is important in waste treatment, and as they are solids their degradation is more complex than for soluble wastes. Knapp's chapter on lignocellulosics also covers a number of points not dealt with elsewhere in the treatise and is of interest to those who are concerned with the biodeterioration of materials, especially the structural material wood.

Wastes come in many forms. Most wastes which have to be dealt with in treatment plants are nutritious to the organisms that are being used to treat the waste, and as a consequence their interaction with receiving waters is generally to stimulate growth rather than inhibit it. However, there are other classes of wastes which need to be dealt with; some are amenable to removal by chemical coagulation and flocculation as described by Gregory. Those wastes, while amenable to biological degradation, in large concentrations can adversely affect growth. These inhibitory or slightly toxic materials must be treated, and require systems slightly different in design and control compared to those used for ordinary sewage. Yet other classes of waste are acutely toxic, cyanide being an obvious example: in general they are not treated biologically and are outside the scope of this volume. Some aspects of treatment and regulation are discussed in the chapters by Melcer.

There are a number of toxic wastes which are within the scope of this volume and these are the materials with long-term (chronic) toxicity. These may be readily treated by microorganisms and, if released into the environment, have no immediately obvious toxic effect, but can cause serious problems in the long term. These have been noted as priority pollutants in the United States by the Environmental Protection Agency and can, in many cases, be successfully treated as Melcer describes. Should they, however, be released, then their effect needs to be considered carefully, especially when they have carcinogenic, mutagenic and teratogenic effects. Parry and Waters discuss the compounds which cause these effects, and how tests for such effects are conducted.

In dealing with actual treatment processes, the most common mode in use is especially that where growth is in a suspended form — especially the activated sludge process and its variants. The basic design equations, oxidation ponds, maturation ponds and other low energy systems are described by Stentiford. These are extremely important in places where there is little opportunity for supervision of a skilled nature, where there is generally a large available area, and especially where the climate is hotter.

In colder climates where land is more at a premium, intense systems are required, and here the activated sludge process merits considerable discussion in several of its variants. Some even more intense systems are often used since these are often possible by improving the aeration efficiency of the activated sludge system by using either pure oxygen or compressed air. These are discussed in Boon and Downing's chapter, which treats aeration in general but concentrates on the more recent forms of high-intensity aeration.

Special attention needs to be paid to the removal of nitrogen, and nitrification-of-ammonia systems are now available which perform these functions in concert with the general removal of carbonaceous material. It is pointed out how these removal systems need to be designed together to give an overall system design, rather than attempting to design each individual removal process as a separate entity.

Fixed film systems in the form of the trickling filter are possibly among the oldest systems ever used, and yet the most modern treatment systems often use organisms immobilized in some form. The rotating biological contactors and fluidized bed systems are especially important and show great promise for the high intensity removal of wastes.

Where wastes are highly concentrated, it is economically appropriate to degrade them anaerobically. Anaerobic decomposition can occur at a much higher intensity than aerobic composition/decomposition, but in general the total treatment of waste by this technique is not possible and polishing by aerobic methods is required. The economics may be favoured by the generation of methane gas, although in many cases the balance between demand and supply is not easy to achieve. Many think it is more important to consider anaerobic digestion as an efficient remover of chemical oxygen demand and to regard methane generation as a possible, but not essential, part of the economics.

Disposal of waste materials on land has long been practised, but only recently has there been any attempt at controlling the decomposition of these wastes in order to achieve economic benefit as part of a rational approach.

Composting is well established as an amateur technology though recent attempts have been made to bring it into a controlled environment, even to the extent of building rotating composters, although the simpler windrows are a most effective means of composting. It is important to develop mathematical models which allow the critical assessment of their performance. Large scale land disposal can lead to waste heaps or, preferably, deposits in beds which can be made to yield considerable quantities of combustible gas in a controlled manner over an extended period of time. Municipal landfills occupy extremely large volumes compared with traditional fermenters or waste treatment systems and, as a result, the quantity of gas generated is of economic significance.

The problems facing a practitioner in biotechnology and waste treatment are similar to those in any other process technology. A plant must be designed and operated to achieve a required performance. The basis for design may result from small scale experiments. The designer wishes to take such information as he has from the small scale to allow him to design a large scale plant. In order to do this he needs to have information as to how the process responds to environmental changes, and to changes in the scale. The designer must understand and predict the influence of the environmental variables, such as mixing, temperature, aeration, *etc.*

In addition, a general process analysis is required to determine the overall performance, and how much material can be expected to be available to catalyse nitrification and denitrification processes. All of these are encompassed in models of the treatment process, and biochemical engineers attempt to take fundamental information from biochemistry and microbiology to create quantitative models of the systems that can then be used for reliable design of treatment plants, interpretation of performance and general trouble-shooting. Comparison between the performance of plants of different designs in an accurate fashion is essential before the economic comparisons and decisions as to the adoption of new technology can be made.

The technology that has been described in this section is right at the forefront of current practice. Not all developments are given equal weight. In the waste treatment industry a wide range of technologies are in use, and it is not always the 'Concorde' technology that is the most appropriate for a particular occasion. In fact, in many cases, advances in use of simpler technology are just as vital and significant as those on the more elaborate and newsworthy ones. Nevertheless, it is clear that developments in anaerobic digestion, toxic waste treatment, control and monitoring, fixed film processes and high intensity aeration systems are areas which we shall see changing relatively rapidly in the near future, and it is intended that these be kept well up to date by periodic reviews. In addition to the advances of technology, there are obvious advantages in increased fundamental understanding. The way in which we look at the processes is continually developing as more knowledge is available and we can develop a more mature approach to interpreting the data that we already have.

The chapters on fundamental considerations of fixed film systems and design equations and activated sludge processes, the production of methane by anaerobic digestion, control and instrumentation, the ecology of microbial treatment systems, are all areas where the authors of this treatise share with the reader the latest advances in technology. They also reflect changes that have occurred in such thinking within the past three years, and thus one can expect further developments in the very near future. Much of these developments can, of course, be related to the recent surge in funding of innovative systems, especially marked in the United States, but also elsewhere where a greater interest in advanced biological treatment of waste has been forthcoming. Possibly it is the legislation of the United States that has stimulated these developments in all countries throughout the world. It is good to find that major contributions to the development of the field have been reported from so many different countries.

41

Stoichiometry and Kinetics of Waste Treatment

R. L. IRVINE
University of Notre Dame, IN, USA

and

J. D. BRYERS
Swiss Institute for Water Resources and Water Pollution (EAWAG), Dübendorf, Switzerland

41.1 INTRODUCTION

Stoichiometry defines the relationship that one chemical species has to another in a reaction. Kinetics involves the rate at which the reaction proceeds and the dependence of that rate upon the concentration of the chemical species and environmental factors such as temperature and pH. Stoichiometry and kinetics are so fundamental to the fields of chemistry and biology that a simple recounting of the titles of the books and journal articles written on the subject would consume the space allocated to this chapter. As a result, this chapter has been limited to a general discussion of stoichiometry as it relates to single and multiple reaction systems and how it can be applied to reactions that are important to researchers and practitioners in the field of biological waste treatment. A brief but formal discussion of kinetics is presented without elaborate detail.

Four key concepts are developed in this chapter. The first is that the fundamental approach to stoichiometry described by Aris (1969) for chemical reactions can be easily extended to biological systems. Irvine (1980) and Irvine *et al.* (1980) established this earlier but without the modifications included herein. The second is that the molar-based approach necessary in so many chemical reaction engineering applications must be replaced by a mass-based approach because of the uncertain chemical composition of both the reactants and products. Herbert (1976) developed this notion rather nicely but restricted his analysis to pure cultures and did not use the more formal approach of Aris (1969). The third is that chemical oxygen demand (COD) be used as the basis of measurement for all reactants and products. The power of this approach was brought to the authors' attention by Gujer (1980). While Gujer (1980) does not fully capture the concepts of system energetics described by Roels (1980), application of his principles reduces each reaction in a multiple reaction system to a simple mass balance with manageable stoichiometric coefficients, while allowing the user to obtain a macroscopic understanding of energetics. The fourth and final concept is that structure must be added to biological systems if the necessary design and operational decisions are to be made in the field of biological waste treatment. As a result, many of the examples used to discuss multiple reaction systems include intracellular components in the reaction set.

Reaction systems covered in this chapter involve aerobic heterotrophs, anoxic heterotrophs (denitrification), anaerobic heterotrophs (anaerobic digestion) and aerobic autotrophs (nitrificaton). The procedures established are independent of the reactor types and configurations selected. That is, concepts developed may be used to analyze both fixed film and suspended growth systems. The rates of formation that are presented herein simply represent one term in a material balance that is used to model a physical system. The chapters which follow describe these models. As a result, no attempt is made in this chapter to develop mathematical models for any biological waste treatment system (*e.g.* activated sludge or trickling filter.)

41.2 A SINGLE REACTION

41.2.1 Overview

Busch (1971) summarized earlier studies involving glucose utilization by mixed cultures in dilute batch systems and found the overall reaction reported below to be representative.

$$24C_6H_{12}O_6 + 59O_2 + 17NH_3 \rightarrow 17C_5H_7NO_2 + 59CO_2 + 110H_2O \qquad (1)$$
$$\text{(glucose)} \qquad\qquad\qquad \text{(cells)}$$

This equation actually represents the net result of many intracellular and extracellular reactions but is presented here in overall form so that principles developed below can be described simply. The molecular formula for cells is obviously approximate and would, of course, depend upon the organism in question and the overall physiological state of the organisms involved.

Equation (1) is on a molar basis. On a mass of glucose basis, this equation can be rewritten as follows:

$$1C_6H_{12}O_2 + 0.44O_2 + 0.07NH_3 \rightarrow 0.44C_5H_7NO_2 + 0.60CO_2 + 0.46H_2O \qquad (2)$$

All stoichiometric coefficients have been expressed in terms of the grams of reactant (*e.g.* oxygen) or product (*e.g.* cells) consumed or produced for each gram of glucose utilized. As written, the total mass of substrates consumed on the left hand side of the reaction must, within round-off error, equal the total mass produced on the right hand side, and each of the stoichiometric coefficients can be thought of as being yield coefficients. For example, 0.44 g of biomass would be produced for each gram of glucose consumed.

Equation (2) can also be rewritten as a carbon atom balance which considers only carbon-containing compounds; here all stoichiometric coefficients are reduced to the basis of a unit of glucose carbon mass consumed.

$$1 \text{ glucose carbon} \rightarrow 0.59 \text{ cell carbon} + 0.41 \text{ CO}_2 \text{ carbon} \qquad (3)$$

In equation (3) the mass of carbon consumed is equal to the mass of carbon produced. Maximum yield on this basis would be 0.59 g of cell carbon produced for each gram of glucose carbon consumed.

A similar relationship can be established in terms of the COD equivalent of each compound or component which changes oxidation state. In equation (1) only nitrogen does not change its oxidation state. The COD mass equivalent of glucose and the assumed molecular formula for cells can be determined directly from equations (4) and (5).

$$1C_6H_{12}O_6 + 1.07O_2 \rightarrow 1.47CO_2 + 0.60H_2O \tag{4}$$

$$1C_5H_7NO_2 + 1.42O_2 \rightarrow 1.95CO_2 + 0.32H_2O + 0.15NH_3 \tag{5}$$

Both of these equations are on a mass basis and represent the mass of oxygen required to completely oxidize a unit mass of either glucose (equation 4) or cells (equation 5) to carbon dioxide and water. Conceptually, -1 g of oxygen is necessary to oxidize (eliminate) 1 g of oxygen itself. As a result, equation (2) can be rewritten in terms of a unit mass of COD equivalents as noted in equation (6).

$$1 \text{ glucose COD} + (-0.41)O_2 \rightarrow 0.59 \text{ cell COD} \tag{6}$$

As before, the total mass of COD utilized on the left hand side of equation (6) equals that produced on the right. The yield in this case is 0.59 g of biomass COD produced for each gram of glucose COD consumed.

The above analysis demonstrates how any biochemical reaction can be written as a separate 'mass balance' using any one of a number of measurement parameters (*e.g.* total mass, carbon or COD).

41.2.2 General Development for a Single Reaction

A reaction written in the general form shown in equation (7) can be readily adapted to a wide variety of biological applications in terms of both mechanism development and kinetic implementation.

$$a_1A_1 + a_2A_2 + \ldots + a_kA_k + \ldots + a_nA_n = 0 \qquad r' \tag{7}$$

where a_i = stoichiometric coefficient for the ith component (moles), A_i = the ith component, A_1 to $A_k = k$ reactants, A_{k+1} to $A_n = n - k$ products, n = total number of components (reactants plus products), r' = rate of reaction (moles volume^{-1} time^{-1}), $a_i < 0$ for reactants, $a_i > 0$ for products, $a_i = 0$ for conservative components.

The value for mechanism development is discussed later under multiple reaction systems. As far as kinetic implementation is concerned, all mass balances, whether they be for suspended growth or fixed film systems, require knowledge of the rate of formation of the ith component, r_{fi}. If the notation defined by equation (7) is used, r_{fi} is defined as shown in equation (8).

$$r_{fi} = a_i r' \tag{8}$$

Because a_i is negative for a reactant, r_{fi} takes on negative values for components that are utilized.

Using the background established up to this point, the molar relationship given in equation (7) can be generalized to fit the mass relationship used in equation (9) as is shown below.

$$(-1)A_1 + (-Y_2)A_2 + \ldots + (-Y_k)A_k + \ldots + (+Y_n)A_n = 0 \qquad r \tag{9}$$

where $Y_i = \dfrac{|a_i|}{|a_1|} \dfrac{MW_i}{MW_1}$

where Y_i = yield of component A_i (mass of component A_i consumed or formed per unit mass of component A_1 consumed), MW_i = molecular weight of component A_i, and r = rate of reaction (mass of component A_1 consumed volume^{-1} time^{-1}).

Written in this form, Y_i is always positive. A negative sign is used for reactants (*e.g.* $-Y_2$ for component A_2) and a positive sign for products.

Note that equation (7) is normalized with respect to component A_1, to produce equation (9). For future reference, the normalizing component will be termed the unit base stoichiometric coefficient and will be discussed in greater detail for multiple reaction systems.

The rate of formation of the ith component in equation (9) is given by

$$r_{fi} = (\pm Y_i)r \tag{10}$$

where the sign depends upon whether A_i is a reactant or product and the units of r_{fi} are given by

$$\frac{\text{mass of component } A_i \text{ consumed or formed}}{\text{mass of component } A_1 \text{ consumed}} \times \frac{\text{mass of component } A_1 \text{ consumed}}{\text{volume} \times \text{time}}$$

$$\text{or } \frac{\text{mass of component } A_i \text{ consumed or formed}}{\text{volume} \times \text{time}}$$

As can be seen from this analysis, the rate of reaction, r, is tied directly to that component which is identified as A_1. While the functional form of r does not depend upon the selection of A_1, the numerical value of the reaction rate constant does.

Each of the equations discussed earlier in Section 41.2.1 can be rewritten in the form represented by equation (9). For example the carbon-based reaction, equation (3), is now revised to read as in equation (11).

$$(-1)G + (+0.59)X + (+0.41)CO_2 = 0\, r \tag{11}$$

where G = glucose carbon (mass), X = cell carbon (mass), CO_2 = carbon dioxide carbon (mass), (-1) = stoichiometric coefficient for G (dimensionless), $(+0.59)$ = stoichiometric coefficient for X (mass X formed per unit mass G consumed), $(+0.41)$ = stoichiometric coefficient for CO_2 (mass CO_2 carbon formed per unit mass G consumed), and r = rate of reaction (mass volume^{-1} time^{-1}). Note that the sum of the stoichiometric coefficients must equal zero when the reaction is written in the form described by equation (11).

Equation (11) also defines a single rate of reaction, r, in terms of the rate of formation of each component as is shown in equation (12).

$$\frac{r_{fG}}{(-1)} = \frac{r_{fX}}{(+0.59)} = \frac{r_{fCO_2}}{(+0.41)} = r \tag{12}$$

Because of this property, the rate of formation of any component is known in terms of that component's stoichiometric coefficient and the overall rate of reaction, r.

41.3   MULTIPLE REACTIONS

41.3.1   Overview

Literally thousands of reactions take place within a biological system. Clearly only a limited number of these can be considered in any reasonable reaction scheme. Those reactions selected must somehow be important in research or in the design and operation of a full-scale treatment facility. A reasonable test for importance might be that omission of one of the reactions would produce sufficient misinformation that either the research findings cannot be sufficiently understood or design and operation are inadequate.

Unfortunately, the number of reactions that should be considered important must be decided on a case by case basis. Most biological reaction schemes typically include reactions for growth and death (or maintenance). A simple example of a reaction scheme containing more than these two reactions would be for a system for which nitrification is taking place. In this case, either one or two additional reactions are usually considered. More complex situations are usually not delineated at the reaction scheme level, often because simple procedures for identifying and quantifying the necessary components and reactions are not generally available. For example, organism selection pressures in staged reactor systems can be markedly different than those in single tank completely mixed systems. In particular, reactions involving intracellular accumulation and utilization of storage products (and the resulting impact on organism selection) are typically far more important in a staged system than in a single tank system and should be introduced into the basic reaction scheme.

The procedures described below can be easily adapted to develop any set of reactions that are felt to be necessary for proper system understanding. The procedures are general and are independent of the type of systems involved whether they be suspended growth or fixed film. The conservation laws used to model the entire system simply include information from the reaction scheme developed.

41.3.2 General Development for Multiple Reactions

The multiple reaction scheme for molar quantities that is equivalent to equation (7) is presented below in equations (13).

$$
\begin{array}{ll}
a_{1,1}A_1 + a_{1,2}A_2 + \ldots + a_{1,n}A_n = 0 & r_{1'} \\
a_{2,1}A_1 + a_{2,2}A_2 + \ldots + a_{2,n}A_n = 0 & r_{2'} \\
\quad\vdots \qquad\quad \vdots \qquad\qquad\quad \vdots & \quad\vdots \\
a_{m,1}A_1 + a_{m,2}A_2 + \ldots + a_{m,n}A_n = 0 & r_{m'}
\end{array}
\tag{13}
$$

where $a_{i,j}$ = stoichiometric coefficient for the jth component in the ith reaction; negative for reactants, positive for products (moles), $r_{i'}$ = rate of the ith reaction (moles volume^{-1} time^{-1}), m = number of reactions, and n = number of components.

Each of these reactions must be transformed to a mass basis using the procedures discussed earlier. Finally, the rate of formation of the ith component, r_{fi}, is given by equation (14).

$$
r_{fi} = \sum_{j=1}^{m} a_{j,i} r'_j
\tag{14}
$$

In the single reaction development, the molar-based reaction was transformed to the unit mass-based reaction by converting the stoichiometric coefficients to yield coefficients by identifying one of the components as the unit base stoichiometric coefficient and normalizing. Conceptually, the selection of the component is arbitrary and left to the discretion of the specific researcher and system under consideration. However, certain selections of the unit base stoichiometric coefficient result in stoichiometric coefficients that have more conventional connotations. For example, the stoichiometric coefficient for X (cell carbon) in equation (11) has units normally attributed to cell yield. If cell carbon had been selected as the normalizing component stoichiometric coefficient for glucose carbon (G) would have been minus the reciprocal of yield and that for X, plus 1.

In the multiple reaction system, each reaction should be treated separately. The particular component (a reactant) selected to have the stoichiometric coefficient of (-1) should be determined for each reaction so that each resulting normalized stoichiometric coefficient has a physical meaning compatible with the user's thinking. The equations (15) represent a unit mass-based reaction scheme for equations (13).

$$
\begin{array}{llllllll}
(-1)A_1 + & \ldots + & (-Y_{1,k})A_k + & (+Y_{1,k+1})A_{k+1} & + & \ldots + & (+Y_{1,n})\ A_n = 0 & r_1 \\
(+Y_{2,1})A_1 + & \ldots + & (-Y_{2,k})A_k + & (-1)A_{k+1} & + & \ldots + & (+Y_{2,n})\ A_n = 0 & r_2 \\
\quad\vdots & & \quad\vdots & \quad\vdots & & & \quad\vdots & \\
(+Y_{m,1})A_1 + & \ldots + & (-Y_{m,k})A_k + & (+Y_{m,k+1})A_{k+1} & + & \ldots + & (-1)\quad A_n = 0 & r_m
\end{array}
\tag{15}
$$

where $Y_{i,j} = \dfrac{|a_{i,j}|}{|a_{i,\alpha}|} \dfrac{MW_j}{MW_\alpha}$

where $Y_{i,j}$ = yield of component A_j in reaction i (mass of component A_j consumed or formed in the ith reaction per unit mass of component A_α consumed in the ith reaction), α = component A_α, that component selected to be the unit base stoichiometric coefficient with a value of (-1) in the ith reaction, and r_i = rate of ith reaction (mass of component A_α consumed in the ith reaction volume^{-1} time^{-1}).

As for the single reaction, $Y_{i,j}$ is always positive with the sign used to differentiate between reactant and product.

The rate of formation of component A_j in the ith reaction, $(r_{fj})_i$, is given by equation (16).

$$
(r_{fj})_i = (+Y_{i,j})\, r_i
\tag{16}
$$

The units of $(r_{fj})_i$ are the mass component A_j consumed or formed in the ith reaction volume^{-1} time^{-1}. As a result, the net amount of component A_j consumed or formed by all the reactions, r_{fj}, is given by equation (17).

$$
r_{fj} = \sum_{i=1}^{m} (r_{fj})_i
\tag{17}
$$

That is, the net rate of formation of any component is equal to the sum of the incremental rates of consumption or production of that component in each of the reactions. If the net rate of formation is negative, that component is being consumed in the system; if positive, the component is being produced.

As a matter of practical importance, each reaction must be isolated as much as possible so that each of the stoichiometric and kinetic coefficients can be intrinsically determined. For example, organisms can be aerated in the absence of extracellular organic substrates so that coefficients for the death reaction can be determined. Unfortunately, complete definition of all coefficients in all reactions is not easy and often forces users to reduce the reaction set to a more 'manageable' number of reactions. By doing this, important information may be lost from any analysis undertaken. In such cases, qualitative information should be obtained from the expanded reaction set using estimates for the 'unknown' coefficients. Such information can then be used in conjunction with results from the reduced reaction set to make decisions.

41.4 FUNCTIONAL FORMS OF REACTION RATE

In general the rate of any reaction is dependent upon the concentration of each of the reactants and each of the products and the system temperature and pressure as is shown in equation (18).

$$r' = r'(T, P, C_1, \ldots, C_k, C_{k+1}, \ldots, C_n) \tag{18}$$

where T = temperature (°C), P = pressure (atmospheres), and C_i = concentration of reactants ($i = 1, \ldots, k$) and products ($i = k+1, \ldots, n$) (mole volume^{-1}).

For systems in which pressure is not a factor and only the forward reaction is of interest, equation (18) reduces to equation (19).

$$r' = r'(T, C_1, \ldots, C_k) \tag{19}$$

For most reactions, the assumption is made that the functional form of the rate expression can be set equal to the product of separate functions, each depending solely on one variable, *i.e.* either temperature or the concentration of a specific reactant as is shown by equation (20).

$$r' = K(T)f_1(C_1). \, . \, .f_k(C_k) \tag{20}$$

where $K(T)$ is the rate constant and depends only on temperature (units depend upon the order of the reaction).

In chemical kinetics (*i.e.* non-biological), the most common functional form for the reaction rate constant, K and f_i are as given in equations (21) and (22).

$$K(T) = A\exp\left(-E_a/RT\right) \tag{21}$$

where A = Arrhenius constant (units depend upon order of reaction), E_a = activation energy (kJ mole^{-1}), R = ideal gas constant (kJ kelvin^{-1} mole^{-1}), and T = absolute temperature (kelvin)

$$f_i(C_i) = C_i^{|a_i|} \tag{22}$$

where $|a_i|$ = absolute value of the stoichiometric coefficient for component A_i. As a result, the rate of a chemical reaction is usually given as

$$r_i' = K_i(T)C_1^{|a_i|} \ldots C_K^{|a_K|} \tag{23}$$

where $|a_1| + |a_2| + \ldots + |a_K|$ is said to be the order of the entire reaction and $|a_i|$ is said to be the order of the reaction with respect to component A_i.

In most cases, the chemical reaction is written such that two reactants at most are involved. Except for the unusual cases of an elementary reaction (*i.e.* one that takes place exactly as described by the reaction), $|a_i|$ is replaced by α_i and α_i must be determined empirically. Equation (24) summarizes a typical rate form for a chemical reaction.

$$r' = K(T)C_1^{\alpha_1}C_2^{\alpha_2} \tag{24}$$

where α_1, α_2 = empirical constants.

Rate forms for biochemical reactions also usually involve a rate constant and the product of two functions, each of which depends on the concentration of one reactant. A typical rate form for a biochemical reaction is given by equation (25).

$$r = K(T)f_1(C_1)f_2(C_2) \tag{25}$$

where C_i now has units of mass volume^{-1} and the prime has been removed from r.

The dependency of the rate constant on temperature is usually written in the form shown in equation (26).

$$K(T) = K(20)\Theta^{(T - 20)} \tag{26}$$

where $K(20)$ = rate constant at 20 °C, Θ = constant determined experimentally, and T = temperature (°C).

As defined by equation (25), components A_1 and A_2 usually represent the rate-limiting substrate (usually the electron donor), S, and cell mass, X, respectively. The dependency on cell mass concentration, x, is often first order. The dependency on the electron donor concentration, C_S, may be either zero order, first order, Monod or one of the many existing forms which can account for the presence of a growth rate inhibitor. The general rate form for biological reactions is given by equation (27).

$$r = K(T)f(C_S)x \tag{27}$$

where $f(C_S) = 1$ (zero order)

or $\quad f(C_S) = C_S$ (first order)

or $\quad f(C_S) = \dfrac{C_S}{K_S + C_S}$ (Monod)

or $\quad f(C_S) = \dfrac{C_S}{C_S + K_S(1 + C_i/K_i)}$ (inhibition)

where K_S = saturation constant (mass volume^{-1}), K_i = inhibition constant (mass volume^{-1}), and C_i = concentration rate of inhibitor (mass volume^{-1}).

Implicit in equation (27) is that the rate of this reaction is zero order with respect to the electron acceptor (*e.g.* oxygen), hydrogen ion concentration (*i.e.* pH), all nutrients, *etc.* Wastewater treatment systems are ideally designed such that these components are either not rate limiting (*e.g.* oxygen and nutrients) or maintained within a range which does not influence the rate of reaction (*e.g.* pH). Otherwise, none of these components can be ignored in the rate expression.

A mixed culture growing on a mixed substrate represents such a complicated system that any of the four functional forms for $f(C_S)$ in equation (27) must be considered as empirical. The Monod form can be nicely developed from a simple one enzyme–one substrate model. Nevertheless, it too must be thought of as an empirical relationship. Upon reflection, the selection of the kinetic expression is not nearly as critical as the selection of the proper number of reactions. This point is emphasized repeatedly later in the chapter.

As a final note, equation (27) should be rearranged to define both the specific metabolic rate, q, and the specific growth rate, μ. If the reaction in question uses an electron donor as the unit base stoichiometric coefficient (see equation 11 for example), q $(=r_{fs}/x)$ and μ $(=r_{fx}/x)$ are defined as shown in equations (28) and (29).

$$q = \frac{-r}{x} = -K(T)f(C_S) \tag{28}$$

$$\mu = \frac{+Y_i r}{x} = +Y_i K(T)f(C_S) \tag{29}$$

where $+Y_i$ = yield coefficient for component A_i where component A_i represents cell mass, r_{fs} = rate of formation of substrate, and r_{fx} = rate of formation of organisms.

In a multiple reaction system the overall specific rate, of course, would be equal to the sum of the specific rates for each reaction.

If $f(C_S)$ takes the Monod form in a multiple reaction system, the specific growth rate for the *i*th reaction, μ_i, is given by equation (30) where the product $Y_{i,j}K_i(T)$ is defined as the maximum specific growth rate, μ_{mi}, for the *i*th reaction.

$$\mu_i = Y_{i,j}K_i(T)\frac{C_S}{K_S + C_S} = \mu_{mi}\frac{C_S}{K_S + C_S} \tag{30}$$

where $K_i(T)$ = rate constant for the *i*th reaction.

The concept of a maximum specific growth rate follows directly from the fact that $C_S/(K_S + C_S)$ numerically reaches a maximum value of 1 when $C_S \gg K_S$.

41.5 SUMMARY IN MATRIX NOTATION

Equations (14) and (16) can be rewritten in terms of a stoichiometric coefficient matrix Y, a component vector A and a rate vector r where

$$Y = \begin{pmatrix} \pm Y_{1,1} & \pm Y_{1,2} & \cdots & \pm Y_{1,n} \\ \pm Y_{2,1} & \pm Y_{2,2} & \cdots & \pm Y_{2,n} \\ & \vdots & & \\ \pm Y_{m,1} & \pm Y_{m,2} & \cdots & \pm Y_{m,n} \end{pmatrix} \tag{31}$$

$$A = \begin{pmatrix} A_1 \\ A_2 \\ \vdots \\ A_n \end{pmatrix} \qquad r = \begin{pmatrix} r_1 \\ r_2 \\ \vdots \\ r_n \end{pmatrix} \tag{32}$$

such that
$$YA = 0 \tag{33}$$

and
$$r_f = Y^T r \tag{34}$$

where r_f is the rate of formation vector and Y^T is the transpose of the Y matrix *i.e.*

$$r_f = \begin{pmatrix} r_{f1} \\ r_{f2} \\ \vdots \\ r_{fm} \end{pmatrix} \tag{35}$$

Actually equation (33) is the matrix notation form of equation (15) and equation (34) is the matrix notation form of equations (16) and (17).

The convenience of matrix notation is obvious. A full set of reactions and rate forms can be developed in a general sense with the specific reactions and rates that are applicable to a particular situation selected on a case by case basis.

41.6 STRUCTURED MODELS

Any reaction scheme which identifies intracellular compounds as members of the component vector A is said to be structured. Examples of components typically included in structured models are ribonucleic acid (RNA), deoxyribonucleic acid (DNA), proteins, glycogen and poly-β-hydroxybutyrate (PHB). All living cells require RNA, DNA and proteins for normal life processes so it should not be surprising that models have been developed that include their interactions. Glycogen and PHB are found as storage reserves in many (but not all) organisms depending upon the growth conditions imposed and the nutrients available. In general the relative distribution of all intracellular compounds in a given organism (*i.e.* the physiological state of that organism) would have a direct and dramatic impact on the organism's morphology and overall ability to compete in the environment imposed. Unfortunately, little information is available at this time on the impact of environmental conditions on the physiological state of the individual organisms that might be present in a mixed culture. As a result, use of such detailed models has been limited to research efforts only.

Because of the potential power of structured models in the design and operation of full-scale biological treatment facilities, these models are included among the reactions described in the next heading. In particular, the classical growth–death reactions are expanded to include intracellular components such as RNA and glycogen.

41.7 IMPORTANT BIOLOGICAL REACTIONS

41.7.1 Overview

Selected aerobic, anoxic and anaerobic reactions involving either heterotrophic or autotrophic organisms are discussed in detail in this section. All reactions will be written first in pictorial form (see equation (36)) and then 'translated' to the form described by equations (15). Except for the anaerobic digestion example presented later, stoichiometric coefficients will take on subscripts in accordance with the component listed in Table 1 with reactions numbered consecutively in order of their discussion in the text. All numerical values given for either stoichiometric or kinetic coefficients are based on the use of COD equivalents for each component. As a result, all stoichiometric coefficients for oxygen are to be shown with a positive sign.

Table 1 Definition of Components Vector **A**

Component[a]	Symbol	Description
A_1	S	Soluble organic substrate considered, for simplicity, as one component
A_2	SS	Suspended solids (also considered as one component)
A_3	N_2	Nitrogen gas
A_4	NH_4-N	Ammonium nitrogen
A_5	NO_2-N	Nitrite nitrogen
A_6	NO_3-N	Nitrate nitrogen
A_7	O_2	Oxygen
A_8	I_{H1}	Intracellular intermediate in X_{H1}
A_9	I_{H2}	Intracellular intermediate in X_{H2}
A_{10}	G_{H1}	Intracellular intermediate in X_{H1}
A_{11}	G_{H2}	Intracellular intermediate in X_{H2}
A_{12}, A_{13}	X_{H1}, X_{H2}	'First two' heterotrophic organisms
A_{14}, A_{15}	X_{N1}, X_{N2}	*Nitrosomonas* spp. and *Nitrobacter* spp.
A_{16}	X_D	Cellular debris
A_{17}	P_1	Extracellular byproduct — any low molecular weight compound
A_{18}	P_2	Extracellular byproduct — a polymer
A_{19}	P	Ortho phosphorus

[a] A_i = component *i* expressed in terms of COD, mass; A'_i = component *i* expressed in terms of nitrogen, mass; A''_i = component *i* expressed in terms of phosphorus, mass.

An example of the procedure to be used is shown in equations (36) and (37) for a structured model comprising five reactions.

In words, the model proposes that substrate is transported through the cell membrane (reaction 1) and is converted into some major intermediate (I_{H1}) which, in turn, serves as the substrate for the production of both a second major intracellular intermediate (G_{H1}, by reaction 2) and a living organism or, strictly speaking, 'the remaining cellular biomass' (X_{H1}, by reaction 3). The living organisms are also produced from G_{H1} (by reaction 4) and decompose to cellular debris (X_D, by reaction 5). The improvement of this structure compared to a simple growth/death reaction system should be as obvious as the inadequacy of equation (36) to capture the full substance of the thousands of reactions that take place in a biological system. The simple inclusion of I_{H1} and G_{H1} into the reaction description, however, allows basic differences between two organisms in a mixed culture to be distinguished in terms of the production of components that influence the organism's ability to respond to rapid changes in environmental conditions and other such characteristics that select for one organism over another in a mixed population.

Pictorial representation

$$\tag{36}$$

Translation for aerobic system (see Table 1)

$$
\begin{aligned}
(-1)A_1 + (+Y_{1.7})A_7 + (+Y_{1.8})A_8 &= 0 & r_1 \\
(-1)A_8 + (+Y_{2.7})A_7 + (+Y_{2.10})A_{10} &= 0 & r_2 \\
(-1)A_8 + (+Y_{3.7})A_7 + (+Y_{3.12})A_{12} &= 0 & r_3 \\
(-1)A_{10} + (+Y_{4.7})A_7 + (+Y_{4.12})A_{12} &= 0 & r_4 \\
(-1)A_{12} + (+Y_{5.7})A_7 + (+Y_{5.16})A_{16} &= 0 & r_5
\end{aligned}
\tag{37}
$$

Some additional characteristics of equations (36) and (37) are worthy of note. In this example, oxygen (component A_7; see Table 1) is shown to be required for each reaction. This, of course, is not a necessary condition. In addition, the two intracellular intermediates, I_{H1} (component A_8) and G_{H1} (component A_{10}), are given two separate roles in this reaction scheme, with that of I_{H1} being somewhat more central than that of G_{H1}, and X_{H1} represents a living cell and all of its intracellular components (including I_{H1} and G_{H1}).

The relative rates of reaction will determine the distribution of each component in the reaction mixture. For example, if r_3 is very much greater than r_2, and r_4 is of the same order as r_2, the fraction of the total mass that is G_{H1} would be quite small.

The relative mass fractions of the individual components can also be adjusted independently of the rates of each of the individual reactions by changing the value of the stoichiometric coefficients. However, since equation (38) below must be maintained for each reaction, the flexibility offered here must be considered wisely.

$$
\sum_{j=1}^{n} (\pm Y_{ij}) = 0, \qquad i = 1, \ldots, m
\tag{38}
$$

where $(\pm Y_{i,a}) = -1$.

For example, if $Y_{2.10}$ were taken as a small number (*e.g.* 0.1 g G_{H1} COD produced per g of I_{H1} COD consumed), a large (and inefficient) mass of oxygen would have to be consumed (*i.e.* $Y_{2.7} = 1 - Y_{2.10}$) in order to produce G_{H1}.

While equations (37) have been written on a COD basis, nitrogen, phosphorus or any other component could have served as the basis for each reaction. Specific examples illustrating nitrogen and phosphorus bases are discussed briefly in the next section.

The procedures described above are repeated in the remainder of this heading for aerobic heterotrophic reactions, nitrification, denitrification and anaerobic digestion. For convenience, however, oxygen will not be shown in the pictorial representation.

41.7.2 Aerobic Heterotrophic Reactions

The simplest reaction set used throughout the biological literature involves the production of cell mass from substrate and the simultaneous loss in cell mass through maintenance (endogenous respiration). The pictorial representation and corresponding translation for aerobic conditions are given in equations (39) and (40) respectively.

$$
S \xrightarrow{\text{(1)}} X_{H1} \xrightarrow{\text{(2)}} X_D
\tag{39}
$$

$$
\begin{aligned}
(-1)A_1 + (+Y_{1.7})A_7 + (+Y_{1.12})A_{12} &= 0 & r_1 \\
(-1)A_{12} + (+Y_{2.7})A_7 + (+Y_{2.16})A_{16} &= 0 & r_2
\end{aligned}
\tag{40}
$$

Equations (40) are on a COD basis. In some cases nitrogen and phosphorus involvement must be considered and the COD basis must be replaced both by a nitrogen-based balance and a phosphorus-based balance. Equations (41) and (42) represent such substitute reactions for equations (40) where the primes (*e.g.* A'_i) stand for nitrogen, and the double primes (*e.g.* A''_i) for phosphorus.

For nitrogen:

$$
\begin{aligned}
(-1)A'_1 + (\pm Y'_{1.4})A'_4 + (+Y'_{1.12})A'_{12} &= 0 & r'_1 \\
(-1)A'_{12} + (+Y'_{2.4})A'_4 + (+Y'_{2.16})A'_{16} &= 0 & r'_2
\end{aligned}
\tag{41}
$$

For phosphorus:

$$
\begin{aligned}
(-1)A''_1 + (\pm Y''_{1.19})A''_{19} + (+Y''_{1.12})A''_{12} &= 0 & r''_1 \\
(-1)A''_{12} + (+Y''_{2.19})A''_{19} + (+Y''_{2.16})A''_{16} &= 0 & r''_2
\end{aligned}
\tag{42}
$$

The plus or minus indicated for nitrogen (component A'_4) and phosphorus (component A''_{19}) allows for the fact that more (the plus sign) or less (the minus sign) nitrogen or phosphorus may be present in the organic substrate (component A'_1) than is necessary for the production of cell mass (component A'_{12}). In both cases the release of nitrogen and phosphorus during 'death' is assumed to be in the inorganic form. This is not a necessary condition.

The conventionally used growth and 'death' reactions described by equation (39) can be expanded to include structure as is shown below.

$$
\begin{array}{c}
\qquad X_{HI} \xrightarrow{\ (2)\ } X_D \\[2pt]
\overset{(1)}{\nearrow} \quad \overset{(4)}{\big\uparrow} \quad \overset{(6)}{\searrow} \\[2pt]
S \xrightarrow{\ (3)\ } I_{HI} \xrightarrow{\ (5)\ } G_{HI}
\end{array}
\qquad (43)
$$

Note that equation (43) differs slightly from the example offered in equation (36) in that this new reaction scheme allows for cell mass production directly from the substrate, S. Reactions, in addition to the two given in equation (40), are depicted in equations (44).

$$
\begin{array}{ll}
(-1)A_1 \ + \ (+Y_{3.7})A_7 \ + \ (+Y_{3.8})A_8 \ = \ 0 & r_3 \\
(-1)A_8 \ + \ (+Y_{4.7})A_7 \ + \ (+Y_{4.12})A_{12} \ = \ 0 & r_4 \\
(-1)A_8 \ + \ (+Y_{5.7})A_7 \ + \ (+Y_{5.10})A_{10} \ = \ 0 & r_5 \\
(-1)A_{10} \ + \ (+Y_{6.7})A_7 \ + \ (+Y_{6.12})A_{12} \ = \ 0 & r_6
\end{array}
\qquad (44)
$$

If necessary, equations (44) could be rewritten in terms of nitrogen or phosphorus or any other element (*e.g.* organic carbon) as was demonstrated by equations (41) and (42).

The structured model described by equation (43) allows the organism system to respond better to sudden changes in the concentration of extracellular substrate and, depending upon the rate of destruction of I_{HI} and G_{HI}, can markedly influence the overall survivability of the culture. Such features are essential to the understanding of transient biological processes, *e.g.* staged or periodically fed reactor systems.

While structure alone can help explain population shifts and stability in a mixed organism system, byproduct synthesis and utilization also increase the potential understanding of system dynamics. Equation (45) adds that detail by including both polymer production, P_2 (to help in floc or biofilm formation), and a second organism, X_{H2}, that grows using a low molecular weight substrate, P_1, with only one major intracellular component, I_{H2}, included.

$$
\begin{array}{c}
\qquad\qquad X_{HI} \xrightarrow{\ (2)\ } X_D \\[2pt]
P_2 \quad \overset{(1)}{\nearrow} \quad \overset{(4)}{\big\uparrow} \quad \overset{(6)}{\searrow} \\[2pt]
\overset{(7)}{\big\uparrow} \quad S \xrightarrow{\ (3)\ } I_{HI} \xrightarrow{\ (5)\ } G_{HI} \\[6pt]
\overset{(8)}{\big\downarrow} \qquad X_{H2} \xrightarrow{\ (12)\ } X_D \\[2pt]
\quad \overset{(9)}{\nearrow} \quad \overset{(11)}{\big\uparrow} \\[2pt]
P_1 \xrightarrow{\ (10)\ } I_{H2}
\end{array}
\qquad (45)
$$

Equations (45) add six reactions to those already included in equations (40) and (44). The additional six reactions are shown below.

$$
\begin{array}{ll}
(-1)A_1 \ + \ (+Y_{7.7})A_7 \ + \ (+Y_{7.18})A_{18} \ = \ 0 & r_7 \\
(-1)A_1 \ + \ (+Y_{8.7})A_7 \ + \ (+Y_{8.17})A_{17} \ = \ 0 & r_8 \\
(-1)A_{17} \ + \ (+Y_{9.7})A_7 \ + \ (+Y_{9.13})A_{13} \ = \ 0 & r_9 \\
(-1)A_{17} \ + \ (+Y_{10.7})A_7 \ + \ (+Y_{10.9})A_9 \ = \ 0 & r_{10} \\
(-1)A_9 \ + \ (+Y_{11.7})A_7 \ + \ (+Y_{11.13})A_{13} \ = \ 0 & r_{11} \\
(-1)A_{13} \ + \ (+Y_{12.7})A_7 \ + \ (+Y_{12.16})A_{16} \ = \ 0 & r_{12}
\end{array}
\qquad (46)
$$

The nitrogen- and phosphorus-based counter sets can also be provided as desired.

Equations (40), (44) and (46) represent a formidable set of 12 reactions that require knowledge of 24 stoichiometric coefficients, 12 reaction rate constants and as many as 10 saturation constants. Obviously considerable research is necessary to develop procedures which can be used to determine many of these coefficients with reasonable ease. Nevertheless, because each reaction is on a COD basis and involves only three components, estimates for most of the stoichiometric coefficients would be relatively simple. In addition, under certain operating conditions only a limited number of these reactions are important and the remaining equations can be eliminated by setting the rate of these reactions equal to zero. As a result, the rate forms can be used to orchestrate the reaction scheme and the reaction scheme can be maintained in 'full form', that is composed of all possible important reactions, independent of the system of interest.

Each of the 12 reactions given by equations (40), (44) and (46) has oxygen (component A_7) shown as a reactant (the plus sign only appears because COD is the basis used for all components). In some instances $Y_{i,7}$ may be approximately zero. Actually the numerical value of $Y_{i,7}$ can be viewed as an estimate of the energy required to carry out the ith reaction, and the product of $Y_{i,7}$ and r_i as an estimate of the rate of energy expended by that reaction. Because only three components are involved in each reaction, the stoichiometric coefficient for the third component in each reaction (oxygen and component α are the other two; see equation 15) is simply equal to 1 $- Y_{i,7}$. As a result, the magnitude of $Y_{i,7}$ also can be used to simulate the efficiency of a biological process by regulating the amount of reactant directed to either storage or byproduct synthesis. If the COD in the electron donor (component α) is not going to oxygen, then it must be going to intracellular reserve (*i.e.* I_{H1} and G_{H1}) or to extracellular components (*i.e.* the byproducts P_1 and P_2). This has a dramatic impact on the notion of the observed yield, Y_{obs} — the net mass of biomass produced per unit mass of substrate consumed. Specifically, the greater the ratio of the rate of formation of oxygen to the rate of formation of substrate, the lower Y_{obs} will be.

As a final note, substrates are generally both soluble, S, and insoluble, SS, and some provision should be made for the conversion of the suspended substrates to a usable soluble form. Actually this is somewhat of an understatement since up to 75% of the COD present in raw domestic wastewater can be in the insoluble form. Solubilization of SS is shown in equations (47) and (48).

$$SS \xrightarrow{(13)} S \tag{47}$$

$$(-1)A_1 + (+Y_{13,7})A_7 + (+Y_{13,2})A_2 = 0 \qquad r_{13} \tag{48}$$

Of course there are likely to be many fractions of suspended solids, each having a different rate of degradation, which may have to be included in the reaction set. Similar considerations should be given to the components which make up the soluble substrate.

The hydrolysis reaction described by equation (48) probably does not include oxygen except in the sense that energy was consumed (*i.e.* oxygen used) in the reactions producing the hydrolytic enzyme which catalyzes this reaction. As a result, a non-zero value for $Y_{13,7}$ is an artificial means for accounting for the energy expended manufacturing the hydrolytic enzyme and not a true representation of the oxygen consumed in the reaction itself.

Equation (48) highlights another interesting problem. When the concentration of cell mass, x, is measured on a dry weight basis, all suspended materials are included. In terms of the individual components used in equations (40), (44), (46) and (48), the concentration of cell mass given in terms of mixed liquor volatile suspended solids is equal to the sum of all of the insoluble components as is shown in equation (49).

$$x = C_2 + C_8 + C_9 + C_{10} + C_{12} + C_{13} + C_{16} + C_{18} \tag{49}$$

Since reaction rates are usually related to x, the obvious assumption that the concentration of true cell mass components (*e.g.* I_{H1}, I_{H2}, G_{H1}, G_{H2}) is either a constant fraction of x or must vary within a reasonably tight range. Adding structure to the reaction set can eliminate this difficulty by making the rate of reaction a function of the actual organism concentration or proper cellular component. Unfortunately the problem is only transferred since the component selected must then be measurable. Note also that the COD equivalent of the cell mass would depend upon the relative distribution of the various important intracellular constituents.

41.7.3　Nitrification

Ammonium nitrogen, either produced during the catabolism of nitrogen-containing organic compounds or present in the raw waste flow, in excess of that incorporated into the cell mass, serves as an electron donor for the autotrophic bacteria, *Nitrosomonas* spp. Nitrite nitrogen produced by the *Nitrosomonas* spp. serves as an electron donor for another autotrophic bacteria, *Nitrobacter* spp., with nitrate nitrogen being the final product of that reaction. The rate of formation for ammonium nitrogen from the nitrogen-based form of equations (40), (44), (46) and (48) times some characteristic volume (*e.g.* the total liquid volume in a completely mixed reactor) represents the net input (mass time^{-1}) for ammonium nitrogen to that characteristic volume from the reactions taking pace. Biological oxidation of this ammonium nitrogen (plus that present in the raw waste) is described by equations (50) and (51).

$$NH_4-N \xrightarrow{(14)} X_{NI} + NO_2-N \atop O_2 \quad \Big|(15) \atop \quad \to X_{N2} + NO_3-N \atop O_2 \tag{50}$$

$$\begin{array}{ll} (-1)A_4 + (+Y_{14.7})A_7 + (+Y_{14.5})A_5 + (+Y_{14.14})A_{14} = 0 & r_{14} \\ (-1)A_5 + (+Y_{15.7})A_7 + (+Y_{15.6})A_6 + (+Y_{15.15})A_{15} = 0 & r_{15} \end{array} \tag{51}$$

Each component in equations (51) is expressed in terms of COD equivalents. These reactions typically do not take place unless the dissolved oxygen concentration is sufficiently high (*e.g.* greater than 0.5 g m^{-3}). Because of the possible confusing nature of this basis for nitrification (*versus*, for example, using nitrogen as the basis), this reaction set is explained further below.

Even though ammonium nitrogen is not oxidized in the COD test, the amount of oxygen required to oxidize nitrogen in the -3 oxidation state to the $+5$ oxidation state in nitrate nitrogen can be calculated from the mole-based equation shown below.

$$NH_4OH + 2O_2 \to HNO_3 + 2H_2O \tag{52}$$

Equation (52) is rewritten on a unit mass basis in equation (53) and on a unit mass of nitrogen basis in equation (54).

$$(-1)NH_4OH + (-1.83)O_2 + (+1.80)HNO_3 + (+1.03)H_2O = 0 \tag{53}$$

$$(-1)NH_4OH \text{ (as N)} + (-4.57)O_2 + (+4.50)HNO_3 \text{ (as N)} + (+2.57)H_2O = 0 \tag{54}$$

Thus, from equation (54) the COD equivalent of NH_4-N is the 4.57 g of oxygen required to oxidize 1 g of ammonium nitrogen completely to nitrate nitrogen. Likewise a COD equivalent of nitrite nitrogen can be shown to be 1.14 g of oxygen per gram NO_2-N oxidized to the $+5$ oxidation state and that of NO_3-N is zero. A simplified procedure for determining the coefficients for equations (51) is described below.

Equations (51) can be rewritten on a mass of nitrogen basis as given by equations (55).

$$\begin{array}{l} (-1)A'_4 + (+Y'_{14.5})A'_5 + (+Y'_{14.14})A'_{14} = 0 \\ (-1)A'_5 + (+Y'_{15.6})A'_6 + (+Y'_{15.15})A'_{15} = 0 \end{array} \tag{55}$$

Recall that the primes designate nitrogen as the basis. Now the COD equivalent of each component can be substituted into equations (55) to obtain, after accounting for the oxygen which must be added, the coefficients for equations (51). An example of this procedure is presented below using numerical values for all of the stoichiometric coefficients.

Step 1. Write nitrogen-based balance:

$$\begin{array}{l} (-1)A'_4 + (+0.98)A'_5 + (+0.02)A'_{14} = 0 \\ (-1)A'_5 + (+0.997)A'_6 + (+0.003)A'_{15} = 0 \end{array} \tag{56}$$

Step 2. Multiply each stoichiometric coefficient by the appropriate COD equivalent and include oxygen in the balance:

$$\begin{array}{l} (-1 \times 4.57)A_4 + (+0.98 \times 1.14)A_5 + (+0.02 \times 16)A_{14} + \text{oxygen} = 0 \\ (-1 \times 1.14)A_5 + (+0.997 \times 0)A_6 + (+0.003 \times 16)A_{15} + \text{oxygen} = 0 \end{array} \tag{57}$$

Note: the COD equivalent of cell mass nitrogen of 16 g of oxygen required to oxidize 1 g of cell mass nitrogen completely to nitrate nitrogen is obtained by combining equation (5) with equation (54).

Step 3. Divide each equation by the absolute value of the stoichiometric coefficient associated with component α (*i.e.* either A_4 or A_5):

$$(-1)A_4 + (+0.24)A_5 + (+0.07)A_{14} + (+Y_{14.7})A_7 = 0$$
$$(-1)A_5 + \qquad (0)A_6 + (+0.05)A_{15} + (+Y_{15.7})A_7 = 0 \qquad (58)$$

Step 4. Calculate $Y_{14.7}[= 1 - (0.24 + 0.07) = 0.69]$ and $Y_{15.7} [= 1 - (0.05) = 0.95]$ (see equation 38):

$$(-1)A_4 + (+0.69)A_7 + (+0.24)A_5 + (+0.07)A_{14} = 0$$
$$(-1)A_5 + (+0.95)A_7 + (+0.05)A_{15} \qquad\qquad = 0 \qquad (59)$$

While these four steps may seem rather tedious, they clearly establish the relationships between a nitrogen and a COD basis. Particular attention should be paid to the COD basis for nitrite nitrogen. As is shown by equations (59), nitrite nitrogen (component A_5) is a reservoir of unreacted COD in exactly the same way as the extracellular byproduct P_1 is in the heterotrophic reaction scheme described in equation (45).

As a final point, equations (51) (or their numerical equivalent, equations 59), should be amended to include endogenous respiration as was done for the reaction 2 in equations (42). This is done in equations (60) below.

$$(-1)A_{14} + (+Y_{16.7})A_7 + (+Y_{16.16})A_{16} = 0 \qquad r_{16}$$
$$(-1)A_{15} + (+Y_{17.7})A_7 + (+Y_{17.16})A_{16} = 0 \qquad r_{17} \qquad (60)$$

In both reactions of equations (60), the final product, cellular debris (component A_{16}), is assumed to be the same.

41.7.4 Denitrification

Denitrification may be either assimilatory or dissimilatory. In assimilatory denitrification, nitrate nitrogen is reduced to ammonium nitrogen which is, in turn, used in cell synthesis (assimilation). In dissimilatory denitrification, nitrate nitrogen is reduced to nitrogen gas. Two major steps are considered of importance in dissimilatory denitrification. The first involves the reduction of nitrate nitrogen to nitrite nitrogen; the second, nitrite nitrogen to nitrogen gas. The second step involves several reactions which will not be considered here. In dissimilatory denitrification neither step will take place unless the dissolved oxygen concentration is sufficiently low (*e.g.* less than 0.5 g m^{-3}), with immeasurable levels preferred. Situations with very low or no dissolved oxygen in the presence of oxidized nitrogen are often referred to as anoxic conditions.

Each of the 12 reactions given in equations (40), (44) and (46) could simply be rewritten with oxygen (component A_7) replaced by nitrate nitrogen (component A_6), and nitrite nitrogen (component A_5) added as a product. A second set of 12 reactions with nitrite nitrogen replacing nitrate nitrogen as an electron acceptor and nitrogen gas (component A_3) replacing nitrite nitrogen as a product would then complete a denitrification reaction scheme. Because ammonium nitrogen is the preferred form of nitrogen in cell synthesis, it should be included in any general set of reactions. In the presence of ammonium nitrogen, oxidized nitrogen will be used strictly as an electron acceptor. In the absence of ammonia nitrogen, the assimilatory denitrification pathway would be used to generate the necessary ammonium nitrogen.

As a final note on denitrification, the subject of COD equivalents must be discussed. The interesting and perhaps confusing aspect of nitrogen is that the COD equivalency depends upon the role that it plays in the reaction. As an electron donor in nitrification, the COD equivalents of nitrite nitrogen and nitrate nitrogen are 1.14 and 0 respectively. In nitrification, the COD equivalents are based on oxidizing nitrogen to the maximum oxidation state of nitrogen ($+5$). In nitrogen's role as an electron acceptor, the basis for calculating COD equivalents is the reduction of nitrogen to the minimum oxidation state for nitrogen (-3 in ammonium nitrogen). Therefore the COD of nitrate nitrogen is -4.57 and that of nitrite nitrogen is -3.43. In this case, the COD equivalent of ammonium nitrogen is zero but the COD equivalent of -1.71 for nitrogen gas nitrogen (component A_3) must also be included. Because the final oxidation state for nitrogen was taken as -3 (actually any consistent basis could have been selected), nitrogen gas represents unreacted oxidizing potential for the nitrite nitrogen.

41.7.5 Anaerobic Digestion

Anaerobic digestion obviously takes place in the absence of all exogenous electron acceptors. Any elecron acceptors present in the feed stream to an anaerobic digestor must be in such low quantities that their use does not seriously change the distribution of organisms necessary to produce the desired end product, methane. Zehnder *et al.* (1982) have provided an excellent review of the ecology of methane formation and Gujer and Zehnder (1983) have summarized this review and other findings into a reaction scheme for anaerobic digestion of particulate organic matter in domestic sludge which includes the interactions of five independent groups of microorganisms. The analysis by Gujer and Zehnder excludes the production of cell mass at each reaction level but a review of their presentation is instructive to the understanding of anaerobic systems and is briefly discussed here.

Rather than introduce a component list similar to that used in Table 1, reactions are written below with abbreviations keyed to the components involved. In addition, numerical values for the stoichiometric coefficients are used instead of the general coefficient, $\pm Y_{i,j}$.

The first reaction involves the hydrolysis of the suspended solids (SS) to proteins (Pr), carbohydrates (C) and lipids (L). This is shown in equation (61).

$$(-1)SS + (+0.21)Pr + (+0.40)C + (0.39)L = 0 \tag{61}$$

In this reaction, 21% of the COD in the SS is converted to proteins, 40% to carbohydrates and 39% to lipids. Because cell mass production is not considered in any of the reactions, the COD of the products will identically equal the COD of the reactant.

The second set of reactions involves the conversion of the protein, carbohydrates and lipids to amino acids plus sugars (AAS) and fatty acids (FA) and is depicted by equations (62).

$$
\begin{aligned}
(-1)Pr + &\quad (+1)AAS &&= 0 \\
(-1)C + &\quad (+1)AAS &&= 0 \\
(-1)L + &\quad (+0.13)AAS + (+0.87)FA &&= 0
\end{aligned}
\tag{62}
$$

Cell mass production in these reactions could be as high as 20% on a COD basis and will, of course, reduce the net amount of extracellular amino acids, sugars and fatty acids produced.

The next set of reactions describes the production of hydrogen gas (H), acetic acid (A) and intermediate byproducts (IBP) such as propionic acid and butyric acid and the conversion of these intermediate byproducts into hydrogen gas and acetic acid. These three reactions are shown in equations (63).

$$
\begin{aligned}
(-1)FA + &\ (+0.32)H + (+0.68)A &&= 0 \\
(-1)AAS + &\ (+0.17)H + (+0.53)A + (+0.30)IBP &&= 0 \\
(-1)IBP + &\ (+0.40)H + (+0.60)A &&= 0
\end{aligned}
\tag{63}
$$

In these reactions, hydrogen gas and acetic acid are the ultimate final products. These two compounds serve as the substrate for two separate groups of microorganisms, one which produces methane gas (M) from hydrogen gas and the other which produces methane gas from acetic acid. These final two reactions are presented in equations (64).

$$
\begin{aligned}
(-1)H + (+1)M &= 0 \\
(-1)A + (+1)M &= 0
\end{aligned}
\tag{64}
$$

Interestingly, approximately 30% of the methane gas produced comes from the conversion of hydrogen gas while the remaining 70% comes from acetic acid, even though over 20 organisms have been identified which produce methane from hydrogen gas and only two have been isolated which produce methane from acetic acid. The limited production of methane from hydrogen gas results from the extreme inhibition of hydrogen production in equations (63) when the hydrogen gas levels exceed 10^{-4} atm.

The overall reaction combining equations (61)–(64) describes the production of 1 g of methane gas COD from 1 g of domestic sludge COD. As noted previously, cell mass production will reduce the net methane production. Gujer and Zehnder (1983) in reviewing work done by others have calculated the cell mass yield (in terms of COD cell mass per unit mass of substrate COD consumed) as approximately 0.15 for carbohydrates and fatty acids, 0.07 for intermediate byproducts and acetic acid and 0.06 for hydrogen gas. Inclusion of these yields into equations (62)–(64) with an estimate of no cell yield from the hydrolysis step (see equation 61) and a yield of 0.15 for proteins, leads to each gram of degradable domestic sludge COD resulting in the production of approximately 0.67 g of methane gas COD and 0.33 g of cell mass COD. The methane gas production, of course, would be increased beyond 0.67 because of the degradation of the cell mass produced.

The above discussion was geared towards the anaerobic digestion of domestic sludge. Anaerobic digestion of industrial wastes would be expected to follow the basic reaction scheme proposed by Gujer and Zehnder (1983) with the following two considerations noted. First, the hydrolysis reactions shown in equation (61) would not be necessary for a soluble waste but must be included to account for methane gas production from the degradation of the biomass produced. Second, a microorganism community will have to be established which carries out reactions to the level noted by equations (62) and (63). That is, since an industrial waste is usually not composed of proteins, carbohydrates and lipids, the microorganism community which converts the organics in an industrial waste to extracellular amino acids, sugars and fatty acids in equations (62) or to the intermediate byproducts shown in equations (63) is likely to be markedly different than that established during the degradation of domestic sludge. This 'upper layer' microorganism community would also be expected to differ from industry to industry. An 'intermediate layer' of microorganisms responsible for the production of hydrogen gas and acetic acid may also depend upon the nature of the industrial waste being treated. Conversely, the 'lower layer' microorganism community involved with the production of methane gas from hydrogen gas and acetic acid must be unaffected by differences in one industrial waste *versus* another. Otherwise methane gas production will not take place. Because of the difficulties associated with the methane gas production during the anaerobic digestion of some industrial wastes, the question of whether or not either of the two organisms which convert acetic acid to methane can survive in some potentially 'toxic' industrial waste environments is raised. Obviously, additional research must be conducted before these questions and many others can be answered.

41.8 SUMMARY

A microorganism is the end result of a myriad of reactions involving a similar number of components. Many microorganisms are involved in the treatment of a given wastewater. A wastewater typically contains a multitude of substrates for these different microorganisms. Obviously all of these complexities cannot be considered in either the design or operation of a biological wastewater treatment facility. Major reactions which describe system behavior can be selected from this vastly complex environment. Even then a quantitative understanding of each of these reactions is difficult.

The methods outlined in this chapter place much greater emphasis on understanding the major reactions involved in a particular biological conversion process than on a particular kinetic expression which describes the process. This was intended. By emphasizing detail at the reaction level, individual kinetic expressions need not be more complicated than the Monod rate form. Thus debates about which rate form to use can be eliminated and replaced by debates concerning which reactions should be included in the reaction scheme. Emphasis was placed on developing the stoichiometric expressions in terms of COD equivalencies with each reaction satisfying a separate 'mass balance'. Even though the COD basis may be initially confusing for reactions involving either the oxidation or reduction of nitrogen, the benefits gained in terms of ease of understanding the reactions and determining the stoichiometric coefficients far outweigh this inconvenience.

41.9 REFERENCES

Aris, R. (1969). *Elementary Chemical Reactor Analysis*. Prentice-Hall, Englewood Cliffs, New Jersey.
Busch, A. W. (1971). *Aerobic Biological Treatment of Wastewaters, Principles and Practice*. Gulf Publishing, Houston, Texas.
Gujer, W. (1980). Personal communication. Swiss Federal Institute of Technology, Dübendorf.
Gujer, W. and A. J. B. Zehnder (1983). Conversion processes in anaerobic digestion. In *Anaerobic Treatment of Wastewater in Fixed Film Reactors*, ed. M. Henze. *Water Sci. Technol.*, **15** (8/9), 127–168.
Herbert, D. (1976). Stoichiometric aspects of microbial growth. In *Proceedings of the 6th International Symposium on Continuous Culture Applications and New Fields*, ed. A. C. R. Dean *et al.*, chap. 1, pp. 1–30. Ellis Horwood, Chichester.
Irvine, R. L. (1980). *Activated Sludge: Stoichiometry, Kinetics and Mass Balances*. ACS Audio Course, cat. no. Z–57. American Chemical Society, Washington, DC.
Irvine, R. L., J. E. Alleman, G. Miller and R. W. Dennis (1980). Stoichiometry and kinetics of biological waste treatment. *J. Water Pollut. Control Fed.*, **52**, 1997–2006.
Roels, J. A. (1980). Application of macroscopic principles to microbial metabolisms: a bioengineering report. *Biotechnol. Bioeng.*, **22**, 2457–2514.
Zehnder, A. J. B., K. Ingvorson and T. Marti (1982). Microbiology of methane bacteria. In *Anaerobic Digestion*, ed. D. E. Hughes *et al.*, pp. 45–68. Elsevier Biomedical Press, Amsterdam.

42

Biochemistry of Waste Treatment

H. A. PAINTER
Water Research Centre, Stevenage, Herts., UK

42.1 INTRODUCTION

The biochemical transformations which occur during wastewater treatment are attributed to microorganisms, mainly bacteria, and have been largely established by inference rather than by direct investigation because of the usually complex nature of wastewaters; sewage, for example, contains thousands of organic compounds at very low concentrations (*e.g.* Painter *et al.*, 1961). Details of metabolic pathways are given in Volume 1 Part 1. It is inferred that the same reactions occur, though at different rates, at low concentrations ($<< 1$ mg l^{-1}) of substrate as occur at much higher concentrations in fermentations and *in vitro* experiments with single or limited numbers of added substrates. Because of the complex multisubstrate/multiorganism situation, conditions sometimes prevail by which inhibition to certain reactions occurs and reactions take place sequentially (or appear to do so), especially if there is competition for cosubstrates, *e.g.* competition for dissolved molecular oxygen between heterotrophs and autotrophic nitrifiers. Some cases of apparent sequential removal of substrate reported in the literature can be explained on kinetic rather than metabolic grounds. Also, because of the relatively low concentration of energy sources and the need for producing effluents from the biological treatment processes containing vanishingly small concentrations for disposal to rivers, *etc.*, the microbial populations sustained in the treatment processes by most wastewaters are necessarily relatively small and the rate of growth is low. These populations contain low proportions of viable cells and in this sense are unlike most populations studied in the laboratory.

The biochemistry of waste treatment obviously depends on the nature of the compounds present. In domestic sewage, it has been shown that about 75% of the organic carbon present (200–300 mg l^{-1} C) can be accounted for by carbohydrates, fats, proteins, amino acids, volatile and higher fatty acids. Other important constituents are anionic (up to 30 mg l^{-1}) and nonionic surfactants (up to 3 mg l^{-1}). As much as two-thirds of the organic matter is present in suspension (Painter *et al.*, 1961).

An important consequence of the low substrate concentrations and the resulting low specific growth rates, as well as the needs of the treatment processes, is that the microbial population grows in flocculent form and it is considered that this is necessarily accompanied by the formation in the sludge of higher than normal concentrations of extracellular and intracellular polymers. Besides these polymers, scores of compounds have been identified by GC–MS in trace amounts in extracts of activated sludge (WRC, unpublished) but it is extremely difficult to decide whether such compounds were present in the wastewater or formed during the process, and they will not be considered further.

Because waste treatment processes are 'open' systems, the microbial population is extremely varied, especially at low growth rates. Consequently, the whole range of catabolic and anabolic reactions (Volume 1, Part 1) is likely to take place. In aerobic treatment processes, the main catabolic reaction will be complete aerobic oxidation with molecular oxygen as the final electron acceptor, while under some circumstances (*e.g.* in contrived anoxic zones for denitrification) anaerobic respiration occurs in which nitrate is the electron acceptor. In anaerobic treatment (not dealt with further here: see Volume 4, Chapters 58–61) the chief reaction is aerobic respiration using inorganic carbon as the electron acceptor after hydrolytic and fermentative reactions. The final product of these reactions, methane, can however be produced by direct fermentation and by reduction of carbon dioxide by hydrogen.

42.2 RESPIRATION — OXYGEN UPTAKE

The measurement of the biochemical uptake of dissolved oxygen is the most usual method of ascertaining whether, and to what extent, a wastewater or chemical compound can be metabolized. A knowledge of the oxygen uptake during the removal of a waste is also important, since aeration in the activated sludge process is a significant fraction of the total cost. Oxygen uptake can be determined by various forms of manometric or electrode respirometer, or by the five-day biochemical oxygen demand (BOD) test. Since the majority of wastewaters, treated or untreated, will eventually be discharged to a river, and since dissolved oxygen is essential for river life, the BOD test was chosen by the Royal Commission on Sewage Disposal (1912) for determining how best to discharge wastes to rivers without seriously depleting the river water of dissolved oxygen. The test is now more often used to assess the degree of removal of degradable substances achieved by aerobic processes.

Whereas the specific endogenous respiration rate of pure cultures of bacteria is an extremely low proportion ($\sim 1\%$) of that with external substrate, the proportion in the case of activated sludge and biological filter film is much higher (~ 10–50%) for two reasons. First, sludge invariably contains insoluble substrates (fatty acids, carbohydrates, proteins) intimately bound to the flocs and these add to the true endogenous rate. Secondly, it is thought that organisms grown at low rates in treatment processes contain more metabolizable 'storage products' than do cells grown at higher rates and again contribute to a falsely high 'endogenous' rate.

As in pure culture, after the expendable cell contents, *i.e.* glycogen, poly-3-hydroxybutyrate, have been oxidized, essential constituents such as proteins and nucleic acids are degraded with the release of ammonia. At some stage the cell loses its viability but continues to respire and is still capable of oxidizing external substrate but without growth; in activated sludge grown under normal operating conditions the proportion of viable heterotrophic cells is low (~ 1–10% of total cells). Finally, complete lysis occurs and cell contents are released causing turbidity and supernatants with relatively high dissolved organic carbon (DOC) and BOD. This sometimes occurs in units with high sludge-retention time (extended aeration). Endogenous metabolism and lysis are important in waste treatment for the assessment of oxygen requirements in the aeration tank, of the amount of excess sludge formed and of the oxygen demand of effluents containing suspended solids when discharged to water courses.

The ratio of carbon dioxide evolved to oxygen taken up, the respiratory quotient (RQ), is constant for a given substrate under constant conditions. The RQ changes throughout the growth phases outlined above but the ratio has not found much application in waste treatment, though the production of carbon dioxide has been used as evidence that substrates, especially insoluble ones, are biodegradable.

The oxygen uptake, expressed as a proportion of either the theoretical, for complete oxidation to carbon dioxide and water, or the chemical oxygen demand (COD), is invariably less than 100%. The actual value varies considerably depending on species composition, nature of the substrates and with conditions of growth (temperature, retention times of sludge and of liquid waste) and is roughly complementary to the cell yield coefficient, Y (weight of cells formed/weight of substrate used). For glucose, the proportion of theoretical oxygen uptake for sludges from a number of sources varied from 8 to 31% (Painter *et al.*, 1968): in general, the longer the sludge retention time ($\equiv$ doubling time) the higher the oxygen uptake.

In adequately aerated, completely mixed tanks, the oxygen uptake rate is fairly uniform throughout the tank, but in plug-flow systems the respiration rate is high at the point of mixing and rapidly falls as the mixed liquor passes along the aeration tank. Since the concentration of individual substrates is low, usually below the corresponding Michaelis constant K_m, the initially high respiration rate is not maintained for long and falls to the previous 'endogenous' rate in a manner roughly parallel to the approximately logarithmic fall in total soluble substrate (COD) concentration.

42.3 DISSOLVED OXYGEN

Molecular oxygen usually serves as a hydrogen or electron acceptor, becoming reduced to water: less commonly, oxygen is transferred directly to the organic molecule. As the concentration of dissolved oxygen is increased, the respiration rate (and growth rate) increases in a hyperbolic manner and asymptotically approaches a maximum value. A concentration is reached, sometimes called the critical concentration, above which little effect is observed. The dependence of rate on dissolved oxygen concentration obeys the Monod–Michaelis–Menten equation and the affinity of bacteria for oxygen may be expressed as the Michaelis or saturation constant K_m (*i.e.* the concentration of dissolved oxygen giving half maximum respiration rate). For pure cultures in dispersed form K_m is very low (~ 0.001 mg l^{-1}) while for activated sludge much higher values are found (0.1–0.2 mg l^{-1}), resulting from the necessity for oxygen to diffuse through the flocs. For $K_m = 0.1$ mg l^{-1}, over 95% of the maximum rate would have been attained at 2 mg l^{-1} dissolved oxygen, so that there is little advantage, kinetically and economically, of operating at concentrations higher than this. The phenomenon of 'critical' concentration is readily demonstrated by examining the respiration curve, obtained using an oxygen electrode, of an enclosed, previously aerated sludge–sewage mixture.

The use of high purity oxygen instead of air in the activated sludge process (see Volume 4, Chapter 50) has the advantage of introducing oxygen more rapidly into the liquid phase. It was originally claimed that the higher concentrations of dissolved oxygen achieved *per se* gave higher rates of substrate removal, but this has been shown not to be so, in accordance with the Michaelis equation. The advantages seem to be largely 'engineering' ones and are still a matter of contention; further comparative tests are needed to show whether any of the possible advantages (*e.g.* lower sludge yields) are biochemical in origin.

42.4 ENZYMES

Enzymes are the ultimate agency by which biochemical reactions are catalysed: free enzymes appear not to be present in activated sludge but to reside within or on the walls of microbial cells. The few recorded attempts to increase the efficiency of the activated sludge process by addition of enzyme preparations have failed.

The dehydrogenase or catalase activity of sludges has been used to investigate possible correlations with sludge activity towards wastewaters but no clear relationships have been found, although sludges with the highest catalase activity purified sewage most rapidly (Sridhar and Pillai, 1966). Dehydrogenase activity increased rapidly from a specific sludge growth rate of 0.02 d^{-1} to 1.4 d^{-1}; thereafter the increase was small up to the highest rate tested, 6.4 d^{-1} (Weddle and Jenkins, 1971). The dehyrogenase activity for viable cells changed much less than the activity per gram of sludge. However, such factors as the presence of inhibitors and the nature of the substrate cause changes in the relation between concentration of dry mass of sludge and enzyme content. The decrease in dehydrogenase activity of sludge has been used to detect the presence of toxic matter in wastewaters and an increase in activity after the addition of a compound to endogenous sludge has been taken to indicate biodegradability of the compound.

Inducible enzymes, such as β-galactosidase and β-galactoside permease, can be inactivated or removed by prolonged aeration, *e.g.* in the contact stabilization process treating wastes containing lactose, and this has led to the suggestion that single stage, complete mixing would be more advantageous, since the sludge would not then undergo a feed–starve cycle and inducible enzymes would be kept at a constant level.

42.5 INHIBITION

Inhibition of the oxidative processes occurs at a number of levels in the metabolic pathways or by acting on the cell wall (See Volume 1); only gross effects will be considered here. Most treatment plants are operated well within their maximum loading capacity, so that minor inhibitions do not significantly lower the quality of the final effluents produced. There can be, however, large differences (2–3 fold range) in the specific respiration rates of activated sludges taken from treatment works operated under roughly the same kinetic conditions and these differences are due to general inhibition from a large number of chemicals. Higher concentrations of inhibitors, such as heavy metals and chlorinated phenols, cause lysis of cells, turbid effluents with high BOD values and finally complete breakdown of the process.

The concentrations necessary to bring about a given degree of inhibition are usually considerably higher in treatment processes than in pure culture because of modifying effects. Chemical reaction (chelation, precipitation) with other sludge constituents and adsorption can reduce the effective concentration of the inhibitor, while acclimatization of the microbial population over days and weeks may enable the inhibitor to be metabolized or allow the development of tolerant strains. As an example *Nitrosomonas* may be quoted (Tomlinson, Boon and Trotman, 1966): in pure culture 0.5 mg l^{-1} thiourea completely inhibits the production of nitrite, while in batch tests with activated sludge 5–10 times as much is necessary. On continuous addition to a laboratory scale sludge plant concentrations of thiourea up to 92 mg l^{-1} could be tolerated if the concentration was increased gradually; the resulting sludge nitrified and not only were organisms present which could degrade thiourea, but a strain of the autotroph was then present which was less susceptible to thiourea. Acclimatization was also observed with dithiooxamide and mercaptobenzothiazole, while streptomycin, which inhibits cell synthesis, had a much greater effect in long-term tests than was predicted from batch tests. Again, as much as 10, 15 and 100 mg l^{-1}, respectively, of copper, chromium and mercury continuously added still permitted nitrification, although in pure culture as little as 0.5, 0.25 and 2.0 mg l^{-1}, respectively of the metals were completely toxic to pure cultures. Also, stoichiometric amounts of copper nullified the effect of thiourea when the mixture was added to sludge; similarly mercury nullified the effect of mercaptobenzothiazole.

The effect of inhibitors is lessened, too, in completely mixed aeration vessels as compared with plug-flow systems and this is particularly important for wastes containing compounds, such as phenol, which though biodegradable are inhibitory at higher concentrations to the species which metabolize them at the lower concentrations.

It is interesting to note that the addition of a chlorophenol to limit the production of 'surplus' activated sludge has been patented.

42.6 NITROGEN METABOLISM

Apart from essential assimilative reactions, nitrogen takes part in fixation, deamination, nitrification and denitrification; the roles of these processes in treatment of wastewaters has been extensively reviewed (Painter, 1970, 1977; Sharma and Ahlert, 1977).

Fixation of dinitrogen by *Azotobacter* and similar organisms shown to be present in activated sludge may play a part in treatment, especially of wastes deficient in nitrogen compounds (BOD:N > 25) but this has not been demonstrated. Such wastes are usually fortified before treatment with ammonium salts, often as phosphate, to reduce the ratio of BOD to nitrogen to 20 or less.

Deamination and other hydrolytic reactions (*e.g.* of urea) release ammonia from nitrogen-containing organic substrates and the released ammonia is used for cell synthesis (amino acids, proteins, nucleic acids). Ammonia is also released during lysis and death of cells and during endogenous metabolism.

Excess ammonia is oxidized by autotrophs to nitrate *via* nitrite (nitrification) if conditions are suitable and nitrate may be used as a source of oxygen (dissimilation) in anaerobic zones or in regions of very low concentration of dissolved oxygen, *e.g.* in settlement tanks, within flocs or filter slimes. Less frequently, nitrate is used a source of nitrogen (assimilation) in the absence of ammonia.

Nitrification is carried out sequentially by two sorts of autotrophs (see Volume 4, Chapter 49) and the specific growth rates (μ) of these species are much lower than those of normal heterotrophs (0.2 d^{-1} at 15 °C against ~5 d^{-1}). Thus, in treatment processes nitrification is more susceptible to general inhibitory situations than is carbonaceous oxidation. There are many specific inhibitors of *Nitrosomonas* (Tomlinson *et al.*, 1966), however, such as thiourea and many other compounds containing N and S which appear to act by competing for essential copper. The addition of allylthiourea may be used to differentiate between oxygen uptake due to carbonaceous oxidation and nitrification in respirometers and to suppress nitrification in the five-day BOD test. It had been thought that organic matter *per se* was inhibitory to nitrification, but it is now realized that the effect is indirect. The addition of extra organic matter (higher loading) without an increase in aeration leads to a lowering of the concentration of dissolved oxygen and, since the nitrifiers have higher K_m values (0.2–0.3 mg l^{-1}) for dissolved oxygen than the heterotrophs (0.1 mg l^{-1}), this results in an inhibition or complete exclusion of nitrification.

During the oxidation of ammonia to nitrite, acid is inevitably produced and this can bring about a lowering of the pH value in treatment processes (see Volume 4, Chapter 49) if the buffering capacity of the waste is insufficient, especially in soft water areas. (Each 1 mg of N oxidized is equivalent to about 7 mg carbonate.) In turn this can lead to inhibition or even complete loss of nitrification, which virtually ceases below pH values of 6–6.5. Although pure cultures of nitrifiers soon lose their activity when starved of substrate, activated sludge has some protective property since its nitrifying activity is retained even under anaerobic conditions in the settlement tanks. *Nitrosomonas* grows faster in pure culture than in activated sludge; Jones and Paskins (1982) have shown that this is not due to a limitation of the carbon source, CO_2. They also showed that an enriched nitrifying culture soon overcame an initial inhibition by high partial pressures of oxygen, encountered in the oxygen activated sludge and the deep-shaft prcesses (see Volume 4, Chapter 50). Thus, lack of nitrification in these processes is probably due to lowering of the pH value, caused by excessive concentrations of carbon dioxide.

Denitrification is the dissimilatory reduction of nitrate and is effected by many species. It occurs fortuitously in settlement tanks causing rising sludge by occlusion of dinitrogen on to flocs and also in microanaerobic zones in flocs and within biological filter film. The process is harnessed to reduce nitrate concentrations at treatment works where the receiving water is used as a potable source and where algal blooms may be a problem in the receiving water. The wastewater, methanol or molasses have been used as carbon sources to reduce the nitrate, either in a second, unaerated reactor or in separate tanks of the main reactor, in which the concentration of dissolved oxygen is < 1 mg l^{-1}. The rate of nitrate removal is linear down to about 5 mg N l^{-1} and the product is usually dinitrogen, though other products (*e.g.* nitric oxide, nitrite) may be formed, depending on the species present and to some extent on the pH value. Since hydroxyl ions are formed on the reduction of nitrate, denitrification usually results in an increase in pH value.

42.7 PHOSPHORUS AND SULFUR

Phosphorus is assimilated as phosphate from a wastewater to be incorporated as nucleic acids, ATP *etc.* and, under some conditions, granules of polyphosphate (volutin) can accumulate within cells (Yall *et al.*, 1972). If the waste is deficient in the element, phosphate must be added to give a ratio of BOD to P of about 100:1; at ratios higher than about 170:1 filamentous organisms (*e.g.* *Sphaerotilus natans*) tend to dominate the population, leading to 'bulking' of activated sludge. It is undesirable to discharge phosphates to rivers, since unwanted algal growth may occur. Many processes have been proposed for the removal of phosphate, which in wastewaters originates mainly from human metabolic activities and detergents; the processes have been reviewed by van Bouwel *et al.* (1979). The so-called 'luxury' microbial uptake of phosphorus has been explained (Menar and Jenkins, 1970) as being largely due to physicochemical reactions leading to the precipitation of phosphates of calcium which become entrapped within flocs. In a sewage containing 10 mg P l^{-1} about 20% was removed by growth of cells and as much as 60% by precipitation as calcium phosphates.

Sulfur is assimilated as sulfate; organic compounds containing SH groups are converted to sulfide, which is rapidly oxidized to sulfate *via* sulfur, thiosulfate and tetrathionate. Organic sulfates and sulfonates (*e.g.* alkylbenzenesulfonates) are hydrolysed to sulfates and sulfites, the latter being rapidly autooxidized to sulfate. Some species appear to oxidize the side chain before desulfonation, but most desulfonate first. Sludges can tolerate concentrations of sulfite up to 1000 mg l^{-1} and sulfide up to 100 mg l^{-1}.

42.8   OTHER ELEMENTS AND GROWTH FACTORS

Many elements other than those already discussed are required for microbial growth and these (K, Na, Ca, Mg, Cu, Zn, *etc.*) are normally present in adequate concentration in most wastewaters. Apart from being required for normal growth processes, some elements (Ca, Mg) are said to be necessary for production of flocs, while imbalances of trace metals are thought to change the sludge flora and possibly cause sludge 'bulking'.

Some individual bacterial species require the presence of one or more organic growth factors, commonly members of the vitamin B group. Yet other species require no external source of the factors, though some grow faster in their presence. Most wastewaters would be expected to contain the B-group vitamins and indeed these have been detected in sewage; there may be relatively few wastes, however, which are deficient. Other unidentified growth factors probably exist, since about 40% of bacterial isolates, obtained from activated sludge on a medium containing hotwater extracts of sludge, could not grow on the most complex synthetic media tested (Prakasam, 1966).

Cooper and Catchpole (1973) and Stafford (1974) have evidence of an unusual stimulatory effect on the oxidation of a number of nitrogen-containing compounds present in coke-oven effluents. The addition of pyruvate, glucose, *p*-aminobenzoic acid or a number of other single chemicals, at 5 mg l^{-1} or less, reduced by 30–66% the minimum retention time needed just to remove all the substrate added to laboratory-scale activated sludge plants. The substrates tested were thiocyanate, cyanide, pyridine and other bases, and ammonia. The mechanisms involved are not known but it was suggested that compounds which produced pyruvate might be effective. This seems unlikely however, since both L- and D-α-alanine and both nicotinic and isonicotinic acids were almost equally effective, reducing the minimum retention time needed to remove all the thiocyanate (500 mg l^{-1} ammonium thiocyanate) from 10 h to 2.5–3.8 h. In some cases, it was possible to withdraw the stimulant without impairing the reaction velocity. The process is patented (British Coke Research Association, 1967, 1969).

42.9   FATE OF INDIVIDUAL CHEMICALS

Whether a given compound is removed from a wastewater on treatment depends on whether an organism is present which can metabolize it within the period of retention in the treatment process. Apart from being metabolized during normal growth, the substance may be removed by cometabolism, or what is called 'fortuitous' metabolism (Knackmuss, 1981), or it may simply be adsorbed physicochemically on to sludge flocs and so removed from the liquid phase. As Alexander (1965) has explained, even though most organic compounds are vulnerable to microbial attack, if the relevant species does not grow at a sufficiently high rate to be maintained in the treatment system, the particular substance will not be degraded. Thus, not all biodegradable substances are treatable under normal conditions. Interest in biodegradability and treatability of chemicals began in the mid-50s on the introduction of synthetic surfactants (detergents) on a wide scale. This was the first time that relatively high concentrations (about 20 mg l^{-1} in sewage) of synthetic compounds had appeared in the aquatic environment. Because the first anionic surfactants (tetrapropylenebenzenesulfonates) could not be fully metabolized in treatment processes (bacteria could not attack the highly-branched side-chain sufficiently rapidly) receiving waters foamed excessively and contained significant concentrations of the surfactant, which was thought to be harmful in other ways. This led to national and international authorities enacting legislation to monitor, and if necessary to control, the introduction of all new chemicals produced in quantities above 1 tonne/year (see Volume 4).

The simplest, but not certain, way to test whether a compound is biodegradable is to follow the disappearance of the substance over a period of, arbitrarily, 28 d in an inorganic medium containing the substance as the only source of carbon. It is usual to add 5–40 mg l^{-1} organic carbon and to inoculate with unacclimatized activated sludge or sewage effluent to give a concentration of 'total count' organisms of 10^2–10^6 ml^{-1}. It has been found that if more than 70% of the dissolved organic carbon is removed in a 10-day period, the substance will normally be removed on treatment. Similarly, if more than 60% of the theoretical oxygen uptake, or carbon dioxide evolution, is observed, the substance is judged treatable: the latter methods are necessary for insoluble substances. If the biodegradability is in doubt, a simulation of a treatment process is carried out: *e.g.* two activated sludge units are operated, one receiving only sewage, or a nutrient solution, and the other treating the test substance in addition. In the agreed international test, observations are carried out over a period of not more than nine weeks. In this way many substances have been shown to be degraded and a smaller number to be nondegraded (*e.g.* Gerike and Fischer, 1979, 1981).

Most synthetic compounds are present in wastewaters at the μg l^{-1} level or less, so that tests should be made at these concentrations for realistic assessments to be made of the kinetics of degradation. Because of obvious analytical difficulties, this point is increasingly being examined by using ^{14}C-labelled compounds (*e.g.* Blanchard *et al.*, 1976; Larson and Games, 1981; Gardner *et al.*, 1982).

There is not enough space here to discuss all groups of compounds so a selection will be made. Analysis of sewage and sewage effluents has shown (Painter *et al.*, 1961) that as expected so-called biological compounds, *i.e.* sugars, amino acids, proteins, fatty acids, are readily degraded by the activated sludge process with >95% removal. This is supported by studies (Chudoba *et al.*, 1968a, 1968b) in which virtually 100% of the single substrate (sugar, amino acid or fatty acid) treated in pilot treatment units was removed. However, soluble organic matter was present in the effluents and this was thought to originate from lysis of cells and other intermediate products.

Insoluble (and slightly soluble) biological compounds are generally removed less rapidly than soluble substrates; Takahashi *et al.* (1969) have shown that the rate of metabolism of suspended solids increased as the particle size decreased. Higher fatty acids (stearic, palmitic) degraded faster than starch, followed by protein and lastly cellulose. While the presence of insoluble substrates did not affect adsorption or oxidation of soluble substrates, the presence of glucose inhibited the degradation of insoluble substrates. Particle size was found to be important in a study (Loehr and Roth, 1968) of the degradability of calcium salts of long-chain fatty acids, and solubility was important in the case of sodium salts; the sodium salts caused a greater uptake of oxygen by activated sludge than did the corresponding calcium salts. Maier (1968) showed that starch was relatively easily degraded by extracellular enzymes to soluble sugars while Banerji *et al.* (1968) found that adsorption of starch was increased by acclimatized sludges and suggested that enzymes attached to the cell wall played a more important role than free, unattached enzymes. Bacteria capable of degrading cellulose and lignin have many times been isolated from activated sludge (see Volume 4, Chapter 46).

Much attention has been paid to surfactants, most recently by Cain (1981). It would seem that, although alkylbenzenesulfonates with relatively unbranched C$_{10}$–C$_{16}$ side chains are readily metabolized with >95% removed on treatment, possibly as much as 20% of the original organic carbon remains in solution in the effluent (Brown *et al.*, 1981). Leidner *et al.* (1976) have identified 11 different sulfophenylcarboxylic acids, in 'die-away' tests and in environmental samples, as intermediates in anionic surfactant metabolism. The extent to which the benzene ring of anionic surfactants is degraded is still debated, but it seems to depend on the ratio of concentration of microorganisms to that of the surfactant (Larson and Payne, 1981). Using GC–MS, Giger *et al.* (1981) have shown that nonylphenols and nonylphenol ethoxylates with one, two and three oxyethylene groups are present in effluents from treatment plants. Giger *et al.* consider the products to be refractory metabolites of certain nonionic surfactants, *i.e.* nonylphenol ethoxylates (APE) containing more than five oxyethylene groups. The refractory nature derives from the extensive branching in the nonyl side-chain. Aliphatic alcohol ethoxylates (AE) are much more readily degraded and not as susceptible to decreased removal at low temperatures (<12 °C), as are the alkylphenol ethoxylates. Kravetz (1981), in reviewing the biodegradation of nonionic surfactants, indicates that AE splits into alkyl and polyoxyethylene (POE) chains followed by rapid oxidation of the alkyl chain and a slower oxidation of the POE chain. By using suitably ^{14}C-labelled linear alcohol ethoxylates, Larson and Games (1981) have shown that neither the alkyl nor ethoxylate chain length had significant effects on the rate or extent of degradation and that the rate was first-order at concentrations between 1 and 100 μg l^{-1}. Degradation of APE is thought to proceed through simultaneous slow oxidation of the alkyl chain, the aromatic ring and the POE chain (Kravetz, 1981). The biodegradation of cationic surfactants based on quaternary ammonium compounds has recently been reviewed by Mackrell and Walker (1978) who found that some short-chain compounds supported bacterial growth, whereas longer-chain members did not. Cationic surfactants are very strongly adsorbed on to sludge, so that there is likely to be only very low concentrations in effluents and special steps have to be taken to demonstrate whether removal is by adsorption or degradation.

The relative degradation rates in batch tests of a large number of aromatic compounds and some aliphatic compounds, including phenols, cresols, xylenols, polyhydric phenols, amino and nitro compounds, have been determined by Pitter *et al.* (1974a, 1974b, 1976). Out of 13 phenolic compounds, only pyrogallol did not degrade; and out of some 20 nitrophenols only 2,4,6-trinitro-phenol and 3,5-dinitrosalicylic acid did not degrade. However, the results are not very helpful in deciding on the treatability of the phenols. Indeed, Hughes and Stafford (1976) in reviewing the complex way in which many aromatic and heterocyclic compounds are metabolized in the treatment of coke-oven effluents are extremely critical of simple batch experiments. Single isolation of the 'occasional' bacteria followed by biochemical tests on cleavage with a particular substrate reveals little of the regulatory mechanisms involved in treating a complex liquor.

Apart from the use of ^{14}C-labelling already mentioned, some other recent examples of data obtained using radiolabelling are: carboxymethylcelluloses were only slowly degraded by activated sludge (20 d in batch tests; Blanchard *et al.*, 1976); 1 mg l^{-1} of ^{14}C-labelled tri-*p*-cresyl phosphate was nearly completely removed by activated sludge, with *p*-hydroxybenzoic acid as an intermediate (Ku and Alvarez, 1982); and ^{14}C-labelled diphenylamine was metabolized *via* aniline, 4-hydroxydiphenylamine and indole, all of which were further rapidly degraded (Gardner *et al.*, 1982). Of great importance to an understanding of the kinetics of trace organic compounds in treatment processes is the work of Alexander (*e.g.* Boethling and Alexander, 1979a, 1979b; Subba-Rao *et al.*, 1982). The degradation was followed of ^{14}C-labelled compounds (glucose, phenol, nitrophenols, 2,4-D, *etc.*) added, at concentrations down to ng l^{-1}, to polluted and unpolluted river and lake waters. The studies have suggested that (1) mineralization of some chemicals at concentrations <1 mg l^{-1} is the result of activities of organisms different from those functioning at higher concentrations or of organisms that metabolize the chemicals but assimilate little or none of the substrate carbon; and (2) the kinetics of mineralization, the capacity of the organisms to assimilate carbon from the substrate and the sensitivity of the population to organic compounds are different at trace levels than at higher concentrations.

42.10 STRUCTURE–ACTIVITY RELATIONSHIPS

As in other biological spheres, relationships have been sought between the molecular structure of chemicals and their biodegradability, but so far no clear cut comprehensive rules have emerged, nor are likely to do so. Alexander (1965) has summarized early work; aliphatic alcohols, aldehydes, monocarboxylic acids and esters are readily degraded by unacclimatized activated sludge; but nitriles, alkanes, ketones, amines, dicarboxylic acids are broken down with difficulty or are toxic. The length of the chain is important; aliphatic hydrocarbons are more readily oxidized the longer the chain, up to C_{12} and the straight chain alkylbenzenesulfonates with C_{10} and C_{12} in the side chain are more easily degraded than those with C_6 or C_8. A high degree of branching confers greater resistance to biological attack; for example, alkanes containing C atoms carrying more than one methyl group, especially on the penultimate C atom, are particularly recalcitrant. The relatively few naturally occurring compounds containing a quaternary C atom are readily degraded, but its presence in the side chain of an aromatic compound endows resistance, although its presence in certain methylated dicarboxylic acids and in pentaerythritol does not greatly hinder degradation, after short acclimatization, while in acids like trimethylacetic acid it does, even after long acclimatization. In aromatic compounds, the introduction of the groupings methyl, chloro, nitro and amino in general make the compound less easily degradable, and *ortho*- and *meta*-substituted isomers are more recalcitrant than the *para*-isomers.

The problem has been more recently reviewed (Howard *et al.*, 1975) and a tentative structure–activity model has been proposed based on results gleaned from the literature on 430 compounds, 70% of which were degradable (Geating, 1981). Stepwise discriminant analysis and ridge progression applied to the data led to degradable and nondegradable coefficients for each substructural molecular fragment (*e.g.* hydroxyl group, amino group), of which only 39 are available at present out of over 200 listed. For a new compound the substructural fragments are identified and the probability of its degradability is calculated from the summation of the degradability and nondegradability coefficients. Apart from the obvious difficulty of interpreting the data in the literature, the author calculated the accuracy of discrimination to be 93% for biodegradable compounds and 70% for nonbiodegradable compounds. About 16% of nondegradable compounds were called degradable by the discriminant equation (false positives) and 2% which were degradable were called nondegradable (false negatives) by the equation.

42.11 MULTISUBSTRATES AND SPECIES INTERACTIONS

Most biochemical studies in this field on topics such as diauxie, sequential removal of substrates, formation of intermediates and polymers, adaptation and enzyme repression have been made using bacterial species derived from sewage or activated sludge, but under conditions far removed from those operating in wastewater treatment (see Volume 4, Chapter 45). The most relevant studies are probably those of Stumm-Zollinger (1968). She grew isolates in pure culture and mixed them after growth to produce mixtures more akin to activated sludge populations than those used by other workers, who inoculated their media with a small inoculum of sewage or activated sludge.

42.11.1 Acclimatization

The ability of activated sludge populations to acclimatize to new substrates is well known and even substrates such as glucose when presented in higher concentrations than normal can cause increases in the specific rate of removal of the substrate. The mechanisms by which these changes occur are not known, but the most common is thought to be changes in species composition rather than enzymic adaptation or mutation. It is probable that no single mechanism operates at any one time and it is likely from the work of Rao and Gaudy (1966) and Prakasam and Dondero (1970), who found that the population of sludge changed even when operating conditions and influent composition were constant, that the relative importance of the various mechanisms changes from time to time. Another possibility is that the newer, more complex organic compounds are degraded by cometabolism or so-called 'fortuitous' metabolism (Knackmuss, 1981).

42.11.2 Multisubstrates

Various pairs of substrates (*e.g.* glucose–fructose, glucose–butyrate, glucose–phenylalanine) exhibited diauxie and sequential removal of substrates in both pure and mixed cultures, derived from activated sludge, at high specific growth rates in both batch and continuous culture (Gaudy *et al.*, 1963; Stumm-Zollinger, 1966). Glucose was oxidized first in all cases, but the phenomenon occurred only with 'young' cells and not with 'aged' cultures equivalent to normal activated sludge. Activated sludge either did not show sequential removal in the above pairs (and also in glucose–acetate, glucose–phenol and glucose–alkylbenzenesulfonate mixtures (Prakasam and Dondero, 1967, 1970; Painter *et al.*, 1968)) or sometimes removed each of the substrates at lower rates than with corresponding single substrates till one had been completely removed (Wuhrmann and Mikx, 1969). Stumm-Zollinger (1968) showed that mixed cultures which exhibited sequential removal usually had a predominance of coliforms or had only a relatively few species. Thus, sequential removal probably does not occur in the treatment of sewage, but could occur in the high rate treatment of wastewaters containing relatively few substrates at relatively high concentrations.

42.11.3 Intermediate and End-products

It is well established that acetate can be detected in continuous cultures of various species growing on well-aerated glucose or glucose–lactate media. Krishnan and Gaudy (1965) demonstrated this, too, for mixed cultures derived from activated sludge, but it is unlikely that acetate would be present in activated sludge at normal rates of treatment. In those cases where acetate has been reported, *e.g.* 160 mg l^{-1} in biological filters (Williams and Taylor, 1968), it has been due to anaerobic zones within the treatment process. However, unidentified organic matter has been found in effluents from activated sludge units treating 'synthetic sewage' with glucose, or other compounds, as the sole source of carbon. Lysis was probably the source of the organic matter but it could also have been intermediate products. Although, in studies on pure single and binary mixtures of five species, it was concluded that competition for the substrate rather than antagonism/synergism accounted for species dominance, Hartmann and Singrün (1968) demonstrated direct stimulations of growth of >100% and inhibitions of about 50% in binary and tertiary mixtures of three species using a 'biocenose simulator', in which the species were separated by membranes.

It has been generally observed that mixtures of species are more effective than single isolates in removing organic matter from wastewaters. This may be due to the mixed species being able to metabolize a wider range of substrates than any single species in the mixture, but synergism might also be playing a part.

42.11.4 Polymers

Bacteria grown under some conditions convert external substrate to intracellular polymers, such as glycogen and poly-3-hydroxybutyrate (PHB), sometimes called storage products. Among conditions known to lead to the formation of these polymers are abnormally high C:N and C:P ratios in the medium or wastewater, and it is possible that other conditions of stress, such as very low substrate concentration or imbalance of metal ions, might encourage their formation. PHB, which occurs only intracellularly, was present in activated sludge from 0.5–1% (Painter *et al.*, 1968) to 12% dry weight (Crabtree *et al.*, 1965). The polymer has been shown to contain not only PHB but also the corresponding valerate and smaller proportions of the C_6 and C_7 compounds (Wallen and Rohwedder, 1974).

The position of polysaccharides is less clear since many species, and activated sludge, produce extracellular polysaccharidic slimes as well as intracellular polymers. The extracellular polymers are thought to be important for the flocculation of bacteria and the incorporation of non-flocculating species and inert finely suspended matter. The polymers usually form macrocapsules but some non-capsule-forming species flocculate by virtue of microgelatinous material on their cell surface. Yet other species form aggregates and chains of cells by rapid synthesis of intracellular PHB.

Novak and Haugan (1981), reviewing the efficiency and specificity of methods of polymer extraction from activated sludge, conclude that both qualitative and quantitative characterization of activated sludge polymers remains difficult. Thus, for example, although up to 15% of activated sludge solids was reported to be polysaccharide (Weddle and Jenkins, 1971), its origin, whether sewage solids, extracellular slime or truly intracellular polysaccharide, is unclear.

Forster (1976) has reviewed flocculation mechanisms and the composition of gels and slimes. One object was to see whether the concentration of some key component could be used as an indicator of the amount of gel present, but so far this has proved impossible. The gelatinous matrix of pure zoogloeal species is by no means uniform: the gel from some species yielded only glucose on hydrolysis, but others yielded many sugars (hexoses, pentoses, aldoses, ketoses) and uronic acids. Mucilages extracted from activated sludge gave a wider variety of sugars on hydrolysis, but not all the sugars and uronic acids appeared in all samples. Further, RNA, DNA, proteins and amino sugars, as well as some 10 metals (Sato and Ose, 1980) were present in sludge extracts. However, since the extracts were made with NaOH, the true extracellular nature of the viscous extract is in much doubt (Novak and Haugan, 1981).

Excess extracellular hydrophilic polysaccharide, for example from zoogloeal species, has been given as a cause of 'bulking' sludge (Pipes, 1969).

42.12 BIOCHEMICAL INDICATORS

Attention has been given to finding a substance(s) in activated sludge which would act as an indicator of biomass or activity, but without much success. For example, Weddle and Jenkins (1971) used ATP as an indicator of oxidative activity in activated sludges grown at various kinetic rates, but concluded that, as with dehydrogenase activity mentioned earlier, ATP gave no better indication of activity than the concentration of mixed liquor suspended solids. A report (Genetelli, 1967) that the DNA content of sludge fell before the sludge volume index (an indication of settling properties) rose, that is the DNA content gave an early warning of the presence of organisms such as *Sphaerotilus nutans*, known to contain lower concentrations of DNA and which are associated with 'bulking', does not seem to have been pursued.

42.13 REFERENCES

Alexander, M. (1965). Biodegradation: problems of molecular recalcitrance and microbial fallibility. *Adv. Appl. Microbiol.*, **7**, 35–80.

Banerji, S. K., B. B. Ewing, R. S. Engelbrecht and R. S. Speece (1968). Kinetics of removal of starch in the activated sludge system. *J. Water Pollut. Control Fed.*, **40**, 161–173.

Blanchard, F. A., I. T. Takahashi and H. C. Alexander (1976). Biodegradability of {^{14}C} methylcellulose by activated sludge. *Appl. Environ. Microbiol.*, **32**, 557–560.

Boethling, R. S. and M. Alexander (1979a). Effect of concentration of organic chemicals on their biodegradation by natural microbial communities. *Appl. Environ. Microbiol.*, **37**, 1211–1216.

Boethling, R. S. and M. Alexander (1979b). Microbial degradation of organic compounds at trace levels. *Environ. Sci. Technol.*, **13**, 989–991.

British Coke Research Association (1967). Improvements in or relating to the treatment of liquid wastes. *Br. Pat.* 1 210 492.

British Coke Research Association (1969). Improvements in or relating to the treatment of liquid wastes. *Br. Pat.* 1 252 460.

Brown, V. M., E. F. King and H. A. Painter (1981). Environmental effects of chemicals. *Report to Department of the Environment 68.M.* Water Research Centre, UK.

Cain, R. B. (1981). Microbial degradation of surfactants and 'builder' components. In *Microbial Degradation of Xenobiotics and Recalcitrant Compounds*, ed. T. Leisinger, R. Hütter, A. M. Cook and J. Nüesch. Academic, London.

Chudoba, J., M. Némec and B. Nemcová (1968a). Residual organic matter in activated-sludge process effluents III. Degradation of saccharides and fats under continuous conditions. *Sb. Vys. Sk. Chem. Technol. Praze, Technol. Vody*, **13**, 27–44.

Chudoba, J., M. Prasil and H. Emmerová (1968b). Residual organic matter in activated-sludge process effluents III. Degradation of amino acids and phenols. *Sb. Vys. Sk. Chem. Technol. Praze, Technol. Vody*, **13**, 45–63.

Cooper, R. L. and J. R. Catchpole (1973). The biological treatment of carbonization effluents. *Water Res.*, **7**, 1137–1153.

Crabtree K., E. McCoy, W. C. Boyle and G. A. Rohlich (1965). Isolation, identification and metabolic role of the sudanophile granules of *Zoogloea ramigera*. *Appl. Microbiol.*, **13**, 218–226.

Forster, C. F. (1976). Bioflocculation in the activated sludge process. *Water SA*, **2**, 119–125.

Gardner, A. M., G. H. Alvarez and Y. Ku (1982). Microbial degradation of ^{14}C-diphenylamine in a laboratory model sewage sludge system. *Bull. Environ. Contam. Toxicol.*, **28**, 91–96.

Gaudy, A. F., K. Komolrit and M. N. Bhatla (1963). Sequential substrate removal in heterogeneous populations. *J. Water Pollut. Control Fed.*, **35**, 903–922.

Geating, J. (1981). Literature study of the biodegradability of chemicals in water. Vol. 1. Biodegradability prediction, advances in and chemical interferences with wastewater treatment. Municipal Environmental Research Laboratory, Cincinnati, OH.

Genetelli, E. J. (1967). DNA and nitrogen relationships in bulking activated sludge. *J. Water Pollut. Control Fed.*, **39**, R32–R44.

Gerike, P. and W. K. Fischer (1979). A correlation study of biodegradability determinations with various chemicals in various tests. *Ecotoxicol. Environ. Safety*, **3**, 159–173.

Gerike, P. and W. K. Fischer (1981). A correlation study of biodegradability determinations with various chemicals in various tests. II Additional results and conclusions. *Ecotoxicol. Environ. Safety*, **5**, 45–55.

Giger, W., E. Stephanou and C. Schaffner (1981). Persistent organic chemicals in sewage effluents. I Identifications of nonylphenols and nonylphenolethoxylates by glass capillary gas chromatography/mass spectrometry. *Chemosphere*, **10**, 1253–1263.

Hartmann, L. and M. E. Singrün (1968). Untersuchung und Auswertung symbiontischer Erscheinungen bei Bakterien mit Hilfe des Biocoenose-Simulators. *Zentralb. Bakteriol.*, **207**, 403–409.

Howard, P. H., J. Saxena, P. R. Durking and L.-T. Ou (1975). Review and evaluation of available techniques for determining persistence and routes of degradation of chemical substances in the environment. US Environmental Protection Agency, Washington, DC.

Hughes, D. E. and D. A. Stafford (1976). The microbiology of the activated-sludge process. *CRC Crit. Rev. Environ. Control*, **6**, 233–257.

Jones, G. L. and A. R. Paskins (1982). Influence of high partial pressure of carbon dioxide and/or oxygen on nitrification. *J. Chem. Technol. Biotechnol.*, **32**, 213–223.

Knackmuss, H.-J. (1981). Degradation of halogenated and sulphonated hydrocarbons. In *Microbial Degradation of Xenobiotics and Recalcitrant Compounds*, ed. T. Leisinger, R. Hütter, A. M. Cook and J. Nüesch. Academic, London.

Kravetz, L. (1981). Biodegradation of nonionic ethoxylates. *J. Am. Oil Chem. Soc.*, **58**, 58A–65A.

Krishnan, P. and A. L. Gaudy (1965). Studies on the response of activated sludge to shock-loadings. *Biotechnol. Bioeng.*, **7**, 455–470.

Ku, Y. and G. H. Alvarez (1982). Biodegradation of ^{14}C-tri-*p*-cresyl phosphate in a laboratory activated sludge system. *Appl. Environ. Microbiol.*, **43**, 619–622.

Larson, R. J. and L. M. Games (1981). Biodegradation of linear alcohol ethoxylates in natural waters. *Environ. Sci. Technol.*, **15**, 1488–1493.

Larson, R. J. and A. G. Payne (1981). Fate of the benzene ring of linear alkylbenzene sulphonate in natural waters. *Appl. Environ. Microbiol.*, **41**, 621–627.

Leidner, H., R. Gloor and K. Wuhrmann (1976). The kinetics of degradation of linear alkyl benzenesulphonate. *Tenside Deterg.*, **13**, 122–130.

Loehr, R. C. and J. C. Roth (1968). Aerobic degradation of long chain fatty acid salts. *J. Water Pollut. Control Fed.*, **40**, R385–R403.

Mackrell, J. A. and J. R. L. Walker (1978). The biodegradation of quaternary ammonium compounds. *Int. Biodeterior. Bull.*, **14**, 77–83.

Maier, W. J. (1968). Model study of colloid removal. *J. Water Pollut. Control Fed.*, **40**, 478–491.

Menar, A. B. and D. Jenkins (1970). Fate of phosphorus in waste treatment processes: enhanced removal of phosphate by activated sludge. *Environ. Sci. Technol.*, **4**, 1115–1121.

Novak, J. T. and B.-E. Haugan (1981). Polymer extraction from activated sludge. *J. Water Pollut. Control Fed.*, **53**, 1420–1424.

Painter, H. A. (1970). A review of literature on inorganic nitrogen metabolism in micro-organisms. *Water Res.*, **4**, 393–450.

Painter, H. A. (1977). Microbial transformations of inorganic nitrogen. *Prog. Water Technol.*, **8**, 3–30.

Painter, H. A., M. Viney and A. Bywaters (1961). Composition of sewage and sewage effluents. *J. Proc. Inst. Sewage Purif.*, **60**, 302–314.

Painter, H. A., R. S. Denton and C. Quarmby (1968). Removal of sugars by activated sludge. *Water Res.*, **2**, 427–447.

Pipes, W. O. (1969). Types of activated sludge which settle poorly. *J. Water Pollut. Control Fed.*, **41**, 714–724.

Pitter, P. (1976). Determination of biological degradability of organic substances. *Water Res.*, **10**, 231–235.

Pitter, P. and L. Richtzová (1974). Relation between molecular structure and biological degradability of organic compounds. IV Biodegradability of aromatic amino compounds. *Sb. Vys. Sk. Chem. Technol. Praze, Technol. Vody*, **19**, 59–75.

Pitter, P. and P. K. Rosolová (1974). Relation between molecular structure and biological degradability of organic compounds. III Biodegradability of aromatic hydroxy derivatives. *Sb. Vys. Sk. Chem. Technol. Praze, Technol. Vody*, **19**, 43–57.

Prakasam, T. B. S. (1966). Enumeration, characterization and adaptation of aerobic heterotrophic bacterial populations of activated sludge. Thesis. Rutgers State Univ. 1966. Univ. Microfilms Inc., Ann Arbor, 1967.

Prakasam, T. B. S. and N. C. Dondero (1967). Aerobic heterotrophic bacterial populations of sewage and activated sludge. III Adaptation in a synthetic waste. *Appl. Microbiol.*, **15**, 1128–1137.

Prakasam, T. B. S. and N. C. Dondero (1970). Aerobic heterotrophic bacterial populations of sewage and activated sludge. IV Adaptation of activated sludge to utilization of aromatic compounds. *Appl. Microbiol.*, **19**, 663–670.

Rao B. S. and A. F. Gaudy (1966). Effects of sludge concentration on various aspect of biological activity in activated sludge. *J. Water Pollut. Control Fed.*, **38**, 794–812.

Royal Commission on Sewage Disposal (1912). 8th Report of the Commission. HMSO, London.

Sato, T. and Y. Ose (1980). Floc-forming substances extracted from activated sludge by sodium hydroxide solution. *Water Res.*, **14**, 333–338.

Sharma, B. and R. C. Ahlert (1977). Nitrification and nitrogen removal. *Water Res.*, **11**, 897–925.

Sridhar, M. K. C. and S. C. Pillai (1966). Enzymes in sewages and sludges. *J. Sci. Ind. Res.*, **25**, 167–171.

Stafford, D. A. (1974). The effects of phenols and pyridines on nitrification in activated sludge. *J. Appl. Bacteriol.*, **37**, 75–82.

Stumm-Zollinger, E. (1966). Effects of inhibition and repression on the utilization of substrates by heterogeneous bacterial communities. *Appl. Microbiol.*, **14**, 654–664.

Stumm-Zollinger, E. (1968). Substrate utilization in heterogeneous bacterial communities. *J. Water Pollut. Control Fed.*, **40**, R213–R229.

Subba-Rao, R. V., H. E. Rubin and M. Alexander (1982). Kinetics and extent of mineralization of organic chemicals at trace levels in freshwater and sewage. *Appl. Environ. Microbiol.*, **43**, 1139–1150.

Takahashi, S., T. Fujita., M. Kato, T. Saiki and M. Maeda (1969). Metabolism of suspended matter in activated sludge treatment. *Proceedings of the 4th International Conference on Water Pollution Research, Prague*, pp. 341–359. Pergamon Press, Oxford.

Tomlinson, T. G., A. G. Boon and C. N. A. Trotman (1966). Inhibition of nitrification in the activated-sludge process of sewage disposal. *J. Appl. Bacteriol.*, **29**, 266–291.

van Bouwel, E. and E. van Vaerenbergh (1979). Possibilities of enhanced phosphate elimination by activated sludge — literature survey. *Trib. CEBEDEAU*, **32**, 145–148.

Wallen, L. C. and W. K. Rohwedder (1974). Poly-β-hydroxyalkanoate from activated sludge. *Environ. Sci. Technol.*, **8**, 576–579.

Weddle, C. L. and D. Jenkins (1971). The viability and activity of activated sludge. *Water Res.*, **5**, 621–640.

Williams, N. V. and H. M. Taylor (1968). The effect of *Psychoda alternata* and *Lumbricillus rivalis* on the efficiency of sewage treatment in percolating filters. *Water Res.*, **2**, 139–150.

Wuhrmann, K. and M. Mikx (1969). In discussion in L. F. Tischler and W. W. Eckenfelder (1969). Linear substrate removal in the activated sludge process. Proceedings of the 4th International Conference on Water Pollution Control, Prague pp. 361–383. Pergamon Press, Oxford.

Yall, I., W. H. Boughton, F. A. Roinestad and N. A. Sinclair (1972). Logical removal of phosphorus. *Prog. Water Technol.*, **1**, 231–241.

43

Precipitation and Coagulation in Waste Treatment

J. GREGORY
University College London, UK

43.1 INTRODUCTION

The success of many water and wastewater treatment processes depends on the conversion of dissolved impurities into insoluble precipitates. This can be achieved in a variety of ways, usually with chemical additives but sometimes by physical or biological means. In some cases, the precipitate may be in a very finely-divided (colloidal) state and the particles may be too small to be conveniently separated. Aggregation of these particles by coagulation (or flocculation) may be necessary for effective removal by processes such as sedimentation and filtration.

In this chapter, a brief outline of the principles involved in precipitation and coagulation will be given, with special reference to some of the more important applications in wastewater treatment. Emphasis will be on inorganic constituents, but clearly the presence of organic substances in wastewaters cannot be ignored, not least because they can have a major influence on the solubility, precipitation and coagulation of inorganic compounds.

As far as possible, we shall treat the phenomena of precipitation and coagulation separately, although in many practical processes they occur together.

43.2 PRECIPITATION

43.2.1 Solubility in Water

Water is a remarkable liquid whose structure and properties have been the subject of a great deal of research in recent years (see *e.g.* Franks, 1979). One unique property is the ability of water to dissolve a very wide range of substances.

The process whereby a solid dissolves in water involves the removal of ions or molecules from close neighbours (*e.g.* in a crystal lattice) and their dispersal throughout the available volume of water. In the case of an ionic solid, such as sodium chloride, oppositely charged ions are held together in a regular three-dimensional array by strong electrostatic forces and these must be overcome if the salt is to dissolve. The work required to overcome the binding energy of the crystal may come from the favourable interactions of the ions with water, generally referred to as 'hydration'. Water molecules are polar and can be attracted quite strongly to ions, especially those which are small and highly charged, to give one or more hydration layers. Once an ion is separated from a crystal lattice and hydrated, electrostatic interactions are considerably diminished partly because the hydration layer prevents the ion from approaching close to other ions and partly because of the high dielectric constant of water.

Overall, the dissolution of a solid depends on the free energy change ΔG associated with the process. This is composed of an enthalpic (heat) term ΔH and an entropy term $T\Delta S$, where T is the absolute temperature and ΔS is the entropy change:

$$\Delta G = \Delta H - T\Delta S \tag{1}$$

The substance can dissolve only if the overall free energy change is negative. This condition could result from a negative enthalpy term or a positive entropy term (so that $-T\Delta S$ would be negative) or both. A negative ΔH term is found when the heat liberated by hydration of the ions exceeds that required to overcome the binding energy of the crystal. In such cases, heat is evolved when the solid dissolves. However, there are many familiar examples (such as sodium nitrate) where a cooling occurs as the solid dissolves in water, indicating a positive ΔH value. In these cases, the entropic term $-T\Delta S$ must be sufficiently large and negative to outweigh the ΔH term. This implies a positive ΔS value and an increased randomness of the system. It might seem self-evident that ions in a solution must be in a more random state than those in a regular crystal lattice, but it must not be forgotten that water itself is a partially structured liquid and that ion hydration involves some ordering of water molecules. Thus, the net entropy change may be either positive or negative.

The overall free energy change ΔG may represent a rather small difference between large, opposing terms and its sign may be changed by relatively small and subtle influences. For this reason, it is not easy to predict the solubility of salts in water, nor to give an entirely satisfactory rationalization of aqueous solubility. In general, ionic solids with a high crystal binding energy and whose ions are not strongly hydrated in water will tend to have rather low solubilities. The attraction between highly-charged ions is much stronger than when singly-charged ions are involved, giving a higher binding energy and lower solubility. Thus, calcium carbonate is relatively insoluble in water because of the strong attraction between Ca^{2+} and CO_3^{2-}, whereas sodium carbonate has a high solubility, by virtue of the lower charge on the cation Na^+. With still more highly charged ions, such as Al^{3+} and PO_4^{3-}, the salts have extremely low solubility. Certain large ions, such as Ag^+, have much greater polarizability than simple cations such as Na^+ and this accounts for the relatively low solubility of salts such as silver chloride.

The solubility of organic substances in water depends on rather different factors to those affecting the solubility of salts. Simple organic compounds, such as hydrocarbons, have very limited solubility in water owing to their hydrophobic character. However, one or more substituent groups such as —OH, —NH$_2$ or —CO$_2$H can greatly increase the solubility because of the formation of hydrogen bonds between these groups and water molecules. For instance, alcohols and, especially, sugars are highly soluble in water by virtue of substituent hydroxyl groups.

The weakly basic amino (—NH$_2$) and weakly acidic carboxylic (—CO$_2$H) groups can ionize in water to form —NH$_3^+$ and —CO$_2^-$ respectively, depending on the pH value of the water. Organic compounds containing these groups are usually much more soluble in the ionized state than when they are unionized. For instance, the class of substances known as 'humic acids', which are very common in natural waters (Schnitzer and Khan, 1972) can be precipitated in acid solutions because the ionization of weakly acidic groups is suppressed under acid conditions. They can also be precipitated by certain metal ions such as Ca^{2+} and Al^{3+} because of the binding of these ions to anionic groups on the humic acid molecule (Hall and Packham, 1965; Mantoura *et al.*, 1978).

Another important class of organic substances is the proteins. Their constituent amino acids can ionize to give either positive or negative groups, depending on the pH of the solution (Figure 1).

Because pH affects the ionization of the two groups in opposite ways, there is a certain pH value at which the net ionization is zero and the protein molecule appears to be uncharged. This is the isoelectric point (i.e.p.), which is characteristic of particular proteins. Most proteins have i.e.p. values in the acid range of pH ($\sim$ 3–6) and precipitation usually occurs at or around the

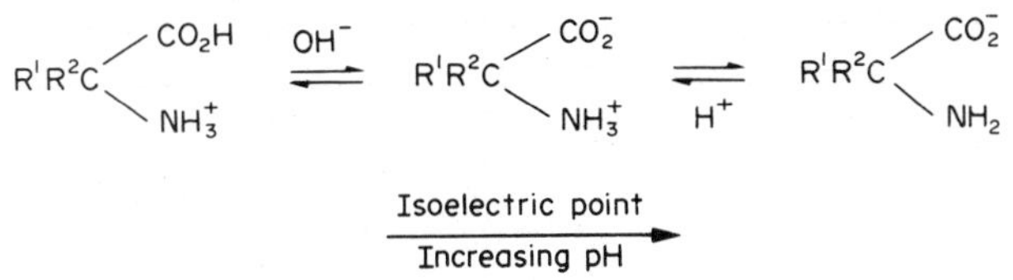

Figure 1 Ionization of amino acids

i.e.p. Many metal ions will also precipitate proteins because of their strong binding to various groups on the protein molecules. A very detailed account of protein solubility and precipitation has been given by Bell *et al.* (1983).

In principle, the experimental measurement of solubility is straightforward. The solid is brought into contact with water under the required conditions and allowed to dissolve until equilibrium is attained between solid and solution. Under equilibrium conditions the solution is said to be saturated and the saturation concentration can be measured by any convenient analytical technique to give the solubility of the substance.

For materials of fairly low solubility in water it is customary to quote the solubility in units such as $mg\,l^{-1}$ (or the SI equivalent $kg\,m^{-3}$). For sparingly soluble ionic solids it is more usual to quote the solubility product, which will be defined in the next section.

43.2.2 Solubility Products

Consider a salt M_yA_z, which dissolves slightly in water to give cations M^{z+} and anions A^{y-}. The dissolution process can be written in the same form as a chemical equilibrium reaction:

$$M_yA_z \text{ (solid)} \rightleftharpoons y\ M^{z+}(aq) + z\ A^{y-}(aq) \tag{2}$$

The term (aq) indicates that the ions are in aqueous solution, but, for convenience, these and (solid) will usuallly be omitted from now on.

By analogy with chemical equilibria, an equilibrium constant can be written for the dissolution reaction:

$$K = \{M^{z+}\}^y \{A^{y-}\}^z / \{M_yA_z\} \tag{3}$$

where the terms in curly brackets represent thermodynamic activities of the species.

Since the solid phase is conventionally specified as the standard state, its activity can be taken as unity. For simplicity, we shall replace the activities of the ions in solution by their molar concentrations $[M^{z+}]$ and $[A^{y-}]$. This is only an acceptable approximation in solutions of very low ionic strength. In waters with fairly high levels of dissolved salts (500 mg l^{-1} or more) it can lead to significant error. However, since our discussion will be largely in qualitative terms and because there are much more significant effects on solubility in wastewater treatment systems, the neglect of activity corrections will be acceptable for the present purposes. The resulting expression gives the solubility product K_s:

$$K_s = [M^{z+}]^y [A^{y-}]^z \tag{4}$$

Tables of solubility products are available in standard texts and a few examples are given in Table 1, taken from the compilation of Snoeyink and Jenkins (1980). Since solubility products are usually very small, it is more convenient to quote values of pK_s $(= -\log_{10}K_s)$.

In principle, for a solution containing the ions M^{z+} and A^{y-}, the product of their concentrations should not exceed the solubility product. If it does, then the solution is supersaturated with respect to the solid M_yA_z and precipitation should occur. If the product of the ion concentrations is less than the solubility product, then the solution is undersaturated and more solid can dissolve. Raising the concentration of one or other ions (or both) can give a supersaturated solution.

Note that in some cases, such as calcium carbonate, the solubility product may differ from one crystalline form to another. Thus, calcite is slightly less soluble than aragonite and might be expected to be the form which precipitates from supersaturated solutions. However, by suitable choice of conditions either calcite or aragonite crystals can be formed (Smallwood, 1977).

Table 1 Solubility Products[a]

Solid	pK_s^b	Solid	pK_s^b
$Al(OH)_3$ (amorphous)	33	$CaHPO_4$	6.66
$Al(OH)_3$ (gibbsite)	36.3	$Ca_3(PO_4)_2$	24.0
$AlPO_4$	21.0	$Ca_5(PO_4)_3OH$ (hydroxyapatite)	55.9
$BaSO_4$	10.0	$Fe(OH)_3$ (amorphous)	38
$CaCO_3$ (calcite)	8.34	$FePO_4$	17.9
$CaCO_3$ (aragonite)	8.22	$Fe_3(PO_4)_2$	33
CaF_2	10.3	$MgCO_3$	5.0
$Ca(OH)_2$	5.3	$Mg(OH)_2$	10.74

[a] All in terms of molar concentrations.   [b] $pK_s = -\log_{10}K_s$.

The relationship between solubility product and solubility is not obvious and will be illustrated with a few examples.

Barium sulfate has a solubility product of about 10^{-10} ($pK_s = 10$). What would be its solubility in pure water? If we suppose x moles of $BaSO_4$ to dissolve in 1 l of water to give a saturated solution, then $[Ba^{2+}] = x$ and $[SO_4^{2-}] = x$ and the solubility product will be $[Ba^{2+}] [SO_4^{2-}] = x^2$. (N.B. since both ions have the same valence their concentrations are not squared, as might be expected from the general formulation of the solubility product in equation 4). So, $x^2 = 10^{-10}$ and, therefore, $x = 10^{-5}$ and the saturation concentration of $BaSO_4$ is 10^{-5} mol l^{-1}. Since the molecuar weight of barium sulfate is about 233 the solubility is about 23 mg l^{-1}.

Now take the case of calcium fluoride in pure water. If x mol l^{-1} of CaF_2 dissolve, then the concentrations of Ca^{2+} and F^- ions will be x and $2x$ respectively. The solubility product is $[Ca^{2+}] [F^-]^2 = 4x^3 \simeq 5 \times 10^{-11}$, so that $x \simeq 3.7 \times 10^{-4}$ mol l^{-1} and the solubility of calcium fluoride is about 29 mg l^{-1}.

The relatively trivial examples above are for substances dissolving in pure water and do not fully bring out the importance of the solubility product concept. Natural waters contain many different ions in solution and these can have a marked effect on solubility. For instance, what would be the solubility of calcium fluoride in a solution which already contained 10^{-2} M Ca^{2+}? If we again assume that x mol l^{-1} of CaF_2 dissolve, then the final concentrations of Ca^{2+} and F^- will be $(10^{-2} + x)$ and $2x$ respectively. Now, as a first guess, it is reasonable to assume that x is very much less than 10^{-2} mol l$^-$ (the previous example shows that this is true, even in pure water). Thus, we can write the solubility product, approximately, as $[Ca^{2+}] [F^-]^2 \simeq 10^{-2}(4x^2) \simeq 5 \times 10^{-11}$. Hence $x \simeq 3.5 \times 10^{-5}$ and the solubility of calcium fluoride would be about 2.8 mg l^{-1}, which is less than 10% of the solubillity in pure water.

The reduction of solubility of a solid caused by the presence of one or other of its constituent ions in water is known as the common ion effect and is of great importance in water chemistry.

An increase in ionic strength which does not involve any common ions usually causes an increase in solubility because activity coefficients are reduced and ion activities become appreciably less than their molar concentrations.

43.2.3 Effect of Hydrolysis on Solubility

The previous examples were chosen because they involved simple ions which are not appreciably hydrolysed in water. In many cases of importance in water and wastewater treatment, solubility depends greatly on the pH value of the water because of interactions involving cations or anions.

The solution of calcium carbonate in water may be represented as:

$$CaCO_3 \text{ (calcite)} \rightleftharpoons Ca^{2+}(aq) + CO_3^{2-}(aq) \qquad pK_s = 8.34$$

Using the solubility product alone the solubility of calcite in pure water could be calculated as 6.8 mg l^{-1}. However, this would be to ignore the equilibrium between carbonate and bicarbonate ions, given by the following dissociation:

$$HCO_3^- \rightleftharpoons CO_3^{2-} + H^+ \qquad pK = 10.3$$

This equilibrium constant, together with the solubility product of calcite and the ionic product of water, $[H^+] [OH^-] = 10^{-14}$, must be used to calculate the solubility of calcite in water. The result turns out to be about 13 mg l^{-1} or about twice the value calculated simply from the solubility product. (Allowance for the presence of ion pairs in solution (Huang, 1975) gives still greater solubility, but will not be considered further here.)

When the pH of a water is adjusted, the equilibrium between carbonate and bicarbonate is modified. As the pH is reduced the proportion of CO_3^{2-} decreases in favour of HCO_3^-. Since calcium bicarbonate is much more soluble in water than calcium carbonate, a reduction in pH effectively increases the solubility of the latter. Conversely, raising the pH of a calcium bicarbonate solution may cause precipitation of calcium carbonate. These effects account for the origin of hard waters, scale formation and the mechanism of lime-soda softening as well as being of enormous importance in controlling the chemistry of natural waters.

The essential point from the above discussion is that the concentration of an ion to be used in solubility product calculations must be that of the particular species concerned, not necessarily the total concentration of the ion added. This can be further illustrated by considering the solubility of hydrous oxides of aluminum and iron, generally given the formulae $Al(OH)_3$ and $Fe(OH)_3$, although these are often amorphous in nature, with rather indeterminate compositions. Aluminum and ferric salts are used extensively in water and wastewater treatment as coagulants and for removal of phosphate and organic colour. Unfortunately, the aqueous chemistry of Al^{3+} and Fe^{3+} is extremely complex and only a very brief summary can be given here. For further details, see Stumm and Morgan (1970) and Snoeyink and Jenkins (1980).

A very important feature of the chemistry of these metals in water is that they undergo a series of hydrolysis reactions, in which water molecules in the hydration shell of an ion are progressively replaced by hydroxyl groups. Thus, for aluminum, we can write the following reaction sequence (in highly simplified form):

$$Al^{3+} \rightleftharpoons Al(OH)^{2+} \rightleftharpoons Al(OH)_2^+ \rightleftharpoons Al(OH)_3 \rightleftharpoons Al(OH)_4^-$$

Increasing pH drives these equilibria to the right. Only in low pH solutions does the unhydrolysed Al^{3+} ion exist to any significant extent. As base is added hydrolysis occurs until 'aluminum hydroxide' $Al(OH)_3$ is precipitated. Further addition of base causes the precipitate to redissolve, as the aluminate ion $Al(OH)_4^-$ is formed.

As well as the simple hydrolysis products listed above, several multinuclear species are known, such as $Al_2(OH)_2^{4+}$ and $Al_{13}(OH)_{34}^{5+}$, but it is thought that these are not true equilibrium species, but kinetic intermediates which are unstable with respect to the solid hydroxides (Hem and Roberson, 1967). However, because of the very slow reactions involved, they can exist in solution for very long periods (up to several weeks) under some conditions. Common constituents of water, especially the sulfate ion, can have a marked effect on the precipitation of aluminum hydroxide (de Hek *et al.*, 1978).

The aqueous chemistry of Fe^{3+} is rather similar to that of Al^{3+} except that there is a lower tendency to form polynuclear hydrolysis products. This makes calculation of the concentrations of the various species somewhat simpler and Figure 2 is a diagram showing the reults of such calculations. This is based on the solubility product for amorphous $Fe(OH)_3$ and the equilibrium constants for the hydrolysis reactions given by Snoeyink and Jenkins (1980). Essentially, the diagram shows the concentrations of the various species in equilibrium with the solid hydroxide as a function of pH. The minimum solubility of Fe^{3+} is seen to occur at about pH 8. Only at low pH values (<2) does the unhydrolysed Fe^{3+} predominate. A similar diagram for Al^{3+} shows a minimum solubility at about pH 6.5 and a rather narrower pH range in which hydroxide precipitation occurs.

It should be pointed out that Fe^{3+} is the oxidized form of iron which usually predominates in waters with dissolved oxygen. However, in anoxic waters, ferrous iron, Fe^{2+}, is the stable form and this precipitates much less readily than Fe^{3+}.

As a final example of the effect of hydrolysis on solubility we shall consider the question of phosphate precipitation. The solubility products of phosphates in Table 1 mainly involve the PO_4^{3-} ion, although this is only one of the possible forms of orthophosphate in water. Orthophosphoric acid, H_3PO_4, dissociates to give $H_2PO_4^-$, HPO_4^{2-} and PO_4^{3-}. The important dissociation equilibria are:

$$H_2PO_4^- \rightleftharpoons HPO_4^{2-} + H^+ \qquad pK_2 = 7.2$$
$$HPO_4^{2-} \rightleftharpoons PO_4^{3-} + H^+ \qquad pK_3 = 12.2$$

(There are also condensed forms of inorganic phosphorus, such as tripolyphosphate, which can be significant in wastewaters, but these will not be considered here.)

The above equilibria indicate that, at neutral pH values, the PO_4^{3-} ion will represent a very small fraction of the total orthophosphate present in solution. Only at pH 12.2 would its concen-

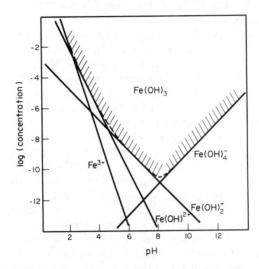

Figure 2 Concentrations of various soluble ferric species in equilibrium with freshly-precipitated $Fe(OH)_3$. The dashed line represents the total concentrations of all species shown, *i.e.* the solubility of $Fe(OH)_3$. Multinuclear hydrolysis products are ignored

tration become equal to that of HPO_4^{2-}. Nevertheless, because of the much lower solubility of its salts, the PO_4^{3-} ion can be important at lower pH values, where HPO_4^{2-} predominates.

To illustrate this point, consider first calcium hydrogen phosphate $CaHPO_4$ and tricalcium phosphate $Ca_3(PO_4)_2$. The solubility products of these solids are given in Table 1 and these values, together with the dissociation constants given above, have been used to construct solubility lines which are shown in Figure 3. These show the total orthophosphate concentration P_T ($= [H_2PO_4^{2-}] + [HPO_4^{2-}] + [PO_4^{3-}]$) in equilibrium with each of these solids as a function of pH. The calcium concentration in solution was chosen as 10^{-3} M (100 mg l^{-1} as $CaCO_3$). The results indicate that at pH values above 8.5, orthophosphate solubility should be controlled by solid tricalcium phosphate, whereas in more acid solutions calcium hydrogen phosphate should be precipitated. In a typical water of pH 8, the total orthophosphate concentration would be about 0.25 mM or about 8 mg l^{-1} as P. (This value is, of course, greatly dependent on the calcium concentration chosen.)

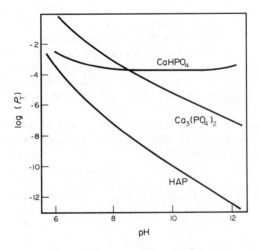

Figure 3 Total orthophosphate concentration P_T in solutions saturated with respect to three different solids : calcium hydrogen phosphate $CaHPO_4$, tricalcium phosphate $Ca(PO_4)_2$ and hydroxyapatite (HAP). Calcium concentration = 10^{-3} M. For any pH value the phosphate concentration in solution should be governed by the solid giving the lowest P_T at that pH

In fact, the previous calculations are rather misleading since they do not consider hydroxyapatite (HAP), which is believed to control the phosphate concentration in natural waters. The solution of this solid can be written as:

$$Ca_5(PO_4)_3OH \rightleftharpoons 5\,Ca^{2+} + 3\,PO_4^{3-} + OH^-$$

The solubility product is given by

$$K_5 = [Ca^{2+}]^5\,[PO_4^{3-}]^3\,[OH^-] = 1.26 \times 10^{-56}$$

For $[Ca^{2+}] = 10^{-3}$ M, as before, the concentration of PO_4^{3-} can be calculated for each pH value and, hence, P_T from the dissociation constants given above. The form of the solubility product for HAP means that activity corrections can be especially significant in this case (Nancollas *et al.*, 1978), and that calculations based on molar concentrations can be very misleading. The results given here would only apply in solutions of very low ionic strength.

The appropriate line for HAP is also given in Figure 3 and shows that for all pH values the phosphate concentration would be much lower than that in equilibrium with either of the other two solids. At pH 8 P_T would be about 5.4×10^{-8} M or about 1.7 μg l^{-1} as P, nearly 5000 times lower than the phosphate concentration in equilibrium with calcium hydrogen phosphate.

These calculations indicate that phosphate concentrations in waters containing moderate amounts of calcium under neutral or alkaline conditions should be extremely low. Furthermore, the addition of calcium in the form of lime should be an effective method of precipitating phosphate, not only because of the increased calcium concentration, but also because of the higher pH.

Aluminum and ferric salts are also used to precipitate phosphate from waters and wastewaters. The solubility products of $AlPO_4$ and $FePO_4$ are given in Table 1, but calculations are complicated by the hydrolysis reactions of Al^{3+} and Fe^{3+} and by the precipitation of the hydrous oxides. As pH increases the phosphate solubility should decrease, because of the increasing proportion of PO_4^{3-}. However, the concentrations of free Al^{3+} and Fe^{3+} decrease as the pH is raised and this causes an increase in phosphate solubility. In the case of aluminum, the minimum phosphate solubility occurs at about pH 5.5, just below the value at which precipitation of amorphous aluminum hydroxide occurs (Snoeyink and Jenkins, 1980).

A further complication is that phosphate readily absorbs on precipitated aluminum hydroxide (Goldschmid and Rubin, 1978) and this might be an effective removal mechanism in practice.

In the use of ferric salts to precipitate phosphate from wastewaters, there may be a problem when the resulting sludges are subjected to highly reducing conditions (as in the sludge digestion process). Under these conditions reduction of iron to the ferrous state should occur and there is a possibility that the precipitated phosphate may redissolve. This does not seem to occur in practice, perhaps due to the formation of the sparingly soluble ferrous phosphate, vivianite (Singer, 1972).

43.2.4 Effect of Complex Formation on Solubility

As well as hydrolysis reactions, metal ions in solution may undergo many interactions with a variety of organic and inorganic constituents to form soluble complexes. Insofar as complex formation reduces the concentration of the free metal ion, the solubility of metal salts can be increased. This is a very large subject with a vast literature (see *e.g.* Turner *et al.*, 1981) and we can do no more than mention a few of the more important aspects.

Among the most significant inorganic ligands (*i.e.* complex-forming agents) are the following common anions: OH^-, F^- Cl^-, SO_4^{2-} and CO_3^{2-}. These can form complexes which may be of predominantly ionic (*e.g.* F^-) or covalent (*e.g.* Cl^-) character. Complexes with OH^- have already been covered in our discussion of metal hydrolysis. Of the remaining anions, many form complexes which significantly affect the speciation of trace metals, for instance about 95% of lead dissolved in fresh water may exist as a carbonate complex (Turner *et al.*, 1981). However, these complexes are not of great significance in wastewater treatment and will not be considered further.

Organic ligands can form very strong complexes with metal ions. These may be of natural original such as the humic acids mentioned earlier, or synthetic chelating agents such as EDTA and NTA which may find their way into wastewater. The effect of this complex formation can be to increase greatly the solubility of metals in water. For instance, Mantoura *et al.* (1978) have calculated that essentially 100% of the copper and mercury in a lakewater were present as humic complexes. The solubilities of iron and aluminum can also be significantly enhanced by humic substances in natural waters.

43.2.5 Formation of Precipitates

When a solid is in contact with an undersaturated solution, the solid will spontaneously dissolve until equilibrium (saturation) is achieved. However, the opposite process, whereby a precipitate forms from a supersaturated solution, may not occur spontaneously. It is generally agreed (see *e.g.* Walton, 1967; Mullin, 1972) that precipitaton must be preceded by nucleation of the solution. This is then followed by crystal growth and, usually, aggregation of crystals.

Nucleation may be either homogeneous (occurring from pure solutions) or heterogeneous (occurring on impurity particles or seed crystals). True homogeneous nucleation is difficult to achieve under laboratory conditions and probably never occurs in practical wastewater treatment processes. Nevertheless, it is worth considering briefly here, since a number of important points emerge.

Homogeneous nucleation involves the formation of a cluster of molecules or ions in the solution which is large enough to initiate crystal growth. Clusters will always be present in a solution, being continually formed and dissolved as a result of random molecular motion. Only when the concentration exceeds the saturation value will there be any chance of the formation of 'critical' clusters or nuclei.

The classical theory of homogeneous nucleation considers the free energy of nucleus formation in terms of two components: (a) the volume free energy due to bond formation, which will always be negative if the solution is supersaturated, and (b) the interfacial free energy per unit area of solid surface, which will always be positive. If r is the radius of a cluster, then the volume free energy will decrease (*i.e.* become more negative) as r^3 and the interfacial free energy will increase as r^2. This meas that the total free energy of nucleus formation will pass through a maximum at some critical cluster size r_c, which represents the minimum size of a stable nucleus. Clusters smaller than the critical size will tend to dissolve, whereas larger ones will tend to grow. The maximum in the free energy curve represents the work required to form the critical nucleus and depends on the supersaturation ratio s.

The rate of nucleation (*i.e.* the number of critical nuclei formed in unit volume and unit time) depends exponentially on the free energy maximum and increases very rapidly with concentration, once a certain critical supersaturation, s^*, is exceeded (see Figure 4). Below this critical value the nucleation rate (and hence the rate of precipitation) can be extremely slow.

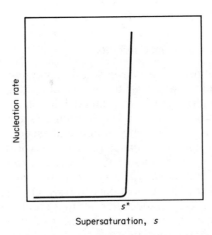

Figure 4 Effect of supersaturation on nucleation rate

Critical supersaturation ratios can be from as high as several hundred for $BaSO_4$ down to about 5 for $AgCl$. The size of the critical nucleus is thought to be around 1–2 nm.

Once the critical supersaturation is exceeded nuclei are formed which begin to grow at a rate which depends either on the diffusion of solute to the surface of the growing crystal or on the rate of incorporation into the crystal lattice. For further details of precipitation rates see Walton (1967) and Nielsen (1981).

If nuclei are formed from a solution only just above the critical supersaturation level, then the concentration of solute will soon drop below the critical value and essentially no new nuclei will be formed. This means that all subsequent growth must occur on the nuclei already present and this can lead to precipitate particles of very uniform size (Overbeek, 1982).

Another mechanism tending to produce uniformity of particle size is the phenomenon of Ostwald ripening, which is a result of the fact that solubility of crystals increases as particle size decreases. Small crystals have a high interfacial area per unit volume and the total interfacial energy of the system can be reduced if large crystals grow at the expense of smaller ones. Once crystals have grown to a size of about 1 μm, Ostwald ripening is of little further significance.

In heterogeneous nucleation crystal growth is initiated on the surface of particles already present (or perhaps on the walls of the container). The particles may be crystals of the same type as those precipitating (seed crystals) or they may be of quite different material. The effectiveness of such particles depends on the extent to which their crystalline structure matches that of the nucleating phase. Recent ideas on heterogeneous nucleation, including kinetic aspects, have been reviewed by Füredi-Milhoffer (1981).

The most important feature of heterogeneous nucleation is that it can occur at much lower supersaturation ratios than those needed for homogeneous nucleation. In principle, growth on seed crystals should occur at a concentration only just above saturation, but the growth can be very slow under such conditions.

For the important case of calcium phosphate precipitation, Nancollas *et al.* (1978) gave some useful examples. For a solution at pH 7.4 with a molar Ca/P ratio of 1.66, they found that spontaneous homogeneous nucleation occurred only at a total calcium concentration of 10 mM or more. The calculated value from the solubility product of hydroxyapatite is only 0.03 mM. Heterogeneous nucleation on impurities such as dust particles occurred at calcium concentrations above about 2.5 mM. Below this value, stable supersaturated solutions could be maintained for days unless specific nucleators (seed crystals) were added.

43.2.6 Nature of Precipitates

After nucleation has occurred, particle growth takes place and, possibly, aggregation of particles to give a precipitate which can be characterized in a number of ways. Crystal form, particle size distribution and particle shape are some of the properties which may be of interest.

It is commonly found that the phase which first precipitates from a supersaturated solution is not the most stable (or least soluble) form, but some metastable intermediate. There are many examples which are relevant to water and wastewater treatment. For instance, hydrous aluminum and ferric oxides precipitate initially as amorphous materials which may undergo slow recrystallization to form more stable products. An amorphous phase is one which has no long-range order, but may consist of small crystallites in a more or less random arrangement. The formation of such phases is favoured at high supersaturations and when the ions are strongly hydrated, leading to a precipitate which is still hydrated. After long periods of ageing (which can be accelerated at high temperatures) these metastable hydrates may be transformed into crystalline oxides such as boehmite ($Al_2O_3 \cdot H_2O$), and haematite (α-Fe_2O_3), but these are unlikely to be of practical significance in treatment processes.

The precipitation of phosphate from solution with calcium should yield hydroxyapatite as the stable phase at all pH values above about 6, below which calcium hydrogen phosphate should precipitate. However, even in alkaline solutions various precursor calcium phosphate phases can precipitate which only slowly transform to HAP (Nancollas *et al.*, 1978).

In water softening, both lime, $Ca(OH)_2$, and soda, Na_2CO_3, may be used, the latter if the water contains significant amounts of non-carbonate hardness. Randtke *et al.* (1982) have reported that very different forms of calcium carbonate are precipitated with the two additives. In the case of lime addition they found small, slowly settling particles consisting almost entirely of calcite. With soda, an amorphous precipitate was formed which settled rapidly and occupied a much larger volume than the sediment formed by lime. The amorphous precipitate was converted fairly rapidly to vaterite (a form of $CaCO_3$) which slowly recrystallized into large calcite crystals with a substantial decrease in sediment volume.

These results were explained in terms of the different degrees of supersaturation in the two cases. Lime is added as a slurry and the particles dissolve fairly slowly, giving only low degrees of supersaturation. Soda dissolves rapidly giving a large increase in carbonate concentration and a sudden sharp rise in supersaturation.

The particle size distribution of precipitates can be very dependent on the nature of the nucleation process and on the concentration of the solution. With heterogeneous nucleation and low degrees of supersaturation fairly compact crystals should form, the number of which depends on the number of nuclei present initially. As the supersaturation ratio is increased dendritic (needle-shaped) crystals may be favoured. When the supersaturation reaches the critical value for homogeneous nucleation, large numbers of new nuclei are formed and the particle size becomes much smaller. Aggregation of particles will be considered separately.

43.2.7 Effect of Impurities on Precipitation

Impurities naturally present in a water, or certain substances deliberately added, can have a profound effect on the precipitation process, quite apart from the effect of impurities on solubility discussed earlier. Many of these effects can be explained in terms of an interference with the normal nucleation and growth stages of precipitate formation. The adsorption of an impurity on the surface of a nucleus or growing crystal may effectively prevent further growth. Certain ionic impurities may be incorporated into the amorphous phase initially precipitated and prevent or retard the conversion to more stable phases. In addition to these effects, certain additives may increase the colloidal stability of the fine precipitate particles and so prevent aggregation (see Section 43.3.1).

The effects of additives on calcium carbonate precipitation have been studied in some detail because of their importance in such areas as scale formation. In wastewater treatment the use of lime to remove phosphate can be more effective if calcium carbonate precipitation is inhibited (Merrill and Jorden, 1975). Reddy (1978) has shown that many wastewater constituents can significantly retard calcite formation, especially organic acids, probably because of their strong adsorption at the calcite–water interface. Condensed phosphates can also be effective and are commonly used to combat scale formation. Small quantities of magnesium have a pronounced inhibitory effect on calcite formation, probably because of the incorporation of Mg^{2+} into calcium carbonate nuclei which reduces their effectiveness.

The precipitation of calcium phosphates can also be inhibited by various additives such as certain tricarboxylic acids (Nancollas *et al.*, 1978), phytate and phosphonate (Koutzoukos *et al.*, 1981). In many of these cases, the inhibitory effect can be interpreted in terms of a Langmuir adsorption isotherm. Feenstra and de Bruyn (1981) have shown that a small amount of magnesium lowers the solubility of amorphous calcium phosphate and retards the formation of hydroxyapatite.

The presence of silica has been shown to have an appreciable influence on the form of iron precipitates produced by the oxidation of ferrous solutions (Robinson *et al.*, 1981). Increasing the silica level caused an increasingly amorphous precipitate and a smaller particle size.

43.3 COAGULATION

43.3.1 Stability of Colloids

Colloidal particles are conventionally defined as having a size in the range 1 nm to 1 μm, although these limits are not rigid. Such small particles do not settle appreciably and can be difficult to remove by filtration. It is often necessary to cause colloidal particles to aggregate (or coagulate) before they can be effectively removed, but the aggregation process can be hindered by repulsion between the particles. The stability of colloids implies their resistance to aggregation and can arise for different reasons.

A basic distinction is traditionally drawn between hydrophilic and hydrophobic colloids. The former are substances which have a high affinity for water but do not form true solutions in the usual sense. Typical hydrophilic colloids are macromolecules such as gelatin and starch which, although soluble in water, have molecules large enough to bring them into the colloidal size range, so that they exhibit properties characteristic of dispersed systems (*e.g.* light scattering). Micelles, which are aggregates of surface-active molecules, can also be regarded as hydrophilic colloids. The essential feature of these systems is that they are thermodynamically stable and are kept in a dispersed state by virtue of their affinity for, and solubility in, water. Aggregation of such colloids would mean the reduction of favourable interactions with water and so does not occur spontaneously. Such systems can be coagulated only by reducing their solubility in water. In the case of proteins, we have already mentioned in Section 43.2.1 that precipitation can be brought about by pH adjustment or by the addition of certain metal ions. Other methods include heating and the addition of large quantities of salts ('salting out') which reduce the solvent power of water for proteins. With hydrophilic colloids such as proteins and other macromolecules, the terms 'precipitation' and 'coagulation' are more or less synonymous.

By contrast, hydrophobic colloids consist of insoluble particles, which happen to exist in a finely-divided state. Because the total interfacial energy of a colloidal dispersion depends on the surface area of the particles, this could be reduced by the formation of aggregates, since there would then be less particle surface in contact with the solution. This implies that hydrophobic colloids cannot be thermodynamically stable and that aggregation should occur spontaneously. In many cases, this does not happen because the particles carry a surface electric charge and repel each other. A colloidal dispersion stabilized in this way is still unstable in a thermodynamic sense, but has a kinetic stability which can keep the particles in a dispersed state virtually indefinitely.

The origin of the surface charge on colloidal particles can vary depending on the nature of the particles (see *e.g.* Shaw, 1980). Ionization of certain groups on the particle surface may be one reason, or a preferential adsorption of anions or cations from solution. Many clay particles are charged by virtue of isomorphous substitution of ions in their crystal structure. Oxide particles, such as Al_2O_3 and the amorphous oxides, have hydroxyl groups bound to their surface which are amphoteric (*i.e.* they can ionize to give either positive or negative surface charge). As pH is increased the negative ionization bcomes more important and at low pH values the surfaces become positively charged. As in the case of proteins, discussed in Section 43.2.1, there is a characteristic pH, the isoelectric point, where the net surface charge is just zero. In the case of Al_2O_3 the i.e.p. is bout 9, so that at pH values below 9 alumina would be positively charged.

In the case of ionic solids, one or other of the constituent ions may have a greater tendency to 'escape' from the crystal lattice, leaving a charged surface. With calcium carbonate, for instance, the carbonate ion dissolves preferentially, so that in a solution containing equal concentrations of Ca^{2+} and CO_3^{2-} (about 7×10^{-5} M at saturation) the solid would be positively charged. Lowering the calcium concentration in solution (and increasing the carbonate concentration to maintain saturation) would cause the solid to become less positive, then uncharged and then increasingly negative. In this case, the isoelectric point can be expressed in terms of calcium concentration and has been found to be about 4×10^{-5} M Ca^{2+}, or pCa = 4.4 (Foxall *et al.*, 1979). The charge on precipitating particles and hence their colloidal stability can thus depend on the ratio of the ion concentrations in the supersaturated solution.

Whatever the origin of the surface charge on colloidal particles, the presence of this charge causes a characteristic distribution of oppositely charged ions (counterions) in the solution around the particle. The particle charge, together with the associated counterion charge, is known as the electrical double layer and detailed accounts of this subject can be found in many texts (*e.g.* Hunter, 1981).

The most important point concerning colloid stability is that the diffuse layer of counterions may not extend very far into the solution and that particles can only experience repulsion when their diffuse layers begin to overlap (see Figure 5). The extent of the diffuse layer depends inversely on the ionic strength of the solution, ranging from about 1 μm in distilled water to less than 1 nm in sea water. In waters with moderate levels of dissolved salts (say 500 mg l^{-1}) the diffuse layer would extend about 5 nm into the solution. This means that charged particles approaching each other in such a water would not experience significant repulsion until the gap between them was less than about 10 nm.

At such small separation distances, another type of interaction can become important. This is the universal van der Waals attraction which operates between all atoms and molecules and gives significant attraction between colloidal particles at close approach. Particles will only be stable if electrical repulsion prevents them coming close enough for van der Waals attraction to become dominant.

The magnitude of the repulsion between particles depends on the ionic strength of the solution and on the charge carried by the particles. Surface charge is not usually measured directly, but the electrical properties of particles can be characterized by their zeta potential which can be measured fairly simply by techniques such as electrophoresis (see Hunter, 1981). Most particles in natural waters and wastewaters are negatively charged with zeta potentials typically in the range −10 to −40 mV.

A quantitative theory of colloid stability in terms of a balance between electrical repulsion and

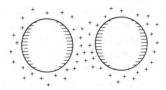

Figure 5 Showing the interaction of two negatively charged particles in water. Each particle is surrounded by a diffuse layer of counterions (cations in this case) and repulsion arises because of the overlap of the diffuse layers

van der Waals attraction was developed some time ago (Verwey and Overbeek, 1948) and has met with some success in model systems. However, in wastewaters there are many complicating factors which may seriously limit the applicability of the theory. For instance, there are innumerable solutes which may adsorb on particles and significantly modify their surface properties. When macromolecules are adsorbed the particles may acquire a hydrophilic character and other types of interaction may have to be considered.

The interaction of particles with adsorbed polymer layers is known as steric stabilization (*e.g.* Tadros, 1982). Significant interaction (usually repulsion) only occurs when the adsorbed layers come into contact, but it can then become very strong. One result of this short-range repulsion is that particles may be held rather weakly by van der Waals attraction and prevented from approaching closer because of the adsorbed layers (see Gregory, 1978). This would lead to relatively weak aggregates which are easily disrupted. Particles in wastewaters are often found to be weakly aggregated.

43.3.2 Coagulation by Salts

According to the simple picture of colloid stability outlined in Section 43.3.1, stable particles are prevented from aggregating by virtue of electrical repulsion between them. By reducing this repulsion the particles become destabilized and aggregation can occur at a rate which depends on the frequency of particle collisions. Collisions may be brought about by the Brownian motion of the particles or, more usually in practice, by particle motion induced by fluid shear, as in stirred tanks, upflow clarifiers or other units (Bratby, 1980).

There is some confusion over the terminology employed in this field. Aggregation of particles is generally referred to as either coagulation or flocculation and these terms are often used interchangeably. However, in the water treatment industry, the term 'coagulation' is usually taken to imply the destabilization of particles by suitable additives and the early phase of aggregation caused primarily by Brownian motion. These processes occur during rapid mixing of chemicals in the water. 'Flocculation' then refers to the subsequent phase of slow stirring in which large aggregates (flocs) are formed as a result of fluid motion.

Colloid scientists often draw a different distinction between the two terms, so that 'coagulation' means aggregation caused by the addition of salts and 'flocculaton' is restricted to the case of aggregation caused by long-chain polymers (see Section 43.3.3.)

It is difficult to adhere rigidly to either of these conventions in a discussion of wastewater treatment and we shall not attempt to do so. The meaning of the terms used should be clear from the context.

Returning to the question of destabilization, there are two practical methods of reducing the repulsion between particles. (a) By increasing the ionic strength of the solution so that the range of repulsion is reduced and particles can approach close enough to be attracted by van der Waals forces. The amount of salt needed to achieve the required double layer compression depends on the valence of the ions, especially that of the counterions, so that for negative particles highly charged cations are most effective. (b) By adding counterions which adsorb specifically on the particles and hence reduce their charge. It is often found that the zeta potential of particles needs to be reduced to about ± 10 mV for adequate destabilization. Specifically-adsorbing counterions should be more effective at lower concentrations than those acting simply through an ionic strength effect. Furthermore, the quantity of adsorbing counterions should depend directly on the particle concentration. There is also the possibility with strongly-adsorbing counterions that excess adsorption could cause charge reversal and restabilization of the particles.

In practice, coagulation of waters and wastewaters probably involves a combination of mechanisms. Typical coagulants are aluminum and ferric salts and lime. The addition of Al^{3+}, Fe^{3+} and Ca^{2+} might seem reasonable in view of the above discussion since they are highly-charged cations and most particles to be coagulated are negatively charged. However, the picture is greatly complicated by a number of factors.

We have seen in Section 43.2.3 that the aqueous chemistry of Al^{3+} and Fe^{3+} is extremely complex. It is known that the hydrolysis products such as $Al(OH)^{2+}$ and $Al_8(OH)_{20}^{4+}$ can be very strongly adsorbed on negative particles and reduce their charge. This can be a major reason for the destabilization of particles, but there is a good deal of evidence to suggest that hydroxide precipitation also plays a part. Precipitates of $Al(OH)_3$ and $Fe(OH)_3$ initially form as very fine colloidal particles which rapidly coagulate to form rather bulky, gelatinous flocs. During this process, particles initially present, after partial destabilization by adsorbed hydrolysis products, can become enmeshed in the growing hydroxide flocs and are carried down with them in sedimentation tanks. This is thought to provide a reasonable explanation of the action of hydrolysing coagulants in water treatment (see Amirtharajah and Mills, 1982, and references therein), but in the more complex systems encountered in wastewater treatment, the picture is still not clear.

In the case of lime addition during sewage treatment, Stones (1981) has shown that the action is not simply due to increased levels of Ca^{2+}, but that positively charged colloidal particles of calcium hydroxide play a part. Precipitation of calcium carbonate and calcium phosphate may also be important in the coagulation process.

The hydrophilic constituents of wastewaters can also be greatly affected by the addition of metal ions, but for different reasons. We have seen in Section 43.2.1 that organic colloids such as humic acids and proteins can bind certain metal ions and that this can reduce their solubility in water. It may be expected, therefore, that precipitation and coagulation of many organic materials will occur during wastewater treatment with metal salts. Adsorption of organics on precipitate particles such as calcite may also be a significant removal mechanism (Randtke *et al.*, 1982).

43.3.3 Polymeric Flocculants

Certain long-chain water-soluble polymers are very effective in aggregating colloidal particles and the mechanism is believed to be a form of 'bridging' between particles by the polymer chains (*e.g.* Gregory, 1982). These polymers adsorb at the particle surface in a way that leaves 'loops' or 'tails' extending into the solution and these sections of adsorbed polymer can attach to unoccupied sites on the surfaces of other particles, thus forming 'bridges' between particles. This process is illustrated schematically in Figure 6. As mentioned previously, aggregation by polymers is sometimes called 'flocculation' to distinguish the bridging mechanism from coagulation caused by the addition of salts. In view of the complexity of the action of some salts, this may not be a worthwhile distinction to make. Nevertheless, we shall adopt this terminology here and refer to the polymers as 'polymeric flocculants'.

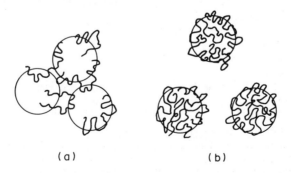

(a)   (b)

Figure 6 Schematic illustration of (a) flocculation and (b) restabilization of particles by adsorbed polymer chains

The nature of bridging flocculation is such that there should be an optimum dosage of polymer to give the best results. This arises because a certain amount of adsorbed polymer is required to give enough links between particles for adequate strength of the aggregates. However, when excess polymer is adsorbed there may be insufficient free particle surface remaining for attachment of polymer chains from other particles (see Figure 6). This can lead to a restabilization of the particles.

Many commercial polymeric flocculants are based on polyacrylamide and can have very high molecular weights (up to several million). Generally, the efficiency of bridging flocculation increases as the polymer molecular weight (or chain length) increases. Flocs produced by these materials can be much larger and stronger than those formed by more conventional agents.

Most polyacrylamides are slightly anionic in character (*i.e.* they have negatively charged groups along the polymer chain) and these can be very effective in some applications. However, in water and wastewater treatment they do not usually work well, except in conjunction with the more traditional aluminum and ferric salts, when they are sometimes called 'coagulant aids'. The reasons may have to do with the negative charge carried by most particles in water and the fact that adsorption of anionic polymers does not readily occur. The hydrous oxides of Al^{3+} and Fe^{3+} may have a slight positive charge, so that they can easily bind anionic polymers. Under certain conditions, the combination of aluminum sulfate and long-chain anionic polymers can give extremely tough flocs and the term 'pelleting flocculation' has been employed (Ide and Kataoka, 1980).

Cationic polymers, being of opposite charge to most wastewater colloids are effective flocculants when used alone. However, in this case, a simple charge neutralization effect may be responsible for their action rather than bridging. In that case, fairly low molecular weight materials might be effective, the important characteristic being the charge density of the cationic polymer. This question has been explored in some detail in dilute model systems (Gregory, 1976), and it was concluded that the behaviour of cationic polymers could be explained largely in terms of charge neutralization, but with some refinements such as 'electrostatic patch' effects.

However, in wastewaters, and especially wastewater sludges, there is much evidence that high molecular weight cationics are more effective than low molecular weight materials, indicating the importance of bridging effects.

When cationic polymers are added to wastewaters they interact strongly with the natural anionic material such as humic acids, forming various complexes and precipitates. This can be an effective means of removing organic material from water (Glaser and Edzwald, 1979), but there is an adverse effect on flocculation (Klute *et al.*, 1979).

43.3.4 Bioflocculation

In biological treatment processes, such as activated sludge, the microorganisms are often aggregated to a large extent, even without chemical additives. It is likely that natural polymers produced by the organisms are at least partly responsible for this 'bioflocculation' (*e.g.* Busch and Stumm, 1968). Extracellular polymers isolated from activated sludge can be effective flocculants for inorganic colloids. However, an extracellular dextran produced during bacterial growth has been shown to prevent flocculation of the bacteria (Harris and Mitchell, 1975).

This is clearly a very important topic in wastewater treatment, but also a very complex one. No further details can be given here.

43.4 REFERENCES

Amirtharajah, A. and K. M. Mills (1982). Rapid-mix design for mechanisms of alum coagulation. *J. Am. Water Works Assoc.*, **74**, 210–216.
Bell, D. J., M. Hoare and P. Dunnill (1982). The formation of protein precipitates and their centrifugal recovery. *Adv. Biochem. Eng. Biotechnol.*, **26**, 1–71.
Bratby, J. (1980). *Coagulation and Flocculation*. Uplands Press, Croydon.
Busch, P.L. and W. Stumm (1968). Chemical interactions in the aggregation of bacteria: bioflocculation in waste treatment. *Environ. Sci. Technol.*, **2**, 49–53.
de Hek, H., R. J. Stol and P. L. de Bruyn (1978). Hydrolysis–precipitation studies of aluminium(III) solutions. 3. The role of the sulphate ion. *J. Colloid Interface Sci.*, **64**, 72–89.
Feenstra, T. P. and P. L. de Bruyn (1981). The Ostwald rule of stages in precipitation from highly supersaturated solutions — a model and its application to the formation of the non-stoichiometric amorphous calcium phosphate precursor phase. *J. Colloid Interface Sci.*, **84**, 66–72.
Foxall, T., G. C. Peterson, H. M. Rendall and A. L. Smith (1979). Charge determination at the calcium salt/aqueous solution interface. *J. Chem. Soc., Faraday Trans. 1*, **75**, 1034–1039.
Franks, F. (1979). *Water — A Comprehensive Treatise*, vol. 6 (and previous volumes). Plenum, New York.
Füredi-Milhofer, H. (1981). Spontaneous precipitation from electrolyte solutions. *Pure Appl. Chem.*, **53**, 2041–2055.

Glaser, H. T. and J. K. Edzwald (1979). Coagulation and direct filtration of humic substances with polyethylenimine. *Environ. Sci. Technol.*, **13**, 299–305.

Goldshmid, T. and A. J. Rubin (1978). Aqueous chemistry and precipitation of aluminium phosphate. In *Chemistry of Wastewater Technology*, ed. A. J. Rubin, pp. 59–80. Ann Arbor Science Publishers, Ann Arbor, MI.

Gregory, J. (1976). The effect of cationic polymers on the colloidal stability of latex particles. *J. Colloid Interface Sci.*, **55**, 35–44.

Gregory, J. (1978). Effects of polymers on colloid stability. In *The Scientific Basis of Flocculation*, ed. K. J. Ives, pp. 101–130. Sijthoff and Noordhoff, Alphen aan den Rijn.

Gregory, J. (1982). Polymeric flocculants. In *Chemistry and Technology of Water Soluble Polymers*, ed. C. A. Finch. Plenum, New York.

Hall, E. S. and R. F. Packham (1965). Coagulation of organic colour with hydrolysing coagulants. *J. Am. Water Works Assoc.*, **57**, 1149–1166.

Harris, R. H. and R. Mitchell (1975). Inhibition of the flocculation of bacteria by polymers. *Water Res.*, **9**, 993–999.

Hem, J. D. and C. E. Roberson (1967). Form and stability of aluminium hydroxide complexes in dilute solutions. *US Geol. Survey Water Supply Pap.* 1827–A.

Huang, C. P. (1975). Ion-pair formation in calcium carbonate equilibria. *Environ. Lett.*, **10**, 319–334.

Hunter, R. J. (1981). *Zeta Potential in Colloid Science*. Academic, London.

Ide, T. and K. Kataoka (1980). A technical innovation in sludge blanket clarifiers. *Filtr. Sep.*, **17**, 152–156.

Klute, R., U. Neis and I. Bantz (1979). Untersuchungen zur Kinetik der Flockung mit Polyelektrolyten in Gegenwart von Huminsäure. *Vom Wasser*, **53**, 203–212.

Koutzoukos, P. G., Z. Amjad and G. H. Nancollas (1981). The influence of phytate and phosphonate on the crystal growth of fluorapatite and hydroxyapatite. *J. Colloid Interface. Sci.*, **83**, 599–605.

Mantoura, R.F.C., A. Dickson and J. P. Riley. (1978). The complexation of metals with humic substances in natural waters. *Estuarine Coastal Mar. Sci.*, **6**, 387–408.

Merrill, D. T. and R. M. Jorden (1975). Lime-induced reactions in municipal wastewaters. *J. Water. Pollut. Control Fed.*, **47**, 2783–2808.

Mullin, J. W. (1972). *Crystallisation*. Butterworth, London.

Nancollas, G. H., M. B. Tomson, G. Battaglia, H. Wawrousek and M. Zuckerman (1978). Precipitation of calcium phosphates: the influence of tricarboxylic acids, magnesium and phosphonate. In *Chemistry of Wastewater Technology*, ed. A. J. Rubin, pp. 17–30. Ann Arbor Science Publishers, Ann Arbor, MI.

Nielsen, A. E. (1981). Theory of electrolyte crystal growth. The parabolic rate law. *Pure Appl. Chem.*, **53**, 2025–2039.

Overbeek, J. T. G. (1982). Monodisperse colloidal systems, fascinating and useful. *Adv. Colloidal Interface Sci.*, **15**, 251–277.

Randtke, S. J., C. E. Thiel, M. Y. Liao and C. N. Yamaya (1982). Removing soluble organic contaminants by lime-softening. *J. Am. Water Works Assoc.*, **74**, 192–202.

Reddy, M. (1978). Kinetic inhibition of calcium carbonate formation by wastewater constituents. In *Chemistry of Wastewater Technology*, ed. A. J. Rubin, pp. 31–58. Ann Arbor Science Publishers, Ann Arbor, MI.

Robinson, R. B., T. Demirel and E. R. Baumann (1981). Identity and character of iron precipitates. *J. Environ. Eng. Div. (Am. Soc. Civ. Eng.)*, **107**, 1211–1227.

Schnitzer, M. and S. U. Khan (1972). *Humic Substances in the Environment*. Dekker, New York.

Shaw, D. J. (1980). *Introduction to Colloid and Surface Chemistry*, 3rd edn. Butterworth, London.

Singer, P. C. (1972). Anaerobic control of phosphate by ferrous iron. *J. Water Pollut. Control Fed.*, **44**, 663–669.

Smallwood, P. V. (1977). Some aspects of the surface chemistry of calcite and aragonite. Part II: Crystal growth. *Colloid Polymer. Sci.*, **255**, 994–1000.

Snoeyink, V. L. and D. Jenkins (1980). *Water Chemistry*. Wiley, New York.

Stones, T. (1981). A compendium of experimental work on chemical precipitation of sewage. *Water Pollut. Control*, **80**, 421–432.

Stumm, W. and J. J. Morgan (1970). *Aquatic Chemistry*. Wiley-Interscience, New York.

Tadros, T. F. (1982). Polymer adsorption and dispersion stability. In *The Effect of Polymers on Dispersion Properties*, ed. T. F. Tadros, pp. 1–38. Academic, London.

Turner, D. R., M. Whitfield and A. G. Dickson (1981). The equilibrium speciation of dissolved components in freshwater and seawater at 25 °C and 1 atm pressure. *Geochim. Cosmochim. Acta*, **45**, 855–881.

Verwey, E. J. W. and J. T. G. Overbeek (1948). *Theory of the Stability of Lyophobic Colloids*. Elsevier, Amsterdam.

Walton, A. G. (1967). *The Formation and Properties of Precipitates*. Interscience, New York.

44

Ecology of Polluted Waters

H. A. HAWKES*
University of Aston in Birmingham, UK

44.1 INTRODUCTION

Pollution, as related to water, has over the years been variously defined depending upon the different interests involved. Earlier definitions, especially by biologists, referred to 'reduction in diversity of aquatic life and eventually destroying the balance of life in the stream' (Patrick, 1953) or 'adverse change in plant and animal communities' (Hawkes, 1957). In cases of legal action, however, ecological change was not, in itself, always acceptable as evidence of pollution. The more practical legal definition — 'the addition of something to the water which changes its natural qualities so that the riparian owner does not get the natural water of the stream transmitted to him' (Wisdom, 1956) — begs the questions (a) what is the natural water of a stream and (b) what are its natural qualities? (Hynes, 1960). Doudoroff and Warren (1957), in attacking the premise of biological assessment of pollution, are critical of ecological change being considered as pollution. Accepting these criticisms of the earlier rather academic definition, most definitions of pollution now refer to the impairment of the uses of the water, *e.g.* 'the addition of substances which directly or indirectly change the nature of the stream to such an extent that its suitability for Man's legitimate uses is impaired' (Hawkes, 1962).

*Present address: Department of Civil Engineering, University of Birmingham, UK

Water-bodies such as lakes and rivers are both a natural resource to satisfy Man's domestic, industrial and agricultural needs and part of his natural environment, providing for his recreational and aesthetic needs. A more recent general definition of pollution by Holdgate (1979) covers both the resource and environmental interests: 'The introduction by Man into the environment of substances or energy liable to cause hazards to human health, harm to living resources and ecological systems, damage to structure or amenity or interference with the legitimate uses of the environment'. In the context of water being both a resource and an environment for Man, water pollution may for our present purpose be defined as:

> The introduction of organisms, substances or energy into water, resulting from Man's activities, which impair the legitimate uses of the water as a natural resource or as a natural environment.

In this chapter on the ecology of polluted waters we shall obviously be more concerned with water as an environment. Nevertheless, many discharges which affect the quality of the water as a resource also induce associated and proportional ecological changes in the aquatic communities. Thus direct biological surveillance of aquatic communities for environmental protection provides a useful indirect monitor of water quality for resource protection.

In providing effective waste-treatment processes to control pollution it is essential that the potential polluting effects of the specific waste are identified in order that the appropriate treatment can be applied. It is the objective of this contribution to identify the several ecological effects of the different pollutants of water, so that methods of treatment and methods of monitoring their effectiveness can be designed.

Water may be polluted either by the introduction of substances or agents quite foreign to the natural environment, *e.g.* synthetic pesticides, or by the occurrence of abnormal levels of natural factors, *e.g.* high phosphate concentrations, elevated temperatures or low oxygen levels. To appreciate the effects of different pollutants it is therefore first necessary to consider the relevant factors affecting communities in their natural aquatic environments. The ecological changes induced by different types of pollution will then be reviewed.

44.2   FACTORS AFFECTING NATURAL AQUATIC ECOSYSTEMS

These may be considered as autecological (physical and chemical factors influencing the distribution and incidence of individual species) and synecological (biotic factors affecting populations, productivity and community structure).

44.2.1   Autecological Factors — Water as a Medium for Life

Several physical properties of water are significant in characterizing the aquatic environment and the forms of life which inhabit it. Compared with the aerial and sub-aerial environments, water provides a much denser medium. The resultant buoyancy makes possible modes of life not possible in other media, *e.g.* plankton, which drift suspended in the water column, and filter-feeders, which strain their food from the water in which it is suspended. Of considerable ecological significance is the elementary fact that the density of water is greatest at 4 °C, above and below which temperature increases or decreases result in a decrease in density, this effect being greatest at high temperatures. This phenomenon results in the stratification of lakes, ice formation on the surface of waters and the induction of convection currents. The surface tension of water makes possible a specialized community of organisms (the neuston) living at the air–water interface. The transparency of water enables light to penetrate to support photosynthesis, enabling plants (producers) to exist to considerable depths. Light, however, is absorbed exponentially, determining the euphotic zone — the depth to which plants and algae can live. The thermal properties of water result in it being more resistant to temperature change than air; the fluctuations in temperature of a body of water are therefore less than those of the atmosphere, thereby providing a more equitable environment. Natural waters differ widely in temperature, and different species are associated with the different temperature regimes over the range 0–88 °C (thermal springs).

Water movements are also ecologically important. Lentic waters (bodies of standing water such as ponds and lakes) and lotic waters (flowing waters such as streams and rivers) are distinguished and support different communities. Different environmental factors are of importance in the two types of habitat. In lentic waters the depth profile, in relation to light penetration and thermal stratification, is a major ecological parameter. In lotic waters the current velocity and related nature of the substratum are important factors determining the nature of the benthic communities. A rapid current over an eroding stony substratum supports quite a different community than does a sluggish current over a depositing muddy substratum, even in the same stretch of river. On the basis of these differences, different river zones may be recognized, the major division of which is into the upper stretches (rhithron) and lowland stretches (potomon).

Chemically, natural water in the aquatic environment is not H_2O but a dilute solution of biogenic materials including nutrient salts and gases. In Nature its composition differs appreciably in different waters. Thus the definition of water purity and water pollution in quantitative chemical terms is difficult. All substances required to support life are soluble in water, both macro-nutrients and micro-nutrients. Of the macro-nutrients, nitrogen and phosphorus are important limiting factors in the aquatic environment. Salinity, the concentration of all ionic constituents (carbonates, sulfates and halides of sodium, potassium, calcium and magnesium) is significant in determining the osmotic pressure of the water. Marine organisms have protoplasm which is isotonic with their external medium. The lower salinity of freshwaters creates problems for freshwater organisms in relation to osmoregulation. The two gases of ecological significance (carbon dioxide and oxygen) behave differently in solution. Carbon dioxide is very soluble in water, entering into chemical reaction to establish an equilibrium which is greatly influenced by pH. Oxygen, in contrast, is poorly soluble in water and its solubility is greatly influenced by temperature. Dissolved oxygen is therefore a major limiting factor in the aquatic environment. For example, compared with 210 ml of oxygen available to aerial organisms in each litre of air there is, even at saturation, in one litre of water approximately 10.2 ml at 0 °C, 7.9 ml at 10 °C and 6.3 ml at 20 °C.

44.2.2 Synecology — Factors Determining Community Structure

All natural aquatic ecosystems involve the cycling of materials through producer, consumer and decomposer populations, thereby trapping light energy entering the hydrosphere which successively supports species populations at different trophic levels (Figure 1). In lentic waters much of the material cycles within the system (a closed ecosystem) although there are both aerial and terrestrial inputs of allochthonous material and loss *via* the outflowing streams. In lotic systems, especially in the rhithron zone, because of the constant loss of material in the current, much of the material is of allochthonous origin, being derived from the terrestrial ecosystems. Lotic ecosystems may therefore be regarded as open ecosystems (Figure 2). Closed lentic ecosystems are therefore relatively independent of allochthonous inputs whereas open lotic systems depend upon it to maintain their existence. As a result, lotic systems tend to be more sensitive to changes in input, such as effluents, and respond sharply. Lentic systems are less sensitive and generally respond more slowly, although the eventual change may be as great and more long-lasting.

44.3 ECOLOGICAL CLASSIFICATION OF WATER POLLUTION

The different types of water pollution have been variously classified (Klein, 1962; Holdgate, 1979). The classification adopted here is based on the ecological factor most affected, although it is appreciated that some discharges affect more than one factor simultaneously. Table 1 presents such a classification and includes examples of pollutants and their sources. Three major groups of factors are recognized: (1) physical, (2) chemical, (3) biotic.

44.4 PHYSICAL POLLUTION

This includes two distinct effects: (1) changes in physical factors of the water itself and (2) changes in the physical nature of the substratum.

44.4.1 Suspended solids, turbidity and colour

These factors all tend to affect the aquatic environment in a similar manner by reducing light penetration and thereby the producer community generally, with consequent effects on foodchains. The sparsity of life in certain streams in Cornwall has been attributed to the turbid conditions caused by china-clay wastes which suppress algal growth (Pentelow, 1949). The suppres-

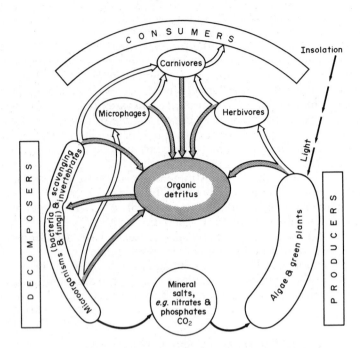

Figure 1 A diagrammatic representation of an ecosystem

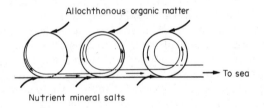

Figure 2 A diagrammatic representation of an 'open' ecosystem

sion of the benthic communities resulted in the reduced availability of fish food. Trout fry were absent from the affected reaches but two-year-old adults were present at about one-seventh the population found in the non-polluted stretches. It was concluded that the fish were not breeding in the polluted waters but migrated into them from non-polluted zones. Many natural rivers carry heavy loads of silt at times of spate. Although fish may survive several hours' or even days' exposure to several thousand mg l^{-1} solids, such conditions should be prevented where good fisheries are to be maintained (Lloyd and Alabaster, 1980).

More important ecologically than the direct effects of suspended solids in the water is the effect the solids have on the benthos when they settle to smother the natural substratum. The deposition on a natural stony substratum of inert solids such as coal dust, gravel washings, metal mining wastes or flocculated metal compounds results in a general reduction in benthos in terms of both numbers of species and populations, thus affecting fish food. Organic solids, however, such as those present in sewage effluents and the effluents from textile, hardboard and paper manufacture, form organic sludges on the river bed. In these a silt community becomes established, dominated by such species as *Chironomus riparius* (bloodworms) and *Asellus aquaticus* (water hoglouse), which often develop high population densities to provide an abundant food supply for those fish which can tolerate the conditions. The most sensitive freshwater habitat in respect of suspended solids is probably the spawning grounds of trout and salmon, where the slightest turbidity in the water or deposition of solids may cause the fish to avoid the area or result in the failure of eggs to develop successfully after they have been laid (Lloyd and Alabaster, 1980).

Colour in river water may result from the discharge of coloured effluents, *e.g.* dye manufacture or textile effluents, or it may appear as the result of the interaction of different effluents in the river. Vegetable tanning effluents mix with ferruginous waters to produce a deep-green to inky-blue

Table 1 A Classification of Types of Water Pollution based on Ecological Factors Involved

Type	Factor	Examples of sources
Physical	Suspended solids	Mining, quarrying, china-clay, pulping
	Turbidity	Most effluents (domestic and industrial)
	Colour	Paint and dye manufacture, textiles, tanneries
	Surface properties	Detergents, oils, synthetic rubber manufacture
	Temperature	Cooling waters, power generation, steel mills
	Radioactivity	Processing nuclear fuels, electricity generation
Chemical	Salinity	Coal and salt mining, oil wells
	pH	Mine waters, most manufacturing industry effluents
	Toxicity	Mining and processing of metals, tanneries, wood pulping, manufacture of dyes, explosives, man-made fibres, paints, pesticides, pharmaceuticals, photographic film, plastics, synthetic rubber
	Deoxygenation	Sulfite wood-pulping liquors, ferruginous mine waters
Biotic	Organic enrichment	Sewage effluents, farm effluents, processing of biological materials: foods, drinks, antibiotics, paper, rubber, textiles, leather, glue, tar
	Inorganic enrichment	Oxidized sewage effluents, agricultural land drainage and run-off, fertilizer manufacture

colour. Although evidently polluted and of reduced quality as a resource, the ecological effects of colour itself are usually minimal compared with other pollutants. Many rivers, such as those draining peat, are naturally highly coloured and yet support a typical biota including trout.

44.4.2 Surface Active Agents

All surface active agents, such as natural soaps or synthetic detergents, affect the physical properties of water in a similar manner by reducing the surface tension and thereby affecting those processes associated with interfaces. However, whereas natural soaps are rapidly degraded and do not therefore create a serious pollutional problem, the earlier synthetic detergents were resistant to biological degradation and their increased use resulted in serious problems on sewage treatment works and in the rivers. Although detergents had been used earlier in industry, especially in the textile industry, creating problems locally, it was their increased use domestically in the period following World War 2 that caused widespread problems. The evident presence of foam, which on some rivers accumulated to cause nuisance, although creating public awareness of the problem, was not the most serious aspect ecologically. The presence of foam would affect the emergence and egg-laying activities of some aquatic insects. The reduced surface tension would also affect the neuston (those organisms associated with the air–water interface).

Of more general ecological significance, however, is the suppression of the rate of stream reaeration caused by the presence of detergents in sewage effluents, thereby retarding self-purification from organic pollution. This effect is greatest in slow-flowing rivers, being less under quiescent conditions and in torrential waters. Detergents are also toxic to fish and other aquatic life at concentrations greater than 5 mg l^{-1} (Manoxol equivalent) (Degens *et al.*, 1950). The toxicity of detergents in sewage effluents was found to be appreciably less, at equivalent Manoxol values, than in pure water (Herbert *et al.*, 1957). The OECD report (Prat and Giraud, 1964) suggested that the hazard to fisheries would be very limited if the concentration of ionic detergents did not exceed 3 mg l^{-1}. Because of the similarity in toxicities of detergents so chemically different as lauryl sulfate and dodecylbenzenesulfonate, Leclerc and Derlaminck (1952) concluded that the toxic process was purely a physical one of reduced surface tension. Kruger (1964) quoted a critical value of 50×10^{-5} N cm^{-1} below which the surface tension was damaging. Bock (1966) showed that at a surface tension of less than 50×10^{-5} N cm^{-1} and at a detergent concentration of 5 mg l^{-1} the epithelial cells of the gills of carp were destroyed and the fish died after several hours exposure. Other workers (Marchetti, 1966) considered that the toxicity was not solely, or even principally, dependent on the reduction in surface tension, but depended on the chemical structure and ionic character of individual detergents.

Detergents are classified on the basis of their electrolytic dissociation as non-ionic, anionic and cationic. The relative proportion of these used in Britain in 1961 was: anionics 92.3%, non-ionics 7.2% and cationics 0.5%. Madai and Lan (1964) investigated the effects of the several different types of detergents on aquatic organisms and concluded that although the cationics were the most harmful, the anionics were more harmful than the non-ionics. Most of the research on detergent pollution has therefore been carried out on the anionic detergents, especially the alkylbenzene-sulfonates (ABS).

In the context of this work it is of interest to note that in the UK the Jephcott Committee, set up to examine the detergent problem, considered alternative suggested approaches. The suggestion that a bacterial flora capable of decomposing the detergents be developed was deemed an unlikely solution and instead the preparation of products more susceptible to biological degradation was pursued. As a result, one such product to be developed was the linear alkylbenzene-sulfonates (LAS) which gradually, between 1958 and 1965, replaced the hard tetrapropylenebenzenesulfonates. Although the advent of the soft detergents resulted in the concentration of detergents in sewage effluents being markedly reduced, it was found that the toxicity of the new soft detergents was greater than that of the hard detergents (Bock, 1964; Liebmann, 1966; Pickering, 1966). With the increasing proportion of soft anionics in use, the non-ionics became more significant in that they stabilized the foam produced by the low concentration of anionics. Besides the continued use of non-ionics in specific industries such as textiles, which resulted in relatively large amounts being discharged locally (Ministry of Housing and Local Government, 1967), there was an increasing use of them in domestic formulations, such as washing-up liquids. As a result, soft non-ionics were also developed and by 1974 most of the non-ionics incorporated into domestic detergents were also soft (Department of the Environment, 1974). As a result of the introduction of these soft detergents, an efficient sewage treatment works can now remove 95% of the detergent present in the sewage compared with some 60–65% removal in the days of the hard detergents. The final report of the Standing Technical Committee on Synthetic Detergents was able to claim: 'The concentration of synthetic surfactants in most, if not all, of United Kingdom waters has fallen to very low levels and gives rise to no problems' (Department of the Environment, 1980). This was accomplished, without legislation, by the voluntary cooperation of industry, regulatory authorities and government departments. The rise and fall of the detergent problem exemplifies the impact of technology on the environment and the use of technology to solve the problems it creates. The environmental problems caused by the increased use of phosphates in detergent-based washing products are dealt with in Section 44.6.2 on eutrophication.

44.4.3 Temperature

The ecological effects of increasing the temperature of natural waters are very complex. The simplest effect is the change in the distribution of species as determined by their temperature requirements. In practice the cold-water stenotherms are to be found in the smaller upland streams which are rarely used as cooling water. The temperature in such streams, however, may be affected by Man's other activities such as deforestation. The elevating of stream temperatures by only a few degrees has seriously affected trout populations in deforested areas in North America (Tarzwell, 1957). In larger lowland rivers, which are used for cooling and therefore receive thermal discharges, the biota is naturally eurythermal being conditioned to a wider range of temperatures and is therefore less affected by temperature increases within limits. Providing the river water temperature does not exceed 30 °C there is not likely to be any drastic direct effect. Temperature has, however, more subtle indirect effects on aquatic ecosystems.

Theoretically elevated temperatures should interfere with the seasonal life cycles of aquatic insects. Recent studies at one site have shown, however, that any such effect was masked by other variables (Langford, 1975). The direct effects of temperature on the growth rate of organisms is determined by the proportion of the increased metabolism that is channelled into synthesis compared with that used in cellular respiration. With limited food or restricted feeding, elevated temperatures could result in reduced growth; with unrestricted feeding, increased growth would probably occur. Heated industrial cooling waters have been used in the culture of marine and freshwater fish and significantly increased growth rates have been achieved. Patalas (1967) found that the primary and secondary productivities of Lichen Lake in Poland, which was used for cooling, were higher than in that of a cold water lake in the same system. The most important indirect effect is probably that on the oxygen depletion resulting in organic pollution, discussed later (Section 44.6.2.2). The effect on toxicity is more complex. Although the rate at which fish die in toxic solutions is increased at higher temperatures, the critical threshold concentration may not be reduced and in the case of some substances, *e.g.* phenol, it may be increased due to the enhanced detoxification processes at higher temperatures (Brown *et al.*, 1967).

44.4.4 Radioactivity

Although in normal circumstances only low-level activity enters inland waters, it is possible for aquatic ecosystems to concentrate this *via* food chains by factors greater than those achieved by dilution in disposal. ^{137}Cs, ^{85}Sr and ^{32}P are probably the most important in this respect. Many algae have been found to be capable of concentrating radioactive materials from low levels in the water. There is a danger, then, of radioactive materials, concentrated by algal blooms from the raw water, being released into supply water at the water works. The deposition of radioactive algae in estuaries as a radioactive sludge is also possible. These possibilities should be considered when determining the permissible levels of activity for discharges to the aquatic environment.

44.5 CHEMICAL POLLUTION

In this section we shall deal with those changes in the chemistry of the water which affect aquatic communities, other than nutritional ones which are the subject of the next section.

44.5.1 Salinity

Freshwater organisms, which have become adapted to living in waters of lower osmotic pressure than that of their body fluids, may be adversely affected by an increase in osmotic pressure resulting from saline discharges. A more important ecological effect of such discharges is probably the change in the proportional ionic composition. In natural waters, salts of Na, Ca, K and Mg are present in such proportions that their individual toxicities are mutually counteracted or antagonized. Discharges which increase the concentration of one ion may result in a toxic solution. Continuous discharges such as those from salt and coal mines and oil wells, which create fairly stable saline conditions, encourage the development of a replacement community of brackish-water species. The brine shrimp, *Artemia salina*, the salt fly, *Ephydra riviparia*, certain rotifers and diatoms have been found to be characteristic of such communities. The WHO limit for drinking water is 200 mg l^{-1} Cl; the use of the former permissible level, under certain conditions, of 600 mg l^{-1} is now discouraged because of the health significance of Na (World Health Organization, 1979). Salinity is not considered a major type of pollution of rivers in Britain in connection with freshwater fisheries.

44.5.2 pH

Within the range 5.0–9.0, pH probably has no significant direct effect on most species, although some taxa are more restricted, *e.g.* snails are usually found in waters with pH above 7.0. Under fairly stable conditions of low pH (5.0) a restricted replacement community becomes established (Harrison and Heukelekian, 1958). More important ecologically than the direct effects of pH is the indirect effect in influencing the toxicity of many substances, especially those in which the toxicity depends upon the degree of dissociation. Ammonia, for example, is 10 times more toxic at pH 8.0 than at 7.0, whereas a nickelocyanide complex was found to be 500 times more toxic to fish at pH 7.0 than at 8.0 (Lloyd and Alabaster, 1980).

An increasing problem of international importance is that of acid rain. Sulfur compounds are released into the atmosphere from volcanoes, the decomposition of sulfur-containing proteins, the burning of fossil fuels and some chemical industrial activities. SO_2 and SO_3 dissolve in rain droplets to form acids which fall as rain, acidifying freshwaters and soils especially in areas of heavy precipitation, often far from the source of emission. This problem is receiving serious attention in Scandinavia, which receives airborne pollution from the industrial nations of Europe to the south.

44.5.3 Toxicity

The overall ecological consequences of toxic pollution on aquatic communities are the reduction in both the number of species and in the total number of individuals, resulting in a reduction in variety and in abundance. Species are eliminated according to their specific tolerance to the toxicant. Different species exhibit different degrees of tolerance to different toxicants. Several species of leech show an unusually high tolerance to DDT. It was found that, in some species at least, this was due to their metabolic ability to detoxify the insecticide by dehydrochlorinating the DDT to the non-toxic DDE (Sawyer, 1947). As the poison becomes diluted or otherwise reduced in concentration downstream of the discharge, there is a successive reappearance of species according to their degree of tolerance to the poison. In some cases the selective elimination of the less tolerant species and the resultant reduction in inter-specific competition or in predation may result in an increase in the population of the more tolerant ones. Toxicity at a level which does not eliminate all life may affect the different trophic levels differentially, resulting in imbalanced community structure.

Toxic effects may be:

(1) Lethal, causing death by direct poisoning.

(2) Sub-lethal, below the level which directly causes death but which may affect growth, reproduction or activity so that the population may ultimately be affected.

(3) Acute, causing an effect (usually death) within a short period.

(4) Chronic, causing an effect (lethal or sub-lethal) over a prolonged period of time.

(5) Accumulative, effect increasing by successive doses.

The acute toxicity of a substance is usually expressed as the LC_{50}, which is that concentration of the substance which kills half the test animals in a specified period of time, *e.g.* 24, 48 or 96 h (Figure 3). Such tests, which determine the lowest concentration of a substance which kills fish in a relatively short period, 1–4 d, are useful in predicting the effects of acute toxicity as produced, for example, by a plug of toxic discharge passing down a river. For many poisons commonly associated with industrial sewage effluents it has been found that the joint toxicity of two or more of them is simply additive (Lloyd, 1961). By expressing the concentration of each poison present as a proportion of its LC_{50} value the combined toxicity can be predicted by addition; if greater than unity the water is likely to be lethal (Brown, 1968).

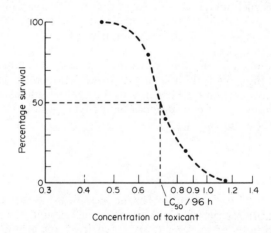

Figure 3 Typical mortality–concentration curve for derivation of LC_{50} values

Although these methods make possible the prediction of acute toxicity where the identity and concentration of the poison are known, the prediction of chronic and sub-lethal effects is far more difficult. Although much is now known about the rates at which fish die in toxic solutions and factors affecting this, little is known about the ecologically significant concentration of toxicants which can be permitted which do not adversely affect a natural fish population continuously exposed to it. Some workers have derived permissible concentrations by applying a factor to the LC_{50} values. A factor of 0.1 has been used for metals (Jones, 1964) and 0.01 for pesticides (Holden, 1964). Such factors, although probably based on sound experience, are no more than intelligent guesses. A more scientifically based value may be derived by plotting the mean periods of survival at a range of concentrations (Figure 4). For many toxicants a curvilinear relationship is exhibited. In such cases the concentration at which the toxicity curve becomes approximately asymptotic to the time axis is the incipient LC_{50} at which the test species could be expected to survive for long periods. Sub-lethal effects may still occur, however, and some poisons exert delayed lethal effects, *e.g.* pesticides and cadmium.

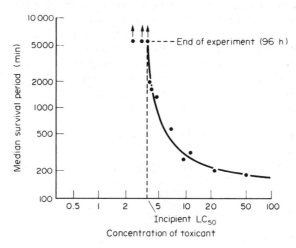

Figure 4   Typical toxicity curve showing survival period *vs.* concentration for derivation of incipient LC_{50} values

Toxicants in the aquatic environment, besides affecting aquatic communities, pollute the water as a resource for public supply. The response of fish to pollutants, such as movement or respiratory activity, is made use of in fish alarm systems to protect water supplies (Morgan, 1978). Public health hazards may arise when toxicants are concentrated *via* aquatic food chains leading to Man. In Japan, thousands of people suffered mercury poisoning after eating fish caught in Minamata Bay into which discharged an effluent containing mercury from a vinyl chloride plant. The disorder, now known as Minamata Disease, involves the progressive loss of coordination, vision, hearing and general intellect. Although the biological surveillance of aquatic communities provides an indication of changes in water quality, care is needed in interpreting ecological changes in terms of water quality for Man's various uses. For example, the presence in very low concentrations of the herbicide TBA (trichlorobenzoic acid) in a river had only slight effects on the benthic macro-invertebrate fauna which was being used to monitor the river-water quality. Nevertheless, water abstracted from the river probably caused the failure of the tomato crop watered with it (Hawkes, 1978). A marked herbicidal effect had in fact been noted downstream of the effluent (the elimination of the water weeds) but the significance of this was not realized.

44.5.4   Deoxygenation

Apart from deoxygenation resulting from organic pollution, outlined in the next section, depletion of oxygen may result from the discharge of industrial effluents containing reducing agents such as sulfites in pulp-mill effluents and ferrous salts in mine waters. Because such effluents are usually associated with other pollutants, the effects due to deoxygenation itself are difficult to establish. However, dissolved oxygen is a major factor determining the distribution of invertebrates and fish in freshwaters. Most species are limited by low oxygen concentration but some invertebrates, such as *Chironomus riparius*, are adversely affected by high oxygen levels and have been found to survive longer at lower oxygen concentrations (Fox and Taylor, 1955). The incipient lethal oxygen level (the lowest oxygen concentration at which a species can survive for a prolonged period of time) is increased at higher temperatures and at higher carbon dioxide concentrations. To permit scope for activity, however, to allow a species to feed, escape predation and reproduce to maintain a population, higher oxygen concentrations are needed. Furthermore, some pelegic organisms, such as fish, are able to select the more suitable sites and leave or avoid less favourable ones. Ultimately therefore the critical oxygen level may be that which allows a species population to remain in a given water.

44.6 BIOTIC EFFECTS

This section deals with the effects of discharges which increase nutrient levels in the water so that the balance of species populations is substantially changed. Organic pollution which results from the discharge of biodegradable organic matter such as sewage and the effluents from the processing of biological materials, *e.g.* foods, textiles and antibiotics, results in an increase in the heterotrophic component of the ecosystem. Other discharges which increase the concentration of plant nutrients, such as nitrogen and phosphorus, in the water enhance the producer component of the ecosystem thus creating an imbalance known as eutrophication (Figure 5). When eutrophication adversely affects the water quality or causes nuisance, it constitutes pollution. In other circumstances, eutrophication may increase fish production in which case it may be beneficial. Eutrophication, in this sense, is not therefore synonymous with pollution.

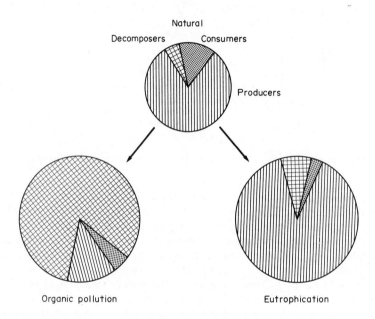

Figure 5   Organic pollution and eutrophication as ecological imbalances

44.6.1 Organic Pollution

The observable effects on the biota associated with organic pollution are the result of a number of interacting factors, the most important of which are: (1) change in nutrient status; (2) change in concentration of the dissolved gases oxygen and carbon dioxide. Sewage effluents, which are the most common cause of organic pollution, contribute additional factors: (1) increase in suspended solids which accumulate on the river bed; (2) toxicity due to ammonia, detergents and possibly toxic metals. Both these factors have been dealt with earlier.

44.6.1.1 Change in nutrient status

The increased organic concentration results in an increase in the heterotrophic microbial populations, both planktonic and benthic. Profuse growths of benthic microorganisms assume macroscopic forms which occur either as white slimes on the surfaces of the substratum or as fluffy fungoid growths extending as long streamers in flowing waters. Such infestations are known as sewage fungus and are considered aesthetically objectionable by some. They are a nuisance to anglers, cause siltation of the eroding substratum in riffles, non-development of trout eggs and, when decomposing, give rise to secondary pollution. Although commonly referred to as sewage fungus, other microorganisms such as bacteria and ciliate protozoa are more commonly involved. From the literature on the subject it would appear that the following are the more commonly recorded taxa in the sewage fungus complex:

Sphaerotilus natans (sheathed filamentous bacteria)
Leptothrix ochracea
Zoogloea spp.
Beggiatoa alba (the sulfur bacterium)
Leptomitus lacteus (a Phycomycetes fungus)
Fusarium aquaeductum (Fungi imperfecti)
Geotrichum candidum
Carchesium polypinum (ciliate protozoon)

The above list does not comprise a typical sewage fungus community but rather a list of organisms, one or more of which provides the basis, both physically and nutritionally, for a heterotrophic biocoenose of other microorganisms, including free-swimming protozoons and diatoms, and invertebrates, such as *Chironomus riparius* (blood worms) and *Asellus aquaticus* (water hoglouse). Much work has been carried out to determine the conditions which encourage these organisms; unfortunately the results are, in some respects, conflicting. This work has been reviewed by Harrison and Heukelekian (1958), who applied the term 'slime infestation', Hawkes (1962) and Curtis (1969). Most work has been reported on the most commonly occurring species, *S. natans*. This requires sources of carbon, nitrogen and phophorus. Many specific compounds, mostly short-chained organic compounds such as sugars and their derivatives, can act as carbon sources. Although an organic source of nitrogen is required for luxuriant growths to develop (a whole range of amino acids is utilized), it has been found that inorganic nitrogen may be used when specific carbon sources such as sucrose, glycerol or succinate are available (Stokes, 1954). Later work (Okrend and Dondero, 1964) demonstrated the need for trace amounts of vitamin B_{12} (cyanocobalamin) when using inorganic nitrogen sources. Phosphorus is also required but apart from vitamin B_{12} no other accessory growth factors have been found necessary. Growth occurs both in soft and hard waters between pH values of 6.8 and 8.0 but is inhibited in saline conditions.

Although increased growths of *Sphaerotilus* occur with increasing nutrient concentrations over the range experienced in rivers, evident growths do not occur in undiluted sewage. This may be due to toxicity but competition with other bacteria may also be important. *Sphaerotilus* has been found to be more efficient in utilizing available nitrogen than other heterotrophic bacteria and thus at low concentrations it successfully competes with other bacteria to utilize the carbon sources available. It would appear that *Sphaerotilus* is physiologically and ecologically well adapted to utilize efficiently organic nutrients from dilute solutions which its attached mode of life enables it to accumulate.

Profuse nuisance-causing growths of *Sphaerotilus* may therefore be produced by the presence of specific nutrients at low concentrations. The growth promoting potential of effluents cannot therefore be assessed in terms of BOD, PV or even organic carbon. Infestations have been reported from rivers polluted with spent sulfite liquors when the BOD was only 0.2 mg l^{-1}. Besides nutrients, dissolved oxygen, temperature and current velocity are all important factors determining the abundance of *Sphaerotilus* in the receiving water. Although it may survive in low oxygen concentrations (Stokes, 1954) it only thrives in moderately to well aerated waters (3–11 mg O_2 l^{-1}) (Curtis, 1969). *Sphaerotilus* growths in rivers are more commonly associated with winter conditions and low temperatures, but this may not be a direct effect of temperature; laboratory data indicate an optimum temperature of 25–30 °C. It may be that the higher metabolic rates at higher temperature result in reduced nutrient levels; Cawley (1958) reported that in the Altamaha river (North America) growths which in the winter at 10 °C extended for 25 km below the polluting effluent were restricted to a distance of only 185 m in the summer at a temperature of 30 °C. It is possible that *Sphaerotilus* is able to grow better than other microorganisms at low temperatures and therefore competes more successfully during the winter. Typical growths only occur in flowing waters, a current velocity in excess of 15 cm s^{-1} being required. The importance of current, clearly evident in field surveys, is well demonstrated by the results from laboratory channel studies reported by Amberg and Cormack (1960). They found that by increasing the velocity from 6 to 18 cm s^{-1} there was a 73% increase in the amount of growth.

Besides *Sphaerotilus*, other microorganisms may form slime infestations on river beds and banks. Of these a true fungus, *Leptomitus lacteus* (*Apodya lactea*), is reported to be more common in the larger polluted continental rivers. Growth of this organism has been associated with the presence of complex organics in dilute solutions. It requires high oxygen concentrations and is favoured by acid conditions (pH 2.5–7.5). Another fungus associated with acid waters is *Fusarium aquaeductum*, which is especially evident as a pinkish growth on weirs. A 17 km stretch of the Otteraa river (Norway) below effluents from a paper and pulp factory and a fibreboard factory was reported to be infested with *Fusarium*, the river bed being covered by a layer of the fungus 1–2 cm thick (Paulson, 1962). Below treated sewage effluents which contain high numbers of bacteria the ciliate protozoon *Carchesium* may dominate the sewage fungus complex. Its stalked branched colonies give it a fungoid appearance although nutritionally it is holozoic, feeding on bacteria.

From these examples of different members of the sewage–fungus complex it will be apparent that just as their nutrient requirements differ so will the method of effluent treatment to control them.

44.6.1.2 *Changes in concentration of dissolved gases*

The increased population and activity of the decomposer community in response to the elevated nutrient level increases the demand for oxygen. This may cause a depletion in the dissolved oxygen in the water to an extent depending upon the concentration of organic nutrients present, as measured by the BOD. As a result of the same decomposing activities the carbon dioxide concentration tends to increase, thus affecting the pH. Most aquatic organisms, including fish, which depend upon the dissolved oxygen for their respiratory requirements, are thus affected to an extent depending upon their tolerance to low oxygen and high carbon dioxide concentrations.

As a result of the changes in nutrient status and in the concentration of dissolved gases the natural benthic community in the river bed is replaced by a decomposer community. This replacement community brings about the progressive stabilization of the organic matter, a process known as self-purification. As the organic matter is degraded the oxygen demand is reduced, thus permitting the dissolved oxygen concentration to recover by re-aeration. This depletion and subsequent recovery of the dissolved oxygen produces the classical oxygen-sag effect (Figure 6). The depth and extent of the sag depends upon the organic load, temperature and current; it is deeper but less extensive at higher temperature.

44.6.1.3 *Self-purification*

Associated with the different ecological conditions at different stages of recovery, different benthic communities become established (Figure 6). These communities form the basis of several biological indicator systems used to assess degrees of pollution. The ecological changes associated with organic pollution may then be regarded as a natural response to a changed nutrient condition and as such is a good example of the homeostatic properties of ecosystems. A river which is capable of reacting in this way may be regarded as ecologically healthy, although some of the practical consequences may be undesirable to Man. Excessive organic loadings may result in total deoxygenation, creating anaerobic conditions suppressing the natural aerobic processes of purification. Although the processes of purification are continued by anaerobic bacteria, the end products such as H_2S create objectionable conditions. Toxic conditions suppress the natural processes of purification.

These natural processes of purification, which occur over several km of river, are those which have been optimized in sewage treatment processes.

44.6.2 Eutrophication

Whereas organic pollution is usually related to rivers and streams, eutrophication has in the past usually been associated with lakes and ponds; it may, however, affect rivers and estuaries. Self-purification of an organically polluted river results in increases in plant nutrients which may give rise to eutrophication. Conversely the ultimate decomposition of the products of excessive primary production in eutrophic waters may create conditions of organic pollution. Thus organic pollution and eutrophication are different phases symptomatic of ecological imbalance.

The eutrophication of lakes is a natural process occurring over prolonged periods of time (thousands of years) caused by the progressive increase in fertility by the trapping of nutrients. This natural process is accelerated by Man's activities; Lake Erie in North America has recently been described as a dying lake, its end being advanced by the discharge into it of sewage and industrial wastes (Stumm and Morgan, 1962). Studies on Lough Neagh in Northern Ireland have shown that there has been a marked increase in the rate of eutrophication over the past 50 years,

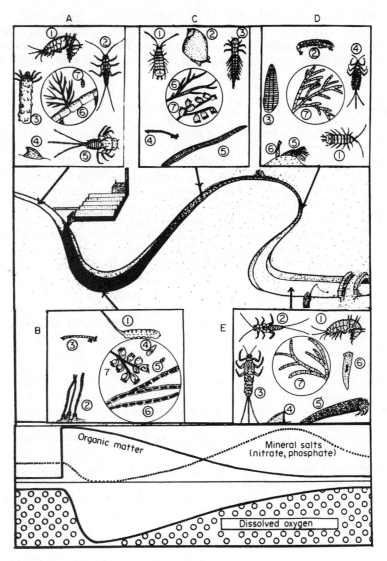

Figure 6 Representative organisms of different stream communities associated with successive stages of self-purification. Key to organisms: (A) (1) *Gammarus pulex* (fresh-water shrimp); (2) *Nemoura* (stone-fly nymph); (3) *Limnophilid caddis*; (4) *Ancylus fluviatilis* (limpet); (5) *Ecdyonurus* (may-fly larva); (6) *Draparnaldia* (green alga); (7) *Cocconeis* (diatom). (B) (1) *Eristalis tenax* (rat-tailed maggot); (2) *Tubifex* (sludge worm); (3) *Chironomus riparius* (blood worm); (4) *Paramecium caudatum*; (5) *Colpidium*; (6) *Sphaerotilus natans* (sewage fungus); (7) *Carchesium* (sewage fungus). (C) (1) *Asellus aquaticus* (water hog-louse); (2) *Lymnaea pereger* (wandering snail); (3) *Sialis lutaria* (alder-fly larva); (4) *Chironomus riparius* (blood worm); (5) *Erpobdella* (leech); (6) *Stigeoclonium* (green alga); (7) *Carchesium* (sewage fungus). (D) (1) *Asellus aquaticus* (water hog-louse); (2) *Hydropsyche* (case-less caddis larva); (3) *Glossiphonia* (leech); (4) *Baetis rhodani* (may-fly nymph); (5 & 6) *Simulium ornatum* (pupa & larva of buffalo gnat); (7) *Cladophora* (blanket weed). (E) (1) *Gammarus pulex* (fresh-water shrimp); (2) *Nemoura* (stone-fly nymph); (3) *Ephemerella* (may-fly nymph); (4) *Ancylus fluviatilis* (limpet); (5) *Stenophylax* (caddis larva); (6) *Dugesia* (flat-worm); (7) *Cladophora* (blanket weed).

recorded by the reduction in diversity of diatom species, changes in dominance and a general increase in diatom production as revealed by a study of the sediment core (Wood and Gibson, 1973). Although both agricultural and domestic sources contributed nutrients to the lake, it was found that 85% of the phosphorus entering during a period April to August (the critical period for phytoplankton growth) was derived from sewage effluents in the catchment (Smith, 1977). The conventional biological oxidation plant, such as the biological filter or the activated sludge process, is designed to remove BOD by biological oxidation and biosynthesis and not to remove plant nutrients such as N and P. By comparing the stoichiometric relationships of C, N and P of domestic sewage and the synthesized biomass it is evident that all the N and P cannot be assimilated in the process and must therefore be discharged in the effluent (Stumm and Morgan, 1962).

The domestic use of detergent washing powders containing polyphosphates as builders doubles the phosphate concentration in domestic sewage effluents (Devey and Harkness, 1973).

The ecological changes of applied significance associated with eutrophication may be considered as: (1) changes in the biota; (2) changes in concentrations of dissolved gases.

44.6.2.1 *Changes in biota*

Successive changes in the nutrient status of the water are associated with changes in the biota. These are usually detrimental to Man's several uses of the lake and water. The onset of eutrophy is characterized by a marked increase in primary productivity, usually in the form of phytoplankton with the frequent occurrence of algal blooms. Although there is an increase in phytoplankton density there is a reduction in species diversity, blue-green algae such as *Oscillatoria rubscens*, *Anabaena*, *Microcystis* and *Aphanizomenon* becoming dominant. There is also a change in the dominant diatoms, species of *Melosira*, *Fragillaria*, *Stephanodiscus* and *Asterionella* replacing such species as *Tabellaria* and *Cyclotella*; the colonial flagellate *Dinobryon* and desmids such as *Staurastrum* appear to decline in abundance under eutrophic conditions. Macro-algae such as *Cladophora* and *Enteromorpha* proliferate and may foul the shores, where they create nuisance as they decay. Associated with the increased primary productivity there is a corresponding increase in secondary productivity in the form of those consumers which can tolerate the conditions and are able to utilize phytoplankton. The death and decay of the increased populations increases the organic load and thereby the bacterial numbers, which in turn support bacteria-feeding animals such as zooplankton and the filter-feeding moss-animalcule *Plumatella*, which foul the bed of some eutrophic lakes and which can be toxic to fish (Jonasson, 1963).

The increased phytoplankton creates turbidity and colour and under summer conditions algal blooms may occur which may render the water toxic to fish and cattle. Illness and lowering of milk yield in cattle have been attributed to their drinking water containing Microcystis. Cases of intestinal disorders and allergic reactions in Man have also been associated with algal blooms. Probably the odours and tastes produced by these organisms deter Man from drinking such waters and thereby limit the number of cases involved. Gorham (1964) considered that toxic algae in waters were more a nuisance and of economic significance than a public health hazard. Eutrophication of waters used as a source for potable supply renders it less acceptable because of the increased bacterial numbers and unpleasant tastes. The removal of the phytoplankton increases the cost of treatment. Examples of nuisance-causing organisms in eutrophic waters are given in Table 2.

Table 2 Problem and Nuisance-causing Algae in Potable Water Supplies

Organism	Problem/Nuisance
Asterionella, *Tabellaria,* *Fragilaria,* *Oscillatoria*	Rapid clogging of sand-filters
Microcystis	Grassy odour; sweet taste
Anabaena, *Aphanizomenon*	Grassy-nasturtium-musty odour; septic odour when abundant
Synura	Cucumber odour; bitter metallic taste
Ceratium	Fishy odour; bitter metallic taste
Cryptomonas	Violet odour; sweet taste
Microcystis, *Aphanizomenon,* *Anabaena*	Toxins resulting in death of fish, birds and mammals
Microcystis	Lowers milk production in cattle

Although the classical accounts of the eutrophication of lakes describe an increase in the littoral macro-vegetation and the final stages involve the invasion of the silted lake by rooted plants to give rise to a marsh or fen community, in some cases submerged macrophytes have been found to be adversely affected by eutrophication. In a study of the Norfolk Broads in England (Mason and Bryant, 1975), it was found that in the nutritionally enriched broads, such as Alderfen Broad, associated with the development of rich algal blooms, the submerged macrophytes were severely restricted consisting mainly of the floating-leaved water lilies. The decline in the submerged macrophytes was attributed to the shading by the phytoplankton and the dense growth of epiphytes on their leaves (Phillips *et al.*, 1978).

In rapidly flowing rivers phytoplankton do not cause problems but under nutrient-enriched conditions, as below a well-oxidized sewage effluent, attached filamentous algae such as *Cladophora* (blanket weed) may develop profuse growths to cause nuisance. In some impounded brackish fenland streams, blooms of the phytoflagellate *Prymnesium parvum* occur which cause fish mortality by the secretion of toxins. With the improvement of sewage effluents, many rivers once infested with growths of sewage fungus, symptomatic of organic pollution, are now affected by dense growths of blanket weed.

44.6.2.2 Changes in concentration of dissolved gases

In the euphotic zone of the lake, *i.e.* the surface layers where light penetration is sufficient to support photosynthesis, the photosynthetic activity of the dense population of phytoplankton by day increases the dissolved oxygen concentration which may at times become supersaturated. By night, however, due to the respiratory demand the oxygen may be depleted thus increasing the magnitude of the diel oxygen fluctuation. Owing to the same activities there is a reverse fluctuation in the carbon dioxide concentration which in turn affects the pH. The increased production results in increased amounts of dead organic matter which sink to the depths of the lake where they decompose, reducing the oxygen concentration in the lower layers. In thermally stratified lakes, in which the warm less-dense waters of the epilimnion become isolated from the cold dense waters of the hypolimnion, the supply of oxygen to meet the need of the decomposers in the profundal zone is cut off and anaerobic conditions become established. Under this condition undesirable chemical and biochemical changes occur, reducing the quality of the water. Methane and sulfides are produced and iron and manganese oxides, precipitated out under aerobic conditions in the epilimnion, are reduced in the anoxic hypolimnion and pass back into solution. In eutrophic lakes and reservoirs used as a source for potable water, the water quality can be seriously affected. Fish are also affected. With some degree of eutrophy, overall fish production is enhanced (Hartmann, 1977), but is reduced under conditions of extreme eutrophy. There is also a change in species dominance, the salmonid sport fish being replaced by coarse fish. In oligotrophic lakes under summer conditions the coldwater-loving salmonids take refuge in the cold waters of the hypolimnion. The depletion of the oxygen in the hypolimnion of eutrophic lakes renders this refuge unsuitable.

In eutrophic rivers also the increase in attached benthic algae, such as *Cladophora*, causes changes in diel fluctuations of dissolved gases. This in turn can affect the animal life in several ways. Stewart (1962) found that by subjecting fish, fed to satiation with earthworms, to wide diel fluctuations in oxygen their growth was impaired. The distribution of *Gammarus pulex* along a length of a stream receiving an organic discharge was accounted for not by the maximum, minimum or mean dissolved oxygen but by the number of hours each night that the oxygen concentration fell below a critical level (Hawkes and Davies, 1971). The significance of the diel oxygen fluctuations on *Gammarus* was later demonstrated by laboratory experiments using an apparatus designed to simulate natural diel oxygen fluctuations (Grant and Hawkes, 1982). The decrease in carbon dioxide due to active photosynthesis in eutrophic rivers can cause a marked increase in pH which may adversely affect fish. In such rivers receiving sewage effluents, ammonia may be present; the increase in pH affects the dissociation of the ammonia: the un-ionized ammonia molecule NH_3 being toxic and the ionized ammonium NH_4^+ having little or no toxicity (Wuhrmann and Woker, 1948), the toxicity is markedly increased.

44.7 RELEVANCE OF BIOTECHNOLOGY IN THE CONTROL OF POLLUTION

The many and diverse ecological effects resulting from the discharge of domestic and industrial wastes into the aquatic environment necessitate the use of equally diverse methods of treatment to prevent pollution. An understanding of the ecology of polluted waters enables appropriate treatment methods to be selected for specific wastes and appropriate tests applied to monitor the effluent. Although standard methods have become established for treating and monitoring the most common discharge (sewage), such methods are not always appropriate for all industrial and agricultural wastes.

Although in this chapter on the ecology of polluted waters we have mainly concerned ourselves with water as an aquatic environment, it is, as stated earlier, also a natural resource of major importance to Man. Treatment processes in pollution prevention must therefore involve both the protection of water as a resource and the protection of the aquatic environment generally. These different objectives are implicit in the UK National Water Council's proposal to classify rivers, for the purpose of establishing discharge consent conditions, by the designation of river water quality objectives based on the uses of the waters involved *and environmental considerations* (National Water Council, 1978). The UK Water Act of 1973, which set up the Water Authorities, charged them with the responsibility of conserving the fresh-water flora and fauna. Although the protection of the aquatic environment safeguards many of Man's uses of the water, the two objectives are not always achieved by the same treatment process. It is of interest to note that most of the pollution prevention treatment processes are designed to protect the aquatic environment rather than the public health quality of the water; sewage effluents are monitored by BOD rather than *E. coli* count.

Of the different types of water pollution considered above, it is naturally those classified as 'biotic' that are most amenable to control by biological treatment methods. In addition, however, some important toxic pollutants are biodegradable and tertiary treatment processes for the removal of fine solids from effluents have been found to depend upon biologically active surfaces for their efficiency (Hawkes, 1983). For monitoring the effects of discharges on the aquatic environment, biological surveillance methods are obviously the most direct means. Biological methods have also been adopted to complement chemical and physical methods by the UK Department of the Environment, for the national river pollution survey of England and Wales, 1980 (National Water Council, 1981).

Ever since the ABC (alum, blood and charcoal) process, patented in 1868, repeated attempts have been made to develop physico-chemical treatment (PCT) methods for treating sewage. The most recent attempt, carried out in the UK under the aegis of the NATO Committee on the Challenge of Modern Society (CCMS), involved the use of purposely designed, continuously operated pilot plants at Coleshill (near Birmingham) and Davyhulme (near Manchester). A subsequent appraisal of PCT (Bayley and Cooper, 1981) concluded that none of the PCT systems currently available were likely to be capable of replacing conventional (biological) treatment of sewage in the UK. The cost of treatment and the energy consumption by PCT would be increased several fold. Apart from specific requirements such as the adsorption of trace organics using activated carbon and the removal of phosphorus by chemical precipitation, biological methods will provide the mainstay of used-water treatment and any advances are likely to be in the field of biotechnology rather than in PCT.

The view has been expressed (La Riviere, 1977) that aerobic oxidative systems of sewage treatment, such as biological filtration and activated sludge, appear to have been developed to a degree of efficiency, in terms of BOD removal, that they are only capable of further marginal improvements. This may be so in respect of the exploitation of the kinetics of BOD removal but further improvements in effluent quality are needed to produce a high-quality effluent suitable for re-use water. Approximately one-third of the water abstracted in England and Wales for public water supply is taken from re-use rivers, *i.e.* rivers which have received effluents; on economic grounds it is likely that such rivers will be used to meet any future increase in demand (Central Water Planning Unit, 1980). The public health aspects of re-using water for public supply are fully appreciated (Fish, 1975). Efficient water treatment practices and appropriate monitoring will ensure the absence of pathogenic organisms and harmful inorganic chemicals. The chronic long-term effects of the presence of tens of thousands of organic compounds is, however, difficult to predict from present knowledge. A review (Central Water Planning Unit, 1976) concluded that most of the synthetic chemicals in river water originated in industrial wastes. Although many of these were considered not removable by normal sewage treatment, others were identified as being potentially biodegradable. The fact that appreciable concentrations of readily biodegradable chemicals often occur in river waters demonstrates the limitations of present biological treatment methods in removing such compounds.

As treatment methods to remove BOD by biological oxidation are perfected, attention will turn to the removal of the other environmental pollutants discussed above, such as N and P. Pollution control is a socio-economic problem and especially in times of economic constraints the economic aspects of treatment processes become paramount. In this situation the low-energy requirements and potential energy source of anaerobic processes make them most attractive, with possible developments of the anaerobic filter and fluidized bed. In the case of the latter the problems of the economics of oxygen transfer would of course be obviated.

In terms of industrial production in the UK, water reclaimed from sewage by biological treatment far exceeds the total of other products processed (Dunhill, 1981). The biological treatment capacity of the water industry is far in excess of all other fermentation industries and the largest reactors (activated sludge plants) are found in sewage treatment. Used-water treatment therefore may be regarded as a major biotechnological industry of importance in safeguarding the nation's water supplies and its aquatic environment.

44.8 REFERENCES

Amberg, H. R. and J. F. Cormack (1960). Factors affecting slime growth in the lower Columbia River and evaluation of some possible control measures. *Pulp. Pap. Can.*, **61**, T70–T80.

Bayley, R. and P. F. Cooper (1981). Physico-chemical treatment—an appraisal. *Notes on Water Research*, No. 26. Water Research Centre, Stevenage, Herts.

Bock, K. J. (1964). Biological properties of surface-active substances. IVth International Congress of Surfactants, 1964. *Manuf. Chem.*, **35**, No. 11.

Bock, K. J. (1966). The effect of raw materials for detergents on fish. *Arch. Fischereiwiss.*, **17**, 68–77.

Brown, V. M. (1968). The calculation of the acute toxicity of mixtures of poisons to rainbow trout. *Water Res.*, **2**, 723–733.

Brown, V. M., D. H. M. Jordan and B. A. Tiller (1967). The effect of temperature on the acute toxicity of phenol to rainbow trout in hard water. *Water Res.*, **1**, 1–9.

Cawley, W. A. (1958). *Sewage Ind. Wastes*, **30**, 1174.

Central Water Planning Unit (1976). *The Significance of Synthetic Chemicals in Rivers used as a Source of Drinking Water.* Tech. Note 14. CWPU, Reading.

Central Water Planning Unit (1980). *Re-use of Water for Potable Supplies.* CWPU Project 4, Final Report. CWPU, Reading.

Curtis, E. J. C. (1969). Sewage fungus: its nature and effects. *Water Res.*, **3**, 289–311.

Degens, P. N., H. Van der Zee, J. D. Kommer and A. H. Kamphuis (1950). Synthetic detergents and sewage processing. V. The effect of synthetic detergents on certain water fauna. *J. Inst. Sewage Purif.*, **1**, 63–68.

Department of the Environment (1974). *Fifteenth Progress Report of the Standing Technical Committee on Synthetic Detergents.* HMSO, London.

Department of the Environment (1980). *Twentieth and Final Report of the Standing Technical Committee on Synthetic Detergents.* HMSO, London.

Devey, D. G. and N. Harkness (1973). The significance of man-made sources of phosphorus: detergents and sewage. *Water Res.*, **7**, 33–54.

Doudoroff, P. and C. E. Warren (1957). Biological indices of water pollution, with special reference to fish populations. In *Biological Problems in Water Pollution*, pp. 144–163. US Department of Health Education and Welfare, Public Health Service, R. A. Taft Sanitary Engineering Center, Cincinnati, OH.

Dunhill, P. (1981). Biotechnology and industry. *Chem. Ind. (London)*, 4 April, 204–217.

Fish, H. (1975) A new look at pollution prevention on lowland rivers. *Water*, No. 3, 2–4.

Fox, H. M. and A. E. R. Taylor (1955). The tolerance of oxygen by aquatic invertebrates. *Proc. R. Soc. London, Ser. B.*, **143**, 214.

Gorham, P. R. (1964). Toxic algae as a public health hazard. *J. Am. Water Works Assoc.*, **56**, 1481.

Grant, I. F. and H. A. Hawkes (1982). The effects of diel oxygen fluctuations on the survival of the freshwater shrimp *Gammarus pulex. Environ. Pollut., Ser. A*, **28**, 53–66.

Harrison, M. E. and H. Heukelekian (1958). Slime infestation — literature review. *Sewage Ind. Wastes*, **30**, 1278–1302.

Hartmann, J. (1977). Fischereiliche veranderungen in kulturbedingt eutrophierenden seen. *Schweiz. Z. Hydrol.*, **39**, 243–254.

Hawkes, H. A. (1957). Biological aspects of river pollution. In *Aspects of River Pollution*, ed. L. Klein, pp. 191–251. Butterworths, London.

Hawkes, H. A. (1962). Biological aspects of river pollution. In *River Pollution II. Causes and Effects*, ed. L. Klein, pp. 311–432. Butterworths, London.

Hawkes, H. A. (1978). River bed animals — tell-tales of pollution. In *Biological Surveillance of River Water Quality*, ed. H. A. Hawkes and J. G. Hughes, pp. 55–57. Proceedings of Section K of the British Association for the Advancement of Science, Aston, 1977.

Hawkes, H. A. (1983). The applied significance of ecological studies of aerobic processes. In *Ecological Aspects of Used-water Treatment. Vol. 3, The Processes and their Ecology*, ed. C. R. Curds and H. A. Hawkes, pp. 173–333. Academic, New York.

Hawkes, H. A. and L. J. Davies (1971). Some effects of organic enrichment on benthic invertebrate communities in stream riffles. In *The Scientific Management of Animal and Plant Communities for Conservation*, ed. E. A. Duffey and A. S. Watt, pp. 271–293. Blackwell, Oxford.

Herbert, D. W. M., G. H. J. Elkins, H. T. Mann and J. Hemens (1957). Toxicity of synthetic detergents to rainbow trout. *Water Waste Treat. J.*, **6**, 394–397.

Holden, A. V. (1964). The possible effects on fish of chemicals used in agriculture. *J. Proc. Inst. Sewage Purif.*, 361–368.

Holdgate, M. W. (1979). *A Perspective of Environmental Pollution.* Cambridge University Press, Cambridge.

Hynes, H. B. N. (1960). *The Biology of Polluted Waters.* Liverpool University Press, Liverpool.

Jonasson, P. J. (1963). The growth of *Plumatella repens* and *P. fungosa* (Bryozoa ectoprocta) in relation to external factors in Danish eutrophic lakes. *Oikos*, **14**, 121.

Jones, J. R. E. (1964). *Fish and River Pollution.* Butterworths, London

Klein, L. (1962). *River Pollution II. Causes and Effects.* Butterworths, London.

Kruger, R. (1964). Recent developments concerning alkylbenzene sulphonate. *Fette Seifen Anstrichm.*, **66**, 217–221 (*Water. Pollut. Abstr.*, 1966, **39**, No. 1138).

Langford, T. E. (1975). The emergence of insects from a British river, warmed by power station cooling water. Part II. The emergence patterns of some species of Ephemeroptera, Trichoptera and Megaloptera in relation to water temperature and river flow, upstream and downstream of the cooling water outfalls. *Hydrobiologia*, **47**, 91–133.

La Riviere, J. W. M. (1977). Microbial ecology of liquid waste treatment. In *Advances in Microbial Ecology*, ed. M. Alexander, Vol. 1, pp. 215–259. Plenum, New York.

Leclerc, E. and F. Derlaminck (1952). Les detergents naturels ou synthetiques et les poissons. *Bull. Trimest. CEBE-DEAU*, **17**, 165–171.

Liebmann, H. (1966). The effects of new detergents on fisheries and on the process of self-purification. *Abwassertechnik*, **17**, No. 1, VII (*Water Pollut. Abstr.*, 1968, **41**, No. 331).

Lloyd, R. (1961). The toxicity of mixtures of zinc and copper sulphates to rainbow trout (*Salmo gairdnerii* Richardson). *Ann. appl. Biol.*, **49**, 535–538.

Lloyd, R. and J. S. Alabaster (1980). *Water Quality Criteria for Freshwater Fish*. Butterworths, London.

Madai, I. and H. A. D. Lan (1964). Concerning the effect of several detergents on freshwater organisms. *Wasser Abwasser*, **4**, 168–183 (*Water Pollut. Abstr.*, 1967, **40**, No. 1035).

Marchetti, R. (1966). Relations between the surface activity, chemical composition and toxicity to aquatic life, of synthetic detergents. In *Pollution of Waters. Some Actual Problems. Assoc. Nat. Res. Tech. Paris*, 185–218 (*Water Pollut. Abstr.*, 1968, **41**, No. 2144).

Mason, C. F. and R. J. Bryant (1975). Changes in the ecology of the Norfolk Broads. *Freshwater Biol.*, **5**, 257–270.

Ministry of Housing and Local Government (1967). *Ninth Progress Report of the Standing Technical Committee on Synthetic Detergents*. HMSO, London.

Morgan, W. S. G. (1978). The use of fish as a biological sensor for toxic compounds in potable water. *Prog. Water Technol.*, **10**, 395–398.

National Water Council. (1978). *River Water Quality — The Next Stage. Review of Discharge Consent Conditions*. NWC Publications, London.

National Water Council. (1981). *River Quality: The 1980 Survey and Future Outlook*. NWC Publications, London.

Okrend, H. and N. C. Dondero (1964). Requirement of *Sphaerotilus* for cyanocobalamin. *J. Bacteriol.*, **87**, 286–292.

Patalas, K. (1967). Original and secondary production of plankton in lake heated by electricity power plant. Summary Reports of 7th Hydrobiologists Congress in Swinoujscie, Sept. 1967.

Patrick, R. (1953). Biological phases of stream pollution. *Proc. Penn. Ac. Sc.*, **27**, 33–36.

Paulson, B. B. (1962). Pollution of the lower reaches of the Otteraa. *Vattenhygien*, **18**, 115 (*Water Pollut. Abstr.*, 1964, **37**, No. 159).

Pentelow, F. T. K. (1949). Fisheries and pollution from china clay works. *Rep. Salm. Freshw. Fish. London*, **31**.

Phillips, G. L., D. Eminson and B. Moss (1978). A mechanism to account for macrophyte decline in progressively eutrophicated freshwaters. *Aquat. Bot.*, **4**, 103–126.

Pickering, Q. H. (1966). Acute toxicity of alkylbenzenesulphonate and linear alkylate sulphonate to the eggs of the fathead minnow, *Pimephales promelas*. *Int. J. Air Water Pollut.*, **10**, 385–391.

Prat, J. and A. Giraud (1964). *The Pollution of Water by Detergents*. Organization for Economic Co-operation and Development, Paris.

Sawyer, R. T. (1947). Leeches (Annelida; Hirudinea). In *Pollution Ecology of Freshwater Invertebrates*, ed. C. W. Hart and S. L. H. Fuller, pp. 81–142. Academic, New York.

Smith, R. V. (1977). Domestic and agricultural contributions to the inputs of phosphorus and nitrogen to Lough Neagh. *Water Res.*, **11**, 453–459.

Stewart, N. E. (1962). The influence of oxygen concentration on the growth of juvenile large mouth bass. M.Sc. Thesis, Oregon State University Library, Corvallis, Oregon (quoted in Doudoroff and Warren, 1965).

Stokes, J. L. (1954). Studies on the filamentous sheathed iron bacterium *Sphaerotilus natans*. *J. Bacteriol.*, **67**, 278–291.

Stumm, W. and J. J. Morgan (1962). Stream pollution by algal nutrients. In *Trans. 12th Conf. Sanit. Eng., Univ. Kansas*, 16–26.

Tarzwell, C. M. (1957). Water quality criteria for aquatic life. *In Biological Problems in Water Pollution*, p. 248. US Department of Health, Education and Welfare, Public Health Service, R. A. Taft Sanitary Engineering Center, Cincinnati, Ohio.

Wisdom, A. S. (1956). *The Law on the Pollution of Waters*. Shaw, London.

Wood, R. B. and C. E. Gibson (1973). Eutrophication and Lough Neagh. *Water Res.* **7**, 173–187.

World Health Organization (1979). *Sodium, Chlorides and Conductivity in Drinking Water*. WHO, Copenhagen.

Wuhrmann, K. and H. Woker (1948). Experimentelle untersuchungen uber die ammoniak- and blausaurevergiftung. *Schweiz. Z. Hydrol.*, **15**, 210–244.

45

Microbiology of Treatment Processes

G. HAMER
Swiss Federal Institute of Technology, Zürich, Switzerland

45.1 INTRODUCTION

In nature, microbiological mechanisms have always been responsible for the degradation of biodegradable waste matter which either inadvertently or by design found its way into the aquatic environment. Self-purification of surface waters is a well known phenomenon. However, over-loading of natural surface waters with biodegradable pollutants as a result of the development of both large centres of population and industrialization, with their attendant discharges of sewage and wastewater, ultimately caused self-purification mechanisms to fail in many regions, Anaerobiosis ensued and surface waters became aesthetically objectionable and totally unsuitable as sources of raw water for drinking water production. Obviously, such problems were most severe at inland locations, coastal locations enjoying the benefit of the much greater potential for discharged waste dilution afforded by the marine environment.

It is important to differentiate clearly between sewage and wastewater disposal on the one hand, and sewage and wastewater treatment on the other. The former involves only a minor element of technology, whilst the latter, depending on the ultimate quality of effluent permitted for discharge, involves a gradation of technology from simple systems that achieve only partial treatment to multistage processes that produce effluents suitable as feed water for drinking water production (Downing, 1972).

The first technique used for sewage treatment at inland locations was introduced in the middle of the 19th century and involved the spreading of settled sewage on land at an appropriate rate so as to permit the oxidation of the biodegradable pollutants present by aerobic soil bacteria, essentially a self-purification mechanism. In this process the application of technology is sufficiently restricted to suggest disposal rather than treatment. However, just before the turn of the century, the first genuinely effective treatment systems began to be introduced. Such systems were either aerobic contact beds or aerobic bacterial beds. Contact beds were operated on the 'fill and draw' principle and comprised a bed of coke breeze which was filled with settled sewage over a period of about two hours, followed by a contact period of about one hour and then drained over a period of five hours, so that three operational cycles were possible per day. The performance of contact beds, although in some cases entirely satisfactory, was frequently variable and the major disadvantages of both intermittent operation and poor response to diurnal variations in sewage flows were soon recognized, and contact beds were rapidly superseded by bacterial beds which provided a continuous flow treatment facility (Reeve, 1973). Such systems when first introduced were operated with continuous dosing to the whole bed, but subsequently, the more effective systems were designed to allow intermittent dosing to each part of the bed by employing moving, usually rotating, distributors. Bacterial beds have variously been described as continuous filters, as trickling filters, as percolating filters or as sprinking filters, but it must be clearly understood that the predominant mechanism of pollutant removal is one of biooxidation by microbes growing as films on the filter packing material rather than any form of straining mechanism as implied in the name filter. Bacterial beds are essentially stationary microbial-film bioreactors, and until the present have provided an effective and economic system for the biological treatment of domestic and municipal sewage and some industrial wastewaters under suitable climatic conditions.

Only one aerobic biological treatment process that has been developed in the interim has challenged and frequently proved more popular than bacterial beds. This is the activated sludge process, with its many variants, that was originally developed by Ardern and Lockett (1914). Activated sludge processes are essentially continuous flow dispersed flocculated growth processes in which the bacterial flocs, described as activated sludge, are separated from the treated effluent by sedimentation in a clarifier and recycled to the aeration tank where they are recontacted with either the sewage or wastewater undergoing treatment, such that a high degree of process intensity is maintained. Activated sludge process aeration tanks are operated either as non-ideal plug-flow bioreactors or as completely mixed bioreactors.

The past decade has seen many proposed innovations in biological sewage and wastewater treatment, some of which might ultimately find widespread application. These include a range of anaerobic treatment processes developed from sludge digestion technology, rotating aerobic microbial film processes and, possibly the process type with the highest potential for the future, hybrid processes of the aerobic fluidized bed type, in which the microbial film grows on solid support particles that are fluidized in a bioreactor, such that the coated support particles are dispersed throughout either the sewage or the wastewater undergoing treatment, in an analogous manner to activated sludge treatment processes. Such processes allow the possibility of controlling the microbial film thickness by appropriate selection of operating parameters.

Essentially, all practical biological sewage and wastewater treatment processes have, irrespective of pertaining treated effluent discharge regulations, to satisfy a number of basic objectives with respect to their operation. These include (i) effective biooxidation of the biodegradable carbonaceous, and frequently nitrogenous, pollutants, such that a minimum quantity of readily dewaterable biomass (sludge) is produced; (ii) adsorption onto and absorption into the sludge of other soluble pollutants and entrapment of particulate pollutants within the microbial flocs or films; and (iii) easy and complete separation of suspended biomass from the treated effluent prior to discharge from the biotreatment process. In this chapter, the impact of microbiology on achieving effective biological sewage and wastewater treatment processes willl be examined.

45.2 CLASSIFICATION OF MICROBES

Essentially, microbial processes can be considered most simply on the basis of their objectives. Sewage and wastewater treatment processes seek to achieve the biodegradation of pollutants, essentially a chemical transformation process and, as such, are processes that do not necessarily require the growth of microbes and, at least theoretically, could be undertaken by either resting microbes or enzymes extracted from microbes. However, developments in biotechnology have not yet permitted the introduction of practical non-growth biotreatment systems, and hence, technologies employing growing cultures to undertake the requisite transformations in biological sewage and wastewater treatment are clearly what must be considered here, although the important requirement to minimize biomass production in such processes cannot be overstressed.

Just as with the classification of all flora and fauna, extensive literature exists concerning the morphology of microbes, but morphological information has only minor impact with respect to the development of an understanding of the mechanisms and functions of mixed microbial populations that develop in biotreatment processes, and in such processes it is the distinction between metabolic characteristics that is most important. However, prior to the elucidation of such characteristics, it is first desirable to summarize the differences between eukaryotic and prokaryotic microbes, as both types are involved in sewage and wastewater treatment, and to introduce the archaebacteria.

45.2.1 Eukaryotes, Prokaryotes and Archaebacteria

Microbes are generally classified as either eukaryotes or prokaryotes on the basis of the complexity of their cellular structure (Gaudy and Gaudy, 1980). Eukaryotic microbes comprise fungi, including yeasts, algae and protozoa, whilst prokaryotic microbes comprise bacteria and cyanobacteria (formally described as blue-green algae). The eukaryote–prokaryote dichotomy has long been accepted as a fundamental concept, but recently it has become evident that a third group of microbes, the archaebacteria, exists (Woese, 1982). As the archaebacteria play important roles in some treatment processes, some details concerning them will also be summarized. In addition to the categories of microbes mentioned above, one also has viruses, which are much simpler in structure and composition and are parasites. However, viruses will not be discussed further, as they probably do not contribute to biodegradation, but it should be mentioned that either the removal or the destruction of mammalian viruses is an important public health requirement of sewage treatment processes and that bacterial viruses (bacteriophages) have potential adverse effects on biotreater performance.

The most immediately obvious difference between eukaryotic and prokaryotic microbes is their size, with the latter, on average, being an order of magnitude smaller than the former. The equivalent spherical diameter of many bacteria is *ca.* 1 μm. However, the fundamental difference between eukaryotes and prokaryotes is one of cellular structure and function. The characteristic organelle of eukaryotic cells is the nucleus, which, together with the several other membrane-bounded organelles present, carries out the essential functions of the cell. In prokaryotes, these same essential functions are generally performed within the cytoplasmic membrane.

The nucleus of eukaryotic cells contains the DNA of the cell and associated proteins that form the chromosomes, and is separated from the cytoplasm by a double porous membrane. The outer nuclear membrane has a complex internal structure, part of which is covered with ribosomes, *i.e.* protein and RNA containing bodies which are the sites of protein synthesis. In aerobic eukaryotes, chemical energy is trapped in the microchondria, organelles bounded by a double membrane and containing an internal membrane structure in which are found both the electron transport system and enzymes associated with respiratory metabolism. Mitochondria are absent in anaerobic eukaryotes. In photosynthetic eukaryotes, *i.e.* green algae, an organelle similar to the mitochondrion, the chloroplast, is found. The chloroplast also contains an internal laminar membrane system housing both the pigments involved in photosynthesis and the photosynthetic electron transport system. Both mitochondria and chloroplasts contain small circular DNA molecules similar to those found in prokaryotes, ribosomes and other components of protein synthesis. Many eukaryotic cells also contain stacked membranes that function in secretion and a variety of types of membrane-bounded vacuoles which may contain substrate, water, storage products or waste products.

All the organelles are contained within the cytoplasm, which comprises an aqueous solution or suspension of the molecules synthesized and utilized by the cell, *i.e.* soluble enzymes, vitamins, amino acids, carbohydrates, lipids, nucleotides, coenzymes and metabolic products and intermediates. The cytoplasm is enclosed in the cytoplasmic membrane, which comprises predominantly phospholipids and proteins and is the effective boundary between the cell and its environment. Generally, additional layers exist outside the cytoplasmic membrane including the cell wall or cell pellicle and a slime layer, but it is the function of the cytoplasmic membrane to transport molecules into and out of the cell with the aid of specific transport proteins associated with the membrane. In both green algae and fungi, the cytoplasmic membrane is surrounded by the cell wall which imparts strength, shape and rigidity to the cell. Protozoa have no true cell wall, although varying degrees of rigidity are imparted to different species by pellicles. For locomotion, eukaryotic microbes generally employ either a single flagellum or a small number of flagella, although one important group of protozoa have large numbers of cilia.

The components of a typical prokaryotic cell are a cell wall, a cytoplasmic membrane, a single molecule of DNA, ribosomes and the cytoplasm. All other components, although they might be present, are essentially dispensable. In the case of the mycoplasmas, even the cell wall is absent. In addition to the absence of complex internal structures, the prokaryotes have a markedly different cell wall structure and components than do eukaryotes. However, the cytoplasmic membrane of prokaryotes is fundamentally similar to that of a eukaryotic cell, *i.e.* a phospholipid double layer with various associated proteins. The variety of proteins in the cytoplasmic membrane is greater in prokaryotes than in eukaryotes because of the significantly greater number of functions performed by the cytoplasmic membrane. As with eukaryotes, the membrane controls the permeability of the cell and in addition to the specific proteins required for the transport of ions and molecules through the membrane, the membrane also contains the enzymes involved in the synthesis of membrane and cell wall components, and is the major site of energy metabolism in the cell. In bacteria, either the electron transport system or the site of photosynthetic energy metabolism is located in the cytoplasmic membrane, and in photosynthetic, aerobic chemoautotrophic and methanotrophic bacteria, the cytoplasmic membrane is extensively folded to provide an extended surface for either photosynthesis or oxidative phosphorylation. In cyanobacteria, the folded membranes that function in photosynthesis are not necessarily continuous with the cytoplasmic membrane.

The ribosomes of prokaryotes are the site of protein synthesis, just as they are in eukaryotes, and although their structure is similar in both types of cells, their size is much reduced in prokaryotes. The genetic information of prokaryotic cells is contained in a DNA molecule that is neither complexed with proteins to form a chromosome, nor separated from the cytoplasm by a membrane, but is present as a single, double-stranded, covalently joined molecule.

In addition to the components of bacterial cells that have already been described, three additional types of internal structure, that are not necessarily essential for the survival of the cell, are found in some bacteria. These structures are several types of storage granules, vesicles and endospores. The most common storage granules in prokaryotes are those for carbon substrate storage and comprise either glycogen or poly-β-hydroxybutyrate. In addition, inorganic phosphate can be stored in polymeric form as volutin granules, and in some specialist sulfur bacteria sulfur granules are found. Of the three types of vesicles found in prokaryotes, gas vesicles, which control the relative density of the cells, are the most important. Spores are restricted to only two genera of bacteria, and will not be discussed further, as they have no role in biotreatment.

Several additional components of bacterial cells are found outside the cell wall. These include capsules and slime layers, flagella, pili (fimbriae) and stalks and holdfasts used for attachment to surfaces. The occurrence of capsules and slime layers is common in prokaryotes. Most bacterial capsules and slimes are polysaccharides, containing either one or several different sugars. Capsular and slime materials bind cells together to form aggregates and, hence, are implicated in microbial floc and film formation.

Many prokaryotes are immotile, but some have either a single flagellum or multiple flagella as their means of locomotion. Bacterial flagella are thin tubular structures composed of spirally arranged protein molecules which arise within the cytoplasmic membrane. Their diameter is much smaller than the diameter of the flagella of eukaryotes. Pili or fimbriae are much shorter, flagella-like structures, that cover the surface of some bacterial species, and although their function is unclear, they probably have a role in cell attachment and adhesion. Stalks and holdfasts are features encountered only in a few bacterial species. Such features are extensions of the cell wall and are involved in the attachment of those bacteria exhibiting such features to surfaces, and, hence, in bacterial film formation.

Archaebacteria form a group of microbes that is distinct from both eukaryotic and prokaryotic microbes and they are neither related to the true bacteria (eubacteria) nor to eukaryotes. They comprise a small, but diverse, collection of phenotypes, and although they have some unique phenotypic characteristics in common, it is difficult to group them convincingly on this basis. Archaebacteria are recognized primarily by genotypic data, but it also seems probable that they differ from the eukaryotes and prokaryotes in significant detail of most of their molecular processes. They comprise three major phenotypes, methanogenic, extreme halophilic and thermoacidophilic, and species from each type are either important in existing wastewater and sludge treatment processes or have significant potential, particularly in thermophilic treatment processes, that can be expected to be developed in the future.

The molecular differences between the archaebacteria and eubacteria are of a comparable nature to those that exist between eukaryotes and prokaryotes, and involve cell wall structure, protein synthesis, ribosomal proteins, membrane lipids, *etc.* as has been discussed in collected papers edited by Kandler (1982).

Irrespective of their diverse structural and molecular characteristics, species of eukaryotic microbes, of prokaryotic microbes and of archaebacteria are found that are classified either as heterotrophs or as autotrophs from the metabolic point of view.

45.2.2 Metabolic Classification

The three nutritional requirements that are quantitatively the most important for microbes are a carbon substrate, an energy source and an electron donor. These, together with the electron acceptor, are governed by the enzymes present in the cell. The carbon source available to microbes can be either organic or inorganic. Those microbes that require organic compounds as either their sole or principal carbon substrate are classified as heterotrophs, whilst those that require inorganic carbonaceous compounds as either their sole or principal carbon substrate are classified as autotrophs. Microbes that utilize light as their energy source are described as phototrophs and those that obtain energy from the oxidation of either organic or inorganic compounds are described as chemotrophs. Microbes that use organic compounds as their source of electrons are desribed as organotrophs, whilst those employing inorganic electron sources are described as lithotrophs.

Such a classification is rather more distinct than that which, in reality, occurs, where overlapping categories are frequently encountered. For example, in a case where light is the energy source, the electron donor can be clearly identified from the energy source, but when an organic chemical energy source is utilized, the same compound is often also utilized as either the sole or principal carbon substrate and in heterotrophs, the situation of combined carbon energy substrates is common. In addition, the versatility of microbial metabolism must also be taken into account, and it is probably more satisfactory, because of this versatility, to think of microbes employing an autotrophic mode of metabolism rather than describing them exclusively as autotrophs. The term mixotroph is used to describe microbes that use alternative energy sources either with equal facility or simultaneously.

The microbes that are classified as phototrophs are, from the eukaryotes, the algae, and from the prokaryotes, the cyanobacteria, the purple non-sulfur bacteria (Rhodospirillaceae), the purple sulfur bacteria (Chromatiaceae) and the green sulfur bacteria (Chlorobiaceae). As far as sewage and wastewater treatment is concerned, two photoautotrophs (photolithotrophs), the green algae and cyanobacteria, have significant roles, but only in either facultative or aerated lagoon treatment systems of the type encountered in tropical and sub-tropical regions, where photosynthesis is important in the provision of oxygen for the aerobic heterotrophic bacteria that biodegrade organic pollutants present in either the sewage or the wastewater undergoing treatment.

The vast majority of microbes are chemotrophs, including all the fungi (and yeasts), all the protozoa, all the archaebacteria and most of the true bacteria. As far as chemoautotrophy (chemolithotrophy) is concerned, a much more restricted group of bacterial species is involved, but most archaebacteria fall into this classification. Protozoa, fungi and yeasts do not exhibit chemoautotrophic metabolism, with the possible exception of methylotrophic yeasts, if one accepts the novel definition that autotrophs are microbes which can synthesize all their cellular constituents from either one or more C_1 compounds (Whittenbury and Kelly, 1977). Most chemoautotrophic bacteria are obligate aerobes and their activities are essential for the cycling of inorganic matter, particularly nitrogen and sulfur, although in sewage and wastewater treatment processes emphasis is, of course, on nitrogen rather than sulfur. Energy sources that can be utilized chemoautotrophically by bacteria include ammonia, nitrite, iron (ferrous), hydrogen sulfide, sulfur and hydrogen.

The nitrifying bacteria are divided into two groups, one that oxidizes ammonia to nitrite and the other that oxidizes nitrite to nitrate. The most widely distributed species in each group, as far as sewage and wastewater treatment is concerned, are *Nitrosomonas* spp. and *Nitrobacter* spp., respectively. Other species exist in both groups, but none are as predominant as the *Nitrosomonas* and *Nitrobacter* spp., and it is these two obligate aerobic, slow growing bacterial species that are the key to effective nitrification in aerobic biotreatment processes. The fact that both species are slow growing and, in addition, tend to be inhibited by biodegradable organic pollutants has caused significant effort to be directed towards elucidation of the nitrification step in biotreatment processes, particularly factors that might enhance efficiency.

Of the other chemoautotrophic microbes present in aerobic biotreatment processes, none have a comparable degree of importance to that of the nitrifying bacteria. Certainly, *Sphaerotilus* spp., which can be classified as iron bacteria, are implicated in sludge bulking and *Alcaligenes* spp., which can be classified as hydrogen bacteria, are frequently encountered, but both species are facultative chemoautotrophs and probably exhibit their chemoheterotrophic modes of metabolism in biotreatment plants. In addition, various species of chemoautotrophic sulfur bacteria are present in biotreaters, but their most predominant role is probably much more implicated in aeration tank concrete degradation than in any positive treatment effect. However, *Thiobacillus denitrificans* might offer interesting possibilities for the denitrification of sewage and wastewater in the future.

In anaerobic biotreatment processes, the role of chemoautotrophic archaebacteria is key in achieving effective mineralization of pollutants. Anaerobic treatment processes have tradition-ally been low rate processes with extended residence times, but recent developments, both with respect to reactor design and a move from the mesophilic to the thermophilic temperature range for process operation, indicate an enhanced potential for rapid treatment. In both mesophilic and thermophilic anaerobic treatment processes, the final step involves the formation of methane from acetate and CO_2/hydrogen. Both routes for methane formation, irrespective of tempera-ture, rely on the activities of methanogenic archaebacteria. The chemoautotrophic step from CO_2/H_2 is undertaken by species that fall in all the three orders of methanogens, *i.e.* Methano-bacteriales, Methanococcales and Methanomicrobiales, whereas acetate utilization seems to be a more restricted characteristic (Zehnder *et al.*, 1982).

The predominant microbes in the two major types of aerobic biotreatment processes, *i.e.* acti-vated sludge type processes, where the biooxidation occurs as a result of the activity of discretely dispersed microbial flocs, and trickling filter (biological bed) processes, where the biooxidation is carried out by attached films of microbes, are chemoheterotrophic bacteria. Fungi have a rela-tively minor role with respect to carbonaceous pollutant biooxidation, but in activated sludge processes are implicated in sludge bulking, *i.e.* in situations where the active biomass tends to float and, therefore, cannot be sedimented for recycling to maintain process intensity. Protozoa play a key role in the performance of activated sludge treatment processes, and this will be dis-cussed subsequently. As far as those microbes that are best suited for effective performance in treatment systems are concerned, it is the ability to grow as either microbial aggregates (either flocs or films) or to be able to become directly associated with such aggregates so as to be either recycled, in the case of activated sludge processes, or retained, in the case of trickling filters, that is most important.

As far as effective dissolved carbonaceous pollutant biodegradation in aerobic treatment pro-cesses is concerned, the predominant microbes are true bacteria of the family Pseudomonadaceae and of the genera *Flavobacterium* and *Alcaligenes*. In spite of the fact that sewage treatment facilities are continuously reinoculated with genera belonging to the family Enterobacteriaceae, which are potential human pathogens and one species particularly, *Escherichia coli*, is used as the indicator of faecal pollution, they account for only a very small fraction of the total bacterial population present in treatment plants. Enterobacteriaceae include, in addition to *E. coli*, bac-teria of the genera *Citrobacter*, *Salmonella*, *Shigella*, *Klebsiella*, *Enterobacter* and *Yersina*, most of which are pathogenic, and in some cases, the causative bacteria of important, water-borne dis-eases, including typhoid and bacillary dysentery.

The two genera of Pseudomonadaceae that are dominant in aerobic sewage and wastewater treatment processes are *Pseudomonas* and *Zoogloea*. *Pseudomonas* spp. are able to use a wide diversity of carbon energy substrates, and this, together with their ability to compete effectively with other bacteria, is undoubtedly responsible for their dominance. *Zoogloea* spp., on the other hand, typically grow embedded in a gelatinous slime and are considered to have a primary func-tion in microbial floc and film formation, although it is now hypothesized that microbial flocs are held together by extracellular fibrils, rather than slime layers.

In addition to those discussed above, many other genera of microbes exist in biotreatment plants. Their specific roles have rarely been identified, but it is important to realize that under mixed culture conditions, primary chemoheterotrophic bacteria frequently become competitive because of correspondingly lower concentrations of associated microbes that modify the microen-vironment but do not use the primary carbonaceous substrates.

Three further groups of microbes warrant discussions because of the important role that they play with respect to biotreatment process performance. One group, the microbes responsible for sludge bulking, *i.e.* the tendency for activated sludge flocs to float, have a negative effect on acti-vated sludge process performance, whilst the two other groups, the denitrifying bacteria and the ciliated protozoa have distinctly positive effects on biotreatment process performance.

Essentially, activated sludge process equipment comprises an aeration tank and a sedimen-tation tank (or clarifier) for separation of the activated sludge from the treated effluent prior to subsequent sludge recycle. With bulking sludge, flotation rather than sedimentation occurs, so that sludge recycle, necessary to maintain process intensity, fails, and, additionally, the sus-pended solid concentration in the treated discharge from the process becomes excessive. Sludge bulking is generally attributed to inbalances between essential nutrients, particularly carbon and nitrogen. The microbes implicated in bulking are sheathed bacteria of the genus *Sphaerotilus*, and fungi of the genus *Geotrichum*.

In any biotreatment plant where nitrification occurs, some degree of denitrification will also occur. Denitrification involves the reduction of nitrate to dinitrogen and, in homogeneous environments, occurs under anoxic conditions. However, in biotreatment systems, heterogeneity is a characteristic, and because of oxygen concentration gradients in heterogenous microbial systems, denitrification occurs even under conditions where oxygen is present in the overall system. Particularly in situations where the hydrological cycle is truncated, denitrification is an important requirement in biotreatment processes. The chemoheterotrophic bacterial species responsible for denitrification in biotreatment processes are usually *Pseudomonas denitrifans*, *Paracoccus denitrificans* and stalked bacteria of the genus *Hyphomicrobium*. The attaching abilities of hyphomicrobia obviously provide them with a competitive advantage in both retained biofilm and recycled floc type systems.

It is clear that the various species of bacteria present are the most important microbes responsible for the biooxidation of biodegradable pollutants in biotreatment processes, even though they frequently function at only a small fraction of their maximum potential biooxidation rate. Hence, until a little more than a decade ago, other microbes present in biotreaters were either ignored or study of their roles was neglected. The most important group of microbes that falls into this category comprises the protozoa, which typically account for some 5% of the suspended solids in the aeration tank. Over 200 species of protozoa have been reported to occur in activated sludge plants, with ciliates being the most dominant both in terms of the number of species present and on the basis of concentration. The most common ciliates found are all Peritrichida, and comprise predominantly *Vorticella* spp., *Opercularia* spp. and *Epistylis* spp., which are all sessile types that are able to attach themselves to sludge flocs.

Protozoa were originally thought to be harmful to the effective functioning of activated sludge processes. However, today, not only is their presence recognized as markedly advantageous, they are also regarded as an indicator of sludge quality; a good quality sludge being one in which ciliated protozoa predominate. Biotreatment plants for the exclusive treatment of industrial wastewaters frequently experience problems with respect to maintaining appropriate populations of protozoa. The role of the protozoa in activated sludge processes involves their ability to feed on discretely dispersed bacteria, which, if not removed, represent both an unacceptable biochemical oxygen demand and an unacceptable suspended solids concentration in the treated discharge from the process. The ability of sessile ciliated species to adhere to bacterial flocs, but not to feed on them, ensures their survival in the process, because they are recycled in the sludge. Protozoa consume discretely dispersed bacteria in both the aeration stage and in the settling stage of activated sludge processes, and the bacteria that they remove probably represent some 10% of the total biomass produced in the process.

45.2.3 Temperature Classification

In addition to classification of microbes on a metabolic basis, they can also be classified on the basis of their optimum temperature for growth and metabolic activity. Temperature is a most important factor affecting both the physiological and biochemical potentials of microbes. The three main temperature ranges used in classification are (i) psychrophilic growth, where the optimum temperature is < 10 °C; (ii) mesophilic growth, where the optimum temperature is between 15 and 40 °C; and (iii) thermophilic growth, where the optimum temperature is > 45 °C. Those microorganisms that can only grow well at temperatures within one of the defined ranges are considered to be obligate with respect to the particular temperature range, whilst those that are able to grow well in an additional temperature range to that which includes their optimum growth temperature are described as facultative with respect to the additional range.

Thermophiles are generally divided into three groups: (i) thermotolerant microbes that grow in the temperature range 40 to 50 °C, a group that includes most facultative thermophiles; (ii) moderately thermophilic microbes that have an optimum temperature range for growth between 50 and 65 °C and a minimum temperature for growth of *ca.* 40 °C; and (iii) extremely thermophilic or caldoactive microbes that have growth optima > 65 °C. The taxonomic diversity of both moderately and extremely thermophilic microbes is restricted when compared with the taxonomic diversity of mesophilic microbes, but it must be stressed that thermophiles are optimally adapted for growth at elevated temperatures rather than struggling to survive at such temperatures. Even so, the extreme thermophiles seem to be relatively fastidious with respect to their growth requirements, and, hence, are probably poorly suited for application in technological processes.

As far as microbial processes are concerned, the effect of temperature on overall reaction rates has received inadequate attention with respect to both process research and process operation. For chemical reactions, an increase in temperature usually enhances the reaction rate and, by analogy, it is generally assumed that a similar effect occurs in the case of microbially mediated reactions, but only within any one of the temperature ranges appropriate for psychrophilic, mesophilic or thermophilic growth. As far as microbes are concerned, they exhibit maximum critical temperatures above which growth ceases within two or three degrees.

Most biological wastewater and sludge treatment processes are operated at mesophilic temperatures, the former processes at temperatures usually close to the minimum and the latter processes at temperatures usually close to the maximum of the mesophilic range.

Unlike many chemical reactions that occur under essentially homogeneous conditions, all microbially mediated processes occur under heterogeneous conditions. The heterogeneous nature of microbial systems will be of considerable significance in any analysis of the effects of temperature on such systems. Further, microbially mediated processes involve a complex sequence of biochemical reactions, none of which may be rate limiting as far as the overall process is concerned. The rate limiting step in most microbial systems is usually a resistance to mass transfer.

As far as microbial systems are concerned, the simplest system that can be devised for analysis is one where a pure monoculture grows, in accordance with Monod type kinetics, as discretely dispersed cells in a defined liquid growth medium, in a completely mixed continuous flow reactor where no wall effects occur, on a single carbon energy substrate that in turn limits growth. For such a system, it is possible to hypothesize concerning the several potential effects of temperature. Clearly identifiable effects include (i) effects on the maximum specific growth rate constant of the cultures; (ii) effect on the affinity of the microorganisms for the growth-limiting substrate; (iii) effect on endogenous metabolism and maintenance requirements; (iv) effect on the specific death and lysis rate constants of the culture. Experimental evidence suggests that, within a specific temperature range, a temperature increase will enhance the maximum specific growth rate constant, the rate of endogenous metabolism and the affinity of the microorganisms for the growth-limiting substrate, but that effects on the specific death and lysis rate constants occur primarily at temperatures above the critical value for growth, *i.e.* outside the optimum range for growth. The overall effect of increasing temperature is, within limits, to increase the maximum potential growth rate of an ideal culture, but with a concomitant decrease in the yield coefficient for the conversion of the limiting carbon energy substrate into microbial biomass.

In process systems, particularly heterogeneous process systems, the effects of temperature frequently stem from changes in physical properties that result from the changes in temperature. Physical properties that are primarily affected in this way in microbiological process systems include gas solubility, liquid phase viscosity and liquid phase diffusivities. The solubility of oxygen is of key importance in aerobic microbial processes, as in such systems oxygen transfer frequently becomes the rate limiting step in the process. For example, in an ideal continuous flow aerobic microbial process operating under carbon energy substrate limitation where the maximum possible rate of supply of oxygen just exceeds the actual rate of demand, an increase in temperature could result in the process becoming oxygen rather than carbon energy substrate limited because of a solubilty dependent reduction in the oxygen transfer rate. In addition, an enhanced oxygen demand resulting from a reduction in the microbial biomass yield coefficient at enhanced temperature will also adversely affect the maintenance of carbon energy substrate limitation. If a similar change and effect were to occur in a wastewater treatment process, biodegradable carbonaceous pollutants would no longer be effectively removed by microbial oxidation from the water undergoing treatment.

Unlike most wastewater treatment systems, sludge treatment systems have liquid phases where viscosity restricts effective mixing in the system such that significant segregation occurs. In single phase liquid systems, any increase in temperature will result in a decrease in viscosity and it is to be expected that similar effects will occur in multiphase systems. In these latter systems, viscosity reduction should enhance the effectiveness of mixing, thereby reducing transfer resistance between phases.

Liquid phase diffusivities will also be enhanced by increases in temperature and in multiphase microbiological process systems; such enhancement can be expected to affect overall process rates favourably by the reduction of the rate limiting resistance for interphase transfer of the overall reaction limiting nutrient.

45.3 MICROBIAL GROWTH THEORY FOR BIOTREATMENT

As has been discussed previously, the dominant microbes that are responsible for the oxidation of biodegradable carbonaceous pollutants in biotreaters are chemoheterotrophic bacteria. Such bacteria reproduce by binary fission, and, hence, it should be expected that the mathematical expressions that describe the growth of individual bacterial species and substrate and nutrient uptake during growth are applicable to the bacterial growth processes that occur in biotreaters, provided limitations with respect to the use of such equations are clearly recognized.

When the necessary requirements for bacterial growth are satisfied, the growth rate of a bacterial culture can be expressed as:

$$dx/dt = \mu x \tag{1}$$

where x is the biomass concentration on a dry weight basis, t is time, and μ is a constant usually described as the specific growth rate constant. The quantitative assessment of growth relative to either substrate or nutrient utilization can be expressed, for substrate utilization, as:

$$Y_{x/s} = - dx/ds \tag{2}$$

where $Y_{x/s}$ is the biomass yield coefficient and s the substrate concentration. For conditions where the yield coefficient remains constant:

$$(x - x_0) = Y_{x/s} (s_0 - s) \tag{3}$$

where subscript 0 represents initial conditions, and no subscript, final conditions. The only other basic relationship needed to describe bacterial growth is the relationship between the specific growth rate constant and the growth limiting substrate concentration. The most widely used relationship between μ and s is that attributed to Monod:

$$\mu = \frac{\mu_m s}{(s + K_s)} \tag{4}$$

where subscript m represents a maximum value and K_s is a saturation constant equal to the value of s when μ is $\mu_m/2$.

Considerable controversy exists concerning the applicability of the Monod relationship to bacterial growth in biotreaters, but most alternative relationships that have been proposed either assume that growth in biotreaters occurs only at very low limiting substrate concentrations and assumes a first-order relationship that virtually coincides with the initial conditions of the μ–s relationship attributed to Monod or seek to allow for diffusional resistances in flocs and films, which should be considered separately.

Equations (1), (3) and (4) can be used to describe the bacterial growth processes in any bioreactor, but the overall performance of the bioreactor depends on the reactor type and mode of operation. As far as aerobic biotreatment is concerned, two principal practical process types exist: activated sludge type processes that incorporate biomass recycle, and trickling filters that incorporate biomass retention to maintain process intensity. Both the recycle and the retention of bacteria in biotreaters results in a significant fraction of the bacteria in the process being non-viable (Jones, 1973), and hence unable to grow. Therefore, for the analysis of bacterial growth in biotreaters, it is essential to consider only the viable biomass present, rather than the total biomass present. Activated sludge processes are continuous flow processes where the bacteria are present as discretely dispersed flocs, and the liquid mixing can range from the completely mixed situation to an approximation of plug-flow, whilst trickling filters are continuous flow systems where the bacteria are present only as attached films and where the liquid is in plug-flow. Equations that describe growth processes in bioreactors must obviously take into account the liquid flow characteristics of the bioreactor and diffusional resistances introduced by the physical structure of the bacterial biomass present.

For an ideal plug-flow reactor the liquid residence time is given by:

$$\tau = V/F = 1/D \tag{5}$$

where τ is the residence time, V the liquid volume of the reactor, F the liquid flow rate and D the dilution rate, a term used in preference to residence time in the microbiological literature. For a completely mixed reactor, where τ is the average residence time, its relationship with V and F is the same as that expressed in equation (5). In a completely mixed dispersed growth bioreactor, the concentration of biomass in the bioreactor outlet will be identical with the concentration of biomass in the bioreactor and the biomass wash-out expression will be:

$$dx/dt = - Dx \tag{6}$$

The growth of a bacterial floc under diffusion limited conditions is frequently described by the cube-root law for growth (Pirt, 1975). However, such an approach fails to account for floc break-up by mechanical shear and resultant floc size distributions. Unfortunately, bacterial floc disruption processes have been inadequately studied and remarkably few data concerning floc size distributions are available so that in order to account for diffusional resistance it becomes necessary to adopt an approach that disregards floc size distribution and structure. One such approach that does not conflict with established bacterial growth theory was proposed by Characklis (1978), and takes into account diffusional resistances in bacterial flocs by considering the limiting substrate flux due to diffusion on the basis of unidimensional transfer, followed by application of the Monod relationship for substrate utilization by the bacteria. The resultant expression for the substrate utilization rate is:

$$r = \frac{\mu_{m}xs/Y_{x/s}a_{f}}{(K_{s} + \mu_{m}x\delta/Y_{x/s}a_{f}\mathcal{D}_{s}) + s} \tag{7}$$

where r is the substrate utilization rate, a_{f} the floc surface area per unit reactor liquid volume, δ the boundary layer thickness and $\mathcal{D}_{s}$ the molecular diffusivity of the substrate in the floc.

45.3.1 Activated Sludge Processes as Bacterial Growth Processes

Obviously, space does not permit the development of bacterial growth theory for each of the several variations of both activated sludge and trickling filter processes and only the completely mixed variant of the activated sludge process will be considered here. A flow diagram for such a process is shown in Figure 1. In Figure 1 F is the liquid flow into the process, s_0 the limiting substrate concentration in the inflow, $\bar{s}$ the steady state limiting substrate concentration in the bioreactor, $\bar{x}$ the steady state bacterial concentration in the bioreactor, F_s the liquid outflow from the bioreactor, a the fraction of F_s recycled with the concentrated biomass, g the concentration factor for biomass in the clarifier, c the fraction of the inlet flow to the process that leaves together with the concentrated waste biomass, and h the concentration of biomass in the clarified discharge from the clarifer. Assuming no accumulation in the process:

$$F_s = F / (1-a) \tag{8}$$

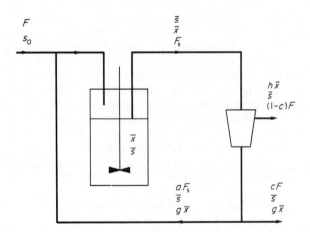

Figure 1 Flow diagram of a completely mixed bioreactor with bacterial biomass (sludge) recycle

In the case of discretely dispersed bacterial growth, Herbert (1961) showed that for steady state conditions in a completely mixed system, by establishing material balance equations for the bioreactor for both the bacterial biomass and the limiting substrate and introducing a steady state form of the Monod relationship, where $\bar{s}$ is substituted for s in equation (4):

$$\bar{s} = \frac{DK_s\,(1-ag)}{\mu_m(1-a)-D(1-ag)} \tag{9}$$

and

$$\bar{x} = \frac{Y_{x/s}\,(s_0-\bar{s})\,(1-a)}{(1-ag)} \tag{10}$$

However, for flocculated growth, as typically occurs in activated sludge processes, Hamer (1983) introduced equation (7) into the original analysis of Herbert (1961), so that:

$$\bar{s} = \frac{D[K_s(ag-1) + \mu_m\delta s_0\,(a-1)/a_f\mathscr{D}]}{\mu_m(a-1)-D(ag-1) + D\mu_m\delta(a-1)/a_f\mathscr{D}} \tag{11}$$

but $\bar{x}$ is still given by equation (10), although the value of $\bar{s}$ to be substituted is that given by equation (11). For the selected values of the constants listed, the discretely dispersed and the flocculant bacterial growth situations are compared in Figure 2.

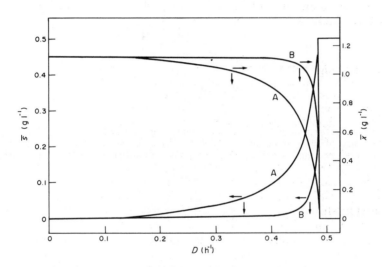

Figure 2 Effect of flocculent growth on the steady state limiting substrate and bacterial biomass concentrations in the bioreactor with respect to dilution rate with bacterial biomass recycle. $\mu_m = 0.108\ \text{h}^{-1}$, $K_y = 0.002\ \text{g l}^{-1}$, $s_0 = 0.5\ \text{g l}^{-1}$, $Y_{x/s} = 0.5$, $\delta = 0.05\ \text{cm}$, $\mathscr{D} = 2 \times 10^{-5}\ \text{cm}^2\ \text{s}^{-1}$, $a_f = 7.5\ \text{cm}^{-1}$, $a = 0.1$, $g = 8$, A = flocculent growth, and B = non-flocculent growth

Clearly, for lower values of D, which would be typical of activated process operating conditions, the difference between the two situations examined is small, but of course no account was taken of the viability of the bacteria in the bioreactor. The impact of biomass recycle on process intensity is clearly evident. To account for partial biomass viability, equations (9) and (11) can be rewritten to incorporate this concept, so that for discretely dispersed growth and biomass recycle:

$$\bar{s} = \frac{DK_s\,(1-ag)}{\mu_m\beta(1-a)-D(1-ag)} \tag{12}$$

and for flocculent growth and biomass recycle:

$$\bar{s} = \frac{D[K_s(ag-1) + \mu_m\delta s_0(a-1)/a_f\mathscr{D}]}{\mu_m\beta(a-1) - D(ag-1) + D\mu_m\,\delta(a-1)/a_f\mathscr{D}} \tag{13}$$

where β is the fractional viability. For both situations the total (viable + non-viable) steady state bacterial biomass concentration, $\bar{x}_t$, is given by:

$$\bar{x}_t = \frac{Y_{x/s}(s_0-\bar{s})\,(1-a)}{(1-ag)} \tag{14}$$

with appropriate substitution for $\bar{s}$.

Assuming the same values for the constants used for Figure 2 and a hypothetical fractional viability, β of 0.1, Figure 3 shows significant deviation in bioreactor performance between the discretely dispersed and the flocculent growth situations at hydraulic dilution rates typical of activated sludge process operation, thereby supporting the argument in favour of using conventional bacterial growth theory as the basis for description of activated sludge process growth kinetics.

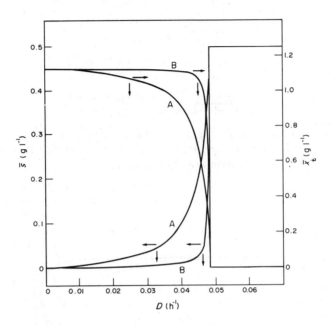

Figure 3 Effect of flocculent growth on the steady state limiting substrate and total bacterial biomass concentrations in the bioreactor with respect to dilution rate with bacterial biomass recycle and a significant level of non-viable cells. $\beta =$ 0.1, other constants as in Figure 2. A = flocculent growth, B = non-flocculent growth

45.3.2 Mixed Bacterial/Mixed Substrate Systems

Activated sludge aeration tanks are bioreactors in which the bacterial population has been subjected to very considerable selection pressures under conditions of bacterial biomass recycle and multiple carbon energy substrate availability. Laboratory continuous culture studies are only rarely concerned with systems employing either biomass recycle or biomass retention. In non-recycle systems, both competition between several bacterial strains for single limiting carbon energy substrates and the utilization of single limiting carbon energy substrates by several bacterial strains have been investigated. However, it seems probable that during operation of activated sludge process aeration tanks the selection pressures occurring will minimize interactions between bacterial strains in spite of continuous reinoculation from the feed and the environment. Competition for each particular carbon energy substrate is unlikely to be significant and stable bacterial associations that are best suited, under the imposed operating conditions, to utilize each individual carbon energy substrate will probably predominate. In fact, a state of neutralism, after a period of competition directly after start-up, can be predicted. The growth rate of stable bacterial associations will depend on the growth rates of the individual component strains. According to Harrison (1978), this can be expressed as:

$$dx/dt = dx_1/dt = dx_2/dt \ldots dx_n/dt = \mu_1 x_2 + \mu_2 x_2 \ldots \mu_n x_n \tag{15}$$

where subscripts $1, 2, \ldots n$ refer to bacterial components of the mixed culture.

For a continuous flow non-interacting multiple bacterial system using separate carbon energy substrates, the change in bacterial concentration can be expressed as:

$$dx/dt = (\mu_1 - D)x_1 + (\mu_2 - D)x_2 \ldots \ldots + (\mu_m - D)x_n \tag{16}$$

The corresponding equation for the utilization of the carbon energy substrates is:

$$\frac{ds}{dt} = D(s_{01} - s_1) - \frac{\mu_1 x_1}{Y_{x/s1}} + D(s_{02} - s_2) - \frac{\mu_2 x_2}{Y_{x/s2}} \ldots \ldots + D(s_{0n} - s_n) - \frac{\mu_n x_n}{Y_{x/sn}} \tag{17}$$

For steady state condition where dx/dt and ds/dt are zero and where the critical dilution rate is not exceeded:

$$D = \mu_1 = \mu_2 = \ldots \ldots \ldots = \mu_n \qquad (18)$$

For steady state operation at dilution rates above the critical value for wash-out of any of the bacterial components, the microbial component in question will be washed-out of the system and the carbon energy substrate and/or the nutrients that it utilized will accumulate. Bacterial biomass recycle eliminates such operating problems and, depending on the recycle ratio employed, the critical dilution rate can be extended so as to allow even very slow growing bacteria to be maintained in the system.

45.3.3 Cooxidation and Cometabolism

In biotreaters, phenomena described variously as cooxidation, cometabolism and fortuitous oxidation are frequently encountered, particularly in systems treating those wastewaters with significant concentrations of synthetic organic chemicals. Essentially, the three terms, which have become virtually synonymous, are all used to describe the oxidation and degradation of non-growth substrates by microbes. Cooxidation was originally defined by Foster (1962) as the phenomenon whereby actively growing microbes oxidize a compound, but do not utilize either carbon or energy derived from the oxidation. Cometabolism was originally defined by Jensen (1963) in order to extend the definition of cooxidation to include reactions such as dehalogenations and excluded an obligate requirement for the simultaneous presence of a growth substrate. However, both definitions have been subjected to criticism on the grounds that they describe metabolic phenomena which can readily be encompassed by existing metabolic nomenclature.

Some microbial oxidation processes that satisfy the definition of cometabolism are merely a reflection of the non-specific nature of monooxygenases, and Stirling and Dalton (1979) have proposed that the transformation of non-growth substrates in the absence of a cosubstrate should simply be referred to as fortuitous activity. In addition, it was also proposed that the term non-growth substrate be used to describe compounds that do not support cellular division, as opposed to an increase in cell mass, as it is probable that some compounds can be assimilated into cellular biosynthetic pathways but are unable to support cellular division. Hence, cometabolism is now defined as the biological transformation of a compound which is unable to support cell replication in the requisite presence of another transformable cosubstrate, thereby clearly differentiating it from fortuitous metabolic events, that results from non-specific monooxygenase activity. It also seems possible that in some cases of cometabolism energy derived from the oxidation of the non-growth substrate can be utilized to fix carbon from the growth substrate.

The capacity for cometabolism and fortuitous activity seems to occur most frequently in hydrocarbon-utilizing bacteria, where both anthropogenic and xenobiotic pollutants are transformed by such mechanisms.

In sewage and wastewater treatment processes, biotransformation of pollutants in the absence of bacterial growth also occurs as a result of the activities of endoenzymes present in dead (non-viable) bacteria as discussed by Jones (1975) and as a result of exoenzymes excreted by viable bacteria, but the magnitude of such effects has yet to be defined.

45.3.4 Endogenous Metabolism and Maintenance

It is well known that when microbial cells are removed from a growing culture and resuspended in fresh growth medium containing no carbon energy source, the cells will for a considerable time take up oxygen and evolve carbon dioxide, consuming their own cell substance in doing so. Thus, the cell biomass decreases so as to give a negative mass growth rate. This activity is described as endogenous metabolism and it has been postulated that it occurs at a constant rate in growing cultures irrespective of the specific growth rate of the culture. Endogenous metabolism occurs as a result of a maintenance energy requirement by the cells. Specific maintenance functions include turn-over of cell materials, osmotic work to maintain concentration gradients between the cell and its exterior, and cell motility.

The effect of endogenous metabolism may be expressed by introducing a modification into the basic equation describing microbial growth, so that:

$$dx = \mu_r x dt - \mu_c x dt \qquad (19)$$

where μ_r is the real specific growth rate constant and μ_c the endogenous rate constant. The observed specific growth rate constant, μ, is given by:

$$\mu = \mu_r - \mu_c \tag{20}$$

When $s = 0$, no real growth occurs, so:

$$\mu = -\mu_c \tag{21}$$

In a steady state continuous culture, the observed growth rate constant, μ, is equal to the dilution rate, D. If $Y'_{x/s}$ is the yield coefficient in the absence of endogenous metabolism, the observed yield coefficient, $Y_{x/s}$, decreases with dilution rate:

$$Y_{x/s} = \frac{Y'_{x/s}D}{(D-\mu_c)} \tag{22}$$

The maintenance coefficient, m, is defined by:

$$m = \mu_c/Y'_{x/s} \tag{23}$$

45.3.5 Death, Lysis and Cryptic Growth

Microorganisms which become incapable of growth under conditions that are normally suitable for growth must be either dead or dormant. The essential difference between the two states is that dormant cells can subsequently regenerate the normal growing (vegetative) state, whereas dead cells cannot. Generally the production of dormant forms is an ordered process with a duration longer than the minimum doubling time of the vegetative form. Bacteria can die as a result of: (i) an adverse physical factor such as high temperature; (ii) the effect of a toxic chemical; (iii) starvation; (iv) an error in autosynthesis: (v) attack by bacteriophages. In a growing culture, neither dormant nor dead cells contribute to the growth of the population and thus behave similarly.

If it is assumed that the death rate is proportional to the total weight of viable biomass present in the culture, x_v, then:

$$dx_v/dt = -kx_v \tag{24}$$

where k is the specific death rate constant. The rate of growth of the viable biomass will be given by:

$$dx_v/dt = (\mu - k)x_v \tag{25}$$

where μ is the specific growth rate constant. The rate of increase in the total biomass, x_t, assuming no lysis, is given by:

$$dx_t/dt = \mu x_v \tag{26}$$

In a chemostat in which cell death occurs, the viable biomass balance, under steady state conditions, is given by:

$$\mu x_v - kx_v - Dx_v = 0 \tag{27}$$

For steady state conditions, the balance for total biomass, x_t, is given by:

$$\mu x_v - Dx_t = 0 \tag{28}$$

In many cases cell death is accompanied by cell lysis, a process in which the cell disintegrates and the contents of the cell enter the growth medium. Many of the components of which the cell consisted are frequently utilized for further cell growth. Growth on such lysis products is described as cryptic growth. Cell lysis can be accounted for in the biomass balance equation for a single-stage chemostat, operating under steady state conditions, by incorporation of a specific lysis rate term in the biomass balance equation:

$$\mu x - k_1 x - Dx = 0 \tag{29}$$

where k_1 is the specific lysis rate constant.

The corresponding material balance equation for lytic product is:

$$k_1 x Y_p - Dp = 0 \qquad (30)$$

where p is the lytic product concentration and Y_p the weight of lytic product produced per unit weight of cells lysed.

Cryptic growth is much more significant in mixed cultures than in pure cultures. In biotreaters, cryptic growth and protozoal predation are both important mechanisms for the reduction of the overall biomass yield coefficient and optimization for the conversion of carbonaceous pollutants into carbon dioxide.

45.3.6 Growth Inhibition

The growth of microbes is frequently affected by the presence of inhibitory compounds in the bioreactor. The presence of inhibitory compounds can be either intentional or unintentional. It is the latter that is of particular interest in the context of biotreatment processes. The unintentional presence of inhibitors in microbial growth systems arises from the presence of inhibitory components in the bioreactor feed, the production of either inhibitory products or intermediates, and the production of inhibitory compounds as a result of cell lysis.

The theory of microbial response to growth inhibitors is based on the kinetics of enzyme inhibition. Essentially two main types of inhibition occur: (i) competitive inhibition, where the inhibitor competes with the growth-limiting substrate for uptake by the microorganism, thus affecting the affinity for the substrate; and (ii) non-competitive inhibition, where the inhibitor is assumed to react with the microorganism at some site other than that for the uptake of the growth-limiting substrate, and does not affect the affinity for the substrate.

For competitive inhibition, the Monod relationship is modified so that:

$$\mu = \frac{\mu_m s}{(s + \alpha K_s)} \qquad (31)$$

where

$$\alpha = \frac{(i + K_i)}{K_i} \qquad (32)$$

and where i is the inhibitor concentration, K_i the inhibition constant and $\alpha > 1$.

For non-competitive inhibition, the Monod relationship is modified so that:

$$\mu = \frac{\mu_m s}{\alpha(s + K_s)} \qquad (33)$$

For a single-stage continuous flow reactor without recycle, in which the specific growth rate of the microorganism is dependent on both the limiting substrate concentration and the concentration of a non-competitive inhibitory product, the expression for the specific growth rate constant will be:

$$\mu = \frac{\mu_m s}{(s + K_s)} \frac{K_p}{(p + K_p)} \qquad (34)$$

where p is the inhibitory product concentration and K_p the product inhibition constant.

45.4 CONCLUDING REMARKS

Microorganisms are a universally available natural resource. However, they can only be effectively harnessed for use in technological processes by those with both perception and the appropriate skills to exploit their physiological and biochemical potential.

The development of microbiology for much of the present century has been characterized by the requirement that microorganisms should be studied and utilized only as pure monocultures and preferably grown only on single carbon energy substrates. This approach presently manifests itself in the significant efforts directed towards the construction, using recombinant DNA techniques, of bacteria with specific metabolic characteristics that allow them to degrade the numerous xenobiotic and recalcitrant compounds that continue to be manufactured. The potential survival characteristics of such 'engineered' bacteria are unlikely to permit their effective use in biotreatment processes. However, relatively stable associations of microbes which exhibit the range of characteristics that are presently being sought from novel 'engineered' strains occur and function effectively in biotreatment processes.

Some 13 years have elapsed since Pirt (1972) made a plea for increased effort on microbiological research oriented towards biotreatment processes. In the meantime, the so-called biotechnological revolution has occurred, but, in spite of this, few major microbiological innovations in biological wastewater treatment technology have occurred. This seems unacceptable considering the size and the importance of the wastewater treatment industry.

45.5　REFERENCES

Ardern, E. and W. T. Lockett (1914). Experiments on the oxidation of sewage without the aid of filters. *J. Soc. Chem. Ind.*, **33** (10), 1–19.

Characklis, W. G. (1978). Microbial reaction rate expressions. *J. Environ. Eng. Div. (Am. Soc. Civ. Eng.)*, **104**, 531–534.

Downing, A. L. (1972). Chemical engineering and the hydrological cycle. *Chem. Eng. (London)*, **260**, 150–157.

Foster, J. W. (1962). Hydrocarbons as substrates for microorganisms. *Antonie van Leeuwenhoek J. Microbiol. Serol.*, **28**, 241–274.

Gaudy, A. F. and E. T. Gaudy, (1980). *Microbiology for Environmental Scientists and Engineers*. McGraw-Hill, New York.

Hamer, G. (1983). The application of continuous culture theory to activated sludge processes. *Chem.-Ing.-Tech.*, **55**, 478–479.

Harrison, D. E. F. (1978). Mixed cultures in industrial fermentation processes. *Adv. Appl. Microbiol.*, **24**, 129–164.

Herbert, D. (1961). A theoretical analysis of continuous culture systems. *S. C. I. Monogr.*, No. 12, 21–53.

Jensen, R. A. (1963). Carbon nutrition of some microorganisms decomposing halogen-substituted aliphatic acids. *Acta Agric. Scand.*, **13**, 404–412.

Jones, G. L. (1973). Bacterial growth kinetics: measurement and significance in the activated sludge process. *Water Res.*, **7**, 1475–1492.

Jones, G. L. (1975). A consolidated approach to activated sludge process design: a microbiological view. *Prog. Water Technol.*, **7**, 199–207.

Kandler, O. (1982). *Archaebacteria*. Gustav Fischer, Stuttgart.

Pirt, S. J. (1972). The microbiological means for the improvement of effluent purification. In *Association of River Authorities Year Book*, pp. 119–123. Association of River Authorities, London.

Pirt, S. J. (1975). *Principles of Microbe and Cell Cultivation*. Blackwell, Oxford.

Reeve, D. A. D. (1973). Secondary treatment. In *Advances in Sewage Treatment*, pp. 27–33. Institute of Civil Engineers, London.

Stirling, D. I. and H. Dalton (1979). The fortuitous oxidation and cometabolism of various carbon compounds by whole-cell suspensions of *Methylococcus capsulatus* (Bath). *FEMS Microbiol. Lett.*, **5**, 315–318.

Whittenbury, R. and D. P. Kelly (1977). Autotrophy: a conceptional phoenix. In *Microbial Energetics*, ed. B. A. Haddock and W. A. Hamilton, pp. 121–149. Cambridge University Press, Cambridge.

Woese, C. R. (1982). Archaebacteria and cellular origins: an overview. *Zentralbl. Bakteriol. Parasitenkd., Infektionskr. Hyg., Abt. 1 Orig. C*, **3**, 1–17

Zehnder, A. J. B., K. Ingvorsen and T. Marti (1982). Microbiology of methane and bacteria. In *Anaerobic Digestion 1981*, ed. D. E. Hughes, pp. 45–68. Elsevier, Amsterdam.

46

Biodegradation of Celluloses and Lignins

J. S. KNAPP
University of Leeds, UK

46.1 INTRODUCTION

Cellulose and lignin are respectively the first and second most abundant single organic compounds in the biosphere (excluding fossil fuels). Both compounds are polymers and are produced by all higher plants. Cellulose is also made by lower plants, some fungi and a few bacteria. Clearly the production of both these compounds is immense and difficult to estimate, but it is likely that about 10^{11} tons of cellulose are synthesized annually (Ratledge, 1977).

Cellulose and lignin make up the highest proportion of dead plant remains which are principally degraded by microorganisms. This degradation is ecologically of the utmost importance since if these materials were not largely broken down there might be a decrease in atmospheric carbon dioxide concentration. Partially degraded cellulose and lignin accumulate in soils as humus and under some conditions give rise to peat, which may eventually be converted to lignite or coal. Human exploitation of plant tissues gives rise to many waste products which have a high cellulose content, *e.g.* waste paper, timber and logging wastes (wood shavings, saw dust, bark), crop residues (notably cereal straws), domestic refuse, sewage solids and farm animal excreta. Humphrey (1975) has estimated that the dry weight of solid organic waste produced annually in the USA is 940×10^6 metric tons; much of it is cellulosic. This waste material poses severe disposal problems to most developed countries and its potential as a valuable raw material for biotechnology is largely unfulfilled.

46.2 STRUCTURE OF CELLULOSE AND PLANT CELL WALLS

Cellulose is a β-1,4-linked linear homopolymer of anhydro-D-glucose, which is generally of very high, but indeterminate, molecular weight. The degree of polymerization (DP) of individual chains is usually in the region of about 1500 to 15 000, giving a molecular weight of *ca.* 2.5×10^5 to 2.5×10^6 and a length of probably up to 7 μm. Usually cellulose molecules are not found individually but are linked together to form microfibrils. The separate molecules are linked by H-bonding into a highly ordered crystalline structure. Some parts of the microfibrils have a less ordered, non-crystalline structure and are referred to as amorphous regions.

Its high molecular weight and ordered tertiary structure make natural cellulose completely insoluble in water. The hydrogen-bonded structure can be disrupted by alkalis and mineral acids which results in swelling of the fibres. Some hot, concentrated mineral acids (notably sulfuric acid) can hydrolyse cellulose to glucose and it can also be hydrolysed by enzymes.

Cellulose is typically found in the walls of plant cells which have secondary thickening. These cell walls also contain pectin, lignin and hemicellulose, a blanket term for plant cell wall polysaccharides other than pectin and cellulose. They are generally much smaller than cellulose and are alkali soluble — examples are xylans, mannans and arabinans.

Cell walls have a primary and a secondary region (itself divided into three layers S_1, S_2 and S_3), and are separated from each other by a middle lamella. It is thought that the cell wall is made up of cellulose fibrils set in a matrix of hemicellulose and encrusted with lignin. The layers of cell walls differ in structure and content: the middle lamella and primary wall have a high lignin and pectin content and relatively low cellulose (20–35%) whereas the thicker secondary cell wall is high in cellulose (35–60%) and hemicellulose and lower in lignin. The primary and secondary wall layers also differ in orientation of the cellulose fibrils. These differences ensure that the various regions of the cell wall differ considerably in terms of their degradation by enzymes. Generally the primary wall and middle lamella are most resistant to cellulases and the S_2 layer is most susceptible. There are also differences of structure between the walls of different types of cells and these too cause differences in degradability.

On average both hard and soft woods contain about 43% cellulose, the rest being made up largely of pectin (1–4%), hemicelluloses (30–40%) and lignins (20–30%). Some special plant fibres may contain much more cellulose, *e.g.* flax or hemp *ca.* 70% and cotton 90%.

Further details of the structure of plant cell walls are given by Norkrans (1967), Jurášek *et al.* (1967) and Jane (1970).

46.3 BIODEGRADATION OF CELLULOSE

46.3.1 General Ecology

Due to the predominance of cellulose in the biosphere its biodegradation is quantitatively important in many natural habitats, *e.g.* soil, leaf litter, fallen wood, water courses and aquatic sediments and also man-made or semi-natural habitats such as compost heaps and anaerobic digesters. Herbivores consume very large quantities of cellulose and its digestion in the gut of such animals is of great importance both to the animals and to man who exploits them. Cellulolytic organisms also cause extensive biodeterioration of materials such as wood, paper and cloths.

Cellulose is susceptible to biodegradation by means of hydrolytic enzymes called cellulases, which convert it to glucose and oligomers thereof. A wide variety of organisms are known to be able to produce cellulases. However, it is generally considered that mammals cannot produce cellulase and there is little information concerning other animals. With most organisms cellulase, if produced, is a major catabolic enzyme utilized to release simple sugars as a growth substrate. The saprophytic microorganisms (including representatives of the fungi, bacteria and protozoa) are pre-eminent as cellulose degraders. A very large number of species and genera have been reported as degrading cellulose, but most of these are limited to particular habitats. Within each major group of microbes cellulolytic activity is shown by many species of different types. Among the cellulolytic bacteria there are Gram-positive and -negative species, aerobes and anaerobes, unicellular and mycelial, immotile, flagellated and gliding bacteria. In the fungi cellulolytic activity is shown by representatives of the phycomycetes, ascomycetes, basidiomycetes, fungi imperfecti and at least one yeast. The cellulolytic ability of protozoa was in doubt for many years due to the difficulty in isolating axenic, bacteria-free cultures. Such cultures have now been isolated and demonstrated to produce carboxymethylcellulase (Yamin and Trager, 1979). It is, however, still possible that *in vivo* some protozoa live in symbiosis with cellulolytic bacteria, relying on bacterial cellulase.

Cellulose is generally degraded extracellularly (except by protozoa) to soluble products which can diffuse to the cellulase-producing cells. These products can of course be taken up by other microbes which are thus dependent on the cellulose degrader; associations of cellulolytic and non-cellulolytic microbes are common in nature. Cellulose is often degraded more rapidly in mixed cultures due to the removal of inhibitory end-products (like cellobiose) and possibly to crossfeeding. In anaerobic habitats satellite microbes (*e.g.* methanogens) may benefit the cellulose degrader by removing inhibitory fermentation products like hydrogen (Chung, 1976). Cellulolytic microbes often grow in close association with their substrate, that is on it or in it. This physical association may be essential if cellulases are cell associated, but will always be of some advantage as it will tend to shorten the distance the product must diffuse.

46.3.2 Degradation of Cellulose in Aerobic Soil and Water

Large amounts of cellulose are found in these and related habitats and a wide variety of microbes are involved in its degradation. The main genera of bacteria are *Cellulomonas*, the non-fruiting gliding bacteria (*Cytophaga* and *Sporocytophaga*), the myxobacteria (*e.g. Angiococcus* and *Polyangium*), some strains of *Pseudomonas* (*Cellvibrio* is now included in this genus) and *Bacillus* and many of the actinomycetes (*e.g. Streptomyces, Micromonospora* and *Thermoactinomyces*). There are a large number of cellulolytic fungi active in soil and leaf litter, among the prominent genera being *Trichoderma, Chaetomium, Fusarium, Aspergillus, Penicillium, Rhizoctonia* and *Verticillium*.

Cellulose decomposition in soil is discussed by Alexander (1977). In aerobic effluent treatment plants bacteria are dominant, notably *Pseudomonas* (*Cellvibrio*) and non-fruiting gliding bacteria (Edberg and Hofsten, 1975). The latter group and the myxobacteria seem to be very important in fresh water habitats and brackish reed swamps respectively (Mullings and Parish, 1981). The fungi are more important in the ordinary soils, leaf litter and fairly dry crop residues. They are also important in cereal straw degradation, as are actinomycetes. Both these groups are involved in aerobic composting of vegetable wastes. In this process there is usually a considerable increase in temperature, up to 60 or 70 °C, which gives rise to successional changes in the microflora, selecting for thermophiles such as *Thermoactinomyces* and the fungi *Chaetomium thermophile* and *Sporotrichum thermophile* (Lacey, 1979).

46.3.3 Biodegradation of Wood

This process is of such economic importance that it has been intensively studied so it is surprising to note that few of the major organisms have been seriously studied with a view to their productive use.

Under normal conditions the main agents of wood decay are fungi, bacteria only being significant in waterlogged timber. Wood-rotting fungi are traditionally split up into five different groups according to their mode of action and effects on the wood structure (Norkrans, 1967; Jurášek *et al.*, 1967; Liese, 1970; Wilcox, 1968).

Blue stain fungi are ascomycetes and fungi imperfecti. They cause little polymer breakdown or physical damage but do grow inside wood cells and cause discolouration.

Members of the ascomycetes and fungi imperfecti are also responsible for soft-rots which attack both hard and soft woods and often occur in situations of high moisture (*e.g.* cooling towers). These fungi often grow within the S_2 layer of the cell wall causing cavity formation. They degrade polysaccharides but generally cause only slight modification to lignin.

The brown-rot fungi are basidiomycetes which preferentially attack soft woods. They have little effect on lignin but extensively degrade hemicellulose and cellulose. Hyphae grow inside the cell lumen and secrete enzymes which usually diffuse through the S_3 layer dissolving the S_2 and S_1 layers before the more lignified S_3, and without affecting the highly lignified compound middle lamella. Some of the best known brown-rotters are the much feared *Merulius lacrymans* (dry rot) and *Coniophora cerabella* (wet rot).

White-rot fungi are also usually basidiomycetes which preferentially degrade lignin and hemicelluloses, only degrading cellulose after prior delignification. The hyphae again grow in the cell lumen and remove the layers of the cell wall in order. It appears that white-rot cellulase cannot attack cellulose without delignification and cannot diffuse through lignified areas. These fungi show preferences for hard over soft woods and include *Fomes annosus* and *Trametes pinni*. The final group of basidiomycetes are the simultaneous-rots which are similar to the foregoing group but degrade all cell wall components simultaneously. An example is *Polyporus versicolor*. Substrate preference is similar to white rots. The different types of attacks on wood are obviously related to the ability to degrade lignin and to the properties of polysaccharases produced.

46.3.4 Anaerobic Habitats

Cellulose degradation is important in three anaerobic habitats: the animal gut, waterlogged soils/aquatic sediments and anaerobic digesters. Strictly anaerobic bacteria are usually the dominant cellulolytic organisms. Anaerobic phycomycete fungi may be found in some animal guts (Orpin, 1981). Some are capable of cellulose degradation; their quantitative significance, however, is unknown. Protozoa also feature in the animal gut and may be involved in cellulose degradation, but seem of no importance in anaerobic digestion (Hobson *et al.*, 1974). Because of its economic importance the microbiology of the rumen is well documented but much less is known concerning the other habitats.

Many herbivorous animals are unable to produce enzymes capable of degrading the structural polysaccharides which make up the bulk of their diet. To overcome this, some of these animals have evolved symbiotic relationships with polysaccharolytic microbes. These symbioses involve the enlargement and morphological alteration of some part of the fore or hind gut to accommodate microbes. This enlargement allows a long detention time, to match the slow degradation of cellulose.

The specified gut regions are highly anaerobic and are populated largely by fermentative microbes. Polysaccharides are hydrolysed to monosaccharides which are then fermented to a mixture of products, notably short chain fatty acids (formic, acetic, propionic and butyric), ethanol, lactate, succinate, hydrogen and carbon dioxide. The gaseous products and also formate and acetate are often converted to methane.

Foregut fermentations are found in mammals in several orders. The best known group is the ruminants (sheep, goats, cattle, antelopes and deer). Other groups include camels, hippopotami, sloths, colobus and proboscis monkeys and macropod marsupials (*e.g.* the kangaroo). In all these animals part of the foregut is hugely enlarged (its contents accounting for 10–20% of total body weight) into a multichambered organ, most of which receives no acid or enzyme secretions but does receive copious amounts of buffered saliva. Not all these systems are fully studied but it is likely that in all of them microbial populations are responsible for the solubilization of polysaccharides and their fermentation to volatile fatty acids. The microflora of the sheep and cattle rumen are now well investigated and understood. The principal cellulolytic bacteria are usually *Bacteroides succinogenes*, *Ruminococcus albus*, *R. flavifaciens*, *Butyrovibrio fibrisolvens* and *Eubacterium cellulosolvens*. There are large numbers of ciliate protozoa in the rumen including several species of entodiniomorphs which ingest cellulose particles and have cellulolytic activity. Their significance in cellulose degradation is unknown however, as is the origin of their cellulase activity (it may be bacterial).

Wood-eating insects such as termites and roaches employ a hind gut fermentation in which flagellate protozoa are prominent. 70–90% of cellulose can be digested in the termite hind gut and at least some of the flagellates are known to produce cellulase in axenic culture (Yamin and Trager, 1979). Cellulolytic bacteria may also be important in termite guts and in the guts of some beetle larvae. Probably the best known hind gut symbiosis is the caecal cellulose digestion which occurs in horses, elephants, rodents and lagomorphs. The cellulolytic microbes found in the caecum are the same as or similar to those in the rumen.

It is interesting that both rumen and caecal fluids normally have very little or no cellulase activity suggesting that the enzyme is always either cell or substrate bound. Cellulolytic rumen bacteria are generally found to grow attached to particles of digesta, and tend to be particularly associated with damaged portions of cell walls and the cut ends of cells. These are areas where more degradable cellulose is present. It also seems that species of bacteria differ in their preferences for certain types of plant cell (Latham *et al.*, 1978).

Cellulolytic clostridia such as *C. cellobioparum* and *C. thermocellum* are found in anaerobic soils and sediments and anaerobic digesters. They also occur in the rumen but are probably of little significance there. Other cellulolytic bacteria have been observed in sediments and digesters but not identified, so their importance is therefore impossible to assess.

For more information on the rumen microbiology the reader is referred to Hungate (1966) and Hobson (1979), for general microbiology and biochemistry of all gut fermentation to Clarke and Bauchop (1977) and on anaerobic digestion to Hobson *et al.* (1974).

46.3.5 Enzymology of Cellulolysis

Cellulose, being an insoluble polymer of very large size, cannot permeate the microbial cell membrane. This means that in fungi and bacteria cellulose must be degraded extracellularly. In ciliate protozoa cellulose particles can be taken up by phagocytosis, so degradation may be intracellular, which obviously reduces competition from other microbes for the released sugars. Fungi and bacteria degrade non-permeating substrates by means of extracellular enzymes which may be truly cell-free or attached to the outer layers (wall or membrane) of the cell (*i.e.* cell associated but extracytoplasmic) (Pollock, 1962).

Truly extracellular enzymes allow the producer to exploit substrate at a distance but increase the risk that the product will be lost. Cell-bound enzymes minimize this risk by reducing the diffusion distance, but limit the amount of substrate attacked at any time. It is desirable to know whether enzymes are cell-free or cell-bound, but this can be difficult to assess as cellulases often have a high affinity for their substrate. Thus if cellulase cannot be found in a culture filtrate it may be either cell- or substrate-bound. This is presumably the reason why some of the most vigorous cellulolytic microbes (*e.g. Myrothecium verrucaria, B. succinogenes* and *Cytophaga*) actually seem to produce little detectable cellulase.

The term cellulase is applied both to single enzymes and to enzyme complexes. The general products are glucose and cellodextrins, notably cellobiose. The best understood cellulose-degrading complexes are those of *Trichoderma reesei* and *T. koningii*, which are unusual in having high activity against crystalline cellulose.

Reese *et al.* (1950) postulated that for a cellulase to attack crystalline cellulose two factors were required, C_1 and C_x. C_x was a carboxymethylcellulase able to attack soluble derivatives and amorphous cellulose, but without activity against crystalline cellulose (*e.g.* cotton). C_1 was a factor which enabled C_x to act on cotton. Its identity and mode of action remained a mystery, although it was thought to possibly be an enzyme responsible for disrupting the structure of crystalline cellulose perhaps by breaking H-bonds. It is now known that the C_1 fraction does have enzyme activity. It is in fact an exo-acting β-1,4-glucanase (often referred to as exocellulase or cellobiohydrolase), which removes cellobiose from the non-reducing end of the polymer. It is now widely (Gong and Tsao, 1979) but not universally (Reese, 1977) accepted that C_1 and cellobiohydrolase are identical. C_x is now more usually called endo-β-1,4-glucanase and is considered to be a randomly acting enzyme, hydrolysing bonds within the polymer chain. The third member of the complex is β-1,4-glucosidase (cellobiase) which hydrolyses cellobiose to glucose.

Some cellulase preparations also have a low molecular weight endoglucanase which forms short fibres from the longer ones present in native cellulose (Halliwell, 1979). Cellobiohydrolase is competitively inhibited by cellobiose and to a lesser extent glucose (Gong and Tsao, 1979) while β-glucosidase is inhibited by glucose (Halliwell, 1979).

In *Trichoderma* it has been shown that the separate components of the cellulase complex are synergistic in action (Wood and McCrae, 1977). β-Glucosidase hydrolyses cellobiose and thus relieves product inhibition of cellobiohydrolase to some extent. Endoglucanase is synergistic with cellobiohydrolase because the former creates new end groups on which the latter can act. Another theory is that action by cellobiohydrolase is essential to prevent the reformation of glycoside bonds in crystalline cellulose which have been broken by endoglucanase and a complex of the two enzymes may be necessary to ensure rapid sequential action (Goksøyr and Eriksen, 1980).

Fractionation of cellulase complexes often seems to produce a multiplicity of endoglucanases (anything from two to five in *T. koningii*, up to 12 in *M. verrucaria*) of different size and often of different carbohydrate content. It has been suggested that these are isoenzymes but there is evidence to suggest that all have been derived from one enzyme, modified by protease activity or by non-specific release from cell binding (Gong and Tsao, 1979).

Many organisms, including most bacteria tested, only seem to produce endoglucanase activity and their culture filtrates have little action against native crystalline cellulose (*e.g.* cotton) although they may be more active against filter paper or avicell. The growth of a culture is not necessarily related to cellulase activity in the culture filtrate. Unfortunately it is often assumed that cellulases from different organisms behave similarly. However this assumption is not valid without detailed fractionation of the preparation and analysis of the modes of action of each component. The ability to degrade any particular type of cellulose does not prove the mode of action of a cellulase complex. Cellulases can be measured in many ways, the most usual of which is to assay reducing sugar production from a suitable substrate. This measures activity of the whole complex. For more details of methods see Goksøyr and Eriksen (1980). Cellulase assay is complicated by the difficulty in achieving substrate saturation and the problem of product inhibition.

The rate of hydrolysis of cellulose declines fairly rapidly with time. The main reason for this decline is product inhibition. Other reasons for the decline in hydrolysis are the early removal of amorphous cellulose (Van Dycke, 1972), resulting in an increase in the proportion of more resistant crystalline cellulose, and the denaturation of adsorbed cellulase (Howell and Mangat, 1978). Mathematical models of cellulose degradation have been reviewed by Lee *et al.* (1980) and Knapp and Howell (1980).

46.3.6 Biosynthesis of Cellulase

Extracellular enzymes are synthesized within the cytoplasm, differing from intracellular enzymes in being synthesized on membrane-bound polyribosomes. Current theory is that these enzymes pass through the cell membrane in linear form and assume their tertiary structure outside the cell. This is known as cotranslational secretion (Davis and Tai, 1980).

Fungi have intracellular membranes, and enzymes may be produced on the rough endoplasmic reticulum and then pass into membrane bound vesicles, which could fuse with the cytoplasmic membrane and release enzymes by exocytosis. There is some evidence for this process in *T. viride* (Chapman and Loewenberg, 1976). This is an extra step in enzyme secretion with further possibilities for control. Otherwise control of the synthesis of extracellular enzymes is similar to that of intracellular enzymes (Watson, 1977). Many, although not all, cellulases appear to be inducible; however, cellulose clearly cannot, at the molecular level, be an inducer. It is likely that the true inducer is a soluble degradation product, probably cellobiose or other cellodextrins. The inducer must therefore be formed by the action of preformed cellulases which will be synthesized at a basal level. It is probable that in *T. reesei*, at least, endo cellulase and cellobiohydrolase are coordinately induced (Gong and Tsao, 1979). In *T. reesei*, cellobiose, as well as inducing, can at high concentration cause repression of cellulase synthesis due to its rapid hydrolysis to glucose, a catabolite repressor. The highest yields of cellulase are obtained by growth on cellulose. Cellobiose and glucose obviously play a key role in the control of cellulose degradation: both are products and inhibitors. Also cellobiose is an inducer and is metabolized to glucose which is a catabolite repressor. Control of cellulase biosynthesis by *T. reesei* is reviewed by Gong and Tsao (1979). In most cellulolytic organisms cellulase synthesis is under dual control of induction and catabolite repression. In some organisms however, *e.g. M. verrucaria* (Hulme and Stranks, 1971) and *Ps. fluorescens* (Suzuki, 1975), it appears to be constitutive and only controlled by catabolite repression. Rather unusually it has been shown that in some brown-rot fungi cellulase synthesis is constitutive and far from causing repression glucose (or another easily metabolized substrate) is actually required for cellulase production (Highley, 1973).

The yields of cellulase from *T. reesei* can be manipulated by altering culture conditions, the type and concentration of cellulose and other nutrients, pH and temperature (see for example Mandels *et al.*, 1975 or Andreotti *et al.*, 1977). Mutation has also been used to obtain hyperproducing strains (Gallo *et al.*, 1978) and also catabolite repression-resistant strains of *T. reesei* (Montenecourt and Eveleigh, 1978) and some bacteria and actinomycetes (Daigneualt-Sylvestre and Kluepfel, 1979). Cellulases have been reviewed recently by Gong and Tsao (1979) and Goksøyr and Eriksen (1980). Halliwell (1979) has reviewed β-glucanases in general and Reese (1977) microbial polysaccharases. The phylogeny of cellulases and other polysaccharases has been discussed by Gooday (1979).

46.3.7 Constraints on Cellulolysis and Pretreatment of Cellulosic Substrates

This aspect of cellulose degradation is of great applied importance from the viewpoint of utilization of natural cellulose as forage for ruminant animals or as a feed stock for biotechnological industry. Cellulose itself is among the least degradable of natural polymers due to its high molecular weight, high degree of structural order, insolubility and low surface area. All these factors serve to limit the availability of glycoside bonds to hydrolytic enzymes. The rate of degradation of cellulose cannot be truly related to concentration in terms of weight/volume but rather must be associated with the surface area. Depending on the type of attack the important surface area will be that available either to enzymes (Stone *et al.*, 1969) or to whole cells or hyphae. Crystalline is more resistant to degradation than amorphous cellulose and the type of crystal structure may also be important. This again is probably related to enzyme accessibility, although there may also be a greater tendency for broken glycoside bonds to reform in crystalline cellulose. Another important factor is the degree of polymerization, the effect of which will depend on the type of cellulase used (Okazaki and Moo-Young, 1978); cellulose of low DP will obviously be more susceptible to exocellulases.

In most situations cellulose does not occur alone but is associated with lignins and hemicelluloses. The presence of lignin is always deleterious to cellulose degradation. The rate of cellulolysis is inversely related to lignin content and is also related to the type of lignin and its association with cellulose. Materials of different origin with similar lignin content may require different degrees of delignification to improve degradability (Millet *et al.*, 1975).

Hemicelluloses are easily biodegraded and should prove little barrier to cellulolysis. However it has been shown that enzymic removal of xylan considerably accelerates cellulolysis in beechwood holocellulose, and that xylanase and cellulase are synergistic on this substrate (Sinner *et al.*, 1976). It is probably desirable that cellulase preparations should also have xylanase activity. Constraints on cellulose degradation are reviewed by Cowling (1975) and Fan *et al.* (1980).

Pretreatments of cellulosic materials to improve degradation include both physical and chemical methods. Many different grinding and milling processes have been tested (Mandels *et al.*, 1974). So far it seems that ball milling is the most effective method as it reduces particle size, thus increasing surface area and accessibility to enzymes and may also reduce crystallinity. Unfortunately ball milling is likely to be far too expensive for commercial use. A new process of two-roll milling has some promise for cellulose pretreatment (Tassinari and Macy, 1977). With most cellulose wastes some size reduction must be accomplished prior to utilization.

The most popular chemical method of pretreatment involves the use of alkalis (ammonia or sodium hydroxide) which have the effect of removing lignin or disrupting its structure, of swelling cellulose fibres to increase their accessible area and possibly of reducing crystallinity. Alkali treatment has been used for about 80 years to improve digestibility of forages to ruminants. It is reasonably cheap and thus may be of use in biotechnological exploitation of cellulose. Pretreatment of natural cellulose to improve degradability has been reviewed by Millet *et al.* (1975) and Dunlap *et al.* (1976).

46.4 LIGNIN BIODEGRADATION

46.4.1 Chemical Structure of Lignin

Lignins are polyaromatic dehydrogenation polymers with a basic repeating phenylpropanoid unit. The monomers are not all linked in the same way: within one type of lignin there are C—C bonds and C—O ether bonds (and sometimes esters). Some monomers may be involved in several linkages and others only one or two. The intermonomer bonds may link the monomers at different positions. The monomers have hydroxy and methoxy substituents, some of which are conserved in the polymer. Overall lignins are rather heterogeneous and much-branched polymers of indefinite structure and size with estimates of molecular weight of lignin derivatives varying from a few thousand to over a million (Goring, 1971).

There are three main types of lignin differing in monomer units and linkage. Gymnosperm lignin is a polymer of coniferyl alcohol, angiosperm lignin is a polymer of coniferyl and sinapyl alcohols and grass lignin is a polymer of coniferyl, sinapyl and *p*-coumaryl alcohols, which involves some ester linkages.

For details of lignin structure the reader is referred to Higuchi (1980) and Sarkanen and Ludwig (1971).

Chemical analysis of lignin is difficult and not very sensitive. Until recently the best assessment of lignin degradation was based on comparative chemical and physical analysis of whole and partly degraded lignins with respect to a variety of chemical parameters, *e.g.* hydroxy, methoxy, carboxyl, carbonyl groups and C, H and O contents (Kirk *et al.*, 1977).

In recent years a sensitive assay for lignin degradation has been developed based on monitoring the production of $^{14}CO_2$ from radiolabelled lignins. Such compounds are obtained by the feeding of labelled precursors to plants or chemical/enzymic synthesis of synthetic lignins (see Crawford *et al.*, 1980).

46.4.2 Microbiology of Lignin Biodegradation

Lignin is highly resistant to biodegradation, much more so than other biopolymers. This is hardly surprising when the structure is considered, as lignin is heterogeneous in bond type and positioning and also most of the bonds are not amenable to hydrolytic cleavage. Added to this lignin is insoluble and difficult to wet.

Lignin undergoes no significant decay in anaerobic environments like the rumen and lake sediments (Zeikus, 1980) and it is likely that molecular O_2 is required for its degradation.

At one time it was considered that the only microbes capable of extensive lignin degradation were the white-rot fungi and related litter degrading basidiomycetes. The ^{14}C-assay has allowed a reappraisal of microbial lignolysis.

Brown-rot fungi cause some modification of lignin, notably extensive demethylation, and possibly limited oxidation. A few species have been shown to mineralize it to CO_2. Soft-rots are preferentially cellulolytic but some can cause degradation of lignin, the extent of which varies from species to species. Bacteria were also considered to be non-lignolytic but to cause slight modification. It is now known that a large number of species can cause degradation, notably species of *Nocardia*, *Streptomyces* and *Bacillus*, but also *Pseudomonas*, *Flavobacterium*, *Aeromonas* and *Xanthomonas*. The rate and extent of degradation are not as great as achieved by white-rot fungi. Other than the wood-rotters, fungi such as strains of *Penicillium*, *Fusarium* and *Aspergillus* have now been shown to degrade lignin. Microbial degradation of lignin is reviewed by Crawford and Crawford (1980), Kirk *et al.* (1980) and Amer and Drew (1980).

Low molecular weight aromatics are often referred to as lignin-related compounds. They may be plant products, fungal secondary metabolites or possibly lignin breakdown products (released by biological or chemical means, *e.g.* pulping). These compounds are fairly simple in structure and are relatively easily biodegraded by a wide range of soil bacteria and fungi. The degradation of this type of compound is reviewed by Cain (1980) and Salkinoja-Salonen and Sundman (1980). Anaerobic degradation of this type of compound has also been demonstrated, the mechanism for which usually involves reduction followed by hydrolysis (Evans, 1977). A strong distinction must be drawn between the ability to degrade lignin and low molecular weight aromatics.

46.4.3 Microbial Physiology and Enzymology of Lignolysis

More information is available on the white-rot fungi than other organisms but this is still very incomplete. It has been shown that for *Phanaerochaete chrysosporium* and *Coriolus versicolor* growth and lignin degradation cannot be supported by lignin alone and that these functions require a cosubstrate, like a carbohydrate. Lignolysis is proportional to the amount of cosubstrate. It seems that lignin degradation fails to produce sufficient energy to support itself let alone growth (Kirk *et al.*, 1976). Similar observations have been made for other, but not all, fungi (Crawford and Crawford, 1980). *P. chrysosporium* requires O_2 for growth and high O_2 for lignolysis, which is also very pH sensitive. The nature of the N-source is unimportant but the concentration is crucial, lignolysis being strongly inhibited by high nitrogen concentration. Lignin degradation begins only after the cessation of linear growth, possibly in response to N-starvation. It has also been suggested that it may be related to secondary metabolism. The presence of lignin is not essential for lignolytic activity to be produced (Kirk, 1980). These findings have been made *in vitro* on artificial media where lignolysis is not essential. This is not the situation *in vivo* where extensive cellulolysis cannot be performed by this organism without prior delignification and where growth on wood is relatively slow and nitrogen will usually be deficient.

Analysis of degraded lignins makes it clear that the attack is oxidative and suggests some of the reactions involved. It is thought that demethylation occurs initially giving rise to dihydroxyphenol moieties which are oxidatively cleaved to give carboxyl-rich residues. These modifications occur while the polymer is still intact. Further oxidative and hydrolytic reactions then degrade the new aliphatic chains. Lignin side chains are also attacked, with hydroxy groups being oxidized to carbonyl and carboxyl groups and β-aryl ether bonds may be cleaved. Some low molecular weight intermediates may be released but not at high concentration. These reactions suggest that lignolysis is not a depolymerization of the normal type (*cf.* cellulose or protein degradation). Lignin undergoes an extensive modification which may ultimately lead to destruction in which the basic units are degraded while still attached to the polymer. Unfortunately the nature of the chemical reactions and their catalysts has yet to be elucidated (Chang *et al.*, 1980; Kirk *et al.*, 1977, Crawford and Crawford, 1980). A similar type of attack on softwood lignin has been proposed for *Streptomyces viridosporus* (Crawford *et al.*, 1982).

Due to its polymeric nature the enzymes involved in lignin degradation must be extracellular. It is however difficult to obtain culture filtrates with appreciable lignolytic activity, the enzymes presumably being cell- or substrate-bound. Lignin can be degraded to some extent by concentrated culture filtrates but the nature of the reactions involved is uncertain (Hall *et al.*, 1980).

It has long been known that extracellular phenol oxidases and laccases can be produced by fungi and the ability to synthesize them has been strongly correlated with lignolytic activity of white-rot fungi. The role of these enzymes in lignolysis however is still unclear. Laccase (EC1.10.3.2) and O_2 have been shown to cause polymerization of lignin and phenolic compounds but also to release low molecular weight compounds from lignin and cause some demethylation and α-carbinol oxidation (Ishihara, 1980). In *P. chrysosporium* (*Sporotrichum pulverulentum*) phenol oxidase has been shown to be essential for lignin degradation and also for growth on wood. Phenol oxidase-less mutants are unable to degrade cellulose or xylan in the presence of phenolic compounds. It may be that this enzyme has a role in detoxifying inhibitory phenolic compounds (Ander and Eriksson, 1976). To summarize, it is generally envisaged that lignin will require mono- and di-oxygenase enzymes to perform demethylation, hydroxylation, ring cleavage and side-chain oxidation. The nature of these enzymes is still uncertain, as is the nature of their cofactors if any are required.

46.5 CELLULOSE IN BIOTECHNOLOGY

The potential of cellulose as a renewable raw material cannot be overestimated. It is produced and wasted annually in huge amounts and has several likely uses if suitable technology can be developed and operated sufficiently cheaply. The main possibilities for its utilization are: (1) production of single-cell protein; (2) conversion to glucose which can be used as such or as a feedstock for other microbial processes, *e.g.* production of fine chemicals like citric acid or glutamate; (3) production of liquid or gaseous fuels and solvents.

At the moment starch or simple sugars are, or can be, used for these processes, some of the products of which could replace fossil fuels. The advantage of using cellulose, which currently has little economic value, is that its use would not compete with human nutrition (*cf.* starch or sugar). Lignin, on the other hand, has little potential for biological exploitation and its presence in wastes is a nuisance. Lignin derivatives in wastewaters are possible pollutants. It is fair to say that the use of cellulose in biotechnology is more potential than actual as critical factors limit cellulose utilization, *viz*: (1) the failure of enzymes to hydrolyse cellulose rapidly and the consequent slow growth of microbes on cellulose; (2) the high cost of pretreatment to improve the degradability of naturally occurring cellulosic substrates.

46.5.1 Single-cell Protein

There have been many accounts of the growth of microbes on cellulose for SCP production using fungi, bacteria and actinomycetes. Knapp and Howell (1980) review many of these processes and list important process parameters. The yield of biomass is generally respectable (from 0.1 to 0.2 kg protein per kg cellulose fed) but due to slow growth productivity is low. The highest values were 0.5 kg protein m^{-3} h^{-1} for a coculture of *Cellulomonas* and *Alcaligenes* (Dunlap, 1975) and 0.34 kg protein m^{-3} h^{-1} for growth of *Thermoactinomyces* (Humphrey *et al.*, 1977). These yields are very low when compared to those obtained in commercial processes which are probably between 2 and 5 kg biomass m^{-3} h^{-1} (corresponding to 0.5/1 to 1.25/2.5 kg protein m^{-3} h^{-1} assuming protein is 25–50% of dry weight). Most of the other suggested processes had productivities of less than 0.1 and many less than 0.05 kg protein m^{-3} h^{-1}. It was noticeable that cocultures of cellulolytic and non-cellulolytic microbes gave increased yield and productivity compared with monocultures (Srinivasan, 1975). SCP production from cellulose has also been reviewed by Callihan and Clemmer (1979).

It has been suggested that an alternative way to produce SCP from cellulose would be to hydrolyse to glucose and use this for SCP production (using established food microbes) (Brown and Fitzpatrick, 1976). This process would involve two or three steps (depending whether acid or enzyme hydrolysis was used) which makes it a less attractive proposition.

Most studies on SCP production have used submerged liquid culture techniques. There is little information on producing SCP from cellulose in solid-state fermentations, although what there is is quite encouraging (Pamment *et al.*, 1978). A full economic comparison of solid-state and submerged fermentation is urgently needed.

It may be that the best way to produce protein from cellulose is to use conventional low technology and feed it to ruminants. Some agricultural wastes, such as cereal straw, are already used and it is likely that more could be used (especially if alkali-treated) without detriment to the live weight gain (Palmer, 1976a, 1976b). Other cellulosic wastes could also be utilized (Dyer *et al.*, 1975).

46.5.2 Saccharification of Cellulose

Much research has been performed on the saccharification of cellulose, almost exclusively using the *T. reesei* cellulase complex.

Cellulose slurries can be converted to glucose syrups and glucose concentrations of up to 30 kg m^{-3} have been produced (Mandels *et al.*, 1978). High percentage conversion can be achieved (*e.g.* 85%, Toyama and Ogawa, 1975); the higher the ratio of cellulase to substrate, the greater the degree of saccharification. The process has been operated at pilot plant level (Nystrom and Andren, 1976). Methods have been devised for continuous (*e.g.* membrane or coarse filtration) (Mandels *et al.*, 1971) or semi-continuous product removal during cellulolysis (Brown and Fitzpatrick, 1976) which overcome to some extent the problem of product inhibition.

The cost of cellulase is high, representing *ca.* 60% of the total cost of saccharification (Wilke *et al.*, 1976). Re-use of enzyme is thus essential and therefore stability is very important. Fortunately *T. reesei* cellulase is very stable and others may also be.

It is obviously important that yields and activities of cellulase preparations be increased further. Mutation and cultural techniques have already proved successful but further study of the physiology of cellulase production and release is required in order to decrease the length of time taken to reach maximum yield. Following growth on cellulose by *T. reesei* the maximum cellulase activity is not obtained until after the cessation of growth (often 2 to 8 days after). The reason for this is not understood although it is probably due to release from cell or substrate binding.

46.6 ADDENDUM

Lignin biodegradation. Recent work (Tien and Kirk, 1983) has demonstrated that the white-rot fungus *Phanerochaete chrysosporium* can produce an extracellular enzyme capable of degrading lignin and lignin-subunit model compounds. The enzyme requires H_2O_2 for its activity.

A mechanism for the mode of action of this enzyme has been suggested (Schoemaker *et al.*, 1985, Harvey *et al.*, 1985). This proposes that the reaction proceeds via a single electron transfer oxidation and thus that the enzyme is a peroxidase rather than an oxygenase.

46.7 REFERENCES

Alexander, M. (1977). *Introduction to Soil Microbiology*, 2nd edn. Wiley, New York.
Amer, G. I. and S. W. Drew (1980). Microbiology of lignin degradation. *Annu. Rep. Ferment. Processes*, **4**, 67–103.
Ander, P. and K.-E. Eriksson (1976). The importance of phenol oxidase activity in lignin degradation by the white-rot fungus *Sporotrichum pulverulentum*. *Arch. Microbiol.*, **109**, 1–8.
Andreotti, R. E., M. Mandels and C. Roche (1977). Effect of some fermentation variables on growth and cellulase production by *Trichoderma* QM9414. In *Proceedings of Bioconversion Symposium*, ed. T. K. Ghose, pp. 249–268. Indian Institute of Technology, Delhi.
Brown, D. E. and S. W. Fitzpatrick (1976). Food from wastepaper. In *Food from Waste*, ed. G. G. Birch, K. J. Parker and J. Worgan, chap. 9, pp. 139–155. Applied Science, New York.
Cain, R. B. (1980). The uptake and catabolism of lignin related aromatic compounds and their regulation in microorganisms. In *Lignin Biodegradation*, ed. T. K. Kirk, T. Higuchi and H.-M. Chang, vol. 1, chap. 2, pp. 21–60. CRC Press, Boca Raton, FL.
Callihan, C. D. and J. E. Clemmer (1979). Biomass from cellulosic materials. In *Economic Microbiology*, ed. A. H. Rose, vol. 4, chap. 9, pp. 271–288. Academic, New York.
Chang, H.-M., C.-L. Chen and T. K. Kirk (1980). The chemistry of lignin degradation by white-rot fungi. In *Lignin Biodegradation*, ed. T. K. Kirk, T. Higuchi and H.-M. Chang, vol. 1, chap. 11, pp. 215–230. CRC Press, Boca Raton, FL.
Chapman, C. M. and J. R. Loewenberg (1976). Ultracytochemistry of cellulase synthesis in the fungus *Trichoderma*. *Plant Physiol.*, **57**, Annual Meeting Supplement, 370.
Chung, K.-T. (1976). Inhibitory effects of H_2 on growth of *Clostridium cellobioparum*. *Appl. Environ. Microbiol.*, **31**, 342–348.
Clarke, R. T. J. and T. Bauchop (eds.) (1977). *Microbial Ecology of the Gut*. Academic, New York.
Cowling, E. B. (1975). Physical and chemical constraints in the hydrolysis of cellulose and lignocellulosic materials. *Biotechnol. Bioeng. Symp.*, **5**, 163–181.

Crawford, D. L. and R. L. Crawford (1980). Microbial degradation of lignin. *Enzyme Microb. Technol.*, **2**, 11–22.
Crawford, R. L., L. E. Robinson and A. M. Cheh (1980). ^{14}C-Labelled lignins as substrates for the study of lignin biodegradation and transformation. In *Lignin Biodegradation*, ed. T. K. Kirk, T. Higuchi and H.-M. Chang, vol. 1, chap. 3, pp. 61–76. CRC Press, Boca Raton, FL.
Crawford, D. L., M. J. Barder, A. L. Pometto, III and R. L. Crawford (1982). Chemistry of softwood lignin degradation by *Streptomyces viridosporus*. *Arch Microbiol.*, **131**, 140–145.
Daigneault-Sylvestre, N. and D. Kluepfel (1979). Method for rapid screening of cellulolytic streptomycetes and their mutants. *Can. J. Microbiol.*, **25**, 858–860.
Davis, B. D. and P.-C. Tai (1980). The mechanism of protein secretion across membranes. *Nature (London)*, **283**, 433–438.
Dunlap, C. E. (1975). Production of single-cell protein from insoluble agricultural waste by mesophiles. In *Single-Cell Protein II*, ed. S. R. Tannenbaum and D. I. C. Wang, chap. 11, pp. 242–262. MIT Press, Cambridge, MA.
Dunlap, C. E., J. Thomson and L. C. Chang (1976). Treatment processes to increase cellulose digestibility. *AIChE Symp. Ser.*, no. 158, **72**, 58–63.
Dyer, I. A., E. Riquelme, L. Baribo and B. Y. Couch (1975). Waste cellulose as an energy source for animal protein production. *World Animal Review*, **15**, 39–43.
Edberg, N. and B. Hofsten (1975). Cellulose degradation in wastewater treatment. *J. Water Pollut. Control Fed.*, **47**, 1012–1020.
Evans, W. C. (1977). Biochemistry of the bacterial catabolism of aromatic compounds in anaerobic environments. *Nature (London)*, **270**, 17–22.
Fan, L. T., Y.-H. Lee and D. H. Beardmore (1980). Major chemical and physical features of cellulosic materials as substrates for enzymic hydrolysis. *Adv. Biochem. Eng.*, **14**, 101–117.
Gallo, B. J., R. Andreotti, C. Roche, D. Ryu and M. Mandels (1978). Cellulase production by a new mutant strain of *Trichoderma reesei*. MCG 77. *Biotechnol. Bioeng. Symp.*, **8**, 89–101.
Goksøyr, J. and J. Eriksen (1980). Cellulases. In *Economic Microbiology*, ed. A. H. Rose, vol. 5, chap. 6, pp. 283–330. Academic, New York.
Gong, C.-S. and G. T. Tsao (1979). Cellulase and biosynthesis regulation. *Annu. Rep. Ferment. Processes*, **3**, 111–140.
Gooday, G. W. (1979). Phylogeny of polysaccharases. In *Microbial Polysaccharides and Polysaccharases*, ed. R. C. W. Berkeley, G. W. Gooday and D. C. Ellwood, chap. 19, pp. 437–460. Academic, New York.
Goring, D. A. I. (1971). Polymer properties of lignin and lignin derivatives. In *Lignins, Occurrence, Formation, Structure and Reactions*, ed. K. V. Sarkanen and C. H. Ludwig, chap. 17, pp. 695–768. Wiley-Interscience, New York.
Hall, P. L., Glasser, W. G. and S. W. Drew (1980). Enzymatic transformations of lignin. In *Lignin Biodegradation*, ed. T. K. Kirk, T. Higuchi and H.-M. Chang, vol. 2, chap. 3, pp. 33–49. CRC Press, Boca Raton, FL.
Halliwell, G. (1979). Microbial β-glucanases. In *Prog. Ind. Microbiol.*, **15**, 1–60.
Harvey, P. J., H. E. Schoemaker, R. M. Bowen and J. M. Palmer (1985). Single electron transfer processes and the reaction mechanism of enzymic degradation of lignin. *FEBS Letters*, **103**, 13–16.
Highley, T. L. (1973). Influence of carbon source on cellulase activity of white-rot and brown-rot fungi. *Wood and Fibre*, **5**, 50–58.
Higuchi, T. (1980). Lignin structure and morphological distribution in plant cell walls. In *Lignin Biodegradation*, ed. T. K. Kirk, T. Higuchi and H.-M. Chang, vol. 1, chap. 1, pp. 1–19. CRC Press, Boca Raton, FL.
Hobson, P. N. (1979). Polysaccharide degradation in the rumen. In *Microbial Polysaccharides and Polysaccharases*, ed. R. C. W. Berkeley, G. W. Gooday and D. C. Ellwood, chap. 16, pp. 377–397. Academic, New York.
Hobson, P. N., S. Bousfield and R. Summers (1974). Anaerobic digestion of organic matter. *Crit. Rev. Environ. Control*, **4**, 131–191.
Howell, J. A. and M. Mangat (1978). Enzyme deactivation during cellulose hydrolysis. *Biotechnol. Bioeng.*, **20**, 847–863.
Hulme, M. A. and D. W. Stranks (1971). Regulation of cellulase production by *Myrothecium verrucaria* grown on noncellulosic substrates. *J. Gen. Microbiol.*, **69**, 145–155.
Humphrey, A. E. (1975). Production outlook and technical feasibility of SCP. In *Single-Cell Protein II*, ed. S. R. Tannenbaum and D. I. C. Wang, chap. 1, pp. 1–23. MIT Press, Cambridge, MA.
Humphrey, A. E., A. Moreira, W. Armiger and D. Zabriske (1977). Production of single-cell protein from cellulose wastes. *Biotechnol. Bioeng. Symp.*, **7**, 45–64.
Hungate, R. E. (1966). *The Rumen and its Microbes*. Academic, New York.
Ishihara, T. (1980). The role of laccase in lignin biodegradation. In *Lignin Biodegradation*, ed. T. K. Kirk, T. Higuchi and H.-M. Chang, vol. 2, chap. 2, pp. 17–31. CRC Press, Boca Raton, FL.
Jane, F. W. (1970). *The Structure of Wood*. Black, London.
Jurášek, L., J. R. Colvin and D. R. Whitaker (1967). Microbiological aspects of the formation and degradation of cellulose fibres. *Adv. Appl. Microbiol.*, **9**, 131–170.
Kirk, T. K. (1980). Physiology of lignin metabolism by white-rot fungi. In *Lignin Biodegradation*, ed. T. K. Kirk, T. Higuchi and H.-M. Chang, vol. 2, chap. 4, pp. 51–63. CRC Press, Boca Raton, FL.
Kirk, T. K., W. J. Connors and J. G. Zeikus (1976). Requirement for a growth substrate during lignin decomposition by two wood-rotting fungi. *Appl. Environ. Microbiol.*, **32**, 192–194.
Kirk, T. K., W. J. Connors and J. G. Zeikus (1977). Advances in understanding the microbiological degradation of lignin. *Recent Adv. Phytochem.*, **11**, 369–394.
Kirk, T. K., T. Higuchi and H.-M. Chang (1980). Lignin biodegradation: summary and perspectives. In *Lignin Biodegradation*, ed. T. K. Kirk, T. Higuchi and H.-M. Chang, vol. 2, chap. 16, pp. 235–243. CRC Press, Boca Raton, FL.
Knapp, J. S. and J. A. Howell (1980). Solid substrate fermentation. In *Topics in Enzyme and Fermentation Biotechnology*, ed. A. Wiseman, vol. 4, chap. 4, pp. 85–143. Ellis Horwood, Chichester.
Lacey, J. (1979). The microflora of straw and its assessment. In *Straw Decay and its Effect on Disposal and Utilisation*, ed. E. Grossbard, pp. 57–64. Wiley, New York.
Latham, M. J., B. E. Brooker, G. L. Pettipher and P. J. Harris (1978). Adhesion of *Bacteroides succinogenes* in pure culture and in the presence of *Rumincoccus flavefaciens* to cell walls in leaves of perennial rye grass (*Lolium perenne*). *Appl. Environ. Microbiol.*, **35**, 1166–1173.

Lee, Y.-H., L. T. Fan and L.-S. Fan (1980). Kinetics of hydrolysis of insoluble cellulose by cellulase. *Adv. Biochem. Eng.*, **17**, 131–168.

Liese, W. (1970). Ultrastructural aspects of woody tissue disintegration. *Annu. Rev. Phytopathol.*, **8**, 231–258.

Mandels, M., J. Kostick and R. Parizek (1971). The use of adsorbed cellulase in the continuous conversion of cellulose to glucose. *J. Polymer Sci., Part C*, **36**, 445–459.

Mandels, M., L. Hontz and J. Nystrom (1974). Enzymatic hydrolysis of waste cellulose. *Biotechnol. Bioeng.*, **16**, 1471–1493.

Mandels, M., D. Sternberg and R. E. Andreotti (1975). Growth and cellulase production by *Trichoderma*. In *Symposium on Enzymatic Hydrolysis of Cellulose*, ed. M. Bailey, T.-M. Enari and M. Linko, pp. 81–109. The Finnish National Fund for Research and Development (SITRA), Helsinki.

Mandels, M., S. Dorval and J. Medeiros (1978). Saccharification of cellulose with *Trichoderma* cellulase. In *Proceedings of the Second Fuels from Biomass Symposium*, p. 627. Wiley, New York.

Millet, B. A., A. J. Baker and L. D. Slatter (1975). Pre-treatment to enhance chemical, enzymic and microbiological attack on cellulosic materials. *Biotechnol. Bioeng. Symp.*, **5**, 193–219.

Montenecourt, B. S. and D. E. Eveleigh (1978). Selective screening methods for isolation of high yielding cellulase mutants of *Trichoderma reesei*. *Adv. Chem. Ser.*, **181**, 289–301.

Mullings, R. and J. H. Parish (1981). Distribution of cellulose degrading gram negative bacteria. In *The Flavobacterium–Cytophaga Group*, ed. H. Reichenbach and O. B. Weeks, pp. 179–187. Verlag Chemie, Weinheim.

Norkrans, B. (1967). Cellulose and cellulolysis. *Adv. Appl. Microbiol.*, **9**, 91–129.

Nystrom, J. M. and R. K. Andren (1976). Pilot plant conversion of cellulose to glucose. *Process Biochem.*, December, 26–34.

Orpin, C. G. (1981). Isolation of a cellulolytic phycomycete fungi from the caecum of the horse. *J. Gen. Microbiol.*, **123**, 287–296.

Okazaki, M. and M. Moo-Young (1978). Kinetics of enzymatic hydrolysis of cellulose: analytical description of a mechanistic model. *Biotechnol. Bioeng.*, **20**, 637–663.

Palmer, F. G. (1976a). The feeding value of straw to ruminants. *ADAS Q. Rev.*, **21**, 220–234.

Palmer, F. G. (1976b). The feeding value and worthwhileness of chemically processed straw for ruminants. *ADAS Q. Rev.*, **22**, 247–266.

Pamment, N., C. W. Robinson, J. Hilton and M. Moo-Young (1978). Solid-state cultivation of *Chaetomium cellulolyticum* on alkali-pretreated sawdust. *Biotechnol. Bioeng.*, **20**, 1735–1744.

Pollock, M. R. (1962). Exoenzymes. In *The Bacteria*, ed. I. C. Gunsalus and R. Y. Stanier, vol. 4, *The Physiology of Growth*, chap. 4, pp. 121–178. Academic, New York.

Ratledge, C. (1977). Fermentation substrates. *Annu. Rep. Ferment. Processes*, **1**, 49–71.

Reese, E. T. (1977). Degradation of polymeric carbohydrates by microbial enzymes. *Recent. Adv. Phytochem.*, **11**, 311–367.

Reese, E. T., R. G. H. Siu and H. S. Levinson (1950). The biological degradation of soluble cellulose derivatives and its relationship to the mechanism of cellulose hydrolysis. *J. Bacteriol.*, **59**, 485–497.

Salkinoja-Salonen, M. and V. Sundman (1980). Regulation and genetics of the biodegradation of lignin derivatives in pulp mill effluents. In *Lignin Biodegradation*, ed. T. K. Kirk, T. Higuchi and H.-M. Chang, vol. 2, chap. 12, pp 180–198. CRC Press, Boca Raton, FL.

Sarkanen, K. V. and C. H. Ludwig (eds.) (1971). *Lignins, Occurrence, Formation, Structure and Reactions*. Wiley-Interscience, New York.

Schoemaker, H. E., P. J. Harvey, R. M. Bowen and J. M. Palmer (1985). On the mechanism of enzymatic lignin breakdown. *FEBS Letters*, **103**, 7–12.

Sinner, M., N. Parameswaran, N. Yamazaki, W. Liese and H. H. Dietrichs (1976). Specific enzymatic degradation of polysaccharides in delignified cell walls. *Appl. Polymer Symp.*, **28**, 993–1024.

Srinivasan, V. R. (1975). Production of Bioproteins from Cellulose. In *Symposium on Enzymatic Hydrolysis of Cellulose*, ed. M. Bailey, T.-M. Enari and M. Linko, pp. 81–109. The Finnish National Fund for Research and Development (SITRA), Helsinki.

Stone, J. E., A. M. Scallan, E. Donefer and E. Ahlgren (1969). Digestibility as a simple function of a molecule of similar size to a cellulase enzyme. *Adv. Chem. Ser.*, **95**, 219–241.

Suzuki, H. (1975). Cellulose formation in *Pseudomonas fluorescens* var. *cellulosa*. In *Symposium on Enzymatic Hydrolysis of Cellulose*, ed. M. Bailey, T.-M. Enari and M. Linko, pp. 151–169. The Finnish National Fund for Research and Development (SITRA), Helsinki.

Tassinari, T. and C. Macy (1977). Differential two speed roll mill pre-treatment of cellulosic materials for enzymatic hydrolysis. *Biotechnol. Bioeng.*, **19**, 1321–1330.

Tien, M. and T. K. Kirk (1983). Lignin-Degrading enzyme from the Hymenomycete *Phanerochaete chrysosporium* Burds. *Science*, **221**, 661–662.

Toyama, N. and K. Ogawa (1975). Sugar production from agricultural woody wastes by saccharification with *Trichoderma viride* cellulase. *Biotechnol. Bioeng. Symp.*, **5**, 225–244.

Van Dycke, B. H., Jr. (1972). *Enzymatic Hydrolysis of Cellulose — A Kinetic Study*. Ph.D Thesis. MIT, Cambridge, MA.

Watson, J. D. (1977). *Molecular Biology of the Gene*, 3rd edn. Benjamin, New York.

Wilcox, W. W. (1968). Changes in wood microstructure through progressive stages of decay. *US Forest Service Research Paper*, FPL 70, July.

Wilke, C. R., R. D. Yang and V. Von Stockar (1976). Preliminary cost analysis for enzymatic hydrolysis of newsprint. *Biotechnol. Bioeng. Symp.*, **6**, 155–176.

Wood, T. M. and S. I. McCrae (1977). The mechanism of cellulase action with particular reference to the C_1 components. In *Proceedings of Bioconversion Symposium*, ed. T. K. Ghose, pp. 111–141. Indian Institute of Technology, Delhi.

Yamin, M. A. and W. Trager (1979). Cellulolytic activity of an axenically-cultivated termite flagellate, *Trichomitopsis termopsidis*. *J. Gen. Microbiol.*, **113**, 417–420.

Zeikus, J. G. (1980). Fate of lignin and related aromatic substrates in anaerobic environments. In *Lignin Biodegradation*, ed. T. K. Kirk, T. Higuchi and H.-M. Chang, vol. 1, chap. 5, pp. 101–109. CRC Press, Boca Raton, FL.

47

Basic Equations and Design of Activated Sludge Processes

E. D. SCHROEDER
University of California, Davis, CA, USA

47.1 INTRODUCTION

The stoichiometry, kinetics and biochemistry associated with the activated sludge process have been discussed in Chapters 41–43 on the chemistry of wastewater treatment. The purpose here is to apply this information to process design. It must be recognized that activated sludge is a relatively old biochemical reaction process and as a result has a large associated literature. Many empirical design parameters have been developed since 1914 when activated sludge processes were introduced and some recognition of these parameters must be made by anyone entering the field.

47.2 PROCESS PARAMETERS

Activated sludge process design is almost always based on three parameters: biochemical oxygen demand (BOD), mixed liquor suspended (or volatile suspended) solids (MLSS or MLVSS) and oxygen uptake rate (q_{O_2}). A fourth parameter, effluent suspended solids concentration, x_e, is very significant, but not directly controllable. As noted in Figure 1, the sedimentation and sludge recycle systems are clearly important components, but neither is a major factor in process design, as will be explained below.

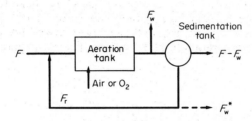

Figure 1   Schematic diagram of a typical activated sludge process

47.2.1   Biochemical Oxygen Demand

In almost all cases, BOD is reported as the five-day (BOD_5), 20 °C value, but for the purpose of process design the ultimate BOD (BOD_u) present is always the appropriate value to use. If a strict proportionality existed between the five-day and ultimate values it would make no difference in practice which value were used, but such a relationship does not exist. Actual measurement of BOD_u is impractical, and the best approximation is obtained through use of either chemical oxygen demand (COD) or the instrumentally obtained total oxygen demand (TOD). Both COD and TOD measurements include refractive material that is difficult to oxidize biologically and which does not enter into the rate relationships and should be subtracted out as described by Busch (1971) and shown in Figure 2.

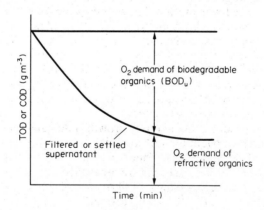

Figure 2   Ultimate BOD as measured by difference in filtered or settled effluent TOD or COD

The method of measuring BOD_u developed by Busch (1971) is called the mass culture T_bOD test and has the advantage of being rapid. Other methods can be used but most require times of the order of days rather than hours.

47.2.2   Mixed Liquor Suspended Solids

Activated sludge is simply a method of developing high cell mass concentrations so that high conversion rates can be obtained. The solids separation and recycle system shown in Figure 1 is the mechanism by which the cell and liquid residence times are separated and allows the development of high cell concentrations. It is important to maintain high cell concentrations because the organic concentration in most wastewaters (particularly municipal) is quite low and the high cell concentration compensates in obtaining satisfactory conversion rates.

The most common measures of cell concentration are MLSS and MLVSS. Both are indirect measures and MLSS is considerably less time consuming to evaluate. The ratio MLVSS/MLSS is not constant and typically ranges between 0.75–0.95 depending on the wastewater and system operating conditions. This is somewhat unfortunate because MLVSS is the more representative value. Under limited operating condition variation and for the same wastewater, MLSS is a satisfactory parameter, however.

A number of other estimates of viable cell concentration have been tried, with the most promising being maximum oxygen uptake rate (Green and Shelef, 1980) and ATP concentration. Maximum oxygen uptake rate can be measured extremely rapidly and could conceivably be a useful method. ATP measurement requires at least 24 h and is no more precise than current methods.

47.2.3 Oxygen Uptake Rate

Oxygen serves as the terminal electron acceptor for the bio-oxidation reactions in activated sludge and must be transferred from the gas phase by some mechanical means. Currently transfer rates of up to 1500–1700 g O_2 m^{-3} d^{-1} are economical using air as the oxygen source. Using pure oxygen the rates are considerably higher but the economically competitive range has not been well defined.

Oxygen transfer rate limitations constrain process design because oxygen uptake rate, BOD removal rate and cell growth rate are stoichiometrically linked. Thus there is a maximum achievable rate that is physically controlled.

47.2.4 Effluent Suspended Solids

As noted above, effluent suspended solids are extremely important in determining process performance but they are not directly controllable. Two general problems exist: (a) hydraulic control of sedimentation and (b) biological control of the culture. Gravity sedimentation is used to separate the MLSS from the treated wastewater (Figure 1) and general design and operating criteria have been empirically developed. A predictive model that relates effluent suspended solids to operating characteristics has not been developed, however. Two types of biological upset occur that result in high effluent solids values. The first, deflocculation, results from high organic loadings that result in high reactor organic concentrations and high growth rates. This problem can be controlled by decreasing the rate of solids wasting and thus forcing the growth rate down. A secondary result is an increase in MLSS and a corresponding increase in q_{O_2}. As long as the maximum q_{O_2} is not exceeded the result is satisfactory.

The second biological problem is filamentous growth. Filamentous bacteria appear to have a competitive advantage under nutrient limited conditions. Examples of the effect of nitrogen limitation are well documented (Kraus, 1965). Chudoba *et al.* (1973), Rensink (1974) and Silverstein (1982) have found that carbon limited cultures are unstable and that long-term carbon limitations can be clearly associated with predominance of filamentous species. These conclusions are supported by the results of Niku *et al.* (1979) who found that completely mixed systems were less stable than nominally plug flow processes.

47.3 KINETIC MODELS

Because the three basic process rates, BOD removal, cell growth and oxygen uptake, are stoichiometrically linked the conceptual models are very similar, as in other microbial processes. A number of kinetic expressions are, or have been, used to describe activated sludge processes, but almost all can be related to the modified Monod model:

$$\mu_g = \mu_m C_s/(K_s + C_s) - \mu_d \tag{1}$$

where μ_g = specific growth rate (t^{-1}), μ_m = maximum specific growth rate (t^{-1}), μ_d = endogenous metabolism rate (t^{-1}), C_s = BOD$_u$ concentration (g m^{-3}) and K_s = saturation constant (g m^{-3}). Typical coefficient values are given in Table 1.

Table 1   Activated Sludge Coefficient Values

Coefficient[a]	Range	Typical
μ_m (d^{-1})	1.5–4	1.6
q_m (d^{-1})	3–5	4
K_s (g m^{-3})	10–50	25
Y (g biomass g^{-1} BOD$_u$)	0.3–0.8	0.4
θ (°C)	1.02–1.15	1.05
D (d^{-1})	1–20	4
θ_c (d)	3–20	5
F/M (d^{-1})	0.05–0.5	0.3
Overflow rate (m d^{-1})	4–40	12
F_r/F	0.2–1.0	0.4

[a] Rates at 20 °C.

BOD removal rate, q_s, is similar to equation (1) but the endogenous metabolism rate is not included:

$$q_s = -q_m C_s/(K_s + C_s) \tag{2}$$

where q_m = maximum BOD$_u$ removal rate (t^{-1}). The ratios μ_m/q_m and μ_g/q_s will be referred to as the yield, Y, and the net yield, Y_G, respectively. Note that mass is the concentration variable in all cases.

47.3.1   Oxygen Uptake Rate

Defining the oxygen uptake rate requires converting the growth rate into units of oxygen equivalents because when organic material (measured as BOD$_u$) is removed, both oxidation and growth take place. The difference between the BOD$_u$ removed and the oxygen equivalents of new cell growth is the oxygen demand:

$$q_{O_2} = q_s + R_{b/x}\, \mu_g \tag{3}$$

where $R_{b/x}$ is the mass of BOD$_u$ per mass of MLSS. Because μ_g is positive and q_s is negative, q_{O_2} is always less than q_s. The value of $R_{b/x}$ varies somewhat with operating conditions but satisfactory estimates can be obtained using the empirical cell formulation $C_5H_7O_2N$, which results in $R_{b/x} = 1.42$ g O$_2$ g^{-1} MLSS.

47.3.2   Limiting Cases

Most activated sludge processes are carbon limited and reported K_s values range between 10 and several hundred g BOD$_u$ m^{-3}. The higher values would appear inaccurate on the basis of pure substrate studies, but three factors may explain why these values were obtained. Firstly, the use of global parameters such as BOD rather than specific species would tend to cloud responses and lead to incorrect rate evaluation. Secondly, refractory organics are often included in rate measurements through the assumption that COD or TOD are accurate estimates of BOD$_u$. As can be seen in Figure 2, this assumption becomes increasingly invalid as reactions progress towards completion. Thirdly, many, if not most, wastewaters contain significant fractions of colloidal and larger biodegradable organics. The complex breakdown of these materials may give a large pseudo-saturation value. Reasonable estimates of K_s that take these factors into account are often in the 20–50 g BOD$_u$ m^{-3} range (Selna and Schroeder, 1978; Grady and Lim, 1981).

Wastewater treatment requirements nearly always specify total effluent BOD$_5$ values below 30 g m^{-3} (approx. 45 g BOD$_u$ m^{-3}), which in most cases corresponds to values of less than 5 g m^{-3} soluble BOD$_5$ (approx. 7.5 g BOD$_u$ m^{-3}), while influent BOD$_u$ values may be anywhere from 100 g BOD$_u$ m^{-3} for a very weak municipal sewage to several thousand g BOD$_u$ m^{-3} for a strong industrial waste. Therefore some activated sludge plants may be carbon limited under all operating conditions and configurations while others may be strongly configuration dependent with respect to the effect of carbon limitation. The results are quite complex. Carbon limitation appears to be one cause of filamentous bulking, as can be deduced from the work of White *et al.* (1980), Chudoba *et al.* (1973), Rensink (1974) and Silverstein (1982). At the other extreme, allowing extended periods of maximum growth rates results in poor flocculation of the culture. Nominally plug flow systems are characterized by concentration gradients along the length of the process and can go from high growth rate (*i.e.* no carbon limitation) to carbon limited conditions if influent BOD values are high enough. The effect on flocculation depends on a balance between the unlimited and limited regions. Systems having relatively weak influents must be designed with control of filamentous bulking in mind. For example, continuous flow stirred tank reactors (CFSTR) are inherently carbon limited systems if effluent requirements are met with a single tank and any tendency for filamentous bulking is accentuated when CFSTRs are used, as shown by Niku *et al.* (1979).

47.3.3 Multiple Limitations

Nutrient limitations in activated sludge are nearly always associated with industrial waste treatment. The most common limiting nutrients are nitrogen, phosphorus and iron which are, not surprisingly, the most common elements in cells after carbon, oxygen and hydrogen. Occasionally other nutrients are limiting in relatively peculiar wastewaters, but the very nature of wastewater disposal tends to minimize such circumstances. Nearly all wastewater streams are contaminated at a number of points before treatment and trace nutrients are part of the added contaminants in most cases.

Kinetic effects of nutrient limitation can be described by double Monod models (DMM) or similar expressions but a better approach is to add excess nutrients to the system. The objective of applying the activated sludge process is to remove organic material and because even strong wastes are dilute by industrial production standards, the cost of adding nutrients is minimal. Because the K_s values for most nutrients are very low ($\ll 1$ g m^{-3}) there is little kinetic effect when concentrations are maintained in the 3–5 g m^{-3} range. Because an excess of nutrients (assumed to be 'kinetic' excess) is a factor in controlling filamentous growth, maintaining maximum rates with respect to nutrients also has a major operating advantage.

Oxygen limitation has been a subject of considerable interest since the introduction of *pure oxygen activated sludge* in the late 1960s. Based on most of the experimental work that has been reported, kinetic limitations do not occur when oxygen concentrations are above 1.0 g m^{-3}, which would imply $K_s < 0.05$ g m^{-3}. Mueller (1966) conducted fundamental studies of oxygen diffusion into floc and concluded that diffusion could be the rate controlling variable in large flocs. A number of investigators have considered oxygen as a limiting nutrient since Mueller's report but results have been inconclusive. Jenkins (1982) has suggested that low dissolved oxygen concentration may be related to filamentous bulking.

Modeling mass transport and reaction in flocs requires a number of simplifying assumptions with a common one being spherical geometry. The scanning electron micrographs of Figure 3 are of typical activated sludge flocs and show considerable deviation from a spherical shape. Highly filamentous flocs are apparently not very 'gelatinous' and would appear to be quite porous. The more desirable 'globular' flocs are also far from spherical, possibly in response to the need to limit transport limitations.

47.3.4 Temperature and pH Effects

Temperature effects are generally considered to follow the Arrhenius relationship between temperatures of 4 and 40 °C:

$$\mu_m = A e^{-E/RT} \tag{4}$$

where A = Arrhenius coefficient (t^{-1}), E = activation energy (J mol^{-1}), R = universal gas constant (J mol^{-1} K^{-1}) and T = temperature (K).

Traditionally, temperature effects in activated sludge are described by a modified form of the Arrhenius equation based on a reference temperature (nearly always 20 °C). Assuming that over the operating temperature range, which is usually much smaller than 40°, the factor $(TT_{20})^{-1}$ is essentially constant, equation (5) results:

$$\mu_{m_T} = \mu_{m_{20}} \theta^{T-20} \tag{5a}$$

$$q_{m_T} = q_{m_{20}} \theta^{T-20} \tag{5b}$$

where θ is a unitless coefficient with an approximate value of 1.05.

Figure 3 Scanning electron micrographs of activated sludge flocs (courtesy of Professor J. A. Silverstein)

Effects of pH variation are very poorly documented and rarely included in design. There is good support for the conclusion that rate and stoichiometric effects are minor in the pH range 6.5–8.5, however. Outside this range rates may or may not drop off drastically depending on the nature of the wastewater. If wastewater characteristics are very restrictive (*e.g.* the organic contents will support a limited population diversity), pH effects may be quite important. Other effects include a tendency for fungi to dominate the population at low pH values. Most wastewaters that are outside the 6.5–8.5 pH range are low in volume however, and pH adjustment is not expensive.

Where wastewaters are poorly buffered, pH problems may occur due to combinations of CO_2 production, nitrification and inadequate mixing. It is generally better to adjust the pH than to try to operate at a value that results in an unstable system.

47.4 STOICHIOMETRY

Activated sludge process design stoichiometry relates to three factors, sludge production, oxygen uptake and in specific cases nutrient requirements. Sludge production, or yield, is usually reported as the maximum value, Y, based on mass of sludge produced per mass of BOD_u removed (in older literature yield may be reported on a BOD_5 basis, however) and defined as the ratio of the maximum rates.

In most cases Y values

$$Y = \mu_m/q_m \tag{6}$$

of about 0.4 are reported. Net yield Y_G, is derived from the ratio of the growth and metabolic rates:

$$Y_G = -\mu_g/q_s \tag{7}$$

$$Y_G = Y/(1 + \mu_d/\mu_g) \tag{8}$$

In the activated sludge literature the inverse of the specific growth rate is often referred to as the *sludge age* or *mean cell (or solids) residence time* and specified as θ_c. It is important to note that these terms are rather crude approximations of the dilution rate/growth rate relationship in chemostats because wastewater treatment systems are rarely steady state processes and are never ideally mixed. Because sludge age is a commonly used design factor it will be discussed in a following section.

47.4.1 Validity of Yield Coefficients

Yield and net yield values often vary from those reported in the literature. Differences in predation and environmental differences that promote production of non-viable biomass are factors that cause the differences or changes in stoichiometry. Values of Y as high as 0.8 have been reported on a BOD_u basis, although this is unusual. It is important to evaluate stoichiometry for given situations however.

Oxygen uptake estimates are based on total BOD removal and organic storage as sludge. Recognition should be made that sludge and cell production are not equivalent. As noted above, the BOD equivalent of activated sludge floc can be approximated as 1.42 g O_2 g^{-1} solids, and therefore the oxygen required, on a rate basis, is:

$$Q_{O_2} = q_{O_2} x \tag{9a}$$

$$= (q_s - 1.42 \, \mu_g) \, x \tag{9b}$$

$$= q_s (1 - 1.42 \, Y_G) \, x \tag{9c}$$

$$= q_s \{1 - 1.42 \, Y/(1 + \mu_d/\mu_g)\} \, x \tag{9d}$$

Conventional oxygen transfer devices can economically provide between 1500 and 2000 g O_2 m^{-3} d^{-1} and this factor is a design constraint on both the biomass concentration, x, and the organic loading rate, as will be shown below.

47.5 LOADING RATES

The first step in activated sludge process design usually is to select a *loading rate*, or loading rate range. The most commonly used loading rate factor is the empirically derived food to microorganisms ratio, **F/M**, which is defined as the ratio of the mass rate of substrate inflow to mass of activated sludge (dry weight basis) in the aeration tank, and therefore has units of t^{-1}:

$$\mathbf{F/M} = FC_{s_i}/Vx \tag{10a}$$

$$= DC_{s_i}/x \tag{10b}$$

where F = volumetric flow rate (m^3 s^{-1}), V = aeration tank volume (m^3), C_{s_i} = influent substrate concentration (g m^{-3}) and x = dry weight biomass concentration (g m^{-3}).

The ratio **F/M** is printed in bold face to distinguish it from the volumetric flow rate, F. This nomenclature problem results from the fact that terminology developed in wastewater treatment is related to civil rather than chemical engineering, where a different notation is used for volumetric flow rate. Although **F/M** is widely used, and as will be shown is crudely related to solids retention time, it is not the best choice for design or operation of activated sludge processes.

Solids retention time (SRT), sludge age and mean cell residence time (θ_c) are one and the same value, but a standard term has not been developed for this loading rate parameter. By definition the sludge age is the average time biological solids remain in the system. Considering a mass balance on the solids in Figure 1 and assuming influent biological solids (as opposed to substrate) are negligible, equation (11) results:

$$V\overline{\mu_g x} = F_w x + (F - F_w)x_e + V\frac{dx}{dt} \tag{11}$$

where $\overline{\mu_g x}$ is the average growth rate in the system and x_e is the effluent solids concentration. At steady state equation (11) reduces to:

$$\mu_g^* = \frac{F_w}{V} + \frac{F - F_w}{V}\frac{x_e}{x} \tag{12}$$

where μ_g^* = average specific growth rate. The inverse of equation (12) is the solids retention time:

$$\theta_c = SRT = \frac{1}{\mu_g} = x/[DF_w \{x_e/F_w + D(x - x_e)/V\}] \tag{13}$$

Quite often the impact of effluent biological solids is ignored and the SRT is approximated as V/F_w. In both equations (12) and 13) the mass of solids in the sedimentation tank and recycle line are assumed neglible.

47.5.1   F/M *vs.* SRT

Effluent substrate concentrations from activated sludge processes are inherently very low and in nearly all cases the approximation $C_{s_i} - C_s \approx C_{s_i}$ can be made. Using this approximation to define the specific substrate removal rates gives

$$Vq_s x = F(C_s - C_{s_i}) \tag{14a}$$

$$\approx - FC_{s_i} \tag{14b}$$

$$q_s^* = - FC_{s_i}/Vx \tag{15a}$$

$$= - DC_{s_i}/x \tag{15b}$$

where q_s^* = average specific substrate removal rate. Comparing equation (15b) to equation (10b) we see that $F/M = |q_s^*|$, and therefore F/M is stoichiometrically related to the sludge age, but the relationship depends heavily on the approximations of equation (14). For this reason F/M is less a suitable parameter than SRT:

$$F/M = \mu_g/Y_G \tag{16a}$$

$$= (Y_G\theta_c)^{-1} \tag{16b}$$

47.5.2   Design Loading Rate Values

In the literature, F/M values are usually reported as g BOD_5 g^{-1} MLSS d^{-1}. Dividing these values by 0.67 gives a good estimate of the F/M based on BOD_u, a more appropriate value, and one that will be used here. The conventional F/M operating range is 0.30–0.75 d^{-1} based on BOD_u. Use of values above 0.75 d^{-1} results in a tendency toward dispersed or non-flocculent growth, while values below 0.30 d^{-1} result in systems that require excessive oxygen due to high endogenous respiration rates and a tendency toward filamentous bulking because of the extreme substrate limiting conditions.

Systems with F/M values above 0.75 d^{-1} are often termed high rate, while those with values below 0.30 d^{-1} are generally referred to as extended aeration systems. Despite operating problems, both types are in use, with extended aeration systems being commonly chosen for small installations and *package* or off-the-shelf plants. When nitrification is required, low F/M values must be used and in such cases extended aeration processes may be used in large installations.

Using equation (16) and suitable coefficient values we can see that SRT values corresponding to 0.30 and 0.75 d^{-1} are approximately 13 and 4 d respectively. Thus systems with SRTs greater than 13 d can be referred to as extended aeration processes and those with residence times less than 4 d can be referred to as high rate systems.

47.6 DESIGN OF CONTINUOUS FLOW PROCESSES

Continuous flow activated sludge has been virtually the only process configuration from the processes inception until very recently. There are four basic modifications of continuous flow processes: conventional or nominal plug flow, continuous flow stirred tank reactor (CFSTR), step feed and contact stabilization. These configurations are shown schematically in Figure 4. Almost without exception design procedures are based on the assumption of steady state conditions.

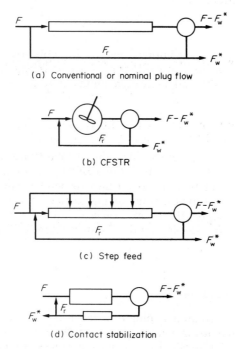

(a) Conventional or nominal plug flow

(b) CFSTR

(c) Step feed

(d) Contact stabilization

Figure 4 Activated sludge process configurations

47.6.1 Design Flows

Design flow rates vary for different portions of the process. Obviously the system must have the hydraulic capacity to convey the maximum instantaneous flow that is expected. Sedimentation tank overflow rates are based on average flow on the maximum day in most cases, as is the aeration system. This means that short periods may occur when design flows are exceeded but there is enough damping in the system to prevent process failure for short periods of overloading. Aeration tank design is based on average flow.

47.6.2 Sedimentation Tanks

Design of sedimentation tanks is based on two factors, overflow rate and solids loading rate. The overflow rate, defined as the flow rate divided by the surface area, is equal to the critical settling velocity under ideal conditions. As noted above, the flow rate used is the average flow on the peak day of the year including recycle. Values used are in the 200–500 m^3 m^{-2} d^{-1} range. Solids loading (dry weight basis) should be in the 3–6 kg m^{-2} h^{-1} range. Tank depths are set to keep horizontal velocities near 30 m h^{-1}, hydraulic detention times below two hours and to provide up to one meter sludge storage. Typical depths are 3.5–5 m. More detailed information is provided in Tchobanoglous and Schroeder (1985).

47.6.3 Sludge Wasting

The schematic diagrams of Figures 1 and 4 show two points of sludge wasting, the first directly from the mixed liquor (F_w) and the second from the settled sludge (F_w^*). In theory there is no difference except that in the latter case a thicker slurry is wasted which would appear to have economic benefit. In practice it is virtually impossible to control the settled sludge concentration and process control using constant SRT or F/M becomes more difficult requiring constant monitoring of this concentration and calculation of the SRT using an on-line computer. In most plants it is possible to waste either a selected solids fraction by wasting all solids from one of the sedimentation tanks or to waste periodically (*e.g.* one hour, three times a day) and this provides positive control based on wasting from the mixed liquor.

47.6.4 Recycle Rates

Because the settled sludge concentration varies considerably (typical values are between 6000 and 12 000 g m^{-3}) there is a need to provide for a variable recycle rate (F_r) also. Recycle ratio (F_r/F) design values of up to 1.0 based on average annual flow are recommended.

47.6.5 Effluent Quality

Most processes are required to meet specific effluent total BOD$_5$ requirements. In most cases these requirements are either 20 or 30 g m^{-3}. On a BOD$_u$ basis the values are 30 and 45 g m^{-3} respectively. Effluent BOD can be partitioned into suspended and filtrate fractions. The suspended fraction results from biological solids being carried over the sedimentation tank weir. Average effluent suspended solids of 20 g m^{-3} or less should be possible in most cases, corresponding to 28 g BOD$_u$ m^{-3} and leaving up to 17 g m^{-3} filtered BOD$_u$ if the discharge requirements is 45 g m^{-3}.

47.6.6 Design of Conventional Processes

Conventional processes are treated as plug flow reactors as stated above. In most cases, conventional processes are sized using loading factors and oxygen transfer rate constraints, but kinetic modelling can be accomplished using mass balances for both substrate and biomass. Because MLSS values are high in all cases (1500–3500 is the typical range) the change in biomass along the tank length is relatively small and the substrate removal rate is not greatly affected by changes in MLSS along the tank length:

$$x = x^* + Y_G(C_s^* - C_s) \tag{17}$$

where x^* = MLSS concentration at the head of the aeration tank, and C_s^* = substrate concentration at the head of the aeration tank.

Both x^* and C^* depend on recycle rate and process performance. Schroeder (1977) suggests the relationships below:

$$x^* = D(1 + F_r/F) \, \text{SRT} - 1 \tag{18}$$

$$C_s^* = \frac{C_{s_i} + (F_r/F) \, C_{s_0}}{1 + F_r/F} \tag{19a}$$

$$C_s^* \approx \frac{C_{s_i}}{1 + F_r/F} \tag{19b}$$

Performing a mass balance on substrate and substituting equations (17), (18) and (19) results in an integrable expression. Integrating along with tank length gives:

$$z = \frac{F}{A}\left(\frac{1 + F_r/F}{q_m}\right)\left[\frac{K_s}{x^* + Y_G C_s^*} \ln\left\{\frac{C_s^*}{x^* C_s}(x^* + Y_G C_s^* - Y_G C_s)\right\} + \frac{1}{Y_G} \ln\left\{\frac{x^* + Y_G(C_s^* - C_s)}{x^*}\right\}\right] \tag{20}$$

where z = distance from tank entrance (m) and A = cross sectional area of tank (m^2).

In most cases C_s is the desired value but the way equation (20) is structured requires working backward from chosen values of C_s. It is also important to note that solving equation (20) requires specification of F_r and SRT as well as having some knowledge of the rate coefficients, flows and BOD of the wastewater. It must also be remembered that all rate coefficients must be temperature corrected.

The most important feature of equation (20) is that it can be used for non-municipal applications. Most municipal plants are designed for hydraulic residence times of four to six hours, but this is only because of the relatively consistent influent BOD concentration of municipal wastewaters. Industrial or industrial/municipal systems often have quite different requirements.

A second useful feature of applying a model such as equation (20) is that once a substrate concentration profile is established, an oxygen uptake rate profile can be predicted. This allows *tapering* the air input, a technique that conserves energy and gives better operating performance.

47.6.7 Continuous Flow Stirred Tank Reactors

Application of CFSTR principles to activated sludge processes began in the mid 1950s. The advantage of large and rapid influent dilution was very attractive where toxic wastes might damage the microbial culture. Experience with CFSTR activated sludge processes led to the conclusion that any decrease in average specific reaction rate could be mitigated by extending the SRT and carrying higher MLSS concentrations. Steady state mass balances result in two algebraic expressions:

$$\mu_g = \frac{\mu_m C_s}{K_s + C_s} - \mu_d = \frac{F_w(x - x_e) - Fx_e}{Vx} \tag{21}$$

$$x = \frac{F}{V} \frac{(C_{s_i} - C_s)(K_s + C_s)}{q_m C_s} \tag{22}$$

In most cases the effect of x_e is small and $\mu_g \approx F_w/V$. The growth rate, or as more commonly used, the SRT, is a design parameter and hence is chosen. Rewriting the modified form of equation (21) gives

$$C_s = \frac{(\mu_d + F_w/V) K_s}{\mu_m - \mu_d - F_w/V} \tag{23}$$

Substituting equation (23) into equation (22):

$$x = \frac{F}{V} \frac{\mu_m}{q_m} \left\{ C_{s_i} \left(\frac{\mu_m}{\mu_d + F_w/V} - 1 \right) - K_s \right\} \tag{24}$$

A good approximation of equation (24) is provided by

$$x = \frac{F}{F_w} \frac{\mu_m}{q_m} C_{s_i} \tag{25}$$

Oxygen uptake rate in CFSTR activated sludge systems can be calculated using equations (2), (9), (23) and (24).

47.6.8 Step Feed Activated Sludge

Step feed processes were developed to mitigate the effect of high organic loadings at the head end of conventional systems. By spreading out the input, maximum oxygen demand rates were decreased and the load was spread more evenly over the aeration tank length. The effect is to approximate more nearly a CFSTR. Design can be accomplished using either a series of PFRs or a series of CFSTRs, as depicted in Figure 5. Because of the relatively short residence times in each segment, either approach is satisfactory. In applying the mass balances it is useful to note that segment volumes (Figure 4) are constant, but flow rates and hydraulic residence times change as flow moves toward the effluent end of the tank.

47.6.9 Contact Stabilization

Contact stabilization is a process configuration (Figure 3d) that incorporates a short contact period between the wastewater and the activated sludge floc followed by separation and separate aeration of the floc. During the contact period (usually 45–60 min), particulate materials are adsorbed on to the floc and easily metabolized materials are removed. Separation of the floc, with the absorbed particulates, results in a concentrated slurry that can be aerated in a smaller volume. Thus the total tankage is smaller than for the other commonly used process schematics. Contact stabilization is often used as a method of expanding capacity without increasing tankage.

There is no way to predict performance of a contact stabilization process other than through pilot plant studies. Niku *et al.* (1979) found the contact stabilization process to be significantly less stable than the other three configurations discussed here. Average effluent concentration values are somewhat higher than conventional processes but can be in the 20–30 g m^{-3} range.

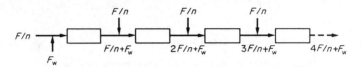

(a) Series of PFRs as a conceptual model of step feed

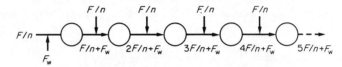

(b) Series of CFSTRs as a conceptual model of step feed

Figure 5   Conceptualization of step feed activated sludge as a series of reactors

47.6.10   Extended Aeration Activated Sludge

Extended aeration is a mode of operation rather than a process configuration, and in fact all of the configurations of Figure 3 can be operated as extended aeration systems. Essentially extended aeration processes are systems with long SRTs (as defined above SRT > 13 d, although that is an arbitrary number). The name extended aeration results from a need to increase tank volume because of oxygen transfer constraints as the SRT is increased. If tank volume is not increased the MLSS values become so high that oxygen demand rates exceed maximum transfer rates. Increasing aeration tank volume to meet these requirements results in longer or extended aeration times. This situation is illustrated in Figure 6.

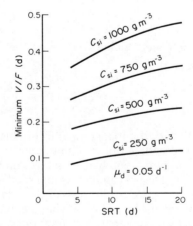

Figure 6   Minimum hydraulic residence times based on a maximum O$_2$ transfer rate of 1500 g m^{-3} d^{-1}

47.6.11   Non Steady-state Effects

Few, if any, real world activated sludge processes receive a steady organic or flow input and, even disregarding temperature change, true steady-state operation would seem to be an unreasonable assumption. Response of activated sludge systems to transients is extremely damped, however, as shown by Selna and Schroeder (1978). Large perturbations in influent characteristics result in small effluent quality responses and when perturbations are of short duration (less than 24 h) have no measurable effect or floc characteristics. Large step changes in feed characteristics can affect performance unless operation is modified, however.

Operating problems, such as filamentous bulking, appear to be a response to instabilities designed into a process as much as transients imposed on the system. Activated sludge processes are usually substrate limited systems and there is an inherent tendency toward filamentous bulking. Influence perturbations may result in bulking by affecting stability. The most appropriate way of stabilizing activated sludge processes is to provide a kinetically unlimited region that favors no filamentous bacterial species. Chudoba *et al.* (1973) suggest use of a *selector*, a small tank at the head end of the activated sludge aeration system which will have high substrate concentration. White *et al.* (1980) used essentially the same concept but further enhanced the kinetic advantage of the non-filamentous bacteria by making the selector anoxic.

47.7 SEQUENCING BATCH REACTORS

The original activated sludge process configuration was batch or fill and draw but operating problems soon forced the conversion to continuous flow systems. Head loss was a consideration, but problems in decanting and turbid effluents were the major factors in the conversion process. Batch processes have three important advantages over continuous flow processes: (1) ideal plug flow and hence maximum reaction rates can be more nearly approximated (using time rather than position as a variable); (2) ideal, quiescent sedimentation can be achieved, a distinct impossibility in continuous flow plants; and (3) flow equalization is an inherent design parameter.

Decanting problems can be solved through use of floating outlets and adjustable weirs (Goronzy, 1979). Reliable systems have been developed and are in use at prototype installations. The effluent turbidity problem was solved by Irvine *et al.* (1979) through use of an anoxic fill. Batch operation, particularly with the anoxic fill, enhances the competitive advantage of flocculent over filamentous bacteria. Thus the configuration confirms to the concepts developed by Chudoba (1973), Rensink (1974) and White *et al.* (1980).

There is an apparent increase in total tankage volume because the large aeration tank is used for settling, but this is at least partially compensated for by faster settling rates (perhaps a factor of 2) and elimination of the sludge recycle system. Typical total hydraulic residence times of less than eight hours for municipal wastes compare favorably with continuous flow systems which typically have residence times of seven to nine hours when sedimentation is included.

47.7.1 Batch Process Design

Experience (Irvine, 1982; Silverstein, 1982) with batch processes using the sequencing mode of operation developed by Irvine *et al.* (1979) has led to the following conclusions.
1. Design flow fill periods of two hours are satisfactory.
2. No mixing is necessary for the first 1.5 h of fill, but about 0.5 h of anoxic mixing is suggested to provide the kinetic enhancement discussed above.
3. Aeration periods as short as one hour are satisfactory, but two to four hours' aeration based on average dry weather flow (ADWF) is appropriate for design.
4. Adequate settling should occur within 30 minutes. Based on ADWF, a one hour settling period should be used for municipal systems.
5. Decant time should be based on decant system design but 30 minutes is satisfactory in most cases.
6. Conventional parameters such as the **F/M** value can be used in process design.

47.8 REFERENCES

Busch, A. W. (1971). *Aerobic Biological Treatment of Wastewaters*. Oligodynamics Press, Houston.
Chudoba, J., V. Ottora and V. Madera (1973). Control of activated sludge filamentous bulking. II. Selection of microorganisms by means of a selecter. *Water Res.*, **7**, 1389.
Goronzy, M. C. (1979). Intermittent operations of extended aeration processes for small systems. *J. Water Pollut. Control Fed.*, **51**, 274.
Grady, C. P. L. Jr. and H. C. Lim (1981). *Biological Wastewater Treatment*. Dekker, San Francisco.
Green, M. and G. Shelef (1980). Sludge viability in a biological reactor. *Water Res.*, **15**, 953.

Irvine, R. L. (1982). *Sequencing Batch Reactors*, USEPA, San Francisco.

Irvine, R. L., G. Miller and A. S. Bhamrah (1979). Sequencing batch treatment of wastewater in rural areas. *J. Water Pollut. Control Fed.*, **51**, 244.

Jenkins D. (1982). Discussion of study of the causes and prevention of sludge bulking in Germany. Cited by F. Wagner in *Bulking of Activated Sludge*, ed. B. Chambers, and E. J. Tomlinson. Ellis Horwood, Chichester.

Kraus, L. S. (1965). Operating practices for activated sludge plants. *J. Water Pollut. Control Fed.*, **37**.

Mueller, J. A. (1966). *Oxygen Diffusion Through a Pure Culture Floc of Zoogloea ramigera*. Ph.D Dissertation, Department of Civil Engineering, University of Wisconsin.

Niku, S., E. D. Schroeder and F. Samaniego (1979). Performance of activated sludge processes and reliability based design. *J. Water Pollut. Control Fed.*, **51**, 2841.

Rensink, J. H. (1974). New approach to preventing bulking sludge. *J. Water Pollut. Control Fed.*, **46**, 1888.

Schroeder, E. D. (1977). *Water and Wastewater Treatment*. McGraw-Hill, New York.

Selna, M. W. and E. D. Schroeder (1978). Response of activated sludge processes of organic transients—kinetics. *J. Water Pollut. Control Fed.*, **50**, 944.

Silverstein, J. A. (1982). *Operational Control of Bioflocculation and Solids Removal in a Batch Wastewater Treatment System*. Ph.D Dissertation, Department of Civil Engineering, University of California, Davis.

Tchobanoglous, G. and E. D. Schroeder (1985). *Water Quality, Characterization, Modeling, Modification*. Addison-Wesley, Reading, MA.

White, M. J. D., E. J. Tomlinson and B. Chambers (1980). The effects of plant configuration on sludge bulking. *Progr. Water Technol.*, **12**, 183.

48

Sedimentation

M. J. D. WHITE
Water Research Centre, Stevenage, Herts., UK
and
D. W. M. JOHNSTONE
Sir William Halcrow and Partners, Swindon, Wilts., UK

48.1 INTRODUCTION

The sedimentation tank in the conventional activated sludge process is required to perform three functions: thickening, clarification and storage.

Thickening is the basic step of concentrating the sludge from the aeration tank and thereby separating from it the treated effluent. In terms of design or operation the thickening function is usually the rate limiting step, and it is the most important function of the three because if it fails the whole activated sludge process may fail.

The clarification function is the settling out of some of the residual fine particles in the treated effluent. If this step is overloaded the final effluent may contain too many suspended solids, but the activated sludge process continues to operate and will for example continue to remove ammonia.

The storage function is necessary to allow for varying input flow, diurnal and storm flows. It would be uneconomic to build sedimentation tanks large enough to cope with the maximum expected peak flow, and in any case at low flow rates large sedimentation capacity and hence long retention times would allow sludge to go anaerobic, leading to floating sludge and encouraging the growth of filamentous or bulking sludge. Sedimentation tanks are therefore designed to cope with a continuous flow of, say, twice the average flow. At higher flows sludge accumulates in the sedimentation tank. Additional capacity (depth) is provided to store the sludge for several hours. If the storage capacity is inadequate there will be intermittent loss of solids, but usually the activated sludge process will recover when the flow falls again.

48.1.1 Design and Operation

Traditionally sedimentation tanks have been designed on the basis of detention time or overflow rate. Although based on the criteria to provide adequate clarification, these methods are essentially rule-of-thumb. They also ignore the thickening function, and so can lead to severe under- or over-sizing. Operation, particularly in the US and UK, has traditionally made use of the Mohlman Sludge Volume Index, SVI (Mohlman, 1934). Many workers have shown that SVI rarely bears any resemblance to the actual settling properties of the sludge in the sedimentation tank (see, for example, Dick and Vesilind, 1969). SVI is therefore an unreliable parameter on which to base operation or design.

In the following sections the theory will be explained to give a rational basis for the design and operation of sedimentation tanks, based on realistic settling characteristics of the sludge.

48.2 THEORY OF THICKENING

There are many theories and empirical expressions for thickening (see Lockyear, 1977) but most of these have been developed for the consolidation of sludges to much greater concentrations than that normally required in the activated sludge process.

A theory that has been found applicable to the activated sludge settlement tank is the mass flux theory, which considers the limitation of the tank in transporting the solids from the feed point to bottom, that is the limitation in the rate of flow of mass per unit area. The theory is described elsewhere (Dick and Ewing, 1967; White, 1975, 1976). Briefly, it considers that the solids in the settlement tank settle by gravity and at the same time are drawn down by the withdrawal of sludge from the bottom of the tank.

The rate of gravity settlement is assumed to be a simple function of the concentration of suspended solids only, and is assumed not to depend on the depth of sludge and hence the weight of any overlying sludge blanket.

An exponential function has been found to be most generally applicable.

$$V = V_0 \exp(-kC) \tag{1}$$

where V is settling velocity, V_0 and k are empirical constants and C is the concentration of solids.

Mass flow rate, F, is the product of settling rate and solids concentration, and for gravity settling is thus $V_0 C \exp(-kC)$ and for sludge withdrawal is $(Q_u/A)C$, where Q_u is the volumetric flow rate of the underflow and A is the cross-sectional area of the settlement tank. Thus overall,

$$F = V_0 C \exp(-kC) + (Q_u/A)C \tag{2}$$

A typical plot of this curve is shown as A in Figure 1.

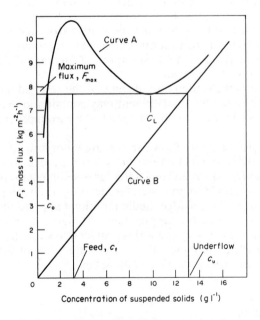

Figure 1 Variation of total mass flux with concentration of suspended solids

The theory assumes that the rate of mass flow, or mass flux, is constant throughout the depth of the tank, despite the increasing solids concentration. That is, in the long term there should be no accumulation of sludge in the tank. Loss of suspended solids in the effluent is assumed to be negligible. By reference to Figure 1 it may be seen that the highest mass flow rate, F_{max}, which can be applied to the settlement tank is that which, if we draw a horizontal line from the feed concentration towards increasing concentration on the right, is able to pass under the minimum on curve A.

Mathematically, F_{max} is therefore given by solving $dF/dC = 0$ when d^2F/dc^2 is positive.

Because of the exponential term in equation (1), F_{max} must be determined graphically, as described by Yoshioka *et al.* (1957), or by an iterative technique (White, 1975).

The rate of sludge return, or the underflow rate, Q_u/A, is given by the slope of line B in Figure 1. As the return rate increases the shape of curve A changes until at a certain critical underflow rate there is a point of inflection and no maximum or minimum, as in Figure 2. Under these conditions the highest mass flow rate is that at the feed point, and is given mathematically by

$$F_{max} = V_f C_f + (Q_u/A)C_f \tag{3}$$

where V_f is the settling velocity of the sludge at the feed solids concentration, C_f.

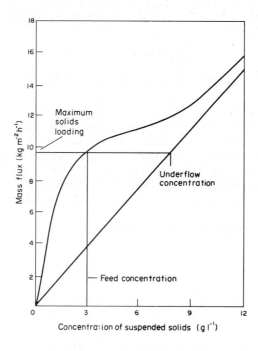

Figure 2 Relation between total mass flux and solids concentration, at an underflow rate above the critical value, for a sludge with poor settling charcteristics

The mass flow rate actually applied to the tanks is F_{appl}.

$$F_{appl} = C_f(Q_0 + Q_u)/A \tag{4}$$

where Q_0 is the volumetric flow rate of incoming sewage. Wastage of surplus sludge is ignored. See Figure 3.

For the final tank not to overflow, the applied solids loading must not exceed the maximum predicted loading:

$$F_{appl} < F_{max} \tag{5}$$

$$C_f(Q_0/A + Q_u/A) < C_f \{V_f + (Q_u/A)\} \tag{6}$$

or

$$Q_0/A < V_f \tag{7}$$

In other words, at high rates of sludge withdrawal the overflow rate must not exceed the settling rate of the sludge in the mixed liquor, which, it may be noted, is one of the criteria used for design of final tanks to ensure that they act as efficient clarifiers.

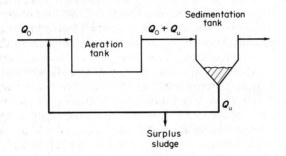

Figure 3   Schematic diagram of activated sludge process

The criterion given by equation (7) also holds below the critical underflow rate if the initial concentration of suspended solids is lower than the value C_0 in Figure 1. This does not appear to occur very frequently, and is only likely to be a problem with sludges which settle very badly.

The concentration of the underflow, C_u, is calculated by balancing the mass flows in and out of the final tank:

$$(Q_0 + Q_u)C_f = Q_u C_u \tag{8}$$

The maximum underflow concentration at any given rate withdrawal is given by

$$C_u = F_{max}A/Q_u \tag{9}$$

The mass flux theory allows us to predict the limiting mass flow rate, or solids loading, if we know the relationship between settling velocity and solids concentration. In the laboratory it is easy, but time consuming, to measure linear settling rates at different concentrations, and hence derive V_0 and k, for example. Operationally it is impractical to keep doing this, and for design it is not possible to predict the relationship from aeration tank design data.

The traditional operation tool, SVI, bears no constant relationship to settling velocity, or to V_0 and k.

Two empirical solutions to this problem are available.

48.2.1   The UK Approach

The disadvantage of the original SVI is that it is influenced both by the presence of the walls of the small measuring cylinder, in which the test is carried out, and by the concentration of solids, especially at high values of the latter.

In the UK White (1975) has advocated the use of a slowly stirred cylinder to overcome the wall effect, and a standard solids concentration, 3.5 g l^{-1}, to overcome the concentration effect. He found that the maximum mass flow rate determined from laboratory measurements of settling velocity correlated well with his Stirred Specific Volume Index, SSVI, measured at 3.5 g l^{-1}, and in turn correlated well with the observed maximum solids loading that could be applied to full-scale settlement tanks.

The maximum allowable mass flow rate, or solids loading, F_{max}, for design or operation, was given empirically by

$$F_{max} = 310(\text{SSVI})^{-0.77} (Q_u/A)^{0.68} \text{ kg m}^{-2}\text{h}^{-1} \tag{10}$$

This relationship has been found to be accurate to $\pm 20\%$.

The applied solids loading, F_{appl}, is simply the flow rate (per unit area) entering the settlement tank multiplied by the mixed liquor suspended solids concentration.

$$F_{appl} = \{(Q_0 + Q_u)/A\}C_f \tag{11}$$

The operator of an activated sludge plant must clearly try to keep the applied loading below the maximum attainable in his settlement tank. For a given return sludge rate he can easily use the above equations, or the nomograph given by White (1976), to predict the maximum MLSS for a given inflow of sewage, or the maximum flow rate of sewage for a given MLSS.

$$(Q_0/A + Q_u/A)C_f < 310(\text{SSVI})^{-0.77} (Q_u/A)^{0.68} \tag{12}$$

In practice the operator will often be able to vary the return sludge rate, Q_u. As this appears on both sides of equation (12), to different powers, it is difficult for him to visualize the effect. For a given sewage flow rate, Q_0, and a given MLSS, we can plot the applied and maximum permissible solids loading against return sludge rate as in Figure 4. It can be seen that as the rate of return is increased the maximum permissible solids loading increases at a greater rate than the applied solids loading. This would imply that when a secondary settlement tank is fully loaded or overloaded an increase in the sludge return rate would alleviate that overloading. However, this is not necessarily always the case. Firstly, if the underflow rate is increased beyond the critical rate then the maximum solids loading possible is not as given in Figure 4, but is also dependent on the concentration of suspended solids in the mixed liquor. Secondly, high rates of sludge return may upset the hydraulic stability of the secondary settlement tank. Thirdly, by operating a plant at high rates of sludge return for long periods the settling properties of the sludge may deteriorate, thus exacerbating the problem (White, 1976).

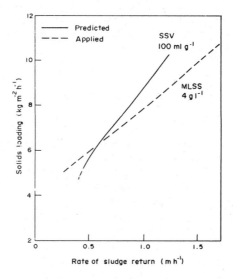

Figure 4   Effect of varying sludge return rates on predicted and applied solids loadings for a flow of settled sewage per unit area of 1 m h^{-1}

Nevertheless varying return sludge rate is a valid means of control and it is convenient to provide the operator with a set of guidelines, as demonstrated by Johnstone *et al.* (1980) and shown in Table 1. This approach has been used widely in the UK.

Although the mass flux theory does not appear to predict the return sludge solids concentration very well the empirical relationship (equation 12) has been found by many workers to predict tank failure quite accurately ($\pm 20\%$). Only in the case of a very poorly settling sludge (SSVI > 200 ml g^{-1}) does the equation appear to overestimate the settling properties of the sludge.

The approach can also be used in design for determining the surface area of the settlement tank. For extensions to existing plants, the SSVI can be measured over a long enough period to give a confident prediction of SSVI values in the future. For a new plant, only rough guidelines for the SSVI can be given at present. A relationship between the SSVI and sludge age, similar to that described by Bisogni and Lawrence (1971), should exist but there are many other variables, including septicity and toxicity of the settled sewage, which have not yet been fully investigated and which influence the shape of the curve. Predictions of SSVI at present are:

(1) For partial-treatment plants with a sludge age less than 1.5 days, where no nitrification is required, the SSVI at 3.5 g l^{-1} may be 80–140 ml g^{-1}.

(2) For plants to give a fully nitrified effluent, with a sludge age greater than 5 days, the SSVI at 3.5 g l^{-1} should be less than 100 ml g^{-1}, falling slightly as the sludge age is increased. A value of SSVI at 3.5 g l^{-1} of 100 ml g^{-1} seems to be a reasonable figure for design purposes for a conventional plant treating domestic sewage. The inclusion of an anoxic denitrification stage before the aeration tank or ensuring a low degree of longititudinal mixing (ensuring plug flow) will increase the confidence of obtaining low values of SSVI.

Table 1   Maximum MLSS Levels to Prevent Sludge Washout

SSVI (at 3.5 g l^{-1}) range	Return sludge flow		Maximum predicted solids handling capacity (kg $m^{-2} h^{-1}$)	Maximum maintainable MLSS levels (g l^{-1}) for a maximum flow of:				
	$\times DWF$	$\dfrac{Q_u}{A}$		2 DWF $\dfrac{Q_0}{A} = 0.5$	2.5 DWF $\dfrac{Q_0}{A} = 0.625$	3 DWF $\dfrac{Q_0}{A} = 0.75$	3.5 DWF $\dfrac{Q_0}{A} = 0.875$	4 DWF $\dfrac{Q_0}{A} = 1.0$
60–79	0.5	.125	2.4	3.75	3.00	2.75	2.40	2.10
80–99	0.5	.125	2.2	3.50	2.75	2.50	2.20	1.90
100–119	0.5	.125	2.0	3.00	2.50	2.25	2.00	1.75
120–139	0.5	.125	1.8	2.75	2.25	2.00	1.80	1.60
60–79	1.0	.25	4.1	5.50	4.75	4.25	3.75	3.25
80–99	1.0	.25	3.8	5.00	4.50	4.00	3.50	3.00
100–119	1.0	.25	3.4	4.50	4.00	3.50	3.00	2.50
120–139	1.0	.25	3.1	4.00	3.50	3.00	2.50	2.00
60–79	1.5	.375	5.5	6.00	5.50	4.75	4.25	4.00
80–99	1.5	.375	5.1	5.50	5.00	4.50	4.00	3.50
100–119	1.5	.375	4.6	5.00	4.50	4.00	3.50	3.25
120–139	1.5	.375	4.1	4.50	4.00	3.50	3.25	3.00
60–79	2.0	.5	6.8	7.00	6.00	5.50	5.0	4.50
80–99	2.0	.5	6.4	6.50	5.50	5.00	4.50	4.00
100–119	2.0	.5	5.7	5.50	5.00	4.50	4.00	3.50
120–139	2.0	.5	5.0	5.00	4.50	4.00	3.50	3.00
60–79	2.5	.625	8.0	7.00	6.50	5.75	5.25	4.75
80–99	2.5	.625	7.4	6.50	6.00	5.25	4.75	4.25
100–119	2.5	.625	6.5	5.75	5.00	4.50	4.25	3.75
120–139	2.5	.625						
60–79	3.0	.75	9.0	7.00	6.50	6.00	5.50	5.00
80–99	3.0	.75	8.3	6.50	6.00	5.50	5.00	4.50
100–119	3.0	.75						
120–139	3.0	.75						

[a] Multiples of DWF (dry weather flow) are given for a particular works, and can easily be derived from Q_u/A for other works.

(3)   For plants operating with sludge ages between 1.5 and 5 days, where nitrification might be incipient or partial, the SSVI at 3.5 g l^{-1} might be 120–160 ml g^{-1}.

All of these SSVI values may be considerably increased where the feed to the aeration tank is septic, or contains a toxic waste, or disproportionate quantities of carbohydrate, N or P, or if there is insufficient oxygen in the aeration tank.

48.2.2   The German Approach

In Germany and many other countries in Europe the dilution sludge index is used. The sludge sample is diluted with final effluent until the settled volume after 30 minutes is about 200 ml. The dilution SVI, I_{sv}, is given by

$$I_{sv} = \frac{\text{settled volume of diluted sludge}}{\text{solids content of diluted sludge}} \tag{13}$$

Since the wall effects of the measuring cylinder are minimal at low solids concentrations and the value of I_{sv} seems to be independent of the original solids content, the shortcomings of the original Mohlman SVI are avoided.

For design the settlement tank area is calculated from overflow rate derived from a new parameter, comparative sludge volume, VS_v, by means of an empirical graph, shown in Figure 5. Comparative sludge volume is the product of the dilution SVI, I_{sv}, and MLSS.

$$VS_v = I_{sv} C_f \tag{14}$$

Guide values for I_{sv} are given, as in Table 2.

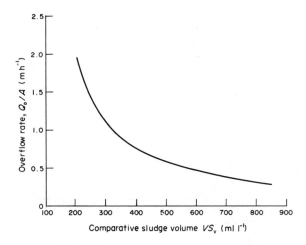

Figure 5 Overflow rate as a function of comparative sludge volume

Table 2 Guide Values for the Dilution SVI

	I_{sv}	(ml g^{-1})
Sludge loading rate (kg BOD kg MLSS^{-1} d^{-1})	$\geqslant 0.05$	$\leqslant 0.05$
Domestic waste water	100	75
Waste water with moderate proportion of organic industrial type	100–150	75–100
Waste water with substantial proportion of organic industrial type	150–200	100–150

A recommendation is also given for the depth in the tank required for the thickening function, an empirical value of $VS_v/1000$ in vertical-flow tanks and $(1 + r)VS_v/1000$ in horizontal-flow tanks, where r is the recycle ratio Q_0/Q_u.

This approach is widely used because it is soundly based on a wealth of operational data, and because the I_{sv} is relatively easy to measure, requiring no special apparatus. On the other hand it is entirely empirical and it does not allow the rate of sludge return to be taken into account.

Recently Koopman and Cadee (1983) have suggested combining the two approaches. They have concluded that dilution SVI was the most useful design variable, having compared it with other settling parameters and total extended filament length. They were able to use I_{sv} in the mass flux theory, and derive a figure similar to Figure 4, shown in Figure 6.

48.3 CLARIFICATION THEORY

When the numbers of particles in the liquid are relatively small, so that they do not affect one another, each particle can be assumed to settle according to Stokes' law. Hazen (1904) first described the theory of sedimentation which equated the settling velocity of the particles to be removed to the upflow or overflow rate (the volumetric inflow rate, Q_0, divided by the cross-sectional area, A). See also Camp (1946).

The situation in a secondary settlement tank is much more complex. It is difficult to estimate the settling velocity of the free-floating particles left after the main blanket has separated, so for design it is usual to use either the settling velocity of the sludge at the feed concentration, or an entirely arbitrary value of 1 m h^{-1}. These approaches probably overestimate the area required for clarification.

In practice enhanced removal can be obtained by natural flocculation in the upward flowing liquid, and if the sludge blanket is above the level of the inlet, it will filter out solids.

On the other hand an effluent poorer than that predicted by Hazen may be obtained if there is

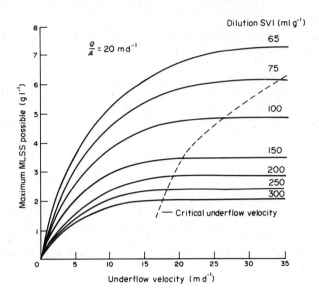

Figure 6 Maximum MLSS as a function of underflow rate and dilution SVI, for an overflow rate of 20 m d^{-1}

too much turbulence in a tank due to poor design of influent or effluent channels. Much work has been done in this area, particularly in the UK and Germany, see for example Price and Clements (1974), Pflanz (1969), Kalbskopf (1974), Resch (1981) and ATV (1981).

A good review is provided by Chapman (1982), who looked at the effect on effluent quality of seven variables: MLSS, sidewater depth, influent flowrate, depth of stilling chamber, rake speed, underflow rate and airflow rate. He found only the first three variables were significant and derived a best-fit equation:

$$C_e = -180.6 + 4.03Cf + 133.24Q_f/A + SWD(90.16 - 62.54Q_f) \quad (15)$$

Even this analysis does not seem to have taken account of the filtering effect of the blanket, the settleability of the sludge particles, or the effects of weir shape, and thus there are still no hard and fast rules for design.

For influent channels the objective is to reduce the incoming kinetic energy as quickly as possible. For rectangular tanks care must be taken to distribute the flow evenly over the whole width of the tank. For circular tanks with a central inlet it is important that the stilling chamber is not too narrow, nor too deep, giving a high downward velocity into the sludge blanket near the sludge outlet. The downward velocity should probably not be greater than 20 m h^{-1}. Sidwick (1977) recommends that the stilling chamber should normally be immersed to one half of the depth at the centre of the tank. The well must also not be too shallow (<1 m), as the incoming kinetic energy will then disturb the surface layers near the effluent channels.

The effluent channels must also be designed to avoid a sudden increase in liquid velocity as the effluent approaches. In the UK a guide value for weir loading is 200 m^3 m^{-1} d^{-1}. In Germany the recommended weir loading is 240 m^3 m^{-1} d^{-1}. They also favour double-sided launders but on the whole it does not seem that these give much better effluents than single-sided launders, and would not justify the extra cost. In the UK the experience is that double-sided launders can reduce the depth available for storage of sludge, and the support structures prevent the use of side-wall scrapers and are difficult to keep clean.

Rectangular tanks are popular in Germany as well, and Kalbskopf (1982) gives a criterion for stable flow that the ratio of depth to length should be at least 1:10 up to about 1:30.

48.4 THEORY OF STORAGE

The mass flux theory of thickening, being a steady-state model, does not allow for accumulation of sludge in the settlement tank. However, it can be seen that when the applied solids loading exceeds the capacity of the tank, sludge will transfer from the aeration tank and accumulate in the settlement tank. The rate of accumulation will be $(F_{appl} - F_{max})$ kg m^{-2} h^{-1}. If an average concentration of the accumulated solids is assumed, the rate of rise of the blanket will be $(F_{appl} - F_{max})/C$ m h^{-1}.

It is difficult to estimate the depth required for storage using the above theory because assumptions have to be made about how long the overload lasts and how $(F_{appl} - F_{max})$ and C change with time.

In the UK storage capacity is not a design parameter, although for many years a rule-of-thumb has been to provide a retention time of 4 hours and experience has shown that storage capacity is normally adequate. In fact it is probably overgenerous.

In Germany an empirical expression for the height of the storage zone has been given as

$$h = \frac{CVI_{sv}}{500A}$$

where C is the difference in MLSS between dry and wet weather, and V is the volume of the aeration tank. The total depth of the tank is made up of the storage zone, the thickening zone (already mentioned), a clear water zone of minimum depth 0.5 m, and a separation zone, which normally has a depth of 1.0 m, but this can be reduced to 0.5 m when the storage zone has a depth greater than 1.0 m. Full details of the German design procedure is given in the ATV Code of Practice (ATV, 1981).

ACKNOWLEDGEMENT

This paper is published by permission of J. L. van der Post, Chief Executive, Water Research Centre. The views expressed are those of the authors and not of their respective employers.

48.5 REFERENCES

ATV (1981). *Guidelines for the sizing of single-stage activated sludge plants for populations above 10,000.* Abwassertechnische Vereinigung e.V. (ATV), Code of Practice A 131.
Bisogni, J. J. and A. W. Lawrence (1971). Relationship between biological solids retention time and settling characteristics of activated sludge. *Water Res.*, **5**, 753.
Camp, T. R. (1946). Sedimentation and the design of settling tanks. *Trans. Am. Soc. Civ. Eng.*, **111**, 895–936.
Chapman, D. T. (1982). The influence of process variables on secondary clarification. *Proc. Water Pollut. Control Fed. Conf.*, *55th*, Missouri.
Dick, R. I. and B. B. Ewing (1967). Evaluation of activated-sludge thickening theories. *J. Sanit. Eng. Div. Am. Soc. Civ. Eng.*, **93**, SA4, 9.
Dick, R. I. and P. A. Vesilind (1969). Sludge Volume Index—what is it? *J. Water Pollut. Control Fed.*, **41**, 1285.
Hazen, A. (1904). On sedimentation. *Trans. Am. Soc. Civ. Eng.*, **53**, 45–88.
Johnstone, D. W. M., M J. Hanbury, A. J. Rachwal and D. J. Critchard (1980). Design and operation of final settlement tanks: use of stirred specific volume index and mass flux theory. *Trib. Cebedeau*, no. 443, **33**, 411–425.
Kalbskopf, K. H. (1974). *Design aspects of biological treatment.* Course 1, IAWPR Workshop, Birmingham.
Kalbskopf, K. H. (1982). Sizing and equipping final clarifiers to ensure satisfactory performance of activated sludge plants. *Gewasser, Wasser, Abwasser*, no. 50. Technical University of Aachen.
Koopman, B. and K. Cadee (1983). Prediction of thickening capacity using diluted sludge volume index. *Water Res.*, in press.
Lockyear, C. F. (1977). *Gravity thickening of biological sludges.* Technical Report TR39. Water Research Centre, Stevenage.
Merkel, W. (1972). Is the Sludge Volume Index a useable parameter for activated sludge. *Korresp. Abwass.*, **19** (8), 171.
Mohlman, F. W. (1934). The sludge index. *Sewage Work Journal*, **6**, 119–122.
Pflanz, P. (1969). Performance of secondary sedimentation basins. In *Advances in Water Pollution Research*, ed. S. H. Jenkins. Pergamon, New York.
Price, G. A. and M. S. Clements (1974). Some lessons from model and full-scale tests in rectangular sedimentation tanks. *Water Pollut. Control*, **73**, 102–113.
Rachwal. A. J., D. W. M. Johnstone, M. J. Hanbury and D. J. Critchard (1982). The application of settleability tests for the control of activated-sludge plants. In *Bulking of Activated Sludge*, ed. B. Chambers and E. J. Tomlinson, pp. 224–243. Ellis Horwood, Chichester.
Resch, H. (1981). New design rules for vertical-flow secondary settlement tanks. *GWF-Wasser/Abwasser*, **H6**, 122, 236–242.
Sidwick, J. M. (1977). *Rationalisation of dimensions and shapes for sewage treatment works construction.* Report 82, Construction Industry Research and Information Association.
Stobbe, G. (1964). Über das Verhalten von belebtem Schlamm in aufsteigende Wasserbewegung. *Veröffentlichung des Institutes für Siedlungswasserwirtschaft*, Heft 18. Technische Hochschule Hannover.
White, M. J. D. (1975). *Settling of activated sludge.* Technical Report TR11. Water Research Centre, Stevenage.

White, M. J. D. (1976). Design and control of secondary settlement tanks. *J. Inst. Water Pollut. Control*, **75**, 459–467.
Yoshioka, N., Y. Holta, S. Tanaka, S. Naito and S. Tsugami (1957). Continuous thickening of homogeneous flocculated slurries. *Chem. Eng. (Tokyo)*, **21**, 66–74.

49

Nitrification in Activated Sludge Processes

E. D. SCHROEDER
University of California, Davis, CA, USA

49.1 INTRODUCTION

Nitrification, the aerobic oxidation of ammonium ion to nitrite and nitrate, has long been observed in Nature. The involvement of bacteria in the process was established by Schloesing and Muntz in 1878. They also noted that the process occurred in two stages: oxidation of ammonium ion to nitrite followed by the oxidation of nitrite to nitrate. Warington (1879) presented supporting evidence for the role of bacteria and also reported an inhibition of nitrification by glucose and that the presence of carbonates was necessary. Ten years later, Winogradsky (1890) isolated nitrifying bacteria in pure culture on a mineral medium. At this point the autotrophic nature of nitrifiers was known, but the inhibition of nitrification by organic material remained a question for many years.

Winogradsky (1890) isolated the two most important nitrifying bacteria in soil, *Nitrosomonas*, which oxidizes ammonium ion to nitrite, and *Nitrobacter*, which oxidizes nitrite to nitrate. Other nitroso bacteria genera that have been isolated are *Nitrosococcus*, *Nitrosospira*, *Nitrosolobus* and *Nitrosovibrio*. Two other genera of autotrophs that oxidize nitrite are *Nitrospina* and *Nitrococcus* (Bremner and Blackmer (1981)). The two organisms isolated by Winogradsky have become models for the overall process and the nitrifying bacteria are usually referred to in the engineering literature as *Nitrosomonas* and *Nitrobacter*.

49.2 STOICHIOMETRY OF NITRIFICATION

The equations describing nitrification and the energy released under optimal conditions were first suggested by Baas-Becking and Parks (1927):

$$NH_4^+ + \tfrac{2}{3}O_2 \rightarrow NO_2^- + H_2O + 2H^+ \tag{1}$$
$$-\Delta G_{298} = 278 \text{ kJ mol}^{-1}$$

$$NO_2^- + \tfrac{1}{2}O_2 \rightarrow NO_3^- \tag{2}$$
$$-\Delta G_{298} = 73 \text{ kJ mol}^{-1}$$

The organisms are not very efficient and approximately 95% of the energy released goes to heat. Yields are estimated to be about 0.15 kg cells per kg ammonia nitrogen oxidized for the first step and about 0.02 kg cells per kg nitrite nitrogen oxidized in the second step. Gujer and Jenkins (1974) have suggested equations (3) and (4) for the coupled oxidation/synthesis reactions:

$$55NH_4^+ + 76O_2 + 109HCO_3^- \rightarrow C_5H_7O_2N + 54NO_2^- + 57H_2O + 104H_2CO_3 \tag{3}$$
$$400NO_2^- + NH_4^+ + 4H_2CO_3 + HCO_3^- + 195O_2 \rightarrow C_5H_7O_2N + 3H_2O + 400NO_3^- \tag{4}$$

Equations (3) and (4) are consistent with the pure culture values of Meyerhoff (1916) with respect to yield. Because cell yield is very low for both reactions it has become common practice to estimate oxygen requirements and H^+ production from equations (1) and (2). Using this method, 4.57 kg O_2 are required to oxidize 1 kg NH_3-N, while inclusion of the cell production results in an estimate of 4.27 kg O_2 required per kg NH_3-N oxidized. The 7% difference is not significant in most cases and provides a small safety factor in design.

49.3 RATE OF NITRIFICATION

The rate of nitrification in pure cultures is dependent upon the concentrations of ammonium ion and nitroso bacteria for the first step and the concentrations of nitro bacteria and nitrite for the second step. Both steps are strongly affected by dissolved oxygen concentration, pH and temperature. Because much of the work carried out on nitrification has been in complicated systems such as soil or sewage treatment, quantitative relationships for these variables are still very approximate. A double Monod model in which nitrogen and oxygen are the limiting concentrations is now commonly used for both reactions:

$$\mu = \mu_m \frac{C_n}{K_n + C_n} \frac{C_o}{K_o + C_o} \tag{5}$$

where μ = specific growth rate (t^{-1}), C_n = nitrogen concentration (g m^{-3}), μ_m = maximum specific growth rate (t^{-1}), C_o = dissolved oxygen concentration (g m^{-3}), K_n = saturation constant for nitrogen (g m^{-3}) and K_o = saturation constant for oxygen (g m^{-3}).

The values of C_n and K_n are of course different for the two steps, as are the maximum specific growth rates. Typical values of these coefficients are given in Table 1. The saturation coefficient for oxygen, K_o, has been somewhat elusive. Current estimates are of the order of 2 g m^{-3} with no separation between the steps. It would be surprising to have the coefficient the same for both steps, but the above value is probably a reasonable approximation in both cases. The high value of K_o relative to K_n is very important where activated sludge type nitrification processes are used because ammonia nitrogen will be present in excess in most cases. This means that oxygen will be the rate limiting variable, at least for the first step.

A second factor to consider in using the coefficients of Table 1 is that they were not obtained in pure culture. This means that, to a certain extent, at least one of the coefficients was obtained by difference. In some cases curve fitting was used to derive the rates. Thus the numbers are at best only reasonable estimates of pure culture values. Of course, nitrification systems are not pure cultures and there is some justification for using 'real world' results. The problems come in extrapolation to new situations, and with nitrification that is a very big problem indeed.

Considering two limiting nutrient concentrations, nitrogen and oxygen, results in Figure 1. The effect of oxygen extends well past 6 g m^{-3} while the effect of nitrogen is primarily at values less than 1 g m^{-3}. In wastewater treatment systems where the nitrifiers are in competition for oxygen with heterotrophs, nitrification essentially stops (or never starts). Operating under very low organic substrate concentrations allows the nitrifiers to compete for oxygen and nitrification proceeds, even at relatively low solids retention times (Poduska and Andrews, 1974). The most

Table 1 Typical Rate Coefficient Values Reported For Nitrifying Organisms[a]

	pH	k_n (d^{-1})	K_n (g N m^{-1})	μ_d (d^{-1})
Ammonia oxidation	7.2	1.5		
	7.8	1.9	0.2–0.4	0.03–0.07
	8.4	0.8		
Nitrate oxidation	6.8	3.3		
	7.8	5.9		
	8.0	5.0	0.2–0.4	0.03–0.07
	8.2	4.8		

[a] Rate values are calculated as g of N oxidized per g nitrifying cells per day. Values are for 20–22 °C.

obvious method of operation is extended aeration (*i.e.* SRT values of 15–20 d), but long hydraulic detention times at low solids retention times will also be satisfactory.

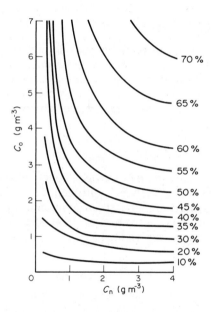

Figure 1 Fraction of maximum nitrifier growth rate obtained as a function of nitrogen and oxygen concentration for $K_n = 0.3$ g m^{-3} and $K_o = 2$ g m^{-3}

In most cases, activated sludge systems are operated at dissolved oxygen concentrations below 2 g m^{-3}. If the K_o value is actually as high as 2 g m^{-3} the maximum growth rates would be less than 45% of the potential values. Raising the rate above this level would require pushing the dissolved oxygen concentration into a region that has generally been considered uneconomical. The upper dissolved oxygen concentration value of 7 g m^{-3} shown in Figure 1 is a practical limit for air systems. Pure oxygen activated sludge systems can exceed this value, but for economic reasons are operated in the 3–4 g m^{-3} range. Thus nitrification systems can be expected to operate in the dissolved oxygen concentration sensitive region in virtually all cases.

A final note on the coefficient values reported in the wastewater treatment literature should be made. Many of the reported values are derived from work with laboratory scale experimental systems. These systems are often operated near dissolved oxygen saturation and without automatic control of recording of the values. This adds to the difficulty of extrapolating from the literature to the field. The reason this situation has occurred is that organic oxidation appears to be far less sensitive to oxygen concentration than nitrification, with saturation being essentially complete at values of 1 g m^{-3} which would indicate K_o of less than 0.1 g m^{-3}.

Why oxygen saturation coefficients for organic and nitrogen oxidation would be an order of magnitude different is unknown, but the fact provides an explanation for the apparent inhibition of nitrification by organic compounds first reported by Warington (1879) and noted by many workers since that time. Heterotrophic bacteria have considerably higher growth rates than the autotrophic nitrifiers and are able to operate at near maximum values even at low dissolved oxygen concentrations. Bulk dissolved oxygen concentration may be in the design range of 1.0–2.0 $g\ m^{-3}$, but microsite values would be much less. Nitrification then ceases until organic oxidation stops. Support for this concept is supplied from studies of nitrification in conventional, nominally plug flow activated sludge processes and in trickling filters. In both cases nitrification follows organic removal.

49.3.1   Maintenance Energy or Endogenous Respiration Rate

Equation (5) does not include a maintenance or endogenous rate factor. Values for μ_d ranging from 0.003 to 0.08 d^{-1} have been reported in the literature. Values used in design are usually similar to those used for heterotrophic organisms: 0.05–0.07 d^{-1}. The modified form of equation (5) is then:

$$\mu = \mu_m \frac{C_n}{K_n + C_n} \frac{C_o}{K_o + C_o} - \mu_d \tag{6}$$

49.3.2   Measuring Nitrifier Concentration

In wastewater treatment systems, nitrifying bacteria do not occur in pure culture and it is not practical to analyze for probable concentrations of these microorganisms. As a result, concentrations are estimated using coefficients from the literature, the nutrient concentrations and the operating set points. This can be done by making a mass balance on nitrogen and solving for the required cell concentration. For the simple example of a continuous flow stirred tank reactor without recycle the resulting equation is:

$$x = \frac{D(C_{ni} - C_n)}{q_n} \tag{7}$$

where D = dilution rate (t^{-1}), C_{ni} = influent nitrogen concentration $(g\ m^{-3})$, C_n = reactor bulk nitrogen concentration $(g\ m^{-3})$ and q_n = specific nitrogen removal rate (t^{-1}).

The nitrogen concentrations used must be either ammonium or nitrite and the resulting cell concentrations will be either nitroso or nitro bacteria. The specific removal rate is stoichiometrically related to the growth rate by the net yield, Y_G:

$$\begin{aligned} q_n &= \mu/Y_G \\ Y_G &= Y/(1 + \mu_d/\mu_g) \end{aligned} \tag{8}$$

The nitrifier concentrations predicted using equation (7) and the coefficient values given above are very small. For example, a feed ammonium concentration of 100 $g\ m^{-3}$ and a detention time of 7 d $(D = 0.14\ d^{-1})$ would result in an effluent ammonium ion concentration of approximately 7.5 $g\ m^{-3}$ and a nitroso bacteria concentration of less than 8 $g\ m^{-3}$. These very low cell concentrations result in potential stability problems, as well as making the modeling process somewhat questionable.

49.3.3   Minimum Solids Retention Time (SRT)

The solids retention time concept has been discussed in previous chapters. In the case of nitrification it can be seen from the rate coefficient values given in Table 1 that theoretical minimum SRT values are of the order of 7 d. In fact voluminous amounts of data exist for excellent nitrification occurring at SRT values as low as 2.5 d (Grady and Lim, 1980). A rational explanation for this phenomenon is that the nitrifier population is attached to reactor walls and on the surface of impellers.

49.3.4   Temperature Effects

As with other biological wastewater treatment processes, the effect of temperature is included in the rate coefficient. Stankewich (1972) has suggested using 15 °C as a reference temperature and the relationship given by equation (9).

$$\mu_{mT} = \mu_{m15} \exp \{b(T-15)\} \tag{9}$$

where μ_{mT} = maximum specific growth rate at $T\,°C\,(t^{-1})$, μ_{m15} = maximum specific growth rate at 15 °C (t^{-1}) and b = empirically derived coefficient. Stankewich found that the values of b were about 0.1 and 0.06 for ammonium and nitrite oxidation, respectively.

49.3.5   pH Effects

The effect of pH on nitrification is twofold. Firstly, the fraction of ammonia present as ammonium ion is strongly pH dependent, as can be seen from equation (10):

$$\text{Fraction NH}_3\text{-N} = \frac{1}{1 + K_{NH_3}\,[H^+]/K_w} \tag{10}$$

The equilibrium coefficient K_{NH_3} and the ionization constant of water, K_w, are both temperature dependent. Consequently the fraction varies with temperature. At temperatures below 40 °C the ammonium ion fraction is greater than 60% for pH values up to 9. At 15 °C, a more normal operating temperature, 77% of the ammonia nitrogen is ionized at pH 9 and 92% at pH 8.5.

The second pH effect is on the nitrifying bacteria. Meyerhoff (1917) reported that *Nitrosomonas* functioned in the pH range 7.6–9.4, with an optimum in the region 8.5–8.8, and that *Nitrobacter* was active in the pH range 5.7–10.5, with an optimum in the range 8.5–9.3. These ranges are not unusual, but the optima are higher than for most species. Srinath *et al.* (1976) reported optimum pH values for *Nitrosomonas* between 7 and 8, and optimum values for *Nitrobacter* between pH 7.7 and 8. The lower values are not readily explainable. Srinath *et al.* were working with heterogeneous cultures and the values they reported may be for other species of nitrifiers or the nitrifiers may have been protected in some manner. Support for the lower values exists in the fact that nitrification readily occurs in biooxidation systems operating at pH values in the 7–8 range. Whether the nitrification rates are optimum with respect to pH is of course unknown.

49.3.6   Inhibition of Nitrification

Inhibition of nitrification by a wide variety of compounds occurs, but the most effective are metal chelaters. Tomlinson *et al.* (1966) studied the effect of 31 compounds and identified concentrations at which approximately 75% inhibition occurred. Of the 31 compounds, eight were inhibitory at concentrations of 0.08 g m^{-3}, while thiosemicarbazide, thioacetamide, sodium methyl dithiocarbamate and methyl isothiocyanate were inhibitory at concentrations less than 1.0 g m^{-3}. Phenol, a compound that commonly occurs in relatively high concentrations, was inhibitory at 5.6 g m^{-3}, and is the only compound on the list likely to occur in any but a very specialized wastewater at inhibitory concentrations.

49.4   ALKALINITY

Alkalinity is important in nitrification because of the strong pH rate relationships described above, and because of the alkalinity destroyed in the first step as shown in equation (1). Two hydrogen ions are produced for each mole of ammonium ion oxidized. Thus two equivalents of alkalinity are destroyed per equivalent of ammonium oxidized. Municipal wastewaters typically contain 30–50 g m^{-3} or 2.1–3.6 eq m^{-3} of total Kjeldahl nitrogen (TKN), most of which is available for oxidation, and therefore nitrification of municipal wastewater will result in the destruction of 4–7 eq m^{-3} of alkalinity. Industrial wastewaters have varying amounts of available nitrogen and a general statement cannot be made, but the concept is the same.

Alkalinity of wastewaters varies considerably, with the most important factor being the water source characteristics. Domestic use usually increases the alkalinity of a water by 2–3 eq m^{-3}. If a water has only 1 or 2 eq m^{-3} of alkalinity to begin with, the total alkalinity of the wastewater will probably be too low and nitrification will be stopped as the pH decreases.

In cases where natural alkalinity of a wastewater is not great enough to prevent an unacceptable drop in pH, chemical addition is the best method of control. Either lime or buffer can be used. Because the acceptable pH range is relatively broad the control mechanisms need not be sophisticated. It is probably better to use a buffer if the chemicals are to be added directly to the reactor. Lime addition to the feed should be satisfactory in well stirred systems, or where the feed pH will not be driven above 9.5.

49.5 DESIGN OF NITRIFICATION SYSTEMS

A number of important design considerations have been discussed above: pH, temperature, alkalinity, dissolved oxygen concentration and nitrogen (ammonium and nitrite). Temperature is generally uncontrollable because of the quantities of wastewater that must be processed, and the nitrogen concentration is given also. The alkalinity pH balance and the dissolved oxygen level can and must be controlled, however. In addition to chemical control, the loading rates and the process configuration are important design variables. Both are coupled to the alkalinity pH balance and dissolved oxygen concentration, as will be explained.

49.5.1 Loading Rates

Both the organic and nitrogen loading rates are significant design variables. The organic loading rate is important because nitrification will not proceed while heterotrophic microorganisms are actively metabolizing organic material. In practice this usually means an organic loading rate (food to microorganisms ratio or **F/M** of less than 0.15 d^{-1}:

$$\mathbf{F/M} = FC_{si}/Vx < 0.15 \ d^{-1} \tag{11}$$

where C_{si} is based on ultimate BOD. It is important to remember that the loading rate is a time averaged parameter and means nothing on an instantaneous basis. Thus **F/M** is independent of process configuration, at least in the sense that it is used in design.

It is conceivable to use kinetic expressions to predict the organic concentration and the resulting oxygen uptake rates in the reactor, but this would be an exercise in futility considering the variability of wastewater characteristics and flow, and the difficulty of establishing both kinetic models and kinetic coefficients.

Nitrogen loading rates are not currently used in design, but useful parameters can quickly be developed. The criteria are stoichiometric nitrogen requirements and oxygen demand for nitrification. The stoichiometric nitrogen requirement is a function of the nitrogen assimilated during growth. Using the empirical cell formulation $C_5H_7O_2N$ we can determine that the assimilative requirement is approximately 0.12 kg N kg^{-1} cells produced. Chapter 47 on activated sludge design equations gives a relationship for the net cell production or yield, Y_G, given above as equation (8). Multiplying the net yield by the assimilative requirement gives the approximate nitrogen requirement, for a given set of operating conditions, as a function of ultimate BOD removed. If the available nitrogen concentration is less than the requirement an unstable system will result, but if the assimilative requirement is exceeded the excess nitrogen is available for nitrification and is a potential source of oxygen demand. Thus nitrogen loading should be stated in terms of excess nitrogen. This term will be defined in terms of mass of excess nitrogen entering the system per unit time and given the symbol L_{n_c}:

$$L_{n_c} = F(C_{ni} - 0.12Y_G(C_{si} - C_{so})) \tag{12}$$

where C_{so} = effluent BOD_u (g m^{-3}).

If $L_{n_c} < 0$, ammonia or nitrate will need to be added to maintain stability, while if $L_{n_c} > 0$ the positive amount will be a potential oxygen demand with a stoichiometric requirement of 4.57 L_{n_c} (g O_2 d^{-1}).

49.5.2 Example: Nitrogen Loading

A wastewater with BOD_u and TKN concentrations of 250 and 60 g m^{-3} respectively has an average flow rate of 0.15 m^3 s^{-1}. If the wastewater is to be treated in an activated sludge system operated at an SRT of 15 d, determine the oxygen requirement for nitrification.

Solution

Assume that the theoretical yield, Y, is 0.4 g cells $g^{-1}BOD_u$ and the rate of endogenous respiration is 0.06 d^{-1}. Using equation (8) the net yield is:

$$Y_G = 0.4/(1 + 0.06(15)) = 0.21 \ \text{g cells} \ g^{-1} \ BOD_u$$

The excess nitrogen is:

$$L_{n_c} = 0.15 \{60 - 0.12(0.21)(250 - C_{so})\}$$

At an SRT of 15 d the effluent BOD_u will be negligible: $C_{so} \approx 0$, and thus

$$L_{n_c} = 8.06 \text{ g s}^{-1} = 696 \text{ kg d}^{-1}$$

Assuming complete nitrification, which like complete BOD_u removal is a good assumption, the stoichiometric oxygen demand is 3180 kg d^{-1}. The oxygen requirement for BOD removal is 2300 kg d^{-1} and therefore nitrification more than doubles the oxygen demand.

It is important to remember that in systems such as the one described in the above example, nitrification will occur unless it is prevented. The simplest ways of preventing nitrification are to limit the oxygen supply and to allow the pH to drop below 6.5.

The rate of oxygen supplied to the reactor(s) in $\text{g m}^{-3} \text{ d}^{-1}$ is an important design number also because the upper limit is approximately $1500–1800 \text{ g m}^{-3} \text{ d}^{-1}$ using conventional equipment. This practical limit is most strongly related to minimum reactor size. If a stirred tank reactor is used the oxygen uptake rate for organic removal is added to that for nitrification, while if a nominally plug flow system is used the carbon removal will occur upstream of the nitrification. Where the heterotrophic and autotrophic reactions are occurring simultaneously the reactor will need to be large enough to allow the nitrifiers to compete for oxygen.

49.5.3 Appropriate Process Configurations

A number of mixed system configurations are in use. The most common type is to use a suspended growth (activated sludge) system for organic removal and an attached growth (trickling filter or rotating biological contractor) for nitrification. This section will not discuss such combinations, but will be limited to the totally suspended growth, activated sludge type configurations. In the United States the most common configuration used in nitrification is the 'one sludge' system. A few 'two sludge' systems have been designed and built and the 'Carousel' process, which was developed in Europe, is coming into use in the United States currently. One of the most exciting new configurations is the sequencing batch reaction (SBR). In all of the activated sludge type systems the nitrifying organisms appear to be attached to floc or on reactor surfaces. The stoichiometry of nitrification, as was discussed above, does not allow the development of high concentrations of cells, even if no wasting is practised. If the MLSS concentration of the systems is very low, an upset will result in a large fraction of the cells being lost. Because nitrifiers grow very slowly, recovery will be very slow also, and the system will be both unreliable and unstable.

49.5.4 Example: Solids Stability of Nitrifying System

Effluent from an activated sludge process is to be nitrified. The flow rate is $0.2 \text{ m}^3 \text{ s}^{-1}$ and the NH_3-N concentration is 35 g m^{-3}. Effluent NH_3-N should not exceed 0.5 g m^{-3}. Determine if the system can be operated as a separate unit.

Solution

The important factors are to determine the average sludge age that must be exceeded to obtain the nitrification and the necessary secondary clarifier efficiency that will maintain the culture.

Several assumptions must be made. Dissolved oxygen concentration = 2.0 g m^{-3}, $\mu_m = 0.4 \text{ d}^{-1}$ for ammonia oxidation and 0.6 d^{-1} for nitrite oxidation, $K_n = 0.3 \text{ g m}^{-3}$ for both reactions, $K_o = \text{g m}^{-3}$ for both reactions, $Y = 0.15$ for ammonia oxidation and 0.06 for nitrite oxidation and $\mu_d = 0.05 \text{ d}^{-1}$ for both reactions. A temperature assumption is implicit in estimating the rate coefficients.

The required average SRT can be determined using equation (6):

$$\mu = \mu_m \{C_n/(K_n + C_n)\} \{C_o/(K_o + C_o)\} - \mu_d$$

For Ammonia Oxidation

$$\mu = 0.4\{0.5/(0.3 + 0.5)\} \{2/(2 + 2)\} - 0.05 = 0.075 = SRT^{-1}$$

The minimum SRT to meet the ammonia standard is then 13.3 d. Using SRT = 13.3 d, equation (8) and the coefficients for nitrite oxidation, a nitrite concentration of 0.21 g m^{-3} can be calculated. This indicates that nearly complete nitrification will occur. The resulting cell production can be estimated using the net yield, Y_G, and the mass nitrogen removals. For the first step, ammonia oxidation:

$$Y_G = 0.15/(1 + 0.05/0.075) = 0.09$$

The mass NH_3-N removal is:

$$M_n = \{0.2 \text{ m}^3 \text{ s}^{-1} (35 - 0.5)\} = 6.9 \text{ g } NH_3\text{-N s}^{-1} = 596 \text{ kg } NH_3\text{-N d}^{-1}$$

Production of *Nitrosomonas* is:

$$M_x = 0.09(6.9) = 0.62 \text{ g s}^{-1}$$

For nitrite oxidation the problem is similar. The net yield is:

$$Y_G = 0.05/(1 + 0.05/0.075) = 0.03$$

The mass NO_2-N removal rate is:

$$M_n \; 0.2 \text{ m}^3 \text{ s}^{-1} (35 - 0.5 - 0.2) = 6.86 \text{ g s}^{-1} = 593 \text{ kg } NO_2\text{-N d}^{-1}$$

Production of *Nitrobacter* is:

$$M_x = 0.03(6.86) = 0.21 \text{ g s}^{-1}$$

The rate of cell loss from the secondary clarifier must be based on a reasonable effluent suspended solids concentration, such as 20 g m^{-3}. This gives a loss rate of 4 g s^{-1}, which is nearly five times the total production of nitroso and nitro bacteria. Thus the system cannot survive as a completely separate unit. A number of options exist, some of which will be discussed below.

49.5.5 One Sludge Systems

The terms 'one sludge' and 'two sludge' came about in the early discussions of possible configurations for nitrifying processes. They are simple methods of describing the configurations shown in Figure 2. In the one sludge system it is assumed that the nitrifying bacteria are attached to the activated sludge floc, or the surfaces of the reactor. The process is really no different from a conventional activated sludge process, except that oxygen transfer requirements must include the nitrogenous demand and pH control must be included in some manner. The fact that some nitrifying systems perform well at SRT values below the minima predicted by literature values is significant, but use of SRT values less than 10 days in design must be considered very risky. Once a process is in operation, lowering the SRT value is possible because all of the physical requirements of a low SRT system will be met by a system designed for a long SRT.

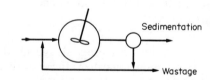

(a) One – sludge nitrification activated sludge

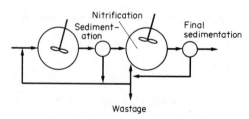

(b) Two – sludge nitrification activated sludge

Figure 2   Schematics of nitrification activated sludge processes

49.5.6 Two Sludge Systems

Two sludge systems are designed to carry out the organic removal and nitrification steps in two separate reactors (Figure 2). In such a system the first reactor can be designed as a low SRT process with the consequent lower oxygen requirements. As suggested by the solids stability example above, the second reactor will be unstable unless a method of increasing the MLSS concentration is applied, however. Increasing the SRT in the second (nitrification) reactor will increase the solids concentration, but stability will remain a problem because recovery from any upset will be very slow. Solids retention times greater than 60 days are usually necessary to increase MLSS concentrations to acceptable values. A second approach is to design the second reactor as a combination aerobic sludge digestor and nitrification unit (Grady and Lim, 1980). In such a system, all, or a portion, of the waste solids from the first reactor are discharged to the second reactor. The nitrifiers will grow as part of the floc and a relatively stable system will result.

49.5.7 Sequencing Batch Reactors (SBRs)

Sequencing batch reactors have been discussed in previous chapters and the fundamentals of the process will not be presented here. A time based schematic of the process is given in Figure 3. The important features of the SBR are:
1. Flow equalization is an inherent part of the fill phase.
2. Sedimentation is truly quiescent.
3. Use of anoxic, no-mix fill greatly reduces the possibility of filamentous bulking.
4. Organic stabilization during fill does not require oxygen.
5. Phase times can be modified as required.

With respect to nitrification, it is the last feature that is important. Aerobic reaction phase times required for organic removal are of the order of 15 minutes to one hour. Nitrification requires extending the aerobic reaction period to 3–4 hours. The system is analogous to the continuous flow single sludge system in many respects. Irvine (1982) has reported essentially complete nitrification in the full scale SBR system at Culver, Indiana, and his results should be typical of other SBR systems.

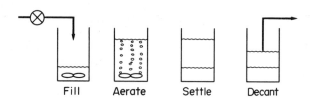

Fill   Aerate   Settle   Decant

Figure 3   Schematic in time of the SBR activated sludge process

49.6 REFERENCES

Baas-Becking, L. G. M. and G. S. Parks (1927). Energy relations in the metabolism of autotrophic bacteria. *Physiol. Rev.*, **7**, 85.
Bremner, J. M. and A. M. Blackmer (1981). Terrestrial Nitrification as a Source of Atmospheric Nitrous Oxide. In *Denitrification, Nitrification and Atmospheric Nitrous Oxide*, ed. C. C. Delwiche. Wiley, New York.
Grady, C. P. L. and H. C. Lim (1980). *Biological Wastewater Treatment*. Dekker, New York.
Gujer, W. and D. Jenkins (1974). *The Contact Stabilization Process. Oxygen and Nitrogen Mass Balances*. SERL Report 74-2, University of California, Berkeley.
Irvine, R. L. (1982). *Full-Scale Study of Sequencing Batch Reactor: Tenth Quarterly Progress Report*. USEPA Grant No. R-806598-01-3, Department of Civil Engineering, University of Notre Dame, Indiana.
Meyerhoff, O. (1916). Untersuchungen uber den atmungsvorgang nitrifizierender bakterien. *Pflug. Arch.*, **164**, 353.
Meyerhoff, O. (1917). Untersuchungen uber den atmungsvorgang nitrifizierender bakterien. IV. Die atmung des nitritbilders und ihre beeniflussung durch chemische substanzen. *Pflug. Arch.*, **166**, 240.
Poduska, R. A. and J. F. Andrews (1974). Dynamics of nitrification in the activated sludge process. In *Proceedings*, *29th Industrial Waste Conference*, Purdue University Engineering Extension Series, No. 145, p.1005. Purdue University, Indiana.
Schloesing, T. and A. Muntz (1878). Recherches sur la nitrification par les ferments organises. *C. R. Acad. Sci.*, **84**, 1018.
Srinath, E. G., R. C. Loehr and T. B. S. Prakasm (1976). Nitrifying organism concentration and activity. *J. Environ. Eng. Div., Am. Soc. Civ. Eng.*, **102**, 449.

Stankewich, M. J. (1972). Biological nitrification with the high purity oxygen process. *Proceedings*, *27th Industrial Waste Conference*, Purdue University Engineering Extension Series, No. 141, p.1. Purdue University, Indiana.

Tomlinson, E. G. *et al.* (1966). Inhibition of nitrification in the activated sludge process of sewage disposal, *J. Appl. Bacteriol.*, **29**, 266.

Warington, R. (1879). On nitrification. *J. Chem. Soc.*, **35**, 429.

Winogradsky, S. (1890). Recherches sur les organisms de la nitrification. *Ann. Inst. Pasteur*, **4**, 213.

50

High Intensity Systems in Activated Sludge Processes

A. L. DOWNING
Binnie and Partners, Consulting Engineers, London, UK
and
A. G. BOON
Water Research Centre, Stevenage, Herts., UK

50.1 INTRODUCTION

The intensity of treatment achievable in an activated-sludge plant is a complex function of the performance of both the aeration units and the settlement tanks. The rate at which polluting matter can be removed in the aeration units depends basically on: the concentration of sludge that can be maintained in the units; the specific growth rate of the organisms within the sludge (*i.e.* their rate of increase in concentration per unit concentration present); and the ability of the aeration equipment to supply oxygen at the rate necessary to meet the respiratory demand of the organisms.

The concentration of sludge that can be maintained in the aeration units and the quality of final effluent produced in respect of suspended solids and their associated BOD are functions of the performance of the settlement tank which in turn are dependent on the influence of conditions in the aeration units on sludge settlability.

In plants equipped with settlement tanks of conventional size the percentage removal of BOD from settled sewage is usually found to vary with sludge loading (the mass of BOD applied in unit time per unit mass of activated sludge in the plant) according to a curve such as that of Figure 1. The implication of this curve is that the maximum BOD load per unit volume of aeration unit, and thus the minimum retention time for production of effluent of given quality, is inversely proportional to concentration of sludge in the aeration units, at least up to the point at which the sludge can no longer be separated satisfactorily in settlement tanks of economical size.

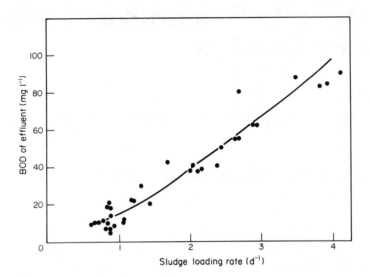

Figure 1   Relation between sludge loading rate of settled domestic sewage (average BOD 250 mg l^{-1}) and BOD of final effluent

The performance of settlement tanks is a function of several variables, including primarily solids loading (mass of suspended solids applied per unit area per unit time), surface overflow rate (volume applied per unit area per unit time), underflow rate (mass of sludge removed per unit area from the bottom of the tanks per unit time) and sludge settlability (usually measured in modern practice as the stirred sludge volume index, SSVI). The magnitudes of operating range of the first three of these variables are normally chosen by designers in the expectation that the SSVI will have an acceptable value. However, no validated method is currently available for predicting SSVI accurately, though many of the factors which influence it have been identified (Chambers and Tomlinson, 1982). One such factor is the degree of turbulence to which the sludge is subjected during aeration (Downing and Wheatland, 1962). If this is excessive the sludge will settle poorly. However, for a given type of aeration system, increasing the rate of solution of oxygen to meet increasing demand from increased organic loading usually involves increasing the degree of turbulence. The adverse effects of excessive turbulence on settlability thus impose restraints on the intensity of treatment that can be achieved using certain types of conventional aeration equipment. Other restraints are imposed by the necessity to prevent excessive foaming and by the physical size of the aerators relative to the size of the aeration tanks.

50.2   OXYGEN REQUIREMENTS

The average rates of solution of oxygen required to satisfy the demand from carbonaceous oxidation processes in activated-sludge plants can be estimated from the empirical relation (Eckenfelder and O'Connor, 1961; Houck and Boon, 1981):

$$R = 0.75D(B_s - B_e) + 0.002C \qquad (1)$$

in which R is the rate of solution of oxygen in mixed liquor (mg l^{-1} h^{-1}); B_s, B_e are the BODs of the wastewater and treated effluent (mg l^{-1}); C is the concentration of activated sludge in the mixed liquor (mg l^{-1}); and D is the dilution rate, *i.e.* the flow of wastewater per unit volume of the aeration units (h^{-1}).

If the plant is designed to achieve full nitrification, then additional oxygen must be supplied at a rate R_n given with adequate accuracy by

$$R_n = 4.5AD \qquad (2)$$

where A is the concentration of ammoniacal nitrogen in the inflowing wastewater (mg l^{-1}).

By combining equations (1) and (2) with the curve of Figure 1, the relations between the rates of solution of oxygen required in nitrifying and non-nitrifying plants, sludge loading and concentration of sludge maintained when treating a settled sewage having a BOD of 200 mg l^{-1} and an ammonia content of 30 mg l^{-1} (roughly typical of many in Europe) are estimated to be as in Figure 2.

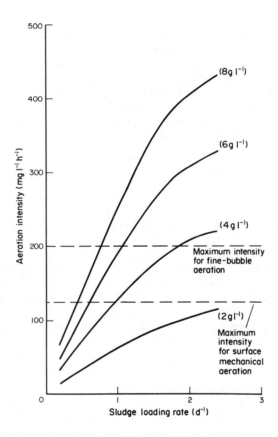

Figure 2 Variation of aeration intensity required to satisfy the BOD of settled sewage (250 mg l^{-1}) as calculated for operation of an activated-sludge plant operated over a range of sludge loadings and at various biomass concentrations (shown in parenthesis by each curve)

Also shown in Figure 2 are the approximate maximum rates of solution normally achievable using commercially available traditional types of mechanical and diffused-air aeration equipment. The oxygenation capacities of such systems fall well below the average demands at high sludge loadings. In practice, demands at any given point in the aeration systems will depend on the extent of longitudinal mixing in the aeration units. In plants in which such mixing is restricted the demand will tend to be higher at the inlet end of the units than the average, and lower towards the outlet. The shortfall between supply and demand would thus be even greater than indicated in Figure 2 at the inlet ends of aeration units equipped with traditional systems.

50.3 METHODS OF INTENSIFICATION

Attempts to intensify activated-sludge processes have sought to bring this about mainly by one or more of the following: increasing rates of solution of oxygen; increasing rates of growth of the organisms within the sludges; improving sludge settlability; and increasing the concentration of sludge that can be maintained under aeration.

Much attention in the 1960s and early 1970s was given to the first of these methods because earlier investigations had revealed that the performances of many older plants were limited by their low oxygenation capacities, and oxygenation capacity could be increased by means such as use of oxygen in place of air, which it was thought might also produce other benefits (see later).

In the last decade much of the impetus has been towards finding not only ways of improving oxygen transfer rates but also of using physical means of retaining sludge in the aeration units in much higher concentration than had hitherto been possible. Broadly, the plants that have been evolved from these efforts can be classified into those using air and those using oxygen for aeration, with each of these categories being subdivisible into those in which the activated sludge is maintained as in traditional plants, without addition of an inert support, and those in which the sludge is largely attached to inert particles. This chapter describes the principal features and performance of such systems.

By using oxygen instead of air, the rate at which oxygen dissolves under similar conditions of temperature and total pressure can be increased at least five-fold. Since the rate of oxygen transfer is proportional to the deficit from the saturation concentration, the difference in rate achievable using oxygen from that using air increases in favour of the former as the concentration approaches the air-saturation value. Among the arguments that have been advanced by many authors for the use of oxygen (particularly Albertsson *et al.*, 1970; Bebin and Renaudin, 1976; Chapman *et al.*, 1976; Jewell *et al.*, 1971; Karnovsky, 1975; McWhirter, 1978) are those for the treatment of wastewaters of high BOD: (i) higher rates of oxygen transfer could be achieved at much lower levels of turbulence than in conventional aeration systems, permitting sludges of higher density and thus improved settlability to be produced, and (ii) higher concentrations of dissolved oxygen could be readily maintained, permitting oxygen to penetrate deeper into the flocs and possibly thereby increasing the activity of those microorganisms to which degradable pollutants could also diffuse and the rate of aerobic digestion of those to which diffusion of pollutants was restricted.

While the ability to achieve higher rates of transfer and higher concentrations of dissolved oxygen at lower levels of turbulence than in traditional aeration systems is unquestioned, the evidence for the other suggested benefits is conflicting. Several studies (Downing and Bayley, 1961; Kalinske, 1971; Mueller *et al.*, 1968; Thabaraj and Gaudy, 1969; Wuhrmann, 1963) have indicated that rates of oxidation of carbonaceous matter and ammonia increase with increasing concentration of dissolved oxygen but only in the range from zero to about respectively 0.5 mg l^{-1} and 2 mg l^{-1}. Chapman *et al.* (1976) and Knudson *et al.* (1982) have suggested that the extent of any benefit from increasing concentration of dissolved oxygen may depend upon the diameter of the activated-sludge flocs, the critical 'diameter' above which beneficial effects are noted being about 400 μm, whereas the average floc diameter of activated sludge has been quoted by Aiba *et al.* (1964), Knudson *et al.* (1982) and Mueller *et al.* (1966) to be less than 200 μm. Wuhrmann (1963) concluded that concentrations of dissolved oxygen of 1.5 to 2.5 mg l^{-1} would adequately keep the flocs aerobic even if the 'diameter' were 400 to 500 μm. We discuss some of the other evidence on these aspects further in subsequent sections dealing with the performance of individual processes.

Several methods have been devised for increasing oxygenation capacities in aerated systems. In some of these diffused-air aeration is combined with mechanical agitation. Another method involves injecting air into mixed liquor circulated through a deep shaft as described by Hines *et al.* (1975), and Bolton and Ousby (1976). The combination of hydrostatic pressure and velocity of circulation (and its associated shearing action) is such as to ensure that the air is almost totally dissolved by the time the liquor reaches the bottom of the shaft.

In considering the relative merits of these newer types of system, costs are obviously important. In the case of oxygen activated-sludge systems, the cost of producing oxygen has particularly important bearing on total costs and in the next section methods of production and use of oxygen are briefly reviewed.

50.4 OXYGENATED PROCESSES

50.4.1 Production of Oxygen

50.4.1.1 Techniques

The cryogenic process, involving liquefaction of air followed by fractional distillation with heat exchange to cool the incoming air with liquid nitrogen, is widely used for the commercial production of oxygen, and plants having a daily production of 1500 t of oxygen of 99.7% purity are in operation. The demand for oxygen at sewage treatment plants is on a smaller scale often falling within the range 1–20 t oxygen daily. For this purpose the process known as 'pressure-swing adsorption' (PSA) has been developed and is described elsewhere by Davis (1972) and Lee (1973). This involves the adsorption of nitrogen from air at a pressure above atmospheric (say, 200–300 kPa) by a zeolite and its subsequent desorption at atmospheric pressure or below; the pressures used differ from one manufacturer to another. The process produces oxygen-enriched air containing between 70 and 95% oxygen, depending on conditions of operation.

Experience, described by Blachford *et al.* (1982), with a full-scale PSA plant at a sewage treatment works in the UK has shown that the plant has operated without serious failure since it was commissioned in August 1976. Nevertheless, to ensure continuity of oxygen supply in the event of mechanical failure in the PSA plant, adequate standby facilities on site are normally provided, including storage capacity for liquid oxygen supplied by tanker.

50.4.1.2 Costs

At production levels of 1–20 t d^{-1} (12–230 g s^{-1}) of oxygen a PSA plant has a lower capital cost than the equivalent cryogenic plant, though it requires more energy for operation (Lee, 1973; Smith and Armand, 1973). At rates above 20 t d^{-1} (230 g s^{-1}) the capital cost of a cryogenic plant is generally less than that of a PSA plant of similar output. The cost of oxygen bought from a commercial supplier depends on the quantity required and on the location to which it has to be delivered. In the UK for daily quantities between 2 and 10 t the price per tonne might be about £45 (January 1982). The cost of providing adequate standby capacity must also be included in the total capital cost.

50.4.1.3 Energy requirements

The energy required to produce a tonne of oxygen is not greatly affected by the scale of production. It is reported (Lee, 1973; Smith and Armand, 1973) to be about 350 kW h t^{-1} (1.26 MJ kg^{-1}) for a cryogenic plant compared with about 600 kW h t^{-1} (2.2 MJ kg^{-1}) for a PSA plant. However, when the demand for oxygen falls below the design output of the PSA plant, the energy required is not reduced proportionally and the equivalent energy may exceed 1000 kW h t^{-1} (3.6 MJ kg^{-1}). New methods of control to avoid excessive use of energy during operation at low rates of output are being developed, but to date these have not reduced the equivalent use of energy below 1000 kW h t^{-1} (3.6 MJ kg^{-1}) when the output was reduced by about 50% (Blachford *et al.*, 1982). Recent experience in the UK with a large plant (16 tonnes oxygen d^{-1}) has shown that energy consumption can be as low as 500 kW h t^{-1} even with a 2:1 turndown of oxygen output.

50.4.2 Dissolving Oxygen

50.4.2.1 Techniques

To avoid wastage of oxygen it must be dissolved efficiently into the liquor of a treatment plant by methods which minimize dilution with or loss into atmospheric air. Several techniques have been developed, including the following: (1) agitation of mixed liquor (biomass plus sewage) with oxygen in an enclosed tank (UNOX, OASES and GIEDEL); (2) dispersion of oxygen as microbubbles at the bottom of an open tank (MAROX); (3) injection of oxygen *via* a venturi into a sidestream of mixed liquor which is returned to a conventionally-aerated tank (VITOX); (4) injection of oxygen to a downflowing stream of sewage the velocity of which is designed to retain the bubbles of gas until they have dissolved (MEGOX and OXITRON).

The names in capital letters and shown in parentheses are registered trade names for the processes by the companies involved.

(i) UNOX

This process was developed in the USA by the Union Carbide Corporation and was the first oxygen activated-sludge process to be used successfully to treat sewage in a full-scale plant, in this case a modified former conventional works at Batavia, New York as described by McWhirter (1978). It was judged in the USA (Albertsson *et al.*, 1970) to have significant economic advantage over a concurrently operated coarse-bubble aeration plant. Since 1972 when the first plant designed from inception to use the UNOX process was built in Japan, a further 64 plants have been constructed throughout the world, treating a variety of wastewaters from domestic and industrial sources. The process (Figure 3) involves agitation of mixed liquor and oxygen (normally by means of a mechanical surface aerator driven through a water seal to prevent loss of oxygen) in an enclosed tank. Often two to four tanks are connected in series. Oxygen enters the first tank, together with settled sewage and recycled activated sludge, at a rate sufficient to maintain the pressure of gas above the liquor at about 20–50 mm water gauge. From the final tank exhaust gas is vented to atmosphere and mixed liquor is discharged to a conventional secondary settlement tank.

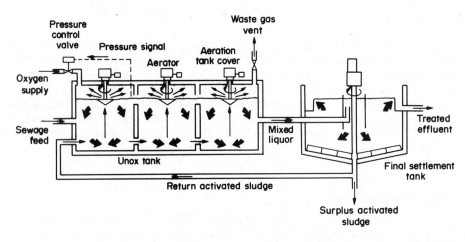

Figure 3   Schematic diagram of a UNOX system (Reproduced by permission of Wimpey-Unox Ltd.)

(ii)   *OASES*

A similar system involving enclosed tanks in which mixed liquor is agitated in an oxygen-rich atmosphere has been developed by Air Products and Chemical Corporation (Robson *et al.*, 1972; Wyatt *et al.*, 1975). The process includes the use of the residual gas (oxygen and CO_2) to strip dissolved nitrogen from the incoming wastewater, thus improving the efficiency of solution of oxygen in the mixed liquor.

(iii)   *MAROX*

The process has been developed in the USA by the FMC Corporation. It has been described and evaluated by Cohen (1973) and Pearlman and Fullerton (1979) and has been further studied in the UK by Wm. E. Farrer Ltd., in cooperation with Severn Trent Water Authority and the Water Research Centre (WRC), on the pilot-scale at the Coleshill Experimental Plant and reported by Clough (1979).

Oxygen is dissolved from microbubbles (50–200 μm) produced by the shearing action of liquor flowing radially across the surface of a rotating disc described by the manufacturers as a 'Rotating Active Diffuser' (RAD). The RAD (Figure 4) consists of a submerged rotating hollow disc, 2.1 m in diameter, suspended on a hollow stainless-steel shaft about 0.9 m above the floor of the aeration tank (Figure 5). A band of porous ceramic diffusers is located in the upper and lower surfaces of the disc, slightly inset from the periphery (Figure 4). Small radial impellers are fitted to the top and bottom of the disc to provide movement of the liquor and thus to produce shear forces across the diffuser to form microbubbles; the action also mixes the liquor to keep the sludge in suspension. The shaft and disc are rotated at 82 to 86 rev min^{-1} through a gearbox by a motor mounted on a bridge over the tank. The production of microbubbles is essential for the effective transfer of oxygen into solution and for the prevention of excessive losses of oxygen from the surface of the liquor. The edge of the disc is tapered to prevent excessive turbulence.

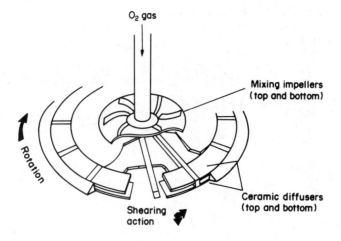

Figure 4 Rotating active diffuser of a MAROX system

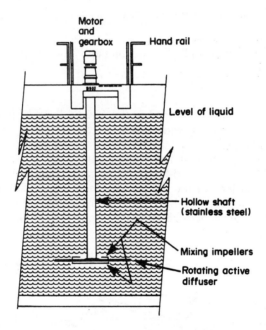

Figure 5 Side elevation of rotating active diffuser in oxygenation tank

(iv) VITOX

This process was developed by BOC (Environmental) Ltd. in the UK, primarily for uprating activated-sludge plants. Its use has been described by Toms and Booth (1982) to supplement existing systems of aeration, usually mechanical.

When the process is used to uprate an activated-sludge plant, settled sewage is pumped through a venturi where gaseous oxygen is injected. The oxygenated sewage, containing a high concentration of dissolved oxygen (about 20–25 mg l^{-1}) and some bubbles of undissolved gas, is then distributed into the liquor at the bottom of the aeration tank by horizontal sparge pipes (Figure 6). Fine bubbles of oxygen created by the high turbulence at the orifice of the sparge pipes are dispersed through the liquor and dissolve rapidly.

Operation of the pump and the oxygen injection system is controlled by a dissolved-oxygen electrode sited below the surface of the liquor. To achieve the maximum and most efficient input

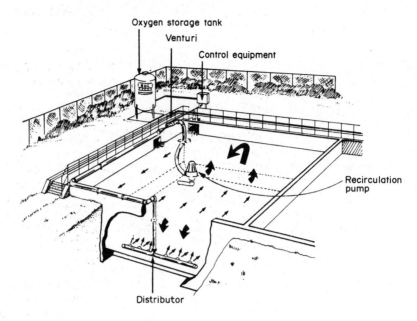

Figure 6 Artist's impression of a VITOX plant (Reproduced by permission of BOC Ltd.)

of oxygen, the system is normally installed at the inlet end of a plug-flow aeration tank where the oxygen demand is at its greatest.

(v) MEGOX

The MEGOX process has been developed in the UK by BOC (Environmental) Ltd. for the partial treatment of wastewaters by providing aerobic conditions for the liquor during its retention in primary settlement tanks.

The process consists of oxygenation in a downflow bubble contactor, followed by reaction and sedimentation in a conventional upward-flow settling tank (Figure 7). Under these conditions bacteria present in the unsettled wastewater will biochemically oxidize some of the pollutants present and are themselves aggregated into flocs. A proportion of the floc is recirculated with incoming wastewater to be reoxygenated and to continue the oxidation process.

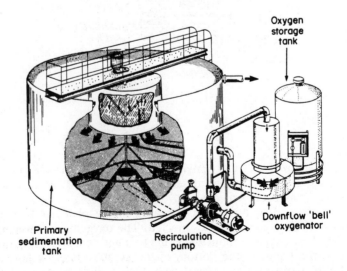

Figure 7 Artist's impression of a MEGOX plant (Reproduced by permission of BOC Ltd.)

The process has been successfully used for partial treatment of industrial wastewaters of high

BOD, and full-scale evaluation is being continued by BOC Ltd. in cooperation with Wessex Water Authority and WRC.

(vi) GIEDAL

The GIEDAL process, developed in France by Degremont and L'Air Liquide, employs oxygen, or oxygen-enriched air, in the activated-sludge process to treat moderately or highly polluted wastewaters.

Mixed liquor is treated in a series of three tanks alternately closed and open to the atmosphere. Oxygen is injected into the closed inlet tank by a mechanical surface aerator. Residual gas (containing about 40–50% oxygen) from the inlet tank enters the gas-space above the liquor in the closed outlet tank, which is also mixed and aerated by a mechanical surface aerator. Exhaust gas is vented to the atmosphere from the outlet tank. In the intermediate tank, which is open to the atmosphere, the liquor can be aerated by submerged or surface aerators to continue biological oxidation and to strip from solution the carbon dioxide produced during biochemical oxidation in the inlet tank. By this means wastewater of high BOD and low alkalinity can be treated without a prolonged reduction in pH value; thus avoiding inhibition of nitrification (see Section 50.4.3.2) and possibly adverse effects on sludge settlability.

(vii) OXITRON

In 1976 in the USA, Dorr-Oliver Inc. purchased a licence for a fluidized-bed process from Ecolotrol Inc. and began development of the OXITRON process to treat wastewaters in a fluidized bed as described by Sutton *et al.* (1981). In this process wastewater is pumped upwards through a bed of particles (in this case sand) at a velocity sufficient to expand the bed to a point at which the particles are in free motion suspended in the flow of liquid. The particles provide a very large surface (per unit volume) upon which the microorganisms grow to develop a concentration approximately an order of magnitude greater than that normally maintained in a conventional activated sludge plant. For sand typically of size 0.2 to 2 mm the concentration of biomass may be up to 18 $g l^{-1}$, although in practice the concentrations are about 10 to 15 $g l^{-1}$. The increased biomass concentration allows the contact period required for treatment to be reduced with a corresponding reduction in the size of the treatment plant.

To satisfy the demand, oxygen is dissolved into the recycled wastewater in a downflow oxygenator. This, it is claimed, will dissolve oxygen at a rate of at least 50 mg l^{-1} in 20 s. A high efficiency of oxygen utilization (about 95%) is also claimed for the process which is shown in outline in Figure 8.

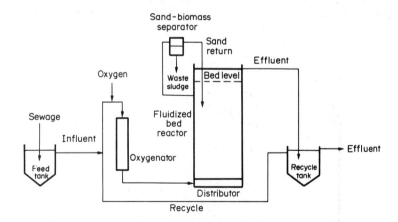

Figure 8 Outline schematic diagram of OXITRON process

A major potential advantage of the process, compared with the conventional activated-sludge process, is the use of a simple device to remove accumulated biomass from the supporting particles without the need for a conventional settlement tank. Potential problems due to poor sludge settlability are thus avoided.

The capital costs of the process are likely to be significantly lower than those of a conventional activated-sludge process designed to achieve the same degree of treatment. However, operating costs are likely to be considerably greater because of the high cost of the energy used to recirculate wastewater through the bed and to separate oxygen from the atmosphere. Estimates made of the difference in the net present values (NPVs) for plants treating flows greater than 30 000 m³ d⁻¹ show that the fluidized-bed process could be competitive with conventional systems. The use of a fluidized bed for denitrification of returned activated sludge, or of nitrified effluent from an oxygenic fluidized bed, was shown by Wheeldon and Bayley (1981) to have potential for economic advantages over conventional alternatives.

The following advantages have been claimed for biological fluidized-bed systems: (i) the high biomass concentration achievable leads to a compact plant which in turn may reduce appreciably the land area required for a treatment works; (ii) secondary clarification may not be needed where effluents containing not more than 30 mg l⁻¹ suspended solids and 20 mg l⁻¹ BOD are required; (iii) a reduction in capital cost is possible because of the smaller size of the plant; (iv) less susceptible than conventional systems to sudden increases in substrate concentration or decreases in temperature, presumably because of the high concentration of biomass in the system; (v) existing overloaded plants can be uprated on the same site by gradual replacement with a smaller fluidized-bed plant having greater treatment capacity; and (vi) thick sludges (4–5% dry solids) may be produced by careful optimization of the sand cleaning/biomass wastage stage.

A full evaluation of this process has recently been completed by WRC using a 500 m³ d⁻¹ pilot plant (Figure 9) at the Coleshill site. Economic analysis by Hoyland and Robinson (1982) of the data obtained indicated that the low capital cost of any oxygenic fluidized bed may at most sites be more than offset by the high costs of operation associated with the supply of oxygen and the electricity required for pumping.

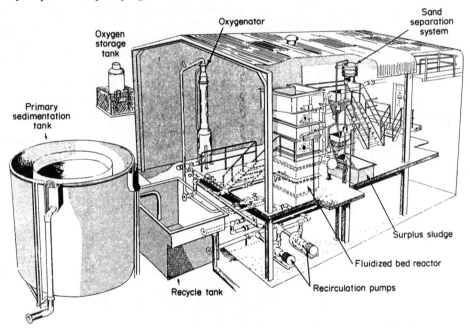

Figure 9   Artist's impression of pilot plant at WRC (Coleshill) to treat 500m³ of sewage daily by the OXITRON process

50.4.2.2 *Energy requirements*

The energy required to dissolve oxygen in water depends largely on the technique used. In the activated-sludge process much energy is also used in achieving mixing and in maintaining the activated sludge in suspension. Data available in the UK indicate that the electrical energy required to dissolve a tonne of commercially produced oxygen ranges from 300 kW h to about 500 kW h (1–1.8 GJ) for agitation in an enclosed tank to about 350 kW h (1.26 GJ) for the side-stream technique.

The total power requirement (for producing 1 tonne of oxygen and dissolving it in the mixed liquor of an activated sludge plant) ranges from 650–1200 kW h (2.3–4.3 GJ). In contrast the total power required to operate conventional aeration equipment using fine-bubble diffused-air is about 450 kW h t⁻¹ (1.6 GJ t⁻¹) oxygen dissolved and is about 600–1000 kW h t⁻¹ (2–3.6 GJ t⁻¹)

using mechanical surface aeration (both operating at zero concentration of dissolved oxygen in the mixed liquor). As much as 1200 kW h t^{-1} (4.3 MJ t^{-1}) may be required for a coarse-bubble aeration plant. To maintain a concentration of 50% of the normal air saturation concentration would, however, increase the total power requirements by only about 5% for an oxygenated plant, compared with as much as 100% in the case of a conventionally aerated plant.

50.4.3 Use of Oxygen in the Activated-sludge Process

50.4.3.1 *Carbonaceous oxidation*

Experiments in pilot plants (Steen and Fuggle, 1975; Wood *et al.*, 1976, Wyatt *et al.*, 1975) and on the full-scale (Blachford *et al.*, 1982) in the UK have shown that there was little difference in the rate of treatment defined as kilograms of BOD removed daily per kilogram of activated sludge [achievable in an oxygenated plant (OP) and in an aerated plant (AP)]. When treating the same sewage at similar sludge loadings there was little difference in the quality of effluents from oxygenated and aerated plants (in terms of BOD and SS) provided that the concentration of dissolved oxygen in the mixed liquor was kept above the rate-limiting value. Benefits claimed from use of oxygen, such as acceptability of increased loadings and increased rate of settlement of sludge, were not always observed by Blachford *et al.* (1982).

The ability of OPs to produce consistently effluents of low BOD (10 mg l^{-1}) and suspended solids (20 mg l^{-1}) has been described by many authors, particularly for full-scale plants operated in the USA (Albertsson *et al.*, 1970; Chapman *et al.*, 1976; McWhirter, 1978), although many of the APs with which the OPs were compared had inadequate aeration capacity and failed to maintain the dissolved oxygen above the rate-limiting value. Most authors confirmed that an OP is able to supply oxygen at high-rates to satisfy the variable demand of a highly-loaded plant, treating wastewaters of high and variable BOD. However, they also found that the rate of treatment was similar provided similar conditions of operation were maintained for both OPs and APs. Some of the other benefits claimed for OPs have been questioned by Kalinske (1971), particularly in respect of advantages of operation at concentrations of dissolved oxygen above 2 mg l^{-1} and of favourable economics. However, most of the other process advantages claimed for the use of oxygen have been substantiated by Blachford *et al.* (1982), Chapman *et al.*, (1976), and Pearlman and Fullerton (1979).

50.4.3.2 *Nitrification*

Difficulties have been encountered in achieving nitrification in oxygenated plants. These are attributed by Blachford *et al.* (1982) and Wyatt *et al.* (1975) to the incomplete stripping of carbon dioxide from the mixed liquor by the relatively small volume of gas ultimately discharged to the atmosphere. Such an effect was confirmed by experiments of Wyatt *et al.* (1975) in the UK with a single complete-mixing OP, which showed that fully nitrified effluent could be produced when alkali (sodium hydroxide) was added to the mixed liquor to prevent the pH value falling below 7.5 but not when the alkali was omitted. Clough (1979) reported that complete nitrification was achieved in a MAROX pilot plant at similar loading rates and at the sludge ages equal to those required for nitrification in an AP operated in parallel and treating similar sewage. To achieve complete nitrification using a UNOX system, two plants have been operated in series, the first stage plant having been designed for high rate oxidation of carbonaceous matter using oxygenation and the second stage plant for conventional aeration to produce a fully nitrified effluent (Steen and Fuggle, 1975). It was possible to achieve 95% removal of ammoniacal nitrogen in a UNOX plant at a sludge loading of 0.25 d^{-1} and with a sludge age of 10 d when the high alkalinity of the sewage (about 350 mg l^{-1} as CaCO$_3$) acted as a buffer stabilizing the pH value (Blachford *et al.*, 1982). By contrast, an AP can consistently produce a fully nitrified effluent in a single-stage process with a sludge age as low as 5 d.

A VITOX system of injecting oxygen into mixed liquor has been installed at a treatment works to uprate the existing aeration capacity by about 25% during the summer months when the population increases. The system is used to raise the daily oxygenation capacity of the existing mechanically aerated plant by up to 5 t, thus enabling the plant to treat about 30 000 m^3 of settled sewage daily to produce consistently a nitrified effluent (Toms and Booth, 1982).

50.4.3.3 Sludge characteristics

Results of experience with a full-scale UNOX plant in the UK treating settled sewage have indicated that settlability of activated sludge was no better than that in an AP operated in parallel (Blachford *et al.*, 1982). For a period of six months severe problems were experienced with sludge bulking in a UNOX plant, the stirred specific volume index (SSVI) increasing to about 600 ml g^{-1}. Increasing the sludge loading and reducing the input flow rate of oxygen to increase the accumulation of carbon dioxide in the liquor reduced the SSVI to 100 ml g^{-1}. In another study (Steen and Fuggle, 1975) the SSVI of sludge from an OP was little affected by changes in sludge loading, whereas the settlability of sludge from an AP treating the same sewage deteriorated when the loading was increased above about 0.3 d^{-1}.

In another experiment with a pilot scale UNOX plant the settling characteristics of sludge were found to be good (Scottish DD, 1977) when treating sewage which contained about 50% of wastewater from distilleries and from yeast production, while an AP fed with similar wastewater failed to perform satisfactorily because of sludge bulking. The SSVI was usually around 50 ml g^{-1} and no tendency towards poor settlability and consequent bulking of the sludge was noticed. Similar experience at other sites has provided justification for construction of UNOX plants at Reading and Tadcaster for treating brewery effluents of high BOD.

The weight of surplus sludge produced varies with the nature of the wastewater being treated. When both types of plant were operated at the same sludge loading (0.3 d^{-1}) treating the same settled sewage in a full-scale plant, the production of sludge from an OP appeared to be about 80% of that from an AP (Blachford *et al.*, 1982). Sludge production is difficult to measure but a difference of this order would have some influence on the total cost of treatment.

Several authors in Europe (Martin, 1978; Staab, 1978; Suominen, 1980) and the USA (Pearlman and Fullerton, 1979; Speece and Malina, 1973; Benefield and Randall, 1979) have reported reduced production of sludge from an OP when compared with an AP.

50.4.3.4 Economics

Data available from the literature (Sidwick *et al.*, 1975) indicate that an OP can have economic advantages over an AP for a large plant (40 000 m^3 d^{-1}) producing a non-nitrified effluent and operating in areas where unit costs of sludge disposal are high.

The selection of a process for a given site will depend on many factors which include the area of land available, the proximity of residential buildings, capital costs, operating costs and performance as well as durability, reliability and versatility of the process. An OP could be cost-competitive with an AP if, by taking advantage of the potential compactness of the former, an area could be used for the works which would reduce substantially the expenditure on sewerage. Capital costs are likely to be similar (Oakley, 1977; Sidwick *et al.*, 1975; Speece and Malina, 1973), particularly for larger works where the increased costs associated with construction of an oxygen-supply plant might be offset by the lower costs of smaller treatment tanks and facilities for sludge disposal. The prime difference in operating cost between an OP and an AP would normally be that of energy used to provide oxygen as already described.

50.4.4 Overall Assessment

The following advantages have been claimed for the use of oxygen activated-sludge processes. Most would depend very little on circumstances such as location or size of the treatment works; they include: (a) the possibility of dissolving oxygen at high rates, leading to the possibility of a smaller plant provided that high MLSS concentration could be maintained; (b) lower production of sludge with possibly better settling characteristics, again depending on high MLSS concentration; (c) reduced possibility of sludge bulking, particularly when treating strong wastes known to produce sludges of poor settlability in a conventional AP such as those from breweries or distilleries; (d) the possibility of treating wastewaters of high BOD in plants smaller than those required for aerated processes; (e) the possibility of reducing aerial nuisance (malodours, aerosols) because only a small volume of gas is discharged to the atmosphere; (f) a consistent quality of effluent and stability of operation of the plant resulting from the ability to control the input of oxygen to cover a wider range of diurnal variations in BOD load; and (g) the possibility of lower overall costs— claims have been made of lower costs when treating dry weather flows exceeding 4500 m^3 d^{-1}.

Similarly, there are disadvantages which apply irrespective of the location and size of the treatment plant: (a) the use of oxygen requires reliance either on fairly complex equipment or on an external supplier; (b) more energy may be consumed in producing and dissolving oxygen than in aeration, although this might be offset by lower costs (including energy costs) for disposal of sludge and construction of the plant; (c) nitrification cannot usually be achieved in a single stage plant without modification of the process because of the effect of carbon dioxide on the pH value of the mixed liquor, particularly in areas where the water is of low alkalinity; (d) the use and storage of oxygen might be considered a hazard, although strict observance of simple safety regulations should reduce the potential risks to a minimum.

It is not possible to generalize about selection, particularly in view of several uncertainties. For the moment each case should be considered on its merits taking account of all the factors listed.

50.5 AERATED PROCESSES

50.5.1 Fine-bubble Diffused-air

Ceramic diffusers have been used successfully to produce fine bubbles (2 to 5 mm diameter) of air at the bottom of aeration tanks in the activated-sludge process for more than 40 years. However, considerable advances have been made in recent years to obtain maximum performance from such a system.

The results of studies carried out in the UK by the WRC (Boon, 1980) and further work undertaken by the US EPA in cooperation with WRC (Houck and Boon, 1981) and others have shown that fine-bubble aeration systems can perform at high efficiency relative to other aeration devices. Data obtained from 19 full-scale treatment plants using fine-bubble aeration have shown that these aeration systems are reliable, have few significant problems, are not expensive to maintain, do not require frequent cleaning, and have a median value of aeration efficiency of 2.75 kg oxygen per kW h (0.76 kJ^{-1}). The aeration efficiency could be significantly affected by the presence of surfactants and if non-biodegradable detergents were present in the sewage the aeration efficiency might be reduced by about 50%. Existing proprietary diffused-air systems have been designed to provide high oxygenation capacities and tests have shown that provided the system is correctly installed, operated and maintained (Table 1) they are greater in efficiency compared with mechanical systems.

Table 1 Aeration Intensity and Efficiency of Various Conventional Systems of Aeration[a]

Aeration system	Maximum aeration intensity (mg l^{-1} h^{-1})	Aeration efficiency (kg kW^{-1} h^{-1})
Diffused air		
Fine bubbles	200	1.5–3.6
Coarse bubbles	100	0.9–1.2
Mechanical surface aerators		
Rotating vertically	125	1.5–2.2
Rotating horizontally	100	1.2–2.4

[a] Measured in tap water (2–8 m deep) containing 5 mg surface active agent per litre. All values measured at maximum deficit of DO. Aeration efficiency calculated from total energy supplied.

To obtain the maximum performance from a fine-bubble aeration system requires correct design of the geometry of the aeration tank and facilities for monitoring and control of DO by varying the flow rate of air. Other factors which should be taken into account to avoid problems with operation of the plant and loss of aeration efficiency are listed below. (i) The minimum flow rate of air at minimum sludge loading should never be lower than the minimum rate specified by the manufacturers if problems of clogging diffusers and inadequate mixing of liquor are to be avoided. (ii) The guidelines specified by the manufacturers for maintenance and cleaning of the equipment should be fully observed. (iii) Growth of slimes on the diffusers, resulting in loss of efficiency through formation of larger bubbles, could be avoided by improved methods of mixing of sewage and recycled sludge at the inlet to reduce high BOD loadings. The provision of anoxic conditions at the inlet of a nitrifying plant, or aeration of recycled sludge in a non-nitrifying plant, would alleviate such problems. (iv) Aeration tanks with 'plug-flow' mixing characteristics tend to produce activated sludge of good settlability compared with sludges grown in 'uniformly-mixed' tanks.

50.5.2 Deep Shaft Process

The Deep Shaft Process is designed to increase the rate of oxygen transfer by increasing the pressure (and hence the saturation concentration of DO at the bottom of the shaft), the contact period of bubbles, and turbulence, without corresponding increases in energy requirement (Bolton and Ousby, 1976; Hines *et al.*, 1975). From earlier research with U-tube aerators, which are similar in operation to a deep shaft, the potential aeration efficiency could be as high as 6 kg kW h^{-1} (1.6 g kJ^{-1}). However, recent results indicate that the efficiency of the deep shaft when treating settled sewage is about 1.5–2 kg kW h^{-1} (0.4–0.55 g kJ^{-1}) with an intensity of about 3000 mg l^{-1} h^{-1} for a shaft 135 m deep.

The plant employs a vertical shaft sunk into the ground, lined with concrete and divided by a vertical partition into a downflow ('downcomer') and upflow ('riser') section. The diagram (Figure 10) of the process shows two concentric tubes in place of the partition and this system may be used in some installations. Recirculation and aeration of the liquor is achieved by 'process' and 'start-up' air. Initially, a large flow of air is introduced into the riser to provide an air-lift to start the circulation. Once the liquor is circulating the process air is turned on and the quantity of air going to the riser is reduced. Process air entering the downcomer is drawn down by the high downward velocity of liquor (1–1.5 m s^{-1}) and is dissolved rapidly in the turbulent flow as the pressure increases, until all the gas is dissolved. In the riser, as the pressure decreases, bubbles of gas (mostly nitrogen and carbon dioxide with perhaps some oxygen) are formed and serve to air-lift the liquor to assist with recirculation. Usually a small proportion (10–30%) of the start-up air must be maintained to ensure that recirculation continues irrespective of the oxygen requirement of the microorganisms, which would normally be largely satisfied by the oxygen provided by the process air. At ground level the shaft terminates in a head tank which serves to permit bubbles to be released to the atmosphere, preventing them from being drawn into the downcomer. Wastewater and recirculated sludge enter the head tank above the downcomer and recirculate in the shaft on average from 20 to 40 times (depending on the BOD of the sewage and the effluent quality required) before passing from the head tank to a degassing system. The treated liquor overflowing from the head tank is supersaturated with nitrogen and carbon dioxide and contains many small bubbles attached to sludge flocs. These gases make it impossible to achieve effective separation of activated sludge by settlement in a conventional tank and in some circumstances (particularly the partial treatment of wastewaters of high BOD) flotation has been used to achieve separation. In the UK the first full-scale plant at Thurrock uses vacuum degassing at a pressure of about 30 kPa (*i.e.* 0.3 atm) to remove excess dissolved gases. The degassed liquor is then settled in a conventional settlement tank and the sludge recycled to the shaft. In an evaluation of the Deep Shaft Process with a pilot plant, Cox *et al.* (1980) developed a double stripping system which involves further aeration of the liquor from the shaft in a plug-flow conventionally-aerated tank before settlement. A similar pilot scale system has been used by Wessex WA to produce consistently a fully nitrified effluent of high quality; an anoxic zone at the inlet of the shaft is provided to utilize nitrate available in the recycled sludge and to improve sludge settlability.

Advantages of the process include: (i) high intensity of oxygen transfer making it possible to treat wastewaters of high BOD in a compact plant; (ii) smaller area of land compared with some other systems, because much of the plant is below ground in a vertical shaft not exceeding 10 m diameter; (iii) fairly high efficiency of aeration compared with some oxygenated processes; (iv) possibility of producing a fully-nitrified effluent in a single stage; (v) fairly good settlability of sludge after degassing (based on existing experience); and (vi) reduced possibility of environmental complications such as odours and aerosols.

The main disadvantage is the high capital cost that may be associated with sinking the shaft, particularly in areas where reinforcing may be needed to avoid damage to the shaft caused by movement of the sub-soil. Some nuisance may be caused by the noise of the compressors (air is required at about 700 kPa for a shaft 130m deep) unless they are contained in a sound-proof building or housed below ground level.

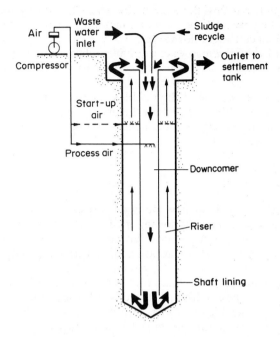

Figure 10 Outline diagram of a Deep Shaft

50.5.3 Captor Process

The Captor Process has been developed in the UK by Simon-Hartley Ltd. from an idea which originated in the Chemical Engineering Department of the University of Manchester Institute of Science and Technology. It is essentially an intensified activated-sludge process in which the biomass (activated sludge) is retained in pieces of porous plastics medium suspended in the aeration tank of an activated-sludge plant. A final settlement tank may not be required where only partial treatment yielding an effluent of high suspended solids content (up to 80 mg l^{-1}) is acceptable.

The biomass support medium is made of a polyether based plastic which is woven to produce a fairly open structure which will retain biomass during aeration and can be squeezed to remove accumulated biomass when required. Each piece of medium is about 25 mm square by about 10 mm thick and about 40 000 are added per cubic metre of aeration tank capacity. The manufacturers claim (Walker and Austin, 1981) that the concentration of biomass in the aeration tank can be increased to at least 8 g l^{-1} (the maximum quoted was 11 g l^{-1}), that is about two to three times the concentration of activated sludge in the aeration tank of a conventional plant. The size of the aeration tank for the Captor Process is therefore reduced *pro rata* to about one half to one third of that required conventionally. To achieve a similar effluent quality the aeration intensity for the Captor Process would have to be two to three times that used conventionally and this may be possible only with a fine-bubble system of aeration.

Another advantage of the process is said to be its ability to yield surplus biomass at the high suspended solids content of 4 to 6%. To obtain this surplus sludge, a proportion of the support medium is removed from the aeration tank, drained, and squeezed between rollers to remove the biomass before the cleaned medium is returned to the aeration tank. If all the biomass were retained by the medium, no final sedimentation stage would be needed. At present there are insufficient data to indicate the effectiveness of the system when treating sewage, but the suspended solids content of the effluent from a pilot plant having no sedimentation stage was generally within the range 20 to 80 mg l^{-1} with an average of about 40 mg l^{-1}. These figures could be reduced by using a final settlement tank to clarify the effluent.

In essence, the process depends on its ability to retain high concentrations of biomass in the medium in an aeration tank in which oxygen is made available at the required rate. A desk study at WRC indicates that the economic viability of the process depends essentially on the durability of the support medium. If 10% of the medium had to be replaced each year, the process could be 20% cheaper (based on a net present value calculated at a discount rate of 5%) than a conventional activated-sludge plant for new works. The benefits of Captor appear to be related to the cost, durability and effectiveness of the support medium and the results of the first full-scale evaluation are awaited with considerable interest.

Laboratory data provided by the manufacturer have indicated that for a given air flow the rate of solution of oxygen from fine bubbles is increased at least three-fold, and from coarse bubbles two-fold, by the introduction of the support medium relative to that in a conventional plant, owing to the increased residence time of the bubbles. These results require confirmation on the full-scale, but indicate considerable potential for the process, particularly for uprating overloaded plants where lack of aeration capacity and the need to retain more biomass in the aeration tank limit the maximum rate of treatment.

50.6 SUMMARY

Activated-sludge processes involving new methods of dissolving oxygen, in some cases accompanied by the use of inert particulate supporting media for maintaining relatively high concentrations of activated sludge in the aeration units, have been introduced in the last decade.

The use of oxygen in place of air for aeration of 'unsupported' sludges has enabled some sewages and industrial wastes to be treated for removal of carbonaceous BOD and suspended solids at a higher rate per unit volume of plant than was possible with the traditional types of aerator available about a decade ago. Although there is still some controversy on the matter, this intensification seems to be realizable for sewage treatment, mainly when higher concentrations of activated sludge can be maintained in the mixed liquor, owing to the improved settlability of sludge in some cases. For industrial wastewaters the intensification is also possible because the high oxygen demands that they generate could not be matched by traditional systems. Other advantages reported for some of such plants is that rates of production of surplus sludge are lower than with traditional types of plant and settlability of sludge is improved.

When the area of land available is adequate for introduction of a traditional plant, oxygenated processes are unlikely to be as cheap to run as the traditional types, because any advantage derived from their smaller size or reduced sludge production would be substantially outweighed by the cost of providing oxygen. An economic advantage might be derived from using an oxygenated process, however, where land was extremely expensive or where a restricted site could be used which would involve much lower costs of sewerage than alternative sites suitable for larger traditional types of plant. Oxygenated systems have been economically justified for supplementing the aeration capacity of existing traditional plants particularly where they have been used intermittently only at times of high loading.

Difficulties have been encountered in achieving nitrification in some single-stage oxygen plants when treating only lightly-buffered wastewaters, mainly because of the relatively low rate of stripping of carbon dioxide from the mixed liquor. Any such difficulties would obviously have an unfavourable impact on the economics of a plant required to produce a nitrified effluent.

Most of the above observations apply qualitatively to oxygen processes in which the sludge is attached to inert particles. However, an additional feature of such plants is that they can be made smaller than traditional plants and indeed than oxygen plants employing unsupported sludges, because by the use of inert supports relatively high concentrations of sludge can be maintained in the aeration units and also because the sludge does not have to be recirculated through large settlement tanks. A reservation to this conclusion is that it does not yet appear to have been conclusively demonstrated that such plants can produce effluent of as high quality as that obtainable by traditional plants and thus the advantages of smaller size may only be achievable over a limited range of effluent qualities.

In the last decade new methods of dissolving oxygen using air have been introduced which enable higher oxygen demands to be met than was possible using traditional types of aerator. There seems to be a possibility that when combined with the use of a supporting medium to increase concentration of activated sludge in the aeration tank, such processes may prove to be cheaper to install and operate than alternatives, especially where high-quality nitrified effluents are required.

50.7 REFERENCES

Aiba, S., S. Kitai and H. Heima (1964). *J. Gen. Appl. Microbiol.*, **10**, 243–256.
Albertsson, J. G., J. R. McWhirter, E. K. Robinson and N. P. Vamldiek (1970). Investigation of the use of high purity

oxygen aeration in the conventional activated-sludge process. *Water Pollution Control Research Series Report No. 17050 DNW 05170*. US Department of the Interior, Federal Water Quality Administration.

Bebin, J. and M. Renaudin (1976). Dairy waste purification by means of highly loaded activated sludge process with pure oxygen supply. *Prog. Water Technol.*, **8**, 301–311.

Benefield, L. D. and C. W. Randall (1979). Air or oxygen activation: verdict still undecided on best system for settleability. I. *Water Sewage Works*, **126** (4), 44–46.

Blachford, A. J., E. M. Tramontini and J. A. Griffiths (1982). Oxygenated process-evaluation at Palmersford. *Water Pollut. Control*, **81**, 1–375.

Bolton, D. H. and J. C. Ousby (1976). The ICI deep shaft effluent treatment process and its potential for large sewage works. *Prog. Water Technol.*, **8**, 256.

Boon, A. G. (1980). Measurement of aerator performance. In *The Profitable Aeration of Waste Water*, pp. 13–28. BHRA Fluid Engineering, Cranfield, Bedford, England.

Chambers, B. and E. J. Tomlinson (1982). *Bulking of Activated Sludge: Preventative and Remedial Methods*. Ellis Horwood, Chichester.

Chapman, T. D., L. C. Matsch and E. H. Zander (1976). Effect of high dissolved oxygen concentration in activated-sludge systems. *J. Water Pollut. Control Fed.*, **48**, 2486–2510.

Clough, G. F. C. (1979). Farrer oxygen activated sludge process pilot plant at Coleshill. In press, to be published by WRC, Stevenage, England.

Cohen, D. B. (1973). A low-cost open-tank pure-oxygen system for high-rate total oxidation. *Proceedings of the 6th International Conference on Water Pollution Control Research, 1972*, pp. 389–403. Pergamon, Oxford.

Cox, G. C., V. H. Lewin, J. T. West, W. J. Brignal, D. L. Redhead, J. G. Roberts, N. K. Shah and C. B. Waller (1980). The use of the deep-shaft process in uprating and extending existing sewage-treatment works. *Water Pollut. Control*, **79**, 70–86.

Davis, J. C. (1972). Oxygen separated from air using molecular sieves. *Chem. Eng. (Albany)*, **79** (23), 88.

Downing, A. L. and R. W. Bayley (1961). Aeration processes for the biological oxidation of waste waters. *Trans. Inst. Chem. Eng.*, **39**, A53–A59.

Downing, A. L. and A. B. Wheatland (1962). Fundamental considerations in biological treatment of effluents. *Trans. Inst. Chem. Eng.*, **40**, 91–103.

Eckenfelder, W. W. and D. J. O'Connor (1961). *Biological Waste Treatment*. Pergamon, New York.

Hines, D. A., M. Bailey, J. C. Ousby and F. C. Roesler (1975). The ICI deep shaft aeration process for effluent treatment. In *Inst. Chem. Eng. Symp. Ser.*, **41**, 01–010.

Houck, D. H. and A. G. Boon (1981). Survey and evaluation of fine-bubble dome-diffuser aeration-equipment. US Environmental Protection Agency, Cincinnati, OH. EPA–600/52–81–222.

Hoyland, G. and P. J. Robinson (1983). Aerobic treatment in 'Oxitron' BFB Plant at Coleshill. *Water Pollut. Control*, **82**, 479–493.

Jewell, W. J., W. W. Eckenfelder and M. E. Cavalier (1971). The use of pure oxygen for the biological treatment of brewery waste waters. *Proceedings of the 26th Industrial Waste Conference, Purdue University*, pp. 487–496. Purdue University, Lafayette, IN.

Kalinske, A. A. (1971). Effect of dissolved oxygen and substrate concentration on the uptake rate of microbial suspensions. *J. Water Pollut. Control Fed.*, **43**, 73–80.

Karnovsky, F (1975). Aeration of waste water with oxygen. *Abwasser Technik*, **26** (3), 18–20.

Knudson, M. K., K. J. Williamson and P. O. Nelson (1982). Influence of dissolved oxygen on substrate utilization kinetics of activated sludge. *J. Water Pollut. Control Fed.*, **54**, 52–60.

Lee, H. (1973). Pressure-swing oxygen—a new source of oxygen for waste water treatment. In *Applications of Commercial Oxygen to Water and Waste Water Systems*, ed. R. E. Speece and J. F. Malina, pp. 16–31. Centre for Research in Water Resources, University of Texas, Austin.

McWhirter, J. R. (1978). *The Use of High-Purity Oxygen in the Activated-Sludge Process*, vol. I and II. CRC Press, Boca Raton, FL.

Martin, P. (1978). Biological and economic factors affecting the use of pure oxygen in aeration systems. *Abwassertechnik*, **29** (3), 30–34.

Mueller, J. A., A. Voelkel and E. W. Lightfoot (1966). Nominal diameter of floc related to oxygen transfer. *J. San. Eng. Div., Am. Soc. Civ. Eng.*, **SA2**, 9–20.

Mueller, J. A., W. C. Boyle and E. N. Lightfoot (1968). Oxygen diffusion through zoogloeal flocs. *Biotechnol. Bioeng.*, **10**, 331–358.

Oakley, H. R. (1977). Pure oxygen in activated sludge. *Prog. Water Technol.*, **8** (6), 135–136.

Parker, D. S. and M. S. Merrill (1976). Oxygen and air activated sludge: another view. *J. Water Pollut. Control Fed.*, **48**, 2511–2528.

Pearlman, S. R. and D. G. Fullerton (1979). Full-scale demonstration of open tank oxygen activated-sludge treatment. US Environmental Protection Agency, Cincinnati, OH. Research Report No. EPA–600/2–79–012.

Robson, C. M. *et al.* (1972). Operational experience of a commercial oxygen activated-sludge plant. Presented at *45th Annual Conference of the Water Pollution Control Federation*, Atlanta, GA.

Scottish Development Department (1977). Pilot plant study of the oxygen activated-sludge process. Engineering Division, Applied Research and Development Report No. ARD5.

Sidwick, J. M., T. P. Lewandowski and K. H. Allum (1975). An economic study of the Unox and conventional aeration systems. *Water Pollut. Control*, **74**, 645–656.

Smith, K. C. and J. W. Armand (1973). Adsorption as a technique for gas separation. *Proceedings of the 'Crystech '73' Symposium*, pp. 101–106.

Speece, R. E. and J. F. Malina (eds) (1973). *Applications of Commercial Oxygen to Water and Wastewater Systems. 6th Water Resources Symposium*. Centre for Research in Water Resources, University of Texas, Austin.

Staab, K. F. (1978). The activated-sludge process at operating pressures in the range 1 to 5 bar. *Wasserwirtschaft*, **68**, 173–179.

Steen, D. and R. W. Fuggle (1975). Unox system pilot plants—two years experience with pure oxygen in the activated-sludge process. *Public Health Eng.*, **3** (16), 91–104.

Suominen, A. (1980). The use of pure oxygen in the activated-sludge process. *Vesitalous*, **21** (1), 14–15.

Sutton, P. M., W. K. Shieh, P. Kos and P. R. Dunning (1981). Dorr-Oliver's Oxitron Systems TM fluidised-bed water and wastewater treatment process. In *Biological Fluidised Bed Treatment of Water and Wastewater*, ed. P. F. Cooper and B. Atkinson, pp. 285–304. Ellis Horwood, Chichester.

Thabaraj, G. J. and A. F. Gaudy (1969). Effect of dissolved-oxygen concentration on the metabolic response of completely-mixed activated sludge. *J. Water Pollut. Control Fed.*, **41**, R322–R335.

Toms, R. G. and M. G. Booth (1982). The use of oxygen in sewage treatment. *Water Pollut. Control*, **81** (2), 151–165.

Walker, I. and E. P. Austin (1981). The use of plastic, porous biomass supports in a pseudo-fluidised bed for effluent treatment. In *Biological Fluidised Bed Treatment of Water and Wastewater*, ed. P. F. Cooper and B. Atkinson, pp. 272–284. Ellis Horwood, Chichester.

Wheeldon, D. H. V. and R. W. Bayley (1981). Economic studies of biological fluidised-beds for wastewater treatment. In *Biological Fluidised Bed Treatment of Water and Wastewater*, ed. P. F. Cooper and B. Atkinson, pp. 306–328. Ellis Horwood, Chichester.

Wood, L. B. *et al.* (1976). The operation of a Simplex activated-sludge pilot plant in an atmosphere of pure oxygen. *Public Health Eng.*, **4**, (2), 36–43.

Wuhrmann, K. (1963). Effect of oxygen tension on biochemical reactions in sewage purification plants. In *Advances in Biological Waste Treatment*, ed. W. W. Eckenfelder and J. McCabe, pp. 27–38. Pergamon, New York.

Wyatt, K. L., P. Brown and F. A. Shabi (1975). Oxidation processes in the activated-sludge process at high dissolved oxygen levels. In *The Application of Chemical Engineering to the Treatment of Sewage and Industrial Liquid Effluents*. *Inst. Chem. Eng. Symp. Ser.*, **41**, E1–E17.

51

Denitrification in Activated Sludge Processes

B. E. JANK
Environment Canada, Wastewater Technology Centre, Burlington, Ontario, Canada

51.1 INTRODUCTION

Biological nitrogen removal represents one of the most cost effective alternatives for nitrogen control in the treatment of municipal and industrial wastewaters. Biological nitrogen removal is essentially a two-step process: nitrification followed by denitrification. In the nitrification step, under aerobic conditions, autotrophic nitrifying organisms oxidize ammonia to nitrate. In the denitrification step nitrate is reduced to molecular nitrogen by heterotrophic organisms in the absence of molecular oxygen.

This chapter describes the basic principles, controlling environmental factors, kinetics and process design procedures for the denitrification phase of integrated suspended growth biological nitrogen removal systems.

51.2 PRINCIPLES OF BIOLOGICAL DENITRIFICATION

Biological denitrification is achieved under anoxic conditions by heterotrophic microorganisms that utilize nitrate or nitrite as a hydrogen acceptor when an organic energy source is available. Ingraham (1981) has reported that numerous genera of bacteria are capable of nitrate respiration. Certain bacteria are capable of completing the entire denitrification pathway (nitrate → nitrite → nitric oxide → nitrous oxide → dinitrogen); others are capable of catalyzing only a portion of the steps of the pathway resulting in partial denitrification. In the activated sludge process a wide variety of common facultative bacteria such as *Pseudomonas*, *Micrococcus*, *Achromobacter*, *Spirillum* and *Bacillus* have been reported to accomplish denitrification (Delwiche, 1956).

Waste Management and Pollution Control

The reduction of nitrate occurs through a series of complicated enzyme catalyzed reactions which can follow either the assimilatory or dissimilatory route. In assimilatory denitrification, nitrate is reduced to ammonia that is used in cell synthesis. In dissimilatory denitrification, in the absence of molecular oxygen, nitrate serves as the hydrogen acceptor in the oxidation of the carbon substrate producing energy for cell growth. The end product of both sets of reactions is principally gaseous nitrogen. A description of the biochemical pathways of denitrification and the physiology of the bacteria that carry out the denitrification reactions has been prepared by Bryan (1981).

Typical reactor configurations for suspended growth processes providing carbon oxidation, nitrification and denitrification are shown in Figure. 1. In the single sludge or separate sludge post-denitrification reactors, nitrate respiration follows carbon oxidation and nitrification. If complete nitrification occurs, the residual organic carbon from the aerobic process will not be of sufficient concentration, nor in a form which is readily usable by the denitrifying population as an organic carbon source. This can either be provided by the endogenous respiration of the sludge, by by-passing part of the influent wastewater to the denitrification reactor, or by the addition of an external carbon source.

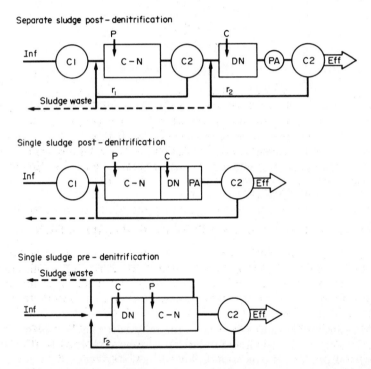

Figure 1 Schematic of nitrogen removal processes: C1 = primary clarifier; C2 = secondary clarifier; C–N = carbon removal plus nitrification; DN = denitrification; Inf = degritted, screened wastewater; Eff = final treated effluent; r = recycle; PA = post aeration; C = external carbon source; P = chemical addition for P removal

In a post-denitrification system relying on endogenous respiration of the biological sludge for nitrate reduction the overall stoichiometric equation can be written:

$$4\,NO_3^- + C_5H_7O_2N \rightarrow 2\,N_2 + 5\,CO_2 + NH_3 + 4\,OH^- \tag{1}$$

On the basis of this equation, 2 mg of biomass will be destroyed per mg of nitrate-nitrogen reduced.

In a post-denitrification reactor relying on methanol as an external carbon source the stoichiometric nitrate reduction equations can be written:

$$\text{Step 1: } 3\,NO_3^- + MeOH \rightarrow 3\,NO_2^- + CO_2 + 2\,H_2O \tag{2}$$

$$\text{Step 2: } 2\,NO_2^- + MeOH \rightarrow N_2 + CO_2 + H_2O + 2\,OH^- \tag{3}$$

$$\text{Overall: } 6\,NO_3^- + 5\,MeOH \rightarrow 3\,N_2 + 5\,CO_2 + 7\,H_2O + 6\,OH^- \tag{4}$$

In addition to the methanol required for nitrate dissimilation, an amount is necessary for bacterial growth. An overall mass balance which considers this requirement is as follows (McCarty, 1973):

$$MeOH + 0.92\,NO_3^- + 0.92\,H^+ \rightarrow 0.06\,C_5H_7O_2N + 0.43\,N_2 + 0.7\,CO_2 + 2.25\,H_2O \qquad (5)$$

On the basis of this equation, 20 mg of NO_3-N would require 19 mg of MeOH-C and would produce 10.5 mg of bacteria. If oxygen is present, an additional amount of methanol will be required for deoxygenation according to:

$$3\,O_2 + 2\,MeOH \rightarrow 2\,CO_2 + 4\,H_2O \qquad (6)$$

Studies involving separate sludge post-denitrification reactors have determined that the methanol requirement and biomass production can be estimated from the following relationship (McCarty *et al.*, 1969):
Methanol requirement:

$$C_m = 2.47\,N_0 + 1.53\,N_1 + 0.87\,D_0 \qquad (7)$$

Biomass production:

$$C_b = 0.53\,N_0 + 0.32\,N_1 + 0.19\,D_0 \qquad (8)$$

where: C_m = required methanol (mg l^{-1}), C_b = biomass production (mg l^{-1}), N_0 = initial NO_3-N concentration (mg l^{-1}), N_1 = initial NO_2-N concentration (mg l^{-1}), and D_0 = the dissolved oxygen concentration in the wastewater, entering the system (mg l^{-1}). A number of researchers (US EPA, 1975) have obtained experimental results which verify that the expressions provide reasonable estimates of methanol requirements and biomass production.

By contrast for pre-denitrification systems no stoichiometric relationships have been developed to define organic carbon requirements for the denitrification reactions. Empirical relationships have been developed and these will be presented in Section 51.6.

51.3 ENVIRONMENTAL FACTORS AFFECTING DENITRIFICATION

Christensen and Harremoës (1977) presented results from numerous authors reporting that nitrate dissimilation is inhibited by molecular oxygen. In both oxygen and nitrate respiration, the same series of reversible enzymatic reactions are involved in the transfer of electrons from organic substrates. Only a different terminal enzyme is required for nitrate respiration and this is formed by bacteria in the absence of molecular oxygen. Molecular oxygen either represses the formation of nitrate reductase or acts as an electron acceptor preventing nitrate reduction (Chang and Morris, 1962). Although denitrification will be observed in systems with a low oxygen tension, a completely anoxic reactor will ensure complete nitrate reduction.

The optimum pH for denitrification is usually reported at conditions near neutrality (Christensen and Harremoës, 1977; US EPA, 1975). Nitrate dissimilation is a basic reaction with hydroxyl ion production increasing the alkalinity. The stoichiometric alkalinity production rate is 3.6 mg of alkalinity as $CaCO_3$ per mg of NO_3-N reduced regardless of whether the carbon source is provided by endogenous respiration of the biological sludge, an external carbon source such as methanol, or the organic carbon in the influent wastewater. When used together with nitrification, the nitrification/denitrification sequence tends to maintain a neutral pH (Barnard, 1974; Sutton *et al.*, 1978b).

Denitrification is reported to occur over a temperature range from 0 to 50 °C with an optimum approaching 40 °C (Christensen and Harremoës, 1977). Figure 2 shows that there is reasonably good agreement concerning the degree of temperature sensitivity over the range normally encountered in municipal wastewater treatment. An extensive list of values for various coefficients normally used to express the reaction rate and temperature dependency in biological systems has been reported by Sutton *et al.* (1975).

In the majority of municipal and industrial waste treatment systems, suspended growth denitrification is associated with nitrification. Considering the sensitivity of the autotrophic nitrifiers to environmental conditions, ensuring that there are no toxicants in the wastewater limiting the nitrification rate will also ensure that the denitrification process will be unaffected.

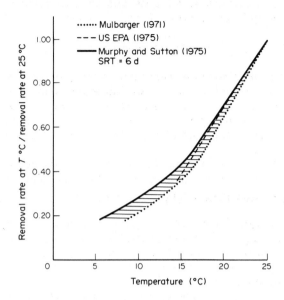

Figure 2　Dependency of denitrification rate on temperature

51.4　KINETICS OF BIOLOGICAL DENITRIFICATION

In the growth of the heterotrophic organisms responsible for denitrification, carbon and nitrate are the important substrates involved in the biological reaction. If the desire is to maximize the nitrate reduction rate, then the denitrification process must be operated under conditions where carbon concentration is not limiting. It has been shown that when methanol is the carbon source, no rate limitation will be observed if the concentration is in excess of that required for nitrate dissimilation and biomass production in accord with equation (7) (McCarty *et al.*, 1969; Dawson and Murphy, 1972; US EPA, 1975). Nitrate reduction will proceed at a constant rate until the carbon and energy source is depleted. At this point the reactor will be operating under carbon limiting conditions and any further nitrate reduction will rely on the organic carbon released during endogenous respiration.

Under non-carbon limiting conditions, little nitrite appearance has been observed and researchers have normally modelled the denitrifier growth rate according to:

$$\mu_{DN} = \mu_{DN}^* \frac{N}{C_N + N} \tag{9}$$

where μ_{DN} = growth rate of denitrifiers (d^{-1}), μ_{DN}^* = maximum growth rate of denitrifiers (d^{-1}), N = nitrate-nitrogen concentration $(mg\ l^{-1})$, and $C_N = N$ at which $\mu_{DN} = 0.5\ \mu_{DN}^*$.

The nitrate reduction can be related to denitrifier growth rate according to:

$$\frac{1}{X_{DN}} \frac{dN}{dt} = \frac{\mu_{DN}}{Y_{DN}} = K_N = K_N^* \frac{N}{C_N + N} \tag{10}$$

where X_{DN} = denitrifier mixed liquor volatile suspended solids concentration $(mg\ l^{-1})$, Y_{DN} = denitrifier yield coefficient $(mg\ mg^{-1})$, K_N = unit denitrification rate $(mg\ mg^{-1}\ d^{-1})$, and K_N^* = maximum denitrification rate $(mg\ mg^{-1}\ d^{-1})$.

At steady state the rate of denitrification in separate sludge systems has been shown to be zero order with respect to nitrate concentration and at a maximum, provided there is an excess of available carbon (Dawson and Murphy, 1972; Sutton *et al.*, 1974). Under these conditions

$$K_N^* = \frac{1}{X_{DN}} \frac{dN}{dt} \tag{11}$$

In single or separate sludge anoxic denitrification reactors, the minimum denitrification rate will be observed under conditions where the endogenous respiration of the biological sludge releases organic carbon which then becomes available for nitrate reduction. Under these conditions the nitrate reduction rate has been reported as zero order with respect to nitrate concentration (Stern and Marais, 1974) and is normally expressed as:

$$\frac{dN}{dt} = K_{NE}X_M \tag{12}$$

where K_{NE} = endogenous denitrification rate (mg mg^{-1} d^{-1}), and X_M = measured total volatile suspended solids concentration (mg l^{-1}) in the denitrification reactor.

Although the coefficients for the kinetic relationships are dependent on the wastewater characteristics, reactor configuration and operating conditions, typical values for denitrification in suspended growth systems have been reported by Sutton *et al.* (1979), Schroeder (1981) and Beccari *et al.* (1983).

51.5 DESIGN PROCEDURES FOR SUSPENDED GROWTH REACTORS

Design of suspended growth biological denitrification systems is normally based on either SRT (solids retention time or sludge age) considerations or the unit denitrification rate approach. The SRT approach is based on the growth rate of the denitrifying organisms, while the unit rate approach is based on the total biomass inventory. The unit denitrification rate is expressed as the mass of equivalent nitrate nitrogen removed per day per mass of total volatile suspended solids. The term 'equivalent nitrate' accounts for the presence of other electron acceptors such as molecular oxygen or nitrite.

51.5.1 SRT Design Approach

In a denitrifying biological system operating at equilibrium conditions, the SRT, a measure of the average retention time of the denitrifiers in the system, and the denitrifier growth rate are related by

$$SRT = \frac{1}{\mu_{DN}} \tag{13}$$

Consequently the growth of denitrifiers is fixed by the SRT of the system. Since the denitrification rate (K_N) is a function of the growth rate (equation 10) then K_N is fixed by the SRT. This relationship between SRT and growth rate of denitrifiers implies a procedure for the design and control of a denitrification system.

For a separate sludge denitrification system (Figure 1), the SRT for the denitrification reactor (SRT$_{DN}$) is defined as the mass of biological solids in the reactor divided by the total rate of loss of biological solids from the system (mass per unit time). For single sludge systems (Figure 1) the SRT for the denitrification reactor is defined as

$$SRT_{DN} = \frac{(V_{DN})(SRT_S)}{V_S} \tag{14}$$

where V_{DN} = the volume of the denitrification reactor (m^3), V_S = the total volume of the suspended growth reactors (m^3), and SRT$_S$ = the total system SRT (d).

Design of systems based on the SRT approach is normally accomplished by utilizing specific growth rate data from the literature or *via* minimum solids retention time data generated from bench-scale testing. To ensure constant denitrification performance a design safety factor should be incorporated. The safety factor is defined (US EPA, 1975) as the design SRT divided by the minimum SRT to achieve denitrification. Application of a safety factor will ensure consistent denitrification. The higher the safety factor the lower will be the effluent nitrate level. From a practical viewpoint, a design safety factor of 2 is normally adequate. This should ensure an effluent NO$_3$-N of less than 0.5 mg l^{-1}.

There is a limited amount of data in the literature regarding design of denitrification systems based on SRT control. In the treatment of municipal wastewater, Sutton *et al.* (1979) determined the minimum SRT necessary to achieve denitrification using the single sludge combined nitrification–denitrification flowsheet. Systems were operated in both the pre- and post-denitrification mode. Carbon sources were methanol for the post-denitrification mode and raw sewage for the pre-denitrification mode. For both systems the minimum anoxic SRT required to achieve complete denitrification was 0.5 to 1.5 d at 24 to 26 °C, 1 to 2 d at 14 to 16 °C and 4 to 5 d at 7 to

8 °C. Results by Sutton *et al.* (1981) from a full scale single sludge pre-denitrification nitrification process configuration, treating wastewater from an organic chemical industry, indicate complete denitrification at 22 to 24 °C with an anoxic SRT of 5 d. Using a similar process configuration, complete denitrification of coke plant wastewaters has been achieved at 20 to 24 °C with an anoxic SRT of 7 d (Bridle *et al.*, 1980).

### 51.5.2	Unit Denitrification Rate Design Approach

The alternative and equally acceptable design methodology is the unit denitrification rate approach. In the design of both municipal (US EPA, 1975; Barnard, 1974) and industrial (Adams and Eckenfelder, 1974; Bridle *et al.*, 1976) reactors for denitrification, the use of a denitrification rate has been the preferred approach.

For separate sludge denitrification reactors operating under non-carbon limiting conditions, the use of a denitrification rate based on the total volatile suspended solids (K_{NV}^*) has been recommended (Christensen and Harremoës, 1977; US EPA, 1975). Assuming that all the volatile solids in the anoxic denitrification reactor are responsible for nitrate reduction, then K_{NV}^* equals K_N^* and equation (11) can be used for sizing the denitrification reactor.

K_N^* can be related to the minimum anoxic SRT required according to:

$$\frac{1}{SRT_{DNM}} = Y_{DN}K_N^* - K_{DN} \tag{15}$$

where SRT_{DNM} = minimum anoxic SRT (d), and K_{DN} = denitrifier decay coefficient (d^{-1}). Equation (15) can be used to calculate SRT given the nitrate removal rate and the net microorganism yield. Wilson *et al.* (1981) used this equation to establish that for a municipal wastewater the SRT for a separate sludge post-denitrification basin should be 18 d at a temperature of 8 °C.

Sutton *et al.* (1978a) developed the following Arrhenius expression for calculating separate sludge denitrification rates at different operating temperatures:

$$K_N^* = 5.42 \times 10^{10}e^{-3660/RT} \tag{16}$$

where R = universal gas constant (kJ kg mol^{-1} °K^{-1}), and T = temperature (°K). This relationship, derived from pseudo steady-state (that is constant flow and varying NO_3-N concentrations) pilot plant data using nitrified municipal wastewater, is shown schematically in Figure 3. Using the same pilot plant data, Sutton *et al.* (1977) have shown that the denitrification rate is not strongly dependent on SRT in the range of 3 to 9 d.

In single sludge systems X_{DN} is not equal to the total mixed liquor volatile suspended solids concentration. For design purposes some estimate of the denitrifier population must be made. The most complex system would be the single sludge system designed for both carbon oxidation and nitrogen control. Under these conditions X_{DN} can be calculated by

$$X_{DN} = \frac{Y_{DN}(N_i - N_e)X}{Y_{DN}(N_i - N_e) + Y_N(A_i - A_e) + Y_C(C_i - C_e)} \tag{17}$$

where $(N_i - N_e)$ = NO_3-N removed (mg l^{-1}), $(A_i - A_e)$ = NH_3-N removed (mg l^{-1}), $(C_i - C_e)$ = TOC removed (mg l^{-1}) in aerobic respiration, X = total mixed liquor volatile suspended solids concentration (mg l^{-1}) in the system, Y_N = nitrifier gross yield (mg VSS per mg NH_3-N removed), and Y_C = heterotrophic gross yield for aerobic respiration (mg VSS per mg TOC removed).

Equation (17) assumes the heterotrophs produced during aerobic respiration are not capable of denitrification. However, it is more likely that a fraction of the heterotrophs initially grown under aerobic conditions are capable of nitrate respiration. Consequently, equation (17) underestimates the denitrifier population.

The design for single sludge systems can, however, be based on total volatile solids, provided the limitations of this approach are realized. It must be emphasized that the denitrification rate will be lower than for separate sludge systems and will be subject to fluctuations dependent on variations in feed composition.

Pilot scale evaluation of single sludge nitrogen removal systems indicated no statistical difference of denitrification rates in the pre- and post-denitrification modes (Sutton *et al.*, 1979). Results from this study are shown in Figure 3. It can be seen that the results compare favourably

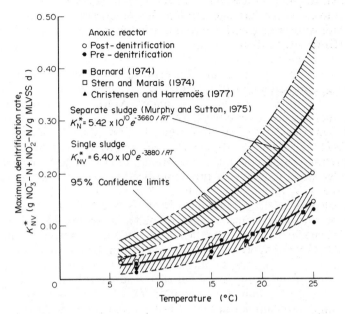

Figure 3 Denitrification rates under non-carbon limiting conditions

with those in the literature and that the rates are roughly 50% of those achieved in separate sludge systems.

51.6 CARBON REQUIREMENTS FOR DENITRIFICATION

51.6.1 Organic Carbon Requirements for Maximum Denitrification Rates

Once the quantity of organic carbon required for bacterial growth and reduction of nitrite, nitrate and oxygen becomes limiting, denitrification will not proceed at the maximum rate. If the chemical structure of the organic carbon source is known one can develop balanced equations which will describe the stoichiometric filtered organic carbon (FOC) requirements for nitrogen removal and deoxygenation. McCarty *et al.* (1969) referred to this as the consumptive ratio, C_r, represented by the relationship:

$$C_r = \frac{\text{Total FOC utilized}}{\text{FOC required for denitrification and deoxygenation}} \qquad (18)$$

If nitrate, nitrite and oxygen are reduced stoichiometrically by organic carbon sources, the C_r should be equal to one. However, a supplemental quantity of organic carbon is required for cell synthesis, and consequently C_r determined experimentally is greater than unity. An organic waste that has a high C_r would tend to generate larger volumes of sludge and would not be desirable as an alternate carbon source for denitrification.

In an extensive assessment of the suitability of industrial waste carbon sources for denitrification, Monteith *et al.* (1980) determined consumptive ratios for industrial effluents containing one organic compound as the major component. The mean C_r for commercial methanol obtained from experiments conducted in batch reactors was 1.65, much higher than an average C_r of 1.3 obtained by McCarty *et al.* (1969) for methanol in semi-continuous experiments. C_r values for other methanol wastes (Monteith *et al.*, 1980) were 1.15 for methanol heads from a paper company, 1.0 for methanol still bottoms from an organic chemical manufacturer and 2.53 for a methyl fuel.

Monteith *et al.* (1980) reported that an alternative method of identifying the effectiveness of an industrial waste as a carbon source was by measuring the substrate consumption ratio, the quantity of filtered COD (FCOD) or FOC consumed per unit of NO_T-N removed. Substrate consumption ratios for the 30 industrial wastes evaluated as alternate carbon sources are presented in

Table 1. Results ranging from 0.71 to 3.7 g FOC per g NO_T-N removed should be compared to the methanol control value of 1.17 g FOC per g NO_T-N removed.

Sutton *et al.* (1979) generated substrate consumption data utilizing the single sludge pilot scale pre- and post-denitrification process configuration treating municipal wastewater. Results from the post-denitrification system (Figure 4), utilizing methanol as the carbon source, revealed that the stoichiometric requirement, of about 3 to 4 g FCOD per g of equivalent NO_3-N reduced, was adequate to ensure complete denitrification. Results from the pre-denitrification mode (Figure 4) indicated that unless approximately 9 g FCOD were available per g of equivalent NO_3-N reduced, the degree of denitrification would be limited. Murphy and Wilson (1981) determined that the 9 g FCOD was comparable to a total COD of 18 g per g of equivalent NO_3-N. Barnard (1973) found the ratio of the COD required to equivalent NO_3-N reduced was 10 in a pre-denitrification reactor. Stern and Marais (1974) and Marsden and Marais (1976), operating bench scale pre-denitrification systems, found the ratio to be greater than 15. They concluded that the reason for the excessive COD requirement in a pre-denitrification reactor is a rapid initial removal of COD by adsorption.

In the pre-denitrification mode the organic carbon requirements to ensure that there was no nitrate or nitrite in the anoxic effluent were determined as a function of the feed carbon to nitrogen ratio. From the data generated by Sutton *et al.* (1979), it was shown that unless the feed COD:N ratio was 13 to 15, oxidized nitrogen was evident in the anoxic reactor. Barnard (1983), stressing the importance of having sufficient carbon for the removal of nitrates, specified that the COD:TKN ratio in the influent should be about 10:1. These results are similar to those reported by Bridle *et al.* (1980) who determined the minimum feed FOC:N ratio required to ensure complete denitrification of coke plant wastewater (pre-denitrification nitrification mode) to be 3.5. Full scale experience at Oswego, New York (Schwinn and Storrier, 1978) and Largo, Florida (Hong *et al.* 1984) has indicated high levels of nitrogen control with influent BOD_5:TKN ratios of 5 to 6:1. Based on the Oswego COD:BOD_5 ratio of 2.2, this represents a COD:TKN ratio of 11 to 13:1, supporting the bench and pilot scale data.

51.6.2 Denitrification Under Carbon Limiting Conditions

When the organics in the raw sewage are not present in sufficient quantity to provide complete denitrification in the anoxic reactor (see Figure 4), complete nitrate removal can be obtained by relying on the endogenous respiration of the biological sludge to complete the reaction. In this case Sutton *et al.* (1979) have shown that the rate of nitrate reduction initially proceeds under non-carbon limiting conditions and then reduces to a rate equivalent to the endogenous rate (see Figure 5). The reduced overall rate of nitrate reduction is equivalent to a reduction in the growth rate of the denitrifiers. Consequently, to obtain complete nitrate reduction under these conditions the SRT in the anoxic reactor of the pre-denitrification nitrification system must be increased. Comparing the non-carbon limited and endogenous nitrate removal rates (see Figure 5), if the anoxic SRT was increased by a factor of 2.5 the resulting SRT should be more than sufficient to obtain complete denitrification.

This was verified by Sutton *et al.* (1979) using continuous results from the pre-denitrification nitrification system under carbon limiting conditions. Complete denitrification was obtained under the increased anoxic SRT conditions at an influent unfiltered COD to TKN ratio as low as nine. Similar findings have been reported by van Vuuren and Wiechers (1977) operating a 130 m^3 per day pre-denitrification nitrification pilot plant in South Africa. The feed to the plant was primary lime treated municipal sewage with a COD to NH_3-N ratio of approximately five. From their data, recorded during operation at the same recycle ratios as employed by Sutton *et al.* (1979), it can be shown that at an initial anoxic SRT of four days, denitrification was not complete in the anoxic reactor. When the anoxic HRT was increased by a factor of 1.8, thereby increasing the SRT, complete denitrification was obtained.

Denitrification rates under non-carbon limiting conditions (see Figure 5) and under conditions where the endogenous respiration of the biological sludge controls the rate of denitrification represent the maximum and minimum values, respectively, attainable in the anoxic reactor(s) of the single sludge systems. The choice of the rate, for design purposes, and the corresponding SRT required for the anoxic reactor will be determined from a knowledge of the organic carbon requirements (see Figure 4) and the influent carbon:nitrogen ratio.

Where endogenous respiration of the biological sludge provided the carbon source the wide variation in the denitrification rates observed by Sutton *et al.* (1979) (see Figure 6) and those

Table 1 Denitrification Rates Using Industrial Wastes as the Carbon Source

Waste	Initial FOC:N	Temperature (°C)	DN rate (d⁻¹)	DN rate relative to MeOH control run on same day	Substrate consumption ratios FCOD consumed (kg) / NO_T–N removed (kg)	FOC consumed (kg) / NO_T–N removed (kg)	Comments
GROUP 1: The following wastes exhibited rates above the 95% confidence interval for methanol:							
Rieder Distillery fusel oils	3.19	20.5	0.331	2.38	2.22	0.77	—
Pea blanchwater (Food Processor 'A')	3.27	18.5	0.261	2.08	5.71		b
Jordan Wines sludge centrate	2.70	20.5	0.207	1.62	7.30	2.28	b
Labatt's Brewery spent grain extract	3.18	20	0.197	2.40	5.48	2.46	—
Molson's Brewery last runnings	2.53	20.5	0.191	1.49	6.67	1.83	b
Molson's Brewery wort	4.29	21	0.187	2.27	6.17	1.35	b
McGuinness Distillers thin stillage	2.71	21	0.184	1.44	6.07	2.18	b
Methanol still bottoms (Org. manuf. 'A')	1.49	20	0.170	0.86	3.66	0.71	b
National Starch process effluent	2.97	18	0.160	1.54	3.26		—
Tomato sludge (Food Processor 'A')	1.72	18	0.160	1.31	2.54	0.80	b
McGuinness Distillers fusel oils	3.17	20	0.159	1.29	5.32	1.46	b
Molson's Brewery beer	4.16	20.5	0.159	1.41	8.57	2.54	b
Du Pont organic acids waste	2.61	21	0.142	1.29	5.14	1.65	c, d
Spent sulfite liquor (Can. Int. Paper)	1.77	19	0.137	1.24	3.94	0.79	b, d
Domtar Packaging whitewater	3.72	21	0.137	2.13	5.74	1.48	—
Vulcan–Cincinnati methyl fuel	4.07	21	0.135	1.22	6.18	1.83	—
Celanese light ends (tray 25)	3.48	21	0.129	1.17	5.23	1.36	—
Methanol heads (Ontario Paper Co.)	1.53	18	0.128	1.06	2.45	0.82	b
Rieder Distillery grape slops	3.21	20	0.125	1.94	5.00	1.42	b
Acetic acid waste (Dow Chemical Co.)	1.76	20	0.123	0.62	3.87	1.71	c
Du Pont high boiling organic waste	2.53	19	0.119	1.07	6.02	1.36	—
McGuinness Distillers light distillate	9.91	20	0.117	1.61	10.16	1.52	—
GROUP 2: The following wastes exhibited rates within the 95% confidence interval for methanol:							
Jordan Wines pomace extract	3.43	19	0.112	1.74	5.69	2.6	b
Millhaven Fibres glycol waste	2.94	20	0.103	1.60	5.98	0.92	e
Methanol control	2.87	20	0.097[a]	—	5.41[a]	1.17[a]	e
Molson's Brewery trub	4.73	20	0.093	1.28	6.40	3.7	e
Isopropanol waste (Norwich)	4.40	20	0.090	1.40	3.64	1.82	c
Gos and Gris cheese whey	2.50	20	0.084	1.31	9.65	0.91	b
GROUP 3: The following wastes exhibited rates below the 95% confidence interval for methanol:							
Domtar Packaging black liquor	2.24	18	0.080	1.25	6.02	1.76	b, d
Waste dextrose (Baxter Travenol Labs)	2.65	20	0.071	0.57	8.19	2.57	b
Formaldehyde Waste (University of Guelph)	3.62	20	0.042	0.37	6.21	1.38	—

[a] Mean value. [b] Waste caused nitrate production. [c] Waste added to TKN concentration. [d] Waste added colour to clarified effluent. [e] DN is mean of 14 runs.

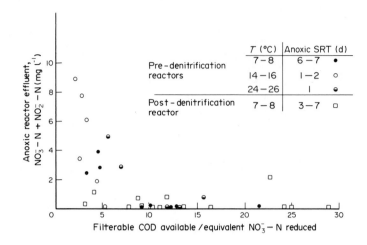

Figure 4   Organic carbon requirements in post- and pre-denitrification reactor

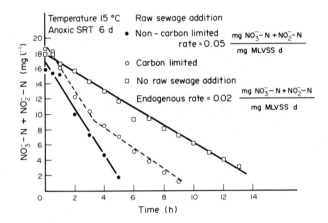

Figure 5   Batch denitrification results under carbon and non-carbon limiting conditions

reported by other authors may be due to the dependence on the system SRT. The observed denitrification rates decreased with increasing SRT. Little or no temperature sensitivity was reported (Sutton *et al.*, 1979) for denitrification proceeding under carbon-limiting conditions.

51.6.3   Alternative Carbon Sources for Denitrification

In the 1970s methanol was commonly suggested as the external carbon source for denitrification. The recent dramatic increase in the price of petrochemical products has all but eliminated its use. Raw sewage or the organics present in industrial wastewaters are probably the only economically viable source of organic carbon.

There has been considerable research involving determination of denitrification rates and substrate consumption ratios for alternative carbon sources. Climenhage and Stelzig (1973) demonstrated that C_1 to C_5 volatile organic acids, generated in the manufacture of nylon intermediates, were effective as the carbon source for denitrification. Rates of 0.36 d^{-1} and 0.1 d^{-1} were reported for single sludge systems at 20 °C and 10 °C, respectively; these are comparable to other results reported for separate sludge systems. Wilson and Newton (1973), utilizing brewery wastes, established denitrification rates of 0.22 to 0.25 d^{-1} at 19 to 24 °C, compared to 0.18 d^{-1} using methanol. Sutton *et al.* (1979) demonstrated that the organics present in raw sewage are

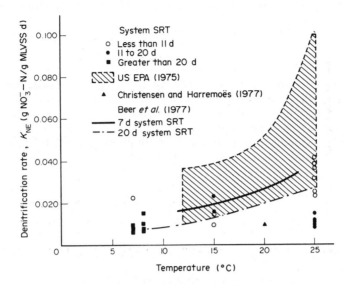

Figure 6 Denitrification rates using endogenous carbon sources

comparable to methanol as carbon sources for denitrification; at an SRT of 6 d and temperature of 7 °C denitrification rates were 0.026 and 0.023 d^{-1}.

Monteith *et al.* (1979) in their evaluation of industrial wastes as carbon sources for denitrification revealed that 27 of the 30 wastes tested exhibited denitrification rates equal to or greater than those observed using methanol (Table 1). Bridle (1982) determined denitrification rates and substrate consumption ratios for the carbon sources evaluated by Skrinde and Bhagat (1982). Results for wine spent concentrate, brewery spent grain extract, starch processing effluent, spent sulfite liquor, the methanol control and whey were comparable to those generated by Monteith *et al.* (1980).

51.7 ACTIVATED SLUDGE PROCESSES INCORPORATING BIOLOGICAL DENITRIFICATION

Integrated biological systems designed for combined nitrogen conversion and removal can be either separate or single sludge systems, operated in either the pre- or post-denitrification mode (Figure 1). Separate sludge systems operate at higher unit removal rates and consequently require lower reactor volumes. This advantage is however normally overshadowed by the advantages attributed to single sludge systems, namely, reduction of clarification requirements, improved sludge properties, reduced need for pH control, reduced aeration requirements, and reduced carbon requirements (pre-denitrification mode).

Wilson *et al.* (1981) undertook a study to review proven methods for biological nitrogen removal from municipal wastewater to select suitable design parameters and to prepare process designs plus capital and operating cost estimates for several process alternatives applicable to the City of Penticton, British Columbia. Process options that were evaluated in detail included separate sludge post-denitrification, single sludge post-denitrification and single sludge pre-denitrification (Figure 1). Process designs were based on achieving a maximum of 5 mg l^{-1} NO$_3$-N and 1 mg l^{-1} TKN in the treated effluent throughout the year at a design flow of 15 500 m^3 d^{-1}. To ensure efficient nitrogen removal at a minimum operating temperature of 8 °C, the design was based on a 20-day SRT for the nitrification basin and a 10-day SRT for the denitrification basin. For Penticton the cost of combined carbon and nitrogen removal by pre-denitrification was found to be the most cost effective nitrogen control option. Land disposal, ammonia stripping and ion exchange proved to be significantly more expensive than the biological methods of nitrogen removal.

Schroeder (1981) has identified alternate system configurations for activated sludge processes providing denitrification. Single tank and multitank configurations similar to those in Figure 1 were described. The oxidation ditch (Barnes *et al.*, 1983) represents one of the single tank systems which can be designed and operated to provide denitrification. The Bardenpho process

(Barnard, 1974) and sequencing batch reactor system (Allmen and Irvine, 1981) are also activated sludge process configurations capable of providing biological denitrification.

51.8 SUMMARY

The design of activated sludge processes providing biological denitrification can be accomplished using either the SRT approach or the unit denitrification rate approach. Considerable research has been conducted to verify that denitrification does not have to be a costly operation using methanol as the carbon source. With the use of waste carbon sources, or the use of the predenitrification flowsheet, total nitrogen control for both municipal and industrial wastewaters can be implemented with minimal increases in operating costs.

51.9 REFERENCES

Adams, C. E., Jr. and W. W. Eckenfelder (1974). *Process Design Techniques for Industrial Waste Treatment.* Enviro Press, Nashville, TN.
Alleman, J. E. and R. L. Irvine (1981). Storage-induced denitrification using sequencing batch reactor operation. *Water Res.*, **14**, 1483–1488.
Barnard, J. L. (1973). Biological denitrification. *Water Pollut. Control*, **72**, 705–720.
Barnard, J. L. (1974). Cut P and N without chemicals. *Water Wastes Eng.*, **11**, 33–36.
Barnard, J. L. (1983). Design consideration regarding phosphate removal in activated sludge plants. *Water Sci. Technol.*, **15**, 319–328.
Barnes, D., C. F. Forster and D. W. M. Johnstone (1983). *Oxidation Ditches in Wastewater Treatment.* Pitman, London.
Beccari, M., R. Passino, R. Ramadori and V. Tandoi (1983). Kinetics of dissimilatory nitrate and nitrite reduction in suspended growth culture. *Water Pollut. Control Fed.*, **55**, 58–64.
Beer, C., J. R. Bergonthal and L. K. Wang (1977). A study of endogenous nitrate respiration of activated sludge. *9th Mid-Atlantic Ind. Waste Conf.*, 207–215.
Bridle, T. R. (1982). Discussion of: Industrial wastes as carbon sources in biological denitrification (by J. R. Skrinde and S. K. Bhagat). *Water Pollut. Control Fed.*, **54**, 1613.
Bridle, T. R., D. C. Climenhage and A. Stelzig (1976). Start-up of a full scale nitrification–denitrification industrial waste treatment plant. *31st Ind. Waste Conf.*, Purdue University, 807–815.
Bridle, T. R., W. K. Bedford and B. E. Jank (1980). Biological nitrogen control of coke plant wastewaters. *Prog. Water Technol,*, **12**, 667–680.
Bryan, B. A. (1981). Physiology and biochemistry of denitrification. In *Denitrification, Nitrification and Atmospheric Nitrous Oxide*, ed. C. C. Delwiche, chap. 4, pp. 67–84. Wiley, Toronto.
Chang, J. P. and J. G. Morris (1962). Studies on the utilization of nitrate by *Micrococcus denitrificans*. *J. Gen. Microbiol.*, **29**, 301–310.
Christensen, M. H. and P. Harremoës (1977). Biological denitrification of sewage: A literature review. *Prog. Water Technol.*, **8**, 509–555.
Climenhage, D. C. and A. Stelzig (1973). Biological processes for nitrogen–BOD removal at Maitland Works, Du Pont of Canada Limited. *Proc. 20th Ont. Ind. Waste Conf.*, 85–98.
Dawson, R. N. and K. L. Murphy (1972). Factors affecting biological denitrification of wastewater. In *Advances in Water Pollution Research*, pp. 671–683. Pergamon, Oxford.
Delwiche, C. C. (1956). Denitrification. In *Symposium on Inorganic Nitrogen Metabolism*, ed. A. B. McElroy and C. D. Glass, pp. 233–256. John Hopkins Press, Baltimore, MD.
Hong, S. N., D. Krichten, A. Best and A. Rachwal (1984). Biological phosphorus and nitrogen removal *via* the A/O process: Recent experience in the United States and United Kingdom. *Water Sci. Technol.*, **16**, 151–172.
Ingraham, J. L. (1981). Microbiology and genetics of denitrifiers. In *Denitrification, Nitrification and Atmospheric Nitrous Oxide*, ed. C. C. Delwiche, chap. 3, pp. 45–65. Wiley, Toronto.
McCarty, P. L. (1973). Nitrification–denitrification by biological treatment. Correspondence Conference on Denitrification of Municipal Wastes, University of Massachusetts.
McCarty, P. L., L. Beck and P. St. Amant (1969). Biological denitrification of wastewaters by addition of organic materials. *24th Ind. Waste Conf.*, Purdue University, 1271–1285.
Marsden, M. and G. V. R. Marais (1976). Role of the primary anoxic reactor in denitrification and biological phosphorus removal. Research Report No. W. 19, Dept. of Civil Eng., University of Cape Town, Republic of South Africa.
Monteith, H. D., T. R. Bridle and P. M. Sutton (1979). Evaluation of industrial waste carbon sources for biological denitrification. Technology Development Report EPS 4-WP-79-9. Wastewater Technology Centre, Burlington, Ontario.
Monteith, H. M., T. R. Bridle and P. M. Sutton (1980). Industrial waste carbon sources for biological denitrification. *Prog. Water Technol.*, **12**, 127–141.
Mulbarger, M. C. (1971). Nitrification and denitrification in activated sludge systems. *Water Pollut. Control Fed.*, **43**, 2059–2070.
Murphy, K. L. and P. M. Sutton (1975). Pilot scale studies on biological denitrification. *Prog. Water Technol.*, **7**, 317–328.
Murphy, K. L. and R. W. Wilson (1981). Evaluation of biological nitrification/denitrification at Penticton, British Columbia. Technology Development Report SCAT 10. Wastewater Technology Centre, Burlington, Ontario.
Schroeder, E. D. (1981). Denitrification in wastewater management. In *Denitrification, Nitrification, and Atmospheric Nitrous Oxide*, ed. C. C. Delwiche, chap. 6, pp. 105–125. Wiley, Toronto.

Schwinn, D. E. and D. F. Storrier (1978). One step for nitrogen removal – a giant step for optimum performance. *Water Wastes Eng.*, **15** (12), 33–37.

Skrinde, J. R. and S. K. Bhagat (1982). Industrial wastes as carbon sources in biological denitrification. *Water Pollut. Control. Fed.*, **54**, 370–377.

Stern, L. B. and G. V. R. Marais (1974). Sewage as an electron donor in biological denitrification. Research Report No. W. 7. Dept. of Civil Eng., University of Cape Town, Republic of South Africa.

Sutton, P. M., K. L. Murphy and R. N. Dawson (1974). Continuous biological denitrification of wastewater. Technology Development Report EPS 4-WP-74-6. Wastewater Technology Centre, Burlington, Ontario.

Sutton, P. M., K. L. Murphy and R. N. Dawson (1975). Low temperature biological denitrification of wastewater. *Water Pollut. Control Fed.*, **47**, 122–134.

Sutton, P. M., K. L. Murphy and B. E. Jank (1977). Nitrogen control: a basis for design with activated sludge systems. *Prog. Water. Technol.*, **8**, 467–481.

Sutton, P. M., K. L. Murphy and B. E. Jank (1978a). Kinetic studies of single sludge nitrogen removal systems. *Prog. Water Technol.*, **10**, 241–253.

Sutton, P. M., K. L. Murphy and B. E. Jank (1978b) Design considerations for integrated nutrient removal systems. *Prog. Water Technol.*, **10**, 469–478.

Sutton, P. M., K. L. Murphy, B. E. Jank and B. A. Monaghan (1979). Single sludge nitrogen removal systems. Canada Ontario Agreement Research Report No. 88. Wastewater Technology Centre, Burlington, Ontario.

Sutton, P. M., T. R. Bridle, W. K. Bedford and J. Arnold (1981). Nitrification and denitrification of an industrial wastewater. *Water Pollut. Control Fed.*, **53**, 176–184.

US EPA (1975). Nitrogen control manual. Office of Technology Transfer, Washington, DC.

van Vuuren, L. R. J. and S. G. Wiechers (1977). Nitrification and denitrification of lime-treated wastewaters. *Prog. Water Technol.*, **10**, 185–198.

Wilson, T. E. and D. Newton (1973). Brewery wastes as a carbon source for denitrification at Tampa, Florida. *28th Ind. Waste Conf.*, Purdue University, 138–149.

Wilson, R. W., K. L. Murphy, P. M. Sutton and S. L. Lackey (1981). Design and cost comparison of biological nitrogen removal systems. *Water Pollut. Control Fed.*, **53**, 1294–1302.

52

Oxidation Ditches, Aerated Lagoons and Waste Stabilization Ponds

E. I. STENTIFORD
University of Leeds, UK

52.1 INTRODUCTION

Wastewater treatment systems not only have to be designed to produce a good quality effluent but also have to do it consistently at minimum cost. In developing countries or remote areas where technical expertise is in limited supply, processes which use sophisticated control systems and involve the maintenance of complicated equipment are not generally suitable. It is for this reason that oxidation ditches, aerated lagoons and waste stabilization ponds have a prominent role in such areas. They have been used extensively under various climatic conditions to treat both domestic and industrial wastewaters producing good effluents consistently. In this section these processes will be examined from both design and operational viewpoints.

52.2 OXIDATION DITCHES

The term oxidation ditch has become widely used to describe the oval channel type activated sludge system shown schematically in Figure 1. In its original form, as developed by Pasveer in Holland (Pasveer, 1959), the ditch was intended to be used for small communities and to operate as an extended aeration system. Since then, this type of ditch configuration has been used frequently for activated sludge reactors operating under other than extended aeration conditions due to the economics of its construction. This section will deal with the ditch operating as an extended aeration system.

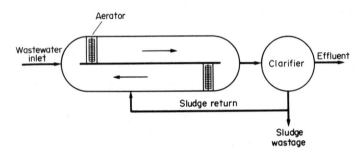

Figure 1 Typical oxidation ditch system with separate clarifier

52.2.1 Extended Aeration

The system of extended aeration operates such that the overall production of excess sludge is kept to a minimum. Consequently no provision is made for primary sedimentation, prior to the oxidation ditch, designed to remove readily settleable solids. In conventional activated sludge systems, 50–70% of the incoming suspended solids are removed in the primary sedimentation tanks. In extended aeration systems all of these solids are fed into the ditch, where a proportion of the organic solids is biodegraded.

Relatively low organic loads are applied to the reactor, which is designed on the basis of relatively long aeration times, *i.e.* long hydraulic retention times. Consequently a considerable degree of auto-oxidation and endogenous respiration takes place with the microorganisms within the ditch. A suggestion by Porges *et al.* (1953), that a system could be operated where net auto-oxidation balanced net synthesis of biomass, prompted much research. Should such a process be possible, then a considerable financial benefit in reduced plant running costs attributable to sludge disposal would result (currently in the UK, sludge disposal costs account for approximately 50% of the total running costs of a sewage treatment works). It was thought that a certain inert fraction of the biomass produced could not be metabolized by the microorganisms present in the aerobic reactor. However, Gaudy *et al.* (1970), working with soluble substrate on a laboratory scale extended aeration system, operated successfully for two years without wasting sludge or losing biochemical efficiency. They observed periodic cycles of biomass build-up followed by a subsequent decrease, but the continual solids accumulation with eventual system failure which was anticipated did not occur. In practice, without primary sedimentation the problem of inert solids accumulation would be much greater due to the amounts of particulate organic and inorganic material in the ditch influent. Possible process pathways producing surplus solids are shown in Figure 2.

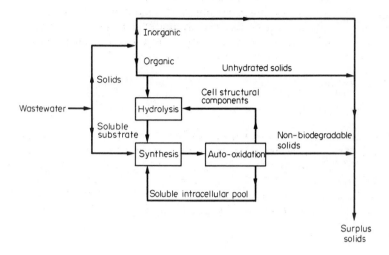

Figure 2 Pathways yielding surplus solids

52.2.2 Modes of Operations

In order for a treatment plant to operate efficiently, the solids present in the ditch have to be separated from the liquid prior to the effluent being discharged. This is normally accomplished by sedimentation either within the ditch itself or in a separate settling tank.

52.2.2.1 Intermittent

The first oxidation ditches of the type developed by Pasveer operated on an intermittent (fill and draw) basis. The type of plant layout used is shown schematically in Figure 3. In this arrangement the provision of a separate sedimenttion tank is not necessary as the ditch acts both as an activated sludge reactor and a settling tank at different stages of the operating cycle.

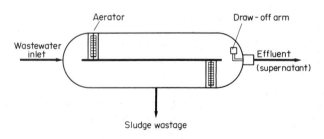

Figure 3 Oxidation ditch: intermittent operation

Goronszy (1979) describes the use of a small ditch operating on an intermittent basis for which some typical operating cycles are shown in Figure 4. The cycle arrangement has to be flexible to obtain the best performance at varying loads and in particular to be able to respond to storm flows. The water level in the ditch varies throughout the cycle so both the aerator and effluent draw-off arrangement should be floating to accommodate this variation.

The liquid in the ditch is moving with a velocity between 0.3 and 0.6 m s^{-1} during the aeration cycle to keep the solids in suspension. In order to slow down the circulation prior to the settlement phase, in many cases the practice is to reverse the aeration for a short period. Wastewater continues to enter the ditch throughout the cycle, so it is good practice to site the supernatant draw-off point at the opposite end of the ditch to the wastewater inlet to avoid turbulence near the outlet during the settling phase. The removal of surplus solids also takes place during the period of settlement.

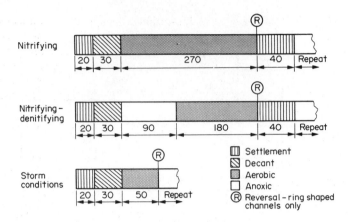

Figure 4   Typical intermittent operating cycles

52.2.2.2   *Continuous*

The use of a sedimentation/clarification unit separate from the oxidation ditch enables it to be operated on a continuous basis. Figure 1 shows the type of layout when the sedimentation tank is kept entirely separate from the ditch structure. The level of the biomass is maintained in the reactor by recirculating sludge settled in the clarifier. The point at which the recirculated sludge enters the ditch has been the subject of much debate, with current thoughts turning towards the effect on sludge bulking (Section 52.2.8). Many ditches currently in operation have the sludge return point sited near the wastewater inlet, which is generally at the opposite end to the outlet weir.

Surplus sludge can be taken from the ditch itself, as with intermittent operation, which gives more precise process control of the concentration of mixed liquor suspended solids (MLSS) than if the sludge is wasted from the return sludge line. The mixed liquor suspended solids concentration in the ditch remains within a fairly narrow range, whereas the sludge solids concentration from the clarifier may vary between 5000 and 15 000 mg l^{-1}. Figure 5, based on the work of Marais and Ekama (1976), shows the variation which can occur.

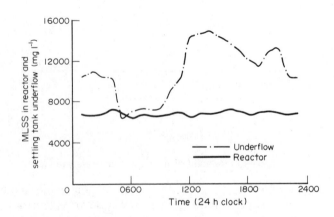

Figure 5   Typical diurnal variations in suspended solids concentration

A plant operator wasting sludge on a volume basis can come closer to removing the required mass of sludge when working with a reasonably steady value for solids concentration which exists in the ditch.

At most treatment works the waste sludge is taken from the clarifier due to economic considerations related to sludge disposal. The disposal costs for a sludge with a high solids content are much less than for a low solids content sludge per unit mass of solids. This is due to the fact that the costs of sludge disposal for pumping, tankering, *etc.*, are related to sludge volume and not to the concentration of solids. Consequently a treatment works manager wants to waste as small a volume of sludge as possible and is not particularly concerned with its solids content. It could be argued that the better process control resulting from wasting sludge from the ditch reduces the net mass of surplus sludge produced. However, it would require more research on this aspect before any works manager would risk increasing his sludge disposal costs by surplussing sludge from the ditch in preference to the clarifier sludge with its much higher suspended solids concentration.

Under an intermittent operating regime the oxidation ditch is suitable for small towns, but on a continuous basis some works have been built to handle wastewater from population equivalents in excess of 100 000.

Several variations on the theme of using a separate clarifier have been developed. Figure 6 shows a split channel system (Water and Waste Treatment, 1981) which effectively incorporates the clarifier into the main body of the ditch. By appropriate use of the channel control gates, two sections of the ditch operate alternately as reactor or clarifier with the effluent and settled sludge being drawn off from the clarification section. On a regular cycle period, very often 4 hours, the control gates change, bringing the settled sludge back into the reactor. Figure 7 shows the use of a circular ditch configuration which has its central area taken up by the clarifier. In both of these variations the intention is to reduce the overall construction costs and improve land usage. There is a certain amount of inflexibility in these systems as each clarification section is dedicated to its own ditch. However, they are intended for smaller flows where only one unit would be required and loss of flexibility is compensated for by the economic advantages.

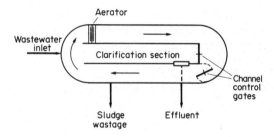

Figure 6 Split channel ditch system

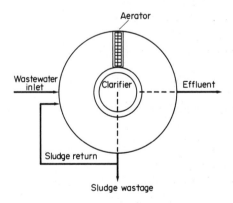

Figure 7 Circular ditch with central clarifier

52.2.3 Design

52.2.3.1 Oxidation ditch

The majority of oxidation ditches which have been constructed to date have been designed on an empirical basis (Denton, 1977). In the UK the loading rates used are based on a 1969 Ministry of Housing and Local Government Technical Memorandum (HMSO, 1969). Table 1 lists these recommendations and shows comparative figures for extended aeration in the United States (Metcalfe and Eddy, 1979) and Australia (Barnes *et al.*, 1981), together with typical values for other activated sludge systems. There is little difference between these design loadings for extended aeration from the different sources when used to treat domestic wastewater. The minimum amount of information required to design on the basis of these loading rates is the mean flow rate and the mean BOD_5 of the wastewater.

Table 1   Design Parameters for Extended Aeration and Other Activated Sludge Systems

Process		Sludge age, θ_c (d)	Food/microorganism (kg BOD_5 (kg MLVSS d)$^{-1}$)[d]	Volumetric loading rate (kg BOD_5 m^{-3} d^{-1})	MLSS (g m^{-3})	Hydraulic retention time (h)	Sludge recycle ratio, F_r/F
Extended	a	—	0.05–0.15[e]	<0.24	2000–6000	24	—
aeration	b	20–30	0.05–0.15	0.1–0.4	3000–6000	18–36	0.75–1.5
	c	20–30	0.04–0.15	0.15–0.4	3000–6000	18–36	0.75–1.5
Conventional		5–15	0.2–0.4	0.3–0.6	1500–3000	4–8	0.25–0.5
Pure oxygen		8–20	0.25–1.0	1.6–4.0	6000–8000	1–3	0.25–0.5
Contact stabilization		5–15	0.2 –0.6	0.9–1.2	1000–3000 (contact unit) 4000–10 000 (stabilization)	0.5–1.0 3–6	0.25–1.0

[a] HMSO, 1969. [b] Metcalfe and Eddy, 1979. [c] Barnes *et al.*, 1981. [d] MLVSS = Mixed Liquor Volatile Suspended Solids. [e] Figures originally for MLSS not MLVSS.

In most design situations a reasonable estimate can be made for both of these parameters. Then an assessment can be made of the likely accuracy of these figures in order to select a suitable volumetric loading rate from the typical range of values used for similar wastewaters. On new developments or where long-term population projections are difficult, then a relatively long hydraulic retention time should be used which with domestic wastewaters will result in a low volumetric loading rate (kg BOD_5 m^{-3} d^{-1}). This design philosophy builds a factor of safety into the system when flow rates and organic strengths are unknown.

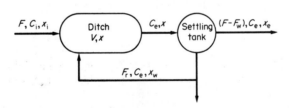

Figure 8   Oxidation ditch system

The ditch volume can be determined for the system shown in Figure 8 from the following relationship:

$$V = \frac{FC_i}{w} \times 10^{-3} \qquad (1)$$

where V = reactor volume (m^3), F = wastewater flow rate (m^3 d^{-1}), C_i = influent BOD_5 concentration (mg l^{-1}) and w = volumetric loading rate (kg BOD_5 m^{-3} d^{-1}).

Possibly due to a great deal of research work on activated sludge using synthetic sewage, which is effectively only a soluble substrate, there is a tendency not to include the contribution from suspended solids in the C_i term. In conventional activated sludge systems, where on average 70% of the solids in the raw wastewater are removed in the primary settling tanks, the influent suspended solids do not exert a great influence on the total substrate. In extended aeration systems, where the raw wastewater is fed directly into the reactor, an allowance for the contribution of the incoming solids, which are potential substrate, should be made. Jenkins and Garrison (1968) propose the use of total influent COD and soluble effluent COD for the system to be described in rational kinetic terms (this appears to ignore the possibility of a chemically oxidizable influent solid being non-biodegradable).

In choosing any loading figures involving substrate concentrations, the designer should understand their basis before using them. The effect of incoming solids on surplus sludge is discussed in Section 52.2.8.

Using the value for V found in equation (1) the hydraulic retention time can be calculated.

$$\theta = \frac{V}{F} \tag{2}$$

where θ = hydraulic retention time for the ditch (d). In the case of domestic wastewater the value of θ should fall within the indicated range for extended aeration shown in Table 1. If the value of θ is outside this range then the value selected for w should be modified until both w and θ are within the normal range. This design method is used if no kinetic characteristics appropriate to the particular wastewater are known.

At the design stage an operating value for mixed liquor volatile suspended solids (MLVSS) can be selected based on the other treatment plants working successfully on a similar wastewater. Using this value the ratio of food (BOD$_5$) to microorganisms can then be calculated:

$$R_{\text{F/M}} = \frac{FC_i}{Vx} \tag{3}$$

where $R_{\text{F/M}}$ = food to microorganism ratio (d^{-1}) and x = concentration of MLVSS in the reactor (mg l^{-1}). If the value of $R_{\text{F/M}}$ lies outside the typical range, then a different operating MLVSS would be selected.

This method of design for extended aeration reactors is based on practical experience rather than extensive research. Equation (1) gives the reactor volume which, if supported by the checks given in equations (2) and (3), is sufficient justification for using this size of reactor if no more detailed data are available on the wastewater characteristics.

A more fundamental approach to reactor design and operation is based on a knowledge of the kinetic constants associated with the wastewater. Certain assumptions have to be made by the designer to complete the design based on this data. In many cases the starting point is to select values for the sludge age, θ_c, and θ which establish the parameters for the biochemical and hydraulic regime of the system, respectively. The expected level of MLVSS can then be estimated for a required effluent standard, C_e, using equation (4):

$$x = \frac{\theta_c}{\theta} \frac{Y(C_i - C_e)}{(1 + k_d\theta_c)} \tag{4}$$

where C_e = effluent BOD$_5$ concentration (mg l^{-1}), Y = maximum yield coefficient (kg biomass produced per kg substrate removed), k_d = endogenous decay coefficient (d^{-1}) and θ_c = sludge age (d).

This value for x is the theoretical operating concentration of MLVSS for a completely mixed reactor with sludge recycle. The racetrack type construction of the ditch often leads designers into thinking of it as a plug flow reactor, but it is a completely mixed system with the wastewater passing around the ditch 50–150 times prior to discharge.

Using the selected value for θ the reactor volume can be found using equation (2). This value for V together with the value for x found from equation (4) can be used in equation (3) to check the food to micoorganism ratio, $R_{\text{F/M}}$.

52.2.3.2 Aeration

There are many methods which can be used for calculating the oxygen required by an activated sludge process. One of the most straightforward (Metcalf and Eddy, 1979; Benefield and Randall, 1980) is based on the change of ultimate biochemical oxygen demand (BOD$_L$) and the amount of surplus biomass produced. If all the substrate was oxidized for energy purposes then

$$\text{Oxygen required} = \text{flow rate} \times \text{change in BOD}_L \tag{5}$$

In practice a proportion of the substrate is used to synthesize biomass.

If it is assumed that the biomass to be utilized can be described by a chemical formula of the type $C_5H_7NO_2$ (Sykes, 1975), then the theoretical oxygen requirement to oxidize this biomass can be calculated:

$$C_5H_7NO_2 + 5O_2 \rightarrow 5CO_2 + 2H_2O + NH_3 \tag{6}$$

$$\begin{array}{cc} 113\,g & 5 \times 32\,g \\ \text{(biomass)} & \text{(oxygen)} \end{array}$$

$$\text{Oxygen requirement} = \frac{5 \times 32}{113} = 1.42 \text{ kg per kg biomass oxidized}$$

Therefore the oxygen required for carbonaceous oxidation by the process shown diagrammatically in Figure 8 can be found as follows:

$$\begin{aligned} \text{Oxygen required} &= \text{total mass of } BOD_L \text{ utilized} - 1.42(\text{wasted biomass}) \\ (\text{kg d}^{-1}) &= RF(C_i - C_e) \times 10^{-3} - 1.42\{F_w x_{wv} + (F - F_w)x_{ev}\} \times 10^{-3} \end{aligned} \tag{7}$$

where F_w = flow rate of waste sludge $(m^3\,d^{-1})$, x_{wv} = concentration of volatile suspended solids (VSS) in the waste sludge $(mg\,l^{-1})$, x_{ev} = concentration of VSS in the effluent $(mg\,l^{-1})$ and R = conversion factor from BOD_5 to BOD_L (typically $BOD_L = 1.5\,BOD_5$, *i.e.* $R = 1.5$).

At the design stage the values of F and C_i should be known to determine the ditch volume as in equation (1). The discharge conditions from the works will control the composition limits for the effluent and hence set a value for C_e. The $F_w x_{wv} + (F-F_w)x_{ev}$ term cannot be determined theoretically since it is dependent upon process operating procedures and the settling qualities of the sludge. However, the total quantity of biomass to be wasted per day can be estimated using the following expression:

$$X_s = Y_{obs}F(C_i - C_e) \times 10^{-3} \tag{8}$$

where X_s = excess biomass produced $(kg\,d^{-1})$ and Y_{obs} = observed net biomass yield $(kg\,(kg\,BOD_5\text{ removed})^{-1})$.

A value for Y_{obs} can be found experimentally or a typical value for similar wastewaters can be used as the basis for an estimate. Equation (7) now becomes:

$$\text{Oxygen required (kg d}^{-1}) = F(C_i - C_e)(R - 1.42Y_{obs}) \times 10^{-3} \tag{9}$$

The relatively high concentration of biodegradable solids which the oxidation ditch receives in comparison to a conventional activated sludge system will have a sizeable effect on the oxygen requirement. An allowance can be made for this by modifying equation (9):

$$\text{Oxygen required (kg d}^{-1}) = F\{(C_i + C_s) - C_e\}\{R - 1.42Y_{obs}\} \times 10^{-3} \tag{10}$$

where C_s = equivalent soluble substrate concentration in terms of BOD_5 to allow for the effect of the influent biodegradable solids $(mg\,l^{-1})$. The effect of biodegradable solids is discussed further in Section 52.2.9.

In addition to carbonaceous oxidation, extended aeration systems achieve a high degree of nitrification for which oxygen is required. The overall nitrification reaction can be represented as follows:

$$NH_4^+ + 2O_2 \rightarrow NO_3^- + 2H^+ + H_2O \tag{11}$$

$$\begin{array}{cc} 14\,g & 2 \times 32\,g \\ (NH_4^+\text{-N}) & \text{(oxygen)} \end{array}$$

$$\text{Oxygen requirement} = \frac{64}{14} = 4.57 \text{ kg } O_2 \text{ per kg } NH_4^+\text{-N oxidized}$$

Therefore the total oxygen requirement for a treatment works design based on full nitrification is given by:

$$\text{Oxygen required (kg d}^{-1}) = F\{(C_i + C_s) - C_e\}\{R - 1.42Y_{obs}\} \times 10^{-3} + 4.57FC_{i,N} \times 10^{-3} \tag{12}$$

where $C_{i,N}$ = concentration of NH_4^+-N in the influent $(mg\,l^{-1})$.

Equation (12) gives a basis from which the overall oxygen requirement can be determined, although very often it is simply taken as 1.25 kg O_2 per kg BOD_5 applied plus 4.5 kg O_2 per kg NH_4^+-N applied. These oxygen requirements are average daily figures and, in order to ensure a sufficient dissolved oxygen level during peak flow periods, extra aeration capacity should be installed. In most cases, two to three times the O_2 requirement calculated from equation (12) is used to estimate the installed aeration capacity.

52.2.4 Aerators

The successful operation of an oxidation ditch depends on keeping the activated sludge flocs in suspension. In order to achieve this and prevent settling, a circulation velocity between 0.3 and 0.6 m s^{-1} is maintained in the ditch. Mechanical aerators are ideally suited to both aerating the wastewater and maintaining an adequate circulation velocity. This is the main reason why diffused air systems are seldom used except as a means of uprating an organically overloaded ditch. The two systems of mechanical aeration in common use are based on either a horizontal shaft (Kessener brush type) or a vertical shaft rotor system (Carrousel process). In both cases oxygen transfer rates in the range 1.4–2.6 kg O$_2$ kW^{-1} h^{-1} can be achieved.

The initial work on ditches by Pasveer used a brush type rotor but its design limited the depth of the ditch to 1.8 metres if adequate aeration and mixing were to be achieved. This depth limitation restricted the flow which the system could handle economically due to the relatively large areas of land required for the ditch to achieve the required volume. Since no larger brush type rotors were available the Carrousel system (Figures 9 and 10) was developed using vertical shaft rotors, which meant that depths up to 4 metres could be used and hence more economic land usage achieved. Development of the larger brush type rotors (Figures 11 and 12) has meant that similar depths of ditch can be used with both systems.

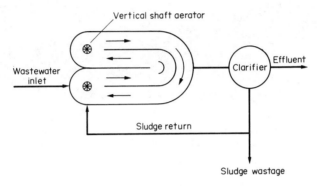

Figure 9 Typical Carrousel system layout

Figure 10 Treatment plant showing Carrousel ditches at various stages of construction (courtesy of Simon-Hartley Ltd., Stoke-on-Trent, UK)

Many comparisons have been made between these two aeration methods (Forster, 1980; Mulready *et al.*, 1982) and each system has its merits. The general view is that they both start as equal

Figure 11 Ditch using Mammoth type horizontal shaft aerator (courtesy of Whitehead and Poole Ltd., Radcliffe, Manchester, UK)

Figure 12 Horizontal shaft aerator (courtesy of Manor Engineering Co. Ltd., Stoke-on-Trent, UK)

alternatives but system costs for a particular site and treatment will probably swing the balance in favour of one of them.

52.2.5 Nitrification

The overall reaction for the conversion of ammonium to nitrate can be represented by equation (11). In the extended aeration oxidation ditch system a high degree of nitrification is possible without sophisticated process control being necessary. A removal of 90% of NH_4^+-N is regularly achieved in oxidation ditch systems of this type due to the following factors.

(i) Available dissolved oxygen. It is the usual practice to maintain concentrations greater than 1 mg l^{-1}, at which level no inhibition of nitrifying bacteria has been found (Wild *et al.*, 1971). Operational levels in the aerobic part of the ditch are generally kept in the 1–2 mg l^{-1} range.

(ii) High MLSS values (4000–6000 mg l^{-1}). Gives a sufficiently high concentration of nitrifiers such that nitrification is not restricted because of low organism concentrations.

(iii) Large sludge age. Allows sufficient time for the nitrifiers, which have relatively low growth rates, to establish themselves.

(iv) Attention to overall reactor pH. In the overall reaction (equation 11), hydrogen ions are released and if insufficient alkalinity is present to buffer the reaction the pH will decrease. As nitrifiers are very sensitive to changes in pH the operator must be ready to look for these changes and control them to maintain reasonable nitrification. The optimum operating pH for nitrification was found by Wild *et al.* (1971) to be 8.4.

52.2.6 Denitrification

Nitrate in a works effluent is a nutrient for algae and as such could be instrumental in producing algal blooms in the receiving water-course. Consequently, it is desirable to keep nitrate levels to a minimum, employing denitrification if necessary. The process of denitrification is carried out by facultative heterotrophic bacteria under anoxic conditions. The overall reaction can be represented as follows, using methanol as a carbon source:

$$6NO_3^- + 5CH_3OH \rightarrow 5CO_2 + 3N_2 + 7H_2O + 6OH^- \qquad (13)$$

In the oxidation ditch it is possible through suitable design of the ditch, and with a reasonable amount of flexibility in the aeration system, to produce anoxic zones. Theoretically it is possible to use a dissolved oxygen measuring probe at the start of the anoxic zone and adjust the aeration to maintain a zero level of dissolved oxygen at the probe. In practice, fouling of the probe can present problems but, notwithstanding, many works successfully carry out denitrification in their ditches, although it is more often established by trial and error rather than precise control.

Some source of carbon is necessary for the reaction described in equation (13) but the purchase of a carbon source, such as methanol, to feed the denitrification process, adds another cost to the wastewater treatment process. This has been found to be unnecessary if some or all of the incoming wastewater is fed directly into the anoxic zone to act as a carbon source. In ditches operating without denitrification the time for the flow to complete one circuit at the mean flow rate is often between 4 and 6 minutes. In order to ensure reasonably successful operation of an anoxic zone a longer circuit time of the order of 10 minutes has been shown to be desirable (Ettlich, 1978). Approximately 5 of these 10 minutes would be taken up by flow through the anoxic region.

Denitrification will carry on in the sludge layer of the settling tank if the conditions are anoxic. This can cause problems if the gas bubbles produced attach themselves to the sludge particles, carrying them to the surface and hence out with the final effluent. This problem can be reduced by having a positive dissolved oxygen carry-over from the ditch and by not allowing the sludge to remain in the settling tank for long periods. Continuous sludge draw-off at a rate appropriate to the settling characteristics of the sludge minimizes this problem associated with anaerobic/anoxic sludges.

52.2.7 Combined Nitrification–Denitrification

There are many activated sludge nitrification–denitrification system layouts used in sewage treatment (Mulbarger, 1971). The majority of these employ separate reactors for the two operations, each with its own specific group of microorganisms. In the oxidation ditch system the same microbiological floc is used for both the aerobic nitrification process and the anoxic dentrification process. A possible mechanism which could explain this is a dynamic aerobic–anoxic zone interface within the sludge floc as shown in Figure 13.

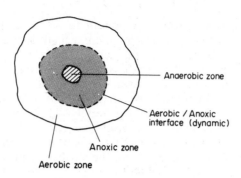

Figure 13   Sludge floc with aerobic/anoxic interface

As the flow passes through the aerobic part of the ditch the available dissolved oxygen is not all consumed at the floc surface and surplus oxygen passes into the floc (Biryukov and Shtoffer (1971) examined the mathematics of oxygen transfer within sludge flocs). This increases the extent of the aerobic zone within the floc, driving the aerobic–anoxic interface towards the anaerobic core. This increases carbonaceous oxidation and nitrification rates as more oxygen becomes available to the facultative aerobes responsible for the processes. As the floc passes into the anoxic section of the ditch the available supply of dissolved oxygen in the fluid external to the floc is reduced and the dissolved oxygen contained in the aerobic zone within the floc is consumed by oxidation processes. This causes the interface to move towards the outer edge of the floc, thereby increasing the size of the anoxic zone within the floc. The rate of denitrification then increases as the anoxic conditions reactivate the denitrifiers, which are facultative anaerobes.

The concentration of dissolved oxygen in a ditch is one of the parameters which the plant operator can control. In this way the position of this interface within the floc can be changed by altering the overall level of dissolved oxygen in the wastewater, with the resulting effect on the degree of nitrification and denitrification which can be carried out. The existence of this interface within the floc could explain why denitrification takes place in ditches where anoxic conditions do not exist in the wastewater. The floc internal anoxic zone can be maintained by just reducing external dissolved oxygen such that the internal floc oxygen consumption rate is greater than the rate at which oxygen can be transferred into the floc.

This dynamic interface is only a suggested mechanism to explain the combined nitrification–denitrification process in the oxidation ditch and further research is needed to establish its feasibility.

52.2.8   Sludge Settling

The secondary sedimentation tank performs two roles: (i) the clarification of effluent prior to discharge and (ii) the thickening of sludge either for return to the activated sludge reactor or for wasting. The overall effectiveness of any wastewater treatment plant is largely dependent on how easily the activated sludge can be removed from the effluent prior to discharge. The effectiveness of sludge thickening affects only the internal process operation and sludge disposal costs.

Bisogni and Lawrence (1971) examined the relationship between the Sludge Volume Index (SVI) and the sludge age for an activated sludge system (Figure 14) using a soluble substrate. Depending on the significance which one attaches to the SVI as a measure of settleability, the indication from their work is that the greater the sludge age the better the sludge settling characteristics. On this basis it is perhaps reasonable to assume that extended aeration, with sludge ages in the 20–30 day range, should produce a readily settled sludge. Barnard (1978) reported that extended aeration plants in South Africa produce flocculent sludges which settle well despite having values of SVI in the 100–200 ml g^{-1} range.

Oxidation ditch sludges appear under normal operating conditions to be no better or worse

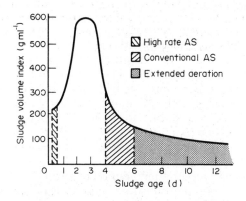

Figure 14 Variation of SVI with sludge age

with regards to settling than sludges from other completely mixed systems. The design of secondary sedimentation tanks should be carried out with the benefit of information from settling tests (Schroeder, 1977) whenever possible. In designing a new works the sludge is not available for such tests until the works is completed and therefore the design of the settling facilities has to be based on surface and solids loading rates. Table 2 gives loading rates for secondary sedimentation tanks which have been used in the past when no information from settling tests has been available. The tendency has been to provide slightly larger secondary sedimentation tanks for extended aeration sludges. The higher MLSS of the influent to the secondary sedimentation tanks, coupled with a lower surface loading rate, gives a very similar solids flux to conventional systems.

Table 2 Typical Design Criteria for Secondary Sedimentation Tanks

Treatment system	Surface loading rate ($m^3 m^{-2} d^{-1}$) Average	Peak	Solids loading rate ($kg m^{-2} h^{-1}$) Average	Peak	Ref.
Conventional	16–32	40–48	3–6	9	a
	—	30–40	—	5	b
Extended aeration	8–16	24–32	1–5	7	a
	—	22	—	5	c

[a] Metcalfe and Eddy, 1979. [b] Barnes *et al.*, 1981. [c] HMSO, 1969.

The problem of sludge bulking in activated sludge systems is currently a very active research area (Chambers and Tomlinson, 1982). Certain recommendations have been made to reduce the chances of bulking occurring and to recognize its onset at an early stage; nonetheless the mechanism itself is still not fully understood. The findings of Pasveer and Heide (1974), that the chances of bulking occurring are much reduced if the return activated sludge floc is introduced into a nutrient rich environment, *i.e.* raw wastewater, have been confirmed at works in the UK. Clearly many factors affect bulking but it seems that the chances of it occurring might be reduced by a design modification using a premix tank as shown in Figure 15. The premix tank should have a hydraulic retention time between 5 and 10 minutes at the mean flow rate.

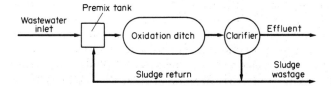

Figure 15 Oxidation ditch system with premix tank

52.2.9   Surplus Sludge Production

It is difficult to find figures in the literature for the actual mass of surplus sludge produced at a treatment works. In recent years, much emphasis has been put on monitoring liquid flows at various points on a works but the usable data on sludge production are minimal. Many works monitor the total volume of sludge wasted, but without knowing the solids concentration any measure of surplus solids is at best a very rough approximation. The problem is complicated further by the way in which various authors express the amount of surplus sludge production. Even if we consider only the cases where BOD is used, then the surplus sludge is either expressed as kg per kg of BOD applied or kg per kg of BOD removed and many workers do not state whether BOD_5 or the ultimate BOD_L is used. Equally important, especially with systems not employing primary sedimentation, does the BOD value include the biodegradable suspended solids or does it only relate to filtrate or supernatant liquid present after a period of settling?

In general terms, for a given soluble substrate concentration the surplus sludge yield from an extended aeration system should be considerably less than for a conventional activated sludge system. In effect, because of the long hydraulic retention times we have an activated sludge system incorporating aerobic digestion of the sludge which should greatly reduce the surplus MLVSS produced (Hartman *et al.*, 1979). This is illustrated in Figure 16, which is based on laboratory work carried out by Sherrard and Schroeder (1973) using COD.

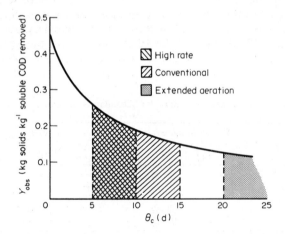

Figure 16   Variation of observed sludge yield with sludge age for different activated sludge systems

In a conventional treatment system which includes suitably designed and effectively operated primary sedimentation tanks, up to 70% of the suspended solids are removed prior to biological treatment. The amount of primary sludge wasted in a conventional works with an average of 70% removal is

$$M_{ps} = Fx_i(0.7) \times 10^{-3} \qquad (14)$$

where M_{ps} = mass of primary sludge wasted (kg d^{-1}) and x_i = concentration of suspended solids in the influent (mg l^{-1}). Typically for a medium strength domestic wastewater, x_i is around 200 mg l^{-1}, which means that $200 \times 0.3 = 60$ mg l^{-1} of suspended solids going into the activated sludge reactor. In an extended aeration system the full 200 mg l^{-1} would go into the oxidation ditch.

The observed yield for a working plant, based on the removal of soluble substrate only, is given by the following relationship:

$$Y'_{obs} = \frac{F_w x_w + (F - F_w)x_e}{F(C_i - C_e)} \qquad (15)$$

where Y'_{obs} = observed sludge yield measured for an actual treatment plant reactor (kg per kg of soluble substrate removed), x_w = concentration of suspended solids in the waste sludge (mg l^{-1}) and x_e = concentration of suspended solids in the final effluent (mg l^{-1}).

On this basis the extended aeration process would appear to produce more surplus sludge relative to conventional activated sludge than we would expect based on laboratory work using only soluble substrate. This is due to the higher level of both biodegradable and non-biodegradable solids entering the extended aeration reactor. The non-biodegradable solids appear unchanged in the surplus sludge and the biodegradable solids become available as an additional substrate. The suspended solids in the influent are composed as follows:

$$\text{TSS} = \text{NVSS} + \text{NDVSS} + \text{DVSS} \tag{16}$$

where TSS = total suspended solids (mg l^{-1}), NVSS = non-volatile suspended solids (mg l^{-1}), NDVSS = non-biodegradable volatile suspended solids (mg l^{-1}) and DVSS = biodegradable volatile suspended solids (mg l^{-1}). It has been proposed (Benefield and Randall, 1980) that the NDVSS concentration in the raw wastewater could be estimated from:

$$\text{NDVSS} = \left(\frac{\text{COD} - \text{BOD}_\text{L}}{\text{COD}}\right) \text{VSS} \tag{17}$$

where COD = COD due to suspended solids (mg l^{-1}), BOD$_\text{L}$ = ultimate BOD due to suspended solids (mg l^{-1}) and VSS = total volatile suspended solids (mg l^{-1}). Using this value for NDVSS, the value for DVSS can be estimated.

The relationship for the surplus sludge from an activated sludge reactor can be expressed as

$$\{(F - F_\text{w})x_\text{e} + F_\text{w}x_\text{w}\} \times 10^{-3} = \text{mass of surplus solids produced in kg d}^{-1}$$
$$= FY_\text{obs}\{(C_\text{i} + C_\text{s}) - C_\text{e}\} \tag{18}$$

where Y_obs = experimentally observed sludge yield (kg per kg of soluble substrate removed). It should be noted that the Y_obs is the yield relating to soluble substrate only and the C_s term is the equivalent soluble substrate attributable to the biodegradable suspended solids in the influent. The equivalent overall yield for a reactor working on a typical wastewater can now be evaluated in terms of soluble substrate removal by combining equations (15) and (18):

$$Y'_\text{obs} = \frac{Y_\text{obs}\{(C_\text{i} + C_\text{s}) - C_\text{e}\} + \text{NDVSS} + \text{NVSS}}{C_\text{i} - C_\text{e}} \tag{19}$$

where Y'_obs = observed sludge yield for the actual wastewater (kg sludge per kg soluble substrate removed). In the case of soluble substrate only, equation (19) reduces to

$$Y'_\text{obs} = Y_\text{obs} \tag{20}$$

The effect of the presence of solids might best be illustrated by an example. In a typical raw wastewater the NVSS fraction accounts for 20–30% of the total suspended solids, with the NDVSS proportion accounting for 20–40% of the total suspended solids. The NDVSS + NVSS fraction can therefore constitute 40–70% of the incoming raw wastewater. The remaining DVSS, which ends up in the activated sludge reactor, makes up the C_s fraction of the substrate. If we consider a wastewater with a soluble COD of 500 mg l^{-1}, TSS 200 mg l^{-1} (NVSS = 0.25 TSS, NDVSS = 0.25 TSS) producing an effluent soluble COD of 50 mg l^{-1} and taking the DVSS to be equivalent to glucose in the reaction described by equation (21), then each mg of DVSS is approximately equivalent to 1 mg COD:

$$\begin{array}{cccc} \text{C}_6\text{H}_{12}\text{O}_6 + & 6\text{O}_2 & \rightarrow 6\text{CO}_2 + 6\text{H}_2\text{O} + \text{energy} & \tag{21} \\ \text{180 g} & \text{192 g} & & \\ \text{(glucose)} & \text{(oxygen)} & & \end{array}$$

In a system incorporating primary sedimentation which removes 70% of TSS and taking a value for Y_obs of 0.3 kg per kg soluble COD removed, then using equation (19):

$$Y'_\text{obs} = \frac{0.3[(500 + 0.5 \times 0.3 \times 200) - 50] + 0.5 \times 200 \times 0.3}{(500 - 50)}$$

$$= 0.39 \text{ kg sludge solids produced per kg soluble COD removed.}$$

For the same wastewater without primary sedimentation, *i.e.* for extended aeration, and using a value for Y_obs of 0.15 kg per kg soluble COD removed, we obtain

$$Y'_\text{obs} = \frac{0.15[(500 + 0.5 \times 200) - 50] + 0.5 \times 200}{500 - 50}$$

$$= 0.41 \text{ kg sludge produced per kg soluble COD removed.}$$

The value for Y'_obs obtained in both cases is very similar despite a considerable difference in Y_obs.

Based on soluble substrate analysis only for the COD removal in the activated sludge reactor, there is no apparent advantage for extended aeration over conventional activated sludge with regard to surplus sludge production. This error is often made by engineers when comparing plant sludge yields, but to make a true comparison either allowance must be made for the extra solids loading on the extended aeration reactor or the sludge from the primary sedimentation tank must be included in the surplus sludge production for a conventional activated sludge system.

The values obtained in practice on various works operating oxidation ditches in the extended aeration mode range from 0.2 to 0.7 kg sludge per kg of BOD_L removed (Denton, 1977). Sludge handling and disposal facilities are a critical part of a treatment works so before using a value for Y_{obs} in design it is essential that its true significance should be appreciated. Some yield coefficient along the lines of Y'_{obs} is more appropriate in a practical situation.

52.2.10 Construction

On smaller works where the land area usage is not critical, then the ditch can be constructed with sloping sides similar to the arrangement shown in Figure 17. The slope of the sides will vary depending on the type of ground conditions but typically would be in the range 1 to 1.5–2. In most cases it is desirable to make the ditch base and sides impermeable, but depending on the site a certain amount of seepage may be permissible. If the ditch is not completely lined then some form of rigid cladding should be used in the following areas: (i) in the vicinity of the aerator to prevent erosion due to the high local turbulence; (ii) around the sides of the ditch at the waterline to prevent erosion due to wave action; (iii) any other area of potential erosion, *e.g.* around the wastewater inlet and sludge return points.

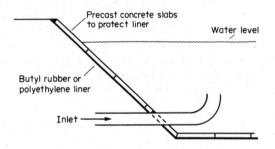

Figure 17   Sectional view of typical ditch construction

Several types of cladding have been used: reinforced concrete (100–150 mm thick layer cast against the trimmed earth face with nominal mesh reinforcement), concrete slabs, gunite, bitumen (known as asphalt in USA), and plastic or rubber sheeting. The selection of cladding material is based on economics and availability. Some form of rigid base is required for the aerators, for which reinforced or mass concrete is generally used. On larger works as the volume of ditch required increases then, in order to optimize land usage, deeper ditches (3–4 metres deep) are used. The deeper ditch is constructed generally in reinforced concrete with vertical side walls. In these cases the central island, which is common on smaller works, is replaced by a single-skin dividing wall. To improve circulation the use of deflector walls at the ends of the ditches as shown in Figure 11 is often recommended. A point frequently forgotten at the design stage is that the plant has to operate efficiently and will require regular maintenance. Access to aerators, valves, outlet weirs, oxygen probes and other equipment associated with ditch operations must be straightforward and safe. Where ditches or ponds are used, a means for completely emptying them should be provided to allow access for maintenance, inspection or desludging.

Sludge return on small works is normally by means of centrifugal pumps operating either continuously or on a timer sequence. The provision of two pumps, duty and standby, is the minimum which should be considered in any design. The simplest and most effective sludge return pump for large works is the archimedian screw type; this will deliver at a fixed head all the sludge flowing to it up to its maximum operating flow rate. In addition, its impeller speed is much less than the equivalent standard centrifugal pump and it is claimed that this causes less structural damage to the return sludge floc and hence improved process performance. However, for a given flow rate the cost of a screw pump (including civil engineering costs) is considerably more than the equivalent centrifugal pump. Once again the designer is faced with a cost/benefit evaluation which varies for each wastewater treatment works.

52.2.11   Process Control

52.2.11.1   Aeration

The control of aeration is necessary on three counts. (i) If the level of residual dissolved oxygen falls too low it can restrict both carbonaceous oxidation and nitrification. (ii) The maintenance of a high level of dissolved oxygen above approximately 2 mg l^{-1} does not improve plant performance but does incur unnecessary power costs. (iii) Anoxic zones or zones with low dissolved oxygen concentrations might be required to ensure adequate denitrification.

The control system can be either manual or automatic depending on the degree of sophistication merited at a particular plant. To ensure any valid form of control, a means of measuring dissolved oxygen must be available and this will either directly control aeration or provide information for manual control. There is no best site for an oxygen probe in the oxidation ditch, but if only one probe is available then a position adjacent to the outlet weir will check that a positive level of dissolved oxygen is carried over into the clarifier (see Section 52.2.6). The best operating dissolved oxygen concentration used for the probe has to be determined by the plant operator as it depends on ditch configuration and wastewater characteristics. The use of more than one probe per ditch refines the possible level of control but increases the amount of equipment to maintain.

There are two favoured methods for adjusting the amount of aeration. (i) Intermittent aerator operation. In this case the aerators are either directly controlled from the oxygen probe or are turned on and off on a time cycle suited to the incoming wastewater. A works operator will very soon get to know the daily flow variations and can set up his cycle timers almost without reference to the oxygen probe readings. (ii) Changing aerator immersion. Fixed aerators are designed in such a way that their aeration rate changes depending upon the amount of immersion. Figure 18 shows the range of this variation. By altering the level of water in the reactor, using an adjustable outlet weir, or in a few plants changing the height of the aerator, the depth of immersion and hence the amount of aeration can be controlled.

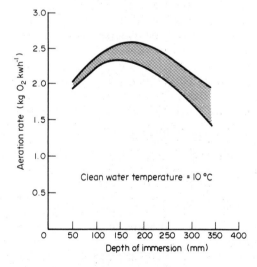

Figure 18   Variation of aeration rate with immersion for mechanical aerators

52.2.11.2   Sludge age

By controlling the sludge age we directly control the level of MLSS in the reactor. In order to control the sludge age it is necessary periodically to waste sludge from the system. This is accomplished either directly from the reactor or from the return sludge line. Figure 5 shows the typical diurnal variation in concentration of suspended solids in the reactor and return sludge line. To achieve steady state conditions a fixed mass of sludge needs to be wasted each day appropriate to the desired sludge age in the plant. By wasting a known volume from the ditch equivalent to $(1/\theta_c)$ times the reactor volume each day, an approximate fixed mass of sludge is wasted since the variation in MLSS is small. This produces a fairly constant θ_c, ignoring the settled solids held in the clarifier. To waste a fixed mass of solids from the sludge return line it is necessary to determine the concentration before wasting the appropriate volume because of the wide variation in concentration. Process control on this basis is far more complex than if wastage takes place directly from the ditch.

The problem from the plant manager's point of view is that sludge wasted from the ditch has a suspended solids concentration in the range 2000–6000 mg l^{-1} whereas the concentration in the sludge return line can be up to 15 000 mg l^{-1}. Therefore for a given mass of solids he may have up to three times the volume of sludge to dispose of if he wastes it from the ditch as opposed to the sludge return line. (This does not take into account any overall reduction in surplus sludge production which may result from better process control achieved by wasting sludge from the ditch.) In most works, therefore, on day-to-day operation the quantity of sludge wastage from the return line is determined based either on a simple settling test carried out on the MLSS in the ditch or for works with basic laboratory facilities the concentration of MLSS is determined. If these tests indicate the level of MLSS is higher than the desired operating concentration, then sludge is wasted accordingly. The settling test, as a measure of the MLSS concentration, is entirely dependent on the settling characteristics of the sludge and consequently as a means of process control this method is inexact. Nonetheless it is used on the majority of small oxidation ditch installations due to its simplicity.

52.3 AERATED LAGOONS

In most cases these are activated sludge units working without sludge return. Essentially they are similar in construction to waste stabilization ponds with the major difference being the method of oxygen supply. In hot climates, algal oxygen is sufficient for pond needs but in temperate climates a supplementary supply is often required. This is normally supplied by a floating mechanical aerator which in addition acts as a mixer. Lagoons are followed by a sedimentation stage to remove the suspended solids prior to discharge of the effluent. Sludge recycle is practised in some lagoon systems, especially in the USA, to improve process performance. In these latter cases the only differences from an oxidation ditch are the somewhat longer hydraulic retention time and the minimum cost earthen basin type of construction.

52.3.1 Design

52.3.1.1 Lagoon

Design has largely been based on hydraulic retention times between 2 and 6 days with typically a value of 4 days being used achieving BOD$_5$ removals greater than 90%. A low operating value for MLSS, in the range 200–4000 mg l^{-1}, exists if the system operates without sludge recycle, but nonetheless these high BOD$_5$ removal rates are routinely attained.

A more rational approach to design has been made by several workers (Mara, 1978) based on a first-order kinetic model for soluble BOD$_5$ related to the type of system shown schematically in Figure 19.

$$C_e = \frac{C_i}{1 + k_T \theta} \tag{22}$$

where k_T = rate constant for BOD$_5$ removal at temperature T °C (d^{-1}).

Provided a value for k_T can be obtained for the particular waste then equation (22) gives a value for θ based on a known C_i and a required C_e. A typical design value for k at 20 °C is in the range 0.4–0.5 d^{-1}, which can be estimated for different operating temperatures from an Arrhenius type equation of the form

$$k_T = k_{20}(1.035)^{T-20} \tag{23}$$

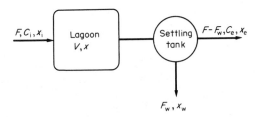

Figure 19 Aerated lagoon system

where k_{20} = BOD$_5$ removal rate constant at 20 °C (d^{-1}). Thirumurthi (1979) also suggests that this value be decreased for low organic loads less then 0.02 kg m^{-3} d^{-1} and hydraulic retention times greater than 10 days.

Determining the operating temperature range expected within a lagoon is essential at the design stage, especially if low temperatures and icing are likely to be encountered. If data are not available from lagoons in the same area or operating under similar climatic conditions, then the temperature can be estimated from a relationship of the form proposed by Mancini and Barnhart (1968):

$$T_{L} = \frac{fAT_{a} + FT_{i}}{fA + F} \tag{24}$$

where T_{L} = lagoon temperature (°C), T_{i} = influent temperature (°C), T_{a} = ambient air temperature (°C), A = surface area of lagoon (m^2), F = wastewater flowrate (m^3 s^{-1}) and f = heat transfer factor (numerical constant which incorporates the effects of aeration, wind and humidity; value used for eastern USA of 0.5). The design should be based on the lowest regularly achieved temperature if guaranteed effluent quality is required.

52.3.1.2 Aerators

To maintain the lagoon as a completely mixed system it is necessary to carry out a certain amount of forced mixing. Floating aerators (Figure 20) are designed to carry out both aeration and mixing. The oxygen requirements can be estimated using equation (12). Commercially available floating surface aerators give oxygen transfer rates in the range 1.25–2.5 kg O$_2$ kWh^{-1}. The installed aerator power includes that required for oxygen transfer at the surface plus an additional amount for mixing for which a figure of 5 W m^{-3} is often used. The longer hydraulic retention times in the lagoon mean that a certain amount of buffering of the peak flows takes place and the installed aeration capacity does not need to be as much in excess of the amount calculated from equation (12) as is necessary with oxidation ditches. If extra aeration capacity is required in a lagoon then it is relatively straightforward to install more floating aerators. Therefore the design calculations in this respect are not as critical as they might be with oxidation ditches and conventional activated sludge systems, where to increase the number of aerators might involve expensive construction work.

52.3.2 Construction

The method of construction, as with waste stabilization ponds, has the emphasis on minimum cost. In this case, with algal growth not being a controlling factor and mechanical mixing being used to keep the activated sludge in suspension, greater depths are used. Values of 3–5 metres are used with embankment slopes in the 1 to 1.5–2 range, which makes a reasonably effective use of land for a given hydraulic retention time.

The operation of the mechanical aerator can cause considerable local scour and some form of liner must be provided, at least in the vicinity of the mixer. The floating aerators may not be finally positioned as intended originally by the designer so a complete lining of the lagoon with butyl rubber, polyethylene or some other material at the construction stage not only prevents infiltration into the surrounding ground but also allows the aerators to be used anywhere within the lagoon without scour damage.

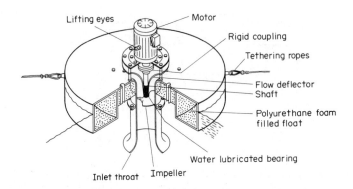

Figure 20   Cut view of floating aerator (courtesy of Satec Ltd., Crewe, UK)

The lagoon is intended to operate as a completely mixed reactor and this is facilitated by using an inlet manifold system to disperse the flow along one side: generally the side opposite the outlet is used. Numerous designs of inlet have been used successfully. One of the simplest is to use a series of upturned pipes as shown in Figure 17, which reduces any shortcircuiting due to high inlet velocities across the lagoon and eliminates the problem of localized bottom scour.

To accommodate the fluctuating levels in the lagoon the outlet is normally of the floating type as used with intermittently operated oxidation ditches. By adjusting the valve of the lagoon outlet the flow rate can be set to suit the final settling basin. The lagoon then also acts as a balancing tank, although provision must be made to limit the bottom and top water levels.

52.4   WASTE STABILIZATION PONDS

Unlike the oxidation ditch, waste stabilization ponds require no energy input in the form of mechanical aeration or mixing. The treatment of the wastewater is carried out naturally by algae and bacteria utilizing solar energy. These processes, although very effective, are not rapid and consequently long hydraulic retention times are needed to achieve a reasonable effluent standard. The low construction and operation costs of ponds, and in particular their suitability for hot climates, have made them a topic of much research and development (Middlebrooks *et al.*, 1978). Approximately 25% of all wastewater treatment plants in the United States use ponds of one form or another, or more often in combination. There are three main types of pond in common use: facultative, maturation and anaerobic.

52.4.1   Facultative Ponds

In the treatment of domestic wastewater the facultative pond is generally the first major process stage after any necessary preliminary treatment. The term facultative is used to describe this type of pond as it contains both aerobic and anaerobic zones. The upper layers of the pond are generally aerobic, with oxygen either being transferred at the pond surface or produced as a result of the photosynthetic activity of algae in the upper 150–300 mm. The importance of algae in this role is shown in Figure 21, which represents the diurnal variation of dissolved oxygen in the pond at different levels. In the absence of algae the pond would soon become largely anaerobic as they provide the majority of the oxygen required by the bacteria *via* photosynthesis.

The concentration of suspended solids in the pond restricts light penetration so the photosynthetically produced oxygen would remain in the upper pond layers without adequate mixing. Mixing not only carries this dissolved oxygen into the lower pond layers but also brings non-motile algae to the surface. The major factor which brings about this mixing is wind action at the surface. Marais and Ekama (1976) reported on the anaerobic pond conditions which resulted from shielding a facultative pond from the wind, thus preventing sufficient mixing. However, recent work by Silva (1982), from which Figure 21 is adapted, shows ponds performing well with apparently very little mixing.

The anaerobic sludge layer which forms in the bottom of the pond readily digests and this sludge digestion means that desludging is required very infrequently, every 10–15 years, provided the temperature in the sludge layer is sufficiently high (>15 °C is suggested by Mara, 1978).

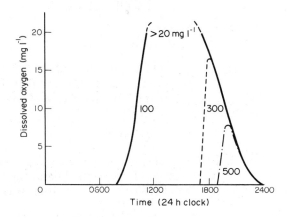

Figure 21 Dissolved oxygen variation with time and depth for a facultative pond (numbers alongside plotted lines show depth below surface in mm)

In a facultative pond the narrow photic zone makes the use of shallow ponds essential to achieve a high algal oxygen production per unit volume. In practice a depth in the range 1–1.5 m is used. Ponds less than 0.8 m deep do not prevent the emergence of vegetation which could then act as a breeding area for mosquitoes. The greater the pond depth the more predominant the anaerobic zone, which is not desirable, and for this reason 1.5 m is generally adopted as the depth limit. The selection of depth as with other design factors is much dependent upon climate.

52.4.2 Maturation Ponds

The role of the maturation pond is essentially that of polishing the effluent from another biological treatment process such as a facultative pond. This polishing role not only refers to soluble organic material but more importantly to pathogen destruction. These ponds are constructed on a similar basis to facultative ponds and are designed to operate wholly aerobically. The essential difference is the shorter retention times, about 7 days, compared with in excess of 30 days for facultative ponds.

52.4.3 Anaerobic Pretreatment Ponds

Anaerobic ponds receive a high organic loading such that the oxygen requirement is greater than the potential photosynthetic oxygen production. These ponds become turbid due to the production of reduced metal sulfides and this restricts algal growth by limiting light penetration. They are used most frequently to pretreat strong wastes such that the organic load on the following aerobic treatment process is reduced, thereby reducing overall pond construction costs.

Incoming organic solids and settleable biomass produced in the pond end up in a sludge layer at the bottom. Provided the temperature in the pond is kept above 15 °C the anaerobic digestion progresses at such a rate that sludge accumulation is not critical. Typically desludging takes place every 3–5 years, the exact period depending on loading rate, temperature and type of wastewater.

52.4.4 Design

52.4.4.1 Facultative ponds

If the pond is completely mixed and we assume the removal of BOD$_5$ follows a first-order relationship along the lines of equation (22), then by rearranging:

$$\theta = \left(\frac{C_i}{C_e} - 1\right)\frac{1}{k_T} \tag{25}$$

Also

$$A = \frac{F\theta}{D_p} \tag{26}$$

where A = pond area (m^2) and D_p = mean pond depth (m). Combining equations (25) and (26):

$$A = \frac{F}{D_p k_T}\left(\frac{C_i}{C_e} - 1\right) \tag{27}$$

The value for A is generally taken as the pond plan area at mid-depth, to allow for the sloping sides. For pond depths in the 1–1.5 m range it has been suggested that the design value for C_e should be 50–70 mg l^{-1}. The value of k_T is about 0.3 d^{-1} at 20 °C and its variation with temperature is described by the relationship (Mara, 1975):

$$k_T = 0.3(1.05)^{T-20} \tag{28}$$

Several empirical methods have been suggested for the design of facultative ponds.

(i) McGarry and Pescod (1970), from a study of operational data, showed that the maximum load which could be applied before the pond became totally anaerobic was:

$$\lambda_m = 11.2(1.054)^T \tag{29}$$

where λ_m = maximum BOD$_5$ loading rate (kg ha^{-1} d^{-1}) and T = temperature (°C).

(ii) In order to use this type of relationship for design Mara (1975) suggested the introduction of a factor of safety to modify equation (29):

$$\lambda_d = 7.5(1.054)^T \tag{30}$$

where λ_d = design BOD$_5$ loading rate (kg ha^{-1} d^{-1}). Once a value for λ_d has been estimated then the pond area is given by:

$$A = \frac{FC_i}{\lambda_d} \times 10 \tag{31}$$

(iii) Hermann and Gloyna (1958) reported that in order to obtain 80–90% BOD$_5$ removal from a wastewater with a BOD$_5$ of 200 mg l^{-1}, the optimum temperature for minimum retention time was 35 °C. They proposed that at any other temperature T the optimum retention time for the same removal criteria was

$$\theta_T = \theta_{35} k_R^{35-T} \tag{32}$$

where θ_T = optimum retention time at temperature T °C (d), θ_{35} = optimum retention time at 35 °C (d) and k_R = rate constant. Since the work was carried out with an influent BOD$_5$ of 200 mg l^{-1}, a correction factor needs to be used if the wastewater to be treated varies from this in order to estimate the pond volume. The pond volume can then be calculated from the following equation:

$$V = \theta_{35} k_R^{35-T} \frac{FC_i}{200} \tag{33}$$

Mara (1976) gives a range of values for θ_{35} from 3.5 to 7 days and it is this large variation which limits the general use of equation (33).

There are many other methods for calculating pond sizes, two of the more widely reported of which were originated by Thirumurthi (1969) and Oswald and Gotaas (1957) and these are based on dispersion and solar energy input, respectively.

It is unlikely that any two of the methods outlined in this section would give the same pond size for the same design parameters, so the selection of the most appropriate method is not straightforward. An advisable line of action at the design stage is to examine the performance of ponds working on a similar wastewater under similar climatic conditions. Information obtained from these existing ponds can then be used to refine any design method.

In laying out the ponds at the design stage it should be noted that a series of ponds is far more efficient than a single pond of the same total volume. Marais (1974) showed that the maximum efficiency for a series of ponds is obtained when the retention time is the same in each pond. However, flexibility must be built into the system such that any pond can be taken out of service without affecting the running of the other ponds in series.

Finney and Middlebrooks (1980) used data from four facultative pond systems to evaluate the three types of design techniques: loading rate and hydraulic retention time methods, empirical design equations and rational design equations. They concluded that none of the methods evaluated was adequate for facultative pond design.

52.4.4.2 Maturation ponds

Marais and Shaw (1961) found that the use of two maturation ponds in series, each with 7 days retention, is sufficient to produce an effluent BOD_5 less than 25 mg l^{-1} for an influent with a BOD_5 around 75 mg l^{-1}. In a series of ponds the number of faecal coliforms in the effluent can be found from the first-order relationship (Mara, 1978):

$$N_e = \frac{N_i}{(1 + k_b\theta_1)(1 + k_b\theta_2)\ldots(1 + k_b\theta_n)} \qquad (34)$$

where N_e = number of faecal coliforms in the effluent, N_i = number of faecal coliforms in the maturation pond influent, k_b = rate constant for faecal coliform removal (d^{-1}) and θ_n = retention time in the nth pond (d).

In some later work, Marais (1974) gave a relationship for k_b within the temperature range 5–21 °C in the form

$$k_{b,T} = 2.6(1.19)^{T-20} \qquad (35)$$

where $k_{b,T}$ = value of k_b at temperature T °C. Mara (1976) proposed a design procedure based on a value for N_i of 4×10^7 faecal coliforms per 100 ml, which for a given design of system can be used to determine N_e. Pond design is then based on a minimum of two ponds with 7 days retention to achieve an effluent BOD_5 of 25 mg l^{-1}, with a check then made on whether N_e is within the permissible limits. If it is too large then the number of ponds in series is increased whilst possibly reducing the retention time in each pond to achieve the desired N_e at minimum cost.

52.4.4.3 Anaerobic ponds

Currently there is no easily applied design system which can be used for anaerobic ponds. In their role as pretreatment ponds for high strength wastewater, the influent they are designed for is likely to be fairly unique. Design is based on data obtained from ponds working on similar wastes under similar climatic conditions. They have been used on a wide range of industrial wastewaters (Benefield and Randall, 1980) under an equally wide range of climatic conditions.

Mara (1975) suggested some conservative design values based on work in Israel, Africa and Australia as follows:

Retention time (d)	BOD_5 reduction (%)
1	50
2.5	60
5	70

The optimum retention time is around 5 days, as for larger values the ponds tend towards being facultative. For times less than five days, problems are likely to be encountered with odour and more rapid sludge build-up. In addition, BOD_5 removal is less and the bacteriological quality of the effluent is not as good.

52.4.5 Construction

There is no localized forced mixing in waste stabilization ponds so the type of lining which is necessary with oxidation ditches and aerated lagoons to reduce erosion is not needed. However, some form of erosion resistant lining might be necessary at the wastewater inlet and at the water-line to prevent wave damage. If ground conditions are such that appreciable seepage can occur, then some form of impermeable lining will be required; polyethylene, butyl rubber, clay and bitumen have all been used for this purpose.

Rectangular shaped ponds are more effective than circular or square designs but flexibility in pond shape means that they are very often formed to suit site geometry and topography. Construction is very similar to the aerated lagoons with the emphasis on low cost and hence earthen basins with side slopes in the 1 to 2–3 range are generally favoured.

52.5 ADVANTAGES AND DISADVANTAGES

The processes considered in this section come at the less sophisticated end of the wastewater treatment process spectrum. The construction costs for each of the three processes are relatively low, which has encouraged their use for a wide range of industrial and domestic wastewaters. The long hydraulic retention times, ranging from 1 to 2 days for oxidation ditches with values sometimes in excess of 50 days for waste stabilization ponds, has made these processes particularly resistant to shock hydraulic loads.

52.3.1 Oxidation Ditches

Ettlich (1978) compared oxidation ditch performance with other activated sludge and attached film treatment systems and for the plants examined he found they consistently produced better results for both TSS and BOD_5 removal. Generally the oxidation ditch can be operated by average personnel to achieve above average performance.

Some specific advantages of ditches are as follows. (i) In its original form for use with relatively small populations, it provides a low cost alternative to conventional activated sludge systems. On larger flows the cost savings, because primary sedimentation tanks are not required with extended aeration, have to be compared with the cost of constructing a ditch with a hydraulic retention time 3–5 times that necessary for a conventional activated sludge reactor. (ii) The overall sludge surplus with a ditch is less than with conventional activated sludge systems, with the sludge itself being more stable due to the longer period of aeration. (iii) 95–99% nitrification can be achieved under normal operating conditions even at low temperatures. (iv) 80% denitrification is possible by setting up anoxic zones within the ditch.

52.5.2 Waste Stabilization Ponds

Waste stabilization ponds require much larger areas of land than other treatment systems and consequently are only used where the cost of land is not a major consideration. Their major advantages are as follows. (i) Treatment costs are low both from the point of view of energy input and the necessity for skilled operators. (ii) Pathogen removal is higher than with most other wastewater treatment systems. (iii) Long hydraulic retention times mean they are able to withstand shock hydraulic loads. (iv) The algae produced in the ponds can be used as a source of protein for use, for instance, in fish farming.

52.5.3 Aerated Lagoons

The aerated lagoon is a treatment system which operationally is between the oxidation ditch and the waste stabilization pond, having some of the advantages of both. In general terms it requires less land than the waste stabilization pond but some form of energy input is necessary to provide adequate aeration.

52.6 REFERENCES

Barnard, J. L. (1978). Solving sludge bulking problems. *Water Pollut. Control*, **77**, 103–106.
Barnes, D., P. J. Bliss, B. W. Gould and H. R. Vallentine (1981). *Water and Wastewater Engineering Systems*. Pitman, London.
Benefield, L. D. and C. W. Randall (1980). *Biological Process Design for Wastewater Treatment*. Prentice Hall, Englewood Cliffs, NJ.
Biryukov, V. V. and L. D. Shtoffer (1971). Effects of stirring on distribution of nutrients and metabolites in bacterial suspension during cultivation. *Appl. Biochem. Microbiol.*, **7**, 9–15.
Bisogni, J. J., Jr. and A. W. Lawrence (1971). Relationship between biological solids retention time and settling characteristics of activated sludge. *Water Res.*, **5**, 753–763.
Chambers, B. and E. J. Tomlinson (1982). *Bulking of Activated Sludge — Preventative and Remedial Methods*. Ellis Horwood, Chichester.
Denton, R. S. (1977). History and development of the oxidation ditch. *Process Biochem.*, **12**, 3–7.
Ettlich, W. F. (1978). *A Comparison of Oxidation Ditch Plants to Competing Processes*. EPA Report No. EPA/600/2–78/051. US Environmental Protection Agency, Washington, DC.
Finney, B. A. and E. J. Middlebrooks (1980). Facultative waste stabilization pond design. *J. Water Pollut. Control Fed.*, **52**, 134–147.
Forster, C. F. (1980). A comparison of the performances achieved by the carrousel and the mammoth rotor versions of the oxidation ditch. *Environ. Technol. Lett.*, **1**, 366–375.
Gaudy, A. F., M. Ramanathan, P. Y. Yang and T. V. De Geare (1970). Studies on the operational stability of the extended aeration process. *J. Water Pollut. Control Fed.*, **42**, 165–179.

Goronszy, M. C. (1979). Intermittent operation of the extended aeration process for small systems. *J. Water Pollut. Control Fed.*, **51**, 274–287.

Hartman, R. B., D. G. Smith, E. R. Bennett and K. D. Linstedt (1979). Sludge stabilisation through aerobic digestion. *J. Water Pollut. Control Fed.*, **51**, 2353–2365.

Hermann, E. R. and E. F. Gloyna (1958). Waste stabilization ponds. *Sewage Ind. Waste*, **30**, 963–975.

HMSO (1969). *Technical Memorandum on Activated Sludge Sewage Treatment Installations Providing for a Long Period of Aeration*. HMSO (Ministry of Housing and Local Government), London.

Jenkins, D. and W. E. Garrison (1968). Control of activated sludge by mean cell residence time. *J. Water Pollut. Control Fed.*, **40**, 1905–1919.

Mancini, J. L. and E. L. Barnhart (1968). Industrial waste treatment in aerated lagoons. In *Advance in Water Quality Improvement*, ed. E. F. Gloyna, and W. W. Eckenfelder. University of Texas Press, Austin.

Mara, D. D. (1975a). *Design Manual for Sewage Lagoons in the Tropics*. East African Literature Bureau, Nairobi.

Mara, D. D. (1975b). Discussion — A note on the design of facultative sewage lagoons. *Water Res.*, **9**, 595–597.

Mara, D. D. (1978). *Sewage Treatment in Hot Climates*. Wiley, New York.

Marais, G. v. R. (1974). Faecal bacterial kinetics in stabilization ponds. *J. Environ. Eng. Div., Am. Soc. Civ. Eng.*, **100**, EEI, 119–139.

Marais, G. v. R. and V. A. Shaw (1961). A rational theory for the design of sewage stabilization ponds in Central and South Africa. *Trans. S. Afr. Inst. Civ. Eng.*, **3**, 205–227.

Marais, G. v. R. and G. Ekama (1976). The activated sludge process: Part I. Steady state behaviour. *Water SA*, **2**, 163–200.

McGarry, M. G. and M. B. Pescod (1970). Stabilization pond design criteria for tropical Asia. In *Proc. 2nd Int. Symp. on Waste Treatment Lagoons*, Kansas City, MO.

Metcalf and Eddy, (1979). *Wastewater Engineering: Treatment, Disposal, Reuse*. McGraw-Hill, New York.

Middlebrooks, E. J., N. B. Jones, J. H. Reynolds and M. F. Torpy (1978). *Lagoon Information Source Book*. Ann Arbor Science, Ann Arbor, MI.

Mulbarger, M. C. (1971). Nitrification and denitrification in activated sludge systems. *J. Water Pollut. Control Fed.*, **43**, 2059–2070.

Mulready, C. J., D. C. Payne and D. P. Watkins (1982). A comparison of the carrousel and Pasveer ditch activated sludge plants. *J. Inst. Water Pollut. Control*, **81**, 297–307.

Oswald, W. J. and H. B. Gotaas (1957). Photosynthesis in sewage treatment. *Trans. Am. Soc. Civ. Eng.*, **122**, 73–105.

Pasveer, A. (1959). A contribution to the development of activated sludge treatment. *J. Proc. Inst. Sewage Purif.*, 436–465.

Pasveer, A. and B. A. Heide (1974). Oxidation ditch: prevention and control of filamentous sludge. H_2O, **7**, 373–377.

Porges, N., L. Jasewicz and S. R. Hoover (1953). Aerobic treatment of dairy wastes. *Appl. Microbiol.*, **1**, 262–270.

Schroeder, E. D. (1977). *Water and Wastewater Treatment*. McGraw-Hill, New York.

Sherrard, J. H. and E. D. Schroeder (1973). Cell yield and growth rate in activated sludge. *J. Water Pollut. Control Fed.*, **45**, 1889–1897.

Silva, S. A. (1982). *On the Treatment of Domestic Sewage in Waste Stabilization Ponds in North East Brazil*. Ph.D. Thesis, University of Dundee, Scotland.

Sykes, R.M. (1975). Theoretical heterotrophic yields. *J. Water Pollut. Control Fed.*, **47**, 591–600.

Thirumurthi, D. (1969). Design principles of waste stabilization ponds. *J. Sanit. Eng. Div., Am. Soc. Civ. Eng.*, **95**, 311–330.

Thirumurthi, D. (1979). Design criteria for aerobic aerated lagoons. *J. Environ. Eng. Div., Am. Soc. Civ. Eng.*, **105**, EEI, 135–148.

Water and Waste Treatment (1981). Split ditch offers big advantages. *Water Waste Treat.*, Dec., 30–33.

Wild, H. E., C. N. Sawyer and T. C. McMahon (1971). Factors affecting nitrification kinetics. *J. Water Pollut. Control Fed.*, **43**, 1845–1854.

53

Management of Toxic Pollutants: Technical Constraints *versus* Legal Requirements

H. MELCER
Environment Canada, Wastewater Technology Centre, Burlington, Ontario, Canada

53.1 INTRODUCTION

Over the past 20 years, rapid industrial development has led to the production of a vast range of synthetic materials. Some of the resulting industrial wastes contain toxic, inhibitory or poorly degradable (biorefractory) compounds. A growing awareness of the potential deleterious environmental impact of these compounds has generated legislative controls over the discharge and disposal of industrial wastes. The execution of such controls can lead to considerable logistic, economic and technical problems, some of which could be avoided with more forethought at the time of formulating regulations and guidelines.

The implementation of the US water pollution control legislation that has been tabled over the past decade may be used to illustrate the type of problems faced by both the regulators and those for whom the legislation is intended. The 1972 amendments to the Federal Water Pollution Control Act represented a radical movement away from controls based on standards related to water quality to technology-based standards. They also recognized the presence of toxic pollutants in the environment and defined a program to minimize their impact. This chapter traces the problems experienced by the regulators and by the pollutors in implementing this legislation. Although toxic pollutants are focussed upon, some background is presented to provide a level of perspective to the overall legislative approach to water pollution control in the US.

53.2 BACKGROUND

During the 1960s and early 1970s there was increasing frustration in the area of US water pollution control policy over the slow pace of clean-up efforts and the inadequacies of existing regulatory programs. Prior to 1972, the central strategy for controlling water pollution was based upon the desired uses of receiving waters and the water quality conditions required to support those uses. Accordingly, State water pollution control agencies applied water quality standards to each stream and, where this standard was not being met, sought to determine the responsible discharger(s) for enforcement action. As increased or new discharges were introduced, it was intended that treatment capabilities be upgraded in order to maintain receiving water quality. This strategy was ineffective due to a number of political, technical and legal weaknesses: stream use designations tailored to protect or attract industrial development, inadequate consideration of the health of aquatic ecosystems, problems of equity between old and new pollution sources and general inattention to non-point discharges (Anon, 1978).

In October 1972, major revisions to the Federal Water Pollution Control Act were embodied in Public Law (PL) 92-500. Section 101 summarized the new order stating: 'The objective of this Act is to restore and maintain the chemical, physical and biological integrity of the Nation's waters'. The Act established national goals of eliminating pollutant discharges by 1985 (zero discharge goal) and achieving an interim water quality level that would protect fish, shellfish and wildlife while providing for recreation in and on the water wherever attainable.

The 1972 amendments changed the course of national water pollution control strategy by dividing discharges into three categories: municipal, industrial and toxic. Unless the quality of the receiving waters dictated more stringent limitations, different general controls were applied to each category based on nationally uniform standards. Municipal discharges from publicly owned treatment works (POTWs) were to meet secondary treatment requirements by July 1, 1977 and municipal best practicable waste treatment technology (BPWTT) by July 1, 1983. Industrial discharges were subject to industrial best practicable technology currently available (BPT) by July 1, 1977 and best available technology economically achievable (BAT) by July 1, 1983. Generally, BPT was accepted to represent secondary treatment standards. BAT represented an upgrading of effluent standards from BPT to tertiary treatment in order to remove nonconventional or toxic pollutants. Toxic effluent standards were to be set by the US Environmental Protection Agency (EPA) for compounds placed on a toxic pollutant list. New source performance standards (NSPS) were to be achieved immediately when a new discharge source commenced operation. All point sources of discharge required national pollutant discharge elimination system (NPDES) permits which were to be renewed and upgraded at least every five years.

In summary, the law distinguished between maximum prohibited and permissible discharge levels by setting pollutant concentration standards for broad classes of industrial and municipal activities. Receiving water standards were substituted by effluent limitations which set end-of-pipe quality standards. At the time, it was widely accepted that water pollution control technology could be forced, through a combination of statutory deadlines and high discharge quality requirements, to provide a much higher level of pollutant removal than had been the case until that time (Anon, 1982). Through deadlines and standards, it was thought that municipal and industrial dischargers would install and refine more economical treatment systems in their own self-interest.

53.3 TECHNOLOGY-BASED STANDARDS

In turning away from water quality to technology-based standards, the Act established pollution removal requirements for manufacturers on a nation-wide basis. In effect, industries were required to apply a base level of control technology regardless of the needs of receiving waters. For the regulatory agencies, this was expected to simplify the administrative task and to prevent the establishment of preferred locations in states anxious to attract industry. The time-consuming burden of proving harmful effects of each plant's discharge was removed. However, the potential for using the assimilative capacity of receiving waters was overlooked. Issue was taken by many industrialists over the under-utilization of this natural phenomenon of biodegradation.

In areas with very sensitive receiving waters, the law allowed the superimposition of water quality based standards where BPT could not achieve the desired water quality based criteria. This tended to place some manufacturers at a disadvantage since they would be required to utilize pollution control technology beyond what had been judged practicable for the major portion of manufacturers in their industrial category.

53.4 EFFLUENT GUIDELINE DEVELOPMENT

The BPT and BAT standards were to be applied in successive five-year periods roughly equivalent to amortization periods. When determining BPT under Section 304(b), EPA was to consider the total cost of the technology in relation to effluent reduction benefits to be achieved. Factors to be examined included industrial plant age, the process used, engineering aspects of proposed control technologies and non-environmental issues such as energy requirements. This was the basis for the development of effluent guidelines. Difficulties were experienced due to the failure to define BPT and BAT for different industrial sectors. While there was general agreement that BPT was to be used to control conventional pollutants (*i.e.* BOD, suspended solids and pH), consensus was harder to achieve in the area of toxic pollutant control.

Guidelines were based upon the operation of 'exemplary' plants in each industrial category. An exemplary plant was one which incorporated the most advanced pollution control facilities in the industry (Minor, 1976). The selection of exemplary plants was controversial. A plant that was outstanding in one type of pollution control, for example, was inadequate in another. Exemplary features from several plants were sometimes combined into a model plant, a practice that was perhaps far removed from actual manufacturing conditions.

A further problem in developing effluent guidelines was the inadequate data base that was then available to assess the performance of water pollution operations on specific waste streams. The technical effort required to analyze fully process technologies, wastewater treatment technologies and their application to specific industrial sectors was underestimated by both industry and EPA.

The cost of treatment was also underestimated. Economists, analyzing the effect of water pollution control costs, limited their analyses to the economic impact on the industry itself. They did not consider the impact of the cumulative costs for all industries on the country. The National Association of Manufacturers (Anon, 1974) indicated that understated costs were based on incomplete analysis of all incurred costs, lack of correction to current levels of costs, ignorance of retrofitting costs (green field situations only were addressed), omission of sludge handling and disposal costs and the underestimation of in-plant control costs which were required for a plant to meet the guidelines.

53.5 RISK ASSESSMENT

Minimizing long-term health risks by removal of suspended toxic pollutants is an important aspect of maintaining high quality receiving waters. Tarlock (1979) defines risk as a future injury which has a low probability of occurrence, but which would have substantial consequences to an individual should it occur. Cancer is regarded as a major public health risk and potentially toxic pollutants are being intensively scrutinized to determine what risks they pose. He postulates that, in environmental law, it has been recognized that proof of risk, as opposed to present harm, is a legitimate basis for banning or limiting the use of a chemical. In view of the highly emotive nature of the cancer issue, scientists have been encouraged to be conservative in estimating risk, no matter what procedure is used. The final use decision is based upon the worst case scenario. This approach puts both the scientist and the regulator decision-makers in a quandary. Given the improved accuracy of modern analytical techniques, many chemicals can be shown to produce some risk. Since it is very difficult to compare the cost of a risk with the benefits of continued use of a chemical, the urge to be conservative causes decision-makers to set high standards in order to minimize the risk potential.

Technology-forcing standards and risk-screening procedures have been criticized for inefficient allocation of resources. In the interest of administrative ease, water pollution control standards make little attempt to relate discharge limitations to the damage caused to society by the discharge and they give dischargers little flexibility in selecting the method of compliance.

Risks are also associated with exposing the aquatic environment to potentially toxic pollutants. Schaumberg (1974) conceptually visualized, by plotting degree of treatment required against net environmental impact, the dilemma that arises when unnecessarily restrictive standards are applied. At some point the net negative environmental impact of the increasing use of energy for treatment exceeds the benefits to be obtained from additional treatment. Garber (1977) documented these observations by case studies at the Hyperion Treatment Plant in Los Angeles. They showed, for example, that physical–chemical treatment would require approximately five times the energy used by the existing activated sludge system.

53.6 TOXIC POLLUTANTS

Section 307 of PL 92-500 required EPA, within a nine-month period, to identify toxic pollutants and issue effluent standards (including prohibition of discharge) for control of those substances on a pollutant-by-pollutant basis rather than by industrial categories. Unfortunately, the toxicological and epidemiological knowledge of the time was inadequate to allow either expeditious identification of toxic pollutants in toxic amounts or timely promulgation of the standards with which to control such substances (Anon, 1982). The generation of scientifically valid data bases for most substances was far more expensive and time-consuming than was anticipated. Results of these studies were debated vigorously and were subject to considerable disagreement; even the term 'toxicity' was open to debate.

During the five-year period following PL 92-500, EPA managed to propose effluent standards for only six toxic pollutants. Several public interest groups including the Natural Resources Defense Council (NRDC) found this rate of progress to be unacceptable. They initiated court action which, in 1976, resulted in a consent decree that became the controlling element of EPA's development of toxic standards (NRDC, 1975). Often referred to as the 'toxics consent decree', it required EPA, by December 1980, to upgrade BAT effluent guidelines, industrial pretreatment standards and new source performance standards by incorporating standards for 65 toxic pollutants (including some categories of substances). These became known as the 'priority pollutants'. The task proved to be beyond the resources at hand.

In 1977 a major revision of PL 92-500 recognized these shortcomings by emphasizing control of toxic substances in a series of amendments that became known as the Clean Water Act, PL 95-217. Three classes of pollutants were created: conventional, nonconventional and toxic. The EPA had until July 1980 to promulgate BAT standards, industry being required to comply by July, 1984. Suspended solids, BOD(5), pH, oils and grease and faecal coliforms were classed as conventional pollutants. Nonconventional pollutants were those substances not defined as either conventional or toxic, for example phosphorus and ammonia. Toxic pollutants included the priority pollutants which EPA extended to a list of 129 compounds.

The listing of the priority pollutants created a formidable problem which remains to the present time. Chemical analysis for the priority pollutants was very difficult. Wastewater characterization had traditionally been carried out in terms of lumped organic parameters such as BOD(5), COD or TOC. The new legislation called for the simultaneous measurement of a range of compounds that occurred at concentrations three orders of magnitude less than the lumped organic parameters. The EPA developed chromatographic methods of analysis (GC, GC/MS and HPLC) (Bishop, 1980; Finnigan *et al.*, 1979; Keith and Telliard, 1979) that were far more sophisticated and costly than most municipalities and industries could accommodate or afford (Vicory and Malina, 1982). There were technical problems, too, with ensuring the accuracy and reproducibility of trace organic data. Extensive quality assurance/quality control programs were required to validate these data.

The use of surrogate parameters (such as suspended solids) or indicator compounds (substances for which reliable information existed and for which a causal relationship between their removal and the removal of treatable pollutants was presumed to exist) was advocated as an alternative to trace organic analysis. In most cases neither surrogates nor indicators could be developed in view of the extensive removal data that were required to establish statistically valid correlations (Washburn *et al.*, 1981).

The EPA has not yet completed the promulgation of BAT effluent standards. Initially industrial dischargers were to have had up to four years to comply. The longer it takes EPA to issue standards, the shorter the time industries will have to meet them before the BAT compliance date of July, 1984. It is doubtful whether this ambitious goal will be met since scientific knowledge about the toxic pollutants to be controlled, about the methods used to analyse them and about the treatment technologies to remove them is limited.

Surveys of municipal (Burns and Roe, 1982; Cohen and Hannah, 1981) and industrial (Tischler and Kocurek, 1982) wastewaters have shown that well designed and operated biological treatment plants can remove the majority of the priority pollutants to very low levels in the final effluent. However, for compounds that are biorefractory, biological treatment results in their accumulation in the biomass, sometimes to levels that are either inhibitory to the biomass (particularly for nitrification systems) or which may render the sludge unsuitable for land disposal. Considerable advances in the technology to control the discharge of toxic pollutants have been stimulated by these laws. For example, the last decade has seen the introduction of powdered activated carbon assisted activated sludge systems, wet oxidation of sludges and hazardous materials, compound-specific ion exchange resins, the expansion of granular activated carbon treatment and the advent of solids retention time control-based biological fixed film reactors such as aerobic or anaerobic fluidized beds, submerged filters and rotating biological contactors.

Also, the legislation has provided for at least two approaches to address the problems of effluent standards development. These are the promotion of innovative technologies and the sponsorship of research into predictive techniques to forecast the potential for toxic behaviour of, and the occurrence of, priority pollutants.

53.7 INNOVATIVE TECHNOLOGY

Under the auspices of PL 95-217, EPA promoted the utilization of innovative waste management schemes in which 'innovative' referred to new and promising technology that was not yet fully proven under the circumstances of its contemplated use (Anon, 1978). In conventional treatment systems, 'innovative' could describe technology that reduced costs by a minimum of 15% or reduced energy consumption by at least 20%. The utilization of fine bubble aeration in activated sludge systems was one such program sponsored by EPA (McCarthy, 1982a). 'Innovative' could also refer to new technology that advanced the state of the art of wastewater treatment. The development of the biological aerated filter (Stensel *et al.*, 1983) and the deep shaft activated sludge process (Weston, 1982) are examples of the latter case. More appropriate to the management of toxic wastes has been the EPA sponsorship of the development of the high-rate anaerobic filter (Heidman, 1982) and the vertical well reactor (McCarthy, 1982b). Although the development of these technologies was directed at the municipal sector, there is considerable potential for applying them to high strength and toxic wastewaters.

53.8 PREDICTION OF TOXIC BEHAVIOUR AND OCCURRENCE OF PRIORITY POLLUTANTS

There are two aspects of toxic behaviour to consider: the prediction of mammalian toxicity of candidate chemicals, and an assessment of the potential for removal of such chemicals in wastewater treatment plants. In the former case, the dependence of biological effects on structural differences between chemical homologues is known as the structure–activity relationship (SAR). A discussion of SAR is beyond the scope of this chapter. In the latter case, there are very few models or data bases which can be used to predict quantitively the degree of removal of these chemicals in wastewater treatment systems. Time-consuming and costly bench- and pilot-scale treatability studies are carried out to determine removal kinetics. One alternative approach proposed by Berkau (1981) was to group chemicals according to the similarity of their behaviour in different wastewater treatment plants. The supposition was that the similarity in response could be correlated with specific physical and chemical characteristics of the chemical. Insufficient data were collected to verify his hypothesis.

This concept was also explored by Strier (1979) who called it the molecular engineering approach. He evaluated a range of physical–chemical and biooxidation properties of the organic priority pollutants to derive theoretical effluent limitations for these compounds in a range of unit treatment processes. In further work, Strier and Gallup (1983) applied the same technique to evaluate the controlling mechanisms for removal of specific organic priority pollutants in the activated sludge process. They showed that by grouping organic priority pollutants into six groups according to solubility and chemical structure similarities, treatability data could be interpreted in terms of the removal mechanisms, adsorption, stripping, and biological or chemical oxidation.

In assessing the potential for occurrence of priority pollutants in raw wastewaters, Wise and Fahrenthold (1981) based their predictions on the feedstock source and the industrial process chemistry. Synthetic routes to the priority pollutants were developed to show the intrinsic potential that certain feedstock–generic process combinations have for the generation of priority pollutants. They characterized the effluents from 172 product/processes of commercial importance at 40 manufacturing sites in the organic chemical and plastics/synthetic fibres industries. Relatively few of the product/processes reviewed were sources of priority pollutants because most did not involve critical precursor/generic process combinations. Priority pollutants and their precursors are simple molecules, derived in most cases from primary feedstocks, and are typically associated with a rather narrow segment near the beginning of synthetic routes to commercial products. Beyond that point, the product molecule or its coproduct impurities become more complex than any of the priority pollutants or their precursors.

These types of study are still in their infancy but offer a less expensive alternative approach to assessing the removal potential of toxic pollutants than conventional treatability studies.

53.9 SUMMARY

The removal of toxic pollutants from wastewater has been found to be a technically challenging task. Experience in implementing the changes to the Federal Water Pollution Control Act has revealed the inherent limitations of a solely water quality-based or a technology-based approach. The scheduling and costs proposed by the Act were unrealistic given the magnitude of the technical difficulties that had to be overcome. Nevertheless, the technology-based effluent guidelines forced the pace of treatment technology development and many controls are now in place. Gradually, accommodations have been made in the Act. The original commitment to technology-based standards for the control of toxic pollutants remains firm but methods of including water quality considerations are being explored.

53.10 REFERENCES

Anon (1974). National pollutant discharge elimination system (NPDES) — an evaluation report. National Association of Manufacturers, Washington, DC.

Anon (1978). A guide to the clean water act. US Environmental Protection Agency. EPA-OPA129/8.

Anon (1982). The clean water act — with amendments. Water Pollution Control Federation, Washington, DC.

Berkau, E. E. (1981). Treatability of organic chemicals. Presented at World Congress of Chemical Engineers, Montreal, Quebec.

Bishop, D. F. (1980). GC/MS methodology for measuring priority organics in municipal wastewater treatment. EPA-600/S2-80-196.

Burns and Roe Industrial Services Corporation (1982). Fate of priority toxic pollutants in publicly-owned treatment works — final report, vol. 1, EPA-440/1-82/303.

Cohen, J. M. and S. A. Hannah (1981). National survey of municipal wastewaters for toxic chemicals. Presented at the 8th US–Japan Conference on Sewage Treatment Technology, Cincinnati, OH.

Finnigan, R. E., D. W. Hoyt and D. E. Smith (1979). Priority pollutants II — cost-effective analysis. *Environ. Sci. Technol.*, **13**, 534–541.

Garber, W. F. (1977). Effluent standards — effect upon designs. *J. Environ. Eng. Div. (Am. Soc. Civ. Eng.)*, **103**, No. EE6, 1115–1127.

Heidman, J. A. (1982). Technology assessment of anaerobic systems for municipal wastewater treatment — anaerobic fluidized bed, anflow. EPA-600/2-82-004.

Keith, L. H. and W. A. Telliard (1979). Priority pollutants I — a perspective review. *Environ. Sci. Technol.*, **13**, 416–423.

McCarthy, J. J. (1982a). Technology assessment of fine bubble aerators. EPA-600/2-82-003.

McCarthy, J. J. (1982b). Technology assessment of the vertical well chemical reactor. EPA-600/2-82-005.

Minor, P. S. (1976). The industry–EPA confrontation — living today with the water pollution control act amendments of 1972. McGraw-Hill, New York.

National Resources Defense Council (NRDC) (1975). M. C. *vs.* Train, 1975 8 ERC, 2120, 2122–29, D.D.C.

Schaumberg, F. D. (1974). Nature, an important factor in management of the total environment. Presented at 7th Biennial Conference, International Association on Water Pollution Research, Paris, France.

Stensel, H. D., K. Lee and H. Melcer (1983). Biological aerated filter full-scale plant investigation. Presented at 1983 Pollution Control Association of Ontario Conference, London, Ontario.

Strier, M. P. (1979). Pollutant treatability — a molecular engineering approach. *Environ. Sci. Technol.*, **14**, 78–81.

Strier, M. P. and J. D. Gallup (1983). Removal pathways and fate of organic priority pollutants in treatment systems — chemical considerations. *Proceedings of the 37th Waste Conference*, Purdue University, W. Lafayette, IN, 813–824. Ann Arbor Science, Ann Arbor, MI.

Tarlock, A. D. (1979). Environmental law — what is it, what it should be. *Environ. Sci. Technol.*, **13**, 1344–1348.

Tischler, L. F. and D. Kocurek (1982). The CMA/EPA five plant study — biological treatment of toxic organic pollutants. Presented at the 55th Water Pollution Control Federation Conference, St. Louis, MO.

Vicory, A. H. and J. F. Malina, Jr. (1982). Apparatus needs and costs for monitoring priority pollutants. *J. Water Pollut. Control Fed.*, **54**, 125–128.

Washburn, J., K. Kirk, G. McDermott and R. O'Dette (1981). Industrial pretreatment needs to be overhauled at the national level. *J. Water Pollut. Control Fed.*, **53**, 1452–1456.

Weston, R. F. (1982). Technology assessment of the deep shaft biological reactor. EPA-600/2-82-002.

Wise, H. E., Jr. and P. D. Fahrenthold (1981). Predicting priority pollutants from petrochemical processes. *Environ. Sci. Technol.*, **15**, 1292–1304.

54
Fundamental Considerations of Fixed Film Systems

W. G. CHARACKLIS, R. BAKKE and M. G. TRULEAR
Montana State University, Bozeman, MT, USA

54.1 INTRODUCTION

The origin of fixed film reactors for wastewater treatment is estimated to have occurred around 1871 (Lohmeyer, 1957). Since that time, fixed film reactors have been widely adopted, although their popularity waned during the period of wide scale adoption of the activated sludge process. Rotating biological contactors (RBC), plastic packing for towers, and the acceptance of fluidized bed reactors have increased the applications of fixed film processes in the last 10 years. An important advantage of the RBC over activated sludge is the lower energy requirement, which has influenced its rapid adoption.

As a result of the new developments, there has been a renewed interest in the modelling of fixed film reactors. The interest in modelling is related to expectations for improved design and operation of the reactors for wastewater treatment. This chapter will deal with phenomenological or mechanistic aspects of fixed film modelling. The intent is to develop models to simulate fixed film reactors under conditions and in applications heretofore unexplored. The oft-used empirical approach can be used with caution for design from pilot data and numerous case histories of this type can be found in the literature (Watt and Cahill, 1980). However, empirical models can rarely provide useful extrapolations for predicting performances of widely disparate systems.

Phenomenological models are developed with attention to the conservation equations derived from thermodynamics and Newton's laws of motion. Using these tools experimental data can yield stoichiometric and kinetic expressions which describe the processes occurring in any fixed film reactor.

Three types of fixed film reactors will be discussed in the succeeding chapters in this section: (1) trickling filters, (2) rotating biological contactors, and (3) fluidized beds. Superficially, these reactor configurations appear quite different. In fact, the fluid flow patterns are significantly different. However, the fundamental processes occurring in the biofilm within these different reactor configurations are essentially the same. The heterogeneous (*i.e.* requiring two phases) reactors contain active biomass immobilized on the wetted surfaces. The immobilized biomass, herein referred to as the *biofilm*, extracts nutrients from the flowing liquid. As noted before, the flow patterns in the various reactor configurations are different and, as a result, fluid shear stress and mass transfer processes are different at the biofilm–fluid interface.

Consequently, a flow diagram of any model for a fixed film reactor would include two major components: (1) a reactor flow model and (2) a biofilm model (Figure 1). The reactor flow model will be different for every different flow configuration. The biofilm model will be the same for all fixed film reactors. The remainder of this chapter will deal with these two model components with emphasis on the biofilm component.

The goal of this chapter is to provide a tool to help the reader formulate questions regarding data presented in succeeding sections. The intent is to stimulate the reader to use this tool as a starting point toward a deeper understanding of the processes occurring within a fixed film reactor which will result in design or operational improvements in a specific wastewater treatment plant.

54.2 SIMPLIFYING ASSUMPTIONS

The growth of even a single population of microorganisms is a very complicated process. The task of modelling such growth requires simplifying assumptions. The assumptions introduce errors since the model will not describe all the various facets of growth. However, the model will describe the processes considered important. Continued experimentation will lead to continued improvement in the accuracy of fixed film mathematical models.

Although many microbial process models make certain simplifying assumptions, it seems important that the more general assumptions be stated explicitly at this point. More specific assumptions will be presented throughout the chapter.

54.2.1 Distribution of States

Biological populations are composed of individual organisms and the individual organisms are physiologically, morphologically and genetically different. More detailed analysis perhaps would reveal even more fundamental distinctions between individuals. In summary, the individuals of a population do not all exist in the same 'state', but rather in a distribution of states.

Our first simplifying assumption is that distribution of states can be ignored and the properties of the culture can be adequately described in terms of a 'typical' individual whose behavior represents an average over the distribution states.

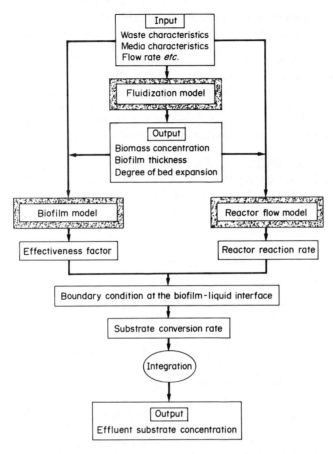

Figure 1 Flow diagram of a model for a fluidized bed biological reactor illustrating interfacing of the biofilm model with the model of the physical characteristics of the reactor

54.2.2 Segregation

In microbial processes, life is segregated into structurally and functionally discrete units, *i.e.* cells. Hence, the number of individual organisms present in a population is as important for the description of the problem as biomass. Despite these considerations, we will sometimes assume that the population is only considered as biomass distributed continuously throughout the reactor. Models based on this assumption are termed nonsegregated or distributed.

If microbial growth is balanced, then increase in biomass and cell number are proportional, which possibly explains some of the success of distributed models in the past. However, microbial growth in biofilms cannot always be considered balanced without introducing large errors.

54.2.3 Structure

Two microorganisms having the same biomass and inhabiting the same environment may, nevertheless, have widely different properties and activities. This is the problem of state again. Even a distributed model should recognize that the average or typical organism can change in response to variations in the environment. This means that parameters other than population number and biomass must be important in describing population behavior.

For most microbial process models, biomass is the sole variable employed for describing the microbial population. Therefore, these models do not recognize any 'structure' in the biotic material. Since this procedure regards organisms and population biomass as featureless, structureless entities, these models are called unstructured. Throughout most of this chapter, unstructured models will be considered.

54.3 FUNDAMENTAL PROCESSES

Development of a systematic understanding of biofilm processes has been limited by the seeming complexity posed by the interaction of several contributing rate processes. The development and persistence of a biofilm may be described as a net result of the following processes: (i) transport of material from the bulk fluid to the wetted surface; materials may be soluble (microbial nutrients both inorganic and organic) and particulate (viable microorganisms, other organic, and inorganic particles); (ii) adhesion, attachment and/or adsorption of soluble and particulate material to the wetted surface or the biofilm; (iii) microbial processes within the biofilm which include growth, product formation, decay or maintenance, and death or lysis; and (iv) detachment or desorption of soluble and particulate material from the biofilm and its reentrainment in the bulk fluid.

These processes are generally occurring simultaneously although their relative rates may be quite different (Figure 2). Any of the processes could provide the rate-limiting step in the overall process of biofilm development or its continued existence.

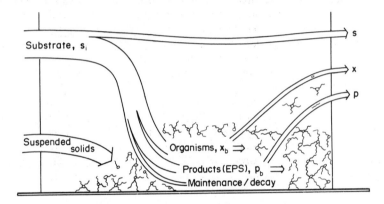

Figure 2 Biofilm development model including transport of soluble and insoluble material to the wetted surface where attachment and microbial transformation processes are occurring. Desorption and detachment of products and cells from the biofilm are also included

54.3.1 Fluid Phase Transport Processes

The rate of transport of materials from the bulk fluid to the biofilm is an important factor contributing to the performance of a fixed film reactor.

54.3.1.1 Soluble materials

The transport of soluble organic and inorganic nutrients, as well as electron donors, can be the rate-controlling step in the removal of these components from the bulk fluid. Transport of soluble materials from the bulk fluid to the biofilm interface can be described adequately by well-tested empirical expressions relating transport rate to fluid flow regime, fluid phase concentration and surface roughness (as reflected by fluid frictional resistance).

54.3.1.2 Particulate materials

Transport of microbial cells, particulate and colloidal substrates, and other relatively inert particles contributes significantly to the development and continued existence of a biofilm. Transport of particles, in contrast to soluble materials, is a more complex phenomenon depending on the properties of the particle (diameter, specific gravity), properties of the fluid, and the fluid flow regime. Equations describing transport of soluble materials are used to describe transport of particulates as well, but mass transfer coefficients cannot be predicted as accurately as those for soluble components.

54.3.2 Interfacial Phenomena

Soluble and particulate materials are transported to the biofilm interface and are immobilized in the biofilm or reentrained in the bulk fluid. The rate of capture, or the 'sticking efficiency', is related to the biofilm surface roughness, its chemical composition and the velocity with which the particle strikes the surface. There is reason to believe that there is significant turnover of microbial cells within the biofilm resulting from the adsorption of cells from the bulk fluid and desorption of cells from the biofilm (Trulear, 1983).

54.3.3 Microbial Processes

The models considered in this chapter will be unstructured for the most part. As a result, we will presume that microorganisms can carry out only four fundamental processes: growth, *i.e.* increase in cellular mass or cell numbers due to reproduction processes; product formation, *i.e.* formation of intracellular or extracellular products which are excreted and attached to the cell or desorbed; maintenance, *i.e.* a process requiring energy to maintain the cellular machinery. Maintenance has not been directly observed but has been used to account for decreases in the observed growth yield, generally at low growth rates; biomass decay, *i.e.* attrition of biomass due to endogenous respiration, autooxidation, and other ill-defined processes. Biomass decay and maintenance are both used to account for the same observations. As a consequence, the two are not used in the same model; and death or lysis, *i.e.* inactivation of cells or their destruction by lysis (can be related to biomass decay).

Growth, product formation and maintenance occur at the expense of substrate and nutrients removed from the wastewater. Therefore, substrate removal rate is the net result of substrate removal for growth, product formation and maintenance. The experimental observation and quantification of the individual process rates is quite difficult. As a consequence, the wastewater treatment literature has generally ignored the fundamental processes choosing to concentrate on the more readily observed quantites such as substrate or oxygen removal rate.

54.3.4 Detachment

Detachment phenomena can be arbitrarily categorized as shearing or sloughing. Shearing refers to continuous removal of small portions of the biofilm which is highly dependent on fluid dynamic conditions. Under these circumstances, rate of detachment increases with increasing biofilm thickness and fluid shear stress at the biofilm–fluid interface (Trulear and Characklis, 1982). Sloughing refers to a random, massive removal of biofilm generally attributed to nutrient or oxygen depletion deep within the biofilm (Howell and Atkinson, 1976) or some dramatic change in the immediate environment of the biofilm (Bakke, 1983). Sloughing is more frequently witnessed with thicker biofilms developed in nutrient-rich environments.

Table 1 is a matrix representation of the fundamental processes and their relationship to the readily observed processes reported regularly in the wastewater treatment literature.

54.4 METHODS

The absence of convenient methods for measuring the biotic and abiotic components has hindered more fundamental process analyses in fixed film reactor systems. Recently, more discriminating methods for such analyses have been developed (Characklis *et al.*, 1982; Trulear, 1983). Material balances and, hence, methods are necessary for quantifying substrate, product and suspended biomass. In addition, a constitutive equation for the biofilm mass is necessary. Therefore, certain properties of the biofilm system must be determined.

54.4.1 Substrate and Product

Biochemical oxygen demand (BOD), chemical oxygen demand (COD) and organic carbon (OC) are nonspecific or colligative measures of organic substrate. BOD is essentially useless for any process analysis because it cannot be used in material balances. COD (in conjunction with oxygen removal rate) and OC can be used in material balances (*i.e.* they are conservative quanti-

Table 1 A Matrix Representation for the Fundamental Microbial Rate Processes

| Process rate / Fundamental process | | Stoichiometry | | | | | | | |
| | | Reactants | | | Products | | | | |
Process	Rate	Substrate S	Nutrient Z	Electron acceptor e	Biomass x_T	x_d	Product p_e	p_i	Metabolite a
Growth	μ	–	–	–	+		+	(+)	+
Maintenance									
Exogenous	m	–	–	–			+		+
Endogenous	k_e		+	–	–	(+)	+	–	+
Product formation	k_p	–	–	–			+	+	+
Death									
Loss of viability	k_d								
Lysis	k_L	(+)	(+)	–	(+)	+			
Observed rate		q_s	q_z	q_e	μ_n		q_p		q_a

q = specific production or removal rate (t^{-1}); μ = specific growth rate or specific biomass production rate (t^{-1}); x_T = total biomass concentration (ml^{-3}); x_d = inert solids concentration (ml^{-3}); p_e = extracellular microbial product concentration (ml^{-3}); p_i = intracellular microbial product concentration (ml^{-3}); s = substrate concentration (ml^{-3}); z = nutrient concentration (ml^{-3}); e = electron acceptor concentration (ml^{-3}); and μ_n = net solids production rate (t^{-1})

ties) but neither can distinguish between substrate and organic microbial products. Wastewaters rarely consist of a single substrate. However, a process analysis with a defined substrate (single or multiple) can provide useful insight into the biofilm processes occurring in wastewater treatment systems. Trulear (1983) and Schaftel (1982) have both used glucose as a substrate in biofilm studies because of the enzymatic analyses available for its specific detection. In addition, they used OC analysis, which measured the sum of the carbon in the glucose as well as the microbial products. The results of their studies are significant and will be discussed in other sections.

54.4.2 Biofilm Properties

Early fixed film studies with wastewater ignored biofilm properties and simply related process characteristics to the wetted surface area covered by the biofilm. As the need became apparent, biofilm thickness and density (dry weight per unit wet volume) were included (*e.g.* Hoehn and Ray, 1973; Trulear and Characklis, 1982). More recently, Trulear (1983) has provided a measure of structure to the biofilm by distinguishing between cellular and extracellular carbon mass within the biofilm. One of his results indicates that the biofilm may consist of as much as 90% extracellular polymer substances (EPS) expressed as organic carbon. Trulear's study was conducted axenically with *Pseudomonas aeruginosa*. Cellular carbon was estimated by acridine orange direct count (AODC) and determination of an average cell volume using epi-illuminated microscopy assuming a specific gravity (1.0) and carbon content (50%) for the cells. The difference between total biofilm organic carbon and cellular biofilm organic carbon is EPS carbon.

54.4.3 Suspended Biomass

Biomass, consisting of cellular mass and insoluble products, is continuously entering the reactor suspended in the influent. However, biomass is also entering the reactor fluid as a result of detachment from the biofilm. The suspended biomass can be classified into cellular and extracellular mass in the same way described above for biofilm mass.

54.5 STOICHIOMETRY

Stoichiometry is the application of the law of conservation of mass and the chemical laws of combining weights to chemical processes. In its broadest sense, stoichiometry is a system of accounting applied to mass and energy participating in a process involving physical and/or chemical change. Stoichiometric calculations permit a surprisingly large amount of information to be obtained from a seemingly small amount of data.

Mass balances for various components across a reactor help define the stoichiometry of the reactions occurring within it. The material balance across a continuous reactor has the following general form:

$$\begin{array}{c} \text{net rate of} \\ \text{accumulation} \\ \text{in reactor} \end{array} = \begin{array}{c} \text{net rate of} \\ \text{transport out} \\ \text{of reactor} \end{array} + \begin{array}{c} \text{rate of} \\ \text{transformation} \\ \text{by reaction} \end{array} \tag{1}$$

The net accumulation of any component in the reactor system can be described mathematically by relating the transport and transformation of its various components through their stoichiometric and rate coefficients. Roels (1980) has presented a systematic method for this type of analysis in homogeneous reactors.

The importance of a systematic mathematical process analysis has become essential in the light of recent experimental observations with fixed film systems. In almost all biofilm studies, with the exception of nitrifying biofilms, the following oversimplified stoichiometry is assumed:

$$\text{substrate} \rightarrow \text{biofilm} \tag{2}$$

The work of Trulear (1983) and Schaftel (1982) clearly indicates that the stoichiometry of the reaction is significantly influenced by other components. A composite of their work would suggest the following stoichiometry:

$$S + E + Z \rightarrow B + PI + PS \tag{3}$$

where S = substrate; E = electron acceptor; Z = other nutrients; B = biofilm mass; PI = insoluble product; and PS = soluble product.

The substrate and electron acceptor may be different in different layers of the biofilm. For example, oxygen may serve as the electron acceptor in the upper layers of a biofilm exposed to aerated media. However, the electron acceptor in the deeper, anaerobic layers may be nitrate, sulfate or an endogenous electron acceptor due to depletion of oxygen by microbial activity as it diffuses through the upper layers of the biofilms. Similarly, the upper layers of biofilm will 'see' the primary substrate while the lower layers may be exposed to products formed in other parts of the biofilm.

54.6 KINETICS

The biofilm is always in a dynamic, nonequilibrium state. Consequently, a stoichiometric analysis alone is not sufficient to model the behavior of a fixed film reactor. The rates of change and the process rates have to be related to the stoichiometry through the mass balance equations. The fundamental process rates discussed in this chapter will include growth, maintenance, product formation, biomass decay, adsorption and detachment. The process rate is characterized by a rate equation which is generally the result of an empirical curve-fitting procedure. The values of the rate coefficients must be determined by experiment, even if the form of the rate expression is suggested by a theoretical analysis or mechanistic model. The determination of the rate expression usually requires a study to determine the influence of concentration followed by the effects of temperature, pH, *etc.* on the reaction rate coefficients.

Rate data for biofilms are best obtained in continuous flow reactors, preferably in a CSTR for which equation (1) applies. Rate of accumulation and rate of transport are rates of change and are measured relatively easily. However, these rates cannot be correlated to fundamental system variables in any meaningful way. Therefore, these rates cannot be extrapolated or used in a predictive manner. The rate of transformation, however, is a process rate which can be used to predict behaviour in systems other than the one in which the data were collected. Generally, process rates can only be inferred from rates of change.

Process rates determine the extent of the process in a reactor with a fixed residence time. Consequently, an estimate of the process rate is necessary for process reactor design. Process rate is perhaps the most important quantity to the reactor design engineer.

The fundamental process rates are difficult to observe directly. The more readily observed process rates are nutrient (substrate) removal rate and electron acceptor (*e.g.* oxygen) removal rate (Table 1). The remainder of this section will discuss observed process rates in terms of the fundamental process rates.

54.6.1 Biomass Production Rate

The overall process of biofilm development is the net result of the fundamental processes of growth, decay, maintenance, product formation, adsorption and detachment.

54.6.1.1 Growth

The specific growth rate, μ, of organisms can be defined as the cellular reproduction rate per total cellular mass per time (t^{-1}) and is, in a biofilm reactor at steady state with negligible suspended activity, measured as (Bakke *et al.*, 1984);

$$\mu = D\frac{xV}{x_bA} \tag{4}$$

where x = suspended cellular biomass concentration (ml^{-3}); and x_b = areal cellular biofilm density (ml^{-2}). Savageau (1969) has shown that growth rate can be related to all of its variables by a power law function. If the system is well defined and accuracy is required, this analysis can be very useful.

A special case of the power function commonly used is the saturation function first presented by Monod (1942) to describe microbial growth. The Monod model assumes that the concentration of one essential nutrient (*i.e.* the substrate) limits the growth of the microorganism as follows:

$$\mu = \frac{\mu_m s}{K_s + s} \tag{5}$$

where μ_m = maximum specific growth rate (t^{-1}); and K_s = saturation coefficient (ml^{-3}). Saturation functions can also describe growth limited by more than one nutrient (Grady and Lim, 1980). The saturation model presumes that the cell is in balanced growth, a condition defined by constant biomass composition and physiological state (Powell *et al.*, 1967). The cells are, in other words, regarded as a homogeneous mass of catalytic agent for substrate conversion. Balanced growth in a chemostat is achieved at steady state. In a batch reactor, balanced growth occurs during exponential growth. In a biofilm, balanced growth cannot be defined in a rigorous manner because of the gradients in concentration, pH, *etc.* that may exist. Modelling unbalanced growth requires structured models.

The Monod model, with its limitation, has been very useful for wastewater treatment despite the variability of wastewater flow rates and composition. One reason for its apparent success may be that steady state may refer to more than 'time-invariant' states. Steady state can refer to a condition in which the derivative with respect to time (net rate of accumulation) is negligible compared to either term on the right hand side of equation (1). In fact, this may be the case of many wastewater treatment processes and may explain the usefulness of the Monod model despite its restrictive underlying assumptions. Low sensitivity requirements may be an additional explanation.

The effects of temperature, pH, substrate quality, *etc.* have to be included in the coefficients K_s and μ_m when applying equation (5).

A general analysis predicting the steady state cellular areal density, x_b, based on the growth rate, μ (equation 4), reveals an inverse proportionality between x_b and μ (Bakke *et al.*, 1984).

54.6.1.2 Product formation

Under balanced growth conditions with only one limiting nutrient, product formation rate has been related to the specific growth rate, μ, and the concentration of organisms, x (Leudeking and Piret, 1959):

$$r_p = k\mu x + k'x \tag{6}$$

where k = growth-associated product formation coefficient $(m_p m_x^{-1})$; and k' = nongrowth-associated product formation coefficient $(m_p m_x^{-1} t^{-1})$.

The first term of the equation refers to product formation related to the metabolic state of the microorganisms. In energy-rich environments, growth rate is high, as is the cellular concentration of metabolic intermediates, which may result in 'leakage' into the bulk fluid. Consequently, the ratio of substrate energy channelled to growth *versus* product formation may be an inverse function of the growth rate.

The second term in equation (6) is independent of metabolic state and is only dependent on biomass concentration. Conceptually, this term may express the result of biomass decay, which has alternatively been modelled as endogenous metabolism, maintenance and/or lysis. The hypothesis is that there is a continuous degradation of cell material resulting in soluble or colloidal organic material which is measured as product. Energy is required to replace and/or 'maintain' the decaying cell components. Trulear (1983) has shown that this simple term can serve as a replacement for biomass decay terms used in other models. The product formation term can be related directly to observations although precision of the coefficents can be a problem (Trulear, 1983). Then k' is a rate coefficient describing the transformation of biomass into soluble organic components. A portion of the soluble organics formed may serve as a secondary substrate for the microorganisms or a source of inorganic nutrients. Figure 3 graphically demonstrates the dependence of r_p on growth rate and biomass concentration as expressed by equation (6).

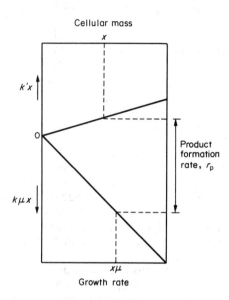

Figure 3 Graphic product formation rate model as described by equation (6)

The product formation rate equation is attractive because it provides a simple method for predicting the soluble organic content of an effluent (substrate and product), a major concern in wastewater treatment. The rate equation (equation 6) may be unable to predict a specific product concentration because the two terms may not express the rate of formation of the same product. A more structured model is required for such predictions.

54.6.1.3 *Extracellular polymer substances (EPS)*

At steady state, the EPS formation rate, r_{EPS}, can be measured as (Bakke *et al.*, 1984):

$$r_{EPS} = D\frac{pV}{p_bA} \tag{7}$$

where p = bulk liquid EPS concentration (ml^{-3}); and p_b = areal biofilm EPS density (ml^{-2}). r_{EPS} has been related to the growth rate, μ, and the areal density of organisms, x_b, using equation (6) (Trulear, 1983).

r_{EPS} variations measured according to equation (7) but not consistent with equation (6) have, however, been observed (Bakke, 1983). Low precision in measured coefficient values (Trulear, 1983) and ignorance of qualitative EPS variations also contribute to the shortcomings of equation (6). The development of a better model in which these limitations have been eliminated depends on increased understanding of the physiological significance of EPS.

54.6.1.4 Net detachment

If the reactor influent contains suspended or particulate material, only net detachment can be discussed. The detachment processes can be classified into two categories: (1) sloughing, which refers to massive removal of large pieces of biofilm which happens at seemingly random intervals; and (2) shearing, which is the continuous removal of microscopic portions of biofilm.

Sloughing is usually observed in nutrient-rich environments and at relatively low Reynolds number. Under these conditions, biofilm thickness can become quite large resulting in oxygen-deficient and/or substrate-deficient layers deep within the biofilm. Howell and Atkinson (1976) have proposed a model which assumes that biomass decay in the deep, substrate-deficient layers of the biofilm causes sloughing.

Bakke (1983) has observed EPS desorption associated with substrate flux changes in the bio-film. For example, an RBC plant may exhibit sloughing if anaerobic digester supernatant is periodically pulsed to the system. Bakke hypothesizes that EPS, involved in diffusion regulation through biofilm density changes during load transitions, are removed from the biofilm by fluid shear forces. The magnitude of the EPS desorption has been correlated to proton flux changes across the cell membrane in symport with the substrate. An electromagnetic feedback diffusion regulation mechanism between the cell membrane and the EPS has been proposed (Bakke, 1983).

Shearing has been observed in relatively oligotrophic environments and at high Reynolds number. Shearing is probably occurring under the same conditions in which sloughing occurs but no such direct observations have been attempted. Shearing is generally attributed to fluid shear forces (Trulear and Characklis, 1982; Zelver, 1979). As long as fluid shear forces prevent the bio-film thickness from exceeding the depth of substrate penetration, detachment will significantly influence biofilm thickness. Hence, a model for predicting biofilm thickness will include a term for biofilm detachment rate which depends on fluid shear stress, σ, biofilm volume, thA, and biofilm density, ϱ:

$$r_d = k_d \sigma thA \varrho \tag{8}$$

where k_d = detachment rate coefficient.

Trulear and Characklis (1982) present data indicating that detachment rate increases with increasing shear stress and increasing biofilm mass. Rittman (1982) has used their data to predict detachment rates in trickling filters and fluidized bed reactors.

54.6.1.5 Biomass decay

Biomass decay is a general term referring to the effects of endogenous metabolism, maintenance, autooxidation, and other ill-defined processes that contribute to the attrition of biomass. Biomass decay rate, like maintenance rate, is proportional to biomass concentration:

$$r_{decay} = k_{de}x \tag{9}$$

where k_{de} = biomass decay rate.

Consequently, biomass decay rate will increase with increasing biofilm thickness and/or density. Bryers and Characklis (1982) report that biomass decay is negligible compared to biomass production processes in thin biofilms. Rittman (1982) has also calculated biomass decay rate in a fixed film system.

54.6.2 Substrate Removal Rate

The overall, r_s, and specific, r_s^*, substrate removal rates for biofilm reactors can be defined as follows (Bakke *et al.*, 1984):

$$r_s = D(s_i - s)\frac{V}{A} \qquad (ml^{-2}t^{-1}) \tag{10}$$

$$r_s^* = D\frac{(s_i - s)}{x_b}\frac{V}{A} \qquad (m_s m_x^{-1}t^{-1}) \tag{11}$$

where D = dilution rate (t^{-1}); s_i = inflow substrate concentration (ml^{-3}); s = bulk liquid substrate concentration (ml^{-3}); V = reactor volume (l^3); A = reactor surface area (l^2); and x_b = biofilm cellular areal density (ml^{-2}).

Microorganisms consume substrate for two basic purposes: (1) to provide energy for various processes such as synthesis of biomass and products as well as maintenance, and (2) to provide materials for synthesis of cell mass and other products. The kinetic expressions describing substrate removal rates by the various fundamental processes will be discussed in the following sections.

54.6.2.1 Growth

The Monod model was developed to describe dispersed microbial growth limited by one essential nutrient. The model has been extended to describe substrate removal by biofilms by presuming it proportional to growth rate through the yield coefficient:

$$r_s = \mu \frac{x}{Y} \tag{12}$$

where Y = yield coefficient ($m_x m_s^{-1}$).

The yield coefficient is a ratio of stoichiometric coefficents describing the biomass formed from the removal of a unit mass of substrate. Intrinsically, this presumes that substrate is only used for biomass formation (*i.e.* synthesis). Trulear (1983) however has measured as much as 90% extracellular organic material in biofilms, formed from primary substrate. Schaftel (1982) documented the formation of soluble product formation by biofilms and their transport into the bulk reactor fluid, even when the biofilm was entirely aerobic.

54.6.2.2 Product formation

Product formation can be accounted for by incorporating equation (6) in the material balance for substrate. The specific substrate removal rate, r_s^*, can be calculated as follows (Trulear, 1983; Bakke *et al.*, 1983):

$$r_s^* = \mu(\frac{1}{Y_{x/s}} + \frac{k}{Y_{p/s}}) + \frac{k'}{Y_{p/s}} \tag{13}$$

where $Y_{x/s}$ = yield of biomass from substrate ($m_x m_s^{-1}$); and $Y_{p/s}$ = yield of products from substrate ($m_p m_s^{-1}$).

Substituting numerical values for the coefficients obtained for *Ps. aeruginosa* grown on dextrose in chemostats (Trulear, 1983) yields:

$$r_s^* = 3.5\mu + 0.6 \tag{14}$$

Data obtained for *Ps. aeruginosa* grown on glucose at steady state in biofilms and dispersed in chemostats correlated well with equation (14). The biofilm cellular areal density, x_b, in these experiments was proportional to the substrate flux into the biofilm, r_s (equation 10), at constant growth rate (Bakke *et al.*, 1984).

Too frequently, substrate removal rate is limited by mass transfer of substrate or oxygen from the bulk fluid to the biofilm or diffusion of substrate or oxygen in the biofilm. Under these circumstances, equation (13) must be modified (see below).

54.6.2.3 Maintenance

The rate of substrate removal for maintenance functions is presumed to be proportional to biomass concentration:

$$r_m = k_m x_b \tag{15}$$

Bryers and Characklis (1981) and Stathopoulos (1981) have estimated maintenance coefficients for biofilms which compare favorably to those determined in dispersed microbial growth systems.

54.6.2.4 Adsorption

Adsorption of particulate substrate, colloidal substrate and microorganisms significantly influences the performance of a fixed film reactor. The rate expressions describing each of these adsorption processes should appear at least twice in a fixed film reactor model. For example, adsorption of particulate substrate will appear as substrate removal in the substrate material balance and will contribute to increased biofilm accumulation in the biofilm constitutive equation. The adsorbed substrate will be degraded with time, at a rate probably controlled by its hydrolysis.

The net result of adsorption/desorption can be described mathematically as follows:

$$r_A = k_A x (1 - \frac{x'}{k'}) \tag{16}$$

Measurements which discriminate between adsorption and degradation of particulate substrate are difficult but have been accomplished (Banerji *et al.*, 1968). Most observations do not distinguish between adsorption and removal since removal is the major concern. However, colloidal substrate may adversely influence removal of soluble substrate (Särner, 1981).

Substrate removal rate may be mass transfer-limited when the substrate is particulate or colloidal. Mass transfer rates may be determined although predicting mass transfer coefficients may be difficult. Substrate degradation rates, however, are relatively slow compared to transport rates.

In summary, substrate removal rate is the net result of four fundamental processes: growth, maintenance/decay, product formation and adsorption. Under different operating conditions, one process may dominate over the others. For example, product formation may be the dominant process in a system with a very low substrate concentration. The plant engineer, ignorant of the fact that the unsatisfactorily high effluent carbon concentration is due to product formation rather than unreacted substrate, mistakenly presumes that increasing retention time will improve effluent quality. In most fixed film research studies related to wastewater treatment, substrate removal attributed to the various fundamental processes is not considered and, as a result, extrapolations of laboratory and pilot scale kinetic studies are often inaccurate and misleading.

54.6.3 Physical Processes Influencing Substrate Removal

Mass transfer, diffusion and detachment all significantly influence the overall performance of a fixed film reactor. The influence of mass transfer and diffusion may be incorporated into the substrate removal rate expression. Net detachment is an additive term in the constitutive equation for the biofilm and also appears in the mass balance for suspended biomass.

54.6.3.1 Mass transfer

The transport of any soluble component from the bulk fluid to the biofilm can be represented by the following:

$$N = k_f(C_b - C_i) \tag{17}$$

where k_f = mass transfer coefficient (lt^{-1}); C_b = bulk fluid concentration (ml^{-3}); and C_i = interfacial concentration (ml^{-3}). This model presumes the two film theory (Sherwood *et al.*, 1975) for interphase transport is valid for the biofilm–fluid system. Empirical relationships are available for calculating the mass transfer coefficient, k_f, as a function of Reynolds number in various geometries (Sherwood *et al.*, 1975). One of the difficulties arises as a result of the nature of the interfacial surface. Most researchers prefer to depict a well-defined interface between the bulk fluid and the biofilm. In fact, some fluid may very well move through the biofilm or portions of the biofilm so that no clear interface exists. In any case, the biofilm does exhibit hydraulic roughness as indicated by the results of Picologlou *et al.* (1980). The biofilm roughness, however, is highly dependent on the microbial species within the biofilm (Trulear and Characklis, 1982). For example, *Sphaerotilus natans* is a filamentous organism which exerts a significant influence on biofilm roughness while *Pseudomonas aeruginosa* exerts little influence on biofilm roughness in its rod-shaped, single cell morphology (Trulear and Characklis, 1982; McCoy and Costerton, 1982).

Process rates in trickling filters, rotating biological contactors and fluidized beds can be rate-limited by transport in the bulk fluid. Detecting this type of rate limitation requires varying the Reynolds number or intrinsic rate coefficients for the processes occurring in the biofilm.

54.6.3.2 *Diffusion in the biofilm*

Rate of electron donor (substrate) or electron acceptor removal may be rate-limited by diffusion of reactants into the biofilm or diffusion of products out of the biofilm. For example, the rate expression for substrate removal may include an effectiveness factor as described by Atkinson and his coworkers (1974) to account for diffusion in the biofilm:

$$N = \frac{\lambda k_1 s x_b}{K_s + S} \tag{18}$$

where λ = effectiveness factor; and k_1 = specific substrate removal rate ($m_s m_x^{-1} t^{-1}$).

Equation (18) is a modified version of the expression for substrate removal in equation (12). Trulear (1983) has modified Atkinson's rate expression to account for product formation in the biofilm as presented in equation (13).

By combining equations (18) and (13), an expression for substrate removal by biofilms is obtained which includes growth, product formation and the influence of diffusion.

Any model of this type suffers due to the difficulty in measuring accurate, meaningful diffusion coefficients in biofilms. Even defining a diffusion coefficient is difficult when a colligative measure of substrate is used. For these cases and other practical applications, the effectiveness factor best serves as a coefficient for modifying the substrate removal rate for the effects of diffusion.

54.6.4 Other Factors Influencing Kinetics

The kinetic expressions presented in the preceding sections are incomplete and suffer from many limitations. From a fundamental standpoint, additional factors may be important and should be considered.

Little is known about microbial population dynamics in a fixed film reactor despite its apparent influence on reactor performance. For example, Michels *et al.* (1979) and Otto *et al.* (1980) have shown that an organism consuming primary substrate may grow 50% faster due to efflux energy generated by organisms consuming a secondary substrate (a metabolite) excreted by the first species. Predation has been studied and has been blamed for total failure of reactors.

The full significance of extracellular polymer substances (EPS) is not known. However, EPS surely influences diffusion rates within the biofilm, adsorption rates within the biofilm and the physiological state of the organisms in the biofilm (Joyce and Dugan, 1970; Klimek and Ollis, 1980; Dudman, 1977).

54.6.5 Summary of Kinetics

Many fundamental processes are recognized as contributing to the overall performance of a fixed film reactor. In some cases, useful, quantitative rate expressions for fundamental processes are not available. A complex model, however, has little immediate relevance to wastewater treatment. Yet continued attention to fundamental aspects of the process may result in improved operation and design in future, more demanding applications.

An illustration may be instructive. Consider the fixed film CSTR in Figure 4. Steady state material balances for substrate, product and suspended cellular biomass are listed, as well as a constitutive equation for the cellular biofilm mass. Easily measured quantities are presented on the left hand side of the equations. The easily measured quantities are also the quantities of most concern to the design and operations engineer. Nevertheless, the right hand sides of the equations indicate the process rates that determine the performance of the system.

The processes discussed above occur in all fixed film reactor systems. The methods used by the observer will determine whether the fundamental processes can be resolved in any given study. For example, if sewage is used as substrate, a colligative measure of substrate is necessary and product formation cannot be considered. Nevertheless, knowing that product may be formed by the biofilm may influence interpretation of the results and subsequent decisions regarding the process.

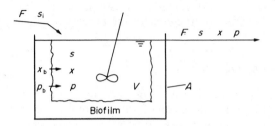

Material balances

Substrate, s

$$\frac{F(s_i - s)}{A} = \frac{r_g}{Y_{x/s}} + \frac{r_p}{Y_{p/s}}$$

Products, p

$$\frac{Fp}{A} = r_p$$

Suspended cellular biomass, x

$$\frac{Fx}{A} = r_d$$

Constitutive equation

Cellular biofilm mass, x_b

$$0 = r_g - r_d$$

where r_g = cellular growth rate $(m_x l^{-2} t^{-1})$
r_p = product formation rate $(m_p l^{-2} t^{-1})$
r_d = cellular detachment rate $(m_x l^{-2} t^{-1})$

Figure 4   A steady state fixed film CSTR analysis

54.7   REACTOR ANALYSIS

Reactor configuration determines the overall liquid phase concentration profiles and transport within the reactor. In addition, of particular importance in fixed film reactors, reactor geometry determines the Reynolds number and fluid shear stress at the biofilm–fluid interface. In this section, liquid phase concentration profiles will be related to reactor configurations of practical interest. Then the flux of components across system boundaries will be related to overall reactor performance.

Two fixed film reactor types will be considered: (1) continuous flow stirred tank reactor (CSTR); and (2) plug flow reactor (PFR). Figure 5 presents schematic diagrams of these two reactors and their characteristic liquid phase concentration profiles. The effluent concentration of a CSTR is the same as the reactor concentration since no spatial gradients exist, *i.e.* the reactor is completely mixed. In the PFR, there is no mixing in the longitudinal direction but mixing is perfect in the radial direction. Consequently, a unit of fluid travels through the reactor as a 'plug'.

Trickling filters without significant recycle flow can be modelled as PFRs while RBCs and fluidized beds are modelled as CSTRs. Any of the CSTRs can be converted to a PFR system by operating several of them in series. A CSTR system can be achieved with a PFR by applying a high recirculation rate. The remainder of this analysis presumes ideal CSTRs and PRFs realizing the limitations imposed by the assumption. At steady state in these reactors, the material balance simplifies to the following:

$$\text{net transport out of reactor} = \text{transformation or reaction} \tag{19}$$

The distinction between the PFR and CSTR occurs on the left hand side of equation (19). Two types of transport are possible: (1) interphase transport, as illustrated by oxygen transfer from air to the wastewater, and (2) bulk transport of materials, as illustration by the input of substrate (dissolved or suspended) in the wastewater influent.

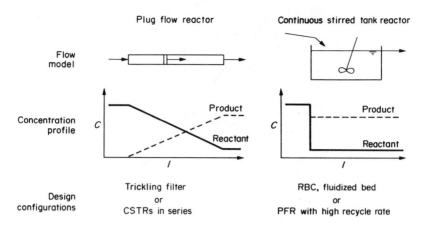

Figure 5 Comparison of PFRs and CSTRs regarding flow scheme, bulk liquid concentration profiles and design configurations

54.7.1 Residence Time

Bulk transport is proportional to the volumetric flow rate across the system boundaries and can be expressed as follows:

$$D = F/V \tag{20}$$

The reactor liquid residence time, τ, is the inverse of D and is useful for comparing reactor configurations. At constant reactor volume and volumetric feed rate, τ required can be calculated from influent and effluent concentrations (C_i and C, respectively) and the observed reaction rate:

$$\text{for a CSTR} \quad \tau_{CSTR} = \frac{C_i - C}{r} \tag{21}$$

$$\text{for a PFR} \quad \tau_{PFR} = \int_{C_i}^{C} \frac{dC}{-r} \tag{22}$$

where r = reaction rate, *e.g.* $r = k_r C^n$ $(ml^{-3}t^{-1})$; k_r = reaction rate coefficient $(l^{3(n-1)}m^{(1-n)}t^{-1})$; n = reaction order (dimensionless); C_i = influent concentration (ml^{-3}); and C = effluent concentration (ml^{-3}). For zero order reaction, $n = 0$ and

$$\tau_{CSTR} = \tau_{PFR} \tag{23a}$$

while for all other reactions, $n > 0$ and

$$\tau_{CSTR} > \tau_{PFR} \tag{23b}$$

Therefore, all non-zero order reactions will require a larger CSTR reactor volume, as compared to a PFR, to achieve the same conversion (Figure 6). This is, however, quite an oversimplification for fixed film biological reactors as discussed below.

Figure 6 illustrates the advantage of the PFR at high reaction orders and when high effluent quality requirements exist. Other aspects, however, must be considered. For example, PFR process performance may be influenced by (1) overloading the influent region and (2) product inhibition at the exit region. Such problems do not exist in a CSTR where dilution of reactants and products occurs (Figure 5). The advantage of dilution is especially apparent in systems experiencing frequent and dramatic shock (transient) loads resulting in nutrient and/or electron acceptor (*e.g.* oxygen) limitations. These types of limitations frequently result in formation of slowly degradable products and contribute to poor effluent quality.

The 'pyramid' mode of wastewater treatment system design and operation illustrates the advantages of a combined CSTR/PFR reactor configuration. The pyramid system consists of CSTRs in series with reactor size decreasing along the reactor train length (Figure 7). Desired

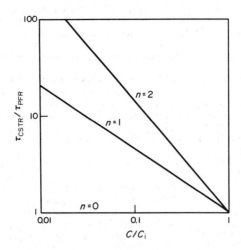

Figure 6 Comparison of reactor detention time, τ, required in a CSTR relative to a PFR to convert reactant C from inflow concentration C_i to a required effluent concentration C for various reaction kinetic orders, n

CSTR volume is commonly obtained by connecting equal size modules in parallel. The first, large CSTR dilutes the substrate while providing high nutrient and electron acceptor transfer capacity thereby ensuring desirable environmental conditions for metabolism. Each succeeding CSTR receives a smaller substrate load and, therefore, requires less dilution, nutrients and electron acceptor. The size of each stage is determined from a process analysis employing principles described in previous sections. The number of stages can be calculated from the same process analysis applied in a PFR system.

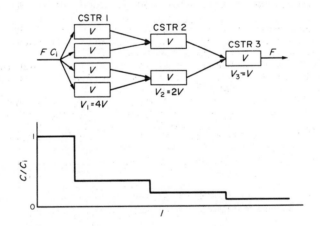

Figure 7 The pyramid mode of wastewater treatment system consisting of CSTRs in series with reactor size decreasing $(V_1 > V_2 > V_3)$ along the reactor train length, l

In summary, the reactor analysis is conceptually relatively simple and is based on traditional techniques of reactor engineering. In conjunction with a thorough kinetic analysis, the reactor may lead to the optimum configuration for the fixed film reactor system.

54.8 REFERENCES

Atkinson, B. (1974). *Biochemical Reactors*. Pion, London.
Bakke, R. (1983). Dynamics of biofilm processes: substrate load variations. Thesis submitted to Montana State University, Bozeman, MT, in partial fulfillment of the degree of Master of Science.
Bakke, R., M. G. Trulear, J. A. Robinson and W. G. Characklis (1984). Activity of Pseudomonas aeruginosa in biofilms: steady state. *Biotechnol. Bioeng.*, **26**, 1418–1424.
Banerji, S. K., B. B. Ewing, R. S. Englebrecht and R. E. Speece (1968). Mechanism of starch removal in the activated sludge process. *J. Water Pollut. Control. Fed.*, **40**, 16–29.

Bryers, J. D. and W. G. Characklis (1981). Early fouling biofilm formation in a turbulent flow system: overall kinetics. *Water Res.*, **15**, 483–491.

Bryers, J. D. and W. G. Characklis (1982). Process governing primary biofilm formation. *Biotechnol. Bioeng.*, **24**, 2451–2476.

Characklis, W. G., J. D. Bryers, M. G. Trulear and N. Zelver (1982). Dynamics of biofilm processes: methods. *Water Res.*, **16**, 1207–1216.

Dudman, W. F. (1977). Surface polysaccharides in natural environments. In *Surface Carbohydrates of the Prokaryotic Cell*, ed. I. W. Sutherland, vol. 1, chap. 9, pp. 357–414. Academic, New York.

Grady, L., Jr. and H. C. Lim (1980). A conceptual model of rotating biological contactor performance. *Proceedings of the First National Symposium/Workshop on Rotating Biological Contactor Technology*, vol. 2, pp. 829–860.

Hoehn, R. C. and A. D. Ray (1973). Effects of thickness on bacterial film. *J. Water Pollut. Control Fed.*, **46**, 2302–2320.

Howell, J. A. and B. Atkinson (1976). Influence of oxygen and substrate concentration on the ideal substrate uptake rate in microbial film fermenters. *Biotechnol. Bioeng.*, **18**, 15–35.

Joyce, H. G. and P. R. Dugan (1970). The role of floc-forming bacteria in BOD removal from waste water. *Dev. Ind. Microbiol.*, **2**, 376–386.

Klimek, J. and D. F. Ollis (1980). Extracellular microbial polysaccharides. *Biotechnol. Bioeng.*, **22**, 2321–2342.

Kornegay, B. H. and J. F. Andrews (1968). Kinetics of fixed-film biological reactors. *J. Water Pollut. Control Fed., Res. Suppl.*, **40**, 460–468.

Lohmeyer, G. T. (1957). Trickling filters and operation tips. *Sewage Ind. Wastes*, **29**, 89–94.

Leudeking, R. and E. L. Piret (1959). A kinetic study of the lactic acid fermentation. *J. Biochem. Microbiol. Technol. Eng.*, **1**, 393–431.

McCoy, W. F. and J. W. Costerton (1982). Growth of sessile *Sphaerotilus natans* in a tubular recycle system. *Appl. Environ. Microbiol.*, **43**, 1490–1494.

Michels, P. A. M., J. P. J. Michels, J. Boonstra and W. N. Konings (1979). Generation of an electrochemical proton gradient in bacteria by the excretion of metabolic end products. *FEMS Microbiol. Lett.*, **5**, 357–364.

Monod, J. (1942). *The Growth of Bacterial Cultures*. Herman, Paris.

Otto, R., J. Hugenholtz, W. N. Konings and A. Veldkamp (1980). Increase of molar growth yield of *Streptococcus cremoris* for lactose as a consequence of lactate consumption by *Pseudomonas stutzeri* in mixed culture. *FEMS Microbiol. Lett.*, **9**, 85–88.

Picologlou, B. F., N. Zelver and W. G. Characklis (1980). Biofilm growth and hydraulic performance. *J. Hydraul. Div., Am. Soc. Civ. Eng.*, **106**, 733–746.

Powell, E. O., C. G. T. Evans, R. E. Strange and D. W. Tempest (eds.) (1967). Editorial comment in *Microbial Physiology and Continuous Culture*, pp. 259–261. HMSO, London.

Rittman, B. E. (1982). The effect of shear stress on biofilm loss rate. *Biotechnol. Bioeng.*, **24**, 501–506.

Roels, J. A. (1980). Application of macroscopic principles to microbial metabolism. *Biotechnol. Bioeng.*, **22**, 2457–2514.

Särner, E. (1981). Removal of dissolved and particulate organic matter in high-rate trickling filters. *Water Res.*, **15**, 671–678.

Savageau, M. A. (1969). Biochemical system analyisis, I and II. *J. Theor. Biol.*, **25**, 365–379.

Schaftel, S. O. (1982). Process of aerobic/anaerobic biofilm development. Thesis submitted to Montana State University, Bozeman, MT, in partial fulfillment of the degree of Master of Science.

Sherwood, T. K., R. L. Pigford and C. R. Wilke (1975). *Mass Transfer*. McGraw-Hill, New York.

Stathopoulos, N. (1981). Influence of temperature on biofilm processes. Thesis submitted to Rice University, Houston, TX, in partial fulfillment of the degree of Master of Science.

Trulear, M. G. (1983). Cellular reproduction and extracellular polymer formation in the development of biofilms. Dissertation submitted to Montana State University, Bozeman, MT, in partial fulfillment of the degree of Doctor of Philosophy.

Trulear, M. G. and W. G. Characklis (1982). Dynamics of biofilm processes. *J. Water Pollut. Control Fed.*, **54**, 1288–1301.

Watt, J. G. and C. J. Cahill (1980). Wastewater treatability studies for a grassroots chemical complex using bench scale rotating biological contactors. *Proceedings of the First National Symposium/Workshop on Rotating Biological Contactor Technology*, vol. 1, pp. 661–690.

Zelver, N. (1979). Biofilm development and associated energy losses in water conduits. Thesis submitted to Rice University, Houston, TX, in partial fulfillment of the degree of Master of Science.

55

High Rate Filters

K. E. PORTER
University of Aston in Birmingham, UK

55.1 INTRODUCTION

High rate filters are used for the treatment of aqueous effluents containing organic pollutants. The polluting material is removed by aerobic microorganisms which grow as a slime on the solid surfaces within the filter, over which the aqueous effluent flows. This mechanism of removing pollution in high rate filters is the same as that which occurs in trickling filters filled with mineral media such as rock or stones. The essential difference between these two devices is that, in high rate filters, the open spaces between the solid surfaces are much larger than those in the traditional rock-filled filter beds. This means that high rate filters may be used at much higher organic loadings without becoming blocked up by the slime or biomass which grows on the solid surfaces. The usual organic loading of 3 kg BOD m^{-3} d^{-1} used with a high rate filter is about 20 times higher than that usually used with a rock- or stone-filled trickling filter.

Filter beds filled with rocks, stones or coke have been used for many years. The size of the pieces of rock or stone is usually between 20 and 100 mm. The earliest high rate filters were developed simply by using much larger pieces of rock so that the void spaces were larger, but this also meant that the solid surface area per unit volume of filter was smaller. To overcome this limitation, specially designed plastic media have been developed so as to provide a combination of high surface area with relatively large spaces between the solid surfaces. All types of plastic media are similar in that the total voidage is large (about 95%) compared with that of stone media (about 25%). Plastic media are also much lighter than mineral media and are usually used in beds more than 3 m deep, depths up to 7 m having been used.

Thus over the last two decades the use of plastic media-filled high rate filters has become more and more common. This chapter describes examples of plastic media and of different designs of high rate filters. A brief review of empirical design methods is then given, and this is followed by a justification for the form of the empirical correlations by an interpretation in terms of diffusion accompanied by a biochemical reaction.

It is convenient to start by first describing how filters and other aeration devices remove dissolved organic pollutants from water.

55.2 AERATION AS A MEANS OF REMOVING WATER POLLUTION

55.2.1 Filters Compared with Other Devices

A general arrangement of almost any aqueous effluent treatment plant is shown in Figure 1. This shows both types of aeration device used in water pollution control, namely the packed filter bed type (which is the subject of this chapter) and the activated sludge type. In the filter type, oxygen is transferred from the air in the void spaces in the media-packed bed to the polluted water which trickles over the media surfaces. The activated sludge type of device contains a deep pool of polluted water. Oxygen is transferred to the water from air which is bubbled or stirred through the pool. Thus both filters and activated sludge devices are methods of aerating water.

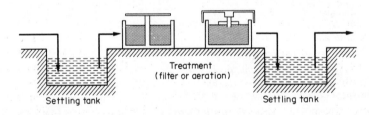

Figure 1   General arrangement of a wastewater treatment plant

The oxygen which is transferred to the water supports living microorganisms which feed on the organic polluting matter. That part of the polluting matter which is dissolved in the water is thus either transformed into the creature which has consumed it (*i.e.* it becomes a solid) or into energy and carbon dioxide.

As shown in Figure 1, solids settling tanks are provided both upstream and downstream of the aeration device. The upstream settling tank removes solids from the water entering the plant. This usually reduces the organic load on the aeration device. The downstream settling tank removes the solids which are produced in the aeration device.

Thus in the aeration device, polluting material which is dissolved in the water is changed into solids which may be settled out. This is how the pollution is removed. It may be noted at this point that if the polluting material is not consumed by microorganisms in an effluent treatment plant, then it will eventually be consumed by similar processes in the rivers, lakes and streams which it pollutes. This results in a removal of oxygen from the water and causes other living things, such as fish, to die. For this reason, the concentration of organic polluting matter is always measured by its 'Biological Oxygen Demand', or BOD.

In filter beds, the microorganisms which consume the organic pollutant grow as a film over the surfaces of the media. This film is also known as 'slime' or generally as 'biomass' or 'bios'. It has considerable importance for understanding how filters work. This is because: (a) the biomass (not the media) removes the BOD; (b) the biomass is continually sloughing, *i.e.* pieces of it break away from the media surface and are washed out of the filter, becoming the solid product which must be settled out; (c) the biomass may grow so thick that the filter becomes completely blocked, circulation of air within the filter is prevented, the bed becomes anaerobic and it no longer works.

55.2.2 The Growth of Biological Film on the Surface of the Media

The film is alive and continually growing. In order for it to stay alive, both food (*i.e.* the polluting material) and oxygen must diffuse through it to reach all its parts. The food and oxygen diffuse into the film through its surface in contact with the water. They thus diffuse through a medium which consumes them, so that the further from the film surface they diffuse, the lower their concentration becomes. This is illustrated in Figure 2.

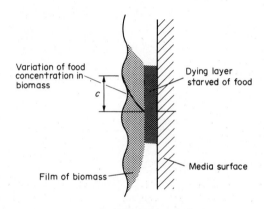

Figure 2 Film of biomass on media surface

At a great enough distance into the film (which is growing thicker all the time) their concentration becomes zero. Microorganisms at greater distances than this become starved and die, and then adhesion to the solid surface is weakened, the film breaks away, and a piece of solid film is washed through the bed of media. New film then grows again on the media surface, getting thicker and thicker until the food (or oxygen) can no longer diffuse right through to those parts near the media surface which die, and the cycle is repeated. This description is somewhat oversimplified. It ignores the effect of snails and worms which graze upon the bios, thus limiting the rate of growth and the way in which living film, if it gets too thick, may be broken away by the weight of the water passing through it. Nevertheless, this oversimplified picture helps to explain why high rate filters are particularly competitive in certain applications.

55.3 APPLICATIONS AND PERFORMANCE

The main applications of high rate filters are as follows. (1) The partial treatment of industrial wastewaters of high BOD concentration. Table 1 gives typical BOD values for various industrial wastes and Table 2 shows typical removal efficiencies. The plastic media Flocor and Biopac, mentioned in Table 2, are described below. (2) The partial treatment of wastewaters which are difficult to treat in an activated-sludge plant, either because of a fluctuating loading or because of the unsatisfactory settling characteristics of the sludge. (3) To increase the capacity of an existing sewage works by providing partial 'roughing' treatment upstream of the main works.

The plastic media used in high rate filters are relatively expensive and they are thus most competitive in applications where the entire surface of the media is covered by a thick growth of biomass. This is achieved by a high organic loading and only partial treatment of the waste. In order to achieve a reasonable thickness of film on the surfaces at the bottom of the filter, it is necessary that some BOD still remains in the liquid at that point.

Table 1   Typical BOD Concentrations for
Various Wastewaters

Type of waste	BOD (mg l^{-1})
Abattoir waste	2000
Oil refinery wastewater	77–200
Brewery waste	600–900
Milk waste	600–1200
Domestic effluent	300

Table 2   Examples of Organic Loadings and Removal
Efficiencies

Type of waste	Medium	Organic load (kg m^{-3} d^{-1})	Percent removal of BOD
Brewery + domestic	Flocor	3.2	60
Abattoir + domestic	Flocor	3.3	65
Abattoir	Flocor	1.5–2.0	60
Poultry processing	Flocor	1.5	75
Industrial + domestic	Biopac	3.0	66
Rubbish tip run off	Biopac	2.3	70
Oil refinery	Biopac 50	2.0–2.5	70–80

55.4   FLOW SHEETS

The flow sheets of Figures 3, 4 and 5 show high rate filters followed by settling tanks. Not shown is the settling tank which usually is installed upstream of the filter. Figures 4 and 5 show the use of filters with recycle of some of the liquid which has passed through the filter back to blend with the liquid fed on to the filter. The need for settling tanks and recycle may be explained as follows.

55.4.1   Settling Tanks

The upstream settling tank removes solids from the water entering the plant. This reduces the organic load on the filter. The downstream settling tank (shown in Figures 3–5) removes solids which are produced in the filter.

55.4.2   Recycle

From the description of the diffusion of food into the slime given above, it may be accepted that the greater the pollutant concentration, the greater the distance into the film it may be expected to diffuse. That is, in general, the higher the BOD of the wastewater, the thicker the film it will support. If the film becomes too thick it will block the filter, thus preventing air from circulating. Anaerobic regions will form and the filter performance will deteriorate. This is known as 'ponding'. Thus it is desirable to limit the concentration of BOD entering the bed, and this can be done by recycling treated water back to dilute the filter feed.

Recycle is also used to improve liquid distribution, particularly with the high BOD wastes. At the same organic load (measured in kg BOD m^{-3} d^{-1}), an increase of BOD concentration means a reduction in hydraulic load (m^3 water m^{-3} d^{-1}).

With recycle, the entering BOD concentration is reduced, the hydraulic load is increased and an improvement both in liquid distribution and in the effectiveness of wetting of the media surfaces is obtained.

55.4.2.1 Single stage without recycle (Figure 3)

This is used for the removal of up to about 70% of the BOD and for low-to-medium strength wastes. For example, for the roughing treatment of domestic sewage (BOD about 300 mg l⁻¹ or 300 g m⁻³).

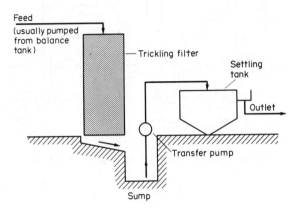

Figure 3 Single-stage filter plant without recycle

55.4.2.2 Single stage with recycle (Figure 4)

This is used for the roughing treatment (up to 70% removal) of more concentrated wastes.

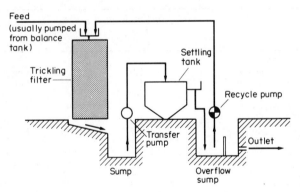

Figure 4 Single-stage filter plant with recycle

55.4.2.3 Two stages with recycle (Figure 5)

This is sometimes used when it is required to remove a large proportion of the polluting BOD from a highly concentrated wastewater. Lower surface area media with large sized void spaces may be used in the first stage to avoid the possibility of ponding. Media of larger surface area per unit volume may be used in the second stage to reduce the BOD to a low level.

55.5 CONSTRUCTION OF HIGH RATE FILTER BEDS

High rate filters are sometimes quite large, as much as 30 m in diameter and 4 m high. Each high rate filter consists of: plastic media, media support and drainage channels, a retaining wall and liquid distributors.

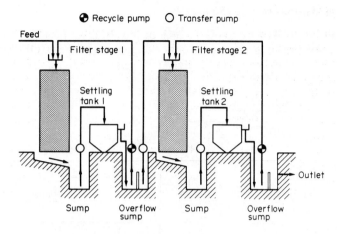

Figure 5   Two-stage filter plant with recycle on both stages

55.5.1   Plastic Media

There are many different shapes of media available from different suppliers. They may be classified in terms of shape as: (a) vertical tube media; (b) vertical sheet media; (c) random media. The properties of various examples of plastic media claimed by the respective manufacturers are given in Table 3.

Table 3   Properties of Plastic Media

	Surface area claimed ($m^2\,m^{-3}$)	Void space (%)	Material
Random Media			
Norton Actifil 90E	101	95	Polypropylene
Norton Actifil 50E	124	92	Polypropylene
Norton Actifil 75	160	92	Polypropylene
MT Filterpak 1127	120	93	Polypropylene
MT Filterpak 1130	190	93	Polypropylene
MT No. 2 Mini Ring	118	93	Polypropylene
MT No. 3 Mini Ring	79	94	Polypropylene
Sheet Media			
ICI Flocor E	90	95	PVC
ICI Flocor M	135	95	PVC
Munters Plasdek B27060	100	95	PVC
Munters Plasdek B19060	140	95	PVC
Munters Plasdek B12060	230	95	PVC
Tube Media			
Cloisonyle	220	94	PVC

55.5.1.1   Vertical tube media

'Cloisonyle' is a vertical tube media, the tubes being placed in the bed with the longitudinal axes vertical. The outside diameter of each tube is 80 mm and the tube is divided internally so as to form a honeycomb structure of 14 small 'tubes' each about 15 mm square. A large surface area is therefore available to support the growth of biomass. The media is usually made of unplasticized PVC and the outer wall thickness is 0.7 mm. The length of the tube is made equal to the height of the bed, which usually varies between 4 and 6 m.

55.5.1.2 Vertical sheet media

There are many different types of sheet media, several of which are referred to in Table 3. Flocor is perhaps the most widely used sheet media and the description of Flocor which follows is typical of sheet media in general. Flocor is made from poly(vinyl chloride) (PVC) sheets which are ridged and corrugated, and this spaces the sheets apart, provides additional surface area and helps distribution of the liquid. The sheets are stuck together to form rectangular blocks, size 1.2 m × 0.6 m × 0.6 m. Flocor may be transported as separate sheets which are made into blocks on site, or alternatively the blocks may be transported complete. The blocks are placed one above the other, making sure that each layer is at right angles to the layer beneath. Bed heights of at least 3 m have been used without intermediate support (Chipperfield, 1967).

The surface area per unit volume of media depends on the spacing of the sheets, which can vary between 40 mm and 60 mm.

55.5.1.3 Random media

A more recent development has been the use of random media such as Actifil and Filterpak. Actifil was developed from Biopac (which was the first random media to be used in the UK) and is typical of random media. The Actifil media elements, made of polypropylene, are right cylinders with perforated walls and both internal and external ribs or fins. These provide additional surface area and slightly space the elements apart, which assists the flow of liquid through the bed. The two most commonly used sizes of Actifil are Actifil 90, each piece of which is 90 mm long and 90 mm in diameter, and Actifil 50, each piece of which is 50 mm long and 50 mm diameter.

Random media are delivered to the site and simply dumped into the filter. There are several efficient ways of quickly filling a large bed. These include using a crane to dump a container load at a time, the use of conveyor belts and pneumatic conveying.

55.5.2 BOD Removal by Different Shapes of Media

Different types of plastic media have been compared at the UK Water Research Centre by Bruce and coworkers (1971, 1973). The work was done in two stages: in the first stage, various types of vertical surface (sheet and tube) media were compared; and in the second stage, sheet media were compared with random media.

Tables 4 and 5 are based on data in the reports on this work and summarize the results obtained. It can be seen that the different shapes of media were chosen so that media of similar surface area could be compared at the same rate of application of sewage. At the same rate of application of sewage per unit surface area of media, the percent removal of BOD was about the same for any of the shapes tested. In other words, the main conclusion from this work was that the efficiency of BOD removal is determined by the surface area per unit volume of the media used.

Table 4 Comparison of Various Sheet and Tube Packing

Type of media	Specific surface area $(m^2 m^{-3})$	Rate of application of sewage $(m^3 m^{-3} d^{-1})$	Percent removal of BOD (average over 12 months)
Cloisonyle	220	12	35.0
Surfpac (crinkle close)	187	12	39.9
Flocor	85	12	21.5
Surfpac (standard)	82	12	23.7
Slag	40	6	21.5

The results in Tables 4 and 5 show a small but consistent deviation from this simple rule in that, at the same rate of application of sewage, the vertical sheet media has a higher efficiency than the vertical tube media (Table 4) and the random media has a higher efficiency than the sheet media

Table 5 Comparative Performance of Random Media with Sheet Media

Type of media	Specific surface area, S (m² m⁻³)	Rate of application of sewage, Q (m³ m⁻³ d⁻¹)	S/Q	Percent removal of BOD
Flucor E	85	4.6	18.5	63.0
Biopac 90	75	4.7	16.0	63.0
Flocor M	137	8.9	15.2	47.6
Biopac 50	124	9.0	13.8	52.8

(Table 5). These differences may be explained in terms of the differing abilities of the media to distribute the liquid and effectively use the available surface. Vertical tube media depend completely on the quality of the initial liquid distribution at the top of the bed. Once liquid has entered a tube it cannot spread out to wet the other tubes. A tube which receives no liquid will remain dry throughout its length. Randomly dumped media spread the liquid in all directions, which helps in achieving efficient utilization of the surfaces. Sheet media can distribute the liquid in one direction and their surface area utilization is therefore expected to be better than that with vertical tube media but not as good as that with random media. This explanation is consistent with the results reported.

55.5.3 Media Supports and Drainage Channels

Vertical sheet media are supported on beams of wood or on concrete piers. For example, it is recommended to support PLASDEK media on wooden beams of 100 mm sides on 400 mm centres, as shown in Figure 6, or on U-shaped tiles fixed in concrete as shown in Figure 7.

Random media are simply dumped on to standard support tiles such as are used to to support the rock beds used in percolating filters (Figure 8).

The treated water leaving the bottom of the media bed must flow out of the filter into a drainage channel without causing a build-up of liquid head within the bed.

Drainage channels are shown in Figures 7 and 8.

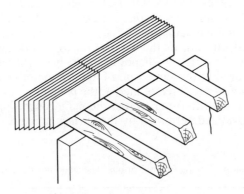

Figure 6 Sheet media supported on wooden beams

55.5.4 Retaining Walls

The plastic media used in high rate filters are very much lighter than mineral media; thus any wall of brick or concrete strong enough to contain mineral media will certainly be strong enough to contain plastic media.

Figure 9 shows a low-cost retaining wall which has been used for both random and sheet media. The media rest against a butyl rubber sheet which is held in place by a mesh of steel rods, which in turn is supported by a steel structure designed to take the forces exerted by the media.

Media made from vertical sheets is theoretically self-supporting and places no lateral forces on the wall, therefore the wall is there to prevent liquid leaving the bed. Random media exert a

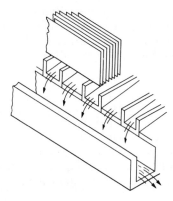

Figure 7 Sheet media supported on concrete tiles

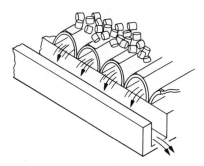

Figure 8 Random media supported on tiles

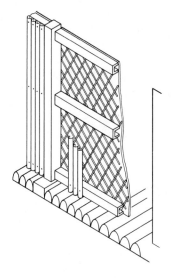

Figure 9 Cost effective retaining wall of butyl rubber and wire mesh

lateral force which may be calculated by conventional methods. Random media are usually used in cylindrical beds where the force to contain the media is supplied by hoop stress in the vessel walls. Sometimes random media are contained within a cylindrical steel tank protected against corrosion by vitreous enamel. This type of tank is often used on farms and is thus easily available.

In calculating forces exerted by the media, due allowance must be made for the weight of the biomass film adhering to the media surfaces. Also, the retaining walls must be designed to be strong enough to withstand wind forces during erection. There is often little difference between the cost of the retaining walls used for different plastic media.

55.5.5 Liquid Distributors

The liquid rates used on high rate filters, although much higher than those used on the traditional mineral media of percolating filter, are still sufficiently low to complicate the design of distributors. Typical liquid flow rates are $1-2 \text{ m}^3$ liquid $\text{m}^{-2} \text{ d}^{-1}$.

Two types of distributor are used: fixed distributors and moving distributors.

55.5.5.1 *Fixed pipe distributors*

A pressure drop of at least 150 mm (water gauge) across the nozzles is required so as to maintain a uniform distribution across the bed. This is the minimum pressure drop which will overcome variation in flow rate due to hydraulic losses and possible out-of-levelness of the distribution system. At the same time, the nozzle diameter should be large enough to avoid the possibility of blocking up by solids in the waste water. These two requirements limit the number of nozzles which may be used; thus it is common for nozzles to be spaced about 1 m apart and for the liquid emerging from each nozzle to be spread out further by causing it to form a spray. Figure 10 shows sprays produced by splash plates fixed beneath liquid distribution nozzles.

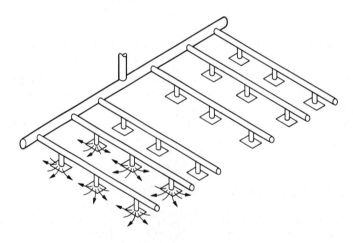

Figure 10   Fixed pipe distributor with splash plates

Fixed distribution systems are usually used on small filters including pilot plants, and on large filter beds, which are rectangular in plan, when the liquid flow rate is high enough. Large circular plant filters are provided with rotating distributors as described below.

55.5.5.2 *Moving distributors*

A typical rotating distributor consists of four radial pipes or troughs, supported on a central bearing, rotating above a cylindrical filter bed which might be 20 m in diameter. Each trough is about 300 mm wide and 600 mm deep. A pipe would be 300 mm in diameter. The polluted water is distributed through a row of holes in the pipe or trough each about 25 mm in diameter. To ensure a uniform distribution there are more holes per length of trough near the retaining wall than near the bearing in the centre of the bed.

With moving distributors, at any particular time only a small part of the upper surface of the bed receives liquid, but as the distributor moves over the bed all parts receive in turn the same amount of liquid. When a rotating distributor with four radial arms is used, each part of the bed receives liquid four times for each revolution of the distributor. Rotating distributors rotate at about two revolutions per minute, depending on the number of radial feed pipes (or troughs).

This means that although the flow of liquid on to any part of the bed is intermittent, the flow per unit length (or per unit area) of the moving distributor is much higher than from a fixed distributor. Thus it is easy to find combinations of head and orifice diameter such that the orifices may be placed much closer together and all of the top of the bed receives the same amount of liquid.

It is claimed that a more effective wetting of the media may be achieved in this way, but the design methods described below, based on uniform flow, are not yet developed to take this into account. Figure 11 shows a rotating distributor designed for a large variation in flow.

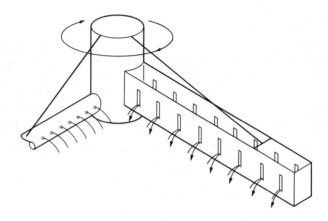

Figure 11   Rotating distributor to cope with a wide variation in flow

Moving distributors may be caused to move by the reaction of the jets of wastewater emerging from the orifices in the distributor. However, the alternative method is to use an electrical motor to move the distributor and of late this has been preferred.

Reciprocating distributors are sometimes provided for rectangular plan beds. These are supported by rails on top of the side walls.

55.6   PRACTICAL DESIGN PROCEDURES

This section presents a review of the methods used to calculate the size and shape of the media bed. Design methods in common use may be classified as: (1) hydraulic volumetric loading methods; (2) organic volumetric loading methods; (3) top surface-area loading irrigation rate methods. Both of the volumetric loading methods predict that a particular duty will require a particular volume of media. The dimensions of the filter required to contain the media are then chosen on the basis of convenience and the costs of the structure and pumping.

The irrigation rate methods (based on liquid flow per unit top surface area of the bed) predict that a smaller volume of media is required in a tall bed of small cross-section and high irrigation rate, than in a shallow bed of large cross section and low irrigation rate. The explanation for this, discussed below, is that at high irrigation rates the surfaces of the media are wet more effectively so that its volumetric performance is improved. The predicted differences in the required media volume are sometimes significant. Thus Kornegay (1975) predicted that 'the filter volume required to reduce an influent concentration of 300 mg l^{-1} to 30 mg l^{-1} (or g m^{-3}) decreases from 2237m^3 to 906m^3 as the hydraulic loading rate is increased from 4.9 to 29.2m^3 m^{-2} d^{-1}'.

Hydraulic loading design procedures are used for low BOD concentration wastes (*e.g.* domestic sewage, BOD say 300 mg l^{-1}, and oil refinery wastewaters, BOD say 70–200 mg l^{-1}). At higher BOD concentrations, say above 400 mg l^{-1} (*e.g.* milk wastes, BOD 600–900 mg l^{-1}) organic design procedures are used.

55.6.1 Hydraulic Volumetric Loading Design Procedures

In hydraulic volumetric loading design procedures, the percentage removal of BOD in the biological filter is assumed to depend only upon the volume of liquid applied to the volume of media in unit time. Data from full scale or pilot plant are correlated by plotting the percentage of BOD removed (or remaining) against Q, the liquid flow per unit volume of media. A typical design curve for this method is shown in Figure 12. The underlying assumption of this method is that the fractional removal of BOD in a given volume of media is independent of concentration. Indeed, this was noted more than 30 years ago by Velz (1948). This implies that the rate processes are 'first order' in BOD; thus a logarithmic relationship between fractional removal and volumetric loading may be derived. For example, Bruce and coworkers (1971, 1973) correlated the results shown in Table 4 for all the different sorts of media using the equation

$$\ln\frac{L_e}{L_i} = -0.034\frac{S}{Q} \tag{1}$$

or, for a particular media, say Flocor E, where $S = 85 \text{ m}^2 \text{ m}^{-3}$

$$\ln\frac{L_e}{L_i} = -\frac{2.89}{Q} \tag{1a}$$

where L_i = inlet BOD (mg l^{-1} or g m^{-3}), L_e = outlet BOD (mg l^{-1} or g m^{-3}), S = specific surface area of media (m^2 m^{-3}) and Q = volumetric loading (m^3 m^{-3} d^{-1}).

To calculate the volume of media required to effect a specified fractional removal of BOD (as L_e/L_i), Q is calculated, and then, for a specified total liquid flow rate, the total media volume may be calculated.

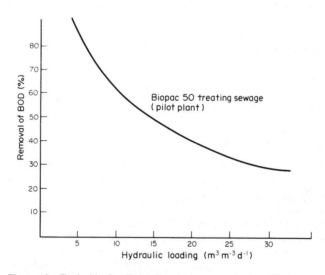

Figure 12   Typical hydraulic loading design curve for low BOD wastes

55.6.2 Organic Volumetric Loading Design Procedures

Organic loading design procedures are similar to hydraulic loading design procedures, except that in the resulting empirical correlations of full scale plant (or pilot plant) data, the percentage removal of BOD is plotted against the volumetric organic load, defined, for example, as kg BOD m^{-3} d^{-1} (see, for example, Chipperfield, 1967). Figure 13 shows an example of an organic loading design curve. The total media volume required is calculated by using the specified fractional removal of BOD to calculate the required volumetric organic load, and then from the total organic load fed to the filter, calculate the volume of media required. (Note that the volumetric organic load is defined as the organic load fed to the filter divided by the media volume.)

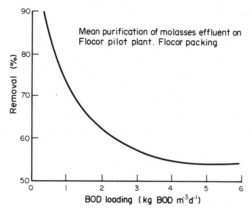

Figure 13 Typical organic loading design curve for high BOD wastes

55.6.3 Top Surface-area Loading Irrigation Rate Methods

An example of an irrigation rate design method is that proposed by Germain (1966). The form of the Germain equation which was recommended by the Norton Company for use with their random Actifil media at BOD concentrations of less than 400 mg l^{-1} is:

$$\ln\frac{L_e}{L_i} = -k\frac{H}{q^n} \tag{2}$$

where H = depth of media bed (m), q = surface irrigation rate (m^3 m^{-2} d^{-1}) and n = media factor (*e.g.* 0.44 for Actifil 90).

The irrigation rate, q, is the rate of application of wastewater per unit area of the top of the bed. It is related to the volumetric loading, Q, by:

$$Q = \frac{q}{H}$$

Thus, the Germain equation may be written

$$\ln\frac{L_e}{L_i} = \frac{k}{Q} q^{(1-n)} \tag{2a}$$

It predicts that a deep-packed bed will require significantly less media than a shallow bed. For example, for Actifil 90 ($n = 0.44$), if at the same total liquid flow rate and fractional removal of BOD the bed cross-sectional area was reduced (and q thus increased) by a factor of 10, then the volumetric hydraulic loading can be increased by a factor of 3.63 (= 10$^{0.56}$). This would require that the bed height be increased by a factor of 2.75, and the media volume be reduced by a factor of 3.63. See also Eckenfelder and Barnhart (1963).

55.7 THEORETICAL JUSTIFICATIONS FOR THE EMPIRICAL DESIGN PROCEDURES

The rate processes which occur inside a high rate filter are complex. Several different types of wastewater are treated in high rate filters as shown in Table 1, which lists, for example, domestic sewage, milk factory wastes and oil refinery wastewater. The feature common to these wastes is that each will feed a slime of microorganisms which is able to attach itself to a solid surface, but different species of microorganisms are associated with each waste. For example, the thick films associated with the high concentration wastes from the food and drink industries are often fungal in nature and may reach thicknesses of 5–8 mm without becoming anaerobic, whereas the film associated with, say, domestic sewage is often bacterial in nature and the living thickness is much less. The situation is made more complicated by the frequent appearance of films of mixed organisms and by changes in type of film and film thickness within the filter.

Practical design methods handle these complexities on the basis that one oil refinery wastewater is much like another, one milk wastewater much like another, and thus practical experience gained in one application may be used in the design of another similar plant to be used for the same waste.

Nevertheless, it is helpful to consider simplified theoretical models of the high rate filter and (at this early stage in their development) use them to put the empirical design methods into a perspective. The theoretical models outlined below are based on the mathematics of diffusion and biochemical reaction within the microbiological slime (biomass). They thus derive expressions for the variation in concentration of food (or oxygen) within the living film and within the high rate filter. In using the theory to explain the design methods, it is assumed that the concentration of polluting organic food is proportional to its BOD, or that any lack of proportionality can be accepted into the correlation coefficients.

55.7.1   Diffusion and Biochemical Reaction

A common starting point for theories of filter performance is the use of an expression for predicting the rate of the biochemical reaction in the film derived from the Monod (1949) equation for predicting the specific growth rate of microorganisms (Bruce and Boon, 1971; Atkinson and Howell, 1975; Porter, 1977; Harremoes, 1977).

The biochemical reaction rate equation which is used has the form

$$R^* = \frac{K_1 C}{K_2 + C} \tag{3}$$

where R = rate of consumption of reactant per unit volume of active biomass per unit time, C = concentration of reactant (*i.e.* pollutant or oxygen) and K_1, K_2 = constants. It should be noted that this form of equation has two limiting solutions. At low concentrations ($K_2 \gg C$) it reduces to

$$R = \left(\frac{K_1}{K_2}\right) C = k_1 C \text{ (a first-order reaction)} \tag{3a}$$

At high concentrations ($K_2 \ll C$) it reduces to

$$R = K_1 = k_0 \text{ (a zero-order reaction)} \tag{3b}$$

For the idealized reaction schemes considered below, in which one reactant is the pollutant and the other oxygen, in principle we may define four limiting cases, *i.e.* for oxygen in excess and pollutant limiting, there are two limiting solutions in terms of pollutant concentration, and for pollutant in excess and oxygen limiting there are two more limiting solutions in terms of oxygen concentration.

The theoretical approaches considered below all lead to equations relating the BOD removal efficiency to hydraulic or organic load which are similar to the empirical design equations. The approaches may be classified as 'the uniform active film approach' and 'the non-uniform active film approach'.

55.7.2   The Uniform Active Film Theory of Filter Performance

The uniform active film approach is that of Kornegay (1975) and assumes the following. (1) At any particular depth in the filter, the concentration of the substrate (*i.e.* pollutant) is constant throughout the active part of the film. (2) The thickness of the active layer is constant throughout the filter and that this active layer thickness is a property of the waste being treated. (3) At all flow rates, all of the media surface is effective, in that it supports biomass with an active layer.

The filter performance equations are derived as follows. Consider the media beneath unit cross-sectional area of the top surface of the bed as shown in Figure 14. The depth of the media is H. We write a mass balance for the removal of pollutant from the water passing through this media as it descends through an increment, dH.

Using Kornegay's assumptions, for a fixed thickness of film t, for media of specific area S, then the volume of active film dV, in incremental height dH, is

$$dV = St \, dH \tag{4}$$

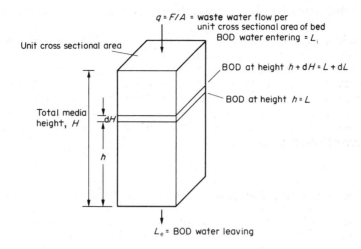

Figure 14 Section through high rate filter showing increment of height (see equations 4–14 and 22–28)

The rate of removal of pollutant in dV is $R\,dV$, or $RSt\,dH$, and this is equal to the change in pollutant concentration in the water, *i.e.*

$$\frac{F}{A}\,dL = RSt\,dH \tag{5}$$

For a situation where oxygen is in excess, and the rate of consumption of pollutant is limited by the pollutant concentration, for low concentrations (*i.e.* low BOD wastes)

$$R = k_1 L \tag{6}$$

and

$$\frac{F}{A}dL = k_1 LSt\,dH \tag{7}$$

which may be integrated to yield

$$k_1\left(\frac{AH}{F}\right)St = \ln\frac{L_i}{L_e} \tag{8}$$

If we define $P_1 = k_1 t$, a constant for the particular waste, then

$$\ln\frac{L_i}{L_e} = \frac{P_1 S}{Q} \tag{9}$$

which is similar to the relationship between fractional BOD removal and hydraulic load given by equation (1) and Figure 12.

Alternatively, we might consider the rate to be limited by the supply of pollutant (oxygen in excess), but the concentration of pollutant to be high enough for the zero-order limit to apply (high BOD wastes), *i.e.*

$$R = k_0 \tag{10}$$

and

$$\frac{F}{A}\,dL = k_0 St\,dH \tag{11}$$

which may be integrated to yield

$$L_i - L_e = k_0 tS\frac{AH}{F} = \frac{P_0 S}{Q} \tag{12}$$

where $P_0 = k_0 t$ and is a constant for a particular waste.

In terms of fractional removal

$$\frac{L_i - L_e}{L_i} = \frac{P_0 S}{QL_i} \tag{13}$$

This is a relationship between fractional removal of BOD and organic load which is similar to that shown in Figure 13.

Kornegay's general equation is

$$K_3 \ln\frac{L_i}{L_e} + (L_i - L_e) = \frac{P_0 S}{Q} \tag{14}$$

55.7.3 The Non-uniform Film Theory of Filter Performance

This approach has been used more recently by Atkinson and Howell (1975), Porter (1977) and Harremoes (1977) and assumes the following. (1) There is a concentration gradient of pollutant and/or oxygen inside the film layer which is a result of the diffusion and reaction processes taking place within it. (2) The thickness of the active layer may depend on the concentration of the pollutants in the water adjacent to the biomass, as well as on the nature of the waste being treated. (3) The proportion of the media surface area which is in contact with the polluted water depends on the liquid irrigation rate.

The calculation procedure accepts that since the concentrations of reactants vary within the film, the rate of removal of pollutant at any point must be calculated from the rate at which it diffuses through the surface of the film. Atkinson and Howell (1975) used the general form of the biochemical reaction rate equation (equation 3) in their work, but not unexpectedly found it necessary to obtain numerical solutions to their equations. These disguise the meaning of the results. Porter (1977) and Harremoes (1977) used the limiting solutions (equations 3a and 3b) and obtained solutions which are more suitable for interpretation. The derivations start from equations, derived from Fick's law of diffusion, which describe diffusion and reaction in an elemental strip of active film. Limiting the analysis to considering only diffusion normal to the film surface, these are:

(a) for a first-order reaction (equation 3a)

$$\frac{\partial C}{\partial \theta} = D \frac{\partial^2 C}{\partial x^2} - k_1 C \tag{15}$$

(b) for a zero-order reaction (equation 3b)

$$\frac{\partial C}{\partial \theta} = D \frac{\partial^2 C}{\partial x^2} - k_0 \tag{16}$$

The solution of these equations depends on the choice of boundary conditions, but in all cases the rate of mass transfer per unit area through the outer surface of the film is given by

$$N = -D\left(\frac{\partial C}{\partial x}\right)_{x=0} \tag{17}$$

55.7.3.1 Boundary conditions

To solve equations (15) and (16), two boundary conditions are required. At the outer surface of the biomass film, assuming the bed as a whole is working at a steady state, it may be assumed that, at any particular depth in the bed, the concentration is constant, that is

$$C = C^* \text{ for } x = 0 \tag{18}$$

If one assumes further that the resistance to mass transfer in the biomass film is much larger than the resistance to mass transfer in the water flowing over the film, then C^* becomes the concentration of pollutant, or oxygen, in the water at the point, *i.e.*

$$C^* = C_D \text{ for } x = 0 \tag{18a}$$

The film thickness is expected to change continually with time due to the sloughing process but, at any instant, most of the mass transfer takes place at points where the film is thick enough for the limiting concentration to be reduced to zero beyond the active surface layer†. The concentration

---

† It is possible to handle the complexity of time dependent sloughing in a theoretical model, but the model is too long for inclusion in this chapter and does not contradict the conclusions obtained in the simpler approach used here.

of the reactant goes down with distance into the film because of the biochemical reaction. The 'infinite thickness' film boundary condition is

$$C = 0 \text{ at } x = \infty \qquad (19)$$

A further simplification may be introduced by noting that (a) the rate of growth of the biomass film is small compared with the rates of diffusion and reaction and (b) the solutions of equations (15) and (16) with boundary conditions (18) and (19) rapidly come to a 'steady state' concentration profile, *i.e.* $(\partial C / \partial \theta)$ is zero, and the 'steady state' solutions are given by

$$\text{from equation (15): } D\frac{\mathrm{d}^2C}{\mathrm{d}x^2} - k_1 c = 0 \qquad (15a)$$

$$\text{and from equation (16): } D\frac{\mathrm{d}^2C}{\mathrm{d}x^2} - k_0 = 0 \qquad (16a)$$

The 'steady state' nature of the concentration profiles means that the rate of mass transfer through the film surface (obtained from applying equation (17) to the solutions) is also constant and given by:

(a) for the first-order reaction (equations 15 and 15a)

$$N = C_D(k_1 D)^{1/2} \qquad (20)$$

(b) for the zero-order reaction (equations 16 and 16a)

$$N = (2Dk_0 C_D)^{1/2} \qquad (21)$$

The concentration term, C_D, in these equations refers to the limiting reactant which may be either the pollutant or oxygen. It is again assumed in what follows that, for rate limitation by pollutant, the concentration C_D may be replaced by L, the BOD.

The rates, N, calculated above are for unit surface area of slime (or biomass) so that (following a similar procedure to that used in equations 4–14 above) the rate of removal in incremental packed height, $\mathrm{d}H$, is $NfS \, \mathrm{d}H$, where f is the fraction of the total surface which is covered by biomass film. So that

$$\frac{F}{A} \mathrm{d}L = NfS \, \mathrm{d}H \qquad (22)$$

For the first-order case, where

$$N = L(k_1 D)^{1/2} \qquad (23)$$

equation (22) on integration yields

$$\ln\frac{L_i}{L_e} = (k_1 D)^{1/2} f\left(\frac{S}{Q}\right) \qquad (24)$$

which is similar in form to equations (1) and (2) and equation (9).

For the zero-order case, where $N = (2Dk_0 L)^{1/2}$, equation (22) becomes

$$\left(\frac{F}{A}\right) \mathrm{d}L = L^{1/2}(2Dk_0)^{1/2} fS \, \mathrm{d}H \qquad (25)$$

which on integration yields

$$L_i^{1/2} - L_e^{1/2} = \left(\frac{Dk_0}{2}\right)^{1/2}\left[\frac{fS}{Q}\right] \qquad (26)$$

which may be rearranged to give the fractional removal of BOD:

$$\frac{L_i - L_e}{L_i} = \left[\left(\frac{Dk_0}{2}\right)^{1/2}\frac{fS}{QL_i^{1/2}}\right]\left[2 - \left(\frac{Dk_0}{2}\right)^{1/2}\frac{fS}{QL_i^{1/2}}\right] \qquad (27)$$

With very high BOD concentrations the active film thickness may be assumed to be constant throughout the filter and controlled by diffusion of oxygen. The BOD removed per unit volume of bed is then considered to be constant and the efficiency of removal is given by

$$\frac{L_i - L_e}{L_i} = K_b f\frac{S}{QL_i} \qquad (28)$$

where K_b is a constant for a particular waste.

Equation (28) is similar in form to equation (13) and to the empirical methods based on organic loading. However, whereas the non-uniform active film theory obtains the 'organic loading' of equation (28) only by assuming oxygen is limiting, the uniform active film theory of equation (13) is derived from BOD limiting with the high-concentration zero-order reaction.

The non-uniform active film approach is well illustrated by equations (26) and (27). This predicts that for high BOD wastes, the filter performance may be correlated in terms of $L_i^{-1/2}$. This is between the empirical correlations for hydraulic load (no dependence on L_i) and that for organic load (fractional removal dependent on L_i^{-1}). Harremoes (1977) showed that the oxidation of glucose and removal of BOD by biofilms can be best interpreted by means of half-order kinetics. He also suggested half-order kinetics applied to his studies of denitrification in submerged downflow filters. Since then other work has supported half-order kinetics and has provided support for the non-uniform active film approach.

55.7.4   The Importance of Irrigation Rate in Effective Wetting of Media

During the development of any of the theoretical models, attention is automatically directed to the term (AH/F) [F/A being the irrigation rate, q] and to that fraction of the surface which is effectively wet, f. At low irrigation rates only a small fraction of the media surface is effectively wet by the polluted water and this fraction increases as the irrigation rate is increased.

Atkinson and Abdul Rahman Ali (1976) used their theory to calculate the effectiveness of wetting of Actifil 50 in a laboratory filter and their results showed the importance of irrigation rate on filter performance. In support of Atkinson's approach, it may be noted that a similar dependence of effective wetted area on liquid rate has been observed for packings used in gas absorption, another type of gas–liquid contacting operation. This explanation in terms of effectiveness of wetting is also consistent with the differences in BOD removal found between the different shapes of media, described above.

Returning to the design equations, it will be noted that in the Germain equation (2a) the term $q^{(1-n)}$ may well reflect the effect of irrigation rate on the wetted area, f. It is also worth noting that Kornegay found his 'constant' changed with irrigation rate, resulting in a more effective use of the media at high irrigation rates. This too may be due to changes in f, the effective wetted area.

55.8   COMMENTS AND CONCLUSIONS

Although not unreasonable explanations for the form of the design methods may be found in the theoretical models described above, there are considerable difficulties in developing the theory to the point where high rate filters may be designed from knowledge of the waste alone. At the present state of the art, design is based on previous experience supplemented by the results of pilot scale tests with the particular wastewater which is to be treated.

Further development of the theoretical models might reasonably be expected to include an allowance not only for the particular mix of microorganisms expected in the biomass slime, but also for any variation in flow rate and concentration in the short (hourly) and longer (monthly) term. Thus there is still a considerable amount to do before the theoretical models become sufficiently accurate for design purposes.

However, high rate filters are similar to many other practical devices (such as distillation columns) in that, while exact theoretical descriptions are of necessity complex, empirical design procedures are both simple and reasonably accurate. Thus inspection of Table 2 shows that for a wide variety of wastewaters, similar removal efficiencies of 60–70% are obtained at similar organic loadings of 2–3 kg BOD m^{-3} d^{-1}.

55.9   REFERENCES

Atkinson, B. and J. A. Howell (1975). Slime holdup, influent B.O.D., and mass transfer in trickling filters. *J. Env. Eng. Div. Am. Soc. Civ. Engrs.*, **101**, EE4, 585.
Atkinson, B. and M. E. Abdul Rahman Ali (1976). Wetted area, slime thickness and liquid-phase mass transfer in packed bed biological film reactors (trickling filters). *Trans. Inst. Chem. Eng.*, **54**, 239.
Bruce, A. M. and A. G. Boon (1971). Aspects of high rate biological treatment of domestic and industrial waste waters. *Water Pollut. Control*, **70**, 487.

Bruce, A. M. and J. C. Merkens (1973). Further studies of partial treatment of sewage by high rate biological filtration. *Water Pollut. Control*, **72**, 499.

Chipperfield, P. N. J. (1967). Performance of plastic filter media in industrial and domestic waste treatment. *J. Water Pollut. Control Fed.*, **39**, 1860.

Eckenfelder, W. W. and E. L. Barnhart (1963). Performance of a high rate trickling filter using selected media. *J. Water Pollut. Control Fed.*, **35**, 1535.

Germain, J. E. (1966). Economical treatment of domestic waste by plastic medium trickling filters. *J. Water Pollut. Control Fed.*, **38**, 192.

Harremoes, P. (1977). Half order reactions in bio-film and filter kinetics. *Vatten*, **2**, 122.

Kornegay, B. H. (1975). The modelling and simulation of fixed bed biological reactors for carbonaceous waste treatment. In *Mathematical Modelling for Water Pollution Control Processes*, p. 271. Ann Arbor Science Publishers, Ann Arbor, MI.

Monod, J. (1949). The growth of bacterial cultures. *Ann. Review Microbiol.*, **3**, 371.

Porter, K. E. (1977). The use of packed beds in water pollution control. *La Tribune du CEBEDEAU*, **30** (398), 30.

Porter, K. E. and E. L. Smith (1979). Plastic-media biological filters. *Water Pollut. Control*, **78**, 371.

Velz, C. J. (1948). A basic law for the performance of biological filters. *Sewage Works J.*, **20**, 607.

56

Rotating Biological Contactors

E. B. PIKE
Water Research Centre, Medmenham, Marlow, Bucks., UK

56.1 NATURE AND DEVELOPMENT OF THE PROCESS

The aerobic rotating biological contactor is a fixed-film reactor in which the microbial community is allowed to develop upon the surfaces provided by a built-up cylindrical rotor which rotates partly immersed in a closely-fitting trough through which the wastewater flows. The rotor serves as a simple aeration device, since re-aeration will take place at the air–liquid interface of the liquid film picked up by the rotor. The dissolved oxygen will then be transferred by diffusion into the biological slime and by mixing into the liquor in the trough. The biological oxidation reactions will occur mainly within the films attached to the rotor surfaces and to a small extent in the liquor of the trough.

The process is not new. Although the first commercial units appeared in Germany from about 1965 (Stengelin), in the USA from 1968 (Allis-Chalmers, succeeded by Autotrol) and in Britain by 1971 (Ames-Crosta Mills 'BioDisc'), the process was invented by Doman in the USA in 1925. Details of these developments are given by Pike *et al.* (1982). The development work conducted for Allis-Chalmers and Autotrol are described in a monograph by Antonie (1976).

Because 'prior art' existed in the process itself, patents have mainly been awarded for improvements or modifications of ancillary features, such as construction of the rotors, or to increase efficiency of aeration, to give greater durability and to reduce construction costs. Thus, the use of discs has been supplemented by assemblages of plastic mouldings giving higher specific surface area, perhaps adapted for drive by coarse-bubble aeration or by rotors of Archimedean-screw form. Some designs of rotor are delivered as 'knocked-down' modules for ease of assembly on site.

The rotating biological contactor must be considered as a biological unit process forming part of a complete wastewater treatment plant (Figure 1). Where treatment is to be provided for populations less than about 500, it is usual to integrate the units into a complete 'package' plant simplifying delivery to the site and installation. A review of the features of package plants available in the UK is given by Pike *et al.* (1981). The process can be adapted for use at large sewage works by installing sufficient rotor modules in series and parallel to provide biological treatment between conventional primary and clarifier treatment stages.

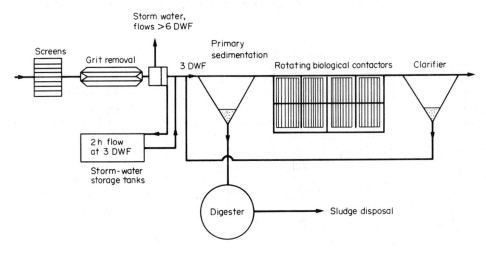

Figure 1 Disposition of treatment units in a sewage works using rotating biological contactors. Maximum rates of flow accepted are shown as multiples of the average dry weather flow (DWF)

The process can also be adapted for treating most biodegradable trade wastewaters, including those containing mineral oils, saline sewage and acid mine-drainage. It has been adapted for carrying out biological nitrification/denitrification (for which the rotors are run completely submerged) and removal of phosphate. References are given by Pike *et al.* (1982). The process is also being developed to give anaerobic treatment of strong organic wastes (Pescod and Norton, 1983).

56.2 MICROBIOLOGY AND KINETICS

56.2.1 Community Structure

The communities of microorganisms and invertebrates inhabiting the films of rotating biological contactors are similar to those found in the activated-sludge process and in the biological percolating filter but with detailed differences. A key monograph on this topic is that of Curds and Hawkes (1975). Depending upon the degree of plug-flow regime in the contactor, the communities will be stratified as shown in Table 1, reflecting the decrease in concentration of pollutants as treatment proceeds and the affinities of the organisms for the available substrate.

Table 1 Changes in Microbial Communities in Films through Successive Stages of the Rotating Biological Contactor

Near inlet	*After partial treatment*	*Near outlet*
Thick filamentous or zoogloeal growth, grey or whitish	Evenly dispersed film, greyish	Sparse film, often patchy, brown
Heterotrophic activity Polyhydroxyalkanoate accumulating in bacterial cells	Heterotrophic activity	Nitrification Grazing of film by invertebrates
Sphaerotilus spp. Other Gram-negative bacteria Flagellate protozoa Amoebae Fungi	Gram-negative bacteria Ciliated protozoa Amoebae	Nitrifying bacteria Rotifers Nematodes Ciliated protozoa Annelid worms Nematode-trapping fungi

The ecology of protozoa within contactors has recently been studied in detail by Madoni and Ghetti (1981) and by Kinner *et al.* (1983). The majority of protozoa inhabit the film and not the liquor. The species range is more restricted than in activated sludge and numbers and diversity are dependent upon organic loading rate and trough retention period. Ciliates and amoebae are both able to ingest bacteria and the former are responsible for clarifying otherwise turbid effluents by removing free bacteria from the liquor. The community is maintained by continuous inoculation by microorganisms and propagules in the sewage and deliberate inoculation with cultures is unnecessary.

56.2.2 Fate of Faecal Bacteria and Pathogens

During aerobic biological treatment, faecal bacteria and pathogens are progressively removed through passage through the plant. Removal is increased when hydraulic loading rates are decreased or when temperature is increased (Pike and Carrington, 1979). In experiments with a 30-population equivalent sized rotating disc plant at a sewage works (Kirk Hammerton, Yorkshire), which was sampled hourly over 24 h on 10 occasions in 1977, the daily input of coliform bacteria to the works was greatest in August and least in February. Although counts in settled sewage varied over 2–3 orders of magnitude over the day, the percentage removal in the plant was unaffected by time of day or count and exceeded 99.5%. The surface hydraulic loading rate was constant ($0.035 \text{ m}^3 \text{ m}^{-2} \text{ d}^{-1}$), but variations in organic loading (1.6–$9.5 \text{ g BOD m}^{-2} \text{ d}^{-1}$) had little effect on removal (Pike and Carrington, 1979).

56.2.3 Factors Affecting Rate of Treatment

56.2.3.1 Oxygen transfer

It is insufficiently appreciated that the rotor is an aerator. It aerates by constantly providing liquid surfaces in contact with air. In doing so, energy is absorbed. The rotor also serves as a mixing device. Re-aeration at the surface of the trough is minimal. It is observed that in plants with compartments or rotors in series nearly anoxic conditions may exist in the inlet region if organic loading rates are high and that the concentration of dissolved oxygen will increase towards the outlet. It is obvious that the rate of treatment is dependent upon the surface area of the rotor per unit volume of trough (*i.e.* on the specific surface area, $\text{m}^2 \text{ m}^{-3}$) and upon the rate of revolution. Experimental determinations of the mass transfer coefficient of oxygen from air to water (K_L) in disc plants were less than predicted by modelling (Bintanja *et al.*, 1975). This departure was greatest (50%) at an intermediate rate of revolution (11 min^{-1} with 0.6 m diameter discs) and was shown to be caused by laminar-flow conditions parallel to the submerged discs, resulting in inefficient mixing (Zeevalkink *et al.*, 1979).

Supplementation of the atmosphere above the rotor with oxygen will increase the rates of removal of BOD and COD and of nitrification (Torpey *et al.*, 1971; Bintanja *et al.*, 1976), demonstrating that oxygen limitation may exist and that the major site of treatment is the film on the rotor.

If organic carbon or other nutrients are not limiting microbial growth, low values of dissolved oxygen will affect the rate of oxidation according to Monod kinetics. For microbial cells, the apparent values of K_s for oxygen are 0.013–0.35 mg l^{-1} (Wimpenny, 1969). Oxygen limitation in rotating biological contactors has not been quantified. Two published experiments give enough data for the value of the Michaelis constant (K_s) to be calculated, namely, for treatment of synthetic sewage (Chesner and Moloff, 1978, their Figure 6), 0.051 and $0.076 \text{ mg O}_2 \text{ l}^{-1}$ and for dairy solids solution (Welch, 1968, Figure 6), 0.30 mg l^{-1}.

These values of K_s are within the range given by Wimpenny, suggesting that diffusion limitations through slime are not serious in comparison with diffusion of oxygen through the microbial cells.

56.2.3.2 Nature of the feed and flow regime

Few microbial substrates in waste treatment are chemically defined and present in low enough concentrations to limit microbial growth and activity. These include oxygen, ammonium ion in nitrification and nitrate or methanol in denitrification. The other condition for Monod kinetics, a defined organism, is probably only met in the case of *Nitrosomonas* and *Nitrobacter* spp. Mixed cultures and stratification of microbial communities together with complexity of flow paths over rotor surfaces imply that theoretical performance models will be complex and, although capable of revealing general principles, will rarely be suitable for predicting performance. The most comprehensive theoretical models are those of Grieves (1972) and of Hansford *et al.* (1978). Recently Harremoës (1983) has advocated the application of biofilm kinetics for describing the oxidation of soluble organic carbon nitrification and denitrification in the rotating biological contactor. This shows that half-order reactions dominate performance because of diffusional limitations and that oxygen is usually the rate-limiting substrate.

Diffusional limitations in the film and in the trough, taken together with substrate limitations when treatment nears completion, imply that the apparent reaction orders will be greater than zero and therefore that an approximately plug-flow regime will provide a greater degree of treatment than with a degree of mixing. The benefits of serial (staged) operation of units or of compartmenting of troughs have been repeatedly noted.

Empirical modelling of performance is carried out by multiple regression analysis or curve-fitting of data obtained from operation of plants under a variety of conditions. This is of direct use in design of plants although general applicability is doubtful unless a sufficiently wide range of variables are studied, the plants are operated properly and the treatability of the sewage can be defined.

Nitrification, as with biological filtration, is observed to occur only in the later stages of treatment when oxidation of organic carbon is substantially completed, presumably because of competition for dissolved oxygen between nitrifiers and heterotrophic bacteria (which are not otherwise antagonistic). Because of the relatively high affinity of *Nitrosomonas* spp. for ammoniacal nitrogen, nitrification rates are largely independent of the concentration of ammoniacal nitrogen above about 5 mg l^{-1}. Below pH 6.5 nitrification is severely inhibited.

56.2.3.3 *Temperature*

Most observers agree that the temperature of the wastewater over the ranges normally experienced (5–20 °C) has relatively little effect upon the rate of removal of BOD compared with its effect on other aerobic processes in which the temperature coefficient (about 1.08 °C^{-1}) normally represents a doubling in rate for a 9–10 °C rise in temperature. In the author's experiments at Kirk Hammerton (Pike *et al.*, 1982) no temperature effect was discernible over the range 7–17 °C, and in the laboratory experiments of Ellis (1976) the temperature coefficient was only 1.006 °C^{-1} for the range 11–18 °C.

The rate of nitrification is, however, greatly affected by temperature (T). In the author's experiments, the temperature coefficient 1.17 °C^{-1} for removal of ammoniacal nitrogen (rate y, g m^{-2} d^{-1}) was shown by

$$y = 1.17(T - 15) \times 0.8120 \tag{1}$$

56.3 PLANT DESIGN AND PERFORMANCE

56.3.1 Construction Features

56.3.1.1 *'Package' plants*

The rotating biological contactor is adaptable to serve very small populations, including single households, up to large cities with 250 000 inhabitants or more. Although the usual arrangement of units at a works is that shown in Figure 1, there is a tendency for the smallest plant, serving populations up to about 500, to be designed as packages containing integrated primary and secondary treatment stages, capable of being delivered to the site by lorry after preparation of the site. The rotating biological contactor has many advantages for small works, particularly those in isolated areas, including those of scenic beauty, those with severe winters and those where low maintenance is required, as with householders, hoteliers, camping sites and at tourist service areas. If properly loaded and given the minimum of attention they will operate without noise, odour or fly nuisance and can be sited to present a low profile to the horizon or even totally below ground level. The packaged unit, being completely contained, is thereby safeguarded from vandalism.

In packaged designs, the primary treatment area and the clarifier are usually situated underneath or alongside the rotor and trough. The primary treatment zone is usually a septic tank, in which settlement, anaerobic digestion of the sludge and storage of sludge take place. It must be properly sized and capable of storing at least three months accumulation of sludge before needing to be emptied. Under current British practice, the volume, V (m^3) should be at least

$$V = 0.18E + 2 \tag{2}$$

where the equivalent population is E units, for domestic housing with a flow of 0.110 m^3 d^{-1} per head (British Standards Institution, 1983). For luxury housing and where 'waste disposal units' (garbage grinders) are used, this is increased to

$$V = 0.25E + 2 \tag{3}$$

In most packaged units the rotor is built-up from discs mounted on a common rotating shaft. Disc diameters are from 0.5–3 m diameter and rotational rates, typically about 3.2 min^{-1} for 3 m diameter discs, are conventionally arranged to give a peripheral velocity of 0.3 m s^{-1}. Spacing of the discs is sufficient to prevent bridging of the discs near the inlet film and is usually about 30 mm. Various materials have been used for fabricating rotors including expanded metal or plastics mesh, high-density polyethylene sheet or moulded expanded polystyrene. There is a growing tendency to use rotors built up from moulded corrugated sheets welded together to form honeycomb structures of higher specific surface area than that attainable with discs.

Usually, rotors are run with 40% of their area immersed. An important consideration, which was often ignored in earlier designs, is the strength of the shaft and drive train and the rigidity of alignment. Power failure, if longer than a few minutes, will cause drainage and drying-out of the exposed film, with considerable unbalance when the rotor is restarted. Similarly, the alignment of the rotor depends upon the rigidity of the tank structures as it is mounted upon them. Operations such as filling the plant at start-up or emptying the primary compartment of sludge can easily distort the structure and cause failure of bearings and drive. Self-aligning bearings should be used.

The clarifier or humus tank may also serve to store sludge, in which case it should have a capacity for at least three months. It is preferable to arrange for secondary sludge to be returned to the primary treatment zone. Under British practice, the minimum volume of the clarifier is recommended as (symbols of equations 2 and 3)

$$V = 0.03E + 1.5 \tag{4}$$

In the 'Biospiral' (Mecana) design, the clarifier is replaced by a cloth-drum microstrainer and the sludge is returned to the primary treatment unit.

It is most important that back-mixing between treatment units should not occur and that the effects of surges of flow, characteristic of small units, should not occur. When the units are packaged, there should be a small head loss, amounting perhaps to only 0.15–0.20 m, across the whole package, achieved by installing weirs between units. The treatment trough should also be compartmented by transverse baffles, with weirs, so that flow regimes approximate to plug-flow. Usually four or five compartments are provided.

Further details of the requirements of package plants are given by Pike *et al.*, 1981.

56.3.1.2 Large plants

In large plants, the rotating biological contactor stage occupies the usual location for biological treatment and comprises multiple combinations of rotors. The layout of large units is discussed in the monograph of Antonie (1976). Practical size limits for individual rotors are 3.7 m diameter (12 ft) and 7.6 m (25 ft) long, with a total surface area of about 9300 m^2 (10^5 ft^2). Compartmenting is provided by arranging rotors in series (normally four), constructed adjacently with shafts parallel to one another, so as to minimize the land area required and the pipework between stages. Larger plants are constructed by serial additions to this basic configuration. There is no interstage settlement, so it is convenient to provide weirs or baffles between successive stages.

Since the surplus floc or sludge from the rotors is large and settles quickly, it is best transported to the clarifiers by gravity, using short runs or by low-speed pumps (*e.g.* Archimedean screws). The rotors should be covered by suitable hemicylindrical enclosures which must be corrosion-proof, since the inside atmosphere will be saturated. Enclosure will prevent heat-loss from the liquor during winter months and safeguard the rotors from destruction by vandalism.

56.3.2 Design and Loading

Mathematical models of performance are able to predict qualitatively the effects of changing features of design and conditions of operation. They have not been considered in any great detail in this chapter, since they cannot be used to design plants to achieve a given performance. In devising them, many assumptions have to be made about kinetics (*e.g.* Monod, Michaelis–Menten, diffusivity of substrate, mass-transfer of oxygen), biomass (concentration, film thickness, specific activity) and limiting substrate (if known, or identifiable). It is not unusual to find that such models have been developed by forcing them to fit experimental data by making suitable adjustments to 'constants'.

Plant manufacturers who have carried out development work are usually able to provide design curves. These have the benefit of being applicable to specific types of wastewater using plants with the particular features being investigated. Their accuracy, however, relies upon the treatability of the wastewater and its temperature being similar to that tested by the manufacturer and upon such imponderables as the efficiency of maintenance of the sewage works and local factors. Examples of such models are those derived by Hartman (1965) and of Antonie (1979).

Most published models (for references, see the analysis of models by Edeline and Vandevenne, 1979) can be reduced to a form in which the fraction of organic matter remaining after treatment is a function (usually exponential) of specific hydraulic loading of surfaces ($m^3 m^{-2} d^{-1}$) and temperature. If the wastewater is of constant treatability and concentration, then the ordinate (Figure 2) can be expressed as concentration of effluent. Figure 2 shows the divergence in predictions of effluent quality produced by five models, which, for similar conditions (BOD, 150 mg l^{-1}, 15 °C), predict effluent BOD levels varying by a range of about 2.5 for specific surface loadings giving full treatment. These models completely underpredict the effluent BOD of the plants examined under carefully controlled conditions by the author's laboratory.

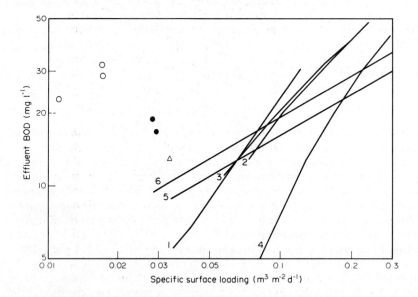

Figure 2 Values of effluent BOD calculated from performance models compared with those from rotating contactors studied by the author's laboratory for different specific surface loading rates. Models are for settled sewage, BOD 150 mg l^{-1}, 15 °C and for four-stage plants, except where given. Models: 1, Antonie (1979), 13 °C; 2, Steels (1974), temperature and number of stages not given; 3, Hansford *et al.* (1978), single stage; 4, Pöpel (1964), single stage; 5, Wu and Smith (1982); 6, Wu and Smith (1982), 7.5 °C. Plants: open circles, four-compartment packages; closed circles, single compartments; open triangle, recommended loading of 5 g m^{-2} d^{-1} (British Standards Institution for settled sewage of 150 mg BOD l^{-1})

Design specifications for the treatment plant and the sizing of units is usually based upon the average organic loading of the rotor surfaces (g BOD m^{-2} d^{-1}) suitable for achieving a given quality of effluent in a specified percentage of examinations. This concept is naturally attractive to the engineer designing the installation, who is concerned with removal of a given daily load, and with manufacturers who will wish to size the plant according to an equivalent population load. It is strictly applicable only when the wastewater strength and composition is constant, because reaction rates, with the exception of nitrification rates, are normally concentration dependent. Inflated claims for loading rates can be ascribed to failure to appreciate this argument.

A number of detailed and independent investigations of rotating biological contactors in the field suggest that when domestic wastewater is being treated in small installations (serving populations below about 1000) the organic loading should not exceed about 5 g BOD m^{-2} d^{-1}, if an effluent standard of 20 mg BOD l^{-1} and 30 mg suspended solids l^{-1} is to be met (Pike *et al.*, 1981; Environment Canada, 1980; British Standards Institution, 1983). For settled sewage with a BOD of 250 mg l^{-1} this loading is equivalent to a specific surface loading rate of 0.02 m^3 m^{-2} d^{-1}.

Observations made by Pike *et al.* (1982) and Environment Canada (1980) suggest that small installations may be unable to achieve the standard for 30 mg suspended solids l^{-1}, because of the presence of colloidal solids in the treated liquor. In this case, tertiary treatment may be advisable, such as microstraining, irrigation of the effluent upon grass plots or lagooning (British Standards Institution, 1983).

56.3.3 Variability in Performance and Effluent Standards

Most design criteria for sewage treatment plants fail to allow for variability in performance and either predict a mean performance, perhaps over 24 h, or merely claim that an effluent of desired quality will be produced consistently, if a given design loading rate is not exceeded. Absolute compliance with a standard is statistically impossible, so a more rational approach is to specify that a given value shall not be exceeded in a given percentage of cases, often 95%. It is possible to calculate the 95 or other percentile value from the mean, if the type of frequency distribution is known.

The principal cause of variability in quality of effluent is the variability in composition and rate of flow of incoming sewage which fluctuates with a regular but complex periodicity over the day and with season, and irregularly with rainfall. Pike (1983) has analysed the features of such variability and the significance of temporal factors and flow regime on the performance of rotating disc plants in the field.

Most sets of data for sewage and effluent composition are distributed approximately log-normally and there is accordingly a strong power law correlation between variance (s^2) and the mean ($\bar{x}$).

$$s^2 = a\bar{x}^b \tag{5}$$

where *a* is a proportionality coefficient and *b* is determined by the type of distribution, being 1 for a normal and 2 for a log-normal distribution. Percentiles are linearly related to the mean (Porter and Boon, 1971).

The study of Pike (1983) showed that different factors were significant in determining variability in effluent quality at the two sites studied. At one works serving a village and a caravan park the effect of treatment was to suppress the effect of diurnal variability on sewage composition and to render seasonal temperature changes the largest factor affecting BOD of the effluent. At the second works a major source of variability in effluent quality, detected by analysis of variance, was thought to have been heavy rainfall on some days. Surprisingly, the main object of this second experiment, to detect a beneficial effect of equalizing the rate of flow over the day, did not yield the expected result. Since the first works did not receive surface drainage water, while at the second flows up to six times the dry-weather value were received by the primary sedimentation units, it would appear that careful design of the inlet works may be more important in controlling fluctuation in effluent quality at small installations than equalizing flows to the rotating biological contactor.

56.4 OPERATION, MAINTENANCE AND RUNNING COSTS

Starting-up of the rotating biological contactor is rapid compared with other biological waste-treatment processes. It can be assisted by filling the trough with dilute and readily degraded waste such as dairy waste or sewage and operating without flow for a few days until film growth is established. The author has noted that a new plant will achieve about 90% of treatment within a fortnight of starting and nitrification within three weeks. In tourist areas and at camp sites a further advantage is that the plant can be left turning during the winter, as long as discs are kept wetted, and the plant will return to full performance in early summer within a few days.

During operation the usual care must be taken with smaller units to ensure that toxic cleaning materials and waste oil are excluded from the drainage system. At catering establishments efficient grease traps must be provided. The manufacturers' instructions for desludging primary tanks and clarifiers must be scrupulously followed, since excessive accumulation of sludge will cause performance to fall rapidly. Manufacturers' estimates of sludge capacity are often optimistic. Care must be taken to avoid damage to the rotors from entry of solid objects into the unit during storms or through vandalism. Apart from desludging the only maintenance normally required is lubrication and adjustment of the drive. Thus the package unit can be specified for isolated communities likely to be cut-off from transport by snow falls in winter.

It is not possible to give relative construction costs for rotating biological contactors, since, even with package units, site work will vary greatly between locations. Similarly, comparative running costs between processes can only usefully be quoted against a base unaffected by inflation or type of energy supply. Gravity and electrically operated processes can be compared in terms of power absorbed by calculating the equivalent head of water (h, m) at the rate of flow being treated (F, m^3 d^{-1}) from the power absorbed (P, kW) as

$$h = 8808PF \qquad (6)$$

Figure 3 shows that the larger rotating biological contactors are more efficient than the smaller and that for populations larger than about 200 they are more efficient than package activated-sludge systems. Since typical strength settled sewage from one person treated fully will require about 46 g of oxygen, if nitrification occurs, it can be calculated that a contactor plant fully treating sewage from 1000 persons would have an efficiency of about 1.8 kg O_2 dissolved kW^{-1}h^{-1}, about the maximum for a diffused-air activated sludge plant working ideally into fully deoxygenated water. Thus, although the rotor system of the biological contactor is basically very simple, it is extremely efficient as an aerator. Although it is less efficient in terms of equivalent energy requirements than passive systems such as the oxidation pond and the biological filter, it is more compact and it may be concluded that compactness in an aerobic process can only be achieved by energy intensiveness. In that respect, the rotating biological contactor achieves a fortunate compromise

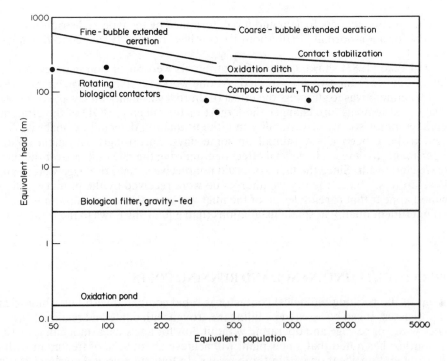

Figure 3 A comparison of power absorbed, expressed as equivalent head of water at flow rate of plant, by rotating contactors, package activated-sludge plants and gravity-operated processes of different sizes giving full treatment. Lines show ranges; package plant data from manufacturers. Circles are for plant investigated by Pike *et al.* (1982).

ACKNOWLEDGEMENT

This chapter is contributed by permission of the Water Research Centre, Environment Directorate.

56.5 REFERENCES

Antonie, R. L. (1976). *Fixed Biological Surfaces — Wastewater Treatment*. CRC Press, Cleveland, OH.

Antonie, R. L. (1979). Applying the rotating biological contactor. *Water Sewage Works*, R 69–R 75.

Bintanja, H. H. T., J. J. V. M. van der Erve and C. Boelhouwer (1975). Oxygen transfer in a rotating disc plant. *Water Res.*, **9**, 1147–1153.

Bintanja, H. H. T., J. J. Brunsmann and C. Boelhouwer (1976). The use of oxygen in a rotating disc process. *Water Res.*, **10**, 561–565.

British Standards Institution (1983). *British Standard Code of Practice for Design and Installation of Small Sewage Treatment Works and Cesspools*. BSI 6297. British Standards Institution, London.

Chesner, W. H. and A. H. Moloff (1978) Biological rotating disc scale-up design: dissolved oxygen effects. *Prog. Water Technol.*, **9**, 811–819.

Curds, C. R. and H. A. Hawkes (eds.) (1975). *Ecological Aspects of Used Water Treatment*, vol. I. Academic, London.

Edeline, F. and L. Vandevenne (1979). Cinetique de l'épuration dans les biodisques. *Trib. CEBEDEAU*, **32**, 3–22.

Ellis, K. V. (1976). A study of rotating-disc treatment units operated at different temperatures. *Water Pollut. Control*, **75**, 73–91.

Environment Canada (1980). *Pilot Plant Studies of Rotating Biological Contactors Treating Municipal Wastewater*. Sewage Collection and Treatment Report SCAT-2. Canada Housing and Mortgage Corporation, Environment Canada, Ottawa.

Grieves, C. G. (1972). *Dynamic and Steady State Models for Rotating Biological Disc Reactor*. PhD Thesis, Clemson University, August 1972.

Hansford, G. S., J. F. Andrews, C. G. Grieves and A. D. Carr (1978). A steady-state model for the rotating biological disc reactor. *Water Res.*, **12**, 855–868.

Harremoës, P. (1983). The application of biofilm kinetics to rotating biological contactors. In *Proceedings EWPCA–IAWPRC Seminar, Rotating Biological Discs*, Fellbach, 6–8 October 1983, pp. 19–39. European Water Pollution Control Association, St Augustin, F. R. Germany.

Hartmann, H. (1965). Der Tauchtropfkörper. *Östereich. Wasserwirtsch.*, **17**, 264–269.

Kinner, N. E., P. L. Bishop and D. Maratea (1983). An evaluation of the bacteria and protozoa inhabiting rotating biological contactor biofilms. In *Proceedings of EWPCA–IAWPRC Seminar, Rotating Biological Discs*, Fellbach, 6–8 October, 1983, pp. 91–107. European Water Pollution Control Association, St Augustin, F. R. Germany.

Madoni, P. and P. F. Ghetti (1981). The structure of ciliated protozoa communities in biological sewage-treatment plants. *Hydrobiologia*, **83**, 207–215.

Pike, E. B. (1983). Analysis of variability in performance of rotating biological contactors at small sewage works and their servicing relationships to standards for effluent quality. In *Proceedings of EWPCA–IAWPRC Seminar, Rotating Biological Discs*, Fellbach, 6–8 October, 1983, pp. 179–192. European Association of Water Pollution Control, St Augustin, F. R. Germany.

Pike, E. B. and E. G. Carrington (1979). The fate of enteric bacteria and pathogens during sewage treatment. In *Biological Indicators of Water Quality*, Newcastle-upon-Tyne, 12–15 September, 1978, ed. A. James and L. M. Evison, paper 20, pp. 1–31. Wiley, Chichester.

Pike, E. B., F. E. Mosey and D. W. Harrington (1981). The use of package plants for treatment of sewage from small communities. In *Proceedings of Symposium, Design and Operation of Small Sewage Works*, Doncaster, 4 November, 1981, paper C, pp. 1–62. Institute of Water Pollution Control, Maidstone.

Pike, E. B., C. H. Carlton-Smith, R. H. Evans, and D. W. Harrington (1982). Performance of rotating biological contactors under field conditions. *Water Pollut. Control*, **8**, 10–27.

Pöpel, F. (1964). Leistung, Berechnung und Gestaltung von Tauchtropfkörperanlagen. *Stuttg. Ber. Siedlungswasserwirtsch.*, **11**. Cited by Edeline and Vandevenne (1979), *q.v.*

Pescod, M. B. and T. S. Norton (1983). Treatment of strong organic wastewater using aerobic and anaerobic packed-cage RBC's. In *Proceedings of EWPCA–IAWPRC Seminar, Rotating Biological Discs*, Fellbach, 6–8 October, 1983, pp. 299–318. European Association of Water Pollution Control, St Augustin, F. R. Germany.

Porter, K. S. and A. G. Boon (1971). Cost of treatment of waste water with particular reference to the river system of the Trent area. In *Symposium on the Trent Research Programme*, Nottingham, 15–16 April, 1971, Paper 3B. Institute of Water Pollution Control, Maidstone.

Steels, I. H. (1974). Design basis for the rotating disc process. *Effluent Water Treat. J.*, **14**, 431–445.

Torpey, W., H. Heukelekian, A. J. Kaplovsky and L. Epstein (1971). Rotating disks with biological growth prepare waste water for disposal or re-use. *J. Water Pollut. Control Fed.*, **43**, 2181–2188.

Welch, F. M. (1968). Preliminary results of a new approach in the aerobic biological treatment of highly concentrated wastes. *Proceedings of the 23rd Industrial Conference, Purdue University*, pp. 428–437. Purdue University, West Lafayette, IN.

Wimpenny, J. W. T. (1969). Oxygen and carbon dioxide as regulators of microbial growth and metabolism. In *Microbial Growth, 19th Symposium of the Society for General Microbiology*, ed. P. Meadow and S. J. Pirt, pp. 161–197. Cambridge University Press, Cambridge.

Wu, Y. C. and E. D. Smith (1982). *J. Environ. Eng. Div., Am. Soc. Civ. Eng.*, **108** (EE3), 578–584.

Zeevalkink, J. A., P. Kelderman, D. C. Visser and C. Boelhouwer (1979). Physical mass transfer in a rotating disc gas-liquid contactor. *Water Res.*, **13**, 913–919.

57

Biological Fluidized Bed Reactors for Treatment of Sewage and Industrial Effluents

P. F. COOPER
Water Research Centre, Stevenage, Herts., UK

57.1 INTRODUCTION

Biological treatment is by far the most commonly used form of sewage treatment. Two distinct types of reactor, classified according to the nature of bacterial growth in them, have been used for the last 70 years. Activated sludge systems are processes in which flocculated bacteria are grown in suspension in aerated tank reactors, whereas systems in which the growth of bacteria occurs on the surfaces of solid media are called attached-growth or fixed-film systems (the traditional plant units are called percolating or biological 'filters'). The Biological Fluidized Bed (BFB) process is an interesting newcomer to the field and is essentially a hybrid of the two processes described above, since it is an attached-growth process in which the bacteria are grown on (fluidized) suspended particles. The basis for the process is the very large surface area/unit volume (specific surface) which is available for growth of very high concentrations of bacteria when small media are suspended in a liquid-fluidized bed. The liquid to be treated is passed up through a bed of small media, such as sand, at a velocity sufficient to cause fluidization of the media. Once fluidized the medium provides a huge specific surface; closely graded cuts of sand particles in the size range 0.3 to 0.9 mm have been used providing a specific surface of the order of 3000 to 4000 m^2 per m^3 of fluidized bed (Jeris *et al.*, 1977). This allows the growth of biomass concentrations of the order of

993

10 to 40 g dry biomass per litre of fluidized bed. This is as high as an order of magnitude greater than that which is possible in activated sludge plants (2 to 5 g l^{-1}) where it is difficult or uneconomic to separate the sludge (biomass) from the effluent if higher concentrations are used. The very high biomass concentrations achieved in BFBs lead to a much higher rate of reaction per unit volume; see Table 1. The rate of reaction per unit weight of biomass remains about the same as in activated sludge processes (Cooper and Wheeldon, 1981). This increase in volumetric treatment rate leads to the following claimed advantages:

(i) Reduction in reactor volumes needed by up to five-fold, hence reduced capital cost.

(ii) The need for secondary clarification is eliminated.

(iii) Since the biomass is grown in high concentrations careful optimization of the sand/biomass separation process can produce a very concentrated sludge (6–7% dry solids) (Cooper *et al.*, 1981). This also results in a reduction in treatment cost.

(iv) Existing overloaded sewage treatment plants can be uprated *in situ* by replacement with BFBs with much higher treatment capacity per unit volume.

(v) (i) and (ii) lead to more compact plants and make the process attractive for industrial effluent treatment plant, hence saving on effluent disposal charges.

Table 1 Process Intensification — A Comparison of BFBs with Conventional Processes

	Biological filters	Organic loading rate[a] (kg BOD m^{-3} d^{-1}) Activated sludge processes	Biological fluidized beds
Carbonaceous oxidation	0.2	1.3	6.0
Nitrification	0.07	0.45	2.0

[a] *Estimate* of the maximum loading rate that could be applied when using a typical UK sewage (250 mg BOD_5 l^{-1}, 150 mg SS l^{-1} and 60 mg Kj–N l^{-1}) and still achieve a good quality effluent. The comparison is expressed in volumetric terms since this is the major difference between activated sludge systems and biological fluidized beds. Had the comparison been made in mass terms the AS and BFB values would have been the same.

A typical arrangement for treatment of sewage in an oxygenic BFB system is shown in Figure 1. One disadvantage of intensifying the rate of treatment (in volumetric terms) is that the rate of dissolved oxygen supply to the reactor must be increased beyond the point at which this can be achieved economically by use of air and hence commercial grade oxygen must be used. Direct injection of oxygen into the fluidized bed has not been used, probably because the turbulent conditions created would cause biomass to be scoured off the particles and prevent its development to high concentration. It has been necessary even with the use of pure oxygen to use a recycle for two complementary purposes: (i) as a diluent for the feed and (ii) as a carrier for the dissolved oxygen. Typically the oxygenator would be designed to achieve a dissolved oxygen concentration of 30 to 50 mg l^{-1} at an operating pressure (in the oxygenator) of 1.5 to 2.0 bar gauge. This pressure can be achieved by building the oxygenator below ground (typically 15 to 20 m deep) in order to take advantage of hydrostatic pressure. Hence in order to supply sufficient oxygen for the complete carbonaceous oxidation of a typical (British) sewage with a carbonaceous BOD_5 of 250 mg l^{-1} the oxygenator would need to have a recycle to feed ratio of 4:1 for a dissolution of 50 mg DO per litre pass.

As the biological growth occurs on the particles their volume increases and their overall density decreases which results in bed expansion. It is thus necessary to remove biomass from the particles in a controlled manner to prevent the bed expanding excessively and causing some of it to flow out of the reactor with the effluent. This is done by withdrawing some of the particles from the top of the bed where they are most thickly coated and detaching the biomass from them by a shear process followed by the separation of the biomass from the particles and finally return of the cleaned particles to the bed where they fall to the bottom for regrowth. Typical operating biomass film thicknesses are in the range 50 to 200 μm. If thick films are allowed to develop, treatment efficiency falls due to the limitations of substrate diffusion and because thickly coated particles can aggregate together into large ineffective clumps, called by some observers 'golf balls' because of their appearance. Another essential feature of BFB design is the distributor which must distribute the liquid evenly across the bed but which must be non-blocking whilst

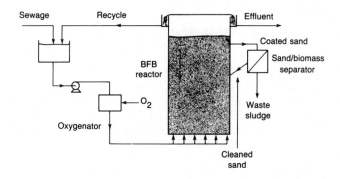

Figure 1 Typical arrangement for oxygenic biological fluidized bed treating sewage

operating at a relatively low pressure-drop. The non-blocking feature may sometimes be a problem to achieve with some systems since biomass will tend to grow on all surfaces associated with the process except where the liquid velocity is very high.

57.2 HISTORICAL BACKGROUND

The BFB process originated from observations made during work to remove dissolved organic compounds from chemically-treated sewage by use of a fluidized bed of activated carbon granules (Weber *et al.*, 1970). Weber and colleagues used additions of sodium nitrate to the feed to prevent the development of anaerobic conditions and production of hydrogen sulfide in the granular carbon beds. They noted that biomass grew on the surfaces of activated carbon particles and that there was a reduction in the nitrate concentration through the bed. In later work (Friedman *et al.*, 1971) they showed that those beds in which biological growth was encouraged by dissolving oxygen in the feed were more effective, in removing organics, than those which were run in the anaerobic mode. This work was followed up by Beer (1970) and Jeris and colleagues (1974) using initially activated carbon and later sand as the media for biological growth. Initially they developed BFBs for nitrate removal from fully-nitrified sewage effluents (Jeris and Owens, 1975). Later they (Jeris *et al.*, 1977) and workers at the United States Environmental Protection Agency's Municipal Engineering Research Laboratory (Oppelt *et al.*, 1978) developed the BFB concept further under oxygenic conditions applying the process to carbonaceous oxidation and nitrification (oxidation of dissolved ammonia) of settled sewage. Jeris and his colleagues obtained patents in the USA, UK and other countries for the use of BFBs in the water and wastewater treatment field (Jeris *et al.*, 1976; Jeris, 1976, 1978). The BFB patents are vested in Ecolotrol Inc. of Bethpage, Long Island, New York, who have exclusively licensed Dorr-Oliver Inc. to exploit the use of BFB processes worldwide using the trade names 'Oxitron' and 'Anitron'.

57.3 PROCESSES

57.3.1 Water Treatment — Nitrification and Denitrification

In 1973 Short (1973, 1975) at the Water Research Centre developed a dilute form of BFB for the oxidation of the low concentration of ammonia (0.5 to 2.0 mg NH_3–N l^{-1}) sometimes found in river water used for potable supply. The idea for the work derived from earlier work done by Pugh (1949) in the 1940s and 1950s in upflow blanket clarifiers in which the blanket was biologically active. In his initial work Short used a blanket of river-silt solids but later realizing the advantage of providing a large specific surface for bacterial growth added fine sand (0.05 to 0.15 mm in size) with dramatic results. This work was developed by Gauntlett in 1979 to overcome another water treatment problem — high concentrations of nitrate in waters being treated for potable use. An anoxic BFB for water denitrification was built at the WRc's Medmenham Laboratory and as a result of this work a small full-scale plant (flow 2500 m^3 d^{-1}) was installed by Anglian Water Authority at Bucklesham Water Works (near Ipswich) in 1982. The plant was supplied by Dorr-Oliver with the design based on the kinetic data derived by WRc.

57.3.2 Sewage Treatment

57.3.2.1 Carbonaceous Oxidation and Nitrification

The earliest work on oxygenic fluidized bed systems for sewage treatment for carbonaceous oxidation was carried out by the United States Environmental Protection Agency (Oppelt *et al.*, 1978) in 1970 to 1973 using a pilot plant comprising eight (254 mm dia) reactors in series with pure oxygen injection into the base of each column. The pilot plant suffered from several problems including blockages in the liquid distribution system and difficulty in obtaining controlled wastage of biomass from the system. These problems were largely overcome in work done by Jeris and his co-workers (1977) at Manhattan College, New York, who progressed from their successful anoxic BFB denitrification work to using a single 0.6 m dia × 4.66 m high reactor for both carbonaceous and nitrogenous oxidation. They used a down-flow bubble-contact oxygenator to achieve dissolution of commercial grade oxygen to achieve the high concentration of oxygen (20–30 mg DO l^{-1}) needed externally to the reactor. Sand/biomass separation was achieved by use of a sand pump and a vibrating sieve to separate the two phases. A BOD reduction from 94 to 15 mg l^{-1} was achieved in only 16 minutes when operating with a BVS* concentration of 14.2 g l^{-1}. This resulted in an F/M ratio of 0.62 kg BOD kg^{-1} BVS d^{-1} and a sludge age of 3.5 d. In a further test aimed at nitrifying a secondary effluent a reduction in ammonia concentration from 19.2 to 0.2 mg NH$_3$–N l^{-1} was achieved in only 10.6 min at a biomass concentration of 8.5 g BVS l^{-1}. The sludge age was estimated to be greater than 15 d.

Dorr-Oliver Inc. since licencing the process from Ecolotrol Inc. have developed the BFB process for sewage treatment and industrial effluent treatment in its own demonstration plants and in cooperation with the Wastewater Technology Centre (WTC) in Canada and the Water Research Centre (WRc) in the UK. In pilot plant work (Nutt *et al.*, 1979, 1980, 1981a; Dearborn, 1981) at WTC Burlington, Ontario, and in a Dorr-Oliver demonstration plant (1.83 m × 1.22 m × 4.27 m high) at Orillia, Ontario, the treatment of settled sewage was examined for both carbonaceous oxidation and ammonia nitrification. The small-scale work was used to define the process response to changes in temperature, hydraulic retention time, recycle ratio, effluent dissolved oxygen concentrations, influent organic concentration and the effect of diurnal flow and load variation. The demonstration plant at Orillia was used to follow up these tests and was used to examine the performance of components in the plant such as the liquid distributor and the oxygenator (Sutton *et al.*, 1979a, 1981). In a similar venture in the UK, the Water Research Centre installed a similar Dorr-Oliver demonstration plant at its Coleshill experimental site. Following on from preliminary small-scale (0.5 m dia × 4.6 m) pilot plant work at the Stevenage Laboratory (Cooper and Wheeldon, 1981, 1982) the Coleshill demonstration plant (1.83 m × 1.22 m × 5.6 m high) was used to examine the carbonaceous oxidation and nitrification of a typical UK-strength sewage (200 mg BOD$_5$ l^{-1}) which is considerably stronger than those examined in the USA and Canadian demonstration plants. The Coleshill pilot plant treated flows of up to 500 m^3 d^{-1} for carbonaceous oxidation. Following this work the demonstration plant was transferred to the Horley STW (Thames Water Authority) (South of London) for use in a joint project by TWA and WRc to investigate the BFB process for high-rate nitrification of effluents from overloaded activated sludge or biological filter works containing 5 to 30 mg NH$_3$–N l^{-1}. This work followed on from tests (Green and Hardy, 1984) done by TWA at Beckton STW (East London) in 0.3 m dia reactors which indicated that it was possible to oxidize economically up to 25 mg NH$_3$–N l^{-1} in a retention of about 25 minutes at a biomass concentration of about 17 g BVS l^{-1}. The Horley plant has a submerged oxygenator (20 m deep) to enable more economical dissolution of oxygen.

The first full-scale BFB for sewage treatment has been installed at Hayward, California by Dorr-Oliver Inc. and should be commissioned in the near future (Sutton, 1983). A demonstration scale BFB has recently been installed at Newark, New Jersey, with EPA funding to demonstrate the treatment of domestic sewage (Sutton, 1983).

In pilot plant at WRc Stevenage Laboratory the process was taken one step further using two reactors linked in series (Cooper and Wheeldon, 1981, 1982; Cooper, 1984). The two reactors, one of which was operated under anoxic conditions and the second which was operated under oxygenic conditions, achieved complete carbonaceous oxidation, nitrification and a 75% reduc-

---

*BVS is the Biomass Volatile Solids concentration measured in the fluidized state and is equivalent to MLVSS for activated sludge systems.

BTS is the Biomass Total Solids concentration measured in the fluidized state and is equivalent to MLSS for activated sludge systems. The relationship between BVS and BTS is generally BVS = 0.8BTS.

tion in effluent nitrate in a combined retention time of 100 minutes. The basic outline of the process is shown in Figure 2. It is an adaptation of the concept used in the modified activated sludge process in which denitrification of effluent takes place in a preliminary anoxic zone. The two-reactor concept has recently been taken a stage further by researchers from the Wastewater Technology Centre in Burlington, Ontario, who used it to treat effluents from coal-carbonization plants (Melcer *et al.*, 1982) which contain high concentrations of ammonia and organic compounds. The process flow-sheet utilizing an anoxic reactor to treat recycled nitrified effluent is particularly suited to minimizing the oxygen input by reusing the oxygen used in producing the nitrate.

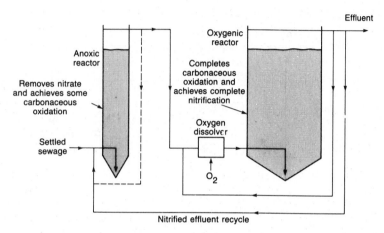

Figure 2 Two-reactor fluidized-bed system for complete sewage treatment including denitrification using the carbon present in settled sewage

Sehic (1978, 1981) attempted to use air rather than commercial oxygen as his oxygen source for carbonaceous oxidation of a settled sewage. He found that it was possible to achieve up to 90% removal of BOD but needed very large recycle ratios (30:1) to dissolve sufficient oxygen.

57.3.3 Industrial Effluent Treatment

57.3.3.1 Denitrification

Francis and colleagues (1977, 1981) at Oak Ridge National Laboratory, Tennessee, extended the use of anoxic BFBs from the relatively low concentrations of nitrate (20 to 40 mg NO_3–N l^{-1}) present in sewage effluents to the much more concentrated wastes (230 to 2300 mg NO_3–N l^{-1}) of the nuclear fuel reprocessing field. A novel feature of the system was the use of a tapered (conical or pyramidal) reactor to produce a reduction in velocity as the liquid passed up through the bed. This was done to prevent the biomass-coated media from being displaced from the reactor by the co-current flow of liquid and the large volumes of nitrogen gas produced. Bosman and colleagues (1978, 1981) at AECI, Johannesburg, South Africa, also applied the anoxic BFB process to the treatment of concentrated nitrate wastes from a chemical complex producing fertilizers and explosives. An interesting feature of this work was the use of a low-value by-product, molasses, as the carbon source for the denitrification process. Miyaji and Kato (1975) used an anoxic fluidized bed of activated sludge pellets (size 1 to 2 mm) to treat a mixture of petrochemical waste-waters. The pellets were grown from the waste to be treated. The nitrate which was present in one waste stream was used to provide the oxygen source needed for the biochemical degradation of another wastewater containing organic matter. Klapwijk (1977, 1981) also used a bed of activated sludge pellets (containing a high percentage of precipitated calcium carbonate) to achieve effluent denitrification.

57.3.3.2 Carbonaceous oxidation and nitrification

Oxygenic BFBs have been used for the carbonaceous oxidation of a concentrated carbohydrate waste (Sutton *et al.*, 1979b), the carbonaceous oxidation and nitrification of a fish-farm effluent

(Goodson, 1978) and for the treatment of strong coal-conversion effluents containing phenols, thiocyanates and hydrogen sulfide (Lee *et al.*, 1979). More recently in Canada the two-reactor (anoxic followed by oxygenic) systems have been applied on pilot scale to the treatment of coal-conversion effluent and steel-works effluents (Nutt *et al.*, 1981b; Melcer *et al.*, 1982).

In the USA Dorr-Oliver Inc. (Dunning, 1980) have designed or installed full-scale or demonstration-scale BFBs for the carbonaceous oxidation of (i) effluents containing paint solvents at the General Motors Plant in Detroit, (ii) brewery wastes, (iii) effluent from a bottling plant, and (iv) synthetic-fibre plant effluent.

57.3.3.3 Anaerobic treatment

Anaerobic BFB treatment has been attempted on pilot- or full-scale with many different wastes. Sometimes a sand medium has been used but in many cases an up-flow sludge blanket reactor using a fluidized blanket of sludge particles has been used (Lettinga *et al.*, 1980).

Some of the wastes treated are as follows: (i) bleaching effluent containing chlorophenols from the pulp and paper industry (Hakulinen and Salkinoja-Salonen, 1981), (ii) diluted cow manure (Jewell, 1981), (iii) soya bean processing effluent (Dunning, 1980), (iv) organic effluent from synthetic-fibre production (Dunning, 1980), (v) effluent from soft-board manufacture (Dunning, 1980), (vi) a dairy wastewater (whey permeate), (vii) detergent (Barnes, 1983a), (viii) blood and tallow from a tannery (Barnes, 1983), (ix) starch wastes (Barnes, 1983), and (x) sugar-beet wastes.

57.4 PROCESS COMPONENTS AND DESIGN CONSIDERATIONS

The following components are likely to be used in all BFBs, whether anoxic, anaerobic or oxygenic: (i) reactor, (ii) media, (iii) distribution system, and (iv) biomass/media separation system.

In addition to the above the oxygenic BFBs will include some form of oxygenator and a dissolved oxygen control system.

57.4.1 Media and Reactor

Mulcahy and La Motta (1978) stated that there are five basic parameters which are directly in the control of the process design engineer who wishes to design a BFB. These are: (a) cross-sectional area of the reactor, (b) operating depth of the bed, (c) density of the medium, (d) mean diameter (and size range) of the medium, and (e) volume of medium.

Another parameter which is indirectly controllable is the biomass concentration. This will be a function of (a) to (e) and the biomass wastage rate which controls the film thickness that is allowed to develop. It is also true to say that the type of treatment and hence the specific bacteria involved will affect the biomass concentration. [For instance it seems to be the case that with the same feed (settled sewage) it is possible to develop biomass concentrations of the order of 30 g BVS l^{-1} under anoxic conditions whereas under oxygenic conditions biomass concentrations of 10–20 g BVS l^{-1} would be the case.] Mulcahy (1978) demonstrated that a film thickness greater than 200 μm was counterproductive because of diffusion limitations. In addition if the film thickness is allowed to expand beyond this it can lead to the effect commonly called 'golf-balling' where thickly-coated particles clump together in aggregates of the order of 50 particles to form balls of 10 to 20 mm in size which can remain in the bed. They are very ineffective and occupy valuable space. They usually result from either too low a sand concentration or an inadequate medium cleaning rate.

The choice of medium is dependent upon a number of factors but the principal one will be with respect to biomass disengagement. If a high-shear stage is to be used to disengage the biomass from the medium then it is likely that a tough medium such as sand would be used. The size and density of the medium must be selected in parallel with the design upflow velocity since these, in conjunction with the calculated required volume, will define the dimensions of the reactor. If for instance it is decided to use a sludge blanket type of reactor in which the biomass is grown on sludge pellets or precipitated calcium carbonate (Klapwijk, 1977, 1981) the fluidized bed will have to be shallow and with a large cross-sectional area to accommodate the low upflow velocity (up to 8 m h^{-1}) and the required retention time.

Sand is the medium most widely used and upflow velocities of the order of 20–45 m h^{-1} can be used for closely-graded sand of an average size 0.45 mm whereas sand with a mean particle size 0.6 mm has been operated in the range 30–70 m h^{-1} (Cooper and Wheeldon, 1981).

As stated above the dimensioning of the reactor is a compromise involving the medium and the reactor cross-sectional area. The reactor volume is dictated by the load to be treated and the rate of treatment, the rate of treatment per unit volume being derived from the biomass concentration and the rate of reaction per unit of biomass. This mass-specific rate of reaction has been shown to be similar to that applying to activated sludge systems (Cooper and Wheeldon, 1981, 1982) and (unless specific BFB pilot plant data exist) may be used as an estimate of the rate of reaction per weight of biomass. Hence in order that the volume of the reactor can be estimated on the basis of the rate of treatment (unit mass) or solids retention time (SRT, Sludge Age) it is necessary to have some estimate of the biomass concentration. The biomass concentration achievable is summarized in broad terms below for sand particles of 0.45 mm (mean size) and upflow velocities of 25–30 m h^{-1}.

Denitrification	20–30 g BVS l^{-1}	(25–40 g BTS l^{-1})
Carbonaceous oxidation	15–20 g BVS l^{-1}	(20–25 g BTS l^{-1})
Nitrification	5–12 g BVS l^{-1}	(6–15 g BTS l^{-1})

Part of the Dorr-Oliver BFB design method has been published in the mathematical model derived by Wen Shieh *et al.* (1979) for predicting the biomass concentration in a BFB reactor. This model is theoretically and empirically based and has been validated in the Dorr-Oliver pilot plants.

When the volume of the reactor has been calculated from the load and the rate of treatment, the aspect ratio involving the cross-sectional area and the height may be determined bearing in mind that, particularly for oxygenic reactors, a recycle flow may be necessary.

57.4.2 Distribution System

An adequate non-blocking system will be needed to distribute the liquid evenly over a relatively large area. A relatively high velocity in the distributor will be needed to prevent growth of biomass on the internal surfaces. In several pilot plants a conical section has been used to provide a turbulent region. Cooper and Wheeldon (1981, 1982) used an inverted cone with the 25 mm feed pipe discharging directly into it. This created a very turbulent spouted bed which occupied about 0.5 m and performed for three years without blockages. Green and Hardy (1984) also used this design successfully following problems with distributor plates and gravel beds. The Dorr-Oliver pilot and demonstration plants have used a similar design in their rectangular reactors. They use a slightly more complicated version which is essentially a number of 12 mm pipes discharging down into inverted rectangular pyramids (furrows) (Dorr-Oliver Co., 1978b). This also produces a vigorous spouted bed and has been successfully operated on pilot plants in the USA, Canada and the UK.

57.4.3 Biomass Media Separation

A number of different methods have been used to separate the media from the biomass to be wasted (Cooper *et al.*, 1981). The most successful methods have involved subjecting the biomass-coated medium to high shear forces by pumping or stirring to break the biomass from the media. The biomass can then be separated from the medium (which is recycled back into the reactor) by either vibrating sieves, sieve-bends or hydrocyclones. It is desirable to keep the process as simple as possible in order to assist equipment reliability. It is possible with careful process optimization to produce sludge concentrations of up to 12% dry solids from anoxic BFBs and up to 6% dry solids from oxygenic BFBs (Cooper *et al.*, 1981). This is significant to the overall cost of the BFB process (Wheeldon and Bayley, 1981).

57.4.4 Oxygen Dissolution

In order to supply oxygen to the reactor at the required rate it is desirable to achieve a high DO concentration in the feed. It is also desirable to minimize the recycle ratio in order to reduce the energy used for pumping.

Down-flow bubble-contact (DBC) oxygenators have been used for many oxygenic BFBs. Jeris *et al.* (1977) used a cruder form of oxygenator (aeration cone) to supply high concentrations to their pilot-scale BFB. This aeration cone was pressurized to improve oxygen solubility. In later work Jeris and Owens (1981) used a U-tube design for supplying high levels of oxygen. The U-tube was built 61 m deep into the ground in order to achieve a dissolved oxygen concentration of 60 mg DO l^{-1}. Dorr–Oliver Inc. (1978b) took this process further producing a more sophisticated bubble-contact tube capable of better oxygen/liquid contact at lower pressures. In demonstration plant tests in the USA and Canada the Dorr-Oliver oxygenator achieved dissolution of 60 mg l^{-1} when at operating pressures up to 2 bar (gauge) against a pressurizing valve. In the WRc pilot plant an oxygentor designed with advice from Dorr-Oliver achieved up to 90 mg DO l^{-1} at a pressure of 2 bar (gauge). The WRc's Coleshill BFB was operated at up to 35 mg DO l^{-1} because it was found that if operated at greater concentrations effervescence of oxygen occurred at the liquid surface. This BFB reactor and its Dorr-Oliver oxygenator have now been moved to Horley STW and the oxygenator is sunk 20 m below ground (with the reactor 4.5 m above ground). Oxygenator tests have shown that dissolved oxygen concentrations of the order of 70 mg DO l^{-1} can be achieved. Figure 3 shows the arrangement of a DBC supplied for the Dorr-Oliver demonstration plants. One feature of this design is that undissolved oxygen gas (bubbles) is separated at the top of the disengagement section whence it returns to the top of the bubble-contact tube by the gas recycle pipe.

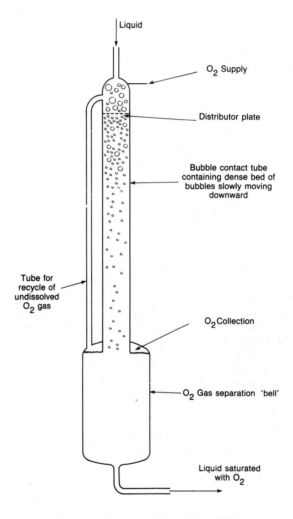

Figure 3   Diagram of Dorr-Oliver DBC

The DBC seems to be capable of achieving greater than 50 mg DO l^{-1} but it should be borne in mind that what is necessary is for all the oxygen to be used up in the passage of the liquid through the reactor. It is likely that some oxygen is released at the base of the reactor as the pressure reduces as the liquid passes upwards but at the same time DO is being used up and the oxygen may redissolve and be utilized. It is desirable to use a below-ground oxygenator in order to achieve the high (hydrostatic) pressure with as little pressure loss as possible (Wheeldon and Bayley, 1981).

If a below-ground oxygenator is used it is possible to utilize the head of liquid coming into the treatment plant. It is possible to create the desired pressure at the bottom of the hole housing the oxygenator without having to pump against a partially closed valve as in the method used in the above-ground oxygenators. If the liquid passing up from the base of the oxygenator is handled gently then the liquid will remain supersaturated and little effervescence will take place.

Very little work has been done using air rather than commercial grade oxygen as the source of oxygen since it has been felt that (i) the recycle ratios needed would make the process uneconomic, (ii) if air were injected directly into the BFB the vigorous conditions created would lead to biomass becoming sheared from the sand particles.

Sehic (1978, 1981) used an external oxygen dissolver (a sieve-plate column) to dissolve oxygen from air but recycle ratios of the order of 30:1 were needed and hence the economics were in doubt.

The Gist-Brocades NV company (1980) in Holland has recently patented a BFB process which indicates that direct air injection into a BFB has been used and that biomass concentrations of the order of 15 g l^{-1} have been achieved. It may be implied from this work that it is possible to select naturally certain strains of common bacteria which can grow in the high-shear conditions created in an aerated reactor.

57.4.5 Start-up of Reactors

Little has been reported in the literature about the start-up or seeding of reactors but it can be said that the start-up of anoxic reactors is relatively easy whereas that for anaerobic reactors is difficult. The start-up of oxygenic reactors is somewhere between the two.

Little difficulty has been reported for start-up of anoxic reactors especially if done under batch-feeding conditions (Cooper and Wheeldon, 1981, 1982). If the reactor and recirculation system is filled with the liquid to be denitrified and the stoichiometric amount of carbon source and recirculated for 1 to 2 days, it is usually found that the dissolved oxygen level reduces to zero on the first day and then the nitrate concentration starts to drop in the second day. The facultative heterotrophic bacteria necessary for the process are in plentiful supply in the environment. It is usually beneficial to change the batch feed after two or three days and after five to seven days a low constant feed can be applied to the system. This feed can be gradually increased and the system should be fully operational in 10 to 14 days. Throughout this period the upflow velocity must be kept constant by adjusting the recycle flow.

It is sometimes possible to seed a BFB for carbonaceous oxidation by direct feeding of the liquid and predissolved oxygen. It is usually beneficial to precede this however by a period of batch feeding (as described above). Cooper and Wheeldon (1981, 1982) experienced problems in starting up an oxygenic BFB (probably caused by an inadequate oxygenator allowing bubbles to pass into the BFB and dislodge biomass) and started the BFB off as an anoxic reactor and then switched to carbonaceous oxidation by merely turning the O$_2$ supply on. Green and Hardy (1984) experienced difficulty in starting off a nitrifying BFB using a well-treated effluent containing ammonia. They found that if they started the bed for carbonaceous oxidation of settled sewage and then gradually reduced the loading rate (SLR or *F/M* ratio) so that nitrifying conditions could be created in the BFB it took about 1 month to reach nitrification.

The start-up of anaerobic BFBs can be difficult since the growth rate of anaerobes is much slower than aerobes. The best advice is probably to seed the bed using an anaerobic sludge from an anaerobic digester.

57.4.6 Control of the BFB

In controlling the BFB reactor several parameters must be carefully monitored in order to ensure satisfactory effluent quality and performance: (i) upflow velocity, (ii) dissolved oxygen concentration, (iii) biomass concentration, and (iv) sand cleaning rate.

Upflow velocity must be controlled within a specific band and if at all possible kept constant. When operating with a varying feed flow rate this control may be exerted by varying the recycle flow rate to operate either at a steady upflow velocity (when the feed flow variation is small) or within a band (when the feed flow variation is wide). The need to maintain dissolved oxygen throughout the reactor may also influence the recycle flow rate.

The dissolved oxygen concentration is usually conveniently controlled by measuring the effluent DO at the outlet of the reactor and then using feed-back control procedures (either two- or three-term control; two-term control is usually sufficient) to change the oxygen input to the oxygenator.

The control of the biomass concentration throughout the reactor is important in order to maintain reactor performance. This is done by taking regular samples (at least once a week) from different depths within reactor. This procedure will indicate whether too much or too little sand cleaning is being practised.

57.5   EXAMPLES OF THE USE OF BFB TREATMENT

Biological fluidized beds are now at the stage where it can be said that sufficient design information and expertise exist to allow the building and operation of a reliable process which will do the treatment required. In the sewage treatment field all the conventional treatment processes (carbonaceous oxidation, nitrification, denitrification and anaerobic digestion) have been pilot tested in BFBs. In addition nitrification and denitrification have been achieved in pilot BFBs for potable water supply. Many industrial effluents have been treated in pilot and full-scale BFBs.

In the late 1970s the equipment and operational problems were thought to be: (a) liquid distribution, (b) sand/biomass separation, (c) suspended solids removal, (d) oxygen dissolution, and (e) rapid start-up (seeding of the bed).

All of these areas have been tackled and seem no longer to pose the problems that they seemed to do at the time of the international conference on BFB treatment in 1980 (Cooper and Atkinson, 1981). Most of the advantages claimed for the BFB process have been borne out in pilot-scale experience. In addition it has been shown that the rates of treatment achieved in BFB reactors are very similar to those achieved in conventional activated sludge processes if the comparison is made on a *mass* basis (Cooper and Wheeldon, 1981). Where the BFB process gains over the conventional process is the higher biomass concentration achieved and consequential reduction in reactor volume needed. For the treatment of sewage for carbonaceous oxidation to achieve a good quality effluent ($BOD_5 < 20$ mg l^{-1} and SS < 30 mg l^{-1}) the loading rate would be about 0.5 kg BOD_5 kg^{-1} BTS d^{-1} (6 kg BOD_5 m^{-3} d^{-1}) resulting in reactors with retention times of 45 to 60 minutes for typical UK strength settled sewage (200 to 250 mg BOD_5 l^{-1}). (The required retention time would drop to the order of 20 to 30 minutes for a typical US sewage which would have a BOD_5 in the range 70 to 120 mg l^{-1}.) This very short contact time is possible because biomass concentrations of the order of 20 g BTS l^{-1} are achievable for carbonaceous oxidation. For combined carbonaceous oxidation and complete nitrification the loading rates would be of the order of 0.15 to 0.20 kg BOD_5 kg^{-1} BTS d^{-1} resulting in a retention time of the order of about 100 minutes (biomass concentration 15–20 g BTS l^{-1}). Table 2 shows the performance of a two-reactor BFB process, of the type shown in Figure 2, with a retention time of 100 minutes (Cooper, 1984). In order to achieve the same performance from a conventional activated sludge process it would be necessary to build sufficient reactor volume to provide a retention time of 6 to 8 hours. The plant from which the data shown in Table 2 were produced did not have a final settling tank.

Nitrification of a well-treated secondary effluent has also been pilot tested in BFBs. In this concept the BFB process was placed at the end of the treatment train to uprate an existing plant which removed all the carbonaceous matter but which was too heavily loaded to allow nitrification to take place. The results of the pilot-scale work done by Thames Water Authority (TWA) in the UK are shown in Table 3 (Green and Hardy, 1984). The table shows the very high nitrification rates achieved which would produce a very compact *add-on* plant needing a retention time of only about 25 minutes to remove ammonia concentrations of about 25 mg NH_3–N l^{-1}. This work has now been taken further by TWA in a joint venture with WRc. The 10 m^3 BFB reactor previously used at the WRc's Coleshill site is now being used at Horley STW (Thames Water Authority) to follow up the primary pilot plant work done by TWA. The results of the large plant (flow 400–900 m^3 d^{-1}) are similar to those of the smaller pilot plant. Once the nitrifying BFB was established very little sand cleaning was necessary because the very low growth rate of the nitrifiers is balanced by the small loss of biomass from the bed. Suspended solids remained very low.

Table 2 Performance of a Two-reactor BFB System for BOD Removal, Nitrification and Denitrification of Settled Sewage (Cooper, 1984)

	BOD_5^a (mg l^{-1})	Suspended solids (mg l^{-1})	Kj–N (mg l^{-1})	NH_3–N (mg l^{-1})	NO_3–N (mg l^{-1})
Settled sewage	221	122	57	37	—
Effluent	9	17	4	<1.0	10.3
Retention time	103 min				
Mean biomass concentration	18 g BTS l^{-1}				
Loading rates					
Mass basis (*F/M* ratio)	0.18 kg BOD kg^{-1} BTS d^{-1}				
Volumetric basis	3.0 kg BOD m^{-3} d^{-1}				

^{a} Nitrification inhibited.

Table 3 Nitrification of Secondary Effluent in a Pilot-Scale BFB (Green and Hardy, 1984)

Test number		1	2	3	4	5	6
Operating period	(d)	28	45	25	32	33	45
Retention time	(min)	23	12	30	16	18	15
Biomass concn.	(g BTS l^{-1})	10.0	14.1	8.0	11.5	16.6	19.2
Feed NH_3–N	(mg l^{-1})	25	32	26	30	32	31
Effl. NH_3–N	(mg l^{-1})	1	7	<0.5	6	4	10
Nitrification rates							
(kg NH_3–N kg^{-1} BTS d^{-1})		0.15	0.15	0.15	0.15	0.13	0.11
(kg NH_3–N m^{-3} d^{-1})		1.5	2.1	1.2	1.7	2.2	2.1

57.6 ECONOMICS OF THE BFB PROCESS

This is a very much neglected area. At the present time there are very few studies relating to the costs of building and operating biological fluidized bed processes.

It may justifiably be argued that it is not possible to give a truly accurate costing for BFB treatment since many of the cost factors are site-specific or depend upon the make-up of the particular effluent or sewage to be treated. It is certainly the case that the cost of any form of effluent treatment is very dependent upon the composition of the feed. Despite these problems it is useful to try to cost BFB treatment by comparison with conventional treatment processes for a given feed composition and flowrate. A number of cost comparisons have been made using UK-strength settled sewage, 200 mg BOD$_5$ l^{-1}, 50 mg Kj-N l^{-1} (Wheeldon and Bayley, 1981; Hoyland and Robinson, 1983).

In their study Wheeldon and Bayley examined a number of process configurations over a range of different flow rates. They costed the following processes in an oxygenic BFB for sewage treatment: (a) carbonaceous oxidation, (b) carbonaceous oxidation and nitrification, and (c) carbonaceous oxidation, nitrification and denitrification in a two-BFB system. For each case they compared the cost of the BFB process with that for building and operating a conventional activated sludge process to produce the same effluent quality. A Net Present Value (NPV) technique was used to cover both capital and operating costs. When this evaluation was done in 1980 the pilot plant work had started at WRc, Stevenage, but the demonstration scale plant (flow 150–400 m^3 d^{-1}) had not been built at Coleshill. When the BFB studies were completed at Coleshill a more limited economic evaluation was done in 1982 (Hoyland and Robinson). In general terms the first evaluations had shown that oxygenic BFB treatment, for carbonaceous oxidation, could show about a 10% saving over conventional activated sludge treatment. It had also clearly pointed out that because of the higher biomass concentration achievable in BFBs the capital cost savings were considerable but the operating costs were significantly greater than those for activated sludge treatment. Pumping and purchase or on-site production of oxygen were shown to be the main operating costs. Taken together the cost of pumping and oxygen could amount to 60% of the total NPV of sewage treatment. The time datum for these estimates was the third quarter (Q3) of 1976. When the demonstration-scale work was completed the cost of carbonaceous oxidation of settled sewage was recalculated and this indicated that BFB treatment could have an

NPV of 15% greater than conventional activated sludge treatment, again using Q3 1976 as the datum. The interest in BFB treatment has recently revived because it has been realized that the 'true' cost of oxygen has reduced since 1976. The cost of tonnage oxygen (in UK) has not increased significantly since Q3 1976 whilst all the other costs involved in the overall treatment cost have at least doubled. Since the cost of oxygen dominates the BFB treatment cost it seems likely that the cost of BFB treatment relative to AS treatment will have reduced. During the BFB treatment conference at UMIST in April 1980 (Cooper and Atkinson, 1981) Dorr-Oliver Inc. staff showed two economic analyses done using the NPV technique but at 1980 *United States prices* (Sutton *et al.*, 1981; Evans, 1981). These evaluations indicated that BFBs would be more economic than activated sludge processes for plants with flows greater than 22 000 m^3 d^{-1}. Their calculations also very clearly showed that *for USA conditions* the cost of pumping and oxygen was much less proportionately than for UK conditions. For carbonaceous oxidation of the same strength sewage as Wheeldon and Bayley in the UK study but using USA costs the Dorr-Oliver study showed the combined cost for energy for pumping and the cost of oxygen only amounted to 35% whereas it had been 60% in the UK study. This was largely attributable to the much higher cost of energy in the UK compared to the USA.

Gregory and Shieham (1981) assessed the cost of using BFB processes for denitrification of water for potable use. They showed that for most circumstances biological dentrification using a BFB was by far the most economic option.

57.7 CONCLUDING REMARKS

Much very sound and elegant research has been done in the past 15 years in taking BFBs to the stage where they can be applied on full-scale in a wide range of processes. BFBs have reached the stage where most of the process problems have been overcome and the hardware for carrying out the process has been proved on pilot-scale. The major stumbling block to their widespread use for sewage and industrial waste treatment is the cost of using oxygen. A secondary problem which appears to put off potential BFB users for sewage treatment is the *apparent* complexity. On first sight they seem complex but if looked at in detail this is not truly the case. Neither of these problems seems to be so significant in the field of industrial effluent treatment where chemical process operators are much more used to dealing with apparently complex processing plants. It is also in the chemical, food and pharmaceutical processing industries where space is at a premium that the extremely compact nature of the BFB process is so attractive. This may explain the much greater acceptance of the BFB technology in the USA for industrial effluent treatment as opposed to its much lesser acceptance for sewage treatment, the area from which the BFB process derived. Of the first 20 large pilot- and full-scale plants sold by Dorr-Oliver in the USA only one was for sewage treatment. To the outsider it would seem that the BFB process would have very great potential for use in the wider biotechnology field of chemical, food, brewing and pharmaceutical high-value products rather than the present more limited role of effluent disposal.

The advantages and disadvantages for BFBs will become clearer when their full-scale performance and economic data become available in the next few years. It already seems likely that the process will find several roles to play in the sewage and effluent treatment field in addition to those already established for potable water treatment.

57.8 REFERENCES

Barnes, D. (1983). Personal communication. University of New South Wales.
Barnes, D., P. J. Bliss, B. Grauer, E. M. Kuo, K. Robins and G. Maclean (1983). Influence of organic shock loads on the performance of an anaerobic fluidised bed system. Paper presented at the 38th Industrial Waste Conference, Purdue University, Indiana, May 1983.
Beer, C. (1970). Evaluation of anaerobic denitrification processes. *J. Sanit. Eng. Div., Am. Soc. Civ. Eng.*, **96** (SA6), 1452–1455.
Bosman J., A. A. Eberhard and C. I. Baskir (1978). Denitrification of a concentrated nitrogenous industrial effluent using a packed column and fluidised bed reactors. *Prog. Water Technol.*, **10** (5/6), 297–308.
Bosman, J. and F. Hendricks (1981). The technology and economics of the treatment of a concentrated nitrogenous industrial effluent by a biological denitrification using a fluidised bed reactor. In *Biological Fluidised Bed Treatment of Water and Wastewater*, ed. P. F. Cooper and B. Atkinson, pp. 222–233. Ellis Horwood, Chichester.
Cooper, P. F. (1984). Biological fluidised treatment of sewage in a two-bed system. Paper presented at the IChemE Symposium, 'Process Engineering Aspects of Immobilised Cell Systems', Manchester, March 1984.
Cooper, P. F. and B. Atkinson (eds.) (1981). *Biological Fluidised Bed Treatment of Water and Wastewater*. Ellis Horwood, Chichester.

Cooper, P. F. and D. H. V. Wheeldon (1981). Complete treatment of sewage in a two-fluidized bed system. In *Biological Fluidised Bed Treatment of Water and Wastewater*, ed. P. F. Cooper and B. Atkinson, pp. 121–144. Ellis Horwood, Chichester.

Cooper, P. F. and D. H. V. Wheeldon (1982). Complete treatment of sewage in a two-fluidized bed system. Part I. *Water Pollut. Control*, **81** (4), 447–462.

Cooper, P. F., D. H. V. Wheeldon, P. E. Ingram-Tedd and D. W. Harrington (1981). Sand/biomass separation with production of a concentrated sludge. In *Biological Fluidised Bed Treatment of Water and Wastewater*, ed. P. F. Cooper and B. Atkinson, pp. 361–367. Ellis Horwood, Chichester.

Dearborn Environmental Consulting Services (1981). Pilot-scale assessment of the biological fluidised bed process for municipal wastewater treatment. Environment Canada, Report SCAT-11.

Dorr-Oliver Co. (1978a). Flow distributor for fluid bed biological reactors. *Br. Pat. Appl.* 909 076.

Dorr-Oliver Co. (1978b). Apparatus and process for dissolution of gases in liquid. *Br. Pat. Appl.* 938 130.

Dunning, P. R. (1980). Biological treatment in fluidised beds. The Dorr-Oliver Oxitron system. Paper presented at a joint meeting of the Institute of Water Pollution Control and the Institution of Public Health Engineers, London, November 1980.

Evans, R. R. (1981). Contribution to discussion. In *Biological Fluidised Bed Treatment of Water and Wastewater*, ed. P. F. Cooper and B. Atkinson, pp. 351–354. Ellis Horwood, Chichester.

Francis, C. W. and C. W. Hancher (1981). Biological denitrification of high-nitrate wastes generated in the nuclear industry. In *Biological Fluidised Bed Treatment of Water and Wastewater*, ed. P. F. Cooper and B. Atkinson, pp. 234–250. Ellis Horwood, Chichester.

Francis, C. W. and C. D. Malone (1977). Anaerobic columnar denitrification of a high-nitrate wastewater. *Prog. Water Technol.*, **8** (4/5), 687.

Friedman, L. D., W. J. Weber, R. Bloom and C. B. Hopkins (1971). *Improving Granular Carbon Treatment*. US Environmental Protection Agency, Water Pollution Control Research Series, 17 020 GDN/07/71.

Gauntlett, R. B. and D. G. Craft (1979). *Biological Removal of Nitrate From River Water*, Technical Report TR 98. Water Research Centre, Medmenham, UK.

Gist-Brocades, NV (1980). *Eur. Pat. Appl.* 80 200 753.3.

Green, M. K. and P. J. Hardy (1984). The development of a high-rate nitrification fluidised bed process. Paper presented to the Institute of Water Pollution Control, London, January 1984.

Gregory, R. and I. Shieham (1981). Biological fluidised bed denitrification of surface water. The economics of a remedy for nitrate in drinking water. In *Biological Fluidised Bed Treatment of Water and Wastewater*, ed. P. F. Cooper and B. Atkinson, pp. 329–350. Ellis Horwood, Chichester.

Goodson, M. (1978). Process and apparatus for biological treatment of wastewater. *Br. Pat. Appl.*

Hakulinen, R. and M. Salkinoja-Salonen (1981). An anaerobic fluidised-bed reactor for the treatment of industrial wastewater containing chlorophenols. In *Biological Fluidised Bed Treatment of Water and Wastewater*, ed. P. F. Cooper and B. Atkinson, pp. 374–382. Ellis Horwood, Chichester.

Hoyland, G. and P. J. Robinson (1983). Aerobic treatment in an OXITRON biological fluidised bed plant at Coleshill. *Water Pollut. Control*, **82** (4), 479–491.

Jeris, J. S., C. Beer and J. A. Mueller (1974). High-rate biological denitrification using a granular fluidised bed. *J. Water Pollut. Control Fed.*, **46** (9), 2118–2128.

Jeris, J. S. and R. W. Owens (1975). Pilot-scale, high-rate biological denitrification. *J. Water Pollut. Control Fed.*, **47** (8), 2043–2057.

Jeris, J. S., C. Beer and J. A. Mueller (1976). Process for the treatment of wastewaters to remove oxidised nitrogen. *Br. Pat.* 1 430 410.

Jeris, J. S. (1976). Treatment of wastewater. *Br. Pat.* 1 433 582.

Jeris, J. S., R. W. Owens, R. Hickey and F. Flood (1977). Biological fluidised-bed treatment for BOD and nitrogen removal. *J. Water Pollut. Control Fed.*, **49** (5), 816–831.

Jeris, J. S. (1978). Apparatus and process for removing ammonia nitrogen from wastewater. *Br. Pat.* 1 520 895.

Jeris, J. S., R. W. Owens and F. Flood (1981). Secondary treatment of municipal wastewater with fluidised bed technology. In *Biological Fluidised Bed Treatment of Water and Wastewater*, ed. P. F. Cooper and B. Atkinson, pp. 112–120. Ellis Horwood, Chichester.

Jewell, W. J., M. S. Switzenbaum and J. W. Morris (1979). Sewage treatment with the anaerobic attached microbial film expanded bed process. Paper presented at the 52nd Water Pollution Control Federation Conference, Houston, Texas, October 1979.

Jewell, W. J. (1981). Development of the attached microbial film expanded-bed process for aerobic and anaerobic waste treatment. In *Biological Fluidised Bed Treatment of Water and Wastewater*, ed. P. F. Cooper and B. Atkinson, pp. 251–267. Ellis Horwood, Chichester.

Klapwijk, A. (1977). New points of view for the removal of organic compounds, nitrogen and phosphate from sewage. H_2O, **10** (9), 208.

Klapwijk, A., H. Smit and A. Moore (1981). Denitrification of domestic wastewater in an up-flow sludge-blanket reactor without carrier material for the biomass. In *Biological Fluidised Bed Treatment of Water and Wastewater*, ed. P. F. Cooper and B. Atkinson, pp. 205–216. Ellis Horwood, Chichester.

Lee, D. D., C. D. Scott and C. W. Hancher (1979). Fluidised bed bio-reactor for coal-conversion effluents. *J. Water Pollut. Control Fed.*, **51** (5), 974–984.

Lettinga, G., A. F. M. Van Velsen, S. W. Hobma, W. de Zeeuw and A. Klapwijk (1980). Use of the upflow sludge blanket (USB) reactor for biological wastewater treatment, especially for anaerobic treatment. *Biotechnol. Bioeng.*, **22**, 699.

Melcer, H., S. G. Nutt, I. Marvan and P. M. Sutton (1982). Combined treatment of coke-plant wastewater and blast furnace blowdown-water in a coupled biological fluidised bed system. Paper presented at the 55th Annual Conference of the Water Pollution Control Federation, St Louis, Missouri, October 1982.

Miyaji, Y. and K. Kato (1975). Biological treatment of industrial wastewater by using nitrate as an oxygen source. *Water Res.*, **9** (1), 95.

Mulcahy, L. T. (1978). *Mathematical Model of the Fluidised Bed Biofilm Reactor*. Ph.D. Thesis, University of Massachusetts.

Mulcahy, L. T. and E. J. La Motta (1978). Mathematical model of the fluidised biofilm reactor. Paper presented at the 51st Annual Conference of the Water Pollution Control Federation, Anaheim, California, October 1978.

Nutt, S. G., J. P. Stephenson and J. H. Pries (1979). Aerobic fluidised bed treatment of municipal wastewater for organic carbon removal. Paper presented at the 52nd Annual Conference of the Water Pollution Control Federation, Houston, Texas, October 1979.

Nutt, S. G., J. P. Stephenson and J. H. Pries (1980). Nitrification kinetics in the biological fluidised bed process. Paper presented at the 53rd Annual Conference of the Water Pollution Control Federation, Las Vegas, Nevada, October 1980.

Nutt, S. G., J. P. Stephenson and J. H. Pries (1981a). Steady and non-steady-state performance of the aerobic (oxygenic) biological fluidised bed. In *Biological Fluidised Bed Treatment of Water and Wastewater*, ed. P. F. Cooper and B. Atkinson, pp. 145–160. Ellis Horwood, Chichester.

Nutt, S. G., H. Melcer and J. H. Pries (1981b). Two-stage biological fluidised bed treatment of coke-plant wastewater for nitrogen control. Paper presented at the 54th Annual Conference of the Water Pollution Control Federation, Detroit, Michigan, October 1981.

Nutt, S. G., H. Melcer., I. J. Marvan and P. M. Sutton (1982). Treatment of coke-plant wastewater in the coupled pre-denitrification–nitrification fluidised bed process. Paper presented at the 37th Industrial Waste Conference, Purdue University, Indiana, May 1982.

Oppelt, E. T., J. M. Smith and W. A. Feige (1978). *Expanded Bed Biological Treatment*. US Environmental Protection Agency, Report No. EPA 600/2-78/-117.

Pugh, N. J. (1949). *J. Inst. Water Eng.*, **3**, 123–149.

Sehic, O. A. (1978). *The Use of Bioactive Fluidised Sand for the Rapid Carbonaceous Oxidation of Domestic Wastewater*. M. Eng. Sc. Thesis, University of Melbourne.

Sehic, O. A. (1981). Fluidised sand recycle reactor for aerobic biological treatment of sewage. In *Biological Fluidised Bed Treatment of Water and Wastewater*, ed. P. F. Cooper and B. Atkinson, pp. 179–189. Ellis Horwood, Chichester.

Short, C. S. (1973). *Removal of Ammonia from River Water*. Technical Paper TP 101. Water Research Association, Medmenham, UK.

Short, C. S. (1975). *Removal of Ammonia from River Water — 2*. Technical Paper TR 3. Water Research Centre, Medmenham, UK.

Sutton, P. M., W. K. Shieh, C. P. Woodcock and R. W. Morton (1979a). Oxitron system fluidised bed wastewater treatment process: development and demonstration studies. Paper presented at the Joint Annual Conference of the Air Pollution Control Association and Pollution Control Association of Ontario, Toronto, April 1979.

Sutton, P. M., D. Langley, K. Warner and K. Park (1979b). Oxitron fluidised bed wastewater treatment system. Application to high-stength industrial wastewaters. Paper presented at the 34th Industrial Waste Conference, Purdue University, Indiana, May 1979.

Sutton, P. M., W. K. Shieh, P. Kos and P. R. Dunning (1981). Dorr-Oliver's Oxitron system fluidised bed water and wastewater treatment process. In *Biological Fluidised Bed Treatment of Water and Wastewater*, ed. P. F. Cooper and B. Atkinson, pp. 285–300. Ellis Horwood, Chichester.

Sutton, P. M. (1983). Dorr-Oliver Inc., Stamford, CT, USA. Personal communication.

Weber, W. J., C. B. Hopkins and R. Bloom (1970). *J. Water Pollut. Control Fed.*, **42** (1), 83–89.

Wen Shieh, W. M., P. M. Sutton and P. Kos (1979). Oxitron system fluidised bed wastewater treatment process. Predicting reactor biomass concentration. Paper presented at the 52nd Annual Water Pollution Control Federation Conference, Houston, Texas, October 1979.

Wheeldon, D. H. V. and R. W. Bayley (1981). Economic studies of biological fluidised beds for wastewater treatment. In *Biological Fluidised Bed Treatment of Water and Wastewater*, ed. P. F. Cooper and B. Atkinson, pp. 306–328. Ellis Howood, Chichester.

58

Conventional Anaerobic Processes and Systems

D. A. STAFFORD
University College Cardiff, UK

58.1 INTRODUCTION

As more people in the developing world have demanded cheaper animal protein of a particular quality, the farming industry has had to adapt to these changing patterns by stressing the virtue of economy of size. This has been developed with a diminishing work force. For example, the UK male agricultural work force fell from 0.61×10^6 in 1947 to 0.31×10^6 in 1967 (Stafford *et al.*, 1981a). When considering pollution problems, the absolute ratio of human to animal population can affect the discharge waste rates of nitrogen and phosphorus from difference countries (Hawkes, 1977) (see Table 1). Thus in Europe, animal livestock contributes several times as much nitrogen and phosphorus as man and, for example, an average pig unit of 300 sows plus followers grown to bacon will produce the equivalent waste of a large town of 10 000–12 000 people. Many such farms exist in the UK and Western Europe. In most developed countries, animals are being produced more intensively in large numbers and on fewer farms (Stafford *et al.*, 1981a). The use of animal wastes as fertilizers has in fact declined over the last 100 years through the use of imported phosphates and nitrogen fertilizers produced using fossil fuel energy. However, with high energy prices and the realization that fertilizer nutrients are concentrated in animal wastes, the trend may well be reversed. The average manure production figures from farm animals are shown in Table 2 (Hawkes, 1977). The amount of water contained in the waste will vary according to local conditions and will also depend on the method of watering and on the food fed to the animals. Wastes are not only produced from animals but also from the storage of plant material, for example during the ensiling process. Thus about 1 tonne of ensiled green crops will produce 260–450 l of effluent (assuming 10–20% dry matter) (Hawkes, 1977).

Table 1 Comparison of Nitrogen and Phosphorus arising from Human and Livestock. Sources in Different OECD Countries[a]

	Nitrogen ($\times 10^3$ tonnes/year)			Phosphorus ($\times 10^3$ tonnes/year)		
	Livestock (A)	*Human (B)*	*A/B*	*Livestock (A)*	*Human (B)*	*A/B*
Austria	265	32	8	50	4	13
Belgium	270	4	7	40	5	8
Denmark	330	21	16	65	2.5	26
France	2015	210	10	270	22.5	11
Germany	1530	240	6	240	29	8
Italy	1060	225	4.5	140	26.5	5.5
Netherlands	375	54	7	55	6.5	9
Norway	145	17	9	20	2	10
Switzerland	175	26	7	25	3	8
United Kingdom	1480	235	6	210	28	8
Average A/B			8			11

[a] Hawkes (1977).

Table 2 Manure Production from Farm Animals (Faeces and Urine)

Parameter	*Poultry*	*Pig*	*Dairy cattle*
Amount produced (kg/day)	0.1	4.5	45
% of live weight	5–6	8–9	9–11
% dry matter	20–30	15–20	10–11
Average dry weight (kg/day)	0.03	0.7	4.5
BOD (kg/day)	0.01	0.2	1.4

58.2 FERTILIZER PRODUCTION

Of course, when determining the possible uses of agricultural wastes, it is vital to determine the nutrient value either as a fertilizer source or as a feed to animals. Table 3 indicates the elemental ranges found in farm manures (Hawkes, 1977). Thus, whether treated in aerobic or anaerobic systems (or not treated), a full analysis of the particular wastes is essential before consideration of the route for utilization.

Table 3 Nutrient Content of Manures

Manure	*Dry matter* (%)	Nutrient (kg/tonne fresh manure)			
		N	*P*	*K*	*Mg*
Cattle	4–23	2.4–6.5	0.4–1.8	2–5.8	0.2–0.6
Pig	5–25	1.6–6.8	0.6–2.1	1.7–3.6	0.3–0.7
Poultry	23–68	9.6–23	2.4–12	3.8–11.6	1.2–2.2

Certainly the fertilizer route offers some amelioration of farming costs in that 41% of UK agricultural costs are attributed to fertilizer costs (Turney, 1980). A study in the US has shown that with two groups (14 each) of farms, one group produced an overall production loss of only 10% when organic fertilizers were used compared with commercial chemical fertilizers (Turney, 1980). The organic farms required 12% more labour but 50% less energy and the net financial returns were about equal.

58.3 PROTEIN PRODUCTION

Whilst the positive aspects of waste utilization are highlighted here, it must be stressed that recycling of these materials provides, to some extent, the solving of an acute disposal problem. The total amount of livestock manure produced in the UK is estimated to be 150×10^6 tonnes. Animal manures are therefore found to be included in the feed of other farm animals, ruminants and non-ruminants. The dry matter digestibility of manure is about 60% in ruminants but only

10–20% in non-ruminants (Stafford *et al.*, 1981b). Feeding manures to ruminants may make a significant contribution to the manure disposal problem. Of course, for some time dry poultry manures (DPM) have been fed to cattle but 'it is impossible to put a figure on the DPM being included in animal feed due to the natural reluctance of those using the product for feeding purposes to divulge figures' (Stafford and Etheridge, 1982). As more people accept the phenomenon, this reluctance will diminish and more organic material will become available from manures for animal feeds. In the US, about 11.6, 2.9 and 15.9 × 10^6 tonnes of manure are available respectively from broilers, turkeys and layers every year (Couch, 1974). It is also stated that about 10% of poultry manure contains about 10% true protein (Couch, 1974). Also in the US, about 1.2 × 10^9 tonnes of cattle manure is calculated to be produced, and cattle manure was estimated to account for 30% of all the solid waste produced in the US (Ettinger and Wade, 1977). It contains twice as much phosphorus as all of the US domestic sewage. More exotic food chains are being devised for utilization of farm wastes and in the US maggot farming may well prove feasible since pupated fly larvae can be fed directly to fish or broilers, or the manure can be treated aerobically or anaerobically and used as a soil conditioner or basis for mushroom farming (Ettinger and Wade, 1977). The manure from 100 000 laying hens is expected to produce between 0.25–0.5 tonnes of pupae meal per day (US Ministry of Agriculture, 1972). It was also stated that up to 80% of sheep food could comprise DPM (US Ministry of Agriculture, 1972). The waste producing DPM will have a potential protein value depending on the presence of non-nitrogenous compounds. The most important of these is uric acid which, like urea, is utilized *via* rumen microorganisms. The uric acid is less soluble, is therefore available more slowly and is thus less susceptible to loss from the system (Oliphant, 1972). Since the DPM contains elements essential to sustain physiological systems, mineral supplements for animal feeds are less needed when DPM is added in the diet. Ruminants are found to be able to utilize the non-protein nitrogenous compounds such as uric acid much more readily than poultry (Warden, 1973).

A very useful application of broiler litter is that of recovery of spoilage tips, especially in mining areas. One such project mixed the litter with Italian ryegrass seed at a rate of 3 tonnes slurry per acre. The litter containing 3.5% N, 1.0% PO_4 and 1.5% K holds moisture well and liberated other essential elements slowly during growth of grass (White, 1970).

Thus the usable products from farm animal wastes major on protein but include contained energy resources such as cellulose, carbohydrates and fats. The refeeding of animal wastes to other farm animals has concentrated on the protein value, the energy portion of the diet deriving from grain fodder in the feed mix.

One of the major products from the bioconversion of animal wastes that is taking on increasing interest is that of methane production using anaerobic digesters. Although the energy resource is a product, others include the valuable fertilizer components, now more readily applied after digestion, and the protein value of the microorganisms released from the digester in the waste liquor. Animal feeding trials will settle the types of animal more suitable for feeding this microbial protein and the percentage of the overall feed can be supplied from the digester residue.

58.4 WHAT ARE ANAEROBIC DIGESTERS?

Digester systems essentially consist of holding tanks of simple to complex design, within which a series of biologial reactions decompose organic materials (present in many wastes) to methane, carbon dioxide, water and a number of other simple chemicals. No one simple design of digester can be considered as ideal, since many factors affect the design and operation. Digester types can be loosely categorized into five types: batch, continuous, high-rate contact, filter and expanded bed (see Figure 1). There are hybrids of all of these and as the technology grows rapidly, more and more variations appear. However, a number of basic designs are appearing to perform well and are commercially viable not only for the manufacturer but also for the operator (Stafford, 1982). This has come about mainly because of the pressure on availability of energy resources and the price of oil. Thus digesters have come into their own as alternative 'in-house' producers of energy. Indeed, some regard digesters as future biological power stations capable of producing methane for a national grid where crops are specially grown for feeding to digesters (Stafford and Hughes, 1981). Such countries that have a high land area to population ratio could be energy self-sufficient in this way, and they include Ireland, Sweden, Finland and New Zealand. It is calculated that in 20–50 years about half of our energy usage will need to be obtained from renewable resources such as green crops (Stafford and Hughes, 1981).

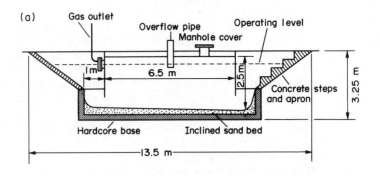

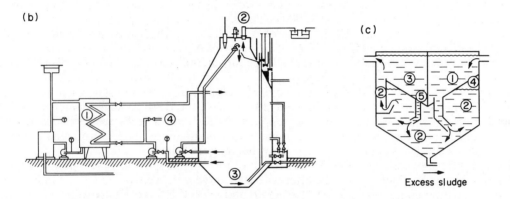

Figure 1 Types of anaerobic digester configurations. (a) Plug flow digester. (b) Classical industrial completely mixed digester: 1, mixed liquor heating system by liquor recirculation; 2, gas outlet; 3, mixed liquor outlet; 4, load outlet. (c) The hydraulic digester — the configuration ensures high hydraulic and organic loading with high biomass yield: 1, feed chamber; 2, main digester tank; 3, waste holding tank; 4, gas space; 5, settlement tube

58.5 WHAT WASTES CAN BE USED?

It is impossible to be exhaustive in listing the potential wastes that could be treated using anaerobic digestion, but which might also produce very good quantities of biogas. The waste characteristics will determine the relative digestibility of the organic materials present, and the pollution load, as measured by the chemical oxygen demand (COD), can be reduced by 20–99%. The higher the percentage reduction in COD the more methane is produced, as a general rule. Thus good reduction in pollution load is usually associated with good biogas production.

58.6 WHAT GOES ON INSIDE A DIGESTER?

The biological working units of anaerobic digesters are microorganisms of different types which play different roles in the overall process. They require to live in the absence of air and normally at a temperature of around 35 °C. Some digester systems operate at 25 °C, others at 55 °C, but they do so for specific applications and not all agree as to the applicability of these. The important point is that the microbes require to have their food (the incoming waste) in a constant supply for them to perform well and it must be readily available to them. Thus elaborate mixing and feeding systems in digesters have been developed. Although many and complex, the transformations that occur within anaerobic digesters are divided into three main stages. The first stage is the microbial hydrolysis of the polymers (if they are present) such as proteins, fats and carbohydrates plus other organics such as aromatic compounds. These polymers are hydrolysed to long chain fatty acids. The second bacterial group degrades these fatty acids to short chain fatty acids such as acetate. This is known as the acidogenic (acid producing) stage (Figure 2).

The methane bacteria are then able to utilize the acetate, hydrogen and carbon dioxide pro-

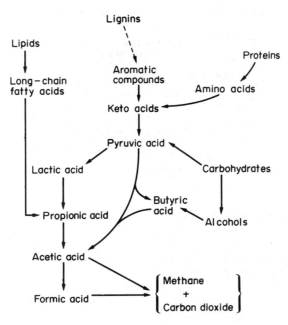

Figure 2 Some of the biochemical pathways during which organic compounds are transformed to methane during anaerobic digestion

duced by the first two stages to produce methane directly. This is known as the methanogenic (methane producing) stage:

$$HCO_2H \rightarrow CO_2 + H_2$$
$$4\,H_2 + CO_2 \rightarrow 2\,H_2O + CH_4$$
$$CH_3CO_2H \rightarrow CH_4 + CO_2$$

These microbial transformations within digester systems all occur at the same time but at different rates and thus the slowest reactions are the rate-limiting ones in the system. It is considered by many that the rate-limiting step is the initial hydrolysis stage, but this will largely depend upon the type of waste chemicals which have been fed to the digester systems. Many research teams are involved in studying the control of the overall process with a view to its improvement and newer digester designs are emerging (Stafford, 1984).

Although one of the more promising potential energy sources from anaerobic digestion is that of farm wastes, these slurries are regarded by farmers often as a necessary evil. However, this tendency to treat farm manures as a waste is changing, if slowly in some minds, and has been given added impetus since the dramatic rises in energy and fertilizer costs (Stafford and Hughes, 1981).

58.7 HANDLING PROBLEMS

The problem of collection and handling has become more acute with the increase in intensive farming methods and increased stringency in water pollution control. The animal waste from modern intensive farms is collected in the form of a slurry or semi-solid with an appropriately designed collection and storage system, thus reducing pollution by natural drainage from the farm. Traditional methods include the addition of bedding material to excreta, to absorb urine and to add carbohydrate to aid composting (Hobson and Shaw, 1973). Table 4 gives some idea of manure production from dairy cattle in the UK.

Pig slurry is normally collected in channels under slats; it will flow slowly and can be spread on land or distributed to a central digester by tanker. Alternatively it may be fed directly *via* a settling tank to a farm based digester.

Much of the manures can go on the land, but again agricultural 'run-off' is a common form of 'point source' pollution of rivers or lakes (Hawkes, 1977). Thus the pressure is to devise some form of treatment of farm wastes which may be less expensive than existing aerobic systems, will be just as efficient in removing pollution loads and which may provide useful energy sources, usable 'on-site' in some instances. From the point of view of the potential for anaerobic digestion

Table 4 Waste Arising from Dairy Cattle in the
UK

Country	Wet weight dairy cattle manure ($\times 10^6$ tonnes/year)
England	48.1
Wales	7.1
Scotland	6.8
N. Ireland	4.1

most farm manures are high in volatile solids (VS), being between 72 and 82%, the lower value for poultry and the higher for beef cattle (Stafford and Etheridge, 1982).

58.8 POTENTIAL GAS PRODUCTION

Loading rates in digesters, in terms of VS added, can vary from 0.7 to 25 kg VS m^{-3} d^{-1} (Stafford and Hughes, 1981). Higher loadings may be possible, but this will depend on whether toxic build up of volatile fatty acids occurs and on the water content of the manure. One of the major problems in the anaerobic digestion of animal wastes is the 'build up' of toxic concentrations of ammonia, produced from the high nitrogen content of the manures. They may be 0.4 g nitrogen per kg animal for cattle, 0.9 for pigs and 0.15 for humans. Toxic concentrations of ammonia on anaerobic digestion are difficult to determine and this will depend upon the hydraulic retention time and water content of the feed (Stafford and Hughes, 1981). One digester system was inhibited at unionized ammonia concentrations of 150 mg l^{-1} whilst another showed successful digestion of swine waste at concentratons of about 5000 mg l^{-1}.

Since animal wastes vary in terms of composition as well as solids content, and since the collection system affects the nature of the waste, each system needs to be evaluated separately before deciding on the type of digester operation. Before evaluating the general merits of the anaerobic digester treatment of animal wastes, it is important to consider the particular behaviour of individual waste types in laboratory, pilot and full-scale systems.

58.9 PIG WASTES — DIGESTION

Early research concentrated on the digestion of swine wastes since it is a strong agricultural waste and thus constitutes a great pollution hazard. The strong pig waste could potentially produce high concentrations of volatile fatty acids as well as high ammonia concentrations (Hobson and Shaw, 1973). It has been shown for example, that loading rates of 2.56 kg VS per m^3 per day at a hydraulic retention time of 14 days produced a fall off in BOD and COD removal with an apparent failure at 3.2 kg VS per m^3 per day. The ammonia concentration reached 1300 mg l^{-1} at maximum loading rate without inhibiting the process. Copper was cited to produce inhibition at 60–85 mg l^{-1} (wet basis), but these concentrations are not often reached (Hobson and Shaw, 1973).

When looking at the figures for pollution reduction when using anaerobic digestion for the treatment of pig waste, the percentage COD removal has ranged from 9.1 to 76% at retention times of from 6 to 30 days. Similar figures for BOD removal were 24.6–87% over a similar range of retention times. The average for both was 42% and 67% respectively. Any digestion operated for the purpose of reducing the pollution load on a system would require further treatment, possibly using a relatively inexpensive trickling filter system. However, one or two pilot sized digesters also produced fairly good quantities of gas in terms of grams per litre per day or litres per litre digester per day. This potential energy bonus could be used for pig-waste treatment, especially if it is coupled with usable gas production. It must be stressed however that in cold climates the energy consumption in winter may be as much as or greater than the total energy recovery (Dahl, 1980). If the good loading rates of about 4 kg VS per m^3 per day at a retention time of 10 days can be improved on, then the gas yielded may be significantly improved upon even if the BOD:COD removal rate is not.

Conventional digesters have been improved upon in recent years in that automated systems are available to farmers to produce a treatment procedure to reduce the pollution load, to produce methane energy and fertilizer for the farms. Such a system has been developed by the author in collaboration with Hamworthy Engineering Ltd., Poole, Dorset, UK (the manufacturers) and Chediston Agri-Systems, Chediston, Suffolk, UK (licensee). A 350 m³ digester is at present treating the waste from 4500 pigs and it is intended also to feed the waste from the digester to 120 cattle. The methane gas is converted to electrical energy using a modified Ford engine and generator set producing 20 kW h. The liquor from the digester is separated through a vibrating screen to produce a liquid stream and a solid residue. The liquid stream is used as irrigation water/fertilizer for arable crops and the solid residue can be used as animal feeds and organic fertilizer.

When considering a choice of digester system the newer types of units developed for smaller farms (up to 600 pigs or 60 cattle) include the plug-flow digester type developed by Professor Jewell of Cornell University, USA (Stafford and Hughes, 1981). These units have been used successfully for cattle digesters up to 30–50 m³. In this laboratory (Cardiff) we have been treating pig waste in conventional mixed systems and also in plug-flow anaerobic digesters. The relative performances are shown in Table 5 where it is shown that gas yields per kilogram of volatile solids added are about one third of the conventional digester, but the temperature of operation is also much lower. It is thus possible that the net gas yields may be comparable since less energy is required to heat the digester and to mix the contents. The pollution reduction is slightly better in the plug-flow due to longer settling times, but the hydraulic retention time is much longer (50 days for the plug flow unit, 15–20 days for conventional systems). For small farmers with low capital cost requirements the plug-flow digester could show advantages. For the farmer with large animal herds a more fully automated system with better gas yields and lower retention times would probably provide the better system.

Table 5 Anaerobic Digestion of Pig Waste: (A) Conventional Digester System; (B) Plug-flow Digester System

	Temp (°C)	Mixing	Volatile fatty acid VFA (mg l⁻¹)	Ammonia (mg l⁻¹)	Gas yield (m³ m⁻³ d⁻¹)	m³ kg⁻¹ VS added	% TS In fluent
(A)	32–35	Intermittent (2 min/ 20 min)	1300–2900	2100–3600	0.40	0.38	3.2
(B)	18–20	No mixing	1000–3500	1100–3500	0.02	0.13	3.2

58.10 AGRICULTURAL CROP WASTES

Since the beginning of the modern energy crisis, which began in 1973 when oil prices were first hoisted to an alarmed world, renewable energy resources have assumed an increased potential in offsetting some of the energy shortfalls. With crop wastes and energy green crops providing a valuable starting material for wet bioconversions, the potential can be put in perspective by noting the productivity on a global basis (Table 6).

Table 6 Breakdown of World's Primary Productivity[a]

Geographical type	Net productivity (%)
Forests and woodlands	44.3
Grasslands	9.7
Cultivated land	5.9
Desert and semi-desert	1.5
Freshwater	3.2
Oceans	35.4

[a] Slesser and Lewis (1979).

Biomass derived from cultivated land represents a small percentage of the total, and the waste plant products an even smaller percentage, nevertheless at a local level and even, in some cases, at a national level, large amounts of waste materials may be produced which could be bioconverted to useful energy resources and other by-products.

Renewable biomass is going to constitute one of the largest reservoirs of potential organic energy since the non-renewable energy sources are being used up rapidly. It is important therefore to highlight the practicalities of using one bioconversion anaerobic digestion system for the treatment of waste materials and to determine the logistics of doing this. It is also important to maximize the energy-yielding potential of the system and to study the most efficient means of using the energy. Economics play an initial role, but local conditions need to be considered such as pollution control, energy availability and the possibility of mixing wastes.

We hope to see more and more of this new biotechnology being applied across a variety of industrial systems.

58.11 HOW CAN THE METHANE VALUES OF WASTES BE DETERMINED?

It is possible to build economic models of digester systems using many waste materials. The most important parameter to determine is the yield of biogas obtainable from a known amount of waste. This can be measured as biogas (m^3) per kilogram organic material added. This must be determined experimentally in laboratory and pilot-scale units. From this information certain other assumptions can be made and yields of gas calculated for certain sizes of digester. Such an exercise was undertaken and the results are shown in Table 7 (Langley, 1980).

Table 7 Summary of Cost Estimates for 'Model' Digestion Processes[a]

Digester capacity (m^3)	30	100	500	2500
Feedstock (tonnes/year)	3	10	50	250
Gross biogas output (m^3/day)	135	450	2250	11 250
Nett energy output (TJ/half year)	0.8	2.6	13	68
Capital cost (£ × 1000)	15	31	88	329
Operating cost (£ × 1000/year)	4.0	8.4	30	119
Nett biogas costs (£/GJ)	4.9	3.2	2.3	1.7
Current costs of				
Natural gas (£/GJ)	1.4–1.9			
Liquid propane (£/GJ)	3.3–3.8			

[a] Langley (1980).

When animal wastes are treated using such treatment processes as anaerobic digestion, the residue provides good fertilizer potential. If this fertilizer is applied to farming land at the end of the main-crop growing season, energy crops may be profitably grown.

58.12 METHANE PRODUCTION FROM ENERGY CROPS

Laboratory studies at University College Cardiff have demonstrated the potential of producing biogas from 'catch' energy crops. Prototype units in New Zealand and the USA are also demonstrating the production of economic quantities of biogas from fuel (energy) crops.

Species already studied in the laboratory as potential 'catch' energy crops include kale, radish, rape and bromus. They were fed to laboratory-scale anaerobic digesters at a solids concentration of about 7% and a volatile solids percentage of 70–80. Radish was the best biogas producer per unit of dry weight added, followed by kale, rape and bromus.

A number of anaerobic digester systems of different designs have been developed to study the most economic method for methane production. Workers at University College Cardiff, UK, in collaboration with the Department of Agriculture, University of Reading, UK, have selected the most useful 'catch' green crops for feeding to the digester systems. In the EEC no such programme exists and it is hoped that a practical system will be developed to produce economic quantities of biogas (Stafford and Etheridge, 1982). It has been calculated by a UK Department of Energy study that biogas could be produced at an economic cost, similar to that from the North Sea. The possible inclusion of other farm wastes will be studied to determine the enhancement of methane production. The process of polymer breakdown (*e.g.* cellulose) will be studied biochemically to determine whether gas production can be improved (Figures 3 and 4).

Thus, by utilizing modern digester technology it will be possible to produce methane energy from farm wastes and even from biomass especially grown for this purpose.

58.13 REFERENCES

Couch, J. R. (1974). Evaluation of poultry manures as a feed ingredient. *Feedstuffs*, **86**, 390–404.

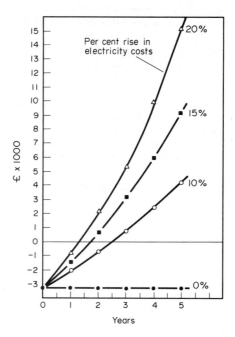

Figure 3 Annual changes in income (from the digestion of the waste from 5000 pigs) in relation to changes in energy costs

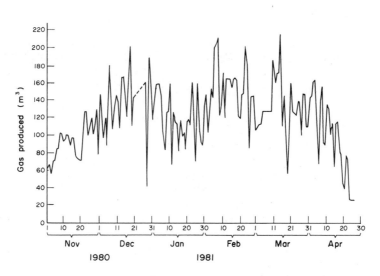

Figure 4 Gas production *versus* time for a 350 m³ farm digester. With a 350 m³ full-scale digester in Suffolk (UK) an attempt was made in the initial design concept to reduce the rate of loss of the solids (containing the active microbes) from the digester. This is done by incorporating a settling tube but low hydraulic retention times are often difficult to achieve. Nonetheless this 350 m³ digester has performed remarkably well. New designs are now being developed to more efficiently retain the active biomass and thus increase performance

Dahl, K. (1980). Anaerobic digestion of Danish cattle waste in a plug flow reactor. In *Anaerobic Digestion — Poster Papers*, ed. D. A. Stafford and B. I. Wheatley. Anaerobic Digestion Scientific Press, Cardiff.

Ettinger, M. B. and L. Wade (1977). Conversion of cattle manure into high quality protein. *26th International Waste Conference*, Purdue, USA.

Hawkes, H. A. (1977). Eutrophication of rivers — effects, causes and control. In *Treatment of Industrial Effluents*, ed. A. G. Callely, C. F. Forster and D. A. Stafford, pp. 159–192. Hodder and Stoughton, London.

Hobson, P. N. and B.G. Shaw (1973). The anaerobic digestion of waste from an intensive pig unit. *Water Res.*, **8**, 437.

Langley, K. F. (1980). Renewable energy through anaerobic digestion. In *Anaerobic Digestion*, ed. D. A. Stafford, B. I. Wheatley and D. E. Hughes. Applied Science, London.

McAllister, J. S. V. (1970). Collection and disposal of farm wastes. *Water Pollut. Control*, **69**, 425.

Oliphant, J. M. (1972). Dried poultry waste and intensive beef. *Agriculture*, **79**, 515–518.

Slesser, M. and C. Lewis (1979). *Biological Energy Resources*. Wiley, New York.

Stafford, D. A. (1982). The effects of mixing and volatile fatty acid concentrations on anaerobic digester performance. *Biomass*, **2**, 43.

Stafford, D. A. (1984). The anaerobic digestion of food processing wastes. *Inst. Chem. Eng. Symp. Ser.*, **84**, 89–108.

Stafford, D. A. and A. P. Etheridge (1982). Farm wastes, energy production and the economics of an anaerobic digester. In *Anaerobic Digestion 1981*, p. 255. Elsevier Biomedical Press, Amsterdam.

Stafford, D. A. and D. E. Hughes (1981). Microbial production of fuels. In *Energy — Present and Future Options*, p.60. Wiley, Chichester.

Stafford, D. A., B. I. Wheatley and D. E. Hughes (1981a). *Anaerobic Digestion*. Applied Science, London.

Stafford, D. A., D. L. Hawkes and R. Horton (1981b). *Methane Production from Waste Organic Matter*. CRC Press, Boca Raton, FL.

Turney, H. (1980). Agricultural wastes as fertilizer. In *Handbook of Organic Waste Conversions*. ed. W. M. Bewick. Van Nostrand Reinhold, Amsterdam.

US Ministry of Agriculture (1972). *Agriculture*, **79**, 98–103.

Warden, W. M. (1973). Dried poultry manure utilization. *Poultry Digestion*, **32**, 344.

White, C. (1970). Broiler litter on Welsh coal tips. *Agriculture*, **77**, 49.

59
Fluidized Bed Anaerobic Reactors

M. S. SWITZENBAUM
University of Massachusetts, Amherst, MA, USA

59.1 HISTORY OF DEVELOPMENT

59.1.1 Introduction

The anaerobic fermentation process has long been recognized as a useful process in wastewater treatment (McCarty, 1981). In the past, however, broad-scale application of this process has been largely with the treatment of municipal sewage sludge and animal residues to achieve waste stabilization and solids reduction. Yet, the anaerobic fermentation process can be used for the treatment of both dilute and concentrated soluble wastes, although it is seldom done in engineering practice. This lack of application is most likely due to the basic lack of understanding of the fundamental concepts associated with anaerobic methane production and the general feeling of unreliability many associate with anaerobic digestion.

Over the more recent past, advances in the basic understanding of the microbiology and biochemistry and advances in the hardware technology have helped to overcome many of the problems associated with anaerobic fermentation (Zeikus, 1980). Along with this, critical energy shortages have provided an additional interest in anaerobic fermentation processes due to their positive energy balances. Among the newer technological advances have been the anaerobic expanded bed and fluidized bed processes. It is the purpose of this paper to describe the anaerobic expanded bed and fluidized bed processes and discuss their application in anaerobic systems.

59.1.2 Development of the Process

The development of both expanded and fluidized beds for wastewater treatment has been reviewed by Cooper and Wheeldon (1980). These systems have been used for denitrification of water and wastewater, aerobic oxidation for BOD removal and nitrification, as well as anaerobic fermentation of wastewater.

The application of this technology for anaerobic treatment was developed in the laboratory of Dr. William J. Jewell of Cornell University. Jewell (1982) originally was looking at a means of optimizing aerobic systems. Work by Jewell and MacKenzie (1972) demonstrated that attached films had twice the removal capacity of suspended systems under comparable conditions. In a subsequent study, Jewell (1974) proposed the attached film expanded bed process as a means of optimizing aerobic treatment systems. This was based on the assumption that large biomass concentrations could be achieved on the large surface area provided by the small sand-sized particles. The small particles, which would be fluidized, would minimize diffusion limitations and eliminate clogging problems.

Beginning in 1974, the major focus on expanded bed development shifted to anaerobic treatment. Leuschner (1976) demonstrated in a short study that the expanded bed was able to treat synthetic sewage at 20 °C, with effluent concentrations reaching 20 mg l^{-1} COD. Jewell *et al.* (1981) conducted a preliminary study with primary effluent from the Ithaca, New York treatment plant. Greater than 70% COD removal efficiencies were obtained at retention times of one hour and greater at 20 °C. A subsequent study (Switzenbaum and Jewell, 1978) was carried out using a synthetic substrate to define the effect of temperature, flow rate, organic volumetric loading rate and influent substrate concentration on process efficiency. The expanded bed was found to be able to achieve high organic removal percentages at low temperatures (10 °C, 20 °C), treating low strength wastes (COD 600 mg l^{-1}) at short detention times (several hours) and at high organic loading rates (up to 8 kg COD m $^{-3}$ d^{-1}).

A subsequent study (Jewell *et al.*, 1981) demonstrated that shock loading (in terms of temperature and loading strength) had relatively little influence on the process. Morris and Jewell (1982) investigated the efficiency of the expanded bed treating particulate wastes and found it to act also as an efficient solids removal system.

Later, the expanded bed was applied to the treatment of higher strength wastes. This will be reviewed in a later section of this chapter.

59.2 REACTOR DESCRIPTION AND PRINCIPLES OF OPERATION

59.2.1 Description

Anaerobic expanded and fluidized beds are similar in concept. Both resemble systems which have been commonly used in chemical engineering processing technology. Generally, they have been applied to gas–solids contacting, mainly for combustion. The exact difference between expanded and fluidized beds is somewhat ambiguous. In many cases, fluidization has been used to refer to more than a doubling in the reactor volume as caused by the high flow rate of gas through the filter composed of small particles. The term 'expanded bed' has been used to designate reactors that have a smaller degree of expansion of the static volume.

The expanded/fluidized bed as applied to wastewater treatment consists of inert sand-sized particles in a column which expand with the upward flow of waste through the column. A schematic of the process is shown in Figure 1. The inert particles act as a support surface for the growth of attached microorganisms. The degree of expansion in these systems is dependent on the type of biological reaction, owing to the fact that biomass grows on the media, and thus decreases the overall density of the particles. Thus higher yielding systems such as aerobic respiration would have a higher degree of expansion (due to thicker biofilms developed) than would an anaerobic fermentation system for a given medium, medium size and superficial velocity. In this discussion, only anaerobic expanded and anaerobic fluidized beds are considered. Both of these systems operate at less than full fluidization or doubling of the reactor volume.

59.2.2 Fluidization Principles

The mechanics of fluidization have been thoroughly described by several investigators (Zenz and Othmer, 1960; Davidson and Harrison, 1971) and will only briefly be reviewed here.

From the point of view of design, two equations are of major importance in describing fluidization in anaerobic fluidized beds. One describes minimum fluidization velocity, the other head loss.

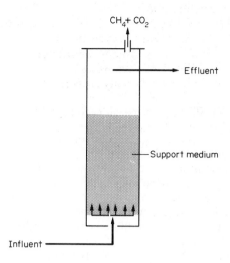

Figure 1 Schematic diagram of an anaerobic fluidized bed reactor

A fluid flowing at low velocities through a porous bed of small particles may not cause the particles to move. However, if the fluid velocity is steadily increased, a point is eventually reached at which the particles no longer remain stable but fluidize under the action of the fluid. At this point, the particles are supported by the fluid. Cleasby and Baumann (1977) have provided a review which characterizes bed expansion and stated that the minimum fluidization velocity can be estimated from the following empirical nonhomogeneous equation:

$$V_{mf} = \frac{0.00381 d^{1.82} \left[\gamma (\gamma_s - \gamma) \right]^{0.94}}{\mu^{0.88}} \tag{1}$$

where V_{mf} = minimum fluidization velocity (m s^{-1}); d = particle diameter (m); γ, γ_s = fluid and particle specific weights (N m^{-3}); and μ = viscosity (N s m^{-2}). The fluidization velocity is of particular importance in anaerobic systems due to the need for recycle to keep the particles in suspension. Higher superficial velocities will of course require more energy.

The head loss or pressure drop within a fluidized bed reactor is also of importance. The head loss may be predicted from the following equation:

$$\Delta P = \frac{L(\gamma_s - \gamma)}{\gamma} (1-\varepsilon) \tag{2}$$

where ΔP = head loss (m); L = bed height (m); and ε = porosity of the fluidized bed.

59.2.3 Operating Conditions

For fluidized beds at a given set of operating conditions, the particle-fluid velocity will determine the concentration of particles, or porosity, within the reactor. The fluid viscosity and density, and the particle size, shape and density will characterize the operating conditions.

A model has been proposed by Shieh *et al.* (1981) for estimating biomass concentration, an important process parameter, in the fluidized bed reactor. A sensitivity analysis was applied to the model to determine the effect of certain variables on biomass concentration. The following general results were shown.

(1) Medium size: for a given film thickness, biomass concentration increases with decreasing medium size until a maximum value is reached. This maximum concentration depends on the superficial upflow velocity and biofilm thickness. It was noted that at smaller media sizes, when a certain critical thickness is reached, carry-over will occur.

(2) Superficial upflow velocity: for a fixed medium size, biomass concentration will decrease with increasing superficial upflow velocity, the rate of change increasing with increasing biofilm thickness.

(3) Biofilm thickness: for a fixed medium size, biomass concentration increases with increasing film thickness until a maximum value is reached. It was noted that for smaller media sizes, thin films offer the advantage of preventing biomass carry-over at higher superficial velocities. Also, the fraction of active biomass, that is the biomass that substrate penetrates through, is greater under identical bulk liquid concentrations with thin films.

(4) Biofilm density: for a fixed film thickness, it was noted that a slight change in biofilm moisture content causes a large change in biomass concentration (inversely related). Moisture content is a function of the wastewater and microbial population that predominates for a particular wastewater and set of environmental conditions.

The information presented by Shieh *et al.* (1981) is of a more general nature, that is not particularly related to anaerobic applications. In regard to anaerobic applications, Heidman (1982) has pointed out that the thin dense films which develop should have little impact on the fluidization characteristics of the bed as a whole. Jewell (1982) has noted that lower density particles, having lower expansion velocities, are more compatible for anaerobic application due to better solids management (that is microbial solids capture within the system) and lower frictional pressure losses. Jewell also pointed out that this would be achieved with relatively small particles, that is diameters between 0.1 and 0.4 mm.

Then, for anaerobic fermentation applications, the fluidized bed should be designed using small, light particles as a support medium. These particles would be easy to expand and would provide a larger surface area to volume ratio, thus allowing higher biomass concentrations. There would exist, however, a minimal particle size, in order to prevent carry out if the overall density of the particles becomes too low. In addition, lower superficial upflow velocities required for lighter particles would result in thicker stagnant liquid layer depths around the bioparticles. This would result in increased diffusional resistances, which in turn could be the overall rate controlling parameter.

Thus, there exist many variables which are interrelated and which must be considered in the proper design and operation of a fluidized bed reactor. These include particle and fluid velocity, fluid viscosity and density, particle size, shape and density. While there exists a large amount of information describing the behavior of gas–solid fluidized beds, liquid–solid systems such as the anaerobic fluidized bed have not been studied as extensively. A proper mathematical model for the anaerobic fluidized bed would have to include a mathematical description of the physical process configuration and a mechanistic model of the biofilm (including intrinsic biological kinetics and mass transport considerations). At the present time, the optimization of these various operating conditions has not been achieved for the design of the anaerobic fluidized bed reactor. The next section of this chapter will review several modeling attempts which have provided valuable insight.

59.2.4 Modeling

A model has been developed by Mulcahy and La Motta (1978) for a denitrification biological fluidized bed reactor. They found that the most significant parameter affecting efficient conversion of substrate was biofilm thickness. It was further determined that biofilm thickness and thus reactor performance could be regulated by specification of five design parameters including (1) expanded height, (2) reactor area perpendicular to flow, (3) support medium density, (4) support medium diameter and (5) total volume of support medium in the reactor.

Shieh (1980) proposed a mathematical model for a fluidized bed reactor operated in a plug flow mode based on the assumption that the substrate conversion reactions following intrinsic zero-order kinetics are limited by the diffusion of substrate within the biofilm and that the substrate penetrates only partially into the biofilm due to internal mass transfer resistance. The solution to the model equations yielded a pseudo rate coefficient which can be maximized by varying media size and density. Excessive bed expansion associated with the use of small media caused a decrease in the biomass concentration and thus the efficiency of the reactor. In addition, excessive media size also decreased the biomass concentration. Lower bed expansions at a given superficial velocity when denser media were used yielded a higher concentration in a reactor. Shieh also indicated that there would exist an optimal biofilm thickness which would result in a maximum substrate conversion rate.

A simplified model for packed-bed biofilm reactors has been developed by Meunier and Williamson (1981a). The model incorporates the concepts previously developed by Williamson and

McCarty (1976) into a simpler framework which can subsequently be used for the engineering design of packed-bed reactors.

Conceptually, the model consists of two basic processes. These include diffusion of substrate from the bulk liquid phase into the biofilm and then conversion at the substrate by the bacteria within the biofilm. The authors verified the model for a number of submerged filter and expanded bed applications (Meunier and Williamson, 1981b). The authors concluded that the expanded bed reactor has a higher removal rate than the submerged filter because of higher surface area per unit volume, lower liquid layer depth and higher maximum substrate utilization rate (Meunier and Williamson, 1981b).

Switzenbaum and Jewell (1978) developed two simplified equations relating process efficiency to two fundamental microbial parameters in an anaerobic fluidized bed reactor, namely the net specific film growth rate and the specific film utilization rate. The equations were used to illustrate the importance of influent substrate concentration and organic loading rate. Data indicate that effluent quality is directly dependent on both influent concentration and organic loading rate. Additionally, at a given organic volumetric loading rate, higher removal efficiencies are achieved with more concentrated wastes.

59.2.5 Design

At the present time there are no deterministic methods available for designing an anaerobic fluidized bed system. As stated previously, empirical methods must be used for design at the present stage of development.

As such, the detention time for a fluidized bed reactor may be calculated as such:

$$\text{detention time (D.T.)} = \frac{\text{medium volume } (V_m)}{\text{waste flow } (Q_o)} \tag{3}$$

and medium volume may be calculated as

$$V_m = (\text{D.T.})(Q_o) = (A_c)(H_b) \tag{4}$$

where A_c = reactor cross-section, and H_b = packed-bed height.

There exists a tradeoff between reactor cross-sectional area (A_c) and packed-bed height (H_b). For fluidization, one would maximize A_c:

$$A_c(\text{max}) = \frac{Q_t}{V_f} \tag{5}$$

where Q_t = total flow rate (waste flow + recycle flow, Q_r) and V_f = fluidization velocity, while for required detention time one would minimize H_b:

$$H_b(\text{min}) = \frac{V_m}{A_c} \tag{6}$$

The design cross-sectional area can be less than the $A_c(\text{max})$, and the design packed-bed height can be greater than $H_b(\text{min})$. Generally, the lowest capital cost is obtained if $(A_c)(H_b) = V_m$.

The fluidized velocity, V_f, will be a function of medium density, ϱ, and medium diameter, d. When expansion is just starting:

$$V_f = \frac{\text{total flow}}{A_c} = \frac{Q_o + Q_r}{A_c} \tag{7}$$

The recycle flow rate, Q_r, is an important parameter in the operation of the anaerobic fluidized bed. Because of the kinetic limitations of the anaerobic methane fermentation process, recycle is necessary to insure adequate contact time along with bed expansion. Recycle will also provide the benefits of maintaining expansion under diurnal flow, equalize loading and provide pH control. Of course, recycle increases Q_t, which results in an increase in pumping energy demand. Pumping energy is also influenced by particle density and packed-bed height.

Finally, one must remember that total systems cost will be the sum of capital and operational costs. Adjustment of Q_r, ϱ and d for minimizing capital costs may be offset by their effect on operational costs.

59.3 CASE STUDIES

There have been several studies using anaerobic fluidized bed reactors. These are summarized in Table 1. As is seen, the process has been tested with domestic sewage, heat treatment liquor and a number of industrial wastes. In general, the performance of these systems on both a laboratory and pilot scale, has been very good. The processes are generally able to operate at high efficiencies and at greater organic volumetric loading rates and lower hydraulic retention times and lower temperatures than other anaerobic processes.

Table 1 Anaerobic Fluidized Bed Studies

Waste	Influent $(g\ l^{-1})^f$	Percent removal	Loading rate $(kg\ COD\ m^{-3}\ d^{-1})$	Temperature ($^{\circ}C$)
Acid whey[a]	50.3–56.1	72.0–83.6	13.4–37.6	35
Acid whey[a]	52.2–55.4	65.2–71.0	15.0–36.8	24
Food processing[a]	7.2–9.4	75–86	3.5–24.1	35
Chemical[a]	12[f]	79–93		
	8.4 (BOD$_5$)	81–98	4.1–27.3	35
Soft drink bottling[a]	6	66–89	4–18.5	35
	3.9 (BOD$_5$)	61–93		
Heat treatment liquor[a]	10	52–75	4.3–21.4	35
	5 (BOD$_5$)	66–95.5		
Whey permeate[b]	6.8	68	8.6–10.4	30–35
Whey permeate[b]	27.3	82	5.3–7.4	30–35
Sewage[c]	0.186	0–81	0.65–35	20
Synthetic sewage[d]	0.2–0.6	24.5–85.3	0.8–43.2	30
	0.2–0.6	38.0–87.8	0.8–43.2	20
	0.2–0.6	35.2–83.2	0.8–43.2	10
Sweet whey[e]	10	36.9–93.1	8.9–60.0	25
Sweet whey[e]	5–20	58.9–92.3	8.2–29.1	35

[a] Hickey and Owens (1981). [b] Sutton and Li (1982). [c] Jewell *et al.* (1981). [d] Switzenbaum and Jewell (1978). [e] Switzenbaum and Danskin (1982). [f] COD unless otherwise stated.

The high performance capability of the anaerobic fluidized bed is due to the successful combination of the anaerobic methane fermentation process, fixed film bioreactors and fluidized bed technology.

Fluidized beds have been widely used in chemical engineering practice, particularly for gas–solids operations, because of the good contact which can be achieved between catalysts and the reactants. More recently, the fluidized bed concept has been applied to biological systems. Both enzymes and whole cells can be immobilized onto water-insoluble materials retaining their catalytic activity, and these materials may then be used in a fluidized bed reactor to insure a large amount of contact between the enzymes and substrate. Atkinson and Davies (1972) proposed a completely mixed microbial film fermenter (CMMFF) based upon the fluidized bed principle for application to continuous operation when using growth-associated systems. Such a system, which would contain microbial mass mainly as surface films, would allow high flow rates with no danger of microbial mass washout. Atkinson and Davies (1972) also pointed out the reduction in diffusional and mass transport limitations in such a system.

The anaerobic fermentation process has long been known in environmental engineering practice. In relation to waste treatment with aerobic processes, the anaerobic process has the advantages that (1) a higher degree of waste stabilization can be achieved; (2) lower amounts of residuals are developed (*i.e.* lower growth yield); (3) lower nutrient additions are necessary due to the lower growth yields; (4) no oxygen is required; and (5) methane, a valuable byproduct, is produced. On the other hand, there are several disadvantages associated with anaerobic processes, such as (1) a temperature requirement of 35 °C; (2) the slow growth rate of the methanogens; and (3) the sensitivity of the methanogens to toxics and shock conditions.

The fluidized bed configuration has been shown to take advantage of the positive aspects associated with anaerobic processes while minimizing the disadvantages (Jewell *et al.*, 1981). The process was able to treat low strength wastes at reduced temperatures and high loading rates. In addition the process was found to adapt well to changes in both temperature and/or loading rates.

Much of the success of the anaerobic fluidized bed is due to the very high solids retention times which can be developed in the process due to the high surface area to volume ratio achieved by the sand-sized particles. However, another significant factor is the reduction in external mass transfer limitations caused by the good mixing in the bulk liquid solution and reduction in diffusional limitations achieved by the thin biofilms and thin stagnant liquid layers around the particles. In fixed film systems, which are heterogeneous systems, mass transfer and diffusional considerations play an important role in the overall reaction rate. The fluidized bed configuration is able to minimize these limitations due to the good mixing and thin films which develop on the support particles. This is especially true in anaerobic systems where the attached films are generally of the order of 10–100 μm.

59.4 APPLICATIONS AND ECONOMIC ASPECTS

59.4.1 Low Strength Wastes

The anaerobic fluidized bed was originally developed for the treatment of low strength wastes, such as municipal domestic wastewater (Switzenbaum and Jewell, 1980; Jewell *et al.*, 1981). The impetus for anaerobic treatment of domestic wastewater is very large. Since the anaerobic fermentation results in a lower cellular yield, less sludge is generated and hence lower sludge handling costs would be possible. In addition, lower energy requirements would result since aeration would not be necessary and methane would be produced as a byproduct. To date, no anaerobic process has been developed for the treatment of domestic wastewater, but the anaerobic fluidized bed shows great promise of achieving this goal (Jewell *et al.*, 1981).

Recently, a preliminary design for an anaerobic fluidized bed system was developed for evaluating the relative merits of this process in greater detail (Owen, 1982). This design was set for the attainment of secondary treatment standards. By doing so, costs and energy consumption estimates can be evaluated in relation to conventional secondary treatment methods such as the activated sludge and trickling filter processes for producing effluents and dewatered sludges of comparable quality.

A summary cost comparison of the three treatment alternatives is shown in Table 2 for two different sized treatment plants. The primary advantage in terms of capital costs for the fluidized bed system are associated with the secondary treatment process itself and with sludge handling unit operations. In terms of sludge processing the anaerobic fluidized bed will produce less sludge than an aerobic system and a more stable sludge because of the high solids retention time inherent to the system. This will result in capital and operation and maintenance costs savings. Another significant saving results from energy costs. The anaerobic fluidized bed was estimated to consume 20 to 48% less energy than comparable aerobic systems. This is due to lesser sludge handling, elimination of aeration and the generation of methane gas.

Table 2 Comparison of Costs of Three Treatment Alternatives for Secondary Wastewater Treatment

	Fluidized bed	Activated sludge	Trickling filter
Capital costs			
3.785 m^3 d^{-1} (1 MGD)	2845	3408	3503
94.625 m^3 d^{-1} (25 MGD)	22 914	25 068	36 424
Operation and maintenance costs			
3.785 m^3 d^{-1} (1 MGD)	140	170	130
94.625 m^3 d^{-1} (25 MGD)	950	1230	990

[a] In $1000.   [b] Owen (1982).

While the results from the analysis presented by Owen (1982) are based on preliminary design information which would require further substantiation, they are significant in demonstrating the potential benefits of municipal wastewater treatment with the anaerobic fluidized bed. Heidman (1982) arrived at a similar conclusion when comparing the anaerobic fluidized bed system with an anaerobic packed-bed system.

59.4.2 High Strength Wastes

As was shown in Table 1, a number of different industrial wastes have been treated successfully on laboratory and pilot scale anaerobic fluidized beds. Switzenbaum (1982) has stated that for high strength wastes, the anaerobic fluidized bed is a more efficient process than anaerobic filters mainly due to the minimization of diffusional limitations which results in higher overall reaction rates and hence smaller required reactor volumes.

At the present time there are several full scale anaerobic fluidized beds which have been constructed. Owens *et al.* (1980) described the use of an anaerobic fluidized bed for the treatment of a soft drinks bottling plant waste. Two 40 m diameter by 10.4 m high reactors were designed to treat approximately 420 kg COD per day. Sutton *et al.* (1982) described the design and construction of an anaerobic fluidized bed for a soy processing waste. The full scale plant consists of four 6.1 m diameter by 12.5 m high reactors with a design capacity of removing 8165 kg BOD_5 per day. Costs for these systems are of the order of \$1 million, but include significant royalty fees due to the proprietary nature of these processes.

59.4.3   Advantages and Disadvantages of the Anaerobic Fluidized Bed

The success of any anaerobic fermentation is dependent on bringing the substrate into contact with the microbial species for a sufficient period of time to allow the reactions to occur. The early anaerobic processes achieved this objective using large reactors and thus long hydraulic retention times (HRT) with long solids retention times (SRT). The anaerobic fluidized process is one means of improving the fermentation technology as it allows the mean cell residence time to be maintained independently of the hydraulic retention time. This allows long SRTs, which are needed for efficient and stable operation, with short HRTs, which are prerequisite for system economy.

An anaerobic fluidized bed may be viewed as a means of realizing the benefits of anaerobic fermentation while overcoming many difficulties historically related to anaerobic fermentation. Among the benefits of the anaerobic fluidized bed are (1) a high treatment efficiency of wastewaters at reduced retention times; (2) lower temperature operation; (3) simplicity in design; (4) good adaptation to changes in feed strength and loading rate; (5) no danger of clogging; (6) small head losses; (7) easier removal or addition of active materials; (8) good hydraulic characteristics (avoidance of short circuiting); (9) high active biomass concentrations; and (10) minimization of diffusional limitations.

One major disadvantage of the system is that recycle of effluent may be necessary to achieve bed expansion and contact time and thus the system would require greater pumping costs. On the other hand, the completely mixed nature of the reactors caused by recirculation ensures better mitigation against toxic shocks and helps to dilute high strength wastes. Another detriment is the slow start-up time associated with anaerobic fixed film systems in general. This could be a particularly harsh problem if the biofilm were lost by some upset during operation, as it would be out of operation for re-establishment of the biofilm. Fortunately, the anaerobic biofilms are quite stable to toxins, particularly with proper acclimation (Speece *et al.*, 1981).

59.5   PROBLEMS AND RESEARCH NEEDS

The anaerobic fluidized bed has been shown to be a capable process for the treatment of high strength and low strength wastes and offers the potential for reducing capital and operating costs. Yet, the system is still in the developmental stage. There are several areas that require further research and development.

One of the more important problems is start-up of the system. It often takes several months to obtain a mature biofilm in an anaerobic fluidized bed. It would be highly desirable to shorten the start-up times. Little is known regarding means of accelerating start-up times.

Heidman (1982) stated that the following questions related to anaerobic fluidized bed technology need to be answered before the design approach can be optimized. These include: reaction kinetics as a function of temperature and reactor response under dynamic loading; optimal reactor depth, medium density and size; need for equalization basins and overall control strategy for adequate bed expansion; net solids production; solids levels attainable in the reactor; biological film properties; effect of biological growth on medium expansion characteristics; solids control strategies in the reactor, if any; need for final clarifiers, particularly with reference to low strength treatment; influence of wastewater sulfate concentration on the desirability and performance of the process; and long-term process stability and reliability at pilot scale.

Thus while the process has shown capability on laboratory and pilot scale testing, much work needs to be accomplished on both fundamental and design levels in order to optimize this technology. Such work is proceeding at this time.

59.6 SUMMARY

The anaerobic fluidized bed process represents a successful combination of the anaerobic fermentation and fluidized bed processes. The process was originally developed for the anaerobic treatment of low strength wastes, such as domestic sewage, but has also been successfully demonstrated for the treatment of high strength nonparticulate wastes.

The process has been shown to be able to realize the benefits of anaerobic treatment, that is high degrees of stabilization, low growth yields and generation of methane. At the same time, many of the detriments historically associated with anaerobic treatment have been minimized with this process. This includes operation at lower temperature, and good process stability in terms of loading and temperature fluctuations.

The system offers great potential for reducing capital and operation and maintenance costs (particularly energy requirements). However, there are several important fundamental and practical design considerations which must be solved before the design of the anaerobic fluidized bed can be optimized.

ACKNOWLEDGEMENTS

Thanks are expressed to the following individuals who helped in preparing this chapter. First, to Mr. Robert Hickey for his valuable insight and suggestions in the preparation of this chapter, especially for his help with the design information. Also, to Mrs. Dorothy Pascoe for typing this chapter.

59.7 REFERENCES

Atkinson, B. and I. J. Davies (1972). The complete mixed microbial film fermenter: A method of overcoming washout in continuous fermentation. *Trans. Inst. Chem. Eng.*, **50**, 208.
Cleasby, J. L. and E. R. Baumann (1977). Backwash of granular filters used in wastewater filtration. *Environmental Protection Technology Series*, EPA-600/2-77-016.
Cooper, P. F. and D. H. V. Wheeldon (1980). Fluidized and expanded bed reactors for wastewater treatment. *Water Pollut. Control*, **79**, 286.
Davidson, J. F. and D. Harrison (1971). *Fluidization*. Academic, New York.
Heidman, J. A. (1982). Technology assessment of anaerobic systems for municipal wastewater treatment: 1. Anaerobic fluidized bed; 2. Anflow. *United States EPA, Research and Development Series*, EPA-600/2-82-004.
Hickey, R. F. and R. W. Owens (1981). Methane generation from high-strength industrial wastes with the anaerobic biological fluidized bed. *Biotechnol. Bioeng. Symp.*, No. 11, 399.
Jewell, W. J. (1974). An optimized biological waste treatment process for oxygen utilization. Paper presented at the 47th Annual Conference of the Water Pollution Control Federation, October 6–11, 1974, Denver, CO.
Jewell, W. J. (1981). Development of the attached microbial film expanded bed process for aerobic and anaerobic waste treatment. In *Biological Fluidized Bed Treatment of Water and Wastewater*, ed. P. F. Cooper and B. Atkinson. Halsted Press, NY.
Jewell, W. J. (1983). Anaerobic attached film expanded bed fundamentals. *Proceedings of the 1st International Conference on Fixed-Film Biological Processes, 1982*, ed. Y. C. Wu *et al.*, p. 17. University of Pittsburg, Pittsburg, PA.
Jewell, W. J. and S. E. MacKenzie (1972). Microbial yield dependence on dissolved oxygen in suspended and attached systems. *Proceedings of the 6th Water Resources Symposium*. University of Texas Press, Austin, TX.
Jewell, W. J., M. S. Switzenbaum and J. W. Morris (1981). Municipal wastewater treatment with the anaerobic attached microbial film expanded bed process. *J. Water Pollut. Control Fed.*, **53**, 482.
Leuschner, A. P. (1976). The feasibility of treating low strength organic wastes with an anaerobic attached film system. Master of Science Thesis, Cornell University, Ithaca, NY.
McCarty, P. L. (1982). One hundred years of anaerobic treatment. In *Anaerobic Digestion 1981*, ed. D. E. Hughes *et al.*, p. 1. Elsevier Biomedical, Amsterdam.
Meunier, A. D. and K. J. Williamson (1981a). Packed bed biofilm reactors: simplified models. *J. Environ. Eng. Div. (Am. Soc. Civ. Eng.)*, **107**, 307.
Meunier, A. D. and K. J. Williamson (1981b). Packed bed biofilm reactors: design. *J. Environ. Eng. Div. (Am. Soc. Civ. Eng.)*, **107**, 319.
Morris, J. W. and W. J. Jewell (1982). Organic particulate removal with the anaerobic attached film expanded bed process. *Proceedings of the 36th Purdue Industrial Waste Conference 1981*, p. 621. Ann Arbor Science, Ann Arbor, MI.
Mulcahy, L. T. and E. J. La Motta (1978). Mathematical model of the fluidized bed biofilm reactor. Report No. Env. E. 59-78-2, Environ. Eng. Prog., Dept. Civ. Eng., University of Massachusetts, Amherst.
Owen, W. F. (1982). *Energy in Wastewater Treatment*. Prentice-Hall, Englewood, NJ.

Owens, R. W., R. F. Hickey, J. S. Jeris and S. Mascitelli (1980). Bottling industry wastewater treatment—pilot to full scale anaerobic fluidized bed. Paper presented at the 53rd Annual Conference of the Water Pollution Control Federation, September 29–October 2, 1980, Las Vegas, NV.

Shieh, W. K. (1980). A suggested kinetic model for the fluidized bed biofilm reactor (FBBR). *Biotechnol. Bioeng.*, **23**, 667.

Shieh, W. K., P. M. Sutton and P. Kos (1981). Predicting reactor biomass concentration in a fluidized-bed system. *J. Water Pollut. Control Fed.*, **53**, 1574.

Speece, R. E., G. F. Parkin, T. Yang and W. Kocher (1981). Methane fermentation toxicity response: contact mode. *Proceedings of the Seminar/Workshop, Anaerobic Filters: An Energy Plus for Wastewater Treatment*, January 9–10, 1980, Howey-in-the-Hills, FL, ANL/CNSV-TM-50, p. 11. Argonne National Laboratory, Argonne, IL.

Sutton, P. M. and A. Li (1982). Anitron System® and Oxitron System® high rate anaerobic and aerobic biological treatment systems for industry. *Proceedings of the 36th Purdue Industrial Waste Conference 1981*, p. 665. Ann Arbor Science, Ann Arbor, MI.

Sutton, P. M., A. Li, R. R. Evans and S. Korchin. (1983). Dorr-Oliver's fixed film and suspended growth anaerobic systems for industrial wastewater treatment and energy recovery. *Proceedings of the 37th Annual Purdue Industrial Waste Conference, 1982*, p. 667. Ann Arbor Science, Ann Arbor, MI.

Switzenbaum, M. S. (1983). A comparison of the anaerobic filter and the anaerobic expanded/fluidized bed processes. *Water Sci. Technol.*, **15** (8/9), 345.

Switzenbaum, M. S. and S. C. Danskin (1982). Anaerobic expanded bed treatment of whey. *Proceedings of the 36th Purdue Industrial Waste Conference 1981*, p. 414. Ann Arbor Science, Ann Arbor, MI.

Switzenbaum, M. S. and W. J. Jewell (1978). The anaerobic attached film expanded bed reactor for the treatment of dilute organic wastes. US Department of Energy Report TID-29398. Available from the National Technical Information Service, US Department of Commerce, Springfield, VA 22161.

Switzenbaum, M. S. and W. J. Jewell (1980). Anaerobic attached film expanded bed reactor treatment. *J. Water Pollut. Control Fed.*, **52**, 1953.

Williamson, K. J. and P. L. McCarty (1976). A model of substrate utilization by bacterial films. *J. Water Pollut. Control Fed.*, **48**, 9.

Zeikus, J. G. (1980). Chemical and fuel production by anaerobic bacteria. *Annu. Rev. Microbiol.*, **34**, 423.

Zenz, F. A. and F. A. Othmer (1960). *Fluidization and Fluid Particle Systems*. Reinhold, New York.

60

Anaerobic Downflow Stationary Fixed Film Reactors

K. J. KENNEDY and L. VAN DEN BERG
National Research Council of Canada, Ottawa, Ontario, Canada

60.1 INTRODUCTION

Downflow stationary fixed film (DSFF) reactors are a relatively recent addition to the family of advanced or second generation anaerobic reactors, all of which are based on retention of the active biomass. The DSFF reactor distinguishes itself from the other advanced reactors by the downflow mode of operation, the architecture of its packing (biofilm support) and the absence or near absence of suspended growth. Retention of biomass is achieved by allowing the microorganisms to grow as a biofilm on the packing surface. These are the factors that give rise to the performance characteristics discussed in this chapter.

In this review we discuss the history and development of the DSFF reactor, the results of intensive research efforts to establish its characteristics and the niche in which this reactor may find application.

60.1.1 History and Development

The development of the DSFF reactor started in 1976 as part of a project to circumvent problems encountered in studying the upflow anaerobic filter (UAF) reactor (Young and McCarty, 1967). These problems included difficulties in assessing the exact role of the packing, plugging of the packing and difficulties in using concentrated wastes and wastes containing large amounts of suspended solids. Interestingly, Lettinga *et al.* (1980, 1982) avoided the first two difficulties by operating without the packing, and as a result developed the upflow anaerobic sludge bed (UASB) reactor. The DSFF reactor was developed to optimize the role of the packing, to avoid plugging or short circuiting and to enable treatment of concentrated wastes and wastes with substantial amounts of suspended solids.

Several reports published before the development of the DSFF reactor (Atkinson and Davies, 1972; Howell and Atkinson, 1976a, 1976b; Howell *et al.*, 1972; Mehta and LeRoux, 1974; Topiwala and Hamer, 1971) discuss the importance of biofilms or wall growth in preventing wash-out of microorganisms in aerobic systems. Prevention of wash-out is of even more significance in anaerobic systems because of the very slow growth rates of the methanogenic bacteria. This was demonstrated by Balmer (1974), Schillinger (1975) and Gill (1978) for completely mixed anaerobic reactors filled with fibre panels.

The role of biofilm development on performance of anaerobic reactors was determined systematically by van den Berg and Lentz (1979, 1980a). By studying first the performance of single tube reactors of varying inside diameters, the effect of reactor surface-to-volume ratio and the role of the tube wall were assessed. The tubes were set vertically to avoid any settling of suspended matter on the exposed walls. Reactors were operated in the downflow mode to allow continuous removal of suspended growth as well as in the more customary upflow mode. This simplified and idealized experimental set-up permitted evaluation of the performance of the film of active biomass which developed on the inside surface of the tubes, in relation to suspended growth.

Once the principle of the DSFF reactor was established, studies were rapidly expanded to evaluate the role of film support properties, the effects of factors such as waste composition and strength, nutrients, temperature, reactor height and width (multichannel), and the robustness of the reactor under adverse operating conditions (hydraulic and organic overloading, intermittent loading and change in waste composition). Large-scale demonstration reactors are presently under construction.

60.1.2 General Description and Principles of Operation

As with other second generation reactor designs, the enhanced performance of anaerobic DSFF reactors (Figure 1) is based on maximizing the biomass concentration in the reactor. By immobilizing the anaerobic bacteria on inert support media (Figure 2) long biomass retention times that are independent of the hydraulic retention time can be achieved. The support media to which the bacteria adhere can be made of various nontoxic, nonbiodegradable materials. The relative advantages and disadvantages of support materials made of plastic, polyvinyl chloride, clay or synthetic fibres are discussed in the section on support media.

Two major design and operational differences make DSFF reactors unique from other second generation anaerobic reactors. DSFF reactors use packing with an oriented geometry that forms vertical channels that run the length of the packing. The upflow anaerobic filter on the other hand usually utilizes a random packed medium such as stones, raschig rings or plastic biorings. The other major difference is the direction of movement of the waste water through the packing. In DSFF reactors waste water enters at the top of the reactor through a submerged distribution device and effluent is removed from the bottom of the tank. In other second generation anaerobic

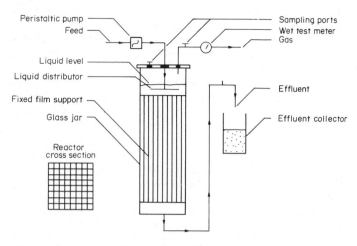

Figure 1 Set-up of multichannel downflow stationary fixed film digester

Figure 2 Electron micrograph of bacteria attached to clay support material

reactors, influent is introduced at the bottom and effluent overflows at the top. As will be shown these two major differences result in major differences in reactor performance.

The downflow mode of operation in combination with the oriented support medium configuration allows utilization of counter current interaction of liquid and evolved digester gas to enhance digester mixing. The liquid volume above and below the packing enhances interchannel mixing while protecting the submerged biofilm and minimizing clogging in the bottom of the reactor. Analysis of the mixing regime in laboratory (unpublished results) and pilot plant DSFF reactors (Hall, 1982) has shown them to behave as completely mixed systems. It is felt that mixing in DSFF reactors is the result of a gas lift pump effect within the reactor channels.

The combination of downflow waste addition and vertically oriented packing also minimizes suspended solids accumulation in the DSFF reactor. This enables DSFF reactors to treat both soluble and insoluble wastes (suspended solids concentrations as high as 4% have been tested). Other second generation reactors are limited to mostly soluble wastes. However, suspended solids lost in the effluent of DSFF reactors include not only untreated waste material but also active suspended biomass. The loss of suspended biomass means that a large majority of the hydrolyzing and acidogenic microbes and almost all of the methanogenic activity is confined to the biofilm. In contrast, biomass profiles in upflow anaerobic filter systems (Young, 1982) have shown it to be a hybrid reactor, a combination of a DSFF reactor and an UASB system. The majority of the reactor biomass in the upflow anaerobic filter (and waste stabilization capacity) is found in the interstitial spaces between the packing material in the lower one third of the reactor.

The performance of DSFF reactors can be discussed in a variety of terms. From a waste treatment point of view, COD removal and stability under practical conditions would be the most significant factors. On the other hand, from an energy point of view, the maximum rate of methane production would be very important. Both these viewpoints are considered in this section.

Loading rates and rates of methane production in DSFF reactors can be expressed in several ways: based on biofilm support surface area, reactor cross-sectional area, reactor void volume, effective void volume, empty bed volume or reactor biomass. Conversion between these expressions is based on the surface-to-volume ratio and biofilm concentration within the reactor. In order to avoid misinterpretation of experimental results it is important to know the various ways that data may be presented. Table 1 summarizes the various notations and variables that may be used.

Table 1 Description of Design Variables

Variable	Description	Unit
V_0	Initial volume of the reactor without support medium below liquid level	m^3
V_L	Initial liquid volume of reactor with support medium below liquid level	m^3
V_B	Volume of reactor support medium ($V_0 = V_L + V_B$)	m^3
V_E	Effective liquid volume below liquid level (volume corrected for biofilm accumulation)	m^3
α	Void factor of the reactor ($V_L = \alpha V_0$)	—
$\theta_0, \theta_L, \theta_E$	Hydraulic retention time based on V_0, V_L, V_E respectively	d
A_0, A_L, A_E	Surface area-to-volume ratio based on V_0, V_L, V_E respectively	$m^2 m^{-3}$
L_0, L_L, L_B	Volumetric organic loading rate based on V_0, V_L, V_E respectively	$kg\, m^{-3}\, d^{-1}$
L'_0, L'_L, L'_E	Surface organic loading rate based on A_0, A_L, A_E respectively	$kg\, m^{-2}\, d^{-1}$
H_0, H_L, H_E	Volumetric hydraulic loading based on V_0, V_L, V_E respectively	$m^3 m^{-3}\, d^{-1}$
H'_0, H'_L, H'_E	Surface hydraulic loading rate based on A_0, A_L, A_E respectively	$m^3 m^{-2}\, d^{-1}$
L_a	Specific loading rate	$kg\, kg^{-1}\, d^{-1}$
v_s	Superficial liquid velocity	$m\, d^{-1}$
G	Volume of gas per unit of time	$m^3 d^{-1}$
G_0, G_L, G_E	Volumetric gas production rate based on V_0, V_L, V_E respectively	$m^3 m^{-3}\, d^{-1}$
G'_0, G'_L, G'_E	Surface gas production rate based on A_0, A_L, A_E respectively	$m^3 m^{-2}\, d^{-1}$

60.2 START-UP

The start-up of anaerobic reactors is generally time consuming and often difficult because of the slow growth rate of acetate-converting methanogens and an often inadequate nutrient supply in wastes. Few systematic studies on start-up are available. Of the four major second generation anaerobic reactors, upflow anaerobic filters and DSFF reactors appear the easiest to start while anaerobic fluidized bed reactors and UASB systems often appear to take somewhat longer. In this section, start-up of DSFF reactors is discussed first in terms of a simple model, then the effects of inoculum, type of support surface, surface-to-volume ratio, temperature, type of waste and nutrient supply are dealt with in more detail.

Four factors have a major effect on the start-up of anaerobic reactors generally and on DSFF reactors especially:

(1) Quality of the inoculum in terms of quantity of slow growing methanogenic microorganisms adaptable to the waste.
(2) Rate of adaptation of these microorganisms to the waste.
(3) Rate of growth of these microorganisms.
(4) Rate of loss of microorganisms in the effluent.

The model is based on these four factors and assumes an adapted inoculum with a growth rate μ (d^{-1}). Additionally, it is assumed that the COD loading rate is proportional to the active biomass concentration X_a, and that acetate conversion is the limiting step. The model leads to the following equation:

$$L_v = \frac{\mu S}{f + \left(\dfrac{\mu S}{a_X X_a} - f\right) e^{-\mu t}}$$ (1)

where L_v = loading rate at time t (kg COD m^{-3} d^{-1}), μ = growth rate of performance limiting methanogens (d^{-1}), X_a = initial active biomass concentration of performance-limiting methanogens (kg m^{-3}), S = substrate concentration in contact with bacteria (kg COD m^{-3}), f = fraction of performance-limiting methanogens lost in the effluent, a_X = specific activity of active biomass (kg COD kg^{-1} d^{-1}) and t = time from inoculation (d).

Using this equation it is possible to compare the start-up of fully mixed reactors without active biomass retention (Figure 3) with reactors with active biomass retention (Figure 4).

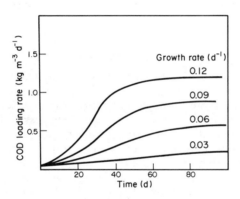

Figure 3 Calculated rates of start-up for a completely mixed anaerobic reactor without biomass retention (waste strength, 10 kg COD m^{-3})

Comparison of Figures 3 and 4 shows that:
(1) The rate of start-up of reactors without biomass retention is markedly dependent on growth rate.
(2) The rate of start-up of reactors with biomass retention is much faster than those without retention and depends markedly on the fraction of bacteria lost in the effluent.
(3) In all cases maximum loading rate is approached gradually as the rate of loss of active biomass in the effluent (decreasing HRT with increasing loading rate) becomes equal to the rate of new biomass formation.
(4) Growth rate is a very important factor in start-up. Anaerobic reactors should therefore, if at all possible, be started at the optimum temperature with a nutritionally balanced waste, and at optimum pH and volatile acid levels.
(5) Since the initial phase of start-up is almost logarithmic (before wash-out becomes significant), initial loading rate and hence inoculum quality is very important.

The model ignores several factors that play a role in normal reactor performance. (1) The growth rate may change during start-up as a result of ecological effects, changes in nutrient content in the reactor liquid, the development of diffusional resistances (fixed films) and changes in mixing (gas production). (2) The specific activity of the bacteria may change, partly as a result of factors already mentioned under (1). (3) The fraction of active biomass retained may change during start-up.

As a result, the model cannot be expected to be predictive, but is useful for delineating some of the factors that play a role in start-up.

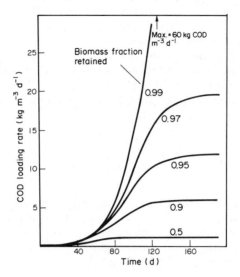

Figure 4 Calculated rates of start-up for a mixed reactor with retention of biomass (waste strength, 10 kg m^{-3}; growth rate, 0.06 d^{-1})

60.2.1 Inoculum

Although little is known about what determines a good inoculum, experience with anaerobic digesters generally has shown that there are large differences in start-up as a result of the inoculum (de Zeeuw and Lettinga, 1980; Lettinga *et al.*, 1980; van den Berg and Kennedy, 1981b; van den Berg and Lentz, 1972, 1980b; Young and McCarty, 1967). With DSFF reactors treating food processing waste similar observations on start-up were made: sewage digester sludge inoculum generally requires a longer time to adapt than inoculum from an active digester treating food processing waste. It is important that the inoculum has a high methanogenic activity (ability to convert acetate, hydrogen and carbon dioxide to methane). The slow growth rate and low yield of methanogens makes it important that they be present in the inoculum in as large a concentration as possible. Where the inoculum is weak, start-up can be improved by increasing the size of the inoculum and by the use of multiple inoculations. In almost all cases start-up is limited by the accumulation of methanogenic bacteria in the biofilm (waste with suspended solids may be limited by liquefaction).

With DSFF reactors, the methanogenic bacteria appear to attach themselves quite quickly to suitable film supports (within a few days) and then grow quite rapidly. On the other hand a large proportion of the non-methanogens appears to maintain themselves quite readily in the bulk of the liquid. Whether or not this partial separation of methanogens and non-methanogens is a factor in the relative ease with which DSFF reactors start in comparison with, for example, UASB reactors (Hall, 1982; Hall and Jovanovic, 1982; van den Berg *et al.*, 1981) is at present a matter of speculation. It is often found that UASB reactors start more readily on volatile acid wastes than on complex waste and this may also indicate that methanogenic bacteria adapt faster when interference from non-methanogens is limited.

60.2.2 Film Support

Film support characteristics and surface-to-volume ratio are major factors in starting a DSFF reactor (Table 2). Both affect the amount of methanogens initially immobilized and hence the initial loading rate.

The effect of film support characteristics is shown in Figure 5. Loading rates are expressed in terms of methane produced per unit surface area, eliminating differences in surface-to-volume effects. Also, the reactors are inoculated with a culture adapted to the waste being treated, and

Table 2 Characteristics of Common Commercially Available Oriented Packings Potentially Useful for DSFF Reactors

Type	Material	Specific surface $(m^2\,m^{-3})$	Porosity (%)
Flocor	PVC foil module	85	98
Surfpac (crinkle close)	Polystyrene sheets (close packed)	187	94
Surfpac (standard)	Polystyrene sheets	82	94
Cloisonyl	Vertical channel, subdivided tubes	220	94
Koro-Z	PVC foil module	210	~96
Vinyl core	PVC foil module	121	~96
Clay	Modular clay blocks fired to 1000 °C	75–140	55–70
Polyester	Needle-punched polyester[a]	75–140	97

[a] Not yet available in a stackable pattern.

with about the same specific methanogenic activity. Glass reactors are slow to start up, presumably because bacteria have difficulty attaching themselves to the smooth inert surface (van den Berg and Lentz, 1979, 1980a). Solid polyvinyl chloride (PVC, used extensively in biological waste water treatment) is substantially better than glass as a film support, but not as good as the fired clay and needle-punched polyester (a felt-like material) (van den Berg and Kennedy, 1981a). Surface roughness appears to play a major role (and in the case of clay, availability of essential minerals due to leaching from the clay) (Murray and van den Berg, 1981, 1982). It should be noted that needle-punched polyester can be considered to have an extremely rough surface.

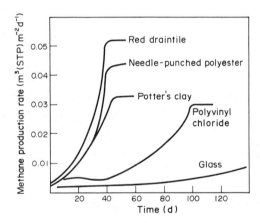

Figure 5 Actual start-up of stationary fixed film reactors (35 °C; bean blanching waste, 10 kg COD m^{-3}; reactor volume, 0.7–1.2l; surface-to-volume ratio, A_0, 100–150 m^2 m^{-3})

The effect of the surface-to-volume ratio is apparent when considering that DSFF reactor performance depends on the total amount of active biomass per unit volume. This is illustrated in Figures 6a and 6b, in which volumetric loading rates L_0 and surface loading rates L'_0 for three reactors with widely differing surface-to-volume ratios are presented. The rate of start-up, when considered on a surface area basis, is the same although rates based on empty bed reactor volume differ widely.

60.2.3 Type of Waste

Since growth rate is such an important factor in start-up, DSFF reactors start up faster when fed a nutritionally adequate waste than when fed a waste that is low in certain nutrients or is slightly toxic. For example, DSFF reactors started up very quickly and consistently on heat-treated liquor obtained from anaerobic sewage sludge (Hall *et al.*, 1981; Hall and Jovanovic, 1982), reasonably quickly on most food processing wastes but very poorly, or not at all, on waste from a chemical industry (unpublished data). It is interesting that UASB reactors do not start up

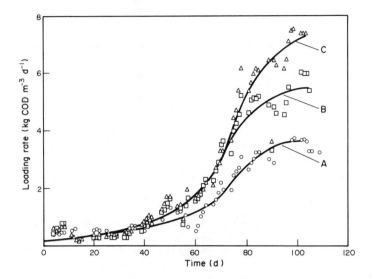

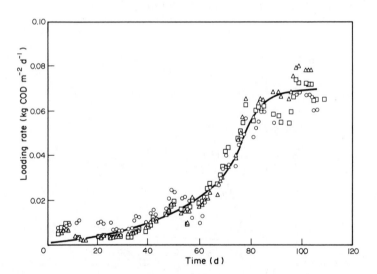

Figure 6   Effect of height on volumetric loading rate, L_0, and surface loading rate, L_0', during start-up of DSFF reactors (bean blanching waste, 10 kg COD m^{-3}; $\bigcirc$, reactor A, 0.50 m tall, surface to volume ratio, A_0, 55 $m^2\,m^{-3}$; $\square$, reactor B, 1.11 m tall, A_0, 83 $m^2\,m^{-3}$; $\triangle$, reactor C, 2.03 m tall, A_0, 94 $m^2\,m^{-3}$.

on heat-treated liquor (Hall, 1982; Hall *et al.*, 1982) nor on chemical industry waste (unpublished data) indicating that fixed film formation in DSFF reactors and sludge granule formation in UASB reactors are not the same process, even though presumably the same or similar bacteria are involved.

60.2.4   Temperature

Temperature affects the rate of start-up markedly. The optimum temperature for starting reactors is 33–37 °C, the optimum growth temperature for mesophilic methanogens (van den Berg, 1977). Another factor that may be important is that the microbiological ecosystems in the reactor will vary with temperature. This is certainly the case at thermophilic temperatures. Also the growth yield factors are higher at lower temperatures (van den Berg, 1977) and this leads to more biomass accumulation (thicker film) at lower temperatures. Consequently the effect of temperature on biofilm development is of major importance.

60.2.5 Trace Elements

Trace elements such as nickel and iron have recently been demonstrated to affect bacterial growth and specific activity (Murray and van den Berg, 1981; Speece, personal communication, 1982). Also as previously mentioned it has been shown for DSFF reactors that leaching of minerals from clay support media affects biofilm development noticeably (Murray and van den Berg, 1982) (Figures 7 and 8). The addition of nickel, cobalt and molybdenum did not affect the rate of start-up, but was important in the maturation of the biofilm (allowed the biofilm to continue to develop).

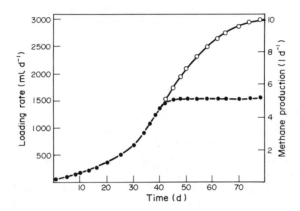

Figure 7 Effect of trace metal addition to bean blanching waste on the development of methanogenic fixed-film reactors, as indicated by waste loading and methane production rates: ●, control; ○, with nickel, cobalt and molybdenum added (waste loading and methane production at the same level for both)

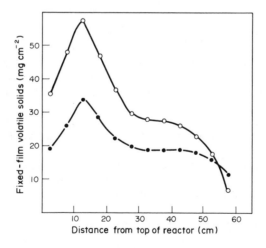

Figure 8 Effect of trace metal addition on the distribution of methanogenic fixed film along the polyester support material of anaerobic reactors treating bean blanching waste: ●, control; ○, with nickel, cobalt and molybdenum added

60.3 STEADY-STATE REACTOR OPERATION

The start-up period of DSFF reactors is followed by a 'steady-state' period, during which the maximum possible loading rate is more or less constant or changes (up or down) relatively slowly. The performance during this 'steady-state' period depends on many factors, which will be discussed. It should be kept in mind, however, that the 'steady-state' condition is arrived at by a combination of several conditions: (1) reduced rate of biofilm formation; (2) reduced specific activity of the biomass in the film (coupled with factors such as accumulation of inert solids, diffusional resistance and change in the composition of the different types of microorganisms present); (3) increased rate of biofilm sloughing.

The maximum possible loading rate may change with time because the three conditions outlined are not constant. Film sloughing, for example, may decrease average film thickness, but at the same time may increase specific activity considerably. As a result, the initial steady state of a DSFF reactor may be different from subsequent steady states (van den Berg *et al.*, 1980). The influence of these conditions varies widely with support material and type of waste. Glass and smooth plastic surfaces result in considerable sloughing of the biofilm. High carbohydrate wastes result in fast rates of biofilm formation (high growth yield coefficient) and thick biofilms that can lead to partial or complete blockage of a channel. Piggery waste, and wastes with a large organic acid component, on the other hand, result in the formation of thin but active biofilms. It should be noted that steady-state operation at below the maximum loading rates is little or not affected by the sloughing of film.

60.3.1 Film Support

The film support affects steady-state performance in addition to start-up (Figure 5). Most if not all of the effect is related to surface roughness: the rougher the surface the greater the average thickness of the biofilm that forms for a particular waste. Generally this leads to higher maximum loading rates (Copp and Kennedy, 1983). Glass and smooth plastic surfaces slough biofilm under the least adverse condition and this limits the average biofilm thickness. Very little sloughing occurs with clay and needle punched polyester and hence maximum surface loading rates for these supports are similar in mature reactors (reactors operated over a long period of time).

Studies with both single tube and multiple channel reactors have shown that the volumetric loading rate L_0 of a reactor is closely related to the amount of available surface (van den Berg and Lentz, 1979; Kennedy and van den Berg, 1982). Channel diameter affects COD removal in that a smaller diameter leads to a reduced volatile suspended solids content in the reactor-mixed liquor. Presumably this reflects the more intimate contact between biofilm and the mixed liquor.

60.3.2 Type of Waste and Waste Strength

A wide variety of wastes have been successfully treated in mature DSFF reactors (Tables 3 and 4). Wastes varied from completely soluble easy-to-treat dilute sugar waste and food processing wastes to concentrated food processing wastes, some with a high sodium and salt content. Some of the latter also had high suspended solid contents. As pointed out earlier, some wastes presented problems in starting up reactors. However, all wastes tested could be treated at a reasonably high loading rate once a reactor had reached steady-state conditions. Maximum loading rates and COD removals depended on both loading rates (hence hydraulic retention time) and type and strength of waste. For relatively dilute and soluble wastes (5–20 kg COD m^{-3}), maximum volumetric loading rates, L_0 are generally over 10 kg m^{-3} d^{-1} with total COD removals between 70% to over 90%. COD removals decrease by about 5% when increasing loading rates from 5 to 15 kg COD m^{-3} d^{-1} (van den Berg and Kennedy, 1982b, 1982c).

At waste strengths below 5 kg COD m^{-3}, the maximum loading rate compatible with a reasonable COD removal decreases. For practical purposes, the lower limit on waste strength is about 2 kg COD m^{-3}, because even at sharply reduced loading rates (L_0 less than 4 kg COD m^{-3} d^{-1}) hydraulic residence times are down to less than 0.5 d and COD removals decrease sharply. For these low strength wastes, the DSFF reactor is at a disadvantage in comparison with UASB, upflow anaerobic filter or fluidized bed reactors. These latter reactors contain more active biomass in the reactor and hence can treat very dilute waste more effectively.

For concentrated wastes (over 20 kg COD m^{-3}) volumetric loading rates L_0 of over 15 kg COD m^{-3} d^{-1} are readily obtained and COD removals are higher than for similar dilute wastes. Removal of COD is affected substantially by the loading rate when the suspended solids in a waste are present in significant quantities as in piggery waste (Kennedy and van den Berg, 1982). The retention time of the suspended solids is too short for extensive liquefaction to take place. That the latter factor depends very much on particle size is indicated by results for synthetic sewage sludge. This waste contained substantial amounts of cellulose in a very fine form and most cellulose was digested even at high loading rates (van den Berg *et al.*, 1980b, 1980c). It is import-

Table 3 Effect of Waste Type on Rates of Methane Production and COD Removal[a]

Waste	Waste strength (kg COD m^{-3})	Suspended COD (% of total)	Feed ratios COD/N	Feed ratios COD/P	Rate of methane production (m^3 (STP) m^{-3} d^{-1})	COD removal (%)
Citric acid waste[b]	3.6	—	90	140	2.6	80
Dairy waste[b]	4.0	<15	20	67	2.5	77
Skim milk waste[b]	4.0	0	20	100	2.3	71
Bean blanching waste[c]	10.0	10–30	25	100	3.0	90
Sugar waste[b]	10.0	0	35	108	3.0	91
Whey[b]	66	< 5	100	100	3.1	95
Heat treated sewage digester sludge liquor (HTL)[c]	10.5	<10	13	100	2.3	70
Chemical industry waste[c]	14	0	5.6	47	2.7	82
Barley stillage waste[c]	53	25	48	—	2.7	82

[a] Mostly soluble wastes, 0.7–1.2 l reactors, operating at 10 kg COD m^{-3} d^{-1} and 35 °C.   [b] Needle-punched polyester support material, $L'_0 = 0.10$ kg COD m^{-2} d^{-1}.   [c] Clay support medium, $L'_0 = 0.072$ kg COD m^{-2} d^{-1}.

ant to bear in mind that the findings discussed in this section are dependent both on the type of waste as well as on the physical characteristics of the particular DSFF reactor used.

Using the concept of solids retention time, a simple equation can be developed that shows the importance of the surface area on the maximum loading rates of DSFF reactors (Henze and Harremoës, 1982).

$$L_0 = \frac{(A_0 X_1 + X_2)/\theta_X}{Y} - L_{0,I} \qquad (2)$$

where A_0 = surface-to-volume ratio in reactor (m^2 m^{-3}), X_1 = biomass concentration in biofilm (kg m^{-2}), X_2 = suspended biomass concentration (kg m^{-3}), θ_X = solids retention time (d), Y = growth yield coefficient (kg kg^{-1} COD), $L_{0,I}$ = load of non-biodegradable suspended organics (kg m^{-3} d^{-1}) and L_0 = maximum volumetric loading rate (kg COD m^{-3} d^{-1}).

Assuming X_2 is negligible in relation to the biomass in the biofilm, equation (1) can be rewritten as

$$L_0 = \frac{(A_0 X_1/\theta_X)}{Y} - L_{0,I} \qquad (3)$$

Although equation (2) does not take into account all parameters (*e.g.* biomass activity) it can be used to show the effect of various other parameters (Figure 9) demonstrating the influence of yield, wastewater composition, surface area and reactor biomass upon maximum obtainable volumetric loading rate.

Wastes with a low yield coefficient (*e.g.* acetic acid) have high volumetric loading rates whereas wastes with a high yield coefficient (carbohydrate wastes) allow only small volumetric loadings. Reactors treating wastes that have low yield coefficients also have an increased proportion of methanogens in the biomass (higher specific activity). As suggested by Henze and Harremoës (1982) the nature of the waste should always be considered when trying to compare tests from different reactor types. Since yield coefficients are not predictable, treatability tests should be conducted on a particular waste in advance of any full-scale design work.

60.3.3 Reactor Size and Configuration

The studies on size and configuration (scale-up) have so far been limited to reactor volumes less than 1.3 m^3 and reactor heights less than 2 m [a full-scale 13 000 m^3 DSFF reactor is now operating in Puerto Rico with stillage waste (reactor height, 42 m; loading rate L_0, 8–12 kg m^{-3} d^{-1}; COD removal, over 70%) (Szendry, 1983)]. Because of the limited nature of these studies only tentative conclusions can be drawn.

Height (0.2 to 1.8 m) appears to have very little if any effect on the maximum surface loading rate, L'_0, of multichannel DSFF reactors (Figure 6). This is surprising since gas bubble mixing should increase with reactor height. However, over the small height range studied these effects may be negligible.

Table 4 Performance of DSFF Reactors

Waste	Waste strength (kg m⁻³)	Suspended solids (% of total)	Feed ratios COD/N	Feed ratios COD/P	Reactor liquid volume (l)	Support material	Reactor temp. (°C)	Volumetric loading rate, L_L (kg COD m⁻³ d⁻¹)	Surface loading[a] rate, L_L (kg COD m⁻² d⁻¹)	COD conversion[b] (%)	Ref.
Bean blanching waste	5.5–22 (TVS)	<1.0	—	—	138	PVC	17 25 35	1.4[c] 2.6 7.5	0.012 0.023 0.067	80 77 75	Stevens and van den Berg (1981)
	10 (COD)	10–30	25	100	0.7	Clay	10 15 25 35	4.2 7.0 12.7 18.4	0.030 0.050 0.091 0.131	88 88 88 88	Kennedy and van den Berg (1981) van den Berg and Lentz (1980a)
Chemical industry waste	14 (COD)	0	5.6	47	0.7	Clay	25 35	14.0 17.9	0.100 0.128	81 84	Kennedy and van den Berg (1981)
Liquor from heat treated sewage digester sludge (HTL)	10.5 (COD)	10	13	100	0.7 650	Clay Clay	35 35	29.0 32.3	0.207 0.461	68–70 72	van den Berg et al. (1980) Hall (1982)
Piggery waste	27–51 (COD)	60–70	13	49	43	Clay	35	4.9 11.7	0.045 0.105	70 43	Kennedy and van den Berg (1982)
Pear peeling waste	110–140 (COD)	40–50	55	275	43	Clay	35	31.7 5.2	0.284 0.046	27 58	van den Berg and Kennedy (1982)
Tomato peeling waste	15 30	<20	54	75	138	PVC	35	7.5 15.3 13.4 13.5	0.067 0.137 0.115 0.115	64 54 53 50	Stevens and van den Berg (1981)

[a] For multichannel DSFF reactor the surface area of the reactor wall is not included in surface area calculations. [b] Chemical oxygen demand, COD. [c] Results based on total volatile solids, TVS.

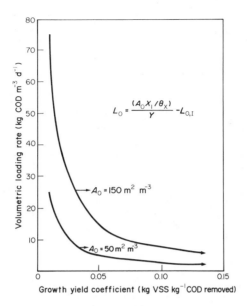

Figure 9 Effect of yield and surface-to-volume ratio on maximum volumetric loading rate, L_0 ($\theta_X = 20$ d, $L_{0,I} = 0$ kg $\mathrm{m^{-3}\,d^{-1}}$, $X_1 = 0.1$ kg VSS $\mathrm{m^{-2}}$)

A minimum channel diameter (minimum distance between opposite walls) is important in avoiding plugging. Particularly at the top of the reactor where fresh waste enters, the biofilm is thicker and can eventually 'choke' the channel (for possible remedies, see below). This minimum distance appears to about 4 cm, indicating that the maximum surface-to-volume ratio of DSFF reactor is about 100 $\mathrm{m^2\,m^{-3}}$, based on packing void volume, or about 70–80 $\mathrm{m^2\,m^{-3}}$ based on total liquid volume.

60.3.4 Temperature

Studies show that DSFF reactors can be operated successfully in the range of 10–55 °C (Kennedy and van den Berg, 1981, 1982a; Kennedy *et al.*, 1981). In the temperature range of 10–35 °C the maximum loading rate of fixed film reactors increases linearly with temperature (Figure 10) but COD removal is not affected. The linear effect may be the result of several factors:

(1) Increase in the biomass yield factors with decreasing temperature (van den Berg, 1978) results in increased biomass in the biofilm.

(2) Methanogenic bacteria that are not in the log phase of growth have activities that increase approximately linearly with temperature (van den Berg, 1977).

The linear effect of temperature in the mesophilic range in DSFF reactors means that fixed film systems can be operated at ambient temperatures with relatively small penalties in performance.

DSFF reactors can also be operated at thermophilic temperatures (55 °C) (Kennedy and van den Berg, 1982). Start-up of thermophilic DSFF reactors is slower than at mesophilic temperatures. Additionally, it has been found that thermophilic systems do not cope well with sudden unexpected changes such as loading rate and substrate composition. It is believed that the low net growth yield factor of thermophilic bacteria (approximately 50% of the yield at 35 °C) is at least in part responsible for the above phenomenon. Another factor may be that the different bacterial species present and active at thermophilic temperatures are not as numerous as at mesophilic temperatures. Zinder and Mah (1979) have also reported that thermophilic bacteria tend to lyse much more easily at thermophilic temperatures and therefore can only survive under conditions of exponential growth.

If changes in operating temperatures of DSFF reactors are necessary, they should be altered slowly (1–2 °C $\mathrm{d^{-1}}$). Sudden changes in temperature tend to inhibit methanogenic bacteria much more than acidogenic bacteria and as a result the buildup of volatile acids can lead to reactor upset.

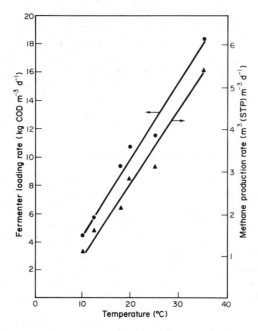

Figure 10 Effect of temperature on loading rate and rate of methane production of a DSFF reactor (bean blanching waste, 10 kg COD m^{-3}; surface-to-volume ratio, A_0, 140 m^2 m^{-3})

60.3.5 Nutrients

Most advanced reactors such as DSFF reactors operate successfully over a wide range of COD–N and COD–P ratios. The generally assumed optimum for COD:N:P is a ratio of 100:5:1. Results (Tables 3 and 4) indicate that DSFF reactors can function with COD:N:P in the range 100:(10–1):(5–1). On the other hand ammonium nitrogen levels of 3500 mg 1^{-1} also are tolerated without adverse affects. The low COD:N:P ratios required for steady-state DSFF reactor operation appear in part to result from recycling of nutrients within the biofilm, although the nature of the wastes also plays an important role. A waste with a low growth yield factor does not need to have as high a COD:N:P ratio as a waste with a larger growth yield factor.

Tests on mineral supplements such as nickel, cobalt and molybdenum (Murray and van den Berg, 1981, 1982) have shown that these elements enhance the development of the anaerobic biofilm with consistently better reactor performance (Figure 7). The effect is the result of a thicker but not more uniform biofilm (3–5 mm) (Figure 8). The thicker biofilm does not seem to be substantially limited by diffusion of substrate into and products out of the biofilm. Under some conditions, iron also affects film performance, but the effect is not as reproducible or clear cut (van den Berg *et al.*, 1980a; Hoban and van den Berg, 1979).

60.3.6 Recirculation

Effluent recycle in DSFF reactors has been found beneficial during start-up and when treating wastes with a high concentration of suspended solids. Recirculation during start-up has a number of advantages.

(1) It increases mixing within the reactor thereby minimizing effects of toxic and inhibitory components in a waste water.

(2) It eliminates the possibility of local high acid concentrations and low pH and makes the alkalinity in the reactor uniform.

(3) It minimizes excess biofilm development in the upper portion of the support medium where fresh substrate enters.

(4) It aids in uniform deposition of inoculum on the support media.

Recirculation is very beneficial for the stabilization of wastes with a high suspended solids content. Settled solids in the bottom of the reactor liquefy very slowly and can eventually lead to blockage problems. By recirculating solids to the top of the reactor they are kept in contact with hydrolytic bacteria and enzymes. It may also be possible to use additional gas mixing or intermittent mechanical mixing in the bottom of the reactor to keep the system mixed.

For mature reactors treating mostly soluble wastes, even when concentrated (80–130 kg COD m^{-3}), recirculation has been found to have little effect on reactor performance. The counter current movement of liquid and biogas within the reactor appears to be sufficient to keep the reactor contents well mixed.

60.4 NON-STEADY-STATE REACTOR OPERATION

Non-steady-state performance of anaerobic reactors is of major importance in waste treatment as mentioned earlier. Food processing plants, for example, often produce widely varying amounts and strengths of waste water during a day. Additionally the composition of the waste may vary considerably from day to day and relatively long shut-down periods may occur. A waste treatment system is expected to cope with all these variations. The DSFF reactor has been extensively tested for non-steady-state performance on a laboratory scale. The ability of the DSFF reactor to withstand, for example, extreme hydraulic overloads is similar to or better than the UASB reactor. The UASB reactor is somewhat more susceptible to upset from hydraulic overloading because the high biogas production rates and liquid flow rates may result in loss of the active biomass.

60.4.1 Hydraulic Overload

Mesophilic DSFF reactors are able to handle extreme hydraulic overloads and recover normal performance within a short period of time (Kennedy and van den Berg, 1981, 1982d). The ability of DSFF reactors to handle this type of shock results from the fact that the bacteria are immobilized within the system and are not lost in the effluent. Waste stabilization decreases with decreasing hydraulic retention times (Figure 11) due to insufficient time for bacteria–substrate contact. Interestingly, methane production during extreme overloading quickly stabilizes at rates much higher than during normal operation. This new pseudo-steady-state biogas production may be an indication of excess methane production capacity by methanogens in the biofilm (Figure 12), better substrate diffusion into the biofilm or a combination of both of these factors.

Hydraulic overloading often leads to sloughing of parts of the biofilm. This usually does not

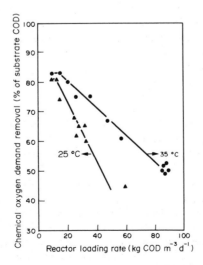

Figure 11 Effect of chemical industry waste overloading at 25 and 35 °C on COD removal (waste strength, 14 kg COD m^{-3}; surface-to-volume ratio, A_0, 140 $m^2 m^{-3}$)

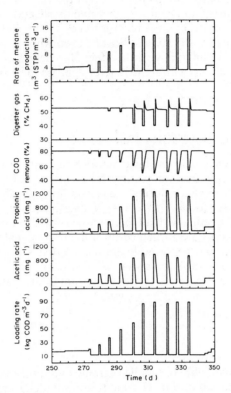

Figure 12 Effect of overloading at 35 °C on several performance characteristics of a stationary fixed film reactor treating chemical industry waste (waste strength, 14 kg COD m^{-3}; surface-to-volume ratio, A_0, 140 m^2 m^{-3})

affect reactor performance drastically and previous performance is quickly restored. It is not clear whether the sloughing of parts of the biofilm is a result of intense gas production in the biofilm or localized high volatile acid levels and low pH in the biofilm (these in themselves can lead to sloughing). It is unlikely that the increased liquid velocities past the film resulting from overloading would play a major role in biofilm sloughing because linear liquid flow velocities are actually very low (usually 0.1–1 cm s^{-1}).

Thermophilic DSFF reactors do not cope with hydraulic overloads as well as their mesophilic counterparts (Duff and Kennedy, 1982). The more limited number of bacterial species presumably present (especially of methanogenic bacteria) may be responsible for the inability of the biofilm to respond quickly to sudden change.

60.4.2 Organic Overload

Mesophilic DSFF reactors can tolerate sudden organic shock loads at constant hydraulic loading (caused by sudden increases in waste strength) (24 hour shock test) and recover normal performance within a few days (Kennedy *et al.*, 1984). Reactors receiving an organic shock load for periods of less than 24 hours recover even sooner. COD removals decrease with increased loading even though the actual amount of waste organics converted and methane gas produced increases. As with hydraulic overloads sloughing of the biofilm also occurs during organic shock loading.

DSFF reactors subjected to organic shock loads from wastes with little buffering capacity do not handle organic shocks well. In this case, high volatile acid levels and low pH cause failure or prolonged recovery periods for the reactors.

Thermophilic DSFF reactors cannot handle sudden organic shocks to the same extent as mesophilic reactors. It would seem again that the thermophilic methanogens are much more sensitive to sudden changes than their mesophilic counterparts.

60.4.3 Starvation

The low decay rates for mesophilic anaerobic methanogenic bacteria allow DSFF reactors to stand idle for weeks or months without a great loss in activity. This is particularly the case when the reactor is allowed to cool to temperatures of 25 °C or lower. In this respect DSFF reactors are similar to upflow filters and USAB reactors, for which it has also been reported that they are able to sit dormant for long periods and regain normal performance within a very short period of time. This ability makes these second generation reactors ideal for wastes that are produced only for a short period of time each month or year (*e.g.* vegetable canning industry).

60.4.4 Changes in Waste Composition

In addition to being able to withstand sudden changes in waste strength, DSFF reactors are able to cope well with changes in waste composition. In many cases, loading rates can be maintained more or less constant, although COD removals change with the composition of the waste. Changes from one food processing waste to another, even when the sodium level changed from 20 to 4000 p.p.m. (Stevens and van den Berg, 1981), and from a sewage waste to a chemical industry waste are possible with little or no setbacks (Figure 13).

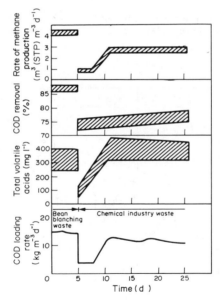

Figure 13 Performance of a fixed film reactor operating on bean blanching waste and subsequently on chemical industry waste (surface-to-volume ratio, A_0, 140 m^2 m^{-3})

60.4.5 Intermittent Loading

A special case of non-steady-state operation involves loading reactors for only a few minutes or hours each day at a high loading rate (calculated for the whole 24 hour period). Tests have shown that the methane production rate (on a 24 hour basis) is usually about the same although it can increase by as much as 30% over continuous loading (due to higher loading rates possible). Under certain conditions, large amounts of hydrogen are also produced (Figure 14). COD removals are lower than for continuous operation (short circuit effect) but operation under these conditions may be advantageous where wastes are handled only once a day (*e.g.* piggery, Figure 15), or where the demand for most of the methane produced occurs during a specific part of the day.

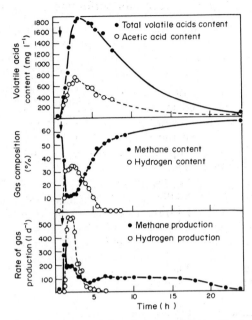

Figure 14　Changes in reactor performance characteristics after a one-daily loading (at arrow) with pear peeling waste (0.545 kg COD; $L_0 = 14$ kg COD m^{-3} d^{-1}; surface-to-volume ratio, A_0, 110 m^2 m^{-3})

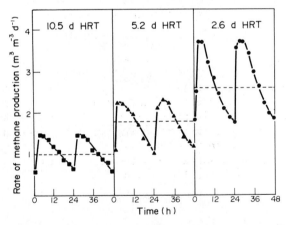

Figure 15　Effect of slug loading on methane production rates of DSFF reactors treating piggery waste at 35 °C: (————) daily average methane production rate; (——————) actual methane production rate (waste strength, 50–60 kg COD m^{-3}; surface-to-volume ratio, A_0, 60 m^2 m^{-3})

60.5　MAINTENANCE OF DSFF REACTORS

Experience with operation of large-scale DSFF reactors is not yet available, and hence only experience with small-scale laboratory and pilot plant reactors can be described. Problem areas with these reactors include foaming, biofilm maintenance and accumulation of undigested suspended solids in the bottom of the reactor when treating wastes with a high suspended solids concentration.

60.5.1　Foaming

Foaming is an intermittent problem that occurs with most anaerobic reactors. The reasons for foaming are often unknown. The effects of foaming can be severe for suspended growth reactors, because of the possibility that biomass may be carried out of the reactor with the foam. However in DSFF reactors, foaming does not affect reactor performance dramatically, but may cause clogging of gas lines.

60.5.2 Biofilm Maintenance

The biofilm in DSFF reactors is not evenly developed over the entire height of the support except for wastes with high suspended solids contents (piggery wastes). In the latter the uniformity in biofilm thickness is believed to result from a combination of the effects of intermittent mixing and the abrasive action between the suspended solid and the biofilm. Non-uniformity is caused by excessive growth near the top of the reactor (see Figure 8). Under certain conditions, this non-uniform growth can cause plugging at the top of the packing.

Several factors determine whether or not plugging will occur: width of the channels, smoothness of the packing (will determine extent of occasional sloughing) and composition of the waste. Carbohydrate wastes are more likely to lead to plugging than fatty acid wastes, because of the higher growth yield coefficients for carbohydrate wastes.

Several methods may have potential for maintaining a reasonably thin biofilm in DSFF reactors.

(1) Organic and hydraulic shock loads: the effect on film sloughing will be greatest near the top of the reactor where the load enters. This method will not be of much use for channels already blocked.

(2) Recirculation of effluent: as already discussed, for wastes with a high suspended solids content, intermittent pumping of liquid from the bottom to the top of the reactor helps maintain a uniform, relatively thin film.

(3) Recirculation of gas: large gas bubbles rising in channels should help the sloughing process and may even open blocked channels.

(4) Reactor configuration: a relatively thin layer of coarse packing on top of the ordinary packing may accumulate the excess biofilm and cope with it.

60.5.3 Solids Accumulation

The downflow method of waste addition and the oriented channel packing enable the DSFF reactor to treat wastes with a relatively high suspended solids component without affecting the biofilm. To prevent accumulation of untreated suspended material in the bottom of the reactor and eventual loss of reactor performance by blockage of the bottom of the packing and plugging the outlets, infrequent mixing from the bottom of the reactor to the top is recommended. This also keeps the solids in suspension so that excess undigested solids are removed with the effluent. Solids recycle will also help maintain a thin active biofilm.

60.6 BIOFILM CHARACTERISTICS

The biofilm generally is a distinct slimy layer on the support material, up to 4 mm thick and grey or black in color. Its thickness varies considerably over short distances (1 cm), but also over the length of the reactor. The volatile suspended solids content of the film (presumably mostly bacteria) is 6–8%, and the specific activity of the volatile suspended solids varies between 0.8 and 2 kg COD kg^{-1} VSS d^{-1}, but is mostly between 1.0 and 1.5 kg COD kg^{-1} VSS d^{-1}.

Biofilm support material affects the film characteristics. Bacteria are known to excrete a glycocalyx that is used to form a bacterial matrix on a support surface. For anaerobic bacteria the effectiveness of this adhesion mechanism works best for materials with a high degree of surface roughness and porosity. Rough surfaces result in the development of more evenly distributed biofilms while smooth surfaces such as glass are characterized by spotty biofilm development. Several other workers using mostly aerobic biofilms have also indicated enhanced biofilm development for rough surfaces (Costerton *et al.*, 1978; Marshall, 1976). Table 2 summarizes the characteristics of various laboratory and commercial support materials, but does not indicate surface roughness.

Improved biofilm development on rough surfaces is partly due to the large initial bacterial deposition on the support material compared to smooth surfaces. Because of the slow net growth

yield and long doubling times for methanogenic bacteria, initial deposition is much more important for start-up of anaerobic processes than for aerobic biofilm processes. It has also been
observed that for support materials on which biofilms develop quickly, the maximum surface
loading rates are usually higher than for materials on which biofilm development is slower (Figure
4).

Difficulties in removing the biofilm intact and in studying it *in situ* have limited the information
available about what is happening in and to the biofilm. Certain observations however, have been
made.

(1) The film is not static. Gas bubbles form both in and on the film and cause film deforma-
    tion. This phenomenon has also been reported by Bochem *et al.* (1982) for granular sludge
    particles.
(2) Pieces of biofilm come off and leave with the effluent from time to time. Occasionally they
    are so active that they rise to the top of the reactor with attached gas bubbles. The bubble
    formation on and in the film may assist sloughing.
(3) Almost all of the methane-forming activity is in the biofilm. At liquid retention times of
    less than one or two days, methanogens are unable to grow in the liquid. Presumably the
    small methanogenic activity of the effluent comes from biofilm sloughing.
(4) Acid-forming activity is present both in the biofilm and in the bulk liquid. The fast growth
    rate of acetogenic bacteria allows them ample time to grow in suspension even at hydraulic
    residence times of less than one day.
(5) Under conditions of high acid levels (pig manure) methanosarcina are present in the bio-
    film in large numbers. To what extent these bacteria are responsible for the conversion of
    acetic acid to methane remains to be determined.
(6) The role of diffusion resistance in the performance of the biofilm is not known. Bubble
    formation inside the film as well as movement of raw waste into the biofilm for conversion
    to organic acids, hydrogen and carbon dioxide (van den Berg and Kennedy, 1982) miti-
    gates against the presence of a large diffusional resistance. Also, specific methanogenic
    activity of biofilm solids (in vial tests) is usually not much different from the activity calcu-
    lated for the biofilm. However, where the film is very thick or includes much inert
    material, film resistance may become important.

Because of the lack of detailed and precise information on the characteristics of anaerobic biofilms, the value of film biokinetics is limited. However, a very good speculative look at the anaerobic biofilm has been presented by Henze and Harremoës (1982).

The biofilm was assumed to have the following characteristics: (i) soluble substrates are transported into the biofilm where the reaction takes place; (ii) the products diffuse out; (iii) the biofilm is a homogeneous mass of active bacteria; and (iv) substrate transport into the biofilm is
governed by molecular diffusion. With these assumptions a simple 0-order or 1'-order kinetic
model of the substrate biofilm interaction can be developed using Fick's equation for molecular
diffusion.

As the biofilm gets thicker its efficiency decreases because substrate cannot penetrate the
whole biofilm depth.

Using known anaerobic kinetic constants (r_x, specific substrate uptake rate, 10 kg COD kg^{-1}
VSS d^{-1}; K_s, half velocity constant for acetate, 100 mg l^{-1}) in their model Henze and Harremoës
(1982) concluded that biofilms are unaffected by molecular diffusion up to thicknesses of 1 mm,
assuming that diffusion is the limiting factor in film activity.

However, if a more realistic value for the specific uptake rate is applied ($r_x = 1.7$ rather than
10.0 kg COD kg^{-1} VSS day^{-1}) even thicker biofilms will be unaffected by diffusional resistance.
Calculations indicate that the biofilm could be up to 2 mm in thickness without adverse effects.
The assumption that thick biofilms can operate unaffected by diffusional resistances is based on
assuming a high half velocity coefficient K_s (100 mg l^{-1}). The K_s value for acetate is 150–900 mg
l^{-1} (Lawrence and McCarty, 1970). If these higher K_s values apply, the model suggests biofilms
could be developed to effective depths that would only be limited by hydraulic shear and organism decay. Additionally, as briefly mentioned earlier, the release of methane gas bubbles from
within the biofilm into the methane-saturated bulk liquid could create micromixing within the
biofilm and minimize the effect of diffusional resistance. This phenomenon may be in part
responsible for the efficiency of DSFF biofilms up to 4 mm in thickness.

For a more detailed evaluation of anaerobic biofilm kinetics the papers by Chian and de Walle
(1977), de Walle and Chian (1977), Rittman and McCarty (1980) and Williamson and McCarty
(1976) should be consulted.

60.7 COMPARISON WITH OTHER ADVANCED REACTORS

Few studies have been made to compare advanced anaerobic reactors. The most significant and well-controlled study is in progress at the Waste Water Technology Center in Burlington, Ontario, Canada (Hall, 1982; Hall and Jovanovic, 1982; Hall *et al.*, 1981). In this study a comparison is made between the DSFF, UASB upflow anaerobic filter and fluidized bed reactor, using at least two wastes and under controlled conditions. Until studies such as this one are finished, definite conclusions cannot be drawn. Nevertheless certain differences between reactors are already apparent (van den Berg and Kennedy, 1982).

(1) Suspended growth reactors (UASB) are potentially capable of higher loading rates and sometimes higher COD conversion than DSFF reactors. The reason is that larger amounts of active biomass can be retained in suspended growth reactors.

(2) DSFF reactors can treat concentrated wastes directly without dilution while other reactors require dilution to low COD levels (10 kg m^{-3}). Dilution is usually obtained by recycling effluent (a process requirement for fluidized and expanded bed reactors).

(3) Anaerobic filters suffer from plugging problems unless very coarse packing with considerable void space as well as a substantial void space under the packing is used. Plugging problems occur because of suspended growth in the bottom one-third of the reactor (Young and Dahab, 1982).

(4) Hydraulic overloading affects DSFF reactors less than suspended growth reactors.

(5) Organic shockloads higher than have been reported for other reactors are not detrimental to DSFF reactors.

(6) Of the advanced reactors only the DSFF reactor appears able to treat wastes with large amounts of suspended material.

(7) COD removals tend to be lower in DSFF reactors because of the absence of a settler. An external settler would improve performance.

60.7.1 Large Scale Application

Research on DSFF reactors has reached the stage where large-scale reactors are now being installed. As noted earlier, one 13 000 m^3 DSFF reactor is in operation in Puerto Rico. In Canada a 500 m^3 DSFF for methane production from pig manure is being designed for construction in 1982/83, while a 1000 m^3 DSFF reactor is being constructed for the treatment of dairy waste. Confirmation of laboratory findings awaits operational experience with these full-scale systems.

60.8 CONCLUSION

The most remarkable feature of the downflow stationary fixed film reactor is its ability to maintain high rates of methane production under adverse conditions. The stationary fixed biofilm appears to provide a high degree of protection for active biomass. As a result, the DSFF reactor can withstand low temperatures, severe and repeated hydraulic overloadings, organic shock loads, sudden changes in waste composition, and starvation, with little or no effect on subsequent performance. Also, high rates of methane production can be obtained while tailoring methane production to energy needs.

60.9 REFERENCES

Atkinson, B. and I. J. Davies (1972). The completely mixed microbial film fermenter — A method of overcoming washout in continuous fermentation. *Trans. Inst. Chem. Eng.*, **50**, 208–216.
Balmer, R. G. (1974). Anaerobic treatment of dairy waste using a fiber wall reactor. M.S. Thesis, Purdue University.
Bochem, H. P., S. M. Schoberth, B. Sprey and P. Wengler (1982). Thermophilic biomethanation of acetic acid: morphology and ultrastructure of a granular consortium. *Can. J. Microbiol.*, **28**, 500–510.
Chian, E. S. K. and F. B. deWalle (1977). Treatment of high strength acidic waste water with a completely mixed anaerobic filter. *Water Res.*, **11** (3), 295–304.
Copp, G. and K. J. Kennedy (1983). Support materials for downflow stationary fixed film (DSFF) methanogenic reactors. *J. Ferment. Technol.*, **61**, 333.
Costerton, J. W., G. G. Geesey and K. J. Cheng (1978). How bacteria stick. *Sci. Am.*, **238** (1), 86.
de Walle, F. B. and E. S. K. Chian (1977). Kinetics of substrate removal in a completely mixed anaerobic filter. *Biotechnol. Bioeng.*, **18**, 1275–1295.

de Zeeuw, W. and G. Lettinga (1980). Acclimation of digested sewage sludge during start-up of an upflow anaerobic sludge blanket (UASB) reactor. In *Proc. Ind. Waste Conf.*, *35th*, *1980*, pp. 39–47. Ann Arbor Science, Michigan.

Duff, S. and K. J. Kennedy (1982). Effect of organic and hydraulic overloading on thermophilic anaerobic fixed film reactors. *Biotechnol. Lett.*, **4**, 815–820.

Gill, M. (1978). Study of the kinetic response of an anaerobic reactor with fibre panels. Ph.D. Thesis, Purdue University.

Hall, E. R. (1982). Biomass retention and mixing characteristics in fixed film and suspended growth anaerobic reactors. In *Anaerobic Treatment of Waste Water in Fixed Film Reactors*, pp. 371–398. IAWPR Specialized Seminar, Copenhagen, Denmark.

Hall, E. R. and M. Jovanovic (1982). Anaerobic treatment of thermal sludge conditioning liquor with fixed-film and suspended growth processes. In *Proc. Ind. Waste Conf.*, *37th*, *1982*, pp. 719–728. Ann Arbor Science, Michigan.

Hall, E. R., B. E. Jank and M. Jovanovic (1981). Energy production from high strength industrial waste water. In *Proc. Bioenergy R & D Seminar*, *3rd*, pp. 125–129. National Research Council of Canada, Ottawa.

Henze, M. and P. Harremoës (1982). In *Anaerobic Treatment of Waste Water in Fixed Film Reactors*, pp. 1–94. IAWPR Specialized Seminar, Copenhagen, Denmark.

Hoban, D. J. and L. van den Berg (1979). Effect of iron on conversion of acetic acid to methane during methanogenic fermentations. *J. Appl. Bacteriol.*, **47**, 153–159.

Howell, J. A. and B. Atkinson (1976a). Influence of oxygen and substrate concentrations on the ideal film thickness and the maximum overall substrate uptake rate in microbial film fermenters. *Biotech. Bioeng.*, **17**, 15–35.

Howell, J. A. and B. Atkinson (1976b). Sloughing of microbial film in trickling filters. *Water Res.*, **10**, 307–315.

Howell, J. A., C. T. Chi and U. Pawlowsky (1972). Effect of wall growth on scale-up problems and dynamic operating characteristics of the biological reactor. *Biotechnol. Bioeng.*, **14**, 253–265.

Kennedy, K. J. and L. van den Berg (1981). Effects of temperature and overloading on the performance of anaerobic fixed film reactors. In *Proc. Ind. Waste Conf.*, *36th*, *1981*, pp. 678–686. Ann Arbor, Michigan.

Kennedy, K. J. and L. van den Berg (1982a). Thermophilic downflow stationary fixed film reactors for methane production from bean blanching waste. *Biotechnol. Lett.*, **4**, 171–176.

Kennedy, K. J. and L. van den Berg (1982b). Continuous vs. slug loading of downflow stationary fixed film reactors digesting piggery wastes. *Biotechnol. Lett.*, **4**, 137–142.

Kennedy, K. J. and L. van den Berg (1982c). Anaerobic digestion of piggery waste using a stationary fixed film reactor. *Agric. Wastes*, **4**, 151–158.

Kennedy, K. J. and L. van den Berg (1982d). Stability and performance of anaerobic fixed film reactors during hydraulic overloading at 10 to 35 °C. *Water Res.*, **16**, 1391–1398.

Kennedy, K. J. and L. van den Berg (1982e). Effect of height on the performance of anaerobic downflow stationary fixed film (DSFF) reactors treating bean blanching waste. In *Proc. Ind. Waste Conf.*, *37th*, *1982*, pp. 71–77. Ann Arbor Science, Michigan.

Kennedy, K. J., L. van den Berg and W. D. Murray (1981). Advanced fixed film reactors for microbial production of methane from waste. In *Proc. World Congr. Chem. Eng.*, *2nd*, *Montreal, Canada*, vol. 1, pp. 317–321.

Kennedy, K. J., M. Muzar and G. Copp (1984). Stability and performance of anaerobic fixed film reactors during organic overloading. *Biotechnol. Bioeng.*, in press.

Lawrence, A. W. and P. L. McCarty (1970). Unified basis for biological treatment design and operation. *J. Sanit. Eng. Div.*, *Am. Soc. Civ. Eng.*, **96**, (SA-3), 757–758.

Lettinga, G., A. F. van Velsen, S. W. Hobma, W. de Zeeuw and A. Klapwijk (1980). Use of the upflow sludge blanket reactor concept for biological wastewater treatment, especially for anaerobic treatment. *Biotechnol. Bioeng.*, **22**, 699–734.

Lettinga, G., S. W. Hobma, L. W. H. Pol, W. de Zeeuw, P. Trin and R. Roersma (1982). Design, operation and economy of anaerobic treatment. In *Anaerobic Treatment of Waste Water in Fixed Film Reactors*, pp. 283–304. IAWPR Specialized Seminar, Copenhagen, Denmark.

Marshall, K. C. (1976). *Interfaces in Microbial Ecology*. Harvard University Press, Cambridge, MA.

Mehta, K. B. and N. W. LeRoux (1974). Effect of wall growth on continuous biological oxidation of ferrous iron. *Biotechnol. Bioeng.*, **16**, 559–563.

Murray, W. D. and L. van den Berg (1981). Effect of support material on the development of microbial fixed film converting acetic acid to methane. *J. Appl. Bacteriol.*, **51**, 257–265.

Murray, W. and L. van den Berg (1982). Effects of nickel, cobalt and molybdenum on the performance of methanogenic fixed film reactors. *Appl. Environ. Microbiol.*, **42**, 502–505.

Rittman, B. E. and P. L. McCarty (1980). Design of fixed-film processes with steady-state-biofilm model. *Progr. Water Technol.*, **12**, 271–281.

Schillinger, M. S. (1975). Treatment and digestion of antibiotic fermentation wastes using a fiber wall bioreactor. M.S. Thesis, Purdue University.

Stevens, T. G. and L. van den Berg (1981). Anaerobic treatment of food processing wastes using a fixed-film reactor. In *Proc. Ind. Waste Conf.*, *36th*, *1981*, pp. 224–233. Ann Arbor Science, Michigan.

Szendry, M. L. (1983). The Bacardi Corporation digestion process for stabilizing rum distilling wastes and producing methane. In *Proc. Energy from Biomass and Wastes Symp.*, *7th*, *1982*, *Buena Vista, FL*, pp. 283–301.

Topiwala, H. H. and G. Hamer (1971). Effect of wall growth in steady-state continuous cultures. *Biotechnol. Bioeng.*, **13**, 919–922.

van den Berg, L. (1977). Effect of temperature on growth and activity of a methanogenic culture utilizing acetate. *Can. J. Microbiol.*, **23**, 898–902.

van den Berg, L. and K. J. Kennedy (1981a). Support materials for stationary fixed film reactors for high-rate methanogenic fermentations. *Biotechnol. Lett.*, **3** (4), 165–170.

van den Berg, L. and K. J. Kennedy (1981b). Potential use of anaerobic processes for industrial waste treatment. In *Proc. Seminar Anaerobic Waste Water Treatment and Energy Recovery*, *Pittsburg, Pa., U.S.A.*

van den Berg, L. and K. J. Kennedy (1982a). Comparison between intermittent and continuous loading of stationary fixed film reactors for methane production from wastes. *J. Chem. Technol. Biotechnol.* **32**, 427–432.

van den Berg, L. and K. J. Kennedy (1982b). Effect of substrate composition on methane production rates of downflow stationary fixed film reactors. In *Proc. Energy from Biomass and Wastes Symp.*, *6th*, *1982*, Buena Vista, FL, pp. 401–423.

van den Berg, L. and K. J. Kennedy (1982c). Dairy waste treatment with anaerobic stationary fixed film reactors. In *Anaerobic Treatment of Waste Water in Fixed Film Reactors*, pp. 229–243. IAWPR Specialized Seminar, Copenhagen, Denmark.

van den Berg, L. and C. P. Lentz (1972). Anaerobic digestion of pear waste: Factors affecting performance. In *Proc. Ind. Waste Conf.*, *27th*, *1972*, pp. 313–323. Ann Arbor Science, Michigan.

van den Berg, L. and C. P. Lentz (1979). Comparison between up and downflow anaerobic fixed film reactors of varying surface-to-volume ratios for the treatment of bean blanching waste. In *Proc. Ind. Waste Conf.*, *34th*, *1979*, pp. 319–325. Ann Arbor Science, Michigan.

van den Berg, L. and C. P. Lentz (1980a). Effects of film area-to-volume ratio, film support, height and direction of flow on performance of methanogenic fixed film reactors. *Proc. U.S. Dep. Energy Workshop/Seminar on Anaerobic Filters*, pp. 1–10.

van den Berg, L. and C. P. Lentz (1980b). Effect of waste, inoculum, and solids retention time on methane production and stability of the anaerobic contact process. *Adv. Biotechnol.*, **2**, 257–262.

van den Berg, L., K. J. Kennedy and M. F. Hamoda (1981). Effect of type of waste on performance of anaerobic fixed film and upflow sludge bed reactors. In *Proc. Ind. Waste Conf.*, *36th*, *1981*, pp. 686–692. Ann Arbor Science, Michigan.

van den Berg, L., K. A. Lamb, W. D. Murray and D. W. Armstrong (1980a). Effects of sulphate, iron and hydrogen on the microbial conversion of acetic acid to methane. *J. Appl. Bacteriol.*, **48**, 437–447.

van den Berg, L., C. P. Lentz and D. W. Armstrong (1980b). Anaerobic waste treatment efficiency comparisons between fixed film reactors, contact digesters and fully mixed, continuously fed digesters. In *Proc. Ind. Waste Conf.*, *35th*, *1980*, pp. 788–793. Ann Arbor Science, Michigan.

van den Berg, L., C. P. Lentz and D. W. Armstrong (1980c). Methane production rates of anaerobic fixed film fermenters as compared to those of anaerobic contact and fully mixed continuous fermenters. *Adv. Biotechnol.*, **2**, 251–256.

Williamson, K. and P. L. McCarty (1976). A model of substrate utilization by bacterial films. *J. Water Pollut. Control Fed.*, **48**, 9–24.

Young, J. C. and M. F. Dahab (1982). Effect of media design on the performance of fixed bed anaerobic filters. In *Anaerobic Treatment of Waste Water in Fixed Film Reactors*, pp. 321–339. IAWPR Specialized Seminar, Copenhagen, Denmark.

Young, J. C. and P. L. McCarty (1967). The anaerobic filter for waste treatment. *Proc. Ind. Waste Conf.*, *22nd*, *1967*, p. 599. Ann Arbor Science, Michigan.

Zinder, S. H. and R. A. Mah (1979). Isolation and characterization of a thermophilic strain of Methanosarcina unable to use H_2–CO_2 for methanogenesis. *Appl. Environ. Microbiol.*, **38**, 996–1008.

61

Methane Production from Anaerobic Digestion

L. VAN DEN BERG, S. J. DUFF and K. J. KENNEDY
National Research Council of Canada, Ottawa, Ontario, Canada

61.1 HISTORY

The production of a burnable gas in ditches and swamps has been recognized for centuries (van Brakel, 1980) but it was not until the nineteenth century that the gas was analyzed as consisting mostly of methane and that the production was demonstrated to be a microbial process decomposing organic material. There has been interesting speculation about the relation between the mythology and legends of fiery dragons on the one hand, and intestinal methane production with accidental ignition on the other.

As far as is known, swamp gas was never collected and used for the production of heat or energy. Presumably this was due to the difficulties of collecting the gas from over a large surface area rather than a lack of recognition of its potential value. The collection and increased use of methane as a fuel started with the development and systematic use of the controlled digestion process as a waste treatment method around the turn of the century.

61.2 CONDITIONS NECESSARY FOR METHANE PRODUCTION

The conditions necessary for microbial production of methane include: (1) the presence of a degradable organic substance; (2) the presence of suitable nutrients; (3) the presence of sufficient water to submerge or dissolve organic substances and to dilute soluble compounds to below toxic levels; (4) the presence of suitable types of bacteria; and (5) a temperature between about 277 and 353 K (4 and 80 °C).

As described in other parts of this chapter most methane is formed from decarboxylation of acetic acid, the remainder being formed by reduction of carbon dioxide using hydrogen. These substrates for methane production are formed by a variety of bacteria, some of which are facultative anaerobes and are able to maintain or restore anaerobic conditions during and after short-term exposure to oxygen.

The relationship between the composition of the organic substrate being degraded and the amount of methane formed has been the subject of some discussion over the years. The following equation is now generally accepted to quantify the relationship:

$$C_nH_aO_b + (n - \frac{a}{4} - \frac{b}{2})\,H_2O \rightarrow (\frac{n}{2} - \frac{a}{8} + \frac{b}{4})\,CO_2 + (\frac{n}{2} + \frac{a}{8} - \frac{b}{4})\,CH_4 \qquad (1)$$

It is clear from this equation that organics high in oxygen (polyacids, sugars) produce less methane than organics low in oxygen (fats). As a result, the parameter total volatile solids (TVS), which is used extensively in wastewater treatment, does not relate well to the amount of methane produced (Figure 1). On the other hand, the chemical oxygen demand (COD) value of a substance or mixture of substances relates stoichiometrically (Henze and Harremoes, 1982; Figure 1, results converted for COD).

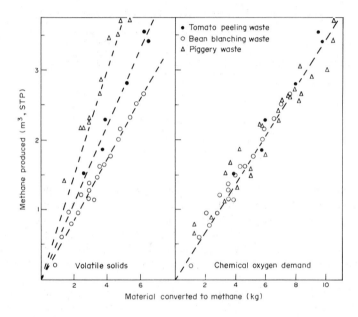

Figure 1 Comparison of amounts of methane produced from volatile solids and from COD for three different wastes (based on laboratory data, from Stevens and van den Berg, 1981 and Kennedy and van den Berg, 1982a)

The relation between COD and methane can be derived from the COD value of methane as determined by complete oxidation:

$$CH_4 + 2\,O_2 \rightarrow CO_2 + 2\,H_2O \qquad (2)$$

In other words, 16.04 g of methane requires 64.00 g of oxygen for complete oxidation. Hence, at 273.15 K and 101.325 kPa (STP—standard temperature and pressure), 0.35 m^3 methane is produced per kg COD removed. The same result is obtained approaching it from the breakdown of organic compounds such as glucose:

$$C_6H_{12}O_6 \rightarrow 3\,CH_4 + 3\,CO_2 \qquad (3)$$

In practice, especially with small laboratory reactors, the measured value is often slightly less (van den Berg and Kennedy, 1982). Reasons for this include difficulty of measuring small gas flows accurately, loss of methane dissolved in the effluent of a digester but not measured as COD, and other imperfections in determining COD values of complex wastes and effluents. Surprisingly, accurate measurements of methane losses in different effluents do not appear to be available. It is usually assumed that the solubility of methane in effluents is the same as in water (0.032 m^3 methane (STP) per m^3 water at 318 K).

61.3 NATURALLY OCCURRING METHANE PRODUCTION

About 5% of the carbon fixed by photosynthesis is converted to methane by microbiological processes. Much of the methane originates from cellulose, a major end-product of photosynthesis

(50 GJ annually). Some of this methane is oxidized to carbon dioxide by methane-oxidizing bacteria and small amounts of methane are stored or trapped in the form of methane hydrates under the extreme pressures found in ocean trenches. Most of the methane is released to the atmosphere, and as a result the atmosphere contains about 1.4 p.p.m. of methane. The 'life expectancy' of a methane molecule in the troposphere or lower stratosphere is about 1.5–7 years before it reacts with products of the ozone cycle in a process called tropospheric oxidation to regenerate carbon dioxide and complete the cycle (Ehhalt, 1974).

Most of the atmospheric methane (90%) originates from ecosystems which exist 1 m above or below the earth's surface (Vogels, 1979). Table 1 lists the major sources of methane as well as the relative amounts contributed by each source. It may be worth noting that the world total annual natural production of methane is almost equal to the annual recovery of fossil methane.

Table 1 Major Sources of Naturally Produced Methane[a]

Source	Amount produced (10^9 kg year^{-1})	Contribution to atmospheric methane (%)
Paddy fields	280	25
Swamps	130–260	45
Ruminants	101	20
Rivers, lakes	1.25–25	3–10
Oceans	5–8	4
Others	15–26	—

[a] Ehhalt (1974).

Although the estimates of methane production from lakes, rivers and oceans might intuitively seem low, it is important to realize that in bodies of water greater than 10 m in depth, bacterial oxidation is sufficient to remove 70–75% of the methane before it reaches the surface. Also, low temperatures in deep waters limit methane production. At a global average temperature of 283 K for fresh water, methanogenesis proceeds at a rate significantly lower than at the optimum 318 K. Moreover, in oceans the presence of sulfate restricts the methane production rate to about one tenth of that in fresh water.

The extent and role of methane production in animals, particularly ruminants, has been studied extensively. In ruminants, 10% of the feed energy may be 'lost' as methane. Most of this methane is formed by reduction of carbon dioxide by hydrogen, both of which are produced as by-products in the digestion of cellulose. Reduction of the methanogenic activity by feeding the antibiotic 'Monensin' indeed increases feed efficiency. Complete elimination of methane production on the other hand might well lead to reduced digestion of forage and loss of feed efficiency.

61.4 MAN-MADE PROCESSES

61.4.1 Methane Production from Sewage, and Agricultural and Industrial Wastes

Methane production from wastes can be approached from two points of view. From a waste treatment point of view, the total amount of methane produced (m^3 m^{-3} waste d^{-1}) is significant because it reflects directly the amount of organic carbon removed as discussed above. From an energy recovery point of view, however, the net amount is significant. Net production is obtained by subtracting the energy required to run the process from the total energy production. As will be shown, process design will vary depending on whether total methane production is the significant one or the net energy production (Rozzi, 1981).

The total amount of methane produced is a function of many factors, such as waste strength, waste composition, residence times (hydraulic and solid), reactor design and temperature. The amount of methane produced generally increases with waste strength, hydraulic and solid retention times, and temperature, because of the longer times organics are exposed to microbial action and, in the case of temperature, the faster rates of microbial action. Soluble substrates can generally be converted at short retention times, while suspended organic material, such as lipids and lignocellulosics, require longer retention times. Lipids are slowly converted, especially when in solidified form, while the breakdown of lignocellulosics is the rate limiting step when they are present in significant amounts. Also, the amount of methane recovered depends on reactor design, such as the degree of mixing, and cell immobilization techniques. A detailed description of these effects has been given earlier in this volume when dealing with anaerobic waste treatment. Also, much useful information has been presented by McFarlane and Pfeffer (1981), Stafford *et al.* (1980), van Brakel (1980) and Wise (1981).

The total volume of methane produced per unit volume of waste, P, and the rate of methane production, R, can readily be calculated using the following expressions:

$$P = (S_o - S_e) Y \tag{4}$$

$$R = \frac{(S_o - S_e) Y}{f_r} \tag{5}$$

where P = methane produced per unit volume of waste added (m^3 m^{-3}); S_o = influent concentration (kg COD m^{-3}); S_e = effluent concentration (kg COD m^{-3}); Y = yield coefficient for methane from COD (m^3 (STP) kg^{-1} COD); R = methane production rate (m^3 (STP) m^{-3} reactor volume d^{-1}); and f_r = hydraulic residence time (d).

These equations assume that the reactor is at steady-state performance and no biomass is accumulating. As pointed out earlier, for practical purposes the yield coefficient ($Y = 0.35$) may have to be reduced slightly.

The simple relationships in equations (4) and (5) make it redundant to provide detailed data on methane production from a variety of wastes. Results given in earlier parts of this volume dealing with waste treatment permit easy calculation of yield and rates of methane production. Table 2 is included to give reference values for yields of methane for some wastes.

Table 2 Methane Production per Head from Human and Animal Wastes

	Organic material produced (kg d^{-1})	Methane produced (m^3 d^{-1})[a]	Conversion (%)
Human			
Sewage	0.1–0.2	0.02–0.03	25–40
Garbage	1.5–3	0.3–0.7	40–60
Chicken	0.01–0.02	0.004–0.008	50–70
Hog	0.3–0.5	0.09–0.14	40–60
Beef steer	3–4.5	0.3–0.6	30–50
Dairy cow	3.5–5.5	0.4–0.7	20–40

[a] At STP.

When net production of methane is the most important factor (reactors specifically designed for energy recovery), other factors may become dominant. Of these, the temperature difference between incoming waste and operating temperature and the net volumetric rate of methane production based on reactor volume are very important. The former can make the net energy production insignificant, the latter affects the capital investment significantly.

A simple equation can demonstrate the importance of substrate temperature. The energy required to heat 1 m^3 of substrate waste to the reactor operating temperature, E_1, is given by:

$$E_1 = S(T_o - T_w) \tag{6}$$

where E_1 = energy required to heat waste to reactor temperature (MJ m^{-3}); S = specific heat (usually close to 4.186 MJ m^{-3} K^{-1}); T_o = operation temperature (K); and T_w = waste temperature (K). (Generally, heat losses from large digesters are relatively small.)

The recoverable energy in the form of methane is derived from equation (4):

$$E_2 = PE_3 \tag{7}$$

where E_2 = recoverable energy (MJ m^{-3}); and E_3 = combustion energy of methane (39 MJ m^{-3}, STP).

The net energy yield therefore is:

$$E_2 - E_1 \text{ (MJ)} = 13.65(S_o - S_e) - 4.186(T_o - T_w) \tag{8}$$

The relationship between $T_o - T_w$, S_o and $S_o - S_e/S_o$, the fraction of COD converted to methane, is demonstrated in Figure 2. In this figure the values of S_o at a given combination of $T_o - T_w$ and $S_o - S_e/S_o$ are the minimum values above which a net methane yield is obtained. For example, with a conversion efficiency of 0.8 (80%) and a reactor temperature of 318 K and waste at 293 K, the minimum value of S_o is 9.4 kg COD m^{-3}. These results clearly show the importance of reducing $T_o - T_w$ to a minimum. A more thorough and detailed description, including the effects of heat losses and other energy requirements, has been given by Rozzi (1981).

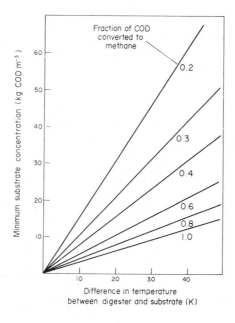

Figure 2 Effect of substrate temperature and fraction of substrate COD converted, $(S_o - S_e)/S_o$, on the minimum substrate concentration required to produce sufficient energy to heat incoming substrate to reactor temperature

The volumetric rate of methane production is a function of reactor design. For example, stirred tank digesters may produce 0.5–1 m^3 STP per m^3 digester per day, while advanced (fixed film, sludge bed or fluidized bed) reactors may produce over 5 m^3 per m^3 digester per day (van den Berg *et al.*, 1982). Assuming the conversion efficiency stays constant with increased rate of gas production (and loading rate) and there is a net methane production according to equation (5), there are advantages of advanced reactors with increased volumetric rates of methane production: (1) reactor size would be substantially smaller, often leading to reduced capital costs (per m^3 of methane produced); and (2) reactor heat losses would be smaller.

It has been well demonstrated, however, that conversion efficiencies decrease with increased loading rates (Henze and Harremoes, 1982; Kennedy and van den Berg, 1982a, 1982b; van den Berg and Kennedy, 1982). This is especially the case for wastes containing lipids and lignocellulosics such as manures and sewage sludge. Under those conditions, increasing loading rates, such as would be possible with advanced reactors, may result in decreasing net energy production (equation 8; Figure 2) and hence may be counterproductive. This is of course most likely where $T_o - T_w$ is large as is often the case in cold climates. In this situation a reactor operating at high rates and waste strengths and at a low temperature would be advantageous.

61.4.2 Methane Production from Energy Crops

Several studies have been made on the conversion of energy crops to methane, but the economics are unfavorable except under special circumstances. This seems reasonable since the digestion process costs have made it mostly uneconomical to produce methane from wastes except where a substantial credit could be assumed for the waste treatment or effluent (Hashimoto and Chen, 1980; Pacific Gas and Electric Co., 1981).

Most of the studies with energy crops have been made with algae, kelp and water hyacinth.

These crops can be grown and harvested quite cheaply in many areas of the world. In some cases they are nuisance crops that have to be removed anyway.

Special land-based energy crops from methane production have been investigated in New Zealand (Stewart, 1980). In most countries, however, agricultural crops are more valuable for food and fiber production or even for the production of fuel alcohol. Crops on non-farm land may also be considered, but harvesting costs are usually high because of the low yield.

Energy crops for anaerobic digestion have to be processed to meet the requirements for digestion outlined earlier. Few major problems have been reported when these conditions have been met.

Conversion efficiencies for crops grown for methane production are generally under 50%, depending on degree of comminution and mixing and the content of cellulosics (Table 3). Salt-water crops tend to have a lower conversion than fresh-water crops or land crops when allowance is made for cellulose or lignocellulose content. Presumably salts from sea water reduce conversion efficiencies.

Table 3 Methane Yields from Energy Crops

Crop	Approximate methane yield $(m^3 kg^{-1})^{a,b}$	Ref.
Oats	0.28	Stewart (1980)
Chopped hay	0.26	Stewart (1980)
Wheat straw	0.22	Stewart (1980)
Grass	0.17	Stewart (1980)
Seaweeds		
Ulva sp.	0.54	Yang (1981)
Euchema sp.	0.14	Yang (1981)
Gracilaria sp.	0.24	Ryther and Hanisak(1981)
Sea kelp	0.19	Fannin *et al.* (1982)
Water hyacinth	0.15–0.2	Chin and Goh (1978)
		De Renzo (1977)
Fresh water algae	0.15–0.2	De Renzo (1977)

[a] At STP. [b] Based on dry weight.

Reactor design is dictated to some extent by crop processing. Substrates usually have a high COD content and high volatile suspended solids content and are therefore not suitable for reactors such as anaerobic filters, sludge bed reactors and expanded or fluidized bed reactors without considerable further processing. If conversion of a large part of the cellulosic contents is required, a stirred tank reactor may be optimum. For faster rates of methane production, downflow stationary fixed film reactors may be preferable.

61.4.3 Gas Utilization

Gas from anaerobic digestion is generally of medium fuel value (15–26 MJ m^{-3}, STP). There are several actual and potential uses for digester gas: (1) direct use of the gas with little or no cleaning, *i.e.* use on or near site for heating purposes, delivery to gas consumer, or generation of electricity on site; (2) on site cleaning to high energy gas for introduction into pipeline systems; (3) conversion to methanol; and (4) condensation to liquid natural gas.

Of the direct uses only on- or near-site combustion for heat production and on-site generation of electrical power have been shown to be successful. Cleaning of the gas to give pipeline quality, conversion to methanol or to liquid natural gas requires large capital investments that are not justified by the size of the resource source or cost of recovery (Schellenbach, 1982).

Cleaning the gas involves removal of water, hydrogen sulfide and carbon dioxide. Water is present in amounts of 2–6% depending on reactor temperature and can readily be removed by cooling and compression. Hydrogen sulfide may be present in trace amounts up to 10% or more depending on the sulfur (especially sulfate) content of the wastes. Where it has to be removed (to prevent corrosion or because of burner emission standards) a part of the energy in the gas may be lost and major costs may be incurred. The sulfide can be removed separately or together with carbon dioxide in a variety of scrubbers (using steel wool or molecular sieves). Since the sulfide is extremely toxic, special precautions are required in handling sulfide-containing gas.

61.5 POTENTIAL OF METHANE AS AN ENERGY SOURCE

Estimates have been made of the amount of methane that can be generated by anaerobic digestion of wastes and energy crops. These estimates vary between less than 1 and up to 5% of a developed country's energy needs. The exact figure depends on whether total or net energy is being considered and whether all wastes and potential crops are included or only those that are economically accessible for treatment. In any case, the contribution of methane to the energy picture is likely to be mostly local, because of the limits on waste and crop transport and the optimum size of digesters.

The economics of methane production from biomass varies from country to country, depending on factors such as energy costs, capital costs, substrate costs and interest of potential operators (van Brakel, 1980; van Buren, 1979). In most developed countries, it may be possible to produce methane at close to the cost to the user of an alternative energy source, if substrate is available at no or negative cost, little or no purification is required (*i.e.* when the gas is mostly used for burning or to run engines) and where gas storage over relatively long periods is not required (Schellenbach, 1982; Slesser and Lewis, 1979). As pointed out, storage and liquefaction require a considerable capital investment, usually not justified by the selling price. As a result there has been little incentive in developed countries to install anaerobic digesters only for the production of methane. Where energy costs are very high or alternative sources of energy are not available, anaerobic digesters can be profitable.

The main incentives for increased production of methane are related to other advantages of anaerobic digestion. First of all anaerobic digestion is a much cheaper method for removing organics from wastewater than aerobic methods and this should lead to acceptance of anaerobic digestion as a method of treatment preceding aerobic treatment. In rural areas, anaerobic digestion of farm wastes and night soil provides a more acceptable odor, and a more acceptable, if not better, fertilizer than the raw wastes, and biogas is a cleaner and less odoriferous source of energy than dried farm wastes. In many areas of the world, anaerobic digestion of farm wastes and night soil also improves hygienic conditions and reduces illness. When all these factors are taken into account, methane production from biomass, particularly waste biomass, must be accorded a rapidly increasing but limited role in the energy picture.

61.6 REFERENCES

Chin, K. K. and T. N. Goh (1978). Bioconversion of solar energy: methane production through water hyacinth. *Proceedings of the 2nd Symposium on Energy from Biomass and Wastes*, pp. 215–228. Institute of Gas Technology, Chicago.
De Renzo, D. J. (1977). *Energy from Bioconversion of Waste Materials*. Noyes Data Corp., Park Ridge, NY.
Ehhalt, D. H. (1974). The atmospheric cycle of methane. In *Microbial Production and Utilization of Gases*, ed. H. G Schlegel, G. Gottschalk and N. Pfennig, pp. 13–22. Akademie der Wissenschaften Zu Göttingen.
Fannin, K. F., V. J. Srivastava and D. P. Chynsweth (1982). Unconventional anaerobic digester designs for improving methane yields from Sea Kelp. *Proceedings of the 6th Symposium on Energy from Biomass and Wastes*, pp. 373–399. Institute of Gas Technology, Chicago.
Hashimoto, A. G. and Y. R. Chen (1980). The overall economics of anaerobic digestion. In *Anaerobic Digestion*, ed. D. A. Stafford and B. I. Wheatley. Applied Science Publishers, London.
Henze, M. and P. Harremoes (1983). Anaerobic treatment of waste waters in fixed film reactors. *Water Sci. Technol.*, **15**, 1–102.
Kennedy, K. J. and L. van den Berg (1982a). Anaerobic digestion of piggery waste using a stationary fixed film reactor. *Agric. Wastes*, **4**, 151–158.
Kennedy, K. J. and L. van den Berg (1982b). Continuous *vs.* slug loading of downflow stationary fixed film reactors digesting piggery wastes. *Biotechnol. Lett.*, **4**, 137–142.
McFarlane, P. N. and J. T. Pfeffer (1981). *Biological Conversion of Biomass to Methane*. Solar Energy Research Institute, Golden, CO, Seri/TR-98357-1.
Pacific Gas and Electric Co. (1981). *Bioconversion Feasibility Study—Cattle Manure to Methane*. Southern California Gas Company.
Rozzi, A. (1981). Energy recovery in anaerobic processes. Paper presented at the Seminar on Anaerobic Waste Water Treatment and Energy Recovery, Pittsburgh, November 3–4. Duncan, Laguese and Associates Inc.
Ryther, J. H. and M. D. Hanisak (1981). Anaerobic digestion and nutrient recycling of small benthic or floating seaweeds. *Proceedings of the 5th Symposium on Energy from Biomass and Wastes*, pp. 383–412. Institute of Gas Technology, Chicago.
Schellenbach, S. (1982). Case study of a farmer owned and operated 1000 head feedlot anaerobic digester. *Proceedings of the 6th Symposium on Energy from Biomass and Wastes*, pp. 545–566. Institute of Gas Technology, Chicago.
Slesser, M. and C. Lewis (1979). *Biological Energy Resources*. E. and F. N. Spon, London.
Stafford, D. A., B. I. Wheatley and D. E. Hughes (eds.) (1980). *Anaerobic Digestion*. Applied Science Publishers, London.
Stevens, T. G. and L. van den Berg (1981). Anaerobic treatment of food processing wastes using a fixed-film reactor. *Proceedings of the 36th Purdue Industrial Wastes Conference*, pp. 224–232. Ann Arbor Science Publishers, Ann Arbor.

Stewart, D. J. (1980). Energy crops to methane. In *Anaerobic Digestion*, ed. D. A. Stafford and B. I. Wheatley. Applied Science Publishers, London.

van Brakel, J. (1980). *The Ignis Fatuus of Biogas*. Delft University Press.

van Buren, A. (ed.) (1979). *A Chinese Biogas Manual*. Intermediate Technology Publications, London.

van den Berg, L. and K. J. Kennedy (1982). Performance characteristics of anaerobic downflow stationary fixed film reactors. *Proceedings of the 1st International Conference on Fixed-Film Biological Processes*, Kings Island, OH, April 20–23.

van den Berg, L., C. P. Lentz and D. W. Armstrong (1982). Methane production rates of anaerobic fixed film fermenters as compared to those of anaerobic contact and fully mixed continuous fermenters. *Adv. Biotechnol.*, **2**, 257–262.

Vogels, G. D. (1979). The global cycle of methane. *Antonie van Leeuwenhoek*, **45**, 347–352.

Wise, D. L. (1981). *Fuel Gas Production from Biomass I, II*. CRC Press, Cleveland, OH.

Yang, P. Y. (1981). Methane fermentation of Hawaiian seaweeds. *Proceedings of the 5th Symposium on Energy from Biomass and Wastes*, pp. 307–358. Institute of Gas Technology, Chicago.

62

Composting

A. J. BIDDLESTONE and K. R. GRAY
University of Birmingham, UK

62.1 INTRODUCTION

Composting is the decomposition of organic matter by a mixed population of microorganisms in a warm, moist, aerobic environment.

Considerable amounts of organic material, produced annually in nature, are eventually degraded by microbial action. This normally takes place slowly on the surface of the ground, at ambient temperature and mainly under aerobic conditions. The natural process of breakdown can be accelerated by gathering the material into heaps to conserve part of the heat of fermentation so that the temperature of the mass rises and faster reaction rates are obtained. This accelerated process is composting.

Wastes amenable to composting vary from the highly heterogeneous organic/inorganic mixture in urban refuse to the reasonably homogeneous farm manures, crop residues and sewage sludges. During the composting process most of the oxygen demand of the wastes is met, the organic materials are converted to more stable products such as the humic acids, and carbon dioxide and water are evolved. An important consideration in increasing agricultural output is raising the level of soil fertility. A method of improving both soil structure and the supply of plant nutrients is the application to the soil of humus, the end product of composting. This is particularly important in arid and semi-arid countries in which the rate of oxidation of soil humus is far higher than in temperate climes.

62.2 THE BASICS OF COMPOSTING

The process of composting involves an interaction between the organic waste, microorganisms, moisture and oxygen. The organic material will normally have an indigenous mixed population of microorganisms derived from the atmosphere, water or soil. When the moisture content of the wastes is brought to a suitable level and the mass aerated, microbial action speeds up. As well as oxygen and moisture, the microorganisms require for their growth and reproduction a source of carbon (the organic waste), macronutrients such as nitrogen, phosphorus and potassium, and certain trace elements. In attacking the organic matter the microorganisms reproduce themselves and liberate carbon dioxide, water, other organic products and energy. Some of the energy is used in metabolism; the remainder is given off as heat. The end product, compost, is made up of the more resistant residues of the organic matter, breakdown products, dead and some living microorganisms, together with products from further chemical reaction between these materials. The overall process is illustrated in Figure 1.

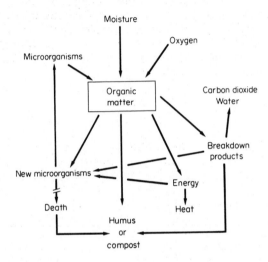

Figure 1 The composting process

62.2.1 Biochemical Aspects

Organic wastes, whether of agricultural, industrial or urban origin, are mixtures of sugars, proteins, fats, hemicelluloses, cellulose, lignin and minerals in a wide variety of concentrations, as shown in Table 1. In animals the composition of manure depends upon the type of animal and its feed. In plants the concentrations depend very much upon the age of the plant as well as upon its type and environment. Fresh green material, such as young grass, contains water-soluble matter, proteins and minerals. As the plant ages, minerals tend to return to the soil and low molecular weight compounds are converted to the higher molecular weight polymers, hemicellulose, cellulose and lignin. Composting is both a breaking-down process and a building-up process. The key point is the cell wall of the microorganism attacking the organic matter. Low molecular weight materials, the water solubles, can pass through the cell wall easily and take part in cell metabolism, providing energy and being built up into larger polymers. The higher molecular weight components of the organic wastes cannot pass through the cell wall and cannot be used without being broken down. In these cases the microorganisms can secrete extracellular enzymes which hydrolyse the polymers into short lengths which are the basic sugar units. Virtually all microorganisms can assimilate the resulting fragments but only a proportion can carry out the hydrolysis.

Some indication of the extent of the biochemical changes taking place is given by the results of Yung Chang (1967) on composting wheat straw amended with ammonium nitrate. The straw had lost over half of its dry weight after 60 days of composting with the majority of the loss in the first 34 days. The loss of total dry weight could be accounted for almost completely by the loss in hemicelluloses and cellulose. Cellulose degradation slowed down during the middle of the cycle whilst the hemicelluloses were broken down fairly steadily. Hemicelluloses are polysaccharides of

Table 1 Composition of Organic Matter

Fraction	% in dry weight Plants	% in dry weight Manures
Hot/cold water solubles: sugars, starches, amino acids, urea and ammonium salts, aliphatic acids	5–30	2–20
Ether/alcohol solubles: fats, oils, waxes and resins	5–15	1–3
Proteins	5–40	5–30
Hemicelluloses	10–30	15–25
Cellulose	15–60	15–30
Lignin	5–30	10–25
Minerals (ash)	1–13	5–20

about 50 to 150 C_5/C_6 sugar units which are fairly susceptible to attack. Cellulose is a much larger polymer of about 1000 to 10 000 glucose units which is significantly more resistant. Lignin consists of a number of aromatic units linked by aliphatic side chains and is extremely resistant to enzyme attack, being the last material degraded in composting. However, during composting the lignin molecule does become modified, losing some methoxy groups and aliphatic side chains and gaining carboxyl and phenolic hydroxy groups (Alexander, 1977).

62.2.2 Microbiology

Composting is a dynamic process brought about by the activities of a succession of mixed microbial populations, each of which is suited to an environment of relatively limited duration. Table 2 lists the types and numbers of organisms normally encountered.

Table 2 Organisms Involved in Composting

	Genus	Numbers per g of moist compost
Microflora	Bacteria	10^8–10^9
	Actinomycetes	10^5–10^8
	Fungi	10^4–10^6
	Algae	10^4
	Viruses	
Microfauna	Protozoa	10^4–10^5
Macroflora	Fungi	
Macrofauna	Mites, springtails, ants, termites, millipedes, centipedes, spiders, beetles, worms	

These organisms represent both the plant and animal kingdoms. Each named microflora is a genus within which are many different species, *e.g.* possibly 2000 of the bacteria and at least 50 of the fungi. Each species can be subdivided according to the temperature ranges of their activity; psychrophiles prefer temperatures below 20 °C, mesophiles 20 to 40 °C and thermophiles above 40 °C. The macroflora and macrofauna which flourish during the final stage of composting are essentially mesophiles.

Although present in vast numbers the bacteria are of very small size and form less than half of the total microbial protoplasm. Some species form endospores which can withstand considerable heat and desiccation.

The actinomycetes develop far more slowly than most bacteria and fungi but become prominent at peak temperature and in the later stages of composting.

The thermophilic fungi are a relatively well-defined group in composting. At least eight species of these fungi have been studied which are capable of growth in the range 40 to 60 °C. They die out above 60 °C, reappearing later as the temperature falls. Studies on the population numbers of bacteria, actinomycetes and fungi during composting have been made by Yung Chang (1967), Yung Chang and Hudson (1967), Hayes and Lim (1979) and Hedger (1972).

Once the compost heap cools from its peak temperature it is accessible to a wide range of soil macrofauna. These feed upon other animals, animal excreta and the plant remains. They normally require well-aerated conditions, adequate moisture and prefer temperatures in the range 7–13 °C. Many of the soil animals make a major contribution to breakdown in the compost heap due to physical maceration; breaking the material into smaller particles exposes greater surface area for subsequent attack by the microflora. They also make a contribution to the mixing of the various constituents.

In temperate climates the earthworm plays a major role in the breakdown processes in the compost heap and in the subsequent incorporation of organic matter into the soil; in arid and semi-arid climates this function is usually undertaken by the termite (Edwards, 1974).

The biochemistry and microbiology of composting are covered in greater detail by Gray *et al*. (1971a).

62.2.3 Temperature–Time Pattern

When organic wastes are gathered into heaps for composting, the insulating effect of the material leads to a conservation of heat and a marked rise in temperature. The subsequent composting process may conveniently be divided into four stages: mesophilic, thermophilic, cooling down and maturing (Figure 2).

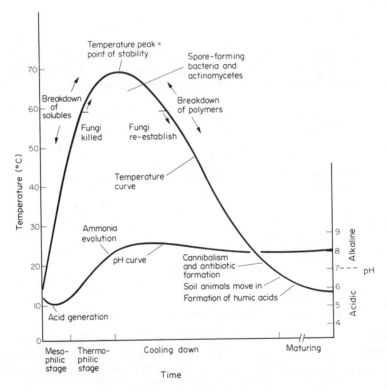

Figure 2 Temperature and pH variations with time

At the start of composting the wastes are at ambient temperature and are slightly acidic. During the mesophilic stage the indigenous microorganisms multiply rapidly, the temperature rises to about 40 °C and the mass becomes increasingly acidic. The temperature continues to rise and the thermophilic strains of organisms take over; the pH turns alkaline as ammonia is liberated during the breakdown of protein molecules. By 60 °C the thermophilic fungi cease activity and the reaction is kept going by the actinomycetes and the spore-forming strains of bacteria. The reaction rate slows down and the temperature peak is reached; the rate of heat generation then becomes equal to the rate of heat loss from the heap surface. This marks the end of the thermophilic stage; the wastes have reached stability at which the easily converted materials, *e.g.* carbohydrates, fats and proteins, have been degraded and most of the oxygen demand met. The wastes are now no longer attractive to flies and vermin and should not give off bad odours; the material may now be put into heaps outside without causing major environmental pollution. The maximum temperature reached depends in the main on the quantity of material being processed.

Following the temperature peak the cooling down stage commences and the pH drops slightly but remains alkaline. When the temperature falls below 60 °C the thermophilic fungi re-invade the mass and together with the actinomycetes attack the long chain polysaccharides, *i.e.* hemicelluloses and cellulose, breaking them down into simple sugars, which may then be utilized by a wide range of microorganisms. Following the breakdown of these polysaccharides the rate of energy liberation becomes very small and the temperature of the mass falls to ambient.

The process now enters the maturing stage in which heat evolution and weight loss are small. The macroflora and macrofauna now invade the heap. As the food supply becomes exhausted antagonism between the microorganisms breaks out and antibiotics are produced. Complex chemical reactions occur between the lignin residues of the original wastes and the proteins from dead microorganisms to form the humic acids. By the end of maturing, which may take several months, the wastes will not heat up on turning nor go anaerobic in storage nor rob nitrogen from the soil on incorporation. The material has become humus or compost.

62.3 PROCESS FACTORS

The composting of organic wastes is a dynamic and complicated ecological process in which temperature, pH and food availability are constantly changing. In consequence, the numbers and species of organisms present also change markedly. The rate of progress towards the mature end product, humus, is dependent on several interrelated process factors. These include particle size, nutrient supply, structural strength of the material, moisture, aeration, agitation, pH and size of heap. It is desirable to adopt the best operating conditions allowed by the economics of the operation. The complexity of the processing plant, and the quality of the final product, will depend upon the type of waste and the level of investment available.

62.3.1 Separation

The major use for compost is as a soil conditioner/fertilizer in agriculture and horticulture. Consequently the compost should have a high organic matter content with a minimum of mineral matter. This is of particular importance when processing urban wastes, the composts from which can contain significant quantities of trace metals such as copper, lead, nickel and zinc. Accordingly, with urban refuse it is desirable to remove as much glass, metal, plastic and debris as is economically possible. There are a variety of devices available for such separations: air classifiers, Rotadisc separators for plastics and ballistic separators for heavy particles. Where sewage sludge is used for composting, it should be mainly from domestic and not industrial sources.

62.3.2 Particle Size

The smaller the size of the particles of waste material, the greater is the surface area exposed to microbial attack, and theoretically, the greater is the rate of composting. However, very small particles pack tightly together, giving material with a high bulk density having narrow pores and channels within it. This restricts the diffusion of air into the mass and carbon dioxide out of it, thereby inhibiting composting. The high bulk density may also cause excessive loads on mechanized turning equipment, especially when the materials are wet.

A compromise on particle size is therefore necessary; the refuse is shredded to approximately 50 mm size for static heaps and windrows. For mechanized plants with agitation and forced aeration, a particle size of 12.5 mm is appropriate. Particle size reduction may be achieved using hammer mills, raspers or by self abrasion in rotary drums.

After composting and maturing the particles of the compost product should be below 2.5 mm screen size.

62.3.3 Nutrients

The microorganisms involved in composting require a source of carbon to provide energy and material for new cells, plus a supply of nitrogen for cell proteins. To a lesser extent there is a requirement for phosphorus, potassium, calcium, sodium, magnesium, sulfur, iron and traces of other elements such as cobalt and zinc. The subject is well discussed by Alexander (1977). In most cases the requirement for these nutrients is adequately met from the original organic wastes; only nitrogen and occasionally phosphorus may need adjustment.

Chemical analysis of microorganisms revealed that on average they contained 50% carbon C, 5% nitrogen N and 0.25–1.0% phosphorus P on a dry weight basis. This gives a C/N ratio of 10:1 and an N/P ratio of 5–20:1. Since approximately 50–60% of the organic carbon in the composting materials is converted to carbon dioxide, an initial C/N ratio of about 25:1 should be optimum if no nitrogen is lost. A higher ratio involves the oxidation of excess carbon, the organisms going through many life cycles to achieve a final C/N ratio of 10:1. With C/N ratios lower than 25:1, as in the case of animal manures and sewage sludges, nitrogen will be lost as ammonia, often in considerable amounts. However, some carbon and nitrogenous compounds are fairly resistant to microbial attack and some nitrogen fixation from the atmosphere may occur during composting. These factors make accurate prediction of initial C/N requirements difficult; in practice a ratio in the range 30–35:1 has proved to be optimum (University of California, 1953). For low initial C/N ratios the loss of nitrogen as ammonia may be suppressed by the addition of extra phosphate.

62.3.4 Other Additives

The use of chemical and herbal additives and the addition of bacterial cultures (inocula) have been claimed, mainly by the additive manufacturers, to increase the rate of composting. Apart from the possible need for extra nitrogen most compostable materials normally contain all the nutrients they require. The situation regarding inocula is less clear. Laboratory scale work by Golueke (1954) indicated that inocula had no effect whereas work on a full size rotary drum by de Bertoldi (1978) showed that the onset of the thermophilic phase could be speeded up by recycling some product compost to the feed.

When composting finely divided organic solids such as sewage sludge and animal manure slurries, bulking agents are normally necessary to ensure an open matrix for air diffusion. Wood chips have been the favoured bulking agent in the Beltsville aerated pile system for sewage sludge (Epstein *et al.*, 1976). Higgins *et al.* (1980) propose pulverized tyres as an alternative; Gray and Biddlestone (1975) used straw in the ARCUB process for manure slurries.

62.3.5 Moisture Content

Water is essential to the composting process for the transport of materials to and from the microorganisms. Below 30% of moisture on a fresh weight basis the biological reactions slow down markedly. If the moisture content is too high the voids within the matrix become waterlogged, limiting access of oxygen to the microorganisms. Some materials, *e.g.* paper, readily lose structural strength when very wet; straws on the other hand can tolerate high moisture content. For refuse the optimum moisture lies in the range 50–60%.

Water is produced during the composting process by microbial action and is lost by evaporation into the air stream. Where forced aeration is applied moisture loss can be excessive and additional water must be supplied to the matrix. Problems of water loss are naturally more severe in hot climates. Urban refuse contains lipids (fats, oils and waxes), which are liquids at composting temperatures. Wiley (1957) has suggested that the total liquid content should be used as a guide, rather than the water content. This is given by

$$\% \text{ liquid} = \frac{100 \times (\% \text{ moisture} + \% \text{ lipids})}{(100 - \% \text{ ash})}$$

62.3.6 Aeration

Oxygen is essential to the metabolism of the aerobic species of microorganisms responsible for composting. Air may be introduced by several methods: by natural gaseous diffusion into the stationary pile, by turning the heap regularly with a machine, or by forced aeration from a fan. Natural diffusion frequently fails to supply adequate oxygen at the start of composting, leading to anaerobic conditions at the heap centre.

Aeration has other functions in the composting process. It removes the carbon dioxide and moisture produced in the microbial reaction and may help to cool the heap by evaporative heat transfer. Oxygen requirements vary throughout the process, being low in the mesophilic stage, increasing to a maximum in the thermophilic stage and decreasing through the cooling down and maturing stages. Wiley (1955) recommended aeration rates to supply 6–19 mg O_2 h^{-1} (g volatile solids in the composting mass)$^{-1}$. It is possible to have too high an aeration rate which can lead to excessive cooling and desiccation of the wastes.

62.3.7 Agitation

Agitation speeds up the composting process by improving aeration and by breaking up large pieces of material thereby exposing fresh surfaces to microbial attack. However too much agitation can lead to excessive cooling and drying of the wastes and shearing of actinomycete and fungal mycelium. Flintoff (1976) considers that turning a windrow heap three or four times should be sufficient while Gray *et al.* (1971b) suggest that in mechanized plants short periods of vigorous agitation should be alternated with periods of no agitation.

62.3.8 pH Control

The pH changes from acid to alkaline during composting as shown in Figure 2. Addition of chemicals to control the pH has generally proved uneconomic in large scale composting plants.

62.3.9 Heat Production and Heap Size

Wiley (1957) studied the heat production when pulverized refuse was composted and concluded that, over 8–10 day cycles, it amounted to approximately 7×10^6 J kg^{-1} of initial volatile solids. Mote and Griffis (1982) determined maximum heat production rates from composting two organic materials obtaining values in the range 20–28 W kg^{-1} of initial dry mass.

By employing well-insulated compost heaps or large masses of wastes, such a heat production can lead to high temperatures in the range 80–90 °C (Spohn, 1970). Most experimental data, however, indicate that at such temperatures the reaction rate is very low. Optimum rates are obtained at temperatures of about 55 °C.

Small masses of material, as in garden compost heaps, have high surface/volume ratios and hence much of the material has to act as insulation. At least 1 tonne of wastes is needed to ensure that a reasonable proportion of the heap reaches a satisfactory temperature. For heaps composting under natural aeration conditions the material should not be piled over 1.5 m high or 2.5 m wide, otherwise diffusion of oxygen to the centre will be impeded. The heap can be elongated into a windrow of any convenient length.

62.3.10 Summary of Ideal Process Conditions

There are now adequate data in the literature on the microbiological, chemical and physical parameters of composting plants. A summary of the optimum values of the important parameters is given in Table 3.

The requirement is to translate these parameters into a low cost but reliable process plant. The complexity of the composting equipment and the degree of approach to the optimum values of the process parameters vary considerably from the simple heap situation to the highly sophisticated mechanized urban plant.

62.4 PRACTICAL PROCESSES

For many centuries composting of organic wastes has been practised by farmers and gardeners in many countries. The outstanding example has been that of the Chinese who have carried out small scale heap composting for some 4000 years (FAO, 1977; King, 1927). Interest in the scientific aspects of composting followed from the visit by King (1927) to the Far East and the subsequent development of the Indore method of heap composting (Howard and Wad, 1931; Howard, 1943).

Table 3 Composting Parameters

Parameter	Value
C:N ratio of feed	30–35:1
C:P ratio of feed	75–150:1
Particle size	12.5 mm for agitated plants and forced aeration 50 mm for windrows, unagitated plants and natural aeration
Moisture content	50–60%
Air flow	0.6–1.8 m^3 air d^{-1} kg^{-1} volatile solids during thermophilic stage, being progressively decreased during cooling down and maturing
Temperature	55 °C
Agitation	Short periods of vigorous agitation, alternating with periods of no agitation which vary in length from minutes in the thermophilic stage to hours during maturing
pH control	Normally none desirable
Heap size	Any length but not over 1.5 m high or 2.5 m wide for heaps and windrows using natural aeration. With forced aeration heap size depends on need to avoid overheating

Recent developments in composting have been in response to the need to deal with the large quantities of urban refuse and sewage sludge and the increasing need to recycle animal manures in modern agriculture.

62.4.1 Materials for Composting

A wide variety of organic materials which are suitable for compost production are produced by human communities and agriculture. Table 4 lists some of these wastes with very approximate values for % nitrogen and C/N ratio.

Table 4 Composition of Materials Suitable for Composting

Material	Nitrogen % dry weight basis	C/N ratio
Urine	15–18	0.8
Dried blood	10–14	3
Nightsoil, dung, sewage sludge	5.5–6.5	6–10
Grass	4	20
Bone meal	4	8
Farm yard manure	2.2	14
Refuse	1.1	34
Wheat straw	0.6	80
Fallen leaves	0.4	45
Sawdust	0.1	500
Paper	0	∞

Typical analyses of urban refuse from cities in different parts of the world are given in Table 5; considerable local variations are to be expected. In refuse only the vegetable/putrescible and paper fractions are of value in compost, the other materials being detritus. In the design of a compost plant it is essential that a thorough analysis be made of the local refuse (Flintoff, 1976; Higgins, 1966).

62.4.2 Large Scale Urban Composting

During the past three decades some 30 different processing schemes have been introduced for composting urban wastes with varying success (Gray *et al.*, 1973). Equipment for feed preparation and compost product finishing are similar in many of these processes. The fermentation

Table 5 Composition of Urban Refuse from Differenct Sources

	% by weight on a fresh weight basis			
	Middle East	Far East	South America	Europe
Vegetable and putrescible	50	75	55	16
Paper	20	2	15	43
Plastics and textiles	10	4	10	7
Glass	2	0	4	10
Metals	10	0	6	10
Miscellaneous	0	7	10	1
Inerts and rubble	8	12	0	13
Approximate moisture content within the total	42	60	45	18

stage, however, has varied widely, being attempted in pits, cells, silos and drums (Brunt *et al.*, 1980; Golueke, 1977; Gotaas, 1956; Haug, 1980; Satriana, 1974). Very recently large scale mechanized plants with capacities in the range 200–500 tonnes of refuse per day have been installed in Scandinavia, Libya and Hong Kong. Some of these plants process refuse alone, others compost sewage sludge with the refuse.

The modern urban composting unit usually comprises refuse storage facilities, feed preparation facilities, a composting stage and final product upgrading (Figure 3).

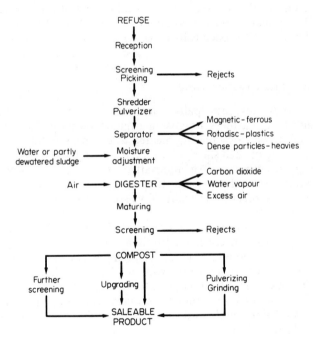

Figure 3 Process flow diagram for urban composting

Refuse collection vehicles discharge their contents into deep hoppers or on to flat concreted areas from where it is conveyed by moving floors, overhead grabs or front-end loaders. The material is then prepared by size reduction, separation of salvageable and unwanted materials, and then adjustment of moisture. Size reduction is achieved by using either wet or dry pulverizers. Wet pulverizers consist of a slightly inclined drum, typically 3 m diameter and up to 10 m long, rotating at 3–10 rev min^{-1} with a capacity of about 10 tonnes h^{-1}. Moistened refuse slowly passes through the drum, tumbles around and gradually breaks down by attrition. Dry pulverizers are normally of the rasping or hammer mill types although ball mills have been used in some recent installations. The power requirements of pulverizers depend mainly upon the final size of the outlet material; for reduction to 50 mm about 8 kWh tonne^{-1} is required whilst reduction to 12.5 mm requires about 20 kWh tonne^{-1}. Following size reduction magnetic separators are used to remove ferrous metals, Rotadisc machines take out sheet plastics and rags, and vibrating screens remove oversize material for reshredding or landfill. Air classifiers are now coming into use for the separation of light from heavy density materials. Prior to composting the moisture content of the material is adjusted, if necessary, by adding water or sewage sludge.

The biological degradation stage is carried out either in windrow heaps, accelerated windrows or mechanized units. In simple windrows the prepared material is placed into long heaps and turned periodically by either a specially designed machine or a front-end loader; composting requires several months. To shorten the composting time to several weeks, accelerated windrows are laid down over forced aeration ducts. Mechanized units involve agitation as well as forced aeration for a period of 2–12 days followed by stockpiling until sufficiently mature. Such mechanized units include automated windrows with turning devices, rotating drums and tower composters with single or multi floors. The product compost is upgraded using trommel or vibratory screens and sometimes by hammer milling to reduce glass and ceramic pieces to harmless, sand-like particles.

62.4.3 Sewage Sludge Composting

Composting is proving to be a useful means of treatment of sewage sludge prior to its use in agriculture or for land reclamation. The sludge is normally partly dewatered to achieve a moisture content in the range 65–85%. It is then mixed with a bulking agent to provide an open matrix facilitating the access of air to the heap. The Beltsville process (Epstein *et al.*, 1976) uses wood chips and forced aeration, the chips being screened out and recycled. Other bulking agents include refuse, straw, shredded tyres and recycled composted sewage sludge (Haug and Davis, 1981; Shea *et al.*, 1979). In many Third World countries nightsoil is generated rather than water borne sewage; Shuval *et al.* (1978) specifically consider nightsoil composting.

62.4.4 Composting of Other Organic Wastes

The composting of farm wastes, principally straw, manures and crop residues, is normally only practised by the minority of farmers practising organic farming. The production of animal manure slurries from intensive stock rearing has led to environmental problems, particularly of smell. The liquid composting of such slurries by forced aeration in enclosed vessels is being practised in various European countries. An alternative approach is conjoint composting of the manure slurries with straw. These farm composting operations have potential for the recovery of low grade heat. The whole subject of agricultural waste composting has recently been reviewed by Gray and Biddlestone (1981).

The composting of garden wastes is widely practised. The quantity of wastes handled is invariably small, 100–500 kg per batch; consequently heat production is low. Considerable care is needed to ensure that most of the wastes reach a sufficient temperature for weed seeds and diseased material to be killed; normally a bin with overhead insulation is necessary (Gray and Biddlestone, 1976). Organic wastes from various food manufacturing operations are also amenable to composting. Large particle materials from many fruit and vegetable canning operations can normally be composted alone while sludges need the addition of a bulking agent such as wood chips (Hyde and Consolazio, 1982).

62.4.5 Reaction Times and Compost Yield

The reaction rate during the three biological stages of composting depends upon the process conditions whereas the final stage of maturing involves complex chemical reactions and appears to be largely independent of these process conditions. Urban waste tipped into heaps without pretreatment or agitation takes 9–12 months to produce rough compost. When piled into windrow heaps with several turnings for agitation it takes about 3 months to reach maturity. In some processes the material is initially treated for 3–5 days in a rotating drum until it reaches the point of stability before it is put into windrows. In the Beltsville process (Epstein *et al.*, 1976; Haug, 1980) with forced aeration but no agitation 3 weeks are required before screening the finely divided sewage sludge compost. Multifloor tower composters with forced aeration and continuous agitation can degrade the material to near the end of stage 3 in about 7 days; maturing may then be required for several weeks, depending on the end use.

The percentage decomposition during composting is normally in the region of 40–60% of the dry organic matter. Yung Chang (1967) found that with straw 50% of the dry weight was lost.

The yield of useful product obtained from a composting reaction depends on several factors. These are the percentage decomposition of the organic matter, the amount of inorganic material in the feed, the final moisture content of the product and the quantity of salvage and unwanted detritus removed during feed preparation and final screening of the product. Typically from 100 tonnes of urban refuse some 40 tonnes of screened product compost can be expected.

62.4.6 Compost Product Composition and Use

The composition of compost products varies widely and reflects mainly the composition of the organic waste used. During the biological breakdown some 40% of the organic matter is oxidized to carbon dioxide and water. Table 6 illustrates the range of compositions normally encountered in composts showing that those prepared from urban wastes tend to be lower in organic matter and the major plant nutrients, N, P and K, than those made from farm/garden wastes. Composts from urban wastes can also contain significant quantities of trace metals such as copper, nickel, lead and zinc; several field trials have examined the take up by crops of these metals from soils treated with such composts (Gray and Biddlestone, 1980).

Table 6 Composition Ranges of Matured Composts

Substance	Composition range % by weight, dry basis
Organic matter	25.0–80.0
Carbon	8.0–50.0
Nitrogen (as N)	0.4–3.5
Phosphorus (as P_2O_5)	0.3–3.5
Potassium (as K_2O)	0.5–1.8
Calcium (as CaO)	7.0–1.5
	← Municipal composts
	Farm/garden composts →

Compost is primarily a soil conditioner due to its organic matter content; the plant nutrients that it contains also give it fertilizer value. The use of compost improves the structure of both clay and sandy soils, increasing the water holding capacity and reducing the susceptibility to wind and water erosion. A major use for urban compost in Europe is in the stabilization of steep vineyard slopes. In many developing countries where chemical fertilizers are expensive or unobtainable, composts prepared from local organic wastes are likely to have a significant effect on increasing food production (Dalzell *et al.*, 1979).

62.5 SUMMARY

Composting is a process of microbial degradation in which organic matter is broken down by a mixed microbial population in a warm, moist, aerobic environment. A wide range of wastes can be converted into useful compost products. This chapter considers the biochemical, microbiological and process factors involved in composting. Practical processes are then considered for the treatment of organic wastes from urban, farm and garden sources. Brief mention is made of compost compositions and uses.

62.6 REFERENCES

Alexander, M. (1977). *Introduction to Soil Microbiology*, 2nd edn. Wiley, New York.
Brunt, L. P., R. B. Dean and O. Tabasaran (1980). *Composting*. World Health Organization, Regional Office for Europe. Report ICP/RCE 401 (C.VII. 2).

Dalzell, H. W., K. R. Gray and A. J. Biddlestone (1979). *Composting in Tropical Agriculture*. International Institute of Biological Husbandry, Stowmarket, UK.

de Bertoldi (1978). Personal communication.

Edwards, C. A. (1974). Macro-arthropods. In *Biology of Plant Litter Decompositon*, ed. C. H. Dickinson and G. J. F. Pugh, vol. 2, chap. 16. Academic, London.

Epstein, E., G. B. Willson, W. D. Burge, D. C. Mullen and N. K. Enkiri (1976). A forced aeration system for composting wastewater sludge. *J. Water Pollut. Control Fed.*, **48** (4), 688–694.

FAO (1977). China: recycling of organic wastes in agriculture. *FAO Soils Bulletin*, no. 40. Food and Agriculture Organization of the United Nations, Rome.

Flintoff, F. (1976). *Management of Solid Wastes in Developing Countries*. World Health Organization Publications, South East Asia Office, New Delhi.

Golueke, C. G. (1954). A critical evaluation of inoculums in composting. *Appl. Microbiol.*, **2**, 45–53.

Golueke, C. G. (1977). *Biological Reclamation of Solid Wastes*. Rodale Press, Emmaus, Pennsylvania.

Gotaas, H. B. (1956). *Composting*, Monograph no. 31. World Health Organization, Geneva.

Gray, K. R., K. Sherman and A. J. Biddlestone (1971a). A review of composting, part 1. *Process Biochem.*, **6** (6), 32–36.

Gray, K. R., K. Sherman and A. J. Biddlestone (1971b). A review of composting, part 2, the practical process. *Process Biochem.*, **6**, (10), 22–28.

Gray, K. R., A. J. Biddlestone and R. Clark (1973). Review of composting, part 3, process and products. *Process Biochem.*, **8** (10), 11–16.

Gray, K. R. and A. J. Biddlestone (1975). New slurry composting process. *Farm Buildings Digest*, **10** (3), 5–6.

Gray, K. R. and A. J. Biddlestone (1976). The garden compost heap. *J. R. Hort. Soc.*, **101**, 540–544, 594–598.

Gray, K. R. and A. J. Biddlestone (1980). Agricultural use of composted town refuse. In *Inorganic Pollution and Agriculture*, pp. 279–305. Ministry of Agriculture, Fisheries and Food, reference book 326. HMSO, London.

Gray, K. R. and A. J. Biddlestone (1981). The composting of agricultural wastes. In *Biological Husbandry*, ed. B. Stonehouse, chap. 6. Butterworths, London.

Haug, R. T. (1980). *Compost Engineering, Principles and Practice*. Ann Arbor Science, Ann Arbor, MI.

Haug, R. T. and B. Davis (1981). Composting results in Los Angeles. *Bio Cycle*, **22** (6), 19–24.

Hayes, W. A. and W. C. Lim (1979). Wheat and rice straw composts and mushroom production. In *Straw Decay and its Effect on Disposal and Utilisation*, ed. E. Grossbard, pp. 85–94. Wiley, New York.

Hedger, J. N. (1972). *The Biology of Fungi in Composts*. Ph. D. Thesis, University of Cambridge.

Higgins, A. E. (1966). *The Analysis of Domestic Refuse*. Institute of Public Cleansing, London.

Higgins, A. J., M. E. Singley, N. Nocitra, K. Callanan, B. Whitson and A. Singh (1980). Shredded rubber tyres as a bulking agent. *Compost Science/Land Utilisation*, **21** (6), 20–23.

Howard, A. (1943). *An Agriculture Testament*. Oxford University Press, London.

Howard, A. and Y. D. Wad (1931). *The Waste Products of Agriculture*. Oxford University Press, London.

Hyde, M. A. and G. A. Consolazio (1982). Composting of food processing waste sludges. *Bio Cycle*, **23** (1), 58–60.

King, F. H (1927). *Farmers of Forty Centuries*. Jonathan Cape, London.

Mote, C. R. and C. L. Griffis (1982). Heat production by composting organic matter. *Agricultural Wastes*, **4**, 65–73.

Satriana, M. J. (1974). *Large-scale Composting*. Noyes Data Corporation, London.

Shea, T. G., J. Braswell and C. S.Coker (1979). Bulking agent selection in sludge composting facility design. *Compost Science/Land Utilisation*, **20** (6), 20–21.

Shuval, H. I., C. G. Gunnerson and D. Julius (1978). *Nightsoil Composting*. P. U. Report no. RES12. The World Bank, Washington, DC.

Spohn, E. (1970). Composting by artificial aeration. *Compost Science*, **11** (3), 22–23.

University of California (1953). *Reclamation of Municipal Refuse by Composting*. Technical Bulletin no. 9, series 37. University of California, Berkeley.

Wiley, J. S. (1955). A preliminary study of high rate composting. *J. Sanit. Eng. Div., Am. Soc. Civ. Eng.*, **81**, 1–27.

Wiley, J. S. (1957). Liquid content of garbage and refuse. *J. Sanit. Eng. Div., Am. Soc. Civ. Eng.*, **83**, 1–11.

Yung Chang (1967). The fungi of wheat straw compost—biochemical and physiological studies. *Trans. Br. Mycol. Soc.*, **50** (4), 667–677.

Yung Chang and H. J. Hudson (1967). The fungi of wheat straw compost—ecological studies. *Trans. Br. Mycol. Soc.*, **50** (4), 649–666.

63

Landfills for Treatment of Solid Wastes

J. F. REES
Harwell Laboratory, Oxon, UK
BioTechnica Ltd, Cardiff, UK*

63.1 INTRODUCTION

An historical review of the fate of domestic refuse (White-Hunt, 1980, 1981b,c) shows that archaeologists have located discrete rubbish heaps (middens) on the sites of Danish, Scandinavian, Hebridean and English New Stone Age Communities. Even in 1900 BC, large, circular pits existed in the city of Knossos for the disposal of rubbish from the Palace area, and much of the Romans' rubbish was discharged into their large sewers. But it seems that this early technology was soon forgotton and in the Middle Ages the favoured dump was the street. The rapid urbanization which was associated with the Industrial Revolution further exacerbated the problem of solid waste disposal. Street dumping was eventually superseded by the more organized collection of refuse and its removal from centres of high population; London's refuse, for example, was transported by barge down stream and deposited on marshland such that by 1930, the filth, smell from burning refuse had become intolerable. This open dumping was further superseded in the mid–late 1930s by the process of controlled tipping, sanitary landfill or landfilling. It was ruled that domestic refuse was to be deposited in a series of layers, each layer not to exceed 6 ft (1.8 m)

*Present address.

in depth, and that all surfaces exposed to the air were to be covered over by 9 in (0.2 m) of earth within 24 h of the refuse being deposited. This practice was designed to minimize (i) landfill fires, (ii) the nuisance from vermin, birds and flies, (iii) windblown litter and (iv) odour. This procedure, by and large, is what is currently recognized by the term landfill. Under these conditions, refuse ferments to yield a mixture of methane and carbon dioxide. The refuse when placed in the landfill is essentially dry with a moisture content ranging from 25–35% wet weight in the UK. When rainfall percolates through the refuse, soluble materials are dissolved and the microbial process of refuse decomposition is stimulated. This yields soluble products which accumulate in the liquid effluent, or leachate. Three major technologies are potentially available for domestic refuse disposal, namely, landfill, incineration and composting. The important role of landfill technology is highlighted in data compiled by Wilson (1981), showing that 89% and 95% by weight of domestic refuse in England and the USA respectively, is landfilled. The remainder is incinerated. In Austria, Belgium, France, the Netherlands, Sweden and West Germany, between 60 and 71% by weight is landfilled. Between 20–30% is incinerated, with composting making a significant contribution in Austria, Belgium and France (11%, 9% and 10% respectively). In Japan, 52% is landfilled and 46% is incinerated, while at the other end of the spectrum, Denmark and Switzerland favour incineration (66% and 70% respectively).

Landfills can be constructed by infilling large holes in the ground, *e.g.* disused ironstone, coal, clay or gravel workings. This is the preferred method in the UK. Currently, in West Germany and the Netherlands, the favoured method is to construct a large, landscaped refuse mound. A combination of both methods is used in the USA and in Canada.

Modern landfill disposal of urban refuse consists typically of a centralized transfer station located in the urban area and served by a large number of refuse collection vehicles. At the station refuse is loaded into larger vehicles or containers for road or rail haulage to a distant landfill, possibly 100 km away. At the landfill, containers are emptied and the refuse is placed in discrete areas, or compacted, by heavy crawler vehicles (Figures 1–3). Thus the landfill is constructed layer by layer.

Figure 1   Containers of London's refuse being lifted from a train at a landfill in Oxfordshire (Photo courtesey of Harwell Laboratory)

Figure 2 Landfill operation at L-Field, Bedfordshire. London's refuse, rail hauled in 15 t containers being landfilled. Refuse is spread and compressed by a compactor vehicle. In the background a crawler vehicle is spreading cover material (Photo courtesey of London Brick Landfill Ltd.)

In some operations, additional processing steps can be included at the transfer station. In an attempt to increase payloads and/or improve the visual appearance of the refuse, the material can either be shredded or baled. The technologies are described in detail by Wilson (1981).

Landfill disposal in disused mineral workings constitutes a very valuable form of land reclamation, particularly in densely populated areas. Potential problems with the technique are (i) pollution of ground and surface waters through injudicious site selection and landfill management (Parker and Williams, 1981); (ii) risk of explosions from landfill gas in nearby properties; and (iii) death of vegetation due to landfill gas. However, together with land reclamation another major benefit is the availability of landfill gas which can be a major local energy source (EMCON Associates, 1981a).

Thus the aim of this chapter is to show how these potential risks and benefits can be reconciled in an environmentally acceptable manner by describing how the processes which occur in a domestic refuse landfill can be controlled.

63.2 DOMESTIC REFUSE — QUANTITIES AND COMPOSITION

Some comparisons of refuse quantities and composition worldwide have been made by Holmes (1982), Wilson (1977), and Bond and Straub (1973). The clear international variation in refuse composition is shown in Table 1, and the chemical composition of some English refuse samples is shown in Table 2.

Some important points to note are (i) the high paper content of American refuse and the correspondingly low vegetable content compared with other Western countries; and (ii) the high vegetable content of Asian refuse and the intermediate composition of Middle Eastern refuse.

The ratio of paper to vegetable wastes in the refuse is a very important factor, affecting the ease with which the methane fermentation can be initiated in the landfill (see Section 63.5).

Figure 3 Compactor vehicle suitable for moving large quantities of refuse quickly (Photo courtesey of Harwell Laboratory)

Table 1 International Variation in Refuse Composition[a,b]

	UK	Asian city	Middle East city	USA
Vegetable	30.6	75	50	20
Paper	31.2	2	16	43
Metals	5.3	0.1	5	7
Glass	3.8	0.2	2	9
Textiles	4.1	3	3	3
Plastics	5.2	1	1	5
Others	19.8	12.7	23	13
Weight per person per day	0.845 kg	0.415 kg	1.060 kg	1.946 kg

[a] Percentage compositions given. [b] Holmes (1982) and Rees (1980a).

Table 2 Chemical Composition of Domestic Refuse Samples from High Wycombe and Chichester, England

Component	High Wycombe (% dry weight)	Chichester (% dry weight)
Cellulose	35.1	37.1
Hemicellulose	1.6	4.4
Lignin	7.6	8.5
Starch	0.3	0.4
Sugars	0.4	0.7
Protein	5.8	5.6
Lipids	< 0.1	< 0.1
Ash	49.1	43.2

A striking feature of the chemical composition of refuse is the low percentage of protein and lipid present, which has led some workers to suggest that the refuse fermentation in landfill is N-limited. Rees (1980) has argued however that landfills are N-sufficient, and probably sufficient in phosphorus as well, although the fermentation of landfill leachate removed from the landfill solids has been shown to be phosphorus-limited (Stegmann and Ehrig, 1980).

63.3 BIODEGRADATION OF REFUSE POLYMERS

Figure 4 summarizes the degradation of refuse polymers in a landfill environment. Although aerobic conditions may prevail transiently or under specialized circumstances in a landfill, anaerobic conditions are rapidly established and methane and carbon dioxide are the final products of carbon metabolism. The landfill methanogenic habitat is a type A system as defined by Wolfe (1979), being similar to aquatic sediments, swamps, bogs, tundra, decaying heartwood and sludge digesters. Early studies on the fate of organic compounds in landfill focussed on the composition of landfill liquids and gases. Volatile fatty acids are major intermediate and often end products of carbon metabolism (Figure 4). Clearly the metabolism of landfilled domestic refuse is dominated by carbohydrate, as protein and lipid account for only 5–10% of the dry weight in the UK.

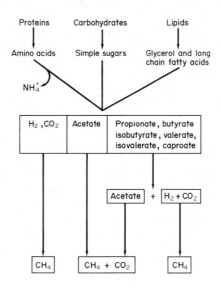

Figure 4 General scheme of landfill reactions

63.3.1 Protein Decomposition

It is very likely that the metabolism of proteins in landfill follows the same course as in anaerobic sewage digestion (Kotze *et al.*, 1968) where hydrolysis to peptides and amino acids occurs. Amino acid deamination leads to the formation of short chain carboxylic acids, CO_2 and NH_3. Amino acid deamination is the sole source of the branched chain, isobutyric and isovaleric acids found in leachates. Typically the concentrations of these two fatty acids are much lower (about 500 mg dm^{-3} in leachates from an experimental 1 year old landfill) than those of the straight chain fatty acids. This is to be expected of a substrate low in protein and rich in carbohydrates. Robinson and Maris (1979) in a review of leachate compositions in the UK, showed that isobutyric and isovaleric acids were present at concentrations ranging from 0 to 150 mg dm^{-3} in leachates. These samples had variously been taken from old sites, diluted with surface water or allowed to percolate through rock formations and been subjected to aerobic degradation. Undiluted liquids from Aveley landfill site, England, a 6 year old landfill actively generating methane, showed the absence of any isovaleric acid and concentrations of isobutyric acid fluctuated between 0–150 mg dm^{-3}, indicating the dynamic production and utilization of these acids. Concentrations of NH_4^+ in liquids from this site range from 600 to 2300 mg dm^{-3}. Concentrations of NH_4^+ from a 1 year old experimental landfill were 700 mg dm^{-3} and rising. The data of Robinson and Maris (1979) showed NH_4^+ concentrations varying between 0 and 1000 mg dm^{-3}. It is difficult to reconcile the presence of high concentrations of NH_4^+ in landfill liquids with the apparent nitrogen-limiting conditions which have been claimed to exist in landfills (Pfeffer, 1974a). Concentrations of NH_4^+ between 1500 and 3000 mg dm^{-3} have been reported to be toxic to anaerobic sewage digestion (McCarty, 1964) but recent work indicates that the methane fermentation can adapt to these high concentrations (Van Velsen, 1979).

63.3.2   Carbohydrate Decomposition

Cellulose is the major carbohydrate found in domestic refuse. It consists of materials of different biodegradabilities, depending on the extent to which it has been processed (Hobson *et al.*, 1974) and also on the degree of crystallinity and lignification (Gaden, 1975). The biodegradability of domestic refuse cellulose is likely to be much greater than that of natural, untreated cellulose. The ratio of cellulose:hemicellulose:lignin in landfilled paper is likely to be close to that of newsprint, *i.e.* 70:15:15 (Grethlein, 1978). Data from Table 2 give ratios of 74:9:17 and 79:4:17. The biodegradability of refuse cellulose derived from paper and food wastes is probably comparable to that found in sewage sludge where the primary structure of cellulose has been destroyed by cooking. However the biodegradibility of garden wastes in landfill is similar to that of the rumen contents where much of the native cellulose is still crystalline and lignified.

Cellulose probably gives rise to glucose and cellobiose as the major products in landfill. These sugars are rapidly fermented to give H_2, CO_2, acetic, propionic, butyric, valeric and caproic acids. Ethanol has only been found under unusual circumstances, and formic, lactic and succinic acids have not been reported. Leachate generated from freshly placed refuse compacted to about 1 t m^{-3} contained the following concentrations of fatty acids: acetic 3800, propionic 1600, butyric 3500, valeric 2100 and caproic 3700 mg dm^{-3}. The same sample contained 145 mg dm^{-3} isobutyric and 70 mg dm^{-3} isovaleric acid derived from protein metabolism.

63.3.3   Methanogenesis and Sulfate Reduction

Hydrogen only appears transiently in landfill, and is prolonged under conditions where the growth of microbes is inhibited by dry conditions established with pulverized fuel ash (PFA), which is a residue from coal burning power stations. The presence of H_2 is also prolonged in wet pulverized refuse where the rapid production of high concentrations of volatile fatty acids (about 40 000 mg total acids dm^{-3}) inhibits methanogenesis and sulfate reduction.

Sulfate concentrations in a domestic refuse landfill are often between 2000 and 3000 mg dm^{-3}, but evidence of good sulfate reduction has been found at Aveley landfill (Rees and Viney, 1982) where concentrations fell as low as 50 mg dm^{-3}. Also at this landfill, fatty acids were virtually undetectable in the leachate and rates of methane production were high (Section 63.5.3).

63.3.4   Leachate Generation

Four principal factors influence the production of leachate from a landfill site. These are (i) the initial refuse moisture content; (ii) the volume of rainfall allowed to enter the landfill; (iii) the volume of other liquids deliberately added to the refuse; and (iv) waste composition and density (Campbell, 1982; Stegmann, 1982; Blakey, 1982). Leachate is often produced before the field capacity of the refuse is reached due to channelling in the heterogeneous mass. The composition of leachate can vary dramatically from one site to another and is influenced by the age of the landfill, and the degree and type of biological activity within it. Table 3 summarises the range of leachate compositions found in practice (Pohland and Engelbrecht, 1976).

The high values would be typical of freshly leached refuse. Migration of water through the fill would favour ready decomposition of the putrescible fraction in particular to give high concentrations of fatty acids with a concomitant low pH. High concentrations of fatty acids might also be accounted for at this early stage in the landfill's history by a poorly developed methanogenic population. Furthermore, the fatty acid residence time in the landfill might also be too short to allow efficient methanogenesis.

The low values would be more typical of leachates from landfills several years old where most of the soluble materials had already been leached from the refuse. Alternatively, leachates containing low total organic carbon (TOC) are to be expected from very reactive landfills where a

Table 3 Range of Chemical Composition of Landfill
Leachate[a]

Component	Quantity ($mg\ dm^{-3}$)
Total organic carbon	256–28 000
pH	3.7–8.5
NH_4 (N)	0–1106
NO_3 (N)	0.2–10.3
PO_4 (P)	6.5–85
SO_4	1–1558
Cl	4.7–2467
K	28–3700
Na	0–7700
Mg	17–15 600
Fe	0–2820
Mn	0.09–125
Ca	50–7200
Cd	0.03–17
Pb	0.10–2
Cu	0–9.9

[a] Pohland and Engelbrecht (1976).

balanced fermentation is occurring. In this situation, fatty acids do not accumulate, but are metabolized to CH_4 and CO_2.

63.3.5 Landfill Microbiology

Many studies of landfill microbes have been dominated by public health considerations (Cromwell, 1965; Peterson, 1971; Engelbrecht and Amirhor, 1975). Attention has been focused on the presence of organisms of faecal origin and viruses in refuse and leachate.

Very few data exist on the number and physiological activities of organisms involved in the decomposition of domestic refuse in landfill. The heterogeneous nature of domestic refuse and the problems associated with sampling have in the past probably discouraged microbial examination. However, recent studies (Jones and Grainger, 1982; Rees *et al.*, 1982, Filip and Küster, 1979) have begun to remedy this situation.

Figure 5 shows the vertical distribution of amylase, protease and cellulase activities throughout the depth of the landfill. It was considered that quantitative measurements of extracellular enzymic activities in landfill sites would be important indicators of microbial activity. Rates of protease and amylase activity increased sharply at the water table. High cellulase activity was detected only in the top *ca*. 1 m of the landfill, although it was evident from the chemical composition of the refuse in the borehole that cellulose decomposition had occurred. The ratio of cellulose:lignin changed from approximately 4.5:1 above the water table to 0.5:1 below it (Rees and Viney, 1982). The possible importance of fungi in cellulose decomposition was indicated by the optimum cellulase activity being observed at pH 5.

Observations on the influence of water on enzymic activity in the landfill and low cellulase activity were confirmed with a series of laboratory experiments using pulverized refuse (Jones and Grainger, 1982). Cellulase, amylase and protease activities were studied over a 6 month period in 3–4 dm^3 samples of anaerobically degrading refuse. Samples were (i) of 65% moisture content and (ii) fully saturated with water. At 65% moisture content all enzymic activities decreased during incubation. Changes in enzymic activities under water-saturated conditions are shown in Figure 6. Protease activity increased three-fold over a 20 day period and amylase activity increased 1000-fold over a 50 day period. After these times the activities decreased. In contrast, cellulase activity decreased rapidly from the beginning of the experiment. The disappearance of cellulase activity was investigated further in view of the importance of cellulosic materials as the major carbohydrate in domestic refuse. In experiments in which a commercial cellulase preparation derived from *Trichoderma reesei* was added to both sterile (autoclaved) and microbially active refuse, the cellulase activity was only recoverable from the sterilized refuse, showing that proteolytic destruction of cellulase is the main cause of loss of activity in this experimental system.

These laboratory experiments essentially confirm the field observation of increasing protease and amylase activities with increasing landfill water content. Low cellulase activity in the field can

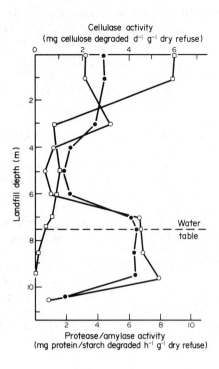

Figure 5   Enzymic activities in Aveley landfill 1980. ○ protease; ● amylase; □ cellulase; borehole temperature 42 °C

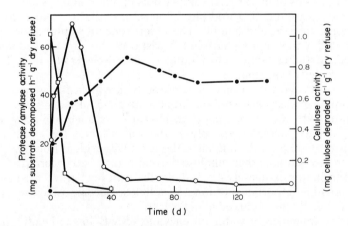

Figure 6   Changes in protease (○); amylase (●); and cellulase (□) activities in pulverized refuse, saturated with water

also be partially accounted for. In a heterogeneous landfill it is not entirely clear how effective proteolytic decomposition of cellulase enzyme will be. Filip and Küster (1979) also showed that proteolytic bacteria were the most prevalent group in the early stages of refuse decomposition in landfills.

Some difficulties associated with enzymic studies are that activities as measured do not accurately reflect rates of starch and protein decomposition in the landfill. Thus the pH optimum of extracted proteases was pH 9 irrespective of the pH of the refuse sample. *In situ* activities where pH values were often below pH 7 would therefore be consistently lower than the value obtained under optimum conditions in the laboratory. Unlike the proteases, optimal pH for amylase varied with refuse pH, thus laboratory data obtained at pH 7 would differ from actual values in the landfill. Similarly, landfill temperatures could well differ from the optimum assay values of 37 °C. Enzymic activities also varied independently of the numbers of proteolytic bacteria. This is not surprising as measured enzyme activities are a function of three different parameters, *viz.* (i) the numbers of microbes; (ii) the specific rate of enzyme production; and (iii) the rate and extent of enzyme inactivation in the refuse. These data underline the problems associated with using bacterial numbers as indicators of bacterial activity, especially where extracellular enzymes are involved.

63.4 FACTORS INFLUENCING REFUSE FERMENTATION

63.4.1 Water Content

The availability of water in any system is measured in terms of water activity (Pirt, 1975). Water plays a major role in hydrolytic reactions, in modifying the conformational structure of macromolecules, *e.g.* enzymes and their activities, in solution of metabolites, and in cell turgidity. Lowering the water activity reduces the specific growth rates of microbes, the degree of reduction often being species-dependent. At low water activities, *Clostridium perfringens* and many silage bacteria exhibited long lag periods (Wujic, 1980). Water also provides the appropriate environment for the mass transfer of nutrients, inhibitors, enzymes and microbes in the landfill environment. Wujic (1980) showed that increasing water activity stimulated rates of methane and carbon dioxide production from manure. Figure 7 summarizes the results of several workers (Merz and Stone, 1968; Merz, 1964; Rovers and Farquhar, 1973; and De Walle and Chian, 1978) who have studied the effects of water content on the stimulation of gas production by domestic refuse. Of all the important physiological groups of bacteria required for complete anaerobic digestion, it is likely that the methanogenic bacteria are most deficient in fresh refuse. The successful growth of these microbes early in the history of the landfill is crucial to the successful operation of a site as a solid phase anaerobic digester.

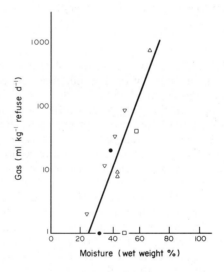

Figure 7 Effect of moisture content on rates of gas production from domestic refuse. Data from Merz and Stone (1968); Merz (1964); Rovers and Farquhar (1973); De Walle and Chian (1978)

Figure 7 shows that the rate of gas production at moisture contents of 30–40% wet weight is less than 10 cm^3 kg^{-1} refuse d^{-1}. In operational landfills in the USA and Europe refuse is often received with moisture contents between 25–35% wet weight, *i.e.* well below the optimum for rapid gas production.

63.4.2 Temperature

Pfeffer (1974) studied the effects of temperature on the anaerobic fermentation of domestic refuse. The optimum mesophilic temperature was found to be 42 °C while the optimum thermophilic temperature was found to be 60 °C. Higher yields and rates of methane were also produced under thermophilic conditions. In some operational landfills, temperatures between 40–45 °C have been sustained consistently (Rees, 1980b) but thermophilic conditions have not been maintained under anaerobic conditions.

63.4.3 Particle Size

The most comprehensive study on the effect of refuse particle size on the rate of decomposition was undertaken by De Walle and Chian (1978). They showed that reducing the particle size from 25 cm to 2.5 cm resulted in a 4.4 times increase in gas production. However, all the gas produced was carbon dioxide. Methane was only produced in experiments using unshredded refuse. Theoretically the shredding of refuse can be expected to improve mixing and nutrient availability in a landfill, and also to increase the substrate surface area available for microbial colonization. In fact, all these phenomena probably do occur in shredded waste giving the faster gas production observed by De Walle and Chian (1978). However, there are problems with initiating methanogenesis in pulverized refuse. This was achieved successfully by Buivid *et al.* (1981) by controlling pH with $CaCO_3$ and inoculating the refuse with a methanogenic sludge from an anaerobic sludge digester. However there are considerable problems with implementing this strategy routinely on a full scale landfill and other, simpler methods are available for initiating methanogenesis in the field. However, the work has provided direct evidence that controlling landfill pH is essential for successful initiation of methanogenesis.

63.4.4 pH

The optimum pH range for the digestion of sewage sludge is 6.8–7.5 but the process still operates successfully at pH 6–8 (Lettinga, 1978), although inhibition of acetate, propionate and butyrate fermentation was observed at pH 6. At Aveley landfill (Rees and Viney, 1982), methane production has been observed at pH values ranging from 5.5 to 8.3. However a pH of 5.5 is still considered very detrimental to the initiation of methanogenesis in the field (Section 63.5).

63.4.5 Refuse Density

Due to the difficulties of measuring the water content of refuse samples, the densities quoted here refer to weights of refuse 'as received' at the landfill, generally containing 25–35% water (of wet weight). Extremes of refuse density, *i.e.* 0.45 t m^{-3} and 1.1 t m^{-3} can significantly modify the type of refuse decomposition process. Loosely placed refuse at a density of 0.35 t m^{-3} allows adequate air ingress into the refuse mass for aerobic processes to be dominant, consequently the major end products are CO_2 and H_2O, and temperatures in excess of 80 °C have been observed. At refuse densities in excess of 0.5 t m^{-3} then the dominant decomposition process becomes fermentative with the appearance of carboxylic acids in leachate and methane in the gas phase.

The influence of refuse density in the range 0.6–1.1 t m^{-3} on refuse decomposition is not clearly understood. Laboratory experiments indicate (Rees and Grainger, 1982) that density can have different effects depending on whether or not water is added to the refuse. With refuse 'as received' (21% water, wet weight), increasing the density from 0.32 t dry refuse m^{-3} to 0.47 t dry refuse m^{-3} stimulated the initial rate of gas production from 401 cm^3 gas kg^{-1} dry refuse d^{-1} to 845 cm^3 gas kg^{-1} dry refuse d^{-1}. Conversely when refuse was placed at densities of 0.2, 0.32 and 0.47 t dry weight m^{-3} and subsequently saturated with water, 2530, 2100 and 1260 cm^3 gas kg^{-1} dry weight d^{-1}, respectively, were obtained. These data were explained in terms of water being squeezed out of the dry refuse at higher densities so increasing the amount of water available to microbes. With the wetted refuse it is likely that high initial densities preclude water admission to the refuse. These conclusions are supported by recent data from large field experiments, which showed that the mass of water absorbed by refuse was inversely proportional to the refuse density (Campbell, 1982).

63.5 CONTROL OF REFUSE FERMENTATION IN THE FIELD

The major objectives of controlling the refuse fermentation in the field are (i) to ensure good initiation of methanogenesis; (ii) to increase the rate of refuse decomposition; and (iii) to improve leachate quality. If these objectives can be met then there is considerable scope for minimizing the potential environmental damage caused by landfill, and for maximizing landfill as an energy source. All of these objectives can be achieved by the control of refuse water content and refuse density. A strategy of controlling landfill pH with $CaCO_3$ addition, adding water and anaerobically digested sewage sludge as an inoculum of methanogenic bacteria, is being pursued by workers in the USA, but results are not yet available. The procedure has proved to be satisfactory on the laboratory scale (Buivid *et al.*, 1981). Apart from the need to regulate the water balance of landfill in order to control microbial processes, there is currently great incentive to control and particularly to minimize the water input to a landfill site so that the volume of polluting leachate which is poduced is as low as possible. Addition of water to most current landfills would not achieve the objectives outlined earlier. Rather, the pollution potential of the landfill would be enhanced. However, Figure 7 indicates that increased water content favours rapid decomposition, so how can this apparent conflict be resolved? Technologies devised to minimize water input to landfill sites can in turn be used in a more sophisticated manner to allow water ingress under controlled and desirable conditions, especially when linked to inoculation procedures.

63.5.1 Control of Landfill Water Content

Major potential sources of water input to operational landfills are (i) groundwater from springs or shallow aquifers; (ii) surface water as streams of varying sizes; and (iii) precipitation as rainfall and/or snow. The adoption of good site selection techniques (Parker and Williams, 1981) and good civil engineering can readily overcome water intrusion from the first two sources. Rainfall intrusion is more difficult to regulate.

Rainfall intrusion occurs in two phases, *i.e.* during the operational life of the site and after completion when the landfill has been adequately covered with a clay seal and soil, and vegetation has been established. Work by Campbell (1982) has shown that at least 55% of incident rainfall will enter operational landfill sites. Between 32 and 39% infiltration can be expected on a completed well-vegetated landfill (Blakey, 1982). Sloping surfaces would further minimize infiltration.

The relatively dry nature of domestic refuse as received at the landfill (25–35% moisture, wet weight) means that there is capacity for further water uptake once refuse is placed in the landfill. The heterogeneous nature of refuse, the unequal distribution of water in a given mass, means that water uptake is often very time-dependent, thus moisture uptake is most favoured when as much refuse as possible is exposed to the moisture and in the absence of fissures. In most landfills this scenario is rarely achieved and consequently much leachate drains to the bottom of a landfill, probably by fissure flow, before the entire landfill site is at field capacity. However some moisture is retained by the refuse prior to leachate generation. This is termed the absorptive capacity. Campbell (1982) has shown that the absorptive capacity of refuse is inversely related to refuse density and that for refuse placed at 0.7 t m^{-3} the absorptive capacity is about 0.09 m^3 water per tonne of refuse.

These data on absorptive capacities and percentage infiltration allow conclusions to be drawn regarding the optimal geometry of the landfill with a particular refuse input, *i.e.* the maximum operational area consistent with minimum leachate generation can be calculated, and this should determine the rate of lateral spread of refuse and the rate at which the refuse should rise vertically.

Campbell (1982) showed that high infiltration rates occurred on his experimental cells despite the presence of a 0.5 m thick layer of clay. Clearly there is considerable misunderstanding of the water-retaining properties of clays within the waste disposal industry. Lake (1981) drew attention to the need to understand the different properties of clays. If clays are not well wetted they cannot be compacted and will contain large voids. Ground desiccation during a British summer can result in cracks ~ 1.5 m deep so clay seals of 2 m are recommended.

Other techniques to minimize water infiltration to landfills are (i) the use of plastic membranes to cover the completed landfill (Emrich and Beck, 1981); and (ii) the use of a urea–formaldehyde foam for intermediate cover during landfill operation (Hofstetter, 1982).

Methods for artificially increasing the moisture content of landfills, particularly in regions of low rainfall, include the addition of industrial liquid wastes to pits excavated in the refuse surface (Wilson, 1979) and the redistribution or recycling of landfill leachate from the base to the top of the landfill (Pohland, 1980).

63.5.2 Control of Refuse Density

Economic considerations dictate that high, in-place refuse densities of *ca.* 1.0 t m^{-3} or higher are desirable. The range of refuse densities available to a landfill operator are about 0.35 t m^{-3} to about 1.1 t m^{-3} when placed, although 0.7–0.9 t m^{-3} is most common. The various operational strategies required to achieve these densities have been described by Campbell and Parker (1980). Essentially, densities of 0.35 t m^{-3} can be achieved by controlled placing of refuse with a mechanical grab and not allowing vehicles to travel on it. Intermediate densities of about 0.7 t m^{-3} can be achieved by pushing refuse over a 2 m deep edge or 'face' and compressing the entire 2 m layer with a vehicle similar to the one shown in Figure 3. Densities of about 1.0 t m^{-3} or more can be achieved using the same compactor vehicle but in a different operational mode. If thin layers of refuse (0.2 m as opposed to 2 m) are compacted then refuse densities in excess of 1.0 t m^{-3} can be achieved. Increasingly this is a favoured landfilling method as the use of landfill space is optimized.

63.5.3   Initiation of Methanogenesis

In most landfill sites methanogenesis is initiated to some degree. However, it is rarely optimized and it often limits the rate of refuse decomposition. At Aveley landfill however (Rees and Viney, 1982) methanogenesis is very well established so that in this landfill the rate-limiting step is polymer hydrolysis. The landfill is characterized by a rising water table. In the water-saturated part of the landfill (Figure 8) high rates of gas production occur and these are associated with < 30 mg total fatty acids dm^{-3} of leachate. Above the water table, the environment is typical of most landfill sites with lower rates of gas production in association with high > 40 000 mg total fatty acids dm^{-3} leachate. Rees (1980b) has calculated that the high rates of methane production in the water-saturated portion of this site generate sufficient heat of reaction to maintain the landfill temperatures between 40 and 45 °C despite freezing ambient temperatures in winter. Thermal buffering was provided by 3–4 m of unsaturated refuse above the bioreactive zone.

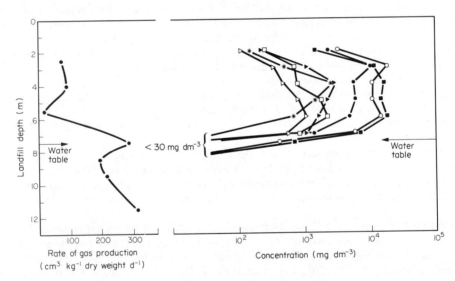

Figure 8   Aveley landfill 1980. Rate of gas production (CH$_4$ + CO$_2$) and distribution of carboxylic acids with depth in the landfill. Acetate ○; propionate ●; isobutyrate □; butyrate ■; isovalerate △; valerate ▲; caproate *

Factors which have been postulated to account for the good initiation of methanogenesis at this site include (i) the intrusion of groundwater into the base of site and (ii) the localized presence of trade and inert waste in the landfill (Rees, 1980a). The physical movement of water through refuse has also been shown to stimulate methane production (Klink and Ham, 1982). Unpublished work in the author's laboratory has also shown that water addition to wet pulverized refuse fails to initiate methanogenesis, but if water is allowed to flow through a mass of refuse then methanogenesis is initiated. At Aveley, the lateral and vertical movement of water has probably contributed towards the establishment of chemical gradients, particularly between areas of readily decomposable refuse and inert materials. Favourable environments for the growth of methanogens would then be established at discrete points within these gradients (Wimpenny *et al.*, 1981; Lovitt and Wimpenny, 1981), where inhibitory concentrations of fatty acid would be reduced.

Stegmann and Ehrig (1982) have demonstrated in the laboratory that methanogenesis can be initiated in pulverized refuse if this material is mixed with refuse previously stabilized by composting (2:1 by dry weight). The compost behaves in a manner analogous to the 'trade' refuse at Aveley and allows for the establishment of chemical gradients and favourable environmental conditions for the growth of methanogenic bacteria.

Composted refuse can be achieved in the field by placing refuse at a low density, between 0.35 and 0.5 t m^{-3} (Rees, 1981). This procedure results in temperatures in excess of 80 °C being generated and the rapid oxidation of fatty acids to give a leachate pH >7 and favourable for the growth of methanogens. Figure 9 compares the quality of leachates obtained from refuse placed at different densities.

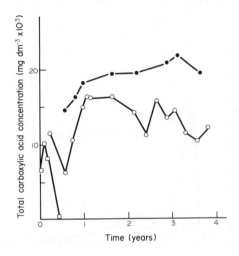

Figure 9 Leachate quality from experimental landfills placed at densities of 0.67 t m^{-3} (○); 0.95 t m^{-3} (●); and 0.35 t m^{-3} (□)

Stegmann and Ehrig (1982) also showed that leachate derived from refuse placed in 0.40 m layers had a lower biological oxygen demand than leachate derived from refuse placed in 2 m thick layers and ascribed this difference to the additional aerobic metabolism which occurred when the refuse was placed in thin layers. It is important in these studies to monitor gas production and leachate quality simultaneously to ascribe fatty acid disappearance to either an aerobic or anaerobic mode.

63.6 GAS PRODUCTION, RECOVERY AND UTILIZATION

63.6.1 Gas Yields

Theoretical estimates of maximum gas yields from domestic refuse have been determined by several workers and are reviewed by EMCON Associates (1981b). Data from six separate studies range from 187 to 424 m^3 gas t^{-1} of refuse (wet weight) with an average of 364 m^3 gas t^{-1}. Rees (1980a) and Ham *et al.* (1979) calculated gas yields of approximately 400 m^3 t^{-1}.

Laboratory scale experiments to measure gas yields from refuse under landfill simulating conditions have been hindered by the difficulties associated with the initiation of methanogenesis (Rees, 1980a). These problems were overcome by Pfeffer (1974b) who fermented refuse as a dilute suspension in a conventional anaerobic digester. Gas yields of 190 m^3 gas t^{-1} refuse (wet weight) were obtained. Buivid *et al.* (1981) measured gas yields in a high solids reactor with pH control (finely divided CaCO$_3$) and inoculation with methanogenic organisms (sewage sludge) and obtained average yields of 215 m^3 t^{-1} dry refuse. Stegmann and Ehrig (1982) overcame the problems of initiation of methane production by mixing domestic refuse with composted refuse and obtained yields of 150–180 m^3 t^{-1} dry weight of refuse.

It becomes more difficult to quantify gas yields from full scale landfills but rates of gas production are high, for example 5×10^4 m^3 landfill gas d^{-1} for the Palos Verdes landfill, CA, and $14 \times$ 10^4 m^3 landfill gas d^{-1} from the C1D landfill, Chicago.

63.6.2   Profiles of Gas Production

Conceptual gas production profiles are shown in Figure 10. Two scenarios are depicted: (i) the conventional landfill, where methanogenesis is suboptimal, moisture contents lie between 30–40% wet weight, and gas production is prolonged, possibly for several decades; and (ii) the controlled landfill where methanogenesis is not the rate-limiting step, moisture contents lie between 50 and 70% wet weight and rapid gas production occurs over a period of about 2–5 years. This profile has been demonstrated on the laboratory scale by Buivid *et al.* (1981) who showed that most of the gas was produced over a 3 month period, with pulverized refuse as substrate.

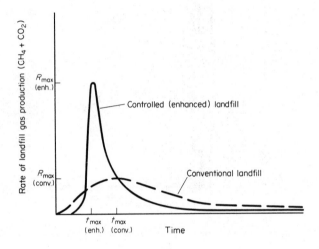

Figure 10   Conceptual profiles of landfill gas production

63.6.3   Gas Recovery

The principles of gas recovery from landfills are reviewed by EMCON Associates (1980). Wells (0.3–1 m diameter) are drilled into the landfill, each well having a radius of influence of about 50 m and these are linked by pipework to a central pumping and transmission station (Figure 11) where as much water vapour as possible is removed from the gas prior to its transmission to the utility. The calorific value of landfill gas is directly related to its methane composition. Landfill gas with a methane content of 50% by volume would thus have a calorific value of 15 kJ m^{-3}. The methane concentration can vary from about 65% for gas drawn from the landfill to 25% CH$_4$ for gas drawn from the landfill perimeter.

63.6.4   Gas Utilization

Three major options exist for the utilization of landfill gas (EMCON Associates, 1980): (i) direct heating; (ii) electricity generation; and (iii) CO$_2$ removal to give 'pipeline' quality gas. Other potential uses include steam generation as a source of heat or for electricity generation, and conversion to methanol.

Historically, CO$_2$ removal to give pipeline quality gas, *e.g.* Mountain View, Palos Verdes and Monterey Park landfills, California, and direct utilization, *e.g.* Ascon and Cinnaminson landfills (EMCON Associates, 1981a), have been popular. More recently increased interest has been shown in electricity generation and many schemes are at an advanced stage of development, *e.g.* Duarte landfill, California, and Braunschweig and Pforzheim landfills, W. Germany (Stegmann, 1981). In the UK landfill gas has been utilized as a fuel in the brickmaking industry.

Figure 11 Pumping station for landfill gas recovery, showing from the right (1) gas pipeline from the landfill; (2) gas compressor; (3) emergency flare stack; (4) refrigerator for water removal; and (5) pipeline to utility (Photo courtesey London Brick Landfill Ltd.)

ACKNOWLEDGEMENTS

The author wishes to acknowledge financial support from the UK Departments of Environment and Energy for his own research work.

63.7 REFERENCES

Blakey, N. C. (1982). Infiltration and absorption of water by domestic wastes in landfills. In *Landfill Leachate Symposium*. Harwell Laboratory, Oxon., UK.

Bond, R. G. and C. P. Straub (1973). *Handbook of Environmental Control*, vol. 2, *Solid Waste*. CRC Press, Cleveland, OH.

Buivid, M. G., D. L. Wise, M. J. Blanchet, E. C. Remedios, B. M. Jenkins, W. F. Boyd and J. G. Pacey (1981). Fuel gas enhancement by controlled landfilling of municipal solid waste. *Resour. Conserv.*, **6**, 3–20.

Campbell, D. J. V. (1982). Absorptive capacity of refuse—Harwell research. In *Landfill Leachate Symposium*. Harwell Laboratory, Oxon., UK.

Campbell, D. J. V. and A. Parker (1980). Density of refuse after deposition using various landfill techniques. *Solid Wastes*, **70**, 435–440.

Cromwell, D. L. (1965). Identification of microbial flora present in landfills. M. Sc. Thesis. West Virginia University.

De Walle, F. B. and E. S. K. Chian (1978). Energy recovery from landfilled solid waste. *Biotechnol. Bioeng. Symp.*, **8**, 317–328.

EMCON Associates (1980). *Methane Generation and Recovery from Landfills*. Ann Arbor Science, Michigan.

EMCON Associates (1981a). State of the art of landfill gas recovery. Report No. ANL/CNSV-TM-85. Argonne National Laboratory, USA.

EMCON Associates (1981b). State of the art of methane gas enhancement in landfills. Report No. ANL/CNSV-23. Argonne National Laboratory, USA.

Emrich, G. H. and W. W. Beck (1981). Top sealing to minimise leachate generation—Status Report in *Proceedings of the 7th Annual Research Symposium on Land Disposal: Hazardous Waste*, USEPA EPA-600/9-81-002b, pp. 292–297.

Engelbrecht, R. S. and P. Amirhor (1975). Biological impact of sanitary landfill leachate on the environment. In *Proceedings of the 2nd National Conference on Complete Water Re-use*, Chicago.

Filip, Z. and E. Küster (1979). Microbial activity and the turnover of organic matter in municipal refuse disposed to landfill. *Eur. J. Appl. Microbiol.*, **7**, 371–379.

Gaden, E. L. (1975). Microbial Cellulolytic system. Summary statement. *Biotechnol. Bioeng. Symp.*, **5**, 161–162.

Grethlein, H. E. (1978). Chemical breakdown of cellulosic materials. *J. Appl. Chem. Biotechnol.*, **28**, 296–308.

Ham, R. K., K. K. Hekimian, S. L. Katten, W. J. Lockman, R. J. Lofy, D. E. McFaddin and E. J. Daly (1979). Recovery, processing and utilization of gas from sanitary landfills. USEPA Report No. 600/2-79-001. Cincinnati, OH.

Hobson, P. N., S. Bousefield and R. Summers (1974). Anaerobic digestion of organic matter. *CRC Crit. Rev. Environ. Control*, **4**, 131–191.

Hofstetter, A. G. (1982). Stahl-und Maschinenbau. CH-3324 Hindelbank, Switzerland.

Holmes, J. R. (1982). Waste management decisions in developing countries. *Chart. Munic. Eng.*, **109**, 11–16.

Jones, K. L. and J. M. Grainger (1983). The application of enzyme activity measurements to a study of protein, starch, and cellulose fermentation in domestic refuse. *Eur. J. Appl. Microbiol. Biotechnol.*, **18**, 181–185.

Klink, R. E. and R. K. Ham (1982). Effects of moisture movements on methane production in solid waste landfill samples. *Resour. Conserv.*, **8**, 29–41.

Kotze, J. P., P. G. Thiel, D. F. Torien, W. H. J. Hatting and M. L. Sieber (1968). A biological and chemical study of several anaerobic digesters. *Water Res.*, **2**, 195–213.

Lake, L. (1981). In discussion. In *Landfill Gas Symposium*. Harwell Laboratory, Oxon., UK.

Lettinga, G. (1978). Feasibility of anaerobic digestion for the purification of waste waters. In *4th European Sewage and Waste Symposium*, EAS Munich.

Lovitt, R. W. and J. W. T. Wimpenny (1981). The gradostat: a bidirectional compound chemostat and its application in microbiological research. *J. Gen. Microbiol.*, **127**, 261–268.

McCarty, P. L. (1964). Anaerobic waste treatment fundamentals. *Public Works*, Sept., 107.

Merz, R. C. (1964). Investigation to determine the quantity and quality of gases produced during refuse decomposition. California State Water Quality Control Board, Sacramento, CA.

Merz, R. C. and R. Stone (1968). Special studies of a sanitary landfill. USPHS Bureau of Solid Waste Management Report EPA-SW8R6-70.

Parker, A. and G. M. Williams (1981). Landfill site selection and operation for municipal and hazardous waste disposal. In *Developmments in Environmental Control and Public Health—2*, ed. A. Porteous, pp. 1–37. Applied Science, Essex.

Peterson, M. L. (1971). Pathogens associated with solid waste processing. USEPA SW-49r.

Pfeffer, J. T. (1974a). Temperature effects on anaerobic fermentation of domestic refuse. *Biotechnol. Bioeng.*, **16**, 771–787.

Pfeffer, J. T. (1974b). Reclamation of energy from organic wastes. USEPA 697/2-74-016.

Pirt, S. J. (1975). *Principles of Microbe and Cell Cultivation*. Blackwell, Oxford.

Pohland, F. G. (1980). Leachate recycle as landfill management option. *J. Environ. Eng. Div. (Am. Soc. Civ. Eng.)*, **EE6**, 1057–1069.

Pohland, F. G. and R. S. Engelbrecht (1976). Impact of sanitary landfills. Report for American Paper Institute, New York.

Rees, J. F. (1980a). The fate of carbon compounds in the landfill disposal of organic matter. *J. Chem. Technol. Biotechnol.*, **30**, 161–175.

Rees, J. F. (1980b). Optimization of methane production and refuse decomposition in landfills by temperature control. *J. Chem. Technol. Biotechnol.*, **30**, 458–465.

Rees, J. F. (1981). Major factors affecting methane production in landfills. In *Landfill Gas Symposium*. Harwell Laboratory, Oxon., UK.

Rees, J. F. and J. M. Grainger (1982). Rubbish dump or fermenter? Prospects for the control of refuse fermentation to methane in landfills. *Process Biochem.*, Nov./Dec., 41–44.

Rees, J. F. and I. Viney (1982). Leachate quality and gas production from a domestic refuse landfill. AERE Harwell Report No. R10328. HMSO, London.

Rees, J. F., K. L. Jones and J. M. Grainger (1982). Untersuchungen über die mikrobiologischen Umsetzungsprocesse in Mülldeponier. In *Proceedings Gas und Wasserhaushalt von mülldeponien*. Braunschweig, Umweltbundesampt, Berlin.

Robinson, H. D. and P. J. Maris (1979). Leachate from domestic waste. Generation, composition and treatment. A review. Technical Report No. 108. Water Research Centre, Medmenham, UK.

Rovers, F. A. and G. J. Farquhar (1973). Infiltration and landfill behaviour. *J. Environ. Eng. Div. (Am. Soc. Civ. Eng.)*, **EE5**, 671–690.

Stegmann, R. (1981). Landfill gas problems. Summary of West German experience. In *Landfill Gas Symposium*. Harwell Laboratory, Oxon., UK.

Stegmann, R. (1982). Absorptive capacity of refuse—West German research. In *Landfill Leachate Symposium*. Harwell Laboratory, Oxon., UK.

Stegmann, R. and H.-J. Ehrig (1980). Operation and design of biological leachate treatment plants. *Prog. Water Technol.*, **12**, 919–947.

Stegmann, R. and H.-J. Ehrig (1982). Enhancement of gas production in sanitary landfill sites. In *Proceedings. Resource Recovery from Municipal, Hazardous and Solid Wastes*, Miami Beach, May 10–12.

Van Velsen, A. F. M. (1979). Adaptation of methanogenic sludge to high ammonia nitrogen concentrations. *Water Res.*, **13**, 995–999.

White-Hunt, K. (1980). Domestic refuse—a brief history. *Solid Wastes*, **70**, 609–615.

White-Hunt, K. (1981b). Domestic refuse—a brief history (part 2). *Solid Wastes*, **71**, 159–166.

White-Hunt, K. (1981c). Domestic refuse—a brief history (part 3). *Solid Wastes*, **71**, 284–292.

Wilson, D. C. (1979). Recent developments in the management of hazardous wastes. In *Developments in Environmental Control and Public Health—1*, ed. A. Porteous, pp. 107–157. Applied Science, London.

Wilson, D. C. (1981). *Waste Management*. Oxford University Press, Oxford.

Wilson, D. G. (1977). *Handbook of Solid Waste Management*. Van Nostrand, New York.

Wimpenny, J. W. T., J. P. Coombs, R. W. Lovitt and S. G. Whittaker (1981). A gel stabilized model ecosystem for investigating microbial growth in spatially ordered solute gradients. *J. Gen. Microbiol.*, **127**, 277–287.

Wolfe, R. S. (1979). Microbial biochemistry of methane. *Int. Rev. Biochem.*, **21**.

Wujic, W. J. (1980). Dry anaerobic fermentation to methane of organic residues. Ph.D. Thesis, Cornell University, Ithaca, NY.

64

Instrumentation for Waste Treatment Processes

R. BRIGGS

Resource Consultants, Cambridge and John Taylor and Sons, London, UK

64.1 INTRODUCTION

The role of instrumentation, control and automation (ICA) in the management, operation and control of sewerage systems and wastewater treatment plants has received world wide consideration over the past decade. The International Association on Water Pollution Research and Control has, for example, held three very useful workshops to discuss the topic and indeed the ICA aspects of river management water resource management, water treatment and distribution also (IAWPR, 1977, 1981) and will be holding a fourth workshop in Houston, Texas, and Denver, Colorado, in the spring of 1985. Recently too in the United Kingdom the National Water Council/Department of the Environment Working Party on Control Systems for the Water Industry (1981) has issued its final report on the topic. This report contains a number of excellent recommendations, for example the need to identify more clearly water industry needs and the need to produce functional specifications, guidelines and codes of practice as quickly as possible.

Also in the report are comments about the need to develop more reliable and robust sensors for the measurement of those variables of most significance in wastewater treatment and control and the requirements for sensors for monitoring equipment status and for determining toxicity and treatability as well as those required for measurement of the more well-known variables such as dissolved oxygen concentrations, sludge solids concentrations, sludge blanket levels, air flows and calorific values of digester gases. Equally important also is the comment that far better liaison is needed between water industry users of instruments and control systems and the manufacturers of them.

64.2   ORGANIZATIONAL AND LOGISTICAL ASPECTS

In order to expedite the implementation of these and other more detailed recommendations a steering committee has been set up in the UK, including the three chairmen of the standing technical committees on water treatment, wastewater treatment and sewers and water mains and representatives of the Department of Industry, the Department of the Environment, the National Water Council, the Water Research Centre, the instrument industry and the Association of Consulting Engineers. This steering group has just published 'Guideline One', a guideline for the water industry upon the purchase of ICA systems and equipment.

In Sweden, too, it has been recognized that proper deployment of ICA systems can reduce operating costs and increase plant reliability. In consequence a three year programme called the Sewage Works Evaluation Programme (SWEP) was initiated in 1980, with broad cooperation between the Swedish National Environmental Protection Board, the Swedish Water and Wastewater Works Association, local treatment plant management and universities. The main objectives of the programme have been stated to be to enhance operational procedures and to increase basic knowledge about sewage treatment, with emphasis on activated sludge systems with chemical precipitation. Forsberg *et al.* (1982) have published details of the Swedish experience to date. In the USA concern has also been expressed about the high investments in wastewater treatment plant, the continuing cost of their maintenance and the generally poor performance obtained from these facilities. Among the many causes cited, poor application of ICA technology ranks fairly high and a test centre is currently being set up under the auspices of the Environmental Protection Agency (EPA) and in collaboration with the Canada Centre for Inland Waters, Burlington, Ontario. This centre is similar in some respects to the SIRA Institute in the UK and will also have amongst its objectives the supervision of and publishing of results obtained from field trials. This aspect of the work is similar to that currently being undertaken by the Water Research Centre at water industry-based evaluation and demonstration facilities (EDF). Two major EDFs are at present being set up, one at a sewage works at Witney in the Thames Water Authority region and a second at a water treatment plant at Eccup in the Yorkshire Water Authority region. The plants will be fitted with a very broad range of instruments, many more than would normally be required for the routine operation and management of the plants. This enhancement by additional instrumentation has increased costs, but conscious of the potential benefits both to the water industry and to the export potential of equipment manufacturers who use the facility, the Department of Industry (DOI) has contributed about £200 000 to the cost of the establishment and initial operation of each of the facilities. These benefits are expected to fall within the following broad areas:

demonstrations of the technical viability of appropriate ICA systems;
defining of operating methods and skills required;
optimization of existing processes utilizing ICA technology;
demonstrations of the value of ICA systems as a management tool, including provision of data for analysis of breakdowns and optimization of maintenance procedures, and use of men and materials;
provision of facilities for determining the efficacy of UK-manufactured ICA equipment in order to provide information both to the water industry and for the manufacturers themselves;
provision of practical demonstrations of appropriate ICA technology to potential overseas clients.

This continually increasing interest internationally in the role of ICA for monitoring and control of wastes and wastewater treatment facilities and for environmental monitoring generally has stimulated the setting up under the auspices of the International Association on Water Pollution

Research and Control (IAWPRC) of a study group on instrumentation for on-line measurement whose main objectives are as follows:

the development of a mechanism for collection, collation and dissemination of pertinent technical information on sensors and supporting systems to the international user community;

the provision of a method of encouraging coordinating international cooperative ventures relevant to those tasks;

identification of variables which need to be measured to monitor and control wastewater treatment plants divided into four groups according to population served (*i.e.* populations less than 5000, between 5000 and 50 000, between 50 000 and 500 000, and greater than 500 000);

the dissemination of information, firstly at the IAWPRC workshops on the ICA, secondly in an annually produced Newsletter, thirdly in review articles published in *Water Research* at appropriate intervals and fourthly by means of short information articles published from time to time in the IAWPRC Newsletter.

Of particular interest are the reports which have been received to date from the Canadian, West German, Swedish and United Kingdom representatives. These have been abstracted to produce a list, reproduced in Table 1, of those variables thought at the present time to require continuous or semicontinuous measurement. These variables have been assigned a level of priority according to the uses to which the results of the measurements are to be put. An indication of instrument availability has also been provided.

64.3 TECHNICAL ASPECTS

Perhaps the most important of the recent technological advances of relevance have been in the microelectronics field: these include massive cost reductions and substantial increases in robustness and reliability, firstly of telemetry and communication systems including the computers and microprocessors themselves, and secondly the sensors, since in some cases these can now be incorporated on to a single 'chip' together with some computing power, thus allowing *in situ* standardization and the provision of a degree of redundancy.

The implications of some of these advances are now discussed in the context of water industry needs and priorities, particularly those associated with the design and operation of large scale wastewater treatment plants.

64.3.1 Needs

Most water authorities, public utilities and other relevant government agencies are committed to a policy of continuing improvements to the environment and to public health services whilst at the same time they are attempting to minimize costs and reduce energy consumption.

If these apparently conflicting goals are to be achieved then management strategies must be improved in all areas and this implies a need for timely and reliable information on all aspects of the water cycle. Matters to consider in this context are the following.

(1) In both developed and developing countries there is a likelihood of an increasing surplus of unskilled labour, permanent shortages of capital for public works and ever-increasing energy costs.

(2) The ability to treat wastewater to a standard sufficient to maintain the quality of the receiving waters downstream of the discharge at a level which is adequate for the use to which the water is to be put. Clearly this will be different for different rivers and different sections of rivers. River sections utilized as a source of drinking water for example must be maintained at a far higher quality than stretches used only for navigation and transport of wastes. In the latter case the quality need only be sufficient for the river section to be acceptable environmentally and give rise to no health hazard, whereas in the case of water used for drinking, for aquaculture and agriculture, in addition to a requirement for acceptable levels of pathogens, there is a need to minimize levels of potentially hazardous chemicals, and because these are expensive to remove in water treatment processes they are perhaps better removed at source. There is also a need to maintain a suitable level of physical and chemical quality: too high a salinity or suspended solids content, for example, could make the water unsuitable for irrigation purposes or for use as a cooling water.

These considerations imply a need for more precise monitoring than hitherto, and modern developments in sensors and systems and the impact of the microcomputer and of developments in large scale integrated circuit technology (LSI) have considerable relevance here.

64.3.2 Automation of Manually Performed Operations

Because of ever-increasing labour costs and the difficulties in the recruitment and maintenance of a suitable workforce there is a clear case for automating as many of the routine manual procedures as is economically practicable. These include the operation of valves and penstocks, cleaning and scraping operations, backwashing of filters and desludging of tanks. The required

Table 1 Variables That Need to be Measured and Availability of the Required Instruments

Availability Code:	(1)	Readily available
	(2)	Sensors available but instruments require further development
	(3)	Experimental prototype stage
	(4)	In development stage
	(5)	Method of measurement being developed or to be developed
Priority Level:	(1)	Measurements that are essential for effective plant operation
	(2)	Measurements required for improving or optimizing plant operation
	(3)	Measurements required for research and development purposes

Variable	Area of application	Availability of equipment	Remarks
Priority level 1	In sewer or at inlet to works	1	Equipment for in-sewer
Liquid flow, level and pressure		3, 4	measurement under development
	Preliminary treatment (screening and grit removal)	1	
	Primary sedimentation tanks	1	
	Secondary settlement tanks	1	Sludge blanket level measurement
	Final effluent channels	1	
	Sludge thickening/ dewatering plant	1	Coagulant control
	Tertiary treatment plants	1	
Temperature	In sewer or at inlet works	1	
	Final effluent channels	1	
	Digesters	1	
Suspended solids (turbidity)	Final effluent channels	1, 2	
Dissolved oxygen	Aeration tanks	1	
pH value	In sewer or at inlet to works	1	
	Sludge thickening/ dewatering plants	1	
pH	Digesters	1	
Priority level 2			
Liquid flow, level and pressure	Biological filtration plants	1	
	Aeration tanks	1	Inlet and outlet
	Recycled and surplus sludge channels	1	
Temperature	Aeration tanks	1	
Suspended solids (turbidity)	In sewer or at inlet to works	1	
	Primary sedimentation tanks	1	
	Biological filtration plants	1	
	Aeration tanks	1, 3	
	Recycled and surplus sludge channels	1, 3	
	Secondary settlement tanks	1, 3	
	Sludge thickening/ dewatering plant	4	
	Digesters	1	
Dissolved oxygen	Final effluent channels	1	
	Supernatant liquor decantation systems	1	

Table 1 (*continued*)

Variable	Area of application	Availability of equipment	Remarks
Dissolved oxygen profile	Aeration tanks	1	Several points in aeration tanks
Nitrate, nitrite and ammonia	Final effluent channels	1	
	Tertiary treatment plants	1	Pure oxygen plants
pH value	Aeration tanks	1	Nitrification, denitrification—several points
Gas flows (air, oxygen, digester gases)	Preliminary treatment (screening and grit removal systems)	1	
	Aeration tanks	1	
	Recycled and surplus sludge channels	1	
Oils	In sewer or at inlet to works and in final effluent	2	
Priority level 3			
Liquid flow, level and pressure	Digester	1	
	Supernatant liquor decantation systems	1	
	Sludge disposal	1	
Temperature	Supernatant liquor decantation systems	1	
Dissolved oxygen	In sewer or at inlet to works	1	
	Primary sedimentation tanks	1	
	Recycled and surplus sludge channels	1	
	Secondary settlement tanks	1	
Organic carbon	In sewer or at inlet to works	2	By combustion or inferred from
	Primary sedimentation tanks	2	UV absorption measurements
	Biological filtration plants	2	
	Final effluent channels	1	
	Tertiary treatment plants	1	
Nitrate, nitrite and ammonia	In sewer or at inlet to works	2	
	Biological filtration plants	2	
	Aeration tanks	2	
	Sludge disposal	1	
SSVI (stirred sludge volume index)	Aeration tanks	2	
	Recycled and surplus sludge channels	2	
pH value	Supernatant liquor decantation system	1	
Phosphate	Final effluent channels	1	
	Tertiary treatment plants	1	
	Sludge disposal	1	
Treatability and toxicity (toxic metals and organics)	In sewer or at inlet to works	4, 5	
	Aeration tanks	4, 5	
	Digesters	1	
	Sludge disposal	1	
Gas flows (air-oxygen, digester gases)	Digesters	1	
	Tertiary treatment plants	1	
Respiration	In sewer or at inlet to works	5	
	Aeration tanks	3	
Oxidation/reduction potential	In sewer or at inlet to works	1	
	Aeration tanks	1	
	Secondary settlement tanks	1	
	Digesters	1	
Gases			
Methane (CH_4)	Digesters	1	
Carbon dioxide (CO_2)	Digesters	1	
Hydrogen sulfide (H_2S)	Digesters	1	
Oxygen (O_2)	Aeration tanks	1	In aeration gas stream

technology is available, and indeed many of the larger works have already automated many of these processes. Equally important is the need to record the status of equipment in order that maintenance may be carried out in the most cost-effective way. This, coupled with the use of standby units where necessary and the centralization of controls, should provide a far greater reliability with, perhaps, lower operating costs than has been the case in the past.

Clearly too, smoother operational procedures and an improved output would result from the automation of those laboratory analytical procedures carried out routinely in the works laboratory which cannot be replaced cost effectively by on-line sensors. These include BOD and COD tests on influents and effluents, determination of suspended solids in settled sewages, mixed liquors, sludges and effluents (although on-line sensors for measurement of suspended-solids concentration can be used in some cases), determination of the concentration of heavy metals and other toxicants in sewage and sludge, the analysis of digester liquor for volatile acids, and measurement of the respiration rate of mixed-liquor samples. Recently, however, methods of measurement of these variables on-line have been developed (Briggs, 1981).

64.3.3 Optimization

The improvements in sensor and system technology described subsequently indicate that varying degrees of optimization may now be possible. For these to be practicable, however, on-line deployment of reliable mathematical models of processes of interest is required.

A mathematical model of the activated-sludge process has been developed by Jones (1975) and has been used to simulate the behaviour of wastewater treatment plant under a wide range of operating conditions. To assess the validity of the model, it has been incorporated in a computer program written specifically to simulate the operation of laboratory-scale pilot plant described by Boon and Burgess (1972). Predictions by the model and data obtained from the plant, in response to diurnal variations in flow, have been compared. Figure 1 shows predictions and observations of effluent quality as determined by the Organic Pollution Monitor described by Briggs *et al.* (1976) and of the concentration of mixed-liquor suspended solids, respectively. In both cases the observed data and simulated curves are well matched and indicate that the model can be calibrated to predict the behaviour of a treatment plant with some degree of confidence.

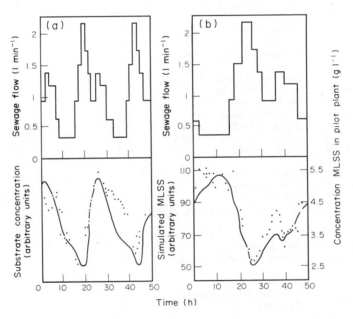

Figure 1 (a) Predicted response to diurnal variations in flow of substrate concentration (*i.e.* organic matter) in effluent and observed data (points) as determined utilizing the Organic Pollution Monitor; (b) predicted and observed levels of suspended solids in mixed liquor (determined instrumentally) as a result of diurnal variation in flow

Additionally the data provide an indication of the likely precision obtainable in this type of application from optically based instruments which can be deployed for measurement of suspended matter and organic matter; these are discussed in more detail subsequently.

The use of this mathematical model for control purposes was investigated by Briggs and Jones (1978). Their work indicated that significant improvements in the day-to-day operation of the activated-sludge process can be achieved by means of accurate measurement of dissolved oxygen profiles in aeration tanks and that sufficiently reliable instrumentation (Briggs, 1981) exists to make this practicable. Further, it was demonstrated that an accurate knowledge of the distribution of suspended solids between effluent, mixed liquor and surplus and returned sludge is essential for fine control of the plant to cope with substantial and prolonged changes in load. In many cases it might be sufficient to determine this by means of sampling and off-line analysis, say, on a daily basis. However, substantial savings might be realized by carrying out these measurements on-line using some of the newer optical and other sensors which have relevance in this context.

Although not strictly necessary for optimization purposes, the ability to continuously measure organic matter, ammoniacal nitrogen and flow in settled sewage and effluents would be extremely useful for monitoring the performance of plants.

Furthermore, since there is a clear trend towards the de-manning of the smaller works, on-line monitoring of relevant variables is clearly worthwhile.

In many cases, too, the costs of sludge treatment and disposal are considerable. Indeed these can be as high as 40% of the total cost of the treatment from smaller works and this implies the need for reliable sludge blanket levels achievable by instrumentally minimizing sludge volumes and optimizing coagulant doses must also be obvious. Instruments of potential value in this context have already been described (Briggs, 1981).

Since disposal of sludges from wastewater treatment plants to land is in many cases the cheapest alternative, there are considerable attractions in using sludge digestion to remove pathogens and there is the additional bonus that the process is a potential source of energy in the form of methane. There is a clear need, therefore, to increase the reliability of the process and as a result reduce the size of the digesters. Sources of digester failure are usually some or all of the following: overloading, poor mixing, accumulation of grit and scum, poisoning by heavy metal, trace organics or detergents.

Again, removal of the problem at source appears to be paramount. However, instrumental methods of monitoring sludges being fed forward to digesters, the degree of mixing of digester contents and the composition of the gases produced have been researched (Briggs, 1981) since in many instances it would be practicable to protect digesters by monitoring the input and diverting sludges that were too toxic, too dilute or too strong for immediate acceptance and altering the feeding arrangements accordingly.

64.4 MEASUREMENT TECHNIQUES

One of the most important, but often unduly weighted, factors when assessing the degree of automation which can be applied to wastewater treatment processes is the commercial availability of the required robust and reliable sensors. In fact, significant improvements have been made in existing sensors in recent years and new ones have been developed, or are being developed at the present time. Many of these advances have been reported at the IAWPR Workshops (1973, 1977, 1981) and elsewhere (Briggs, 1981, 1983). They include the production of improved dissolved oxygen electrodes and suspended solids meters, development of methods of measurement of organic matter, ammonia and nitrate ion, improved methods of flow and level measurement, methods of measurement of sludge level, sludge solids concentration and sludge dewaterability, measurement of the amount of film in biological filters and methods of measurement of heavy metals and other toxicants. Certain aspects of sensor technology will be described subsequently.

64.4.1 Sensor Logistics

Prolonged exposure of sensors to industrial and domestic wastewater and to the products of the various treatment processes inevitably leads to a degradation in performance and reliability, particularly in the case of those sensors which have evolved from a laboratory instrument which was designed to measure the same or a related variable but under favourable environmental conditions. Such sensors are frequently unreliable when deployed in situations that are not climatically well controlled, where gross fouling and/or growth of biological films and slimes is likely to occur and where skilled human supervision is not readily available.

Various remedies exist but in general they add considerably to the cost and complexity of the instrumentation. They range from good housekeeping practices to a redesign of the whole system. The more important of the practical remedies include: sensor modification or replacement with an alternative which, amongst other things, utilizes conservatively rated components; careful design of the hydraulic system to minimize fouling; provision of automatic cleaning and calibration facilities; provision of an adequate level of maintenance; protection from adverse environmental effects. Perhaps even more important is the use of a telemetry system to provide a 'real time' indication of sensor and system performance. Additionally the use of the telemetry system and a centrally located computer or, alternatively, a locally based microprocessor system enables the deployment of such techniques as majority polling or the use of a dual sensor system such as that developed initially to protect intakes to water treatment plants where the maximum reliability is required (Briggs, 1981). With this system two identical sensors are deployed, one on stream and the other on standby (immersed in a calibrating solution containing a biocide).

By automatically transposing the functions of the sensors, both on a time basis and on receipt of data indicating that a sensor reading has deviated significantly from past data trends, it is possible to detect a sudden and serious change in sample quality on the one hand or a sensor failure on the other. In the latter case the good sensor is left on stream permanently until the other is repaired or replaced. Heslop (1981) stated that a system of this type installed on the River Wear to protect the intake to Lumley Water Treatment Works has operated reliably and continuously with a 100% data return for at least 18 months and recent unpublished reports have indicated that the system has continued to perform reliably.

64.4.2   Existing Sensors

Since the viability of any monitoring and control system depends primarily on the commercial availability of robust and reliable sensors for essential variables, progress with the development of sensors is now discussed.

64.4.2.1   Flow

In open channels, conventional techniques based on level measurement are common but have limitations in unattended operation when applied to liquids containing high concentrations of suspended and/or floating matter. Remote level sensing using pressure or ultrasonic transducers gives better results for some applications and both ultrasonic and optical devices have been used successfully for measurement of sludge blanket level. Magnetic flow meters are being used increasingly to measure flows of heavily polluted liquids and slurries in pipes. There are instances, however, where large variations occur, and the accuracy of measurement at the lower end of the scale suffers because of zero drift. It has been suggested that magnetic flow meters with smaller internal diameters than the adjoining pipework could be used for the following reasons: (1) less susceptibility to zero drift, (2) during periods where the main pipework may not be completely full, adequate measurements could still be made in the reduced section, provided the meter was correctly located, (3) since the throughput would necessarily be higher, self-cleaning of the electrodes mounted on the inner wall would be more effective. However, these advantages must be carefully weighed against the increase in line pressure and the possibility of blockage under high flow conditions. A technique based on measurement of turbulence-induced electrical noise at on-line ultrasonic sensors has been shown to give data proportional to the solids concentration. Mass and volumetric flow rates can therefore be determined simultaneously on sludges at various stages in the process (Balachandran and Briggs, 1981).

64.4.2.2   Temperature (−10 to +40 ± 0.5 °C)

Temperature measurement is seldom a problem. Platinum resistance thermometers give the best results, but in many cases the long-term stability of thermistors is adequate.

64.4.2.3 *Dissolved oxygen*

Dissolved oxygen is measured at 0 to 100% ± 1% or 1 to 200% ± 2% saturation, and galvanic or polarographic dissolved-oxygen probes will work reliably in flowing samples. However, both rely on the diffusion of gaseous oxygen through a hydrophobic membrane (usually polythene or Teflon), and in polluted waters it is necessary to inhibit the formation of algal growths and bacterial slimes on the membrane and to calibrate the probes regularly. To avoid interferences from gaseous hydrogen sulfide it is possibly to deploy a silver/silver sulfide salt as an electrolyte and thus avoid the possibility of sensitivity loss by poisoning.

The theoretical basis for operation of the probes follows from the cathode reaction:

$$O_2 + 2H_2O + 4e \rightarrow 4OH^- \tag{1}$$

Steady state output current (I) is given by

$$I = nFA(P_m/b)a_{O_2} \tag{2}$$

where n = the number of electrons involved in the cell reaction, F = Faraday's constant, A = the area of membrane exposed both to the sample and to the cathode, P_m = the membrane permeability coefficient, b = the membrane thickness, and a_{O_2} = the activity of the oxygen molecules. The activity, a_{O_2}, is given by

$$a_{O_2} = \exp(K_s I_s C) \tag{3}$$

where K_s = the salting-out coefficient, I_s = the ionic strength of the water being monitored and C = the concentration of dissolved oxygen (mg dm^3).

Temperature dependence is given by

$$I_T = Ke^{-J/T} \tag{4}$$

where J = 2400 K for polythene. This yields a temperature coefficient of about 6% °C^{-1}. Methods of temperature compensation utilizing the thermistors have been described in detail by Briggs and Viney (1964).

A useful innovation devised by Mancy and Schmid (1970) employs a pulsed voltammetric system of increased sensitivity and flow independence (*e.g.* it can be used in microbial films and static solutions). Its performance is governed by

$$I_t = \frac{nFAD}{t} \times 0.5a_{O_2} \tag{5}$$

where n = the number of electrons involved in the cell reaction, F = Faraday's constant, A = the area of the cathode, t = the time in seconds, D = the diffusion coefficient and a_{O_2} = the activity of oxygen molecule a.

The temperature coefficient is approximately 0.02% °C^{-1}. More recently, Leeds and Northrup Ltd have developed a maintenance-free probe based on a coulometric operating principle. The probe contains three electrodes: an anode or oxygen generator, a cathode at which oxygen reduction occurs and a reference electrode. As is the case with other systems, the electrolyte is contained by a polythene or Teflon membrane but this is protected by a second robust silicone rubber membrane which is far more permeable to oxygen.

Although meant to be maintenance-free, it still seems vital to use even these probes in a system which provides automatic cleaning, at least daily, and some form of automatic calibration.

64.4.2.4 *Oxidized nitrogen (0 to 50 mg N l^{-1} ± 5% of reading)*

Apart from the automatic wet-chemistry analyser approach, which is not generally recommended for unattended monitoring, the only technique known to be in common usage is the deployment of a 'specific ion monitor' fitted with a nitrate ion-selective liquid ion-exchange electrode. The electrode performance is affected by interference from other ions in solution, in particular perchlorate, iodide, permanganate, thiocyanate and zinc. In addition, the algicide Panacide, which could in other circumstances be used to inhibit algal growth, also has an adverse effect on electrode performance.

However, the specific ion monitor has the facility of mixing the sample with a reagent, in this case a total ionic strength adjustment buffer (TISAB) to stabilize the ionic strength and to reduce the pH value below 4.5 to eliminate carbonate and bicarbonate interference. The means of thermal stabilization of the sample and electrode assembly, and automatic calibration and standardization are also built into the monitor and, provided that the sample does not contain a high concentration of suspended solids, the 3 mm bore pipework and peristaltic pumps will operate effectively for up to a week unattended, thus providing a measure of nitrate ion concentration in the sample in the range 1.4–140 mg N l^{-1} with a precision of approximately ±10% of reading (Briggs and Melbourne, 1972).

64.4.2.5   Ammoniacal nitrogen (0 to 5, 10 or 50 mg N l^{-1} ± 5% of reading)

A specific ion monitor similar to that described for the measurement of nitrate ion may be used for the determination of total ammonia (as NH_3) in the sample. In this case the electrode consists of a glass pH electrode situated behind a thin gas-permeable hydrophobic membrane, with a small quantity of ammonium chloride solution in contact with the electrode tip and the membrane. When the probe is immersed in a sample containing free ammonia, the latter diffuses through the membrane until the partial pressure of ammonia is equalized on both sides. Thus, as the ammonia concentration in the sample changes, the pH electrode detects the change in hydrogen ion concentration and the probe as a whole has a Nernstian response to ammonia. The probe exhibits considerable resistance to contamination by dissolved ions and gases. Carbon dioxide, hydrogen sulfide and sulfur dioxide which might be expected to interfere do not affect the performance at the high pH value at which measurement of free ammonia is made. In practice the reagent added to the sample consists of a mixture of sodium hydroxide solution to increase the pH value to 11–12 and EDTA to minimize the deposition of hardness.

The response time of the probe varies from a few seconds above 1 mg l^{-1} to several minutes at 0.1 mg l^{-1}, and when used in the specific ion monitor it will operate unattended in samples containing low concentrations in the range 0.4–140 mg N/l^{-1} with a precision of approximately ± 10% of reading (Briggs and Melbourne, 1972).

64.4.2.6   Organic matter (0 to 10 or 0 to 100 mg C l^{-1} ± 5% of reading)

Ultraviolet absorption at 254 nm has been shown to correlate well with TOC for a wide variety of samples ranging from settled sewage and effluents to raw and treated river waters (Briggs *et al.*, 1976) and an instrument has been developed for water quality monitoring applications and for control of water and wastewater treatment processes. When the 'Organic Pollution Monitor' first became commercially available, difficulties were encountered because of drifts in the intensities of light emitted in the ultraviolet and visible parts of the spectrum (the latter is used to correct for suspended solids which may be present in the sample). This drawback has now been overcome by the provision of a thermal stabilization unit around the base of the UV lamp. After initial settling down, which usually takes several hours after initial switch on, the monitor will operate satisfactorily for at least a week without any attention.

64.4.2.7   Suspended matter (5 to 500 to 5000 mg l^{-1} ± 5% of reading)

Strictly speaking, this parameter can only be determined gravimetrically by filtration or centrifugation of a known volume of sample followed by drying and weighing the residue. However, optical techniques have been used for some time to measure turbidity which may be related to the suspended solids concentration provided that variations in the nature of the particulates in suspension, their shape and size distribution are known. Turbidimeters fall into two basic categories, *i.e.* absorptiometers, which measure the absorption of light through a sample, and nephelometers, which measure the amount of light scattered at one or more angles to the incident beam. There is now a very wide range of absorptiometers and nephelometers commercially available; these satisfy most operation requirements including the monitoring of ultrapure waters, river waters, effluents, sludges and slurries, whose concentration is measured in % rather than mg l^{-1}. It should be possible therefore, by careful selection, to find a turbidimeter suitable for almost any river water quality monitoring application. Some of the better turbidimeters have built-in compensation for fouling of optical surfaces and/or mechanical means of cleaning these surfaces, and in some instances the optical elements are not in contact with the sample at all. These can be calibrated by means of the Formazin standard which has been shown to be reproducible between samples and stable after several months of storage in recommended conditions.

When obtaining a practical correlation, however, it is necessary to take into account the effect of flow on particle size distribution since this will in general cause a reduction in sensitivity at higher solids concentration. Provided adequate cleaning and calibration procedures are used and the hydraulic circuit is designed to avoid trapping gas bubbles, and sediment settlement, data of the required precision can be obtained (Briggs *et al.*, 1967).

64.4.2.8 Chloride

Chloride-ion concentration is not normally monitored in rivers, unless pollution by industrial wastes or saline is suspected, in which case a specific-ion monitor similar to those described for nitrate and ammonia may be used. For this there are basically two types of chloride ion-selective electrode, one of which utilizes a solid-state membrane and the other a liquid ion-exchange membrane with different extents of interference susceptibilities. Buffering is only necessary when interfering ions are present, and it should be possible to monitor chloride-ion concentration in the range 0.2–350 mg $Cl\,l^{-1}$ with $\pm 10\%$ precision for up to one week's unattended operation.

64.4.2.9 Hardness

There are several divalent cation-selective electrodes commercially available, some of which may be used to monitor temporary and/or permanent hardness measured at 10 to 1000 mg l^{-1} as $CaCO_3 \pm 5\%$ of reading. Interference may be caused by copper(II), zinc, iron(II), nickel, strontium and barium ions.

The value of hardness measurement in on-line monitoring applications in river systems must be questionable, because the hardness of a natural water in an area does not normally vary significantly. Hence, this measurement is more likely to be used in the control of water treatment processes or other industrial applications.

64.4.2.10 pH value

A large selection of glass sensing and reference electrodes are now commercially available. To be effective, the reference electrode, which ideally should be of the double-junction type, should not be electrically grounded or present a low impedance to ground. An earthed guard electrode would be placed around the pH probe, and the glass/reference electrodes connected to a dual-input stage-balanced differential amplifier with a very high input impedance (10^{13} Ω). This overcomes the drift problems and provided that periodic mechanical (brush), ultrasonic or chemical (algicide and bacteriocide) cleaning techniques are also used, it is possible to monitor pH in the range 1–12 with a precision of ± 0.2 for periods of up to one week's unattended operation. The interval between servicing may be extended considerably if low and high calibration solutions are fed to the probe on a daily basis and used to correct the instrument's calibration automatically.

64.4.2.11 Dissolved carbon dioxide

Several probes are commercially available where the dissolved CO_2 diffuses through a hydrophobic membrane into a filling solution and causes a change in pH value which is detected by a conventional pH electrode situated within the membrane. The sample must be buffered to below pH 4, and there is interference from volatile acids. However, the value of this determinand is questionable in other than specialized applications, such as the study of the intrinsic toxicity of CO_2 to fish and the additional toxicity of ammonia in the presence of CO_2.

64.4.3 New Developments in Sensor Technology

As stated earlier, the solutions currently deployed to achieve high performance reliability are costly. Also, that which is required in many cases is 'fit and forget' instrumentation, particularly for deployment in remote and inaccessible locations or where there is a lack of adequately trained maintenance staff. Furthermore, where intrinsically safe instrumentation is required, for example, for monitoring flow and level in sewers or for analysing digester gases, the additional costs associated with bringing standard instrumentation up to BASEEFA ZONE ONE Standard are also considerable.

Techniques which are relevant here include the use where practicable of non-contact or non-intrusive sensors which are robust and reliable and also the use of passive voltage-free sensors. Additionally, when the sensor must contact the sample, for example, in the case of some water quality sensors, extremely low voltage and low current devices such as chemically sensitive field effect transistors (CHEMFETs) can be used.

Also relevant are the ever increasing reliability of microelectronics, including memory devices and microprocessors, and their ready availability and ever decreasing costs. This makes such techniques as majority polling, pattern recognition and the use of relative measurements rather than absolutes practicable in many cases.

Some examples of the techniques outlined above are therefore now discussed.

64.4.3.1 Flow measurement

Conventional instrumentation for flow measurements, such as level recording in association with weirs and flumes for measurements in open channels, propeller-type meters, Doppler meters, magnetic meters, Pitot and Dahl tubes for measurement of flows in carrying clean water under pipe full conditions, has been discussed in considerable detail elsewhere (Medlock, 1982) and will not be discussed further here. However, perhaps the most difficult of the Water Authority application areas is that of flow measurement in sewers and the following details apply.

The suggested functional specification for measurement of flows in sewers and effluent channels provided by WRC is reproduced in Table 2.

Table 2   Functional Specification for Flow Measuring Equipment for Use in Effluent Channels and Sewers

	Permanent	*Survey*	*Portable*
(1) Velocity range (m s^{-1})	0.33	0.15–6	0.15–6
(2) Error			
(a) systematic	±5–10%	±5–10%	±5–10%
(b) random	±2%	±4%	±1%
at 95% confidence level			
(% of reading)			
(3) Sensor remote or in liquid	Either	Either but remote preferable	In liquid
(4) Maintenance	Design for ease of maintenance	Design for ease of maintenance	Design for ease of maintenance
(5) Power	Mains	Battery (7 days supply for operation every 5 min)	
(6) Output	On-line/logger/chart	Logger	Analogue/digital/meter/logger
(7) Recording interval	5–60 min	1–15 min	Continuous
(8) Environment			
(a) sensor	Withstand a 3m head of water (IP68)	Withstand a 3 m head of water (IP 68)	Withstand a 3 m head of water (IP 68)
(b) recorder	Stormproof	1 m head of water for 24 h (IP68)	Stormproof (IP63)
(c) temperature (°C)	0–50	0–50	0–50
(d) atmosphere	Withstand corrosion and be electrically safe to BASEEFA Zone 1 standards		
(9) Manhole/pipe/channel	Channel/pipe	Pipe/channel/manhole	Channel/pipe/manhole

As a result of a considerable number of survey and measurement exercises, the Water Research Centre (WRC) concluded that a survey package developed by the Golden River Company Limited could (after certain modifications now incorporated) have the required precision, robustness and reliability.

The Doppler principle was chosen for the velocity monitor in preference to the electromagnetic method because: (a) it is more robust; (b) it consumes less power; and (c) it measures velocity at a position less influenced by the head of the sensor.

Depth measurement is carried out by means of a pressure transducer. The device selected was originally developed for borehole work but has been modified to have a suitable operating range.

Although in general pressure transducers exhibit some long term drift, it was reasoned that provided check measurements were carried out weekly, the device would be suitable and certainly the best available for measurements of water depth in an approximately horizontal pipe.

Both the velocity and depth sensor are mounted in a single streamlined housing and attached to a ring which can be inserted into the pipe from a manhole. (In general flow conditions are far better in hydraulic terms a short distance up a pipe rather than in the manhole itself.) The sensing part of the pressure transducer is located under the wedge-shaped housing in order to minimize velocity head effects.

As a result of repeated calibrations carried out on a number of meters both in the Hydraulic Research Laboratory's test tank and *in situ*, WRC have deduced the following relationship:

$$v = \text{velocity} = \left[0.93 + \frac{0.16 \times \text{depth}}{\text{diameter}}\right]$$

for depths up to 1 m, and are confident that an overall precision of ±5% of reading is achievable in the field up to depths considerably greater than 1 m.

Flow gauging instruments based on the Faraday effect are well known in pipe flow measurement applications, but only comparatively recently has the technique been applied to rivers, open channels and pipes such as sewers which can be partially or completely full. Herschy and Newman (1982) have provided a full description of the technique.

In the system now available commercially from Sarasota Automation, a coil is installed below the river bed, above a pipe or bridged across a channel or pipe width. Electrodes set into opposite banks of the channel or as loops on opposite sides of the pipe sense the potential induced by the water flowing through the magnetic field. Control equipment drives an electric current through the coil and measures the induced potential, which is proportional to the average velocity of the water. This, combined with a measurement of cross-sectional area, gives a measure of quantity flow (discharge). For measurement of flows, culverts and reasonably small rivers, it is practicable to electrically insulate the channel or pipe from the water and thus avoid problems due to short-circuiting in the bed or surroundings.

Results to date from a number of test sites, both sewers and open channels, have in all cases produced flow data with a precision better than ±5% of reading throughout the range of flows experienced and the equipment is able to detect direction of flow also when this is required.

Auto and cross-correlation flow meters. Methods of measurement of mass and volumetric flows of domestic and industrial water and wastes and of sewage and sewage and waterworks sludges using auto and cross-correlation techniques have been described by Briggs *et al.* (1974) and Balachandran and Briggs (1982).

The method is based on measurement of turbulence-induced electrical noise in the outputs of a wide range of sensors such as thermal and electrical conductivity detectors, optical and ultrasonic sensors, and is fairly well known. A measure of time of travel of a disturbance pattern is obtained by cross-correlating decoupled and normalized 'noise'-induced signals obtained from two sets of sensors in series, and applying a variable time delay to the signal obtained from the first. The time delay required to maximize the cross-correlation signal provides a measure of the mean velocity, and the percentage depth of 'noise'-induced modulation at one of the pairs of sensors provides a measure of the concentration of particulates to the fluid.

Cross-correlation techniques applied to pairs of ultrasonic sensors placed in the wake of a vortex shedder have also been utilized to increase the effective range of vortex shedding meters, and a promising technique for measuring flows in water mains and in pipes to domestic and industrial premises has been described by Coulthard and Briggs (1981).

Until recently, however, the cost and complication of the required electronics have inhibited commercial exploitation of these techniques, but it is expected that recent advances in microelectronics and microcomputers will alter the situation in the near future.

Additionally, two other possibilities are being researched. The first is the use of collimated γ-radiation from a sealed source to provide 'clip-on' fit-and-forget sensors for use in inaccessible areas. The other is the use of fibre optics, both as the telemetry link and as the sensor itself. This is particularly attractive for use in hazardous areas since the sensor is passive and voltage- and current-free. These and other optical techniques have been described in detail elsewhere (Briggs *et al.*, 1981). However, the following examples may be used as an illustration of the principles to be deployed.

   (1) In the case of depth measurement it would be practicable to mount a mirror in such a way that it would reproduce the deflections of a diaphragm and to use a twin optical fibre link (one of which was close-coupled to a solid-state laser source such as gallium arsenide and

which would be pulsed at a suitable frequency and the other to a receiver) to measure the deflection of the diaphragm to whatever precision was required.

(2) For velocity measurement it would seem practicable to terminate twin optical fibres at various points around the circumference of the pipe, to use the same method of energization for one fibre and to pick up the Doppler (back scatter) signal from the other. Cleaning of optical surfaces could be achieved chemically on an intermittent basis.

64.4.3.2 Metals

Van Duin and de Kreuk (1979) have described a continuous electrochemical analyser for monitoring zinc, cadmium, lead and copper, the limits of detection being of the order of 0.1 mg l^{-1}. The instrument, a pulsed differential anodic-stripping voltammeter with a hanging mercury-drop electrode, is capable of unattended operation for up to one month. Sample treatment comprises UV irradiation and acidification, heating to 95 °C to liberate the metals, cooling and filtration. Before the sample is introduced into the sample cell its pH value is adjusted to about 5 and dissolved oxygen is removed.

A rather different approach, based on coulometry utilizing two platinum electrodes, has been reported by Cnobloch *et al.* (1980). The metals Ag, Pb, Cu, Co, Fe, Cd, Zn and Ni have been detected at levels below 1 mg dm^{-3}. The instrument has been successfully deployed for eight months at the inlet of a municipal wastewater treatment plant and for nine months at the outlet of a neutralization plant of a metal-plating works. Sample preparation, particularly in the case of sewage samples and sludges, presented problems but a system of heating with sulfuric acid, followed by buffering to pH 3.5, and filtering using a plate filter was found to be adequate.

An exciting approach to cost-effective metal analysis currently being investigated is production of an array of sensitive MOSFETs on a single waterproof chip with the exception of the gates, to which will be bonded soluble phosphate or borate glass loaded with appropriate concentrations of metal oxide, and which will be exposed to the sample. One of the gates will be arranged to be selective to a reference ion which can be added to the sample in a variety of ways, including contact with soluble glass. Using a microprocessor to implement a matrix inversion procedure, the resultant simultaneous Nernstian equations can be solved; a single chip could then be utilized to monitor a wide range of metal ions.

So far, discrete ion-selective electrodes have been fabricated for copper and for iron and give a normal Nernstian response when coupled with a conventional reference electrode. Additionally, research into the deposition of the glasses on to MOSFET gates, and into the waterproofing of the device for direct immersion in water samples, looks promising.

64.4.3.3 Trace organics

In general, the wide range of compounds encountered, the fact that they are usually present in microgram quantities, and the uncertainty in respect of their potentially hazardous nature imply a need for sampling followed by sophisticated analytical techniques. Gas chromatography/mass spectrometry, for example, is often required to obtain qualitative and quantitative data of a sufficiently reliable nature to establish trends. However, again for the specific purpose of protecting intakes and indeed in the context of assessing the efficacy of a particular treatment process, the ability to monitor trace organics continuously is clearly desirable.

A number of possibilities are being considered at present. The first of these is the development of a robust GC/MS system for on-line use, such as that currently deployed in aircraft in air pollution studies (Evans and Arnold, 1975); a major difficulty, now solved commercially, is removal of water from the sample. A possibility is to strip volatiles, adsorb them on to and then desorb them from a continuous filament of activated-carbon fibre or to measure in the head space using gas-sensitive semiconductors.

Additionally, recent sensor developments brought about by advances in microelectronics and membrane technology appear to show considerable promise. A typical example is the use of polymers, phthalocyanines and other organic films, deposited either on MOSFET gates or on resistive elements, to measure volatile organics initially in the head space above a sample and subsequently by direct immersion. Research on the development of special glasses, lipid membranes and captive-enzyme systems also has considerable relevance. Phenol electrodes based on enzyme deposited on an oxygen electrode and glucose electrodes utilizing oxygen electrodes and glucose oxidase are already available in small numbers.

64.4.3.4 Oil

By far the greatest number of reported pollution incidents is due to oil spills. A number of devices are currently available for monitoring oil. Surface films are reasonably easily detected by utilizing the change in refractive index/measurement of UV fluorescence, or by a novel device (Malz, 1970) comprising a polythene disc on which was mounted a number of conducting sectors. The disc is slowly rotated in the vertical plane partially submerged in the water. The conductivity between sectors is measured and the disc is wiped clean between measurements. Since oil has a far greater affinity for polythene than has water, a significant change in conductivity results if the disc passes through an oil film.

This basic principle is now utilized in a commercially available instrument but in this case the oil adhering to the disc is determined by means of UV fluorescence measurement.

Dispersed oil is of far greater significance, however, since it can readily enter intakes and seriously contaminate treatment works and reservoirs even though the intake is located below the surface. Probably the more reliable methods of detection of significant amounts of oil would be to use either the Malz or similar commercial development within the monitoring station, or alternatively a recently available instrument from Fisher Controls, based on solvent extraction followed by infrared analysis.

64.4.3.5 Measurement of plant and equipment status

Again one of the more useful of the possibilities likely to become practicable in the near future is that of determining plant and equipment status by means of pattern recognition techniques utilizing non-contact sensors, for example, the use of an array of hydrophones to 'listen' to machinery and storage of the noise spectrum in memory and comparison of current patterns with past patterns in order to obtain an early indication of faults leading to excessive or different vibration.

Interestingly, too, levels and flows in inaccessible places can be monitored utilizing collimated electromagnetic radiation obtained from robust solid-state sensors, thus making maintenance less frequent and far easier to carry out.

64.4.4 Telemetry and Communication Systems

It has been stated by White (1981) that although the UK is highly industrialized, 95% of its population being served by main drainage, yet of the 7800 sewage works, less than 200 serve populations of more than 50 000 and less than 90 serve populations greater than 100 000, and that this is probably not untypical of other industralized countries. Clearly, therefore, the bulk of the sewage treated is by means of large numbers of widely distributed works serving populations between 5000 and 50 000 and this dispersion enables the best use to be made of the recovery capacity of rivers. It is not likely, therefore, that there will be any significant concentration of sewage treatment in the near future. Although this concentration of sewage treatment at a few large works has the benefit of scale and the ability to provide skilled supervision, it has disadvantages now well recognized, where populations are not themselves well concentrated. These disadvantages include the concentration of discharge of treated sewage effluent at a few points and consequently the need for an average higher standard of treatment, and also the deterioration of raw sewage due to the time taken to bring it to the works.

The result of all this, coupled with the fact that both skilled and unskilled employees are increasingly less likely to accept unsocial hours and dirty work, has been a trend to an increasing use of unmanned works from which performance and equipment status data are relayed to appropriate regional works where skilled staff reside. Increasingly, therefore, Water Authority managements, encouraged by the rapidly reducing costs of microelectronic equipment and associated computing power, are installing wide ranging telemetry and communications systems. A typical schematic of a data transmission and communication system suitable for a large multifunctional division of a Regional Water Authority is depicted in Figure 2. Here several modes of data transmission have been utilized and existing equipment and communication methods integrated with new ones. An indication of data flow routes is provided in Figure 3 and a typical data base configuration in Figure 4.

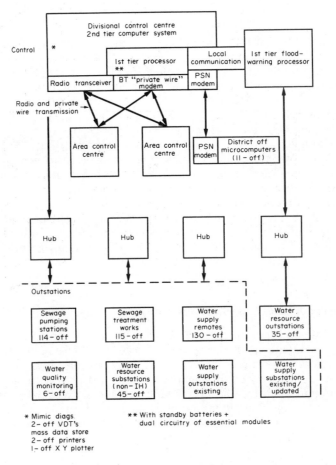

Figure 2   Typical schematic of a data transmission and communication system suitable for use by a large multifactional division of a Regional Water Authority

Implicit in such schemes is the need for local closed loop automation of essential aspects of water and wastewater treatment works and of pumping stations with the attendant requirement for robust and reliable equipment for measurement and storage of analogue information and for telemetry of alarm data and equipment status inspection. Additionally, adequate standby equipment and machinery must be provided.

In justifying such schemes, therefore, the costs of the equipment and its operation and maintenance must be offset by savings in energy, reduction in tariff charges and reduced manning levels both for operation and maintenance. Additionally, the value of providing a better service to the public needs to be quantified.

64.5   SUMMARY AND CONCLUSIONS

The basic requirements for instruments and systems for monitoring and control of wastewater treatment plants and in water pollution control generally have been discussed, and the organizational and logistical aspects of applying instrumentation, control and automation (ICA) in a cost-effective and sensible manner have been explored.

Existing and developing instruments and systems have been reviewed and so too have current applications on a worldwide basis. Lack of space has prevented much by way of instrument design construction and instrument performance details being given but adequate references to other publications have been provided.

Perhaps the most interesting conclusions to draw are that, compared with other industries,

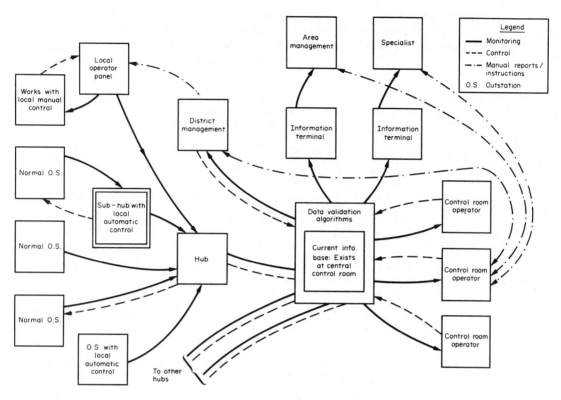

Figure 3 Typical data flow schematic

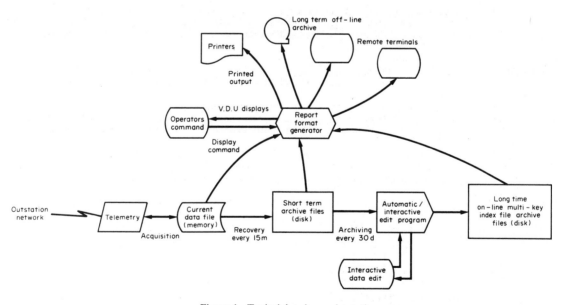

Figure 4 Typical data base schematic

water industries have not hitherto made anything like as much use of ICA technology as have, for example, the mining, steel, oil, gas, petrochemical, pharmaceutical and food industries. However, as discussed earlier the changing economic situation, the lack of capital for public works, ever increasing energy costs and different attitudes to labour, coupled with the substantially lower costs of basic sensors, telemetry and data processing equipment, have already brought about significant changes and the UK Steering Group on ICA has played a significant role here.

Finally, it must be emphasized that the momentum of this change will soon be lost unless the instrument industry concentrates on the production of low cost, robust and reliable sensors and support systems which can be easily maintained on a modular replacement basis.

64.6 REFERENCES

Andres, J. F., R. Briggs and S. H. Jenkins. (1974). *Progress in Water Technology*, vol. 6. Pergamon, Oxford.

Balachandran, W. and R. Briggs (1981). Ultrasonic sensors for monitoring flows and sludge solids concentrations in water and wastewater treatment. *Water Sci. Technol.*, **13** (11), 581–588.

Boon, A. G. and Burgess (1972). Effects of diurnal variations in flow of settled sewage on the performance of high-rate activated-sludge plants. *Water Pollut. Control*, **71** (5), 493.

Briggs, R. (1981). Water quality monitoring and control. In *Developments in Environmental Control and Public Health*, ed. A. Porteus, vol. 2, p. 155. Applied Science Publishers, Barking, UK.

Briggs, R. (1983). On-line analytical instrumentation in the water industry —problems and practical solutions. *Measurement and Control*, **16** (3), 94–97; **16** (4), 151–155.

Briggs, R. and G. Jones (1978). Aspects of automation of the activated sludge process. *Water Pollut. Control*, **77** (4), 439–451.

Briggs, R. and K. V. Melbourne. (1972). Ion-selective electrodes in water quality monitoring. Paper presented at a Conference on Air and Water Pollution, pp. 37–64. Institution of Mechanical Engineers, London.

Briggs, R. and G. Olsson (1982). On-line measurement, data processing and digital control. In *Proceedings of IWPC Symposium: Data Processing in the Water Industry*, Bournemouth, April 1982.

Briggs, R., K. V. Melbourne and G. E. Eden (1967). The monitoring of water quality. In *River Management*, ed. P. C. G. Isaac, pp. 38–55. Maclaren, London.

Briggs, R., C. N. Wormold, M. S. Beck and A. Cornish (1974). Sludge solids concentration and velocity flow measurement using electrical noise technique. *Prog. Water Technol.*, **6**, 114–123.

Briggs, R., J. W. Schofield and P. A. Gorton (1976). Instrumental methods of monitoring organic pollution. *Water Pollut. Control*, **75** (1), 47–57.

Briggs, R., A. R. Farmer and M. Anson (1981). *Proceedings of Symposium on Optical Sensors and Optical Techniques in Instrumentation*. Institute of Measurement and Control, London.

Cnobloch, H., W. Kellerman, H. Kulh, H. Nischik, K. Pantel and H. Poppa (1980). Continuous monitoring of heavy metals in industrial wastewaters. *Anal. Chim. Acta*, **114**, 303–310.

Coulthard, J. and R. Briggs (1981). Suggestions for a low cost method of metering domestic and industrial water supplies based on auto-correlation of vortex induced ultrasonic modulation. *Water Sci. Technol.*, **13**, 311.

Evans, J. E. and J. T. Arnold (1975). Monitoring organic vapors. *Environ. Sci. Technol.*, **9**, 1134–1138.

Forsberg, C., B. Hawerman and B. Hultman (1982). Experience from 10 years advanced wastewater treatment. *Water Sci. Technol.*, in press.

Herschy, R. W. and J. D. Newman (1982). The measurement of open channel flow by the electromagnetic gauge. In *Advances in Hydrometry* (Proceedings of the Exeter Symposium, July 1982). IAHS Publ. No. 134. IAHS.

Heslop, R. T. (1981). The River Wear telemetry scheme and control system. Paper presented at IPHE Symposium on Control and Telemetry, Exeter, January 1981.

IAWPR (1977). *Prog. Water Technol.*, **9** (5/6).

IAWPR (1981). *Water Sci. Technol.*, **13**, (8–12).

Jones, G. L. (1975). A consolidated approach to activated sludge process design: a microbiological view. *Prog. Water Technol.*, **7** (1), 199–208.

Maltz, F. (1970). Emschergenossenschaft und Lipperverband. Personal communication.

Mancy, K. H. and M. Schmid (1970). Dissolved oxygen measurement in natural waters and wastes by pulsed voltmetric membrane electrodes. *Progress Report*. Environmental Chemistry Laboratory, School of Public Health, University of Michigan, Ann Arbor, MI.

Medlock, R. S. (1982). The techniques of flow measurement. *Measurement Control*, **15** (12), 458–462.

Van Duin, P. J. and C. W. de Kreuk (1979). A continuous metal analyser: instrument and application. In *Proceedings of International Conference on Management and Control of Heavy Metals in the Environment*, London, September 1979, pp. 412–415. CEP Consultants, Edinburgh.

White, S. F. (1981). Requirement and application of control and automation systems in waste water treatment — a critical view. *Water Sci. Technol.*, **13** (8), 41–48.

65

Control Strategies for the Activated Sludge Process

G. OLSSON
Lund Institute of Technology, Sweden

65.1 INTRODUCTION

Even though wastewater treatment processes are highly dynamic, most current design and operation methods are based on the assumption of steady state. Disturbances connected to the influent flow rate, concentration and composition have to be met by instrumentation and control, so that operating costs can be reduced and plant reliability can be increased.

The dynamics of a biological wastewater treatment system includes phenomena ranging in time scale from minutes to months, therefore control actions have to be made in relevant time scales. Experiences of process identification of the activated sludge process including aerator and clarifier dynamics will be discussed.

Estimation of organic and nitrogenous load will be discussed. Self-tuning control of the dissolved oxygen concentration can improve plant economy and will also be the basis for the estimation of respiration.

The dynamics of the clarifier has to be understood, since it is closely related to influent pump control. Sludge inventory control by recycling or step feed can be approached in a systematic manner, so that the coupling between aerator and settler dynamics is taken into account.

65.2 DYNAMICS OF WASTEWATER TREATMENT SYSTEMS

There are several inherent dynamical problems in wastewater treatment. It is well known by plant operators that wastewater treatment plants are seldom in steady state. Despite this fact traditional design methods are based on static criteria. Consequently a plant will behave poorly if the variations in influent concentration or flow rate are too large, even if the average values are within the design limits (Olsson, 1977).

Influent disturbances to a wastewater treatment plant may vary significantly, both in amplitude and in frequency. In a concentrated sewer network flow variations are often large, whereas they are attenuated in more widely distributed networks. Rainstorms can cause shock loads that upset the settler and clarifier systems and this can seriously affect the whole plant system. Lower frequency variations, say on a weekly basis, must also be considered. However, a more flexible piping system may often allow the plant to be adjusted to different loading conditions, *e.g.* seasonal variations. During low load periods, for example, some aerators may be shut off, whilst still leaving all settlers in operation. Such a flexibility can both create a more balanced system and decrease the operating costs.

Concentration disturbances vary in amplitude and frequency. In addition to diurnal variations in BOD, several types of additional loads may appear. The operations required to meet these disturbances have to be flexible. One condition demanding such a flexibility is periodic overloading. In addition to causing an increase in effluent BOD it frequently gives rise to bulking sludge and then in turn increases effluent suspended solids, which causes a significant deterioration in performance. The organization of the plant to enable it to cope with such a load change will in general be quite different to that required to deal with a BOD shock load.

The possibility of successful plant control in the presence of significant disturbances is coupled to the probability of detecting them in time. In order to detect a shock load the sampling rate for measurements must be sufficiently high to make sure that the pulse does not pass undetected. Each analytical test has a finite cost. Therefore the cost of measurements at high sampling rates must be directly compared with the cost penalty for not detecting the disturbances. In several cases the situation can be significantly improved if upstream information is obtained (*i.e.* feedforward information). This is particularly useful in dealing with industrial effluents which often give rise to pulselike disturbances. Sometimes the discharge may be buffered in order to arrive at the plant during low load periods. Sometimes an early warning can be utilized to avoid a major plant upset. For example, pretreatment such as neutralization can be performed. Also, step feed or other flow distributions can be automaticallly put into action.

The need for automatic detection of large disturbances is even more significant for small and unmanned plants. These may be designed to operate satisfactorily during normal diurnal variations provided adequate flow balancing is provided. Certain discrete events, however, may upset the process and demand the attention of an operator. Remote supervision by telemetering of essential plant performance data can be very cost-effective in these circumstances since its use enables anticipatory, preventive and planned maintenance of control procedures to be carried out (Briggs and Olsson, 1982).

65.3 ACTIVATED SLUDGE PROCESS DYNAMICS

The activated sludge process is recognized as the most common and major unit process for the reduction of organic waste. Its dynamical properties have been studied extensively during the last decade and several modelling studies have been presented and documented, *e.g.* in three international symposia on instrumentation and control (IAWPR, 1973, 1977, 1981). A comprehensive overview of activated sludge models is found in Dold *et al.*, 1980.

A fully structured model of an activated sludge system is very complex. A large number of substrates and organism species as well as settling and hydraulic properties have to be included. The most important features of a structured model include the degradation of degradable pollutants, containing both organic carbon, phosphorus and nitrogen; cell growth and basal metabolism; the oxygen requirements of the system; the flow regime in the aeration basin; the representation of the performance of the settler and clarifier; and the effect of secondary parameters, such as temperature, pH and toxic or inhibitory substances.

To verify a fully structured model is an awkward task, and only verifications of details of the models have been published, *e.g.* Dold and Marais (1985). Process identification has been a useful tool (see Olsson and Hansson, 1976a; Beck, 1981).

The goal for the reactor operation is of course to degrade the degradable pollutants. However, the operation must be such that organisms with the preferred floc formation are produced, thus giving desired clarification and thickening properties. Otherwise the operation of the system will fail, even if the degradation is efficient.

Research on reactor dynamics has advanced considerably, and the models are getting quite structured and detailed. The reactor models, however, are not matched by a corresponding understanding of settlers. These models are still incomplete and lack a full understanding of the underlying mechanisms. Reliable dynamic models of activated sludge systems cannot be obtained until the interaction between aerator and settler operations is well understood.

65.4 RESPIRATION AND DISSOLVED OXYGEN DYNAMICS AND CONTROL

The DO concentration is an essential variable of the activated sludge process. It has a significant influence on both the plant operation economy and on the biological activity, and consequently on the quality of the effluent water.

65.4.1 Basic DO Dynamics

The DO concentration of an aerator is determined by several physical and biological phenomena such as: hydraulic transport of oxygen in to and out of the aerator; mass transfer of gaseous oxygen into dissolved oxygen; oxygen utilization due to cell synthesis of zoogleal heterotrophic organisms or nitrifying organisms, such as *Nitrosomonas* and *Nitrobacter*; and oxygen utilization due to decay of organisms.

A detailed derivation of the equations can be found in Olsson and Andrews (1978). In a dispersed plug flow reactor the resulting DO dynamics can be described by:

$$\frac{\partial c}{\partial t} = E\frac{\partial^2 c}{\partial z^2} - v\frac{\partial c}{\partial z} + k_La(c_s - c) - R \tag{1}$$

where z = length along the reactor; $c\,(z,t)$ = dissolved oxygen concentration; c_s = dissolved oxygen saturation concentration; k_La = oxygen mass transfer rate; E = dispersion coefficient; v = stream velocity; and R = oxygen uptake rate (respiration rate)

The oxygen uptake rate (R) can be represented by:

$$R = \alpha_1\mu_x(c,s)x_z + \alpha_2\mu_{ns}(c,s_{NH_4})x_{ns} + \alpha_3\mu_{nb}(c,s_{NO_2})x_{nb} + \tag{2}$$
$$\alpha_4 k_{dx}x_z + \alpha_5 k_{dns}x_{ns} + \alpha_6 k_{dnb}x_{nb}$$

where α_i = stoichiometric constants; $\mu_x, \mu_{ns}, \mu_{nb}$ = specific growth rates for zoogleal, *Nitrosomonas* and *Nitrobacter* organisms respectively; k_{dx}, k_{dns}, k_{dnb} = decay rates of zoogleal, *Nitrosomonas* and *Nitrobacter* organisms respectively; x_z, x_{ns}, x_{nb} = concentrations of zoogleal, *Nitrosomonas* and *Nitrobacter* organisms respectively; s = concentration of soluble organic substrate; and s_{NH_4}, s_{NO_2} = concentration of ammonium and nitrite nitrogen respectively. All concentrations are functions of both space and time.

The phenomena influencing the DO concentration have widely different time scales. Consequently DO control actions can be considered in different time scales. In a short time frame (fractions of hours) the air flow rate can counteract disturbances. In a longer time scale (several hours) the air flow distribution along the reactor may be changed thus influencing the activity along the reactor (see Olsson and Andrews, 1978; Olsson and Andrews, 1981a, 1981b). In an even longer time scale the set-point may be varied in order to favour the growth of certain organisms, and consequently influence settleability and clarifying properties of the sludge. For the discussion the DO dynamics for a complete mix reactor is given and (1) is simplified to

$$\frac{dc}{dt} = \frac{Q}{V}[c_{in} - (1 + r)c] + k_La(c_s - c) - R \tag{3}$$

where Q = influent flow rate; V = reactor volume; c_{in} = influent DO concentration; R = return sludge flow rate = rQ.

65.4.2 Oxygen Uptake Rate

The oxygen uptake rate or respiration (R) has been recognized for a long time as a relevant measure of organic load of the activated sludge system (Stenstrom and Andrews, 1979). Knowledge of R along with the measurement of mixed liquor volatile suspended solids defines the value of SCOUR, the specific oxygen uptake rate. To measure or estimate SCOUR is a crucial first step in sludge inventory control. With a respirometer the oxygen uptake rate can be measured directly. However, as part of a dissolved oxygen control system it can be estimated. The obvious way to estimate R on-line is by considering the DO mass balance in the aerator as long as $k_L a$ is known (equations 1 and 3; see Brouzes, 1968). In practice, however, $k_L a$ is not known and may vary gradually with time, so it has to be estimated together with R.

On-line estimation of $k_L a$ is demonstrated and the simultaneous estimation of $k_L a$ and R is discussed. In non-homogeneous reactor systems R can be estimated using the concentration profile of dissolved oxygen (Olsson and Andrews, 1981b).

65.4.3 Relation Between Respiration and Air Flow Rate

The oxygen mass transfer coefficient $k_L a$ is a function of the quality of the wastewater as well as of the air flow rate and the mixing of the water. It depends approximately linearly on the air flow rate (F_{air}). For each aerator an empirical relationship can be established around an operating point:

$$k_L a = a_1 F_{air} + a_2 \tag{4}$$

where the coefficients a_1 and a_2 have to be estimated on-line or off-line.
Assuming that a control system keeps the DO concentration constant, *i.e.*, $c = c_0$, then R is directly related to the air flow rate. Consider the steady state DO mass balance:

$$0 \approx \frac{dc_0}{dt} = \frac{Q}{V}(c_{in} - (1 + r)c_0) + (a_1 F_{air} + a_2)(c_s - c_0) - R \tag{5}$$

The first term of the right hand side can often be neglected in comparison with the other terms. Consequently, if F_{air} does not saturate, then

$$R = b_1 F_{air} + b_2 \tag{6}$$

where $b_1 = a_1(c_s - c_0)$; $b_2 = a_2(c_s - c_0)$.

65.4.4 Identification of the Oxygen Transfer Rate

The oxygen mass transfer rate can be estimated on-line or off-line by identification of the DO concentration dynamics. This is made possible by one simple observation. The oxygen uptake rate is independent of the DO concentration as long as the DO concentration is large enough. If R is constant, then a change of the air flow rate will create an approximate first order system response of the DO concentration. The time constant of the dynamics is given by the linearized DO mass balance (equations 3 and 4)

$$\frac{d\Delta c}{dt} = -\frac{Q}{V}(1 + r)\Delta c - (a_1 F_{air} + a_2)_0 \Delta c + (c_s - c_0)a_1 \Delta F_{air} \tag{7}$$

where subscript 0 denotes steady state values. The two unknown parameters a_1 and a_2 are readily identified. When a linearized model is used, the validity of the linearization is limited if the air flow changes are large.

The oxygen mass transfer rate was identified, using maximum likelihood identification in a series of experiments reported in Olsson and Hansson (1976a, 1976b). Other methods are reported by Holmberg (1981). The experiences gained from these experiments were: $k_L a$ may be largely different in different parts of an aerator; and $k_L a$ varied only little with time, even over a period of one year. Figure 1 shows the kind of estimate that can be achieved.

It is interesting to update $k_L a$ and R recursively. The reason is that both parameters are time varying. Since both parameters are unknown they have to be estimated simultaneously and under closed loop DO control. This is not a trivial task. A change in any of the parameters will be noticed *via* the DO concentration. Consequently the parameter influences cannot be distinguished from each other without further information.

It is known that $k_L a$ varies slowly with time and it can be assumed to be constant on an hourly time scale. On the other hand, R varies more rapidly. It is the R variations that are the main reason for the control of the air flow.

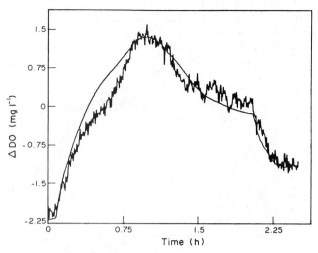

Figure 1 Off-line estimation of $k_L a$ in a linearized model. Model and experimental DO concentrations are shown

The trick is now to separate the estimations of the two parameters and the following procedure is proposed. If R is constant, then $k_L a$ can be estimated. Usually such a period of at least one hour can be found during the day in a typical plant. It can be verified afterwards whether or not R was constant. Then, under closed loop DO control the air flow rate would be almost constant. To estimate $k_L a$ with a constant R the air flow will be disturbed manually as indicated in the previous section. The a_1 and a_2 parameters are then estimated according to equation (7). A typical $k_L a$ estimation may take place in an hour, since the time constant is of the order 10–20 minutes. Given $k_L a$, R can be directly calculated from the DO mass balance equation.

Experimental data from the Käppala treatment plant at Lidingö outside Stockholm have been used to update $k_L a$ recursively. An extended least squares method has been used to produce the result in Figure 2.

New methods for the simultaneous estimation of $k_L a$ and R have been developed recently (Holmberg and Olsson, 1985).

65.4.5 Estimation of the Oxygen Uptake Rate Using Concentration Profiles

The DO profile has been recognized to be a good measure of the organic load to a non-homogeneous reactor system (see Olsson and Andrews, 1978, 1981b). Figure 3 indicates a steady state profile for a plug flow reactor with uniform air flow distribution along the tank, where it is assumed that only organic removal takes place. Close to the inlet the substrate concentration is high. Consequently the growth is large, and the DO concentration will be correspondingly low. As the growth goes to completion the DO uptake rate will decrease, which allows the DO concentration to assume the typical profile shown.

Consider one subreactor, i, and the estimation of the local value of the respiration. The DO mass balance of the ith subreactor is:

$$\frac{dc_i}{dt} = \frac{Q}{V_i}(c_{i-1} - c_i) + k_L a(c_s - c_i) - R_i \tag{8}$$

If the DO concentrations of two neighbouring subreactors and the oxygen transfer rate are known, then the local respiration is estimated in a similar manner to that of the complete mix reactor.

For a system of reactors in series the knowledge of the DO concentration profile along the system can be used favourably. Consider the DO concentration difference between two consecutive subreactors:

$$\delta c_i = c_i - c_{i-1} \tag{9}$$

Subtract the DO balance equations from reactors i and $(i-1)$, and assume the volumes equal, then:

$$\frac{d\delta c_i}{dt} = \frac{Q}{V}(\delta c_{i-1} - \delta c_i) - k_L a(\delta c_i) - \delta R_i \tag{10}$$

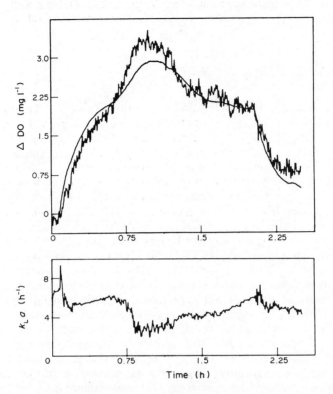

Figure 2   On-line estimation of $k_L a$ in a linearized model of DO dynamics. Model and experimental DO are shown in the upper curve and the parameter estimate in the lower curve

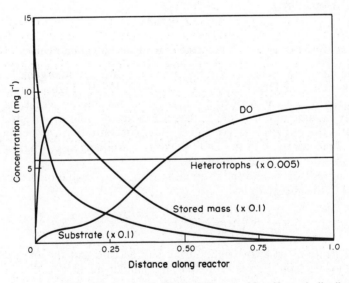

Figure 3   A typical DO concentration profile in an aerator with uniform air distribution

Ideally, if the growth has been completed, the difference in respiration, δR_i, is small. This is directly reflected in the slope of the DO concentration, according to equation (10); the slope ideally would approach a horizontal value towards the outlet.

Consider the outlet region. First, we assume that only organic removal takes place. At the outlet the substrate concentration (s_i) is small, which makes the growth first order:

$$\mu_{xi} x_{zi} = \text{constant} \times s_i = \alpha s_i \qquad (11)$$

Moreover, assume that there is no DO limitation at the outlet, and the sludge concentration x along the tank is constant, *i.e.* $x_i = x_{i-1}$. Assume steady state. From equations (10) and (11) it is found that

$$0 = \frac{Q}{V}(\delta c_{i-1} - \delta c_i) - k_L a(\delta c_i) + \alpha(s_{i-1} - s_i) \tag{12}$$

The first term in equation (12) is negligible compared to the second one. The equation thus shows that if the DO slope is small close to the outlet, then the growth is close to completion. The fact that the DO slope is proportional to the substrate utilization rate along the reactor has to be understood as an approximation. Equation (12) can only be used as a qualitative indicator if the reaction has gone to completion and not as an estimator of the reaction rate (Olsson and Andrews, 1981b). The reason is that the slope can only be calculated with a limited accuracy. Close to the outlet, when it is small, the relative accuracy is of course poor.

The DO profile can, however, be used for estimation of the reaction rate at another location of the reactor, close to the breakpoint of the DO concentration profile towards the centre of the reactor. Here several interesting features can be observed. Simple analysis shows that the DO sensitivity for load changes has its maximum close to this location (Olsson and Andrews, 1978). Moreover it is reasonable to assume that there is no DO limitation for the growth. The growth is significantly larger than the decay component.

Consider the slope of the respiration at this location. It is assumed that the substrate has been reduced sufficiently to give a first order growth. This gives

$$\delta R_i = k_1[\mu_{xi} x_i - \mu_{x,i-1} x_{i-1}] + k_2 k_{dx}[x_i - x_{i-1}]$$

since $x_i \approx x_{i-1}$ the last term can be neglected and

$$\delta R_i = \text{constant} \times x_i(s_i - s_{i-1}) \tag{13}$$

The DO slope is seen to be proportional to the substrate slope, which indicates the growth. As the slopes are larger in this part of the aerator the relative accuracy of the calculation is more satisfactory.

In a fully nitrifying reactor the profile looks similar to the previous description. The components that make up the profile are, however, quite different. The fact that heterotrophic and nitrogenous organism growth rates are very different is used to distinguish between the phenomena. If nitrification occurs, the organic removal is completed long before the nitrogenous growth, *i.e.* already in the first part of the reactor. The characteristic slope towards the outlet therefore is determined only by the completion of the *Nitrosomonas* and the *Nitrobacter* growths (see Olsson, 1980). In a system with uniform air flow distribution the slope of the DO is a simple indicator of the growth. If the air flow is non-uniform, then the slope of the respiration profile has to be used instead. The analysis is extended in a straight-forward manner.

65.4.6 Dissolved Oxygen Control

The control of DO as a physical variable does not require any in-depth knowledge of the microbial dynamics. Naturally the proper actuators and sensors are assumed to be available. Despite the relatively simple dynamics of the DO the control task may suffer serious difficulties. An illustrative study is documented in Wells (1979) and shows that digital control is superior to manual control, from both an economic and a qualitative point of view, mainly because a digital controller can compensate for time delays of the system.

The problem of finding the right DO set-point has been discussed elsewhere (Olsson, 1979; Olsson *et al.*, 1985). The traditional assumption of a constant DO set-point may not be valid. Instead there are indications that some periodic set-point change may influence the process favourably. One example is the use of anoxic zones in the inlet area of the aerator (Dold *et al.*, 1980; Marais and Ekama, 1983). Nitrates can be used as a replacement for oxygen. This contributes not only to energy saving but to a more favourable water quality.

There may be several kinds of disturbances to the DO concentration, such as hydraulic disturbance of the influent wastewater (Q) or the recycle flow rate (r) (min and h); step feed changes influencing the activity distribution along the reactor (h); organic (*s*) or ammonia (s_{NH_4}) disturbances (min and h); toxic substances or changing environmental conditions. If a toxic material enters the reactor, organisms will be killed. This is modelled by increasing the k_d coefficients. A pH change may temporarily change the maximum specific growth rates (min or h); temperature changes are mostly noted as seasonal variations. Several parameters are affected by temperature, such as $k_L a$, c_s, growth rates and decay rates, particularly for nitrification. Recycle streams from sludge handling units also influence the DO concentration.

The discussion on DO concentration profiles has shown that DO control can not be based on only one DO sensor at the outlet. Two sensors are needed not only to find a proper DO setpoint, but also to avoid process failures. An operational experience can illustrate this. A conventional aerator was controlled based on one sensor at the outlet. Due to clogging of the diffusers close to the inlet, more air was forced into the tail end area of the reactor. This increased the DO at the outlet, and consequently the controller decreased the total air flow. This made the clogging at the head end even worse. Thus the control system contributed to making the situation worse.

Conventional PID controllers have been used with favourable results in several plants for DO control. Time delays will make analog control less successful, and digital control can compensate for the dead-times. However, since parameters like $k_L a$ vary with time, the tuning of the controller may have to be changed regularly. That is the first reason for using self-tuning controllers for the DO control (Åström, 1980). The second reason has already been discussed. The estimation of the $k_L a$ and respiration rate will be made as part of the self-tuning control of the DO. The value of these estimates can be used for other loops of the sludge inventory control, discussed later in this chapter. Experiments are being conducted at the Käppala wastewater treatment plant to control the DO with self-tuning controllers. Some results have been published in Olsson *et al.* (1985).

65.5 CLARIFIER DYNAMICS

The final settler of an activated sludge plant is crucial for the whole operation of the system. Plant operators are well aware of the problems created by hydraulic shocks to the clarifier, but the modelling of its dynamical properties is still not highly developed.

In many plants hydraulic disturbances are created internally. If the primary pumping is not smooth, *i.e.* if pumps are switched onto the system in too large steps, the hydraulic wave is propagated through the plant. This often creates a large overshoot of suspended solids concentration in the effluent. It will take several hours before the effluent concentration will decrease to its normal value again. In the meantime a large amount of sludge has been lost from the system. This leads to not only a deterioration of the effluent quality but also a loss of viable bacteria.

In cooperation with the Wastewater Technology Centre, Burlington, Ontario, Canada, a series of dynamic experiments have been performed in a half scale activated sludge system (Olsson and Chapman, 1985). From the experiments dynamic models have been derived that describe the relation between the clarifier influent load and effluent suspended solids concentration. The models are generalizations of the empirical static relations derived by Pflanz (1969).

In the work it was attempted to obtain dynamic models with a minimum number of parameters, but still with a physical interpretation. All observed phenomena of the clarifier can be included in second order dynamic models. There is one model type for flow rate increases and another one for flow rate decreases. The reason is connected with the turbulence created by pumping changes. The turbulence can be influenced directly only at flow rate increases.

Figure 4 demonstrates a typical experiment showing the clarifier influent flow rate and the effluent suspended solids concentration. A linear model, even of high order, gives quite a poor adjustment to the experiment. Instead one model for flow increases and another one for flow decreases are identified. They are combined, and can give a satisfactory adjustment. Note that the effluent suspended solids concentration depends not only on the flow rate but also on its rate of change.

A clarifier typically demonstrates time-variable behaviour. Thus not even very complex models are valid for a long time. It is natural to update the clarifier parameters recursively, based on measurements of the effluent suspended solids concentration and the influent load. From an operational point of view parameter changes rather than effluent concentration changes are relevant indicators of quality changes of the sludge.

One apparent way to influence the clarifier performance on a short time scale is by primary

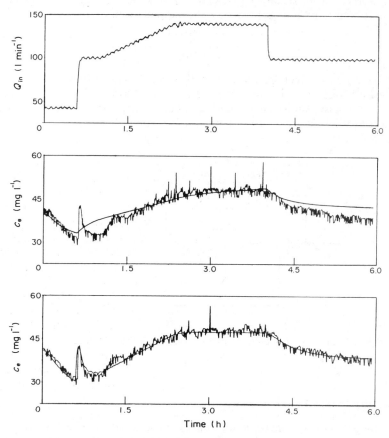

Figure 4 Dynamical experiment of a secondary settler. The upper curve is the influent rate. The middle curves show one time-invariant linear model and the experimental output. The lower curve shows a simulation of combined models. The experiment was performed at the Wastewater Technology Centre, Burlington, Ontario, Canada

pumping control. Typically the total water volume to be pumped for the next few hours may be predicted. The clarifier model can be used to optimize the pumping. The criterion is that the total amount of suspension in the effluent should be minimized. Work is in progress to derive such control schemes.

65.6 SLUDGE INVENTORY CONTROL

There are basically three control variables for the sludge inventory in an activated sludge system. The waste activated sludge (WAS) flow rate controls the total sludge mass in the system. The sludge distribution within the system is then controlled with the step feed flow distribution or the return activated sludge (RAS) flow rate. The former can dynamically redistribute the sludge within the aerator while the latter can shuffle sludge between the settler and the aerator.

65.6.1 Waste Sludge Flow Rate Control

The WAS flow rate is used to control the average solids inventory in the aerator–settler system. With the sludge wastage the sludge retention time (SRT) can be kept at a desired level. The SRT does not account for the dynamics of the system. If the total solids mass balance is set up for the system, then the dynamical changes of the total sludge mass due to the WAS flow rate can be calculated. Analysis shows that the approximate time constant of the total sludge system is about the same as the SRT (see Olsson, 1983). This verifies the practical experience that a change from one SRT level to another will take several sludge ages. Traditionally the wastage rate is calculated through a steady-state calculus. Instead, based on estimates of the time-varying total sludge content, a more relevant feedforward–feedback scheme of WAS control can be defined (Figure 5). The feedforward link is a quasi-steady-state calculation:

$$F_w = \left(\frac{\bar{X}_T}{(\theta_x)_d} - \bar{F}\bar{x}_e \right)(\bar{x}_w)^{-1} \tag{14}$$

where the bars indicate average values and F_w = waste sludge flow rate; X_T = total sludge mass; $(\theta_x)_d$ = desired SRT; F = influent flow rate; x_e = clarifier effluent suspended solids concentration; and x_w = waste sludge suspended solids concentration. The feedback can be realized by a traditional controller. More details are given in Olsson (1983).

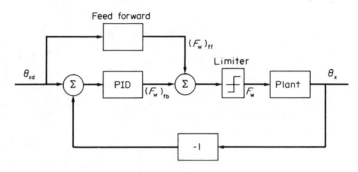

Figure 5   Block diagram of the feedforward–feedback structure of the waste sludge flow control

65.6.2   Return Sludge Flow Control

Return sludge control is the topic for many contradictory conclusions. In practice the two most common controls are either constant flow rate or influent flow proportional control. None of these control schemes take into consideration concentration variations in the aerator, the settler or changing settling conditions. The main difficulties may be summarized as follows. An increase in recycle flow rate may increase the sludge content of the aerator, and then the dissolved BOD may be decreased due to more activity. The increased hydraulic load, however, may influence the clarifier efficiency, so that what is gained in dissolved BOD is lost by an increase in particulate BOD in the effluent.

A more systematic approach to return sludge control has been made. Here only some main ideas are given (see Olsson, 1983 for details).

The recycle flow rate can only redistribute the sludge content between the settler and the aerator, but cannot influence the average sludge mass of the system.

First the criterion for control is defined. It is suggested that the recycle should be controlled so as to maintain a given specific oxygen uptake rate setpoint ($SCOUR_{sp}$). This in turn will define a desired sludge mass in the aerator (X_{ad}):

$$X_{ad} = \frac{R}{V\,(SCOUR_{sp})} \tag{15}$$

where V is the aerator volume and R the estimated value of the oxygen uptake rate. A solids mass balance around the aerator will give the feedforward recycle flow rate $(F_r)_{ff}$. The recycle is also controlled by a feedback loop, as shown in Figure 6 where the correction term $(F_r)_{fb}$ is calculated.

The constraints in recycle control are crucial and not trivial. They are summarized as follows in terms of limits of the total RAS flow F_r. (i) sludge blanket constraints: to keep the sludge blanket between limits, $\alpha_1 < F_r < \beta_1$; (ii) sludge dilution: a too high recycle rate will dilute the sludge. The dry mass flow rate has to increase for a hydraulic increase, *i.e.*

$$\Delta(F_r x_r)/\Delta F_r > 0$$

(iii) dissolved oxygen depletion in the settler or storage: the hold-up time of sludge in the settler must be small enough, *i.e.* $F_r > \beta_2$; and (iv) hydraulic load on clarifier: any increase in the hydraulic load to the clarifier may cause losses in suspended solids, *i.e.* $F_r < \beta_3$.

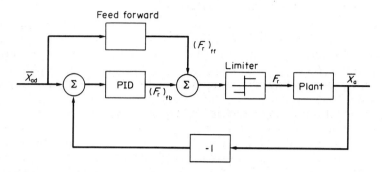

Figure 6 Block diagram of the feedforward–feedback structure of return sludge control

To check the constraints the sludge inventory has to be checked and suspended solids concentrations have to be known in the recycle stream, in the effluent, and in the aerator. Moreover the sludge blanket has to be known at least approximately.

The proposed control for the recycle flow rate can be seen as the generalization of previously suggested schemes. Special cases are found in Petersack and Smith (1975), Tanuma (1980) and Ching (1980).

There may be a possibility that due to the combination of constraints there is no control authority left. Therefore the potential for recycle control may be quite limited. This has been verified by simulation of a number of control schemes (Holmberg, 1984).

65.6.3 Sludge Distribution Control by Step Feed Flow Rate

The use of step feed for dynamical control is seldom practised, despite its many interesting possibilities. Its use should be carefully evaluated. Basically, what happens during step feed control is that sludge is transported to different parts of the aerator. This gives a control to take care of large hydraulic or organic shock loads or toxic spills. It can also be used to prevent process failures due to poorly settling sludge or bulking sludge. This was observed a long time ago by Torpey (1948).

As a criterion for choosing the step feed control the SCOUR can be used. This has been practised on a laboratory scale (Yust *et al.*, 1981) and further consequences are explored in Olsson (1983).

A criterion based on a desired SCOUR value will give satisfactory step feed control for toxic and non-toxic disturbances. For a special case with two influent rate ports an analysis is presented in Olsson (1983) to show how a combination of feedforward and feedback step feed control can be implemented. The time scale of step feed control is of the order of several hours. This means that step feed control can often be implemented manually, at least for feasibility studies. However, the flow rates in different influent ports are seldom measured automatically. This of course makes an accurate step feed control difficult. The control, however, can be performed sometimes with quite gross measures of the flow rates (*i.e.* zero, half and full) giving reasonable results.

65.7 CONTROL BY OPERATIONAL STATES

The lack of adequate sensors is and will be a major problem in wastewater treatment control. Even if new sensors are developed still only a small fraction of all components can be measured. Moreover, some measurements are very difficult to perform automatically. No sensor can replace an operator's judgements of patterns, odours or colours. Therefore any control system should be constructed in such a way that maximum use is made of automatic measurements. Any additional information from laboratory tests or manual observations can be added in order to make more advanced or more relevant control actions. This is the background to 'operational state' control.

Control actions in a complex system like an activated sludge plant will in fact be determined in two steps. In the first one the combination of control variables based on qualitative measures of the operational conditions of the process are determined. In the second step the actual quantitative control actions are calculated. The idea of such a 'human operator' model was suggested independently by Beck (1977) and Gillblad and Olsson (1977), and was further explored by Tong *et al.* (1980).

To give an example, completely different operational conditions can be characterized from influent load disturbances. Both decreasing organic load and a toxic load to a plant will result in an increasing DO concentration. In both cases a decreasing specific oxygen uptake rate (SCOUR) has occurred. As the toxic pulse cannot be counteracted by simply increasing the air flow rate the DO measurement typically has to be combined with additional information before a proper control action can be taken.

An operational state is a detectable process condition. The source of the information is irrelevant. It may come from an on-line instrument, from a laboratory test or from operator observations. It is of course of particular interest to examine which operational conditions are automatically observable from on-line instruments. Some examples of operational states are: poor sedimentation; too low SCOUR in the aerator; full/empty sludge buffer; and toxic spill.

Thus the operational states are discrete in nature. There may be just a few of them defined. As the knowledge of a specific plant is increased the states may be more refined to allow a more sophisticated detection. The determination of operational states is made up from logical decisions, and at this level there may be a sequence table of questions and answers of the type yes/no. Some of the answers are given automatically by the on-line instruments, others have to be given by the operator directly from the console. In case no answer is given the computer has to act in the best possible way, given the available information. This is a decision under uncertainty.

The concept of operational state gives a good method of communication between the operator and the process in a natural language. The operational state is diagnostic information to the operator. Because of the interaction between the operator and the computer the human operator can implement his own experience in the programs. The operational states aim to reflect the 'human operator model' or the operator's mental model of the plant behaviour.

The detection of an operational state only gives qualitative information, *e.g.* which control variables to use and which set-points to choose. On the next level there are the actual controllers, like DO or sludge inventory controllers. They will decide by how much actual control variables are going to be changed.

65.8 CONCLUSIONS

Several factors have contributed to making control of wastewater treatment plants more interesting. On the one hand there are several factors that will force more control into treatment plants. The increasing operating costs due to energy, chemicals and personnel motivate more automation. The increasing quality demands on effluent water will also motivate better control and operation. On the other hand there are several promising developments that will stimulate better operation. The process knowledge is increasing significantly. Methods in control, such as digital control methods, process identification, self-tuning regulators, *etc.*, are applicable to this field. Computers are becoming more available at plants, thus increasing the flexibility. More sensors have been developed and the general competence of operators and designers in the areas of dynamics, computers and control will be gradually improved.

Some control schemes have been illustrated in this chapter. It is not intended to be a complete survey of possible control schemes in a full scale plant. An important aspect that is not discussed is the coupling between individual sub-processes. Only the aerator–settler coupling has been mentioned. Other important couplings are those between sludge handling and the water treatment, different recirculation patterns of, for example, chemical sludge or reject water.

The coupling between the sewer network and the plant has to be better understood and operated in order to compensate for perhaps the most difficult disturbances of all, hydraulic changes. Finally, the operation of a plant naturally affects the receiving water. Consequently the quality of the recipient ought to govern the specifications for the plant operation.

ACKNOWLEDGEMENTS

Several people have inspired the ideas presented here. A long cooperation with Professor John F. Andrews, Rice University, Houston, Texas has had great influence on the author. The experiments on the clarifier were performed by D. Chapman, Wastewater Technology Centre,

Burlington, Ontario. Some of the simulations and identifications were made by A. Wikström, Lund. The work has been supported by the Swedish Board for Technical Development under contract 82-3352.

65.9 REFERENCES

Åström, K. J. (1980). Self-tuning regulators—Design principles and applications. In *Applications of Adaptive Control*, ed. K. S. Narendra and R. V. Monopoli. Academic, New York.

Beck, M. B. (1977). Modeling and control in practice. *Prog. Water Technol.*, **9**, 557–564.

Beck, M. B. (1981). Operational estimation and prediction of nitrification dynamics in the activated sludge process. *Water Res.*, **15**, 1313–1330.

Briggs R. and G. Olsson (1982). On-line measurement, data processing and digital control. *Symposium on Processing in the Water Industry, Bournemouth, England. Journal of the Institute of Water Pollution Control*.

Brouzes, P. (1968). Automated activated sludge plants with respiratory metabolism control. *4th International Conference on Water Pollution Control Research*. IAWPRC, London.

Ching, D. L. (1980). Process performance based on control strategies. *Adv. Instrum.*, **35**, 427–452.

Dold, P. and G. von R. Marais (1985). Evaluation of the general activated sludge model incorporating modifications proposed by the IAWPRC task group. *Specialized seminar, Modelling of Biological wastewater treatment, Copenhagen*. IAWPRC, London.

Dold, P. L., G. A. Ekama and G. von R. Marais (1980). A general model for the activated sludge process. *Prog. Water Technol.*, **12**, 47–77.

Gillblad, T. and G. Olsson (1977). Computer control of a medium-sized activated sludge plant. *Prog. Water Technol.*, **9**, 427–433.

Holmberg, A. (1981). Microprocess-based estimation of oxygen utilization in the activated sludge wastewater treatment process. *Int. J. Syst. Sci.*, **12**, 703–718.

Holmberg, U. (1984). Simulation of the dynamics of activated sludge systems (in Swedish), MS thesis, Report TFRT 5318, Dept. of Automatic Control, Lund Inst. of Technology, Lund, Sweden.

Holmberg, U. and G. Olsson (1985). Simultaneous on-line estimation of oxygen transfer rate and respiration. Paper, *1st IFAC Symposium on Modelling and Control of Biotechnological Processes, Noordwijkerhout, Holland*.

IAWPR (1973). International workshop on instrumentation control and automation for wastewater treatment systems, London and Paris. *Prog. Water Technol.*, **6**.

IAWPR (1977). International workshop on instrumentation and control for water and wastewater treatment and transport systems, London and Stockholm. *Prog. Water Technol.*, **9** (5/6).

IAWPR (1981). International workshop on practical experience of control and automation in wastewater treatment and water resources management, Munich and Rome. *Water Sci. Technol.*, **13** (8–12).

Olsson, G. (1977). State of the art in sewage treatment plant control. *AIChE Symp Ser.*, **72**, 52–76.

Olsson, G. (1979). Automatic control in wastewater treatment plats. Invited paper, International Environment Colloquium, Liege, Belgium, May 1979.

Olsson, G. (1980). Modeling and control of the activated sludge process. Invited paper, *Am. Chem. Soc. Symp. Ser.*, **124**, 367–382.

Olsson, G. (1983). A unified approach to sludge inventory control. Dept. of Automatic Control, Lund Inst. of Technology, Lund, Sweden.

Olsson, G. and J. F. Andrews (1978). The dissolved oxygen profile — a valuable tool for the control of the activated sludge process. *Water Res.*, **12**, 985–1004.

Olsson, G. and J. F. Andrews (1981a). Dissolved oxygen control in the activated sludge process. *Water Sci. Technol.*, **13**, 341–347.

Olsson, G. and J. F. Andrews (1981b). Estimation of loading changes for the nitrifying activated sludge process. *Proc. IFAC World Congress, Kyoto*. Pergamon Press, Oxford.

Olsson, G. and O. Hansson (1976a). Stochastic modeling and computer control of a full scale wastewater treatment plant. *Proceedings of the Symposium on Instrument Models in Air and Water Pollution*. The Institute of Measurement and Control, London.

Olsson, G. and O. Hansson (1976b). Modeling and identification of an activated sludge process. *Proceedings of the 4th IFAC Symposium on Identification and System Parameter Estimation, Tbilisi, USSR*. Pergamon Press, Oxford.

Olsson, G. and D. Chapman (1985). Modelling the dynamics of clarifier behaviour in activated sludge systems. *Instrumentation and control of water and wastewater treatment and transport systems*, IAWPRC, 405–412. Pergamon Press, Oxford.

Olsson, G., L. Rundqwist, L. Eriksson and L. Hall (1985). Self tuning control of the dissolved oxygen concentration in activated sludge systems. *Instrumentation and control of water and wastewater treatment and transport systems*, IAWPRC, 473–480. Pergamon Press, Oxford.

Petersack, J. and R. Smith (1975). Advanced automatic control strategies for the activated sludge treatment process, EPA Technical Series EPA-670/2-75-039.

Pflanz, P. (1969). Performance of activated sludge secondary sedimentation basins. In *Advances in Water Pollution Research*, ed. S. H. Jenkins. Pergamon, Oxford.

Srenstrom, M. and J. F. Andrews (1979). Real-time control of activated sludge process. *J. Environ. Eng. Div. (Am. Soc. Civ. Eng.)*, **EE2**, 245–260.

Tanuma, M. (1980). Water quality management in a wastewater treatment. *Proceedings of a Task Force in Real Time Water Quality Management*, ed. M. B. Beck. IIASA, Laxenburg, Austria.

Tong, R. M., M. B. Beck and A. Latten (1980). Fuzzy control of the activated sludge wastewater treatment process. *Automatica*, **16**, 695–701.

Torpey, W. N. (1948). Practical results of step aeration. *Sewage Works J.*, **20**, 781–788.

Wells, C. H. (1979). Computer control of fully nitrifying activated sludge processes. *Instrum. Technol.*, April, 32–36.

Yust, L. J., J. P. Stephenson and K. L Murphy (1981). Dynamic step feed control for organic carbon removal in a suspended growth system. *Water Sci. Technol.*, **13**, 341.

66

Computer Implementation of Control and Monitoring of Waste Treatment

G. L. JONES
Water Research Centre, Stevenage, Herts., UK

66.1 INTRODUCTION

Treatment of wastewaters in purpose built biological reactors has been known for about 100 years. In that time two basic types of reactor have been used. The earliest was the fixed-film reactor, typified by the percolating filter, and later the stirred-tank reactor was introduced (Ardern and Lockett, 1914). The fixed-film reactor has recently been subject to considerable development with the introduction of the fluidized bed and other cell immobilization techniques. Prior to these changes there was little that could be done with the typical percolating filter installation to modify the performance. Indeed percolating filters have a reputation for robustness which could well be a consequence of the fact that operating conditions are virtually fixed at the design stage and it is extremely difficult for the operator to make any changes, particularly to the quantity of active biomass present in the system.

For the activated sludge process there has been little change in operational practice since its inception (Jones, 1976). Modern examples, like the ICI Deep Shaft, are recognizably activated sludge systems. The constraints on operation are still the requirement to reduce the concentration of biochemical oxygen demand (BOD), whether it be exerted by soluble biochemically oxidizable material, or by the respiration of freely suspended bacteria and protozoa. To achieve satisfactory standards the solutes contained within the wastewater must be utilized with the minimum conversion to biomass and any biomass formed must be readily separable from the bulk of the liquid waste. Essentially wastewater treatment is an exercise in concentration, and the crucial stage in the activated sludge system is the process of sedimentation. Without an adequate sedimentation stage neither the solids nor the solutes in the final effluent can be reduced to a sufficiently low level to make the process worth carrying out.

Early efforts at automatic control of activated sludge were directed towards controlling the level of solids within the aeration stage. Mohlmann (1917) showed the dependence of sludge yield on the rate at which sludge was wasted, solids production increasing as the rate of wastage was increased. Most plants are operated on the basis of maintaining a given level of suspended solids within the aeration stage, and the wastage rate is adjusted, if possible, to provide the required concentration.

Garrett (1958), using the work of Monod (1942) as a theoretical basis, demonstrated that automatic control of the sludge level was possible if biomass was wasted at a fixed and constant rate from the aeration tank. This technique ensures that the mean value of the specific growth rate of the solids, and hence the concentration of the substrates, is affected only by the rate of flow of the waste from the aeration tank and by the volume of the aeration tank, provided that the capacity of the final sedimentation tank is not exceeded (Jones, 1973).

It has been argued that the performance of an activated sludge plant may be controlled by adjusting the concentration of solids within the aeration tank so as to maintain a constant ratio between the solids and the available substrate. Many models have been produced which propose operational strategies to achieve this objective. Most suggest controlling the rate of recycle from the sedimentation tank specifically to counter the variability in effluent quality that occurs because of diurnal variations in flow and strength of the incoming waste. While theoretically this may be possible, many such treatments ignore the hydraulic effects of the final sedimentation tank, which, under certain circumstances, may cause performance to deteriorate. The limitations of this technique were demonstrated on both theoretical (Jones, 1977) and on practical grounds (Cashion *et al.*, 1977), while from a survey of the literature, and from simulations, Andrews and Stenstrom (1979) concluded that for real-time control of a treatment plant such methods were of little use.

A major effort has been made in recent years to improve not only the transfer of oxygen into wastewater treatment systems but also to control the level of oxygen within the aeration stage so as to make maximum use of that oxygen. Supply of oxygen to the microorganisms responsible for the purification process is one of the critical conditions for the performance of a plant. Increasing the rate of supply permits the maintenance of a higher concentration of microorganisms, thereby reducing the size of plant needed to achieve a given degree of purification. An early, quite simple method of controlling the level of dissolved oxygen in a pilot plant by switching between two alternate rates of oxygen transfer was described by Briggs *et al.* (1967). With the advent of small computers more sophisticated methods of controlling oxygen supply are now being investigated. For example, programmable logic controllers are in use on a full scale plant to investigate the extent to which energy savings may be made by controlling the level of dissolved oxygen (Chambers, 1982), while Olsson and Andrews (1978) have for a number of years been investigating sophisticated methods for using measurements of the respiration rate of the microorganisms within the reactor to provide inputs to an algorithm developed to maintain the most economic use of oxygen.

66.2 OPERATIONAL CONSTRAINTS

Of the fermentation-based industries that of wastewater treatment is unique in that in most instances, with the possible exception of some industrial waste treatment systems, the operator has no control over either the quantity or the quality of the feedstock to the plant. In addition, particularly in the case of sewage treatment, the concentration of solutes in the waste will be low. Again, unlike other fermentation processes, there is no end product; the objective is to remove as much biochemically oxidizable material as possible from the waste, and at the same time to minimize the conversion of that material to biomass. In that way the problems associated with the disposal of surplus solids are limited.

In order to effect some control over the performance of the plant, the operator must be able to adjust one or more of the operating parameters and have enough information on which to base his decision. For the activated sludge system it may be possible to adjust the rate of recycle of the suspended solids from the sedimentation tank to the aeration tank. Further adjustments to conditions may be possible by altering the rate of wastage of sludge from the system and by changes in the rate of aeration.

Fixed-film systems present different problems. The classic percolating filter type of plant, which is used extensively in sewage treatment, allows for little or even no automatic adjustment. More recent developments of fixed-film technology, *e.g.* the biological fluidized bed, do permit some degree of control. The rates of recycle and upflow velocities through the bed are controllable, and it should also be possible to achieve some control of the amount of active biomass present (Atkinson, 1974), which is by no means true of the conventional percolating filter.

66.3 MODELS OF TREATMENT SYSTEMS

Models of wastewater treatment systems have largely been confined to the activated sludge type. Such models are simpler to develop than those for fixed-film reactors, since it is not essential to take into account diffusion of oxygen and solutes through the biological film in order to construct a usable model.

Kinetic models of activated sludge are of two main types. The most used, and arguably the most successful, have been the pragmatic models which described the overall performance of a plant in terms of a non-specific parameter such as the BOD. In most instances these models have been based on first-order kinetics which give an adequate description of the removal of BOD (Tucek *et al.*, 1971), and have been of most use in the design stages for activated sludge systems. However, such models are not well suited to control purposes, whether for theoretical study or for actual control.

With the development of the chemostat and the general acceptance of the work of Monod (1942) on the kinetics of the growth of bacteria, there has been an explosion in the development of deterministic models of the activated sludge process. The first successful example of such a model was that described by Downing *et al.* (1964) for the case of nitrification. Many of these models, particularly in the early stages of their development, concerned themselves solely with the biological processes occurring within the aeration tank. When used to assess possible strategies in the control of the activated sludge process they could produce misleading results, since no account was taken of the hydraulic effects of the final sedimentation tank on the overall performance of the system. The models also tended to overestimate the concentration of microorganisms within the system, which led to some unreal estimates of the kinetic constants for substrate removal (Jones, 1973). This problem was overcome by the development of structured models, which paved the way for far more realistic simulations of the process.

'Black-box' or statistical models have also been produced. Typical of these is the model developed by Beck (1981) for the prediction of nitrification at the Whittlingham sewage works in Norwich. While not of immediate use in increasing the understanding of the fundamental principles of wastewater treatment, such models have a role to play in any proposed control system.

66.3.1 Use of Mathematical Models

To date, the major use of mathematical models of treatment plants has been for the design, *e.g.* STOM (WRC, 1981), and for guidance as to the most suitable method of operating a treatment plant. In this way a considerable number of attempts at defining operational strategies have been made, although not all of these attempts may be considered to have been an unqualified success (Andrews and Stenstrom, 1979).

At all times there are two factors to be considered when making use of a mathematical model of the activated sludge process. Firstly the biology of the process must be considered. The objective of wastewater treatment is to reduce the BOD of the wastewater to such a level that the effect it has on the receiving environment is negligible. By definition this means that the environmental conditions within the treatment plant will be unsuited to the vigorous support of bacterial growth. It follows therefore that the population of microorganisms within a system producing a high quality effluent will be severely substrate-limited, and will be of low viability (Pike and Carrington, 1972).

The relatively poor physiological state of the microorganisms within the system has serious consequences for the behaviour of the plant when it is presented with changes in operating conditions. Because of the low percentage viability of the population, any growth response will be both slow and small in comparison with the total amount of solids present. Thus an increase in load on the plant will stimulate neither a large, nor even a rapid, increase in growth. Since the organisms present in the plant are substrate-limited, an increase in the concentration of their substrate should initiate an immediate increase in activity as predicted by Michaelis–Menten kinetics.

The response of the system to an increase in load brought about by an increase in the flow rate may therefore be expected to produce more marked fluctuations in the quality of the final effluent than if the increased load were due to an increase in concentration. Figure 1 shows the changes to be expected for a doubling in load predicted by the WRC model of the activated sludge process for just this situation. Provided that the supply of oxygen to the organisms is not limiting, then one would expect that the increase in concentration of substrate in the effluent will be less when the load is doubled by an increase in concentration than by an equivalent doubling brought about by an increase in flow rate.

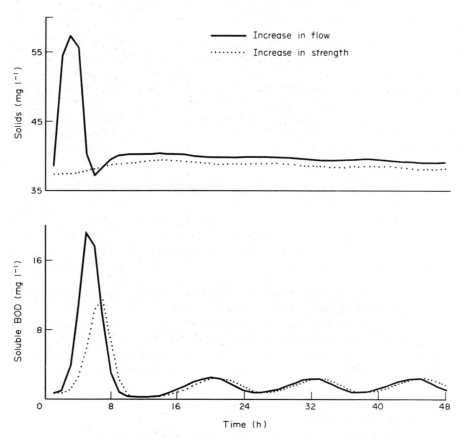

Figure 1   Comparison of the predicted effects of a doubling in load brought about by increasing the rate of flow, or the concentration of substrate, on the soluble BOD and the suspended solids in the final effluent

66.4   PLANT CONTROL

Equation (1) describes the removal of substrate (s) by a growing population of microorganisms. X is the concentration of cells present, Ks is the concentration of the limiting substrate, S, at which the specific growth rate is half the maximum specific growth rate μ_m, and Y is the mass of cells produced per unit unit mass of substrate consumed.

$$\mathrm{d}s/\mathrm{d}t \ = \ -\mu_m XS/Y(Ks \ + \ S) \tag{1}$$

Consideration of this equation indicates that in a succesful plant producing an effluent with a very low concentration of soluble BOD, the overall rate of removal of substrate is governed by the concentration of microorganisms present. For a conventional activated sludge system there is a limit to the level of suspended solids that may be maintained within the system. Constraints on having a very high level of suspended solids are the capacity of the final sedimentation tank, the settlability of the sludge, and the rate of transfer of oxygen into the system. In experimental plants using conventional aeration domes, suspended solids levels of up to 10 g l^{-1} were achieved, together with very high rates of treatment (Boon and Burgess, 1972). Such a plant could be expected to achieve a rate of treatment some three times that of one using a more normal level of suspended solids. The introduction of the fluidized bed has allowed even greater concentrations of biomass to be brought into contact with the wastewater. Cell concentrations in excess of 50 g l^{-1} have been reported, which has allowed the size of the reactor to be reduced considerably.

Following from equation (1), the influence of aeration period might also be expected to have some impact on the performance of the plant. One of the simplest equations describing the rate of growth of microorganisms in a biological reactor with recycle was developed by Herbert (1961).

$$\mu = D[1 - a(b - 1)] \tag{2}$$

This equation expresses the specific growth rate, μ, and hence the activity of the organisms, in terms of the flow rate of the wastewater and the volume of the aeration tank, D, the ratio of the recycle flow of the solids and the rate of flow of the influent, a, and the concentration ratio between the solids in the recycled sludge and in the aeration tank, b. The equation obscures the interdependence of the recycle ratio and the concentration ratio of the solids. It implies that a and b are independent variables, which they are not, and μ is extremely sensitive to small changes in the value of b. For example, assuming values of 0.1 h^{-1} for D, and 1 for a, a concentration of 3000 mg l^{-1} for the solids in the aeration tank, and 5900 mg l^{-1} for the recycled solids gives a value for b and μ of 1.96667 and 0.0033 h^{-1} respectively. If the concentration of returned suspended solids were in fact 5950 mg l^{-1}, a difference not detectable by the standard method of analysis, and which would in any case be obscured by sampling error, then the values for b and μ respectively are 1.98333 and 0.0016 h^{-1}. Thus an error of measurement of only 50 mg l^{-1} in 5900 mg l^{-1} leads to an over- or under-estimate of the activity of the microorganisms of about 100%.

The effect of the interdependence of the concentration of the solids in the aeration stage, recycle, influent flow rate and the rate of recycle is shown in Table 1. The data in the table were obtained from an analytical solution of the WRC activated sludge model (Jones and Paskins, 1979). The table shows how the degree of nitrification for a plant operated at a constant flow rate is affected by the rate of recycle. This effect arises because, with the volumetric wastage of sludge from the underflow kept constant, the concentration of solids in the underflow decreases as the recycle ratio increases. Thus, although the aeration period is decreased, and the product of solids concentration and aeration time declines, the overall wastage of solids from the system is reduced, which allows the nitrifying bacteria to become established.

Table 1 Effect of Recycle Ratio on Sludge Age and Nitrification[a]

Recycle ratio	Aeration period (h)	MLSS (mg l^{-1})	RSS (mg l^{-1})	BOD (mg l^{-1})	Ammonia (mg l^{-1})	Sludge age (d)
0.75	3.4	3202	7224	18.3	45.0	4.5
1.0	3.0	3394	6607	18.3	2.5	4.7
1.25	2.7	3503	6162	18.3	2.2	4.9

[a] Retention time (1/D) = 6 h.

66.4.1 Control with Recycle

As has been indicated earlier many proposed control strategies have been based on the control of the rate of recycle of suspended solids from the final sedimentation tank. Table 1 shows that if excess solids are wasted from the underflow from the final sedimentation tank, some control of the specific growth rate can be obtained by adjusting the rate of recycle. Figure 2 shows simulations in which the effect of both increased and decreased rates of recycle in response to a 50% increase in substrate concentration are compared with taking no action at all. In each case the response to the increased loading persisted long after the original disturbance had passed. Adjustment of the recycle rate, together with an immediate return to the original value after the disturbance, had a deleterious effect whether the recycle rate were increased or decreased. In each case the peak concentration of BOD in the final effluent was greater than if no action were taken at all. Decreasing the rate of recycle delayed the appearance of the substrate peak in the final effluent. Some improvement over taking no action could be obtained if the recycle rate was returned to its original value over a period of time.

A possibly important effect of the rate of recycle is the influence it may have on the distribution

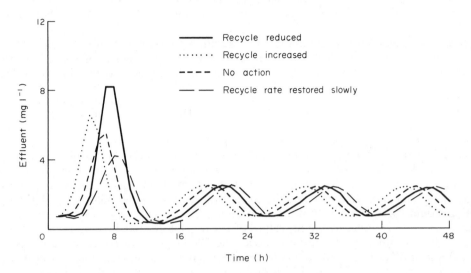

Figure 2 Comparison of the predicted effects of changes in the rate of recycle of sludge in response to an increase in the flow rate of sewage

of effluent quality from a plant treating a wastewater subject to diurnal flow. Figure 3 shows the predicted distribution in effluent quality obtained using alternative strategies for controlling the rate of recycle. The four situations shown are: (i) constant recycle ratio; (ii) constant recycle rate; (iii) constant aeration period; and (iv) as (iii) but with maximum recycle ratio limited to 1.0. The main effects were observed on the samples having the lowest BOD. A constant recycle ratio provided the lowest BOD, while a constant recycle rate produced a more compact distribution pattern.

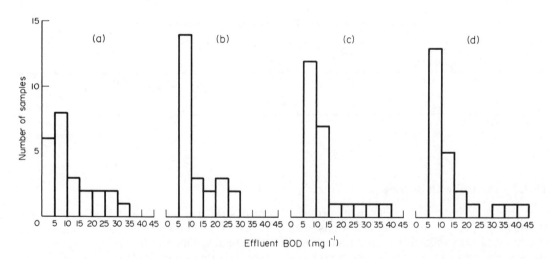

Figure 3 Predicted effect of alternative recycle regimes on the distribution of effluent BOD over a 24 hour period: (a) constant recycle ratio; (b) constant recycle rate; (c) constant aeration period; (d) constant aeration period but with maximum recycle ratio limited to 1.0

Control of activated sludge systems by simple on/off changes to the rate of flow of some of the process streams appears not to be a very satisfactory method. The high 'inertia' of the process leads to the system over-compensating for any hydraulic change that is imposed on it. When treating a waste of variable flow and strength, it is possible to keep the effluent quality within a reasonable standard by maintaining the recycle rate and the wastage rate of solids at a suitable constant value.

66.4.2   Control of Oxygen

The major possibilities of direct control are associated with the use of oxygen. An indication of the possible savings when treating a wastewater which shows diurnal variation in flow is illustrated in Figure 4. Typically, an activated sludge reactor is designed to restrict the degree of longitudinal mixing. Consequently there will be a concentration gradient along its length, and the oxygen demand will decrease as the waste passes along the tank. Many plants make no provision for this change, hence the concentration of dissolved oxygen will rise rapidly, particularly at low rates of flow, towards the end of the tank. This may be compensated for by reducing the rate of oxygen transfer in the later stages of the aeration tank. Tapered aeration, whereby the rate of oxygen transfer towards the end of the tank is reduced, by reducing the number of aeration domes for example, will save some oxygen, although it is not very flexible. More useful is some form of direct control of the oxygen transfer rate. Simulations are shown for the variation in the concentration of dissolved oxygen in the final effluent for a typical diurnal variation in flow. Three situations are shown: (i) no control—constant rate of supply at all times; (ii) tapered aeration—reduced oxygenation towards the end of the tank; and (iii) a simple proportional control which increases or decreases the rate of supply of oxygen by 10% depending upon whether the level of dissolved oxygen at that stage of the tank is above or below the set level. Even with such a simple control algorithm, a considerable reduction in the waste of oxygen is achieved, although a more sophisticated algorithm would be desirable to enable a slightly higher level of oxygen to be maintained at the higher loadings.

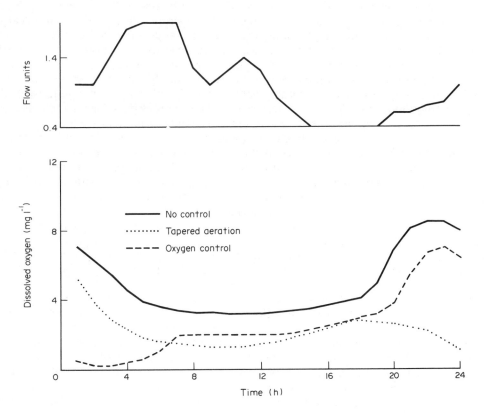

Figure 4   Predicted concentration of dissolved oxygen in the final effluent without control, using tapered aeration, and with direct control for a plant receiving a diurnal flow of wastewater

66.5   USE OF COMPUTERS

Computers have invaded all walks of life, and there are few industrial processes to which attempts have not been made to apply them. With the decline in real cost of considerable computing power this trend may be expected to continue. That computing power is becoming cheaper does not however absolve the user from considering what the most useful applications of that computer should be. In wastewater treatment there are many instances where computers have been used to assist in the management and control of plants, for example Cotton (1981), Chappell (1981) and Lund (1981). In some cases these have been restricted to monitoring and controlling sludge levels and the opening and closing of valves, the more mundane operations of the plant being readily amenable to automation. Of more complexity, and possibly of less easily demonstrable advantage, is the use of computers to control the biological performance of the plant.

66.5.1 Process Control

It is clear that the biology of the process is such that there may not be much, if any, advantage to be obtained by attempting to use a computer to influence the biological process. The use of BOD, a biological parameter, as the criterion of effective performance ensures that the treatment plant is designed and operated with the optimum environment for the production of an effluent low in both solutes and suspended solids. Pike and Carrington (1972) showed that under these conditions the majority of the microorganisms responsible for the removal of BOD are moribund. Such a population is probably fragile and will be incapable of a rapid positive response to a change in conditions.

Simulations with mathematical models show how readily the system may be made to oscillate. This is a biological phenomenon which is also influenced by the performance of the final sedimentation tank. Any change in operating conditions, whether it be one of flow or in strength of the waste being treated, if it is large enough to influence the operation of the plant will, because of the inertia of the system, remain visible long after the original disturbance has passed. There is sufficient evidence, both from simulation and from practical experience (*e.g.* Garrett, 1958; Cashion *et al.*, 1977), to suggest that the best strategy for optimizing the biological performance of a reactor is to adjust the operating conditions only very rarely, and then only in response to long-term changes in performance. Typically activated sludge plants are operated, especially if nitrification is required, at sludge ages in excess of 5 days. This implies a mean cell generation time of some 3.5 days. Using the criteria of workers in the field of continuous culture that in order to establish the existence of a steady state one should monitor the culture for a minimum period of two to three generation times, it is obvious that for the activated sludge process any change in performance due to an alteration of operating conditions will be impossible to detect with any certainty within 7 to 10 days. Since most sewage treatment plants operate with an input varying during the course of 24 hours, it becomes an academic exercise to determine whether a change in operating conditions in response to a short-term change has had the desired effect or not.

Alteration of operating conditions which influence the biological processes occurring within the system should be made with caution. The objective should be to operate the plant within a desirable operating window. Provided that the effluent meets the required criteria for BOD and suspended solids, the operator can be sure that the environmental conditions within the plant are satisfactory.

A major saving in the operation of activated sludge can be obtained from the control of dissolved oxygen. Particularly at times of low flow the amount of oxygen wasted can be considerable. Some saving can be made by use of tapered aeration, although this is relatively crude and inflexible. The best control can be obtained by monitoring the dissolved oxygen level at a number of locations along the length of the aeration tank, and adjusting the rate of aeration according to demand. For a technique like this there appears to be no advantage in using the data obtained to control the aeration rate by way of a model of bacterial respiration; simple proportional control, or, perhaps better still, a three-term PID controller would be suitable. This approach is currently in use by WRC at the Rye Meads Works of the Thames Water Authority using programmable logic controllers.

66.5.2 Data Logging

That biological control of a treatment plant is best left to the microorganisms within the plant does not mean there is no place for a computer on a wastewater treatment plant. Provision of a computer allows a number of things to be done more effectively. There are many cases where the computer has been used to control the mechanical operation of the plant, and to provide assistance in plant maintenance schedules. Certainly there is an excellent case for controlling the concentration of dissolved oxygen within the system. This is the one substrate for the microorganisms that the operator can ensure does not become limiting, but at the same time waste is to be avoided. Data collection and presentation is another area where the computer would come into its own.

For the plant operator to make a rational assessment of whether or not changes to operating conditions should be made, he needs to consider a large quantity of data and facts. It is here that the computer can make a useful contribution. Given that the plant is adequately instrumented, and that all necessary manual analysis is entered into the computer, software can be written that will present to the operator an unbiased summary of the performance of the plant at the current time. In addition to this he can interrogate the machine for historical data which will tell him whether the current performance, and the current conditions, are at all unusual. With an adequate model of the system, the computer can then be asked to suggest any possible changes that may be necessary or desirable.

66.5.3   Sensor Control

Any control system is only as good as the information that it receives. A major role for a computer in a wastewater treatment plant must be not only the monitoring and logging of data, but also in the field of quality control. The wastewater treatment environment is well known for its hostility to instrumentation. Rates of flow tend to be slow and biological growths abound, both factors which tend to reduce the reliability of the sensors used. One advantage that the process control engineer has with a wastewater treatment plant is the leisurely rate at which changes occur. Diurnal variations tend to be regular; a few 24 hour surveys of flow and strength soon reveal a general pattern which, unless there are any major shifts in population such as may occur at holiday resorts, will show only minor variations, with possibly longer term seasonal variations. The changes that occur are not abrupt as some buffering will occur in the sewerage system. Also, as simulations show, the response of an activated sludge plant to a change in conditions is far from rapid.

Figure 4 illustrates this concept quite well. The simulations of oxygen control were carried out using a simple proportional control algorithm which used the concentration of dissolved oxygen before each integration period as the reference point. The minimum length of the integration period was about 10 minutes, while the maximum was set at 1 hour. Even with sampling periods as long as this the system was quite capable of reducing considerably the concentration of dissolved oxygen in the final effluent. This lack of sensitivity by the biological system to external influences has important consequences for the choice and operation of the sensors used in the system. To meet with the rigours of the environment the sensors need to be robust, and provision needs to be made for regular cleaning and calibration checks. It is in this area that the computer can be of great use to the operator. As has been seen above, sampling of the system need not be carried out very frequently. Without the need for continuous data acquisition, the sensors can be removed from the hostile environment for long perods, during which time the computer can be used to carry out reliability and calibration checks, so that the acquired data has the maximum reliability, and any sensor faults can be flagged immediately.

66.5.4   Telemetry

A potential role for the computer in wastewater treatment is the monitoring of remote sites *via* the public telephone network. A system for monitoring river quality data for inlet protection has proved feasible at both the Thames Water Authority and The Sunderland and South Shields Water Company (Briggs, 1979). With regular, but not necessarily frequent, interrogation of judiciously sited sensors a single computer could be used not only to monitor but also possibly to control a number of small, normally unattended, treatment works.

66.6   SUMMARY

For the wastewater treatment industry the computer, together with suitable mathematical models of the many treatment processes involved, could be very effective. As far as the activated sludge process is concerned, direct computer control of the biological process, other than the control of the level of dissolved oxygen, does not appear to be justified. However, there is a need to have more effective control over the complete performance of treatment plants, not just the biological stage, and for this the computer needs to be an integral part of the overall process design.

Given that a computer is available with suitable sensors and control signals then the operator has available a whole range of facilities. These include: (i) optimization of plant performance; (ii) efficient use of aeration equipment; (iii) automatic calibration of sensors; (iv) automation of routine tasks; (v) detection of equipment failure; (vi) control of maintenance schedules; (vii) quality control; (viii) supervision of remote plant; (ix) ready availability of historical data; and (x) report generation. None of these facilities individually would necessarily justify the use of a computer. Taken together however they increase the possibility of making the most effective use of the equipment available to produce effluents of the required quality and consistency.

66.7 REFERENCES

Andrews, J. F. and M. K. Stenstrom (1979). Real-time control of activated sludge process. *J. Environ. Eng. Div. (Am. Soc. Civ. Eng.).*, **105**, (EE2), 245–260.

Ardern, E. and W. T. Lockett (1914). Experiments on the oxidation of sewage without the aid of filters. *J. Soc. Chem. Ind.*, **33**, 523–539.

Atkinson, B. (1974). *Biochemical Reactors*, chap. 7. Pion Press, London.

Beck, M. B. (1981). Operational estimation and prediction of nitrification dynamics in the activated sludge process. *Water Res.*, **15**, 1313–1330.

Boon, A. G. and D. Burgess (1972). Effects of diurnal variations in flow of settled sewage on the performance of high-rate activated sludge plants. *J. Inst. Water Pollut. Control*, **5**, 493–522.

Briggs, R. (1979). The continuous monitoring of raw water intakes. *J. Inst. Water Eng.*, **33**, 84–94.

Briggs, R., K. Jones and A. B. Oaten (1967). Monitoring and automatic control of dissolved oxygen level in activated sludge plants. Effluent and Water Treatment Convention, London.

Cashion, B. S., T. M. Keinath and W. W. Schuk (1977). Evaluation of instantaneous F M control strategies for the activated sludge process. *Prog. Water Technol.*, **9**, 593–596.

Chambers, B. (1982). Energy saving—optimisation of fine bubble aeration. *WRC Preliminary Report.*

Chappell, T. E. (1981). A computer-based centralized control and supervisory system for an effluent treatment plant. *Chem. Eng. (London)*, **371/2**, 377–380.

Cotton, P. (1981). Automation at Norwich sewage treatment works: five years experience. *J. Inst. Water Pollut. Control*, **80**, 243–259.

Downing, A. L., H. A. Painter and G. Knowles (1964). Nitrification in the activated sludge process. *J. Inst. Sewage Purif.*, Part 2, 130–158.

Garrett, M. T. Jr. (1958). Hydraulic control of activated sludge growth rate. *Sewage Ind. Wastes*, **30**, 253–261.

Herbert, D. (1961). A theoretical analysis of continuous culture systems. Continuous culture of microorganisms. *SCI Monogr.*, **12**, 21–53.

Jones, G. L. (1973). Bacterial growth kinetics: measurement and significance in the activated sludge process. *Water Res.*, **7**, 1475–1492.

Jones, G. L. (1976). Microbiology and activated sludge. *Process Biochem.*, Jan/Feb, 3–5, 24.

Jones, G. L. (1977). Mathematical model of activated sludge. *Prog. Water Technol.*, **9**, 571–577.

Jones, G. L. and A. R. Paskins (1979). Mathematical model of activated sludge—a comparison of predictions and experimental results. *WRC Laboratory Report 1002.*

Lund, D. E. (1981). Empire controlled by computer. *Water/Eng. Management (USA)*, **128** (6), 30–33.

Mohlmann, F. W. (1917). The activated sludge method of sewage treatment. *Water Survey Ser. Illinois Univ.*, **14**, 75–113.

Monod, J. (1942). *Recherches sur la Croissance des Cultures Bacteriennes*. Hermann, Paris.

Olsson, G. and J. F. Andrews (1978). The dissolved oxygen profile—a valuable tool for control of activated sludge. *Water Res.*, **12**, 985–1004.

Pike, E. B. and E. G. Carrington (1972). Recent developments in the study of bacteria in the activated sludge process. *J. Inst. Water Pollut. Control*, **71**, 583–605.

Tucek, F., J. Chudoba and V. Madera (1971). Unified basis for design of biological aerobic treatment process. *Water Res.*, **5**, 647–680.

WRC (1981). WRC/CIRIA Sewage treatment optimisation model (STOM). *WRC Technical Report 144.*

67

Practices in Activated Sludge Process Control

J. P. STEPHENSON
Environment Canada, Wastewater Technology Centre, Burlington, Ontario, Canada

67.1 INTRODUCTION

During the past few decades many concerns have been raised about water quality management, and significant resources have been directed towards the construction of municipal and industrial wastewater collection and treatment systems. More recently, inadequacies in the performance of these systems have become apparent. This led in the US to a survey to ascertain the factors limiting municipal wastewater treatment plant performance (Hegg *et al.*, 1978). These investigators concluded that, of the ten major factors leading to inadequate performance, the two most significant factors were the application of concepts and testing to operational process control, and of process control testing procedures. To some extent these faults can be diminished by the application of on-line instrumentation and automated control procedures to upgrade operations.

Hegg *et al.* (1978) also found that plant structural and mechanical designs were not always conducive to good operational control. The lack of plant flexibility severely limited the degree of control that could be attained in these plants. Without structural and mechanical flexibility, control schemes can be too constrained to be effective.

Faults in the design and operation of treatment plants are interactive and must be addressed simultaneously by those who design, regulate and operate these plants. Beck (1977a) emphasized the importance of effective communication between all of the many disciplines which impact upon the design, the operation and the regulatory responsibilities. Beck (1977a) indicated a reluctance upon the part of control engineers to move from theory to practice and a reluctance upon the part of wastewater treatment design engineers and operational personnel to acquaint themselves sufficiently with control engineering. There is no doubt that, with the pressing day-to-day demands within each industry, there is difficulty in achieving these goals. In wastewater engineering from the regulatory to the design and operating teams, it is safe to surmise that a miniscule portion of these teams have sufficient expertise in basic control theory let alone acquaintance with the more sophisticated and potentially rewarding branches of control. We can hope that the inroads of microprocessors in wastewater management will introduce automated control and on-line estimation principles to more wastewater treatment engineers. Even today, it is appropriate to recount one of Beck's statements about the lack of communication between the important groups: the consequences of this lack of communication are that we may not be making the most of our opportunities to apply new (and old) control techniques in practice (Beck, 1977a). Except for those examples of effective control which have been demonstrated and reported, the vast majority of water pollution control plants have developed without the attention to automatic sensor-based process control which they deserve.

In most plants the principal operational control methods are governed by time-consuming and time-delayed laboratory-generated analytical results and by human observation of process performance which is indispensible if performed well. Supporting these approaches to operational control are the concepts that wastewater treatment processes are slow acting and self-controlling (von Buchstab, 1983). In practice, many aspects of biological wastewater treatment either are fast acting and have little possibility of being controlled manually, or are so slow acting that more productive management through automated control techniques is appropriate.

Fortunately, there is a growing tendency to explore automatic control possibilities in biological wastewater treatment to improve performance and to reduce energy consumption and operating costs. Three major workshops (IAWPR, 1974, 1977, 1981) have addressed this area during the past decade. More recently, administrative calls have been made for improvements in the operation of wastewater treatment works (WPCF, 1983). To emphasize the need for improvements in operational control, a recent review of 30 municipal wastewater treatment facilities in the Great Lakes Region identified that less than one-third of the plants complied with their regulatory requirements in terms of five-day biochemical oxygen demand (BOD_5), total suspended solids and total phosphorus (IJC, 1983). Amongst other factors, the report claimed that the operators of these facilities had a lack of understanding of fundamental principles of solids inventory control in biological treatment systems. This provides indication that considerably more effort needs to be expended to solve operational problems in existing facilities before large capital investments are made to expand these facilities. Considered in the proper context and within the constraints of plant design, instrument-based operational control has a significant role to play to improve plant operations (White, 1981; DOE and NWC, 1981). At the outset, it is important to realize that automation will not correct for design deficiencies.

67.2 ON-LINE MEASUREMENTS IN WASTEWATER TREATMENT

Many on-line instruments are deemed to be readily available for monitoring wastewater treatment plants (DOE and NWC, 1981). However, instrumentation actually in use is weighted towards the detection and measurement of liquid level and fluid flow rate. This is not surprising since it reflects the importance of quantity measurement in wastewater treatment. Holmström and Hultman (1981) reviewed the usage of wastewater instrumentation distributed in 30 plants in Sweden. They reported that 77.5% of the instruments were used for level and flow rate measurement; only 22.5% of the instrumentation was of an analytical type. The US Environmental Protection Agency (EPA, 1977) surveyed the on-line instrumentation at 50 wastewater treatment plants that were known to be at least partly automated. This survey, based upon the report by Molvar *et al.* (1976), showed that even in these plants only 29% of the installed instrumentation was analytical, the balance being flow related.

The Swedish and US studies summarized the performance of the installed instrumentation. The track records of the instruments showed that 30 to 40% of them were less than good or satisfactory. Usually, adverse process conditions or in the US case poor instrument maintenance were the causes of poor performance. In the US survey, 31% of instruments and/or associated automatic control loops were abandoned because of inadequacy in the instrumentation or its usage.

The concern that acceptable instrumentation be selected or developed has led to the establishment of a commercial instrument testing service for the water and wastewater industry (Garrett *et al.*, 1982). The aims of this service are to generate instrument test protocols, to oversee scientific testing at established laboratories and to transfer the findings to the users. The dissatisfaction with instrument performance and usage has been echoed by numerous writers including White (1981), Bonnick and Sidwick (1981), von Buchstab (1983), Garrett *et al.* (1982) and Schuk *et al.*

(1981). Clearly, instrumentation limitations need to be solved before we can expect widespread acceptance of more advanced analytical equipment in wastewater treatment than exists now.

The reviews by Holmström and Hultman (1981) and Molvar *et al.* (1976) do suggest that many instruments are used satisfactorily for monitoring and control (Figure 1). In general, the less complex the instrumentation, the higher the likelihood that successful usage is experienced. Stephenson *et al.* (1981a, 1983) evaluated several analytical instruments for activated sludge monitoring including pH, dissolved oxygen, filterable organic carbon, and high- and low-range suspended solids analyzers. Their result showed that the instrument accuracies and precisions were sufficiently reliable for continuous monitoring and control functions if adequate installation, maintenance and calibration efforts were made.

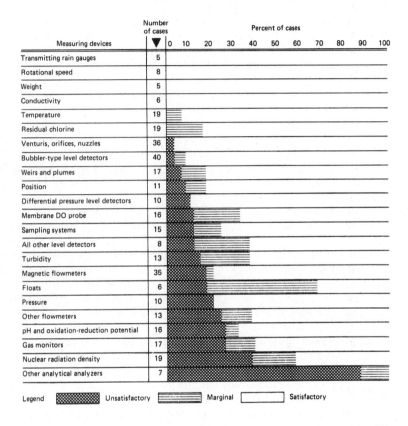

Figure 1   Performance summary of measuring devices in wastewater treatment facilities (EPA, 1977)

A key factor to consider is the degree of accuracy actually required for practical application of instrumentation in wastewater treatment. To improve plant operational consistency in a realistic sense, relaxing instrument accuracy, if necessary, and improving instrument reliability would help ensure the industry's acceptance of continuous instrumentation in practice. Although instruments may not meet the degree of accuracy and precision stated by the manufacturer, they may be adequate for monitoring and controlling plant operation to achieve consistent results. White (1981) indicated that instrument accuracy of say ±20% is adequate if it is consistent. Russell (1980) gave a critical example connected with flow rate measurements in which he stated that measurements could have greater than 25% error in any case.

To stand as an example, Figure 2 gives an impression of the degree of accuracy and precision of a continuous monitoring suspended solids monitor used in return activated sludge (Stephenson *et al.*, 1981a, 1983). For these data, the instrument response was attributed with all of the observed error, but it was apparent that a significant yet unmeasured component of error resulted from the process fluctuations which made manual sampling and gravimetric reference analysis difficult, *e.g.* the clarifier rake drive produced a sinusoidal fluctuation in the underflow solids concentration with period of one-half the revolution period of the rake (Stephenson *et al.*, 1983). This example of the difficulty in accurate manual sampling of process streams underlines the need for consistent continuous measuring hardware that can compensate for manually unobservable process phenomena with the highest justifiable accuracy.

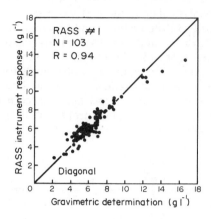

Figure 2   Performance of a continuous monitoring photometric suspended solids monitor for return activated sludge (RASS)

67.3   PREREQUISITE MEASUREMENTS IN ACTIVATED SLUDGE WASTEWATER TREATMENT

Usually only a few measurements are routinely used by plant operating staff to monitor activated sludge process operations to meet regulatory requirements. In practice, flow rate and liquid levels are usually the only continuous measurements obtained. Other conventional measurements are manually performed usually on grab or composite samples and include settleability, suspended solids, alkalinity, pH, temperature, dissolved oxygen, chemical and biochemical oxygen demand, organic carbon, nitrogen and phosphorus to name most. In smaller facilities the more complex analyses are frequently performed at a central laboratory on an infrequent basis. The sampling techniques and their inherent time delays are not conducive to detection of dynamic disturbance nor to effective control.

For activated sludge process operation and control, the prerequisite measurements include flow rates, activated sludge settleability, suspended solids concentration and dissolved oxygen concentration. One of the first duties of plant operating staff is to ensure that process failure does not occur (Beck, 1981; Andrews *et al.*, 1980). In most cases this means that operating staff must ensure that sludge does not escape in the effluent (Keinath *et al.*, 1977). The traditional method used to control plant solids inventory is to measure the activated sludge settleability through, say, a 30 minute settling test. The plant operator then may use the settled volume to determine two control possibilities: return activated sludge flow rate, or waste activated sludge flow rate. Less frequently, the secondary clarifier sludge blanket level is measured to determine the desired return activated sludge flow rate. However, it is known that settleability tests can lead to the choice of the wrong control action, for example, under sludge bulking conditions. The more reliable method recommended to control plant solids inventory has been control of the growth rate or the solids retention time (SRT) (Garrett, 1958; Lawrence and McCarty, 1970). These methods require only the measurement of suspended solids and flow rate. However, like F/M control, they apply only to steady-state operating conditions and do not account for dynamic variations within the operation of the plant (Strenstrom and Andrews, 1979).

In aerobic wastewater treatment the dissolved oxygen concentration is a necessary measurement to ensure that sufficient aeration occurs. Inadequate dissolved oxygen concentration can lead to treatment problems such as bulking or inhibition of nitrification (EPA, 1975; Palm *et al.*, 1980; Sezgin *et al.*, 1978), while excessive aeration is costly from the energy consumption viewpoint (Andersson, 1979; Sørensen, 1979; Flanagan and Bracken, 1977) and can lead to settleability problems because of floc shear (Parker *et al.*, 1971). Unfortunately, DO measurements are rarely performed continuously, at least in Canada. Due to the time-varying nature of the dissolved oxygen concentration, point measurements in space and time do not guarantee that the concentration is continuously maintained within a desirable range.

Because the loading to most wastewater treatment plants is never at steady state, there are compelling reasons for the industry to adopt continuous on-line measuring principles to effectively monitor and control the process. Not all measurements are easily automated, *e.g.* BOD$_5$, and digital microprocessor capabilities should be used effectively to extract the maximum amount of needed information from the readily available measurements (Olsson, 1981a). To extract this information requires careful consideration of the instrumentation, the process dynamics and the intricacies of process models. The value of this information will be the estimation of process variables which cannot be measured directly.

Besides the prerequisite measurements discussed above, other measurements are sometimes necessary in the day-to-day operation of the activated sludge process, *e.g.* pH. Depending upon the wastewater characteristics and the application, these other measurements may or may not be prerequisite to successful operation.

67.4 ACTIVATED SLUDGE WASTEWATER TREATMENT PROCESS CONTROL

67.4.1 Manipulable Variables

The hydraulic and organic loading disturbances influencing a plant generally are uncontrollable (Beck, 1977b; Niku *et al.*, 1981). Municipally operated systems experience diurnal loads upon which random shock loads are superimposed. In industrial systems, similar loading disturbances are experienced.

For the activated sludge process there are a limited number of manipulable variables in comparison to the number of measureable variables. Four major manipulable variables are available for control of the activated sludge process and should be properly considered at the treatment plant design stage (Olsson and Andrews, 1977):

> wastewater feed distribution between and within aeration vessels,
> air flow rate and air flow distribution,
> waste activated sludge flow rate,
> return activated sludge flow rate.

To manipulate these variables requires flexible process hardware such as variable output devices, *e.g.* multiple speed pumps and blowers. Unfortunately, many treatment plants do not have the hardware flexibility required to provide for control (Hegg *et al.*, 1978; Flanagan and Bracken, 1977). Plant hardware is commonly oversized for the maximum anticipated load (Barnhart, 1985). Partly this inflexibility results from process design information that has been based upon peak requirements from experience or from steady-state solutions to the available empirical models (Lawrence and McCarty, 1970; Niku *et al.*, 1981) instead of an analysis of the loading and process dynamics (Beck, 1977b; Olsson, 1976, 1981b; Strenstrom and Andrews, 1979) which provide a more complete description of the process control requirements.

67.4.2 Pump Station Control

Hobson (1982) examined the impact of pumping practice upon energy consumption for pumping in rising mains. He concluded that significant energy savings (62.5%) could be gained by more rational design of pump stations and controllers to provide optimum operation for dry weather flow (DWF) conditions rather than for wet weather flows. Moreover, the instantaneous pumping rate can be significantly reduced, thereby reducing the repeated hydraulic shocks from pumps designed to deliver the peak wet weather flow at say 6 DWF. Owen (1982) considered further the design of pump stations and control hardware for energy conservation including the selection of pumps, motors and variable speed drives.

Wet weather flows and flow disturbances can cause severe problems in the settlement of activated sludge, particularly under bulking sludge conditions (Tomlinson, 1982). Wet weather pumping rates used during dry weather conditions, however short the duration, exacerbate settleability and clarification problems (Chapman, 1983; Collins and Crosby, 1980; Sørensen, 1980; Wilson and Polkowski, 1981). Data from Wilson and Polkowski (1981) showed secondary clarifier hydraulic disturbances varying from 0.21 to 0.33 m^3 s^{-1} within a 600 s time interval. Even within a 600 s interval, these authors showed significant flow disturbances to the clarifier with the same order of magnitude as the previously mentioned value with a period of about 60 s and a deviation from the mean flow of ±20%. Flow to the plant was reported to be delivered by constant speed pumps operated intermittently. Figure 3 depicts the flow rate measured at the secondary clarifier of a typical activated sludge plant. Flow was delivered by intermittently operated lift pumps controlled by liquid level and sized to handle the wet weather flow. These data show that, in practice, activated sludge plants frequently receive large disturbances of varying frequencies and amplitude.

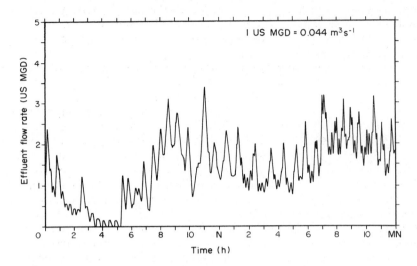

Figure 3　Diurnal flow characteristics in municipal wastewater treatment

Chapman (1983) investigated the impact of hydraulic disturbances upon secondary clarification and effluent suspended solids. He showed that the effluent suspended solids concentration responded directly to disturbances in clarifier feed flow rate. The response to a step increase in feed flow rate was rapid with a first-order time constant of about 10^3 s and an average increase in effluent solids of about 10 mg l^{-1}. He observed that rapid feed flow disturbances tended to flush unsettled debris collected around the clarifier weir into the effluent for a temporary but significant increase in suspended solids above the final steady-state value. Decreases in flow rate were accompanied by reductions in effluent solids, but of a slower dynamic nature.

The foregoing discussion, especially in connection with clarification of effluent suspended solids and fluctuations in sludge blanket height, points to the need for improved control of wastewater transport to treatment systems over and above the need for energy savings. The control philosophies for fluid pumping are well established and design engineers should be more motivated to consider the effects of disregarding the control possibilities. Indeed, in many cases it would appear that the elementary forms of control usually practised (*i.e.* on/off limit control) tend to amplify the frequency and amplitude of the hydraulic disturbances influencing a plant. Fortunately, there is evidence that more advanced designs of flow regulation systems are being implemented in practice. Brueck *et al.* (1981) reported the results of a sophisticated monitoring and control system which forecast diurnal DWF and used the combined sewer system to equalize and control flows to the plant. During rainstorms, wastewater overflows to the receiving stream were minimized while using the sewer storage capacity to limit the flow rate to the treatment plant to an acceptable level.

67.4.3　Dissolved Oxygen Concentration Control

The aerobic biological treatment of wastewater depends upon the transfer of gaseous oxygen to the dissolved phase. A certain minimum dissolved oxygen concentration (C) must be maintained within the aeration tank for satisfactory treatment. In particular, the nitrification process is sensitive to low concentrations of dissolved oxygen (EPA, 1975). However, the energy intensity of the process mitigates against the maintenance of a surplus dissolved oxygen concentration. In activated sludge systems, aeration typically consumes 60 to 80% of the total plant energy demand. Flanagan and Bracken (1977) reported the basic techniques to be used and the advantages to be gained from automatic aeration control, although actual applications had been made as early as 1963 in Colombes, France (Brouzes, 1969).

The primary objectives of aeration control are to satisfy the oxygen demand of the respiring biomass and to minimize energy consumption. Secondary objectives, much less frequently applied but important in their own right, include on-line estimation of the oxygen transfer coefficient, respiration rate, substrate removal rate and excess sludge wasting requirements. In all likelihood, the infrequent application of the secondary objectives is because many activated sludge plants still do not practise the primary objectives. However, the energy crisis of the last decade has recreated widespread interest in aeration control and the resulting energy savings (Aarinen *et al.*, 1979; Andersson, 1979; Flanagan and Bracken, 1977; Nielsen *et al.*, 1981; Olsson and Andrews, 1978; Sørensen, 1979, 1981; Schlegel and Lohmann, 1981; Stephenson *et al.*, 1981b; Tanuma *et al.*, 1981; Usrael, 1980).

The potential energy savings through aeration control are of the order of 30%. Andersson (1979) reviewed eight full scale activated sludge systems in Denmark and showed energy savings ranging from 2.5 to 60% through aeration and solids inventory control. For 12 plants examined by Flanagan and Bracken (1977) in the US there was an average 33% improvement in BOD removed per kWh of energy supplied for aeration. Caution must be exercised in designing energy efficient aeration control systems since the potential energy savings may not be realized due to electrical and mechanical system efficiency losses when blowers are turned down. Aarinen *et al.* (1979) calculated potential shaft energy savings of 30 to 35% through aeration control; however, the actual wire savings were only 6 to 10% due to the compressor DC-motor speed controller losses. Wood and Ross (1983) enumerated practical considerations in applying controllers to electrical motors to ensure energy savings.

Several investigations have established that activated sludge settleability improves in plants practising automatic dissolved oxygen control. Wells (1979) reported that filamentous organisms were minimized in the process train of a full scale facility having dissolved oxygen control compared to a parallel train under manual control. If the automatic control was disabled, the settleability deteriorated. Further support to the concept that dissolved oxygen concentration, in combination with organic loading, influences settleability was provided by Palm *et al.* (1980). The conjecture that dissolved oxygen concentration or the rate of change of concentration has a bearing upon settleability has to be tempered by the knowledge that diverse factors can affect settleability (Chambers and Tomlinson, 1982). The consensus, however, is that dissolved oxygen should not become a limiting concentration in activated sludge wastewater treatment.

The largest proportion of control strategies for dissolved oxygen concentration has been derived either as continuous proportion-integral (P-I) feedback controllers (Aarinen *et al.*, 1979; Flanagan and Bracken, 1977; Holmberg, 1981a, 1981b, 1982a; Petersack and Smith, 1975; Iwaki *et al.*, 1977; Ohto *et al.*, 1977; Sørensen, 1979; Stephenson *et al.*, 1981b) or as on/off limit controllers *via* multi-speed motors or multiple blowers (Andersson, 1979; Flanagan and Bracken, 1977; Schlegel and Lohman, 1981). Figure 4 shows the results of a P-I control strategy in a diffused air activated sludge plant. Figure 5 shows results from an on/off control strategy using multi-speed and multi-unit positive displacement blowers. Most studies have shown that a feedback loop responds well to loading disturbances. Aarinen *et al.* (1979) reported that a fast acting feedback P-I controller maintained the dissolved oxygen concentration at a setpoint value in a completely mixed basin subjected to step changes in wastewater flow rate. Stephenson *et al.* (1983) found a 20-fold to 100-fold reduction in dissolved oxygen concentration variance using a feedback P-I controller in a pilot-scale completely mixed aeration basin. Full scale success of feedback P-I control of dissolved oxygen concentration has been amply demonstrated (Ohto *et al.*, 1977; Olsson and Hansson, 1976). Boon and Chambers (1985) implemented a sophisticated aeration control system in a pseudo-plug flow aeration system in which the distribution of air along the length of the aeration tank was manipulated to maintain setpoint concentrations at four locations in a nitrifying system and at five locations in a non-nitrifying system. Individual dissolved oxygen concentrations were governed by local programmable logic controllers. They found that load increases in concentration did not necessarily create excessive oxygen demand at the aeration tank inlet, but that the demand was exerted further downstream due to the more bioresistant components present. Transient shifts in the oxygenation profile in a plug-flow aeration tank suggest that air flow distribution control is a necessary requirement in aeration control systems. This requirement is over and above the design of tapered aeration systems which can only exactly match the demand under a specific steady-state situation.

Practical considerations for blower control are important: anti-surge control in centrifugal systems, minimized start/stop, blowoff control during startup, maintenance of electro/mechanical efficiency to name a few (Dobrowolski and Mendelsohn, 1977; Flanagan and Bracken, 1977).

Some investigators have noted the need for dead-time compensation (Wells, 1979; Aarinen *et al*, 1979). Aarinen *et al.* (1979) used a Dahlin controller for aeration control and realized a closer regulation towards the setpoint compared to a discretized P-I controller; both controllers gave

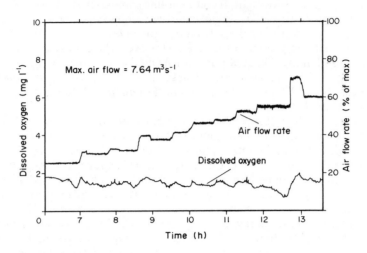

Figure 4   Performance of continuous P-I regulator for dissolved oxygen control

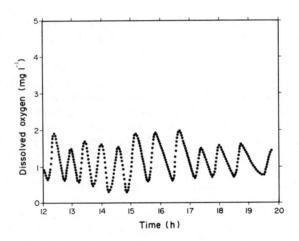

Figure 5   Response of dissolved oxygen to on/off controller adjustments

satisfactory results. Wells (1979) used a 600 s dead-time compensation within a dissolved oxygen control loop.

In some cirumstances feedforward/feedback dissolved oxygen control has been suggested (Arthur, 1980; Schegel and Lohmann, 1981; Flanagan and Bracken, 1977). Clifft and Andrews (1981a) proposed the use of the wastewater flow rate signal to modulate aeration rate in small (<44 l s^{-1}) activated sludge plants through continuous or on/off control. They predicted energy savings between 20 and 30% using this control procedure. There are several reasons to be cautious about the use of flow rate to control aeration: small treatment plants usually have significant and rapid flow rate disturbances from on/off pump stations; the point of flow rate measurement is usually removed from the aeration vessel and there are significant delays in the propagation of flow through a plant and dampening of the flow disturbances; the effects of spills or rain water dilution would be undetected. Clifft and Andrews (1981a) recognized the limitations to feedforward control alone and provided several suggestions to minimize deviations from desired behaviour including programmed control based on analysis of historical data. Stephenson *et al.* (1983) noted that aeration demands decreased during weekends even when diurnal weekend flow rates were identical to those during weekdays having higher organic loading. Therefore, flow proportional control should be trimmed through feedback action, if used at all. Feedforward control is unlikely necessary in most cases if well-tuned feedback control is practised.

Less attention has been given to the importance of providing the necessary flexibility and turndown capability in the aeration, electrical and mechanical hardware to achieve the primary objectives. Many aeration designs are based upon the requirements to provide oxygen at a mass rate equivalent to the peak mass BOD loading rate which occurs less than 10% of the time (Barnhart, 1985; Olsson, 1980a). This often has resulted in oversized blowers and aerators with insufficient turndown capability. Moreover, recent studies have shown that the BOD mass loading rate is not exerted immediately in activated sludge systems, but that peaks are significantly damped out in terms of aeration requirements and partially oxidized under subsequently lower organic loading conditions (Clifft and Andrews, 1981a, 1981b; Sørensen, 1979, 1981; Stephenson *et al.*, 1983). Gujer (1981) showed that the removed particulate organics exert a damped oxygen demand dependent on sludge age and the ratio of particulate to soluble BOD. These factors lend a new dimension to the proper sizing of aeration hardware for operation and control purposes. Ignorance of these effects is likely responsible for much of the overdesign observed in installed systems. Data describing the damping effect or aeration control results suggest that turndown ratios from 1.5:1 (Gujer, 1981) to 5:1 (Usrael, 1980) should be available for the operation and control of municipal activated sludge systems.

67.4.4 Estimation of Process Parameters and State Variables Through Dissolved Oxygen Control

The use of microprocessors for control has opened new possibilities for the monitoring and operation of activated sludge systems. At the same time, the entry of computer control specialists into the field has meant that these possibilities are being realized in practice to a greater extent.

In aeration control, the well-known dynamics of reaeration in respiring systems are identified by:

$$dC/dt = \alpha k_L a (C_s - C) - r$$

where C = dissolved O_2 concentration, t = time, $\alpha = (k_L a)_{process}/(k_L a)_{tap\ water}$, $k_L a$ = volumetric O_2 mass transfer coefficient, C_s = saturation value of C, and r = volumetric respiration rate. This relationship has been used to provide controller design information and to provide for on-line calculation of two important reaeration parameters, $\alpha k_L a$ and r.

Brouzes (1969, 1979) used measurements of air flow rate and dissolved oxygen concentration under control conditions to calculate the *in situ* respiration rate of the microorganisms. From fundamental mass balances, Brouzes calculated the growth rate and the required wasting rate of the microorganisms. Furthermore, he estimated the organic loading and removal within the plant. The strategy has been used in several full scale plants in France.

Olsson and Hansson (1976), using digital control computing capabilities in a 1.3 m³ s⁻¹ activated sludge system, showed that the process volumetric oxygen mass transfer coefficient could be estimated recursively on-line using a maximum likelihood algorithm. The noise characteristics of the process and the sampling and control interval required for control were established. Control constraints included allowable rates of change and limits of the air flow rate, and of the developed pressure of the blowers. They determined the mass transfer coefficient at two dissolved oxygen probe locations on-line within a 100 m long, step-loaded aeration tank based upon disturbances in the air flow rate. The estimates showed that the transfer coefficient could be updated recursively. The technique indicated widely different values of $\alpha k_L a$ at the two locations (4.3 h⁻¹ and 2.1 h⁻¹, respectively). Since the lower $\alpha k_L a$ value was nearer the aeration tank outlet, Olsson and Hansson (1976) postulated that the air flow rate was less at that point either from diffuser clogging or from an unintentionally different diffuser submergence depth. This type of information, based upon on-line results from control applications, has value beyond the energy savings achieved by control. For example, for Olsson and Hansson's data for the same organic loading, at least twice as much delivered energy would be required at the control location having the $\alpha k_L a$ value of 2.1 h⁻¹. By early and accurate detection of process failures such as diffuser clogging, significant operating and maintenance cost savings can be realized. It is hoped that, based upon results like these, greater understanding of the possibilities of on-line estimation will be gained by the designers of treatment plant control systems and by the regulatory agencies that influence and fund them. The applications then can become more widespread.

Gillblad and Olsson (1977) reported the use of on-line computation capability in a 0.9 m³ s⁻¹ activated sludge system to control the dissolved oxygen concentration to a setpoint value established by sludge activity calculations derived from oxygen mass balances. The specific carbonaceous oxygen utilization rate (SCOUR) was calculated on-line and used to govern the return activated sludge flow rate. The plant operators used the indicated dissolved oxygen profile to adjust the waste activated sludge rate or the wastewater feed distribution.

The placement of dissolved oxygen sensors for control purposes is a frequent question. In completely mixed systems, probe positioning depends upon a location where sufficient velocity passes the probe to avoid diffusion limitations. In dispersed plug-flow reactors, Olsson (1980a, 1980b) and Olsson and Andrews (1981a, 1981b) indicate that one oxygen electrode is insufficient for control purposes or even for determination of a dissolved oxygen setpoint. They suggest a minimum of two, but preferably three, oxygen electrodes near the reactor outlet to determine the shape of the dissolved oxygen profile. The dissolved oxygen profile yielded information about the degree of completion of the biochemical reactions for carbonaceous and nitrogenous components, the required dissolved oxygen setpoint and the air flow rate required for control.

One drawback in the original scheme for on-line estimation of respiration rate used by Brouzes (1969, 1979) is the hazard that the oxygen transfer coefficient will vary in time and that the oxygen concentration time derivative will be non-zero. This could result from several factors including presence of surfactants, clogging of diffusers or even manual manipulation of air flow distribution by an operator. As mentioned, Olsson and Hansson (1976) overcame this limitation through on-line estimation of the overall oxygen transfer coefficient. In Finland, Holmberg (1981a, 1981b, 1981c, 1982a) and Holmberg and Ranta (1982) recommended recursive techniques to estimate the oxygen transfer coefficient and the respiration rate in a $0.004 \text{ m}^3 \text{ s}^{-1}$ activated sludge system. They concluded that the oxygen transfer parameters were estimable during large changes of the dissolved oxygen setpoint. Further, using state estimation, Holmberg and Ranta (1982) calculated the growth rate of the microorganisms in the activated sludge process from the directly measurable process variables and the estimated oxygen utilization rate. During a 24 h diurnal loading period the growth rate varied from about 3 to $30 \text{ g m}^{-3} \text{ h}^{-1}$. From this an accurate wasting rate was calculated. The influent organic loading and sludge concentration also were predicted on-line.

Stephenson *et al.* (1981b, 1983) used the non-recursive technique established by Brouzes (1969, 1979) to estimate the oxygen utilization rate in a pilot scale activated sludge process. The waste activated sludge mass was controlled to maintain a 24 h setpoint oxygen utilization rate. Compared to a parallel plant having instantaneous sludge age control to a stable setpoint value (3 to 5 d), the respiration rate-controlled system experienced excursions in the waste sludge rate and thus the instantaneous SRT (1 to 20 d) due to organic loading disturbances. The daily equilibrium SRT remained within about ±0.5 d of the median SRT in the respiratory-controlled system. There was no deterioration in the effluent quality of the respiratory-controlled system. The effluent suspended solids concentration in the respiratory-controlled system was slightly better during a 47 d period than the sludge age-controlled system and the SVI was significantly lower.

The application of these more advanced operational management tools has tremendous potential in the wastewater industry. More effort will be required by the control industry to ensure that, where feasible, the techniques are put into practice. A good rapport must be maintained between the various disciplines involved to ensure this. The trend is in this direction (Cook and Marsili-Libelli, 1981; Goodwin *et al.*, 1981).

67.4.5 Recycle and Feed Distribution Control

Significant attention has been directed to solids management within activated sludge systems. The steady-state formulations relating recycle ratio to the attainable biomass concentrations in the reactor, *e.g.* Lawrence and McCarty (1970), demonstrated the importance of recycle flow rate as a design variable. The use of recycle flow rate as a control variable for unsteady-state conditions, manipulated by some control law, is not as clear. Current evidence suggests that recycle flow rate has poor control authority and, if anything, its manipulation serves to deteriorate effluent quality.

Keinath and Cashion (1980) determined that clarification efficiencies were reduced slightly by the hydraulic transients imposed by recycle control. Holmberg (1982a) examined recycle control in proportion to feed flow rate and determined that this control philosophy has an effect opposite to the one desired, *i.e.* sludge would accumulate in the settler rather than in the aeration tank. From experiments Holmberg (1982a) determined that constant recycle flow rate control was preferred. During an instrument evaluation program, Stephenson *et al.* (1983) measured reactor and return activated sludge suspended solids concentrations on-line in parallel plants having diurnal hydraulic and organic loadings (Figure 6). In one plant the recycle flow rate was held constant and equal to the average daily feed flow rate; in the other plant the return flow rate was controlled to 100% of the feed flow rate. There was little difference in the reactor solids concentration in either of the plants during a 24 h period. The concentration was essentially constant in the reactor with variable recycle rate, whereas only a small change of about ±5% was observed in the other plant between the peak and minimum flow rates. With constant recycle flow rate, the return solids concentration was observed to vary with the clarifier solids loading. This indicates that the aerator/clarifier/thickener can be self-regulating in terms of solids distribution within the limits of the thickening constraints of the settler.

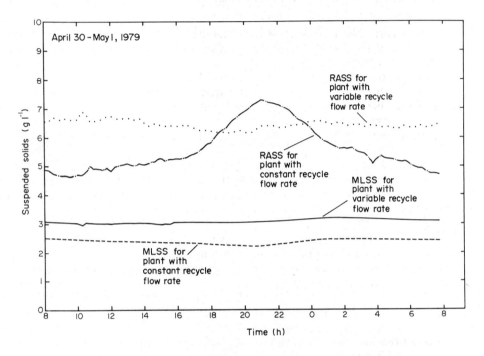

Figure 6 Comparison of return activated sludge and mixed liquor (MLSS) suspended solids concentration with and without recycle control

Olsson (1981a, 1981b) pointed out that there is much controversy about the selection of return activated sludge flow rate as a manipulable variable for control. This doubtlessly will continue until a systematic approach (Olsson, 1981b) is identified that ties together the constraints of clarifier operation with the overall plant operation.

To overcome the limitations of hydraulic transients imposed by variable recycle flow rate control, the use of dynamic step feed control to match the organic loading to the biomass has been proposed and evaluated (Andrews, 1976; Keinath, 1981). Yust *et al.* (1981) adapted the technique proposed by Andrews to control SCOUR by manipulating the wastewater feed distribution. SCOUR was calculated on-line in the third compartment of a three-tanks-in-series pilot plant using the technique proposed by Brouzes (1969) and dividing the volumetric rate by the compartment solids concentration. The control objective was to maintain a constant growth rate through SCOUR control in the final compartment. SCOUR control was achieved within the physical constraints imposed by the process. At high organic loadings the limited results suggested that the filterable effluent organic carbon concentration and its variability may have been reduced compared to a parallel completely mixed system. However, the effluent suspended solids concentration was significantly higher in the SCOUR-controlled case. It was felt that this resulted from higher air flow shear rates in the step feed system (Yust, 1981). Results were similar to those of Keinath and Cashion (1980).

Hruschka and Hegemann (1981) evaluated a modified step feed approach to control F/M ratio in a full scale activated sludge plant in Germany. Aeration tanks were taken in and out of service based on respiration rate measurements. Effluent quality was maintained and 30 to 40% aeration energy savings were achieved.

67.5 SUMMARY

In activated sludge systems, the major control areas are aeration and solids inventory. Effective pump station control is important to minimize plant hydraulic disturbances and reduce the level and variability of plant effluent suspended solids.

There are many applications and possibilities for instrument-based control in wastewater treatment. The applications have not been exhausted, but significantly more communication needs to be exercised between the parties involved in treatment plant design and those knowledgeable in control. Automation cannot rectify plant design deficiencies. Regardless of the type of control strategy chosen, the control objective cannot be reached if the plant is too conservatively designed or inflexible, nor can a control strategy necessarily correct for a chronically overloaded situation.

Better links need to be made between the groups influencing the design of wastewater treatment plants and the experts in control. This will help to ensure that control and estimation possibilities are realized in practice.

67.6 REFERENCES

Aarinen, R., J. Tirkkonen and A. Halme (1979). Experiences on instrumentation and control of activated sludge plants — a microprocessor application. In *A Link Between Science and Applications of Automatic Control*, ed. A. Niemi, B. Wahlstrom and J. Virkunnen, pp. 255–262. Pergamon, Oxford.

Andersson, L. G. (1979). Energy savings at wastewater treatment plants. Report to the Commissioner of the European Communities and the Danish Council of Technology, Water Quality Institute, DK-2970, Hørsholm, Denmark.

Andrews, J. F. (1976). Specific oxygen utilization rate for control of the activated sludge process. *Prog. Water Technol.*, **8** (6), 451–460.

Andrews, J. F., P. E. Sørensen and M. T. Garrett (1980). Control of nitrification in the oxygen activated sludge process. *Prog. Water Technol.*, **12** (5), 497–519.

Arthur, R. M. (1980). On-line measurements improve activated sludge process control. *Intech*, **27** (9), 103–107.

Barnhart, E. L. (1985). An overview of oxygen-transfer systems. *Proceedings: Seminar Workshop on Aeration System Design, Testing, Operation and Control*, Report EPA 600/9-85-005. US Environmental Protection Agency, Cincinnati, OH.

Beck, M. B. (1977a). Critical assessment of present-day attitudes towards control systems in water and wastewater management. *Prog. Water Technol.*, **9** (5/6), 13–15.

Beck, M. B. (1977b). Modelling and control in practice. *Prog. Water Technol.*, **9** (5/6), 557–564.

Beck, M. B. (1981). *Operational Water Quality Management: Beyond Planning and Design*. Executive Report 7. International Institute for Applied Systems Analysis (IIASA), A-2361, Laxenburg, Austria.

Bonnick, A. S. and J. M. Sidwick (1981). Instrumentation, control and automation—the choices. *Water Sci. Technol.*, **13** (8), 35–40.

Boon, A. G. and B. Chambers (1985). Design protocol for aeration systems—U.K. perspective. *Proceedings: Seminar Workshop on Aeration System Design, Testing, Operation and Control*, Report EPA 600/9-85-005. US Environmental Protection Agency, Cincinnati, OH.

Brouzes, P. (1969). Automated activated sludge plants with respiratory metabolism control. In *Advances in Water Pollution Research*, ed. S. H. Jenkins, pp. 385–396. Pergamon, Oxford.

Brouzes, P. H. (1979). Monitoring and controlling activate sludge plants in connection with aeration. *Prog. Water Technol.*, **11** (3), 193–200.

Brueck, T. M., D. I. Knudsen and D. F. Peterson (1981). Automatic computer-based control of a combined sewer system. *Water Sci. Technol.*, **13** (8) 103–109.

Chambers, B. and E. J. Tomlinson (1982). *Bulking of Activated Sludge — Preventative and Remedial Methods*. Ellis Horwood, Chichester.

Chapman, D. T. (1983). *The Influence of Dynamic Loads and Process Variables on Secondary Clarification*. Ph.D. Thesis, University of Alberta, Edmonton.

Clifft, R. C. and J. F. Andrews (1981a). Aeration control for reducing energy consumption in smalll activated sludge plants. *Water Sci. Technol.*, **13** (10), 371–379.

Clifft, R. C. and J. F. Andrews (1981b). Predicting the dynamics of oxygen utilization in the activated sludge process. *J. Water Pollut. Control Fed.*, **53**, 1219–1232.

Collins, M. A. and R. M. Crosby (1980). Impact of flow variation on secondary clarifier performance. Paper presented at the 53rd Annual Conference of the Water Pollution Control Federation, Las Vegas, September 28–October 3.

Cook, S. and S. Marsili-Libelli (1981). Estimation and control problems in activated sludge processes. *Water Sci. Technol.*, **13** (11/12), 737–742.

Dobrowolski, F. J. and J. Mendelsohn (1977). Automated pump and blower operation for reliability, cost reduction and energy conservation. *Prog. Water. Technol.*, **9** (5/6), 405–408.

DOE and NWC (1981). Working party on control systems for the water industry. Final Report Standing Technical Committee on Wastewater Treatment, National Water Council, London, UK, May.

EPA (1975). Process design manual for nitrogen control. In *Technology Transfer*. US Environmental Protection Agency, Cincinnati, OH.

EPA (1977). Current status of research in automation of wastewater treatment in the United States. In *Technology Transfer*. US Environmental Protection Agency, Cincinnati, OH.

Flanagan, M. J. and B. D. Bracken (1977). Design procedures for dissolved oxygen control of activated sludge processes. Report EPA-600/2-77-032. US Environmental Protection Agency, Cincinnati, OH.

Garrett, M. T. (1958). Hydraulic control of activated sludge growth rate. *Sewage Ind. Wastes*, **30** (3), 253–261.

Garrett, M. T., J. Andrews, I. J. Kugelman and W. W. Schuk (1982). Instrument testing center proposed. *APWA Reporter*, March, 6–10.

Gillblad, T. and G. Olsson (1977). Computer control of a medium sized activated sludge plant. *Prog. Water Technol.*, **9** (5/6), 427–433.

Goodwin, G. C., B. McInnis and R. S. Long (1981). *Adaptive Control Algorithm for Wastewater Treatment and pH Neutralization*. Technical Report EE8112. Department of Electrical and Computer Enginering, University of Newcastle, NSW 2308, Australia.

Gujer, W. (1981). The effect of particulate organic matter on activated sludge yield and oxygen requirement. *Prog. Water Technol.*, **13** (4/5), 79–95.

Hegg, B. A., K. L. Rakness and J. R. Schultz (1978). Evaluation of operation and maintenance factors limiting municipal wastewater treatment plant performance. Part 1. *J. Water Pollut. Control. Fed.*, **50**, 491–426.

Hobson, J. A. (1982). Energy-saving: pumping of sewage. Paper presented at Seminar on Low-cost and Low-energy Wastewater Treatment and Recycling, Madrid, Spain, October 25–29.

Holmberg, A. (1981a). Microprocessor-based estimation of oxygen utilization in the activated sludge wastewater treatment process. *Int. J. Systems Sci.*, **12**, 703–718.

Holmberg, A. (1981b). A systems engineering approach to biotechnical processes — experiences of modelling, estimation and control methods. *Acta Polytechnica Scandinavica*. Mathematics and Computer Science Series No. 33. Systems Theory Laboratory, Helsinki University of Technology, SF-02150 Espoo 15, Finland.

Holmberg, A. (1981c). Modelling of the activated sludge process for state estimation and control. Report B62. Systems Theory Laboratory, Helsinki University of Technology, SF-02150, Espoo 15, Finland.

Holmberg, A. (1982a). Modelling of the activated sludge process for microprocessor-based state estimation and control. *Water Res.*, **16**, 1233–1246.

Holmberg, A. (1982b). A microprocessor based estimation and control system for the activated sludge process. Presented at 1st IFAC Workshop on Modelling and Control of Biotechnical Processes, Helsinki.

Holmberg, A. and J. Ranta (1982). Procedure for parameter and state estimation of microbial growth process models. *Automatica*, **18** (2), 181–193.

Holmström, H. and B. Hultman (1981). Improved operation of municipal wastewater treatment plants in Sweden by use of instrumentation and automation. *Water Sci. Technol.*, **13** (8), 7–13.

Hruschka, H. and W. Hegemann (1981). Investigation on constant F/M ratio by adaption of aeration tank volume. *Water Sci. Technol.*, **13** (8), 495–499.

IAWPR (1974). Instrumentation, control and automation for wastewater treatment systems. *Prog. Water. Technol.*, **6**, 570 pp.

IAWPR (1977). International workshop on instrumentation and control for water and wastewater treatment and transport systems. *Prog. Water Technol.*, **9** (5/6), 646 pp.

IAWPR (1981). Practical experiences of control and automation in wastewater treatment and water resources management. *Water Sci. Technol.*, **13** (8–12), 845 pp.

IJC (1983). Detailed review of thirty municipal wastewater treatment facilities in the Great Lakes Basin. Report prepared for Work Group III, Municipal Abatement Task Force, International Joint Commission, Windsor, Ontario.

Iwaki, H., T. Ohto, S. Nogita and S. Kato (1977). Preliminary study of dissolved oxygen control of a diffused air aeration plant. *Prog. Water Technol.*, **9** (5/6), 393–397.

Keinath, T. M. (1981). Solids inventory control in the activated sludge process. *Wat. Sci. Tech.*, **13** (10), 413–419.

Keinath, T. M. and B. S. Cashion (1980). Control strategies for the activated sludge proces. Report EPA-600/2-80-131. US Environmental Protection Agency, Cincinnati, OH.

Keinath, T. M., M. D. Ryckman, C. H. Dana and D. A. Hofer (1977). Activated sludge—unified system design and operation. *J. Environ. Eng. Div., Am. Soc. Civ. Eng.*, **103**, 829–849.

Lawrence, A. W. and P. L. McCarty (1970). Unified basis for biological treatment design and operation. *J. Sanit. Eng. Div., Am. Soc. Civ. Eng.*, **96**, 757–778.

Molvar, A., J. Roesler, R. Wise and R. Babcock (1976). Instrumentation and automation experiences in wastewater treatment facilities. Report EPA-600/2-76-298. US Environmental Protection Agency, Cincinnati, OH.

Nielsen, M. K., O. Persson and M. Kümmel (1981). Computer control of nitrifying and denitrifying activated sludge process. *Water Sci. Technol.*, **13** (9), 285–291.

Niku, S., E. D. Schroeder, G. Tchobanoglous and F. J. Samaniego (1981). Performance of activated sludge processes: reliability, stability and variability. Report EPA-600/52-81-227. US Environmental Protection Agency, Cincinnati, OH.

Ohto, T., S. Nogita, H. Iwaki and S. Kato (1977). Experiences of dissolved oxygen control of a diffused air aeration plant. *Prog. Water Technol.*, **9** (5/6), 399–403.

Olsson, G. (1976). State of the art in sewage treatment plant control. *AIChE Symp. Ser.*, **72**, No. 159, 52–76.

Olsson, G. (1980a). Automatic control in wastewater treatment plants. *Trib. CEBEDEAU*, **33**, no. 436, 121–130.

Olsson, G. (1980b). Some new results on activated sludge control based on dissolved oxygen profiles. In *Proc. Joint Automatic Control Conf., San Francisco*.

Olsson, G. (1981a). Control and automation in wastewater treatment — advances, unsolved problems and challenges. *Water Sci. Technol.*, **13** (11/12), 826–831.

Olsson, G. (1981b). A unified approach to sludge inventory control. Manuscript. Department of Automatic Control, Lund Institute of Technology, 22007 Lund, Sweden.

Olsson, G. and J. F. Andrews (1977). Estimation and control of biological activity in the activated sludge process using dissolved oxygen measurements. Presented at the IFAC Symposium on Environmental Systems Planning, Design and Control, August 1–5, 1977, Kyoto, Japan.

Olsson, G. and J. F. Andrews (1978). The dissolved oxygen profile, a valuable tool for control of the activated sludge process. *Water Res.*, **12**, 985–1004.

Olsson, G. and J. F. Andrews (1981a). Dissolved oxygen control in the activated sludge process. *Water Sci. Technol.*, **13** (10), 341–347.

Olsson, G. and J. F. Andrews (1981b). Estimation of loading changes for the nitrifying activated sludge process. Paper presented at the IFAC World Congress, Kyoto, Japan.

Olsson, G. and O. Hansson (1976). Stochastic modelling and computer control of a full scale wastewater treatment plant. In *Proc. Symp. Systems Models Air Water Pollut.* Institute of Measurement and Control, London.

Owen, W. F. (1982). *Energy in Wastewater Treatment.* Prentice-Hall, Englewood Cliffs, NJ.

Palm, J. C., D. Jenkins and D. S. Parker (1980). The relationship between organic loading, dissolved oxygen concentration and sludge settleability in the completely mixed activated sludge process. *J. Water Pollut. Control Fed.*, **52**, 2484–2506.

Parker, D. S., W. J. Kaufman and D. Jenkins (1971). Physical conditioning of activated sludge floc. *J. Water Pollut. Control Fed.*, **43**, 1817–1833.

Petersack, J. F. and R. G. Smith (1975). Advanced automatic control strategies for the activated sludge treatment process. Report EPA-670/2-75-039, NTIS PB-246 726, May. US Environmental Protection Agency, Cincinnati, OH.

Russell, D. L. (1980). Monitoring and sampling liquid effluents. *Chem. Eng. (NY)*, **97** (21), 108–120.

Schlegel, S. and J. Lohmann (1981). Control of dissolved oxygen in activated sludge plants. *Water Sci. Technol.*, **13** (9), 225–232.

Schuk, W. W., I. J. Kugelman and W. A. Rosenkranz (1981). Procedures for the development and evaluation of sensors to be used in monitoring wastewater. *Water Sci. Technol.*, **13** (9), 133–138.

Sezgin, M., D. Jenkins and D. S. Parker (1978). A unified theory of activated sludge bulking. *J. Water Pollut. Control Fed.*, **50**, 362–381.

Sørensen, P. E. (1979). Pilot scale evaluation of control schemes for the activated sludge process. Contributions from the Water Quality Institute, No. 1, 2970 Hørsholm, Denmark.

Sørensen, P. E. (1980). Plant optimization for effluent quality and energy control purposes. Paper presented at the workshop — Automation of the Activated Sludge Process, held at Wastewater Technology Centre, Burlington, Canada, June 27.

Sørensen, P. E. (1981). Method of evaluating the possible energy savings by implementing DO-control of fine bubble diffusors. *Water Sci. Technol.*, **13** (9), 263–268.

Stenstrom, M. K. and J. F. Andrews (1979). Real-time control of activated sludge process. *J. Environ. Eng. Div., Am. Soc. Civ. Eng.*, **105**, 245–260.

Stephenson, J. P., E. G. Luxon, B. A. Monaghan and R. G. Gillespie (1981a). Evaluation of instruments for continuous activated sludge monitoring. *Water Sci. Technol.*, **13** (11/12), 721–728.

Stephenson, J. P., B. A. Monaghan and P. J. Laughton (1981b). Automatic control of solids retention time and dissolved oxygen in the activated sludge process. *Water Sci. Technol.*, **13** (11/12), 751–758.

Stephenson, J. P., B. A. Monaghan and L. J. Yust (1983). Pilot scale investigation of computerized control of the activated sludge process. Report SCAT-12. Environmental Protection Service, Environment Canada, Ottawa.

Tanuma, R., O. Shimizu, K. Takeda and N. Tanji (1981). Dissolved oxygen control using aeration exhaust gas. *Water Sci. Technol.*, **13** (9), 183–188.

Tomlinson, E. J. (1982). The emergence of the bulking problem and the current situation in the UK. In *Bulking of Activated Sludge — Preventative and Remedial Methods*, ed. B. Chambers and E. J. Tomlinson, pp. 17–23. Ellis Horwood, Chichester.

Usrael, G. (1980). Régulation automatique de la fourniture d'oxygène dans une installation d'activation en vue de l'élimination de l'azote. *Trib. CEBEDEAU*, **33**, no. 436, 141–145.

von Buchstab, V. (1983). Treatment plants enter computer immersion phase. *Water Pollut. Control (Canada)*, **121** (2), 18–22.

Wells, C. H. (1979). Computer control of fully nitrifying activated sludge processes. *Instrument. Technol.*, **26** (4), 32–36.

White, S. F. (1981). Requirements and application of control and automation systems in wastewater treatment — a critical view. *Water Sci. Technol.*, **13** (8), 41–48.

Wilson, B. D. and L. B. Polkowski (1981). Evaluation of operational control strategies for final clarifiers. Presented at the 54th Annual Conference of the Water Pollution Control Federation, Detroit, Michigan, October 4–9.

Wood, J. A. and R. Ross (1983). Electrical considerations for pumping stations. Presented at the 10th Annual B. C. Water and Waste Association Conference, Harrison Hot Springs, British Columbia, April 24–27.

WPCF (1983). A decade of progress – America's quest for clean water. *Highlights*, **20**, April, 8 pp.

Yust, L. J. (1981). Dynamic real-time control of the activated sludge process using step feed. M. Eng. Thesis, McMaster University, Hamilton, Canada.

Yust, L. J., J. P. Stephenson and K. L. Murphy (1981). Dynamic step feed control for organic carbon removal in a suspended growth system. *Water Sci. Technol.*, **13** (11/12), 729–736.

Appendix 1: Glossary of Terms

Because of the broad multidisciplinary nature of biotechnology, both beginners and specialists in this field may find this glossary useful. It covers terms often used in the relevant areas of the biological, chemical and engineering sciences. It is not intended to be exhaustive. The material was generated primarily from four sources: 'Commercial Biotechnology', Office of Technology Assessment, Congress of the US (1984); 'Advances in Biotechnology', M Moo-Young *et al.* (1981); *Pure and Applied Chemistry*, Vol. 54, No. 9, pp. 1743–1749 (1982); and 'Dictionary of Biochemistry', J. Stenish (1975).

Acclimatization: The biological process whereby an organism adapts to a new environment. For example, it describes the process of developing microorganisms that degrade toxic wastes in the environment.

Activated sludge: Biological growth that occurs in aerobic, organic-containing systems. These growths develop into suspensions that settle and possess clarification and oxidative properties.

Activation energy: The difference in energy between that of the activated complex and that of the reactants; the energy that must be supplied to the reactants before they can undergo transformation to products.

Active immunity: Disease resistance in a person or animal due to antibody production after exposure to a microbial antigen following disease, inapparent infection or inoculation. Active immunity is usually long-lasting.

Active transport: The movement of a solute across a biological membrane such that the movement is directed against the concentration gradient and requires the expenditure of energy.

Activity: A measure of the effective concentration of an enzyme, drug, hormone or some other substance. It is also the product of the molar concentration of an ionic solute and its activity coefficient.

Adsorption: The taking up of molecules of gases, dissolved substances or liquids at the surfaces of solids or liquids with which they are in contact.

Aerobic: Living or acting only in the presence of free-form oxygen, as in air.

Affinity chromatography: The use of compounds, such as antibodies, bound to an immobile matrix to 'capture' other compounds as a highly specific means of separation and purification.

Airlift fermenter: Vessel in which a bioconversion process takes place; the sparged gas is the only source of agitation. The presence of a draft tube inside a fermenter distinguishes this type of fermenter from a bubble column.

Alga (pl. **algae**): A chlorophyll-containing, photosynthetic protist; algae are unicellular or multicellular, generally aquatic and either eukaryotic or prokaryotic.

Amino acids: The building blocks of proteins. There are 20 common amino acids.

Amino acid sequence: The linear order of amino acids in a protein.

Amylase: An enzyme that catalyzes the hydrolysis of starch.

Anabolism: The phase of intermediary metabolism that encompasses the biosynthetic and energy-requiring reactions whereby cell components are produced. Also, the cellular assimilation of macromolecules and complex substances from low-molecular weight precursors (*cf.* Catabolism).

Anaerobic: Living or acting in the absence of free-form oxygen.

Anaerobic digestion: The energy-yielding metabolic breakdown of organic compounds by microorganisms that generally proceeds under anaerobic conditions and with the evolution of gas. The term is most often used to describe the reduction of waste sludges to less solid mass and for methane gas production.

Antibiotic: A specific type of chemical substance that is administered to fight infections, usually bacterial infections, in humans or animals. Many antibiotics are produced by using microorganisms; others are produced synthetically.

Antibody: A protein (immunoglobulin) produced by humans or higher animals in response to

1145

exposure to a specific antigen and characterized by specific reactivity with its complementary antigen (see also Monoclonal antibodies).

Antigen: A substance, usually a protein or carbohydrate, which, when introduced into the body of a human or higher animal, stimulates the production of an antibody that will react specifically with it.

Antiserum: Blood serum containing antibodies from animals that have been inoculated with an antigen. When administered to other animals or humans, antiserum produces passive immunity.

Aromatic compound: A compound containing a benzene ring. Many speciality and commodity chemicals are aromatic compounds.

Ascites: Liquid accumulations in the peritoneal cavity. Used as a method for producing monoclonal antibodies.

Asepsis: The prevention of access of microorganisms causing disease, decay or putrefaction to the site of a potential infection.

Aseptic: Of, or pertaining to, asepsis. In directed fermentation processes, the exclusion of unwanted (contaminating) organisms.

Assay: A technique that measures a biological response.

Attenuated vaccine: Whole, pathogenic organisms that are treated with chemical, radioactive, or other means to render them incapable of producing infection. Attenuated vaccines are injected into the body, which then produces protective antibodies against the pathogen to protect against disease.

Autolysis: The self-destruction of a cell as a result of the action of its own hydrolytic enzymes.

Autotrophic: Capable of self-nourishment (opposed to heterotrophic).

Axial dispersion: Mixing along the flow path of fluids during processing as in a bioreactor.

Bacteria: Any of a large group of microscopic organisms having round, rodlike, spiral or filamentous unicellular or noncellular bodies that are often aggregated into colonies, are enclosed by a cell wall or membrane, and lack fully differentiated nuclei. Bacteria may exist as free-living organisms in soil, water, organic matter, or as parasites in the live bodies of plants and animals.

Bacteriophage (or **phage**)/**bacterial virus**: A virus that multiplies in bacteria. Bacteriophage lambda is commonly used as a vector in rDNA experiments.

Batch processing: A method of processing in which a bioreactor, for example, is loaded with raw materials and microorganisms, and the process is run to completion, at which time products are removed (*cf.* Continuous processing).

Binary fission: Asexual division in which a cell divides into two, approximately equal, parts, as in the growth method of some single cells.

Biocatalyst: An enzyme, in cell-free or whole-cell forms, that plays a fundamental role in living organisms or industrially by activating or accelerating a process.

Biochemical: Characterized by, produced by, or involving chemical reactions in living organisms; a product produced by chemical reactions in living organisms.

Biochip: An electronic device that uses biological molecules as the framework for molecules which act as semiconductors and functions as an integrated circuit.

Bioconversion: A chemical conversion using a biocatalyst.

Biodegradation: The breakdown of substances by biological agents, especially microbes.

Biological oxygen demand (BOD): The oxygen used in meeting the metabolic needs of aerobic organisms in water containing organic compounds.

Biological response modifier: Generic term for hormones, neuroactive compounds and immunoactive compounds that act at the cellular level; many are possible targets for production with biotechnology.

Biologics: Vaccines, therapeutic serums, toxoids, antitoxins and analogous biological products used to induce immunity to infectious diseases or harmful substances of biological origin.

Biomass: Organic matter of biological origin such as microbial and plant material.

Biooxidation: Oxidation (the loss of electrons) catalyzed by a biocatalyst.

Biopolymers: Naturally occurring macromolecules that include proteins, nucleic acids and polysaccharides.

Bioprocess: Any process that uses complete living cells or their components (*e.g.* enzymes, chloroplasts) to effect desired physical or chemical changes.

Bioreactor: Vessel in which a bioprocess takes place; examples include fermenter, enzyme reactor.

Biosensor: A device, usually electronic, that uses biological molecules to detect specific compounds.

Biosurfactant: A compound produced by living organisms that helps solubilize compounds such as organic molecules (*e.g.* oil and tar) by reducing surface tension between the compound and liquid.

Biosynthesis: Production, by synthesis or degradation, of a chemical compound by a living organism.

Biotechnology: Use of biological agents or materials to produce goods or services for industry, trade and commerce; a multidisciplinary field.

Bubble column: A gas–liquid contacting vessel in which sparged gas is the only source of agitation.

Bubbly flow: Type of two-phase flow in which the gas phase is distributed in the liquid in the form of bubbles whose dimensions are small compared to the characteristic dimension of the flow cross-section, as in certain designs of bioreactor geometrics.

Budding: A form of asexual reproduction typical of yeast, in which a new cell is formed as an outgrowth from the parent cell.

Buffer: A solution containing a mixture of a weak acid and its conjugate weak base that is capable of resisting substantial changes in pH upon the addition of small amounts of acidic or basic substances, as used for some fermentation media.

Bulking: Increase in volumetric solids as in certain waste treatment bioreactors, which limits the weight concentration that a clarifier can handle.

Callus: An undifferentiated cluster of plant cells that is a first step in regeneration of plants from tissue culture.

Carbohydrate: An aldehyde or a ketone derivative of a polyhydroxy alcohol that is synthesized by living cells. Carbohydrates may be classified either on the basis of their size into mono-, oligo- and poly-saccharides, or on the basis of their functional group into aldehyde or ketone derivatives.

Carboxylation: The addition of an organic acid group (COOH) to a molecule.

Catabolism: The phase of intermediary metabolism that encompasses the degradative and energy-yielding reactions whereby nutrients are metabolized. Also, the cellular breakdown of complex substances and macromolecules to low-molecular weight compounds (*cf.* Anabolism).

Catalysis: A modification, especially an increase, in the rate of a chemical reaction induced by a material (*e.g.* enzyme) that is chemically unchanged at the end of the reaction.

Catalyst: A substance that induces catalysis; an agent that enables a chemical reaction to proceed under milder conditions (*e.g.* at a lower temperature) than otherwise possible. Biological catalysts are enzymes; some nonbiological catalysts include metallic complexes.

Cell: The smallest structural unit of living matter capable of functioning independently; a microscopic mass of protoplasm surrounded by a semipermeable membrane, usually including one or more nuclei and various nonliving products, capable alone, or interacting with other cells, of performing all the fundamental functions of life.

Cell culture: The *in vitro* growth of cells usually isolated from a mixture of organisms. These cells are usually of one type.

Cell differentiation: The process whereby descendants of a common parental cell achieve and maintain specialization of structure and function.

Cell fusion: Formation of a single hybrid cell with nuclei and cytoplasm from different cells (as in 'cell fusion technology'; see also Hybridoma.)

Cell line: Cells that acquire the ability to multiply indefinitely *in vitro* (especially in plant cell tissues).

Cellulase: The enzyme that digests cellulose to sugars.

Cellulose: A polymer of six carbon sugars found in all plant matter; the most abundant biological compound on earth.

Chemical clarification: Characterization of a wastewater process involving distinct operations: coagulation, flocculation and sedimentation.

Chemostat: An apparatus for maintaining microorganisms in the (exponential) phase of growth over prolonged periods of time. This is achieved by the continuous addition of fresh medium and the continuous removal of effluent, so that the volume of the growing culture remains constant.

Chemostat selection: Screening process used to identify microorganisms with desired properties, such as microorganisms that degrade toxic chemicals (see Acclimatization).

Chemotherapeutic agent: A chemical that interferes with the growth of either microorganisms or cancer cells at concentrations at which it is tolerated by the host cells.

Chemotherapy: The treatment of a disease by means of chemotherapeutic agents.

Chitin: A homopolysaccharide of *N*-acetyl-D-glucosamine that is a major constituent of the hard, horny exoskeleton of insects and crustaceans.

Chlorophyll: The green pigment that occurs in plants and functions in photosynthesis by absorbing and utilizing the radiant energy of the sun.

Chloroplasts: Cellular organelles where photosynthesis occurs.

Chromosome: A structure in the nucleus of eukaryotic cells that consists of one or more large double-helical DNA molecules that are associated with RNA and histones; the DNA of the chromosome contains the genes and functions in the storage and in the transmission of the genetic information of the organism.

Chromatography: A process of separating gases, liquids or solids in a mixture or solution by adsorption as the mixture or solution flows over the adsorbent medium, often in a column. The substances are separated because of their differing chemical interaction with the adsorbent medium.

Chromosomes: The rodlike structures of a cell's nucleus that store and transmit genetic information; the physical structures that contain genes. Chromosomes are composed mostly of DNA and protein and contain most of the cell's DNA. Each species has a characteristic number of chromosomes.

Clinical trial: One of the final stages in the collection of data for drug approval where the drug is tested in humans.

Clone: A group of genetically identical cells or organisms produced asexually from a common ancestor.

Cloning: The amplification of segments of DNA, usually genes.

Coagulation: The process whereby chemicals are added to wastewater resulting in a reduction of the forces tending to keep suspended particles apart.

Coagulation–flocculation aids: Materials used in relatively small concentrations which are added either to the coagulation and/or flocculation basins and may be classified as: oxidants, such as chlorine or ozone; weighting agents, such as bentomite clay; activated silica; and polyelectrolytes.

Coding sequence: The region of a gene (DNA) that encodes the amino acid sequence of a protein.

Codon: The sequence of three adjacent nucleotides that occurs in messenger RNA (mRNA) and that functions as a coding unit for a specific amino acid in protein synthesis. The codon determines which amino acid will be incorporated into the protein at a particular position in the polypeptide chain.

Coefficient of thermal conductivity: A physical parameter characterizing intensity of heat conduction in a substance; it is numerically equal to the conductive heat flux density due to a temperature gradient of unity.

Coenzyme: The organic molecule that functions as a cofactor of an enzyme.

Cofactor: The nonprotein component that is required by an enzyme for its activity. The cofactor may be either a metal ion (activator) or an organic molecule (coenzyme) and it may be attached either loosely or tightly to the enzyme; a tightly attached cofactor is known as a prosthetic group.

Colony: A group of contiguous cells that grow in or on a solid medium and are derived from a single cell.

Complementary DNA (cDNA): DNA that is complementary to messenger RNA; used for cloning or as a probe in DNA hybridization studies.

Compulsory licensing: Laws that require the licensing of patents, presumably to ensure early application of a technology and to diffuse control over a technology.

Constitutive enzyme: An enzyme that is present in a given cell in nearly constant amounts regardless of the composition of either the tissue or the medium in which the cell is contained.

Conjugation: The covalent or noncovalent combination of a large molecule, such as a protein or a bile acid, with another molecule. Also, the alternating sequence of single and double bonds in a molecule. Also, the genetic recombination in bacteria and in other unicellular organisms that resemble sexual reproduction and that entails a transfer of DNA between two cells of opposite mating type which are associated side by side.

Continuous processing: Method of processing in which raw materials are supplied and products are removed continuously, at volumetrically equal rates (*cf.* Batch processing).

Continuum: A medium whose discrete heterogeneous structure can be neglected, as in certain fluid flow problems.

Convective mass transfer: Mass transfer produced by simultaneous convection and molecular diffusion. The term is usually used to describe mass transfer associated with fluid flow and involves the mass transfer between a moving fluid and a boundary surface or between two immiscible moving fluids.

Convective transfer: The transfer of mass, heat or momentum in a medium with a nonhomogeneous distribution of velocity, temperature, or concentration; it is accompanied by the displacement of macroscopic elements through the medium.

Cosmid: A DNA cloning vector consisting of plasmid and phage sequences.

Corporate venture capital: Capital provided by major corporations exclusively for high-risk investments.

Crabtree effect: The inhibition of oxygen consumption in cellular respiration that is produced by increasing concentrations of glucose (see also Pasteur effect).

Critical dilution rate: Dilution rate at which wash-out conditions of cells in a bioreactor occurs (see Chemostat).

Cross flow filtration: Method of operating a filtration device whereby the processed material prevents undue build-up of filtered material on filter.

Crystalloid: A noncolloidal low-molecular weight substance.

Culture medium: Any nutrient system for the artificial cultivation of bacteria or other cells; usually a complex mixture of organic and inorganic materials.

Cyclic batch culture: Method of operating a bioreactor whereby a fill-and-dump approach retains enough biocatalyst to avoid need for re-inoculation. In cell cultures, relatively insignificant growth between fill and dump stages occurs.

Cytoplasm: The 'liquid' portion of a cell outside and surrounding the nucleus.

Cytotoxic: Damaging to cells.

Declining phase (or **Death phase**): The phase of growth of a culture of cells that follows the stationary phase and during which there is a decrease in the number (or the mass) of the cells.

Denitrification: The formation of molecular nitrogen from nitrate by way of nitrite.

Deoxyribonucleic acid (DNA): A linear polymer, made up of deoxyribonucleotide repeating units, that is the carrier of genetic information; present in chromosomes and chromosomal material of cell organelles such as mitochondria and chloroplasts, and also present in some viruses. The genetic material found in all living organisms. Every inherited characteristic has its origin somewhere in the code of each individual's DNA.

Diagnostic products: Products that recognize molecules associated with disease or other biological conditions and are used to diagnose these conditions.

Dialysis: The separation of macromolecules from ions and low-molecular weight compounds by means of a semipermeable membrane that is impermeable to (colloidal) macromolecules but is freely permeable to crystalloids and liquid medium.

Dicots (dicotyledons): Plants with two first embryonic leaves and nonparallel veined mature leaves. Examples are soybean and most flowering plants.

Diffusion boundary layer: Characterized by a transverse concentration gradient of a given component in a mixture; the effect of the gradient produces a transverse (mass transfer) of this component.

Diffusion coefficient: A physical parameter that appears as a proportionality coefficient with the gradient of concentration of a specified component in the equation which establishes the dependence of the mass diffusion flux density of the given component on the concentration gradients of all the components in the mixture. 'Self-diffusion coefficient' denotes a physical parameter which characterizes the diffusion of some molecules in the same medium with respect to others for a single-component medium.

Diffusional mass flux: Mass flux due to molecular diffusion.

Dilution rate: Reciprocal of the residence time of a culture in a bioreactor; given by the flow rate divided by bioreactor volume.

Dimensional analysis: Method of determining the number and structure of dimensionless groups consisting of variables essential to a given process on the basis of a comparison of the dimensions of these variables.

Diploidy (or **Diploid state**): The chromosome state in which each of the various chromosomes, except the sex chromosome, is represented twice.

Disclosure requirements: A patent requirement for adequate public disclosure of an invention that enables other people to build and use the invention without 'undue' experimentation.

Distal: Remote from a particular location or from a point of attachment.

DNA: Deoxyribonucleic acid (see above).

DNA base pair: A pair of DNA nucleotide bases. Nucleotide bases pair across the double helix in a very specific way: adenine can only pair with thymine; cytosine can only pair with guanine.

DNA probe: A sequence of DNA that is used to detect the presence of a particular nucleotide sequence.

DNA sequence: The order of nucleotide bases in the DNA helix; the DNA sequence is essential to the storage of genetic information.

DNA synthesis: The synthesis of DNA in the laboratory by the sequential addition of nucleotide bases.

Doubling time: The observed time required for a cell population to double in either the number of cells or the cell mass; it is equal to the generation time only if all the cells in the population are capable of doubling, have the same generation time, and do not undergo lysis (see Generation time).

Downstream processing: After bioconversion of materials in a bioreactor, the separation and purification of the product(s).

Drug: Any chemical compound that may be administered to humans or animals as an aid in the treatment of disease.

Dry weight: The weight of a sample from which liquid has been removed, usually by drying of filtered material.

Eddy mass diffusivity: A quantity characterizing the intensity of turbulent mass transfer of a particular component, as in intensely mixed bioreactors.

Electrophoresis: The movement of charged particles through a stationary liquid under the influence of an electric field. Electrophoresis is a tool for the separation of particles in both preparative and analytical studies of macromolecules. Separation is achieved primarily on the basis of the charge on the particles and to a lesser extent on the basis of the size and shape of the particles. Potentially useful in downstream processing.

Elution: The removal of adsorbed material from an adsorbent, such as the removal of a product from an enzyme bound in a chromatography column.

Emulsification: The process of making lipids, oils, fats, more soluble in water.

Endoenzyme: An enzyme that acts at random in cleaving molecules of substrate.

Endorphins: Opiate-like, naturally occurring peptides with a variety of analgesic effects throughout the endocrine and nervous systems.

Enrichment culture: A culture used for the selection of specific strains of an organism from among a mixture; such a culture favors the growth of the desired strain under the conditions used.

Enzyme: Any of a group of catalytic proteins that are produced by living cells and that mediate and promote the chemical processes of life without themselves being altered or destroyed.

Enzyme induction: The process whereby an inducible enzyme is synthesized in response to an inducer. The inducer combines with a repressor and thereby prevents the blocking of an operator by the repressor.

***Escherichia coli* (*E. coli*)**: A species of bacteria that inhabits the intestinal tract of most vertebrates. Some strains are pathogenic to humans and animals. Many nonpathogenic strains are used experimentally as hosts for rDNA.

Eukaryote: A cell or organism with membrane-bound, structurally discrete nuclei and well-developed cell organelles. Eukaryotes include all organisms except viruses, bacteria and blue-green algae (*cf*. Prokaryote).

Exoenzyme: An enzyme that acts by cleaving the ends of molecular chains in a substrate.

Exponential growth: The growth of cells in which the number of cells (or the cell mass) increases exponentially.

Export controls: Laws that restrict technology transfer and trade for reasons of national security, foreign policy or economic policy.

Facultative: Capable of living under more than one set of conditions, usually with respect to aerobic or anaerobic conditions (*e.g.* a facultative anaerobe is an organism or a cell that can grow either in the absence, or in the presence, of molecular oxygen).

Fatty acids: Organic acids with long carbon chains. Fatty acids are abundant in cell membranes and are widely used as industrial emulsifiers.

Fed-batch culture: As in 'cyclic batch culture' except that significant changes in the medium (*e.g.* cell growth) occur during the addition and/or removal of materials from bioreactor.

Feedback inhibition: A negative feedback mechanism in which a product of an enzymatic reaction inhibits the activity of an enzyme that functions in the synthesis of this product.

Feedstocks: Raw materials used for the production of chemicals.

Fermentation: A bioprocess. Fermentation is carried out in bioreactors and is used in various industrial processes for the manufacture of products such as antibiotics, alcohols, acids and vaccines by the action of living organisms (strictly speaking, anaerobically).

Film boiling: Boiling in which a continuous film of vapor that collapses periodically into the bulk of the liquid is formed on the heated surface, as in certain evaporation processes.

Flagellum: (pl. **flagella**): A threadlike, cellular extension that functions in the locomotion of bacterial cells and of unicellular eukaryotic organisms.

Flavin adenine dinucleotide (FAD): The flavin nucleotide, riboflavin adenosine diphosphate, which is a coenzyme form of the vitamin riboflavin, and which functions in dehydrogenation reactions catalyzed by flavoproteins.

Flocculating agent: A reagent added to a dispersion of solids in a liquid to bring together the fine particles into larger masses.

Flocculation: The agglomeration of suspended material to form particles that will settle by gravity, as in the 'tertiary' treatment of waste materials.

Food additive (or **Food ingredient**): A substance that becomes a component of food or affects the characteristics of food and, as such, is regulated, *e.g.* by the US Food and Drug Administration.

Forced convection: Motion of fluid elements induced by external forces, *e.g.* in a bioreactor by a mechanical stirrer.

Free convection: Motion of fluid elements induced by 'natural' forces, *e.g.* by density differences caused by concentration or temperature gradients.

Free-living organism: An organism that does not depend on other organisms for survival.

Fruiting body: A mass of vegetative cells which swarm together at the same stage of growth.

Fungus: Any of a major group of saprophytic and parasitic plants that lack chlorophyll, including molds, rusts, mildews, smuts and mushrooms.

Gamma globulin (GG): A protein component of blood that contains antibodies and confers passive immunity.

Gene: The basic unit of heredity; an ordered sequence of nucleotide bases, comprising a segment of DNA. A gene contains the sequence of DNA that encodes one polypeptide chain (*via* RNA).

Gene amplification: In biotechnology, an increase in gene number for a certain protein so that the protein is produced at elevated levels.

Gene expression: The mechanism whereby the genetic directions in any particular cell are decoded and processed into the final functioning product, usually a protein (see also Transcription and Translation).

Gene transfer: The use of genetic or physical manipulation to introduce foreign genes into host cells to achieve desired characteristics in progeny.

Generation time: The time required by a cell for the completion of one cycle of growth (see also Doubling time).

Genetic engineering: Loose term used to describe any gene manipulative technique, especially recombinant DNA techniques.

Genome: The genetic endowment of an organism or individual.

Genus: A taxonomic category that includes groups of closely related species.

Germ cell: The male and female reproductive cells; egg and sperm.

Germplasm: The total genetic variability available to a species.

Glycoproteins: Proteins with attached sugar groups.

Glycoside: A mixed acetal (or ketal) derived from the cyclic hemiacetal (or hemiketal) form of an aldose (or a ketose); a compound formed by replacing the hydrogen or the hydroxyl group of the anomeric carbon of the carbohydrate with an alkyl or aryl group.

Glycosylation: The attachment of sugar groups to a molecule, such as a protein.

Gram negative: Designating a bacterium that does not retain the initial Gram stain but retains the counterstain. Gram-negative bacteria possess a relatively thin cell wall that is not readily digested by the enzyme lysozyme, and in which the peptidoglycan layer is covered with lipopolysaccharide.

Gram positive: Designating a bacterium that retains the initial Gram stain and is not stained by the counterstain. Gram-positive bacteria generally possess a relatively thick and rigid cell wall that is readily digested by the enzyme lysozyme, and that consists of a layer of peptidoglycan.

Gram stain: A set of two stains (chemicals) that are used to stain bacteria; the staining depends on the composition and the structure of the bacterial cell wall.

Growth hormone (GH): A group of peptides involved in regulating growth in higher animals.

Heat flux: The quantity of heat that passes through an arbitrary surface per unit time.

Heat flux density: Heat flux per unit area.

Heat transfer: Spontaneous irreversible process of heat transmission in a space with a nonisothermal temperature field, as in the cooling or heating of bioreactors and ancilliary equipment.

Hemicellulose: A polymer of D-xylose that contains side chains of other sugars and that serves to cement plant cellulose fibers together.

Herbicide: An agent (*e.g.* a chemical) used to destroy or inhibit plant growth; specifically, a selective weed killer that is not injurious to crop plants.

Heterofermentative lactic acid bacteria: Lactic acid bacteria that produce in fermentation less than 1.8 moles of lactic acid per mole of glucose; in addition to lactic acid, these organisms produce ethanol, acetate, glycerol, mannitol and carbon dioxide (see also Homofermentative lactic acid bacteria).

Heterotroph: A cell or organism that requires a variety of carbon-containing compounds from animals and plants as its source of carbon, and that synthesizes all of its carbon-containing biomolecules from these compounds and from small inorganic molecules.

Heterotrophic: Pertaining to a regulatory enzyme in which the effector is a metabolite other than the substance of the enzyme.

High performance liquid chromatography (HPLC): A recently developed type of chromatography that is potentially important in downstream processing of bioreactor products.

Histone: A basic, globular, and simple protein that is characterized by its high content of arginine and lysine. Histones are found in association with nucleic acids in the nuclei of many eukaryotic cells.

Homofermentative lactic acid bacteria: Lactic acid bacteria that produce 1.8–2.0 moles of lactic acid per mole of glucose during fermentation.

Hormone: A chemical messenger found in the circulation of higher organisms that transmits regulatory messages to cells.

Host: A cell whose metabolism is used for growth and reproduction of a virus, plasmid or other form of foreign DNA.

Host–vector system: Compatible combinations of host (*e.g.* bacterium) and vector (*e.g.* plasmid) that allow stable introduction of foreign DNA into cells.

Hybrid: The offspring of two genetically dissimilar parents (*e.g.* a new variety of plant or animal that results from cross-breeding two different existing varieties; a cell derived from two different cultured cell lines that have fused).

Hybridization: The act or process of producing hybrids.

Hybridoma: Product of fusion between myeloma cell (which divides continuously in culture and is 'immortal') and lymphocyte (antibody-producing cell); the resulting cell grows in culture and produces monoclonal antibodies.

Hybridoma technology: See Monoclonal antibody technology.

Hydrolysis: Chemical reaction involving addition of water to break bonds.

Hydroxylation: Chemical reaction involving the addition of hydroxyl (OH) groups to chemical compounds.

Hypha (pl. hyphae): The filamentous and branched tube that forms the network which contains the cytoplasm of the mycelium of a fungus.

Immobilized enzyme or cell techniques: Techniques used for the fixation of enzymes or cells on to solid supports. Immobilized cells and enzymes are used in continuous bioprocessing in bioreactors and upstream or downstream processing of materials.

Immune response: The reaction of an organism to invasion by a foreign substance. Immune responses are often complex, and may involve the production of antibodies from special cells (lymphocytes), as well as the removal of the foreign substance by other cells.

Immunization: The administration of an antigen to an animal organism to stimulate the production of antibodies by that organism. Also, the administration of antigens, antibodies or lymphocytes to an animal organism to produce the corresponding active, passive or adoptive immunity.

Immunoassay: The use of antibodies to identify and quantify substances. The binding of antibodies to antigen, the substance being measured, is often followed by tracers such as radioisotopes.

Immunogenic: Capable of causing an immune response (see also Antigen).

Immunotoxin: A molecule attached to an antibody capable of killing cells that display the antigen to which the antibody binds.

Inducible enzyme: An enzyme that is normally either absent from a cell or present in very small amounts, but that is synthesized in appreciable amounts in response to an inducer in the process medium.

Interface: The boundary between two phases.

Interferons (Ifns): A class of glycoproteins (proteins with sugar groups attached at specific locations) important in immune function and thought to inhibit viral infections.

In vitro: Literally, in glass; pertaining to a biological reaction taking place in an artificial apparatus; sometimes used to include the growth of cells from multicellular organisms under cell culture conditions. *In vitro* diagnostic products are products used to diagnose disease outside of the body after a sample has been taken from the body.

In vivo: Literally, in life; pertaining to a biological reaction taking place in a living cell or organism. *In vivo* products are products used within the body.

Ionic strength: A measure of the ionic concentration of a solution.

Laminar flow: Fluid motion in which the existence of steady fluid particle trajectories can exist; in processing equipment, it represents relatively low levels of mixing intensities.

Isoelectric pH (isoelectric point): The pH at which a molecule has a net zero charge; the pH at which the molecule has an equal number of positive and negative charges, which includes those due to any ions bound by the molecule.

Isoelectrophoretic pH (isoelectrophoretic point): The pH at which the electrophoretic mobility of a protein is zero; this pH may coincide with the theoretical isoelectric pH of the protein, depending on the surface structure of the protein, the ionic strength, and the nature of the ionic double layer around the protein.

Lag phase: That phase of growth of a cell that precedes the exponential phase and during which there is only little or no growth.

Leaching: The removal of a soluble compound such as an ore from a solid mixture by washing or percolating.

Lignocellulose: The composition of woody biomass, including lignin and cellulose.

Lignolytic: Pertaining to the breakdown of lignin.

Lime: Various natural forms of the chemical compound CaO, as in hydrated lime and dolomitic lime.

Linker: A small fragment of synthetic DNA that has a restriction site useful for gene cloning, which is used for joining DNA strands together.

Lipase: An enzyme that catalyzes the hydrolysis of fats.

Lipids: A large, varied class of water-insoluble fat-based organic molecules; includes steroids, fatty acids, prostaglandins, terpenes and waxes.

Lipopolysaccharide: A water-soluble lipid–polysaccharide complex.

Liposome transfer: The process of enclosing biological compounds inside a lipid membrane and allowing the complex to be taken up by a cell.

Lymphocytes: Specialized white blood cells involved in the immune response; B lymphocytes produce antibodies.

Lymphokines: Proteins that mediate interactions among lymphocytes and are vital to proper immune function.

Lyophilization: The removal of water under vacuum from a frozen sample; a relatively gentle process in which water sublimes.

Lysis: The rupture and dissolution of cells.

Mass exchange: Mass transfer across an interface or a permeable wall (membrane) between two phases.

Mass flux: The mass of a given mixture component passing per unit time across any surface.

Mass flux density: Mass flux per unit area of surface.

Mass transfer: Spontaneous irreversible process of transfer of mass of a given component in a space with a nonhomogenous field of the chemical potential of the component. In the simplest case, the driving force is the difference in concentration (in liquids) or partial pressure (in gases) of the component. Other physical quantities, *e.g.* temperature difference (thermal diffusion), can also induce mass transfer.

Mass transfer coefficient: A quantity characterizing the intensity of mass transfer; it is numerically equal to the ratio of the mass flux to the difference of its mass fractions. For the case of mass transfer between a liquid medium and a gas, the mass fraction of a given component in the liquid is determined by phase equilibrium parameters (distribution coefficient) with allowance, if necessary, for resistance to transfer at the phase boundary *per se*.

Mass velocity: Mass flow rate across a unit area perpendicular to the direction of the velocity vector.

Mesophile: An organism that grows at moderate temperatures in the range 20–45 °C, and that has an optimum growth temperature in the range 30–39 °C.

Mesophilic: Of, or pertaining to, mesophiles.

Messenger RNA (mRNA): RNA that serves as the template for protein synthesis; it carries the transcribed genetic code from the DNA to the protein synthesizing complex to direct protein synthesis.

Metabolism: The physical and chemical processes by which chemical components are synthesized into complex elements, complex substances are transformed into simpler ones, and energy is made available for use by an organism.

Metabolite: Any reactant, intermediate or product in the reactions of metabolism.

Metallothioneins: Proteins, found in higher organisms, that have a high affinity for heavy metals.

Methanogens: Bacteria that produce methane as a metabolic product.

Microorganisms: Microscopic living entities; microorganisms can be viruses, prokaryotes (*e.g.* bacteria) or eukaryotes (*e.g.* fungi). Also referred to as microbes.

Microencapsulation: The process of surrounding cells with a permeable membrane.

Mitochondrion (pl. **mitochondria**): A subcellular organelle in aerobic eukaryotic cells that is the site of cellular respiration and that carries out the reactions of the citric acid cycle, electron transport, and oxidative phosphorylation. Mitochondria contain DNA and ribosomes, carry out protein synthesis, and are capable of self-replication.

Mixed culture: Culture containing two or more types of microorganisms.

Molecular diffusion: Mass transfer resulting from thermal motion. Concentration diffusion refers to molecular diffusion resulting from a nonhomogenous distribution of concentrations of components of a mixture.

Monoclonal antibodies (MAbs): Homogeneous antibodies derived from a single clone of cells; MAbs recognize only one chemical structure. MAbs are useful in a variety of industrial and medical capacities since they are easily produced in large quantities and have remarkable specificity.

Monoclonal antibody technology: The use of hybridomas that produce monoclonal antibodies for a variety of purposes. Hybridomas are maintained in cell culture or, on a larger scale, as tumors (ascites) in mice. Also referred to as 'hybridoma' technology.

Monocots (monocotyledons): Plants with single first embryonic leaves, parallel-veined leaves, and simple stems and roots. Examples are cereal grains such as corn, wheat, rye, barley and rice.

Monosaccharide: A polyhydroxy alcohol containing either an aldehyde or a ketone group; a simple sugar.

Multigenic: A trait specialized by several genes.

Multi-phase medium: A medium consisting of two or more single-phase portions with physical properties changing discontinuously (stepwise) at the boundaries of the medium, as in a gas–liquid dispersion used in aerobic fermentations.

Mutagenesis: The induction of mutation in the genetic material of an organism; researchers may use physical or chemical means to cause mutations that improve the production of capabilities of organisms.

Mutagen: An agent that causes mutation.

Mutant: An organism with one or more DNA mutations, making its genetic function or structure different from that of a corresponding wild-type organism.

Mutation: A permanent change in a DNA sequence.

Mycelium (pl. **mycelia**): The vegetative structure of a fungus that consists of a multinucleate mass of cytoplasm, enclosed within a branched network of filamentous tubes known as hyphae.

Myeloma: Antibody-producing tumor cells.

Myeloma cell line: Myeloma cells established in culture.

Natural convection: Free motion due to gravitational forces in a system with a non-homogeneous density distribution (see also Free convection).

Neurotransmitters: Small molecules found at nerve junctions that transmit signals across those junctions.

Newtonian fluid: A fluid, the viscosity of which is independent of the rate and/or duration of shear.

NIH Guidelines: Guidelines, established by the US National Institutes of Health, on the safety of research involving recombinant DNA.

Nitrogen fixation: The conversion of atmospheric nitrogen gas to a chemically combined form, ammonia (NH_3), which is essential to growth. Only a limited number of microorganisms can fix nitrogen.

Nodule: The anatomical part of a plant root in which nitrogen-fixing bacteria are maintained in a symbiotic relationship with the plant.

Nodulins: Proteins, possibly enzymes, present in nodules; function unknown.

Non-Newtonian fluid: A fluid, the viscosity of which depends on the rate and/or duration of shear.

Nucleate boiling: Boiling in which vapor is generated in the form of periodically produced and growing discrete bubbles.

Nucleic acids: Macromolecules composed of sequences of nucleotide bases. There are two kinds of nucleic acids: DNA, which contains the sugar deoxyribose, and RNA, which contains the sugar ribose.

Nucleoside: A glycoside composed of D-ribose or 2-deoxy-D-ribose and either a purine or a pyrimidine.

Nucleotide base: A structural unit of nucleic acid. The bases present in DNA are adenine, cytosine, guanine and thymine. In RNA, uracil substitutes for thymine.

Nucleus: In the biological sciences, a relatively large spherical body inside a cell that contains the chromosomes.

Oligomer: A molecule that consists of two or more monomers linked together, covalently or non-covalently.

Oligonucleotides: Short segments of DNA or RNA.

Optical density (absorbance): A measure of the light absorbed by a solution.

Organelle: A specialized part of a cell that conducts certain functions. Examples are nuclei, chloroplasts and mitochondria, which contain most of the genetic material, conduct photosynthesis and provide energy, respectively.

Organic compounds: Molecules that contain carbon.

Organic micropollutant: Low molecular weight organic compounds considered hazardous to humans or the environment.

Osmosis: The movement of water or another solvent across a semipermeable membrane from a region of low solute concentration to one of a higher solute concentration.

Osmotic pressure: The pressure that causes water or another solvent to move in osmosis from a solution having a low solute concentration to one having a high solute concentration; it is equal to the hydrostatic pressure that has to be applied to the more concentrated solution to prevent the movement of water (solvent) into it.

Oxidation ponds: Quiescent earthen basins that provide sufficient hydraulic hold-up time for the natural processes to effect removal and stabilization of organic matter.

Oxygen transfer rate (OTR): Mass transfer for oxygen solute as in fermentation medium.

Parasite: An organism that lives in or upon another organism from which it derives some or all of its nutrients.

Passive immunity: Disease resistance in a person or animal due to the injection of antibodies from another person or animal. Passive immunity is usually short-lasting (*cf.* Active immunity).

Pasteur effect: The inhibition of glycolysis and the decrease of lactic acid accumulation that is produced by increasing concentrations of oxygen.

Patent: A limiting property right granted to inventors by a government allowing the inventor of a new invention the right to exclude all others from making, using or selling the invention unless specifically approved by the inventor, for a specified time period in return for full disclosure by the inventor about the invention.

Pathogen: A disease-producing agent, usually restricted to a living agent such as a bacterium or virus.

Pectin: A polysaccharide that occurs in fruits and that consists of a form of pectic acid in which many of its carboxyl groups have been methylated.

Peptide: A linear polymer of amino acids. A polymer of numerous amino acids is called a *polypeptide*. Polypeptides may be grouped by function, such as 'neuroactive' polypeptides.

Permease: An enzyme that is instrumental in transporting material across a biological membrane or within a biological fluid.

Pharmaceuticals: Products intended for use in humans, as well as *in vitro* applications to humans, including drugs, vaccines, diagnostics and biological response modifiers.

Photorespiration: Reaction in plants that competes with the photosynthetic process. Instead of fixing CO_2, RuBPCase can utilize oxygen, which results in a net loss of fixed CO_2.

Photosynthesis: The reaction carried out by plants where carbon dioxide from the atmosphere is fixed into sugars in the presence of sunlight; the transformation of solar energy into biological energy.

Plasma: The liquid (noncellular) fraction of blood. In vertebrates, it contains many important proteins (*e.g.* fibrinogen, responsible for clotting).

Plasmid: An extrachromosomal, self-replicating, circular segment of DNA; plasmids (and some viruses) are used as 'vectors' for cloning DNA in bacterial 'host' cells.

Plug flow: Flow of materials in which there is no mixing in the direction of flow (see Axial dispersion).

Polymer: A linear or branched molecule of repeating subunits.

Polypeptide: A long peptide, which consists of amino acids.

Polysaccharide: A polymer of sugars.

Pool boiling: Boiling with convective (free) motion in a liquid volume whose dimensions in all directions are large compared to the breakaway diameter of the bubble.

Primary metabolite: Metabolite that is required for the function of the organism's life support system.

Proinsulin: A precursor protein of insulin.

Prokaryote: A cell or organism lacking membrane-bound, structurally discrete nuclei and organelles. Prokaryotes include bacteria and the blue-green algae. (*cf.* eukaryote).

Promoter: A DNA sequence in front of a gene that controls the initiation of 'transcription' (see below).

Prophylaxis: Prevention of disease.

Protease: Protein-digesting enzyme.

Protein: A polypeptide consisting of amino acids. In their biologically active states, proteins function as catalysts in metabolism and, to some extent, as structural elements of cells and tissues.

Protist: A unicellular or multicellular organism that lacks the tissue differentiation and the elaborate organization that is characteristic of plants and animals.

Protoplast fusion: The joining of two cells in the laboratory to achieve desired results, such as increased viability of antibiotic-producing cells.

Protozoa: Diverse forms of eukaryotic microorganisms; structure varies from simple single cells to colonial forms; some protozoa are pathogenic.

Psychrophile: An organism that grows at low temperatures in the range of 0–25 °C, and that has an optimum growth temperature in the range 20–25 °C.

Pure culture: A culture containing only microorganisms from one species.

Pyrogenicity: The tendency for some bacterial cells or parts of cells to cause inflammatory reactions in the body, which may detract from their usefulness as pharmaceutical products.

Recarbonation: Unit water treatment process in which carbon dioxide is added to a lime-treated water. Basic purpose is the downward adjustment of the pH of the water.

Recombinant DNA (rDNA): The hybrid DNA produced by joining pieces of DNA from different organisms together *in vitro* (*i.e.* in an artificial apparatus).

Recombinant DNA technology: The use of recombinant DNA for a specific purpose, such as the formation of a product or the study of a gene.

Recombination: Formation of a new association of genes or DNA sequences from different parental origins.

Reducing sugar: A sugar that will reduce certain inorganic ions in solution, such as the copper(II) ions of Fehling's or Benedict's reagent; the reducing property of the sugar is due to its aldehyde or potential aldehyde group.

Regeneration: In biological sciences, the laboratory process of growing a whole plant from a single cell or small clump of cells.

Regulatory sequence: A DNA sequence involved in regulating the expression of a gene.

Repressible enzyme: An enzyme, the synthesis of which is decreased when the intracellular concentration of specific metabolites reaches a certain level.

Resistance gene: Gene that provides resistance to an environmental stress such as an antibiotic or other chemical compound.

Respiration: The cellular oxidative reactions of metabolism, particularly the terminal steps, by which nutrients are broken down; the reactions which require oxygen as the terminal electron acceptor, produce carbon dioxide as a waste product, and yield utilizable energy.

Restriction enzymes: Enzymes that cut DNA at specific DNA sequences.

Ribosome: One of a large number of subcellular, nucleoprotein particles that are composed of approximately equal amounts of RNA and protein and that are the sites of protein synthesis in the cell.

Ribosomal RNA: The RNA that is linked noncovalently to the ribosomal proteins in the two ribosomal subunits and that constitutes about 80% of the total cellular RNA.

RNA: Ribonucleic acid (see also Messenger RNA).

RuBPCase (ribulosebisphosphate carboxylase): An enzyme that catalyzes the critical step of the photosynthetic CO_2 cycle.

Salting out: The decrease in the solubility of a protein that is produced in solutions of high ionic strength by an increase of the concentrations of neutral salts.

Saccharification: The degradation of polysaccharides to sugars.

Scale-up: The transition of a process from research laboratory bench scale to engineering pilot plant or industrial scale.

Secondary metabolite: Metabolite that is not required by the producing organism for its life-support system.

Semiconductor: A material such as silicon or germanium with electrical conductivities intermediate between good conductors such as copper wire and insulators such as glass.

Shake flask: A laboratory flask for culturing microorganisms in a shaker–incubator which provides mixing and aeration.

Shear rate: The variation in velocity within a flowing material, as in a bioreactor.

Single cell protein (SCP): Cells, or protein extracts, of microorganisms grown in large quantities for use as human or animal protein supplements. A misnomer for multicellular SCP products.

Single-phase medium: Continuous single- or multi-component medium whose properties in space can vary in a continuous manner with no phase boundaries, *e.g.* a gas or a liquid.

Slaking: Process of adding water to quicklime, or recalcined lime, to produce a slurry of hydrated lime.

Slant culture: A culture grown in a tube that contains a solid nutrient medium which was solidified while the tube was kept in a slanted position.

Slimes: Aggregations of microbial cells that pose environmental and industrial problems; may be amenable to biological control.

Sludge: Precipitated solid matter produced by water and sewage treatment or industrial problems; may be amenable to biological control.

Slug flow: Type of two-phase flow in which the gas phase flows in the form of large bubbles whose transverse dimensions are commensurate with the characteristic dimension of the flow cross section as in some pipeline operations.

Somaclonal variation: Genetic variation produced from the culture of plant cells from a pure breeding strain; the source of the variation is not known.

Species: A taxonomic subdivision of a genus. A group of closely related, morphologically similar individuals which actually or potentially interbreed.

Specific growth rate: The rate of growth of a population of microorganisms, per unit mass of cells.

Spectrometer: An instrument used for analyzing the structure of compounds on the basis of their light-absorbing properties.

Spheroplast: A bacterial cell that is largely, but not entirely, freed of its cell wall.

Spore: A dormant cellular form, derived from a bacterial or a fungal cell, that is devoid of metabolic activity and that can give rise to a vegetative cell upon germination; it is dehydrated and can survive for prolonged periods of time under drastic environmental conditions.

Starch: The major form of storage carbohydrates in plants. It is a homopolysaccharide, composed of D-glucose units, that occurs in two forms: amylose, which consists of straight chains, and in which the glucose residues are linked by means of alpha(1–4) glycosidic bonds; and amylopectin, which consists of branched chains, and in which the glucose residues are linked by means of both alpha(1–4) and alpha(1–6) glycosidic bonds.

Stationary phase: The phase of growth of a culture of microorganisms that follows the exponential phase and in which there is little or no growth.

Sterile: Free from viable microorganisms.

Sterilization: The complete destruction of all viable microorganisms in a material by physical and/or chemical means.

Steroid: A group of organic compounds, some of which act as hormones to stimulate cell growth in higher animals and humans.

Stirred tank bioreactor: Agitated vessel in which a bioprocess takes place; mixing is provided by the mechanical action of an impeller/agitator.

Storage protein genes: Genes coding for the major proteins found in plant seeds.

Strain: A group of organisms of the same species having distinctive characteristics but not usually considered a separate breed or variety. A genetically homogeneous population of organisms at a subspecies level that can be differentiated by a biochemical, pathogenic or other taxonomic feature.

Substrate: A substance acted upon, for example, by an enzyme.

Subunit vaccine: A vaccine that contains only portions of a surface molecule of a pathogen. Subunit vaccines can be prepared by using rDNA technology to produce all or part of the surface protein molecule or by artificial (chemical) synthesis of short peptides.

Surfactant: A substance that alters the surface tension of a liquid, generally lowering it; detergents and soaps are typical examples.

Symbiont: An organism living in symbiosis, usually the smaller member of a symbiotic pair of dissimilar size.

Symbiosis: In the biological sciences, the living together of two dissimilar organisms in mutually beneficial relationships.

Synchronous growth: Growth in which all of the cells are at the same stage in cell division at any given time.

Tangential flow filtration: See Cross-flow filtration.

Taxis: The movement of an organism in response to a stimulus.

Taxonomy: The scientific classification of plants and animals that is based on their natural relationships; includes the systematic grouping, ordering and naming of the organisms.

Therapeutics: Pharmaceutical products used in the treatment of disease.

Thermal diffusivity: Numerically equal to the ratio of the coefficient of thermal conductivity to the volumetric specific heat of a substance (see also Heat transfer).

Thermophile: An organism that grows at high temperatures in the range 45–70 °C (or higher temperatures) and that has an optimum growth temperature in the range 50–55 °C.

Thermophilic: Heat loving. Usually refers to microorganisms that are capable of surviving at elevated temperatures; this capability may make them more compatible with industrial biotechnology schemes.

Thermotolerant: Capable of withstanding relatively high temperatures (45–70 °C).

Thrombolytic enzymes: Enzymes such as streptokinase and urokinase that initiate the dissolution of blood clots.

Ti plasmid: Plasmid from *Agrobacterium tumefacciens*, used as a plant vector.

Tissue plasminogen activator (TPA): A hormone that selectively dissolves blood clots that cause heart attacks and strokes.

Totipotency: The capacity of a higher organism cell to differentiate into an entire organism. A totipotent cell contains all the genetic information necessary for complete development.

Toxicity: The ability of a substance to produce a harmful effect on an organism by physical contact, ingestion or inhalation.

Toxin: A substance, produced in some cases by disease-causing microorganisms, which is toxic to other living organisms.

Toxoid: Detoxified toxin, but with antigenic properties intact.

Transcription: The synthesis of messenger RNA on a DNA template; the resulting RNA sequence is complementary to the DNA sequence. This is the first step in gene expression (see also Translation).

Transfer RNA (tRNA): A low-molecular weight RNA molecule, containing about 70–80 nucleotides, that binds an amino acid and transfers it to the ribosomes for incorporation into a polypeptide chain during translation.

Transformation: In the biological sciences, the introduction of new genetic information into a cell using naked DNA.

Transistor: An active component of an electrical circuit consisting of semiconductor material to which at least three electrical contacts are made so that it acts as an amplifier, detector or switch.

Translation: In the biological sciences, the process in which the genetic code contained in the nucleotide base sequence of messenger RNA directs the synthesis of a specific order of amino acids to produce a protein. This is the second step in gene expression (see also Transcription).

Transposable element: Segment of DNA which moves from one location to another among or within chromosomes in possibly a predetermined fashion, causing genetic change; may be useful as a vector for manipulating DNA.

Turbulent flow: In the engineering sciences, fluid motion with particle trajectories varying chaotically (randomly) with time; irregular fluctuations of velocity, pressure and other parameters, non-uniformly distributed in the flow; indicates relatively high levels of mixing (*cf.* Laminar flow).

Vaccine: A suspension of attenuated or killed bacteria or viruses, or portions thereof, injected to produce active immunity (see also Subunit vaccine).

Vector: DNA molecule used to introduce foreign DNA into host cells. Vectors include plasmids, bacteriophages (virus) and other forms of DNA. A vector must be capable of replicating autonomously and must have cloning sites for the introduction of foreign DNA.

Viable: Describing a cell or an organism that is alive and capable of reproduction.

Viable count: The number of viable cells in a culture of microorganisms.

Virus: Any of a large group of submicroscopic agents infecting plants, animals and bacteria and unable to reproduce outside the tissues of the host. A fully formed virus consists of nucleic acid (DNA or RNA) surrounded by a protein or protein and lipid coat.

Viscosity: A measure of a liquid's resistance to flow.

Volatile fatty acids (VFAs): Mixture of acids, primarily acetic and propionic, produced by acidogenic microorganisms during anaerobic digestion.

Volatile organic compounds (VOCs): Group of toxic compounds found in ground water and that pose environmental hazards; their destruction during water purification may be done biologically.

Wash out: Condition at which the critical dilution rate is exceeded in a chemostat.

Wild-type: The most frequently encountered phenotype in natural breeding populations.

Yeast: A fungus of the family Saccharomycetacea that is used especially in the making of alcoholic liquors and fodder yeast for animal feeds, and as leavening in baking. Yeasts are also commonly used in bioprocesses.

Yield: For a general chemical reaction: the weight of product obtained divided by the theoretical amount expected. Also, for the isolation of an enzyme: the total activity at a given step in the isolation divided by the total activity at a reference step.

Appendix 2:
Nomenclature Guidelines

Provisional list of symbols and units recommended for use in biotechnology by the IUPAC Commission on Biotechnology as reported in *Pure Appl. Chem.*, **54**, 1743–1749 (1982). This list is intended for use in conjunction with other recommendations on symbols, in particular *Manual of Symbols and Terminology for Physicochemical Quantities and Units* (Pergamon, 1979) and 'Letter Symbols for Chemical Engineering', *Chem. Eng. Prog.*, 73–80 (1978).

A2.1 GENERAL CONCEPTS

	Symbol	*SI units*	*Other units*
Activation energy	E	J mol^{-1}	cal mol^{-1}
for growth	E_g		
for death	E_d		
Area dimensions			
area per unit volume	a	m^{-1}	cm^{-1}
Linear dimension			
impeller diameter	D_i	m	m
tank diameter	D_t	m	m
liquid depth	D_l	m	m
width of baffle	D_b	m	m
Pressure	p	Pa	atm, bar
denote partial pressure with appropriate subscript, *e.g.* P_{O_2} for partial pressure of oxygen			
Ratio, in general	R		
for stoichiometric mass ratio, *e.g.* mass of substrate A consumed per mass of substrate B consumed	$R_{A/B}$		
for stoichiometric molar ratio, *e.g.* mole of substrate A consumed per mole of B consumed	$R_{MA/B}$		
Temperature			
absolute	T	K	K
general	t, T	°C	°C
Time	t	s	min, h
identify specific time periods by appropriate subscripts, *e.g.* t_d for			

	Symbol	SI units	Other units
doubling time, t_1 for lag time, and t_r for replacement or mean residence time			

Volume dimensions

volume	V	m^3, L	L

identify by subscript, *e.g.* V_1 for volume of stage 1, *etc.*

Yield, general mass ratio expressing output over input	Y		

without further definition, Y refers to the mass conversion ratio in terms of g dry weight biomass per g mass of substrate used. It should be further defined by subscripts to denote other ratios, *e.g.* $Y_{P/S}$ and $Y_{P/X}$ for g mass of product per g mass of substrate and per g dry weight of biomass, respectively

Yield, growth mass ratio corrected for maintenance, where　　　　　Y_G

$$\frac{1}{Y} = \frac{1}{Y_G} + \frac{m}{\mu}$$

or

$$q_s = \frac{\mu}{Y_G} + m$$

Yield, molar growth	Y_{GM}	kg mol^{-1}	g mol^{-1}

kg biomass formed per mole of mass used, or further defined as above to denote other molar yields

A2.2　CONCENTRATIONS AND AMOUNTS

Concentration

Biomass*

total mass (dry wt. basis)	x	kg	g
mass concentration (dry wt. basis)	X	kg m^{-3}	g L^{-1}
volume fraction	ϕ		
total number	N		
number concentration	n	m^{-3}	
Substrate concentration mass or moles per unit volume	C_S	kg m^{-3}, kmol m^{-3}	mg L^{-1}, mmol L^{-1}
Product concentration mass or moles per unit volume	C_P	kg m^{-3}, kmol m^{-3}	mg L^{-1}, mmol L^{-1}
Gas hold-up volume of gas per volume of dispersion	ε_G		

* Note: because of the difficulty in expressing biomass (cells) in molar terms, a separate symbol (other than C) is recommended.

	Symbol	SI units	Other units
Inhibitor concentration mass or moles per unit volume	C_i	kg m^{-3}, kmol m^{-3}	mg L^{-1}, mmol L^{-1}
Inhibitor constant dissociation constant of inhibitor–biomass complex	K_i	kg m^{-3}	g L^{-1}
Saturation constant as in the growth rate expression $\mu = \mu_m C_S/(K_s + C_S)$	K_s	kmol m^{-3}	g L^{-1}, mmol L^{-1}
Total amount, *e.g.* mass or moles	C	kg, kmol	g, mol

A2.3 INTENSIVE PROPERTIES

	Symbol	SI units	Other units
Density, mass	ϱ	kg m^{-3}	g L^{-1}
Diffusivity, molecular, volumetric	D_v	m^2 s^{-1}	cm^2 s^{-1}
Enthalpy, mass, of growth	H_X	J kg^{-1}	J g^{-1}
heat produced per unit of dry weight biomass formed			
Enthalpy, molar, of substrate consumption or of product formation	H_S, H_P	J mol^{-1}	J mol^{-1}
Vapor pressure denote with appropriate subscript, *e.g.* $p_i^* =$ vapor pressure of material i	p^*	Pa	atm, bar
Viscosity, absolute	μ	Pa s	poise
Viscosity, kinematic	v	m^2 s^{-1}	cm^2 s^{-1}

A2.4 RATE CONCEPTS

	Symbol	SI units	Other units
Death rate, specific $\delta = -(dn/dt)/n$	δ	s^{-1}	s^{-1}, h^{-1}
Dilution rate volume flow rate/culture volume	D	s^{-1}	h^{-1}, d^{-1}
Dilution rate, critical value at which biomass washout occurs in continuous flow culture	D_c	s^{-1}	h^{-1}, d^{-1}
Doubling time, biomass $t_d = (\ln 2)/\mu$	t_d	s	min, h
Flow rate, volumetric identify stream by appropriate subscript, *e.g.* A for air, G for gas, L for liquid, *etc.*	F	m^3 s^{-1}	L h^{-1}
Growth rate, colony radial rate of extension of biomass colony on a surface	K_r	m s^{-1}	μm h^{-1}
Growth rate, maximum specific	μ_m	s^{-1}	h^{-1}, d^{-1}
Growth rate, specific $\mu = (dx/dt)/x$	μ	s^{-1}	h^{-1}, d^{-1}

	Symbol	SI units	Other units
Heat transfer coefficient			
individual	h	$\text{W m}^{-2}\,\text{K}^{-1}$	$\text{cal h}^{-1}\,\text{cm}^{-2}\,°\text{C}^{-1}$
overall	U	$\text{W m}^{-2}\,\text{K}^{-1}$	$\text{cal h}^{-1}\,\text{cm}^{-2}\,°\text{C}^{-1}$
Maintenance coefficient, substrate or non-growth term associated with substrate consumption as defined in yield relationship (see yield term)	m	s^{-1}	h^{-1}
Mass transfer coefficient (molar basis)			
Individual, area basis	k	$\text{kmol m}^{-2}\,\text{s}^{-1}$ (driving force)$^{-1}$	$\text{gmol h}^{-1}\,\text{cm}^{-2}$ (driving force)$^{-1}$
gas film	k_G	$\text{kmol m}^{-2}\,\text{s}^{-1}\,\text{kPa}^{-1}$	"
liquid film	k_L	m s^{-1}	"
Overall, area basis	K	"	"
gas film	K_G	"	"
liquid film	K_L	"	"
Individual, volumetric basis	ka	$\text{kmol m}^{-3}\,\text{s}^{-1}\,\text{kPa}^{-1}$	h^{-1}
gas film	$k_\text{G}a$	"	h^{-1}
liquid film	$k_\text{L}a$	s^{-1}	h^{-1}
Metabolic rate, maximum specific	q_m	s^{-1}	h^{-1}
Metabolic rate, specific	q	s^{-1}	h^{-1}

$q = (\mathrm{d}C/\mathrm{d}t)/X$
where C may be a substrate or product mass concentration. Subscripts may further define the rates, *e.g.* q_S, q_P, q_{O_2}, which are substrate utilization, product formation, and oxygen uptake rates, respectively

	Symbol	SI units	Other units
Mutation rate	w	s^{-1}	h^{-1}
Power	P	W	W
Productivity, mass concentration rate basis, use appropriate subscripts, *e.g.* r_X for biomass productivity and r_P for product productivity	r	$\text{kg m}^{-3}\,\text{s}^{-1}$	$\text{kg m}^{-3}\,\text{h}^{-1}$
Revolutions per unit time or stirring speed	N	s^{-1}	s^{-1}
Velocity	V	m s^{-1}	cm s^{-1}

V_s for superficial gas velocity $= F_\text{G}/\pi D_\text{t}^2$

V_i for impeller tip velocity $= \pi N D_\text{i}$

Cumulative Subject Index

Bubbles
  large
    oxygen transfer coefficients and, **2**, 25–26
  size, **2**, 275
    distribution, sparger design and, **2**, 274
    oxygen transfer coefficients and, **2**, 24
  small
    oxygen transfer coefficients and, **2**, 24–25
  swarms
    oxygen transfer coefficients and, **2**, 24
Budapest Treaty on International Recognition of the
    Deposit of Microorganisms for the Purposes of
    Patent Procedure, **4**, 497
Buffering capacity
  microorganisms and, **1**, 268
Buffers
  in media
    for cell culture growth, **1**, 314
  in microbial cultures, **1**, 268
  pH control by, **1**, 268
Buildings
  design
    for containment of biohazards, **4**, 468–472
Bulgarican
  fermented dairy products
    shelf-life, **3**, 577
Bulk density
  particulate solids, **2**, 268
  solids, **2**, 259
Bulk handling equipment, **2**, 261
Bulking agents
  in composting, **4**, 1064
Bulk materials
  storage and handling of, **1**, 133
Bulk mixing
  bioreactors, **2**, 71
Buoyancy
  water
    life and, **4**, 802
Burkitt's lymphoma, **4**, 578
Butadiene
  acetone–butanol fermentation and, **3**, 915
  preparation, **3**, 942
  production
    from ethanol, **3**, 905
2,3-Butadione — *see* Diacetyl
1,3-Butanediol
  in affinity chromatography
    diglycidyl ether, **2**, 512
2,3-Butanediol, **3**, 933–944
  analysis
    by liquid chromatography, **3**, 941
  bacterial oxidation, **3**, 938
  distillation, **3**, 935
  formation from acetoin
    kinetics, **3**, 939
  microbial overproduction of, **1**, 610
    pH and, **1**, 611
  physical properties, **3**, 935
  production
    biochemistry, **3**, 937–941
    by *Enterobacter aerogenes*, **1**, 371
  reactions, **3**, 941
  recovery, **3**, 943
  stereochemistry, **3**, 934
2,3-Butanediol dehydrogenases, **3**, 938
Butanol
  extraction
    from fermentation products, **3**, 926
  fermentation, **3**, 915–929
    capital costs, **3**, 921
    economics, **3**, 922
    equipment costs, **3**, 921

income summary, **3**, 922
  production, **3**, 916
  production costs, **3**, 922
fermentation production
  petrochemicals and, **3**, 924
  regulation, **3**, 925
manufacture, **3**, 916
microbial overproduction of
  pH and, **1**, 612
on-line electrode measurement
  microbial growth rate measurement and, **4**, 319
production, **3**, 753
  by *Clostridium acetobutylicum*, **1**, 629
  by fermentation, **3**, 732
  by glucose fermentation, **1**, 369
  by microorganisms, **3**, 661
tolerance
  in *Clostridium acetobutylicum*, **3**, 927
toxicity
  to *Clostridium acetobutylicum*, **3**, 926, 927
  to *Clostridium butylicum*, **3**, 927
uses, **3**, 916
n-Butanol
  production
    bioprocesses, **4**, 615
2-Butene
  preparation, **3**, 942
Butikacin
  synthesis, **3**, 155
Butirosin
  biosynthesis, **3**, 72, 73
  fermentation
    phosphatase, **3**, 78
  formation
    alanine, **3**, 79
  structure, **3**, 139
Butirosin B
  synthesis, **3**, 155
Buttermilk
  consumption, **3**, 568
  cultured
    manufacture, **3**, 578
  flavor
    defects, **3**, 579
  nutritional composition, **3**, 588
  texture
    defects, **3**, 579
Butyl acetate, **3**, 30
  uses, **3**, 916
Butyl butyrate
  butanol extraction by
    fermentation products, **3**, 926
Butylene
  as substrate
    for bacterial biomass, **3**, 466
Butyramide
  utilization
    by *Pseudomonas* spp., **1**, 556
Butyrate, γ-amino-
  transport of
    ions and, **1**, 146
Butyric acid, **3**, 731–754
  bacterial formation, **3**, 737–739
  bacterial producers, **3**, 734
  chemical industry feedstock, **3**, 753
  coproduction
    with acetic and propionic acid, **3**, 743
  esters
    uses, **3**, 752
  fermentation production
    regulation, **3**, 925
  industrial production, **3**, 739–751
  isobutyl ester

in artificial cells, **4**, 57
Glutamine synthetase
adenylylation system, **4**, 85
nitrogen source repression and, **1**, 622
Glutamine synthetase–glutamate synthase
in ammonia assimilation, **4**, 80
Glutaraldehyde
carboxypeptidase A immobilization and, **2**, 199
collagen systems
cell immobilization, **2**, 211
in enzyme immobilization
on amine carriers, **2**, 197
glucose isomerase
immobilization, **3**, 847
Glutaric acid, α-keto-
production
in glutamic acid fermentation, **3**, 595
Glutaric acid, 2-oxo-
overproduction
by *Klebsiella aerogenes*, **1**, 607
Glutathione
in yeast cells
carbon dioxide retention and, **3**, 453
Gluten
in dough
carbon dioxide retention and, **3**, 453
Glycerol
in beer
haze and, **3**, 367
lactopalmitate
uses, **3**, 772
lactostearate
uses, **3**, 772
production
in glucose fermentation, **1**, 366
recovery from ethanol distillation, **3**, 899
from soybean oil, **3**, 410
Glycerol, triacyl-
biosynthesis, **1**, 475
temperature and, **1**, 470
nomenclature, **1**, 460
production
by bacteria, **3**, 988
by yeasts and molds, **3**, 933–995
from yeasts, **1**, 464
Glycerophosphates
in nutritional requirements
of microorganisms, **1**, 117
Glycerophospholipids
biosynthesis
temperature and, **1**, 470
Glycine, glycyl-
uptake of
by *Escherichia coli*, **1**, 142
Glycine decarboxylase
nitrogen source repression and, **1**, 622
Glycogen
brewer's yeast and, **3**, 362
in cyanobacteria, **1**, 581
dissimilation, **3**, 363
formation
in wastewaters, **4**, 782
microbial overproduction of, **1**, 607
Glycogen phosphorylase *b*
purification
hydrophobic chromatography, **2**, 529
Glycol, 2,3-butylene-
production
redox potential and, **2**, 140
Glycol, polyoxypropylene-
diether
in citric acid fermentation, **3**, 672
Glycolipids

in bacteria, **1**, 463
microbial production, **3**, 990
nomenclature, **1**, 461
from yeasts, **1**, 465
Glycolysis
in Embden–Meyerhof pathway, **1**, 358
in Entner–Douderoff pathway, **1**, 358
meat
tenderness, **3**, 1047
in phosphoketolase pathway, **1**, 358
rate
control of, **1**, 371
Glycolytic pathways
in lactic streptococci metabolism, **3**, 513
Glycophase G
for molecular sieve chromatography, **2**, 495
Glycophorin
purification
by hydrophobic chromatography, **2**, 532
Glycoproteins
in beer
foam stability and, **3**, 370
Glycosidases
glycoside mutagenicity and, **4**, 454
Glycosides
mutagenicity
comutagens, **4**, 453
Glycosides, amino-
ion exchange recovery of, **2**, 476
Glyoxylate cycle, **1**, 348
Gold
accumulation
in *Bacillus subtilis* cell walls, **4**, 239
cathodes
for dissolved oxygen probes, **4**, 372
recovery
biotechnology and, **4**, 219
Golgi apparatus
enzyme secretion and, **1**, 594
Gomphidiaceae
occurrence, **4**, 116
systematics, **4**, 115
Gonatobotrys spp.
gluconic acid production by, **3**, 684
Gonobactin
production, **3**, 281
Good Laboratory Practices Regulations, **4**, 626
Good Manufacturing Practice Regulations, **4**, 626
drugs, **4**, 593
regulations, **4**, 593
Gougerotin, **3**, 120
Governmental programs
for biotechnology
development and applications, **4**, 695–715
Government regulations
biohazardous materials, **4**, 609–631
desirable elements, **4**, 588
recombinant DNA, **4**, 587–604
Grading
cheese, **3**, 546
Graft *versus* host disease
chemotherapy
monoclonal antibodies and, **4**, 37
Graft *versus* host rejection, **4**, 45
Gramicidin, **3**, 121
in antibiotic peptide formation, **3**, 109
microbial overproduction of
pH and, **1**, 611
Gramicidin S, **3**, 121
overproduction, **1**, 610
production, **3**, 109, 110
Granulation
enzymes

γ-Octalactone
 microbial, **3**, 968
Octane enhancer
 ethanol as, **3**, 863
Octapeptins, **3**, 125
Octapines
 plasmid-induced synthesis, **4**, 189
1-Octen-3-ol
 mushroom aroma, **3**, 977
Oestrogens
 synthesis
  from diosgenin, **4**, 190
Oil
 aromatic
  biological activity, **4**, 297
 biological activity
  under anaerobic conditions, **4**, 296
 measurement
  in wastewater treatment, **4**, 1103
 recovery
  microbially enhanced, **4**, 295–303
  microbially enhanced, applications, **4**, 302
  xanthan gum in, **4**, 615
 reservoirs
  microbially enhanced recovery, **4**, 295, 296
Oil reservoirs
 bacteria in
  indigenous, **4**, 300
Oils
 interesterification
  lipases, **3**, 832
 from microorganisms, **3**, 983–1001
 plant seed
  fatty acyl composition, **3**, 987
 prices, **3**, 984, 986
 production, **3**, 985
 uses, **3**, 984, 986
 yeast growth on, **3**, 998
Oilseed rape
 black leg tolerant plants
  cell selection systems, **4**, 163
 somaclonal variants
  enhanced tolerance to pathogenic diseases, **4**, 158
Oil wells
 bore plugging
  microbially enhanced, **4**, 297
Oleaginous microorganisms, **1**, 467; **3**, 984–997
 definition, **3**, 987
Olefins
 fuels, **3**, 753
 production, **3**, 732
Oleic acid
 biosynthesis
  biotin in, **3**, 596
 in wort
  high gravity, **3**, 373
Oligogalacturonases
 in *Aspergillus niger*, **3**, 820
Olivanic acid, **3**, 64
 discovery, **3**, 50
Olivomycin
 structure, **3**, 266
One sludge system
 nitrification, **4**, 878
Onion
 flavor
  from plant tissue culture, **3**, 978
On-line measurements
 in wastewater treatment, **4**, 1132–1134
On/off control
 in fermentation, **2**, 158
Opercularia spp.
 in treatment of wastewater, **4**, 825

Operons, **1**, 619
 in enzyme synthesis, **1**, 595
Operophthera brumata
 EPV, **3**, 950
Ophiobolus herpotrichus
 21-hydroxylation by
  of steroids, **3**, 304
Opiates
 as plant metabolites, **4**, 190
Opines
 plasmid-induced synthesis, **4**, 189
Opportunistic pathogens
 health hazards, **4**, 563–564
Optical sensors
 for agitation measurement
  in fermentation process control, **2**, 154
Optimization
 dynamic
  in fermentation systems, **2**, 171
 of steady state
  in fermentation systems, **2**, 171
Oral contraceptives
 synthesis
  from diosgenin, **4**, 190
Orchard grass
 acid hydrolysates
  fermentation, **3**, 742
Ore bodies
 fracturing, **4**, 216
Ores
 beneficiation, **4**, 219
Organelle fusion, **4**, 189
Organelles
 microbial sensors, **4**, 418
Organelle transplantation, **4**, 189
Organic carbon
 biomass
  microbial growth rate and, **4**, 314
Organic matter
 measurement
  wastewater treatment, **4**, 1098
Organic overload
 downflow stationary fixed film reactors, **4**, 1042
Organic pollution monitor, **4**, 1094, 1098
Organic volumetric loading design
 high rate filters, **4**, 974
Organic wastes
 composition, **4**, 1061
 composting, **4**, 1060
Organoleptic evaluation
 of microbial flavors, **3**, 978
Organoleptic properties
 of beer, **3**, 343–346
Organotrophs
 in treatment of wastewater, **4**, 823
Ornamented vesicles
 Gigaspora spp., **4**, 130
Ortho fission pathways, **1**, 493
Oscillatoria limnetica
 hydrogen conversion by, **1**, 449
 photoassimilation in, **1**, 574
Oscillatoria rubscens
 in polluted waters, **4**, 814
Osmolarity
 microbial overproduction and, **1**, 606
Osmotic pressure
 dough ingredients and, **3**, 451
 water salinity and, **4**, 807
Osmotic shock
 in cell wall disruption, **2**, 314
Osmotolerance
 of yeasts
  in ethanol production, **3**, 875

in fermentation process control, **2**, 152
conditioning amplifiers
in fermentation process control, **2**, 153
transmission
in fermentation process control, **2**, 153
Signal sequence
function, **1**, 591
structure, **1**, 591
translocation, **1**, 591
Silane, γ-aminopropyltriethoxy-
as carrier
in enzyme immobilization, **2**, 197
Silanes
as carriers
in enzyme immobilization, **2**, 197
Silica
iron precipitation and, **4**, 794
for molecular sieve chromatography, **2**, 494, 495
porous
in affinity chromatography, **2**, 510
surface-modified
molecular sieve chromatography, **2**, 494
Silica hydrogels
in beer
stability and, **3**, 367
Silicalite
in fatty acid recovery, **3**, 751
Silicates
biodegradation, **4**, 202
degradation, **4**, 203
minerals
biodegradation, **4**, 207
Silicon
in nutritional requirements
of algae, **1**, 123
Silicone oil
artificial cells, **4**, 66
Silicone rubber
in enzyme immobilization
by entrapment, **2**, 199
Silicone rubber microspheres
artificial cells, **4**, 66
Silkworms
diseases of, **3**, 950
Silkworm sugar peptide, **3**, 127
Silver
accumulation by microorganisms, **4**, 236
cathodes
for dissolved oxygen probes, **4**, 372
intracellular accumulation, **4**, 241
recovery
biotechnology and, **4**, 219
Silver salts
solubility
in water, **4**, 786
Simulation models
continuum
of ion exchange of antibiotics, **2**, 477–479
Simultaneous adaptation, **1**, 496
Single cell protein, **3**, 1000
acceptance of, **4**, 689–691
as food, **4**, 673–683
for animal feeds
acceptance of, **4**, 685–693
bacterial, **3**, 463
as animal feed, **3**, 474
functional properties, **3**, 477
bacterial enrichment for, **1**, 31
from cellulose, **4**, 843–844
centrifugation, **2**, 341
development
organizations involved in, **4**, 686
digestibility and protein level of, **4**, 681

fermentation contaminants in, **4**, 687
fermentation residues in, **4**, 687
fermenters for, **1**, 325
as food, **4**, 681
growth of
reduction and, **1**, 230
manufacture
in West Germany, **4**, 707
workers health and, **4**, 692–693
mixed cultures and, **1**, 287
nature of, **4**, 687–688
nucleic acids in, **4**, 679–680
nutritional value of, **4**, 680–681, 691–692
production, **1**, 11, 379, 401
large scale, **1**, 130
from methanol, **1**, 614
from whey, **3**, 555
safety of, **4**, 685–688
testing of
guidelines for, **4**, 685–687
toxicity, **4**, 674
toxicology of, **4**, 688–689
Single-stage air lift reactors
with inner loop channel, **2**, 112–113
with outer loop, **2**, 111–112
with split cylinder, **2**, 112
Single-stage air lift tower loop reactors, **2**, 110–113
with draught tubes, **2**, 110–111
Single-stage tower reactors, **2**, 109–110
Sintering
in membrane preparation, **2**, 386
Siomycin, **3**, 127, 234
Sirodesmins, **3**, 127
Sisomicin
biosynthesis, **3**, 74
discovery, **3**, 138
formation
alanine, **3**, 79
production
new strains for, **3**, 153
structure, **3**, 139
5-*epi*-Sisomicin
market, **3**, 159
production, **3**, 155
Sisomicin, 1-*N*-ethyl-
synthesis, **3**, 155
Sister chromatid exchange
genotoxic agents
testing, **4**, 449
Sitosterol
degradation of
by fermentation, **3**, 306
Size
particulate solids, **2**, 268
Size exclusion chromatography, **2**, 539
Skim milk
homogenization
for cheese making, **3**, 531
nutritional composition, **3**, 588
Skin
microbial spoilage
health hazards, **4**, 562
Slaframine
mycotoxicoses, **4**, 550
Sleeping pills
detoxification
artificial cells, **4**, 61
Slide culture
biological enumeration techniques, **4**, 312
Slime
in high rate filters, **4**, 964
Sloping plane biological film fermenter, **2**, 85
Sloughing